Handbook of SCHEDULING

Algorithms, Models, and Performance Analysis

CHAPMAN & HALL/CRC COMPUTER and INFORMATION SCIENCE SERIES

Series Editor: Sartaj Sahni

PUBLISHED TITLES

HANDBOOK OF SCHEDULING: ALGORITHMS, MODELS, AND PERFORMANCE ANALYSIS
Joseph Y-T. Leung

FORTHCOMING TITLES

HANDBOOK OF COMPUTATIONAL MOLECULAR BIOLOGY
Srinivas Aluru

HANDBOOK OF ALGORITHMS FOR WIRELESS AND MOBILE NETWORKS AND COMPUTING
Azzedine Boukerche

DISTRIBUTED SENSOR NETWORKS
S. Sitharama Iyengar and Richard R. Brooks

SPECULATIVE EXECUTION IN HIGH PERFORMANCE COMPUTER ARCHITECTURES
David Kaeli and Pen-Chung Yew

HANDBOOK OF DATA STRUCTURES AND APPLICATIONS
Dinesh P. Mehta and Sartaj Sahni

HANDBOOK OF BIOINSPIRED ALGORITHMS AND APPLICATIONS
Stephan Olariu and Albert Y. Zomaya

HANDBOOK OF DATA MINING
Sanjay Ranka

THE PRACTICAL HANDBOOK OF INTERNET COMPUTING
Munindar P. Singh

SCALABLE AND SECURE INTERNET SERVICE AND ARCHITECTURE
Cheng Zhong Xu

CHAPMAN & HALL/CRC COMPUTER and INFORMATION SCIENCE SERIES

Handbook of SCHEDULING

Algorithms, Models, and Performance Analysis

Edited by
Joseph Y-T. Leung

CHAPMAN & HALL/CRC

A CRC Press Company

Boca Raton London New York Washington, D.C.

Library of Congress Cataloging-in-Publication Data

Catalog record is available from the Library of Congress

International Standard Book Number 1-58488-397-9
Printed in the United States of America 1 2 3 4 5 6 7 8 9 0
Printed on acid-free paper

Dedication

To my wife Maria

Preface

Scheduling is a form of decision-making that plays an important role in many disciplines. It is concerned with the allocation of scarce resources to activities with the objective of optimizing one or more performance measures. Depending on the situation, resources and activities can take on many different forms. Resources may be nurses in a hospital, bus drivers, machines in an assembly plant, CPUs, mechanics in an automobile repair shop, etc. Activities may be operations in a manufacturing process, duties of nurses in a hospital, executions of computer programs, car repairs in an automobile repair shop, and so on. There are also many different performance measures to optimize. One objective may be the minimization of the mean flow time, while another objective may be the minimization of the number of jobs completed after their due dates.

Scheduling has been studied intensively for more than 50 years, by researchers in management, industrial engineering, operations research, and computer science. There is now an astounding body of knowledge in this field. This book is the first handbook on scheduling. It is intended to provide a comprehensive coverage of the most *advanced* and *timely* topics in scheduling. A major goal of this project is to bring together researchers in the above disciplines in order to facilitate cross fertilization. The authors and topics chosen cut across all these disciplines.

I would like to thank Sartaj Sahni for inviting me to edit this handbook. I am grateful to all the authors and co-authors (more than 90 in total) who took time from their busy schedules to contribute to this handbook. Without their efforts, this handbook would not have been possible. Edmund Burke and Michael Pinedo have given me valuable advice in picking topics and authors. Helena Redshaw and Jessica Vakili at CRC Press have done a superb job in managing the project.

I would like to thank Ed Coffman for teaching me scheduling theory when I was a graduate student at Penn State. My wife, Maria, gave me encouragement and strong support for this project.

This work was supported in part by the Federal Aviation Administration (FAA) and in part by the National Science Foundation (NSF). Findings contained herein are not necessarily those of the FAA or NSF.

Preface

The Editor

Joseph Y-T. Leung, Ph.D., is Distinguished Professor of Computer Science in New Jersey Institute of Technology. He received his B.A. in Mathematics from Southern Illinois University at Carbondale and his Ph.D. in Computer Science from the Pennsylvania State University. Since receiving his Ph.D., he has taught at Virginia Tech, Northwestern University, University of Texas at Dallas, University of Nebraska at Lincoln, and New Jersey Institute of Technology. He has been chairman at University of Nebraska at Lincoln and New Jersey Institute of Technology.

Dr. Leung is a member of ACM and a senior member of IEEE. His research interests include scheduling theory, computational complexity, discrete optimization, real-time systems, and operating systems. His research has been supported by NSF, ONR, FAA, and Texas Instruments.

Contributors

Richa Agarwal
Georgia Institute of Technology
Department of Industrial & Systems Engineering
Atlanta, Georgia

Sanjay L. Ahire
University of Dayton
Department of MIS, OM, and DS
Dayton, Ohio

Ravindra K. Ahuja
University of Florida
Department of Industrial & Systems Engineering
Gainesville, Florida

Yalçin Akçay
Koç University
Istanbul, Turkey

James Anderson
University of North Carolina
Department of Computer Science
Chapel Hill, North Carolina

Hakan Aydin
George Mason University
Department of Computer Science
Fairfax, Virginia

Anantaram Balakrishnan
University of Texas
Austin, Texas

Philippe Baptiste
CNRS, Ecole Polytechnique
Palaiseau, France

Sanjoy Baruah
University of North Carolina
Department of Computer Science
Chapel Hill, North Carolina

Jacek Błażewicz
Poznań University of Technology
Institute of Computing Science
Poznań, Poland

N. Brauner
IMAG
Grenoble, France

R. P. Brazile
University of North Texas
Department of Computer Science & Engineering
Denton, Texas

Peter Brucker
University of Osnabrück
Department of Mathematics
Osnabrück, Germany

Edmund K. Burke
University of Nottingham
School of Computer Science
Nottingham, United Kingdom

Marco Caccamo
University of Illinois
Department of Computer Science
Urbana, Illinois

Xiaoqiang Cai
Chinese University of Hong Kong
Department of Systems Engineering & Engineering Management
Shatin, Hong Kong

Jacques Carlier
Compiègne University of Technology
Compiègne, France

John Carpenter
University of North Carolina
Department of Computer Science
Chapel Hill, North Carolina

Xiuli Chao
North Carolina State University
Department of Industrial Engineering
Raleigh, North Carolina

Chandra Chekuri
Bell Laboratories
Murray Hill, New Jersey

Bo Chen
University of Warwick
Warwick Business School
Coventry, United Kingdom

Deji Chen
Fisher-Rosemount Systems, Inc.
Austin, Texas

Artur Czumaj
New Jersey Institute of Technology
Department of Computer Science
Newark, New Jersey

Patrick De Causmaecker
KaHo Sint-Lieven
Department of Industrial Engineering
Gent, Belgium

Sudarshan K. Dhall
University of Oklahoma
School of Computer Science
Norman, Oklahoma

Maciej Drozdowski
Poznań University of Technology
Institute of Computing Science
Poznań, Poland

Pierre-François Dutot
CNRS
Lab Informatique et Distribution
Montbonnot, France

Kelly Easton
Kansas State University
School of Industrial & Manufacturing Systems Engineering
Manhattan, Kansas

G. Finke
IMAG
Grenoble, France

Shelby Funk
University of North Carolina
Department of Computer Science
Chapel Hill, North Carolina

Karsten Gentner
University of Karlsruhe
Institute of Economic Theory and Operations Research
Karlsruhe, Germany

Teofilo F. Gonzalez
University of California
Department of Computer Science
Santa Barbara, California

Joël Goossens
Université Libre de Brussels
Department of Data Processing
Brussels, Belgium

Valery S. Gordon
National Academy of Sciences of Belarus
United Institute of Informatics Problems
Minsk, Belarus

Michael F. Gorman
University of Dayton
Department of MIS, OM, and DS
Dayton, Ohio

Kevin I-J. Ho
Chun Shan Medical University
Department of Information Management
Taiwan, China

Dorit Hochbaum
University of California
Haas School of Business, and Department of Industrial Engineering & Operations Research
Berkeley, California

Philip Holman
University of North Carolina
Department of Computer Science
Chapel Hill, North Carolina

H. Hoogeveen
Utrecht University
Department of Computer Science
Utrecht, Netherlands

Antoine Jouglet
CNRS
Compiègne, France

Joanna Józefowska
Poznań University of Technology
Institute of Computing Science
Poznań, Poland

Philip Kaminsky
University of California
Department of Industrial Engineering & Operations Research
Berkeley, California

John J. Kanet
University of Dayton
Department of MIS, OM and DS
Dayton, Ohio

Hans Kellerer
University of Graz
Institute for Statistics & Operations Research
Graz, Austria

Sanjeev Khanna
University of Pennsylvania
Department of Computer & Information Science
Philadelphia, Pennsylvania

Young Man Kim
Kookmin University
School of Computer Science
Seoul, South Korea

Gilad Koren
Bar-Ilan University
Computer Science Department
Ramat-Gan, Israel

Wieslaw Kubiak
Memorial University of Newfoundland
Faculty of Business Administration
St. John's, Canada

Raymond S.K. Kwan
University of Leeds
School of Computing
Leeds, United Kingdom

Ten H. Lai
The Ohio State University
Department of Computer & Information Science
Columbus, Ohio

Gilbert Laporte
University of Montreal
Centre of Research in Transportation
Montreal, Canada

Chung-Yee Lee
Hong Kong University of Science & Technology
Department of Industrial Engineering & Engineering Management
Kowloon, Hong Kong

Joseph Y-T. Leung
New Jersey Institute of Technology
Department of Computer Science
Newark, New Jersey

Rami Melhem
University of Pittsburgh
Department of Computer Science
Pittsburgh, Pennsylvania

Aloysius K. Mok
University of Texas
Department of Computer Science
Austin, Texas

Daniel Mosse
University of Pittsburgh
Department of Computer Science
Pittsburgh, Pennsylvania

Grégory Mounié
ENSIMAG
Lab Informatique et Distribution
Montbonnot, France

George Nemhauser
Georgia Institute of Technology
School of Industrial & Systems Engineering
Atlanta, Georgia

Klaus Neumann
University of Karlsruhe
Institute for Economic Theory and Operations Research
Karlsruhe, Germany

Laurent Péridy
West Catholic University
Applied Mathematics Institute
Angers, France

Sanja Petrovic
University of Nottingham
School of Computer Science
Nottingham, United Kingdom

Michael Pinedo
New York University
Department of Operations Management
New York, New York

Eric Pinson
West Catholic University
Applied Mathematics Institute
Angers, France

Jean-Marie Proth
INRIA-Lorraine
SAGEP Project
Metz, France

Kirk Pruhs
University of Pittsburgh
Computer Science Department
Pittsburgh, Pennsylvania

Xiangtong Qi
Hong Kong University of Science and Technology
Department of Industrial Engineering and Engineering Management
Kowloon, Hong Kong

David Rivreau
West Catholic University
Applied Mathematics Institute
Angers, France

Sartaj Sahni
University of Florida
Department of Computer & Information Science & Engineering
Gainesville, Florida

Christoph Schwindt
University of Karlsruhe
Institute for Economic Theory & Operations Research
Karlsruhe, Germany

Jay Sethuraman
Columbia University
Department of Industrial Engineering & Operations Research
New York, New York

Jiří Sgall
Mathematical Institute, AS CR
Prague, Czech Republic

Lui Sha
University of Illinois
Department of Computer Science
Urbana, Illinois

Dennis E. Shasha
New York University
Department of Computer Science
Courant Institute of Mathematical Sciences
New York, New York

Zuo-Jun Shen
University of Florida
Department of Industrial & Systems Engineering
Gainesville, Florida

Anand Srinivasan
Microsoft Corporation
Windows Embedded Team
Redmond, Washington

Vitaly A. Strusevich
University of Greenwich
School of Computing and Mathematical Sciences
London, United Kingdom

K. M. Swigger
University of North Texas
Department of Computer Science & Engineering
Denton, Texas

Alexander Thomasian
New Jersey Institute of Technology
Department of Computer Science
Newark, New Jersey

Vadim G. Timkovsky
CGI Group Inc.
McMaster University
Algorithms Research Group
Ontario, Canada

Eric Torng
Michigan State University
Department of Computer Science and Engineering
East Lansing, Michigan

Norbert Trautmann
University of Karlsruhe
Institute for Economic Theory and Operations Research
Karlsruhe, Germany

Michael Trick
Carnegie Mellon University
Graduate School of Industrial Administration
Pittsburgh, Pennsylvania

Denis Trystram
Institut National Polytechnique
Lab ID-IMAG
Grenoble, France

George L. Vairaktarakis
Case Western Reserve University
Department of Operations
Cleveland, Ohio

Marjan van den Akker
Utrecht University
Department of Computer Science
Utrecht, Netherlands

Greet Vanden Berghe
KaHo Sint-Lieven
Department of Industrial Engineering
Gent, Belgium

Jan Węglarz
Poznań University of Technology
Institute of Computing Science
Poznań, Poland

Susan Xu
Penn State University
Department of Supply Chain and Information Systems
University Park, Pennsylvania

Jian Yang
New Jersey Institute of Technology
Department of Industrial & Manufacturing Engineering
Newark, New Jersey

G. Young
California State Polytechnic University
Department of Computer Science
Pomona, California

Gang Yu
University of Texas
Department of Management Science & Information Systems
Austin, Texas

Xian Zhou
The Hong Kong Polytechnic University
Department of Applied Mathematics
Kowloon, Hong Kong

Contents

Part I: Introduction

Part II: Classical Scheduling Problems

Part III: Other Scheduling Models

Part IV: Real-Time Scheduling

Part V: Stochastic Scheduling and Queueing Networks

Part VI: Applications

I

Introduction

1

Introduction and Notation

1.1 Introduction

1

Introduction and Notation

Joseph Y-T. Leung
New Jersey Institute of Technology

1.1 Introduction

Scheduling is concerned with the allocation of scarce resources to activities with the objective of optimizing one or more performance measures. Depending on the situation, resources and activities can take on many different forms. Resources may be machines in an assembly plant, CPU, memory and I/O devices in a computer system, runways at an airport, mechanics in an automobile repair shop, etc. Activities may be various operations in a manufacturing process, execution of a computer program, landings and take-offs at an airport, car repairs in an automobile repair shop, and so on. There are also many different performance measures to optimize. One objective may be the minimization of the makespan, while another objective may be the minimization of the number of late jobs.

The study of scheduling dates back to 1950s. Researchers in operations research, industrial engineering, and management were faced with the problem of managing various activities occurring in a workshop. Good scheduling algorithms can lower the production cost in a manufacturing process, enabling the company to stay competitive. Beginning in the late 1960s, computer scientists also encountered scheduling problems in the development of operating systems. Back in those days, computational resources (such as CPU, memory and I/O devices) were scarce. Efficient utilization of these scare resources can lower the cost of executing computer programs. This provided an economic reason for the study of scheduling.

The scheduling problems studied in the 1950s were relatively simple. A number of efficient algorithms have been developed to provide optimal solutions. Most notable are the work by Jackson [1, 2], Johnson [3], and Smith [4]. As time went by, the problems encountered became more sophisticated, and researchers were unable to develop efficient algorithms for them. Most researchers tried to develop efficient branch-and-bound methods that are essentially exponential-time algorithms. With the advent of complexity theory [5–7], researchers began to realize that many of these problems may be inherently difficult to solve. In the 1970s, many scheduling problems were shown to be NP-hard [8, 9–11].

In the 1980s, several different directions were pursued in academia and industry. One direction was the development and analysis of approximation algorithms. Another direction was the increasing attention paid to stochastic scheduling problems. From then on, research in scheduling theory took off by leaps and bounds. After almost 50 years, there is now an astounding body of knowledge in this field.

This book is the first handbook in scheduling. It is intended to provide a comprehensive coverage of the most *advanced* and *timely* topics in scheduling. A major goal is to bring together researchers in computer

1-58488-397-9/$0.00+$1.50

science, industrial engineering, operations research, and management science so that cross fertilization can be facilitated. The authors and topics chosen cut across all of these disciplines.

1.2 Overview of the Book

The book comprises six major parts, each of which has several chapters.

Part I presents introductory materials and notation. Chapter 1 gives an overview of the book and the $\alpha|\beta|\gamma$ notation for classical scheduling problems. Chapter 2 is a tutorial on complexity theory. It is included for those readers who are unfamiliar with the theory of NP-completeness and NP-hardness. Complexity theory plays an important role in scheduling theory. Anyone who wants to engage in theoretical scheduling research should be proficient in this topic. Chapter 3 describes some of the basic scheduling algorithms for classical scheduling problems. They include Hu's, Coffman-Graham, LPT, McNaughton's, and Muntz-Coffman algorithms for makespan minimization; SPT, Ratio, Baker's, Generalized Baker's, Smith's, and Generalized Smith's rules for the minimization of total (weighted) completion time; algorithms for dual objectives (makespan and total completion time); EDD, Lawler's, and Horn's algorithms for the minimization of maximum lateness; Hodgson-Moore algorithm for minimizing the number of late jobs; Lawler's pseudo-polynomial algorithm for minimizing the total tardiness.

Part II is devoted to classical scheduling problems. These problems are among the first studied by scheduling theorists, and for which the 3-field notation ($\alpha|\beta|\gamma$) was introduced for classification.

Chapters 4 to 7 deal with job shop, flow shop, open shop, and cycle shop, respectively. Job shop problems are among the most difficult scheduling problems. There was an instance of job shop with 10 machines and 10 jobs that was not solved for a very long time. Exact solutions are obtained by enumerative search. Chapter 4 gives a concise survey of elimination rules and extensions that are one of the most powerful tools for enumerative search designed in the last two decades. Hybrid flow shops are flow shops where each stage consists of parallel and identical machines. Chapter 5 describes a number of approximation algorithms for two-stage flexible hybrid flow shops with the objective of minimizing the makespan. Open shops are like flow shops, except that the order of processing on the various machines is immaterial. Chapter 6 discusses the complexity of generating exact and approximate solutions for both nonpreemptive and preemptive schedules, under several classical objective functions. Cycle shops are like job shops, except that each job passes through the same route on the machines. Chapter 7 gives polynomial-time and pseudo-polynomial algorithms for cycle shops, as well as NP-hardness results and approximation algorithms.

Chapter 8 shows a connection between an NP-hard preemptive scheduling problem on parallel and identical machines with the corresponding problem in a job shop or open shop environment for a set of chains of equal-processing-time jobs. The author shows that a number of NP-hardness proofs for parallel and identical machines can be used to show the NP-hardness of the corresponding problem in a job shop or open shop.

Chapters 9 to 13 cover the five major objective functions in classical scheduling theory: makespan, maximum lateness, total weighted completion time, total weighted number of late jobs, and total weighted tardiness. Chapter 9 discusses the makespan objective on parallel and identical machines. The author presents polynomial solvability and approximability, enumerative algorithm, and polynomial-time approximations under this framework. Chapter 10 deals with the topic of minimizing maximum lateness on parallel and identical machines. Complexity results and exact and approximation algorithms are given for nonpreemptive and preemptive jobs, as well as jobs with precedence constraints. Chapter 11 gives a comprehensive review of recently developed approximation algorithms and approximation schemes for minimizing the total weighted completion time on parallel and identical machines. The model includes jobs with release dates and/or precedence constraints. Chapter 12 gives a survey of the problem of minimizing the total weighted number of late jobs. The chapter concentrates mostly on exact algorithms and their correctness proofs. Total tardiness is among the most difficult objective functions to solve, even for a single machine. Chapter 13 gives branch-and-bound algorithms for minimizing the total weighted tardiness on one machine, where jobs are nonpreemptible and have release dates (but not precedence constraints).

Many NP-hard scheduling problems become solvable in polynomial time when the jobs have identical processing times. Chapter 14 gives polynomial-time algorithms for several of these cases, concentrating on one machine as well as parallel and identical machines' environments.

The scheduling problems dealt in the above-mentioned chapters are all offline deterministic scheduling problems. This means that the jobs' characteristics are known to the decision maker before a schedule is constructed. In contrast, online scheduling restricts the decision maker to schedule jobs based on the currently available information. In particular, the jobs' characteristics are not known until they arrive. Chapter 15 surveys the literature in online scheduling.

A number of approximation algorithms for scheduling problems have been developed that are based on linear programming. The basic idea is to formulate the scheduling problem as an integer programming problem, solve the underlying linear programming relaxation to obtain an optimal fractional solution, and then round the fractional solution to a feasible integer solution in such a way that the error can be bounded. Chapter 16 describes this technique as applied to the problem of minimizing the total weighted completion time on unrelated machines.

Part III is devoted to scheduling models that are different from the classical scheduling models. Some of these problems come from applications in computer science and some from the operations research and management community.

Chapter 17 discusses the master-slave scheduling model. In this model, each job consists of three stages and processed in the same order: preprocessing, slave processing, and postprocessing. The preprocessing and postprocessing of a job are done on a master machine (which is limited in quantity), while the slave processing is done on a slave machine (which is unlimited in quantity). Chapter 17 gives NP-hardness results, polynomial-time algorithms, and approximation algorithms for makespan minimization.

Local area networks (LAN) and wide area networks (WAN) have been the two most studied networks in the literature. With the proliferation of hand-held computers, Bluetooth network is gaining importance. Bluetooth networks are networks that have an even smaller distance than LANs. Chapter 18 discusses scheduling problems that arise in Bluetooth networks.

Suppose a manufacturer needs to produce d_i units of a certain product for customer i, $1 \leq i \leq n$. Assume that each unit takes one unit of time to produce. The total time taken to satisfy all customers is $D = \sum d_j$. If we produce all units for a customer before we produce for the next customer, then the last customer will have to wait for a long time. Fair sequences are those such that each customer would ideally receive $(d_j/D)t$ units at time t. Chapter 19 gives a review of fair sequences. Note that fair sequences are related to Pfair scheduling in Chapter 27 and fair scheduling of real-time tasks on multiprocessors in Chapter 30.

In scheduling problems with due date-related objectives, the due date of a job is given *a priori* and the scheduler needs to schedule jobs with the given due dates. In modern day manufacturing operations, the manufacturer can negotiate due dates with customers. If the due date is too short, the manufacturer runs the risk of missing the due date. On the other hand, if the due date is too long, the manufacturer runs the risk of loosing the customer. Thus, due date assignment and scheduling should be integrated to make better decisions. Chapters 20 and 21 discuss due date assignment problems.

In classical scheduling problems, machines are assumed to be continuously available for processing. In practice, machines may become unavailable for processing due to maintenance or breakdowns. Chapter 22 describes scheduling problems with availability constraints, concentrating on NP-hardness results and approximation algorithms.

So far we have assumed that a job only needs a machine for processing without any additional resources. For certain applications, we may need additional resources, such as disk drives, memory, and tape drives, etc. Chapters 23 and 24 present scheduling problems with resource constraints. Chapter 23 discusses discrete resources, while Chapter 24 discusses continuous resources.

In classical scheduling theory, we assume that each job is processed by one machine at a time. With the advent of parallel algorithms, this assumption is no longer valid. It is now possible to process a job with several machines simultaneously so as to reduce the time needed to complete the job. Chapters 25 and 26 deal with this model. Chapter 25 gives complexity results and exact algorithms, while Chapter 26 presents approximation algorithms.

Part IV is devoted to scheduling problems that arise in real-time systems. Real-time systems are those that control real-time processes. As such, the primary concern is to meet hard deadline constraints, while the secondary concern is to maximize machine utilization. Real-time systems will be even more important in the future, as computers are used more often to control our daily appliances.

Chapter 27 surveys the pinwheel scheduling problem, which is motivated by the following application. Suppose we have n satellites and one receiver in a ground station. When satellite j wants to send information to the ground, it will repeatedly send the same information in a_j consecutive time slots, after which it will cease to send that piece of information. The receiver in the ground station must reserve one time slot for satellite j during those a_j consecutive time slots, or else the information is lost. Information is sent by the satellites dynamically. How do we schedule the receiver to serve the n satellites so that no information is ever lost? The question is equivalent to the following: Is it possible to write an infinite sequence of integers, drawn from the set $\{1, 2, \ldots, n\}$, so that each integer j, $1 \leq j \leq n$, appears at least once in any a_j consecutive positions? The answer, of course, depends on the values of a_j. Sufficient conditions and algorithms to construct a schedule are presented in Chapter 27.

In the last two decades, a lot of attention has been paid to the following scheduling problem. There are n periodic, real-time jobs. Each job i has an initial start time s_i, a computation time c_i, a relative deadline d_i, and a period p_i. Job i initially makes a request for execution at time s_i, and thereafter at times $s_i + kp_i$, $k = 1, 2, \ldots$. Each request for execution requires c_i time units and it must finish its execution within d_i time units from the time the request is made. Given $m \geq 1$ machines, is it possible to schedule the requests of these jobs so that the deadline of each request is met? Chapter 28 surveys the current state of the art of this scheduling problem.

Chapter 29 discusses an important issue in the scheduling of periodic, real-time jobs — a high-priority job is blocked by a low-priority job due to priority inversion. This can occur when a low-priority job gains access to shared data, which will not be released by the job until it is finished; in other words, the low-priority job cannot be preempted while it is holding the shared data. Chapter 29 discusses some solutions to this problem.

Chapter 30 presents Pfair scheduling algorithms for real-time jobs. Pfair algorithms produce schedules in which jobs are executed at a steady rate. This is similar to fair sequences in Chapter 19, except that the jobs are periodic, real-time jobs.

Chapter 31 discusses several approaches in scheduling periodic, real-time jobs on parallel and identical machines. One possibility is to partition the jobs so that each partition is assigned to a single machine. Another possibility is to treat the machines as a pool and allocate upon demand. Chapter 31 compares several approaches in terms of the effectiveness of optimal algorithms with each approach.

Chapter 32 describes several approximation algorithms for partitioning a set of periodic, real-time jobs into a minimum number of partitions so that each partition can be feasibly scheduled on one machine. Worst-case analyses of these algorithms are also presented.

When a real-time system is overloaded, some time-critical jobs will surely miss their deadlines. Assuming that each time-critical job will earn a value if it is completed on time, how do we maximize the total value? Chapter 33 presents several algorithms, analyzes their competitive ratios, and gives lower bounds for any competitive ratios. Note that this problem is equivalent to online scheduling of independent jobs with the goal of minimizing the weighted number of late jobs.

One way to cope with an overloaded system is to completely abandon a job that cannot meet its deadline. Another way is to execute less of each job with the hope that more jobs can meet their deadlines. This model is called the imprecise computation model. In this model each job i has a minimum execution time $\min_i$ and a maximum execution time $\max_i$, and the job is expected to execute α_i time units, $\min_i \leq \alpha_i \leq \max_i$. If job i executes less than $\max_i$ time units, then it incurs a cost equal to $\max_i - \alpha_i$. The objective is to find a schedule that minimizes the total (weighted) cost or the maximum (weighted) cost. Chapter 34 presents algorithms that minimize total weighted cost, and Chapter 35 presents algorithms that minimize maximum weighted cost as well as dual criteria (total weighted cost and maximum weighted cost). Chapter 36 studies the same problem with arbitrary cost functions. It is noted there that this problem has some connections with power-aware scheduling.

Chapter 37 presents routing problems of real-time messages on a network. A set of n messages reside at various nodes in the network. Each message M_i has a release time r_i and a deadline d_i. The message is to be routed from its origin node to its destination node. Both online and offline routing are discussed. NP-hardness results and optimal algorithms are presented.

Part V is devoted to stochastic scheduling and queueing networks. The chapters in this part differ from the previous chapters in that the characteristics of the jobs (such as processing times and arrival times) are not deterministic; instead, they are governed by some probability distribution functions.

Chapter 38 compares the three classes of scheduling: offline deterministic scheduling, stochastic scheduling, and online deterministic scheduling. The author points out the similarities and differences among these three classes.

Chapter 39 deals with the earliness and tardiness penalties. In Just-in-Time (JIT) systems, a job should be completed close to its due date. In other words, a job should not be completed too early or too late. This is particularly important for products that are perishable, such as fresh vegetables and fish. Harvesting is another activity that should be completed close to its due date. The authors studied this problem under the stochastic setting, comparing the results with the deterministic counterparts.

The methods to solve queueing network problems can be classified into exact solution methods and approximation solution method. Chapter 40 reviews the latest developments in queueing networks with exact solutions. The author presents sufficient conditions for the network to possess a product-form solution, and in some cases necessary conditions are also presented.

Chapter 41 studies disk scheduling problems. Magnetic disks are based on technology developed 50 years ago. There have been tremendous advances in magnetic recording density resulting in disks whose capacity is several hundred gigabytes, but the mechanical nature of disk access remains a serious bottleneck. This chapter presents scheduling techniques to improve the performance of disk access.

The Internet has become an indispensable part of our life. Millions of messages are sent over the Internet everyday. Globally managing traffic in such a large-scale communication network is almost impossible. In the absence of global control, it is typically assumed in traffic modeling that the network users follow the most rational approach; i.e., they behave selfishly to optimize their own individual welfare. Under these assumptions, the routing process should arrive into a Nash equilibrium. It is well known that Nash equilibria do not always optimize the overall performance of the system. Chapter 42 reviews the analysis of the coordination ratio, which is the ratio of the worst possible Nash equilibrium and the overall optimum.

Part VI is devoted to applications. There are chapters that discuss scheduling problems that arise in the airline industry, process industry, hospitals, transportation industry, and educational institutions.

Suppose you are running a professional training firm. Your firm offers a set of training programs, with each program yielding a different payoff. Each employee can teach a subset of the training programs. Client requests arrive dynamically, and the firm must decide whether to accept the request, and if so which instructor to assign to the training program(s). The goal of the decision maker is to maximize the expected payoff by intelligently utilizing the limited resources to meet the stochastic demand for the training programs. Chapter 43 describes a formulation of this problem as a stochastic dynamic program and proposes solution methods for some special cases.

Constructing timetables of work for personnel in healthcare institutions is a highly constrained and difficult problem to solve. Chapter 44 presents an overview of the algorithms that underpin a commercial nurse rostering decision support system that is in use in over 40 hospitals in Belgium.

University timetabling problems can be classified into two main categories: course and examination timetabling. Chapter 45 discusses the constraints for each of them and provides an overview of some recent research advances made by the authors and members of their research team.

Chapter 46 describes a solution method for assigning teachers to classes. The authors have developed a system (GATES) that schedules incoming and outgoing airline flights to gates at the JFK airport in New

York City. Using the GATES framework, the authors continue with its applications to the new domain of assigning teachers to classes.

Chapter 47 provides an introduction to constraint programming (CP), focusing on its application to production scheduling. The authors provide several examples of classes of scheduling problems that lend themselves to this approach and that are either impossible to formulate, using conventional Operations Research methods or are clumsy to do so.

Chapter 48 discusses batch scheduling problems in the process industry (e.g., chemical, pharmaceutical, or metal casting industries), which consist of scheduling batches on processing units (e.g., reactors, heaters, dryers, filters, or agitators) such that a time-based objective function (e.g., makespan, maximum lateness, or weighted earliness plus tardiness) is minimized.

The classical vehicle routing problem is known to be NP-hard. Many different heuristics have been proposed in the past. Chapter 49 surveys most of these methods and proposes a new heuristic, called *Very Large Scale Neighborhood Search*, for the problem. Computational tests indicate that the proposed heuristic is competitive with the best local search methods.

Being in a time-sensitive and mission-critical business, the airline industry bumps from the left to the right into all sorts of scheduling problems. Chapter 50 discusses the challenges posed by aircraft scheduling, crew scheduling, manpower scheduling, and other long-term business planning and real-time operational problems that involve scheduling.

Chapter 51 discusses bus and train driver scheduling. Driver wages represent a big percentage, about 45 percent for the bus sector in the U.K., of the running costs of transport operations. Efficient scheduling of drivers is vital to the survival of transport operators. This chapter describes several approaches that have been successful in solving these problems.

Sports scheduling is interesting from both a practical and theoretical standpoint. Chapter 52 surveys the current body of sports scheduling literature covering a period of time from the early 1970s to the present day. While the emphasis is on *Single Round Robin Tournament Problem* and *Double Round Robin Tournament Problem*, the chapter also discusses *Balanced Tournament Design Problem* and *Bipartite Tournament Problem*.

1.3 Notation

In all of the scheduling problems considered in this book, the number of jobs (n) and machines (m) are assumed to be finite. Usually, the subscript j refers to a job and the subscript i refers to a machine. The following data are associated with job j:

Processing Time (p_{ij}) — If job j requires processing on machine i, then p_{ij} represents the processing time of job j on machine i. The subscript i is omitted if job j is only to be processed on one machine (any machine).

Release Date (r_j) — The release date r_j of job j is the time the job arrives at the system, which is the earliest time at which job j can start its processing.

Due Date (d_j) — The due date d_j of job j represents the date the job is expected to complete. Completion of a job after its due date is allowed, but it will incur a cost.

Deadline ($\bar{d}_j$) — The deadline $\bar{d}_j$ of job j represents the hard deadline that the job must respect; i.e., job j must be completed by $\bar{d}_j$.

Weight (w_j) — The weight w_j of job j reflects the importance of the job.

Graham et al. [12] introduced the $\alpha|\beta|\gamma$ notation to classify scheduling problems. The α field describes the machine environment and contains a single entry. The β field provides details of job characteristics and scheduling constraints. It may contain multiple entries or no entry at all. The γ field contains the objective function to optimize. It usually contains a single entry.

The possible machine environment in the α field are as follows:

Single Machine (1) — There is only one machine in the system. This case is a special case of all other more complicated machine environments.

Parallel and Identical Machines (*Pm*) — There are m identical machines in parallel. In the remainder of this section, if m is omitted, it means that the number of machines is arbitrary; i.e., the number of machines will be specified as a parameter in the input. Each job j requires a single operation and may be processed on any one of the m machines.

Uniform Machines (*Qm*) — There are m machines in parallel, but the machines have different speeds. Machine i, $1 \leq i \leq m$, has speed s_i. The time p_{ij} that job j spends on machine i is equal to p_j/s_i, assuming that job j is completely processed on machine i.

Unrelated Machines (*Rm*) — There are m machines in parallel, but each machine can process the jobs at a different speed. Machine i can process job j at speed s_{ij}. The time p_{ij} that job j spends on machine i is equal to p_j/s_{ij}, assuming that job j is completely processed on machine i.

Job Shop (*Jm*) — In a job shop with m machines, each job has its own predetermined route to follow. It may visit some machines more than once and it may not visit some machines at all.

Flow Shop (*Fm*) — In a flow shop with m machines, the machines are linearly ordered and the jobs all follow the same route (from the first machine to the last machine).

Open Shop (*Om*) — In an open shop with m machines, each job needs to be processed exactly once on each of the machines. But the order of processing is immaterial.

The job characteristics and scheduling constraints specified in the β field may contain multiple entries. The possible entries are β_1, β_2, β_3, β_4, β_5, β_6, β_7, β_8.

Preemptions (*pmtn*) — Jobs can be preempted and later resumed possibly on a different machine. If preemptions are allowed, $pmtn$ is included in the β field, otherwise, it is not included in the β field.

No-Wait (*nwt*) — The no-wait constraint is for flow shops only. Jobs are not allowed to wait between two successive machines. If nwt is not specified in the β field, waiting is allowed between two successive machines.

Precedence Constraints (*prec*) — The precedence constraints specify the scheduling constraints of the jobs, in the sense that certain jobs must be completed before certain other jobs can start processing. The most general form of precedence constraints, denoted by *prec*, is represented by a directed acyclic graph, where each vertex represents a job and job i precedes job j if there is a directed arc from i to j. If each job has at most one predecessor and at most one successor, the constraints are referred to as *chains*. If each job has at most one successor, the constraints are referred to as an *intree*. If each job has at most one predecessor, the constraints are referred to as an *outtree*. If *prec* is not specified in the β field, the jobs are not subject to precedence constraints.

Release Dates (r_j) — The release date r_j of job j is the earliest time at which job j can begin processing. If this symbol is not present, then the processing of job j may start at any time.

Restrictions on the Number of Jobs (*nbr*) — If this symbol is present, then the number of jobs is restricted; e.g., $nbr = 5$ means that there are at most five jobs to be processed. If this symbol is not present, then the number of jobs is unrestricted and is given as an input parameter n.

Restrictions on the Number of Operations in Jobs (n_j) — This subfield is only applicable to job shops. If this symbol is present, then the number of operations of each job is restricted; e.g., $n_j = 4$ means that each job is limited to at most four operations. If this symbol is not present, then the number of operations is unrestricted.

Restrictions on the Processing Times (p_j) — If this symbol is present, then the processing time of each job is restricted; e.g., $p_j = p$ means that each job's processing time is p units. If this symbol is not present, then the processing time is not restricted.

Deadlines ($\bar{d}_j$) — If this symbol is present, then each job j must be completed by its deadline $\bar{d}_j$. If the symbol is not present, then the jobs are not subject to deadline constraints.

The objective to be minimized is always a function of the completion times of the jobs. With respect to a schedule, let C_j denote the completion time of job j. The *lateness* of job j is defined as

$$L_j = C_j - d_j$$

The *tardiness* of job j is defined as

$$T_j = \max(L_j, 0)$$

The *unit penalty* of job j is defined as $U_j = 1$ if $C_j > d_j$; otherwise, $U_j = 0$.

The objective functions to be minimized are as follows:

Makespan ($C_{\max}$) — The makespan is defined as $\max(C_1, \ldots, C_n)$.

Maximum Lateness ($L_{\max}$) — The maximum lateness is defined as $\max(L_1, \ldots, L_n)$.

Total Weighted Completion Time ($\sum w_j C_j$) — The total (unweighted) completion time is denoted by $\sum C_j$.

Total Weighted Tardiness ($\sum w_j T_j$) — The total (unweighted) tardiness is denoted by $\sum T_j$.

Weighted Number of Tardy Jobs ($\sum w_j U_j$) — The total (unweighted) number of tardy jobs is denoted by $\sum U_j$.

For more information about the $\alpha|\beta|\gamma$ classification, the reader is referred to the website *www.mathematik.uni-osnabrueck.de/research/OR/class/*

Acknowledgment

This work is supported in part by the NSF Grant DMI-0300156.

References

[1] J. R. Jackson, Scheduling a production line to minimize maximum tardiness, Research Report 43, Management Science Research Project, University of California, Los Angeles, 1955.

[2] J. R. Jackson, An extension of Johnson's results on job lot scheduling, *Naval Research Logistics Quarterly*, 3, 201–203, 1956.

[3] S. M. Johnson, Optimal two and three-stage production schedules with setup times included, *Naval Research Logistics Quarterly*, 1, 61–67, 1954.

[4] W. E. Smith, Various optimizers for single stage production, *Naval Research Logistics Quarterly*, 3, 59–66, 1956.

[5] S. A. Cook, The complexity of theorem-proving procedures, in *Procedings of the 3rd Annual ACM Symposium on Theory of Computing*, Association for Computing Machinery, New York, 1971, pp. 151–158.

[6] M. R. Garey and D. S. Johnson, *Computers and Intractability: A Guide to the Theory of NP-Completeness*, W. H. Freeman, New York, 1979.

[7] R. M. Karp, Reducibility among combinatorial problems, in R. E. Miller and J. W. Thatcher (eds), *Complexity of Computer Computations*, Plenum Press, New York, 1972, pp. 85–103.

[8] P. Brucker, *Scheduling Algorithms*, 3rd ed., Springer-Verlag, New York, 2001.

[9] J. K. Lenstra and A. H. G. Rinnooy Kan, Computational complexity of scheduling under precedence constraints, *Operations Research*, 26, 22–35, 1978.

[10] J. K. Lenstra, A. H. G. Rinnooy Kan, and P. Brucker, Complexity of machine scheduling problems, *Annals of Discrete Mathematics*, 1, 343–362, 1977.
[11] M. Pinedo, *Scheduling: Theory, Algorithms, and Systems*, 2nd ed., Prentice Hall, New Jersey, 2002.
[12] R. L. Graham, E. L. Lawler, J. K. Lenstra, and A. H. G. Rinnooy Kan, Optimization and approximation in deterministic sequencing and scheduling: A survey, *Annals of Discrete Mathematics*, 5, 287–326, 1979.

2

A Tutorial on Complexity

Joseph Y-T. Leung
New Jersey Institute of Technology

2.1 Introduction

Complexity theory is an important tool in scheduling research. When we are confronted with a new scheduling problem, the very first thing we try is to develop efficient algorithms for solving the problem. Unfortunately, very often, we could not come up with any algorithm more efficient than essentially an enumerative search, even though a considerable amount of time had been spent on the problem. In situations like this, the theory of NP-hardness may be useful to pinpoint that no efficient algorithms could possibly exist for the problem in hand. Therefore, knowledge of NP-hardness is absolutely essential for anyone interested in scheduling research.

In this chapter, we shall give a tutorial on the theory of NP-hardness. No knowledge of this subject is assumed on the reader. We begin with a discussion of time complexity of an algorithm in Section 2.2. We then give the notion of polynomial reduction in Section 2.3. Section 2.4 gives the formal definition of NP-completeness and NP-hardness. Pseudo-polynomial algorithms and strong NP-hardness will be presented in Section 2.5. Finally, we discuss polynomial-time approximation schemes (PTAS) and fully polynomial-time approximation schemes (FPTAS) and their relations with strong NP-hardness in Section 2.6.

The reader is referred to the excellent book by Garey and Johnson [1] for an outstanding treatment of this subject. A comprehensive list of NP-hard scheduling problems can be found on the website www.mathematik.uni-osnabrueck.de/research/OR/class/.

2.2 Time Complexity of Algorithms

The running time of an algorithm is measured by the number of basic steps it takes. Computers can only perform a simple operation in one step, such as adding two numbers, deciding if one number is larger than or equal to another, moving a fixed amount of information from one memory cell to another, or reading a

1-58488-397-9/$0.00+$1.50

fixed amount of information from external media into memory. Computers cannot, in one step, add two vectors of numbers, where the dimension of the vectors is unbounded. To add two vectors of numbers with dimension n, we need n basic steps to accomplish this.

We measure the running time of an algorithm as a function of the size of the input. This is reasonable since we expect the algorithm to take longer time when the input size grows larger. Let us illustrate the process of analyzing the running time of an algorithm by means of a simple example. Shown below is an algorithm that implements bubble sort. Step 1 reads n, the number of numbers to be sorted, and Step 2 reads the n numbers into the array A. Step 3 to 5 sort the numbers in ascending order. Finally, Step 6 prints the numbers in sorted order.

2.2.1 Bubble Sort

```
1. Read n;
2. For i = 1 to n do { Read A(i); }
3. For i = 1 to n − 1 do
   {
4.    For j = 1 to n − i do
      {
5.       If A(j) > A(j + 1) then {temp ← A(j); A(j) ← A(j + 1); A(j + 1) ← temp;}
      }
   }
6. For i = 1 to n do { Print A(i); }
```

Step 1 takes c_1 basic steps, where c_1 is a constant that is dependent on the machine, but independent of the input size. Step 2 takes $c_2 n$ basic steps, where c_2 is a constant dependent on the machine only. Step 5 takes c_3 basic steps each time it is executed, where c_3 is a constant dependent on the machine only. However, Step 5 is nested inside a double loop given by Steps 3 and 4. We can calculate the number of times Step 5 is executed as follows. The outer loop in Step 3 is executed $n-1$ times. In the ith iteration of Step 3, Step 4 is executed exactly $n-i$ times. Thus, the number of times Step 5 is executed is

$$\sum_{i=1}^{n-1}(n-i) = \frac{n(n-1)}{2}$$

Therefore, Step 5 takes a total of $c_3 n(n-1)/2$ basic steps. Finally, Step 6 takes $c_4 n$ basic steps, where c_4 is a constant dependent on the machine only. Adding them together, the running time of the algorithm, $T(n)$, is

$$T(n) = c_1 + c_2 n + \frac{c_3 n(n-1)}{2} + c_4 n$$

In practice, it is not necessary, or desirable, to get such a detailed function for $T(n)$. Very often, we are only interested in the growth rate of $T(n)$. We can see from the above function that $T(n)$ is dominated by the n^2 term. Thus, we say that $T(n)$ is $O(n^2)$, ignoring the constants and any terms that grow slower than n^2. Formally, we say that a function $f(n)$ is $O(g(n))$ if there are constants c and n' such that $f(n) \leq cg(n)$ for all $n \geq n'$.

In the remainder of this chapter, we will be talking about the running time of an algorithm in terms of its growth rate $O(\cdot)$ only. Suppose an algorithm $\mathcal{A}$ has running time $T(n) = O(g(n))$. We say that $\mathcal{A}$ is a polynomial-time algorithm if $g(n)$ is a polynomial function of n; otherwise, it is an exponential-time algorithm. For example, if $T(n) = O(n^{100})$, then $\mathcal{A}$ is a polynomial-time algorithm. On the other hand, if $T(n) = O(2^n)$, then $\mathcal{A}$ is an exponential-time algorithm.

Since exponential functions grow much faster than polynomial functions, it is clearly more desirable to have polynomial-time algorithms than exponential-time algorithms. Indeed, exponential-time algorithms are not practical, except for small-size problems. To see this, consider an algorithm $\mathcal{A}$ with running time

$T(n) = O(2^n)$. The fastest computer known today executes one trillion (10^{12}) instructions per second. If $n = 100$, the algorithm will take more than 30 billion years using the fastest computer! This is clearly infeasible since nobody lives long enough to see the algorithm terminates.

We say that a problem is *tractable* if there is a polynomial-time algorithm for it; otherwise, it is *intractable*. The theory of NP-hardness suggests that there is a large class of problems, namely, the NP-hard problems, that *may* be intractable. We emphasize the words "may be" since it is still an open question whether the NP-hard problems can be solved in polynomial time. However, there are circumstantial evidence suggesting that they are intractable. Notice that we are only making a distinction between polynomial time and exponential time. This is reasonable since exponential functions grow much faster than polynomial functions, regardless of the degree of the polynomial.

Before we leave this section, we should revisit the issue of "the size of the input." How do we define "the size of the input"? The official definition is the number of "symbols" (drawn from a fixed set of symbols) necessary to represent the input. This definition still leaves a lot of room for disagreement. Let us illustrate this by means of the bubble sort algorithm given above. Most people would agree that n, the number of numbers to be sorted, should be part of the size of the input. But what about the numbers themselves? If we assume that each number can fit into a computer word (which has a fixed size), then the number of symbols necessary to represent each number is bounded above by a constant. Under this assumption, we can say that the size of the input is $O(n)$. If this assumption is not valid, then we have to take into account the representation of the numbers. Suppose a is the magnitude of the largest number out of the n numbers. If we represent each number as a binary number (base 2), then we can say that the size of the input is $O(n \log a)$. On the other hand, if we represent each number as a unary number (base 1), then the size of the input becomes $O(na)$. Thus, the size of the input can differ greatly, depending on the assumptions you make. Since the running time of an algorithm is a function of the size of the input, they differ greatly as well. In particular, a polynomial-time algorithm with respect to one measure of the size of the input may become an exponential-time algorithm with respect to another. For example, a polynomial-time algorithm with respect to $O(na)$ may in fact be an exponential-time algorithm with respect to $O(n \log a)$.

In our analysis of the running time of bubble sort, we have implicitly assumed that each integer fits into a computer word. If this assumption is not valid, the running time of the algorithm should be $T(n) = O(n^2 \log a)$.

For scheduling problems, we usually assume that the number of jobs, n, and the number of machines, m, should be part of the size of the input. Precedence constraint poses no problem, since there are at most $O(n^2)$ precedence relations for n jobs. What about processing times, due dates, weights, etc.? They can be represented by binary numbers or unary numbers, and the two representations can affect the complexity of the problem. As we shall see later in the chapter, there are scheduling problems that are NP-hard with respect to binary encodings but not unary encodings. We say that these problems are NP-hard in the *ordinary* sense. On the other hand, there are scheduling problems that are NP-hard with respect to unary encodings. We say that these problems are NP-hard in the *strong* sense.

The above is just a rule of thumb. There are always exceptions to this rule. For example, consider the problem of scheduling a set of chains of unit-length jobs to minimize $C_{\max}$. Suppose there are k chains, $C_1, C_2, \ldots, C_k$, with n_j jobs in chain C_j, $1 \le j \le k$. Clearly, the number of jobs, n, is $n = \sum n_j$. According to the above, the size of the input should be at least proportional to n. However, some authors insist that each n_j should be encoded in binary and hence the size of the input should be proportional to $\sum \log n_j$. Consequently, a polynomial-time algorithm with respect to n becomes an exponential-time algorithm with respect to $\sum \log n_j$. Thus, when we study the complexity of a problem, we should bear in mind the encoding scheme we use for the problem.

2.3 Polynomial Reduction

Central to the theory of NP-hardness is the notion of polynomial reduction. Before we get to this topic, we want to differentiate between decision problems and optimization problems. Consider the following three problems.

2.3.1 Partition

Given a list $A = (a_1, a_2, \ldots, a_n)$ of n integers, can A be partitioned into A_1 and A_2 such that $\sum_{a_j \in A_1} a_j = \sum_{a_j \in A_2} a_j = \frac{1}{2} \sum a_j$?

2.3.2 Traveling Salesman Optimization

Given n cities, $c_1, c_2, \ldots, c_n$, and a distance function $d(i, j)$ for every pair of cities c_i and $c_j (d(i, j) = d(j, i))$, find a tour of the n cities so that the total distance of the tour is minimum. That is, find a permutation $\sigma = (i_1, i_2, \ldots, i_n)$ such that $\sum_{j=1}^{n-1} d(i_j, i_{j+1}) + d(i_n, i_1)$ is minimum.

2.3.3 0/1-Knapsack Optimization

Given a set U of n items, $U = \{u_1, u_2, \ldots, u_n\}$, with each item u_j having a size s_j and a value v_j, and a knapsack with size K, find a subset $U' \subseteq U$ such that all the items in U' can be packed into the knapsack and such that the total value of the items in U' is maximum.

The first problem, Partition, is a decision problem. It has only "Yes" or "No" answer. The second problem, Traveling Salesman Optimization, is a minimization problem. It seeks a tour such that the total distance of the tour is minimum. The third problem, 0/1-Knapsack Optimization, is a maximization problem. It seeks a packing of a subset of the items such that the total value of the items packed is maximum.

All optimization (minimization or maximization) problems can be converted into a corresponding decision problem by providing an additional parameter ω, and simply asking whether there is a feasible solution such that the cost of the solution is $\leq$ (or $\geq$ in case of a maximization problem) ω. For example, the above optimization problems can be converted into the following decision problems.

2.3.4 Traveling Salesman Decision

Given n cities, $c_1, c_2, \ldots, c_n$, a distance function $d(i, j)$ for every pair of cities c_i and $c_j (d(i, j) = d(j, i))$, and a bound B, is there a tour of the n cities so that the total distance of the tour is less than or equal to B? That is, is there a permutation $\sigma = (i_1, i_2, \ldots, i_n)$ such that $\sum_{j=1}^{n-1} d(i_j, i_{j+1}) + d(i_n, i_1) \leq B$?

2.3.5 0/1-Knapsack Decision

Given a set U of n items, $U = \{u_1, u_2, \ldots, u_n\}$, with each item u_j having a size s_j and a value v_j, a knapsack with size K, and a bound B, is there a subset $U' \subseteq U$ such that $\sum_{u_j \in U'} s_j \leq K$ and $\sum_{u_j \in U'} v_j \geq B$?

It turns out that the theory of NP-hardness applies to decision problems only. Since almost all of the scheduling problems are optimization problems, it seems that the theory of NP-hardness is of little use in scheduling theory. Fortunately, as far as polynomial-time hierarchy is concerned, the complexity of an optimization problem is closely related to the complexity of its corresponding decision problem. That is, an optimization problem is solvable in polynomial time if and only if its corresponding decision problem is solvable in polynomial time. To see this, let us first assume that an optimization problem can be solved in polynomial time. We can solve its corresponding decision problem by simply finding an optimal solution and comparing its objective value against the given bound. Conversely, if we can solve the decision problem, we can solve the optimization problem by conducting a binary search in the interval bounded by a lower bound (LB) and an upper bound (UB) of its optimal value. For most scheduling problems, the objective functions are integer-valued, and LB and UB have values at most a polynomial function of its input parameters. Let the length of the interval between LB and UB be l. In $O(\log l)$ iterations, the binary search will converge to the optimal value. Thus, if the decision problem can be solved in polynomial time, then the algorithm of finding an optimal value also runs in polynomial time, since $\log l$ is bounded above by a polynomial function of the size of its input.

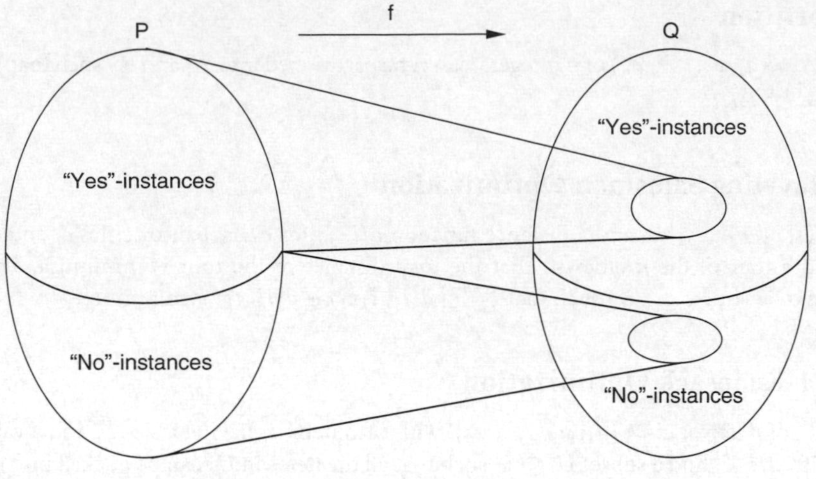

FIGURE 2.1 Illustrating polynomial reducibility.

Because of the relationship between the complexity of an optimization problem and its corresponding decision problem, from now on we shall concentrate only on the complexity of decision problems. Recall that decision problems have only "Yes" or "No" answer. We say that an instance I is a "Yes"-instance if I has a "Yes" answer; otherwise, I is a "No"-instance.

Central to the theory of NP-hardness is the notion of polynomial reducibility. Let P and Q be two decision problems. We say that P is *polynomially reducible* (or simply *reducible*) to Q, denoted by $P \propto Q$, if there is a function f that maps *every* instance I_P of P into an instance I_Q of Q such that I_P is a Yes-instance if and only if I_Q is a Yes-instance. Further, f can be computed in polynomial time.

Figure 2.1 depicts the function f. Notice that f does not have to be one-to-one or onto. Also, f maps an instance I_P of P without knowing whether I_P is a Yes-instance or No-instance. That is, the status of I_P is unknown to f. All that is required is that Yes-instances are mapped to Yes-instances and No-instances are mapped to No-instances. From the definition, it is clear that $P \propto Q$ does not imply that $Q \propto P$. Further, reducibility is transitive, i.e., if $P \propto Q$ and $Q \propto R$, then $P \propto R$.

Theorem 2.1

Suppose we have two decision problems P and Q such that $P \propto Q$. If Q is solvable in polynomial time, then P is also solvable in polynomial time. Equivalently, if P cannot be solved in polynomial time, then Q cannot be solved in polynomial time.

Proof

Since $P \propto Q$, we can solve P indirectly through Q. Given an instance I_P of P, we use the function f to map it into an instance I_Q of Q. This mapping takes polynomial time, by definition. Since Q can be solved in polynomial time, we can decide whether I_Q is a Yes-instance. But I_Q is a Yes-instance if and only if I_P is a Yes-instance. So we can decide if I_P is a Yes-instance in polynomial time. □

We shall show several reductions in the remainder of this section. As we shall see later, sometimes we can reduce a problem P from one domain to another problem Q in a totally different domain. For example, a problem in logic may be reducible to a graph problem.

Theorem 2.2

The Partition problem is reducible to the 0/1-Knapsack Decision problem.

Proof

Let $A = (a_1, a_2, \ldots, a_n)$ be a given instance of Partition. We create an instance of 0/1-Knapsack Decision as follows. Let there be n items, $U = \{u_1, u_2, \ldots, u_n\}$, with u_j having a size $s_j = a_j$ and a value $v_j = a_j$. In essence, each item u_j corresponds to the integer a_j in the instance of Partition. The knapsack size K and the bound B are chosen to be $K = B = \frac{1}{2}\sum a_j$. It is clear that the mapping can be done in polynomial time.

It remains to be shown that the given instance of Partition is a Yes-instance if and only if the constructed instance of 0/1-Knapsack is a Yes-instance. Suppose $I' \subseteq \{1, 2, \ldots, n\}$ is an index set such that $\sum_{i \in I'} a_i = \frac{1}{2}\sum a_j$. Then $U' = \{u_i \mid i \in I'\}$ forms a solution for the instance of 0/1-Knapsack Decision, since $\sum_{i \in I'} s_i = \frac{1}{2}\sum a_j = K = B = \sum_{i \in I'} v_i$. Conversely, if there is an index set $I' \subseteq \{1, 2, \ldots, n\}$ such that $U' = \{u_i \mid i \in I'\}$ forms a solution for the instance of 0/1-Knapsack Decision, then $A' = \{a_i \mid i \in I'\}$ forms a solution for Partition. This is because $\sum_{i \in I'} s_i = \sum_{i \in I'} a_i \leq K = \frac{1}{2}\sum a_j$ and $\sum_{i \in I'} v_i = \sum_{i \in I'} a_i \geq B = \frac{1}{2}\sum a_j$. For both inequalities to hold, we must have $\sum_{i \in I'} a_i = \frac{1}{2}\sum a_j$. □

In the above proof, we have shown how to obtain a solution for Partition from a solution for 0/1-Knapsack Decision. This is the characteristic of reduction. In essence, we can solve Partition indirectly via 0/1-Knapsack Decision.

Theorem 2.3

The Partition problem is reducible to the decision version of $P2 \,||\, C_{\max}$.

Proof

Let $A = (a_1, a_2, \ldots, a_n)$ be a given instance of Partition. We create an instance of the decision version of $P2 \,||\, C_{\max}$ as follows. Let there be n jobs, with job i having processing time a_i. In essence, each job corresponds to an integer in A. Let the bound B be $\frac{1}{2}\sum a_j$. Clearly, the mapping can be done in polynomial time. It is easy to see that there is a partition of A if and only if there is a schedule with makespan no larger than B. □

In the above reduction, we create a job with processing time equal to an integer in the instance of the Partition problem. The given integers can be partitioned into two equal groups if and only if the jobs can be scheduled on two parallel and identical machines with makespan equal to one half of the total processing time.

Theorem 2.4

The 0/1-Knapsack Decision problem is reducible to the decision version of $1 \mid d_j = d \mid \sum w_j U_j$.

Proof

Let $U = \{u_1, u_2, \ldots, u_n\}$, K, and B be a given instance of 0/1-Knapsack Decision, where u_j has a size s_j and a value v_j. We create an instance of the decision version of $1 \mid d_j = d \mid \sum w_j U_j$ as follows. For each item u_j in U, we create a job j with processing time $p_j = s_j$ and weight $w_j = v_j$. The jobs have a common due date $d = K$. The threshold ω for the decision version of $1 \mid d_j = d \mid \sum w_j U_j$ is $\omega = \sum v_j - B$.

Suppose the given instance of 0/1-Knapsack Decision is a Yes-instance. Let $I' \subseteq \{1, 2, \ldots, n\}$ be the index set such that $\sum_{j \in I'} s_j \leq K$ and $\sum_{j \in I'} v_j \geq B$. Then I' is a subset of jobs that can be scheduled on time, since $\sum_{j \in I'} p_j \leq K = d$. The total weight of all the jobs in I' is $\sum_{j \in I'} w_j = \sum_{j \in I'} v_j \geq B$. Hence the total weight of all the tardy jobs is less than or equal to $\sum v_j - B$. Thus, the constructed instance of the decision version of $1 \mid d_j = d \mid \sum w_j U_j$ is a Yes-instance.

Conversely, if the constructed instance of the decision version of $1 \mid d_j = d \mid \sum w_j U_j$ is a Yes-instance, then there is a subset $I' \subseteq \{1, 2, \ldots, n\}$ of on-time jobs with total weight greater than or equal to $\sum v_j - \omega \geq \sum v_j - \sum v_j + B = B$. The set $U' = \{u_i \mid i \in I'\}$ forms a solution to the instance of the 0/1-Knapsack Decision problem. □

In the above reduction, we create, for each item u_j in the 0/1-Knapsack Decision, a job with a processing time equal to the size of u_j and a weight equal to the value of u_j. We make the knapsack size K to be the common due date of all the jobs. The idea is that if an item is packed into the knapsack, then the corresponding job is an on-time job; otherwise, it is a tardy job. Thus, there is a packing into the knapsack with value greater than or equal to B if and only if there is a schedule with the total weight of all the tardy jobs less than or equal to $\sum v_j - B$.

Before we proceed further, we need to define several decision problems.

Hamiltonian Circuit. Given an undirected graph $G = (V, E)$, is there a circuit that goes through each vertex in G exactly once?

3-Dimensional Matching. Let $A = \{a_1, a_2, \ldots, a_q\}$, $B = \{b_1, b_2, \ldots, b_q\}$, and $C = \{c_1, c_2, \ldots, c_q\}$ be three disjoint sets of q elements each. Let $T = \{t_1, t_2, \ldots, t_l\}$ be a set of triples such that each t_j consists of one element from A, one element from B, and one element from C. Is there a subset $T' \subseteq T$ such that every element in A, B, and C appears in exactly one triple in T'?

Deadline Scheduling. Given one machine and a set of n jobs, with each job j having a processing time p_j, a release time r_j, and a deadline $\bar{d}_j$, is there a nonpreemptive schedule of the n jobs such that each job is executed within its executable interval $[r_j, \bar{d}_j]$?

Theorem 2.5

The Hamiltonian Circuit problem is reducible to the Traveling Salesman Decision problem.

Proof

Let $G = (V, E)$ be an instance of Hamiltonian Circuit, where V consists of n vertexes. We construct an instance of Traveling Salesman Decision as follows. For each vertex v_i in V, we create a city c_i. The distance function $d(i, j)$ is defined as follows: $d(i, j) = 1$ if $(v_i, v_j) \in E$; otherwise, $d(i, j) = 2$.
We choose B to be n.

Suppose $(v_{i_1}, v_{i_2}, \ldots, v_{i_n})$ is a Hamiltonian Circuit. Then $(c_{i_1}, c_{i_2}, \ldots, c_{i_n})$ is a tour of the n cities with total distance equal to n; i.e., $\sum_{j=1}^{n-1} d(i_j, i_{j+1}) + d(i_n, i_1) = n = B$. Thus, the constructed instance of Traveling Salesman Decision is a Yes-instance.

Conversely, suppose $(c_{i_1}, c_{i_2}, \ldots, c_{i_n})$ is a tour of the n cities with total distance less than or equal to B. Then the distance between any pair of adjacent cities is exactly 1, since the total distance is the sum of n distances, and the smallest value of the distance function is 1. By the definition of the distance function, if $d(i_j, i_{j+1}) = 1$, then $(v_{i_j}, v_{i_{j+1}}) \in E$. Thus, $(v_{i_1}, v_{i_2}, \ldots, v_{i_n})$ is a Hamiltonian Circuit. □

The idea in the above reduction is to create a city for each vertex in G. We define the distance function in such a way that the distance between two cities is smaller if their corresponding vertexes are adjacent in G than if they are not. In our reduction we use the values 1 and 2, respectively, but other values will work too, as long as they satisfy the above condition. We choose the distance bound B in such a way that there is a tour with total distance less than or equal to B if and only if there is a Hamiltonian Circuit in G. For our choice of distance values of 1 and 2, the choice of B equal to n (which is n times the smaller value of the distance function) will work.

Theorem 2.6

The Partition problem is reducible to the Deadline Scheduling problem.

Proof

Let $A = (a_1, a_2, \ldots, a_n)$ be an instance of Partition. We create an instance of Deadline Scheduling as follows. There will be $n+1$ jobs. The first n jobs are called "Partition" jobs and the last job is called the "Divider" job. For each job j, $1 \leq j \leq n$, $p_j = a_j$, $r_j = 0$, and $\bar{d}_j = \sum a_j + 1$. For job $n+1$, $p_{n+1} = 1$, $r_{n+1} = \frac{1}{2}\sum a_j$, and $\bar{d}_{n+1} = r_{n+1} + 1$.

Suppose the given instance of Partition has a solution. Let $I' \subseteq \{1, 2, \ldots, n\}$ be an index set such that $\sum_{i \in I'} a_i = \frac{1}{2}\sum a_j$. We schedule the jobs in the index set from time 0 until time $\frac{1}{2}\sum a_j$ in any order. We then schedule job $n+1$ from time $\frac{1}{2}\sum a_j$ until time $\frac{1}{2}\sum a_j + 1$. Finally, we schedule the remaining Partition jobs from time $\frac{1}{2}\sum a_j + 1$ onward, in any order. It is easy to see that all jobs can meet their deadline.

Conversely, if there is a schedule such that every job is executed in its executable interval, then the Divider job must be scheduled in the time interval $[\frac{1}{2}\sum a_j, \frac{1}{2}\sum a_j + 1]$. The timeline is now divided into two disjoint intervals – $[0, \frac{1}{2}\sum a_j]$ and $[\frac{1}{2}\sum a_j + 1, \sum a_j + 1]$ – into which the Partition jobs are scheduled. Clearly, there must be a partition of A. □

The idea behind the above reduction is to create a Divider job with a very tight executable interval ($[\frac{1}{2}\sum a_j, \frac{1}{2}\sum a_j + 1]$). Because of the tightness of the interval, the Divider job must be scheduled entirely in its executable interval. This means that the timeline is divided into two disjoint intervals, each of which has length exactly $\frac{1}{2}\sum a_j$. Since the Partition jobs are scheduled in these two intervals, there is a feasible schedule if and only if there is a partition.

Theorem 2.7

The 3-Dimensional Matching problem is reducible to the decision version of $R \,||\, C_{\max}$.

Proof

Let $A = \{a_1, a_2, \ldots, a_q\}$, $B = \{b_1, b_2, \ldots, b_q\}$, $C = \{c_1, c_2, \ldots, c_q\}$, and $T = \{t_1, t_2, \ldots, t_l\}$ be a given instance of 3-Dimensional Matching. We construct an instance of the decision version of $R \,||\, C_{\max}$ as follows. Let there be l machines and $3q + (l - q)$ jobs. For each $1 \leq j \leq l$, machine j corresponds to the triple t_j. The first $3q$ jobs correspond to the elements in A, B, and C. For each $1 \leq i \leq q$, job i (resp. $q+i$ and $2q+i$) corresponds to the element a_i (resp. b_i and c_i). The last $l - q$ jobs are dummy jobs. For each $3q+1 \leq i \leq 3q+(l-q)$, the processing time of job i on *any* machine is 3 units. In other words, the dummy jobs have processing time 3 units on any machine. For each $1 \leq i \leq 3q$, job i has processing time 1 unit on machine j if the element corresponding to job i is in the triple t_j; otherwise, it has processing time 2 units. The threshold ω for the decision version of $R \,||\, C_{\max}$ is $\omega = 3$.

Suppose $T' = \{t_{i_1}, t_{i_2}, \ldots, t_{i_q}\}$ is a matching. Then we can schedule the first $3q$ jobs on machines $i_1, i_2, \ldots, i_q$. In particular, the three jobs that correspond to the three elements in t_{i_j} will be scheduled on machine i_j. The finishing time of each of these q machines is 3. The dummy jobs will be scheduled on the remaining machines, one job per machine. Again, the finishing time of each of these machines is 3. Thus, there is a schedule with $C_{\max} = 3 = \omega$.

Conversely, if there is a schedule with $C_{\max} \leq \omega$, then the makespan of the schedule must be exactly 3 (since the dummy jobs have processing time 3 units on any machine). Each of the dummy jobs must be scheduled one job per machine; otherwise, the makespan will be larger than ω. This leaves q machines to schedule the first $3q$ jobs. These q machines must also finish at time 3, which implies that each job scheduled on these machines must have processing time 1 unit. But this means that the triples corresponding to these q machines must be a matching, by the definition of the processing time of the first $3q$ jobs. □

The idea in the above reduction is to create a machine for each triple. We add $l - q$ dummy jobs to *jam* up $l - q$ machines. This leaves q machines to schedule other jobs. We then create one job for each element in A, B, and C. Each of these jobs has a smaller processing time (1 unit) if it were scheduled on the machine corresponding to the triple that contains the element to which the job corresponds; otherwise, it will have a larger processing time (2 units). Therefore, there is a schedule with $C_{\max} = 3$ if and only if there is a matching.

Using the same idea, we can prove the following theorem.

Theorem 2.8

The 3-Dimensional Matching problem is reducible to the decision version of $R \mid \bar{d}_j = d \mid \sum C_j$.

Proof

Let $A = \{a_1, a_2, \ldots, a_q\}$, $B = \{b_1, b_2, \ldots, b_q\}$, $C = \{c_1, c_2, \ldots, c_q\}$, and $T = \{t_1, t_2, \ldots, t_l\}$ be a given instance of 3-Dimensional Matching. We construct an instance of the decision version of $R \mid \bar{d}_j = d \mid \sum C_j$ as follows. Let there be l machines and $3q + (l - q)$ jobs. For each $1 \leq j \leq l$, machine j corresponds to the triple t_j. The first $3q$ jobs correspond to the elements in A, B, and C. For each $1 \leq i \leq q$, job i (resp. $q + i$ and $2q + i$) corresponds to the element a_i (resp. b_i and c_i). The last $l - q$ jobs are dummy jobs. For each $3q + 1 \leq i \leq 3q + (l - q)$, the processing time of job i on *any* machine is 6 units. For each $1 \leq i \leq 3q$, job i has processing time 1 unit on machine j if the element corresponding to job i is in the triple t_j; otherwise, it has processing time 2 units. All jobs have deadline $d = 6$. The threshold ω for the decision version of $R \mid \bar{d}_j = d \mid \sum C_j$ is $\omega = 6l$.

Notice that there is always a schedule that meets the deadline of every job, regardless of whether there is a matching or not. It is easy to see that there is a schedule with $\sum C_j = \omega$ if and only if there is a matching. □

2.4 NP-Completeness and NP-Hardness

To define NP-completeness, we need to define the NP-class first. NP refers to the class of decision problems which have "succinct" certificates that can be verified in polynomial time. By "succinct" certificates, we mean certificates whose size is bounded by a polynomial function of the size of the input. Let us look at some examples. Consider the Partition problem. While it takes an enormous amount of time to decide if A can be partitioned into two equal groups, it is relatively easy to check if a given partition will do the job. That is, if a partition A_1 (a certificate) is presented to you, you can quickly check if $\sum_{a_i \in A_1} = \frac{1}{2} \sum a_j$. Furthermore, if A is a Yes-instance, then there must be a succinct certificate showing that A is a Yes-instance, and that certificate can be checked in polynomial time. Notice that if A is a No-instance, there is no succinct certificate showing that A is a No-instance, and no certificate can be checked in polynomial time. So there is a certain asymmetry in the NP-class between Yes-instances and No-instances. By definition, Partition is in the NP-class.

Consider the Traveling Salesman Decision problem. If a tour is presented to you, it is relatively easy to check if the tour has total distance less than or equal to the given bound. Furthermore, if the given instance of the Traveling Salesman Decision problem is a Yes-instance, there would be a tour showing that the instance is a Yes-instance. Thus, Traveling Salesman Decision is in the NP-class. Similarly, it is easy to see that 0/1-Knapsack Decision, Hamiltonian Circuit, 3-Dimensional Matching, Deadline Scheduling, as well as the decision versions of most of the scheduling problems are in the NP-class.

A decision problem P is said to be *NP-complete* if (1) P is in the NP-class and (2) *all* problems in the NP-class are reducible to P. A problem Q is said to be *NP-hard* if it satisfies (2) only; i.e., all problems in the NP-class are reducible to Q.

Suppose P and Q are both NP-complete problems. Then $P \propto Q$ and $Q \propto P$. This is because since Q is in the NP-class and since P is NP-complete, we have $Q \propto P$ (all problems in the NP-class are reducible to P). Similarly, since P is in the NP-class and since Q is NP-complete, we have $P \propto Q$. Thus, any two NP-complete problems are reducible to each other. By our comments earlier, either all NP-complete problems are solvable in polynomial time, or none of them. Today, thousands of problems have been shown to be NP-complete, and none of them have been shown to be solvable in polynomial time. It is widely conjectured that NP-complete problems cannot be solved in polynomial time, although no proof has been given yet.

To show a problem to be NP-complete, one needs to show that *all* problems in the NP-class are reducible to it. Since there are infinite number of problems in the NP-class, it is not clear how one can prove any problem to be NP-complete. Fortunately, Cook [2] in 1971 gave a proof that the Satisfiability problem (see the definition below) is NP-complete, by giving a generic reduction from Turing machines to Satisfiability. From the Satisfiability problem, we can show other problems to be NP-complete by reducing it to the target problems. Because reducibility is transitive, this is tantamount to showing that all problems in the NP-class are reducible to the target problems. Starting from Satisfiability, Karp [3] in 1972 showed a large number of combinatorial problems to be NP-complete.

Satisfiability. Given n Boolean variables, $U = \{u_1, u_2, \ldots, u_n\}$, and a set of m clauses, $C = \{c_1, c_2, \ldots, c_m\}$, where each clause c_j is a disjunction (or) of some elements in U or its complement (negation), is there an assignment of truth values to the Boolean variables so that every clause is simultaneously true?

Garey and Johnson [1] gave six basic NP-complete problems that are quite often used to show other problems to be NP-complete. Besides Hamiltonian Circuit, Partition, and 3-Dimensional Matching, the list includes 3-Satisfiability, Vertex Cover, and Clique (their definitions are given below). They [1] also gave a list of several hundred NP-complete problems, which are very valuable in proving other problems to be NP-complete.

3-Satisfiability. Same as Satisfiability, except that each clause is restricted to have exactly three literals (i.e., three elements from U or their complements).

Vertex Cover. Given an undirected graph $G = (V, E)$ and an integer $J \leq |V|$, is there a vertex cover of size less than or equal to J? That is, is there a subset $V' \subseteq V$ such that $|V'| \leq J$ and such that every edge has at least one vertex in V'?

Clique. Given an undirected graph $G = (V, E)$ and an integer $K \leq |V|$, is there a clique of size K or more? That is, is there a subset $V' \subseteq V$ such that $|V'| \geq K$ and such that the subgraph induced by V' is a complete graph?

Before we leave this section, we note that if the decision version of an optimization problem P is NP-complete, then we say that P is NP-hard. The reason why P is only NP-hard (but not NP-complete) is rather technical, and its explanation is beyond the scope of this chapter.

2.5 Pseudo-Polynomial Algorithms and Strong NP-Hardness

We begin this section by giving a dynamic programming algorithm for the 0/1-Knapsack Optimization problem. Let K and $U = \{u_1, u_2, \ldots, u_n\}$ be a given instance of the 0/1-Knapsack Optimization problem, where each u_j has a size s_j and a value v_j. Our goal is to maximize the total value of the items that can be packed into the knapsack whose size is K.

We can solve the above problem by constructing a table $R(i, j)$, $1 \leq i \leq n$ and $0 \leq j \leq V$, where $V = \sum v_j$. Stored in $R(i, j)$ is the smallest total size of a subset $U' \subseteq \{u_1, u_2, \ldots, u_i\}$ of items such that the total value of the items in U' is exactly j. If it is impossible to find a subset U' with total value exactly j, then $R(i, j)$ is set to be ∞.

The table can be computed row by row, from the first row until the nth row. The first row can be computed easily. $R(1, 0) = 0$ and $R(1, v_1) = s_1$; all other entries in the first row are set to ∞. Suppose we have computed the first $i - 1$ rows. We compute the ith row as follows:

$$R(i, j) = \min\{R(i-1, j), R(i-1, j-v_j) + s_j\}$$

The time needed to compute the entire table is $O(nV)$, since each entry can be computed in constant time. The maximum total value of the items that can be packed into the knapsack is k, where k is the largest integer such that $R(n, k) \leq K$. We can obtain the set U' by storing a pointer in each table entry, which shows from where the current entry is obtained. For example, if $R(i, j) = R(i-1, j)$, then we store a pointer in $R(i, j)$ pointing at $R(i-1, j)$ (which means that the item u_i is not in U'). On the other hand, if $R(i, j) = R(i-1, j-v_j) + s_j$, then we store a pointer in $R(i, j)$ pointing at $R(i-1, j-v_j)$ (which means that the item u_i is in U').

We have just shown that the 0/1-Knapsack Optimization problem can be solved in $O(nV)$ time, where $V = \sum v_j$. If v_j's are represented by unary numbers, then $O(nV)$ is a polynomial function of the size of the input and hence the above algorithm is qualified to be a polynomial-time algorithm. But we have just shown that the 0/1-Knapsack Optimization problem is NP-hard, and presumably NP-hard problems do not admit any polynomial-time algorithms. Is there any inconsistency in the theory of NP-hardness? The answer is "no." The 0/1-Knapsack Optimization problem was shown to be NP-hard only under the assumption that the numbers s_j and v_j are represented by binary numbers. It was not shown to be NP-hard if the numbers are represented by unary numbers. Thus, it is entirely possible that a problem is NP-hard under the binary encoding scheme, but solvable in polynomial time under the unary encoding scheme. An algorithm that runs in polynomial time with respect to the unary encoding scheme is called a *pseudo-polynomial* algorithm. A problem that is NP-hard with respect to the binary encoding scheme but not the unary encoding scheme is said to be NP-hard in the *ordinary* sense. A problem that is NP-hard with respect to the unary encoding scheme is said to be NP-hard in the *strong* sense. Similarly, we can define NP-complete problems in the ordinary sense and NP-complete problems in the strong sense.

We have just shown that 0/1-Knapsack Optimization can be solved by a pseudo-polynomial algorithm, even though it is NP-hard (in the ordinary sense). Are there any other NP-hard or NP-complete problems that can be solved by a pseudo-polynomial algorithm? More importantly, are there NP-hard or NP-complete problems that cannot be solved by a pseudo-polynomial algorithm (assuming NP-complete problems cannot be solved in polynomial time)? It turns out that Partition, $P2 \,||\, C_{\max}$, $1 \,|\, d_j = d \,|\, \sum w_j U_j$ all admit a pseudo-polynomial algorithm, while Traveling Salesman Optimization, Hamiltonian Circuit, 3-Dimensional Matching, Deadline Scheduling, $R \,||\, C_{\max}$, and $R \,|\, \bar{d}_j = d \,|\, \sum C_j$ do not admit any pseudo-polynomial algorithm. The key in identifying those problems that cannot be solved by a pseudo-polynomial algorithm is the notion of NP-hardness (or NP-completeness) in the strong sense, which we will explain in the remainder of this section.

Let Q be a decision problem. Associated with each instance I of Q are two measures, *SIZE*(I) and *MAX*(I). *SIZE*(I) is the number of symbols necessary to represent I, while *MAX*(I) is the *magnitude* of the largest number in I. Notice that if numbers are represented in binary in I, then *MAX*(I) could be an exponential function of *SIZE*(I). We say that Q is a *number* problem if *MAX*(I) is *not* bounded above by *any* polynomial function of *SIZE*(I). Clearly, Partition, Traveling Salesman Decision, 0/1-Knapsack Decision, the decision version of $P2 \,||\, C_{\max}$, the decision version of $1 \,|\, d_j = d \,|\, \sum w_j U_j$, Deadline Scheduling, the decision version of $R \,||\, C_{\max}$, and the decision version of $R \,|\, \bar{d}_j = d \,|\, \sum C_j$ are all number problems, while Hamiltonian Circuit, 3-Dimensional Matching, Satisfiability, 3-Satisfiability, Vertex Cover, and Clique are not.

Let $p(\cdot)$ be a polynomial function. The problem Q_p denotes the subproblem of Q, where all instances I of Q_p satisfy the condition that *MAX*$(I) \leq p($*SIZE*$(I))$. Notice that Q_p is *not* a number problem even if Q is a number problem. We say that Q is NP-complete in the strong sense if Q_p is NP-complete for some polynomial $p(\cdot)$. An optimization problem is NP-hard in the strong sense if its corresponding decision problem is NP-complete in the strong sense.

If Q is not a number problem and Q is NP-complete, then by definition Q is NP-complete in the strong sense. Thus, Hamiltonian Circuit, 3-Dimensional Matching, Satisfiability, 3-Satisfiability, Vertex Cover, and Clique are all NP-complete in the strong sense, since they are all NP-complete [1]. On the other hand, if Q is a number problem and Q is NP-complete, then Q may or may not be NP-complete in the strong sense. Among all the number problems that are NP-complete, Traveling Salesman Decision, Deadline Scheduling, the decision version of $R \,||\, C_{\max}$, and the decision version of $R \,|\, \bar{d}_j = d \,|\, \sum C_j$ are NP-complete in the strong sense, while Partition, 0/1-Knapsack Decision, the decision version of $P2 \,||\, C_{\max}$, and the decision version of $1 \,|\, d_j = d \,|\, \sum w_j U_j$ are not (they are NP-complete in the ordinary sense since each of these problems has a pseudo-polynomial algorithm).

How does one prove that a number problem Q is NP-complete in the strong sense? We start from a known strongly NP-complete problem P and show a *pseudo-polynomial reduction* from P to Q. A pseudo-polynomial reduction is defined exactly as polynomial reduction given in Section 2.3, except that $MAX(I_Q)$ further satisfies the condition that $MAX(I_Q) \le p(MAX(I_P), SIZE(I_P))$ for some polynomial $p(\cdot)$.

Let us examine the proof of Theorem 2.5. Hamiltonian Circuit is known to be NP-complete in the strong sense. The reduction defines the distance function to have value 1 or 2. Clearly, this is a pseudo-polynomial reduction. Thus, Traveling Salesman Decision is NP-complete in the strong sense. 3-Dimensional Matching is known to be NP-complete in the strong sense. The reductions given in Theorems 2.7 and 2.8 are also pseudo-polynomial reductions. Thus, the decision version of $R \,||\, C_{\max}$ and $R \mid \bar{d}_j = d \mid \sum C_j$ are NP-complete in the strong sense.

In Chapter 3 of the book by Garey and Johnson [1], the author gave a reduction from 3-Dimensional Matching to Partition. The reduction given there was not a pseudo-polynomial reduction. Thus, Partition was not shown to be NP-complete in the strong sense, even though 3-Dimensional Matching is NP-complete in the strong sense.

We have said earlier that Deadline Scheduling is NP-complete in the strong sense. Yet the proof that Deadline Scheduling is NP-complete is by a reduction from Partition (see Theorem 2.6), which is not NP-complete in the strong sense. As it turns out, Deadline Scheduling can be shown to be NP-complete in the strong sense by a pseudo-polynomial reduction from the strongly NP-complete 3-Partition problem.

3-Partition. Given a list $A = (a_1, a_2, \ldots, a_{3m})$ of $3m$ positive integers such that $\sum a_j = mB$, $\frac{1}{4}B < a_j < \frac{1}{2}B$ for each $1 \le j \le 3m$, is there a partition of A into $A_1, A_2, \ldots, A_m$ such that $\sum_{a_j \in A_i} a_j = B$ for each $1 \le i \le m$?

Theorem 2.9

3-Partition is reducible to Deadline Scheduling.

Proof

Given an instance $A = (a_1, a_2, \ldots, a_{3m})$ of 3-Partition, we construct an instance of Deadline Scheduling as follows. There will be $4m - 1$ jobs. The first $3m$ jobs are Partition jobs. For each $1 \le j \le 3m$, job j has processing time a_j units, release time 0, and deadline $mB + (m - 1)$. The last $m - 1$ jobs are Divider jobs. For each $3m + 1 \le j \le 4m - 1$, job j has processing time 1 unit, release time $(j - 3m)B + (j - 3m - 1)$, and deadline $(j - 3m)(B + 1)$.

The $m - 1$ Divider jobs divide the timeline into m intervals into which the Partition jobs are scheduled. The length of each of these intervals is exactly B. Thus, there is a feasible schedule if and only if there is a 3-Partition. □

It is clear that the above reduction is a pseudo-polynomial reduction.

2.6 PTAS and FPTAS

One way to cope with NP-hard problems is to design approximation algorithms that run fast (polynomial time), even though they may not always yield an optimal solution. The success of an approximation algorithm is measured by both the running time and the quality of solutions obtained by the algorithm vs. those obtained by an optimization algorithm. In this section we will talk about approximation algorithm, polynomial-time approximation schemes (PTAS), and fully polynomial-time approximation schemes (FPTAS).

We will use 0/1-Knapsack Optimization as an example to illustrate the ideas. There is a fast algorithm that always generates at least 50% of the total value obtained by an optimization algorithm, and the algorithm runs in $O(n \log n)$ time. The algorithm is called the Density-Decreasing-Greedy (DDG) algorithm and it works as follows. Sort the items in descending order of the ratios of value vs. size. Let $L = (u_1, u_2, \ldots, u_n)$ be the sorted list such that $v_1/s_1 \geq v_2/s_2 \geq \cdots \geq v_n/s_n$. Scanning L from left to right, pack each item into the knapsack if there is enough capacity to accommodate the item. Let v' be the total value of items packed in the knapsack. Let v'' be the value obtained by merely packing the item with the largest value into the knapsack; i.e., $v'' = \max\{v_j\}$. If $v' > v''$, then output the first solution; otherwise, output the second solution.

Theorem 2.10

For any instance I of the 0/1-Knapsack Optimization problem, let DDG(I) and OPT(I) denote the total values obtained by the DDG and optimization algorithms, respectively. Then we have

$$\frac{OPT(I)}{DDG(I)} \leq 2$$

Moreover, there are instances such that the ratio can approach 2 arbitrarily closely.

We shall omit the proof of Theorem 2.10; it can be found in Ref. [1]. DDG is an approximation algorithm that gives a worst-case bound of 2. One wonders if there are approximation algorithms that approximate arbitrarily closely to the optimal solution. The answer is "yes." Sahni [4] gave a family of algorithms A_k that, for each integer $k \geq 1$, gives a worst-case bound of $1 + 1/k$; i.e., for each instance I, we have

$$\frac{OPT(I)}{A_k(I)} \leq 1 + \frac{1}{k}$$

By choosing k large enough, we can approximate arbitrarily closely to the optimal solution.

For each $k \geq 1$, the algorithm A_k works as follows. Try all possible subsets of k or fewer items as an initial set of items, and then pack, if possible, the remaining items in descending order of the ratios of value vs. size. Output the best of all possible solutions. The algorithm runs in $O(n^{k+1})$ time, which is polynomial time for each fixed k. Notice that while the family of algorithms A_k has the desirable effect that it can approximate arbitrarily closely to the optimal solution, it has the undesirable effect that k appears in the exponent of the running time. Thus, for large k, the algorithm becomes impractical. We call a family of approximation algorithms with this kind of characteristics *polynomial-time approximation scheme*.

It would be nice if we have an approximation scheme whose running time is a polynomial function of both n and k. Such an approximation scheme is called a *fully polynomial-time approximation scheme*. Indeed, for 0/1-Knapsack Optimization, there is a FPTAS due to Ibarra and Kim [5]. The idea of the method of Ibarra and Kim is to scale down the value of each item, use the pseudo-polynomial algorithm given in Section 2.5 to obtain an exact solution for the scaled down version, and then output the items obtained in the exact solution. The net effect of scaling down the value of each item is to reduce the running time of the algorithm in such a way that accuracy loss is limited. Most FPTAS reported in the literature exploit pseudo-polynomial algorithms in this manner. Thus, one of the significances of developing pseudo-polynomial algorithms is that they can be converted into FPTAS.

The pseudo-polynomial algorithm given in Section 2.5 runs in $O(nV)$ time, where $V = \sum v_j$. If $v = \max\{v_j\}$, then the running time becomes $O(n^2 v)$. Let $U = \{u_1, u_2, \ldots, u_n\}$ be a set of n items, with each item u_j having a size s_j and a value v_j. Let I denote this instance and let I' denote the instance obtained from I by replacing the value of each item by $v'_j = \lfloor v_j/K \rfloor$, where $K = v/(k+1)n$. We then apply the pseudo-polynomial algorithm to I', and the resulting solution (i.e., the subset of items packed in the knapsack) will be used as the solution for I. The running time of the algorithm is $O(n^2 v/K) = O((k+1)n^3)$, which satisfies the timing requirement of FPTAS.

It remains to be shown that the worst-case bound is no more than $1 + 1/k$. Observe that

$$A_k(I) \geq OPT(I) - Kn = OPT(I) - \frac{v}{(k+1)}$$

and since $OPT(I) \geq v$, we have

$$\frac{OPT(I)}{A_k(I)} \leq \frac{A_k(I) + \dfrac{v}{(k+1)}}{A_k(I)} \leq 1 + \frac{\dfrac{v}{(k+1)}}{v - \dfrac{v}{(k+1)}} = 1 + \frac{1}{k}$$

Theorem 2.11

There is a family of algorithms A_k that runs in time $O((k+1)n^3)$ such that for every instance I of 0/1-Knapsack Optimization, we have

$$\frac{A_k(I)}{OPT(I)} \leq 1 + \frac{1}{k}$$

As it turns out, the existence of a pseudo-polynomial algorithm is closely tied to whether the problem is NP-hard in the strong sense.

Theorem 2.12

If an optimization problem Q is NP-hard in the strong sense, then there cannot be any pseudo-polynomial algorithms for Q unless all NP-complete problems can be solved in polynomial time.

The proof of the above theorem can be found in Ref. [1].

Acknowledgment

This work is supported in part by the NSF Grant DMI-0300156.

References

[1] S. A. Cook, The complexity of theorem proving procedures, *Procedings of the 3rd Annual ACM Symposium on Theory of Computing*, Association for Computing Machinery, New York, 1971, pp. 151–158.

[2] M. R. Garey and D. S. Johnson, *Computers and Intractability: A Guide to the Theory of NP-completeness*, W. H. Freeman, New York, 1979.

[3] R. M. Karp, Reducibility among combinatorial problems, in R. E. Miller and J. W. Thatcher (eds), *Complexity of Computer Computations*, Plenum Press, New York, 1972, pp. 85–103.

[4] S. Sahni, Approximate algorithms for the 0/1 knapsack problem, *Journal of the ACM*, 22, 115–124, 1975.

[5] O. H. Ibarra and C. E. Kim, Fast approximation algorithms for the knapsack and sum of subset problems, *Journal of the ACM*, 22, 463–468, 1975.

3

Some Basic Scheduling Algorithms

Joseph Y-T. Leung
New Jersey Institute of Technology

3.1 Introduction

In this chapter we shall review some of the most well-known and basic scheduling algorithms. We shall concentrate on identical and parallel machines only. The performance metrics we shall be dealing with are $C_{\max}$, $\sum C_j$, $L_{\max}$, $\sum U_j$, and $\sum T_j$. These will be presented in Sections 3.2 to 3.7.

3.2 The Makespan Objective

We first consider nonpreemptive scheduling. The problem $P2 \mid\mid C_{\max}$ can be shown to be NP-hard in the ordinary sense by a simple reduction from the Partition problem (see Chapter 2 for a definition of the Partition problem). Thus, there is no hope for any polynomial-time algorithms (unless $P = \text{NP}$) if we allow the processing times to be arbitrary.

The problem becomes easier if the processing times are identical. $P \mid p_j = p, \mathit{intree} \mid C_{\max}$ and $P \mid p_j = p, \mathit{outtree} \mid C_{\max}$ are both solvable in polynomial time [1], so is $P2 \mid p_j = p, \mathit{prec} \mid C_{\max}$ [2]. On the other hand, $P \mid p_j = p, \mathit{prec} \mid C_{\max}$ has been shown to be NP-hard in the strong sense [3]; it is still an open question whether $Pm \mid p_j = p, \mathit{prec} \mid C_{\max}$ can be solved in polynomial time for each fixed $m \geq 3$. In the following we shall describe the polynomial-time algorithms for equal-length jobs.

The set of jobs and precedence constraint are described by a directed acyclic graph $G = (V, A)$, where V is the set of vertexes and A is the set of arcs. Each job i is represented by a vertex in V. There is a directed arc from i to j (written as $i \rightarrow j$) if job i is an *immediate predecessor* of job j, i.e., job j cannot start until job i has finished. We assume that G has no transitive arcs, i.e., no arc $i \rightarrow j$ whenever there are two arcs $i \rightarrow k$ and $k \rightarrow j$ in G for some $k \neq i, j$. We assume that the processing time of each job is one unit, whatever that unit represents. Figure 3.1 shows a graph G with 19 jobs along with their precedence constraint.

1-58488-397-9/$0.00+$1.50

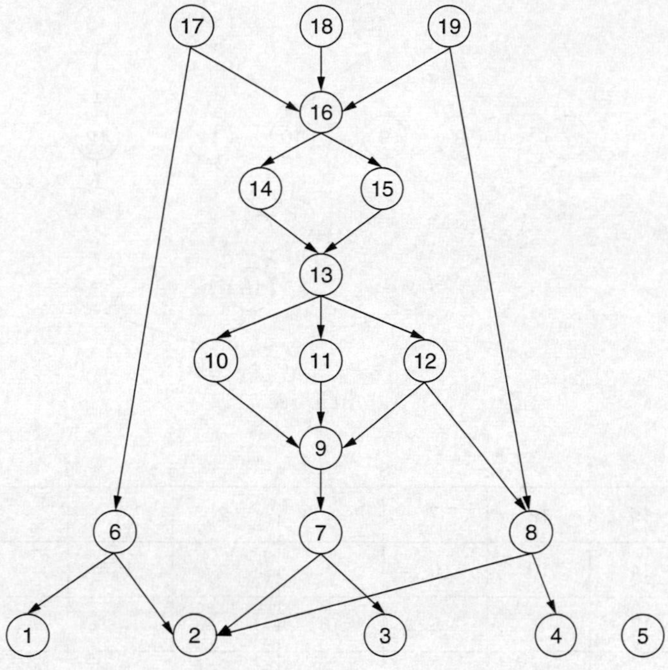

FIGURE 3.1 An example set of jobs with precedence constraint.

Job i is the *predecessor* of job j if there is a path from i to j. *Successor* and *immediate successor* are similarly defined. G is an *intree* if every vertex has one immediate successor, except the *root* that has no immediate successor. G is an *outtree* if every vertex has one immediate predecessor, except the *root* that has no immediate predecessor. Reversing the direction of the arcs converts an intree into an outtree, and vice versa. For the C_{max} objective, an optimal algorithm for intrees can be used as an optimal algorithm for outtrees, and vice versa. We note that this may not be true for objectives other than C_{max}.

Each vertex in G can be assigned a level as follows. All vertexes without any immediate successors are at level 1. For all other vertexes, the level of the vertex is one plus the maximum level of its immediate successors. The optimal algorithm for intrees is to assign the ready job that is at the highest level among all ready jobs. In the literature this algorithm is called highest-level-first or critical-path or Hu's algorithm [1].

Hu's Algorithm. Whenever a machine is free for assignment, assign that ready job (i.e., the one all of whose immediate predecessors have already finished execution), which is at the highest level among all ready jobs. Ties can be broken in an arbitrary manner.

Figure 3.2 shows an intree and a schedule produced by Hu's algorithm on four machines. Notice that the root is always (the only job) executed in the last time unit in any schedule. Since we need to finish all nonroot jobs before we can start the root, the schedule produced by Hu's algorithm must be optimal if there is no machine idle time before the second to last time unit. Also, since each job has at most one immediate successor, the number of ready jobs is nonincreasing in time, and hence the number of jobs executed in each time unit is also nonincreasing in time.

Theorem 3.1

Hu's algorithm is optimal for $P \mid p_j = p, \text{intree} \mid C_{max}$.

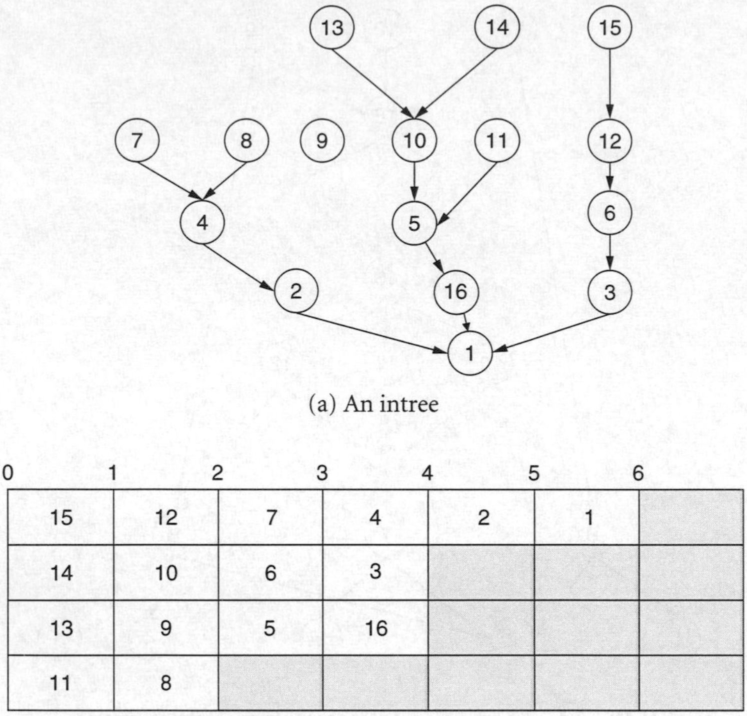

(a) An intree

(b) Hu's schedule on four machines

FIGURE 3.2 Illustrating Hu's algorithm.

Proof

We shall prove the theorem by induction on the height h of the intree, which is defined as the maximum level of any vertex in the intree. The basis case, $h = 1$ or 2, is obvious. Assuming that the theorem is true for all $h < k$, we shall prove that it is true for $h = k$.

The inductive step is proved by contradiction. Let T be an intree with height k such that the Hu's schedule S is not optimal. Let S have makespan ω. Let the optimal makespan be ω_O. Since S is not optimal, we have $\omega_O < \omega$. Also, since S is not optimal, there must be a time t, $t \leq \omega - 2$, such that there is a machine idle in the time slot $[t-1, t]$. Since T is an intree, we must have an idle machine in every time slot after t. We can also conclude that there is a chain of jobs executing from $t - 1$ until ω, i.e., the root (a level-1 job) executed in the time slot $[\omega - 1, \omega]$, a level-2 job executed in the time slot $[\omega - 2, \omega - 1]$, and so on.

Consider a new intree T' obtained from T by deleting all level-2 jobs and making all level-3 jobs to be the immediate predecessors of the root. Clearly, the height of T' is $k - 1$. Furthermore, the level of every vertex in T', except the root, is exactly one less than that in T. It can be shown that the Hu's schedule S' for T' is exactly the schedule S, except that all level-2 jobs are replaced by idle times and the root is scheduled one time unit earlier. Thus, $\omega' = \omega - 1$. On the other hand, the optimal schedule length ω'_O satisfies $\omega'_O \leq \omega_O - 1$. Thus, $\omega'_O < \omega'$, and hence T' also violates the theorem. But T' has height $k - 1$, contradicting our assumption that Hu's algorithm is optimal for all intrees with height less than k. □

Hu's algorithm admits an $O(n \log n)$-time implementation.

The algorithm that solves $P2 \mid p_j = p, prec \mid C_{\max}$ is the well-known Coffman-Graham algorithm [2]. It works by first assigning a label to each job, which is a function of the labels of its immediate successors. It then schedules jobs according to their labels, highest label first.

Let $N = (n_1, n_2, \ldots, n_t)$ and $N' = (n'_1, n'_2, \ldots, n'_{t'})$ be two decreasing sequences of integers with $t \leq t'$. We say that N is *lexicographically smaller than* N' (denoted by $N \prec N'$) if either (1) there is a $k \leq t$ such that $n_i = n'_i$ for all $1 \leq i < k$ and $n_k < n'_k$ or (2) $n_i = n'_i$ for all $1 \leq i \leq t$ and $t < t'$. We say that N is *lexicographically smaller than or equal to* N' (denoted by $N \preceq N'$) if N is either lexicograhically smaller than or equal to N'. The labeling algorithm is described as follows.

Coffman-Graham Labeling Algorithm. Let there be k jobs with no immediate successors. Assign in any order the integers $1, 2, \ldots, k$ to these k jobs. Suppose we have already assigned the first j integers. Let J' be the set of jobs all of whose immediate successors have already been assigned a label. For each job i in J', let $N(i)$ denote the decreasing sequence of integers formed from the labels of its immediate successors. Assign the integer $j+1$ to the job i such that $N(i) \preceq N(k)$ for all jobs k in J'. Repeat this process until every job has been assigned a label.

We note that jobs at a higher level are always assigned higher labels than jobs at a lower level. When two jobs are at the same level, they are differentiated by the labels of their immediate successors.

Coffman-Graham Algorithm. Assign labels to the jobs by the above algorithm. Schedule the jobs as follows: Whenever a machine is free for assignment, assign that ready job that has the highest label among all ready jobs.

Figure 3.3(a) shows a labeling by the Coffman-Graham Labeling Algorithm. The label of each job is the integer next to the circle that represents the job. The integer inside each circle is the job index.

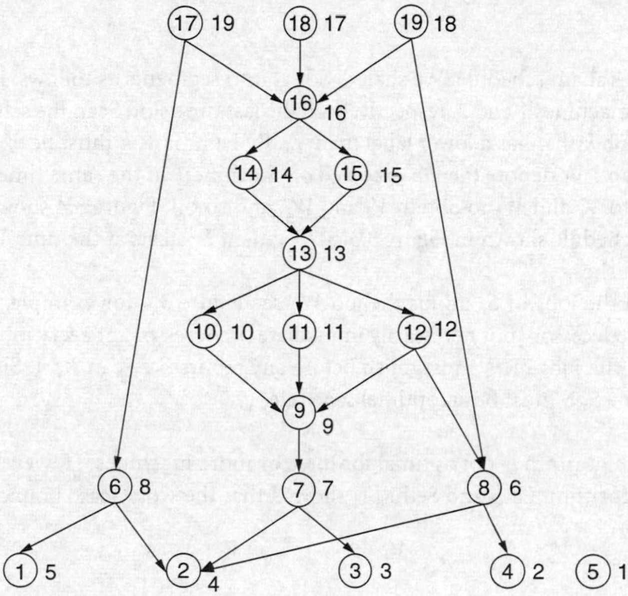

(a) Label assigned by Coffman-Graham labeling algorithm for the jobs in Figure 3.1.

0	1	2	3	4	5	6	7	8	9	10
17	18	16	15	13	12	10	9	7	2	
19	6	1	14	5	11	8	4		3	

(b) Schedule produced by Coffman-Graham algorithm

FIGURE 3.3 Illustrating Coffman-Graham algorithm.

Figure 3.3(b) shows a schedule produced by the Coffman-Graham algorithm. With appropriate data structure, Coffman-Graham algorithm can be implemented to run in $O(|V| + |A|)$ time.

Since each job has unit processing time, both machines become available for assignment simultaneously. We make the convention that machine 1 is always assigned before machine 2. If a machine is idle in a time slot, we say that it is executing a job with label zero.

Lemma 3.1

Let S be a Coffman-Graham schedule and let job j be executed on machine 1 in the time slot $[t, t+1]$. If k is any other job executed on any machine in the time slot $[t', t'+1]$, where $t' \geq t$, then the label of j is higher than the label of k.

Proof

Consider all the jobs that have not been executed at time t. Let job i be the one that has the highest label. Job i must be ready for execution since its predecessors must have higher labels. By our convention, job i is assigned to machine 1. Therefore, it has a higher label than any jobs that are assigned after it. □

Theorem 3.2

Coffman-Graham algorithm is optimal for $P2 \mid p_j = p, prec \mid C_{\max}$.

Proof

Let S be a Coffman-Graham schedule. We shall divide S into segments as follows. Let V_0 and W_0 denote the jobs executed on machines 1 and 2, respectively, in the last time slot. Scan the schedule backward until we first encounter a job k that has a lower label than V_0. By Lemma 1, k must be executed on machine 2. Let W_1 denote job k and V_1 denote the job executed on machine 1 in the same time slot as k. Repeat this process with respect to V_1 and W_1 to obtain V_2 and W_2, and so on. Figure 3.4 shows the locations of the V_i's and W_i's in the schedule shown in Figure 3.3(b). Segment S_i starts at the time V_{i+1} finishes and ends at the time that V_i finishes.

Define F_i to be all the jobs in S_i minus the job W_i, see Figure 3.4 for example. It can be shown that every job in F_i is a predecessor (not necessarily immediate predecessor) of every job in F_{i-1}. Thus, in any schedule whatsoever, the jobs in F_i must finish before any job in F_{i-1} can start. Since the jobs in F_i are scheduled optimally in S, S must be an optimal schedule. □

Coffman-Graham algorithm is not optimal for three or more machines. However, we can still use it as an approximation algorithm. Lam and Sethi [4] showed that the worst-case bound (compared with the

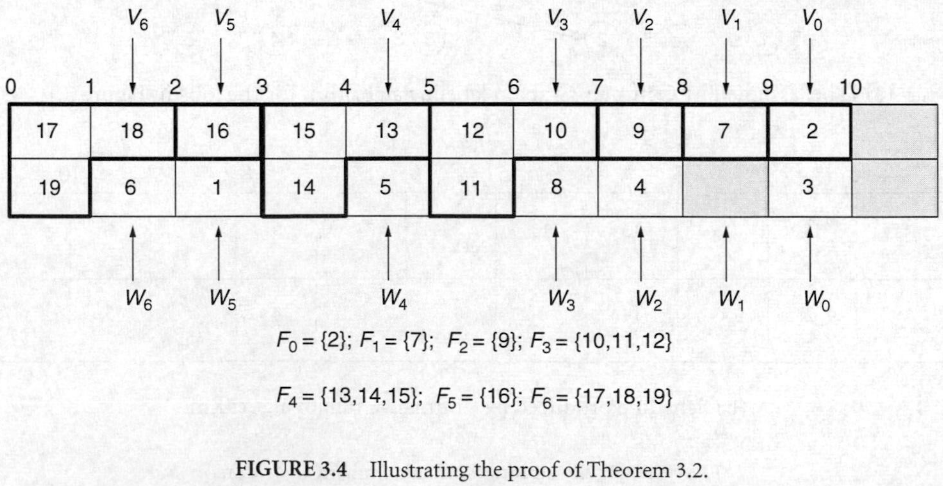

FIGURE 3.4 Illustrating the proof of Theorem 3.2.

optimal schedule length) is no more than $2 - 2/m$, where m is the number of machines. Moreover, the bound is tight.

$P \mid\mid C_{\max}$ can be shown to be NP-hard in the strong sense by a simple reduction from the 3-Partition problem (see Chapter 2 for a definition of the 3-Partition problem). Thus, unless $P = \text{NP}$, there is no polynomial-time algorithm to solve this problem. However, there is an effective approximation algorithm for it. This algorithm is the first approximation algorithm reported in the literature [5,6] whose worst-case performance was successfully analyzed.

Largest-Processing-Time (LPT) Rule. Whenever a machine is free for assignment, assign that job with the largest processing time among all unassigned jobs. Ties can be broken arbitrarily.

The worst-case bound of LPT is $4/3 - 1/3m$, which is a tight bound. We refer the readers to Graham [5,6] for the proof.

We now consider preemptive scheduling. Unlike $P \mid\mid C_{\max}$, $P \mid pmtn \mid C_{\max}$ can be solved in linear time by the well-known McNaughton's wrap-around rule [7].

McNaughton's Wrap-Around Rule. Compute $D = \max\{\max\{p_j\}, (1/m)\sum p_j\}$. Assign the jobs in any order from time 0 until time D on machine 1. If a job's processing extends beyond time D, preempt the job at time D, and continue its processing on machine 2, starting at time 0. Repeat this process until all jobs are assigned.

Theorem 3.3

McNaughton's wrap-around rule is optimal for $P \mid pmtn \mid C_{\max}$.

Proof

It is clear that D is a lower bound for the optimal schedule length. If we can show that the wrap-around rule can always generate a feasible schedule in the time interval $[0, D]$, then the schedule must be optimal. First, no job can have overlaps, i.e., simultaneously executing on more than one machine. This is because $D \geq \max\{p_j\}$. Second, there is enough capacity in the time interval $[0, D]$ to schedule all the jobs, since $mD \geq \sum p_j$. Thus, the wrap-around rule can always generate a feasible schedule. □

It turns out that the wrap-around rule can be modified to handle online situations where the jobs are released at unpredictable times and we do not have *a priori* knowledge about the times jobs are released. Consider the following modified wrap-around rule.

Modified Wrap-Around Rule

1. Let J be the set of n jobs sorted in descending order of processing times, i.e., $p_1 \geq p_2 \geq \cdots \geq p_n$. Let there be m machines.
2. If $\max\{p_j\} \leq (1/m)\sum p_j$, then schedule all jobs in J by McNaughton's wrap-around rule.
3. Schedule the longest job on one machine and delete the job from J. Decrement m by 1 and go to step 2.

The only difference between the modified wrap-around rule and McNaughton's wrap-around rule is observed when $\max\{p_j\} > (1/m)\sum p_j$. In this case, McNaughton's wrap-around rule schedules many machines up to time $\max\{p_j\}$, while the modified wrap-around rule schedules one machine up to that time (i.e., the longest job). Figure 3.5 shows the schedules generated by the two different rules for a set of independent jobs.

Hong-Leung Algorithm. Whenever new jobs are released, schedule the new jobs along with the unexecuted portions of the remaining jobs by the modified wrap-around rule.

J_i	1	2	3	4	5	6	7	8	9	10
p_i	11	10	4	3	2	2	2	1	1	1

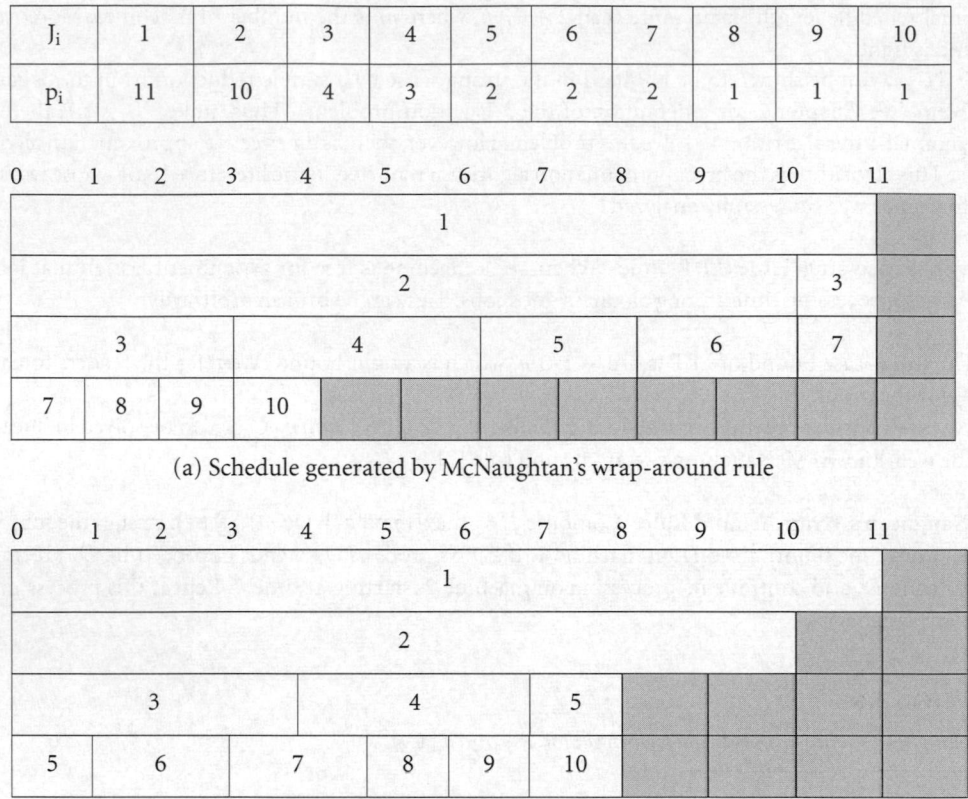

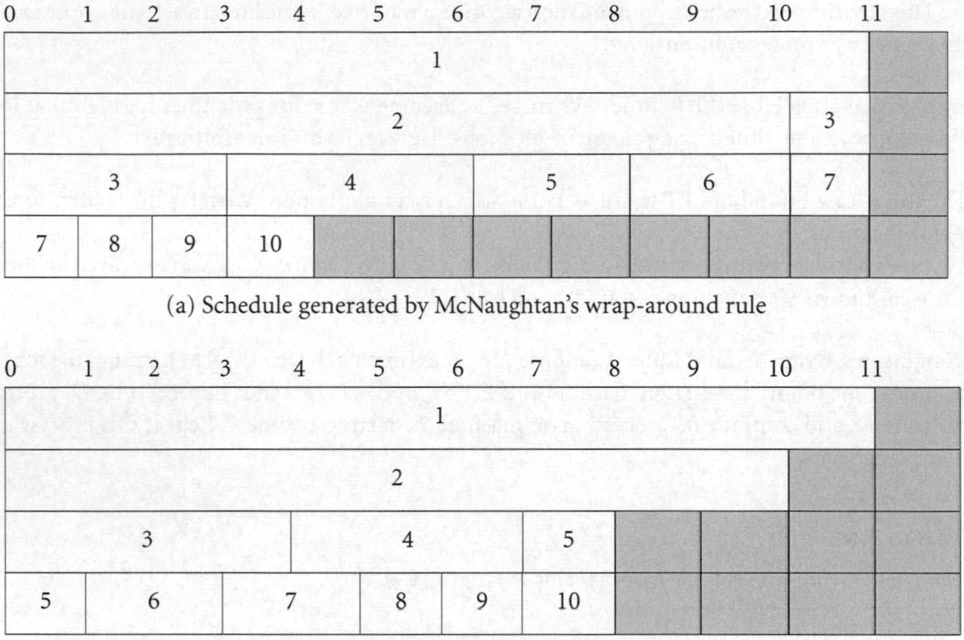

(a) Schedule generated by McNaughtan's wrap-around rule

(b) Schedule generated by modified wrap-around rule

FIGURE 3.5 Illustrating the two wrap-around rules.

Hong and Leung [8] showed that scheduling jobs by the modified wrap-around rule is optimal for $P \mid pmtn, r_j \mid C_{\max}$. Because of space limitation, we will omit the proof here.

Theorem 3.4

Hong-Leung algorithm is optimal for the online version of $P \mid pmtn, r_j \mid C_{\max}$.

When precedence constraint is added to the jobs, the problem can be solved optimally for intrees or two machines by the well-known Muntz-Coffman algorithm [9,10]. Muntz-Coffman algorithm is described in terms of machine sharing, which allows a fraction β, $0 < \beta \leq 1$, of a machine to be assigned to a job. A machine-sharing schedule can be converted to a preemptive schedule without increasing the schedule length. Let S be a machine-sharing schedule and let $t_0 < t_1 < \cdots < t_k$ be the time instants where the machine assignment changed in S. We will convert each segment $[t_{i-1}, t_i]$, $1 \leq i \leq k$, of S into a preemptive schedule. For each segment, compute for each job the processing done in the segment and then use McNaughton's wrap-around rule to schedule the jobs. The result is a preemptive schedule with the same schedule length as S.

Like Hu's algorithm, Muntz-Coffman algorithm is also a highest-level-first algorithm. When jobs have arbitrary processing times, we need to generalize the definition of level as follows. The level of any job that does not have any immediate successors is its own processing time. For any other job, its level is its own processing time plus the maximum level of its immediate successors.

Muntz-Coffman Algorithm. Assign one machine each to the jobs at the highest level. If there is a tie among y jobs for the last x $(x < y)$ machines, then assign x/y of a machine to each of these y jobs.

M = 3

(a) Schedule generated by Muntz-Coffman algorithm

0	3/2	5/2	10/3	14/3	17/3	20/3	23/3
8 ($\beta=1$)	8 ($\beta=1$)	1 ($\beta=3/5$)	3 ($\beta=3/4$)	5 ($\beta=1$)	7 ($\beta=1$)	9 ($\beta=1$)	
2 ($\beta=2/3$)	1 ($\beta=1/2$)	2 ($\beta=3/5$)	4 ($\beta=3/4$)	6 ($\beta=1$)	8 ($\beta=1$)		
4 ($\beta=2/3$)	2 ($\beta=1/2$)	4 ($\beta=3/5$)	6 ($\beta=3/4$)	8 ($\beta=1$)			
6 ($\beta=2/3$)	4 ($\beta=1/2$)	6 ($\beta=3/5$)	8 ($\beta=3/4$)				
	6 ($\beta=1/2$)	8 ($\beta=3/5$)					

(b) Preemptive schedule

Time markers: 0, 1, 3/2, 2, 5/2, 3, 10/3, 4, 14/3, 17/3, 20/3, 22/3

Machine 1	8	8	1	2	3	4	5	7	9
Machine 2	2	4	1	2	2 4 6	4	6	6	8
Machine 3	4	6	4	6	6	8	6	8	

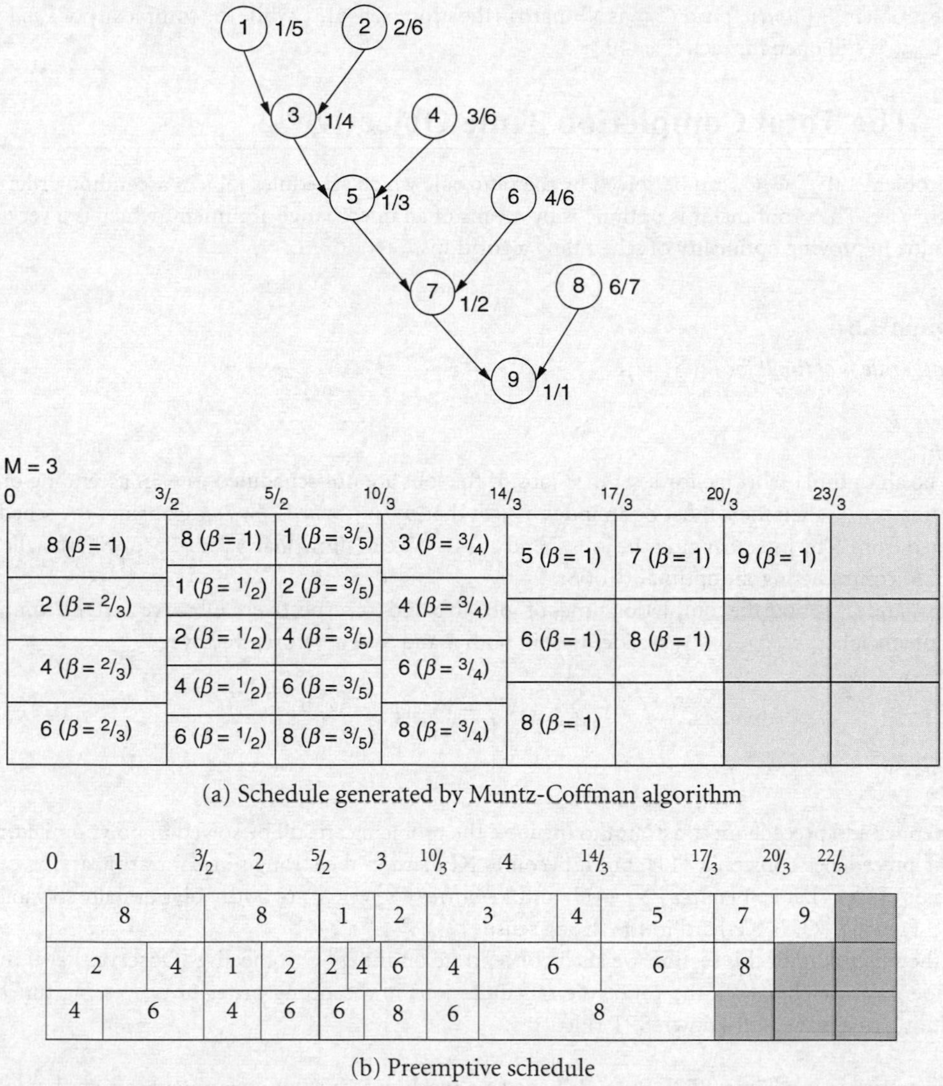

FIGURE 3.6 Illustrating Muntz-Coffman algorithm.

Reassign the machines to the unexecuted portion of G according to the above rule, whenever either of the two events described below occurs:

Event 1. A job is completed.
Event 2. We reach a point where, if we were to continue the current assignment, we would be executing a job at a lower level at a faster rate than other jobs at a higher level.

Figure 3.6 shows a set of jobs along with the precedence constraint. The number α/β right next to each vertex denotes the processing time and level of the job, respectively. Figure 3.6(a) shows the schedule generated by Muntz-Coffman algorithm and Figure 3.6(b) shows the schedule converted to a preemptive schedule. The proof of the following theorem can be found in Muntz and Coffman, Jr. [9, 10].

Theorem 3.5

Muntz-Coffman algorithm is optimal for $P \mid pmtn,\ intrees \mid C_{\max}$ and $P2 \mid pmtn,\ prec \mid C_{\max}$.

The problem $P \mid pmtn, prec \mid C_{\max}$ is NP-hard in the strong sense [3], while the complexity of $Pm \mid pmtn, prec \mid C_{\max}$ is still open for each fixed $m \geq 3$.

3.3 The Total Completion Time Objective

The problem $1 \mid\mid \sum w_j C_j$ can be solved by the ratio rule which schedules jobs in ascending order of the ratios p_i/w_i. The proof that it is optimal is by means of an interchange argument, which is a very useful technique in proving optimality of scheduling algorithms.

Theorem 3.6

The ratio rule is optimal for $1 \mid\mid \sum w_j C_j$.

Proof

Let S be an optimal schedule for a set of n jobs. If the jobs are not scheduled in S in ascending order of the ratios p_i/w_i, then there must be an index j such that $p_j/w_j > p_{j+1}/w_{j+1}$. Consider the schedule S' obtained from S by interchanging the jobs j and $j+1$. We can show that $\sum w_j C_j$ for S' is smaller than that of S, contradicting the optimality of S.

Let C_i and C_i' denote the completion times of job i in S and S', respectively. Observe that the completion times of any job k, $k \neq j, j+1$, are identical in both S and S'. Therefore, we have

$$\sum w_j C_j - \sum w_j C_j' = w_{j+1} p_j - w_j p_{j+1} > 0$$

since $p_j/w_j > p_{j+1}/w_{j+1}$. □

When we add precedence constraint to the jobs, the problem can still be solved in polynomial time for treelike precedence constraint [11], but it becomes NP-hard in the strong sense for arbitrary precedence constraint [12]. That is, $1 \mid intree \mid \sum w_j C_j$ and $1 \mid outtree \mid \sum w_j C_j$ are both polynomially solvable [11], while $1 \mid prec \mid \sum C_j$ is NP-hard in the strong sense [12].

In the remainder of this section we shall concentrate on the $\sum C_j$ objective. Observe that if the jobs have identical weights, then the ratio rule schedules jobs in ascending order of processing times. This scheduling rule is the well-known SPT rule.

Smallest-Processing-Time (SPT) Rule. Whenever a machine is free for assignment, assign that job with the smallest processing time among all unassigned jobs. Ties can be broken arbitrarily.

The proof that SPT is optimal for $1 \mid\mid \sum C_j$ follows directly from Theorem 3.6. Besides this indirect proof, there are two direct proofs that show the optimality of SPT. The first one is by means of the concept of *strong dominance*. A schedule S is said to be *strongly dominant* if at any time instant t, the number of jobs completed in S is at least as large as any other schedule S'. A moment of reflection shows that any strongly dominant schedule must be optimal with respect to the $\sum C_j$ objective. It is easy to see that SPT schedules are strongly dominant schedules.

The second proof is by means of the cost function itself. Let jobs $1, 2, \ldots, n$ be scheduled in this order in S. The cost function $\sum C_j$ can be expressed as

$$\sum C_j = np_1 + (n-1)p_2 + (n-2)p_3 + \cdots + 2p_{n-1} + p_n$$

From the above cost function, we can see that the processing time of the first job contributes n times to the cost function, the processing time of the second job contributes $n-1$ times, and so on. In other words, the processing time of a job contributes to the cost function by x number of times, where x is one plus the number of jobs that follow it. To minimize the cost function, it is clear that we should put the smallest job in the first position (because it has the largest multiplier), the second smallest job in the second position, and so on.

The SPT rule is also optimal for $P \mid\mid \sum C_j$; the proof of its optimality follows the same line as above. Since McNaughton [7] has shown that preemption cannot reduce $\sum C_j$ for a set of independent jobs, SPT is optimal for $P \mid pmtn \mid \sum C_j$ as well.

Theorem 3.7

The SPT rule is optimal for $P \mid\mid \sum C_j$ and $P \mid pmtn \mid \sum C_j$.

If we introduce ready times into the problem, we can still solve the problem in polynomial time if preemption is allowed, but it becomes NP-hard in the ordinary sense for nonpreemptive scheduling. That is, $1 \mid r_j \mid \sum C_j$ is NP-hard in the ordinary sense [13], but $1 \mid pmtn, r_j \mid \sum C_j$ is solvable in polynomial time [14]. The algorithm that solves $1 \mid pmtn, r_j \mid \sum C_j$ is actually the preemptive version of the SPT rule.

Baker's Rule. At any moment of time, schedule that ready job with the smallest remaining processing time.

Next, consider a set of jobs each of which has a deadline to meet. All jobs are ready at time 0. We seek a schedule such that all deadlines are met and such that $\sum C_j$ is minimized. In the $\alpha \mid \beta \mid \gamma$ notation, this problem is denoted as $1 \mid \bar{d}_j \mid \sum C_j$. This problem can be solved by the Smith's Rule which schedules the jobs backward [15].

Smith's Rule. Let $t = \sum p_j$. From among those jobs that can complete at time t (i.e., $\bar{d}_j \geq t$), choose the one with the largest processing time to complete at time t. This leaves a problem instance with one fewer jobs to which the same rule applies.

Theorem 3.8

Baker's rule is optimal for $1 \mid pmtn, r_j \mid \sum C_j$ and Smith's rule is optimal for $1 \mid \bar{d}_j \mid \sum C_j$ and $1 \mid pmtn, \bar{d}_j \mid \sum C_j$.

Preemption is not necessary when all jobs are ready at the same time. Thus, Smith's rule solves $1 \mid pmtn, \bar{d}_j \mid \sum C_j$ as well.

Smith's rule exploits the fact that for one machine the last job must finish at time $t = \sum p_j$. For parallel machines, it is not clear when the last job finishes on each machine in an optimal schedule. Therefore, the idea of scheduling jobs backward cannot be extended to parallel machines. For parallel machines, we need an algorithm that schedules jobs from the front. For one machine, it is not difficult to design such an algorithm.

We start out with a SPT order. Let $L = (1, 2, \ldots, n)$ be the list of jobs sorted in ascending order of processing times and in ascending order of deadlines in case of identical processing times. If there is no deadline violation when jobs are scheduled in this order, then the schedule is already optimal. Typically, however, there are some deadline violations and we need to reorder the jobs so that a feasible schedule can be obtained. The job reordering follows as much a SPT-like order as possible. Starting with job 1, we check if the jobs that follow job 1 can be feasibly scheduled. This can be determined by scheduling the remaining jobs by the *Earliest Deadline* (*EDD*) rule. If the EDD schedule is feasible, then we fix job 1 at the first position and move on to the next job in the list. On the other hand, if the EDD schedule is infeasible, let d be the first deadline violation in the EDD schedule. We move all the jobs whose deadline is less than or equal to d ahead of job 1. In other words, job 1 has been moved down the list (which may or may not be its final position). The jobs that were moved ahead are then sorted in ascending order of processing times and in ascending order of deadlines in case of identical processing times, and the first job is then considered by the same process as we did for job 1. This process is repeated until the position of every job is fixed.

It is easy to see that the schedule produced by the above algorithm is identical to the one produced by Smith's rule. This idea of scheduling jobs from the front has been exploited by Leung and Pinedo [16] who developed a polynomial-time algorithm for $P \mid pmtn,\ \bar{d}_j \mid \sum C_j$. Of course, for parallel machines, the feasibility test is a bit more complicated.

The problems $1 \mid pmtn,\ r_j \mid \sum C_j$ and $1 \mid pmtn,\ \bar{d}_j \mid \sum C_j$ are both solvable in polynomial time. One wonders about the complexity of the problem if both the ready times (r_j) and deadlines ($\bar{d}_j$) are in the problem instance, i.e., the problem $1 \mid pmtn,\ r_j, \bar{d}_j \mid \sum C_j$. Unfortunately, Du and Leung [17] showed that it is NP-hard in the ordinary sense. Thus, as far as polynomial solvability is concerned, we cannot have both parameters in the problem instance. If we restrict ourselves to have ready times only, then the problem is NP-hard in the ordinary sense for every $m \geq 2$ [18], i.e., $Pm \mid pmtn,\ r_j \mid \sum C_j$ is NP-hard in the ordinary sense for every $m \geq 2$. On the other hand, if we restrict ourselves to have deadlines only, then the problem can be solved in polynomial time for arbitrary number of machines [16], i.e., $P \mid pmtn,\ \bar{d}_j \mid \sum C_j$ is polynomially solvable. There seems to be no symmetry between the ready times and deadlines.

Du and Leung [17] also generalized Baker's rule and Smith's rule to solve special cases of the problem $1 \mid pmtn,\ r_j, \bar{d}_j \mid \sum C_j$. Let us start with the generalized Baker's rule first. Recall that Baker's rule makes scheduling decisions only at the time a job completes or at the time new jobs are released. At each decision point, Baker's rule chooses the available job with the smallest remaining processing time.

Generalized Baker's Rule. Call an available job j *eligible* if it is feasible to complete j prior to any of the other available jobs. At each decision point, choose to process an eligible job with the smallest remaining processing time. Break ties between smallest eligible jobs by choosing the one with the earliest possible deadline.

Recall that Smith's rule determines which job to complete last. Once that job is determined, it leaves a problem instance with one less job which is recursively scheduled by the same rule. The generalized Smith's rule operates in the same spirit as Smith's rule.

Generalized Smith's Rule. Schedule the jobs by the EDD with the following refinement. When a choice must be made between jobs with the same deadline, choose to process any job, other than the one with the largest remaining processing time. (Break ties between jobs with the largest remaining processing time by choosing to process a job with the earliest possible ready time.) Suppose k is the job that completes last in the EDD schedule. Job k may be preempted several times and divides the schedule into several segments. Recursively schedule the jobs in each segment by the same rule.

With appropriate data structure, generalized Baker's rule and generalized Smith's rule can both be implemented to run in $O(n^2)$ time. Both rules can solve a much broader class of problem instances than the ones they generalized. A triple of jobs $(i,\ j,\ k)$ is called an *obstruction* if $r_i, r_k < r_j < d_i < d_j, d_k$ and $p_j < p_i, p_k$.

Theorem 3.9

Both generalized Baker's rule and generalized Smith's rule are optimal for $1 \mid pmtn,\ r_j,\ \bar{d}_j \mid \sum C_j$ if the set of jobs does not contain an obstruction.

An algorithm recognizing problem instances with no obstruction can be implemented to run in $O(n^2)$ time. Note that problem instances without an obstruction include those for which (1) the intervals $[r_j, d_j]$, $j = 1, 2, \ldots, n$, are nested; i.e., there is no pair of indexes i and j such that $r_i < r_j < d_i < d_j$; (2) release times and deadlines are oppositely ordered, i.e., there is a numbering of jobs such that $r_1 \leq r_2 \leq \cdots \leq r_n$ and $d_1 \geq d_2 \geq \cdots \geq d_n$; (3) processing times and deadlines are similarly ordered, i.e., there is a numbering of jobs such that $p_1 \leq p_2 \leq \cdots \leq p_n$ and $d_1 \leq d_2 \leq \cdots \leq d_n$; (4) processing times and release times are similarly ordered; and (5) processing times are identical.

3.4 Dual Objectives: Makespan and Total Completion Time

In the last two sections we considered the $C_{\max}$ and $\sum C_j$ objectives separately. In this section, we consider scheduling problems that minimize both objectives. A schedule that simultaneously minimizes both $C_{\max}$ and $\sum C_j$ is called an *ideal* schedule. Ideal schedules exist for certain sets of jobs. In nonpreemptive scheduling, it is possible to have an ideal schedule for two machines and equal-length jobs. Indeed, Coffman-Graham algorithm generates an ideal schedule. It is also possible to have ideal schedules for equal-length jobs with precedence constraint in the form of outtree. Brucker et al. [19] recently gave an algorithm that generates an ideal schedule for this case. On the other hand, it can be shown that there is no ideal schedule for equal-length jobs on three or more machines, even when the precedence constraint is in the form of intree. In preemptive scheduling, Coffman et al. [20] recently gave an algorithm to generate ideal schedules for two machines and equal-length jobs. The idea is to convert a Coffman-Graham schedule into a preemptive schedule in such a way that $C_{\max}$ and $\sum C_j$ are both minimized.

Since ideal schedules exist under very restrictive conditions, it makes sense to consider dual objectives. For example, we can consider the problem of minimizing $C_{\max}$ subject to the condition that $\sum C_j$ is minimum; we denote this problem by $\alpha \mid \beta \mid C_{\max}(\sum C_j)$. Or, we can consider the problem of minimizing $\sum C_j$ subject to the condition that $C_{\max}$ is minimized; this problem will be denoted by $\alpha \mid \beta \mid \sum C_j(C_{\max})$.

For nonpreemptive scheduling, $P2 \mid\mid C_{\max}(\sum C_j)$ is NP-hard in the ordinary sense, as the next theorem shows.

Theorem 3.10

$P2 \mid\mid C_{\max}(\sum C_j)$ is NP-hard in the ordinary sense.

Proof

We shall reduce the Even–Odd Partition problem (see Chapter 2 for a definition of the Even–Odd Partition problem) to $P2 \mid\mid C_{\max}(\sum C_j)$. Let $a_1 < a_2 < \cdots < a_{2n}$ be an instance of Even–Odd Partition and let $B = \frac{1}{2}\sum a_j$. Create $2n$ jobs, where job j has processing time $p_j = a_j$. Let the threshold for $C_{\max}$ be B.

A schedule with minimum $\sum C_j$ is one that assigns jobs $2i - 1$ and $2i$, $1 \le i \le n$, in any order on machines 1 and 2, one per machine. This schedule will have $C_{\max} \le B$ if and only if there is an Even–Odd Partition. □

Coffman and Sethi [21] proposed an approximation algorithm for $P \mid\mid C_{\max}(\sum C_j)$ with a worst-case bound of $(5m-4)/(4m-3)$, which is a tight bound. Eck and Pinedo [22] proposed a better algorithm for two machines with a worst-case bound of 28/27.

While $P2 \mid\mid C_{\max}(\sum C_j)$ is NP-hard, $P \mid pmtn \mid C_{\max}(\sum C_j)$ is solvable in polynomial time. Leung and Young [23] gave a polynomial-time algorithm which is described as follows.

Leung-Young Algorithm

1. Let there be n jobs and m machines. Assume n is a multiple of m. Otherwise, we can create additional dummy jobs with zero processing times. Sort the jobs in ascending order of processing times, i.e., $p_1 \le p_2 \le \cdots \le p_n$.
2. Put aside the longest m jobs (i.e., jobs $n, n-1, \ldots, n-m+1$) and schedule the remaining jobs by strict SPT rule. That is, jobs $1, m+1, 2m+1, \ldots$ are assigned to machine 1; jobs $2, m+2, 2m+2, \ldots$ are assigned to machine 2; and so on. Let S be the schedule and let the finishing times of machine i, $1 \le i \le m$, be denoted by f_i.
3. Compute the optimal schedule length D as follows. For each i, $1 \le i \le m$, let $t_i = (1/i)(\sum_{j=1}^{i} f_j + \sum_{j=1}^{i} p_{n-j+1})$. Let $D = \max\{t_i\}$.
4. Schedule the jobs $n, n-1, \ldots, n-m+1$, in that order on the m machines, one job per machine. The scheduling is done by one of the three rules given below. Suppose we are scheduling job j, $n-m+1 \le j \le n$.

a. If $p_j \leq D - f_m$, schedule job j completely on machine m. Delete machine m and decrement m by 1. Reindex the machines so that they are in ascending order of finishing times.
b. If there is a machine i such that $p_j = D - f_i$, schedule job j completely on machine i. Delete machine i and decrement m by 1. Reindex the machines so that they are in ascending order of finishing times.
c. If Rules (a) and (b) do not apply, there must be a machine i such that $p_j < D - f_i$ and $p_j > D - f_{i+1}$. Schedule $D - f_{i+1}$ amount of job j on machine $i+1$ and the remaining amount on machine i. Delete machine $i + 1$, update the finishing time of machine i, and decrement m by 1. Reindex the machines so that they are in ascending order of finishing times.

The proof that the above algorithm is optimal is rather lengthy. We refer the interested readers to Ref. [23] for a detailed proof.

Theorem 3.11

Leung-Young algorithm is optimal for $P \mid pmtn \mid C_{\max}(\sum C_j)$.

We now turn our attention to the problem of minimizing $\sum C_j$ subject to the condition that $C_{\max}$ is minimum. We shall restrict our attention to preemptive scheduling only, since it is NP-hard to find a minimum-length nonpreemptive schedule. The problem $P \mid pmtn \mid \sum C_j(C_{\max})$ can be solved in polynomial time. First, the optimal schedule length D is given by $D = \max\{\max\{p_j\}, (1/m)\sum p_j\}$. Next, we assign a deadline D to each job and seek a schedule that meets the common deadline and minimizes $\sum C_j$. Recently, Leung and Pinedo [16] gave a polynomial-time algorithm to solve $P \mid pmtn, \bar{d}_j \mid \sum C_j$, where $\bar{d}_j$ is the deadline of job j. Thus, we can apply the algorithm of Leung and Pinedo [16] to find an optimal schedule. Alternatively, we can use the algorithm of Gonzalez [24] to solve the problem as well.

Theorem 3.12

The problem $P \mid pmtn \mid \sum C_j(C_{\max})$ *can be solved in polynomial time.*

We note that $Q \mid pmtn \mid \sum C_j(C_{\max})$ can still be solved in polynomial time by using the algorithm of Gonzalez for uniform machines [25]. As well, McCormick and Pinedo [26] have given an algorithm to compute the Pareto curve for $\sum C_j$ and $C_{\max}$ on uniform machines.

3.5 The Maximum Lateness Objective

The problem $1 \mid\mid L_{\max}$ can be solved by the earliest due date (EDD) rule. The EDD rule schedules at each instant of time the job with the earliest due date; ties can be broken arbitrarily. Since preemption is unnecessary for jobs with identical ready times, the EDD rule solves $1 \mid pmtn \mid L_{\max}$ as well. The preemptive version of the EDD rule is optimal for $1 \mid pmtn,\ r_j \mid L_{\max}$. Unfortunately, $1 \mid r_j \mid L_{\max}$ is NP-hard in the strong sense, since there is a simple reduction from the 3-partition problem to it.

If we add precedence constraint to the jobs, we can still solve the problem in polynomial time. In fact, there is a polynomial-time algorithm due to Lawler [27] to solve a more general problem. Suppose each job j is subject to a cost function $f_j(C_j)$ and our goal is to find a schedule with the minimum $\max\{f_j(C_j)\}$. This can be solved as follows. Let $t = \sum p_j$. From among all the jobs with no immediate successor, choose the job j such that $f_j(t)$ is the smallest and schedule job j to complete at time t; ties can be broken arbitrarily. Delete job j from G and recursively apply the same rule to the remaining jobs.

We can apply the above algorithm to solve $1 \mid prec \mid L_{\max}$. For each job j, let $f_j(C_j)$ be defined as $f_j(C_j) = C_j - d_j$. Clearly, the above algorithm will find a schedule with the minimum $L_{\max}$. We can also use the above algorithm to solve $1 \mid prec, \bar{d}_j \mid L_{\max}$. In this problem, each job j has a deadline $\bar{d}_j$, in addition to its due date d_j. Our goal is to find a schedule such that all deadlines are met and such that $L_{\max}$ is minimized. For this case, we let $f_j(C_j) = C_j - d_j$ if $C_j \leq \bar{d}_j$; otherwise, $C_j = \infty$.

For parallel machines, $P2 \,||\, L_{\max}$ is NP-hard in the ordinary sense, while $P \mid pmtn, r_j \mid L_{\max}$ is solvable in polynomial time. That $P2 \,||\, L_{\max}$ is NP-hard follows from the observation that $P2 \,||\, C_{\max}$ is a special case of $P2 \,||\, L_{\max}$ (by letting each job to have a due date equal to zero) and that $P2 \,||\, C_{\max}$ is NP-hard in the ordinary sense. The problem $P \mid pmtn, r_j \mid L_{\max}$ can be solved as follows. Parametrize on the maximum lateness. Assuming $L_{\max} = z$, create for all jobs deadlines $\bar{d}_j = d_j + z$. We now check if there is a feasible schedule such that every job is executed between its ready time and deadline. The optimal value of $L_{\max}$ can be obtained by a binary search of z in a range between the lower and upper bounds of $L_{\max}$.

The problem of determining whether there is a schedule such that each job is executed between its ready time and deadline is called *Deadline Scheduling*. This problem can be solved by a network flow approach [28]. Let $b_0 < b_1 < \cdots < b_k$ be the distinct values of $\{r_1, \bar{d}_1, r_2, \bar{d}_2, \ldots, r_n, \bar{d}_n\}$. These $k+1$ values divide the time frame into k intervals: $u_i = [b_{i-1}, b_i]$, $1 \leq i \leq k$. We construct a network with source S_1 and sink S_2. For each job i, $1 \leq i \leq n$, create a job vertex J_i, and for each interval u_j, $1 \leq j \leq k$, create an interval vertex I_j. There is an arc from S_1 to each job vertex J_i. The capacity of the arc is p_i. For each job vertex J_i, there is an arc to each of the interval vertex I_j in which job i can execute in the interval u_j. The capacity of the arc is the length l_j of the interval u_j (i.e., $l_j = b_j - b_{j-1}$). Finally, there is an arc from each interval vertex I_j to S_2, and the capacity of the arc is $m \times l_j$. It is clear that there is a feasible schedule if and only if the maximum flow from S_1 to S_2 is $\sum p_j$.

Deadline scheduling can be solved by a simpler algorithm due to Sahni [29] if the job's ready times are identical, say $r_j = 0$ for all j. The algorithm schedules the jobs in ascending order of due dates. Suppose we are scheduling job j. Let the machines be sorted in descending order of finishing times and let f_i denote the finishing time of machine i. Job j will be scheduled by one of the following four rules:

Rule 1. If $p_j < d_j - f_1$, then schedule job j completely on machine 1.
Rule 2. If $p_j > d_j - f_m$, then print "Infeasible" and stop.
Rule 3. If $p_j = d_j - f_i$ for some i, then schedule job j completely on machine i.
Rule 4. If none of the above three rules apply, then there must be an index i such that $p_j > d_j - f_i$ and $p_j < d_j - f_{i+1}$. Schedule $d_j - f_i$ amount of job j on machine i and the remaining amount on machine $i+1$.

We note that Hong and Leung [30] have given another algorithm for jobs whose executable intervals are nested. That is, for any pair of jobs i and j, either $[r_i, \bar{d}_i]$ and $[r_j, \bar{d}_j]$ are disjoint, or one is contained in the other.

We can also solve the problem of minimizing $\sum C_j$ subject to the condition that $L_{\max}$ is minimum, i.e., $P \mid pmtn \mid \sum C_j(L_{\max})$. We first find the minimum $L_{\max}$, say z^*, by the above method. We then create for each job j a deadline $\bar{d}_j = d_j + z^*$. Finally, we use the algorithm of Leung and Pinedo [16] to solve $P \mid pmtn, \bar{d}_j \mid \sum C_j$.

3.6 The Number of Late Jobs Objective

The problem $1 \mid pmtn, r_j \mid \sum w_j U_j$ can be solved by a dynamic programming algorithm due to Lawler [31]. The running time of the algorithm is $O(n^3 W^3)$, where $W = \sum w_j$. The running time becomes polynomial when the objective is to minimize $\sum U_j$, since $W = n$ in this case. Unfortunately, Du et al. [32] showed that $Pm \mid pmtn, r_j \mid \sum U_j$ is NP-hard in the ordinary sense for every fixed $m \geq 2$. Thus, there is no hope to obtain a polynomial-time algorithm on parallel machines if the jobs have arbitrary ready times.

When the jobs have identical ready times, the problem can be solved in polynomial time on a fixed number of machines, even for uniform machines. Lawler [33] showed that $Qm \mid pmtn \mid \sum w_j U_j$ can be solved in $O(n^2 W^2)$ time for $m = 2$ and in $O(n^{3m-5} W^2)$ time for $m > 2$. Thus, $Qm \mid pmtn \mid \sum U_j$ can be solved in $O(n^4)$ time for $m = 2$ and in $O(n^{3m-3})$ time for $m > 2$. This is the best one can hope for since $P \mid pmtn \mid \sum U_j$ has been shown to be NP-hard in the ordinary sense [34].

The problem is even more difficult for nonpreemptive scheduling. $1 \mid d_j = d \mid \sum w_j U_j$ is NP-hard in the ordinary sense, as it is equivalent to the 0/1-knapsack problem (see Chapter 2 for a definition of the 0/1-knapsack problem). $P2 \mid d_j = d \mid \sum U_j$ is also NP-hard in the ordinary sense, by a simple reduction from the partition problem. On the other hand, $1 \mid\mid \sum U_j$ can be solved in $O(n \log n)$ time by the well-known Hodgson-Moore algorithm [35]. Since preemption is unnecessary when all jobs have identical ready times, Hodgson-Moore algorithm solves $1 \mid pmtn \mid \sum U_j$ as well. In the following we shall describe this algorithm in detail.

Hodgson-Moore Algorithm. Schedule jobs in ascending order of due dates. If in scheduling job j we encounter a due date violation, then delete from among the jobs in the schedule (including job j) the one with the largest processing time. Ties are broken arbitrarily. All deleted jobs are scheduled at the end, after the on-time jobs.

Theorem 3.13

Hodgson-Moore algorithm is optimal for $1 \mid\mid \sum U_j$ *and* $1 \mid pmtn \mid \sum U_j$.

Proof

The idea is to show that at each stage of the algorithm, it schedules the maximum number of on-time jobs and maintains the smallest total processing time of all on-time jobs. This assertion can be proved by induction. □

3.7 The Total Tardiness Objective

Minimizing $\sum T_j$ is computationally the most difficult among all scheduling problems. Du and Leung [36] showed that $1 \mid\mid \sum T_j$ is NP-hard in the ordinary sense. Notice that $1 \mid\mid \sum T_j$ is a very restricted version of the problem; it does not have ready times and precedence constraint. The only parameter that is variable is the processing time of the jobs. Thus, as far as polynomial-time algorithms are concerned, we cannot have arbitrary processing times in the problem instance, unless $P = \text{NP}$.

The problem $1 \mid p_j = p \mid \sum T_j$ can be solved in $O(n \log n)$ time. An optimal schedule can be obtained by sequencing the jobs in ascending order of due dates. Thus, it appears to be easier when the jobs have identical processing times. Polynomial solvability can still be maintained if we introduce ready times to the problem. Baptiste [37] gave a polynomial-time algorithm for $1 \mid p_j = p, r_j \mid \sum T_j$; see also [38]. On the other hand, we loose polynomial solvability if we add precedence constraint to the problem, as Leung and Young [39] showed that $1 \mid p_j = 1, chains \mid \sum T_j$ is NP-hard in the strong sense.

Lawler [40] gave a pseudo-polynomial algorithm for $1 \mid\mid \sum T_j$. His algorithm forms the basis of a fully polynomial approximation scheme for the problem [41]. In the following we shall describe this algorithm in detail.

Lawler's pseudo-polynomial algorithm is based on the next two lemmas. The proof of the next lemma is not difficult and can be found in Refs. [42, 43].

Lemma 3.2

If $p_j \leq p_k$ *and* $d_j \leq d_k$, *then there exists an optimal sequence in which job* j *is scheduled before job* k.

In the following lemma, the sensitivity of an optimal sequence to the due dates is considered. Two problem instances are considered, both of which have n jobs with processing times $p_1, p_2, \ldots, p_n$. The first instance has due dates $d_1, d_2, \ldots, d_n$. Let C'_k be the *latest* possible completion time of job k in *any* optimal sequence for this instance. The second instance has due dates the same as before, except that $d'_k = \max(d_k, C'_k)$.

Lemma 3.3

Any sequence that is optimal for the second instance is optimal for the first instance as well.

The proof of Lemma 3.3 can be found in Refs. [42, 43]. Suppose that the jobs are indexed such that $d_1 \leq d_2 \leq \cdots \leq d_n$, and $p_k = \max\{p_j\}$. That is, the job with the kth smallest due date has the largest processing time. From Lemma 3.2, it follows that there exists an optimal sequence in which jobs $1, 2, \ldots, k-1$ all appear, in some order, before job k. Let C'_k be the latest possible completion time of job k in any optimal sequence. If $d_l \leq C'_k < d_{l+1}, l \geq k$, then from Lemma 3.3, we can modify the due date of job k to $d'_k = C'_k$ without changing the problem; i.e., an optimal sequence for the new instance is also an optimal sequence for the original instance. From Lemma 3.2, it follows that there exists an optimal sequence in which jobs $1, 2, \ldots, k-1, k+1, \ldots, l$ all appear, in some order, before job k. Unfortunately, we do not know the value of C'_k, which means that we do not know the value of l. Fortunately, the possible values for l are $k, k+1, k+2, \ldots, n$, and so we can try all possibilities and choose the best of all schedules.

This suggests a dynamic programming algorithm. We assume that the jobs are indexed in ascending order of due dates. The heart of the algorithm is a recursive procedure SEQUENCE(t, I), where t is a time instant and I is a set of job indices. SEQUENCE(t, I) returns an optimal sequence of the jobs in I, starting at time t.

SEQUENCE(t, I)

```
If I = ∅, then σ* = empty sequence
Else
{
Let i_1 < i_2 < ··· < i_r be the jobs in I;
Find i_k with p_{i_k} = max{p_i | i ∈ I};
f* = ∞;
For j = k To r Do
     {
     I_1 = {i_v | 1 ≤ v ≤ j and v ≠ k}; t_1 = t;
     σ_1 = SEQUENCE(t_1, I_1);
     I_2 = {i_v | j < v ≤ r}; t_2 = t + Σ_{v=1}^{j} p_{i_v};
     σ_2 = SEQUENCE(t_2, I_2);
     σ = σ_1 || i_k || σ_2;
     Calculate the objective value of f(σ, t) for σ;
     If f(σ, t) < f* then
          {
          σ* = σ;
          f* = f(σ, t);
          }
     }
}
Return(σ*);
```

The algorithm calls SEQUENCE(t, I) with $t = 0$ and $I = \{1, 2, \ldots, n\}$. It can be shown that the running time of the algorithm is $O(n^3 p)$, where n is the number of jobs and $p = \sum p_j$.

Acknowledgment

This work is supported in part by the NSF Grant DMI-0300156.

References

[1] T. C. Hu, Parallel sequencing and assembly line problems, *Oper. Res.*, 9, 841–848, 1961.

[2] E. G. Coffman, Jr. and R. L. Graham, Optimal scheduling for two-processor systems, *Acta Informatica*, 1, 200–213, 1972.

[3] J. D. Ullman, NP-complete scheduling problems, *J. Comp. Syst. Sci.*, 10, 384–393, 1975.

[4] S. Lam and R. Sethi, Worst-case analysis of two scheduling algorithms, *SIAM J. Comp.*, 6, 518–536, 1977.

[5] R. L. Graham, Bounds for certain multiprocessing anomalies, *Bell Syst. Tech. J.*, 45, 1563–1581, 1966.

[6] R. L. Graham, Bounds on multiprocessing timing anomalies, *SIAM J. Appl. Math.*, 17, 263–269, 1969.

[7] R. McNaughton, Scheduling with deadlines and loss functions, *Manag. Sci.*, 6, 1–12, 1959.

[8] K. S. Hong and J. Y.-T. Leung, On-line scheduling of real-time tasks, *IEEE Trans. Comp.*, C-41, 1326–1331, 1992.

[9] R. R. Muntz and E. G. Coffman, Jr., Optimal preemptive scheduling on two-processor systems, *IEEE Trans. Comp.*, C-18, 1014–1020, 1969.

[10] R. R. Muntz and E. G. Coffman, Jr., Preemptive scheduling of real-time tasks on multiprocessor systems, *J. ACM*, 17, 324–338, 1970.

[11] W. A. Horn, Single-machine job sequencing with treelike precedence ordering and linear delay penalties, *SIAM J. Appl. Math.*, 23, 189–202, 1972.

[12] E. L. Lawler, Sequencing jobs to minimize total weighted completion time subject to precedence constraints, *Ann. Discrete Math.*, 2, 75–90, 1978.

[13] J. K. Lenstra, *Seq. Enumerative Met.*, Mathematical Centre Tracts 69, Mathematisch Centrum, Amsterdam, 1977.

[14] K. R. Baker, *Introduction to Sequencing and Scheduling*, Wiley, New York, 1974.

[15] W. E. Smith, Various optimizers for single stage production, *Naval Res. Log.*, 3, 59–66, 1956.

[16] J. Y.-T. Leung and M. L. Pinedo, Minimizing total completion time on parallel machines with deadline constraints, *SIAM J. Comp.*, to appear.

[17] J. Du and J. Y.-T. Leung, Minimizing mean flow time with release time and deadline constraints, *J. Algor.*, 14, 45–68, 1993.

[18] J. Du, J. Y.-T. Leung, and G. H. Young, Minimizing mean flow time with release time constraint, *Theor. Comp. Sci.*, 75, 347–355, 1990.

[19] P. Brucker, J. Hurink, and S. Knust, A polynomial algorithm for $P \mid p_j = 1, r_j, outtree \mid \sum C_j$, *Math. Meth. Oper. Res.*, 56, 407–412, 2003.

[20] E. G. Coffman, Jr., J. Sethuraman, and V. G. Timkovsky, Ideal preemptive schedules on two processors, *Acta Informatica*, to appear.

[21] E. G. Coffman, Jr. and R. Sethi, Algorithm minimizing mean flow time: schedule-length properties, *Acta Informatica*, 6, 1–14, 1976.

[22] B. Eck and M. Pinedo, On the minimization of the makespan subject to flow time optimality, *Oper. Res.*, 41, 797–800, 1993.

[23] J. Y.-T. Leung and G. H. Young, Minimizing schedule length subject to minimum flow time, *SIAM J. Comp.*, 18, 314–326, 1989.

[24] T. Gonzalez, Minimizing the mean and maximum finishing time on identical processors, Technical Report CS-78-15, Computer Science Dept., Pennsylvania State University, University Park, PA, 1978.

[25] T. Gonzalez, Minimizing the mean and maximum finishing time on uniform processors, Technical Report CS-78-22, Computer Science Dept., Pennsylvania State University, University Park, PA, 1978.

[26] S. T. McCormick and M. L. Pinedo, Scheduling n independent jobs on m uniform machines with both flow time and makespan objectives: a parametric analysis, *ORSA J. Comp.*, 7, 63–77, 1995.

[27] E. L. Lawler, Optimal sequencing of a single machine subject to precedence constraints, *Manag. Sci.*, 19, 544–546, 1973.

[28] W. A. Horn, Some simple scheduling algorithms, *Naval Res. Logist. Quart.*, 21, 177–185, 1974.

[29] S. Sahni, Preemptive scheduling with due dates, *Oper. Res.*, 27, 925–934, 1979.

[30] K. S. Hong and J. Y.-T. Leung, Preemptive scheduling with release times and deadlines, *J. Real-Time Syst.*, 1, 265–281, 1989.

[31] E. L. Lawler, A dynamic programming algorithm for preemptive scheduling of a single machine to minimize the number of late jobs, *Ann. Oper. Res.*, 26, 125–133, 1990.

[32] J. Du, J. Y.-T. Leung, and C. S. Wong, Minimizing the number of late jobs with release time constraint, *J. Combin. Math. Combin. Comput.*, 11, 97–107, 1992.

[33] E. L. Lawler, Preemptive scheduling of uniform parallel machines to minimize the weighted number of late jobs, Report BW 105, *Mathematisch Centrum*, Amsterdam, 1979.

[34] E. L. Lawler, Recent results in the theory of machine scheduling, in A. Bachem, M. Grotschel, and B. Kortel (eds.), *Mathematical Programming: The State of the Art*, Springer, 1982.

[35] J. M. Moore, An *n* job, one machine sequencing algorithm for minimizing the number of late jobs, *Manag. Sci.*, 15, 102–109, 1968.

[36] J. Du and J. Y.-T. Leung, Minimizing total tardiness on one processor is NP-hard, *Math. Operat. Res.*, 15, 483–495, 1990.

[37] Ph. Baptiste, Scheduling equal-length jobs on identical parallel machines, *Discr. Appl. Math.*, 103, 21–32, 2000.

[38] Ph. Baptiste and P. Brucker, Scheduling equal processing time jobs, in J. Y.-T. Leung (ed.), *Handbook of Scheduling Algorithms, Models and Performance Analysis*, CRC Press, Boca Raton, FL, 2004.

[39] J. Y.-T. Leung and G. H. Young, Minimizing total tardiness on a single machine with precedence constraint, *ORSA J. Comp.*, 2, 346–352, 1990.

[40] E. L. Lawler, A 'pseudopolynomial' algorithm for sequencing jobs to minimize total tardiness, *Ann. Discr. Math.*, 1, 331–342, 1977.

[41] E. L. Lawler, A fully polynomial approximation scheme for the total tardiness problem, *Oper. Res. Lett.*, 1, 207–208, 1982.

[42] P. Brucker, *Scheduling Algorithms*, Springer-Verlag, New York, 2001.

[43] M. L. Pinedo, *Scheduling Theory Algorithms, and Systems*, Prentice Hall, New Jersey, 2002.

II

Classical Scheduling Problems

4

Elimination Rules for Job-Shop Scheduling Problem: Overview and Extensions

Jacques Carlier
Compiègne University of Technology

Laurent Péridy
West Catholic University

Eric Pinson
West Catholic University

David Rivreau
West Catholic University

4.1 Introduction

4.1.1 Problem Statement

In the job shop scheduling problem a set J of n jobs $J_1, J_2, \ldots, J_n$ has to be processed on a set M of m different machines $M_1, M_2, \ldots, M_m$. Each job J_j consists of a sequence of m_j operations $O_{j1}, O_{j2}, \ldots, O_{jm_j}$ that must be scheduled in this order. Moreover, each operation needs to be processed only on a specific machine among the m available ones. Preemption is not allowed and machines can handle at most one operation at a time. Operation O_{jk} has a fixed processing time p_{jk}. The objective is to find an operating sequence for each machine to minimize the makespan $C_{\max} = \max_{j=1,n} C_j$, where C_j denotes the completion time of the last operation of job $J_j (j = 1, \ldots, n)$.

The problem is modeled by a disjunctive graph $\mathcal{G} = (G, D)$ (Figure 4.1) [1], where $G = (X, U)$ is a conjunctive graph associated with the job sequences and D is a set of disjunctions. A disjunction

1-58488-397-9/\$0.00+\$1.50

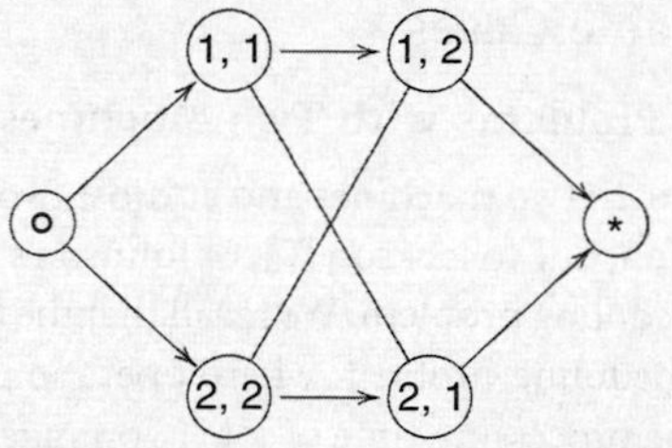

FIGURE 4.1 Disjunctive graph: two jobs, $J_1 = \{(1, 1), (1, 2)\}$ and $J_2 = \{(2, 2), (2, 1)\}$ on two machines.

$[i, j] = \{(i, j), (j, i)\}$ is associated with each pair of operations processed by the same machine. Two dummy operations $\circ$ and $\star$ which represent respectively the beginning and the end of the schedule are included in the graph G. A schedule on the disjunctive graph $\mathcal{G} = (G, D)$ is a set of starting times $T = \{t_i \mid i \in X\}$ such that

– the conjunctive constraints are satisfied:

$$t_j - t_i \geq p_i, \quad \forall (i, j) \in U$$

– the disjunctive constraints are satisfied:

$$t_j - t_i \geq p_i \vee t_i - t_j \geq p_j, \quad \forall (i, j) \in D$$

This formulation can be used to compute heads and tails for operations. The head (release date), r_j, is the length of the longest path, from node $\circ$ to the node associated with operation j. This head is clearly a lower bound on the starting time of the given operation. Analogously, the length of the longest path from node j to node $\star$ is referred to as the tail, q_j. If an upper bound *UB* is known, a due date, d_j, can be calculated for operation j by setting $d_j = UB - q_j$.

A mixed integer formulation can also be derived by introducing a binary variable x_{ij} for each disjunction, which specifies whether operation i precedes operation j or not (M is a large integer).

$$\begin{aligned}
&\min C_{\max} \\
&\text{s.t. } \forall i \in X, t_i \geq 0 \\
&\quad\ \ \forall i \in X, t_i + p_i \leq C_{\max} \\
&\quad\ \ \forall (i, j) \in U, t_i + p_i \leq t_j \\
&\quad\ \ \forall [i, j] \in D, t_i + p_i - M(1 - x_{ij}) \leq t_j \\
&\quad\ \ \forall [i, j] \in D, t_j + p_j - Mx_{ij} \leq t_i \\
&\quad\ \ \forall [i, j] \in D, x_{ij} \in \{0, 1\}
\end{aligned}$$

4.2 A Brief State of the Art

Only few special cases of this problem can be solved in polynomial time, and they are briefly reviewed in the next section. Problems with two machines and $m_j \leq 2 (j = 1, \ldots, n)$, and problems with three machines and $m_j \leq 2 (j = 1, \ldots, n)$ are $\mathcal{NP}$-hard [2,3]. Problems with three machines and unit processing times have been proved to be strongly $\mathcal{NP}$-Hard [4]. The job shop problem with three jobs is $\mathcal{NP}$-hard [5]. Note that the previous results hold even if preemption is allowed [3]. Finally, we indicate that determining the existence of a schedule with an upper bound of four is $\mathcal{NP}$-complete even when all operations have unit processing times [6].

4.2.1 Polynomially Solvable Cases

4.2.1.1 Job Shop Scheduling Problems with Two Machines

The job shop scheduling problem with two machines and at most two operations per job can be solved in $\mathcal{O}(n \log n)$ time by a simple extension, due to Jackson [7], of Johnson's algorithm [8] which was introduced for the two-machine flow shop scheduling problem. We recall that the flow-shop scheduling problem is the particular case of the job shop scheduling problem where operation sequences do not depend on jobs. Let p_{i1}(respectively p_{i2}) denote the processing time of job J_i on machine M_1 (respectively M_2), and m_{i1} (respectively m_{i2}) the index of the first (respectively second) machine assigned to job J_i in its job sequence.

Proposition 4.1 [Johnson's rule]

If $\min(p_{i1}, p_{j2}) \leq \min(p_{i2}, p_{j1})$ *then there exists an optimal schedule for the two-machine flow shop scheduling problem in which job* J_i *precedes job* J_j.

Proposition 4.2 [Jackson's rule]

Partition the set of jobs into the following four groups:

$$J_{12} = \{J_j \mid m_j = 2, m_{1j} = 1, m_{2j} = 2\} \qquad J_{21} = \{J_j \mid m_j = 2, m_{1j} = 2, m_{2j} = 1\}$$
$$J_1 = \{J_j \mid m_j = 1, m_{1j} = 1\} \qquad J_2 = \{J_j \mid m_j = 1, m_{1j} = 2\}$$

An optimal schedule for the job shop scheduling problem with two machines and $m_j \leq 2 (j = 1, \ldots, n)$ corresponds to the following processing orders:

– on machine M_1, $J_{12} - J_1 - J_{21}$
– on machine M_2, $J_{21} - J_2 - J_{12}$

where the jobs in J_{12} and J_{21} are ordered according to Johnson's rule, and the jobs in J_1 and J_2 are in an arbitrary order.

Note that the two-machine job shop scheduling problem with unit processing times can be solved in $\mathcal{O}(\sum_{j=1,\ldots,n} m_j)$ time [9].

4.2.1.2 Job Shop Scheduling Problem with Two Jobs

The job shop scheduling problem with two jobs can also be solved in polynomial time [10,11]. Brucker proposes an $\mathcal{O}(m_1 m_2 \log m_1 m_2)$ time algorithm achieving this goal. After Akers, he shows that the problem can be formulated as a shortest-path problem in the plane with rectangular objects as obstacles. These obstacles correspond to a common machine requirement of both jobs according to their job sequence. More precisely, denote by m_{1j}(respectively m_{2j}) the index of the jth machine assigned to job J_1 (respectively J_2) in its job sequence, and by p_{1j} (respectively p_{2j}) the corresponding processing times. If $m_{1i} = m_{2j} (i = 1, \ldots, m_1, j = 1, \ldots, m_2)$, we create an obstacle with south-west coordinates $(\sum_{k=1,\ldots,i-1} p_{1k}, \sum_{k=1,\ldots,j-1} p_{2k})$ and length p_{1i} (respectively p_{2j}) on the x axis (respectively y axis). It can be proved that any feasible path starts at the origin O with coordinates $(0, 0)$ and ends at F with coordinates $(\sum_{k=1,\ldots,m_1} p_{1k}, \sum_{k=1,\ldots,m_2} p_{2k})$, going either diagonally or in parallel to the axes and avoiding the interior of the obstacles. This shortest-path problem can be reduced to the problem of finding a shortest path from O to F in a specific network that can be constructed in $\mathcal{O}(p \log p)$ steps, where p is the number of obstacles. Moreover, Brucker shows that the shortest path on this network can be computed in $\mathcal{O}(p)$ steps (notice that $p = \mathcal{O}(m_1 m_2)$) (see Figure 4.2).

4.2.2 Exact Methods

Balas [12] presents one of the first branch and bound method for the job shop scheduling problem, based on critical operations on the disjunctive graph. McMahon and Florian [13] present one of the first

Job sequence				
J_1	$(m_1, 3)$	$(m_2, 2)$	$(m_1, 3)$	$(m_3, 5)$
J_2	$(m_1, 3)$	$(m_3, 2)$	$(m_2, 4)$	

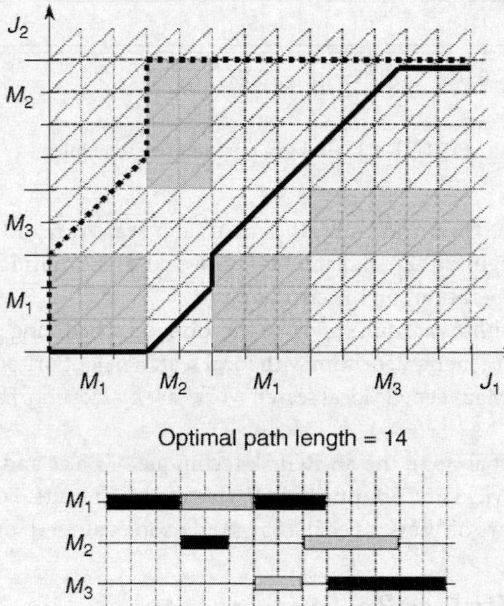

FIGURE 4.2 Job shop scheduling problem with two jobs.

successful exact method by determining the critical job and all the jobs with a due date greater than the critical job. Following the works of Lageweg et al. [14] and Barker and McMahon [15], Carlier and Pinson [16] solve the famous ten jobs and ten machines scheduling problem proposed by Fisher and Thompson [17] and remained open for over 20 years. They develop the concept of immediate selections and use Jackson's Preemptive Schedule as lower bound. In Refs. [18,19], they improve their work by introducing new immediate selections and new algorithms with lower complexity. Many works are derived from [16]: Applegate and Cook [20], Brucker et al. [21] (branching based on the critical block scheme of Grabowski et al. [22]), Perregard and Clausen [23] (parallel search strategy), Caseau and Laburthe [24] (interval approach), Baptiste and Le Pape [25] (constraint programming approach), Martin and Shmoys [26], Brucker and Brinkotter [27](see survey of Jain and Meeran [28]).

4.2.3 Heuristic Methods

Lots of heuristic methods are proposed in the literature to solve job shop scheduling problems such as priority dispatching rules, simulated annealing, tabu search, or genetic algorithm.

The most well-known and comprehensive survey of priority dispatching rules for scheduling problems is by Panwalkar and Iskander [29] where 113 rules are presented. More recently, Blackstone et al. [30], Haupt [31], Bhaskaran and Pinedo [32], and Chang et al. [33] provided extended discussions on this topic.

One of the most famous heuristic method is probably the *shifting bottleneck procedure* developed by Adams et al. [34]. The main idea is to optimally sequence one by one the machines, using Carlier's algorithm [35] for the one machine scheduling problem. At each step, the sequence and the heads and tails of the last optimized machine are used to reoptimize the bottleneck machine. In 1998, Balas and Vazacopoulos [36] improved the method and obtained very good results.

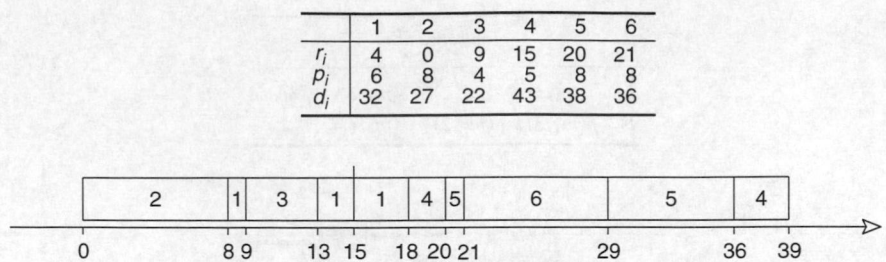

	1	2	3	4	5	6
r_i	4	0	9	15	20	21
p_i	6	8	4	5	8	8
d_i	32	27	22	43	38	36

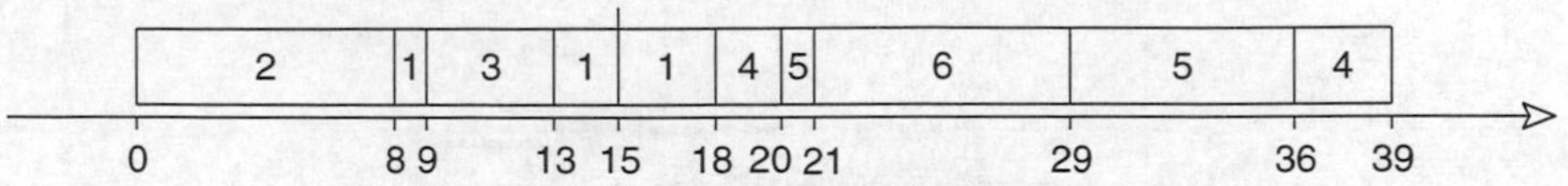

FIGURE 4.3 Jackson's preemptive schedule.

Derived from the works of Glover [37–39], the main tabu search methods are proposed by Laguna et al. [40,41], Barnes and Laguna [42], Taillard [43], Nowicki and Smutnicki [44]. They are based on insertion or swap moves of operations on the critical path.

The results of genetic algorithms are quite poor on the job shop scheduling problem [45–48]. The superiority of genetic local search (genetic algorithm with local search neighborhood structures) is highlighted by Della Croce et al. [49,50]. Main genetic local search works are by Dorndorf and Pesch [51] and Mattfeld [52].

As indicated by Vaessens et al. [53], the SB-RGLSk technique of Balas and Vazacopoulos [36] and the tabu search algorithm of Nowicki and Smutnicki [44] give the best results. For extended discussions, the reader is referred to the surveys of Vaessens et al. [53], Anderson et al. [54], or Jain and Meeran [28].

4.3 Elimination Rules: Basic Concepts

4.3.1 Introduction

Elimination rules appear of prime interest in B&B methods for efficiently pruning the associated search tree. Almost all of them focus on a subset I of operations that have to be processed on a given processor by characterizing particular partial operating sequences that cannot lead to a global feasible solution for the job shop problem. So, in some cases, it is possible either to select immediately some disjunctive constraints (immediate selections) or to narrow some processing intervals of operations (adjustment of time windows). This section deals with basic elimination rules and related adjustments proposed for the job shop problem during the last two decades. Some of these are based on Jackson's Preemptive Scheduling problem we present now.

Let us consider a subet I of operations that have to be processed on the same machine. Jackson's preemptive schedule (*JPS*) (Figure 4.3) is the list schedule associated with the earliest due date (EDD) priority dispatching rule. From time 0 to the end of the schedule, we schedule at each time instant — the part of available operation with minimal due date. Using heap structures, *JPS* can be computed in $\mathcal{O}(n \log n)$ steps. Its makespan is a lower bound of the job shop scheduling problem [35].

We will aso use the following notations:

$$\forall J \subset I, r(J) = \min_{j \in J} r_j$$

$$\forall J \subset I, p(J) = \sum_{j \in J} p_j$$

$$\forall J \subset I, d(J) = \max_{j \in J} d_j$$

4.3.2 Immediate Selections on Disjunctions

A first and simple way for designing elimination rules for disjunctive problems is to focus on disjunctions.

Immediate Selections on Disjunctions [55]

Let i and j be two operations of I.

If $r_j + p_j + p_i > d_i$, then $(i \rightarrow j)$ in any solution, and we can set

$$d_i = \min(d_i, d_j - p_j)$$
$$r_j = \max(r_j, r_i + p_i)$$

Carlier and Pinson in [19] have proposed an $\mathcal{O}(n \log n)$ time algorithm, computing all the immediate selections on disjunctions.

4.3.3 Immediate Selections on Ascendant/Descendant Sets

Carlier and Pinson extended the previous concept to the case of the relative positioning of an operation i and a subset of operations $J \subset I$ $(i \notin J)$. For this purpose, they identify three possible cases:

– if

$$r_i + p_i + p(J) > d(J) \tag{4.1}$$

then Job i cannot be scheduled *before* the set J

– if

$$r(J) + p_i + p(J) > d(J) \tag{4.2}$$

then Job i cannot be scheduled *in the middle* of the set J

– if

$$r(J) + p(J) + p_i > d_i \tag{4.3}$$

then Job i cannot be scheduled *after* the set J

By definition, J is called an ascendant set of operation i if relations (4.1) and (4.2) are satisfied. Symmetrically we can define the notion of descendant set, on the basis of relations (4.2) and (4.3). We obtain the following.

Immediate Selection on Ascendant Set [16]

Let $i \in I$ and $J \subseteq I \setminus \{i\}$.

If $r(J \cup \{i\}) + p_i + p(J) > d(J)$, then $(J \rightarrow i)$ in any solution, and we can set (Figure 4.4).

$$r_i = \max(r_i, \max_{J' \subset J}(r(J') + p(J')))$$

Immediate Selection on Descendant Set [16]

Let $i \in I$ and $J \subseteq I \setminus \{i\}$.

If $r(J) + p_i + p(J) > d(J \cup \{i\})$, then $(i \rightarrow J)$, in any solution, and we have

$$d_i = \min(d_i, \min_{J' \subset J}(d(J') - p(J')))$$

Clearly, if J is an ascendant set of operation i, then i is an output of $J \cup \{i\}$. Denoted by p_j^+ the processed time of operation $j \in J$ after time instant r_i in *JPS*, and C_j its completion time in *JPS*, Carlier and Pinson proved that finding the optimal adjustment of r_i is equivalent to looking for a nonempty subset $K^+ \subset \{j \in I \mid p_j^+ > 0\}$, satisfying

$$r_i + p_i + p^+(K^+) > d(K^+) \tag{4.4}$$

	1	2	3	4	5	6
r_i	4	0	9	15	20	21
p_i	6	8	4	5	8	8
d_i	32	27	22	43	38	36

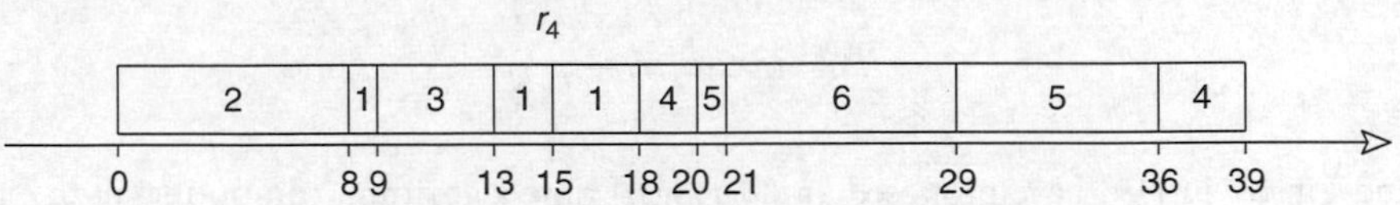

If we apply the previous result with $i = 4$, we obtain:

$$r_4 + p_4^+ + p_1^+ + p_5^+ + p_6^+ = 15 + 5 + 3 + 8 + 8 > \max(d_1, d_5, d_6)$$

So $K_4^+ = \{1, 5, 6\}$ and we can adjust r_4 by setting $r_4 = \max\{C_1, C_5, C_6\} = 36$

FIGURE 4.4 Immediate selection on ascendant set.

Obviously if such a set exists, there is a maximal set K^* satisfying (4.4), and we have the optimal adjustment

$$r_i = \max(r_i, \max_{j \in K^*} C_j)$$

Figure 4.4 shows a simple application of this result.

An $\mathcal{O}(n \log n)$ time algorithm performing all the immediate selections on ascendant/descendant sets can be found in Ref. [19].

4.3.4 Not-First/Not-Last Rules

If relation (4.2) is not satisfied simultaneously with one of the relations (4.1) or (4.3), no immediate selection can be deduced. For instance, if (4.1) is verified, the only information we can obtain is that i is not an input of $J \cup \{i\}$. Thus, job i cannot be scheduled entirely before the set of operations J, which implies that at least one job in J must be scheduled before job i in any feasible solution. Of course, we get a symmetric result by considering the case where only relation (4.3) is satisfied.

Not-First Adjustments [55,56]

Let $i \in I$ and $J \subseteq I \setminus \{i\}$.

If $r_i + p_i + p(J) > d(J)$, then we can adjust r_i by setting

$$r_i = \max(r_i, \min_{j \in J}\{r_j + p_j\})$$

Not-Last Adjustments [55,56]

Let $i \in I$ and $J \subseteq I \setminus \{i\}$.

If $r(J) + p(J) + p_i > d_i$, then we can adjust d_i by setting

$$d_i = \min(d_i, \max_{j \in J}\{d_j - p_j\})$$

4.3.5 r-Sets Conditions

Dewess [57] proposes to examine the feasible *permutations* set of r jobs, in order to identify immediate selections. The notation of this section comes from the paper of Brucker et al. [21].

Let us consider a subset of jobs J, $|J| = r$, and $\pi = (\pi(1), \ldots, \pi(r))$ a permutation of the set J. The sequence π is called *feasible* if, for any selected disjunction $(i \rightarrow j)$, we have $\pi^{-1}(i) < \pi^{-1}(j)$.

Moreover, we define $\delta(\pi)$ by

$$\delta(\pi) = \min_{1 \leq i \leq j \leq r} \left\{ d_{\pi(j)} - r_{\pi(i)} - \sum_{k=i}^{j} p_{\pi(k)} \right\}$$

It can be easily seen that if $\delta(\pi)$ is strictly negative, then the sequence π is not feasible.

Let i and j be two jobs of the set J. The pair (i, j) is called *J-derivable* if and only if, for any feasible sequence in which the job j is scheduled before the job i, we have $\delta(\pi) < 0$.

r-Set Conditions [57]
Let i, j be two jobs of the set $J \subseteq I$.
If (i, j) is J-derivable, then $(i \to j)$ in any solution.

The case $r = 2$ complies with the immediate selections on disjunctions. The studies realized by Dewess show that beyond the case $r = 3$, the time consumption to compute the immediate selections becomes too large in relation to obtained adjustments.

Brucker et al. [21] propose the following particular version of the 3-set conditions based on the assumption that all the immediate selections on disjunctions (2-set conditions) and all the associated adjustments have been performed.

3-Set Conditions [21]
Let i, j and k be three jobs of the set I.
If $r_j + p_j + p_i + p_k > d_k$, $r_j + p_j + p_k + p_i > d_i$, and $r_k + p_k + p_j + p_i > d_i$,
then $(i \to j)$ in any solution.

Likewise, Brucker et al. exhibit an $\mathcal{O}(n^2)$ time algorithm that strictly dominates the 3-set conditions. By using this rule, the search tree in a branch and bound is reduced by 10% in average with a similar increase of the computing time.

4.4 Extensions

4.4.1 Introduction

As described in the previous section, most of the basic elimination rules of the job shop problem focus on the relative positioning of one operation and a subset of operations. Obviously more complex configurations can be considered in order to capture more local information on the problem. In this section, we introduce a general framework matching this goal, the so-called Set Partition Selections Problem, and generalizing the basic elimination rules. Next, a particular extension of the not-first/not-last (NF-NL) rule is proposed.

4.4.2 The Set Partition Selections Problem (SPSP)

Beyond the study of the relative positioning of two distinct operations or one operation and a set of distinct operations to each other, we can naturally intend to get information relying on the relative positioning of two sets of operations. Let L and H be two disjoint subsets of operations in I. A relative positioning of operations of H in relation with those in L leads to a partition of L in three subsets B, M, and A:

- B is the set of operations scheduled before the set H
- M is the set of operations scheduled in the middle of the set H
- A is the set of operations scheduled after the set H

Let P_L denote the collection of all the possible 3-partitions (B, M, A) from set L. Define by

$$P_{L,H} = \{(B, M, A) \in P_L \mid \max\{r(H), r(B) + p(B)\} + p(H) + p(M) \leq \min\{d(H), d(A) - p(A)\}\}$$

we get the following.

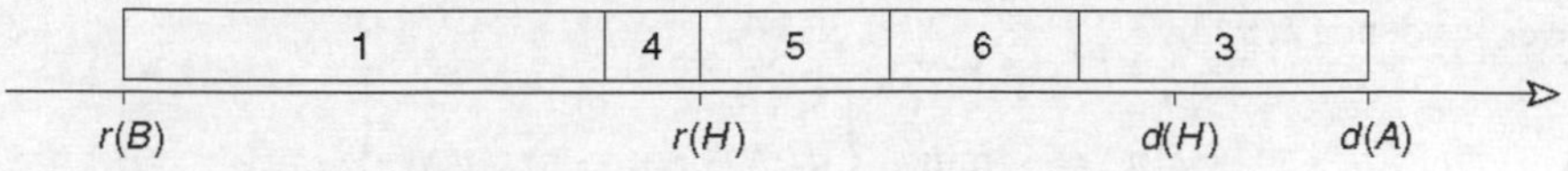

FIGURE 4.5 Relative positioning of L in relation to H.

Set Partition Rule [58]
Let L and H be two subsets of I such that $L \cap H = \varnothing$ and $l \in L$.
If $\forall(B, M, A) \in P_{L,H}$, $l \in B$, then $(l \rightarrow H)$ in any solution.
If $\forall(B, M, A) \in P_{L,H}$, $l \in A$, then $(H \rightarrow l)$ in any solution.

As pointed out in the introduction, immediate selections on disjunctions and on ascendant/descendant sets, as well as the immediate selections on 3-set conditions can be formulated as particular cases of this general rule.

– Immediate selections on disjunctions comply with the particular case

$$L = \{i\} \quad \text{and} \quad H = \{j\}$$

– Immediate selections on sets comply with the particular case

$$L = \{i\} \quad \text{and} \quad H \subseteq I \setminus \{i\}$$

– 3-set conditions comply with the particular case

$$L = \{ik\} \quad \text{and} \quad H = \{j\}$$

Unfortunately, but not really surprisingly, such a general framework cannot operate in practice on hard instances because of the following complexity result.

Proposition 4.3
Let L and H be two fixed sets and $l \in L$. Establishing by Set Partition Selection the immediate selection $(H \rightarrow l)$ is a co–NP complete problem.

Several attempts focusing on well suited relaxations on SPS rule can be found in Ref. [58]. Nevertheless, the deduced information is in average quite disappointing in comparison with the required computational effort.

Example 4.1
Let us consider the following example:

	1	2	3	4	5	6
r_i	0	1	3	0	6	7
p_i	5	1	3	1	2	2
d_i	15	15	13	13	11	11

Let $L = \{1, 3, 4\}$ and $H = \{5, 6\}$.
Figure 4.5 complies with the set partition $(B, M, A) = (\{1, 4\}, \varnothing, \{3\})$.
The application of the rules from Section 4.3 does not allow immediate selections to be found between the jobs in L and the set of jobs H.[1] But, an accurate analysis of the set partitions shows that it is not

[1]In fact, classical immediate selections do not find any selection.

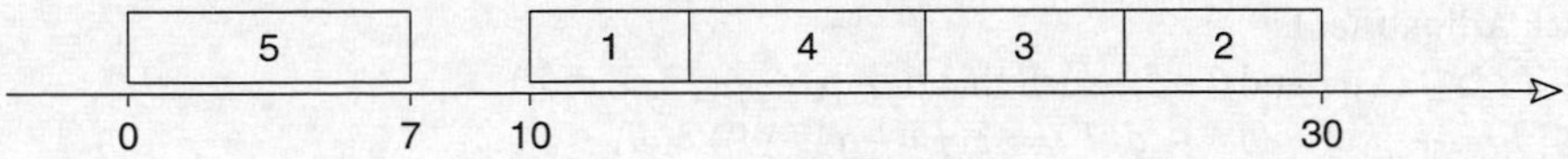

FIGURE 4.6 Operation 1 is started at 10.

possible to schedule job 4 before the set H. Indeed, the jobs 1 and 3 cannot be scheduled simultaneously before or after the set H. So, we have two cases:

- either job 1 is scheduled before the set H, and job 3 has to be scheduled after set H
- or job 1 is scheduled after the set H, and job 3 has to be scheduled before the set H

In both cases, we observe that job 4 cannot be scheduled in the middle of or after the set of jobs H. So, in any solution, job 4 is scheduled before set H and we have a set partition selection.

4.4.3 Stack Adjustments

In contrast with the previous elimination rules that try to detect immediate selections to deduce adjustments, stack adjustments attempt to directly narrow the time windows of jobs.

Example 4.2

For clarity purpose, let us consider the following example:

	1	2	3	4	5
r_i	9	3	1	0	0
p_i	4	5	5	6	7
d_i	30	21	26	21	28

As it can be observed, there is no immediate selection, since permutations [3, 2, 1, 4, 5], [3, 4, 1, 2, 5], [4, 2, 1, 3, 5] and [5, 2, 4, 3, 1] are all feasible.

Let us determine if operation 1 can start at time 10 of its domain $D_1 = [9, 26]$. Operations from $J = \{2, 3, 4, 5\}$ represent a total of 23 units of processing times that must be performed either before or after operation 1. If we do not interrupt operations before 1, we can easily check that at most 7 units of processing times from J can be processed before time 10 (see Figure 4.6). This entails a minimum of 16 units from set J to be scheduled after operation 1. So, under the assumption that operation 1 starts at time 10, it can be deduced that the set J completes its execution at least at time $t = 10+4+16 = 30$. Considering that no operation of J can finish its processing after time 28, then we can invalidate the assumption that 1 can be started at time 10 (see Figure 4.6). This argument still holds for time $t = 9$. Consequently, we have $D_1 \leftarrow D_1 \backslash \{9, 10\} = [11, 26]$ and therefore we can set r_1 to 11.

In general, if stack adjustments can adjust heads and tails, they will try before all to reduce the domain of feasible starting times of a given operation i.

The following proposition gives a formal statement of stack adjustments.

Stack Adjustments [59]

Let $i \in I$, $J \subseteq I \backslash \{i\}$ and $\theta \in D_i$ such that $r(J) + p(J) > \theta$.
If $\forall J' \subseteq J, r(J') + p(J') \leq \theta \Rightarrow \theta + p_i + p(J \backslash J') > d(J)$,
then θ is a forbidden starting time for operation i ($D_i \leftarrow D_i \backslash \{\theta\}$).

The proof of this proposition is obvious. Indeed, the stack adjustment condition just states that it is not possible to split set J around operation i, if this operation is started at time θ.

The stack adjustment condition is not symmetrical. The starting time of part J' is set to $r(J')$, whereas the finishing time of part $J' \backslash J$ is overestimated: $d(J)$ (instead of $d(J \backslash J')$). Consequently, a dual version of these adjustments can be expressed as follows.

Dual Stack Adjustments

Let $i \in I$, $J \subseteq I \backslash \{i\}$ and $\theta \in D_i$ such that $\theta + p_i + p(J) > d(J)$.
If $\forall J' \subseteq J, \theta + p_i + p(J') \leq d(J') \Rightarrow r(J) + p(J \backslash J') > \theta$,
then θ is a forbidden starting date for operation i ($D_i \leftarrow D_i \backslash \{\theta\}$).

Anything that applies to stack adjustments can easily be transferred to dual stack adjustments. Therefore, in the remainder of this section, we will focus on primal stack adjustments.

To compute them, we will use the following function:

$$\gamma_J(\theta) = \begin{cases} \max_{J' \subseteq J} \{p(J') \,|\, r(J') + p(J') \leq \theta\} & \text{if } \min_{j \in J}(r_j + p_j) \leq \theta \\ 0 & \text{otherwise} \end{cases}$$

The quantity $\gamma_J(\theta)$ represents an evaluation of the maximum amount of processing times of operations from J that can be performed before θ. We can notice that the case $\gamma_J(\theta) = 0$ is similar to a not-first condition. Stack adjustments can be reformulated as follows.

Forbidden Starting Times

Let $J \subseteq I \backslash \{i\}$.
If $\gamma_J(\theta) < p(J)$ and $\theta + p_i + p(J) - \gamma_J(\theta) > d(J)$,
then θ is a forbidden starting time for operation i.

Quantities $\gamma_J(\theta)$ can easily be computed by using a dynamic programming approach. Moreover, it can be shown that stack adjustments can be made in $O(n^3 \cdot T)$ for all operations where $T = d(I) - r(I)$. Unfortunately, finding a polynomial algorithm for that kind of adjustment is unlikely, as it is shown by the following proposition.

Proposition 4.4

Let $i \in I$, $J \subseteq I \backslash \{i\}$ and $\theta \in D_i$ such that $r(J) + p(J) > \theta$. Finding stack adjusment $D_i \leftarrow D_i \backslash \{\theta\}$ is a co−NP complete problem.

However, a last comment should be added on function γ_J. Since γ_J is clearly a knapsack-like function, it finally appears to be a step function, as can be seen in Figure 4.7 for Example 4.2.

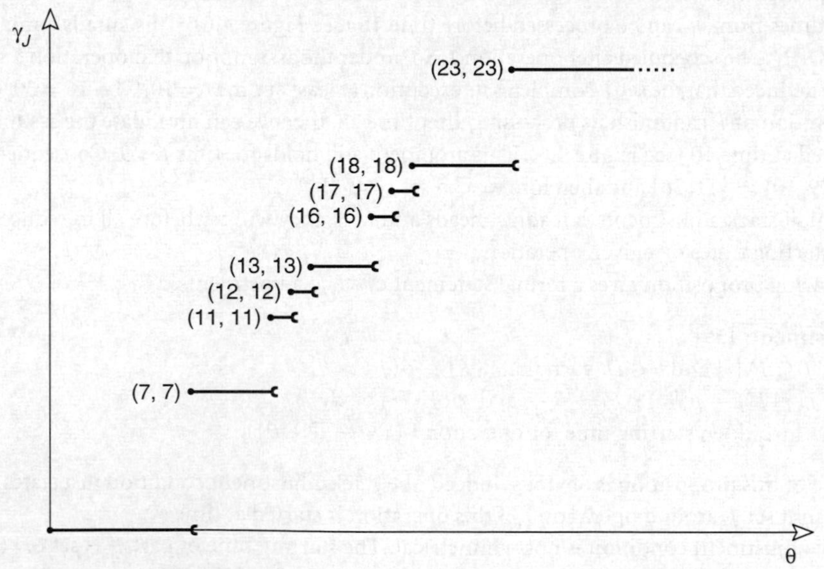

FIGURE 4.7 γ_J function for $J = \{1, 2, 3, 4\}$.

As a result, γ_J function can be fully defined on some critical pair $(\theta_l, \gamma_J(\theta_l))$. This simple statement leads to two main issues. First, this property entitles us to use state space reduction techniques, in order to speed up recurrence relations. Moreover, this critical pair $(\theta_l, \gamma_J(\theta_l))$ allows us to directly remove intervals inside domains more than single values.

4.5 Local Shaving

Carlier and Pinson [19], and Peridy [58] in parallel with Martin and Shmoys [26] introduce the concept of global operations. Global operations use a satisfiability test to determine whether or not there exists a solution that satisfies a given condition *cond*. If no solution satisfies *cond*, then $\neg$ *cond* is satisfied in any solution. Notice that this concept is called *shaving* by Martin and Shmoys [26].

Let us consider the time window associated with an operation i, and a time instant α of its domain. Clearly, if we prove that there is no solution for which the starting time of operation i is strictly lower than that of operation α, then we can adjust the release date of i to α. Carlier, Pinson and Peridy propose to use immediate selections to test the aforementioned configuration by fixing $d_i \leftarrow \alpha + p_i - 1$, and applying a dichotomic search to compute the value α realizing the best adjustment. By using propagation on the conjunctive component, the global operations — or shaving — present the advantage to treat simultaneously all the interacting cliques of disjunction. The objective of this chapter is the study of the local operations, we use this principle only at the local level (only interactions of disjunctions on the same clique during the dichotomic search).

4.5.1 Local Shaving on Sets

We call Local Shaving on Sets the local operation obtained by applying the previous principle at a local level and by using the immediate selections on sets as a satisfiability test.

Local Shaving (Primal Version)

Let $i \in K, \alpha \in]r_i, d_i - p_i]$.
If immediate selections show there exists no solution satisfying $(t_i < \alpha)$,
then we can set: $r_i \leftarrow \alpha$.

Local Shaving (Dual Version)

Let $i \in K, \alpha \in [r_i, d_i - p_i[$.
If immediate selections show there exists no solution satisfying $(t_i > \alpha)$,
then we can set: $d_i \leftarrow \alpha + p_i$.

The dichotomic search applied on all the time-windows is performed in $\mathcal{O}(n \log T)$ time, where $T = d(J) - r(J)$, which leads to an overall time complexity of $\mathcal{O}(n^2 \log n \log T)$. If this complexity is not strongly polynomial, it remains nevertheless smaller than the complexity of the set partition problem (cf. Section 4.4.1) and of stack adjustments (cf. Section 4.4.3).

4.5.2 Exact Adjustments

The exact adjustments operate according to the same principle, but by applying an *exact resolution* for the satisfiability test. Clearly, this elimination rule allows us to obtain the maximal adjustments from only $1 \mid r_i \mid L_{\max}$ problems. It was used by Martin and Shmoys [26].

Exact Adjustments (Primal Version)

Let $i \in K, \alpha \in]r_i, d_i - p_i[$.
If there is no feasible solution satisfying $(t_i < \alpha)$,
then we can set: $r_i \leftarrow \alpha$.

Exact Adjustments (Dual Version)

Let $i \in K, \alpha \in]r_i, d_i - p_i[$.
If there is no feasible solution satisfying $(t_i > \alpha)$,
then we can set: $d_i \leftarrow \alpha + p_i$.

The counterpart of the optimality of these adjustments is its exponential computational time (because it is necessary to solve a set of $\mathcal{NP}$-Hard problems $1 \mid r_i \mid L_{\max}$). However, the exact solving of the single machine scheduling problem can be performed in a very effective way by using the branch and bound method proposed by Carlier [35].

The value $\alpha^\star$ of the best adjustment can be obtained by a dichotomic search on the interval $[r_i, d_i - p_i]$. The value $\alpha^\star$ is the smallest value satisfying the satisfiability test. From this value, the release date adjustment is deduced, $r_i \leftarrow \alpha^\star - p_i + 1$.

Finally we observe an interesting property of the exact adjustments. Contrary to the other classic adjustments and whatever is the order considered to realize the exact adjustments, the algorithm converges in one run. To prove this, it is sufficient to consider the value $\alpha^\star$ realizing an adjustment. For this value, we know explicitly a feasible solution for which the operation i is shifted on $\alpha^\star$. So, it is a feasible starting date for operation i.

4.6 Computational Experiments

Most of the results given in this section are reported in Ref. [59].

4.6.1 One Machine Scheduling Problem

In order to evaluate the immediate effect of local operations on disjunctive problems, we compare the different rules on the one machine scheduling probem $1 \mid r_i, q_i \mid C_{\max}$.

The one-machine problem generator of Carlier [35] was used to create a set of 420 instances. This generator admits for parameters n, $p_{\max}$, and k: n indicates the number of operations, $p_{\max}$ is the maximal duration for an operation, and k adjusts the dispersal of release dates and tails. For a fixed combination of parameters, the creation of a set of n operations is made by uniform generation in $[1, p_{\max}]$ of the processing times, and in $[0, \frac{1}{50}\, n\, k\, p_{\max}]$ of release dates and tails.

As mentioned in the first section, the tail q_i of an operation represents the minimal distance between the completion time of an operation and the total duration C of the schedule. To compute the due dates of the operations, the duration of the schedule was arbitrarily set to the minimal feasible value $C^\star$, given by the application of the B&B algorithm of Carlier [35].

The values of parameters used for the creation of the problems were $p_{\max} = 99$, $n \in \{10, 20, 50, 100, 200, 500\}$, and $k \in \{1, 5, 10, 15, 20, 25, 30\}$. Finally, a set of 10 instances was generated for each of the 42 combinations.

The fact that exact adjustments find maximal adjustments of time windows was used to estimate the reduction given by the other elimination rules. Our main criterion is the sum of the sizes of operation time windows, $S = \sum_{i=1}^{n}(d_i - r_i - p_i)$.

We report in Table 4.1 the average quantity of information which can be obtained for each instance. This value is equal to $(S - S^\star)/S$ where S is the sum of the time windows sizes before adjustments and $S^\star$ is the optimal value.

On this first table, it should be pointed out that part of reduction of time windows sharply decreases with the number of operations of the problem. This is not completely suprising since the "degree of freedom" on the optimal solution should increase due to the increase of the number of operations as well as to a global increase of the size of time windows. One observes as well that the increase of information is maximum for medium values of the parameter k (k in interval $[5, 20]$).

We have mentioned that the quantity $S - S^\star$ is the maximum of reduction of time windows that can be reached. Let S_R be the sum of the sizes of windows obtained after application of an elimination rule R.

TABLE 4.1 Average Part of Reduction Reached

	n						
k	10	20	50	100	200	500	Total
1	25.61%	13.73%	4.79%	2.09%	0.76%	0.36%	7.89%
5	41.13%	18.31%	6.57%	2.51%	1.25%	0.54%	11.72%
10	41.23%	22.11%	9.70%	5.32%	2.29%	0.85%	13.58%
15	43.38%	20.26%	8.34%	0.19%	0.03%	0%	12.03%
20	42.58%	18.36%	0.28%	0.05%	0.02%	0%	10.22%
25	29.28%	5.06%	0.28%	0.04%	0.01%	0%	5.78%
30	8.94%	1.59%	0.13%	0.03%	0.01%	0%	1.78%
Total	33.17%	14.20%	4.30%	1.46%	0.62%	0.25%	9%

TABLE 4.2 Percentage of Reduction Given by Classical Elimination Rules

		n						
k	Rule	10	20	50	100	200	500	Total
1	Disj.	0%	0%	0%	0%	0%	0%	0%
	NF-NL	59.68%	44.93%	28.69%	15.37%	24.24%	3.95%	29.48%
	A/D Set	83.77%	83.30%	92.27%	87.20%	77.80%	99.10%	87.24%
5	Disj.	0%	0%	0%	0%	0%	0%	0%
	NF-NL	43.20%	21.94%	10.59%	5.96%	2.51%	1.16%	14.23%
	A/D Set	94.11%	95.14%	97.27%	98.19%	99.04%	99.77%	97.25%
10	Disj.	5.29%	0%	0%	0%	0%	0%	0.88%
	NF-NL	46.58%	17.98%	6.03%	3.10%	1.53%	0.74%	12.66%
	A/D Set	92.76%	94.75%	98.43%	99.14%	99.49%	99.72%	97.38%
15	Disj.	26%	47.02%	73.16%	100%	100%	100%	74.36%
	NF-NL	54.39%	54.47%	74.31%	100%	100%	100%	80.53%
	A/D Set	86.57%	96.57%	97.58%	94.08%	98.66%	92.70%	94.36%
20	Disj.	49.40%	65.31%	99.24%	100%	100%	100%	85.66%
	NF-NL	63.56%	74.19%	93.60%	100%	100%	100%	88.56%
	A/D Set	95.03%	89.88%	98.86%	96.60%	99.74%	100%	96.69%
25	Disj.	79.79%	95.77%	96.08%	100%	100%	100%	95.27%
	NF-NL	91.01%	94.10%	96.08%	100%	100%	100%	96.86%
	A/D Set	97.54%	94.96%	96.93%	99.77%	98.92%	100%	98.02%
30	Disj.	91.49%	98.76%	100%	100%	99.78%	100%	98.34%
	NF-NL	87.22%	99.23%	100%	95.25%	99.78%	100%	96.91%
	A/D Set	94.53%	99.47%	99.90%	99.79%	99.89%	99.97%	98.93%
Total Disj.		35.99%	43.84%	52.64%	57.14%	57.11%	57.14%	50.64%
Total NF-NL		63.66%	58.12%	58.47%	59.96%	61.15%	57.98%	59.89%
Total A/D Set		92.04%	93.44%	97.32%	96.39%	96.22%	98.75%	95.69%

The quantity $(S - S_R)/(S - S^{\star})$ allows us to estimate the part of information obtained by the rule R over the maximal quantity that can be obtained. Table 4.2 presents the average part obtained by the classic adjustments for every combination of parameters.

We can see that immediate selection on ascendant/descendant sets (A/D sets) clearly outperforms the other classical elimination rules. Not-first/not-last adjustments seem also to surpass immediate selections on disjunctions (Disj.). It can be observed as well that the more the quantity of information to be collected is weak (see Table 4.2), the more it is easy to obtain. In that case, it appears to be essentially composed by immediate selections on disjunctions that can not be detected A/D sets.

Finally, Table 4.3 allows us to compare the local shaving on sets (Local Shav.) and the combination of the previous classical elimination rules (Cla. Adj.). The difference does not seem to be very significant.

TABLE 4.3 Comparison of Strongly Polynomial Elimination Rules / Local Shaving

		n						
k	Rule	10	20	50	100	200	500	Total
1	Cla. Adj.	99.48%	99.03%	97.73%	93.65%	96.34%	99.68%	97.65%
	Local Shav.	100%	99.97%	99.83%	99.64%	99.77%	99.98%	99.86%
5	Cla. Adj.	99.64%	98.33%	98.68%	99.58%	99.84%	99.93%	99.33%
	Local Shav.	99.92%	99.35%	99.67%	99.87%	99.89%	99.97%	99.78%
10	Cla. Adj.	97.70%	97.56%	99.29%	99.74%	99.85%	99.90%	99.01%
	Local Shav.	98.78%	98.39%	99.62%	99.91%	99.92%	99.95%	99.43%
15	Cla. Adj.	97.29%	97.95%	99.93%	100%	100%	100%	99.19%
	Local Shav.	98.87%	99.28%	100%	100%	100%	100%	99.69%
20	Cla. Adj.	98.53%	96.47%	100%	100%	100%	100%	99.17%
	Local Shav.	98.57%	98.27%	100%	100%	100%	100%	99.47%
25	Cla. Adj.	100%	96.53%	100%	100%	100%	100%	99.42%
	Local Shav.	100%	100%	100%	100%	100%	100%	100%
30	Cla. Adj.	98.07%	100%	100%	100%	99.89%	100%	99.66%
	Local Shav.	100%	100%	100%	100%	100%	100%	100%
Total Cla.		98.67%	97.98%	99.37%	99%	99.42%	99.93%	99.06%
Total Shav.		99.45%	99.32%	99.87%	99.92%	99.94%	99.99%	99.75%

TABLE 4.4 Single Elimination Rules

	Disj.		NF-NL		A/D Set	
Instance	Nodes	*T* (s.)	Nodes	*T* (s.)	Nodes	*T* (s.)
FT10	—	—	—	—	—	—
ABZ5	—	—	—	—	—	—
ABZ6	7 635	6	7 893	16	1 912	3
La16	1 135	1	1 453	3	415	1
La17	1 970	1	1 247	2	152	0
La18	5 492	5	3 745	8	1 738	3
La19	—	—	—	—	—	—
La20	—	—	—	—	—	—
ORB02	—	—	—	—	—	—
ORB04	—	—	—	—	—	—
ORB07	—	—	—	—	9 665	16
ORB08	—	—	—	—	225	0
ORB09	1 925	2	6 539	12	477	1
ORB10	—	—	—	—	3 859	6
Moyenne	3 631.4	3	4 175.4	8.2	2 305.4	3.75

4.6.2 Job Shop Scheduling Problem

Table 4.4 shows that immediate selections on sets are the most efficient ones. They allow us to solve more problems and faster. Among the other immediate selections the difference is small but the immediate selections on disjunctions solve the problems faster than NF-NL rules.

When we combine two rules (see Table 4.5), immediate selections on sets remain the most efficient rule, although the difference between its association with immediate selections on disjunction or with NF-NL rules is not significant.

Finally, Table 4.6 compares polynomial adjustments, local shaving and exact adjustments (Exact Adjust.). The nonpolynomial adjustments have the same behavior, they solve all the problems. But exact adjustments are faster, while polynomial adjustments leave one open problem, although they are very fast and seem to be the good trade-off between time and efficiency.

TABLE 4.5 Multiple Elimination Rules

	Disj. & NF-NL		Disj. & A/D Set		NF-NL & A/D Set	
Instance	Nodes	*T* (s.)	Nodes	*T* (s.)	Nodes	*T* (s.)
FT10	—	—	7 969	15	8 183	29
ABZ5	—	—	6 041	13	6 543	25
ABZ6	2 472	6	743	1	928	3
La16	251	1	227	1	230	1
La17	607	1	187	0	119	1
La18	1 634	4	670	2	696	2
La19	12 343	29	6 732	13	6 826	26
La20	—	—	8 958	19	8 362	32
ORB02	—	—	—	—	—	—
ORB04	—	—	—	—	—	—
ORB07	—	—	5 962	12	5 363	19
ORB08	1 526	3	189	0	149	0
ORB09	1 352	3	352	3	350	2
ORB10	8 616	21	2 348	5	2 103	8
Moyenne	3 600.1	8.5	3 364.8	6.8	3 321	12.33

TABLE 4.6 Polynomial Adjustments, Local Shaving and Exact Adjustments

	Disj. & NF-NL & A/D Set		Local Shaving		Exact Adjust.	
Instance	Nodes	*T* (s.)	Nodes	*T* (s.)	Nodes	*T* (s.)
FT10	6 139	22	4 835	155	4 675	60
ABZ5	5 577	21	4 591	156	4 632	50
ABZ6	746	3	383	11	387	4
La16	206	1	187	5	185	2
La17	119	1	106	3	106	1
La18	614	2	466	15	472	5
La19	5 814	22	4 330	151	4 282	45
La20	6 996	27	5 751	199	5 905	69
ORB02	—	—	11 097	365	11 047	123
ORB04	9 768	32	7 310	227	7 378	80
ORB07	4 605	16	3 760	116	3 701	43
ORB08	137	0	117	4	134	2
ORB09	342	2	280	9	286	3
ORB10	1 897	7	1 392	48	1 366	19
Moyenne	3 304.6	12	3 186.1	104.57	3 182.6	36.14

4.7 Conclusion

The progress of solving exactly the job-shop problem during the last two decades are mainly due to the development of elimination rules. Indeed these rules permit us to narrow efficiently the intervals of starting times of operations. We have shown in this chapter, that all classical elimination rules are useful, but the most efficient ones are those based on ascendant sets and descendant sets. The question for the future is — how to go further? Peridy [58] suggests to split the interval associated with an operation into several intervals and to study the incompatibility graph between pair of operations of the corresponding intervals. By using the latter approach, a proof of optimality for MT10 was obtained without backtrack. So our opinion is that it is a very promising way of research for the coming decade.

References

[1] B. Roy and B. Sussman. Les problemes d'ordonnancement avec contraintes disjonctives. Technical report, Note DS no 9 bis, SEMA, Paris, 1964.

[2] J.K. Lenstra, A.H.G. Rinnooy Kan, and P. Brucker. Complexity of machine scheduling problems. *Ann. Discrete Math.*, 1:343–362, 1977.

[3] T. Gonzalez and S. Sahni. Flow shop and job shop schedules: complexity and approximation. *Oper. Res.*, 26:36–52, 1978.

[4] J.K. Lenstra and A.H.G. Rinnooy Kan. Computational complexity of discrete optimization problems. *Ann. Discrete Math.*, 4:121–140, 1979.

[5] Y.N. Sotskov. The complexity of scheduling problems with two and three jobs. *Eur. J. Oper. Res.*, 53:326–336, 1991.

[6] D.P. Williamson, L.A. Hall, J.A. Hoogeveen, C.A.J. Hurkens, J.K. Lenstra, S.V. Sevast'janov, and D.B. Shmoys. Short shop schedules. *Oper. Res.*, 45(2):288–294, 1997.

[7] J.R. Jackson. Scheduling a production line to minimize maximum tardiness. Technical Report 43, University of California, Los Angeles, 1955.

[8] S.M. Johnson. Optimal two- and three-stage production schedules with setup times included. *Naval Res. Log. Q.*, 1:61–68, 1954.

[9] I. Adiri and N. Hefetz. An efficient optimal algorithm for the two-machines unit-time job shop schedule length problem. *Math. Oper. Res.*, 7:354–360, 1982.

[10] S.B. Akers. A graphical approach to the production scheduling problem. *Oper. Res.*, 4:244–245, 1956.

[11] P. Brucker. An efficient algorithm for the job-shop problem with two jobs. *Comp.*, 40:353–359, 1988.

[12] E. Balas. Machine scheduling via disjunctive graphs: implicit enumeration algorithm. *Oper. Res.*, 17:941–957, 1969.

[13] G.B. McMahon and M. Florian. On scheduling with ready times and due dates to minimize maximum lateness. *Oper. Res.*, 23(3):475–482, 1975.

[14] B.J. Lageweg, J.K. Lenstra, and A.H.G. Rinnooy Kan. Job shop scheduling by implicit enumeration. *Manag. Sci.*, 24(4):441–450, 1977.

[15] J.R. Barker and G.B. McMahon. Scheduling in general job shop. *Manag. Sci.*, 31(5):594–598, 1985.

[16] J. Carlier and E. Pinson. An algorithm for solving the job-shop problem. *Manag. Sci.*, 35(2):165–176, February 1989.

[17] H. Fisher and G.L. Thompson. *Industrial Scheduling*, Probabilistic learning combinations of local job-shop scheduling rules, pp. 225–251. Prentice Hall, 1963.

[18] J. Carlier and E. Pinson. A practical use of Jackson's preemptive schedule for solving the job-shop problem. *Ann. Oper. Res.*, 26:269–287, 1990.

[19] J. Carlier and E. Pinson. Adjustment of heads and tails for the job-shop problem. *Eur. J. Oper. Res.*, 78:146–161, 1994.

[20] D. Applegate and W. Cook. A computational study of the job-shop scheduling problem. *ORSA J. Comp.*, 3:149–156, 1991.

[21] P. Brucker, B. Jurisch, and A. Kramer. The job-shop problem and immediate selections. *Ann. Oper. Res.*, 50, 1994.

[22] J. Grabowski, E. Nowicki, and S. Zdralka. A block approach for single machine scheduling with release dates and due dates. *Eur. J. Oper. Res.*, 26:278–285, 1986.

[23] M. Perregard and J. Clausen. Parallel branch and bound methods for the job shop scheduling problem. *Ann. Oper. Res.*, 83:137–160, 1998.

[24] Y. Caseau and F. Laburthe. Disjunctive scheduling with task intervals. Technical report, LIENS Technical Report 95-25, Ecole Normale Superieure Paris, France, 1995.

[25] Ph. Baptiste and C. Le Pape. Edge-finding constraint propagation algorithms for disjunctive and cumulative scheduling. In *Proc. 15th Workshop U.K. Plan. Spec. Int. Gr.*, Liverpool, United Kingdom, 1996.

[26] P. Martin and D.B. Shmoys. A new approach to computing optimal schedules for the job-shop scheduling problem. In *Proc. 5th Int. IPCO Conf.*, pp. 389–403, 1996.
[27] W. Brinkkotter and P. Brucker. Solving open benchmark problems for the job shop problem. *J. Sched.*, 4:53–64, 2001.
[28] A.S. Jain and S. Meeran. Deterministic job shop scheduling: Past, present, future. *Eur. J. Oper. Res.*, 113:390–434, 1999.
[29] S.S. Panwalkar and W. Iskander. A survey of scheduling rules. *Oper. Res.*, 25(1):45–61, 1977.
[30] J.H. Blackstone, D.T. Philipps, K. Hogg, G.L. Bhaskaran, and M. Pinedo. A state of the art survey on dispatching rules for manufacturing job shop operations. *Intl. J. Prod. Res.*, 20:27–45, 1982.
[31] R. Haupt. A survey of priority rule based scheduling. *Oper. Res. Spect.*, 11:3–16, 1989.
[32] K. Bhaskaran and M. Pinedo. *Handbook of Industrial Engineering*, Dispatching. John Wiley & Sons, 1991.
[33] Y.L. Chang, T. Sueyoshi, and R.S. Sullivan. Ranking dispatching rules by data envelopment analysis in a job shop environment. *IIE Trans.*, 28(8):631–642, 1996.
[34] J. Adams, E. Balas, and D. Zawack. The shifting bottleneck procedure for job shop scheduling. *Manag. Sci.*, 34:391–401, 1988.
[35] J. Carlier. The one-machine sequencing problem. *Eur. J. Oper. Res.*, 11:42–47, 1982.
[36] E. Balas and A. Vazacopoulos. Guided local search with shifting bottleneck for job shop scheduling. *Manag. Sci.*, 44:262–275, 1998.
[37] F. Glover. Tabu search - part i. *ORSA J. Comp.*, 1(3):190–206, 1989.
[38] F. Glover. Tabu search - part ii. *ORSA J. Comp.*, 2(1):4–23, 1990.
[39] F. Glover and M. Laguna. *Tabu Search.* Kluwer Academic, Norwell, MA, 1997.
[40] M. Laguna, J.M. Barnes, and F. Glover. Tabu search methods for a single machine scheduling problem. *J. Intell. Manufac.*, 2:63–74, 1991.
[41] M. Laguna and F. Glover. Integrating target analysis and tabu search for improved scheduling systems. *Expert Syst. Appli.*, 6:287–297, 1993.
[42] J.W. Barnes and M. Laguna. A tabu search experience in production scheduling. *Ann. Oper. Res.*, 41:141–156, 1993.
[43] E. Taillard. Parallel taboo search techniques for the job shop scheduling problem. *ORSA J. Comp.*, 16(2):108–117, 1994.
[44] E. Nowicki and G. Smutnicki. A fast taboo search algorithm for the job shop scheduling problem. *Manag. Sci.*, 42(6):797–813, 1996.
[45] R. Nakano and T. Yamada. Conventional genetic algorithm for job shop problems. In M.K. Kenneth and L.B. Booker (editors), *Proc. 4th Int. Conf. Gen. Algor. Their Appl.*, pp. 474–479, San Diego, U.S. 1991.
[46] C. Biewirth. A generalized permutation approach to job shop scheduling with genetic algorithms. *Oper. Res. Spect.*, 17(2, 3):87–92, 1995.
[47] C. Biewirth, D.C. Mattfeld, and H. Kopfer. On permutation representation for scheduling problems. In H.M. Voigt et al. (editor), *PPSN'IV Parallel Problem Solving from Nature*, pp. 310–318, Springer-Verlag, Berlin, Heidelberg, Germany, 1996.
[48] G. Shi. A genetic algorithm applied to a classic job shop scheduling problem. *Intl. J. Syst. Sci.*, 28(1):25–32, 1997.
[49] F. Della Croce, R. Tadei, and R. Rolando. Solving real world project scheduling problem with a genetic approach. *Belgian J. Oper. Res.*, 33(1, 2):65–78, 1994.
[50] F. Della Croce, R. Tadei, and G. Volat. A genetic algorithm for the job shop problem. *Comp. Oper. Res.*, 22(1):15–24, 1995.
[51] U. Dorndorf and E. Pesch. Evolution based learning in a job shop scheduling environment. *Comp. Oper. Res.*, 22(1):25–40, 1995.
[52] D.C. Mattfeld. Evolutionary search and the job shop: investigations on genetic algorithms for production scheduling. *Physica-Verlag*, 1996.

[53] R.J.M. Vaessens, E.H.L. Aarts, and J.K. Lenstra. Job shop scheduling by local search. *J. Comp.*, 8(3):302–137, 1996.

[54] E.J. Anderson, C.A. Glass, and C.N. Potts. *Local Search*, Machine scheduling. Discrete mathematics and optimization. John Wiley & Sons, 1997.

[55] J. Carlier. Ordonnancement ffntraintes disjonctives. These de 3[e] cycle, Paris VI, juin 1975.

[56] J. Erschler, F. Roubellat, and J.P. Vernhes. Finding some essential characteristics of the feasible solutions for a scheduling problem. *Oper. Res.*, 24:774–783, 1976.

[57] G. Dewess. Ein existenzsatz fur packungsprobleme mit konsequenzen fur die berechnung optimaler maschinenbelegungsplne. Technical report, Universitat Leipzig, 1991.

[58] L. Péridy. *Le problème de job-shop: arbitrages et ajustements.* PhD thesis, Université de Technologie de Compiègne, 1996.

[59] D. Rivreau. *Problèmes d'ordonnancement disjonctifs: règles d'élimination et bornes inférieures.* PhD thesis, University of Technology of Compiègne, 1999.

5

Flexible Hybrid Flowshops

George Vairaktarakis
Case Western Reserve University

5.1 Introduction and Literature Review

The study of scheduling problems in flexible manufacturing and assembly systems has attracted significant attention (see Herrmann and Lee, 1992; Lee et al., 1993; Sriskandarajah and Sethi, 1989; Wittrock, 1988) due to the importance of such systems for small-to-medium batch manufacturing. Expanding this research, in this chapter we address a couple of flexible variations of the well-known hybrid flowshop system (HFS).

The hybrid flowshop consists of two or more flowshop stages each of which consists of multiple identical processors operated in parallel. The two modes of flexibility considered in this chapter are (i) processing flexibility and (ii) routing flexibility. Most of our analysis is limited to two-stage hybrid flowshops. Throughout this chapter the objective is minimization of makespan.

The HFS with processing flexibility will be referred to as *flexible hybrid flowshop system* or FHFS in our further discussions. A job can be either completely processed at one of the stages or its processing splits in a prespecified fashion between the two stages. For the latter case the routing is in only one direction, from the upstream stage to the downstream stage, which preserves the routing structure of a traditionally defined flowshop (see Baker, 1993; Pinedo, 2001). The stages can consist of a single or multiple machines. In the latter case we assume that the machines in a stage are operated as parallel identical machines. We use the notation FHFS_{m_1,m_2} to denote a FHFS with m_1 machines at stage 1 and m_2 machines at stage 2. We assume the presence of adequate storage space between the two stages, and no preemption allowed for the tasks. Let us compare the HFS and FHFS systems. The HFS environments are multistage production environments with multiple identical parallel machines at each stage. The flexibility of these environments is derived from the ability to process a job on any one of the parallel machines at a stage (processing flexibility within a production stage), while FHFS has additional flexibility as a consequence of the fact

1-58488-397-9/$0.00+$1.50

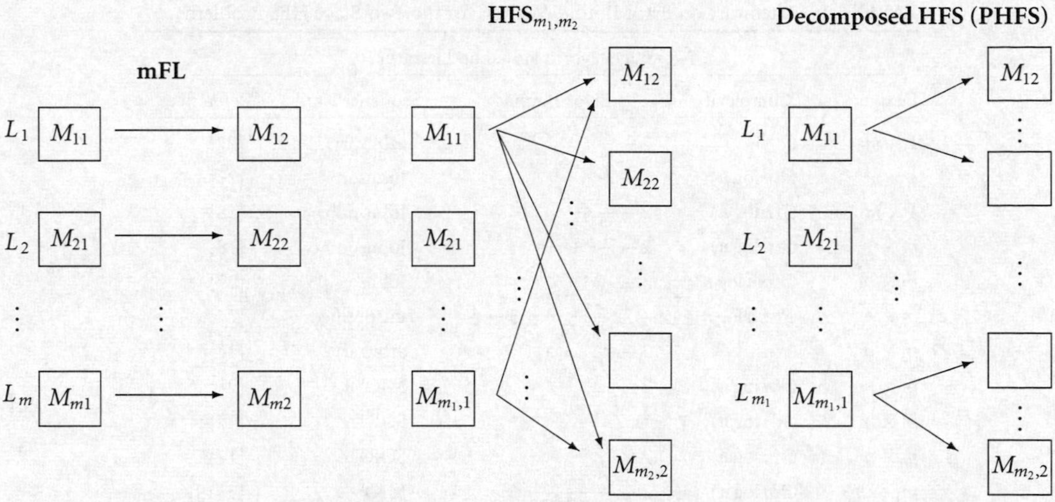

FIGURE 5.1 Hybrid flowshops with different routing structures.

that the job, or a part of it, could be processed on any one of the production stages (processing flexibility across production stages). The FHFS$_{m_1,m_2}$ can thus be viewed as a generalization of the HFS. The common characteristic of the HFS and FHFS is the preservation of the strict routing structure of the traditional flowshop environment, with parts being routed from an upstream stage to a downstream stage.

The second variation of HFS addressed in this chapter is a system with limited routing flexibility compared to FHFS. The HFS allows complete routing flexibility between processors of consecutive production stages. Such flexibility results in high throughput rates that come at the expense of sophisticated material handling systems, such as automated guided vehicles, automated transfer lines, etc. Such handling systems usually require major investment, both in capital and management technology. Let HFS$_{m_1,m_2}$ denote the HFS system with m_k identical processors in stage $k = 1, 2$ and without loss of generality $m_1 \leq m_2$. We will present a design called *Parallel HFS* (PHFS) where the HFS$_{m_1,m_2}$ is decomposed into smaller HFS sub-designs operated in parallel. Each HFS sub-design requires unidirectional routing and results to substantial reduction in routing complexity, savings in material handling costs and simplified management. More specifically, PHFS is composed of m_1 independent sub-designs each of which is a hybrid flowshop of the form HFS$_{1,k}$ where k is an appropriate integer (see Figure 5.1). This decomposition of HFS$_{m_1,m_2}$ provides unidirectional routing from the only stage-1 machine to the k stage-2 machines.

In Figure 5.1 we include the simplest possible PHFS system. It consists of m two-machine flowshops (or flowlines) operated in parallel and is denoted by mFL (see Figure 5.1). In-depth study of the mFL system is conducted later in this chapter before the PHFS design is addressed. Next we review the literature on each of the two flexible variations of HFS.

5.1.1 HFS with Processing Flexibility

Flexible hybrid flowshops are encountered in printed circuit manufacturing environments. During the production of printed circuit boards (PCBs), automated pick-and-place machines are used to place components on the surface of the PCB. Each machine can dispense many different components, thus allowing either the complete component placement process of a PCB to take place at one machine, or to be split among machines. In some consumer electronics factories the component placement process of the PCB production is configured as a two-stage HFS with identical component feeders for all the machines in each stage.

The hybrid flowshop HFS$_{m_1,m_2}$ with m_k machines in stage $k = 1, 2$ is a special case of FHFS and is encountered in many industrial sectors including semiconductor manufacturing (see Herrmann and Lee, 1992; Lee and Herrmann, 1993), the glass container industry (see Paul, 1979), cable manufacturing

TABLE 5.1 Literature on Error Bound Analysis for the Two-Stage HFS Problem

Two-Stage Hybrid Flowshop Literature				
Design	Complexity	Error Bound	Sequence Used	Reference
$1 \times m$	$\mathcal{O}(n)$	$3 - \frac{1}{m}$	arbitrary	[27]
$1 \times m$	$\mathcal{O}(n \log n)$	$3 - \frac{3}{m} + \frac{1}{m^2}$	Johnson	[27]
$1 \times m$	$\mathcal{O}(n \log n)$	$2 - \frac{1}{m}$	Johnson[a]	[18]
$m \times 1$	$\mathcal{O}(n \log n)$	$2 - \frac{1}{m}$	Johnson[b]	[18]
$m \times m$	$\mathcal{O}(mn \log n)$	$3 - \frac{1}{m}$	Johnson	[27]
$m \times m$	$\mathcal{O}(mn \log n)$	$\frac{7}{3} - \frac{2}{3m} \leq r \leq 3 - \frac{1}{m}$	unspecified	[27]
$m \times m$	$\mathcal{O}(n)$	$3 - \frac{1}{m}$	arbitrary	[17]
$m \times m$	$\mathcal{O}(n \log n)$	2	MJO[c]	[2]
$m \times m$	$\mathcal{O}(n \log n)$	$\frac{5}{2}$	SORTA[d]	[17]
$m \times m$	$\mathcal{O}(n \log n)$	2	SORTB[e]	[17]
$m_1 \times m_2$	$\mathcal{O}(n \log n)$	$2 - \frac{1}{\max\{m_1, m_2\}}$	MJO[f]	[3, 19]

[a] Johnson with respect to $(a_i, b_i/m)$.
[b] Johnson with respect to $(a_i/m, b_i)$.
[c] MJO is the Modified Johnson's order; see Buten and Shen (1973).
[d] SORTA is the shortest processing time order with respect to a_i.
[e] SORTB is the shortest processing time order with respect to b_i.
[f] Johnson with respect to $(a_i/m_1, b_i/m_2)$.

(see Narasimhan and Panwalker, 1984) and others. Let $J_i = (a_i, b_i)$ for $1 \leq i \leq n$ be a set of jobs where task a_i (b_i) requires processing in stage 1 (stage 2) for a_i (b_i) units of time. The HFS design is a generalization of the m parallel identical machine environment (Pm) where it is assumed that $b_i = 0$ for every $J_i \in J$. The Pm environment has been well studied over the last 25 years by several researchers. A review including most results in this area is given by Cheng and Sin (1990).

Several special cases of the two-stage hybrid flowshop problem have been considered in the literature. Arthanary and Ramaswamy (1971) developed a branch-and-bound algorithm for $\text{HFS}_{m,1}$, which is inefficient for problem sizes of more than ten jobs. Several heuristics have been proposed for $\text{FS}_{m,1}$ and $\text{HFS}_{1,m}$ by Gupta (1988), Gupta and Tunc (1991), Rao (1970), and others. These heuristics are evaluated by establishing a lower bound and then computing the average relative gap of the heuristic solution from the lower bound. However, the lower bounds that have been proposed for $\text{HFS}_{m,1}$ and $\text{HFS}_{1,m}$ are either trivial or loose.

Heuristics accompanied by worst-case error bound analysis are also found in literature. We tabulate these results in Table 5.1. In every case we mention the complexity, the worst-case error bound, the sequence used by the heuristic as well as the appropriate reference.

Hoogeveen et al. (1996) proved that $\text{HFS}_{2,1}$ is strongly $\mathcal{NP}$-complete and hence so is $\text{FHFS}_{2,1}$. Guinet and Solomon (1996) compared the performance of several heuristics on hybrid flowshops with three or five stages and makespan or maximum tardiness objectives. For the makespan objective they found that the best among the heuristics considered in their study exhibits a relative deviation (from the lower bound) of about 8% on the average. Gupta et al. (1997) developed a branch-and-bound algorithm for $\text{HFS}_{m,1}$, and presented computational results on problems with up to 20 jobs within 25 to 30 sec on an IBM 3090 computer. Solomon et al. (1996) presented five heuristics, and experimented with problems of size 50, 150, and 300 jobs. The reported relative deviations from their lower bounds range from 0.75% to 4.27% (on the average) depending on the heuristic and the range of the task processing times.

5.1.2 HFS with Routing Flexibility

The decomposition of HFS_{m_1,m_2} into the PHFS system requires a simpler but equally important design — the m parallel flowline design — denoted by mFL (Figure 5.1). The mFL design is a special case of PHFS, and has appeared in He et al. (1996). In this, the authors consider a design that consists of several flowlines (i.e., traditional flowshops with a single processor per stage) operated in parallel, motivated by an application from the glass industry. They consider several product types, setup times between different types, and no-wait in process. The latter constraint renders this problem very different than mFL.

The mFL problem offers an alternative design for flexible flowshops with limited routing flexibility. In mFL, the production system is decomposed into m independent cells, each of which can be managed independently. This approach is in line with principles of cellular manufacturing that gained popularity the last three decades. Admittedly, the mFL design will incur a loss in throughput performance as compared to an equivalent design that allows free routing between stages.

When $m_1 = m_2 = m$, the HFS_{m_1,m_2} system has the same number and layout of machines as the mFL design. The outline of the rest of this chapter is as follows. System FHFS is studied in Section 5.2. In Section 5.3 we study the PHFS system. In Section 5.4 we summarize our computational findings and draw managerial insights. We conclude the chapter in Section 5.5.

5.2 Hybrid Flowshops with Processing Flexibility

A formal definition of the $\text{FHFS}_{1,1}$ problem is provided in Section 5.2.1 along with properties of optimal $\text{FHFS}_{1,1}$ schedules. An optimal dynamic program is presented for $\text{FHFS}_{1,1}$ in Section 5.2.2. This dynamic program is extended to the case of stage dependent processing times in Section 5.2.3. In Section 5.2.4 we present a heuristic for HFS_{m_1,m_2} with good error bound performance. This heuristic is extended for the k-stage HFS in Section 5.2.5. The FHFS_{m_1,m_2} system with stage dependent processing times is considered in Section 5.2.6, where a heuristic is presented along with its error bound. This heuristic is improved for the special case $\text{FHFS}_{m,m}$ in Section 5.2.7.

5.2.1 Optimal Properties for the Two-Machine FHFS

Problem $\text{FHFS}_{1,1}$ is formally stated as follows. A set of jobs $J = \{J_1, J_2, \ldots, J_n\}$ is given and every job J_i consists of two tasks with processing time requirements a_i and b_i. We will use a_i, b_i to denote both the tasks and their processing requirements. All jobs are assumed to be available at time zero and no preemption is allowed. The jobs in J are to be processed in a two-stage flowshop with the following processing flexibility. Every job J_i can be processed entirely on the upstream machine M_1 for $p_i = a_i + b_i$ periods, or entirely in the downstream machine M_2 for p_i periods, or task a_i is processed on M_1 and task b_i is processed on M_2. To maximize throughput, as well as machine utilization, we are interested in sequencing the jobs in J so that the resulting schedule on the two-stage FHFS has minimum makespan.

FHFS represents a generalization of the traditional flowshop F2, and the parallel identical machine shop P2. It is an environment with three job classes: class V_1 of jobs requires processing only at the upstream stage, class V_2 of jobs requires processing only at the downstream stage and class V_3 of jobs requires processing at both stages. It is easy to observe that the parallel machine environment is an environment with classes V_1 and V_2 of jobs only, and the traditional flowshop with class V_3 only.

Observe that when $b_i = 0$ for all $J_i \in J$, our flexible hybrid flowshop problem reduces to the problem of minimizing makespan on two identical parallel machines, which is known to be ordinary $\mathcal{NP}$-complete (see Garey and Johnson, 1979). Therefore our problem, since it contains the above problem as a special case, is also $\mathcal{NP}$-complete.

We identify a number of properties of optimal $\text{FHFS}_{1,1}$ schedules. First, given a schedule S for the $\text{FHFS}_{1,1}$ we partition the set of jobs J into the three classes listed below:

V_1: The jobs in J for which both tasks are processed on M_1.
V_2: The jobs in J for which both tasks are processed on M_2.
V_3: The jobs in J where task a_i is processed on M_1 and task b_i is processed on M_2.

The following properties hold for the optimal makespan schedule of $\text{FHFS}_{1,1}$.

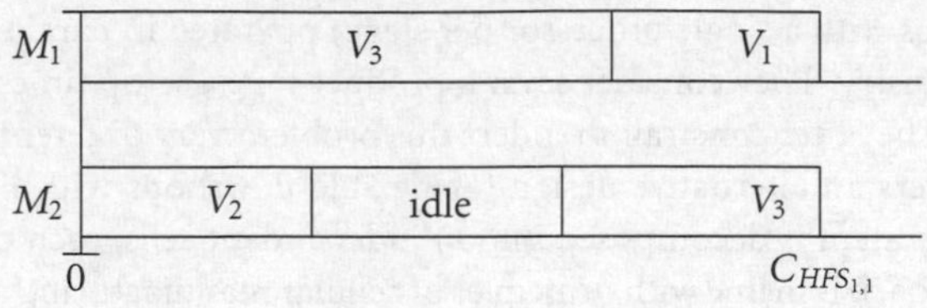

FIGURE 5.2 The structure of an optimal solution for the $FHFS_{1,1}$.

Properties [16]:

There exists an optimal schedule where

1. all jobs in V_1 follow the jobs in V_3
2. all jobs in V_2 precede the jobs in V_3
3. all jobs in V_3 are in Johnson's order
4. there is no idle time on M_1

Properties 1 to 3 are verified either by straightforward job interchange arguments or by observing that the $FHFS_{1,1}$ is a special case of the two-machine job shop (J_2). In the job shop we have four classes of jobs: classes V_1, V_2, V_3, and an additional class V_4 of jobs whose task a_i is processed on machine M_2 and task b_i is processed on M_1. Property 4 is motivated by the makespan objective. For a derivation of Properties 1 to 4 see Jackson, 1956. When the jobs belonging to the classes V_1–V_4 are prespecified, then $J_2//C_{\max}$ can be solved by Johnson's algorithm in $\mathcal{O}(n \log n)$ time (see Johnson, 1954). However, for $FHFS_{1,1}$ the jobs in each class must be specified by the solution procedure.

Since there may be idle time on M_2 we can always shift jobs in V_2 to the left so that the resulting schedule processes all jobs in V_2 contiguously starting at time zero. Therefore, any idle time on M_2 is induced by jobs in V_3 only. Also, we can always start the last b-task of V_3, say b_l, as soon as possible, preceded by the remaining b tasks in V_3 with no idle time in between. Hence we will assume without loss of generality that in $FHFS_{1,1}$, idle time can occur between the processing of classes V_2 and V_3 in M_2, or at the end of processing in M_2 or M_1. Clearly, the above shifting of jobs can be done so that neither the makespan is increased nor the flowshop constraints are violated. Combining properties 1 to 4 with the previous observations, we conclude that there exists an optimal solution for $FHFS_{1,1}$ which has the structure depicted in Figure 5.2.

In the next subsection we use the above properties to develop a pseudopolynomial dynamic programming algorithm for the $FHFS_{1,1}$ problem.

5.2.2 Optimal Dynamic Program for the Two-Machine Flowshop

Assume that the set $J = \{J_1, J_2, \ldots, J_n\}$ is ordered according to Johnson's order. We use the term *Johnson's Order* to refer to a sequence of jobs that has the property that *Job* J_i *precedes job* J_j *if* $\min\{a_i, b_j\} \leq \min\{a_j, b_i\}$ (see Johnson, 1954). Let

$$P_i = \sum_{j=1}^{i} p_j \quad \text{for } i = 1, 2, \ldots, n$$

and $f_i(I, t, S)$ = the makespan of an optimal schedule whose total processing time on M_1 of jobs in $V_3 \cap \{J_1, J_2, \ldots, J_i\}$ is t, the idle time at the end of job processing on M_1 is I, and the idle time at the end of processing on M_2 is S.

By definition, $I \cdot S = 0$ since the makespan value is attained on at least one of M_1 and M_2. More specifically, if the makespan is attained on M_1 we have $I = 0$, and if it is attained on M_2 we have $S = 0$.

In the above definition of $f_i(I,t,S)$, the variables t, I, S take values on the interval $[0, P_n]$. These observations indicate that the state space of the dynamic program (DP) to calculate $f_i(\cdot,\cdot,\cdot)$ is $\mathcal{O}(nP_n^2)$.

The following DP is based on the fact that in an optimal schedule, J_i $(1 \leq i \leq n)$ belongs to precisely one of V_1, V_2, V_3, and the optimal makespan is attained on at least one of M_1 and M_2; thus producing six possible cases.

Definition 5.1

Let C_{vm} be the case where J_i is allocated to V_v and the makespan of the optimal schedule is attained on machine M_m, for $v = 1,2,3$, and $m = 1,2$.

For each C_{vm}, we depict in Figure 5.3 all alternative schedule configurations for the job set $\{J_1, J_2, \ldots, J_{i-1}\}$ that can result to C_{vm} after the insertion of J_i. We use the notation $V_k^i = V_k \cap \{J_1, J_2, \ldots, J_i\}$ for $k = 1,2,3$, and $1 \leq i \leq n$.

Evidently, there are two alternative schedule configurations for $\{J_1, J_2, \ldots, J_{i-1}\}$ and C_{11}. Configuration C_{11} corresponds to the case where prior to scheduling J_i the optimal makespan is attained on M_2 and configuration C_{11} corresponds to the case where prior to scheduling J_i the optimal makespan is attained on M_1. Similarly, there are two subcases for C_{22}, C_{31}, and C_{32}. Note that it is impossible to start with a schedule that attains its makespan on M_1, and make it attain its makespan on M_2 by inserting a job on M_1. Hence there is a unique schedule configuration for C_{12} and similarly for C_{21}.

The six cases C_{vm}, $v = 1,2,3$, and $m = 1,2$, motivate the following recurrence relation.

5.2.2.1 Recurrence Relation

Let $J_1, J_2, \ldots, J_n$ be the set of jobs ordered according to Johnson's order. Then,

$$f_i(0,t,S) = \min \begin{cases} C_{11}: f_{i-1}((p_i - S)^+, t, (S - p_i)^+) + \min\{p_i, S\} \\ C_{21}: \begin{cases} f_{i-1}(0,t,S) & \text{if } 2f_{i-1}(0,t,S) - P_i - S \geq 0 \\ \min_{0 \leq y \leq S+p_i} f_{i-1}(0,t,S+p_i-y) & \text{otherwise} \end{cases} \\ C_{31}: \min \begin{cases} f_{i-1}((a_i - b_i - S)^+, t - a_i, (b_i - a_i + S)^+) + \min\{b_i + S, a_i\} \\ \min_{0 \leq I' < a_i - b_i - S}\{f_{i-1}(I', t - a_i, 0) - I'\} + a_i \\ \min_{S' > b_i + S - a_i} f_{i-1}(0, t - a_i, S') + a_i \end{cases} \end{cases}$$

$$f_i(I,t,0) = \min \begin{cases} C_{12}: f_{i-1}(I + p_i, t, 0) \\ C_{22}: \begin{cases} \min_{0 \leq S' \leq p_i - I} f_{i-1}(0,t,S') + I & \text{if } 2f_{i-1}(0,t,S') = P_i - I \\ \min_{0 \leq I' \leq I} f_{i-1}(I',t,0) + \max\{P_i - 2f_{i-1}(I',t,0) + I', 0\} \end{cases} \\ C_{32}: \min \begin{cases} f_{i-1}((a_i + I - b_i)^+, t - a_i, (b_i - a_i - I)^+) + b_i \\ \min_{0 \leq I' < a_i + I - b_i} \{f_{i-1}(I', t - a_i, 0) - I'\} + a_i + I \\ \min_{S' > b_i - a_i - I} f_{i-1}(0, t - a_i, S') + a_i + I \end{cases} \end{cases}$$

5.2.2.2 Boundary Conditions

$$f_1(I,t,S) = \begin{cases} p_1 & \text{if } (I,t,S) \in \{(0,0,p_1), (p_1,0,0), (b_1,a_1,0)\} \\ \infty & \text{otherwise} \end{cases}$$

5.2.2.3 Optimal Solution

Let f_n^* denote the optimal solution of DP with the above boundary conditions. Then

$$f_n^* = \min_{I,t,S} f_n(I,t,S)$$

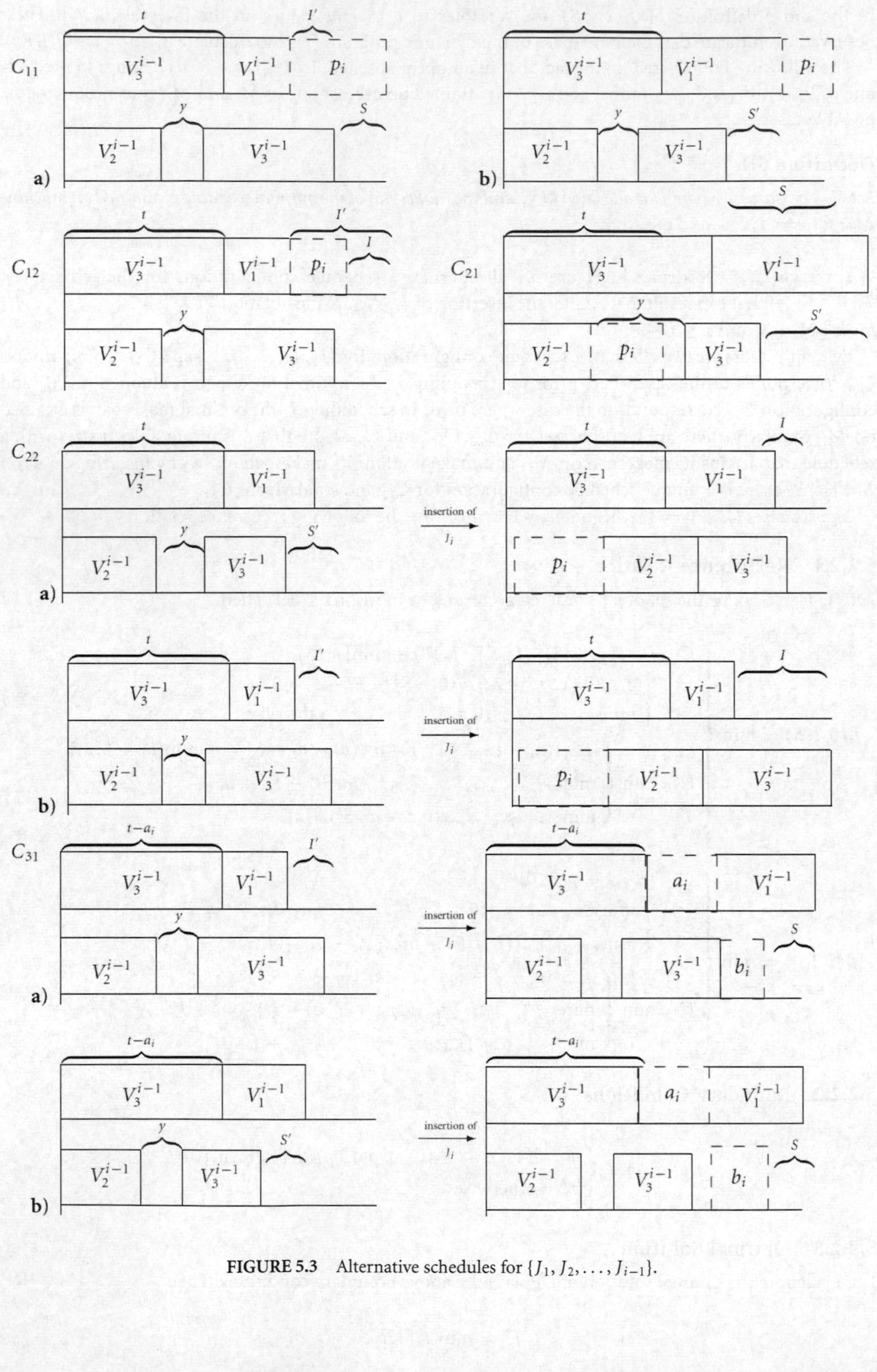

FIGURE 5.3 Alternative schedules for $\{J_1, J_2, \ldots, J_{i-1}\}$.

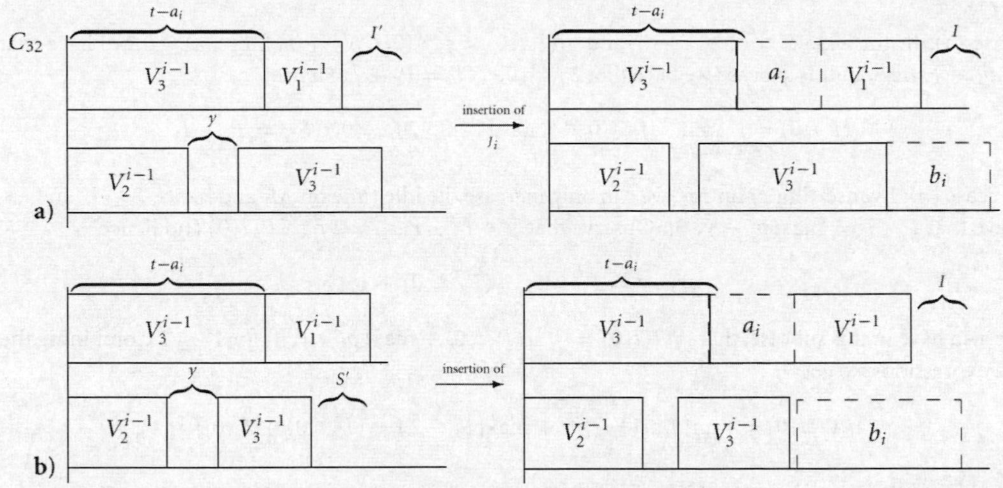

FIGURE 5.3 *(Continued)*

Theorem 5.1

The recurrence relation for $f_i(I, t, S)$, along with the above boundary conditions, produce the optimal makespan value [16].

Proof

Throughout this proof we refer to Figure 5.3. We denote by I', S' and y the idle times of the schedule of the first $i - 1$ jobs in the set J. I' is the idle time at the end of the schedule on M_1, S' is the idle time at the end of the schedule on M_2, and y is the idle time between the sets V_2^{i-1} and V_3^{i-1} on M_2. Next, we examine the cases one by one.

Case C_{11}

Subcase (a): The insertion of J_i in V_1 will affect the makespan only if $p_i \geq I'$. Then, the resulting schedule will have idle time S on M_2, where $p_i = I' + S$. Thus, $f_i(0, t, S) = f_{i-1}(p_i - S, t, 0) + S$. Note that in this case the constraint $p_i \geq I'$ holds because $I' = p_i - S \leq p_i$.

Subcase (b): The insertion of J_i in V_1 increases the makespan by p_i, and $S = S' + p_i$. Since $S' \geq 0$, this subcase holds only if $p_i \leq S$. Then, $f_i(0, t, S) = f_{i-1}(0, t, S - p_i) + p_i$. Note that the value of $f_{i-1}(0, t, S - p_i)$ can be finite only if $S - p_i \geq 0$, and hence the condition $S \geq p_i$ is satisfied.

Combining subcases (a) and (b) of C_{11} we get

$$f_i(0, t, S) = f_{i-1}((p_i - S)^+, t, (S - p_i)^+) + \min\{p_i, S\}$$

Case C_{12}

Note that $I' = I + p_i$ and hence $f_i(I, t, 0) = f_{i-1}(I + p_i, t, 0)$.

Case C_{21}

For the idle time y we have that $y = 2f_{i-1}(0, t, S') - P_{i-1} - S'$. If $p_i \leq y$, then $S' = S$ and $f_i(0, t, S) = f_{i-1}(0, t, S)$. Hence,

$$f_i(0, t, S) = f_{i-1}(0, t, S) \quad \text{if} \quad p_i \leq 2f_{i-1}(0, t, S) - P_{i-1} - S$$

If $y < p_i$ then $f_i(0, t, S) = f_{i-1}(0, t, S')$ and $S' = S + p_i - y$. Since $S' \geq 0$, we have $0 \leq y \leq S + p_i$. Thus,

$$f_i(0, t, S) = \min_{0 \leq y \leq S + p_i} f_{i-1}(0, t, S + p_i - y)$$

The above two formulas provide the desired result for C_{21}.

Case C_{22}

Subcase (a): In this case $S' = p_i - I - y$ and $f_i(I, t, 0) = f_{i-1}(0, t, S') + I$. Since $y \geq 0$, we have that $S' \leq p_i - I$. Also, in this subcase we have that $2 f_{i-1}(0, t, S') = P_i - I$. Hence,

$$f_i(I, t, 0) = \min_{0 \leq S' \leq p_i - I} f_{i-1}(0, t, S') + I \quad \text{if} \quad 2 f_{i-1}(0, t, S') = P_i - I$$

Subcase (b): By inserting p_i on M_2, we can only increase the idle time on M_1 and hence $I \geq I'$. In fact, we have that $I - I' = \max\{p_i - y, 0\}$. Observe that $y + I' + P_{i-1} = 2 f_{i-1}(I', t, 0)$ and hence

$$p_i - y = P_i - 2 f_{i-1}(I', t, 0) + I'$$

We also have in this subcase, that $f_i(I, t, 0) = f_{i-1}(I', t, 0) + \max\{p_i - y, 0\}$ for $I' \leq I$ Combining the above expressions we get

$$f_i(I, t, 0) = \min_{I' \leq I}\{f_{i-1}(I', t, 0) + \max\{P_i - 2 f_{i-1}(I', t, 0) + I', 0\}\}$$

Combining the two subcases of C_{22}, we get the desired result.

Case C_{31}

Subcase (a): In this subcase it holds that $a_i \geq I' + b_i + S$. Depending upon whether the insertion of a_i induces idle time on M_2, we distinguish the following two subcases:

i. In this subcase the insertion of a_i into V_3^i does not increase the idle time on M_2, and $a_i = I' + b_i + S$. Since $I' \geq 0$, this subcase holds only if $a_i \geq b_i + S$. Thus, we have that

$$f_i(0, t, S) = f_{i-1}(a_i - b_i - S, t - a_i, 0) + b_i + S, \quad \text{for} \quad a_i \geq b_i + S$$

Note that $f_{i-1}(a_i - b_i - S, t - a_i, 0)$ cannot be finite unless $a_i \geq b_i + S$, and hence this condition is already accounted for by the recurrence relation and the boundary conditions.

ii. In this subcase the insertion of a_i into V_3^i increases the idle time on M_2, i.e., $a_i > I' + b_i + S$. Then, it holds that

$$b_i + S = \sum_{j \in V_1^{i-1}} p_j = f_{i-1}(I', t - a_i, 0) - (t - a_i) - I'$$

or equivalently

$$t = a_i - b_i - S - I' + f_{i-1}(I', t - a_i, 0)$$

After the insertion of J_i, the makespan becomes

$$f_i(0, t, S) = t + b_i + S = \min_{0 \leq I' < a_i - b_i - S} f_{i-1}(I', t - a_i, 0) - I' + a_i$$

Subcase (b): As in subcase (a) we further distinguish two subcases:

i. In this subcase the insertion of a_i into V_3^i does not increase the idle time on M_2, and $a_i = b_i + S - S'$. Since $S' \geq 0$, we have that $a_i \leq b_i + S$ for the case to hold. Thus, we have that

$$f_i(0, t, S) = f_{i-1}(0, t - a_i, b_i + S - a_i) + a_i, \quad \text{for} \quad S \geq a_i - b_i$$

Note that $f_{i-1}(0, t - a_i, b_i + S - a_i)$ cannot be finite unless $b_i + S - a_i \geq 0$, or $S \geq a_i - b_i$.

ii. In this subcase the insertion of a_i into V_3^i increases the idle time on M_2, i.e., $a_i > b_i + S - S'$. Then, it holds that

$$b_i + S = \sum_{j \in V_1^{i-1}} p_j = f_{i-1}(0, t - a_i, S') - (t - a_i)$$

or equivalently

$$t = a_i - b_i - S + f_{i-1}(0, t - a_i, S')$$

When this case holds, the makespan is

$$f_i(0, t, S) = t + b_i + S = \min_{a_i > b_i + S - S'} f_{i-1}(0, t - a_i, S') + a_i$$

The formulas for the subcases (a.i) and (b.i) can be combined into

$$f_i(0, t, S) = f_{i-1}((a_i - b_i - S)^+, t - a_i, (b_i - a_i + S)^+) + \min\{b_i + S, a_i\}$$

Hence, the recurrence relation for C_{31} is

$$f_i(0, t, S) = \min \begin{cases} f_{i-1}((a_i - b_i - S)^+, t - a_i, (b_i - a_i + S)^+) + \min\{b_i + S, a_i\} \\ \min_{I' < a_i - b_i - S}\{f_{i-1}(I', t - a_i, 0) - I'\} + a_i \\ \min_{a_i > b_i + S - S'} f_{i-1}(0, t - a_i, S') + a_i \end{cases}$$

Case C_{32}

Subcase (a). In this subcase it holds that $a_i \geq I' - I + b_i$. Depending upon whether the insertion of a_i induces idle time on M_2, we distinguish the following two subcases:

i. In this subcase the insertion of a_i into V_3^i does not increase the idle time on M_2, and $a_i = I' - I + b_i$. Since $I' \geq 0$, this subcase holds only if $b_i \leq a_i + I$. Thus, we have that

$$f_i(0, t, S) = f_{i-1}(a_i + I - b_i, t - a_i, 0) + b_i \quad \text{for} \quad I \geq b_i - a_i$$

Clearly, $f_{i-1}(a_i + I - b_i, t - a_i, 0)$ cannot be finite unless $I \geq b_i - a_i$.

ii. In this subcase the insertion of a_i into V_3^i increases the idle time on M_2, i.e., $a_i > I' - I + b_i$. Then, it holds that

$$b_i = \sum_{j \in V_1^{i-1}} p_j + I = f_{i-1}(I', t - a_i, 0) - (t - a_i) - I' + I$$

or equivalently

$$t + b_i = f_{i-1}(I', t - a_i, 0) + a_i - I' + I$$

After the insertion of J_i, the makespan becomes

$$f_i(I, t, 0) = t + b_i = \min_{0 \leq I' < I + a_i - b_i} f_{i-1}(I', t - a_i, 0) + a_i - I' + I$$

Subcase (b): As in subcase (a) we further distinguish two subcases.

i. In this subcase the insertion of a_i into V_3^i does not increase the idle time on M_2, and $b_i = a_i + I + S'$. Since $S' \geq 0$, we have that $b_i \geq a_i + I$ for the case to hold, and

$$f_i(I, t, 0) = f_{i-1}(0, t - a_i, b_i - a_i - I) + b_i$$

ii. In this subcase the insertion of a_i into V_3^i increases the idle time on M_2 and $b_i < a_i + I + S'$. Then, it holds that

$$b_i = \sum_{j \in V_1^{i-1}} p_j + I = f_{i-1}(0, t - a_i, S') - (t - a_i) + I$$

or equivalently

$$t + b_i = f_{i-1}(0, t - a_i, S') + a_i + I$$

After the insertion of J_i, the makespan becomes

$$f_i(I,t,0) = t + b_i = \min_{S' > b_i - a_i - I} f_{i-1}(0, t - a_i, S') + a_i + I$$

We combine the cases $a_i + I \geq b_i$ and $a_i + I \leq b_i$ further into

$$f_i(I,t,0) = \min \begin{cases} f_{i-1}((a_i + I - b_i)^+, t - a_i, (b_i - a_i - I)^+) + b_i \\ \min_{0 \leq I' < a_i + I - b_i} \{f_{i-1}(I', t - a_i, 0) - I'\} + a_i + I \\ \min_{S' > b_i - a_i - I} f_{i-1}(0, t - a_i, S') + a_i + I \end{cases}$$

Collecting the recurrence relations for all possible configurations of Figure 5.3 we get the recurrence relation for DP. This completes the proof of the theorem. □

As indicated earlier, the state space of DP is $\mathcal{O}(nP_n^2)$. It is easy to check that the effort required at every iteration of the DP is order $\mathcal{O}(P_n)$. Therefore, the complexity of the DP is $\mathcal{O}(nP_n^3)$. In the following subsection we extend DP to the case of stage dependent processing times.

5.2.3 Stage Dependent Processing Times

The dynamic program DP′ below is an adaptation of DP for the $\text{FHFS}_{1,1}$ with stage dependent processing times. We use the following additional notation.

a_{ik}, b_{ik}: the processing time required by the first and second task of J_i, respectively, if it is processed on M_k, $k = 1, 2$.

$$p_{ik} = a_{ik} + b_{ik}, \quad k = 1, 2$$

We refer to the dynamic programming algorithm for the $\text{FHFS}_{1,1}$ problem with stage dependent processing times as DP′, which is described as follows. Assume that the set $J = \{J_1, J_2, \ldots, J_n\}$ is ordered according to Johnson's order with respect to (a_{i1}, b_{i2}). Define the quantity:

$f_i(I, t, Q, S)$ = the makespan of an optimal schedule whose total processing time on M_1 of jobs in $V_3 \cap \{J_1, J_2, \ldots, J_i\}$ is t, the idle time at the end of job processing on M_1 is I, the idle time at the end of processing on M_2 is S, and the total processing time required by $J_1, J_2, \ldots, J_i$ is Q.

5.2.3.1 Recurrence Relation

$$f_i(0,t,Q,S) = \min \begin{cases} C_{11}: f_{i-1}((p_{i1} - S)^+, t, Q - p_{i1}, (S - p_{i1})^+) + \min\{p_{i1}, S\} \\ C_{21}: \begin{cases} f_{i-1}(0, t, Q - p_{i2}, S) & \text{if } 2f_{i-1}(0,t,Q - p_{i2}, S) - Q - S \geq 0 \\ \min_{0 \leq u \leq S + p_{i2}} f_{i-1}(0, t, Q - p_{i2}, S + p_{i2} - y) & \text{otherwise} \end{cases} \\ C_{31}: \min \begin{cases} f_{i-1}((a_{i1} - b_{i2} - S)^+, t - a_{i1}, Q - a_{i1} - b_{i2}, (b_{i2} - a_{i1} + S)^+) + \min\{b_{i2} + S, a_{i1}\} \\ \min_{0 \leq I' < a_{i1} - b_{i2} - S} \{f_{i-1}(I', t - a_{i1}, Q - a_{i1} - b_{i2}, 0) - I'\} + a_{i1} \\ \min_{S' > b_{i2} + S - a_{i1}} f_{i-1}(0, t - a_{i1}, Q - a_{i1} - b_{i2}, S') + a_{i1} \end{cases} \end{cases}$$

$$f_i(I,t,Q,0) = \min \begin{cases} C_{12}: f_{i-1}(I + p_{i1}, t, Q - p_{i1}, 0) \\ C_{22}: \begin{cases} \min_{0 \leq S' \leq p_{i2} - I} f_{i-1}(0, t, Q - p_{i2}, S') + I & \text{if } 2f_{i-1}(0, t, Q - p_{i2}, S') = Q - I \\ \min_{0 \leq I' \leq I} f_{i-1}(I', t, Q - p_{i2}, 0) + \max\{Q - 2f_{i-1}(I', t, Q - p_{i2}, 0) + I', 0\} \end{cases} \\ C_{32}: \min \begin{cases} f_{i-1}((a_{i1} - b_{i2} + I)^+, t - a_{i1}, Q - a_{i1} - b_{i2}, (b_{i2} - a_{i1} - I)^+) + b_{i2} \\ \min_{0 \leq I' < a_{i1} + I - b_{i2}} \{f_{i-1}(I', t - a_{i1}, Q - a_{i1} - b_{i2}, 0) - I'\} + a_{i1} + I \\ \min_{S' > b_{i2} - a_{i1} - I} f_{i-1}(0, t - a_{i1}, Q - a_{i1} - b_{i2}, S') + a_{i1} + I \end{cases} \end{cases}$$

5.2.3.2 Boundary Conditions

$$f_1(I,t,Q,S) = \begin{cases} p_{11} & \text{if } (I,t,Q,S) = (0,0,p_{11},p_{11}) \\ p_{12} & \text{if } (I,t,Q,S) = (p_{12},0,p_{12},0) \\ a_{11}+b_{12} & \text{if } (I,t,Q,S) = (b_{12},a_{11},a_{11}+b_{12},0) \\ \infty & \text{otherwise} \end{cases}$$

5.2.3.3 Optimal Solution

Let f_n^* denote the optimal solution of DP′ with the above boundary conditions. Then

$$f_n^* = \min_{I,t,Q,S} f_n(I,t,Q,S)$$

Theorem 5.2

The recurrence relation for $f_i(I,t,Q,S)$, along with the above boundary conditions, produce the optimal makespan value for the $FHFS_{1,1}$ problem with stage dependent processing times [16].

Proof

We use a logic similar to that of Theorem 5.1. In light of the additional notation for stage dependent processing times, the following changes are made:

a_i is replaced by a_{i1} for those subcases that a_i is scheduled on M_1 in Figure 5.3.
b_i is replaced by b_{i2} for those subcases that b_i is scheduled on M_2 in Figure 5.3.
p_i is replaced by p_{ik} for those subcases that p_i is scheduled on M_k, k=1, 2 in Figure 5.3.
P_i is replaced by Q.
P_{i-1} is replaced by $Q - a - b$ where a, b are the processing times of the first and second task of J_i, respectively, for each particular subcase.

It is easy to check that the above changes along with the insertion of the new parameter Q into the definition of f_i, transform the recurrence relation of DP into that of DP′. This completes the proof of the theorem. □

Due to the addition of the parameter Q, the state space of DP′ is $\mathcal{O}(nP_n'^3)$, where $P_n' = \sum_i(\max\{a_{i1}, a_{i2}\} + \max\{b_{i1}, b_{i2}\})$. It is easy to check that the effort required at every iteration of the DP′ has order $\mathcal{O}(P_n')$. Therefore, the complexity of DP′ is $\mathcal{O}(nP_n'^4)$, while its state space is $\mathcal{O}(nP_n'^3)$.

The algorithm DP′ can be easily modified to ensure (if desired by the user) that J_i belongs to a specific job set among V_1, V_2, V_3. For instance, we can force J_i to be included in V_1 or V_2 by eliminating from the definition of $f_i(I,t,Q,S)$ the branches C_{31} and C_{32}. Algorithm DP′ is used later to develop a heuristic for $FHFS_{m_1,m_2}$ with stage dependent processing times.

5.2.4 HFS with Two Stages

In this subsection, we focus attention to the HFS_{m_1,m_2} design which is the special case of $FHFS_{m_1,m_2}$ where $V_1 = V_2 = \emptyset$. The theory developed will be extended to $HFS_{m_1,m_2,\ldots,m_k}$ design that consists of k stages and stage k consists of m_l parallel identical machines $1 \le l \le k$. Since HFS_{m_1,m_2} is a hybrid between the multiple parallel identical machine environment *Pm* and the traditional two-machine flowshop $FS_{1,1}$, we are motivated to develop heuristic algorithms that are hybrids of existing good algorithms for these two environments. Minimizing the makespan in $HFS_{1,1}$ is solved in $\mathcal{O}(n \log n)$ time by Johnson's algorithm, which we refer to as JA (see Johnson, 1954).

Most successful heuristics for *Pm* utilize an ordering S of the jobs along with the *first available machine* (FAM) rule. According to the FAM rule, the job to be scheduled next in the *Pm* environment, is assigned to the first machine that becomes available, i.e., the machine that finishes first the job (if any) previously assigned to it. Depending on the order S, the above heuristic produces different solutions.

The heuristic presented below for HFS_{m_1,m_2} uses the mirror image of the FAM rule which we call *last busy machine* (LBM) rule. This rule will be used to assign jobs to the second stage machines. Given a constant $T > 0$ and an ordering S of jobs where each job j has a specified processing time p_j, a description of the rule is given next.

LBM Rule:

1. Set $t_m = T$ for $m = 1, \dots, m_2$.
2. Let j be the last unscheduled job of S and m a machine with largest t_m. Schedule job j on machine m to finish at time t_m.
3. Set $t_m = t_m - p_j$ and $S = S - \{j\}$. If $S \neq \emptyset$ then go to Step 2, else stop.

Remark: The value of t_m is the time that machine m becomes busy. In Step 2 we assign the job j to the machine with largest t_m, i.e., the last busy machine. Hence we call this rule the last busy machine rule. Also, note that the value of T is only a reference point and has no effect on the allocation of jobs to machines. With this background we can present the following heuristic which uses the Johnson's order in conjunction with the FAM and LBM rules.

Heuristic H:

1. Apply JA with respect to the processing times $\{(\frac{1}{m_1}a_i, \frac{1}{m_2}b_i) : i = 1, 2, \dots, n\}$. Let S be the resulting sequence.
2. Apply the FAM rule on the stage-1 tasks of the sequence S.
3. Apply the LBM rule on the stage-2 tasks of the sequence S.
4. On each stage-2 machine M_m, reorder the tasks assigned to M_m (during Step 3) according to nonincreasing completion times of the corresponding stage 1 tasks. Let S_m be the resulting order for $m = 1, 2, \dots, m_2$.
5. On each stage-2 machine M_m, schedule the tasks in S_m in this order, as soon as possible.

At step 1 of H a sequence S of jobs is produced, at step 2 an assignment of $(i, 1)$ tasks on the first stage machines is made and at Step 3 to Step 5 the tasks of Stage 2 are scheduled. In particular, Step 3 determines which tasks will be processed by each stage-2 machine, Step 4 determines the order of stage-2 tasks within a stage-2 machine and Step 5 proceeds with the scheduling of the stage-2 tasks on stage-2 machines. Since JA requires $\mathcal{O}(n \log n)$ time, this is also the computational effort required by H.

Example

Consider the case $m_1 = m_2 = 2$ and a set of five jobs with processing time requirements $(1, 3)$, $(1, 4)$, $(2, 4)$, $(3, 6)$, and $(6, 2)$. The jobs are already written according to Johnson's order. The Gantt chart of an application of H on this example is given in Figure 5.4.

The next lemma will be used in finding the error bound of H. For this lemma we introduce the following *auxiliary* problem. Replace the first stage machines $M_{11}, M_{21}, \dots, M_{m_1,1}$ by a dummy machine M_1 and replace the processing time requirement of a_i units on one of $M_{11}, M_{21}, \dots, M_{m_1,1}$ by the requirement of $\frac{1}{m_1}a_i$ units on M_1. Similarly, replace the second stage machines $M_{12}, M_{22}, \dots, M_{m_2,2}$ by a dummy machine M_2 and replace the processing time requirement of b_i units on one of $M_{1,2}, M_{2,2}, \dots, M_{m_2,2}$ by the requirement of $\frac{1}{m_2}b_i$ units on M_2. This way we define an auxiliary two-machine flowshop problem on M_1, M_2. We refer to this problem as AFS.

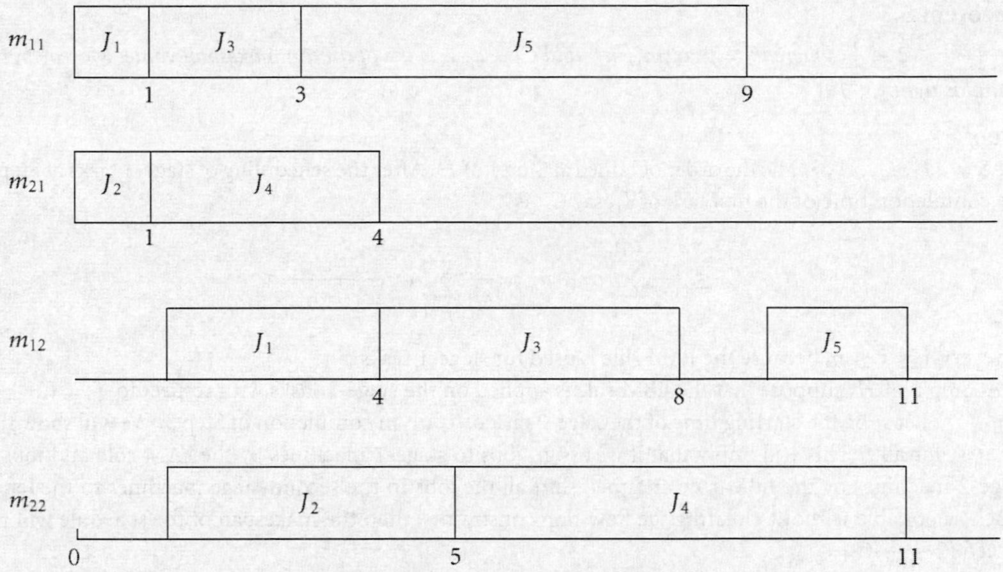

FIGURE 5.4 The Gantt chart of an application of H.

Lemma 5.1

Let C_{LB} be the completion time found by JA for the auxiliary problem. Then $C_{\text{LB}} \leq C_{\text{HFS}_{m_1,m_2}}$ where $C_{\text{HFS}_{m_1,m_2}}$ is the optimum makespan value.

Proof

Consider an optimal solution S^* for HFS_{m_1,m_2}. Let $S = \{J_1, J_2, \ldots, J_n\}$ be the set of jobs ordered in nondecreasing order of completion times of the stage-1 tasks in S^*. Consider the partial schedule of S^* consisting of the first i stage-1 tasks of S and the last $n - i + 1$ stage-2 tasks of S. Since in S^* the last $n - i + 1$ stage-2 tasks of S start no earlier than the completion time of the first i stage-1 tasks of S (due to the flowshop constraints), we have that

$$C_{\text{HFS}_{m_1,m_2}} \geq \frac{1}{m_1} \sum_{j=1}^{i} a_j + \frac{1}{m_2} \sum_{j=i}^{n} b_j \quad \text{for every } 1 \leq i \leq n$$

For the AFS problem (where the processing time requirements for J_i are $\frac{1}{m_1} a_i, \frac{1}{m_2} b_i$), schedule the jobs according to sequence S. Let C_S be the resulting makespan. Then, it is clear that

$$C_S = \frac{1}{m_1} \sum_{j=1}^{i_0} a_j + \frac{1}{m_2} \sum_{j=i_0}^{n} b_j$$

where J_{i_0} is the last job whose stage-2 task starts immediately after the completion of the corresponding stage-1 task (note that such a task always exists; since J_1 satisfies this property).

Combining the last two expressions we get that $C_S \leq C_{\text{HFS}_{m_1,m_2}}$. However, the sequence S is not necessarily optimal for the AFS problem and hence $C_{\text{LB}} \leq C_S$. The last two relations establish that $C_{\text{LB}} \leq C_{\text{HFS}_{m_1,m_2}}$. This completes the proof of the lemma. □

The following theorem establishes the worst case performance of H. Let C_H be the makespan value obtained by H.

Theorem 5.3

$\frac{C_H}{C_{\text{HFS}m_1,m_2}} \leq 2 - \frac{1}{m}$, *where* $m = \max\{m_1, m_2\}$ *and* $C_{\text{HFS}m_1,m_2}$ *is the optimum makespan value. Moreover, this bound is tight* [3, 19].

Proof

Let $S = \{J_1, J_2, \ldots, J_n\}$ be the order obtained at Step 1 of H. After the scheduling of stage-1 tasks at Step 2, the completion time of the first task of J_i, say C_i, is

$$C_i \leq \frac{1}{m_1}\sum_{j=1}^{i-1} a_j + a_i = \frac{1}{m_1}\sum_{j=1}^{i} a_j + \frac{m_1 - 1}{m_1} a_i$$

for every $1 \leq i \leq n$, because the FAM rule is used for stage-1 tasks.

At Step 3 of H, suppose that the LBM rule is applied on the stage-2 tasks with respect to $T = (2 - \frac{1}{m})$ $C_{\text{HFS}_{m_1,m_2}}$. Let s_i be the starting time of the stage-2 task of J_i upon completion of Step 3. We will show that $C_i \leq s_i$ for all i. This will imply that if we assign jobs to stage-1 machines by the FAM rule and jobs to stage-2 machines by the LBM rule and then shift all the jobs in the second-stage machines to the left as much as possible without violating the flowshop constraints, then the makespan of the schedule will not exceed $(2 - \frac{1}{m})C_{\text{HFS}_{m_1,m_2}}$.

To see that $s_i \geq C_i$ note that for every $1 \leq i \leq n$

$$s_i \geq T - \frac{1}{m_2}\sum_{j=i}^{n} b_j - \frac{m_2 - 1}{m_2} b_i$$

for every $1 \leq i \leq n$, because the LBM rule is used for stage-2 tasks. Then,

$$s_i - C_i \geq \left(2 - \frac{1}{m}\right) C_{\text{HFS}_{m_1,m_2}} - \frac{1}{m_2}\sum_{j=i}^{n} b_j - \frac{m_2 - 1}{m_2} b_i - \frac{1}{m_1}\sum_{j=1}^{i} a_j - \frac{m_1 - 1}{m_1} a_i$$

Note that

$$\frac{1}{m_1}\sum_{j=1}^{i} a_j + \frac{1}{m_2}\sum_{j=i}^{n} b_j \leq C_{\text{LB}} \leq C_{\text{HFS}_{m_1,m_2}} \quad \text{for every } 1 \leq i \leq n$$

as we showed in the proof of Lemma 5.1. Also,

$$\frac{m_1 - 1}{m_1} a_i + \frac{m_2 - 1}{m_2} b_i \leq \frac{m - 1}{m}(a_i + b_i) \quad \text{where } m = \max\{m_1, m_2\}$$

$$\leq \frac{m - 1}{m} C_{\text{HFS}_{m_1,m_2}}$$

Hence,

$$s_i - C_i \geq T - \left(1 + \frac{m - 1}{m}\right) C_{\text{HFS}_{m_1,m_2}} = 0$$

Therefore the schedule produced by concatenating, as described above, the partial schedules of Step 2 and Step 3 has a makespan no greater than $(2 - \frac{1}{m})C_{\text{HFS}_{m_1,m_2}}$. It is easy to check that Step 4 and Step 5 can only improve the makespan performance of the above schedule. This completes the upper bound performance of H.

To see that the bound of $2 - \frac{1}{m}$ is tight for H, consider the case where $b_j = 0$ for all jobs. Then the heuristic H reduces to the RDM list scheduling heuristic studied by Graham (1966), which has a tight worst-case error bound of $2 - \frac{1}{m_1}$. In a problem where $b_j = 0$ for all jobs and $m_1 \geq m_2$ the error bound of $2 - \frac{1}{m}$ is tight for the heuristic H. This completes the proof of the theorem. □

5.2.5 HFS with k Stages

In what follows we describe a heuristic algorithm H' for the k-stage hybrid flowshop $\text{HFS}_{m_1,m_2,\ldots,m_k}$. The heuristic H' utilizes H and assumes that the number k of stages is even (otherwise we can introduce a dummy stage with zero machines).

Heuristic H':

1. Apply H on stages $2r-1, 2r$ and let S_r be the resulting schedule, for $r = 1, 2, \ldots, \frac{k}{2}$.
2. Concatenate the schedules S_r, for $r = 1, 2, \ldots, \frac{k}{2}$. Eliminate all idle time between stage-1 tasks. Eliminate unnecessary idle time between stage-2 tasks.

The complexity of H' is $\mathcal{O}(kn \log n)$. The next theorem finds a bound to the error of H'.

Theorem 5.4

$\frac{C_{H'}}{C_{\text{HFS}_{m_1,m_2,\ldots,m_k}}} \leq k - \frac{1}{\max\{m_1,m_2\}} - \frac{1}{\max\{m_3,m_4\}} - \cdots - \frac{1}{\max\{m_{k-1},m_k\}}$ *where $C_{\text{HFS}_{m_1,m_2,\ldots,m_k}}$ is the optimum makespan value* [19].

Proof

Let C_r^* be the optimal makespan for stages $2r-1$, $2r$. Then, from Theorem 5.3 we have

$$C_{H'} \leq \left(2 - \frac{1}{\max\{m_1, m_2\}}\right)C_1^* + \left(2 - \frac{1}{\max\{m_3, m_4\}}\right)C_2^* + \cdots + \left(2 - \frac{1}{\max\{m_{k-1}, m_k\}}\right)C_{k/2}^*.$$

Clearly, $C_r^* \leq C_{\text{HFS}_{m_1,m_2,\ldots,m_k}}$ for $r = 1, 2, \ldots, \frac{k}{2}$. Then,

$$\frac{C_{H'}}{C_{\text{HFS}_{m_1,m_2,\ldots,m_k}}} \leq k - \frac{1}{\max\{m_1, m_2\}} - \frac{1}{\max\{m_3, m_4\}} - \cdots - \frac{1}{\max\{m_{k-1}, m_k\}}$$

This completes the proof of the theorem. □

The above theorem is linked with existing theory for the flowshop problem as follows. If $m_1 = m_2 = \cdots = m_k = 1$ then the hybrid flowshop reduces to the traditional k-machine flowshop. In this case,

$$\frac{C_{H'}}{C_{\text{HFS}_{1,1,\ldots,1}}} \leq \left\lceil \frac{k}{2} \right\rceil$$

The latter bound coincides with the bound of Gonzalez and Sahni (1978), which, to the best of our knowledge, is the best existing error bound for the k-machine flowshop problem.

Several cases where the error bound of H' is tight are known. The tightness of the k-machine flowshop for $k = 3, 4$ is given in Gonzalez and Sahni (1978). In case that the hybrid flowshop consists of only two stages, the tight examples given in the previous section provide tight examples for the hybrid flowshop as well. If the hybrid flowshop consists of one stage only, then H' reduces to RDM, which as we mentioned earlier has a tight error bound of $2 - \frac{1}{m_1}$.

5.2.6 Heuristic for FHFS_{m_1,m_2} and Stage Dependent Processing Times

We now study the flexible hybrid flowshop with m_1 identical machines at stage 1 and m_2 identical machines at stage 2 where the machines in different stages may require different amounts of time to process each task. To avoid additional notation, we will continue to use FHFS_{m_1,m_2} to denote the flexible hybrid flowshop with m_1, m_2 machines at stage 1 and stage 2, respectively, even for the case of stage-dependent processing times. Next we present and analyze a heuristic algorithm for FHFS_{m_1,m_2}. In Section 5.2.7, we consider the important special case of FHFS_{m_1,m_2} where $m_1 = m_2 = m$.

In this subsection we develop a heuristic algorithm for FHFS$_{m_1,m_2}$ with good worst-case error bound. The heuristic uses DP′, the FAM and LBM rules. Algorithm DP′ will be used to allocate tasks to the sets V_1, V_2, V_3: the FAM rule will be used to schedule tasks to stage-1 machines, and the LBM rule to schedule tasks to the stage-2 machines.

Heuristic H_D:

- Let $p_L = \max_i \min\{p_{i1}, p_{i2}, a_{i1} + b_{i2}\}$, and $p_H = \max_i \max\{p_{i1}, p_{i2}, a_{i1} + b_{i2}\}$.
- **For** every integer $p \in [p_L, p_H]$, **do** the following:
  1. Apply DP′ with respect to the processing times $\{(\frac{1}{m_1}a_{i1}, \frac{1}{m_2}a_{i2}, \frac{1}{m_1}b_{i1}, \frac{1}{m_2}b_{i2}) : i = 1, 2, \ldots, n\}$; where the definition of $f_i(I, t, Q, S)$ allows J_i to be included in V_1, V_2, or V_3; only if $p_{i1} \leq p$, $p_{i2} \leq p$, or $a_{i1} + b_{i2} \leq p$ respectively. Let S_p and C_p be the resulting sequence and associated objective function value.
  2. Apply the FAM rule on the stage-1 tasks of the sequence S_p.
  3. Apply the LBM rule on the stage-2 tasks of the sequence S_p.
  4. On each stage-2 machine, schedule the p_{i2} jobs to start as soon as possible, and the b_{i2} tasks according to the first-come-first-serve rule. Let $S(p)$ and $C(p)$ be the resulting schedule and makespan respectively.
- Let $C_{H_D} = C(p^*)$ be the smallest among the $C(p)$ values, and S_{H_D} the associated schedule.

The for-loop in the above heuristic is taken over all possible values of p, where p is the maximum amount of processing allowed for each job to spend on FHFS$_{m_1,m_2}$. For each such value of p, a schedule with makespan $C(p)$ is generated. The best among these schedules is the output S_{H_D} of H_D. In H_D, the for-loop is used to facilitate the worst case performance result presented next. In this, the smallest among the objective function values C_p at Step 1 is shown to be a lower bound of $C_{\text{FHFS}_{m_1,m_2}}$ — the optimal makespan for FHFS$_{m_1,m_2}$. However, to ensure that $C(p) \leq (2 - \frac{1}{\max\{m_1,m_2\}})C_{\text{FHFS}m_1,m_2}$, it is required that $p \leq C_{\text{FHFS}m_1,m_2}$. This latter inequality is ensured by the for-loop.

An observation for heuristic H_D follows. Consider an arbitrary machine M_{j2} of stage 2. Then, Step 4 of H_D dictates that the b_{i2}-tasks of jobs in V_3 that are allocated to M_{j2} should be processed in nondecreasing order of completion times of the corresponding a_{i1}-tasks. This is referred to as the first-come-first-serve rule. A straightforward interchange argument shows that this rule is optimal for a given allocation of stage 1 tasks to machines. Since DP′ requires $\mathcal{O}(nP_n'^4)$ time, the computational effort required by H_D is $\mathcal{O}(nP_n'^4 p_H)$. The following theorem establishes the worst case performance of H_D.

Let S^* be an optimal schedule for FHFS$_{m_1,m_2}$, and p^* the maximum processing requirement by any of the jobs in S^*. Then, $\max_i \min\{p_{i1}, p_{i2}, a_{i1} + b_{i2}\} \leq p^* \leq \max_i \max\{p_{i1}, p_{i2}, a_{i1} + b_{i2}\}$. Consider the auxiliary FHFS$_{1,1}$ problem (say FAFS) which is constructed as follows. Replace the stage-1 machines $M_{11}, M_{21}, \ldots, M_{m_1,1}$ by a dummy machine M_1 and the stage-2 machines $M_{12}, M_{22}, \ldots, M_{m_2,2}$ by a dummy machine M_2. Also, replace the processing requirements a_{i1}, a_{i2}, b_{i1}, and b_{i2} by a_{i1}/m_1, a_{i2}/m_2, b_{i1}/m_1, and b_{i2}/m_2, respectively. The following lemma shows that the makespan of DP′ subject to the constraint that no job requires more than p^* units of processing (as required in Step 1 of H_D for $p = p^*$), is a lower bound of $C_{\text{FHFS}_{m_1,m_2}}$.

Lemma 5.2

Let C_{p^*} be the completion time found by DP′ for $p = p^*$ for the auxiliary problem FAFS. *Then,* $C_{p^*} \leq C_{\text{FHFS}_{m_1,m_2}}$.

Proof

From S^* we construct a schedule for FAFS as follows. Order the stage-1 tasks of S^* (these are p_{i1} tasks that are processed exclusively by a stage-1 machine, and a_{i1} tasks of jobs whose b_{i2} is processed by a stage-2 machine) in nondecreasing order of completion times; let S_1 be the resulting order. Then, schedule the tasks in S_1, on machine M_1 of FAFS, in the order S_1.

Symmetrically, order the stage-2 tasks of S^* (i.e., p_{i2} and b_{i2} tasks) in nondecreasing order of completion times; let S_2 be the resulting order. Then, schedule the tasks in S_2 on machine M_2 of FAFS, in the order S_2, so that the last task on M_2 completes at time $C_{\text{FHFS}_{m_1,m_2}}$. Let S_{FAFS} be the resulting schedule for FAFS.

Due to the order S_1, the fact that FHFS_{m_1,m_2} has m_k machines in stage $k = 1, 2$, and that the stage 1 (stage 2) processing requirements in FAFS are $(1/m_1)$th $[(1/m_2)$ th$]$ of the corresponding requirements in S^*, the completion time of every job in S_1 has completion time in S_{FAFS} no greater than the corresponding completion time on S^*. By symmetry, the starting time of every task in S_2, starts in S^* no later than it starts in S_{FAFS}.

Therefore, S_{FAFS} is a feasible schedule for FAFS that has makespan $C_{\text{FAFS}} \leq C_{\text{FHFS}_{m_1,m_2}}$. However, the sequence S_{FAFS} is not necessarily optimal for the FAFS problem and hence $C_{p^*} \leq C_{\text{FAFS}}$. The last two inequalities establish that $C_{p^*} \leq C_{\text{FHFS}_{m_1,m_2}}$. This completes the proof of the lemma. □

Theorem 5.5

$\frac{C_{H_D}}{C_{\text{FHFS}m_1,m_2}} \leq 2 - \frac{1}{\max\{m_1,m_2\}}$ *where* C_{H_D} *is the makespan value obtained by* H_D, *and this bound is tight* [16].

Proof

Let S^* be an optimal schedule for FHFS_{m_1,m_2}, and p^* the maximum processing requirement by any of the jobs in S^*. Let $S(p^*)$ be the schedule produced by H_D when $p = p^*$, and $C(p^*)$ the corresponding makespan. Also, let V_1, V_2 be the set of jobs that are processed in $S(p^*)$ exclusively in stage 1 and stage 2, respectively, and V_3 the set of jobs whose processing is split between the two stages. First we will prove that $C(p^*) \leq (2 - \frac{1}{\max\{m_1,m_2\}})C_{\text{FHFS}_{m_1,m_2}}$, where $m = \max\{m_1, m_2\}$. Recall that, if $C(p)$ denotes the makespan of $S(p)$ at Step 4 of H_D, we have that $C_{H_D} \leq C(p)$ for every trial value of p, and hence the claim would mean that $C_{H_D} \leq (2 - \frac{1}{m_2})C_{\text{FHFS}_{m_1,m_2}}$.

Therefore, it remains to show that $C(p^*) \leq (2 - \frac{1}{m_2})C_{\text{FHFS}_{m_1,m_2}}$. We distinguish three cases depending on the task that attains $C(p^*)$; namely, the case where $C(p^*)$ is attained by a p_{i1}: $J_i \in V_1$, or a p_{i2}: $J_i \in V_2$, or a b_{i2}: $J_i \in V_3$.

Let S_1 be the nondecreasing order of completion times of stage-1 tasks in $S(p^*)$, i.e., p_{i1} tasks that are processed exclusively on the stage-1 machines and a_{i1} tasks that require processing of b_{i2} on stage-2. Let also S_2 be the nonincreasing order of start times of stage-2 tasks in $S(p^*)$, i.e., p_{i2} tasks that are processed exclusively on stage-2 and b_{i2} tasks.

Case 1: $C(p^*)$ is attained by a job in V_1.

Since the FAM rule is used at Step 2 of H_D for the stage-1 jobs, we have that

$$C(p^*) \leq \frac{1}{m_1}\left(\sum_{J_i \in V_1} p_{i1} + \sum_{J_i \in V_3} a_{i1}\right) + \frac{m_1 - 1}{m_1} \max\{p_{i1}: J_i \in V_1,\ a_{i1}: J_i \in V_3\}$$

Note that all stage-1 jobs in S_1 are processed by the M_1 machine of the FAFS problem, and hence $\frac{1}{m_1}(\sum_{J_i \in V_1} p_{i1} + \sum_{J_i \in V_3} a_{i1}) \leq C_{p^*}$. Hence,

$$C(p^*) \leq C_{p^*} + \frac{m_1 - 1}{m_1} \max\{p_{i1}, a_{i1} \in S_1\}$$

From Lemma 5.2 we have that $C_{p^*} \leq C_{\text{FHFS}_{m_1,m_2}}$. Also, by choice of p^*, we have that $\max\{p_{i1}: J_i \in V_1, a_{i1}: J_i \in V_3\} \leq C_{\text{FHFS}_{m_1,m_2}}$. Hence, we have that $C(p^*) \leq (2 - \frac{1}{m_1})C_{\text{FHFS}_{m_1,m_2}}$ and the claim holds.

Case 2: $C(p^*)$ is attained by a job in V_2.

This case realizes only if $V_3 = \emptyset$ (see Figure 5.3) and hence the proof is similar to Case 1.

Case 3: $C(p^*)$ is attained by a job in V_3.

Let $M_{m_0,2}$ be a stage-2 machine that attains $C(p^*)$. If there is no idle time on $M_{m_0,2}$, then a proof similar to Case 1 applies. In this case

$$C(p^*) \le \frac{1}{m_2}\left(\sum_{J_i \in V_2} p_{i2} + \sum_{J_i \in V_2} b_{i2}\right) + \frac{m_2 - 1}{m_2}\max\{p_{i2}\colon J_i \in V_2,\ b_{i2}\colon J_i \in V_3\} \le \left(2 - \frac{1}{m_2}\right) C_{\mathrm{FHFS}_{m_1,m_2}}$$

Else, suppose that there is nonzero idle time on $M_{m_0,2}$ and all tasks allocated on $M_{m_0,2}$ start as early as possible. Also, suppose that $V_3 = \{J'_1, J'_2, \ldots, J'_k\}$ where $J'_i = (a'_{i1}, b'_{i2})$ for $1 \le i \le k \le n$. Let J'_{i_0} be the last job whose b-task starts in $M_{m_0,2}$ immediately after the corresponding a-task. Such a job exists since otherwise all jobs allocated to $M_{m_0,2}$ could be shifted to the left so that the idle time is eliminated from $M_{m_0,2}$.

Since the FAM rule is used for the tasks in S_1, the completion time of $a'_{i_0,1}$ is

$$C_{i_0} \le \frac{1}{m_1}\sum_{i=1}^{i_0} a'_{i1} + \frac{m_1 - 1}{m_1} a'_{i_0,1}$$

At Step 3 of H_D the LBM rule is applied on the tasks of S_2. Hence, if s_{i_0} is the starting time of the task $b'_{i_0,2}$ we have that

$$C(p^*) - s_{i_0} \le \frac{1}{m_2}\sum_{i=i_0}^{k} b'_{i2} + \frac{m_2 - 1}{m_2} b'_{i_0,2}$$

Since $C_{i_0} \le s_{i_0}$, we have that

$$C(p^*) \le \frac{1}{m_2}\sum_{i=i_0}^{k} b'_{i2} + \frac{m_2 - 1}{m_2} b'_{i_0,2} + \frac{1}{m_1}\sum_{i=1}^{i_0} a'_{i1} + \frac{m_1 - 1}{m_1} a'_{i_0,1}$$

Note that

$$\frac{1}{m_1}\sum_{i=1}^{i_0} a'_{i1} + \frac{1}{m_2}\sum_{i=i_0}^{k} b'_{i2} \le C_{p^*} \le C_{\mathrm{FHFS}_{m_1,m_2}}$$

because the tasks a'_{i1}/m_1 for $i \le i_0$ are processed on M_1 of FAFS prior to the tasks b'_{i2}/m_2 for $i \ge i_0$ that are processed on M_2 of FAFS. Also,

$$\frac{m_1 - 1}{m_1} a'_{i_0,1} + \frac{m_2 - 1}{m_2} b'_{i_0,2} \le \frac{\max\{m_1, m_2\} - 1}{\max\{m_1, m_2\}} C_{\mathrm{FHFS}_{m_1,m_2}}$$

since by choice of p^*, we have that $a'_{i1} + b'_{i2} \le p^* \le C_{\mathrm{FHFS}_{m_1,m_2}}$. Hence,

$$C(p^*) \le \left(2 - \frac{1}{\max\{m_1, m_2\}}\right) C_{\mathrm{FHFS}_{m_1,m_2}}$$

This completes the proof of the error bound. To see that the bound is tight, assume that $m_1 \ge m_2$. Then, consider an instance with n-jobs where $b_{i1} = 0$ and $b_{i2} = \infty$ for every job J_i. Then, all jobs will be entirely processed at stage 1, and the H_D heuristic reduces to the RDM list scheduling heuristic studied by Graham (1966) which has a tight worst-case error bound of $2 - \frac{1}{m_1}$. This completes the proof of the theorem. □

Lemma 5.2 shows that the makespan of DP′ for the auxiliary problem FAFS, when $p = p^*$, is a lower bound on $C_{\mathrm{FHFS}_{m_1,m_2}}$. This observation is used in the proof of the corollary below to identify a lower bound for FHFS_{m_1,m_2} that is useful for computational purposes (as opposed to C_{p^*} which is subject to knowing p^*).

Let C_{p_H} be the optimal function value of DP′ when there is no constraint on the maximum processing length p allowed for a job J_i. Effectively, every J_i is allowed to be included in either V_1, V_2 or V_3.

Corollary 5.1

C_{p_H} is a lower bound for $C_{\text{FHFS}_{m_1,m_2}}$ [16].

Proof

By definition of p_H the constraint $p < p_H$ is satisfied for every $p \in [p_L, p_H]$ and hence $C_{p_H} \leq C_p$ for every $p \in [p_L, p_H]$. Therefore,

$$C_{p_H} = \min_p C_p \leq C_{p^*} \leq C_{\text{FHFS}_{m_1,m_2}}$$

the last inequality stemming from Lemma 5.2. This completes the proof of the corollary. □

The following corollary indicates that, when $C_{p_H} \geq p_H$, the error bound of Theorem 5.5 holds for the schedule $S(p_H)$. In this case the for-loop of heuristic H_D can be eliminated by a single iteration — the one for $p = p_H$ — thus bringing the complexity of H_D down to $\mathcal{O}(nP_n'^4)$. In most practical situations the if-condition holds true unless there is a super large job that is longer than all other jobs combined. Then, a single iteration of H_D is needed and the complexity of H_D is $\mathcal{O}(nP_n'^4)$.

Corollary 5.2

If $C_{p_H} \geq p_H$, then $\frac{C(p_H)}{C_{\text{FHFS}m_1,m_2}} \leq 2 - \frac{1}{\max\{m_1,m_2\}}$ [16].

Proof

For the reasons indicated in Corollary 5.1, we have that

$$C_{p_L} \geq C_{p_L+1} \geq \cdots \geq C_{p_H}$$

and therefore, $\max_i \max\{p_{i1}, p_{i2}, a_{i1} + b_{i2}\} \leq C_{p_H} \leq C_{p^*} \leq C_{\text{FAFS}_{m_1,m_2}}$ the latter inequality coming from Lemma 5.2 (again, p^* denotes the longest processing required by any job in an optimal schedule S^* for FHFS_{m_1,m_2}).

In Theorem 5.5, the proof of the inequality

$$C(p^*) \leq C_{\text{FHFS}_{m_1,m_2}} + \frac{\max\{m_1, m_2\} - 1}{\max\{m_1, m_2\}} \max_i \max\{p_{i1}, p_{i2}, a_{i1} + b_{i2}\}$$

carries over if p^* is replaced by p_H and S_{p^*} by S_{p_H}. Due to the condition $\max_i \max\{p_{i1}, p_{i2}, a_{i1} + b_{i2}\} \leq C_{p_H} \leq C_{\text{FHFS}_{m_1,m_2}}$, we have $C_{p_H} \leq (2 - \frac{1}{\max\{m_1,m_2\}})C_{\text{FHFS}_{m_1,m_2}}$. This completes the proof of the corollary. □

5.2.7 The Special Case $\text{FHFS}_{m,m}$

In this subsection, we consider the $\text{FHFS}_{m,m}$ system with $a_{i1} = a_{i2}$ and $b_{i1} = b_{i2}$ for $i = 1, 2, \ldots, n$ (i.e., the case of stage-independent processing times). Consider the following heuristic.

Heuristic H_D':

1. Apply DP with respect to the processing times $\{(\frac{1}{m}a_i, \frac{1}{m}b_i) : i = 1, 2, \ldots, n\}$. Let S be the resulting sequence.
2. Apply the FAM rule on the stage-1 tasks of the sequence S.

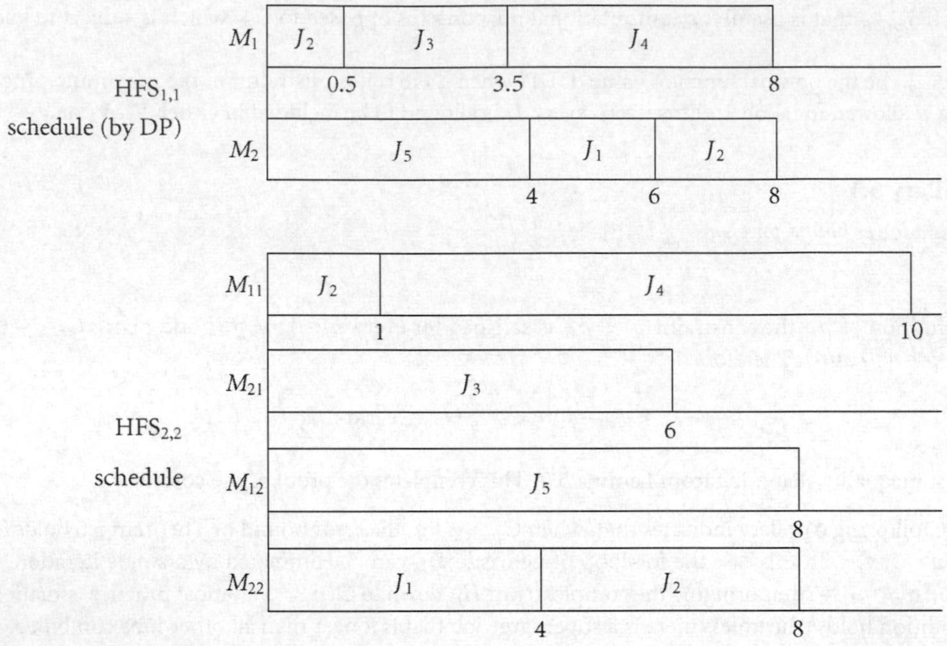

FIGURE 5.5 An application of H'.

3. Apply the LBM rule on the stage-2 tasks of the sequence S.
4. On each stage-2 machine, schedule the $p_i = a_i + b_i$ tasks to start as soon as possible, and the b_i tasks according to the first-come-first-serve rule.

Heuristic $H_D{}'$ is a special case of H_D where the for-loop is eliminated. In this case algorithm DP$'$ can be simplified to DP. As a straightforward consequence of Theorem 5.5, and the fact that $a_i + b_i \leq C_{\text{FAFS}_{m,m}}$ for $i = 1, 2, \ldots, n$, the simplified heuristic $H_D{}'$ has a worst-case error bound of $2 - \frac{1}{m}$. Since DP requires $\mathcal{O}(nP_n^3)$ time, the computational effort required by $H_D{}'$ is also $\mathcal{O}(nP_n^3)$.

Example

Consider the case $m = 2$ and a set of five jobs with processing time requirements $J_1 = (1, 3)$, $J_2 = (1, 4)$, $J_3 = (2, 4)$, $J_4 = (3, 6)$, and $J_5 = (6, 2)$. The jobs are already numbered according to Johnson's order. The Gantt chart of an optimal schedule generated by DP with respect to the processing times $\{(\frac{1}{2}a_i, \frac{1}{2}b_i) : i = 1, 2, 3, 4, 5\}$, as well as the associated schedule for FAFS$_{2,2}$ generated by H_D, are given in Figure 5.5.

At Step 1 we get $C_{\text{DP}} = 8$. At Step 2 the jobs $a_2 = 1$, $p_4 = 9$ are allocated onto M_{11} and $p_3 = 6$ is allocated onto M_{21}. At Step 3 jobs $b_2 = 4$, $p_1 = 4$ are allocated onto M_{22} and $p_5 = 8$ is allocated onto M_{12}. Since b_2 already starts on M_{22} as soon as possible, Step 4 cannot improve the schedule. Evidently, $C_{H_D{}'} = 10$.

5.3 Hybrid Flowshops with Routing Flexibility

In order to present the decomposition of HFS$_{m_1,m_2}$ into the PHFS system, we first study a simpler but equally important design; the m parallel flowlines denoted by mFL (see Figure 5.1). To develop algorithms for this system we first study the case of $m = 2$ in Subsection 5.3.1 where we present an optimal dynamic programming algorithm. This algorithm is used in Subsection 5.3.2 to develop heuristics for mFL. The resulting heuristics are adapted in Subsection 5.3.3 for PHFS.

5.3.1 The 2*FL* Problem

Problem $2FL$ is stated formally as follows. A set of jobs $J = \{J_1, J_2, \ldots, J_n\}$ is given and every job J_i consists of two tasks, with processing time requirements a_i and b_i. We will use a_i, b_i to denote both the tasks and the requirements of job J_i. All jobs are assumed to be available at time zero and no preemption is allowed for the tasks. Each job in J must be processed exclusively by one of two available flowlines L_1, L_2, where each flowline is a two-machine flowshop; see Figure 5.1. To maximize throughput, as well as the machine utilization of the 2*FL* system, we are interested in scheduling the jobs in J to the two flowlines L_1, L_2, so that the resulting schedule minimizes makespan. Hence, the 2*FL* problem is equivalent to partitioning the job set J into two subsets of jobs, say I_1 and I_2, and then dedicate the flowline L_k to the subset I_k, $k = 1, 2$. After resolving this assignment problem, scheduling on L_1 and L_2 is a simple task involving Johnson's algorithm for minimizing makespan on a two-machine flowshop (see Johnson, 1954).

When $b_i = 0$ for all $J_i \in J$, our 2*FL* problem reduces to the problem of minimizing makespan on two identical parallel machines. This scheduling problem is known to be ordinary $\mathcal{NP}$-complete (see Garey and Johnson, 1979), and therefore our problem, since it contains the above problem as a special case, is $\mathcal{NP}$-complete as well.

Assume that the set $J = \{J_1, J_2, \ldots, J_n\}$ is ordered according to Johnson's order. Define the quantities:

$$p_i = a_i + b_i \qquad A_i = \sum_{j=1}^{i} a_j$$

$f_i(I, S_1, S_2)$ = the optimal makespan value of 2*FL* for the jobs $J_1, J_2, \ldots, J_i$, when the amount of idle time on M_{11} is I, and the idle times after the last job of M_{12} and M_{22} are S_1 and S_2, respectively.

By definition, $S_1 \cdot S_2 = 0$ since the makespan value is attained on at least one of M_{12} and M_{22}. More specifically, if the makespan is attained on M_{12} we have $S_1 = 0$, and if it is attained on M_{22} we have $S_2 = 0$. Also, in the above definition of $f_i(I, S_1, S_2)$, the variables I, S_1, S_2 take values from the interval $[0, P_n]$, where $P_n = \sum_i p_i$. These observations indicate that the state space of the dynamic program for 2*FL* (abbreviated by DPF) to calculate $f_i(\cdot, \cdot, \cdot)$ is $\mathcal{O}(nP_n^2)$.

The following DPF algorithm is based on the fact that the optimal makespan $f_i(I, S_1, S_2)$ is attained on at least one of M_{12} and M_{22} before the scheduling of J_i, and on at least one of M_{12} and M_{22} after the scheduling of J_i, thus producing four possible combinations.

Definition 5.2

Let C_{kr} be the combination where the value $f_i(I, S_1, S_2)$ is attained by L_r after the scheduling of J_i, and by L_k prior to scheduling J_i, $k, r \in \{1, 2\}$.

For each C_{kr}, we depict in Figure 5.6 the alternative schedule configurations for the job set $\{J_1, J_2, \ldots, J_{i-1}\}$, that can result to C_{kr} after the insertion of J_i.

Evidently, there are two alternative schedule configurations for C_{11}. In C_{11}(a), J_i is assigned to L_1; and in C_{11}(b), J_i is assigned to L_2 (in Figure 5.6 the dotted boxes indicate job J_i). Similarly, there are two configurations for C_{22}. In C_{21}, the makespan is attained on L_2 prior to inserting J_i, and hence J_i *must* be inserted into L_1 if the makespan is to be attained on L_1 after the insertion. Hence, there is a single configuration for C_{12}. Similarly, there is a unique configuration for C_{21}. The four C_{kr} combinations motivate the following recurrence relation.

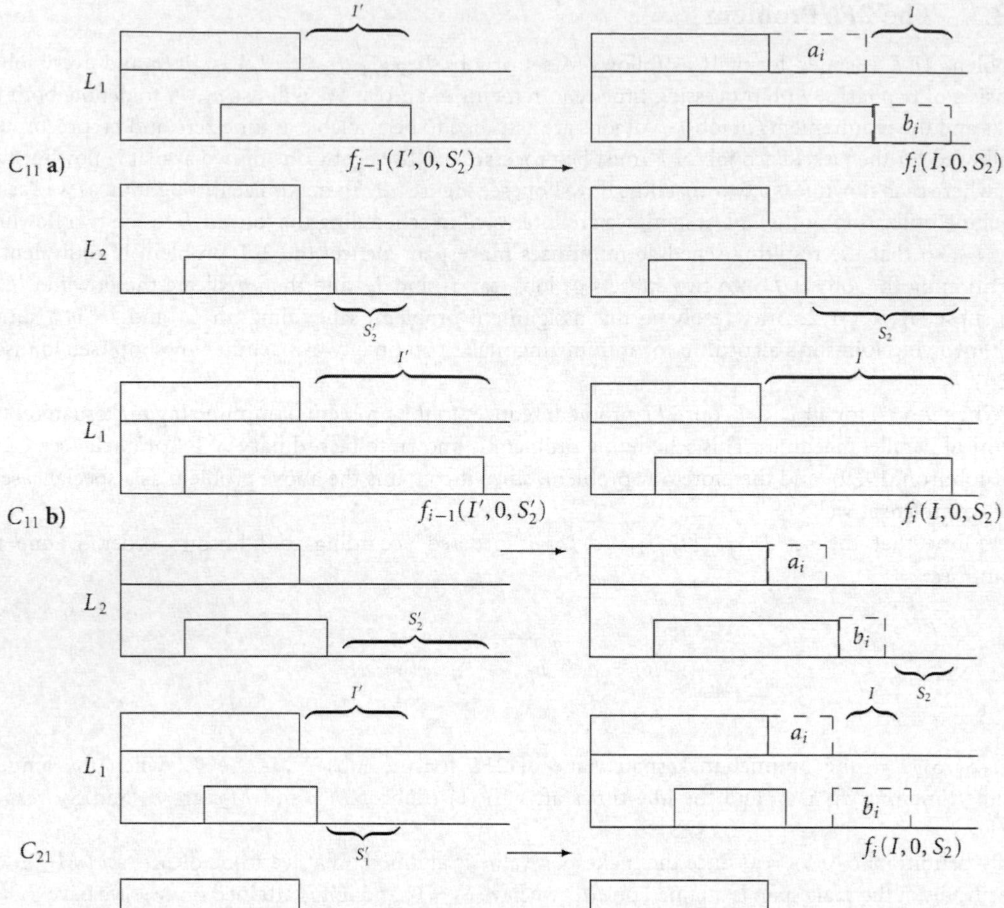

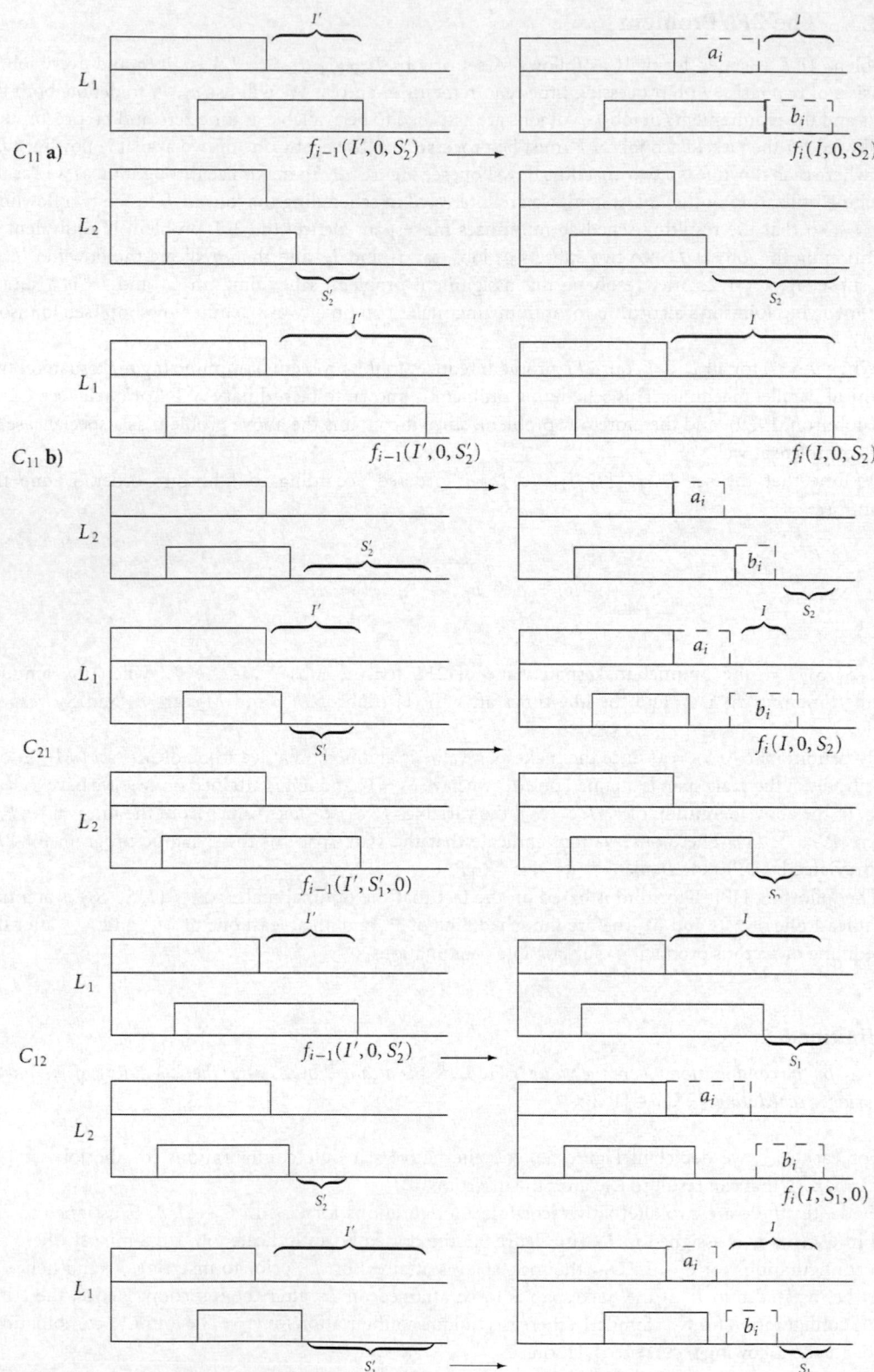

FIGURE 5.6 Alternative schedules for $\{J_1, J_2, \ldots, J_{i-1}\}$.

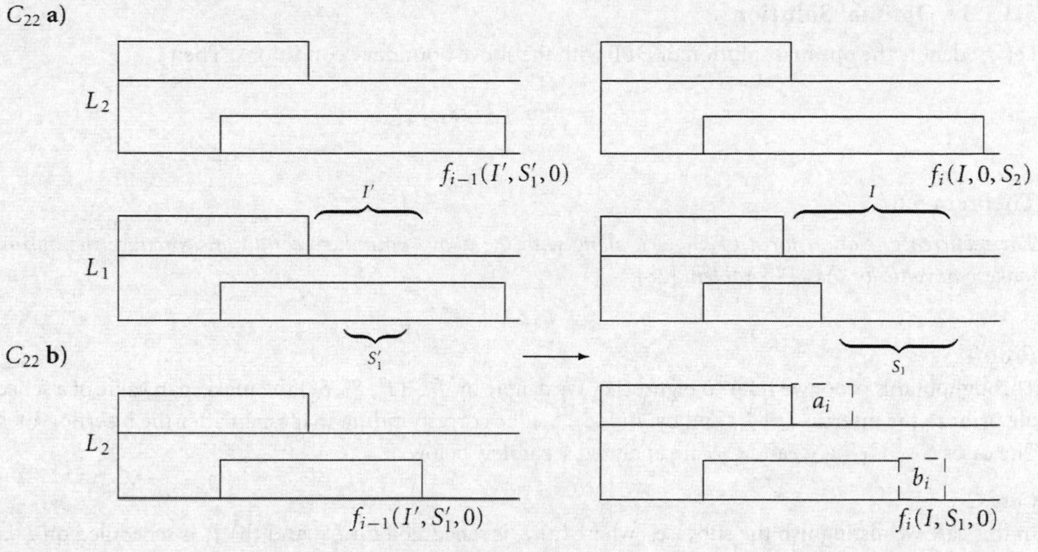

FIGURE 5.6 *(Continued)*

5.3.1.1 Recurrence Relation

Let $J_1, J_2, \ldots, J_n$ be the set of jobs ordered according to Johnson's order. Then,

$$f_i(I,0,S_2)$$

$$= \min \begin{cases} C_{11}\colon \min \begin{cases} \min_{I' \le a_i}\{f_{i-1}(I',0,S_2+I'-p_i) - I'\} + p_i & \text{if } I = b_i \\ f_{i-1}(I-b_i+a_i,0,S_2-b_i) + b_i & \text{if } I > b_i \\ \min_{S_2' \ge S_2}\{f_{i-1}(I,0,S_2') : S_2' = S_2 + b_i + (A_i - 2f_{i-1}(I,0,S_2') + I + S_2')^+\} \end{cases} \\ \\ C_{21}\colon \begin{cases} f_{i-1}(I-S_2+a_i, b_i-S_2, 0) + S_2 & \text{if } I > b_i \\ \min_{0 \le S_1' \le a_i+b_i} f_{i-1}(p_i - S_2, S_1', 0) + S_2 & \text{if } I = b_i \end{cases} \end{cases}$$

$$f_i(I,S_1,0)$$

$$= \min \begin{cases} C_{12}\colon \min \begin{cases} \min_{0 \le S_2' \le b_i} f_{i-1}(I-S_1,0,S_2') + S_1 \\ \min_{0 \le S_2' \le a_i+b_i-S_1} f_{i-1}(I-S_1,0,S_2') + S_1 & \text{if } A_i + I - S_1 + S_2' > 2f_{i-1}(I-S_1,0,S_2') \\ \quad \text{and } (A_i + I - S_1 + S_2' - 2f_{i-1}(I-S_1,0,S_2'))^+ + b_i = S_1 + S_2' \end{cases} \\ \\ C_{22}\colon \begin{cases} \min_{S_1' \ge I} f_{i-1}(I+a_i,S_1',0) & \text{if } I = S_1 + b_i \\ f_{i-1}(I+a_i, S_1+b_i, 0) & \text{if } I > S_1 + b_i \\ \min_{0 \le y \le a_i}\{f_{i-1}(I-b_i-y, S_1-b_i-y, 0) + y\} + b_i & \text{if } y = (A_i - 2f_{i-1}(I-b_i-y, S_1-b_i-y, 0) + I')^+ \end{cases} \end{cases}$$

5.3.1.2 Boundary Conditions

$$f_1(I,S_1,S_2) = \begin{cases} p_1 & \text{if } (I,S_1,S_2) = (b_1,0,p_1) \\ p_1 & \text{if } (I,S_1,S_2) = (p_1,p_1,0) \\ \infty & \text{otherwise} \end{cases}$$

5.3.1.3 Optimal Solution

Let f_n^* denote the optimal solution of DPF with the above boundary conditions. Then

$$f_n^* = \min_{I,S_1,S_2} f_n(I, S_1, S_2)$$

Theorem 5.6

The recurrence relation for $f_i(I, S_1, S_2)$, along with the above boundary conditions, produce the optimal makespan value for the 2FL problem [28].

Proof

Throughout this proof we refer to Figure 5.6. We denote by $f_{i-1}(I', S_1', S_2')$ the makespan value of a schedule prior to the insertion of J_i, and by $f_i(I, S_1, S_2)$ the corresponding makespan after the insertion of J_i. The cases C_{11}, C_{12}, C_{21}, and C_{22} are analyzed separately below.

Case C_{11}

In this case we distinguish the subcases, where (a) J_i is scheduled on L_1, and (b) J_i is scheduled on L_2.

Subcase (a): In this subcase we have that $I \geq b_i$. If $I = b_i$, then we must have $a_i \geq I'$. Then the idle time on M_{22} increases by $b_i + a_i - I'$ after the insertion of J_i, and hence $S_2 = S_2' + p_i - I'$. Also, the makespan of M_{12} increases by $p_i - I'$. Hence,

$$f(I, 0, S_2) = \min_{I' \leq a_i} f_{i-1}(I', 0, S_2 - p_i + I') + p_i - I' \quad \text{if} \quad I = b_i$$

On the other hand, if $I > b_i$, then we must have that $a_i < I'$. More specifically, in this case we have that $I' = I - b_i + a_i$. Also, the makespan of M_{22} increases by b_i and hence $S_2' = S_2 - b_i$. Therefore,

$$f(I, 0, S_2) = f_{i-1}(I - b_i + a_i, 0, S_2 - b_i) + b_i \quad \text{if} \quad I > b_i$$

Subcase (b): When J_i is scheduled on L_2, we have that $I' = I$ and

$$f_i(I, 0, S_2) = \min_{S_2' \geq S_2} f_{i-1}(I, 0, S_2')$$

where

$$S_2' = S_2 + b_i + (A_i - 2f_{i-1}(I, 0, S_2') + I + S_2')^+$$

The term $(A_i - 2f_{i-1}(I, 0, S_2') + I + S_2')^+$ indicates the idle time induced to M_{22} by the scheduling of a_i on M_{12}.

Combining the two subcases of C_{11} we get the recurrence relation:

$$f(I, 0, S_2) = \min \begin{cases} \min_{I' \leq a_i}\{f_{i-1}(I', 0, S_2 + I' - p_i) - I'\} + p_i & \text{if } I = b_i \\ f_{i-1}(I - b_i + a_i, 0, S_2 - b_i) + b_i & \text{if } I > b_i \\ \min_{S_2' \geq S_2}\{f_{i-1}(I, 0, S_2') : S_2' = S_2 + b_i + (A_i - 2f_{i-1}(I, 0, S_2') + I + S_2')^+\} \end{cases}$$

Case C_{21}

This case holds only if

$$\max\{f_{i-1}(I', S_1', 0) - I' + a_i, f_{i-1}(I', S_1', 0) - S_1'\} + b_i \geq f_{i-1}(I', S_1', 0) \tag{5.1}$$

or equivalently,

$$\max\{p_i - I', b_i - S_1'\} \geq 0$$

We distinguish two subcases for C_{21}, namely (a) $I > b_i$ and (b)$I = b_i$. In (a), we have that $b_i = S'_1 + S_2$, and $I + a_i = I' + S_2$. Hence,

$$f_i(I, 0, S_2) = f_{i-1}(I + a_i - S_2, b_i - S_2, 0) + S_2 \quad \text{if} \quad I > b_i$$

In subcase (a), relation (5.1) is equivalent to $\max\{S_2 - I + b_i, S_2\} \geq 0$, which holds true because $S_2 \geq 0$ by definition.

In subcase (b) the idle time inserted on M_{21} after the insertion of J_i is $a_i - (I' - S'_1)$, and hence $S'_1 + S_2 = b_i + a_i - (I' - S'_1)$ or $I' = p_i - S_2$. Also, it is true that $S'_1 \leq a_i + b_i$ and hence

$$f_i(b_i, 0, S_2) = \min_{0 \leq S'_1 \leq a_i + b_i} f_{i-1}(p_i - S_2, S'_1, 0) + S_2 \quad \text{if} \quad I = b_i$$

In this subcase, relation (5.1) is equivalent to $\max\{S_2, b_i - S'_1\} \geq 0$ which holds always true.

Combining the recurrence relations for subcases (a) and (b), we get

$$f_i(I, 0, S_2) \begin{cases} f_{i-1}(I - S_2 + a_i, b_i - S_2, 0) + S_2 & \text{if } I > b_i \\ \min_{0 \leq S'_1 \leq a_i + b_i} f_{i-1}(p_i - S_2, S'_1, 0) + S_2 & \text{if } I = b_i \end{cases}$$

Case C_{12}

This case holds only iff

$$\max\{A_{i-1} - (f_{i-1}(I', 0, S'_2) - I') + a_i, f_{i-1}(I', 0, S'_2) - S'_2\} + b_i \geq f_{i-1}(I', 0, S'_2) \tag{5.2}$$

Hence, we distinguish the subcases where

(a) $A_i - f_{i-1}(I', 0, S'_2) + I' \leq f_{i-1}(I', 0, S'_2) - S'_2$
(b) $A_i - f_{i-1}(I', 0, S'_2) + I' > f_{i-1}(I', 0, S'_2) - S'_2$

In (a), the insertion of a_i into L_2 does not increase the idle time on M_{22}. In this subcase relation (5.2) is equivalent to $b_i \geq S'_2$. We have that $b_i = S_1 + S'_2$, and hence condition (5.2) is verified. Also, we have that $I = I' + S_1$. Hence,

$$f_i(I, S_1, 0) = \min_{0 \leq S'_2 \leq b_i} f_{i-1}(I - S_1, 0, S'_2) + S_1$$

In (b), the insertion of a_i into L_2 increases the idle time on M_{22}. Since $I = I' + S_1$, the additional idle time on M_{22} is given by $(A_i + I - S_1 + S'_2 - 2 f_{i-1}(I - S_1, 0, S'_2))^+$. To ensure that only the right combination of values of S_1, S'_2, and I are considered by the recurrence relation, we need to ensure that

$$(A_i + I - S_1 + S'_2 - 2 f_{i-1}(I - S_1, 0, S'_2))^+ + b_i = S_1 + S'_2$$

which indicates that the newly inserted idle time on M_{22} plus b_i equal $S_1 + S'_2$ (see Figure 5.6). With the above observations we get that

$$f_i(I, S_1, 0) = \min_{0 \leq S'_2 \leq a_i + b_i - S_1} f_{i-1}(I - S_1, 0, S'_2) + S_1$$

given that $(A_i + I - S_1 + S'_2 - 2 f_{i-1}(I - S_1, 0, S'_2))^+ + b_i = S_1 + S'_2$ and $A_i + I - S_1 + S'_2 > 2 f_{i-1}(I - S_1, 0, S'_2)$.

The following comment should be made for the range of values of S'_2. Note that $S_1 + S'_2 \leq a_i + b_i$ and hence $S'_2 \leq a_i + b_i - S_1$. Combining (a) and (b) we get the desired recurrence relation for C_{12}.

Case C_{22}

In this case we distinguish the subcases, where (a) J_i is scheduled on L_1, and (b) J_i is scheduled on L_2.

Subcase (a): This subcase holds iff after the insertion of J_i, M_{22} finishes after M_{21}, i.e.,

$$\max\{f_{i-1}(I', S'_1, 0) + a_i - I', f_{i-1}(I', S'_1, 0) - S'_1\} + b_i + S_1 = f_{i-1}(I', S'_1, 0)$$

or equivalently

$$\max\{a_i + b_i - I', b_i - S_1'\} = -S_1 \tag{5.3}$$

In this subcase we have that $I' = I + a_i$, and hence the flowshop condition (5.3) becomes

$$\min\{I, S_1'\} = S_1 + b_i \tag{5.4}$$

Hence,

$$S_1 + b_i = \begin{cases} I & \text{if } I \leq S_1' \\ S_1' & \text{if } I > S_1' \end{cases}$$

Observing that $f_i(I, S_1, 0) = f_{i-1}(I', S_1', 0)$ in this case, we get the recurrence relation

$$f_i(I, S_1, 0) = \min \begin{cases} \min_{S_1' \geq I} f_{i-1}(I + a_i, S_1', 0) & \text{if } I = S_1 + b_i \\ f_{i-1}(I + a_i, S_1 + b_i, 0) & \text{if } I > S_1' = S_1 + b_i \end{cases}$$

Subcase (b): The insertion of J_i into L_2, induces $(A_i - 2f_{i-1}(I', S_1', 0) + I')^+$ units of idle time on M_{22}, and the makespan of M_{22} increases by $b_i + (A_i - 2f_{i-1}(I', S_1', 0) + I')^+$. Define

$$y = (A_i - 2f_{i-1}(I', S_1', 0) + I')^+$$

Then, we have that $I = I' + y + b_i$, and $S_1 = S_1' + y + b_i$. Also, observe that the idle time y cannot exceed a_i. Hence,

$$f_i(I, S_1, 0) = \min_{0 \leq y \leq a_i} \{f_{i-1}(I - b_i - y, S_1 - b_i - y, 0) + y\} + b_i$$

Combining the recurrence relations for the above four cases, we get the recurrence relation stated in the theorem. This completes the proof of the theorem. □

5.3.1.4 Complexity of DPF Algorithm

As indicated earlier, the state space of DPF is $\mathcal{O}(nP_n^2)$. It is easy to check that the effort required at every iteration of DPF is of order $\mathcal{O}(P_n)$. Therefore, the complexity of DPF is $\mathcal{O}(nP_n^3)$.

5.3.2 The *m*FL Problem

In this subsection we develop lower bound and heuristic algorithms for the *m*FL problem. These algorithms have been used in Vairaktarakis and Elhafsi (2000) in computational experiments.

5.3.2.1 Lower Bounds

Given the set J of jobs, we can construct an auxiliary two-machine flowshop problem (AFS) for *m*FL, by replacing a_i by $\frac{1}{m}a_i$, and b_i by $\frac{1}{m}b_i$. Hence, the AFS problem is a makespan problem on a single flowline. Note that the AFS subproblem is identical to the one defined in Section 5.2.4. Let C_{AFS} be the optimal makespan value obtained by the application of Johnson's algorithm on AFS (see Johnson, 1954). Then, if $C_{m\text{FL}}$ denotes the optimal makespan value for *m*FL, we have the following result.

Lemma 5.3

$C_{\text{AFS}} \leq C_{\text{mFL}}$.

Proof

Let S^* be an optimal schedule for *m*FL. Let S_1 be the nondecreasing order of completion times of a_i tasks in S^*. We can use the order S_1 to generate a schedule S for AFS as follows.

Schedule the a_i/m tasks on the upstream machine of AFS according to the order S_1. Then, schedule the b_i/m tasks on the downstream machine of AFS according to the order S_1. The schedule S constructed in this way is not necessarily optimal for AFS and in general it is not a permutation schedule. Let C_S denote the makespan of the schedule S constructed above. Without loss of generality, we reorder the jobs so that $S_1 = \{a_1, a_2, \ldots, a_n\}$. Then, due to the order S_1 we have that the task a_i/m completes in S at time

$$C_i' = \frac{1}{m}\sum_{j\leq i} a_j \leq C_i$$

where C_i is the completion time of a_i on S^*.

For the b_i/m tasks we can assume that they are scheduled contiguously so that the last task finishes at time $C_{m\text{FL}}$. Then, the task b_i/m starts in S at time

$$s_i' = C_{m\text{FL}} - \frac{1}{m}\sum_{j\geq i} b_j$$

while b_i starts in S^* at time $s_i \leq s_i'$, due to the order S_1 and the fact that all b_j tasks with $j \geq i$ start in S^* after a_i.

Therefore, for every $1 \leq i \leq n$ we have that $C_i' \leq C_i$, and $s_i \leq s_i'$. These relations mean that the flowshop constraints for S are satisfied when the last b_i/m task of AFS is scheduled to finish at time $C_{m\text{FL}}$. Hence, $C_S \leq C_{m\text{FL}}$. However, the makespan value C_{AFS} produced by Johnson's algorithm is optimal for the AFS problem and hence $C_{\text{AFS}} \leq C_S \leq C_{m\text{FL}}$. This completes the proof of the lemma. □

The above lemma is used to develop a better lower bound for mFL, when m is a power of 2, i.e., $m = 2^k$ for $k \geq 1$. As we will see, the case $m = 2^k$ is not restrictive at all, since with minor additional computational effort we can transform problems with $2^{k-1} < m < 2^k$ to equivalent problems with $m = 2^k$.

5.3.2.2 The Case $m = 2^k$

Let AFL be the auxiliary problem on two flowlines (*2FL*) where a_i is replaced by $2a_i/m$, and b_i is replaced by $2b_i/m$. Let us denote by C_{LB} the makespan value obtained by the application of DPF on AFL. Then, we have the following result.

Theorem 5.7

$C_{\text{LB}} \leq C_{m\text{FL}}$ [28].

Proof

Let S^* be an optimal schedule for mFL. Apply Lemma 5.3 on the first $m/2$ flowlines of the mFL environment. Then, the auxiliary flowshop problem, say AFS_1, is the two-machine flowshop where the processing times of all tasks assigned to the first $m/2$ flowlines of mFL are multiplied by 2. Equivalently, (a_i, b_i) is replaced in AFS_1 by $(2a_i/m, 2b_i/m)$ if J_i is assigned to one of the first $m/2$ flowlines of mFL in S^*. Let C_{AFS_1} be the makespan obtained by the JA for the AFS_1 problem. Then, by Lemma 5.3 we have that $C_{\text{AFS}_1} \leq C_{m\text{FL}}$ since $C_{m\text{FL}}$ is the makespan of S^*. Similarly, we define the auxiliary flowshop problem AFS_2 for the last $m/2$ flowlines of mFL. Let C_{AFS_2} be the resulting makespan value. Then, $C_{\text{AFS}_2} \leq C_{m\text{FL}}$.

Note that the schedules obtained by the AFS_1 and AFS_2 problems, provide a feasible solution for the 2FL problem with processing time requirements of $(2a_i/m, 2b_i/m)$, $1 \leq i \leq n$. However, this solution is not necessarily optimal for this 2FL problem. By Theorem 5.6, the optimal makespan value for the 2FL auxiliary problem equals C_{LB}, and hence $C_{\text{LB}} \leq \max\{C_{\text{AFS}_1}, C_{\text{AFS}_2}\} \leq C_{m\text{FL}}$. This completes the proof of the theorem. □

5.3.2.3 The Case $m \neq 2^k$

Consider the case where $2^{k-1} < m < 2^k$. Then, the mFL problem can be transformed into an equivalent (2^k)FL problem by adding to the job set J, $2^k - m$ dummy jobs with processing requirements $(0, B)$, and $2^k - m$ dummy jobs with processing requirements $(B, 0)$, where $B = C_{m\text{FL}}$. Then, an optimal schedule

S^{2^k} for $(2^k)FL$ induces an optimal schedule for mFL by disregarding the $2^k - m$ flowlines of S^{2^k} that are assigned to process the $2(2^k - m)$ dummy jobs. Since the optimal makespan $C_{m\text{FL}}$ is unknown, we can perform bisection search on B, in the range

$$B \in \left[\frac{1}{m}\sum_{i=1}^{n} p_i, \sum_{i=1}^{n} p_i\right]$$

In this construction, the optimal makespan of the mFL problem, is the least value of B for which $C_{m\text{FL}} = B$.

We can use the above observation to adapt Theorem 5.7 to the case where $m \neq 2^k$, by applying our DPF algorithm to the AFL problem on the revised job set that except for $(2a_i/m, 2b_i/m)$, $1 \leq i \leq n$, includes $2^k - m$ dummy jobs with processing requirements $(0, B')$ and $2^k - m$ dummy jobs with processing requirements $(B', 0)$. In this case we have that

$$B' \in \left[\frac{2}{m^2}\sum_{i=1}^{n} p_i, \frac{2}{m}\sum_{i=1}^{n} p_i\right]$$

and the optimal value for B' is the least value for which $C_{\text{LB}} = B'$. Since the computational effort required by DPF is $\mathcal{O}(nP_n^3)$, the described bisection search scheme can provide a lower bound to the mFL problem for $m \neq 2^k$ in $\mathcal{O}(nP_n^3 \log P_n)$ time.

5.3.2.4 Heuristic Algorithms

Several heuristics are developed in Vairaktarakis and Elhafsi (2000) for problem mFL. Below we describe the one that exhibited the best performance. We refer to this heuristic as H_F. The tree T of Figure 5.7 facilitates the description of H_F.

At level 0 we apply DPF with respect to the processing times $(2a_i/m, 2b_i/m)$. Algorithm DPF partitions the job set J in two subsets: the jobs that are going to be processed by L_1 and the jobs that are going to be

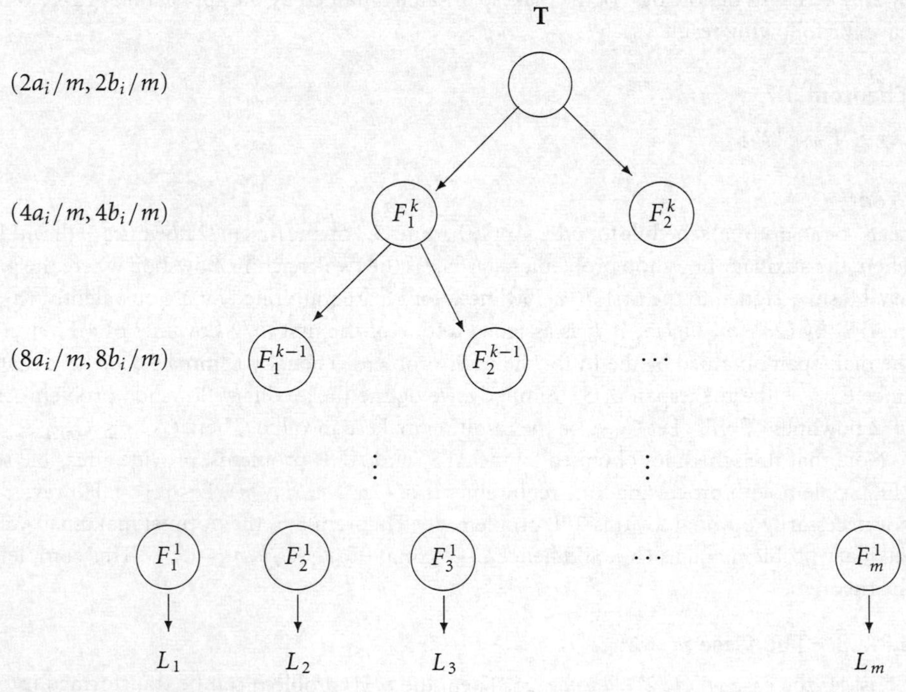

FIGURE 5.7 The tree structure of heuristic H_F.

processed by L_2 (recall that the DPF solves optimally the 2FL problem). The jobs allocated to L_1 correspond to the jobs that will be processed on the upper half ($m/2$ machines) of mFL; we refer to this subproblem as F_1^k, where $k = \log_2 m$. The jobs allocated to L_2 are the jobs that will be processed on the lower half of flowlines in mFL; we refer to this subproblem as F_2^k. At level 1, we apply the DPF algorithm on the subproblem F_1^k, thus allocating jobs to the first and second quarter of machines of mFL. Similarly, the application of DPF on F_1^k allocates jobs to the third and fourth quarter of machines of mFL. In every level of the tree T of Figure 5.7 we indicate the processing times utilized by DPF. The leaves of T indicate the allocation of jobs to single flowlines and hence the DPF algorithm is utilized only by the nodes in levels 0 through $k-1$ of T. A schedule is provided for each of the m flowlines represented by leaves, from the DPF applications in the previous level. There are $2^k - 1$ or $\mathcal{O}(m)$ nodes in T that are not leaves. Since DPF is applied once for each such node and since each application of DPF requires $\mathcal{O}(nP_n^3)$ time, the complexity of H_F is $\mathcal{O}(mnP_n^3)$.

Having described algorithms for problem mFL, we proceed in the next Subsection with adaptations that decompose HFS_{m_1,m_2} to flowline-like systems.

5.3.3 Decomposing Hybrid Flowshops to Flowline-Like Designs

We consider the decomposition of HFS_{m_1,m_2} into m_1 smaller independent units (we assume $m_1 \leq m_2$) each of which is a hybrid flowshop of the form $\text{HFS}_{1,k}$, where k is an appropriately selected number (see Figure 5.1). The assumption $m_1 \leq m_2$ is not restrictive since the case $m_1 > m_2$ is symmetric and hence all the results developed in this chapter still apply.

Intuitively it is beneficial to distribute the workload of HFS_{m_1,m_2} so that each machine can handle about the same workload. This is in line with similar conclusions for scheduling environments where balanced designs result to improved throughput performance. For this reason, we assume that the number k of machines in $\text{HFS}_{1,k}$ is either $\lceil \frac{m_2}{m_1} \rceil$ or $\lceil \frac{m_2}{m_1} \rceil - 1$ (although our approach is applicable to any arbitrary decomposition). To further simplify our exposition, we assume that m_2 is an integral multiple of m_1, i.e., $m_2 = km_1$. If a HFS_{m_1,m_2} design does not possess this property, we can introduce $km_1 - m_2$ dummy stage-2 machines along with $km_1 - m_2$ dummy jobs with processing time requirements $(0, B)$, where B is a trial makespan value. Bisection search on B can be used to identify the least value of B for which there exists a feasible decomposition of HFS_{m_1,m_2} into m_1 units of the form $\text{HFS}_{1,\frac{m_2}{m_1}}$, so that the resulting makespan value equals B. Moreover, without loss of generality we retain our assumption that m_1 is a power of 2. We refer to the decomposition problem as PHFS since it decomposes the HFS design to parallel HFS sub-designs.

The algorithms developed in Vairaktarakis and Elhafsi (2000) for the mFL problem were extended to the PHFS. Here we present the adaptation of H_F to PHFS.

5.3.3.1 Heuristic for PHFS

1. Apply heuristic H_F to the m_1FL problem with respect to the processing times $(a_i, \frac{m_2}{m_1} b_i)$. Let A_r be the set of jobs assigned to L_r, $1 \leq r \leq m_1$.
2. For $r = 1$ to m_1 do apply algorithm H to schedule the jobs in A_r to $\text{HFS}_{1,\frac{m_2}{m_1}}$.

At Step 1 of the above algorithm we use H_F to partition the job set J into m_1 parts $A_1, A_2, \ldots, A_{m_1}$. For each $1 \leq r \leq m_1$, we use heuristic H to schedule the jobs in A_r to a $\text{HFS}_{1,\frac{m_2}{m_1}}$ sub-design. Let C_{PHFS} be the optimal makespan value for the PHFS problem. Then, the suboptimality of the heuristics for PHFS are attributed to two sources. Namely, suboptimality of the heuristic H_F for the m_1FL problem, and suboptimality of heuristic H for the $\text{HFS}_{1,\frac{m_2}{m_1}}$ problem. The worst case error bound for the latter source of suboptimality is $2 - \frac{m_1}{m_2}$. Let PFL be the auxiliary problem on two flowlines (2FL) where a_i is replaced by $2a_i/m_1$, and b_i is replaced by $2b_i/m_2$. Then, a proof similar to that of Theorem 5.7 verifies the following result.

Theorem 5.8

$C_{\text{PFL}} \leq C_{\text{PHFS}}$ [28].

This completes the presentation of results for problems PHFS and FHFS. In the next Section we present managerial insights stemming from the computational experiments performed in Kouvelis and Vairaktarakis (1998) and Vairaktarakis and Ehafsi (2000).

5.4 Managerial Insights

System $FHFS_{m_1,m_2}$ is related to the hybrid flowshop HFS_{m_1,m_2} and the $m_1 + m_2$ parallel identical machine environment $P(m_1 + m_2)$. In HFS_{m_1,m_2}, stage-1 machines process a_i tasks only, and stage-2 machines process b_i tasks only. In $P(m_1 + m_2)$ any of the $m_1 + m_2$ machines can process task $p_i = a_i + b_i, 1 \leq i \leq n$. Below we present findings reported in Kouvelis and Vairaktarakis (1998) comparing the performance of $FHFS_{m_1,m_2}$ with HFS_{m_1,m_2} and $P(m_1 + m_2)$.

Experiments were performed with the designs $HFS_{2,3}$, $HFS_{3,3}$, $HFS_{3,2}$, $HFS_{2,4}$, $HFS_{2,8}$, and $HFS_{4,8}$; problem sizes with 20, 30, 40, and 50 jobs; and processing time ratios 2 : 3, 3 : 3, and 3 : 2. The ratio 2 : 3, for instance, means that the processing time of a_i is a randomly selected integer from a uniform distribution on [1,20], and the processing time of b_i from a uniform distribution on [1,30]. In each problem instance the makespan C_H of HFS_{m_1,m_2} is computed by means of algorithm H. The makespan of $P(m_1 + m_2)$ is computed using an optimal dynamic programming algorithm (see Cheng and Sin, 1990).

The experiment in Kouvelis and Vairaktarakis (1998) indicates that the makespan performance of the $P(m_1 + m_2)$ environment is very close to the performance of $FHFS_{m_1,m_2}$ and the overall average performance gap equals 1.33%. Similarly, it is reported that the difference in makespan performance between $FHFS_{m_1.m_2}$ and HFS_{m_1,m_2} compiles an overall average relative gap of 11.76%. This shows that investing in processing flexibility offers significant benefits in throughput performance. Moreover, it is found that these performance gaps increase with the total number of machines in the system. For instance, when $m_1 + m_2 = 6$ the average performance gap is 12.66% while the average gap when $m_1 + m_2 = 5$ is 11.16%. Average performance gaps decrease as the number of jobs increase. This completes our report on the comparison between FHFS and related HFS and parallel identical machine systems.

Next we summarize the findings reported in Vairaktarakis and Elhafsi (2000) for the PHFS system. The authors experimented with problem instances with $n = 20, 30, 40$, and 50 jobs; and the designs $HFS_{2,2}$, $HFS_{2,4}$, $HFS_{2,8}$, $HFS_{4,4}$, $HFS_{4,8}$, and $HFS_{4,16}$. The $HFS_{2,4}$ design, for instance, is decomposed to two independent $HFS_{1,2}$ systems and similarly for the remaining HFS_{m_1,m_2} systems. In this experiment, for every randomly generated problem the C_{PHFS} and C_H makespan values were obtained for the same scenario of processing times. These processing times were uniformly selected from the range [1,20]. The results reported indicate that the throughput performance difference between PHFS and HFS_{m_1,m_2} is surprisingly small and becomes negligible as the number of jobs increases. In particular, for $n = 50$ the average throughput difference between the values C_{PHFS} and C_H is 1.33%. The corresponding gap for $n = 20$ is 4.06%. This indicates that, it may be marginally beneficial to use the HFS_{m_1,m_2} system for batches of 20 or fewer jobs. For larger problem sizes it appears that HFS systems are harder to justify compared to PHFS. Another trend reported in this experiment is that the performance gaps increase with the total number of machines in the system, $m_1 + m_2$.

Overall, the experiments presented in Vairaktarakis and Elhafsi (2000) indicate that investing in sophisticated material control structures and multidirectional routing offers minute throughput increases compared to the unidirectional routing structure proposed here that appropriately decomposes the machines to independent production cells. Combining the managerial insights described above we conclude that the performance of $FHFS_{m_1,m_2}$ is superior to $P(m_1 + m_2)$ which is superior to HFS_{m_1,m_2}, which in turn is superior to the corresponding PHFS. System $FHFS_{m_1,m_2}$ is the most flexible because it consists of $m_1 + m_2$ machines all of which can process both a_i and b_i tasks and it can split jobs of length $p_i = a_i + b_i$ into pairs of flowshop operations (a_i, b_i). Evidently, job splitting is hard to justify, especially because $FHFS_{m_1,m_2}$ has a much more complicated routing structure than $P(m_1 + m_2)$. On the other hand, both of these systems exhibit significantly faster throughput than HFS_{m_1,m_2}. Therefore, processing flexibility yields significant benefits. On the other hand, routing flexibility does not add significant benefits. This is evident when comparing HFS_{m_1,m_2} with PHFS and when comparing FHFS_{m_1,m_2} with $P(m_1 + m_2)$.

5.5 Conclusion

In this chapter we presented results for FHFS; a hybrid flowshop with processing flexibility across stages. This system subsumes the well studied HFS system and incorporates processing flexibility across production stages as well as stage dependent processing times. The resulting system is quite general. Then, we presented results for PHFS which possesses very simple routing structure. For all these systems we presented optimal and heuristic algorithms and reported managerial insights drawn from computational experiments. We found that processing flexibility yields significant throughput improvements while routing flexibility does not, despite the additional responsibility to manage complex routing. This research demonstrates the importance of studying flexibility modes before investments are made. Alternative modes of flexibility include multiprocessor task processing, different job-splitting protocols, lot streaming, and others. Comparison of competing designs is a fruitful research direction.

References

[1] Baker, K.R. (1993). *Elem. Seq. Sched.*, Dartmouth, Hanover, NH 03755.

[2] Buten, R.E. and V.Y. Shen (1973). A Scheduling Model for Computer Systems with Two Classes of Processors. *Sagamore Comp. Conf. Parall. Proc.* 130–138.

[3] Chen, B. (1994). *Scheduling Multiprocessor Flow Shops*. In Du, D.-Z. and Sun, J. (Editors); *New Adv. Opt. Approx.*, Kluwer, Dordrecht, 1–8.

[4] Cheng, T.C.E. and C.C.S. Sin (1990). A State-of-the-Art Review of Parallel-Machine Scheduling Research. *Eur. J. Oper. Res.*, 47:271–192.

[5] Garey, M.R. and D.S. Johnson (1979). *Comp. Intract.* W.H. Freeman, San Francisco, CA.

[6] Gonzalez, T. and S. Sahni (1978). Flowshop and Jobshop Schedules; Complexity and Approximation. *Oper. Res.*, 26:36–52.

[7] Guinet, A. and M.M. Solomon (1996). Scheduling Hybrid Flowshops to Minimize Maximum Tardiness or Maximum Completion Time. *Intl. J. Prod. Res.*, 34:1643–1654.

[8] Gupta, J.N.D. (1988). Two Stage Hybrid Flowshop Scheduling Problem. *J. Oper. Res. Soc.*, 39(4):359–364.

[9] Gupta, J.N.D., A.M.A. Hariri and C.N. Potts (1997). Scheduling a Two-Stage Hybrid Flowshop with Parallel Machines at the First Stage. *Ann. Oper. Res.*, series on *Math. Ind. Syst.* 69:171–191.

[10] Gupta, J.N.D. and E.A. Tunc (1991). Schedules for a Two Stage Hybrid Flowshop with Parallel Machines at the Second Stage. *Intl. J. Prod. Res.*, 29(7):1489–1502.

[11] He, D.W., A. Kusiak and A. Artiba (1996). A Scheduling Problem in Glass Manufacturing, *IIE Transactions* 28: 129–139.

[12] Herrmann, J.W. and C.-Y. Lee (1992). Three-Machine Look-Ahead Scheduling Problems. *Res. Rep.* 92-23. Department of Industrial and Systems Engineering, University of Florida.

[13] Hoogeveen, J.A., J.K. Lenstra and B. Veltman (1996). Preemptive Scheduling in a Two-Stage Multiprocessor Flowshop is $\mathcal{NP}$-hard. *Euro. J. Oper. Res.*, 89:172–175.

[14] Jackson, J.R. (1956). An Extension of Johnson's Results on Job Lot Scheduling. *Nav. Res. Log. Quart.*, 3:201–203.

[15] Johnson, S.M. (1954). Optimal Two- and Three-Stage Production Schedules with Setup Times Included. *Nav. Res. Log. Quart.*, 1:61–68.

[16] Kouvelis P. and G. Vairaktarakis (1998). Flowshops with Processing Flexibility Across Production Stages. *IIE Transactions*, 30(8):735–746.

[17] Langston M.A. (1978). Interstage Transportation Planning in the Deterministic Flow-Shop Environment. *Oper. Res.*, 35(4):556–564.

[18] Lee, C.-Y. and G.L. Vairaktarakis (1993). Design for Schedulability: Look-Behind and Look-Ahead Flowshops. *Research Report* 93-23. Department of Industrial and Systems Engineering, University of Florida.

[19] Lee, C.-Y. and G. Vairaktarakis (1994). Minimizing Makespan in Hybrid Flowshops. *Oper. Res. Lett.*, 16:149–158.

[20] Lee, C.-Y. and J.W. Herrmann (1993). A Three-Machine Scheduling Problem with Look-Behind Characteristics. *Res. Rep.* 93-11. Department of Industrial and Systems Engineering, University of Florida.

[21] Lee, C.-Y., T.C.E. Cheng and B.M.T. Lin (1993). Minimizing the Makespan in the 3-Machine Assembly-Type Flowshop Scheduling Problem. *Manag. Sci.*, 39(5):616–625.

[22] Narasimhan, S.L. and S.S. Panwalker (1984). Scheduling in a Two-Stage Manufacturing Process. *Intl. J. Prod. Res.*, 22:555–564.

[23] Paul, R.J. (1979). A Production Scheduling Problem in the Glass-Container Industry. *Oper. Res.*, 22:290–302.

[24] Pinedo, M. (2001). Scheduling: Theory, *Algor. Sys.* Prentice Hall, Englewood CliffS, NJ.

[25] Rao T.B.K. (1970). Sequencing in the Order A, B with Multiplicity of Machines for a Single Operation. *Opsearch*, 7:135–144.

[26] Solomon, M.M., P.K. Kedia and A. Dussauchoy (1996). A Computational Study of Heuristics for Two-Stage Flexible Flowshops. *Intl. J. Prod. Res.*, 34:1399–1416.

[27] Sriskandarajah, C. and S.P. Sethi (1989). Scheduling Algorithms for Flexible Flowshops: Worst and Average Case Performance. *Euro. J. Oper. Res.*, 43:143–160.

[28] Vairaktarakis, G. and M. Elhafsi (2000). Efficient Decompositions of Hybrid Flowshops for Simplified Routing. *IIE Trans.*, 32(8):687–699.

[29] Wittrock, R.J. (1988). An Adaptable Scheduling Algorithm for Flexible Flow Lines. *Oper. Res.*, 36:445–453.

[30] Blazewicz, J., K. Ecker, G. Schmidt and J. Weglartz (1993). *Sched. Comp. Manuf. Sys.* Springer Verlag, Berlin, Germany.

[31] Graham, R.L. (1966). Bounds For Certain Multiprocessing Anomalies. *Bell Sys. Tech. J.*, 45: 1563–1581.

6

Open Shop Scheduling

Teofilo F. Gonzalez
University of California

6.1 Introduction

The open shop scheduling problem consists of m machines denoted by $M_1, M_2, \ldots, M_m$ that perform different tasks. There are n jobs $(J_1, J_2, \ldots, J_n)$ each of which consists of m tasks. The jth task of job J_i is denoted by $T_{i,j}$ and it must be processed by machine M_j for $p_{i,j} \geq 0$ time units. The total processing time for job J_i is $p_i = \sum_j p_{i,j}$, the processing requirement for machine M_j is $m_j = \sum_i p_{i,j}$, and we define $h = \max\{p_i, m_j\}$. The scheduling restrictions in an open shop problem are as follows:

1. Each machine may process at most one task at a time.
2. Each job may be processed by at most one machine at a time.
3. Each task $T_{i,j}$ must be processed for $p_{i,j}$ time units by machine M_j.

Clearly, the finishing time of every open shop schedule must be at least h. The main difference between the flow shop and the open shop problems is that in the former problem the tasks for each job need to be processed in order, i.e., one may not begin processing task $T_{i,j}$ until task $T_{i,j-1}$ has been completed for $1 < j \leq m$. In the open shop problem the order in which tasks are processed is immaterial. In the job shop problem jobs may have any number of tasks, rather than just m. Each task of each job is assigned to one of the machines rather than assigning the jth task to machine M_j as in the flow shop and open shop problems. But, the order in which tasks must be processed in the job shop problem is sequential as in the flow shop problem. One may think of the open shop problem as the flow shop problem with the added flexibility that the order in which tasks are processed is immaterial.

Gonzalez and Sahni [1] introduced the open shop scheduling problem back in 1974 to model several real-world applications that did not quite fit under the flow shop model. They developed a linear-time algorithm for the two machine makespan nonpreemptive as well as the preemptive scheduling problems ($O2 \,||\, C_{\max}$, and $O2 \,|\, pmtn \,|\, C_{\max}$). This result compares favorably to Johnson's two machine flow shop algorithm that takes $O(n \log n)$ time. Gonzalez and Sahni [1] also showed that for three or more machines the nonpreemptive open shop problem ($O3 \,||\, C_{\max}$) is NP-hard. Their main result was two efficient algorithms for the preemptive version of the makespan open shop problem ($O \,|\, pmtn \,|\, C_{\max}$). Since 1974 hundreds of open shop papers have been published in all sorts of conference proceedings and journals. The

1-58488-397-9/$0.00+$1.50

minimum makespan open shop preemptive scheduling problem has found applications in many different fields of study, which is one of the reasons for the popularity of the open shop problem.

The most interesting application of the open shop preemptive scheduling problem is in scheduling theory where the problem naturally arises as a subproblem in the solution of other scheduling problems. A typical application arises when one solves (optimally or suboptimally) a scheduling problem via Linear Programming (LP). In the first step one defines a set of intervals, and a set of LP problems defines the amount of time each job is to be scheduled in each machine in each time interval. Once the set of LP programs are solved, we are left with a set of one or more open shop makespan problems. When the resulting open shop problem is such that preemptions are allowed, one can use the algorithm in Ref. [1] to construct the final schedule. Lawler and Labetoulle [2] were the first to use the open shop problem this way. Since then, it has become common practice. The most interesting use of this approach is given by Queyranne and Sviridenko [3] to generate suboptimal solutions to open shop preemptive scheduling problems with various objective functions as well as for interesting generalizations of the open shop problem.

Another application arises in the area of Satellite-Switched Time-Division Multiple Access (SS/TDMA) [4] where information has to be interchanged between multiple land sites using a multibeam satellite. The scheduling of the communications has been modeled by an open shop problem [4]. The open shop problem also arises in the scheduling and wavelength assignment (SWA) problem in optical networks that are based on the wavelength-division-multiplexing (WDM) technology [5]. Wang and Sahni [6] also use the open shop problem for routing in OTIS (optical transpose interconnect system) optoelectronic computers to find efficiently permutation routings. For mesh computers with row and column buses, the open shop problem is used for routing packets [7]. Even when routing in heterogeneous networks, the open shop problem has been used to model communications schedules [8]. Iyengar and Chakrabarty [9] used the makespan open shop preemptive scheduling problem for system-on-a-chip (SOC) testing. The computational techniques behind the makespan open shop preemptive scheduling algorithm in Ref. [1] have been applied to the solution of stochastic switching problems [10].

For the *multimessage multicasting*, Gonzalez [11–13] has developed efficient offline and online approximation algorithms not only for the fully connected networks but also for Benes-type of networks capable of replicating data and realizing all permutations. A class of approximation algorithms for the multimessage multicasting problem generate solutions by solving a couple of problems, one of which is the *multimessage unicasting* problem [12]. Gonzalez [11] has shown that the multimessage unicasting problem is equivalent to the unit-processing time makespan open shop scheduling problem [12], which can be solved by the algorithms given in Ref. [1]. The multimessage unicasting problem is also known as the h-relations problem [14] and the $(h - h)$-routing request problem.

The open shop problem is a generalization of the bipartite graph edge coloring problem. A *graph* consists of a set of V vertices and a set of edges E. A graph is said to be *bipartite* when the set of vertices can be partitioned into two sets A and B such that every edge is incident to a vertex in set A and to a vertex in set B. The *bipartite graph edge coloring problem* consists of assigning a color to each edge in the graph in such a way that no two edges incident upon the same vertex are assigned the same color and the total number of different colors utilized is least possible. The open shop problem in which all the $p_{i,j}$ values are 0 or 1 is called the *unit-processing time* open shop problem. This open shop problem with the objective function of minimizing the makespan $(O \mid p_{i,j} \in \{0, 1\} \mid C_{\max})$ corresponds to the bipartite graph edge coloring problem. To see this, map the set of vertices A to the set of jobs and the set of vertices B to the set of machines. An edge from a vertex in set A to a vertex in set B represents a task with unit processing time. Each color represents a time unit. The coloring rules guarantee that an edge coloring for the graph corresponds to the unit-processing time open shop schedule. The makespan or finishing time corresponds to the number of different colors used to color the bipartite graph.

When there are multiple edges between at least one pair of nodes, the edge coloring of bipartite graphs problem is called the *bipartite multigraph edge coloring* problem. As pointed out in Ref. [4], this problem can be solved by the constructive proof of Egerváry [15], which uses König-Hall theorem [16–18]. A more general version of this problem has been defined over an edge-weighted (positive real values) bipartite graph, and the problem is to find a set of matchings $M_1, M_2, \ldots, M_m$ and positive real-valued weights $w_1, w_2, \ldots, w_m$ such that the bipartite graph is equal to the sum of the weighted matchings. This problem corresponds to the problem solved by the Birkhoff–von Neumann theorem, which establishes that a doubly stochastic matrix (i.e., a square nonnegative real matrix with all lines (rows and columns) equal to one) is a convex combination of permutation matrices. Berge [19] presents a graph-theory based proof for this theorem and then points out: "The proof illustrates the value of the tool provided by the theory of graphs, the direct proof of the theorem of Birkhoff and von Neumann is very much longer."

The timetable problem [20,21] is a generalization of the open shop scheduling problem. The professors are the machines, the jobs are the classes, and the objective is to find times at which the professors can instruct their classes without any professor teaching more than one class at a time and any class meeting with more than one professor at a time. In addition, the classical timetable problem includes constraints where professors or classes cannot meet during certain time periods.

In Section 6.2 we discuss the open shop problem with the objective function of minimizing the makespan. We outline a linear-time algorithm for the two machine problem, as well as polynomial-time algorithms for the general preemptive version of the problem. For the nonpreemptive version of the problem we present mainly NP-hard results and approximation algorithms. We also discuss the problem of generating suboptimal solutions to these problems, as well as to the distributed version of the problem. Section 6.3 covers the open shop problem with the objective function of minimizing the mean flow time. We discuss the NP-hardness results for these problems as well as several approximation algorithms. In Section 6.4 we briefly discuss the open shop problem under various objective functions as well as generalization of the basic open shop problem.

6.2 Minimum Makespan Problems

In this section we discuss the open shop problem with the objective function of minimizing the makespan. We outline a linear-time algorithm for the two machine problem, as well as polynomial-time algorithms for the general preemptive version of the problem. For the nonpreemptive version of the problem we discuss mainly NP-hard results. We also discuss the problem of generating suboptimal solutions to these problems as well as to the distributed version of the problem.

6.2.1 Two Machines

Gonzalez and Sahni [1] developed a very clever algorithm to construct a minimum makespan schedule for the two machine nonpreemptive version of the problem ($O \,||\, C_{\max}$). This algorithm also generates an optimal preemptive schedule ($O2 \,|\, pmtn \,|\, C_{\max}$). Let us now outline a variation of this algorithm.

We represent each job by a pair of positive integers whose first component is the processing time on machine M_1 and the second component is the processing time on machine M_2 for the job. We partition these pairs into two groups: A and B. Group A contains all the tuples whose first component is greater or equal to the second, and group B contains all the tuples whose second component is larger than the first one. We will only discuss the case when both sets are nonempty, because the other cases are similar. The group A is represented as a sequence by $A_1, A_2, \ldots, A_R$ and B is represented by the sequence $B_1, B_2, \ldots, B_L$. The processing time on machine M_j for tuple A_i is denoted by $A_i(j)$. Similarly, we define $B_i(j)$. From our definitions we know that $A_i(1) \geq A_i(2)$, and $B_j(1) < B_j(2)$. We assume without loss of generality that A_R is the job in A with largest processing time on machine M_1, i.e., $A_R(1)$ is largest. Similarly, B_L is the job in B with largest processing time on machine M_2.

Now construct the schedule that processes jobs in the order $A_1, A_2, \ldots, A_R$ as shown in Figure 6.1(b). For each job the task on machine M_1 is scheduled immediately after the completion of the previous task (if any) on machine M_1, and the task on machine M_2 is processed at the earliest possible time after the completion of the job's task on machine M_1. Clearly, the only idle time between tasks is on machine M_2. Since $A_R(1)$ is the largest value, then $A_R(1) \geq A_{R-1}(2)$. Therefore, the task on machine M_2 for job A_R

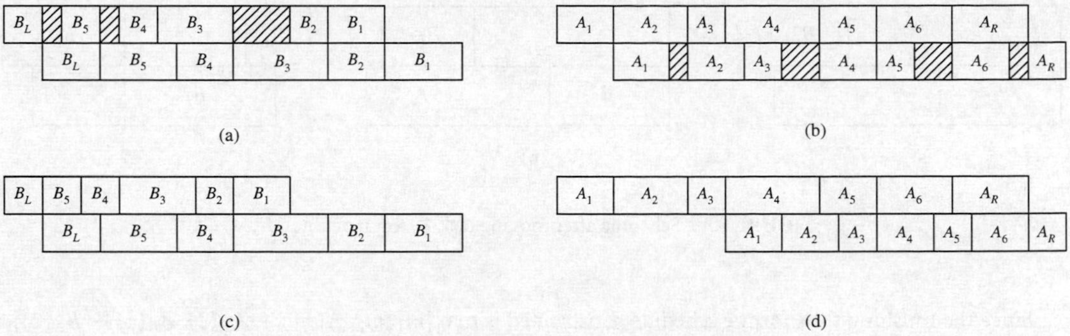

FIGURE 6.1 Schedule for the jobs in A and B.

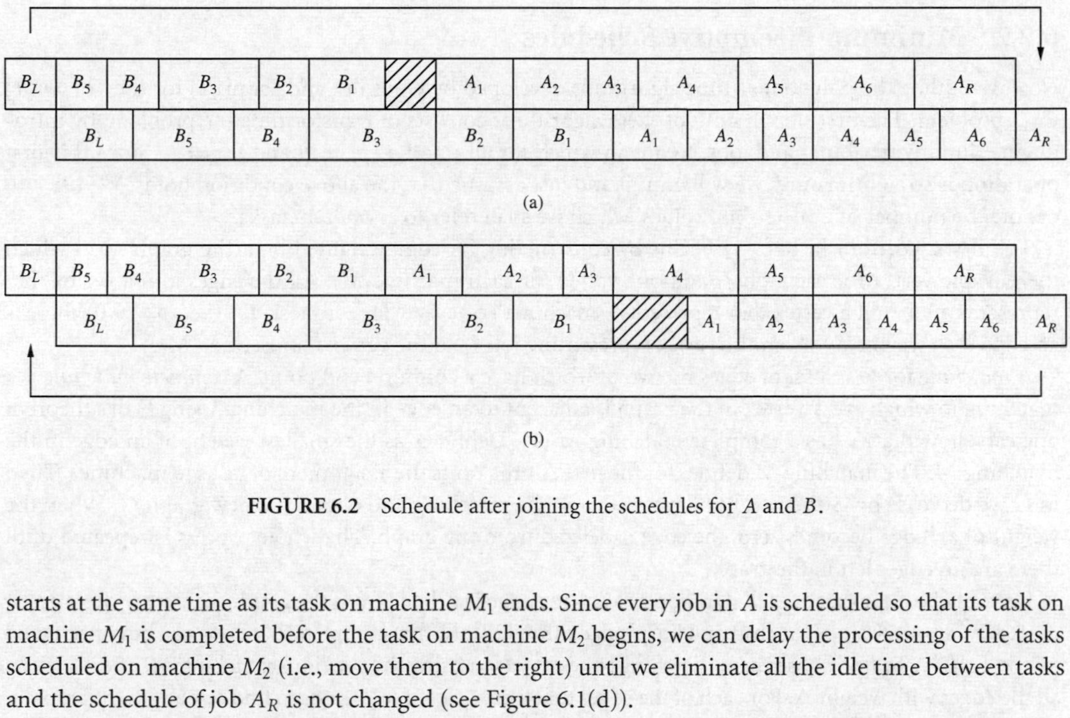

FIGURE 6.2 Schedule after joining the schedules for A and B.

starts at the same time as its task on machine M_1 ends. Since every job in A is scheduled so that its task on machine M_1 is completed before the task on machine M_2 begins, we can delay the processing of the tasks scheduled on machine M_2 (i.e., move them to the right) until we eliminate all the idle time between tasks and the schedule of job A_R is not changed (see Figure 6.1(d)).

We obtain a similar schedule for B (see Figure 6.1(a)). Note that in this case the idle time between tasks is only on machine M_1, and the tasks are processed in the order $B_1, B_2, \ldots, B_L$ but from right to left. In this case the tasks scheduled on machine M_1 are moved to the left to eliminate idle time between tasks (see Figure 6.1(c)).

Now we concatenate the schedule for B and A in such a way that there is no idle time between the schedules on machine M_1 (Figure 6.2(a)), or machine M_2 (Figure 6.2(b)), or both (either of the figures without idle time between the schedules on both machines).

If one ends up with the schedule in Figure 6.2(a), we move job B_L on machine M_1 from the front of the schedule to the back. We then push all the tasks to the left until the block of idle time between the schedules is eliminated (Figure 6.3(a) or 6.3(b)), or until job B_L starts on machine M_1 at the time when it finished on machine M_2 (Figure 6.3(c)). In the former case the total makespan is equal to $m_1 = \sum A_i(1) + \sum B_j(1)$ or $m_2 = \sum A_i(2) + \sum B_j(2)$, and in the latter case it is given by $B_L(1) + B_L(2)$. On the other hand, if we end up with the schedule in Figure 6.2(b), a similar operation is performed, but now we move job A_R on machine M_2. The resulting schedules are similar to those in Figure 6.3 except that job A_R replaces job B_L and the machines are interchanged. Clearly the time complexity of the algorithm is $O(n)$.

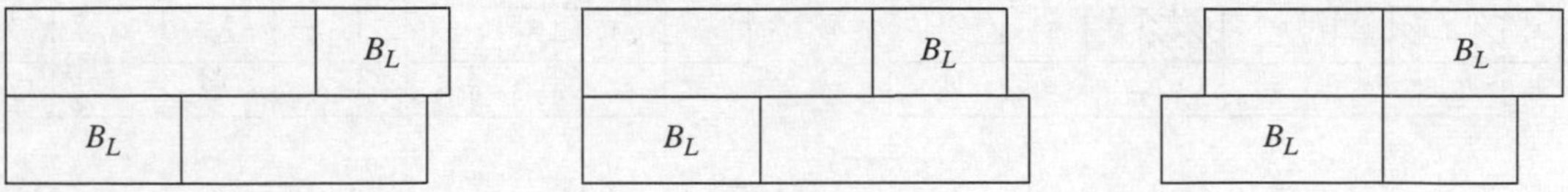

(a) (b) (c)

FIGURE 6.3 Schedule after moving task B_L on machine M_1.

Since the finishing time for the schedule constructed is $\max\{m_1, m_2, A_R(1) + A_R(2), B_L(1) + B_L(2)\}$, which is simply h, there are no preemptions in the schedule, and every schedule must have finishing time greater than or equal to h, it then follows that the schedule constructed by our procedure is a minimum makespan nonpreemptive preemptive schedule, i.e., it solves the $O \,||\, C_{\max}$ and $O \,|\, pmtn \,|\, C_{\max}$ problems.

6.2.2 Minimum Preemptive Schedules

Now we outline the polynomial-time algorithms developed by Gonzalez and Sahni [1] for the $O \,|\, pmtn \,|\, C_{\max}$ problem. The first step in both of these algorithms consists of transforming the problem, by introducing dummy machines and jobs, to one in which for all i and j, $p_i = h$ and $m_j = h$. Since this first operation is straightforward, we will omit it and just assume that the above condition holds. We use r to denote the number of nonzero $p_{i,j}$ values, which we shall refer to as *nonzero* tasks.

The first algorithm in Ref. [1] begins by constructing an edge-weighted bipartite graph G in which there is one vertex for each job J_i and one vertex for each machine M_j. All the edges join a vertex that represents a job and a vertex that represents a machine. For every nonzero task $T_{i,j}$ (i.e., $p_{i,j} > 0$) there is an edge with weight $p_{i,j}$ from the vertex representing job J_i to the vertex representing M_j.

A *matching* for G is a set of edges no two of which have a common end point. A *complete matching* is a matching in which every vertex in the graph is adjacent to an edge in the matching. Using Hall's theorem one can show that G has a complete matching M [1]. Define Δ as the smallest weight of an edge in the matching M. The matching M defines for the first Δ time units the assignment of tasks to machines. Then in G we decrease by Δ the weight of every edge in the matching and obtain the new graph G. When the weight of an edge becomes zero, the edge is deleted from the graph. The whole process is repeated until there are no edges left in the graph.

It should be clear that at each iteration at least one edge is deleted from the graph. Therefore there are at most r iterations. Using Hopcroft and Karp's algorithm [22] the first complete matching can be constructed in $r(n+m)^{0.5}$. To find subsequent matchings one resorts to using the previous matching M, after deleting all the edges with weight Δ. For each of the edges that were deleted one needs to find one augmenting path. An *augmenting (or alternating) path* relative to a matching M is a simple path with an odd number of edges such that the ith edge in the path, for i odd, is in the graph but not in the matching, and for i even, it is an edge in the matching. The "even" edges are represented by the set M' and the "odd" ones are represented by the set of edges M_N. If we delete from M all the edges in M' and then we add all the edges in M_N, we obtain a new matching that contains one more edge than the matching we had before the operation. The reader is referred to Gonzalez and Sahni [1] for the precise definitions and procedures to find augmenting paths. The time required to find an augmenting path is $O(r)$ and it is obtained via breadth first search. So, if the complete matching M had l edges with weight Δ, then one needs to construct l augmenting paths to find the next complete matching. Therefore this technique can be implemented to take $O(r^2)$ time. The above approach is essentially the typical constructive proof for the Birkoff–von Neumann theorem [18] and Egerváry theorem [15]. As pointed out by Berge [19], the proof of the Birkoff–von Neumann theorem when viewing the problem as a graph problem is much simpler than the original one.

The second algorithm given by Gonzalez and Sahni [1] takes $O(r(\min\{r, m^2\} + m \log n))$ time. The algorithm is designed for the case when $m < n$, and it is better than the first one when $r > m^2$. As in the first algorithm, the first step adds dummy machines and jobs so that $p_i = m_j = h$ for all i and j. Then all the dummy jobs are deleted. As a result of this operation some jobs will have total processing time less than h. The jobs with $p_i = h$ are called *critical* and those with $p_i < h$ are called *noncritical*. All the machines are called *critical* because $m_j = h$ for all j. The strategy behind the algorithm is similar to that in the previous algorithm. The main difference is that the cost of an augmenting path is m^2 rather than r. In the first step we find a matching I that includes all critical jobs and machines. We build this matching by finding and then using an augmenting path for each critical job, and then finding and using an augmenting path for each critical machine that is not in the matching. We define the slack time s_i of a job as $h - c - g_i$, where c is the current time (this is just the makespan of the schedule so far constructed), and g_i is the remaining processing time for the job. Initially $g_i = p_i$ and $c = 0$. A noncritical job becomes critical when its slack time s_i becomes zero. Remember that at every point in time t all jobs that are critical at time t must be scheduled, i.e., must be in the matching I at time t. Therefore the minimum slack time for the jobs that are not in the matching at time t is greater than zero. We define Δ as the minimum of the remaining processing time of a task represented by an edge in the matching I and the minimum slack time of the jobs that are not in the matching. The minimum slack time can be computed in $O(\log n)$ time by storing all the slack times of the jobs that are not in the matching in a balanced tree. Then we generate the schedule implied by the matching for the next Δ time units. We update the slack times for the noncritical jobs in the matching, which takes $O(m)$ time. We delete all the noncritical jobs from the matching, and delete all the tasks with remaining processing time zero. Then we add to the resulting matching any critical jobs that are not in the matching. Each of these additions are carried out by finding and then using an augmenting path that takes $O(m^2)$ time to construct, as shown in Ref. [1]. Now we add to the matching any of the noncritical jobs that were just deleted. For these jobs we do not find an augmenting path, we just add them if the corresponding machine for the task that was part of the previous matching is not covered by the current matching. Then through the augmenting path technique we add all the critical machines that are not in the matching and we end up with another matching that includes all the critical jobs and critical machines. The whole process is repeated until all the tasks have been scheduled for the appropriate amount of time. Readers interested in the complete details (which are complex) are referred to Gonzalez and Sahni [1]. The total number of iterations (matchings) constructed is at most $r + m$ because at each iteration at least one nonzero task has been scheduled for its full processing time, or a noncritical job becomes critical. Since $r \geq m$ and the cost of each augmenting path is at most m^2, all of these operations take $O(r \min\{r, m^2\})$ time. The other factor in the time complexity bound, $r \cdot m \log n$, originates for the operation of updating the slack time in the balanced binary search tree of at most m tasks at each iteration. This time also includes the time to find the smallest slack time of a task that is not in the matching. Therefore, the overall time complexity bound for the second algorithm is $O(r(\min\{r, m^2\} + m \log n))$. The total number of preemptions introduced by the algorithm is at most $r(m - 1) + m^2$ preemptions.

The algorithm for the minimum makespan open shop preemptive scheduling problem by Vairaktarakis and Sahni [23] has the same worst case time complexity; however, it is in general faster and requires less space than the previous two algorithms. This is achieved by not introducing dummy jobs and dummy machines, and only critical jobs and machines are required to be in the matching at each step. This eliminates the overhead of dealing with the additional jobs and machines. The added constraints of *load balancing* as well as *maintenance* are incorporated into the algorithms given in Ref. [23]. The former constraints balance the number of busy machines throughout the schedule, and the latter constraints deal with machine maintenance with limited personnel. The algorithms in Ref. [23] use linear programming, max flow as well as additional graph theory properties of bipartite matchings.

The algorithms in Refs. [1,23] are in general much more efficient than the algorithms that were developed for the corresponding bipartite graph and multigraph edge coloring problems as well as for the doubly stochastic matrices.

As we mentioned before, Gonzalez [11] has established the equivalence of the makespan open shop preemptive scheduling problem and the multimessage unicasting problem. In this problem each processor must send equal size messages to other processors over a fully connected network. The scheduling rules are that no processor may send more than one message at a time and no processor may receive more than one message at a time. The objective is to find a schedule with the least total completion time.

In the distributed version of the multimessage unicasting problem every processor only knows the messages it will be sending and does not know what other processors have to do. Though every processor may send or receive at most d messages. When this is translated back into the open shop problem we need to introduce n users each of which is trying to process its corresponding job on the machines without knowing what the other users are trying to do. At each time unit each user decides which (if any) of its task he/she will attempt to get processed by the corresponding machine. If more than one user attempts to use the same machine, then the machine will be idle and both users will be informed that their tasks were not processed at that time. Gereb-Graus and Tsantilas [24] presented distributed algorithms with $\Theta(d + \log n \log\log n)$ expected communication steps. The multimessage unicasting and multicasting problems with forwarding have been studied in the context of optical-communication parallel computers [14,24–26]. Forwarding means that a message does not need to be sent directly, the message may be sent through another machine. In Section 6.4 we explain what forwarding means in the context of open shop scheduling.

6.2.3 Limiting the Number of Machines, Jobs or Tasks

First let us consider the open shop preemptive scheduling problem when either the number of jobs or machines is not part of the problem input, e.g., the number of machines is at most 20 (or any other constant) but the number of jobs may be any arbitrary number, or vice versa. Gonzalez [27] shows that these problems can be solved in linear time by using a technique called *combine-and-conquer*. The main idea is that any subset of jobs or machines whose total processing time is at most h can be combined into a super-job or a super-machine. From a solution to the super-problem one can easily solve the original problem. Obviously this combine-and-conquer approach is not recursive, it is applied just once. The selection of which jobs to combine is determined by solving an instance of the bin-packing problem. The *bin-packing* problem consists of packing into the least number of bins with capacity h a set of objects whose size corresponds to the job processing times $p_1, p_2, \ldots, p_1$. If $k = \min\{n, m\}$, then $\sum p_i \leq k \cdot h$. There are simple linear-time algorithms that pack all of these objects in at most $2k - 1$ bins each of size h. The same approach is used for the machines. Therefore we will end up with a problem instance that has at most $2k - 1$ (super) jobs and $2k - 1$ (super) machines. Since there is a fixed number of jobs or machines (independent of the input), the resulting problem has a number of jobs and machines that is bounded by a constant independent of the input. Any of the algorithms in the previous subsection can be used to solve this reduced size problem and the solution can be easily used to obtain a solution to the original problem. The above procedure can be easily implemented to take $O(n + m)$ time.

Let us now discuss algorithms that perform very well when every job has very few tasks, and every machine needs to process very few tasks. From our previous discussion this problem may be viewed as the multigraph edge coloring problem. Gabow and Kariv's [28] algorithm to color the edges of a bipartite multigraph takes time $O(\min\ \{m \log^2 n, n^2 \log n\})$, where n is the number of nodes in the graph and m is the number of edges. Cole and Hopcroft [29] developed a faster algorithm for the same problem with time complexity bounded by $O(m \log n)$. Cole and Hopcroft's [29] algorithm uses the combine-and-conquer approach [27] as well as the idea of finding a matching with only critical jobs as in the algorithms in the previous subsection [1]. These algorithms are the fastest ones when the degree of the multigraph is small. When the edge multiplicity is large, it is better to use Gonzalez and Sahni's [1] open shop algorithms, because multiple edges are treated as a single weighted edge and the schedule for a whole interval may be generated at each iteration, rather than one for one time unit.

6.2.4 Nonpreemptive Schedules

To establish that the minimum makespan open shop nonpreemptive scheduling problem is NP-hard, Gonzalez and Sahni [1] reduced the partition problem to three machine problem instances ($O3 \,||\, C_{\max}$). Given n objects denoted by $a_1, a_2, \ldots, a_n$ and a size or weight function $s \colon a \to I^+$ the *partition* problem is to determine whether or not the set A can be partitioned into two sets, A_1 and A_2, such that the sum of the weight (or size) of the objects in each set is equal to $T/2$, where $T = \sum s(a_i)$.

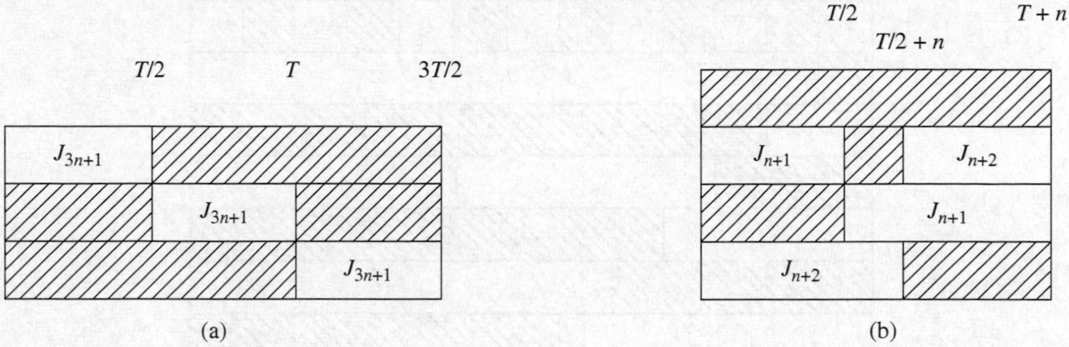

FIGURE 6.4 Architecture of reductions from partition.

The reduction to the three machine nonpreemptive open shop makespan problem ($O3 || C_{\max}$) constructs an instance from partition with $3n + 1$ jobs and three machines. The last job, J_{3n+1}, has processing time on each machine of $T/2$ time units. Therefore, if there exists a schedule with makespan equal to $3T/2$, then job J_{3n+1} has to be processed without interruptions. Since the schedule cannot have preemptions, it must be that on one of the machines the task from J_{3n+1} is processed from time $T/2$ to time T (see Figure 6.4(a)). That leaves two disjoint blocks of idle time each of length $T/2$ on one machine. These blocks will be used to process a set of n tasks whose processing time corresponds to the size of the objects in the instance of partition. Therefore, a schedule with finishing time at most $3T/2$ exists iff the instance of partition we start from has a partition for set A. To make sure that such a set of tasks exists, we need to introduce $3n$ jobs as follows. For $1 \leq j \leq 3$, the jth set of n jobs have only nonzero tasks on machine M_j and their processing time for the ith job on machine M_j is $s(a_i)$. Therefore, the open shop makespan decision problem is NP-hard even when $m = 3$ [1].

All the jobs in the above construction have one nonzero task except for the last job that has nonzero processing time on the three machines. Does the problem remain NP-hard even when each job has at most two nonzero tasks? Gonzalez and Sahni [1] addressed this problem and showed that it remains NP-hard when $m \geq 4$. The reduction is similar in nature to the previous one. The difference is that now there are two jobs with processing requirements such that in every schedule with makespan at most $T + n$, they leave a block of idle time from time $T/2$ to $T/2 + n$ on machine M_2, and M_1 is not utilized (see Figure 6.4(b)). A set on n jobs is introduced each with processing time of 1 unit on machine M_2 and a processing time corresponding to the size of an object in the partition problem on machine M_1. By scaling up the size of the objects we can guarantee that if there is partition with total size between $T/2$ and $T/2 + n$, then there is also a partition of size $T/2$. This will guarantee that a schedule with finishing time at most $T + n$ exists iff we start with a yes-instance of partition. Therefore, the minimum makespan open shop scheduling problem is NP-hard even when every job has at most two nonzero tasks and $m = 4$ [1].

The above reductions do not establish that the open shop problem is NP-hard in the strong sense [30], i.e., the problem is not shown to be NP-hard when the sum of the task times is bounded by a polynomial of n and m. This is because the partition problem is not NP-complete in the strong sense unless $P =$ NP. However, Lenstra [31,32] has shown that the open shop problem is NP-hard in the strong sense for an arbitrary number of machines. To show that this problem is NP-hard in the strong sense we reduce the 3-partition problem to it. In the 3-partition problem we are given m objects and a size or weight function defined as in the partition problem. The problem is to decide if the set of objects can be partitioned in $m/3$ subsets such that the sum of the size of the objects in each subset is identical. Figure 6.5 gives the architecture of the reduction for the case when $m = 18$. The main idea is to introduce a set of jobs that no matter how they are assigned in a schedule with certain finishing time, there will be $m/3$ equally sized blocks of idle time on machine M_1 whose total size corresponds to the sum of the size of the objects

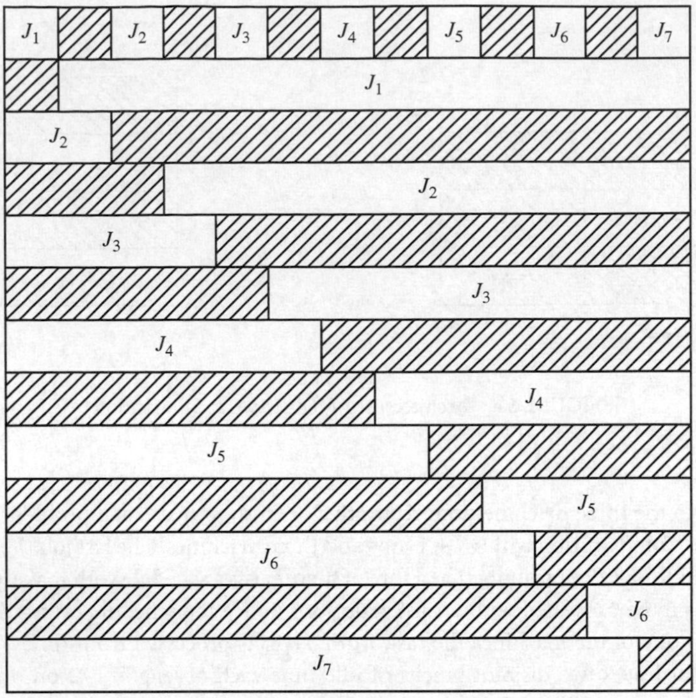

FIGURE 6.5 Architecture of reduction from 3-partition.

in the instance of 3-partition we start from. Without further discussion of the details we claim that the minimum makespan open shop nonpreemptive scheduling problem ($O \mid\mid C_{\max}$) is NP-hard in the strong sense [31].

Another interesting restricted open shop problem is one in which all the nonzero tasks have the same processing time (or unit processing time, $O \mid p_{i,j} \in \{0, 1\} \mid C_{\max}$). For this case it is simple to see that the preemptive scheduling algorithms presented in the previous section generates a minimum makespan schedule.

There is a very simple approximation algorithm for the makespan open shop nonpreemptive scheduling problem ($O \mid\mid C_{\max}$). This algorithm was developed and analyzed by Racsmány (see [33]), but a better approximation bound for the algorithm is given by Shmoys, Stein, and Wein [33]. The scheduling strategy mimics Graham's well-known list scheduling algorithm. Whenever a machine becomes available we assign to each available machine a task which has not yet been processed on that machine that belongs to a job that is not currently being processed by another machine. This very simple procedure generates a schedule with finishing time, which is at most twice the length of an optimal schedule. To see this, consider a job that finishes at the latest time in the schedule just constructed. Let us say it is job J_i and the latest time it is being processed is on machine M_j. The finishing time on machine M_j (which is the finishing time of the schedule) is equal to the processing time demands on that machine, which we have previously defined to be at most h plus the total idle time on machine M_j. Now, when there is idle time on machine M_j it must be that a task of job J_i was being processed by another machine, as otherwise the scheduling algorithm should have assigned the task of job J_i to machine M_j. Therefore the total idle time on machine M_j is less than the total processing time for job J_i, which is equal to p_i. Since by definition $p_i \leq h$, it then follows that the algorithm generates a schedule with finishing time at most $2h$. Clearly, every schedule must have finishing time at least h. So it follows that the schedule generated by the above procedure has makespan at most two times the optimal makespan.

Williamson et al. [34] showed that the problem of generating schedules within 5/4 times the length of the optimal schedule for the open shop nonpreemptive scheduling problem is NP-hard. They also present a polynomial-time algorithm for the case when for all i and j, $p_{i,j} \in \{0, 1, 2, 3\}$, and there is an optimal schedule with finishing time at most 3.

6.3 Minimum Mean Flow Time or Minsum Problems

In this section we discuss the open shop problem with the objective function of minimizing the mean flow time. Since the minimum mean flow time, $\sum C_i/n$, objective function is equivalent to the minsum one, $\sum C_i$, we use these terms interchangeably. We discuss the NP-hardness results for these problems as well as several approximation algorithms.

Achugbue and Chin [35] have established that the minsum problem is NP-complete in the strong sense even for two machines $(O2 \,||\, \sum C_i)$. Their reduction is quite complex and it is similar in architecture to that for the flow shop by Garey, Johnson, and Sethi [36]. Liu and Bulfin [37] showed that the problem is NP-hard in the strong sense for three machines when preemptions are allowed by using a reduction from 3-partition $(O3 \,|\, pmtn \,|\, \sum C_i)$. Subsequently Du and Leung [38] showed that the minsum open shop preemptive scheduling problem is NP-hard (in the normal sense) even when there are only two machines $(O2 \,|\, pmtn \,|\, \sum C_i)$ using a reduction from a restricted version of partition.

The simplest version of the minsum open shop problem is when all the nonzero tasks have equal processing times $(O \,|\, p_{i,j} \in \{0, 1\} \,|\, \sum C_i$ and $O \,|\, p_{i,j} \in \{0, 1\}; pmtn \,|\, \sum C_i)$. Gonzalez [39] showed that this problem is NP-hard in both the preemptive and nonpreemptive mode. Since the flavor of this reduction is quite different from most other reductions, we explore it in more detail. The reduction is from the graph coloring problem. The *graph coloring* problem is given an undirected graph assign a color to each vertex in the graph in such a way that the least number of colors is used and no two adjacent vertices are assigned the same color. The reduction from the graph coloring problem to the open shop problem is as follows. The set of jobs represents nodes, edges, and node-edge pairs in the graph, and the machines represent nodes and edges in the graph. Time is partitioned into three different intervals, each corresponds to one color. The node jobs force the corresponding node-edge job to be confined to one of the three time intervals. The edge jobs are introduced to simplify the accounting of the objective function value. The node-edge jobs are defined in such a way that two jobs that represent adjacent nodes in the graph must be scheduled in different time intervals. The jobs and machines are defined in such a way that if the graph we start from is three colorable, then one-third of the jobs finish at time 5, another third finish at time 10, and the remaining third finish at time 15, with a total mean flow time of 10. When the graph is not three colorable, then all schedules have mean flow time greater than 10. The whole reduction is quite complex, so readers interested in additional details are referred to Ref. [39]. The reduction does not work for the makespan problem $(O \,|\, p_{i,j} \in \{0, 1\} \,|\, C_{\max})$ because for graphs that are not three colorable there are schedules with finishing time equal to 15, though it will not be the case that one-third of the jobs finish at time 5 and the next third of the jobs finish at time 10. But the reduction does work for the minimum makespan open ship no-wait scheduling problem. By *no-wait* scheduling we mean that all the tasks from a job must be executed contiguously.

Achugbue and Chin [35] showed that any schedule that does not have idle time on all the machines at the same time for the $O \,||\, \sum C_j$ problem has total completion time that is at most n times the optimal one. They also showed [35] that the simple SPT scheduling rule guarantees that the total completion time of the schedules generated is no more than m times the optimal one. Hoogeveen, Schuurman, and Woeginger [40] showed that generating near optimal solutions for the $O \,||\, \sum C_j$ problem is as difficult (computationally) as generating an optimal solution. The approximation bound for which the problem is NP-hard is somewhere in the $1 + 10^{-5}$ range. They showed that this problem is APX-complete, which means that if the problem has a polynomial-time approximation scheme, then $P = \text{NP}$.

Queyranne and Sviridenko [3] developed a quite sophisticated approximation algorithm for the open shop preemptive problem (as well as for more general versions of the problem) under different objective functions. The approximation bound is $(2+\epsilon)$. The idea is to have an interval-indexed formulation, which may be viewed as a set of intervals defined in terms of ϵ and another small constant δ over which an LP problem is defined. From the solution to the LP problem they construct a schedule using the algorithm for $O \mid pmtn \mid C_{\max}$ given in Ref. [1]. Then randomization is used through Schultz and Skutella's slow-motion algorithm with the factor β being randomly chosen with certain properties. After this process there is a derandomization step. The analysis of this process is quite complex. They showed that this technique cannot generate solutions with an approximation factor better than 2. However, the algorithm works for generalized versions of the open shop. For example, when a task may be processed on several machines at different speeds, or the objective function includes the finishing time of tasks rather than just the finishing time of the jobs.

6.4 Related Objective Functions

In this section we discuss briefly the open shop problem under various objective functions as well as generalization of the basic open shop problem.

The open shop problem has been studied under other classical objective functions as well as other restrictions. All of these results are very important, but for brevity we cannot possibly discuss all of them. Chapter 9 and Chapter 10 discuss algorithms for other objective functions, and there are several lists of current results for open shop problems available on the Internet (e.g., Ref. [41]). We will just point to some polynomial-time algorithms that have received attention in the past: $O \mid pmtn; r_i \mid L_{\max}$ by Cho and Sahni [42], $O \mid p_{ij} = 1; intree \mid L_{\max}$ by Brucker [43], $O2 \mid p_{ij} = 1; prec \mid \sum C_i$ by Coffman et al. [44], and $O \mid p_{ij} = 1 \mid \sum T_i$ by Liu and Bulfin [37]. Recent NP-hard results for restricted open shop problems are given by Timkovsky [45].

Vairaktarakis and Sahni [23] also present algorithms for some very interesting extensions of the open shop problem. One of these problems is called the *generalized open shop* problem that allows multiple copies of the same machine, but the same scheduling constraints remain. They also define the *flexible open shop*, where a machine is allowed to perform different tasks, not just one as in the open shop. The algorithms in Ref. [23] use linear programming, max flow, as well as additional graph theory properties of bipartite matchings.

The multimessage multicasting problem may be viewed as a more general open shop problem. In this generalization, each task of a job consists of a subset of subtasks each of which is to be processed by a different machine. But the processing of the subtasks of a task may be concurrent. This problem has been shown to be NP-hard even when all subtasks have unit-processing time [11]. However, Gonzalez [11,12] has developed efficient approximation algorithms for this problem. The most efficient ones use message forwarding. That means that a message that needs to be sent from processor i to processor j may be sent indirectly. For example, first it may be sent from processor i to processor l, and then from processor l to processor j. When we translate forwarding to the multimessage unicasting problem, which is equivalent to the open shop problem, we obtain another new version of the open shop problem that allows for solutions whose scheduling has an added flexibility. The added flexibility is that if job J_i needs to be processed by machine M_j, it can be replaced by job J_i that needs to be processed by machine M_l and then job J_l needs to be processed by machine M_j, provided that the second job is performed after the first one. Another important point is that the resulting open shop that needs to be solved is such that some tasks of some jobs may be processed concurrently when forwarding is allowed. This added scheduling flexibility simplifies the scheduling problem, except when each message is to be delivered to just one processor.

In the distributed version of the multimessage multicasting problem every processor only knows the messages it must send. Gonzalez [13] has developed approximation algorithms for this problem, which reduce the problem to the solution of two problems, one of which is the multimessage unicasting problem with forwarding, and in the previous paragraph we explained how forwarding is translated into the open shop problem. In a previous section we discussed the meaning of the distributed version of the problem in the context of open shop scheduling.

The distributed version of the multimessage unicasting problem with forwarding (which corresponds to a form of open shop problem) has been studied in the context of optical-communication parallel computers [14,24–26]. Valiant [26] presented a distributed algorithm with $O(d + \log n)$ total expected communication cost. The algorithm is based in part on the algorithm by Anderson and Miller [25]. The communication time is optimal, within a constant factor, when $d = \Omega(\log n)$, and Gereb-Graus and Tsantilas [24] raised the question as to whether a faster algorithm for $d = o(\log n)$ exists. This question was answered in part by Goldberg, Jerrum, Leighton, and Rao [14] who show all communication can take place in $O(d + \log\log n)$ communication steps with high probability, i.e., if $d < \log n$, then the failure probability can be made as small as n^{α} for any constant α.

6.5 Discussion

The decision version of all the NP-hard scheduling problems that were discussed in this chapter can be shown to be NP-complete. For the preemptive scheduling problems one needs to establish the minimum number of preemptions needed by an optimal solution.

Even though the open shop problem is relatively young, there have been several hundred papers dealing with this problem. The main popularity of the problem is that it models a large number of real-world problems. The algorithms are quite interesting and many of NP-hard reductions are quite complex. The main invariant of all of the work is that as we explore more of the open shop problem we find more interesting versions, generalizations, as well as applications of the problem.

References

[1] Gonzalez, T.F. and Sahni, S., Open shop scheduling to minimize finish time, *J. ACM,* 23, 665, 1976.

[2] Lawler, E.L. and Labetoulle, J., On preemptive scheduling of unrelated parallel processors by linear programming, *J. ACM,* 25, 612, 1978.

[3] Queyranne, M. and Sviridenko, M., A $(2+\epsilon)$-approximation algorithm for generalized preemptive open shop problem with minsum criteria, *J. Algor.,* 45, 202, 2002.

[4] Dell'Amico, M. and Martello, S., Open shop, satellite communication and a Theorem by Egerváry, *Oper. Res. Lett.,* 18, 207, 1996.

[5] Bampis, E. and Rouskas, G.N., The scheduling and wavelength assignment problem in optical WDM networks, *IEEE/OSA J. Lightwave Technol.,* 20, 782, 2002.

[6] Wang, C.F. and Sahni, S., OTIS optoelectronic computers, in Li, K. and Zheng, S.Q. (eds.), *Parallel Computing Using Optical Interconnections,* Kluwer, 1998, pp. 99–116.

[7] Suel, T., Permutation routing and sorting on meshes with row and column buses, *Parallel Proc. Lett.,* 5, 63, 1995.

[8] Bhat, P.B., Prasanna, V.K., and Raghavendra, C.S., Block-cyclic redistribution over heterogeneous networks, *Cluster Comp.,* 3, 25, 2000.

[9] Iyengar, V. and Chakrabarty, K., System-on-a-chip test scheduling with precedence relationships, preemption, and power constraints, *IEEE CAD,* 21, 1088, 2002.

[10] Altman, E., Liu, Z., and Righter, R., Scheduling of an input-queued switch to achieve maximal throughput, *Probab. Eng. Inf. Sci.,* 14, 327, 2000.

[11] Gonzalez, T.F., Complexity and approximations for multimessage multicasting, *J. Par. Dist. Comput.,* 55, 215, 1998.

[12] Gonzalez, T.F., Simple multimessage multicasting approximation algorithms with forwarding, *Algorithmica,* 29, 511, 2001.

[13] Gonzalez, T.F., Distributed multimessage multicasting, *J. Interconnection Networks,* 1, 303, 2000.

[14] Goldberg, L.A., Jerrum, M., Leighton, F.T., and Rao, S., Doubly logarithmic communication algorithms for optical-communication parallel computers, *SIAM J. Comput.,* 26, 1100, 1997.

[15] Egerváry, E., Matrixok kombinatorius tulajdonságairol, *Matematikai és Fizikai Lapok,* 38, 16, 1931. (English translation by Kuhn, H.W., On combinatorial properties of matrices, *Logistic Papers,* 11, 1, 1955, George Washington University.)

[16] König, D., Graphos és matrixok, *Matematikai és Fizikai Lapok*, 38, 116, 1931.

[17] Hall, M., *Combinatorial Theory*, Blaisdell, Waltham, MA, 1967.

[18] Birkoff, G., Tres observaciones sobre el algebra lineal, Revista Facultad de Ciencias Exactas, Puras y Aplicadas, Universidad Nacional de Tucuman, Series A (Mathematicas y Ciencias Teoricas), 5, 147, 1946.

[19] Berge, C., *The Theory of Graphs and it Applications*, Wiley, 1962.

[20] Gotlieb, C.C., The construction of class-teacher timetables, *Proc. IFIP Congress*, 1962, pp. 73–77.

[21] Even, S., Itai, A., and Shamir, A., On the complexity of timetable and multicommodity flow problems, *SIAM J. Comput.*, 5, 691, 1976.

[22] Hopcroft, J. and Karp, R.M., An $n^{2.5}$ algorithm for maximum matchings in bipartite graphs, *SIAM J. Comput.*, 2, 225, 1973.

[23] Vairaktarakis, G. and Sahni, S., Dual criteria preemptive open shop problems with minimum finish time, *Naval Res. Logistics*, 42, 103, 1995.

[24] Gereb-Graus, M. and Tsantilas, T., Efficient optical communication in parallel computers, *Proc. 4th ACM Symp. Parallel Algor. Architect.*, ACM, New York, Vol. 41, 1992.

[25] Anderson, R.J. and Miller, G.L., Optical communications for pointer based algorithms, TRCS CRI 88 – 14, USC, Los Angeles, 1988.

[26] Valiant, L.G., General purpose parallel architectures, in van Leeuwen, J. (ed.), *Handbook of Theoretical Computer Science*, Elsevier, New York, Chap. 18, 1990, p. 967.

[27] Gonzalez, T.F., A note on open shop preemptive schedules, *IEEE Trans. Comput.*, 28, 782, 1979.

[28] Gabow, H., and Kariv, O., Algorithms for edge coloring bipartite graphs and multigraphs, *SIAM J. Comput.*, 11, 117, 1982.

[29] Cole, R. and Hopcroft, J., On edge coloring bipartite graphs, *SIAM J. Comput.*, 11, 540, 1982.

[30] Garey, M.R. and Johnson, D.S., *Computers and Intractability: A Guide to the Theory of NP-Completeness*, W. H. Freeman and Company, New York, 1979.

[31] Lenstra, J.K. (unpublished).

[32] Lawler, E.L., Lenstra, J.K., Rinnooy Kan, A.H.G., and Shmoys, D.B., Sequencing and scheduling: algorithms and complexity, in Graves, S.C., Rinnooy Kan, A.H.G., and Zipkin, P.H. (eds.), *Handbooks in Operations Research and Management Science, Vol. 4: Logistics of Production and Inventory*, North-Holland, 1993.

[33] Shmoys, D.B., Stein, C., and Wein, J., Improved approximation algorithms for shop scheduling problems, *SIAM J. Comput.*, 23, 617, 1994.

[34] Williamson, D.P., Hall, L.A., Hoogeveen, J.A., Hurkens, C.A.J., Lenstra, J.K., Sevastianov, S.V., and Shmoys, D.B., Short shop schedules, *Oper. Res.*, 45, 288, 1997.

[35] Achugbue, J.O. and Chin, F.Y., Scheduling the open shop to minimize mean flow time, *SIAM J. Comput.*, 11, 709, 1982.

[36] Garey, M.R., Johnson, D.S., and Sethi, R., The complexity of flowshop and jobshop scheduling, *Math. Oper. Res.*, 1, 117, 1976.

[37] Liu, C.Y. and Bulfin, R.L., On the complexity of preemptive open shop scheduling problems, *Oper. Res. Lett.*, 4, 71, 1985.

[38] Du, J. and Leung, J.Y.-T., Minimizing mean flow time in two-machine open shops and flow shops, *J. Algor.*, 14, 24, 1993.

[39] Gonzalez, T.F., Unit execution time shop problems, *Math. Oper. Res.*, 7, 57, 1982.

[40] Hoogeveen, H., Schuurman, P., and Woeginger, G.J., Nonapproximability results for scheduling problems with minsum criteria, *INFORMS J. Comput.*, 13, 157, 2001.

[41] Brucker, P., Hurink, J., and Jurisch, J., Operations Research: complexity results of scheduling problems, www.mathematik.uni-osnabrueck.de/research/OR/class/.

[42] Cho, Y. and Sahni, S., Preemptive scheduling of independent jobs with release and due times on open, flow and job shops, *Oper. Res.*, 29, 511, 1981.

[43] Brucker, P., Jurisch, B., and Jurisch, M., Open shop problems with unit time operations, *Z. Oper. Res.*, 37, 59, 1993.

[44] Coffman, Jr., E.G. and Timkovsky, V.G., Ideal two-machine schedules of jobs with unit-execution-time operations, *Proc. 8th Inter. Workshop on Proj. Manag. Sched.*, Valencia, Spain, April 2002.
[45] Timkovsky, V.G., Identical parallel machines vs. unit-time shops, preemptions vs. chains, and other offsets in scheduling complexity, Technical Report, Star Data Systems, Inc. 1998.
[46] Sahni, S. and Gonzalez, T.F., P-complete approximation problems, *J. Assoc. Comput. Machinery*, 23, 555, 1976.

7

Cycle Shop Scheduling

Vadim G. Timkovsky
CGI Group Inc. & McMaster University

7.1 Introduction

A *cycle shop* was introduced in scheduling theory by Degtiarev and Timkovsky [1] (see also References [2–5]) as a general model of machine environments related to periodic processes in a *job shop*. A cycle shop can be defined as the special case of a job shop and the extension of a *flow shop* where all jobs have the same sequence of operations on the machines, but in contrast to a flow shop, some operations can be repeated on some machines a number of times, and this number can differ from one machine to another.

Cycle shop scheduling problems arose from the VLSI technology research that was done in the mid-1970s in the Soviet electronic industry. The term "cycle" was derived from images of technological aesthetics: the sequence of operations on the machines in a cycle shop can be nicely depicted by a spiral *cyclogram*. See Figure 7.1 for an illustration. VLSI technology designers used this type of cyclogram for presenting the technological route in a flexible production line that was set up for mass manufacturing of certain types of microchips. In contrast to the hypothetical example given in Figure 7.1, cyclograms of real technological routes contained several hundreds of operations involving several dozens of machines.

1-58488-397-9/$0.00+$1.50

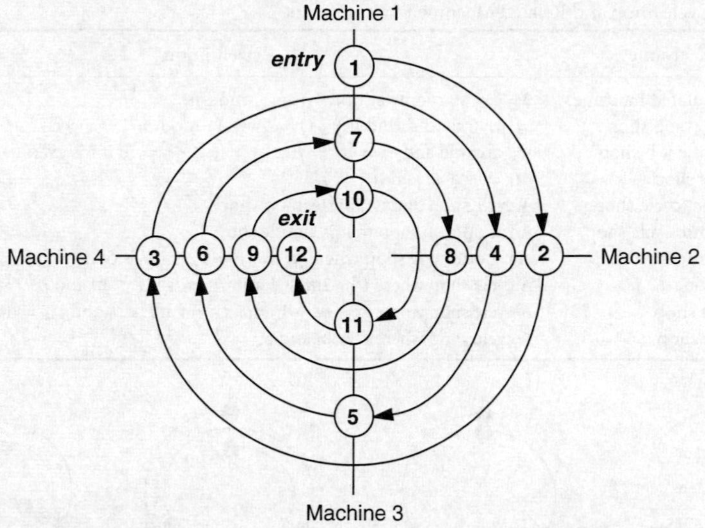

FIGURE 7.1 A cyclogram for a four-machine cycle shop with 12-operation jobs: Machine 1 processes the 1st, 7th, and 10th operation; Machine 2 processes the 2nd, 4th, and 8th operation; Machine 3 processes the 5th and 11th operation; and Machine 4 processes the 3rd, 6th, 9th, and 12th operation of each job. (From Middendorf, M. and Timkovsky, V.G., On scheduling cycle shops: classification, complexity and approximation, *Journal of Scheduling*, 5:135–169, 2002. © John Wiley & Sons. With permission.)

Cycle shops also have purely theoretical origins. Attempts to extend flow shop scheduling theory to the case where a flow shop is complemented by a transport robot as an additional machine led to a special cycle shop in which only one machine (the robot) repeats operations. Thus, the *flow line with an operator* of Livshits et al. [6] is a special cycle shop. Other special cases of cycle shops were considered later under different names, such as the *reentrant flow shop* of Graves et al. [7], Morton and Pentico [8], the *V-shop* of Lev and Adiri [9], the *robotic flow shop* of Asfahl [10], the *periodic job shop* of Serafini and Ukovich [11], the *repetitive flow shop* of Ramazani and Younis [12], the *cyclic job shop* of Roundy [13], the *robotic cell* of Sethi et al. [14], the *loop reentrant flow shop* of Gupta [15], the *chain-reentrant shop* of Wang et al. [16], the *cyclic robotic flow shop* of Kats and Levner [17] and the *flow shop with transportation times and a single robot* of Hurink and Knust [18]. Crama et al. [19] survey recent papers on periodic scheduling in cycle shops that are robotic flow shops. Serafini and Ukovich [20] describe a cycle shop that appeared as a *traffic light scheduling model*.

This chapter presents the fundamentals of cycle shop scheduling by giving a classification of cycle shops, describing periodic and nonperiodic cycle shop scheduling problems, basic exact and approximation algorithms for them and related complexity results.

7.2 Cycle Shop and Related Machine Environments

Complementing the $\alpha|\beta|\gamma$ notation given in Chapter 1, we will use the extended machine environment field $\alpha = \alpha_0\alpha_1\alpha_2$. Before giving definitions we need to remember that $\circ$ denotes the empty symbol; α_1 and α_2 define the machine environment type and the machine number, respectively; $\alpha_1 = P$ means *identical parallel machines*, where each job J_j consists of a single operation that can be processed on any machine during time p_j; and that $\alpha_1 = J$ means a *job shop*, where each job J_j consists of a chain of operations $O_{1j}, O_{2j}, \ldots, O_{m_j j}$, where O_{ij} has to be processed on a specified machine $M_{\mu_{ij}}$ during time p_{ij} starting on or after the completion of $O_{i-1,j}$ and $\mu_{i-1,j} \neq \mu_{ij}$ for $i > 1$. Other definitions are given in Table 7.1 and illustrated in Figure 7.2.

Jobs in a job shop are *identical* if $m_j = \ell$ and $\mu_{ij} = \mu_i$ and $p_{ij} = p_i$. Thus, a job shop with identical jobs is a special case of a cycle shop. Observe that a robotic job shop is constructed from a job shop by inserting an additional operation on an additional common machine between each two consecutive

TABLE 7.1 Cycle Shop and Related Machine Environments

α_1	Name	Definition
P mod	Modulated machines	M_i can start only at times $(i-1) \bmod m$
J per	Perfect job shop	$m_j = 0 \bmod m$ and $\mu_{ij} - 1 = (i-1) \bmod m$
RJ	Robotic job shop	m_j are odd and $m = \mu_i \neq \mu_{i-1,j} \neq \mu_{i+1,j} \neq m$ if i is even
C	Cycle shop	$m_j = \ell \geq m$ and $\mu_{ij} = \mu_i$
C per	Perfect cycle shop	A cycle shop that is a perfect job shop
RC	Robotic cycle shop	A robotic job shop that is a cycle shop
RF	Robotic flow shop	A robotic cycle shop where $\mu_i = (i+1)/2$ if i is odd
V	V-shop	A cycle shop where $\ell = 2m - 1$ and $\mu_i = i$ if $i \leq m$ and $\mu_i = 2m - i$ if $i > m$
L	Loop shop	A cycle shop where $\ell = m + 1$ and $\mu_i = i$ if $i < \ell$ and $\mu_\ell = 1$
F	Flow shop	A cycle shop where $\ell = m$ and $\mu_i = i$

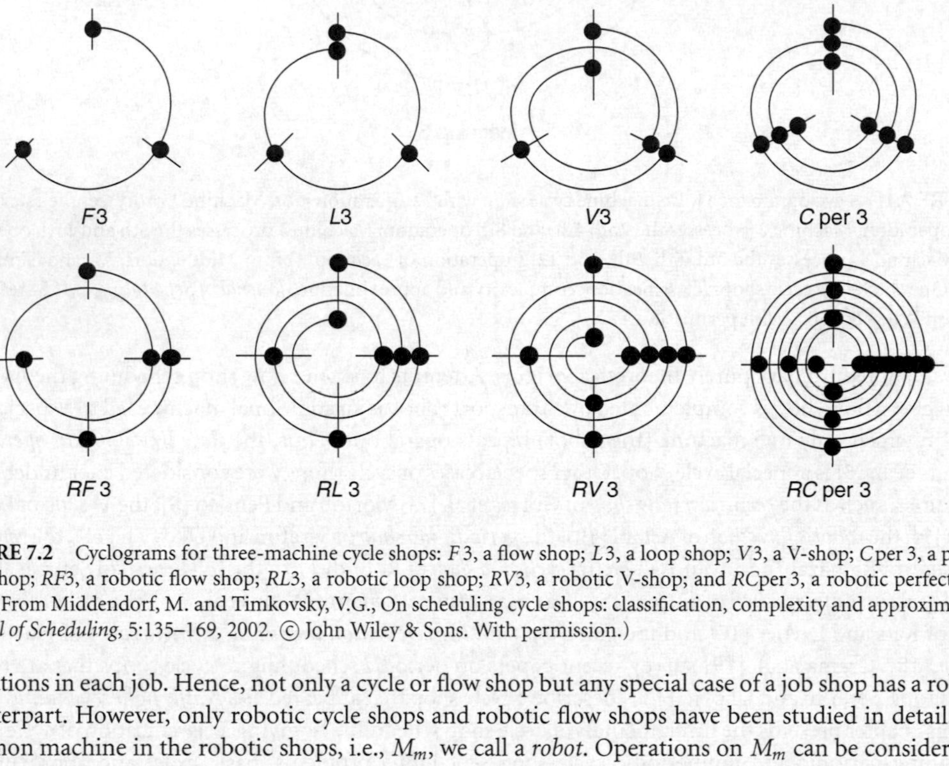

FIGURE 7.2 Cyclograms for three-machine cycle shops: $F3$, a flow shop; $L3$, a loop shop; $V3$, a V-shop; Cper 3, a perfect cycle shop; $RF3$, a robotic flow shop; $RL3$, a robotic loop shop; $RV3$, a robotic V-shop; and RCper 3, a robotic perfect cycle shop. (From Middendorf, M. and Timkovsky, V.G., On scheduling cycle shops: classification, complexity and approximation, *Journal of Scheduling*, 5:135–169, 2002. © John Wiley & Sons. With permission.)

operations in each job. Hence, not only a cycle or flow shop but any special case of a job shop has a robotic counterpart. However, only robotic cycle shops and robotic flow shops have been studied in detail. The common machine in the robotic shops, i.e., M_m, we call a *robot*. Operations on M_m can be considered as the transportation of parts (related to the jobs) between the other machines. The empty robot in all our considerations can move instantly, and all related problems can be studied without consideration of robot moves. We do not consider here the case where the empty robot does not move instantly, or the related problems that require a consideration of robot moves. You can find comprehensive information about such problems in several papers [14,19,21–24].

In robotic shops, p_{ej} and p_{oj} will denote the processing times only of even operations $O_{ej}, e = 2, 4, \ldots, m_j - 1$, on M_m, and odd operations $O_{oj}, o = 1, 3, \ldots, m_j$ on $M_1, M_2, \ldots, M_{m-1}$, respectively. In addition, we assume that a robot loads and unloads parts instantly as well because nonzero loading and unloading times can be included in processing times p_{oj} or transportation times p_{ej}.

Cycle Shop Machine Number	
α_2	Definition
$\circ$	Machine number is not fixed
m	Machine number is fixed at $m+1$ for robotic shops or m otherwise

Let M_k be the machine in a cycle shop that processes ν_k operations in a job. Then define ν_k to be the *multiplicity* of M_k. We call M_k a *unimachine* or a *multimachine* if $\nu_k = 1$ or $\nu_k > 1$, respectively. Define the number of multimachines, and $\nu = \max_{1 \le k \le m} \nu_k$ to be the *rank* and the *multiplicity* of the cycle shop, respectively.

Rank		Multiplicity	
α_0	Definition	ϵ	Definition
$\circ$	Rank is not fixed	$\circ$	Multiplicity is not fixed
ρ	Rank is fixed and equal ρ	ν	Multiplicity is fixed and equal ν

For example, a flow shop is of rank 0 and multiplicity 1; a loop shop is of rank 1 and multiplicity 2; a robotic flow shop is of rank 1 and multiplicity $m - 1$; and a two-machine cycle shop with four or more operations in jobs is of rank 2. A two-machine robotic flow shop is a three-machine flow shop. A three-machine robotic flow shop is a cycle shop of rank 1 and multiplicity 2, i.e., the simplest robotic flow shop is not a flow shop. A V-shop is of rank $m - 1$ and multiplicity 2. A loop shop or a robotic flow shop has only one multimachine. In a similar way, a V-shop has only one unimachine. Notice that a two-machine loop shop, a two-machine V-shop and a two-machine cycle shop with three operations in jobs represent the same cycle shop machine environment that differs minimally from a two-machine flow shop. Further, we consider only cycle shops of positive rank, i.e., which are not flow shops.

7.3 Periodic Schedules

Let σ be a schedule for a given finite job set $\mathcal{J}$ with n jobs, and let *unlimited breeding* of σ by an infinite number of shifts by P units in time produce a feasible infinite schedule τ. Then we call the schedule τ *infinitely n-job periodic* with *period P*, the schedule σ a *generator*, and the job set $\mathcal{J}$ a *generating job set* of τ. To stress that τ is generated by σ we will use the denotation τ_σ. We obtain the definition of an *N-finitely n-job periodic* schedule if we use *limited breeding* of σ by a finite number N of shifts by P units in time. In this chapter, we discuss primarily infinitely single-job periodic schedules and n-finitely single-job periodic schedules. In what follows, if the number of jobs in $\mathcal{J}$ is clear from the context, the term "n-job" will be omitted. Additionally, if not stated otherwise, the terms "periodic" and "finitely periodic" will mean "infinitely single-job periodic" and "n-finitely single-job periodic," respectively.

Define a *periodic problem* to be a problem of finding a periodic schedule. Define a *flow time* of the periodic schedule τ_σ to be the maximum completion time in σ. One of the main points in our approach is the consideration of any periodic problem as a nonperiodic problem with an additional constraint.

Let O_{ij} have m_{ij} nonpreemptive parts in σ, and let O_{ijk} be the kth nonpreemptive part of O_{ij} in σ. Then O_{ij} can be considered as a chain $O_{ij1} O_{ij2} \dots O_{ijm_{ij}}$, where $m_{ij} = 1$ and $O_{ij} = O_{ij1}$ if O_{ij} is nonpreemptive. Let S_{ijk}, C_{ijk}, and p_{ijk} be the start time, the completion time, and the length of O_{ijk} in σ, respectively. Thus, $C_{ijk} - S_{ijk} = p_{ijk}$ and $p_{ij1} + p_{ij2} + \dots + p_{ijm_{ij}} = p_{ij}$. Obviously, τ is a feasible periodic schedule if and only if the parts O_{ijk} scheduled on $M_\imath$ do not overlap modulo P. So we have the following *periodicity constraint:*

$$(i, j, k) \neq (i', j', k') \quad \text{and} \quad \mu_i = \mu_{i'} \Longrightarrow [S_{ijk}, C_{ijk}] \cap [S_{i'j'k'}, C_{i'j'k'}] = \emptyset \bmod P \tag{7.1}$$

Any problem of finding a periodic schedule, therefore, is a problem of finding a nonperiodic schedule satisfying Constraint (7.1). Hence, any nonperiodic problem is a relaxation of its periodic counterpart. Obviously, $P \ge \max_{1 \le \imath \le m} \sum_{\imath = \mu_{ij}} p_{ij}$. We call a problem *nonperiodic* if it is free of Constraint (7.1).

Let $s_\imath$ and $c_\imath$ be the earliest start time and the latest completion time on $M_\imath$ in σ. Thus, $B_\imath = c_\imath - s_\imath$ is the *busy time* of $M_\imath$ in σ. Define $B_{\max}$ to be the maximum busy time in σ, i.e., $B_{\max} = \max_{1 \le \imath \le m} B_\imath$. We call a periodic schedule τ_σ with period P *trivially periodic* if $B_{\max} \le P$. It is easy to see that this inequality implies Constraint (7.1), and a minimum period trivially periodic schedule τ_σ is a result of the concatenation of copies of σ, where $B_{\max} = P$. Hence, finding a trivially periodic schedule of minimum period reduces to finding a nonperiodic schedule of minimum maximum busy time. Note that the maximum busy time in robotic shops is always reached on M_m, i.e., on the robot, and is called the robot *cycle time*. Since all the events in the sequence ε can be taken modulo P, we can get the analogous sequence of *events modulo P* in the interval $[0, P-1]$.

Periodic problems in this chapter are based on the above periodicity concept. We avoid considering periodic problems on identical parallel machines [25,26] and in *recurrent job shops* [27] with *potential* or *conjunctive constraints* [27,28] originating from pipeline computing [29].

In addition to the minimization criteria traditionally used in nonperiodic scheduling (see Chapter 1) we will also use the following criteria in periodic scheduling:

Additional Criteria for Periodic Schedules

γ	Definition
P	Period
$P * C_{\max}$	Pareto criterion on P and $C_{\max}$
$B_{\max}$	Maximum busy time

7.4 Assumptions, Notation, and Terminology

All numerical parameters are assumed to be positive and integral. If a problem involves only one job, we will omit the job index in denotations of operations and processing times. So, if $n = 1$, then we will write O_i and p_i instead of O_{ij} and p_{ij}.

Following a commonly accepted assumption [30], we ignore logarithms of numerical parameters in estimations of problem sizes and time complexity bounds, considering them as constants. For example, any instance of $L \mid p_{ij} = 1 \mid C_{\max}$ can be defined by two integers, m and n, so a compact binary code of the problem is of length $\log m + \log n$. Since we ignore logarithms, the problem size is a constant. Analogously, any instance of $C \mid p_{ij} = 1 \mid C_{\max}$ can be defined by integers m, n, ℓ, and ℓ integers mapping operations to machines. Thus, a compact binary code of the problem is of length $\log m + \log n + \log \ell + \ell \log m$. Hence, the problem size is ℓ.

We assume that $\gamma \in \{P, P * C_{\max}\}$ implies a periodic problem. Denotations of nonperiodic problems with $\gamma = B_{\max}$ will also denote the related trivially periodic problems, since they are equivalent. In periodic problems with $\gamma \notin \{P, P*C_{\max}, B_{\max}\}$, the period P will be considered to be given. To distinguish periodic problems with $\gamma \notin \{P, P * C_{\max}, B_{\max}\}$ from their nonperiodic counterparts, we will use the following accents above the last letter in the denotations of their machine environments.

Special Periodic Machine Environments

α_1	Definition
$R\tilde{F}$	Periodic robotic flow shop with $\gamma \notin \{P, P * C_{\max}, B_{\max}\}$
$\tilde{C}$	Periodic cycle shop with $\gamma \notin \{P, P * C_{\max}, B_{\max}\}$
$\dot{C}$	Finitely periodic cycle shop

For example, $\tilde{C} \| C_{\max}$ and $\dot{C} \| C_{\max}$ denote the infinitely and finitely periodic counterparts of the nonperiodic problem $C \| C_{\max}$, while $C \| P$, $C \| P * C_{\max}$, $C \| B_{\max}$, and $RF \| P$ denote infinitely periodic problems.

Several authors use the term "cyclic scheduling" instead of "periodic scheduling" and "cycle time" instead of "period." The latter two terms, however, are not synonyms in our classification because they mean $B_{\max}$ and P, respectively. In several papers, cycle time has been denoted as C_t, but we use $B_{\max}$ instead, to avoid an association with completion time or period. We avoid the term "reentrant flow shop" because one can find different definitions of it in the literature that actually represent (in our terminology) a perfect cycle shop, or a loop shop, or a general cycle shop where zero processing times are allowed. Notice that the classification of cycle shops given here makes sense only for positive processing times.

7.5 Structural Relationship

As we noticed before, $L2\|\gamma$, $V2\|\gamma$, and $C2 \mid \ell = 3 \mid \gamma$ are identical problems. This obviously remains true even in the case of precedence and release-date constraints. There exist other structural relations among special cycle shops that are worth noting. Thus, $RF\alpha_2 \mid \beta \mid \gamma$ and $RC\alpha_2 \mid \beta \mid \gamma$ are special cases of $1C\alpha_2 \mid \beta \mid \gamma$ and $C\alpha_2 \mid \beta \mid \gamma$, respectively. Note that a single-job job shop is a cycle shop, i.e., $C\alpha_2 \mid \beta_{1-4}, n = 1, \beta_{6-7} \mid \gamma$ and $J\alpha_2 \mid \beta_{1-4}, n = 1, \beta_{6-7} \mid \gamma$ are identical. The same is true in the periodic case. Since the maximum busy time and the maximum completion time are always reached on M_1 in loop shops, $L\alpha_2 \mid \beta \mid B_{\max}$ and $L\alpha_2 \mid \beta \mid C_{\max}$ are also identical.

The following relationship is not as evident. Define a *flow component* of a cycle shop to be any maximal subsequence of adjacent unimachines in the sequence $M_{\mu_1} M_{\mu_2} \ldots M_{\mu_\ell}$. Define the *flow size* of a cycle shop to be the length of its longest flow component.

Let A be $R\tilde{F} \mid n = 1, p_o = q_{\mu_o}, p_e = 1 \mid C_{\max}$; B be $1\tilde{C} \mid n = 1, p_i = 1 \mid C_{\max}$; and let P_{B} and f_{B} be the given period and the flow size of the cycle shop in B, respectively. Then A $\propto$ B, but B $\propto$ A only if $P_{\mathsf{B}} \geq f_{\mathsf{B}}$.

The proof is as follows. Let P_{A} denote the period in A, and let ν_{A} and ν_{B} denote multiplicities of the cycle shops in A and B, respectively. Obviously, $P_{\mathsf{A}} \geq \max\{\nu_{\mathsf{A}}, q_{\max}\}$, where $q_{\max} = \max_\imath q_\imath$, and $P_{\mathsf{B}} \geq \nu_{\mathsf{B}}$. Without loss of generality, we assume that M_1 is the multimachine in both problems.

For the reduction A $\propto$ B, replace each unimachine in an instance of A with processing time q by q unimachines with unit processing time and put $P_{\mathsf{B}} = P_{\mathsf{A}}$. This produces an instance of B that obviously has the same minimum $C_{\max}$.

If $P_{\mathsf{B}} \geq f_{\mathsf{B}}$, the reduction B $\propto$ A can be obtained by replacing each flow component of length q in an instance of B by one unimachine with processing time q and putting $P_{\mathsf{A}} = P_{\mathsf{B}}$. If $\mu_1 = 1$ or $\mu_\ell = 1$ in B, add one artificial starting or finishing unit-time operation on one or two artificial unimachines, respectively. Depending on zero, one or two artificial unimachines are added, the minimum $C_{\max}$ for the constructed instance of A will equal, or exceed by one or two, the minimum $C_{\max}$ for the instance of B. Since $f_{\mathsf{B}} = q_{\max}$ and $P_{\mathsf{B}} \geq \nu_{\mathsf{B}} = \nu_{\mathsf{A}}$, we have $P_{\mathsf{A}} \geq \max\{\nu_{\mathsf{A}}, q_{\max}\}$.

7.6 Traffic Light Scheduling

The simplest scheduling model of a traffic light controlling the intersection of two streets is C per $2 \mid$ *no wait*, $n = 2$, $p_{i+1,1} = p_{i2} \mid B_{\max}$. The parameters and related constraints in this model should be interpreted as follows:

Perfectness: Since the cycle shop is perfect, the traffic light goes through the same number of green–yellow–red time cycles for each street.

Machines: M_1 is the "green–yellow safety red-light" machine that allows traffic and sets a safety period while the red light is on for both streets; M_2 is the "red-light" machine that stops traffic.

No-wait requirement: Traffic lights are not permitted to be dark.

Jobs: Each job represents a traffic light activity on a single street.

Number of operations in jobs: ℓ is the number of cycles the traffic light is set up on. Assuming that the length of the green–yellow–red time cycle is one minute and that the traffic light activity must be repeated each 24 hs, $\ell = 1440$.

Operations: $O_{2l-1,j}$, where $l = 1, 2, \ldots, \ell/2$, $j \in \{1, 2\}$ (odd operations), implement the lth green–yellow safety red-light cycle on the jth street; $O_{2l,j}, l = 1, 2, \ldots, \ell/2$, (even operations) implement the lth red-light cycle on the jth street.

Processing times of odd operations: $p_{2l-1,j}, l = 1, 2, \ldots, \ell/2$, has three consecutive components which are $g_{2l-1,j}$, green-light time; y, yellow-light time; and s, safety red-light time. Thus, $p_{2l-1,j} = g_{2l-1,j} + y + s$.

Processing-time constraints: $p_{i+1,1} = p_{i2}$, $i = 1, 2, \ldots, \ell - 1$, (hedging condition) prohibits traffic on both streets at the same time; if there is the requirement $p_{ij} = p$, then the streets have the same

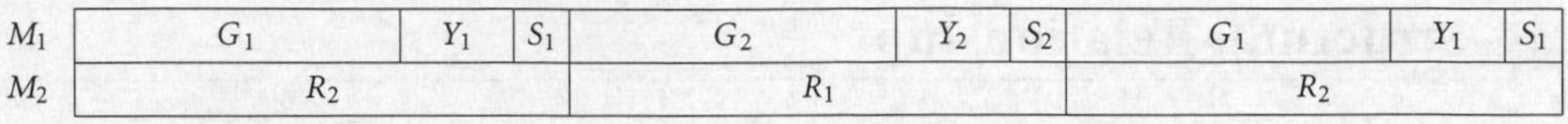

FIGURE 7.3 Three cycles of a traffic-light schedule, where G_j, Y_j, R_j, and S_j denote green-, yellow-, red-light, and safety red-light signals, respectively, on the jth street.

cycle frequency $1/p$; if there is the requirement $p_{ij} \in \{p, q\}$, then the streets have different cycle frequencies $1/p$ and $1/q$.

Minimization criterion: It is easy to see that for any instance of the problem, there exist only two trivially found optimal schedules with minimum

$$B_{\max} = \sum_{l=1}^{\ell/2} (p_{2l-1,1} + p_{2l-1,2})$$

that convert to each other by transposition of the jobs. Figure 7.3 shows a fragment of the traffic-light schedule.

7.7 Tractable Cycle Shops

7.7.1 $C \mid p_{ij} = 1 \mid C_{\max}$, $C2 \mid p_{ij} = q_{\mu_i} \mid C_{\max}$

An $O(\ell)$ algorithm for $C \mid p_{ij} = 1 \mid C_{\max}$, whose size is ℓ, follows from the following recursion relation. Let $F_i(k)$ denote the finishing time of M_k in an optimal schedule of the first k operations of all n jobs. Setting $F_{\mu_0}(0) = \infty$ and $F_i(0) = 0$ for $i = 1, 2, \ldots, m$, it is easy to verify that

$$F_{\mu_{k+1}}(k+1) = \max\left\{F_{\mu_k}(k) + 1, F_{\mu_{k+1}}(k) + n\right\}, \quad k = 0, 1, 2, \ldots, \ell - 1$$

where $F_{\mu_\ell}(\ell)$ is obviously minimum $C_{\max}$. In accordance with this equation, blocks of operations $O_{k1}, O_{k2}, \ldots, O_{kn}$ must be scheduled without interruptions in the order $k = 1, 2, \ldots, \ell$ as early as possible [31]. We will call the schedule this algorithm produces an *operation-block* schedule. A simple example is given in Figure 7.4.

The algorithm simultaneously solves $\dot{C} \mid p_{ij} = 1 \mid P * C_{\max}$ producing a finitely periodic schedule with period one. $C2 \mid p_{ij} = q_{\mu_i} \mid C_{\max}$ can be solved in constant time by a similar recursion relation [31]. Note that $J3 \mid p_{ij} = 1 \mid C_{\max}$ is NP-hard [32].

7.7.2 $C2 \mid r_j, p_{ij} = 1 \mid C_{\max}$, $C2 \mid p_{ij} = 1 \mid \sum U_j$

The size of $C2 \mid r_j, p_{ij} = 1 \mid C_{\max}$ is n, which is the size of $J2 \mid r_j, p_{ij} = 1 \mid C_{\max}$ as well. Since there are $O(n^2)$ algorithms known for the job shop problem [33,34], the cycle shop problem can be solved in polynomial time by the same algorithms. Analogously, $C2 \mid p_{ij} = 1 \mid \sum U_j$ can be solved in polynomial time $O(n^7)$ by using a job shop scheduling algorithm for $J2 \mid p_{ij} = 1 \mid \sum U_j$ [35].

$C2 \mid r_j, p_{ij} = 1 \mid L_{\max}$, $J2 \mid r_j, p_{ij} = 1 \mid L_{\max}$, and $L2 \mid r_j, p_{ij} = 1 \mid \sum U_j$ remain open, while $C2 \mid$ *chains*, $p_{ij} = 1 \mid \sum U_j$ is NP-hard because the flow shop counterpart is so [36].

7.7.3 $C2 \mid p_{ij} = 1 \mid \sum C_j$

The *Earliest-Completion-First* (ECF) algorithm [37,38] solves $J2 \mid p_{ij} = 1 \mid \sum C_j$, whose size is n, in polynomial time $O(n \log n)$. This time, however, is pseudopolynomial for $C2 \mid p_{ij} = 1 \mid \sum C_j$ because the size of the cycle shop problem is constant.

To explain how to get a polynomial-time solution, let us consider the ECF algorithm in detail. The algorithm splits each job into two uninterrupted parts, where the first part includes only the first operation,

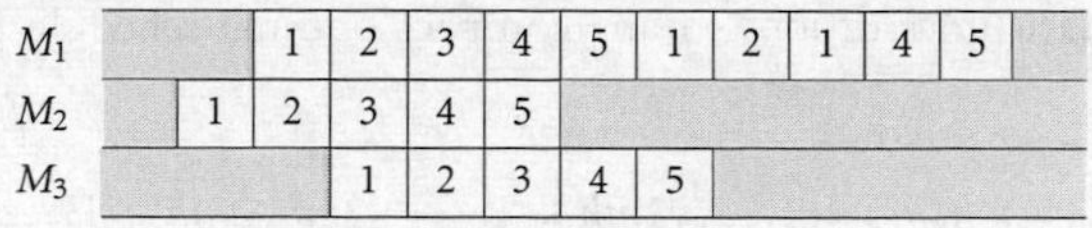

$$\begin{array}{lll} F_1(1) = 0, & F_2(1) = 5, & F_3(1) = 0, \\ F_1(2) = 6, & F_2(2) = 5, & F_3(2) = 0, \\ F_1(3) = 6, & F_2(3) = 5, & F_3(3) = 7, \\ F_1(4) = 11, & F_2(4) = 5, & F_3(4) = 7. \end{array}$$

FIGURE 7.4 An operation-block schedule for the instance of $C3|p_{ij} = 1|C_{\max}$ with $n = 5, \ell = 4, \mu_1 = 2, \mu_2 = 1, \mu_3 = 3, \mu_4 = 1$.

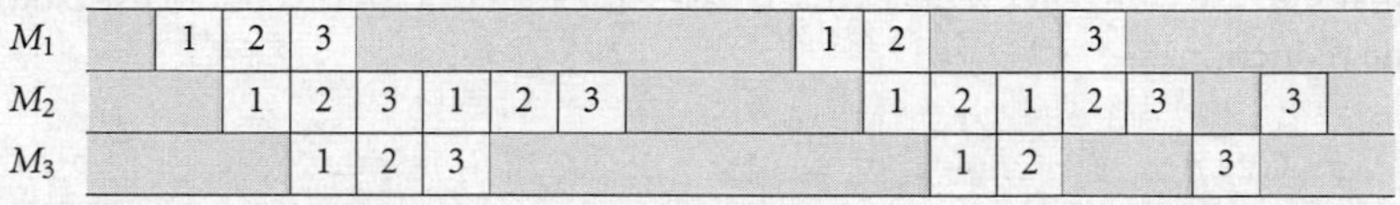

FIGURE 7.5 An operation-block schedule and an optimal schedule with total completion times $18 = 5 + 6 + 7$ and $17 = 4 + 5 + 8$, respectively, for the instance of $C3 \mid p_{ij} = 1 \mid \sum C_j$ with $n = 3, \ell = 4, \mu_1 = 1, \mu_2 = 2, \mu_3 = 3, \mu_4 = 2$.

and the second part includes the remaining operations. Then it schedules the parts such that every next job completion time happens as early as possible.

It is easy to check that the ECF algorithm does not interrupt jobs of any instance of $C2 \mid p_{ij} = 1 \mid \sum C_j$ since the jobs in this problem are identical. Thus, an ECF schedule for the cycle shop can be defined by start times of the jobs. It is also easy to check that the start times in the ECF schedule are

$$\begin{array}{llllllllll} 0, & 1, & \ell, & \ell+1, & 2\ell, & 2\ell+1, & 3\ell, & 3\ell+1,\ldots & \text{if } \ell \text{ is even, or} \\ 0, & 1, & \ell+1, & \ell+2, & 2\ell+2, & 2\ell+3, & 3\ell+3, & 3\ell+4,\ldots & \text{if } \ell \text{ is odd} \end{array}$$

In both cases, the sequence consists of n integers. Thus, an ECF schedule for the cycle shop can be presented by evident recursion formulas that require constant space requirements.

$C2 \mid r_j, p_{ij} = 1 \mid \sum C_j$ and $C \mid p_{ij} = 1 \mid \sum C_j$ remain open problems. Figure 7.5 shows that an operation-block schedule is not optimal for the latter problem. Also note that any algorithm solving $C \mid p_{ij} = 1 \mid \sum C_j$ in time polynomial in n is of exponential time because the problem size is ℓ.

7.7.4 $C2 \mid$ *no wait*, $p_{ij} = 1 \mid \sum f_j$

Since an ECF schedule for $C2 \mid p_{ij} = 1 \mid \sum C_j$ does not contain interrupted jobs, it is an optimal schedule for $C2 \mid$ *no wait*, $p_{ij} = 1 \mid \sum C_j$ as well. Therefore, the no-wait problem can also be solved in constant time. Now we show that the extension to the case with the arbitrary total-cost criterion $\sum f_j$ can actually be solved in polynomial time $O(n^3)$ if the cost function f_j is specified by a constant-space and constant-time oracle for every $j = 1, 2, \ldots, n$. Since the space $O(n)$ is needed to specify the criterion, the problem size is n. Thus, the two-machine no-wait unit-time cycle shop scheduling problem with any criterion from the scheduling classification can be solved in polynomial time.

The case with odd ℓ actually reduces to the case with even ℓ. Let us show that the problem with even ℓ reduces to an assignment problem. Note that there exist the following two evident isomorphisms (cf. Chapter 17):

$$\begin{aligned} C2 \mid \beta_{2-5}, \ell = 2h, p_{ij} = 1 \mid \gamma \quad &\approx \quad C\,\mathrm{per}\,2 \mid \beta_{2-5}, \ell = 2h, p_{ij} = 1 \mid \gamma \\ P\,\mathrm{mod}\,2 \mid \beta_{2-5}, p_j = 2h \mid \gamma \quad &\approx \quad C\,\mathrm{per}\,2 \mid \beta_{2-5}, \ell = 2h, p_{ij} = 1 \mid \gamma \end{aligned} \tag{7.2}$$

But $P\,\mathrm{mod}\,m \mid p_j = mh \mid \sum f_j$, can be solved by an assignment procedure as follows. Since f_j are nondecreasing functions, we can consider only compact optimal schedules, where each machine has no idling. Hence, the start times of jobs in any compact optimal schedule for $P\,\text{mod}\,|\,p_j = mh\,|\,\gamma$ are

$$0, 1, \ldots, m-1, \qquad mh, mh+1, \ldots, 2mh-1, \qquad 2mh, 2mh+1, \ldots, 3mh-1, \qquad \ldots$$

which we denote as $S_1, S_2, \ldots, S_n$, respectively. They are all different, and therefore the problem is to find a minimum-total-cost assignment of n jobs to n start times, where the assignment of J_j to S_k costs

$$f_j(S_k + mh)$$

But the assignment problem can be solved in polynomial time $O(n^3)$ (see [39]).

To show that the case with odd ℓ reduces to the case with even ℓ, it is sufficient to use Isomorphism (7.2) and prove the reduction

$$C2\,|\,\beta_{2-5}, \ell = 2h-1, p_{ij} = 1\,|\,\gamma \quad \propto \quad C2\,|\,\beta_{2-5}, \ell = 2h, p_{ij} = 1\,|\,\gamma \tag{7.3}$$

Let I be an instance of the left problem. Construct the instance I' of the right problem from I by appending one operation to the beginning of each job and increasing all due dates (if they are given) by one. It is easy to make sure that each compact optimal schedule for I can be obtained from a compact optimal schedule for I' by deleting all appended operations and shifting time back by one unit.

$C2\,|\,\textit{no wait, chains}, p_{ij} = 1\,|\sum f_j$ (see Chapter 17) and $J2\,|\,\textit{no wait}, p_{ij} = 1\,|\sum f_j$ [5,31,40] are already NP-hard, while $C2\,|\,\textit{no wait}, r_j, p_{ij} = 1\,|\sum C_j$ remains open. Notice that Kravchenko's $O(n^6)$ algorithm [41] for $J2\,|\,\textit{no wait}, p_{ij} = 1\,|\sum C_j$ does not solve $C2\,|\,\textit{no wait}, p_{ij} = 1\,|\sum C_j$ in polynomial time because the size of the cycle shop problem is constant.

7.7.5 $C2\,|\,prec, r_j, p_{ij} = 1\,|\,L_{max}$, $L2\,|\,prec, r_j, p_{ij} = 1\,|\,L_{max}$

The cycle shop problem and the loop-shop problem, whose size is n, can be solved in pseudopolynomial time $O(\ell^2 n^2)$ and polynomial time $O(n^2)$, respectively.

Reduction (7.3) and a *chaining* technique [36] increasing the number of jobs in ℓ/m times give the following reduction, which is a trivial isomorphism for the criterion $\gamma = C_{max}$:

$$C\,\text{per}\,\alpha_2\,|\,\beta_{2-5}, \ell = mh, p_{ij} = 1\,|\,\gamma \quad \propto \quad F\alpha_2\,|\,\beta_2, \beta_3 \cdot \textit{chains}, \beta_{4-5}, p_{ij} = 1\,|\,\gamma \tag{7.4}$$

where $\gamma \neq \sum C_j$. Each instance of the former problem can be converted into an instance of the latter by considering jobs with processing time mh as chains of h jobs with processing time m. If J_j is a job with release date r_j, due date d_j and weight w_j of an instance of the former problem, then the new h jobs in the related chain have a common release date r_j, due dates $D, D, \ldots, D, d_j$, where D is a large number bounded by a polynomial in the problem size, and weights $0, 0, \ldots, 0, w_j$, respectively. The reduction does not hold for $\gamma = \sum C_j$ due to the absence of weights.

Reductions (7.3) and (7.4) allow us to solve $C2\,|\,prec, r_j, p_{ij} = 1\,|\,L_{max}$ by an $O(n^2)$ algorithm for $F2\,|\,prec, r_j, p_{ij} = 1\,|\,L_{max}$ [42]. Since reduction (7.4) increases the number of jobs in ℓ/m times, the cycle shop problem can be solved in pseudopolynomial time $O(\ell^2 n^2)$. This time bound, however, is $O(n^2)$ for $L2\,|\,prec, r_j, p_{ij} = 1\,|\,L_{max}$ because $\ell = 3$ in this case. Thus, though $C2\,|\,prec, r_j, p_{ij} = 1\,|\,L_{max}$ is not strongly NP-hard (unless $P = \text{NP}$), it remains open.

7.7.6 $C\,|\,prec, r_j\,|\,P$

There is a simple polynomial time algorithm even for $J\,|\,prec, r_j\,|\,P$ with a time requirement proportional to the total number of operations in the jobs. Remember that the flow time is not restricted in this problem, and the algorithm freely spends it. The first three steps solve the problem without regard to release dates. The fourth step turns the schedule into one that satisfies given release dates.

Step 1. Construct the precedence relation R for the total set of operations in the jobs and find a linear order including R, i.e., a list L of operations consistent with R.

Step 2. Set P to be the maximum total processing time on the machines, i.e.,

$$P = \max_{1 \le \imath \le m} \sum_{\imath = \mu_{ij}} p_{ij}$$

and divide a time interval of length PT, where

$$T = \sum_{j=1}^{n} m_j$$

i.e., T is the total number of operations in jobs, into T subintervals of length P. Each subinterval will include only one operation.

Step 3. Let the first $j - 1$ operations of the list L be already included in the schedule, and let them occupy the total processing times $T_\imath$ on the machines $M_\imath$, $\imath = 1, 2, \ldots, m$. Let the jth operation in L require the machine M_k. Then assign its start time on M_k to be T_k as a local time inside the jth subinterval.

Step 4. Assuming that the local start times of operations in the subintervals are fixed, move the subintervals apart so that the release dates of the jobs are satisfied and the distance between any two adjacent subintervals is divisible by P.

The algorithm obviously finds a minimum-period schedule because P is the maximum total processing time on the machines, a lower bound for periods. Special cases of $J \mid prec, r_j \mid P$ that cannot be solved in polynomial time by this algorithm are $J2 \mid prec, r_j, p_{ij} = 1 \mid P$ and even $J2 \mid p_{ij} = 1 \mid P$ because the size of these problems includes the number of operations in jobs under logarithms, i.e., the algorithm runs only in pseudopolynomial time. We suggest that these special cases can also be solved in polynomial time but they require special consideration like $J2 \mid p_{ij} = 1 \mid C_{\max}$ and $J2 \mid r_j, p_{ij} = 1 \mid C_{\max}$ (see [33]). On the other hand, in terms of the periodic scheduling model described in [43], there is an extension of $J \mid prec, r_j \mid P$ (capturing parallel machines, for example) that can still be solved in polynomial time [44].

7.7.7 $1C \mid$ *no wait*, $n = 1, p_i = q_{\mu_i}/q \mid P$, $C \mid$ *no wait*, $n = 1, p_i = 1 \mid P$

The no-wait constraint makes cycle shops nontrivial even with $n = 1$. Note that the sizes of these problems are ν and ℓ, respectively, and that $\nu < \ell$, $m < \ell$.

Let us take $1C \mid no\ wait, n = 1, p_i = q_{\mu_i}/q \mid P$. Since all processing times are divisible by q we will consider an equivalent form of the problem, where $q = 1$, dividing all processing times by q. Without loss of generality we assume that the multimachine in this problem is M_1. Thus, all operations on M_1 are of unit processing time, and the condition that start times on M_1 should be different modulo P becomes necessary for any feasible schedule.

Let x_i be the the start time of the ith operation on M_1, and let $a_i - 1$ be the total processing time of all operations between the ith and $i + 1$st operations on M_1, if $i > 0$; or all operations before the first operation on M_1, if $i = 0$; or all operations after the last operation on M_1, if $i = \nu$. Set

$$x_1 = 0, \qquad x_i = \sum_{j=1}^{i-1} a_j \tag{7.5}$$

for $i = 2, 3, \ldots, \nu$. Thus, the problem is to find minimum P such that $x_1, x_2, \ldots, x_\nu$ are all different modulo P. It is clear that for a fixed P this condition can be checked in time $O(\nu)$. Remember that

$$P \ge Q = \max_{1 \le k \le m} \nu_k q_k$$

and note that, due to the no-wait requirement, the flow time is fixed and equal to

$$C_{\max} = -1 + \sum_{j=0}^{\nu} a_j = \sum_{k=1}^{m} \nu_k q_k$$

Thus, to solve the problem we need to check whether $x_1, x_2, \ldots, x_\nu$ are different modulo P for each period in the interval $[Q, C_{\max} - 1]$. This takes pseudopolynomial time

$$O\left(\nu\left[\sum_{k=1}^{m} \nu_k q_k - \max_{1\le k\le m} \nu_k q_k\right]\right)$$

which is $O\left(\nu^2 m q_{\max}\right)$.

Similar considerations lead to a polynomial-time algorithm for $C \mid no\ wait, n = 1, p_i = 1 \mid P$. Let $x_1 < x_2 < \cdots < x_\tau$ be the start times of all operations $O_{s_1}, O_{s_2}, \ldots, O_{s_\tau}$ in a single job on all multimachines. Thus, τ is the total multiplicity of the multimachines. Let $a_i - 1$ be the total processing time of all operations between O_{s_i} and $O_{s_{i+1}}$ if $i > 0$ or all operations before O_{s_1} if $i = 0$ or all operations after O_{s_τ} if $i = \tau$. Define I_k to be the set of indices of operations on the kth multimachine, $k = 1, 2, \ldots, \tau$. So, $I_1, I_2, \ldots, I_\rho$ is a partition of $\{1, 2, \ldots, \tau\}$.

Set the start times by Equalities (7.5), where $i = 2, 3, \ldots, \tau$. Then the problem is to find minimum P such that start times x_i, $i \in I_k$, are different modulo P for each $k = 1, 2, \ldots, \rho$. For a fixed P this can be checked in time $O(\tau)$. Since we consider the case with unit processing times, $C_{\max} = \ell$ and $Q = \nu$. Hence, the total time complexity is $O(\tau[\ell - \nu]) = O(\ell^2 - \nu\ell)$.

7.7.8 $\tilde{V} \mid n = 1, p_i = 1 \mid C_{\max}$, $V \mid n = 1, p_i = 1 \mid P{*}C_{\max}$

The size of each of these problems is constant because an instance of the former problem is defined by two integers m and P while an instance of the latter problem is defined by only one integer m. The problems have almost trivial solutions that can also be found in constant time. Since the problem multiplicity is two in both cases, $P \ge 2$ is a necessary and sufficient condition for their existence.

If P is odd in an instance of $\tilde{V} \mid n = 1, p_i = 1 \mid C_{\max}$, then there exists an optimal no-wait schedule with $C_{\max} = m - 1$. If P is even, then there is an optimal schedule with $C_{\max} = m$, where only O_{m+1} waits a single time unit. Such a schedule with $P = 2$ is obviously a solution to $V \mid n = 1, p_i = 1 \mid P * C_{\max}$. Figure 7.6 gives a four-machine example. By doubling the processing time of O_m in the schedule we can see that $V \mid n = 1, p_{i \ne m} = 1, p_m = 2 \mid P * C_{\max}$ has the same solution but without waiting jobs (see Figure 7.7).

7.7.9 $\tilde{C}$ per $\mid n = 1, p_i = 1 \mid C_{\max}$

The size of the problem is constant because only three numbers, m, ν, and P, are necessary to store. Note that $\ell = m\nu$ here. It is easy to see that the problem is equivalent to the single-machine problem of finding a minimum-length schedule of ν unit-time jobs whose start times are different modulo P and distinguishable by at least m.

M_1	1		2		3		**4**	1	5	2	6	3	7	**4**
M_2		1		2		3	1	**4**	2	5	3	6	**4**	7
M_3			1		2	1	3	2	**4**	3	5	**4**	6	5
M_4				1		2		3		**4**		5		6

FIGURE 7.6 The optimal schedule for $V4 \mid n = 1, p_i = 1 \mid P * C_{\max}$.

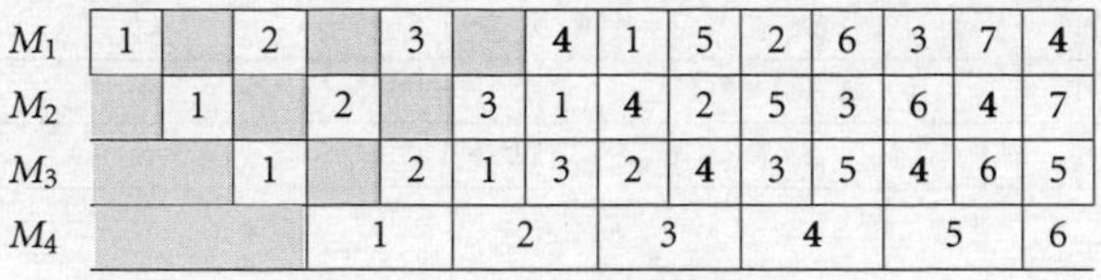

FIGURE 7.7 The optimal no-wait schedule of V-shape jobs for $V4 \mid n = 1, p_{i \neq m} = 1, p_m = 2 \mid P * C_{\max}$.

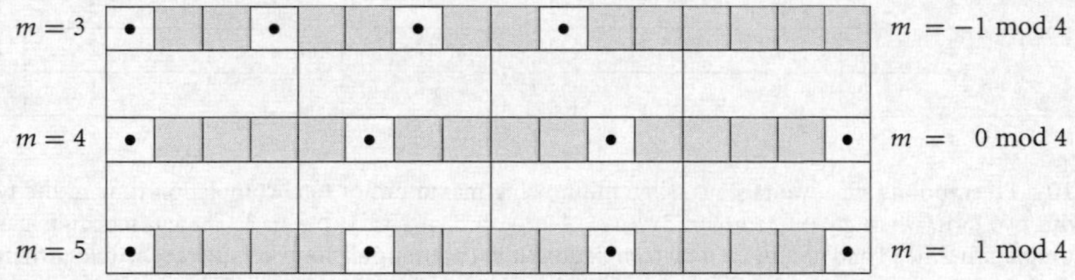

FIGURE 7.8 Solutions to the three instances of the single-machine version of C per $m \mid n = 1, p_i = 1 \mid C_{\max}$ with $P = 4$ and $m = 3, 4, 5$.

Formally, the problem is given positive integers m and P finding integers $x_1, x_2, \ldots, x_n$ with minimum x_n such that

$$x_1 = 0, \quad i > 1 \Rightarrow x_i - x_{i-1} \geq m, \quad i > j \Rightarrow x_i - x_j \neq 0 \bmod P.$$

If $m = 1 \bmod P$ or $m = -1 \bmod P$, then there exists an obvious no-wait optimal schedule. If $m = 0 \bmod P$, then it is clear that a no-wait optimal schedule does not exist and there exists an obvious optimal schedule with $x_i = x_{i-1} + 1$. See Figure 7.8 for an example.

Dual conditions $P = -1, 0, 1 \bmod m$ also lead to trivial solutions. In general, however, the problem remains open because it is not clear whether there exists a solution that can be specified and found in constant time in the case where $m \neq -1, 0, 1 \bmod P$.

7.8 Intractable Cycle Shops

7.8.1 $L2 \| C_{\max}$, $L2 \mid pmtn \mid C_{\max}$

The simplest cycle shop scheduling problem of minimizing maximum completion time, where processing times of jobs are not restricted, is already NP-hard even in a two-machine loop shop [9,31]. Remember that a two-machine loop shop and a two-machine V-shop are one and the same minimal cycle shop.

Let us show a reduction from the *partition problem* [30]: given positive integers $a_1, a_2, \ldots, a_k$ with $\sum_{j=1}^{k} a_j = 2b$, does there exist a set $S \subset \{1, 2, \ldots, k\}$ with $\sum_{j \in S} a_j = b$?

In the nonpreemptive case, for any instance I of the partition problem we create the instance I' of the decision version of $L2 \| C_{\max}$ with

$$n = k + 1, \qquad p_{1j} = p_{3j} = 1, \qquad p_{2j} = ka_j, \qquad j = 1, 2, \ldots, k$$
$$p_{1,k+1} = p_{3,k+1} = kb, \qquad p_{2,k+1} = k$$

and the upper bound $2k(b+1)$ on the schedule length. If the answer to I is positive, then I' has a schedule of length $2k(b + 1)$ without idle time on M_1. Figure 7.9 presents such a schedule. If the answer to I is negative, then all schedules for I' are of length more than $2k(b + 1)$ because M_1 will be idle at least one time unit in any schedule.

	$|S|$	kb	k	kb	$|\bar{S}|$
M_1	S	$k+1$	$S+\bar{S}$	$k+1$	\bar{S}
M_2		S	$k+1$	\bar{S}	

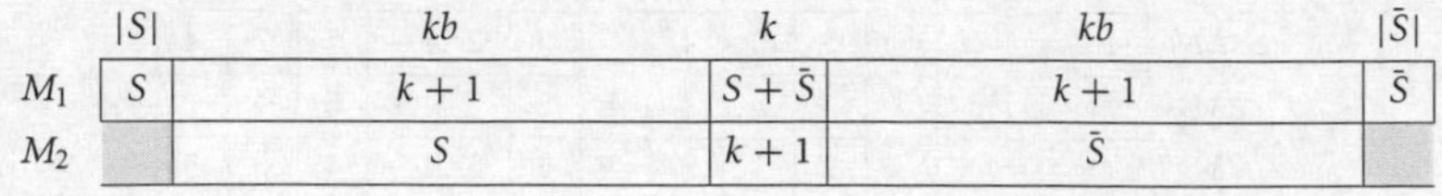

FIGURE 7.9 The schedule for the instance I' of $L2\|C_{\max}$ if the answer to the instance I of the partition problem is positive, where $\bar{S} = \{1, 2, \ldots, k\} - S$.

	3	1	3	3	1	1	3	1	3	1
M_1	2	1	2		1	1	2		2	1
M_2		2	1			1	2	1		

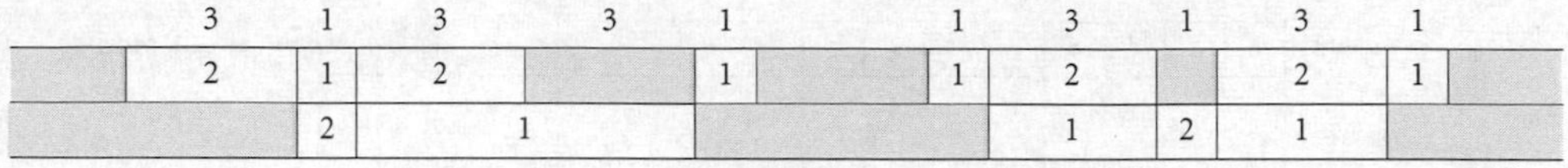

FIGURE 7.10 Preemptions are advantageous when minimizing maximum or total completion time in the two-machine loop shop with two jobs, where $p_{11} = 1$, $p_{21} = 3$, $p_{31} = 1$, $p_{12} = 3$, $p_{22} = 1$, $p_{32} = 3$. The left schedule is of minimum maximum completion time 11 and minimum total completion time 18 among all nonpreemptive schedules, whereas the right schedule is of minimum maximum completion time 9 and minimum total completion time 17 among all preemptive schedules.

	$|S|$	kb	$3kb$	k	$3kb$	kb	$|\bar{S}|$
M_1	S	$k+1$	$k+2$	$S+\bar{S}$	$k+2$	$k+1$	\bar{S}
M_2		S	$k+1$	$k+2$	$k+1$	\bar{S}	

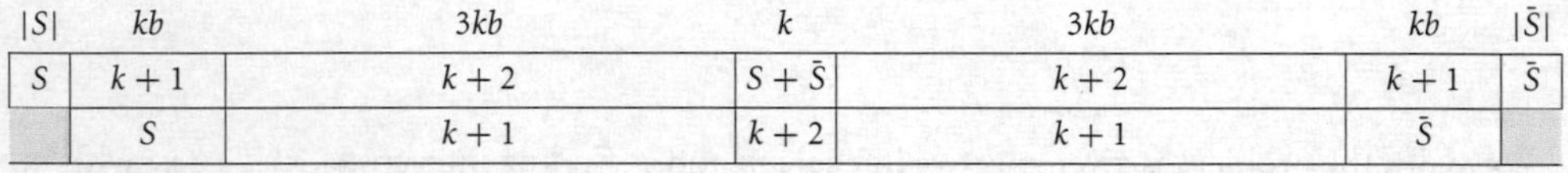

FIGURE 7.11 The schedule for the instance I' of $L2 \mid pmtn \mid C_{\max}$ if the answer to the instance I of the partition problem is positive, where $\bar{S} = \{1, 2, \ldots, k\} - S$.

To describe a similar reduction in the preemptive case, we first observe that preemptions in a two-machine loop shop can be advantageous. Figure 7.10 gives a simple example. The reduction remains almost the same. We only need to replace the $(k+1)$th job by two jobs that will generate a preemptive fragment similar to the right schedule in Figure 7.10. The other k jobs remain the same.

In the preemptive case, for any instance I of the partition problem we create the instance I' of the decision version of $L2 \mid pmtn \mid C_{\max}$ with

$$n = k+2, \qquad p_{1j} = p_{3j} = 1, \qquad p_{2j} = ka_j, \qquad j = 1, 2, \ldots, k$$
$$p_{1,k+1} = p_{3,k+1} = kb, \qquad p_{2,k+1} = 6kb$$
$$p_{1,k+2} = p_{3,k+1} = 3kb, \qquad p_{2,k+2} = k$$

and the upper bound $k(8b+1)$ on the schedule length. If the answer to I is positive, then I' has a schedule of length $k(8b+1)$ without idle time on M_1. Figure 7.11 presents such a schedule. If the answer to I is negative, then all schedules for I' are of length more than $k(8b+1)$ because M_1 will be idle at least one time unit in any schedule.

Note that it is still unknown whether the problems can be solved in pseudopolynomial time or are strongly NP-hard. It is only known that $L2 \mid r_j \mid C_{\max}$ is strongly NP-hard because there is a reduction to this problem from the 3-*partition problem* [31].

7.8.2 $L2\|\sum C_j$, $L2 \mid pmtn \mid \sum C_j$, $L3\|C_{\max}$, $L3 \mid pmtn \mid C_{\max}$

The NP-hardness proofs for $L2\|C_{\max}$ and $L2 \mid pmtn \mid C_{\max}$ can easily be converted to NP-hardness proofs for minimizing $\sum C_j$ by adding a "large" number of "long" jobs to the instance I' such that they can be presented by a code polynomially bounded on k. These jobs must be processed last in schedules with smaller values of $\sum C_j$. For example, we can add $2k$ jobs with operations of processing time $3kb$. Since the

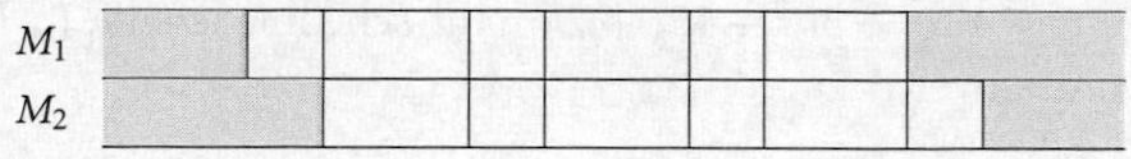

FIGURE 7.12 String denotation of this schedule is ${}^{121212}_{212121} = [{}^{1}][{}^{21}_{21}]^2[{}^{2}_{2}][{}_{1}]$.

partition problem is ordinarily NP-hard we obtain the NP-hardness of the problems, leaving the question of whether they can be solved in pseudopolynomial time or are strongly NP-hard.

The problems, however, are strongly NP-hard indeed because there can be shown reductions from $F2\|\sum C_j$ and $F2\,|\,pmtn\,|\sum C_j$, which are proved to be strongly NP-hard in [45] and [46], respectively.

Let I be an instance of one of the flow shop problems with n jobs and upper bound u on total completion time. We can construct the instance I' of the related loop-shop problem from I by transposition of M_1 and M_2, increasing all time parameters by a "large enough" integral factor a, adding a unit-time operation on M_1 to the beginning of each job, and setting the upper bound $au + n$. Then it is not hard to show that, for appropriately chosen a, a schedule for I with total completion time at most u exists if and only if there exists a schedule for I' with total completion time at most $au + n$. For example, we can take $a = p - 1$, where p is minimum among processing times of first operations of jobs in I.

The strong NP-hardness of $L3\|C_{\max}$ and $L3\,|\,pmtn\,|\,C_{\max}$ can be shown by similar reductions from $F3\|C_{\max}$ and $F3\,|\,pmtn\,|\,C_{\max}$ that are known to be strongly NP-hard problems.[1]

7.8.3 $C2\,|\,p_{ij} \in \{1,2\}\,|\,C_{\max}$

Since cycle shops have an equal number of operations in their jobs, the strong NP-hardness proof for this problem strengthens the strong NP-hardness result for the job-shop counterpart $J2\,|\,p_{ij} \in \{1,2\}\,|\,C_{\max}$. As in [32], we use a reduction from the 3-*partition problem* [30]: given positive integers $a_1, a_2, \ldots, a_{3t}, b$ with $\frac{b}{4} < a_j < \frac{b}{2}$ for all $j = 1, 2, \ldots, 3t$ and $\sum_{j=1}^{3t} a_j = bt$ does there exist a partition of $\{1, 2, \ldots, 3t\}$ into t three-element blocks $B_1, B_2, \ldots, B_t$ such that $\sum_{j \in B_k} a_j = b$ for every $k = 1, 2, \ldots, t$?

It will be convenient to specify schedules as strings in the six-symbol alphabet

$$\left\{ {}^{1}, {}^{2}, {}_{1}, {}_{2}, {}^{1}_{1}, {}^{2}_{2} \right\}$$

where the symbols present single or two parallel operations with processing times 1 or 2 on M_1 and M_2. For any string s and positive integer k, we assume that

$$[s]^k = s\,[s]^{k-1}, \quad \text{where} \quad [s]^1 = [s] = s, \quad [s]^0 = \circ$$

As before, $\circ$ denotes the empty symbol. See Figure 7.12 for an example. Jobs will be specified by their no-wait schedules.

Let I be an instance of the 3-partition problem, and let us create the instance I' of $C2\,|\,p_{ij} \in \{1,2\}\,|\,C_{\max}$ with $n = 3t + 1$ jobs, each of which has $\ell = 2bt + 12t + 2$ operations:

$$J_j = \left[{}^{2}_{\;\;1}\right]\left[{}^{2}_{\;\;2}\right]^{a_j}\left[{}^{1}_{\;\;2}\right]\left[{}^{1}_{\;\;1}\right]^{bt+6t-a_j-1}, \quad j = 1, 2, \ldots, 3t$$

$$J_{3t+1} = \left[{}^{1}\right]\left[\left[{}_{2}{}^{1}\right]\left[{}_{2}\right]^{3}\left[{}^{2}_{\;\;1}\right]^{b}\left[{}_{1}{}^{2}\right]^{3}\right]^{t}\left[{}_{1}\right]$$

---

[1] The original proofs of the strong NP-hardness of the three-machine flow shops (cf. [45,47,48]) use operations with zero processing times. Thus, the proofs establish the strong NP-hardness of the three-machine flow shops with missing operations indeed. However, as observed in [31], the three-machine flow shops with positive processing times remain strongly NP-hard. Related discussions can be found in [49].

and upper bound $u = 3bt^2 + 18t^2 + 3bt + 15t + 2$ on the schedule length. In J_j we distinguish the *head* and the *tail*, which are, respectively,

$$H_j = \begin{bmatrix} 2 & \\ & 1 \end{bmatrix} \begin{bmatrix} 2 & \\ & 2 \end{bmatrix}^{a_j} \begin{bmatrix} 1 & \\ & 2 \end{bmatrix} \quad \text{and} \quad T_j = \begin{bmatrix} 1 & \\ & 1 \end{bmatrix}^{bt+6t-a_j-1}$$

If I has the positive answer, then there exists a schedule

$$S = \begin{bmatrix} \mathbf{1} \\ \end{bmatrix} \left[\begin{bmatrix} 2 & \mathbf{1} \\ \mathbf{2} & 1 \end{bmatrix}^3 \begin{bmatrix} 2 & \mathbf{2} \\ \mathbf{2} & 2 \end{bmatrix}^b \begin{bmatrix} 1 & \mathbf{2} \\ \mathbf{1} & 2 \end{bmatrix}^3 \right]^t \begin{bmatrix} 1 \\ \mathbf{1} \end{bmatrix} \begin{bmatrix} 1 \\ 1 \end{bmatrix}^{3bt^2+18t^2-bt-3t-1} \begin{bmatrix} \\ 1 \end{bmatrix}$$

for I' with length u, where J_{3t+1} (shown in bold) is processed without interruption. In S we also distinguish the *head* and the *tail*, which are, respectively,

$$H = \begin{bmatrix} \mathbf{1} \\ \end{bmatrix} \left[\begin{bmatrix} 2 & \mathbf{1} \\ \mathbf{2} & 1 \end{bmatrix}^3 \begin{bmatrix} 2 & \mathbf{2} \\ \mathbf{2} & 2 \end{bmatrix}^b \begin{bmatrix} 1 & \mathbf{2} \\ \mathbf{1} & 2 \end{bmatrix}^3 \right]^t \begin{bmatrix} \\ \mathbf{1} \end{bmatrix} \quad \text{and} \quad T = \begin{bmatrix} 1 \\ \end{bmatrix} \begin{bmatrix} 1 \\ 1 \end{bmatrix}^{3bt^2+18t^2-bt-3t-1} \begin{bmatrix} \\ 1 \end{bmatrix}$$

For every $k = 1, 2, \ldots, t$, the three $\{H_j : j \in B_k\}$ is appropriately inserted in time interval $[1 + (4b + 18)(k - 1), 1 + (4b + 18)k]$ in H. This is possible because $\sum_{j \in B_k} a_j = b$. Obviously, H can be specified in time $O(t)$.

The tails T_j, $j = 1, 2, \ldots, 3t$, generate an instance I^* of $J2 \mid p_{ij} = 1 \mid C_{\max}$ with jobs of even length, where each job starts on M_1 and finishes on M_2. The tail T is an optimal schedule for I^* with length $3bt^2 + 18t^2 - bt - 3t + 1$. To ensure that such a schedule exists, we only need to note that since $\frac{b}{4} < a_j$, the length of the longest job in I^* is shorter than half the total length of jobs in I^*, i.e.,

$$\begin{aligned} \max_{1 \le j \le 3t} \{2bt + 12t - 2a_j - 2\} &< 2bt + 12t - \frac{b}{2} - 2 \\ &< \frac{\sum_{j=1}^{3t} (2bt + 12t - 2a_j - 2)}{2} = 3bt^2 + 18t^2 - bt - 3t \end{aligned}$$

for $b > 0$ and $t > 0$. As was shown in [50] (see also [5]), this condition means that there exists an optimal schedule for I^* with at most one interrupted job, no interior idle time units and, therefore, of length $3bt^2 + 18t^2 - bt - 3t + 1$. Besides, $J2 \mid p_{ij} = 1 \mid C_{\max}$ can be solved in time $O(n)$ [50], and therefore, T can be specified in time $O(t)$. Thus, S can be specified in polynomial time.

Now let us show that the existence of a schedule S of length u for I' implies a positive answer for I. Note that S has only two idle time units that are in time intervals $[0, 1]$ and $[u - 1, u]$ on M_2 and M_1, respectively. The first operation on M_1 in S is $O_{1,3t+1}$. Otherwise, S is of length more than u because the first operations of the other jobs are of length 2, and the total processing time of all operations on M_2 is $u - 1$. The first operation on M_2 in S can only be $O_{2,3t+1}$ that starts at time 1. The second operation on M_1 can only be O_{1j_1} with $1 \le j_1 \le t$. The third operation on M_1 can only be $O_{3,3t+1}$; otherwise it is O_{1j_2} with $1 \le j_2 \le t$ and $j_1 \ne j_2$. But the processing time of O_{1j_2} is 2, then we would have an idle time unit in $[3, 5]$. The second operation on M_2 can only be O_{2j_1}. Continuing the same reasoning, we can conclude that the first seven operations on M_1 and the first six operations on M_2 are scheduled in S, as shown in Figure 7.13.

By time 10, only four jobs, $J_{3t+1}, J_{j_1}, J_{j_2}, J_{j_3}$, appear in S, where $j_1, j_2, j_3 \in \{1, 2, \ldots, t\}$ and $j_1 \ne j_2 \ne j_3 \ne j_1$. Only operations of these jobs can appear in S in the time interval $[10, 4b + 10]$ as shown in

M_1	$3t+1$	j_1	$3t+1$	j_2	$3t+1$	j_3	$3t+1$
M_2		$3t+1$	j_1	$3t+1$	j_2	$3t+1$	j_3

FIGURE 7.13 The schedule $[^1][^{21}_{21}]^3$ of the first seven operations of J_{3t+1} and the first two operations of $J_{j_1}, J_{j_2}, J_{j_3}$ in the time interval $[0, 10]$ of the schedule S.

M_1	j_1	$3t+1$	...		j_3	$3t+1$
M_2	$3t+1$	j_1		...	$3t+1$	j_3

FIGURE 7.14 The schedule $[^{22}_{22}]^b$ of the $8, 9, \ldots, (2b+7)$th operations of J_{3t+1} and the $3, 4, \ldots, (a_j+2)$th operations of $J_{j_1}, J_{j_2}, J_{j_3}$ in the time interval $[10, 4b+10]$ of the schedule S.

M_1	j_1	$3t+1$		j_2	$3t+1$		j_3	$3t+1$
M_2	$3t+1$	j_1		$3t+1$	j_2		$3t+1$	j_3

FIGURE 7.15 The schedule $[^{12}_{12}]^3$ of the $2b+8, 2b+9, \ldots, 2b+13$th operations of J_{3t+1} and the last two operations of $J_{j_1}, J_{j_2}, J_{j_3}$ in the time interval $[4b+10, 4b+19]$ of the schedule S.

Figure 7.14; otherwise a fifth job would start in this interval, and hence, after the completion of the second operation of this fifth job, S would contain an idle time unit on M_2. Hence, $a_{j_1} + a_{j_2} + a_{j_3} = b$.

In the time interval $[4b+10, 4b+19]$, the jobs $J_{j_1}, J_{j_2}, J_{j_3}$ must be completed as shown in Figure 7.15; otherwise at least one time unit on M_2 would be idle. Thus, the time interval $[1, 4b+19]$ in S defines the first three $B_1 = \{j_1, j_2, j_3\}$. Proceeding by induction, we can determine that the time interval $[1 + (4b + 18)(k-1), 1 + (4b+18)k]$ in S defines the kth three B_k for every $k = 1, 2, \ldots, t$. Thus, the existence of a schedule S of length u for I' implies a positive answer for I.

7.8.4 $R\tilde{F} \mid n = 1, p_o = q_{\mu_o}, p_e = 1 \mid C_{\max}$, $1\tilde{C} \mid n = 1, p_i = 1 \mid C_{\max}$

The robotic flow shop $R\tilde{F} \mid n = 1, p_o = q_{\mu_o}, p_e = 1 \mid C_{\max}$ is proved to be NP-hard in [23] by a reduction from the 3-*satisfiability problem* [30]. The reduction is rather cumbersome, so we refer the reader to [23] for details. As we have showed in Section 7.5, the robotic flow shop reduces to $1\tilde{C} \mid n = 1, p_i = 1 \mid C_{\max}$. Therefore, the latter problem is also NP-hard.

7.9 Approximations for Intractable Cycle Shops

7.9.1 $L2\|C_{\max}$, $L2 \mid pmtn \mid C_{\max}$

$L2\|C_{\max}$ has an elegant $O(n \log n)$ approximation algorithm [15,16] with performance ratio 3/2 that runs twice the Johnson algorithm [51] for $F2\|C_{\max}$.

To present the result we need to observe that, as well as $F2\|C_{\max}$ and $F3\|C_{\max}$ [52], $L2\|C_{\max}$ also has an optimal *permutation schedule*, i.e., the sets of the first operations, the second operations and the third operations of jobs are processed on M_1, M_2, and M_1 in the same sequence, respectively. Further, there are optimal schedules where M_1 starts processing third operations only after finishing all the first operations. These two facts can be easily verified by contradiction. The problem therefore reduces to finding a permutation $J_{j_1} J_{j_2} \ldots J_{j_n}$ of given n jobs and scheduling as soon as possible successively

$$\begin{array}{llll} O_{1j_1}, O_{1j_2}, \ldots, O_{1j_n} & \text{on} & M_1, & \text{then} \\ O_{2j_1}, O_{2j_2}, \ldots, O_{2j_n} & \text{on} & M_2, & \text{and then} \\ O_{3j_1}, O_{3j_2}, \ldots, O_{3j_n} & \text{on} & M_1, & \text{after the completion of } O_{1j_n} \end{array}$$

Let I, I^{12}, and I^{23} be an instance of $L2\|C_{\max}$ and the two instances of $F2\|C_{\max}$ that are created by deleting the third operations or the first operations of jobs in I, respectively, let $C^*_{\max}$, $C^{12}_{\max}$, and $C^{23}_{\max}$ denote the minimum schedule length for I, I^{12}, and I^{23}, respectively, and let $C^{JJ}_{\max}$ denote the length of the shorter schedule between the two schedules for I with the Johnson permutations for I^{12} and I^{23}. Then we have

$$C^{JJ}_{\max} \leq C^{12}_{\max} + \sum_{j=1}^{n} p_{3j}, \qquad C^{JJ}_{\max} \leq C^{23}_{\max} + \sum_{j=1}^{n} p_{1j}$$

Since $C^{12}_{\max}$ and $C^{23}_{\max}$ are obvious lower bounds for $C^{*}_{\max}$, we have

$$C^{JJ}_{\max} \leq C^{*}_{\max} + \sum_{j=1}^{n} p_{3j}, \qquad C^{JJ}_{\max} \leq C^{*}_{\max} + \sum_{j=1}^{n} p_{1j}$$

Summing up the equations and using third obvious lower bound $\sum_{j=1}^{n} p_{1j} + \sum_{j=1}^{n} p_{3j}$ we eventually have

$$2C^{JJ}_{\max} \leq 2C^{*}_{\max} + \sum_{j=1}^{n} p_{1j} + \sum_{j=1}^{n} p_{3j} \leq 3C^{*}_{\max}$$

A better $O(n \log n)$ approximation algorithm has been found [53] with performance ratio 4/3 that creates 12 schedules and chooses the best among them. The proof of this bound is rather long. Note that $J2\|C_{\max}$ is known to be solved in polynomial time only with a performance ratio more than two [54].

$L2 \mid pmtn \mid C_{\max}$ can also be solved in polynomial time with performance ratio 4/3 because there is an $O(n^2)$ approximation algorithm with this ratio even for $J2 \mid pmtn \mid C_{\max}$ [55]. The algorithm strengthens a similar result with performance ratio 3/2 [56].

7.9.2 $\tilde{C} \mid n = 1, p_i = 1 \mid C_{\max}$, $C \mid n = 1, p_i = 1 \mid P{*}C_{\max}$

The cycle shop $\tilde{C} \mid n = 1, p_i = 1 \mid C_{\max}$ can be easily solved in polynomial time with absolute error at most $\frac{\ell}{2}(\nu^2 - \nu)$ as follows. Let $x_i, i = 1, 2, \ldots, \ell$, denote the start time of O_i. Set $x_1 = 0$. Then the problem is to find integers $x_2, x_3, \ldots, x_\ell$ with minimum x_ℓ such that for every $k = 2, 3, \ldots, m$ integers x_i with $\mu_i = k$ are all different modulo P, and $x_i \geq x_{i-1} + 1$ for $i = 2, 3, \ldots, \ell$. Obviously, the problem has a solution if and only if $P \geq \nu = \max_{1 \leq k \leq m} \nu_k$.

The *Nearest-Window-First* (NWF) algorithm [1,4,31] maintains a current set R_k of residuals modulo P engaged by start times on M_k for every $k = 1, 2, \ldots, m$. After assigning all start times, R_k will contain ν_k residuals modulo P. The NWF algorithm initially sets $x_0 = -1$ and $R_k = \emptyset$ for $k = 1, 2, \ldots, m$. At the ith step, $i = 1, 2, \ldots, \ell$, the algorithm "moves" O_i to the "nearest window", i.e., it sets x_i to be the smallest integer

$$x_i \geq x_{i-1} + 1, \quad \text{where } x_i \bmod P \notin R_{\mu_i}$$

and puts $x_i \bmod P$ in R_{μ_i}. The time complexity of the algorithm is clearly $O(\nu\ell)$.

Taking into account that in the worst case R_k is an interval of integers, we can conclude that the total lost in flow time is

$$\begin{aligned}\sum_{i=1}^{\ell}(x_i - x_{i-1} - 1) &\leq \sum_{k=1}^{m}(1 + 2 + \cdots + \nu_k - 1)\\ &= \sum_{\nu_k > 1}\frac{1}{2}(\nu_k^2 - \nu_k) \leq \frac{\rho}{2}(\nu^2 - \nu)\end{aligned}$$

Thus, $1\tilde{C} \mid n = 1, p_i = 1 \mid C_{\max}$ can be solved in polynomial time with absolute error at most $\frac{1}{2}(\nu^2 - \nu)$.

A more sophisticated $O(\nu^2)$ algorithm for the latter problem with absolute error at most $\frac{1}{2}(\nu^2 - \frac{1}{14}\nu^{\frac{3}{2}})$ has been proposed in [23].

Since the NWF algorithm can find a periodic schedule with period $P = \nu$ and periodic schedules do not exist if $P < \nu$, the algorithm also finds an optimal solution to $C \mid n = 1, p_i = 1 \mid P$ and an approximate solution to $C \mid n = 1, p_i = 1 \mid P * C_{\max}$ with the same absolute error bounds as those for $\tilde{C} \mid n = 1, p_i = 1 \mid C_{\max}$.

7.9.3 $\tilde{C} \mid n = 1, p_i = q_{\mu_i} \mid C_{\max}$, $C \mid n = 1, p_i = q_{\mu_i} \mid P{*}C_{\max}$, $C \mid p_i = q_{\mu_i} \mid C_{\max}$

Obviously, $\tilde{C} \mid n = 1, p_i = q_{\mu_i} \mid C_{\max}$ has a solution if and only if $P \geq Q = \max_{1 \leq k \leq m} \nu_k q_k$. As before, let $x_i, i = 1, 2, \ldots, \ell$, denote the start time of O_i. Set $x_1 = 0$. Then the problem is to find $x_2, x_3, \ldots, x_\ell$ with minimum x_ℓ such that

$$i > 1 \quad \Rightarrow \quad x_i - x_{i-1} \geq q_{\mu_{i-1}} \tag{7.6}$$

$$\mu_i = \mu_j = k \quad \text{and} \quad i > j \quad \Rightarrow \quad (x_i - x_j) \bmod P \geq q_k \tag{7.7}$$

To apply the NWF algorithm to the problem, let us define an *extended multiplicity* and an *extended machine processing time* of M_k to be

$$V_k = \left\lfloor \frac{P}{q_k} \right\rfloor \quad \text{and} \quad Q_k = \frac{P}{V_k}$$

respectively. The NWF algorithm maintains a current set R_k of residuals modulo V_k engaged by start times on M_k for every $k = 1, 2, \ldots, m$. After assigning all start times, R_k will contain ν_k residuals modulo V_k. The NWF algorithm initially sets $x_0 = \mu_0 = q_0 = 0$ and $R_k = \emptyset$ for $k = 1, 2, \ldots, m$. At the ith step, $i = 1, 2, \ldots, \ell$, the algorithm checks whether O_i is the first operation on M_{μ_i}. If it is, the algorithm sets

$$x_i = x_{i-1} + q_{\mu_{i-1}}$$

and memorizes engagement time $y_{\mu_i} = x_i$ of M_{μ_i}. If it is not, then the algorithm "moves" O_i to the "nearest window," i.e., it sets x_i to be the minimum number such that

$$x_i - x_{i-1} \geq q_{\mu_{i-1}}$$

$$Q_{\mu_i}(z_i - 1) \quad \leq \quad x_i - y_{\mu_i} \leq Q_{\mu_i} z_i - q_{\mu_i}, \text{ where } z_i \notin R_{\mu_i} \tag{7.8}$$

and puts $z_i \bmod V_{\mu_i}$ in R_{μ_i}. The time complexity is obviously $O(\nu\ell)$. Figure 7.16 gives an example with $P = Q$.

It is clear that the algorithm satisfies Condition (7.6). It is easy to verify that Condition (7.7) is also satisfied because the fact that for any $k = 1, 2, \ldots, m$ we have $P = Q_k V_k$ and the set R_k contains ν_k residuals modulo V_k implies that, in accordance with Condition (7.8), O_i and O_j appear in different intervals of length $Q_k \geq q_k$ modulo P if $\mu_i = \mu_j = k$ and $i > j$. And, since Condition (7.8) has two inequalities, M_k contributes maximum lost $2Q_k(1 + 2 + \cdots + \nu_k - 1)$ into the absolute error bound

$$\sum_{\nu_k > 1} Q_k(\nu_k^2 - \nu_k) \leq \rho(\nu^2 - \nu) q_{\max} \tag{7.9}$$

Since the NWF algorithm can find a periodic schedule with period $P = Q$ and periodic schedules do not exist if $P < Q$, it also finds an optimal solution to $C \mid n = 1, p_i = q_{\mu_i} \mid P$ and an approximate solution to $C \mid n = 1, p_i = q_{\mu_i} \mid P * C_{\max}$ with the same absolute error bound as that for $\tilde{C} \mid n = 1, p_i = q_{\mu_i} \mid C_{\max}$.

Assigning Residuals to Time Windows

R_1	0			0			0			0	1			0	1	2		0	1	2	3	0	
R_2		0			0			0	1		0		1		0		1		0		1		

Schedule for Operations

M_1	1		2		3		4		5		6		7
M_2	*1*		*2*		*3*		*4*		*5*		*6*		
M_3		1		2		3		4		5		6	
M_2		*1*		*2*		*3*		*4*		*5*			
M_4			1		2		3		4				
M_1				1		2		3					
M_5				1		2		3					
M_1					1		2						
M_6					1		2						
M_1						1							

Schedule for Machines

M_1	1		2		3		4	1		5	2	1		6	3	2	1	7
M_2		*1*		2	*1*	3		2	4	*3*	5		4	6		5		
M_3			1		2			3			4			5			6	
M_4					1			2			3			4				
M_5						1			2			3						
M_6							1				2							

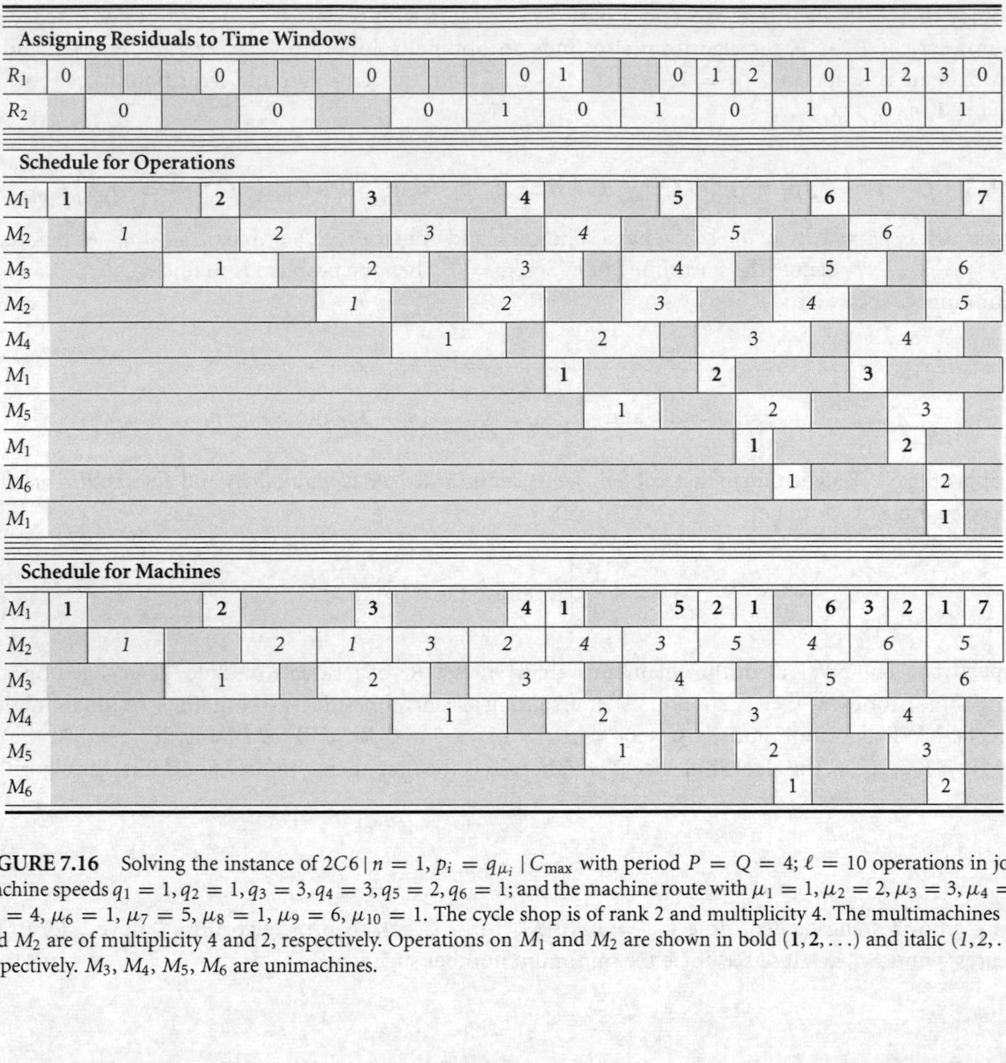

FIGURE 7.16 Solving the instance of $2C6\,|\,n=1,\,p_i=q_{\mu_i}\,|\,C_{\max}$ with period $P=Q=4$; $\ell=10$ operations in jobs; machine speeds $q_1=1,q_2=1,q_3=3,q_4=3,q_5=2,q_6=1$; and the machine route with $\mu_1=1,\mu_2=2,\mu_3=3,\mu_4=2$, $\mu_5=4,\mu_6=1,\mu_7=5,\mu_8=1,\mu_9=6,\mu_{10}=1$. The cycle shop is of rank 2 and multiplicity 4. The multimachines M_1 and M_2 are of multiplicity 4 and 2, respectively. Operations on M_1 and M_2 are shown in bold ($\mathbf{1,2,\ldots}$) and italic ($1,2,\ldots$), respectively. M_3, M_4, M_5, M_6 are unimachines.

It is also evident that the breeding of an NWF schedule n times with period Q produces a schedule for the related instance of $C\,|\,p_i=q_{\mu_i}\,|\,C_{\max}$. Absolute error bound (7.9) remains the same because the breeding minimally increases the schedule length by $Q(n-1)$. More than that, the following performance ratio, where C_{\max}^{NWF} and C_{\max}^* denote the length of an NWF schedule and the minimum length, respectively, shows that the breeding reaches an asymptotically optimal solution on the number of jobs:

$$
\begin{aligned}
\frac{C_{\max}^{\text{NWF}}}{C_{\max}^*} &\leq \frac{\sum_{k=1}^{m} q_k v_k + Q(n-1) + \sum_{v_k>1} Q_k(v_k^2 - v_k)}{\sum_{k=1}^{m} q_k v_k + Q(n-1)} = 1 + \frac{\sum_{v_k>1} Q_k(v_k^2 - v_k)}{\sum_{k=1}^{m} q_k v_k + Q(n-1)} \\
&\leq 1 + \frac{(v-1)\sum_{v_k>1} Q_k v_k}{\sum_{k=1}^{m} q_k v_k + Q(n-1)} \leq 1 + \frac{Q\rho(v-1)}{\left(\sum_{k=1}^{m} q_k v_k - Q\right) + Qn} \\
&\leq 1 + \frac{\rho(v-1)}{n}
\end{aligned}
$$

Since ρ and v are not dependent on n, where $\rho \leq m$ and $v \leq \ell/2$, we have an unlimited approach to the optimum while unlimited increasing of n.

7.10 Conclusion

In addition to the open cycle shop scheduling problems mentioned above, we can indicate a few other intriguing problems.

As we have shown in Section 7.7, $C2 \mid chains, p_{ij} = 1 \mid C_{\max}$ can be solved in pseudopolynomial time. The special case of the problem with even ℓ, i.e., C per $2 \mid chains, p_{ij} = 1 \mid C_{\max}$, can be solved in polynomial time by job shop scheduling algorithms for $J2 \mid p_{ij} = 1 \mid C_{\max}$ [50] (see also [33,34]). Therefore, the obstacle is the case with odd ℓ. The cycle shop $C2 \mid chains, p_{ij} = 1 \mid \sum C_j$ is also open. It is unknown, however, whether C per $2 \mid chains, p_{ij} = 1 \mid \sum C_j$, can be solved in polynomial time by job shop scheduling algorithms for $J2 \mid p_{ij} = 1 \mid \sum C_j$ [37,38]. It is important to point out that the size of these problems is the number of chains. Hence, algorithms that are polynomial in the number of jobs are of exponential time.

Another interesting open problem is $C \mid p_{ij} = q_{\mu_i} \mid C_{\max}$. As we have shown in Section 7.9, a finitely periodic schedule for this problem that is asymptotically optimal on the number of jobs can be found in polynomial time. However, finding an optimal infinitely periodic schedule even for $1\tilde{C} \mid n = 1, p_i = 1 \mid C_{\max}$ is a strongly NP-hard problem [23]. It is also interesting to characterize the set of instances of $C \mid p_{ij} = q_{\mu_i} \mid C_{\max}$ that have optimal finitely periodic schedules. Note that all instances of $C \mid p_{ij} = 1 \mid C_{\max}$ have optimal finitely periodic schedules (see Section 7.7), but there are instances of $C2 \mid p_{ij} = q_{\mu_i} \mid C_{\max}$ that do not have optimal finitely periodic schedules [31]. Other open cycle shop scheduling problems can be found in [23,31].

Performance guarantee approximations for cycle shop scheduling problems, especially periodic problems, are not well studied. In all likelihood, this chapter includes all known results in this area.

References

[1] Yu. I. Degtiarev and V. G. Timkovsky. On a model of optimal planning systems of flow type. In *Technique of Communication Means*, number 1 in Automation Control Systems, pages 69–77. Ministry of Communication Means of the USSR, Moscow, 1976. (in Russian)

[2] V. G. Timkovsky. On transition processes in systems of flow type. In *Technique of Communication Means*, number 1(3) in Automation Control Systems, pages 46–49. Ministry of Communication Means of the USSR, Moscow, 1977. (in Russian)

[3] V. G. Timkovsky. An approximation for the cycle-shop scheduling problem. *Economics and Mathematical Methods*, 22(1):171–174, 1986. (in Russian)

[4] V. G. Timkovsky. *Discrete Mathematics in the World of Machines and Parts: Introduction to Mathematical Modelling for Discrete Manufacturing Problems*. Nauka, Moscow, 1992. (in Russian).

[5] V. G. Timkovsky. Is a unit-time job shop not easier than identical parallel machines? *Discrete Applied Mathematics*, 85:149–162, 1998.

[6] E. M. Livshits, Z. N. Mikhailetsky, and E. V. Chervyakov. A scheduling problem in automated flow line with an automated operator. *Computational Mathematics and Computerized Systems*, 5:151–155, 1974. (in Russian)

[7] S. C. Graves, H. C. Meal, D. Stefek, and A. H. Zeghmi. Scheduling of reentrant flow shops. *Journal of Operations Management*, 3(4):197–207, 1983.

[8] T. E. Morton and D. W. Pentico. *Heuristic Scheduling Systems*. John Wiley, New York, 1993.

[9] V. Lev and I. Adiri. V-shop scheduling. *European Journal of Operational Research*, 18(1):51–56, 1984.

[10] C. R. Asfahl. *Robots and Manufacturing Automation*. John Wiley, New York, 1985.

[11] P. Serafini and W. Ukovich. A mathematical model for periodic scheduling problems. *SIAM Journal on Discrete Mathematics*, 2:550–581, 1989.

[12] R. Ramazini and N. Younis. Repetitive pure flowshop problem: a permutation approach. *Computers and Industrial Engineering*, 24(1):125–129, 1993.

[13] R. Roundy. Cyclic schedules for job shops with identical jobs. *Mathematics of Operations Research*, 17:842–865, 1992.

[14] S. P. Sethi, C. Sriskandarajah, G. Sorger, J. Blazewicz, and W. Kubiak. Sequencing of parts and robot moves in a robotic cell. *International Journal of FMS*, 4:331–358, 1992.

[15] J. N. D. Gupta. Two-stage reentrant flowshop problem with repeated processing at the first stage. Technical report, Department of Management, Ball State University, Muncie, IN, 1993.

[16] M. Y. Wang, S. P. Sethi, and S. L. van de Velde. Minimizing makespan in a class of reentrant shops. *Operations Research*, 4:702–712, 1997.

[17] V. Kats and E. Levner. A strongly polynomial algorithm for no-wait cyclic robotic flowshop scheduling. *Operations Research Letters*, 21:159–164, 1997.

[18] J. Hurink and S. Knust. Flow-shop problems with transportation times and a single robot. *Discrete Applied Mathematics*, 112:199–216, 2001.

[19] Y. Crama, V. Kats, J. van de Klundert, and E. Levner. Cyclic scheduling in robotic flowshops. *Annals of Operations Research*, 96:97–124, 2000.

[20] P. Serafini and W. Ukovich. A mathematical model for the fixed-time traffic control problem. *European Journal of Operational Research*, 41:1–14, 1989.

[21] N. G. Hall, H. Kamoun, and C. Sriskandarajah. Scheduling in robotic cells: classification, two and three machine cells. *Operations Research*, 45:421–439, 1997.

[22] I. N. K. Abadi, N. G. Hall, and C. Sriskandarajah. Minimizing cycle time in a blocking flowshop. *Operations Research*, 48:177–180, 2000.

[23] M. Middendorf and V. G. Timkovsky. On scheduling cycle shops: classification, complexity and approximation. *Journal of Scheduling*, 5:135–169, 2002.

[24] A. Che, C. Chu, and E. Levner. A polynomial algorithm for 2-degree cyclic robot scheduling. *European Journal of Operational Research*, 145(1):31–44, February 2003.

[25] C. Hannen and A. Munier. A study of the cyclic scheduling problem on parallel processors. *Discrete Applied Mathematics*, 57:167–192, 1995.

[26] A. Munier. The complexity of a cyclic scheduling problem with identical machines and precedence constraints. *European Journal of Operational Research*, 91:471–480, 1996.

[27] C. Hannen. Study of an NP-hard cyclic scheduling problem: the periodic recurrent job shop. *European Journal of Operational Research*, 72:82–101, 1994.

[28] J. Carlier and Ph. Chrétienne. On the dominance of k-periodic schedules in cyclic scheduling. In *Proc. 2nd International Workshop on Project Management and Scheduling*, pages 181–187. Compiègne, France, 1990.

[29] P. M. Kogge. *The Architecture of Pipelined Computers*. McGraw-Hill, New York, 1981.

[30] M. R. Garey and D. S. Johnson. *Computers and Intractability: A Guide to the Theory of NP-Completeness*. Freeman, San Francisco, 1979.

[31] V. G. Timkovsky. On the complexity of scheduling an arbitrary system. *Soviet Journal of Computer and System Sciences*, (5):46–52, 1985.

[32] J. K. Lenstra and A. H. G. Rinnooy Kan. Computational complexity of discrete optimization problems. In *Annals of Discrete Mathematics*, volume 4, pages 121–140. North Holland, 1979.

[33] V. G. Timkovsky. A polynomial-time algorithm for the two-machine unit-time release-date job-shop schedule-length problem. *Discrete Applied Mathematics*, 77:185–200, 1997.

[34] V. G. Timkovsky. How to make Brucker's algorithm polynomial: scheduling by periodization. *International Journal of Mathematical Algorithms*, 2:325–360, 2001.

[35] S. A. Kravchenko. Minimizing the number of late jobs for the two-machine unit-time job-shop scheduling problem. *Discrete Applied Mathematics*, 98(3):209–217, 1999.

[36] P. Brucker and S. Knust. Complexity results for single-machine problems with positive finish-start time lags. *Computing*, 63:299–316, 1999.

[37] S. A. Kravchenko. Minimizing total completion time in two-machine job shops with unit processing times. Technical Report 4, Institute of Technical Cybernetics, Minsk, Belorus, 1994. (in Russian).

[38] W. Kubiak and V. G. Timkovsky. A polynomial-time algorithm for total completion time minimization in two-machine job-shop with unit-time operations. *European Journal of Operational Research*, 94:310–320, 1996.

[39] D. Gusfield. Design (with analysis) of efficient algorithms. In E. G. Coffman, Jr., J. K. Lenstra, and A. H. G. Rinnooy Kan (editors), *Computing*, volume 3 of *Handbooks in Operations Research and Management Science*, chapter 8, pages 375–453. Elsevier Science Publishers B. V., North-Holland, Amsterdam, 1992.

[40] C. Sriskandarajah and P. Ladet. Some no-wait job scheduling problems: complexity results. *European Journal of Operational Research*, 24:424–438, 1986.

[41] S. A. Kravchenko. A polynomial time algorithm for a two-machine no-wait job-shop scheduling problem. *European Journal of Operational Research*, 106:101–107, 1998.

[42] J. Bruno, J. W. Jones, and K. So. Deterministic scheduling with pipelined processors. *IEEE Transactions on Computing*, C-29:308–316, 1980.

[43] C. Hannen and A. Munier. Cyclic scheduling problems: an overview. In Ph. Chrétienne, E. G. Coffman, Jr., J. K. Lenstra, and Z. Liu (editors), *Scheduling Theory and Applications*, chapter 4. John Wiley, New York, 1996.

[44] A. Munier. Résolution d'un problème d'ordonnancement cyclique à itérations indépendantes et contraintes de ressources. *RAIRO Recherche Opérationnelle*, 25:161–182, 1991.

[45] M. R. Garey, D. S. Johnson, and R. Sethi. The complexity of flowshop and jobshop scheduling. *Mathematics of Operations Research*, 1:117–129, 1976.

[46] J. Du and J.Y.-T. Leung. Minimizing mean flow time in two-machine open shops and flow shops. *Journal of Algorithms*, 14:24–44, 1993.

[47] T. Gonzalez and S. Sahni. Flow shop and job shop schedules: complexity and approximation. *Operations Research*, 26:36–52, 1978.

[48] Y. Cho and S. Sahni. Preemptive scheduling of independent jobs with release and due times on open, flow and job shops. *Operations Research*, 29:511–522, 1981.

[49] V. S. Tanaev, Yu. N. Sotskov, and V. A. Strusevich. *Scheduling Theory: Multi-Stage Systems*. Nauka, Moscow, 1989. (in Russian). English translation: Kluwer, Dordrecht, 1994.

[50] V. G. Timkovsky. Polynomial-time algorithm for the Lenstra–Rinnooy Kan two-machine scheduling problem. *Kibernetika*, (2):109–111, 1985. (in Russian).

[51] S. M. Johnson. Optimal two- and three-machine production schedules with set up times included. *Naval Research Logistics Quarterly*, 1:61–68, 1954.

[52] R. W. Conway, W. L. Maxwell, and L. W. Miller. *Theory of Scheduling*. Addison-Wesley, Reading, Massachusetts, 1967.

[53] I. G. Drobouchevitch and V. A. Strusevich. A heuristic algorithm for two-machine reentrant shop scheduling. *Annals of Operations Research*, 86:417–439, 1999.

[54] D. B. Shmoys, C. Stein, and J. Wein. Improved approximation algorithms for shop scheduling problems. *SIAM Journal on Computing*, 23:617–632, 1994.

[55] J. Han, J. Wen, and G. Zhang. A new approximation algorithm for UET-scheduling with chain-type precedence constraints. *Computers and Operations Research*, 25(9):767–771, 1998.

[56] S. V. Sevastianov and G. J. Woeginger. Makespan minimization in preemptive two machine job shops. *Computing*, 60:73–79, 1998.

8

Reducibility among Scheduling Classes

Vadim G. Timkovsky
CGI Group Inc. & McMaster University

8.1 Introduction

A *scheduling class* can be defined informally as a general scheduling problem that includes many special scheduling problems with a common property. For example, we can speak about the class of problems on parallel machines, where machines are interchangeable; the class of shop scheduling problems, where machines are specific; or the class of problems with precedence delays, where a precedence of jobs enforces waiting time between them. We say that a scheduling class A reduces to a scheduling class B if there exists a mapping of every problem a in A to a problem b in B such that a reduces to b. If A is a representative class in a certain sense, then we say that the mapping is a *mass reduction*. We consider a scheduling class to be representative if it includes problems with many criteria, precedence, and release-date constraints; and the mass reductions we discuss in this chapter, as a rule, do not change criteria or cancel the constraints.

In terms of the theory of categories and functors, the mass reductions are functors connecting scheduling categories. Thus, mass reductions convert studying problems in one class to using approaches known in another class. The complexity status of many scheduling problems has been established by polynomial or pseudopolynomial mass reductions, since they pass complexity results from one class to another. Thus, mass reductions have proven to be a powerful tool for studying the complexity of scheduling problems. Special interest has recently been paid to mass reductions that are equivalences or even isomorphisms, since they exhibit a similarity between scheduling problems that were originally treated as quite distinct.

Strusevich [1] (see also [2]) and, independently, Liu and Bulfin [3] were the first to observe that the simplest problems of scheduling jobs with processing times m and integer preemptions on m identical parallel machines (with equal release dates and without precedence constraints) are equivalent to

1-58488-397-9/$0.00+$1.50

m-machine unit-time no-no-wait open-shop scheduling problems. Bräsel, Kluge, and Werner [4] consider such an equivalence between problems with tree-like precedence constraints. Further, we call it a *preemptive no-no-wait* equivalence. Kubiak, Sriskandarajah, and Zaras [5] show that the *nonpreemptive no-wait* counterpart, which is even more transparent, also takes place. Brucker, Jurisch, and Jurisch [6] generalize these results and establish that both equivalences, no-no-wait preemptive and no-wait nonpreemptive, take place in the general case involving precedence constraints, release dates and all regular criteria. A key element of these equivalences is *latinization* of the parallel-machine schedules, i.e., their transformation into Latin rectangles by machine permutations in each time unit. Bräsel [7,8] was the first to notice the isomorphism between unit-time open-shop schedules and Latin rectangles.

There exists a reduction of all scheduling problems whose decision versions are in the class NP [9] to their *expansions* whose optimal schedules contain only divisible by m durations of uninterrupted parts of operations [10]. The expansions are trivially obtained by multiplying all time parameters by a large enough multiple of m that is bounded by a polynomial in the problem size. The latinization of such expanded schedules leads to mass reductions to unit-time open shops. Expanded schedules can then be converted into *modulated* schedules where every uninterrupted part of each operation starts at time $i - 1$ modulo m if the part is run on the ith machine. The key observation here is that modulated schedules are isomorphic to schedules in special unit-time job shops which we call *perfect*. For example, flow shops are the simplest perfect job shops. Thus, the expansion and the modulation lead to mass reductions to unit-time job shops and flow shops. For example, modulated schedules of jobs with processing times m on m parallel machines are isomorphic to m-machine unit-time flow-shop schedules.

Reductions from parallel machines to unit-time shops are called *shoppings*, while the inverse reductions, i.e., from unit-time shops to parallel machines, are called *parallelings*.

The following two mass reductions connecting the class of parallel-machine problems with the class of unit-time open-shop problems are not equivalences [11]. One of them, *open-shopping*, maps preemptive problems on identical parallel machines with arbitrary processing time jobs to unit-time open-shop problems. The other, *open-shop paralleling*, maps unit-time open-shop problems to nonpreemptive problems on identical parallel machines with unit-time jobs. Both reductions take place for all classical criteria except the total completion time. It is also shown that open shopping cannot be extended to this criterion unless $P = \mathrm{NP}$.

The relationship between parallel-machine problems and job-shop or flow-shop problems is less symmetrical since the latter problems, as has been observed through the history of scheduling theory, appear to be harder. The one-way mass reductions found from parallel-machine problems to shop scheduling problems prove this observation theoretically. The most general among them is *job-shopping* [10] that maps *preemptive/nonpreemptive* problems of scheduling jobs with arbitrary processing times on m identical parallel machines to m-machine *no-no-wait/no-wait* unit-time job-shop scheduling problems. The reduction takes place for all classical criteria and any precedence and release-date constraints. In the case of two machines and the maximum completion time criterion, inverse reductions have been found: Kubiak, Sethi, and Sriskandarajah [12] describe *no-no-wait preemptive* paralleling, and the result of Kubiak [13] can be interpreted as *no-wait nonpreemptive* paralleling. Job shopping of parallel-machine problems with equal-processing-time jobs is *cycle shopping*, i.e., a mass reduction to unit-time cycle shops [14]. Complemented by a chaining technique job shopping leads to flow shopping [15].

Brucker and Knust [15] consider *precedence delaying*, i.e., mass reductions of parallel-machine problems to single-machine problems with unit-time jobs under precedence delays. Bruno, Jones, and So [16], Schäffer and Simons [17], and Finta and Liu [18] consider the inverse reductions, i.e., *precedence-delay paralleling*.

Lawler [19] observes that, in the case of identical parallel machines, polynomial-time algorithms known for some nonpreemptive problems of scheduling jobs with unit processing times have counterparts for the related preemptive problems of scheduling jobs with arbitrary processing times. Such a relationship motivates *preemption chaining* [11] that changes preemptive jobs with arbitrary processing times into chains of nonpreemptive jobs with unit processing times in the same machine environment. As well as open shopping and flow shopping, preemption chaining is proved to be not extendible to the case of the total completion time criterion unless $P = \text{NP}$. In addition, preemption chaining is proved to be valid for identical parallel machines, uniform parallel machines and job shops, but not extendible to flow shops and open shops for the same reason. The proof uses the fact that the nonpreemptive problems of scheduling jobs with unit processing times and chain-like precedence constraints can be solved in polynomial time, while the related preemptive problems of scheduling jobs with arbitrary processing times are NP-hard.

Finding polynomial-time algorithms for problems with chain-like precedence constraints is complicated for the following reasons. Compact encoding schemes known for unit-time problems with treelike or general precedence constraints imply the problem size to be polynomial in the number of jobs. However, the restricted cases with chain-like precedence constraints, equal release dates, and the maximum or total completion time criterion allow its *own compact encoding*, where the problem size is polynomial in the number of chains. In application to such cases, the polynomial-time algorithms known for the problems with treelike or general precedence constraints run in exponential time because they remain polynomial in the number of jobs. The same observation can be made regarding extensions with different release dates, if the number of different release dates is polynomial in the number of chains. Thus, problems with chain-like precedence constraints require special consideration for finding polynomial-time algorithms for them. These algorithms have usually been tracked down.

The remainder of this chapter is devoted to a detailed discussion of the mass reductions mentioned above. We will follow the interior logic of the results rather than their historical sequence.

8.2 Preliminaries

8.2.1 Assumptions

8.2.1.1 Scheduling Data

All input data of problems in this chapter is considered to be integer. We call a schedule *divisible by a rational number* if the length of the time interval between any two events in it (start times, completion times, preemption times) is a multiple of the number. An *integer schedule* is a schedule divisible by 1. A *schedule of a fixed fractionality* is a schedule divisible by a fixed fraction. Unless stated otherwise, we consider further only integer schedules.

For any sequence of parameters $a_1, a_2, \ldots, a_n$, we set

$$a_{\min} = \min_{1 \le j \le n} a_j, \qquad a_{\max} = \max_{1 \le j \le n} a_j$$

8.2.1.2 Problem Size

Following a commonly accepted simplifying technique [9], we ignore logarithms of input data (replacing them by constants) in evaluations of problem sizes everywhere in the chapter except in Section 8.3.1, where a detailed consideration is necessary. For example, the size of the problem of finding a minimum number among integers $a_1, a_2, \ldots, a_{n^2+n}$ is $n^2 + n$ regardless of the appearance of $\log a_1 + \log a_2 + \cdots + \log a_{n^2+n}$ bits in the binary code of the input data.[1] Thus, any algorithm for the problem is polynomial if its time complexity has an upper bound polynomial in n only. If an upper bound is polynomial in n and at least one of the integers, say a_1, then the algorithm is pseudopolynomial.

A problem A is a *subproblem* or a *special case* of a problem B if the set of instances of A is a subset of the set of instances of B. It is important to note that, since a subproblem can have its own compact encoding scheme, the instances of the subproblem can have much shorter own codes than those for the original problem. A subproblem A of a problem B causes a *size collapse* if the size of B is not polynomially bounded by the size of A. Then a polynomial algorithm for B does not necessarily imply a polynomial algorithm

[1] Note that any polynomial in n can be assigned to be the problem size [9].

TABLE 8.1 Marking the Complexity Status of Problems.

Mark	Meaning
✓	Polynomially solvable
✓?	Pseudopolynomially solvable, but neither polynomial solvability nor NP-hardness is established
?	The complexity status is open
†?	NP-hard, but neither pseudopolynomial solvability nor strong NP-hardness is established
†	Ordinarily NP-hard, i.e., pseudopolynomially solvable and NP-hard
††	Strongly NP-hard

The bold marks **✓**, **†**, and **††** in further tables point to the complexity results that are originally obtained as examples of the mass reductions in the tables.

for A.[2] For example, finding a minimum number among integers $a_1, a_2, \ldots, a_{n^2+n}$, where $a_i = n - i$ for $i = 1, 2, \ldots, n^2 + n$, is a special case of the former problem. Since we know the formula for a_i, any instance of this special case can be specified by only one number n. Hence, the size of the special case is 1. Thus, an algorithm enumerating $n^2 + n$ numbers is not polynomial for it, but an algorithm that only outputs the number $-n^2$ is.

8.2.1.3 Denotations for Scheduling Classes and Problems

The denotation $\alpha \mid \beta \mid \gamma$ [22], where the parameters α, β, and γ specify machine environments, job characteristics, and minimization criteria, is understood as the universal (most general) scheduling class. If some or all of these parameters are specified, then the related denotation defines a particular scheduling class (subclass) or a problem, respectively. For example, $\alpha \mid \beta \mid C_{\max}$ is the class of scheduling problems with the maximum completion time criterion, $P3 \mid \beta \mid C_{\max}$ is the subclass whose machine environment comprises three identical parallel machines, while $P3 \mid \text{pmtn, prec}, p_j = 1 \mid C_{\max}$ is the problem in the subclass of finding a preemptive schedule of unit-time jobs with arbitrary precedence constraints. Thus, the absence or presence of Greek letters in the denotation means that a problem in a class is specified or unspecified, respectively. We assume that the decision versions of preemptive scheduling problems in the universal class belong to NP [9]. [3] It is easy to check that the decision versions of all nonpreemptive scheduling problems in this chapter are in NP. To indicate the complexity status of scheduling problems, we use the marking symbols listed in Table 8.1.

8.2.2 Special Machine Environments

Together with the classical machine environments $\alpha = \alpha_1\alpha_2$, where $\alpha_1 \in \{1, P, Q, O, F, J\}$ specifies the machine environment type among *single machine, identical parallel machines, uniform parallel machines, open shop, flow shop*, and *job shop*, respectively, and α_2 specifies the machine number,[4] we also consider the following special types.

[2] One may come across the erroneous opinion, ignoring the size collapse effect, that a polynomial algorithm for a problem implies polynomial algorithms for all its special cases. This implies $P = NP$ because there exist problems solvable in polynomial time that have NP-hard subproblems. For example, let us consider a set of $n!$ directed graphs $G_1, G_2, \ldots, G_{n!}$ each of which has n vertices. Using these graphs as components, Shafransky [20] constructs the garland by adding new vertices $u_0, u_1, \ldots, u_{n!}$ and arcs $u_{k-1} \to v_k$ and $v_k \to u_k$ for every vertex v_k in G_k, $k = 1, 2, \ldots, n!$. The problem is to find a Hamiltonian path in Shafransky's garland from u_0 to $u_{n!}$. Since the problem size is $n!$, it can be trivially solved in polynomial time $O(n!^2)$ by exhaustion of all $n!$ sequences of vertices in each component. The subproblem where all $n!$ components of the garland are identical has size n and is equivalent to finding a Hamiltonian path in a graph with n vertices. The latter problem is known to be NP-hard [21].

[3] In other words, we assume that the NP-preemption hypothesis [23] is true.

[4] See Chapter 1.

8.2.2.1 Modulated Machine Environment

The denotation $\alpha_1 \bmod \alpha_2$ represents a *modulated machine environment*, where the machines M_i start operations or their uninterrupted parts only at times $t = (i-1) \bmod m$ for all $i = 1, 2, \ldots, m$. Schedules in modulated machine environments are called *modulated schedules* or *schedules modulo m.*

8.2.2.2 Special Job Shops

The denotation J per α_2 presents a *perfect job shop*, where the operations O_{ij} are to be processed on the machines $M_{\mu_{ij}}$ with $\mu_{ij} - 1 = (i-1) \bmod m$. A *cycle shop*, $C\alpha_2$, is a job shop where $\mu_{ij} = \mu_i$. All jobs pass through the same machine route as in flow shops, but some of the machines can be repeated. A *perfect cycle shop*,[5] C per α_2, is a perfect job shop that is a cycle shop. An *open job shop*, $\mathcal{O}\alpha_2$, can be defined as the relaxation of a job shop where precedence relation between operations inside jobs is cancelled; note that a classical open shop can be called an *open flow shop* because it can also be defined as the same relaxation of a flow shop.

8.2.3 Special Job Characteristics

In this chapter, the job characteristics field β contains seven classical subfields in the following order: β_1, the preemption switch; β_2, the no-wait switch (in shops only); β_3, the precedence-relation classifier; β_4, the release-date assigner; β_5, the job-number fixer, β_6, the operation-number fixer (for shops only); and β_7, the processing-time equalizer. We use all traditional job characteristics (see Chapter 1) and the following modifications.

8.2.3.1 Restricted Preemptions and No-Wait Constraints

Together with the classical preemption and the no-wait constraint, we consider the *p-preemption* and the *p-no-wait constraint,* denoting them in problem denotations as p-pmtn and p-no-wait. These mean that preemptions are allowed, but the duration of each nonpreemptive part of every operation is divisible by a positive integer p; and, respectively, interruptions of jobs between operations are allowed but every uninterrupted part of each job contains the number of operations divisible by p. Preemptive schedules having only p-preemptions are called *p-preemptive*; schedules observing the p-no-wait constraint are called *p-no-wait.*

8.2.3.2 Expanded Time Parameters and Series of Constraints

We write $p\beta_4$, $p\beta_6$, and $p\beta_7$, if the release dates, the numbers of operations in jobs and processing times, respectively, are divisible by p. We also use the abbreviations

$$\beta_{v_1 v_2 \ldots v_k} = \beta_{v_1}, \beta_{v_2}, \ldots, \beta_{v_k}, \qquad p\beta_{v_1 v_2 \ldots v_k} = p\beta_{v_1}, p\beta_{v_2}, \ldots, p\beta_{v_k}$$

where $v_1, v_2, \ldots, v_k \in \{4, 6, 7\}$.

8.2.3.3 Redundancy and Dominance

We say that a job characteristic is *redundant* or *dominant* if its cancellation or imposition, respectively, does not worsen the optimal value of a given criterion. For example, we will touch on *preemption redundancy* and *no-wait dominance.*

8.2.3.4 Magnification of Precedence and Release-Date Constraints

A *homeomorph* of a precedence relation between jobs, by definition, is obtained by replacing the jobs by chains of new jobs. The denotation chain-β_3 will stand for a *homeomorph* of the precedence relation of the type β_3. Thus,

$$\text{chain-}\circ = \text{chains}, \quad \text{chain-chains} = \text{chains}, \quad \text{chain-tree} = \text{tree}, \quad \text{chain-prec} = \text{prec}$$

[5] See Chapter 7.

where $\circ$ denotes the empty precedence relation. In the case of the chain-like precedence relation, if jobs inside each chain have a common release date, i.e., the release dates are specified just for chains, they will be called *chain release dates* and denoted as c_k.

If the job J_j precedes the job J_k, then the *precedence delay* ℓ between the jobs means that J_k cannot start earlier than ℓ time units after the completion time of J_j. We use the denotation $\beta_3(\ell)$ instead of β_3 (assuming that $\circ(l) = \circ$) if any precedence implies the *precedence delay* ℓ.

The denotation r_j-β_4 imposes unequal release dates meaning that r_j-$\circ = r_j$ and r_j-$r_j = r_j$, where $\circ$ denotes equal release dates.

8.2.4 Criteria Features

8.2.4.1 Regular and Classical Criteria

We consider in this chapter only *regular* criteria, i.e., the criteria that are nondecreasing functions of job completion times. Hence, any scheduling problem can be understood as a problem of finding an *active schedule*, where no job or its part can be shifted to the left without making the schedule infeasible. Regular criteria are also necessary to ensure that optimal schedules exist and avoid anomalies like infinite number of preemptions [24]. Moreover, if not stated otherwise, we mean only the classical criteria $C_{\max}$, $L_{\max}$, $\sum C_j$, $\sum w_j C_j$, $\sum T_j$, $\sum w_j T_j$, $\sum U_j$, and $\sum w_j U_j$.

8.2.4.2 Proportion and Permanence Criteria

Any criterion γ can be presented as a function of certain time parameters, $t_1, t_2, \ldots, t_k$, and job completion times, $C_1, C_2, \ldots, C_n$. Thus, $\gamma = f(t_1, t_2, \ldots, t_k, C_1, C_2, \ldots, C_n)$. Let

$$\gamma(a) = f(at_1, at_2, \ldots, at_k, aC_1, aC_2 \ldots, aC_n)$$

for a positive time-scale factor a. We will distinguish

- *proportion criteria,* for which $\gamma(a) = a\gamma(1)$
- *permanence criteria,* for which $\gamma(a) = \gamma(1)$

Thus, proportion criteria "repeat" and permanence criteria "ignore" time scales. It is easy to check that all classical criteria are of only these two types. For example, $\sum T_j$ is a proportion criterion, since $T_j = \max\{C_j - d_j, 0\}$ and

$$\sum_{j=1}^{n} \max\{aC_j - ad_j, 0\} = a \sum_{j=1}^{n} \max\{C_j - d_j, 0\}$$

Thus, $C_{\max}$, $L_{\max}$, $\sum C_j$, $\sum w_j C_j$, $\sum T_j$, and $\sum w_j T_j$ are proportion criteria, and $\sum U_j$ and $\sum w_j U_j$ are permanence criteria. Of course, it is easy to design a regular criterion that is neither proportion nor permanence.

8.2.5 Reduction Types

8.2.5.1 Turing's and Karp's Reductions

We use the following two well-known types of reductions. Let A denote a minimization problem, and let A' denote its decision version: Given an instance I of A and a constant k, does there exist a solution to I with criteria value at most k? Thus, the pair (I, k) is an instance of A' if I is an instance of A.

Turing's Reduction: A problem A reduces to a problem B, denoted by $A \propto_T B$, if there is an algorithm a solving A by using (possibly many times) an algorithm b solving B as a subroutine. The reduction is polynomial if, under the condition that the algorithm b works instantaneously, the algorithm a is polynomial.

Karp's Reduction: A problem A reduces to a problem B, denoted by $A \propto_K B$, if there is an algorithm a transforming any instance of A' to an instance of B' such that the instances have the same answer. The reduction is polynomial if the algorithm a is polynomial.

It is important to emphasize that, since the algorithm a should take into account both encoding schemes for the problems A and B, the reductions cannot be polynomial if the size of B is not polynomially bounded by the size of A. The reductions are pseudopolynomial if the algorithm a is pseudopolynomial, and there are polynomial dependencies between the sizes of A and B and between the maximal numerical parameters of A and B [9].

Obviously, Karp's reduction implies Turing's reduction, i.e., $A \propto_K B \Rightarrow A \propto_T B$. It is not known, however, that the converse is also true, because it is not clear that all applications of the algorithm b in Turing's reduction can be done as a single transformation in Karp's reduction. So far, nobody has proven that it can or cannot be done. If a particular Turing's reduction uses the algorithm b only once, then the related Karp's reduction is evident.

We will construct chains of reductions mixing both types following the most convenient way or the original style of the results. Such a mixture, however, is not misleading because any chain of reductions of mixed types implies the related chain of only Turing's reductions. Therefore, we omit the lower index T or K and write simply $A \propto B$ because the reduction type will be clear from the context.

If $\mathcal{A}$ and $\mathcal{B}$ are scheduling classes, then $\mathcal{A} \propto \mathcal{B}$ will mean that any problem in $\mathcal{A}$ reduces to a problem in $\mathcal{B}$. Obviously, reductions generate a transitive relation between scheduling classes or problems.

If the problem A is a special case of the problem B, then there exists a trivial reduction $A \propto B$, which is called an *inclusion* of A into B. It is clear that an inclusion is not a polynomial reduction if A causes a size collapse of B.

8.2.5.2 Size Inflation and Properness

Let A and B be problems with sizes x and y. We say that a reduction $A \propto B$ is *size inflating* if there is no polynomial p such that $y \leq p(x)$. Thus, a size-inflating reduction is not polynomial. Each time we obtain a polynomial reduction $A \propto B'$, where B' is a subproblem of B that causes a size collapse, we actually obtain a size inflating reduction $A \propto B$. For example, we can obtain a reduction

$$P \mid \text{chains}, p_j = 1 \mid C_{\max} \;\propto\; P \mid \text{tree}, p_j = 1 \mid C_{\max}$$

by adding a new job that precedes all the original jobs. The reduction is size-inflating because the size of the former problem is the number of chains while the size of the latter problem is the number of jobs that is not polynomially bounded by the number of chains. Hence, Hu's algorithm [25] that schedules trees in time linear in the number of jobs is not polynomial for the problem of scheduling chains. A polynomial algorithm of scheduling chains that iteratively uses McNaughton's wrap-around rule can be found in Ref. [26].

Since a tree-like precedence relation is a special case of an arbitrary precedence relation, we have the inclusion

$$P2 \mid \text{tree}, p_j = 1 \mid C_{\max} \;\propto\; P2 \mid \text{prec}, p_j = 1 \mid C_{\max}$$

that, in contrast to the previous reduction, is not size-inflating because the size of both the problems is the number of jobs. Hence, the Coffman-Graham algorithm [27] that solves the problem with arbitrary precedence constraints in time quadratic in the number of jobs also solves in polynomial time the problem with tree-like precedence constraints.

Let us define a *job model* to be the set of values of the fields: β_3, the type of precedence constraints; β_4, the type of release dates; and γ, the criterion. We call a reduction *proper* if it does not change the job model. For example, the last two reductions are not proper because they change the type of precedence constraints, while the inclusions

$$1 \mid p_j = 1 \mid C_{\max} \;\propto\; 1 \| C_{\max} \quad \text{and} \quad F2 \mid \text{chains}, r_j \mid C_{\max} \;\propto\; J2 \mid \text{chains}, r_j \mid C_{\max}$$

are proper. Proper reductions thus can change only the machine environment or the constraint on processing times. Hence, proper reductions can be size inflating, like the latter two, only at the expense of changing the machine environment or relaxing the constraint on processing times. Note also that the inclusions

$$1 \mid r_j, p_j = 1 \mid C_{\max} \propto 1 \mid r_j \mid C_{\max} \quad \text{and} \quad J \text{ per } 2 \mid p_{ij} = 1 \mid C_{\max} \propto J2 \mid p_{ij} = 1 \mid C_{\max}$$

are proper polynomial reduction, while the inclusions

$$1 \mid p_j = p \mid C_{\max} \propto 1 \| C_{\max} \quad \text{and} \quad J \text{ per } 3 \mid p_{ij} = 1 \mid C_{\max} \propto J3 \mid p_{ij} = 1 \mid C_{\max}$$

are size inflating proper reductions, which are pseudopolynomial because they become polynomial if the job lengths are bounded by a polynomial in n.

8.2.5.3 Equivalence and Isomorphism

An *equivalence* $A \sim B$ means $A \propto B$ and $B \propto A$. As a reflexive, symmetric, and transitive relation, an equivalence generates *equivalence classes* on the set of problems. We consider further only polynomial equivalences. For example, NP-complete problems are equivalent, so they generate the equivalence class NPC [9].

Two problems that can be solved by a common polynomial algorithm are trivially equivalent. So, we can define an *equivalence class on an algorithm* to be the set of problems that can be solved in polynomial time by the algorithm. For example, problems in

$$P\alpha_2 \mid \text{pmtn} \mid C_{\max}, \quad \text{where } \alpha_2 \in \{1, 2, \dots, m, \circ\}$$

belong to the equivalence class on McNaughton's rule for $P \mid \text{pmtn} \mid C_{\max}$ [28]; problems in

$$P2 \mid \beta_3, p_j = 1 \mid \gamma \quad \text{and} \quad 1 \mid \beta_3(1), p_{ij} = 1 \mid \gamma, \quad \text{where } \beta_3 \in \{\circ, \text{tree}, \text{prec}\}, \ \gamma \in \{C_{\max}, \sum C_j\}$$

belong to the equivalence class on the Coffman-Graham algorithm for $P2 \mid \text{prec}, p_j = 1 \mid C_{\max}$ [27] (cf. [23] and Section 8.5.5).

An *isomorphism* $A \approx B$ is an equivalence that sets a bijection $f : A \to B$ such that for any instance $I \in A$ there is a bijection $g : S_I \to S_{f(I)}$, where S_I and $S_{f(I)}$ are the sets of solutions to I and $f(I)$, respectively. Thus, we have the hierarchy:

$$A \approx B \ \Rightarrow \ A \sim B \ \Rightarrow \ A \propto B$$

To show that a reduction $A \propto D$ results from a chain of reductions $A \approx B \propto C \sim D$ we draw the diagram

$$\begin{array}{ccc} B & \overset{\mathsf{b}}{\propto} & C \\ \wr\wr\ \mathsf{a} & & \wr\ \mathsf{c} \\ A & \propto & D \end{array}$$

where a, b, c denote references to the reductions. The isomorphism $A \approx A$ is a *tautology*.

8.2.6 Some Fundamental Reductions and Equivalences

8.2.6.1 Reductions of Classical Criteria

We write $\gamma_1 \to \gamma_2$ if $\alpha \mid \beta \mid \gamma_1 \propto \alpha \mid \beta \mid \gamma_2$. The following diagram is well-known (cf. [29,30]):

$$\begin{array}{ccccccc} & & \sum w_j U_j & \leftarrow \sum U_j & \sum C_j & \rightarrow & \sum w_j C_j \\ & & & \uparrow & \downarrow & & \downarrow \\ C_{\max} & \rightarrow & & L_{\max} & \rightarrow \sum T_j & \rightarrow & \sum w_j T_j \end{array}$$

Replacing the criteria in accordance with this diagram gives another set of size inflating reductions. For example, the inclusion $1 \mid p_j = 1 \mid \sum C_j \propto 1 \mid p_j = 1 \mid \sum w_j C_j$ is size inflating, because the problem sizes are 1 and n, respectively. However, the inclusions

$$1\,|\,r_j, p_j=1\,|\,\sum C_j \;\propto\; 1\,|\,r_j, p_j=1\,|\,\sum w_jC_j \quad \text{and} \quad 1\,\|\,\sum C_j \;\propto\; 1\,\|\,\sum w_jC_j$$

are polynomial reductions, because the size of each of the four problems is n.

8.2.6.2 Reduction Symbols

Up to now we have been using the symbol $\propto$ to denote any reduction. Further, it will denote only a polynomial reduction. The symbol $\not\propto$ will denote only a pseudopolynomial reduction. If not stated otherwise, the term "reduction" means a polynomial reduction.

8.2.6.3 Time Reversal

It is clear that any feasible schedule observing given precedence constraints can be transformed into a feasible schedule observing the inverse precedence constraints by listing all the events in reverse order. Besides, any optimal schedule for an instance of a problem with the criterion $L_{\max}$ remains optimal after increasing or decreasing all due dates by a common constant.

Hence, any schedule of jobs with release dates r_j, precedence constraints, and minimum value t of $C_{\max}$ in the time interval $[r_{\min} = 0, t]$, is the reverse of a schedule of jobs with release dates 0, the inverse precedence constraints, due dates $t - r_j$, and minimum value 0 of $L_{\max}$.

Conversely, any schedule of jobs with release dates 0, precedence constraints, and minimum value l of $L_{\max}$ in the time interval $[0, t]$, is the reverse of a schedule of jobs with release dates $t - d_j - l$, the inverse precedence constraints, and minimum value t of $C_{\max}$. Thus, we have proved the isomorphism

$$\alpha\,|\,\beta_{123567}, r_j\,|\,C_{\max} \;\approx\; \alpha\,|\,\beta_{123567}\;|\;L_{\max}$$

Similar reasoning that changes the roles of release dates and due dates gives the time reversal

$$\alpha\,|\,\beta_{12567}, \text{intree}, r_j\,|\,L_{\max} \;\approx\; \alpha\,|\,\beta_{12567}, \text{outtree}, r_j\,|\,L_{\max}$$

It is clear that this isomorphism can be extended to an arbitrary precedence relation since any precedence relation is invertible. Such an extension, however, is a tautology.

8.2.6.4 Shopping Squares

In Section 8.5.2, we will establish the *job-shopping square:*

$$\begin{array}{ccc}
P\ \text{mod}\,\alpha_2\,|\,m\text{-pmtn}, \beta_{345}, m\beta_7\,|\,\gamma & \overset{(8.23)}{\approx} & J\ \text{per}\,\alpha_2\,|\,m\text{-no-wait}, \beta_{345}, m\beta_6, p_{ij}=1\,|\,\gamma \\
\wr\ (8.22) & & \wr\ (8.25) \\
P\ \text{mod}\,\alpha_2\,|\,\text{pmtn}, \beta_{345}, m\beta_7\,|\,\gamma & \sim & J\ \text{per}\,\alpha_2\,|\,\beta_{345}, m\beta_6, p_{ij}=1\,|\,\gamma
\end{array}$$

The vertical equivalences state that m-preemptions are enough to obtain optimal preemptive schedules for modulated identical parallel machines and that the m-no-wait requirement does not worsen optimal nonpreemptive schedules in unit-time perfect job shops. The isomorphism proves that any preemptive problem of scheduling jobs with processing times divisible by m on m modulated identical parallel machines is just another form of a perfect unit-time job-shop problem of scheduling jobs whose number of operations is divisible by m.

Note that m-preemptions of jobs with processing times m is obviously redundant, and a flow shop is the minimal perfect job shop. Therefore, as a special case, we have the *flow-shopping square:*

$$\begin{array}{ccc}
P\ \text{mod}\,\alpha_2\,|\,\beta_{345}, p_j=m\,|\,\gamma & \approx & F\alpha_2\,|\,m\text{-no-wait}, \beta_{345}, p_{ij}=1\,|\,\gamma \\
\wr & & \wr \\
P\ \text{mod}\,\alpha_2\,|\,\text{pmtn}, \beta_{345}, p_j=m\,|\,\gamma & \sim & F\alpha_2\,|\,\beta_{345}, p_{ij}=1\,|\,\gamma
\end{array}$$

The equivalence on the main diagonal shows that any nonpreemptive problem of scheduling jobs with processing times m on m modulated identical parallel machines is just another form of an m-machine unit-time flow-shop scheduling problem.

8.2.6.5 Preemption Redundancy

Preemption redundancy results can be presented as equivalences between preemptive problems and nonpreemptive problems. Table 8.2 presents all preemption redundancy results that we are aware of. Note that the well-known Equivalence I in the table has the following extension to the many-machine case. A function F on real n-vectors is *quasiconcave* if $0 \leq \lambda \leq 1$ implies

$$F(\lambda x + [1-\lambda]y) \geq \min\{F(x), F(y)\}$$

for any x and y. If the inequality is true only for x and y with $y = x + \alpha e$, where $\alpha > 0$, and e runs through all n-vectors with components $-1, 0, 1$, then F is an *e-quasiconcave* function. Thus, quasiconcave functions are all e-quasiconcave. Tanaev [31] (cf. [32]) shows that preemptions in $P \mid \text{pmtn} \mid \gamma$ are redundant if the criterion γ is a nondecreasing e-quasiconcave function of job completion times.

In what follows, we cross out the abbreviation pmtn to emphasize that preemptions in a preemptive problem are redundant. For example, we write $P \mid$ ~~pmtn~~ $\mid \sum w_j C_j$ to refer to the preemption redundancy result A in Table 8.2.

TABLE 8.2 Preemption Redundancy Results

A	$P \mid \text{pmtn} \mid \sum w_j C_j$	$\sim$	$P \| \sum w_j C_j$
B	$P \mid \text{pmtn, chains} \mid \sum w_j C_j$	$\sim$	$P \mid \text{chains} \mid \sum w_j C_j$
C	$P \mid \text{pmtn}, r_j, \text{outtree}, p_j = 1 \mid \sum C_j$	$\sim$	$P \mid r_j, \text{outtree}, p_j = 1 \mid \sum C_j$
D	$P \mid \text{pmtn, intree}, p_j = 1 \mid \sum w_j U_j$	$\sim$	$P \mid \text{intree}, p_j = 1 \mid \sum w_j U_j$
E	$P \mid \text{pmtn}, r_j, p_j = 1 \mid \sum w_j U_j$	$\sim$	$P \mid r_j, p_j = 1 \mid \sum w_j U_j$
F	$P \mid \text{pmtn}, r_j, p_j = 1 \mid \sum w_j T_j$	$\sim$	$P \mid r_j, p_j = 1 \mid \sum w_j T_j$
G	$P \mid \text{pmtn, prec}, r_j, n \leq m \mid \gamma$	$\sim$	$P \mid \text{prec}, r_j, n \leq m \mid \gamma$
H	$P\ \text{mod} \mid \text{pmtn, prec}, r_j, p_j = m \mid \gamma$	$\sim$	$P\ \text{mod} \mid \text{prec}, r_j, p_j = m \mid \gamma$
I	$1 \mid \text{pmtn} \mid \gamma$	$\sim$	$1 \| \gamma$
J	$1 \mid \text{pmtn, prec}, r_j, p_j = 1 \mid \gamma$	$\sim$	$1 \mid \text{prec}, r_j, p_j = 1 \mid \gamma$
K	$1 \mid \text{pmtn}, r_j \mid C_{\max}$	$\sim$	$1 \mid r_j \mid C_{\max}$
L	$J \mid \text{pmtn, prec}, r_j, p_{ij} = 1 \mid \gamma$	$\sim$	$J \mid \text{prec}, r_j, p_{ij} = 1 \mid \gamma$

Equivalence A is McNaughton's theorem [28]. Equivalence B was obtained by Du, Leung, and Young [33] as an extension of McNaughton's theorem to the case of chain-like precedence constraints. Equivalence C is a conjecture of Baptiste and Timkovsky [34] that has been proved by Brucker, Hurink, and Knust [35]. Note that these results are maximal in the sense that preemptions become advantageous for any extensions of these problems [34]. Equivalences D and E are due to Brucker, Heitman, and Hurink [36]. Equivalence F, obtained by Baptiste [37], strengthens the preemption redundancy results obtained by Brucker, Heitman, and Hurink [36] for $P \mid \text{pmtn}, p_j = 1 \mid \sum T_j$ and $P2 \mid \text{pmtn}, r_j, p_j = 1 \mid \sum w_j T_j$. Equivalence G follows from the obvious observation that a number of jobs being started on separate machines can be processed without preemptions. Equivalence H is an example of the left vertical equivalence in the flow-shopping square. Equivalence I is proved in the book of Conway, Maxwell, and Miller [38]. Gordon [39] (cf. [32]) strengthens the result showing that preemptions in $1 \mid \text{pmtn, prec}, r_j \mid \gamma$ can be advantageous only at release dates. Equivalences J, K, and L can be trivially proved by contradiction using the exchange argument (cf. [36,38]).

8.2.6.6 Parallel Increase

A *parallel increase* is a reduction that increases the number of parallel machines by one. Although such reductions are traditionally treated as "minor troubles," some of them are quite far from trivial exercises. We consider only the case of identical parallel machines, giving only descriptions of the reductions and ideas underlying their proofs. Details are left to the reader.

It is easy to see that a parallel increase,

$$Pm \,|\, \beta_{134} \,|\, \gamma \propto Pm{+}1 \,|\, \beta_{134} \,|\, \gamma \tag{8.1}$$

is always proper. Let us show how it can be obtained for different criteria. We assume that there exists an algorithm A for the problem with $m+1$ machines that produces an optimal schedule S convertible in polynomial time into an optimal schedule where a job with maximum processing time appears on only one machine. If this condition is not satisfied, then the reduction may not be polynomial, in general.

Let us take an instance I of a scheduling problem on m identical parallel machines of scheduling n jobs with processing times p_j, release dates r_j, due dates d_j, and weights w_j. Without loss of generality, we consider that $r_j = 0$, $d_j = 0$, and $w_j = 1$ if release dates, due dates, and weights are not specified, respectively. Then we extend the instance I by adding a new machine M_{m+1} and a new job J_{n+1} with release date 0 and set

$$p = r_{\max} + \sum_{j=1}^{n} p_j$$

The idea of the reduction is to complete the definition of the extended instance (specifying the processing time p_{n+1} and possibly other parameters) such that the algorithm A applied to it must schedule J_{n+1} in the time interval $[0, p_{n+1}]$. Using the assumption that the algorithm A exists, we can transform an optimal schedule for the extended instance in polynomial time into an optimal schedule where one of the machines contains only J_{n+1}. Hence, the rest of the schedule must define an optimal schedule on m machines for I.

Now we describe how to complete the definition of the extended instance and how to obtain the schedule S for different classical criteria. The following two criteria require the bisection method:

$\gamma = C_{\max}$: Let p_{n+1} be variable in the time interval $[r_{\max}, p]$. The bisection method applied to the time interval with the algorithm A finds the minimum value of p_{n+1} that is the minimum value of $C_{\max}$ in the schedule S.

$\gamma = L_{\max}$: Set $p_{n+1} = p$ leaving d_{n+1} variable in the time interval $[p + d_{\min}, p + d_{\max}]$. The bisection method applied to the time interval with the algorithm A finds the minimum value d^*_{n+1} of d_{n+1} such that $p_{n+1} - d^*_{n+1}$ is the minimum value of $L_{\max}$ in the schedule S.

For the following criteria, only a single application of the algorithm A gives the schedule S.

$\gamma = \sum w_j C_j$: Set $p_{n+1} = p$, $w_{n+1} = 1$ and replace weights w_j by $2pw_j$ for $j = 1, 2, \ldots, n$

$\gamma = \sum w_j U_j$: Set $p_{n+1} = p$, $w_{n+1} = 2nw_{\max}$, $d_{n+1} = p$

$\gamma = \sum w_j T_j$: Set $p_{n+1} = p$, $w_{n+1} = 2nw_{\max}$, $d_{n+1} = p$

The original jobs (in the first case) and the new job (in the other two cases) get heavy weighting to ensure that a schedule starting the new job later than at time 0 is worse than any other schedule. Hence the new job starts at time 0 in any optimal schedule.

Since the extended instance can obviously be constructed in polynomial time, an instantaneous work of an algorithm solving the extended instance implies a polynomial algorithm for the original instance. Hence, the described parallel increase is polynomial.

TABLE 8.3 Parallel Increases: Set I

[43–45]	✓	$1 \mid$ ~~pmtn~~, chains, $p_j = 1 \mid \sum w_j C_j$	$\propto$	$P2 \mid$ ~~pmtn~~, chains, $p_j = 1 \mid \sum w_j C_j$	††	[26]
[43–45]	✓	$1 \mid$ chains, $p_j = 1 \mid \sum w_j C_j$	$\propto$	$P2 \mid$ chains, $p_j = 1 \mid \sum w_j C_j$	††	[26]
[21,46]	†	$1 \mid$ ~~pmtn~~ $\mid \sum w_j U_j$	$\propto$	$P2 \mid$ pmtn $\mid \sum w_j U_j$	†	
[42,47]	††	$1 \mid$ ~~pmtn~~ $\mid \sum w_j T_j$	$\propto$	$P2 \mid$ pmtn $\mid \sum w_j T_j$	††	
[42,47]	††	$1 \| \sum w_j T_j$	$\propto$	$P2 \| \sum w_j T_j$	††	
[48]	††	$1 \mid$ ~~pmtn~~, chains, $p_j = 1 \mid \sum U_j$	$\propto$	$P2 \mid$ pmtn, chains, $p_j = 1 \mid \sum U_j$	††	[49]
[50]	††	$1 \mid$ ~~pmtn~~, chains, $p_j = 1 \mid \sum T_j$	$\propto$	$P2 \mid$ pmtn, chains, $p_j = 1 \mid \sum T_j$	††	
[48]	††	$1 \mid$ chains, $p_j = 1 \mid \sum U_j$	$\propto$	$P2 \mid$ chains, $p_j = 1 \mid \sum U_j$	††	
[50]	††	$1 \mid$ chains, $p_j = 1 \mid \sum T_j$	$\propto$	$P2 \mid$ chains, $p_j = 1 \mid \sum T_j$	††	
[51]	††	$1 \mid$ pmtn, $r_j \mid \sum w_j C_j$	$\propto$	$P2 \mid$ pmtn, $r_j \mid \sum w_j C_j$	††	
[47]	††	$1 \mid r_j \mid \sum C_j$	$\propto$	$P2 \mid r_j \mid \sum C_j$	††	
[47]	††	$1 \mid r_j \mid L_{\max}$	$\propto$	$P2 \mid r_j \mid L_{\max}$	††	

The extensions of the first two single-machine problems to the case with tree-like precedence constraints and arbitrary processing times can also be solved in polynomial time [43–45]. A pseudopolynomial algorithm for $Qm \mid \text{pmtn} \mid \sum w_j U_j$ can be found in Refs. [46].

An parallel increase that is always proper for $\gamma \in \{\sum C_j, \sum U_j, \sum T_j\}$ is not known. There are known only

$$Pm \mid \beta_{134} \mid \gamma \;\propto\; Pm{+}1 \mid \beta_{14}, \text{chain-}\beta_3 \mid \gamma \tag{8.2}$$

$$Pm \mid \beta_{34} \mid \gamma \;\propto\; Pm{+}1 \mid \beta_3, r_j\text{-}\beta_4 \mid \gamma \tag{8.3}$$

Reduction (8.2) is known for all three criteria, and it is proper only if $\beta_3 \in \{\text{chains, tree, prec}\}$. It can be obtained by adding to I a single chain of $2p$ unit-time jobs. Note that this operation makes the reduction pseudopolynomial unless we deal with unit-time jobs. We also assign due dates $1, 2, \ldots, 2p$ for these jobs if $\gamma \in \{\sum U_j, \sum T_j\}$. The idea of the reduction is based on the observation that the chain is long enough to guarantee that any schedule, where the chain is processed with delays, is worse than the best among the schedules, where the chain is processed without delays. Hence any optimal schedule for the instance I complemented by the chain includes it without delays. Using the algorithm A we can place the whole chain on one machine, and hence the residual schedule will define an optimal schedule for I.

Reduction (8.3) is known for $\gamma \in \{\sum C_j, \sum T_j\}$, and it is proper only if $\beta_4 = r_j$. We consider only the case where $m = 1$ and $\gamma = \sum C_j$ because the other cases can be managed analogously. Following the idea of Shafransky [40], let us add a second machine and a job J_{n+1} with processing time $2np$ and release date 0 and replace release dates r_j by $np + r_j$, $j = 1, 2, \ldots, n$. A key point of this reduction is that any active schedule that starts J_{n+1} at time 0 has total completion time less than $3np$, but any active schedule that starts J_{n+1} at time np or later (and then both machines are idle in the interval $[0, np]$) has total completion time at least $3np$. Hence, an optimal schedule for the extended instance starts J_{n+1} at time 0, and therefore the original jobs must appear on one machine.

Table 8.3 presents Set I of parallel increases of types (8.1), (8.2), and (8.3) that can be obtained by the techniques described above. We refer to this table further to point to the NP-hardness of the problems on two identical parallel machines in it. Table 8.4 presents Set II of parallel increases of the same types; it is not known, however, whether they can be obtained without using Cook's theorem [9] or the fact that the problems belong to an equivalence class on a polynomial algorithm. The most interesting open question related to parallel increases is

$$1 \mid \sout{\text{pmtn}} \mid \sum T_j \overset{?}{\propto} P2 \mid \text{pmtn} \mid \sum T_j$$

The single-machine problem is proved to be ordinarily NP-hard [41,42], but the complexity status of the two-machine problem remains open.

In conclusion of this section, we observe that all parallel increases of Set I add at least one job to the problem. It is interesting to question about the existence of polynomial parallel increases that save

TABLE 8.4 Parallel Increases: Set II

[28]	✓	$1 \mid$ ~~pmtn~~ $\mid \sum C_j$	$\sim$	$P2 \mid$ ~~pmtn~~ $\mid \sum C_j$	✓	[28]
[53,54]	✓	$1 \mid$ ~~pmtn~~ $\mid \sum U_j$	$\sim$	$P2 \mid \text{pmtn} \mid \sum U_j$	✓	[46,55]
[56]	✓	$1 \mid \text{pmtn}, r_j \mid \sum C_j$	$\propto$	$P2 \mid \text{pmtn}, r_j \mid \sum C_j$	†?	[57]
[58]	✓	$1 \mid \text{pmtn}, r_j \mid \sum U_j$	$\propto$	$P2 \mid \text{pmtn}, r_j \mid \sum U_j$	†?	[59]
[41]	†?	$1 \mid \text{pmtn}, r_j \mid \sum T_j$	$\sim$	$P2 \mid \text{pmtn}, r_j \mid \sum T_j$	†?	[57]
[28]	✓	$1 \| \sum C_j$	$\sim$	$P2 \| \sum C_j$	✓	[28]
[53,54]	✓	$1 \| \sum U_j$	$\propto$	$P2 \| \sum U_j$	†	[47,60]
[41,42]	†	$1 \| \sum T_j$	$\sim$	$P2 \| \sum T_j$	†	[47]
[47]	†?	$1 \mid r_j \mid \sum U_j$	$\sim$	$P2 \mid r_j \mid \sum U_j$	†?	[47]

The equivalences between $P2 \mid$ ~~pmtn~~ $\mid \sum C_j$, $P2\|\sum C_j$, $P2 \mid \text{pmtn} \mid \sum U_j$, and their single-machine counterparts follow from the membership of the problems in the equivalence class on McNaughton's algorithm for $P \| \sum C_j$ and Lawler's algorithm for $Qm \mid \text{pmtn} \mid \sum U_j$, respectively. The other equivalences follow from the membership of the decision versions of the problems in NPC. The reductions follow from Cook's theorem [9]. Note that $Pm \mid \text{pmtn}, r_j \mid \sum C_j$ [33] and $Pm \mid \text{pmtn}, r_j \mid \sum U_j$ [59] are proved to be NP-hard for any fixed positive integer $m \geq 2$ without using a parallel increase. NP-hardness proofs for $P2\|\sum U_j$ and $P2\|\sum T_j$ can be obtained by the same reduction from the partition problem [9] with integers a_j as for $P2\|C_{\max}$ [47] by setting a common due date $\sum a_j/2$. Pseudopolynomial algorithms are presented in Refs. [42,60].

the number of jobs. They exist indeed and imply that the problems to which they are applicable can be solved in polynomial time. They show it is enough to note that problems of scheduling n jobs on m identical parallel machines with $n \leq m$ can be trivially solved in polynomial time since there exist optimal nonpreemptive schedules where every machine runs at most one job (cf. Caption to Table 8.2). Thus, if we find a polynomial parallel increase of a problem that save the number of jobs, then, applying it at most n times, we can reduce the problem to its counterpart with $n \leq m$.

The polynomial increases saving the number of jobs are known, however, only for the problems that can be solved in polynomial time for an unrestricted number of machines. For example, due to McNaughton's algorithm [28] for $P \mid \text{pmtn} \mid C_{\max}$,

$$1 \mid \cancel{\text{pmtn}} \mid C_{\max} \sim P2 \mid \text{pmtn} \mid C_{\max} \sim \cdots \sim Pn{-}1 \mid \text{pmtn} \mid C_{\max} \sim Pn \mid \cancel{\text{pmtn}} \mid C_{\max}$$

8.3 Preemptions vs. Chains

8.3.1 Expansion

8.3.1.1 Properties of Polynomial Expansions

The main technique used in almost all mass reductions is an *expansion* of scheduling problems by increasing all time parameters by a common factor such that optimal schedules become integer or divisible by m, i.e., the number of machines. Since we are interested in obtaining polynomial reductions, we consider only *polynomial expansions*, i.e., those where the factor size is polynomially bounded in the problem size. Polynomial expansions are trivially attainable in the nonpreemptive case by factor m, since $\log m$ is obviously polynomially bounded in the problem size. The preemptive case is not trivial because it is not known, *a priori*, how small nonpreemptive parts of jobs can appear in optimal preemptive schedules.

The *extended fractionality conjecture* implies that for any regular criterion and integer data, there exists an optimal preemptive schedule of fractionality $1/m$. Hence, if the conjecture is true, then we could use factor m or m^2 to attain integer or divisible by m optimal preemptive schedules, respectively.

However, even the *fractionality conjecture*[6] remains unverified. Therefore to obtain polynomial reductions in the preemptive case, we consider only preemptive problems whose decision versions belong to NP. As we show, the preemptive problems have polynomial expansions with integer, or divisible by m, optimal preemptive schedules.

Let I be an instance of a scheduling problem in the class $\alpha \mid \beta \mid \gamma$. Then the pair (I, k) is the related instance of the decision version of the scheduling problem with upper bound k for γ values. Since we consider only decision problems belonging to NP, we assume that for any instance (I, k) with a positive answer there exists a feasible schedule with a binary code ℓ whose length, $\log \ell$, is polynomial in the size of (I, k). Thus, all numerical data specifying the schedule is presented by at most $\log \ell$ fractions with denominators at most ℓ. Therefore

1. The data become integer after increasing by a common factor at most $\ell^{\log \ell}$.
2. The data differ by at least $1/\ell^2$.

8.3.1.2 Expansion Isomorphisms

For positive integer a and an instance I, let aI denote the *expanded* instance obtained by increasing processing times, release dates, and due dates in I by multiple factor a, and let $a\sigma$ denote a feasible schedule for aI. Then, the schedule $a\sigma$ can be obtained from σ by stretching it a times. Obviously, $a\sigma$ is a feasible schedule for aI and all completion times C_j in σ become aC_j in $a\sigma$. Let $[\sigma, I]$ denote the γ value of σ for I. Then it is easy to see that

$$a[\sigma, I] = [a\sigma, aI] \quad \text{for proportion criteria } \gamma \tag{8.4}$$

$$[\sigma, I] = [a\sigma, aI] \quad \text{for permanence criteria } \gamma \tag{8.5}$$

Let p and q be positive integers at most ℓ. The following isomorphisms, which we call *preemptive* and *nonpreemptive expansions*, are valid for all proportion and permanence criteria:

$$\alpha \mid \text{pmtn}, \beta_{234567} \mid \gamma \approx \alpha \mid p\text{-pmtn}, \beta_{2356}, pq\beta_{47} \mid \gamma \tag{8.6}$$

$$\alpha \mid \beta_{234567} \mid \gamma \approx \alpha \mid \beta_{2356}, q\beta_{47} \mid \gamma \tag{8.7}$$

The two factors p and q are introduced to make the scale of preemptions smaller than that of processing times; we will need this feature as a link to open shops. Expansion (8.7) is trivial. To prove Expansion (8.6), we set

$$a = pqd, \qquad I' = aI, \qquad k' = \begin{cases} ak & \text{for proportion criteria } \gamma \\ k & \text{for permanence criteria } \gamma \end{cases}$$

where d is the least common denominator of the fractions specifying a feasible schedule for (I, k). By Property 1, all the lengths of nonpreemptive parts of operations in σ become integer and divisible by pq in $a\sigma$. Since Equalities (8.4) and (8.5) hold, a positive answer for (I, k) means a positive answer for (I', k') and vice versa. Besides, since $p \le \ell$, $q \le \ell$ and $d \le \ell^{\log \ell}$, we have

$$\log a = \log p + \log q + \log d \le 2\log \ell + (\log \ell)^2$$

[6]The fractionality conjecture is implicitly stated in the paper of Muntz and Coffman [52] for the unit-time scheduling problem $P \mid \text{pmtn}, \text{prec}, p_j = 1 \mid C_{\max}$. It was extended to all classical scheduling problems with regular criteria and integer data in the paper of Coffman, Sethuraman, and Timkovsky [23]. The extended fractionality conjecture was motivated by the observation that all preemptive problems with regular criteria and integer data that are known to be solved in polynomial time have optimal preemptive schedules of fractionality $1/m$; and all known NP-hardness proofs for preemptive problems with regular criteria and integer data use only preemptive schedules of fractionality $1/m$. Thus, it is unknown whether NP-hard preemptive problems and preemptive problems whose complexity status is unknown have optimal schedules of fractionality $1/m$. Note that there exist preemptive problems with irregular criteria, whose optimal schedules have an unrestricted number of preemptions, and hence are of an unrestricted fractionality [24].

i.e., $\log a$ is bounded by a polynomial in $\log \ell$. Hence (I', k') can be constructed in time polynomial in the size of (I, k).

8.3.2 Preemption Chaining

8.3.2.1 Operation Chaining Technique

The *operation chaining* technique replaces operations or single-operation jobs by chains of new jobs. It can be described in detail as follows: Let a job J_j with release date r_j, due date d_j, and weight w_j be the chain of operations

$$O_{1j} \rightarrow O_{2j} \rightarrow \cdots \rightarrow O_{m_j j}$$

where the operation O_{ij} has processing time pqq_{ij}. Then we transform O_{ij} into the chain of single-operation jobs

$$J_{ij1} \rightarrow J_{ij2} \rightarrow \cdots \rightarrow J_{ijq_{ij}}$$

with equal processing times pq, equal release dates r_j, equal due dates d_j, and weights

$$w_{ijk} = \begin{cases} w_j & \text{if } i = m_j \text{ and } k = q_{ij} \\ 0 & \text{otherwise} \end{cases}$$

If O_{ij} is assigned to be processed on a machine, then the jobs J_{ijk}, $k = 1, 2, \ldots, q_{ij}$ are assigned to be processed on the machine. If some data are not specified for the job J_j, then the related data will not be specified for the chain of the new jobs. For example, release dates and due dates of jobs can be missed in problems with the criterion $C_{\max}$. The transformation can obviously be done in pseudopolynomial time $O(\sum_{j=1}^{n} \sum_{i=1}^{m_j} q_{ij})$, which is polynomial if the parameters q_{ij} are bounded by a polynomial in the problem size.

8.3.2.2 Preemption Chaining Diagram

It is easy to check that the operation chaining technique does not change the machine environment if it is identical parallel machines, uniform parallel machines, a job shop, or an open job shop. However, it transforms flow shops and open shops to job shops and open job shops, respectively. Besides, the original job and the chain of new jobs will contribute the same value to any classical proportion or permanence criterion, with the exception of $\sum C_j$, because job weights are not specified in this case. Thus, if $\alpha_1 \notin \{F, O\}$ and $\gamma \neq \sum C_j$, then the operation chaining technique gives the reduction

$$\alpha_1\alpha_2 \mid p\text{-pmtn}, \beta_3, pq\beta_{47} \mid \gamma \;\propto\; \alpha_1\alpha_2 \mid p\text{-pmtn}, \text{chain-}\beta_3, pq\beta_4, p_{ij} = pq \mid \gamma \tag{8.8}$$

Note that the equality $q = 1$ induces the *preemption elimination* effect, since

$$\alpha_1\alpha_2 \mid p\text{-pmtn}, \beta_{356}, p\beta_4, p_{ij} = p \mid \gamma \approx \alpha_1\alpha_2 \mid \beta_{356}, p\beta_4, p_{ij} = p \mid \gamma \tag{8.9}$$

The composition of Expansion (8.6), Chaining (8.8), and Elimination (8.9) with $p = 1$, $q = 1$ gives the following diagram:

$$\begin{array}{ccc} \alpha_1\alpha_2 \mid 1\text{-pmtn}, \beta_{2347} \mid \gamma & \overset{(8.8)}{\propto} & \alpha_1\alpha_2 \mid 1\text{-pmtn}, \text{chain-}\beta_3, \beta_{24}, p_j = 1 \mid \gamma \\ \wr\wr\ (8.6) & & \wr\wr\ (8.9) \\ \alpha_1\alpha_2 \mid \text{pmtn}, \beta_{2347} \mid \gamma & \propto & \alpha_1\alpha_2 \mid \text{chain-}\beta_3, \beta_{24}, p_{ij} = 1 \mid \gamma \end{array}$$

The lower reduction is *preemption chaining* that holds for $\alpha_1 \notin \{F, O\}$ and $\gamma \neq \sum C_j$. Preemption chaining is a polynomial reduction if the operation chaining technique works in polynomial time. Table 8.5 presents examples of preemption chaining.

TABLE 8.5 Reductions by Preemption Chaining

[28]	✓	$P \mid \text{pmtn} \mid C_{\max}$	$\propto$	$P \mid \text{chains}, p_j = 1 \mid C_{\max}$	✓	[26]
[63,64]	✓	$P \mid \text{pmtn}, r_j \mid C_{\max}$	$\propto$	$P \mid \text{chains}, r_j, p_j = 1 \mid C_{\max}$	✓	[65]
[64,66]	✓	$P \mid \text{pmtn, tree} \mid C_{\max}$	$\propto$	$P \mid \text{tree}, p_j = 1 \mid C_{\max}$	✓	[25]
[19]	✓	$P \mid \text{pmtn, outtree}, r_j \mid C_{\max}$	$\propto$	$P \mid \text{outtree}, r_j, p_j = 1 \mid C_{\max}$	✓	[65]
[19]	✓	$P \mid \text{pmtn, intree} \mid L_{\max}$	$\propto$	$P \mid \text{intree}, p_j = 1 \mid L_{\max}$	✓	[65]
[19]	✓	$P2 \mid \text{pmtn, prec}, r_j \mid L_{\max}$	$\propto$	$P2 \mid \text{prec}, r_j, p_j = 1 \mid L_{\max}$	✓	[67]
[68]	✓	$Q \mid \text{pmtn, chains} \mid C_{\max}$	$\propto$	$Q \mid \text{chains}, p_j = 1 \mid C_{\max}$	††	[69]
[53,54]	✓	$1 \mid$ ~~pmtn~~ $\mid \sum U_j$	$\propto$	$1 \mid \text{chains}, p_j = 1 \mid \sum U_j$	††	[48]
[21,55]	†	$1 \mid$ ~~pmtn~~ $\mid \sum w_j U_j$	$\propto$	$1 \mid \text{chains}, p_j = 1 \mid \sum w_j U_j$	††	[48]
[41,42]	†	$1 \mid$ ~~pmtn~~ $\mid \sum T_j$	$\propto$	$1 \mid \text{chains}, p_j = 1 \mid \sum T_j$	††	[50]
[51]	††	$1 \mid \text{pmtn}, r_j \mid \sum w_j C_j$	$\propto$	$1 \mid \text{chains}, r_j, p_j = 1 \mid \sum w_j C_j$	††	[48]
[70]	††	$P \mid \text{pmtn, intree}, r_j, p_j = 1 \mid C_{\max}$	$\propto$	$P \mid \text{intree}, r_j, p_j = 1 \mid C_{\max}$	††	[65]
[70]	††	$P \mid \text{pmtn, outtree}, p_j = 1 \mid L_{\max}$	$\propto$	$P \mid \text{outtree}, p_j = 1 \mid L_{\max}$	††	[65]
[71]	††	$P \mid \text{pmtn, prec}, p_j = 1 \mid C_{\max}$	$\propto$	$P \mid \text{prec}, p_j = 1 \mid C_{\max}$	††	[72]
[62]	††	$J2 \mid \text{pmtn} \mid C_{\max}$	$\propto$	$J2 \mid \text{chains}, p_{ij} = 1 \mid C_{\max}$	††	[73]
Table 8.3	††	$P2 \mid \text{pmtn, chains}, p_j = 1 \mid \sum U_j$	$\propto$	$P2 \mid \text{chains}, p_j = 1 \mid \sum U_j$	††	Table 8.3
Table 8.3	††	$P2 \mid \text{pmtn, chains}, p_j = 1 \mid \sum T_j$	$\propto$	$P2 \mid \text{chains}, p_j = 1 \mid \sum T_j$	††	Table 8.3
[33]	††	$P2 \mid$ ~~pmtn~~, chains $\mid \sum w_j C_j$	$\propto$	$P2 \mid \text{chains}, p_j = 1 \mid \sum w_j C_j$	††	[26]
[70]	††	$O2 \mid \text{pmtn, chains} \mid C_{\max}$	$\propto$	$O2 \mid \text{chains}, p_{ij} = 1 \mid C_{\max}$	††	[26]
[70]	††	$O2 \mid \text{pmtn} \mid \sum w_j C_j$	$\propto$	$O2 \mid \text{chains}, p_{ij} = 1 \mid \sum w_j C_j$	††	[26]

The strong NP-hardness of the last three problems in the left column follows from a trivial polynomial reduction from $P2 \mid$ ~~pmtn~~, chains $\mid \sum C_j$ [33], and obvious pseudopolynomial reductions from $O2 \mid \text{pmtn, chains} \mid C_{\max}$ [70], and $O2 \mid \text{pmtn} \mid \sum w_j C_j$ [70], respectively. Lawler [19] shows that the polynomial algorithms of Brucker, Garey, and Johnson [65] and of Garey and Johnson [67] for the nonpreemptive problems with unit-time jobs on identical parallel machines can be adjusted to solve the related preemptive problems with arbitrary processing times on identical and even uniform parallel machines. Pseudopolynomial algorithms for the preemptive single-machine problems can be found in Ref. [42,55].

Table 8.6 shows that preemption chaining cannot be extended in general to cases where $\alpha_1 \in \{F, O\}$ or $\gamma = \sum C_j$, unless $P = \text{NP}$. The open questions in preemption chaining are

$$P\alpha_2 \mid \cancel{\text{pmtn}} \mid \sum C_j \overset{?}{\propto} P\alpha_2 \mid \text{chains}, p_j = 1 \mid \sum C_j \tag{8.10}$$

$$P\alpha_2 \mid \text{pmtn}, r_j \mid \sum C_j \overset{?}{\propto} P\alpha_2 \mid \text{chains}, r_j, p_j = 1 \mid \sum C_j \tag{8.11}$$

$$O\alpha_2 \mid \text{pmtn} \mid \sum C_j \overset{?}{\propto} O\alpha_2 \mid \text{chains}, p_{ij} = 1 \mid \sum C_j \tag{8.12}$$

$$C\alpha_2 \mid \text{pmtn} \mid \sum C_j \overset{?}{\propto} C\alpha_2 \mid \text{chains}, p_{ij} = 1 \mid \sum C_j \tag{8.13}$$

An answer to Question (8.10) is not known even in the single-machine case. Since the parallel-machine problems and open-shop problems to the right can be solved in polynomial time [26], positive answers to Questions (8.11) and (8.12) would give pseudopolynomial algorithms for the preemptive problems to the left whose two-machine special cases $P2 \mid \text{pmtn}, r_j \mid \sum C_j$ [57], and $O2 \mid \text{pmtn} \mid \sum C_j$ [61] are proved to be NP-hard but remain open for the strong NP-hardness. As to Question (8.13), we can say only that $C2 \mid \text{pmtn} \mid \sum C_j$ is strongly NP-hard even with three-operation jobs (see Chapter 7), but the complexity status of $C2 \mid \text{chains}, p_{ij} = 1 \mid \sum C_j$ remains open.

Note that preemption chaining with $\gamma = \sum C_j$ is possible. For example, the reduction

$$J2 \mid \text{pmtn} \mid \sum C_j \propto J2 \mid \text{chains}, p_j = 1 \mid \sum C_j$$

follows from Cook's theorem because both problems are strongly NP-hard [10,62]. Other examples of this kind have not been found.

TABLE 8.6 Counterexamples to Preemption Chaining with $\alpha_1 \in \{F, O\}$, $\gamma = \sum C_j$

[33]	††	$P2 \mid$ ~~pmtn~~ $, \text{chains} \mid \sum C_j$	$P2 \mid \text{chains}, p_j = 1 \mid \sum C_j$	✓	[26]
[61]	††	$F2 \mid \text{pmtn} \mid \sum C_j$	$F2 \mid \text{chains}, p_j = 1 \mid \sum C_j$	✓	[26]
[70]	††	$F2 \mid \text{pmtn, chains} \mid C_{\max}$	$F2 \mid \text{chains}, p_j = 1 \mid C_{\max}$	✓	[26]
[62,74]	††	$F2 \mid \text{pmtn}, r_j \mid C_{\max}$	$F2 \mid \text{chains}, r_j, p_j = 1 \mid C_{\max}$	✓	[26]
[62,74]	††	$F2 \mid \text{pmtn} \mid L_{\max}$	$F2 \mid \text{chains}, p_j = 1 \mid L_{\max}$	✓	[16]
[62,74]	††	$F3 \mid \text{pmtn} \mid C_{\max}$	$F3 \mid \text{chains}, p_j = 1 \mid C_{\max}$	✓	[26]
[70]	††	$O2 \mid \text{pmtn, chains} \mid C_{\max}$	$O2 \mid \text{chains}, p_j = 1 \mid C_{\max}$	✓	[26]
[75]	††	$O2 \mid \text{pmtn}, r_j \mid \sum C_j$	$O2 \mid \text{chains}, r_j, p_j = 1 \mid \sum C_j$	✓	[76]
[77]	††	$O3 \mid \text{pmtn} \mid \sum C_j$	$O3 \mid \text{chains}, p_j = 1 \mid \sum C_j$	✓	[26]

As shown in the papers cited in the right column, even more general problems than those indicated can be solved in polynomial time.

8.4 Parallel Machines vs. Open Shops

8.4.1 Latinization

8.4.1.1 Partition Matrices and Latin Rectangles

Any preemptive schedule of n jobs with processing times m and maximum completion time k on m identical parallel machines can be presented by an $[m, n, k]$ *partition matrix,* i.e., a matrix with m rows, k columns, and elements $1, 2, \ldots, n$, where $m \leq k$, such that every element appears in m columns, at most once in each column. Any m-machine unit-time open-shop schedule with n jobs and maximum completion time k can be presented by an $[m, n, k]$ *latin rectangle,* i.e., a matrix with m rows, k columns, and elements $1, 2, \ldots, n$, where $m \leq k$, such that every element appears only once in each row and at most once in each column.

Obviously, any $[m, n, k]$ latin rectangle is an $[m, n, k]$ partition matrix, and hence any m-machine unit-time open-shop schedule with n jobs and maximum completion time k is isomorphic to a preemptive schedule of n jobs with processing times m and maximum completion time k on m identical parallel machines. The inverse assertion is obviously not true; however, any $[m, n, k]$ partition matrix can be transformed into an $[m, n, k]$ latin rectangle by permutations of elements in each column. Such a transformation or its scheduling counterpart we call a *latinization.*

8.4.1.2 Edge Coloring in Bipartite Graphs

The fact that the latinization is always possible follows from a classical result on *edge-coloring* of bipartite graphs. We say that the edges of a graph are correctly colored if every two edges with a common end have different colors. We consider only bipartite graphs whose set of vertices can be divided into two parts such that no part contains both ends of an edge. König's theorem (cf. [78,79]) states that the minimum number of colors needed for correct coloring edges of a bipartite graph equals its maximum vertex degree.

Let $\{1, 2, \ldots, n\}$ and $\{1, 2, \ldots, k\}$ be parts of the bipartite graph $G[n, k]$ where the edge (j, t) exists if and only if element j appears in the tth column of the $[m, n, k]$ partition matrix, i.e., the job J_j is processed in the tth time unit of the parallel-machine schedules defined by the matrix. Since every element appears in m columns and every column contains at most m elements, the maximum vertex degree in $G[n, k]$ is m. Hence, by König's theorem, m is the minimum number of colors in edge coloring.

Let us identify m colors with the rows of the $[m, n, k]$ partition matrix. Since every two edges with a common end have different colors, edge coloring defines permutations of elements in the columns of the $[m, n, k]$ partition matrix that transforms it into an $[m, n, k]$ latin rectangle; and the set of edge colorings by m colors of the graph $G[n, k]$ is in a one-to-one correspondence with the set of $[m, n, k]$ latin rectangles

Partition Matrix

	4	2		2
1	1	4	1	3
2	3	3		4

Latin Rectangle

1	3	4		2
	4	2	1	3
2	1	3		4

Edge coloring

Job-Time (Element-Column) — Machine (Row)

Edge	Color
1–1	1
1–2	3
1–4	2
2–1	3
2–3	2
2–5	1
3–2	1
3–3	3
3–5	2
4–2	2
4–3	1
4–5	3

FIGURE 8.1 The $[3, 4, 5]$ partition matrix presents a schedule with maximum completion time 5 for a four-job instance of a problem in the class $P3 \mid 1\text{-pmtn}, \beta_3, \beta_4, p_j = 3 \mid \gamma$. The $[3, 4, 5]$ latin rectangle presents the corresponding schedule for the corresponding instance of the equivalent problem in the class $O3 \mid \beta_3, \beta_4, p_{ij} = 1 \mid \gamma$. The table presents the corresponding edge coloring of the related bipartite graph $G[4, 5]$ with three colors.

that can be obtained from the $[m, n, k]$ partition matrix by permutations of elements in its columns. Figure 8.1 gives a simple example.

8.4.1.3 Latinization by Edge-Coloring Algorithms

An edge-coloring algorithm directly follows from the classical proof of König's theorem using the fact that all paths (including cycles) in a bipartite graph are even, i.e., contain even number of edges. Let m be the maximum vertex degree in a bipartite graph.

If $m = 1$, then edge-coloring is trivial. If $m = 2$, then we go over every path of the graph coloring the edges by simply alternating the two colors. Since every path is even, every two edges with a common end are colored differently. If $m \geq 3$, then we color edges of the graph in any order using a new color only if it is necessary. During this process we maintain the set of colors χ_v for every vertex v that have been used for coloring edges incident to v. It is clear that we can continue unless we meet an uncolored edge (j, t) with

$$|\chi_j \cup \chi_t| = m \quad \text{and} \quad |(\chi_j \cup \chi_t) - (\chi_j \cap \chi_t)| \geq 2$$

If this happens, we choose any two colors a and b in $(\chi_j \cup \chi_t) - (\chi_j \cap \chi_t)$ and consider the subgraph generated by (j, t) and all of the edges with colors a and b. Obviously, the maximum vertex degree in the subgraph is at most two, i.e., it consists of separate paths of even length. Hence, we can recolor the edges of the subgraph by alternating a and b. Since the edges of the residual graph are not colored by a and b, we include (j, t) in edge coloring. This algorithm requires roughly a quadratic time in the number of edges, i.e., mn in our case.[7]

[7]Gabow and Kariv [80] and Cole and Hopcroft [81] propose more efficient edge-coloring algorithms for bipartite graphs with time complexities $O(mn \log^2 mn)$ and $O(mn \log(m + n))$, respectively. Using the model of Liu and Bulfin [3] based on the assignment problem, Brucker, Jurisch, and Jurisch [6] proposed another version of latinization by an assignment algorithm whose time complexity is $O(mn^2)$.

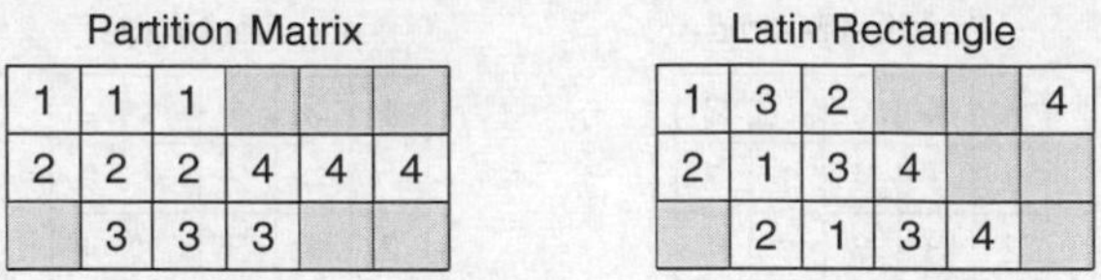

FIGURE 8.2 The $[3, 4, 6]$ partition matrix presents a nonpreemptive schedule with maximum completion time 6 for a four-job instance of a problem in the class $P3 \mid \beta_3, \beta_4, p_j = 3 \mid \gamma$. The cyclic shift transforms the schedule into a $[3, 4, 6]$ latin rectangle that presents the corresponding no-wait schedule for the corresponding instance of the equivalent problem in the class $O3 \mid$ no-wait, $\beta_3, \beta_4, p_{ij} = 1 \mid \gamma$.

Note that latinization algorithms with time complexity polynomial in mn are of pseudopolynomial time if the size of the equivalent problems is n. Hence, the latinization can be done in polynomial time by edge-coloring algorithms only if $m < n$ or m is fixed. We will see further how to manage the case where m is not fixed and $m \geq n$.

8.4.1.4 Latinization by Cyclic Shift

The latinization also establishes an equivalence between nonpreemptive problems of scheduling jobs with processing times m on m identical parallel machines and m-machine unit-time no-wait open-shop scheduling problems. In this case, however, $[m, n, k]$ partition matrices can be transformed into $[m, n, k]$ latin rectangles even more simply by a *cyclic shift* of elements inside columns (which means a cyclic permutation of the machines inside time units).[8]

Specifically, the cyclic shift moves the element of an $[m, n, k]$ partition matrix in row i and column t from row i to row $i + t \bmod m$ without changing the column. In other words, it moves the part of a job that is assigned to be processed in a schedule in the time unit $[t, t + 1]$ from the machine M_i to the machine $M_{i+t \bmod m}$. Informally speaking, the cyclic shift uniformly "twists" a schedule. See Figure 8.2 for an example.

Since a nonpreemptive schedule of jobs with processing times m on m identical parallel machines complemented by the formula $(i, t) \mapsto (i + t \bmod m, t)$ defines an m-machine unit-time no-wait open-shop schedule, the latinization in the nonpreemptive case can be trivially done in constant time.

Note that the latinization of preemptive parallel-machine problems of size n in the case where m is not fixed and $m \geq n$ can also be done in constant time because, in accordance with Equivalence G in Table 8.2, preemptions are redundant if $m \geq n$.

8.4.1.5 Latinization Equivalences

Thus, for any criterion γ, we have proved the following two equivalences that we call the *preemptive no-no-wait latinization* and the *nonpreemptive no-wait latinization*:

$$P\alpha_2 \mid 1\text{-pmtn}, \beta_{345}, p_j = m \mid \gamma \sim O\alpha_2 \mid \beta_{345}, p_{ij} = 1 \mid \gamma \tag{8.14}$$

$$P\alpha_2 \mid \beta_{345}, p_j = m \mid \gamma \sim O\alpha_2 \mid \text{no-wait}, \beta_{345}, p_{ij} = 1 \mid \gamma \tag{8.15}$$

Table 8.7 presents examples of the latinizations.

[8] Such a transformation was considered by Adiri and Amit [82] and Kubiak, Sriskandarajah, and Zaras [5] for solving the simplest unit-time open-shop scheduling problems.

TABLE 8.7 Examples of Latinizations

(a) Equivalences by preemptive no-no-wait latinization						
[28]	✓	P | 1-~~pmtn~~, $p_j = m \mid \sum C_j$	~	$O \mid p_{ij} = 1 \mid \sum C_j$	✓	[82]
[28]	✓	P | 1-~~pmtn~~, $p_j = m \mid \sum w_j C_j$	~	$O \mid p_{ij} = 1 \mid \sum w_j C_j$	✓	[1]
[3]	✓	P | 1-~~pmtn~~, $p_j = m \mid \sum U_j$	~	$O \mid p_{ij} = 1 \mid \sum U_j$	✓	[3]
[3]	✓	P | 1-pmtn, $p_j = m \mid \sum T_j$	~	$O \mid p_{ij} = 1 \mid \sum T_j$	✓	[3]
[26]	✓	P | 1-~~pmtn~~, chains, $p_j = m \mid \sum C_j$	~	O | chains, $p_{ij} = 1 \mid \sum C_j$	✓	[26]
[25]	✓	P | 1-~~pmtn~~, outtree, $p_j = m \mid \sum C_j$	~	O | outtree, $p_{ij} = 1 \mid \sum C_j$	✓	[4]
[52]	✓	$P2$ | 1-pmtn, prec, $p_j = 2 \mid C_{\max}$	~	$O2$ | prec, $p_{ij} = 1 \mid C_{\max}$	✓	[23]
[23]	✓	$P2$ | 1-pmtn, prec, $p_j = 2 \mid \sum C_j$	~	$O2$ | prec, $p_{ij} = 1 \mid \sum C_j$	✓	[23]
[76]	✓	$P2$ | 1-pmtn, outtree, $r_j, p_j = 2 \mid \sum C_j$	~	$O2 \mid r_j$, outtree, $p_{ij} = 1 \mid \sum C_j$	✓	[76]
[83]	✓	Pm | 1-pmtn, $r_j, p_j = m \mid \sum C_j$	~	$Om \mid r_j, p_{ij} = 1 \mid \sum C_j$	✓	[83]
[84]	✓	Pm | 1-pmtn, $r_j, p_j = m \mid \sum w_j U_j$	~	$Om \mid r_j, p_{ij} = 1 \mid \sum w_j U_j$	✓	[84]
[85]	††	P | 1-pmtn, $r_j, p_j = m \mid \sum U_j$	~	$O \mid r_j, p_{ij} = 1 \mid \sum U_j$	††	[85]
(b) Equivalences by nonpreemptive no-wait latinization						
[5]	✓	$P \mid p_j = m \mid \sum w_j U_j$	~	O | no-wait, $p_{ij} = 1 \mid \sum w_j U_j$	✓	[5]
[86]	✓	$P \mid p_j = m \mid \sum w_j T_j$	~	O | no-wait, $p_{ij} = 1 \mid \sum w_j T_j$	✓	[6]
[25]	✓	P | tree, $p_j = m \mid C_{\max}$	~	O | no-wait, tree, $p_{ij} = 1 \mid C_{\max}$	✓	[6]
[65]	✓	P | outtree, $r_j, p_j = m \mid C_{\max}$	~	O | no-wait, outtree, $r_j, p_{ij} = 1 \mid C_{\max}$	✓	[6]
[65]	✓	P | intree, $p_j = m \mid L_{\max}$	~	O | no-wait, intree, $p_{ij} = 1 \mid L_{\max}$	✓	[6]
[26]	✓	P | chains, $p_j = m \mid C_{\max}$	~	O | no-wait, chains, $p_{ij} = 1 \mid C_{\max}$	✓	[26]
[26]	✓	$P2$ | chains, $c_k, p_j = 2 \mid C_{\max}$	~	$O2$ | no-wait, chains, $c_k, p_{ij} = 1 \mid C_{\max}$	✓	[26]
[26]	✓	$P2$ | chains, $c_k, p_j = 2 \mid \sum C_j$	~	$O2$ | no-wait, chains, $c_k, p_{ij} = 1 \mid \sum C_j$	✓	[26]
[27]	✓	$P2$ | prec, $p_j = 2 \mid \sum C_j$	~	$O2$ | no-wait, prec, $p_{ij} = 1 \mid \sum C_j$	✓	[26]
[87]	✓	$Pm \mid r_j, p_j = m \mid \sum w_j C_j$	~	Om | no-wait, $r_j, p_{ij} = 1 \mid \sum w_j C_j$	✓	[26]
[87]	✓	$Pm \mid r_j, p_j = m \mid \sum T_j$	~	Om | no-wait, $r_j, p_{ij} = 1 \mid \sum T_j$	✓	[26]
[88]	✓	$P \mid r_j, p_j = m \mid \sum C_j$	~	O | no-wait, $r_j, p_{ij} = 1 \mid \sum C_j$	✓	[26]
[88]	✓	$P \mid r_j, p_j = m \mid L_{\max}$	~	O | no-wait, $r_j, p_{ij} = 1 \mid L_{\max}$	✓	[89]
[72]	††	P | prec, $p_j = m \mid C_{\max}$	~	O | no-wait, prec, $p_{ij} = 1 \mid C_{\max}$	††	[6]
[90]	††	P | prec, $p_j = m \mid \sum C_j$	~	O | no-wait, prec, $p_{ij} = 1 \mid \sum C_j$	††	[6]
[65]	††	P | intree, $r_j, p_j = m \mid C_{\max}$	~	O | no-wait, intree, $r_j, p_{ij} = 1 \mid C_{\max}$	††	[6]
[65]	††	P | outtree, $p_j = m \mid L_{\max}$	~	O | no-wait, outtree, $p_{ij} = 1 \mid L_{\max}$	††	[6]
[26]	††	$P2$ | chains, $p_j = 2 \mid \sum w_j C_j$	~	$O2$ | no-wait, chains, $p_{ij} = 1 \mid \sum w_j C_j$	††	[26]
Table 8.3	††	$P2$ | chains, $p_j = 2 \mid \sum U_j$	~	$O2$ | no-wait, chains, $p_{ij} = 1 \mid \sum U_j$	††	[26]
Table 8.3	††	$P2$ | chains, $p_j = 2 \mid \sum T_j$	~	$O2$ | no-wait, chains, $p_{ij} = 1 \mid \sum T_j$	††	[26]

The last three problems on two identical parallel machines are isomorphic to their unit-time counterparts. Hence, the strong NP-hardness of the first follows from Table 8.5, and the strong NP-hardness of the other two follows from Table 8.3.

It is important to note that integer preemptions are essential in Equivalence (8.14) and cannot be replaced by arbitrary preemptions in the reduction from unit-time open shops to identical parallel machines. See Figure 8.3 for an example. The inverse reduction, as we show in the section devoted to open shopping, can be extended to arbitrary preemptions if $\gamma \neq \sum C_j$ at the expense of additional chain-like precedence constraints.

Summing up the results of this section, we emphasize that the latinization algorithms presented here for preemptive parallel-machine problems are polynomial only if the problem size is polynomial in n. If the problem size is not polynomial in n and preemptions are not redundant, then the algorithms are not polynomial due to a size collapse.

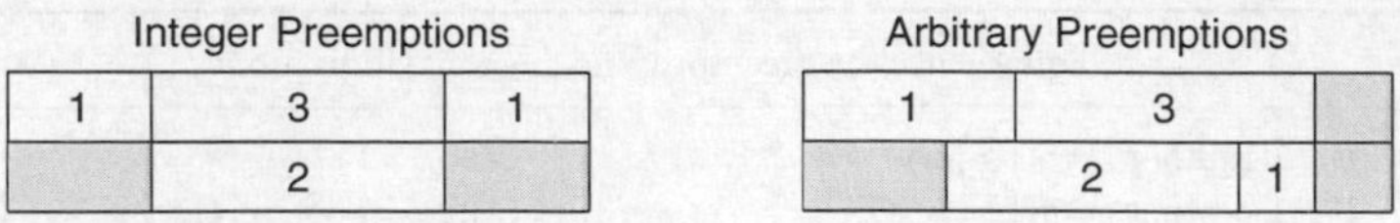

FIGURE 8.3 An optimal schedule for the instance of $P2 \mid 1\text{-pmtn}, p_j = 2 \mid C_{\max}$ with three jobs and release dates 0, 1, 1 is of length 4, but an optimal schedule for the same instance of $P2 \mid \text{pmtn}, p_j = 2 \mid C_{\max}$ is of length 7/2.

Among problems in the class $P\alpha_2 \mid 1\text{-pmtn}, \beta_{345}, p_j = m \mid \gamma$ there are only six problems whose size is not polynomial in n. These are

$$P\alpha_2 \mid 1\text{-pmtn}, p_j = m \mid C_{\max}, \quad P\alpha_2 \mid 1\text{-pmtn}, p_j = m \mid \sum C_j$$
$$P\alpha_2 \mid 1\text{-pmtn, chains}, p_j = m \mid C_{\max}, \quad P\alpha_2 \mid 1\text{-pmtn, chains}, p_j = m \mid \sum C_j$$
$$P\alpha_2 \mid 1\text{-pmtn, chains}, c_k, p_j = m \mid C_{\max}, \quad P\alpha_2 \mid 1\text{-pmtn, chains}, c_k, p_j = m \mid \sum C_j$$

The size of the problems without precedence constraints is 1. The size of the problems with chain-like precedence constraints is the number of chains. Obtaining the latinization for these problems in polynomial time requires finding special algorithms. The cases with equal release dates and the total completion time criterion, however, are easily manageable since preemptions, by Reduction A and B in Table 8.2, are redundant. Therefore, the latinizations

$$P\alpha_2 \mid 1\text{-pmtn}, p_j = m \mid \sum C_j \sim O\alpha_2 \mid p_{ij} = 1 \mid \sum C_j$$
$$P\alpha_2 \mid 1\text{-pmtn, chains}, p_j = m \mid \sum C_j \sim O\alpha_2 \mid \text{chains}, p_{ij} = 1 \mid \sum C_j$$

can be done in constant time by the cyclic shift. The latinizations with the maximum completion time criterion

$$P\alpha_2 \mid 1\text{-pmtn}, p_j = m \mid C_{\max} \sim O\alpha_2 \mid p_{ij} = 1 \mid C_{\max}$$
$$P\alpha_2 \mid 1\text{-pmtn, chains}, p_j = m \mid C_{\max} \sim O\alpha_2 \mid \text{chains}, p_{ij} = 1 \mid C_{\max}$$

can also be done in constant time because the problems in the first pair are equivalent to $F\alpha_2 \mid p_{ij} = 1 \mid C_{\max}$, and the problems in the second pair are equivalent to $F\alpha_2 \mid \text{chains}, p_{ij} = 1 \mid C_{\max}$.

Let us consider an instance of the flow-shop problem without precedence constraints. It has a trivial optimal schedule of length $m + n - 1$. It is easy to see that the schedule can be transformed into an optimal schedule of length $\max\{m, n\}$ for the related instance of the open-shop problem (as well as the parallel-machine problem) by shifting the content of the time interval $[\max\{m, n\}, m + n - 1]$ to the time interval $[0, m + n - 1 - \max\{m, n\}]$. Figure 8.4 gives two examples.

There exists a similar transformation of optimal schedules of flow-shop problems with chain-like precedence constraints. It requires only a polynomial algorithm of scheduling chains in unit-time flow shops. Refer to [26] for details.

Thus, Equivalence (8.14) is proved to be of polynomial time except for the following two questionable reductions:

$$O\alpha_2 \mid \text{chains}, c_k, p_{ij} = 1 \mid C_{\max} \overset{?}{\propto} P\alpha_2 \mid 1\text{-pmtn, chains}, c_k, p_j = m \mid C_{\max}$$
$$O\alpha_2 \mid \text{chains}, c_k, p_{ij} = 1 \mid \sum C_j \overset{?}{\propto} P\alpha_2 \mid 1\text{-pmtn, chains}, c_k, p_j = m \mid \sum C_j$$

It is known that the parallel-machine problems can be solved in polynomial time [26], while the open-shop problems remain open.

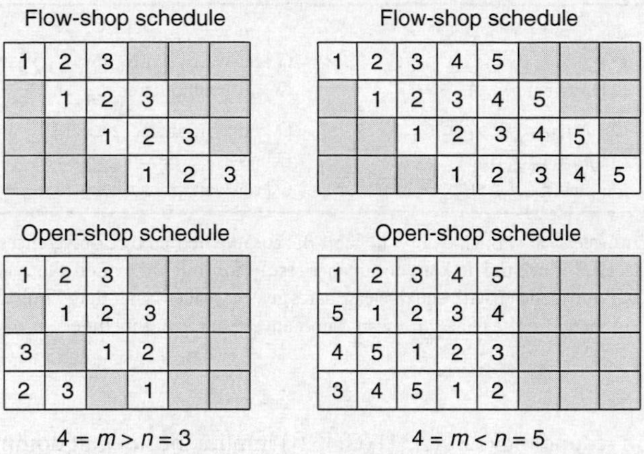

FIGURE 8.4 Transformations of an optimal unit-time flow-shop schedule to an optimal unit-time open-shop schedule for the instances with three and five jobs.

8.4.2 No-Wait Dominance

We say that a no-wait constraint is dominant for a shop scheduling problem if it does not worsen optimal schedules for this problem. Each case of preemption redundancy in a preemptive parallel-machine problem with $p_j = m$ can be converted to the related no-wait dominance case in the related unit-time open shop by using Equivalences (8.14) and (8.15). Note that the redundancy of arbitrary preemptions implies the redundancy of integer preemptions.

8.4.2.1 By McNaughton's Theorem

Equivalence A in Table 8.2 applied to the problem of scheduling jobs with processing times m and complemented by Equivalences (8.14) and (8.15) gives the following diagram:

$$\begin{array}{ccc} P \mid 1\text{-pmtn}, p_j = m \mid \sum w_j C_j & \overset{A}{\sim} & P \mid p_j = m \mid \sum w_j C_j \\ \wr\ (8.14) & & \wr\ (8.15) \\ O \mid p_{ij} = 1 \mid \sum w_j C_j & \sim & O \mid \text{no-wait}, p_{ij} = 1 \mid \sum w_j C_j \end{array}$$

The lower equivalence here proves the no-wait dominance in unit-time open shops with no precedence constraints, equal release dates and the total weighted completion time criterion.

8.4.2.2 By Other Preemption Redundancy Results

Parallel-machine problems with $p_j = 1$ and equal release dates being expanded by factor m also produce no-wait dominance results in related unit-time open shops. For example, Equivalence C in Table 8.2 gives the following diagram:

$$\begin{array}{ccc} P \mid 1\text{-pmtn, outtree}, p_j = m \mid \sum C_j & \overset{C}{\sim} & P \mid \text{outtree}, p_j = m \mid \sum C_j \\ \wr\ (8.14) & & \wr\ (8.15) \\ O \mid \text{outtree}, p_{ij} = 1 \mid \sum C_j & \sim & O \mid \text{no-wait, outtree}, p_{ij} = 1 \mid \sum C_j \end{array}$$

All the no-wait dominance results for regular criteria we are aware of are presented in Table 8.8. Equivalences H, I, and J in Table 8.2 do not give no-wait dominance results because open shops have at least two machines.

TABLE 8.8 No-Wait Dominance Results

a	$O \mid p_{ij}=1 \mid \sum w_j C_j$	$\sim$	$O \mid \text{no-wait}, p_{ij}=1 \mid \sum w_j C_j$
b	$O \mid \text{chains}, p_{ij}=1 \mid \sum w_j C_j$	$\sim$	$O \mid \text{no-wait, chains}, p_{ij}=1 \mid \sum w_j C_j$
c	$O \mid \text{outtree}, p_{ij}=1 \mid \sum C_j$	$\sim$	$O \mid \text{no-wait, outtree}, p_{ij}=1 \mid \sum C_j$
d	$O \mid \text{intree}, p_{ij}=1 \mid \sum w_j U_j$	$\sim$	$O \mid \text{no-wait, intree}, p_{ij}=1 \mid \sum w_j U_j$
f	$O \mid p_{ij}=1 \mid \sum w_j T_j$	$\sim$	$O \mid \text{no-wait}, p_{ij}=1 \mid \sum w_j T_j$
g	$O \mid \text{prec}, r_j, n \le m, p_{ij}=1 \mid \gamma$	$\sim$	$O \mid \text{no-wait, prec}, r_j, n \le m, p_{ij}=1 \mid \gamma$

Equivalences A, B, C, D, F, G in Table 8.2 complemented by Equivalences 8.14 and 8.15 produce the no-wait dominance results for unit-time open shops a, b, c, d, f, g. Equivalence E with equal release dates gives a weaker result than Equivalence D, and therefore the related no-wait dominance result is not included.

Tanaev's preemption redundancy result [31] (cf. [32]) implies the no-wait dominance in $O \mid p_{ij}=1 \mid \gamma$, i.e.,

$$O \mid p_{ij}=1 \mid \gamma \sim O \mid \text{no-wait}, p_{ij}=1 \mid \gamma$$

if the criterion γ is a nondecreasing e-quasiconcave function of job completion times (see Section 8.2.6.5).

8.4.3 Open Shopping and Open-Shop Paralleling

8.4.3.1 Open Shopping Diagram

The composition of Expansion (8.6) and Chaining (8.8) with $\alpha_1 = P$, $p = 1$, and $q = m$; and Latinization (8.14), gives the following diagram:

$$\begin{array}{ccc} P\alpha_2 \mid 1\text{-pmtn}, \beta_3, m\beta_{47} \mid \gamma & \overset{(8.8)}{\propto} & P\alpha_2 \mid 1\text{-pmtn, chain-}\beta_3, m\beta_4, p_j=m \mid \gamma \\ \wr\wr\ (8.6) & & \wr\ (8.14) \\ P\alpha_2 \mid \text{pmtn}, \beta_{347} \mid \gamma & \propto & O\alpha_2 \mid \text{chain-}\beta_3, m\beta_4, p_{ij}=1 \mid \gamma \end{array}$$

The lower reduction here is *open shopping.* Since the operation chaining technique does not work for the total completion time criterion, open shopping is valid for all proportion and permanence criteria except $\gamma = \sum C_j$. Table 8.9 presents examples of open shopping.

As with preemption chaining, open shopping cannot be extended in general to the case $\gamma = \sum C_j$, unless P=NP, because $P2 \mid$ ~~pmtn~~, chains $\mid \sum C_j$ is strongly NP-hard [33] but $O2 \mid \text{chains}, p_{ij} = 1 \mid \sum C_j$ can be solved in polynomial time [26].

Note that a polynomial algorithm for $O2 \mid \text{chains}, p_{ij}=1 \mid \sum C_j$ is essential since polynomial algorithms for $O2 \mid \text{outtree}, r_j, p_{ij} = 1 \mid \sum C_j$ [76], and $O2 \mid \text{prec}, p_{ij} = 1 \mid \sum C_j$ [23] do not solve it in polynomial time. The open questions in open shopping are

$$P\alpha_2 \mid \sout{\text{pmtn}} \mid \sum C_j \overset{?}{\propto} O\alpha_2 \mid \text{chains}, p_j=1 \mid \sum C_j \tag{8.16}$$

$$P\alpha_2 \mid \text{pmtn}, r_j \mid \sum C_j \overset{?}{\propto} O\alpha_2 \mid \text{chains}, r_j, p_j=1 \mid \sum C_j \tag{8.17}$$

Since $O2 \mid \text{chains}, r_j, p_j = 1 \mid \sum C_j$ [26], and even $O2 \mid \text{outtree}, r_j, p_j = 1 \mid \sum C_j$ [76] can be solved in polynomial time, a positive answer to Question (8.17) in the two-machine case would give a pseudopolynomial algorithm for $P2 \mid \text{pmtn}, r_j \mid \sum C_j$, which is proved to be NP-hard [57] but remains open for the strong NP-hardness.

TABLE 8.9 Reductions by Open Shopping

[28]	✓	$P\,\|\,\text{pmtn}\,\|\,C_{\max}$	$\propto\!\!\!\mid$	$O\,\|\,\text{chains}, p_j=1\,\|\,C_{\max}$	✓	[26]
[63,64]	✓	$P\,\|\,\text{pmtn}, r_j\,\|\,C_{\max}$	$\propto\!\!\!\mid$	$O\,\|\,\text{chains}, r_j, p_j=1\,\|\,C_{\max}$	✓	[6]
[64,66]	✓	$P\,\|\,\text{pmtn, tree}\,\|\,C_{\max}$	$\propto\!\!\!\mid$	$O\,\|\,\text{tree}, p_j=1\,\|\,C_{\max}$	✓	[8]
[19]	✓	$P\,\|\,\text{pmtn, outtree}, r_j\,\|\,C_{\max}$	$\propto\!\!\!\mid$	$O\,\|\,\text{outtree}, r_j, p_j=1\,\|\,C_{\max}$	✓	[6]
[19]	✓	$P\,\|\,\text{pmtn, intree}\,\|\,L_{\max}$	$\propto\!\!\!\mid$	$O\,\|\,\text{intree}, p_j=1\,\|\,L_{\max}$	✓	[6]
[19]	✓	$P2\,\|\,\text{pmtn, prec}, r_j\,\|\,L_{\max}$	$\propto\!\!\!\mid$	$O2\,\|\,\text{prec}, r_j, p_j=1\,\|\,L_{\max}$	✓	[6]
Table 8.3	††	$P2\,\|\,\text{pmtn, chains}, p_j=1\,\|\sum U_j$	$\propto$	$O2\,\|\,\text{chains}, p_{ij}=1\,\|\sum U_j$	††	[26]
Table 8.3	††	$P2\,\|\,\text{pmtn, chains}, p_j=1\,\|\sum T_j$	$\propto$	$O2\,\|\,\text{chains}, p_{ij}=1\,\|\sum T_j$	††	[26]
[33]	††	$P2\,\|$ ~~pmtn~~ $,\text{chains}\,\|\sum w_jC_j$	$\propto\!\!\!\mid$	$O2\,\|\,\text{chains}, p_{ij}=1\,\|\sum w_jC_j$	††	[26]
[70]	††	$P\,\|\,\text{pmtn, intree}, r_j, p_j=1\,\|\,C_{\max}$	$\propto$	$O\,\|\,\text{intree}, r_j, p_{ij}=1\,\|\,C_{\max}$	††	[26]
[70]	††	$P\,\|\,\text{pmtn, outtree}, p_j=1\,\|\,L_{\max}$	$\propto$	$O\,\|\,\text{outtree}, p_{ij}=1\,\|\,L_{\max}$	††	[26]
[71]	††	$P\,\|\,\text{pmtn, prec}, p_j=1\,\|\,C_{\max}$	$\propto$	$O\,\|\,\text{prec}, p_{ij}=1\,\|\,C_{\max}$	††	[26]

The strong NP-hardness of $P2\,|$ ~~pmtn~~, chains $|\sum w_jC_j$ follows from a trivial reduction from $P2\,|$ ~~pmtn~~, chains $|\sum C_j$ that is proved to be strongly NP-hard [33].

TABLE 8.10 Reductions by Open-Shop Paralleling

[26]	✓	$O\,\|\,p_{ij}=1\,\|\,C_{\max}$	$\propto$	$P\,\|\,\text{chains}, p_j=1\,\|\,C_{\max}$	✓	[26]
[26]	✓	$O\,\|\,\text{chains}, p_{ij}=1\,\|\,C_{\max}$	$\propto$	$P\,\|\,\text{chains}, p_j=1\,\|\,C_{\max}$	✓	[26]
[7]	✓	$O\,\|\,r_j, p_{ij}=1\,\|\,C_{\max}$	$\propto$	$P\,\|\,\text{chains}, r_j, p_j=1\,\|\,C_{\max}$	✓	[65]
[8]	✓	$O\,\|\,\text{tree}, p_{ij}=1\,\|\,C_{\max}$	$\propto$	$P\,\|\,\text{tree}, p_j=1\,\|\,C_{\max}$	✓	[25]
[6]	✓	$O\,\|\,\text{outtree}, r_j, p_{ij}=1\,\|\,C_{\max}$	$\propto$	$P\,\|\,\text{outtree}, r_j, p_j=1\,\|\,C_{\max}$	✓	[65]
[49]	✓	$O\,\|\,\text{chains}, r_j, p_{ij}=1\,\|\,L_{\max}$	$\propto$	$P\,\|\,\text{chains}, r_j, p_j=1\,\|\,L_{\max}$	✓	[91]
[6]	✓	$O\,\|\,\text{intree}, p_{ij}=1\,\|\,L_{\max}$	$\propto$	$P\,\|\,\text{intree}, p_j=1\,\|\,L_{\max}$	✓	[65]
[6]	✓	$O2\,\|\,\text{prec}, r_j, p_{ij}=1\,\|\,L_{\max}$	$\propto$	$P2\,\|\,\text{prec}, r_j, p_j=1\,\|\,L_{\max}$	✓	[67]
[26]	††	$O\,\|\,\text{prec}, p_{ij}=1\,\|\,C_{\max}$	$\propto$	$P\,\|\,\text{prec}, p_j=1\,\|\,C_{\max}$	††	[72]
[26]	††	$O\,\|\,\text{intree}, r_j, p_{ij}=1\,\|\,C_{\max}$	$\propto$	$P\,\|\,\text{intree}, r_j, p_j=1\,\|\,C_{\max}$	††	[65]
[26]	††	$O\,\|\,\text{outtree}, p_{ij}=1\,\|\,L_{\max}$	$\propto$	$P\,\|\,\text{outtree}, p_j=1\,\|\,L_{\max}$	††	[65]
[26]	††	$O2\,\|\,\text{chains}, p_{ij}=1\,\|\sum w_jC_j$	$\propto$	$P2\,\|\,\text{chains}, p_j=1\,\|\sum w_jC_j$	††	[26]
[26]	††	$O2\,\|\,\text{chains}, p_{ij}=1\,\|\sum U_j$	$\propto$	$P2\,\|\,\text{chains}, p_j=1\,\|\sum U_j$	††	Table 8.3
[26]	††	$O2\,\|\,\text{chains}, p_{ij}=1\,\|\sum T_j$	$\propto$	$P2\,\|\,\text{chains}, p_j=1\,\|\sum T_j$	††	Table 8.3

The unit-time open-shop problems can be solved by open-shop paralleling in polynomial time by the polynomial algorithms for the related parallel-machine problems.

8.4.3.2 Open-Shop Paralleling Diagram

The composition of Latinization (8.14), Chaining (8.8) with $\alpha_1 = P$, $pq = 1$, $q_j = m$, and Elimination (8.9) with $\alpha_1 = P$, $p = 1$, gives an inverse to open shopping:

$$\begin{array}{ccc} P\alpha_2\,|\,\text{1-pmtn}, \beta_{34}, p_j=m\,|\,\gamma & \overset{(8.8)}{\propto} & P\alpha_2\,|\,\text{1-pmtn, chain-}\beta_3, \beta_4, p_j=1\,|\,\gamma \\ \wr\ (8.14) & & \wr\wr\ (8.9) \\ O\alpha_2\,|\,\beta_{34}, p_j=1\,|\,\gamma & \propto & P\alpha_2\,|\,\text{chain-}\beta_3, \beta_4, p_{ij}=1\,|\,\gamma \end{array}$$

The lower reduction we call *open-shop paralleling*. Since parallel-machine with problems $p_j = m$ are not trivial only if $m \leq n$, arguments similar to those for latinization, show that Chaining (8.8) with $\alpha_1 = P$, $pq = 1$, $q_j = m$, and therefore, open-shop paralleling can be done in polynomial time. Table 8.10 presents examples of open-shop paralleling. Since the diagram involves the operation chaining technique, the reduction is valid for $\gamma \neq \sum C_j$.

However, in contrast to open shopping, it is not known whether open-shop paralleling can be extended to the case $\gamma = \sum C_j$, since counterexamples have not been found.

Note that preemption chaining of identical parallel machines can be obtained by composition of open shopping and open-shop paralleling. For example,

$$P \mid \text{pmtn} \mid C_{\max} \propto O \mid \text{chains}, p_{ij}=1 \mid C_{\max} \propto P \mid \text{chains}, p_j=1 \mid C_{\max}$$

The nonpreemptive no-wait latinization of the last problem expanded m times gives the following new no-wait dominance result for unit-time open shops:

$$O \mid \text{chains}, p_{ij}=1 \mid C_{\max} \propto O \mid \text{no-wait, chains}, p_{ij}=1 \mid C_{\max}$$

It is clear that Table 8.10 and the nonpreemptive no-wait part of Table 8.7 generate other no-wait dominance results.

8.5 Parallel Machines vs. Job Shops

8.5.1 Modulation

8.5.1.1 Linear Transformations

Let I be an instance of $\alpha \mid \beta \mid \gamma$, and let σ be a feasible schedule for I. Without loss of generality, we assume that $d_j = 0$ and $w_j = 1$ for $j = 1, 2, \ldots, n$ if γ does not include due dates and γ is not a weighted criterion, respectively. Set $w = w_1 + w_2 + \cdots + w_n$.

For positive integers a and b, let us construct the instance $aI+b$ from I by replacing p_{ij}, r_j, d_j by $ap_{ij}, ar_j, ad_j + b$ and the schedule $a\sigma$ by stretching σ a times. For positive integers $b_j \leq b$, define $a\sigma + b$ to be a feasible schedule for aI with completion times $aC_j + b_j$. Obviously, $a\sigma + b$ is a feasible schedule for $aI+b$.

Let, as before, $[\sigma, I]$ denote the γ value of σ. Then it is easy to check that, for the proportion criteria γ, we have

$$a[\sigma, I] = [a\sigma, aI] \leq [a\sigma + b, aI] \leq [a\sigma, aI] + bw \tag{8.18}$$

Now let $b > 0$ and $a/b > \ell^2$. Then from Property 2 we have

$$\begin{array}{ccccc} C_j - d_j > 0 \Leftrightarrow C_j - d_j & \geq 1/\ell^2 & & & \\ & \Updownarrow & & & \\ & a(C_j - d_j) \geq a/\ell^2 & \Leftrightarrow & aC_j - (ad_j + b) > 0 & \\ & & & \Updownarrow & \\ & & & (aC_j + b_j) - (ad_j + b) > 0 & \end{array}$$

This means that the sets of late jobs in σ for I, in $a\sigma$ for aI and in $a\sigma + b$ for $aI + b$ are the same. Therefore, for permanence criteria γ, we have

$$[\sigma, I] = [a\sigma, aI] = [a\sigma + b, aI + b] \tag{8.19}$$

8.5.1.2 Shift Procedure

Now we describe a *shift procedure* that transforms σ into a feasible schedule modulo m for I that is denoted as $\sigma \bmod m$.

Define $\mathcal{P}$ to be the set of all nonpreemptive parts of operations of jobs in σ. Introduce a precedence relation $\prec$ on $\mathcal{P}$ as follows. Set $x \prec y$ if and only if

1. x and y belong to J_j and J_k, respectively, such that J_j precedes J_k in accordance with given precedence constraints; or

2. x and y are parts of O_{ij} and O_{lj}, respectively, where $i < l$; or
3. x and y are parts of one operation, and y must be processed after x; or
4. x and y are scheduled in σ on one machine that processes y after x.

Let Min $\mathcal{P}$ be the set of minimal parts in $\mathcal{P}$ in accordance with $\prec$, Rest $\mathcal{P} = \mathcal{P} -$ Min $\mathcal{P}$, $\text{Near}_i\, t = \min\{u : u \geq t \wedge u = (i-1) \bmod m\}$, i.e., it is the earliest time modulo m on the machine M_i that is not earlier than t, and let Machine x be the index of the machine on which x is scheduled in σ. Let s_x and t_x denote the start times of x in σ and $\sigma \bmod m$, respectively. The following program presents the shift procedure:

while $\mathcal{P} \neq \emptyset$ **do**
  for $x \in$ Min $\mathcal{P}$ **do**
    $i \leftarrow$ Machine x
    $t_x \leftarrow \text{Near}_i\, s_x$
  $d \leftarrow \{t_x - s_x : x \in \text{Min}\, \mathcal{P}\}$
  for $x \in$ Rest $\mathcal{P}$ **do**
    $s_x \leftarrow s_x + d$
  $\mathcal{P} \leftarrow$ Rest $\mathcal{P}$

Informally, it moves the uninterrupted parts of σ to the right so that they attain the earliest possible start times $(i-1) \bmod m$ without violation of $\prec$. Since $d < m$, the procedure increases the completion time of every job by at most mg, where $g = |\mathcal{P}|$. Thus, $a\sigma \bmod m$ is an example of $a\sigma + mg$.

8.5.1.3 Modulation Reductions

Now we are ready to prove the *preemptive* and *nonpreemptive modulations*

$$\alpha \mid m\text{-pmtn}, \beta_{356}, m\beta_{47} \mid \gamma \;\propto\; \alpha_1 \bmod \alpha_2 \mid m\text{-pmtn}, \beta_{356}, m\beta_{47} \mid \gamma \tag{8.20}$$

$$\alpha \mid \beta_{356}, m\beta_{47} \mid \gamma \;\propto\; \alpha_1 \bmod \alpha_2 \mid \beta_{356}, m\beta_{47} \mid \gamma \tag{8.21}$$

To prove the case with proportion criteria, we take $a = (mgw+1)\ell^2, b = mg$ and construct the modulated instance (I', k') with $I' = aI$, $k' = ak + bw$. If σ is a feasible schedule for I, then, by Inequality (8.18), $a\sigma \bmod m$ is a feasible schedule for I':

$$[\sigma, I] \leq k \;\Rightarrow\; [a\sigma \bmod m, aI] = [a\sigma + b, aI] \leq a[\sigma, I] + bw \leq ak + bw = k'$$

On the other hand, let $[\sigma, I] > k$ for any feasible schedule σ for I. From Equality (8.18) and Property 2 we have: $[a\sigma, aI] = a[\sigma, I] \geq ak + a/\ell^2 = ak + bw + 1$ for any feasible schedule $a\sigma$ for aI. Using Equality (8.18) again, we obtain

$$[\sigma, I] > k \;\Rightarrow\; [a\sigma \bmod m, aI] = [a\sigma + b, aI] \geq a[\sigma, I] \geq ak + bw + 1 > k'$$

To prove the case with permanence criteria, we take the same a and b as those in the proof for proportion criteria and set $I' = aI + bw$, $k' = k$. Note that the inequality $a/b > \ell^2$ is satisfied. The Equality (8.19) shows that the answers for (I, k) and (I', k') are the same.

Informally, Modulations (8.20) and (8.21) say that the proportion and the permanence criteria remain insensitive to relatively small time shifts of operations when stretching schedules out in time by large amounts. On the other hand, the stretching can be small enough to save a polynomial length of schedule codes.

8.5.1.4 Dominance of *m*-Preemptions

We observe that m-preemptions are dominant in preemptive problems of scheduling jobs with processing times divisible by m on m modulated identical parallel machines, i.e.,

$$P \bmod \alpha_2 \mid m\text{-pmtn}, \beta_{345}, m\beta_7 \mid \gamma \sim P \bmod \alpha_2 \mid \text{pmtn}, \beta_{345}, m\beta_7 \mid \gamma \quad (8.22)$$

To prove this fact it is enough to notice that, if a job in a preemptive schedule on modulated identical parallel machines is scheduled on M_i and preempted at time t, then the time interval $[t, u]$, where u is the earliest time modulo m after time t, remains idle because M_i, by definition, starts jobs or their parts only at times $(i-1) \bmod m$. Hence, the portion of the remaining job of duration $u - t$ can be brought to the interval $[t, u]$ on M_i without increasing the completion times of all the jobs. Recursively applying this procedure, we can obtain a nonpreemptive schedule that is not worse than the original preemptive schedule.

8.5.2 Perfect Job Shop

8.5.2.1 Overlap Property

We consider only perfect[9] unit-time job-shop problems with the requirement that the durations of (maximal) uninterrupted parts of jobs in the schedule sought are divisible by m. As we show later, this requirement is actually redundant in finding active schedules.

Thus, every uninterrupted part of a job in a perfect unit-time job-shop schedule has the first operation on M_1 and the last operation on M_m. To determine such a schedule it is sufficient to indicate only start times and durations of uninterrupted parts of jobs because unit-time operations inside the parts are periodically distributed among the machines.

The *overlap property* of perfect unit-time job shops can be formulated as follows. Let I be an instance of a perfect unit-time job-shop problem, and let σ be a feasible schedule for I. Then the overlap of time intervals of two uninterrupted parts of jobs with start times s and t implies $s \neq t \bmod m$. The property trivially follows from contradiction because $s = t \bmod m$ implies that every machine must process two operations in each time unit of the overlap.

8.5.2.2 Job-Shopping Isomorphisms

Now we are ready to prove the *job shopping isomorphisms* that reveal links between modulated parallel machines and perfect job shops:

$$P \bmod \alpha_2 \mid m\text{-pmtn}, \beta_{345}, m\beta_7 \mid \gamma \approx J \text{ per } \alpha_2 \mid m\text{-no-wait}, \beta_{345}, m\beta_6, p_{ij}{=}1 \mid \gamma \quad (8.23)$$

$$P \bmod \alpha_2 \mid \beta_{345}, m\beta_7 \mid \gamma \approx J \text{ per } \alpha_2 \mid \text{no-wait}, \beta_{345}, m\beta_6, p_{ij}{=}1 \mid \gamma \quad (8.24)$$

Let I be an instance of the problem on modulated parallel machines. Then we construct the instance I' of the perfect unit-time job-shop problem by replacing every single-operation job J_j in I by a multi-operation job J'_j with unit-time operations $O_{1j}, \ldots, O_{mq_j,j}$ for all $j = 1, \ldots, n$. The replacement establishes a natural one-to-one correspondence between nonpreemptive parts of J_j and uninterrupted parts of J'_j of the same duration.

If σ is a feasible schedule for I, then it can be transformed into a feasible schedule σ' for I' by translating each nonpreemptive part of J_j in σ into the corresponding uninterrupted part of J'_j and assigning it the

[9] In the paper where perfect job shops were introduced [10], the term "periodic" was used instead of the term "perfect" because a route in a perfect job shop periodically changes the machines. We use the latter term in order to distinguish the structural periodicity related to the way of changing machines in routes from the time periodicity related to periodic schedules. The term "periodic" is commonly accepted now in literature devoted to rapidly growing theory of periodic schedules.

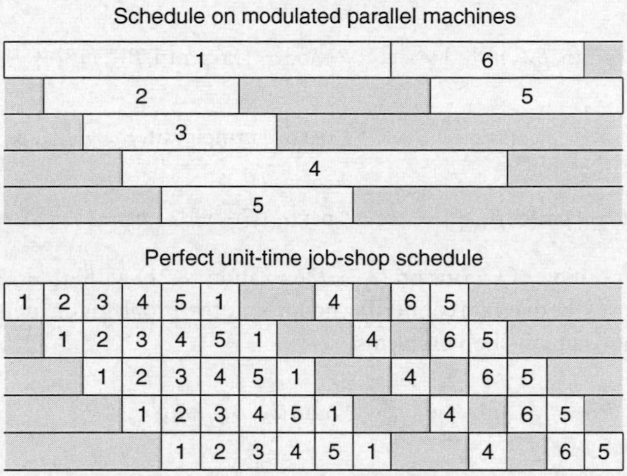

FIGURE 8.5 A schedule of six single-operation jobs with processing times 10, 5, 5, 10, 10, 5, respectively, on five modulated parallel machines and the isomorphic perfect unit-time five-machine job-shop schedule.

same start time. Since start times on different machines in σ are not comparable modulo m, start times of any two overlapping uninterrupted parts are also not comparable modulo m. Therefore, σ' is feasible for I'.

Analogously, any feasible schedule σ' for I' can be transformed into a feasible schedule σ for I by translating uninterrupted parts of σ' into the corresponding nonpreemptive parts of σ and assigning them the same start times t on $M_{t \bmod m+1}$. Thus, start times on M_i are $(i-1) \bmod m$. Therefore, any two overlapping nonpreemptive parts appear on different machines. Hence, σ is a feasible schedule for I.

The transformations of σ into σ' and vice versa do not change the completion times of the jobs, and therefore γ values for σ and σ' are equal. Figure 8.5 illustrates the job-shopping isomorphism.

8.5.3 Job Shopping

8.5.3.1 *m*-No-Wait Dominance

The following equivalence reveals the m-no-wait dominance in perfect unit-time job shops:

$$J \text{ per } \alpha_2 | m\text{-no-wait}, \beta_{345}, m\beta_6, p_{ij}{=}1 \,|\, \gamma \sim J \text{ per } \alpha_2 \,|\, \beta_{345}, m\beta_6, p_{ij}{=}1 \,|\, \gamma \tag{8.25}$$

We can obtain the proof by showing that any feasible perfect no-no-wait unit-time job-shop schedule can be transformed into a feasible perfect m-no-wait unit-time job-shop schedule without increasing job completion times. Looking through the former schedule from left to right, find the first maximal uninterrupted part of length not divisible by m. Since it is the first, it should be the chain of operations $O_{i+1,j}, \ldots, O_{i+k,j}$ on the machines $M_1, \ldots, M_k$, where $k < m$. If $O_{i+k,j}$ occupies time unit $[t-1, t]$, then time unit $[t, t+1]$ on M_{k+1} is idle, and therefore $O_{i+k+1,j}$ occupies on M_{k+1} a time unit after time $t+1$. Let us move $O_{i+k+1,j}$ to the left in time unit $[t, t+1]$. Since this operation is inside a job, the move does not violate the precedence relation or release dates, and the completion times are not increased. Obviously, repeating this procedure gives the necessary transformation.

8.5.3.2 Job-Shopping Diagrams

Isomorphism (8.23) complemented by Dominances (8.22) and (8.25) produces the job-shopping squire we mentioned in Section 8.2.5. The composition of Expansion (8.6); Modulation (8.20) with $\alpha_1 = P$, $p = 1, q = m$; Isomorphism (8.23); and Dominance (8.25) gives the diagram connecting preemptive problems on identical parallel machines to perfect unit-time no-no-wait job-shop problems:

$$\begin{array}{ccc} P\alpha_2 \mid m\text{-pmtn}, \beta_{35}, m\beta_{47} \mid \gamma & \overset{(8.20)}{\propto} & P \bmod \alpha_2 \mid m\text{-pmtn}, \beta_{35}, m\beta_{47} \mid \gamma \\ & & \wr\wr\ (8.23) \\ \wr\wr\ (8.6) & & J \text{ per}\, \alpha_2 \mid m\text{-no-wait}, \beta_{35}, m\beta_{46}, p_{ij}=1 \mid \gamma \\ & & \wr\ (8.25) \\ P\alpha_2 \mid \text{pmtn}, \beta_{3457} \mid \gamma & \propto & J \text{ per}\, \alpha_2 \mid \beta_{35}, m\beta_{46}, p_{ij}=1 \mid \gamma \end{array} \tag{8.26}$$

Analogously, the composition of Expansion (8.7), Modulation (8.21) with $\alpha_1 = P$, $p = 1$, $q = m$, and Isomorphism (8.24) gives the diagram connecting nonpreemptive problems on identical parallel machines to perfect unit-time no-wait job-shop problems:

$$\begin{array}{ccc} P\alpha_2 \mid \beta_{35}, m\beta_{47} \mid \gamma & \overset{(8.21)}{\propto} & P \bmod \alpha_2 \mid \beta_{35}, m\beta_{47} \mid \gamma \\ \wr\wr\ (8.7) & & \wr\wr\ (8.24) \\ P\alpha_2 \mid \beta_{3457} \mid \gamma & \propto & J \text{ per}\, \alpha_2 \mid \text{no-wait}, \beta_{35}, m\beta_{46}, p_{ij}=1 \mid \gamma \end{array} \tag{8.27}$$

Since a perfect job shop is a special case of a job shop, Reductions (8.26) and (8.27) can be extended by the trivial inclusion into job shops producing the *preemptive no-no-wait* and the *nonpreemptive no-wait* reductions, respectively, of *job shopping:*

$$P\alpha_2 \mid \text{pmtn}, \beta_{3457} \mid \gamma \propto J\alpha_2 \mid \beta_{3456}, p_{ij}=1 \mid \gamma$$

$$P\alpha_2 \mid \beta_{3457} \mid \gamma \propto J\alpha_2 \mid \text{no-wait}, \beta_{3456}, p_{ij}=1 \mid \gamma$$

It is important to note that Reductions (8.26) and (8.27) are polynomial, while the job-shopping reductions are pseudopolynomial because perfect job-shop problems reduce to job-shop problems in pseudopolynomial time. Obviously, two-machine job shopping is a polynomial reduction. Table 8.11 shows examples of job shopping.

8.5.3.3 Job-Shop Paralleling

Only two reductions that parallel job shops are known. These are

$$J2 \mid p_{ij}=1 \mid C_{\max} \propto P2 \mid \text{pmtn} \mid C_{\max}$$

$$J2 \mid \text{no-wait}, p_{ij}=1 \mid C_{\max} \propto P2 \| C_{\max}$$

The first reduction is obtained by Kubiak, Sethi, and Sriskandarajah [12] by using as a subroutine McNaughton's wrap-around rule applied to $P2 \mid \text{pmtn} \mid C_{\max}$. The second reduction is obtained by Kubiak [13] by the adjustment of a pseudopolynomial algorithm for the partition problem [9], i.e., the decision version of $P2\|C_{\max}$.

8.5.3.4 Cycle Shopping

Job shopping conserves the number of jobs, translating the processing times p_j of jobs in problems on identical parallel machines into the numbers m_j of operations of jobs in perfect unit-time job-shop problems such that $m_j = ap_j$ for a common positive integer a divisible by m. Thus, job shopping does not change the relative length of jobs. Besides, a perfect unit-time job-shop problem with equal number of operations in jobs is obviously a unit-time cycle shop problem. Therefore, job shopping applied to problems on identical parallel machines with equal processing times does *cycle shopping*:

$$P\alpha_2 \mid \text{pmtn}, \beta_{345}, p_j=p \mid \gamma \propto C\alpha_2 \mid \beta_{345}, p_{ij}=1 \mid \gamma$$

$$P\alpha_2 \mid \beta_{345}, p_j=p \mid \gamma \propto C\alpha_2 \mid \text{no-wait}, \beta_{345}, p_{ij}=1 \mid \gamma$$

TABLE 8.11 Examples of Job Shopping

(a) Reductions by preemptive no-no-wait job shopping						
[28]	✓	$P2 \mid \text{pmtn} \mid C_{\max}$	$\propto$	$J2 \mid p_{ij}=1 \mid C_{\max}$	✓	[12,73]
[63,64]	✓	$P2 \mid \text{pmtn}, r_j \mid C_{\max}$	$\propto$	$J2 \mid r_j, p_{ij}=1 \mid C_{\max}$	✓	[92]
[19]	✓	$P2 \mid \text{pmtn}, r_j \mid L_{\max}$	$\propto$	$J2 \mid r_j, p_{ij}=1 \mid L_{\max}$	?	[93]
[38]	✓	$P2 \mid$ ~~pmtn~~ $\mid \sum C_j$	$\propto$	$J2 \mid p_{ij}=1 \mid \sum C_j$	✓	[94,95]
[55,96]	✓	$P2 \mid \text{pmtn} \mid \sum U_j$	$\propto$	$J2 \mid p_{ij}=1 \mid \sum U_j$	✓	[97]
	?	$P2 \mid \text{pmtn} \mid \sum T_j$	$\propto$	$J2 \mid p_{ij}=1 \mid \sum T_j$	?	
[98]	✓?	$P \mid \text{pmtn, prec}, r_j, n=k \mid \sum w_j U_j$	$\not\propto$	$J \mid \text{prec}, r_j, n=k, p_{ij}=1 \mid \sum w_j U_j$	✓	[49,98]
[98]	✓?	$P \mid \text{pmtn, prec}, r_j, n=k \mid \sum w_j T_j$	$\not\propto$	$J \mid \text{prec}, r_j, n=k, p_{ij}=1 \mid \sum w_j T_j$	✓	[49,98]
Table 8.3	†	$P2 \mid$ ~~pmtn~~ $\mid \sum w_j U_j$	$\propto$	$J2 \mid p_{ij}=1 \mid \sum w_j U_j$	†	[10,97]
Table 8.3	††	$P2 \mid$ ~~pmtn~~ $\mid \sum w_j T_j$	$\propto$	$J2 \mid p_{ij}=1 \mid \sum w_j T_j$	††	[10]
Table 8.3	††	$P2 \mid \text{pmtn}, r_j \mid \sum w_j C_j$	$\propto$	$J2 \mid r_j, p_{ij}=1 \mid \sum w_j C_j$	††	[10]
[28,99]	†	$P2 \mid$ ~~pmtn~~ $\mid \sum w_j C_j$	$\propto$	$J2 \mid p_{ij}=1 \mid \sum w_j C_j$	†?	[10]
[57]	†?	$P2 \mid \text{pmtn}, r_j \mid \sum C_j$	$\propto$	$J2 \mid r_j, p_{ij}=1 \mid \sum C_j$	†?	[10]
[59]	†?	$P2 \mid \text{pmtn}, r_j \mid \sum U_j$	$\propto$	$J2 \mid r_j, p_{ij}=1 \mid \sum U_j$	†?	[10]
[33]	††	$P2 \mid$ ~~pmtn~~ , chains $\mid \sum C_j$	$\propto$	$J2 \mid \text{chains}, p_{ij}=1 \mid \sum C_j$	††	[10]
(b) Reductions by nonpreemptive no-wait job shopping						
[98]	✓	$P \mid \text{prec}, r_j, n=k \mid \sum w_j U_j$	$\not\propto$	$J \mid \text{no-wait, prec}, r_j, n=k, p_{ij}=1 \mid \sum w_j U_j$	✓	[49,98]
[98]	✓	$P \mid \text{prec}, r_j, n=k \mid \sum w_j T_j$	$\not\propto$	$J \mid \text{no-wait, prec}, r_j, n=k, p_{ij}=1 \mid \sum w_j T_j$	✓	[49,98]
[38]	✓	$P2 \| \sum C_j$	$\propto$	$J2 \mid \text{no-wait}, p_{ij}=1 \mid \sum C_j$	✓	[100]
[47]	†?	$P2 \| \sum T_j$	$\propto$	$J2 \mid \text{no-wait}, p_{ij}=1 \mid \sum T_j$	†?	[10]
[28,99]	†	$P2 \| \sum w_j C_j$	$\propto$	$J2 \mid \text{no-wait}, p_{ij}=1 \mid \sum w_j C_j$	†?	[10]
[9,47]	†	$P2 \| C_{\max}$	$\propto$	$J2 \mid \text{no-wait}, p_{ij}=1 \mid C_{\max}$	†	[13,73]
[47,60]	†	$P3 \| C_{\max}$	$\not\propto$	$J3 \mid \text{no-wait}, p_{ij}=1 \mid C_{\max}$	††	[101]
Table 8.3	††	$P2 \mid r_j \mid L_{\max}$	$\propto$	$J2 \mid \text{no-wait}, r_j, p_{ij}=1 \mid L_{\max}$	††	[10]
Table 8.3	††	$P2 \mid r_j \mid \sum C_j$	$\propto$	$J2 \mid \text{no-wait}, r_j, p_{ij}=1 \mid \sum C_j$	††	[10]
Table 8.3	††	$P2 \| \sum w_j T_j$	$\propto$	$J2 \mid \text{no-wait}, p_{ij}=1 \mid \sum w_j T_j$	††	[10]
[33]	††	$P2 \mid \text{chains} \mid C_{\max}$	$\propto$	$J2 \mid \text{no-wait, chains}, p_{ij}=1 \mid C_{\max}$	††	[10]
[33]	††	$P2 \mid \text{chains} \mid \sum C_j$	$\propto$	$J2 \mid \text{no-wait, chains}, p_{ij}=1 \mid \sum C_j$	††	[10]

See the comment on the NP-hardness of $P2\|\sum T_j$ in the caption to Table 8.4. Related pseudopolynomial algorithms can be found in References [9,13,28,42,46,60,97]. The preemptive problems with a fixed number of jobs on identical parallel machines are proved to be solvable in pseudopolynomial time only under the conjecture that the NP-preemption hypothesis [23] is true.

Table 8.12 shows examples of cycle shopping. Since a cycle shop is a special case of a job shop and a generalization of a flow shop, there exists the related intermediate special case between the job-shopping and flow-shopping squares, which we call the *cycle-shopping square:*

$$
\begin{array}{ccc}
P \bmod \alpha_2 \mid m\text{-pmtn}, \beta_{345}, p_j = pm \mid \gamma & \approx & C \operatorname{per} \alpha_2 \mid m\text{-no-wait}, \beta_{345}, m_j = pm, p_{ij} = 1 \mid \gamma \\
\wr & & \wr \\
P \bmod \alpha_2 \mid \text{pmtn}, \beta_{345}, p_j = pm \mid \gamma & \sim & C \operatorname{per} \alpha_2 \mid \beta_{345}, m_j = pm, p_{ij} = 1 \mid \gamma
\end{array}
$$

Note that cycle shopping is just the contraction of job shopping on the class of parallel-machine problems with equal processing times. Like job shopping, it is a polynomial reduction in the two-machine case. Since job shopping produces job-shop problems where the number of operations in jobs exceeds m, a further

TABLE 8.12 Examples of Cycle Shopping

		(a) Reductions by preemptive no-no-wait cycle shopping				
[28]	✓	$P \mid \text{pmtn}, p_j = p \mid C_{\max}$	$\propto_*$	$C \mid p_{ij} = 1 \mid C_{\max}$	✓	[73]
[102]	†?	$P \mid \text{pmtn}, p_j = p \mid \sum w_j U_j$	$\propto_*$	$C \mid p_{ij} = 1 \mid \sum w_j U_j$	?	
[85]	††	$P \mid \text{pmtn}, r_j, p_j = m \mid \sum U_j$	$\propto_*$	$C \mid r_j, p_{ij} = 1 \mid \sum U_j$	††	[14]
[103]	††	$P \mid \text{pmtn}, r_j, p_j = p \mid \sum w_j C_j$	$\propto_*$	$C \mid rj, p_{ij} = 1 \mid \sum w_j C_j$	††	[14]
		(b) Reductions by nonpreemptive no-wait cycle shopping				
[86]	✓	$P2 \mid p_j = p \mid \sum w_j U_j$	$\propto$	$C2 \mid \text{no-wait}, p_{ij} = 1 \mid \sum w_j U_j$	✓	[14]
[86]	✓	$P2 \mid p_j = p \mid \sum w_j T_j$	$\propto$	$C2 \mid \text{no-wait}, p_{ij} = 1 \mid \sum w_j T_j$	✓	[14]
[70]	††	$P \mid \text{intree}, r_j, p_j = 1 \mid \sum C_j$	$\propto_*$	$C \mid \text{no-wait, intree}, r_j, p_{ij} = 1 \mid \sum C_j$	††	[14]
[65]	††	$P \mid \text{intree}, r_j, p_j = 1 \mid C_{\max}$	$\propto_*$	$C \mid \text{no-wait, intree}, r_j, p_{ij} = 1 \mid C_{\max}$	††	[14]
[65]	††	$P \mid \text{outtree}, p_j = 1 \mid L_{\max}$	$\propto_*$	$C \mid \text{no-wait, outtree}, p_{ij} = 1 \mid L_{\max}$	††	[14]
[72]	††	$P \mid \text{prec}, p_j = 1 \mid C_{\max}$	$\propto_*$	$C \mid \text{no-wait, prec}, p_{ij} = 1 \mid C_{\max}$	††	[14]
[90]	††	$P \mid \text{prec}, p_j = 1 \mid \sum C_j$	$\propto_*$	$C \mid \text{no-wait, prec}, p_{ij} = 1 \mid \sum C_j$	††	[14]
[26]	††	$P2 \mid \text{chains}, p_j = 1 \mid \sum w_j C_j$	$\propto$	$C2 \mid \text{no-wait, chains}, p_{ij} = 1 \mid \sum w_j C_j$	††	[14]
Table 8.3	††	$P2 \mid \text{chains}, p_j = 1 \mid \sum U_j$	$\propto$	$C2 \mid \text{no-wait, chains}, p_{ij} = 1 \mid \sum U_j$	††	[14]
Table 8.3	††	$P2 \mid \text{chains}, p_j = 1 \mid \sum T_j$	$\propto$	$C2 \mid \text{no-wait, chains}, p_{ij} = 1 \mid \sum T_j$	††	[14]

Note that $Pm \mid \text{pmtn}, p_j = p \mid \sum w_j U_j$ can be solved in polynomial time [49].

contraction does not reach a special case of job shopping for flow shops. In the next section, however, we show that preemptive no-no-wait job shopping can be "continued" to a similar mass reduction for flow shops by magnification of precedence constraints.

8.5.4 Flow Shopping

8.5.4.1 Job Chaining Technique

For *flow shopping* we need to use a technique slightly different from the operation chaining described in Section 8.3.2. The *job chaining* technique decomposes multioperation jobs into chains of smaller multioperation jobs that save the chains of operations in the original jobs. It can be described in detail as follows. Let a job J_j with release date r_j, due date d_j, and weight w_j be the chain of unit-time operations

$$O_{1j} \to O_{2j} \to \cdots \to O_{m_j,j},$$

where $m_j = mq_j$. Then we transform J_j into the chain of m-operation jobs

$$J_{j1} \to J_{j2} \to \cdots \to J_{jq_j}$$

where J_{jk}, $k = 1, 2, \ldots, q_j$, is the chain of operations

$$O_{(k-1)m+1,j} \to O_{(k-1)m+2,j} \to \cdots \to O_{km,j}$$

with release date r_j, due date d_j, and weight

$$w_{jk} = \begin{cases} w_j & \text{if } k = q_j \\ 0 & \text{otherwise} \end{cases}$$

TABLE 8.13 Reductions by Flow Shopping

[28]	✓	$P \mid pmtn \mid C_{max}$	$\propto$	$F \mid chains, p_j = 1 \mid C_{max}$	✓	[26]
[63,64]	✓	$P \mid pmtn, r_j \mid C_{max}$	$\propto$	$F \mid chains, r_j, p_j = 1 \mid C_{max}$	✓	[16]
[64,66]	✓	$P \mid pmtn, tree \mid C_{max}$	$\propto$	$F \mid tree, p_j = 1 \mid C_{max}$	✓	[16]
[19]	✓	$P \mid pmtn, outtree, r_j \mid C_{max}$	$\propto$	$F \mid outtree, r_j, p_j = 1 \mid C_{max}$	✓	[16]
[19]	✓	$P \mid pmtn, intree \mid L_{max}$	$\propto$	$F \mid intree, p_j = 1 \mid L_{max}$	✓	[16]
[19]	✓	$P2 \mid pmtn, prec, r_j \mid L_{max}$	$\propto$	$F2 \mid prec, r_j, p_j = 1 \mid L_{max}$	✓	[16]
Table 8.3	††	$P2 \mid pmtn, chains, p_j = 1 \mid \sum U_j$	$\propto$	$F2 \mid chains, p_{ij} = 1 \mid \sum U_j$	††	[15]
Table 8.3	††	$P2 \mid pmtn, chains, p_j = 1 \mid \sum T_j$	$\propto$	$F2 \mid chains, p_{ij} = 1 \mid \sum T_j$	††	[15]
[33]	††	$P2 \mid \cancel{pmtn}, chains \mid \sum C_j$	$\propto$	$F2 \mid chains, p_{ij} = 1 \mid \sum w_j C_j$	††	[2]
[70]	††	$P \mid pmtn, intree, r_j, p_j = 1 \mid C_{max}$	$\propto$	$F \mid intree, r_j, p_{ij} = 1 \mid C_{max}$	††	[15]
[70]	††	$P \mid pmtn, outtree, p_j = 1 \mid L_{max}$	$\propto$	$F \mid outtree, p_{ij} = 1 \mid L_{max}$	††	[15]
[71]	††	$P \mid pmtn, prec, p_j = 1 \mid C_{max}$	$\propto$	$F \mid prec, p_{ij} = 1 \mid C_{max}$	††	[104]

The strong NP-hardness of $P2 \mid \cancel{pmtn}, chains \mid \sum w_j C_j$ follows from a trivial reduction from $P2 \mid \cancel{pmtn}, chains \mid \sum C_j$ that is proved to be strongly NP-hard [33].

The machine assignment to the operations remains the same. The decomposition can obviously be done in pseudopolynomial time $O(\sum_{j=1}^{n} q_j)$, which is polynomial if the parameters q_j are bounded by a polynomial in the problem size.

8.5.4.2 Flow-Shopping Diagram

Thus, the job chaining technique gives the pseudopolynomial reduction

$$J \text{ per } \alpha_2 \mid \beta_{345}, m\beta_6, p_{ij} = 1 \mid \gamma \propto F\alpha_2 \mid \text{chain-}\beta_3, \beta_4, p_{ij} = 1 \mid \gamma \tag{8.28}$$

for all criteria except $\gamma = \sum C_j$ since the job chaining technique, like the operation chaining technique, does not work for this criterion. Extending Reduction (8.26) by Chaining (8.28) we obtain the diagram

$$\begin{array}{ccc} & J \text{ per } \alpha_2 \mid \beta_{34}, p_{ij} = 1 \mid \gamma & \\ \swarrow (8.26) & & \searrow (8.28) \\ P\alpha_2 \mid pmtn, \beta_{34} \mid \gamma & \propto & F\alpha_2 \mid \text{chain-}\beta_3, \beta_4, p_{ij} = 1 \mid \gamma \end{array} \tag{8.29}$$

The lower reduction here is *flow shopping*.[10] Table 8.13 shows examples of flow shopping.

Like preemption chaining and open shopping, flow shopping is pseudopolynomial. It cannot be extended in general to the case $\gamma = \sum C_j$, unless P = NP, because $P2 \mid \cancel{pmtn}, chains \mid \sum C_j$ is strongly NP-hard [33] but $F2 \mid chains, p_{ij} = 1 \mid \sum C_j$ can be solved in polynomial time [26].

Note that a polynomial algorithm for $F2 \mid chains, p_{ij} = 1 \mid \sum C_j$ is essential since polynomial algorithms for $F \mid outtree, r_j, p_{ij} = 1 \mid \sum C_j$, and $F2 \mid prec, p_{ij} = 1 \mid \sum C_j$ [15] do not solve it in polynomial time. The open questions in flow shopping are

$$P\alpha_2 \mid \cancel{pmtn} \mid \sum C_j \overset{?}{\propto} F\alpha_2 \mid chains, p_j = 1 \mid \sum C_j \tag{8.30}$$

$$P\alpha_2 \mid \cancel{pmtn}, r_j \mid \sum C_j \overset{?}{\propto} F\alpha_2 \mid chains, r_j, p_j = 1 \mid \sum C_j \tag{8.31}$$

[10] Brucker and Knust [15] proved this reduction in a different way, using the modulated version of Reduction (8.8).

Since $F2\,|\,\text{chains}, r_j, p_j = 1\,|\,\sum C_j$ [26], and even $F2\,|\,\text{outtree}, r_j, p_j = 1\,|\,\sum C_j$ [15] can be solved in polynomial time, a positive answer to Question 8.31 in the two-machine case would give a pseudopolynomial algorithm for $P2\,|\,\sout{\text{pmtn}}, r_j\,|\,\sum C_j$, which is proved to be NP-hard [57] but remains open for the strong NP-hardness.

Note that flow shopping maps problems in the same way as open shopping does. Table 8.13 and the counterexample to flow shopping with $\gamma = \sum C_j$ can be obtained from Table 8.9 and the counterexample to open shopping with $\gamma = \sum C_j$ only by changing the references in the right columns and the machine-environment symbol from O to F.

8.5.4.3 Flow-Shopping Triangle

The main observation here is that an m-no-wait schedule S on the machines $M_1, M_2, \ldots, M_m$ for an instance I of the m-machine unit-time flow-shop scheduling problem can easily be transformed into a schedule S' for the related instance $I - (m-1)$ of the single-machine problem of scheduling unit-time jobs with equal precedence delays $m-1$ by deleting the part of S on the machines $M_2, M_3, \ldots, M_m$. Conversely, a schedule S' can be extended to a schedule S by adding the machines $M_2, M_3, \ldots, M_m$ and uninterrupted chains of $m-1$ unit-time operations on the machines right after the completion of jobs in S' on M_1.

The transformation changes every job completion time by $m-1$. Therefore, the optimal values of the criteria

$$C_{\max}, \quad \sum C_j, \quad \sum w_j C_j$$

change by constants $m-1$, $(m-1)m$, and $(m-1)\sum w_j$, respectively. While the optimal values of the criteria

$$L_{\max}, \quad \sum T_j, \quad \sum U_j, \quad \sum w_j T_j, \quad \sum w_j U_j$$

do not change. Thus, the m-machine unit-time m-no-wait flow-shop scheduling problem and the single-machine problem of scheduling unit-time jobs with equal precedence delays $m-1$ are isomorphic. Taking into account the right vertical equivalence and the equivalence on the main diagonal in the flow-shopping square (see Section 8.2.5), we obtain the *flow-shopping triangle*

$$\begin{array}{ccc} Fm\,|\,\beta_{345}, & p_{ij} = 1\,|\,\gamma & \\ \wr & \wr & \\ 1\,|\,\beta_3(m-1), \beta_{45}, p_j = 1\,|\,\gamma & \sim & P \bmod m\,|\,\beta_{345}, p_j = m\,|\,\gamma \end{array} \tag{8.32}$$

This shows that unit-time single-machine scheduling problems with precedence delays $m-1$ and nonpreemptive problems of scheduling jobs with processing times m on m modulated identical parallel machines are just two other forms of unit-time flow-shop scheduling problems. Figure 8.6 gives an example of related schedules for the three classes. Because of the left side of the flow-shopping triangle, all the results in the next section with unit processing times can be reformulated in terms of unit-time flow shops.

8.5.4.4 Flow-Shop Paralleling

All known flow-shop paralleling reductions can be obtained from those in Table 8.14 in the next section by changing the single-machine precedence-delay problems into the equivalent unit-time flow-shop problems.

TABLE 8.14 Reductions by Precedence-Delay Paralleling

[17,18]	✓	$1 \mid \text{prec}(1), p_j = 1 \mid C_{\max}$	$\propto$	$P2 \mid \text{prec}, p_j = 1 \mid C_{\max}$	✓	[27]
[15]	✓	$1 \mid \text{prec}(1), p_j = 1 \mid \sum C_j$	$\propto$	$P2 \mid \text{prec}, p_j = 1 \mid \sum C_j$	✓	[23,27]
[16]	✓	$1 \mid \text{prec}(1), r_j, p_j = 1 \mid L_{\max}$	$\propto$	$P2 \mid \text{prec}, r_j, p_j = 1 \mid L_{\max}$	✓	[67]
[16]	✓	$1 \mid \text{outtree}(m-1), r_j, p_j = 1 \mid C_{\max}$	$\propto$	$P \mid \text{outtree}, r_j, p_j = 1 \mid C_{\max}$	✓	[65]
[16]	✓	$1 \mid \text{tree}(m-1), p_j = 1 \mid C_{\max}$	$\propto$	$P \mid \text{tree}, p_j = 1 \mid C_{\max}$	✓	[25]
[26]	✓	$1 \mid \text{chains}(m-1), p_j = 1 \mid C_{\max}$	$\propto$	$P \mid \text{chains}, p_j = 1 \mid C_{\max}$	✓	[26]
[26]	✓	$1 \mid \text{chains}(m-1), p_j = 1 \mid \sum C_j$	$\propto$	$P \mid \text{chains}, p_j = 1 \mid \sum C_j$	✓	[26]

The reductions are obtained in the papers cited in the left column. These papers show that the algorithms on identical parallel machines proposed in the papers in the right column also solve the related single-machine unit-time problems with equal precedence delays. By the flow-shopping triangle 8.32, the problems in the left column can be replaced by the equivalent flow-shop problems or problems on modulated parallel machines, which can also be solved in polynomial time.

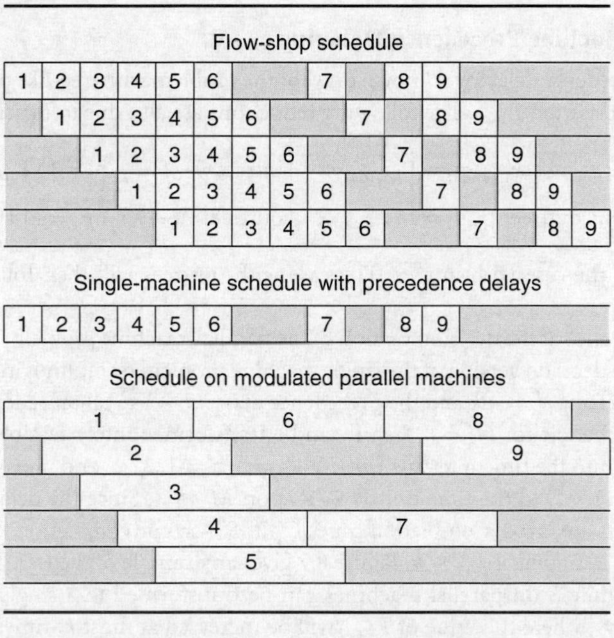

FIGURE 8.6 Three schedules for the instance with five machines, nine jobs and the precedence constraints 1→6, 3→7, 6→8, 5→8: a unit-time flow-shop schedule; the corresponding single-machine schedule with precedence delays 4; and the corresponding schedule on modulated parallel machines.

8.5.5 Precedence Delaying

8.5.5.1 Single-Machine Precedence Delaying Diagram

Applying parallel increase (8.1), or chain parallel increase (8.2) in the case of the criteria $\sum C_j$ and $\sum U_j$, to preemptive single-machine problems ℓ times, flow shopping (8.29) with $\alpha_2 = m = \ell + 1$ and the left side of the flow-shopping triangle (8.32), we obtain the diagram

$$\begin{array}{ccc} P\ell+1|\text{pmtn}, \beta_{34} \mid \gamma & \overset{(8.29)}{\propto} & F\ell+1 \mid \text{chain-}\beta_3, \beta_4, p_{ij}=1 \mid \gamma \\ \ell \text{ times } \propto \; (8.1) \text{ or } (8.2) & & \wr \; (8.32) \\ 1 \mid \text{pmtn}, \beta_{34} \mid \gamma & \propto & 1 \mid \text{chain-}\beta_3(\ell), \beta_4, p_j=1 \mid \gamma \end{array} \qquad (8.33)$$

The lower reduction, which we call *preemptive precedence delaying*, takes place for all classical criteria except $\gamma = \sum C_j$. Like preemption chaining, open shopping, and flow shopping, preemptive precedence delaying is pseudopolynomial and cannot be extended in general to the case $\gamma = \sum C_j$, unless P = NP, because $1 \mid \text{pmtn}, \text{chains}, r_j \mid \sum C_j$ is strongly NP-hard [70] but $1 \mid \text{chains}(1), r_j, p_{ij} = 1 \mid \sum C_j$ and even the extension of this problem to tree-like precedence constraints can be solved in polynomial time [15,26].

Preemptive precedence delaying becomes polynomial under the same condition as flow shopping does. The only open question regarding preemptive precedence delaying is

$$1 \mid \cancel{\text{pmtn}} \mid \sum C_j \overset{?}{\propto} 1 \mid \text{chains}(\ell), p_j = 1 \mid \sum C_j \tag{8.34}$$

for a fixed delay ℓ. Brucker and Knust [15] propose *nonpreemptive precedence delaying*

$$1 \mid \beta_{347} \mid \gamma \propto 1 \mid \text{chain-}\beta_3(\ell), \beta_{47} \mid \gamma$$

We refer the reader to [15] for the proof of this reduction.

8.5.5.2 Parallel-Machine Precedence Delaying

Parallel-machine precedence delaying is known only for the problem with tree-like precedence constraints, release dates and the criterion $L_{\max}$. The following reductions are also due to Brucker and Knust [15]:

$$Pm \mid \text{intree}, r_j, p_j = 1 \mid L_{\max} \propto 1 \mid \text{intree}(m-1), r_j, p_j = 1 \mid L_{\max}$$

$$Pm \mid \text{outtree}, r_j, p_j = 1 \mid L_{\max} \propto 1 \mid \text{outtree}(m-1), r_j, p_j = 1 \mid L_{\max}$$

Let us consider only the case with intrees. The case with outtrees will then follow from the time reverse 8.2.6.3.

Let (I, k) be an instance of the decision version of the parallel-machine problem. Then we construct the instance (I', k') of the decision version of the single machine problem by multiplying release dates and due dates in I by m and setting $k' = mk$ and the precedence delay $m - 1$ to all precedences in the intree in I.

If S' is a feasible schedule for (I', k'), then it can be transformed into a feasible schedule S for (I, k) as follows. We cut S' into the time intervals $[m(t-1), mt]$, $t = 1, 2, \ldots$ and move the content of the ith time unit in $[m(t-1), mt]$ to the time unit $[t-1, t]$ on M_i in S. Since the delay is at least $m - 1$, all precedence constraints are saved. Completion times C'_j in S' become completion times $C_j = \lceil C'_j / m \rceil$ in S. Therefore, $L'_{\max} \leq k'$ implies $L_{\max} \leq k$. Figure 8.7 gives an example of the transformation.

Conversely, a schedule S on parallel machines can be transformed to a single-machine schedule S' with precedence delays, where the value of $L_{\max}$ will be increased at most m times. This can be done by

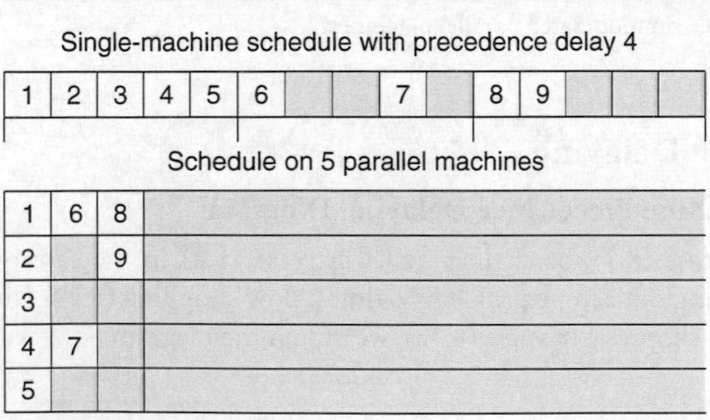

FIGURE 8.7 A single-machine schedule with precedence delay 4 and the related schedule on five parallel machines for the instance with nine jobs and the precedence constraints 1→7, 2→7, 3→7, 4→7, 5→8, 6→9.

reassigning the jobs to the machines in each time unit in S such that J_j and J_k appear in the neighbouring time units $[t-1, t]$ and $[t, t+1]$ on M_h and M_i, respectively, if and only if

$$J_j \prec J_k \Rightarrow h \leq i \tag{8.35}$$

This can always be done because we deal with intree-like precedence constraints. Executing the described transformation of S' to S in the reverse order we obtain a schedule S' with precedence delays. It is not hard to make sure that Condition (8.35) guarantees precedence delays at least $m-1$. Since $C'_j \leq mC_j$, $L_{\max} \leq k$ implies $L'_{\max} \leq k'$.

8.5.5.3 Precedence-Delay Paralleling

Precedence-delay paralleling reductions are listed in Table 8.14. They have been obtained by observations that algorithms originally designed for problems on identical parallel machines also solve the related single-machine unit-time problems with equal precedence delays.

8.6 Concluding Remarks

The most tangible complexity result of mass reductions is the mass proof of the solvability in polynomial time or the NP-hardness of many unit-time scheduling problems. For example, the strong NP-hardness of $P2 \mid \text{chains}, p_j = 1 \mid \sum w_j C_j$ was proved by the preemption chaining of $P2 \mid \cancel{\text{pmtn}}, \text{chains} \mid \sum C_j$; the proof of solvability in polynomial time for $O \mid \text{no-wait, tree}, p_{ij} = \mid C_{\max}$ was obtained by nonpreemptive no-wait latinization of $P \mid \text{tree}, p_j = 1 \mid C_{\max}$. The area of unit-time scheduling has gained the most benefit from these mass reductions.

Not only complexity issues become interesting in light of the mass reductions. They also turn to be useful for a deeper understanding of the relationship between scheduling problems having different combinatorial structures. The shopping squares and the flow-shopping triangle represent bright examples here. Also, the mass reductions reveal the unexpectedly subtle nature of the criterion $\sum C_j$, which in many cases proves to be unmanageable. Thus, only job shopping (and its narrowed version, cycle shopping), latinization, and nonpreemptive precedence delaying are possible for this criterion because they do not use a chaining technique, and hence they are proper reductions. The other mass reductions are not proper, since they depend on a chaining technique. Preemption chaining, open shopping, flow shopping, and preemptive precedence delaying with the exception of the unclear cases (8.10), (8.16), (8.30), and (8.34) do not exist unless P = NP. These questions generate the following four-question diagram:

$$\begin{array}{ccc}
O2 \mid \text{chains}, p_{ij} = \mid \sum C_j & \sim & P2 \mid \text{chains}, p_j = 1 \mid \sum C_j \\
\not\propto ? & \not\propto ? & \wr \\
P2 \mid \cancel{\text{pmtn}} \mid \sum C_j & \overset{?}{\propto} & F2 \mid \text{chains}, p_{ij} = \mid \sum C_j \\
\wr & & \wr \\
1 \mid \cancel{\text{pmtn}} \mid \sum C_j & \overset{?}{\propto} & 1 \mid \text{chains}(1), p_j = 1 \mid \sum C_j
\end{array}$$

Thus, a positive answer to at least one question in the diagram gives positive answers to all four. Another open issue is open-shop paralleling for $\sum C_j$. Studying the relationship between solutions of problems like $P2 \mid 1\text{-pmtn} \mid \sum C_j$ and $P2 \mid \text{chains}, p_j = 1 \mid \sum C_j$ would probably throw some light on the possibility of open-shop paralleling with the criterion $\sum C_j$ and, in general, the properties of this criterion related to chaining techniques.

To justify the potential of extending the mass reductions to criteria more general than classical, we can emphasize that the basic reductions we used are shown to work as follows: modulation, for all the classical criteria; expansion, for all proportion and all permanence criteria; chaining, for all the classical criteria except $\sum C_j$; and latinization, for any regular criteria. Job shopping, for example, can be extended to a criterion that is more general than a proportion criterion [10].

Extensions of the mass reductions involving the class with identical parallel machines to the class with uniform parallel machines would bring interesting results. For example, since any machine environment has a modulated counterpart, we can consider the class $Q \bmod \alpha_2 \mid \beta \mid \gamma$, assuming that the machines M_i increase processing times by q_i times. We can also introduce an extension of the perfect job shop, denoted by $K \operatorname{per} \alpha_2 \mid \beta \mid \gamma$, that increases the number of operations in uninterrupted parts of jobs by q_i times if they start at times $(i-1) \bmod m$. Then it is easy to show the isomorphisms

$$Q \bmod \alpha_2 \mid m\text{-pmtn}, \beta_{345}, m\beta_7 \mid \gamma \approx K \operatorname{per} \alpha_2 \mid m\text{-no-wait}, \beta_{345}, m\beta_6, p_{ij} = \mid \gamma$$

$$Q \bmod \alpha_2 \mid \beta_{345}, m\beta_7 \mid \gamma \approx K \operatorname{per} \alpha_2 \mid \text{no-wait}, \beta_{345}, m\beta_6, p_{ij} = \mid \gamma$$

The machine environment $K \operatorname{per} \alpha_2$ is a special case of the machine environment, denoted by K, where the route through the machines for an uninterrupted part of a job depends on the time it starts; a job and a time unit specify a fixed route through the machines and processing times on them.[11] Thus, we can obtain the reductions

$$Q\alpha_2 \mid \text{pmtn}, \beta_{3457} \mid \gamma \propto K\alpha_2 \mid \beta_{3456}, p_{ij} = \mid \gamma$$

$$Q\alpha_2 \mid \beta_{3457} \mid \gamma \propto K\alpha_2 \mid \text{no-wait}, \beta_{3456}, p_{ij} = \mid \gamma$$

The scheduling class associated with the machine environment K possesses an enormously powerful structure. It covers, for example, the class of scheduling problems with limited machine availability [105,106], since a time unit can specify no route. Nonetheless, we believe there exist nontrivial problems in this class that can still be solved in polynomial time.

In conclusion, we pose $K2 \mid p_{ij} = \mid C_{\max}$ as an example of such a nontrivial problem. Let us consider a time horizon h and jobs J_j, $j = 1, 2, \ldots, n$, that have to be processed on two machines, M_1 and M_2, as follows: Each pair (j, t), where $t = 0, 1, \ldots, h-1$, specifies the route

$$\rho_{jt} = M_{\mu_{1jt}} M_{\mu_{2jt}} \ldots M_{\mu_{m_{jt}jt}}$$

of m_{jt} unit-time operations, where $\mu_{ijt} \neq \mu_{i+1,jt}$ for $i = 1, 2, \ldots, m_{jt} - 1$ and

$$\sum_{j=1}^{n} \max_{1 \le t \le h} m_{jt} \le h$$

The last condition guarantees that a solution exists. Note that the routes in this problem differ by only the first machines and lengths. The problem size, therefore, is nh. So, if J_j starts at time t, then it follows the route ρ_{jt}. The problem is to find a minimum-length schedule. If $\rho_{j0} = \rho_{j1} = \ldots = \rho_{j,h-1}$ for every $j = 1, 2, \ldots, n$, then we have the classical unit-time job-shop problem $J2 \mid p_{ij} = \mid C_{\max}$, which is solvable in polynomial time (see Table 8.11).

References

[1] V. A. Strusevich. Minimizing the sum of completion times for the open shop. *Vestnik BGU, Series 1*, (1):44–46, 1988. (In Russian.)

[2] V. S. Tanaev, Yu. N. Sotskov, and V. A. Strusevich. *Scheduling Theory: Multi-Stage Systems*. Nauka, Moscow, 1989. (In Russian); Kluwer, Dordrecht, 1994. (English translation)

[3] C. Y. Liu and R. L. Bulfin. Scheduling open shops with unit execution times to minimize functions of due dates. *Operations Research*, 36:553–559, 1988.

[11] In music, a key is a scale of musical notes that starts on one specific note. This allegory motivates to call the machine environment, which is denoted by K, a *key shop*. Thus, a key shop can be thought as a generalization of a job shop where the "scale" of routes changes in time.

[4] H. Bräsel, D. Kluge, and F. Werner. A polynomial time algorithm for an open shop problem with unit processing times and tree constraints. *Discrete Applied Mathematics*, 59:11–21, 1995.
[5] W. Kubiak, C. Sriskandarajah, and K. Zaras. A note on the complexity of open shop scheduling problems. *INFOR*, 29:284–294, 1991.
[6] P. Brucker, B. Jurisch, and M. Jurisch. Open shop problems with unit time operations. *ZOR - Methods and Models of Operations Research*, 37:59–73, 1993.
[7] H. Bräsel. *Lateinische Rechtecke und Maschinenbelegung*. Dissertation B, TU, Magdeburg, 1990.
[8] H. Bräsel, D. Kluge, and F. Werner. A polynomial algorithm for the $[n/m/0, t_{ij} = 1, \mathrm{tree}/C_{\max}]$ open shop problem. *European Journal of Operational Research*, 72:125–134, 1994.
[9] M. R. Garey and D. S. Johnson. *Computers and Intractability: A Guide to the Theory of NP-Completeness*. Freeman, San Francisco, 1979.
[10] V. G. Timkovsky. Is a unit-time job shop not easier than identical parallel machines? *Discrete Applied Mathematics*, 85:149–162, 1998.
[11] V. G. Timkovsky. Scheduling unit-time operation jobs on identical parallel machines and in a flow shop: complexity and correlation. Technical Report No. 98–03 4, Department of Computer Science and Systems, McMaster University, Hamilton, Ontario, 1998.
[12] W. Kubiak, S. Sethi, and C. Sriskandarajah. An efficient algorithm for a job shop problem. *Annals of Operations Research*, 57:203–216, 1995.
[13] W. Kubiak. A pseudopolynomial algorithm for a two-machine no-wait job shop problem. *European Journal of Operational Research*, 43:267–270, 1989.
[14] M. Middendorf and V. G. Timkovsky. On scheduling cycle shops: classification, complexity and approximation. *Journal of Scheduling*, 5:135–169, 2002.
[15] P. Brucker and S. Knust. Complexity results for single-machine problems with positive finish-start time lags. *Computing*, 63:299–316, 1999.
[16] J. L. Bruno, J. W. Jones, and K. So. Deterministic scheduling with pipelined processors. *IEEE Transactions on Computers*, C-29:308–316, 1980.
[17] A. Schäffer and B. Simons. Computing the bump number with techniques from two-processor scheduling. *Order*, 5:131–141, 1988.
[18] L. Finta and Z. Liu. Single machine scheduling subject to precedence delays. *Discrete Applied Mathematics*, 70:247–266, 1996.
[19] E. L. Lawler. Preemptive scheduling of precedence-constrained jobs on parallel machines. In M. A. H. Dempster, J. K. Lenstra, and A. H. G. Rinnooy Kan (editors), *Deterministic and Stochastic Scheduling*, pages 101–123. Reidel, Dordrecht, 1982.
[20] V. S. Tanaev, M. Y. Kovalev, and Y. M. Shafransky. *Scheduling Theory: Group Technologies*. Institute of Engineering Cybernetics, Minsk, 1998. (In Russian)
[21] R. M. Karp. Reducibility among combinatorial problems. In R. E. Miller and J. W. Thatcher, editors, *Complexity of Computer Computations*, pages 85–103. Plenum Press, New York, 1972.
[22] R. L. Graham, E. L. Lawler, J. K. Lenstra, and A. H. G. Rinnooy Kan. Optimization and approximation in deterministic sequencing and scheduling: a survey. *Annals of Discrete Mathematics*, 5:287–326, 1979.
[23] E. G. Coffman, Jr., J. Sethuraman, and V. G. Timkovsky. Ideal preemptive schedules on two processors. Technical report, Columbia University, NY, 2002. *Acta Informatica*, 39:597–612, 2003.
[24] Ph. Baptiste, J. Carlier, A. Kononov, M. Queyranne, S. Sevastianov, and M. Sviridenko. Structural properties of preemptive schedules. (Unpublished)
[25] T.C. Hu. Parallel sequencing and assembly line problems. *Operations Research*, 9:841–848, 1961.
[26] V. G. Timkovsky. Identical parallel machines vs. unit-time shops and preemptions vs. chains in scheduling complexity. *European Journal of Operational Research*, 149(2):355–376, 2003.
[27] E. G. Coffman, Jr. and R. L. Graham. Optimal scheduling for two-processor systems. *Acta Informatica*, 1:200–213, 1972.

[28] R. McNaughton. Scheduling with deadlines and loss functions. *Management Science*, 6:1–12, 1959.

[29] P. Brucker. *Scheduling Algorithms*. Springer, Heidelberg, second edition, 1998.

[30] M. Pinedo. *Scheduling: Theory, Algorithms, and Systems*. Prentice Hall, NJ, second edition, 2002.

[31] V. S. Tanaev. Preemptions in deterministic scheduling systems with parallel identical machines. *Vestsi Akademii Nauk BSSR. Ser. fizika-matematychnykh navuk*, (6):44–48, 1973. (In Russian)

[32] V. S. Tanaev, V. S. Gordon, and Y. M. Shafransky. *Scheduling Theory: Single-Stage Systems*. Nauka, Moscow, 1984. (In Russian); Kluwer, Dordrecht, 1994. (English translation)

[33] J. Du, J. Y.-T. Leung, and G. H. Young. Scheduling chain-structured tasks to minimize makespan and mean flow time. *Information and Computation*, 92:219–236, 1991.

[34] Ph. Baptiste and V. G. Timkovsky. On preemption redundancy in scheduling unit processing time jobs on two parallel machines. *Operations Research Letters*, 28:205–212, 2001.

[35] P. Brucker, J. Hurink, and S. Knust. A polynomial algorithm for $P|p_j = 1, r_j, \text{outtree}|\sum C_j$. *Mathematical Methods of Operations Research*, 56:407–412, 2003.

[36] P. Brucker, S. Heitman, and J. Hurink. How useful are preemptive schedules. *Operations Research Letters*, 31:129–136, 2003.

[37] Ph. Baptiste. On preemption redundancy. Technical report, IBM Watson Research Center, Yorktown Heights, NY, 2002.

[38] R. W. Conway, W. L. Maxwell, and L. W. Miller. *Theory of Scheduling*. Addison-Wesley, Reading, Massachusetts, 1967.

[39] V. S. Gordon. The deterministic single-stage scheduling with preemptions. In *Computers in Engineering*, pages 30–38. Institute of Engineering Cybernetics, Minsk, June 1973. (In Russian)

[40] Y. M. Shafransky. Private communication. 2003.

[41] J. Du and J. Y.-T. Leung. Minimizing total tardiness on one machine is NP-hard. *Mathematics of Operations Research*, 15(3):483–495, 1990.

[42] E. L. Lawler. A 'pseudopolynomial' algorithm for sequencing jobs to minimize total tardiness. *Annals of Discrete Mathematics*, 1:331–342, 1977.

[43] W. A. Horn. Single-machine job sequencing with treelike precedence ordering and linear delay penalties. *SIAM Journal on Applied Mathematics*, 23:189–202, 1972.

[44] D. Adolphson and T. C. Hu. Optimal linear ordering. *SIAM Journal on Applied Mathematics*, 25:403–423, 1973.

[45] J. B. Sidney. Decomposition algorithms for single-machine sequencing with precedence relations and deferral costs. *Operations Research*, 23:283–298, 1975.

[46] E. L. Lawler. Preemptive scheduling of uniform parallel machines to minimize the weighted number of late jobs. Report bw 105, Centre for Mathematics and Computer Science, Amsterdam, 1979.

[47] J. K. Lenstra, A. H. G. Rinnooy Kan, and P. Brucker. Complexity of machine scheduling problems. *Annals of Discrete Mathematics*, 1:343–362, 1977.

[48] J. K. Lenstra and A. H. G. Rinnooy Kan. Complexity results for scheduling chains on a single machine. *European Journal of Operational Research*, 4:270–275, 1980.

[49] P. Baptiste, P. Brucker, S. Knust, and V. Timkovsky. Fourteen notes on equal-processing-time scheduling. Technical report, University of Obsanbrück, Osnabrück, 2002.

[50] J. Y.-T. Leung and G. H. Young. Minimizing total tardiness on a single machine with precedence constraints. *ORSA Journal on Computing*, 2:346–352, 1990.

[51] J. Labetoulle, E. L. Lawler, J. K. Lenstra, and A. H. G. Rinnooy Kan. Preemptive scheduling of uniform machines subject to release dates. In W. R. Pulleyblank (editor), *Progress in Combinatorial Optimization*, pages 245–261. Academic Press, 1984.

[52] R. R. Muntz and E. G. Coffman, Jr. Optimal preemptive scheduling on two-processor systems. *IEEE Transactions on Computers*, C-18:1014–1020, 1969.

[53] J. M. Moore. An n job, one machine sequencing algorithm for minimizing the number of late jobs. *Management Science*, 15:102–109, 1968.

[54] W. L. Maxwell. On sequencing *n* jobs on one machine to minimize the number of late jobs. *Management Science*, 16:295–297, 1970.

[55] E. L. Lawler and C. U. Martel. Preemptive scheduling of two uniform machines to minimize the number of late jobs. *Operations Research*, 37:314–318, 1989.

[56] K. R. Baker. *Introduction to Sequencing and Scheduling*. John Wiley, New York, 1974.

[57] J. Du, J. Y.-T. Leung, and G. H. Young. Minimizing mean flow time with release time constraint. *Theoretical Computer Science*, 75:347–355, 1990.

[58] E. L. Lawler. A dynamic programming algorithm for preemptive scheduling of a single machine to minimize the number of late jobs. *Annals of Operations Research*, 26:125–133, 1990.

[59] J. Du, J. Y.-T. Leung, and C. S. Wong. Minimizing the number of late jobs with release time constraints. *Journal of Combinatorial Mathematics and Combinatorial Computing*, 11:97–107, 1992.

[60] E. L. Lawler, J. K. Lenstra, A. H. G. Rinnooy Kan, and D. B. Shmoys. Sequencing and scheduling: algorithms and complexity. In S. Graves, A. H. G. Rinnooy Kan, and P. Zipkin (editors), *Logistics of Production and Inventory*, volume 4 of *Handbooks in Operations Research and Management Science*, chapter 9, pages 445–522. North Holland, Amsterdam, 1993.

[61] J. Du and J. Y.-T. Leung. Minimizing mean flow time in two-machine open shops and flow shops. *Journal of Algorithms*, 14:24–44, 1993.

[62] T. Gonzalez and S. Sahni. Flow shop and job shop schedules: complexity and approximation. *Operations Research*, 26:36–52, 1978.

[63] W. A. Horn. Some simple scheduling algorithms. *Naval Research Logistics Quarterly*, 21:177–185, 1974.

[64] T. Gonzalez and D. B. Johnson. A new algorithm for preemptive scheduling of trees. *Journal of the ACM*, 27:287–312, 1980.

[65] P. Brucker, M. R. Garey, and D. S. Johnson. Scheduling equal-length tasks under tree-like precedence constraints to minimize maximum lateness. *Mathematics of Operations Research*, 2:275–284, 1977.

[66] R. R. Muntz and E. G. Coffman, Jr. Preemptive scheduling of real time tasks on multiprocessor systems. *Journal of the ACM*, 17:324–338, 1970.

[67] M. R. Garey and D. S. Johnson. Two-processor scheduling with start-times and deadlines. *SIAM Journal on Computing*, 6:416–426, 1977.

[68] E. C. Horvath, S. Lam, and R. Sethi. A level algorithm for preemptive scheduling. *Journal of the ACM*, 24:32–43, 1977.

[69] W. Kubiak. Optimal scheduling of unit-time tasks on two uniform processors under tree-like precedence constraints. *ZOR — Methods and Models of Operations Research*, 33:423–437, 1989.

[70] J. K. Lenstra. Private communication. 2001.

[71] J. D. Ullman. Complexity of sequencing problems. In E. G. Coffman, Jr. editor, *Computer and Job-Shop Scheduling Theory*, pages 139–164. John Wiley, New York, 1976.

[72] J. D. Ullman. NP-complete scheduling problems. *Journal of Computer and System Sciences*, 10:384–393, 1975.

[73] V. G. Timkovsky. A polynomial algorithm for the Lenstra-Rinnooy Kan scheduling problem on two machines. *Kibernetika*, (2):109–111, 1985. (In Russian)

[74] Y. Cho and S. Sahni. Preemptive scheduling of independent jobs with release and due dates on open, flow and job shops. *Operations Research*, 29:511–522, 1981.

[75] A. A. Gladky. A two-machine preemptive openshop scheduling problem: an elementary proof of NP-completeness. *European Journal of Operational Research*, 103:113–116, 1997.

[76] I. Lushchakova. Two machine preemptive scheduling problem with release dates, equal processing times and precedence constraints. Technical report, Belorussian State University, Minsk, Belarus, 2001.

[77] C. Y. Liu and R. L. Bulfin. On the complexity of preemptive open shop scheduling problems. *Operations Research Letters*, 4:71–74, 1985.

[78] C. Berge. *The Theory of Graphs and its Applications*. John Wiley, New York, 1962.

[79] J. A. Bondy and U. S. R. Murty. *Graph Theory and Applications*. Macmillan, 1976.

[80] H. N. Gabov and O. Kariv. Algorithms for edge coloring bipartite graphs and multigraphs. *SIAM Journal on Computing*, 11:117–129, 1982.

[81] R. Cole and J. Hopcroft. On edge coloring bipartite graphs. *SIAM Journal on Computing*, 11:540–546, 1982.

[82] I. Adiri and N. Amit. Openshop and flowshop scheduling to minimize sum of completion times. *Computers & Operations Research*, 11(3):275–289, 1984.

[83] T. Tautenhahn and G. J. Woeginger. Minimizing the total completion time in a unit-time open shop with release times. *Operations Research Letters*, 20:207–212, 1997.

[84] Ph. Baptiste.On minimizing the weighted number of late jobs in unit execution time open-shops. *European Journal of Operational Research*, 149:344–354, 2003.

[85] S. A. Kravchenko. On the complexity of minimizing the number of late jobs in unit time open shop. *Discrete Applied Mathematics*, 100:127–132, 1999.

[86] M. I. Dessouky, B. J. Lageweg, J. K. Lenstra, and S. L. van de Velde. Scheduling identical jobs on uniform parallel machines. *Statistica Neerlandica*, 44:115–123, 1990.

[87] Ph. Baptiste. Scheduling equal-length jobs on identical parallel machines. *Discrete Applied Mathematics*, 103:21–32, 2000.

[88] B. Simons. Multiprocessor scheduling of unit-time jobs with arbitrary release times and deadlines. *SIAM Journal on Computing*, 12:294–299, 1983.

[89] B. Simons and M. Warmuth. A fast algorithm for multiprocessor scheduling of unit-length jobs. *SIAM Journal on Computing*, 18:690–710, 1989.

[90] J. K. Lenstra and A. H. G. Rinnooy Kan. Complexity of scheduling under precedence constraints. *Operations Research*, 26:22–35, 1978.

[91] M. Dror, W. Kubiak, and P. Dell'Olmo. 'Strong'-'weak' chain constrained scheduling. *Ricerca Operativa*, 27:35–49, 1998.

[92] V. G. Timkovsky. A polynomial-time algorithm for the two-machine unit-time release-date job-shop schedule-length problem. *Discrete Applied Mathematics*, 77:185–200, 1997.

[93] V. G. Timkovsky. How to make Brucker's algorithm polynomial: scheduling by periodization. *International Journal of Mathematical Algorithms*, 2:325–360, 2001.

[94] S. A. Kravchenko. Minimizing total completion time in two-machine job shops with unit processing times. Technical Report 4, Institute of Technical Cybernetics, Minsk, Belorus, 1994. (In Russian)

[95] W. Kubiak and V. G. Timkovsky. A polynomial-time algorithm for total completion time minimization in two-machine job-shop with unit-time operations. *European Journal of Operational Research*, 94:310–320, 1996.

[96] E. L. Lawler. Recent results in theory of machine scheduling. In A. Bachem, M. Grötschel, and B. Korte (editors), *Mathematical Programming: the State of the Art — Bonn 1982*, pages 202–234. Springer, Berlin, 1983.

[97] S. A. Kravchenko. Minimizing the number of late jobs for the two-machine unit-time job-shop scheduling problem. *Discrete Applied Mathematics*, 98(3):209–217, 1999.

[98] M. Middendorf and V. G. Timkovsky. Transversal graphs for partially ordered sets: sequencing, merging and scheduling problems. *Journal of Combinatorial Optimization*, 3:417–435, 1999.

[99] J. L. Bruno, E. G. Coffman, Jr., and R. Sethi. Scheduling independent tasks to reduce mean finishing time. *Communications of the ACM*, 17:382–387, 1974.

[100] S. A. Kravchenko. A polynomial time algorithm for a two-machine no-wait job-shop scheduling problem. *European Journal of Operational Research*, 106:101–107, 1998.

[101] C. Sriskandarajah and P. Ladet. Some no-wait jobs scheduling problems: complexity results. *European Journal of Operational Research*, 24:424–438, 1986.

[102] P. Brucker and S. Kravchenko. Preemption can make parallel machine scheduling problems hard. Technical report, University of Osnabrück, Osnabrück, 1999.

[103] J. Y.-T. Leung and G. H. Young. Preemptive scheduling to minimize mean weighted flow time. *Information Processing Letters*, 34:47–50, 1990.

[104] J. Y.-T. Leung, O. Vornberger, and J. D. Witthoff. On some variants of the bandwidth minimization problem. *SIAM Journal on Computing*, 13:650–667, 1984.

[105] C.-Y. Lee. Two-machine flowshop scheduling with availability constraints. *European Journal of Operational Research*, 114:420–429, 1999.

[106] G. Schmidt. Scheduling with limited machine availability. *European Journal of Operational Research*, 121(1):1–15, 2000.

9

Parallel Scheduling for Early Completion

Bo Chen
University of Warwick

9.1 Introduction

This chapter assumes knowledge of some basic scheduling terminology provided in Part I of this book. We are concerned in this chapter with classical deterministic parallel scheduling with the objective of minimizing the makespan. Both preemptive and nonpreemptive schedules are considered. There can be precedence relationships between jobs. Excluded from consideration in this chapter are parallel scheduling models of other typical job characteristics, online and randomized algorithms, all of which are covered in separate chapters of this book. We will address the following aspects of the scheduling problems:

- Polynomial solvability and approximability
- Enumerative algorithms
- Polynomial-time approximations

Although we also consider scheduling jobs of different release dates, we will mainly concentrate on models of equal job release dates. The reason is as follows. If all jobs are not released at the same time, then the scheduling problems of minimizing the makespan can be viewed as special cases of the corresponding problems of minimizing maximum lateness, which is dealt with in a separate chapter. To see this, note that due to schedule symmetry on the time axis, there is an equivalence relationship (with respect to the objective value) between scheduling of the former objective and that of the latter: To a schedule S of instance I with makespan $C_{\max}$ corresponds a schedule S' of instance I' with maximum lateness $C_{\max}$, where in instance I' all jobs are released at time 0, the due date and start time of job j are respectively $d'_j = -r_j$ and $S'_j = C_{\max} - C_j$, where r_j and C_j are the release date and completion time of job j in schedule S, respectively.

Also note that minimization of makespan is a special case of minimization of the maximum lateness. The reader may find some further results in the separate chapter dealing with the latter objective function.

1-58488-397-9/$0.00+$1.50

We will mainly use worst-case guarantee as algorithm performance indicator. As a result, a rich set of literature on local search algorithms, including meta-heuristics, are not considered here. This chapter is a more focused and updated study on the relevant scheduling models than those treated by Chen, Potts, and Woeginger [1].

The rest of the chapter is organized as follows. First, we present in the next section some important algorithms and basic relationships, which not only have significant influences on parallel scheduling, but on shop scheduling as well. Then in Sections 9.3, 9.4, and 9.5 we consider respectively preemptive, nonpreemptive scheduling of independent jobs and scheduling with precedence constraints.

9.2 Some Basic Algorithms and Relationships

9.2.1 List Scheduling

As a simplest combinatorial algorithm, *list scheduling* (LS) has been well known for almost half a century. In this algorithm, jobs are fed from a pre-specified list and, whenever a machine becomes idle, the first *available* job on the list is scheduled and removed from the list, where the availability of a job means that the job has been released and, if there are precedence constraints, all its predecessors have already been processed.

In its simplicity and the fact that any optimal schedule can be constructed by LS with an appropriately chosen list, LS is by far the most popular scheduling approach. Moreover, since LS requires no knowledge of unscheduled jobs as well as of all jobs currently being processed, it is very powerful in online scheduling and especially so in online nonclairvoyance scheduling, in which it remains a dominant heuristic. (See Chapter 15 for more details.)

9.2.2 LPT and SPT

If the job list in LS can be sorted in order of nonincreasing processing times before they are scheduled, then the resulting algorithm is known as *largest processing time first* (LPT). The minor extra computing in sorting the jobs significantly improves the worst-case performance of the algorithm. Since LPT was first proposed by Graham [2], it has become the touchstone for the design of efficient algorithms, both offline and online. Moreover, LPT has also found applications in combinatorial optimization.

If preemption is allowed, then LPT becomes LRPT — *largest remaining processing time first*: At all times, a job with largest remaining processing time among all available jobs is processed, preempt a job whenever necessary.

On the other hand, if the pre-specified job list in LS is sorted in order of nondecreasing processing times, then the resulting algorithm is known as *shortest processing time first* (SPT). In contrast to LPT, applications of SPT in scheduling are primarily for minimizing the average job completion time. For example, problems $1 || \sum w_j C_j$ and $1 | r_j, pmtn | \sum C_j$ are solved to optimality with two SPT-based LS algorithms [3, 4].

9.2.3 HLF and Critical Path

In the presence of precedence relations between jobs, Hu's level algorithm [5] for scheduling UET jobs (i.e., jobs of unit processing time), known as *highest level first* (HLF), is fundamental. This algorithm is based on the notion of *critical path* (CP) in network optimization and always schedules a job that heads the longest current chain of unscheduled jobs of arbitrary lengths.

9.2.4 Relationships

Let us consider two basic relationships in parallel scheduling to minimize the makespan, one between optimal preemptive schedules and optimal nonpreemptive schedules, while the other between schedules of independent jobs of arbitrary processing times and schedules of UET jobs of precedence constraints.

It is evident that preemption can reduce the makespan of a schedule, although such time saving might incur a cost for making the preemptions in practice. But how efficient can preemption be in reducing the makespan? Given a system of arbitrarily constrained jobs, Liu [6] conjectures that the makespan of any optimal nonpreemptive schedule is at most $2 - 2/(m+1)$ times that of an optimal preemptive schedule, where m is the number of parallel identical machines. This conjecture is proved by Hong and Leung [7] for two special cases: (i) of UET jobs and (ii) of *in-tree* precedence constraints, where every job has at most one immediate successor. In the case of (ii), Hong and Leung show that any CP schedule (as a nonpreemptive schedule not necessarily optimal) has already satisfied the conjectured upper bound. Note that the special case of (ii) can be extended to the case of independent jobs. As a result, any LPT schedule (as a CP schedule) satisfies the conjectured bound. This result is also obtained recently by Braun and Schmidt [8] in a different way. Furthermore, for both cases of (i) and (ii) above, the bound of $2 - 2/(m+1)$ holds true for any optimal (nonpreemptive) LS schedules, since Hong and Leung show that unforced idle time in any optimal nonpreemptive schedule is not beneficial in reducing the makespan.

Liu's conjecture is proved by Coffman and Garey [9] if $m = 2$, although it is still open in general if $m \geq 3$. Recently, Braun and Schmidt [8] investigated the benefit of *each* additional preemption. They showed that, if the jobs are independent, the makespan of any schedule that is optimal among all schedules of at most k preemptions is at most $2 - 2/[m/(k+1)+1]$ times the makespan of an optimal preemptive schedule. Note that when $k = 0$ it reduces to a special case for Liu's conjecture.

On the other hand, complexity similarity between scheduling independent preemptable jobs of arbitrary processing times and scheduling nonpreemptable UET jobs of precedence constraints has recently led to an identification of a more formal relationship between the two classes of problems. In his *Preemption-Chaining Theorem*, Timkovsky [10] establishes that scheduling problems of the former class are reducible to the scheduling problems of the latter with the precedence constraints being chain-like. For more details, the reader is referred to Chapter 8 which deals with reducibility of problem classes.

9.3 Preemptive Scheduling

Scheduling with preemption allowed greatly mitigates the difficulty of solving a scheduling problem, although each preemption may incur a cost in practice. For almost all scheduling problems, the preemptive version is no harder than its nonpreemptive counterpart, with only a few exceptions ($J2 \mid n = 3 \mid C_{\max}$ [11, 12], $P \mid p_j = p \mid \sum w_j U_j$ [13], and $R \mid\mid \sum C_j$ [14]). The following observation can help understand this phenomenon. If we consider a scheduling problem as a zero-one integer program with decision variables indicating the assignment of jobs (or their operations) to machines and time slots, then preemption allows a variable to take a value between zero and one, thereby indicating the proportion of the job or its operation to be assigned. From this perspective, preemption is a vital element for polynomial solvability for many scheduling problems.

Due to the reducibility mentioned at the end of the previous section to scheduling nonpreemptable UET jobs of precedence constraints, which we will consider in Section 9.5, we mention here only a few basic results of scheduling independent preemptable jobs of arbitrary processing times.

It is not surprising that the first paper in parallel machine scheduling deals with preemption. In his paper, McNaughton [15] uses a simple wrap-around procedure, which runs in $O(n)$ time and generates at most $m-1$ preemptions, to solve problem $P \mid pmtn \mid C_{\max}$ to optimality. The basic idea behind this is quite simple: First, calculate a lower bound on the value of an optimal schedule and then construct a schedule that matches the bound. Note that in this approach, the lower bound is used as a deadline for all jobs. Once a complete schedule is constructed in this way, it is automatically optimal. The lower bound used in this approach is the maximum of the average machine workload and the processing time of the longest job.

McNaughton's approach has been successfully applied to other preemptive scheduling problems. Following a generalization of McNaughton's lower bound to the case of uniform machines by Liu and Yang [16] after the machine speeds are taken into account, Horvath, Lam, and Sethi [17] suggest the *largest remaining processing time on fastest machines* (LRPT-FM) rule. At each time point, the job with the most remaining processing time is scheduled on the fastest available machine; thereby they solve problem $Q \mid pmtn \mid C_{\max}$ in $O(mn^2)$ time. As a variant of LPT for preemptive scheduling, the LRPT-FM rule is also an application of the notion of critical path as is HLF, which we mentioned in the previous section. A less intuitive and more case-based algorithm is developed by Gonzalez and Sahni [18], which significantly reduces both the running time to $O(n + m \log m)$ and the number of preemptions.

Further down the road when the machines are unrelated, more expensive approaches are proposed. At the cost of solving a linear program, in which a decision variable represents the total time spent by each job on each machine, Lawler and Labetoulle [19] provide a general lower bound on the optimal makespan for problem $R \mid pmtn \mid C_{\max}$, and then solve it to optimality as an instance of problem $O \mid pmtn \mid C_{\max}$.

9.4 Nonpreemptive Scheduling

One of the fundamental problems in scheduling theory is to schedule independent nonpreemptable jobs onto parallel machines to minimize the makespan. This general problem is difficult to solve to optimality, as two most basic models $P2 \mid\mid C_{\max}$ and $P \mid\mid C_{\max}$ are already NP-hard and strongly NP-hard, respectively (Garey and Johnson [20, 21]). In fact, it is shown by Lenstra, Shmoys, and Tardos [22] that, unless $P = \mathrm{NP}$, it is not at all easier even to approximate problem $R \mid\mid C_{\max}$ within a factor better than 3/2.

9.4.1 Enumerative Algorithms

It is immediately clear that problem $P \mid\mid C_{\max}$ allows for a very simple dynamic algorithm with all possible job assignments taken into account. A far more efficient branch and bound algorithm is proposed by Dell'Amico and Martello [23]. If the jobs are in LPT order, then a trivial lower bound is given by $\max\{\sum_{j=1}^{n} p_j/m, p_1, p_m + p_{m+1}\}$. Based on bin-packing arguments (see Section 9.4.2 for more details), they have developed a procedure to improve iteratively this lower bound, while their branching rule assigns a longest unscheduled job to a machine. Computational results show that the algorithm can solve instances with up to 10,000 jobs.

For the more general problem $R \mid\mid C_{\max}$, branch and bound algorithms are proposed by Van de Velde [24] and by Martello, Soumis, and Toth [25]. The former is based on surrogate relaxation and duality and the latter on Largrangian relaxations and additive techniques. Computational results demonstrate the superiority of the algorithm of Martello, Soumis, and Toth over Van de Velde's algorithm. Recently, Mokotoff and Chrétienne [26] introduce a cutting plane technique, which is based on a newly identified valid inequality of the convex hull of the problem instance. Their algorithm includes a preprocessing phase to compute an upper bound with LS heuristic and a lower bound obtained from the preemptive relaxation. Computational results show that the algorithm gives an optimal solution for almost all tested cases within the fixed time and memory limits.

9.4.2 Approximation Algorithms

9.4.2.1 LS-Based Techniques

In the first paper on the worst-case analysis of scheduling algorithms, Graham [27] studies algorithm LS. He shows that for problem $P \mid\mid C_{\max}$, LS is $(2 - 1/m)$-approximate by observing that the makespan of any LS schedule is at most $2 - 1/m$ times the lower bound of optimum — the maximum of the length of a largest job and average machine workload. Note that this simple lower bound has been used to construct an optimal preemptive schedule (see the previous section for details).

The worst case arises in the above estimate on LS performance when the last job has the longest length. This gives rise to an improved algorithm LPT [2] by first sorting the jobs in a better order in an attempt to avoid the worst case. LPT has improved the worst-case performance ratio to $4/3 - 1/(3m)$. Moreover, with a further observation that the worst case happens in LPT actually when the terminating machine processes only three jobs, it is later proved by Coffman and Sethi [28] and Chen [29] that LPT schedule is actually asymptotically optimal with the increase of the number k of jobs on the terminating machine. In fact, the makespan of any such LPT schedule is at most $(k+1)/k - 1/(km)$ times the optimum.

Another variant of LS, as suggested by Graham, is to schedule the k longest jobs optimally, and then apply LS to the list of remaining jobs. This gives a ratio guarantee of $1 + (1 - 1/m)/(1 + \lfloor k/m \rfloor)$. Therefore, for fixed m, a family of these algorithms for different values of k constitute a polynomial-time approximation scheme (PTAS), although the running time of $O(n^{km})$ is huge. Ibarra and Kim [30] show that LPT is no more than $1 + 2(m-1)/n$ away from the optimum makespan, if $n \geq 2(m-1)\pi$, where $\pi = \max_j p_j / \min_j p_j$.

If machines have different speeds, then Morrison [31] shows that LPT for problem $Q \,||\, C_{\max}$ has a worst-case ratio of $\max\{\sigma/2, 2\}$, where $\sigma = \max_i s_i / \min_i s_i$. A modified LPT algorithm, which assigns the current job to the machine on which it will finish first, is shown by Gonzales, Ibarra, and Sahni [32] to improve the ratio guarantee to $2 - 2/(m+1)$. Subsequently, this guarantee is improved (except for $m \leq 3$) to 19/12 by Dobson [33] and Friesen [34].

9.4.2.2 Bin-Packing Techniques

A second main approximation approach is to use the dual form of the problem, known as *bin-packing*. In a bin-packing problem, a number of items of various sizes are to be packed into a minimum number of bins of a common given capacity. Note that problems of scheduling to minimize the makespan and of bin-packing share the same decision version. Naturally, bin-packing techniques are considered for scheduling.

Based on this principle of duality, Coffman, Garey, and Johnson [35] proposed an algorithm for problem $P \,||\, C_{\max}$, called *Multifit* (MF), to find by binary search the minimum capacity of the m bins into which the n items can be packed by an efficient packing heuristic known as *first-fit decreasing* (FFD). In the FFD heuristic, each iteration packs the largest remaining item into the first bin into which it fits. By using the technique of weight functions in bin-packing, they prove that MF has a ratio guarantee of $\rho + 2^{-k}$, where $\rho \leq 1.22$ and k denotes the number of binary search iterations. Subsequently, by refining the approach of weight functions, Friesen [36] improves the bound of ρ to 1.2, and Yue [37] further to 13/11, which is shown tight. At the expense of a larger running time, the MF algorithm is refined by Friesen and Langston [38] to achieve a slightly better worst-case ratio of $72/61 + 2^{-k}$.

A similar principle of duality between scheduling and bin-packing is considered by Hochbaum and Shmoys [39]. Given the machine capacity d, a *ρ-dual approximation* algorithm ($\rho > 1$) produces a job packing that uses at most the minimum number of machines of capacity d at the expense of possible capacity violation by no more than $(\rho - 1)d$. Using a family of dual approximation algorithms, Hochbaum and Shmoys provide a PTAS for problem $P \,||\, C_{\max}$.

The MultiFit approach and dual approximation approach are extended to scheduling of uniform machines. When MF is applied to the case of uniform machines, the corresponding bin-packing problem is of bins of different capacities. Friesen and Langston [40] show that the ρ in the corresponding worst-case ratio $\rho + 2^{-k}$ of MF is between 1.341 and 1.40, which is later improved to 1.38 by Chen [41]. On the other hand, extension of the dual approximation approach to uniform machines by Hochbaum and Shmoys has led to a PTAS for problem $Q \,||\, C_{\max}$ [42].

9.4.2.3 Linear Programming Techniques

Extensive research has appeared in the literature on computing near-optimal solutions for scheduling models by rounding optimal solutions to linear programming relaxations. Such techniques are not only for minimization of makespan. There are two main general ways to exploit the optimal solution to a linear programming relaxation. It is used either to guide the assignment, deterministic or random, of jobs to machines or to derive job priorities to be used in constructing the schedule.

Extending the optimal solution to a linear programming relaxation by an enumerative process, Potts [43] obtains a 2-approximation algorithm for problem $R \,||\, C_{\max}$ when m is fixed. Lenstra, Shmoys, and Tardos [22] extend Potts' approach by first establishing that the fractional solution to the linear programming relaxation can be rounded to a good integral approximation in polynomial time, thereby obviating the need for enumeration and removing the exponential dependence on m, and then deriving a 2-approximation algorithm even when m is part of the input.

This approach is further extended to accommodate a more general objective criterion. Shmoys and Tardos [44] introduce a stronger rounding technique than that of Lenstra, Shmoys, and Tardos to develop a polynomial algorithm that can find a schedule with mean job completion time M and makespan at most $2T$, if a schedule with mean job completion time at most M and makespan at most T exists.

9.4.2.4 Other Techniques

In approximation of NP-hard problems, a fully polynomial-time approximation scheme (FPTAS) is the strongest possible result that one can hope for unless, of course, $P = \text{NP}$. It seems that all FPTASs are based on *dynamic programming* (DP) formulations, which always find an optimal solution, though not necessarily in polynomial time. As a first FPTAS, Sahni [45] develops a DP-based approach for approximating problem $Pm \,||\, C_{\max}$, in which the state variables at each stage i form a set $S^{(i)}$ of $(m-1)$-tuples, representing the total workloads of the m machines. The cardinality $|\, S^{(i)} \,|$ of each such set is controlled by rounding the input data of the instance with use of an *interval partitioning* method. In a similar vein, Horowitz and Sahni [46] derive an FPTAS for problems $Qm \,||\, C_{\max}$ and $Rm \,||\, C_{\max}$.

In addition to DP formulation, semidefinite programming techniques have recently also played a role in scheduling approximation. The reader is referred to a separate chapter for more details.

9.5 Scheduling with Precedence Constraints

9.5.1 For UET Jobs

Presence of general precedence relationships dramatically increases the difficulty of any scheduling problems. To appreciate the effect of such presence, let us first simplify other elements of scheduling and assume that all jobs are of unit length. Then all such problems are NP-hard if the number of machines is part of the input, since this is already the case for problem $P \,|\, prec, p_j = 1 \,|\, C_{\max}$, as shown by Lenstra and Rinnooy Kan [47]. Actually, they show that it is NP-complete even to approximate the solution of problem $P \,|\, prec, p_j = 1 \,|\, C_{\max}$ within a factor better than 4/3.

However, this stagnant situation improves immediately if either the form of the precedence relations is relaxed or if the number of machines is fixed. As one of the earliest addressing scheduling of precedence constraints, Hu's algorithm HLF [5] (see Section 9.2) solves problem $P \,|\, tree, p_j = 1 \,|\, C_{\max}$ to optimality in polynomial time. The HLF approach has been generalized and various algorithms have been designed for a number of special cases of problem $Pm \,|\, prec, p_j = 1 \,|\, C_{\max}$, which include the case where the precedence graph is an *opposing forest*, that is, the disjoint union of an in-forest where each job has at most one immediate successor and an out-forest where each job has at most one immediate predecessor (Garey, Johnson, Tarjan, and Yannakakis [48]).

Similarly, the formulation by Fujii, Kasami, and Ninomiya [49] of problem $P2 \,|\, prec, p_j = 1 \,|\, C_{\max}$ as a maximum cardinality matching problem, which is solvable in $O(n^3)$ time, provides the groundwork for the development of a series of improved algorithms. Moreover, polynomial solvability has been extended to *simultaneously* minimizing $\sum C_j$ (Coffman and Graham [50] and, if preemption is allowed, Coffman, Sethuraman, and Timkovsky [51]) and to including job release dates and due dates (Garey and Johnson [52, 53]).

Some of the efficient algorithms for problem $P2 \,|\, prec, p_j = 1 \,|\, C_{\max}$ are adapted to the more general problem $P \,|\, prec, p_j = 1 \,|\, C_{\max}$, and are shown to have some good worst-case guarantees. Recently, by replacing the precedence constraints with release dates and due dates, tight lower bounds are derived by Baev, Meleis, and Eichenberger [54] for $P \,|\, prec, p_j = 1 \,|\, C_{\max}$. These bounds are probably tighter than known lower bounds and empirical experiments demonstrate that over more than 90% of time they lead to the optimal values over a synthetic benchmark.

Research on the case where the machines have different speeds has so far been limited. Using as a guide solutions to the two relaxed problems $Q2 \,|\, pmtn \,|\, C_{\max}$ and $Q2 \,|\, p_j = 1 \,|\, C_{\max}$, Brucker, Hurink, and

Kubiak [55] solve the problem $Q2 \mid chains, p_j = 1 \mid C_{\max}$ in linear time with respect to the number of chains. However, the complexities of two slightly more difficult problems $Q2 \mid tree, p_j = 1 \mid C_{\max}$ and $Q3 \mid chains, p_j = 1 \mid C_{\max}$ are still open.

We remark that the aforementioned HLF-based algorithms run in time polynomial in the number of jobs, but exponential in the problem size when it comes to the chainlike precedence constraints if a succinct coding scheme is used.

9.5.2 For Jobs of General Lengths

It is interesting to observe that research on parallel scheduling with precedence constraints has been mainly concentrated on the preemptive version.

Due to reducibility property mentioned in Section 9.2, the polynomial solvability picture for scheduling jobs with arbitrary processing times under preemption is very similar to that for the nonpreemptive scheduling of UET jobs. This similarity suggests a close relationship between these two models, as Lawler [56] observes as a result of deriving polynomial algorithms for a series of counterparts in the former model of well-solvable problems in the latter model.

In parallel to the complexity results for scheduling nonpreemptable UET jobs, problem $P \mid pmtn, prec \mid C_{\max}$ is shown to be NP-hard by Ullman [57], whereas problems $P \mid pmtn, tree \mid C_{\max}$ and $Q2 \mid pmtn, prec \mid C_{\max}$ are both polynomially solvable by the algorithms of Muntz and Coffman [58, 59] and Horvath, Lam, and Sethi [17], respectively.

Interestingly, the algorithm of Muntz and Coffman [58, 59] also focuses on the critical path of the precedence graph, as is the case in Hu's algorithm HLF for scheduling UET jobs. However, Gonzalez and Johnson [60] use a totally different LS approach for solving problem $P \mid pmtn, tree \mid C_{\max}$. Their algorithm segregates the jobs into two classes. In one class there is what can be termed the "backbone" of the problem, a superset of those jobs whose start and finish times are fixed in any optimal schedule. The other jobs can, in general, be scheduled with some freedom. Their algorithm runs in $O(n \log m)$ time. In the same spirit as their LRPT-FM algorithm for problem $Q \mid pmtn \mid C_{\max}$ (see Section 9.3), Horvath, Lam, and Sethi [17] solve problem $Q2 \mid pmtn, prec \mid C_{\max}$ in $O(mn^2)$ time.

Lam and Sethi [61] adapt the algorithm of Muntz and Coffman for problem $P \mid pmtn, prec \mid C_{\max}$ and show that it has a worst-case ratio of $2 - 2/m$. Similarly, in the case of uniform machines, Horvath, Lam, and Sethi [17] prove that this algorithm has a ratio guarantee of $\sqrt{3m/2}$, which is tight up to a constant factor. To improve the Muntz–Coffman algorithm, Jaffe [62] suggests scheduling jobs without unforced idleness by always using fastest machine available. He proves that this improves the ratio guarantee to $\sqrt{m} + 1/2$.

While also considering multiprocessor jobs, Błażewicz and Liu [63] study problems $P \mid chains \mid C_{\max}$ and $P \mid chains, r_j \mid C_{\max}$. As a variant of LPT and HLF, their algorithms run in $O(n \log n)$ time and in $O(n^2)$ time, respectively, and are still applicable if preemption is allowed. Their work represents one of a few results dealing with nonpreemptable jobs of arbitrary lengths and of precedence constraints. The same remark applies here as the one made at the end of Section 9.5.1.

In another recent study dealing with nonpreemptable jobs, Hurink and Knust [64] present complexity results that have influence on the strength of LS algorithm. They show that, with sequence-dependent setup times, dominant schedules for problem $P \mid prec, s_{ij} \mid C_{\max}$ cannot be calculated efficiently with LS techniques.

References

[1] Chen, B., Potts, C.N., and Woeginger, G.J., A review of machine scheduling: complexity, algorithms and approximability, in: Du, D.-Z. and Pardalos, P.M. (eds.), *Handbook of Combinatorial Optimization*, Kluwer, Boston, 1998, 21–169.

[2] Graham, R.L., Bounds on multiprocessing timing anomalies, *SIAM Journal on Applied Mathematics* 17 (1969), 416–429.

[3] Smith, W.E., Various optimizers for single-stage production, *Naval Research Logistics Quarterly* 3 (1956), 59–66.

[4] Schrage, L., A proof of the shortest remaining processing time processing discipline, *Operations Research* 16 (1968), 687–690.

[5] Hu, T.C., Parallel sequencing and assembly line problems, *Operations Research* 9 (1961), 841–848.

[6] Liu, C.L., Optimal scheduling on multi-processor computing systems, *Proceedings of the 13th Annual Symposium on Switching and Automatic Theory*, IEEE Computer Society, Los Alamitos, CA, 1972, 155–160.

[7] Hong, K.S. and Leung, J.Y.-T., Some results on Liu's conjecture, *SIAM Journal on Discrete Mathematics* 5 (1992), 500–523.

[8] Braun, O. and Schmidt, G., Parallel processor scheduling with limited number of preemptions, *SIAM Journal on Computing* 62 (2003), 671–680.

[9] Coffman, Jr., E.G. and Garey, M.R., Proof of the 4/3 conjecture for preemptive vs. nonpreemptive two-processor scheduling, *Journal of the ACM* 20 (1993), 991–1018.

[10] Timkovsky, V.G., Identical parallel machines vs. unit-time shops and preemptions vs. chains in scheduling complexity, *European Journal of Operational Research* 149 (2003), 355–376.

[11] Brucker, P., Kravchenko, S.A., and Sotskov, Y.N., Preemptive job-shop scheduling problems with a fixed number of jobs, *Mathematical Methods of Operations Research* 49 (1999), 41–76.

[12] Kravchenko, S.A. and Sotskov, Y.N., Optimal makespan schedule for three jobs on two machines, *Mathematical Methods of Operations Research* 43 (1996), 233–238.

[13] Brucker, P. and Kravchenko, S.A., Preemption can make parallel machine scheduling problems hard, *Osnabruecker Schriften zur Mathematik*, Reihe P (Preprints), Heft 211 (1999).

[14] Sitters, R., Two NP-hardness results for preemptive minsum scheduling of unrelated parallel machines, in: Aardal, K. and Gerards, B. (eds.), *Integer Programming and Combinatorial Optimization*, LNCS 2081 (2001), 396–405.

[15] McNaughton, R., Scheduling with deadlines and loss functions, *Management Science* 6 (1959), 1–12.

[16] Liu, J.W.S. and Yang, A., Optimal scheduling of independent tasks on heterogeneous computing systems, *Proceedings of the ACM Annual Conference*, 1974, 38–45.

[17] Horvath, E.C., Lam, S., and Sethi, R., A level algorithm for preemptive scheduling, *Journal of the ACM* 24 (1977), 32–43.

[18] Gonzalez, T. and Sahni, S., Preemptive scheduling of uniform processor systems, *Journal of the ACM* 25 (1978), 92–101.

[19] Lawler, E.L. and Labetoulle, J., On preemptive scheduling of unrelated parallel processors by linear programming, *Journal of the ACM* 25 (1978), 612–619.

[20] Garey, M.R. and Johnson, D.S., *Computers and Intractability: A Guide to the Theory of NP-Completeness*, Freeman, San Francisco, 1979.

[21] Garey, M.R. and Johnson, D.S., Strong NP-completeness results: motivation, examples and implications, *Journal of the ACM* 25 (1978), 499–508.

[22] Lenstra, J.K., Shmoys, D.B., and Tardos, É., Approximation algorithms for scheduling unrelated parallel machines, *Mathematical Programming* 46 (1990), 259–271.

[23] Dell'Amico, M. and Martello, S., Optimal scheduling of tasks on identical parallel processors, *ORSA Journal on Computing* 7 (1995), 191–200.

[24] Van de Velde, S.L., Duality-based algorithms for scheduling unrelated parallel machines, *ORSA Journal on Computing* 5 (1993), 192–205.

[25] Martello, S., Soumis, F., and Toth, P., Exact and approximation algorithms for makespan minimization on unrelated parallel machines, *Discrete Applied Mathematics* 75 (1997), 169–188.

[26] Mokotoff, E. and Chrétienne, P., A cutting plane algorithm for the unrelated parallel machine scheduling problem, *European Journal of Operational Research* 141 (2002), 515–525.

[27] Graham, R.L., Bounds for certain multiprocessing anomalies, *Bell System Technical Journal* 45 (1966), 1563–1581.

[28] Coffman, Jr., E.G. and Sethi, R., A generalized bound on LPT sequencing, *Revue Française d' Automatique Informatique, Recherche Operationnelle,* Supplément au 10 (1976), 17–25.

[29] Chen, B., A note on LPT scheduling, *Operations Research Letters* 14 (1993), 139–142.

[30] Ibarra, O.H. and Kim, C.E., Heuristic algorithms for scheduling independent tasks on nonidentical processors, *Journal of the ACM* 24 (1977), 280–289.

[31] Morrison, J.F., A note on LPT scheduling, *Operations Research Letters* 7 (1988), 77–79.

[32] Gonzalez, T., Ibarra, O.H., and Sahni, S., Bounds for LPT schedules on uniform processors, *SIAM Journal on Computing* 6 (1977), 155–166.

[33] Dobson, G., Scheduling independent tasks on uniform processors, *SIAM Journal on Computing* 13 (1984), 705–716.

[34] Friesen, D.K., Tighter bounds for LPT scheduling on uniform processors, *SIAM Journal on Computing* 16 (1987), 554–660.

[35] Coffman, Jr., E.G., Garey, M.R., and Johnson, D.S., An application of bin-packing to multiprocessor scheduling, *SIAM Journal on Computing* 7 (1978), 1–17.

[36] Friesen, D.K., Tighter bounds for the multifit processor scheduling algorithm, *SIAM Journal on Computing* 13 (1984), 170–181.

[37] Yue, M., On the exact upper bound for the multifit processor scheduling algorithms, *Annals of Operations Research* 24 (1990), 233–259.

[38] Friesen, D.K. and Langston, M.A., Evaluation of a MULTIFIT-based scheduling algorithm, *Journal of Algorithms* 7 (1986), 35–59.

[39] Hochbaum, D.S. and Shmoys, D.B., Using dual approximation algorithms for scheduling problems: Theoretical and practical results, *Journal of the ACM* 34 (1987), 144–162.

[40] Friesen, D.K. and Langston, M.A., Bounds for multifit scheduling on uniform processors, *SIAM Journal on Computing* 12 (1983), 60–70.

[41] Chen, B., Tighter bounds for MULTIFIT scheduling on uniform processors, *Discrete Applied Mathematics* 31 (1991), 227–260.

[42] Hochbaum, D.S. and Shmoys, D.B., A polynomial approximation scheme for machine scheduling on uniform processors: Using the dual approximating approach, *SIAM Journal on Computing* 17 (1988), 539–551.

[43] Potts, C.N., Analysis of a linear programming heuristic for scheduling unrelated parallel machines, *Discrete Applied Mathematics* 10 (1985), 155–164.

[44] Shmoys, D.B. and Tardos, É., An approximation algorithm for the generalized assignment problem, *Mathematical Programming* 62 (1993), 461–474.

[45] Sahni, S., Algorithms for scheduling independent tasks, *Journal of the ACM* 23 (1976), 116–127.

[46] Horowitz, E. and Sahni, S., Exact and approximate algorithms for scheduling nonidentical processors, *Journal of the ACM* 23 (1976), 317–327.

[47] Lenstra, J.K. and Rinnooy Kan, A.H.G., Complexity of scheduling under precedence constraints, *Operations Research* 6 (1978), 22–35.

[48] Garey, M.R., Johnson, D.S., Tarjan, R.E., and Yannakakis, M., Scheduling opposing forests, *SIAM Journal on Algebraic Discrete Mathematics* 4 (1983), 72–93.

[49] Fujii, M., Kasami, T., and Ninomiya, K., Optimal sequencing of two equivalent processors, *SIAM Journal on Applied Mathematics* 17 (1969), 784–789. Erratum: *SIAM Journal on Applied Mathematics* 20 (1971), 141.

[50] Coffman, Jr., E.G. and Graham, R.L., Optimal scheduling for two-processor systems, *Acta Informatica* 1 (1972), 200–213.

[51] Coffman, Jr., E.G., Sethuaraman, J., and Timkovsky, V.G., Ideal preemptive schedules on two processors, *Acta Informatica* (to be published).

[52] Garey, M.R. and Johnson, D.S., Scheduling tasks with nonuniform deadlines on two processors, *Journal of the ACM* 23 (1976), 461–467.

[53] Garey, M.R. and Johnson, D.S., Two-processor scheduling with start-times and deadlines, *SIAM Journal on Computing* 6 (1977), 416–426.

[54] Baev, I.D., Meleis, W.M., and Eichenberger, A., Lower bounds on precedence-constrained scheduling for parallel processors, *Information Processing Letters* 83 (2002), 27–32.

[55] Brucker, P., Hurink, J., and Kubiak, W., Scheduling identical jobs with chain precedence constraints on two uniform machines, *Mathematical Methods of Operations Research* 49 (1999), 211–219.

[56] Lawler, E.L., Preemptive scheduling of precedence-constrained jobs on parallel machines, in: Dempster, M.A.H., Lenstra, J.K., and Rinnooy Kan, A.H.G. (eds.), *Deterministic Stochastic Scheduling*, Reidel, Dordrecht, 1982, 101–123.

[57] Ullman, J.D., Complexity of sequencing problems, in: Coffman, Jr., E.G. (ed.), *Computer and Job-Shop Scheduling Theory*, Wiley, New York, 1976, 139–164.

[58] Muntz, R.R. and Coffman, Jr., E.G., Optimal scheduling on two-processor systems, *IEEE Transactions on Computing* C-18 (1969), 1014–1020.

[59] Muntz, R.R. and Coffman, Jr., E.G., Preemptive scheduling of real tasks on multiprocessor systems, *Journal of the ACM* 17 (1970), 324–338.

[60] Gonzalez, T. and Johnson, D.B., A new algorithm for preemptive scheduling of trees, *Journal of the ACM* 27 (1980), 287–312.

[61] Lam, S. and Sethi, R., Worst case analysis of two scheduling algorithms, *SIAM Journal on Computing* 6 (1977), 518–536.

[62] Jaffe, J.M., An analysis of preemptive multiprocessor job scheduling, *Mathematics of Operations Research* 5 (1980), 415–521.

[63] Błażewicz, J. and Liu, Z., Linear and quadratic algorithms for scheduling chains and opposite chains, *European Journal of Operational Research* 137 (2002), 248–264.

[64] Hurink, J. and Knust, S., List scheduling in a parallel machine environment with precedence constraints and setup times, *Operations Research Letters* 29 (2001), 231–239.

10 Minimizing the Maximum Lateness

Hans Kellerer
Universität Graz

10.1 Introduction

In this chapter we treat deterministic scheduling problems with the objective of minimizing the maximum lateness. Formally, we have given a set N of n jobs, each job j with processing time p_j and due date d_j, $j = 1, \dots, n$, and one or m parallel machines. For a given schedule the *lateness* L_j of job j is defined as $L_j := C_j - d_j$, that is, the (positive or negative) time difference between the completion time and the due date. The objective is to minimize the *maximum lateness*

$$L_{\max} := \max_{1 \le j \le n} L_j$$

We get an equivalent scheduling problem if instead of a due date each job j has a non-negative *delivery time* q_j. The delivery time can, for example, be interpreted as a transportation time to the customer. For a job j the *delivery-completion time* is defined as $L_j := C_j + q_j$. The objective is to minimize the *maximum delivery-completion time* $L_{\max} := \max_{1 \le j \le n} C_j + q_j$.

The equivalence between the two objectives can easily be obtained by subtracting a sufficiently large constant from each due date, thus allowing negative due dates with $d_j = -q_j$. We do not explicitly distinguish between these models and use always the expressions L_j and $L_{\max}$, independently of whether due dates or delivery times are considered, and speak usually of lateness in both cases.

Due to the equivalence optimal solution sets for both objectives are identical. Things become more complicated if we consider the relative performance guarantee of approximation algorithms. The maximum lateness of an optimal schedule can be zero or even negative. Thus, it makes no sense to study the worst-case behavior for the due date criterion. This is not the case for delivery times. Therefore, it is possible to speak of due dates in the context of exact algorithms, but we will definitely work with delivery times for approximation algorithms.

1-58488-397-9/$0.00+$1.50

In the presence of job release dates r_j *and* delivery times q_j, the maximum lateness problem on a single machine can also be modelled as a three-machine flow shop problem. Machines 1 and 3 are nonbottleneck machines, while machine 2 is a bottleneck machine processing only one job at a time. Jobs have to be processed on machines 1, 2, and 3 (in that order). Job j spends time r_j on machine 1, p_j on machine 2, and d_j on machine 3. The objective is to minimize the maximum makespan. In this context, the release dates are often denoted as *heads*, the processing times as *bodies* and the delivery times as *tails*.

Due to *forward–backward symmetry*, for each problem with release dates and delivery times there is an equivalent *inverse problem* by interchanging release dates and delivery times for each job and possibly inverting existing precedence relations. The inverse problem has the same optimal solution value and has for single machine problems the additional property that a job sequence $j_1, \ldots, j_n$ is optimal for the original problem if and only if the inverted sequence $j_n, \ldots, j_1$ is optimal for the inverse problem.

Note that especially the problem $P \mid r_j \mid C_{\max}$ of minimizing the makespan on parallel machines with release dates can be formulated as a maximum lateness problem $P \mid \cdot \mid L_{\max}$, by turning an instance of jobs j with release dates r_j into an instance without release dates and due dates $d_j = -r_j$.

Deadlines $\bar{d}_j$ are different from *due dates* d_j in the sense that no job j is allowed to be processed after its deadline $\bar{d}_j$. Hence, problems with deadlines are always feasibility problems and for a given instance with deadlines it is not clear whether a feasible schedule exists or not.

We address also generalizations of the lateness criterion, where the lateness L_j is replaced by a (usually nondecreasing) cost function $h_j(t)$. This function $h_j(t)$ represents the cost of j that is incurred by the completion of job j at time t. The problem is to find a sequence that minimizes the maximum $h_{\max}$ of the incurred costs.

Both preemptive and nonpreemptive scheduling problems and also precedence relations are included in this chapter on minimizing the maximum lateness. We do not cover shop problems, online scheduling models, availability constraints, or batching problems. For the first three problem classes, we refer to the corresponding chapters of this book. For results on scheduling problems with batching decisions, we recommend the excellent survey by Potts and Kovalyov [1].

In the description of the algorithms, we concentrate more on algorithms for polynomially solvable problems and on the worst-case analysis of approximation algorithms for $\mathcal{NP}$-hard problems than on results for branch-and-bound algorithms or local search techniques. Some basic results will also be described in detail.

Let us also mention the survey on scheduling problems by Chen, Potts, and Woeginger [2], which gives a very good overview on the literature related to scheduling theory.

The rest of this chapter is organized as follows: In Section 10.2, we treat single machine problems without release dates. Section 10.3 is devoted to single machine problems with release dates. We finish with algorithms for parallel machines in Section 10.4.

10.2 Single Machine Problems without Release Dates

It is well known that the basic problem $1 \mid \cdot \mid L_{\max}$ of minimizing the maximum lateness on a single machine can be solved in $O(n \log n)$ time by sorting the jobs in nondecreasing order of their due dates

$$d_1 \leq d_2 \leq \cdots \leq d_n \tag{10.1}$$

(or equivalently in nonincreasing order of their delivery times) according to Jackson [3] back in 1955. This method of sequencing is called the *earliest due date (EDD) rule* or *Jackson's rule*. Note that Jackson's rule gives also the optimal schedule for the problem of minimizing the *maximum tardiness* $T_{\max} = \max\{0, L_{\max}\}$ on a single machine $1 \mid \cdot \mid T_{\max}$.

The proof of Jackson's rule can be done by a simple *neighborhood exchange argument*. Assume there is a schedule S in which the jobs are not sorted according to Jackson's rule. Then there are two jobs i and j, job j directly processed after i, with $d_i > d_j$. Consider the schedule S' obtained by interchanging i and j. Since the completion times of the other jobs do not change by this operation, we only have to compare the values L_i and L_j of jobs i and j. W.l.o.g. assume that the starting time of job i is equal to zero in schedule S. Then the maximum lateness of the subschedule of S consisting of i and j is equal to $L_{\max} = \max\{p_i - d_i, p_i + p_j - d_j\}$. The maximum lateness of the corresponding subschedule of S' is $L'_{\max} = \max\{p_j - d_j, p_i + p_j - d_i\}$. It can be easily seen that the former expression is not smaller than the latter. Thus, the maximum lateness of the new schedule S' is decreasing . Repeating this process iteratively, we obtain an optimal schedule in which the jobs are sorted in EDD order.

If each job has additionally a deadline $\bar{d}_j$, then the problem $1 \mid \bar{d}_j \mid L_{\max}$ can still be solved optimally in polynomial time by applying Jackson's rule to the problem with the new due dates $d'_j := \min\{\bar{d}_j, d_j + K\}$ embedded in a bisection search on the value K.

Single machine problems without release dates can be solved in polynomial time for much more general objective functions. In 1973, Lawler [4] presented an $O(n^2)$ algorithm for general nondecreasing cost functions h_j of job j even in the presence of precedence relations, that is, for problem $1 \mid prec \mid h_{\max}$. For any subset $\tilde{N} \subseteq N$, let $p(\tilde{N})$ denote the total processing time of the jobs in $\tilde{N}$. Starting with completion time $p(N) = \sum_{j=1}^{n} p_j$ the jobs are assigned backward to the machine and iteratively each time a job with minimum cost and no successors is chosen. Lawler's algorithm can formally be described as follows.

Algorithm of Lawler

1. Initialization: Set $N' := N$, let N^s denote the jobs of N' without successors and set $j := 0$.
2. Set $i := \arg\min\{h_\ell(p(N')) \mid \ell \in N^s\}$. Process job i such that it finishes at time $p(N')$. Set $N' := N' \setminus \{i\}$ and $j := j + 1$. Modify N^s to represent the new set of jobs to be scheduled.
3. If $j = n$, output the resulting schedule and stop. Otherwise, go to Step 2.

We repeat the proof of Lawler due to its simplicity.

Theorem 10.1

The algorithm of Lawler is optimal for problem $1 \mid prec \mid h_{\max}$.

Proof

Assume that the schedule obtained by Lawler's algorithm is not optimal. Let S^* denote an optimal schedule with maximal j, where j is the first iteration in Lawler's algorithm such that it chooses a job i different from the job i^* in this optimal schedule S^*. Due to the definition of j, job i must be processed before i^* in schedule S^*. We obtain a new schedule S' from S^* by inserting job i directly after job i^* and leaving the other jobs unchanged. Schedule S' is still feasible since job i is processed at the same position by Lawler's algorithm. The only job which completes later now is job i, so the other jobs cannot increase the maximum cost. But by definition of Lawler's algorithm, the completion time cost of job i is in S' not larger than the cost of job i^* in S^*. Thus, the new schedule S' is still optimal, a contradiction to the maximality of j. □

For the special case of minimizing the maximum weighted tardiness on a single machine Hochbaum and Shamir [5] found an $O(n \log^2 n)$ algorithm. Their result has been improved by Fields and Frederickson [6] to an $O(h + n \log n)$ algorithm for the problem of minimizing the maximum weighted tardiness subject to h precedence constraints. Both algorithms use methods from computational geometry. An algorithm with the same running time $O(h + n \log n)$ is given by Sourd and Nuijten [7] for the problem with deadlines $1 \mid \bar{d}_j, prec \mid h_{\max}$.

10.3 Single Machine Problems with Release Dates

The existence of release dates makes the problem of minimizing the maximum lateness much more difficult. According to Lenstra, Rinnooy Kan, and Brucker [8] problem $1 \mid r_j \mid L_{\max}$ is strongly $\mathcal{NP}$-hard by reduction from 3-Partition.

10.3.1 Polynomial Algorithms

There are only a few special cases in the presence of release dates for which polynomial algorithms exist. Horn [9] proves that the problems $1 \mid r_j, p_j = 1 \mid L_{\max}$ and $1 \mid r_j, pmtn \mid L_{\max}$ can be solved by the *extended Jackson's rule* in $O(n \log n)$ time. Every time when the machine is available, schedule the job with smallest due date. In case that preemption is allowed and a job is released with a smaller due date than that of the job currently processed, preempt the job and continue with the new released job. At most $n - 1$ preemptions occur.

Frederickson [10] gives an algorithm for $1 \mid r_j, p_j = 1 \mid L_{\max}$ with improved running time $O(n)$. Simons [11] considers the generalization to equal processing times of arbitrary length. She gives an $O(n^2 \log n)$ algorithm for finding a feasible schedule with respect to release dates r_j and deadlines $\bar{d}_j$, that is, for looking for a feasible solution for problem $1 \mid r_j, p_j = p, \bar{d}_j \mid$. An alternative algorithm with the same time complexity has been obtained by Carlier [12]. The feasibility results of Simons and Carlier have been improved by Garey, Johnson, Simons and Tarjan [13] to an algorithm with running time in $O(n \log n)$. Bisection search over the possible values for $L_{\max}$ turns these algorithms into polynomial time algorithms for $1 \mid r_j, p_j = p \mid L_{\max}$. For details on these algorithms we refer to the chapter on equal processing time jobs.

The problems above can also be solved in polynomial time in the presence of precedence constraints as observed by Lageweg, Lenstra, and Rinnooy Kan [14]. This is done by an appropriate *modification of the due dates* and then applying the extended Jackson's rule to the instance without precedence constraints. Consider, for example, the problem $1 \mid r_j, p_j = 1, prec \mid L_{\max}$ and assume job i must precede job j according to the precedence constraints. In the case $r_i \leq r_j$ and $d_i \leq d_j$, the extended Jackson's rule will automatically process job i before job j. (If Jackson's rule can choose among more than one job with the same due date, jobs shall be preferred whose predecessors have all been scheduled already.) If $r_i > r_j$, redefine $r_j := r_i$ and analogously if $d_i > d_j$, set $d_i := d_j$. Jackson's rule will again deliver an optimal solution without violating the precedence constraints. Monma [15] even presented a $O(n + h)$ algorithm for problem $1 \mid p_j = 1, prec \mid L_{\max}$ with h precedence constraints.

A generalization of Lawler's algorithm described in Section 10.2 has been independently found by Baker et al. [16] and Gordon and Tanaev [17] for solving the preemptive problem $1 \mid r_j, pmtn, prec \mid h_{\max}$ with h_j nondecreasing cost functions. An optimal schedule is found in $O(n^2)$ time with at most $n - 1$ preemptions. Based on the ideas in Ref. [16] Sourd and Nuijten [7] solve the problem with deadlines $1 \mid r_j, \bar{d}_j, pmtn, prec \mid L_{\max}$ (or equivalently $1 \mid r_j, \bar{d}_j, p_j = 1, prec \mid L_{\max}$) subject to h precedence constraints in $O(h + n \log n)$ time. Note that this result gives an $O(n \log n)$ algorithm for $1 \mid r_j, \bar{d}_j, pmtn \mid L_{\max}$.

10.3.2 Approximation Algorithms

We will describe in this section approximation algorithms for $1 \mid r_j \mid L_{\max}$. We will speak now of delivery times instead of due dates to allow a meaningful discussion of the worst-case behavior. In this context, $L_{\max}(OPT)$ shall denote the maximum lateness (i.e., the maximum delivery-completion time) of an optimal schedule.

Most of the approximation algorithms are based on variations of the extended Jackson's rule, which we will also call briefly *Schrage*, since the extended Jackson's rule was introduced by Schrage [18]. We analyze first properties of Schrage schedules. W.l.o.g. the jobs shall be indexed such that Schrage yields the sequence $(1, \dots, n)$. The job c, which attains the maximum lateness in the Schrage schedule, is called the *critical job*. Then we have

$$L_{\max} = r_a + \sum_{j=a}^{c} p_j + q_c \tag{10.2}$$

Job a is the first job so that there is no idle time between the processing of jobs a and c, that is, either there is idle time before a or a is the first job to be scheduled. The sequence of jobs $a, a+1, \dots, c$ forms a block and is called the *critical sequence* with the corresponding set $\Lambda := \{a, a+1, \dots, c\}$. It is obvious that all jobs j in the critical sequence have release dates $r_j \geq r_a$.

For any set $A \subseteq N$ of jobs, the inequality

$$L_{\max}(OPT) \geq LB(A) := \min_{j \in A} r_j + p(A) + \min_{j \in A} q_j \tag{10.3}$$

yields a useful lower bound $LB(A)$ for the maximum lateness in an optimal schedule. Especially, for the critical sequence (10.3) transforms into

$$L_{\max}(OPT) \geq LB(\Lambda) = r_a + \sum_{j=a}^{c} p_j + \min_{j \in \Lambda} q_j \tag{10.4}$$

Subtracting (10.4) from (10.2) we get the following upper bound for the absolute error in terms of the delivery time of the critical item

$$L_{\max} - L_{\max}(OPT) \leq q_c - \min_{j \in \Lambda} q_j \tag{10.5}$$

Moreover, Schrage is optimal if $q_c \leq q_j$ for all $j \in \Lambda$. Otherwise, there is a job in Λ with smaller delivery time than q_c. The last job f in the critical sequence with the property $q_f < q_c$ is called the *interference job*. Let $\Lambda_f := \{f+1, \ldots, c\}$ be the jobs in Λ processed after the interference job f. Clearly, $q_j \geq q_c > q_f$ and $r_j > s_f$ hold for all $j \in \Lambda_f$, where s_f denotes the starting time of the interference job f. Inequality (10.3) applied to Λ_f gives the lower bound

$$L_{\max}(OPT) \geq LB(\Lambda_f) = \min_{j \in \Lambda_f} r_j + p(\Lambda_f) + \min_{j \in \Lambda_f} q_j > s_f + p(\Lambda_f) + q_c \tag{10.6}$$

Since there is no idle time during the execution of the jobs of the critical sequence, the maximum lateness of the Schrage schedule is $L_{\max} = s_f + p_f + p(\Lambda_f) + q_c$. Subtracting this equation from (10.6) we get the following upper bound for the absolute error in terms of the processing time p_f of the interference job f

$$L_{\max} - L_{\max}(OPT) < p_f \tag{10.7}$$

Using that $L_{\max}(OPT) \geq p_f$ and applying, for example, inequality (10.7) shows that Schrage yields a relative performance guarantee of 2. This was first observed by Kise, Ibaraki, and Mine [19]. The same performance guarantee holds even for simple list scheduling, that is, processing the first job which is available from an arbitrary list. Unfortunately, this bound is tight, which can be seen from a simple instance with two jobs characterized by $r_1 = 0$, $p_1 = M$, $q_1 = 0$, $r_2 = 1$, $p_2 = 1$, $q_2 = M$ with M a sufficiently large number.

Potts [20] improves the extended Jackson's rule by running Schrage at most n times to slightly varied instances. If there is a interference job f in the current Schrage sequence, f is forced to be processed after the critical item c in the next iteration by setting $r_f := r_c$. The algorithm stops if there is no interference job in the current iteration or n iterations have been performed. Finally, the best solution is selected. The algorithm of Potts runs in $O(n^2 \log n)$ time.

Theorem 10.2

The algorithm of Potts has a relative performance guarantee of 3/2.

Proof

If all processing times are less than or equal to $L_{\max}(OPT)/2$, the bound follows directly from (10.7) with the schedule of the first iteration. Hence, we may assume that there is a unique job u with $p_u > L_{\max}(OPT)/2$.

We distinguish two cases. First assume that all precedence constraints, induced by the algorithm of Potts by changing the release dates, are consistent with an optimal sequence. If a job other than u is interference job, the claim follows again from (10.7). Job u can be interference job at most $n-1$ times.

Thus, the algorithm terminates with no interference job before the n iterations are finished. But the heuristic sequence is optimal, if the algorithm terminates with no interference job.

Now assume there is no optimal sequence which is consistent with the subsequently added precedence constraints. Consider the first time that the interference job f is forced to be scheduled after the critical job c, whereas in the optimal sequence f precedes c. From (10.5) and (10.7) we conclude $q_c > L_{\max}(OPT)/2$ and $p_f > L_{\max}(OPT)/2$, respectively. But this is a contradiction to $L_{\max}(OPT) \geq r_f + p_f + p_c + q_c$. □

A slight modification of the delivery times guarantees that the algorithm of Potts has the same worst-case performance even for precedence constraints as shown by Hall and Shmoys [21]. Assume job i must precede job j. If $r_i > r_j$, set $r_j := r_i$ and analogously if $q_i < q_j + p_j$, set $q_i := q_j + p_j$ before the start of the algorithm. Whenever the release date r_j of an interference job f is set to $r_f := r_c$ in an iteration of the algorithm, the release dates of all jobs that must be preceded by f, are also set to r_c.

A more efficient 3/2-approximation algorithm with $O(n \log n)$ running time is presented by Nowicki and Smutnicki [22] by using the fact that Schrage yields already a performance ratio of 3/2 if the interference job f has processing time $p_f \leq p(N)/2$. Otherwise, a schedule is constructed by first assigning the jobs of $A := \{j \in N \mid j \neq f, r_j \leq q_j\}$ in nondecreasing order of release dates, then the interference job f, and then the jobs of $B := \{j \in N \mid j \neq f, r_j > q_j\}$ in nonincreasing order of delivery times. The better of this schedule and the Schrage schedule is selected.

A performance guarantee of 4/3 was obtained by Hall and Shmoys [21] by performing the algorithm of Potts for the original *and* the inverse problem and taking the best solution. In the same paper Hall and Shmoys develop two polynomial time approximation schemes (PTAS) for $1 \mid r_j \mid L_{\max}$. The first algorithm uses that a polynomial time dynamic programming algorithm exists when there are only a constant number of release dates and the processing times are encoded in unary. The second algorithm divides the jobs into large and small jobs with a constant number of large jobs. For every configuration of the large jobs several ways of fitting in the small jobs are tried. Finally, the best solution is selected. A third scheme due to Lawler is described by Hall in [23]. More sophisticated preprocessing algorithms than simple modification of delivery times are used by Hall and Shmoys in Ref. [24] to extend the scheme for $1 \mid r_j \mid L_{\max}$ to a PTAS for the problem with precedence constraints $1 \mid r_j, prec \mid L_{\max}$.

Hall [25] shows that the algorithm of Potts has a performance guarantee of 2 for general cost functions $h_j(t)$, which fulfill the following properties:

- The h_j are *subadditive*, that is, $h_j(t_1 + t_2) \leq h_j(t_1) + h_j(t_2)$ holds for every $t_1, t_2 \geq 0$ and $j = 1, \ldots, n$. Examples for subadditive functions are concave functions h_j with $h_j(0) \geq 0$.
- For every $i, j, 1 \leq i, j \leq n$, if $h_i(t) \geq h_j(t)$ holds for some $t \geq 0$, then $h_i(t) \geq h_j(t)$ holds for every $t \geq 0$. This property requires that the functions do not cross each other, which induces an ordering on the jobs.

Also for these cost functions the algorithm of Potts can be extended to precedence constraints.

10.3.3 Enumerative Algorithms

There are various enumerative algorithms for $1 \mid r_j \mid L_{\max}$, we will not give a complete list of the existing literature. The first branch-and-bound algorithm is due to Baker and Sue [26]. They compute lower bounds obtained by solving the preemptive problem $1 \mid r_j, pmtn \mid L_{\max}$ using the extended Jackson's rule. The strength of the lower bound obtained from preemption can be seen from the fact that the optimal solution $L_{\max}(pmtn)$ of $1 \mid r_j, pmtn \mid L_{\max}$ is identical to the maximum of the lower bounds $LB(A)$ over all sets $A \subseteq N$, as defined in (10.3), that is,

$$L_{\max}(OPT) \geq L_{\max}(pmtn) = \max_{A \subseteq N} LB(A) \tag{10.8}$$

This was observed by Carlier [27]. Note that by (10.8) $\max_{A \subseteq N} LB(A)$ can be calculated in $O(n \log n)$ time.

The algorithm of Baker and Su is improved by McMahon and Florian [28] presenting a branch-and-bound algorithm that is based on applying the extended Jackson's rule at each node of the enumeration tree. Lageweg, Lenstra, and Rinnooy Kan [14] refine the approach by McMahon and Florian by taking into account the inverse problems.

Carlier [27] applies also the extended Jackson's rule to develop a branch-and-bound algorithm. The node of the branch-and-bound tree associated with a single machine problem is branched depending on whether the interference job f shall be executed before the set Λ_f (see Section 10.3.2) or not. The lower bound is computed as the maximum of the lower bounds $LB(\Lambda_f)$ and $LB(\Lambda_f \cup \{f\})$ obtained from (10.3). His method is especially useful for problems with a large number of items. Similar methods are applied by Larson, Dessouky, and Devor [29]. Grabowski, Nowicki, and Zdrzalka [30] use the so-called *block approach* for solving $1 \mid r_j \mid L_{\max}$. This block approach was applied by the authors to solve flow shop problems optimally.

Most of the procedures above work also when precedence constraints are included. A branch-and-bound algorithm for the problem with precedence constraints and nondecreasing cost functions $1 \mid r_j, prec \mid h_{\max}$ is given by Zdrzalka and Grabowski [31] by extending the ideas used for $1 \mid r_j, prec \mid L_{\max}$.

Tinhofer and Farnbacher [32] developed a new lower bound LB_S for $1 \mid r_j \mid L_{\max}$: For $A \subseteq N$ being an arbitrary set of jobs and $a \in A$ let C_{Aa} denote the maximum delivery–completion time in a Schrage schedule of job set $A \setminus a$. Then, the lower bound LB_S with

$$L_{\max}(OPT) \geq LB_S := \max_{A \subseteq N} \min \{C_{Aa} + p_a + q_a \mid a \in A\} \tag{10.9}$$

can be calculated in $O(n^2)$ time. Moreover, the bound LB_S dominates the bound $L_{\max}(pmtn)$ from (10.8) obtained by the relaxation to preemptive schedules.

10.4 Parallel Machine Problems

In this section we study scheduling problems for m parallel machines with the objective of minimizing the maximum lateness. We will present algorithms for problems with preemptions or jobs with equal processing times, and approximation and enumerative algorithms for nonpreemptive problems with general-length jobs.

10.4.1 Problems with Preemption

Parallel machine problems can often be formulated as linear integer programs with decision variables $x_{ij} \in \{0, 1\}$, which decide to which machine i a job j is assigned. Since preemption can be considered as "relaxing" these variables to nonnegative variables $x_{ij} \geq 0$, the IP-formulation of the scheduling problem often turns into a linear program. Thus, it is not surprising that most of the preemptive problems can be solved in polynomial time (as long as precedence constraints are not involved).

Starting with parallel, identical machines Horn [9] gives an $O(n^2)$ algorithm for solving $P \mid pmtn \mid L_{\max}$ (or equivalently $P \mid r_j, pmtn \mid C_{\max}$). Horn generalizes the classical approach of McNaughton [33] for $P \mid pmtn \mid C_{\max}$ who first constructs a lower bound on the makespan of the optimal schedule and then finds a schedule with makespan equal to that bound. Gonzalez and Johnson [34] give an algorithm for $P \mid pmtn \mid L_{\max}$ with reduced running time in $O(nm)$. An algorithm of Sahni [35] generates an optimal schedule for the pure feasibility problem with deadlines $P \mid \bar{d}_j, pmtn \mid \cdot$ in $O(n \log mn)$ time with at most $n - 2$ preemptions.

Baptiste [36] proposes a necessary and sufficient condition for the existence of a feasible solution for $P \mid \bar{d}_j, pmtn \mid \cdot$ which can be stated as follows. There is a feasible schedule for $P \mid \bar{d}_j, pmtn \mid \cdot$ if and only if $p_j \leq \bar{d}_j$ and

$$\sum_{i=1}^{n} \max\{0, p_i - \max\{0, \bar{d}_i - \bar{d}_j\}\} \leq m\bar{d}_j$$

hold for all $j = 1, \ldots, n$. This allows him to construct an $O(n \log n)$ algorithm for the feasibility problem and an algorithm with the same time complexity for $P \mid pmtn \mid L_{\max}$.

If there are also release dates, Horn [9] formulates the feasibility problem $P \mid r_j, \bar{d}_j, pmtn \mid \cdot$ as a network flow problem which could be solved in $O(n^3)$ time. Labetoulle et al. [37] embed Horn's approach into a binary search and show that this results in an algorithm for $P \mid r_j, pmtn \mid L_{\max}$ with running time in $O(n^3 \min\{n^2, \log n + \log \max_j p_j\})$.

The feasibility problem for uniform machines with deadlines $Q \mid \bar{d}_j, pmtn \mid \cdot$ has been studied by Sahni and Cho [38]. Their algorithm runs in $O(n \log n + mn)$ time and generates in the worst-case at most $O(mn)$ preemptions. The same result is obtained by Labetoulle et al. [37]. In the presence of release dates Martel [39] presents an $O(m^2 n^4 + n^5)$ algorithm with $O(mn)$ preemptions for the feasibility problem $Q \mid r_j, \bar{d}_j, pmtn \mid$. The algorithm uses a generalization of network flow techniques and can be adapted to $Q \mid r_j, pmtn \mid L_{\max}$ by using bisection search over the possible values for $L_{\max}$. Federgruen and Groenevelt [40] show that a feasible schedule for $Q \mid r_j, \bar{d}_j, pmtn \mid L_{\max}$ can be obtained by determining the maximum flow in a network. The running time reduces to $O(nt^3)$ operations when t is the number of different machine speeds. An adaption to the maximum lateness criterion is also described.

The problem for unrelated machines $R \mid r_j, pmtn \mid L_{\max}$ is solved by Lawler and Labetoulle [41] in polynomial time by formulating the problem as a linear programming problem. It is shown that no more than $O(m^2 n)$ preemptions are necessary. Sourd and Nuijten [7] extend the linear programming approach of Lawler and Labetoulle to prove that the problem with deadlines $R \mid r_j, \bar{d}_j, pmtn \mid L_{\max}$ can also be solved in polynomial time.

For precedence constraints everything becomes immediately difficult. Ullman [42] shows that the basic problem $P \mid p_j = 1, pmtn, prec \mid C_{\max}$ is $\mathcal{NP}$-hard. There are not many problems with precedence relations for which polynomial time algorithms exist. We want to mention the paper by Lawler [43] which contains an $O(n^2)$ algorithms for $P \mid pmtn, intree \mid L_{\max}$, whereas the problem $P \mid pmtn, outtree \mid L_{\max}$ is $\mathcal{NP}$-hard. For the problem with two uniform machines $Q2 \mid pmtn, intree \mid L_{\max}$ he found also an $O(n^2)$ algorithm. In the presence of release dates the running time for optimally solving problem $Q2 \mid r_j, pmtn, intree \mid L_{\max}$ increases to $O(n^6)$.

10.4.2 Problems with Equal Processing Times

Unless there are no precedence constraints or release dates, problems with unit-length jobs are fairly simple. The problem $P \mid p_j = 1 \mid L_{\max}$ is trivially solved by the EDD rule. For uniform machines the problem $Q \mid p_j = 1 \mid h_{\max}$, with h_j nondecreasing cost functions, is treated as follows: First, it can be observed that there exists an optimal schedule in which the jobs are processed in the time periods with the n earliest completion times. Then, the jobs are assigned backwards analogously to the algorithm of Lawler described in Section 10.2. This yields an optimal schedule in $O(n^2)$ time. For $Q \mid p_j = 1 \mid L_{\max}$ the running time reduces to $O(n \log n)$. A description of this algorithm can be found in Lawler et al. [44] or in Dessouky et al. [45].

For release dates the results are usually given for feasibility problems with deadlines but can be easily converted into results for minimizing the maximum lateness. The problem of scheduling unit-length jobs with release dates and deadlines $P \mid r_j, \bar{d}_j, p_j = 1 \mid \cdot$ is shown to be solvable in $O(n \log n)$ time by Frederickson [10]. An algorithm with the same time complexity $O(n \log n)$ is given by Sourds and Nuijten [7] for $P \mid r_j, \bar{d}_j, p_j = 1 \mid L_{\max}$. For the problem of finding a feasible schedule with equal processing times $P \mid r_j, \bar{d}_j, p_j = p \mid \cdot$ Simons [46] gives an $O(n^3 \log \log n)$ algorithm. Simons and Warmuth [47] improve the running time of this algorithm to $O(mn^2)$ by doing some preprocessing before the jobs are actually scheduled.

The complexity status of the problem with uniform machines $Q \mid r_j, p_j = 1 \mid L_{\max}$ is still an open question. Dessouky [48] compares six simple heuristic solution procedures for $Q \mid r_j, p_j = 1 \mid L_{\max}$ and uses them to develop a branch-and-bound procedure for this problem.

Lenstra and Rinnooy Kan [49] show that even the problem of deciding whether there is a feasible schedule of makespan at most three for the problem $P \mid p_j = 1, prec \mid C_{\max}$ is already $\mathcal{NP}$-complete. Thus, there is no polynomial approximation algorithm for $P \mid p_j = 1, prec \mid C_{\max}$ with performance ratio better than 4/3 unless $\mathcal{P} = \mathcal{NP}$ holds. For more results on equal processing times and precedence relations we refer to the chapter dedicated to equal processing times.

10.4.3 Approximation Algorithms for Nonpreemptive Problems with General-Length Jobs

There is not much hope to find polynomial time algorithms for parallel machines and jobs with general lengths. Even the fundamental problem of minimizing the makespan on two identical machines $P2 \mid \cdot \mid C_{\max}$ is $\mathcal{NP}$-hard and the problem $P \mid \cdot \mid C_{\max}$ is strongly $\mathcal{NP}$-hard [50]. This excludes the existence of a fully polynomial time approximation scheme (FPTAS) if the number of machines is part of the input. For unrelated machines the situation is even worse. Lenstra, Shmoys, and Tardos [51] show that there is no approximation algorithm for $R \mid \cdot \mid C_{\max}$ with performance ratio better than 3/2 unless $\mathcal{P} = \mathcal{NP}$ holds.

For a fixed number of machines pseudopolynomial algorithms have been found by Lawler and Moore [52] for the problem with unrelated machines $Rm \mid \cdot \mid L_{\max}$ applying dynamic programming techniques. Based on the dynamic programming algorithms by Lawler and Moore, Horowitz and Sahni [53] use interval partitioning of the state space to convert the dynamic program into an FPTAS for $Rm \mid \cdot \mid C_{\max}$. Unfortunately, they could not generalize this result to the maximum lateness criterion. Only results for minimizing the *maximum flow time*

$$F_{\max} := \max_{1 \le j \le n} (C_j - r_j)$$

is known. By definition, the flow time F_j is a special case of the lateness L_j with $d_j := r_j$. Recently, Bansal [54] shows that there is a PTAS for the problem with a fixed number of machines $Rm \mid \cdot \mid F_{\max}$ which runs in $n^{O(m/\varepsilon)}$ time.

From now on we assume that the number of machines is part of the input. Woeginger [55] analyzes the worst-case behavior of a mixture of list scheduling and the EDD rule for the problem $P \mid \cdot \mid L_{\max}$ with parallel, identical machines and delivery times. His algorithm has a performance ratio of $2 - 2/(m+1)$ and runs in $O(n \log n)$ time. Hall and Shmoys [56] manage to show the existence of a PTAS for $P \mid \cdot \mid L_{\max}$. Their result is a common generalization of the PTAS of the same authors for the single machine problem $1 \mid r_j \mid L_{\max}$ and a result of Hochbaum and Shmoys [57] for $P \mid \cdot \mid C_{\max}$, and uses the notion of an *outline scheme*. Hall and Shmoys also prove in the same paper that the following simple list scheduling algorithm is a 2-approximation algorithm: Whenever a machine becomes idle, we choose any released job with all of its predecessors having been finished processing.

For the problem with uniform machines $Q \mid \cdot \mid L_{\max}$ Koulamas and Kyparisis [58] study the worst-case behavior of an extension of Jackson's rule to a uniform parallel machine setting. They show that the algorithm yields a maximum lateness, which does not exceed the optimal value by more than the maximum job processing time.

10.4.4 Enumerative Algorithms for Nonpreemptive Problems with General-Length Jobs

There are only a few results concerning enumerative algorithms for problems on parallel machines. Carlier [59] considers a generalization of the extended Jackson's rule to the problem $P \mid r_j \mid L_{\max}$: A branching scheme is proposed by associating with each job an interval of time during which it has to be processed. To branch, the interval for a particular job is divided into two smaller ones. Two lower bound $G(A)$ and $G'(A)$ are presented for any set $A \subseteq N$ of jobs. The first lower bound

$$G(A) := \min_{j \in A} r_j + p(A)/m + \min_{j \in A} q_j \tag{10.10}$$

is an obvious generalization of the lower bound (10.3) for the single machine case. Note that the bound $\max_{A \subseteq N} G(A)$ can be found in $O(n \log n)$ time. Now assume that A has at least m elements and that

$r_1, \ldots, r_m$ are the m largest release dates of A and $q_1, \ldots, q_m$ the m largest delivery times of A, respectively. Then, the second bound $G'(A)$ is given as

$$G'(A) := \frac{1}{m}\left(\sum_{j=1}^{m} r_j + p(A) + \sum_{j=1}^{m} q_j\right) \tag{10.11}$$

and can be computed in linear time.

Recall that the solution algorithm of Labetoulle et al. [37] for the preemptive problem $P \mid r_j, pmtn \mid L_{\max}$ requires a running time of $O(n^3 \min\{n^2, \log n + \log \max_j p_j\})$. This time complexity forbids an extensive use of the preemptive relaxation in enumerative algorithms for $P \mid r_j \mid L_{\max}$. Carlier and Pinson [60] introduce the so-called *Jackson's pseudo-preemptive schedule*, where a job is allowed to be processed on more than one machine at a time. A pseudo-preemptive schedule is not as strong as a preemptive schedule but can be calculated in $O(n \log n + nm \log m)$ time.

Haouari and Gharbi [61] introduce a new concept of semipreemptive scheduling and show how it can be used to derive a maximum-flow-based lower bound for $P \mid r_j \mid L_{\max}$. The new bound dominates the preemptive lower bound, and thus also the pseudo-preemptive schedules of Carlier and Pinson. It can be computed in $O(n^3(\log n + \log \max_j p_j))$ time. Gharbi and Haouari [62] also develop branch-and-bound algorithms for $P \mid r_j \mid L_{\max}$. They solve instances with up to 300 jobs in a moderate CPU time.

References

[1] C.N. Potts and M.Y. Kovalyov. Scheduling with batching: a review. *European Journal of Operational Research*, 120:228–249, 2000.

[2] B. Chen, C. Potts, and G. Woeginger. A review of machine scheduling: Complexity, algorithms and approximability. In D.Z. Du and P. Pardalos, (eds.), *Handbook of Combinatorial Optimization*, Kluwer, 1998, pp. 21–169.

[3] J.R. Jackson. Scheduling a production line to minimize maximum tardiness. Technical Report 43, University of California, Los Angeles, 1955.

[4] E.L. Lawler. Optimal sequencing of a single machine subject to precedence constraints. *Management Science*, 19:544–546, 1973.

[5] D. Hochbaum and R. Shamir. An $O(n \log^2 n)$ algorithm for the maximum weighted tardiness problem. *Information Processing Letters*, 31:215–219, 1989.

[6] M.C. Fields and G.N. Frederickson. A faster algorithm for the maximum weighted tardiness problem. *Information Processing Letters*, 36:39–44, 1990.

[7] F. Sourd and W. Nuijten. Scheduling with tails and deadlines. *Journal of Scheduling*, 4:105–121, 2001.

[8] J.K. Lenstra, A.H.G. Rinnooy Kan, and P. Brucker. Complexity of machine scheduling problems. *Annals of Operations Research*, 1:342–362, 1977.

[9] W.A. Horn. Some simple scheduling algorithms. *Naval Research Logisitcs Quarterly*, 21:177–185, 1974.

[10] G.N. Frederickson. Scheduling unit time tasks with integer release times and deadlines. *Information Processing Letters*, 16:171–173, 1983.

[11] B.B. Simons. A fast algorithm for single processor scheduling. In *Proceedings of the 19th Annual Symposium on Foundations of Computer Science*, 1978, pp. 246–252.

[12] J. Carlier. Problème à une machine dans le cas où les taches ont des durées égales. Technical report, Université Paris VI, Paris, 1979.

[13] M.R. Garey, D.S. Johnson, B.B. Simons, and R.E. Tarjan. Scheduling unit-time tasks with arbitrary release times and deadlines. *SIAM Journal on Computing*, 10:256–269, 1981.

[14] B.J. Lageweg, J.K. Lenstra, and A.H.G. Rinnooy Kan. Minimizing maximum lateness on one machine: computational experience and some applications. *Statistica Neerlandica*, 30:25–41, 1976.

[15] C.L. Monma. Linear time algorithms for scheduling on parallel processors. *Operations Research*, 30:116–124, 1982.
[16] K.R. Baker, E.L. Lawler, J.K. Lenstra, and A.H.G. Rinnooy Kan. Preemptive scheduling of a single machine to minimize maximum cost subject to release dates and precedence constraints. *Operations Research*, 31:381–386, 1983.
[17] V. Gordon and V.S. Tanaev. On minimax problems of scheduling theory for a single machine. *Vetsi Akademii Navuk BSSR*, 1983, pp. 3–9. [in Russian] in Russian.
[18] L. Schrage. Obtaining optimal solutions to resource constrained network scheduling problems. unpublished manuscript, 1971.
[19] H. Kise, T. Ibaraki, and H. Mine. Performance analysis of six approximation algorithms for the one-machine maximum lateness scheduling problem with ready times. *Journal of the Operational Research Society of Japan*, 22:205–224, 1979.
[20] C. Potts. Analysis of a heuristic for one machine sequencing with release dates and delivery times. *Operations Research*, 28:1436–1441, 1980.
[21] L. Hall and D. Shmoys. Jackson's rule for single-machine scheduling: making a good heuristic better. *Mathematics of Operations Research*, 17:22–35, 1992.
[22] E. Nowicki and C. Smutnicki. An approximation algorithm for single-machine scheduling with release times and delivery times. *Discrete Applied Mathematics*, 48:69–79, 1994.
[23] L. Hall. Approximation algorithms for scheduling. In D. Hochbaum, (ed.), *Approximation Algorithms for NP-hard Problems*, PWS Publishing Co., 1997, pp. 1–45.
[24] L. Hall and D. Shmoys. Near-optimal sequencing with precedence constraints. In *Proceedings of the 1st Integer Programming and Combinatorial Optimization Conference (IPCO)*, 1990, pp. 249–260.
[25] L. Hall. A note on generalizing the maximum lateness criterion for scheduling. *Discrete Applied Mathematics*, 47:129–137, 1993.
[26] K.R. Baker and Z.-S. Su. Sequencing with due-dates and early start times to minimize maximum tardiness. *Naval Research Logistics Quarterly*, 21:171–176, 1974.
[27] J. Carlier. The one-machine sequencing problem. *European Journal of Operational Research*, 11:42–47, 1982.
[28] G.B. McMahon and M. Florian. On scheduling with ready times and due dates to minimize maximum lateness. *Operations Research*, 23:475–482, 1975.
[29] R.E. Larson, M.I. Dessouky, and R.E. Devor. A forward-backward procedure for the single machine problem to minimize the maximum lateness. *IIE Transactions*, 17:252–260, 1985.
[30] J. Grabowski, E. Nowicki, and S. Zdrzalka. A block approach for single-machine scheduling with release dates and due dates. *European Journal of Operational Research*, 26:278–285, 1986.
[31] S. Zdrzalka and J. Grabowski. An algorithm for single machine sequencing with release dates to minimize maximum cost. *Discrete Applied Mathematics*, 23:73–89, 1989.
[32] G. Tinhofer and E. Farnbacher. A new lower bound for the makespan of a single machine scheduling problem. In P. Kall, (ed.), *System Modelling and Optimization*, *Lecture Notes in Control and Information Science*, Springer, 1992, pp. 209–218.
[33] R. McNaughton. Scheduling with deadlines and loss functions. *Management Science*, 12:1–12, 1959.
[34] T. Gonzalez and D.B. Johnson. A new algorithm for preemptive scheduling of trees. *Journal of the Association of Computing Machinery*, 27:287–312, 1980.
[35] S. Sahni. Preemptive scheduling with due dates. *Operations Research*, 27:925–934, 1979.
[36] P. Baptiste. Preemptive scheduling of identical machines. Technical Report 2000/314, Université Technologie de Compiègne, 2000.
[37] J. Labetoulle, E.L. Lawler, J.K. Lenstra, and A.H.G. Rinnooy Kan. Preemptive scheduling of uniform machines subject to release dates. In W.R. Pulleyblank, (ed.), *Progress in Combinatorial Optimization*, Academic Press, 1984, pp. 245–261.
[38] S. Sahni and Y. Cho. Scheduling independent tasks with due times on a uniform processor system. *Journal of the Association of Computing Machinery*, 27:550–563, 1980.

[39] C. Martel. Preemptive scheduling with release dates, deadlines, and due dates. *Journal of the Association of Computing Machinery*, 29:812–829, 1982.

[40] A. Federgruen and H. Groenevelt. Preemptive scheduling of uniform machines by ordinary network flow techniques. *Management Science*, 32:341–349, 1986.

[41] E.L. Lawler and J. Labetoulle. On preemptive scheduling of unrelated parallel processors by linear programming. *Journal of the Association of Computing Machinery*, 25:612–619, 1978.

[42] J.D. Ullman. NP-complete scheduling problems. *Journal of Computing and System Sciences*, 10:384–393, 1975.

[43] E.L. Lawler. Preemptive scheduling of precedence-constrained jobs on parallel machines. In M.A.H. Dempster, J.K. Lenstra, and A.H.G. Rinnooy Kan, (eds.), *Deterministic and Stochastic Scheduling*, Reidel, 1982, pp. 101–123.

[44] E.L. Lawler, J.K. Lenstra, A.H.G. Rinnooy Kan, and D. Shmoys. Sequencing and scheduling: algorithms and complexity. In S. Graves, A.H.G. Rinnooy Kan, and P. Zipkin, (eds.), *Handbooks in Operations Research and Management Science*, North Holland, 1993, pp. 445–522.

[45] M.I. Dessouky, B.J. Lageweg, J.K. Lenstra, and S.L. van de Velde. Scheduling identical jobs on uniform parallel machines. *Statistica Neerlandica*, 44:115–123, 1990.

[46] B.B. Simons. Multiprocessor scheduling of unit-time jobs with arbitrary release times and deadlines. *SIAM Journal on Computing*, 12:294–299, 1983.

[47] B.B. Simons and M.K. Warmuth. A fast algorithm for multiprocessor scheduling of unit-length jobs. *SIAM Journal on Computing*, 18:690–710, 1989.

[48] M.M. Dessouky. Scheduling identical jobs with unequal ready times on uniform parallel machines to minimize the maximum lateness. *Computers and Industrial Engineering*, 34:793–806, 1998.

[49] J.K. Lenstra and A.H.G. Rinnooy Kan. Complexity of scheduling under precedence constraints. *Operations Research*, 26:22–35, 1978.

[50] M.R. Garey and D.S. Johnson. *Computers and Intractability: a Guide to the Theory of NP-Completeness*. Freeman, 1979.

[51] J.K. Lenstra, D. Shmoys, and E. Tardos. Approximation algorithms for scheduling unrelated parallel machines. *Mathematical Programming*, 46:259–271, 1990.

[52] E.L. Lawler and J.M. Moore. A functional equation and its application to resource allocation and sequencing problems. *Management Science*, 16:77–84, 1969.

[53] E. Horowitz and S. Sahni. Exact and approximate algorithms for scheduling nonidentical processors. *Journal of the Association of Computing Machinery*, 23:317–327, 1976.

[54] N. Bansal. Approximation schemes for flow time on multiple machines. Technical report, Carnegie Mellon University, 2003.

[55] G.J. Woeginger. Heuristics for parallel machine scheduling with delivery times. *Acta Informatica*, 31:503–512, 1994.

[56] L. Hall and D. Shmoys. Approximation schemes for constrained scheduling problems. In *Proceedings of the 30th Symposium on Foundations of Computer Science*, 1989, pp. 134–139.

[57] D. Hochbaum and D. Shmoys. Using dual approximation algorithms for scheduling problems: theoretical and practical results. *Journal of the Association of Computing Machinery*, 34:144–162, 1987.

[58] C. Koulamas and G.J. Kyparisis. Scheduling on uniform parallel machines to minimize maximum lateness. *Operations Research Letters*, 26:175–179, 2000.

[59] J. Carlier. Scheduling jobs with release dates and tails on identical machines to minimize the makespan. *European Journal of Operational Research*, 29:298–306, 1987.

[60] J. Carlier and E. Pinson. Jackson's pseudo preemptive schedule for the $Pm \mid r_j, q_i \mid C_{\max}$ scheduling problem. *Annals of Operations Research*, 83:41–58, 1998.

[61] M. Haouari and A. Gharbi. An improved max-flow-based lower bound for minimizing maximum lateness on identical parallel machines. *Operations Research Letters*, 31:49–52, 2003.

[62] A. Gharbi and M. Haouari. Minimizing makespan on parallel machines subject to release dates and delivery times. *Journal of Scheduling*, 5:329–355, 2002.

11

Approximation Algorithms for Minimizing Average Weighted Completion Time

Chandra Chekuri
Bell Labs

Sanjeev Khanna
University of Pennsylvania

11.1 Introduction

In this chapter we survey approximation algorithms for scheduling to minimize average (weighted) completion time (equivalently sum of (weighted) completion times). We are given n jobs $J_1, \ldots, J_n$, where each job J_j has a positive weight w_j. We denote by C_j the completion time of job j in a given schedule. The objective in the problems we consider is to find a schedule to minimize $\sum_j w_j C_j$ (average weighted completion time). The most basic problem in this context is to minimize $\sum_j C_j$ on a single machine with job j having a processing time p_j, and all jobs are available at time 0, in other words, the problem $1 \,||\, \sum_j C_j$. It is easy to see that ordering jobs by the SPT rule (shortest processing time first) gives an optimal schedule. A slight generalization with jobs having weights, $1 \,||\, \sum_j w_j C_j$, also has a simple optimality rule first stated by Smith [1] (known as Smith's rule): schedule jobs in nondecreasing order of the ratio p_j / w_j.

Earlier work on minimizing weighted completion times focussed on identifying polynomial time solvable cases by generalizing Smith's rule. Algorithms were found for solving the problem on a single machine when jobs are allowed to have some restricted class of precedence constraints such as in- and out-tree precedences and series-parallel precedence constraints [2–4]. Sydney [4] unified these results by showing the applicability of Smith's rule to *decomposable* classes of precedence constraints. We are interested in a more

1-58488-397-9/\$0.00+\$1.50

general setting with release dates, arbitrary precedence constraints, and multiple machines, any of which make the problem NP-hard [5]. Thus, we will consider approximation algorithms. Online algorithms for some of these problems will be considered elsewhere in the book.

The problem of minimizing makespan ($\max_j C_j$) is an extremely well studied measure in scheduling and starting with Graham's seminal paper on multiprocessor scheduling [6], approximation algorithms have been studied for several variants. However, even though weighted completion time is closely related to makespan (generalizes makespan when precedence constraints are present), it is only recently that approximation algorithms have been designed for this objective. The work of Phillips, Stein, and Wein [7] was the first to explore weighted completion time in detail from the approximation algorithms point of view, in particular for problems with release dates. A variety of algorithms and techniques were developed in the same work; in particular simple ordering rules based on preemptive schedules and LP relaxations were shown to be effective in obtaining near-optimal schedules. Following [7], scheduling to minimize weighted completion time received substantial attention. Polyhedral formulations for single machine problems have been an active area of research and a variety of results are known from the 80s and early 90s. Building on some of the ideas in [7] and insight from polyhedral formulations, Hall, Shmoys, and Wein [8], and independently Schulz [9], developed the first constant factor approximation algorithms for minimizing weighted completion time with precedence constraints. Subsequently, many improvements, generalizations, and combinatorial methods were developed in a number of papers [10–21]. Hoogeveen, Schuurman, and Woeginger [22] systematically studied in-approximability results and established APX-hardness for a number of problems. More recently, polynomial time approximation schemes (PTASes) were obtained for several variants [20,23] based on enumeration and dynamic programming based ideas. Together, these results give us a fairly comprehensive picture of the approximability of scheduling to minimize weighted completion time. Figure 11.1 tabulates the known results and in Section 11.5 we mention several as yet unresolved questions.

This survey aims to describe in detail selected algorithms that illuminate some of the useful and interesting ideas for scheduling to minimize weighted completion time. It is infeasible to do justice to all the known literature given the nature of the survey and the space constraints. We hope that we have captured several central ideas and algorithms that would enable the reader to get both an overview of the known results and some of the more advanced technical ideas. In Section 11.5 we give pointers to material that we would have liked to include but could not.

The rest of the survey is organized as follows. We have broadly partitioned the algorithms into three categories. In Section 11.2 we describe approximation algorithms when jobs have precedence constraints.

Problem	Approx. Ratio	Ref.	Inapproximability
$1 \| \| \sum_i w_j C_j$	1	[1]	
$1 \| r_j, pmtn \| \sum_j C_j$	1		
$P \| pmtn \| \sum_j C_j$	1	[25, 26]	
$P \| r_j \| \sum_j w_j C_j$	PTAS	[20]	NP-hard
$Q \| r_j \| \sum_j w_j C_j$	PTAS	[27]	NP-hard
$Rm \| r_j \| \sum_j w_j C_j$	PTAS	[20]	NP-hard
$R \| \| \sum_j w_j C_j$	3/2	[28]	APX-hard [22]
$R \| r_j \| \sum_j w_j C_j$	2	[28]	APX-hard [22]
$1 \| prec, sp \| \sum_j w_j C_j$	1	[3]	
$1 \| prec \| \sum_j w_j C_j$	2	[29]	NP-hard
$1 \| prec, r_j \| \sum_j w_j C_j$	$(e + \in)$	[30]	NP-hard
$P \| prec, r_j \| \sum_j w_j C_j$	4	[16]	4/3 [31]
$Q \| prec, r_j \| \sum_j w_j C_j$	$O(\log m)$	[15]	4/3 [31]
$O \| r_j \| \sum_j w_j C_j$	5.83	[21]	APX-hard [22]
$J \| r_{hj} \| \sum_j w_j C_j$	$O((\frac{\log m\mu}{\log \log m\mu})^2)$	[21]	APX-hard [22]

FIGURE 11.1 Complexity of minimizing average completion time.

Linear programming formulations have played an important role in algorithms for this case and we describe several formulations and how they have been used to develop constant factor approximation algorithms. We also show how simpler combinatorial methods can be used to obtain similar ratios. The latter algorithms have the advantage of being more efficient than the LP based methods as well as providing useful structural insight into the problem. In Section 11.3 we describe an algorithm for unrelated machines. The algorithm illustrates the need for a time indexed formulation as well as the usefulness of randomized rounding. Finally, in Section 11.4 we describe a framework that has led to polynomial time approximation schemes for a variety of problems with release dates. In particular we present a PTAS for scheduling on identical parallel machines. Section 11.5 concludes with discussion, pointers to related work, and open problems that remain.

Applications and Motivation: Makespan of a schedule measures the time by which all the jobs in the schedule finish. However, when jobs are independent and competing for the same resource, a more natural measure of performance is the *average* completion time of jobs. This motivated early work for this measure including the work of Smith [1]. More recently, an application to instruction scheduling in VLIW processors is presented in the work of Chekuri et al. [24]. In this application the weight of a job (instruction) is derived from profile information and indicates the likelihood of the program block terminating at that job. This is a compelling application for scheduling jobs with precedence constraints on a complex machine environment to minimize sum of weighted completion times.

In addition to applications, minimizing weighted completion time presents interesting theoretical problems in finding orderings of posets and has spurred work on polyhedral formulations as well as combinatorial methods. There is an intimate relation between minimizing makespan and minimizing weighted completion time and exploring this has improved our understanding of both.

Techniques: The algorithms we present, except for the approximation schemes in Section 11.4 have an underlying theme. They obtain an ordering of the jobs in one form or the other: via LP relaxations or preemptive schedules or based on decomposition methods. The ordering is converted into a feasible schedule by one of several variants of list scheduling, sometimes based on randomization. Typically, the bounds are shown by a job by job analysis. Although the above paradigm broadly unifies many of the algorithms, subtle and nontrivial problem specific ideas are necessary to obtain the best results. Finally, approximation schemes require a different approach, that of dynamic programming based enumeration over a structured space of schedules.

11.2 Algorithms for Problems with Precedence Constraints

In this section we describe algorithms when jobs have precedence constraints. We note that in this case minimizing makespan is a special case of minimizing weighted completion time. Add a dummy job that is preceded by all the other jobs, the dummy job has weight 1 and the rest have weight 0. The first constant factor approximation algorithms for minimizing completion time when jobs have precedence constraints were obtained via linear programming relaxations. For single machine and identical parallel machines, combinatorial algorithms have been developed although the ratios obtained are some what worse than the ones obtainable by LP methods. However, the combinatorial methods, in addition to yielding efficient algorithms, have provided structural insight, have been useful in extending algorithms to some stochastic settings [32], and in obtaining improved algorithms for special classes of precedence constraints [33,34]. However, for complex machine environments (related or unrelated machines) the only nontrivial algorithms we know are based on LP methods.

11.2.1 LP-Based Algorithms

A variety of LP formulations have been explored for minimizing weighted completion time, in particular for the single machine problem. We mention some of the work here: Balas [35], Wolsey [36], Dyer and Wolsey [37], Queyranne [38], Queyranne and Wang [39], Lasserre and Queyranne [40], Sousa and Wolsey [41], von Arnim and Schulz [42], Crama and Spieksma [43], Van den Akker, Van Hoesel, and Savelsbergh [44], Van den Akker [45], Van den Akker, Hurkens, and Savelsbergh [46], von Arnim, Schrader, and Wang [47]. The goal of this line of work was to examine the strength of various formulations in obtaining exact solutions via branch and bound and branch and cut methods. More recently these formulations have been used to obtain provably good approximation algorithms, starting with the work of Phillips, Stein, and Wein [7] and subsequently Hall, Shmoys, and Wein [8] and Schulz [9], and many others. Below we describe algorithms for the single machine and identical parallel machines.

11.2.1.1 Single Machine Scheduling

We start by describing and analyzing the completion time formulation for the problem $1 \mid r_j, prec \mid \sum_j w_j C_j$. We then describe two other formulations, the time indexed formulation and the linear ordering formulation.

Completion Time Formulation: The completion time formulation is based on the work of Queyranne. The formulation has variables C_j, $j = 1, \ldots, n$. C_j indicates the completion time of job j in the schedule. The formulation is the following.

$$\min \sum_{j=1}^{n} w_j C_j$$

subject to

$$C_j \geq r_j + p_j \qquad j = 1, \ldots, n \tag{11.1}$$

$$C_k \geq C_j + p_k \qquad \text{if } j \prec k \tag{11.2}$$

$$\sum_{j \in S} p_j C_j \geq \frac{1}{2}\left[\left(\sum_{j \in S} p_j\right)^2 + \sum_{j \in S} p_j^2\right] \qquad S \subseteq N \tag{11.3}$$

The nontrivial constraint in the above formulation is Equation (11.3). We justify the constraint as follows. Consider any valid schedule for the jobs. Without loss of generality, for ease of notation, assume that the jobs in S are $\{1, 2, \ldots, |S|\}$ and that they are scheduled in this order in the schedule. Then it follows that for $j \in S$, $C_j \geq \sum_{k \leq j} p_k$, hence $p_j C_j \geq p_j \sum_{k \leq j} p_k$. Summing over all $j \in S$ and simple algebra results in the inequality (11.3). It is easy to see that the inequality is agnostic to the ordering of S in the schedule and hence holds for all orderings. Note that Equation (11.3) generates an exponential number of constraints. Queyranne [38] has shown a polynomial time separation oracle for these constraints and hence the above formulation can be solved in polynomial time by the ellipsoid method.

We now state an important lemma regarding the formulation.

Lemma 11.1

Let $\bar{C}$ be a feasible solution to the completion time formulation and without loss of generality let $\bar{C}_1 \leq \bar{C}_2 \leq \cdots \leq \bar{C}_n$. Then, the following is true for $j = 1, \ldots, n$.

$$\bar{C}_j \geq \frac{1}{2} \sum_{k=1}^{j} p_k$$

Proof

We have that $\bar{C}_k \leq \bar{C}_j$ for $k = 1, \ldots, j$, hence $\sum_{k=1}^{j} p_k \bar{C}_k \leq (\sum_{k=1}^{j} p_k)\bar{C}_j$. Consider the set $S = \{1, 2, \ldots, j\}$ in Equation (11.3). It follows that

$$\left(\sum_{k=1}^{j} p_k\right)\bar{C}_j \geq \sum_{k=1}^{j} p_k \bar{C}_k$$
$$\geq \frac{1}{2}\left[\left(\sum_{k=1}^{j} p_k\right)^2 + \sum_{k=1}^{j} p_k^2\right]$$
$$\geq \frac{1}{2}\left(\sum_{k=1}^{j} p_k\right)^2$$

The lemma follows. □

We now analyze the performance of the algorithm Schedule-by-$\bar{C}$ which obtains a valid schedule as follows. First, the above LP is solved to obtain an optimal solution $\bar{C}$. Without loss of generality assume that $\bar{C}_1 \leq \bar{C}_2 \leq \cdots \leq \bar{C}_n$. The algorithm then schedules jobs in order of nondecreasing $\bar{C}_j$ inserting idle time as necessary if the release date r_j is greater than the completion time of the $(j-1)$th job.

Lemma 11.2

Let $\tilde{C}_j$ denote the completion time of j in the schedule produced by Schedule-by-$\bar{C}$. Then, for $j = 1, \ldots, n$,

$$\tilde{C}_j \leq \max_{k=1}^{j} r_k + 2\bar{C}_j.$$

Hence $\tilde{C}_j \leq 3\bar{C}_j$.

Proof

Fix a job j and let $S = \{1, 2, \ldots, j\}$. In the schedule produced by Schedule-by-$\bar{C}_j$, it is clear that there is no idle time in the interval $[\max_{k=1}^{j} r_k, \tilde{C}_j]$ since all jobs in S are released by $\max_{k=1}^{j} r_k$. From the ordering rule of the algorithm it follows that $\tilde{C}_j \leq \max_{k=1}^{j} r_k + p(S)$, and using Lemma 11.1, we conclude that $\tilde{C}_j \leq \max_{k=1}^{j} r_k + 2\bar{C}_j$. Finally, we observe that Equation (11.1) implies that $\bar{C}_k \geq r_k$ for $k = 1, \ldots, n$, and since $\bar{C}_j \geq \bar{C}_k$, $k = 1, \ldots, j$, we have that $\bar{C}_j \geq \max_{k=1}^{j} r_k$. Thus $\tilde{C}_j \leq 3\bar{C}_j$. □

The following theorem follows from the lemma above.

Theorem 11.1

Schedule-by-$\bar{C}$ gives a 2-approximation for $1 \mid prec \mid \sum_j w_j C_j$ and a 3-approximation for $1 \mid prec, r_j \mid \sum_j w_j C_j$.

Proof

In the problem $1 \mid prec \mid \sum_j w_j C_j$, $r_j = 0$ for all j. Hence, from Lemma 11.2 we obtain that $\tilde{C}_j \leq 2\bar{C}_j$, hence $\sum_j w_j \tilde{C}_j \leq 2\sum_j w_j \bar{C}_j$. The quantity $\sum_j w_j \bar{C}_j$ is the optimum value of the LP relaxation and hence is a lower bound on the integral optimum solution. This proves the 2-approximation.

For the problem $1 \mid prec, r_j \mid \sum_j w_j C_j$, we use the fact that $\tilde{C}_j \leq 3\bar{C}_j$ from Lemma 11.2 to obtain the 3-approximation. □

The analysis can be refined to obtain a $(2 - \frac{2}{n+1})$-approximation for $1 \mid prec \mid \sum_j w_j C_j$, see Schulz [9] for details.

Time-Indexed Formulation: The time-indexed formulation for $1 \mid prec \mid \sum_j w_j C_j$ was introduced by Dyer and Wolsey [37]. The size of the formulation is pseudo-polynomial; however, as Hall et al. [29] have shown, an approximate polynomial size formulation can be derived from it. In the time-indexed formulation an upper bound on the schedule length, T, is used. In the case of $1 \mid prec \mid \sum_j w_j C_j$ it is easy to see that $T = \sum_j p_j$ since in absence of release dates, idle time can be eliminated from any schedule. For each job j in $1, 2, \ldots, n$ and time t in $1, \ldots, T$ there is a variable x_{jt} that is 1 if j completes processing at time t. The formulation below is for $1 \mid prec \mid \sum_j w_j C_j$.

$$\min \sum_{j=1}^{n} w_j \sum_{t=1}^{T} t x_{jt}$$

subject to

$$\sum_{t=1}^{T} x_{jt} = 1 \qquad j = 1, \ldots, n \tag{11.4}$$

$$\sum_{s=1}^{t} x_{js} \geq \sum_{s=1}^{t+p_k} x_{ks} \quad \text{if } j \prec k, t = p_j, \ldots, T - p_k \tag{11.5}$$

$$\sum_{j=1}^{n} \sum_{s=t}^{\min\{t+p_j-1, T\}} x_{js} \leq 1 \qquad t = 1, \ldots, T \tag{11.6}$$

$$x_{jt} \geq 0 \qquad j = 1, \ldots, n,\ t = 1, \ldots, T \tag{11.7}$$

$$x_{jt} = 0 \qquad t = 1, \ldots, r_j + p_j - 1 \tag{11.8}$$

Linear Ordering Formulation of Potts: We describe the formulation of Potts [48] that uses linear ordering variables δ_{ij} for $i \neq j$. The variable δ_{ij} is 1 if i is completed before j in the schedule, and is 0 otherwise. The formulation below is for $1 \mid prec \mid \sum_j w_j C_j$.

$$\min \sum_{j=1}^{n} w_j C_j$$

subject to

$$C_j = p_j + \sum_{i=1}^{n} p_i \delta_{ij} \qquad j = 1, \ldots, n \tag{11.9}$$

$$\delta_{ij} + \delta_{ji} = 1 \qquad i \neq j \tag{11.10}$$

$$\delta_{ij} + \delta_{jk} + \delta_{ki} \leq 2 \qquad i \prec j \prec k \text{ or } k \prec j \prec i \tag{11.11}$$

$$\delta_{ij} = 1 \qquad i \prec j \tag{11.12}$$

$$\delta_{ij} \geq 0 \qquad i \neq j \tag{11.13}$$

Both of the above formulations can be easily extended to handle release dates, that is for $1 \mid r_j, prec \mid \sum_j w_j C_j$. Schulz [9] has shown that the completion time formulation is no stronger than both the time-indexed formulation and the linear ordering formulation. Thus, it follows that both these formulations can also be used to obtain Theorem 11.1. The advantage of the linear ordering formulation is that it is polynomial sized. Chudak and Hochbaum [19] have shown that the linear ordering formulation can be rewritten with only two variables per inequality, this allows it to be solved by using a minimum cut computation in an associated graph. For the problem $1 \mid r_j, prec \mid \sum_j w_j C_j$, Schulz and Skutella [30] obtain a ratio of $(e + \epsilon)$, where e is the base of the natural logarithm, this improves upon the 3-approximation that we presented. For this, they use the idea of ordering jobs by their α-completion times in a solution to the time-indexed formulation for the problem, where α is picked according to a probability distribution. See discussion in Section 11.5.

All the formulations for the single machine problem $1 \mid prec \mid \sum_j w_j C_j$ have an integrality gap of 2. While it is easy to construct gaps for the completion time and time-indexed formulations [29], it is some what nontrivial to do so for the linear ordering formulation. In [13] an example using bipartite strong expander graphs is used to show a gap of 2 for all the known formulations.

11.2.1.2 Identical Parallel Machines

The completion time formulation has been extended to the problem $P \mid prec, r_j \mid \sum_j w_j C_j$ by Hall et al. [29] as follows. The constraint (11.3) is replaced by the following where m is the number of machines.

$$\sum_{j \in S} p_j C_j \geq \frac{1}{2m}\left(\sum_{j \in S} p_j\right)^2 + \frac{1}{2}\sum_{j \in S} p_j^2 \quad S \subseteq N \tag{11.14}$$

For the validity of the above constraint, see Ref. [29]. Once again we can order jobs by their completion times $\bar{C}$ in the LP relaxation. The following lemma can be shown with an analysis similar to that of Lemma 11.1.

Lemma 11.3

Let $\bar{C}$ be a feasible solution to the completion time formulation for parallel machines and without loss of generality let $\bar{C}_1 \leq \bar{C}_2 \leq \ldots \bar{C}_n$. Then, the following is true for $j = 1, \ldots, n$.

$$\bar{C}_j \geq \frac{1}{2m}\sum_{k=1}^{j} p_k$$

The following is also an easy lemma to prove from the constraints.

Lemma 11.4

For any sequence of jobs $j_1, j_2, \ldots, j_k$ such that $j_1 \prec j_2 \prec \cdots \prec j_k$, the following holds:

$$\bar{C}_{j_k} \geq \max_{i=1}^{k}\left(r_{j_i} + \sum_{\ell=i}^{k} p_{j_\ell}\right)$$

It is however not straight forward to generalize Schedule-by-$\bar{C}$ to the case of parallel machines. There is a tradeoff between utilizing the resources of all machines and giving priority to the ordering of the jobs. Two list scheduling algorithms are natural. The first is the greedy algorithm of Graham [6] for minimizing makespan on parallel machines. In Graham's list scheduling, we are given an ordering of the jobs that is consistent with precedence constraints. A job j is *ready* at time t if all its predecessors are completed by t and $t \geq r_j$. If a machine is free, the algorithm schedules the earliest job in the given ordering that is ready. Using Lemmas 11.3 and 11.4 the following theorem can be shown for jobs with unit processing times. The analysis is very similar to Graham's analysis for minimizing makespan. The advantage of equal processing times is that jobs that are scheduled out of order do not delay earlier jobs in the order. For the same reason the theorem below also holds if jobs can be preempted. For more details see Hall et al. [29].

Theorem 11.2

For the problems $P \mid prec, r_j, p_j = 1 \mid \sum_j w_j C_j$ and $P \mid prec, r_j, pmtn \mid \sum_j w_j C_j$, Graham's list scheduling with the ordering according to $\bar{C}$ yields a 3-approximation algorithm.

Even though Graham's list scheduling is effective for the problems in the above theorem, it can produce schedules as much as an $\Omega(m)$ factor away from the optimum for the problem $P \mid prec \mid \sum_j w_j C_j$ [16]. The second list scheduling is one that strictly schedules in order, in spirit akin to Schedule-by-$\bar{C}$. It requires some care to define the precise generalization to multiple machines. Below we describe the variant from Munier, Queyranne, and Schulz [16].

List Scheduling:

1. The list $L = (\ell(1), \ell(2), \ldots, \ell(n))$ is given.
2. Initially all machines are empty. For $h = 1, \ldots, m$, set machine completion time $\gamma_h = 0$.
3. For $k = 1, \ldots, n$ do:
   a. Let job $j = \ell(k)$, its start time $S_j = \max(\{C_i : i \prec j\}, \min\{\gamma_h : h = 1, \ldots, m\})$ and its completion time $C_j = S_j + p_j$.
   b. Assign j to a machine h such that $\gamma_h \leq S_j$. Update $\gamma_h = C_j$.

Notice that the above algorithm schedules strictly in order of the list. It can be shown that using the above algorithm with the ordering provided by $\bar{C}$ also leads to a poor schedule [16]. Using a deceptively simple modification, Munier et al. show that if the list is obtained by ordering jobs in nondecreasing order of their LP *midpoints* (the LP midpoint of job j, $\bar{M}_j$, is defined as $(\bar{C}_j - p_j/2)$), then the list schedule results in a 4-approximation! We now prove this result. For job j we denote by $\tilde{S}_j$ the start time of j in the schedule constructed. The main result from [16] is the following.

Theorem 11.3

Let $\bar{C}$ and $\bar{M}$ denote the LP completion time and midpoint vectors of the jobs. Let $\tilde{S}$ denote the vector of start times in the feasible schedule constructed by List Scheduling. Then for $j = 1, \ldots, n$, $\tilde{S}_j \leq 4\bar{M}_j$ and hence $\tilde{C}_j \leq 4\bar{C}_j$.

The following lemmas follow easily from the modified constraint (11.14).

Lemma 11.5

Let $\bar{C}$ be a feasible solution to the completion time formulation for parallel machines and without loss of generality let $\bar{M}_1 \leq \bar{M}_2 \leq \ldots \bar{M}_n$. Then, the following is true for $j = 1, \ldots, n$.

$$\bar{M}_j \geq \frac{1}{2m} \sum_{k=1}^{j-1} p_k$$

Lemma 11.6

For any sequence of jobs $j_1, j_2, \ldots, j_k$ such that $j_1 \prec j_2 \prec \cdots \prec j_k$, the following holds.

$$\bar{M}_{j_k} - \bar{M}_{j_1} \geq \frac{1}{2} \sum_{\ell=1}^{k} p_{j_\ell}$$

We will use $[j]$ to refer to the set $\{1, \ldots, j\}$. Precedence constraints between jobs induce a directed acyclic graph $G = ([n], A)$ where $(i, j) \in A$ if and only if $i \prec j$. Given the schedule produced by the list scheduling algorithm we define for each j the graph $G^j = ([j], A^j)$ where

$$A^j = \{(k, \ell) \in A : k, \ell \in [j] \text{ and } \tilde{C}_\ell = \tilde{C}_k + p_\ell\}$$

Thus A^j is a subset of the precedence constraints in $[j]$ that are *tight* in the schedule. Fix a job j and consider the state of the schedule when j is scheduled by the algorithm. The time interval $[0, \tilde{S}_j)$ can be partitioned into those intervals in which all machines are busy processing jobs and those in which at least one machine is idle. Let μ be the total time in which all machines are busy and λ the rest. To prove Theorem 11.3 it is sufficient to show that $\mu \leq 2\bar{M}_j$ and $\lambda \leq 2\bar{M}_j$. Since no job $i > j$ is considered for scheduling by the time j is scheduled, $\mu \leq \frac{1}{m} \sum_{k \leq j-1} p_k$; by Lemma 11.5 it follows that $\mu \leq 2\bar{M}_j$.

Now we show that $\lambda \leq 2\bar{M}_j$. The nonbusy intervals in $[0, \tilde{S}_j]$ can be partitioned into maximal intervals $(a_1, b_1), (a_2, b_2), \ldots, (a_q, b_q)$ such that $0 \leq a_1$, and for $h = 2, \ldots, q$, $b_{h-1} < a_h < b_h$, and $b_q \leq \tilde{S}_j$. For ease of notation define $b_0 = 0$ and $a_{q+1} = \infty$. We now prove the following crucial lemma.

Lemma 11.7

Let $k \in [j]$ be a job such that $\tilde{S}_k \in [b_h, a_{h+1}]$ for some $h \in 1, \ldots, q$. Let $v_1, v_2, \ldots, v_s = k$ be a maximal path in A^j that ends in k. Then, there is an index $g \leq h$ such that $\tilde{S}_{v_1} \in [b_g, a_{g+1}]$ and if $g > 0$ then there exists a job $\ell \neq k$ such that $\bar{M}_\ell \leq \bar{M}_k$ and $\tilde{S}_\ell = b_g$. Further

$$\bar{M}_k - \bar{M}_{v_1} \geq \max\left[\frac{1}{2}(b_h - a_{g+1}), 0\right] \tag{11.15}$$

Proof

Since the path is maximal, the job v_1 has no precedence constraints preventing it from being scheduled earlier than $\tilde{S}_{v_1}$. Hence, it must be that $\tilde{S}_{v_1}$ belongs to some busy interval $[b_g, a_{g+1}]$. Suppose $g > 0$. Since $[a_g, b_g]$ is a maximal busy interval some job ℓ starts at b_g. We claim that $\bar{M}_\ell \leq \bar{M}_{v_1}$. If not, v_1 would have been considered earlier than ℓ for scheduling and again by maximality of the path, v_1 would have started strictly before b_g, a contradiction.

Now we prove Equation (11.15). If $g = h$, then $b_h - a_{g+1} < 0$ and the equation is trivial. Suppose $g < h$, then because from the definition of edges in A^j, we have that

$$b_h - a_{g+1} \leq \tilde{S}_{v_s} - \tilde{S}_{v_1} = \sum_{i=1}^{s-1} \tilde{S}_{v_{i+1}} - \tilde{S}_{v_i} = \sum_{i=1}^{s-1} p_{v_i} \tag{11.16}$$

Since $(v_i, v_{i+1}) \in A^j$, from the LP constraints (11.2) we have the following for $i = 1, \ldots, s-1$.

$$\bar{M}_{v_{i+1}} \geq \bar{M}_{v_i} + \frac{1}{2}(p_{v_i} + p_{v_{i+1}}) \tag{11.17}$$

Putting together the above two equations,

$$\bar{M}_{v_s} - \bar{M}_{v_1} = \bar{M}_k - \bar{M}_{v_1} \geq \frac{1}{2}\sum_{i=1}^{s-1} p_{v_i} \geq \frac{1}{2}(b_h - a_{g+1}) \tag{11.18}$$

Now we are ready to prove Theorem 11.3. Using Lemma 11.7 we create a sequence of indices $q = i_1 > i_2 > \cdots > i_r = 0$. With each index i_ℓ we associate a job $x(i_\ell)$. The sequence is constructed as follows. We let $x(i_1) = x(q)$ be a job such that $\tilde{S}_{x(q)} = b_q$, such a job must exist. Let $v_1, v_2, \ldots, v_s = x(q)$ be a maximal path that ends in $x(q)$. If $v_1 \in [0, a_1]$ we stop and set $i_2 = i_r = 0$. Otherwise, from Lemma 11.7 there is a job ℓ such that $\tilde{S}_\ell = b_g$ for some $g \leq q$. We claim that $g < q$ since otherwise i_1 would have been scheduled earlier than b_q. We set $i_2 = g$ and $x(i_2) = \ell$. From Equation (11.15) $\bar{M}_{x(q)} - \bar{M}_{v_1} \geq \frac{1}{2}(b_q - a_{g+1})$. Since $\bar{M}_{v_1} \geq \bar{M}_\ell$, we also have that $\bar{M}_{x(q)} - \bar{M}_{x(g)} \geq \frac{1}{2}(b_q - a_{g+1})$, in other words $\bar{M}_{x(i_1)} - \bar{M}_{x(i_2)} \geq \frac{1}{2}(b_{i_1} - a_{i_2+1})$. We continue the process with i_2 to obtain i_3 and so on until $\tilde{S}_{i_r} \in [0, a_1]$. It is clear that the process terminates since the maximal path length at each step is of length at least 2. Using the same reasoning as above we have the following equation for $k = 1, \ldots, r-1$.

$$\bar{M}_{x(i_k)} - \bar{M}_{x(i_{k+1})} \geq \frac{1}{2}(b_{i_k} - a_{i_k+1}) \tag{11.19}$$

Adding up all the above equations for $k = 1, \ldots, r-1$ yields

$$\bar{M}_{x(i_1)} - \bar{M}_{x(i_r)} \geq \sum_{k=1}^{r-1} \frac{1}{2}(b_{i_k} - a_{i_k+1}) \tag{11.20}$$

It is easy to see from the construction that every idle interval $[a_h, b_h]$ is contained in one of the intervals $[a_{i_k+1}, b_{i_k}]$ for some k. Hence, it follows that

$$\lambda = \sum_{h=1}^{q}(b_h - a_h) \leq \sum_{k=1}^{r-1}(b_{i_k} - a_{i_k+1}) \leq 2(\bar{M}_{x(i_1)} - \bar{M}_{x(i_r)}) \leq 2\bar{M}_j \tag{11.21}$$

This finishes the proof of Theorem 11.3. □

It is instructive for the reader to understand why the above proof fails if we try to use an ordering based on either the completion times or the start times ($\bar{S}_j = \bar{C}_j - p_j$). In [16] examples are given that show that the approximation ratio provided by the algorithm is tight (factor of 4) and that the LP integrality gap is at least 3.

The algorithm and the bound generalize in a straight forward fashion to the case where there are delays associated with the precedence constraints, that is, for all $i \prec j$ there is a delay d_{ij} which indicates the time that j has to wait to start after i completes.

11.2.2 Combinatorial Algorithms

We have seen LP based methods to minimize weighted completion time when jobs have precedence constraints and/or release dates. Now we describe combinatorial methods based on structural insights into the problem. Our first algorithm is for the single machine problem $1 \,|\, prec \,|\, \sum_j w_j C_j$. For identical parallel machines we give an algorithm that takes an approximately good single machine schedule and converts into an approximate schedule on parallel machines.

11.2.2.1 One-Machine Scheduling with Precedence Constraints

Polynomial time solvable cases of $1 \,|\, prec \,|\, \sum_j w_j C_j$ include precedence constraints induced by forests (in and out forests) and generalized series-parallel graphs [2–4]. These special cases have been solved by combinatorial methods (in fact by $O(n \log n)$ time algorithms).

We now describe a simple combinatorial 2-approximation algorithm for the single machine problem with general precedence constraints. This algorithm was obtained independently by Chekuri and Motwani [13], and Margot, Queyranne, and Wang [14]. Chudak and Hochbaum [19] derived a linear programming relaxation for $1 \,|\, prec \,|\, \sum_j w_j C_j$ that uses only two variables per inequality, using the linear ordering formulation of Potts [48] described earlier. Such formulations can be solved by a minimum cut computation and this also leads to a combinatorial approximation algorithm. However, the running time obtained is worse than that of the algorithm we present below [13,14] by a factor of n.

Let $G = (V, E)$ denote the precedence graph where V is the set of jobs. We will use jobs and vertices interchangeably. We say that i precedes j, denoted by $i \prec j$, if and only if there is a path from i to j in G. For any vertex $i \in V$, let G_i denote the subgraph of G induced by the set of vertices preceding i.

Definition 11.1

The rank *of a job J_i, denoted by q_i, is defined as $q_i = p_i / w_i$. Similarly, the* rank *of a set of jobs A denoted by $q(A)$ is defined as $q(A) = p(A)/w(A)$, where $p(A) = \sum_{J_i \in A} p_i$ and $w(A) = \sum_{J_i \in A} w_i$.*

Definition 11.2

A subdag G' of G is said to be precedence closed *if for every job $J_i \in G'$, G_i is a subgraph of G'.*

The rank of a graph is simply the rank of its node set.

Definition 11.3

We define G^* to be a precedence-closed subgraph of G of minimum rank, i.e., among all precedence-closed subgraphs of G, G^* is of minimum rank.

Note that G^* could be the entire graph G.

A Characterization of the Optimal Schedule: Smith's rule for a set of independent jobs states that there is an optimal schedule that schedules jobs in nondecreasing order of their ranks. We generalize this rule for the case of precedence constraints in a natural way. A version of the following theorem was proved by Sydney [4]. The proof relies on a careful exchange argument ([13] also provides a proof).

Definition 11.4

A segment *in a schedule S is any set of jobs that are scheduled consecutively in S.*

Theorem 11.4 (Sydney)

There exists an optimal sequential schedule where the optimal schedule for G^* occurs as a segment that starts at time zero.

Note that when G^* is the same as G this theorem does not help in reducing the problem.

A 2-approximation Theorem 11.4 suggests the following natural algorithm. Given G, compute G^* and schedule G^* and $G - G^*$ recursively. It is not *a priori* clear that G^* can be computed in polynomial time, however this is indeed the case. The other important issue is to handle the case when G cannot be decomposed because G^* is the same as G. We have to settle for an approximation in this case, for otherwise we would have a polynomial time algorithm to compute the optimal schedule.

The following lemma establishes a strong lower bound on the optimal schedule value when $G^* = G$.

Lemma 11.8

If G^* is the same as G, OPT $\geq w(G)p(G)/2$.

Proof

Let $\alpha = q(G)$. Let S be an optimal schedule for G. Without loss of generality assume that the ordering of the jobs in S is $J_1, J_2, \ldots, J_n$. For any j, $1 \leq j \leq n$, observe that $C_j = \sum_{1 \leq i \leq j} p_i \geq \alpha \sum_{1 \leq i \leq j} w_i$. This is because the set of jobs $J_1, J_2, \ldots, J_j$ form a precedence closed subdag, and from our assumption on G^* it follows that $\sum_{i \leq j} p_j / \sum_{i \leq j} w_i \geq \alpha$. We bound the value of the optimal schedule as follows.

$$
\begin{aligned}
\text{OPT} &= \sum_{j=1}^{n} w_j C_j \geq \sum_{j=1}^{n} w_j \sum_{i=1}^{j} \alpha w_i \\
&= \alpha \left(\sum_{j=1}^{n} w_j^2 + \sum_{1 \leq i < j \leq n} w_i w_j \right) = \alpha \left(\left(\sum_{j=1}^{n} w_j \right)^2 - \sum_{1 \leq i < j \leq n} w_i w_j \right) \\
&\geq \alpha \left(w(G)^2 - w(G)^2/2 \right) \\
&= \alpha w(G)^2/2 = w(G)p(G)/2
\end{aligned}
$$

The last equality is true because $q(G^*) = q(G) = p(G)/w(G) = \alpha$. □

The following lemma is straight forward.

Lemma 11.9

Any feasible schedule with no idle time has a weighted completion time of at most $w(G)p(G)$.

Theorem 11.5

If G^* for a graph can be computed in time $O(T(n))$, then there is a 2-approximation algorithm for computing the minimum weighted completion time schedule that runs in time $O(nT(n))$.

Proof

Given G, we compute G^* in time $O(T(n))$. If G^* is the same as G we schedule G arbitrarily and Lemmas 11.8 and 11.9 guarantee that we have a 2-approximation. If G^* is a proper subdag we recurse on G^* and $G - G^*$. From Theorem 11.4 we have $\text{OPT}(G) = \text{OPT}(G^*) + p(G^*)w(G - G^*) + \text{OPT}(G - G^*)$. Inductively if we have 2-approximate solutions for G^* and $G - G^*$ it is clear that we can combine them to get a 2-approximation of the overall schedule. Now we establish the running time bound. We observe that $(G^*)^* = G^*$, therefore it suffices to recurse only on $G - G^*$. It follows that we make at most n calls to the routine to compute G^* and the bound on the running time of the procedure follows. □

An algorithm to compute G^* using a parametric minimum cut computation in an associated graph is presented in Lawler's book [49]. The associated graph is dense and has $\Omega(n^2)$ edges. Applying known algorithms for parametric minimum cut, G^* can be computed in strongly polynomial time of $O(n^3)$ [50] or in $O(n^{8/3} \log nU)$ time [51], where $U = \max_{i=1}^{n}(p_i + w_i)$.

11.2.2.2 A General Conversion Algorithm

In this section we describe a technique of Chekuri et al. [11] to obtain parallel machine schedules from one-machine schedules that works even when jobs have precedence constraints and release dates. Given an average weighted completion time scheduling problem, we show that if we can approximate the one-machine preemptive variant, then we can also approximate the m-machine nonpreemptive variant, with a slight degradation in the quality of approximation.

We use the superscript m to denote the number of machines; thus, X^m denotes a schedule for m machines, C^m denotes the sum of weighted completion time of X^m, and C_j^m denotes the completion time of job j under schedule X^m. The subscript OPT refers to an optimal schedule; thus, an optimal schedule is denoted by X_{OPT}^m, and its weighted completion time is denoted by C_{OPT}^m. For a set of jobs A, $p(A)$ denotes the sum of processing times of jobs in A.

Definition 11.5

For any vertex j, recursively define the quantity κ_j as follows. For a vertex j with no predecessors $\kappa_j = p_j + r_j$. Otherwise, define $\kappa_j = p_j + \max\{\max_{i \prec j} \kappa_i, r_j\}$. Any path P_{ij} from i to j where $p(P_{ij}) = \kappa_j$ is referred to as a critical path to j.

We now describe the Delay-List algorithm. Given a one-machine schedule which is a ρ-approximation, Delay-List produces a schedule for $m \geq 2$ machines whose value is within a factor $(k_1\rho + k_2)$ of the optimal m-machine schedule, where k_1 and k_2 are small constants. We will describe a variant of this scheduling algorithm which yields $k_1 = (1 + \beta)$ and $k_2 = (1 + 1/\beta)$ for any $\beta > 0$.

The main idea is as follows. The one-machine schedule taken as a list (jobs in order of their completion times in the schedule) provides some priority information on which jobs to schedule earlier. However, when trying to convert the one-machine schedule into an m-machine one, precedence constraints prevent complete parallelization. Thus, we may have to execute jobs out-of-order from the list to benefit from parallelism. If all p_i are identical (say 1), we can afford to use Graham's list scheduling. If there is an idle machine and we schedule some available job on it, it is not going to delay jobs which become available soon, since it completes in one time unit. On the other hand, if not all p_i's are the same, a job could keep a machine busy, delaying more profitable jobs that become available soon. At the same time, we cannot afford to keep machines idle. We strike a balance between the two extremes: schedule a job out-of-order only if there has been enough idle time already to justify scheduling it. To measure whether there has been enough idle time, we introduce a charging scheme.

Assume, for ease of exposition, that all processing times are integers and that time is discrete. This restriction can be removed without much difficulty and we use it only in the interests of clarity and intuition. A job is *ready* if it has been released and all its predecessors are done. The time at which job j is ready in a schedule X is denoted by q_j^X and time at which it starts is denoted by S_j^X.

We use X^m to denote the m-machine schedule that our algorithm constructs and for ease of notation the superscript m will be used in place of X^m to refer to quantities of interest in this schedule. Let $\beta > 0$ be some constant. At each discrete time step t, the algorithm applies one of the following three cases:

1. *There is an idle machine M and the first job j on the list is ready at time t* — schedule j on M and charge all uncharged idle time in the interval (q_j^m, S_j^m) to j.
2. *There is an idle machine — and the first job j in the list is not ready at time t but there is another ready job on the list* — focusing on the job k which is the first in the list among the ready jobs, schedule it if there is at least βp_k *uncharged* idle time among all machines, and charge βp_k idle time to k.
3. *There is no idle time or the above two cases do not apply* — do not schedule any job, merely increment t.

Definition 11.6

A job is said to be scheduled in order *if it is scheduled when it is at the head of the list. Otherwise it is said to be scheduled* out of order. *The set of jobs which are scheduled before a job j but which come later in the list than j is denoted by O_j. The set of jobs which come after j in the list is denoted by A_j and those which come before j by B_j (includes j).*

Definition 11.7

For each job i, define a path $P_i' = j_1, j_2, \ldots, j_\ell$, with $j_\ell = i$ with respect to the schedule X^m as follows. The job j_k is the predecessor of j_{k+1} with the largest completion time (in X^m) among all the predecessors of j_{k+1} such that $C_{j_k}^m \geq r_{j_{k+1}}$; ties are broken arbitrarily. j_1 is the job where this process terminates when there are no predecessors which satisfy the above condition. The jobs in P_i' define a disjoint set of time intervals $(0, r_{j_1}], (S_{j_1}^m, C_{j_1}^m], \ldots, (S_{j_\ell}^m, C_{j_\ell}^m]$ in the schedule. Let κ_i' denote the sum of the lengths of the intervals.

Fact 11.1 $\kappa_i' \leq \kappa_i$.

Fact 11.2 *The idle time charged to each job i is less than or equal to βp_i.*

Proof

The fact is clear if idle time is charged to i according to case 2 in the description of our algorithm. Suppose case 1 applies to i. Since i was ready at q_i^m and was not scheduled according to case 2 earlier, the idle time in the interval (q_i^m, S_i^m) that is charged to i is less than βp_i. We remark that the algorithm with discrete time units might charge more idle time due to integrality of the time unit. However, that is easily fixed in the continuous case where we schedule i at the first time instant when at least βp_i units of uncharged idle time have accumulated. □

A crucial feature of the algorithm is that when it schedules jobs, it considers only the first job in the list that is ready, even if there is enough idle time for other ready jobs that are later in the list. The proof of the following lemma makes use of this feature.

Lemma 11.10

For every job i, there is no uncharged idle time in the time interval (q_i^m, S_i^m), and furthermore all the idle time is charged only to jobs in B_i.

Proof

By the preceding remarks, it is clear that no job in A_i is started in the time interval (q_i^m, S_i^m) since i was ready at q_i^m. From this we can conclude that there is is no idle time charged to jobs in A_i in that time interval. Since i is ready at q_i^m and was not scheduled before S_i^m, from cases 1 and 2 in the description of our algorithm there cannot be any uncharged idle time. □

The following lemma shows that for any job i, the algorithm does not schedule too many jobs from A_i before scheduling i itself.

Lemma 11.11

For every job i, the total idle time charged to jobs in A_i, in the interval $(0, S_i^m)$, is bounded by $m(\kappa_i' - p_i)$. It follows that $p(O_i) \leq m(\kappa_i' - p_i)/\beta \leq m(\kappa_i - p_i)/\beta$.

Proof

Consider a job j_k in P_i'. The job j_{k+1} is ready to be scheduled at the completion of j_k, that is $q_{j_{k+1}}^m = C_{j_k}^m$. From Lemma 11.10, it follows that in the time interval between $(C_{j_k}^m, S_{j_{k+1}}^m)$ there is no idle time charged to jobs in $A_{j_{k+1}}$. Since $A_{j_{k+1}} \supset A_i$ it follows that all the idle time for jobs in A_i has to be accumulated in the intersection between $(0, S_i^m)$ and the time intervals defined by P_i'. This quantity is clearly bounded by $m(\kappa_i' - p_i)$. The second part follows since the total processing time of the jobs in O_i is bounded by $1/\beta$ times the total idle time that can be charged to jobs in A_i (recall that $O_i \subseteq A_i$). □

Theorem 11.6

Let X^m be the schedule produced by the algorithm DELAY LIST *using a list X^1. Then for each job i, $C_i^m \leq (1+\beta)p(B_i)/m + (1+1/\beta)\kappa_i' - p_i/\beta$.*

Proof

Consider a job i. We can split the time interval $(0, C_i^m)$ into two disjoint sets of time intervals T_1 and T_2 as follows. The set T_1 consists of all the disjoint time intervals defined by P_i'. The set T_2 consists of the time intervals obtained by removing the intervals in T_1 from $(0, C_i^m)$. Let t_1 and t_2 be the sum of the times of the intervals in T_1 and T_2, respectively. From the definition of T_1, it follows that $t_1 = \kappa_i' \leq \kappa_i$. From Lemma 11.10, in the time intervals of T_2, all the idle time is either charged to jobs in B_i and, the only jobs which run are from $B_i \cup O_i$. From Fact 11.2, the idle time charged to jobs in B_i is bounded by $\beta p(B_i)$. Therefore, the time t_2 is bounded by $(\beta p(B_i) + p(B_i) + p(O_i))/m$. Using Lemma 11.11 we see that $t_1 + t_2$ is bounded by $(1+\beta)p(B_i)/m + (1+1/\beta)\kappa_i' - p_i/\beta$. □

One-Machine Relaxation: In order to use Delay List, we will need to start with a one machine schedule. The following two lemmas provide lower bounds on the optimal m-machine schedule in terms of the optimal one-machine schedule. This one-machine schedule can be either preemptive or nonpreemptive, the bounds hold in either case.

Lemma 11.12

$C^m_{\text{OPT}} \geq C^1_{\text{OPT}}/m.$

Proof

Given a schedule X^m on m machines with total weighted completion time C^m, we will construct a one-machine schedule X^1 with total weighted completion time at most mC^m as follows. Order the jobs according to their completion times in X^m with the jobs completing early coming earlier in the ordering. This ordering is our schedule X^1. Note that there could be idle time in the schedule due to release dates. If $i \prec j$ then $C^m_i \leq S^m_j \leq C^m_j$ which implies that there will be no precedence violations in X^1. We claim that $C^1_i \leq mC^m_i$ for every job i. Let P be the sum of the processing times of all the jobs which finish before i (including i) in X^m. Let I be the total idle time in the schedule X^m before C^m_i. It is easy to see that $mC^m_i \geq P + I$. We claim that $C^1_i \leq P + I$. The idle time in the schedule X^1 can be charged to idle time in the schedule X^m and P is the sum of all jobs which come before i in X^1. This implies the desired result.

□

Lemma 11.13

$C^m_{\text{OPT}} \geq \sum_i w_i \kappa_i = C^\infty_{\text{OPT}}.$

Proof

The length of the critical path κ_i, is an obvious lower bound on the completion time C^m_i of job i. Summing up over all jobs gives the first inequality. It is also easy to see that if the number of machines is unbounded that every job i can be scheduled at the earliest time it is available and will finish by κ_i yielding the equality.

□

Obtaining Generic *m*-Machine Schedules: In this section we derive our main theorem relating m-machine schedules to one-machine schedules.

We begin with a corollary to Theorem 11.6.

Corollary 11.1

Let X^m be the schedule produced by the algorithm Delay-List using a one-machine schedule X^1 as the list. Then for each job i, $C^m_i \leq (1+\beta)C^1_i/m + (1 + 1/\beta)\kappa_i$.

Proof

Since all jobs in B_i come before i in the one-machine schedule, it follows that $p(B_i) \leq C^1_i$. Plugging this and Fact 11.1 into the bound in Theorem 11.6, we conclude that $C^m_i \leq (1+\beta)C^1_i/m + (1+1/\beta)\kappa_i$. □

Theorem 11.7

Given an instance I of scheduling to minimize sum of weighted completion times and a one-machine schedule for I that is within a factor ρ of an optimal one-machine schedule, Delay-List gives a m-machine schedule for I that is within a factor $(1+\beta)\rho + (1+1/\beta)$ of an optimal m-machine schedule. Further, Delay-List can be implemented in $O(n \log n)$ time.

Proof

Let X^1 be a schedule which is within a factor ρ of the optimal one-machine schedule. Then $C^1 = \sum_i w_i C_i^1 \leq \rho C_{\text{OPT}}^1$. By Corollary 11.1, the schedule created by the algorithm Delay-List satisfies,

$$\begin{aligned} C^m &= \sum_i w_i C_i^m \\ &\leq \sum_i w_i \left[(1+\beta)\frac{C_i^1}{m} + \left(1+\frac{1}{\beta}\right)\kappa_i \right] \\ &= \frac{1+\beta}{m}\sum_i w_i C_i^1 + \left(1+\frac{1}{\beta}\right)\sum_i w_i \kappa_i \end{aligned}$$

From Lemmas 11.12 and 11.13 it follows that

$$\begin{aligned} C^m &\leq \frac{(1+\beta)\rho C_{\text{OPT}}^1}{m} + \left(1+\frac{1}{\beta}\right) C_{\text{OPT}}^\infty \\ &\leq \left[(1+\beta)\rho + \left(1+\frac{1}{\beta}\right) \right] C_{\text{OPT}}^m \end{aligned}$$

The running time bound on Delay-List can be easily seen from the description of the algorithm. □

There is an interesting property of the conversion algorithm that is useful in its applications and worth pointing out explicitly. We explain it via an example. Suppose we want to compute an m-machine schedule with release dates and precedence constraints. From Theorem 11.7 it would appear that we need to compute a one-machine schedule for the problem that has both precedence constraints and release dates. However, we can completely ignore the release dates in computing the one-machine schedule X^1. This follows from a careful examination of the upper bound proved in Theorem 11.6 and the proof of Theorem 11.7. This is useful since the approximation ratio for the problem $1|prec|\sum_j w_j C_j$ is 2, while it is $(e+\epsilon)$ for $1\,|\,prec, r_j\,|\,\sum_j w_j C_j$ [30]. In another example, the problem $1\,||\,\sum_j w_j C_j$ has a very simple polynomial time algorithm using Smith's ratio rule while $1\,|\,r_j\,|\,\sum_j w_j C_j$ is NP-hard. Thus release dates play a role only in the conversion algorithm and not in the single machine schedule. A similar claim can be made when there are delays between jobs. In this setting a positive delay d_{ij} between jobs i and j indicates that i is a predecessor of j and that j cannot start until d_{ij} time units after i completes. Delay-List and its analysis can be generalized to handle delays, and obtain the same results as those in Theorem 11.6 and Theorem 11.7. The only change required is in the definition of ready time of a job which now depends also on the delay after a predecessor finishes. As with release dates we can ignore the delay values (not the precedence constraints implied by them though) in computing the single machine schedule.

11.3 Unrelated Machines

In this section we present LP-based approximation algorithms for $R\,||\,\sum_j w_j C_j$ and $R\,|\,r_j\,|\,\sum_j w_j C_j$ from [17]. For each job j and machine i, p_{ij} denotes the processing time of j on i. Let $T = \max_{j\in J} r_j + \sum_{j\in J} \max_i p_{ij}$ denote an upper bound on the schedule length. The algorithms are based on the following time indexed relaxation which is pseudo-polynomial in the input size. The variable y_{ijt} indicates the fraction of the time interval $[t, t+1)$ that j is processed on i.

$$\min \sum_j w_j C_j$$

subject to

$$\sum_{i=1}^{m}\sum_{t=r_j}^{T}\frac{y_{ijt}}{p_{ij}} = 1 \qquad \text{for all } j \tag{11.22}$$

$$\sum_{j\in J} y_{ijt} \le 1 \qquad \text{for all } i \text{ and } t \tag{11.23}$$

$$C_j \ge \sum_{i=1}^{m}\sum_{t=r_j}^{T}\left[\frac{y_{ijt}}{p_{ij}}\left(t+\frac{1}{2}\right)+\frac{1}{2}y_{ijt}\right] \qquad \text{for all } j \tag{11.24}$$

$$y_{ijt} \ge 0 \qquad \text{for all } i,\ j, \text{ and } t \tag{11.25}$$

Equation (11.22) states that job j is processed completely. Equation (11.23) ensures that machine i can process one unit of work in each unit time interval. To understand Equation (11.24) consider the ideal situation, when a job j is processed on a single machine i continuously from time h to $h + p_{ij}$. Then, it can be verified easily that Equation (11.24) gives the exact completion time for j. In other circumstances, it can be easily checked that Equation (11.24) provides a lower bound on C_j. Note that the formulation remains a relaxation even if y_{ijt} are constrained to be binary.

We now describe a randomized rounding algorithm, Rand-Round, from [17] that uses an optimal solution to the above LP.

1. Let $\bar{C}$ be an optimal solution to the LP.
2. Assign each job j, *independently*, to machine-time pair (i, t), where the probability of j being assigned to (i, t) is exactly y_{ijt}/p_{ij}. Let (i_j, t_j) be the chosen pair for j.
3. On each machine i, schedule the jobs assigned to i nonpreemptively in order of t_j (ties broken *randomly*).

Let $\tilde{C}_j$ be the completion time of j produced by the above algorithm. Note that $\tilde{C}_j$ is a random variable. The analysis rests on the following lemma.

Lemma 11.14

The expected completion time of j, $E[\tilde{C}_j]$ is upper bounded by the following:

$$E[\tilde{C}_j] \le 2\sum_{i=1}^{m}\sum_{t=r_j}^{T}\frac{y_{ijt}}{p_{ij}}\left(t+\frac{1}{2}\right)+\sum_{i=1}^{m}\sum_{t=r_j}^{T} y_{ijt}$$

Proof

Consider a fixed job j and let t' be the earliest time such that there is no idle time in the interval $[t', \tilde{C}_j)$ on machine i_j. Let A_j be the set of jobs that are scheduled on i_j in this interval. It follows from the definition of t' and A_j that

$$\tilde{C}_j = t' + \sum_{k\in A_j} p_{ik}$$

It is easy to see that $t' \le t_j$ for, otherwise we would have scheduled j earlier. Since i_j was busy in $[t', \tilde{C}_j)$, we can upper bound $\tilde{C}$ as

$$\tilde{C}_j \le t_j + \sum_{k\in A_j} p_{ik}$$

To bound $E\,[\tilde{C}_j]$ we analyze the conditional expectation $E_{\,i,t}[\tilde{C}_j]$ under the event that j is assigned to the machine–time pair (i,t). It follows from the previous equation that

$$\begin{aligned} E_{\,i,t}[\tilde{C}_j] &\leq t + E_{\,i,t}\left[\sum_{k\in A_j} p_{ik}\right] \\ &\leq t + p_{ij} + \sum_{k\neq j} p_{ik}\mathrm{Pr}_{i,t}[k \text{ scheduled on } i \text{ before } j] \\ &= t + p_{ij} + \sum_{k\neq j} p_{ik}\left(\sum_{s=r_k}^{t-1}\frac{y_{iks}}{p_{ik}} + \frac{1}{2}\frac{y_{ikt}}{p_{ik}}\right) \\ &\leq t + p_{ij} + \left(t+\frac{1}{2}\right) \leq 2\left(t+\frac{1}{2}\right) + p_{ij} \end{aligned}$$

In the penultimate line above the factor of $\frac{1}{2}$ comes from the fact that the algorithm breaks ties randomly. In the last inequality above we use the LP constraints given in Equation (11.24). Now we can obtain the unconditional expectation as

$$\begin{aligned} E\,[\tilde{C}_j] &= \sum_{i=1}^{m}\sum_{t=r_j}^{T} \Pr[j \text{ assigned to } (i,t)]\cdot E_{i,t}[\tilde{C}] \\ &= \sum_{i=1}^{m}\sum_{t=r_j}^{T}\frac{y_{ijt}}{p_{ij}} E_{\,i,t}[\tilde{C}_j] \;\leq\; 2\sum_{i=1}^{m}\sum_{t=r_j}^{T}\frac{y_{ijt}}{p_{ij}}\left(t+\frac{1}{2}\right) + \sum_{i=1}^{m}\sum_{t=r_j}^{T} y_{ijt} \end{aligned}$$

This gives us the desired equation. □

Combining Lemma 11.14 and the constraint (11.24) it follows that $E\,[\tilde{C}_j] \leq 2\bar{C}_j$. By linearity of expectation we obtain the following.

Theorem 11.8

For instances of $R\,|\,r_j\,|\sum_j w_jC_j$, Rand-Round produces a solution whose expected value is within twice the value of the optimum solution the LP.

If there are no release dates, that is for the problem $R\,||\sum_j w_jC_j$ the integrality gap can be improved to 3/2 as follows. First, we can strengthen the LP by adding the following additional constraints:

$$C_j \geq \sum_{i=1}^{m}\sum_{t=r_j}^{T} y_{ijt} \quad \text{for all } j \tag{11.26}$$

When all release dates are 0, we can also strengthen Lemma 11.14 to show that

$$E\,[\tilde{C}_j] \leq \sum_{i=1}^{m}\sum_{t=r_j}^{T}\frac{y_{ijt}}{p_{ij}}\left(t+\frac{1}{2}\right) + \sum_{i=1}^{m}\sum_{t=r_j}^{T} y_{ijt}$$

The above equation comes about by observing in the proof of Lemma 11.14 that $t' = 0$ and $\tilde{C}_j = \sum_{k\in A_j} p_{ik}$ which in turn leads to $E_{\,i,t}[\tilde{C}_j] \leq p_{ij} + (t+\frac{1}{2})$.

We leave it as an exercise to the reader to prove the validity of (11.26) and put the above facts together to show the following theorem.

Theorem 11.9

For instances of $R\,||\sum_j w_jC_j$, Rand-Round produces a solution whose expected value is within 3/2 times the value of an optimum solution to the LP with the strengthened inequalities (11.26).

In [17], the integrality gaps in the above two theorems are shown to be tight. However, as we point out in Section 11.5, the integrality gaps of the LP when strengthened with an additional inequality is unknown. Rand-Round can be derandomized using the method of conditional probabilities. We refer the reader to [17] for more details. As we mentioned earlier, the LP is pseudo-polynomial in the input size. The formulation can be modified to obtain a polynomial sized LP while losing only a factor of $(1+\epsilon)$ in the objective function value. This idea was originally used by Hall, Shmoys, and Wein [8] and details in the context of the above results can be found in [17]. Using the modified LP and the Rand-Round algorithm results in approximation ratios of $(2+\epsilon)$ and $(3/2+\epsilon)$ for $R \mid r_j \mid \sum_j w_j C_j$ and $R \mid\mid \sum_j w_j C_j$, respectively. The latter result for $R \mid\mid \sum_j w_j C_j$ was independently observed by Chudak [18]. Using convex quadratic formulations, Skutella [28] improved the ratios to 2 and 3/2, respectively.

11.4 Approximation Schemes for Problems with Release Dates

The algorithms considered thus far share the basic paradigm of solving a relaxation of the completion time problem and use information from the relaxation to obtain an ordering on the jobs and/or an assignment of the jobs to machines. Even though these ideas yield constant factor approximation bounds, there are fundamental barriers to turning these approaches into approximation schemes. In particular, there are provable constant factor gaps between the objective value of the relaxation and the average completion time of the optimal schedule. Thus these approaches cannot directly yield approximation schemes.

In this section we develop a framework to obtain approximation schemes when jobs have release dates. In particular we present an approximation scheme for the problem $P \mid r_j \mid \sum_j w_j C_j$. The framework essentially consists of a dynamic programming-based search in a carefully chosen subset of all possible schedules. Given any $\epsilon > 0$, we will show that we can identify a subset of schedules such that each schedule in the subset has a nice structure and the subset contains a schedule that is $(1+\epsilon)$-approximate. We then use dynamic programming to explore this structured space and find the best possible schedule. There are two key components for implementing this framework: (i) input transformations, and (ii) dynamic programming. Input transformations add structure to a given arbitrary instance while dynamic programming helps us efficiently enumerate over the space of near-optimal schedules for a given structured instance. Each transformation on the input instance as well as optimality relaxation in the dynamic programming potentially increases the objective function value by a $1+O(\epsilon)$ factor, so we can perform a constant number of them while still staying within a $1+O(\epsilon)$ factor of the original optimum. When we describe such a transformation, we shall say it produces $1+O(\epsilon)$ *loss*.

We next develop each of these components in detail. The ideas described below apply to both $P \mid r_j \mid \sum w_j C_j$ and $P \mid r_j, pmtn \mid \sum w_j C_j$. In describing the dynamic programming, we will primarily focus on the nonpreemptive case, and toward the end mention the modifications needed for the preemptive case. To simplify notation we will assume throughout that $1/\epsilon$ is an integer and that $\epsilon \leq 1/4$). We use C_j and S_j to denote the completion and start time respectively of job j, and OPT to denote the objective value of some fixed optimal schedule.

We note that the framework that we describe can be extended with further ideas to obtain approximation schemes for the problems $Rm \mid r_j \mid \sum_j w_j C_j$ [20] and $Q \mid r_j \mid \sum_j w_j C_j$ [27].

11.4.1 Input Transformations

The goal of the input transformations is to impose a certain structure on the input instance that will facilitate efficient search by dynamic programming. In order to analyze the effect of each transformation, we show that an optimal schedule can be modified with a small increase in objective value so as to conform to the modified input structure. We will apply two transformations. The first transformation is a standard discretization of the input instance by *geometric rounding*. The second transformation, referred to as *job shifting*, is a new transformation that ensures that at each release date, only a small number of jobs arrive.

11.4.1.1 Geometric Rounding

Our first transformation reduces the number of distinct processing times and release dates in a given input instance.

Lemma 11.15

With $1+\epsilon$ loss, we can assume that all processing times and release dates are integer powers of $1+\epsilon$.

Proof

We transform the given instance in two steps. First, we multiply every release date and processing time by $1+\epsilon$; this increases the objective by the same amount (we are simply changing time units). Then we *decrease* each date and time to the next *lower* integer power of $1+\epsilon$ (which is still greater than the original value). This can only decrease the objective function value. □

Notation: For an arbitrary integer x, we define $R_x := (1+\epsilon)^x$. As a result of Lemma 11.15, we can assume that all release dates are of the form R_x for some integer x. We partition the time interval $[0, \infty)$ into disjoint intervals of the form $I_x := [R_x, R_{x+1})$ (Lemma 11.16 below ensures that no jobs are released at time 0). We will use I_x to refer to both the interval and the size $(R_{x+1} - R_x)$ of the interval. We will often use the fact that $I_x = \epsilon R_x$, i. e., the length of an interval is ϵ times its start time.

11.4.1.2 Job Shifting

The goal of job shifting is to ensure that for every job, the difference between its release time and completion time is only a small number of intervals, say $f(1/\epsilon)$, for some function f. It is easy to see intuitively that this property can be helpful in efficiently implementing a dynamic programming approach. An input instance can violate this property in two ways. First, a job released at some time R_x can be arbitrarily larger than $(1+\epsilon)^x$ and hence must be alive for an arbitrarily large number of intervals. Second, many jobs can be simultaneously released at some time R_x, no schedule can finish all of the jobs in a fixed number of intervals following R_x. The first of these two problems is easily fixed by suitably delaying the release date of each job. The second problem is more difficult to fix and it requires statically pruning and delaying jobs at each release date. This shifting of jobs is a critical ingredient of our approach.

We start by classifying each job as either small or large, and handle them differently. We say that a job is *small* if its size is less than ϵ^2 times the size of the interval in which it arrives (i.e., $p_x < \epsilon^2 I_x$), and *large* otherwise. The lemma below shows that we can always ensure that no job is arbitrarily large compared to the size of the interval in which it is released.

Lemma 11.16

With $1+\epsilon$ loss, we can enforce $r_j \geq \epsilon p_j$ for all jobs j.

Proof

Increase all processing times by a factor of $(1+\epsilon)$; thus a job j with original processing time p_j, now has processing time $(1+\epsilon)p_j$. As noted in Lemma 11.15, this increases the optimal objective value by a factor of at most $(1+\epsilon)$. Now consider an optimal schedule σ for this modified instance. For each job j, ignore the first ϵp_j units of its processing in the schedule σ. All remaining processing of job j now occurs at time ϵp_j or later, and moreover, p_j units of job j still get processed. The lemma follows. □

An immediate corollary is that in nonpreemptive schedules, a job never crosses many intervals.

Corollary 11.2

In any nonpreemptive schedule, each job crosses at most $s := \lceil \log_{1+\epsilon}(1+\frac{1}{\epsilon}) \rceil$ intervals.

Proof

Suppose job j starts in interval $I_x = [R_x, R_{x+1})$. Since $R_x \geq r_j \geq \epsilon p_j$ (Lemma 11.16), we have $I_x = \epsilon R_x \geq \epsilon^2 p_j$. The s intervals following x sum in size to $I_x/\epsilon^2 \geq p_j$. □

Lemma 11.17

There exists a $(1 + O(\epsilon))$–approximate schedule such that the following hold:

1. *for any two small jobs j and k with $r_j \leq r_k$, $\frac{p_j}{w_j} \leq \frac{p_k}{w_k}$ and $j < k$, $x(j) \leq x(k)$ holds, where $x(j)$ and $x(k)$ are the intervals in which j and k are scheduled*
2. *each small job finishes within the interval that it is started*

Proof

Fix an optimal schedule. Let s_x be the time allocated by this schedule to executing small jobs within interval I_x. We can assume that the small jobs are executed contiguously within each interval — this increases the schedule value by at most a $(1 + \epsilon)$-factor. We now create a new schedule to satisfy the first property in the statement of the lemma as follows. We first describe an assignment of small jobs to intervals without specifying the precise schedule of the jobs within an interval. The assignment is done as follows. Suppose we have assigned small jobs to intervals I_1 to I_{x-1}. Now we describe the assignment of jobs to I_x. Let A_x be set of all small jobs that have been released by I_x but have not been assigned to any of the intervals I_1 to I_{x-1}. For job j in A_x consider the tuple $(p_j/w_j, r_j, j)$. We order jobs in A_x in nondecreasing order of their tuples, using the dictionary ordering. We assign jobs in this order to I_x until the total processing time of the jobs assigned to I_x just exceeds s_x. Since each job in A_x is small, the total processing assigned to I_x is no more than $s_x + \epsilon I_x$. We claim that the total weight of jobs assigned to intervals I_1 to I_x by the above procedure dominates the weight of small jobs completed by the optimal schedule in intervals I_1 to I_x. This is relatively easy to verify and we leave it as an exercise to the reader. Also, observe that for each x, small jobs assigned to I_x by the above procedure can be scheduled in I_x using the space left by the optimal schedule, provided we stretch I_x by a $(1 + \epsilon)$-factor. Stretching the intervals by a $(1 + \epsilon)$-factor increases the objective function value by at most a $(1 + \epsilon)$-factor. The assignment procedure satisfies the first desired property. The second property also can be easily accomplished by expanding each interval by a $(1 + \epsilon)$-factor. □

As a result of Lemma 11.17, we can order all small jobs released at R_x according to their ratio $\frac{p_j}{w_j}$ and consider them for scheduling only in that order. Let T_x and H_x denote the small and large jobs released at R_x. Note that in this section small means an ϵ^2 fraction of the interval. Let $p(S)$ denote the sum of the processing times of the jobs in set S. The next lemma says that any input instance I can be modified with $1 + \epsilon$ loss to an instance I' so that the total size of the small and large jobs released at any release date R_x is $O(mI_x)$. This lemma plays an important role in our implementation of the dynamic programming framework since it allows us to represent compactly information about unfinished jobs as we move from one block to the next.

Lemma 11.18

(Job Shifting) An instance can be modified with $1 + O(\epsilon)$ loss to an instance I' such that the following conditions hold.

- *$p(T'_x) \leq (1 + \epsilon)mI_x$ for all x*
- *The number of distinct job sizes in H'_x is at most $\lfloor 1 + 4\log_{1+\epsilon} \frac{1}{\epsilon} \rfloor$*
- *The number of jobs of each distinct size in H'_x is at most $\frac{m}{\epsilon^2}$*

Proof

Consider the input instance I. The total processing time available in interval I_x is mI_x. Order the small jobs in T_x by nondecreasing ratios $\frac{p_j}{w_j}$ and pick jobs according to this order until the processing time of jobs picked would exceed $(1+\epsilon)mI_x$ if one more job is added. Picking jobs according to this order is justified by Lemma 11.17. The remaining jobs, which are released at R_x but cannot be processed in I_x, can safely be moved to the next release date R_{x+1}.

For each job j in H_x, Lemma 11.16 yields $R_x \geq \epsilon p_j$. On the other hand, since j is large we get $p_j \geq \epsilon^2 I_x = \epsilon^3 R_x$. Since all job sizes are powers of $1+\epsilon$, the number of distinct job sizes in H_x is as claimed. For a particular size that is large for I_x, we can order jobs by nonincreasing weights. The number of jobs of each size class that can be executed in the current interval is limited to $\frac{mI_x}{\epsilon^3 R_x} = \frac{m}{\epsilon^2}$. □

11.4.2 Dynamic Programming

The basic idea of the dynamic programming is to decompose the time horizon into a sequence of *blocks*. A block is a set of $s = \lceil \log_{1+\epsilon}(1+\frac{1}{\epsilon}) \rceil$ consecutive intervals. Note that $s \leq \frac{1}{\epsilon^2}$. Let $\mathcal{B}_0, \mathcal{B}_1, \ldots, \mathcal{B}_\ell$ be the partition of the time interval $[\min_j r_j, D)$ into blocks where D is an upper bound on the schedule makespan say, $(\sum_j p_j + \max_j r_j)$. Our goal is to do dynamic programming with blocks as units. There is interaction between blocks since jobs from an earlier block can cross into the current block. However, by the choice of the block size and Corollary 11.2, no job crosses an entire block in any nonpreemptive schedule. In other words, jobs that start in $\mathcal{B}_i$ finish either in $\mathcal{B}_i$ or $\mathcal{B}_{i+1}$. A *frontier* describes the potential ways that jobs in one block finish in the next. An incoming frontier for a block $\mathcal{B}_i$ specifies for each machine the time at which the crossing job from $\mathcal{B}_{i-1}$ finishes on that machine.

Lemma 11.19

There exists a $(1+\epsilon)$-approximate schedule which considers only $(m+1)^{s/\epsilon}$ feasible frontiers between any two blocks.

Proof

By Lemma 11.18 we can restrict attention to schedules in which small jobs never cross an interval. Each block consists of a fixed number s of intervals. Fix an optimal schedule and consider any machine in a block $\mathcal{B}_i$. A large job j continuing from the preceding block finishes in one of the s intervals of block $\mathcal{B}_i$ which we denote by $I_{x(j)}$. We can round up C_j to C'_j where $C'_j = R_{x(j)} + i\epsilon I_{x(j)}$ for some integer $0 \leq i \leq \frac{1}{\epsilon} - 1$. This will increase the schedule value by only a $1+\epsilon$ factor. Thus, we can restrict the completion times of crossing jobs to $\frac{s}{\epsilon}$ discrete time instants. Each machine realizes one of these possibilities. A frontier can thus be described as a tuple $(m_1, \ldots, m_{s/\epsilon})$ where m_i is the number of machines with crossing jobs finishing at the ith discrete time instant. Therefore, there are at most $(m+1)^{s/\epsilon}$ frontiers to consider. □

Let $\mathcal{F}$ denote the possible set of frontiers between blocks. The high level idea behind the dynamic programming is now easy to describe. The dynamic programming table entry $O(i, F, U)$ stores the minimum weighted completion time achievable by starting the set U of jobs before the end of block $\mathcal{B}_i$ while leaving a frontier of $F \in \mathcal{F}$ for block $\mathcal{B}_{i+1}$. Given all the table entries for some i, the values for $i+1$ can be computed as follows. Let $W(i, F_1, F_2, V)$ be the minimum weighted completion time achievable by scheduling the set of jobs V in block $\mathcal{B}_i$, with F_1 as the incoming frontier from block $\mathcal{B}_{i-1}$ and F_2 the outgoing frontier to block $\mathcal{B}_{i+1}$. We obtain the following equation:

$$O(i+1, F, U) = \min_{F' \in \mathcal{F}, V \subset U} \left[O(i, F', V) + W(i+1, F', F, U-V) \right]$$

There are two difficulties in implementing the dynamic programming. First, we cannot maintain the table entries for each possible subset of jobs in polynomial time. Therefore, we need to show the existence of approximate schedules that have compact representations for the set of subsets of jobs remaining after each block. Second, we need a procedure that computes the quantity $W(i, F_1, F_2, V)$. In what follows, we describe how these elements can be efficiently implemented for the parallel machine case. We focus on the nonpreemptive case and later describe the necessary modifications for handling the preemptive case. Below, we give a lemma that bounds the duration for which a job can remain unprocessed after its release.

Lemma 11.20

There exists a $[1+O(\epsilon)]$*–approximate schedule in which every job finishes within* $O(\frac{1}{\epsilon^2})$ *blocks after it is released.*

Proof

Consider some fixed optimal schedule. For each block $\mathcal{B}_i$, let A_i denote the set of jobs released in $\mathcal{B}_i$ which are not finished in the optimal schedule by the end of $\mathcal{B}_{\ell_i}$, where $\ell_i = i + \frac{1}{\epsilon^2}$. For each i, we alter the schedule by scheduling all the jobs in A_i in $\mathcal{B}_{\ell_i}$. Let I_x be the smallest interval in block $\mathcal{B}_{\ell_i}$. Using Lemma 11.18, the total volume of jobs released in $\mathcal{B}_i$ can be verified to be at most $\epsilon m I_x$, in other words $p(A_i) \le \epsilon m I_x$. Further, for every job $j \in A$, $p_j \le \epsilon^4 I_x$ holds. To schedule jobs of A_i in $\mathcal{B}_{\ell_i}$, we group jobs of A_i into units each with volume between ϵI_x and $(\epsilon + \epsilon^4)I_x$. From the bound on $p(A_i)$ the number of units is at most m. We assign each unit to an exclusive machine and schedule the unit on the machine as soon as the crossing job on that machine from $\mathcal{B}_{\ell_i - 1}$ finishes. We shift the jobs on the machine to accommodate the extra volume. Clearly, the completion times of jobs in A_i have only decreased. It is easy to see that expanding intervals by a $(1+O(\epsilon))$-factor can accommodate the increase in the completion times of jobs in $\mathcal{B}_i$ for all i. The new schedule satisfies the desired property. □

11.4.3 Nonpreemptive Scheduling on Parallel Machines

We now describe how to implement the two main components of the dynamic program: Compact representation of job subsets and scheduling within a block. We use time-stretching to show existence of schedules for which both tasks can be efficiently performed.

11.4.3.1 Compact Representation of Job Subsets

Recall that H_x and T_x denote the large and small jobs released at R_x. Let A_{xi} and B_{xi} denote the set of large and small jobs released at R_x that are scheduled in block $\mathcal{B}_i$. Let U_{xi} and V_{xi} denote the set of large and small jobs among jobs released at R_x that remain *after* block $\mathcal{B}_i$. Our goal is to show that there exist $(1+\epsilon)$ — approximate schedules with compact representations for these sets. Let $b(x)$ denote the block containing the interval I_x.

Large Jobs: Consider large jobs released in interval I_x. By Lemma 11.18, there are $O(\frac{1}{\epsilon}\log\frac{1}{\epsilon})$ distinct size classes in H_x. Within each size class, we assume that the jobs are ordered in nonincreasing order of weight. For each block $\mathcal{B}_i$, we specify the set U_{xi} by simply indicating for each size class, the smallest indexed job that remains to be scheduled. Since within each size class, any optimal schedule executes the jobs in nonincreasing weight order, this completely specifies the set A_{xi}. The total number of choices for U_{xi} are $(m/\epsilon^2)^{O(\log(1/\epsilon)/\epsilon)}$ by Lemma 11.18. Moreover, for any block $\mathcal{B}_i$, by Lemma 11.20, there are $O(s/\epsilon^2)$ release dates before $\mathcal{B}_i$ whose jobs can be still alive. Thus, there are only $(m/\epsilon^2)^{O((\log(1/\epsilon)/\epsilon)(s/\epsilon^2))}$ possible sets of large jobs that can be alive at the end of any block $\mathcal{B}_i$.

Small Jobs: The approach described above is not efficient for maintaining a description of unprocessed small jobs since the number of small jobs arriving at any release date can be arbitrarily large. We will instead focus on maintaining information about the unprocessed volume of small jobs.

Recall that for each release date R_x, we order the set T_x using Smith's ratio rule. Without any loss of generality, assume that the jobs in T_x are $\{1, 2, \ldots, \alpha\}$, indexed in nondecreasing order of p/w ratio. Let j_1 be the least index such that the total processing time of the first j_1 jobs exceeds $\epsilon^2 I_x$, denote this set of jobs by $J_1(x)$. Let j_2 be the least index such that the total processing time of the next $j_2 - j_1$ jobs exceeds

$\epsilon^2 I_x$, denote this set of jobs by $J_2(x)$. We continue in this manner to construct sets $J_1(x), J_2(x), \ldots, J_\ell(x)$, where each set $J_i(x)$, $1 \leq i < \ell$ contains a processing volume of at least $\epsilon^2 I_x$, and at most $2\epsilon^2 I_x$. The last set $J_\ell(x)$ contains at most $2\epsilon^2 I_x$ processing volume. By Lemma 11.18, we know that $\ell \leq 2m$. Our main observation is as follows.

Lemma 11.21

There exists a $(1 + O(\epsilon))$-approximate schedule in which for every release date R_x, all jobs in a set $J_i(x)$ are scheduled in the same interval.

Proof

By Lemma 11.17, we know that there exits a $(1+\epsilon)$-approximate schedule in which all small jobs arriving at any release date are executed in accordance with Smith's ratio rule. We start with such a schedule, say σ, and show that by stretching each interval by a $(1 + 2\epsilon)$ factor, we can modify σ to satisfy the property indicated in the Lemma. In the modified schedule, no job is scheduled in a later interval than it is scheduled in σ. Thus the modification only increases the objective function value by $(1 + 2\epsilon)$.

Fix a set $J_i(x)$ and let I_y be the first interval in which a job from $J_i(x)$ is scheduled in σ. Let M be a machine on which this event occurs. If all jobs in $J_i(x)$ are scheduled in I_y, the property above is satisfied. Otherwise, we know that no jobs from any set $J_{i'}(x)$ for $i' > i$ were scheduled in this interval. We now schedule all jobs in $J_i(x)$ on machine M in this interval. The total additional processing load added is at most $2\epsilon^2 I_x$. Adding over all intervals I_x with $x < y$, we can bound the additional processing overhead on machine M in interval I_y to be at most $2\epsilon I_y$. Thus, this additional load is accommodated by the stretching described above. □

Lemma 11.21 above says that for each block $\mathcal{B}_i$, we can specify the set V_{xi} by simply indicating the smallest index i such that $J_i(x)$ remains to be scheduled. This completely specifies the set B_{xi}. The total number of choices for V_{xi} are (m/ϵ^2) by Lemma 11.18. Moreover, for any block $\mathcal{B}_i$, by Lemma 11.20, there are $O(s/\epsilon^2)$ release dates before $\mathcal{B}_i$ whose jobs can be still alive. Thus, there are only $(m/\epsilon^2)^{O(s/\epsilon^2)}$ possible sets of small jobs that can be alive at the end of any block $\mathcal{B}_i$.

We summarize our considerations and results of this subsection in the following lemma:

Lemma 11.22

There is a $(1 + O(\epsilon))$–approximate schedule $\mathcal{S}$ such that for each block $\mathcal{B}_i$ the following is true:

- *There are $k = (\frac{m}{\epsilon^2})^{O(1/\epsilon^9)}$ sets $G_i^1, \ldots, G_i^k$ that can be constructed in polynomial time*
- *G_i, the set of jobs remaining in $\mathcal{S}$ after block $\mathcal{B}_i$, is one of $\{G_i^1, \ldots, G_i^k\}$*

11.4.3.2 Scheduling Jobs Within a Block

We now describe how to compute $W(i, F_1, F_2, V)$. Since this is itself an NP-hard problem we settle for a relaxation. A $1 + \epsilon$ decision procedure for computing $W(i, F_1, F_2, V)$ outputs a schedule that is within $1 + \epsilon$ of $W(i, F_1, F_2, V)$ and shifts the frontier F_2 by at most a $1 + \epsilon$ factor. Clearly such a procedure suffices in order to compute a $(1 + O(\epsilon))$-optimal solution to the dynamic program given above. We now describe a $1 + \epsilon$ decision procedure that runs in polynomial time for each fixed ϵ.

For the purposes of scheduling jobs in the block, we partition the job set V into *big* and *tiny* as follows. Let I_x be the smallest interval in $\mathcal{B}_i$. We call a job $j \in V$ big if $p_j \geq \epsilon I_x$, otherwise we call it tiny. Note that this definition differs from the earlier definition of large and small. Since the block consists of s intervals, for any big job $j \in V$ and interval I_y in $\mathcal{B}_i$ we have the property that $p_j \geq \epsilon^2 I_y$.

Our objective is to enumerate over all potential schedules of big jobs. In particular, we restrict ourselves to schedules where, in each interval I_x, a big job starts only at one of the $\frac{1}{\epsilon^3}$ times specified by $R_x + i\epsilon^3 I_x$, for $i = 0, \ldots, \frac{1}{\epsilon^3} - 1$. Furthermore, in our enumeration of big job schedules we will only specify the size and the start times of the big jobs scheduled. This is sufficient information to reconstruct their schedule: whenever we have two jobs of same size available, we always schedule the one with the larger weight first. With these restrictions, the schedule of big jobs on a machine within a block is completely determined by three things: its incoming frontier, its outgoing frontier, and the sizes of the big jobs started at each of the discrete time units in each of the s intervals. We estimate the number of different possibilities for each machine. The number of discrete times where a big job can be scheduled is $\frac{s}{\epsilon^3} \leq 1/\epsilon^5$. The number of big job sizes in a block can be seen to be $O(\log \frac{1}{\epsilon})$. Hence, the total number of possibilities is $(O(\log \frac{1}{\epsilon}))^{1/\epsilon^5}$ which is upper bounded by $k = 2^{O(1/\epsilon^6)}$. Thus, the configurations of all machines are from one of $(m+1)^k$ possibilities. Out of these we consider only those that are compatible with the incoming and outgoing frontiers F_1 and F_2 and have a feasible schedule for the big jobs in V. Both conditions can be checked in a straightforward way.

We schedule the tiny jobs in a greedy fashion in the spaces left by the big jobs. For each interval I_x, all the big jobs that start and finish in I_x can be assumed to execute contiguously at the earliest available time in the interval. We increase each of the spaces by a $1 + \epsilon$ factor to accommodate all the tiny jobs. The scheduling of tiny jobs is similar to that described in the proof of Lemma 11.17. In each interval I_x in $\mathcal{B}_i$, we order the tiny jobs in V that are released by I_x and not yet scheduled, in nondecreasing order of p_j/w_j. The jobs are scheduled in this order within the spaces left by the big jobs until there is no more space left in interval I_x. Then we proceed to I_{x+1}. This procedure can be accomplished in polynomial time in the number of tiny jobs in V, but can also be done in polynomial in m by grouping tiny jobs into units of size ϵI_x. We omit details. The scheduling of tiny jobs has to be repeated with each possibility of scheduling large jobs. We thus obtain the following.

Lemma 11.23

There is a $1 + \epsilon$ decision procedure to compute $W(i, F_1, F_2, V)$ that runs in time $(m + k)^k$ where $k = 2^{O(1/\epsilon^6)}$.

We remark that the running time of the procedure can be improved by doing dynamic programming between intervals of the block instead of brute force enumeration of all big job schedules. The improved running time will be $m^{\text{poly}(1/\epsilon)}$. We omit details here and state the result.

Theorem 11.10

There is a PTAS for $P \,|\, r_j \,|\, \sum w_j C_j$ that constructs a $(1 + \epsilon)$–approximation in time $O(n \log n + n (m+1)^{poly(1/\epsilon)})$.

The number of potential blocks for the dynamic programming is $O(\log D)$, where D is an upper bound on the schedule makespan. However, there are only $O(n/\epsilon^3)$ interesting blocks since each job j finishes by r_j/ϵ^4.

11.4.4 Preemptive Scheduling on Parallel Machines

In the preemptive case, several computational aspects of the preceding algorithm can be simplified, leading to an approximation scheme with a better running time. Specifically, since large jobs can be executed fractionally, we do not need to keep track of the frontier formed by the crossing jobs. Moreover, we can do dynamic programming directly with intervals instead of blocks and an approximate schedule can be specified by the fractions of jobs that are processed in any interval. This significantly reduces the amount of enumeration needed in the dynamic programming. For instance, since there are no release dates within an interval, we can use McNaughton's wrap around rule [25] to compute a preemptive schedule with optimal makespan in $O(n)$ time. Thus if we knew the job fragments that execute within an interval, they can be efficiently scheduled. We omit here the various technical details involved and summarize below the running time of our approximation scheme.

Theorem 11.11

There is a PTAS for $P \mid r_j, pmtn \mid \sum w_j C_j$ *that constructs a* $(1+\epsilon)$*–approximation in time* $O(n \log n + n (m+1)^{poly(1/\epsilon)})$.

11.5 Conclusions and Open Problems

We have surveyed approximation algorithms for minimizing sum of weighted completion times under a variety of constraints and machine environments. We have chosen to present a few algorithms and ideas in detail rather than giving an overview of all the known results. Below we mention several related ideas and results that we would have liked to cover but had to omit due to space constraints.

- Before [20] obtained PTASes for problems with release dates, constant factor approximation algorithms were known that had ratios better than the ones we presented in Section 11.2. They are still the algorithms of choice for simplicity and efficiency. We refer the reader to [11] for a $(\frac{e}{e-1})$-approximation for $1 \mid r_j \mid \sum_j C_j$, to [52] for a 1.6853-approximation for $1|r_j|\sum_j w_j C_j$, and to [17] for a 2-approximation to $P \mid r_j \mid \sum_j w_j C_j$. These algorithms have the additional advantage of being online. The underlying technical idea is to order the jobs by their α_j-completion times in a time-indexed relaxation for the problem where the α_j are chosen from an appropriately chosen probability distribution. The notion of α-completion times was used in [8] and the power of randomization to improve the ratios was first demonstrated independently in [11] and [12]. This technique also improves the approximation ratio for $1 \mid r_j, prec \mid \sum_j w_j C_j$ from 3 to $(e + \epsilon)$ [30].
- As we mentioned in a few places, makespan and weighted completion time are related in several ways. With precedence constraints, makespan is a special case of weighted completion time. However, (approximation) algorithms for minimizing makespan can be used to obtain approximation algorithms for minimizing weighted completion time. We mention two techniques in this regard. The first is dual-approximation based approach for online algorithms developed in [8]. The second approach which is more general and suited for offline algorithms is the one of Queyranne and Sviridenko [21] where approximation algorithms for makespan that are based on simple lower bounds are translated into approximation algorithms for weighted completion time. They use this approach to obtain algorithms for shop scheduling problem with minsum criteria.
- As indicated earlier, time-indexed formulations are pseudo-polynomial in size of the input. To obtain polynomial time algorithms, it is necessary to reduce their size. Hall et al. [8,29] accomplish this by partitioning time into geometrically increasing intervals: for parameter $\epsilon > 0$, the intervals are of the form $[(1+\epsilon)^{\ell}, (1+\epsilon)^{\ell+1})$ for $0 \le \ell \le \lceil \log_{1+\epsilon} T \rceil$. The formulation has variables $x_{j\ell}$ to indicate if job j finishes in interval ℓ. The interval indexed formulation is particularly useful in obtaining an $O(\log m)$-approximation for the problem $Q \mid r_j, prec \mid \sum_j w_j C_j$ [15]. Interval indexed formulations allow a more direct way to use a makespan algorithm to obtain an algorithm for minimizing weighted completion times. For example, the algorithms in [21] which are based on completion time variables do not explicitly use interval indexed formulations. However, the notion of geometrically increasing intervals is used in the rounding phase. We also mention that this relation between makespan and weighted completion time shows the existence of schedules that are simultaneously good for both measures. For a more formal treatment of the models where such results are possible and for the best known tradeoffs, see [53–55].
- We did not detail in-approximability results in this survey except to mention the known results in the table in Figure 11.1. Hoogeveen, Schuurman, and Woeginger [22] derived most of the known nontrivial results, establishing APX-hardness for several variants. The reductions are, in spirit, similar to the earlier known in-approximability results for minimizing makespan, although there are technical differences.

This survey has focussed primarily on recent work for minimizing weighted completion time. Some prior surveys on scheduling provide pointers to earlier work. These include the surveys of Graham et al. [56] from 1979 and that of Lawler et al. [5] from 1993. More recent surveys of interest are by Hall [57] and by Karger, Stein, and Wein [58]. We list below several interesting open problems in the context of scheduling to minimize weighted completion time.

1. Approximability of the single machine problem $1 \mid prec \mid \sum_j w_j C_j$. The approximation ratio we have is 2 via both LP based methods and combinatorial methods. On the other hand we do not know if the problem is APX-hard. Closing this gap is perhaps the most interesting open problem in this area. All the known LP formulations have an integrality gap of 2 as shown in [13]. Woeginger [33] has given an approximation preserving reduction from arbitrary instances to a restricted class of instances used in [13] to establish integrality gaps.
2. Improved approximation ratio for $P \mid prec \mid \sum_j w_j C_j$. Munier et al. [16], as described in Section 11.2.1.2, obtain a 4-approximation. However, even for the LP relaxation they use, the best lower bound on the integrality gap is 3. In terms of in-approximability, there is no evidence that the problem is harder to approximate than the makespan problem $P \mid prec \mid C_{\max}$ for which we have a 2-approximation and an in-approximability ratio of 4/3.
3. Improved approximation for $R \mid\mid \sum_j w_j C_j$ and $R \mid r_j \mid \sum_j w_j C_j$. The integrality gap of the LP for $R \mid r_j \mid \sum_j w_j C_j$, as presented in Section 11.3, is 2, and even with the additional constraint (11.26), the gap is 3/2 for $R \mid\mid \sum_j w_j C_j$. We can further strengthen the LP by adding the following set of constraints.

$$\sum_{i=1}^{m} y_{ijt} \leq 1 \quad \text{for all jobs } j \text{, and } t = 0, \ldots, T \tag{11.27}$$

   The integrality gap of the LP strengthened with the above inequality is unknown for both the problems and it would be interesting to see if improved approximation ratios can be obtained by using it. This problem is suggested in [17].

There are several open problems for minimizing makespan when jobs have precedence constraints that translate into open problem for minimizing weighted completion time. We refer the reader to the list of open problems compiled by Schuurman and Woeginger [59] for some of them.

References

[1] W. E. Smith. Various optimizers for single-stage production. *Naval Research Logistics Quarterly*, 3:59–66, 1956.

[2] D. Adolphson. Single machine job sequencing with precedence constraints. *SIAM Journal on Computing*, 6:40–54, 1977.

[3] E. L. Lawler. Sequencing jobs to minimize total weighted completion time. *Annals of Discrete Mathematics*, 2:75–90, 1978.

[4] J. Sydney. Decomposition algorithms for single-machine sequencing with precedence relations and deferral costs. *Operations Research*, 23(2):283–298, 1975.

[5] E. L. Lawler, J. K. Lenstra, A. H. G. Rinnooy Kan, and D. B. Shmoys. Sequencing and scheduling: algorithms and complexity. In S. C. Graves et al. (eds.), *Handbooks in OR & MS*, Elsevier Science Publishers, 4:445–522, 1993.

[6] R. L. Graham. Bounds for certain multiprocessor anomalies. *Bell System Technical Journal*, 45:1563–81, 1966.

[7] C. Phillips, C. Stein, and J. Wein. Minimizing average completion time in the presence of release dates. *Mathematical Programming B*, 82:199–223, 1998.

[8] L. A. Hall, D. B. Shmoys, and J. Wein. Scheduling to minimize average completion time: offline and online algorithms. In *Proceedings of the 7th ACM-SIAM Symposium on Discrete Algorithms*, 142–151, 1996.

[9] A. S. Schulz. Scheduling to minimize total weighted completion time: performance guarantees of lp based heuristics and lower bounds. In *Proceedings of IPCO*, 301–315, 1996.

[10] S. Chakrabarti, C. A. Phillips, A. S. Schulz, D. B. Shmoys, C. Stein, and J. Wein. Improved scheduling algorithms for minsum criteria. In *Proceedings of the 23rd ICALP*, 1996.

[11] C. Chekuri, R. Motwani, B. Natarajan, and C. Stein. Approximation techniques for average completion time scheduling. In *SIAM Journal on Computing*, 31(1): 146–166, 2000.

[12] M. X. Goemans. Improved approximation algorithms for scheduling with release dates. In *Proceedings of the 8th ACM-SIAM Symposium on Discrete Algorithms*, 591–598, 1997.

[13] C. Chekuri and R. Motwani. Precedence constrained scheduling to minimize weighted completion time on a single machine. *Discrete Applied Mathematics*, 98(1-2): 29–38, 1999.

[14] F. Margot, M. Queyranne, and Y. Wang. Decompositions, network flows, and a precedence constrained single machine scheduling problem. *Operations Research*, 2000.

[15] F. Chudak and D. Shmoys. Approximation algorithms for precedence-constrained scheduling problems on parallel machines that run at different speeds. *Journal of Algorithms*, 30:323–343, 1999.

[16] A. Munier, M. Queyranne, and A. S. Schulz. Approximation bounds for a general class of precedence constrained parallel machine scheduling problems. In *Proceedings of IPCO*, 367–382, 1998.

[17] A. Schulz and M. Skutella. Scheduling unrelated machines by randomized rounding. In *SIAM Journal on Discrete Mathematics*, 15(4): 450–469, 2002.

[18] F. Chudak. A min-sum 3/2-approximation algorithm for scheduling unrelated parallel machines. *Journal of Scheduling*, 2:73–77, 1999.

[19] F. Chudak and D. Hochbaum. A half-integral linear programming relaxation for scheduling precedence-constrained jobs on a single machine. *Operations Research Letters*, 25: 1999, 199–204.

[20] F. Afrati, E. Bampis, C. Chekuri, D. Karger, C. Kenyon, S. Khanna, I. Milis, M. Queyranne, M. Skutella, C. Stein, and M. Sviridenko. Approximation schemes for minimizing average weighted completion time with release dates. In *Proceedings of FOCS*, 1999.

[21] M. Queyranne and M. Sviridenko. Approximation algorithms for shop scheduling problems with minsum objective. *Journal of Scheduling*, 5: 287–305, 2002.

[22] J. A. Hoogeveen, P. Schuurman, and G. J. Woeginger. Non-approximability results for scheduling problems with minsum criteria. In *Proceedings of IPCO*, 353–366, 1998.

[23] M. Skutella and G. J. Woeginger. A PTAS for minimizing the weighted sum of job completion times on parallel machines. In *Proceedings of the 31st Annual ACM Symposium on Theory of Computing*, 400–407, 1999.

[24] C. Chekuri, R. Johnson, R. Motwani, B. K. Natarajan, B. R. Rau, and M. Schlansker. Profile-driven instruction level parallel scheduling with applications to super blocks. In *Proceedings of the 29th Annual International Symposium on Microarchitecture (MICRO-29)*, 58–67, 1996.

[25] R. McNaughton. Scheduling with deadlines and loss functions. *Management Science*, 6:1–12, 1959.

[26] W. Horn. Minimizing average flowtime with parallel machines. *Operations Research*, 21:846–847, 1973.

[27] C. Chekuri and S. Khanna. A PTAS for minimizing weighted completion time on uniformly related machines. In *Proceedings of ICALP*, 2001.

[28] M. Skutella. Convex quadratic and semidefinite programming relaxations in scheduling. *Journal of the ACM*, 48(2):206–242, 2001.

[29] L. A. Hall, A. S. Schulz, D. B. Shmoys, and J. Wein. Scheduling to minimize average completion time: offline and online algorithms. *Mathematics of Operations Research*, 22:513–544, 1997.

[30] A. Schulz and M. Skutella. Random-based scheduling: new approximations and LP lower bounds. In *Proceedings of RANDOM*, 1997.

[31] J. K. Lenstra and A. H. G. Rinnooy Kan. Complexity of scheduling under precedence constraints. *Operations Research*, 26:22–35, 1978.

[32] M. Skutella and M. Uetz. Scheduling precedence-constrained jobs with stochastic processing times on parallel machines. In *Proceedings of SODA*, 589–590, 2001.

[33] G. Woeginger. On the approximability of average completion time scheduling under precedence constraints. In *Proceedings of ICALP*, 887–897, 2001.

[34] S. Kolliopoulos and G. Steiner. Partially-ordered knapsack and applications to scheduling. In *Proceedings of ESA*, 2002.

[35] E. Balas. On the facial structure of scheduling polyhedra. *Mathematical Programming Studies*, 24: 179–218, 1985

[36] L. Wolsey. Mixed integer programming formulations for production planning and scheduling problems. *Invited Talk at the 12th International Symposium on Mathematical Programming*, MIT, Cambridge.

[37] M. Dyer and L. Wolsey. Formulating the single machine sequencing problem with release dates as a mixed integer program. *Discrete Applied Mathematics*, 26:255–270, 1990.

[38] M. Queyranne. Structure of a simple scheduling polyhedron. *Mathematical Programming*, 58: 263–285, 1993.

[39] M. Queyranne and Y. Wang. Single-machine scheduling polyhedra with precedence constraints. *Mathematics of Operations Research*, 16(1):1–20, 1991.

[40] J.-B Lasserre and M. Queyranne. Generic scheduling polyhedra and a new mixed-integer formulation for single-machine scheduling. In *Proceedings of IPCO*, 136–149, 1992.

[41] J. Sousa and L. Wolsey. A time-indexed formulation of nonpreemptive single-machine scheduling problems. *Mathematical Programming*, 54:353–367, 1992.

[42] A. von Arnim and A. Schulz. Facets of the generalized permutahedron of a poset. *Discrete Applied Mathematics*, 72: 179–192, 1997.

[43] Y. Crama and C. Spieksma. Scheduling jobs of equal length: complexity, facets, and computational results. In *Proceedings of IPCO*, 277–291, 1995.

[44] J. Van den Akker, C. Van Hoesel, and M. Savelsbergh. Facet-inducing inequalities for single-machine scheduling problems. Memorandum COSOR 93-27, Eindhoven University of Technology, Eindhoven, The Netherlands, 1993.

[45] J. Van den Akker. LP-based solution methods for single-machine scheduling problems. PhD Thesis, Eindhoven University of Technology, Eindhoven, The Netherlands, 1994.

[46] J. Van den Akker, C. Hurkens, and M. Savelsbergh. A time-indexed formulation for single-machine scheduling problems: branch and cut. Preprint, 1995.

[47] A. von Arnim, R. Schrader, and Y. Wang. The permutahedron of N-sparse posets. Preprint, 1996.

[48] C. N. Potts. An algorithm for the single machine sequencing problem with precedence constraints. *Mathematical Programming Studies*, 13:78–87, 1980.

[49] E. L. Lawler. *Combinatorial Optimization*. Holt, Rinehart, and Winston 1976.

[50] G. Gallo, M. D. Grigoriadis, and R. Tarjan. A fast parametric maximum flow algorithm and applications. *SIAM Journal on Computing*, 18:30–55, 1989.

[51] A. V. Goldberg and S. Rao. Beyond the flow decomposition barrier. *Proceedings of the 38th Annual IEEE Symposium on Foundations of Computer Science*, 2–11, 1997.

[52] M. Goemans, M. Queyranne, A. Schulz, M. Skutella, and Y. Wang. Single machine scheduling with release dates. *SIAM Journal on Discrete Mathematics*, 15:165–192, 2002.

[53] C. Stein and J. Wein. On the existence of schedules that are near-optimal for both makespan and total weighted completion time. *OR Letters*, 21: 1997.

[54] J. Aslam, A. Rasala, C. Stein, and N. Young. Improved bicriteria existence theorems for scheduling problems. In *Proceedings of SODA*, 846–847, 1999.

[55] A. Rasala, C. Stein, E. Torng, and P. Uthaisombut. Existence theorems, lower bounds and algorithms for scheduling to meet two objectives. In *Proceedings of SODA*, 723–731, 2002.

[56] R. L. Graham, E. L. Lawler, J. K. Lenstra, and A. H. G., Rinnooy Kan. Optimization and approximation in deterministic sequencing and scheduling: a survey. *Annals of Discrete Mathematics*, 5:287–326, 1979.

[57] L. Hall. Approximation algorithms for scheduling. In D. Hochbaum (ed.), *Approximation Algorithms for NP-Hard Problems*, PWS Publishing, 1995.

[58] D. Karger, C. Stein, and J. Wein. Scheduling algorithms. In M. Atallah (ed.), *Algorithms and Theory of Computation Handbook*, CRC Press, 1998.

[59] P. Schuurman and G. Woeginger. Polynomial time approximation algorithms for machine scheduling: ten open problems. *Journal of Scheduling*, 2(5):203–213, 2000.

[60] N. Alon, Y. Azar, G. J. Woeginger, and T. Yadid. Approximation schemes for scheduling on parallel machines. *Journal of Scheduling*, 1:55–66, 1998.

[61] K. R. Baker. *Introduction to Sequencing and Scheduling*. Wiley, 1974.

[62] J. L. Bruno, E. G. Coffman, and R. Sethi. Scheduling independent tasks to reduce mean finishing time. *Communications of the ACM*, 17:382–387, 1974.

[63] E. Horowitz and S. Sahni. Exact and approximate algorithms for scheduling nonidentical processors. *Journal of the Association for Computing Machinery*, 23:317–327, 1976.

[64] J. K. Lenstra, A. H. G. Rinnooy Kan, and P. Brucker. Complexity of machine scheduling problems. *Annals of Discrete Mathematics*, 1:343–362, 1977.

[65] M. W. P. Savelsbergh, R. N. Uma, and J. Wein. An experimental study of LP-based scheduling heuristics. In *Proceedings of the 9th ACM-SIAM Symposium on Discrete Algorithms*, 453–461, 1998.

12

Minimizing the Number of Tardy Jobs

Marjan van den Akker
Utrecht University

Han Hoogeveen
Utrecht University

12.1 Introduction

In this chapter we consider a problem faced by many students: after having had a great time during the semester spent on other activities, examination time has (suddenly) come close, and hence there is not enough time for thoroughly preparing all of the exams. Since a time-machine has not been invented yet, there is no alternative but to concentrate on a number of courses, such that as many as possible of the exams are passed. The problem is of course: which ones?

Moore (1968) was the first one to study this problem in the context of machine scheduling. In this case, we are given a single machine (corresponding to the student), on which we have to execute a given set of jobs (corresponding to the preparations of the exams). The machine is assumed to be continuously available from time zero onward, and it can perform at most one job at a time. The machine has to execute n jobs, denoted by $J_1, \ldots, J_n$. Performing task J_j requires a period of uninterrupted length p_j (the (estimated) time to prepare the exam), and the execution of this task is preferably finished by its due date d_j (the start of the examination). If job J_j is finished after its due date, then it is marked as late. The objective is to minimize the (weighted) number of late jobs. In the standard problem considered by Moore it does not matter at which time a late job is finished, late jobs are skipped altogether, and the machine only carries out the jobs that will finish on time.

Moore shows that the unweighted case, in which each job is equally important, can be solved in $O(n \log n)$ time by an algorithm that since then is known as Moore-Hodgson's algorithm. The weighted case, where w_j expresses the importance or the reward of executing job J_j $(j = 1, \ldots, n)$, can be solved in $O(n \sum p_j)$ time by dynamic programming (Lawler and Moore, 1969). Karp (1972) shows that pseudo-polynomial running time is unavoidable for this problem (unless $\mathcal{P} = \mathcal{NP}$) by establishing $\mathcal{NP}$-hardness in the ordinary sense, even if all due dates are equal.

Throughout this paper we use the three-field notation scheme by Graham et al., (1979) to denote the problems. We further use the standard scheduling notation: for each job J_j (or simply job j), we use p_j to denote the processing time, d_j to denote the due date of J_j, w_j to denote the weight (if no weights have been specified, then we put all weights equal to 1), r_j to denote the release date of job j (if not specified,

1-58488-397-9/$0.00+$1.50

then we assume that $r_j = 0$, at which time the machine becomes available), $\bar{d}_j$ to denote the deadline of job j (if this has not been specified, then $\bar{d}_j = +\infty$), and $C_j(\sigma)$ to denote the completion time of job j in schedule σ (and we use simply C_j if it is clear to which schedule we are referring). To denote the objective function we introduce the indicator variable U_j, which gets the value 1 if job j is late, that is, if $C_j > d_j$, and 0, otherwise. Hence, the number of late jobs is then equal to $\sum U_j$.

The remainder of the paper is organized as follows. In Section 12.2 we review the problem of minimizing the number of late jobs. Here we state Moore–Hodgson's algorithm and prove its correctness by interpreting it as a dynamic programming algorithm. In Section 12.3 we discuss the weighted case. We first show how to solve the problem through dynamic programming, where we present both Lawler and Moore's algorithm, which runs in $O(n \sum p_j)$ time and space, and an extension of the dynamic programming algorithm of Section 12.2 that runs in $O(n \sum w_j)$ time and space. We then present Karp's $\mathcal{NP}$-hardness proof, which shows that pseudo-polynomial running times are unavoidable. In Section 12.4 we add release dates, which imply a lower bound on the start time. In case of arbitrary release dates, the resulting problem is $\mathcal{NP}$-hard in the strong sense, but when the release dates and due dates are similarly ordered, then the problem is solvable in $O(n \log n)$ time by an algorithm due to Lawler (-). In Section 12.5 we introduce deadlines, which decree an upper bound on the completion time; in this case, we assume that the machine must execute the late jobs as well. We present Lawler's proof (Lawler (-)) that the problem with deadlines is $\mathcal{NP}$-hard in the ordinary sense. In Section 12.6 we consider the problem with parallel machines and show how to solve this using column generation; this approach is due to Van den Akker, Hoogeveen, and Van de Velde (1999) and Chen and Powell (1999).

12.2 Minimizing the Number of Late Jobs on a Single Machine

In this section, we review the problem described by Moore (1968). Using the standard notation we denote this problem by $1 || \sum U_j$. We first state Moore–Hodgson's algorithm without a proof, and then solve the problem again from scratch. Fortunately, this detour of an $O(n^2)$ dynamic programming algorithm will bring us back to Moore–Hodgson's $O(n \log n)$ algorithm. We developed this approach ourselves, and only found out recently that it has been described in an unpublished paper by Lawler (-) for the more complicated case described in Section 12.4.

Moore–Hodgson's Algorithm

Step 1. Set σ equal to the Earliest Due Date (EDD)-schedule for all jobs $J_1, \ldots, J_n$.

Step 2. If each job in σ is on time, then this subschedule followed by the remaining jobs in any order is optimal.

Step 3. Find the first job in σ that is late; let this be job J_j. Find the largest job from the set containing J_j and all its predecessors in σ and remove it from σ. Remove the idle time from the resulting schedule by shifting jobs forward; call this new schedule σ, and go to Step 2.

We apply this algorithm to the following 5-job instance, the data of which are shown in Table 12.1. Due to the numbering of the jobs, σ is equal to J_1, J_2, J_3, J_4, J_5. The first job in σ that is late is J_3. The longest job in the subschedule J_1, J_2, J_3 is J_1, which hence is removed. Therefore, σ becomes equal to J_2, J_3, J_4, J_5, and J_4 is the first job in σ that is late. The longest job in J_2, J_3, J_4 is J_4, which gets removed. Now σ is equal to J_2, J_3, J_5, and each job in σ is on time, which means that the solution in which J_2, J_3, J_5 are on time and J_1, J_4 are late is optimal.

TABLE 12.1 Due Dates and Processing Times

	J_1	J_2	J_3	J_4	J_5
d_j	6	7	8	9	11
p_j	4	3	2	5	6

The optimality proof of Moore–Hodgson's algorithm is rather complicated. We therefore show its correctness by solving the problem from scratch. We start with some simple, well-known observations.

Observation 12.1 *There exists an optimal solution with the following properties:*

1. *The on time jobs precede all late jobs*
2. *The on time jobs are executed in EDD-order, that is, in order of nondecreasing due date*

Proof

The proof of the first part is straightforward: Suppose that an on time job is preceded by a late job. Moving this late job to the end of the schedule and shifting forward all the jobs currently succeeding it will not decrease the quality of the schedule. Repetition of this argument shows the correctness of the first property.

The proof of the second property follows from the result by Jackson (1955), who showed that executing the jobs in EDD-order minimizes the maximum lateness, where the lateness of a job j is defined as $C_j - d_j$ (this maximum lateness is at most equal to zero if all jobs are on time). □

Hence, instead of specifying the whole schedule, we can limit ourselves to specifying the set E (for early) containing the on time jobs, as it can be checked whether the jobs in E can be all on time together by putting them in EDD-order. A set E of jobs that are selected to be on time is called *feasible* if none of them is late when executed in EDD-order. The goal is therefore to find a feasible set E of maximum cardinality. From now on, we will use the following notation. By $|Q|$ and $p(Q)$ we denote the number and total processing time of the jobs in a given set Q, respectively. Since the EDD-order is crucial in the design of the algorithm, we assume from now on that the jobs are numbered such that

$$d_1 \leq d_2 \leq \cdots \leq d_n$$

We want to solve the problem by applying dynamic programming. For this we need the following *dominance rule.*

Dominance Rule 12.1 *Let E^1 and E^2 be two feasible subsets of the tasks $\{J_1, \ldots, J_j\}$ with $|E^1| = |E^2|$. If $p(E^1) < p(E^2)$, then any solution E with $E \cap \{J_1, \ldots, J_j\} = E^2$ can be ignored.*

Proof

Let E correspond to an optimal solution, and suppose that $E \cap \{J_1, \ldots, J_j\} = E^2$. We will show that replacing the jobs in E^2 by the jobs in E^1 yields a feasible subset $\bar{E}$ of $\{J_1, \ldots, J_n\}$; since $|\bar{E}| = |E|$, the subset $\bar{E}$ must then correspond to an optimal solution, too.

To show that $\bar{E}$ is a feasible subset, we must show that in the EDD-schedule of $\bar{E}$ all jobs are on time. Let σ and π denote the EDD-schedule of the jobs in $\bar{E}$ and E, respectively. Due to the numbering of the jobs, we know that the jobs in E^1 precede the remaining jobs of $\bar{E}$ in σ; as E^1 is a feasible subset, the jobs in E^1 are on time in σ. The remaining jobs in $\bar{E}$ start $p(E^2) - p(E^1) > 0$ time units earlier in σ than in π, and hence these jobs are on time as well. □

As a consequence of the dominance rule, the only feasible subset of $\{J_1, \ldots, J_j\}$ with cardinality k that is of interest is the one with minimum total processing time. We define mc_j (of maximum cardinality), for $j = 1, \ldots, n$, as the maximum cardinality of a feasible subset of the jobs $\{J_1, \ldots, J_j\}$; the value of mc_j will be determined through the algorithm. Since mc_j denotes the maximum number of on time jobs out of the first j, the value of the optimum solution is equal to $n - mc_n$. We further use $E_j^*(k)$ $(j = 1, \ldots, n; k = 0, \ldots, mc_j)$ to denote a feasible subset containing exactly k jobs from $\{J_1, \ldots, J_j\}$ with minimum total processing time. Our dynamic programming algorithm works as follows. We add the jobs one by one in EDD-order. For each combination (j, k), where j $(j = 1, \ldots, n)$ refers to the

TABLE 12.2 Result of the Dynamic Programming Algorithm

	$j = 0$	$j = 1$	$j = 2$	$j = 3$	$j = 4$	$j = 5$
$k = 0$	0	0	0	0	0	0
$k = 1$	∞	4	3	2	2	2
$k = 2$	∞	∞	7	5	5	5
$k = 3$	∞	∞	∞	∞	∞	11

number of jobs that have been considered and k ($k = 0, \ldots, mc_j$) denotes the number of on time jobs, we introduce a state-variable $f_j(k)$ with value equal to $p(E_j^*(k))$. As an initialization, we define $mc_0 = 0$ and put $f_j(k) = 0$ if $j = k = 0$ and $f_j(k) = \infty$, otherwise. Suppose that we have determined the values $f_j(k)$ for a given j and all $k = 0, \ldots, mc_j$. We first determine mc_{j+1}. Obviously, mc_{j+1} can only become equal to $mc_j + 1$ if it is possible that J_{j+1} is on time together with mc_j jobs from $\{J_1, \ldots, J_j\}$, and it follows immediately from the proof of Observation 12.1 that to achieve this (if possible) we must pick the subset $E_j^*(mc_j)$ for the mc_j jobs from $\{J_1, \ldots, J_j\}$. Since J_{j+1} succeeds the jobs in $E_j^*(mc_j)$, it can be on time only if $p(E_j^*(mc_j)) + p_{j+1} \leq d_{j+1}$, which condition is equal to $f_j(mc_j) + p_{j+1} \leq d_{j+1}$ by definition of $f_j(k)$. Therefore, we find

$$\begin{aligned} mc_{j+1} &= mc_j + 1 \quad \text{if } f_j(mc_j) + p_{j+1} \leq d_{j+1} \\ mc_{j+1} &= mc_j \quad\quad\; \text{otherwise} \end{aligned}$$

Now we come to the process of finding the correct values of the state-variables $f_{j+1}(k)$, for $k = 0, \ldots, mc_{j+1}$. We simply put $f_{j+1}(0) = 0$ and determine $f_{j+1}(k)$ ($k = 1, \ldots, mc_{j+1}$) through the recurrence relation

$$f_{j+1}(k) = \min\{f_j(k), f_j(k-1) + p_{j+1}\} \tag{12.1}$$

Assuming the correctness of our approach (which we shall show below), we can compute the values $f_j(k)$ in $O(n^2)$ time altogether, from which we immediately determine the minimum number of late jobs as $(n - mc_n)$, whereas the optimum schedule can be determined through backtracking. If we apply it to the example in Table 12.1, we find the values displayed in Table 12.2 (in which we have omitted the lines for $k = 4$ and $k = 5$, since these have ∞ in each field). From Table 12.2 we find that $mc_5 = 3$, which implies that in the optimum solution two jobs are late; through backtracking, we find that the jobs J_2, J_3, J_5 are on time and the jobs J_1, J_4 are late. What results is to show the correctness of the dynamic programming algorithm.

Lemma 12.1

The values $f_j(k)$ ($j = 1, \ldots, n; k = 0, \ldots, n$) computed through the recurrence relation are equal to $p(E_j^*(k))$.

Proof

Because of the initialization, we find $f_j(k) = \infty$ if $k > mc_j$, which is the correct value by definition. We use induction to prove the correctness for all j and $k \leq mc_j$. For $j = 0$ and $k = 0$, correctness holds because of the initialization. Suppose that the values $f_j(k)$ are correct for all $j = 0, \ldots, m$ and $k = 0, \ldots, mc_m$ for some value $m \geq 0$. We will show correctness of $f_{m+1}(k)$, where k is any value from $\{1, \ldots, mc_{m+1}\}$ (the case $k = 0$ being trivial). If $k \leq mc_m$, then there are two possibilities: we add job J_{m+1} to the set of on time jobs, or we mark it as late. In the latter case, we find a minimum total processing time of $f_m(k)$. If we include J_{m+1} in the set of on time jobs, then we have to select a feasible subset from the jobs $\{J_1, \ldots, J_m\}$ containing $k - 1$ jobs, and we know that the best one has total processing time $f_m(k-1)$.

Job J_{m+1} does not necessarily have to be on time when placed after this subset of $k-1$ jobs, but it will be if $f_m(k-1)+p_{m+1} < f_m(k)$, since $f_m(k)$ is equal to the completion time of the last job in the set of on time jobs, which amounts to no more than d_m and $d_m \leq d_{m+1}$ by definition. The recurrence relation selects the smaller of these two values, which implies the correctness of $f_{m+1}(k)$.

If $k = mc_{m+1} = mc_m + 1$, then we need to add job J_{m+1} to the set of on time jobs to get the desired number, as we have indicated above. In that case, the check performed when computing mc_{m+1} guarantees that J_{m+1} is on time when it starts at time $f_m(mc_m)$, and the correctness of the value $f_m(mc_m)$ settles the proof of the correctness of $f_{m+1}(mc_{m+1})$. □

We will show that by looking at the dynamic programming algorithm in the right way we can obtain Moore-Hodgson's algorithm. Here we need the close relation between the sets $E_j^*(k)$ and $E_j^*(k-1)$ expressed in the following lemma.

Lemma 12.2

A set $E_j^*(k-1)$ is obtained from the set $E_j^*(k)$ by removing a task from $E_j^*(k)$ with largest processing time.

Proof

Suppose that the lemma does not hold for some combination of j and k. Let J_i be a task in $E_j^*(k)$ with maximum processing time. According to our assumption, $p(E_j^*(k-1)) < p(E_j^*(k)) - p_i$. Determine J_q as the job with maximum due date that belongs to $E_j^*(k)$ but not to $E_j^*(k-1)$. Now consider the subset $\bar{E}_j^*(k) = E_j^*(k-1) \cup \{J_q\}$; we will prove that this subset is feasible, which contradicts the optimality of $E_j^*(k)$, since $p(E_j^*(k)) > p(E_j^*(k-1)) + p_i \geq p(E_j^*(k-1)) + p_q = p(\bar{E}_j^*(k))$. We have to check the EDD-schedule, which we denote by σ, for the tasks in $\bar{E}_j^*(k)$, which is obtained from the EDD-schedule for the jobs in $E_j^*(k-1)$ by inserting J_q at the correct spot. Due to the feasibility of $E_j^*(k-1)$, the only jobs that may be late are J_q and the jobs after J_q in $E_j^*(k-1)$. Due to the choice of J_q, we have that J_q and all jobs following J_q in the EDD-schedule of $E_j^*(k)$, which we denote by π, are also present in σ, but they are started at least $p(E_j^*(k)) - p(\bar{E}_j^*(k)) > 0$ time units earlier now, which implies that these jobs are on time in σ as well. The only jobs that still have to be checked are the jobs in σ following J_q that are not present in $E_j^*(k)$. Let J_l be any such job, and let J_k be the last job in σ before J_l that is also present in π; such a job J_k definitely exists, because of the presence of J_q. Since J_k precedes J_l in σ, we have that $d_k \leq d_l$. Moreover, we have that $C_k(\pi) > C_l(\sigma)$, since all jobs following J_k in π are also present in σ and $C_{\max}(\pi) > C_{\max}(\sigma)$. Putting things together, we find that $C_l(\sigma) < C_k(\pi) \leq d_k \leq d_l$, which implies that J_l is on time as well. □

As a consequence, if we use a state-variable $g_j(k)$ to denote the set $E_j^*(k)$ (and not just its total processing time), then we can determine $g_j(k)$ from $g_j(mc_j)$, for $k = 0, \ldots, mc_j$, by simply choosing the k smallest jobs. Hence, we then only need to compute and store $g_j(mc_j)$, for $j = 1, \ldots, n$, to find the optimum solution. We can update the recurrence relation correspondingly: If $mc_{j+1} = mc_j + 1$, then we simply put $g_{j+1}(mc_{j+1}) \leftarrow g_j(mc_j) \cup \{J_{j+1}\}$. If on the other hand $mc_{j+1} = mc_j$, then we must choose between the sets $g_j(mc_j)$ and $g_j(mc_j - 1) \cup \{J_{j+1}\}$ as the new set $g_{j+1}(mc_{j+1})$, where the choice depends on the total processing time of the two sets. Because of Lemma 12.2, we know that these two sets both contain $g_j(mc_j - 1)$ as a subset, which implies that there is only one job that makes the difference. Therefore, we have to compare the processing time of the longest job in $g_j(mc_j)$ to p_{j+1}. But this boils down to finding the largest job, say J_i, in the set $E_j^*(mc_j) \cup \{J_{j+1}\}$ and setting $E_{j+1}^*(mc_{j+1})$ equal to $E_j^*(mc_j) \cup \{J_{j+1}\} \setminus \{J_i\}$. Summarizing, we get the following:

- If $p(E_j^*(mc_j)) + p_{j+1} \leq d_{j+1}$, then $E_{j+1}^*(mc_{j+1}) \leftarrow E_j^*(mc_j) \cup \{J_{j+1}\}$
- If $p(E_j^*(mc_j)) + p_{j+1} > d_{j+1}$, then find the longest job in $E_j^*(mc_j) \cup \{J_{j+1}\}$ and set $E_{j+1}^*(mc_{j+1})$ equal to $E_j^*(mc_j) \cup \{J_{j+1}\}$ minus this job

But this is exactly Moore-Hodgson's algorithm, which we defined above. Hence, we have proven the following theorem.

Theorem 12.1

Moore-Hodgson's algorithm solves the problem $1\,||\sum U_j$.

We want to conclude this section by discussing a special case of the problem $1\,||\sum U_j$ in which the job set has been partitioned into important and less important jobs; the objective is to complete as many important jobs as possible on time, and in addition to have as many jobs as possible from the less important category on time.

To solve this variant we apply dynamic programming like described above. In this case we need state-variables that measure the number of jobs from the first and the number of jobs from the second category out of the set $\{J_1, \ldots, J_j\}$ that are on time; the state-variable $f_j(k_1, k_2)$ then has value equal to the minimum total processing time possible for a feasible subset of the jobs $\{J_1, \ldots, J_j\}$ that contains exactly k_1 important jobs and k_2 less important jobs. This can be implemented in a straightforward manner to run in $O(n^3)$ time.

Note that, if it is feasible to complete all of the important jobs on time, then we can find an optimal solution to the problem above, that is, a solution that completes all important jobs on time and as many of the less important jobs as possible, in $O(n \log n)$ time by adjusting Step 3 in Moore-Hodgson's algorithm: If J_j is the first tardy job in σ, then remove from the set containing J_j and all its predecessors in σ the longest of the less important jobs. We leave the proof of this statement to the reader.

12.3 Minimizing the Weighted Number of Tardy Jobs

In many applications it is necessary to differentiate between the jobs. This can be done by attaching a positive, integral weight w_j to job J_j, for $j = 1, \ldots, n$, which leads to the problem $1\,||\sum w_j U_j$. This problem has been studied by Lawler and Moore (1969), who show that it can be solved in $O(n \sum p_j)$ time and space through dynamic programming. The dynamic programming algorithm is based on Observation 12.1; it uses state-variables $f_j(t)$ ($j = 0, \ldots, n; t = 0, \ldots, \sum_{i=1}^{j} p_i$), where $f_j(t)$ is equal to the value of the optimum solution for the jobs $J_1, \ldots, J_j$ subject to the constraint that the total processing time of the on time jobs amounts to t.[1] The initialization is standard: We put $f_j(t) = 0$ if $j = t = 0$ and ∞, otherwise. Given $f_j(t)$ for $t = 0, 1, \ldots, \sum_{i=1}^{j} p_i$, we compute $f_{j+1}(t)$ ($t = 0, 1, \ldots, \sum_{i=1}^{j+1} p_i$) through the following recurrence relation

$$f_{j+1}(t) = \begin{cases} \infty & \text{if } t > d_{j+1} \\ \min\{f_j(t) + w_{j+1}, f_j(t - p_{j+1})\} & \text{otherwise} \end{cases}$$

Assuming that this procedure returns the correct values for the state-variables, which we will show below, we find that $f_n(t)$ ($t = 0, \ldots, \sum_{i=1}^{n} p_i$) gives the value of the optimum solution subject to the constraint that the total processing time of the on time jobs amounts to t. Therefore, we can find the optimum by determining the minimum of the $f_n(t)$ values over all values of t. Since each of the $O(n \sum p_j)$ state-variables is computed in constant time, the algorithm can be implemented to run in $O(n \sum p_j)$ time and space.

Lemma 12.3

The state-variables $f_j(t)$ ($j = 0, \ldots, n; t = 0, \ldots, \sum_{i=1}^{j} p_i$) *get the right value.*

[1] In the paper by Lawler and Moore, the total processing time should amount to *no more than* t; the current definition makes the recurrence relation simpler.

Proof

We prove this lemma by induction. Suppose that the values $f_j(t)$ are correct for $t = 0, 1, \ldots, \sum_{i=1}^{j} p_i$; this is obviously true for $j = 0$. We will now show that the recurrence relation assigns the correct values to $f_{j+1}(t)$. First, suppose that $t > d_{j+1}$. In that case, the last job in the on time set completes after its due date, which implies infeasibility, and hence $+\infty$ is the correct value then. If $t \leq d_{j+1}$, then one possibility is to add J_{j+1} to the tardy set, which implies that the interval $[0, t]$ must be filled with jobs from $\{J_1, \ldots, J_j\}$; this gives the first term in the minimand. The other possibility is to put J_{j+1} in the on time set, which implies that only the interval $[0, t - p_{j+1}]$ is left for the jobs in $\{J_1, \ldots, J_j\}$; this leads to the second term in the minimand. Choosing the better one of these two options yields the correct value of $f(j, t)$. □

The implementation of the dynamic programming algorithm can obviously be improved by restricting the domain of t in $f_j(t)$ to be no more than $\min\{\sum_{i=1}^{j} p_i, d_j\}$, since the remaining $f_j(t)$ values have been initialized at $+\infty$ already.

The dynamic programming algorithm described for the problem $1 || \sum U_j$ can be adapted to solve the problem $1 || \sum w_j U_j$ in a straightforward manner. Recall that the state-variable $f_j(k)$ $(j = 0, \ldots, n;$ $k = 0, \ldots, j)$ was defined as the minimum total processing time over all feasible subsets of $\{J_1, \ldots, J_j\}$ consisting of exactly k elements. When we change this such that $f_j(k)$ is equal to the minimum total processing time over all feasible subsets of $\{J_1, \ldots, J_j\}$ that have total weight equal to k (which is equal to the cardinality in the unit-weight case), then we can use it to solve the problem $1 || \sum w_j U_j$, since the whole analysis of Section 12.2 goes through. We only need to change the recurrence relation to

$$f_{j+1}(k) = \min\{f_j(k), f_j(k - w_{j+1}) + p_{j+1}\}$$

through this, we can compute the correct values of the state-variables $f_j(k)$ $(j = 0, \ldots, n; k = 0, 1, \ldots, \sum_{i=1}^{j} w_i)$ in constant time. Hence, this dynamic programming algorithm can be implemented to run in $O(n \sum w_i)$ time and space. The value of the optimum solution is then equal to $\sum_{i=1}^{n} w_i - k^*$, where k^* corresponds to the maximum value of k such that $f_n(k) < +\infty$; the corresponding schedule is determined by backtracking.

Both dynamic programming algorithms above have a pseudo-polynomial running time. Karp (1972) showed that this is inevitable, unless $\mathcal{P} = \mathcal{NP}$, by showing that the problem $1 || \sum w_j U_j$ is $\mathcal{NP}$-hard in the ordinary sense, even if all due dates are equal; this was presumably the first $\mathcal{NP}$-hardness proof formulated for a scheduling problem. The reduction is from the Partition problem, which is known to be $\mathcal{NP}$-hard in the ordinary sense. Partition is defined as follows.

Partition Given n positive integers $a_1, a_2, \ldots, a_n$ with sum equal to $2A$, does there exist a subset S of the index set $\{1, \ldots, n\}$ such that

$$\sum_{i \in S} a_i = A$$

Given any instance of Partition, we construct the following instance of the decision variant of our problem. We let each integer a_i $(i = 1, \ldots, n)$ correspond to a job J_i with processing time $p_i = a_i$, weight $w_i = a_i$, and due date $d_i = A$. The decision variant of the problem of $1 || \sum w_j U_j$ is: Does the instance defined above have a solution with value no more than A? We denote this problem by Decision Variant of Weighted Number of Late Jobs problem (DVWNLJ).

Lemma 12.4

The answer to the instance of Partition is "yes" if and only if the answer to DVWNLJ *is "yes."*

Proof

First suppose that the answer to Partition is "yes"; let S be a subset of $\{1, \ldots, n\}$ such that $\sum_{i \in S} a_i = A$. It is easily verified that we then can construct a "yes" solution of DVWNLJ by putting job J_i $(i = 1, \ldots, n)$ in the on time set if $i \in S$ and in the tardy set if $i \notin S$.

Conversely, suppose that we have a "yes" solution of DVWNLJ; let S be the index set containing the indices of the jobs in the on time set. Since each job has due date equal to A, we know that $\sum_{i \in S} p_i = \sum_{i \in S} a_i \leq A$. We further know that, since the total weight of all jobs equals $2A$ and the total weight of the jobs in the tardy set is no more than A, the total weight of the jobs in the on time set is at least equal to A, that is, $\sum_{i \in S} w_i = \sum_{i \in S} a_i \geq A$. Combining these two inequalities implies that $\sum_{i \in S} a_i = A$, and hence S constitutes a "yes" solution to Partition. □

Theorem 12.2

The problem $1 \,||\, \sum w_j U_j$ *is* $\mathcal{NP}$*-hard in the ordinary sense.*

12.4 Scheduling with Release Dates

In this section we assume that each job has a *release date,* which we denote by r_j $(j = 1, \ldots, n)$; this implies that job J_j $(j = 1, \ldots, n)$ cannot be started before time r_j. Note that if no release date has been specified for some job J_j, then we can simply put $r_j = 0$, as the machine is not assumed to be available before this time. Using standard notation, we denote this problem by $1 \,|\, r_j \,|\, \sum U_j$.

Unfortunately, the problem of deciding whether there exists a solution of $1 \,|\, r_j \,|\, \sum U_j$ without tardy jobs is identical to the problem of deciding whether there exists a schedule for this instance with $L_{\max} \leq 0$. Since the latter problem is known to be $\mathcal{NP}$-hard in the strong sense (Lenstra, Rinnooy Kan, and Brucker (1977)), the general $1 \,|\, r_j \,|\, \sum U_j$ problem is $\mathcal{NP}$-hard in the strong sense as well.

Kise, Iberaki, and Mine (1978) consider the special case in which the release dates and due dates are similarly ordered, that is, the jobs can be renumbered such that

$$r_1 \leq r_2 \leq \cdots \leq r_n \quad \text{and} \quad d_1 \leq d_2 \leq \cdots \leq d_n$$

Since this order is essential, we assume that the jobs have been renumbered according to this order. Kise et al. show that this problem can be solved in $O(n^2)$ time. This was improved by Lawler (-), who presented an $O(n \log n)$ algorithm. Here we discuss Lawler's algorithm, which is based on the dynamic programming formulation of Section 12.2.

Since the release dates and due dates are ordered similarly, Observation 12.1 is still valid. Therefore, given a subset of jobs, we can easily check if this subset is feasible, that is, if all jobs in it can be on time, by putting the jobs in EDD-order. Analogous to our definition of the state-variable $f_j(k)$ in Section 12.2, we define $f_j(k)$ here as the earliest completion time of the last job in a feasible subset of $\{J_1, \ldots, J_j\}$ in which exactly k jobs are on time; because of the release dates, $f_j(k)$ is not necessarily equal to the total processing time of the jobs in this subset. Since the dominance rule of Section 12.2 remains valid, we can solve this special case of $1|r_j|\sum U_j$ through dynamic programming. The initialization is standard: $f_j(k) = 0$ if $j = k = 0$ and $+\infty$, otherwise. Denoting the maximum cardinality of the on time set of $\{J_1, \ldots, J_j\}$ by mc_j again, we find that

$$\begin{aligned} mc_{j+1} &= mc_j + 1 \quad \text{if } \max\{f_j(mc_j), r_{j+1}\} + p_{j+1} \leq d_{j+1} \\ mc_{j+1} &= mc_j \quad \text{otherwise} \end{aligned}$$

where we use $mc_0 = 0$. The process of updating the state-variables is essentially the same as in Section 12.2. We simply put $f_{j+1}(0) = 0$ and determine $f_{j+1}(k)$ $(k = 1, \ldots, mc_{j+1})$ through the recurrence relation

$$f_{j+1}(k) = \min\{f_j(k), \max\{f_j(k-1), r_{j+1}\} + p_{j+1}\}$$

The correctness proof is straightforward. This algorithm can be implemented to run in $O(n^2)$. To speed it up to run in $O(n \log n)$ time, we have to establish the relation between $f_j(k)$ and $f_j(k-1)$. Let $S_j(k)$ be the set that corresponds to $f_j(k)$. Because of the release dates, we cannot simply remove the longest job from $S_j(k)$ anymore to determine $S_j(k-1)$, but we need a more involved approach. We start with the following well-known lemma.

Lemma 12.5

For any subset S of $\{J_1, \ldots, J_n\}$ we find the minimum times at which the first j ($j = 1, \ldots, |S|$) jobs have been completed through the preemptive shortest remaining processing time (PSRPT) rule, which is defined as follows: at any time, execute the available job with shortest remaining processing time, where ties are broken by choosing the job with smallest index.

Proof

This follows by applying an easy interchange argument. □

We will need the PSRPT-rule in the following situation. Suppose that we are given two sets of jobs $S \cup \{J_0\}$ and $S \cup \{J_1\}$, where in both schedules determined by the PSRPT-rule for these two job sets all jobs from S are completed before J_0 and J_1, respectively. If we want to choose which job of J_0 and J_1 to add to S such that the resulting set of jobs is completed as soon as possible, then we can decide this by computing the PSRPT-schedule for the set $S \cup \{J_0, J_1\}$ and choosing the job that is not completed last. The correctness of this statement follows from the observation that the jobs in S are completed before the other two jobs and from the fact that the PSRPT-rule completes the first j ($j = 1, \ldots, |S| + 2$) jobs as soon as possible, and hence also the first $|S| + 1$ jobs.

Lemma 12.6

Let $S_j(k)$ denote the subset of $\{J_1, \ldots, J_j\}$ that corresponds to the state-variable $f_j(k)$ ($j = 1, \cdots, n$; $k = 0, \ldots, mc_j$). We have that

$$\emptyset = S_j(0) \subset S_j(1) \subset \cdots \subset S_j(mc_j) \quad \textit{for } j = 1, \ldots, n$$

Proof

We prove the lemma by induction. For $j = 1$ it is clearly true. Suppose that it holds for $j = l - 1$; we will now show that it also holds for $j = l$. Consider the dynamic programming algorithm when computing state-variables $f_l(k)$ ($k = 0, \ldots, mc_{l-1}$) in relation to the corresponding sets $S_l(k)$: We have to choose between the sets $S_{l-1}(k)$ and $S_{l-1}(k-1) \cup J_l$. Suppose that k_0 is the smallest value of k for which we decide to include J_l. If we choose to include J_l in all of the sets $S_l(k)$, for $k = k_0 + 1, k_0 + 2, \ldots, mc_l$, then we are done, since in that case we have

$$S_l(k-1) = S_{l-1}(k-1) \subset S_{l-1}(k) = S_l(k), \quad \forall k = 0, \ldots, k_0 - 1$$

$$S_l(k_0 - 1) = S_{l-1}(k_0 - 1) \subset S_{l-1}(k_0 - 1) \cup \{J_l\} = S_l(k_0)$$

$$S_l(k-1) = S_{l-1}(k-2) \cup \{J_l\} \subset S_{l-1}(k-1) \cup \{J_l\} = S_l(k), \quad \forall k = k_0 + 1, \ldots, mc_{l-1}$$

and finally, if $mc_l = mc_{l-1} + 1$, then we find that $S_l(mc_l) = S_l(mc_l - 1) \cup \{J_l\}$. The crucial part is hence to show that J_l is put into all sets $S_l(k)$, for $k = k_0 + 1, k_0 + 2, \ldots, mc_l$. Here, we need Lemma 12.5. As remarked above, the choice between the sets $S_{l-1}(k)$ and $S_{l-1}(k-1) \cup J_l$ for any relevant value of k can be made by applying the PSRPT-rule to the set $S_{l-1}(k) \cup J_l$ and remove the job that is completed last (or, equivalently, keep the k jobs that are completed first). Since all jobs are present at time r_l, the job that will be completed last is the job with largest remaining processing time at time r_l (where ties are broken in favor of a smaller index). When determining $S_l(k_0)$, we see that J_l is not removed, but some other job, say J_h, which implies that the processing time of job J_h remaining at time r_l was more than p_l, because of the tie-breaker. If we subsequently determine $S_l(k_0+1)$ by applying the PSRPT-rule to the set $S_{l-1}(k_0+1) \cup J_l$, then we see that the processing time of J_h remaining at time r_l is the same as it was in the PSRPT-schedule for the set $S_{l-1}(k_0)$, and hence, J_l is preferred to J_h, which implies that J_l will be included (remark that J_h will be completed at the (k_0+1)th place and the job belonging to $S_{l-1}(k_0+1)$ but not to $S_{l-1}(k_0)$ will be removed). This argument can be repeated to show that J_l will get included in all sets $S_l(k)$, $k = k_0, k_0+1, \ldots, mc_l$. □

The proof in combination with the PSRPT-rule leads to the following algorithm, which can easily be implemented in $O(n \log n)$ time. We first give the algorithm in pseudo-code, and we discuss it afterward.

Lawler's Algorithm

Step 1. $j \leftarrow 0$; $S \leftarrow \emptyset$.
Step 2. $j \leftarrow j+1$. If $j = n+1$, then go to Step 6.
Step 3. Compute the PSRPT-schedule for the jobs J_i with $i \in S$ for the interval $[0, r_j]$. Determine the processing time p_i' remaining at time r_j for each job J_i with $i \in S$.
Step 4. Check if $r_j + \sum_{i \in S} p_i' + p_j \leq d_j$. If yes, then $S \leftarrow S \cup \{j\}$, and go to Step 2.
Step 5. Determine $h \leftarrow \arg\max_{i \in S} p_i'$; in case of a tie, choose h as large as possible. If $p_h' > p_j$, then $S \leftarrow S \cup \{j\} \setminus \{h\}$. Go to Step 2.
Step 6. S contains the set of on time jobs.

The idea behind the algorithm is the following. We add the jobs in EDD-order. The set S denotes the current set of on time jobs. When adding J_j, we check whether we can add it to the on time set; this is done in Step 4. If not, then we find the job with the largest processing time remaining at time r_j (job J_h in the algorithm) and compare p_h' to p_j. If $p_h' > p_j$, then we replace h by j in the set S; otherwise, we keep S the same (Step 5). The process of determining the p_i' values for all $i \in S$ takes place in Step 3. Obviously, it is not necessary to compute the PSRPT-schedule for the jobs in S over and over again, since the part until r_{j-1} is still the same as before (if job J_h is processed somewhere in this schedule and h gets removed from S, then we simply replace it with idle time, since clearly none of the other jobs in S was then available; this does not alter the values of the processing time remaining at time r_{j-1} of the other jobs). If we compute the remaining processing times p_i' ($i \in S$) in an incremental way, then we can implement the algorithm to run in $O(n)$ time after the jobs have been sorted.

We conclude this section by applying the above algorithm to the following 7-job instance, which is due to Lawler (-). The data are shown in Table 12.3. We use here the incremental way to determine the p_i' values for all jobs J_i with $i \in S$.

In iteration 1, we put $S \leftarrow \{1\}$; $p_1' = 3$.

In iteration 2, we see that J_1 is executed in the interval $[0, 2]$; hence, p_1' is reduced to 1. We cannot add J_2 to S, and hence we compare p_2 to p_1'. Since p_1' is smaller, we keep $S = \{1\}$.

In iteration 3, we compute the PSRPT-schedule for the interval $[2, 4]$. We see that J_1 is completed at time 3, and there are no more jobs in S. Hence, we find $p_1' \leftarrow 0$. When we add J_3 to S, then it is on time; therefore, we put $S \leftarrow \{1, 3\}$.

In iteration 4, we compute the PSRPT-schedule for the interval $[4, 5]$; we put $p_3' \leftarrow 3$. There is room to add J_4 to the on time set, and therefore we put $S \leftarrow \{1, 3, 4\}$.

In iteration 5, we compute the PSRPT-schedule for the interval $[5, 6]$; we put $p_4' \leftarrow 0$. There is room to add J_5 to the on time set; hence, we put $S \leftarrow \{1, 3, 4, 5\}$.

TABLE 12.3 Release Dates, Due Dates, and Processing Times

	J_1	J_2	J_3	J_4	J_5	J_6	J_7
r_j	0	2	4	5	6	7	8
d_j	4	4	9	9	12	12	13
p_j	3	2	4	1	3	2	1

In iteration 6, we compute the PSRPT-schedule for the interval $[6, 7]$; we put $p'_3 \leftarrow 2$ (the tie-breaker in the PSRPT-schedule chooses the job with smallest index). Adding J_6 to the on time set is not feasible. We see that $p'_5 = 3$ is the largest remaining processing time in S, which is more than p_6. Hence, we put $S \leftarrow \{1, 3, 4, 6\}$.

In iteration 7, we compute the PSRPT-schedule for the interval $[7, 8]$; we put $p'_3 \leftarrow 1$. There is room to add J_7 to the on time set; hence, we put $S \leftarrow \{1, 3, 4, 6, 7\}$, which is the solution to this instance.

12.5 Scheduling with Deadlines

In this section we discuss the effect of adding *deadlines* to the problem $1 \,||\, \sum U_j$. The deadline of job J_j is denoted by $\bar{d}_j$ $(j = 1, \ldots, n)$, and it indicates that job J_j must be completed by time $\bar{d}_j$. Obviously, this only makes a difference if we are forced to execute the tardy jobs as well; otherwise, we can just adjust the due date of J_j such that it becomes equal to $\min\{d_j, \bar{d}_j\}$. Using standard notation, we denote this problem by $1 \,|\, \bar{d}_j \,|\, \sum U_j$.

Lawler (-) shows that this problem is $\mathcal{NP}$-hard in the ordinary sense by a reduction from Partition; we present this proof below. Since no pseudo-polynomial algorithm is known, this $\mathcal{NP}$-hardness proof does not fully fix the computational complexity of the problem. Another open problem is the complexity of the special case in which the due dates and deadlines are related; an example of such a special case is the problem of minimizing the number of tardy jobs subject to the constraint that the lateness of a tardy job should not exceed a given threshold.

Consider the general instance of Partition defined in Section 12.3. We define the following special instance of $1 \,|\, \bar{d}_j \,|\, \sum U_j$. There are $4n$ jobs, the data of which are found in Table 12.4. Here $M = 3A$ and

$$P_{i-1} = \sum_{k=1}^{n} (2^{n+k-1} + 2^{k-1}) + \sum_{k=1}^{i-1} (2^{n+k-1} + 2^{k-1})$$
$$= 2^n(2^n - 1) + 2^n - 1 + 2^n(2^{i-1} - 1) + 2^{i-1} - 1$$

The decision variant that we will show $\mathcal{NP}$-hardness of is defined as: Does there exist a feasible schedule for the instance of the problem $1 \,|\, \bar{d}_j \,|\, \sum U_j$ defined above in which there are no more than $2n$ tardy jobs? We will denote this problem as Decision Variant of the Number of Late Jobs problem With Deadlines (DVNLJWD).

The reduction defined above looks rather complicated. The idea behind it is that, for each $i = 1, \ldots, n$, the jobs $\{i, 3n + i\}$ and $\{n + i, 2n + i\}$ will form two pairs, one pair of which goes in the on time set, whereas the other pair goes into the tardy set. We determine the set $S \subset \{1, \ldots, n\}$ by putting $i \in S$ if the pair $\{i, 3n + i\}$ goes into the on time set, for $i = 1, \ldots, n$. Because of the structure of the due dates and deadlines, a schedule leading to "yes" for DVNLJWD will then have the following form: first the on time job from the pair $\{J_i, J_{n+i}\}$ $(i = 1, \ldots, n)$ followed by the on time job from the pair $\{J_{2n+i}, J_{3n+i}\}$ $(i = 1, \ldots, n)$; these jobs are followed by the tardy jobs in order of earliest deadline. There are two critical

TABLE 12.4 Data for the Reduction

j	p_j	d_j	$\bar{d}_j$
i $(i = 1, \ldots, n)$	$2^{i-1}M + a_i$	$(2^i - 1)M + A$	$(P_{i-1} + 2^{i-1})M - A$
$n + i$ $(i = 1, \ldots, n)$	$2^{i-1}M$	$(2^i - 1)M + A$	$(P_{i-1} + 2^{i-1} + 2^{n+i-1})M - A$
$2n + i$ $(i = 1, \ldots, n - 1)$	$2^{n+i-1}M$	$(2^{n+i} - 1)M + A$	$(P_{i-1} + 2^{n+i-1})M - A$
$3n$	$2^{2n-1}M$	$(2^{2n} - 1)M - A$	$(P_{n-1} + 2^{2n-1})M - A$
$3n + i$ $(i = 1, \ldots, n - 1)$	$2^{n+i-1}M - 2a_i$	$(2^{n+i} - 1)M + A$	$(P_{i-1} + 2^{i-1} + 2^{n+i-1})M - A$
$4n$	$2^{2n-1}M - 2a_n$	$(2^{2n} - 1)M - A$	$(P_{n-1} + 2^{n-1} + 2^{2n-1})M - A$

points in this schedule. The first one is the completion time of the on time job from the pair $\{J_n, J_{2n}\}$, which is equal to $(2^n - 1)M + \sum_{i \in S} a_i$, whereas its due date is equal to $(2^n - 1)M + A$; hence, it is on time only if $\sum_{i \in S} a_i \leq A$. The second critical point is the completion time of the on time job from the pair $\{J_{3n}, J_{4n}\}$, which is equal to $(2^{2n} - 1)M - \sum_{i \in S} a_i$. Since the due date of jobs $3n$ and $4n$ is equal to $(2^{2n} - 1)M - A$, and one of these two jobs must be on time, we need $\sum_{i \in S} a_i \geq A$. Combining both inequalities, we find that $\sum_{i \in S} a_i = A$.

Using the description of a "yes" solution above, we can easily verify that if the subset S of $\{1, \ldots, n\}$ corresponds to a "yes" solution of Partition, then the answer to DVNLJWD is "yes" as well.

Conversely, suppose that the answer to DVNLJWD is "yes." We are done if we are able to show the special structure of this solution, which is that, for each $i = 1, \ldots, n$, the on time set contains, either the jobs $\{i, 3n + i\}$, or the jobs $\{n + i, 2n + i\}$. We start with the following observations.

Observation 12.2 *Any feasible schedule for* DVNLJWD *with no more than $2n$ tardy jobs possesses the following properties:*

1. *At most one of the jobs from $\{J_i, J_{n+i}\}$ is on time, for each $i = 1, \ldots, n$.*
2. *At most one of the jobs from $\{J_{2n+i}, J_{3n+i}\}$ is on time, for each $i = 1, \ldots, n$.*
3. *There are exactly $2n$ on time jobs.*
4. *At most one of the jobs from $\{J_{n+i}, J_{3n+i}\}$ is on time, for each $i = 1, \ldots, n$.*

Proof

The first property holds, since $p_i + p_{n+i} > 2^i M > (2^i - 1)M + A = d_i = d_{n+i}$ for $i = 1, \ldots, n$, where $M = 3A$. A similar proof shows the second property, and the third property follows immediately from the combination of the first two properties and the assumption that there are at least $2n$ on time jobs. Hence, we have exactly one job from $\{J_i, J_{n+i}\}$ and one job from $\{J_{2n+i}, J_{3n+i}\}$ in the on time set, for each $i = 1, \ldots, n$.

The proof of the fourth property follows from a contradiction. Suppose that for some i both J_{n+i} and J_{3n+i} belong to the on time set. A quick computation shows that in any feasible schedule with exactly $2n$ jobs on time we must first execute the on time jobs, and then the remaining jobs in order of earliest deadline. If we compute the completion time of job J_{2n+i} in such a schedule, then we find that it is more than $P_{i-1}M - 4A + 2^{i-1}M + 2^{n+i-1}M \geq (P_{i-1} + 2^{n+i-1})M + (M - 4A) \geq \bar{d}_{2n+i}$, since $M = 3A$. □

Property 4 implies that if J_{3n+i} $(i = 1, \ldots, n)$ belongs to the on time set, then J_{n+i} does not belong to the on time set, and hence Properties 1 and 3 imply that J_i belongs to the on time set. What is left to show is that if J_i $(i = 1, \ldots, n)$ belongs to the on time set, then J_{3n+i} belongs to the on time set as well. This we prove in the next lemma.

Lemma 12.7

Consider any feasible schedule for DVNLJWD *with $2n$ tardy jobs. If J_i $(i = 1, \ldots, n)$ belongs to the on time set, then J_{3n+i} belongs to the on time set, too.*

Proof

Determine the index set S_0 as the subset of $\{1, \ldots, n\}$ that contains the indices of the jobs J_i $(i = 1, \ldots, n)$ that belong to the on time set. Similarly, determine S_3 such that it contains the indices of the jobs J_{3n+i} $(i = 1, \ldots, n)$ in the on time set. We have shown above that $S_3 \subset S_0$.

Recall that in any "yes" schedule for DVNLJWD we have to execute first the on time job from the pair $\{J_i, J_{n+i}\}$ $(i = 1, \ldots, n)$, then the on time job from the pair $\{J_{2n+i}, J_{3n+i}\}$ $(i = 1, \ldots, n)$, and finally the tardy jobs in order of earliest deadline. As observed in the description of the idea behind the reduction, we need that $\sum_{i \in S_0} a_i \leq A$ to be sure that the job from the pair $\{J_n, J_{2n}\}$ that belongs to the on time set meets its due date. Similarly, we need that

$$\sum_{i \in S_0} a_i - 2 \sum_{i \in S_3} a_i \leq -A$$

Adding up the inequality $\sum_{i \in S_0} a_i \leq A$ yields the inequality $\sum_{i \in S_0} a_i \leq \sum_{i \in S_3} a_i$. Since $S_3 \subset S_0$ and all a_i values are nonnegative, this can occur only if $S_0 = S_3$. □

By now, we have shown that the structure claimed in the sketch of the idea behind the reduction is necessary for a feasible schedule with no more than $2n$ tardy jobs. Therefore, we have proven the following theorem.

Theorem 12.3
The problem $1 \mid \bar{d}_j \mid \sum U_j$ *is* $\mathcal{NP}$*-hard in the ordinary sense.*

12.6 Scheduling on *m* Parallel Machines

In this section we consider the problem of minimizing the weighted number of tardy jobs, where we have m parallel, identical machines available for executing the jobs. This implies that we have the additional freedom of deciding which machine is going to execute a job; since the machines are identical, the processing time is independent of the choice of machine. We denote machine i by M_i $(i = 1, \ldots, m)$; as before, each machine is continuously available from time zero onward and can handle no more than one job at a time. Using standard notation, we denote the problem by $P \mid\mid \sum w_j U_j$.

This problem can be solved through a generalization of the dynamic programming algorithm of Section 12.3, since the properties stated in Observation 12.1 remain true for each machine. We need state-variables $f_j(t_1, t_2, \ldots, t_m)$, which have value equal to the value of the optimum solution for the jobs $J_1, \ldots, J_j$ subject to the constraint that the total processing time of the on time jobs on machine M_i amounts to t_i, for $i = 1, \ldots, m$. This leads to an algorithm that is easily implemented to run in time $O(n(\sum p_j)^m)$ and space. Especially the space requirement becomes a problem when m increases.

We describe in this section how this problem can alternatively be solved through the method of *column generation*. It was first applied by Van den Akker, Hoogeveen, and Van de Velde (1999) and Chen and Powell (1999); we describe the method by Van den Akker et al., which is more efficient in the branch-and-bound phase.

The crucial step is to view the problem $P \mid\mid \sum w_j U_j$ as a *partitioning* problem. Looking at any feasible schedule, we see that each of the m machines executes a subset of the on time jobs in EDD-order, possibly followed by some tardy jobs. Hence, we need to partition the jobs into m subsets of on time jobs, and one subset containing the tardy jobs; we need only one subset of tardy jobs, since these can be divided over the machines in any way, if we need to execute these at all. Obviously, we require each of the m subsets of on time jobs to be *feasible*, that is, all of the jobs in a subset should be on time when executed in EDD-order.

We apply this partitioning idea by formulating the problem as a set covering problem with an exponential number of binary variables, n covering constraints, and a single side constraint. We let each feasible subset correspond to a *machine schedule*, which we define as a string of jobs that can be assigned together to any single machine, such that each job in it is on time when this string of jobs is scheduled in EDD-order. Let a_{js} be a constant that is equal to 1 if job J_j is included in machine schedule s and 0 otherwise. Accordingly, the column $(a_{1s}, \ldots, a_{ns})^T$ represents the jobs in machine schedule s. Instead of minimizing the weighted number of tardy jobs, we look at the equivalent problem of maximizing the weighted number of on time jobs. Accordingly, we define the value c_s of machine scheduling s as the sum of the weights of the jobs included in it, that is, $c_s = \sum_{j=1}^{n} a_{js} w_j$. Let S be the set containing all machine schedules. We introduce variables x_s $(s = 1, \ldots, |S|)$ that assume value 1 if machine schedule s is selected and 0 otherwise. The problem is then to select at most m machine schedules, such that each job occurs in at most one machine schedule and such that the total value is maximum. Mathematically, the problem is then to determine values x_s that maximize

$$\sum_{s \in S} c_s x_s$$

subject to

$$\sum_{s \in S} x_s \leq m \tag{12.2}$$

$$\sum_{s \in S} a_{js} x_s \leq 1, \quad \text{for each } j = 1, \ldots, n \tag{12.3}$$

$$x_s \in \{0, 1\}, \quad \text{for each } s \in S \tag{12.4}$$

Condition (12.2) and the integrality condition (12.4) ensures that at most m machine schedules are selected. Condition (12.3) ensures that each job is executed at most once.

To obtain an upper bound that we can use in a branch-and-bound procedure, we solve the linear programming relaxation of this integer linear programming formulation. We obtain the LP-relaxation by replacing the constraints $x_s \in \{0, 1\}$ with $x_s \geq 0$, as conditions (12.3) prohibit x_s from exceeding the value of 1. Because of the large number of eligible machine schedules, which in general is too big to enumerate them all, we use the technique of column generation to solve the LP-relaxation, which method considers the feasible columns implicitly. Starting with a restricted linear programming problem in which only a subset $\bar{S}$ of the variables is available, the column generation method solves the LP-relaxation by adding new columns that may improve the solution value, if the optimal solution has not been determined yet; these new columns are obtained by solving the so-called *pricing problem.*

When we have solved the LP-relaxation for a restricted set of variables, we find the values of the dual variables λ_0, which corresponds to conditions (12.2), and λ_j ($j = 1, \ldots, n$), where λ_j corresponds to the jth one of the constraints 12.3. It is well known from the theory of linear programming (see, e.g., Bazaraa, Jarvis, and Sherali (1990)) that the *reduced cost* of any variable x_s is then equal to

$$c'_s = c_s - \lambda_0 - \sum_{j=1}^{n} \lambda_j a_{js} = \sum_{j=1}^{n} w_j a_{js} - \sum_{j=1}^{n} \lambda_j a_{js} - \lambda_0 = \sum_{j=1}^{n} (w_j - \lambda_j) a_{js} - \lambda_0$$

and the current solution, in which only a subset of the variables have been taken into account, is guaranteedly optimal only if $c'_s \leq 0$ for each $s \in S$. The pricing problem now boils down to finding the variable x_s with corresponding column $(a_{1s}, \ldots, a_{ns})^T$ for which c'_s is maximum. If this maximum has value smaller than or equal to zero, then we have solved the LP-relaxation to optimality; if it is positive, then we add this variable to the LP-relaxation. Since the term $-\lambda_0$ is an unavoidable constant, we ignore it, and hence we want to find the column $(a_{1s}, \ldots, a_{ns})^T$ that maximizes

$$\sum_{j=1}^{n} (w_j - \lambda_j) a_{js}$$

subject to the constraint that it corresponds to a feasible machine schedule. But this results in a familiar problem: We have to maximize the weighted number of on time jobs, where the weight of job J_j ($j = 1, \ldots, n$) is equal to $(w_j - \lambda_j)$. We solve this problem through the algorithm of Lawler and Moore (1969) described in Section 12.3; the alternative algorithm described in Section 12.3 is not well-suited for this purpose here, since the weights are not integral anymore.

After we have solved the LP-relaxation, we have found a strong upper bound that can be applied in a branch-and-bound procedure if necessary. We have to be careful when choosing a branching strategy in case of column generation: it is well-known that the branching strategy of fixing a variable x_s at either zero or one does not work in combination with column generation, as the pricing algorithm may come up with this column again, even though we fixed the variable at zero. Our initial branching strategy is based upon splitting the set of possible completion times.

Suppose that we have found an optimal solution to the LP-relaxation; let S^* denote the set of columns that have been (partially) selected, that is, for each $s \in S^*$ we have $x_s > 0$. For each machine schedule $s \in S^*$ we compute $C_j(s)$ as the completion time of job J_j $(j = 1, \ldots, n)$ if $a_{js} = 1$. We distinguish between the case that for each job J_j $(j = 1, \ldots, n)$ the completion time $C_j(s)$ is the same for each machine schedule $s \in S^*$ it occurs in, and the opposite case. We first address the latter case. Suppose that job J_j is a so-called *fractional* job, that is, a job for which $C_j(s)$ is not the same for each $s \in S^*$. We further define

$$\bar{C}_j \leftarrow \min\{C_j(s) \,|\, s \in S^*\}$$

We design a binary branch-and-bound tree for which in each node we first identify the fractional job with smallest index, and, if any, then create two descendant nodes: One for the condition that $C_j \leq \bar{C}_j$ and one for the condition that $C_j \geq \bar{C}_j + 1$. The first condition essentially specifies a deadline for J_j by which it must be completed. The second condition specifies a release date $\bar{C}_j + 1 - p_j$ before which J_j cannot be started. Given such a deadline or release date, we first separate the current set of columns accordingly. When we need to solve the pricing problem in a node for which deadlines or release dates have been specified, we can incorporate these bounds in the pricing algorithm. For example, when we are given the additional constraint that $C_j \leq \bar{C}_j$, then we have to find the machine schedule that maximizes the total weight (based on weights $w'_j = (w_j - \lambda_j)$) of the set of on time jobs, where we can only select feasible machine schedules that possess the properties of Observation 12.1 in which job J_j can be on time only if it is completed by time $\bar{C}_j$. As this problem is equivalent to minimizing the total weight of the tardy jobs, we can solve it through Lawler and Moore's algorithm by adjusting the recurrence relation to

$$f_j(t) = \begin{cases} \infty & \text{if } t > d_j \\ \min\{f_{j-1}(t) + w'_j, f_{j-1}(t - p_j)\} & \text{if } t \leq \bar{C}_j \\ f_{j-1}(t) + w'_j & \text{if } t > \bar{C}_j \end{cases}$$

Hence, this branching strategy does not complicate the pricing algorithm, which is a major advantage.

After having solved this problem through branch-and-bound, we find the optimal solution of the LP-relaxation in which the completion time of job J_j $(j = 1, \ldots, n)$ is the same in each column $s \in S^*$ that contains job J_j; since this is an essential property of any integral solution, the value of the LP-relaxation is still an upper bound, and if this solution is integral, then we have found an optimum solution. If it is not integral, but we have that

$$\sum_{s \in S^*} a_{js} x_s \in \{0, 1\}, \qquad \text{for each } j = 1, \ldots, n$$

then we can find an optimal solution in the following way. We put all jobs J_j with $\sum_{s \in S^*} a_{js} x_s = 1$ in the on time set, which we denote by E and all the remaining jobs in the tardy set T. We construct a schedule σ for the on time jobs by executing each of the jobs $J_j \in E$ in the interval $[C_j - p_j, C_j]$, where C_j is the common completion time in the columns in S^* containing J_j. This leads to the following theorem.

Theorem 12.4

Suppose that the fractional solution satisfies the following properties

1. *For each job j $(j = 1, \ldots, n)$, we either have $\sum_{s \in S^*} a_{js} x_s = 1$, in which case we put $j \in E$, or $\sum_{s \in S^*} a_{js} x_s = 0$, in which case we put $j \in T$.*
2. *For each $j \in E$, we have that the completion time of job J_j is equal to C_j in each machine schedule $s \in S^*$ containing J_j.*

Then the schedule σ obtained by processing J_j in the time interval $[C_j - p_j, C_j]$ for all $j \in E$ is feasible and has value equal to the value of the fraction solution.

Proof

The schedule σ in which job J_j $(j \in E)$ is processed from time $C_j - p_j$ to time C_j is feasible if and only if at most m jobs are processed at the same time and no job starts before time zero. The second condition is obviously satisfied, since the C_j values originate from feasible machine schedules. The first constraint is satisfied if we show that at most m jobs are started at time zero and that the number of jobs started at any point in time $t \in [1, T]$ is no more than the number of jobs completed at that point in time, where T denotes the latest point in time that a job is started. We first prove that at most m jobs are started at time 0. We define the set Z such that it contains the indices of all jobs in σ that are started at time zero; we need to prove that $|Z| \leq m$. Since each job that starts at time zero in any column $s \in S^*$ is included in Z, we know that $\sum_{j \in Z} a_{js} = 1$ for each $s \in S^*$. Hence, we find that

$$|Z| = \sum_{j \in Z} 1 = \sum_{j \in Z} \left(\sum_{s \in S^*} a_{js} x_s \right) = \sum_{s \in S^*} \left(\sum_{j \in Z} a_{js} \right) x_s = \sum_{s \in S^*} x_s \leq m$$

because of condition (12.2). Hence, we know that at most m jobs are started at time zero. As to the final part of the feasibility proof, let $A(t) \subseteq S^*$ be the set of all machine schedules in which at least one job starts at time t; similarly, let $B(t) \subseteq S^*$ be the set of all machine schedules in which at least one job completes at time t. As $C_j(s) = C_j$, for any machine schedule containing J_j, the number of jobs started at time t is equal to $\sum_{s \in A(t)} x_s^*$; similarly, the number of jobs completed at time t is equal to $\sum_{s \in B(t)} x_s^*$. Since each machine schedule s is constructed such that there is no idle time between the jobs, a job in s can start at time t only if some other job in s is completed at time t. Hence, $A(t) \subset B(t)$, which means that the indicated schedule is feasible.

The value of the fractional solution is equal to

$$\sum_{s \in S^*} c_s x_s = \sum_{s \in S^*} \left(\sum_{j=1}^{n} a_{js} w_j \right) x_s = \sum_{j=1}^{n} w_j \sum_{s \in S^*} a_{js} x_s = \sum_{j \in E} w_j$$

where the last inequality follows from the definition of E and T. But this is equal to the weighted number of on time jobs in σ. □

If for some j we have that $0 < \sum_{s \in S^*} a_{js} x_s < 1$, then we cannot apply the theorem, and we have to branch further. In this case, we create two nodes, where we assume in node 1 that $\sum_{s \in S^*} a_{js} x_s = 0$ and in node 2 we assume that $\sum_{s \in S^*} a_{js} x_s = 1$; this is equivalent to putting J_j tardy (node 1) or on time (node 2). Note that putting J_j into the tardy set simply means that we ignore J_j in this node; putting J_j into the on time set does not burden the pricing problem, and hence the whole procedure goes through. There is one major difference, though, in comparison with the situation before: due to our second branching strategy, we require some jobs to be present in the on time schedule, in which they have to meet a given deadline and possibly a release date (because of the first part of the branching strategy). Therefore, we may run the risk that in our current node no feasible schedule exists for the jobs that have been put in the on time set by the second part of the branching procedure in which they meet the release dates and the deadlines that have been issued by the first part of the branch-and-bound procedure. The problem of finding out if there exists a feasible solution is of course $\mathcal{NP}$-complete. We therefore check a necessary condition for the existence of a feasible solution, where we only need to consider the jobs that must be present in the on-time schedule — if it fails this check, then we may prune the corresponding node and backtrack. This necessary condition proceeds by transforming the feasibility problem of determining whether there exists a feasible schedule that meets all release dates and deadlines into an optimization problem. We replace each deadline $\bar{d}_j$ by a due date d_j of equal value, and consider the problem of minimizing the maximum lateness $L_{\max}$, which is defined as $\max_{1 \leq j \leq n}(C_j - d_j)$. Because $d_j = \bar{d}_j$, the constraint $C_j \leq \bar{d}_j$ is equivalent to the constraint $C_j - d_j \leq 0$. Hence, the answer to the feasibility problem is "yes" if and only if there exists a feasible schedule with $L_{\max} \leq 0$. The optimization problem is $\mathcal{NP}$-hard, but several good lower bounds that are quickly computed are known (see for instance Vandevelde, Hoogeveen, Hurkens, and Lenstra (2003)).

A second problem that can occur when we include the constraint $\sum_{s \in S^*} a_{js} x_s = 1$ is that we must be more careful about having a subset $\bar{S}$ of the columns that constitutes a feasible solution (we did not have this problem before, since we could just put $x_s = 0$ to find a feasible solution). One way is to work around this problem in the following way. We first remove the infeasible columns that are not part of the current solution to the linear programming relaxation. As to the infeasible columns that are currently part of the linear programming solution, we decrease their costs with some big value, say M, such that the current solution value decreases to a value smaller than the incumbent lower bound (which can be the value of any feasible solution, even zero). Using this trick, we can proceed with the column generation algorithm to solve the current instance of the problem, because we have ensured that there exists at least one feasible solution.

References

[1] J.M. van den Akker, J.A. Hoogeveen, and S.L. van de Velde (1999). Parallel machine schedule by column generation. *Operations Research 47*, 862–872.

[2] M.S. Bazaraa, J.J. Jarvis, and H.D. Sherali (1990). *Linear Programming and Network Flows*. Wiley, New York.

[3] Z.-L. Chen and W.B. Powell (1999). Solving parallel machine scheduling problems by column generation. *INFORMS Journal on Computing 11*, 78–94.

[4] R.L. Graham, E.L. Lawler, J.K. Lenstra, and A.H.G. Rinnooy Kan (1979). Optimization and approximation in deterministic sequencing and scheduling: a survey. *Annals of Discrete Mathematics* 5, 287–326.

[5] J.R. Jackson (1955). Scheduling a production line to minimize maximum tardiness. Research Report 43, Management Science Research Project, University of California, Los Angeles.

[6] H. Kise, T. Iberaki, and H. Mine (1978). A solvable case of the one-machine scheduling problem with ready and due times. *Operations Research 26*, 121–126.

[7] R.M. Karp (1972). Reducibility among combinatorial problems, in: R.E. Miller and J.W. Thatcher (eds.), (1972). *Complexity of Computer Computations*, Plenum Press, New York, pp. 85–103.

[8] E.L. Lawler (-). Scheduling a single machine to minimize the number of late jobs. Unpublished manuscript.

[9] E.L. Lawler and J.M. Moore (1969). A functional equation and its application to resource allocation and sequencing problems. *Management Science 16*, 77–84.

[10] J.K. Lenstra, A.H.G. Rinnooy Kan, and P. Brucker (1977). Complexity of machine scheduling problems. *Annals of Discrete Mathematics 1*, 343–362.

[11] J.M. Moore (1968). An *n* job, one machine sequencing algorithm for minimizing the number of late jobs. *Management Science 15*, 102–109.

[12] A.M.G. Vandevelde, J.A. Hoogeveen, C.A.J. Hurkens, and J.K. Lenstra (2003). Lower bounds for the head-body-tail problem on parallel machines: a computational study of the multiprocessor flow shop. To appear in *Informs Journal on Computing*.

13

Branch-and-Bound Algorithms for Total Weighted Tardiness

Antoine Jouglet
Compiègne University of Technology

Philippe Baptiste
CNRS

Jacques Carlier
Compiègne University of Technology

13.1 Introduction

In this chapter we consider the situation where a set $N = \{J_1, \ldots, J_n\}$ of n jobs has to be processed by a single machine and where the objective is to minimize the total weighted tardiness. Associated with each job J_i, are a release date r_i, a processing time p_i, a due date d_i, and a weight w_i. All data are nonnegative integers. A job cannot start before its release date, preemption is not allowed, and only one job at a time can be scheduled on the machine. The tardiness of J_i is defined as $T_i = \max(0, C_i - d_i)$, where C_i denotes the completion time of job J_i. The problem is to find a feasible schedule with minimum total weighted tardiness $\sum w_i T_i$. Note that the problems of minimizing total tardiness ($\sum T_i$), total weighted completion time ($\sum w_i C_i$), and total completion time ($\sum C_i$) are special cases of this problem. Hence, all our results apply for these problems. All these problems, denoted as $1 \mid r_i \mid \sum w_i T_i$, $1 \mid r_i \mid \sum T_i$, $1 \mid r_i \mid \sum w_i C_i$, and $1 \mid r_i \mid \sum C_i$ are known to be NP-hard in the strong sense [1,2].

A lot of research has been carried on the unweighted total tardiness problem with equal release dates $1 \mid\mid \sum T_i$. Powerful dominance rules have been introduced by Emmons [3]. Lawler [1] has proposed a dynamic programming algorithm that solves the problem in pseudo-polynomial time. Finally, Du and

1-58488-397-9/$0.00+$1.50

Leung have shown that the problem is NP-hard [4]. Most of the exact methods for solving $1 \mid\mid \sum T_i$ strongly rely on Emmons' dominance rules. Potts and Van Wassenhove [5], Chang et al. [6], and Szwarc et al. [7], have developed branch-and-bound methods using the Emmons rules coupled with the decomposition rule of Lawler [1] together with some other elimination rules. The best results have been obtained by Szwarc, Della Croce, and Grosso [7,8] with a branch-and-bound method that efficiently handles instances with up to 500 jobs. The weighted problem $1 \mid\mid \sum w_i T_i$ is strongly NP-hard [1]. For this problem, Rinnooy Kan et al. [9] and Rachamadugu [10] extended the Emmons rules [3]. Rachamadugu [10] identifies a condition characterizing adjacent jobs in an optimal sequence. Most of the exact approaches have been tested by Abdul-Razacq, Potts, and Van Wassenhove [11] which have provided a survey of dynamic programming and branch-and-bound algorithms. Recently, Akturk and Yildrim [12] have proposed a new dominance rule and a lower bounding scheme for this problem.

There are less results on the $1 \mid r_i \mid \sum T_i$ problem. Chu and Portmann [13] have introduced a sufficient condition for local optimality which allows them to build a dominant subset of schedules. Chu [14] has also proposed a branch-and-bound method using some efficient dominance rules. This method handles instances with up to 30 jobs for the hardest instances and with up to 230 jobs for the easiest ones. More recently, Baptiste, Carlier, and Jouglet [15] have described a new lower bound and some dominance rules which are used in a branch-and-bound procedure which handles instances with up to 50 jobs for the hardest instances and 500 jobs for the easiest ones. For the $1 \mid r_i \mid \sum w_i T_i$ problem, Akturk and Ozdemir [16] have proposed a sufficient condition for local optimality which is used to improve heuristic algorithms. This rule is then used with a generalization of the Chu's dominance rules to the weighted case in a branch-and-bound algorithm [17]. This branch-and-bound method handles instances with up to 20 jobs.

For the total completion time problem, in the case of identical release dates, both the unweighted and the weighted problems can easily be solved polynomially in $O(n \log n)$ time by applying the weighted shortest processing time WSPT priority rule, which is also called the Smith's rule [18] : it consists of sequencing the jobs in nondecreasing order of their ratio of processing time to weight. For the unweighted problem with release dates, several researchers have proven some dominance properties and proposed a number of algorithms [19–21]. Chu [22,23] has proved several dominance properties and has provided a branch-and-bound algorithm. Chand, Traub, and Uzsoy used a decomposition approach to improve branch-and-bound algorithms [24]. Among the exact methods, the most efficient algorithms [22,24] can handle instances with up to 100 jobs. The weighted case with release dates $1 \mid r_i \mid \sum w_i C_i$ is NP-hard in the strong sense [2], even when preemption is allowed [25]. Several dominance rules and branch-and-bound algorithms have been proposed [26–29]. To our knowledge, the best results are obtained by Belouadah, Posner, and Potts with a branch-and-bound algorithm which has been tested on instances involving up to 50 jobs.

The aim of this chapter is to propose an efficient branch-and-bound method to solve the problem of minimizing total weighted tardiness with arbitrary release dates. We describe in Section 13.2 the general framework of our branch-and-bound procedure. We then describe some techniques which allow us to improve the behavior of the branch-and-bound method. In Section 13.3, we describe the notion of "better" sequences that allows us to compare two partial sequences of the same set of jobs. On the basis of this notion we describe a no-good recording technique in Section 13.4. We then provide in Section 13.6, some local dominance properties that allow us to define dominant subsets of schedules. We will see that each node of the search tree is characterized by a partial sequence of jobs scheduled at the beginning of the schedule and a set of unscheduled jobs. In Section 13.7, we provide a dominance rule relying on the partial sequence of scheduled jobs. We then provide in Section 13.8, two dominance rules relying on the set of unscheduled jobs. We show in Section 13.9 that these dominance rules dominate, generalize, and include a lot of dominance rules of the literature [14,17,22,27]. Finally, we describe in Section 13.10, a "Look — Ahead" propagation rule relying on lower bounds of the problem. The no-good recording technique, the dominance rules, and the look-ahead propagation rule have been integrated into the branch-and-bound procedure and have been experimentally tested in Section 13.11. This shows the efficiency of our techniques to drastically reduce both the search space and the computational time. Moreover, our results outperform the best known approaches not only for total weighted tardiness but also for total (weighted) completion time and total tardiness: for the $1 \mid r_j \mid \sum w_i T_i$, $1 \mid r_j \mid \sum T_i$, $1 \mid r_j \mid \sum w_i C_i$ and $1 \mid r_j \mid \sum C_i$ problems, the branch-and-bound procedures can handle, respectively, up to 35, 50, 90, and 120 jobs for the hardest instances instead of, respectively, 20, 30, 50, and 100 jobs.

13.2 Branch-and-Bound Procedure

The aim of this section is to propose an efficient branch-and-bound method with constraint propagation to solve the one machine total weighted tardiness problem with release dates.

In constraint programming terms, a variable C_i representing the completion time of J_i is associated with each job. The criterion is an additional variable $\overline{WT}$ that is constrained to be equal to $\sum_i w_i \max(0, C_i - d_i)$. Arc-B-Consistency (see, for instance, [30]) is used to propagate this constraint. It ensures that when a schedule has been found, the value of $\overline{WT}$ is actually the total weighted tardiness of the schedule.

To find an optimal solution, we *solve successive variants of the decision problem*. At each iteration, we try to improve the best known solution and thus we add an additional constraint stating that $\overline{WT}$ is lower than or equal to the best solution minus one. Each time, the search resumes at the last visited node of the previous iteration. It now remains to show how to solve the decision variant of the problem. We rely on the *Edge-Finding* branching scheme (see, for instance, Carlier [31]). It consists of ordering jobs on the machine ("edges" in a graph representing the possible orderings of jobs): at each node, a set of jobs is selected and, for each job J_i belonging to this set, a new branch is created where job J_i is constrained to be first (or last) among the jobs in this set. Thus, rather than searching for the starting times of jobs, we look for a *sequence* of jobs. This sequence is built both from the beginning and from the end of the schedule by using a *depth first strategy*. Throughout the search tree, we dynamically maintain several sets of jobs that represent the current state of the schedule (see Figure 13.1).

1. P is the sequence of the jobs scheduled at the beginning.
2. Q is the sequence of the jobs scheduled at the end.
3. NS is the set of unscheduled jobs that have to be sequenced between P and Q.
4. $PF \subseteq NS$ (Possible First) is the set of jobs which can be scheduled immediately after P.
5. $PL \subseteq NS$ (Possible Last) is the set of jobs which can be scheduled immediately before Q.

At each node of the search tree, a job J_i is chosen among those in PF and it is scheduled immediately after P. Upon backtracking, this job is removed from PF. Of course, if NS is empty, then a solution has been found and we can iterate to the next decision problem. If $NS \neq \emptyset$ while PF or PL is empty, then a backtrack occurs. Note that set PL is only used in the following two ways. At first, if filtering set PL leads to a set which is empty, then a backtrack occurs. Moreover, if filtering set PL leads to a set which contains only one job, then this job is immediately scheduled at the end of the schedule, i.e., at the beginning of sequence Q.

From now on, in aim to simplify the presentation we suppose that the release dates of the jobs are adjusted at each node of the search tree according to the completion time of the last job in the partial sequence P, i.e., for all jobs $J_i \in NS$, we have $r_i = \max(r_i, C_{\max}(P))$, in which $C_{\max}(P)$ is the completion time of sequence P.

Due to our branching scheme, jobs are sequenced from left to right, so it may happen that at some node of the search tree, all jobs of NS have the same release date (the completion time of the last job in P). Moreover, if the greatest due date is greater or equal to this release date, then all unscheduled jobs are

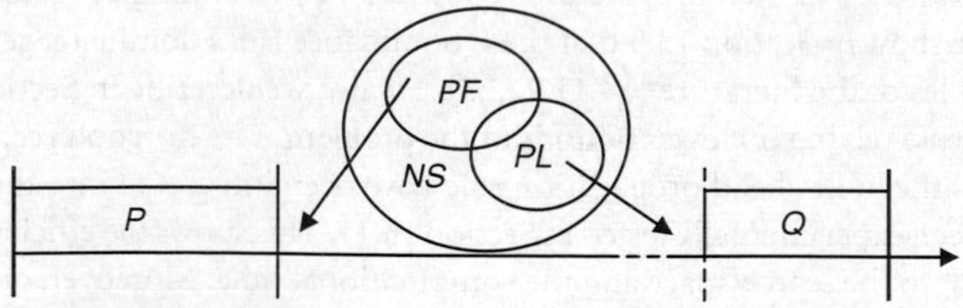

FIGURE 13.1 A node of the search tree.

late. As a result finding a schedule of the remaining unscheduled jobs to minimize total weighted tardiness is equivalent to finding a schedule of the remaining unscheduled jobs to minimizing total weighted completion time where all release dates are equal. In such a case, to improve the behavior of the branch-and-bound method, we can apply the WSPT rule to optimally complete the schedule [17]. Moreover, if all weights are equal and if all release dates are reached, then finding a schedule of the remaining unscheduled jobs to minimize total weighted tardiness is equivalent to the $1 \,||\, \sum T_i$ problem, and we apply the dynamic programming algorithm of Lawler [1]. Finally, for the case in which all due dates are equal to zero ($1 \,|\, r_i \,|\, \sum w_i C_i$ and $1 \,|\, r_i \,|\, \sum w_i C_i$), we use the forward sequencing branching rule with release dates adjustments of Belouadah, Posner, and Potts [26].

At each node of the search tree, we compute a lower bound of the cost of scheduling remaining unscheduled jobs. If this lower bound is greater than the upper bound $\overline{WT}$ of the problem decision, then a backtrack occurs. For that, we use several lower bounds in the literature according to the criterion we want to minimize. For the $1 \,|\, r_i \,|\, \sum w_i T_i$ and $1 \,|\, r_i \,|\, \sum w_i C_i$ problems, we use the lower bound of Belouadah, Posner and Potts [26] relying on job splitting. For the $1 \,|\, r_i \,|\, \sum C_i$ problem we use the SRPT rule. Finally, for the $1 \,|\, r_i \,|\, \sum T_i$ problem, we use the lower bound of Baptiste, Carlier, and Jouglet [15] using the propagation of Generalized Emmons Rules.

13.3 "Better" Sequences

In this section we introduce the notion of "better" sequences which allows us to compare two partial sequences σ_1 and σ_2 of the same set of jobs (i.e., σ_1 is a permutation of σ_2). Informally speaking, we say that a sequence σ_1 is "better" than a sequence σ_2, if σ_2 can be replaced advantageously by σ_1 in any feasible schedule which starts with the sequence σ_2, i.e., σ_2 is dominated by σ_1. Let $\bar{\sigma}_1$ be the set of jobs which do not belong to σ_1. Note that we have $\bar{\sigma}_1 = \{J_l \in N \,|\, J_l \notin \sigma_1\} = \{J_l \in N \,|\, J_l \notin \sigma_2\}$. Let $WT(\sigma)$ and $C_{\max}(\sigma)$ be, respectively, the total weighted tardiness and the completion time of the schedule associated with the sequence σ in which jobs belonging to sequence σ are scheduled as soon as possible (i.e., the active schedule associated with sequence σ). Now, let us examine under which conditions sequence σ_1 is "as good as" sequence σ_2, i.e., under which conditions it is possible to replace sequence σ_2 by sequence σ_1 in any feasible schedule.

1. If $C_{\max}(\sigma_1) \leq C_{\max}(\sigma_2)$ and $WT(\sigma_1) \leq WT(\sigma_2)$, then we can replace σ_2 by σ_1 in any feasible schedule. So, sequence σ_1 is at least "as good as" sequence σ_2.
2. Now, assume that $C_{\max}(\sigma_1) > C_{\max}(\sigma_2)$. Let $r_{\min} = \min_{\{J_l \in \bar{\sigma}_1\}} r_l$ be the smallest release date of jobs belonging to set $\bar{\sigma}_1$. If we replace σ_2 by σ_1 in a feasible schedule, all jobs in $\bar{\sigma}_1$ have to be shifted at most: $\max(C_{\max}(\sigma_1), r_{\min}) - \max(C_{\max}(\sigma_2), r_{\min})$ time units. So, the additional cost for jobs in $\bar{\sigma}_1$ is at most $(\max(C_{\max}(\sigma_1), r_{\min}) - \max(C_{\max}(\sigma_2), r_{\min})) \sum_{\{J_l \in \bar{\sigma}_1\}} w_l$. Hence, we can say that σ_1 is at least "as good as" σ_2 if $WT(\sigma_1) + (\max(C_{\max}(\sigma_1), r_{\min}) - \max(C_{\max}(\sigma_2), r_{\min})) \sum_{\{J_l \in \bar{\sigma}_1\}} w_l \leq WT(\sigma_2)$.

Now, we can define the notion of "better" sequence.

Definition 13.1

We say that sequence σ_1 is "better" than sequence σ_2 if sequence σ_1 is at least "as good as" sequence σ_2, and if, either (1) sequence σ_2 is not at least "as good as" sequence σ_1 or (2) if sequence σ_1 is lexicographically smaller than σ_2.

13.4 No-Good Recording

The no-good recording technique [32,33] records throughout the search tree, some situations that do not lead to a feasible (nondominated) solution. Given a node ν of the search tree, let P^{ν} be the value taken by sequence P at node ν. Whenever a backtrack occurs, the node ν is stored in a no-good list. We only save the following data:

1. the set of jobs belonging to sequence P^ν
2. the completion time $C_{\max}(P^\nu)$ of sequence P^ν
3. the total weighted tardiness $WT(P^\nu)$ associated with sequence P^ν

Now assume that we encounter another node μ such that sequence P^ν is a permutation of sequence P^μ, i.e., the set of jobs belonging to sequence P^μ is exactly the set of jobs belonging to sequence P^ν. If sequence P^ν is a "better" sequence (see Definition 13.1) than sequence P^μ, then μ is dominated by ν.

We use this information to filter the set of *PF* jobs. Suppose that scheduling a job J_i just after the current sequence P leads to a node μ which is dominated by a node ν which has been stored in the no-good list. The sequence $P \mid J_i$ is then dominated and we can remove J_i from set *PF*: $PF \leftarrow PF/\{J_i\}$.

With a view to using efficiently the no-good list, we use an Hash-Table. Most of the time, it allows us to obtain a pertinent state of the list in $O(n \log n)$. A state of the list consists of a set of jobs sorted in Lexicographic order coded with a list of pointers, and a list of couples $(WT(P^\kappa), C_{\max}(P^\kappa))$, also coded with a list of pointers, which fit to the different states of nodes κ which have ever been visited with this set of jobs. Assume that we encounter a node μ such that the set of jobs belonging to sequence P^μ is the same as a state of the no-good list. Sorting jobs in Lexicographic order runs in $O(n \log n)$ time. Computing an index in the Hash-Table of this set of jobs runs in $O(n)$ time. Since jobs are sorted in Lexicographic order, comparing two sets of jobs runs in $O(n)$ time. If an associated couple $(WT(P^\kappa), C_{\max}(P^\kappa))$ of this state is such that its cost $WT(P^\kappa)$ and its completion time $C_{\max}(P^\kappa)$ are, respectively, greater than or equal to those of sequence P^μ, then this couple is replaced by the cost $WT(P^\mu)$ and the completion time $C_{\max}(P^\mu)$ of sequence P^μ.

13.5 Dominance of Active Schedules

In this section we recall that active schedules are dominant for the total weighted tardiness problem. An *active schedule* S is such that no job can be completed earlier in S without delaying another job. An objective function which is monotone with respect to all variables C_i is called a *regular criterion*. It has been shown that the subset of active schedules is dominant for regular criteria [34,35]. Since the criterion of the minimization of total weighted tardiness (and its special cases) is a regular criterion, the following result is immediately obtained.

Theorem 13.1 [34,35]

Let J_k be a job belonging to NS. If $r_k \geq \min_{\{J_i \in NS\}}\{r_i + p_i\}$, then the sequence $P \mid J_k$ is dominated.

If J_k is such a job, then we can remove it from the set of *PF* jobs, i.e., $PF \leftarrow PF/\{J_k\}$. Indeed, if we schedule J_k in first position, it is then possible to obtain a better schedule by inserting a job J_j such that $r_j + p_j = \min_{\{J_i \in NS\}}\{r_i + p_i\}$, without delaying other jobs. In order to obtain active schedules, we assume that at each node of the search tree, the set *PF* is filtered such that, for all jobs $J_j \in PF$, we have $r_j < \min_{\{J_i \in NS\}}\{r_i + p_i\}$. Note that we cannot use this dominance rule to filter set *PL* since we do not know the start time of jobs belonging to this set.

13.6 Local Dominance Properties

Following the works of Chu on the total tardiness and the total completion time criteria [13,23], we provide a necessary and sufficient condition for local optimality and we define *dominant subsets of schedules* on the basis of this condition.

Chu [13,14,22,23] has described a sufficient condition for local optimality for the total tardiness criterion, and a necessary and sufficient condition for the total completion time criteria. By *local optimality*, Chu means *the optimality of two adjacent jobs* in a given schedule, without taking into account the jobs following them, and the jobs preceding them. This is similar to considering the one machine two jobs problem. Suppose that we have to schedule two jobs J_j and J_k on a machine available at time t. Let $WT_{jk}(t)$ be the sum of weighted tardiness of jobs J_j and J_k obtained by scheduling J_j before J_k at time t: $WT_{jk}(t) = w_j \max[0, \max(r_j, t) + p_j - d_j] + w_k \max\{0, \max[\max(r_j, t) + p_j, \max(r_k, t)] + p_k - d_k\}$. Then we have the following theorem whose proof is obvious.

Theorem 13.2

Considering a scheduling problem involving only two jobs J_j and J_k which have to be scheduled on a machine available after time t. Scheduling J_j before J_k is optimal if and only if $WT_{jk}(t) - WT_{kj}(t) \leq 0$.

This condition enables us to build a dominant subset to minimize total weighted tardiness. Let $\Delta_j(S)$ be the completion time of the job immediately preceding job J_j in a schedule S. If J_j is the first job of the schedule S, then $\Delta_j(S) = -\infty$.

Definition 13.2

Consider a given active schedule S. If any couple of adjacent jobs J_j and J_k (J_j followed by J_k) satisfies at least one of the following conditions:

1. $\max(r_j, \Delta_j(S)) < \max(r_k, \Delta_j(S))$
2. $WT_{jk}(\Delta_j(S)) - WT_{kj}(\Delta_j(S)) \leq 0$

then S is said to be LO-Active (Locally Optimal Active Schedule).

According to the previous definition, Theorem 13.2 leads to the following theorem.

Theorem 13.3

All the optimal active schedules of the total weighted tardiness one-machine problem are LO-Active.

Proof

Assume that there exists an optimal active schedule S which is not *LO*-Active. It means that there is at least one pair of adjacent jobs J_j and J_k (J_j followed by J_k), such that $WT_{jk}(\Delta_j(S)) - WT_{kj}(\Delta_j(S)) > 0$ and $\max(r_j, \Delta_j) \geq \max(r_k, \Delta_j)$. We construct another schedule S' by interchanging J_j and J_k without moving any other job. This interchange does not delay jobs after J_k since $\max(r_k, \Delta_j) \leq \max(r_j, \Delta_j)$. Only the tardiness of jobs J_j and J_k are changed. It is then clear that $WT(S) - WT(S') = WT_{jk}(t) - WT_{kj}(t) > 0$. It means that interchanging jobs J_j and J_k decreases strictly the total weighted tardiness and that S' is strictly better than S. It contradicts the assumption that S is an optimal schedule. □

Not all *LO*-Active schedules are optimal and it is possible to remove some schedules which are dominated from the set of *LO*-Active schedules. The idea is to remove schedules S in which there exist useless idle times, i.e., in which there exist two adjacent jobs J_j and J_k (job J_j before job J_k), such that $WT_{jk}(\Delta_j(S)) - WT_{kj}(\Delta_j(S)) = 0$ and $\max(r_j, \Delta_j(S)) > \max(r_k, \Delta_j(S))$.

Definition 13.3

Consider a given active schedule S. If any pair of adjacent jobs J_j and J_k (J_j followed by J_k) satisfies at least one of the following conditions:

1. $WT_{jk}(\Delta_j(S)) - WT_{kj}(\Delta_j(S)) < 0$
2. $WT_{jk}(\Delta_j(S)) - WT_{kj}(\Delta_j(S)) = 0$ *and* $\max(r_j, \Delta_j(S)) \leq \max(r_k, \Delta_j(S))$
3. $WT_{jk}(\Delta_j(S)) - WT_{kj}(\Delta_j(S)) > 0$ *and* $\max(r_j, \Delta_j(S)) < \max(r_k, \Delta_j(S))$

then S is said to be LOWS-Active (Locally Optimal Well Sorted Active Schedule).

According to the previous definition, we establish the following theorem.

Theorem 13.4

The subset of the LOWS-Active schedules is dominant for the one machine total weighted tardiness problem.

Proof

Consider an optimal *LO*-Active schedule S which is not *LOWS*-Active. It means that there is at least one pair of adjacent jobs J_j and J_k (J_j followed by J_k), such that $WT_{jk}(\Delta_j(S)) - WT_{kj}(\Delta_j(S)) \leq 0$ or $\max(r_j, \Delta_j) < \max(r_k, \Delta_j)$, and such that none of the three conditions of a *LOWS*-Active schedule is satisfied. Condition (1) implies that $WT_{jk}(\Delta_j(S)) - WT_{kj}(\Delta_j(S)) \geq 0$, else schedule S should be *LOWS*-Active.

First, assume that $WT_{jk}(\Delta_j(S)) - WT_{kj}(\Delta_j(S)) = 0$. From condition (2), it follows that we have $\max(r_j, \Delta_j(S)) > \max(r_k, \Delta_j(S))$. Now assume that $WT_{jk}(\Delta_j(S)) - WT_{kj}(\Delta_j(S)) > 0$. From condition (3), it follows that $\max(r_j, \Delta_j(S)) \geq \max(r_k, \Delta_j(S))$. Note that, in all cases, $\max(r_j, \Delta_j(S)) \geq \max(r_k, \Delta_j(S))$.

We construct another schedule S' by interchanging J_j and J_k without moving any other job. This interchange does not delay the jobs after J_k since $\max(r_k, \Delta_j) \leq \max(r_j, \Delta_j)$. Only the tardiness of jobs J_j and J_k are changed. It is then clear that $WT(S) - WT(S') = WT_{jk}(t) - WT_{kj}(t) = 0$. It means that interchanging jobs J_j and J_k does not increase the total weighted tardiness and that schedule S' is at least as good as schedule S. We can iterate this process until we obtain a *LOWS*-Active schedule as good as schedule S. □

We can use this local dominance in our forward sequencing rule with the following theorem.

Theorem 13.5

Let J_j be the last job of a partial sequence P and let J_k be a job such that $J_k \in PF$. If $\max(r_k, \Delta_j) \leq \max(r_j, \Delta_j)$ and $WT_{kj}(\Delta_j(S)) - WT_{jk}(\Delta_j(S)) \leq 0$ (with at least one strict inequality or with J_j lexicographically smaller than J_k), then the sequence $P \mid J_k$ is dominated.

Proof

If $\max(r_k, \Delta_j) \leq \max(r_j, \Delta_j)$ and $WT_{kj}(\Delta_j(S)) - WT_{jk}(\Delta_j(S)) \leq 0$ (with at least one strict inequality), and if J_k is scheduled just after J_j, then we can interchange J_j and J_k to obtain a better schedule. □

In this case, we can remove J_k from set *PF* of possible first jobs: $PF \leftarrow PF/\{J_k\}$. Note that this dominance rule dominates the dominance rules described by Akturk and Ozdemir [16] and Chu [13,23]. In the next section we show that this theorem is dominated by a more general dominance rule.

13.7 Dominance Rule Relying on Scheduled Jobs

The local dominance rule can be used to filter set *PF* according to the last job of P. Now, we consider *dominance properties that take into account the complete partial sequence* P of scheduled jobs. Suppose we try to schedule a job J_i just after P. Let $P \mid J_i$ be this sequence. Informally speaking, our most basic rule states that if the sequence $P \mid J_i$ can be improved, then it is dominated, so J_i can be removed from set *PF*.

Theorem 13.6

Let J_i be a job belonging to PF. If there exists a permutation which is "better" than $P \mid J_i$, then sequence $P \mid J_i$ is dominated.

Proof

Assume there exists a sequence σ that is "better" than sequence $P \mid J_i$, then we can replace sequence $P \mid J_i$ by sequence σ in any feasible schedule, so $P \mid J_i$ is dominated. □

See Section 13.3 for the notion of "better" sequences. If scheduling J_i after P leads to a dominated schedule, then J_i can be removed from the set *PF*. To compare two sequences, we just have to build the schedules associated with them and this can be done in linear time. We use several ways to find better permutations π of $P \mid J_i$.

1. At first, we can enumerate all permutations π that are identical to $P \mid J_i$ except for the last k jobs, where k is a fixed number. When k is large, we have a great reduction of the search space but this takes a lot of time. Experimentally, $k = 6$ seems to be a good trade-off between the reduction of the search tree and the time spent in the enumeration.
2. We can also enumerate the permutations π that are obtained from $P \mid J_i$ by inserting J_i somewhere inside P or by interchanging J_i with another job of P. Since there are $O(n)$ such permutations and since comparing two sequences can be done in linear time, the algorithm runs in $O(n^2)$ time for a given job J_i. In aim to improve the quality of sequences built along the algorithm, we try to ensure that the local optimality is satisfied between each pair of adjacent jobs: all along the construction of a sequence, two adjacent jobs can be interchanged to try to obtain *LOWS*-Active schedules. Note that such an interchange runs in $O(1)$ time.
3. Finally, we use the list of no-good sequences which have been recorded with the no-good recording technique (see Section 13.4).

Note that this list is not exhaustive and that some other imaginative ways can be found to improve the filtering of the set *PF*. Note that trying an interchange with the last job of sequence P dominates Theorem 13.5.

13.8 Dominance Rules Relying on Unscheduled Jobs

In the previous section we have described dominance rules which *take into account the jobs which have already been scheduled* (jobs belonging to the sequence P). Now we describe two new dominance rules which take into account the jobs which have not been scheduled (jobs belonging to set *NS*).

Let J_j and J_k be two jobs such that $J_j \in NS$ and $J_k \in PF$. The idea is to examine under which conditions J_j dominates J_k in the first position. By dominance in the first position, we mean that, if J_k is scheduled just after the sequence P, and if we complete the sequence $P \mid J_k$ by a sequence Z including job J_j, then it will be always "better" to interchange J_j and J_k or to insert J_j just before J_k, i.e., the sequence $P \mid J_k \mid Z$ is dominated for all sequences Z, and job J_k can be removed from the set *PF* of possible first jobs. We show how we can obtain an evaluation of the minimum gain of interchanging jobs J_j and J_k, or of inserting job J_j before job J_k, and that, for all sequences Z, if we schedule J_k in first position. We also show that this interchange or this insertion does not increase total weighted tardiness under some conditions. We can prove that the deduced dominance rules dominate or are equivalent to most of the dominance rules described for the studied criteria [14,17,22,27].

In Section 13.8.1 we describe the dominance rule obtained by using interchanges of jobs. Next we describe in Section 13.8.2, the dominance rule obtained by using insertions of jobs. Finally, we introduce in Section 13.8.3, an algorithm in $O(n^2)$ that shows how we can apply these dominance rules to filter the set *PF* of possible first jobs.

Recall that the release dates of jobs have been adjusted according to the partial sequence P of scheduled jobs. We assume that r_k and r_j are lower than $\min_{\{J_i \in NS\}}\{r_i + p_i\}$ else J_k should not be first according to Theorem 13.1. In particular, we have $r_k < r_j + p_j$ and $r_j < r_k + p_k$.

13.8.1 Dominance Rule Based on Interchanges of Jobs

In this section we describe a dominance in the first position *based on interchange of jobs*. Let J_j and J_k be two jobs such that $J_j \in NS$ and $J_k \in PF$. We examine under which conditions it will be always "better" to interchange jobs J_j and J_k, if J_k is scheduled just after the sequence P. This dominance rule is valid only if job J_j has a heavier weight than job J_k. From now on, we assume that $w_j \geq w_k$. Suppose that job J_k is

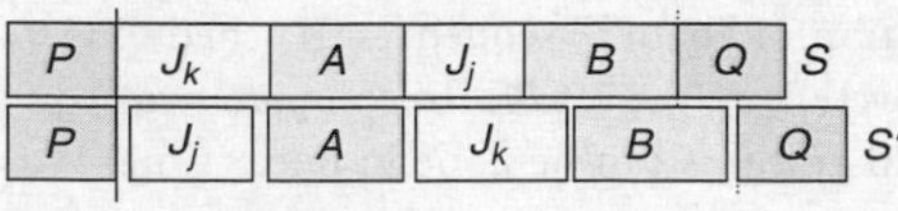

FIGURE 13.2 Interchanging J_k and J_j.

scheduled in the first position just after the current sequence P. Assume that the current node of the search tree can be extended to a feasible schedule S. Let A and B be respectively the sequence of jobs scheduled between jobs J_k and J_j, and the sequence of jobs scheduled between job J_j and sequence Q in schedule S. Suppose now that we construct another schedule S' by interchanging jobs J_j and J_k (see Figure 13.2). To show that it will be always "better" to interchange jobs J_j and J_k, we have to prove that interchanging jobs J_j and J_k does not increase total weighted tardiness. For that, it is sufficient to prove that the difference between the total weighted tardiness of S and of S' is positive, i.e., $WT(S) - WT(S') \geq 0$. The gain of interchanging jobs J_j and J_k is equal to $WT_k(S) + WT_j(S) - WT_j(S') - WT_k(S')$. Note that this gain can be negative. During the interchange, some other jobs may have been shifted to the right. In such a case, we must evaluate the maximum cost to schedule these jobs later.

To begin with, we focus on the interchange of jobs J_j and J_k, and we evaluate $WT_k(S) + WT_j(S) - WT_j(S') - WT_k(S')$. In schedule S, job J_k is completed at time $C_k(S) = r_k + p_k$ and job J_j starts at time $t_j(S)$. In S', job J_j is completed at time $C_j(S') = r_j + p_j$ and job J_k starts at time $t_k(S')$. *Let τ_1 be the difference between the start time of job J_k in S' and $t_j(S)$*, i.e., $\tau_1 = t_k(S') - t_j(S)$. Note that if J_k starts before $t_j(S)$ in S', then τ_1 is negative, else τ_1 is positive. We have exactly $WT_k(S)+WT_j(S)-WT_j(S')-WT_k(S') = w_k \max(0, r_k + p_k - d_k) - w_j \max(0, r_j + p_j - d_j) + w_j \max(0, t_j(S) + p_j - d_j) - w_k \max(0, t_j(S) + \tau_1 + p_k - d_k)$.

In several cases, the interchange leads to a schedule in which some jobs are scheduled later than in the initial schedule. Suppose that during the interchange, some other jobs have been shifted to the right. These jobs can be either jobs belonging to set $OJ = NS/\{J_j, J_k\}$ (other jobs) of not scheduled jobs or jobs belonging to the partial sequence Q.

Let τ_2 be the maximum shift to the right of the jobs belonging to set OJ. Note that value τ_2 is always greater than or equal to 0. If no job is scheduled later, then shift τ_2 is equal to 0. The jobs belonging to sequence A have been scheduled from at most $\max(0, C_j(S') - C_k(S))$ later. The jobs belonging to sequence B have been scheduled from at most $\max(0, C_k(S') - C_j(S))$ later. *Note that we do not know if a job $J_i \in OJ$ belongs to sequence A or to sequence B, but we know that* $OJ = A \cup B$. It follows that at most all jobs in set OJ are shifted to the right from at most $\tau_2 = \max(0, C_j(S') - C_k(S), C_k(S') - C_j(S))$. The completion time of each job $J_i \in OJ$ is increased by at most τ_2 and its weighted tardiness by at most $w_i\tau_2$. Hence, the additional cost is therefore at most $\tau_2 \sum_{\{J_l \in OJ\}} w_l$.

Let τ_3 be the maximum shift to the right of jobs belonging to sequence Q. Likewise, τ_3 is always greater than or equal to 0. Note that jobs of sequence Q are scheduled later only if the jobs of sequence B are scheduled later. In this case, the shift is the same as the jobs of sequence B. Hence, $\tau_3 = \max(0, C_k(S') - C_j(S))$.

These evaluations allow us to obtain a lower bound of the gain of interchanging jobs J_k and J_j, if job J_k is scheduled in the first position, and if job J_j is scheduled at time $t_j(S)$. For given τ_1, τ_2 and τ_3 we define the function Γ_{jk}, for two jobs J_j and J_k and a time t as:

$$\begin{aligned}\Gamma_{jk}(t, \tau_1, \tau_2, \tau_3) = {} & w_j \max(0, t + p_j - d_j) - w_k \max(0, t + \tau_1 + p_k - d_k)\\ & + w_k \max(0, r_k + p_k - d_k) - w_j \max(0, r_j + p_j - d_j)\\ & - \tau_2 \sum_{\{J_l \in OJ\}} w_l - \tau_3 \sum_{\{J_l \in Q\}} w_l \end{aligned} \tag{13.1}$$

Note that $WT(S) - WT(S') \geq \Gamma_{jk}(t_j(S), \tau_1, \tau_2, \tau_3)$. The value $\Gamma_{jk}(t_j(S), \tau_1, \tau_2, \tau_3)$ can therefore be seen as a lower bound of the gain of interchanging jobs J_k and J_j if job J_k is scheduled in the first position and job J_j is scheduled at time $t_j(S)$.

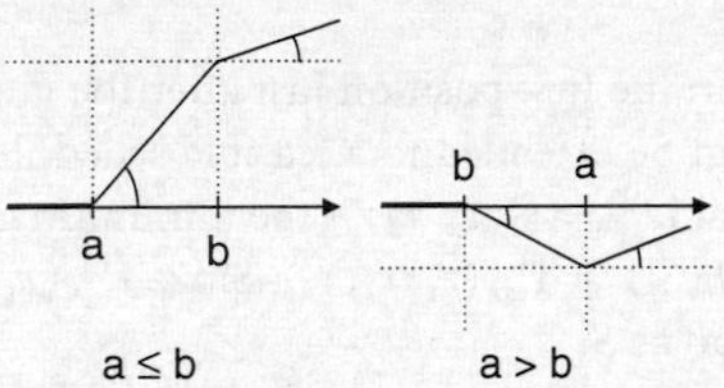

FIGURE 13.3 Lemma 13.1.

Note that for given τ_1, τ_2, and τ_3, function Γ_{jk} is only dependent of the time t. Now, we show that function Γ_{jk} is nondecreasing on $[\max(d_j - p_j, r_k + p_k), +\infty]$ and has a global minimum at time $t = \max(d_j - p_j, r_k + p_k)$. This global minimum allows us to obtain a minimum gain of interchanging jobs J_j and J_k if J_k is scheduled in the first position.

It is easy to see that function Γ_{jk} can be expressed as a function $f(t) = \alpha \max(0, t - a) - \beta \max(0, t - b) + C$, where $\alpha = w_j$, $\beta = w_k$, $a = d_j - p_j$, $b = d_k - p_k - \tau_1$, and $C = -\tau_2 \sum_{\{J_l \in O'\}} w_l - \tau_3 \sum_{\{J_l \in Q\}} w_l$. We introduce the following lemma which allows us to study function Γ_{jk}.

Lemma 13.1

Let $f : t \to \alpha \max(0, t - a) - \beta \max(0, t - b) + C$ be a function defined on $\mathbb{R}^+$ for which $a, b, C \in \mathbb{R}$ and $\alpha, \beta \in \mathbb{R}^{+*}$. If $\alpha \geq \beta$, then the function f has a global minimum at time $t = a$ and is nondecreasing on interval $[a; +\infty]$.

Proof

First assume that $a \leq b$ (see Figure 13.3). On interval $[-\infty, a]$, function f is constant. On interval $[a, b]$, function f is strictly increasing. Finally, on interval $[b, +\infty]$, if $\alpha > \beta$, then function f is strictly increasing, and if $\alpha = \beta$, then function f is constant. Since $f(t) = f(a)$ on interval $[-\infty, a]$ and since function f is nondecreasing on interval $[a, +\infty]$, we have a global minimum of $f(t)$ at time $t = a$.

Now, assume that $a > b$ (see Figure 13.3). On interval $[-\infty, b]$, function f is constant. On interval $[b, a]$, function f is strictly decreasing. Finally, on interval $[a, +\infty]$, if $\alpha > \beta$, then function f is strictly increasing, and if $\alpha = \beta$ then function f is constant. Since function f is nonincreasing on $[-\infty, a]$ and nondecreasing on $[a, +\infty]$, we have a global minimum of $f(t)$ at time $t = a$. In all cases, we obtain a global minimum of $f(t)$ at time $t = a$. □

We have $w_j \geq w_k$ and $t_j(S) \geq r_k + p_k$. Following Lemma 13.1, function Γ_{jk} is nondecreasing on $[\max(d_j - p_j, r_k + p_k), +\infty]$ and has a global minimum at time $t = \max(d_j - p_j, r_k + p_k)$. We define this global minimum as

$$\hat{\Gamma}_{jk}(\tau_1, \tau_2, \tau_3) = \Gamma_{jk}(\max(d_j - p_j, r_k + p_k), \tau_1, \tau_2, \tau_3) \tag{13.2}$$

Note that $\hat{\Gamma}_{jk}$ is not a function of time t, but only of τ_1, τ_2, and τ_3. Informally speaking, $\hat{\Gamma}_{jk}$ can be seen as the minimum gain obtained by interchanging jobs J_j and J_k, if J_k is scheduled in first position. Now, it is clear that the condition $w_j \geq w_k$, which is necessary for the validity of the rule established in this section, comes from the condition $\alpha \geq \beta$ from Lemma 13.1, which allows us to obtain this global minimum. Now we can introduce a global theorem, which shows under which conditions a job dominates another one in the first position.

Theorem 13.7

Let J_j and J_k be two jobs such that $J_j \in NS$, $J_k \in PF$, and $w_j \geq w_k$. If computing values τ_1, τ_2, and τ_3 according to job J_j, job J_k, and set NS leads to $\hat{\Gamma}_{jk}(\tau_1, \tau_2, \tau_3) \geq 0$, then J_j dominates J_k in the first position.

Proof

Suppose that job J_k is scheduled in the first position just after the current sequence P. Assume that the current node of the search tree can be extended to a feasible schedule S. Construct another schedule S' by interchanging J_j and J_k. The value $\hat{\Gamma}_{jk}(\tau_1, \tau_2, \tau_3)$ is the minimum gain obtained by interchanging jobs J_j and J_k. We have $WT(S) - WT(S') \geq \hat{\Gamma}_{jk}(\tau_1, \tau_2, \tau_3)$. Since $\hat{\Gamma}_{jk}(\tau_1, \tau_2, \tau_3) \geq 0$, we thus obtain a new schedule S' which is at least as good as S. □

Note that interchanging jobs J_k and J_j leads either to a "better" schedule when we can have a strict inequality ($WT(S) - WT(S') > 0$), or to a schedule which is at least as good as the initial schedule, when we can have an equality ($WT(S) - WT(S') \geq 0$). In the latter case, we will say that J_j dominates J_k in the first position if J_j is lexicographically smaller than J_k.

13.8.1.1 Different Cases of Interchange

In aim to obtain a minimum gain $\hat{\Gamma}_{jk}(\tau_1, \tau_2, \tau_3)$ of interchanging jobs J_k and J_j, it now remains to show how we can compute τ_1, τ_2, and τ_3. We propose an *exhaustive list of the cases of interchange* which has been derived to take advantage of several situations. This list is summarized in the Tree from Figure 13.4, in the cases from Figure 13.5 and in the Table 13.1. Now, we show in each case how the *shifts* τ_1, τ_2, *and* τ_3 *are evaluated.* In each case, we suppose that job J_k is scheduled in the first position just after the current partial sequence P. We assume that the current node of the search tree can be extended to a feasible schedule S which begins with the partial sequence P and ends with the partial sequence Q and that sequence A, sequence B, set OJ, and schedule S' are obtained as described previously. Recall also that $\tau_1 = t_k(S') - t_j(S)$, $\tau_2 = \max(0, C_j(S') - C_k(S), C_k(S') - C_j(S))$, and $\tau_3 = \max(0, C_k(S') - C_j(S))$. Note that each case is illustrated in Figure 13.5.

- *Case 1.* $r_j + p_j \leq r_k + p_k$ and $p_j < p_k$. The completion times of jobs belonging to sequence A can remain the same because $r_j + p_j \leq r_k + p_k$. Hence, job J_k can start at $t_k(S') = t_j(S)$, and we have $\tau_1 = 0$. Nevertheless, jobs belonging to sequence $B \mid Q$ are scheduled from at most $p_k - p_j$ later. These jobs are at most all jobs belonging to set $OJ \cup Q$. Hence, we have $\tau_2 = \tau_3 = p_k - p_j$.

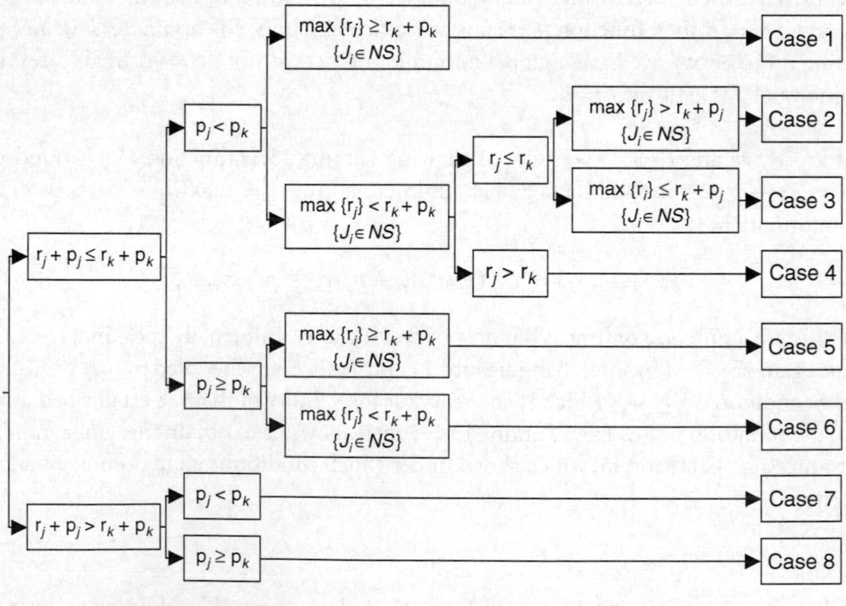

FIGURE 13.4 Tree of the cases of interchange.

TABLE 13.1 Cases of Interchange

Case	Values of τ_1, τ_2, τ_3
1	$\tau_1 = 0, \tau_2 = \tau_3 = p_k - p_j$
2	$\tau_1 = \max_{\{J_i \in NS\}}\{r_i\} - r_k - p_k, \tau_2 = \tau_3 = \max_{\{J_i \in NS\}}\{r_i\} - r_k - p_j$
3	$\tau_1 = \max(r_j + p_j, \max_{\{J_i \in NS\}}\{r_i\}) - r_k - p_k, \tau_2 = \tau_3 = 0$
4	$\tau_1 = \max(r_j + p_j, \max_{\{J_i \in NS\}}\{r_i\}) - r_k - p_k, \tau_2 = \tau_3 = \max(r_j + p_j, \max_{\{J_i \in NS\}}\{r_i\}) - r_k - p_j$
5	$\tau_1 = \tau_2 = \tau_3 = 0$
6	$\tau_1 = \max(r_j + p_j, \max_{\{J_i \in NS\}}\{r_i\}) - r_k - p_k, \tau_2 = \tau_3 = 0$
7	$\tau_1 = r_j + p_j - r_k - p_k, \tau_2 = \tau_3 = r_j - r_k$
8	$\tau_1 = r_j + p_j - r_k - p_k, \tau_2 = r_j + p_j - r_k - p_k, \tau_3 = \max(0, r_j - r_k)$

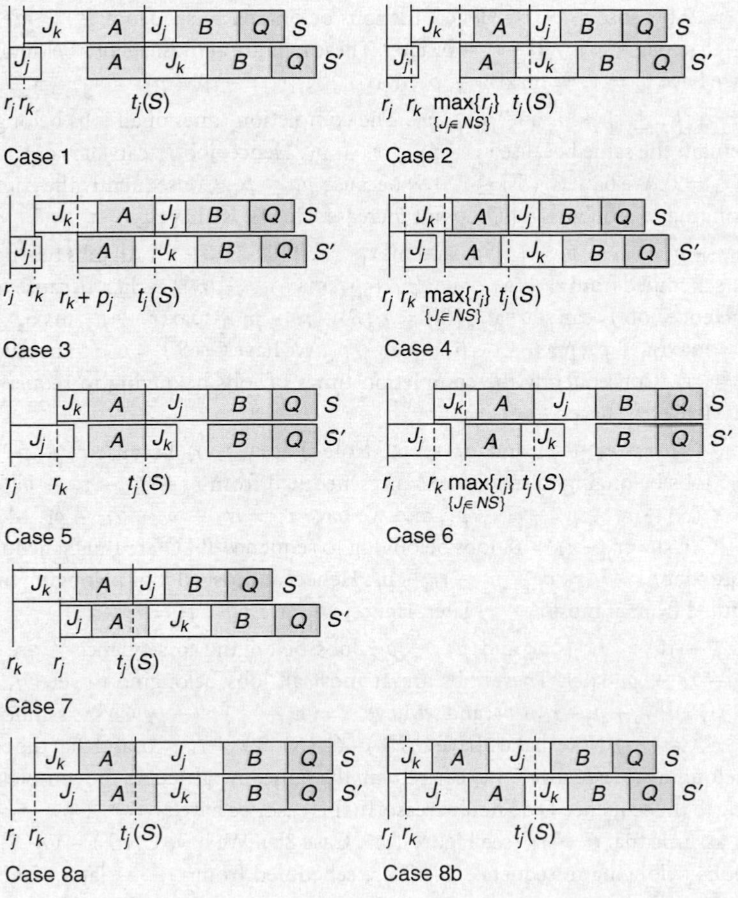

FIGURE 13.5 Cases of interchange.

- *Case 2.* $r_j + p_j \leq r_k + p_k$, $p_j < p_k$, $r_k + p_j < \max_{\{J_i \in NS\}}\{r_i\} < r_k + p_k$, and $r_j \leq r_k$. Note that $C_j(S') = r_j + p_j < \max_{\{J_i \in NS\}}\{r_i\}$ because $r_k + p_j < \max_{\{J_i \in NS\}}\{r_i\}$ and $r_j \leq r_k$. All jobs belonging to sequence A can be scheduled from $r_k + p_k - \max_{\{J_i \in NS\}}\{r_i\}$ earlier because $\max_{\{J_i \in NS\}}\{r_i\} < r_k + p_k$. Hence, job J_k can start at $t_k(S') = t_j(S) - r_k - p_k + \max_{\{J_i \in NS\}}\{r_i\}$ and we have $\tau_1 = \max_{\{J_i \in NS\}}\{r_i\} - r_k - p_k$. The jobs belonging to sequence $B \mid Q$ are scheduled from $C_k(S') - C_j(S) = \max_{\{J_i \in NS\}}\{r_i\} - r_k - p_j > 0$ later. These jobs are at most all jobs belonging to set $OJ \cup Q$. Hence, we have $\tau_2 = \tau_3 = \max_{\{J_i \in NS\}}\{r_i\} - r_k - p_j$.

- *Case 3.* $r_j + p_j \le r_k + p_k$, $p_j < p_k$, $r_j \le r_k$, and $\max_{\{J_i \in NS\}}\{r_i\} \le r_k + p_j$. Note that $C_j(S') = r_j + p_j \le r_k + p_j$ because $r_j \le r_k$. All jobs belonging to sequence A can be scheduled from $r_k + p_k - \max(r_j + p_j, \max_{\{J_i \in NS\}}\{r_i\})$ earlier, because $\max_{\{J_i \in NS\}}\{r_i\} \le r_k + p_j \le r_k + p_k$. Hence, job J_k can start at time $t_k(S') = t_j(S) - r_k - p_k + \max(r_j + p_j, \max_{\{J_i \in NS\}}\{r_i\})$, and we have $\tau_1 = \max(r_j + p_j, \max_{\{J_i \in NS\}}\{r_i\}) - r_k - p_k$. Since $r_k + p_j \ge \max_{\{J_i \in NS\}}\{r_i\}$, we have $r_k + p_k - \max(r_j + p_j, \max_{\{J_i \in NS\}}\{r_i\}) \ge r_k + p_k - (r_k + p_j) = p_k - p_j$, and the completion times of jobs belonging to sequence $B \mid Q$ can remain the same because $C_k(S') \le t_j(S) - (p_k - p_j) + p_k = t_j(S) + p_j = C_j(S)$. Hence, we have $\tau_2 = \tau_3 = 0$.
- *Case 4.* $r_j + p_j \le r_k + p_k$, $p_j < p_k$, $\max_{J_i \in NS}\{r_i\} < r_k + p_k$, and $r_j > r_k$. All jobs belonging to sequence A can be scheduled from $r_k + p_k - \max(r_j + p_j, \max_{\{J_i \in NS\}}\{r_i\})$ earlier. Hence, job J_k can start at time $t_k(S') = t_j(S) - r_k - p_k + \max(r_j + p_j, \max_{\{J_i \in NS\}}\{r_i\})$, and we have $\tau_1 = \max(r_j + p_j, \max_{\{J_i \in NS\}}\{r_i\}) - r_k - p_k$. Note that $C_k(S') - C_j(S) = \max(r_j + p_j, \max_{\{J_i \in NS\}}\{r_i\}) - r_k - p_j > 0$ because $r_j > r_k$. Hence, the jobs belonging to sequence $B \mid Q$ are scheduled from $\max(r_j + p_j, \max_{\{J_i \in NS\}}\{r_i\}) - r_k - p_j$ later. These jobs are at most all jobs belonging to set $OJ \cup Q$. Hence, we have $\tau_2 = \tau_3 = \max(r_j + p_j, \max_{\{J_i \in NS\}}\{r_i\}) - r_k - p_j$.
- *Case 5.* $r_j + p_j \le r_k + p_k$ and $p_j \ge p_k$. The completion times of all jobs belonging to sequence A can remain the same because $r_j + p_j \le r_k + p_k$. Hence, job J_k can start at $t_k(S') = t_j(S)$, and we have $\tau_1 = 0$. We have $C_k(S') \le C_j(S)$ because $p_k \le p_j$. Consequently the completion times of jobs belonging to sequence $B \mid Q$ do not increase. Hence, we have $\tau_2 = \tau_3 = 0$.
- *Case 6.* $r_j + p_j \le r_k + p_k$, $p_j \ge p_k$ and $\max_{\{J_i \in NS\}}\{r_i\} < r_k + p_k$. All jobs belonging to sequence A can be scheduled from $r_k + p_k - \max(r_j + p_j, \max_{\{J_i \in NS\}}\{r_i\})$ earlier, because $\max_{\{J_i \in NS\}}\{r_i\} < r_k + p_k$. Hence, job J_k can start at $t_k(S') = t_j(S) - r_k - p_k + \max(r_j + p_j, \max_{\{J_i \in NS\}}\{r_i\})$, and we have $\tau_1 = \max(r_j + p_j, \max_{\{J_i \in NS\}}\{r_i\}) - r_k - p_k$. We have $C_k(S') \le C_j(S)$ because $t_k(S') \le t_j(S)$ and $p_k \le p_j$. Consequently the completion times of jobs belonging to sequence $B \mid Q$ do not increase. Hence, we have $\tau_2 = \tau_3 = 0$.
- *Case 7.* $r_j + p_j > r_k + p_k$ and $p_j < p_k$. Note that $r_j > r_k$ because $r_j + p_j > r_k + p_k$ and $p_j < p_k$. Jobs belonging to sequence A are scheduled from $r_j + p_j - r_k - p_k$ later. J_k starts at $t_k(S') = t_j(S) + r_j + p_j - r_k - p_k$, and we have $\tau_1 = r_j + p_j - r_k - p_k$. Moreover, we have $C_k(S') - C_j(S) = r_j - r_k > 0$. Jobs belonging to sequence $B \mid Q$ are thus scheduled from $r_j - r_k$ later. Note that $r_j - r_k > r_j + p_j - r_k - p_k$. Hence, at most all jobs belonging to set $OJ \cup Q$ can be scheduled from at most $r_j - r_k$ later. Hence, we have $\tau_2 = \tau_3 = r_j - r_k$.
- *Case 8.* $r_j + p_j > r_k + p_k$ and $p_j \ge p_k$. Jobs belonging to sequence A are scheduled from $r_j + p_j - r_k - p_k$ later. These jobs are at most all jobs belonging to set OJ. Job J_k starts at $t_k(S') = t_j(S) + r_j + p_j - r_k - p_k$, and we have $\tau_1 = r_j + p_j - r_k - p_k$. First, assume that $r_j \le r_k$ (see Figure 13.5, Case 8a). Note that we have $C_k(S') - C_j(S) = r_j - r_k \le 0$, and that the completion times of jobs belonging to sequence $B \mid Q$ can remain the same. In particular, the completion times of jobs belonging to the sequence Q do not increase. In this case, we have $\tau_2 = r_j + p_j - r_k - p_k$ and $\tau_3 = 0$.

  Now, assume that $r_j > r_k$ (see Figure 13.5, Case 8b). We have $C_k(S') - C_j(S) = r_j - r_k > 0$. Hence, jobs belonging to sequence $B \mid Q$ are scheduled from $r_j - r_k$ later. Note that $r_j - r_k < r_j + p_j - r_k - p_k$ because $p_j > p_k$. In particular, jobs belonging to the sequence Q are scheduled from $r_j - r_k$ later, whereas jobs belonging to set OJ can be scheduled from $r_j + p_j - r_k - p_k$ later. Hence, we have $\tau_2 = r_j + p_j - r_k - p_k$ and $\tau_3 = r_j - r_k$.

13.8.2 Dominance Rule Based on Insertions of Jobs

In this section we describe a dominance in the first position *based on insertion of a job.* Let J_j and J_k be two jobs such that $J_j \in NS$ and $J_k \in PF$. We examine under which conditions it will be always "better" to insert job J_j before job J_k, if job J_k is scheduled just after the partial sequence P. We show how it is possible to derive a new dominance rule "in the first position". Fortunately, for this dominance rule, there is *no condition on weights*, i.e., weight w_j can be lower than weight w_k.

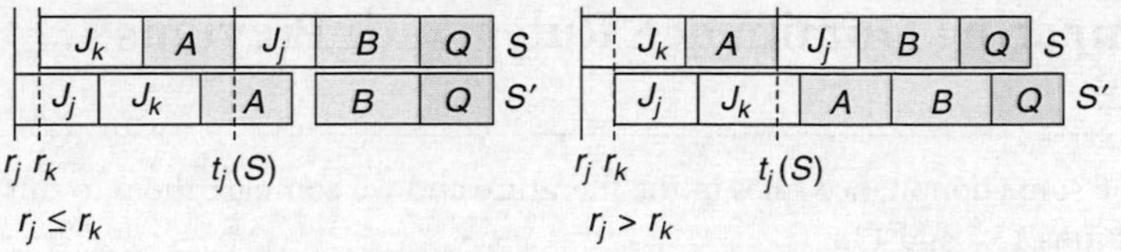

FIGURE 13.6 Cases of insertions.

Theorem 13.8

Let J_j and J_k be two jobs such that $J_j \in NS$ and $J_k \in PF$. If $\Gamma_{jk}(r_k + p_k, r_j + p_j - r_k - p_k, r_j + p_j - r_k, \max(0, r_j - r_k)) \geq 0$, then job J_j dominates job J_k in the first position.

Proof

Suppose that job J_k is scheduled in the first position just after the current sequence P. Assume that the current node of the search tree can be extended to a feasible schedule S. Construct another schedule S' by inserting J_j just before J_k (see Figure 13.6). To show that it will be always "better" to insert job J_j before job J_k, we have to prove that it does not increase the total weighted tardiness. For that, it is sufficient to prove that $WT(S) - WT(S') \geq 0$.

To begin with, we try to evaluate $WT_k(S) + WT_j(S) - WT_j(S') - WT_k(S')$ which is the gain of inserting job J_j before job J_k. Note that this gain can be negative. In schedule S, job J_k is completed at time $C_k(S) = r_k + p_k$, and job J_j is starting at $t_j(S)$. In schedule S', job J_j is completed at time $C_j(S') = r_j + p_j$, and job J_k at time $C_k(S') = r_j + p_j + p_k$. Hence, we have $WT_k(S) + WT_j(S) - WT_k(S') - WT_j(S') = w_k \max(0, r_k + p_k - d_k) + w_j \max(0, t_j(S) + p_j - d_j) - w_j \max(0, r_j + p_j - d_j) - w_k \max(0, r_j + p_j + p_k - d_k)$. Note that $t_j(S) \geq r_k + p_k$, and that $w_j \max(0, t + p_j - d_j)$ is a function in t which is not decreasing on $[r_k + p_k, +\infty]$. Hence, we have $WT_k(S) + WT_j(S) - WT_k(S') - WT_j(S') \geq w_k \max(0, r_k + p_k - d_k) + w_j \max(0, r_k + p_k + p_j - d_j) - w_j \max(0, r_j + p_j - d_j) - w_k \max(0, r_j + p_j + p_k - d_k)$.

During the insertion, some other jobs have been shifted to the right. In such a case, we must evaluate the maximum cost to schedule these jobs later. The jobs of sequence A are scheduled from $r_j + p_j + p_k - r_k - p_k = r_j + p_j - r_k$ later, and are at most all jobs belonging to OJ. Let J_u be the last job of sequence A. Note that $C_u(S) \leq t_j(S) = C_j(S) - p_j$. We have $C_u(S') = C_u(S) + r_j + p_j - r_k \leq C_j(S) + r_j - r_k$. If $r_j \leq r_k$, then the completion times of jobs scheduled after J_u in S' do not increase. In particular, the tardiness of jobs belonging to sequence Q do not increase. If $r_j > r_k$, then jobs scheduled after J_u in S' are scheduled from at most $r_j - r_k$ later. In particular, the total weighted tardiness of jobs belonging to sequence Q increase from at most $(r_j - rj) \sum_{\{J_l \in Q\}} w_l$. Note that $r_j - r_k < r_j + p_j - r_k$.

Finally, we have $WT(S) - WT(S') \geq w_k \max(0, r_k + p_k - d_k) + w_j \max(0, r_k + p_k + p_j - d_j) - w_j \max(0, r_j + p_j - d_j) - w_k \max(0, r_j + p_j + p_k - d_k) - (r_j + p_j - r_k) \sum_{\{J_l \in OJ\}} w_l - \max(0, r_j - r_k) \sum_{\{J_l \in Q\}} w_l \geq 0$. Note that we can express that by using function Γ_{jk}. Thus, we have $WT(S) - WT(S') \geq \Gamma_{jk}(r_k + p_k, r_j + p_j - r_k - p_k, r_j + p_j - r_k, \max(0, r_j - r_k)) \geq 0$. We thus obtain a new schedule S' which is at least as good as S. □

13.8.3 Applying Dominance Rules

Dominance rules described in Sections 13.8.1 and 13.8.2, can be propagated in $O(n^2)$ time, at each node of the search tree. For each job $J_k \in PF$, we search a job $J_j \in NS$ such that computing τ_1, τ_2, and τ_3 with $\hat{\Gamma}_{jk}(\tau_1, \tau_2, \tau_3)) \geq 0$ or $\Gamma_{jk}(r_k + p_k, r_j + p_j - r_k - p_k, r_j + p_j - r_k, \max(0, r_j - r_k)) \geq 0$. If such a job J_j exists, then J_k is removed from PF. Since computing τ_1, τ_2, τ_3, $\hat{\Gamma}_{jk}$, and Γ_{jk} runs in 0(1) time, and since we have to try at most $|NS| - 1$ jobs J_j for each job J_k, then the whole algorithm runs in $O(n^2)$ time.

13.9 Comparing Dominance Rules with Previous Ones

Now, we describe some dominance rules in the literature and we compare them to our dominance rules described in Sections 13.7 and 13.8.

The two first dominance rules (Theorems 13.9 and 13.10), have been first described by Chu [14] for the total tardiness criterion, and then generalized by Akturk and Ozdemir for the weighted case. Note that Theorem 13.10 has also been described for the total weighted completion time by Bianco and Ricciardelli. The three next dominance rules (Theorems 13.11 to 13.13), which have been described by Bianco and Ricciardelli, are only valid for the total weighted completion time criterion. For Theorem 13.13, we assume that the current partial sequence P is such that $P = (P' \mid J_k \mid J_y)$, i.e., jobs J_k and J_y are the last two jobs of the partial sequence P. Consider a job J_j belonging to set *PF* of jobs which can be scheduled just after sequence P. Theorem 13.13 gives sufficient condition for sequence $(P' \mid J_j, J_y, J_k)$ to dominate sequence $(P' \mid J_k, J_y, J_j)$. Finally, the two last dominance rules (Theorems 13.14 and 13.15) have been provided by Chu [23], and are only valid for the total completion time criterion. Recall that $\Delta_k(\sigma)$ denotes the completion time of job preceding job J_k in a sequence σ.

Theorem 13.9 [14,17]

Let J_j and J_k be two jobs belonging to NS. If $w_j \geq w_k$, $p_j \geq p_k$, $r_j + p_j \leq r_k + p_k$, and $d_j \leq d_k$, then job J_j dominates job J_k in the first position.

Theorem 13.10 [14,17,27]

If $\max_{J_i \in NS}\{r_i\} \leq \min_{J_i \in NS}\{r_i + p_i\}$ (it is said that the jobs are "dense"), and J_j and J_k are two jobs belonging to NS such that $r_j \leq r_k$, $d_j \leq d_k$, $w_j \geq w_k$, and $r_j + p_j \leq r_k + p_k$, then job J_j dominates job J_k in the first position.

Theorem 13.11 [27]

Let J_j and J_k be two jobs such that $J_j \in NS$, $J_k \in PF$. If $w_j \geq w_k$, $r_j + p_j \leq r_k + p_k$ and $w_j(r_k + p_k - r_j - p_j) + w_j p_j - w_k p_k \geq (p_k - p_j) \sum_{\{J_l \in \{J_l \in NS/\{J_j, J_k\} \cup Q\}} w_l$, then J_j dominates J_k in the first position.

Theorem 13.12 [27]

Let J_j and J_k be two jobs such that $J_j \in NS$, $J_k \in PF$. If $w_j \geq w_k$, $r_j + p_j \geq r_k + p_k$, and $w_j p_j - w_k p_k \geq (r_j + p_j - r_k - p_k) \sum_{\{J_l \in NS \cup Q\}} w_l + \max(0, p_k - p_j) \sum_{\{J_l \in NS/\{J_j, J_k\} \cup Q\}} w_l$, then J_j dominates J_k in the first position.

Theorem 13.13 [27]

Given a partial sequence $P = (P' \mid J_k \mid J_y)$, and given a job J_j belonging to set PF. If sequences (J_j, J_y, J_k) and (J_k, J_y, J_j) have no inserted idle times and $(r_k + p_k - r_j - p_j)(w_j + w_k + w_y) + (w_j - w_k)(p_j + p_k + p_y) + w_k p_j - w_j p_k \geq \max(0, r_j - r_k) \sum_{\{J_l \in NS/\{J_j\}\}} w_l$, then sequence $(P' \mid J_j, J_y, J_k)$ dominates sequence $(P' \mid J_k, J_y, J_j)$.

Theorem 13.14 [23]

Given a job J_j belonging to set PF. If there exists a job $J_k \in P$ such that $\max(r_j, \Delta_k(P)) + p_j \leq \max(r_k, \Delta_k(P)) + p_k$ and $\max(r_j, \Delta_k(P)) - \max(r_k, \Delta_k(P)) \leq (p_j - p_k)(|NS| - 1)$, then sequence $P \mid J_j$ is dominated.

TABLE 13.2 Example in which Theorem 13.7 Dominates Theorem 13.9 [14,17]

J_i	w_i	r_i	p_i	d_i
J_1	2	0	6	8
J_2	1	3	3	7
J_3	1	1	5	17
J_4	1	6	7	15

Theorem 13.15 [23]

Given a job J_j belonging to set PF. If there exists a job $J_k \in P$ in the ith position in sequence P such that $p_j \geq p_k$ and $p_j - p_k \geq (\max(r_j, \Delta_k(P)) + p_j - \max(r_k, \Delta_k(P)) - p_k)(|P| - i + 2)$, then sequence $P|J_j$ is dominated.

We can prove that our dominance rules described in Sections 13.7 and 13.8 dominate, generalize, and include all the previous dominance rules. This leads to the following propositions.

Proposition 13.1

Theorem 13.7 dominates Theorems 13.9 to 13.12.

Proposition 13.2

Theorem 13.6, finding better sequences with interchanges and insertions dominates Theorems 13.13 to 13.15.

Theorems 13.9 to 13.12 are all dominated by Theorem 13.7, since these theorems are different combinations of Cases 1 to 8. For each of these theorems, it is sufficient to prove that the theorem is a combination of special cases of Theorem 13.7, and that the corresponding situation leads always to a value of $\hat{\Gamma}_{ij}$ which is nonnegative. Theorem 13.7 allows us to deal with a lot of cases that are not taken into account by the theorems of the literature. For instance, Theorem 13.7 is more general than Theorem 13.9 because there is no condition on due dates. We provide an example of a scheduling situation in which Theorem 13.7 detects that J_1 dominates J_2 in the first position, while Theorem 13.9 does not detect it (see Table 13.2). We have $w_1 \geq w_2$, $r_1 + p_1 \leq r_2 + p_2$, $p_1 \geq p_2$, and $\hat{\Gamma}_{1,2}(0, 0, 0) = 4 > 0$; and according to Theorem 13.7, job J_1 dominates J_2 in first position. We have $d_1 > d_2$ and Theorem 13.9 cannot be applied.

It is also clear that Theorem 13.13 is a special case of Theorem 13.6 in which a "better" sequence is obtained by interchanging a job $J_j \in PF$ with the last but one job of the partial sequence P. Finally, Theorems 13.14 and 13.15 identify special cases for which Theorem 13.6 using interchanges or insertions always finds that job J_j dominates J_k in the first position. Note that complete proofs of Propositions 13.1 and 13.2 can be found in [36].

13.10 Look-Ahead Propagation Rule

To take advantage of lower bounds, we use a kind of look-ahead technique to test whether a job J_i can be removed of the set *PF* of possible first jobs: suppose that job J_i is sequenced immediately after P, and a lower bound (depending on the studied criteria) of the new scheduling situation is computed. If this lower bound is greater than $\bar{F}$, then J_i cannot be first and it is removed from *PF*. A symmetric rule is used for the set *PL*.

Note that the complexity of propagating this rule for all possible first and possible last jobs depends on the lower bound which is used for the propagation. The better the lower bound, the more the sets *PF* and *PL*, and thus the search space are reduced. Nevertheless, the best lower bounds are generally of high complexity and take a lot of time to be computed. We have therefore to pay attention to the lower bound which is used in the propagation. Experimental results have to be performed to find a lower bound which gives the best trade-off between the reduction of the search tree and the time spent in its computation.

Actually, we use the same lower bounds introduced in Section 13.2 for the branch-and-bound procedure.

13.11 Experimental Results

All techniques presented in this chapter have been incorporated into a branch-and-bound method implemented on top of Ilog Solver and Ilog Scheduler. All experimental results have been computed on a PC Dell Latitude 650 MHz running Windows 98. In aim to compare our results with the relevant previous works, the instances have been generated with corresponding schemes of the literature.

For the $1 \mid r_i \mid \sum T_i$ and the $1 \mid r_i \mid \sum w_i T_i$, we use the schemes of Chu [14] and Akturk and Ozdemir [17]. Each instance is generated randomly from uniform distributions of r_i, p_i, and d_i. The distributions of p_i are always between 1 and 10. The distributions of r_i and d_i depend on 2 parameters: α and β. For each job J_i, r_i is generated from the distribution $[0, \alpha \sum p_i]$ and $d_i - (r_i + p_i)$ is generated from the distribution $[0, \beta \sum p_i]$. In the weighted case, the distributions of w_i are always between 1 and 10. Four values for α and three values for β were combined to produce 12 instances sets, each containing 10 instances of n jobs with $n \in \{10, 20, 30, 40, 50, \ldots, 150\}$ jobs. We then obtain 120 instances for each size n.

For the $1 \mid r_i \mid \sum C_i$ and the $1 \mid r_i \mid \sum w_i C_i$, the instances are generated using the same scheme as the test problems of Hariri and Potts [28], and Belouadah, Posner, and Potts [26]. For each job J_i, we generate a processing time p_i from the uniform distribution [11,24] and a weight w_i from the uniform distribution [11,24] for the weighted case. For a size $n = \{10, 20, \ldots, 150\}$ of problem, an integer release date r_i for each job J_i was generated from the uniform distribution $[0, 50.5nR]$, where R controls the range of the distribution. For each selected value of n, five problems were generated for each of the R values 0.2, 0.4, 0.6, 0.8, 1.0, 1.25, 1.5, 1.75, 2.0, and 3.0 producing 50 problems for each value of n.

13.11.1 Dominance Rules, No-Good Recording, and Look-Ahead

In Table 13.3, we show the efficiency of Dominance Rules, no-good recording technique, and Look-Ahead propagation rule which have been described previously. For that, the branch-and-bound procedure is executed with different combinations of techniques, a time limit of 3600 seconds has been fixed.

For each criterion and for each n, we report the average number of fails ("Bck."), and the average computation time in seconds over all the generated instances. Each time, we first report the results obtained with the dominance properties used in the previous works of the literature (Akturk and Ozdemir [17] for $1 \mid r_i \mid \sum w_i T_i$, Chu [14,22] for $1 \mid r_i \mid \sum T_i$ and $1 \mid r_i \mid \sum C_i$, and Belouadah, Posner, and Potts [26] for $1 \mid r_i \mid \sum w_i C_i$). Note that for some values of n, we do not report results ("—") since several instances are not solved within the time limit of 3600 seconds. We then report, the results when dominance rules of the literature are replaced with our dominance rules described in Sections 13.7 and 13.8 (DR). Next, we report results obtained when we add the no-good recording (NGR) technique described in Section 13.4. Finally, we report results obtained when we add the Look-Ahead propagation rule (LA) described in Section 13.10. To complete the results, each technique is then removed one by one in order to see the real contribution of each of one to reduce the search space or the computational time. For the $1 \mid r_i \mid \sum T_i$ problem, since the Look-Ahead propagation rule seems to be very costly from a computational point of view, we remove it to finish the tests.

We can see that all our techniques are useful to reduce the search space and the computational time. We can see that our dominance rules are more efficient than the dominance rules of the literature. However, the Look-Ahead technique seems to be very costly in terms of CPU time compared to the corresponding reduction of the search tree. This is due to the relatively high complexity of the lower bound that is used several times in the Look-Ahead. We tried to use some weaker lower bound but it does not reduce the

TABLE 13.3 Efficiency of Dominance Properties, No-Good Recording, and Look-Ahead Propagation Rule

				$n = 15$		$n = 15$		$n = 15$	
				Bck.	CPU	Bck.	CPU	Bck.	CPU
	Akturk and Ozdemir			2378	1.23	—	—	—	—
	DR			40	0.11	273	0.80	1464	4.73
$1\|r_i\|\sum w_i T_i$	DR	+NGR		27	0.11	173	0.66	695	2.95
	DR	+NGR	+LA	16	0.09	95	0.48	505	2.39
		+NGR	+LA	62	0.19	536	1.47	4559	23.23
	DR		+LA	19	0.10	118	0.55	727	3.32
				$n = 20$		$n = 30$		$n = 40$	
				Bck.	CPU	Bck.	CPU	Bck.	CPU
	Chu			99	0.11	8469	12	384628	1211
	DR			44	0.13	353	1.98	3194	33.6
$1\|r_i\|\sum T_i$	DR	+NGR		39	0.15	297	1.79	2824	31.8
	DR	+NGR	+LA	12	0.15	103	2.10	1051	88.8
		+NGR		96	0.15	2336	6.15	27029	172.8
				$n = 40$		$n = 50$		$n = 60$	
				Bck.	CPU	Bck.	CPU	Bck.	CPU
	Belouadah et al.			4338	7.91	57704	130	—	—
	DR			182	1.07	841	7.45	4716	47.4
$1\|r_i\|\sum w_i C_i$	DR	+NGR		108	0.79	497	5.34	1339	17.49
	DR	+NGR	+LA	43	0.66	143	3.75	477	13.64
		+NGR	+LA	103	0.89	342	6.80	1259	28.96
	DR		+LA	60	0.78	203	4.86	1262	32.28
				$n = 40$		$n = 50$		$n = 60$	
				Bck.	CPU	Bck.	CPU	Bck.	CPU
	Chu			329	0.35	832	1.37	23697	46
	DR			58	0.23	156	0.82	838	5.70
$1\|r_i\|\sum C_i$	DR	+NGR		34	0.15	86	0.53	274	2.33
	DR	+NGR	+LA	14	0.16	44	0.51	125	2.27
		+NGR	+LA	25	0.11	86	0.48	314	3.10
	DR		+LA	17	0.16	59	0.62	240	4.05

search space. We can see that the dominance rules seem to be the more efficient technique to reduce both the search space and the computational time. Moreover, our techniques seem to be very efficient compared with the ones described in the literature.

13.11.2 Main Results

The results obtained with the version of the algorithm that incorporates the best combination of our techniques are presented in Table 13.4. A time limit of 3600 seconds has been fixed. For each criterion and for several values of n, we provide the minimum (min.), the average (av.), and the maximum (max.) number of fails (bck.); and the minimum, the average, and the maximum computation time in seconds over the generated instances. The numbers in parentheses are the number of instances which are not solved within the time limit. To compute the averages for these instances, we then use the lower bound of the real number of backtracks and the real computation time, by replacing these values with the current values at 3600 sec.

For the $1\,|\,r_i\,|\sum w_i T_i$ problem, all instances are solved within the time limit for up to 30 jobs. For $n = 35$, five instances cannot be solved within the time limit. These instances belong to the set of instances generated with ($\alpha = 0.5, \beta = 0.5$). As noticed earlier by Chu [14] for the total tardiness

TABLE 13.4 Main Results of the Algorithm Incorporating the Best Combination of Techniques

$1 \mid r_i \mid \sum w_i T_i$			$1 \mid r_i \mid \sum T_i$		
n	Bck.	CPU (s)	n	Bck.	CPU (s)
10	4	0,03	10	1	0,02
15	16	0,09	20	12	0,05
20	95	0,49	30	52	0,29
25	505	2,37	40	313	3,44
30	4255	26	50	1307	21
35	18192	283 (5)			
$1 \mid r_i \mid \sum w_i C_i$			$1 \mid r_i \mid \sum C_i$		
n	Bck.	CPU (s)	n	Bck.	CPU (s)
10	1	0,03	10	0	0,03
20	5	0,05	20	6	0,06
30	22	0,25	30	15	0,17
40	43	0,65	40	37	0,54
50	143	3,90	50	124	2,37
60	477	14	60	269	7
70	556	23	70	323	10
80	1007	43	80	484	19
90	3559	179	90	1175	57
100	7363	416 (2)	100	2093	118
			110	3046	223 (1)
			120	9415	566 (4)

problem, instances generated according to this particular combination seem to be *hard* to solve in practice. From this table, we can remark that the *hardness* increases very quickly with n. In practice most of the instances are solved within a few seconds with up to 30 jobs and our results compare well to those of Akturk and Ozdemir [17]. For instance, the average number of fails for the combination ($\alpha = 0.5, \beta = 0.5$) with $n = 20$ jobs was greater than one million in [17], whereas this number is now lower than 700. Moreover, Akturk and Ozdemir solve all instances with up to 15 jobs, whereas this number is now 30.

For the $1 \mid r_i \mid \sum T_i$ problem, all instances are solved within the time limit for up to 50 jobs. For $n = 60$, several instances cannot be solved within the time limit. These instances belong to the set of instances generated with ($\alpha = 0.5, \beta = 0.5$). In practice most of the instances are solved within 30 seconds and our results compare well to those of [14]. For instance, the average number of fails for the combination ($\alpha = 0.5, \beta = 0.5$) was greater than 36000 in [14], whereas this number is now lower than 300. Moreover, Chu solves all instances with up to 30 jobs, whereas this number is now 50.

For the $1 \mid r_i \mid \sum w_i C_i$ problem, all instances are solved within the time limit of 3600 seconds for up to 90 jobs. For $n = 100$, two instances have not been solved within the time limit. It is a little tricky to compare our results to those of [26], since Belouadah, Posner, and Potts have only tested their branch-and-bound procedure with up to 50 jobs. Nevertheless, our branch-and-bound procedure using only the techniques described in [26], cannot solve all instances with $n = 60$ jobs within the time limit (see Table 13.3). Moreover, the average number of fails for $n = 50$ was greater than 14000 in [26], whereas this number is now lower than 150.

For the $1 \mid r_i \mid \sum C_i$ problem, all instances are solved within the time limit of one hour for up to 100 jobs. Only one instance of 110 jobs is not solved within the time limit. It is also a little tricky to compare our results to those of [22,24], since they use another branch-and-bound method and other techniques and dominance properties which are specific to their branch-and-bound procedure. Therefore, our results only slightly improve these previous results.

13.12 Conclusion

We have described new dominance rules, a new propagation technique, and a new no-good recording technique for the $1 \mid r_i \mid \sum w_i T_i$ problem. Computational results show the efficiency of our techniques to drastically reduce both the search space and the computational time. Moreover, the proposed approach outperforms the best known procedures for the $1 \mid r_i \mid \sum w_i T_i$ problem and its special cases ($1 \mid r_i \mid \sum T_i$, $1 \mid r_i \mid \sum w_i C_i$, $1 \mid r_i \mid \sum C_i$). We hope to improve our results with new lower bounds and new dominance rules. In particular, no specific lower bound has been found for the $1 \mid r_i \mid \sum w_i T_i$ problem and developing such a lower bound should improve the results for this criterion. Note that the described Dominance Rules, the no-good recording, and the Look-Ahead propagation rule are original and powerful generic techniques which can be adapted to a lot of scheduling problems in which other objective functions and other constraints can be taken into account.

References

[1] E.L. Lawler. A pseudo-polynomial algorithm for sequencing jobs to minimize total tardiness. *Annals of Discrete Mathematics*, 1:331–342, 1977.

[2] A.H.G. Rinnooy Kan. *Machine Sequencing Problem: Classification, Complexity and Computation*. Nijhoff, The Hague, 1976.

[3] H. Emmons. One-machine sequencing to minimize certain functions of job tardiness. *Operations Research*, 17:701–715, 1969.

[4] J. Du and J.Y.T. Leung. Minimizing total tardiness on one processor is np-hard. *Mathematics of Operations Research*, 15:483–495, 1990.

[5] C.N. Potts and L.N. Van Wassenhove. A decomposition algorithm for the single machine total tardiness problem. *Operations Research Letters*, 26:177–182, 1982.

[6] S. Chang, Q. Lu, G. Tang, and W. Yu. On decomposition of the total tardiness problem. *Operations Research Letters*, 17:221–229, 1995.

[7] W. Szwarc, F. Della Croce, and A. Grosso. Solution of the single machine total tardiness problem. *Journal of Scheduling*, 2:55–71, 1999.

[8] W. Szwarc, A. Grosso, and F. Della Croce. Algorithmic paradoxes of the single machine total tardiness problem. *Journal of Scheduling*, 4:93–104, 2001.

[9] A.H.G. Rinnooy Kan, B.J Lageweg, and J.K. Lenstra. Minimizing total costs in one-machine scheduling. *Operation Research*, 23:908–927, 1975.

[10] R.M.V. Rachamadugu. A note on weighted tardiness problem. *Operations Research*, 35:450–452, 1987.

[11] T.S. Abdul-Razacq, C.N. Potts, and L.N. Van Wassenhove. A survey of algorithms for the single machine total weighted tardiness scheduling problem. *Discrete Applied Mathematics*, 26:235–253, 1990.

[12] M.S. Akturk and M.B. Yildrim. A new lower bounding scheme for the total weighted tardiness problem. *Computers and Operations Research*, 25:265–278, 1998.

[13] C. Chu and M.C. Portmann. Some new efficient methods to solve the $n \mid 1 \mid r_i \mid \sum T_i$ Scheduling problem. *European Journal of Operational Research*, 58:404–413, 1991.

[14] C. Chu. A branch and bound algorithm to minimize total tardiness with different release dates. *Naval Research Logistics*, 39:265–283, 1992.

[15] P. Baptiste, J. Carlier, and A. Jouglet. A branch and bound procedure to minimize total tardiness on one machine with arbitrary release dates. *European Journal of Operational Research*, 2002. (Unpublished)

[16] M.S. Akturk and D. Ozdemir. A new dominance rule to minimize total weighted tardiness with unequal release dates. *European Journal of Operational Research*, 135:394–412, 2001.

[17] M.S. Akturk and D. Ozdemir. An exact approach to minimizing total weighted tardiness with release dates. *IIE Transactions*, 32:1091–1101, 2000.

[18] W.E. Smith. Various optimizers for single stage production. *Naval Research Logistics Quarterly*, 3:59–66, 1956.
[19] R. Chandra. On $n/1/\bar{F}$ dynamic determistic systems. *Naval Research Logistics*, 26:537–544, 1979.
[20] M.I. Dessouky and D.S. Deogun. Sequencing jobs with unequal ready times to minimize mean flow time. *SIAM Journal of Computing*, 10:192–202, 1981.
[21] D.S. Deogun. On scheduling with ready times to minimize mean flow time. *Computer Journal*, 26:320–328, 1983.
[22] C. Chu. A branch and bound algorithm to minimize total flow time with unequal release dates. *Naval Research Logistics*, 39:859–875, 1991.
[23] C. Chu. Efficient heuristics to minimize total flow time with release dates. *Operations Research Letters*, 12:321–330, 1992.
[24] S. Chand, R. Traub, and R. Uzsoy. Single-machine scheduling with dynamic arrivals: Decomposition results and an improved algorithm. *Naval Research Logistics*, 43:709–716, 1996.
[25] J. Labetoulle, E.L. Lawler, J.K. Lenstra, and A.H.G Rinnooy Kan. Preemptive scheduling of uniform machines subject to release dates. pages 245–261, 1984. in W.R. Pulleyblank (ed.), *Progress in Combinatorial Optimization*. Academic Press, New York.
[26] H. Belouadah, M.E. Posner, and C.N. Potts. Scheduling with release dates on a single machine to minimize total weighted completion time. *Discrete Applied Mathematics*, 36:213–231, 1992.
[27] L. Bianco and S. Ricciardelli. Scheduling of a single machine to minimize total weighted completion time subject to release dates. *Naval Research Logistics*, 29:151–167, 1982.
[28] A.M.A Hariri and C.N. Potts. An algorithm for single machine sequencing with release dates to minimize total weighted completion time. *Discrete Applied Mathematics*, 5:99–109, 1983.
[29] G. Rinaldi and A. Sassano. On a job scheduling problem with different ready times: Some properties and a new algorithm to determine the optimal solution. *Operations Research*, 1977. Rapporto dell'Ist. di Automatica dell'Universita di Roma e del C.S.S.C.C.A.-C.N.R.R, Report R.77-24.
[30] O. Lhomme. Consistency techniques for numeric csps. *Thirteenth International Joint Conference on Artificial Intelligence*, Chambéry, France, 1993.
[31] J. Carlier. Ordonnancements à contraintes disjonctives. *RAIRO*, 12:333–351, 1978.
[32] G. Verfaillie and T. Schiex. Solution reuse in dynamic constraint satisfaction problems. *AAAI'94*, Seattle, WA, U.S., 307–312, 1994.
[33] G. Verfaillie and T. Schiex. Maintien de solutions dans les problèmes dynamiques de satisfaction de contraintes : bilan de quelques approches. *Revue d'Intelligence Artificielle*, 9(3), 1995.
[34] K.R. Baker. *Introduction to Sequencing and Scheduling*. John Wiley and Sons, 1974.
[35] R.W. Conway, W.C. Maxwell, and L.W. Miller. *Theory of Scheduling*. Addison Wesley, Reading, MA, 1967.
[36] A. Jouglet. *Ordonnancer une Machine pour Minimiser la Somme des Coûts — The One Machine Total Cost Sequencing Problem*. PhD thesis, Université de Technologie de Compiègne, Compiègne, France, 2002.

14

Scheduling Equal Processing Time Jobs

Philippe Baptiste
CNRS

Peter Brucker
Universität Osnabrück

14.1 Single Machine Problems

14.1.1 Problem $1 \mid p_j = 1; r_j \mid \sum f_j$

We assume that the objective function has the form $\sum_{j=1}^{n} f_j(C_j)$, where $f_j(C_j)$ is a nondecreasing function of the finishing time C_j of job j $(j = 1, \ldots, n)$.

Due to the monotonicity of the functions f_j the jobs should be scheduled as early as possible. The n earliest time slots t_j for scheduling all n jobs may be calculated using the following algorithm, in which we assume that the jobs are enumerated such that

$$r_1 \leq r_2 \leq \cdots \leq r_n$$

Algorithm Time Slots

1. $t_1 := r_1$;
2. FOR $i := 2$ TO n DO
   $t_i := \max\{r_i, t_{i-1} + 1\}$

One can show by simple induction that an optimal schedule exists in which a job starts at each of the times t_i. Thus the problem can be formulated by a bipartite min-cost assignment problem, in which jobs

1-58488-397-9/$0.00+$1.50

$j = 1, \dots, n$ are to be assigned to time slots $I_i = [t_i, t_i + 1]$ $(i = 1, \dots, n)$. In the corresponding bipartite graph there is an arc (j, i) with cost $f_j(t_i + 1)$ for each i with $t_i \geq r_j$.

14.1.2 Problem $1 \mid p_j = p, r_j \mid \sum f_j$

In the following, we show that single machine problems with identical processing times can be solved polynomially for a large class of so-called *ordered objective functions.* All these results are extended to parallel machines problems in Section 14.2.9.

Definition 14.1

F is an ordered objective function if and only if:

1. *F is a sum function, i.e.* $F = \sum f_j(C_j)$.
2. *F is regular, i.e.,* $\forall j$, f_j *is nondecreasing.*
3. f_j *is constant after a time point* δ_j*. i.e.,* $\forall t > \delta_j$, $f_j(t) = \omega_j$.
4. $\forall j < k$, $\delta_j \leq \delta_k$ *and* $t \mapsto (f_j - f_k)(t)$ *is nondecreasing over* $[0, \delta_j]$.

It is easy to verify that the weighted number of late jobs, $\sum w_i U_i$, is an ordered objective function if we assume that $d_1 \leq d_2 \leq \cdots \leq d_n$. On the contrary, $\sum w_i C_i$ with $w_i \geq 0$ and $\sum T_i$ are not ordered objective functions. However, conditions 1 and 2 of Definition 14.1 hold for these functions and jobs can be renumbered to meet condition 4.

We show how a function such as $\sum w_i C_i$ or $\sum T_i$ can be modified, without changing the optimum value, to become an ordered objective function: Consider a large time point T and alter the functions f_i after T so that $\forall t \geq T$, $f_i(t) = M$, where M is another large value. If T and M are large enough, the optimum of the modified problem is also the optimum of the original one. Moreover, the modified functions are ordered objective functions.

From now on, we restrict our study to ordered objective functions. We have $\delta_1 \leq \cdots \leq \delta_n$. By analogy with due date scheduling, we say that a job is late when it is completed after δ_i and that it is on-time otherwise. The "late" cost is ω_i and the early cost is time-dependent. Notice that a late job can be scheduled arbitrarily late.

We first introduce a dominance property on starting times and we describe the variables of the dynamic programming algorithm which solves these problems and a proposition that links the variables together. Finally, the dynamic programming algorithm itself is described.

14.1.2.1 Dominance Property

Proposition 14.1 provides a characterization of the time points at which jobs start and end on active schedules (a schedule is called active if it is not possible to schedule jobs earlier without violating some constraint).

Proposition 14.1

In active schedules, start and completion times belong to $\Theta = \{t \mid \exists r_i, \exists l \in \{0, \dots, n\}, t = r_i + lp\}$.

Proof

Consider an active schedule and a job k. Let t be the largest time point, before the start time of k, at which the machine is idle immediately before t. If t is not a release date, the job scheduled immediately after t could be scheduled earlier and thus the schedule would not be active. Therefore, t is then a release date, say r_i. Between r_i and the starting time of k, l jobs execute $(0 \leq l \leq n - 1)$. Hence the starting time and the ending time of k belong to Θ. □

Since active schedules are dominant for regular objective functions, we will restrict our search to schedules where starting times belong to Θ.

14.1.2.2 Variables Definition and Decomposition Scheme

Recall that late jobs can be scheduled arbitrarily late. Hence, there is no point to assign a starting time to late jobs. In the following, given a set of jobs O and a schedule $\mathcal{H}$ of some of these jobs (some jobs are not scheduled), the "cost" of $\mathcal{H}$ with respect to O is

- the sum, over all jobs $i \in O$ scheduled in $\mathcal{H}$, of $f_i(C_i)$
- plus the sum, over all other jobs $i \in O$ not scheduled in $\mathcal{H}$, of ω_i

For any integer $k \leq n$, let $U_k(s, e)$ be the set of jobs whose index is lower than or equal to k and whose release date is in the interval $[s, e)$. Let $F_k(s, e)$ be the minimal cost over all feasible schedules $\mathcal{H}$ of the jobs in $U_k(s - p, e)$ such that

1. the machine is not used before s on $\mathcal{H}$
2. the machine is not used after e on $\mathcal{H}$
3. starting times of jobs on $\mathcal{H}$ belong to Θ

$F_k(s, e)$ is always defined since an empty schedule (i.e., a schedule where all jobs of $U_k(s - p, e)$ are late) is a feasible schedule and meets Constraints 1, 2 and 3. Notice that given our definition, $F_0(s, e)$ is equal to 0.

Proposition 14.2

Let $k \in [1, n]$ and let $[s, e)$ be any time interval. If $r_k \notin [s - p, e)$, $F_k(s, e) = F_{k-1}(s, e)$. If $r_k \in [s - p, e)$, then $F_k(s, e)$ is equal to $\min(\omega_k + F_{k-1}(s, e), F'_k(s, e))$*; where*

$$F'_k(s, e) = \min_{\substack{\max(s, r_k) \leq t_k \\ t_k + p \leq \min(e, \delta_k) \\ t_k \in \Theta}} F_{k-1}(s, t_k) + F_{k-1}(t_k + p, e) + f_k(t_k + p)$$

If $F'_k(s, e)$ is undefined, assume that $F'_k(s, e) = \infty$.

Proof

If $r_k \notin [s - p, e)$ the proposition holds because $U_k(s - p, e) = U_{k-1}(s - p, e)$. Now, assume that $r_k \in [s - p, e)$.

We first prove that $\omega_k + F_{k-1}(s, e) \geq F_k(s, e)$ and that $F'_k(s, e) \geq F_k(s, e)$.

- The schedule that realizes $F_{k-1}(s, e)$ and where k is late is feasible for the set of jobs $U_k(s - p, e)$. The cost of this schedule is $\omega_k + F_{k-1}(s, e)$. So, we have $\omega_k + F_{k-1}(s, e) \geq F_k(s, e)$
- Assume that $F'_k(s, e)$ is finite. According to the definition of $F'_k(s, e)$, there exists a time point t_k such that (1) $t_k \geq \max(s, r_k)$, (2) $t_k \leq \min(e, \delta_k) - p$ and (3) $F'_k(s, e) = F_{k-1}(s, t_k) + F_{k-1}(t_k, e) + f_k(t_k + p)$. Let $\mathcal{H}_1$ and $\mathcal{H}_2$ be two schedules that realize $F_{k-1}(s, t_k)$ and $F_{k-1}(t_k + p, e)$, respectively. Notice that any job in $U_{k-1}(s - p, e)$ is either late or scheduled in $\mathcal{H}_1$ or in $\mathcal{H}_2$. Consider the schedule $\mathcal{H}$ built as follows: schedule k at time t_k and "add" $\mathcal{H}_1$ and $\mathcal{H}_2$. Given the definition of F_{k-1}, $\mathcal{H}$ is a feasible schedule of $U_k(s - p, e)$, *i.e.*, jobs do not overlap in time and start (after their release dates) at time points that obviously belong to Θ. On top of that, $\mathcal{H}$ is idle before s and after e. Since $t_k + p \leq \delta_k$, k is on-time and thus, the cost of $\mathcal{H}$ is exactly $F'_k(s, e)$. Hence, $F'_k(s, e) \geq F_k(s, e)$.

We now prove that $\min(\omega_k + F_{k-1}(s, e), F'_k(s, e)) \leq F_k(s, e)$. Consider a schedule that realizes $F_k(s, e)$ and let O be the set of jobs scheduled in this schedule (jobs in $O \setminus U_k(s - p, e)$ are late). Among all schedules (1) that realize $F_k(s, e)$ and (2) in which the same set of jobs O are scheduled, consider the schedule $\mathcal{H}$ that lexicographically minimizes the vector made of completion times of jobs in O. (The completion time of the job in O, with the smallest index, is minimum, then the completion time of the job in O, with the second smallest index, etc.) The job k is either late or on-time on $\mathcal{H}$.

- If k is late then $\mathcal{H}$ is also a feasible schedule for $U_{k-1}(s-p,e)$ and its cost is exactly $F_k(s,e) - \omega_k$ (the late cost ω_k is removed since k is not considered any longer). Thus, $F_k(s,e) - \omega_k \geq F_{k-1}(s,e)$ and our claim holds.
- Now assume that k is on-time and let $t_k \in \Theta$ be the starting time of k in $\mathcal{H}$. Note that $t_k + p \leq \min(e, \delta_k)$ and $\max(s, r_k) \leq t_k$. In the following, we show that this starting time allows us to "partition" all jobs.

  Suppose that there is a job i with $r_i \leq t_k$ that is executed on-time after k, at time t_i in $\mathcal{H}$ $(t_k \leq t_i)$. Let $\mathcal{H}'$ be the schedule obtained from $\mathcal{H}$ by exchanging the jobs i and k. The exchange is valid because $r_i \leq t_k$ and moreover, both jobs remain on-time because $\delta_i \leq \delta_k$. The relative cost Δ of the exchange is

$$\begin{aligned}\Delta &= f_i(t_k+p) + f_k(t_i+p) - (f_i(t_i+p) + f_k(t_k+p))\\ &= (f_i - f_k)(t_k+p) - (f_i - f_k)(t_i+p)\end{aligned}$$

  Note that $t_i + p \leq \delta_i$ because i is on-time. Moreover, $i < k$ and thus $t \mapsto (f_i - f_k)(t)$ is nondecreasing over $[0, \delta_i]$. Hence $\Delta \leq 0$. Given the definition of $F_k(s,e)$, Δ cannot be strictly negative (this would contradict the fact that $\mathcal{H}$ realizes the minimum). Hence $\mathcal{H}$ and $\mathcal{H}'$ have the same cost. However, $\mathcal{H}'$ is better than $\mathcal{H}$ for the lexicographical order. This contradicts our hypothesis on $\mathcal{H}$. *Hence, all jobs with a release date lower than or equal to t_k are late or are scheduled before t_k.*

  Now, let us decompose $\mathcal{H}$. Let $\mathcal{H}_1$ and $\mathcal{H}_2$ denote the left and the right parts of $\mathcal{H}$ (before t_k and after $t_k + p$). The cost of $\mathcal{H}$ is the sum of the costs of $\mathcal{H}_1$ and $\mathcal{H}_2$ plus $f_k(t_k+p)$. Since all jobs with a release date lower than or equal to t_k are late or are scheduled before t_k, the cost of $\mathcal{H}_1$ is greater than $F_{k-1}(s, t_k)$ and the cost of $\mathcal{H}_2$ is greater than $F_{k-1}(t_k+p, e)$. Hence, $F_k(s,e) \geq F_{k-1}(s,t_k) + F_{k-1}(t_k+p,e) + f_k(t_k+p) \geq F_k'(s,e)$. □

14.1.2.3 A Dynamic Programming Algorithm

Active schedules are dominant so, the optimum is exactly $F_n(\min_{t\in\Theta} t, \max_{t\in\Theta} t)$. Thanks to Proposition 14.2, we have a straight dynamic programming algorithm to reach the optimum. The relevant values for s and e are exactly those in Θ. The values of $F_k(s,e)$ are stored in a multi-dimensional array of size $O(n^5)$ (n possible values for k, n^2 possible values both for s and e). Our algorithm works as follows

- In the initialization phase, $F_0(s,e)$ is set to 0 for any values s, e in Θ $(s \leq e)$.
- We then iterate from $k = 1$ to $k = n$. Each time, F_k is computed for all the possible values of the parameters thanks to the formula of Proposition 14.2, and to the values of F_{k-1} computed at the previous step.

The initialization phase runs in $O(n^4)$ because the size of Θ is upper bounded by n^2. Afterward, for each value of k, $O(n^4)$ values of $F_k(s,e)$ have to be computed. For each of them, a maximum among $O(n^2)$ terms is computed (because there are $O(n^2)$ possible values for $t_k \in \Theta$). This leads to an overall time complexity of $O(n^7)$. A rough analysis of the space complexity leads to an $O(n^5)$ bound but since, at each step of the outer loop on k, one only needs the values of F computed at the previous step $(k-1)$, the algorithm can be implemented with two arrays of $O(n^4)$ size: One for the current values of F and one for the previous values of F. (To build the optimal schedule, all values of $F_k(s,e)$ have to be kept; hence the initial $O(n^5)$ bound applies.)

14.1.3 Problem $1 \mid p_j = p; prec; r_j \mid \sum C_j$

If $i \rightarrow j$ and $r_i + p > r_j$ then job j cannot start before $r_j' := r_i + p$. Thus, we may replace r_j by r_j'. The release times can be modified in a systematic way such that for the new release times r_j' the conditions

$$r_j' \geq r_i' + p > r_i' \quad \text{if } i \rightarrow j$$

are satisfied (cf. Brucker [2001], p.63).

The following algorithm calculates the job starting times S_j of an optimal schedule.

Algorithm $1 \mid p_j = p; prec; r_j \mid \sum C_j$

1. Calculate the modified release times r'_j
2. Enumerate the job such that $r'_1 \leq r'_2 \leq \cdots \leq r'_n$
3. $S_1 := r'_1$
4. FOR $j = 2$ TO n DO
   $S_j := \max\{S_{j-1} + p, r'_j\}$

It is not difficult to prove that a schedule S constructed by this algorithm must be optimal. First note that a schedule satisfies the original release times if and only if it satisfies the modified release times. Thus, S is feasible. Now, define a block as a maximal subschedule without idle times. Optimality can be shown by induction on the number of blocks in the schedule without machine idle time. Assume that S decomposes into blocks $B_1, \ldots, B_r$ where B_{i-1} is scheduled before B_i. Each block B_i starts at a release time r'_j of a job j in B_i. B_r is optimally scheduled because the jobs in B_r are scheduled without idle times and all release times of jobs in B_r are greater or equal to the starting time of B_r. By induction jobs in $\bigcup_{\nu=1}^{r-1} B_\nu$ are scheduled optimally as well.

14.1.4 Problem $1 \mid p_j = p, prec, pmtn, r_j \mid \sum Cj$

As we show, there is no advantage to preemptions in this problem. Thus it can be solved in polynomial time by the algorithm for $1 \mid p_j = p, prec, r_j \mid \sum Cj$.

Let t be the earliest preemption time, and let j be the job interrupted at time t, started at time S_j and completed at time C_j in a preemptive schedule. Let $\mathcal{K}$ and k denote the set and number of jobs, respectively, that are started not earlier than t and completed earlier than C_j.

We can reconstruct the schedule in the time interval $[t, C_j]$ by moving all parts of j into the time interval $[t, S_j + p]$, so that j becomes uninterrupted, and placing all parts of other jobs in $[t, C_j]$ without changing their order into the time interval $[S_j + p, C_j]$. The reconstruction obviously observes precedence constraints and release dates.

One can make sure that the decrease of completion time of j is at least kp and that the increase of total completion time of jobs in $\mathcal{K}$ is at most $k(p - t + S_j)$. Thus, the reconstruction decreases the total completion time by at least $k(t - S_j)$.

Recursively applying the reconstruction procedure we obtain a nonpreemptive schedule that is not worse than the original preemptive schedule.

14.1.5 Problem $1 \mid p_j = p, pmtn, r_j \mid \sum w_j U_j$

$1 \mid pmtn, r_j \mid \sum w_j U_j$ is NP-hard but can be solved in pseudo-polynomial time by Lawler's algorithm [1] whose time and space complexities are respectively $O(nk^2W^2)$ and $O(k^2W)$, where k is the number of distinct release dates and where W is the sum of the weights of the jobs. If weights are equal the problem obviously becomes strongly polynomial.

In the following, we describe a dynamic programming from [2] that solves the weighted problem when processing times are equal. We assume that due dates are sorted in nondecreasing order, i.e., $d_1 \leq d_2 \leq \cdots \leq d_n$.

14.1.5.1 Problem Reformulation

Let Z be a subset of jobs and let Jackson Preemptive Schedule (JPS) be the preemptive schedule of these jobs associated to the EDD dispatching rule: Whenever the machine is free and at least one job is available, schedule the job i for which d_i is the smallest. If a job j becomes available while i is in process, stop i and start j if d_j is strictly smaller than d_i; otherwise continue i. Jackson Preemptive Schedule has several interesting properties (e.g., [3]). In particular, if a job is scheduled on JPS after its due date, there is no preemptive schedule of Z where all jobs are on-time. In the following, Z is said to be feasible if and only if all jobs are on-time in its JPS. Given this definition, searching for a schedule on which the weighted number of late jobs is minimal, reduces to finding a set of jobs whose weight is maximal and whose JPS is feasible.

14.1.5.2 Dominance Property

Proposition 14.3

For any subset of jobs Z, the start and end times of the jobs on the JPS of Z belong to the set Θ. $\Theta = \{t \mid \exists r_i, \exists l \in \{0, \ldots, n\}, t = r_i + lp\}$.

Proof

We first prove that the end time of a job on the JPS of Z belongs to Θ. Let k be any job and let s and e be respectively its start and end times on JPS. Let t be the minimal time point such that between t and s JPS is never idle. Because of the structure of JPS, t is a release date, say r_x. The jobs that execute (even partially) between s and e do not execute before s nor after e (because Jackson Preemptive schedule is based upon the EDD rule). Thus $e - s$ is a multiple of p. Two cases can occur

- Either k causes an interruption and hence $s = r_k$.
- Or k does not cause any interruption and hence the jobs that execute between r_x and s, are fully scheduled in this interval. Consequently, $s - t$ is a multiple of p.

In both cases, there is a release date r_y (either r_k or r_x) such that between r_y and e, JPS is never idle and such that e is equal to r_y plus a multiple of p. On top of that, the distance between r_y and e is not greater than np (because JPS is not idle). Hence, $e \in \Theta$.

Now consider the start time of any job on JPS. This time point is either the release date of the job or is equal to the end time of the "previous" one. Thus, start times also belong to Θ. □

In the following, we note $t_1 < t_2 < \cdots < t_q$ the distinct time-points in Θ. Recall that $q \leq n^2$.

Definition 14.2

For any time points t_u, t_v in Θ with $u < v$ and for any integer value k such that $1 \leq k \leq n$

- *Let $U_k(t_u, t_v) = \{i \mid i \leq k, t_u \leq r_i < t_v\}$.*
- *For any $m \leq n$, let $W_k(t_u, t_v, m)$ be the maximal weight of a subset $Z \subseteq U_k(t_u, t_v)$ of m jobs such that, the JPS of Z is feasible and ends before t_v. If there is no such subset, $W_k(t_u, t_v, m)$ is set to $-\infty$.*

Proposition 14.4

For any time points $t_u, t_v \in \Theta$ with $u < v$ and any integer values k and m such that $1 < k \leq n$ and $1 \leq m \leq n$, $W_k(t_u, t_v, m)$ can be computed as follows.

If $r_k \notin [t_u, t_v)$, $W_k(t_u, t_v, m) = W_{k-1}(t_u, t_v, m)$. Otherwise, $W_k(t_u, t_v, m) = \max(W_{k-1}(t_u, t_v, m), W')$ where W' is the maximum of

$$W_{k-1}(t_u, t_x, m_1) + W_{k-1}(t_x, t_y, m_2) + W_{k-1}(t_y, t_v, m_3) + w_k$$

under the constraints

$$\begin{cases} t_x, t_y \in \Theta \\ \max(r_k, t_u) \leq t_x < t_y \leq \min(d_k, t_v) \\ m_1 + m_2 + m_3 = m - 1 \\ p(m_2 + 1) = t_y - t_x \end{cases}$$

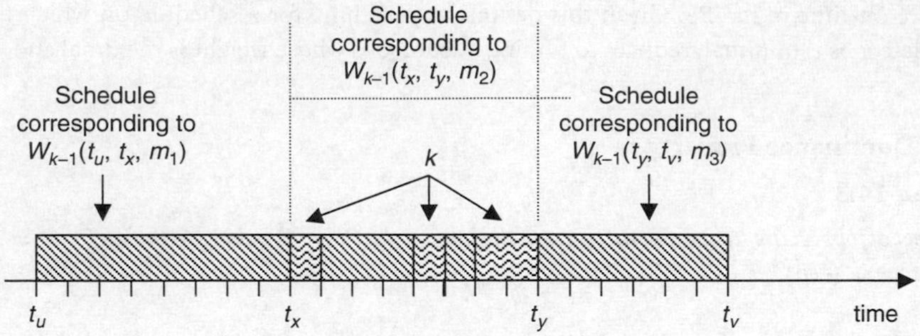

FIGURE 14.1 Preemptive schedule of k.

Proof

It is obvious that if $r_k \notin [t_u, t_v)$, $W_k(t_u, t_v, m)$ is equal to $W_{k-1}(t_u, t_v, m)$. In the following, we suppose that $r_k \in [t_u, t_v)$.

We first prove that $\max(W_{k-1}(t_u, t_v, m), W') \leq W_k(t_u, t_v, m)$.

- *Case 1.* $W_{k-1}(t_u, t_v, m) \geq W'$. Since $U_{k-1}(t_u, t_v) \subseteq U_k(t_u, t_v)$, we have

$$\max(W_{k-1}(t_u, t_v, m), W') = W_{k-1}(t_u, t_v, m) \leq W_k(t_u, t_v, m)$$

- *Case 2.* $W' > W_{k-1}(t_u, t_v, m)$. There exist $t_x \in \Theta, t_y \in \Theta$ and 3 integers m_1, m_2, m_3 such that

$$\begin{cases} \max(r_k, t_u) \leq t_x < t_y \leq \min(d_k, t_v) \\ m_1 + m_2 + m_3 = m - 1 \\ p(m_2 + 1) = t_y - t_x \\ W' = W_{k-1}(t_u, t_x, m_1) + W_{k-1}(t_x, t_y, m_2) + W_{k-1}(t_y, t_v, m_3) + w_k \end{cases}$$

Obviously, the subsets $U_{k-1}(t_u, t_x)$, $U_{k-1}(t_x, t_y)$ and $U_{k-1}(t_y, t_v)$ do not intersect. Thus, the JPS schedules of the subsets that realize $W_{k-1}(t_u, t_x, m_1)$, $W_{k-1}(t_x, t_y, m_2)$ and $W_{k-1}(t_y, t_v, m_3)$, put one after another define a valid overall schedule of a set of $m-1$ jobs taken in $U_{k-1}(t_u, t_v)$. Moreover, between t_x and t_y there is enough space to schedule k since m_2 jobs in $U_{k-1}(t_x, t_y)$ are scheduled and since $p(m_2 + 1) = t_y - t_x$ (see Figure 14.1). As a consequence, $W' \leq W_k(t_u, t_v, m)$.

Now, we prove that $W_k(t_u, t_v, m) \leq \max(W_{k-1}(t_u, t_v, m), W')$. We only consider the case where $W_k(t_u, t_v, m)$ is finite otherwise the claim holds. Consider a set Z that realizes $W_k(t_u, t_v, m)$. If k does not belong to Z then $W_k(t_u, t_v, m) = W_{k-1}(t_u, t_v, m) \leq \max(W_{k-1}(t_u, t_v, m), W')$. Suppose now that $k \in Z$. Let t_x and t_y be the start and end times of k on the JPS of Z. Thanks to Proposition 14.3, we know that $t_x \in \Theta$ and $t_y \in \Theta$. We also have $\max(r_k, t_u) \leq t_x < t_y \leq \min(d_k, t_v)$. Let Z_1, Z_2, Z_3 be the partition of $Z - \{k\}$ into the jobs that have a release date between t_u and t_x, t_x and t_y, and t_y and t_v. Because of the structure of JPS (k is the job whose due date is maximal), all jobs in Z_1 are completed before t_x. Moreover, all jobs in Z_2 start after t_x and are completed before t_y, and all jobs in Z_3 are completed before t_v. On top of that, $p(|Z_2| + 1) = t_y - t_x$ because k is also scheduled between t_x and t_y. Moreover, we have $|Z_1| + |Z_2| + |Z_3| + 1 = m$. Finally the weight of Z_1 is not greater than $W_{k-1}(t_u, t_x, |Z_1|)$, the weight of Z_2 is not greater than $W_{k-1}(t_x, t_y, |Z_2|)$ and the weight of Z_3 is not greater than $W_{k-1}(t_y, t_v, |Z_3|)$. This leads to $W_k(t_u, t_v, m) \leq \max(W_{k-1}(t_u, t_v, m), W')$. □

Our dynamic programming algorithm relies on the above proposition. The values of $W_k(t_u, t_v, m)$ are stored in a multi-dimensional array of size $O(n^6)$ (n possible values for k, n^2 possible values for t_u, n^2 possible values for t_u, and n possible values for m).

- In the initialization phase the value of $W_1(t_u, t_v, m)$ is set to w_1 if $m = 1$ and if p is not greater than $\min(d_1, t_v) - \max(r_1, t_u)$ and to $-\infty$ otherwise.
- We then iterate from $k = 2$ to $k = n$. Each time, W_k is computed for all the possible values of the parameters thanks to the formula of Proposition 14.4 and to the values of W_{k-1} computed at the previous step.

The maximum weighted number of on-time jobs is equal to

$$\max(W_n(t_0, t_q, 1), W_n(t_0, t_q, 2), \ldots, W_n(t_0, t_q, n))$$

The overall complexity of the algorithm is $O(n^5)$ for the initialization phase. For each value of k, $O(n^5)$ values of W_k have to be computed. For each of them, a maximum among $O(n^4)$ terms has to be computed (for given values of t_x, m_1 and m_2, there is only one possible value for both t_y and m_3). This leads to an overall time complexity of $O(n^{10})$. A rough analysis of the space complexity leads to an $O(n^6)$ bound but since, at each step of the outer loop on k, one only needs the values of W computed at the previous step $(k - 1)$, the algorithm can be implemented with two arrays of $O(n^5)$ size (one for the current values of W and one for the previous value of W). Relying on similar decomposition schemes, a more complex algorithm running in $O(n^4)$ is described in [4].

14.1.6 Problem $1 \mid p_j = p, prec, r_j \mid L_{\max}$

We describe a dynamic programming algorithm to solve this problem. A much more efficient (and much more complex) algorithm has been proposed by Simons [5].

Without loss of generality, we assume that release dates and due dates are adjusted to precedence constraints by changing r_k to $\max\{r_j + p, r_k\}$ and d_j to $\min\{d_j, d_k - p\}$ when $j \rightarrow k$. It is easy to see that the adjusted release dates are met in any feasible schedule and that the maximum lateness does not change. In the following, we assume that due dates are sorted in nondecreasing order, i.e., $d_1 \leq d_2 \leq \cdots \leq d_n$.

Note that in active schedules, start and completion times belong to the set

$$\Theta = \{t \mid \exists r_i, \exists l \in \{0, \ldots, n\}, t = r_i + lp\}$$

Now consider the subproblem defined by five parameters $k, r_{\min}, r_{\max}, s, e$ ($r_{\min} \leq r_{\max}$ and $s \leq e$) in which we look for an optimal schedule of the jobs

$$\{i \leq k \mid r_i \in [r_{\min}, r_{\max}]\}$$

in which jobs cannot start before s and are completed before or at e. Let $L_{\max}(k, r_{\min}, r_{\max}, s, e)$ be the corresponding minimum maximal lateness. Note that

- we look for $L_{\max}(n, \min_i r_i, \max_i r_i, \min \Theta, \max \Theta\}$
- if $\{i \leq k \mid r_i \in [r_{\min}, r_{\max}]\} = \emptyset$, then $L_{\max}(k, r_{\min}, r_{\max}, s, e) = 0$
- given our initial remark on active schedules, we can restrict the search to subproblems where s and e belong to Θ

Now we present the fundamental recursion formula. First, if $r_k \notin [r_{\min}, r_{\max}]$ then

$$L_{\max}(k, r_{\min}, r_{\max}, s, e) = L_{\max}(k - 1, r_{\min}, r_{\max}, s, e)$$

Now assume that $r_k \in [r_{\min}, r_{\max}]$ and among optimal schedules, consider one where the starting time $t \in \Theta \cap [s, e - p]$ of job k is maximal. Remark that jobs i scheduled after k are such that (1) $d_i \leq d_k$ (because $i < k$) and (2) $r_i > t$ (otherwise assume that, among such jobs, r_i is minimal and exchange i and k; this is valid because k has no successor in the subproblem and because i has no predecessor scheduled after k and the exchange does not decrease the objective function while increasing the starting time of k). Hence, we can decompose the schedule in two subschedules before and after k. More formally, $L_{\max}(k, r_{\min}, r_{\max}, s, e)$ is the minimum over $t \in \Theta \cap [s, e - p]$ of

$$\max(L_{\max}(k-1, r_{\min}, \max\{r_{\min}\} \cup \{r_u \mid r_u \leq t\}, s, t)$$

$$L_{\max}(k-1, \min\{r_{\max}\} \cup \{r_u \mid r_u > t\}, r_{\max}, t+p, e))$$

Note that there are $O(n^2)$ values in Θ hence, there are $O(n^7)$ subproblems to consider. Each time, we have to test all values $\Theta \cap [s, e - p]$ and to compute the above expression. Since $\max\{r_u \mid r_u \leq t\}$ and $\min\{r_u \mid r_u > t\}$ can be precomputed, the algorithm runs in $O(n^9)$ time.

14.2 Identical Parallel Machine Problems

14.2.1 Problem $P \mid p_j = 1, r_j \mid \sum f_j$

Like problem $1 \mid p_j = 1; r_j \mid \sum f_j$ also the corresponding problem with m identical machines can be formulated as min-cost assignment problem. Again, we assume that the functions f_j are nondecreasing functions of the finishing times C_j of jobs j $(j = 1, \ldots, n)$. One has to assign to each job j a machine M_i and a time $t \geq r_j$ such that j is processed in $[t, t+1]$ on M_i. n relevant machine–time pairs (μ_l, t_l) $(l = 1, \ldots, n)$ are calculated by the following **Algorithm Machine–Time Pairs.** Job j can be assigned to (μ_l, t_l) with costs $f_j \cdot (t_l + 1)$ if and only if $t_l \geq r_j$.

For calculating the machine–time pairs (μ_l, t_l) we assume that the jobs are enumerated such that $r_1 \leq r_2 \leq \cdots \leq r_n$. The idea of the algorithm is to set $t = r_1$ and to collect in a set I all jobs j with $r_j = t$. If $h := |I| \leq m$ then h machine–time pairs (M_i, t) $(i = 1, \ldots, h)$ are built, t is replaced by the smallest $r_j > t$, and I is replaced by the set of jobs released at the new t-value. Otherwise m machine–time pairs (M_i, t) $(i = 1, \ldots, m)$ are built, m jobs are eliminated from I, t is replaced by $t + 1$, and jobs possibly released at the new t-value are added to I. The process is continued until all jobs have been added to I and released.

Algorithm Machine–Time Pairs

```
j := 1;
K := 0;
I := ∅;
WHILE j ≤ n DO
    BEGIN
    t := r_j;
    WHILE r_j = t AND j ≤ n DO
        BEGIN I := I ∪ {j}; j := j + 1 END;
    WHILE I ≠ ∅ DO
        BEGIN
        Create s := min{|I|, m} machine-time pairs (μ_{K+i}, t_{K+i})
        with μ_{K+i} := M_i and t_{K+i} := t (i = 1, ... , s) and delete
        s jobs from I;
        K := K + s;
        t := t + 1;
        WHILE r_j = t AND j ≤ n DO
            BEGIN I := I ∪ {j}; j := j + 1 END;
        END
    END
```

14.2.2 Premption Redundancy for Unit Processing Times

When optimal nonpreemptive schedules are also optimal in the preemptive case, preemptions are said to be redundant. This is not always the case and the following well-known examples show that preemptions are advantageous for $P2 \mid p_j = 1, pmtn \mid C_{\max}$ and $1 \mid pmtn, r_i \mid \sum C_i$.

Example 14.1

$P2 \mid p_j = 1, pmtn \mid C_{\max}$. Completion time for all nonpreemptive two-machine schedules of three jobs is at least 2. But there exists a preemptive two-machine schedule of the jobs with maximum completion time 1.5. In schedules of this example and next example, 1, 2, 3 and $\cdot$ denote half a unit of job 1, job 2, job 3, and machine idling, respectively.

$$\begin{array}{ll} M_1 & 1133\cdots\cdot \\ M_2 & 22\cdots\cdots \\ & C_{\max}=2 \end{array} \qquad \begin{array}{ll} M_1 & 113\cdots\cdots \\ M_2 & 322\cdots\cdot \\ & C_{\max}=1.5 \end{array}$$

Example 14.2

$1 \mid pmtn, r_i \mid \sum C_i$. Given two jobs 1 ($r_1 = 0, p_1 = 3$) and 2 ($r_2 = 1, p_2 = 1$), the total completion time for all nonpreemptive one–machine schedules is at least 7 but there is a preemptive one–machine schedule of the jobs with total completion time 6.

$$\begin{array}{ll} M_1 & \cdot\cdot 22111111\cdots\cdot \\ & \sum C_i=7 \end{array} \qquad \begin{array}{ll} M_1 & 11221111\cdots\cdot\cdot \\ & \sum C_i=6 \end{array}$$

In a single machine schedule (with no release dates), preemptions are redundant for all classical scheduling criteria, even with arbitrary precedence constraints. This is also true if we have release dates and if all processing times are unitary. The situation is much more complex in the case of parallel-machine scheduling. McNaughton's theorem [6] states that preemptions in scheduling arbitrary processing time jobs on identical parallel machines to minimize the total weighted completion time are redundant. Du, Leung, and Young [7] strengthened this result proving that preemptions are redundant even in $P \mid pmtn, chains \mid \sum w_i C_i$.

There are known simple counterexamples showing that extensions of McNaughton's theorem to other criteria or more general precedence constraints such as intrees or outtrees, or different release dates of jobs, or different speeds of machines, are not true even for equal weights of jobs [7–9]. Recently, a lot of work has been carried out on the special case where all jobs have unit processing times. Baptiste and Timkovsky [10] have provided some simple examples that show that preemptions are also advantageous in $P2 \mid p_j = 1, pmtn, intree \mid \sum C_j$, $P2 \mid p_j = 1, pmtn, outtree \mid \sum w_j C_j$ and $Q2 \mid p_j = 1, pmtn \mid \sum C_j$. In the same paper, it is shown that preemptions in $P2 \mid p_j = 1, pmtn, outtree, r_j \mid \sum C_j$ are redundant. This latter result has been extended to any number of machines by Brucker, Hurink, and Knust [11].

We show that preemptions are redundant for any sum objective function $\sum_{j=1}^{n} f_j$ when the f_j functions are nondecreasing and linear between consecutive integer time points. Since $P \mid p_j = 1, r_j \mid \sum f_j$ can be solved in polynomial time for such functions (see Section 14.2.1), the corresponding preemptive problem is also solvable in polynomial time.

Proof

Assume that we are given an optimal preemptive schedule and let $\mathcal{T}$ denote the set of all integer time points between the minimum release date and the maximum completion time. We build a bipartite graph $G = (\{1, \ldots, n\} \cup \mathcal{T}, E)$ as follows: Given a job i and a time point $t \in \mathcal{T}$, (i, t) belongs to E if and only if $r_i \leq t < C_i$. All edges in E have a unit *capacity*. Now add a source vertex σ, link it to each job i with a *capacity* p_i and add a sink vertex ν linked from each time point in $\mathcal{T}$ with *capacity* m. Finally, we introduce some *costs* on the edges defined as follows. They are all equal to 0 except for one edge per job: The cost of the edge between a job i and the time point $\lfloor C_i \rfloor$ equals the slope of f_i in the interval $[\lfloor C_i \rfloor, \lfloor C_i \rfloor + 1]$.

From the initial schedule, we can easily build a flow ϕ of capacity $\sum p_i$ in this network (the flow through edge (i, t) is the amount of time during which i is processed in $[t, t+1)$). Because of our hypothesis on the f_i functions,

$$\sum f_i(C_i) \geq \sum f_i(\lfloor C_i \rfloor) + \text{cost of } \phi$$

Now consider the maximal flow with minimum cost $\bar{\phi}$. It is better than ϕ hence,

$$\sum f_i(C_i) \geq \sum f_i(\lfloor C_i \rfloor) + \text{cost of } \bar{\phi}$$

Since $\bar{\phi}$ is an integer flow, we immediatly have a preemptive schedule with integer preemptions. Moreover, its cost is not larger than $f_i(\lfloor C_i \rfloor) + \text{cost of } \bar{\phi}$, hence it is optimal. □

Note that the results presented above have been extended in [12] where it is shown that, for arbitrary jobs and any regular objective function, preemptions only occur at integer time points.

14.2.3 Problem $P \mid p_j = p, pmtn \mid \sum f_j$

Brucker, Heitmann, and Hurink [13] have shown that preemptions are redundant in $P \mid p_j = p, pmtn \mid \sum T_j$. But $P \mid p_j = p \mid \sum T_j$ can be solved in polynomial time by a simple reduction to the assignment Problem 14.2.1. We show that an extension of the problem to a more general criterion, where preemptions may be advantageous, can also be solved in polynomial time.

Let f_j be convex nondecreasing functions such that they are defined on a time space $\mathcal{T}_{pmtn}$ polynomially bounded in n, and the differences $f_i - f_j$ are all monotone functions (i.e., nonincreasing or nondecreasing functions). Without loss of generality, we can assume that the jobs are in a linear order where for each pair of jobs i and j the functions $f_i - f_j$ are nondecreasing if $i < j$. It is shown in [14] that there is an optimal schedule where $i \leq j \Rightarrow C_i \leq C_j$.

Let us consider completion times C_j for all $j = 1, 2, \ldots, n$ as deadlines. It is known [15] that a feasible schedule for the decision problem $P \mid pmtn, D_j \mid -$ with deadlines C_j exists if and only if

$$\forall j : \sum_{i=1}^{n} \max\{0, p_i - \max\{0, C_i - C_j\}\} \leq mC_j \quad \text{and} \quad C_j \geq p_j$$

Under the conditions $p_j = p$ and $i \leq j \Rightarrow C_i \leq C_j$ this predicate is

$$\forall j : \sum_{i=j+1}^{n} \max\{0, p - C_i + C_j\} \leq mC_j - jp \quad \text{and} \quad C_j \geq p$$

Introducing additional variables

$$X_{ij} = \max\{0, p - C_i + C_j\}$$

for all $i = 2, 3, \ldots, n$ and $j = 1, 2, \ldots, n$ we finally obtain the problem of minimizing the *convex-separable* objective function

$$\sum_{j=1}^{n} f_j(C_j)$$

under the linear constraints

$$\forall j : \sum_{i=j+1}^{n} X_{ij} \leq mC_j - jp \quad \text{and} \quad C_j \geq p$$

$$\forall i : \forall j : X_{ij} \geq p - C_i + C_j \qquad \text{and} \quad X_{ij} \geq 0$$

The problem reduces to a linear programming problem in time polynomial in the total number of *break points* of the functions f_j [16]. In our case, the number of break points of each of the n functions is at most $|\mathcal{T}_{pmtn}|$. Hence, the problem can be solved in time polynomial in $n \cdot |\mathcal{T}_{pmtn}|$.

Once the optimal completion times are known, the optimal schedule can be produced by Sahni's algorithm [17].

14.2.4 Problem $P|p_j = 1, chains, r_j|\sum f_j$

Dror et al. [18] established that $P \mid p_j = 1, chains, r_j \mid L_{\max}$ can be solved in polynomial time. We show that there exists a similar polynomial-time solution to minimizing $\sum f_j$ if the cost functions of chained jobs are interchangeable, i.e.,

$$j \to k \;\Rightarrow\; f_j(x) + f_k(y) = f_k(x) + f_j(y) \quad \text{for all } x \neq y$$

If chained jobs have identical cost functions, i.e., $j \to k \Rightarrow f_j = f_k$, then they are obviously interchangeable. Thus, $P \mid p_j = 1, chains, r_j \mid \sum w_j U_j$ and $P \mid p_j = 1, chains, r_j \mid \sum w_j T_j$ can be solved in polynomial time if chained jobs have identical weights and due dates.

Without loss of generality, we assume that release dates are adjusted to precedence constraints by changing r_k to $\max\{r_j + 1, r_k\}$ for each immediate precedence $j \to k$. Hence, we can assume that release dates are increasing along each chain of precedence constraints.

Let us create a network with a source S, job vertices j, $1 \le j \le n$, chain-time vertices (c, t) for each chain c and integer time point $t \in \mathcal{T}$, time vertices $t \in \mathcal{T}$ and a sink T. Since the number of chains is at most n and $|\mathcal{T}| = O(n^2)$, the number of vertices in the network is $O(n^3)$. The arcs in the network will be the arcs in all possible paths

$$S \to j \to (c, t) \to t \to T$$

under the condition that arcs $j \to (c, t)$ exist if and only if $r_j \le t$ and $j \in c$. We assign capacity m to the arcs $t \to T$ and capacity 1 to the other arcs, cost $f_j(t + 1)$ to the arcs $j \to (c, t)$ and cost 0 to the other arcs, supply n to S and demand n to T.

It is well known that the quantities of a minimum-cost flow running through the arcs are integer if all arcs in the network have integer capacities. Thus, the unit flow through the arc $j \to (c, t)$ with unit capacity models an assignment of the job j belonging to the chain c to be processed in the time unit $[t, t + 1]$ with $r_j \le t$. The flow of value at most m through the arc $t \to T$ with capacity m models an assignment of at most m jobs to be processed in the time unit $[t, t + 1]$.

The network-flow model observes all precedence constraints with the exception of possibly those between jobs i and j that start at time $\max\{r_i, r_j\}$ or later. However, transposing such jobs we can obtain a schedule that observes all precedence constraints in time $O(n^2)$. Since the cost functions of chained jobs are interchangeable, the total cost remains the same after the transposition. The time complexity of solving the problem is $O(n^9)$ because finding a minimum-cost flow in a network takes at most cubic time in the number of its vertices.

Note that $P \mid p_j = 1, chains, r_j \mid L_{\max}$ reduces to a binary search on

$$\Delta = D_1 - d_1 = D_2 - d_2 = \ldots = D_n - d_n$$

for finding a maximal flow (of quantity n) in the same network, where the arc condition $r_j \le t$ is replaced by $r_j \le t < D_j$.

14.2.5 Problem $P \mid p_j = 1; intree \mid L_{\max}$

The procedure which solves this problem has two steps. In the first step, the due dates of the jobs are modified in such a way that they are consistent with the precedence constraints.

In the second step, jobs are scheduled in an order of nondecreasing modified due dates.

The idea of the due date modification procedure is to replace d_i by $\min\{d_i, d_j - 1\}$ whenever $i \rightarrow j$. This is done in a systematic way going from the roots (vertices i with no successor) to the leaves (vertices i with no predecessor) of the intree. After modifying the due date d_i, we eliminate i from the intree.

Implementing this procedure in an appropriate way yields an $O(n)$-algorithm if applied to intrees.

We denote the modified due dates by d'_j. Note that $d'_i < d'_j$ whenever $i \rightarrow j$. Furthermore, the following proposition holds.

Proposition 14.5

A schedule has no late jobs with respect to the original due dates d_j if and only if it has no late jobs with respect to the modified due dates.

Proof

Because $d'_j \leq d_j$ for all jobs j, a schedule without late jobs with respect to the d'_j-values has no late jobs with respect to the d_j-values.

To prove the other direction, assume w.l.o.g. that $n, n-1, \ldots, 1$ is the order in which the due dates are modified. Consider a schedule with finishing times $C_1, \ldots, C_n$ satisfying $C_j \leq d_j$ for $j = 1, \ldots, n$. Then $C_n \leq d_n = d'_n$. If for some $1 < r \leq n$ we have $C_j \leq d'_j$ for $j = r, \ldots, n$, and there exists a job $i \in \{r, \ldots, n\}$ with $s(r-1) = i$ (where $s(r-1)$ denotes the successor of $r-1$) then we have $C_{r-1} \leq \min\{d_{r-1}, d'_i - 1\} = d'_{r-1}$. □

In the second step the jobs are scheduled sequentially in order of nondecreasing modified due dates. This is done by scheduling each job at the earliest available starting time, i.e. the earliest time at which less than m jobs are scheduled to start and all predecessors of the job have been completed. A more precise description is given by the following algorithm. We assume that the jobs are numbered in such a way that $d'_1 \leq d'_2 \leq \ldots, d'_n$. Furthermore, F denotes the earliest time at which a machine is available, and $r(j)$ is the latest finishing time of a predecessor of job j. $n(t)$ counts the number of jobs scheduled at time t and $x(j)$ is the starting time of job j. As before, $s(j)$ is the successor of j.

Algorithm $P \mid p_j = 1; intree \mid L_{\max}$

```
1. F := 0;
2. FOR j := 1 TO n DO r(j) := 0;
3. FOR t := 0 TO n DO n(t) := 0;
4. FOR j := 1 TO n DO
       BEGIN
5.     t := max{r(j), F};
6.     x(j) := t;
7.     n(t) := n(t) + 1;
8.     IF n(t) = m THEN F := t + 1;
9.     i := s(j);
10.    r(i) := max{r(i), t + 1}
       END
```

The schedule constructed by this algorithm has the important property that the number of jobs scheduled at any time is never less than the number scheduled at a later time. This can be seen as follows. Suppose k jobs are scheduled to start at a time t and at least $k+1$ jobs are scheduled to start at time $t+1$. Since the procedure schedules jobs at the earliest available starting time and less than m jobs are scheduled at time t, the $k+1$ jobs scheduled at time $t+1$ must each have an immediate predecessor scheduled to start at time t. This is impossible because, due to the intree structure, each job starting at time t has at the most one successor.

The running time of Algorithm $P \mid p_j = 1; intree \mid L_{\max}$ is $O(n)$. Thus, problem $P \mid p_j = 1; intree \mid L_{\max}$ can be solved in $O(n \log n)$ time.

We still have to prove that Algorithm $P \mid p_j = 1; intree \mid L_{\max}$ is correct.

Proposition 14.6

If there exists a schedule in which no job is late, then a schedule constructed by Algorithm $P \mid p_j = 1; intree \mid L_{\max}$ has this property.

Proof

Assume that there is a late job in the schedule $x(1), \ldots, x(n)$ constructed by the algorithm. Then there is also a late job with respect to the modified due dates. Consider the smallest i with $x(i) + 1 > d'_i$.

Let $t < d'_i$ be the largest integer with the property that $|\{j \mid x(j) = t, d'_j \leq d'_i\}| < m$.

Such a t exists because otherwise md'_i jobs j with $d'_j \leq d'_i$ are scheduled before d'_i. Job i does not belong to this set because $x(i) + 1 > d'_i$. This means that at least $md'_i + 1$ jobs must be scheduled in the time interval $[0, d'_i]$ if no job is late. This is a contradiction.

Each job j with $d'_j \leq d'_i$ and $x(j) > t$ must have a (not necessarily immediate) predecessor starting at time t. Now we consider two cases.

Case 1. $\boldsymbol{t = d'_i - 1}$

We have $x(i) > d'_i - 1 = t$. Thus, a predecessor k of i must start at time t and finish at time d'_i. Because $d'_k \leq d'_i - 1 < d'_i = x(k) + 1$, job k is late, too. However, this is a contradiction to the minimality of i.

Case 2. $\boldsymbol{t < d'_i - 1}$

Exactly m jobs j with $d'_j \leq d'_i$ start at time $t + 1$, each of them having a predecessor starting at time t. Due to the intree structure, all these predecessors must be different. Furthermore, if k is such a predecessor of a job j, then $d'_k \leq d'_j - 1 < d'_j \leq d'_i$ which contradicts the definition of t.

Proposition 14.7

The algorithm $P \mid p_j = 1; intree \mid L_{\max}$ is correct.

Proof

Let $L^*_{\max}$ be the optimal solution value. Then there exists a schedule satisfying

$$\max_{j=1}^{n}\{C_j - d_j\} \leq L^*_{\max} \tag{14.1}$$

which is equivalent to

$$C_j \leq d_j + L^*_{\max} \quad \text{for } j = 1, \ldots, n \tag{14.2}$$

Due to Proposition 14.6, a schedule S constructed by the algorithm for the due dates $d_j + L^*_{\max}$ satisfies (14.2) or equivalently (14.1). Thus it is optimal. However, S is identical to a schedule constructed by the algorithm for the due dates d_j because $(d_j + L^*_{\max})' = d'_j + L^*_{\max}$ for $j = 1, \ldots, n$. □

The complexity of problem $P \mid p_j = 1; intree \mid L_{\max}$ can be reduced to $O(n)$ by using hash techniques (see [19]).

$P \mid p_j = p; intree \mid L_{\max}$ can be solved with the same complexity because by scaling with factor $1/p$ this problem is transformed into the corresponding equivalent problems with $p_j = 1$. Symmetrically to $P \mid p_j = p; intree \mid L_{\max}$, problem $P \mid p_j = p; outtree; r_i \mid C_{\max}$ can be solved in linear time.

14.2.6 Problem $P \mid p_j = 1; outtree; r_j \mid \sum C_j$

We assume that the release dates r_j are integer and compatible with the outtree precedence constraints, i.e., $r_i + 1 \leq r_j$ for all precedences $i \to j$.

In the following we will consider two relaxations of problem $P \mid p_j = 1; outtree; r_j \mid \sum C_j$. In the first we relax all precedence constraints, i.e., we consider problem $P \mid p_j = 1, r_j \mid \sum C_j$. Let $\tilde{r}_1 < \tilde{r}_2 < \cdots < \tilde{r}_k$

be the different release dates and define $\tilde{r}_{k+1} := \infty$. Denote by S_ν the set of jobs with release date $\tilde{r}_\nu$. An optimal schedule for this problem can be constructed by the following algorithm.

Algorithm $P \mid p_j = 1; r_j \mid \sum C_j$

```
1. S := Ø;
2. FOR ν := 1 TO k DO BEGIN
3.     t := r̃_ν; S := S ∪ S_ν;
4.     WHILE t < r̃_{ν+1} and S ≠ Ø DO BEGIN
5.         m_t := min{m, |S|};
6.         Schedule a set S_t ⊆ S of m_t jobs at time t;
7.         S := S \ S_t; t := t + 1;
       END
   END
```

In the for-loop of this algorithm jobs are scheduled in the time periods $[\tilde{r}_\nu, \tilde{r}_{\nu+1}]$. This is done by first adding the set S_ν to the set of jobs which are available at time $t = \tilde{r}_\nu$ and then scheduling a maximal number of jobs from S in each of the next time periods. For all time periods t, where m_t is not defined by the algorithm let $m_t := 0$.

In the second relaxation we replace the number m of machines by the number of jobs n. In this relaxation we can schedule all jobs from the set S_ν at time $\tilde{r}_\nu$ $(\nu = 1, \ldots, k)$. The resulting schedule is feasible (since the release dates are compatible with the precedences) and certainly optimal for the problem with n machines. Let

$$\hat{m}_t := \begin{cases} |S_\nu|, & \text{if } t = \tilde{r}_\nu \\ 0, & \text{otherwise} \end{cases}$$

To construct an optimal schedule for the original problem we transform the schedule of the second relaxation in which $\hat{m}_t$ jobs are scheduled in each time period $[t, t+1]$ into a feasible schedule where exactly m_t jobs are scheduled in period t (without violating the precedence constraints). The resulting schedule must be optimal for the original problem since it has the same objective value as the optimal schedule for the first relaxation.

This transformation is done by iteratively moving jobs from left to right. During the transformation the following invariance properties are satisfied:

(i) The outtree precedences are respected.
(ii) $\sum_{\nu \le \tau} \hat{m}_\nu \ge \sum_{\nu \le \tau} m_\nu$ holds for all time indices τ.

Clearly, these invariance properties hold at the beginning.

Assume that $\hat{m}_\nu < m_\nu$ for some index ν holds. Otherwise, $n = \sum_\nu \hat{m}_\nu \ge \sum_\nu m_\nu = n$ implies $\hat{m}_\nu = m_\nu$ for all ν and we are finished. Let t be the smallest time index with $\hat{m}_t < m_t \le m$. Due to property (ii) we must have an index $t' < t$ with $\hat{m}_{t'} > m_{t'}$ and $\hat{m}_\tau = m_\tau$ for $t' < \tau < t$. Furthermore, due to the definition of m_t we have $m_\tau = m$ for $t' < \tau < t$ which implies

$$\hat{m}_{t'} \ge \hat{m}_{t'+1} = \hat{m}_{t'+2} = \ldots = \hat{m}_{t-1} > \hat{m}_t$$

We successively will move a job from time period $[\tau - 1, \tau]$ to $[\tau, \tau + 1]$ for $\tau = t, t-1, \ldots, t'+1$ such that property (i) keeps satisfied. This can be established as follows.

Since in $[\tau - 1, \tau]$ at least one more job is scheduled than in $[\tau, \tau + 1]$ and each job has at most one predecessor, we can always find in the interval $[\tau - 1, \tau]$ a job without successors in $[\tau, \tau + 1]$ which can be moved one time unit to the right. Also property (ii) will be satisfied for t' because $\sum_{\nu < t'} \hat{m}_\nu \ge \sum_{\nu < t} m_\nu$ and $\hat{m}_\nu > m_\nu$. Thus, after the move of one job from $[t', t'+1]$ to $[t'+1, t'+2]$ inequality (ii) still holds for $\tau = t'$. It also holds for all $\tau > t'$ because by the transformation the $\hat{m}_\nu$ values do not decrease for $\nu > t'$.

In one step of the transformation the value of $\max\{0, m_\nu - \hat{m}_\nu\}$ is decreased by one for $\nu = t$ and kept unchanged for all other ν. Thus, $\sum_\nu \max\{0, m_\nu - \hat{m}_\nu\}$ is decreased by one in each step. Due to the fact that at the beginning $\sum_\nu \max\{0, m_\nu - \hat{m}_\nu\} \leq \sum_\nu m_\nu = n$, after at most n iterations we reach an optimal schedule for the original problem.

Since each iteration can be implemented in $O(n)$ time, the overall complexity to solve problem $P \mid p_j = 1; outtree; r_j \mid \sum C_j$ is $O(n^2)$.

Example 14.3

Consider an instance of $P \mid p_j = 1; outtree; r_j, \mid \sum C_j$ with $m = 3$ machines, $n = 21$ jobs and the following outtree precedences and release dates:

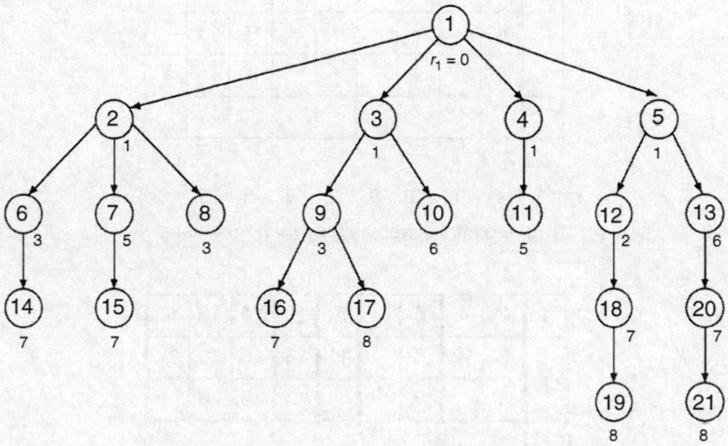

Applying *Algorithm* $P \mid p_j = 1; r_j \mid \sum C_j$ to this instance produces the schedule shown in Figure 14.2(a) in which the precedences $5 \rightarrow 12, 18 \rightarrow 19$ and $20 \rightarrow 21$ are violated. An optimal schedule for $m = n$ machines is presented in Figure 14.2(b) and an optimal transformed schedule can be found in Figure 14.2(c). In this transformation job 4 is moved from the interval $[1, 2]$ to the interval $[2, 3]$, jobs 17, 21 are moved from $[8, 9]$ to $[9, 10]$ and jobs 16, 20 are moved from $[7, 8]$ to $[8, 9]$ (cf. Figure 14.2(b) and (c)).

14.2.7 Problem $P \mid p_j = 1; pmtn; outtree; r_j \mid \sum C_j$

In this section we show that for problem $P \mid p_j = 1; pmtn; outtree; r_j \mid \sum C_j$ an optimal schedule exists in which preemption is not necessary. Thus, the algorithm derived in the previous section also solves problem $P \mid p_j = 1; pmtn; outtree; r_j \mid \sum C_j$.

Our claim is a consequence of the following facts:

1. The optimal solution value for the relaxation $P \mid p_j = 1; pmtn; r_j \mid \sum C_j$ is a lower bound for the corresponding problem $P \mid p_j = 1; pmtn; outtree; r_j \mid \sum C_j$.
2. Corresponding instances of $P \mid p_j = 1; pmtn; r_j \mid \sum C_j$ and $P \mid p_j = 1; r_j \mid \sum C_j$ have the same optimal solution value.
3. Given an instance I of $P \mid p_j = 1; outtree; r_j \mid \sum C_j$ where the release dates are compatible with the outtree-precedences, an optimal solution of the related instance of $P \mid p_j = 1; r_j \mid \sum C_j$ with the same release dates can be transformed into a feasible solution of I without changing the objective value.

Clearly, (1) holds (3) has been shown in the previous section. To prove (2) we decompose the schedule constructed by Algorithm $P \mid p_j = 1, r_j \mid \sum C_j$ (or the corresponding schedule for $P \mid p_j = 1; outtree; r_j \mid \sum C_j$) into *blocks*. These blocks are defined as follows.

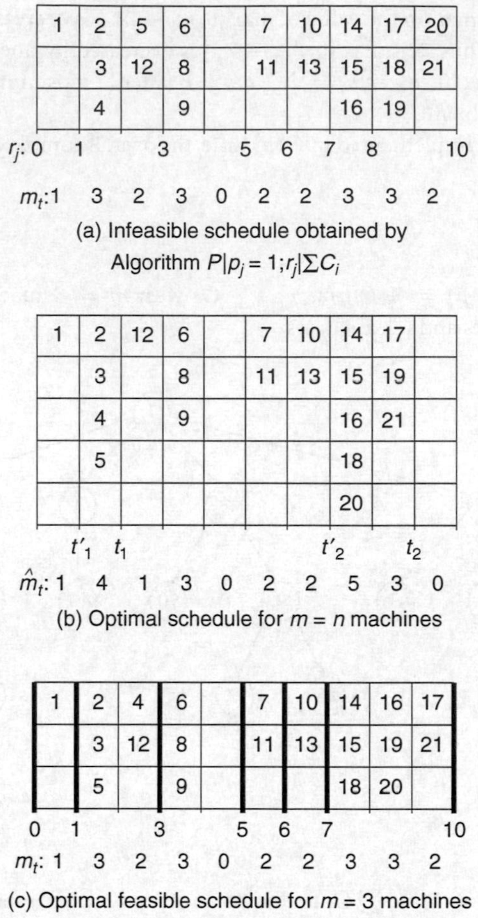

(a) Infeasible schedule obtained by Algorithm P|pj = 1;rj|ΣCi
(b) Optimal schedule for m = n machines
(c) Optimal feasible schedule for m = 3 machines

FIGURE 14.2 Schedules for problem $P \mid p_j = 1; outtree; r_j \mid \sum C_j$.

Let $i_0 = 1 < i_1 < \cdots < i_q$ be indices such that for $\nu = 0, \ldots, q-1$, $i_{\nu+1}$ is the first index greater than i_ν such that

- algorithm $P \mid p_j = 1; r_j \mid \sum C_j$ schedules all jobs j with $\tilde{r}_{i_\nu} \le r_j < \tilde{r}_{i_{\nu+1}}$ within the interval $[\tilde{r}_{i_\nu}, \tilde{r}_{i_{\nu+1}})$, and
- at least one machine is idle for at least one time unit in $[\tilde{r}_{i_\nu}, \tilde{r}_{i_{\nu+1}}]$.

Note that jobs of block ν cannot be scheduled before the corresponding interval $[\tilde{r}_{i_\nu}, \tilde{r}_{i_{\nu+1}})$ since their release dates are greater or equal to $\tilde{r}_{i_\nu}$. The partial schedule corresponding to the interval $[\tilde{r}_{i_\nu}, \tilde{r}_{i_{\nu+1}})$ is called a block. In the example in Figure 14.2(c) we have six blocks in the intervals $[0,1)$,$[1,3)$,$[3,5)$,$[5,6)$, $[6,7)$ and $[7,10)$.

Now (2) can be proved by induction on the number of blocks. If only one block exists, then consider the relaxed problem $P \mid p_j = 1; pmtn \mid \sum C_j$ in which all release dates are ignored. McNaughton [11] has shown that for this problem without release dates preemption is redundant. The resulting schedule fits completely in the time interval corresponding to the block. Furthermore, an optimal solution of $P \mid p_j = 1 \mid \sum C_j$ has the same structure as the solution provided by Algorithm $P \mid p_j = 1; r_j \mid \sum C_j$ which is feasible for $P \mid p_j = 1; pmtn; r_j \mid \sum C_j$. Thus, the latter solution must be optimal for $P \mid p_j = 1; pmtn; r_j \mid \sum C_j$.

Assume that (2) holds for an instance with $k-1$ blocks and consider an instance with k blocks. Consider the relaxation of this problem in which the first block and the remaining $k-1$ blocks are optimized separately. By induction preemption is redundant for each of these subproblems and the corresponding schedules fit into their time intervals. Thus, both schedules do not overlap, i.e., the composed schedule is feasible for the problem with k blocks. Consequently, the composed schedule is an optimal nonpreemptive schedule for the preemptive problem.

14.2.8 Problem $Pm \mid p_j = 1, intree \mid \sum Cj$

Using the fact that the precedence relation is an intree, it is easy to make sure by contradiction that the number of busy machines is not increasing in time in any optimal schedule. Hence, an optimal schedule defines the time unit $[t, t+1]$ that contains a set L with less than m independent jobs such that either $t = 0$ or the time unit $[t-1, t]$, where $t > 0$, contains m jobs.

It is evident that L is the set of leaves of the subtree S scheduled in $[t, C_{\max}]$, and the set F of other jobs is a forest with mt jobs scheduled in $[0, t]$. Let C_j denote the completion time of j in an optimal schedule. Then the total completion time of the schedule can be written as

$$m(1 + 2 + \cdots + t) + \sum_{j \in S} C_j$$

Note that any active schedule for S is optimal because it requires less than m machines. Once L is given, F and S can be obviously restored in linear time. The subschedule for F and the time point t can also be found in linear time by Hu's algorithm [20].

To find a set L we need to check all $O(n^{m-1})$ subsets with less than m jobs. Since m is fixed, we obtain a solution in polynomial time.

14.2.9 Problem $Pm \mid p_j = p, r_j \mid \sum f_j$

In this section, we show that, in the nonpreemptive case, most of the results described in Section 14.1.2 can be extended to the situation where m parallel identical machines are available.

We show that, for any ordered objective function F (Definition 14.1), $Pm \mid p_j = p, r_j \mid F$ can be solved in polynomial time.

We extend a dominance property initially introduced for the single machine case and we describe the variables of the dynamic programming algorithm and the Proposition that links them together. Finally, the dynamic programming algorithm itself is described.

14.2.9.1 Dominance Property

As for the one-machine case, we can restrict our search to active schedules and it is easy to see that in active schedules, start and completion times belong to

$$\Theta = \{t \mid \exists r_i, \exists l \in \{0, \ldots, n\}, t = r_i + lp\}$$

14.2.9.2 Variables Definition and Decomposition Scheme

Recall that late jobs can be scheduled arbitrarily late. Hence, there is no point to assign a starting time to late jobs. As in the single machine case, when we refer to a schedule $\mathcal{H}$ of some subset of jobs O, it does not mean that all jobs in O are in the schedule (the on-time jobs only are in the schedule). So the cost of $\mathcal{H}$ with respect to O is

- The sum, over all jobs $i \in O$ scheduled in $\mathcal{H}$, of $f_i(C_i)$
- Plus the sum, over all other jobs $i \in O$ not scheduled in $\mathcal{H}$, of ω_i

Roughly speaking, the decomposition scheme for the single machine case (Proposition 14.2) relies on a particular time point $t_k \in [s, e)$ that allows us to split the problem into two subproblems (between s and

t_k and between t_k and e). The same scheme is kept here; however, s and e are replaced by two vectors σ and ϵ that represent some resource profiles. We introduce more formally this notion.

Definition 14.3

A resource profile ξ is a vector $(\xi_1, \xi_2, \cdots, \xi_m)$ such that $\xi_1 \leq \xi_2 \leq \cdots \leq \xi_m$ and $\xi_m - \xi_1 \leq p$. In the following, Ξ denotes the set of resource profiles ξ such that $\forall i, \xi_i \in \Theta$.

Intuitive meaning of resource profiles. A resource profile $\xi = (\xi_1, \xi_2, \ldots, \xi_m)$ represents the state of the resource at some time point ξ_1 (in practice this time point always matches the starting time or the completion time of a job). If q machines are idle at time ξ_1 then the q first components of the resource profile equal ξ_1. The other components are the completion times of the $m-q$ jobs that are being processed at time ξ_1. Since the processing time of jobs is p, we have $\forall i, \xi_i - \xi_1 \leq p$. The state of the resource is considered from a global point of view and thus, the ith component of the resource profile is not systematically related to the ith machine. In the following, some "left" resource profiles σ are used to state that no machine is available before σ_1, one machine is available between σ_1 and σ_2; two machines are available between σ_2 and σ_3, etc. Conversely, some "right" resource profiles ϵ are used to state that no machine is available after ϵ_m, one machine is available between ϵ_{m-1} and ϵ_m; two machines are available between ϵ_{m-2} and ϵ_{m-1}, etc. The resource profiles σ and ϵ allow us to determine the exact amount of resource that is available at each time point in $[\sigma_1, \epsilon_m]$.

Definition 14.4

Given two resource profiles ξ and μ, $\xi \ll \mu$ if and only if for any index i in $\{1, \ldots, m\}$, $\xi_i \leq \mu_i$.

Notice that $\ll$ defines a partial order on resource profiles. We introduce a technical proposition that will be used later on.

Proposition 14.8

Given two resource profiles ξ and μ, $\xi \ll \mu$ if and only if for any time point t, $|\{i : t < \xi_i\}| + |\{i : \mu_i \leq t\}| \leq m$.

Proof

Sufficient Condition. (By contradiction.) Let k be the first index such that $\mu_k < \xi_k$. Let us compute $|\{i : t < \xi_i\}| + |\{i : \mu_i \leq t\}|$ for $t = \mu_k$. It is equal to $|\{i : \mu_k < \xi_i\}|$ plus $|\{i : \mu_i \leq \mu_k\}|$; which is greater than or equal to $(m - k + 1) + k > m$. This contradicts our hypothesis.

Necessary Condition. Let t be any time point. In the following, $\delta(P)$ denotes the binary variable that equals 1 if the condition P holds, 0 otherwise. Notice that for any value of i, $\delta(t < \xi_i) + \delta(\mu_i \leq t) \leq 1$ because $\xi_i \leq \mu_i$. Hence,

$$|\{i : t < \xi_i\}| + |\{i : \mu_i \leq t\}| = \sum_{i=1}^{m} (\delta(t < \xi_i) + \delta(\mu_i \leq t)) \leq m$$

We now define the variables of the dynamic programming algorithm. For any integer $k \leq n$, for any resource profiles σ and ϵ ($\sigma \ll \epsilon$), let $F_k(\sigma, \epsilon)$ be the minimal cost over all feasible schedules of the jobs in $U_k(\sigma_m - p, \epsilon_1)$ such that

- starting times belong to Θ
- the number of machines available at time t to schedule the jobs in the set $U_k(\sigma_m - p, \epsilon_1)$ is $m - |\{i : t < \sigma_i\}| - |\{i : \epsilon_i \leq t\}|$

Notice that given our definition, $F_0(\sigma, \epsilon)$ is equal to 0. □

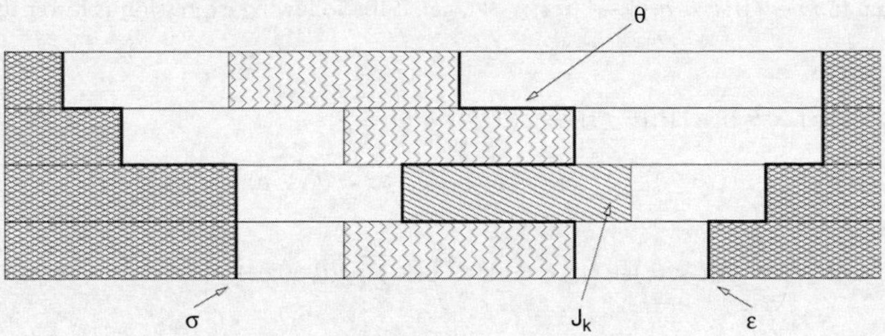

FIGURE 14.3 Resource profiles.

Proposition 14.9

Let $k \in [1, n]$ and let $\sigma \ll \epsilon$ be two resource profiles. If $r_k \notin [\sigma_m - p, \epsilon_1)$, $F_k(\sigma, \epsilon) = F_{k-1}(\sigma, \epsilon)$. If $r_k \in [\sigma_m - p, \epsilon_1)$, $F_k(\sigma, \epsilon)$ is equal to $\min(\omega_k + F_{k-1}(\sigma, \epsilon), F'_k(\sigma, \epsilon))$; where

$$F'_k(\sigma,\epsilon) = \min_{\substack{\theta \in \Xi \\ r_k \le \theta_1 \\ \sigma \ll \theta \\ \theta' = (\theta_2, \ldots, \theta_m, \theta_1 + p) \\ \theta' \ll \epsilon}} F_{k-1}(\sigma,\theta) + F_{k-1}(\theta',\epsilon) + f_k(\theta_1 + p)$$

(If $F'_k(\sigma, \epsilon)$ is undefined, assume that $F'_k(\sigma, \epsilon) = \infty$.)

Notice that in the above formula the value θ' is derived from θ. Figure 14.3 provides an illustration of this proposition. Proposition 14.9 basically states that the optimum schedule for k, σ, ϵ can be computed by trying all possible resource profiles θ of Ξ that are "between" the resource profiles σ and ϵ. For each candidate resource profile θ, the job k starts at θ_1.

Proof

First notice that, θ' is a resource profile because $\theta \in \Xi$. Hence the use of $F_{k-1}(\theta', \epsilon)$ is correct. If $r_k \notin [\sigma_m - p, \epsilon_1)$ the proposition obviously holds because $U_k(\sigma_m - p, \epsilon_1) = U_{k-1}(\sigma_m - p, \epsilon_1)$. We now consider the case where $r_k \in [\sigma_m - p, \epsilon_1)$.

We first prove that $\omega_k + F_{k-1}(\sigma, \epsilon) \geq F_k(\sigma, \epsilon)$ and that $F'_k(\sigma, \epsilon) \geq F_k(\sigma, \epsilon)$. The first claim is obvious (consider the schedule that realizes $F_{k-1}(\sigma, \epsilon)$, "add" k and put it late). Now, let $\theta \in \Xi$ be the resource profile that realizes $F'_k(\sigma, \epsilon)$. There is a schedule $\mathcal{H}_1$ that realizes $F_{k-1}(\sigma, \theta)$ and a schedule $\mathcal{H}_2$ that realizes $F_{k-1}(\theta', \epsilon)$. Notice that any job in $U_{k-1}(\sigma_m - p, \epsilon_1)$ is either late or scheduled in $\mathcal{H}_1$ or in $\mathcal{H}_2$. Consider the schedule $\mathcal{H}$ built as follows: schedule k at time θ_1 and all other jobs in $U_k(\sigma_m - p, \epsilon_1)$ at the time they were scheduled on $\mathcal{H}_1$ or on $\mathcal{H}_2$. Let us prove that $\mathcal{H}$ is a feasible schedule of $U_k(\sigma_m - p, \epsilon_1)$. Since $r_k \leq \theta_1$ and since $\mathcal{H}_1$ and $\mathcal{H}_2$ are feasible, all jobs are scheduled after their release date. Moreover, we claim that the m-machine constraint holds.

Let t be any time point. The number of machines used by the jobs scheduled on $\mathcal{H}_1$ is upper bounded by $m - |\{i : t < \sigma_i\}| - |\{i : \theta_i \leq t\}|$. The number of machines used by the jobs scheduled on $\mathcal{H}_2$ is upper bounded by $m - |, \{i : t < \theta'_i\}| - |\{i : \epsilon_i \leq t\}|$. Finally, k uses a machine at time t if and only if $t \in [\theta_1, \theta_1 + p)$. The resource constraint is satisfied at time t if the sum of the upper bounds is lower than or equal to $m - |\{i : t < \sigma_i\}| - |\{i : \epsilon_i \leq t\}|$, i.e., if the following expression is lower than or equal to 0.

$$\begin{aligned}
&m - |\{i : t < \theta'_i\}| - |\{i : \theta_i \leq t\}| + \delta(\theta_1 \leq t < \theta_1 + p) \\
&= m - \sum_{i=1}^{m} \delta(t < \theta'_i) - \sum_{i=1}^{m} \delta(\theta_i \leq t) + \delta(\theta_1 \leq t < \theta_1 + p) \\
&= m - \left(\sum_{i=2}^{m} \delta(t < \theta_i) + \delta(t < \theta_1 + p) \right) - \sum_{i=1}^{m} \delta(\theta_i \leq t) + \delta(\theta_1 \leq t < \theta_1 + p) \\
&= m - \sum_{i=2}^{m} (\delta(t < \theta_i) + \delta(\theta_i \leq t)) - \delta(t < \theta_1 + p) - \delta(\theta_1 \leq t) + \delta(\theta_1 \leq t < \theta_1 + p) \\
&= 1 - \delta(t < \theta_1 + p) - \delta(\theta_1 \leq t) + \delta(\theta_1 \leq t < \theta_1 + p) \leq 0
\end{aligned}$$

We have proven that $\mathcal{H}$ is a feasible schedule of $U_{k-1}(\sigma_m - p, \epsilon_1)$. On top of that, starting times obviously belong to Θ. The cost of $\mathcal{H}$ is exactly $F'_k(\sigma, \epsilon)$ and hence, $F'_k(\sigma, \epsilon) \geq F_k(\sigma, \epsilon)$. *We now prove that* $\min(\omega_k + F_{k-1}(\sigma, \epsilon), F'_k(\sigma, \epsilon)) \leq F_k(\sigma, \epsilon)$. Consider a schedule that realizes $F_k(\sigma, \epsilon)$ and let O be the set of jobs scheduled in this schedule (jobs in $O - U_k(s - p, e)$ are late). Among all schedules (1) that realize $F_k(\sigma, \epsilon)$ and (2) in which the same set of jobs O are scheduled, consider the schedule $\mathcal{H}$ that lexicographically minimizes the vector made of completion times of jobs in O. (The completion time of the job in O, with the smallest index, is minimum, then the completion time of the job in O, with the second smallest index, etc.) The job k is either late or on-time on $\mathcal{H}$.

If k is late then $\mathcal{H}$ is also a schedule of $U_{k-1}(\sigma_m - p, \epsilon_1)$ that could fit between σ and ϵ. Its cost is exactly $F_k(\sigma, \epsilon) - \omega_k$ (the late cost ω_k is removed since k is not considered any longer). Hence $F_{k-1}(\sigma, \epsilon) \leq F_k(\sigma, \epsilon)$.

Now assume that k is on-time and let t_k be its starting time on $\mathcal{H}$. The proof works as follows. We first show that on $\mathcal{H}$, jobs with a release date lower than or equal to t_k are either late or start before or at t_k. We then exhibit a resource profile $\theta \in \Xi$ such that (1) $\theta_1 = t_k$, (2) $\sigma \ll \theta$, (3) $\theta' = (\theta_2, \ldots, \theta_m, \theta_1 + p) \ll \epsilon$. We then conclude the proof.

In $\mathcal{H}$, jobs i with $r_i \leq t_k$ are late or start before or at t_k. Suppose that there is an on-time job i with $r_i \leq t_k$ that is executed at time $t_i > t_k$ on $\mathcal{H}$. Let $\mathcal{H}'$ be the schedule obtained from $\mathcal{H}$ by exchanging the jobs i and k. $\mathcal{H}'$ is better than $\mathcal{H}$ (*cf.* proof of the second part of Proposition 14.2). This contradicts our hypothesis on $\mathcal{H}$.

Definition of θ. Let τ the vector be built component per component as follows: The first component of τ is t_k, the time at which k starts. The following components of τ are the end times on $\mathcal{H}$ of the jobs (except k) that start before or at t_k and end strictly after t_k. The following components are the values σ_i that are strictly greater than t_k. Since the resource constraint holds at time t for $\mathcal{H}$, it is easy to prove that the dimension of τ is lower than or equal to m. The vector τ is extended to a vector τ' of dimension m by adding a sufficient number of times a component t_k. Let θ be the vector obtained from τ' by sorting its components in increasing order.

θ belongs to Ξ. Consider a component θ_j of θ. Either it is the end time of a job and then $t_k < \theta_j \leq t_k + p$ or it is a $\sigma_i > t_k$ value (and then $t_k < \theta_j \leq t_k + p$ otherwise t_k would be strictly lower than σ_1 and hence no machine would be available at time t_k) or it is equal to t_k. Hence, all components belong to the interval $[t_k, t_k + p]$, as a consequence $\theta_m - \theta_1 \leq p$. We have proven that θ is a resource profile. It is also easy to verify that all components of θ belong to Θ and hence $\theta \in \Xi$. On top of that, it is obvious that $\theta_1 = t_k$. The proof that $\sigma \ll \theta$ is also immediate given the definition of θ.

$\theta' = (\theta_2, \ldots, \theta_m, \theta_1 + p) \ll \epsilon$. The fact that θ' is a resource profile comes immediately from $\theta \in \Xi$. Suppose that the relation $\theta' \ll \epsilon$ does not hold. Then, according to Proposition 14.8, there is a time point t such that $|\{i : t < \theta'_i\}| + |\{i : \epsilon_i \leq t\}| > m$. Recall that $\epsilon_1 \geq t_k$ otherwise no machine would be available at the time point where k ends. Hence, if $t < t_k$, $|\{i : \epsilon_i \leq t\}| = 0$ and consequently, $|\{i : t < \theta'_i\}| > m$; which contradicts the fact that θ' is a vector of dimension m. As a consequence, we have $t_k \leq t$. Let O be the set

of jobs that start before or at t_k and end strictly after t_k. The components of θ', are either the completion times of the jobs in O or the σ_i values that are strictly greater than t_k or are equal to t_k. Hence, the number of components of θ' that are strictly greater than t is equal to the sum of (1) the number of jobs in O that end strictly after time t and of (2) the number of components of σ_i that are strictly greater than t. Since $t_k \leq t$, the jobs in O all start before t. Hence, the total number of jobs N_t that start before or at t and end strictly after t plus the number of components of σ_i that are strictly greater than t, is greater than or equal to $|\{i : t < \theta'_i\}|$. Hence, $|\{i : t < \theta'_i\}| + |\{i : \epsilon_i \leq t\}| > m$ leads to, $N_t + |\{i : t < \sigma_i\}| + |\{i : \epsilon_i \leq t\}| > m$. This contradicts the fact that the resource constraint is met at time t on $\mathcal{H}$.

Conclusion. The cost of $\mathcal{H}$, restricted to the jobs with a release date lower than or equal to t_k (except k) is not lower than $F_{k-1}(\sigma, \theta)$. Similarly, the cost of $\mathcal{H}$ restricted to the jobs with a release date greater than t_k is not lower than $F_{k-1}(\theta', \epsilon)$. Hence, the total cost of $\mathcal{H}$ ($F_k(\sigma, \epsilon)$), is greater than or equal to $F_{k-1}(\sigma, \theta) + F_{k-1}(\theta', \epsilon) + f_k(\theta_1 + p)$.

14.2.9.3 A Dynamic Programming Algorithm

The optimum is exactly $F_n((\min_{t \in \Theta} t, \ldots, \min_{t \in \Theta} t), (\max_{t \in \Theta} t, \ldots, \max_{t \in \Theta} t))$. Thanks to Proposition 14.9, we have a straight dynamic programming algorithm to compute this value. The relevant values for σ and ϵ are exactly the vectors in Ξ. We claim that there are $O(n^2 n^{m-1}) = O(n^{m+1})$ relevant resource profiles. Indeed, there are n^2 possible values for the first component and once it is fixed there are only n possible choices for the $m-1$ remaining ones (because of the structure of Θ and because the difference between the m-th component and the first one is upper bounded by p). This means that there are $O(n^{2m+2})$ relevant pairs (σ, ϵ). The values of $F_k(\sigma, \epsilon)$ are stored in a multidimensional array of size $O(n^{2m+3})$ (n possible values for k, n^{m+1} possible values for σ and n^{m+1} possible values for ϵ). Our algorithm then works as follows.

- In the initialization phase, $F_0(\sigma, \epsilon)$ is set to 0 for any values σ, ϵ in Ξ such that $\sigma \ll \epsilon$
- We then iterate from $k = 1$ to $k = n$. Each time, F_k is computed for all the possible values of the parameters thanks to the formula of Proposition 14.9, and to the values of F_{k-1} computed at the previous step.

Before analyzing the complexity of the overall algorithm, remark that one can generate easily all possible resource profiles θ between σ and ϵ (i.e., $\sigma \ll \theta \ll \epsilon$) in $O(n^{m+1})$ steps. Indeed, there are $O(n^2)$ possible values $\theta_1 \in \Theta \cap [\sigma_1, \epsilon_1]$. The other components of θ belong to $\Theta \cap [\theta_1, \theta_1 + p]$. There are only $O(n)$ values in this set. Components θ_i are generated, one after another; each time a test verifying that $\sigma_i \leq \theta_i \leq \epsilon_i$ is being performed in constant time.

In the initialization phase, $O(n^{2m+2})$ pairs ($\sigma \ll \epsilon$) are generated. Afterward, for each value of k, $O(n^{2m+2})$ values of F_k have to be computed. For each of them, $O(n^{m+1})$ resource profiles θ are generated with $\sigma \ll \theta \ll \epsilon$. A minimum among $O(n^{m+1})$ terms is computed. This leads to an overall time complexity of $O(n^{3m+4})$. A rough analysis of the space complexity leads to an $O(n^{2m+3})$ bound but since, at each step of the outer loop on k, one only needs the values of F computed at the previous step ($k-1$), the algorithm can be implemented with two arrays of $O(n^{2m+2})$ size: one for the current values of F and one for the previous values of F.

Notice that we can perform a backward computation on the values $F_k(\sigma, \epsilon)$ to recover the optimum schedule. Indeed, for all relevant values of k, σ and ϵ, it is easy to identify which resource profile θ is the best. Since the starting time of k is the first component θ_1 of θ, we can recover the starting times of all jobs at once.

14.2.10 Problem $P2 \mid p_j = p; pmtn; r_j \mid \sum C_j$

Assume that the jobs $j = 1, \ldots, n$ are indexed such that

$$r_1 \leq r_2 \leq \cdots \leq r_n$$

and let

$$R_1 < R_2 < \cdots < R_z$$

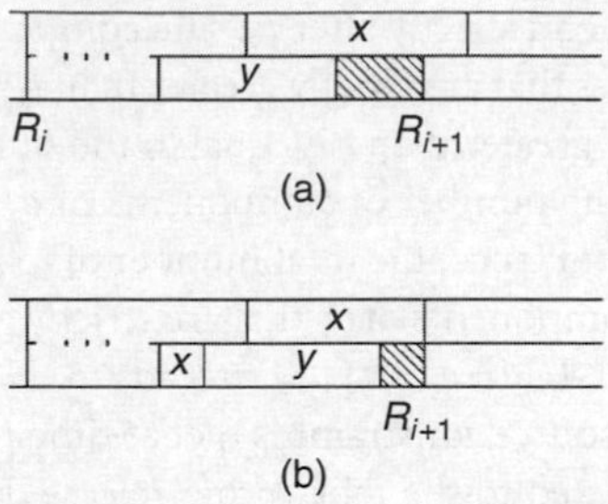

FIGURE 14.4 Eliminating idle time.

be all distinct release times. Additionally we set $R_{z+1} = \infty$. By $\mathcal{J}_i$ we denote the set of all jobs with release time R_i. The idea of the algorithm which solves problem $P2 \mid p_j = p; pmtn; r_j \mid \sum C_j$ is to create for each interval $[R_i, R_{i+1}]$ $(i = 1, \ldots, z)$ a subschedule B_i called block. Blocks are created in increasing order of block indices by following the SPT-rule as much as possible, subject to the constraint that the amount of processing done in the block is maximized. Preemption is used within a block only when it is possible to increase the amount of processing done without increasing the total flow time $\sum_{j=1}^{n} C_j$. Such a situation is shown in Figure 14.4(a). Figure 14.4(b) shows how to eliminate idle time without increasing the total flow time. The total flow time does not change because job y is shifted right by the same amount as job x is moved to the left.

A more precise description is given by the following.

Algorithm $P2 \mid p_j = p; pmtn; r_j \mid \sum C_j$

$\mathcal{J}_1 = \{j \mid r_j = R_1\}$;

FOR $i = 1$ to z DO

BEGIN

Schedule jobs in $\mathcal{J}_i$ by the SPT-rule until either all jobs have been assigned or the next job starts at or after time R_{i+1}, whichever occurs first, creating a schedule S_i;

$\mathcal{J}_{i+1} = \{j \mid r_j = R_{i+1}\}$;

Case 1. IF all jobs in $\mathcal{J}_i$ have been assigned and they are all completed by time R_{i+1} in S_i THEN $B_i := S_i$;

Case 2. IF there is no idle time in S_i in the time interval $[R_i, R_{i+1}]$ then preempt, if necessary, the jobs at time R_{i+1} and pass all uncompleted parts of jobs in $\mathcal{J}_i$ to the set $\mathcal{J}_{i+1}$. Set B_i to be equal to the subschedule of S_i in the time interval $[R_i, R_{i+1}]$;

Case 3. IF there are some idle machine times in S_i in the time interval $[R_i, R_{i+1}]$ and there is a job x not completed by time R_{i+1}, THEN modify S_i as follows. Let y be the last job completed on the machine other than the one which processed x. Shift parts of the processing of x to precede y until: (a) all idle machine times are removed (which implies that y is completed at time R_{i+1}) or (b) x is completed at time R_{i+1} or (c) x is started at time R_i and processed without interruption, whichever comes first. IF x is not completed by time R_{i+1} in the modified schedule, THEN preempt it at time R_{i+1} and pass the uncompleted part to the set $\mathcal{J}_{i+1}$. Set B_i to be equal to the subschedule of the modified schedule in $[R_i, R_{i+1}]$

END

j	1	2	3	4	5	6	7	8	9	10	11	12	13
r_j	0	0	0	16	16	16	16	33	40	40	40	55	55

$P_j = 10$

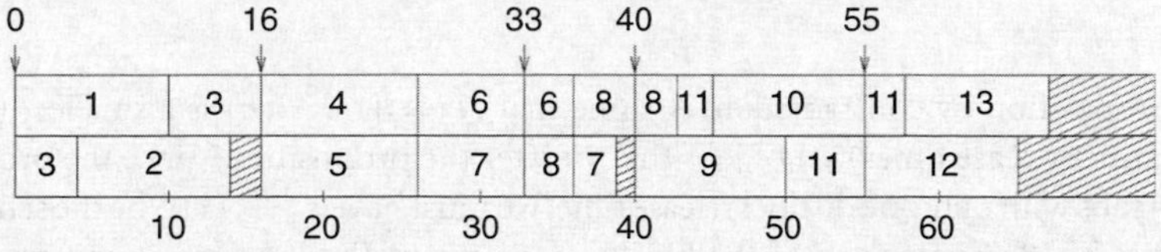

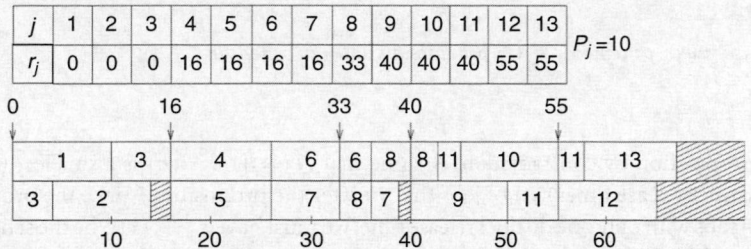

FIGURE 14.5 Example for algorithm $P2\,|\,p_j = p; pmtn; r_j\,|\sum C_j$.

Depending on the cases different scheduling rules are applied. For $i = 1, 2, 3$, Rule i denotes the rule applied in connection with Case i.

The algorithm is illustrated by the example shown in Figure 14.5. The first block B_1 is scheduled in $[0, 16]$ according to Case 3b. Block B_2 is a Case 2 block scheduled in $[16, 33]$ followed by Case 3c Block B_3 scheduled in $[33, 40]$. B_4 is a Case 3a block scheduled in $[40, 55]$. Finally B_5 is a Case 1 block scheduled in $[55, \infty]$.

It is easy to see that the algorithm can be implemented to run in $O(n \log n)$ time. Sorting the jobs in ascending order of their release times takes $O(n \log n)$ time. Scheduling the jobs takes linear time. Note, that even though the algorithm may produce preempted jobs, the number of preemption is bounded by $2z \leq 2n$. Thus, the scheduling of jobs can be done in time providing an overall complexity of $O(n \log n)$.

For the correctness proof of the algorithm we use the following.

Proposition 14.10

A schedule S constructed by algorithm $P2\,|, p_j = p; pmtn; r_j\,|\sum C_j$ has the property that the amount of processing done within each block B_i is maximum among all preemptive schedules under the condition that this property holds for all previous blocks.

Proof

One can prove the proposition by induction by showing that for each block $B_i (i = 1, \ldots, z - 1)$ one of the following statements, which are clearly equivalent, holds

1. The total amount of processing done within B_i is maximum.
2. The total remaining processing time passed to B_{i+1} is minimum.
3. Non-unnecessary idle time occurs within B_i.

If B_i was scheduled by Rule 1 or Rule 2, then there is no possibility to increase the amount of processing done within B_i. If B_i was scheduled by Rule 3, then there was some idle machine times in the interval $[R_i, R_{i+1}]$ and the job x not completed by time R_{i+1} was shifted before the job y completed last on the machine that contained idle times until

a. all idle machine times are removed
b. x is complete by time R_{i+1}
c. x is started at time R_i

whichever occurs first. If (a) occurs first then there is no idle machine time within B_i. If (b) occurs first, then the remaining processing time passed to B_{i+1} is zero. Finally, if (c) occurs first, then x must have been released at time R_i and is continuously assigned to a machine throughout B_i. Clearly, no schedule can process more of x within $[R_i, R_{i+1}]$ than B_i. All jobs j scheduled after R_{i+1} have a release time $r_j \geq R_{i+1}$. Thus, nonunnecessary idle machine time occurs within B_i, and Proposition 14.10 holds.

Proposition 14.11

Algorithm $P2 \mid p_j = p; pmtn; r_j \mid \sum C_j$ *provides an optimal schedule.*

Proof

We prove the proposition by contradiction. Assume that $\mathcal{J} = \{1, \dots, n\}$ is a smallest (in terms of the number z of distinct release times) set of jobs that violates the proposition. Since the proposition clearly holds for a set of jobs with only one distinct release time, we must have $z \geq 2$. Let S be the schedule produced by the algorithm for $\mathcal{J}$. Then none of the blocks of S, except the last one, is constructed by Rule 1. For if block B_i with $i < z$ were constructed by Rule 1 then the schedule splits into the independent schedules $B_1, \dots, B_i$ and $B_{i+1}, \dots, B_z$. One of them must violate the proposition. Furthermore, we may assume that none of the blocks of S is scheduled by Rule 2. For if block B_i were scheduled in this way, then we could construct a smaller set of jobs $\mathcal{J}'$ from $\mathcal{J}$ by letting the jobs released at time R_{i+1} be released at time R_i. The schedule produced by the algorithm for $\mathcal{J}'$ is the same as S, while an optimal schedule for $\mathcal{J}'$ must have an optimal $\sum C_j$ value no larger than the optimal $\sum C_j$ value for $\mathcal{J}$. Thus, $\mathcal{J}'$ is a smaller set of jobs violating the proposition, contradicting our assumption that $\mathcal{J}$ is a smallest such set.

Therefore, all blocks in S, except the last one are scheduled by Rule 3, and the last block in S is scheduled by Rule 1. Let S_0 denote an optimal schedule for $\mathcal{J}$. Since S is not optimal, we have

$$\sum_{j=1}^{n} f_j(S_0) < \sum_{j=1}^{n} f_j(S) \tag{14.3}$$

where $f_j(S_0)$ and $f_j(S)$ denote the finishing time of job j in S_0 and S, respectively.

Let $k, k+1, \dots, n$ $(1 < k \leq n)$ be the jobs released at time R_z. Assume for the moment

$$\sum_{j=1}^{k-1} f_j(S_0) < \sum_{j=1}^{k-1} f_j(S) \tag{14.4}$$

Consider the new set of jobs $\mathcal{J}'$ that contains only the jobs $1, 2, \dots, k-1$. Let S' denote the schedule produced by Algorithm $P2 \mid p_j = p; pmtn; r_j \mid \sum C_j$ for $\mathcal{J}'$. Then blocks 1 to $z-2$ of S' are identical to blocks 1 to $z-2$ of S, and block $z-1$ of S' is scheduled by Rule 1 while block $z-1$ of S is scheduled by Rule 3. However, block $z-1$ of S' differs from block $z-1$ of S only if the processing of $k-1$ is shifted to precede the processing of $k-2$. Letting x be the amount shifted we have $f_{k-2}(S) = f_{k-2}(S') + x$ and $f_{k-1}(S) = f_{k-1}(S') - x$. Thus

$$\sum_{j=1}^{k-1} f_j(S) = \sum_{j=1}^{k-1} f_j(S') \tag{14.5}$$

If we denote by S'_0 an optimal schedule for $\mathcal{J}'$, we have from (14.4) and (14.5).

$$\sum_{j=1}^{k-1} f_j(S'_0) \leq \sum_{j=1}^{k-1} f_j(S_0) < \sum_{j=1}^{k-1} f_j(S) = \sum_{j=1}^{k-1} f_j(S') \tag{14.6}$$

But (14.6) shows that $\mathcal{J}'$ is a smaller set of jobs violating the proposition which is a contradiction. Therefore, (14.4) must be false and we have

$$\sum_{j=1}^{k-1} f_j(S_0) \geq \sum_{j=1}^{k-1} f_j(S) \tag{14.7}$$

From (14.3) and (14.7) we have

$$\sum_{j=k}^{n} f_j(S_0) < \sum_{j=k}^{n} f_j(S)$$

which is a contradiction because the partial schedule of S for the jobs $k, k+1, \ldots, n$ must be optimal for the following reasons:

- Job $k-1$ is the only job, if any, which passes processing time from block $z-1$ to block z. The amount of passed processing time is minimal.
- Block z is an SPT-schedule for jobs $k, k+1, \ldots, n$ together with the shortened job $k-1$ passed to block z under the condition that all release times are equal to R_z. This is an optimal nonpreemptive schedule.
- For $P \mid pmtn \mid \sum C_j$ a nonpreemptive schedule is optimal due to a result of McNaughton. □

14.2.11 Problem $P2 \mid p_j = 1; prec; r_j \mid L_{\max}$

To solve this problem it is sufficient to solve its decision version: given a threshold value L^* find a feasible schedule (given by the finishing times C_j of all jobs $j = 1, \ldots, n$) such that the maximum lateness $\max_{j=1}^{n}\{C_j - d_j\}$ is bounded by L^*. Because $C_j - d_j \leq L^*$ is equivalent to $C_j \leq \overline{d}_j := L^* + d_j$ this problem is equivalent to the **time window problem** $P2 \mid p_j = 1; prec; r_j; d_j \mid$, which can be stated as follows.

Find starting times S_j of all jobs such that

- for all $t \in [0, \infty]$ the number of jobs j with $S_j = t$ is at most 2
- $S_i + p_i \leq S_j$ whenever $i \rightarrow j$
- $r_j \leq S_j \leq d_j - 1$ for all jobs j

We will present an $O(n^3)$ algorithm which solves the time window problem. Thus, problem $P2 \mid p_j = 1; prec; r_j \mid L_{\max}$ can be solved with time complexity $O(n^3 \log n)$ by applying binary search.

The algorithm has two main steps. In the first step a priority list $L = (j_1, j_2, \ldots, j_n)$ for the jobs is calculated. In the second step L is used to construct a corresponding list schedule in the following intuitive manner: Initially, both machines are idle. Starting with $t = \min r_j$, at any time t at which a machine is idle the list L is scanned from the beginning to select the first job k which may be validly executed, i.e., $r_k \leq t$ and all predecessors of k have been completed. If such a job exists it is scheduled on the idle machine starting at time t. Otherwise, t is increased to

$$\max\{t+1, \min\{r_j \mid j \text{ is not scheduled }\}\}$$

By using an appropriate data structure the list schedule can be constructed in time $O(n^2)$.

The list L is constructed by first calculating modified deadlines having the property that a valid schedule meets all the modified deadlines if and only if it meets all the original deadlines. In the second step jobs are ordered according to nondecreasing modified deadlines.

In order to state the proposition on which our deadline modifications are based, we require a few preliminary definitions. For any job i and integers r, d satisfying $r_i \leq r \leq d_i \leq d$, we define $S(i, r, d)$ to be the set of all jobs $j \neq i$ which have $d_j \leq d$ and either are successors of i or have $r_j \geq r$. Let $N(i, r, d)$ denote the number of jobs in $S(i, r, d)$. We use $\lceil x \rceil$ to denote the least integer no less than x.

Proposition 14.12

For any job i and integers r, d satisfying $r_i \leq r \leq d_i \leq d$, if $N(i, r, d) \geq 2(d - r)$, then i must be completed by time $d - \lceil N(i, r, d)/2 \rceil$ in any valid schedule that meets all job deadlines.

Proof

Suppose $N(i, r, d) \geq 2(d - r)$ and consider a valid schedule $S = (S_i)$. Then we divide the proof into two cases.

Case 1: $N(i,r,d) > 2(d-r)$
Since all jobs in $S(i,r,d)$ must be completed by time d and there are only two machines, operating nonpreemptively, there must be some job $j \in S(i,r,d)$ for which $S_j \leq d - \lceil N(i,r,d)/2 \rceil < r$. Since $r_j \leq S_j < r$, the definitions of $S(i,r,d)$ implies that j must be a successor of i. Hence i must be completed when j starts at time S_j and the desired result follows.

Case 2: $N(i,r,d) = 2(d-r)$
Then we have $\lceil (N(i,r,d)+1)/2 \rceil = \lceil N(i,r,d)/2 \rceil + 1$. We parallel the previous argument using $S = S(i,r,d) \cup \{i\}$ instead of $S(i,r,d)$. Since all jobs in S must be completed by time d and there are only two machines, operating nonpreemptively, there must be some jobs $j \in S$ for which $S_j \leq d - \lceil (N(i,r,d)+1)/2 \rceil = d - \lceil N(i,r,d)/2 \rceil - 1$. Since $r_j \leq S_j < r$ the definition of $S(i,r,d)$ implies that either $j = i$ or j is a successor of i. In either case we have

$$S_i + 1 \leq S_j + 1 \leq r = d - \lceil N(i,r,d)/2 \rceil$$

as desired.

If $N(i,r,d) \geq 2(d-r)$ and $d - \lceil N(i,r,d))/2 \rceil < d_i$ then we may set d_i equal to $d - \lceil N(i,r,d))/2 \rceil$. Such modifications may be performed repeatedly until either no further modifications are possible or we have $d_i < r_i + 1$ for some job i. In this case no valid schedule can possibly meet all the deadlines. Later we shall describe an algorithm for performing these successive modifications in an organized and efficient manner.

Motivated by the preceding discussion, we will call deadlines *consistent* whenever the following two conditions hold for every job i:

- $d_i \geq r_i + 1$
- for every pair of integers r, d satisfying $r_i \leq r \leq d_i \leq d$, if $N(i,r,d) \geq 2(d-r)$, then $d_i \leq d - \lceil N(i,r,d)/2 \rceil$

The following properties of consistent deadlines are useful. □

Property 1
If the deadlines are consistent then $i \rightarrow j$ implies $d_i < d_j$.

Proof
Assume that $i \rightarrow j$ and $d_i \geq d_j$ holds. Then job j belongs to $S(i, d_i, d_i)$ and hence $N(i, d_i, d_i) > 0 = 2(d_i - d_i)$. However, the second consistency property requires $d_i \leq d_i - \lceil N(i, d_i, d_i)/2 \rceil \leq d_i - 1$ which is a contradiction. □

Property 2
If the deadlines are consistent, then $r \leq d$ implies

$$|\{i \mid r \leq r_i \text{ and } d_i \leq d\}| \leq 2(d-r)$$

Proof
Suppose that for some $r \leq d$ the set $S = \{i \mid r \leq r_i \text{ and } d_i \leq d\}$ would satisfy $|S| > 2(d-r)$. Let k be a job in S having the earliest release time. Then $S \backslash \{k\} \subseteq S(k, r_k, d)$ and hence $N(k, r_k, d) \geq |S| - 1 \geq 2(d - r) \geq 2(d - r_k)$. The second consistency property then requires that

$$d_k \leq d - \lceil N(i, r_k, d)/2 \rceil \leq r_k$$

Thus, the first consistency property implies the contradiction $d_k \leq r_k < r_k + 1 \leq d_k$. □

We are now prepared to prove the main result.

Proposition 14.13

Assume that jobs are indexed such that $d_i \leq d_{i+1}$ for $i = 1, \ldots, n-1$ holds. If the deadlines are consistent then the valid schedule defined by the list

$$L = \{1, 2, \ldots, n\}$$

meets all the job deadlines.

Proof

Suppose that the valid schedule (S_i) constructed from L fails to meet the deadlines. Since all jobs have unit processing times and all start-times and deadlines are integers the algorithm provides integer starting times S_i. Let j be a job with minimum S_j-value which fails to meet its deadline, i.e., $S_j + 1 > d_j$ and hence by integrality $S_j \geq d_j$. Let r be the greatest integer time, $0 \leq r \leq S_j$, for which the set $P(r) = \{i \mid S_i = r - 1 \text{ and } d_i \leq d_j\}$ satisfies $| P(r) | < 2$. Defining $S = \{i \mid r \leq S_i < S_j\} \cup \{j\}$, we observe that $| S | = 2(S_j - r) + 1$ and each $i \in S$ has $d_i \leq d_j$ by definition of r. Now we divide the proof into two cases.

***Case 1:* $| P(r) | = 0$**

None of the jobs in S were ready to begin execution at time $r - 1$. Furthermore, due to Property 1, any jobs that were executed at time $r - 1$ could not have been predecessors of any job in S. Therefore, every job in S must have a release time exceeding $r - 1$ and hence, by integrality, has a release time r or greater. This implies that

$$S \subseteq \{i \mid r \leq r_i \text{ and } d_i \leq d_j\}$$

from which it follows that

$$| \{i \mid r \leq r_i \text{ and } d_i \leq d_j\} | \geq | S | = 2(S_j - r) + 1 \geq 2(d_j - r) + 1 > 2(d_j - r)$$

However, the last inequality contradicts Property 2.

***Case 2:* $| P(r) | = 1$**

Let k be the single job in $P(r)$. Then similar to Case 1 every task in S either has a release time at least r or is a successor of k. This implies that $S \subseteq S(k, r, d_j)$ from which it follows that

$$N(k, r, d_j) \geq | S | = 2(S_j - r) + 1 \geq 2(d_j - r) + 1 > 2(d_j - r)$$

From the second consistency property we then must have

$$d_k \leq d_j - \lceil N(k, r, d_j)/2 \rceil \leq d_j - (d_j - r + 1) = r - 1$$

Therefore k failed to meet its deadline, contradicting the choice of k. □

The basic algorithm, which determines whether a valid schedule meeting all deadlines exists, and if so, constructs one, can be summarized as follows:

1. Successively modify the deadlines using Proposition 14.12 until either some $r_i \geq d_i$ or the deadlines are consistent. If some $r_i \geq d_i$ then report that no schedule exists and halt.
2. Form a priority list L by sorting the jobs in order of nondecreasing modified deadlines.
3. Compute a valid schedule defined by the priority list L.

By Proposition 14.13 a list schedule constructed in this way meets all job deadlines.

Step 2 can be done in $O(n \log n)$ while Step 3 can be implemented in time $O(n^2)$. Next, we will present an algorithm which calculates the modified deadlines and checks consistency. It can be implemented in such a way that its time complexity is $O(n^3)$.

14.2.11.1 A Deadline Modification Algorithm

Assume that the jobs are indexed such that $r_1 \leq r_2 \leq \cdots \leq r_n$. Then, a suitable deadline modification algorithm can be structured as three nested loops, each selecting successive values of one of the three parameters i, r and d. The outer loop (Statements 1 to 17 in the following description) selects values of d in decreasing order. For each d, the next loop (Statements 6 to 16) selects values of i in increasing order, skipping those values of i with $d_i > d$. Finally, for fixed i and d, the inner loop (Statements 11 to 16) selects appropriate values of r in increasing order and modifies d_i, if required by the value of $\lceil N(i, r, d)/2 \rceil$.

Garey and Johnson [21] have shown that for fixed i and d, the inner loop need only consider those integers r, $r_i \leq r \leq d_i$, which are job release times or d_i itself. Furthermore, the d-values can be restricted to original or modified d_i-values. If, after considering all values i and r for some d, no task remains with deadline d then no valid schedule exists.

The following algorithm describes the details of the deadline modification procedure.

Algorithm Deadline Modification

```
1.   d := 1 + max{d_i | i = 1, ..., n};
2.   WHILE a job i with d_i < d exists DO
         BEGIN
3.       IF d_j <= r_j for some job j THEN HALT (no schedule exists)
4.       ELSE
             BEGIN
5.           Set d to the largest job deadline less than d;
6.           Set i to the least job deadline d_i with d_i <= d;
7.           i := i - 1;
8.           WHILE a job j with j > i and d_j <= d exists DO
                 BEGIN
9.               Let j be the least index greater than i with d_j <= d;
10.              i := j;
11.              k := i;
12.              WHILE r_k < d_i DO
                     BEGIN
13.                  UPDATE (i, r_k, d);
14.                  k := k + 1;
                     END;
15.              r_k := min{r_k, d_i};
16.              UPDATE (i, r_k, d);
                 END;
             END;
17.      IF no job j with d_j = d exists THEN HALT (no schedule exists)
         END
```

In this algorithm UPDATE (i, r, d) is the following procedure.

UPDATE (i, r, d)

1. IF $N(i, r, d) \geq 2(d - r)$ AND $d_i > d - \lceil N(i, r, d)/2 \rceil$ THEN
   $d_i := d - \lceil N(i, r, d)/2 \rceil$;
2. FOR ALL predecessors j of i with $d_j > d_i$ DO $d_j := d_i$.

Garey and Johnson [21] have proved that the Algorithm Deadline Modification correctly calculates consistent deadlines, if a feasible schedule exists. Furthermore, they have shown that it can be implemented to run in time $O(n^3)$.

14.2.12 $P \mid p_j = p; r_j \mid \sum C_j$

An optimal schedule can be constructed by scheduling jobs j in an order of nondecreasing release times at the earliest time $t \geq r_j$ at which j can be scheduled on some machine.

To prove that the scheduling rule provides an optimal schedule assume that the jobs are indexed in such a way that $r_1 \leq r_2 \leq \cdots \leq r_n$. Let $j - 1 < n$ be the largest index with the property that an optimal schedule S^* exists which coincides with the schedule constructed by our rule up to job $j - 1$. Let t be the earliest time at which job j can be scheduled as the next job on some machine, say M_i. Let k be the job scheduled next after t at M_i. Such a job exists because in S^* the starting time of j is greater than or equal to t. Thus, in S^* job j could be moved to M_i starting at time t. Let S_k be the starting time of job k and S_j be the starting time of job j in S^*.

If $S_j \geq S_k$ then $r_j \leq r_k \leq S_k \leq S_j$ and we can swap j and k. This contradicts the maximality of $j - 1$.

If $S_j < S_k$ then $t \leq S_j < S_k$ and M_i is idle between t and S_k. let $M_l \neq M_i$ be the machine on which job j is processed. Then we can swap the jobs scheduled on M_l from time S_j with the jobs scheduled on M_i from time S_j and then move job j on M_i to the left providing again a contradiction.

Notice that when applying the scheduling rule it is not possible that a job j will be scheduled in an idle period between two jobs k and l scheduled on the same machine without changing the position of these jobs. This follows from the fact that $r_j \geq \max\{r_k, r_l\}$. Thus, using a priority queue with elements (M_i, t_i), where t_i is the currently latest finishing time of a job scheduled on M_i, it is possible to construct an $O(n \log n)$-algorithm.

14.2.13 Problem $P \mid p_j = p; r_j \mid L_{\max}$

If $p = 1$ then we get an optimal schedule by scheduling available jobs in an order of nondecreasing due dates. More specifically, if at current time t not all machines are occupied and there is an unscheduled job j with $r_j \leq t$ then we schedule such a job with the smallest due date.

Scheduling available jobs with smallest due dates fails to provide an optimal schedule if all jobs have a processing time $p \neq 1$ as shown in Figure 14.6.

To solve problem $P \mid p_j = p; r_j \mid L_{\max}$ we consider the problem $P \mid p_j = p; r_j; d_j \mid -$ of finding a schedule such that each job j has to be processed within a time window $[r_j, d_j]$ if such a schedule exists. To solve $P \mid p_j = p; r_j \mid L_{\max}$ binary search can be applied.

First we introduce the notion of a cyclic list schedule. A list is a permutation $\pi : \pi(1), \pi(2), \ldots, \pi(n)$ of all jobs. A corresponding cyclic list schedule is constructed by the following algorithm. In connection with this algorithm, it is convenient to number the machines from 0 to $m - 1$. The starting time of job j in this schedule is denoted by $x(j)$, $h(j)$ denotes the machine on which j is to be processed, and $t(i)$ is the finishing time of the last job on machine i.

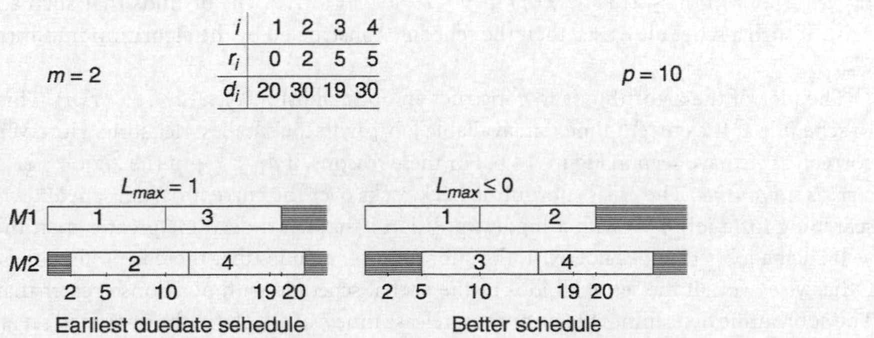

FIGURE 14.6 Earliest due date schedule in not optimal for problem $P \mid p_j = p; r_j \mid L_{\max}$.

Algorithm Cyclic List Schedule

1. FOR $i := 0$ TO $m - 1$ DO $t(i) := 0$;
2. FOR $j := 1$ TO n DO
   BEGIN
3. Schedule job j on machine $h(j) := j(mod\ m)$ at time
   $x(j) := \max\{t(h(j)), r_j\}$;
4. $t(h(j)) := x(j) + p$
   END

The following proposition shows that we only need to consider cyclic list schedules if we want to solve a problem of the form $P \mid p_j = p; r_j; d_j \mid f$ with regular objective function f.

Proposition 14.14

Let f be a regular objective function and assume that $P \mid p_j = p; r_j; d_j \mid f$ has a feasible solution. Then there always exists a cyclic list schedule which is optimal.

Proof

We will show that any feasible solution $(x(j), h(j))$ can be transformed into a feasible cyclic list schedule without increasing the objective function value. Such a transformation is done in two steps. The first step changes the machine on which the jobs are scheduled as follows.

Consider a permutation π with

$$x(\pi(1)) \leq x(\pi(2)) \leq \cdots \leq x(\pi(n))$$

Then we schedule job $\pi(k)$ on machine $k(mod\ m)$. The corresponding schedule has no overlapping jobs on the same machine. This can be seen as follows. Assume that two jobs $\pi(i_0)$ and $\pi(i_1)$, with $i_0 = hm+k$ and $i_1 = lm + k$ where $l > h$, overlap in an interval I. We have $x(\pi(i_0)) \leq x(\pi(i)) \leq x(\pi(i_1))$ for all $i_0 \leq i \leq i_1$ and the processing time of all jobs is equal to p. Therefore, all jobs $\pi(i)$ $(i_0 \leq i \leq i_1)$ are processed during the interval I. This contradicts the feasibility of $(x(i), h(i))$ because there are at least $m + 1$ jobs $\pi(i)$ $(i_0 \leq i \leq i_1)$.

In the second step, the new schedule is transformed into the list schedule which corresponds to π by decreasing the starting times $x(\pi(i))$ of all jobs. Thus, the regular objective function does not increase during the transformation. □

We will present Simon's algorithm which constructs a schedule $(x(j), h(j))$ respecting all time windows $[r_j, d_j]$ (i.e. with $r_j \leq x(j) < x(j) + p \leq d_j$ for $i = 1, \ldots, n$) or finds that such a schedule does not exist. If such a schedule exists, then the schedule constructed by the algorithm minimizes $\sum C_j$ as well as $C_{\max}$.

The idea of the algorithm is to construct an optimal list $\pi(1), \pi(2), \ldots, \pi(n)$. This is done by trying to schedule at the current time t an available job j with the smallest deadline. However, this is not always correct, as we have seen in Figure 14.6. For these reasons, if $d_j < t + p$ (i.e., if job j is late) we have to call a *crisis subroutine*. The crisis subroutine backtracks over the current partial schedule $\pi(1), \ldots, \pi(k-1)$ searching for a job $\pi(i)$ with a highest position i that has a deadline greater than that of the crisis job j. If such a job $\pi(i)$ does not exist, the subroutine concludes that there is no feasible schedule and halts. Otherwise, we call the set of all jobs in the partial schedule with positions greater than i a *restricted set*. The subroutine determines the minimum release time r of all jobs in the restricted set and creates a *barrier* (i, r). This barrier is an additional restriction to the starting time of jobs scheduled in positions $k \geq i$. It is added to a barrier list which is used to calculate the current scheduling time. Finally, $\pi(i)$ and all jobs in the restricted set are eliminated from the partial schedule and we continue with the partial schedule $\pi(1), \ldots, \pi(i-1)$.

If $d_j \geq t + p$, then job j is scheduled at time t on machine $h(j) = j (mod\ m)$.

Details are given below. U is the set of unscheduled jobs, the barrier list contains all barriers, and $\pi(i)$ denotes the i-th job currently scheduled for $i = 1, 2, \ldots, k-1$.

Algorithm $P \mid p_j = p; r_j; d_j \mid \sum C_j, C_{max}$

```
1. Initialize;
2. WHILE there are unscheduled jobs DO
      BEGIN
3.    Calculate current time t;
4.    Find unscheduled job j available at time t
      with smallest due date;
5.    If d_j ≥ t + p THEN schedule job j
      ELSE
6.         crisis (j)
      END
```

We now describe the **initialize, calculate current time t, schedule job j** and **crisis** (j) modules. The most important module is **crisis (j)**, which resolves a crisis with job j.

Crisis (j)

IF there exists an index $1 \leq v \leq k-1$ with $d_{\pi(v)} > d_j$ THEN
BEGIN
Calculate largest index $1 \leq i \leq k-1$ with $d_{\pi(i)} > d_j$;
$r := \min(\{r_{\pi(v)} \mid v = i+1, \ldots, k-1\} \cup \{r_j\})$;
Add (i, r) to *barrierlist*;
Add jobs $\pi(i), \pi(i+1), \ldots, \pi(k-1)$ to U;
$k := i$
END
ELSE HALT (There exists no feasible schedule)

Initialize

barrierlist $:= \phi; k := 1; U := \{1, \ldots, n\}$

Schedule job j

$x(j) := t$;
$h(j) := k(mod\ m)$;
$U := U \backslash \{j\}$;
$t(h(j)) := t + p$;
$\pi(k) := j$;
$k := k + 1$

Calculate current time t

IF $1 \leq k \leq m$ THEN $t_1 := 0$ ELSE $t_1 := t(k\ (mod\ m))$;
$t_2 := \min\{r_j \mid j$ is an unscheduled job $\}$;
$t_3 := \max(\{r \mid (i, r)$ is a barrier $; 1 \leq i \leq k\} \cup \{0\})$;
$t := \max\{t_1, t_2, t_3\}$

Simons [22] has shown that Algorithm $P \mid p_j = p; r_j; d_j \mid \sum C_j, C_{\max}$ provides a feasible schedule for the time-window problem in time $O(n^3 \log\log n)$ if appropriate data structures are used. Furthermore, the algorithm minimizes simultaneously both objective functions $\sum C_j$ and $C_{\max}$ in that case.

14.3 Summary of Complexity Results

In this section we will summarize the results in this chapter and include the corresponding references. We also list the scheduling problems which can be solved polynomially even with arbitrary processing times and possibly uniform or unrelated machines. Additionally, NP-hardness results will be listed for problems with constant processing times.

Table 14.1 contains the hardest single machine problems which are known to be polynomially solvable.

Table 14.2 contains single machine problems with equal processing times which have been shown to be NP-hard (even in the strong sense which is indicated by "$*$").

Complexity results for parallel machine problem without preemptions and with preemptions are listed in separate tables.

The hardest nonpreemptive scheduling problems with identical, uniform, or related machines can be found in Table 14.3.

Table 14.4 shows nonpreemptive parallel machine problems with equal processing times which are NP-hard.

Tables 14.5 and 14.6 show corresponding results for parallel machine scheduling problems with preemptions. Comparing Tables 14.3 and 14.5 it can be seen that many problems which are polynomially solvable if the processing times are equal have a polynomially solvable counterparts with arbitrary processing times if preemption is allowed. However, this is not the rule. A surprising counterexample can be found in Tables 14.3 and 14.6. $P \mid p_j = p \mid \sum w_j U_j$ is polynomially solvable whereas the corresponding preemptive problem is NP-hard.

Lists of open problems can be found under
http://www.mathematik.uni-osnabrueck.de/research/OR/class

TABLE 14.1 Polynomial Solvable Single Machine Problems

$1 \mid prec; r_j \mid C_{max}$	Lawler [23]
$1 \mid p_j = p; prec; r_j \mid L_{max}$	Simons [5]
$1 \mid prec; r_j; pmtn \mid L_{max}$	Baker et al. [24]
$1 \mid p_j = p; prec; r_j \mid \sum C_j$	Simons [22]
$1 \mid p_j = p; prec; pmtn; r_j \mid \sum C_j$	Baptiste et al. [25]
$1 \mid r_j; pmtn \mid \sum C_j$	Baker [26]
$1 \mid p_j = p; r_j \mid \sum w_j C_j$	Baptiste [27]
$1 \mid sp - graph \mid \sum w_j C_j$	Lawler [28]
$1 \mid r_j; pmtn \mid \sum U_j$	Lawler [1]
$1 \mid p_j = p; r_j \mid \sum w_j U_j$	Baptiste [2]
$1 \mid p_j = p; pmtn; r_j \mid \sum w_j U_j$	Baptiste [2]
$1 \mid p_j = p; r_j \mid \sum T_j$	Baptiste [27]
$1 \mid p_j = 1; r_j \mid \sum w_j T_j$	Assignment-problem

TABLE 14.2 NP-Hard Single Machine Problems

$*$	$1 \mid p_j = 1; chains; r_j \mid \sum w_j C_j$	Lenstra and Rinnooy Kan [29]
$*$	$1 \mid p_j = 1; prec \mid \sum w_j C_j$	Lawler [28], Lenstra and Rinnooy Kan [30]
$*$	$1 \mid p_j = 1; chains \mid \sum U_j$	Lenstra and Rinnooy Kan [29]
$*$	$1 \mid p_j = 1; chains \mid \sum T_j$	Leung and Young [31]

TABLE 14.3 Polynomially Solvable Parallel Machine Problems without Preemption

$P \mid p_j = p; outtree; r_j \mid C_{max}$	Brucker et al. [32]
$P \mid p_j = p; tree \mid C_{max}$	Hu [20], Davida and Linton [33]
$Q \mid p_j = p; r_j \mid C_{max}$	Dessouky et al. [34]
$Q2 \mid p_j = p; chains \mid C_{max}$	Brucker et al. [35]
$P \mid p_j = 1; chains; r_j \mid L_{max}$	Baptiste et al. [25]
$P \mid p_j = p; intree \mid L_{max}$	Brucker et al. [32], Monma [19]
$P \mid p_j = p; r_j \mid L_{max}$	Simons [5]
$P2 \mid p_j = p; prec \mid L_{max}$	Garey and Johnson [36]
$P2 \mid p_j = 1; prec; r_j \mid L_{max}$	Garey and Johnson [21]
$P \mid p_j = 1; outtree; r_j \mid \sum C_j$	Brucker et al. [11]
$P \mid p_j = p; outtree \mid \sum C_j$	Hu [20]
$P \mid p_j = p; r_j \mid \sum C_j$	Simons [22]
$P2 \mid p_j = p; prec \mid \sum C_j$	Coffman and Graham [37]
$Pm \mid p_j = p; intree \mid \sum C_j$	Baptiste et al. [25]
$Pm \mid p_j = p; tree \mid \sum C_j$	Baptiste et al. [25], Hu [20]
$Qm \mid p_j = p; r_j \mid \sum C_j$	Dessouky et al. [34]
$R \mid\mid \sum C_j$	Horn [38], Bruno et al. [39]
$Pm \mid p_j = p; r_j \mid \sum w_j C_j$	Baptiste [27]
$P \mid p_j = 1; r_j \mid \sum w_j U_j$	Networkflow-problem
$Pm \mid p_j = p; r_j \mid \sum w_j U_j$	Baptiste et al. [25]
$Q \mid p_j = p \mid \sum w_j U_j$	Assignment-problem
$Pm \mid p_j = p; r_j \mid \sum T_j$	Baptiste [27]
$P \mid p_j = 1; r_j \mid \sum w_j T_j$	Networkflow-problem
$Q \mid p_j = p \mid \sum w_j T_j$	Assignment-problem

TABLE 14.4 Nonpreemptive Parallel Machine Problems with Equal Processing Times Which Are *NP*-Hard

*	$P \mid p_j = 1; intree; r_j \mid C_{max}$	Brucker et al. [32]
*	$P \mid p_j = 1; prec \mid C_{max}$	Ullman [40]
*	$P2 \mid chains \mid C_{max}$	Kubiak [41]
*	$P \mid p_j = 1; outtree \mid L_{max}$	Brucker et al. [32]
*	$P \mid p_j = 1; intree; r_j \mid \sum C_j$	Lenstra [42]
*	$P \mid p_j = 1; prec \mid \sum C_j$	Lenstra and Rinnooy Kan [30]
*	$P2 \mid p_j = 1; chains \mid \sum w_j C_j$	Timkovsky [43]
*	$P2 \mid p_j = 1; chains \mid \sum U_j$	Single-machine problem
*	$P2 \mid p_j = 1; chains \mid \sum T_j$	Single-machine problem

TABLE 14.5 Polynomially Solvable Parallel Machine Problems with Preemptions

$P \mid outtree; pmtn; r_j \mid C_{max}$	Lawler [44]
$P \mid tree; pmtn \mid C_{max}$	Muntz and Coffman [45], Gonzalez and Johnson [46]
$Q \mid chains; pmtn \mid C_{max}$	Horvath et al. [47]
$P \mid intree; pmtn \mid L_{max}$	Monma [19], Garey and Johnson [48], Lawler [44]
$R \mid pmtn; r_j \mid L_{max}$	Lawler and Labetoulle [49]
$Q2 \mid prec; pmtn; r_j \mid L_{max}$	Lawler [44]
$P \mid p_j = 1; outtree; pmtn; r_j \mid \sum C_j$	Brucker et al. [11]
$P \mid p_j = p; outtree; pmtn \mid \sum C_j$	Brucker et al. [11]
$P2 \mid p_j = 1; prec; pmtn \mid \sum C_j$	Coffman et al. [50]
$P2 \mid p_j = p; pmtn; r_j \mid \sum C_j$	Herrbach and Leung [51]
$Q \mid pmtn \mid \sum C_j$	Labetoulle et al. [52]
$P \mid p_j = p; pmtn \mid \sum w_j C_j$	McNaughton [6]
$Q \mid p_j = p; pmtn \mid \sum U_j$	Baptiste et al. [25]
$P \mid p_j = 1; pmtn; r_j \mid \sum w_j U_j$	Brucker et al. [13]
$Pm \mid p_j = p; pmtn \mid \sum w_j U_j$	Baptiste at al. [25]
$P \mid p_j = p; pmtn \mid \sum T_j$	Baptiste et al. [25]
$P \mid p_j = 1; pmtn; r_j \mid \sum w_j T_j$	Baptiste et al. [12]

TABLE 14.6 Preemptive Parallel Machine Problems with Equal Processing Times Which Are *NP*-Hard

*	$P \mid p_j = 1; prec; pmtn \mid C_{max}$	Ullman
*	$P \mid p_j = p; pmtn; r_j \mid \sum w_j C_j$	Leung and Young [53]
*	$P2 \mid p_j = 1; chains; pmtn \mid \sum w_j C_j$	Timkovsky [43], Du et al. [7]
*	$P2 \mid p_j = 1; chains; pmtn \mid \sum U_j$	Baptiste et al. [25]
	$P \mid p_j = p; pmtn \mid \sum w_j U_j$	Brucker and Kravchenko [54]

References

[1] E. L. Lawler. A dynamic programming algorithm for preemptive scheduling of a single machine to minimize the number of late jobs. *Annals of Operations Research*, 26:125–133, 1990.

[2] Ph. Baptiste. Polynomial time algorithms for minimizing the weighted number of late jobs on a single machine when processing times are equal. *Journal of Scheduling*, 2:245–252, 1999.

[3] J. Carlier. *Problèmes d'Ordonnancement à Contraintes de Ressources : Algorithmes et Complexité*. Thèse de Doctorat d'Etat, Université Paris VI, 1984.

[4] Ph. Baptiste, M. Chrobak, Ch. Dürr, W. Jawor, and N. Vakhania. Preemptive scheduling of equal-length jobs to maximize weighted throughput. Submitted to Operations Research Letters.

[5] B. Simons. A fast algorithm for single processor scheduling. *FOCS*, 1978.

[6] R. McNaughton. Scheduling with deadlines and loss functions. *Management Science*, 6:1–12, 1959.

[7] J. Du, J. Y.-T. Leung, and G. H. Young. Scheduling chain-structured tasks to minimize makespan and mean flow time. *Information and Computation*, 92: 219–236, 1991.

[8] B. Chen, C. N. Potts, and G. J. Woeginger. A review of machine scheduling: complexity, algorithms and approximability. In *D.-Z. Du and P. M. Pardalos (eds.), Handbook of Combinatorial Optimization*, Kluwer, 1998

[9] E. L. Lawler, J. K. Lenstra, A. H. G. Rinnooy Kan, and D. B. Shmoys. Sequencing and scheduling: algorithms and complexity. *Handbook on Operations Research;* and S. C. Graves, A. H. G. Rinnooy Kan, and P. Zipkin (eds.), *Management Science*, Elsevier, 1993.

[10] Ph. Baptiste and V. Timkovsky. On preemption redundancy in scheduling unit processing time jobs on two parallel machines. *Operations Research Letters*, 28:205–212, 2001.

[11] P. Brucker, J. Hurink, and S. Knust. A polynomial algorithm for $P \mid p_j = 1, r_j, outtree \mid \sum C_j$. *Technical Report, University of Osnabrueck*, Germany, 2001.

[12] Ph. Baptiste, J. Carlier, A. Kononov, M. Queyranne, S. Sevastianov, and M. Sviridenko. Structural properties of preemptive schedules. IBM Research Report.

[13] P. Brucker, S. Heitmann, and J. Hurink. How useful are preemptive schedules? *Operations Research Letters*, 31:129–136, 2003.

[14] Ph. Baptiste. Preemptive scheduling of identical machines. *Research Report 314*, University of Technology of Compiègne, 2000.

[15] W. Horn. Some simple scheduling problems. *Naval Research Logistics Quarterly*, 21:177–185, 1974.

[16] G. B. Dantzig. *Linear Programming and Extensions*, 11th printing, Princeton University Press, Princeton, NJ, 1998.

[17] S. Sahni. Preemptive scheduling with due dates. *Operations Research*, 27:925–934, 1979.

[18] M. Dror, W. Kubiak, and P. Dell'Olmo. *'Strong'-'Weak' Chain Constrained Scheduling*. Ricerca Operativa 27, 1998.

[19] C. L. Monma. Linear-time algorithms for scheduling on parallel processors, *Operations Research*, 30: 116–124, 1982.

[20] T. C. Hu. Parallel sequencing and assembly line problems. *Operations Research*, 9:841–848, 1961.

[21] M. R. Garey and D. S. Johnson. Two-processor scheduling with start-times and deadlines. *SIAM Journal on Computing*, 6: 416–426, 1977.

[22] B. Simons. Multiprocessor scheduling of unit-time jobs with arbitrary release times and deadlines. *SIAM Journal of Computing*, 12:294–299, 1983.

[23] E. L. Lawler. Optimal sequencing of a single machine subject to precedence constraints, *Management Science*, 19:544–546, 1973.

[24] K. R. Baker, E. L. Lawler, J. K. Lenstra, and A. H. G. Rinnooy Kan. Preemptive scheduling of a single machine to minimize maximum cost subject to release states and precedence constraints, *Operations Research*, 26: 111–120, 1983.

[25] Ph. Baptiste, P. Brucker, S. Knust, and V. G. Timkovsky. Fourteen notes on equal-processing-time scheduling. Submitted.

[26] K. R. Baker. *Introduction to Sequencing and Scheduling*. John Wiley & Sons, New York, 1974.

[27] Ph. Baptiste. Scheduling equal-length jobs on identical parallel machines. *Discrete Applied Mathematics*, 103:21–32, 2000.

[28] E. L. Lawler. Sequencing jobs to minimize total weighted completion time subject to precedence constraints, *Annals of Discrete Mathematics*, 2:75–90, 1978.

[29] J. K. Lenstra and A. H. G. Rinnooy Kan. Complexity results for scheduling chains on a single machine. *European Journal of Operation Research* 4, 270–275, 1980.

[30] J. K. Lenstra and A. H. G. Rinnooy Kan. Complexity of scheduling under precedence constraints, *Operations Research*, 26: 22–35, 1978.

[31] J. Y.-T. Leung and G. H. Young. Minimizing total tardiness on a single machine with precedence constraints. *ORSA Journal on Computing*, 2: 346–352, 1990.

[32] P. Brucker, M. R. Garey, and D. S. Johnson. Scheduling equal-length tasks under tree-like precedence constraints to minimize maximum lateness. *Mathematics of Operations Research*, 2: 275–284, 1977.

[33] G. I. Davida and D. J. Linton. A new algorithm for the scheduling of tree structured tasks, *Proceedings of Conference on Information, Science and System*. Baltimore, MD, 543–548, 1976.

[34] M. I. Dessouky, B. J. Lageweg, J. K. Lenstra, and S. L. van de Velde. Scheduling identical jobs on uniform parallel machines. *Statistica Neerlandica* 44: 115–123, 1990.

[35] P. Brucker, J. Hurink, and W. Kubiak. Scheduling identical jobs with chain precedence constraints on two uniform machines: *Mathematical Methods of Operations Research*, 49: 211–219.

[36] M. R. Garey and D. S. Johnson. Scheduling tasks with nonuniform deadlines on two processors. *Journal of the Association for Computing Machinery*, 23: 461–467, 1976.

[37] E. G. Coffman, Jr. and R. L. Graham. Optimal scheduling for two-processor systems, *Acta Informatica*, 1: 200–213, 1972.

[38] W. Horn. Minimizing average flow time with parallel machines. *Operations Research*, 21: 846–847, 1973.

[39] J. L. Bruno, E. G. Coffman, and R. Sethi. Scheduling independent tasks to reduce mean finishing time, *Communications of the ACM*, 17: 382–387, 1974.

[40] J. D. Ullman. NP-complete scheduling problems, *Journal of Computing System Science* 10: 384–393, 1975.

[41] W. Kubiak. Exact and approximate algorithms for scheduling unit time tasks with tree-like precedence constraints, Abstracts EURO IX - TIMS XXVIII Paris, 1988.

[42] J. K. Lenstra. Unpublished.

[43] V. G. Timkovsky. Identical parallel machines vs. unit-time shops, preemptions vs. chains, and other offsets in scheduling complexity, Technical Report, Department of Computer Science and Systems, McMaster University, Hamilton, 1998.

[44] E. L. Lawler. Preemptive scheduling of precedence-constrained jobs on parallel machines, in Dempster et al.: *Deterministic and Stochastic Scheduling*, Reidel, Dordrecht, 1982.

[45] R. R. Muntz and E. G. Coffman, Jr. Optimal preemptive scheduling on two-processor systems, *IEEE Transactions on Computers* C-18, 1014–1020, 1969.

[46] T. Gonzalez, D. S. Johnson. A new algorithm for preemptive scheduling of trees. *Journal of the Association for Computing Machinery*, 27: 287–312, 1980.

[47] E. C. Horvath, S. Lam, and R. Sethi. A level algorithm for preemptive scheduling. *Journal of the Association for Computing Machinery*, 24: 32–43, 1977.

[48] M. R. Garey and D. S. Johnson. *Computers and Intractability. A Guide to the Theory of NP-Completeness*. W. H. Freeman & Co., 1979.

[49] E. L. Lawler and J. Labetoulle. On preemptive scheduling of unrelated parallel processors by linear programming. *Journal of ACM*, 25:612–619, 1978.

[50] E. G. Coffman, J. Sethuraman, and V. G. Timkovsky. Ideal preemptive schedules on two processors, Technical Report, Columbia University, 2002.

[51] L. A. Herrbach and J. Y.-T. Leung. Preemptive scheduling of equal length jobs on two machines to minimize mean flow time. *Operations Research*, 38:487–494, 1990.

[52] J. Labetoulle, E. L. Lawler, J. K. Lenstra, and A. H. G. Rinnooy Kan. Preemptive scheduling of uniform machines subject to release dates. *Progress in Combinatorial Optimization*, 245–261, 1984.

[53] J. Y.-T. Leung and G. H. Young. Preemptive scheduling to minimize mean weighted flow time. *Information Processing Letters*, 34: 47–50, 1990.

[54] P. Brucker and S. A. Kravchenko. Preemption can make parallel machine scheduling problems hard. *Osnabruecker Schriften zur Mathematik*, Reihe P, Nr. 211.

[55] P. Brucker. *Scheduling Algorithms*. Springer Lehrbuch, 2001.

[56] J. Carlier. Problème à une machine et algorithmes polynômiaux. *QUESTIO*, 5(4):219–228, 1981.

[57] R. L. Graham, E. L. Lawler, J. K. Lenstra, and A. H. G. Rinnooy Kan. Optimization and approximation in deterministic sequencing and scheduling: a survey. *Annals of Discrete Mathematics*, 5:287–326, 1979 .

[58] M. R. Garey, D. S. Johnson, B. B. Simons, and R. E. Tarjan. Scheduling unit-time tasks with arbitrary release times and deadlines. *SIAM Journal of Computing*, 10:256–269, 1981.

[59] J. K. Lenstra, A. H. G. Rinnooy Kan, and P. Brucker. Complexity of machine scheduling problems. *Annals of Discrete Mathematics*, 1:343–362, 1977.

15

Online Scheduling

15.1 Introduction

15

Online Scheduling

Kirk Pruhs*
University of Pittsburgh

Jiří Sgall†
Mathematical Institute, ASCR

Eric Torng‡
Michigan State University

15.1 Introduction

In this chapter, we summarize research efforts on several different problems that fall under the rubric of online scheduling. In online scheduling, the scheduler receives jobs that arrive over time, and generally must schedule the jobs without any knowledge of the future. The lack of knowledge of the future precludes the scheduler from guaranteeing optimal schedules. Thus much research has been focused on finding scheduling algorithms that guarantee schedules that are in some way not too far from optimal.

We focus on problems that arise within the ubiquitous client-server setting. In a client-server system, there are many clients and one server (or perhaps a few servers). Clients submit requests for service to the server(s) over time. In the language of scheduling, a server is a processor, and a request is a job. Applications that motivate the research we survey include multiuser operating systems such as Unix and Windows, web servers, database servers, name servers, and load balancers sitting in front of server farms.

*Supported in part by NSF grant CCR-0098752, NSF grant ANI-0123705, NSF grant ANI-0325353, and a grant from the United States Air Force.

†Partially supported by Institute for Theoretical Computer Science, Prague (project LN00A056 of MŠMT ČR), grant 201/01/1195 of GA ČR, grant A1019901 of GA AV ČR, and cooperative grant KONTAKT-ME476/CCR-9988360-001 from MŠMT ČR and NSF.

‡Supported in part by NSF grant CCR-9701679, NSF grant CCR-0105283, and NSF grant EIA-0219229.

1-58488-397-9/$0.00+$1.50

The area of online scheduling is much too large for a chapter sized unabridged survey. Our goal is to highlight the critical ideas and techniques that have driven much of the recent research and to focus attention on the open problems that appear to be most interesting.

15.1.1 Online Paradigms and Notation

The idea behind an online algorithm is that the algorithm does not have access to the entire input instance as it makes its decisions. For a thorough introduction to online algorithms that extends beyond online scheduling algorithms, please refer to the book by Borodin and El-Yaniv [1] and the collection of surveys [2]. In scheduling, we model a range of different environments which differ in the way the information is released. These are discussed below, as well as some additional notations specific to online problems that we use to supplement the standard three-field notation introduced in Chapter 1.

In the *online-time* paradigm, the scheduler must decide at each time t which job to run at time t. Problems within the *online-time* model typically have release dates, and the scheduler is not aware of the existence of a job until its release date. Once a job is released, we assume that the scheduler learns the processing time of a job. For example, a web server serving static documents might reasonably be modeled by the *online-time* model since the web server can know the size of the requested file. In contrast, in the *online-time-nclv* model, the scheduler is given no information about the processing time of a job at its release date. For example, the process scheduling component of an operating system is better modeled by the *online-time-nclv* model than the *online-time* model since the operating system typically will not know the execution time of a process. This lack of knowledge of the processing time is called nonclairvoyance.

If preemption is not allowed for problems in either the *online-time* or *online-time-nclv* model, and jobs can have arbitrary processing time, then there is usually a trivial example that shows that any online scheduler will produce schedules that are far from optimal. This is why server systems, such as operating systems and web servers, generally allow preemption. Thus most research in online scheduling assumes preemption unless all jobs have similar processing times (as could be the case for a name server, for example). In the online setting, there is another possibility, which is meaningless for offline algorithms. Namely, a running job can be stopped and later *restarted* from the beginning on the same or different machine(s). Thus in order to finish, a job has to be assigned to the same machine(s) for its whole running time without an interruption; in the offline case this is equivalent to nonpreemptive scheduling as the unfinished parts of a job can simply be removed from the schedule. This possibility will be denoted by *pmtn-restart* in the middle (job) field of the three-field notation.

In the *online-list* paradigm, the jobs are ordered in a list/sequence. As soon as the job is presented, we know all its characteristics, including the processing time. The job has to be assigned to some machine and time slots (consistent with the restrictions of the given problem) before the next job is seen. The scheduling algorithm cannot change this assignment once it has been made. In the *online-list* model, in contrast to the *online-time* and *online-time-nclv* models, the time between when jobs are assigned is irrelevant or meaningless. The *online-list* model might be an appropriate model for a load balancer sitting in front of a server farm.

The notation *online-time*, *online-time-nclv* or *online-list* will be included in the job field of the three-field notation. So, for example, $1 \mid \textit{online-time}, r_j, \textit{pmtn} \mid \sum F_j$ represents the problem of minimizing total flow time on identical machines in the *online-time* model with preemption, which models the problem faced by a web server. And $P \mid \textit{online-list} \mid C_{\max}$ is the problem of minimizing the makespan on identical machines when jobs are presented one by one.

15.1.2 Competitive Analysis

Given that an online algorithm has only partial knowledge of the input instance, for most problems, no online algorithm can produce an optimal solution for all input instances. Probably the most obvious method for evaluating the worst-case performance of an algorithm is the worst-case relative error between the quality of the computed solution for an instance and the quality of the corresponding optimal solution. For example, this is the standard technique for evaluating polynomial-time approximation algorithms for NP-hard problems. In the context of online algorithms, this method is called *competitive analysis* [3,4]. Let $f(A, I)$ denote the objective value of the schedule produced by algorithm A on input instance I where A could be an online or offline algorithm and f be an objective value that we are trying to minimize such as makespan or total flow time. We say that an online algorithm A is c-competitive if $f(A, I) \leq c \cdot f(OPT, I) + b$ for any input instance I for a fixed constant b where *OPT* is the optimal offline scheduling algorithm for this problem. For most of the problems we consider, we can ignore the additive constant b. This follows from the fact that scheduling problems are typically scalable; by scaling all the jobs so that the objective is arbitrarily large, the possible benefit of the additive constant disappears. The competitive ratio of algorithm A, denoted c_A, is the infimum of c such that A is c-competitive.

The goal in any problem is to find an algorithm with a competitive ratio as small as possible. Ideally, this competitive ratio should be a constant independent of any parameter of the input instance such as the number of jobs faced, but we shall see this is not always possible.

15.1.3 Worst-Case Analysis and Other Alternatives

Competitive analysis allows us to prove lower bounds using the so-called adversary method. This means that a malicious omnipotent adversary uses the partial schedule generated by the online algorithm to decide what further jobs should be generated. If the algorithms considered are deterministic, this process can be simulated beforehand, and thus it provides a lower bound on the competitive ratio.

For many scheduling problems, worst case competitive analysis gives quite strong lower bounds. For example, for the problem of $1 \mid$ *online-time-nclv*, r_j, *pmtn* $\mid \sum F_j$, the competitive ratio of every deterministic algorithm is $\Omega(n^{1/3})$, see [5]. Consider for the moment the possibility that we have an $O(n^{1/3})$-competitive algorithm A for this problem. In absence of other information, this is positive evidence of the superiority of A to other possible algorithms with higher competitive ratios. However, given the magnitude of the $O(n^{1/3})$ guarantee on relative error, it is probable that an operating system designer would not take this as strong evidence to adopt A. Such situations have led to the development of many alternative techniques for analyzing online algorithms.

15.1.3.1 Randomized Algorithms

One standard alternative is to consider randomized algorithms that make random choices as they construct a schedule. We say that a randomized algorithm A is c-competitive if $E[f(A, I)] \leq c \cdot f(OPT, I)$ for all input instances I where $E[f(A, i)]$ is the expected cost of algorithm A on input instance I. This corresponds to the so-called oblivious adversary in online algorithms terminology [1,6]. An oblivious adversary has to commit to an input instance *a priori* without any knowledge of the random events internal to the algorithm. Intuitively, this takes away the "unfair" power of the adversary to completely predict the behavior of the algorithm. The assumption of an oblivious adversary is appropriate for scheduling problems where the scheduling decisions do not affect future input. Even in situations where the oblivious adversary assumption is not fully justified, such an analysis might still provide new insights. For some online scheduling problems, the use of randomized algorithms dramatically decreases the competitive ratio. For example, the competitive ratio for $1 \mid$ *online-time-nclv*, r_j, *pmtn* $\mid \sum F_j$ drops from $\Omega(n^{1/3})$ to $\Theta(\log n)$ when one allows randomized algorithms against an oblivious adversary [7].

The most common technique for proving a lower bound on the competitive ratio for any randomized algorithm against an oblivious adversary is Yao's technique. In Yao's technique you lower bound the expected competitive ratio of any deterministic algorithm on an input distribution of your choosing. Generally this expected lower bound for deterministic algorithms also then lower bounds the competitive ratio for any randomized algorithm against an oblivious adversary. However, there are some cases, particularly for maximization problems, where one needs to be a bit careful in applying this technique. For more information see [1].

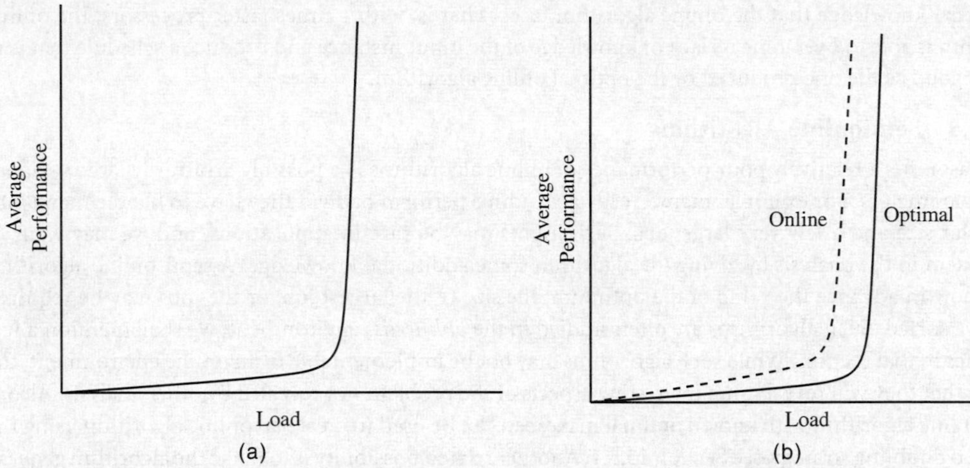

FIGURE 15.1 (a) Standard performance curve and (b) the worst possible performance curve of an s-speed c-competitive online algorithm.

15.1.3.2 Resource Augmentation

Another alternative that has proven especially useful in the context of online scheduling is resource augmentation. The recent popularity of resource augmentation analysis of scheduling problems emanates from a paper by Kalyanasundaram and Pruhs [8]. The term *resource augmentation*, and the associated terminology we use, was introduced by Phillips et al. [9]. In this model, we augment the online algorithm with extra resources in the form of faster processors or extra processors. For now, we focus on faster processors as most resource augmentation results utilize faster processors. Let A_s denote an algorithm that works with processors of speed s where $s \geq 1$. We say that an online algorithm A is an s-speed c-competitive algorithm if $f(A_s, I) \leq c \cdot f(OPT_1, I)$ for all input instances I.

Research with resource augmentation has focused on two primary goals. The first focuses on minimizing the speed subject to the constraint that the competitive ratio is $O(1)$. To understand this goal, we first need to understand how client-server systems typically behave. Figure 15.1(a) depicts "typical" average performance curves for client-server systems. That is, the average performance at loads below capacity is good, and the average performance above capacity is intolerable. So, in some sense, one can specify the performance of such a system by simply giving the value of the capacity of the system. Note that formally defining load (or capacity) is not easy, but load generally reflects the size of jobs and their rate of arrival over time. We next need to understand what it means to have s-speed processors. An alternative interpretation is that the jobs created have processing time p_j/s. That is, we can interpret the load as shrinking by a factor of s. This means that an s-speed c-competitive algorithm A performs at most c times worse than the optimal performance on inputs with s times higher load. Assuming that the performance does correspond to the curves in Figure 15.1(b) and is either good or intolerable, a modest c times either good or intolerable still gives you quite good or intolerable. So an s-speed c-competitive algorithm should perform reasonably well up to load $1/s$ of the capacity of the system as long as c is of modest size. Thus an ideal resource augmentation result would be to prove an algorithm is $(1+\epsilon)$-speed $O(1)$-competitive.

We call a $(1+\epsilon)$-speed $O(1)$-competitive algorithm *almost fully scalable* since it should perform well up to almost the peak capacity of the system. For many scheduling problems, there are almost fully scalable algorithms even though there are no $O(1)$-competitive algorithms. The intuition behind this is that if a system's load is near its capacity, then the scheduler has no time to recover from even small mistakes. Note many of the strong lower bounds for online scheduling problems utilize input instances where the load is essentially the capacity of the system.

The second goal is to find s-speed 1-competitive algorithms for these problems for as small a value of s as possible. The intuition behind these results is that s represents the tradeoff between extra resources and the partial knowledge that the online algorithm faces. That is, with s times faster processors, the online algorithm is able to overcome its lack of knowledge of the input instance and produce a schedule that is at least as good as the one produced by the optimal offline algorithm.

15.1.3.3 Semionline Algorithms

One reason for a relatively poor performance of online algorithms is a possibly arbitrarily large variance of job parameters. For example, many greedy algorithms perform badly if they have to handle many jobs of similar size and a few very large jobs. Such inputs may be rare in applications, and we may want to avoid them in the analysis by giving the algorithm some additional knowledge. A semi-online algorithm may know in advance the value of the optimum, the size of the largest job, or the jobs may be required to arrive sorted. Such algorithms are often studied in the *online-list* environment; we shall mention a few examples in that section. While such algorithms may not be implementable in an online environment, the hope is that they will reveal some interesting aspects of the problem not revealed by other analysis. Also, a semi-online algorithm with known optimal makespan can be used to create an online algorithm using the so-called doubling strategy, see Section 15.3.4. Another related possibility is to make the algorithm general, but study the dependence of the competitive ratio on the variance of some parameter (for example, the ratio of the largest and the smallest processing times).

15.1.3.4 Average-Case Analysis

Average-case analysis of algorithms is desirable if we have a reasonable approximation of what the input distribution should be. For some client server systems, this is known. For example, traffic for a web server is often modeled using a Poisson distribution for job arrivals and independent identical Zipf distributions for job lengths [10].

While there are many alternative analysis options, we note that worst-case analysis of online algorithms is of fundamental importance. In addition to the standard arguments in its favor (guarantee under any circumstances, etc.), in many online systems, positive feedback appears and thus bad situations may happen more often than one would expect. For example, in many embedded scheduling systems, a request not serviced sufficiently quickly may be reissued.

15.1.4 History

Many natural heuristics for scheduling are in fact online algorithms; thus some of the early scheduling literature prove bounds on the performance of online algorithms, in current terminology.

The first proof of competitiveness of an online algorithm for a scheduling problem, and perhaps for any problem, was given by Graham in 1966 [11]. It is quite remarkable that this happened at about the same time as Edmonds discovered his famous polynomial time algorithm for matching in graphs, long before notions like polynomial time and NP-hard problems were standard.

Graham [11] studied a simple deterministic greedy algorithm, now commonly called *list scheduling*, for $P \mid\mid C_{\max}$. Each job is scheduled to a machine with currently the smallest load (total size of jobs assigned to it). Graham proved that the job arrival order can change the resulting makespan by a factor of at most $2 - 1/m$ and that this bound is the best possible. Since this algorithm and a slightly refined analysis works in all three online environments we consider, even with release times and precedence constraints if jobs arrive over time, we get a $(2 - 1/m)$-competitive online algorithm for $Pm \mid \textit{online-list}, r_j \mid C_{\max}$ and $Pm \mid \textit{online-time-nclv}, \textit{prec}, r_j \mid C_{\max}$.

Graham [11] even considered the case when the number of machines changes, again giving tight bounds for this algorithm. Today we may view this as a result on resource augmentation. In the follow-up paper [12], Graham shows that the factor of $2 - 1/m$ decreases to $4/3 - 1/(3m)$ if we require the jobs to arrive in a sequence sorted according to nonincreasing processing times. Thus this is a semi-online algorithm for $Pm \mid \textit{online-list} \mid C_{\max}$.

Two other early papers that contain results about online scheduling algorithms are [13,14]. The first one gives an optimal (1-competitive) algorithm for $P \mid \textit{online-time}, \textit{pmtn} \mid C_{\max}$ and explicitly mentions that the algorithm is online. (Actually, the algorithm is not quite online, since it assumes that at any time, the next release time of a job is known. This additional assumption was later removed in [15].) The second paper is, to our best knowledge, the first one that states explicitly a lower bound on the performance ratio of any online algorithm for some scheduling problem, namely the bound of $\Omega(\sqrt{m})$ for $Qm \mid online\text{-}time\text{-}nclv \mid C_{\max}$; the paper even suggests that restarts may be helpful, a conjecture later proven to be correct [16].

Around 1990 new results were discovered concerning many variants of online scheduling, both old and new. The development over the last 15 years is the main topic of this chapter.

15.1.5 Organization of the Chapter

In Section 15.2, we focus on the problem of scheduling jobs that arrive over time, both clairvoyantly and nonclairvoyantly. We focus our efforts primarily on minimizing total flow time and related objective functions, but we also briefly discuss related results on other objective functions. In Section 15.3, we focus on the problem of scheduling jobs one by one. As nonclairvoyance does not really make sense in this model, we cover only clairvoyant algorithms. Most research in this area has focused on minimizing the makespan.

15.1.6 Related Areas

As we stated earlier, we do not provide a complete overview of all research in online scheduling. We admit that our choices are necessarily idiosyncratic.

One particularly interesting and important class of problems that we do not cover is that of real-time scheduling where jobs have deadlines. This topic is covered in Chapter 33 of this book.

We do not cover any of results for minimizing total completion time or total weighted completion time. Some discussion of offline algorithms for these problems can be found in Chapter 13 of this book. Note several of the techniques used for offline algorithms can be adapted to construct online algorithms. See also the surveys [17,53].

There seem to be only a limited amount of results concerning online shop scheduling, we refer to the survey [17] for some references.

We also do not cover any average-case analysis results in this chapter. More information can be found in part V of this book and in particular in Chapter 38.

15.2 Jobs that Arrive Over Time

In this section we discuss online scheduling problems in the job model *online-time*.

15.2.1 Standard Algorithms

Most results in the literature end up analyzing one of a handful of standard algorithms. We now introduce these algorithms, along with their standard abbreviations and some brief comments. The standard clairvoyant algorithms are:

SRPT The algorithm Shortest Remaining Processing Time always runs the job with the least remaining work. It is well known that SRPT is optimal for average flow time on one processor.

FIFO The algorithm First in First Out always runs the job with the earliest release time. It is well known that FIFO is optimal for maximum flow time on one processor. FIFO is also called *First Come First Served* in the literature.

SJF The algorithm Shortest Job First always runs the job with the least initial work. For resource augmentation analysis results, SJF is often easier to analyze than SRPT.

HDF The algorithm Highest Density First always runs the job with the highest density, which is the weight of the job divided by the initial work of the job.

The standard nonclairvoyant algorithms are:

RR The algorithm Round Robin devotes an equal amount of processing resources to all jobs. An understanding of RR is important because it is the underlying scheduling algorithm for many technologies. For example, the congestion control protocol within the ubiquitous TCP Internet protocol can be viewed as scheduling connections through a single bottleneck using RR. This algorithm is also called Processor Sharing, or Equi-Partition.

SETF The algorithm Shortest Elapsed Time First devotes all the resources to the job that has been processed the least. In the case of ties, this amounts to RR on the jobs that have been processed the least. While RR perhaps most intuitively captures the notion of fairness, SETF can be seen as fair in an affirmative action sense of fairness.

MLF The algorithm Multilevel Feedback can be viewed as mimicking SETF, while keeping the number of preemptions per job to be logarithmic. In most real systems, preemptions take a nontrivial amount of time. In MLF, there are a collection $Q_0, Q_1, \ldots$ of queues. There is a target processing time T_i associated with each queue. Typically, $T_i = 2^{i+1}$, but some results require more slowly growing targets, e.g., $T_i = (1+\epsilon)^{i+1}$. Each job J_j gets processed for $T_i - T_{i-1}$ units of time while in queue Q_i before being promoted to the next queue, Q_{i+1}. MLF maintains the invariant that it is always running the job in the front of the lowest nonempty queue.

It is of natural interest to ask about the scheduling algorithms used by current server technology. Unfortunately, because of the messiness of real software, it is often debatable what the best abstraction of the implemented algorithm is. Let us give a couple of examples. Currently the most commonly used web server software is Apache. The underlying scheduler for the Apache is usually described to be FIFO. But it would probably be more accurate to say that threads are allocated to pending requests on a FIFO basis, and then threads are scheduled using another algorithm. Often it is reported that the underlying process scheduling algorithm for the Unix and Windows NT operating systems is MLF. But in Unix and NT there are only a fixed number of queues, where the lowest priority queue may be scheduled using RR. Thus whether the underlying scheduler is best viewed as MLF or RR depends on the relationship between the job sizes and the largest job quantum.

15.2.2 Objective Functions

Perhaps the most intuitive measure of quality of service (QoS) received by an individual job J_i is the flow time $F_i = C_i - r_i$. The terminology is not standard, and flow time is also called response time, wait time, latency, etc. Another intuitive QoS measure for a job J_i is the *stretch* $S_i = (C_i - r_i)/p_i$. If a job has stretch s, then it appears to the client that it received dedicated service from a speed $1/s$ processor. One motivation for considering stretch is that a human user may have some feeling for the work of a job. For example, in the setting of a web server, the user may have some knowledge about the size of the requested document (for example the user may know that video documents are generally larger than text documents) and may be willing to tolerate a larger response time for larger documents.

In order to get a QoS measure for a collection of jobs, one needs to combine the QoS measures for the individual jobs in some way. Almost all the literature uses an l_p norm for some $1 \le p \le \infty$. For example, the l_p norm of the flow times is $(\sum_{i=1}^{n} F_i^p)^{1/p}$.

By far the most commonly used QoS measure in the computer systems literature is average flow time, that is, the l_1 norm of flow times. Schedules with good average QoS may still provide very bad QoS to some small number of jobs. In the computer systems literature these jobs are said to be starved. Some scheduling systems such as process scheduling in Unix have mechanisms to try to prevent starvation. To measure how well a schedule avoids starvation, one may use the l_∞ norm. However, an optimal schedule under the l_∞ norm may provide relatively lousy service to the majority of jobs. A common compromise in many settings is the l_p norm, for something like $p = 2$ or $p = 3$. The l_p, $1 < p < \infty$, objective function still considers the average in the sense that it takes into account all values, but because x^p is a strictly convex function of x, the l_p norm more severely penalizes outliers than the standard l_1 norm.

15.2.3 Notation and Analysis Techniques

If A is an algorithm and I is an input instance, then $A(I)$ will refer to the schedule that A outputs on I. If F is an objective function and S is a schedule, then $F(S)$ is the value of the objective function on that schedule. *OPT* generally refers to the optimal schedule.

A common difficulty in presenting scheduling results is keeping the notational complexity manageable. One technique to achieve this, which we will adopt, is to drop notation when it is understood. For example, A may refer to the schedule that algorithm A produces on some understood input, as well the value of some understood objective function on that schedule.

The majority of analyses in this area use local competitiveness. Let $A(t)$ be the rate at which the objective function is increasing at time t for the scheduling algorithm A. Generally, $A(t)$ has a nice physical interpretation. For example, if the objective function is total flow time, then $A(t)$ is the number of unfinished jobs at time t.

A local competitiveness argument has the following form. The argument starts by fixing an arbitrary time t. The argument then shows that $A(t) \leq c \cdot OPT(t)$, where $OPT(t)$ is the schedule that minimizes the rate of the increase of the objective function for the specific time t under consideration. So if the objective function was average flow time, $OPT(t)$ would be the schedule that minimizes the number of unfinished jobs at time t. Note that this gives the adversary some advantage since she is not constrained to have any consistency between the schedules $OPT(t)$ and $OPT(t')$ for two different times. But the compensation for the human prover is that the structure of the schedules $OPT(t)$ are often simpler to deal with than globally optimal schedules. It then immediately follows that A is c-competitive in the global sense since

$$A = \int_0^\infty A(t)\,dt \leq \int_0^\infty c \cdot OPT(t)\,dt \leq c \cdot OPT$$

The condition that $A(t) \leq c \cdot OPT(t)$, is called *local c-competitiveness*. Note that in any local c-competitiveness argument, it must be the case that $c \geq 1$. To see this consider an instance consisting of just one job.

Sometimes a local competitiveness argument is not possible because no matter what the online algorithm does, the adversary can get in the lead for at least a short period of time. In general, if a local competitiveness argument is not possible, there is usually a relatively straightforward instance that formally demonstrates this. Also in general, arguments that do not use local competitiveness are more complicated. We will particularly emphasize the analyses in the literature that do not use local competitiveness.

Often our competitive ratios will depend on some parameter. Following standard convention we generally use n to denote the number of jobs, and m to denote the number of processors. We use $P_{\max}$ to denote the maximum processing time of any job, $P_{\min}$ to denote the minimum processing time of any job, and use P to denote $P_{\max}/P_{\min}$.

In the context of resource augmentation results, or other results where there are variable speed processors, the processing time of a job is not fixed. So it no longer makes sense to call p_i processing time. Instead, p_i is usually referred to as the *work* of a job. The time that a job is processed is then its work divided by the average speed at which it is processed.

15.2.4 Clairvoyant Scheduling to Minimize Average/Maximum Flow/Stretch

In this subsection, we focus on the *online-time* model where the scheduling algorithms know the processing requirements of jobs as soon as they are released. We first cover results for the total flow time and total stretch objective functions that measure average response time for clients. We then discuss results for the max flow and max stretch objective functions that measure server fairness to outlier jobs. In both subsections, we first summarize the results and then highlight some of the key proofs behind these results. We assume that the algorithms are preemptive except when considering maximum flow.

15.2.4.1 Total Flow Time and Total Stretch

For both single machine scheduling and parallel machine scheduling, SRPT has, within constant factors, the best possible competitive ratio of any online algorithm for minimizing both total flow time and total stretch. On a single machine, SRPT is an optimal algorithm for minimizing total flow time, i.e., $1 \mid online\text{-}time, pmtn, r_j \mid \sum F_j$. On parallel machines, it is $\Theta(\min(\log P, \log n/m))$-competitive for minimizing total flow time, i.e., $P \mid online\text{-}time, pmtn, r_j \mid \sum F_j$, and this is known to be optimal within constant factors [18]. The lower bound applies to randomized as well as deterministic online algorithms. A simpler analysis of SRPT's performance for minimizing flow time is available in [19]. Applying resource augmentation to $P \mid online\text{-}time, pmtn, r_j \mid \sum F_j$, Phillips et al. showed that SRPT is a $(2 - 1/m)$-speed 1-competitive algorithm for minimizing total flow time [9]. McCullough and Torng improved this result by showing that SRPT is an s-speed $1/s$-competitive for minimizing total flow time for $s \geq 2 - 1/m$ [20]. That is, SRPT "optimally" uses its faster processors. Meanwhile, SRPT is 2-competitive for minimizing total stretch on a single machine, $1 \mid online\text{-}time, pmtn, r_j \mid \sum S_j$, it is 14-competitive for minimizing total stretch on parallel machines, $P \mid online\text{-}time, pmtn, r_j \mid \sum S_j$, and no 1-competitive online algorithm exists for minimizing total stretch on a single machine or parallel machines [21].

While SRPT has essentially the best possible competitive ratio, it utilizes both preemptions and job migrations to achieve its performance. Awerbuch et al. developed an algorithm without job migration (each job is processed on only one machine) that is $(O(\min(\log P, \log n))$-competitive with respect to total flow time [22] and 37-competitive with respect to total stretch [23]. Chekuri, Khanna, and Zhu developed a related algorithm without migration that is $(O(\min(\log P, \log n/m))$-competitive with respect to total flow time and 17.32-competitive with respect to total stretch [24]. If migration is allowed, a modified algorithm is 9.82-competitive for minimizing total stretch [24]. While these algorithms never migrate jobs, they do hold some jobs in a central pool after their release date. Avrahami and Azar developed an algorithm without migration with immediate dispatch (each job is assigned to a machine upon its release) that is $O(\min(\log P, \log n))$-competitive for minimizing total flow time [25]. Chekuri, Khanna, and Kumar have shown that Avrahami and Azar's immediate dispatch algorithm is almost fully scalable for minimizing both total flow time and total stretch [26]. (Table 15.1)

Difference Between One Machine and Parallel Machines

We now examine why online algorithms can do well for minimizing total flow time on a single machine but cannot do well for minimizing total flow time on parallel machines. The key observation is that for any time t on a single machine, SRPT has completed as many jobs as any other algorithm.

The following notation will be used throughout this subsection. Let $A(t)$ be both the set of unfinished jobs for algorithm A applied to input instance I at time t as well as the number of jobs in this set. The specific meaning should be clear from context. Furthermore, let $A_j(t)$ be the jth smallest job in $A(t)$ as

TABLE 15.1 Summary of Results for Minimizing Total Flow Time and Total Stretch for Single Processor and Parallel Machines

	Total Flow Time		Total Stretch	
Algorithm	Uniprocessor	Parallel Machines	Uniprocessor	Parallel Machines
Best upper bound	1	$\Theta(\min(\log P, \log n/m))$	2	9.82
Best lower bound	1	$\Theta(\min(\log P, \log n/m))$	1.036	1.093
SRPT	1	$\Theta(\min(\log P, \log n/m))$	2	14
Speed-s SRPT $s \geq 2 - 1/m$	$1/s$	$1/s$	—	—
No migration		$\Theta(\min(\log P, \log n/m))$		17.32
Immediate dispatch		$\Theta(\min(\log P, \log n))$		—
Speed-$(1+\epsilon)$ immediate dispatch		$O(1 + 1/\epsilon)$		$O(1 + 1/\epsilon)$

well as that job's remaining processing time, and let $A^j(t)$ be the jth largest job in $A(t)$ as well as that job's remaining processing time. Again, the specific meaning should be clear from context.

Lemma 15.1

Consider any input instance I, and schedule S, and any time t. When we consider a single machine environment, $SRPT(t) \leq S(t)$.

To prove this, we typically prove a stronger result first.

Lemma 15.2

Consider any input instance I, any schedule S that never idles the machine unnecessarily, any time t, and any integer $k \geq 0$. When we consider a single machine environment, $\sum_{j=1}^{k} SRPT^j(t) \geq \sum_{j=1}^{k} S^j(t)$.

Lemma 15.2 implies Lemma 15.1. Consider any schedule S for some input instance I at an arbitrary time t. We first observe that $\sum_{j=1}^{SRPT(t)} SRPT^j(t) = \sum_{j=1}^{S(t)} S^j(t)$ since neither algorithm ever unnecessarily idles the processor. Suppose $S(t) < SRPT(t)$ and let $y = S(t)$. Then it must be the case that $\sum_{j=1}^{y} S^j(t) > SRPT^j(t)$. Since this contradicts Lemma 15.2, it follows that $S(t) \geq SRPT(t)$.

Note that the number of unfinished jobs is the weight of the schedule at any time, and thus $SRPT$ is locally 1-competitive on a single machine. No comparable guarantee can be made in the parallel machine environment, even for randomized algorithms. In particular, no online algorithm can be locally c-competitive for any constant c as demonstrated by the following proof from Leonardi and Raz [18].

Theorem 15.1

Any randomized online algorithm for $P \mid online\text{-}time, pmtn, r_j \mid \sum F_j$ is $\Omega(\log P)$-competitive. Likewise, any such algorithm is $\Omega(\log(n/m))$-competitive.

Proof

We focus on deterministic algorithms but the argument is essentially unchanged for randomized algorithms. The lower bound is obtained by considering a family of input instances composed of repeating phases P_i for $i \geq 0$ followed eventually by a stream of short jobs. The first phase P_0 is organized as follows. At time 0, release a collection of $m/2$ long jobs of size P. At times 0 through $P/2 - 1$, release a collection of m short jobs of size 1. One algorithm A_1 finishes all the short jobs by time $P/2$ by devoting all m machines to the short jobs from time 0 to time $P/2$. A second algorithm A_2 finishes all jobs by time P by devoting $m/2$ of the machines to the long jobs and the other $m/2$ machines to the short jobs.

If an online algorithm A is not locally $(\log P)$-competitive with A_1 at time $P/2$, we introduce m jobs of size 1 for P^2 time units starting at time $P/2$. It can be easily seen that the competitive ratio of any such algorithm A will be $\Omega(\log P)$. Since $n = O(mP^2)$, it follows that the algorithm will also be $\Omega(\log(n/m))$.

Thus, we can focus our attention on algorithms that are locally $(\log P)$-competitive with algorithm A_1 at time $P/2$. Since $A_1(P/2) = m/2$, this means that A must finish all but roughly $m \log P$ short jobs by time $P/2$. This means that at most $m \log P$ time units can be devoted to the long jobs before time $P/2$. If P is sufficiently larger than m, this means that the $m/2$ long jobs will still have remaining processing times of essentially $P/2$ at time P.

Phase P_i for $i \geq 1$ has an initial release time of $R_i = 2P - P/2^{i-1}$ and a "halfway" time of $H_i = R_i + P/2^{i-2}$. The $m/2$ long jobs of length $P/2^i$ are released at time R_i while the short jobs of length 1 are released at times R_i through $H_i - 1$. As before, there is an algorithm A_1 that finishes all the short jobs of phase P_i by time H_i while another algorithm A_2 finishes all the jobs of phase P_i by R_{i+1}. Similar to the analysis of phase P_0, any online algorithm that is not locally $(\log P)$-competitive with algorithm A_1 at time H_i has a competitive ratio of $\Omega(\log P)$ and $\Omega(\log(n/m))$. Thus, we restrict our attention to online algorithms that are locally $(\log P)$-competitive with the algorithm A_1 that finishes all short jobs of phase P_i by time H_i. After phase $P_{\log P-1}$, such an online algorithm will not have finished any of its long jobs from any phase. Thus, the online algorithm will have $m/2 \log P$ jobs left while *OPT* will have no jobs left. We now introduce a stream of m jobs of length 1 for P^2 time units, and these algorithms also have competitive ratios of $\Omega(\log P)$ and $\Omega(\log(n/m))$, and the result follows. □

Structure of SRPT's Extra Jobs

From the lower bound argument above, we see that unlike in the single machine case, SRPT can idle some of the parallel machines unnecessarily leading to situations where SRPT has many more unfinished jobs than some other schedule for the same input instance. However, Leonardi and Raz were able to show that while SRPT can have arbitrarily more extra jobs, there is a structure to these extra jobs. This leads to an upper bound on SRPT's flow time. We present a brief analysis of SRPT utilizing crucial ideas from [18] and [21].

The concept of volume [18] captures the total remaining work that needs to be completed at any time. For any schedule S, any input instance I, any time t, let the volume $\text{vol}(S,t) = \sum_j S_j(t)$ be the sum of remaining processing times of jobs in $S(t)$. Let the volume difference $V'(S,t) = \text{vol}(SRPT,t) - \text{vol}(S,t)$ be the difference in volume between SRPT's schedule and any other schedule S. We will be interested in focusing on some restricted subsets of jobs when looking at volumes and volume differences. Let $\text{vol}(S,t,x)$ be the sum of remaining processing times in $S(t)$ when restricted to jobs of size at most x, and let $V'(S,t,x) = \text{vol}(SRPT,t,x) - \text{vol}(S,t,x)$.

The following proof from [21] provides a bound on $V'(S,t,x)$.

Lemma 15.3

For any time t, any input instance I, any real x, and any schedule S, $V'(S,t,x) \leq mx$.

Proof

Suppose there are at most m jobs with remaining processing times at most x in $SRPT(t)$. Then clearly $\text{vol}(SRPT,t,x) \leq mx$ and the result follows. Thus assume there are more than m jobs with remaining processing times at most x in $SRPT(t)$.

Let t' be the last moment before time t where fewer than m jobs of remaining processing time at most x exist in SRPT's schedule. To simplify the proof description, we also use t' to denote the moment immediately after t' (i.e., the moment where there are now more than m jobs with remaining processing time at most x in $SRPT(t')$). If no such time exists, $t' = 0$. Clearly, during the interval (t',t), SRPT will devote all m machines to jobs with remaining processing times at most x, so we can bound $V'(S,t,x)$ by $V'(S,t',x)$.

We now analyze $V'(S,t',x)$. There were $y \leq m$ jobs of size at most x at time t'. These contribute at most yx work to $\text{vol}(SRPT,t',x)$. New jobs with processing times at most x might arrive, but these jobs will contribute to both $\text{vol}(SRPT,t',x)$ and $\text{vol}(S,t',x)$, so they do not affect $V'(S,t',x)$. Finally, the $m-y$ machines not working on jobs with remaining processing times at most x may create $m-y$ jobs with remaining processing times of x. No other jobs of size at most x can be created at time t' and the result follows. □

Applying a result from [21], we derive the following characterization of the extra jobs in SRPT's schedule. We first consider the case where S has finished all jobs at time t.

Lemma 15.4

For any input instance I, for any schedule S, any time t where $S(t) = 0$, and any $i \leq SRPT(t) - 2m$, $SRPT_{2m+i}(t) \geq P_{\min}(m/(m-1))^i$ for $i \geq 0$, and the sum of these $2m+i$ smallest jobs in $SRPT(t)$ is at least $mP_{\min}(m/(m-1))^i$.

Proof

We prove this result by induction on i. We first show the base case for $i = 0$. SRPT can have at most m jobs with remaining processing time less than $P_{\min}$ as such a job will never be preempted by a newly arriving job. The next m smallest jobs in SRPT's schedule must have size at least $P_{\min}$ and the base case follows. We now assume the result holds for some n and show that it applies for $n+1$.

By the induction hypothesis, we know that the sum of the $2m+n$ smallest jobs in $SRPT(t) \geq mP_{\min}(m/(m-1))^n$. Let y denote the size of the $(2m+n+1)$st smallest job in $SRPT(t)$. From Lemma 15.3, we have that $V'(S,t,y) \leq my$. Since $\text{vol}(S,t)=0$, this means $\text{vol}(SRPT,t,y) \leq my$. However, we know that $\text{vol}(SRPT,t,y) \geq mP_{\min}(m/(m-1))^n + y$. Thus, we derive that $my \geq mP_{\min}(m/(m-1))^n + y$ which means that $y = SRPT_{2m+n+1}(t) \geq P_{\min}(m/(m-1))^{n+1}$ completing the first part of the induction. Adding this lower bound on $SRPT_{2m+n+1}(t)$ with the lower bound on the sum of the $2m+n$ smallest jobs in $SRPT(t)$ completes the second part of the induction and the result follows. □

We can extend this result and eliminate the restriction that $S(t)=0$ as follows.

Lemma 15.5

For any input instance I, for any schedule S, any time t, and any $i \leq SRPT(t) - 2m - S(t)$, $SRPT_{2m+i+S(t)}(t) \geq P_{\min}(m/(m-1))^i$.

Proof

The key observation is that an unfinished job $j \in S(t)$ of size z cannot increase the number of jobs in $SRPT(t)$ by more than one, and this job must have size at least z. Consider for example the job $S_1(t)$ of size z. Despite the existence of job $S_1(t)$, $\text{vol}(S,t,y)=0$ for all $y<z$, and thus $\text{vol}(SRPT,t,y) \leq my$. The fact that $\text{vol}(S,t,z)=z$ (assuming no other jobs of size z are in $S(t)$) implies $\text{vol}(SRPT,t,z) \leq mz+z$ which allows the addition of one job of size at least z to $SRPT(t)$ in addition to the jobs generated by the argument of Lemma 15.4. □

With this result, we can now derive the following bound on $SRPT(t)$.

Theorem 15.2

For any input instance I, any schedule S, and any time t, $SRPT(t) \leq S(t) + m(2+\ln P)$.

Proof

Applying Lemma 15.5, we have that $SRPT(t) \leq S(t) + 2m + \log_{m/(m-1)} P$. Now $(m/(m-1))^m = (1+1/(m-1))^m \geq e$ for $m \geq 2$. Thus, $\log_{m/(m-1)} P \leq m \ln P$ and the result follows. □

To derive the upper bound on SRPT's flow time, we first observe that the contribution of jobs in $SRPT(t)$ that correspond to jobs in $S(t)$ to SRPT's flow time is at most the total flow time incurred by $S(I)$. We now divide time into two categories: intervals where all m machines are busy and intervals where some machines are idle. SRPT can only have extra jobs during busy times. If we focus only on the active jobs during these busy times, their contribution to SRPT's flow time is at most the sum of processing times of jobs, and this is clearly a lower bound on the optimal flow time. Note there are m active jobs at any time during these busy intervals. Thus, the at most $O(m \log P)$ extra jobs in SRPT's schedule during these busy intervals is at most $O(\log P)$ more than the m active jobs, and it then follows that SRPT is $O(\log P)$-competitive for the problem of minimizing total flow time, $P \mid$ *online-time, pmtn*, $r_j \mid \sum F_j$. To prove that SRPT is $O(\log(n/m))$-competitive requires more sophisticated arguments which we omit.

Eliminating Migration and Immediate Dispatch

The key idea in algorithms that eliminate migration is the idea of classifying jobs by size [22,24,25]. In [22], jobs are classified as follows: a job j whose remaining processing time is in $[2^k, 2^{k+1})$ is in class k for $-\infty < k < \infty$. Note that jobs change classes as they execute. This class definition reflects the structure of extra jobs in SRPT's schedule first observed in [18].

The algorithm A uses the following data structures to organize the jobs. There is a central pool containing jobs not yet assigned to any machine. With each machine, we associate a stack to hold jobs currently assigned to that machine.

The algorithm A works as follows. Each machine processes the job at the top of its stack. When a new job arrives, the algorithm looks for a machine that is idle or currently processing a job of a higher class than the new job. If it finds one, the new job is pushed into that machine's stack and its processing begins. Otherwise, the job enters the central pool. Finally, if a job is completed on some machine, the algorithm compares the job at the top of the stack of that machine with the minimum class of any job in the pool. If the minimum in the pool is smaller than the class of the job on top of the stack, then any job in the pool of that minimum class is then pushed onto that stack. Using ideas similar to those used in [18], they derive the following result.

Lemma 15.6

For any input instance I, for any schedule S, and for any time t when all m machines are busy, $A(t) \leq 2S(t) + mO(\log P)$.

Again, this leads to the result that the algorithm is $O(\log P)$-competitive. With more work, this algorithm can be shown to be $O(\log n)$-competitive, slightly worse than $O(\log(n/m))$-competitive.

New algorithms proposed in [24] achieve the same bounds as SRPT within constant factors for minimizing total flow time by modifying the class definition from [22]. A job is now assigned to class k if its *original* processing time is in the range $[2^k, 2^{k+1})$. Thus, the class of a job does not change as it executes. This simplifies the analysis of their algorithm, particularly when considering total stretch. Furthermore, their simpler analysis allows the optimization of the constant used to define classes (the definitions above use constant 2).

A new algorithm that dispenses with the central pool of unassigned jobs was proposed in [25]. That is, each job is immediately assigned to a machine, and there is no migration of jobs. They show that this algorithm is $O(\min(\log P, \log n))$-competitive for minimizing total flow time on parallel machines. This algorithm uses the class definition of [24]. When a job j of class k arrives, it is assigned to the machine that has been assigned the minimum total processing time of jobs of class k so far. That is, Graham's List Scheduling rule is used to assign jobs to machines within each class of jobs. Note that this assignment rule ignores information such as what is the current load on each machine or which jobs in the specified class have actually been processed or completed at the current time. Each machine then implements the SRPT algorithm which is optimal for scheduling jobs on a single machine to minimize flow time. However, to simplify the analysis, they analyze a modified version of this algorithm that uses SJF on each machine instead.

Here are a few of the key observations in the analysis of this algorithm [25]. The first fact is that the difference in total volume of jobs of any class k assigned to any two machines by any time t is at most 2^{k+1}, the size of the largest job in class k, since they use greedy List Scheduling. This implies that difference in the total volume of work processed by any time t of jobs in class at most k on any two machines is at most 2^{k+2}. Combining these two observations implies that the difference in unfinished work from jobs of class at most k at any time t on any two machines is at most 2^{k+3}. With these facts, [25] are able to apply many of the arguments used in the analysis of other algorithms without migration to prove the flow time bound for their algorithm.

Resource Augmentation Results

With sufficiently faster processors, [9] showed that SRPT will never have extra jobs. Specifically, they extended Graham's analysis of List Scheduling [11] to show that any s-speed algorithm where $s \geq 2 - 1/m$ that never idles a machine when jobs are available always completes as much work by any time t as any 1-speed algorithm on the same input instance. Adding to this the greedy nature of SRPT, their analysis shows that speed-$(2 - 1/m)$ SRPT is locally 1-competitive.

This result has recently been improved to show that SRPT is an s-speed $1/s$-competitive algorithm for $s \geq 2 - 1/m$ [20]. The analysis in [20] uses some new ideas to prove a competitiveness bound smaller than 1. First is the idea that s-speed processors can be approximated by multiplying release dates by a factor of s. In [9,20], the resulting input instance is called a stretched input instance. The key observation is that an algorithm on a stretched input instance will incur a flow time exactly s times larger than the same algorithm using s-speed processors on the original input instance. Thus, they need only show that SRPT on an s-stretched input instance does as well as the optimal algorithm does on the original input instance to prove the $1/s$ bound for s-speed SRPT. This introduces a complication as they need to compare SRPT on a stretched input instance to the optimal algorithm on the original input instance, and thus jobs are released at different times for the two algorithms. They overcome this difficulty by introducing a proxy algorithm for the original input instance. This proxy algorithm will in some cases produce schedules that are not legal. This is acceptable since the proxy algorithm is used for analysis purposes only. However, they do need to introduce a charging scheme to handle cases when the schedule is not legal. They then show that the proxy algorithm is locally 1-competitive and that the proxy algorithm incurs a flow time on the original input instance that is at least as large as the flow time incurred by SRPT on the s-stretched input instance. This argument does not use local competitiveness but rather a structural relationship between the proxy schedule and the SRPT schedule on the stretched input instance.

Chekuri, Khanna, and Kumar have shown that the immediate dispatch algorithm of Avrahami and Azar is almost fully scalable for total flow time and total stretch [26]. This result also applies when the algorithm is given extra machines instead of faster machines, and the result extends to show that the algorithm is almost fully scalable for l_p norms of flow and stretch for all $p \geq 1$. Their analysis builds upon Bansal and Pruhs' analysis of SJF and SRPT for minimizing l_p norms of flow and stretch on a single machine [27]. These results are discussed in more detail in Section 15.2.5.

A few open questions remain regarding resource augmentation and minimizing total flow time. While we now know that there is an almost fully scalable algorithm for minimizing total flow time on parallel machines, no such analysis is known for SRPT.

Open Problem 15.1

For the problem $P \mid online\text{-}time, pmtm, r_j \mid \sum F_j$, is SRPT almost fully scalable?

Furthermore, from [9], we know that SRPT is at least as good as optimal when given speed-$(2 - 1/m)$ machines and that no speed-$(22/21 - \epsilon)$ 1-competitive algorithm exists for minimizing total flow time on parallel machines for $m \geq 2$ and $\epsilon > 0$.

Open Problem 15.2

For the problem $P \mid online\text{-}time, pmtm, r_j \mid \sum F_j$, what is the minimum speed s such that there exists an s-speed 1-competitive algorithm, and what is the corresponding algorithm?

Both of these questions can be extended to all l_p norms of flow and stretch.

Difference Between Total Stretch and Total Flow Time

At first glance, it may seem surprising that there exist algorithms with constant approximation factors for total stretch on parallel machines but not total flow time. This discrepancy is explained by considering the structure of extra jobs for SRPT and the fact that the total stretch objective function weights jobs by the inverse of their original processing times. For example, while $SRPT(t) - S(t)$ can be unbounded, there can only be a relatively few extra jobs with small remaining processing times in SRPT's schedule at any time. In particular, the large jobs add a negligible amount to the total weight of jobs at any given moment. This property is exploited more explicitly in the algorithms that use job classifications.

For example, consider the algorithm of [24] and consider the jobs on the stack of any machine of their algorithm. Suppose the job that is currently executing is from class k. The original processing times of the remaining jobs on the stack for that machine are at least $2^{k+1}, 2^{k+2}, 2^{k+3}, \ldots$, and their weights sum to at most $1/2^k$. This is at most twice the weight of the job currently executing and thus the increase in total stretch can be charged to this currently executing job. Handling jobs in the central pool is more complicated and we ignore these details.

On the other hand, we observe that no online algorithm can be optimal for minimizing total stretch on a single machine while there does exist an optimal online algorithm, namely SRPT, for minimizing total flow time on a single machine. The lower bound example below shows that we can create a situation where it is optimal to prioritize one job j_1 over a second job j_2 in some cases while in other cases, it is optimal to prioritize job j_2 over job j_1.

Lemma 15.7

No online algorithm can be better than 1.036-competitive for the problem of minimizing total stretch on a single machine [21].

Proof

Consider an adversary strategy using at most three jobs of sizes q, m, and s where $q > m > s$. Under the first scenario, the job of size q is released at time 0, and the job of size m is released at time $q - k$ for some $k \leq m$, and the third job is never released. Under the second scenario, the third job of size s is released at time q. The adversary makes its decision on which scenario to implement based on the online algorithm's decisions up to time q.

The optimal strategy for the first scenario is to run the second job as soon as it arrives. The optimal strategy for the second scenario is to finish the first job first, run the third job as soon as it arrives, and then finish the second job. Clearly no online algorithm can do both. Using a proper choice of q, m, s, and k, the bound of 1.036 follows. □

More Detailed Analysis of SRPT for Total Stretch [21]

The analysis of SRPT for total stretch on a single machine utilizes a matching property between the jobs waiting in SRPT's queue at any time t and the jobs in any other schedule S's queue at time t.

Lemma 15.8

For any input instance I, for any schedule S, and any time t, and any $k > 1$, $SRPT_k(t) \geq S_{k-1}(t)$.

Proof

Suppose this is not true at some time t. Let $k > 1$ be the smallest integer such that the relationship does not hold. Let $b = SRPT_k(t)$. It follows that the number of jobs in $SRPT(t)$ of size at most b is at least k, while the number of jobs in $S(t)$ of size at most b is at most $k - 2$. Furthermore, given the definition of k, we have that $SRPT_j(t) \geq S_{j-1}(t)$ for $1 < j < k$. Thus, $\text{vol}(SRPT, t, b) - \text{vol}(S, t, b) \geq b + SRPT_1(t)$ which means that $V'(S, t, b) > b$. This is a contradiction since Lemma 15.3 implies that $V'(S, t, b) \leq b$, and the result follows. □

With this matching property, Muthukrishnan et al. [21] bound the amount that SRPT's waiting jobs contribute to SRPT's total stretch by the total stretch incurred by any other algorithm. They then observe that the total stretch incurred by SRPT's active job over time is exactly n which is a lower bound on the optimal total stretch, and the factor of 2 result follows.

In the parallel machine case, there is the extra complication that SRPT has extra jobs. However, given the structural property observed earlier, Muthukrishnan et al. [21] are able to derive a similar mapping of some of SRPT's waiting jobs to at least as small unfinished jobs for schedule S. The unmapped jobs for SRPT then obey the structure observed earlier and their total contribution to total stretch can be bounded by a constant times the optimal total stretch.

15.2.4.2 Maximum Flow Time and Maximum Stretch

While SRPT and the related algorithms perform well for client jobs on average, these algorithms do have the undesirable property of starving some jobs in order to service most jobs well. For example, consider an input instance on a single machine where a job with processing time 2 is released at time 0 and jobs with processing time 1 are released at unit intervals starting at time 0 and ending at time x. SRPT will always process the jobs with processing time 1 delaying the job with processing time 2 until the end of the long stream of jobs, so its flow time will be $x + 3$. An alternative algorithm would schedule the job with processing time 2 first, and then schedule the jobs with processing time 1 in order of their arrival. The flow time of the job with processing time 2 will be 2 while the flow time of all jobs with processing time 1 will be 3. As we can make x as large as we desire, this shows that for the $F_{\max}$ or $S_{\max}$ objective functions on a single machine, i.e., $1 \mid$ *online-time, pmtn*, $r_j \mid F_{\max}$ and $1 \mid$ *online-time, pmtn*, $r_j \mid S_{\max}$, SRPT is $\Omega(n)$-competitive.

Different algorithms are needed to provide good guarantees for maximum flow and maximum stretch. The best results known for these objective functions come from Bender, Chakrabarti, and Muthukrishnan [28] and Bender, Muthukrishnan, and Rajaraman [29]. For maximum flow, [28] show that FIFO is $(3 - 2/m)$-competitive for $P \mid$ *online-time, pmtn*, $r_j \mid F_{\max}$ and provide a lower bound of 4/3 for any nonpreemptive algorithm for $m \geq 2$. (The paper claims a lower bound of 3/2, but the proof seems to work only for a 4/3 lower bound.) The maximum stretch objective function turns out to be harder to minimize than maximum flow. This stands as an interesting contrast to the case of total stretch and total flow time where, in the parallel machine environment, there exist constant competitive algorithms for total stretch but no constant competitive algorithms for total flow time, even with preemption. For maximum stretch on a single machine, $1 \mid$ *online-time, pmtn*, $r_j \mid S_{\max}$, [28] provides an algorithm that is $O(P^{1/2})$-competitive based on the earliest deadline first (EDF) real-time scheduling algorithm, and they provide a lower bound of $\Omega(P^{1/3})$ on the competitive ratio of any online algorithm for maximum stretch. A simpler and more efficient algorithm to achieve the $O(P^{1/2})$ bound is given in [29]. No results are known for maximum stretch on parallel machines, $P \mid$ *online-time, pmtn*, $r_j \mid S_{\max}$. These results are summarized in Table 15.2.

Maximum Flow

The fact that FIFO, a nonpreemptive algorithm, is constant competitive for minimizing maximum flow shows how this problem is quite different than that of minimizing total flow time. We provide below a proof that FIFO is optimal for the single machine environment.

Theorem 15.3

FIFO is an optimal algorithm for minimizing maximum flow time on a single machine, $1 \mid$ online-time, pmtn, $r_j \mid F_{\max}$ [28].

Proof

Without loss of generality, we consider only input instances I such that there is no idle time in *FIFO*(I). Consider any such input instance I and a job j such that F_j is maximized in *FIFO*(I). This means that from time r_j to time C_j, FIFO is working only on jobs that had release times at most r_j. Since FIFO is not

TABLE 15.2 Summary of Results for Minimizing Max Flow Time and Max Stretch for Single Processor and Parallel Machines

	Max Flow Time		Max Stretch	
Algorithm	Uniprocessor	Parallel Machines	Uniprocessor	Parallel Machines
Best upper bound	1	$3 - 2/m$	$O(P^{1/2})$	—
Nonpreemptive lower bound	1	4/3	$\Omega(P)$	$\Omega(P)$
Preemptive lower bound	1	—	$\Omega(P^{1/3})$	—

idle prior to r_j, it is not possible to finish all the jobs released prior to r_j plus job j any earlier than C_j. Thus, in any schedule for input instance I, some job released at time no later than r_j must complete no earlier than C_j, and the result follows. □

When we consider parallel machines, it is no longer true that FIFO is not idle prior to the release time of the job with maximum flow time in *FIFO*(I). Thus, FIFO is not optimal for the parallel machine environment, but it is still constant competitive.

The only lower bound known for this problem is 4/3 for $m = 2$ for nonpreemptive algorithms (see [28]). At time 0, two jobs with processing time 3 are released. If the algorithm starts both jobs by time 1, a job with processing time 6 is released at time 1; otherwise, no more jobs are released. In the first case, the optimal $F_{\max}$ is 6 while the online algorithm's $F_{\max}$ is at least 8. In the second case, the optimal $F_{\max}$ is 3 while the online algorithm's $F_{\max}$ is at least 4.

Maximum Stretch

Theorem 15.4

On a single machine, no preemptive online algorithm is $P^{1/3}/2$-competitive for minimizing maximum stretch, $1 \mid \textit{online-time}, \textit{pmtn}, r_j \mid S_{\max}$ [28].

Proof

Consider the following input instance. Two jobs with length P are released at time 0. Meanwhile, jobs of size $k = P^{2/3} - 1$ are released at times $2P - k, 2P, \ldots, P^{4/3} - k$. To simplify the proof, we assume that $y = P^{4/3} - 2P$ is an integral multiple of k.

An optimal schedule for minimizing maximum stretch for this input is FIFO which results in a maximum stretch of 2. Thus, for an online algorithm to be $P^{1/3}/2$-competitive, the first two jobs must be completed by time $P^{4/3}$ giving them a stretch of $P^{1/3}$. This means one of the length k jobs cannot complete before $P^{4/3} + k$.

Now suppose that jobs of length 1 arrive every unit of time starting at time $P^{4/3}$ and ending at time $2P^{4/3} - k - 1$. Either one of the length 1 jobs finishes at essentially $2P^{4/3}$ or one of the length k jobs finishes then. In the first case, the maximum stretch will then be at least $k + 1 = P^{2/3}/2$ while in the second case, the maximum stretch will be essentially $P^{4/3}/k > P^{2/3}/2$. Meanwhile, the optimal algorithm schedules the jobs of size 1 and size k as they arrive and finishes one of the jobs of size P at time $2P^{4/3}$. The result then follows. □

One algorithm for minimizing maximum stretch uses ideas from real-time scheduling. Suppose the online algorithm somehow knew in advance what the maximum stretch S^* for the input instance would be. It could then treat each job j as if it had a deadline of $r_j + S^* p_j$ and use algorithms from real-time scheduling to attempt to meet all deadlines. One such online algorithm is earliest deadline first (EDF) that prioritizes available jobs by their deadlines breaking ties arbitrarily. EDF is known to legally complete all jobs by their deadlines on a single machine if it is possible to do so. By the definition of maximum stretch, it clearly is possible to schedule all jobs such that they end by $r_j + S^* p_j$. Thus, EDF armed with knowledge of the maximum stretch of the input instance is an optimal online algorithm.

Unfortunately, the online algorithm cannot possibly know the maximum stretch ahead of time. Instead, the best that any online algorithm can do is compute what the maximum stretch of an input instance would be if no more jobs arrive. This algorithm, stretch-so-far [28] has a further refinement of overestimating the maximum stretch computed so far by setting a job's deadline to be $r_j + \alpha S^* p_j$ where $\alpha \geq 1$. Choosing an appropriate value of α is critical to minimizing maximum stretch. Also note that the deadlines will change as S^* is refined. Stretch-so-far with $\alpha = 1$ is P-competitive for this problem [28]. If α is instead chosen to be $O(P^{1/2})$, the algorithm is then $O(P^{1/2})$-competitive. Note that the $P^{1/2}$ used here is based on the jobs seen so far, so this is an online algorithm. Constant competitive algorithms exist if there are only two distinct job lengths [28].

While stretch-so-far achieves a competitive ratio of $O(P^{1/2})$, it has the disadvantage of requiring $\Omega(n^2)$ processing per job arrival. In particular, the optimal maximum stretch must be recalculated on each job arrival requiring the algorithm to remember all jobs seen so far at any point in time during its execution. A simpler greedy strategy that achieves the same competitive ratio is proposed in [29]. Suppose that $P_{\min} = 1$ and $P = P_{\max}$ is known to the online algorithm. Their algorithm computes a pseudostretch for each available job at any time t that is $(t - r_j)/P^{1/2}$ if $1 \leq p_j \leq P^{1/2}$ and is $(t - r_j)/P$ if $P^{1/2} < p_j \leq P$. That is, they replace the p_j in the denominator by $P^{1/2}$ if the job is small and P if the job is larger. This algorithm is $O(P^{1/2})$-competitive. However, it assumes *a priori* knowledge of $P_{\min}$ and $P_{\max}$. To make this more online, they assume that the algorithm knows the minimum job size 1 in advance, and they use the largest job seen so far as their estimate for P, recalculating as needed when new jobs arrive.

15.2.5 l_p Norms of Flow and Stretch

In this section, we consider the problems of minimizing the l_p norms of flow times and stretch. We discuss both clairvoyant and nonclairvoyant algorithms. Recall that the motivation for considering the l_p, $1 < p < \infty$, norms of flow and stretch was that they represent some compromise between optimizing for the worst case QoS and the average QoS. Although most of the results we give below also hold when $p = 1$ and when $p = \infty$. These results generalize both the average and maximal flow and stretch, since the total flow time or stretch is then the l_1 norm while maximum flow time or stretch is the l_∞ norm.

Let us initially focus on one machine. The study of l_p norms of flow and stretch was initiated by Bansal and Pruhs [27]. Bansal and Pruhs [27] show that are no $n^{o(1)}$-competitive online clairvoyant scheduling algorithms for any l_p norm, $1 < p < \infty$ of either flow or stretch. This is a bit surprising, at least for flow time, as there are optimal online algorithms, SRPT and FIFO, for the l_1 and l_∞ norms of flow time.

Theorem 15.5

For the problems $1 \mid$ *online-time, pmtn,* $r_j \mid (\sum F_j^p)^{1/p}$ *and* $1 \mid$ *online-time, pmtn,* $r_j \mid (\sum S_j^p)^{1/p}$, $1 < p < \infty$, *the competitive ratio of any randomized algorithm A against an oblivious adversary is* $n^{\Omega(1)}$.

Proof

We only give lower bound proofs for flow norms, and only for deterministic algorithms. It is easy to extend the lower bound to randomized algorithms using Yao's technique. The input is parameterized by integers L, $\alpha = (p+1)/(p-1)$, and $\beta = 2$. A long job of size L arrives at time 0. From 0 to time until time $L^\alpha - 1$ a job of size 1 arrives every unit of time.

In the case that A does not finish the long job by time L^α then this is the whole input. Then $F^p(A)$ is at least the flow of the long job, which is at least $L^{\alpha p}$. In this case the adversary could first process the long job and then process the unit jobs. Hence, $F^p(Opt) = O(L^p + L^\alpha \cdot L^p) = O(L^{\alpha+p})$. The competitive ratio is then $\Omega(L^{\alpha p - \alpha - p})$, which is $\Omega(L)$ by our choice of α.

Now consider the case that A finishes the long job by time L^α. In this case $L^{\alpha+\beta}$ short jobs of length $1/L^\beta$ arrive every $1/L^\beta$ time units from time L^α until $2L^\alpha - 1/L^\beta$. One strategy for the adversary is to finish all jobs, except for the long job, when they are released. Then $F^p(Opt) = O(L^\alpha \cdot 1^p + L^{\alpha+\beta} \cdot (1/L^\beta)^p + L^{\alpha p})$. It is obvious that the dominant term is $L^{\alpha p}$, and hence $F^p(Opt) = O(L^{\alpha p})$. Now consider the subcase that A has at least $L/2$ unit jobs unfinished by time $3L^\alpha/2$. Since these unfinished unit jobs must have been delayed by at least $L^\alpha/2$, $F^p(A) = \Omega(L \cdot L^{\alpha p})$. Clearly in this subcase the competitive ratio is $\Omega(L)$. Alternatively, consider the subcase that A has finished at least $L/2$ unit jobs by time $3L^\alpha/2$. Then A has at least $L^{\alpha+\beta}/2$ released, and unfinished, small jobs at time $3L^\alpha/2$. By the convexity of F^p, the optimal strategy for A from time $3L^\alpha/2$ onwards is to delay each small job by the same amount. Thus A delays $L^{\alpha+\beta}/2$ short jobs by at least $L/2$. Hence in this case, $F^p(A) = \Omega(L^{\alpha+\beta} \cdot L^p)$. This gives a competitive ratio of $\Omega(L^{\alpha+\beta+p-\alpha p})$, which by the choice of β is $\Omega(L)$. □

This negative result motivated Bansal and Pruhs [27] to fall back to resource augmentation analysis. They showed that the standard clairvoyant algorithms SJF and SRPT are almost fully scalable for l_p norms of flow and stretch. They showed that SETF and MLF are almost fully scalable for flow objective functions, but not for stretch objective functions. In contrast, RR is not almost fully scalable even for flow objective functions. This is a bit surprising as starvation avoidance is an often cited reason for adopting RR.

While the analysis of SJF used a local competitive argument, the analysis of SRPT was not strictly a local competitiveness argument as a newly released job J_i is not counted until time $r_i + \Theta(p_i)$. The analysis of SETF and MLF used the same method that Bansal et al. [30] used to analyze nonclairvoyant average stretch. We shall sketch the analysis of SJF.

Theorem 15.6

For the problems $1 \mid online\text{-}time, pmtn, r_j \mid (\sum F_j^p)^{1/p}$, *and* $1 \mid online\text{-}time, pmtn, r_j \mid (\sum S_j^p)^{1/p}$, *SJF is* $(1+\epsilon)$*-speed* $O(1/\epsilon)$*-competitive.*

Proof

The proof is by local competitiveness on the objective function $\sum F_j^p$. Let $U(SJF, t)$ and $U(Opt, t)$ denote the unfinished jobs at time t in SJF and Opt respectively, and $\mathcal{D} = U(SJF, t) - U(Opt, t)$. Let $Age^p(X, t)$ denote the sum over all jobs $J_i \in X$ of $(t - r_i)^{p-1}$. Note that $A(t)$, the rate of increase of the objective function for algorithm A, is the sum over all $J_i \in U(A, t)$ of $Age^p(U(A, t), t)$. That is, $F^p(A) = p \int_t Age^p(U(A, t), t)\, dt$.

Thus to have a local competitiveness argument, it is sufficient to establish that

$$Age^p(\mathcal{D}, t) \leq O(1/\epsilon^p) Age^p(U(Opt, t), t)$$

This is established in the following manner. Let $V(t, \alpha)$ denote the aggregate unfinished work at time t among those jobs J_j that satisfy the conditions in the list α. Let $1, \ldots, k$ denote the indices of jobs in $\mathcal{D}$ such that $p_1 \leq p_2 \ldots \leq p_k$. Consider the jobs in $\mathcal{D}$ in the order in which they are indexed. Assume that we are considering job J_i. One can allocate to J_i an $\epsilon p_i/4(1+\epsilon)$ amount of work from $V(t, \{J_j \in U(Opt, t), r_j \leq t - \epsilon(t - r_i)/(4(1+\epsilon)), p_j \leq p_i\})$ that was previously not allocated to a lower indexed job in $\mathcal{D}$. This establishes $O(1/\epsilon^p)$ local competitiveness for F^p for the following reasons. The total unfinished work in each $J_j \in U(Opt, t)$ is associated with $O(1/\epsilon)$ longer jobs in $\mathcal{D}$. Since the jobs J_j are $\Omega(\epsilon)$ as old as J_i, the contribution to $Age^p(U(Opt, t), t)$ for J_j is $\Omega(\epsilon^{p-1})$ as large as the contribution of J_i to $Age^p(U(SJF, t), t)$. Using the same reasoning, and the fact that $p_j \leq p_i$, one establishes local competitiveness for S^p. □

Chekuri, Khanna, and Kumar [26] show how to combine immediate dispatching algorithm of Avrahami and Azar [25] with a scheduling policy such as SJF to obtain an almost fully scalable algorithm for l_p norms of flow and stretch on multiple machines. The analysis is essentially a local competitive argument similar to Bansal and Pruhs' analysis of SJF and SRPT [27].

15.2.6 Weighted Flow Time

In the online weighted flow time problem, each job J_i has an associated positive weight w_i that is revealed to the clairvoyant scheduler at the release time of J_i. The objective function is $\sum w_i F_i$. If all $w_i = 1$ then the objective function is total flow time, and if all $w_i = 1/p_i$ then the objective function is total stretch. Some systems, such as the Unix operating system, allows different processes to have different priorities. In Unix, users can use the `nice` command to set the priority of their jobs. Weights provide a way that a system might implement priorities. For the moment let us focus on one machine.

Becchetti et al. [31] show that besides being a sufficient condition, local c-competitiveness is a necessary condition for an algorithm to be c-competitive.

Theorem 15.7

Every c-competitive deterministic algorithm A for $1 \mid \text{online-time}, r_j, pmtn \mid \sum w_j F_j$ *must be locally c-competitive.*

Proof

Suppose there is a time t where $A(t) > cOPT(t)$. The adversary can punish the online algorithm by bringing in a stream of dense short jobs with the following properties. The density of the jobs in the stream is large enough so that the optimal strategy for all algorithms is to run each of these jobs immediately. At the same time, the weight of the jobs in the stream can be made small enough so the contribution of the stream jobs to the total weighted flow time is arbitrarily close to 0 when the stream jobs are run immediately. The stream is made long enough so that the ratio of A's total flow time when compared to the adversary's total flow time is arbitrarily close to $A(t)/OPT(t)$.

At first glance, it may seem impossible for the stream of jobs to be dense enough to warrant running immediately yet have low enough weight that they contribute almost nothing to the total weighted flow time. This phenomenon becomes clearer when we consider the following example. Consider a job with weight x, processing time 1, and thus density x released at some time t and compare this to a stream of x jobs of weight 1, processing time $1/x$, and thus density also x released at times $t, t+1/x, \ldots, t+(x-1)/x$. If both are processed immediately, the job of weight x will contribute x to the total weighted flow time while the stream of weight 1 jobs will contribute only 1 to the total weighted flow time. On the other hand, if both are delayed by 1 time unit and then processed at time $t+1$, then the increase in total weighted flow time due to the delay for both cases will be exactly x. What we see is that the stream of density x jobs incurs the same delay cost as the one job with weight x, but the actual processing costs are vastly different. Finally observe that we can push this to the extreme where the stream of density x jobs have arbitrarily small weight $\epsilon > 0$ and processing times ϵ/x. If the stream jobs are processed immediately, they add only ϵ to the total weighted flow time. If the stream jobs are delayed by 1 time unit and then processed at time $t+1$, the increase in weighted flow time will still be exactly x. □

The following instance shows that the obvious greedy algorithms have high competitive ratios. Consider the following set of jobs, released at time 0: One job of weight k, length k^2, and hence density $1/k$, and k^3 jobs of weight 1, length 1, and hence density 1. Two natural schedules are: (1) first run the low density job followed by the k^3 high density jobs, and (2) first run the k^3 high density jobs followed by the low density job. It is easy to see that the first algorithm is not constant locally competitive at time k^3, and that the second algorithm is not constant locally competitive at time k^3+k^2-1. In fact, what the online algorithm should do in this instance is to first run k^3-k of the high density jobs, then run the low density job, and then finish with the remaining k high density jobs. This instance demonstrates that the scheduler has to balance between delaying low density jobs, and delaying low weight jobs.

Using this intuition, Chekuri, Khanna, and Zhu gave an $O(\log^2 P)$-competitive algorithm for a single machine [24]. This algorithm is semi-online; it needs *a priori* knowledge of P. The algorithm partitions jobs based on approximate weights and on approximate densities. It then considers the weight classes from largest to smallest. Assume it is considering weight class w. It runs the densest job J_i from weight class w if and only if the total weight of jobs, with weight $<w$ and density greater than the density of J_i, is less than w. Otherwise it is safe for the algorithm to proceed to lower weight and higher density jobs. The analysis is a rather complicated local competitiveness argument.

If all jobs have the same weight, or if all jobs have the same processing time, or if all jobs have the same density, then $O(1)$-competitiveness is easy. This leads to three obvious algorithms. The algorithm partitions the jobs based on approximate weight or length, or density. Some job in the partition with maximum total weight is run. All jobs within a partition are run using the $O(1)$-competitive algorithm for same weight/length/density jobs. Intuitively, all of these algorithms should have competitive ratios that are linear in the number of partitions, or equivalently, logarithmic in the range of possible weights/lengths/densities. Bansal and Dhamdhere [32] proved this for the version of the algorithm where you partition based on the weight. The analysis is a local competitiveness argument that is a variation on the local competitiveness argument for SRPT.

Perhaps the most intellectually intriguing open question in online scheduling in the *online-time* model, with a flow or stretch objective function, is the following.

Open Problem 15.3

For the problem $1 \mid$ *online-time, pmtn,* $r_j \mid \sum w_j F_j$, *is there an* $O(1)$*-competitive clairvoyant algorithm?*

Several positive results have been developed for weighted flow time problems using resource augmentation. Phillips et al. [9] showed that an algorithm they named Preemptively-Schedule-By-Halves is 2-speed 1-competitive algorithm for minimizing total weighted flow time on a single machine. Becchetti et al. [31] observed that the analysis of Phillips et al. [9] also applied to HDF. Furthermore, using a more direct local competitiveness argument, Becchetti et al. [31] showed that HDF is $(1+\epsilon)$-speed $(1+1/\epsilon)$-competitive on a single machine.

Bansal and Pruhs then consider the problem of minimizing the weighted l_p norms of flow time [33]. They show that HDF is almost fully scalable for the problem $1 \mid$ *online-time, pmtn,* $r_j \mid (\sum w_j F_j^p)^{1/p}$. They then consider the obvious generalization, Weighted SETF (WSETF), of the nonclairvoyant algorithm SETF. WSETF operates as follows. For a job J_i, let $x_i(t)$ denote the amount of work done on that job by time t. Amongst jobs with the smallest $x_i(t)/w_i$, WSETF splits the processor proportionally to weights of the jobs. They show that WSETF is almost fully scalable for the problem $1 \mid$ *online-time-nclv, pmtn,* $r_j \mid (\sum w_j F_j^p)^{1/p}$. The analysis of HDF and WSETF are similar to the analysis of SJF and SETF in [27].

For the parallel machine setting, only a few results are known. Chekuri, Khanna, and Zhu [24] give a lower bound on the competitive ratio of any algorithm of $\Omega(\min(\sqrt{P}, \sqrt{W}, (n/m)^{1/4})$ for the problem $P \mid$ *online-time, pmtn,* $r_j \mid \sum w_j F_j$ where W is the largest weight. Becchetti et al. [31] show that HDF is a $(2+\epsilon)$-speed $O(1)$-competitive algorithm for the same problem.

15.2.7 Semiclairvoyant Scheduling for Average Flow/Stretch

The concept of semiclairvoyant scheduling was introduced by Bender, Muthukrishnan, and Rajaraman [29]. A semiclairvoyant algorithm only has approximate knowledge about processing times. A *strong semiclairvoyant* algorithm knows a constant approximation of the remaining processing time of a job, and a *weak semiclairvoyant* algorithm knows only a constant approximation of the original processing time of a job. While there may be some practical application for these results, for example, a web server serving dynamic documents may only be able to estimate the size of a resulting document as it dynamically constructs the document, the main motivation seems to be that such results may then be used as subroutines in other algorithms that round the processing times of jobs. Rounding processing times often seems to make the development of an algorithm or analysis simpler.

For the parallel machine setting, both the strong and weak semiclairvoyant models are not significantly different than the clairvoyant setting when considering the total flow time and total stretch objective functions because of the classification nature of the clairvoyant nonmigratory algorithms developed earlier. For example, the algorithm of Awerbuch et al. [22] that classifies jobs according to their remaining processing times can be adapted to be a strong semiclairvoyant algorithm for minimizing total flow time and total stretch while the algorithms of Chekuri et al. [24] and Avrahami and Azar [25] that classify jobs according to their initial processing times can be adapted to be weak semiclairvoyant algorithms for minimizing total flow time or stretch. Thus, we focus on the uniprocessor setting for the remainder of this subsection.

Let us first consider strong semiclairvoyant algorithms on a single machine. The most obvious algorithm is to run the job that appears to have the least remaining processing time. Bender et al. [29] show that this algorithm is $O(1)$-competitive with respect to average stretch, but is only $\Theta(\log P)$-competitive with respect to average flow time. They then give modified algorithm that is $O(1)$-competitive with respect to average flow time. The main idea behind this algorithm is that if there is a choice between two jobs with similar remaining processing times, then the algorithm should favor the job whose initial processing time was less.

We now consider weak semiclairvoyant algorithms. Bender et al. [29] show that the obvious generalization of SJF is $O(1)$-competitive with respect to average stretch. Bender, Muthukrishnan, and Rajaraman [29] also proposed an algorithm for average flow time. The basic idea of this algorithm is to run the apparent shortest job first, except in one special case. This special case is that if the job J_i with the apparent least original processing time has not been run at all, and the job J_j with the second least apparent original processing time has been partially run, and there are no other jobs with comparable apparent original processing times, then J_j is run instead of J_i.

Becchetti et al. [34] showed that this algorithm is in fact $O(1)$-competitive with respect to average flow time. A simpler analysis was developed by Nikhil Bansal. Bansal's analysis was a variation of the local competitiveness analysis of SRPT. At any time, order both the online algorithms jobs and the adversary's jobs by increasing remaining processing time. Then Bansal shows that the following invariant always holds: The total work contained in the online algorithm's jobs up to the kth unexecuted job, is at least the total work in the adversary's first k jobs. It is straightforward to observe that this invariant implies $O(1)$-competitiveness.

Becchetti et al. [34] show that there is no weak semiclairvoyant algorithm that can be simultaneously $O(1)$-competitive with respect to average flow time and average stretch. This is in contrast to the clairvoyant setting where SRPT is $O(1)$-competitive with respect to both objective functions.

15.2.8 Nonclairvoyant Scheduling to Minimize Average/Maximum Flow/Stretch

In the *online-time-nclv* model, the nonclairvoyant scheduler is given no information about the processing time of a job. For example, the process scheduling component of an operating system is best viewed as being nonclairvoyant in that the operating system in general does not know the execution times of the various client processes.

15.2.8.1 Maximum and Average Flow Time on One Machine

If the objective function is minimizing maximum flow, then a nonclairvoyant scheduler can still be optimal since FIFO does not require knowledge of the processing times of the jobs. The situation for average flow looks more bleak. The optimal algorithm is SRPT. However, from the nonclairvoyant scheduler's point of view, any job might conceivably be the one with shortest remaining processing time. In the absence of any information about remaining processing times, the most obvious nonclairvoyant algorithm is probably RR. Motwani, Philips and Torng [5] show that RR is 2-competitive in the case that all jobs are released at time 0. However, Matsumoto [35], and independently Motwani, Philips, and Torng [5] showed that the competitive ratio for RR is $\Omega(n/\log n)$ in the case of release dates. Kalyanasundaram and Pruhs [8] noted that a variation of this lower bound instance shows that modest resource augmentation is not enough to allow RR to be $O(1)$-competitive.

Theorem 15.8

For the problem $1 \mid$ online-time-nclv, pmtn, $r_j \mid \sum F_j$, the competitive ratio of RR is at least $\Omega(n/\log n)$, and the competitive ratio of RR with speed s, $1 < s < 2$ processor is at least $\Omega(n^{2-s})$.

Proof

Let $s = 1 + \epsilon$. We divide time into stages. Let the ith stage, $i \geq 0$ start at time t_i. We let $t_0 = 0$, and $t_1 = 1 + \epsilon$. There are two jobs of length $(1 + \epsilon)$ released at time t_0, and one job is released at each time t_i, $i \geq 1$, with length $p(i)$ that is exactly the same length as RR has left on each of the previous jobs. In order to guarantee that the adversary can finish the job released at time t_{i-1} by time t_i, $i \geq 2$, we let $t_i = t_{i-1} + p(i - 1)$. Observe that during the interval $[t_{i-1}, t_i]$, RR executes each of the $i + 1$ jobs for

$p(i-1)/(i+1)$ units of time. Since RR also uses a $1+\epsilon$ speed processor, the work done on a job during that interval is $(1+\epsilon)p(i-1)/(i+1)$. Therefore, we get the recurrence

$$p(i) = p(i-1) - \frac{(1+\epsilon)p(i-1)}{i+1} = \left(\frac{i-\epsilon}{i+1}\right)p(i-1)$$

The total flow time for the adversary is then $\Theta(\sum_{i=1}^{n} 1/(i-\epsilon)^{1+\epsilon})$, which is a convergent sum. The total flow time for RR is then $\Theta(\sum_{i=1}^{n} i/(i-\epsilon)^{1+\epsilon})$, which is $\Theta(n^{1-\epsilon})$. The result then follows. □

More generally, Motwani, Phillips and Torng [5] showed that the competitive ratio of every deterministic nonclairvoyant algorithm for average flow is $\Omega(n^{1/3})$. Thus one cannot get a strong positive result for deterministic nonclairvoyant algorithms using standard competitive analysis. This construction is the basis for most general lower bound proofs on average flow time.

Theorem 15.9

For the problem $1 \mid online\text{-}time\text{-}nclv, pmtn, r_j \mid \sum F_j$, *the competitive ratio of every deterministic algorithm is* $\Omega(n^{1/3})$.

Proof

We present an adversary strategy which works in two stages. In the first stage, the adversary releases k jobs at time 0 and lets the algorithm A schedule them for k time units. The adversary ensures that the remaining processing time of each job at time k for A is $1/(k-1)$ while OPT has 1 unfinished job at time k. The second stage starts at time k, when the adversary releases a job of length $1/(k-1)$ every $1/(k-1)$ time units apart, until time k^2. No matter what A does after time k, A has a total flow time of $\Omega(k^3)$ while *OPT* has total flow time $O(k^2)$. □

This strong lower bound motivated Kalyanasundaram and Pruhs [8] to propose resource augmentation analysis as a standard method of analysis. Notice that in this lower bound example the load after time k is 1. Furthermore, if the nonclairvoyant scheduler had a slightly faster processor, then it would not be behind at time k. Kalyanasundaram and Pruhs [8] showed that SETF is almost fully scalable. In [8] the algorithm SETF was called Balance.

Theorem 15.10

For the problem $1 \mid online\text{-}time\text{-}nclv, pmtn, r_j \mid \sum F_j$, *SETF is* $(1+\epsilon)$*-speed* $(1+1/\epsilon)$*-competitive.*

Proof

We give the intuition here using the borrow technique introduced in [8]. The proof is a local competitiveness argument. That is, at any particular time t, SETF does not have too many more unfinished jobs than the adversary. Let J_i and J_j be jobs such that SETF is running J_j during the time interval $[a,b] \subseteq [r_i, C_i]$. The adversary may then do $(b-a)$ units of work on J_i during time $[a,b]$. We think of this as the adversary borrowing $(b-a)$ units of work from J_j to give to J_i. Borrowing can also be transitive; J_i can borrow from J_j which can borrow from J_k, etc. This borrowing might be advantageous to the adversary if J_i is almost finished by SETF. Let $w_i(t)$ be the remaining unfinished work on job J_i for SETF at time t. If the adversary is going to finish a job J_i, then it must arrange for J_i to borrow at least $\epsilon w_t(i)$ units of work since the adversary's processor is ϵ slower than the processor used by SETF. However, the description of SETF ensures that this time can only come from jobs that SETF ran for less time than SETF ran J_i. Hence, we would expect that each job that the adversary borrows time from can only be used to finish $1/\epsilon$ jobs that SETF has not finished. □

Berman and Coulston [36] improved this result for larger speeds. They showed that SETF is s-speed $2/s$-competitive for total flow time for $s \geq 2$. They showed inductively that for each job that is added to a schedule, the increased cost that $SETF_s$ pays is at most $2/s$ the cost that the adversary pays.

Turning to randomized algorithms, Motwani, Philips, and Torng [5] showed that the competitive ratio for total flow time of every randomized algorithm against an oblivious adversary is $\Omega(\log n)$. Kalyanasundaram and Pruhs [37] noted that this argument can be modified to give a lower bound of $\Omega(P)$. Recall that an oblivious adversary must fix the input a priori.

Theorem 15.11

For the problem $1 \mid online\text{-}time\text{-}nclv, pmtn, r_j \mid \sum F_j$, the competitive ratio of every randomized algorithm against an oblivious adversary is $\Omega(\log n)$.

Proof

We use Yao's technique and prove a lower bound on the competitive ratio of deterministic algorithms on a particular input distribution.

The jobs are released in two phases. In the first phase, at time 0, k jobs are released whose sizes are independently drawn from the exponential distribution with mean 1. The scheduling algorithm is then allowed to run until time $k - 2k^{3/4}$. Because the expected remaining work of an unfinished job is independent of how long it has been executed, the state of the unfinished jobs for the nonclairvoyant algorithm at time $k - 2k^{3/4}$ is the same for all nonclairvoyant algorithms. Hence, at the end of the first stage, the nonclairvoyant algorithm has remaining $\Omega(k^{3/4})$ jobs with remaining work at least 1. With high probability, the adversary scheduler can set aside $k^{3/4}/\log k$ jobs of size at least $(\log k)/4$, and finish all other jobs by the end of the first phase.

The second phase consists of releasing a job of size 1 at each time unit, for a total of k^2 time units. Clearly the nonclairvoyant algorithm should execute these jobs before the large jobs it has remaining from the first phase, for an expected total flow time of $\Omega(k^{2.75})$. The adversary executes the second phase jobs as they arrive, and lastly schedules the set-aside jobs. The expected total waiting time of the adversary algorithm is $O(k^{2.75}/\log k)$. □

It is not at all obvious what strategy a randomized algorithm should adopt in order to obtain a logarithmic competitive ratio. One benefit of resource augmentation analysis of deterministic algorithms is that the analysis can suggest a randomized strategy. This problem is an example of this phenomenon. Let us reflect on Kalyanasundaram and Pruhs' resource augmentation analysis of SETF [8] for a moment. One sees that to argue that SETF is locally $O(c \cdot d)$-competitive, it is sufficient to argue the following property holds:

> For an at least $1/c$ fraction of SETF's unfinished jobs, it is the case that they have at least $1/d$ of their original processing time left unfinished.

This suggests finding a randomized algorithm that favors newly released jobs (like SETF does) and that guarantees the above property holds with high probability. This line of reasoning led Kalyanasundaram and Pruhs [37] to propose the algorithm RMLF. RMLF is identical to MLF except that the target of each job in queue Q_i is 2^{i+1} minus an exponentially distributed independent random variable.

We now give some intuition why this approach should give poly-logarithmic competitiveness. Assume that at time 0 the adversary releases a collection of n jobs of length $2^i + x$. The adversary is hoping that at the first time that all remaining unfinished jobs for RMLF are all in Q_{i+1} that the following holds: it is the case that RMLF will have $\omega(1)$ jobs in Q_{i+1} and that these jobs are almost all finished. For example, if RMLF uniformly at random selected the target between 2^i and 2^{i+1}, then by picking $x = 2^i/\sqrt{n}$, the adversary could expect that RMLF has $\sqrt{n}$ jobs with at most a $1/\sqrt{n}$ of their initial processing time left. By bringing in a stream of small jobs, the adversary could then push the competitive ratio up to $\Omega(n^\epsilon)$. So RMLF wants that the number of targets set to $X - 2^i/\log n$ should be a constant fraction of the number of targets set to X. By setting the targets randomly in this way, you expect that a constant fraction of the jobs have $1/\log n$ of their original processing time left. Two more points need to be made. First, this argument is not valid if x is very small, that is if the jobs have very little processing time left on the job when it reaches Q_i. However, in this case, each job is finished in Q_i and does not reach Q_{i+1} with very high probability.

Second, in order to turn this into a formal proof, you need to have a high probability argument, which adds another factor of $\log n$ to the calculated competitive ratio. This argument can be formalized to show that RMLF is $O(\log^2 n)$-competitive for the problem $1 \mid$ *online-time-nclv, pmtn*, $r_j \mid \sum F_j$.

This $O(\log^2 n)$ analysis can be improved. Kalyanasundaram and Pruhs [37] showed that RMLF is $\Theta(\log n \log\log n)$ against an adversary that at all times knows the outcome of all of the random events internal to RMLF up until that time. This accounts for the possibility of inputs where future jobs may depend on the past schedule. Becchetti and Leonardi [7] improved upon this analysis to obtain a tight analysis of RMLF.

Theorem 15.12

For the problem $1 \mid$ *online-time-nclv, pmtn*, $r_j \mid \sum F_j$, *RMLF is* $O(\log n)$*-competitive against an oblivious adversary.*

Note that if the target for jobs in queue Q_i is c^i, then MLF is c-speed $O(1)$-competitive. In particular, if $c = 1+\epsilon$, MLF devolves into SETF and is also almost fully scalable. These facts (SETF/MLF is almost fully scalable, and RMLF is optimally competitive amongst randomized algorithms) provide strong support for the adoption of MLF for process scheduling within an operating system.

Open Problem 15.4

Obtain a tight bound on the competitive ratio of deterministic algorithms for the problem $1 \mid$ *online-time-nclv, pmtn*, $r_j \mid \sum F_j$. *On one hand, given that there is a high,* $\Omega(n^{1/3})$, *lower bound on the competitive ratio, this may seem to be only of academic interest. On the other hand, this is arguably the most basic problem in nonclairvoyant scheduling, and it is quite unsatisfactory that a tighter bound is not known.*

15.2.8.2 Maximum and Average Stretch on One Machine

Kalyanasundaram and Pruhs [38] observed that the competitive ratio for maximum stretch is $\Omega(n)$ for nonclairvoyant algorithms and also that resource augmentation is of minimal help. For average stretch, it is easy to see that the competitive ratio for nonclairvoyant algorithms is $\Omega(n)$ and $\Omega(P)$. However, Bansal et al. [30] show that a moderately positive result can be obtained using resource augmentation.

Theorem 15.13

For the problem $1 \mid$ *online-time-nclv, pmtn*, $r_j \mid \sum S_j$, *MLF is an* $O(1)$*-speed* $O(\log^2 P)$*-competitive algorithm.*

Proof

It is easy to see that one cannot prove $MLF(J) = O(OPT(J))$ using local competitiveness (even if MLF has O(1) faster processor). To see this consider the case of a single unit length job and a small number of long jobs released at time 0. One can verify that every nonclairvoyant algorithm will be $\Omega(P)$ locally competitive at say time 2.

To show that $MLF(J) = O(\log^2 P) \cdot OPT(J)$, there are two main ideas in the proof. The first main idea was to show that $MLF(J) = O(SJF(L))$, where J is the original input, and L is some other input derived from J. In this modified instance L, each job J_i in J is replaced by a collection of jobs with geometrically increasing work, with aggregate work p_i, and with release date r_i. The idea is that at any particular time, MLF has the original job J_i in the jth queue if and only if SJF finished the $j-1$ shortest jobs in L corresponding to J_i.

To show $MLF(J) = O(SJF(L))$, Bansal et al. [30] introduce an auxiliary objective function, called inverse work, that can be used to show local competitiveness. Let $w_i(t)$ be the amount of work done on job i by time t. Then the inverse work for a job is $\int_{r_j}^{C_j} 1/w_j(t)\,dt$. Clearly the inverse work of a job is greater than its stretch $\int_{r_j}^{C_j} 1/p_j\,dt$. Hence, $MLF(J) \le MLF'(J)$, where $MLF'(J)$ is total inverse work. Then the authors show that by local competitiveness that $MLF'(J) = O(SJF(L))$. Applying Becchetti et al. [31] analysis of HDF, Bansal et al. conclude that $SJF(L)$ with a slightly faster processor has total stretch $O(OPT(L))$.

To finish the proof, one needs to upper bound $OPT(L)$ by $O(\log^2 P) \cdot OPT(J)$. The second main idea was the method used to relate $OPT(L)$ and $OPT(J)$. Given the schedule $OPT(J)$, one can construct a schedule for L (which is a union of geometrically scaled copies of J) in the following way. Take the schedule $OPT(J)$ and consider suitably scaled down copies of this schedule. These schedules are thought of running a scaled down copy of J with some carefully chosen processor speed. Executing all these schedules simultaneously can be thought of as a schedule for L. Bansal et al. [30] show that this scaling can be done in such a way that the total additional speed required is $O(1)$ and the total stretch for L is $O(\log^2 P) \cdot OPT(J)$. □

For the problem $1 \mid$ *online-time-nclv, pmtn*, $r_j \mid \sum S_j$, Bansal et al. [30] also shows that every $O(1)$-speed algorithm has a competitive ratio of $\Omega(\log P)$. If all release dates are zero, Bansal et al. [30] gives an $O(\log P)$-competitive algorithm, and prove a general $\Omega(\log n)$ lower bound on the competitive ratio.

15.2.8.3 Average Flow Time on Parallel Machines

An immediate question that one has to ask when formalizing a scheduling problem on parallel machines is whether a single job can simultaneously run on multiple machines. In some settings this may not be possible; in other settings this may be possible but the speed-up that one obtains may vary. Thus one can get myriad different scheduling problems on parallel machines depending on what one assumes. A very general model is to assume that each job has a speed-up function that specifies how much the job is sped up when assigned to multiple machines. More formally, a *speed-up function* $\Gamma(s)$ measures the rate at which work is finished on the job if s processing resources (say s processors) are given to the job.

The simplest speed-up model is the *fully parallelizable* where $\Gamma(s) = s$. Fully parallelizable work has the property that if you devote twice as many resources to the work, it completes at twice the rate. The normal assumption in the uniprocessor scheduling literature is that all work is fully parallelizable. In the uniprocessor setting, this means that if you devote a fraction f of a single processor to a job, you will complete work at rate f instead of rate 1. To simplify notation, we will often use the word parallel instead of fully parallelizable when there is no possibility of ambiguity.

The normal multiprocessor setting can be modeled by the speed-up function $\Gamma(s) = s$ for $s \leq 1$ and $\Gamma(s) = 1$ for $s > 1$. That is, a job is fully parallelizable on one processor, but assigning the job to multiple processors does not help.

In any real application, speed-up functions will be sublinear and nondecreasing. A speed-up function is sublinear if doubling the number of processors at most doubles the rate at which work is completed on the job. A speed-up function is nondecreasing if increasing the number of processors does not decrease the rate at which work is completed on the job. One can also generalize this so that jobs are made of phases, each with their own speed-up function. We will use the notation sc_j in the job field of the three-field scheduling notation to denote parallel machines with job phases that have speed-up curves that are sublinear and nondecreasing.

We typically assume that a nonclairvoyant scheduling algorithm does not know the speed-up function of any job. Given how little knowledge a nonclairvoyant scheduler has in this setting, there are few natural algorithms to consider. The obvious ones to analyze are SETF and RR. Edmonds showed that SETF is not a good algorithm when jobs are not fully parallelizable [39].

Theorem 15.14

The deterministic and randomized versions of SETF are not s-speed $O(1)$-competitive if the speed-up curves of jobs are not fully parallelizable no matter how large s is [39].

Furthermore, in a remarkable analysis, Edmonds showed that RR is $(2+\epsilon)$-speed $O(1+1/\epsilon)$-competitive for jobs with phases that have speed-up functions that are sublinear and nondecreasing [39]. His result extends with slightly weaker bounds to the case where RR is given extra machines instead of faster machines.

Theorem 15.15

For the problem $P \mid$ online-time-nclv, pmtn, $r_j, sc_j \mid \sum F_j$, RR is $(2+\epsilon)$-speed $O(1+1/\epsilon)$-competitive.

One obvious difficulty in constructing an $O(1)$-speed $O(1)$-competitiveness analysis for RR is that, as the example lower bound instance in Theorem 15.8 shows, one cannot use a local competitiveness argument. One of Edmonds' insights was the identification of an appropriate potential function so that one could prove local competitiveness in an amortized sense. Arguably another insight was that analysis of RR for $1 \mid$ *online-time-nclv, pmtn*, $r_j \mid \sum F_j$, seems to require the introduction of speed-up curves. It is at least of academic interest whether there is an analysis of RR for $1 \mid$ *online-time-nclv, pmtn*, $r_j \mid \sum F_j$ that does not require the generalization to speed-up curves. Edmonds' analysis is too involved to give in its entirety here. We shall instead focus on the intuition that the proof gives about why RR performs reasonably well.

A key step in the proof is the introduction of the *constant* speed-up curve where $\Gamma(s) = c$ for all $s \geq 0$ and some constant $c > 0$. Devoting additional processing resources to constant jobs does not result in any faster processing of these jobs. In fact constant jobs complete at the same rate even if they are not run. The motivation for defining the constant speed-up curve is its utility for analytic purposes, not as a model of real job behavior. Note, Edmonds uses the term *sequential* instead of constant in [39].

With this definition of constant jobs, Edmonds transforms each possible input into a canonical input that is streamlined. An input is *streamlined* if: (1) every phase is either fully parallelizable or constant, and (2) the adversary is able to execute each job at its maximum possible speed. This implies that at any one time, the adversary has only one parallel job phase to which it is allocating all of its resources. The idea of this transformation is that if RR is devoting more resources to some work than the adversary, it is to the adversary's advantage to make this work be constant work that completes at the rate that the adversary was originally processing that work. In contrast, if the adversary is devoting more resources to a job than is RR, and the adversary has no other unfinished jobs, then it is to the adversary's advantage to make this work to be fully parallelizable. If the adversary has several unfinished jobs, then the transformation is only slightly more involved; each bit of work is replaced by a constant phase, followed by a parallel phase, followed by a constant phase. As a consequence of this transformation, you get that the adversary is never behind RR on any job. Given that the input is streamlined, we can for simplicity assume that RR has one processor of speed $s = 2 + \epsilon$ and OPT has one processor of speed 1.

We now turn to the potential function Φ, which is defined to be the work that has not been completed by RR but that has been completed by the adversary. Then $\Phi(t)$ is the rate of change of Φ at time t. Edmonds then proves local competitiveness using this potential function. That is, he shows that at all times t, $RR_s(t) \leq O(1+1/\epsilon)OPT(t) + \Phi(t)$.

We now give the intuition behind the Edmonds' proof from [39]. Let l_t be the number of constant jobs at time t. Note that l_t is the same for all schedules. RR devotes at most $s/(l_t+1)$ of its speed to the unique fully parallelizable job that the adversary is working on at time t. To ensure that RR falls further behind on this job, l_t must be at least s or else RR may complete as much work on the parallel job as OPT does at time t. On the other hand, the adversary does not want l_t to be too large as the adversary must also pay this cost.

The key observation is that as the fully parallelizable work on which RR is behind builds up, RR self-adjusts by devoting more resources to this parallel work. Let m_t be the number jobs with parallel work for RR at time t. Note that m_t can be larger than 1 since RR is behind the adversary in some jobs. RR devotes $s/(l_t + m_t)$ of its s speed to each of the $l_t + m_t$ jobs it sees. Hence, RR completes fully parallelizable work at a rate of $s \cdot m_t/(l_t + m_t)$. Since the adversary works at unit rate on the fully parallelizable work, RR falls behind on this work at a rate of at most $1 - s \cdot m_t/(l_t + m_t)$. The steady state is when this rate is 0, that is, when $m_t = l_t/(s-1)$.

In this steady state, the competitive ratio is then at most $(l_t + l_t/(s-1))/(l_t+1) \leq s/(s-1)$. Intuitively, RR tries to move to this steady state. To see this consider that RR is either above or below this steady state. If $m_t < l_t/(s-1)$ then more fully parallelizable work is being released than RR is completing, and hence RR is falling further behind and the potential function increases. The potential function increase is compensated by the fact that RR's flow time is increasing at a slower rate than it is at steady state. On the other hand, if $m_t > l_t/(s-1)$ then RR must be catching up to the adversary in terms of uncompleted parallel work. In this case, the decrease in the potential function must pay the additional increase in flow time that RR has to pay for being behind.

Note that the speed s has to be at least $2+\epsilon$ in order for the potential function to decrease quickly enough. A simple instance that shows that speed 2 is required is n jobs, with equal processing time, that all arrive at time 0. In this case RR needs speed at least 2 so that it is always $O(1)$-competitive in terms of the number of unfinished jobs.

The obvious and interesting open question is the following.

Open Problem 15.5

Is there an almost fully scalable algorithm in the case of sublinear and nondecreasing speed-up functions when the objective function is total flow time?

Edmonds, Datta, and Dymond [40] extend Edmonds' analysis of RR to Internet TCP protocol. Becchetti and Leonardi [7] extend their analysis of RMLF to show that it is $O(\log n \log(n/m))$-competitive on m machines under the assumption that jobs may not be simultaneously run on multiple machines.

15.2.9 Multicast Pull Scheduling for Average Flow

In a multicast/broadcast system, when the server sends a requested page/item, all outstanding client requests to this page are satisfied by this multicast. The system may use broadcast because the underlying physical network provides broadcast as the basic form of communication (e.g., if the network is wireless or the whole system is on a LAN). Multicast may also arise in a wired network as a method to provide scalable data dissemination. One commercial example of a multicast-pull client-server system is Hughes' DirecPC system. In the DirecPC system the clients request documents via a low bandwidth dial-up connection, and the documents are broadcast via high bandwidth satellite to all clients. In this section we will restrict our attention to the case that the objective function is total flow time. We use the notation B in the machine field of the 3-field scheduling notation to denote broadcast, or more precisely, multicast pull.

While this problem is interesting in its own right, it is also interesting because of its connection to weighted flow time and its surprising connection to scheduling jobs with speed-up functions. We first explain why this problem generalizes weighted flow time. If one restricts the instances in multicast pull scheduling such that for each page, all requests for that page arrive at the same time, then the multicast pull scheduling problem and the weighted flow scheduling problem are identical. Here the number of requests that arrive for the page is the weight.

At first glance, it seems that the only difficulty the scheduler faces is how to favor both shorter pages as well as more popular pages. However, the situation is more complicated than this. Consider the case where all pages have the same size. The obvious algorithm to consider is Most Requests First (MRF) that broadcasts the page with the most outstanding requests thus generalizing the HDF weighted scheduling algorithm. At first, one might even be tempted to think that MRF is optimal. Kalyanasundaram, Pruhs, and Velauthapillai [41] showed that MRF is not $O(1)$-speed $O(1)$-competitive.

Lemma 15.9

For the problem $B \mid online\text{-}time, pmtn, r_j, p_j = 1 \mid \sum F_j$, the algorithm MRF is not $O(1)$-speed $O(1)$-competitive.

Proof

Assume that MRF has an $s = O(1)$ speed processor. Let $k = n^2$. At time 0, the adversary requests pages $P_1, \ldots, P_{n-s}$ once each, and requests pages $P_{n-s+1}, \ldots, P_n$ twice each. At each time t, $1 \le t \le k$, the adversary requests pages $P_{n-s+1}, \ldots, P_n$ twice each.

For all times $t \in [1, k]$, MRF will broadcast pages $P_{n-s+1}, \ldots, P_n$. Only after time k will MRF finally broadcast pages $P_1, \ldots, P_{n-s}$. Since the initial requests to pages $P_1, \ldots, P_{n-s}$ are not satisfied by MRF during the first k time units, the total flow time for MRF is $\Omega(nk)$, which is $\Omega(n^3)$ since $k = n^2$.

On the other hand, for time $1 \leq i \leq n - s$, the adversary broadcasts page P_i. From this time on, the adversary broadcasts pages $P_{n-s+1}, \ldots, P_n$ in a round robin fashion from time $(n - s) + 1$ to time k. Each of the $O(ns)$ requests made before time $n - s$ is satisfied within n time units, and each of the $O(ks)$ requests made after time $n - s$ is satisfied within s time units. Hence, the total flow time for the adversary is $O(sn^2 + ks^2)$, which is $O(n^2)$ since $k = n^2$ and $s = O(1)$. Therefore, the competitive ratio for MRF is $\Omega(n)$. □

The lower bound instance in Lemma 15.9 shows that the online scheduler has to be concerned with how to best aggregate jobs. Without knowledge of the future or resource augmentation, this turns out to be impossible. Kalyanasundaram, Pruhs, and Velauthapillai [41] show that no $O(1)$-competitive algorithm exists even in the case of unit pages if preemption is not allowed. Edmonds and Pruhs [42] extend the lower bound to the case that preemption is allowed.

Lemma 15.10

For the problem $B \mid$ online-time, pmtn, $r_j, p_j = 1 \mid \sum F_j$, the competitive ratio of every randomized online algorithm A against an oblivious adversary is $\Omega(n)$ where n is the number of different pages.

Proof

We give only the deterministic lower bound proof. This can be generalized to a lower bound for randomized algorithms using Yao's technique. At time 0, every page is requested once. Then no pages are requested until time $n/2$. From time 1 until time $n/2$, the adversary broadcasts the $n/2$ pages not broadcasted by A by time $n/2$. At time $n/2$, the adversary requests all of the pages previously broadcasted by A. Note that there are at most $n/2$ such pages and they were not previously broadcasted by the adversary. No more pages are requested until time n. After the broadcast at time n, the adversary has satisfied all of the requests to date, while A has at least $n/2$ unsatisfied requests. At each time t, for t from n to $k = n^2$, the adversary requests the page broadcasted by A at time $t - 1$. Hence, at each time in $[n, k]$, A has $n/2 + 1$ unsatisfied requests. At each time $t \in [n + 1, k + 1]$, the adversary can satisfy the request at time $t - 1$. Hence, the adversary has at most 1 unsatisfied request at each time $t \in [n + 1, k]$. Hence, the total flow time for the adversary is $O(n^2 + k)$, and the total flow time for A is $\Omega(nk)$. □

Before considering upper bounds, we need to note that several reasonable models are possible depending on what one assumes about the capabilities of the server and the clients to send and receive segments of the pages out of order. For example, it is not clear whether a client that requests a large page, in the middle of the broadcast, will need the whole page rebroadcast, or only the first half. For example, in a protocol, like the http protocol, where the content is identified only in the header, rebroadcast would be required. Pruhs and Uthaisombut [43] compare the optimal schedules, under various objective functions, in the different models. They show that allowing the server to send segments out of order is of no real benefit. On the other hand, they show that the ability of the clients to receive data out of order can drastically improve the average flow time, but not the maximum flow time. Further they show that a speed 2 server can compensate for clients not be able to receive pages out of order.

The general lower bound in Lemma 15.10 actually contains the key insight that ties multicast pull scheduling to scheduling with speed-up curves and thus suggests a possible algorithm. After the online algorithm has performed a significant amount of work on a page that was requested by a single client, the adversary can again direct another client to request that page. The online algorithm must service this second request as well. In contrast, the optimal schedule knows not to initially give any resources to the first request because the broadcast for the second request simultaneously services the first. Thus, even though the online algorithm devotes a lot of resources to the first request and the optimal algorithm devotes no resources to the first request, it completes under both at about the same time. In this regard, the work associated with the first request can be thought of as "constant." This suggests that the real difficulty of broadcast scheduling is that the adversary can force some of the work to have a constant speed-up curve.

Formalizing this intuition, Edmonds and Pruhs [42] give a method to convert any nonclairvoyant unicast scheduling algorithm A to a multicast scheduling algorithm B under the assumption that the clients must receive all pages in order. A unicast algorithm can only answer one request at a time. All the standard algorithms listed in Section 15.2.1 are unicast algorithms. Edmonds and Pruhs [42] show that if A works well when jobs can have parallel and constant phases, then B works well if it is given twice the resources. The basic idea is that B simulates A, creating a separate job for each request, and then the amount of time that B broadcasts a page is equal to the amount of time that A runs the corresponding jobs. More formally, if A is an s-speed c-competitive unicast algorithm, then its counterpart, algorithm B, is a $2s$-speed c-competitive multicast algorithm. In the reduction, each request in the multicast pull problem is replaced by a job whose work is constant up until the time that either the adversary starts working on the job or the online algorithm finishes the job. After that time, the work of the replacement job is parallel. The amount of parallel work is such that A will complete a request exactly when B completes the corresponding job.

Using RR for algorithm A, one obtains an algorithm, called BEQUI in [42], that broadcasts each page at a rate proportional to the number of outstanding requests. Using Edmonds' analysis of RR for jobs with speed-up functions, one gets the following result.

Theorem 15.16

For the problem $B \mid$ online-time-nclv, pmtn, $r_j \mid \sum F_j$, under the assumption that all pages must be received in order, BEQUI is $(4+\epsilon)$-speed $O(1+1/\epsilon)$-competitive.

Note that BEQUI preempts even unit sized jobs. Edmonds and Pruhs also give a $(4+\epsilon)$-speed $O(1+1/\epsilon)$-competitive algorithm BEQUI-EDF for the problem $B \mid$ *online-time*, $r_j, p_j = 1 \mid \sum F_j$. The idea of the algorithm is to simulate BEQUI to give a deadline for each request of the release time of that job plus some constant times the flow time of the job in BEQUI's schedule. BEQUI-EDF then runs the job with the Earliest Deadline First.

For the problem $B \mid$ *online-time*, $r_j, p_j = 1 \mid \sum F_j$, the most popular algorithm in the computer systems literature is Longest Wait First (LWF). LWF always services the page for which the aggregate waiting times of the outstanding requests for that page is maximized. In the natural setting where for each page, the request arrival times have a Poisson distribution, LWF broadcasts each page with frequency roughly proportional to the square root of the page's arrival rate, which is essentially optimal. Edmonds and Pruhs [44] provide an analysis of LWF. They show that LWF is 6-speed $O(1)$-competitive, but is not almost fully scalable. It is not too difficult to see that there is no possibility of proving such a result using local competitiveness. The rather complicated analysis given by Edmonds and Pruhs [44] compares the total cost of LWF to the total cost of the adversary.

The obvious interesting open question follows.

Open Problem 15.6

For the problems $B \mid$ online-time, pmtn, $r_j \mid \sum F_j$, $B \mid$ online-time-nclv, pmtn, $r_j \mid \sum F_j$, $B \mid$ online-time, $r_j, p_j = 1 \mid \sum F_j$, is there an almost fully scalable algorithm? For the problems $B \mid$ online-time, pmtn, $r_j \mid \sum F_j$, $B \mid$ online-time-nclv, pmtn, $r_j \mid \sum F_j$, one should consider both the version where the client has to receive the page in order, and the version where the client can receive the page out of order.

Bartal and Muthukrishnan [45] stated that FIFO is 2-competitive when the objective is minimizing maximum flow time under the assumption clients may receive documents out of order.

15.2.10 Nonclairvoyant Scheduling to Minimize Makespan

A general reduction theorem from [16] shows that in any variant of scheduling in *online-time-nclv* environment with makespan objective, any batch-style σ-competitive algorithm can be converted into a 2σ-competitive algorithm in a corresponding variant which in addition allows release times. In [46] it is proved that for a certain class of algorithms the competitive ratio is increased only by additive 1, instead of the factor of 2 in the previous reduction; this class of algorithms includes all algorithms that use a greedy approach similar to List Scheduling. The intuition beyond these reductions is that if the release times are fixed, the optimal algorithm cannot do much before the last release time. In fact, if the online algorithm would know which job is the last one, it could wait until its release, then use the batch-style algorithm once, and achieve the competitive ratio of $\sigma + 1$ easily. These reductions are completely satisfactory if we are interested only in the asymptotic behavior of the competitive ratio. However, if the competitive ratio is a constant, we may be interested in a tighter result.

In the basic model where the only characteristic of a job is the running time, there is not much we can do if we do not know it. For the basic problem $P \mid \textit{online-time-nclv} \mid C_{\max}$, no deterministic algorithm is better than $2 - 1/m$, i.e., better than List Scheduling, and randomization does not help much, as the lower bound is $(2 - O(1/\sqrt{m}))$ [16]. In Section 15.1.4 we mentioned that List Scheduling is $(2 - 1/m)$-competitive even for $P \mid \textit{online-time-nclv}, \textit{prec}, r_j \mid C_{\max}$. Hence we do not lose anything in the competitive ratio by allowing release times, unlike in the general reductions above.

15.2.10.1 Different Speeds

Here we consider both uniformly related machines and unrelated machines. In the case of related machines, the speed of each machine is the same for all jobs and given in advance. For unrelated machines, the speeds are different for each job, and we assume that the speed for each job on each machine is known when the job is released. Only the running time is not known (i.e., for each job we know the relative speeds of machines).

If no preemptions are allowed, even for uniformly related machines, $Qm \mid \textit{online-time-nclv} \mid C_{\max}$, a simple example shows that no algorithm is better than $\Omega(\sqrt{m})$-competitive [14]. A matching, $O(\sqrt{m})$-competitive, algorithm is known even for unrelated machines, $Rm \mid \textit{online-time-nclv} \mid C_{\max}$, see [14]. This is not very satisfactory, as a trivial greedy algorithm is m-competitive even for $Rm \mid \textit{online-time-nclv}, \textit{prec}, r_j \mid C_{\max}$, see [47].

However, for related machines, allowing preemptions or even only restarts helps significantly. In this case we can use a variant of a doubling method to convert an arbitrary offline algorithm into an online algorithm. Since we do not know the running times, we guess that all jobs have some chosen running time, then run the appropriate schedule. If any job is not finished in the guessed time, we stop it, double its estimate, and repeat the procedure for all such jobs. This method, together with additional improvements, yields an $O(\log m)$-competitive algorithm for uniformly related machines with restarts, $Qm \mid \textit{online-time-nclv}, \textit{pmtn-restart}, r_j \mid C_{\max}$ [16]. A matching lower bound shows that this is optimal even for $Qm \mid \textit{online-time-nclv}, \textit{pmtn} \mid C_{\max}$ [16].

15.2.10.2 Parallel Jobs

In this variant, each job is characterized by its running time and the number of identical machines (processors) it requests. This is denoted by $size_j$ in the middle field of the three-field notation. While the running times are unknown, the number of machines a job requests is known as soon as the job becomes available. We consider two variants according to how strict the request is. In the first, the jobs are nonmalleable, which means that they have to be scheduled on the requested number of machines. On the other hand, malleable jobs may be scheduled on fewer machines, at the cost of increasing the processing time. Most of the time we consider ideally malleable jobs. Using the terminology of speed-up curves, such jobs are fully parallelizable up to the requested number of machines. That is, scheduling on $q' < q$ machines takes time pq/q' instead of the original processing time p.

Consider the simplest greedy approach for batch-style algorithms: whenever there are sufficiently many machines idle, we schedule some job on as many machines as it requests. This leads to $(2-1/m)$-competitive algorithm, regardless of the rule by which we choose the job to be scheduled (note that here we have a meaningful choice, as we know how many machines each job requests), even with release times, i.e., for $P \mid online\text{-}time\text{-}nclv, size_j, r_j \mid C_{\max}$ [48]. This is optimal, as the basic model corresponds to the special case when each job requests only one machine. Moreover, this algorithm works even for nonmalleable jobs.

If we allow precedence constraints, $P \mid online\text{-}time\text{-}nclv, size_j, prec \mid C_{\max}$, no reasonable online algorithm exists for nonmalleable parallel jobs. For deterministic algorithms there is a lower bound of m on the competitive ratio (a trivial greedy algorithm matches this) [49], and for randomized algorithms there is a lower bound of $m/2$ [50].

In contrast, with ideally malleable jobs, $P \mid online\text{-}time\text{-}nclv, size_j, prec \mid C_{\max}$ allows a constant competitive ratio. The optimal competitive ratio for deterministic algorithms for $P \mid online\text{-}time\text{-}nclv, prec \mid C_{\max}$ is $1 + \phi \approx 2.6180$ [49]. The optimal strategy is again greedy, with the following rules for using malleability of the jobs: (i) If there is an available job requesting q machines and q machines are idle, schedule this job on q machines. (ii) Otherwise, if less than m/ϕ machines are busy and some job is available (requesting more machines), schedule it on all available machines. Note that this algorithm uses malleability only for large jobs. Accordingly, if there is an upper bound on the number of machines a job can use, we can get better algorithms and also algorithms for nonmalleable jobs. The tight trade-offs are given in [49].

In practice, it is perhaps not realistic to assume that any subset of machines can be used for a parallel job. A model which takes into account a particular network topology of the parallel machine was considered in [49,51,52] (without precedence constraints, with precedence constraints, and randomized algorithms, respectively). In this model, if the underlying network is, for example, a mesh (two-dimensional array), each job requests a rectangular subset of processors with given dimensions. Perhaps the most interesting results in this area concern the power of randomization. For $Pm \mid online\text{-}time\text{-}nclv, size_j \mid C_{\max}$ with the mesh restriction, no deterministic algorithm has a constant competitive ratio, the tight bound is $\Theta(\sqrt{\log\log m})$; on the other hand, there exists a randomized $O(1)$-competitive algorithm. This randomized algorithm is based on random sampling, and it is one of the first results showing the power of randomization in the area of online algorithms. In contrast, if we allow precedence constraints, there are lower bounds showing that randomization does not change the competitive ratio significantly. For a more complete survey of results for various topologies (linear array, hypercube, mesh), see [53].

15.3 Scheduling Jobs One by One

This paradigm corresponds most closely to the standard model of request sequences in competitive analysis. It can be formulated in the language of online load balancing as the case where the jobs are permanent and the load is their only parameter corresponding to our processing time. Consequently, there are many results on load balancing that extend the basic results on online scheduling in a different direction. As a reference in this area, we recommend the survey [54].

In this paradigm, we do not allow release times and precedence constraints, as these restrictions appear to be unnatural with scheduling jobs one by one. In most of the variants, it is also sufficient to assign each job to some machine(s) for some length of time, but it is not necessary to specify the actual time slot(s). In other words, it is not necessary or useful to introduce idle time on any machine. An exception is the area of preemptive scheduling on related machines where introducing idle times seems to be very useful.

We first give the results considering minimizing the makespan; only in Sections 15.3.8 and 15.3.9 we do briefly mention results for other objective functions, namely minimizing the l_p norm and the total completion time.

15.3.1 The Basic Model

We start by studying the basic parallel machine scheduling problem $P \mid online\text{-}list \mid C_{\max}$. This is probably the most extensively studied online scheduling problem, yet many questions remain open. In this section, we are interested in deterministic algorithms.

We have m machines and a sequence of jobs characterized by their processing times. The jobs are presented one by one, and we have to schedule each job to a single machine before we see the next one. There are no additional constraints, preemption is not allowed, all the machines have the same speed, and the objective function is the makespan.

The greedy List Scheduling algorithm schedules each arriving job on a least loaded machine. From Graham's analysis [11], it follows that the competitive ratio of List Scheduling is $2 - 1/m$. This is provably the best possible for $m = 2$ and $m = 3$ [55], but for larger m it is possible to develop better algorithms.

From the analysis of List Scheduling, it is clear what is the main issue in designing algorithms better than List Scheduling. If all machines have equal loads and a job with long processing time is presented, we may create a schedule which is almost twice as long as the optimal one. This is a problem if the scheduled jobs are sufficiently small, and the optimal schedule can distribute them evenly on $m - 1$ machines in parallel with the last long job on the remaining machine. Thus, to achieve better results, we have to create some imbalance and keep some machines lightly loaded in preparation for large jobs that have not yet arrived.

To design a good algorithm, current results use two different approaches. One is to schedule each job on one of the two currently least loaded machines [56,57]. This gives better results than List Scheduling for any $m \geq 4$, and achieves the currently best upper bounds for small m. However, for large m, the competitive ratio still approaches 2. The difficulty is that this approach only ensures that there is one lightly loaded machine. Thus, after many small jobs and two long jobs, we get a long schedule and the competitive ratio is at least $2 - 2/m$.

To keep the competitive ratio bounded away from 2 even for large m, it is necessary to keep some constant fraction of machines lightly loaded. Such an algorithm was first developed in [58]. Later better algorithms based on this idea were designed in [59–61] to give the currently best upper bounds for large m. The analysis of all these algorithms is relatively complicated. However, at a basic level, all of these algorithms use the following three lower bounds on the optimal makespan: (i) the total processing time of all jobs divided by m, (ii) the largest processing time of any job, and (iii) the sum of the mth largest and $(m + 1)$st largest processing times. Recently it has been shown that we cannot prove any deterministic algorithm has a better competitive ratio than 1.919 using only these lower bounds on the optimal makespan [62]. It has been conjectured that an improved algorithm and an improved analysis might by using additional lower bounds on the optimal makespan that include the size of the $(2m + 1)$st largest processing time, or in general the $(km + 1)$st processing time for $k \geq 2$. However, we still cannot prove any deterministic algorithm has a better competitive ratio than 1.917 using these additional lower bounds on the optimal makespan [62].

The lower bounds for this problem are typically proven by explicitly giving a hard sequence of jobs. The observation that List Scheduling is optimal for $m = 2, 3$ is due to [55]. For $m = 4$, the lower bound is $\sqrt{3} \approx 1.7321$, see [63,64]. The other lower bounds for small m are from [57]. The lower bounds for large m were gradually improved in [60,63,65].

The current state of our knowledge is summarized in Table 15.3. For comparison we include also the competitive ratio of List Scheduling. (See Section 15.3.2 for a discussion of results for randomized algorithms and Section 15.3.3 for preemptive scheduling.)

TABLE 15.3 Current Bounds for $Pm \mid$ *online-list* $\mid C_{\max}$ and $Pm \mid$ *online-list, pmtn* $\mid C_{\max}$

	Deterministic			Randomized		Preemptive
m	LS	Upper Bound	Lower Bound	Upper Bound	Lower Bound	Upper and Lower Bound
2	1.5000	1.5000	1.5000	1.3333	1.3333	1.3333
3	1.6666	1.6666	1.6666	1.5373	>1.4210	1.4210
4	1.7500	1.7333	1.7321	1.6567	1.4628	1.4628
5	1.8000	1.7708	1.7462	1.7338	1.4873	1.4873
6	1.8333	1.8000	1.7730	1.7829	1.5035	1.5035
7	1.8571	1.8229	1.7910	1.8168	1.5149	1.5149
∞	2.0000	1.9230	1.8800	1.9160	1.5819	1.5819

15.3.2 Randomized Algorithms

Much less is known about randomized algorithms for the basic model $P \mid online\text{-}list \mid C_{\max}$ studied in Section 15.3.1. We only know an optimal randomized algorithm for the case $m = 2$. A 4/3-competitive randomized algorithm and a matching lower bound for two machines, $P2 \mid online\text{-}list \mid C_{\max}$, was presented in [58].

First we show that this is best possible. Consider a sequence of three jobs with processing times 1, 1, and 2. After the first two jobs, the optimal makespan is 1, so the expected makespan of the online algorithm has to be at most 4/3. This means that after the first two jobs, the expected load of the less loaded machine is at least 2/3, and after the third job, even if it is always scheduled on the smaller machine, the expected makespan is at least $2/3 + 2 = 8/3$. Since the optimum is 2, the algorithm cannot be better that 4/3-competitive.

In the proof, we can replace the first two jobs by an arbitrary sequence of jobs with total processing time 2. Hence the proof actually shows that in any 4/3-competitive algorithm, the expected load of the more loaded machine has to be at least twice as much as the expected load of the other machine at all times. This has to be tight whenever we can partition the jobs into two sets with exactly the same sum of processing times. The most natural way to design an algorithm with this in mind is to keep the desired ratio of expected loads at all times. It turns out this works, with some additional considerations for large jobs [53,58].

The idea of the lower bound for two machines can be extended to an arbitrary number of machines [66,67]. This leads to a lower bound of $1/(1-(1-1/m)^m)$, which approaches $e/(e-1) \approx 1.5819$ for large m and increases with increasing m. This lower bound shows that for m machines, the expected loads should be in geometric sequence with the ratio $m : (m-1)$, if the machines are always ordered so that their loads are nondecreasing. (For example, for $m = 3$, the ratio of loads is $4 : 6 : 9$.) An algorithm based on this invariant would be a natural generalization of the optimal algorithm for two machines from [58]; it would also follow the suggestion from [68] (see Section 15.3.3). However, it is impossible to always maintain this ratio of expected loads. For three machines, $P3 \mid online\text{-}list \mid C_{\max}$, we know that this lower bound of 27/19 is not tight [69]. More precisely, we know that for some $\varepsilon > 0$, there is no $(27/19+\varepsilon)$-competitive algorithm, but the value of ε is very small and not explicit in [69].

New randomized algorithms for small m were developed in [70,71]. They are provably better than any deterministic algorithm for $m = 3, 4, 5$ and better than the currently best deterministic algorithm for $m = 6, 7$. They always assign the new job on one of the two least loaded machines, similar to the deterministic algorithms for small m from [56,57]. Consequently, the competitive ratio approaches two as m grows.

Another observation is that any randomized algorithm that never assigns jobs to the most loaded machine is at best 1.5-competitive. Consider a sequence of two jobs with processing time 1 and $m-1$ jobs with processing time 2. The first two jobs are assigned to two distinct machines due to the restriction of the algorithm. After the remaining jobs, the makespan is at least 3, while the optimum is 2.

Recently a new 1.916-competitive randomized algorithm for any number of machines, $P \mid online\text{-}list \mid C_{\max}$, was given in [62]. It is interesting that this algorithm as well as the algorithms for small m from [71] are *barely random*, i.e., need only a finite number of random bits (or different schedules) independent of the number of jobs. In contrast to this, the optimal algorithm for $m = 2$ from [58] needs to maintain an increasing collection of schedules; their number is linear in the number of jobs. Note also that 1.916 is better than the best possible deterministic competitive ratio provable using "standard" lower bounds on the optimal makespan.

To summarize, we have the optimal randomized algorithm for $m = 2$, a significant improvement over the deterministic algorithms for small m and a tiny improvement over the deterministic algorithms for large m. See Table 15.3.

15.3.3 Preemptive Scheduling

Next we consider the preemptive version of the problem, $P \mid online\text{-}list, pmtn \mid C_{\max}$. Each job may be assigned to one or more machines and time slots (the time slots have to be disjoint, of course), and this assignment has to be determined completely as soon as the job is presented. In this model the offline case is easily solved, and the optimal makespan is the maximum of the maximal processing time and the sum of the processing times divided by m (i.e., the average load of a machine), see Chapter 3.

It is easy to see that the lower bounds from Section 15.3.2 hold in this model, too, as they only use the arguments about expected load (with the exception of the improved bound for three machines). This again leads to a lower bound of $1/(1-(1-1/m)^m)$, which approaches $e/(e-1) \approx 1.5819$ for large m, valid even for randomized algorithms [68]. As it turns out, there exists a deterministic algorithm matching this lower bound. It essentially tries to preserve the invariant that the expected loads are in geometric sequence with the ratio $m:(m-1)$ with some special considerations for large jobs [68].

Thus, in this model, both deterministic and randomized cases are completely solved, giving the same bounds as the randomized lower bounds in Table 15.3. Moreover, we know that randomization does not help. This agrees with the intuition. In the basic model, randomization can serve us to spread the load of a job among more machines, but we still have the problem that the individual configurations cannot look exactly as we would like. With preemption, we can maintain the ideal configuration by spreading the loads as we wish among the m machines. Thus, preemption is more powerful than randomization.

15.3.4 Semionline Algorithms with Known Optimum and Doubling Strategies

Assuming that the algorithm knows the optimum value of the objective function is perhaps not realistic from a practical viewpoint. However, as the following theorem shows, such a semionline algorithm can be used as a building block for an online algorithm for the same problem. Instead of a known optimal makespan, we use an estimate and double it whenever it turns out that the estimate was too small.

Theorem 15.17

Suppose that for some scheduling problem in the online-list environment with the objective to minimize makespan there exists an R-competitive semionline algorithm if the optimum is known. Then for the same problem there exists both a deterministic online algorithm with competitive ratio $4R$ and a randomized online algorithm with competitive ratio $eR < 2.7183R$.

Proof

Let G_0 be the value of the optimal schedule considering only the first job of the sequence and let OPT be the optimal makespan on the whole instance. Let A_G denote the semionline algorithm provided with the information that G is the optimal makespan. First note that if $G \geq OPT$, then A_G always produces a schedule with makespan at most RG: the sequence can be appended with jobs that increase the makespan to exactly G and on this appended sequence the algorithm guarantees not to schedule any job after time RG.

The deterministic online algorithm computes G_0 and sets $G := G_0$ upon the arrival of the first job. Then it runs the algorithm A_G modified so that the jobs are scheduled in time interval $[RG, 2RG)$ instead of $[0, RG)$. If A_G fails to schedule the next job, the online algorithm sets $G := 2G$ and starts the algorithm A_G with the new value of G. The intervals in which the algorithm A_G schedules for different values of G are disjoint and thus the algorithm is well defined. The value of G can be increased only when $G < OPT$, by the property of the semionline algorithm mentioned above. Thus at the end of the algorithm we have $G \leq 2OPT$ and the final makespan is at most $2RG \leq 4R \cdot OPT$ and the algorithm is $4R$-competitive.

The randomized online algorithm computes G_0 and sets $G := G_0 e^z$, where z is a random variable uniformly distributed in $[0, 1)$ and e is the base of natural logarithms. Then it runs the algorithm A_G modified so that the jobs are scheduled in the time interval $[RG \cdot 1/(e-1), RG \cdot e/(e-1))$ instead of $[0, RG)$. If A_G fails to schedule the next job, the online algorithm sets $G := eG$ and starts the algorithm A_G with the new value of G. Again, the intervals in which the algorithm A_G schedules for different values of G are disjoint and the value of G can be increased only when $G < OPT$. Thus at the end of the algorithm G is at most $G' = G_0 \cdot e^{k+z}$ where k is the smallest integer such that this value is at least OPT. This is equivalent to saying that $x = \ln(G'/OPT)$ is the fractional part of $y = \ln(G_0/OPT) + k + z$. Since z is uniform in $[0, 1)$, k is an integer and $\ln(G_0/OPT)$ is a constant, x is also uniformly distributed in $[0, 1)$. The expected value of the final makespan is at most $\text{Exp}[RG' \cdot e/(e-1)] = \text{Exp}[Re^x \cdot OPT \cdot e/(e-1)] = \text{Exp}[e^x] \cdot R \cdot OPT \cdot e/(e-1) = eR \cdot OPT$ and the algorithm is eR-competitive. □

A doubling strategy similar to this theorem is a very common tool in computer science. In the area of online algorithms, it leads to optimal algorithms for search on a line (also known as cow-path problem) and its generalizations, both for deterministic and randomized algorithms, see [72–74]. In the context of online scheduling, it was used the first time in [16,75], see Section 15.2.10.1 and Section 15.3.5. In some cases, to get currently best results, this method may need some refinements, however, the basic idea of multiplying the estimate by a fixed constant as well as type of distribution used for the initial guess of a randomized algorithm is always the same.

If the optimum is known, the problem $P \mid$ *online-list* $\mid C_{\max}$ is also studied as so-called online bin-stretching. We know that the jobs fit into some number of bins of some height, and we ask how much we need to "stretch" the bins to fit the jobs online. For two machines, there exists a 4/3-competitive algorithm and this is tight. For more machines a 1.625-competitive algorithm is presented in [76]. Of course, in this case, doubling algorithms are not useful as other algorithms perform better.

For uniformly related machines nonpreemptive scheduling, $Q \mid$ *online-list* $\mid C_{\max}$, scheduling a job on the slowest machine that completes the job by the time equal to twice the optimal makespan is a 2-competitive semionline algorithm [75]. For preemptive scheduling on related machines, $Q \mid$ *online-list, pmtn* $\mid C_{\max}$, we can even produce an optimal schedule if the optimal makespan is known; a 1-competitive semionline algorithm is given in [77] for two machines and in [78] for any number of machines.

15.3.5 Different Speeds

For uniformly related machines, most results are based on the doubling strategy from Section 15.3.4 or its variants. For nonpreemptive scheduling, $Q \mid$ *online-list* $\mid C_{\max}$, a simple doubling strategy leads to a constant competitive ratio [75]. The competitive ratio can be improved by using more sophisticated analysis of doubling strategies. The current best algorithms are $3 + \sqrt{8} \approx 5.828$-competitive deterministic and 4.311-competitive randomized [79]. For an alternative very nice presentation see [80]. The lower bounds are 2.438 for deterministic algorithms [79] and 2 for randomized algorithms [81].

For uniformly related machines preemptive scheduling, $Q \mid$ *online-list, pmtn* $\mid C_{\max}$, we already mentioned that it is possible to design an optimal (1-competitive) semi-online algorithm if the optimal makespan is known in advance. Thus, by Theorem 15.17, this yields 4-competitive deterministic and 2.7183-competitive randomized algorithms [78]. The lower bound is 2 — both for deterministic and randomized algorithms [81].

For unrelated machines, $R \mid$ *online-list* $\mid C_{\max}$, it is possible to obtain $O(\log m)$-competitive deterministic algorithm [75,82]. A matching lower bound of $\Omega(\log m)$ holds both for deterministic and randomized algorithms even in the special case of the so-called restricted assignment, where each job specifies a set of machines on which it may be processed (it is processed infinitely slowly on the others) and besides this restriction all the machines have the same speed [83]. The lower bound also works for $R \mid$ *online-list, pmtn* $\mid C_{\max}$.

It is interesting that both for related and unrelated machines, the optimal algorithms are asymptotically better than List Scheduling. Here List Scheduling is modified so that the next job is always scheduled so that it will finish as early as possible (for the case of identical speed this is clearly equivalent to the more usual formulation that the next job is scheduled on the machine with the smallest load). For unrelated machines, $R \mid$ *online-list* $\mid C_{\max}$, the competitive ratio of List Scheduling is exactly m [75]. For related machines, $Q \mid$ *online-list* $\mid C_{\max}$ and $Q \mid$ *online-list, pmtn* $\mid C_{\max}$, the competitive ratio of List Scheduling is asymptotically $\Theta(\log m)$ [75,78,84] (the lower bound, the upper bound, and the preemptive case, respectively). The exact competitive ratio for $m = 2$ is ϕ and for $3 \le m \le 6$ it is equal to $1 + \sqrt{(m-1)/2}$ [84]; moreover for $m = 2, 3$ it can be checked easily that there is no better deterministic algorithm.

For two machines, $Q2 \mid$ *online-list* $\mid C_{\max}$ and $Q2 \mid$ *online-list, pmtn* $\mid C_{\max}$, we are able to analyze the situation further, depending on the speeds [85]. We first consider the nonpreemptive problem.

TABLE 15.4 Current Bounds for $Qm \mid online\text{-}list \mid C_{\max}$ and $Qm \mid online\text{-}list, pmtn \mid C_{\max}$

		Deterministic			Randomized	
Preemption	m	LS	Upper Bound	Lower Bound	Upper Bound	Lower Bound
Nonpreemptive	2	1.618	1.618	1.618	1.528	1.500
	∞	$\Theta(\log m)$	5.828	2.438	4.311	2.000
Preemptive	2	1.500	1.333	1.333	1.333	1.333
	∞	$\Theta(\log m)$	4.000	2.000	2.718	2.000

Suppose that the speeds of the two machines are 1 and $s \geq 1$. It is easy to see that List Scheduling is the best deterministic online algorithm for any choice of s. For $s \leq \phi$ the competitive ratio is $1+s/(s+1)$, increasing from 3/2 to ϕ. For $s \geq \phi$ the competitive ratio is $1 + 1/s$, decreasing from ϕ to 1; this is the same as for the algorithm which puts all the jobs on the faster machine. It turns out that this is also the best possible randomized algorithm for $s \geq 2$. On the other hand, for any $s < 2$, randomized algorithms are better than deterministic ones, and the overall upper bound is 1.5278. The competitive ratio of the optimal deterministic preemptive algorithm is better than the competitive ratio of the optimal nonpreemptive randomized algorithm for any $s \geq 1$. Furthermore, the worst case is the identical machine case when $s = 1$. In contrast, without preemption, the worst competitive ratio (both deterministic and randomized) is achieved for some $s > 1$ [85,86].

The current bounds for scheduling on uniformly related machines are summarized in Table 15.4.

15.3.6 Semionline Algorithms

In addition to algorithms that know the optimum which we discussed in Section 15.3.4, the most commonly studied semionline variant is the one where the jobs arrive sorted according to their processing times. In case of the makespan objective, the jobs are sorted largest first, i.e., by nonincreasing processing time, to improve the performance.

When the jobs are sorted, the greedy online algorithm List Scheduling becomes the so-called LPT (Largest Processing Time first) semionline algorithm. We already mentioned that for $P \mid online\text{-}list \mid C_{\max}$, the competitive ratio of LPT is $4/3 - 1/(3m)$, see [12]. For related machines the competitive ratio of LPT is a small constant, unlike List Scheduling which is only $\Theta(\log m)$-competitive. For $Q \mid online\text{-}list \mid C_{\max}$, the competitive ratio of LPT is between 1.52 and 1.66 [87]; a better upper bound of 1.58 is claimed in [88], but the proof appears to be incomplete. For $Q2 \mid online\text{-}list \mid C_{\max}$, the complete analysis of the dependence of the competitive ratio on the speed ratio was given in [89]. For $Q \mid online\text{-}list, pmtn \mid C_{\max}$, the competitive ratio of LPT is 2, see [78].

The semionline case of $P \mid online\text{-}list \mid C_{\max}$ and $P \mid online\text{-}list, pmtn \mid C_{\max}$ was further studied in [90]. It turns out that for $P2 \mid online\text{-}list \mid C_{\max}$, LPT is an optimal deterministic algorithm. For randomized algorithms a better competitive ratio of 8/7 is possible and optimal. For $P \mid online\text{-}list, pmtn \mid C_{\max}$, the optimal competitive ratio is $(1+\sqrt{3})/2 \approx 1.336$; this is surprisingly higher than the performance of LPT in the nonpreemptive case. The semionline case of $Q2 \mid online\text{-}list \mid C_{\max}$ and $Q2 \mid online\text{-}list, pmtn \mid C_{\max}$ was completely analyzed in [91,92].

15.3.7 Scheduling with Rejections

In this version, jobs may be rejected at a certain penalty. Each job is characterized by the processing time and the penalty. A job can either be rejected, in which case its penalty is paid, or scheduled on one of the machines, in which case its processing time contributes to the completion time of that machine (as usual). The objective is to minimize the makespan of the schedule for accepted jobs plus the sum of the penalties of all rejected jobs. Again, there are no additional constraints and all the machines have the same speed.

The main goal of an online algorithm is to choose the correct balance between the penalties of the rejected jobs and the increase in the makespan for the accepted jobs. At the beginning, it might have to reject some jobs if the penalty for their rejection is small compared to their processing time. However, at some point, it would have been better to schedule some of the previously rejected jobs since the increase in the makespan due to scheduling those jobs in parallel is less than the total penalty incurred.

We first look at deterministic algorithms in the case when preemption is not allowed [93]. At first it would seem that a good algorithm has to do well both in deciding which jobs to accept, and on which machines to schedule the accepted jobs. However, it turns out that after the right decision is made about rejections, it is sufficient to schedule the accepted jobs using List Scheduling. This is certainly surprising, as we know that without rejections, List Scheduling is not optimal. Thus, it is natural to expect that any algorithm for scheduling with rejections would benefit from using a better algorithm for scheduling the accepted jobs.

We can solve this problem optimally for $m = 2$ and for unbounded m; the competitive ratios are ϕ and $1 + \phi$, respectively. However, the best competitive ratio for fixed $m \geq 3$ is not known. It certainly tends to $1 + \phi$, which is the optimum for unbounded m, but the rate of convergence is not clear. While the upper bound is $1 + \phi - 1/m$ (i.e., the same rate of convergence as for List Scheduling), the lower bound is only $1 + \phi - 1/O(\log m)$.

The lower bounds for small m from [93] work also for preemptive deterministic algorithms, but for large m yield only a lower bound of 2. An improved algorithm for deterministic preemptive scheduling was designed in [94]. It achieves competitive ratio 2.3875 for all m. An interesting question is whether a better than 2-competitive algorithm can be found for $m = 3$: we know several different 2-competitive algorithms even without preemption, but the lower bound does not match this barrier.

Randomized algorithms for this problem, both with and without preemption, were designed in [94–96]. No algorithms better than the deterministic ones are known for large m. The lower bounds for randomized scheduling without rejection (Table 15.3) clearly apply here (set the penalties infinitely large), and no better lower bounds are known.

The results are summarized in Table 15.5. The deterministic lower bounds apply both for algorithms with and without preemption, with the exception of arbitrary m where the lower bound is only 2 with preemption.

15.3.8 Minimizing the l_p Norm

Here we minimize the l_p norm of the vector of the loads of machines, instead of the makespan, which is equivalent to the l_∞ norm. Of special interest is the Euclidean l_2 norm, the square root of the sum of squares of loads, which has a natural interpretation in load balancing [54,97]. For identical machines, a convexity argument implies that if all the machine loads are equal, the schedule is optimal. Thus, similar to measuring makespan, this performance measure quantifies how well we can approximate this ideal schedule; however note that a single overloaded machine has a much smaller effect on the l_p objective.

Minimizing the l_2 norm on identical machines was studied in [98]. List Scheduling is $\sqrt{4/3}$-competitive, and this is optimal. The performance of List Scheduling is not monotone in the number of machines. It is equal to $\sqrt{4/3}$ only for m divisible by 3; otherwise it is strictly better. More surprisingly, there exists an algorithm which is for sufficiently large m better than $\sqrt{4/3} - \delta$ for some $\delta > 0$. Since the lower bound

TABLE 15.5 Current Bounds for Algorithms Scheduling Jobs One by One with Possible Rejection

m	Deterministic Lower Bounds	Deterministic Upper Bounds		Randomized Upper Bounds	
		Nonpreemptive	Preemptive	Nonpreemptive	Preemptive
2	$\phi \approx 1.6180$	ϕ	ϕ	1.5000	1.5000
3	1.8392	2.0000	2.0000	1.8358	1.7774
4	1.9276	2.1514	2.0995	2.0544	2.0227
5	1.9660	2.2434	2.1581	2.1521	2.0941
∞	$1 + \phi \approx 2.6180$	$1 + \phi$	2.3875	—	—

of $\sqrt{4/3}$ holds for $m = 3$, this means that the optimal competitive ratio is also not monotone in m. This is perhaps the most interesting feature of these results: for the basic problem $Pm \mid \mid C_{\max}$ we often expect, based on the current results, that the competitive ratio will increase with the number of machines m; this intuition thus fails at least for a slightly different objective function. For a general p, the same approach leads also to an algorithm better than List Scheduling for large m.

For unrelated machines, [97] gives a simple greedy algorithm with a competitive ratio $1 + \sqrt{2}$ for the l_2 norm and $O(p)$ for a general l_p norm. In contrast to makespan, the competitive ratio is a constant that does not depend on the number of machines or jobs.

15.3.9 Minimizing the Total Completion Time

In this variant it is necessary to use idle times, as we have to finish the jobs with short processing times first to minimize the total completion time. Even on a single machine, $1 \mid \textit{online-list} \mid \sum C_j$, it is hard to design a good algorithm and the competitive ratio depends on the number of jobs logarithmically. More precisely, there exists a deterministic $(\log n)^{1+\varepsilon}$-competitive algorithm on a single machine without preemptions, but no $\log n$-competitive algorithm exists even if preemption is allowed [99].

15.3.10 Open Problems

Randomized algorithms. We still understand very little about the power of randomization in this online paradigm, despite some recent progress. In particular, for the basic problem $P \mid \textit{online-list} \mid C_{\max}$, the lower bound is 1.581 while the best algorithm is 1.916-competitive; this gap is quite large compared to the case of deterministic algorithms. It is reasonable to expect that improvements of the algorithm are more likely, but the lower bound of [69] for $P3 \mid \textit{online-list} \mid C_{\max}$ indicates some possibility of improving the lower bound as well.

Preemptive scheduling on related machines. In the offline case, $Q \mid \textit{pmtn} \mid C_{\max}$ we understand preemptive scheduling very well. The optimum is easy to calculate and the structure of optimal schedules is well understood [100,101]. In the online identical machine case, $P \mid \textit{online-list, pmtn} \mid C_{\max}$ we have a similar complete understanding of the optimal online algorithm. Despite some effort, the case of online scheduling on uniformly related machines, $Q \mid \textit{online-list, pmtn} \mid C_{\max}$ remains open. Our intuition is that, similarly as for $P \mid \textit{online-list, pmtn} \mid C_{\max}$, randomization should not help and thus the deterministic 4-competitive algorithm can be improved.

Acknowledgments

We are grateful to many colleagues for useful comments, pointers to the literature, and manuscripts. Without them this survey could not possibly cover as many results as it does.

References

[1] A. Borodin and R. El-Yaniv. *On-line Computation and Competitive Analysis*. Cambridge University Press, 1998.

[2] A. Fiat and G. J. Woeginger (editors). *On-line Algorithms: The State of the Art*. Springer, 1998.

[3] A. Karlin, M. Manasse, L. Rudolph, and D. Sleator. Competitive snoopy caching. *Algorithmica*, 3:79–119, 1988.

[4] D. Sleator and R. E. Tarjan. Amortized efficiency of list update and paging rules. *Communications of the ACM*, 28:202–208, 1985.

[5] R. Motwani, S. Phillips, and E. Torng. Non-clairvoyant scheduling. *Theoretical Computer Science*, 130:17–47, 1994.

[6] S. Ben-David, A. Borodin, R. M. Karp, G. Tardos, and A. Widgerson. On the power of randomization in on-line algorithms. In *Proc. 22nd Symp. Theory of Computing (STOC)*, pp. 379–386. ACM, 1990.

[7] L. Becchetti and S. Leonardi. Non-clairvoyant scheduling to minimize the average flow time on single and parallel machines. In *Proc. 33rd Symp. on Theory of Computing (STOC)*, pp. 94–103. ACM, 2001.

[8] B. Kalyanasundaram and K. Pruhs. Speed is as powerful as clairvoyance. *Journal of the ACM*, 47:214–221, 2000.

[9] C. Phillips, C. Stein, E. Torng, and J. Wein. Optimal time-critical scheduling via resource augmentation. *Algorithmica*, pp. 163–200, 2002.

[10] R. Càceres, F. Douglis, A. Feldmann, G. Glass, and M. Rabinovich. Web proxy caching: The devil is in the details. In *Proc. ACM SIGMETRICS Workshop on Internet Server Performance*, 1998.

[11] R. L. Graham. Bounds for certain multiprocessing anomalies. *Bell System Technical Journal*, 45:1563–1581, 1966.

[12] R. L. Graham. Bounds on multiprocessing timing anomalies. *SIAM Journal on Applied Mathematics*, 17:263–269, 1969.

[13] S. Sahni and Y. Cho. Nearly on line scheduling of a uniform processor system with release times. *SIAM Journal on Computing*, 8:275–285, 1979.

[14] E. Davis and J. M. Jaffe. Algorithms for scheduling tasks on unrelated processors. *Journal of the ACM*, 28:721–736, 1981.

[15] K. S. Hong and J. Y.-T. Leung. On-line scheduling of real-time tasks. *IEEE Transactions on Computing*, 41:1326–1331, 1992.

[16] D. B. Shmoys, J. Wein, and D. P. Williamson. Scheduling parallel machines on-line. *SIAM Journal on Computing*, 24:1313–1331, 1995.

[17] B. Chen, C. N. Potts, and G. J. Woeginger. A review of machine scheduling: Complexity, algorithms and approximability. In D.-Z. Du and P. M. Pardalos (editors), *Handbook of Combinatorial Optimization*, volume 3, pp. 21–169. Kluewer, 1998.

[18] S. Leonardi and D. Raz. Approximating total flow time on parallel machines. In *Proceedings of 29th Symposium on Theory of Computing (STOC)*, pp. 110–119. ACM, 1997.

[19] S. Leonardi. A simpler proof of preemptive flow-time approximation. In *Approximation and on-line Algorithms*, Lecture Notes in Computer Science. Springer, 2003.

[20] J. McCullough and E. Torng. SRPT optimally uses faster machines to minimize flow time. In *Proceedings of 15th Symposium on Discrete Algorithms (SODA)*. ACM/SIAM, 2004.

[21] S. Muthukrishnan, R. Rajaraman, A. Shaheen, and J. E. Gehrke. On-line scheduling to minimize average stretch. In *Proceedings of 40th Symposium on Foundations of Computer Science (FOCS)*, pp. 433–443. IEEE, 1999.

[22] B. Awerbuch, Y. Azar, S. Leonardi, and O. Regev. Minimizing the flow time without migration. *SIAM Journal on Computing*, 31:1370–1382, 2001.

[23] L. Becchetti, S. Leonardi, and S. Muthukrishnan. Scheduling to minimize average stretch without migration. In *Proceedings of 11th Symposium on Discrete Algorithms (SODA)*, pp. 548–557. ACM/SIAM, 2000.

[24] C. Chekuri, S. Khanna, and A. Zhu. Algorithms for weighted flow time. In *Proceedings of 33rd Symposium on Theory of Computing (STOC)*, pp. 84–93. ACM, 2001.

[25] N. Avrahami and Y. Azar. Minimizing total flow time and total completion time with immediate dispatching. In *Proceedings of 15th Symposium on Parallel Algorithms and Architectures (SPAA)*, pp. 11–18. ACM, 2003.

[26] C. Chekuri, S. Khanna, and A. Kumar. Multiprocessor scheduling to minimize l_p norms of flow and stretch. Manuscript, 2003.

[27] N. Bansal and K. Pruhs. Server scheduling in the L_p norm: A rising tide lifts all boats. In *Proceedings of 35th Symposium on Theory of Computing (STOC)*, pp. 242–250. ACM, 2003.

[28] M. A. Bender, S. Chakrabarti, and S. Muthukrishnan. Flow and stretch metrics for scheduling continuous job streams. In *Proceedings of 9th Symposium on Discrete Algorithms (SODA)*, pp. 270–279. ACM/SIAM, 1998.

[29] M. A. Bender, S. Muthukrishnan, and R. Rajaraman. Improved algorithms for stretch scheduling. In *Proceedings of 13th Symposium on Discrete Algorithms (SODA)*, pp. 762–771. ACM/SIAM, 2002.

[30] N. Bansal, K. Dhamdhere, J. Konemann, and A. Sinha. Non-clairvoyant scheduling for mean slowdown. In *Proceedings of 20th Symposium on Theoretical Aspects of Computer Science (STACS)*, volume 2607 of *Lecture Notes in Computer Science*, pp. 260–270. Springer, 2003.

[31] L. Becchetti, S. Leonardi, A. Marchetti-Spaccamela, and K. Pruhs. On-line weighted flow time and deadline scheduling. In *RANDOM-APPROX*, volume 2129 of *Lecture Notes in Computer Science*, pp. 36–47. Springer, 2001.

[32] N. Bansal and K. Dhamdhere. Minimizing weighted flow time. In *Proceedings of 14th Symposium on Discrete Algorithms (SODA)*, pp. 508–516. ACM/SIAM, 2003.

[33] N. Bansal and K. Pruhs. Server scheduling in the weighted l_p norm. Manuscript, 2003.

[34] L. Becchetti, S. Leonardi, A. Marchetti-Spaccamela, and K. Pruhs. Semiclairvoyant scheduling. In *Proceedings of 11th European Symposium on Algorithms (ESA)*, volume 2832 of *Lecture Notes in Computer Science*. Springer, 2003.

[35] T. Matsumoto. Competitive analysis of the Round Robin algorithm. In *Proceedings of 3rd International Symposium on Algorithms and Computation (ISAAC)*, volume 650 of *Lecture Notes in Computer Science*, pp. 71–77. Springer, 1992.

[36] P. Berman and C. Coulston. Speed is more powerful than clairvoyance. *Nordic Journal of Computing*, 6(2):181–193, 1999.

[37] B. Kalyanasundaram and K. Pruhs. Minimizing flow time nonclairvoyantly. *Journal of the ACM*, 50:551–567, 2003.

[38] B. Kalyanasundaram and K. Pruhs. Fault-tolerant scheduling. In *Proceedings of 26th Symposium on Theory of Computing (STOC)*, pp. 115–124. ACM, 1994.

[39] J. Edmonds. Scheduling in the dark. *Theoretical Computer Science*, 235:109–141, 2000.

[40] J. Edmonds, S. Datta, and P. W. Dymond. Tcp is competitive against a limited adversary. In *Proceedings of 15th Symposium on Parallel Algorithms and Architectures (SPAA)*, pp. 174–183. ACM, 2003.

[41] B. Kalyanasundaram, K. R. Pruhs, and M. Velauthapillai. Scheduling broadcasts in wireless networks. *Journal of Scheduling*, 4:339–354, 2001.

[42] J. Edmonds and K. Pruhs. Multicast pull scheduling: when fairness is fine. *Algorithmica*, 36:315–330, 2003.

[43] K. Pruhs and P. Uthaisombut. A comparison of multicast pull models. In *Proceedings of 10th European Symposium on Algorithms (ESA)*, volume 2461 of *Lecture Notes in Computer Science*, pp. 808–819. Springer, 2002.

[44] J. Edmonds and K. Pruhs. A maiden analysis of longest wait first. In *Proceedings of 15th Symposium on Discrete Algorithms (SODA)*. ACM/SIAM, 2004.

[45] Y. Bartal and S. Muthukrishnan. Minimizing maximum response time in scheduling broadcasts. In *Proceedings of 11th Symposium on Discrete Algorithms (SODA)*, pp. 558–559. ACM/SIAM, 2000.

[46] A. Feldmann, B. Maggs, J. Sgall, D. Sleator, and A. Tomkins. Competitive analysis of call admission algorithms that allow delay. Technical Report CMU-CS-95-102, Carnegie-Mellon University, 1995.

[47] L. Epstein. A note on on-line scheduling with precedence constraints on identical machines. *Information Processing Letters*, 76:149–153, 2000.

[48] E. Naroska and U. Schwiegelshohn. On an on-line scheduling problem for parallel jobs. *Information Processing Letters*, 81:297–304, 2002.

[49] A. Feldmann, M.-Y. Kao, J. Sgall, and S.-H. Teng. Optimal on-line scheduling of parallel jobs with dependencies. *Journal of Combinatorial Optimization*, 1:393–411, 1998.

[50] J. Sgall. *On-line Scheduling on Parallel Machines*. PhD thesis, Technical Report CMU-CS-94-144, Carnegie-Mellon University, Pittsburgh, PA, 1994.

[51] A. Feldmann, J. Sgall, and S.-H. Teng. Dynamic scheduling on parallel machines. *Theoretical Computer Science*, 130:49–72, 1994.

[52] J. Sgall. Randomized on-line scheduling of parallel jobs. *Journal of Algorithms*, 21:149–175, 1996.

[53] J. Sgall. on-line scheduling. In A. Fiat and G. J. Woeginger (editors), *On-line Algorithms: The State of the Art*, pp. 196–231. Springer, 1998.

[54] Y. Azar. On-line load balancing. In A. Fiat and G. J. Woeginger (editors), *On-line Algorithms: The State of the Art*, pp. 178–195. Springer, New York, 1998.

[55] U. Faigle, W. Kern, and G. Turán. On the performane of on-line algorithms for partition problems. *Acta Cybernetica*, 9:107–119, 1989.

[56] G. Galambos and G. J. Woeginger. An on-line scheduling heuristic with better worst case ratio than Graham's list scheduling. *SIAM Journal on Computing*, 22:349–355, 1993.

[57] B. Chen, A. van Vliet, and G. J. Woeginger. New lower and upper bounds for on-line scheduling. *Operations Research Letters*, 16:221–230, 1994.

[58] Y. Bartal, A. Fiat, H. Karloff, and R. Vohra. New algorithms for an ancient scheduling problem. *Journal Computer Systems Science*, 51:359–366, 1995.

[59] D. R. Karger, S. J. Phillips, and E. Torng. A better algorithm for an ancient scheduling problem. *Journal of Algorithms*, 20:400–430, 1996.

[60] S. Albers. Better bounds for on-line scheduling. *SIAM Journal on Computing*, 29:459–473, 1999.

[61] R. Fleischer and M. Wahl. On-line scheduling revisited. *Journal of Scheduling*, 3:343–353, 2000.

[62] S. Albers. On randomized on-line scheduling. In *Proceedings of 34th Symposium on Theory of Computing (STOC)*, pp. 134–143. ACM, 2002.

[63] J. F. Rudin III. *Improved Bound for the On-line Scheduling Problem*. PhD thesis, The University of Texas at Dallas, 2001.

[64] J. F. Rudin III and R. Chandrasekaran. Improved bound for the on-line scheduling problem. *SIAM Journal on Computing*, 32:717–735, 2003.

[65] Y. Bartal, H. Karloff, and Y. Rabani. A better lower bound for on-line scheduling. *Information Processing Letters*, 50:113–116, 1994.

[66] B. Chen, A. van Vliet, and G. J. Woeginger. Lower bounds for randomized on-line scheduling. *Information Processing Letters*, 51:219–222, 1994.

[67] J. Sgall. A lower bound for randomized on-line multiprocessor scheduling. *Information Processing Letters*, 63:51–55, 1997.

[68] B. Chen, A. van Vliet, and G. J. Woeginger. An optimal algorithm for preemptive on-line scheduling. *Operations Research Letters*, 18:127–131, 1995.

[69] T. Tichý. Randomized on-line scheduling on 3 processors. *Operations Research Letters*, 32:152–158, 2002.

[70] S. S. Seiden. Randomized on-line multiprocessor scheduling. *Algorithmica*, 28:173–216, 2000.

[71] S. Seiden. Barely random algorithms for multiprocessor scheduling. *Journal of Scheduling*, 6:309–334, 2003.

[72] R. Baeza-Yates, J. Culberson, and G. Rawlins. Searching in the plane. *Information and Computation*, 106:234–252, 1993.

[73] M.-Y. Kao, J. H. Reif, and S. R. Tate. Searching in an unknown environment: An optimal randomized algorithm for the cow-path problem. *Information and Computation*, 131:63–79, 1996.

[74] M.-Y. Kao, Y. Ma, M. Sipser, and Y. Yin. Optimal constructions of hybrid algorithms. *Journal of Algorithms*, 29:142–164, 1998.

[75] J. Aspnes, Y. Azar, A. Fiat, S. Plotkin, and O. Waarts. On-line load balancing with applications to machine scheduling and virtual circuit routing. *Journal of the ACM*, 44:486–504, 1997.

[76] Y. Azar and O. Regev. On-line bin stretching. *Theoretical Computer Science*, 268:17–41, 2001.

[77] L. Epstein. Bin stretching revisited. *Acta Informatica*, 39:97–117, 2003.

[78] T. Ebenlendr and J. Sgall. Optimal and on-line preemptive scheduling on uniformly related machines. To appear in *Proc. 21st Symp. on Theoretical Aspects of Computer Science (STACS)*, Lecture Notes in Computer Science, Springer-Verlag, 2004.

[79] P. Berman, M. Charikar, and M. Karpinski. On-line load balancing for related machines. *Journal of Algorithms*, 35:108–121, 2000.

[80] A. Bar-Noy, A. Freund, and J. Naor. New algorithms for related machines with temporary jobs. *Journal of Scheduling*, 3:259–272, 2000.

[81] L. Epstein and J. Sgall. A lower bound for on-line scheduling on uniformly related machines. *Operations Research Letters*, 26(1):17–22, 2000.

[82] S. Leonardi and A. Marchetti-Spaccamela. On-line resource management with applications to routing and scheduling. *Algorithmica*, 24:29–49, 1999.

[83] Y. Azar, J. Naor, and R. Rom. The competitiveness of on-line assignments. *Journal of Algorithms*, 18:221–237, 1995.

[84] Y. Cho and S. Sahni. Bounds for list schedules on uniform processors. *SIAM Journal on Computing*, 9:91–103, 1980.

[85] L. Epstein, J. Noga, S. S. Seiden, J. Sgall, and G. J. Woeginger. Randomized on-line scheduling for two related machines. *Journal of Scheduling*, 4:71–92, 2001.

[86] J. Wen and D. Du. Preemptive on-line scheduling for two uniform processors. *Operations Research Letters*, 23:113–116, 1998.

[87] D. K. Friesen. Tighter bounds for LPT scheduling on uniform processors. *SIAM Journal on Computing*, 16:554–560, 1987.

[88] G. Dobson. Scheduling independent tasks on uniform processors. *SIAM Journal on Computing*, 13:705–716, 1984.

[89] P. Mireault, J. B. Orlin, and R. V. Vohra. A parametric worst case analysis of the LPT heuristic for two uniform machines. *Operations Research*, 45:116–125, 1997.

[90] S. Seiden, J. Sgall, and G. J. Woeginger. Semi-on-line scheduling with decreasing job sizes. *Operations Research Letters*, 27:215–221, 2000.

[91] L. Epstein and L. M. Favrholdt. Optimal non-preemptive semi-on-line scheduling on two related machines. In *Proceedings of 27th Symposium on Mathematical Foundations of Computer Science (MFCS)*, volume 2420 of *Lecture Notes in Computer Science*, pp. 245–256. Springer, 2002.

[92] L. Epstein and L. M. Favrholdt. Optimal preemptive semi-on-line scheduling to minimize makespan on two related machines. *Operations Research Letters*, 30:269–275, 2002.

[93] Y. Bartal, S. Leonardi, A. Marchetti-Spaccamela, J. Sgall, and L. Stougie. Multiprocessor scheduling with rejection. *SIAM Journal on Discrete Mathematics*, 13:64–78, 2000.

[94] S. S. Seiden. Preemptive multiprocessor scheduling with rejection. *Theoretical Computer Science*, 262:437–458, 2001.

[95] S. S. Seiden. *Randomization in On-line Computation*. PhD thesis, University of California, Irvine, 1997.

[96] S. S. Seiden. More multiprocessor scheduling with rejection. Technical Report Woe-16, TU-Graz, 1997.

[97] B. Awerbuch, Y. Azar, E. F. Grove, M.-Y. Kao, P. Krishnan, and J. S. Vitter. Load balancing in the l_p norm. In *Proceedings of 36th Symposium Foundations of Computer Science (FOCS)*, pp. 383–391. IEEE, 1995.

[98] A. Avidor, Y. Azar, and J. Sgall. Ancient and new algorithms for load balancing in the l_p norm. *Algorithmica*. 29(3):422–441, 2001.

[99] A. Fiat and G. J. Woeginger. On-line scheduling on a single machine: Minimizing the total completion time. *Acta Informatica*, 36:287–293, 1999.

[100] E. Horwath, E. C. Lam, and R. Sethi. A level algorithm for preemptive scheduling. *Journal of the ACM*, 24:32–43, 1977.

[101] T. F. Gonzales and S. Sahni. Preemptive scheduling of uniform processor systems. *Journal of the ACM*, 25:92–101, 1978.

16

Convex Quadratic Relaxations in Scheduling

Jay Sethuraman
Columbia University

16.1 Introduction

Scheduling problems have been at the heart of several important developments in the fields of combinatorial optimization, approximation algorithms, and complexity theory. Over the last four decades, research on scheduling theory has resulted in new tools and techniques that have been useful in larger contexts. At the same time, advances in these areas have resulted in an improved understanding of scheduling problems. The theory of NP-completeness helped distinguish relatively easy scheduling problems from difficult ones. Attempts to find reasonable solutions to these difficult problems gave birth to the field of approximation algorithms, which focuses on finding efficient algorithms for NP-hard (or harder) problems with a provable performance guarantee. While the earliest approximation algorithms were combinatorial in nature, the last fifteen years have resulted in an improved understanding of the power of mathematical programming methods in designing approximation algorithms.

Linear programming (LP) has proved to be an invaluable tool in the design and analysis of approximation algorithms. The typical steps in designing an LP-based approximation algorithm are as follows: first, we formulate the problem under consideration as an integer programming problem; second, we solve the underlying linear programming relaxation of the problem to obtain an optimal fractional solution, whose cost is obviously a lower bound on the achievable optimal cost; we then "round" the fractional solution to a feasible integer solution in a systematic way; and finally, we bound the "error" incurred in this rounding process. LP-based methods have been used to design improved approximation algorithms for an impressive array of problems arising in diverse settings. Moreover, these techniques have resulted in a significantly better understanding of the problem itself, often leading to a better combinatorial approximation algorithm. The attractiveness of LP as a tool in designing (approximation) algorithms is due to several factors: LPs can be solved efficiently, both in theory and practice; they admit an elegant duality theory,

1-58488-397-9/$0.00+$1.50

which is often the key to analyzing the proposed (approximation) algorithm; and they arise naturally as relaxations of many optimization problems.

Given the success of LP-based methods, it seems natural to ask if more general mathematical programming methods can be used to design approximation algorithms for difficult optimization problems. After all, convex programming problems admit efficient algorithms (in theory). Moreover, semidefinite programming methods were used by Lovász [1] to find the Shannon capacity of a graph and Grötschel et al. [2] to design a polynomial-time algorithm for finding a maximum independent set in a perfect graph. However, the power of convex (specifically, semidefinite) programming methods was illustrated by Goemans and Williamson [3] in their pioneering work on the MAXCUT problem. Since then, researchers have investigated the use of semidefinite programming in various problems such as graph coloring [4] and betweenness [5], and graph bisection [6].

Scope and Organization. Our brief survey chapter is on the role of convex relaxations in scheduling. Specifically, we consider the problem of scheduling unrelated machines with the objective of minimizing weighted completion time, and several natural variants of this problem. We show that this problem leads to a natural integer convex programming problem. Relaxing the integrality constraints leads to a convex programming problem, which can be solved efficiently. We describe a natural rounding algorithm that converts this "fractional" schedule into an "integral" one, and analyze its performance.

Credits. Since this chapter is of an expository nature, we strive to keep the bibliographic references in the text to a minimum. The first convex relaxations for scheduling unrelated machines in the absence of release dates was proposed and analyzed independently by Skutella [7] and Sethuraman and Squillante [8,9]; this is discussed in Section 16.2.1. Skutella [7] further presented improved results for the two machine case based on a semidefinite programming relaxation (not discussed in this article). In a subsequent paper, Skutella [12] considered the problem of scheduling unrelated machines in the presence of machine-dependent release dates: this is discussed in Section 16.3.1. The preemptive versions of these problems were treated by Skutella [10]; these ideas are discussed in Section 16.2.2 and Section 16.3.1 respectively. The basic problem with cardinality constraints was considered by Ageev and Sviridenko [11] and forms the subject of Section 16.4. The presentation here draws heavily from Skutella [12], which summarizes his work on the problem.

16.2 Scheduling Unrelated Machines

We consider various versions of the problem of scheduling a set of n jobs on m unrelated machines. Job j can be processed on machine i at or after its (machine-dependent) release date r_{ij}, and requires p_{ij} time units. We assume that the r_{ij} and p_{ij} are positive integers. A job can be processed by at most one machine at any time, and each machine can process at most one job at a time. If preemption is not allowed, any job once started must be run to completion; if preemption is allowed, jobs may be interrupted during their execution, and resumed later on any machine. The objective is to find a schedule that minimizes the sum of weighted completion times, where job j is assumed to have a non-negative weight w_j. Since this problem is NP-hard (except for some trivial special cases), our goal is to find a near-optimal schedule with a provable approximation guarantee. Following the standard classification scheme of Graham et al. [13], the problems we consider are $R|r_{ij}|\sum w_j C_j$ and $R|r_{ij}, \text{pmtn}\,|\sum w_j C_j$.

The parallel-machine scheduling problem just described involves two decisions: first, the "assignment" (or "routing") decision for each job; and second, the sequencing decision for each machine given the set of jobs it processes. This viewpoint naturally leads to a nonlinear programming formulation on binary variables, which can be reformulated as a convex integer program. Relaxations of this latter problem form the basis of the approximation algorithms discussed here.

The rest of this section is organized as follows: we begin with the problem of scheduling unrelated machines without release dates, the simplest (and chronologically the first) problem on which many of the key ideas are illustrated. We then discuss the case in which preemption is allowed.

16.2.1 $R || \sum w_j C_j$

Recall that we view the scheduling problem as a two-stage problem of first assigning jobs to machines optimally, and then optimally sequencing the jobs in each machine. This latter problem is a $1 || \sum w_j C_j$ scheduling problem, which can be solved optimally using Smith's rule [14]: the jobs assigned to machine i are scheduled in decreasing order of w_j/p_{ij}, with ties broken arbitrarily. For convenience, we assume that machine i has an ordering $<_i$ of all the jobs such that $j <_i k$ if $w_j/p_{ij} > w_k/p_{ik}$. (Thus, ties are resolved consistent with the ordering $<_i$.) This observation reduces the scheduling problem to the assignment problem, which we formulate using the indicator variables x_{ij}, representing whether or not job j is assigned to machine i. The nonlinear integer program we thus obtain is

$$\text{IQP} \qquad \text{Min} \sum_j w_j \left\{ \sum_i x_{ij} \left(p_{ij} + \sum_{k:k<_i j} x_{ik} p_{ik} \right) \right\}$$

subject to:

$$\sum_i x_{ij} = 1, \quad \forall j$$

$$x_{ij} \in \{0, 1\}, \quad \forall i, j$$

The two sets of constraints ensure that each job is assigned to exactly one machine. Thus, of the m terms in the expression within the braces in the objective function, exactly one — the one corresponding to the machine to which j is assigned — will be nonzero, and this nonzero expression is exactly the sum of job j's processing time and the processing times of all higher priority jobs running on the same machine as j. The correctness of the formulation IQP is immediate.

The formulation using assignment variables suggests a natural decomposition of the given parallel machine scheduling problem into a set of single machine scheduling problems. Therefore, it is convenient to rewrite the objective function of IQP from a "machine" point of view. Specifically, by interchanging the first two summations, the objective function can be rewritten as $\sum_i \Phi(i)$, where

$$\Phi(i) \equiv \sum_j w_j x_{ij} \left(p_{ij} + \sum_{k:k<_i j} x_{ik} p_{ik} \right)$$

is the cost incurred by machine i. Note that $\Phi(i)$ is a function of the n variables $x_{i1}, x_{i2}, \ldots, x_{in}$. To check if it is convex, we compute its Hessian $D(i) = (d_{jk})$, which is an $n \times n$ matrix. Relabel the jobs, if necessary, so that $1 <_i 2 <_i, \ldots, <_i n$. Then, d_{jk} is exactly the coeficient of the term $x_{ij} x_{ik}$ in the expression for $\Phi(i)$; this is $w_k p_{ij}$, if $j < k$, $w_j p_{ik}$, if $j > k$, but zero, if $j = k$. Thus,

$$D(i) = \begin{bmatrix} 0 & w_2 p_{i1} & w_3 p_{i1} & \cdots & w_{n-1} p_{i1} & w_n p_{i1} \\ w_2 p_{i1} & 0 & w_3 p_{i2} & \cdots & w_{n-1} p_{i2} & w_n p_{i2} \\ w_3 p_{i1} & w_3 p_{i2} & 0 & \cdots & w_{n-1} p_{i3} & w_n p_{i3} \\ \vdots & \vdots & \vdots & \ddots & \vdots & \vdots \\ w_{n-1} p_{i1} & w_{n-1} p_{i2} & w_{n-1} p_{i3} & \cdots & 0 & w_n p_{in-1} \\ w_n p_{i1} & w_n p_{i2} & w_n p_{i3} & \cdots & w_n p_{in-1} & 0 \end{bmatrix} \tag{16.1}$$

Moreover, letting $x(i) = (x_{i1}, x_{i2}, \ldots, x_{in})$ and $c(i) = (w_1 p_{i1}, w_2 p_{i2}, \ldots, w_n p_{in})$, we can rewrite $\Phi(i)$ as

$$\Phi(i) = c(i)^T x(i) + \frac{1}{2} x(i)^T D(i) x(i) \tag{16.2}$$

Clearly, $\Phi(i)$ is convex if and only if its Hessian $D(i)$, shown in Equation (16.1), is positive semidefinite. Recall that a matrix is positive semidefinite if and only if all its principal minors have nonnegative determinants. As the submatrix

$$\begin{bmatrix} 0 & w_2 p_{i1} \\ w_2 p_{i1} & 0 \end{bmatrix}$$

has a negative determinant (when $w_2, p_{i1} > 0$), $D(i)$ is not necessarily positive semidefinite. Even though $D(i)$ fails to be positive semidefinite, its form begs for considering the related matrix

$$\hat{D}(i) = \begin{bmatrix} w_1 p_{i1} & w_2 p_{i1} & w_3 p_{i1} & \cdots & w_{n-1} p_{i1} & w_n p_{i1} \\ w_2 p_{i1} & w_2 p_{i2} & w_3 p_{i2} & \cdots & w_{n-1} p_{i2} & w_n p_{i2} \\ w_3 p_{i1} & w_3 p_{i2} & w_3 p_{i3} & \cdots & w_{n-1} p_{i3} & w_n p_{i3} \\ \vdots & \vdots & \vdots & \ddots & \vdots & \vdots \\ w_{n-1} p_{i1} & w_{n-1} p_{i2} & w_{n-1} p_{i3} & \cdots & w_{n-1} p_{in-1} & w_n p_{in-1} \\ w_n p_{i1} & w_n p_{i2} & w_n p_{i3} & \cdots & w_n p_{in-1} & w_n p_{in} \end{bmatrix} \tag{16.3}$$

obtained by adding the terms that are "obviously" missing from the diagonal.

Lemma 16.1

The matrix $\hat{D}(i)$ is positive semidefinite.

Proof

We show that every principal submatrix has a nonnegative determinant. Consider the rth principal submatrix. Divide the (j, k)th entry by $p_{ij} p_{ik}$ to get the matrix

$$\begin{bmatrix} w_1/p_{i1} & w_2/p_{i2} & \cdots & w_r/p_{ir} \\ w_2/p_{i2} & w_2/p_{i2} & \cdots & w_r/p_{ir} \\ \vdots & \vdots & \ddots & \vdots \\ w_r/p_{ir} & w_r/p_{ir} & \cdots & w_r/p_{ir} \end{bmatrix}$$

Subtracting the second row from the first, the third row from the second, etc., we obtain a lower triangular matrix, with the jth diagonal entry being $w_j/p_{ij} - w_{j+1}/p_{i,j+1}$, which, by our labeling of the jobs, is nonnegative. The lemma follows by noting that all the operations transforming $\hat{D}(i)$ to the lower triangular matrix preserve the sign of the determinant. □

Lemma 16.1 motivates us to consider working with $\hat{D}(i)$ instead of $D(i)$, so we write $\Phi(i)$ as

$$\Phi(i) = c(i)^T x(i) + \frac{1}{2} x(i)^T \hat{D}(i) x(i) - \frac{1}{2} \sum_{j=1}^{n} w_j p_{ij} x_{ij}^2 \tag{16.4}$$

As $x_{ij} \in \{0, 1\}$, $x_{ij}^2 = x_{ij}$, so $\sum_{j=1}^{n} w_j p_{ij} x_{ij}^2 = \sum_{j=1}^{n} w_j p_{ij} x_{ij} = c(i)^T x(i)$, yielding

$$\Phi(i) = \frac{1}{2}\left(c(i)^T x(i) + \frac{1}{2} x(i)^T \hat{D}(i) x(i) \right) \tag{16.5}$$

This discussion leads to a reformulation of the scheduling problem as the following convex integer programming problem:

$$(ICQP) \qquad \text{Min} \sum_i \frac{1}{2}(c(i)^T x(i) + x(i)^T \hat{D}(i) x(i))$$

$$\text{subject to:}$$

$$\sum_i x_{ij} = 1, \quad \forall j$$

$$x_{ij} \in \{0, 1\}, \quad \forall i, j$$

We thus obtain two nonlinear integer programming formulations of the scheduling problem: the formulation IQP has a nonconvex objective, whereas the formulation ICQP has a convex objective. Of course, these integer programming problems are hard to solve, so it is natural to investigate their (fractional) relaxations.

First we consider the formulation (IQP), whose fractional relaxation is

$$\text{(QP)} \qquad \text{Min} \sum_j w_j \left\{ \sum_i x_{ij} \left(p_{ij} + \sum_{k:k<_i j} x_{ik} p_{ik} \right) \right\}$$

subject to:

$$\sum_i x_{ij} = 1, \quad \forall j$$

$$x_{ij} \geq 0, \quad \forall i, j$$

Let x^* be an optimal solution to this problem. Our goal is to now convert this fractional assignment to an integral one. Consider the (random) assignment in which job j is sent to machine i with probability x^*_{ij}, independent of all other jobs. Since $\sum_i x^*_{ij} = 1$, every job is assigned to some machine. Moreover, the expected completion time of job j is exactly

$$\sum_i x^*_{ij} \left(p_{ij} + \sum_{k:k<_i j} x^*_{ik} p_{ik} \right)$$

Therefore, the expected cost of this random schedule is exactly the cost of the fractional solution x^*. Standard techniques can be used to derandomize this scheme, and thus to find an optimal schedule. Thus, finding an optimal (fractional) solution to QP is no easier than solving the scheduling problem $R \,||\, \sum_j w_j C_j$. In other words, given an optimal (possibly fractional) solution to QP, it is easy to find an optimal integer solution; however, finding an optimal solution to QP is difficult.

We next turn to the formulation ICQP and its fractional relaxation

$$\text{(CQP)} \qquad \text{Min} \sum_i \frac{1}{2} [c(i)^T x(i) + x(i)^T \hat{D}(i) x(i)]$$

subject to:

$$\sum_i x_{ij} = 1, \quad \forall j$$

$$x_{ij} \geq 0, \quad \forall i, j$$

Recall that ICQP was obtained from IQP by judiciously replacing some of the x_{ij} by x^2_{ij}. This transformation is valid if $x_{ij} \in \{0, 1\}$, but not if $x_{ij} \in (0, 1)$. In the latter case, $x^2_{ij} < x_{ij}$, so the objective function of CQP underestimates the true optimal cost. In this sense, CQP is a true relaxation of the problem $R \,||\, \sum_j w_j C_j$. Fortunately, CQP is a simple complex quadratic programming problem subject to linear constraints, and so can be solved efficiently in polynomial time. Thus, the situation here is in sharp contrast to that we had earlier: the fractional relaxation CQP is easy to solve, but there could be fractional solutions with cost strictly below that of an optimal integer solution. The task now is to somehow recover a good schedule from an optimal solution to CQP. The randomized assignment scheme described earlier is a natural candidate, and we analyze this next.

Let x^* be an optimal solution to CQP, with $\Phi^*(i)$ representing the cost incurred by machine i in this relaxation. Consider the (random) schedule obtained by assigning, as before, job j to machine i with probability x^*_{ij}, independent of all other jobs. The expected cost incurred by machine i in the schedule thus obtained, $E[\Phi(i)]$, is

$$\sum_j w_j x^*_{ij} \left(p_{ij} + \sum_{k:k<_i j} x^*_{ik} p_{ik} \right)$$

which, by Equation (16.4), is

$$c(i)^T x^*(i) + \frac{1}{2} x^*(i)^T \hat{D}(i) x(i) - \frac{1}{2} \sum_{j=1}^{n} w_j p_{ij} (x^*_{ij})^2$$

Simplifying this expression, we have

$$\begin{aligned} E[\Phi(i)] &\le \frac{1}{2} c(i)^T x^*(i) + \frac{1}{2} x^*(i)^T \hat{D}(i) x(i) + \frac{1}{2} c(i)^T x^*(i) \\ &= \Phi^*(i) + \frac{1}{2} c(i)^T x^*(i) \qquad (16.6) \\ &\le 2\Phi^*(i) \qquad (16.7) \end{aligned}$$

where expression (16.7) follows expression (16.6) because $c(i)^T x^*(i)/2$ is a lower bound on $\Phi^*(i)$.

We have thus shown that the expected cost incurred by any machine i is at most twice its cost in the (fractional) relaxation CQP. Can this be improved? One possibility is to use the

$$\Phi^*(i) \ge c(i)^T x^*(i)$$

instead of the weaker

$$\Phi^*(i) \ge \frac{1}{2} c(i)^T x^*(i)$$

After all, $\Phi^*(i)$ is a lower bound on the cost incurred by machine i, and it seems natural to expect it to be at least as much as the sum of the weighted *processing times* of the jobs assigned to it. This may not be true because we are dealing with a fractional relaxation, which allows a job to be split across multiple machines. For instance, consider a single job with $w = 1$ and $p_i = 1$ on two machines, the optimal schedule splits this job into two pieces, each piece processed by a different machine. It is easy to check that $\Phi^*(1) = \Phi^*(2) = 3/8$, whereas the total weighted processing time on each of the machines is $1/2 > 3/8$. Notice, however, that $\Phi(i)$ must be at least $c(i)^T x(i)$ for all (integral) feasible schedules of the original problem, so we could explicitly enforce this inequality as a constraint. We are thus led to the following stronger formulation of the problem $R \,||\, \sum w_j C_j$:

$$\begin{aligned} \text{(SCQP)} \qquad & \text{Min} \sum_i \Phi(i) \\ & \text{subject to:} \\ & \Phi(i) \ge \frac{1}{2}\left(c(i)^T x(i) + x(i)^T \hat{D}(i) x(i)\right), \quad \forall i \\ & \Phi(i) \ge c(i)^T x(i), \quad \forall i \\ & \sum_i x_{ij} = 1, \quad \forall j \\ & x_{ij} \ge 0, \quad \forall i, j \end{aligned}$$

Given an optimal solution to SCQP, we can again interpret the x^*_{ij} as assignment probabilities. Assigning job j to machine i with probability x^*_{ij}, and sequencing the set of jobs assigned to any particular machine i using Smith's rule, we obtain a (random) schedule in which the expected cost of machine i is within 3/2 of its cost in the relaxation SCQP (see Expression (16.6) and Expression (16.7)). Since this is true for every machine, the total expected cost of the schedule obtained is at most 3/2 times the cost of the relaxed problem SCQP. Thus, we have a randomized 3/2-approximation algorithm, assuming we can find an optimal solution to SCQP.

The formulation SCQP is no longer a simple quadratic program, but is still a convex programming problem, for which a solution within an additive error of any $\epsilon > 0$ can be obtained in polynomial-time.

This, when combined with the randomized rounding scheme, gives us a randomized $3/2+\epsilon$ approximation algorithm. As we deal with integer data (the optimal cost is also an integer), choosing any $\epsilon < 1/3$ and derandomizing the rounding algorithm using standard techniques (see [11]) yields a schedule with cost at most 3/2 of the optimal cost. We thus have the following result, whose proof follows from our discussion so far.

Theorem 16.1

Given an instance of $R \,||\, \sum w_j C_j$, let x^* be a near-optimal solution to SCQP within an additive error of 1/3. Consider the (random) assignment obtained by assigning job j to machine i with probability x^*_{ij}, and let each machine sequence its job according to Smith's rule. Let π be the resulting (random) schedule, Z^π its cost, and let Z^* be the optimal cost. Finally, let $\hat{\pi}$ be the schedule computed by derandomizing the random assignment. Then (a) $E[Z^\pi] \leq (3/2)Z^* + 1/3$; and (b) $E[Z^{\hat{\pi}}] \leq (3/2)Z^*$.

16.2.2 $R \,|\, \text{pmtn} \,|\, \sum w_j C_j$

The preemptive version of the problem of scheduling unrelated machines can be treated using similar techniques. However, we need to use a slightly different reasoning to get a valid relaxation. Recall that the contribution of job j to the cost of machine i in a nonpreemptive schedule in terms of the assignment variables is given by

$$w_j x_{ij}\left(p_{ij} + \sum_{k:k<_i j} x_{ik} p_{ik}\right)$$

This is zero if j is not assigned to i; otherwise, this is simply the weighted completion time of job j.

Consider the preemptive version of the same problem. It is easy to see that this expression is no longer correct. To derive a valid expression, we consider a simple charging scheme based on another interpretation of the objective function: assume that each unit of job j in the system incurs a cost of w_j per unit time. If x_{ij} is the portion of job j assigned to machine i, then, machine i will only be responsible for the cost incurred by these x_{ij} units of job j; the amount of job j at machine i is exactly x_{ij} in the time interval $[0, \sum_{k:k<_i j} x_{ik} p_{ik}]$, and decreases linearly to zero in the time interval $[\sum_{k:k<_i j} x_{ik} p_{ik}, x_{ij} p_{ij} + \sum_{k:k<_i j} x_{ik} p_{ik}]$. Thus, the cost incurred by the (portions of the) jobs in machine i, $\Phi(i)$, is

$$\Phi(i) \geq \sum_j w_j x_{ij}\left(\frac{x_{ij}}{2} p_{ij} + \sum_{k:k<_i j} x_{ik} p_{ik}\right)$$

This leads us to the following convex programming problem:

$$\text{(SCPP)} \qquad \text{Min} \sum_i \Phi(i)$$

subject to:

$$\Phi(i) \geq \frac{1}{2}(x(i)^T \hat{D}(i) x(i)), \quad \forall i$$

$$\Phi(i) \geq c(i)^T x(i), \quad \forall i$$

$$\sum_i x_{ij} = 1, \quad \forall j$$

$$x_{ij} \geq 0, \quad \forall i, j$$

The only difference from the nonpreemptive case is that the terms $c(i)^T x(i)/2$ are missing from the first set of constraints. We can solve this problem (SCPP) to obtain a near optimal solution x^*. We can think of allocating job-fractions according to x^*, but we will then have the difficult task of making sure no job is run simultaneously on multiple machines. Since our approach is to reduce everything to dealing with independent single-machine problems, we do not attempt to implement this fractional solution x^* directly. Rather, we resort to finding a nonpreemptive schedule $\hat{x}$ by applying randomized rounding on x^*. The schedule $\hat{x}$ has expected cost that is at most

$$\sum_i c(i)^T x(i) + \frac{1}{2}(x(i)^T \hat{D}(i) x(i)),$$

which is within twice the cost of SCPP. We thus have the following result.

Theorem 16.2

Given an instance of $R \mid \text{pmtn} \mid \sum w_j C_j$, *let* x^* *be a near-optimal solution to SCPP within an additive error of 1/3. Consider the (random) assignment obtained by assigning job* j *to machine* i *with probability* x^*_{ij}, *and let each machine sequence its job according to Smith's rule. Let* π *be the resulting (random) schedule,* Z^π *its cost, and let* Z^* *be the optimal cost. Finally, let* $\hat{\pi}$ *be the schedule computed by derandomizing the random assignment. Then (a)* $E[Z^\pi] \geq 2Z^* + 1/3$, *and (b)* $E[Z\hat{\pi}] \geq 2Z^*$.

In addition, we also get a bound on the *power of preemption.* Since we find a nonpreemptive schedule within a factor of 2 of the best possible preemptive schedule, we can also conclude that by allowing preemption we can reduce the sum of weighted completion times by at most a factor of 2.

16.3 Scheduling Unrelated Machines with Release Dates

We now turn to the problem of scheduling unrelated machines when the jobs have (possibly machine-dependent) release dates. Specifically, a job j can be processed on machine i only at or after time r_{ij} . As before, we first consider the nonpreemptive version followed by the preemptive one.

16.3.1 $R \mid r_{ij} \mid \sum w_j C_j$

Recall that our basic approach is to think of the scheduling problem as consisting of two subproblems: first, the routing or assignment of jobs to machines; and second, the sequencing of the jobs assigned to each machine. The latter is simply a set of m independent single machine scheduling problems, each of which can be solved to optimality by Smith's rule if all the jobs are released at the same time. We would of course like to pursue this approach for the problem with release dates. To do so, however, we need to overcome two difficulties: first, the single machine scheduling problem $1 \mid r_j \mid \sum w_j C_j$ is itself NP-hard [15], so the single machine sequencing problem is itself nontrivial; and second, even if we did know the optimal sequencing rule for all subsets of jobs assigned to a given machine, the resulting cost, as a function of the assignment variables, may not be "convex" (or easily "convexifiable"). Thus an optimal solution to $1 \mid r_j \mid \sum w_j C_j$ may actually not be helpful to our approach. However, we do know that scheduling according to Smith's rule results in a cost function that can be convexified easily. Therefore, we would like the sequencing rule to be "similar" to Smith's rule.

For the single machine scheduling problem with release dates, when is Smith's rule optimal? Fix a machine i, and suppose machine i has an ordering $<_i$ of all the jobs such that $j <_i k$ if $w_j/p_{ij} > w_k/p_{ik}$. Clearly, if the jobs are released in the order of their priority, Smith's rule is optimal, i.e., if $r_{ij} \leq r_{ik}$ for $j <_i k$, then scheduling j before k is optimal. Thus, if $j <_i k$, and both are assigned to machine i and are available for processing at any time t, then we can schedule j before k in an optimal schedule. Therefore, in any interval of time during which no job is released, we may assume that jobs are scheduled according to Smith's rule in an optimal schedule. This motivates us to work with intervals of time during which no job is released; these are called *slots*.

We work with a fixed machine i; without loss of generality, we assume that jobs are labeled so that $1 <_i 2 \cdots <_i n$. Let $\rho_{i,1} < \rho_{i,2} < \cdots < \rho_{i,q_i}$ be an ordering of the distinct release dates in the set $\{r_{i1}, r_{i2}, \ldots, r_{in}\}$. The interval $[\rho_{i,k}, \rho_{i,k+1}]$ will be called slot (i,k) or simply slot k if i is evident. Job j is eligible for slot (i,k) if and only if $r_{ij} \le \rho_{i,k}$. By our discussion, jobs scheduled within a slot are sequenced according to Smith's rule. So it is tempting to treat the scheduling problem as one of allocating jobs to slots (instead of machines), with slot (i,k) "available for processing jobs only in the interval $[\rho_{i,k}, \rho_{i,k+1})$. If we do so, note that we need to allow "fractional" jobs to be assigned to slots; otherwise we implicitly impose an additional constraint, namely that no job straddles any slot, which is simply absent from the original problem. Because we eventually plan to relax the assignment constraints anyway, and because the overall effect of allowing this flexibility is to underestimate the cost, this does not pose a problem. Let $\Phi(i,k)$ be the cost incurred by slot k of machine i, and let $\Phi(i) = \sum_k \Phi(i,k)$ be the cost incurred by machine i. Clearly,

$$\Phi(i,k) = \sum_j w_j x_{ikj} \left(\rho_{i,k} + p_{ij} + \sum_{l: l <_i j} x_{ikl} p_{il} \right)$$

where x_{ikj} is the portion of job j assigned to slot (i,k). We are thus led to the following nonlinear programming problem:

$$\text{(QPR)} \qquad \text{Min} \sum_{i,k} \sum_j w_j x_{ikj} \left(\rho_{i,k} + p_{ij} + \sum_{l: l <_i j} x_{ikl} p_{il} \right)$$

subject to:

$$\sum_{i,k} x_{ikj} = 1, \quad \forall j$$

$$\sum_j x_{ikj} p_{ij} \le \rho_{i,k+1} - \rho_{i,k}, \quad \forall i, k$$

$$x_{ikj} = 0, \quad \text{if } r_{ij} < \rho_{i,k}, \quad \forall j$$

$$x_{ikj} \ge 0, \quad \forall i, k, j$$

We note that given any feasible schedule to the original problem, we can set the assignment variable x_{ikj} to be the fraction of job j processed in slot (i,k). The cost of processing a job by a machine (i.e., its weighted completion time) is computed for these different pieces (processed in different slots) independently. This, and the additional flexibility that each slot has to reorder its assigned jobs, results in an underestimate of the actual weighted completion time of the discrete schedule. Thus, QPR is a relaxation of the problem $R \mid r_{ij} \mid \sum w_j C_j$.

As before, the objective function $\Phi(i,k)$ is not necessarily convex. However, the same trick can be used to make this a convex function: we replace

$$w_j p_{ij} x_{ikj}$$

in $\Phi(i,k)$ by

$$w_j p_{ij} \left(\frac{x_{ikj}}{2} + \frac{x_{ikj}^2}{2} \right)$$

which has the effect of further underestimating the actual objective value, causing it to remain a valid relaxation. This convex relaxation is formally stated below:

$$\text{(CQPR)}\qquad \text{Min} \sum_{i,k} \sum_{j} \left\{ w_j \rho_{i,k} x_{ikj} + w_j p_{ij} \frac{x_{ikj}}{2} + w_j x_{ikj} \left(\frac{p_{ij} x_{ikj}}{2} + \sum_{l:l <_i j} x_{ikl} p_{il} \right) \right\}$$

subject to:

$$\sum_{i,k} x_{ikj} = 1, \quad \forall j$$

$$\sum_{j} x_{ikj} p_{ij} \le \rho_{i,k+1} - \rho_{i,k}, \quad \forall i, k$$

$$x_{ikj} = 0, \quad \text{if } r_{ij} < \rho_{i,k}, \quad \forall j$$

$$x_{ikj} \ge 0, \quad \forall i, k, j$$

An optimal solution x^* to CQPR can be found in polynomial time. As before, we interpret x^*_{ijk} as the probability with which job j is assigned to slot (i, k), thus we consider each job j and independently assign it to a slot (i, k) with probability x^*_{ikj}. Let T_{ik} be the set of jobs assigned to slot (i, k). A machine i processes its assigned jobs in the order of their slots, sequencing the jobs within a slot according to $<_i$. Thus machine i processes all the jobs in T_{i1} first, then the jobs in T_{i2}, etc. We next prove that the cost of this schedule is at most twice the optimal cost.

Theorem 16.3

Let x^* be an optimal solution to CQPR. Assign jobs to slots independently, with job j assigned to slot k of machine i with probability x^*_{ikj}. Sequence the jobs at any machine in order of the slots, and sequence the jobs within a slot according to Smith's rule. The cost of the resulting schedule is at most twice the optimal cost.

Proof

Let $T_{i,k}$ be the set of jobs assigned to slot k of machine i. We shall show that the expected cost of the jobs in $T_{i,k}$ is at most $2\hat{\Phi}^*(i, k)$. Let $t_{i,k}$ be the epoch at which the first job from the set $T_{i,k}$ is processed. Let $\hat{x}_{ikj}$ be the indicator function of the resulting (random) schedule. Then, the expected cost of the jobs in $T_{i,k}$ is clearly

$$E \left[\sum_j w_j \hat{x}_{ikj} \left(t_{i,k} + p_{ij} + \sum_{l:l <_i j} p_{il} \hat{x}_{ikl} \right) \right]$$

which equals

$$\sum_j w_j E[\hat{x}_{ikj} t_{i,k}] + \sum_j w_j E \left[\hat{x}_{ikj} \left(p_{ij} + \sum_{l:l <_i j} p_{il} \hat{x}_{ikl} \right) \right] \tag{16.8}$$

Fix a job j. The term corresponding to j in the second expectation is simply

$$w_j x^*_{ikj} \left(p_{ij} + \sum_{l:l <_i j} p_{il} x^*_{ikl} \right)$$

Using an argument similar to the one in Section 16.2.1, this expression can be seen to be at most twice

$$w_j p_{ij} \frac{x^*_{ikj}}{2} + w_j x_{ikj} \left(\frac{p_{ij} x^*_{ikj}}{2} + \sum_{l:l <_i j} x^*_{ikl} p_{il} \right)$$

which is identical to the last two terms in the objective function of the relaxation CQPR. So to prove the theorem, it is enough to show that

$$E[\hat{x}_{ikj} t_{i,k}] \leq 2x^*_{ikj} \rho_{i,k}$$

for any job j and any slot (i, k). Since $\hat{x}_{ikj} \in \{0, 1\}$

$$E[\hat{x}_{ikj} t_{i,k}] = \text{Prob}[\hat{x}_{ikj} = 1] E[t_{i,k} | \hat{x}_{ikj} = 1]$$

This latter expectation can be found by elementary reasoning. A job from $T_{i,k}$ will be processed as soon as at least one of the jobs in $T_{i,k}$ is released, and all of the jobs in the sets $T_{i,1}, T_{i,2}, \ldots, T_{i,k-1}$ are processed. Thus,

$$\begin{aligned} E[t_{i,k} \mid \hat{x}_{ikj} = 1] &\leq E\left[\rho_{i,k} + \sum_{l=1}^{k-1} \sum_j \hat{x}_{ilj} p_{il}\right] \\ &= \rho_{i,k} + \sum_{l=1}^{k-1} E\left[\sum_j \hat{x}_{ilj} p_{il}\right] \\ &= \rho_{i,k} + \sum_{l=1}^{k-1} (\rho_{i,l+1} - \rho_{i,l}) \\ &\leq 2\rho_{i,k} \end{aligned}$$

where the penultimate equation follows from a constraint of CQPR. Thus,

$$E[\hat{x}_{ikj} t_{i,k}] = \text{Prob}[\hat{x}_{ikj} = 1] E[t_{i,k} \mid \hat{x}_{ikj} = 1] \leq 2x^*_{ikj} \rho_{i,k}$$

and the proof is complete. □

16.3.2 $R \mid r_{ij}, \text{pmtn} \mid \sum w_j C_j$

By now all the key ideas are in place: the result here is mainly a matter of combining the ideas of Section 16.2.2 and Section 16.3.1, so we merely sketch the details.

As in $R \mid \text{pmtn} \mid \sum w_j C_j$, we first derive a lower bound on the weighted completion time of job j. The analogous result here (proof omitted) is

$$\sum_j w_j C_j \geq \sum_j w_j \sum_k x_{ikj} \left(\rho_{ik} + \frac{x_{ikj}}{2} p_{ij} + \sum_{l:l<_i j} x_{ikl} p_{il} \right)$$

The convex relaxation now becomes

$$\begin{aligned} \text{(CQPPR)} \qquad & \text{Min} \sum_{i,k} \sum_j \left\{ w_j \rho_{i,k} x_{ikj} + w_j x_{ikj} \left(\frac{p_{ij} x_{ikj}}{2} + \sum_{l:l<_i j} x_{ikl} p_{il} \right) \right\} \\ & \text{subject to:} \\ & \sum_{i,k} x_{ikj} = 1, \quad \forall j \\ & \sum_j x_{ikj} p_{ij} \leq \rho_{i,k+1} - \rho_{i,k}, \quad \forall i, k \\ & x_{ikj} = 0, \quad \text{if } r_{ij} < \rho_{i,k}, \ \forall j \\ & x_{ikj} \geq 0, \quad \forall i, k, j \end{aligned}$$

Again, the main difference from the nonpreemptive version is the lack of the term $w_j p_{ij} x_{ikj}/2$ in the objective function of the relaxation. Using the same trick of Section 16.2.1, we can improve this relaxation by requiring $\sum_{i,k,j} w_j p_{ij} x_{ikj}$ to be a lower bound on the optimal cost. This strengthened formulation can be solved in polynomial time. Applying randomized rounding on an optimal solution x^* of CQPPR gives us a schedule whose cost is within a factor of 3 of the best possible, this also yields a bound on the power of preemption in this setting.

16.4 Cardinality Constraints

In all of the models considered so far, we find a good fractional solution of an appropriate relaxation and then use randomized rounding to find a (random) integer solution whose expected cost is within a (small) constant factor of the optimal cost. In this section we consider similar scheduling problems except that we have additional constraints on the number of jobs any given machine can process. In the presence of such cardinality constraints, the randomized rounding algorithm may not even produce a *feasible* solution, so we have to rely on a different set of techniques. Our purpose here is to illustrate a deterministic rounding algorithm — called *pipage* rounding — that can be used in such situations. For concreteness we focus only on the problem $R \,||\, \sum w_j C_j$ with additional cardinality constraints. The model with release dates can be dealt with in a similar fashion.

Let n_i be the maximum number of jobs that machine i can process. Then the problem $R \,||\, \sum w_j C_j$ can be formulated as the following integer programming problem.

$$\text{(IQPC)} \qquad \text{Min} \sum_j w_j \left\{ \sum_i x_{ij} \left(p_{ij} + \sum_{k:k<_i j} x_{ik} p_{ik} \right) \right\}$$

subject to:

$$\sum_i x_{ij} = 1, \quad \forall j$$

$$\sum_j x_{ij} \leq n_i, \quad \forall i$$

$$x_{ij} \in \{0, 1\} \quad \forall i, j$$

Fix a machine i, and relabel the jobs so that $1 <_i 2 <_i \cdots <_i n$. Let $c(i)$ be the n-vector $(w_1 p_{i1}, w_2 p_{i2}, \ldots, w_n p_{in})$, and let $x(i)$ be the n-vector $(x_{i1}, x_{i2}, \ldots, x_{in})$. As before, let

$$\hat{D}(i) = \begin{bmatrix} w_1 p_{i1} & w_2 p_{i1} & w_3 p_{i1} & \cdots & w_{n-1} p_{i1} & w_n p_{i1} \\ w_2 p_{i1} & w_2 p_{i2} & w_3 p_{i2} & \cdots & w_{n-1} p_{i2} & w_n p_{i2} \\ w_3 p_{i1} & w_3 p_{i2} & w_3 p_{i3} & \cdots & w_{n-1} p_{i3} & w_n p_{i3} \\ \vdots & \vdots & \vdots & \ddots & \vdots & \vdots \\ w_{n-1} p_{i1} & w_{n-1} p_{i2} & w_{n-1} p_{i3} & \cdots & w_{n-1} p_{in-1} & w_n p_{in-1} \\ w_n p_{i1} & w_n p_{i2} & w_n p_{i3} & \cdots & w_n p_{in-1} & w_n p_{in} \end{bmatrix}$$

The following convex programming problem is a relaxation of the problem of scheduling unrelated machines with cardinality constraints; if the $x_{ij} \in \{0, 1\}$, it is an exact formulation. (As discussed earlier in the context of the problem $R\|\sum w_jC_j$, we add an additional set of constraints, which is redundant for the integer program but strengthens the relaxation.)

$$\text{(SCPC)} \qquad \text{Min} \sum_i \Phi(i)$$

subject to:

$$\Phi(i) \geq \frac{1}{2}(c(i)^T x(i) + x(i)^T \hat{D}(i)x(i)), \quad \forall i$$

$$\Phi(i) \geq c(i)^T x(i), \quad \forall i$$

$$\sum_i x_{ij} = 1, \quad \forall j$$

$$\sum_j x_{ij} \leq n_i, \quad \forall i$$

$$x_{ij} \geq 0, \quad \forall i, j$$

Suppose x^* is an optimal solution to SCPC. A simple randomized rounding argument, similar to the one used in Section 16.2.1, shows that the expected cost of the (random) schedule found is at most 3/2 of the cost of SCPC. However, the schedule found may violate the cardinality constraint at some machine. Moreover, there does not seem to be any simple way to satisfy the cardinality constraints if we resort to randomized rounding. As we discussed earlier, we address this problem by relying on a deterministic rounding algorithm called *pipage* rounding.

Pipage rounding is a deterministic rounding algorithm that is useful in handling cardinality constraints, budget constraints, etc. The main idea is to start with a fractional solution and to gradually and systematically reduce the number of fractional components, while always "improving" the objective value.

Recall that our proof of the 3/2-approximation algorithm for the problem $R\,||\sum w_jC_j$ proceeded by finding an optimal solution x^* to *SCQC* and showing that its cost, which is

$$F(x^*) \equiv \sum_j w_j \left\{ \sum_i x^*_{ij} \left(p_{ij} + \sum_{k:k<_i j} x^*_{ik} p_{ik} \right) \right\}$$

is within 3/2 of the optimal cost of SCQC. We shall now round this fractional solution deterministically to obtain a solution $\hat{x}$ with $F(\hat{x}) \leq F(x^*)$, so the cost of $\hat{x}$ will be within 3/2 of the optimal.

Consider the bipartite graph with m nodes — one for each machine — on one side, and n nodes — one for each job — on the other. Let the weight of edge (i, j) be x^*_{ij}. The assignment constraints imply that the degree of every job-node is exactly 1; the cardinality constraints imply that the sum of weights of the edges incident to machine-node i is at most n_i. Since our goal is to reduce the number of fractional components of x^*, we look at the subgraph defined by the fractional x^*_{ij} variables. We shall call this subgraph $G(x^*)$.

Suppose $G(x^*)$ contains a cycle $C = (e_1, e_2, \ldots, e_{2l})$. Consider the sets $M_1 = \{e_1, e_3, \ldots, e_{2l-1}\}$, and $M_2 = \{e_2, e4, \ldots, e_{2l}\}$. Let

$$\epsilon_1 = \min\{\min_{e \in M_1}(1 - x_e), \min_{e \in M_2} x_e)\}$$

and

$$\epsilon_2 = \min\{\min_{e \in M_1} x_e, \min_{e \in M_2}(1 - x_e)\}$$

There are two natural ways to modify the weights on this cycle so as to reduce the number of fractional components and preserve the node degrees: increase the weights of the edges in M_1 by ϵ_1 and decrease the weights of the edges in M_2 by the same amount; or decrease the weights of the edges in M_1 by ϵ_2 and increase the weights of the edges in M_2 by the same amount. Let $x^*_{\epsilon 1}$ and $x^*_{\epsilon 2}$ be the solutions obtained. We shall now argue that

$$\min\{F(x^*_{\epsilon_1}), F(x^*_{\epsilon_2})\} \leq F(x^*)$$

We do this by considering how $F(x^*)$ changes when the x variables are altered along a cycle as suggested earlier. Observe that the solutions we obtain — $x^*_{\epsilon_1}$ and $x^*_{\epsilon_2}$ — are essentially one dimensional modifications of x^*: we arrive at these by increasing $x^*_{e_1}$ by ϵ_1 and decreasing $x^*_{e_1}$ by ϵ_2, respectively. This motivates us to study $F(x^*_\epsilon)$ where $\epsilon \in [-\epsilon_2, \epsilon_1]$ is the amount added to $x^*_{e_1}$. Since the x variables change only along the cycle C, every machine-node in C will have the weights of one of its edges increased by ϵ and the weights of another of its edges decreased by ϵ; from the form of $F(x)$, it is clear that $F(\cdot)$ is a concave function of ϵ, so achieves its minimum at one of the extreme points of the feasible region. Therefore

$$\min\{F(x^*_{\epsilon_1}), F(x^*_{\epsilon_2})\} \leq F(x^*)$$

If $G(x^*)$ does not contain a cycle, it must be a collection of trees. So we pick any path connecting two nodes of degree 1, and consider the obvious two modifications along this path: we find ϵ_1 and ϵ_2 as before by looking at alternate edges along the path. The degrees of all intermediate nodes are preserved. Although the degrees of the endpoints are not preserved, the new degrees obtained in $x^*_{\epsilon_1}$ and $x^*_{\epsilon_2}$ satisfy both the assignment constraints and the cardinality constraints. The assignment constraints hold in these two new solutions because no job-node can have degree 1 (as every job is completely assigned), thus any path has to be from a machine-node to another machine-node. Since the n_i are integers, and since no edge's weight is increased beyond the next integer value, the cardinality constraints cannot be violated for any machine i in $x^*_{\epsilon_1}$ and $x^*_{\epsilon_2}$. It is clear again that $F(x^*_\epsilon)$ is concave in ϵ for $\epsilon \in [-\epsilon_2, \epsilon_1]$, so it will attain its minimum at one of the endpoints.

Thus, in either case we can "round" x^* into another feasible solution x^*_ϵ such that $F(x^*_\epsilon) \leq F(x^*)$ and x^*_ϵ has fewer fractional components than x^*. Continuing this process, we eventually reach an integer solution $\hat{x}$, which is the required near-optimal solution. (Clearly, the number of steps to reach this integer solution is bounded by the number of fractional components of x^*, which we know is small.)

An analogous argument can be shown to hold for the more general problem $R \mid r_{ij} \mid \sum w_j C_j$, the details are left to the interested reader.

16.5 Conclusions

In this chapter we saw how the problem of scheduling unrelated machines (and several variations) can be reduced to solving an assignment problem followed by a sequencing problem. This point-of-view resulted in a natural convex relaxation, which can be solved efficiently. We also saw how fractional solutions can be rounded to obtain integer solutions whose performance can be effectively bounded.

While the work discussed here appears to be very natural, such an approach was not pursued until 5 years ago. Perhaps these ideas can be pushed further, both to obtain improved guarantees and to broaden their applicability. As for the former, almost all of our analysis can be shown to be tight, so to obtain improved guarantees, we may have to obtain improved relaxations. As for the latter, while no significant advances have been made in this direction since the appearance of the original papers, we remain optimistic. It will be interesting to see new applications of such ideas, both in scheduling and, more broadly, in the field of approximation algorithms.

Acknowledgments

I thank Akshay-Kumar Katta, Joseph Leung, and Martin Skutella for their comments on an earlier version of this manuscript.

References

[1] L. Lovasz. On the shannon capacity of a graph. *IEEE Trans. Info. Theo.*, 25:1–7, 1979.

[2] M. Groetschel, L. Lovasz, and A. Schrijver. The ellipsoid method and its consequences in combinatorial optimization. *Combinatorica*, 1:169–197, 1981.

[3] M. Goemans and D. P. Williamson. Improved approximation algorithms for maximum cut and satisfiability problems using semidefinite programming. *J. ACM*, 42:1115–1145, 1995.

[4] D. R. Karger, R. Motwani, and Madhu Sudan. Approximate graph coloring by semidefinite programming. In *IEEE Symposium on Foundations of Computer Science*, pp. 2–13, 1994.

[5] B. Chor and M. Sudan. A geometric approach to betweenness. *SIAM J. Disc. Math.*, 11:511–523, 1998.

[6] Y. Ye. A 0.699 approximation algorithm for Max-Bisection, *Mathematical Programming*, 90(1):101–111, 2001.

[7] M. Skutella. Semidefinite relaxations for parallel machine scheduling. In *IEEE Symp. Foundations of Computer Science*, pp. 472–481, 1998.

[8] J. Sethuraman and M. S. Squillante. Optimal scheduling of multiclass parallel machines. In *Proceding of 10th ACM-SIAM Symposium on Discrete Algorithms (SODA)*, pp. 963–964, 1999.

[9] J. Sethuraman and M. S. Squillante. Optimal stochastic scheduling in multiclass parallel queues. In *ACM Sigmetrics Conference on Measurement and Modeling of Computer Systems*, pp. 93–102, 1999.

[10] M. Skutella. Convex quadratic programming relaxations for network scheduling problems. In *European Symposium on Algorithms*, pp. 127–138, 1999.

[11] A. Ageev and M. Sviridenko. Pipage rounding: a new method of constructing algorithms with proven performance guarantee. *J. Combinat. Opt.* (Unpublished)

[12] M. Skutella. Convex quadratic and semidefinite programming relaxations in scheduling. *J. ACM*, 48(2):206–242, 2001.

[13] R. L. Graham, E. Lawler, J. K. Lenstra, and A. H. G. Rinooy Kan. Optimization and aaproximation in deterministic sequencing and scheduling: A survey. *Annals of Discrete Mathematics*, 5, 287–326, 1979.

[14] W. E. Smith. Various optimizers for single-stage production. *Nav. Res. Log. Q.*, 3:59–66, 1956.

[15] J. K. Lenstra, A. H. G. Rinnooy Kan, and P. Brucker. Complexity of machine scheduling problems. *Ann. Disc. Math.*, 1:343–362, 1977.

III

Other Scheduling Models

III

Other Scheduling Models

17

The Master-Slave Scheduling Model

Sartaj Sahni
University of Florida

George Vairaktarakis
Case Western Reserve University

17.1 Introduction

The master-slave scheduling model, which was introduced by Sahni [1], involves two sets of processors — the master processors that are responsible for pre- and postprocessing of work orders, and the slave processors that are responsible for the actual execution of the orders. The number of slave processors is no less than the number of work orders. Applications of this model include parallel computing, semiconductor testing and problems in transportation as will be described shortly.

First, we give a brief description of the model. A set of jobs is to be processed by a system of master and slave processors. Each job has three tasks associated with it. The first is a preprocessing task, the second is a slave task, and the third a postprocessing task. The tasks of each job are to be performed in the order: preprocessing, slave, postprocessing. Let a_i, b_i, and c_i, respectively, denote the preprocessing, slave, and postprocessing tasks (and task times) of job i. All task times are assumed to be greater than zero (i.e., $a_i > 0$, $b_i > 0$, and $c_i > 0$, for all i). The available processors are divided into two categories: master and slave. If n denotes the number of jobs, then no schedule can use more than n slaves. Hence, we may assume that there are exactly n slaves. The *makespan* or *finish time* of a schedule is the earliest time at which all tasks have been completed.

Figure 17.1 (a) shows a possible schedule for the case when $n = 2$, $(a_1, b_1, c_1) = (2, 6, 1)$, and $(a_2, b_2, c_2) = (1, 2, 3)$. In this schedule, the preprocessing of job 1 is handled first by the master; all other tasks begin at the earliest possible time. M denotes the master processor and S_1 and S_2 denote the slaves. The finish time is 9. The schedule that results when the master preprocesses job 2 first and all other tasks begin at the earliest possible time is shown in Figure 17.1 (b). This has a finish time of 10.

Let us examine the schedules of Figure 17.1. Notice that in both schedules, once the processing of a job begins, the job is processed continuously until completion. Schedules with this property are said to have *no-wait-in-process*. In industrial applications, one may impose this requirement on a schedule.

1-58488-397-9/$0.00+$1.50

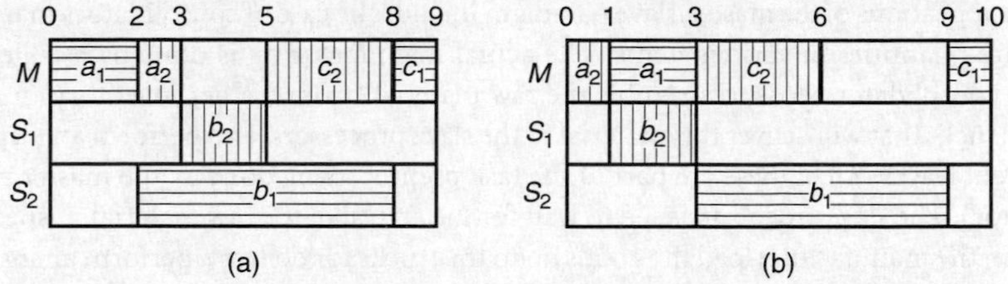

FIGURE 17.1 Example schedules.

Another interesting feature of the schedules of Figure 17.1 is that in one the postprocessing is done in the reverse order of the preprocessing while in the other the pre- and postprocessing orders are the same. In some settings, we may require that schedules satisfy one order or the other. For example, reverse order could simplify the postprocessing if a stack is used, by the master, to maintain a record of jobs in process. Similarly, if the master uses a queue to maintain this information, we might require that the postprocessing be done in the same relative order as the preprocessing. Another discipline that might be imposed on the master is to complete all the preprocessing tasks before beginning the first postprocessing task. Both of the schedules of Figure 17.1 obey this discipline.

Several applications of the master-slave model are found in parallel computer scheduling. A common parallel programming paradigm involves the use of a single main computational thread that employs the fork and join operations to spawn parallel tasks/threads and then to synchronize following the completion of these tasks. The fork operation involves the passing of varying amounts of data to remote processors that will execute the spawned threads (we assume that each spawned thread will be executed on a different processor). These processors will, in turn, return the results to the main thread. So, associated with each of the spawned threads, we have three amounts of work:

1. Preprocessing by main thread. This is the work needed to initiate the thread. It includes the effort expended in collecting the data needed by the remote processor (in case of a distributed memory environment); overheads involved in transmitting this data to the remote processor, and so on.
2. Work done in the thread. This includes the computational activity assigned to the remote processor, the work this processor must do to receive the data and send back the results, and the transmission times in receiving and sending.
3. Postprocessing by the main thread. This represents the effort expended in receiving the answers and performing any postprocessing on them.

Since the different threads may execute very different pieces of code, the relative values of the amounts of work involved in preprocessing, in thread execution, and in postprocessing can vary widely from thread to thread.

Certain semiconductor testing operations also utilize the master-slave paradigm. In the case of burn-in operations, chips are subject to thermal stress for an extended period of time in order to bring out latent defects leading to infant mortality that might otherwise surface in the operating environment. The thermal stressing is accomplished by maintaining the oven at a constant temperature while powering up the chip. The burn-in times for each chip are specified by the customer for whom it is made and it is thus fixed *a priori*. After the initial burn-in operation each chip cools off for a specified amount of time that depends on the length and intensity of the initial burn-in period. After cooling, each chip is subject to a final burn-in operation (see [2] for a more detailed description of semiconductor burn-in operations). In this application the burn-in oven corresponds to the master processor, the two burn-in tasks correspond to pre- and postprocessing and the cooling period corresponds to the slave task. Since the burn-in operations are near the end of the production process, scheduling is critical in determining on-time delivery and output performance for the entire company.

Industrial applications of the master-slave paradigm include the case of consolidators that receive orders to manufacture quantities of various items. The actual manufacturing is done by a collection of slave agencies. The consolidator needs to assemble the raw material (from his/her inventory) needed for each task, load the trucks that will deliver this material to the slave processors, and perform an inspection before the consignment leaves. All of these are part of the task preprocessing done by the master processor (i.e., the consolidator). The slave processors need to wait for the arrival of the raw material, inspect the received goods, perform the manufacture, load the goods on to the trucks for delivery, perform an inspection as the trucks are leaving. These activities together with the delay involved in getting the trucks to their destination (i.e., the consolidator) represent the slave work. When the finished goods arrive at the consolidator, they are inspected and inventoried. This represents the postprocessing.

Suppose that all the raw material is loaded on a single truck and that the slaves are uniformly spaced. Whenever the truck stops, it has to wait at the slave location while the material for that location is unloaded and checked. This constitutes the preprocessing. When the truck returns to pick up the finished goods, it must again wait to load and check. This constitutes the postprocessing. If the truck route is circular, then the pre- and postprocessing orders are the same. If the route is linear, then the postprocessing is done when the truck is returning to its point of origin and so is done in the reverse order of preprocessing. In both cases, all preprocessing tasks are done before the first postprocessing task.

In certain maintenance/repair environments, the maintenance manager examines the maintenance tasks to be performed and writes up a formal work order for each and prepares the task for maintenance; the work orders are executed by different maintenance crews that are dispatched following the receipt of the work order; upon completion, the maintenance manager inspects the completed work and signs an acceptance document.

It is easy to see that the examples cited earlier for single master systems generalize to multiple master systems. For example, we may have a computational resource that is comprised of a large number of processors. This resource is shared by several host computers whose function is to obtain the data and code for each job (say from a disk) and to store the results on a disk or to print the results out. For each job, the actual computation is done on a single processor of the shared computational resource. Each job has a preprocessing task (gather the data and code needed), a postprocessing task (output the results), and a slave task (computation). Assuming that the total number of jobs is no more than the number of processors in the shared computational resource, the problem of scheduling the jobs can be modeled as a multiple master scheduling problem. In this application, it is required that for each job, the pre- and postprocessing tasks be done by the same master. This is referred to as *restricted multiple master* scheduling.

If the consolidator example is generalized to include several consolidators, then the resulting scheduling problem may be modeled as a restricted multiple master system. On the other hand if there is a single consolidator with multiple trucks and each truck has its own crew for loading, inspecting, etc., then the scheduling problem can be modeled as a multiple master system (each truck and crew define one master) in which the master that preprocesses job i (i.e., the truck that delivers the raw material for the job) need not be the same as the one that postprocesses job i (i.e., the truck that brings back the finished goods corresponding to this job).

It is interesting to note that the master-slave scheduling model may be regarded as a variant of the job shop as described below:

1. The job shop has two classes of machines: master and slave.
2. There is exactly one master machine and the number of slave machines equals the number of jobs.
3. Each job has three tasks to be done in order; the first and third on the master and the second on a slave.

The two-machine flowshop model with transfer lags (2FTL) is a close relative to the master-slave model. In this model the preprocessing task has to be processed by the upstream machine, followed by a waiting period known as transfer lag, followed by the postprocessing task at the downstream machine.

Special cases of this model are among the first problems considered in scheduling theory; see [3–5]. In [6], the problem of finding minimum makespan schedules for 2FTL was shown to be strongly NP-hard. Further results on 2FTL may be found in [7]. The problem of scheduling single machines with time lags and two tasks per job is identical to the single-master master-slave model. Since the former problem is strongly NP-hard [8], the single master problem is also strongly NP-hard.

The client-server models studied in [9–11], for example, are related, yet different from the master-slave model described here. In the client-server model of [10], for example, there is one server (equivalent to a master processor) and m (slave) processors. Each of the n jobs to be performed has a preprocessing (say, read a file from the server), slave (say, perform some computation), and postprocessing (say, write the results back to the server) task. Although the preprocessing and postprocessing tasks keep the server busy while they are performed, these tasks also keep a slave processor busy. When $n \leq m$, the client-server model is equivalent to the single master master-slave model (note that in the master-slave model, $n \leq m$ by definition). When $n \leq m$, the involvement of both a slave and the master in each pre- and postprocessing task in the client-server model does not materially affect scheduling as the slave would otherwise (i.e., in the master-slave model) be idle. The model of [10] permits slave processors to have buffers (useful when $n > m$) in which preprocessed jobs may be held and also includes a variant in which each slave task has a designated slave processor on which it is to be performed.

Section 17.2 focuses on single-master master-slave problems. Section 17.3 considers the case when the master-slave system has multiple masters. In Section 17.4 we consider problems where the pre- and postprocessing tasks are executed by different groups of master processors.

17.2 Single-Master Master-Slave Systems

17.2.1 Unconstrained MFT

The unconstrained minimum finish time problem (UMFT) is NP-hard [8]. In this section, we investigate heuristic algorithms that have good worst case performance. If S is an unconstrained schedule, then a straightforward interchange argument shows that we may rearrange the master tasks so that all preprocessing tasks complete before any postprocessing task starts. Such a rearrangement can be done without increasing the makespan of the schedule. Further, the rearranged schedule has no preemptions. We may shift the a tasks in the rearranged schedule left so as to start at time 0 and complete at time $\sum a_i$ and the b tasks may be shifted left so as to begin as soon as their corresponding a tasks complete. The c tasks may be ordered to begin in the same order as the b tasks complete. None of these rearrangement operations affects the makespan of S. With this as motivation, we define a *canonical schedule* to be one which satisfies the following properties:

1. There are no preemptions.
2. The a tasks begin on the master at time 0 and complete at time $\sum a_i$.
3. The b tasks begin as soon as their corresponding a tasks complete.
4. The c tasks are done in the same order as the b tasks complete and as soon as possible.

It is evident that for every unconstrained schedule S, there is a corresponding canonical schedule with better or the same makespan. So, in the remainder of this section we limit ourselves to canonical schedules. Note that a canonical schedule is completely specified by giving the relative order in which the preprocessing tasks are done. As a result, such a schedule is defined by a permutation that gives the relative order in which the preprocessing tasks are done. We will use the terminology i follows (precedes) j to mean i comes after (before) j in the permutation that defines the schedule.

The next theorem finds the worst case performance of an arbitrary canonical schedule S. Let C^S be the makespan of the canonical schedule S and C^* the optimal makespan of UMFT.

Theorem 17.1 [12]

For any canonical schedule S, $\frac{C^S}{C^*} \leq 2$ and the bound is tight.

Proof

If $C^S = \sum_i (a_i + c_i)$ then S is optimal and the error bound of 2 is valid. Else, $C^S > \sum_i (a_i + c_i)$ in which case there exists idle time on the master processor. Since S is canonical, this idle time will have to precede one or more postprocessing tasks. Let c_{i_0} be the last postprocessing task in S that starts immediately after its corresponding slave task b_{i_0}. Since there is idle time on the master, such an i_0 exists. Then,

$$C^S = \sum_{i \text{ precedes } i_0} a_i + (a_{i_0} + b_{i_0} + c_{i_0}) + \sum_{i \text{ follows } i_0} c_i \leq 2C^*$$

since $a_{i_0} + b_{i_0} + c_{i_0} \leq C^*$ and $\sum_i (a_i + c_i) \leq C^*$.

To see that the error bound is tight consider an instance with $k+1$ jobs, where k is an arbitrary positive integer. The first k jobs have processing requirements $(1, \epsilon, \epsilon)$ while the $(k+1)$-st job has requirements $(\epsilon, k, \epsilon), \epsilon < 1/k$. The schedule S that processes $a_{k+1} = \epsilon$ last among all preprocessing tasks has makespan $C^S = 2k + 2\epsilon$. The schedule S^* that processes a_{k+1} first among all preprocessing tasks has makespan $C^* = k + (k+2)\epsilon$ and hence $\frac{C^S}{C^*} \to 2$ as $\epsilon \to 0$. □

In what follows we present a heuristic whose error bound is $\frac{3}{2}$.

Heuristic H

Step 1. Let $S_1 = \{i : a_i \leq c_i\}$ and $S_2 = \{i : a_i > c_i\}$.
Step 2. Reorder the jobs in S_1 according to nondecreasing order of b_i.
Step 3. Reorder the jobs in S_2 according to nonincreasing order of b_i.
Step 4. Generate the canonical schedule in which the a tasks of S_1 precede those of S_2.

The complexity of heuristic H is readily seen to be $O(n \log n)$. Let C^H be the makespan of the schedule generated by the above heuristic.

Theorem 17.2 [12]

$\frac{C^H}{C^*} \leq \frac{3}{2}$ *and the bound is tight.*

Proof

Let S^* be an optimal schedule for UMFT with makespan C^*. Based on the processing requirements (a_i, b_i, c_i) of job i, we define an auxiliary problem P' with processing requirements (a'_i, b'_i, c'_i) defined as follows:

$$a'_i = \begin{cases} 0 & \text{if } a_i \leq c_i \\ a_i & \text{otherwise} \end{cases} \qquad b'_i = b_i \qquad c'_i = \begin{cases} 0 & \text{if } c_i < a_i \\ c_i & \text{otherwise} \end{cases}$$

Note that P' is not a legal instance of UMFT as it contains tasks whose processing requirement is zero. However, this does not affect the validity of our proof.

In P', all preprocessing tasks in S_1 are zero and hence they can precede all nonzero preprocessing tasks (i.e., the preprocessing tasks of S_2). Similarly, all postprocessing tasks in S_2 are zero and hence they can follow all nonzero postprocessing tasks (i.e., the postprocessing tasks of S_1). Also, in P' every job has either $a'_i = 0$ or $c'_i = 0$.

A straightforward interchange argument shows that there exists an optimal schedule for P' where all postprocessing tasks for which $a'_i = 0$ are ordered in nondecreasing order of b_i. Similarly, all preprocessing tasks with $c'_i = 0$ are ordered in nonincreasing order of b_i. Therefore, an optimal sequence S' for P' looks as in Figure 17.2.

Note that S' is the schedule generated by step 4 of H if applied on P'. Let C' be the makespan of S'. By optimality of S' we have that $C' \leq C^*$. From the schedule S' for P' we generate a schedule S^H for the original problem (where the processing requirements are (a_i, b_i, c_i)) by appending the tasks a_i; $i \in S_1$ in the beginning of S' and the tasks c_i; $i \in S_2$ at the end of S'. Note that the resulting schedule S^H is feasible for the original data because S' is feasible for the modified data and $b' = b_i$. It is easy to check that S^H is the schedule generated by H for the input data (a_i, b_i, c_i) $i = 1, 2, \ldots, n$.

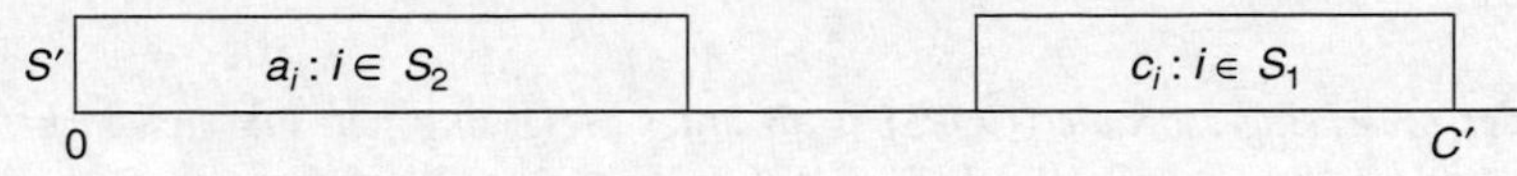

FIGURE 17.2 An optimal sequence for P'.

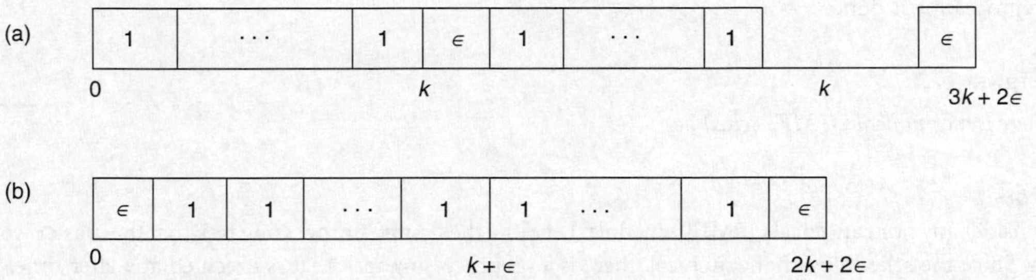

FIGURE 17.3 The bound of $\frac{3}{2}$ is tight.

Let C^H be the makespan of S^H. Then, by construction

$$C^H = C' + \sum_{i \in S_1} a_i + \sum_{i \in S_2} c_i \leq C^* + \frac{1}{2}\sum_{i \in S_1}(a_i + c_i) + \frac{1}{2}\sum_{i \in S_2}(a_i + c_i)$$
$$= C^* + \frac{1}{2}\sum_i (a_i + c_i) \leq \frac{3}{2} C^*$$

since $\sum_i (a_i + c_i) \leq C^*$.

To see that the bound of $\frac{3}{2}$ is tight, consider an instance that consists of $k + 1$ jobs, where k is an arbitrary positive integer. The first k jobs have processing requirements $(1, \epsilon, 1)$ while the $(k + 1)$-st job has requirements $(\epsilon, 2k, \epsilon)$. For this instance, we have $S_2 = \emptyset$ and H produces the canonical schedule of Figure 17.3 (a). In this the preprocessing tasks of jobs 1 through k are done first, in any order, and then that of job $k + 1$ is done. The makespan is $3k + 2\epsilon$.

An optimal solution with makespan $2k + 2\epsilon$ is depicted in Figure 17.3 (b) and hence

$$\frac{C^H}{C^*} = \frac{3k + 2\epsilon}{2k + 2\epsilon} \to \frac{3}{2}$$

as $\epsilon \to 0$. This completes the proof of the theorem. □

17.2.2 Same Pre- and Postprocessing Orders

In this section, we develop an $O(n \log n)$ algorithm to construct an order preserving minimum finish time (OPMFT) schedule. Without loss of generality, we place the following restrictions on schedules we consider in this section:

R1: The schedules are nonpremptive.
R2: Slave tasks begin as soon as their corresponding preprocessing tasks are complete.
R3: Each postprocessing task begins as soon after the completion of its slave task as is consistent with the order preserving constraint.

First, we establish some properties of order preserving schedules that satisfy these assumptions.

Definition 17.1

A canonical order preserving schedule (COPS) is an order preserving schedule in which (a) the master processor completes the preprocessing tasks of all jobs before beginning any of the postprocessing tasks, and (b) the preprocessing tasks begin at time zero and complete at time $\sum_{i=1}^{n} a_i$.

Because of restrictions R1 – R3, every COPS is uniquely described by providing the order in which the preprocessing is done.

Lemma 17.1

There is a canonical OPMFT schedule.

Proof

Consider any noncanonical OPMFT schedule. Let c_j be the first postprocessing task that the master works on. Since the schedule is noncanonical, there is a preprocessing task that is executed at a later time. Let a_i be the first of these. Slide a_i to the left so that it begins just after the preprocessing task (if any) that immediately precedes c_j (if there is no such task preceding c_j, then slide a_i left so as to start at time 0). Slide the postprocessing tasks beginning with c_j and ending at the postprocessing task that immediately preceded a_i (before it was moved) rightward by a_i units. Slide the slave and postprocessing tasks left so as to satisfy restrictions R2 and R3. The result is another OPMFT schedule that is closer to canonical form. By repeating this transormation at most $n - 1$ times we can obtain a canonical OPMFT schedule. □

Lemma 17.2

If $a_i = c_i$, $1 \leq i \leq n$, *then every COPS is an OPMFT schedule.*

Proof

Because of the preceding lemma, it is sufficient to show that all COPS have the same length. Each COPS is uniquely identified by the order in which the preprocessing tasks are executed. We shall show that exchanging two adjacent jobs in this ordering does not increase the schedule length. Since we can go from one permutation to any other via a finite sequence of adjacent exchanges, it follows that no matter what the preprocessing order, canonical schedules have the same finish time when jobs have equal pre- and postprocessing times. Hence, all COPS are OMFT schedules.

Consider two jobs j and $j+1$ that are adjacent in the preprocessing order (Figure 17.4). Let t_j and t_{j+1}, respectively, be the times at which the master begins tasks c_j and c_{j+1}. Slide job $j+1$ left by a_j so that all its tasks begin a_j units earlier than before, slide tasks a_j and b_j right by a_{j+1} units so that they begin a_{j+1} units later than before, and move task c_j so that it begins just after c_{j+1} finishes. As a result, task c_{j+1} now begins at $t_{j+1} - a_j = t_{j+1} - c_j \geq t_j$. Hence, the rescheduling of job $j+1$ does not result in the master working on two or more jobs simultaneously. In addition, the postprocessing of job $j+1$ does not begin until after its slave task is complete. The postprocessing of task c_j now begins at $t_{j+1} - a_j + c_{j+1} = t_{j+1} - c_j + a_{j+1} \geq t_j + a_{j+1}$ which is greater than or equal to the time at which the slave finishes b_j. Task c_j finishes at $t_{j+1} + a_{j+1}$. Hence, the schedule for the remaining jobs is unchanged. □

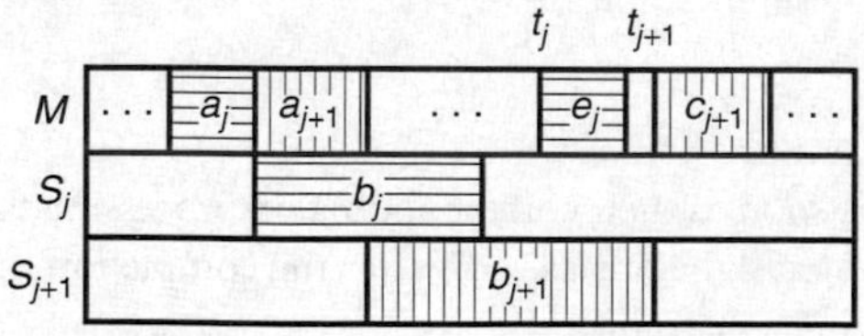

FIGURE 17.4 Figure for Lemma 17.2.

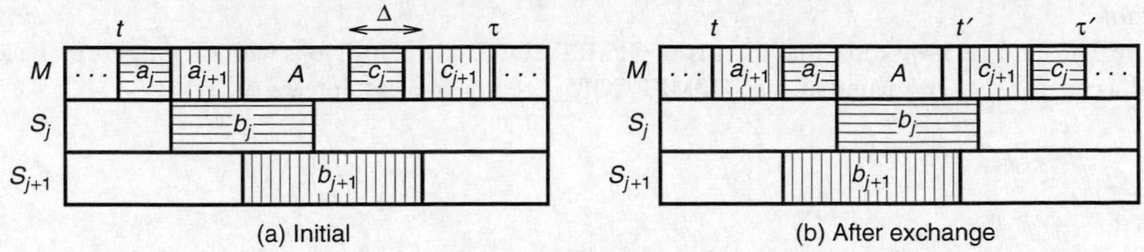

FIGURE 17.5 Figure for Lemma 17.3.

Lemma 17.3

Consider the COPS defined by some permutation σ. Assume that job j is preprocessed immediately before job $j+1$ (i.e., j immediately precedes $j+1$ in σ). If $c_j \leq a_j$ and $c_{j+1} \geq a_{j+1}$, then the schedule length (i.e., its finish time) is no less than that of the COPS obtained by interchanging j and $j+1$ in σ.

Proof

A diagram of the schedule with job j immediately preceding job $j+1$ is shown in Figure 17.5 (a). In this figure, t is the time at which the preprocessing of job j starts, A is the elapsed time between the completion of task a_{j+1} and the start of the postprocessing of job j (note that $A \geq \sum_{k \text{ follows } j+1} a_k + \sum_{k \text{ precedes } j} c_k$), $\Delta > 0$ is the time between the start of c_j and c_{j+1}, and τ is the time at which c_{j+1} completes.

Let σ' be the permutation obtained by interchanging jobs j and $j+1$ in σ. The schedule corresponding to σ' is shown in Figure 17.5 (b). Let t' and τ', respectively, be the times at which c_{j+1} and c_j finish in this schedule. If $\Delta \geq a_j$, then $t' \leq \tau - a_j$. Also, from Figure 17.5 (a), we observe that $b_j \leq a_{j+1} + A \leq c_{j+1} + A$. So, b_j finishes by t' in Figure 17.5 (b). Hence, $\tau' = t' + c_j \leq \tau - a_j + c_j \leq \tau$. As a result, the postprocessing tasks of the remaining jobs can be done so as to complete at or before their completion times in σ and the interchanging of j and $j+1$ does not increase the schedule length.

If $\Delta < a_j$, then c_{j+1} starts at time $t + a_{j+1} + a_j + A$ in σ'. So, $t' = t + a_{j+1} + a_j + A + c_{j+1}$. The time at which b_j finishes in σ' is $t + a_{j+1} + a_j + b_j \leq t + 2a_{j+1} + a_j + A \leq t + a_{j+1} + a_j + A + c_{j+1} = t'$. So, c_j finishes at $t' + c_j = t + a_{j+1} + a_j + A + c_{j+1} + c_j \leq \tau$. Consequently, the OPS defined by σ' has a finish time that is $\leq$ that of the OPS defined by σ. □

Theorem 17.3 [1]

There is an OPMFT schedule which is a COPS in which the preprocessing order satisfies the following:

1. *Jobs with $c_j > a_j$ come first.*
2. *Those with $c_j = a_j$ come next.*
3. *Those with $c_j < a_j$ come last.*

Proof

Immediate consequence of Lemma 17.3. □

Lemma 17.4

Let σ define an OPMFT COPS that satisfies Theorem 17.3. Its length is unaffected by the relative order of jobs with $a_j = c_j$.

Proof

Follows from Lemma 17.3. □

Lemma 17.5

There is an OPMFT COPS in which all jobs with $c_j > a_j$ are at the left end in nondecreasing order of $a_j + b_j$.

Proof

From Theorem 17.3, we know that there is an OPMFT COPS in which all jobs with $c_j > a_j$ are at the left end. Let σ $(1, 2, \ldots, n)$ define such an OPMFT COPS. Let j be the least integer such that

1. $a_j + b_j > a_{j+1} + b_{j+1}$
2. $c_j > a_j$
3. $c_{j+1} > a_{j+1}$

If there is no such j, then the lemma is established. So, assume that such a j exists. Figure 17.5 (a) shows the relevant part of the schedule. A denotes the time span between the finish of task a_{j+1} and the finish of the task that immediately precedes c_j (in the figure, this happens to coincide with the start of c_j). Figure 17.5 (b) shows the relevant part of the schedule, σ', that results from interchanging the jobs j and $j+1$. We shall show that $\tau' \leq \tau$. As a result, the finish time of σ' is no more than that of σ. So, σ' is also an OPMFT schedule. By repeated application of this exchange process, σ is transformed into an OPMFT that satisfies the lemma.

Case (a): $\boldsymbol{b_j \leq a_{j+1} + A}$ **and** $\boldsymbol{b_{j+1} \leq A + a_j}$

Now, $b_j < c_{j+1} + A$ and $b_{j+1} < A + c_j$. So, $\tau = t + a_j + a_{j+1} + A + c_j + c_{j+1} = \tau'$.

Case (b): $\boldsymbol{b_j \leq a_{j+1} + A}$ **and** $\boldsymbol{b_{j+1} > A + a_j}$

The conditions for this case imply that $A + a_j + b_j < A + a_{j+1} + b_{j+1}$ or $a_j + b_j < a_{j+1} + b_{j+1}$ which contradicts the assumption on j. Hence, this case cannot arise.

Case (c): $\boldsymbol{b_j > a_{j+1} + A}$ **and** $\boldsymbol{b_{j+1} \leq A + a_j}$

Since, $c_j > a_j, b_{j+1} < A+c_j, \tau = t+a_j+b_j+c_j+c_{j+1}$, and $\tau' = t+a_{j+1}+a_j+\max\{A+c_{j+1}, b_j\}+c_j$. For τ' to be $\leq \tau$, we need

$$b_j + c_{j+1} \geq a_{j+1} + \max\{A + c_{j+1}, b_j\}$$

So, if $b_j \geq A + c_{j+1}$, we need $b_j + c_{j+1} \geq a_{j+1} + b_j$ or $c_{j+1} \geq a_{j+1}$. This is true by choice of j. If $b_j < A+c_{j+1}$, we need $b_j+c_{j+1} \geq a_{j+1}+A+c_{j+1}$ or $b_j \geq a_{j+1}+A$. This is part of the assumption for this case.

Case (d): $\boldsymbol{b_j > a_{j+1} + A}$ **and** $\boldsymbol{b_{j+1} > A + a_j}$

This time, $\tau = t + a_j + \max\{b_j + c_j, a_{j+1} + b_{j+1}\} + c_{j+1} = t + \max\{a_j + b_j + c_j + c_{j+1}, a_j + a_{j+1} + b_{j+1} + c_{j+1}\}$, and $\tau' = t + a_{j+1} + \max\{b_{j+1} + c_{j+1}, a_j + b_j\} + c_j = t + \max\{a_{j+1} + b_{j+1} + c_{j+1} + c_j, a_j + b_j + c_j + a_{j+1}\}$. Since, $a_j + b_j > a_{j+1} + b_{j+1}$, $a_j + b_j + c_j + c_{j+1} > a_{j+1} + b_{j+1} + c_{j+1} + c_j$. Also, since $c_{j+1} > a_{j+1}$, $a_j + b_j + c_j + c_{j+1} > a_j + b_j + c_j + a_{j+1}$. Hence, $\tau > \tau'$. □

Lemma 17.6

There is an OPMFT COPS in which all jobs with $c_j < a_j$ are at the right end in nonincreasing order of $b_j + c_j$.

Proof

Similar to that of Lemma 17.5. □

Theorem 17.4 [1]

There is an OPMFT COPS in which the preprocessing order satisfies the following:

1. *Jobs with $c_j > a_j$ come first and in nondecreasing order of $a_j + b_j$.*
2. *Those with $c_j = a_j$ come next in any order.*
3. *Those with $c_j < a_j$ come last and in nonincreasing order of $b_j + c_j$.*

Proof

This follows from Theorem 17.3 and the fact that the proofs of Lemmas 17.4, 17.5, 17.6 are local to the portion of the schedule they are applied to. □

Theorem 17.4 results in the simple $O(n \log n)$ algorithm given below to find a preprocessing order that defines a COPS which is an OPMFT schedule. We shall call this algorithm OOPS(1).

Step 1. Partition the jobs into three sets L, M, and R such that $L = \{j \mid c_j > a_j\}$, $M = \{j \mid c_j = a_j\}$, and $R = \{c_j < a_j\}$.
Step 2. Sort the jobs in L such that $a_j + b_j \leq a_{j+1} + b_{j+1}$. Let $\bar{L}$ be the resulting ordered sequence.
Step 3. Sort the jobs in R such that $b_j + c_j \geq b_{j+1} + c_{j+1}$. Let $\bar{R}$ be the resulting ordered sequence.
Step 4. The preprocessing order for the COPS is: $\bar{L}$ followed by the jobs in M in any order followed by $\bar{R}$.

17.2.3 Reverse Order Postprocessing

For any given preprocessing permutation, σ, we can construct a reverse-order schedule as below:

1. The master preprocesses the n jobs in the order σ.
2. Slave i begins the slave processing of job i as soon as the master completes its preprocessing.
3. The master begins the postprocessing of the last job (say k) in σ as soon as its slave task is complete.
4. The master begins the postprocessing of job $j \neq k$ at the later of the two times (a) when it has finished the postprocessing of the succesor of j in σ, and (b) when slave j has finished b_j.

Schedules constructed in the above manner will be referred to as *canonical reverse order schedules* (CROS). Given a preprocessing permutation σ, the corresponding CROS is unique. It is easy to establish that every minimum finish-time reverse order (ROMFT) schedule is a CROS. So, we can limit ourselves to finding a minimum finish-time CROS.

Lemma 17.7

Let $\sigma = (1, 2, \ldots, n)$ be a preprocessing permutation. Let $j < n$ be such that $b_j < b_{j+1}$. Let σ' be obtained from σ by interchanging jobs j and $j+1$. Let τ and τ', respectively, be the finish times of the CROSs S and S' corresponding to σ and σ'. $\tau' \leq \tau$.

Proof

Let t be the time at which job $j+2$ finishes in S and S'. If $j = n-1$, let $t = 0$. Let s_j (s'_j) be the time at which task b_j finishes in S (S'). Let s_{j+1} and s'_{j+1} be similarly defined. From the definition of a CROS, it follows that:

$$s_j = \sum_1^j a_k + b_j \qquad s_{j+1} = \sum_1^{j+1} a_k + b_{j+1} \tag{17.1}$$

$$s'_j = \sum_1^{j+1} a_k + b_j \qquad s'_{j+1} = \sum_1^{j+1} a_k - a_j + b_{j+1} \tag{17.2}$$

Let q (q') be the time at which c_j (c_{j+1}) finishes in σ (σ'). It is sufficient to show that $q' \leq q$. We see that

$$\begin{aligned} q &= \max\{\max\{t, s_{j+1}\} + c_{j+1}, s_j\} + c_j \\ &= \max\{t + c_j + c_{j+1}, s_{j+1} + c_j + c_{j+1}, s_j + c_j\} \end{aligned} \tag{17.3}$$

and

$$\begin{aligned} q' &= \max\{\max\{t, s'_j\} + c_j, s'_{j+1}\} + c_{j+1} \\ &= \max\{t + c_j + c_{j+1}, s'_j + c_j + c_{j+1}, s'_{j+1} + c_{j+1}\} \end{aligned} \tag{17.4}$$

From Equation (17.1) to Equation (17.3), and the inequality $b_j < b_{j+1}$, we obtain

$$\begin{aligned} s'_j + c_j + c_{j+1} &= s_{j+1} + b_j - b_{j+1} + c_j + c_{j+1} \\ &< s_{j+1} + c_j + c_{j+1} \\ &\leq q \end{aligned} \tag{17.5}$$

and

$$\begin{aligned} s'_{j+1} + c_{j+1} &= s_{j+1} - a_j + c_{j+1} \\ &< s_{j+1} + c_{j+1} \\ &< s_{j+1} + c_{j+1} + c_j \\ &\leq q \end{aligned} \tag{17.6}$$

From Equation (17.3) to Equation (17.6), it follows that $q' \leq q$. □

Theorem 17.5 [1]

The CROS defined by the ordering $b_1 \geq b_2 \geq \cdots \geq b_n$ is an ROMFT schedule.

Proof

Follows from Lemma 17.7. □

Using Theorem 17.5, one readily obtains an $O(n \log n)$ algorithm to construct an ROMFT schedule. This algorithm is called *OROS*(1).

17.2.4 No Wait in Process

In this section we consider the following single-master master-slave scheduling problems:

1. [MFTNW] Minimize finish time subject to the no-wait-in-process constraint.
2. [OP-MFTNW] This is the order-preserving version of MFTNW. That is, minimize finish time subject to the no-wait-in-process and order-preserving constraints.
3. [RO-MFTNW] Minimize finish time subject to the no-wait-in-process and reverse-order constraints.

The first two of the above problems are NP-hard, while the third admits a polynomial time solution. Our NP-hard proofs use the subset sum problem which is known to be NP-hard [13]. This problem is defined below:

Input: A collection of positive integers x_i, $1 \leq i \leq n$ and a positive integer M.
Output: "Yes" iff there is a subset with sum exactly equal to M.

Theorem 17.6 [1]

MFTNW is NP-hard.

Proof

From any instance of the subset sum problem, we may construct an equivalent instance of MFTNW as below:

$$\begin{aligned} a_i = c_i = x_i/2, b_i &= \epsilon, 1 \leq i \leq n \\ a_{n+1} = c_{n+1} = S - M + 1, b_{n+1} &= M + n\epsilon \\ a_{n+2} = c_{n+2} = M + 1, b_{n+2} &= S - M + n\epsilon \end{aligned}$$

where S is the sum of the x_i's and $0 < \epsilon < 1/n$.

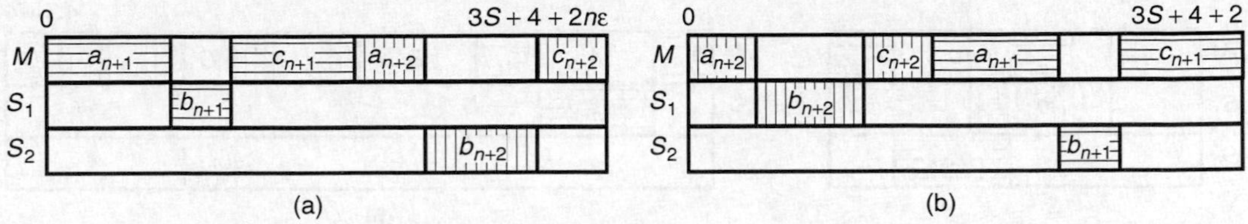

FIGURE 17.6 Templates for NP-hard proof.

In the no wait case, the master processor cannot preempt any job as such a preemption would violate the no wait constraint. In the preceding section, we remarked that there is no advantage to preemptions on slave processors. So, we may assume nonpreemptive schedules. Since $a_{n+1} = c_{n+1} > b_{n+2}$, the preprocessing and/or postprocessing tasks of job $n+1$ cannot be done while a slave is working on job $n+2$. Similarly, the preprocessing and/or postprocessing tasks of job $n+2$ cannot be overlapped with the slave task of job $n+1$. Hence, every no wait schedule has a finish time f that is at least the sum of the task times of these two jobs. That is,

$$f \geq a_{n+1} + b_{n+1} + c_{n+1} + a_{n+2} + b_{n+2} + c_{n+2} = 3S + 4 + 2n\epsilon$$

There are exactly two templates for schedules with this length. One has job $n+1$ processed before job $n+2$ and the other has $n+2$ preceding $n+1$ (see Figure 17.6).

To complete the schedule using either of the templates and not exceed the finish time of $3S+4+2n\epsilon$, some of the remaining jobs must fully overlap with b_{n+1} and the remainder with b_{n+2}. For this, the sum of the first group's task times cannot exceed $b_{n+1} = M + n\epsilon$ and the sum of second group's task times cannot exceed $b_{n+2} = S - M + n\epsilon$. Since the sum of the task times for the remaining jobs is $S + n\epsilon$ and $\epsilon < 1/n$, the only way to accomplish this is when there is a subset of the x_i's that sums to M. Hence, MFTNW is NP-hard. □

Theorem 17.7 [1]

OP-MFTNW is NP-hard.

Proof

The construction is similar to that of Theorem 17.6. The task times for the $n+2$ jobs are:

$$a_i = c_i = x_i, \quad b_i = S - x_i + 1, \quad 1 \leq i \leq n$$
$$a_{n+1} = c_{n+1} = S - M + 1, \quad b_{n+1} = M$$
$$a_{n+2} = c_{n+2} = M + 1, \quad b_{n+2} = S - M$$

The finish time is at least the sum of the master processor task times. So,

$$f \geq \sum a_i + \sum c_i = 4S + 4$$

It is easy to see that there is an order preserving no wait schedule with length $4S+4$ whenever there is a subset of the s_i's that sums to M. We shall show that whenever there is a schedule with this length, there is a subset that sums to M.

As in the previous proof, the tasks of jobs $n+1$ and $n+2$ cannot overlap. So, jobs $n+1$ and $n+2$ are done in sequence. Suppose that job $n+1$ is done before $n+2$ (the case $n+2$ before $n+1$ is similar). Since the sum of the task times for these two jobs is $3S+4$, the only way to finish processing by time $4S+4$ is for the master processor to be busy throughout the time the slaves are working on tasks b_{n+1} and b_{n+2} and for task a_{n+1} to begin by time S. The first requirement means that there are only S other time units when the master can work on the remaining S units of pre- and postprocessing needed by jobs $1, \ldots, n$. There are three cases to consider.

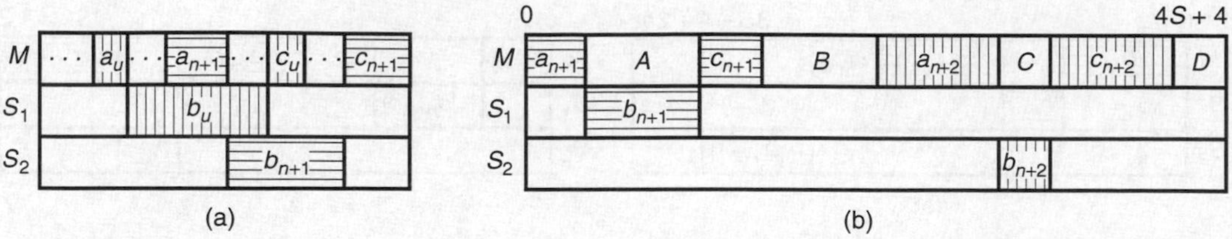

FIGURE 17.7 Templates for order preserving NP-hard proof.

Case 1. *There is at least one job whose preprocessing is done before a_{n+1} and whose postprocessing is done after a_{n+1}.* Let u, $1 \leq u \leq n$, be the first such job. The postprocessing of this job must be done while a slave is working on b_{n+1} as $a_{n+1} + b_{n+1} = S + 1 > b_u = S - x_u + 1$. Hence, we have the situation shown in Figure 17.7 (a).

The tasks (if any) scheduled between a_{n+1} and c_u, must be preprocessing tasks. To see this, note that to schedule a postprocessing task here, the corresponding preprocessing task must have been scheduled either before a_u (in which case u is not the first job with preprocessing before a_{n+1} and postprocessing after a_{n+1}), or in between a_u and a_{n+1} (in which case the order requirement is violated as the postprocessing of this job precedes that of u), or between a_{n+1} and c_u (which is not possible as the sum of task lengths for each of jobs $1, \ldots, n$ exceeds $S + 1$ which in turn is larger than b_{n+1}).

The tasks scheduled between a_u and a_{n+1} are either postprocessing tasks of jobs started before a_u or preprocessing tasks of jobs that will finish after c_u (because of the order requirement). Hence, the tasks beginning with a_u and ending just before c_u that are processed by the master correspond to different jobs. The total amount of time from the beginning of a_u to the start of c_u is $a_u + b_u = S + 1$. Subtracting a_{n+1} from this leaves us with M units of time, all of which must be utilized by the master in order for the schedule to complete by $4S + 4$. This can happen iff there is a subset of the x_i's that sums to M.

Case 2. *There is at least one job whose pre- and postprocessing are done before a_{n+1}.* Let u be one such job. Since the sum of the task lengths of u is $S + x_u + 1$, task a_{n+1} cannot begin until $S + x_u + 1$ and so the schedule cannot complete by $4S + 4$. Therefore, this case is not possible.

Case 3. *Task a_{n+1} is the first task scheduled.* Figure 17.7 (b) shows the scheduling template for this case. For the schedule length to be $4S + 4$, the total time represented by the regions A, B, C, and D must be $2S$. The master processor cannot be idle in any of these regions as the amount of pre- and postprocessing not scheduled in Figure 17.7 (b) is exactly $2S$. Because of the order constraint, in region A, we can schedule only the preprocessing of some subset of the jobs $1, \ldots, n$. Hence, there needs to be a subset of the x_i's that sums to $b_{n+1} = M$.

Hence, OP-MFTNW is NP-hard. □

Note that when there is no ordering constraint between pre- and postprocessing and also when these two orders are required to be the same, there is always at least one feasible solution (i.e., process the jobs in sequence using any permutation). When the postprocessing order is required to be the reverse of the preprocessing order and no wait is permitted in process, then the processing of the jobs must be fully nested. That is, the processing of the jth scheduled job must begin and end while a slave is working on the $(j-1)$th job. As a result, if jobs are preprocessed in the order $1, 2, \ldots, n$, then the following must be true:

$$b_i \geq a_{i+1} + b_{i+1} + c_{i+1}, \quad 1 \leq i < n \tag{17.7}$$

Since the a_j's and c_j's are positive, it follows that

$$b_1 > b_2 > \cdots > b_n \tag{17.8}$$

The preceding inequality implies a unique ordering of the jobs. The algorithm to determine feasibilty, as well as a feasible schedule that minimizes both the finish and mean finish times is [1]:

1. (**Verify Equation 17.8**) Sort the jobs into decreasing order of b_j's. If such an ordering does not exist, there is no feasible schedule. In this case, terminate.
2. (**Verify Equation 17.7**) For $i = 1, \ldots, n-1$, verify that $b_i \geq a_{i+1} + b_{i+1} + c_{i+1}$. If there is an i for which this is not true, then there is no feasible schedule. In this case, terminate.
3. The minimum finish time and mean finish time schedule is obtained by preprocessing the jobs in the order determined in step 1.

The complexity of the above algorithm is readily seen to be $O(n \log n)$.

17.3 Multiple Master Systems

For master-slave systems with multiple master processors we can distinguish two classes of problems. In the first class we require both pre- and postprocessing tasks to be processed by the same processor; we shall refer to such systems as *restricted multiple master systems*. In the second class we allow the pre- and postprocessing tasks of each job to be processed by different processors; we shall refer to such systems as *unrestricted multiple master systems*. Both modes of operation are applicable in semiconductor testing in the presence of multiple burn-in ovens.

For unrestricted multiple master systems we need to be careful about the definition of order-preserving and reverse-order schedules as the pre- and postprocessing tasks of a job may be done by different master processors.

Definition 17.2

For multiple master processor systems we shall say that a schedule is order preserving iff for every pair of jobs i and j such that the preprocessing of i begins before the preprocessing of j, the postprocessing of i completes before or at the same time as the postprocessing of j.

Definition 17.3

For multiple master processor systems we shall say that a schedule is a reverse order schedule iff for every pair of jobs i and j such that the preprocessing of i begins before the preprocessing of j, the postprocessing of i completes after or at the same time as the postprocessing of j.

A versatile heuristic, referred to as *general*, that obtains multimaster schedules with an error bound of at most 2 is developed in Section 17.3.1. For the case of reverse order sequencing a heuristic with worst case error bound $2 - \frac{1}{m}$ (m is the number of master processors) is presented in Section 17.3.2.

17.3.1 A General Heuristic

The heuristic *general* may be used for both restricted and unrestricted systems as well as when constraints are placed between the orders in which the pre- and postprocessing tasks are executed. Before presenting this heuristic, we define the *first available machine* (FAM) rule. In this, jobs are assigned to master processors one-at-a-time. Each job has a time t_i associated with it and the jobs are considered in a given order σ. When a job is considered, it is assigned to the master on which the sum of the times of already assigned jobs is the least (ties are broken arbitrarily).

17.3.1.1 Heuristic *general(m)*

Step 1. For each job, let $t_i = a_i + c_i$. Sort the jobs so that $t_1 \geq t_2 \geq \cdots \geq t_n$.

Step 2. Consider the jobs in this order and use the FAM rule to assign jobs to masters.

Step 3. On each master, schedule the preprocessing tasks in any order from time 0 to time T, where T is the sum of the preprocessing tasks of the jobs assigned to this master. The slave tasks are scheduled to begin as soon as their corresponding preprocessing tasks are complete. The postprocessing tasks are scheduled to begin as soon after the completion of their slave tasks as is feasible.

The heuristic *general*(m) constructs schedules with the property that each job's pre- and postprocessing tasks are done by the same master. Hence, the schedules are feasible for both the restricted and unrestricted master models. The complexity of the heuristic is readily seen to be $O(n \log n)$.

Let $C^{general}$ be the makespan of the schedule generated by heuristic *general*. Let C^*_{UMFT} and C^*_{RMFT}, respectively, be the makespans of the optimal unrestricted and restricted master system schedules.

Theorem 17.8 [12]

$C^{general}/C^*_{UMFT} \leq 2$ and $C^{general}/C^*_{RMFT} \leq 2$.

Proof

Since $C^*_{UMFT} \leq C^*_{RMFT}$, it is sufficient to show that $C^{general}/C^*_{UMFT} \leq 2$. Assume that on the kth master the last postprocessing task completes at time $C^{general}$. If there is no idle time on this master, then from step 2 it follows that

$$C^{general} \leq \frac{1}{m}\sum_{i=1}^{l-1}(a_i + c_i) + (a_l + c_l) \leq \frac{1}{m}\sum_{i=1}^{n}(a_i + c_i) + \frac{m-1}{m}(a_l + c_l)$$

where l is the last job assigned to master k by the FAM rule. Since, $C^*_{UMFT} \geq \frac{1}{m}\sum_{i=1}^{n}(a_i + c_i)$ and $C^*_{UMFT} \geq a_l + c_l$, we get

$$C^{general} \leq C^*_{UMFT} + \frac{m-1}{m}C^*_{UMFT} = \left(2 - \frac{1}{m}\right)C^*_{UMFT}$$

or $C^{general}/C^*_{UMFT} \leq 2 - \frac{1}{m}$.

If the kth master has idle time, then from step 3 it follows that there is a job q scheduled on this master such that the master is busy from time 0 to the start of b_q and again from the finish of b_q to time $C^{general}$. Let Q be the set of jobs assigned to this master in step 2.

$$C^{general} \leq \sum_{i \in Q}(a_i + c_i) + b_q = \sum_{i \in Q}(a_i + c_i) - (a_q + c_q) + (a_q + b_q + c_q)$$

From step 2, it follows that $\sum_{i \in Q}(a_i + c_i) \leq \frac{1}{m}\sum_{i=1}^{n}(a_i + c_i) + \frac{m-1}{m}(a_l + c_l)$, where l is the last job assigned to the master in step 2. Because of the ordering of step 1, $a_l + c_l \leq a_q + c_q$. Hence,

$$C^{general} \leq \frac{1}{m}\sum_{i=1}^{n}(a_i + c_i) + (a_q + b_q + c_q)$$

Each term on the right hand side of the above inequality is easily seen to be no more than C^*_{UMFT}. Hence, $C^{general} \leq 2C^*_{UMFT}$.

Combining the bounds for the two cases, we get $C^{general}/C^*_{UMFT} \leq 2$. □

To see that the bound of 2 is tight, consider the $n(m-1)+2$ job instance in which the first job's pre-, slave, and postprocessing tasks are given by $(n-\epsilon, \epsilon, \epsilon/2)$, the next $n(m-1)$ job task times are $(1/2, \epsilon, 1/2)$ and the last job has times (ϵ, n, ϵ). Here, $0 < \epsilon < 1/2$. The jobs have been given in the order produced in step 1. The heuristic assigns jobs 1 and $n(m-1)+2$ to master 1. The remaining jobs are distributed evenly across the remaining masters. If in step 3, the first master is scheduled to process a_1 first, then $C^{general} = 2n + \epsilon$. However, $C^*_{UMFT} = C^*_{RMFT} = n + 2.5\epsilon$. The ratio approaches 2 as $\epsilon \to 0$.

Heuristic *general* may be used to obtain order preserving and reverse order schedules by modifying step 3 to produce such schedules. In fact, since optimal single master order preserving and reverse order schedules can be obtained in polynomial time ([1]), step 3 can generate optimal schedules using the jobs assigned to each master. Since the proof of Theorem 17.8 does not rely on how the schedule is constructed in step 3, the error bound of 2 applies even for the case of order preserving and reverse order schedules.

17.3.2 Restricted Reverse Order Schedules

In this subsection we develop an approximation algorithm for restricted multiple master systems in which each master processor is required to process its postprocessing tasks in an order that is the reverse of the order in which it processes its preprocessing tasks. This problem is abbreviated as $ROS(m)$ (reverse order scheduling with m masters). The $OROS(1)$ algorithm of Section 17.2.3 solves optimally the $ROS(1)$ problem.

The approximation algorithm, *Heuristic* $ROS(m)$, given below obtains schedules with an error bound no more than $2 - 1/m$.

17.3.2.1 Heuristic *ROS(m)*

Step 1. Sort the jobs so that $b_1 \geq b_2 \geq \cdots \geq b_n$.

Step 2. Consider the jobs in this order and use the FAM rule to assign jobs to masters using $t_i = a_i + c_i$.

Step 3. On each master, schedule the preprocessing tasks in the order the jobs were assigned to the master. Schedule the postprocessing tasks in the reverse order and to begin as soon as possible after all preprocessing tasks complete.

Note that in step 1, we obtain the ordering needed to construct an $OROS(1)$ for the n jobs and that in step 3 the jobs assigned to each master are scheduled to form an $OROS(1)$ for that master. The complexity of $ROS(m)$ is easily seen to be $O(n \log n)$. To establish the error bound, we need to first establish two other results. This is done in Lemmas 17.8 and 17.9. The error bound itself is established in Theorem 17.9.

Let I, I', and I'' be three sets of jobs. $I = \{(a_i, b_i, c_i) \mid 1 \leq i \leq n\}$, I' has n jobs defined by $a_i' = c_i' = (a_i + c_i)/2$ and $b_i' = b_i$, and I'' has n jobs defined by $a_i'' = c_i'' = (a_i + c_i)/(2m)$ and $b_i' = b_i$. Let $C_I^*(m)$, $C_{I'}^*(m)$, and $C_{I''}^*(m)$, respectively, denote the makespans of the $OROS(m)$ for I, I', and I''.

Lemma 17.8

$C_I^*(m) = C_{I'}^*(m)$ for all m.

Proof

Let the optimal schedules for I and I' be $S_I^*(m)$ and $S_{I'}^*(m)$, respectively. In $S_I^*(m)$, consider a job k with $a_k \neq c_k$. Let p be the master processor on which job k is scheduled in $S_I^*(m)$. If $a_k < c_k$, then increase the time for which the preprocessing of k is scheduled to $a_k' = (a_k + c_k)/2$ and reduce the time for which its postprocessing is scheduled to $c_k' = (a_k + c_k)/2$. This will require us to shift right by $a_k' - a_k$ all tasks of jobs whose preprocessing is scheduled after the preprocessing of job k on master p and also the slave and postprocessing tasks of job k. This transformation does not increase the schedule length. A similar transformation can be made when $a_k > c_k$. By applying this transformation to all jobs with $a_i \neq c_i$, we transform $S_I^*(m)$ into a feasible reverse order m master schedule for I' without increasing the schedule length. So, $C_{I'}^*(m) \leq C_I^*(m)$.

Using a reverse transformation, we can transform $S_{I'}^*(m)$ into a feasible reverse order schedule for I without increasing the schedule length. So, $C_I^*(m) \leq C_{I'}^*(m)$. Hence, $C_I^*(m) = C_{I'}^*(m)$. □

Lemma 17.9

$C_{I''}^*(1) \leq C_I^*(m)$ for all m.

Proof

From Lemma 17.8, it follows that it is sufficient to show that $C_{I''}^*(1) \leq C_{I'}^*(m)$. In $S_{I'}^*(m)$, we may assume that the preprocessing tasks on each master are scheduled continuously (i.e., with no idle time) from time zero to the time the last preprocessing task on that master completes (this may require us to shift some preprocessing tasks to the left). Also, we may assume that the postprocessing tasks are scheduled continuously from the start of the first postprocessing task on the master to time $C_{I'}^*(m)$ (this may require us to shift some postprocessing tasks to the right). Let F_i and B_i, respectively, denote the finish time of a_i'

and the start time of c'_i in $S^*_{I'}(m)$. Since $a'_i = c'_i = \frac{1}{2}(a_i + c_i)$, it follows that $F_i = C^*_{I'}(m) - B_i$ for all i. Assume that the jobs are numbered so that $F_i \leq F_{i+1}, 1 \leq i < n$. Hence, $B_i \geq B_{i+1}, 1 \leq i < n$.

Since $F_i \leq F_{i+1}, 1 \leq i < n, mF_i \geq \sum_{j=1}^{i} a'_j$ or $F_i \geq (\sum_{j=1}^{i} a'_j)/m$. So, $C^*_{I'}(m) - B_i \geq (\sum_{j=1}^{i} a'_j)/m = (\sum_{j=1}^{i} c'_j)/m$. Since, $C^*_{I'}(m) \geq F_i + b'_i + (C^*_{I'}(m) - B_i) \geq (\sum_{j=1}^{i}(a'_j + c'_j))/m + b'_i$, for all i, we get

$$C^*_{I'}(m) \geq \max_i \left\{ \left(\sum_{j=1}^{i} (a'_j + c'_j) \right) \Big/ m + b'_i \right\}$$

Now, consider the reverse order schedule $S_{I''}(1)$ obtained by scheduling the preprocessing tasks of I'' in the order 1, 2, ..., n. In this schedule, we may assume that the preprocessing tasks are scheduled continuously and the slave and postprocessing tasks are scheduled as early as is feasible. Let $C_{I''}(1)$ be its makespan. Clearly, $C^*_{I''}(1) \leq C_{I''}(1)$. If there is no idle time on the master, then $C_{I''} = \sum_{j=1}^{n}(a''_j + c''_j) = (\sum_{j=1}^{n}(a'_j + c'_j))/m < \sum_{j=1}^{n}(a'_j + c'_j)/m + b'_n \leq \max_i\{(\sum_{j=1}^{i}(a'_j + c'_j))/m + b'_i\} \leq C^*_{I'}(m)$.

If there is idle time on the master, then there is a k such that $C_{I''} = \sum_{j=1}^{k}(a''_j + c''_j) + b''_k = (\sum_{j=1}^{k}(a'_j + c'_j))/m + b'_k \leq \max_i\{(\sum_{j=1}^{i}(a'_j + c'_j))/m + b'_i\} \leq C^*_{I'}(m)$.

So, $C^*_{I''}(1) \leq C_{I''}(1) \leq C^*_{I'}(m) = C^*_I(m)$. □

Effectively, Lemmas 17.8 and 17.9 show that the makespan of the schedule produced by *OROS*(1) on I'' is a lower bound on the optimal makespan value for *ROS*(m) which is denoted by $C^*_I(m)$. The following theorem makes use of this result.

Theorem 17.9 [12]

Let $C^{ROS}_I(m)$ be the makespan of the schedule generated by ROS(m) on instance I. $C^{ROS}_I(m)/C^*_I(m) \leq 2 - 1/m$ and this bound is tight.

Proof

Assume that the jobs are numbered so that $b_i \geq b_{i+1}, 1 \leq i < n$. Using the transformations of Lemma 17.8, we can transform $S^{ROS}_I(m)$ into a schedule for I' that has the same makespan. Let this schedule be $S_{I'}(m)$. In this schedule, we may assume that the preprocessing tasks on each master are scheduled continuously from time zero to the time the last preprocessing task on that master completes and that the postprocessing tasks are scheduled continuously from the start of the first postprocessing task on the master to time $C^{ROS}_I(m)$. Let F_i and B_i, respectively, denote the finish time of a'_i and the start time of c'_i in $S^{ROS}_{I'}(m)$. From the application of the FAM rule in step 2 of *ROS*(m), it follows that

$$F_i \leq \frac{1}{m}\sum_{j=1}^{i-1} a'_j + a'_i = \frac{1}{m}\sum_{j=1}^{i} a'_j + \frac{m-1}{m} a'_i$$

and

$$B_i \geq C^{ROS}_I(m) - \left[\frac{1}{m}\sum_{j=1}^{i} c'_j + \frac{m-1}{m} c'_i \right]$$

If there is no idle time on at least one of the masters, then let Q be the set of jobs processed by any master, say k, with no idle time. If l is the last job preprocessed on the kth master, it follows that $C^{ROS}_I(m) = \sum_{i \in Q}(a'_i + c'_i) \leq \frac{1}{m}\sum_{i=1}^{n} a'_j + \frac{m-1}{m} a'_l + \frac{1}{m}\sum_{i=1}^{n} c'_j + \frac{m-1}{m} c'_l = \frac{1}{m}\sum_{i=1}^{n}(a'_j + c'_j) + \frac{m-1}{m}(a'_l + c'_l)$ $\leq C^*_{I'}(m) + \frac{m-1}{m} C^*_{I'}(m) = C^*_I(m) + \frac{m-1}{m} C^*_I(m)$.

If all masters have idle time, then $C^{ROS}_I(m) = \max_i\{F_i + b'_i + C^{ROS}_I(m) - B_i\} \leq \max_i\{\frac{1}{m}\sum_{j=1}^{i}(a'_j + c'_j) + \frac{m-1}{m}(a'_i + c'_i) + b'_i\}$. Now consider the instance I''. In *OROS*(1), the preprocessing tasks of I'' are scheduled in the order $1, 2, \ldots, n$ because $b_i \geq b_{i+1}$. As a result, $C^*_{I''}(1) \geq \max_i\{\frac{1}{m}\sum_{j=1}^{i}(a'_j + c'_j) + b_i\}$.

From Lemma 17.9, we know that $C_I^*(m) \geq C_{I''}^*(1)$. So, $C_I^{ROS}(m) \leq \max_i\{\frac{1}{m}\sum_{j=1}^{i}(a'_j + c'_j) + \frac{m-1}{m}(a'_i + c'_i) + b'_i\} \leq C_{I''}^*(m) + \frac{m-1}{m}C_{I''}^*(m) \leq C_I^*(m) + \frac{m-1}{m}C_I^*(m)$.

Hence, $C_I^{ROS}(m)/C_I^*(m) \leq 2 - 1/m$.

To see that this error bound is tight, consider the m master instance I with $n = m(m-1)+1$ jobs. The first $m(m-1)$ of these jobs have $(a_i, b_i, c_i) = (1, \epsilon, \epsilon)$, where ϵ is a small number. The last job has $(a_n, b_n, c_n) = (m, \epsilon/2, \epsilon)$. The job numbering corresponds to that produced in step 1 of *ROS*(m). In step 2, each master is assigned $m-1$ of the first $m(m-1)$ jobs and one of them gets job n in addition. The optimal schedule for this master has makespan $2m - 1 + \epsilon/2 + m\epsilon$. The *OROS*$(m)$ schedule assigns job n alone to one of the m masters and distributes the remaining jobs equally among the remaining $m-1$ masters. So, $C_I^*(m) = m + (m+1)\epsilon$. So, $C_I^{ROS}(m)/C_I^*(m) = (2m - 1 + \epsilon/2 + m\epsilon)/(m + (m+1)\epsilon)$ which tends to $2 - 1/m$ (from below) as $\epsilon \to 0$. □

The above result has some similarities with the problem of minimizing makespan in a two-stage hybrid flowshop considered in [14]. In this problem preprocessing tasks are executed on the machines of stage 1, postprocessing tasks are executed on the machines of stage 2, and the slave tasks are null. A heuristic with bound $2 - \frac{1}{m}$ was developed for that problem as well.

17.4 Distinct Pre- and Postprocessing Masters

In this section, we consider a variant of the basic master-slave system. In this variant, there are two types of master processors–those that perform the preprocessing tasks and those that perform the postprocessing tasks. This variant is a 3-stage flowshop in which the first and third stage consist of master processors while the second stage consists of slave processors. A set of jobs each consisting of three tasks is to be processed by the 3-stage master-slave system. The first task of each job is referred to as preprocessing master task, the second is called slave task and the third is referred to as postprocessing master task.

A special case of the master-slave variant just described where each of the master stages consists of a single machine was among the first problems ever studied in scheduling theory; see [3–5]. In that research the objective is to find an order preserving minimum makespan sequence. Instead of slave tasks the above literature considers for each job a transfer lag i.e., a minimum amount of time that must elapse from the completion of a preprocessing task to the initiation of the corresponding postprocessing task.

The problems to be considered in this section are as follows:

OPS(m_1, m_2). The problem to find a minimum makespan order preserving schedule for the master-slave variant with m_1 preprocessors and m_2 postprocessors

ROS(m_1, m_2). The problem to find a minimum makespan reverse order schedule for the master-slave variant with m_1 preprocessors and m_2 postprocessors

UMFT(m_1, m_2). The problem to find a minimum makespan schedule for the master-slave variant with m_1 preprocessors and m_2 postprocessors with no order restriction

An optimal algorithm *OPS*$(1, 1)$ has been developed in [4], and is outlined as follows:

1. Jobs with $c_j \geq a_j$ come first in nondecreasing order of $a_j + b_j$.
2. Jobs with $c_j < a_j$ come last in nonincreasing order of $b_j + c_j$.
3. Generate the order preserving schedule whose preprocessing tasks are ordered according to steps 1 and 2.

Note the similarity between this algorithm and the algorithm *OOPS*(1) of Section 17.2.2. Clearly *OOPS*$(1, 1)$ has complexity $\mathcal{O}(n \log n)$.

A simple interchange argument shows that there always exists an optimal schedule for *ROS*$(1, 1)$ in which all preprocessing tasks are executed prior to the initiation of any postprocessing task. Then, since the first postprocessing task follows the last preprocessing task, the optimal makespan value is the same as if both pre- and postprocessing tasks were processed by the same processor. Hence, *ROS*$(1, 1)$ is equivalent to problem *ROS*(1).

UMFT(1, 1) has been shown to be strongly $\mathcal{NP}$-complete in [6]. An interesting special case of this problem is when all slave tasks are null (i.e., $b_i = 0$ for $1 \leq i \leq n$). In this case *UMFT*(1, 1) reduces to the traditional 2-stage flowshop which is solved by Johnson's algorithm [15], while $UMFT(m_1, m_2)$ reduces to the *2-stage hybrid flowshop*; see [14].

In the following we present a mirror image of the FAM rule and call it *last busy machine* rule (**LBM**). This rule will be used later to assign postprocessing tasks. Given a constant $T > 0$ and an ordering S of postprocessing tasks a description of the rule is given next.

LBM rule

1. Set $f_m := T$ for every postprocessor m, $m = 1, \ldots, m_2$.
2. Let c_j be the last unscheduled task of S and m a postprocessor with largest f_m. Schedule task c_j on the m-th postprocessor to finish at time f_m.
3. Set $f_m := f_m - c_j$ and $S := S - \{c_j\}$. If $S \neq \emptyset$ then goto step 2 else Stop.

Remark The value of f_m is the time that machine m becomes busy. In step 2 we assign c_j to a postprocessor with largest f_m, i.e., the last busy postprocessor. Hence we call this rule the last busy machine rule. Also, note that the value of T is only a reference point and has no effect on the allocation of postprocessing tasks to the postprocessors.

We can now describe a generic heuristic H for the $OPS(m_1, m_2)$, $ROS(m_1, m_2)$, and $UMFT(m_1, m_2)$ problems. Appropriate changes at step 1 and step 4 of H, produce the heuristics H_{OP}, H_{RO} and H_U for the above three problems, respectively.

Heuristic H

1. Specify a sequence S of the jobs.
2. Apply the FAM rule on the preprocessing tasks of S.
3. Apply the LBM rule on the postprocessing tasks of S.
4. Rearrange some of the tasks of the resulting schedule.

At step 1 of H a sequence S of tasks is produced. As we will see, all of these sequences are variants of the Johnson sequence for the traditional 2-machine flowshop. At step 2 (3) an assignment of preprocessing (postprocessing) tasks is made on the preprocessors (postprocessors). At step 4 of our heuristics we specify which tasks need to be shifted and how.

17.4.1 Order Preserving Scheduling – *OOPS*(m_1,m_2)

OPS(1,1) is solved in $\mathcal{O}(n \log n)$ time using the algorithm *OOPS*(1,1) described in Section 17.4. For more than one processor in either the pre- or postprocessing stage the problem $OPS(m_1, m_2)$ is $\mathcal{NP}$-hard since the problem of minimizing makespan on parallel identical machines is a special case (i.e., $b_i = c_i = 0$) that is known to be $\mathcal{NP}$-hard; see [13]. This observation motivates the development of heuristics which have good performance. In what follows we present a heuristic H_{OP} with worst case error bound $2 - \frac{1}{m}$, where $m = \max\{m_1, m_2\}$. H_{OP} has the general format of the heuristic H presented earlier and the steps 1 and 4 of H are replaced by the following.

17.4.1.1 Heuristic H_{OP}

Step 1. Apply *OOPS*(1, 1) with respect to the processing times $\{J_i = (\frac{1}{m_1}a_i, b_i, \frac{1}{m_2}c_i) : i = 1, 2, \ldots, n\}$. Let S be the resulting sequence.

Step 4. Start all postprocessing tasks earlier by the maximum amount of time t that does not violate the order and flowshop constraints.

Note that in the resulting sequence S, the order of all pre- and postprocessing tasks is the same. Also, at step 4 we shift <u>all</u> postprocessing tasks t units earlier, and hence the nondecreasing order of completion times after the shift of step 4 remains the same as it was at the end of step 3. As a result, the order preserving property is retained. Since $OOPS(1, 1)$ requires $\mathcal{O}(n \log n)$ time, this is also the computational complexity required by H_{OP}.

The next lemma will be used to find the error bound of H_{OP}. For this lemma we introduce the following *auxiliary* problem. Replace the m_1 parallel identical preprocessors by a single preprocessor, replace the m_2 parallel identical postprocessors by a single postprocessor and replace the processing time requirements (a_i, b_i, c_i) of J_i by the requirement $(\frac{1}{m_1} a_i, b_i, \frac{1}{m_2} c_i)$. This way we define an auxiliary problem on a single flowline. We will refer to this problem as **AMS.**

Lemma 17.10 [16]

Let C_{LB} be the completion time of the solution found by $OOPS(1, 1)$ on the auxiliary problem AMS. Then, $C_{LB} \leq C_{OPS(m_1,m_2)}$, where $C_{OPS(m_1,m_2)}$ denotes the optimal makespan value of $OPS(m_1, m_2)$.

Proof

Consider an optimal solution S^* for $OPS(m_1, m_2)$. Let $S = \{J_1, J_2, \ldots, J_n\}$ be the set of jobs ordered in nondecreasing order of completion times of the slave tasks in S^*. Consider the partial schedule of S^* consisting of the first i preprocessing tasks of S, the first i slave tasks of S, and the last $n - i + 1$ postprocessing tasks of S. Since in S^* the last $n - i + 1$ postprocessing tasks of S start no earlier than the completion time of the first i slave tasks of S (due to the master-slave flowshop constraints), we have that

$$C_{OPS(m_1,m_2)} \geq \frac{1}{m_1} \sum_{j=1}^{i} a_j + b_i + \frac{1}{m_2} \sum_{j=i}^{n} c_j \quad \text{for every } 1 \leq i \leq n$$

For the AMS problem (where the processing time requirements for J_i are $(\frac{1}{m_1} a_i, b_i, \frac{1}{m_2} c_i)$), schedule the jobs according to sequence S. Let C_S be the resulting makespan. Then, it is clear that

$$C_S = \frac{1}{m_1} \sum_{j=1}^{i_0} a_j + b_{i_0} + \frac{1}{m_2} \sum_{j=i_0}^{n} c_j$$

where J_{i_0} is the last job whose postprocessing task starts immediately after the completion of the corresponding slave task (note that such a task always exists; since J_1 satisfies this property).

Combining the last two expressions we get that $C_S \leq C_{OPS(m_1,m_2)}$. However, the sequence S is not necessarily optimal for the AMS problem and hence $C_{LB} \leq C_S$. The last two relations establish that $C_{LB} \leq C_{OPS(m_1,m_2)}$. This completes the proof of the lemma. □

The following theorem establishes the worst case performance of H_{OP}. Let $C_{H_{OP}}$ be the makespan value obtained by H_{OP}.

Theorem 17.10 [16]

$\frac{C_{H_{OP}}}{C_{OPS(m_1,m_2)}} \leq 2 - \frac{1}{m}$ *where* $m := \max\{m_1, m_2\}$, *and the bound is tight.*

Proof

Let $S = \{J_1, J_2, \ldots, J_n\}$ be the order obtained at step 1 of H_{OP}. To see that the sequence produced by H_{OP} is order preserving, note that the nondecreasing order of starting times provided by the FAM rule at step 2 of H_{OP} coincides with the nondecreasing order of completion times provided by the LBM rule at step 3 of H_{OP}.

For the rest of the proof assume that the jobs are indexed in the nondecreasing order of completion times of slave tasks. After the scheduling of the preprocessing tasks at step 2, the completion time of the slave task b_i, say C_i, is

$$C_i \le \frac{1}{m_1}\sum_{j=1}^{i-1} a_j + a_i + b_i = \frac{1}{m_1}\sum_{j=1}^{i} a_j + \frac{m_1-1}{m_1}a_i + b_i$$

for every $1 \le i \le n$, because the FAM rule is used for preprocessing tasks.

At step 3 of H_{OP}, the LBM rule is applied on the postprocessing tasks. Let us consider the postprocessing tasks that follow J_i in S. The elapsed time between the earliest starting time of these tasks and the latest finishing time of these tasks (which is $C_{H_{OP}}$) does not exceed

$$\frac{1}{m_2}\sum_{j=i}^{n} c_j + \frac{m_2-1}{m_2}c_i$$

for every $1 \le i \le n$, because the LBM rule is used for postprocessing tasks.

Note that the schedule produced by H_{OP} has all the postprocessors finish at time $C_{H_{OP}}$. Also, at step 4 of our heuristic we ensure that it is not possible to left-shift the postprocessing tasks any further. As a result there must exist a critical job J_{i_0} whose b_{i_0} task starts immediately after a_{i_0} and finishes immediately before c_{i_0} commences. The postprocessor on which c_{i_0} is allocated, finishes at time $C_{H_{OP}}$ (as any other postprocessor) and therefore

$$C_{H_{OP}} \le \frac{1}{m_1}\sum_{j=1}^{i_0} a_j + \frac{m_1-1}{m_1}a_{i_0} + b_{i_0} + \frac{1}{m_2}\sum_{j=i_0}^{n} c_j + \frac{m_2-1}{m_2}c_{i_0} \tag{17.9}$$

Note that

$$C_{LB} \ge \frac{1}{m_1}\sum_{j=1}^{i_0} a_j + b_{i_0} + \frac{1}{m_2}\sum_{j=i_0}^{n} c_j \tag{17.10}$$

due to the flowshop constraints of the auxiliary problem AMS. Also, it is clear that

$$\frac{m_1-1}{m_1}a_{i_0} + \frac{m_2-1}{m_2}c_{i_0} \le \frac{m-1}{m}C_{OPS(m_1,m_2)}$$

where $m := \max\{m_1, m_2\}$.

Therefore, the schedule produced by steps 2, 3, and 4 of H_{OP} has a makespan no greater than $(2-\frac{1}{m})C_{OPS(m_1,m_2)}$. This completes the proof for the worst case performance of H_{OP}.

To see that the bound of $2-\frac{1}{m}$ is tight for H_{OP}, consider the case where $b_i = c_i = 0$ for all jobs. Then the heuristic H_{OP} reduces to the RDM (for random) list scheduling heuristic studied in [17] which has a tight worst case error bound of $2-\frac{1}{m}$. This completes the proof of the theorem. □

Example 17.1

To illustrate H_{OP}, consider a 5-job example with processing requirements $J_1 = (1,2,3)$, $J_2 = (1,2,4)$, $J_3 = (2,3,4)$, $J_4 = (3,4,6)$ and $J_5 = (6,1,2)$, and the 2-stage flowshop environment with $m_1 = m_2 = 2$. These jobs are already indexed in the order resulting from step 1 of H_{OP}. The Gantt chart of the application of H_{OP} on this 5-job example is given in Figure 17.8. Note that all postprocessors finish at time $C_{OP} = 14$ and that the only critical job is J_4. Namely, the postprocessing tasks of J_1, J_2, J_3, and J_5 could start earlier but this would violate the order preserving requirement.

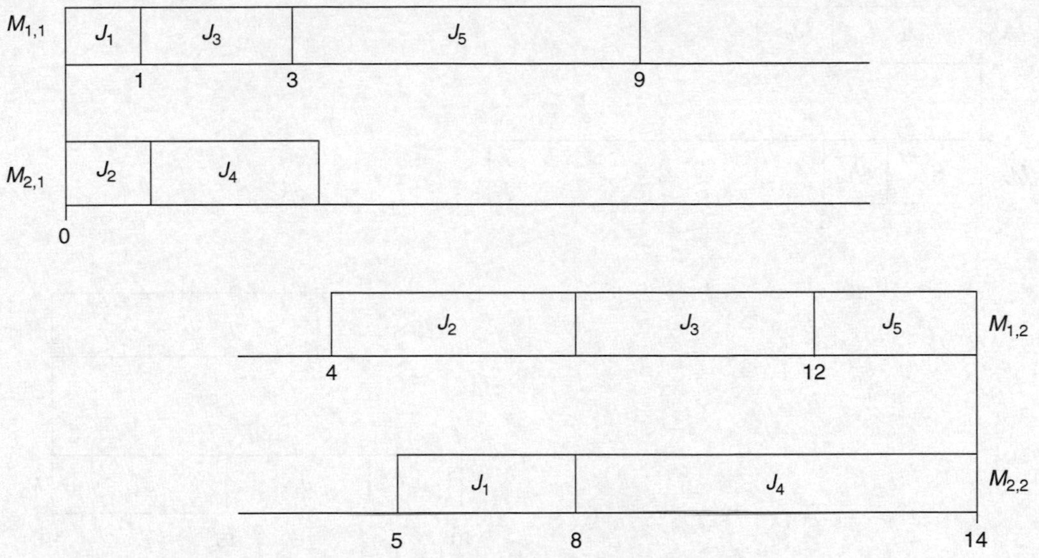

FIGURE 17.8 An application of H_{OP}.

17.4.2 Reverse Ordering – $ROS(m_1, m_2)$

Finding a reverse order sequence in $ROS(1, 1)$ is solved in $\mathcal{O}(n \log n)$ time using the algorithm $OROS(1, 1)$ described in Section 17.4. For more than one processor in either the pre- or postprocessing stage the problem $ROS(m_1, m_2)$ is $\mathcal{NP}$-hard since the problem of minimizing makespan on parallel identical machines is a special case (i.e., $b_i = c_i = 0$) known to be $\mathcal{NP}$-hard; see [13]. In this section we develop a heuristic algorithm for the problem to minimize the makespan of $ROS(m_1, m_2)$. This heuristic too, follows the general format of H, except that steps 1 and 4 are replaced by the following.

17.4.2.1 Heuristic H_{RO}

Step 1. Apply $OROS(1, 1)$ with respect to the processing times $\{J_i = (\frac{1}{m_1}a_i, b_i, \frac{1}{m_2}c_i) : i = 1, 2, \ldots, n\}$. Let S be the resulting sequence.

Step 4. Start all postprocessing tasks earlier by the maximum amount of time t that does not violate the order and flowshop constraints.

In S, the order of preprocessing tasks is reverse to the order of postprocessing tasks. Therefore, at step 3 the heuristic H_{RO} applies LBM on the postprocessing tasks in reverse S order. At step 4 we shift all postprocessing tasks t units earlier and therefore the nonincreasing order of completion times at the end of step 3 does not change at step 4. Since $OROS(1, 1)$ requires $\mathcal{O}(n \log n)$ time, this is also the computational complexity required by H_{RO}. Consider the auxiliary flowshop problem (AMS) defined in the previous subsection.

Lemma 17.11 [16]

Let C_{LB} be the completion time of the solution found by $OROS(1, 1)$ on the auxiliary problem AMS. Then, $C_{LB} \leq C_{ROS(m_1,m_2)}$, where $C_{ROS(m_1,m_2)}$ denotes the optimal makespan value of $ROS(m_1, m_2)$.

Proof

The proof of this lemma is similar to the proof of Lemma 17.10 [16]. □

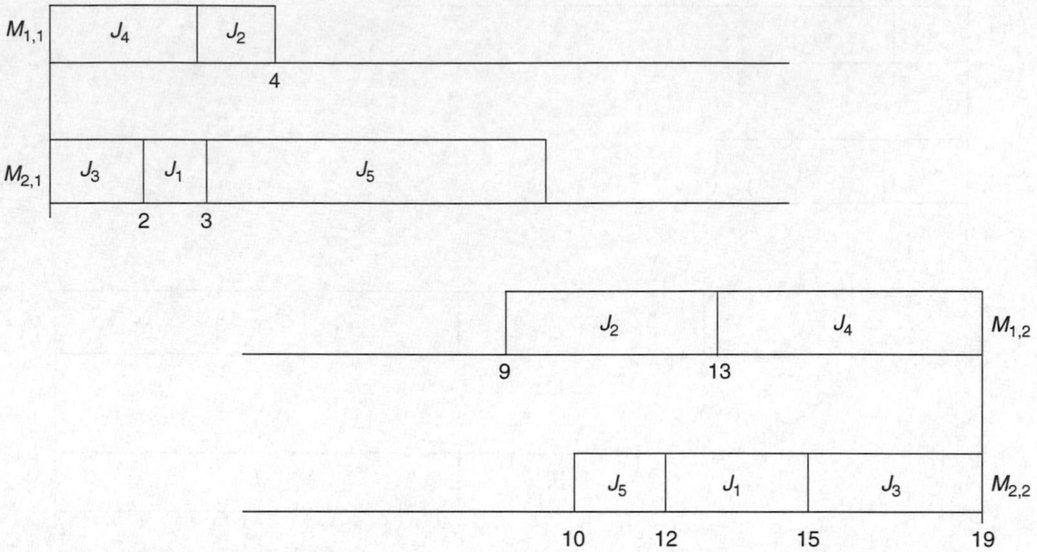

FIGURE 17.9 An application of H_{RO}.

The following theorem establishes the worst case performance of H_{RO}. Let $C_{H_{RO}}$ be the makespan value obtained by H_{RO}.

Theorem 17.11 [16]

$\frac{C_{H_{RO}}}{C_{ROS(m_1,m_2)}} \leq 2 - \frac{1}{m}$ *where* $m := \max\{m_1, m_2\}$, *and the bound is tight.*

Proof

The proof of this theorem is similar to that of Theorem 17.10. In this case inequality (17.9) gets replaced by

$$C_{H_{RO}} \leq \frac{1}{m_1}\sum_{j=1}^{i_0} a_j + \frac{m_1 - 1}{m_1} a_{i_0} + b_{i_0} + \frac{1}{m_2}\sum_{j=1}^{i_0} c_j + \frac{m_2 - 1}{m_2} c_{i_0} \tag{17.11}$$

and inequality (17.10) gets replaced by

$$C_{LB} \geq \frac{1}{m_1}\sum_{j=1}^{i_0} a_j + b_{i_0} + \frac{1}{m_2}\sum_{j=1}^{i_0} c_j \tag{17.12}$$

□

Example 17.2

To illustrate H_{RO}, consider the 5-job instance of Example 17.1. The order produced by step 1 of H_{RO} is J_4, J_3, J_1, J_2, J_5. Note that the reverse order used by step 3 is J_5, J_2, J_1, J_3, J_4. The Gantt chart of the application of H_{RO} on this 5-job example is given in Figure 17.9. Note that all postprocessors finish at time $C_{RO} = 19$ and that the only critical job is J_5. More specifically, the postprocessing tasks of J_1, J_2, J_3, and J_4 could start earlier but this would violate the reverse order constraint.

17.4.3 Unconstrained Ordering — $UMFT(m_1, m_2)$

In this section we investigate the problem $UMFT(m_1, m_2)$ where no order (such as order preserving and reverse orders) constraint is imposed on the pre- and postprocessing tasks. As we mentioned, this problem is $\mathcal{NP}$-hard in the strong sense even if $m_1 = m_2 = 1$. This is in contrast to the polynomial time algorithms $OOPS(1, 1)$ and $OROS(1, 1)$ indicated earlier. In light of the complexity status of $UMFT(m_1, m_2)$, we turn our attention to simple heuristic algorithms that perform well in both the worst case and in practice.

In what follows we present a heuristic H_U with a worst case error bound of 2. H_U utilizes the format of the heuristic H. Also, it utilizes Johnson's algorithm (JA) which is known to minimize makespan in a simple two-stage flowshop (see [15]). The JA algorithm follows immediately from the rule: *Job J_i precedes job J_j if* $\min\{a_i, c_j\} \leq \min\{a_j, c_i\}$.

17.4.3.1 Heuristic H_U

Step 1. Apply JA with respect to the processing times $\{J_i = (\frac{1}{m_1}a_i, \frac{1}{m_2}c_i) : i = 1, 2, \ldots, n\}$. Let S be the resulting sequence.

Step 4. Schedule the slave tasks to start as soon as the corresponding preprocessing tasks complete. Then, schedule the postprocessing tasks to start as soon as possible.

Since JA requires $\mathcal{O}(n \log n)$ time, this is also the computational complexity of H_U. To compute the worst case error bound of H_U we first make the following observations. Consider any instance of $UMFT(m_1, m_2)$. From that instance we can construct an instance in which $b_i = 0$ for all $1 \leq i \leq n$. The flowshop environment that consists of two multistation stages and has null slave tasks is known in the literature as *hybrid flowshop* and is denoted by $HFS(m_1, m_2)$. As a result, if $C^*_{HFS(m_1,m_2)}$ is the optimal makespan of $HFS(m_1, m_2)$ for the data set $(a_i, c_i) : 1 \leq i \leq n$ and $C^*_{UMFT(m_1,m_2)}$ is the optimal makespan of $UMFT(m_1, m_2)$ for the data set $(a_i, b_i, c_i) : 1 \leq i \leq n$, we have that

$$C^*_{HFS(m_1,m_2)} \leq C^*_{UMFT(m_1,m_2)}$$

Furthermore, in [14] it is shown that the makespan C_{JA} obtained by JA when applied to the data $(\frac{1}{m_1}a_i, \frac{1}{m_2}c_i)$ is a lower bound to $C^*_{HFS(m_1,m_2)}$. Therefore, $C_{JA} \leq C^*_{UMFT(m_1,m_2)}$. With these observations we proceed with the following proof. Let S_{H_U} and C_{H_U} be the schedule and makespan respectively obtained by H_U.

Theorem 17.12 [16]

$\frac{C_{H_U}}{C^*_{UMFT(m_1,m_2)}} \leq 2$ *and the bound is tight.*

Proof

An argument similar to that of Theorem 17.10 shows that

$$C_{H_U} \leq \frac{1}{m_1}\sum_{j=1}^{i_0} a_j + \frac{m_1 - 1}{m_1}a_{i_0} + b_{i_0} + \frac{1}{m_2}\sum_{j=i_0}^{n} c_j + \frac{m_2 - 1}{m_2}c_{i_0}$$

where J_{i_0} is the last job whose b_{i_0} task starts immediately after a_{i_0} and finishes immediately before c_{i_0}.

Note that

$$C_{JA} \geq \frac{1}{m_1}\sum_{j=1}^{i_0} a_j + \frac{1}{m_2}\sum_{j=i_0}^{n} c_j$$

Hence,

$$C_{H_U} \leq C_{JA} + \frac{m_1 - 1}{m_1}a_{i_0} + b_{i_0} + \frac{1}{m_2}c_{i_0} \leq 2C^*_{UMFT(m_1,m_2)}$$

since by the above observation $C_{JA} \leq C^*_{UMFT(m_1,m_2)}$ and the fact that $C^*_{UMFT(m_1,m_2)} \geq \frac{m_1-1}{m_1}a_{i_0} + b_{i_0} + \frac{1}{m_2}c_{i_0}$ for $1 \leq i_0 \leq n$. This completes the first part of the theorem.

To see that the bound of 2 is tight, consider an instance of $m_1 n$ jobs as follows. The processing requirements of J_1 are $(1, n-1, 0)$ and the requirements of all the remaining $m_1 n - 1$ jobs are $(1, 0, 0)$. Any sequence qualifies as Johnson's sequence with respect to $(\frac{1}{m_1}a_i, \frac{1}{m_2}c_i)$. If J_1 is the first job in the sequence, the heuristic H_U will produce a schedule with makespan $C^* = n$. On the other hand, if J_1 is the last job in the sequence, H_U will produce a schedule with makespan $C_{H_U} = 2n - 1$. The ratio $\frac{C_{H_U}}{C^*}$ approaches 2 from below as n increases. This completes the proof of the theorem. □

Note that at step 4, the algorithm H_U applies the **FCFS** rule (i.e., each c_i task is scheduled as soon after the completion of b_i as possible), however, on the same postprocessor. Hence, at the end of step 4 each c-task is scheduled at the processor decided by step 3 of the algorithm. A straightforward interchange argument shows that it is always advantageous to schedule the postprocessing tasks according to the FCFS rule irrespective of the postprocessor assignments of step 3. As a result, step 4 of H_U can be replaced by

Step 4′. Schedule the postprocessing tasks of S according to FCFS across processors.

The original exposition of H_U by using step 4 rather than 4′ simplified the proof of Theorem 17.12. Replacing 4 by 4′ does not improve the worst case error bound of 2, since the instance provided to prove the tightness part of Theorem 17.12 has $c_i = 0$ for all jobs J_i.

An extensive computational experiment is reported in [16] on the performance of the heuristics H_{OP}, H_{RO}, and H_U. It is found that all three heuristics exhibit near optimal performance, the relative gaps of the heuristics decrease as the duration of slave tasks increases, the average relative gaps increase with the sum of the number of preprocessors plus postprocessors and they increase with the preprocessing workload. Also, it is found that the relative gaps decrease as the ratio $\frac{n}{m}$ increases; the heuristics perform well for $\frac{n}{m} > 8$.

17.5 Conclusion

In this chapter we introduced various applications of the master-slave paradigm under three different configurations for the relative order of pre- and postprocessing tasks. For each of the configurations we developed heuristic procedures that enjoy small worst-case error bounds. Throughout this chapter the objective has been to minimize makespan. An appealing characteristic of the master-slave system considered is that it includes several production environments (such as the parallel identical machines, where $b_i = c_i = 0$, and the hybrid flowshop, where $b_i = 0$) as special cases of it. An interesting future research direction may be to consider objectives other than makespan. The flowtime objective is of particular interest. Future directions also include the identification of other variations of the master-slave model that are of practical relevance.

Acknowledgment

The work of Sartaj Sahni was supported, in part, by the National Science Foundation under grant CCR-9912395.

References

[1] S. Sahni, Scheduling master-slave multiprocessor systems, *IEEE Transactions on Computers*, 45(10), 1195–1199, 1996.

[2] C.-Y. Lee, R. Uzsoy and L.A.M. Vega, Efficient algorithms for scheduling semiconductor burn-in operations, *Operations Research*, 40(4), 764–775, 1992.

[3] S.M. Johnson, Discussion: Sequencing n jobs on two machines with arbitrary time lags, *Management Science*, 5, 299–303, 1959.

[4] L.G. Mitten, Sequencing n jobs on two machines with arbitrary time lags, *Management Science*, 5, 293–298, 1959.

[5] W. Szwarc, On some sequencing problems, *Naval Research Logistics Quarterly*, 15, 127–155, 1968.

[6] R. Graham, E. Lawler, J. Lenstra, and A. Rinnooy Kan, Optimization and approximation in deterministic sequencing and scheduling: A survey, *Annals of Discrete Mathematics*, 5, 287–326, 1979.

[7] M. Dell'Amico, Shop problems with two machine and time lags, *Operations Research*, (in press).

[8] W. Kern and W. Nawijn, Scheduling multi-operation jobs with time lags on a single machine, University of Twente, 1993.

[9] M. Aref and M. Tayyib, Lana-Match algorithm: A parallel version of the Rete-Match algorithm, *Parallel Computing*, 24, 763–775, 1998.

[10] J. Blazewicz, P. Dell'Olmo, and M. Drozdowski, Scheduling of client-server applications, *International Transactions in Operational Research*, 6, 345–363, 1999.

[11] A. Jean-Marie, S. Lefebvre-Barboux, and Z. Liu, An analytical approach to the performance evaluation of master-slave computational models, *Parallel Computing*, 24, 841–862, 1998.

[12] S. Sahni and G. Vairaktarakis, The master-slave paradigm in parallel computer and industrial settings, *Journal of Global Optimization*, 9, 357–377, 1996.

[13] M. Garey and D. Johnson, *Computers and Intractability: A Guide to the Theory of NP-Completeness*, W. H. Freeman & Co., New York, 1979.

[14] C.-Y. Lee and G.L. Vairaktarakis, Minimizing makespan in hybrid flowshops, *Operations Research Letters*, 16, 149–158, 1994.

[15] S.M. Johnson, Optimal two and three-stage production schedules with setup times included, *Naval Research Logistics Quarterly*, 1, 61–68, 1954.

[16] G. Vairaktarakis, Analysis of algorithms for master-slave systems, *IIE Transactions*, 29(11), 939–949, 1997.

[17] R.L. Graham, Bounds for certain multiprocessing anomalies, *Bell System Technical Journal*, 45, 1563–1581, 1966.

[18] P. Krueger, T. Lai, and V. Dixit-Radiya, Job scheduling is more important than processor allocation for hypercube computers, *IEEE Transactions on Parallel and Distributed Systems*, 5(5), 488–497, 1994.

18

Scheduling in Bluetooth Networks

Young Man Kim
Kookmin University

Ten H. Lai
The Ohio State University

18.1 Introduction

Bluetooth [1–3] is a new radio interface standard that provides means to interconnect mobile electronic devices into a personal area ad hoc network (PAN). The devices include cell phones, laptops, headphones, GPS navigators, palm pilots, beepers, portable scanners, etc., in addition to access points to Internet, sensors and actuators. When users walk into a new environment like a conference room, business office, hospital, or home, they might want to quickly become aware of what services are provided in it and how to use them; for example, to exchange real-time multimedia data, to browse web pages, to control room temperatures, and to adjust the lighting.

Thus, a PAN infrastructure based on Bluetooth should provide many different communication services [4,5] such as Internet access, real-time monitoring and control over sensor and actuators, multimedia stream service, etc.

Bluetooth is a short-range radio technology operating in the unlicensed industrial–scientific–medical (ISM) band at 2.45 GHz. A frequency hop transceiver is applied to combat interference and fading. Two or more nodes sharing the same channel form a *piconet*, where one unit acts as a *master* and the other units up to seven act as *slaves*. Within a piconet, the channel is shared using a slotted Time-Division Duplex (TDD) scheme. The master polls the slaves to exchange data.

1-58488-397-9/$0.00+$1.50

Scatternet is established by linking several piconets together in an ad hoc fashion to yield a global wireless ad hoc network in a restricted space. A *bridge* in scatternet delivers interpiconet messages between two neighboring piconets. Since each piconet operates in a unique frequency sequence determined by its own master, a bridge should know the exact instant of polling from each master in advance for an efficient data exchange, while the master can schedule, in any order, the communication between the master and the pure slaves.

A number of researchers have addressed the issues of scheduling in Bluetooth. Most of these studies have been restricted, however, to piconet scheduling, where the fundamental question is the polling discipline used by the piconet master to poll its slaves. These algorithms are often referred to as intrapiconet scheduling schemes. In [6], the authors assume a simple round robin polling scheme and investigate queueing delays in master and slave units depending on the length of the Bluetooth packets used. In [7], the authors analyze and compare the behavior of the three different polling algorithms. They conclude that the simple round robin scheme may perform poorly in Bluetooth systems. Instead, they propose a scheme called Fair Exhaustive Polling that demonstrates a good preformance. Similar conclusions are drawn in [8] in which it is argued that the traditional round robin scheme may result in waste and unfairness. They also propose three enhanced versions of round robin scheme, Exhaustive Round Robin (ERR), Limited Round Robin (LRR), and Limited and Weighted Round Robin(LWRR) schemes, that show better throughtput and less delay. In [9,10] the authors concentrate on scheduling policies designed with the aim of low power consumption. A number of scheduling policies are proposed which exploit either *park* or *sniff* low power modes in Bluetooth.

While piconet can be scheduled in any possible order by the master, there exist two constraints in scatternet scheduling. First one is *node capacity condition*: the sum of the scheduled bandwidths of individual links incident on a node cannot exceed the maximum node bandwidth. Second is *link exclusion condition*: two links incident on a node cannot be scheduled simultaneously. The scheduling problem here is augmented by the need to coordinate the presence of bridges such that timing mismatches are avoided. Also, the slot boundaries of different piconets do not match in general. This is called *phase difference*.

In [11] an interpiconet scheduler is analyzed. They have shown that constructing a perfect link schedule that maximizes total throughput in a Bluetooth scatternet is an NP-hard problem [12] even if scheduling is performed by a central entity. The authors also propose a scheduling algorithm referred to as Distributed Scatternet Scheduling Algorithm(DSSA), which falls in the category of distributed, hard coordination schemes. Although this scheme solves the required bandwidth requirement, the algorithm is complex. Recently, Rácz et al. [13] have proposed a pseudorandom scatternet scheduling algorithm that is based on checkpoints determined by random function. The exchange of user data starts at any checkpoint and continues until there is no data to transfer or next checkpoint is encountered. This scheme is so simple to meet one of the design goal for Bluetooth scheduling algorithm. In [14], this idea is polished finely and proposed as a desired Bluetooth standard called *JUMP mode*. The rendezvous point (conceptually same as checkpoint) may be periodic without loss of merit as proposed in [15–17] where they proposed several schemes that somewhat accomplish both fairness and performance.

In a piconet, the master polls the slaves according to a piconet scheduling scheme so as to maximize the throughput, minimize the delay, or provide a fairness over all slaves as piconet design goals. The design goal for scatternet scheduler includes the QoS setup service in addition to the above goals for piconet.

The algorithm simplicity is important in the Bluetooth environment where the available resources are often restricted in energy and storage size. In Bluetooth network, different applications and protocols place different demands on the link. For example, a file transfer application requires a reliable data transmission. On the other hand, in application transferring compressed video or audio stream, small percentage of data loss is tolerable unless the delay and jitter are so high. Sensor and control data are still another class of traffic where data should arrive at the destination within a fixed time and message loss rate should be minimized.

Therefore, it is necessary for a successful delivery of QoS in Bluetooth scatternet to consider both QoS-sensitive traffic and QoS-unconscious traffic simultaneously in one global scheduling scheme. The

QoS-sensitive traffic requires QoS-aware scheduling scheme to provide a proper QoS service. On the other hand, the simple algorithm like checkpoint based algorithm is suitable for the QoS-unconscious traffic. Thus, both schemes are highly desirable to be implemented together to provide a comprehensive scheduling service for all kinds of data traffic.

This chapter is organized as follows. In Section 18.2, we present piconet scheduling schemes proposed in [7,8]. In Section 18.3, checkpoint based scatternet scheduling algorithm [13,15] is introduced. Then, in Section 18.4, a QoS-aware scatternet algorithm is proposed. The complexity of this algorithm can be reduced by adopting the distributed protocol that is used as an integral part of DSSA scheduling scheme [11] presented in Section 18.5. Also, the hybrid scheme, which combines the QoS-aware scatternet algorithm with checkpoint based scheduling algorithm, is presented in Section 18.6 so that QoS-aware time slot assignment is supplemented with dynamic load allotment to meet the whole spectrum of various traffic requests. Finally, we conclude in Section 18.7.

18.2 Piconet Scheduling Schemes

The problem of finding an efficient scheduling algorithm for piconet is quite similar to the classical optimization problem of polling schemes for which theoretical results are available in the literature [18–20]. In [18], it is proved that, if the same traffic is offered to all queues (symmetric system) and all queue-lengths are known to the server, the optimal polling scheme is exhaustive polling scheme that keeps polling a queue until there is no more data in it and the queue with the longest waiting data becomes the next polling queue.

In [19], a subset of polling schemes with a fairness constraint is considered: each queue must be visited once per cycle but the visiting order may be dynamically changed. The optimal polling policy with symmetric system is the one that enforces a decreasing queue-length order at the beginning of each cycle. However, the optimality of this scheme with the general traffic has not been proved.

The Bluetooth operation model differs from that of the classical schemes mainly for two reasons. After the transmission of a packet from the master to a slave, there is always a chance for the slave to transmit a packet to the master; i.e., the visits of the server to the master-to-slave and the slave-to-master queues are strictly related. Moreover, since the Bluetooth scheduling procedure is managed by the master, the knowledge of the queue status is only partial. The master cannot have an updated information on the status of the slave-to-master queues.

The practical polling schemes proposed in [8] are described as follows:

Pure Round Robin (PRR): A fixed cyclic order is defined, and a single chance to transmit is given to each master–slave queue pair according to the cyclic order. This scheme is not exhaustive.

Exhaustive Round Robin (ERR): A fixed order is defined as with PRR but the scheme is exhaustive and the master does not switch to the next pair until the master and slave queues are empty.

Exhaustive Pseudocyclic Master queue length (EPM): A dynamic cyclic order is defined at the beginning of each cycle according to a decreasing master-to-slave queue length order.

Limited Round Robin (LRR): The operation rule is the same as that of ERR except that the maximum number of transmission per cycle allows to get a limit on the cycle length and to avoid the unbounded cycle length according to the traffic characteristics.

Limited and Weighted Round Robin (LWRR): The operation rule is the same as that of ERR except that each slave is assigned a weight, maximum priority (MP), at the beginning. Each time a slave is polled without data exchange between master and slave, the weight of the slave is reduced by 1. Otherwise, it is increased by 1 up to MP. For the slave with the weight W, the next polling cycle is (MP-W)th cycle from the current one. Thus, when MP is four and the weight of a slave is 2, the slave must wait polling until two cycles are passed and the third one arrives.

PRR is fair among all schemes with the expense that the probability of polling without data transmission is the highest. On the other hand, ERR is near the optimal in delay performance. However, it has *capture effect*; the cycle length can vary without bound and according to the traffic characteristics. Thus, ERR is unfair scheme. To improve the fairness of ERR without reducing the performance, LRR is devised.

The performance of LRR is improved in LWRR by reducing the rate of visits to queues which have been found empty in the last visits and should have a lower probability to be the longer queues.

In [8], several simulation senarios have been executed under IP datagram traffic and TCP connection-oriented traffic. With the traffic up to 0.85, PRR shows the worst performance. The delay difference between ERR and EPM is negligible and the curve of ERR is slightly lower. However, with the traffic over 0.85, PRR shows the lower delay than ERR. LRR is close to ERR under low traffic load and close to PRR under high traffic load. LWRR is the improved version of LRR for the connection-oriented traffic and the performance is good in all conditions and usually very close to that of the ERR (PRR) scheme under low (high) traffic and removes the unfair channel capture problem.

In [7], the author proposes another good scheme, called fair exhaustive polling (FEP) scheme, that works like ERR (PRR) under the low (high) traffic load. The main idea behind FEP is to poll slaves that probably have nothing to send as seldom as possible. The slaves belong to one of two complementary states, the *active state* and the *inactive state*. A polling cycle starts with the master moving all slaves to the active state, and then begins one of possibly several polling subcycles. In a polling subcycle all active slaves are polled once in a round robin fashion. The master performs the task of packet scheduling for both the downlink and uplink flows, however, the master has only limited knowledge of the arrival processes at the slaves. This means that the scheduling of the uplink flows has to be based on the feedback it gets when polling the slaves. Based on this feedback and the current state of the master output queues, slaves are moved between the different states. A slave is moved from the active state to the inactive state when both of the following conditions are fulfilled: (i) the slave has no information to send, and (ii) the master has no information to send to the specific slave.

A slave is moved to the active state when the master has information to send to it. This is an iterative process that continues until the active state is emptied (the exhaustive part of the algorithm) and when it is, a new polling cycle starts. Also this algorithm will favor slaves generating packets at maximum rate and therefore an additional parameter is added, which limits the polling interval of any slave to some predetermined maximum time. For the algorithm this means that a slave, whose maximum polling interval timer has expired, is moved to the active state and therefore will be polled in the next polling subcycle.

FEP behaves, asymptotically, as an exhaustive scheduler at low loads and as a round robin scheduler at high loads. This is a very attractive feature since the mean delay for the exhaustive scheduler is less than or equal to the mean delay for the round robin scheduler for low traffic, whereas the opposite holds for heavy traffic. According to the simulation, FEP experiences lower delays than both the exhaustive and the round robin scheduler at all loads.

18.3 Soft Coordinated Scheduling Scheme for Scatternet

This and the following two sections present three categories of scatternet scheduling schemes. Bluetooth is a short-range radio technology operating in the unlicensed ISM band using a frequency hopping scheme. Bluetooth units are organized into piconets. There is one Bluetooth device in each piconet that acts as the master, which can have any number of slaves out of which up to seven can be active simultaneously. The communication within a piconet is organized by the master which polls each slave according to some polling scheme. A slave is only allowed to transmit in a slave-to-master slot if it has been polled by the master in the previous master-to-slave slot.

A Bluetooth unit can participate in more than one piconet at any time but it can be a master in only one piconet. A unit that participates in multiple piconets can serve as a bridge, thus allowing the piconets to form a larger network. We define *bridging degree* as the number of piconets a bridging node is member of. A set of piconets that are all interconnected by such bridging units is referred to as a *scatternet* network. Since a Bluetooth unit can transmit or receive in only one piconet at a time, bridging units must switch between piconets on a time division basis. Due to the fact that different piconets are not synchronized in time a bridging unit necessarily loses some time while switching from one piconet to the other. Furthermore, the temporal unavailability of bridging nodes in the different piconets makes it difficult to coordinate the communication with them, which impacts throughput and can be an important performance constraint in building scatternets.

There are two important phenomena that can reduce the efficiency of the polling-based communication in Bluetooth scatternets:

1. Slaves that have no data to transmit may be unnecessarily polled, while other slaves with data to transmit may have to wait to be polled.
2. At the time of an expected poll, one of the nodes of a master–slave node pair may not be present in the piconet (the slave that is being polled is not listening or the master that is expected to poll is not polling).

The first problem applies to polling-based schemes in general, while the second one is specific to the Bluetooth environment. In the previous section, LWRR and FEP significantly reduce the performance degradation by the first problem. On the other hand, in order to improve the efficiency of interpiconet communication the scheduling algorithm has to coordinate the presence of bridging nodes in the different piconets such that the effect of the second phenomenon is minimized.

However, the scheduling of interpiconet communication expands to a scatternet wide coordination problem. Each node that has more than one Bluetooth link has to schedule the order in which it communicates with its respective neighbors. A node with multiple Bluetooth links can be either a piconet master or a bridging node or both. The scheduling order of two nodes will mutually depend on each other if they have a direct Bluetooth link in which case they have to schedule the communication on their common link for the same time slots. This necessitates some coordination between the respective schedulers.

For instance in Figure 18.1, the scheduling order of node C and the scheduling order of its bridging neighbors, a, c, and e mutually depend on each other, while the bridging nodes further effect nodes A, B, and D as well. Furthermore, the possible loops in a scatternet (e.g., C-a-B-c) make it even more complicated to resolve scheduling conflicts.

In case of bursty traffic in the scatternet the scheduling problem is further augmented by the need to adjust scheduling order in response to dynamic variation of traffic intensity. In a bursty traffic environment it is desirable that a node spends most of its time on those links that have a backlogged burst of data.

One way to address the coordination problem of interpiconet scheduling is to explicitly allocate, in advance, time slots for communication in each pair of nodes. Such a *hard coordination* approach eliminates ambiguity with regard to a node's presence in piconets, but it implies a complex, scatternet wide coordinational problem and requires explicit signalling between nodes of a scatternet. In the case of bursty traffic, hard coordination scheme generates a significant computation and signaling overhead as the communication slots have to be reallocated in response to changes in traffic intensity and each time when a new connection is established or released. However, in the case of continuous traffic

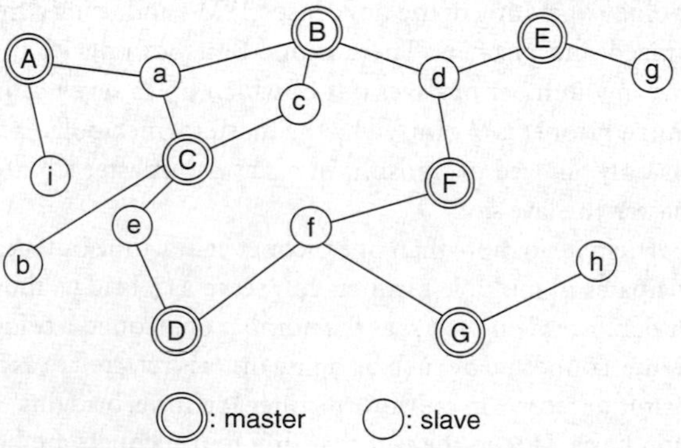

FIGURE 18.1 Example scatternet.

like multimedia, the slot allocation period is long and it is possible to check and adjust the performance measures like delay and jitter that are fundamental requirements for multimedia. Furthermore, the computational complexity can be reduced by adopting the distributed algorithm. In the next two sections, a hard coordinated scheduling scheme and a distributed scatternet scheduling algorithm are presented.

In *soft coordination* scheme, nodes decide their presence in piconets based on local information. By nature, soft coordination schemes cannot guarantee conflict-free participation of bridging nodes in the different piconets, however, they have a significantly reduced complexity. In this section, we present one soft coordination scheme called *pseudorandom coordinated scatternet scheduling* (PCSS) algorithm [13]. In PCSS algorithm, coordination is achieved by implicit rules in the communication without the need of exchanging explicit control information. The low complexity of the algorithm and its conformance to the current Bluetooth specification allow easy implementation and deployment.

The PCSS algorithm consists of two key components, *checkpoints* mechanism and *dynamic adjustment* of checking intensity. We explain the whole operations of PCSS as follows.

Coordination in the PCSS algorithm is achieved by the unique pseudorandom sequence of checkpoints that is specific to each master–slave node pair and by implicit information exchange between peer devices. A checkpoint is a designed Bluetooth frame. The activity of being present at a checkpoint is referred to as *to check*. A master node actively checks its slave by sending a packet to the slave at the corresponding checkpoint and waiting for a response from the slave. The slave nodes passively checks its master by listening to the master at the checkpoint and sending a response packet in case of being addressed.

The expected behavior of nodes is that they show up at each checkpoint on all of their links and checks their peers for available user data. The exchange of user data packets started at a checkpoint can be continued in the slots following the checkpoint. A node remains active on the current link until there is user data in either the master-to-slave or slave-to-master directions or until it has to leave for a next checkpoint on one of its other links. In the PCSS scheme we exploit the concept of randomness in assigning the position of checkpoints, which excludes the possibility that checkpoints on different links of a node will collide systematically, thus giving the node an equal chance to visit all of its checkpoints.

The pseudorandom procedure is similar to the one used to derive the pseudorandom frequency hopping sequence. In particular, the PCSS scheme assigns the positions of checkpoints on a given link following a pseudorandom sequence that is generated based on the Bluetooth clock of the master and the MAC address of the slave. This scheme guarantees that the same pseudorandom sequence will be generated by both nodes of a master–slave pair, while the sequences belonging to the different node pairs will be different. Figure 18.2 shows an example for the pseudorandom arrangement of checkpoints in case of a node pair A and B. The length of the current base checking interval is denoted by $T_{\text{check}}^{(i)}$ and the current checking intensity is defined accordingly as $1/T_{\text{check}}^{(i)}$. There is one checkpoint within each base checking interval and the position of the checkpoint within this window is changing from one time window to the other in a pseudorandom manner.

Since the pseudorandom sequence is different from one link to another, checkpoints on different links of a node will collide only occasionally. In case of collision the node can attend only one of the colliding

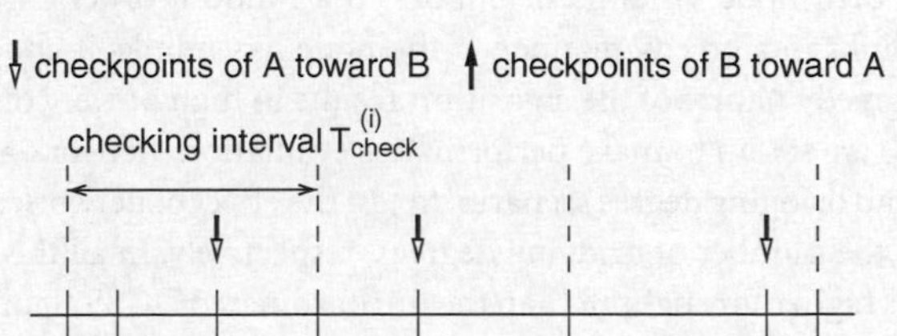

FIGURE 18.2 Example scatternet.

checkpoints, which implies that the corresponding neighbors have to be prepared for a nonpresent peer. That is, the master might not poll and the slave might not listen at a checkpoint. We note that a collision occurs either if there are more than one checkpoint scheduled for the same time slot or if the checkpoints are so close to each other that a packet transmission started at the first checkpoint necessarily overlaps the second one. Furthermore, if the colliding checkpoints belong to links in different piconets, the necessary time to perform the switching must be also taken into account.

During the communication there is the possibility to increase or decrease the intensity of checkpoints depending on the amount of user data to be transmitted and on the available capacity of the node. According to the PCSS algorithm a node performs certain traffic measurements at the checkpoints and increases or decreases the current checking intensity based on these measurements. Since nodes decide independently about the current checking intensity without explicit coordination, two nodes on a given link may select different base checking periods. In order to ensure that two nodes with different checking intensities on the same link can still communicate we require the pseudorandom generation of checkpoints to be such that the set of checkpoint positions at a lower checking intensity is a subset of checkpoint positions at any higher checking intensities.

To evaluate the PCSS scheme a reference scheme called *ideal coordinated scatternet scheduler* (ICSS) algorithm is prepared as follows. In the ICSS algorithm, a node has the following extra information about its neighbors, which represents the idealized property of the algorithm:

1. A node is aware of the already prescheduled transmissions of its neighbors.
2. A node is aware of the content of the transmission buffers of its neighbors.

According to the ICSS algorithm each node maintains a *scheduling list*, which contains the already prescheduled tasks of the node. A task always corresponds to one packet pair exchange with a given peer of the node. Knowing the scheduling list of the neighbors allows the node to schedule communication with its neighbors without overlapping their other communication, such that the capacity of the nodes is utilized as much as possible. Furthermore, being aware of the content of the transmission buffers of neighbors eliminates the inefficiencies of the polling-based scheme, since there will be no unnecessary polls and the system will be work-conserving.

In the scheduling of a node there is at most one packet pair exchange scheduled in relation to each of its peers, provided that there is a Bluetooth packet carrying user data either in the transmission buffer of the node or in the transmission buffer of the peer or in both. After completing a packet exchange on a given link the two nodes schedule the next packet exchange, provided that there is user data to be transmitted in at least one of the directions. If there is user data in only one of the directions, a POLL or NULL packet is assumed for the reverse direction depending on whether it is the master-to-slave or slave-to-master direction, respectively. The new task is fitted into the scheduling lists of the nodes using the first fitting strategy. According to this strategy the task is fitted into the first time interval that is available in both of the scheduling lists and that is long enough to accommodate the new task. Note that the algorithm strives for maximal utilization of node capacity by trying to fill in the unused gaps in the scheduling lists.

Another reference scheme for the evaluation of the PCSS scheme is *uncoordinated greedy scatternet scheduler* (UGSS) algorithm. In the UGSS algorithm, Bluetooth nodes do not attempt to coordinate their meeting points; instead each node visits its neighbors in a random order. Nodes switch continuously among their Bluetooth links in a greedy manner. If the node has n links it chooses each of them with a probability of $1/n$. The greedy nature of the algorithm results in high power consumption of Bluetooth.

In [13] three scenarios are set up to make performance evaluation: network access point scenario, forwarding hops scenario, and bridging degree scenario, to test the effect of network access points, the number of forwarding hops, and the number of bridging degrees, respectively. In all the senarios investigated it is found that PCSS achieves higher throughput than the uncoordinated UGSS algorithm. Moreover, with the traffic dependent meeting point intensity adjustments, the throughput and power measures of PCSS quite closely match the results of the ideal ICSS algorithm. At the same time PCSS consumes approximately the same amount of power as the ideal scheduler to achieve the same throughput, which is significantly less than the power consumption of UGSS algorithm.

18.4 Hard Coordinated Central Scheduling Scheme for Scatternet

In this section, a hard coordinated central scheduling algorithm is introduced. The distributed version of the hard scheduling algorithm is also introduced in the next section. The algorithm has three design goals: (1) The computation time must be polynomial in the number of nodes and links, (2) the solution must be perfect to assign any feasible traffic load request successfully, and (3) each individual link load is evenly distributed along the frame interval so that QoS delay and jitter are minimized.

The proposed algorithm is based on two principles: (1) The scheduling process is based on divide and conquer method, and (2) each intermediate link load is evenly partitioned into two subintervals to yield the optimal QoS performance unless the even partitioning is impossible in which the unresolved single request is randomly assigned into either interval.

In the following subsections, the necessary terminology and problem definition are presented, the perfect scheduling algorithm for bipartite scatternet is proposed, its QoS performance is analyzed, and the scheduling algorithm for general scatternet is introduced.

18.4.1 Perfect Scheduling Problem for Bipartite Scatternet

We distinguish between four types of nodes in a scatternet: pure masters, pure slaves, master bridges, and slave bridges. Pure master or slave is a master or a slave that belongs to one exclusive piconet. Thus, scatternet needs the existence of bridges that connect multiple piconets. Master bridge has two roles: it acts as the master in one piconet and the slave in the other piconet(s). Slave bridge always operates as a slave in each piconet. Let M_p be the set of all pure masters; M_b, the set of all master bridges; S_p, the set of pure slaves; and S_b, the set of all slave bridges. These four sets are pairwise disjoint. Also, for convenience, let $M = M_p \cup M_b$ and $S = S_p \cup S_b$. We represent the topology of a scatternet with an adjacency matrix $A(M, S \cup M_b)$ such that $A(i, j) = 1$, where $i \in M$ and $j \in S \cup M_b$, if j is an active slave of i. Each row in A describes the structure of a piconet and thus has at most seven nonzero entries. Each column of A shows piconet interconnection structure at a slave or master bridge.

A scatternet is said to be *bipartite* if it contains no master bridges (i.e., $M_b = \emptyset$). Figure 18.3 depicts a bipartite scatternet consisting of eight pure masters and nine slaves. A *general* or *nonbipartite* scatternet is one that contains at least one master bridge.

In Bluetooth, the time line is divided into slots, each 625 μs long. Basic communication between a master and its slaves consists of two consecutive slots, the first one for polling and the next for response. Thus, a couple of slots compose a basic unit of internodal communication. From now on, we use the notation "slot" to actually represent two consecutive slots.

In a scatternet, each link is associated with a *load* or *bandwidth* request, which is expressed in terms of required slots. Thus, for $i \in M$ and $j \in S \cup M_b$, $L(i, j)$ denotes the number of slots that link (i, j) requires

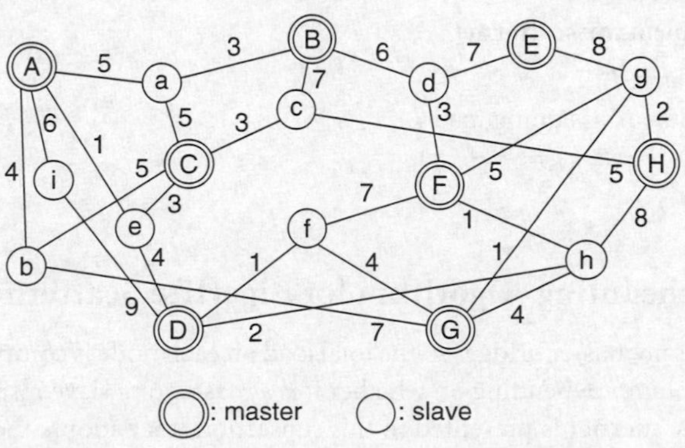

FIGURE 18.3 A bipartite graph with eight masters and nine slaves.

(master) \ (slave)	a	b	c	d	e	f	g	h	i	
A	5	4			1				6	16
B	3		7	6						16
C	5	5	3		3					16
D					4	1		2	9	16
E				7			8			15
F				3		7	5	1		16
G		7				4	1	4		16
H			5				2	8		15
	13	16	15	16	8	12	16	15	15	T(L, x)

(x: master or slave)

FIGURE 18.4 Load matrix L and total load $T(L, x)$.

in each frame of 2^n slots. L, defined on $M \times (S \cup M_b)$, is referred to as a *load matrix*, which indicates the load of each link on the scatternet.

Given a load matrix $L(M, S \cup M_b)$, let $L(i, *) = \sum_j L(i, j)$, and $L(*, j) = \sum_i L(i, j)$. For a pure master $i \in M_p$, the total load on i is $L(i, *)$. Similarly, the total load on a pure slave or slave bridge $j \in S$ is $L(*, j)$. However, the total load on a master bridge $x \in M_b$ is $L(x, *) + L(*, x)$. The total load of a node x in the load matrix L is denoted as $T(L, x)$. Notice that, for a master or slave x in a bipartite scatternet, $T(L, x)$ is equal to $L(x, *)$ or $L(*, x)$, respectively. Figure 18.4 shows load matrix L and total load of node $x \in M \cup S$, $T(L, x)$, on the scatternet of Figure 18.3.

A *feasible* load matrix L must satisfy, at minimum, the constraint $T(L, x) \leq 2^n$ for all nodes x. For example, load matrix L in Figure 18.4 is feasible in case of $n \geq 4$. It is, however, not clear whether this condition is sufficient for a load matrix to be feasible.

A schedule F for a scatternet over a period of 2^n slots is a function $F(i, j, k)$, $i \in M$, $j \in S \cup M_b$, $k \in [0..2^n - 1]$, where $F(i, j, k) = 1$ if link (i, j) is allocated at slot k; $F(i, j, k) = 0$, otherwise. $F(i, j, *)$ gives the schedule of link (i, j) over a time period of 2^n slots, while $F(*, *, k)$ depicts the slot assignment over the whole scatternet at time k. *Scheduled load* $S(x, k)$ of node x at slot k denotes total loads of node x in slot assignment $F(*, *, k)$. For a pure master $i \in M_p$, scheduled load $S(i, k)$ is $\sum_j F(i, j, k)$. Similarly, the scheduled load on a slave $j \in S$ is $\sum_i F(i, j, k)$. However, on a master bridge $x \in M_b$, it is $\sum_i F(i, x, k) + \sum_j F(x, j, k)$. Schedule F is *feasible* iff scheduled load $S(x, k)$ is at most one for all nodes $x \in M \cup S$ and all slots $k \in [0..2^n - 1]$.

Now we define the *Perfect Scheduling Problem for Scatternet* as follows.

Perfect scheduling problem for scatternet

Input: A feasible load matrix L.

Output: A feasible schedule F assigning load L perfectly, i.e., $\sum_k S(x, k) = T(L, x)$, $\forall x \in M \cup S$.

18.4.2 Perfect Scheduling Algorithm for Bipartite Scatternet

A bipartite scatternet has no master bridge. So the total load on each node is computed by summing up the corresponding row or column depending on whether it is a master or a slave respectively. The scheduling algorithm for bipartite scatternet is presented in this subsection, that adopts the methodology of divide and conquer. Given the initial load matrix L satisfying the constraint $T(L, x) \leq 2^n, \forall x$, the algorithm generates two load matrices L_1 and L_2 satisfying $T(L_i, x) \leq 2^{n-1}, i = 1$ or 2. In general, given a load

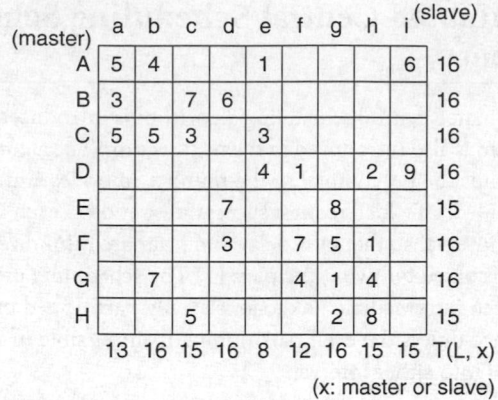

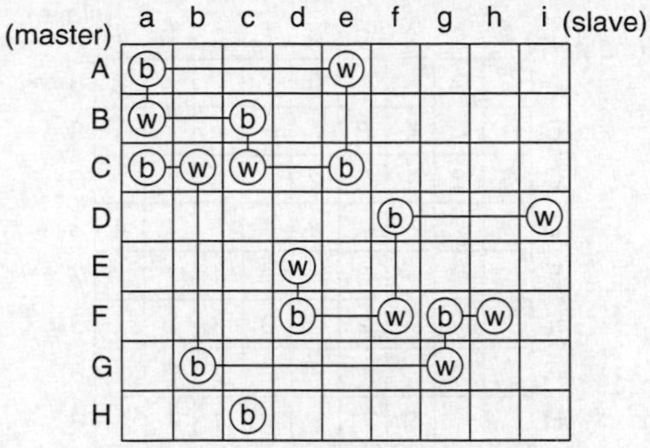

FIGURE 18.5 0/1 matrix A and graph G.

matrix L satsfying $T(L,x) \leq 2^k$ at level k such that $n \geq k \geq 1$, the algorithm partitions it into two load matrices L_1 and L_2 each satisfying the constraint $T(L_i,x) \leq 2^{k-1}$, where $i = 1$ or 2.

In other words, this process of dividing load matrix evenly is repeated recursively until the upper bound reaches 2^0 where total load on each node is at most one, implying that no contention exists. Thus L at the last recursion is always a feasible assignment in itself. By assigning L into $F(*,*,l)$ for all time slices $l, 0 \leq l \leq 2^n - 1$, a feasible schedule F is generated and satisfies perfect condition $\sum_k S(x,k) = T(L,x), \forall x \in M \cup S$, since no parts of the initial load matrix entries are allowed to drop in the algorithm.

The proposed scheduling algorithm *Bluetooth_Scheduling* implements the above process by calling procedure *Divide_Load* that recursively calls itself until the set of feasible slot assignments $F(*,*,l)$ are produced. The actual load division is done by another procedure *Load_Partition.*

Consider a particular case that Divide_Load (L,k,l) calls Load_Partition(L,L_1,L_2). An even entry in L is divided by two and each half is set into the same entries of L_1 and L_2. Division of odd entry in L is more complex. First, the odd entry is decremented by one and then it is divided by two. The computation result is set at the same entries of L_1 and L_2, as an intermediate value. Now, the residual value can be allotted either into L_1 or L_2. All the residual values to be assigned further are represented in 0/1 matrix A. For example, Figure 18.5 depicts A derived from load matrix L in Figure 18.4.

For a fair division of residual load, the nonzero entries at each row of A are grouped into pairs with at most one possible unpaired entry. It is repeated for each column in A. Let the resulting graph be G. Later we prove that G is always a bipartite graph for any given bipartite scatternet and load matrix. Figure 18.5 shows G computed from L in Figure 18.4, in which there is one even cycle, two linear sequences, and one isolated entry. Then, each nonzero entry of A (i.e., the vertices of G) is colored either with *black (b)* or *white (w)* so that no two directly linked entries have the same color; this rule is the key to the even partitioning of A into L_1 and L_2, i.e., $T(L_1,x) \leq 2^{k-1}$ and $T(L_2,x) \leq 2^{k-1}$ for a given load L satisfying $T(L,x) \leq 2^k$. Finally, black or white entry in A is allotted into L_1 or L_2, respectively. For the load matrix L in Figure 18.4 and the entries in A colored like Figure 18.5, the output load matrices L_1 and L_2 are computed by Load_Partition(L,L_1,L_2), as shown in Figure 18.6 and Figure 18.7.

Algorithm Bluetooth_Scheduling
Input: Load matrix L
Output: A feasible schedule F
Statements:

  call *Divide_Load*$(L,n,0)$

Procedure Divide_Load(L,k,l)
  if $k = 0$ **then**
    $F(i,j,l) := L(i,j) \quad \forall i,j$

(master) \ (slave)	a	b	c	d	e	f	g	h	i	T(L, x)
A	3	2			0				3	8
B	1		4	3						8
C	3	2	1		2					8
D					2	1		1	4	8
E				3			4			7
F				2		3	3	0		8
G		4				2	0	2		8
H			3				1	4		8
T(L, x)	7	8	8	8	4	6	8	7	7	

(x: master or slave)

FIGURE 18.6 Load matrix L_1.

(master) \ (slave)	a	b	c	d	e	f	g	h	i	T(L, x)
A	2	2			1				3	8
B	2		3	3						8
C	2	3	2		1					8
D					2	0		1	5	8
E				4			4			8
F				1		4	2	1		8
G		3				2	1	2		8
H			2				1	4		7
T(L, x)	6	8	7	8	4	6	8	8	8	

(x: master or slave)

FIGURE 18.7 Load matrix L_2.

else
  call *Load_Partition*(L, L_1, L_2)
  call Divide_Load$(L_1, k-1, 2l)$
  call Divide_Load$(L_2, k-1, 2l+1)$

Procedure Load_Partition(L, L_1, L_2)
Input: Load matrix L
Output: Load matrices L_1 and L_2 such that $|L_1(i,j) - L_2(i,j)| \leq 1$ and $|T(L_1,x) - T(L_2,x)| \leq 1$ for all links (i,j) and all nodes x.

1. For each entry $L(i,j)$, $i \in M$, $j \in S$, let

$$L_1(i,j) := L_2(i,j) := \lfloor L(i,j)/2 \rfloor$$
$$A(i,j) := L(i,j) \bmod 2$$

   A is a 0/1 matrix.
2. For each row in A, group the nonzero entries into pairs with at most one possible unpaired entry; do the same for each column in A. Let the resulting graph be G. (We will show that G is a bipartite graph.)

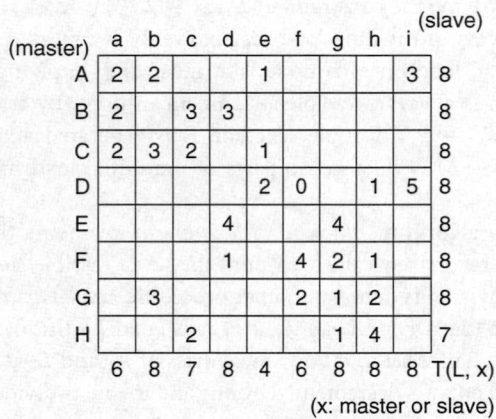

3. Color the nonzero entries of A (i.e., the vertices of G) with *black* or *white* so that no two directly linked entries have the same color.
4. For each nonzero entry $A(i, j)$, if it is *black*, increment $L_1(i, j)$ by 1, else increment $L_2(i, j)$ by 1.

The following lemmas and theorem show some graphical and scheduling properties of bipartite scatternet and Algorithm Bluetooth_Scheduling, and prove that the proposed algorithm solves perfect scheduling problem for bipartite scatternet.

Lemma 18.1

If the scatternet is bipartite, then it has no cycle of odd length.

Proof

Proof by contradiction. Suppose that there is an odd cycle in a bipartite scatternet. A bipartite scatternet has no master bridge. Thus, there is only master–slave or slave–master link in bipartite scatternet and, thus, the sequence of nodes in the odd cycle can be enumerated alternatively like master followed by slave followed by master followed by slave and so on. However, if the first node in the cycle would be master (slave), then the last node in the sequence must be master (slave) because of the odd length of the cycle. Since the last and first node are also connected by a link, there exists a master–master (slave–slave) link contrary to the previous fact. □

Lemma 18.2

Graph G, that is produced by step 2 of Procedure Load_Partition, is a bipartite graph.

Proof

Notice that each nonzero entry $A(i, j)$ has at most two links incident on itself, one horizontal and the other vertical. The number of possible types of isolated graph components consisting of such entries is three: isolated node, linear sequence, and simple cycle. Furthermore, the cycle always has an even number of vertices, since the odd cycle implies that some vertex must have two identical type of links, e.g., two horizontal (vertical) links, that do not exist in G. It is easy to observe that all the above types of graphs can be arranged as bipartite graphs. □

Lemma 18.3

If the given scatternet is bipartite and the input matrix L to Procedure Load_Partition satisfies $T(L, x) \leq 2^k$, then the output matrices, L_1 and L_2, satisfy $T(L_i, x) \leq 2^{k-1}$, $i = 1$ or 2.

Proof

Suppose that input matrix L has the property $T(L, x) \leq 2^k$. If the output matrices L_1 and L_2 satisfy the relation $|T(L_1, x) - T(L_2, x)| \leq 1$, then it is evident that $T(L_i, x) \leq 2^{k-1}$, $i = 1$ or 2. Thus, it is enough to prove that Procedure Load_Partition generates two load matrices L_1 and L_2 satisfying $|T(L_1, x) - T(L_2, x)| \leq 1$. Notice that $T(L, x)$ is either $L(x, *)$ or $L(*, x)$ for a master or slave x, respectively. Consider only the case of master x. Similar reasoning can be applied for the proof of the other case and we skip it here. Remind that $L(x, *)$ is $\sum_j L(x, j)$. In the procedure, each entry $L(x, j)$ is evenly divided into $L_1(x, j)$ and $L_2(x, j)$ except the indivisible value one in each odd entry of $L(x, *)$. Such entries are denoted as 1 at the same location in 0/1 matrix A. Then, nonzero entires of A are paired and connected by horizontal links. According to Lemma 18.2, G is a bipartite graph where it is always possible to color the vertices of G with black or white so that no two directly linked entries have the same color (Step 3). Since black and white vertices increment $L_1(x, *)$ and $L_2(x, *)$ respectively by one, and there is at most one entry remaining without pairing in the row, it is true that $|L_1(x, *) - L_2(x, *)| = |T(L_1, x) - T(L_2, x)| \leq 1$. □

Theorem 18.1

For bipartite scatternets, a load matrix L is feasible iff $T(L, x) \leq 2^n$ for every node x. If L is feasible, then Algorithm Bluetooth_Scheduling will produce a perfect, feasible schedule.

Proof

Suppose L is feasible or $T(L, x) \leq 2^n$ for every node x. We consider the case that x is master. The other case is similar to the following proof and will be omitted here. According to Algorithm Bluetooth_Scheduling and Lemma 18.3, load matrix L' at level k, having the property $T(L', x) \leq 2^k$, is evenly partitioned into two matrices L'_1 and L'_2, having the properties $T(L'_i, x) \leq 2^{k-1}, i = 1$ or 2. By induction, load matrix L'' at last level $k = 0$ has the property $L''(x, j) \leq 1, j \in S$. Remember that $F(x, j, l) = L''(x, j), \forall x, j, l$. Since $S(x, l) = \sum_j F(x, j, l) = \sum_j L(x, j) \leq 1$ and initial load is preserved at all levels of load matrices, i.e., $\sum_k S(x, k) = T(L, x), \forall x \in M \cup S$, the schedule produced by Algorithm Bluetooth_Scheduling is feasible and perfect. □

Notice that the algorithm distributes the original link loads evenly over the time period of 2^n slots so that the generated schedule has regular distribution, yielding tight bounds of the delay and jitter in addition to perfect allocation of the required bandwidth. Thus, the proposed algorithm realizes QoS-aware scatternet scheduling. In the next subsection, QoS analysis will be presented to figure out the delay and jitter quantitatively.

18.4.3 QoS Performance Analysis

18.4.3.1 Bandwidth

One of the most essential performance measures for QoS is the *bandwidth*, the amount of data transferred per unit time. Since the scheduling algorithm proposed in the previous subsection is reservation-based, they are suitable for multimedia and real-time traffic to obtain the guaranteed bandwidths along the communication routes. Once the algorithm generates a feasible schedule from the set of link load requests, the assigned bandwidths are exclusively used for the requesting applications. In case of bipartite scatternet (i.e., having no master bridge), the algorithm yields a perfect, feasible schedule. Therefore, the feasible schedule always guarantees the exact amounts of bandwidth to the requesting applications.

18.4.3.2 Delay and Jitter

Another two fundamental measures of QoS property are *delay (D)* and *jitter (J)*. Multimedia and real-time data are time-critical and hence they should be delivered within a bounded time (delay) and distributed within a range of arrival time (jitter). Notice that, in the previous algorithm, link load is evenly partitioned into two at each recursion until the schedule is produced. Thus, the delay and jitter are expected to be tightly bounded.

There are three kinds of delays: *minimum* ($D_{\min}$), *average* (D_{avg}), and *maximum* ($D_{\max}$) delays. Among them, average delay is the most important. The *lower* and *upper* jitters, J_l and J_u, denote the lower and the upper deviation ranges of delay from the average, i.e., $J_l = D_{\text{avg}} - D_{\min}$ and $J_u = D_{\max} - D_{\text{avg}}$. Since J_u is much more critical to the application than J_l, J_u is used to represent the jitter.

Let t_s denote one physical slot time in Bluetooth, that is 1.25 msec. On the other hand, each logical slot used in the algorithm consists of a constant number of consecutive physical slots, v, to reduce the bandwidth loss due to phase difference. Then, the logical slot time, t_v, is $v \cdot t_s$. Remind that $L(i, j)$ is the initial link load on edge (i, j) that will be scheduled into 2^n number of consecutive slots. For example, if some multimedia application needs 128 kbps of bandwidth, 128 kbps/1 Mbps $= \frac{1}{8}$ of total bandwidth, i.e., 2^{n-3} slots out of 2^n, should be scheduled for it.

We analyze a simple case where the initial load $L(i, j)$ on link (i, j) is $2^m, 0 \leq m \leq n$. The general case can be analyzed in a similar fashion, and will be skipped for brevity. Since the algorithm partitions the load evenly into two subloads in the lower level, from level n down to $(n - m)$, each slot span of size 2^{n-m} at level $(n - m)$ that is called *basic span* of $L(i, j)$ loaded with 2^m, contains exactly one request for link (i, j). $S(m)$ denotes this basic span size, 2^{n-m}. Then, $S(m)t_v$ equals the physical time of one basic span.

In the scatternet analysis, it is desirable to derive the global performance measures over a route. Let $D(h)$ and $J(h)$ be the total delay and jitter along the route of h hops. First, we consider the special case of $h = 1$. The general case will be covered later. For simplicity, assume that the scheduled slot is distributed with uniform probability along the basic span. Then, $D_{\min}(1) = t_v$, $D_{\text{avg}}(1) = 0.5S(m)t_v$, and $D_{\max}(1) = S(m)t_v$. According to the definition of jitter, $J_l(1) = (0.5S(m) - 1)t_v$ and $J_u(1) = 0.5S(m)t_v$.

Now, we consider the general case of $h > 1$. There are two different message management schemes adoptable in the intermediate nodes. In the *pipelining* scheme, a message arriving from uplink node waits until the next basic span starts even though a free slot is available later in the current span. Thus a message always advances one hop per basic span. The delay and jitter under the pipelining scheme are identified with superscript p: $D^p_{\min}(h) = ((h-1)S(m)+1)t^v$, $D^p_{\text{avg}}(h) = (h-0.5)S(m)t_v$, and $D^p_{\max}(h) = hS(m)t_v$. Also, according to the definion of jitter, $J^p_l(h) = (0.5S(m) - 1)t_v$ and $J^p_u(h) = 0.5S(m)t_v$. Notice that, in the pipelining scheme, the jitter is independent of the number of hops h.

The second message management scheme is the *pass-through* scheme in which a message can pass through any number of hops during one basic span, if there is no other message in the buffer waiting for the available slot. Otherwise, the new message yields the slot access to the old message in the buffer. Suppose that there is no buffered message along the route to simplify the analysis. Notice that this assumption is optimistic and the real performance measure exists somewhere between the result derived under the pass-through scheme and that under pipelining.

The location variable x_k, $1 \le k \le h$, denotes the allocated slot location at the kth hop in a basic span. Let the basic span be normalized between 0.0 and 1.0 and assume that it contains many slots, i.e., $2^{n-m} \gg 1$ such that x_k is supposed to be uniformly and continuously distributed with uniform probability in the range [0.0, 1.0]. Then, the case that a message at the source advances k hops during one basic span, occurs when the following relation is true: $x_1 < x_2 < \cdots < x_k \ge x_{k+1}$. The instant probability of such a case is represented as $(1 - x_1)(1 - x_2)\cdots(1 - x_{k-1})x_k$, for which $(1 - x_p)$ denotes the probability that x_{p+1} is greater than x_p, and x_{k+1} is the probability that x_{k+1} is no greater than x_k. The average probability of k hops advance per basic span, P_k, is

$$P_k = \int_0^1 \int_{x_1}^1 \cdots \int_{x_{k-1}}^1 (1 - x_1)(1 - x_2)\cdots(1 - x_{k-1})x_k \, dx_k dx_{k-1} \cdots dx_1$$

In the above, P_k is decomposed into two terms, $P_k = P_k^1 + P_k^2$, defined as follows:

$$\begin{aligned} P_k^1 &= \int_0^1 \int_{x_1}^1 \cdots \int_{x_{k-1}}^1 (1 - x_1)(1 - x_2)\cdots(1 - x_{k-1})(1 - x_k) \, dx_k dx_{k-1} \cdots dx_1 \\ &= \frac{1}{2^k k!} \\ P_k^2 &= \int_0^1 \int_{x_1}^1 \cdots \int_{x_{k-1}}^1 (1 - x_1)(1 - x_2)\cdots(1 - x_{k-1})1 \, dx_k dx_{k-1} \cdots dx_1 \\ &= \frac{2^k k!}{(2k)!} \end{aligned}$$

L_k denotes the average location of x_k multiplied with P_k, in which k hops of advance occurs:

$$L_k = \int_0^1 \int_{x_1}^1 \cdots \int_{x_{k-1}}^1 (1 - x_1)(1 - x_2)\cdots(1 - x_{k-1})(x_k)^2 \, dx_k dx_{k-1} \cdots dx_1$$

L_k can be represented by P_k^1 and P_k^2:

$$L_k = P_{k+1}^2 + P_k^2 - 2P_k^1$$

H_{avg}, the average number of hops for a message to advance, is shown within four digits precision:

$$H_{\text{avg}} = \sum_{k=1}^{\infty} kP_k \approx 1.0868$$

L_{avg}, the average stop location of a message in a basic span, is shown within four digits precision:

$$L_{\text{avg}} = \sum_{k=1}^{\infty} L_k \approx 0.5239$$

Using H_{avg} and L_{avg}, we denote the delay and the jitter for the pass-through scheme with superscript t, as follows:

$$D^t_{\min}(h) = ht_v$$

$$D^t_{\text{avg}}(h) \approx \left(\frac{h}{H_{\text{avg}}} - 1 + L_{\text{avg}}\right) S(m)t_v$$

$$\approx (0.9201h - 0.4761)S(m)t_v$$

$$D^t_{\max}(h) = hS(m)t_v$$

$$J^t_l(h) \approx ((0.9201h - 0.4761)S(m) - h)t_v$$

$$J^t_u(h) \approx (0.0799h + 0.4761)S(m)t_v$$

With a route of 5 hops, $D^t_{\text{avg}}(5)$ is reduced 8.3% in comparison with $D^p_{\text{avg}}(5)$. On the contrary, $J^t_u(5)$ increases 75% compared to $J^p_u(5)$. Notice that there is a trade-off on delay and jitter between the two schemes.

On the other hand, regardless of the message scheme applied, both the delay and the jitter are proportionally dependent on the logical slot time t_v, the basic span $S(m)$, and the number of hops h. As the logical slot time is also proportional to the number of physical slots v, if v decreases then the delay and jitter become shorter and smaller, yielding better performance. However, the link bandwidth utility shrinks on the contrary. Remember that basic time span $S(m)$ is 2^{n-m}. Since the requested link bandwidth $L(i, j)$ is $\frac{1}{S(m)}$, m is dependent on the bandwidth request. Therefore, the bigger the bandwidth request, the smaller the basic time span, and the better the QoS performance.

We examine the analysis results further using some numerical examples. Suppose $L(i, j)$ is 128 kbps. It is $\frac{1}{8}$ of the total bandwidth, 1 Mbps. Because of phase difference, practical minimum value of v is 2, and $\frac{1}{4}$ of total slots (i.e., $m = n - 2$) should be assigned to that request, 50% of which is lost according to asynchronism between two piconets. Moreover, $S(m) = 4$ and $t_v = 2.5$ msec. Thus, the delay and jitter for the requested bandwidth of 2×64 kbps $= 128$ kbps are $D^p_{\text{avg}}(h) = 10(h - 0.5)$ msec and $J^p_u(h) = 5$ msec. If the pass-through scheme is used, these measures are $D^t_{\text{avg}}(h) = 10(0.9201h - 0.4761)$ msec and $J^t_u(h) = 10(0.0799h + 0.4761)$ msec. If we use larger value for v, say 8, the bandwidth loss is reduced from 50% to 12.5%. However, the QoS performance becomes worse: $D^p_{\text{avg}}(h) = 80(h - 0.5)$ msec, $J^p_u(h) = 40$ msec, $D^t_{\text{avg}}(h) = 80(0.9201h - 0.4761)$ msec, and $J^t_u(h) = 80(0.0799h + 0.4761)$ msec.

18.4.4 Scheduling Algorithm for General Scatternet

Master bridge is allowed in the general scatternet so that a master may be enrolled in a number of piconets; it plays a master in one piconet and the bridge (i.e., slave) in the others.

Figure 18.8 shows an example of general scatternet with nine masters and six slaves in which there are five master bridges (B, E, G, H, and I). Notice that there exist several odd-cycles: e.g., (A, B, b) and (B, E, f, F, b). The existence of odd-cycle in the graph results in the fact that there is no perfect scheduling solution for the general scatternet. We will demonstrate this by depicting example configuration in this subsection. Then, *Subperfect Scheduling Problem for General Scatternet*, where a load matrix is partially scheduled, is defined. Finally, one particular algorithm solving this problem is presented.

18.4.4.1 Nonexistence of Perfect Scheduling Algorithm for General Scatternet

When a scatternet allows master bridge as an integral part, the perfect scheduling problem has no solution. For example, suppose that there are three nodes, x, y, and z, in scatternet; x, y, and z are fully connected master bridges. In other words, three nodes form an odd cycle of length three. Figure 18.9 depicts this configuration.

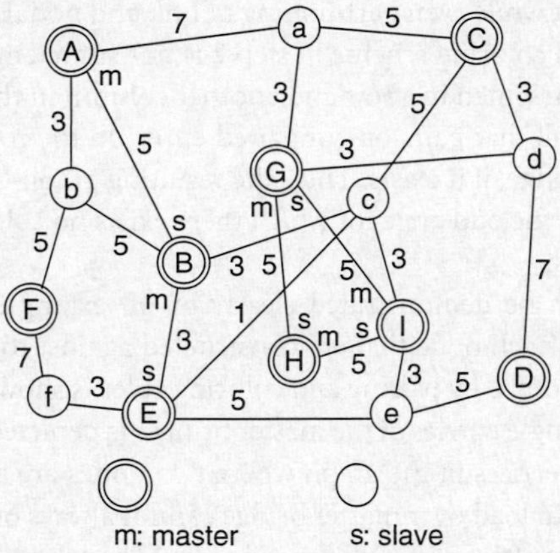

FIGURE 18.8 A general scatternet with nine masters and six slaves.

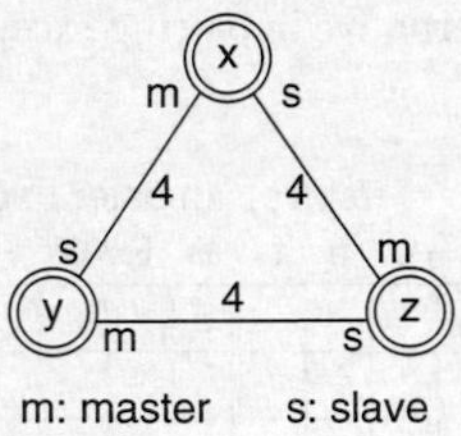

FIGURE 18.9 A general scatternet with an odd cycle of three nodes.

Consider the case that each link load is 2^2 and the maximum node bandwidth is 2^3 or $n = 3$. The load matrix L is feasible by definition. However, no matter how the load would be assigned to slots, at most $8(=2^3)$ slots out of the total load requests, $12(=3 \cdot 2^2)$, could be scheduled, i.e., 50% of slot assignment failure, because any slot assignment on a link does not allow the assignments of the other two links in any slot instant and feasible schedule.

Although the existence of odd cycle in the scatternet prevents any solution for the perfect scheduling problem, there exists a solution if we sacrifice some of the requested link loads. Since a master bridge creates an odd cycle in the topology, if scheduling loss is inevitable, it would be applied to the links incident on the master bridges. The following one defines a new subperfect scheduling problem.

Subperfect scheduling problem for general scatternet

Input: A load matrix L satisfying the constraint

$T(L, x) \leq 2^n$ for all nodes x.

Output: A feasible schedule F such that

$$\sum_{l=0}^{2^n-1} F(i, j, l) \begin{cases} = L(i, j), & \text{if } i \in M_p \text{ and } j \in S \\ \geq \frac{1}{2} L(i, j), & \text{otherwise} \end{cases}$$

18.4.4.2 Subperfect Scheduling Algorithm for General Scatternet

In the previous subsection, we showed that there exists some configuration of general scatternet and its load matrix in which it is impossible to produce a perfect feasible schedule. In particular, an odd cycle in the scatternet creates such a problem. Let us explain this with reference to Algorithm *Bluetooth_Scheduling*. Step 2 in Procedure *Load_Partition* is not sufficient because of the existence of a master bridge that creates some odd cycle in the scatternet while even partitioning of link and nodal load is necessary for the perfect schedule. However, pairing and coloring scheme in step 2 is not sufficient for even partitioning since the load of any master bridge is distributed in a row and another column in the load matrix L. Thus, we have to introduce a third *inclined* link that pairs one unpaired entry on the row with another unpaired entry on the column of the master bridge, if it exists. Thus, the resulting graph G with inclined links is no more bipartite due to the existence of the odd cycle for which there exists no coloring scheme for two entries on any link to have different colors.

This informal statement can be demonstrated clearly by an example. Figure 18.10 shows the load matrix L and its total loads for each node that are constructed against the scatternet in Figure 18.8. The corresponding graph derived from L by pairing and coloring rules is shown in Figure 18.11. In the figure, the inclined link between two single entries of the master bridges is depicted by a dotted line. There are two odd cycles consisting of three vertices in the graph where two nodes are assigned with the same color so that even partitioning of the total load over master bridge cannot always be feasible. In fact, the odd cycles in the figure can be avoided if the alternative node pairing would be selected. For example, if entry (A, b) is paired with (A, a) instead of (A, B), then the odd cycle $(A, b) - (A, B) - (B, b)$ that is replaced by a linear sequence $(A, B) - (B, b) - (A, b) - (A, a) - (C, a) - (C, c) - (B, c) - (B, E)$, disappears in the graph. However, repairing does not always work to remove odd cycle. If there were no link like (A, a), (B, c), and (B, E), then there would be no other alternative in pairing except the odd cycle $(A, b) - (A, B) - (B, b)$.

(master)	a	b	c	d	e	f	B	E	G	H	I	
	(slave)						(master bridge)					
A	7	3					5					15
B		5	3					3				11 + (5) = 16
C	5		5	3								13
D				7	5							12
E			1		5	3						9 + (3) = 12
F		5				7						12
G	3			3						5		11 + (5) = 16
H											5	5 + (5) = 10
I			3		3				5			11 + (5) = 16
	15	13	12	13	13	10	(5)	(3)	(5)	(5)	(5)	T(L, x)

( ): link load for master bridge x: master or slave

FIGURE 18.10 Load matrix L and total load $T(L, x)$ for the general scatternet.

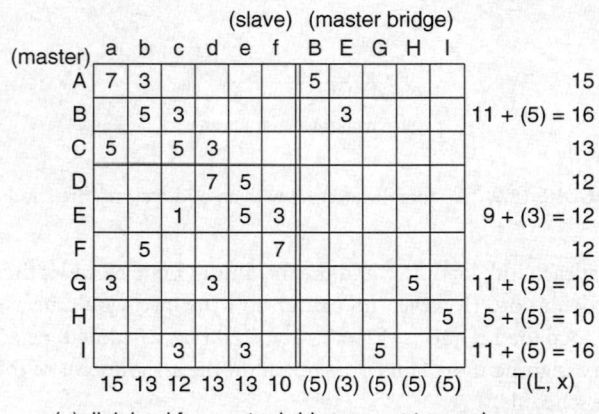

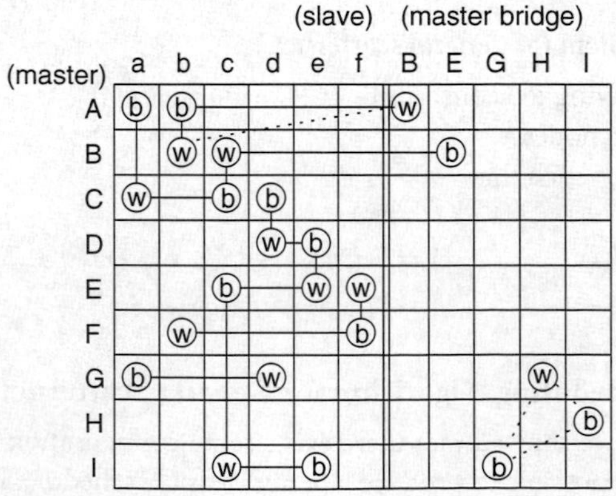

FIGURE 18.11 0/1 matrix A and graph G for the general scatternet.

The following algorithm *Bluetooth_Scheduling_General* and two procedures, *Divide_Load_General* and *Load_Partition_General*, are basically identical to *Bluetooth_Scheduling* and the corresponding procedures except that rules for pairing with inclined link and entry coloring for odd cycle are added. Furthermore, when the final load at step $k = 0$ is to be transformed into schedule F, if the total load at a node is greater than one, that is at most two and will be proved later, it will be reduced to one before copying it into schedule F.

Algorithm Bluetooth_Scheduling_General
Input: Load matrix L
Output: A feasible schedule F such that

$$\sum_{l=0}^{2^n-1} F(i,j,l) \begin{cases} =L(i,j), & \text{if } i \in M_p \text{ and } j \in S \\ \geq \frac{1}{2} \cdot L(i,j) & \text{otherwise} \end{cases}$$

Procedure:
call *Divide_Load_General*$(L, n, 0)$

Procedure Divide_Load_General(L, k, l)
if $k = 0$ **then**
adjust L such that, if $T(0, x) = 2$, removes
one from $L(x, *)$ or $L(*, x)$ in a fair way
$F(i, j, l) := L(i, j) \forall i, j$
else
call *Load_Partition_General*(L, L_1, L_2)
call Divide_Load_General$(L_1, k-1, 2l)$
call Divide_Load_General$(L_2, k-1, 2l+1)$

We define two predicates $BC(L,k)$ and $EC(L_1, L_2)$ as follows. $BC(L,k)$ will be proved to be the invariant in the algorithm and $EC(L_1, L_2)$ is necessary to make the link load to be evenly distributed into the slots.

$$\begin{aligned} BC(L,k) &\equiv T(L,x) \leq 2^k + 1, \forall x \in M \cup S \\ EC(L_1, L_2) &\equiv |L_1(i,j) - L_2(i,j)| \leq 1 \\ &\quad \wedge |T(L_1, x) - T(L_2, x)| \leq 1, \forall i, j, x \end{aligned}$$

Procedure Load_Partition_General(L, L_1, L_2)
Input: Load matrix L such that $BC(L,k)$ is true where L is a load matrix at step k
Output: Load matrices L_1 and L_2 such that $BC(L_1, k-1)$, $BC(L_2, k-1)$, and $EC(L_1, L_2)$ are true.

1. For each entry $L(i, j), i \in M, j \in S$, let

$$\begin{aligned} L_1(i,j) &:= L_2(i,j) := \lfloor L(i,j)/2 \rfloor \\ A(i,j) &:= L(i,j) \bmod 2 \end{aligned}$$

   A is a 0/1 matrix.
2. For each row in A, group the nonzero entries into pairs with at most one possible unpaired entry, by connecting with a horizontal link; do the same for each column in A, by connecting with a vertical link. Then, for each master bridge whose load is presented by one row as piconet master and another column as bridging slave interconnecting two piconets, group an unpaired nonzero row entry with another unpaired nonzero column entry, if it exists, by connecting with an inclined link. Let the resulting graph be G. (Note that G is not a bipartite graph.)

3. Color the nonzero entries of A (i.e., the vertices of G) with *black* or *white* so that no two directly linked entries have the same color except that two entries on any one inclined link in each odd cycle are identically colored.
4. For each nonzero entry $A(i, j)$, if it is *black*, increment $L_1(i, j)$ by 1, else increment $L_2(i, j)$ by 1.

The following theorem proves two facts: first, a link load over no master bridge is perfectly scheduled, and second, at least half of the load over master bridge can be guaranteed to be scheduled successfully.

Theorem 18.2

For general scatternets, if L is feasible, then Algorithm Bluetooth_Scheduling_General will produce a feasible schedule having two properties about the link (i, j): (i) if i and j are pure master and pure slave, respectively, the total load assigned in F(i,j,∗) is the same as the link load $L(i, j)$, (ii) otherwise, the total assigned load of bridge x in the schedule is no less than one half of $T(*, x)$ in the initial load L.

Proof

If there is no odd cycle in graph G at all steps k and time slices l, there is no inclined link in G and thus a perfect schedule S is produced by Theorem 18.1. Therefore, we analyze the effect of some existing odd cycle in G at step k and time slice l. Suppose that predicate $BC(L, k)$ for input load L is true. Initially, it is true because initial load L at step n is feasible. We show that predicates $BC(L_1, x)$ and $BC(L_2, x)$ for loads L_1 and L_2 computed by Load_Partition_General remain true. For each master bridge x in M_b, x may be included in at most one odd cycle, since unpaired entries for x are at most two according to the pairing rule. In such case, $T(L, x)$ must be even and $|T(L_1, x) - T(L_2, x)| = 2$. Otherwise, $|T(L_1, x) - T(L_2, x)| \leq 1$. In the first case, because $T(L, x) \leq 2^k + 1$ by assumption and $T(L, x)$ is even, the larger and smaller values between $T(L_1, x)$ and $T(L_2, x)$ are no greater than $2^k/2+1 = 2^{k-1}+1$ and $2^k/2-1 = 2^{k-1}-1$, respectively. In the other case, since $|T(L_1, x) - T(L_2, x)| \leq 1$, the larger and smaller values between $T(L_1, x)$ and $T(L_2, x)$ are no greater than $2^{k-1} + 1$ and 2^{k-1}, respectively. Thus, on all occasions, predicates $BC(L_1, x)$ and $BC(L_2, x)$ are true. By induction, predicate $BC(L, x)$ remains true over all nodes x, steps k and time slices l. Moreover, when in the final step or $k = 0$, $BC(L, x) \leq 2$. That is, at worst a half of loads, i.e., 1 out of 2 can be scheduled feasibly. Otherwise, the full load could be assigned into the schedule successfully. □

18.5 Hard Coordinated Distributed Scheduling Scheme for Scatternet

In this section, a hard coordinated distributed scheduling algorithm is introduced. The algorithm has three design goals: (1) The number of messages and its computation time must be polynomial in the number of nodes, (2) the solution must be easy to implement by a distributed algorithm, and (3) it must remain correct in a mobile environment.

The proposed algorithm is based on two principles: (1) The scheduling process, based on polling, is managed by the masters of the scatternet (robustness), and (2) the reduction of the algorithm's input, from considering all nodes by each node, to that where each master is only considering its slaves (ease of implementation).

The outline of the distributed solution can be given as follows. It is assumed that nodes have distinct identities and are aware of the identities and traffic requirements of their neighbors. On entering the algorithm each master needs the permission of all its neighbors to schedule its piconet. Permission is granted to the neighboring master with the highest ID among those neighboring masters that have not yet scheduled their piconets. Permissions are passed in messages together with a set of restrictions specifying which frames that are not possible to allocate due to previous assignments performed by other neighboring masters. After receiving the permission rights from all neighbors, the master assigns timeslots to its slaves. This is done by allocating timeslots for a slave from those timeslots that are known not to be previously assigned, by some other master or the node itself, for that particular slave. The algorithm terminates when all masters have scheduled their piconets.

The algorithm has the following properties: (1) The information propagation is always limited to a one-hop radius. (2) Due to the locality property, the adaptation of slot assignment triggered by topological changes can also be executed locally. This permits fast network adaptation and concurrent execution of multiple changes. (3) The logical cycle can be reduced to the maximum slot number selected in the network. To reduce the size of the cycle the nodes need to agree on the new size and on the time in which the new cycle should go into effect simultaneously at all nodes.

18.5.1 Network Operational Assumptions

In this section, the following network operational assumptions are given. (1) The nodes have distinct identities and know the identities of their neighbors, (2) the network topology does not change during the algorithm execution, and (3) each node is aware of changes in its set of neighbors due to its own movement and loss of contact with its neighbors due to other reasons.

18.5.2 The Distributed Scatternet Scheduling Algorithm

In this subsection, the Distributed Scatternet Scheduling Algorithm (DSSA) is presented. DSSA is distributed in the sense that it requires no centrally stored information, no global knowledge of the topology and there is no special central station. The algorithm is continuously adaptive to topological changes and can be executed in parallel.

Before the proposed algorithm is formally presented, the following *functionality sets* are defined, as seen from a master's perspective regarding the slaves in its piconet.

Set 1: Slaves that are masters in other piconets.
Set 2: Slaves that are active in several piconets, i.e., bridges, but not masters.
Set 3: Slaves that are only active in the master's piconet.

The algorithm is message driven and four types of messages are used. On entering the algorithm each node has the following lists:

1. *LOCAL.Neighbors* which contains the identities, functionalities and degrees of all its neighbors. It also contains two flags for every neighbor, j: *PERMISSION.Received*, indicating whether a PERMISSION has been received from j and *Piconet_Scheduled*, indicating whether the slot assignment has been accomplished for j's piconet.
2. *LOCAL.Schedule* which lists its status, busy or idle, during each slot in the scheduling frame.
3. *NEIGHBOR.Schedules* stores the LOCAL.Schedule for each neighbor and is used to update the LOCAL.Schedule and, in case the node is a master, to construct the LOCAL.Schedule. The node also has a *LOCAL.ID*, a *LOCAL.Degree*, and a *LOCAL.Functionality*.

A node starts participating in the algorithm either on receiving a *WAKE* message from an upper layer algorithm (protocol) or by the first reception of some message sent by another node executing the algorithm. Each slave entering the algorithm sends a LOCAL(LOCAL.Degree, LOCAL.Functionality) message to all neighboring masters. Each master entering the algorithm sends a WAKE message to each slave in its slave list and the slaves respond with the same LOCAL message as in the previous phrase. This information is used by the master to fill in the degree and functionality in LOCAL.Neighbors. On completion of this list, the master is waiting for the proper conditions for scheduling its piconet. At a master this condition is obtained when it has received a *PERMISSION* message from all neighboring units, i.e., when the flag PERMISSION_Received is raised for all its neighbors. PERMISSION messages are sent together with a set of restrictions(LOCAL.Schedule) specifying frames that the master is not allowed to assign itself for this particular node.

The master collects these restrictions in NEIGHBOR.Schedules, initially empty, and they are used by the master when constructing its schedule. The PERMISSION message is sent to the node with the highest ID in LOCAL.Neighbors, i.e., in its network locality, whose Piconet_Scheduled flag is false. In general, a master that has received the PERMISSION message from all its neighbors can schedule its piconet, taking the accumulated restrictions into account when doing so. After the schedule has been completed, a message *SCHEDULE*(LOCAL.Schedule) is transmitted to all slaves in the piconet. Upon reception of the SCHEDULE message the LOCAL.Schedule and LOCAL.Neighbors.Piconet_Scheduled are updated and the PERMISSION message is sent to the next master in line that has not scheduled its piconet, if any.

It is easy to see that we now can form conflict free schedule based on the information given in the different lists mentioned above. The following rules are used by the masters to find their LOCAL.Schedules. The slaves are first prioritized according to their functionality set belonging (set one has the highest priority). Within a functionality set, slaves are prioritized according to their degrees, i.e., the slave with the highest degree within the functionality set has the highest priority and so on. Going through all slaves (according to their priorities) in descending order, the $\lambda_{i,j}$ first frames, i.e., the frames with lowest frame number, which can be found in master i's LOCAL.Schedule that does not conflict with j's LOCAL.Schedule (stored in NEIGHBORS.Schedules in the master), are assigned to slave j.

The reason for prioritizing the slaves is that it minimizes the impact of switching overhead by first scheduling the switching nodes and thereafter letting the nonswitching nodes, so to speak, "fill in the gaps." By allocating the first nonconflicting frames encountered, it is possible to find maximal transmission sets, i.e., to maximize the spatial reuse. Notice that the selected frames are the final decision unless topological changes occur in the node's locality.

18.5.3 Adapting to Topological Changes

When topological changes occur, changes in the schedules may be necessary. Assume that a node whose set of neighbors has been altered is aware of such an alteration. A topological change occurs when (1) nodes move to a new location, (2) nodes are deleted, (3) nodes are added, and (4) nodes alter their bandwidth requirements. For a uniform and simple treatment of all possible changes and their combinations we use a model, which defines a location change as a pair of events consisting of a node disconnection and connection from/to all its neighbors that are affected by the change. Disconnection and connection need not come in pairs and a single disconnection or connection is also allowed.

The algorithm is extended to deal with a disconnection. The adaption process here is based on the claim that disconnection information is needed only by neighboring nodes of the disconnected node. Only these nodes need to free the frames taken up by the disconnecting node. After learning of a disconnection, the masters alter its LOCAL.Schedule and NEIGHBOR.Schedules accordingly.

One way to achieve the connection process is to cause a WAKE event in every node which notices a change in its set of neighbors or in the link requirements, and to execute the algorithm in the previous subsection. However, the schedule changes can often be restricted to a locality smaller than the whole network saving control messages and computation time.

For any $u \in V$, define the *master set*, $Ms(i)$, as the set of masters of the piconets, where u is one of its active members. Let G be a network graph for which the DSSA has been invoked and terminated, and in which a topological change has occured. Let $V' \subseteq V$ be the set of nodes which caused the topological change. Clearly every master not in $Ms(V')$ can use the same schedule as was assigned to it before the topological change.

Instead of a WAKE/LOCAL message being sent by every waking master/slave, WAKE messages will only be sent by the masters in $Ms(V')$. LOCAL messages will not be used by the slaves to wake up masters since all topological changes within the piconet will be noticed by the master of the piconet. This is so, since the master, according to our assumptions, will notice any change in its set of neighbors and since every bandwidth alteration has to be negotiated with the master. This change in the waking process will enable nodes to stop the scheduling process from covering the whole scatternet.

Notice that the update procedure can also be used to generate the initial schedule. This observation corresponds with the intuition which views network initialization as an instance of a topological update.

18.6 QoS-Guaranteed Global Scheduling Scheme

In this section, a traffic model for scatternet, having a vast spectrum of QoS requests, is presented. Then, as an integrated scheme to meet such needs successfully, a global hybrid scheduling scheme, which combines the hard scheduling algorithm with the soft coordinated one, is proposed that is suitable for the real scatternet traffic.

18.6.1 Scatternet Traffic Model

There are two classes of connection-oriented traffic [21] in multimedia-supporting networks: constant bit rate (CBR) traffic and variable bit rate (VBR) traffic. Many multimedia transmission standards (e.g., PCM, G.721, EVRC) support CBR traffic with guaranteed bandwidth, bounded delay, and bounded jitter. Some other multimedia standards such as MPEG4, H.263, and PNG, employ VBR traffic with variable bandwidth, bounded delay, and bounded jitter. In VBR traffic, a minimum bandwidth must be guaranteed for successful multimedia operation. Internet applications like Web browser are a representative source creating CBR and VBR traffic. Another source of CBR mode can be found in monitoring and control application based on networked sensors and actuators. Inherently, such data is time critical and should be delivered within a bounded time and data loss must be minimized for a successful mission.

Thus, in Bluetooth scatternet environment, we identify three different traffic requests. First, CBR traffic is necessary to support multimedia and real-time sensor/actuator messages. CBR traffic schedule should satisfy two QoS conditions: (1) The schedule contains a proper number of slots providing the required bandwidth, and (2) scheduled slots are distributed evenly to meet delay and jitter requirements.

Second, VBR traffic supports multimedia and connection-oriented Internet applications. Scatternet management protocol messages are another example of VBR traffic. VBR traffic requirement has the same conditions as CBR traffic requirement except that, initially, the minimum bandwidth required by the application should be statically scheduled as a guarantee and the additional bandwidth is dynamically assigned by another scheme.

Finally, there is *statistical bursty rate* (SBR) traffic in which message traffic occurs at random, bursty phase, and large delay/jitter is often tolerable. File transfer is one example of SBR traffic. Since the due time does not exist, the still available bandwidth after CBR and VBR traffic assignment is assigned for SBR traffic.

In the next subsection, we present several rules that combine QoS-guaranteed scheduling algorithm with the dynamic checkpoint scheduling scheme that assigns the bursty load request to the remaining slots on-the-fly so as to yield the maximum bandwidth utility.

18.6.2 Traffic-Model Based Global Scheduling Scheme

The global scheduling scheme should deal with traffic property, message priority, and local polling harmoniously. In real applications, there exist several classes of traffic having different QoS requirements. We give a set of rules for the practical global scheduling scheme, that can deal with various QoS requirements, priority, and bursty traffic scheduling, as follows.

Global scheduling scheme

Rule 1: Bandwidth division
Total bandwidth for CBR and VBR traffic should be upper bounded to leave some bandwidth for SBR, protocol messages, and local traffic, depending upon the scatternet topology and the current applications in the network.

Rule 2: Instant rescheduling
The actual traffic is statically distributed even in the case of CBR traffic because of mutual interference among independent traffic flows along routing paths and local delay in task scheduling. Thus, when data

to be transfered on a link is less than the size of one logical slot, the unused physical slots, if it exists, should be rescheduled for the other purpose: for example, local traffic usage by switching to local polling. In the reverse case that local traffic data is not too many to utilize the whole allotted slots, the free local slots should be converted into the global slots between piconets. One possible scheme is introduced here. First, the logical slot is magnified by 20 to 40% so that it is used both for interpiconet and local traffic. Let us explain the remaining part of the scheme using an example. Assume that one logical slot has ten physical slots, so that eight out of ten are intended for master-bridge message transfer and two remaining ones are reserved for local communication. There are four cases depending on the current number of messages in the buffer. When the number of messages for global traffic and local traffic are both less than eight and two slots, respectively, the switching between master-bridge and master-slave occurs between the eighth and the ninth physical slots, as the original assignment does. However, if one type of traffic messages are bigger than the scheduled slots and the other messages are less than the corresponding slots, for example, if there are five global and four local messages in the buffer at the beginning of a logical slot, the clock switching occurs at the boundary of sixth and seventh slots so as to four slots would be supplied for local message transfer. The final case that both global and local messages are larger than the available slots is considered in the next rule.

Rule 3: Resolution for bandwidth competition

There are two alternatives for resolving bandwidth competition between global and local messages that are both larger than their own available slots. The first one is keeping the assigned boundary between global and local slots without reallocation. According to message priority, messages with highest priorities are selected up to the available number of slots and transfered. In general, the management protocol message has the highest priority. Then, time-critical, real-time message, e.g., emergency or control information in sensor network, is assigned with the next highest priority. Multimedia data may follow these. The remaining data will have the lowest priority. The second alternative is changing the boundary according to priority. For example, suppose that high-priority global and local messages are five and three, in addition to the low-priority global and local messages four and seven. One possible modified boundary would be between seventh and eighth slots so that three high-priority local messages are guaranteed to be transfered. Another proper boundary would be sixth and seventh so that at least one slot for low-priority global and local messages could be delivered. There exist a lot of variations among which one will be selected according to the requirements of the scatternet environment or the current applications.

Rule 4: Dynamic slot assignment

In the previous two rules, rescheduling scheme between global (master-bridge) and local (master-slave) traffic is explained. However, there exists more flexible, dynamic slot assignment scheme. Suppose that the system has low load matrix so that there may be free slot(s) between scheduled slots. These free slots can be assigned on-the-fly by the consent of both parties. A typical senario is the following. Suppose that a master starts communication with a bridge at the beginning of a logical slot. Initially, scheduling-related information, like the message sizes waiting in the buffer and the available number of free slots following the current one, is exchanged. Then, according to a common rule, both nodes could allot the following free slots dynamically, if it exists. For example, suppose that three and two consecutive logical slots are free at the master and the bridge, respectively. If there are a lot of messages waiting in the buffers, the nodes would make three consecutive logical slots for the message transfer. That is, in the global scheduling scheme, we can adopt the checkpoint algorithm to supplement the QoS-aware algorithm to meet the various QoS requirements and to maximize the bandwidth utility by the dynamic slot assignment simultaneously.

18.7 Conclusion

In this chapter, the Bluetooth scheduling schemes are studied. First, the piconet scheduling schemes proposed in [7,8] are introduced. Then, checkpoint based scatternet scheduling algorithm [13,15] is introduced. This algorithm is suitable for the QoS-unconscious application. On the other hand, a QoS-aware scatternet algorithm is proposed to meet the QoS request from the multimedia and real-time applications. The complexity of this algorithm can be reduced by adopting the distributed scheduling protocol proposed in [11]. Finally, the hybrid scheme, which combines the QoS-aware scatternet algorithm with the checkpoint based scheduling algorithm, is proposed so that the QoS-aware time slot assignment is supplemented with the dynamic load allotment to meet the whole spectrum of various traffic requests.

References

[1] The Bluetooth SIG, The Bluetooth Core and Profile Specifications, *http://www.bluetooth.com*.

[2] C. Bisdikian, An Overview of the Bluetooth Wireless Technology, *IEEE Communications Magazine*, pp. 86–94, Dec. 2001.

[3] J. Bray and C. F. Sturman, *Bluetooth: Connect without Cables*, Prentice Hall, New York, 2000.

[4] R. Kapoor, et al., Multimedia Support over Bluetooth Piconets, *ACM Workshop on Wireless Mobile Internet*, pp. 50–55, July 2001.

[5] M. Albrecht, et al., IP Services over Bluetooth: Leading the Way to a New Mobility, *Conference on Local Computer Networks*, pp. 2–11, Oct. 1999.

[6] P. Johansson, N. Johansson, U. Körner, J. Elgg, and G. Svennarp, Short Range Radio Based Ad Hoc Networking: Performance and Properties, *Proceedings of ICC'99*, Vancouver, 1999.

[7] N. Johansson, et al., Performance Evaluation of Scheduling Algorithms for Bluetooth, *Broadband Communications: Convergence of Network Technologies*, Kluwer Academic Publishers, pp. 139–150, June 2000.

[8] A. Capone, M. Gerla, and R. Kapoor, Efficient Polling Schemes for Bluetooth Picocells, *IEEE ICC'01*, pp. 1990–1994, June 2001.

[9] M. Kalia, D. Bansal, and R. Shorey, MAC Scheduling Policies for Power Optimization in Bluetooth: A Master Driven TDD Pico-Cellular Wireless System, *IEEE Vehicular Technology Conference 2000*, Tokyo, 2000.

[10] M. Kalia, S. Garg, and R. Shorey, Efficient Policies for Increasing Capacity in Bluetooth: An Indoor Pico-Cellular Wireless System, *IEEE Vehicular Technology Conference 2000*, Tokyo, 2000.

[11] N. Johansson, U. Körner, and L. Tassiulas, A Distributed Scheduling Algorithm for a Bluetooth Scatternet, *International Teletraffic Congress*, pp. 61–72, Sept. 2001.

[12] M. L. Garey and D. S. Johnson, *Computers and Intractability: A Guide to the Theory of NP-Completeness*, W. H. Freeman, San Francisco, 1979.

[13] A. Rácz, et al., A Pseudo Random Coordinated Scheduling Algorithm for Bluetooth Scatternets, *MobiHOC'01*, pp. 193–203, 2001.

[14] N. Johansson, et al., JUMP Mode — A Dynamic Window-Based Scheduling Framework for Bluetooth Scatternets, *MobiHOC'01*, pp. 204–211, 2001.

[15] P. Johansson, R. Kapoor, M. Kazantzidis, and M. Gerla, Rendezvous Scheduling in Bluetooth Scatternets, *ICC, 2002*.

[16] M. Kazantzidis and M. Gerla, On the Impact of Inter-Piconet Scheduling in Bluetooth Scatternets, *Internet Workshop on Wired/Wireless Internet Communications*, Las Vegas, June 2002.

[17] S. Baatz, M. Frank, C. Kühl, P. Martini, and C. Scholz, Bluetooth Scatternets: An Enhanced Adaptive Scheduling Scheme, *INFOCOM 2002*, New York, June 2002.

[18] Z. Liu, P. Nain, and D. Towsley, On Optimal Polling Policies, *Queueing Systems*, vol. 11, pp. 59–83, 1992.

[19] O. Fabian and H. Levy, Polling System Optimization through Dynamic Routing Policies, *IEEE INFOCOM '93*, vol. 1, pp. 194–200.

[20] H. Takagi, *Analysis of Polling Systems*, MIT Press, 1986.

[21] A. Tanenbaum, *Computer Networks*, Prentice Hall PTR, 3rd ed., 1996.

19

Fair Sequences

Wieslaw Kubiak
Memorial University of Newfoundland

19.1 Introduction

The idea of *fair* sequences seemed to have been started on an assembly line, in fact on a mixed-model, just-in-time assembly line at Toyota. There sequences of different models to produce were sought to smooth out the usage rate of all parts and the production rate of all different models, and consequently to reduce shortages, on one hand, and excessive inventories, on the other, Monden [1]. It is worth noticing that Tijdeman [2] worked on a similar problem, called the Chairmen Assignment Problem, in 1980. It has then been realized at Toyota that a sequence with different models evenly spread throughout the sequence meets these requirements much better than a *batch* sequence that keeps all copies of the same model as close to each other as possible. The latter, a hallmark of *mass* production, was being replaced by the former, a hallmark of mass *customization*. Miltenburg [3] formulated the problem as a nonlinear integer programming problem with the objective to minimize the total *deviation* of actual model production from the desired quantity of model production determined by demands $d_1, \ldots, d_n$ for models $1, \ldots, n$, respectively. This original model, referred to as the *product rate variation* (PRV) problem, was subsequently extended to multilevel assembly systems where the *derived* demands for the outputs of *supply chain* downstream processes were also factored in the sequence of final product assembly, Miltenburg and Goldstein [4], Kubiak, Steiner and Yeomans [5]. Drexl and Kimms [6] study the PRV with temporal capacity constraints of the assembly systems.

Miltenburg [3], and Inman and Bulfin [7] presented a number of heuristic algorithms for the original PRV problem, and Kubiak and Sethi [8,9] demonstrated that the problem can be reduced to the assignment problem, and thus efficiently solved. Interestingly, when studying these heuristics, Bautista, Companys, and Corominas [10] discovered that one of them is a simple extension of the *Hamilton apportionment* and thus fails to build optimal sequences as it suffers from the infamous *paradox of Alabama*, well-known phenomenon in the theory and practice of *apportionment*, see Balinski and Young [11]. This observation

1-58488-397-9/$0.00+$1.50

prompted Balinski and Shahidi [12] to propose a simple approach to the product rate variation problem *via* axiomatics.

On another side of the spectrum, computer science seemed to have independently and apparently unknowingly stepped into *fair* sequencing idea when Waldspurger and Wheil [13] proposed a *stride* scheduling approach to resource allocation in operating systems, the Internet and networks. We argue in this chapter that their stride scheduling approach is simply the Thomas *Jefferson's method* of the apportionment, see Balinski and Shahidi [12], and Balinski and Young [11] for the definition of the method.

Altman, Gaujal, and Hordijk [14] have recently asked the following question which echoes both product rate variation problem and the stride scheduling: Is it possible to construct an infinite sequence over n letters where each letter is distributed as evenly as possible and appears with a given rate? The sequences proving positive answer to this question provide optimal routings for tasks to resources and minimize expected average workload in event graphs. Altman, Gaujal, and Hordijk propose *balanced words* as the answer. In balanced words, any two subsequences of the same length must have more or less the same number of occurrences of each letter, more precisely, these numbers for each letter must not differ by more than 1. Though *balanced* words answer the question positively, in many cases it may prove *impossible* to obtain a balanced word for given letter rates. In fact, the famous Fraenkel's conjecture, see Tijdeman [15], states that for each $n > 2$ there is only *one* set of *distinct* rates for which a balanced word exists. However, the question whether other sequences, besides balanced words, would also provide optimal routings seems still open. Interestingly, a special case of Fraenkel's conjecture has been formulated by Brauner and Crama [16] for solutions to the product rate variation problem with small maximum deviations, that is less than $\frac{1}{2}$.

All these practical and theoretical problems clearly revolve around an elusive concept of *fair* sequences, see Balinski and Young [11] who used the term *fair representation* in the context of the apportionment problem. This chapter will argue that the concept fair sequences is *fundamental* for a growing number of applications in manufacturing, hard real-time systems, computer and operating systems, the Internet, and networks. Having said that we also realize that one, mathematically sound, and universally accepted definition of fairness in resource allocation and scheduling has not yet emerged, if it ever will. This of course does not prevent us from investigating and tracking different models of fair sequences and mapping links between them. Needless to say, we wish for a comprehensive review in this chapter, the goal which is even more elusive than the concept of fair sequences itself.

We start by defining fair sequences as the solutions to the PRV problem which will be formally defined in Section 19.2. Next, Section 19.3 presents the reduction of the PRV problem to the assignment problem, and thus provides an efficient algorithm for the former. This section also presents the main properties of optimal sequences for the PRV problem as well as an efficient method for solving the min-max deviation problem. Section 19.4 shows main properties of min-max optimal sequences and their applications to the generalized pinwheel scheduling and the periodic scheduling. Section 19.5 discusses the Fraenkel's conjecture for balanced words as well as a solution of the small deviations conjecture. Section 19.6 focuses on special balanced words called constant gap words or exact coverings. Section 19.7 introduces the stride scheduling, shows that the stride scheduling is a parametric method of apportionment and discusses main properties of stride scheduling as well as two main performance metrics suggested for stride scheduling, the throughput error and the response time variability. Finally, Section 19.8 proposes a new model of fair sequences, peer-to-peer fair sequences and concludes the chapter.

19.2 The Product Rate Variation Problem

We study the following optimization problem. Given n products (models) $1, \ldots, i, \ldots, n$, n positive integers (*demands*) $d_1, \ldots, d_i, \ldots, d_n$, and n convex and symmetric functions $f_1, \ldots, f_i, \ldots, f_n$ of a single variable, called deviation, all assuming minimum 0 at 0. Find a sequence $S = s_1 \ldots s_D$, $D = \sum_{i=1}^{n} d_i$, of products $1, \ldots, i, \ldots, n$, where product i occurs exactly d_i times that minimizes

$$F(S) = \sum_{i=1}^{n} \sum_{k=1}^{D} f_i(x(S)_{ik} - r_i k)$$

where $x(S)_{ik}$ = the number of product i occurrences (*copies*) in the prefix $s_1 \ldots s_k$, we shall refer to this prefix as k-prefix, $k = 1, \ldots, D$, and $r_i = d_i/D$, $i = 1, \ldots, n$. The problem is fundamental in flexible just-in-time production systems, where sequences, we refer to them as *JIT* sequences, that make these systems tick must keep the actual, equal $x(S)_{ik}$, and the desired, equal $r_i k$, production levels of each product i as close to each other as possible all the time, Monden [1], Miltenburg [3], Vollman, Berry, and Wybark [17], and Groenevelt [18]. This problem is known as the *product rate variation (PRV)* problem in the literature, see Kubiak [19], Bautista, Companys and Corominas [10,20], and Balinski and Shahidi [12]. It will be a point of departure for our discussion in this chapter. Therefore, in the next section, we describe its reduction to the assignment problem which will allow us to efficiently solve it as well as a number of closely related problems to optimality. In our subsequent discussions we often use the notation x_{ik} instead of $x(S)_{ik}$ for convenience whenever this simplification does not introduce confusion.

19.3 Ideal Positions and Reduction to Assignment Problem

We begin this section with an example to provide an insight into the reduction of the *PRV* problem to the assignment problem introduced by Kubiak and Sethi [8,9]. In this example we consider product i with demand $d_i = 3$, $D = 17$, and the absolute deviation function $f_i(x_{ik} - r_i k) = |x_{ik} - kr_i|$. We first observe that since the variable x_{ik} can take on only values $0, 1, \ldots, d_i$, then by replacing it by its $d_i + 1$ possible values we obtain $d_i + 1$ functions $|0 - kr_i|, |1 - kr_i|, \ldots, |d_i - kr_i|$ of a single variable k that assumes integer values $0, 1, \ldots, D$. The graphs of these functions in the interval $[0, D]$ are shown in Figure 19.1. These graphs are all we need to explain the definition of the cost coefficients in the assignment problem. The idea is as follows. Ideally, product i should be produced in positions 3, 9, and 15, marked by the small circles on the horizontal time line in Figure 19.1. Notice that 3 is the smallest integer greater than the crossing point of $|0 - kr_i|$ and $|1 - kr_i|$, 9 is the smallest integer greater than the crossing point of $|1 - kr_i|$ and $|2 - kr_i|$, and 15 is the smallest integer greater than the crossing point of $|2 - kr_i|$ and $|3 - kr_i|$. If the three copies of product i are sequenced in positions 3, 9, and 15, respectively, then product i will contribute the $\inf_j |j - kr_i|$ at $k = 0, 1, \ldots, D$, i.e., the "lower envelope" of the set of functions $|0 - kr_i|, |1 - kr_i|, \ldots, |d_i - kr_i|$ at $k = 0, 1, \ldots, D$, to the total cost of the solution. However, i must compete with other products for the three ideal positions and therefore we must be able to calculate costs

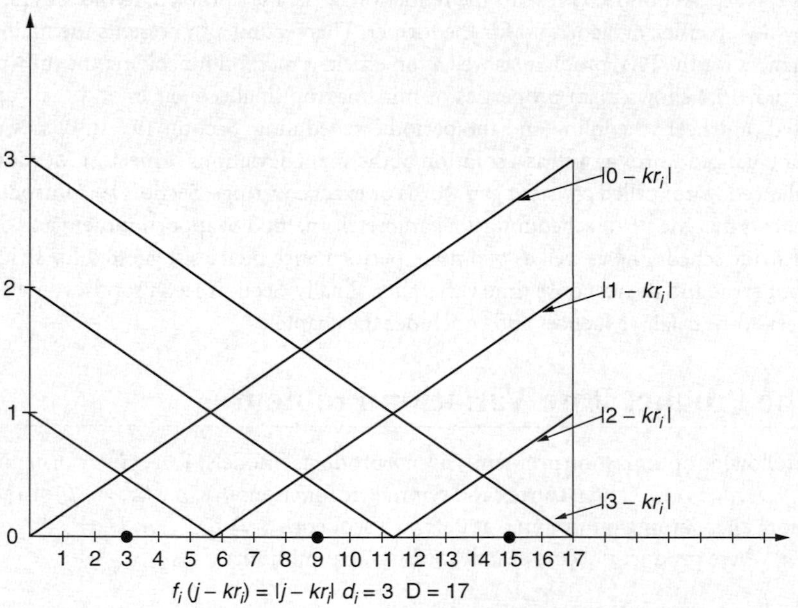

FIGURE 19.1 The ideal positions 3, 9, and 15 and the level curves $|j - kr_i|$ for copies $j = 1, 2$, and 3.

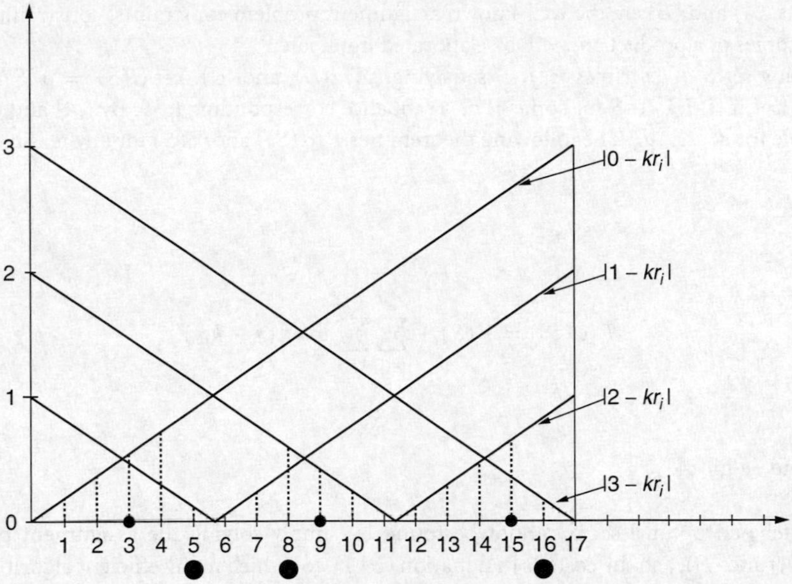

FIGURE 19.2 Cost calculation for a solution with copies of product i in positions 5, 8, and 16 instead of the ideal 3, 9, and 15.

of deviating from these positions if necessary. For instance, suppose that i ends up in positions 5, 8, and 16 instead, see Figure 19.2. Then, the *additional* cost is incurred in 3 and 4 since the first copy of i is too *late*, i.e., the cost is charged according to $|0 - r_i k|$ until 5 when the solution switches to $|1 - r_i k|$ and therefore the differences $|0 - r_i k| - |1 - r_i k|$ for k=3 and 4 add to the cost of the ideal solution that would switch from $|0 - r_i k|$ to $|1 - r_i k|$ earlier, at 3. Then, the cost is charged according to $|1 - r_i k|$ from 5 until it switches to $|2 - r_i k|$ at 8, this adds additional cost in 8 since the second copy is too *early*. Finally, the additional cost is charged in 15 since the third copy is in position 16, which is too *late*.

We are now ready to formally introduce the reduction. Let $X = \{(i, j, k) | i = 1, \ldots, n;\ j = 1, \ldots, d_i;\ k = 1, \ldots, D\}$. Following Kubiak and Sethi [8,9], define cost $C^i_{jk} \geq 0$ for $(i, j, k) \in X$ as follows:

$$C^i_{jk} = \begin{cases} \sum_{l=k}^{Z^i_j - 1} \psi^i_{jl}, & \text{if } k < Z^i_j \\ 0, & \text{if } k = Z^i_j \\ \sum_{l=Z^i_j}^{k-1} \psi^i_{jl}, & \text{if } k > Z^i_j \end{cases} \tag{19.1}$$

where for symmetric functions f_i, $Z^i_j = \lceil \frac{2j-1}{2r_i} \rceil$ is called the *ideal position* for the jth copy of product i, and

$$\begin{aligned} \psi^i_{jl} &= | f_i(j - lr_i) - f_i(j - 1 - lr_i)| \\ &= \begin{cases} f_i(j - lr_i) - f_i(j - 1 - lr_i), & \text{if } l < Z^i_j \\ f_i(j - 1 - lr_i) - f_i(j - lr_i), & \text{if } l \geq Z^i_j \end{cases} \end{aligned} \tag{19.2}$$

Notice that the point $\frac{2j-1}{2r_i}$ is the crossing point of $f_i(j - 1 - kr_i)$ and $f_i(j - kr_i)$, $j = 1, \ldots, d_i$.

Let $S \subseteq X$, we define $V(S) = \sum_{(i,j,k)\in S} C^i_{jk}$, and call S *feasible* if it satisfies the following three constraints:

($\mathcal{A}$) For each k, $k = 1, \ldots, D$, there is exactly one pair (i, j), $i = 1, \ldots, n$; $j = 1, \ldots, d_i$ such that $(i, j, k) \in S$.

($\mathcal{B}$) For each pair (i, j), $i = 1, \ldots, n$; $j = 1, \ldots, d_i$, there is exactly one k, $k = 1, \ldots, D$, such that $(i, j, k) \in S$.

($\mathcal{C}$) If $(i, j, k), (i, j', k') \in S$ and $k < k'$, then $j < j'$.

Constraints ($\mathcal{A}$) and ($\mathcal{B}$) are the well-known assignment problem constraints, constraint ($\mathcal{C}$) imposes an order on copies of a product and will be elaborated upon later.

Consider any set S of D triples (i, j, k) satisfying ($\mathcal{A}$), ($\mathcal{B}$), and ($\mathcal{C}$). Let $\alpha(S) = \alpha(S)_1, \ldots, \alpha(S)_D$, where $\alpha(S)_k = i$ if $(i, j, k) \in S$ for some j, be a sequence corresponding to S. By ($\mathcal{A}$) and ($\mathcal{B}$) sequence $\alpha(S)$ is feasible for $d_1, \ldots, d_n$. The following theorem ties $F(\alpha(S))$ and $V(S)$ for any feasible S.

Theorem 19.1

We have

$$F(\alpha(S)) = V(S) + \sum_{i=1}^{n} \sum_{k=1}^{D} \inf_j f_i(j - kr_i) \tag{19.3}$$

Proof

See Kubiak and Sethi [8]. □

Unfortunately, an optimal set S cannot be found by simply solving the assignment problem with constraints ($\mathcal{A}$) and ($\mathcal{B}$), and the costs as in Equation (19.1), for which many efficient algorithms exist, see for example Kuhn [21]. The reason is constraint ($\mathcal{C}$), which is not of the assignment type. Informally, ($\mathcal{C}$) ties up copy j of a product with the jth ideal position for the product and it is necessary for Theorem 19.1 to hold. In other words, for a set S satisfying ($\mathcal{A}$) and ($\mathcal{B}$) but not ($\mathcal{C}$) we may generally have inequality in Equation (19.3). However, the following theorem remedies this problem.

Theorem 19.2

If S satisfies ($\mathcal{A}$) and ($\mathcal{B}$), then S' satisfying ($\mathcal{A}$), ($\mathcal{B}$), and ($\mathcal{C}$), and such that

$$V(S) \geq V(S')$$

can be constructed in $O(D)$ steps. Furthermore, each product occupies the same positions in $\alpha(S')$ as it does in $\alpha(S)$.

Proof

See Kubiak and Sethi [8]. □

The set of optimal solutions $\mathcal{S}^*$ includes *cyclic* solutions for symmetric functions f_i if the greatest common divisor of demands $d_1, \ldots, d_n$, denoted by $\gcd(d_1, \ldots, d_n)$, is greater than 1. This follows from the following theorem.

Theorem 19.3

Let β be an optimal sequence for $d_1, \ldots, d_n$. Then β^m, $m \geq 1$, is optimal for $md_1, \ldots, md_n$.

Proof

See Kubiak [22]. □

Moreover, if $\alpha \in \mathcal{S}^*$, then $\alpha^R \in \mathcal{S}^*$, where α^R is a mirror reflection of α.

This just presented approach to solving the *PRV* problem, proposed by Kubiak and Sethi [8,9], can be *readily* modified to solve the min-max deviation problem, where the objective function is as follows:

$$H(S) = \min \max_{i,k} f_i(x(S)_{ik} - r_i k)$$

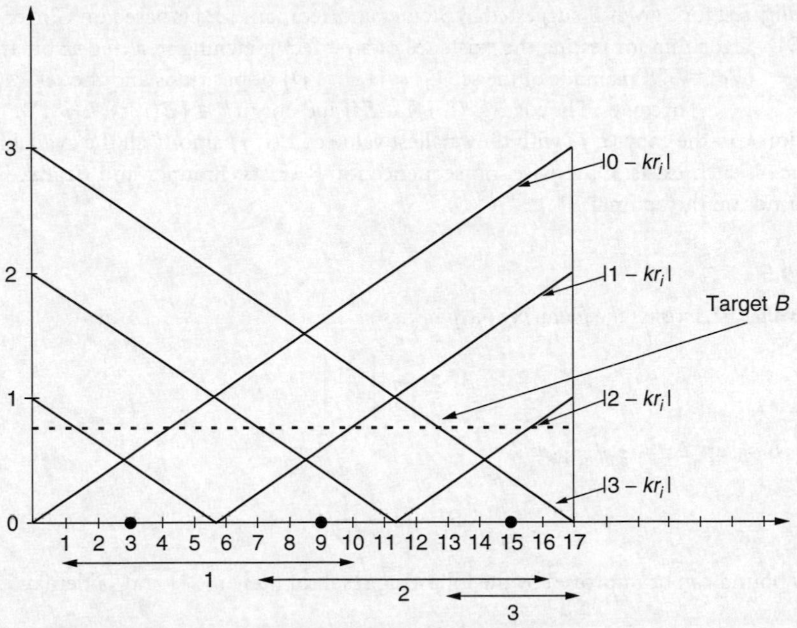

FIGURE 19.3 The earliest and latest positions for copies 1, 2, and 3 of product i and bound B on the maximum deviation.

This was done by Steiner and Yeomans [23] for the case with the same absolute deviation function, $f_i(x_{ik} - r_i k) = |x_{ik} - r_i k|$, for all products, however, their approach can easily be extended to other symmetric and convex functions of the deviation. We shall explain their idea of the solution with the same example as used in Figure 19.1 and Figure 19.2 in our earlier discussion of the *PRV* problem. Suppose, that we wish to test if there exists S with maximum deviation $H(S) \leq B$, where B is a given upper bound imposed *a priori* on the maximum deviation. We could use the graph in Figure 19.1, draw a horizontal line at the distance B above the horizontal time line as in Figure 19.3, and then learn from the graph how far copies of product i are allowed to deviate from their ideal positions in order *not* to violate the bound B imposed on the deviation. Figure 19.3 shows that product i will not violate B as long as its first copy is somewhere between 1 and 10 inclusive, copy 2 somewhere between 7 and 16 inclusive, and copy 3 somewhere between 13 and 17 inclusive. For instance, copy 2 sequenced either before 7 or after 16 would result in the deviation above B along $|2 - r_i|$. Consequently, the two crossing points of line B and $|j - r_i k|$ determine the *earliest* and the *latest* positions for copy j to be sequenced. We assume that the earliest position is 1 and the latest is D if the corresponding crossing points do not exist in the interval $[1, D]$. The earliest and latest positions can be easily calculated as shown by Steiner and Yeomans [23]. Brauner and Crama [16] propose the following formulae for calculating this positions.

Theorem 19.4

If a sequence S with maximum absolute deviation not exceeding B exists, then copy j of product i, $i = 1, \ldots, n$ and $j = 1, \ldots, d_i$ occupies a position in the interval $[E(i, j), L(i, j)]$, where

$$E(i, j) = \left\lceil \frac{j - B}{r_i} \right\rceil$$

and

$$L(i, j) = \left\lfloor \frac{j - 1 + B}{r_i} + 1 \right\rfloor$$

The *feasibility* test for a given B suggested by Steiner and Yeomans [23] is based on Glover's [24] earliest due date (EDD) algorithm for testing the existence of a perfect matching in a *convex* bipartite graph G. The graph $G = (V_1 \cup V_2, \mathcal{E})$ is made of the set $V_1 = \{1, \dots, D\}$ of positions and the set $V_2 = \{(i, j) | i = 1, \dots, n; j = 1, \dots, d_i\}$ of copies. The edge $(k, (i, j)) \in \mathcal{E}$ if and only if $k \in [E(i, j), L(i, j)]$. The algorithm assigns position k to the copy (i, j) with the smallest value of $L(i, j)$ among all the available copies with $(k, (i, j)) \in \mathcal{E}$, if such exists. Otherwise, no sequence for B exists. Brauner and Crama [16] show the following bounds on the optimal B^*.

Theorem 19.5

The optimal value B^* satisfies the following inequalities

$$B^* \geq \frac{1}{\Delta_i} \left\lfloor \frac{\Delta_i}{2} \right\rfloor$$

for $i = 1, \dots, n$, where $\Delta_i = \frac{D}{gcd(d_i, D)}$ and

$$B^* \leq 1 - \frac{1}{D}$$

The upper bound can be improved by the following result of Meijer [25] and Tijdeman [2].

Theorem 19.6

Let λ_{ij} be a double sequence of nonnegative numbers such that $\sum_{1 \leq i \leq k} \lambda_{ik} = 1$ for $k = 1, \dots$. For an infinite sequence S in $\{1, \dots, n\}$ let x_{ik} be the number of i's in the k-prefix of S. Then there exists a sequence S in $\{1, \dots, n\}$ such that

$$\max_{i,k} \left| \sum_{1 \leq j \leq k} \lambda_{ij} - x_{ik} \right| \leq 1 - \frac{1}{2(n-1)}$$

Let us define $\lambda_{ik} = r_i = \frac{d_i}{D}$ for $k = 1, \dots$. Then, this theorem ensures the existence of an infinite sequence S such that

$$\max_{i,k} | kr_i - x_{ik} | \leq 1 - \frac{1}{2(n-1)}$$

To ensure that the required number of copies of each product is in D-prefix of S, consider the D-prefix of S and suppose that there is i with $x_{iD} > d_i$. Then, there is j with $x_{jD} < d_j$. It can be easily checked that replacing the last i in the D-prefix by j does not increase the absolute maximum deviation for the D-prefix. Therefore, we can readily obtain a D-prefix where each i occurs exactly d_i times and with maximum deviation not exceeding $1 - \frac{1}{2(n-1)}$. We consequently have the following stronger upper bound.

Theorem 19.7

The optimal value B^* satisfies the following inequality

$$B^* \leq 1 - \max\left\{ \frac{1}{D}, \frac{1}{2(n-1)} \right\}$$

Though, $D \geq 2(n-1)$ most often in practice. It is obviously possible that $D < 2(n-1)$, for instance when $d_i = 1$ for all i and $n > 2$. We refer the reader to Tijdeman [2] for details of the algorithm to generate a sequence with $B \leq 1 - \frac{1}{2(n-1)}$.

Finally, Theorem 19.7 along with the fact that the product DB^* is integer, see Steiner and Yeomans [23], allow the binary search to find the optimum B^* and the corresponding matching by doing $O(\log D)$ tests. Other efficient algorithms based on the reduction to the bottleneck assignment problem were proposed by Kubiak [19], Bautista, Companys, and Corominas [20], and Moreno [26].

We shall study solutions to the maximum absolute deviation problem in the subsequent sections, where we will refer to these solutions as simply min-max optimal solutions and to the problem itself as simply min-max problem. In fact our discussion will equally well apply to any solution with $B < 1$, not necessarily optimal, which by Theorem 19.7 always exists.

19.4 Properties and Applications of Min-Max Optimal Solutions

We shall discuss some properties of min-max optimal sequences in this section. Also, we show some application of min-max optimal solutions to the generalized pinwheel sequencing problem and to the periodic scheduling in hard real-time environments.

19.4.1 Properties

Let $d_1, \ldots, d_n$ be an instance of the min-max problem. In this and following sections we shall use the term letter (or symbol) instead of product to make our discussion more general, that is going beyond just-in-time assembly line sequences. Let $r_i = \frac{d_i}{D} = \frac{a_i}{b_i}$, where a_i and b_i are relatively prime, be the rate for letter i. We shall consider an infinite periodic sequence $s = S^* = SS\ldots$ with period S obtained by the min-max EDD algorithm, for instance, described in Section 19.3 with a given bound B. First, we observe that copy $j+1$ must follow copy j, $j = 1, \ldots$, in any S with $B < 1$. We have the following lemma.

Lemma 19.1

For $B < 1$ and $k = 1, 2, \ldots$ $j = 1, \ldots$, we have

$$L(i, j) \leq E(i, j+1)$$

Proof

By Theorem 19.4

$$L(i, j) = \left\lfloor \frac{j-1+B}{r_i} + 1 \right\rfloor$$

and

$$E(i, j+1) = \left\lceil \frac{j+1-B}{r_i} \right\rceil = \left\lceil \frac{j-1+B}{r_i} + \frac{2(1-B)}{r_i} \right\rceil$$

If $\frac{j-1+B}{r_i}$ is an integer, then

$$L(i, j) = \frac{j-1+B}{r_i} + 1 \leq \frac{j-1+B}{r_i} + \left\lceil \frac{2(1-B)}{r_i} \right\rceil = E(i, j+1)$$

since $\frac{2(1-B)}{r_i} > 0$ for $B < 1$. If $\frac{j-1+B}{r_i}$ is not an integer, then

$$L(i, j) = \left\lceil \frac{j-1+B}{r_i} \right\rceil \leq \left\lceil \frac{j-1+B}{r_i} + \frac{2(1-B)}{r_i} \right\rceil = E(i, j+1)$$

since again $\frac{2(1-B)}{r_i} > 0$ for $B < 1$. Thus, the lemma holds. □

Consider a sequence s with maximum deviation $B < 1$, we show that any its subsequence w with at least ka_i i's is not *longer* than $kb_i + 1$, $k = 1, 2, \ldots$. The length of w equals the number of letters in w. Let the first i in subsequence w be copy $j+1$, and the last copy of i in subsequence w be copy l. Obviously, $l \geq j + ka_i$, otherwise there would be less than ka_i i's in subsequence w. In s, copy $j+1$ cannot be in any position prior to $E(i, j+1)$ and copy $j + ka_i$ cannot be in any position higher than $L(i, j + ka_i)$. However, we have the following upper bound on the difference between the latter and the former in s.

Lemma 19.2

For $B < 1$ and $k = 1, 2, \ldots$ $j = 0, 1, \ldots$, we have

$$L(i, j + ka_i) - E(i, j+1) \le kb_i$$

Proof

By Theorem 19.4

$$L(i, j + ka_i) = \left\lfloor \frac{j + ka_i - 1 + B}{r_i} + 1 \right\rfloor$$

and

$$E(i, j+1) = \left\lceil \frac{j + 1 - B}{r_i} \right\rceil$$

Thus,

$$L(i, j + ka_i) - E(i, j+1) \le \frac{ka_i}{r_i} + 1 + \frac{2B - 2}{r_i}$$

By assumption $\frac{2B-2}{r_i} < 0$ and $r_i = \frac{a_i}{b_i}$. Furthermore, the left-hand side of the inequality is an integer, thus the lemma holds. □

Now, consider a sequence s again with maximum deviation $B < 1$, we show that any subsequence w of s with at least $ka_i + 2$ i's is not *shorter* than $kb_i + 1$, $k = 1, 2, \ldots$. Let the first i in subsequence w be copy j and the last copy of i in subsequence w be copy l. Obviously, $l \ge j + ka_i + 1$, otherwise there would be less than $ka_i + 2$ i's in subsequence w. In s, copy j cannot be in any position higher than $L(i, j)$ and copy $ka_i + 1$ cannot be in any position prior to $E(i, j + ka_i + 1)$. However, we have the following lower bound on the difference between the latter and the former in s.

Lemma 19.3

For $B < 1$ and $k = 1, 2, \ldots, j = 1, \ldots$, we have

$$E(i, j + ka_i + 1) - L(i, j) \ge kb_i$$

Proof

By Theorem 19.4

$$E(i, j + ka_i + 1) = \left\lceil \frac{j + ka_i + 1 - B}{r_i} \right\rceil$$

and

$$L(i, j) = \left\lfloor \frac{j - 1 + B}{r_i} + 1 \right\rfloor$$

Thus,

$$E(i, j + ka_i + 1) - L(i, j) \ge \frac{ka_i}{r_i} - 1 + \frac{2 - 2B}{r_i}$$

By assumption, $\frac{2-2B}{r_i} > 0$ and $r_i = \frac{a_i}{b_i}$. Furthermore, the left hand side of the inequality is an integer, thus, the lemma holds. □

We use the two lemma to characterize distribution of letter i in s with $B < 1$.

Theorem 19.8

Let $s_{j+1} \dots s_{j+b_i}$ be any subsequence of b_i consecutive letters of $s = S^*$, with S obtained by the min-max algorithm with $B < 1$. Then, letter i occurs either $a_i - 1$, or a_i, or $a_i + 1$ times in the subsequence. Furthermore, if subsequences $s_{j+1} \dots s_{j+b_i}$ and $s_{j+1+kb_i} \dots s_{j+(k+1)b_i}$, for some $k \geq 1$ have $a_i - 1$ letter i occurrences each, then $k \geq 2$ and there are exactly $(k-1)a_i + 1$ i's in the sequence $s_{j+1+b_i} \dots s_{j+kb_i}$.

Proof

Consider subsequence $w = s_{j+1} \dots s_{j+b_i}$, $j \geq 0$. We first show that there are at least $a_i - 1$ i's in w. This claim obviously holds for $a_i = 1$. Thus, let $a_i \geq 2$, and assume that the number of i's in w is less than $a_i - 1$. If there is no i in $s_1 \dots s_j$, then there is $l > j + b_i + 1$ such that the sequence $s_1 \dots s_l$ has exactly a_i i's but this contradicts the fact that the sequence with a_i i's is not longer than $b_i + 1$. Now, if there is an i in $s_1 \dots s_j$, then there are $k \leq j$ and $l > j + b_i$ such that the sequence $s_k \dots s_l$ has exactly a_i i's. However, $l - k + 1 > b_i + 1$ which again contradicts the fact that the sequence with a_i i's is not longer than $b_i + 1$. Therefore, there is at least $a_i - 1$ i's in $s_{j+1} \dots s_{j+b_i}$. Assume now, that there are at least $a_i + 2$ i's in w. This, however, again leads to a contradiction since the sequence has b_i letters only and thus it cannot include as many as $a_i + 2$ i's, any subsequence with at least $a_i + 2$ i's must be at least $b_i + 1$ long. Therefore, the only possible numbers of i's in $s_{j+1} \dots s_{j+b_i}$ are $a_i - 1$, a_i, and $a_i + 1$. This completes the proof of the first part of the theorem.

Now, let $s_{j+1} \dots s_{j+b_i}$ and $s_{j+kb_i+1} \dots s_{j+(k+1)b_i}$ for $k \geq 1$ be two sequences with $a_i - 1$ i's. Assume that each sequence $s_{j+lb_i+1} \dots s_{j+(l+1)b_i}$ $1 < l < k$, if any, in between the two has exactly a_i i's. If there is no i in $s_1 \dots s_j$, then sequence $s_1 \dots s_{j+(k+1)b_i}$ has no more than $(k+1)a_i - 2$ i's and there is $l > j + (k+1)b_i + 1$ such that the sequence $s_1 \dots s_l$ has exactly $(k+1)a_i$ i's but this contradicts the fact that the sequence with ka_i i's is not longer than $kb_i + 1$. If there is an i in $s_1 \dots s_j$, then there are $h \leq j$ and $l > j + (k+1)b_i + 1$ such that the sequence $s_h \dots s_l$ has exactly $(k+1)a_i$ i's. However, $l - h + 1 > (k+1)b_i + 1$, which again contradicts the fact that the sequence with ka_i i's is not longer than $kb_i + 1$. Then in both cases, $k \geq 2$ and there must be a sequence $s_{j+lb_i+1} \dots s_{j+(l+1)b_i}$ $1 < l < k$ with $a_i + 1$ i's.

To complete the proof we observe that there is a positive integer α_i such that $\alpha_i b_i = D$ and $\alpha_i a_i = d_i$. Consequently, there are $\alpha_i a_i$ i's in $s_1 \dots s_D$. For any j, $j = 0, \dots, b_i - 1$, consider sequences $S_k = s_{j+kb_i+1} \dots s_{j+(k+1)b_i}$ for $k = 0, \dots, \alpha_i - 2$, and $S_{\alpha_i - 1} = s_{j+(\alpha_i-1)b_i+1} \dots s_D s_1 \dots s_j$. Let $S_{j_0}, \dots, S_{j_m}$ for $j_0 < \cdots < j_m$ be all the sequences with $a_i - 1$ i's. Then there must be *exactly* one sequence S_j with $j_{k \bmod (m+1)} < j < j_{k \bmod (m+1)+1}$ with $a_i + 1$ i's. This claim follows from the fact that there must be at *least* one such sequence, which we have already shown, and the fact that there cannot be more than one since the number of i's is $\alpha_i a_i$. □

For review of other interesting properties see Jost [27].

19.4.2 Applications

We now show two applications of the min-max sequences with $B < 1$, which by Theorem 19.7 always exists, to the generalized pinwheel scheduling problem, Baruah and Lin [28], and the periodic scheduling problem, Liu and Layland [29]. We begin with the former. Let $(a_1, b_1), \dots, (a_n, b_n)$ be n pairs of positive integers. Following Baruah and Lin we have the following definition.

Definition 19.1 (Generalized Pinwheel Schedule)

A generalized pinwheel schedule *on alphabet* $\{1, 2, \dots, n\}$ *is an infinite sequence* $s = s_1 s_2 \dots$ *such that*

1. $s_j \in \{1, 2, \dots, n\}$ *for all* $j \in \mathbb{N}$
2. *each* $i \in \{1, 2, \dots, n\}$ *occurs at least* a_i *times in any subsequence* σ *consisting of* b_i *consecutive elements of* s

The schedule for $a_1 = a_2 = \cdots = a_n = 1$ is referred to simply as a pinwheel schedule. We have the following sufficient condition for the existence of the generalized pinwheel schedule.

Theorem 19.9

If $\sum_{1\le i\le n} \frac{a_i}{b_i} + \frac{1}{b_i} \le 1$, *then there is a generalized pinwheel schedule for pairs* $(a_1, b_1), \ldots, (a_n, b_n)$. *The schedule can be found by the min-max algorithm with* $B < 1$.

Proof

Let $(a_1, b_1), \ldots, (a_n, b_n)$ be an instance of the generalized pinwheel scheduling problem such that $\sum_{1\le i\le n} \frac{a_i}{b_i} + \frac{1}{b_i} \le 1$. Define $d_i = \frac{L(a_i+1)}{b_i}$, for $i = 1, \ldots, n$, where L =lcm $(b_1, \ldots, b_n)$ is the least common multiple of the numbers $b_1, \ldots, b_n$. Then, $\sum_{i=1}^{n} d_i \le L$, and if $\sum_{i=1}^{n} d_i < L$, then define $d_{n+1} = L - \sum_{i=1}^{n} d_i$. Theorem 19.8 ensures that the min-max algorithm with $B < 1$ when applied to the instance including ratios $\frac{d_i}{L} = \frac{a_i+1}{b_i}$, for $i = 1, \ldots, n$, will deliver a sequence with at least $(a_i + 1) - 1 = a_i$ occurrences of i in any subsequence of b_i consecutive letters, and therefore a generalized pinwheel schedule for $(a_1, b_1), \ldots, (a_n, b_n)$. □

A similar result was independently obtained by Baruah and Lin [28]. Notice that as a corollary from Theorem 19.9 we have that there always exist a pinwheel schedule for $(1, b_1), \ldots, (1, b_n)$ as long as $\sum_{1\le i\le n} \frac{1}{b_i} \le \frac{1}{2}$, see also Holte et al. [30].

Finally, we discus the hard real-time periodic scheduling. Following Liu and Layland [29] we consider n independent, preemptive, periodic jobs $1, \ldots, n$ with their request periods being $T_1, \ldots, T_n$ and their run-times being $C_1, \ldots, C_n$. The execution of the k-th request of task i, which occurs at moment $(k-1)T_i$, must finish by moment kT_i when the next request for the task is being made. Missing a deadline is fatal to the system, therefore the deadlines $T_i, 2T_i, \ldots$ are considered hard for job i. All numbers are positive integers and $C_i \le T_i$ for $i = 1, \ldots, n$. We need to find an infinite sequence $s = s_1 \ldots$ on the alphabet $\{1, 2, \ldots, n\}$ such that i occurs at exactly C_i times in each subsequence $s_{(k-1)T_i+1} \ldots s_{kT_i}$ for $k = 1, \ldots$ and $i = 1, \ldots, n$. We shall refer to the sequence as the *periodic schedule*. We have the following theorem, Kubiak [31].

Theorem 19.10

Any solution to the min-max deviation problem with ratios $r_i = \frac{C_i}{T_i}$, $i = 1, \ldots, n$, $\sum r_i \le 1$, *and* $B < 1$ *is a periodic schedule.*

19.5 Small Deviations, Balanced Words and Fraenkel's Conjecture

This section will study optimal solutions of the min-max problem with small, that is *less* than $\frac{1}{2}$, maximum deviations. In their 2001 report, Brauner and Crama [16] conjectured that the only standard instance of the min-max problem with $n > 2$ and maximum deviation *less* than $\frac{1}{2}$ is made up of the first n non-negative powers of 2. More precisely, let us call an instance *standard* if $0 < d_1 \le d_2 \le \cdots \le d_n$, $n \ge 2$, and the greatest common divisor of $d_1, d_2, \ldots, d_n$, and D is 1, i.e., $\gcd(d_1, \ldots, d_n, D) = 1$. Their conjecture states the following.

Conjecture 19.1

For $n > 2$, *a standard instance* $(d_1, \ldots, d_n)$ *of the min-max absolute deviation sequencing problem has optimal value,* B^*, *less than* $\frac{1}{2}$ *if and only if* $d_i = 2^{i-1}$ *for* $i = 1, 2, \ldots, n$, *and* $B^* = \frac{2^{n-1}-1}{2^n-1}$.

Or, equivalently, it can be stated as follows.

Conjecture 19.2 (Competition-Free Instances)

For $n \geq 3$, all D numbers $\lceil \frac{2j-1}{2r_i} \rceil$ where $j = 1, \ldots, d_i$ and $i = 1, \ldots, n$ are different if and only if $r_i = \frac{2^{i-1}}{2^n-1}$.

Brauner and Crama [16] showed that the conjecture holds for $n \leq 6$. Subsequently, Kubiak [32], and Brauner, Jost, and Kubiak [33] have shown that the conjecture holds for any $n > 2$. The former proof is *geometric* and exploits a natural *symmetry* of regular polygons inscribed in a circle of circumference D. The latter is based on a mathematically elegant concept of *balanced* words, which we now explore in more detail.

Definition 19.2 (Balanced Word)

A balanced word on alphabet $\{1, 2, \ldots, n\}$ is an infinite sequence $s = s_1 s_2 \ldots$ such that

1. *$s_j \in \{1, 2, \ldots, n\}$ for all $j \in \mathbb{N}$*
2. *if σ_1 and σ_2 are two subsequences consisting of t consecutive elements of s ($t \in \mathbb{N}$), then the number of occurrences of the letter i in σ_1 and σ_2 differs by at most 1, for all $i = 1, 2, \ldots, n$*

The *balanced* words have appeared in many contexts in mathematics, see Tijdeman [15], and Vuillon [17], and have recently been proven optimal for the expected average workload minimization for routing customers in *events graphs*, a subset of Petri nets, by Altman, Gaujal, and Hordijk [14]. The main problem in practice, however, is to construct a balanced word, if any exists, where each letter occurs with a given rate, [14]. For instance, losses of packets are allowed in the asynchronous transfer mode (ATM) networks, however, the rate of these losses must not exceed a given rate called cell loss ratio (CLR), see Altman et al. [14] for more detailed discussion of this issue, or the system administrator may specify rates (called proportional preferences) for each file transfer protocol on the NeST storage appliance, see Bent et al. [34].

As balanced sequences seem to come close to the *ideal* of being *fair*, since they keep any letter as evenly distributed in the sequence as possible, they are certainly sought for in practice. The *unpleasant* surprise comes from the fact that the chance of finding a balanced word for most of rates *may* be rather slim. In fact, the Fraenkel's conjecture claims that for any $n > 2$ there is exactly one set of n *distinct* rates for which a balanced word can be found, see Tijdeman [15] and Vuillon [43] for excellent reviews of the state-of-the-art for the Fraenkel's conjecture, which has been proven for $n = 2, 3, 4, 5, 6$ but remains open for $n > 6$. Here are details of this conjecture.

For a finite word S on alphabet $\{1, 2, \ldots, n\}$, let us denote by $|S|$ (or D) the length of S and by $|S|_i$ (or d_i) the number of occurrences of the letter i in S. Also let us define the rate r_i of the letter i as the fraction $r_i = \frac{|S|_i}{|S|} = \frac{d_i}{D}$. We assume $r_1 \leq \cdots \leq r_n$. For a finite word S, let $S^* = SS\ldots$ be the infinite repetition of S. An infinite word s is called *periodic* if $s = S^*$ for some finite word S. An infinite word s is called *periodic, balanced* word if s is balanced and $s = S^*$ for some finite word S. A finite word S is called *symmetric* if $S = S^R$, where S^R is a mirror reflection of S, then, S is obviously a *palindrome*. An infinite word s will be called *periodic, symmetric* and *balanced* word if s is balance and $s = S^*$ for some finite symmetric word S. Tijdeman [35] and Altman, Gaujal, and Hordijk [14] show that the Fraenkel's conjecture is equivalent to the following conjecture for balanced words.

Conjecture 19.3 (Fraenkel's Conjecture for Periodic, Balanced Words)

There exists a periodic, balanced word on $n \geq 3$ letters with rates $r_1 < r_2 < \cdots < r_n$ if and only if $r_i = \frac{2^{i-1}}{2^n-1}$.

Though this conjecture remains open, a simpler one for periodic, symmetric and balanced words has recently been proven by Brauner, Jost, and Kubiak [33].

Theorem 19.11 (Fraenkel's Symmetric Case)

There exists a periodic, symmetric and balanced word on $n \geq 3$ letters with rates $r_1 < r_2 < \cdots < r_n$, if and only if the rates verify $r_i = \frac{2^{i-1}}{2^n-1}$.

Consequently, the small deviations conjecture holds true, and we have the following theorem.

Theorem 19.12

For $n > 2$, a standard instance $(d_1, \ldots, d_n)$ of the min-max absolute deviation sequencing problem has optimal value, B^*, less than $\frac{1}{2}$ if and only if $d_i = 2^{i-1}$ for $i = 1, 2, \ldots, n$, and $B^* = \frac{2^{n-1}-1}{2^n-1}$.

While for any $n \geq 3$ there is only one standard instance with maximum deviation less than $\frac{1}{2}$ the number of standard instances with maximum deviation less than $\frac{1}{2}$ for $n = 2$ is *infinite*. We have the following result from Kubiak [32].

Theorem 19.13

$B^* < \frac{1}{2}$ if and only if one of demands d_1 or d_2 is odd and the other even.

19.6 Constant Gap Words

This section will discuss special balanced words called *constant gap* words, Altman, Gaujal, and Hordijk [14]. Following them we define the constant gap words as follows. A periodic infinite word $s = S^*$ where each i is separated from the next i by a constant number of letters is referred to as a constant gap word. These words are balanced and have a perfectly *fair* distribution of each letter i, which occurs at positions $f_i + k\frac{D}{d_i}$ where f_i is the position of the first occurrence of the letter i in s (or equivalently in S) and $k = 0, 1, 2, \ldots$. Obviously, $1 \leq f_i \leq \frac{D}{d_i}$ and d_i divides D (we denote this fact by $d_i \mid D$) for $i = 1, \ldots, n$. Therefore, the set $\{(f_1 - 1, \frac{D}{d_1}), \ldots, (f_n - 1, \frac{D}{d_n})\}$ of n ordered pairs is an *exact covering sequence*. An exact covering sequence is a set $\{(a_1, b_1), \ldots, (a_n, b_n)\}$ of ordered pairs of nonnegative integers with the property that for every nonnegative integer n there is one and only one i, $i = 1, \ldots, n$, such that n is congruent to a_i mod b_i, i.e., $n \equiv a_i \bmod b_i$, see Wilf [36]. This observation leads to an interesting application of generating functions, Wilf [36], which states that it is *impossible* to obtain a constant gap word for $d_1, \ldots, d_n$ which are all *distinct*, see Newman [37], Tijdeman [15], and Altman, Gaujal, and Hordijk [14], where the following lemma is proven.

Lemma 19.4

If set $\{(a_1, b_1), \ldots, (a_n, b_n)\}$ with $b_1 \leq \cdots \leq b_n$ is an exact covering sequence, then $b_{n-1} = b_n$.

For any n there is a finite number of possible ratios $(r_1 = \frac{d_1}{D}, \ldots, r_n = \frac{d_n}{D})$ for which constant gap words are possible, Altman, Gaujal, and Hordijk [14]. They also show that for $n = 3$ there are only two possibilities $(\frac{1}{3}, \frac{1}{3}, \frac{1}{3})$ and $(\frac{1}{4}, \frac{1}{4}, \frac{1}{2})$, for $n = 4$ there are only four $(\frac{1}{4}, \frac{1}{4}, \frac{1}{4}, \frac{1}{4})$, $(\frac{1}{8}, \frac{1}{8}, \frac{1}{4}, \frac{1}{2})$, $(\frac{1}{6}, \frac{1}{6}, \frac{1}{6}, \frac{1}{2})$, and $(\frac{1}{6}, \frac{1}{6}, \frac{1}{3}, \frac{1}{3})$. The following theorem, see Exercise 25 on p 165 in Wilf [36], provides a polynomial time, in D and n, test for the existence of a constant gap word with the first letter i in position f_i, $i = 1, \ldots, n$.

Theorem 19.14

For $\{(a_1, b_1), \ldots, (a_n, b_n)\}$ to be an exact covering sequence it is necessary and sufficient that each number $0, 1, \ldots, lcm(b_1, \ldots, b_n)$ is congruent to a_i mod b_i for exactly one i.

The test proceeds as follows. It first checks if $d_i | D$ for each $i = 1, \ldots, n$. If not, then obviously no constant gap word exists for $d_1, \ldots, d_n$ regardless of the position each letter occurs first in. Otherwise, we continue knowing that the least common multiple of $\frac{D}{d_1}, \ldots, \frac{D}{d_n}$, that is lcm $(\frac{D}{d_1}, \ldots, \frac{D}{d_n})$, does not exceed D. Next, we apply Theorem 19.14 to $\{(f_1 - 1, \frac{D}{d_1}), \ldots, (f_n - 1, \frac{D}{d_n})\}$. The conditions of the theorem can be checked by doing $(\text{lcm}(d_1, \ldots, d_n) + 1)n \leq (D + 1)n$ divisions. Therefore, the whole test requires doing no more than $n + (D + 1)n$ divisions.

Finally, the decision problem, we refer to this problem as the constant gap problem, as to whether there is a constant gap word for $d_1, \ldots, d_n$, which is equivalent to the decision problem as to whether there are positive integers $f_1, \ldots, f_n$ such that $\{(f_1 - 1, \frac{D}{d_1}), \ldots, (f_n - 1, \frac{D}{d_n})\}$ is an exact covering, appears open as to its computational complexity. Though the latter looks similar to the Periodic Maintenance Scheduling Problem, Anily, Glass, and Hassin [38], which can be stated as follows.

Definition 19.3 (Periodic Maintenance Scheduling Problem)
Given positive integers $l_1 \leq \cdots \leq l_n$, the maintenance intervals for machines $1, \ldots, n$ respectively, such that $\sum_{i=1}^{n} \frac{1}{l_i} \leq 1$, are there nonnegative integers $a_1, \ldots, a_n$ such that the sets $A_i = \{a_i + kl_i | k = 0, 1, \ldots\}$ for $i = 1, \ldots, n$ are disjoint, i.e., no two machines require maintenance in the same time slot?

Notice that if $\sum_{i=1}^{n} \frac{1}{l_i} < 1$, then regardless of what $a_1, \ldots, a_n$ are the set $\{(a_1, l_1), \ldots, (a_n, l_n)\}$ is *not* exact covering. The NP-hardness of the Periodic Maintenance Scheduling Problem, proved by Bar-Noy et al. [39], does not imply the NP-hardness of the constant gap problem as the proof strongly relies on the fact that $\sum_{i=1}^{n} \frac{1}{l_i} < 1$ and all $l_1, \ldots, l_n$ are distinct. However, notice that by Lemma 19.4 we must have $l_{n-1} = l_n$ in any exact covering. Consequently, filling the gap between 1 and $\sum_{i=1}^{n} \frac{1}{l_i} < 1$ necessary for exact covering through a polynomial (or even pseudoplolynomial) transformation does not seem trivial and remains open.

19.7 Stride Scheduling

Stride scheduling was introduced by Waldspurger and Weihl in their 1995 report [13] as a universal scheduling paradigm. The stride scheduling allows clients to buy, sell and trade tickets, the number of tickets acquired by a client then determines the rate at which the client will receive the resources needed for its job, which competes for resources with jobs of other clients. Stride scheduling has since been used for scheduling resource allocations, in the Linux kernel, [13], in network routers by the Click modular router, Kohler et al. [40], and in storage appliances by the NeST software-only storage appliance, Bent et al. [34]. In this section, we discuss main properties of stride schedules and their metrics.

19.7.1 Properties

The stride scheduling is a deterministic scheduling technique where each of n clients, $i = 1, \ldots, n$, is first issued a positive integer number of *tickets* so that client i receives d_i tickets. The total number of tickets issued equals $D = \sum_{1 \leq i \leq n} d_i$. The resources are then allocated in discrete time slices $1, 2, \ldots$, called *quanta*. The client i^* to be allocated resources in quantum $k + 1$, $k = 0, 1, 2, \ldots$, is calculated as follows

$$i^* = \arg\min_{i} \left\{ \frac{stride_1 \cdot (x_{ik} + 1)}{d_i} \right\} \tag{19.4}$$

where x_{ik} is client's i number of allocations that it received during the first k allocations, and $stride_1$ is an integer constant. We assume $x_{i0} = 0$ for $i = 1, \ldots, n$, and ties are broken using the ascending order of the client's *strides* equal $\frac{stride_1}{d_i}$ for $i = 1, \ldots, n$. The constant $stride_1$ is *only* used to obtain high-precision fixed-point integer representation of the strides $\frac{stride_1}{d_i}$ for individual clients, see [13] for details. However, it is obvious from (19.4) that the sequence of allocations produced by the stride scheduling is *independent* of the value of $stride_1$. Therefore, from now on, we shall assume without loss of generality that $stride_1 = 1$, and that i^* is calculated, equivalently, as follows:

$$i^* = \arg\max_i \left\{ \frac{d_i}{x_{ik} + 1} \right\} \tag{19.5}$$

In our discussion we shall consider Equation (19.5) as a simple method of producing an infinite sequence $s = s_1 \ldots$ on a finite alphabet $\{1, \ldots, n\}$, where s_q is the client that is allocated resources in quantum q. We shall study the sequences produced by stride scheduling and propose possible improvements over the stride scheduling. The crucial observation is that the stride scheduling sequence given by (19.5) is a *parametric method* studied in the axiomatic theory of *apportionment*, see Balinski and Shahidi [12]. The parametric method ϕ^δ based on $0 \leq \delta \leq 1$ is defined recursively as follows: begin with the empty sequence, if the sequence is built up to k and has clients' allocations x_{ik}, then allocate quant $k + 1$ to client i^* where

$$i^* = \arg\max_i \left\{ \frac{d_i}{x_{ik} + \delta} \right\} \tag{19.6}$$

Obviously, we have $\delta = 1$ for the stride scheduling. In fact, the stride scheduling is the Thomas *Jefferson's method* of the apportionment, see Balinski and Shahidi [12] and Balinski and Young [11], which is well known, Balinski and Shahidi [12], to optimize

$$\min_k \max_i \left\{ \frac{x_{ik}}{d_i} \right\} \tag{19.7}$$

Any parametric method, and thus the stride scheduling:

P1 is *cyclic and exact.*
P2 is *anonymous.*
P3 respects *priorities.*

Following Balinski and Shahidi [12], we shall now define and briefly discuss these concepts. Let us consider an infinite sequence $s = s_1 s_2 \ldots$. Consider sequences $c_k = s_{kD+1} \ldots s_{(k+1)D}$ for $k = 0, \ldots$. If *all* of them are equal, i.e., $c = c_0 = c_1 = \ldots$ than we call sequence s cyclic and sequence c its cycle. If cycle c includes exactly d_i client's i allocations, then s is called *cyclic and exact.* The method of generating a sequence s is called cyclic and exact if it always produces sequences that are both cyclic and exact. Now, consider a permutation π of clients $i = 1, \ldots, n$. Consider a cyclic and exact sequence s with cycle $c = s_1 \ldots s_D$ and an exact sequence $\pi(s)$ with cycle $\pi(s) = \pi(s_1) \ldots \pi(s_D)$. The method that produces s for clients $1, 2, \ldots, n$ and $\pi(s)$ for clients $\pi(1), \ldots, \pi(n)$ is called *anonymous.* Finally, consider clients i and j, define $\frac{d_i}{d_j}$ as the relative priority of i with respect to j. Now, suppose that at some point client i increases its number of tickets relative to client j, i.e., the new ratio $\frac{d'_i}{d'_j} > \frac{d_i}{d_j}$, where clients i and j now have d'_i and d'_j tickets, respectively. Then the reasonable expectation, according to the axiomatic approach to the apportionment problem, is that from that point on if client's j allocations are advanced, then client's i allocations are *not* delayed. If a method always meets this expectation, then we say that it *respects priorities,* Balinski and Young [11]. Formally, see for instance Balinski and Shahidi [12], the definition is as follows.

Let s be a sequence produced by a parametric method for clients $1, \ldots, n$ with tickets $d_1, \ldots, d_n$. Denote by x_{iq} the number of client's i allocations in the prefix $s_1, \ldots, s_q$ of s. Let p be a sequence produced by the same parametric method for another set of clients $1, \ldots, m$ with tickets $t_1, \ldots, t_m$, respectively. Denote by y_{iq} the number of client's i allocations in the prefix $p_1 \ldots p_q$ of p. Consider clients i, j, k and l, $\frac{d_i}{d_j} \geq \frac{t_k}{t_l}$. The method *respects priorities* if

(i) for $\frac{d_i}{d_j} > \frac{t_k}{t_l}$, $x_{iq} < y_{kq'}$ implies $x_{jq} \leq y_{lq'}$
(ii) for $\frac{d_i}{d_j} = \frac{t_k}{t_l}$, if $x_{iq} < y_{kq'}$ and $x_{jq} > y_{lq'}$, then exchanging a single j for i in the prefix $s_1 \ldots s_q$ of s gives another sequence produced by the method

The following theorem follows immediately from Theorems 19.1 and 19.2 in Balinski and Shahidi [12].

Theorem 19.15

Any parametric method is anonymous, cyclic and exact, and respects priorities.

Consequently, we have

Corollary 19.1

The stride scheduling is anonymous, cyclic and exact, and respects priorities.

By setting δ to 0 the parametric method advances the allocations for the low-throughput clients (i.e., with small number of tickets or, equivalently, long strides), whereas by setting δ to 1 the parametric method advances the allocations for the high-throughput (i.e., with large number of tickets or, equivalently, short strides) clients, Balinski and Rachev [41]. The latter obviously holds for the stride scheduling. Consequently, neither of these parametric methods is *symmetric*. A method is symmetric if with each cycle c it also generates (by breaking ties differently) its mirror reflection c^R. Balinski and Shahidi [12] prove that the only *symmetric* parametric method is the one with $\delta = \frac{1}{2}$.

As we have shown, the notion of stride scheduling can naturally be extended to exactly match the parametric methods of apportionment with different values of δ. Therefore, we shall introduce the term δ-stride scheduling, $0 \leq \delta \leq 1$, as the one that corresponds to the parametric method with δ. We shall now proceed to the analysis of two metrics for stride scheduling: throughput error and response time variability.

19.7.2 Throughput Error

Waldspurger and Weihl [13] suggest *throughput error (accuracy)* as an important performance metric for the stride scheduling. Their definition of the throughput error coincides with our definition of the maximum absolute deviation introduced in Section 19.3. Waldspurger and Weihl realize that their *basic* 1-stride scheduling may produce sequences with large absolute throughput error. We now know that this is because a parametric method with $\delta = 1$ always advances the allocations of the high-throughput clients. Waldspurger and Weihl also point out that a similar behavior, resulting in large throughput error, has been exhibited by similar to the stride scheduling rate-based network flow control algorithms. To illustrate this problem with parametric methods consider an instance with 101 clients and the following ticket allocations $d_1 = 100$, $d_2 = \cdots = d_{101} = 1$. The 1-stride algorithm would result in a cycle where client 1 receives the first 100 quanta followed by the remaining 100 clients receiving 1 quantum each. Clearly the throughput error of this cycle is 50. Notice that since $x_{1k} = 100$ for $k = 100$ and $r_1 = \frac{1}{2}$, $|x_{1k} - kr_1| = 50$ for $k = 100$. The $\frac{1}{2}$-stride scheduling would reduce this error twice. It would result in a cycle where client 1 receives the first 50 quanta followed by the remaining 100 clients receiving 1 quantum each, and followed by 50 quanta allocated to client 1. Clearly, the throughput error of this cycle is 25. Notice that since $x_{1k} = 50$ for $k = 50$ and $r_1 = \frac{1}{2}$, then $|x_{1k} - kr_1| = 25$ for $k = 50$. However, it follows from Theorem 19.7 that the optimal throughput error does not exceed $1 - \frac{1}{200} = \frac{199}{200}$, a cycle that attains this throughput error allocates all *odd* quanta between 1 and 200 to client 1, and all *even* quanta to the remaining clients in an arbitrary way. In fact, an optimal cycle results in a throughput error of $\frac{198}{200}$. This optimal cycle allocates all *odd* quanta between 1 and 100, and all *even* quanta between 101 and 200 to client 1 and all the remaining quanta to the remaining clients in an arbitrary way. Notice that a more sophisticated *hierarchical* stride scheduling, we refer the reader to [13] for details, is able to reduce the throughput error for the instance discussed above to 4.5 only. We close this section with the following theorem.

Theorem 19.16

The Earliest Due Date algorithm with due dates defined as in Theorem 19.4 minimizes the throughput error.

It is an interesting question whether there always exists a throughput error optimal solution that also respects priorities, see axiom **P3**. The positive answer to this question would show that it is actually *not* necessary to compromise the throughput error in order to generate sequences with the desired properties **P1**, **P2**, and **P3** of the parametric methods for apportionment. The negative answer would prove that respecting priorities, a key axiom in the axiomatic approach to the apportionment problem, can only be achieved at the cost of increased throughput error. Unfortunately, the latter holds, which is a consequence of the famous theorem of *impossibility* shown by Balinski and Young [11]; see also Balinski and Shahida [12].

Theorem 19.17

Respecting priorities and meeting $\lfloor kr_i \rfloor \leq x_{ik} \lceil kr_i \rceil$ *for all* k *and* i *are incompatible.*

However, by Theorem 19.7, $B^* < 1$, and consequently $\lfloor kr_i \rfloor \leq x^*_{ik} \leq \lceil kr_i \rceil$ for all k and i in any solution x^* minimizing the throughput error, which by Theorem 19.17 is incompatible with axiom **P3**.

19.7.3 Response Time Variability

Another performance metrics suggested by [13] is *response time variability*. The response time is defined as the number of quanta between a client's two consecutive quantum allocations plus one. Since the quantum duration is fixed this definition is equivalent to the one given by Waldspurger and Weihl [13] who define it as the elapsed time from a client's completion of one quantum up to and including its completion of next. A natural measure of response time variability for a client is the *variance* of its response time. For cyclic sequences this variance for client i can be defined as the variance of response time for the first $d_i + 1$ allocations. More formally, let t_j be the number of quanta between the completion of the jth allocation and the completion of the $(j+1)$st, $j = 1, \ldots, d_i$. Then, the response time variability for client i is

$$d_i \text{Var}_i = \sum_{1 \leq j \leq d_i} (t_i - \bar{t})^2$$

We would like to find a sequence that minimizes the total variance

$$V = d_1 \text{Var}_1 + \cdots + d_n \text{Var}_n \tag{19.8}$$

of all clients. Actually, the problem is as follows: Given that clients $1, \ldots, n$ were issued $d_1, \ldots, d_n$ tickets, respectively, find a cyclic sequence with client i allocated exactly d_i quanta in a cycle that minimizes response time variability defined in Equation (19.8). Notice that the average response time $\bar{t}$ is constant and equal to $\frac{D}{d_i}$ for client i. Unfortunately, the problem of minimizing the response time variability is computationally more difficult than the problem of minimizing throughput error. We have the following theorem proven by Kubiak and Moreno [42] by reduction from the Periodic Maintenance Scheduling Problem shown NP-complete by Bar-Noy et al. [39].

Theorem 19.18

The problem of minimizing the response time variability is NP-hard.

However, it is open whether the problem of minimizing the response time variability is NP-hard in the strong sense.

We now show that $V = 0$ is impossible for instances where no two clients are issued the *same* number of tickets. We have the following theorem.

Theorem 19.19

If $d_1 < \cdots < d_n$ *for* $n > 1$, *then* $V > 0$.

Proof

If $d_i \nmid D$ for some i, then the theorem holds since the average distance for i equal $\frac{D}{d_i}$ is not an integer and all response times are integer. Otherwise, $d_i \mid D$ for each i. By contradiction, if $V = 0$, then for each i all response times are equal to the average response time $\frac{D}{d_i}$. Consequently, there are non-negative integers $a_1, \ldots, a_n$ such that $(a_1, \frac{D}{d_1}), \ldots, (a_n, \frac{D}{d_n})$ is an exact covering sequence. However, by Lemma 19.4, then $\frac{D}{d_{n-1}} = \frac{D}{d_n}$ and consequently $d_{n-1} = d_n$, which leads to a contradiction. This proves the theorem. □

The throughput error and the response time variability are two *independent* metrics. Except for the case of two clients, Kubiak and Moreno [42], we cannot optimize them both at the same time. For example, for the instance $d_1 = 100$, $d_2 = \cdots = d_{101} = 1$ the solution given in the last section with maximum deviation $\frac{199}{200}$ minimizes the response time variability since the solution results in $V = 0$, notice that all consecutive client 1 allocations are at a distance 2 in this solution. On the other hand no optimal solution, which has the maximum deviation $\frac{198}{200}$, results in minimum response time variability since the solution must have exactly two consecutive allocations for client i next to each other, i.e., at a distance 1, and all remaining consecutive allocations separated by some other client allocation(s), i.e., at a distance at least 2. Consequently, its response time variability is $V > 0$.

19.8 Peer-to-Peer Fair Sequences

We close this chapter with a brief discussion of a promising model of fair sequences. First, we give some motivation behind the model using a simple example of a stride scheduling problem, though the model itself is general. Consider two clients a and b. Suppose client a obtains three tickets and client b obtains six, then b expects advancing its task at a pace which is twice the pace of a. Thus, if clients a and b were the only two competing for the shared resources, then the infinite cyclic sequence $(abb)(abb)a\ldots$, with the cycle abb, would be *fair* for both and neither would have a *casus* for complaining since out of any three consecutive quanta two are allocated to b, with six tickets, and one to a, with three tickets. Notice that $\frac{3}{6} = \frac{1}{2}$ and consequently $1 + 2 = 3$ is the smallest number of quanta to consider for clients a and b in any *reasonable*, at least mathematically, discussion of what constitute a fair allocation of quanta for the couple. Any smaller number of quanta would obviously be *biased* toward one of the clients and the $\frac{1}{2}$ ratio could not then be achieved. However, if another client, say c, joins the couple with four tickets in their competition for the shared resource, then the sequence $(cbabcbabcbabc)\ldots$, with the maximum deviation $B = \frac{9}{13}$, for the three becomes $(bab)(bab)(bab)\ldots$, for a and b, which ensures that out of any three consecutive quantum allocations two are made to b and one to a. This should certainly be *fair* for the two. Also, for clients a and c it becomes $(cacacac)\ldots$, which ensures that out of any seven consecutive quanta 3 are allocated to a and 4 to c. Again, it is fair to both a and c as their ticket ratio is $\frac{3}{4}$, and $3 + 4 = 7$ is the length of the shortest cycle where this ratio is achievable. Finally, for clients b and c the sequence becomes $(cbbcbbcbbc)\ldots$, which means that out of any five consecutive quanta client b is allocated at least 3, but sometimes 4. This could make client c feel that it does not receive its fair share of allocations with respect to client c as it has as many as $\frac{2}{3}$ of the number of tickets client b has, though it sometimes gets only $\frac{1}{5}$ of consecutive 5 quanta. This potential perception of *unfairness* can be avoided by the sequence $bcbacbbcabcba\ldots$, with the maximum deviation $B = \frac{11}{13}$, as it becomes $bcbcbbcbcb\ldots$, for clients b and c. Thus, out of any consecutive 5 quanta, 3 are allocated to b and 2 to c. Furthermore, the sequence becomes $bbabbabba\ldots$, for clients a and b, which ensures that out of any three consecutive quantum allocations two are made to b and one to a. The last sequence, that is $(bcbacbbcabcba)\ldots$, has an obvious advantage of being *peer-to-peer* fair though it is not optimal from the maximum deviation stand point. The sequence $(cbabcbabcbabc)\ldots$ we began with is not peer-to-peer fair but has lower maximum deviation of $\frac{9}{13}$. The question then is: can we have a peer-to-peer fair sequence for a, b, c which at the same time minimizes maximum deviation? In our example of clients a, b, and c the answer is positive, the sequence $(bcabcbabcbacb)\ldots$ minimizes maximum deviation, its optimal value is $\frac{7}{13}$, and is peer-to-peer fair. Notice that this sequence is not a balanced word since it includes subsequences bb with two b's and ca with no b. In fact, since the Fraenkel's conjecture holds for $n \leq 6$, see Tijdeman [35], no balanced word for clients a, b, c with tickets 3, 6, 4 respectively, is possible.

We have the following definition.

Definition 19.4 (Peer-to-Peer Fair Sequencing)

Given clients $1, \cdots, n$ with tickets $d_1 \leq \ldots \leq d_n$, respectively. Define the peer-to-peer ratio for clients i and j, $i < j$, as $f_{ij} = \frac{d_i}{d_j} = \frac{\alpha_{ij}}{\beta_{ij}}$, where α_{ij} and β_{ij} are relatively prime. Find an infinite periodic sequence

SS..., with period S which minimizes maximum deviation, and such that client i occurs exactly d_i times in S and for each couple i, j, i < j, of clients any subsequence of $S_{ij} S_{ij} \ldots$ with $\alpha_{ij} + \beta_{ij}$ clients has exactly α_{ij} client i allocations and exactly β_{ij} client j allocations. Sequence S_{ij} is obtained from S by deleting all clients except i and j, whose relative positions remain as in the original sequence S.

It seems tempting to conjecture that for any instance of the peer-to-peer fair sequencing problem there always exists solution that minimizes maximum deviation. Though this may hold for many instances, an instance with three clients a, b and c and their tickets 3, 5 and 15, respectively, makes a simple counterexample to the conjecture. However, the sequences for these clients (*cbcaccbcccabcccbccacbcc*)... is peer-to-peer fair though it has maximum deviation $B = \frac{19}{24}$ while the optimal maximum deviation is $B^* = \frac{14}{23}$.

19.9 Conclusions

This chapter has reviewed a number of fair sequences' models: just-in-time sequences, balanced words, constant gap words, sequences generated by stride scheduling, and peer-to-peer fair sequences. It has shown that the just-in-time model of fair sequences, though originally conceived for sequencing mixed-model, just-in-time systems, is fundamental to resource allocation problems in a large number of diverse environments. These include operating systems, the Internet, hard real-time systems, multiprocessors and communication networks. The chapter has pointed out many of these applications and reviewed some of them in detail, namely, the generalized pinwheel scheduling and the periodic scheduling. Consequently, the chapter has demonstrated that the just-in-time model of fair sequencing provides a common framework for fair scheduling problems in many environments. The chapter has also reviewed results on the recently introduced *small deviation* conjecture which links fair sequences with the *Fraenkel's* conjecture. Last, but not least, the chapter has discussed the axiomatic approach to modeling fair sequences. This approach results from the fundamental work done on the *apportionment* problem. Finally, the chapter has presented a number of open and challenging problems in this new, practically important, and relatively unexplored area of research reaching far beyond scheduling.

Acknowledgment

This research has been supported by the Natural Sciences and Engineering Research Council of Canada grant OPG0105675. It was completed when the author was visiting the Department of Mathematics of the Ecole Polytechnique Fédérale de Lausanne; the support of the EPFL is gratefully acknowledged.

References

[1] Y. Monden (1983) *Toyota Production Systems* Industrial Engineering and Management Press, Norcross, GA.

[2] R. Tijdeman (1980), The chairman assignment problem, *Discrete Mathematics* **32**, 323–330.

[3] J.G. Miltenburg (1989) Level schedules for mixed-model assembly lines in just-in-time production systems, *Management Science* **35**, 192–207.

[4] J.G. Miltenburg and T. Goldstein (1991) Developing production schedules which balance par usage and smooth production loads in just-in-time production systems, *Naval Research Logistics* **38**, 893–910.

[5] W. Kubiak, G. Steiner, and S. Yeomans (1997) Optimal level schedules in mixed-model, muli-level just-in-time assembly systems, *Annals of Operations Research* **69**, 241–259.

[6] A. Drexl and A. Kimms (2001) Sequencing JIT mixed-model assembly lines under station-load and part-usage constraints, *Management Science*, **47**, 480–491.

[7] R. Inman and R. Bulfin (1991) Sequencing JIT mixed-model assembly lines, *Management Science* **37**, 901–904.

[8] W. Kubiak and S.P. Sethi (1994) Optimal just-in-time schedules for flexible transfer lines, *The International Journal of Flexible Manufacturing Systems* **6**, 137–154.

[9] W. Kubiak and S.P. Sethi (1991) A note on level schedules for mixed-model assembly lines in just-in-time production systems, *Management Science* **6**, 137–154.

[10] J. Bautista, R. Companys, and A. Corominas (1996) A note on the relation between the product rate variation (PRV) and the apportionment problem, *Journal of Operational Research Society* **47**, 1410–1414.

[11] M. Balinski and H.P. Young (1982) *Fair Representation: Meeting the Ideal of One Man, One Vote*, Yale University Press, New Haven, CT.

[12] M. Balinski and N. Shahidi (1998) A simple approach to the product rate variation problem via axiomatics, *Operations Research Letters* **22**, 129–135.

[13] C.A. Waldspurger and W.E. Weihl (1995) Stride scheduling: deterministic proportional-share resource management. Technical Report MIT/LCS/TM-528, Massachusetts Institute of Technology, MIT Laboratory for Computer Science, June 1995.

[14] E. Altman, B. Gaujal, and A. Hordijk (2000), Balanced sequences and optimal routing, *Journal of ACM* **47(4)**, 754–775.

[15] R. Tijdeman (2000), Exact covers of balanced sequences and Fraenkel's conjecture, in: *Algebraic Number Theory and Diophantine Analysis*, F. Halter-Koch and R. F. Tichy (eds.), Walter de Gruyter, Berlin, New York, 2000, 467–483.

[16] N. Brauner and Y. Crama (2001) Facts and questions about the maximum deviation just-in-time scheduling problem, Research Report G.E.M.M.E. No 0104, University of Liège, Liège.

[17] L. Vuillon (2003) Balanced words, Rapports de Recherche 2003-006, LIAFA CNRS, Université Paris 7.

[18] H. Groenevelt (1993) *The Just-in-Time Systems*, in: A.H.G. Rinnooy Kan, and P.H. Zipkin, *Handbooks in Operations research and Management Science* Vol. 4 S.C. Graves, North Holland.

[19] W. Kubiak (1993) Minimizing variation of production rates in just-in-time systems: a survey, *European Journal of Operational Research* **66**, 259–271.

[20] J. Bautista, R. Companys, and A. Corominas (1997) Modelling and solving the production rate-variation problem, *TOP* **5**, 221–239.

[21] H.W. Kuhn (1955) The Hungarian method for the asignment problem, *Naval Research Logistics Quarterly* **2**, 83–97.

[22] W. Kubiak (2003) Cyclic just-in-time sequences are optimal, *Journal of Global Optimization*, **27**, 333–347.

[23] G. Steiner and S. Yeomans (1993) Level schedules for mixed-model, just-in-time production processes, *Management Science* **39**, 401–418.

[24] F. Glover (1967) Maximum matching in a convex bipartite graph, *Naval Research Logistics Quarterly* **4**, 313–316.

[25] H.G. Meijer (1973) On a distribution problem in finite sets, *Nederlands Akademie Wetenschappen Indag. Math.* **35**, 9–17.

[26] N. Moreno Palli (2002) Solving the product rate variation problem (PRVP) of large dimensions as an assignment problem, Doctoral Thesis, Department D'Organitzacio D'Empreses, UPC, Barcelona.

[27] V. Jost (2003) Deux problemes d'approximation Diophantine: Le patage proportionnel en nombres entries et Les pavages equilibres de Z, DEA ROCO, Laboratoire Leibniz-IMAG.

[28] S. K. Baruah and S-S. Lin (1998) Pfair scheduling of generalized pinwheel task systems, *IEEE, Transactions on Computers* **47**, 812–816.

[29] C.L. Liu and J.W. Layland (1973) Scheduling algorithm for multiprogramming in a hard-real-time environment, *Journal of ACM* **20**, 46–61.

[30] R. Holte, A. Mok, L. Rosier, I. Tulchinsky, and D. Varvel (1989) The pinwheel: a real-time scheduling problem, *Proceedings of the 22nd Hawaii International Conference on System Science*, 693–702.

[31] W. Kubiak (2003) The Liu-Layland problem revisited, MAPSP 2003, Aussois, France, 74–75.

[32] W. Kubiak (2003) On small deviations conjecture, *Bulletin of the Polish Academy of Sciences* **51**, 189–203.

[33] N. Brauner, V. Jost, and W. Kubiak (2002) On symmetric Fraenkel's and small deviations conjectures. Les cahiers du Laboratoire Leibniz-IMAG, no 54, Grenoble, France.

[34] J. Bent, V. Venkateshwaran, N. LeRoy, A. Roy, J. Stanley, A.C. Arpaci-Dusseau, R.H. Arapaci-Dusseau, and M. Livny (2002) Flexibility, manageability, and performance in a grid storage appliance. In Proceedings of the Eleventh IEEE Symposium on High Performance Distributed Computing, Edinburgh, Scotland, July 2002.

[35] R. Tijdeman (2000) Fraenkel's conjecture for six sequences, *Discrete Mathematics*, 2000, 222, 223–234.

[36] H.S. Wilf (1994) *Genratingfunctionology*, Academic Press, 2nd ed., 1994.

[37] M. Newman (1971) Roots of unity and covering sets, *Mathematics Annals* **191**, 279–282.

[38] S. Anily, C.A. Glass, and R. Hassin (1998), The scheduling of maintenance service, *Discrete Applied Mathematics* **82**, 27–42.

[39] A. Bar-Noy, R. Bhatia, J. Naor, and B. Schieber (1998) Minimizing service and operation costs of periodic scheduling. In *Ninth Annual ACM-SIAM Symposium on Discrete Algorithms SODA, 1998*, 11–20.

[40] E. Kohler, R. Morris, B. Chen, J. Jannotti, and M.H. Kaashoek (2000) The click modular router, *ACM Transactions on Computer Systems* **18**, 263–279.

[41] M. Balinski and S.T. Rachev (1997) Rounding proportions: methods of rounding, *Mathematical Scientis* **22**, 1–26.

[42] W. Kubiak and N. Moreno Palli (2003) Towards the theory of stride scheduling, *Optimization Days 2003*, Montreal, Canada.

[43] T.E. Vollman, W.L. Berry, and D.C. Wybark (1992) *Manufacturing Planning and Control Systems*, 3rd ed., IRWIN.

20

Due Date Quotation Models and Algorithms

Philip Kaminsky
University of California

Dorit Hochbaum
University of California

20.1 Introduction

When firms operate in a make-to-order environment, they must set due dates (or lead times) which are both relatively soon in the future and can be met reliably in order to compete effectively. This can be a difficult task, since there is clearly an inherent trade-off between short due dates, and due dates that can be easily met. Nevertheless, the vast majority of due date scheduling research assumes that due dates for individual jobs are exogenously determined. Typically, scheduling models that involve due dates focus on sequencing jobs at various stations in order to optimize some measure of the ability to meet the given due dates. However, in practice, firms need an effective approach for quoting due dates and for sequencing jobs to meet these due dates. In this chapter, we consider a variety of models that contain elements of this important and practical problem, which is often known as the due date quotation and scheduling problem, or the due date management problem.

In this chapter, we focus on papers that contain *analytical results*, and describe the algorithms and results presented in those papers in some detail. We do not discuss simulation-based research, or papers that focus on industrial applications rather than theory. We will follow many of the conventions of traditional scheduling theory, and assume our reader is familiar with basic scheduling concepts. For a comprehensive description of due date–related papers, including simulation-based research and descriptions of industrial applications, see Keskinocak and Tayur [1]. We also refer the reader to Cheng and Gupta [2], an earlier survey of this area.

1-58488-397-9/$0.00+$1.50

20.2 Overview

Most of the models discussed in this chapter contain two elements: a due date setting element, and a sequencing element. Since capacity is inherently limited in scheduling models, it is frequently impossible to set ideal due dates, and to sequence jobs so that they complete processing precisely at these ideal due dates. Indeed, the interplay between sequencing jobs to meet due dates, and setting due dates so that sequencing is possible, makes these problems very difficult (see Figure 20.1).

Ideally, of course, the sequencing and due date quotation problems will be solved simultaneously—rules or algorithms are developed that both quote due dates, and suggest an effective sequence. Unfortunately, in many cases, solving these problems simultaneously is difficult or impossible. In these cases, researchers turn either to *sequenced-based models*, where some sequence or sequencing rule is selected, and then due dates are optimized based on this sequence or rule, or *due date–based models* in which the due date is first assigned, and then the sequence is set based on this due date. Frequently, the model or analysis approach dictates this choice. In queuing models, for example, the analysis frequently requires a sequencing rule (such as first come, first served) to be selected, and then due dates to be set based on this rule. Indeed, one can argue that in general the sequence-based approach makes more sense, since the due date depends on the available capacity, which is directly dependent upon how jobs are sequenced.

Much of the notation used in the chapter will be introduced as needed. Some of the notation, however, is fairly standard, and is introduced below. For job i:

1. For any problem, there are N jobs and M machines. Where appropriate, N and M also refer to the set of jobs and the set of machines, respectively. If the number of machines is not discussed, it is assumed to be a single machine.
2. r_i represents the release time, or availability for processing, of the job.
3. p_i represents the processing time of the job. If the job is processed on more than one machine, p_i^m represents the processing time of job i on machine m.
4. Given a schedule, C_i represents the completion time of the job in that schedule.
5. Given a due date quotation approach, d_i is the due date of job i in that schedule. If the model under consideration is a *common due date* model, then d represents the common due date.
6. Given C_i and d_i, E_i represents the earliness of job i, $\max\{d_i - C_i, 0\}$.
7. Given C_i and d_i, T_i represents the tardiness of job i, $\max\{C_i - d_i, 0\}$.
8. The *quoted lead time* of a job is the time between its release and its due date, $d_i - r_i$. Sometimes models will be expressed in terms of quoted lead times rather than quoted due dates. The *flow time* of a job is the actual time between its release time and its completion, $C_i - r_i$.
9. Given a sequence of jobs, job $j_{[i]}$ is the ith job in the sequence, with processing time $p_{[i]}$, release time $r_{[i]}$, etc.

The models considered in this chapter for the most part follow standard scheduling convention. We consider single machine models, parallel machine models, job shops, and flow shops, both dynamic (that is, jobs have different release or available times) and static (that is, all jobs are available at the start of the horizon.) Some due date quotation models do not restrict the quoted due dates. In other words, any due date can be quoted for any job. Some models are so-called *common-due date* models. In these models, a single due date must be quoted for all of the jobs.

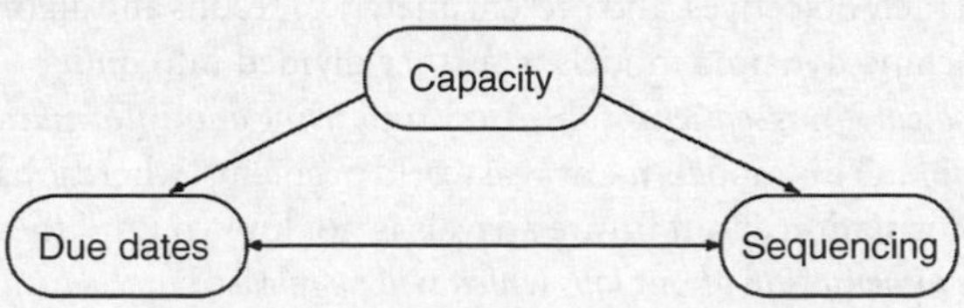

FIGURE 20.1 Sequencing vs. due date quotation.

Because it is sometimes impractical to quote due dates from an unrestricted set of possible due dates, researchers have considered a variety of problems in which the class of possible due dates is limited. In general, this research involves proposing a simple due date setting rule, and then attempting to optimize the *parameters* of that rule in order to achieve some objective. Three types of due date setting rules are commonly used:

1. CON: Jobs are given constant lead times, so that for job j, $d_j = r_j + \gamma$. Note that for a static problem (with all release times equal), this is equivalent to a common due date.
2. SLK: Jobs are given lead times that reflect equal slacks, so that for job j, $d_j = r_j + p_j + \beta$.
3. TWK: Jobs are assigned lead times proportional to their lengths (or their *Total WorK*), so that $d_j = r_j + \alpha p_j$.

A variety of different sequencing and scheduling rules have been employed for these types of models. Some standard dispatch rules include:

1. Shortest Processing Time (SPT): Jobs are sequenced in nondecreasing order of processing times.
2. Longest Processing Time (LPT): Jobs are sequenced in nonincreasing order of processing times.
3. Weighted Shortest Processing Time (WSPT) and Weighted Longest Processing Time (WLPT): Job i has an associated weight w_i. For WSPT, jobs are sequenced in nondecreasing order of p_i/w_i; for WLPT, jobs are sequenced in nonincreasing order of the same ratio.
4. Earliest Due Date (EDD): Jobs are sequenced in nondecreasing order of due dates. We observe that for due date quotation problems, sequencing jobs EDD in some sense removes a degree of freedom from the optimizer, since instead of making sequencing and due date quotation decisions, for an EDD problem, the due date quotation directly implies a sequence.
5. Shortest Processing Time among Available jobs (SPTA): In a dynamic model, each time a job completes processing, the next job to be processed is the shortest job in the set of released but not yet processed jobs.
6. Preemptive SPT (PSPT): In a dynamic model, each time a job is released, the current job will be stopped, and the newly released job will be processed, if the remaining processing time of the currently processing job is longer than the processing time of the newly released job. When a job completes processing, the job with shortest remaining processing time is processed.
7. Preemptive EDD (PEDD): In a dynamic model, each time a job is released, the current job will be stopped, and the newly released job will be processed, if the newly released job has an earlier due date than the currently processing job. When a job completes processing, the remaining job with the earliest due date will be processed.

Researchers have considered a variety of objectives for due date quotation models. Many of them involve functions of the quoted due date or due dates, and the earliness and tardiness of sequenced jobs. In addition, some models feature reliability constraints. For example, some models feature a 100% reliability constraint, which requires each job to complete processing by its quoted due date. Some models feature probabilistic reliability constraints, which limit the probability that a job will exceed its quoted due date. Some reliability constraints limit the fraction of jobs that can be tardy, or the the total amount of tardiness.

The remainder of this chapter is organized as follows. Each section considers a class of models: single machine common due date models, single machine static distinct due date models, single machine dynamic models, parallel machine models, and jobshop and flowshop models. Within each section, we introduce a variety of models, present their objectives, and present analytical results and algorithms from the literature. The section on single machine dynamic models is further divided into *online* and *offline* models. In this context, *online scheduling algorithms sequence jobs at any time using only information pertaining to jobs which have been released by that time.* This models many real world problems, where job information is not known until a job arrives, and information about future arrivals is not known until these jobs arrive. In contrast, *offline algorithms may use information about jobs which will be released in the future to make sequencing and due date quotation decisions.* The online section is further divided into subsections featuring probabilistic analysis of heuristics, worst-case analysis of heuristics, and queuing-theory based analysis of related models.

We conclude with a discussion of some models that do not fit these categories, and a short discussion of research opportunities in the area of due date quotation.

20.3 Single Machine Static Common Due Date Models

In the models considered in this section, all jobs are assumed to be available at the start of the scheduling horizon (a static problem, with $r_i = 0$ for all jobs). Jobs must be processed sequentially on a single machine, and processing times are deterministic (with a few exceptions, described below) and known at the start of the scheduling horizon. We use d to represent the common due date.

The most frequently explored objective for this class of models is a function of the weighted sum of the due date, earliness and tardiness over all jobs. Each of these three components is given a weight that can differ by job, so that the overall objective is thus $\sum_{i=1}^{N}(\pi_i^d d + \pi_i^e E_i + \pi_i^t T_i)$ where $\pi_i^d, \pi_i^e, \pi_i^t$, are the due date, earliness, and tardiness weights associated with job i respectively. In standard three-field scheduling notation, the model can be expressed as $1 \mid d^{opt} \mid \sum_{i=1}^{N}(\pi_i^d d + \pi_i^e E_i + \pi_i^t T_i)$, where the notation d^{opt} is used to indicate that the due date is determined within the model, and not exogenously assigned.

Baker and Scudder [3] and Quaddus [4] analyzed the structure of this model. Observe that if

$$\sum_{i=1}^{N} \pi_i^d \geq \sum_{i=1}^{N} \pi_i^t \tag{20.1}$$

then the optimal common due date $d^* = 0$. To see this, notice that for any sequence, increasing the due date from 0 will increase due date costs more than it decreases tardiness costs if condition (20.1) is met. In addition, any increase in due dates can only increase earliness cost. If condition (20.1) is met, all of the jobs will be tardy, so the total tardiness is minimized by sequencing jobs in nondecreasing order of p_i/π_i^t. If condition (20.1) is not met, it is not difficult to show that for any given sequence, the optimal schedule involves no inserted idle time, and the optimal due date must be equal to the completion time of one of the jobs.

To see this, observe that if there is any idle time in the schedule, either the job immediately preceding the idle time is early, and could be shifted later, decreasing the total earliness penalty; or the job immediately after the idle is tardy, and could be shifted earlier, decreasing the total tardiness. By repeatedly applying this observation, any schedule with inserted idleness could be converted to a schedule without inserted idleness with a lower objective function value. Suppose that jobs are contiguously scheduled, but that the first job does not start processing at time 0, and instead starts processing at time T. If the starting time of each of the jobs is decreased by T, and the due date is decreased by T, then earliness and tardiness costs will not change, but the due date cost will decrease by $\sum_{i=1}^{N} \pi_i^d T$.

Now, suppose that in such a schedule, the due date d does not coincide with the completion time of one of the jobs. If $d < C_{[1]}$, then no jobs are early and d can be increased by δ to $C_{[1]}$. In this case, the objective function will decrease by at least $\delta(\sum_{i=1}^{n} \pi_i^t - \sum_{i=1}^{n} \pi_i^d)$, and since condition (20.1) is not met, this is a positive quantity. If $d > C_{[N]}$, then no jobs are tardy and d can be decreased to $C_{[N]}$. In this case, both earliness and due date costs will decrease, and there will still be no tardiness costs, so the objective decreases.

Finally, suppose that for some job in position i, $1 \leq i \leq N-1$, $C_{[i]} < d < C_{[i+1]}$. Let F represent the objective function value given d, and let $x = d - C_{[i]}$ and $y = C_{[i+1]} - d$. Clearly, both $x > 0$ and $y > 0$. If the due date is changed to $C_{[i]}$, the new objective F_i will equal

$$F_i = F + x\left(\sum_{j=1}^{N}(\pi_i^t - \pi_i^d) - \sum_{j=1}^{i}(\pi_i^e + \pi_i^t)\right)$$

Similarly, if the due date is changed to $C_{[i+1]}$, the new objective F_{i+1} will equal

$$F_{i+1} = F - y\left(\sum_{j=1}^{N}(\pi_i^t - \pi_i^d) - \sum_{j=1}^{i}(\pi_i^e + \pi_i^t)\right)$$

Clearly, if $\sum_{j=1}^{N}(\pi_i^t - \pi_i^d) - \sum_{j=1}^{i}(\pi_i^e + \pi_i^t)$ is positive, then $F_{i+1} < F$; and if it is negative, then $F_i < F$.

Now, consider a given sequence, and two adjacent jobs, $[j-1]$ and $[j]$. Following Baker and Scudder [3], we will compare two schedules: S, in which $d = C_{[j-1]}$; and S', in which $d' = C_{[j]}$. The objective function can be written as

$$f(S,d) = \sum_{i=1}^{N} \pi^d_{[i]} d + \sum_{i=1}^{j-1} \pi^e_{[i]}(d - C_{[i]}) + \sum_{i=1}^{N} \pi^t_{[i]}(C_{[i]} - d)$$

Now, observing that

$$C_{[i]} = \sum_{k=1}^{i} p_{[k]}$$

and

$$d = \sum_{k=1}^{j-1} p_{[k]}$$

and substituting into the objective function, we get

$$f(S,d) = \sum_{k=1}^{j-1} p_{[k]} \left(\sum_{i=1}^{k-1} \pi^e_{[i]} + \sum_{i=1}^{N} \pi^d_{[i]} \right) + \sum_{k=j}^{N} p_{[k]} \left(\sum_{i=k}^{N} \pi^t_{[i]} \right)$$

Letting $G(S,S') = f(S,d) - f(S',d')$, we get

$$G(S,S') = p_{[j]} \sum_{k=1}^{j-1} \pi^e_{[k]} + p_{[j]} \sum_{k=1}^{N} \pi^d_{[k]} - p_{[j]} \sum_{k=j}^{N} \pi^t_{[i]}$$

Observe that S' has a better objective value than S, and the due date should be later than $C_{[j-1]}$, if

$$\sum_{k=1}^{j-1} (\pi^e_{[k]} + \pi^t_{[k]}) < \sum_{k=1}^{N} (\pi^t_{[k]} - \pi^d_{[k]})$$

and that the due date should be no later than $C_{[j]}$ if the reverse is true. Therefore, for any given sequence, we can conclude that the optimal due date $d = C_{[r]}$, where r is the smallest integer for which

$$\sum_{k=1}^{r} (\pi^e_{[k]} + \pi^t_{[k]}) \geq \sum_{k=1}^{N} (\pi^t_{[k]} - \pi^d_{[k]}) \tag{20.2}$$

Quaddus [4] provides an alternative proof of this result using duality theory.

Note that for any sequence, the jobs will be partitioned into two sets: one of on-time jobs and one of tardy jobs. Using a simple adjacent pairwise interchange proof, it can be shown that the on-time jobs are scheduled WLPT (in nonincreasing order of p_i/π^e_i), and the tardy jobs are scheduled WSPT (in nondecreasing order of p_i/π^t_i). This is known as a V-shaped schedule, and this type of schedule is optimal for many related common due date problems. (See Figure 20.2 for an example of a V-shaped sequence.) For example, Raghavachari [5] uses an interchange argument to prove that the optimal sequence of jobs around a common due date must be V-shaped when the objective is to minimize the sum of deviations around the due date (in other words, the problem described above, with $\pi_d = 0$ and $\pi_e = \pi_t$).

Hall and Posner [6] prove that $1 \mid d^{opt} \mid \sum_{i=1}^{N} (\pi^d_i d + \pi^e_i E_i + \pi^t_i T_i)$ is NP-hard. Baker and Scudder [3] propose an optimization procedure to find the optimal sequence that involves enumerating V-shaped sequences and determining due dates and thus objective values as described above.

Panwalker, Smith, and Seidmann [7] consider a special case of the model described above, where earliness, tardiness, and the (single) due date are given weights that do not differ by job. The overall objective is thus $\sum_{i=1}^{N} (\pi^d d + \pi^e E_i + \pi^t T_i)$.

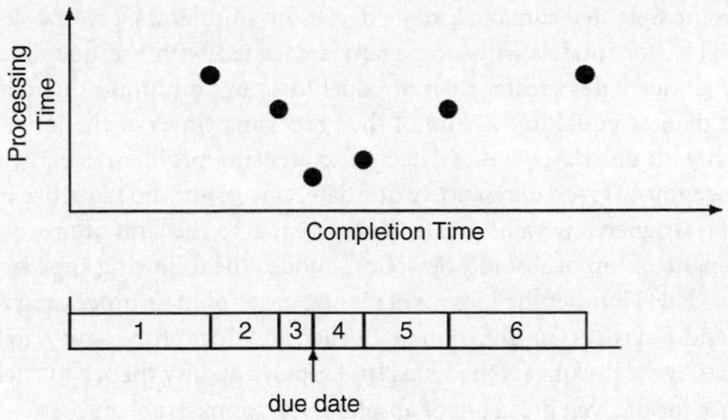

FIGURE 20.2 A V-shaped schedule.

As observed, if $\pi^d \geq \pi^t$, then $d^* = 0$ and it is optimal to sequence jobs in SPT order, and for any sequence there is an optimal d value equal to the completion times of one of the jobs.

For this version of the model, Equation (20.2) can be simplified so that for any specified sequence, there is an optimal due date $C_{[r]}$, where

$$r = \left\lceil N \frac{\pi^t - \pi^d}{\pi^e + \pi^t} \right\rceil \tag{20.3}$$

Next, observe that given a sequence of jobs, the objective function can be rewritten

$$\sum_{j=1}^{r} (n\pi^d + (j-1)\pi^e) p_{[j]} + \sum_{j=r+1}^{N} \pi^t (N+1-j) p_{[j]} = \sum_{j=1}^{N} \Gamma_j p_{[j]}$$

where

$$\Gamma_j = \begin{cases} n\pi^d + (j-1)\pi^e & \text{if } j \leq r \\ \pi^t(n+1-j) & \text{otherwise} \end{cases}$$

Furthermore, observe that this problem can be solved optimally by sequencing jobs so that the smallest value of Γ is matched with the largest processing time, the next smallest value of Γ is matched with the next largest processing time, etc. This suggests the following optimal solution procedure: Determine r using Equation (20.3); if this quantity is not greater than 0, $d^* = 0$, and the SPT sequence is optimal; otherwise, match processing times with Γ values as described above (that is, match the smallest value of Γ with the largest processing time, etc.) and sequence jobs in order of Γ indices; finally, set the due date $d^* = p_{[1]} + p_{[2]} + \cdots + p_{[r]}$.

It is interesting to note that Γ_i is increasing as i increases from 1 to r, and decreasing as i increases from $r+1$ to n. Thus, processing times are decreasing as i increases from 1 to r; and increasing as i increases from $r+1$ to N (assuming ties are broken appropriately). Thus the optimal schedule is LPT until the rth job, and then SPT — this approach finds a V-shaped schedule.

Cheng [8] provides an interesting alternative proof of this result utilizing constrained convex programming theory.

With slight modifications, Panwalker, Smith, and Seidmann [7] extend this result when there is an additional term in the objective, representing weighted flow time, that is, $1 \mid d^{opt} \mid \sum_{i=1}^{N} (\pi^d d + \pi^e E_i + \pi^t T_i + \pi^f F)$, where $F = \sum_{i=1}^{n} C_{[i]}$.

Other authors, including Kanet [9] and Quaddus [10], consider an even more simplified version of the original model, with no due date penalty, and equal earliness and tardiness weights: $\pi_d = 0$ and $\pi_e = \pi_t$.

This is also known as the weighted sum of absolute deviations problem. Of course, as observed by Bagchi, Chan, and Sullivan [11], for models with no penalty associated with the due date, the objective value will be the same for all due dates greater than or equal to some minimum due date. Furthermore, this due date will be less than or equal to the sum of the processing times of the jobs. Therefore, if there is no penalty associated with due dates, a due date and sequencing problem is equivalent to a sequencing problem with an exogenously given *unrestrictive* due date, since once the sequence is determined the due date can arbitrarily be assigned any value greater than or equal to the sum of processing times. Kanet [9] shows that for this weighted sum of absolute deviations model, the following approach leads to an optimal sequence, given a due date: Number the jobs in increasing order of their processing times. Assign the jobs alternately to sets A and B. Process set B first in LPT order, and then process set A in SPT sequence, where the first job in A starts at the due date. Quaddus [10] employs duality theory to characterize the optimal due date and sequence for the weighted sum of absolute deviations problem.

Bagchi, Chan, and Sullivan [11] consider the weighted sum of squared deviation problem, $1 \mid d^{opt} \mid \sum_{i \in N} \pi_e E_i^2 + \pi_t T_i^2$. Many of the properties described above hold. Unfortunately, *there is not necessarily an optimal schedule in which the completion times of one of the jobs coincides with the due date.* However, Bagchi, Chan, and Sullivan [11] characterize the optimal due date for any given sequence using first order conditions:

$$d^* = \frac{\pi^e \sum_{i \in N: C_i < d^*} C_i + \pi^t \sum_{i \in N: C_i > d^*} C_i}{\pi^e \mid i \in N : C_i < d^* \mid + \pi^t \mid i \in N: C_i > d^* \mid}$$

They propose an iterative procedure to determine the due date, and a branch-and-bound procedure to find the optimal sequence.

Cheng [12] considers a related model, the weighted common due date problem. In this model, $\pi_i^d = 0$, and the earliness and tardiness penalties are identical to each other, but *differ for each job*, so that $\pi_i^e = \pi_i^t$. As before, a V-shaped schedule is optimal for this model. Cheng [12] proves that the optimal due date for a given sequence can be found using the approach discussed in Equation (20.2), which in this case simplifies to finding r such that the optimal due date coincides with the completion time of the rth job in the sequence, where r is determined as follows:

$$\sum_{i=1}^{r-1} \pi_i^e < \frac{\sum_{i=1}^{N} \pi_i^e}{2}$$

$$\sum_{i=1}^{r} \pi_i^e \geq \frac{\sum_{i=1}^{N} \pi_i^e}{2}$$

Cheng [12] proposes an (exponential) algorithm based on partially enumerating possible sequences using these observations.

This last model was generalized by Cheng [13], who proposes a model with a due date related penalty, and a lateness penalty, leading to the following objective:

$$\sum_{i=1}^{N} \pi^d d + \pi^i \mid C_i - d \mid^m$$

where m is some given integer parameter. Cheng [13] identifies some necessary conditions for optimality, and an iterative procedure to find the optimal due date for this problem (although a procedure to find the optimal sequence is not known).

Cheng [14] considers the SLK rule in relationship to a version of this model, where the objective involves minimizing $\pi_\beta \beta + \max_{i \in N} T_i$. For this problem, it is well known that EDD is the optimal sequence.

Writing the objective function as a function of β, Cheng [14] observes that the optimal β is as follows (corrected in Gordon [15]):

$$\beta = \begin{cases} x \in [C_{[N-1]}, \infty) & \text{if } \pi_\beta = 0 \\ C_{[N-1]} & \text{if } 0 < \pi_\beta < 1 \\ x \in [0, C_{[N-1]}] & \text{if } \pi_\beta = 1 \\ 0 & \text{if } \pi_\beta > 1 \end{cases}$$

20.4 Single Machine Distinct Due Date Static Models

Of course, in many realistic problems, each job can be assigned a distinct due date. In this model, we review a variety of single machine static models with distinct due dates.

Seidmann, Panwalker, and Smith [16] consider a multiple due date assignment model where the objective is a function of earliness, tardiness, and length of lead time. Each of these three components is given a weight that does not differ by job, and the authors introduce the concept of excessive lead time, so that if some time A is considered a reasonable lead time, the lead time penalty π_l is multiplied by the excess lead time $L_i = \max(d_i - A, 0)$. The overall objective is thus $\sum_{i \in N} \pi^l L_i + \pi^e E_i + \pi^t T_i$.

Seidmann et al. [16] show that this problem, $1 \mid d_i^{opt} \mid \sum_{i \in N} \pi^l L_i + \pi^e E_i + \pi^t T_i$, can be solved optimally as follows: if $\pi^l s \pi^t$, sequence jobs in SPT order, and set due dates equal to completion times of jobs. Otherwise, for each job, set due dates equal to the minimum of A, and the completion time.

This result follows from the fact that the SPT sequence minimizes the sum of completion times, and that there is no benefit to assigning due dates later than completion times. Thus, the only trade-off is between lead time penalty and completion time penalty. If lead time penalty is greater, it makes sense to assign a due date equal to the reasonable lead time to a job, whereas if tardiness penalty is greater, it makes sense to assign a due date equal to the completion time of the job. Seidmann et al. [16] utilize a simple interchange argument to formally prove this result.

The majority of single machine distinct due date static model research involves optimizing the parameters of one of the due date setting rules described in Section 20.2, CON, SLK, and TWK. Note that for static models, CON is equivalent to a common due date model, and recall that the parameters for these three rules are γ, β, and α, respectively.

Karacapilidis and Pappis [17] note an interesting relationship between the CON and SLK versions of the single machine static due date quotation problem with the objective of minimizing weighted earliness and tardiness. For the CON problem, which is equivalent to the static due date problem described above, with $\pi^d = 0$, we have the following objective:

$$\sum_{i \in N} \pi_i^e E_i + \pi_i^t T_i = \sum_{i \in N} \pi_i^e [C_i - d]^+ + \pi_i^t [d - C_i]^+$$

Furthermore, the following relationship holds:

$$d + T_i - E_i = C_i$$

On the other hand, for the SLK version of the problem, we have $\sum_{i \in N} \pi_i^e E_i + \pi_i^t T_i = \sum_{i \in N} \pi_i^e [C_i - p_i - \beta]^+ + \pi_i^t [\beta + p_i - C_i]^+$. Note that $C_i - p_i = W_i$, the waiting time of job i, and that the following relationship holds:

$$\beta + T_i - E_i = W_i$$

Thus, for any sequence, the mathematical program for the two problems is equivalent, except that W_i replaces C_i in the SLK version, and $W_i \geq 0 \, \forall i$ in the SLK problem, whereas $C_i \geq 0 \, \forall i$ in the CON problem.

Furthermore, the optimal sequence and due date for the CON problem can always be shifted to the right. This implies that given an optimal solution to one of the problems, we can find an optimal solution to that problem which is feasible and optimal for the other problem. Karacapilidis and Pappis [17] use this observation to develop algorithms that find the complete set of optimal sequences for both problems.

Baker and Bertrand [18] consider CON, SLK, and TWK for single machine static models with the objective of minimizing the sum of assigned due dates subject to the constraint that no job can finish later than its assigned due date (three 100% reliable single machine static models: $(1 \mid d_i^{opt} \mid \sum_{i \in N} r_i + \gamma)$, $(1 \mid d_i^{opt} \mid \sum_{i \in N} r_i + p_i + \beta)$, and $(1 \mid d_i^{opt} \mid \sum_{i \in N} r_i + \alpha p_i))$. Of course, for this model, the optimal schedule is easy to determine — schedule jobs in SPT order, and assign due dates equal to their completion times. Nevertheless, in practice, rules such as these might be useful.

First, observe that in this case, for any set of due dates, EDD will minimize the objective, so it is sufficient to assume that whatever the due date assignment parameters, the sequence will be EDD. For the CON version of the problem, clearly all due dates will be equal, so in order to meet the due date,

$$d_i = d = \gamma = \sum_{i=1}^{N} p_i$$

For the SLK version of the problem, the EDD sequence is equal to the SPT sequence. In order for the last job to finish on time, assume that jobs are numbered in SPT order, and observe that

$$\beta = \sum_{i=1}^{N-1} p_i$$

Finally, for the TWK rule, EDD is again equivalent to SPT. The minimum value of α that ensures that due dates will be met in the SPT sequence can be determined as follows. Observe that

$$\alpha p_i \geq C_i$$

for

$$\alpha = \max_{1 \leq i \leq N} C_i / p_i$$

And thus,

$$\alpha = \max_{1 \leq i \leq N} \frac{\sum_{j=1}^{i} p_j}{p_i}$$

Also, consider the special case of these problems when all processing times are equal. Baker and Bertrand [18] observe that if all jobs have the same length, all of the approaches yield the same objective, the ratio of the optimal CON, SLK, or TWK objective to the optimal solution to the problem (the one arrived at using SPT) is

$$\frac{Z^H}{Z^*} = \frac{2N}{N+1}$$

Qi and Tu [19] also consider the 100% reliable SLK model, but with two different objectives: minimizing the sum of a monotonically increasing function of lateness $\left(\sum_{i=1}^{N} g(d_i - C_i)\right)$, and minimizing the total weighted earliness $\left(\sum_{i=1}^{N} w_i(d_i - C_i)\right)$. It is easy to see that exactly one job (the final job in the sequence) will be on time, and all other jobs will be early. It can be shown using an interchange argument that for the first problem, all early jobs are sequenced in LPT (longest to shortest) order, and for the second problem, all early jobs are sequenced in nondecreasing order of p_i / w_i.

For the first objective, by comparing the cost of a schedule with an arbitrary job scheduled as the on-time job with the cost of a schedule with the longest job scheduled as an on-time job, Qi and Tu [19] show that there is an optimal schedule in which the longest job is the on-time job. Thus, the first problem can be solved by putting the longest job last, scheduling the remaining jobs LPT, and setting β such that the due date of the final job is equal to its completion time.

The total weighted earliness problem can be solved by trying each of the jobs in the on-time position, scheduling the rest of the jobs in nondecreasing order of p_i/w_i, and finding the best sequence. β is once again set so that the due date of the final job is equal to its completion time.

Gordon and Strusevich [20] present a more efficient approach for solving this problem, and extend these results by providing structural results as well as efficient algorithms for the total weighted exponential earliness objective $\left(\sum_{i=1}^{N} w_i \exp\left(d_i - C_i\right)\right)$, as well as problems with precedence constraints.

Cheng [21] considers the same single machine dynamic distinct due date model with the TWK rule, and with the objective of minimizing total squared lateness, $\sum_{i=1}^{N}(C_i - d_i)^2$. By differentiating the objective function, the optimal value of the multiplier α for a given sequence can be seen to be

$$\alpha = \frac{\sum_{i=1}^{N} p_{[i]} \sum_{j=1}^{i} p_{[j]}}{\sum_{i=1}^{N} p_{[i]}^{2}}$$

Cheng [21] shows that the optimal value of α given above is in fact independent of the sequence of jobs, and constant for a given set of processing times, by using an interchange argument. Thus, the objective function value can be written as

$$\sum_{i=1}^{N}\left(\sum_{j=1}^{i} p_{[j]}\right)^2 + \left(\alpha \sum_{j=1}^{N} p_{[j]}\right)^2 - 2\alpha \sum_{i=1}^{N}\sum_{j=1}^{i} p_{[j]}$$

Furthermore, Cheng [21] demonstrates using interchange arguments that the second and third terms of this expression are constant, and that the first term is minimized by sequencing jobs in SPT order.

Cheng [22] extends this model to the case in which processing times are independent random variables from the same distribution family (where the processing time of job i has a mean μ_i and a standard deviation σ_i), and the objective is to minimize the expected squared lateness. Using an analogous approach to that of Cheng [21], in Cheng [22] it is shown that if the random variables have known means and the same coefficient of variation,

$$\alpha = \frac{\sum_{i=1}^{N}\left(\sigma_{[i]}^2 + \mu_{[i]} \sum_{j=1}^{i} \mu_{[j]}\right)}{\sum_{i=1}^{N}\left(\mu_{[i]}^2 + \sigma_{[i]}^2\right)}$$

where this value is independent of the sequence of jobs. Also, if the variances of processing times are monotonic functions of the means, then the shortest expected processing time sequence is optimal.

Cheng [12] considers a similar model, but employs the so called TWK-P rule, so that $d_i = \alpha p_i^m$, where m is a problem parameter. For this problem, it is necessary to explicitly prohibit inserted idleness. For a given sequence, this problem can be written as an LP, and using duality theory, Cheng [12] characterizes the optimal α value.

20.5 Single Machine Dynamic Models

In many models, some jobs are not available to be processed at the start of the time horizon. In this section, we consider single machine models in which jobs have associated release times, and cannot be processed before these times. We first consider offline models, and then online models. For online models, we consider worst case and probabilistic analysis of algorithms, and then queueing models.

20.5.1 Offline Single Machine Dynamic Models

Baker and Bertrand [18] consider the CON, SLK, and TWK rules for single machine dynamic models with preemption, and the objective of minimizing the sum of assigned due dates subject to the constraint that no job can finish later than its assigned due date (three 100% reliable single machine dynamic preemption models: $(1 \mid pmtn, r_i, d_i^{opt} \mid \sum_{i \in N} r_i + \gamma)$, $(1 \mid pmtn, r_i, d_i^{opt} \mid \sum_{i \in N} r_i + p_i + \beta)$, and $(1 \mid pmtn, r_i, d_i^{opt} \mid \sum_{i \in N} r_i + \alpha p_i))$. For these problems, the EDD sequence is optimal once due dates have been determined.

For the CON rule, the optimal solution can be found recursively by scheduling jobs in order of release times, and then determining the lead time as follows:

$$\gamma = \max_{1 \le i \le N} C_i - r_i$$

Similarly, for the SLK rule, all jobs have the same allowed flow time, so jobs will optimally be sequenced preemptively, in order of $r_j + p_j$ among available jobs. That is, each time a job is released, the uncompleted, released job with minimum $r_j + p_j$ will be processed, even if this means interrupting the currently processing job. Once the sequence is determined, the flow time can be determined:

$$\beta = \max_{1 \le i \le N} C_i - r_i - p_i$$

The TWK rule is more complex, as the EDD sequence cannot be determined before α is assigned. Baker and Bertrand [18] point out that given a sequence,

$$\alpha = \max_{1 \le i \le N} \frac{C_i - r_i}{p_i}$$

Thus, by applying an algorithm for minimizing the maximum cost for a single machine problem with preemption after expressing the cost as a function of α (Baker et al. [23]), the optimal sequence can be found.

Gordon [15] considers the SLK rule for a single machine dynamic model with preemption and the objective of minimizing the maximum tardiness plus a penalty associated with the slack $(1 \mid pmtn, r_i, d_i^{opt} \mid \pi_\beta \beta + \max_{i \in N} T_i)$. It is well known that for a given set of due dates, maximum tardiness is minimized by scheduling jobs in preemptive EDD order. Also, clearly the EDD sequence is independent of the value of β. If we assign $\beta = 0$ and find the job j^* with maximum tardiness if jobs are sequenced preemptive EDD, clearly no job will have larger tardiness for other positive values of β. Gordon [15] expresses the objective function in terms of this job j^*, and calculates its minimum as follows:

$$\beta = \begin{cases} x \in [C_{j^*} - r_{j^*} - p_{j^*}, \infty) & \text{if } \pi_\beta = 0 \\ C_{j^*} - r_{j^*} - p_{j^*} & \text{if } 0 < \pi_\beta < 1 \\ x \in [0, C_{j^*} - r_{j^*} - p_{j^*}] & \text{if } \pi_\beta = 1 \\ 0 & \text{if } \pi_\beta > 1 \end{cases}$$

The results in Gordon [15] are actually more general than this, as they allow for precedence constraints among the jobs. To modify this approach for precedence constraints, an $O(n^2)$ algorithm for minimizing $1/pmtn, prec, r_i/f_{\max}$ (Baker et al. [24], Gordon and Tanaev [25]) is used to sequence jobs, and an analogous approach is used to find j^*. Given j^*, the optimal value β is as described above.

Cheng and Gordon [26] extend this approach for single machine dynamic models with preemption and precedence constraints to two extensions of the due date assignment rules: PPW (process time plus wait), an extension and combination of TWK and SLK where the due date $d_i = \alpha p_i + \beta$; and TWK-power, an extension of TWK, where $d_i = \alpha p_i^\beta$.

20.5.2 Online Single Machine Dynamic Models

20.5.2.1 Probabilistic Analysis

Kaminsky and Lee [27] consider a model in which a set of jobs must be processed on a single machine. Each job has an associated processing time and release time, and no job can be processed before its release time. At its release time, each job is assigned a due date. When a job is released, its processing time is revealed, and its due date must be quoted. In this model, all due dates are met, with the objective of minimizing average quoted due date. In [27], they present three online heuristics for this model, which they call First Come First Serve Quotation (FCFSQ), Sequence/Slack I (SSI), and Sequence/Slack II (SSII).

In the FCFSQ heuristic, jobs are sequenced in order of their release, so that accurate due dates are easy to quote. In the SSI and SSII heuristics, a two-phase approach is used. First, the newly released job is inserted in the queue of jobs waiting to be processed in a position as close as possible to its position in an SPT ordering of the queue, without violating any previously assigned due dates. If there were no new arrivals, the completion time of the new job could now be quoted exactly. However, there may be future arrivals that would ideally be sequenced ahead of the newly inserted job. Thus, some slack is added to the projected completion time of this job. In SSI, the slack assigned is roughly proportional to the length of the newly inserted job. SSII attempts to estimate the possible waiting time for a new job after it is inserted by estimating the number and length of jobs that will arrive after the new job arrives and before it is processed, but are shorter than it.

To analyze this model, Kaminsky and Lee [27] use probabilistic analysis, and in particular, asymptotic probabilistic analysis of the model and heuristics. In this type of analysis, a sequence of randomly generated deterministic instances of the problem is considered, and the objective values resulting from applying the heuristics to these instances as the size of the instances (the number of jobs) grows to infinity is characterized. For the probabilistic analysis, they generate problem instances as follows: the processing times are drawn from independent identical distributions bounded above by some constant, with expected value *EP*. Release times are determined by generating interarrival times drawn from identical independent distributions bounded above by some constant, with expected value *ET*. Processing times are assumed to be independent of interarrival times.

They consider two sets of randomly generated problem instances. If problem instances are generated from distributions such that $EP < ET$, they demonstrate that each of our heuristics is asymptotically optimal, as the number of jobs tends to infinity. For problem instances generated from distributions such that $ET < EP$, they prove that SSII is asymptotically optimal, as the number of jobs tends to infinity.

20.5.2.2 Worst Case Analysis

Keskinocak, Ravi, and Tayur [28] consider a single machine model in which each job j has a release time r_j, each job has the same processing time p, each job has the same acceptable maximum lead time l, and each job has the same penalty (lost revenue) per unit time the order is delayed before its processing starts w. The objective is thus to quote a lead time d_j in order to maximize revenue, where revenue for a particular job j, $R(d_j)$, is

$$R(d_j) = \begin{cases} (l - d_j)w & \text{if } d_j < l \\ 0 & \text{otherwise} \end{cases}$$

Keskinocak, Ravi, and Tayur [28] consider a variety of versions of this problem, both with and without due date quotation. For the due date quotation versions of this problem, they consider a 100% reliable problem, where due dates must be quoted immediately when jobs are released, and the jobs have to start processing before the quoted lead time. In the delayed due date quotation version of the problem, 100% reliable due dates must still be quoted, but they can be quoted within q time units after the order arrives, where $q < l$. Several heuristics are proposed for these models, and to analyze the heuristics, a technique known as competitive analysis is employed. In this approach, an online algorithm is compared to an optimal offline algorithm. If Z^*_{offline} is the optimal offline solution objective value, and Z^{online} is the online solution generated by a heuristic, then the online heuristic is called *c-competitive* if for any instance,

$$Z^{\text{online}} \leq c Z^*_{\text{offline}} + a$$

where a is some constant independent of the instance.

Keskinocak, Ravi, and Tayur [28] propose an algorithm they call **Q-FRAC**: Select a value of α such that $0 < \alpha < 1$. At time t, schedule each order to the earliest available position only if a revenue of at least αl can be obtained, and reject all other orders that arrive at time t. Use the scheduled start time to quote a lead time. They prove that if $\alpha = 0.618$, then Q-FRAC has a competitive ratio less than or equal to $1/\alpha = 1.618$.

For the delayed quotation version of the problem, Keskinocak, Ravi, and Tayer [28] assume that $q = (1 - \lambda)(l - 1)$, and consider the case when $p = 1$. They propose an algorithm call **Q-HRR**: At time t, from the set of orders available for scheduling, choose the one with the largest remaining revenue and process it. Quote the appropriate lead time. Reject all orders with remaining revenue $\leq \lambda l$. They prove that for this problem, there is an online quotation algorithm with a competitive ratio at most $\min\{1.619, 1/(1 - \lambda^2)\}$. This can be achieved by using Q-FRAC if $\lambda > 0.618$, and QHRR if $\lambda \leq 0.618$.

Keskinocak, Ravi, and Tayur [28] develop related results for more complicated models, where there are two types of customers: an urgent type who needs the product immediately, and a normal type who can accept a longer lead time.

20.5.2.3 Queuing Models

Various researchers have utilized queuing theory to analyze the single machine due date quotation problem. Typically, researchers assume some sort of sequencing rule, and then optimize a parameter of some lead time quotation rule, which may or may not depend on the state of the system when the job arrives.

Seidmann and Smith [29] consider due date quotation in a G/G/1 queuing model with CON due date assignment. They present a method to find the optimal due date assignment with the objective of minimizing total expected cost, where β is the constant lead time, and the cost of a job is the sum of three monotonically increasing strictly convex functions of $[\beta - A]^+$, job earliness, and job tardiness, respectively (represented by $C_d()$, $C_e()$, and $C_t()$). Note that the quantity $[\beta - A]^+$ penalizes a constant lead time greater than some constant A – lead times less than that quantity incur no cost penalty. Seidmann and Smith [29] assume jobs are sequenced EDD (or equivalently, First Come First Serve, or in order of arrival).

Let θ be a random variable representing the time that a job spends in between when it arrives and when it departs, and let $f(\theta)$ represent the probability density function of θ, where $f(\theta) > 0, 0 < \theta < \infty$. The distribution of θ is assumed to be common to all jobs. The three components of the cost function can therefore be expressed as follows:

$$\Pi_d(\beta) = \begin{cases} 0 & \text{if } \theta \leq A \\ C_d(\theta - A) & \text{if } \theta > A \end{cases}$$

$$\Pi_e(\theta, \beta) = \begin{cases} C_e(\beta - \theta) & \text{if } \theta < \beta \\ 0 & \text{if } \theta \geq \beta \end{cases}$$

$$\Pi_t(\theta, \beta) = \begin{cases} 0 & \text{if } \theta \leq \beta \\ C_t(\theta - \beta) & \text{if } \theta > \beta \end{cases}$$

Then, the expected value of the total cost $TC(\theta, \beta)$ is

$$E[TC(\theta,\beta)] = \begin{cases} \int_0^\infty (\Pi_e(\theta,\beta) + \Pi_t(\theta,\beta)) f(\theta) d\theta & \text{if } \theta \leq A \\ \int_0^\infty (\Pi_d(\beta) + \Pi_e(\theta,\beta) + \Pi_t(\theta,\beta)) f(\theta) d\theta & \text{if } \theta > A \end{cases} \tag{20.4}$$

Thus, the expected cost function has to be investigated over two intervals defined by A. Now, let $\tilde{\beta}_I$ be the unconstrained minimum of the first line of Equation (20.4), and let $\tilde{\beta}_{II}$ be the unconstrained minimum of the second line. By exploring the structure of the two parts of the $E[TC(\theta, \beta)]$, noting that both parts of the function are individually strictly convex, and exploring the relationship between the two parts of the function, Seidmann and Smith [29] show that the optimal value of β, β^*, can be characterized as follows:

1. If $\tilde{\beta}_I < A$, then $\beta^* = \tilde{\beta}_I$
2. If $\tilde{\beta}_{II} < A < \tilde{\beta}_I$, then $\beta^* = A$
3. If $A < \tilde{\beta}_{II}$, then $\beta^* = \tilde{\beta}_{II}$

Note that these are all of the possible cases.

Seidmann and Smith [29] then consider a linear cost function. Utilizing the notation of previous sections of this chapter, the components of this function can be expressed as

$$\Pi_d(\beta) = \begin{cases} 0 & \text{if } \theta \leq A \\ \pi^l(\theta - A) & \text{if } \theta > A \end{cases}$$

$$\Pi_e(\theta,\beta) = \begin{cases} \pi^e(\beta - \theta) & \text{if } \theta < \beta \\ 0 & \text{if } \theta \geq \beta \end{cases}$$

$$\Pi_t(\theta,\beta) = \begin{cases} 0 & \text{if } \theta \leq \beta \\ \pi^t(\theta - \beta) & \text{if } \theta > \beta \end{cases}$$

Integrating to find expected values as discussed above, Seidmann and Smith [29] take derivatives of the two parts of the cost function, and find the unconstrained minima meet the following conditions, where $F(\theta)$ is the cumulative distribution function of θ:

$$F(\tilde{\beta}_I) = \frac{\pi^t}{\pi^e + \pi^t}$$

$$F(\tilde{\beta}_{II}) = \frac{\pi^t - \pi^d}{\pi^e + \pi^t}$$

This suggests the following algorithm to minimize this function:

1. Check if $\pi^e = 0$. If yes, go to step 3. If no, go to step 2.
2. Check if $F(A) > \frac{\pi^t}{\pi^e+\pi^t}$. If yes, set $\beta^* = \tilde{\beta}$, where $F(\tilde{\beta}) = \frac{\pi^t}{\pi^e+\pi^t}$. If no, go to step 3.
3. Check if $F(A) < \frac{\pi^t-\pi^d}{\pi^e+\pi^t}$. If yes, set $\beta^* = \tilde{\beta}$, where $F(\tilde{\beta}) = \frac{\pi^t-\pi^d}{\pi^e+\pi^t}$. If no, set $\beta^* = A$.

Dellaert [30] considers a queuing model of lead time quotation that incorporates setups. Arrivals are assumed to be Poisson, and setup and service are exponential, with different rates. Lead times (that is, time until job completion) are quoted to customers as they arrive, and customers can leave the system if the quoted lead time is too large. If the lead time is greater than some parameter $d_{\max}$, customers will definitely leave the system. Otherwise, they will stay in the system with the following probability: $1 - d/d_{\max}$. A setup is charged each time the processor switches from being idle to processing jobs. However, no setup is charged between consecutively processed jobs without idle time. The objective is to minimize expected cost, and the cost per job is a sum of π^e per time unit for jobs that complete processing ahead of their due date, π^t per time unit for jobs that complete processing after their due date, and s per setup. Jobs are assumed to be processed FCFS, and the following rule is used to schedule production: every time production is started, all available orders will be processed. However, once production has stopped, no production takes place until m jobs are available to be processed, where the value for m is determined simultaneously with the due date decision. Dellaert [30] characterizes the distribution of time that jobs spend in the system as a function of m, and uses the approach of Seidmann and Smith [29] to determine the optimal constant lead time. The expected cost is compared for a variety of values of m. Dellaert [30] also shows that expected cost performance can be dramatically improved by quoting state dependent lead times, where the state of the system is the number of jobs waiting to be processed, and whether the server is down, or in setup or processing model.

Duenyas and Hopp [31] consider a model similar to the model of Dellaert [30], except that there is no penalty for completing jobs ahead of their due date, and each job generates the same net revenue, so that the objective is to maximize the expected net revenue. Also, the acceptance probability is more general than in Dellaert [30]; the probability that a customer places an order is only assumed to a decreasing function of the quoted lead time. Duenyas and Hopp [31] first focus on an M/M/1 queue, and FCFS dispatch rules. The state of the system at any arrival time is characterized by k, the number of jobs in the system. Duenyas and Hopp [31] characterize a number of properties of this system. For example, there is some $\bar{k}$, such that for $k \geq \bar{k}$, it is optimal to reject arriving customers (that is, to quote a due date greater than A, the maximum acceptable due date). For $k < \bar{k}$, the optimal policy is a set of lead times β_k such that arrivals to the system at state k are quoted lead time β_k. Duenyas and Hopp [31] prove that β_k is increasing in k. They also show that in this model, regardless of due date quotation rule, it is optimal to sequence jobs in EDD order.

Duenyas [32] extends these results to a similar model with multiple customer classes. There are n different classes of customers, each demanding the same product, but with different preferences for lead time (in other words, different state dependent probabilities of placing orders for a given lead time), different net revenue per job, and Poisson arrival processes with different rates. However, the tardiness penalty and the processing time distribution is the same for each class. As in Duenyas and Hopp [31], a policy is a set of lead times β_k^i such that arrivals to the system of class i at state k are quoted lead time β_k^i. Duenyas [32] shows that for each class i, β_k^i is increasing in k, and that for any pair of classes i and j such that (1) the revenue of i is greater than the revenue of j, (2) the probability of a customer of type i placing an order for a given lead time is less than the probability of a customer of type j placing an order for the same lead time, and (3), customers of type i are more sensitive to changes in lead time than customers of type j, for any k, the optimal quoted lead time $\beta_k^i \leq \beta_k^j$. Duenyas [32] also shows that in this model, regardless of due date quotation rule, it is optimal to sequence jobs in EDD order.

A variety of researchers have also explored due date quotation with some constraint on expected fraction of orders that are allowed to be tardy. The majority of these are simulation-based, employing dispatch rules that are in some cases based on sophisticated analysis of the time jobs will spend in the system under various scheduling disciplines (for example, Wein [33] – see Keskinocak and Tayur [1] for a complete survey).

Spearman and Zhang [34] consider a general queuing system with a single job class and multiple stages. Although this is a very difficult problem for which to obtain analytical results, Spearman and Zhang [34] focus on two performance measures, and obtain some structural results. In particular, they focus on objectives that minimize the average quoted lead times of jobs subject to either (problem 1) a constraint on the fraction of jobs that exceed the quoted lead time, or (problem 2) the average tardiness of jobs. They further assume that both the steady state distribution of the number and location of jobs in the system is known, that the time remaining to process those jobs is known upon job arrival, and that flow time distributions of jobs given the number and location of jobs in the system and the time remaining to process those jobs, is known. By determining optimality conditions for the lead time quotation problems, they discover some interesting behavior of the system. For example, for problem 2, the mean flow time of jobs is increasing as the number of jobs in the system increases. However, in many cases, there exists a number of jobs in the system such that if a customer arrives when there are more than that number of jobs in the system, it is optimal to quote a lead time of zero, even though there is no likelihood of completing the job immediately. For problem 2, however, both mean flow time and optimal quoted lead time increase with the number of jobs in the system. They argue that this implies that problem 1 leads to less ethical due date quotation practices than problem 2.

When the goal is to quote a lead time that meets a service objective independent of cost, Hopp and Roof Sturgis [35] observe that for an M/G/1 queuing system, the distribution of flow time given n jobs in the system, T_n, equals

$$P(T_n \leq t) = 1 - e^{-\lambda t} - \frac{e^{-\lambda t}(\lambda t)^1}{1!} - \cdots - \frac{e^{-\lambda t}(\lambda t)^{n-1}}{(n-1)!}$$

Thus, if α represents the user specified target level, and if $P(T_n \leq \beta_n) = \alpha$, then if job i sees n jobs in the system including itself, the quoted lead time should be β_n. Hopp and Roof Sturgis [35] use this observation to develop heuristics when the flow time distribution is not known.

So and Song [36] consider an M/M/1 queuing system, and determine both a single price, and a single quoted lead time, for all customers. In this model, demand is a function of both price and lead time, following the Cobb-Douglas demand function, so that if $D(p, \beta)$ is the demand rate as a function of price p and lead time β, then

$$D(p, \beta) = -k_1 p^{-k_2} \beta^{-k_3}$$

where k_1, k_2, and k_3 are positive constants, representing respectively level of potential demand, price elasticity, and delivery-time guarantee elasticity. Furthermore, for an M/M/1 queue with service rate μ, assuming a FCFS scheduling discipline, the delivery time reliability, or probability that the time spent in the system is less than some quantity x, $R(p, x)$, can be explicitly calculated, so that

$$1 - R(p, x) = \exp\{-(\mu - D(p, x))x\}$$

So and Song [36] also define c, the cost per unit, and α, the desired delivery level (exogenously determined), and then make the following substitution: $k = -\ln(1 - \alpha)$. This enables them to state the problem in the following way:

$$\begin{aligned} &\text{maximize } (p - c)k_1 p^{-k_2} \beta^{-k_3} \\ &\text{subject to } (\mu - k_1 p^{-k_2} \beta^{-k_3})\beta \geq k, \quad p, \beta \geq 0 \end{aligned}$$

Unfortunately, the objective function is not jointly concave, so there is no straightforward approach to solving this problem. Nevertheless, So and Song [36] analyze the structure of the objective and feasible region, leading to a variety of interesting observations about this model, including

1. Firms with lower cost should select a lower price and a longer delivery lead time than firms with higher cost.
2. If a firm desires a higher service level, it should quote a longer delivery lead time, and reduce its prices.
3. All things being equal, there is a larger profit loss in promising a shorter than optimal delivery lead time than in promising a longer than optimal one, providing they deviate from optimal by the same amount.
4. All things being equal, it is more important for firms with higher unit operating costs to set an optimal lead time than it is for firms with lower unit costs.

Palaka, Erlebacher, and Kropp [37] consider the same model, except that demand is a linear function of lead time:

$$D(p, \beta) = k_1 - k_2 p - k_3 \beta$$

and the model explicitly considers holding costs (h per unit per unit time), and tardiness costs (π_l per unit per unit time). Thus, letting μ be the service rate and λ be the arrival rate, and observing that the probability that the firm does not meet the quoted lead time β is $e^{-(\mu-\lambda)\beta}$ and that $\frac{1}{\mu-\lambda}$ is the expected lateness, the objective function of their model is

$$\lambda(p-c) - \frac{h\lambda}{\mu-\lambda} - \frac{\pi_l\lambda}{\mu-\lambda}e^{-(\mu-\lambda)\beta}$$

Palaka, Erlebacher, and Kropp [37] show that if the cubic equation

$$(k_1 - ck_2 - 2\lambda)(\mu-\lambda)^2 = G\mu$$

where

$$G = k_3 \log\beta + hk_2 + \frac{\pi_l k_2}{x}$$

and

$$x = \max\left\{\frac{1}{1-s}, \frac{k_2\pi_l}{k_3}\right\}$$

has a root on the interval $[0,\mu]$ then the optimal lead time β^* is given as follows:

$$\beta^* = \frac{\log x}{\mu-\lambda^*}$$

where λ^* is the root of the cubic equation above on the interval $[0,\mu]$, and the optimal price can be determined using the following relationship:

$$p^* = \frac{k_1 - \lambda^* - k_3\beta^*}{k_2}$$

Plambeck [38] considers an exponential single server queue with two classes of customers that differ in price and delay sensitivity, patient customers and impatient customers. The arrival rates of the two classes of customers are linear functions of price and quoted lead time. The objective is to maximize profit, subject to an asymptotic constraint on lead time performance, which roughly says the likelihood of actual time in the system exceeding quoted lead time is relatively small. Plambeck [38] develops a simple policy: promise immediate delivery to impatient customers and charge them more, and for the patient customers, quote a lead time proportional to the queue length when they arrive at the system. Always process the jobs of impatient customers when they are available, and within classes, process jobs FCFS. This policy is shown to be asymptotically optimal for the system. In addition, Plambeck [38] modifies the policy to make it "incentive compatible" when the class of arriving customers is unknown, and show that this modified policy is asymptotically optimal among all incentive compatible policies.

20.6 Parallel Machine Models

Cheng [39] considers the parallel machine common due date model, where the objective is a function of earliness, tardiness, and the (single) due date. Each of these three components is given a weight that does not differ by job. The overall objective is thus $\sum_{i=1}^{N} \pi^d d + \pi^e E^i + \pi^t T_i$, and the model is therefore $Pm \,|\, d^{opt} \,|\, \sum_{i\in N} \pi^d d + \pi^e E_i + \pi^t T_i$.

Cheng [39] generalizes Equation (20.3), and by taking the derivative of the cost function with respect to the due date, concludes that there is an optimal due date $C_{[r]}$, where $r = \lceil N\frac{\pi^t-\pi^d}{\pi^e+\pi^t}\rceil$.

Of course, this observation is not immediately useful, as the sequence and assignment of machines is not obvious. De, Ghosh, and Wells [40] observe that although the sequences on each machine should not be interrupted by idle times after they start, they should start at different times. Once jobs are assigned to machines, the optimal machine value r can be determined as described earlier for the single machine case, and thus for a given d^*, the starting time of the sequence can easily be determined.

Unfortunately, the problem is still NP-hard, but De, Ghosh, and Wells [40] propose an optimal algorithm based on enumerating machine assignments and V-shaped schedules.

Sundararaghavan and Ahmed [41] generalize Kanet's [9] results for the common due date weighted sum of absolute deviations problem to the parallel machine case. Recall that in this case, any due date larger than a minimum will optimize the problem. Sundararaghavan and Ahmed [41] generalizes the optimal scheduling algorithm of Kanet [9] as follows.

First, observe that the algorithm of Kanet [9] should be used to schedule jobs on each of the parallel machines once jobs are allocated to machines. Furthermore, it is easy to show (by contradiction) that in any optimal schedule, for any two machines i and j, if n_i is the number of jobs assigned to machine i and n_j is the number of jobs assigned to machine j, then

$$| n_i - n_j | \leq 1$$

Now, to optimally allocate jobs to machines, jobs should be assigned from largest to smallest. Arbitrarily sequence the machines, and assign the m largest jobs one to each machine.

If there are fewer than m jobs remaining, assign one job each to a different machine. If there are between m and $2m - 1$ jobs remaining, assign the next m jobs one to each machine, and then assign the remaining jobs each to a different machine. If there are $2m$ or more jobs remaining, assign two jobs each to every machine, observe the number of jobs remaining to be assigned, and repeat the approach described in this paragraph.

Once jobs are assigned, sequence and schedule the jobs on each machine using Kanet's [9] approach.

Sundararaghavan and Ahmed [41] sketch a proof that this algorithm solves this problem by expressing the problem as an integer programming problem, and observing intuitively that this approach will minimize the objective function and meet the constraints.

20.7 Jobshop and Flowshop Models

Observe that the queuing-based approach of Seidmann and Smith [29] described in Section 20.5.2.3 is not restricted to a single machine — indeed, this approach only requires the distribution of time that jobs will spend in the system. Shanthikumar and Sumita [42] extend this approach to a variety of job-shop models. They consider models of job shops with one machine in each center, where the following assumptions are made:

1. Jobs arrivals to the system form a Poisson process.
2. Each job consists of a series of operations, each performed by only one machine.
3. The processing times of all jobs at a specific machine are finite, iid, and can be determined before processing starts.
4. Jobs can wait between machines.
5. Jobs are processed without preemption.
6. Machines are continuously available.
7. Jobs are processed on only one machine at a time.

Specifically, Shanthikumar and Sumita [42] consider an open queuing model of a job shop, with M machines. Jobs arrive at the system according to a Poisson process, and each job must first be processed on machine i with some specified probability q_i. Sequencing at machines is according to one of a variety of possible dispatch rules, including FCFS, SPT, and a variety of rules based on separating jobs into different priority classes, and then applying FCFS or SPT within the classes. When a job completes processing at its first machine i, it proceeds to machine j with probability p_{ij}. Note that $p_{ii} = 0 \ \forall i \in M$, and after completing processing, a job departs from the shop with probability

$$1 - \sum_{i=1}^{M} p_{ij}$$

Clearly, this approach can be used to model a variety of shop structures. For example, for a flowshop, $q_1 = 1, q_i = 0, 1 < i \leq M, p_{i,i+1} = 1, 1 \leq i \leq M - 1$, and all other probabilities are zero.

Shanthikumar and Sumita [42] develop approximations for the distribution of time that jobs spend in the system, by extending the results of Shanthikumar and Buzacott [43] for analysis of open queuing systems (who in turn extended the seminal results of Jackson [44]), and Shanthikumar and Buzacott [45], who develop approximations for the mean and standard deviation of time that jobs spend in the system. In particular, Shanthikumar and Sumita [42] characterize the expected value of N_i, the number of times a job will return to machine i, as well as the covariance $\text{Cov}(N_i, N_j)$, along with the expected value and variance of S_i, the service time of an arbitrary job at machine i. They then calculate a quantity called the service index of the job shop I_s,

$$I_s = \frac{\sum_{i=1}^{M} E(N_i)\text{Var}(S_I) + \sum_{i=1}^{M}\sum_{j=1}^{M} \text{Cov}(N_i, N_j)E(S_i)E(S_j)}{\left\{\sum_{i=1}^{M} E(N_i)E(S_i)\right\}^2}$$

Shanthikumar and Sumita [42] show that if $I_s \cong 1$, the distribution of time spent in the shop is closely approximated by an exponential distribution, if $I_s \ll 1$, the distribution is best approximated by a generalized Erlang distribution, and if $I_s \gg 1$, the distribution is best approximated by a hyperexponential distribution. They then apply the results of Seidmann and Smith [29] to quote CON due dates to minimize the expected cost of the system.

20.8 Other Models

There are a variety of other papers that include elements of lead time quotation and sequencing, along with other problem features. For example, Easton and Moodie [46] develop a model of appropriate bidding strategies for make-to-order firms, where a job bid consists of a price and a promised delivery lead time. This model accounts for *contingent orders* — other outstanding bids placed by same firm.

Elhafsi and Rolland [47] consider the problem faced by a firm that must assign incoming orders to a set of workstations, each of which can process all of the orders, but at different rates. Processing times and machine availability is random, and the authors develop a tool that helps management determine the cost and completion times for a variety of different assignments of jobs to machines. In this way, the tool can be used to estimate a reliable lead time, and to estimate the cost, and thus help with negotiating of pricing, for different lead times. Several models are presented, and a variety of solution approaches are developed for these models.

A variety of authors explore the relationship between inventory levels and quoted lead times in assemble-to-order systems. In these systems, components are produced and held in inventory, and arriving orders are filled by assembling various combinations of components held in inventory. The service level in these systems is usually modeled as the fraction of orders that are filled within the target lead time. Clearly, for a given target service level, there is a trade-off between component inventory levels and quoted lead time. For details, see Glasserman and Wang [48], Lu, Song, and Yao [49], and the references therein.

20.9 Conclusions

Although due date quotation research has been ongoing since at least the 1970s and significant advances have been made, analytical results have for the most part been limited to relatively simple models. In contrast to simulation-based research, much of the analytical research has focused on static models, common due date models, single machine models, and simple queuing systems with Poisson arrivals and exponential service times. In addition, many researchers focus on systems with simple due date quotation rules. In many cases, these models and rules do not sufficiently capture the important characteristics of real-world systems. Consequently, there are many interesting opportunities to develop advanced modeling and analysis techniques that capture more of the system features frequently seen in practice, such as many machines and jobs, a variety of operating characteristics, and complex arrival and production processes.

References

[1] Keskinocak, P. and S. Tayur (2003), Due date management policies. In D. Simchi-Levi, D. Wu, and M. Shen (editors), *Sin Supply Chain Analysis in the eBusiness Era*, Kluwer.

[2] Cheng, T.C.E. and M.C. Gupta (1989), Survey of scheduling research involving due date determination decisions. *European Journal of Operational Resaerch* **38**, pp. 156–166.

[3] Baker, K.R. and G. Scudder (1989), On the assignment of optimal due dates. *Journal of the Operational Research Society* **40**(1), pp. 93–95.

[4] Quaddus, M.A.(1987), A generalized model of optimal due date assigment by linear programming. *Journal of the Operational Research Society* **38**(4), pp. 353–359.

[5] Raghavachari, M. (1986), A V-shape property of optimal schedule of jobs about a common due date. *European Journal of Operational Research* **23**, pp. 401–402.

[6] Hall, N.G. and M.E. Posner (1991), Earliness-tardiness scheduling problems I: weighted deviation of completion times about a common due date. *Operations Research* **39**, pp. 836–846.

[7] Panwalker, S.S., M.L. Smith, and A. Seidmann (1982), Common due date assignment to minimize total penalty for the one machine scheduling problem. *Operations Research* **30**(2), pp. 391–399.

[8] Cheng, T.C.E. (1988), An alternative proof of optimality for the common due date assignment problem. *European Journal of Operational Research* **37**, pp. 250–253.

[9] Kanet, J.J. (1981), Minimizing the average deviation of job completion times about a common due date. *Naval Research Logistics Quarterly* **28**, pp. 642–651.

[10] Quaddus, M.A. (1987), On the duality approach to optimal due date determination and sequencing in a job shop. *Engineering Optimization* **10**, pp. 271–278.

[11] Bagchi, U. Y.-L. Chan, and R. Sullivan (1987), Minimizing absolute and squared deviations of completion times with different earliness and tardiness penalties and a common due date. *Naval Research Logistics* **34**, pp. 739–751.

[12] Cheng, T.C.E. (1987), An algorithm for the CON due date determination and sequencing problem. *Computers and Operations Research* **14**(6), pp. 537–542.

[13] Cheng, T.C.E. (1989), On a generalized optimal common due date assignment problem. *Engineering Optimization* **15**, pp. 113–119.

[14] Cheng, T.C.E. (1989), Optimal assignment of slack due dates and sequencing in a single-machine shop. *Applied Mathematics Letters* **2**(4), pp. 333–335.

[15] Gordon, V.S. (1993), A note on optimal assignment of slack due dates in single machine scheduling. *European Journal of Operational Research* **70**, pp. 311–315.

[16] Seidmann, A, S.S. Panwalker, and M.L. Smith (1981), Optimal assignment of due dates for a single processor scheduling problem. *International Journal of Production Research* **19**, pp. 393–399.

[17] Karacapilidis, N.I. and C.P. Pappis (1995), Form similarties of the CON and SLK due date determination methods. *Journal of the Operational Research Society* **46**(6), pp. 762–770.

[18] Baker, K.R. and J.W.M. Bertrand (1981), A comparison of due date selection rules. *AIIE Transactions* **13**(2), pp. 123–131.

[19] Qi, X. and F-S Tu (1998), Scheduling a single machine to minimize earliness penalties subject to the SLK due date determination method. *European Journal of Operational Research* **105**, pp. 502–508.

[20] Gordon, V.S. and V.A. Strusevich (1998), Earliness penalties on a single machine subject to precendence constraints: SLK due date assignment. *Computers & Operations Research* **26**, pp. 157–177.

[21] Cheng, T.C.E. (1984), Optimal due date determination and sequencing of n jobs on a single machine. *Journal of the Operational Research Society* **35**(5), pp. 433–437.

[22] Cheng, T.C.E. (1988), Optimal due date assignment for a single machine problem with random processing times. *International Journal of Production Research* **17**(8), pp. 1139–1144.

[23] Baker, K.R., E.L. Lawler, J.K. Lenstra, and A.H.G. Rinnooy Kan (1980), Preemptive Scheduling of a Single Machine to Minimize Maximum Cost Subject to Release Dates and Precedence Constraints. Report BW 128, Mathematisch Centrum, Amsterdam.

[24] Baker, K.R., E.L. Lawler, J.K.Lenstra, and A.H.G.Rinnooy Kan (1983), Preemptive scheduling of a single machine to minimize maximum cost subject to release dates and precedence constraints. *Operations Research* **31**, pp. 381–386.
[25] Gordon, V.S. and V.S. Tanaev (1983), On minimax single machine scheduling problems. *Transactions of the Academie of Sciences of the BSSR* **3**, pp. 3–9. (in Russian)
[26] Cheng, T.C.E. and V.S.Gordon (1994), Optimal assignment of due dates for preemptive single-machine scheduling. *Mathematical and Computer Modelling* **20**(2), pp. 33–40.
[27] Kaminsky, P.M. and Z-H Lee (2003), Asymptotically Optimal Algorithms for Reliable Due Date Scheduling. Working paper, University of California, Berkeley.
[28] Keskinocak, P., R. Ravi, and S. Tayur (2001), Scheduling and reliable lead-time quotation for orders with availability intervals and lead-time sensitive revenues. *Management Science* **47**(2), pp. 264–279.
[29] Seidmann, A. and M.L. Smith (1981), Due date assignment for production systems. *Management Science* **27**(5), pp. 571–581.
[30] Dellaert, N. (1991), Due date setting and production control. *International Journal of Production Economics* **23**, pp. 59–67.
[31] Duenyas, I. and W.J. Hopp (1995), Quoting customer lead times. *Management Science* **41**(1), pp. 43–57.
[32] Duenyas, I. (1995), Single facility due date setting multiple customer classes. *Management Science* **41**(4), pp. 608–619.
[33] Wein, L.M. (1991), Due date setting and priority sequencing in a multiclass M/G/I queue. *Management Science* **37**(7), pp. 834–850.
[34] Spearman, M.L. and R.Q. Zhang (1999), Optimal lead time policies. *Management Science* **45**(2), pp. 290–295.
[35] Hopp, W.J. and M.L. Roof Sturgis (2000), Quoting manufacturing due dates subject to a service level constraint. *IIE Transactions* **32**, pp. 771–784.
[36] So, K.C. and J-S Song (1998), Price, delivery time guarantees, and capacity selection. *European Journal of Operational Research* **111**, pp. 28–49.
[37] Palaka, K., S. Erlebacher, and D.H. Kropp (1998), Lead time setting, capacity utilization, and pricing decisions under lead-time dependent demand. *IIE Transactions* **30**, pp. 151–163.
[38] Plambeck, E. (2000), Pricing, Leadtime Quotation and Scheduling in a Queue with Heterogeneous Customers. Working paper, Stanford University.
[39] Cheng, T.C.E. (1989), A heuristic for common due date assignment and job scheduling on parallel machines. *Journal of the Operational Research Society* **40**(12), pp. 1129–1135.
[40] De, Prabuddha, J.B. Ghosh, and C.E.Wells (1991), On the multiple-machine extension to a common due date assignment and scheduling problem. *Journal of the Operational Research Society* **42**(5), pp. 419–422.
[41] Sundararaghavan, P.S. and M. Ahmed (1984), Minimizing the sum of absolute lateness in single-machine and multimachine scheduling. *Naval Research Logistics Quarterly* **31**, pp. 325–333.
[42] Shanthikumar, J.G. and U. Sumita (1988), Approximations for the time spent in a dynamic job shop, with applications to due date assignment. *International Journal of Production Research* **26**(8), pp. 1329–1352.
[43] Shanthikumar, J.G. and J.A. Buzacott (1981), Open queueing network models of dynamic job shops. *International Journal of Production Research* **19**(3), pp. 255–266.
[44] Jackson, J.R. (1963), Job shop-like queuing systems. *Management Science* **10**, pp. 131–142.
[45] Shanthikumar, J.G. and J.A. Buzacott (1984), The time spent in a dynamic job shop. *European-Journal of Operational Research* **17**, pp. 215–226.
[46] Easton, F.F. and D.R. Moodie (1999), Pricing and lead time decisions for make-to-order firms with contingent orders. *European Journal of Operational Research* **116**, pp. 305–318.
[47] Elhafsi, M. and E. Rolland (1999), Negotiating price/delivery date in a stochastic manufacturing environment. *IIE Transactions* **31**, pp. 255–270.

[48] Glasserman, P. and Y. Wang. (1998), Leadtime-inventory trade-offs in assemble-to-order systems. *Operations Research* **46**(6), pp. 858–871.

[49] Lu, Yingdong, J-S Song, and D.D. Yao (2003), Order fill rate, leadtime variability, and advance demand information in an assemble-to-order system. *Operations Research* **51**(2), pp. 292–308.

[50] Cheng, T.C.E. (1987), Optimal total-work-content-power due date determination and sequencing. *Computers and Mathematical Applications* **148**, pp. 579–582.

21 Scheduling with Due Date Assignment

Valery S. Gordon
National Academy of Sciences of Belarus

Jean-Marie Proth
INRIA-Lorraine

Vitaly A. Strusevich
University of Greenwich

21.1 Introduction

The introduction of just-in-time concepts in inventory management according to which jobs are to be completed as close to their due dates as possible, has stimulated research on the scheduling problems with due date assignment and nonregular objective functions dependent on earliness and tardiness costs. In the relevant models, the decision maker not only takes sequencing and scheduling decisions, but also determines the optimal values of controllable parameters that specify the due dates. The controllable parameters usually contribute into the objective function, therefore the optimal choice of their values reflects a possible cooperation between various participants of a production process or a supply chain (a supplier–manufacturer link or a manufacturer–customer link).

In this review, we aim at providing a unified framework of the due date assignment in the deterministic scheduling problems. We consider only the static production settings in which a fixed set of jobs is available for processing, as opposed to the dynamic production settings where jobs continuously arrive in the system and should be scheduled online. We restrict our consideration to the analytical methods, leaving simulation techniques beyond the scope of this review. The survey papers by Cheng and Gupta [1], Baker and Scudder [2], Gordon, Proth, and Chu [3,4] contain extensive lists of references for the scheduling problems with due date assignment.

In the literature, three major due date assignment models can be found:

1. Common due date assignment, in which a due date common to all jobs has to be determined.
2. Assignment of due dates dependent on the processing times (assignment policies with common slack due dates, total-work-content or processing-plus-wait due dates).
3. Assignment of due dates from a given set of values to the jobs (e.g., depending on the positions of the jobs in a schedule).

The subsequent sections of this chapter address these three modes of assignment, respectively. Most of the discussed results are related either to a single machine or to parallel machine models. This is

1-58488-397-9/$0.00+$1.50

due to the fact that the due date assignment problems for more complicated types of production environment, e.g., multistage or shop scheduling systems, have been studied rather scarcely.

The following notation is used in the rest of the chapter. Let $N = \{1, 2, \ldots, n\}$ be a set of jobs to be processed with given release dates r_j and processing times that are denoted either by p_j in the case of single-operation model (e.g., a single machine) or by p_{ij} in the case of multioperation models with m machines, $i = 1, \ldots, m$, $j = 1, \ldots, n$. For a certain schedule s, let d_j, C_j, E_j, and T_j denote respectively the due date, the completion time, the earliness, and the tardiness of job j, where

$$E_j = \max\{0, d_j - C_j\}, \qquad T_j = \max\{0, C_j - d_j\} \tag{21.1}$$

Lateness L_j of job j is defined as $L_j = C_j - d_j$. Hence, $E_j = \max\{0, -L_j\}$ and $T_j = \max\{0, L_j\}$.

21.2 Common Due Date

In this section, we discuss the simplest due date assignment model in which the jobs receive the same due date, *common* to all jobs, i.e., $d_j = d$, $j = 1, \ldots, n$. This method of due date assignment is also known in scheduling literature as common due date or *CON model*, where CON stands for *constant* flow allowance. Such a model corresponds to a system in which, for some reason (appointment, technical constraints, etc.), several tasks are to be completed at the same time: for instance, in a shop several jobs form an order by a single customer, or the components of the product should be ready by the time of assembly. In chemical and food production, the common due date model applies if some of the involved substances or components have a limited life span (a "best before" time), and that determines a due date for the whole mixture or the final product.

In the three-field notation for scheduling problems, we write $d_j := d$ in the middle position to refer to the scheduling problems in which a common due date has to be *assigned*, to be distinguished from notation $d_j = d$ which means that the common due date is *fixed*, i.e., known in advance and not assigned.

21.2.1 Single Machine

Consider first the problem of scheduling n simultaneously available jobs on a single machine without preemption. Let $\sigma = ([1], [2], \ldots, [n])$ be an arbitrary sequence of the jobs, where $[j]$ is the jth job in σ.

21.2.1.1 Panwalkar–Smith–Seidmann Problem

Historically, problem $1 \mid d_j := d \mid \sum(\alpha E_j + \beta T_j + \gamma d)$ with $\alpha, \beta, \gamma \geq 0$ considered by Panwalkar, Smith, and Seidmann [5] is one of the first due date assignment scheduling problems studied in the literature. In this problem, it is required to find both a sequence σ^* of jobs and a common due date d^* which deliver the minimum value to the function

$$f(d, \sigma) = \sum_{j=1}^{n} (\alpha E_j + \beta T_j + \gamma d) \tag{21.2}$$

Consideration in [5] is based on the results of Kanet [6] for minimizing the absolute lateness with respect to a given common due date. Many useful properties of problem $1 \mid d_j := d \mid \sum(\alpha E_j + \beta T_j + \gamma d)$ established in [5] are often observed for other due date assignment problems.

It can be seen that for the optimal due date d^* the inequality $d^* \leq \sum_{j=1}^{n} p_j$ holds. Given a sequence σ of jobs, it suffices to consider schedules in which the jobs are processed according to σ starting at time zero and without intermediate idle time. Furthermore, if $\gamma \geq \beta$, an optimal sequence σ^* follows SPT order and $d^* = 0$.

Assume now that $\gamma < \beta$.

Property 21.1

For problem $1 \,|\, d_j := d \,|\, \sum(\alpha E_j + \beta T_j + \gamma d)$ *with* $\gamma < \beta$*, given any fixed sequence* σ*, there exists an optimal value of* d *that is equal to the completion time of one of the jobs.*

Indeed, if for a schedule s associated with sequence σ we have that $d < C_{[1]}$, we can decrease the total penalty (21.2) by $n(\gamma - \beta)(C_{[1]} - d)$ by increasing the due date to $C_{[1]}$. If $C_{[j]} < d < C_{[j+1]}$, $j = 1, \ldots, n-1$, with $x = d - C_{[j]}$ and $y = C_{[j+1]} - d$, we can set $d = C_{[j+1]}$ and decrease the total penalty by $y(n(\beta-\gamma) - j(\alpha+\beta))$ (if $n(\beta-\gamma) - j(\alpha+\beta) \geq 0$), or set $d = C_{[j]}$ and decrease the penalty by $-x(n(\beta-\gamma) - j(\alpha+\beta))$ (if $n(\beta-\gamma) - j(\alpha+\beta) < 0$).

Property 21.2

For problem $1 \,|\, d_j := d \,|\, \sum(\alpha E_j + \beta T_j + \gamma d)$ *with* $\gamma < \beta$*, given a fixed sequence* σ*, there exists an optimal due date equal to* $C_{[K]}$*, where* K *is the smallest integer greater than or equal to* $n(\beta-\gamma)/(\alpha+\beta)$*, i.e.,* $K = \lceil n(\beta-\gamma)/(\alpha+\beta) \rceil$*.*

Indeed, if an optimal due date coincides with the completion time of the job in position K, then the right shift of the due date from $C_{[K]}$ to $C_{[K]}+\varepsilon$ causes an increase of the total penalty by $\varepsilon(n\gamma + K\alpha - (n-K)\beta) \geq 0$ and we have $K \geq n(\beta-\gamma)/(\alpha+\beta)$. The left shift of the due date results in $K - 1 \leq n(\beta-\gamma)/(\alpha+\beta)$. Since K is an integer, Property 21.2 follows immediately.

Thus, the total penalty is equal to $f(C_{[K]}, \sigma)$ and, since $C_{[K]} = p_{[1]} + p_{[2]} + \cdots + p_{[K]}$, it follows that

$$f(C_{[K]}, \sigma) = \sum_{j=1}^{K} (n\gamma + (j-1)\alpha) p_{[j]} + \sum_{j=K+1}^{n} \beta(n+1-j) p_{[j]} = \sum_{j=1}^{n} a_j p_{[j]}$$

where a_j is a positional penalty (or weight) of job $[j]$ and is equal to $n\gamma + (j-1)\alpha$ for $j \leq K$ and to $\beta(n+1-j)$ for $j > K$. To find an optimal sequence σ^*, we need to minimize $\sum_{j=1}^{n} a_j p_{[j]}$ over all sequences σ. This can be done in $O(n \log n)$ time by matching the smallest positional penalty to the largest processing time, the second smallest positional penalty to the second largest processing time, and so on. Thus, the problem can be solved by the following procedure.

Algorithm PSS (Panwalkar, Smith, and Seidmann [5])

1. Compute $K = \lceil n(\beta-\gamma)/(\alpha+\beta) \rceil$.
2. Compute $a_j = n\gamma + (j-1)\alpha$ for $j \leq K$ and $a_j = \beta(n+1-j)$ for $j > K$.
3. Find an optimal sequence σ^* by matching the positional weights to the processing times.
4. Set d^* equal to the completion time of the K–th job in sequence σ^*.

In the corresponding optimal schedule s^*, the jobs are processed starting at time zero in accordance with σ^* without intermediate idle time. The sequence generated by Algorithm PSS is V-shaped. A *V-shaped* sequence is such that a subset of jobs sorted in nonincreasing order of p_j (LPT order) is followed by the remaining jobs in nondecreasing order of p_j (SPT order).

If $K = n(\beta-\gamma)/(\alpha+\beta)$, it is not necessary to constrain d^* to be equal to one of the job completion times [7,8]. The optimal due date d^* can be set equal to any value in the interval $[C_{[K]}, C_{[K+1]})$, so that there are infinitely many optimal due dates.

If the jobs become available at their release dates r_j, we obtain problem $1 \,|\, r_j, d_j := d \,|\, \sum(\alpha E_j + \beta T_j + \gamma d)$ considered by Cheng, Chen, and Shakhlevich [9]. This problem is strongly NP-hard even if $\alpha = 0$ since optimal sequencing of late jobs is equivalent to problem $1 \,|\, r_j \,|\, \sum C_j$, which is strongly NP-hard [10]. If $\beta = 0$, then $d^* = 0$ and any feasible schedule is optimal. If $\gamma = 0$, an optimal sequence can be found by Algorithm PSS, and the corresponding schedule starts at time $R = \max_{j \in N}\{r_j\}$. For a fixed sequence of jobs, a polynomial-time algorithm is given that finds the optimal due date and determines an optimal schedule. If all processing times are equal, then an optimal schedule corresponds to the sequence in which the jobs arrive. Approximation algorithms are proposed for the general case (the worst-case ratio is $1 + R$) and for several special cases (e.g., for $\alpha = 0$ the worst-case ratio is 2).

21.2.1.2 Mean Absolute Deviation Problem

If $\alpha = \beta = 1$ and $\gamma = 0$ in (21.2), we arrive at the problem of minimizing the mean absolute lateness with the objective function $\sum_{j=1}^{n}(E_j + T_j) = \sum_{j=1}^{n}|C_j - d|$ or, equivalently, at the problem of minimizing the Mean Absolute Deviation (MAD) of completion times about a common due date. From Property 21.2, an optimal due date in this case is $d^* = C_{[K]}$, where $K = n/2$ if n is even, and $K = (n+1)/2$ if n is odd. For MAD problem, the positional penalty a_j for the job in position j is equal to $j-1$ if $j \leq K$ and to $n+1-j$ if $j > K$. Therefore, the penalties are the same for positions 2 and n, for 3 and $n-1$, for 4 and $n-2$, and so on. In this case, there are two ways of matching the positional weights to the processing times and hence of assigning jobs to the positions of an optimal sequence. Thus, there may be different optimal due dates and different optimal sequences, see [11–13]. If all processing times are distinct, the total number of optimal sequences is 2^r, where $r = n/2$ if n is even, and $r = (n-1)/2$ if n is odd. Obviously, increasing d^* by a given value $\varepsilon > 0$ and increasing the starting times of the jobs in the optimal sequence by the same value ε leads to another optimal solution.

Finding the smallest value of the optimal due date in MAD problem is considered by Bagchi, Chang, and Sullivan [14]. Assuming that the jobs are numbered in SPT order, we define the ordered sets A and B as follows: $B = \{n, n-2, \ldots, 1\}$, $A = \{2, 4, \ldots, n-1\}$ if n is odd, and $B = \{n, n-2, \ldots, 2\}$, $A = \{1, 3, \ldots, n-1\}$ if n is even. Let $\Delta = \sum_{j \in B} p_j$. Then the minimal value of the optimal due date d^* is equal to Δ. In other words, this is the earliest completion time of the Kth job in an optimal schedule and the minimum sum of the processing times of nontardy jobs.

The complexity of MAD problem changes drastically if an acceptable value of the due date cannot be greater than D such that $D < \Delta$. In this case, Properties 21.1 and 21.2 do not hold any longer, although there still exists a V-shaped optimal sequence. Such a problem is called the *restricted* common due date problem as opposed to the *unrestricted* problems in which the due date d can take any value.

The restricted MAD problem is NP-hard even with a fixed common due date d, see [15,16]. For this problem, there exists an optimal schedule in which either some job completes at the due date, or the schedule starts at time zero. An optimal schedule may contain a *straddling* job, which starts before d and is completed after d.

21.2.1.3 Weighted Sum of Absolute Deviations Problem

If in (21.2) $\gamma = 0$ but α and β are not equal to 1, we arrive at the problem of minimizing the weighted sum of absolute deviations (WSAD) of the completion times with respect to a common due date: $\alpha \sum_{j \in E}|C_j - d| + \beta \sum_{j \in T}|C_j - d| = \sum_{j=1}^{n}(\alpha E_j + \beta T_j)$, where $E = \{j \mid C_j \leq d\}$ is a set of jobs that are not tardy and $T = \{j \mid C_j > d\}$ is a set of jobs that are tardy.

Similar to MAD problem, WSAD problem may have different optimal due dates, and an infinite number of optimal solutions can be obtained by increasing the starting times of the jobs and the optimal due date by the same value. An $O(n \log n)$ algorithm different from the matching procedure of Algorithm PSS is proposed in [14]. This algorithm yields a unique sequence with the smallest due date value among the optimal due dates.

Similar to the restricted MAD problem, the restricted WSAD problem is NP-hard.

21.2.1.4 Total Weighted Earliness and Tardiness Problem

The total weighted earliness and tardiness (TWET) problem is a natural generalization of MAD and WSAD problems: find a sequence σ^* and a common due date d^* that minimize the objective function $\sum_{j=1}^{n}(\alpha_j E_j + \beta_j T_j)$, where α_j and β_j are the unit earliness and tardiness penalties (weights) for job j, respectively. In the symmetric TWET problem, we have $\alpha_j = \beta_j$, $j = 1, \ldots, n$.

For the fixed common due date, the symmetric TWET problem is NP-hard even in the unrestricted setting [17], but solvable by pseudopolynomial-time dynamic programming (DP) algorithms, see [17–19]. The problem also admits a fully polynomial-time approximation scheme (FPTAS) [20].

For the asymmetric TWET problem with due date assignment, the following property holds, see [11,21,22].

Property 21.3

For the unrestricted TWET problem, there exists an optimal schedule s^* such that:

(a) *s^* is V-shaped in the sense that the jobs in set E (nontardy jobs) are sequenced in nonincreasing order of the ratio p_j/α_j and are followed by the jobs of set T (tardy jobs) in nondecreasing order of the ratio p_j/β_j*
(b) *in s^* the optimal due date coincides with the completion time of the last nontardy job in an optimal sequence*
(c) *the inequality $\sum_{j\in T}\beta_j \le \sum_{j\in E}\alpha_j$ holds*

Property 21.3(a) is valid also for the restricted problem (excluding the straddling job). However, Properties 21.3(b) and 21.3(c) do not hold in the restricted case.

For the unrestricted symmetric TWET problem, De, Ghosh, and Wells [18] develop $O(n\sum p_j)$ and $O(n\sum \alpha_j)$ DP algorithms. Cheng [23] presents an $O(n^{1/2}2^n)$ local search algorithm, while Hao et al. [24] design a tabu search algorithm. For $\alpha_j = \beta_j = \lambda p_j$, $\lambda > 0$, $j = 1,\ldots,n$, Karacapilidis and Pappis [25] present an algorithm for finding $r!(n-r)!$ alternative optimal sequences, where r is the position of the job with the completion time equal to the optimal due date.

For the unrestricted asymmetric TWET problem, branching procedures [26,27] and heuristic algorithms [26–28] are proposed. A problem generator for obtaining the benchmark instances of TWET problem (by changing a parameter to get more or less restrictive common due date) is developed in [29].

21.2.1.5 Almost Common Due Date

Several problems with different due dates appear to be close to the common due date problems. Hoogeveen and van de Velde [30] propose an $O(n^2)$ DP algorithm for solving the unrestricted WSAD problem with almost equal due dates, i.e., the due dates of the jobs are different but lie between a large constant d and $d+p_j$ for each job j. Seidmann, Panwalkar, and Smith [31] consider a problem in which job j can be assigned any due date d_j, but there is a given lead time D. The problem is to find the optimal due dates d_j^*, $j = 1,\ldots,n$, and an optimal sequence σ^* of jobs that minimize the total penalty $\sum_{j=1}^{n}(\alpha E_j+\beta T_j+\gamma \max\{0, d_j - D\})$, where E_j and T_j are given by (21.1) and $\gamma > 0$. In an optimal schedule, the jobs are processed in the order of their numbering by SPT rule, and the optimal due dates are given either by $d_j^* = \sum_{i=1}^{j} p_i$ if $\gamma \le \beta$, or by $d_j^* = \min\{D, \sum_{i=1}^{j} p_i\}$, otherwise. Cheng [32] considers a due date d_j for job j to be defined as $d_j = S_j+d$, where S_j is the starting time of job j, and d is a constant flow allowance. The problem consists in finding a sequence $\sigma = ([1],[2],\ldots,[n])$ and a due date allowance that minimize $\sum_{j=1}^{n}(\alpha_{[j]}\,|C_{[j]} - d_{[j]}| + \gamma d)$, where $0 \le \alpha_j \le 1$, $\sum_{j=1}^{n}\alpha_j = 1$. A linear programming (LP) formulation is proposed for this problem and, by considering the dual problem, it is shown that the optimal flow allowance is independent of the job sequence and equal either to one of the job processing times if $0 \le \gamma < 1/n$ or to $d^* = 0$ if $\gamma \ge 1/n$.

21.2.1.6 The Weighted Number of Late Jobs

Another problem that involves the lead time D is formulated by De, Ghosh, and Wells [33]: find a due date d, common to all jobs, and a sequence σ that minimize the objective function $\sum_{j\in T} w_j + \gamma \max\{0, d - D\}$, where $\gamma > 0$, $w_j > 0$, $j = 1,\ldots,n$. The order in which the jobs are sequenced in each of the sets E (nontardy jobs) and T is immaterial for this problem. The problem is called restricted if $D < M$ and unrestricted otherwise, where M is the sum of the processing times of the jobs j with $w_j/p_j > \gamma$. A polynomial-time algorithm is proposed for the unrestricted problem while the restricted problem is shown to be NP-hard. The latter problem is formulated as a knapsack problem, and a heuristic based on the continuous relaxation of the knapsack problem is presented.

Cheng [34] considers the due date assignment problem to minimize the objective function $\sum_{j=1}^{n}(wU_j + \gamma d)$, where w is the penalty cost for finishing a job late, and $U_j = 1$ if $C_j > d$ or $U_j = 0$, otherwise. An SPT sequence is shown to be optimal while the optimal due date coincides with the completion time of Kth job such that the inequality $p_K \leq w/n\gamma \leq p_{K+1}$ holds. The due date assignment problem with a more general objective function $\sum_{j\in T} w_j + \sum_{j\in E} h(d - C_j) + g(d)$ is considered by Kahlbacher and Cheng [35]. Here, g and h are nondecreasing functions, $w_j > 0$, $j = 1, \ldots, n$. The problem is shown to be NP-hard even if (a) $h(x) = 0$ and $g(d)$ is an arbitrary function, or (b) $g(d) = 0$ and $h(x)$ is an arbitrary function. Polynomial-time algorithms are proposed for the following cases: (a) $w_j = w$, $j = 1, \ldots, n$ (an $O(n \log n)$ algorithm), (b) $h(x) = 0$ and $g(d) = \gamma d$, $\gamma \geq 0$ (an $O(n)$ algorithm), (c) both $h(x)$ and $g(d)$ are linear (an $O(n^4)$ algorithm).

A modification of TWET problem with the objective function $\sum_{j=1}^{n}(\alpha_j E_j + \beta_j T_j + w_j U_j)$ that incorporates an additional fixed cost for each tardy job is considered in [36]. Under the assumption that $p_i/\alpha_i < p_j/\alpha_j$ implies $p_i/\beta_i < p_j/\beta_j$ for all i and j, pseudopolynomial-time DP algorithms are proposed for both unrestricted and restricted variants of the problem. For the unrestricted problem with constant weights ($\alpha_j = \alpha$, $\beta_j = \beta$, and $w_j = w$ for all j), an $O(n^2)$ time algorithm is proposed.

21.2.1.7 Controllable Processing Times

In many scheduling applications, the processing times are not fixed but can be chosen from a given interval. In scheduling problems with controllable processing times, for each job j we are given a 'standard' processing time $\bar{p}_j$ that can be crashed down to some actual processing time p_j, $\underline{p}_j \leq p_j \leq \bar{p}_j$, where $\underline{p}_j$ denotes the minimum acceptable processing time. This incurs the additional cost $\lambda_j x_j$, where $x_j = \bar{p}_j - p_j$ is the compression amount of job j.

Cheng, Oguz, and Qi [37] generalize the results of [5] to the case of controllable processing times and solve the problem of minimizing the function $\sum_{j=1}^{n}(\alpha E_j + \beta T_j + \gamma d + \lambda_j x_j)$ by formulating it as an assignment problem.

WSAD problem with controllable processing times is considered by Panwalkar and Rajagopalan [38]. The objective is to find the actual processing times, a sequence of jobs, and the smallest due date that minimize the function $\sum_{j=1}^{n}(\alpha E_j + \beta T_j + \lambda_j x_j)$. It is shown that there exists an optimal schedule without partial compression (with actual processing times p_j equal either to $\underline{p}_j$ or to $\bar{p}_j$). The problem is formulated as an assignment problem and solved in $O(n^3)$ time. Biskup and Cheng [39] consider the problem of minimizing the function $\sum_{j=1}^{n}(\alpha E_j + \beta T_j + \theta C_j + \lambda_j x_j)$ with $\theta \geq 0$. Again the problem is reduced to the assignment problem and solved in polynomial time.

Biskup and Jahnke [40] study the model with jointly reducible processing times, i.e., $p_j = x\bar{p}_j$ for all jobs, $0 \leq x \leq x_{\max} < 1$. They consider the problem of minimizing total penalty (21.2) with an extra cost $k(x)$ added to the objective function, where $k(x)$ is a monotone increasing function. Since the processing times of all jobs are reduced by the same proportion, Algorithm PSS leads to the same job sequence, and the new objective function is $\sum_{j=1}^{n} a_j(1-x)p_{[j]} = (1-x)f^*$, where f^* is the objective function value for the problem with standard processing times, and a_j are the positional weights. The designed procedure finds an optimal sequence by algorithm PSS and then determines the optimal reduction of processing times and calculates the optimal due date. A similar approach is applied to minimization of the function $\sum_{j=1}^{n}(wU_j + \gamma d)$, where w is the penalty cost for finishing a job late. These results are further extended to the situation in which $p_j = \bar{p}_j - xv_j$ for all jobs, where $0 \leq x \leq x_{\max} < 1$ and v_j is a given job compression rate [41]; the corresponding algorithms require $O(n^2 \log n)$ time.

The introduction of learning effects to job processing times is considered by Biskup [42] for the problem of minimizing $\sum_{j=1}^{n}(\alpha E_j + \beta T_j + \theta C_j)$. It is assumed that standard processing time p_j of job j is incurred if the job is scheduled first in a sequence. If the job is scheduled in position $r > 1$, its processing time is smaller than p_j because of the learning effect, and is set equal to $p_{jr} = p_j r^l$. Here, $l \leq 0$ is the learning index, defined as the logarithm to the base of 2 of a given learning rate. It is shown that the problem can be solved in $O(n^3)$ time in two steps: first, as an assignment problem which leads to an optimal sequence of the jobs, and then as a problem of finding the optimal due date similarly to the problem without learning effects (since introduction of these effects does not influence the number of early and tardy jobs). To formulate the assignment problem, the binary variables x_{jr} are introduced such that $x_{jr} = 1$ if job j is scheduled in position r, and $x_{jr} = 0$, otherwise. A similar approach is applied by Mosheiov [43] to minimization of the objective function (21.2) with the same learning effect. Mosheiov and Sidney [44] discuss another type of learning effect ($p_{jr} = p_j r^{l_j}$ with a job-dependent negative index l_j) and apply it in a similar way to minimization of (21.2).

21.2.1.8 Mean Squared Deviation and Completion Time Variance Problems

The problem of minimizing the mean squared deviation (MSD) of job completion times about a common due date addresses the situation when only small deviations of the completion times from the due date are acceptable. The objective function $MSD(d,\sigma) = \sum_{j=1}^{n}(C_j - d)^2 = \sum_{j=1}^{n}(E_j^2 + T_j^2)$, being a quadratic penalty function, penalizes larger deviations at a higher rate.

The following property holds for the unrestricted MSD problem [45].

Property 21.4

The optimal due date that minimizes $MSD(d,\sigma)$ for any given sequence σ is equal to the mean completion time $\bar{C} = \frac{1}{n}\sum_{j=1}^{n} C_j$.

As a result, the unrestricted MSD problem is equivalent to the problem of minimizing $\sum_{j=1}^{n}(C_j - \bar{C})^2$, that is, to the completion time variance (CTV) problem considered first in [46]. Since CTV problem is NP-hard [47], the unrestricted MSD problem is also NP-hard. For CTV problem, pseudopolynomial-time DP algorithms are developed in [48,49], while Cai [50] presents a FPTAS. A genetic algorithm is designed in [51], and a family of heuristics for solving CTV problem and the symmetric TWET problem is presented in [19]. Similar to the unrestricted MAD problem, the unrestricted MSD problem is invariant to any feasible shift of a schedule together with the due date. An approach to finding the smallest optimal due date is presented in [45].

Enumerative procedures for the restricted MSD problem and for the restricted problem to minimize the weighted sum of squared deviations $\sum_{j=1}^{n}(\alpha E_j^2 + \beta T_j^2)$ are given in [45] and [14], respectively. The enumerative procedures in [52,53] avoid the unnecessary assumption that a schedule should begin at time zero. For the above problems (as well as for the restricted MAD and WSAD problems), De, Ghosh, and Wells [54] provide a solution methodology of pseudopolynomial time complexity based on DP algorithms. An $O(n\sum p_j)$ algorithm for the restricted variant of the problem with a more general objective function $\sum_{j=1}^{n}(\alpha E_j^c + \beta T_j^c)$ for an arbitrary positive c is presented in [55,56].

21.2.1.9 Arbitrary Earliness--Tardiness Penalties

Cai, Lum, and Chan [57] consider the due date assignment problem to minimize the objective function $\sum_{j\in E'}(e_j + \alpha_j h(L_j)) + \sum_{j\in T}(t_j + \beta_j g(L_j))$, where g and h are arbitrary nondecreasing functions, L_j is the lateness of job j, and $E' = \{j \mid C_j < d\}$ is the set of early jobs. Fixed costs $e_j \geq 0$ and $t_j \geq 0$, respectively, for an early and a tardy job j are introduced into the objective function. Let $S_j = C_j - p_j$ denote the starting time of job j. A schedule is said to be *W-shaped* if $p_j \geq p_k$ for $C_j < C_k < d$, and $p_j \leq p_k$ for $d \leq S_j < S_k$, and moreover, it is possible to have a straddling job u with $S_u < d \leq C_u$ such that $p_u > p_{u'}$ and $p_u > p_{u''}$, where jobs u' and u'' are sequenced immediately before and after job u, respectively. Under the agreeable weight condition (i.e., if $p_i > p_j$ implies that $\alpha_i \leq \alpha_j$ and $\beta_i \leq \beta_j$ for all i, j), the problem has a W-shaped optimum. An $O(n^2\sum p_j)$ DP algorithm is presented that finds an optimal solution in the case of agreeable weights.

For the special case of the problem with $e_j = t_j = 0, \alpha_j = \beta_j = 1, j = 1,\ldots,n$, Kahlbacher [56] gives a pseudopolynomial-time algorithm, while Federgruen and Mosheiov [58] develop a greedy heuristic which requires $O(n^3)$ elementary operations and evaluations of functions g and h. If a nondecreasing function $\gamma(d)$ is added to the objective function, a pseudopolynomial-time algorithm is proposed in [59].

21.2.1.10 Maximum Weighted Absolute Lateness Problem

The maximum weighted absolute lateness (MWAL) problem consists in finding a due date d^* and an associated sequence σ^* that minimize the objective function $\max_{1 \leq j \leq n}\{w_j \mid C_j - d \mid\}$. This problem is NP-hard even for a fixed d, as proved in [60], and polynomially solvable in the case of unit weights [61].

Li and Cheng [60] give a polynomial-time algorithm that solves MWAL problem for a given sequence σ. Let d' be the best possible due date for a given sequence. There exist an early job i and a tardy job k such that $w_i(d' - C_i) = w_k(C_k - d') = \max_{1 \leq j \leq n}\{w_j \mid C_j - d' \mid\}$ since, otherwise, it would exist an $\varepsilon > 0$ such that either $d' + \varepsilon$ or $d' - \varepsilon$ is a better due date. From $w_k(C_k - d') = w_i(d' - C_i)$, it follows that $d' = (w_i C_i + w_k C_k)/(w_i + w_k)$. Comparing $d(i,k) = (w_i C_i + w_k C_k)/(w_i + w_k)$ for all possible i and k, we can find the value of d' in $O(n^2)$ time.

21.2.1.11 Batching

Many problems arising in industry combine three types of decisions: scheduling, batching, and due date assignment. Consider a common due date problem in which the completed jobs have to be delivered to the customer in batches. The jobs with $C_j \leq d$ (early jobs) are delivered at time d in one batch without any delivery cost. Other jobs are delivered in tardy batches with the delivery dates equal to the largest completion time among the jobs in the batch. Each tardy batch incurs a fixed delivery cost θ. Let D_j be the delivery date of job j, and α (or β, respectively) be the lateness (or tardiness) penalty weight, assuming that $\alpha \leq \beta$. Earliness and tardiness of job j are defined, respectively, as $E_j = D_j - C_j$ and $T_j = D_j - d$. Thus, the penalty for an early job j is $\alpha(d - C_j)$. Note that a tardy job may incur both tardiness and earliness penalties if it is not delivered immediately upon its completion. The problem consists in finding a due date d^*, a job sequence σ^*, and a partition of the set of jobs into batches which jointly minimize the objective function $f(\sigma, d, l) = \gamma d + \theta l + \sum_{j=1}^{n}(\alpha E_j + \beta T_j)$, where l is the number of tardy batches. Chen [62] presents an $O(n^5)$ DP algorithm for the unrestricted variant of the problem. A pseudopolynomial-time algorithm for the restricted variant of the problem (with a fixed due date $d < \sum_{j=1}^{n} p_j$ and $\gamma = 0$) is given in [63].

Cheng and Kovalyov [64] consider a problem with f families of jobs, where each family h consists of $q_h \geq 1$ identical jobs with the same processing time $p_h \geq 0$ and the same weight $w_h \geq 0$. Each family can be partitioned into batches with contiguously scheduled jobs, and a setup time s_h is required if a batch of family h follows a batch of another family. The machine cannot process any job while a setup is performed. The problem consists in finding a common due date d, the number of jobs in each batch and a processing order of the batches that minimize $\sum_{h=1}^{f} w_h U_h' + \gamma \max\{0, d - D\}$. Here, U_h' is the number of tardy jobs of family h, and D is the lead time. The problem is shown to be NP-hard in each of the following cases: (a) $w_h = 1$, $p_h = 1$; (b) $w_h = 1$, $s_h = s$; (c) $q_h = 1$, $s_h = 0$; and (d) $q_h = 1$, $p_h = 0$, $h = 1, \ldots, f$. Two DP pseudopolynomial algorithms and a FPTAS are presented. In the following cases the problem is solvable in $O(f \log f)$ time: (a) $w_h = 1$, $q_h = q$; (b) $w_h = 1$, $p_h = p$, $s_h = s$; and (c) $q_h = q$, $p_h = p$, $s_h = s$, $h = 1, \ldots, f$. The problem with $s_h = s$ and $p_h = p$ is shown to be NP-hard [65].

In the batching model with *family setup times*, n jobs are in advance split into f families, and the setup time s_h is required in order to process a job of family h after a job of some other family. For the unrestricted TWET problem with family setup times, Azizoglu and Webster [66] present a branch-and-bound algorithm effective for instances up to 15 jobs and a beam search heuristic, i.e., a truncated branch-and-bound procedure in which the choice of a beam width (that is, a limited number of nodes selected for branching) provides a trade-off between the computation time needed to obtain a solution and its quality. Chen, Li, and Tang [67] consider the due date assignment, family setup times problem to minimize $\alpha \sum_{j=1}^{n} U_j + \gamma d$, where $U_j = 1$ if job j is late and $U_j = 0$, otherwise. They propose polynomial-time algorithms for two versions of the problem: each family is processed as a whole batch, or families are allowed to be split into subbatches.

21.2.1.12 Common Due Window

If a job completes close enough to the due date, it is reasonable to assume that no penalty should be incurred. These situations are handled by introducing a common due window, also known as a tolerance interval. It can be specified either by a pair of the earliest and the latest due dates as its end points, or by a completion time deviation allowance e around a common due date d. Under the assumption that the deviation of the job completion time from the due date is small enough so that at most one job can avoid the penalty (i.e., $2e < \min\{p_j \mid 1 \leq j \leq n\}$), Cheng [68] proposes an $O(n \log n)$ algorithm for finding a due date d^* and a sequence of jobs that minimize $\sum_{j=1}^{n}(E_j U(E_j - e) + T_j U(T_j - e))$, where $U(x-e) = 0$ if $x \leq e$, and $U(x-e) = 1$ otherwise. If n is odd, then $d^* \in [C_{[(n+1)/2]} - e, C_{[(n+1)/2]} + e)]$, while for n even either $d^* = C_{[n/2]} + e$ or $d^* = C_{[(n/2)+1]} - e$, see [69].

For MAD problem with an arbitrary fixed tolerance e, polynomial-time algorithms are designed in [70,71]. Unlike [68], for the model studied in [70], the earliness and tardiness penalties are calculated with respect to the end points of the tolerance interval rather than with respect to the due date. A more general case of MAD problem with nonidentical tolerance intervals for earliness and tardiness is considered in [71]. A similar version of the TWET due window assignment problem is studied by Baker and Scudder [2] who assume that the completion time deviation allowances may be different for different jobs and the due date tolerance interval is relatively small compared to the processing times of the jobs.

If the location of the common due window is a decision variable, $O(n \log n)$ algorithms are known for WSAD problem [72] and for its modification with the additional due window location penalty [73]. A generalized version of the latter problem with controllable processing times to determine both the location and the size of the due window is solvable in $O(n^3)$ time [74].

For the due window location assignment TWET problem, a branch-and-bound algorithm effective for instances up to 25 jobs is proposed in [75]. If the ratios of the processing times to the earliness–tardiness penalties are agreeable, a pseudopolynomial DP algorithm is presented in [76] for the problem with the objective function that includes the sum of weighted earliness-tardiness, the weighted number of early and tardy jobs and the common due window location penalty.

Mosheiov [77] considers the problem of finding the starting time d and the length l of the due window that minimize $\max\{\alpha \max_{1 \leq j \leq n} E_j, \beta \max_{1 \leq j \leq n} T_j, \gamma d, \delta l\}$. The optimal values of d and l are found by simple calculation after constructing the optimal schedule which starts with the largest job at time zero, and has no intermediate idle time (the order of the rest of the jobs is immaterial).

21.2.2 Parallel Machines

We now pass to the problems of scheduling n jobs on m parallel machines. All jobs are available for processing immediately and preemption is not allowed. Let p_{ij} be the processing time of job j on machine i, with $p_{ij} = p_j$ for identical machines, and $p_{ij} = p_j / v_i$ for uniform machines, where v_i is the speed of machine i.

The parallel identical machine version of Panwalkar–Smith–Seidman problem, that is, the problem of finding a schedule s^* and a common due date d^* that minimize the function $f(d,s) = \sum_{j=1}^{n}(\alpha E_j + \beta T_j + \gamma d)$ becomes NP-hard even if $m = 2$, see [78–80]. This is not surprising since the case $\alpha = \beta = 0$ corresponds to the minimum makespan problem $P2| |C_{\max}$. De, Ghosh, and Wells [80] show that the problem is strongly NP-hard for an arbitrary m, and solvable in pseudopolynomial time for a fixed m. The following property is valid for this problem [8,12,80].

Property 21.5

For problem $P \mid d_j := d \mid \sum(\alpha E_j + \beta T_j + \gamma d)$, there exists an optimal schedule with no intermediate idle time on any machine, and $C_j = d^*$ for at least one job j. Moreover, on each machine, the sequence of jobs is V-shaped about the due date, i.e., the early jobs are processed in LPT order and the tardy jobs in SPT order.

Given a schedule s, there may be several due dates that deliver minimum to the objective function. Let $d^*(s)$ denote the smallest among these optimal due dates. This value can be easily computed using the following property similar to Property 21.2 for the single machine case. For a schedule s, let $C_{[K]}$ be the Kth smallest completion time, n_0 be the number of jobs completed at time $C_{[K]}$, and $l = n(\beta - \gamma)/(\alpha + \beta)$.

Property 21.6

For problem $P \mid d_j := d \mid \sum(\alpha E_j + \beta T_j + \gamma d)$, in any given schedule s, we have $d^*(s) = C_{[K]}$, where K satisfies the inequality $l \leq K \leq l + n_0$.

Problem $P \mid d_j := d \mid \sum(\alpha E_j + \beta T_j + \gamma d)$ becomes polynomially solvable if either $p_j = p$, $j = 1, \ldots, n$, (an $O(1)$ algorithm is given in [79]) or $\gamma \geq \beta$ (the problem reduces to $P \mid \mid \sum C_j$). If $\gamma = 0$, the problem becomes WSAD problem, which is shown to be polynomially solvable by Emmons [81]. For identical machines, an $O(n \log n)$ algorithm generates an optimal due date and allows finding alternative optima. For uniform machines, an $O(n \log n)$ algorithm produces a unique optimal solution with the common due date equal to $\max_{1 \leq i \leq m}\{\sum_{j=1}^{n'_i} p_{[ij]}/v_i\}$. Here, $[ij]$ denotes the jth early job on machine i in the optimal schedule, and n'_i is the number of early jobs on machine i. Notice that Emmons was the first who studied due date assignment problems with parallel machines.

For unrelated parallel machines, an $O(n^3)$ algorithm for finding an optimal due date and the associated optimal schedule for $\alpha = \beta = 1$ (MAD problem) is developed in [82] by providing reduction to the transportation problem. This approach can be derived from the algorithm for WSAD problem (without due date assignment) described in [83]. By extending the results in [38], problem $R \mid d_j := d \mid \sum(\alpha E_j + \beta T_j)$ with controllable processing times is also reduced to the transportation problem in [84].

As seen earlier, the problem of finding the optimal due date differs from the problem of finding the smallest value among several optimal due dates. Unlike in the single machine case, these problems differ in complexity in the case of parallel machines. For WSAD problem with the identical parallel machines, finding the smallest optimal due date and the associated schedule is NP-hard even if $m = 2$, and strongly NP-hard for a variable m, see [80]. For MAD problem, a heuristic procedure that is aimed at decreasing the optimal due date and minimizing the makespan among optimal schedules is developed in [82].

For problem $P \mid d_j := d \mid \sum(\alpha E_j + \beta T_j + \gamma d)$, an efficient heuristic is described in [8], however, each machine is required to begin processing at time zero and no idle time is allowed. The NP-hardness of the problem with this requirement is shown in [78]. De, Ghosh, and Wells [85] argue that the zero-start requirement is not needed and characterize some necessary conditions for an optimal schedule. Diamond and Cheng [86] alter the heuristic employed in [8], apply it to the problem without the zero-start requirement and show that it is asymptotically optimal as n approaches infinity while the parameters m, α, β are kept constant. Xiao and Li [87] demonstrate that the worst-case behavior of the heuristic in [86] can be arbitrarily bad, and present an extension of this heuristic that guarantees a worst-case ratio of 3 (reduced to 2 for $\alpha = 0$). For the case of fixed m, a FPTAS is developed [87] using a rounding DP technique. A heuristic procedure for problem $R \mid d_j := d \mid \sum(\alpha E_j + \beta T_j + \gamma d)$ is described in [88]. Biskup and Cheng [89] present an NP-hardness proof for problem $P \mid d_j := d \mid \sum(\alpha E_j + \beta T_j + \theta C_j)$, develop a heuristic algorithm, and show that the problem with equal processing times is polynomially solvable.

A version of WSAD problem for parallel identical machines with the objective $\delta m + \sum_{j=1}^{n}(\alpha E_j + \beta T_j)$, where δ is the penalty for activating a machine, is studied in [90]. In this problem, it is required to determine the number m of active machines, and, for each machine, to find a common due date for the jobs assigned to that machine. An $O(n^2)$ algorithm is based on the idea of matching the positional penalties with the processing times, see Algorithm PSS. In the case of multiple optimal due dates, the algorithm is aimed at minimizing the largest common due date. The algorithm can be extended to the case of uniform machines. The case of the due date that is common for all machines is also discussed.

The symmetric TWET problem $P \mid d_j := d \mid \sum \alpha_j(E_j + T_j)$, which is NP-hard in the ordinary sense in the single machine case, is strongly NP-hard for parallel identical machines in both restricted and unrestricted cases, as proved by Webster [91].

Kahlbacher and Cheng [35] consider problem $P \mid d_j := d \mid \sum_{j \in E} h(d - C_j) + \sum_{j \in T} w_j + g(d)$, where $w_j > 0$ is a weight associated with job j, $j = 1, \ldots, n$; g and h are real-valued, nondecreasing functions such that $g(0) = h(0) = 0$. They propose polynomial-time algorithms for a linear function $h(x)$ and $g(d) = 0$ (the running time is $O(n \log n)$ if w_j are equal, and $O(n^4)$ if w_j are different). The problem is shown to be NP-hard for (a) an externally given due date even if $h(x) = 0$ and all $w_j = 1$ (it may be also regarded as the restricted due date assignment problem), (b) linear $g(d)$ and $h(x) = 0$, and (c) arbitrary $h(x)$ and $g(d) = 0$. The two latter cases are NP-hard even if $w_j = w$.

Problems $P \mid d_j := d \mid \max_{1\leq j\leq n}\{E_j, T_j\}$ and $P \mid d_j := d \mid \max_{1\leq j\leq n} w_j \mid C_j - d \mid$ are shown to be NP-hard [79] and strongly NP-hard [60], respectively. For problem $P \mid d_j := d \mid \max\{\alpha \max_{1\leq j\leq n} E_j, \beta \max_{1\leq j\leq n} T_j, \gamma d\}$, Mosheiov [92] presents a two-step asymptotically optimal heuristic algorithm. In the first step, the jobs are scheduled by an LPT-based procedure, and in the second step, an optimal due date for the obtained schedule is computed by finding the points of intersection of the linear functions. The heuristic is adapted for the problem with due window assignment in [77].

21.3 Due Dates Depending on Processing Times

In this section, we consider various models of due date assignment in which the due dates for the jobs depend on their processing times or release dates. That is not the case in the CON model. Below, we mainly concentrate on the relevant models for single machine scheduling.

In the *SLK due date assignment* model (where SLK stands for *slack*), a flow allowance that reflects equal waiting time or a slack, denoted by q, is assigned to the jobs. Thus, in the general case $d_j = r_j + p_j + q$, or, if all jobs are available simultaneously at time zero, $d_j = p_j + q$, $j = 1, \ldots, n$. In this model, the goal is to find an optimal value of the slack q and an associated optimal schedule with respect to a criterion to be optimized. The situation with an additive common slack arises, for instance, when all jobs have to be delivered to the same customer, or for some reason it is necessary to show 'consistent' decision behavior.

In the *TWK due date assignment* (where TWK stands for *total work content*), the due dates are equal to a multiple of the job processing times, i.e., $d_j = kp_j$. If jobs have release dates r_j, the due dates may be set as $d_j = r_j + kp_j$. The optimal due date assignment consists in finding the value of the common multiplier $k > 0$ and the corresponding schedule that minimize the objective function under consideration. Sometimes, the TWK method includes also a nonnegative exponent c which can be either given or found as a decision parameter, and the due dates are set as $d_j = kp_j^c$ or $d_j = r_j + kp_j^c$ (the so-called TWK-power due dates).

The *PPW due date assignment* (PPW stands for *Processing-Plus-Waiting*) combines the CON, SLK, and TWK due date assignment rules. In this model, the due dates are set as either $d_j = r_j + kp_j + q$ or $d_j = kp_j + q$. Given an objective function, the optimal due date assignment consists in finding the optimal values of the common multiplier k, slack allowance q and the related schedule.

The SLK, TWK, and PPW models reflect the situation when the production facility is itself responsible for assigning due dates, and the decision-maker assigns due dates by estimating job flow times.

21.3.1 SLK Model

For scheduling problems with the SLK due date assignment, in the middle position of the three-field classification notation we write $d_j := r_j + p_j + q$ or $d_j := p_j + q$.

21.3.1.1 Maximum Tardiness Problem

For problem $1 \mid pmtn, prec, r_j, d_j := r_j + p_j + q \mid \gamma q + T_{\max}$ (where $\gamma \geq 0$ is the cost per time unit of slack q, and $T_{\max} = \max_{1\leq j\leq n} T_j$), a schedule that minimizes $\gamma q + T_{\max}$ for any given q can be obtained by an $O(n^2)$ algorithm proposed for problem $1 \mid pmtn, prec, r_j \mid f_{\max}$ by Gordon and Tanaev [93] and Baker et al. [94] (see also [95] for a similar procedure). We refer to this algorithm as Algorithm GT-BLLR. Gordon [96] shows that the schedule obtained by applying Algorithm GT-BLLR is optimal independent of the value of q, and that the optimal slack q^* is calculated as either $q^* = C_l - r_l - p_l$ if $0 < \gamma < 1$, or $q^* = 0$ if $\gamma > 1$. Moreover, if $\gamma = 0$, then q^* may take any value that satisfies the inequality $q^* \geq C_l - r_l - p_l$, while $0 \leq q^* \leq C_l - r_l - p_l$ if $\gamma = 1$. Here, l is the job with the maximal tardiness in the optimal schedule. Note that if a precedence graph is a tree, the optimal solution can be obtained in $O(n \log n)$ time. A special case of the problem with simultaneously available independent jobs has been considered in [97,98].

In this case, the optimal schedule is defined by an SPT sequence σ_{SPT}, and the optimal value of q is obtained by replacing $C_l - r_l - p_l$ with $C_{[n-1]}$ in the above formulas, where $C_{[n-1]}$ is the completion time of the $(n-1)$-th job in σ_{SPT} (see also Section 20.3 in Chapter 20).

21.3.1.2 TWET and MAD Problems

Consider the symmetric TWET problem $1 \mid d_j := p_j + q \mid \sum \alpha_j (E_j + T_j)$, where $\alpha_j > 0$ is a weight of job j. Notice that the objective function can be rewritten as $\sum_{j=1}^{n} \alpha_j \mid C_j - d_j \mid$. For TWET problem, it has been observed in [13,28], that CON and SLK due date assignment models have similar properties that can be derived from an analogy in their LP formulations. Indeed, let $[j]$ be the j-th job in sequence σ, and d be a common due date. Then, the LP formulation for the CON model is given by

$$\begin{aligned} \text{minimize} \quad & \sum_{j=1}^{n} \alpha_{[j]} \mid C_{[j]} - d \mid = \sum_{j=1}^{n} \alpha_{[j]} (E_{[j]} + T_{[j]}) \\ \text{subject to} \quad & d + T_{[j]} - E_{[j]} = C_{[j]}, \ j = 1, \ldots, n \\ & d, E_{[j]}, T_{[j]} \geq 0, \ j = 1, \ldots, n \end{aligned}$$

where $E_{[j]} = d - C_{[j]}$, $T_{[j]} = C_{[j]} - d$, and these variables cannot be positive simultaneously. Thus, for any job, only one of the variables $E_{[j]}$ and $T_{[j]}$ can be basic. Let $W_{[j]} = C_{[j]} - p_{[j]} = C_{[j-1]}$ denote the waiting time for job $[j]$ in σ, $j = 2, 3, \ldots, n$, and $W_{[1]} = 0$. Then, an LP formulation for the SLK method is given by

$$\begin{aligned} \text{minimize} \quad & \sum_{j=1}^{n} \alpha_{[j]} \mid W_{[j]} - d \mid = \sum_{j=1}^{n} \alpha_{[j]} (E_{[j]} + T_{[j]}) \\ \text{subject to} \quad & q + T_{[j]} - E_{[j]} = W_{[j]}, \quad j = 1, \ldots, n \\ & d, E_{[j]}, T_{[j]} \geq 0, \quad j = 1, \ldots, n \end{aligned}$$

where $E_{[j]} = q - W_{[j]}$, $T_{[j]} = W_{[j]} - q$. Here, as above, for any job only one of $E_{[j]}$ and $T_{[j]}$ is greater than zero.

Gupta, Bector, and Gupta [28] use the structural similarity between these LP formulations to derive the following property, analogous to Property 21.3.

Property 21.7

For problem $1 \mid d_j := p_j + q \mid \sum \alpha_j (E_j + T_j)$, there exists an optimal schedule s^* such that

1. *s^* is V-shaped, i.e., the jobs of set E (nontardy jobs) are sequenced in nonincreasing order of the ratio p_j/α_j and are followed by the jobs of set T (tardy jobs) in nondecreasing order of this ratio*
2. *the optimal slack q^* coincides with the waiting time of the last nontardy job*
3. *the inequality $\sum_{j \in T} \alpha_j \leq \sum_{j \in E} \alpha_j$ holds*

Moreover, Gupta, Bector, and Gupta [28] present a polynomial-time algorithm for the case of $\alpha_j = 1$, $j = 1, \ldots, n$, (the mean absolute deviation, MAD, problem) that is similar to the algorithm known for the CON due dates (see Section 21.2.1.2), and a heuristic approach in the case of different α_j. Notice that in the case of general due dates, problem $1 \mid\mid \sum (E_j + T_j)$ without due date assignment is NP-hard [99].

For MAD problem, an algorithm for finding an optimal slack and the set of all optimal sequences is given in [13,100] (each optimal sequence is found in $O(n \log n)$ time). The algorithm essentially uses similarity between the CON and the SLK models of due date assignment.

The restricted variant of MAD problem is shown to be NP-hard in [101]. This is a problem with $q < M$, where $M = p_1 + p_3 + \cdots + p_{n-1}$ if n is even, and $M = p_2 + p_4 + \cdots + p_{n-1}$ if n is odd, assuming that the jobs are taken in SPT order. The result follows from [15].

For a version of the asymmetric TWET problem, namely, for problem $1\,|\,d_j := p_j+q\,|\,\sum(\alpha_j E_j+\beta_j T_j)$ with $\alpha_j = \lambda p_j^a, \beta_j = \lambda p_j^b$ (where $\lambda > 0$, and parameters a and b are nonnegative integers), the branch-and-bound algorithms for solving the problem under certain conditions on a and b are developed in [102,103]. If an optimal solution is not unique, the set of all alternative optimal sequences can be found.

21.3.1.3 Other Problems with SLK and CON Similarity

The problem formulated in [5] for the common due date, is considered in [104] for the SLK model: find a slack q and a sequence σ that minimize the objective function $\sum_{j=1}^{n}(\alpha E_j + \beta T_j + \gamma q)$. Similar to the CON model, the optimal sequence is obtained by determining the positional weights and matching them with the processing times as in algorithm PSS. The optimal value of q is equal to $W_{[K]} = C_{[K]} - p_{[K]}$, where $K = \lceil n(\beta - \gamma)/(\alpha + \beta)\rceil$.

Kahlbacher [56] has proved the following general result on similarity between the SLK and the CON models.

Property 21.8

The problem to minimize the objective function $\sum_{j=1}^{n} g(C_j - d_j)$ for the CON due date assignment model with due dates $d_j = d$ for all jobs j, and the problem to minimize the objective function $\sum_{j=1}^{n} h(C_j - d_j)$ for the SLK model with due dates $d_j = p_j + q$ for all jobs j are equivalently solvable under the following conditions on the functions g and h and on the due dates: $q = \sum_{j=1}^{n} p_j - d$; $h(x) = g(-x)$, and $g(x)$ is a unimodal, real valued function such that (i) $g(0) = 0$, (ii) $g(x_1) \geq g(x_2)$ for all $x_1 \leq x_2 \leq 0$, and (iii) $g(x_1) \leq g(x_2)$ for all $0 \leq x_1 \leq x_2$.

It can be shown that the objective functions for MAD, WSAD, and MSD problems (see Section 21.2.1) satisfy these conditions. As follows from Property 21.8 and the results of Kubiak [47] for the CON due dates, MSD problem with SLK due dates is NP-hard.

21.3.1.4 Earliness Penalties with No Tardy Jobs

Consider problem $1\,|\,prec, d_j := p_j + q, C_j \leq d_j\,|\,\varphi(F,q)$, where the objective function $\varphi(F,q)$ is an arbitrary nondecreasing function in both arguments and $F = F(E_1, E_2, \ldots, E_n)$ is a nondecreasing function with respect to each of its arguments. Here, a schedule is feasible if no job is tardy and the sequence of jobs respects the precedence constraints. Function F represents penalties for the early completion of the jobs and can be viewed as the holding cost function.

Gordon and Strusevich [105] propose an algorithm which provides a general scheme for solving the problem. The time complexity depends on the structure of precedence constraints and on the form of function F. For arbitrary precedence constraints and $F = \sum_{j=1}^{n} E_j$, the problem is shown to be NP-hard in the strong sense. For series-parallel precedence constraints and $F = \sum_{j=1}^{n}\alpha_j E_j$ or $F = \sum_{j=1}^{n}\alpha_j \exp(\gamma E_j)$ with $\gamma \neq 0$ and $\alpha_j > 0$, the running time is $O(n^2 \log n)$. For the same functions F and independent jobs, the running time reduces to $O(n \log n)$, which improves an $O(n^2)$ algorithm from [106] designed for minimizing $F = \sum_{j=1}^{n}\alpha_j E_j$.

An $O(n \log n)$ algorithm for problem $1\,|\,d_j := p_j + q, C_j \leq d_j\,|\,\sum g(E_j)$, where g is a nondecreasing function is described in [106] (see also Section 20.4 in Chapter 20).

21.3.2 TWK Model

For scheduling problems with the TWK due date assignment, in the middle position of the three-field classification notation we write $d_j := kp_j$ (or $d_j := kp_j^c$ in the case of TWK-power due dates).

21.3.2.1 Maximum Tardiness Problem

Cheng [107] considers problem $1\,|\,d_j := kp_j\,|\,\Phi(k) + \alpha T_{\max}$, where $\alpha > 0$ and $\Phi(k)$ is a nondecreasing convex function, and shows that an optimal sequence follows the earliest due date (EDD) order.

Problem $1 \mid pmtn, prec, r_j, d_j := kp_j^c \mid \Phi(k,c) + \alpha T_{\max}$, where $\Phi(k,c)$ is a nondecreasing convex function of k and c with $\Phi(0,0) = 0$, is considered by Cheng and Gordon [108]. The following property allows finding an optimal schedule in $O(n^2)$ time.

Property 21.9

For problem $1 \mid pmtn, prec, r_j, d_j := kp_j^c \mid \Phi(k,c) + \alpha T_{\max}$, the schedule found by Algorithm GT-BLLR is optimal independent of the values of k and c.

According to this property, the completion times in the optimal schedule depend only on r_j, p_j, and the precedence constraints. However, the values of k and c determine which of the jobs has the maximal tardiness in this schedule.

21.3.2.2 Minsum Problems

For problem $1 \mid d_j := kp_j \mid \sum L_j^2$, it is shown that an SPT sequence σ_{SPT} is optimal and the optimal value of k is given by $k^* = (\sum_{j=1}^n p_{[j]} \sum_{i=1}^j p_{[i]}) / \sum_{j=1}^n p_{[j]}^2$, where $[j]$ is the j-th job in σ_{SPT}, see [109,110]. Sequence σ_{SPT} is also optimal for problem $1 \mid d_j := kp_j^c \mid (1/n) \sum C_j + \sum L_j^2$ (where $c \geq 1$ is given) considered in [111,112]. For finding an optimal multiplier k^*, any nonlinear search method can be used.

For problem $1 \mid d_j := kp_j \mid \sum \alpha_j(E_j + T_j)$ with no machine idle time, given a sequence of jobs, the optimal value of k is equal to the ratio of C_j/p_j for exactly one job j, see [28]. Problem $1 \mid d_j := kp_j \mid \gamma k + \sum \alpha_j(E_j + T_j)$, where γ is the cost per unit value of k, is NP-hard [113]. However, if $\gamma = 0$ and machine idle time is allowed, it becomes solvable in $O(n \log n)$ time even for the asymmetric objective function $\sum_{j=1}^n (\alpha_j E_j + \beta_j T_j)$. As described in [113], the value of k can be chosen large enough without increasing penalty, and the problem can be decomposed into subproblems of scheduling jobs with equal processing times about common due dates to be solved by the algorithm from [16] for the CON due dates.

Problem $1 \mid d_j := kp_j^c \mid \gamma k + \sum(E_j + T_j)$ with a given sequence of jobs and a fixed $c \geq 1$ (i.e., k is the only decision variable) can be formulated as an LP problem and solved by an $O(n^2)$ algorithm [114] or by an improved $O(n \log n)$ algorithm [115].

21.3.3 PPW Model

For scheduling problems with the PPW due date assignment, in the middle position of the three-field classification notation, we write $d_j := kp_j + q$.

21.3.3.1 Maximum Tardiness and Lateness Problems

A property similar to Property 21.9 is observed by Cheng and Gordon [108] for problem $1 \mid pmtn, prec, r_j, d_j := kp_j + q \mid \Phi(k,q) + \alpha T_{\max}$ (where $\Phi(k,q)$ is a nondecreasing convex function of k and q, such that $\Phi(0,0) = 0$): a schedule found by Algorithm GT-BLLR is optimal independent of the values of k and q. Any convex optimization technique can be used for finding optimal values of k and q. A simple graphical approach is described in [108] for $\Phi(k,q) = \Phi_1(k) + \gamma q, \gamma \geq 0$, where $\Phi_1(k)$ is a nondecreasing convex function with $\Phi_1(0) = 0$.

For problem $1 \mid d_j := kp_j + q \mid \max_{1 \leq j \leq n} |L_j|$, Kahlbacher and Cheng [116] observe that an SPT sequence is optimal if $k \geq 0$, and the problem of finding the optimal values of parameters k and q is reduced to an LP problem with three variables and $2n$ constraints. A simple graphical approach to finding the optimal parameters k and q in $O(n \log n)$ time is presented in [117].

21.3.3.2 MSD Problem

Problem $1 \mid d_j := kp_j + q \mid (1/n) \sum (C_j - d_j)^2$ of minimizing the mean squared deviation (MSD) with respect to the due dates is considered in [116] under the assumption of no machine idle times. Using basic results from linear regression analysis, it is shown that, for a given sequence of jobs, the optimal values of parameters k and q are determined by $k^* = (\sum_{j=1}^n (C_j - \bar{C})(p_j - \bar{p})) / \sum_{j=1}^n (p_j - \bar{p})^2$ and $d^* = \bar{C} - k^* \bar{p}$, where $\bar{p} = \frac{1}{n}\sum_{j=1}^{n} p_j$ is the mean processing time, and $\bar{C} = \frac{1}{n}\sum_{j=1}^{n} C_j$ is the mean completion time. The combined sequencing and due date assignment problem is NP-hard. An optimal schedule is defined by a V-shaped sequence, and an $O(n \log n)$ heuristic algorithm is developed for its finding.

Assume that the jobs are numbered according to an SPT sequence σ_{SPT}. Lee [118] shows that σ_{SPT} is optimal for MSD problem, provided that $-kp_1 \leq q \leq (p_1+p_2)/2$, and is asymptotically optimal, provided that (i) $q > (p_1+p_2)/2$, and (ii) there is an equal increment between the successive values of the processing times. If $f(\sigma^*)$ is an optimal value of MSD and $f(\sigma_{SPT})$ is the value of MSD for sequence σ_{SPT}, then for $q > (p_1 + p_2)/2$ an absolute error of the SPT heuristic is given by $f(\sigma_{SPT}) - f(\sigma^*) \leq n^2 q(p_n - p_1)/2$.

21.4 Positional and Other Due Dates

In this section, we review scheduling problems in which the values of due dates are given but are not associated with specific jobs. For the related models, the process of due date assignment consists in assigning the due dates to the jobs and finding an optimal schedule.

21.4.1 Positional Due Dates

Hall [119] has introduced the *generalized* due dates that are specified according to the position in which a job is completed, rather than to the identity of that specific job. For these due dates, k jobs should be completed by the kth due date, i.e., the job that is completed first is assigned the earliest due date, the job completed second is assigned the second earliest due date, and so on. We refer to these due dates as the *positional* due dates. Hall [119] and Hall, Sethi, and Sriskandarajah [120] describe several applications of positional due date problems in the petrochemical industry, in public utility planning, survey design, and flexible manufacturing. A comprehensive review of scheduling problems with positional due dates can be found in [119] and [120] (see also [4]), therefore here we only summarize the results on the complexity of the relevant problems in Table 21.1, where notation $d_{[j]} = d_j^p$ denotes the positional due dates. Additionally, we mention that an enumerative algorithm for problem $1 \,|\, pmtn, r_j, d_{[j]} = d_j^p \,|\, \sum U_j$ is given in [122], while an approximation algorithm with a guaranteed worst-case performance for problem $1 \,|\, d_{[j]} = d_j^p \,|\, (L_{\max} - L_{\min})$ is presented in [123].

21.4.2 Unit Interval Assignment

Given the set of due dates and, independently, the set of jobs, the due dates can be assigned to the jobs using some rule without taking into account the positions of the jobs in the schedule. For job j, we denote such due dates by d_j^a. Additionally, if the release dates are also assignable, we use notation r_j^a.

Agnetis et al. [125] and Gordon and Kubiak [126,127] consider the model of due date assignment in which the scheduling horizon $[0, n]$ is divided into n unit intervals. The two end points of each interval correspond to a release date and a due date to be assigned to a job, respectively. The model corresponds to sequencing problems arising in the management of paced-flowlines, that is production lines where jobs are released at constant time intervals. Each job j has rational processing time p_j and can be either short (if $p_j < 1$) or long (if $p_j > 1$), assuming that $\sum_{j=1}^{n} p_j = n$. The jobs with $p_j = 1$ can be disregarded without loss of generality. Let $[u]$ be the job in the uth position (in the preemptive case, the job with the uth earliest starting time) in a schedule. In problem $1 \,|\, r_{[u]} := u - 1,\ d_{[u]} := u,\ C_{\max} = n \,|\, \sum w_j T_j$, it is required to find a schedule that completes at time n and minimizes the total weighted tardiness, provided that the job in the uth position is assigned the release date $u-1$ and the due date u, $u = 1, \ldots, n$. The problem is proved to be NP-hard in the strong sense, and a branch-and-bound algorithm effective for instances up to 18 jobs is developed in [125] . Problems $1 \,|\, r_{[u]} := u - 1,\ d_{[u]} := u,\ C_{\max} = n \,|\, \sum U_j$ and $1 \,|\, r_j^a := u - 1,\ d_j^a := u,\ C_{\max} = n \,|\, \sum U_j$ differ in the following: in the latter problem, the release date $u - 1$ and the due date u can be assigned to any job, provided that no two jobs have equal release or due dates; in other words, the release date of the job in the $(u - 1)$th position may be greater than the release date of the job in the uth position in a schedule. Both problems are shown to be strongly NP-hard in [126,127]. The preemptive

TABLE 21.1 Positional Due Date Assignment and Scheduling

Problem	Complexity	Reference
$1 \mid d_{[j]} = d_j^p \mid \sum T_j$	$O(n \log n)$	[119]
$1 \mid d_{[j]} = d_j^p \mid \sum U_j$	$O(n \log n)$	[119]
$1 \mid d_{[j]} = d_j^p \mid \sum w_j T_j$	NP-hard	[121]
$1 \mid d_{[j]} = d_j^p \mid \sum w_j U_j$	NP-hard	[119]
$1 \mid r_j, d_{[j]} = d_j^p \mid L_{\max}$	Strongly NP-hard	[120]
$1 \mid r_j, d_{[j]} = d_j^p \mid \sum T_j$	Strongly NP-hard	[119]
$1 \mid r_j, d_{[j]} = d_j^p \mid \sum U_j$	Strongly NP-hard	[120]
$1 \mid pmtn, r_j, d_{[j]} = d_j^p \mid L_{\max}$	$O(n \log n)$	[120]
$1 \mid pmtn, r_j, d_{[j]} = d_j^p \mid \sum T_j$	$O(n \log n)$	[122]
$1 \mid pmtn, r_j, d_{[j]} = d_j^p \mid \sum U_j$	$O(n \log n)$	[122]
$1 \mid prec, p_j = 1, d_{[j]} = d_j^p \mid \sum T_j$	$O(n^2)$	[120]
$1 \mid prec, p_j = 1, d_{[j]} = d_j^p \mid \sum U_j$	$O(n^2)$	[120]
$1 \mid chain, d_{[j]} = d_j^p \mid L_{\max}$	Strongly NP-hard	[121]
$1 \mid chain, pmtn, r_j, d_{[j]} = d_j^p \mid L_{\max}$	Strongly NP-hard	[121]
$1 \mid d_{[j]} = d_j^p \mid L_{\max} - L_{\min}$	Strongly NP-hard	[123]
$1 \mid d_{[j]} = d_j^p \mid \max_j\{E_j, T_j\}$	Strongly NP-hard	[124]
$P \mid pmtn, d_{[j]} = d_j^p \mid L_{\max}$	$O(nm)$	[120]
$P2 \mid d_{[j]} = d_j^p \mid F \in \{L_{\max}, \sum U_j\}$	NP-hard	[119]
$O2 \mid d_{[j]} = d_j^p \mid F \in \{L_{\max}, \sum U_j, \sum T_j\}$	Strongly NP-hard	[120]
$F2 \mid d_{[j]} = d_j^p \mid F \in \{L_{\max}, \sum U_j, \sum T_j\}$	Strongly NP-hard	[120]
$J2 \mid d_{[j]} = d_j^p \mid F \in \{L_{\max}, \sum U_j, \sum T_j\}$	Strongly NP-hard	[120]
$O2 \mid pmtn, d_{[j]} = d_j^p \mid F \in \{L_{\max}, \sum U_j, \sum T_j\}$	Strongly NP-hard	[120]
$F2 \mid pmtn, d_{[j]} = d_j^p \mid F \in \{L_{\max}, \sum U_j, \sum T_j\}$	Strongly NP-hard	[120]
$J2 \mid pmtn, d_{[j]} = d_j^p \mid F \in \{L_{\max}, \sum U_j, \sum T_j\}$	Strongly NP-hard	[120]

counterparts of these problems, i.e., problems $1 \mid pmtn, r_{[u]} := u - 1, d_{[u]} := u, C_{\max} = n \mid \sum w_j U_j$ and $1 \mid pmtn, r_j^a := u - 1, d_j^a := u, C_{\max} = n \mid \sum w_j U_j$, are polynomially solvable [126,127]: two $O(n)$ algorithms (that differ in the number of preemptions) are proposed which are applicable to each of the problems.

21.4.3 Arbitrary Assignment Rule

Qi, Yu, and Bard [128] extend the equal interval model studied in [125–127] to allow arbitrary release dates and due dates. In the middle field of the scheduling classification notation, we write $d_j := d_j^a$ and $r_j := r_j^a$ to denote the fact that the due dates and the release dates have to be assigned to jobs from a given set of values. Given a set of n positive integers $d_1 \le d_2 \le \cdots \le d_n$ corresponding to the due dates, and the processing times $p_1 \le p_2 \le \cdots \le p_n$ numbered in SPT order, the problem is to assign a due date and a position in the schedule to each job so that the objective function is minimized. The SPT-EDD sequence is constructed by assigning the due dates to the jobs in nondecreasing order according to their positions in the SPT sequence, that is, the assigned due date d_j^a of job j is equal to d_j. As shown in [128], the SPT-EDD sequence is optimal with respect to the objective functions $L_{\max} = \max_{1 \le j \le n}\{C_j - d_j^a\}$ and $\sum_{j=1}^{n} T_j$.

The following results are also due to [128]. Each of the problems $1 \mid d_j := d_j^a \mid \sum w_j T_j$, $1 \mid d_j := d_j^a \mid \sum w_j(E_j + T_j)$, $1 \mid r_j, d_j := d_j^a \mid L_{\max}$, $1 \mid r_j := r_j^a, d_j := d_j^a \mid L_{\max}$, and $1 \mid pmtn, r_j := r_j^a, d_j := d_j^a \mid L_{\max}$ is NP-hard in the strong sense. Problem $1 \mid d_j := d_j^a \mid \sum(E_j + T_j)$ is NP-hard but is open with respect to the NP-hardness in the strong sense. Problem $1 \mid d_j := d_j^a \mid \sum w_j U_j$ is NP-hard and is solvable by a pseudopolynomial-time DP algorithm that requires $O(n^2 d_n)$ time. An $O(n^2)$ algorithm is given for problem $1 \mid pmtn, r_j, d_j := d_j^a \mid \sum U_j$, while each of the problems $1 \mid d_j := d_j^a \mid \sum U_j$ and $1 \mid pmtn, r_j, d_j := d_j^a \mid F$ with $F \in \{L_{\max}, \sum T_j\}$ is solvable in $O(n \log n)$ time.

Qi, Yu, and Bard [128] consider also problems in which a job that is assigned due date d_j^a is also assigned the weight w_j^a. Notice that, unlike the traditional model, here the weight is associated with the due date rather than with the job. By constructing an appropriate assignment problem, problems $1 \mid d_j := d_j^a \mid \sum w_j^a T_j$ and $1 \mid d_j := d_j^a \mid \sum w_j^a U_j$ are shown to be polynomially solvable, while problem $1 \mid d_j := d_j^a \mid \sum w_j^a (E_j + T_j)$ is NP-hard in the strong sense.

Acknowledgment

This work is supported in part by INTAS (Projects 00-217 and 03-51-5501). The first author is also partly supported by ISTC and is grateful to The Royal Society and INRIA for supporting his visits to the UK and France during the preparation of this review.

References

[1] Cheng, T.C.E. and Gupta, M.C., Survey of scheduling research involving due date determination decisions, *Eur. J. Oper. Res.*, 38, 156–166, 1989.

[2] Baker, K.R. and Scudder, G.D., Sequencing with earliness and tardiness penalties: a review, *Oper. Res.*, 38, 22–36, 1990.

[3] Gordon, V.S., Proth, J.-M., and Chu, C., A survey of the state-of-the-art of common due date assignment and scheduling, *Eur. J. Oper. Res.*, 139, 1–25, 2002.

[4] Gordon, V.S., Proth, J.-M., and Chu, C., Due date assignment and scheduling: SLK, TWK and other due date assignment models, *Prod. Plan. Control*, 13, 117–132, 2002.

[5] Panwalkar, S.S., Smith, M.L., and Seidmann, A., Common due date assignment to minimize total penalty for the one machine scheduling problem, *Oper. Res.*, 30, 391–399, 1982.

[6] Kanet, J.J., Minimizing the average deviation of job completion times about a common due date, *Naval Res. Logist. Quart.*, 28, 643–651, 1981.

[7] Cheng, T.C.E., An alternative proof of optimality for the common due date assignment problem, *Eur. J. Oper. Res.*, 37, 250–253, 1988. Corrigendum: *Eur. J. Oper. Res.*, 38, 259, 1989.

[8] Cheng, T.C.E., A heuristic for common due date assignment and job scheduling on parallel machines, *J. Oper. Res. Soc.*, 40, 1129–1135, 1989.

[9] Cheng, T.C.E., Chen, Z.-L., and Shakhlevich, N.V., Common due date assignment and scheduling with ready times, *Comput. Oper. Res.*, 29, 1957–1967, 2002.

[10] Lenstra, J.K., Rinnooy Kan, A.H.G., and Brucker, P., Complexity of machine scheduling problems, *Ann. Discr. Math.*, 1, 343–362, 1977.

[11] Bagchi, U., Sullivan, R.S., and Chang, Y.L., Minimizing mean absolute deviation of completion times about a common due date, *Naval Res. Logist.*, 33, 227–240, 1986.

[12] Hall, N.G., Single- and multiple-processor models for minimizing completion time variance, *Naval Res. Logist. Quart.*, 33, 49–54, 1986.

[13] Karacapilidis, N.I. and Pappis, C.P., Form similarities of the CON and SLK due date determination methods, *J. Oper. Res. Soc.*, 46, 762–770, 1995.

[14] Bagchi, U., Chang, Y.L., and Sullivan, R.S., Minimizing absolute and squared deviations of completion times with different earliness and tardiness penalties and a common due date, *Naval Res. Logist.*, 34, 739–751, 1987.

[15] Hall, N.G., Kubiak, W., and Sethi, S.P., Earliness-tardiness scheduling problems, II: Deviation of completion times about a restrictive common due date, *Oper. Res.*, 39, 847–856, 1991.

[16] Hoogeveen, J.A. and van de Velde, S.L., Scheduling around a small common due date, *Eur. J. Oper. Res.*, 55, 237–242, 1991.

[17] Hall, N.G. and Posner, M.E., Earliness-tardiness scheduling problems, I: Weighted deviation of completion times about a common due date, *Oper. Res.*, 39, 836–846, 1991.

[18] De, P., Ghosh, J.B., and Wells, C.E., CON due date determination and sequencing, *Comput. Oper. Res.*, 17, 333–342, 1990.

[19] Jurisch, B., Kubiak, W., and Jozefowska, J., Algorithms for minclique scheduling problems, *Discr. Appl. Math.*, 72, 115–139, 1997.

[20] Kovalyov, M.Y. and Kubiak, W., A fully polynomial approximation scheme for the weighted earliness-tardiness problem, *Oper. Res.*, 47, 757–761, 1999.

[21] Cheng, T.C.E., A duality approach to optimal due date determination, *Eng. Optimiz.*, 9, 127–130, 1985.

[22] Cheng, T.C.E., An algorithm for the CON due date determination and sequencing problem, *Comput. Oper. Res.*, 14, 537–542, 1987.

[23] Cheng, T.C.E., A note on a partial search algorithm for the single-machine optimal common due date assignment and sequencing problem, *Comput. Oper. Res.*, 17, 321–324, 1990.

[24] Hao, Q., Yang, Z., Wang, D., and Li, Z., Common due date determination and sequencing using tabu search, *Comput. Oper. Res.*, 23, 409–417, 1996.

[25] Karacapilidis, N.I. and Pappis, C.P., Optimization algorithms for a class of single machine scheduling problems using due date determination methods, *Yugosl. J. Oper. Res.*, 5, 289–297, 1995.

[26] Dileepan, P., Common due date scheduling problem with separate earliness and tardiness penalties, *Comput. Oper. Res.*, 20, 179–181, 1993.

[27] De, P., Ghosh, J.B., and Wells, C.E., Solving a generalized model for CON due date assignment and sequencing, *Int. J. Prod. Econ.*, 34, 179–185, 1994.

[28] Gupta, Y.P., Bector, C.R., and Gupta, M.C., Optimal schedule on a single machine using various due date determination methods, *Computers in Industry*, 15, 245–253, 1990.

[29] Biskup, D. and Feldmann, M., Benchmarks for scheduling on a single-machine against restrictive and unrestrictive common due dates, *Comput. Oper. Res.*, 28, 787–801, 2001.

[30] Hoogeveen, J.A. and van de Velde, S.L., Earliness-tardiness scheduling around almost equal due date, *INFORMS J. Comput.*, 9, 92–99, 1997.

[31] Seidmann, A., Panwalkar, S.S., and Smith, M.L., Optimal assignment of due dates for a single processor scheduling problem, *Int. J. Prod. Res.*, 19, 393-399, 1981.

[32] Cheng, T.C.E., Optimal constant due date determination and sequencing of *n* jobs on a single machine, *Int. J. Prod. Econ.*, 22, 259–261, 1991.

[33] De, P., Ghosh, J.B., and Wells, C.E., Optimal delivery time quotation and order sequencing, *Decis. Sci.*, 22, 379–390, 1991.

[34] Cheng, T.C.E., Common due date assignment and scheduling for a single processor to minimize the number of tardy jobs, *Eng. Optimiz.*, 16, 129–136, 1990.

[35] Kahlbacher, H.G. and Cheng, T.C.E., Parallel machine scheduling to minimize costs for earliness and number of tardy jobs, *Discr. Appl. Math.*, 47, 139–164, 1993.

[36] Lee, C.Y., Danusaputro, S.L., and Lin, C.-S., Minimizing weighted number of tardy jobs and weighted earliness-tardiness penalties about a common due date, *Comput. Oper. Res.*, 18, 379–389, 1991.

[37] Cheng, T.C.E., Oguz, C., and Qi, X.D., Due date assignment and single machine scheduling with compressible processing times, *Int. J. Prod. Econ.*, 43, 29–35, 1996.

[38] Panwalkar, S.S. and Rajagopalan, R., Single machine sequencing with controllable processing times, *Eur. J. Oper. Res.*, 59, 298-302, 1992.

[39] Biskup, D. and Cheng, T.C.E., Single-machine scheduling with controllable processing times and earliness, tardiness and completion time penalties, *Eng. Optimiz.*, 31, 329–336, 1999.

[40] Biskup, D. and Jahnke, H., Common due date assignment for scheduling on a single machine with jointly reducible processing times, *Int. J. Prod. Econ.*, 69, 317–322, 2001.

[41] Ng, C.T.D., Cheng, T.C.E., Kovalyov, M.Y., and Lam, S.S., Single machine scheduling with a variable common due date and resource-dependent processing times, *Comput. Oper. Res.*, 30, 1173–1185, 2003.

[42] Biskup, D., Single-machine scheduling with learning considerations, *Eur. J. Oper. Res.*, 115, 173–178, 1999.

[43] Mosheiov, G., Scheduling problems with a learning effect, *Eur. J. Oper. Res.*, 132, 687–693, 2001.

[44] Mosheiov, G. and Sidney, J.B., Scheduling with general job-dependent learning curves, *Eur. J. Oper. Res.*, 147, 665–670, 2003.

[45] Bagchi, U., Sullivan, R.S., and Chang, Y.L., Minimizing mean squared deviation of completion times about a common due date, *Manag. Sci.*, 33, 894–906, 1987.
[46] Merten, A.G. and Muller, M.E., Variance minimization in a single machine sequencing problems, *Manag. Sci.*, 18, 518–528, 1972.
[47] Kubiak, W., Completion time variance minimization on a single machine is difficult, *Oper. Res. Lett.*, 14, 49–59, 1993.
[48] De, P., Ghosh, J.B., and Wells, C.E., On the minimization of completion time variance with a bicriteria extension, *Oper. Res.*, 40, 1148–1155, 1992.
[49] Kubiak, W., New results on the completion time variance minimization, *Discr. Appl. Math.*, 58, 157–168, 1995.
[50] Cai, X., Minimization of agreeably weighted variance in single machine systems, *Eur. J. Oper. Res.*, 85, 576–592, 1995.
[51] Gupta, M.C., Gupta, Y.P., and Kumar, A., Minimizing flow time variance in a single machine system using genetic algorithms, *Eur. J. Oper. Res.*, 70, 289–303, 1993.
[52] De, P., Ghosh, J.B., and Wells, C.E., A note on the minimization of mean squared deviation of completion times about a common due date, *Manag. Sci.*, 35, 1143–1147, 1989.
[53] De, P., Ghosh, J.B., and Wells, C.E., Scheduling about a common due date with earliness and tardiness penalties, *Comput. Oper. Res.*, 17, 231–241, 1990.
[54] De, P., Ghosh, J.B., and Wells, C.E., On the general solution for a class of early/tardy problems, *Comput. Oper. Res.*, 20, 141–149, 1993.
[55] Kahlbacher, H.G., SWEAT — a program for a scheduling problem with earliness and tardiness penalties, *Eur. J. Oper. Res.*, 43, 111–112, 1989.
[56] Kahlbacher, H.G., Scheduling with monotonous earliness and tardiness penalties, *Eur. J. Oper. Res.*, 64, 258–277, 1993.
[57] Cai, X., Lum, V.Y.S., and Chan, J.M.T., Scheduling about a common due date with job-dependent asymmetric earliness and tardiness penalties, *Eur. J. Oper. Res.*, 98, 154–168, 1997.
[58] Federgruen, A. and Mosheiov, G., Greedy heuristics for single-machine scheduling problems with general earliness and tardiness costs, *Oper. Res. Lett.*, 16, 199–208, 1994.
[59] Federgruen, A. and Mosheiov, G., Simultaneous optimization of efficiency and performance balance measures in single-machine scheduling problems, *Naval Res. Logist.*, 40, 951–970, 1993.
[60] Li, C.-L. and Cheng, T.C.E., The parallel machine min-max weighted absolute lateness scheduling problem, *Naval Res. Logist.*, 41, 33–46, 1994.
[61] Cheng, T.C.E., Minimizing the maximum deviation of job completion time about a common due date, *Comput. Math. Applic.*, 14, 279–283, 1987.
[62] Chen, Z.-L., Scheduling and common due date assignment with earliness-tardiness penalties and batch delivery costs, *Eur. J. Oper. Res.*, 93, 49–60, 1996.
[63] Herrmann, J.W. and Lee, C.-Y., On scheduling to minimize earliness-tardiness and batch delivery costs with a common due date, *Eur. J. Oper. Res.*, 70, 272–288, 1993.
[64] Cheng, T.C.E. and Kovalyov, M.Y., Batch scheduling and common due date assignment on a single machine, *Discr. Appl. Math.*, 70, 231–245, 1996.
[65] Kovalyov, M.Y., Batch scheduling and common due date assignment problem: An NP-hard case, *Discr. Appl. Math.*, 80, 251–254, 1997.
[66] Azizoglu, M. and Webster, S., Scheduling job families about an unrestricted common due date on a single machine, *Int. J. Prod. Res.*, 35, 1321–1330, 1997.
[67] Chen, D., Li, S., and Tang, G., Single machine scheduling with common due date assignment in a group technology environment, *Math. Comput. Modelling*, 25, 81–90, 1997.
[68] Cheng, T.C.E., Optimal common due date with limited completion time deviation, *Comput. Oper. Res.*, 15, 91–96, 1988.
[69] Dickman, B., Wilamowsky, Y., and Epstein, S., Optimal common due date with limited completion time, *Comput. Oper. Res.*, 18, 125–127, 1991.

[70] Weng, M.X. and Ventura, J.A., Scheduling about a large common due date with tolerance to minimize mean absolute deviation of completion times, *Naval Res. Logist.*, 41, 843–851, 1994.

[71] Wilamowsky, Y., Epstein, S., and Dickman, B., Optimal common due date completion time tolerance, *Comput. Oper. Res.*, 23, 1203–1210, 1996.

[72] Krämer, F.-J. and Lee, C.-Y., Common due-window scheduling, *Prod. Oper. Manag.*, 2, 262–275, 1993.

[73] Liman, S.D., Panwalkar, S.S., and Thongmee, S., Determination of common due window location in a single machine scheduling problem, *Eur. J. Oper. Res.*, 93, 68–74, 1996.

[74] Liman, S.D., Panwalkar, S.S., and Thongmee, S., A single machine scheduling problem with common due window and controllable processing times, *Ann. Oper. Res.* 70, 145–154, 1997.

[75] Azizoglu, M. and Webster, S., Scheduling about an unrestricted common due window with arbitrary earliness/tardiness penalty rates, *IIE Trans.*, 29, 1001–1006, 1997.

[76] Yeung, W.K., Oguz, C., and Cheng, T.C.E., Single-machine scheduling with a common due window, *Comput. Oper. Res.*, 28, 157–175, 2001.

[77] Mosheiov, G., A due-window determination in minmax scheduling problems, *INFOR*, 39, 107–123, 2001.

[78] Cheng, T.C.E. and Kahlbacher, H.G., The parallel machine common due date assignment and scheduling problem is NP-hard, *Asia Pacif. J. Oper. Res.*, 9, 235–238, 1992.

[79] Cheng, T.C.E. and Chen, Z.-L., Parallel-machine scheduling problems with earliness and tardiness penalties, *J. Oper. Res. Soc.*, 45, 685–695, 1994.

[80] De, P., Ghosh, J.B., and Wells, C.E., Due date assignment and early/tardy scheduling on identical parallel machines, *Naval Res. Logist.*, 41, 17–32, 1994.

[81] Emmons, H., Scheduling to a common due date on parallel uniform processors, *Naval Res. Logist. Quart.*, 34, 803–810, 1987.

[82] Alidaee, B. and Panwalkar, S.S., Single stage minimum absolute lateness problem with a common due date on non-identical machines, *J. Oper. Res. Soc.*, 44, 29–36, 1993.

[83] Kubiak, W., Lou, S., and Sethi, S., Equivalence of mean flow time problems and mean absolute deviation problems, *Oper. Res. Lett.*, 9, 371–374, 1990.

[84] Alidaee, B. and Ahmadian, A., Two parallel machine sequencing problems involving controllable job processing times, *Eur. J. Oper. Res.*, 70, 335–341, 1993.

[85] De, P., Ghosh, J.B., and Wells, C.E., On the multiple-machine extension to a common due date assignment and scheduling problem, *J. Oper. Res. Soc.*, 42, 419–422, 1991.

[86] Diamond, J.E. and Cheng, T.C.E., Error bound for common due date assignment and job scheduling on parallel machines, *IIE Trans.*, 32, 445–448, 2000.

[87] Xiao, W.-Q. and Li, C.-L., Approximation algorithms for common due date assignment and job scheduling on parallel machines, *IIE Trans.*, 34, 467–477, 2002.

[88] Adamopoulos, G.I. and Pappis, C.P., Scheduling under a common due date on parallel unrelated machines, *Eur. J. Oper. Res.*, 105, 494–501, 1998.

[89] Biskup, D. and Cheng, T.C.E., Mulitple-machine scheduling with earliness, tardiness and completion time penalties, *Comput. Oper. Res.*, 26, 45–57, 1999.

[90] Panwalkar, S.S. and Liman, S.D., Single operations earliness-tardiness scheduling with machine activation costs, *IIE Trans.*, 34, 509–513, 2002.

[91] Webster, S. T., The complexity of scheduling job families about a common due date, *Oper. Res. Lett.*, 20, 65–74, 1997.

[92] Mosheiov, G., A common due date assignment problem on parallel identical machines, *Comput. Oper. Res.*, 28, 719–732, 2001.

[93] Gordon, V.S. and Tanaev, V.S., On minmax single machine scheduling problems, *Izvestiya Akademii Nauk BSSR. Ser. Fiz.-Mat. Nauk*, 3, 3-9, 1983. (In Russian)

[94] Baker, K.R., Lawler, E.L., Lenstra J.K., and Rinnooy Kan, A.H.G., Preemptive scheduling of a single machine to minimize maximum cost subject to release dates and precedence constraints, *Oper. Res.*, 31, 381–386, 1983.

[95] Blazewicz, J., Scheduling dependent tasks with different arrival times to meet deadlines, In *Modelling and Performance Evaluation of Computer Systems*, Gelenbe, E. and Beilner, H. (editors), North-Holland, Amsterdam, 1976, 57–65.
[96] Gordon, V.S., A note on optimal assignment of slack due dates in single-machine scheduling, *Eur. J. Oper. Res.*, 70, 311–315, 1993.
[97] Cheng, T.C.E., Optimal assignment of slack due dates and sequencing in a single-machine shop, *Appl. Math. Lett.*, 2, 333–335, 1989.
[98] Alidaee, B., Optimal assignment of slack due dates and sequencing in a single machine shop, *Appl. Math. Lett.*, 4, 9–11, 1991.
[99] Garey, M.R., Tarjan, R.E., and Wilfong, G.T., One-processor scheduling with symmetric earliness and tardiness penalties, *Math. Oper. Res.*, 13, 330–348, 1988.
[100] Karacapilidis, N.I. and Pappis, C.P., Optimal due date determination and sequencing of n jobs on a single machine using the SLK method, *Comput. Ind.*, 21, 335–339, 1993.
[101] Oguz, C. and Dinger, C., Single machine earliness-tardiness scheduling problems using the equal-slack rule, *J. Oper. Res. Soc.*, 45, 589–594, 1994.
[102] Pappis, C.P. and Adamopoulos, G.I., Scheduling under the due date criterion with varying penalties for lateness, *Yugosl. J. Oper. Res.*, 3, 189–198, 1993.
[103] Adamopoulos, G.I. and Pappis, C.P., Scheduling with different, job-dependent earliness and tardiness penalties using SLK method, *Eur. J. Oper. Res.*, 88, 336–344, 1996.
[104] Adamopoulos, G.I. and Pappis, C.P., Single machine scheduling with flow allowances, *J. Oper. Res. Soc.*, 47, 1280–1285, 1996.
[105] Gordon, V.S. and Strusevich, V.A., Earliness penalties on a single machine subject to precedence constraints: SLK due date assignment, *Comput. Oper. Res.*, 26, 157–177, 1999.
[106] Qi, X. and Tu, F.S., Scheduling a single machine to minimize earliness penalties subject to SLK due date determination method, *Eur. J. Oper. Res.*, 105, 502–508, 1998.
[107] Cheng, T.C.E., Optimal assignment of total-work-content due dates and sequencing in a single machine shop, *J. Oper. Res. Soc.*, 42, 177–181, 1991.
[108] Cheng, T.C.E. and Gordon, V.S., Optimal assignment of due dates for preemptive single-machine scheduling. *Math. Comput. Modelling*, 20, 33–40, 1994.
[109] Cheng, T.C.E., Optimal due date determination and sequencing of n jobs on a single machine, *J. Oper. Res. Soc.*, 35, 433–437, 1984.
[110] Mutlu, O., Comments on optimal due date determination and sequencing of n jobs on a single machine, *J. Oper. Res. Soc.*, 44, 1062, 1993.
[111] Cheng, T.C.E., Optimal TWK-power due date determination and sequencing, *Int. J. Syst. Sci.*, 18, 1–7, 1987.
[112] Cheng, T.C.E. and Li, S., Some observations and extensions of the optimal TWK-power due date determination and sequencing problem, *Comput. Math. Applic.*, 17, 1103–1107, 1989.
[113] Chu, C. and Gordon, V., TWK due date determination and scheduling: NP-hardness and polynomially solvable case, *Int. J. Mathl. Algorithms*, 2, 251–267, 2001.
[114] Cheng, T.C.E., Optimal total-work-content-power due date determination and sequencing, *Comput. Math. Applic.*, 14, 579–582, 1987.
[115] van de Velde, S.L., A simpler and faster algorithm for optimal total-work-content-power due date determination, *Math. Comput. Modelling*, 13, 81–83, 1990.
[116] Kahlbacher, H.G. and Cheng, T.C.E., Processing-plus-wait due dates in single-machine scheduling, *J. Opt. Theory Appl.*, 85, 163–186, 1995.
[117] Cheng, T.C.E. and Kovalyov, M.Y., Complexity of parallel machine scheduling with processing-plus-wait due dates to minimize maximum absolute lateness, *Eur. J. Oper. Res.*, 114, 403–410, 1999.
[118] Lee, I.-S., A worst-case performance of the shortest-processing-time heuristic for single machine scheduling, *J. Oper. Res. Soc.*, 42, 895–901, 1991.
[119] Hall, N.G., Scheduling problems with generalized due dates, *IIE Trans.*, 18, 220–222, 1986.

[120] Hall, N.G., Sethi, S.P., and Sriskandarajah, C., On the complexity of generalized due date scheduling problems, *Eur. J. Oper. Res.*, 51, 100–109, 1991.

[121] Sriskandarajah, C., A note on the generalized due date scheduling problems, *Naval Res. Logist.*, 37, 587–597, 1990.

[122] Chu, C., A new class of scheduling criteria and their optimization, *RAIRO Rech. Oper.*, 30, 171–189, 1996.

[123] Tanaka, K. and Vlach, M., Single machine scheduling to minimize the maximum lateness with both specific and generalized due dates, *IEICE Trans. Fundl. Electron., Commun. Comput. Sci.*, E80-A, 557–563, 1997.

[124] Tanaka, K. and Vlach, M., Minimizing maximum absolute lateness and range of lateness under generalized due dates on a single machine, *Ann. Oper. Res.*, 86, 507–526, 1999.

[125] Agnetis, A., Macchiaroli, R., Pacciarelli, D., and Rossi F., Assigning jobs to time frames on a single machine to minimize total tardiness, *IIE Trans.*, 29, 965–976, 1997.

[126] Gordon, V.S. and Kubiak, W., Minimizing the weighted number of late jobs with release and due date assignment, In *Industrial Scheduling of Robots and Flexible Manufacturing Systems*, Levner, E. (editor), CTEH Press, 1996, 127–130.

[127] Gordon, V.S. and Kubiak, W., Single machine scheduling with release and due date assignment to minimize the weighted number of late jobs, *Inform. Proc. Lett.* , 68, 153–159, 1998.

[128] Qi, X., Yu, G., and Bard, J.F., Single machine scheduling with assignable due dates, *Discrete Appl. Math.*, 122, 211–233, 2002.

22

Machine Scheduling with Availability Constraints

Chung-Yee Lee
The Hong Kong University of Science and Technology

22.1 Introduction

Most of the literature on scheduling assumes that the machines are available at all times. However, due to various reasons, machines may not be always available in many realistic situations. Under such circumstances, special consideration is needed in order to obtain optimal solutions. Note that machine scheduling is usually done on a rolling horizon basis. At the beginning of the scheduling period, machines may continue processing some jobs scheduled in the previous time horizon and hence are not available at the beginning of the current horizon. It is also possible that certain important jobs may have been promised during the previous planning horizon. Hence, some machines have been committed to process the promised jobs in certain periods of the current horizon and will not be available in those periods. In such cases, the optimal machine scheduling policy can be different from that for classical problems.

Another example is the case in which a machine is not available due to maintenance. In such a case, the question involved is whether the maintenance decision was done separately or jointly with the job scheduling decision. If the maintenance decision is made separately in advance, the nonavailability intervals will then become a given parameter or a constraint while we plan the job scheduling. That is, while we

1-58488-397-9/$0.00+$1.50

are doing the job scheduling, the machine unavailability period is given. In our experience with industry applications, especially in the semiconductor industry, it is not uncommon to observe an operational machine in an idle state waiting for maintenance while jobs are waiting to be processed. This is due to the lack of coordination between operators (or production planning personnel) and maintenance personnel. Clearly, this lack of coordination may make the whole system ineffective even if the job scheduling itself is done optimally. On the other hand, if the maintenance decision is made jointly with the job scheduling, the system will be more effective and the problem is actually more interesting. In such a case, the machine nonavailability interval is a decision variable. Most literature studies problems of machine scheduling given nonavailability intervals. Due to the popularity of supply chain management, logistics scheduling that integrates maintenance and job-scheduling decisions jointly has recently become more important.

Another example of machine unavailability can be found during machine breakdowns. Machine scheduling problems with potential breakdown considerations have been studied intensively in the literature. An interesting yet little studied area is the case in which the machine has partially failed and is working at a less-efficient speed. The question then is whether we should stop the machine now and repair it or repair it later. If we do not repair it, the machine can still operate yet at a less-efficient speed. Furthermore, the full breakdown of the machine can happen any time. Clearly, after the repair activity, the machine will be back to normal speed. These characteristics make the problem interesting and important, yet very few researchers have studied this problem.

The machine scheduling with availability constraints is thus an important topic in scheduling (see, e.g., Blazewicz et al., 2001; Pinedo, 2002) and has attracted much attention recently (see the surveys by Lee et al., 1997; Schmidt, 2000). The nonavailability consideration adds complexity to any scheduling problem. This chapter addresses particularly this issue and will focus on the problems that are motivated by logistics applications. For survey of problems motivated by other applications, please see Schmidt (2000).

22.2 Notation and Problem Definition

We are given m machines and n jobs with the following notation.

J_i	Job $i, i = 1, \ldots, n$
p_i	Processing time for J_i
d_i	Due date for J_i
C_i	The completion time for J_i
w_i	Weight for J_i (Hence, $\sum w_i C_i$ is the total weighted completion time.)
L_i	Lateness for $J_i, L_i = C_i - d_i$
$C_{\max}$	Makespan $= \text{Max}\{C_i, i = 1, \ldots, n\}$
C^*	Optimal makespan
$L_{\max}$	$= \text{Max}\{L_i, i = 1, \ldots, n\}$
U_i	$= 1$, if $L_i > 0$, and 0 otherwise (Hence, ΣU_i denotes the total number of tardy jobs.)
M_j	Machine $j, j = 1, \ldots, m$
s_{ji}, t_{ji}	the ith nonavailability interval for $M_j$is $[s_{ji}, t_{ji}]$ for all j, where $0 \leq s_{ji} \leq t_{ji}$

For the problem with at most one nonavailability interval in each machine, we will simply use $[s_j, t_j]$ to denote the nonavailability interval for M_j. Furthermore, for a single machine problem with only one nonavailability interval, $[s, t]$ will be used.

In this chapter we focus on problems in which job preemption is not allowed. There are three types of machine unavailability (Lee, 1996, 1999) discussed in the literature.

Resumable: A machine is called *resumable* if a job that cannot be finished before a down period of a machine can be continued without any penalty after the machine becomes available again.

Nonresumable: A machine is called *nonresumable* if the job that cannot be completed before a period of machine nonavailability must be totally restarted rather than continuing after the machine is brought back on line.

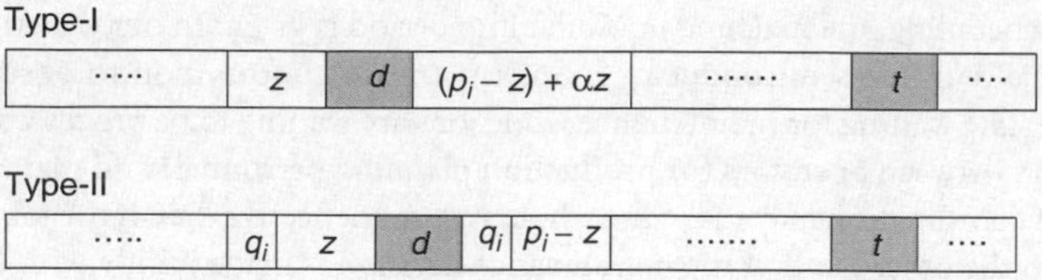

FIGURE 22.1 The semiresumable case.

Semiresumable: A machine is called *semiresumable* if the nonfinished job before a period of machine nonavailability must be partially restarted. There are two types of *semiresumabililty*: In Type-I, in addition to processing the nonfinished part, the machine needs to process extra work that is proportional to the finished part of that job. In Type-II, if a job is not processed to completion before the machine is stopped for maintenance, an additional setup is necessary when the processing is resumed. Figure 22.1 depicts the two types. In the Type-I figure, z is the finished part of job i before the nonavailability interval, d, and $0 \leq \alpha \leq 1$. In the Type-II figure, q_i is the setup time of job i.

We use r-a, *nr-a*, *sr*1-*a*, and *sr*2-*a* to denote a resumable, nonresumable, Type-I semiresumable, and Type-II semiresumable availability constraints, respectively. It is important to note that the Type-I semiresumable case becomes a resumable case when $\alpha = 0$ and a nonresumable case when $\alpha = 1$. Also, the Type-II semiresumable case becomes a nonresumable case if $z = 0$ and a resumable case if $q_i = 0$. Hence, both resumable and nonresumable cases are special cases of the semiresumable case.

In this chapter, we follow the notation of Pinedo (2002) and the additional notation introduced by Lee (1997). This notation allows us to define a scheduling problem using three fields, $\alpha/\beta/\gamma$. The α field denotes the machine environment. The β field gives problem characteristics and constraints. The γ field contains the performance measure to be minimized. For example, $Pm/r\text{-}a/C_{\max}$ and $Pm/nr\text{-}a/C_{\max}$ denote m parallel machines scheduling by minimizing the makespan with resumable and nonresumable availability constraints, respectively, and $F2/r\text{-}a/C_{\max}$ denotes two-machine flow shop scheduling by minimizing the makespan with a resumable availability constraint.

22.3 Problems with Nonavailability Intervals at the Beginning of Planning Horizon ($s_j = 0$ for all j)

Lee (1991) studies a problem in which there are m parallel identical machines to process n jobs by minimizing the makespan, yet each machine may not be available at time zero ($s_j = 0$ for all j). It was shown that the makespan obtained by applying the classical longest processing time first (LPT) algorithm to this problem, denoted as C_{LPT}, satisfies $C_{\text{LPT}} \leq [3/2 - 1/(2m)]C^*$. A modified LPT (MLPT) algorithm was then developed to solve the problem with performance $C_{\text{MLPT}} \leq (4/3)C^*$, where C_{MLPT} is the makespan obtained by applying the MLPT algorithm to this problem. This MLPT algorithm first treats the nonavailability interval at each machine as the processing time of an artificial job and applies the LPT algorithm to assign jobs to machines while interchanging between jobs at different machines, if necessary, to prevent the situation that there are more than one artificial jobs at any machine. Finally, the artificial job in each machine is sequenced first in that machine. Kellerer (1998) studies the same problem and provides a 5/4-approximation algorithm. In the same settings, yet with two machine open shop problem, it is solvable in $O(n)$ time (Lu and Posner, 1993).

For the problem of minimizing the total completion time, $\sum C_i$, the shortest processing time first (SPT) algorithm is still optimal (Kaspi and Montruil, 1988). Also, for $F2//C_{\max}$ with $s_j = 0$ for all j, Johnson's algorithm is still optimal (Lee, 1997). For the same problem with no wait constraint, it is solvable in $O(n \log n)$ time (Kubzin and Strusevich, 2002).

22.4 Problems with Nonavailability Intervals During the Planning Horizon ($s_j > 0$)

22.4.1 One Machine Problems

22.4.1.1 The Resumable Case

Lee (1996) studies a single machine problem with different performance measures assuming that there is only one nonavailability interval. This study shows that an arbitrary sequence is optimal for $1/r\text{-}a/C_{\max}$, the SPT sequence is optimal for $1/r\text{-}a/\sum C_i$, and the earliest due date first (EDD) sequence solves $1/r\text{-}a/C_{\max}$ optimally. That is, the algorithms are the same as those for the classical problems without availability constraints. For $1/r\text{-}a/\sum U_i$, the Moore-Hodgson Algorithm with slight modification is still optimal. However, for $1/r\text{-}a/\sum w_i C_i$, the weighted shortest processing time first (WSPT) algorithm is not optimal unless there is one job, J_i, with a completion time equal to the starting time of the nonavailability interval ($C_i = s$) in the WSPT solution. This problem is NP-hard even if $w_i = p_i$ for all i. Furthermore, $F_w(\text{WSPT})/F_w^*$ can be arbitrarily large even if $w_i = p_i$ for all i, where $F_w(\text{WSPT})$ is the weighted total completion time obtained by applying WSPT to the problem and F_w^* is the optimal weighted total completion time. Another heuristic approach was developed to solve the problem with an error-bound analysis. Dynamic programming was also proposed to solve the problem optimally and hence it was deemed NP-hard in the ordinary sense (Lee, 1996).

22.4.1.2 The Nonresumable Case

On the other hand, for the nonresumable case, Adiri et al. (1989) and Lee and Liman (1991) show that $1/nr\text{-}a/\sum C_i$ is NP-hard. Lee and Liman (1991) prove that applying SPT to $1/nr\text{-}a/\sum C_i$ will have $F_{\text{SPT}}/F^* \leq 9/7$, where F_{SPT} is the total completion time by applying SPT to the problem and F^* is the optimal total completion time. Lee (1996) shows that $1/nr\text{-}a/C_{\max}$ is NP-hard. It is also noticed in the paper that the problem becomes NP-hard in the strong sense if there are multiple nonavailability intervals as the problem can be transformed from the Three-Partition problem. Since $1/nr\text{-}a/C_{\max}$ is NP-hard it implies that $1/nr\text{-}a/L_{\max}$ and $1/nr\text{-}a/\sum U_i$ are also NP-hard. It is shown that if we apply the Moore-Hodgson Algorithm to solve $1/nr\text{-}a/\sum U_i$ then the maximum deviation of the number of tardy from optimal is one. The error of applying EDD to solve $1/nr\text{-}a/L_{\max}$ is at most $p_{\max}$ where $p_{\max} = \max\{p_i : i = 1, \ldots, n\}$. It is clear that $1/nr\text{-}a/\sum w_i C_i$ is NP-hard and $F_w(\text{WSPT})/F_w^*$ can be arbitrarily large even if $w_i = p_i$ for all i. Leon and Wu (1992) study one-machine scheduling with multiple nonavailability intervals with ready-time constraints on the jobs. The objective is to minimize the maximum lateness. They propose a branch and bound algorithm to solve the problem. Saddfi et al. (2001) provide a 20/17-approximation algorithm for problem $1/nr\text{-}a/\sum C_i$.

22.4.1.3 The Semiresumable Case

Since the nonresumable case is a special case of the semiresumable case, both types of semiresumable problems ($1/sr1\text{-}a/C_{\max}$, $1/sr2\text{-}a/C_{\max}$, $1/sr1\text{-}a/\sum C_i$, $1/sr2\text{-}a/\sum C_i$, $1/sr1\text{-}a/L_{\max}$, $1/sr2\text{-}a/L_{\max}$, $1/sr1\text{-}a/\sum U_i$ and $1/sr2\text{-}a/\sum U_i$) are all NP-hard.

22.4.2 Parallel Machine Problems

22.4.2.1 The Resumable Case

It is well known that the classical parallel machine scheduling problem without an availability constraint, ($Pm//C_{\max}$), is NP-hard. Clearly, $Pm/r\text{-}a/C_{\max}$ and $Pm/nr\text{-}a/C_{\max}$ are both NP-hard. Lee (1996) studies the problem in which each machine has at most one nonavailability interval and shows that if $s_j > 0$ for all j then C_{LPT}/C^* can be arbitrarily large even for the two-machine problem. This implies that no polynomial approximation scheme exists unless $P = \text{NP}$. Lee (1996) noted that in the classical LPT algorithm, assigning jobs one by one to the minimum loaded machine is aiming for the job to be finished as early as possible. Even though in the classical problem ($Pm//C_{\max}$) these two goals are equivalent, there may be different results if we apply the following two versions of LPT to our $Pm/r\text{-}a/C_{\max}$.

LPT1: Assigning a job to the least loaded machine.

LPT2: Assigning a job on the top of the list to a machine such that the finishing time of that job is minimized.

Lee (1996) shows that C_{LPT1}/C^* can be arbitrarily large even for $P2/r\text{-}a/C_{\max}$. However, $C_{\text{LPT2}}/C^* \leq 3/2 - 1/2m$, assuming that there is at least one machine always available. Since $1/r\text{-}a/\sum w_i C_i$ is NP-hard, $P2/r\text{-}a/\sum w_i C_i$ is also NP-hard. A pseudo-polynomial dynamic programming algorithm is then provided to solve $P2/r\text{-}a/\sum w_i C_i$ optimally.

22.4.2.2 The Nonresumable Case

For $P2/nr\text{-}a/\sum C_i$ with $s_1 = t_2 = \infty$ (machine 1 is always available while machine 2 is not available from s_2 to infinity), Lee and Liman (1993) show that the problem is NP-hard. They also provide an SPT-based heuristic with $F_H/F^* \leq 3/2$. Mosheviov (1994) studies $Pm/nr\text{-}a/\sum C_i$ under the condition that M_j is available only in the interval $[x_j, y_j]$ and shows that SPT is asymptotically optimal as the number of jobs approaches infinity. For $Pm/nr\text{-}a/C_{\max}$, Lee (1996) shows that $C_{LS}/C^* \leq m$ under the assumption that at least one machine is always available ($s_j = \infty$ for some j), where LS denotes the List Scheduling, which is defined as "Given an arbitrary order of jobs, assign the job to the machine such that the job can be finished as early as possible." He also shows that $C_{\text{LPT}}/C^* \leq (m+1)/2$ and the bound is tight.

22.4.3 Flow Shops

In the past few years, a significant amount of literature has been devoted to the study of two-machine flow shop and open shop problems with availability constraints. Since most problems are NP-hard, and some are even NP-hard in the strong sense, studies have developed heuristics with error-bound analysis or have developed fully polynomial approximation schemes to solve the problems.

22.4.3.1 The Resumable Case

Lee (1997) shows that both $F2/r\text{-}a(M_1)/C_{\max}$ and $F2/r\text{-}a(M_2)/C_{\max}$ are NP-hard, where $r\text{-}a(M_1)$ and $r\text{-}a(M_2)$ indicate that there is only one nonavailability interval located in machine 1 and machine 2, respectively. Lee (1997) also shows that for $F2/r\text{-}a(M_1)/C_{\max}$, we have $C_{JA}/C^* \leq 2$ and the bound is tight. An improved heuristic (call it H1) is then provided with $C_{H1}/C^* \leq 3/2$. Interestingly, for $F2/r\text{-}a(M_2)/C_{\max}$, $C_{JA}/C^* \leq 3/2$ and there exists an example with $C_{JA}/C^* \to 3/2$ as $n \to \infty$. Lee (1997) also provides an improved heuristic (call it $H2$) with $C_{H2}/C^* \leq 4/3$. Lee (1997) notes that the problem is irreversible, an important characteristic that is distinct from the classical flow shop problem. Later, Cheng and Wang (2000) provide an algorithm for $F2/r\text{-}a(M_1)/C_{\max}$ with an improved performance, $C_H/C^* \leq 4/3$. Kubiak et al. (2002) show that no polynomial-time algorithm with a fixed worst-case performance ratio exists unless $P = NP$, provided that in the flow shop problem there are either at least two intervals on the second machine or at least one interval on each machine. A branch and bound algorithm is provided to solve the problem with multiple nonavailability intervals. Blazewicz et al. (2001) provide constructive and local search based heuristic algorithms for $F2/r\text{-}a/C_{\max}$. Braun (2002) and Braun et al. (2002) derive sufficient conditions for the optimality of Johnson's permutation in the case of one or more nonavailability intervals. They show that usually Johnson's permutation remains optimal in the case of nonavailability intervals.

22.4.3.2 The Nonresumable and Semiresumable Cases

Lee (1999) studies $F2/sr1\text{-}a/C_{\max}$ where an availability constraint is imposed only on one machine as well as being imposed on both machines. The problem is clearly NP-hard because a special case, $1/nr\text{-}a/C_{\max}$, is already NP-hard. Furthermore, if we allow multiple nonavailability intervals, then the problem is strongly NP-hard as its special case, the one machine problem, is already strongly NP-hard (Lee, 1996). Lee (1999) also provides a pseudo-polynomial dynamic programming algorithm to solve the problem $F2/sr1\text{-}a(M_1)/C_{\max}$ optimally in which the semiresumable availability constraint is imposed on machine 1. If we apply Johnson's algorithm to sequence jobs for $F2/sr1\text{-}a(M_1)/C_{\max}$,then $C_{JA} \leq 2C^*$ and the bound is tight. It is interesting to note that the tight error bound is independent of the value of α. Thus, both the resumable and nonresumable cases have the same error bound. On the other hand, if we apply Johnson's algorithm to sequence jobs for $F2/sr1\text{-}a(M_2)/C_{\max}$ then $C_{JA}/C^* \leq \max\{3/2, 1+\alpha\}$ and the bound is tight. That is, if we apply Johnson's algorithm to sequence jobs for $F2/nr\text{-}a(M_2)/C_{\max}$, then $C_{JA}/C^* \leq 2$ and the bound is tight. An improved algorithm, call it H, is then developed with $C_H/C^* \leq 3/2$. For the problem, where an availability constraint is imposed on both machines, $F2/sr1\text{-}a(M_1, M_2)/C_{\max}$, the problem is NP-hard even if $s_1 = s_2 = s$, and $t_1 = t_2 = t$. Nonetheless, under such a special case, $s_1 = s_2 = s$, and $t_1 = t_2 = t$, we have $C_{JA}/C^* \leq 1 + \alpha$. That is, Johnson's algorithm is optimal for $F2/r\text{-}a(M_1, M_2)/C_{\max}$, if $s_1 = s_2 = s$, and $t_1 = t_2 = t$. For the general problem $F2/r\text{-}a(M_1, M_2)/C_{\max}$, Lee (1999) shows that C_{JA}/C^* can be arbitrarily large. This also implies that no polynomial approximation scheme exists unless $P = \text{NP}$.

22.4.3.3 A Flow Shop with No-wait

Espinouse et al. (1999, 2001) show that $F2/nw, nr\text{-}a/C_{\max}$ is NP-hard, where *nw* denotes the no-waiting constraint. This problem is NP-hard in the strong sense if there is more than one nonavailability interval at one machine. The theorems and the corresponding proofs are not surprising as they hold even for the one-machine problem (see Lee, 1996). For the problem with a single nonavailability interval, Espinouse et al. (1999, 2001) provide a heuristic with $C_H/C^* \leq 2$. Wang and Chen (2001) improve this result to $C_H/C^* \leq 5/3$. Cheng and Liu (2003) present a polynomial time approximation scheme for the same problem yet machine 1 and machine 2 have overlapping nonavailability intervals or only one machine has an nonavailability interval. Kubzin and Strusevich (2002) show that $F2/nw, nr\text{-}a/C_{\max}$ $F2/nw, r\text{-}a/C_{\max}$ and $F2/nw, sr1\text{-}a/C_{\max}$ are all NP-hard. They provide a heuristic for all three types of problems with $C_H/C^* \leq 3/2$. They also provide a heuristic particularly for $F2/nw, r\text{-}a/C_{\max}$ with $C_H/C^* \leq 4/3$, which is a better error bound.

22.4.4 An Open Shop

The two-machine open shop problem to minimize the makespan with a single nonavailability interval and a resumable case is shown to be NP-hard by Breit et al. (2001). They also provide a heuristic with $C_H/C^* \leq 4/3$. A pseudo-polynomial algorithm has been provided by Lorigeon et al. (2001) to solve this problem optimally. Similarly to the flow shop, Briet (2000) shows that no polynomial-time algorithm with a fixed worst-case performance ratio exists unless $P = \text{NP}$, provided that in the open shop problem there are at least two intervals on one machine and at least one interval on the other machine. Kubzin et al. (2003) provide a polynomial approximation scheme for problems under each of the following cases: (i) one nonavailability interval on each machine and (ii) several nonavailability intervals on one machine. For the nonresumable case, Breit et al. (2002) show that for the problem with one nonavailability interval in one machine and two nonavailability intervals on the other machine, no polynomial approximation scheme exists unless $P = \text{NP}$. They also provide a heuristic with $C_H/C^* \leq 2$ when there is one nonavailability interval on each machine and another heuristic with $C_H/C^* \leq 4/3$ for the problem with only one nonavailability interval.

22.5 Problems with Nonavailability Intervals as Decision Variables

22.5.1 One-Machine Problems

If the starting time of the nonavailability interval is a decision variable, i.e., it is decided jointly with the job scheduling, Qi et al. (1999) show that the problem is NP-hard in the strong sense if there are multiple such intervals. They also present a branch and bound algorithm for solving the problem under the nonresumable assumption. Grave and Lee (1999) and Grave (1998) study the problem under the Type-II semiresumable assumption. The Type-II semiresumable assumption is motivated by industrial application, in which a job not processed to completion before the machine is stopped for maintenance requires additional setup when the processing is resumed. Furthermore, they observe that the maintenance is usually required after certain periods of time that are usually larger than the planning horizon of shop floor scheduling. Hence, there are at most one or two machines down for maintenance during a planning horizon. The problem of greatest interest is the case when a certain period of time has passed and, hence, in the current planning horizon the maintenance activity must be implemented. In the problem they studied, the time period between two consecutive maintenance periods cannot exceed a fixed number, say T. That is, the maintenance must be performed within a fixed period, T, and the time for the maintenance is a decision variable.

Property 22.1

(Graves and Lee, 1999) There exists an optimal solution such that no more than $2m$-2 maintenance periods are required in any $1/sr2\text{-}a/$ problems with regular performance measures if $\sum_{i=1}^{n}(q_i + p_i) \leq mT$, for some positive integer m, where q_i is the setup time of job i.

Hence, for any problem with $\sum_{i=1}^{n}(q_i + p_i) \leq 2T$, there are at most two maintenance periods required. They study the problem with the objective of minimizing the total weighted job completion times, $1/sr2\text{-}a/\sum w_i C_i$, and minimizing the maximum lateness, $1/sr2\text{-}a/L_{\max}$. In both cases, they study two scenarios concerning the planning horizon. When the planning horizon is not short in relation to T the problem with either objective function is NP-hard, and they present pseudo-polynomial time dynamic programming algorithms for both objective functions. For example, if $\sum_{i=1}^{n}(q_i + p_i) \leq 2T$, and the problems are denoted as $1/sr2\text{-}a, 2T/\sum w_i C_i$ and $1/sr2\text{-}a, 2T/L_{\max}$, both problems are NP-hard. In the second scenario, the planning horizon is short in relation to T. However, part of the period T may have elapsed before any jobs are scheduled in this planning horizon, and the remaining time before the next maintenance activity is shorter than the current planning horizon. This is denoted as $v \leq T'$. Thus, at least one maintenance activity must be scheduled in this planning horizon. They show that the problem of minimizing the total weighted completion times ($1/2sr\text{-}a, v \leq T'/\sum w_i C_i$) in this scenario is NP-hard, while the SPT rule and the EDD rule are optimal for the total completion time problem, $1/2sr\text{-}a, v \leq T'/\sum C_i$, and the maximum lateness problem, $1/2sr\text{-}a, v \leq T'/L_{\max}$, respectively.

22.5.2 Parallel Machine Problems

Lee and Chen (2000) study the problem of processing a set of jobs on several parallel machines in which each machine must be maintained once during the planning horizon. Their objective is to schedule maintenance activities and jobs so that the total weighted completion time of all jobs is minimized. Two cases are demonstrated in this paper. In the first case, there are sufficient resources such that different machines can be maintained simultaneously if necessary (hence, the machines are treated independently). The problem is denoted as $Pm/\,nr\text{-}a, ind\,/\sum C_j$. In the second case, only one machine can be maintained at any given time, (hence machines are dependent) and the problem is denoted as $Pm\,/\,nr\text{-}a, dep\,/\sum C_j$. In such a case, they assume that $T \geq mt$; otherwise, there is simply no feasible schedule. They first show that, even when all the jobs have the same weight, both cases of the problem are NP-hard even though the corresponding classical problem, the one without maintenance, is polynomially solvable. They then provide a pseudo-polynomial algorithm to solve both problems. Hence, both problems are NP-hard in the ordinary sense when the number of machines is fixed. Since the complexity is very high (even though it is pseudo-polynomial), they then propose branch and bound algorithms based on the *column generation* approach for solving both cases of the problem. The algorithms are capable of solving medium-size problems to optimality within a reasonable computational time. Their approach can also solve the general problem when at most k machines can be maintained simultaneously ($k \leq m$, and m is the total number of machines).

22.6 Problems with Machine Speed Changes after Maintenance or Repair Activities

Most of the literature assumes that after the nonavailability interval (maintenance or breakdown-repair activity) the machine speed remains the same. Motivated by a problem commonly found in electronic assembly lines, Lee and Leon (2001) study a job scheduling problem with maintenance activity on a single machine, in which the maintenance activity will change the production rate of the equipment under consideration (or equivalently, change the job processing times). They call such maintenance activity a *rate-modifying* activity. Thus, the processing times of jobs depend on whether the job is processed before or after the rate-modifying activity. Clearly, rate-modifying activities fall under a certain class of maintenance activities and the machine is unavailable during that activity. This type of problem occurs often in the surface-mount technology lines of electronic assembly systems, in which some of the pick-and-place nozzles may not be working, which results in the machine running at a less-efficient speed. In this situation, a decision must be made about whether to stop the machine and fix the problem, or simply to continue producing at an inferior production rate. The decisions under consideration are when to schedule the rate-modifying activity and the sequence of jobs to optimize some performance measure. This section summarizes some results from Lee and Leon (2001).

Corresponding to each job i and a fixed known processing time p_i, there is a constant, $\alpha_i > 0$, *called the modifying rate*. If job i is processed before the rate-modifying activity, its processing time is p_i. On the other hand, if job i is processed after the rate-modifying activity, then the processing time is $\alpha_i p_i$. Note that α_i can be greater or less than one although, in most practical problems, $\alpha_i < 1$. We use rm to denote the rate-modifying activity and assume that t is its duration length. Hence, $1/rm/C_{\max}$, $1/rm/\sum C_i$, $1/rm/\sum w_i C_i$, and $1/rm/L_{\max}$ denote the single machine scheduling problem with the rate-modifying activity with the objective of minimizing $C_{\max}$, $\sum C_j$, $\sum w_j C_j$, and $L_{\max}$, respectively. Note that for all these problems, it is optimal to start the rate-modifying activity only after the completion of some job or at the beginning of the sequence. That is, we will never start the rate-modifying activity during the processing of a job.

For the $1/rm/C_{\max}$ problem, whether or not we should have a rate-modifying activity during the planning horizon depends on the trade-off between the saving and delay caused by the rate-modifying activity. If we let $s = \sum_{i \in V} p_i(1-\alpha_i)$, where $V = \{i: \alpha_i < 1\}$, then it is optimal to sequence arbitrarily all jobs without the rate-modifying activity when $s \leq t$. Otherwise, it is optimal to sequence arbitrarily all jobs not in V, followed by a rate-modifying activity, then sequence arbitrarily all jobs in V.

For the $1/rm/\sum C_i$ problem, if we fix the position of the rate-modifying activity, say immediately after the completion of the job at the kth position, it is easy to calculate the contribution of a ith position job to the objective function, $\sum C_i$. The objective function can be expressed as follows:

$$Z(k) = \sum_{i=1}^{k} (n-i+1) p_{[i]} + (n-k)t + \sum_{i=k+1}^{n} (n-i+1)\,\alpha_{[i]}\, p_{[i]}, \quad k = 1, \ldots, n-1$$

$$Z(0) = nt + \sum_{i=1}^{n} (n-i+1)\,\alpha_{[i]}\, p_{[i]}$$

$$Z(n) = \sum_{i=1}^{n} (n-i+1) p_{[i]}$$

If we define

$$B(k) = \sum_{i=1}^{k} (n-i+1)p_{[i]} + \sum_{i=k+1}^{n} (n-i+1)\alpha_{[i]}\, p_{[i]}$$

then $Z(k) = B(k) + (n-k)t$. For a given k, minimizing $Z(k)$ is equivalent to minimizing $B(k)$. The contribution of the ith job to $B(k)$ is $(n-i+1)p_{[i]}$ or $(n-i+1)\alpha_{[i]}p_{[i]}$ depending on whether $i \le k$ or $i > k$, respectively. This problem can be solved by a weighted-bipartite matching problem.

For the $1/rm/\sum w_i C_i$ problem, a dynamic programming algorithm has been developed to solve the problem in pseudo-polynomial time under the agreeable assumption that $p_i < p_j$ implies $\alpha_i p_i \le \alpha_j p_j$. It is an open question whether the general problem is NP-hard.

For the $1/rm/L_{\max}$ problem, it is proved that the EDD sequence is optimal if $\alpha_i < 1$ for all i. For the general α_i problem, it is shown that the problem is NP-hard and a pseudo-polynomial algorithm to solve the problem optimally is provided.

22.7 Problems with Machine Breakdown and Repair Activities

Note that, in the problem studied by Lee and Leon (2001), it was assumed that if the production planner continues to run the machine without fixing it, the machine breakdown will not happen. However, in a realistic case, it is possible that the machine breakdown can happen at any time and will have to be repaired immediately. That is, the nonavailability can be due to machine breakdown in addition to the maintenance activity. Studies of machine scheduling with potential breakdown includes Glazebrook (1984, 1987, 1991), Adiri et al. (1989), Pinedo and Rammouz (1988), Birge and Glazebrook (1988), Birge et al. (1990), Frostig (1991), and Albers and Schmidt (2001). Very few of these studies discussed the case in which the machine has partially failed and maintenance will change the machine speed. Lee and Lin (2001) study the problem in which maintenance and *repair* activities can occur and will change the machine speed. When a machine is running at less than an efficient speed, a production planner can decide to stop the machine and fix it or wait and fix it later. If the choice is made to continue running the machine without fixing it, it is possible that the machine can breakdown and immediate repair will be required, where the time to breakdown is assumed to be a random variable following certain distribution. Both maintenance and repair activities can change the machine speed from a sub-normal production rate to a normal one. Hence, Lee and Lin (2001) call these *rate-modifying activities* and denote them as *rms*. The purpose is to sequence jobs and schedule maintenance activity simultaneously to optimize regular performance measures. This section uses some results from Lee and Lin (2001).

Both *resumable and nonresumable* cases are considered by Lee and Lin (2001) and the expected makespan ($E[C_{\max}]$), the total expected completion time ($\sum E[C_j]$) and the maximum expected lateness ($\max E[L_j]$) are used as the objective functions. They assume that the processing times of a job before and after the rate-modifying activity are known. However, the machine can be down at any time. The time that the machine will be down is stochastic while the repair time is fixed. The repair time is assumed to be longer than the maintenance time. In case the machine is down before the maintenance, then, after the repair activity, the production rate of the machine will be back to normal. The decisions under consideration are when to schedule the maintenance and the sequence of jobs to optimize some regular performance measure. Several interesting results are shown.

The *static list policy* was used to analyze the problem. The decision to insert maintenance activity at a certain point in the process will not change during the processing unless the machine is broken before we reach the insertion point. In such a case, we will not insert the maintenance activity. Furthermore, it is assumed that the repair or maintenance activity will bring the machine to its normal speed *until the end of the planning horizon* and that the machine breakdown will not happen during the remainder of the planning horizon, as a shop floor scheduling problem planning horizon is usually short and it is very rare to have the machine break down again after repairs or maintenance activities are implemented.

The $1/rms,r\text{-}a/\mathrm{E}[C_{\max}]$, $1/rms,r\text{-}a/\sum\mathrm{E}[C_j]$ and $1/rms,r\text{-}a/\max\mathrm{E}[L_j]$ problems for resumable cases, as well as the $1/rms,nr\text{-}a/\mathrm{E}[C_{\max}]$, $1/rms,nr\text{-}a/\sum\mathrm{E}[C_j]$, and $1/rms,nr\text{-}a/\max\mathrm{E}[L_j]$ problems for nonresumable cases are studied in the paper. In both cases, it was assumed that $\alpha_i = \alpha$ for all i.

For the resumable cases, results obtained are similar to those in the classical problems. For example, for the $1/rms,r\text{-}a/\sum E[C_i]$ problem, it is optimal to sequence jobs in the SPT order. Hence, we can calculate the total expected cost of inserting the maintenance at the kth position for $k = 1, \ldots, n+1$. The optimal solution is the one with the minimal total expected cost. Similarly, for $1 \mid rms, r\text{-}a \mid \max\mathrm{E}[L_i]$, it is optimal to sequence jobs in the EDD order.

On the other hand, results obtained for the nonresumable case are more interesting. For example, (i) for the problem of minimizing the makespan, if we decide not to do maintenance activity, then it is optimal to sequence jobs in the SPT order when the distribution function is concave and in the LPT order when the distribution function is convex up to the total sum of processing times; (ii) for the problem of minimizing the total expected completion times, it is optimal to sequence jobs in the SPT order when the distribution function is concave, (iii) for the problem of minimizing the total expected lateness, when the problem satisfies the agreeable assumption, (i.e., if $d_i < d_j$, then $p_i \le p_j$ for all $1 \le i, j \le n$), it is optimal to sequence jobs in the EDD order when the distribution function is concave. (A distribution function F is said to be concave on $[0, \alpha]$ if the function F is continuous and satisfies $F(x+s) - F(x) \ge F(y+s) - F(y)$ for each x, y, s such that $0 \le x < y, s > 0$, and $y+s \le \alpha$, $F(\alpha) = 1$ and $F(x) = 0$ for $x < 0$. Also, a distribution function F is said to be convex on $[0, \alpha]$ if the function is continuous, $F(\alpha) = 1$, $F(x) < 1$ for $x < \alpha$, $F(x) = 0$ for $x < 0$, and $F(x+s) - F(x) \le F(y+s) - F(y)$ for each x, y, s such that $0 \le x < y, s > 0$, and $y+s \le \alpha$.)

22.8 Conclusions and Future Research

Machine scheduling with availability constraints has attracted much attention in the scheduling field recently. Most works with deterministic models assume that the nonavailability intervals are given and then focus on the job scheduling. The works with stochastic models focus on job scheduling with potential uncertain machine breakdown. Here are some potential research topics.

(i) Problems with unavailability periods at the beginning of periods. In many industrial settings, the assignment of job scheduling in the previous horizon may not be suitable for the current horizon due to urgent job requirements and special setup times. In such cases, it may be necessary to move jobs (already assigned) from one machine to another one. This involves the possible extra sequence-dependent setup times plus transportation time and costs.

(ii) As mentioned above, it will be more effective and more interesting to have the maintenance decision made jointly with the job scheduling. In such cases, the machine nonavailability interval is a decision variable. Due to the popularity of supply chain management, logistics scheduling involving integrated maintenance and job-scheduling decisions has become more important. This is a relatively less studied area.

(iii) Blazewicz et al. (2000) study the multiprocessor tasks scheduling problem with availability constraints. Schmidt (2000) study online scheduling policy for the machine scheduling problem with availability constraints. These two areas are also relatively less studied.

(iv) Semiresumable problems, especially Type-II, are also less studied. This type of unavailability constraint is popular in industrial settings and is worth more study.

(v) When facing breakdown disruptions, the original scheduling plans are rarely executed as smoothly as anticipated. As mentioned above, most literature assumes that the nonavailability intervals are given in advance while doing the job scheduling. Research seldom discusses the postdisruption remedy policy. That is, how do we reschedule in a short time after the disruption such that the deviation from the original plan is minimized. Qi et al. (2001) study a problem in the one machine environment. Lee and Yu (2003) study multiple machine problems under various scenarios. Their model also includes transportation costs for moving jobs from disrupted machines to nondisrupted machines and the deviation (from the original plan) cost. More research is needed in this area.

References

Adiri, I., J. Bruno, E. Frostig, and A. H. G. Rinnooy Kan, Single Machine Flow-Time Scheduling with a Single Breakdown, *Acta Informatica*, **26**, (1989), pp. 679–696.

Albers, S. and G. Schmidt, Scheduling with Unexpected Machine Breakdowns, *Discrete Applied Mathematics*, **110**, (2001), pp. 85–99.

Bean, J.C., J.R. Birge, J. Mittenthal, and C.E. Noon, Matchup Scheduling with Multiple Resources, Release Dates and Disruptions, *Operations Research*, **39**, (1991), pp. 470–483.

Birge, J. and K.D. Glazebrook, Assessing the Effects of Machine Breakdowns in Stochastic Scheduling, *Operations Research Letters*, **7**, (1988), pp. 267–271.

Birge, J., J.B.G. Frenk, J. Mittenthal, and A.H.G. Rinnooy Kan, Single-Machine Scheduling Subject to Stochastic Breakdowns, *Naval Research Logistics*, **37**, (1990), pp. 661–677.

Blazewicz, J., J. Breit, P. Formanowicz, W. Kubiak, and G. Schmidt, Heuristic Algorithms for the Two-Machine Flowshop with Limited Machine Availability, *Omega*, **29**, (2001), pp. 599–608.

Blazewicz, J., P. Dell'Olmo, M. Drozdowski, and P. Maczka, Scheduling Multiprocessor Tasks on Parallel Processors with Limited Availability, *European Journal of Operational Research*, **149**, (2003), pp. 377–389.

Blazewicz, J., M. Drozdowskiu, P. Formanowicz, W. Kubiak, and G. Schmidt, Scheduling Preemptable Tasks on Parallel Processors with Limited Availability, *Parallel Computing*, **26**, (2000), pp. 1195–1211.

Blazewicz J., K. Ecker, G. Schmidt, and J. Weglarz, *Scheduling Computer and Manufacturing Processes*, 2nd Ed., (2001), Springer-Verlag, Berlin.

Braun, O., Scheduling with Limited Available Processors and with Limited Number of Preemptions, PhD thesis (in German), (2002), Saarland University, Saarbruecken, Germany.

Braun, O., T.-C. Lai., G., Schmidt, and Y.N. Sotskov, Stability of Johnson's Schedule with Respect to Limited Machine Availability, *International Journal of Production Research*, 40, (2002), pp. 4381–4400.

Breit J., *Heuristische Ablaufplanungsverfahren fur Flowshops und Openshops mit beschrankt verfugbaren Prozessoren*. Ph.D. Thesis, (2000), University of Sarrland, Sarrland.

Breit J., G. Schmidt and V. Strusevich, Two-Machine Open Shop Scheduling with an Availability Constraint, *Operations Research Letters*, **29**, (2001), pp. 65–77.

Breit J., G. Schmidt and V. Strusevich, NonPreemptive Two-Machine Open Shop Scheduling with Nonavailability Constraints, Department of Information and Technology Managemnt, Sarrland University, 2002.

Cheng, T.C.E. and Z. Liu, Approximability of Two-Machine Flowshop Scheduling with Availability Constraints, *Operations Research Letters*, **31**, (2003), pp. 319–322.

Cheng, T.C.E., and G. Wang, An Improved Heuristic for Two-Machine Flowshop Scheduling with an Availability Constraint, *Operations Research Letters*, **26**, (2000), pp. 223–229.

Espinouse, M.-L., P. Formanowicz and B. Penz, Minimizing the Makespan in the Two Machine No Wait Flow Shop with Limited Machine Availability, *Computers Industrial Engineering*, **37**, (1999), pp. 497–500.

Espinouse, M.-L., P. Formanowicz and B. Penz, Complexity Results on and Approximation Algorithms for the Two Machine No Wait Flow Shop with Limited Machine Availability, *Journal of Operational Research Society*, **52**, (2001), pp. 116–121.

Frostig, E., A Note on Stochastic Scheduling on a Single Machine Scheduling Subject to Breakdown — The Preemptive Repeat Model, *Probability in the Engineering and Informational Sciences*, **5**, (1991), pp. 349–354.

Glazebrook, K.D., Scheduling Stochastic Jobs on a Single Machine Subject to Breakdowns, *Naval Research Logistics Quarterly*, **31**, (1984), pp. 251–264.

Glazebrook, K.D., Evaluating the Effects of Machine Breakdowns in Stochastic Scheduling Problems, *Naval Research Logistics*, **34**, (1987), pp. 319–335.

Glazebrook, K.D., On Nonpreemptive Policies for Stochastic Single Machine Scheduling with Breakdowns, *Probability in the Engineering and Informational Sciences*, **5**, (1991), pp. 77–87.

Graves, G.H., *Application of the Genetic Algorithm for Global Scheduling and a Single Machine Scheduling Problem with Periodic Maintenance and Semiresumable Jobs.* Master of Science Thesis, Department of Industrial Engineering, Texas A&M University, College Station, TX, 1998.

Graves, G.H. and C.-Y. Lee, Scheduling Maintenance and Semiresumable Jobs on a Single Machine, *Naval Research Logistics*, **46**, (1999), pp. 845–863.

Kelleler, H. Algorithms for Multiprocessor Scheduling with Machine Release Dates, *IIE Transactions*, **31**, (1998), pp. 991–999.

Kubiak, W., J. Blazewicz, P. Formanowicz, J. Breit, and G. Schmidt, Two-Machine Flow Shops with Limited Machine Availability, *European Journal of Operational Research*, **136**, (2002), pp. 528–540.

Kubzin, M.A., V. Strusevich, J. Breit and G. Schmidt, Polynomial-Time Approximation Schemes for the Open Shop Scheduling Problem with Nonavailability Constraints, School of Computing and Mathematical Science, University of Greenwich. Paper 02/IM/100, 2002.

Kubzin, M.A. and V. Strusevich, Two-Machine Flow Shop No-Wait Scheduling with a Nonavailability Interval, School of Computing and Mathematical Science, University of Greenwich. 2002.

Lee, C.-Y., Parallel Machines Scheduling with Nonsimultaneous Machine Available Time, *Discrete Applied Mathematics*, **30**, (1991), pp. 53–61.

Lee, C.-Y., Machine Scheduling with an Availability Constraint, *Journal of Global Optimization*; Special Issue on Optimization on Scheduling Applications, **9**, (1996), pp. 395–416.

Lee, C.-Y., Minimizing the Makespan in the Two-Machine Flowshop Scheduling Problem with an Availability Constraint, *Operations Research Letters*, **20**, (1997), pp. 129–139.

Lee, C.-Y., Two-Machine Flowshop Scheduling with Availability Constraints, *European Journal of Operational Research*, **114**, (1999), pp. 420–429.

Lee, C.-Y. and Z.L. Chen, Scheduling of Jobs and Maintenance Activities on Parallel Machines, *Naval Research Logistics*, **47**, (2000), pp. 145–165.

Lee, C.-Y., L. Lei, and M. Pinedo, Current Trend in Deterministic Scheduling, *Annals of Operations Research*, **70**, (1997), pp. 1–42.

Lee, C.-Y. and J. Leon, Machine Scheduling with A Rate-Modifying Activity, *European Journal of Operational Research*, **128,** (2001), pp. 119–128.

Lee, C.-Y. and S.D., Liman, Single Machine Flow-Time Scheduling With Scheduled Maintenance, *Acta Informatica*, **29**, (1992), pp. 375–382.

Lee, C.-Y. and Liman, S.D. Capacitated Two-Parallel Machines Scheduling to Minimize Sum of Job Completion Times, *Discrete Applied Mathematics*, **41**, (1993), pp. 211–222.

Lee, C.-Y. and C.-S. Lin, Machine Scheduling with Maintenance and Repair Rate-Modifying Activities, *European Journal of Operational Research*, **135**, (2001), pp. 491–513.

Lee, C.-Y. and G. Yu, Logistics Scheduling Under Disruptions, Submitted for publication (Working paper, Department of Industrial Engineering and Engineering Management, The Hong King University of Science and Technology, Hong Kong 2003).

Leon, V.J. and S.D. Wu, On Scheduling with Ready-Times, Due-Dates and Vacations, *Naval Research Logistics*, **39**, (1992), pp. 53–65.

Lu, L. and M.E. Posner, An NP-hard Open Shop Scheduling Problem with Polynomial Average Time Complexity, *Mathematical Operations Research*, **18**, (1993), pp. 12–38.

Mosheiov, G., Minimizing the Sum of Job Completion Times on Capacitated Parallel Machines, *Mathematical Computing Modelling*, **20**, (1994), pp. 91–99.

Pinedo, M.L., *Scheduling: Theory, Algorithms and Systems*, 2nd Ed., (2002), Prentice-Hall, Englewood Cliffs, NJ.

Pinedo, M.L. and E. Rammouz, A Note on Stochastic Scheduling Machine Subject to Breakdown and Repair, *Probability in the Engineering and Informational Sciences*, **2**, (1988), pp. 41–49.

Qi, X., T. Chen and F. Tu, Scheduling the Maintenance on a Single Machine. Journal of the Operational Research Society, **50,**(1999), pp. 1071–1078.

Qi, X., J.F. Bard, and G. Yu, Disruption Management for Machine Scheduling, Working paper, Department of Management Science and Information Systems, (2002), College of Business Administration, The University of Texas, Austin, TX.

Saddfi, C., J. Blazewicz, P. Formanowicz, B. Penz, and C.Rapine. A Better Approximation Algorithm for the Single Machine Total Completion Time Scheduling Problem with Availability Constraints. International Conference on Industrial Engineering and Production Management, Quebeck, Canda, 2001.

Sanlaville, E. and G. Schmidt, Machine Scheduling with Availability Constraints, *Acta Informatica*, **5**, (1998), pp. 795–811.

Schmidt, G., Scheduling on Semi-identical Processors. *ZOR — Zeitschrift fur Operations Research*, **28**, (1984), pp. 153–162.

Schmidt, G., Scheduling Independent Tasks with Deadlines on Semi-identical Processors, *Journal of Operational Research Society*, **39**, (1988), pp. 271–277.

Schmidt, G., Scheduling with Limited Machine Availability, *European Journal of Operational Research*, **121**, (2000), pp. 1–15.

Wang, G. and T.C.E. Cheng, Heurisitcs for Two-Machine No-wait Flowshop Scheduling with Availability Constraint, *Information Process Letters*, **80**, (2001), pp. 305–309.

23

Scheduling with Discrete Resource Constraints*

J. Błażewicz
Technical University of Poznan

N. Brauner
IMAG, Grenoble

G. Finke
IMAG, Grenoble

23.1 Introduction

In this chapter we will consider the problem of *scheduling* tasks on *processors (machines)* under *discrete resource constraints.* The additional resources may stand for memory, channels, reentrant procedures in case of computer systems or for tools, fixtures, automated guided vehicles, and robots in case of manufacturing systems. Depending on their nature and usage, resources can be characterized by *classes, categories,* and *types.*

A division of resources into classes depends on the fact whether or not the resources are needed together with a processor (machine) during the processing of a given task set. If the answer is *yes,* then the resources will be called *processing resources.* Examples of these resources are memory, reentrant procedure, tools, and fixtures. If the answer is *no,* i.e., the resource is needed either *before* the processing of a task on a processor or *after* it, then the resources will be called *input-output resources* (or *i/o-resources*).

The classification into *categories* will concern two points of view. First, we differentiate three categories of resources from the viewpoint of resource constraints. We will call a resource *renewable,* if only its total usage, i.e., temporary availability at every moment, is constrained (in other words this resource, once used, may be used again after being released from a task). A resource is called *nonrenewable,* if only its total consumption, i.e., integral availability up to any given moment, is constrained (in other words this resource once used by some task cannot be assigned to any other task). A resource is called *doubly constrained,* if both total usage and total consumption are constrained. Second, we distinguish two resource categories from the viewpoint of resource divisibility: *discrete* (i.e., discretely divisible) and *continuous* (i.e., continuously divisible) resources. In other words, by a discrete resource we will understand the resource that can be allocated to tasks in discrete amounts from a given finite set of possible allocations, which in particular

*The work has been supported by the KBN grant 4T11C03925.

1-58488-397-9/$0.00+$1.50

may consist of one element only. Continuous resources, on the other hand, can be allocated in arbitrary, *a priori* unknown, amounts from a given interval.

The classification into *types* takes into account only the functions resources fulfill: resources of the same type are assumed to fulfill the same functions.

In this chapter only discrete resources will be considered, while Chapter 24 is devoted to the analysis of continuous resources. Section 23.2 will be concerned with processing resources that are discrete and renewable. We start with a complexity analysis of the basic scheduling problems on parallel processors. Then in Section 23.3, a management problem for processing resources will be considered. Section 23.4 is dealing with several cases of input-output resources.

23.2 Scheduling with Processing Resources

As we said this section will be devoted to the analysis of basic cases of scheduling with processing resources that are discrete and renewable. Thus, we may assume that s types of additional resources $R_1, R_2, \dots, R_s$ are available in $m_1, m_2, \dots, m_s$ units, respectively. Each task T_j requires for its processing one processor and certain fixed amounts of additional resources specified by the resource requirement vector $\mathbf{R}(T_j) = [R_1(T_j), R_2(T_j), \dots, R_s(T_j)]$, where $R_l(T_j)(0 \leq R_l(T_j) \leq m_l)$, $l = 1, 2, \dots, s$, denotes the number of units of resource R_l required for the processing of T_j. We will assume here that all required resources are granted to a task before its processing begins or resumes (in the case of preemptive scheduling), and they are returned by the task after its completion or in the case of its preemption.

Before discussing basic results in that area we would like to complement the notation scheme introduced in Chapter 1 of the Handbook that describes additional resources. In fact, they are denoted by parameter $\beta_2 \epsilon \{\emptyset, res\ \lambda\sigma\rho\}$, where [1]

$\beta_2 = \emptyset$: no resource constraints.

$\beta_2 = res\ \lambda\sigma\rho$: there are specified resource constraints; $\lambda, \sigma, \rho \epsilon \{\cdot, k\}$ denote respectively the number of resource types, resource limits, and resource requirements. If $\lambda, \sigma, \rho = \cdot$, then the number of resource types, resource limits, and resource requirements are, respectively, arbitrary, and if $\lambda, \sigma, \rho = k$, then, respectively, the number of resource types is equal to k; each resource is available in the system in the amount of k units and the resource requirements of each task are at most equal to k units.

At this point we would also like to present possible transformations among scheduling problems Π that differ only by their resource requirements (see Figure 23.1). In this figure six basic resource requirements are presented. All but two of these transformations are quite obvious. Transformation $\Pi(res \cdots) \propto \Pi(res\, 1 \cdot \cdot)$ has been proved for the case of saturation of machines and additional resources [2] and will not be presented

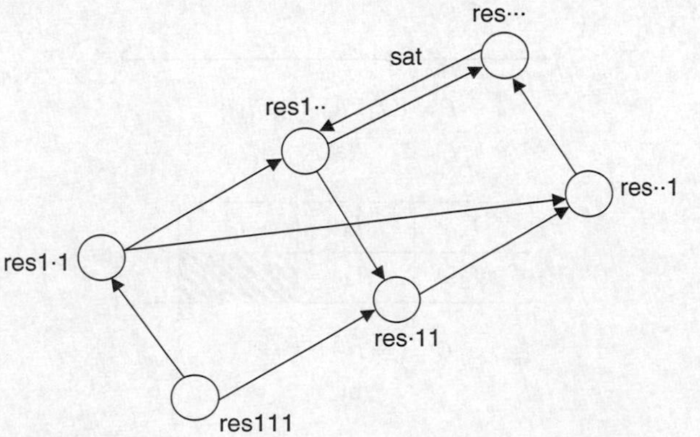

FIGURE 23.1 Polynomial transformations among resource constrained scheduling problems.

here. The second, $\Pi(res\ 1 \cdot\cdot) \propto \Pi(res \cdot 11)$, has been proved in [3]. To sketch its proof, for a given instance of the first problem we construct a corresponding instance of the second problem by assuming the parameters are all the same, except resource constraints. Then for each pair T_i, T_j such that $R_1(T_i)+R_1(T_j) > m_1$ (in the first problem), resource R_{ij} available in the amount of one unit is defined in the second problem. Tasks T_i, T_j require a unit of R_{ij}, while other tasks do not require this resource. It follows that $R_1(T_i) + R_1(T_j) \leq m_1$ in the first problem if and only if for each resource R_k, $R_k(T_i) + R_k(T_j) \leq 1$ in the second problem.

We will now pass to the presentation of some important results obtained for the above model of resource constrained scheduling. Space limitations prohibit us even from only quoting all these results; however, an extensive survey may be found in [4]. As an example, we chose the problem of scheduling tasks on parallel identical processors to minimize schedule length. Basic algorithms in this area will be presented.

Let us first consider the case of independent tasks and nonpreemptive scheduling. Let us start with problem $P2|res \cdots, p_j = 1|C_{\max}$. The problem of scheduling unit-length tasks on two processors with arbitrary resource constraints and requirements can be solved optimally by the following algorithm.

Algorithm 23.1 *Algorithm by Garey and Johnson for* $P2|res \cdots, p_j = 1|C_{\max}$ [2].

begin

Construct an n-node (undirected) graph G with each node labelled as a distinct task and with an edge joining T_i to T_j if and only if $R_l(T_i) + R_l(T_j) \leq m_l$, $l = 1, 2, \ldots, s$;

Find a maximum matching F of graph G;

Put the minimal value of schedule length $C^*_{\max} = n - |F|$;

Process in parallel the pairs of tasks joined by the edges comprising set F;

Process other tasks individually;

end;

Notice that the key idea here is the correspondence between maximum matching in a graph displaying resource constraints and the minimum-length schedule. The complexity of the above algorithm clearly depends on the complexity of the algorithm determining the maximum matching. There are several algorithms for finding it, the complexity of the most efficient by Kariv and Even [5] being

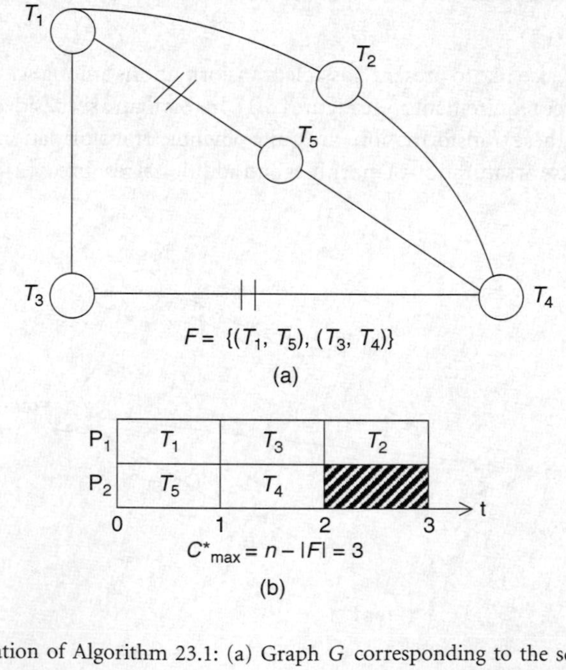

FIGURE 23.2 An application of Algorithm 23.1: (a) Graph G corresponding to the scheduling problem, (b) An optimal schedule.

$O(n^{2.5})$. An example of the application of this algorithm is given in Figure 23.2 where it is assumed that $n = 5, m = 2, s = 2, m_1 = 3, m_2 = 2, \mathbf{R}(T_1) = [1,1], \mathbf{R}(T_2) = [1,2], \mathbf{R}(T_3) = [2,1], \mathbf{R}(T_4) = [1,1]$, and $\mathbf{R}(T_5) = [2,1]$.

An even faster algorithm can be found if we restrict ourselves to the one-resource case. It is not hard to see that in this case an optimal schedule will be produced by ordering tasks in nonincreasing order of their resource requirements and assigning tasks in that order to the first free processor on which a given task can be processed because of resource constraints. Thus, problem $P2 \mid res\ 1 \cdot\cdot, p_j = 1 | C_{\max}$ can be solved in $O(n \log n)$ time.

If in the last problem tasks are allowed only for 0-1 resource requirements, the problem can be solved in $O(n)$ time even for arbitrary ready times and an arbitrary number of machines, by first assigning tasks with unit resource requirements up to m_1 in each slot, and then filling these slots with tasks having zero resource requirements [6].

Let us now consider problem $P \mid res\ sor, p_j = 1 | C_{\max}$. When the numbers of resource types, resource limits, and resource requirements are fixed (i.e., constrained by positive integers s, o, r, respectively), problem $P \mid res\ sor, p_j = 1 | C_{\max}$ is still solvable in linear time, even for an arbitrary number of processors [7]. We describe this approach below, since it has more general application. Depending on the resource requirement vector $[R_1(T_j), R_2(T_j), \ldots, R_s(T_j)] \in \{0, 1, \ldots, r\}^s$, the tasks can be distributed among a sufficiently large (and fixed) number of classes. For each possible resource requirement vector we define one such class. The correspondence between the resource requirement vectors and the classes will be described by a 1-1 function $f \colon \{0, 1, \ldots, r\}^s \to \{1, 2, \ldots, k\}$, where k is the number of different possible resource requirement vectors, i.e., $k = (r+1)^s$. For a given instance, let n_i denote the number of tasks belonging to the ith class, $i = 1, 2, \ldots, k$. Thus, all the tasks of class i have the same resource requirement $f^{-1}(i)$. Observe, that most of the input information describing an instance of a problem $P | res\ sor, p_j = 1 | C_{\max}$ is given by the resource requirements of n given tasks (we bypass for the moment the number m of processors, the number s of additional resources and resource limits o). This input may now be replaced by the vector $\mathbf{v} = (v_1, v_2, \ldots, v_k) \in IN_o^k$, where v_i is the number of tasks having resource requirements equal to $f^{-1}(i), i = 1, 2, \ldots, k$. Of course, the sum of the components of this vector is equal to the number of tasks, i.e., $\sum_{i=1}^{k} v_i = n$.

We now introduce some definitions useful in the following discussion. An *elementary instance* of $P \mid res\ sor, p_j = 1 | C_{\max}$ is defined as a sequence $\mathbf{R}(T_1), \mathbf{R}(T_2), \ldots, \mathbf{R}(T_u)$, where each $\mathbf{R}(T_i) \in \{1, 2, \ldots, r\}^s - [0, 0, \ldots, 0]$, with properties $u \le m$ and $\sum_{i=1}^{u} \mathbf{R}(T_i) \le (o, o, \ldots, o)$. Note that the minimal schedule length of an elementary instance is always equal to 1. An *elementary vector* is vector $\mathbf{v} \in IN_o^k$ which corresponds to an elementary instance. If we calculate the number L of different elementary instances, we see that L cannot be greater than $(o+1)^{(r+1)s-1}$; however, in practice L will be much smaller than this upper bound. Denote the elementary vectors (in any order) by $\mathbf{b}_1, \mathbf{b}_2, \ldots, \mathbf{b}_L$.

We observe two facts. First, any input $\mathbf{R}(T_1), \mathbf{R}(T_2), \ldots, \mathbf{R}(T_n)$ can be considered as a union of elementary instances. This is because any input consisting of one task is elementary. Second, each schedule is also constructed from elementary instances, since all the tasks which are executed at the same time form an elementary instance.

Now, taking into account the fact that the minimal length of a schedule for any elementary instance is equal to one, we may formulate the original problem as that of finding a decomposition of a given instance into the minimal number of elementary instances. One may easily see that this is equivalent to finding a decomposition of the vector $\mathbf{v} = (v_1, v_2, \ldots, v_k) \in IN_o^k$ into a linear combination of elementary vectors $\mathbf{b}_1, \mathbf{b}_2, \ldots, \mathbf{b}_L$, for which the sum of coefficients is minimal: Find $e_1, e_2, \ldots, e_L \in IN_o^k$ such that $\sum_{i=1}^{L} e_i \mathbf{b}_i = \mathbf{v}$ and $\sum_{i=1}^{L} e_i$ is minimal.

Thus, we have obtained an integer linear programming problem, which in the general case would be NP-hard. Fortunately, in our case the number of variables L is fixed. It follows that we can apply a result due to Lenstra [8] which states that the integer linear programming problem with a fixed number of variables can be solved in polynomial time depending on both, the number of constraints of the integer linear programming problem and $\log a$, but not on the number of variables, where a is the maximum of all the coefficients in the linear integer programming problem. Thus, the complexity of the problem is $O(2^{L^2}(k \log a)^{cL})$ for some constant c. In our case the complexity of that algorithm is $O(2^L(k\log n)^{cL}) < O(n)$. Since the time needed to construct the data for this integer programming problem is $O(2^s(L+\log n)) = O(\log n)$, we conclude that the problem $P\,|\,res\;sor,\,p_j = 1|C_{\max}$ can be solved in linear time.

Let us consider now problem $Pm\,|\,res\;sor\,|C_{\max}$. We will generalize the above considerations for the case of nonunit processing times and tasks belonging to a fixed number k of classes only. That is, the set of tasks may be divided into k classes and all the tasks belonging to the same class have the same processing and resource requirements. If the number of processors m is fixed, then the following algorithm, based on dynamic programming, has been proposed by Błażewicz et al. [9]. A schedule will be built step by step. In every step one task at a time is assigned to a processor. All these assignments obey the following rule: If task T_i is assigned after task T_j, then the starting time of T_i is not earlier than the starting time of T_j. At every moment an assignment of processors and resources to tasks is described by a *state of the assignment process*. For any state a *set of decisions* is given each of which transforms this state into another state. A *value of each decision* will reflect the length of a partial schedule defined by a given state to which this decision led. Below, this method will be described in more detail.

The state of the assignment process is described by an $m \times k$ matrix $\mathbf{X}$, and vectors $\mathbf{Y}$ and $\mathbf{Z}$. Matrix $\mathbf{X}$ reflects numbers of tasks from particular classes already assigned to particular processors. Thus, the maximum number of each entry may be equal to n. Vector $\mathbf{Y}$ has k entries, each of which represents the number of tasks from a given class not yet assigned. Finally, vector $\mathbf{Z}$ has m entries and they represent classes which recently assigned tasks (to particular processors) belong to.

The initial state is that for which matrices $\mathbf{X}$ and $\mathbf{Z}$ have all entries equal to 0 and $\mathbf{Y}$ has entries equal to the numbers of tasks in the particular classes in a given instance.

Let S be a state defined by $\mathbf{X}$, $\mathbf{Y}$, and $\mathbf{Z}$. Then, there is a decision leading to state S' consisting of $\mathbf{X}'$, $\mathbf{Y}'$, and $\mathbf{Z}'$ if and only if

$$\exists t \in \{1,\ldots,k\} \quad \text{such that } Y_t > 0 \tag{23.1}$$

$$|M| = 1 \tag{23.2}$$

where M is any subset of

$$F = \left\{ i \,\Big|\, \sum_{1\le j\le k} X_{ij}\,p_j = \min_{1\le g\le m} \left\{ \sum_{1\le j\le k} X_{gj}\,p_j \right\} \right\}$$

and finally

$$R_l(T_t) \le m_l - \sum_{1\le j\le k} R_l(T_j)|\{g\,|\,Z_g = j\}| \quad l = 1,2,\ldots,s \tag{23.3}$$

where this new state is defined by the following matrices:

$$X'_{ij} = \begin{cases} X_{ij}+1 & \text{if } i \in M \text{ and } j = t \\ X_{ij} & \text{otherwise} \end{cases}$$

$$Y'_j = \begin{cases} Y_j - 1 & \text{if } j = t \\ Y_j & \text{otherwise} \end{cases} \tag{23.4}$$

$$Z'_i = \begin{cases} t & \text{if } i \in M \\ Z_i & \text{otherwise} \end{cases}$$

In other words, a task from class t may be assigned to processor P_i, if this class is nonempty (inequality (23.1) is fulfilled), there is at least one free processor (Equation (23.2)), and resource requirements of this task are satisfied (Equation (23.3)).

If one (or more) conditions (23.1) to (23.3) are not satisfied, then no task can be assigned at this moment. Thus, one must simulate an assignment of an idle-time task. This is done by assuming the following new state S'':

$$X''_{ij} = \begin{cases} X_{ij} & \text{if } i \notin F \\ X_{hj} & \text{otherwise} \end{cases}$$

$$\mathbf{Y}'' = \mathbf{Y} \tag{23.5}$$

$$Z''_i = \begin{cases} Z_i & \text{if } i \notin F \\ 0 & \text{otherwise} \end{cases}$$

where h is one of these $g, 1 \leq g \leq m$, for which

$$\sum_{1\leq j\leq k} X_{gj} p_j = \min_{i\notin F, 1\leq i\leq m} \left\{ \sum_{1\leq j\leq k} X_{ij} p_j \right\}$$

This means that the above decisions lead to state S'' which repeats a pattern of assignment for processor P_h, i.e., one which will be free as the first from among those which are now busy.

A decision leading from state S to S' has its value equal to

$$\max_{1\leq i\leq m} \left\{ \sum_{1\leq j\leq k} X_{ij} p_j \right\} \tag{23.6}$$

This value, of course, is equal to a temporary schedule length.

The final state is that for which matrices **Y** and **Z** have all entries equal to 0. An optimal schedule is then constructed by starting from the final state and moving back, state by state, to the initial state. If there is a number of decisions leading to a given state, then we choose the one having the least value to move back along it. More clearly, if state S follows immediately S', and S (S' respectively) consists of matrices $\mathbf{X}, \mathbf{Y}, \mathbf{Z}(\mathbf{X}', \mathbf{Y}', \mathbf{Z}'$ respectively), then this decision corresponds to assigning a task from **Y** to **Y**′ at the time

$$\min_{1\leq i\leq m} \left\{ \sum_{1\leq j\leq k} X_{ij} p_j \right\}$$

The time complexity of this algorithm clearly depends on the product of the number of states and the maximum number of decisions which can be taken at the states of the algorithm. A careful analysis shows that this complexity can be bounded by $O(n^{k(m+1)})$; thus, for fixed numbers of task classes k and of processors m, it is polynomial in the number of tasks.

Let us note that another dynamic programming approach has been described in [9] in which the number of processors is not restricted, but a fixed upper bound on task processing times p is specified. In this case the time complexity of the algorithm is $O(n^{k(p+1)})$.

Let us now pass to problem $P \mid res \cdots, p_j = 1 | C_{\max}$. It follows that when we consider the non-preemptive case of scheduling of unit length tasks we have five polynomial-time algorithms and this is probably as much as we can get in this area, since other problems of nonpreemptive scheduling under resource constraints have been proved to be NP-hard. Let us mention the parameters that have an influence on the hardness of the problem. First, different ready times cause the strong NP-hardness of the problem even for two processors and very simple resource requirements, i.e., problem $P2|\ res\ 1\cdot\cdot, r_j, p_j = 1|C_{\max}$ is already strongly NP-hard [3] (From Figure 23.1 we see that problem $P2|res\ \cdot 11, r_j, p_j = 1|C_{\max}$ is strongly NP-hard as well). Second, an increase of the number of processors from two to three results in the strong NP-hardness of the problem. That is, problem $P3|\ res1\cdot\cdot, r_j, p_j = 1|C_{\max}$ is strongly NP-hard as proved by Garey and Johnson [2]. (Note that this is the famous 3-Partition problem, the first strongly NP-hard problem.) Again from Figure 23.1 we conclude that problem $P3|\ res\ \cdot 11, r_j, p_j = 1|C_{\max}$ is NP-hard in the strong sense. Finally, even the simplest precedence constraints result in the NP-hardness of the scheduling problem, i.e., $P2|\,res\ 111, chains, p_j = 1|C_{\max}$ is NP-hard in the strong sense [1]. Because all these problems are NP-hard, there is a need to work out approximation algorithms. We quote some of the results. Most of the algorithms considered here are list scheduling algorithms which differ from each other by the ordering of tasks on the list. We mention three approximation algorithms analyzed for the problem. (Let us note that the resource constrained scheduling for unit task processing times is equivalent to a variant of the bin packing problem in which the number of items per bin is restricted to m. On the other hand, several other approximation algorithms have been analyzed for the general bin packing problem and the interested reader is referred to [10] for an excellent survey of the results obtained in this area.)

1. *First fit (FF).* Each task is assigned to the earliest time slot in such a way that no resource and processor limits are violated.
2. *First fit decreasing (FFD).* A variant of the first algorithm applied to a list ordered in nonincreasing order of $R_{\max}(T_j)$, where $R_{\max}(T_j) = \max\{R_l(T_j)/m_l | 1 \le l \le s\}$.
3. *Iterated lowest fit decreasing (ILFD — applies for $s = 1$ and $p_j = 1$ only).* Order tasks as in the FFD algorithm. Put C as a lower bound on $C^*_{\max}$. Place T_j in the first time slot and proceed through the list of tasks, placing T_j in a time slot for which the total resource requirement of tasks already assigned is minimum. If we ever reach a point where T_j cannot be assigned to any of C slots, we halt the iteration, increase C by 1, and start over.

   Below we will present the main known bounds for the case $m < n$. In [11] several bounds have been established. Let us start with the problem $P|\,res\ 1\cdot\cdot, p_j = 1|C_{\max}$ for which the three above-mentioned algorithms have the following bounds:

$$\frac{27}{10} - \frac{\lceil 37 \rceil}{10m} < R^{\infty}_{FF} < \frac{27}{10} - \frac{24}{10m} \qquad R^{\infty}_{FFD} = 2 - \frac{2}{m} \qquad R_{ILFD} \le 2$$

We see that the use of an ordered list improves the bound by about 30%. Let us also mention here that problem $P|\,res\cdots, p_j = 1|C_{\max}$ can be solved by the approximation algorithm based on the two-machine aggregation approach by Röck and Schmidt [12]. The worst case behavior of this algorithm is $R = \frac{\lceil m \rceil}{2}$.

Let us now describe problem $P|res\cdots|C_{\max}$. For arbitrary processing times some other bounds have been established. For problem $P|\,res\cdots|C_{\max}$ the first fit algorithm has been analyzed by Garey and Graham [13]:

$$R^{\infty}_{FF} = \min\left\{\frac{m+1}{2}, s + 2 - \frac{2s+1}{m}\right\}$$

Finally, when dependent tasks are considered, the first fit algorithm has been evaluated for problem $P|\,res\cdots, prec\,|\,C_{\max}$ by the same authors: $R^{\infty}_{FF} = m$. Unfortunately, no results are reported on the probabilistic analysis of approximation algorithms for resource constrained scheduling.

Now, let us consider problem $P|\,pmtn, res\ 1{\cdot}1\,|C_{\max}$. It can be solved via a modification of McNaughton's rule [14] by taking

$$C^*_{\max} = \max\left\{\max_j\{p_j\}, \sum_{j=1}^{n} p_j/m, \sum_{T_j \in Z_R} p_j/m_1\right\}$$

as the minimum schedule length, where Z_R is the set of tasks for which $R_1(T_j) = 1$. The tasks are scheduled as in McNaughton's algorithm (cf. Chapter 3), the tasks from Z_R being scheduled first. The complexity of the algorithm is obviously $O(n)$.

Let us consider now problem $P2|\,pmtn, res\cdots|C_{\max}$. This problem can be solved via a transformation into the transportation problem [1].

Without loss of generality we may assume that task T_j, $j = 1, 2, \ldots, n$, spends exactly $p_j/2$ time units on each of two processors. Let (T_j, T_i), $j \ne i$, denote a resource feasible task pair, i.e., a pair for which $R_l(T_j) + R_l(T_i) \le m_l, l = 1, 2, \ldots, s$. Let Z be the set of all resource feasible pairs of tasks. Z also includes all pairs of the type (T_j, T_{n+1}), $j = 1, 2, \ldots, n$, where T_{n+1} is an idle time (dummy) task. Now, we may construct a transportation network. Let $n+1$ sender nodes correspond to the $n+1$ tasks (including the idle time task) which are processed on processor P_1 and let $n+1$ receiver nodes correspond to the $n+1$ tasks processed on processor P_2. Stocks and requirements of nodes corresponding to T_j, $j = 1, 2, \ldots, n$, are equal to $p_j/2$, since the amount of time each task spends on each processor is equal to $p_j/2$. The stock and the requirements of two nodes corresponding to T_{n+1} are equal to $\sum_{j=1}^{n} p_j/2$, since these are the maximum amounts of time each processor may be idle. Then, we draw directed arcs (T_j, T_i) and (T_i, T_j) if and only if $(T_j, T_i) \in Z$, to express the possibility of processing tasks T_j and T_i in parallel on processors P_1 and P_2. In addition we draw an arc (T_{n+1}, T_{n+1}). Then we assign for each pair $(T_j, T_i) \in Z$ a cost associated with arcs (T_j, T_i) and (T_i, T_j) equal to 1, and a cost associated with arc (T_{n+1}, T_{n+1}) equal to 0. (This is because an interval with idle times on both processors does not lengthen the schedule.) Now, it is quite clear that the solution of the corresponding transportation problem, i.e., the set of arc flows $\{x_{ji}^*\}$, is simply the set of the numbers of time units during which corresponding pairs of tasks are processed (T_j being processed on P_1 and T_i on P_2).

The complexity of the above algorithm is $O(n^4 n) = O(n^5 n)$ since this is the complexity of finding a minimum cost flow in a network, with the number of vertices equal to $O(n)$.

We will now consider problem $Pm \mid pmtn, res \cdots \mid C_{\max}$. This problem can still be solved in polynomial time via the linear programming approach [15] where the notion of a *resource feasible set* is used. By the latter we mean the set of tasks which can be simultaneously processed because of resource limits (including processor limit).

Finally, we mention that for problem $P \mid pmtn, res\, 1 \cdot\cdot \mid C_{\max}$, the approximation algorithms *FF* and *FFD* had been analyzed by Krause et al. [11]:

$$R_{FF}^{\infty} = 3 - \frac{3}{m} \qquad R_{FFD}^{\infty} = 3 - \frac{3}{m}$$

Surprisingly, the use of an ordered list does not improve the bound.

23.3 Management of Processing Resources

Section 23.2 describes the classical theory of scheduling with processing resources. Notice that these resources are assigned to the tasks, but nothing is said about the details of this resource allocation. This theory is quite sufficient as long as only few resources are concerned. For instance, a human operator who is required as an additional resource to supervise the processing at a semiautomatic machine fits well into the scope of this theory. However, whenever many resources are concerned, one is faced with the problem of resource allocation and management integrated into the job-machine schedule. Problems of this kind occur in modern manufacturing systems [16]. We shall concentrate in this section on one such case.

We consider a single numerically controlled machine (NC-machine) where the job is the production of a mechanical part. The part is obtained on a single machine by cutting elements from a piece of raw material, by means of many different cutting tools. Each category of tools represents one resource. All tools required for a job have to be loaded into the tool magazine of the machine before the processing and have to stay there until the completion of the job. Clearly, tools are typical processing resources. Tool magazines may have a capacity for 20 to 130 tools and the variety of tools (resources) may exceed 1000. These tools are to be kept in a general storage area and there is a steady flow of new tools to the magazine whenever a new job (part) is sequenced. This problem has two ingredients: the job scheduling and the tool management problem. For each new job, it is necessary to choose which tools should be unloaded to make room for the requested tools. The objective is to find the job sequence that minimizes the total number of tool switches (*tool-switching problem*). The two problems are, of course, related. It has been shown that this problem is already NP-hard for a magazine capacity of two tools [17]. In fact, there is a reduction from the Hamiltonian path problem in an edge-graph to the given scheduling problem with tool management. If the job sequence is given (the *offline* version), then the tool switching problem has simple optimal solutions. We shall describe this optimal tool management in terms of so-called *k-server problems*. It is shown that there is a surprising analogy between the tool switching problem and the *paging problem* in computer memory systems [18,19]. For a long time, the two communities (computer science and operations research) seem to have been unaware of these similarities.

Server problems are well known in connection with computer memory allocation problems [20,21]. The k-server problem may be stated as follows. Let a complete graph with n vertices, numbered $1, 2, \ldots, n$, be given together with a collection of k mobile servers residing on k of the vertices. A server can be moved to another vertex in response to a request for service. Moving a server from vertex u to vertex v results in the cost $C_{uv}(C_{uu} = 0)$. The objective is to minimize the total cost that has to be paid in response to a sequence of requests. Our generalization concerns the way these requests are made. In the general k-server problem, the servers have to be moved in response to a sequence of N bulk requests $(B_1, \ldots, B_N)$. A bulk request is a subset of vertices $B_i \in \{1, \ldots, n\}$ of size $|B_i| \leq k$, which correspond to customers that demand to be served together. It means that one cannot move a server to a requested vertex and then use this server again to move to another vertex if it belongs to the same bulk. Note, however, that requests belonging to the same bulk can be processed sequentially in any order since all the corresponding vertices are covered before a new bulk is to be handled. We call this general problem the *bulk request k-server problem*. The standard case, where all $C_{uv} = 1, u \neq v, (C_{uu} = 0)$, is called *uniform*.

An example of the bulk request k-server problem is given in Figure 23.3: $n = 6$, $k = 3$ and the bulk request (2, 3, 4) is satisfied at the cost $C_{62} + C_{53}$.

There are two main application areas for server problems in computer and manufacturing systems. The classical case refers to two-level computer memory systems. A computer has information stored in n pages of memory, k of which are *fast-access* and the remaining $n - k$ are *slow-access* memory. The n pages correspond to the n vertices of a complete graph and a page in the fast-access memory refers to a vertex being served. Let C_{uv} be the cost of exchanging page u for page v in the fast-access memory (i.e., inserting page v and eliminating page u from the fast-access memory). Performing a task (running a program), requires a specific sequence of pages $(B_1, \ldots, B_N)$. If a requested page is in the slow-access memory, it has to be exchanged for a page in the fast-access memory before it can be used. The goal is to minimize the total cost spent to exchange the required pages. Since the pages are requested one at a time, one has $|B_i| = 1$ for all i. This paging problem has been widely studied, for example [22].

Let us now turn to the second application area, the tool-switching problem. This replacement problem can be modeled as a bulk request k-server problem in the following way. We have a two-level tooling

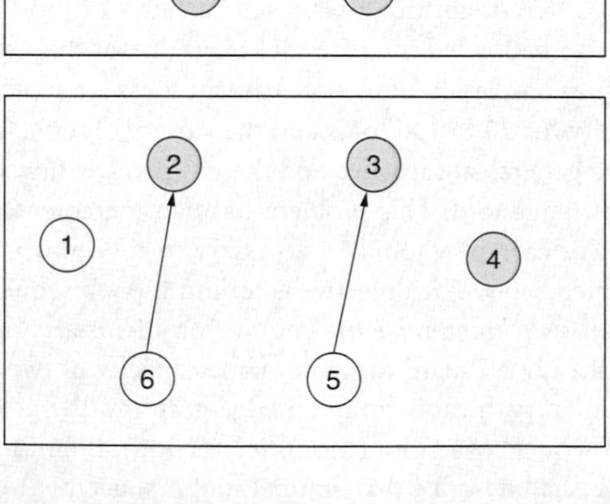

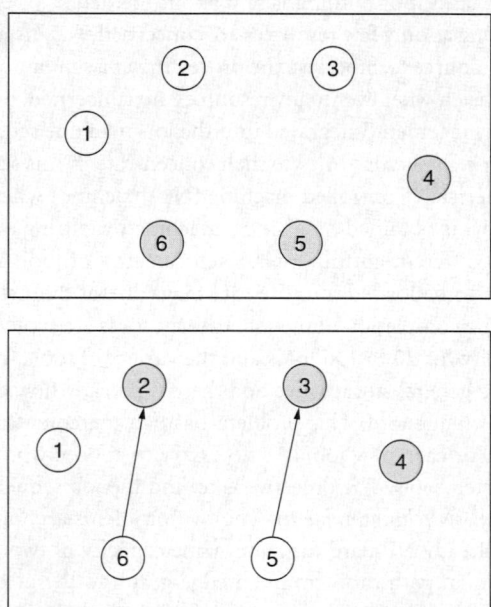

FIGURE 23.3 Servers at 4, 5, 6 and a bulk request for (2, 3, 4).

system consisting of the tool magazine of an NC-machine which has room for k tools, and a main tooling reservoir which can contain $n - k$ tools. A given collection of parts has to be produced on this flexible machine. All tools requested for a part (job) must be present in the tool magazine of the machine, during the processing of the job (processing resources). Some tools may have to be transferred from the main tooling reservoir to the machine, and if no place is available in the magazine, unused tools have to be unloaded and transferred back to the main tool storage area. The choice of these tools to be unloaded depends on the next jobs to be processed. Each loading-unloading operation is time consuming and may require a sizeable fraction of the current machining time. We may then identify the bulk B_i with the set of tools needed to process job i, and the problem is to find a tool replacement strategy in order to minimize the total number of tool switches. This defines the uniform case where we assume that each tool switching takes the same amount of time. This occurs frequently, especially if an automatic tool handling system is present. This uniform tool switching problem was introduced by Tang and Denardo [23], but no link was made to the paging problem. The case $|B_i| = 1$ for each i is of course the same as the paging problem. In an offline situation, the parts to be produced (pages to be used) as well as their production sequence on the machine (page request sequence) are given.

The optimal replacement strategy was first described for computer systems [22,24]. Only about 20 years later, the optimal strategy for the bulk switching problem in manufacturing systems was given [23], which is, as explained, the generalization to bulk sizes exceeding one. The optimal rule is called *Keep Tools Needed Soonest* (*KTNS*) which gives in terms of k-servers:

(i) Move a server to an uncovered vertex only if the service is requested.
(ii) Uncover the vertices for which the next request is farthest in the future (i.e., belongs to the bulk that occurs as late as possible, or never again, in the sequence).

As an illustration, consider the tool-job incidence matrix in Figure 23.4(a), and the optimal tool replacement strategy in Figure 23.4(b). Here it is assumed that the magazine capacity is four and the jobs are processed in natural order $(1, \ldots, 10)$. We see that 14 tool switches are required.

Bulk requests cannot simply be handled as a sequence of individual requests, as the following small example in Figure 23.5 demonstrates. Let us set $n = 3$ and $k = 2$. For the six single requests in Figure 23.5(a), the server would have to be moved only once as indicated.

If we group the requests into the three bulks (Figure 23.5(b)), two moves are required and in addition the previous positions of the servers are no longer feasible for these bulks.

It appears that the extension to bulk requests is a real generalization. But in fact both problems are equivalent if one applies the following transformation [19]: Double the length of the request sequence by replacing a bulk request $B = (b_1, b_2, \ldots, b_p)$ by the sequence $b_1\ b_2 \ldots b_p b_1 b_2 \ldots b_p$. Then apply the

tool \ Job	1	2	3	4	5	6	7	8	9	10
1	1	1					1			1
2			1							1
3		1				1		1		
4	1									1
5		1					1	1	1	
6			1		1					
7			1	1			1		1	
8	1		1					1		
9	1						1			

(a)

1	1 -----	2	2	2	2 -----	1 -----	3	3 -----	1
4 -----	3 -----	6	6	6 -----	3 -----	5	5	5 -----	2
8	8	8	8	8	8 -----	9 -----	8	8 -----	4
9 -----	5 -----	7	7	7	7	7	7	7	7

(b)

FIGURE 23.4 Tool management: (a) Tool-job incidence matrix, (b) Optimal tool loading (14 tool switches).

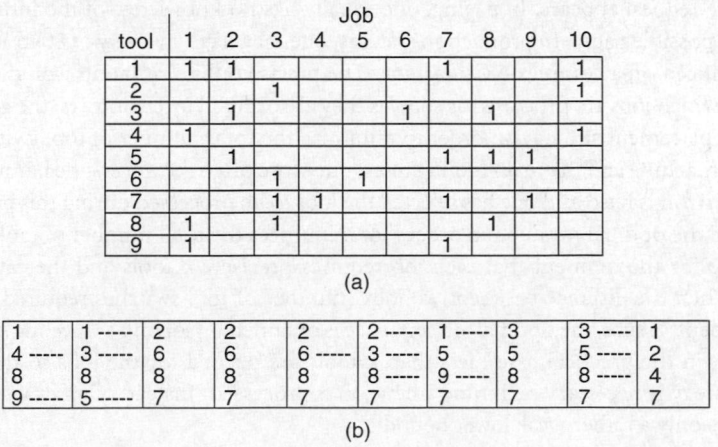

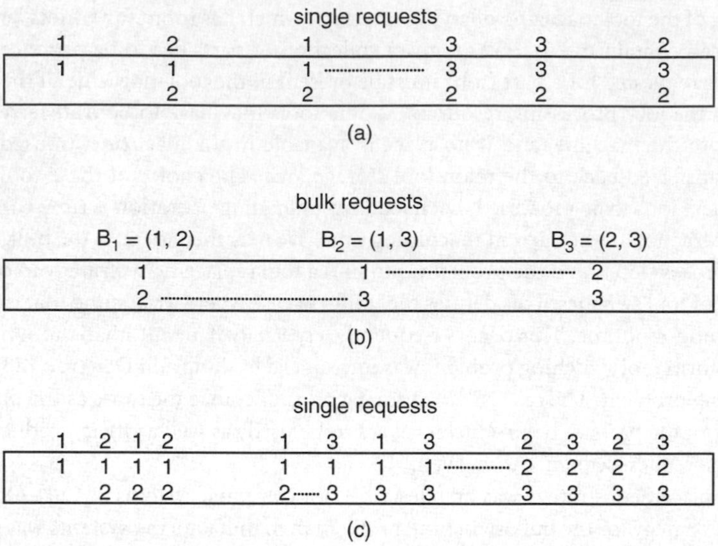

FIGURE 23.5 Example of a 2-server problem: (a) Sequence of single requests, (b) Bulk requests, (c) Transformation to single requests.

optimal replacement strategy to the augmented sequence for the individual requests. After the doubled sequence for each bulk has been treated, all vertices of this bulk are covered. This is true, since the bulk requests are the nearest in the sequence if one processes the first part of the bulk sequence (no movements occur when treating the second part of the bulk sequence). The total cost, the number of exchanges, remains the same (Figure 23.5(c)).

Occasionally, tool switchings are weighted because the time needed for loading and unloading depends on the type of tool. For this nonuniform case with general transition costs C_{uv}, the goal is to minimize the sum of the switching weights for the given job sequence. Also this problem has a polynomial time solution. It is modeled in the form of a network flow problem, as shown in [25] for the paging problem and in [18] for the bulk problem.

So far, we have only considered the offline version of the problem, where the entire request sequence is revealed to the decisionmaker before any switches are made. In the *online* version, the decisionmaker has no information about the request sequence (and the bulk cardinalities) and he must serve each request before any future request appears. In paging, one usually has no knowledge of the future because of the huge number of possible pages. In production systems, one may have a variety of 80 to 100 different parts that can be produced on a complex NC-machine. The precise online situation would correspond to an environment in which jobs are processed as soon as they are ordered by customers: the goal is then to find an online tool replacement strategy, in order to minimize the total number of tool switches between the magazine of the machine and the main tooling reservoir. More often, however, one has partial knowledge: A production horizon is fixed and one has selected the jobs to be processed during this period. One would like to determine the optimal production order that minimizes the total number of tool switches.

Let us assume for the moment that each job requires precisely k tools and the capacity of the tool magazine is k. Then the distance between two jobs (number of tool switches required) is fixed and not sequence dependent. Hence, the optimal job sequence is obtained by solving a traveling salesman problem (TSP). However, in the practical cases, less than k tools are needed for the jobs so that an online tool replacement strategy is necessary each time a new job is processed. Instead of an exact distance between two jobs, one has only a rather weak lower bound:

$$LB(i, j) = \max\{0, |B_i| + |B_j| - |B_i \cap B_j| - k\}$$

jobs

3	4	5	8	2	6	7	9	1	10
2	2	2	3	3	3	9	9	9	9
6	6	6	5	5	5	5	5	4	4
7	7	7	7	7	7	7	7	8	2
8	8	8	8	1	1	1	1	1	1

FIGURE 23.6 Optimal job sequence (7 tool switches).

Again, the solution of the Hamiltonian path of minimum length gives the job sequence. The results are poor since the *LB* approximation is too optimistic. Improvements to TSP related methods have been developed [17,18]. The following two-phase heuristic has been proposed with acceptable results. Firstly, a grouping strategy is applied which consists of gathering several jobs that use common tools, until a new fictitious job is created that nearly fills the whole tool magazine. This grouping stage is repeated until no more jobs can be added without exceeding the tool capacity of the machine. In this way, we obtain sets of tasks for which it is easier to approximate the number of tool changes: The lower bound *LB* between these sets of fictitious tasks nearly gives the exact number of switches. Afterward the new jobs (and the possible single remaining jobs) are scheduled according to one of the well-known TSP heuristics and at the end the KTNS strategy may be applied.

In a way it would be more appropriate to develop a heuristic that constructs the sequence job by job and then applies a tool management strategy each time a new job is scheduled. As explained earlier, this is in a way an online situation, because tools have to be unloaded without knowledge of the order of the jobs to be processed next. The difficulty is exactly this "tool unloading criterion." For instance, one may unload the tools that are the least frequently used by the remaining jobs. Such methods have failed so far. However, we have established earlier the equivalence of paging and tool-switching problems. For paging, one knows the best online strategy in form of the so-called *Partitioning Algorithm* [21,26]. A strongly H_k-competitive randomized algorithm is given for the uniform k-server problem, where H_k is the kth harmonic number ($H_k = 1 + 1/2 + \cdots + 1/k$). In [19], this theory is used to develop an efficient tooling algorithm: One schedules as the next job the one that fits the best (the least number of tools to be inserted) and applies the Partitioning algorithm as the tool management strategy. The optimal schedule for the example (Figure 23.4(a)) is displayed in Figure 23.6. The job sequence is the permutation (3, 4, 5, 8, 2, 6, 7, 9, 1, 10) with seven tool switches.

23.4 Scheduling with Input/Output Resources

Input/Output resources are required by a job at the beginning or at the end of the job processing. During the execution of the job, the resource may carry out (process) other tasks in the system. In [27] and the therein cited literature, a theory of parallel machines with a common server (the input resource) is given. The server is required to set up the job on a machine and can handle at most one job at a time.

An example with six jobs on three parallel machines is displayed in Figure 23.7. The notation for this problem is $P3, S1||C_{\max}$ where $S1$ indicates the presence of a single server. The processing and setup times are as follows: $p = [4, 1, 2, 2, 1, 1]$ and $s = [2, 2, 3, 1, 1, 2]$. The given solution is optimal since $C_{\max} = \sum s_j + \min p_j$.

Numerous complexity results are given in [27]. For instance, polynomially solvable are the problems $P, S1 \mid p_j = 1, r_j, s_j = s \mid f$ for the performance criteria $f \in \{C_{\max}, \sum w_j C_j, \sum w_j U_j, \sum T_j\}$. NP-hard are the problems $P2, S1|s_j = 1|f$ for $f \in \{C_{\max}, L_{\max}, \sum C_j, \sum T_j\}$. The following problems have an unknown complexity status: $P, S1|s_j = 1, p_j = p, r_j|L_{\max}$ or $\sum w_j C_j$.

For these scheduling problems with a common server, the resource is reserved for a setup, but nothing is detailed about the current displacements of the server. In important practical problems, the i/o-resources are delivery and pickup resources. For instance, automated guided vehicles (AGV) have to be routed so

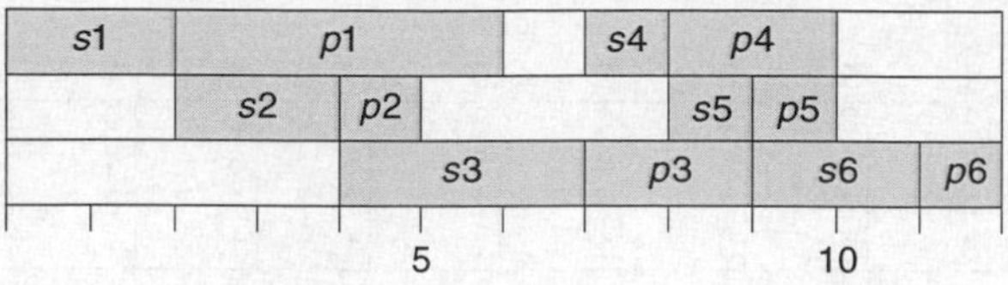

FIGURE 23.7 Parallel machine schedule with a common server.

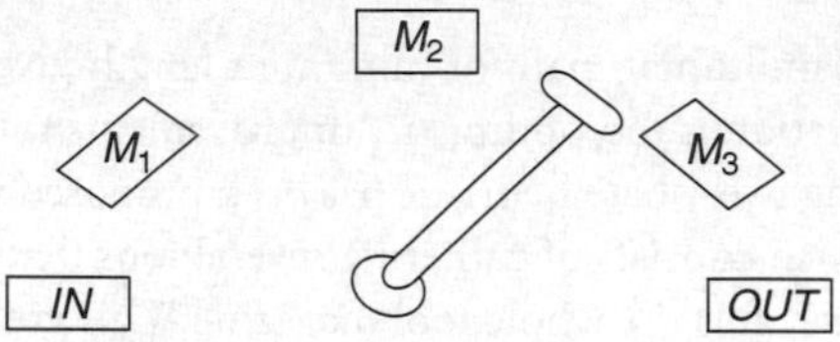

FIGURE 23.8 Robotic cell with $m = 3$ machines.

that they are at the right time at the right place to handle a job. One is therefore confronted with complex routing or traffic problems. Very often, heuristics are developed using dispatching rules (send the closest vehicle, the least frequently used, etc.) and they are tested in simulation studies.

We want to describe in detail a case of an i/o-resource for which many theoretical results exist, namely the so-called *robotic cells*. Robotic cells consist of m machines arranged in a circular layout and served by a single central robot (the i/o-resource). They were first introduced by [28] and studied by [29]. In [28], a line for machining castings for truck differential assemblies is described in the form of a three-machine robotic cell where a robot has to transfer heavy mechanical parts between large machines. The system contains a conveyor belt for the incoming parts and another one for the outgoing parts. In this particular system the robot is not able to traverse the conveyor. Therefore, the movement of the robot from the output to the input station has to traverse the entire cell (Figure 23.8).

The robot may have unit capacity, as will be the case in our model, or one may have two-unit robots [30]. The robotic cells produce, in general, mechanical parts that may wait at the machines as long as required before being picked up by the robot. There is a vast literature on the more restricted case, the so-called hoist scheduling problem, where the waiting times at the machines (the chemical baths) are limited.

The original application has the form of a flow shop to produce a large number of castings. It is known that the robotic scheduling problem is already NP-hard for a flow shop with $m \geq 3$ machines and two or more different part types [31]. It remains the interesting case of the m-machine robotic cell in which one wants to produce identical parts. We shall concentrate on this case. Then, the problem reduces to finding the optimal strategy for the robot moves in order to obtain the maximal throughput rate for this unique part. A survey on general robotic cells can be found in [32].

The m machines of a robotic cell are denoted by $M_1, M_2, \ldots, M_m$ and we add two auxiliary machines, M_0 for the input station IN and M_{m+1} for the output station OUT. The robotic cell represents a flow-shop with the central robot as the i/o-resource. The raw material for the parts to be produced is available in unlimited quantity at M_0. The central robot can handle a single unit at a time. A part is picked up at M_0 and transferred in succession to $M_1, M_2, \ldots, M_m$, where it is machined in this order until it finally reaches the output station M_{m+1}. At M_{m+1}, the finished parts can be stored in unlimited amounts. We focus on the classical case as in [29], where the machines $M_1, M_2, \ldots, M_m$ are without buffer facility. In this case, the robot has to be empty whenever it wants to pick up a part at $M_h (h = 0, 1, \ldots, m)$. In order to travel between two machines, the robot takes the shortest path on the circle. Consequently, the distances are additive.

We consider cyclic robot moves for the production process of the parts and define a *k-cycle* as a production cycle of exactly k parts. It can be described as a sequence of robot moves, where exactly k parts enter the system at M_0, k parts leave the system at M_{m+1} and each time the robot executes the k-cycle, the system returns to the same state, i.e., the same machines are loaded, the same machines are empty and the robot returns to the starting position. Having restored the same cell configuration permits to repeat the same moves of the k-cycle. Note that we are not, for each repetition of the k-cycle, assuming the same remaining processing times for the parts on the machines. We only return to the same part-machine incidence vector. To describe k-cycles we use the concept of activities [33]. The activity $A_h(h = 0, 1, \ldots, m)$ consists of the following sequence:

- The idle robot takes a part from M_h.
- The robot travels with this part from M_h to M_{h+1}.
- The robot loads this part onto M_{h+1}.

Note, that many sequences are not feasible, e.g., $(\ldots A_0, A_0 \ldots)$ since the robot carries a part to M_1 which is occupied. In [33], the authors characterize the k-cycles as follows: A *k-cycle* C_k is a sequence of activities, in which each activity occurs exactly k times and between two consecutive (in a cyclic sense) occurrences of $A_h(h = 1, 2, \ldots, m-1)$ there is exactly one occurrence of A_{h-1} and exactly one occurrence of A_{h+1}.

We represent a k-cycle C_k as in Figure 23.9. The vertical axis represents the cell. The graph indicates the position of the robot in the cell during the cycle. The dashed lines are the empty robot moves and the plain lines are the loaded robot moves.

In the *state graph* G_m, associated with the m-machine robotic cells, each vertex is an m-vector which represents a state of the cell (empty machines and loaded machines): The ith component is 0 if machine M_i is empty and 1 otherwise. The arcs represent the activities of the robot to go from one state to another. The state graph G_3 is given in Figure 23.10. In G_3, the vertex 100 represents the state of the line with M_1 occupied and M_2 and M_3 empty. To move from state 100 to state 010, the robot has to take a part form M_1 and to put it onto M_2, thus executing A_1.

Each k-cycle corresponds to a cycle of length $k(m+1)$ in the graph and conversely. For instance, the 1-cycle $(A_0A_3A_1A_2)$ corresponds to the sequence of vertices 001, 101, 100, 010 and the 2-cycle

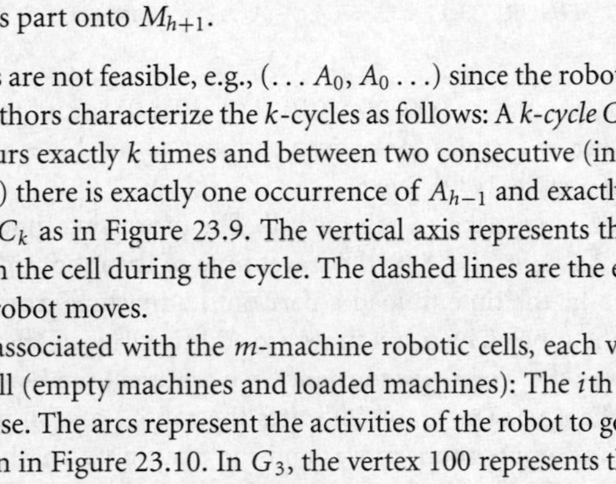

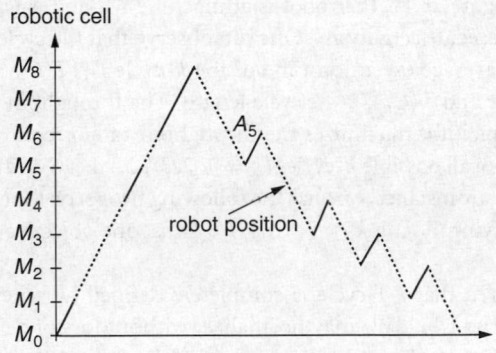

FIGURE 23.9 The 1-cycle $\pi = (A_0, A_4, A_6, A_7, A_5, A_3, A_2, A_1)$.

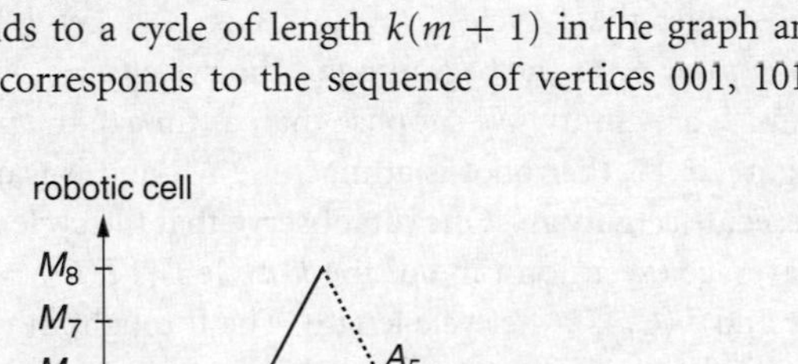

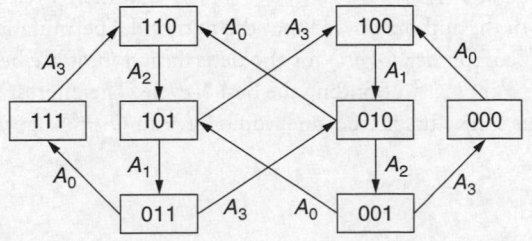

FIGURE 23.10 The state graph G_3.

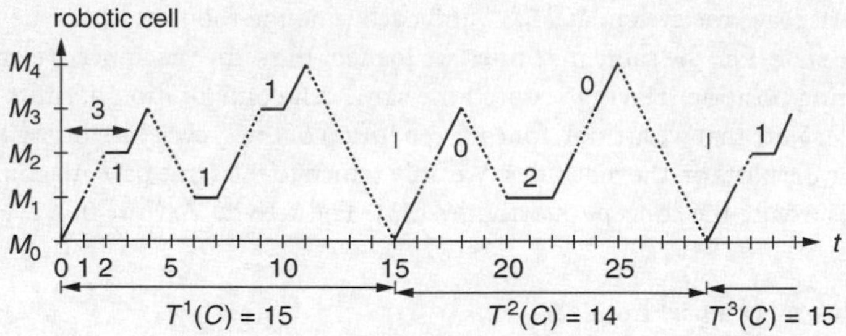

FIGURE 23.11 $C = (A_0 A_2 A_1 A_3)$ for instance I.

$(A_0 A_1 A_0 A_2 A_1 A_3 A_2 A_3)$ corresponds to 000, 100, 010, 110, 101, 011, 010, 001. Using the state graphs, one can calculate the number of k-cycles for a given number of machines m. For instance, in a four-machine cell, the number of 5-cycles is 1 073 696.

Consider an instance I of an m-machine robotic cell. The processing times of the parts are given by p_h on machine $M_h (h = 1, 2, \ldots, m)$. Let δ be the travel time of the robot (idle or loaded) between two consecutive machines. Let ε be the time to load a part onto a machine from the robot or to unload a part from a machine onto the robot. The travel times are additive. Hence, the trip of the idle robot from M_h to $M_{h'} (h \neq h')$ takes $|h - h'|\delta$. For a given instance, we represent the k-cycles as in Figure 23.11: the horizontal axis represents the time. The graph indicates the robot position in the cell according to the time. The dashed lines are the empty robot moves and the plain lines are the loaded robot moves, the loading/unloading processes or the waiting times of the robot at the machines. Let us illustrate the concept with an example.

In a three-machine cell, consider the 1-cycle $C = (A_0 A_2 A_1 A_3)$. Let I be the following instance: $\delta = 1; \varepsilon = 0; p_1 = 6; p_2 = 9; p_3 = 6$. At the beginning of the cycle $C = (A_0 A_2 A_1 A_3)$, machine M_2 is loaded and machines M_1 and M_3 are empty. We suppose that, at time 0, the part has been on M_2 for six time units. At time 9, on Figure 23.11, the robot is at machine M_3 and is waiting one time unit for the part to be ready in order to execute activity A_3. One can observe that the cycle does not repeat identically. Let $T(C_k)$ be the long run average execution time of the k-cycle $C_k (T(C) = 14,5$ for the instance I). We call $T(C_k)$ the cycle time and $T(C_k)/k$ the cycle length. The throughput rate is defined by $k/T(C_k)$. Thus the k-cycle C_k is optimal if it maximizes the throughput rate or equivalently minimizes the cycle length $T(C_k)/k$ over the set of all possible k-cycles ($k = 1, 2, 3, \ldots$). Let S and S' be two sets of production cycles. S *dominates* S' if, for any instance, one has the following property: for every k'-cycle $C_{k'}$ of S', there exists a k-cycle C_k in S verifying $T(C_k)/k \leq T(C_{k'})/k'$. Thus, one objective is to identify dominant sets of production cycles.

In [29], the authors proved that a 1-cycle is completely defined by a permutation of the activities. Hence, the number of 1-cycles is $m!$. We may normalize, without loss of generality, the sequences π of activities and start with activity A_0. The 1-cycles are then of the form $\pi = (A_0, A_{i1}, A_{i2}, \ldots, A_{im})$ where $(i_1, i_2, \ldots, i_m)$ is a permutation of $\{1, 2, \ldots, m\}$. Let us consider the 1-cycles π belonging to the set of pyramidal permutations. We call $\pi = (A_0, A_{i1}, A_{i2}, \ldots, A_{im})$ *pyramidal* if there is an index p such that $1 \leq i_1 < \cdots < i_p = m$ and $m > i_{p+1} > \cdots > i_m \geq 1$. Note that the 1-cycle π in Figure 23.9 is a pyramidal permutation. In [34], the authors proved that the pyramidal permutations dominate the 1-cycles. They give an algorithm of complexity $O(m^3)$ for the determination of the best pyramidal permutation which is therefore also the complexity of finding the best 1-cycle. The interest in 1-cycles is motivated by the conjecture below. If this conjecture is true, one would have an $O(m^3)$ algorithm to obtain the optimal cycle.

One-Cycle Conjecture [29]

The set of 1-cycles dominates the set of all production cycles. It means that the maximum throughput rate over all sequences of finite cyclic robot moves is obtained by executing a 1-cycle.

Cell types	Dominance of 1-cycles and performance factor λ						Calculation of best 1-cycle
	$m = 2$	$m = 3$	$m = 4$, $k = 2$	$m = 4$, $k \geq 3$	$5 \leq m \leq 14$	$m \geq 15$	
Euclidean	True	False: $\lambda \leq 4$					NP-hard [37]
General additive	True [33,35]		False [36]: $\lambda \leq 2$				$O(m^3)$ [34]
Regular	True [36]			False [36]: $\lambda \leq 1.5$			$O(m^3)$ [34]
Regular balanced	True [38]					?	Constant [37]

FIGURE 23.12 An overview of results in robotic cells.

Figure 23.12 summarizes the status of the one-cycle conjecture concerning general and specialized cell configurations. So far, we have considered the so-called *regular case*: the machines are equidistant (δ) with constant loading and unloading times (ε). A further specialization is the *regular balanced case* (all $p_i = p$). In *general additive cells* we allow variable distances δ_i between M_i and M_{i+1} and loading/unloading times ε_i^l and ε_i^u. One may also generalize the travel times for the robot, δ_{ij}, between the machines M_i and M_j. For the so-called *Euclidean case,* we impose the symmetry ($\delta_{ij} = \delta_{ji}$) and the triangle inequality ($\delta_{ij} + \delta_{jk} \geq \delta_{ik}$).

Figure 23.12 provides all known results on the dominance of 1-cycles (one-cycle conjecture). Note, that the dominance of 1-cycles varies with the cell type. In Figure 23.12, we also include the complexity of finding the best 1-cycle as well as the performance factor λ of 1-cycles (ratio of the best 1-cycle to the optimal k-cycle).

Other classes of travel times have also been studied: *circular cells* with coinciding input and output machines ($\delta_{ij} = \min(|i - j|, m + 1 - |i - j|)\delta$, where δ is the travel time between two consecutive machines), *constant cells* [39] where all inter-machine travel times are equal ($\delta_{ij} = \delta$).

It would be interesting to settle the 1-cycle conjecture for regular balanced cells with $m \geq 15$ machines. Of course, the most challenging open problem that remains is to find the optimal cycle in a robotic cell whenever the 1-cycle conjecture is false.

References

[1] Błażewicz, J., Lenstra, J.K., and Rinnooy Kan, A.H.G., Scheduling subject to resource constraints: classification and complexity, *Discrete Appl. Math.*, 5, 11, 1983.

[2] Garey, M.R. and Johnson, D.S., Complexity results for multiprocessor scheduling under resource constraints, *SIAM J.Comput.*, 4, 397, 1975.

[3] Błażewicz, J., Barcelo, J., Kubiak, W., and Röck, H., Scheduling tasks on two processors with deadlines and additional resources, *Euro. J.Oper. Res.*, 26, 364, 1986.

[4] Błażewicz, J., Cellary, W., Słowiński, R., and Węglarz, J., *Scheduling under Resource Constraints: Deterministic Models*, J.Baltzer, Basel, 1986.

[5] Kariv, O. and Even, S., An $O(n^2)$ algorithm for maximum matching in general graphs, in *Proc. 16*th *Ann. IEEE Symp. on Found. Comp. Sci.*, 1975, 100.

[6] Błażewicz, J., Complexity of computer scheduling algorithms under resource constraints, in *Proc. I Meeting AFCET — SMF on Appl. Mathe.*, Palaiseau, 1978, 169.

[7] Błażewicz, J. and Ecker, K., A linear time algorithm for restricted bin packing and scheduling problems, *Oper. Res. Lett.*, 2, 80, 1983.

[8] Lenstra, H.W., Jr., Integer programming with a fixed number of variables, *Math. Oper. Res.*, 8, 538, 1983.

[9] Błażewicz, J., Kubiak, W., and Szwarcfiter, J., Scheduling independent fixed-type tasks, in Słowiński, R. and Węglarz, J., (eds.), *Advances in Project Scheduling*, Elsevier, Amsterdam, 1989, 225.

[10] Coffman, E.G., Jr., Garey, M.R., and Johnson, D.S., Approximation algorithms for bin-packing — an updated survey, in Ausiello, G., Lucertini, M., and Serafini, P. (eds.), *Algorithms Design for Computer System Design*, Springer, Vienna, 1984, 49.

[11] Krause, K.L., Shen, V.Y., and Schwetman, H.D., Analysis of several task-scheduling algorithms for a model of multiprogramming computer systems, *J. Assoc. Comput. Mach.*, 22, 522, 1975. Erratum: *J. Assoc. Comput. Mach.*, 24, 527, 1977.

[12] Röck, H. and Schmidt, G., Machine aggregation heuristics in shop scheduling, *Methods Oper. Res.*, 45, 303, 1983.

[13] Garey, M.R. and Graham, R.L., Bounds for multiprocessor scheduling with resource constraints, *SIAM J. Comput.*, 4, 187, 1975.

[14] McNaughton, R., Scheduling with deadlines and loss functions, *Manage. Sci.*, 12, 1, 1959.

[15] Węglarz, J., Błażewicz, J., Cellary, W. and Słowiński, R., An automatic revised simplex method for constrained resource network scheduling, *ACM Trans. Math. Software*, 3, 295, 1977.

[16] Błażewicz, J. and Finke, G., Scheduling with resource management in manufacturing systems, *Euro. J. Oper. Res.*, 76, 1, 1994.

[17] Crama, Y., Kolen, A.W.J., Oerlemans, A.G., and Spieksma, F.C.R., Minimizing the number of tool switches on a flexible machine, *Int. J. Flexible Manufac. Syst.*, 6 (1), 33, 1994.

[18] Privault, C. and Finke, G., Modelling a tool switching problem on a single NC-machine, *J. Intell. Manufac.*, 6, 87, 1995.

[19] Privault, C., and Finke, G., *k*-server problems with bulk requests: an application to tool switching in manufacturing, *Annals Oper. Res.*, 96, 255, 2000.

[20] Manasse, M.S., McGeogh, L.A., and Sleater, D.D., Competitive algorithms for server problems, *J. Algor.*, 11, 208, 1990.

[21] McGeogh, L.A. and Sleater, D.D., A strongly competitive randomized paging algorithm, *Algorithmica*, 6, 816, 1991.

[22] Belady, L.A., A study of replacement algorithms for virtual storage computers, *IBM Syst. J.*, 5, 78, 1966.

[23] Tang, C.S. and Denardo, E.V., Models arising from a flexible manufacturing machine, part 1: Minimization of the number of tool switches, *Oper. Res.*, 36, 767, 1988.

[24] Mattson, R., Gecsei, J., Slutz, D.R., and Traiger, I.L., Evaluation techniques for storage hierarchies, *IBM Sys. J.*, 2, 78, 1970.

[25] Chrobak, M., Karloff, H., Payne, T., and Vishwanathan, S., New results on server problems, *SIAM J. Discrete Math.*, 4 (2), 172, 1991.

[26] Fiat, A., Karp, R.M., Luby, M., McGeogh, L.A., Sleater, D.D., and Young, N.E., Competitive paging algorithms, *J. Algor.*, 12, 685, 1991.

[27] Brucker, P., Dhaenens-Flipo, C., Knust, S., Kravchenko, S.A., and Werner, F., Complexity results for parallel machine problems with a single server, *J. Scheduling*, 5, 429, 2002.

[28] Asfahl, C. R., *Robots and Manufacturing Automation*, John Wiley & Sons, New York, NY, 1985.

[29] Sethi, S.P., Sriskandarajah, C., Sorger, G., Blazewicz, J., and Kubiak, W., Sequencing of parts and robot moves in a robotic cell, *Int. J. FMS*, 4, 331, 1992.

[30] Su, Q. and Chen, F. F., Optimally sequencing of double-gripper gantry robot moves in tightly-coupled serial production systems, Technical Report, Florida International University, 1995.

[31] Hall, N. G., Kamoun, H., and Sriskandarajah, C., Scheduling in robotic cells: Classification, two and three machine cells, *Oper. Res.*, 45, 421, 1997.

[32] Crama, Y., Kats, V., van de Klundert, J., and Levner, E., Cyclic scheduling in robotic flowshops, *Annals Oper. Res. Math. Indus. Sys.* 96, 97, 2000.

[33] Crama, Y. and van de Klundert, J., Cyclic scheduling in 3-machine robotic flow shops, *J. Scheduling*, 2, 35, 1999.

[34] Crama, Y. and van de Klundert, J., Cyclic scheduling of identical parts in a robotic cell, *Oper. Res.*, 45, 952, 1997.

[35] Brauner, N. and Finke, G., On a conjecture about robotic cells: new simplified proof for the three-machine case, *J. Info. Sys. Oper. Res. — INFOR*, 37, 20, 1999.

[36] Brauner, N. and Finke, G., Cycles and permutations in robotic cells, *Math. Comput Model.*, 34, 565, 2001.

[37] Brauner, N., Finke, G., and Kubiak, W., Complexity of one-cycle robotic flow-shops, *J. Scheduling*, 6, 355, 2003.

[38] Brauner, N., Crama, Y., and Finke G., One-unit production cycles in balanced robotic cells, in *Proc. IEPM'01, Int. Conf. Ind. Eng. Product. Manag.*, Quebec City, 2001, 508.

[39] Dawande, M., Sriskandarajah, C., and Sethi, S., On throughput maximization in constant travel-time robotic cells, *Manufac. Serv. Oper. Manag.*, 4(4), 296, 2003.

24

Scheduling with Resource Constraints — Continuous Resources

Joanna Józefowska
Poznań University of Technology

Jan Węglarz
Poznań University of Technology

24.1 Introduction

In the classical, discrete model of scheduling problems on parallel machines with resource constraints, the resources are allotted to jobs in amounts (numbers of units) from given finite sets only. In consequence, each job is characterized by a vector of processing times representing all its alternative execution modes. Within this model a number of results are known in the literature, concerning the computational complexity and exact and approximation algorithms (see, e.g., Błażewicz et al. [1]). However, in many practical situations additional resources can be allotted to jobs in amounts (unknown in advance) from given intervals. In such situations we speak about continuously divisible or simply continuous resources. An example of such a situation may be the execution of jobs on parallel processors driven by a common (electric, hydraulic, pneumatic) power source, e.g., commonly supplied grinding or mixing machines, electrolytic tanks or refueling terminals. Another example concerning the forging process in steel plants was considered by Janiak [2]. Forgings are preheated by gas up to an appropriate temperature in forge furnaces. Gas flow intensity, limited for the whole battery of forge furnaces, is a continuous resource. Also in computer systems multiple processors may share a common primary memory, as pointed out in Węglarz [3]. If it is a paged-virtual memory system and the number of pages runs into hundreds, primary memory can be treated as a continuous resource. On the other hand, in scalable (SPP) and massively parallel (MPP) systems with hundreds or even thousands of processors, processors themselves can be considered as a continuous resource.

Two models of a job are proposed in the literature in the context of continuous resources: processing rate vs. resource amount and processing time vs. resource amount. In this chapter we examine the first model

1-58488-397-9/$0.00+$1.50

which is more general. The goal is to find a sequence of n jobs on m machines and a continuous resource allocation which optimizes a given schedule performance measure. It is assumed that each machine can process at most one job at a time, and that all jobs and machines are simultaneously available at the start of the process. Since machines may be considered as a discrete resource, these problems are called discrete-continuous scheduling problems. Notice that discrete-continuous problems may be also defined in the context of project scheduling. Basic results for discrete-continuous project scheduling problems may be found in Józefowska et al. [4].

The presented methodology is not restricted to the class of problems described above and can be generalized in a number of directions. Moreover, it should be emphasized that some properties proved for the discrete-continuous model can be successfully applied to solving classical discrete scheduling problems, where job performance modes (resource allocations) are known in advance.

In Section 24.2, a general formulation of the considered class of problems is defined; while in Section 24.3, the general solution approach is characterized. In Section 24.4 to Section 24.6 the problems concerning makespan, maximum lateness, and mean flow time minimization are discussed, respectively.

24.2 Problem Formulation

Let us consider n nonpreemptable and independent jobs. Each job $i, i = 1, 2, \ldots, n$, requires for its processing at time t a machine $j, j = 1, 2, \ldots, m$, from a set of identical, parallel machines, and an amount (unknown in advance) of a continuous resource, $u_i(t) > 0, \sum_{i=1}^{n} u_i(t) = 1$ for every t (we assume without loss of generality that the total available resource amount equals 1).

The processing rate of job i is described by the following equation [5]:

$$\dot{x}_i(t) = \frac{dx_i(t)}{dt} = f_i[u_i(t)], \quad x_i(0) = 0, \; x_i(C_i) = \tilde{x}_i \tag{24.1}$$

where

$x_i(t)$ = state of job i at time t

f_i = continuous, nondecreasing function, where $f_i(0) = 0$

$u_i(t) \in [0, 1]$ = amount of the continuous resource allotted to job i at time t

C_i = (unknown in advance) completion time of job i

$\tilde{x}_i$ = final state or processing demand of job i

State $x_i\ (t)$ of job i at time t is an objective measure of work related to the processing of job i up to time t. It may denote, for example, the number of man-hours already spent on processing of job i, the number of standard instructions in processing of computer program i and so on. The problem is to find a sequence of jobs on machines, and, simultaneously, a continuous resource allocation, which minimize the considered optimization criterion Q, e.g., makespan, maximum lateness or mean flow time. The continuous resource allocation is defined by a piece-wise continuous, nonnegative vector function $\underline{u}^*(t) = [u_1^*(t), u_2^*(t), \ldots, u_n^*(t)]$ whose values $\underline{u}^* = (u_1^*, u_2^*, \ldots, u_n^*)$ are (continuous) resource allocations corresponding to Q^*, the minimal value of Q.

Finally, we can define a discrete-continuous scheduling problem π as a sequence of parameters $n, m, f_1, f_2, \ldots, f_n, \tilde{x}_1, \tilde{x}_2, \ldots, \tilde{x}_n$, and a criterion Q. Notice that the other parameters of π, in particular release dates (equal to zero), precedence relation (empty), and set U (defined by $\sum_{i=1}^{n} u_i \leq 1$) are fixed. The proposed methodology, however, can be easily adopted to handle more general formulations.

24.3 Solution Approach

Let us notice that in the case of $n \leq m$, the problem is to find an optimal continuous resource allocation. This problem has been studied in a number of papers (see Węglarz [6] for a survey). Below some basic results will be recalled which will be useful in further considerations. Notice that one can assume that $n = m$ since for $n < m$, at least $m - n$ machines remain idle.

Let $U \in \mathbf{R}^n$ denote the set of all points $\underline{u}, u_i \geq 0, i = 1, 2, \ldots, n$, satisfying the relation

$$\sum_{i=1}^{n} u_i \leq 1 \tag{24.2}$$

and V the set defined as follows:

$$\underline{v} \in V \Leftrightarrow \underline{u} \in U \quad \text{where } v_i = f_i(u_i) \tag{24.3}$$

where f_i are the functions in (24.1). Of course, sets U and V are the sets of all feasible resource allocations in the system of coordinates $\underline{u}$ and $\underline{v}$, respectively.

Optimal schedule length and corresponding continuous resource allocation can be calculated using the following result, proved by Węglarz [7].

Theorem 24.1

The minimum makespan as a function of final states of jobs, $\underline{\tilde{x}} = (\tilde{x}_1, \tilde{x}_2, \ldots, \tilde{x}_n)$, *can always be given by*

$$M^*(\underline{\tilde{x}}) = \min\{M > 0: \underline{\tilde{x}}/M \in \text{co}V\}$$

where co$V$ *is the convex hull of* V. $M^*(\underline{\tilde{x}})$ *is always a convex function.*

In other words, this result says that resource allocation $\underline{v}^*$ (in the system of coordinates $\underline{v}$) determined by the intersection point of the straight line with parametric equations

$$v_i = \tilde{x}_i/M, \quad i = 1, 2, \ldots, n \tag{24.4}$$

and the boundary of co$V$, always defines M^*. Below, two particular results following directly from Theorem 24.1 are presented.

Corollary 24.1

Assume that f_i *are concave,* $i = 1, 2, \ldots, n$. *Then the makespan is minimized by the fully parallel processing of all jobs using the following resource amounts:*

$$u_i^* = f_i^{-1}(\tilde{x}_i/M^*), \quad i = 1, 2, \ldots, n \tag{24.5}$$

where M^* is the (unique) positive root of the equation

$$\sum_{i=1}^{n} f_i^{-1}(\tilde{x}_i/M) = 1 \tag{24.6}$$

Corollary 24.2

For the n *job and* m *machine scheduling problem with uniform machines and* $f_i \leq c_i u_i$, $c_i = f_i$ *(1),* $i = 1, 2, \ldots, n$, *the makespan is minimized by scheduling all the jobs on the fastest machine.*

Corollary 24.1 identifies very important cases in which an optimal resource allocation, i.e., a makespan optimal schedule for $n = m$, can be found analytically. Generally speaking, these are the cases when Equation (24.6) can be solved analytically. From among these the ones in which (24.6) is an algebraic equation of an order ≤ 4 are of special importance. This is, for example, the case of $f_i = c_i u_i^{1/\alpha_i}$, $\alpha_i \in \{1, 2, 3, 4\}, i = 1, 2, \ldots, n$. Notice that using these functions we can model job processing rates in a variety of practical problems, e.g., those arising in multiprocessor scheduling with memory allocation [8].

Let us now consider the case of $n > m$. First, notice that Corollary 24.2 holds also for $n > m$ and thus the case of the functions $f_i \leq c_i u_i$, $c_i = f_i$ (1), $i = 1, 2, \ldots, n$, will be excluded from further considerations.

Next, let us divide a feasible schedule (i.e., a solution of a discrete-continuous problem) into $p \leq n$ intervals, defined by the completion times of the consecutive jobs.

Let $M_k, k = 1, 2, \ldots, p$, denote the length of the kth interval, and Z_k denote the combination of jobs executed in this interval. Thus a feasible sequence S of combinations $Z_k, k = 1, 2, \ldots, p$, is associated with each feasible schedule. Feasibility of such sequence requires, in addition to the number of elements in each combination restricted by m, that each job appears in at least one combination and that the nonpreemptability of each job is guaranteed. The last condition means that each job appears in one or in consecutive combinations in S. Furthermore, the processing demand of each job can be divided into parts $\tilde{x}_{ik} \geq 0$ corresponding to particular combinations. For a given feasible sequence S one can find an optimal division of processing demands of jobs, $\tilde{x}_i, i = 1, 2, \ldots, n$, among combinations in S, i.e., a division which leads to an optimal schedule from among all feasible schedules generated by S. To this end a nonlinear programming problem can be formulated in which adequate function of the minimum-length intervals (i.e., parts of a feasible schedule) generated by the consecutive combinations in S, as functions of the $\tilde{x}_{ik}$'s, is minimized subject to the constraint that each job has to be completed. To define these functions we use Theorem 24.1 applied to each interval $k, k = 1, 2, \ldots, p$. In this way the properties of optimal resource allocations for $n = m$ are fully utilized, simplifying the resulting optimization problem. In particular, for concave $f_i, i = 1, 2, \ldots, n$, these functions are obtained by solving Equation (24.6) written for each interval. Notice also that for concave $f_i, i = 1, 2, \ldots, n$, it is sufficient to consider feasible schedules in which the resource allocation among jobs remains constant in each interval $k, k = 1, 2, \ldots, p$ (see Corollary 24.1). Of course, knowing a division of $\tilde{x}_i$'s one can easily calculate the corresponding resource allocations.

The above reasoning is based, from the discrete side, on feasible sequences, and from the continuous side on divisions of $\tilde{x}_i$'s. Thus schedules will be defined in terms of these two components.

To formalize the above idea some additional definitions and notations are introduced.

Definition 24.1

A feasible sequence S for an instance I is a sequence $Z_1, Z_2, \ldots, Z_p$, $p \leq n$, of subsets of jobs such that

1. *each of them contains at most m jobs*
2. *each job appears in at least one Z_k*
3. *all Z_k's containing the same job must be consecutively indexed*

Conditions (1) and (2) are obvious, condition (3) concerns nonpreemptability. The cardinality of the set of all feasible sequences grows exponentially with the number of jobs. A concept of a potentially optimal set is defined below.

Definition 24.2

A potentially optimal set (POS) is a set of feasible sequences S, which contains at least one sequence corresponding to an optimal schedule.

Assume that a feasible sequence S is given and that the problem is to find an optimal continuous resource allocation. Formalizing the idea described at the beginning of this section, let $M_k^*(\{\tilde{x}_{ik}\}_{i \in Z_k})$ be the minimum length of a part of a schedule generated by $Z_k \in S$, as a function of $\{\tilde{x}_{ik}\}_{i \in Z_k}$, where $\tilde{x}_{ik}$ is the part of job i processed in combination Z_k. Let K_i be the set of all indices of Z_k's such that $i \in Z_k$. The following mathematical programming problem is obtained to find an optimal continuous resource allocation.

Problem P

$$\text{Minimize} \quad F(\{M_k\{\tilde{x}_{ik}\}_{i \in Z_k}\}_{k=1}^{p}) \tag{24.7}$$

$$\text{subject to} \quad \sum_{k \in K_i} \tilde{x}_{ik} = \tilde{x}_i, \quad i = 1, 2, \ldots, n \tag{24.8}$$

$$\tilde{x}_{ik} \geq 0, \quad i = 1, 2, \ldots, n; \; k \in K_i \tag{24.9}$$

where F is a function depending on Q, and $M_k^*(\{\tilde{x}_{ik}\}_{i\in Z_k})$ is calculated using Theorem 24.1 applied to jobs in $Z_k, k = 1, 2, \ldots, p$.

Of course constraints (24.8) correspond to the condition of realizing processing demands of all jobs. The function F is usually a convex function (particularly for the makespan, maximum lateness, and mean flow time minimization). Thus our problem is to minimize a convex function subject to linear constraints.

The above reasoning can be significantly simplified for concave $f_i, i = 1, 2, \ldots, n$, the case which is most important in practice. First of all, in this case, on the basis of Corollary 24.1, one can calculate $M_k^*(\{\tilde{x}_{ik}\}_{i\in Z_k})$ in Problem P as the unique positive root of the equation

$$\sum_{i\in Z_k} f_i^{-1}\left(\frac{\tilde{x}_{ik}}{M_k}\right) = 1 \tag{24.10}$$

which can be solved analytically for some important cases.

In the sequel we restrict ourselves to the case of concave f_i's. Notice that in order to find an optimal schedule for I, one must solve, in general, Problem P for all feasible sequences S in the POS and find the best solution. Unfortunately, the cardinality of POS defined for the optimization criteria considered below grows exponentially with the number of jobs n. However, heuristic search can be performed on the space of feasible sequences. Using this concept, simulated annealing, tabu search and genetic algorithms were proposed for this class of problems. The presented methodology can be easily extended to consider precedence constraints between jobs. It is enough to add one condition to the definition of the feasible sequence, namely, that the precedence constraints between jobs have to be satisfied. This condition limits the search space of the metaheuristics; however, the problem becomes more difficult because avoiding unfeasible solutions requires additional operations in the algorithm. For that reason tabu search, the most effective heuristic for independent jobs, was outperformed by simulated annealing in case of precedence related jobs.

24.4 Makespan Minimization

Let us start the analysis of properties of discrete-continuous scheduling problems from the minimization of makespan, i.e., $\max\{C_i\}$, which is the completion time of the last job scheduled. In this section we will present properties of optimal schedules as well as some special cases, where optimal schedules can be found effectively. Finally, heuristic approaches will be described.

24.4.1 Properties of Optimal Schedules

For concave $f_i, i = 1, 2, \ldots, n$, the following result about the structure of the POS can be proved [9].

Lemma 24.1

If $f_i, i = 1, 2, \ldots, n$, are concave then a set of all feasible sequences consisting of $p = n - m + 1$, m-element combinations is a POS.

Although, the cardinality of such POS grows exponentially with the number of jobs, the property shown in Lemma 24.1 can be used to prove further properties of optimal schedules.

Problem P(M)

$$\text{Minimize} \quad \sum_{k=1}^{n-m+1} M_k^*(\{\tilde{x}_{ik}\}_{i\in Z_k}) \tag{24.11}$$

$$\text{subject to} \quad \sum_{k\in K_i} \tilde{x}_{ik} = \tilde{x}_i, \quad i = 1, 2, \ldots, n \tag{24.12}$$

$$\tilde{x}_{ik} \geq 0, \qquad i = 1, 2, \ldots, n;\ k \in K_i \tag{24.13}$$

where $M_k^*(\{\tilde{x}_{ik}\}_{i\in Z_k})$ is calculated using Theorem 24.1 applied to jobs in $Z_k, k = 1, 2, \ldots, n - m + 1$.

24.4.2 Special Cases

In some special cases the problem of continuous resource allocation becomes easier to solve. The following results were proved in [9].

Proposition 24.1

Consider a discrete-continuous scheduling problem with $f_i = f, i = 1, 2, \ldots, n$, being concave. Then for a given S such that $|Z_k| = m$ for each k, an optimal continuous resource allocation can be obtained by solving the following equation:

$$\sum_{j=1}^{m} f^{-1}\left(\frac{\tilde{x}_j}{M}\right) = 1 \tag{24.14}$$

where $\tilde{x}_j = \sum_{i \in J_j} \tilde{x}_i$, and J_j denotes the set of jobs appearing in S on machine j.

Proposition 24.2

Consider a discrete-continuous scheduling problem with $f_i = c_i u_i^{1/\alpha}, \alpha > 1, c_i > 0, i = 1, 2, \ldots, n$. Then in an optimal schedule, each job is processed using a constant resource amount.

More detailed analysis of special cases was presented in [10,11]. The main results are recalled below.

Proposition 24.3

Consider a discrete-continuous scheduling problem with $f_i = u_i^{1/\alpha_i}, \alpha_i \geq 1, i = 1, 2, \ldots, n$, and the number of jobs with $\alpha_i \geq 2$ greater than $m - 1$. Then in an optimal schedule at most one job with $\alpha_i = 1$ occurs in each combination of the corresponding feasible sequence. Let us notice that in the case of $f_i = u_i^{1/a}$, $\alpha > 1, i = 1, 2, \ldots, n$, the objective function is of the following form:

$$\text{Minimize} \quad \sum_{k=1}^{n-m+1} \left(\sum_{i \in Z_k} \tilde{x}_{ik}^{a}\right)^{1/a} \tag{24.15}$$

This form of the objective function leads to the following corollary.

Corollary 24.3

Consider the discrete-continuous scheduling problem with $f_i = u_i^{1/\alpha}, \alpha > 1, i = 1, 2, \ldots, n$. Then for a given assignment of jobs to machines the length of an optimal schedule for a given feasible sequence can be calculated as:

$$M^* = \left(\sum_{j=1}^{m}\left(\sum_{i \in J_j} \tilde{x}_i\right)^{a}\right)^{1/a} \tag{24.16}$$

where J_j denotes the set of jobs scheduled on machine j.

In the case of $f_i = u_i^{1/a_i}, \alpha_i \in \{1, 2\}, i = 1, 2, \ldots, n$, two sets of jobs will be considered. The first one consists of jobs with $f_i = u_i$ (i.e., $\alpha_i = 1$) and the second one consists of jobs with $f_i = u_i^{1/2}$ (i.e., $\alpha_i = 2$). These sets will be denoted as A_1 and A_2, respectively. Let n_1 be the cardinality of set A_1 and n_2 the cardinality of set A_2. Of course, $n = n_1 + n_2$. Then the following four subcases can be considered.

In the case of $n_1 = n, n_2 = 0$, all jobs have processing rates of the form $f_i = u_i$. It is easy to see from (24.16) that for such processing rates any assignment of jobs to machines gives the same value of the makespan.

It follows from (24.15) that for functions $f_i = u_i^{1/2}$ the objective function in Problem P has the form:

$$\text{Minimize} \quad \sum_{k=1}^{n-m+1} \sqrt{\sum_{i \in Z_k} \tilde{x}_{ik}^2} \tag{24.17}$$

Thus when there are no jobs with $f_i = u_i (n_1 = 0$ and $n_2 = n)$, an optimal schedule for a given feasible sequence can be found by applying Corollary 24.3.

In the case when the number of jobs with $f_i = u_i^{1/2}$ is less than the number of machines ($n_1 > 0$ and $n_2 < m$), an optimal schedule can be obtained immediately from Proposition 24.3. Each job from set A_2 is then scheduled on a different machine and all jobs from set A_1 are scheduled on another machine. Notice that in this case $m - n_2 - 1$ machines remain idle.

In the case of $f_i = u_i^{1/a_i}$ where $a_i \in \{1, 2\}, i = 1, 2, \ldots, n$ ($n_1 > 0$ and $n_2 \geq m$), Equation (24.6) has the following form:

$$\sum_{i \in Z_k^2} \left(\frac{\tilde{x}_{ik}}{M_k} \right)^2 + \sum_{i \in Z_k^1} \frac{\tilde{x}_{ik}}{M_k} = 1$$

where $Z_k^1 = Z_k \cap A_1$ and $Z_k^2 = Z_k \cap A_2$.

Consequently, the length of an optimal schedule for a given feasible sequence can be calculated as

$$M^* = \frac{1}{2} \left(\sum_{i \in A_1} \tilde{x}_i' + \sqrt{\sum_{j=1}^{m} \left(\sum_{i \in J_j} \tilde{x}_i' \right)^2} \right) \tag{24.18}$$

where $\tilde{x}_i' = \begin{cases} \tilde{x}_i & \text{if } a_i = 1 \\ 2\tilde{x}_i & \text{if } a_i = 2 \end{cases}$

Let us remind that Equation (24.18) is useful if and only if at most one job with $f_i = u_i^{1/a_i}$ and $\alpha_i = 1$ is processed at a time. We have shown that if $f_i = u_i^{1/a_i}$, $\alpha_i \in \{1, 2\}, i = 1, 2, \ldots, n$, the continuous resource allocation for a given feasible sequence can be found analytically.

24.4.3 Metaheuristic Approaches

A natural application of the search methods follows from the general approach described in Section 24.2. Instead of enumerating all the potentially optimal sequences, it is possible to search over the set of potentially feasible sequences using simulated annealing, tabu search, or genetic algorithms. This approach was examined in [12–14]. In all the proposed algorithms a solution is defined by a feasible sequence since for each feasible sequence an optimal continuous resource allocation can be found by solving Problem P.

24.4.3.1 Simulated Annealing

In the simulated annealing algorithm a feasible sequence is represented by two n-element sequences. The first one is a permutation of job indices and describes the order of jobs. The second one consists of the indices of machines on which the jobs from the first sequence will be executed.

The initial solution is generated by setting all the jobs in an ascending order of job indices and assigning each job to a machine according to the following rule:

$$\text{machine\_index} = ((i - 1) \bmod m) + 1$$

A neighbor of a current solution may be generated in two ways, either by a small perturbation in the sequence of jobs, or by a small change in the sequence of machines. In the latter case, an element from the sequence of machines is randomly chosen and replaced by another random integer from the interval $[1, m]$.

The change in the sequence of jobs, called the *shift neighborhood* is defined by removing a randomly chosen job from one position and putting it into another randomly chosen position. The value of the objective function of a feasible solution is defined as the minimal schedule length for the corresponding feasible sequence. A simple cooling strategy was implemented, the details can be found in [14]. The algorithm stops after a fixed number of iterations.

24.4.3.2 Tabu Search

A feasible solution for TS is a feasible sequence consisting of $n - m + 1$, m-element combinations of jobs. It is worth noticing that in such a feasible solution every combination $Z_k, k = 2, \dots, n - m + 1$, differs from the previous one by exactly one job. A starting solution is generated in two steps. In the first step successive jobs in particular combinations are generated randomly but a job is accepted only if it does not violate feasibility of a sequence. In the second step a feasible solution generated randomly is transformed according to the vector of processing demands in such a way that the job which occurs the largest number of times is replaced by the job with the largest processing demand and so on. In consequence, a job with a larger processing demand appears in a larger number of combinations. Intuitively, it should lead to better schedules.

The value of the objective function of a feasible solution is defined as the minimal schedule length for the corresponding feasible sequence. Various methods of the tabu list management have been tested in [15].

Consider a feasible solution consisting of $(n-m+1)m$ positions. Each job occurs in at least one position. A neighbor of a current solution is obtained by replacing a job in a chosen position by another job. A job may be replaced only if it occurs more than once in the feasible solution (every job has to be executed) and only in the first or last combination in that it occurs (nonpreemptability). It is easy to notice that in order to avoid repetitions of combinations in a sequence, a job in its first combination has to be replaced by a job from the next combination and a job in its last combination by a job from the previous one. A neighborhood consists of feasible solutions obtained by performing all such replacements. The algorithm stops after a fixed number of iterations.

24.4.3.3 Genetic Algorithms

In the implementation proposed in [12] each individual consists of two chromosomes. The first one defines the order in which jobs are executed, and thus it consists of n unique job indices. Since the jobs are available at the start of the process, the first m jobs from this chromosome form the first combination of jobs on machines. The occupation of machines by subsequent jobs (the next combinations of jobs on machines) is defined by the second chromosome.

The second chromosome consists of $n - m$ elements, each one representing an index of the machine on which jobs are exchanged. In every combination, except the first one, an index of the job executed on a released machine is replaced by the index of the first job from the first chromosome which has not been started yet. An initial population is generated randomly and consists of feasible solutions. The fitness of an individual depends on the minimum schedule length for the relevant feasible sequence. The shorter schedule length, the higher fitness of the individual. As a selection operator, a linear ranking with elitism has been selected.

The recombination operators replace randomly chosen individuals (so-called parent individuals) by offspring individuals. The number of individuals that undergo recombination depends on the recombination parameter setting. There are four recombination operators, especially defined for the assumed representation of a feasible sequence. Two of them operate on a single parent individual (mutation type) and the remaining two work on pairs of parent individuals (crossover type).

The first mutation type operator, called *mutation I,* chooses randomly a position in the second chromosome and replaces it by another one, randomly chosen from the range $[1, m]$.

The second mutation type operator, called *mutation II,* exchanges a random number of randomly generated job indices. In other words, a number of job indices which will be exchanged (from the range $[2, n]$), and then these indices, have to be generated randomly. Of course, this operator concerns only the first chromosome in the selected individual. As a result of mutation II we obtain a new order of jobs.

Two crossover operators were considered. The first one, the *head crossover,* operates on the first chromosome only. This operator uses two crossing points. First, the segments between the crossing points are copied into offspring individuals. Next, starting from the second crossing point of one parent, the jobs from the other parent are copied in the same order, omitting symbols already present. Reaching the end of the string, copying is continued from the first place of the string.

The second crossover operator, the *tail crossover*, operates on the second chromosome only. It cuts the second chromosomes of both parent individuals at the same randomly chosen cut point and exchanges the resulting segments. The population size has been set at 50. The algorithm terminates after examining at least 1000 solutions.

24.4.4 Heuristic Allocation of the Continuous Resource

The results of the computational experiments [12–14] showed that although solutions generated by the heuristics were very close to optimal ones, the processing time was, in general case, not acceptable. The reason was that solving Problem P in each iteration is very time consuming.

It appeared clear that in practical applications the continuous resource allocation has to be done heuristically, at least during the search process. Two heuristics proposed in [16] are recalled below.

24.4.4.1 Heuristic H1m

Step 1. Define the remaining part r_i of the processing demand of each job $i, i = 1, 2, \ldots, n$, as $y_i = \tilde{x}_i$.
Step 2. Calculate the length of the kth interval $M_k, k = 1, 2, \ldots, n - m$, as

$$M_k = \frac{y_i}{f_i(1/m)}$$

where y_i is the remaining part of the processing demand of job i completed in combination Z_k.
Calculate the new y_i of each uncompleted job i as $y_i := y_i - M_k f_i(1/m)$; if $y_i \leq 0$, then $y_i := 0$.
Step 3. Calculate the length of the $(n - m + 1)$th interval as

$$M_{n-m+1} = \max_{i \in Z_{n-m+1}} \left\{ \frac{y_i}{f_i(1/m)} \right\}$$

Step 4. Calculate the value of the makespan as

$$M = \sum_{k=1}^{n-m+1} M_k$$

In contrast to the simple heuristic H1m, the second heuristic does not assume a fixed continuous resource allocation but calculates the allocation on the basis of the machine load, i.e., on the sum of the processing demands of jobs assigned to each machine.

24.4.4.2 Heuristic HCRA

Step 1. Calculate the makespan lower bound as

$$M_{LB} = \frac{\sum_{i=1}^{n} \frac{\tilde{x}_i}{f_i(1/m)}}{m}$$

Step 2. Calculate the average processing speed of each machine $j, j = 1, 2, \ldots, m$, as

$$MAPS_j = \frac{\sum_{i \in J_j} \frac{\tilde{x}_i}{f_i(1/m)}}{M_{LB}}$$

where J_j is the set of jobs processed on machine j.

Step 3. Define the remaining part y_i of the processing demand of each job i, $i = 1, 2, \dots, n$, as $y_i = \tilde{x}_i$.
Step 4. Calculate the length of the kth interval $M_k, k = 1, 2, \dots, n - m$, as

$$M_k = \frac{y_i}{f_i(u_{jk})}$$

where y_i is the remaining part of the processing demand of job i completed in combination Z_k on machine j and u_{jk} is given by

$$u_{jk} = \frac{f_i^{-1}(MAPS_j)}{\sum_{i \in Z_k} f_i^{-1}(MAPS_j)} \qquad j = 1, 2, \dots, m; \; k = 1, 2, \dots, n - m; \; i \in Z_k$$

Calculate the new y_i of each uncompleted job i as $y_i := y_i - M_k f_i(u_{jk})$; if $y_i \leq 0$, then $y_i := 0$.
Step 5. Calculate the length of the $(n - m + 1)$th interval M_{n-m+1} as the unique positive root of the equation

$$\sum_{i \in Z_{n-m+1}} f_i^{-1}(y_i / M_{n-m+1}) = 1$$

Step 6. Calculate the value of the makespan as

$$M = \sum_{k=1}^{n-m+1} M_k$$

Of course, it should be emphasized that solving the equation in step 5 may not be quite simple but in many cases it can be solved analytically.

These heuristics were included in the tabu search procedure. The computational experiments showed that solving the mathematical programming problem only once, for the best feasible sequence found, allows shortening the processing time by 1000 times, while the solutions found using the HCRA heuristics remain within 10% from optimum.

24.5 Maximum Lateness Minimization

Assume that each job has a given due date $d_i, i = 1, 2, \dots, n$, and it is required that all the jobs have to be completed before their due dates. Maximum lateness is defined as $L_{\max} = \max\{C_i - d_i\}$.

24.5.1 Properties of Optimal Schedules

Let us first consider a discrete-continuous scheduling problem with $f_i \leq c_i u_i, c_i = f_i(1), i = 1, 2, \dots, n$. It can be proved that for any instance of this problem an optimal schedule can be found in polynomial time [17].

Proposition 24.4

Consider a discrete-continuous scheduling problem with $f_i \leq c_i u_i, c_i = f_i(1)$. The optimum schedule is obtained by scheduling all the jobs on one machine according to their nondecreasing due dates and allotting to each job the total available amount of the continuous resource.

It is easy to see that the complexity of the solution procedure is $O(n \log n)$. Moreover, it is easy to see that Proposition 24.4 holds also for uniform machines if we schedule jobs on the fastest one. In the sequel, we assume concave functions f_i $i = 1, 2, \ldots, n$. The objective function can be calculated as follows:

$$L_{\max} = \max \left\{ \sum_{k \leq l_i} M_k^* - d_i \right\}_{i=1}^{n} \tag{24.19}$$

where l_i is the index of the last combination in which job i is processed, and M_k^* is the minimum length of the kth interval which can be calculated as the unique positive root of the equation

$$\sum_{i \in Z_k} f_i^{-1}\left(\frac{\tilde{x}_{ik}}{M_k}\right) = 1 \quad \text{for } k = 1, 2, \ldots, p$$

If we assume that the index of a job corresponds to the number of the last combination in which this job occurs in the feasible sequence S, the problem can be formulated as follows:

Problem P(L)

$$\text{Minimize} \quad z \tag{24.20}$$

$$\text{subject to} \quad \sum_{k \in K_i} \tilde{x}_{ik} = \tilde{x}_i, \qquad i = 1, 2, \ldots, n \tag{24.21}$$

$$\sum_{i \leq k} M_i^* - d_k \leq z, \quad k = 1, 2, \ldots, n \tag{24.22}$$

$$\tilde{x}_{ik} \geq 0, \qquad i = 1, 2, \ldots, n, \; k \in K_i \tag{24.23}$$

The most general POS that can be defined for the considered criterion consists of n combinations such that the first $n - m + 1$ combinations consist of exactly m elements and the combinations $n - m + 2$, $n - m + 3, \ldots, n - 1, n$ contain $m - 1, m - 2, \ldots, 2$, and 1 element, respectively.

24.5.2 Special Cases

There are some special cases among instances with "loose" due dates for which the optimum schedule can be found in polynomial time. Let us number jobs according to their nondecreasing due dates.

Proposition 24.5

Consider discrete-continuous problem with $(d_i - d_{i-1}) \cdot f_i(1) \geq \tilde{x}_i, i = 2, 3, \ldots, n$. Then the optimal schedule is obtained by scheduling all the jobs on one machine according to their increasing numbers and allotting to each job the total available amount of the continuous resource.

Proof of Proposition 24.5 can be found in [17].

24.5.3 Heuristic Approaches

Let us consider a heuristic algorithm based on the idea of sequencing the jobs in nondecreasing order of their due dates. The first m jobs are executed in parallel using the optimal amount of the continuous resource. After the last job in this subset is completed, the next m jobs are scheduled in a similar way. It is easy to see that idle time occurs on some machines, so they are not properly utilized. Thus, an improved version of this algorithm was proposed in [18], where the first m jobs are executed with optimal resource allocation until the first job is completed. Then the processing demands corresponding to unexecuted parts of the remaining $m - 1$ jobs are calculated. The next job on the list is added to this set and the procedure is repeated. The schedule obtained in this way does not contain idle time, however, it is not an optimal schedule. A feasible sequence corresponding to the solution found by this algorithm is of the form: $\{1, 2, 3, \ldots, m\}, \{2, 3, \ldots, m, m + 1\}, \ldots, \{n - m + 1, n - m + 2, \ldots, n - 1, n\}, \ldots, \{n - 1, n\}, \{n\}$. Thus, solving Problem P(L) for this feasible sequence will result in a schedule at least as good as before.

This observation has been used to propose other heuristic algorithms for the maximum lateness minimization problem [17]. In the first algorithm, the jobs are ordered according to nondecreasing order of their due dates. The first m jobs constitute Z_1. Then, the job with the earliest due date is replaced by the next job in the sequence and so Z_2 is created. This step is repeated until the feasible sequence is ready (i.e., contains all the jobs). For this sequence the Problem P(L) is solved to find the resource allocation. Another algorithm takes into account the processing demands and processing rates of the jobs. It is also based on the earliest due date (EDD) order. A predefined amount of the continuous resource ρ is allotted to each job and preliminary processing times are calculated according to the formula $p_i = \tilde{x}_i / f_i(\rho)$. A preliminary schedule is constructed using the EDD rule for the problem with preliminary processing times. A feasible sequence following from the obtained schedule is built and the Problem P(L) is solved for this sequence. It is easy to see that different rules of the preliminary resource allocation may be used, e.g., $\rho = 1/m, \rho = 1/n$ or $\rho_i = 1/\alpha_i, i = 1, 2, \ldots, n$. Depending on the rule, variations of this algorithm may be implemented. Other algorithms are based on the LPT or SPT rules, where processing demands are used instead of processing times. The preliminary schedule is constructed using the LPT or SPT rules, the corresponding feasible sequence is built and Problem P(L) is solved.

Of course, another heuristic approach is to apply metaheuristics (simulated annealing, tabu search or genetic algorithms) as for the makespan criterion. However, the mathematical programming problem is now more difficult to solve and computational times of the heuristics increase compared to makespan minimization problem.

Computational results reported in [17] show that the algorithms perform well, finding good solutions and using much less computational time than metaheuristics.

24.6 Mean Flow Time Minimization

It is worth recalling that for classical scheduling problems, without additional constraints, the problem of makespan minimization is NP-hard, while the mean flow time minimal schedule can be found effectively using the SPT algorithm. It is interesting that for the discrete-continuous problem similar relation holds. We will show in the next paragraph that for a class of processing rate functions a mean flow time optimal schedule can be found in polynomial time. Some interesting properties of optimal resource allocation, as well as feasible sequences will also be presented.

24.6.1 Properties of Optimal Schedules

It follows from Corollary 24.2 that if $f_i, i = 1, \ldots, n$ are convex functions, an optimal schedule can be found even for uniform machines by scheduling jobs on the fastest machine in nondecreasing order of their processing demands $\tilde{x}_i$. Further on we will consider concave functions f_i.

It is easy to see that feasible sequences consist of n combinations. Let us now formulate the Problem P for the mean flow time minimization.

Problem P(F)

$$\text{Minimize} \quad \frac{1}{n}\sum_{k=1}^{n}(n-k+1)M_k^*(\{\tilde{x}_{ik}\}_{i\in Z_k}) \tag{24.24}$$

$$\text{subject to} \quad \sum_{k\in K_i}\tilde{x}_{ik} = \tilde{x}_i, \qquad i = 1, 2, \ldots, n \tag{24.25}$$

$$\tilde{x}_{ik} \geq 0, \qquad i = 1, 2, \ldots, n;\ k \in K_i \tag{24.26}$$

where $M_k^*(\{\tilde{x}_{ik}\}_{i\in Z_k})$ is calculated using Theorem 24.1 applied to jobs in $Z_k, k = 1, 2, \ldots, n-m+1$.

It was shown in [19] that the optimal resource allocation with respect to the mean flow time minimization does not depend on the values of processing demands. The optimal amount of the continuous resource allotted to job i in interval k depends only on the processing rate f_i and the number of interval k. Below we will recall in a more explicit formulation two theorems proved in [19].

Theorem 24.2

For a discrete-continuous scheduling problem with concave processing rate functions f_i, $i = 1, \ldots, n$, having continuous first derivatives, an optimal resource allocation with respect to the mean completion time fulfils the following equations:

$$\frac{df_i(u_{ik})}{du_{ik}} = \frac{\beta_k}{(-\lambda_i)}, \quad k = 1, \ldots, n, \ i \in Z_k \tag{24.27}$$

$$\sum_{i \in Z_k} (-\lambda_i) f_i(u_{ik}) = n - k + 1, \quad k = 1, \ldots, n, \ i \in Z_k \tag{24.28}$$

where $\beta_k = \rho_k / M_k, \lambda_i$ and ρ_k are Lagrange multipliers corresponding to constraints (24.4) and (24.5).

Theorem 24.3

For a discrete-continuous scheduling problem with processing rates of jobs $f_i = c_i f(u_i)$ an optimal schedule with respect to the mean flow time is obtained for the following feasible sequence: $S = \{1, 2, \ldots, m\}$, $\{2, 3, \ldots, m+1\}, \ldots, \{n-m+1, n-m+2, \ldots, n\}, \{n-m+2, n-m+3, \ldots, n\}, \ldots, \{n\}$, where jobs are numbered in nondecreasing order of their processing demands.

24.6.2 Special Cases

The above theorems were used [20] to prove additional properties of an optimal resource allocation for power processing rate functions $f_i(u_i) = u_i^{1/\alpha_i}, \alpha_i \geq 1$.

Theorem 24.4

For a discrete-continuous scheduling problem with processing rates of jobs $f_i(u_i) = c_i u_i^{1/\alpha_i}$, $\alpha_i \geq 1$, an optimal resource allocation (for a given feasible sequence) with respect to the mean flow time requires finding roots of a polynomial of order at most max$\{\alpha_i\}$. Thus for $\alpha_i \in \{1, 2, 3, 4\}$ the optimal resource allocation can be found analytically.

Corollary 24.4

For a discrete-continuous scheduling problem with processing rates of jobs $f_i(u_i) = c_i u_i^{1/\alpha}$, $\alpha_i \in \{1, 2, 3, 4\}$, an optimal schedule with respect to the mean flow time can be found in polynomial time.

Corollary 24.5

For a discrete-continuous scheduling problem with the processing rate functions of jobs $f_i(u_i) = c_i u_i^{1/2}$, $m = 2$, and a feasible sequence S defined above, an optimal resource allocation with respect to the mean flow time can be calculated using the following formulas:

$$u_{kk} = \frac{n - k + 2}{2(n - k + 1)} \tag{24.29}$$

$$u_{k+1k} = \frac{n - k}{2(n - k + 1)} \tag{24.30}$$

where u_{kk} denotes the amount of the continuous resource allotted to a job being completed in the kth interval.

Theorem 24.5

For a discrete-continuous scheduling problem with processing rates of jobs $f_i(u_i) = c_i u_i^{1/\alpha_i}$, $\alpha_i \geq 1$, *there exists a mean flow time optimal schedule such that at most one job with* $\alpha_i = 1$ *is scheduled in each interval* $[C_{i-1}, C_i], i = 1, \ldots, n, C_0 = 0$. *Moreover, if* $\alpha_i = \alpha_j$ *and* $\tilde{x}_i \leq \tilde{x}_j$, *then* $C_i \leq C_j$.

The above theorems allow improving the efficiency of the local search algorithms developed for the considered problem either by removing the time-consuming step of solving the mathematical programming problem (24.24)–(24.26) or by reducing the search space.

Although many interesting properties of discrete-continuous scheduling problems are already known, there are still some open questions, especially the structure of optimal feasible sequences for the mean flow time minimization problem with arbitrary concave processing rate functions.

Acknowledgment

This research has been supported by the KBN grant No. 8T11F 001 22.

References

[1] Błażewicz J., Ecker K.H., Pesch E., Schmidt G., and Węglarz J., *Scheduling Computer and Manufacturing Processes*, 2nd ed., Springer-Verlag, Berlin, 2001.

[2] Janiak A., Scheduling and resource allocation problems in some flow type manufacturing processes, In Fandel G. and Zapfel G. (editors), *Modern Production Concepts*, Springer-Verlag, Berlin, 404, 1991.

[3] Węglarz J., Multiprocessor scheduling with memory allocation — a deterministic approach, *IEEE Trans. Comput.*, C-29, 703, 1980.

[4] Józefowska J., Mika M., Różycki R., Waligóra G., and Węglarz J., Project scheduling under discrete and continuous resources, In Węglarz J. (editor), *Project Scheduling: Recent Models, Algorithms and Applications*, Kluwer, Dordrecht, 289, 1999.

[5] Burkov V.N., Optimal project control, in *Proc. 4th IFAC Congress*, Warszawa, 46, 1969.

[6] Węglarz J., Modelling and control of dynamic resource allocation project scheduling systems, In Tzafestas S.G. (editor), *Optimization and Control of Dynamic Operational Research Models*, North Holland, Amsterdam, 105, 1982.

[7] Węglarz J., Time-optimal control of resource allocation in a complex of operations framework, *IEEE Trans. Systems, Man and Cybernetics*, SMC-6, 783, 1976.

[8] Belady L.A. and Kuehner C.J., Dynamic space sharing in computer systems, *Comm. ACM*, 12, 282, 1968.

[9] Józefowska J. and Węglarz J, On a methodology for discrete-continuous scheduling, *Europ. J. Opnl. Res.*, 107, 338, 1998.

[10] Józefowska J., Mika M., Różycki R., Waligóra G., and Węglarz J., Discrete-continuous scheduling with identical processing rates of jobs, *Found. Comput. Decision Sci.*, 22, 279, 1997.

[11] Józefowska J., Mika M., Różycki R., Waligóra G., and Węglarz J., Discrete-continuous scheduling to minimize makespan for power processing rates of jobs, *Discr. Appl. Math.*, 94, 263, 1999.

[12] Józefowska J., Różycki R., and Węglarz J., A genetic algorithm for discrete-continuous scheduling problems, In Pearson, D.W., Steele, N.C., and Albrecht, R.F. (editors), *Artificial Neural Nets and Genetic Algorithms*, Springer-Verlag, Berlin, 273, 1995.

[13] Józefowska J., Waligóra G., and Węglarz J., A tabu search algorithm for discrete-continuous scheduling problems, In Rayward-Smith V. J., Osman I. H., Reeves C. R., and Smith G. D., (editors), *Modern Heuristic Search Methods*, John Wiley & Sons, Chichester, 169, 1996.

[14] Józefowska J., Mika M., Różycki R., Waligóra G., and Węglarz J., Local search metaheuristics for discrete-continuous scheduling problems, *Europ. J. Opnl. Res.*, 107, 354, 1998.

[15] Józefowska J., Waligóra G., and Węglarz J., Tabu list management methods for a discrete-continuous scheduling problem, *Europ. J. Opnl. Res.*, 137, 288, 2002.
[16] Józefowska J., Mika M., Różycki R., Waligóra G., and Węglarz J., A heuristic approach to allocating the continuous resource in discrete-continuous scheduling problems to minimize the makespan, *J. Scheduling*, 5, 487, 2002.
[17] Józefowska J., Mika M., Różycki R., Waligóra G., and Węglarz J., Discrete-continuous scheduling to minimize maximum lateness, In Domek S., Emirsajlow Z., and Kaszyńki R. (editors), *Proc. MMAR'97 — 4th International Symposium on Methods and Models in Automation and Robotics*, Wyd. Uczelniane Pol. Szczecińskiej, Szczecin, 947, 1997.
[18] Janiak A. and Przysada J., Parallel machine scheduling with resource allocation-minimization of maximum lateness, manuscript, 1997. (Unpublished)
[19] Józefowska J. and Węglarz J., Discrete-continuous scheduling problems — mean completion time results, *Europ. J. Opnl. Res.*, 94, 302, 1996.
[20] Józefowska J. and Węglarz J., New results for discrete-continuous mean flow time scheduling problems, *Proc. of the Eight International Workshop on Project Management and Scheduling*, Valencia, Spain, 217, 2002.

25 Scheduling Parallel Tasks—Algorithms and Complexity

25.1 Introduction

25

Scheduling Parallel Tasks — Algorithms and Complexity

M. Drozdowski
Poznań University of Technology

25.1 Introduction

Parallel tasks can be executed on more than one processor at the same time. The execution time of a parallel task is a function of the number of assigned processors. The actual number of processors used by a parallel task is selected by a scheduling algorithm. The idea of simultaneous use of processors by the tasks is relatively new in the scheduling theory because one of the fundamental assumptions in the classic deterministic scheduling theory was to execute a task on one processor only [1–3]. With the emergence of new production, communication, and parallel computing systems this requirement has become, in many cases, obsolete and unfounded.

The concept of simultaneous use of the processors appeared in various environments. This resulted in many special cases of parallel tasks, and many naming conventions. For example, *multiprocessor tasks* simultaneously require a number of processors which are fixed and known a priori. Multiprocessor tasks were called *concurrent* [4], *rigid* [5,6], or *parallel* [7] tasks. The following names have been used to denote parallel tasks or their special types: *moldable tasks* [5,6,8], *multiversion tasks* [9], *malleable tasks* [8,10]. Differences in the definition of the parallel task model result from various assumptions on: the relation between the execution time and the number of assigned processors, changing the number of used processors while executing a task, allowing for migration of a task between the processors, setting an upper limit on the number of usable processors. We present and discuss these assumptions in Section 25.2.

1-58488-397-9/$0.00+$1.50

We study deterministic scheduling on parallel processors (the reader is kindly referred to the initial chapters of this book for the basic definitions.), i.e., the processors able to execute any task. Deterministic scheduling problems are not restricted to static schedules because schedules can be built on-line, as long as the information the algorithms need is provided. This is a common misconception that schedules for deterministic scheduling problems are static and must be constructed off-line because of the required information, and the prohibitive computational complexity.

It is worth noting, that a different class of multiprocessor tasks exists which require a set of dedicated processors. Scheduling multiprocessor tasks on dedicated processors is not a subject of this chapter. We do not present approximation algorithms for scheduling parallel tasks here, because it is the subject of the next chapter.

The rest of this chapter is organized as follows: In Section 25.2 the model of parallel tasks is justified by practical applications. The problem of scheduling parallel tasks is formulated in Section 25.3. Basic algorithms, and complexity results on scheduling various classes of parallel tasks are presented in Sections 25.4, 25.5, and 25.6. Some earlier surveys on deterministic scheduling of parallel tasks can be found in [11–15].

25.2 Motivation of the Parallel Task Model

In this section we discuss the motivation behind the parallel task model and its assumptions. Applications of this model will be presented.

25.2.1 Parallel Applications

Parallel computer applications are an established form of parallel tasks. The studies on the workload of massively parallel processing (MPP) computer systems (see [16] for a review) have been conducted on the basis of the user surveys [5,6], and accounting logs [17]. According to the current practice, at the submission time a user selects a number of processors on which the application will run. By the selection of the processor number a task, which is parallel by its nature, is turned to a multiprocessor task. The costs of migration of the parts of a parallel application are high both for the parallel algorithm and for the operating system. Therefore, the selected number of processors is not changed during the execution of the application. The investigations show that more than 95% of the massively parallel applications are parallel tasks in the above sense.

To the surprise of the researchers it has been discovered that majority of the parallel applications ($\approx$70–80%) can be executed on a number of processors which is a power of 2. This phenomenon has been ascribed to the behavioral inertia of the users or some bias in the design of the batch queue submission interfaces. It may have also deeper justification: solving the packing problem on the element sizes that are multiples of each other is computationally easy [18]. This artifact of the parallel task sizes should be, and is, welcome by the designers of the scheduling algorithms for the operating systems and batch queuing systems. A good example of exploiting this feature are buddy systems used for processor allocation [19–22]. In the buddy allocation systems processors are grouped into clusters of the power-of-two sizes. Usually, the information about the available clusters is held in a balanced binary tree structure. A request for processors is extended to the nearest power of two and assigned to a processor cluster of that size.

In the above discussion a parallel application was represented by a rectangle in a time $\times$ processors space. It is reasonable representation for the batch queues of the MPP systems. However, parallel applications often have complex internal structure. *Parallelism profile* is a function of the number of used processors over time. As the parallelism profiles show (see, e.g., [23,24]) the number of used processors is not constant over time. Applications with the above features were called *evolving tasks* [8]. Therefore, a representation by a rectangle in a time $\times$ processors space may lead to a significant loss of the utilization of a parallel system if the application is assigned the maximum number of processors it can exploit. On the other hand, assigning to few processors (e.g., one) neglects any potential gain from parallelism. In parallel computer systems with shared memory a different scheduling paradigm is used. A parallel application creates threads as needed. Each thread is able to use one processor. A single queue of threads is maintained by the operating system. The threads eligible for execution are activated by, e.g., round-robin scheduling algorithm. Such an organization is advantageous for achieving high utilization of the parallel computer system. It is the case because threads performing lengthy I/O operations, internal communications, or using operating system services can be suspended, while threads ready to run can be activated. The operating system decides which threads to run and when. The threads of a parallel application are not guaranteed to be run in parallel in the real-time. The shape of parallel application threads running in time × processors space is not necessarily rectangular. For the sake of defining a scheduling problem we can infer from the above description that in some computer systems the number of processors used over time may be changing.

So far we demonstrated parallel computer applications where many processors simultaneously *can* be used in parallel. There is also evidence that parallel application should or even must be executed in parallel. In the real-time parallel applications many processors must be used simultaneously to obtain sufficient computing speed. For example, it is the case of real-time applications processing video stream [25,26]. Let us consider two more examples related to shared-memory multiprocessor systems with time sharing. Imagine a parallel application which uses many threads accessing a critical section guarded by a lock. The lock must be acquired by threads before using the critical section. A thread which captured the lock may lose its processor due to the time-sharing scheduling discipline. Then, other threads must wait for the release of the lock until the next time quantum in which the thread holding the lock is running. Thus, a thread which is holding the lock should not be descheduled when other threads of the same application are running. Consequently, the threads of the same application should be run in parallel in real-time. Consider the second example: The threads of the same application communicate with each other but are run in different time quanta. A message must wait until the receiver obtains its time quantum and is running. A response must also wait to be received until the first thread is started and running. In the above two examples the advance in computation depends on the speed of switching threads on the processors rather than on the speeds of the processors or the communication medium. To alleviate this drawback a modification in the time-sharing discipline called *coscheduling* or *gang scheduling* [27–30] has been invented. Coscheduling consists in granting the processors to the threads of the same application in the same time quantum. One more fundamental reason for running parallel applications on some number of processors in parallel is the lack of the information on the internal structure of the parallel application. As already mentioned the internal structure of a parallel applications may be complex and diversified. Only the creator of the application, or the compiler, have sufficient understanding of the application internals to efficiently schedule it. Therefore, such information is hard to obtain for the operating system, and difficult to deal with for the computational complexity reasons. Hence, for the operating system, or batch queue management system, it is rational to assume that a parallel application is a rectangle in time × processor space.

One more important characteristic of a parallel application is the dependence of processing time on the number of assigned processors. Many models of the parallel application execution time have been proposed [31–33]. Though no model is widely accepted, most of them assume that either parallelism costs, and adding more processors returns diminishing reductions of the execution time, or that a parallel application processing time decreases proportionately to the number of used processors. In other words, the processing speed is either a concave or a linear increasing function of the number of assigned processors. In the latter case a parallel application is said to be *work preserving* or that it has a linear speedup.

25.2.2 Reliable Computing

Due to the nature of the tasks being performed, many computer systems require increased reliability. This can be the case of control systems in factory, avionics, automotive, medical and life support systems. It is a goal of the fault-tolerant system to detect errors, and mask or recover from errors.

In the software systems the required reliability can be obtained by executing redundant replicas of software components and comparing their outputs in order to determine the correct result. The outputs can be compared by voting or some more sophisticated method of achieving a consensus in a distributed system. Such systems have been considered, e.g., in [9,28,34–36]. The copies of a computer program must work in parallel on different processors in order to guarantee the required reliability. Furthermore, the redundant copies must visit the same states in about the same time to maintain replica determinism [36]. Thus, the replicated copies of a reliable software system constitute a multiprocessor task.

Reliability of the software systems can be achieved not only by redundant copies of the same code, but also by running in parallel different codes solving the same problem [9,37,38]. In other words, diverse design methods are used to create multiple versions of the procedures solving the same problem. Each of the versions has a different execution time, and different reliability measure. The role of the scheduling algorithm is to select the versions to be run in parallel in order to meet timing constraints and to maximize some reliability measure. Tasks of this kind were called *multiversion tasks* in [9]. If we disregard the nonstandard objective functions based on reliability measures, then multiversion tasks are parallel tasks with some predetermined discrete execution time function depending on the number of used versions, and consequently, the number of used processors.

Electronic devices need to be tested to verify correctness of their operation. The testing may have various forms and may take place in various moments of the apparatus life cycle. In the production phase a device can be tested by two processors: One processor stimulates the tested device, the other processor collects the output signals and verifies their correctness [39]. VLSI chips and electronic devices have built-in self-tests [40] which are often run at the boot up time. During such a test one functional unit tests some other functional unit. If the results of the test are correct, both units together can be used to test other more complex structures. In multiprocessor computer systems one processor may be tested by some other processor [41]. These operations require at least two functional units or processors simultaneously which is another example of a multiprocessor task.

25.2.3 Communication

Parallel tasks appear in one more form in the dynamic bandwidth allocation problem [42–44]. The bandwidth allocation problem arises in communication media that have guaranteed bandwidth policy (e.g., ATM [45]). Applications, such as the ones delivering compressed video, require certain level of quality of service (QoS) in the form of guaranteed bandwidth. The applications submit requests for communication channel bandwidth. The scheduling algorithm has to determine the assignment of bandwidth for a given interval to fulfill the submitted requirements. Bandwidth plays the role of processors here. Thus, bandwidth allocation problem can be seen as the assignment of application requests in the bandwidth × time space.

Very often the transmitter and the receiver have some buffers for smoothing the source bitrate variations [46–48]. Such buffers allow for periods of momentary transmission speed variations. At the low level of communication stack the information stream (e.g., video stream) is converted into packets. The amount of the information sent over some period of time must be big enough to satisfy the quality requirements. From this low level point of view the momentary bandwidth usage may be variable as long as the bandwidth requirement accumulated over a longer period of time is satisfied. Thus, the shape of the area in the bandwidth × time space assigned to satisfy some request can change over time. It is not necessarily a rectangle in a bandwidth × time space. Furthermore, preempting the communication and changing the bandwidth assignment is possible and can be profitable both for the utilization of the channel [42], and for the number of transferred messages [43].

Another application for multiprocessor task can be scheduling of file transfers [49]. A file transfer requires at least two processing elements simultaneously: the sender and the receiver. Simultaneous transfers on multiple buses can also be considered as multiprocessor tasks [50].

25.2.4 Other Applications

Parallel tasks have been mentioned also in the context of other applications.

In the textile industry loom scheduling problem arises [51]. The looms can be considered as processors in this problem. The tasks are the orders for products to be woven. Each task can be associated to a certain subset of the looms on which it can be processed. A special feature of this problem is that jobs can be *split* into parts processed independently on the admissible machines. Preemption is also allowed. Thus, the area occupied by a task in the processors × time space can be arbitrary provided that the required quantity of the product is woven.

In berth allocation problems, vessels may occupy several berths simultaneously [52] during loading and unloading. The loading/unloading operation is a task, while the berths are processors here. Clearly, this is an example of a multiprocessor task scheduling problem.

In the semiconductor industry, complex modular projects are assigned to a team of people that must work simultaneously [52]. This must be the case also in other kind of complex projects and activities involving workforce management, e.g., in engineering, agriculture, forestry.

From the above examples we can conclude that there are several classes of parallel tasks. These classes differ in the fixed or flexible number of processors that can be used, the admissible shape of the area in the processor × time space, or the form of the processing time dependence on the number of used processors.

Let us finish this section with an observation that the model of parallel, or multiprocessor tasks has strong ties with other combinatorial optimization problems, such as two-dimensional bin-packing, dynamic memory allocation, dynamic spectrum allocation [53,54], scheduling with resource constraints [55,56]. The legitimacy of such similarities often depends on minor but crucial differences in the definitions of the problems.

In the further discussion we stick to a parallel application metaphor of the parallel task.

25.3 Problem Formulation and Notation

In this section we define the parallel task scheduling problem. Specialized subtypes of parallel task model are defined first. In our naming convention for the parallel task types we follow the conventions from [8,57,58]. The notation introduced in the initial chapters of this book is adjusted to the needs of the considered problems. Finally, the $\alpha \mid \beta \mid \gamma$ notation is extended to represent the parallel task scheduling problems.

Let $\mathcal{P} = \{P_1, \ldots, P_m\}$ denote the set of available processors, and $\mathcal{T} = \{T_1, \ldots, T_n\}$ the set of tasks to be executed.

By *parallel tasks* we mean the tasks that are executed by some number of processors used in parallel. The various types of parallel tasks are characterized by the different ways of using the processors. The mutual relations between the parallel task classes are shown in Figure 25.1(a). The following task types are specializations of the parallel task model:

Multiprocessor tasks are the tasks which require a single fixed number of processors simultaneously. For task T_j the number of required processors will be called task *size* and denoted by $size_j$. The execution time of a multiprocessor task T_j is given, and will be denoted by p_j. Set $\mathcal{T}$ of multiprocessor tasks comprises subsets $\mathcal{T}^1, \ldots, \mathcal{T}^m$ of tasks with sizes equal to $1, \ldots, m$, respectively. Let $n_i = |\mathcal{T}^i|$. Thus, $n = \sum_{i=1}^{m} n_i$. A further specialization within the multiprocessor task class are the tasks which require a number of processors which is a power of 2. Multiprocessor tasks constitute a simple case of parallel tasks, and are useful in determining the computational complexity status of the parallel task scheduling problems.

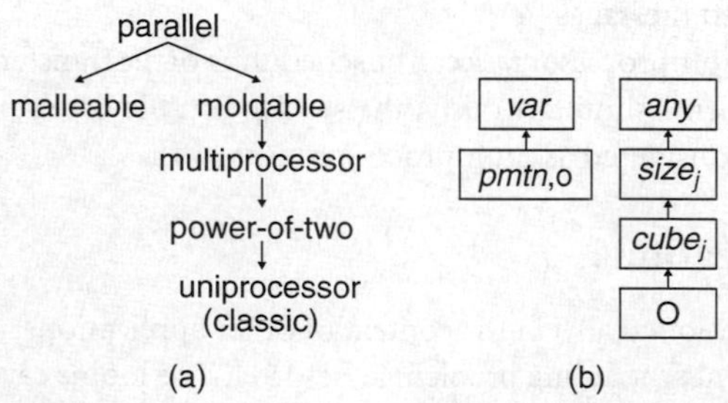

FIGURE 25.1 (a) Relation between the parallel task types, (b) Polynomial time transformations between the parallel task types.

Moldable tasks can be executed on several processors in parallel. The actual number of used processors is determined before starting the execution of a task, and remains unchanged until the task's completion. The execution time of moldable task T_j will be denoted by $p_j(q)$, where $q \in \{1, \ldots, m\}$ is the number of used processors. $p_j(q') = \infty$ means that T_j may not be executed on q' processors.

In the case of *malleable tasks* the number of assigned processors may change during the execution of a task. Thus, malleable tasks can accommodate new processors or release some processors on request of the system. This assumption has several consequences. First there must exist some infrastructure for migrating pieces of the tasks between the processors which are received from or returned to the computer system. If task migration is possible, then also task preemption should be easy as a prerequisite for the migration. Hence, malleable tasks can be considered both migratable and preemptable. Second, malleable tasks are not necessarily rectangles in the processor × time space, because the number of used processors changes over time. This feature distinguishes malleable tasks from multiprocessor, and moldable tasks. Finally, there must be some method of assessing the progress in executing a task because a summation of the lengths of intervals when T_j is executed is not sufficient. Any feasible schedule is a sequence of l intervals in which the number of processors assigned to a task does not change. Interval k ($k = 1, \ldots, l$) determines its length L_k, and the number of processors q_{kj} occupied by task T_j. Let $p_j(q)$ be the processing time of a malleable task T_j, if executed continuously on $q \in \{1, \ldots, m\}$ processors. In any feasible schedule, $\sum_{k=1}^{l} \frac{L_k}{p_j(q_{kj})} = 1$, for task $T_j \in \mathcal{T}$. In the case of work preserving malleable task T_j processing time on q processors is $p_j(q) = p_j/q$, where p_j is the amount of work that must be performed on T_j. A work preserving task is considered as finished when the area occupied by T_j in processor × time space is equal to p_j. Note that malleable tasks are not the same as a chain of multiprocessor tasks with different sizes, though schedules may look similar. It is because in the case of malleable tasks the number of processors changes by the decision of the scheduling algorithm. In the case of chains of multiprocessor tasks the number of processors used by each task in the chain is given in the instance data.

We will denote by δ_j an upper bound on the number of processors that may be used by task T_j. δ_j may be an upper limit of the number of processors that can be effectively exploited or a maximum bandwidth allowed for communication T_j.

Speedup $S_j(q) = \frac{p_j(1)}{p_j(q)}$ is a measure of a parallel application performance. *Linear speedup* means that $p_j(q) = p_j(1)/q$, and a moldable or a malleable task T_j is *work preserving*. *Sublinear speedup* means that $p_j(q) > p_j(1)/q$ and using additional processors returns diminishing reductions of the execution time.

In our definition of the parallel task scheduling problem we inherit all the properties typical of deterministic scheduling problems as defined in the initial chapters of this book. Beyond the notation specific for parallel tasks we will denote by c_j, the completion time; d_j, the due date; $\tilde{d}_j$, the hard deadline; r_j, the ready time; w_j, the weight of task $T_j \in \mathcal{T}$. $T_i \prec T_j$ will mean that task T_i must be completed before task T_j. The optimality criteria will be denoted as follows: $C_{\max}$, schedule length (makespan); $L_{\max}$, maximum lateness; $\sum c_j$, sum of completion times (which is equivalent to the mean flow time); $\sum w_j c_j$, weighted sum of completion times (which is equivalent to the mean weighted flow time).

The standard $\alpha \mid \beta \mid \gamma$ deterministic scheduling problem notation must be extended to accommodate specific features of parallel tasks. It is hard to devise a notation that is simple, intuitive, and consistent because properties of parallel tasks are not mutually independent. We adapt and extend the notation introduced in [15] and in [13]. On the basis of the applications presented in Section 25.2 and the above definitions, we infer that two features distinguish parallel tasks: simultaneous use of many processors, and dependence of processing time on the number of used processors. To denote various types of parallel tasks we will use the following special symbols in the β field:

- $size_j$ — to denote that tasks are multiprocessor;
- $cube_j$ — to denote that tasks are multiprocessor and require a number of processors which is a power-of-two;
- *any* — to denote that tasks are moldable;
- *var* — to denote malleable tasks. Observe that the number of processors assigned to a malleable task over the course of the task's execution is *var*iable. For this property task preemption is a necessity,

and the word *var* implies *pmtn*. Therefore, we will skip *pmtn* in the notation of malleable task scheduling problems.

The relation between the processing time and the number of assigned processors will be expressed as follows:

- $p_j(q) = p_j/q$ denotes that tasks are *work preserving*, and execution time decreases proportionately to the number of assigned processors q.
- $p_j(q) = f(q)$ denotes that dependence of the processing time on the number of assigned processors q is determined by some particular continuous function $f(q)$. For example, $p_j(q) = \frac{p_j}{q^\alpha}$, where $0 < \alpha < 1$.
- The lack of the above symbols will mean that processing time $p_j(q)$ is an arbitrary discrete function of the number of assigned processors q.

We will also use symbol δ_j in the β field, to express the fact that the number of processors that can be used is limited from above. Symbol δ_j is absent in the case of uniprocessor (classic) tasks, while in the case of parallel tasks its absence denotes a lack of the limit on the number of usable processors.

Polynomial time transformations exist between scheduling problems for various types of parallel tasks. The existence of such transformations is demonstrated in Figure 25.1(b). Symbol $\circ$ signifies the lack of any specific parallel task symbols in the $\alpha \mid \beta \mid \gamma$ notation. It denotes uniprocessor tasks, i.e., the classic tasks requiring only one processor, and equivalently, $\delta_j = 1$ for all $T_j \in \mathcal{T}$. Each arrow shows the existence of polynomial time transformation between the problems differing in exactly the parallel task type. The relations of this kind are useful in deriving the complexity status of a more general problem from the complexity status of a simpler problem. For example, an **NP**-hardness of some multiprocessor task scheduling problem implies **NP**-hardness of the moldable task scheduling problem with the same remaining features of the problems. Furthermore, **NP**-hardness of uniprocessor task preemptive scheduling problem implies **NP**-hardness of an equivalent malleable scheduling problem. For the sake of conciseness we will write **NPh** and s**NPh** to denote **NP**-hardness and strong **NP**-hardness of some scheduling problems, respectively.

25.4 Multiprocessor Tasks

In this section we present complexity status and algorithms specific for multiprocessor task scheduling problems. For the sake of conciseness we will write k-tasks about the tasks with $size_j = k$.

25.4.1 Schedule Length and Maximum Lateness Criteria

25.4.1.1 Nonpreemptive Schedules, Arbitrary Processing Times

For the schedule length criterion, arbitrary execution times of the tasks, nonpreemptive multiprocessor task scheduling problem is **NPh** because classic problem $P2 \,||\, C_{\max}$ is **NPh**. The multiprocessor task scheduling problem is s**NPh** starting from $m = 5$ processors, i.e., $P5 \mid size_j \mid C_{\max}$ [58]. For $m = 2, 3$ this problem is solvable in pseudopolynomial time [58]. We write about it in the next section. To our best knowledge whether problem $P4 \mid size_j \mid C_{\max}$ is s**NPh** remains an unanswered question. Adding precedence constraints causes that even the problem with $m = 2$ processors $P2 \mid size_j, chain \mid C_{\max}$ is s**NPh** [58]. The problems with maximum lateness criterion are s**NPh** even for $m = 2$ processors, i.e., problem $P2 \mid size_j \mid L_{\max}$ is s**NPh** [52]. Since the schedules for problems with due dates and $L_{\max}$ criterion read backwards are schedules for problems with ready times and $C_{\max}$ criterion, the above result implies that also problem $P \mid size_j, r_j \mid C_{\max}$ is s**NPh**.

25.4.1.2 Nonpreemptive Schedules, Unit Execution Times (UET)

Restricting execution times to unit often makes scheduling problems computationally simpler. Yet, problem $P \mid size_j, p_j = 1 \mid C_{\max}$ is also s**NPh** which was shown in [4] (transformation from $P \,||\, C_{\max}$) and independently in [57] (transformation from 3-Partition).

Problem $Pm \mid size_j, p_j = 1 \mid C_{\max}$ with fixed number of processors can be solved in $O(n)$ time by the use of integer linear programming with a fixed number of variables as proposed in [57]. In any schedule for $Pm \mid size_j, p_j = 1 \mid C_{\max}$ tasks executed in some unit of time form a *processor feasible set*, i.e., a set of tasks which require no more than m tasks. The number of processor feasible sets is the mth Bell number ω_m. Since m is fixed, the number of processor feasible sets ω_m is fixed. Let $\overline{b_i}$ be a vector representing processor feasible set i. Component k of $\overline{b_i}$ is the number of tasks with size equal to k in processor feasible set i. Problem $Pm \mid size_j, p_j = 1 \mid C_{\max}$ consists in decomposing vector $\overline{n} = (n_1, n_2, \ldots, n_m)$ into a minimum number of vectors $\overline{b_1}, \ldots, \overline{b_{\omega_m}}$. This can be expressed as an integer linear program

$$\text{Minimize} \quad \sum_{i=1}^{\omega_m} x_i \tag{25.1}$$

$$\text{subject to} \quad \sum_{i=1}^{\omega_m} x_i \overline{b_i} = \overline{n} \tag{25.2}$$

where x_i is the number of processor feasible set i instances. Integer linear program (25.1)–(25.2) can be solved in $O(n)$ time because the number of variables is fixed [57].

Scheduling multiprocessor tasks with precedence constraints is **sNPh** for $P3 \mid size_j, p_j = 1, chain \mid C_{\max}$, which was shown by transformation from $P \mid prec, p_j = 1 \mid C_{\max}$ in [4], and independently by transformation from 3-Partition in [59]. As the problem with $m = 3$ is **sNPh**, reducing m may result in some polynomially solvable cases. Indeed, problem $P2 \mid size_j, p_j = 1, prec \mid C_{\max}$ can be solved in polynomial time in the way proposed in [4]: First all the transitive arcs are added in the graph of precedence constraints. Then, the 2-tasks are removed, and a schedule for the remaining 1-tasks is built using the algorithm for $P2 \mid p_j = 1, prec \mid C_{\max}$ [60]. Finally, the 2-tasks are reinserted into the schedule at the earliest possible time moment. The transitive closure can be found in time $O(n^{2.376})$ [61]. However, construction of a complete transitive closure is not necessary here. It was proposed in [62] to use a partial transitive closure containing the original arcs of the precedence graph and arcs bypassing the 2-tasks. Such a partial transitive closure can be built in time $O(nk)$ using depth-first search from each node of the task precedence graph where k is the number of arcs in the graph. Consequently, this method can be applied in $O(n^2 + \min\{nk, n^{2.376}\})$ time.

Monotone chains and uniform chains of UET tasks were studied in [59]. Monotone chains consist of two subchains: the leading chain is a sequence of Δ-tasks, and the trailing chain is a sequence of 1-tasks. Uniform chains consist of the same size tasks. $O(n \log n)$ algorithms have been given in [59] for monotone chains with $\Delta > m/2$, and uniform chains with either Δ-tasks or 1-tasks. Note that these algorithms are pseudopolynomial because the length of a string encoding the problem instance depends on the numbers of Δ-tasks and 1-tasks in each of the streams. Thus, the length of the string encoding problem instance is proportional to $\max\{\log n_\Delta, \log n_1\}N$, where N is the number of strings. A polynomial time algorithm with complexity $O(N^2)$ was given in [63].

The next step in delineating the border between the polynomially solvable problems and the hard problems is to consider due dates and ready times. Problem $P2 \mid size_j, p_j = 1, chain, r_j \mid C_{\max}$, i.e., scheduling UET multiprocessor tasks on two processors with precedence constraints and ready times is **sNPh** [64]. Consequently, problem $P2 \mid size_j, p_j = 1, chain \mid L_{\max}$ is also **sNPh**. The complexity status of problem $P2 \mid size_j, p_j = 1, r_j \mid L_{\max}$ remained unknown for several years. This problem has been shown to be polynomially solvable in [65]. Two algorithms testing feasibility for a given value of $L_{\max}$ have been proposed for this problem. The first one is based on solving a linear program for a preemptive version of the problem, and then converting it to a feasible nonpreemptive schedule for UET tasks. The second algorithm builds a feasible schedule, if there is any, in $O(n^4)$ time.

Problem $Pm \mid r_j, p_j = 1, size_j \mid C_{\max}$ has been studied in [66]. An algorithm based on dynamic programming has been proposed. The time complexity of the algorithm is $O(n^{3m})$. Thus, problems $Pm \mid r_j, p_j = 1, size_j \mid C_{\max}$ and equivalently $Pm \mid p_j = 1, size_j \mid L_{\max}$ are solvable in polynomial time provided m is fixed. The dynamic program proposed in [66] can be adapted to solve problem $Pm \mid r_j, p_j = 1, size_j \mid \sum c_j$ as well.

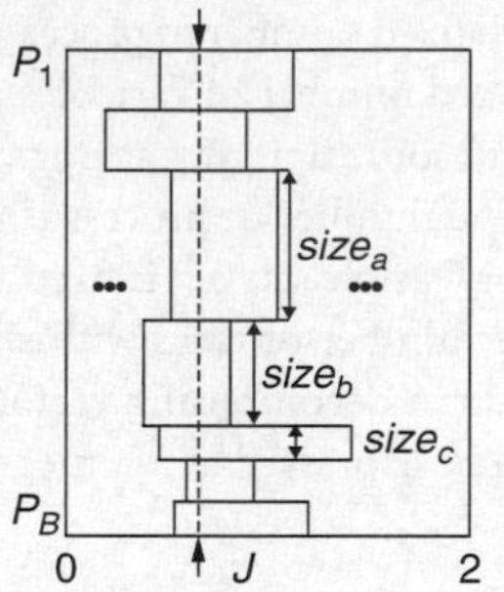

FIGURE 25.2 In the transformation from Partition to $P\,|\,size_j, pmtn\,|\,C_{\max}$ in any time instant task sizes must sum up to $m = B$.

25.4.1.3 Preemptive Schedules

In this section we proceed from the more general problems to more specialized cases.

Problem $P\,|\,pmtn\,|\,C_{\max}$ is solvable in polynomial time [67]. Unfortunately, problem $P\,|\,size_j, pmtn\,|\,C_{\max}$ of constructing preemptive schedules for multiprocessor tasks is **NP**-hard for arbitrary number of processors [68]. The proof of **NP**-hardness is based on the polynomial time transformation from the Partition [69] to the decision version of $P\,|\,size_j, pmtn\,|\,C_{\max}$. The idea of the proof is as follows. Suppose $a_1, \ldots, a_q$ are sizes of elements in the Partition, and $\sum_{i=1}^{q} a_i = 2B$. After setting $n = q, m = B, size_i = a_i, p_i = 1$ for $i = 1, \ldots, n$, we realize that a schedule with length $C_{\max} = 2$ may not have any idle time. This means that the sizes of tasks executed in parallel in any time instant satisfy $m = \sum_{i \in J} size_i = \sum_{i \in J} a_i = B$, where J is the set of tasks executed in parallel. Thus, the answer to Partition must also be positive (see Figure 25.2).

For fixed number of processors m, problem $Pm\,|\,size_j, pmtn,\,|\,C_{\max}$ can be solved in polynomial time [57] by the so-called *one-stage* approach proposed in [70]. This method is based on the generation of all processor feasible sets. Let M be the number of processor feasible sets constructed on the basis of tasks in $\mathcal{T}$, and let x_i denote the duration of executing processor feasible set i. If Q_j is the set of processor feasible sets comprising task T_j then problem $Pm\,|\,size_j, pmtn,\,|\,C_{\max}$ can be solved by the following linear program (observe similarity to (25.1)–(25.2)):

$$\text{Minimize} \quad \sum_{i=1}^{M} x_i \tag{25.3}$$

$$\text{subject to} \quad \sum_{i \in Q_j} x_i = p_j \quad \text{for } j = 1, \ldots, n \tag{25.4}$$

The number of constraints is n, and the number of variables is equal to the number of processor feasible sets. The number of processor feasible sets is $O(n^m)$ [57]. Thus, $Pm\,|\,size_j, pmtn,\,|\,C_{\max}$ can be solved in polynomial time provided m is fixed. This approach has been extended to scheduling multiprocessor tasks with maximum lateness criterion [71], and later, also tasks with ready times, executed on processors with windows of limited availability, i.e., to problem $P, win\,|\,size_j, pmtn, r_j\,|\,L_{\max}$, where *win* denotes changing processor availability pattern [72].

Problem $Pm\,|\,size_j, pmtn,\,|\,C_{\max}$ has been further studied in [73], and an algorithm with time complexity $O(n)$ has been proposed. The method proposed in [73] exploits the ideas that can be traced back to [74]. Let us assume that tasks are ordered according to decreasing sizes, i.e., $size_1 \geq size_2 \geq \cdots \geq size_n$, and tasks of certain size l are ordered according to decreasing processing times, i.e., $p_{a_l+1} \geq p_{a_l+2} \geq \cdots \geq p_{a_l+n_l}$, where $a_l + 1 = \sum_{i=l+1}^{m} n_i + 1$ is the index of the first task with size l. l-tasks must be executed on l processors in parallel which we will call l-stripes. Let PC_i^l be a length of the ith longest l-stripe in a feasible schedule. A necessary and sufficient condition for the existence of a feasible schedule is [74,75]:

$$\sum_{i=1}^{j} PC_i^l \geq \sum_{i=a_l+1}^{a_l+j} p_i \quad \text{for} \quad j = 1, \ldots, \left\lfloor \frac{m}{l} \right\rfloor - 1 \tag{25.5}$$

$$\sum_{i=1}^{\lfloor \frac{m}{l} \rfloor} PC_i^l \geq \sum_{i=a_l+1}^{a_l+n_l} p_i \tag{25.6}$$

If inequalities (25.5)–(25.6) are satisfied, then it is possible to schedule l-tasks from the longest to the shortest one without violating (25.5)–(25.6) [74]. Let us note that this concept has also been used in [76–78] (we write about this in the following). Thus, a feasible schedule must comprise processor feasible sets executed long enough to satisfy (25.5)–(25.6) for all sizes $l = 1, \ldots, m$. This requirement can be transformed to a linear program:

$$\text{Minimize} \sum_{i=1}^{M} x_i \tag{25.7}$$

$$\text{s.t.} \sum_{i=1}^{M} \min\{b_{il}, j\} x_i \geq \sum_{i=a_l+1}^{a_l+j} p_i \quad \forall l = 1, \ldots, m, \quad j = 1, \ldots, \left\lfloor \frac{m}{l} \right\rfloor - 1 \tag{25.8}$$

$$\sum_{i=1}^{M} b_{il} x_i \geq \sum_{i=a_l+1}^{a_l+n_l} p_i \qquad \forall l = 1, \ldots, m, \tag{25.9}$$

where x_i is duration of executing processor feasible set i, $\overline{b_i} = [b_{i1}, \ldots, b_{im}]$ is a vector representing processor feasible set i, component b_{il} of $\overline{b_i}$ is the number of l-stripes in processor feasible set i, and $a_l + 1 = \sum_{i=l+1}^{m} n_i + 1$ is the index of the first (and the longest) l-task. Linear program (25.7)–(25.9) still has the number of variables exponential in m. However, its structure does not depend on n. Given an optimum solution of (25.7)–(25.9), a feasible schedule can be constructed in $O(n)$ time. It was shown in [73], that linear program (25.7)–(25.9) can be solved in time polynomially depending on m. Thus, if we disregard the cost of ordering tasks according to their sizes and processing times, and the cost of formulating (25.7)–(25.9) (because it depends on m which is constant), then problem $Pm \,|\, size_j, pmtn \,|\, C_{\max}$ can be solved in $O(n)$ time.

The above algorithms for $Pm \,|\, size_j, pmtn \,|\, C_{\max}$ seem to have rather theoretical meaning because their computational cost, though polynomial or even linear in n, is still very big. For multiprocessor task executed on power-of-two processors, i.e., according to model *cube*, polynomial algorithms have been proposed in [78–81].

An algorithm testing feasibility of the given instance of problem $P \,|\, cube_j, pmtn \,|\, C_{\max}$ for schedule length C has been studied in [80]. The proposed algorithm builds schedules with processor allocation profile which is *stair-like*. In a stair-like profile (1) each processor P_j is busy before some time $f(P_j)$ and idle after $f(P_j)$, (2) f is nonincreasing function of processor number. In other words, a stair-like schedule consists of a number of steps (cf. Figure 25.3(a)). Tasks are scheduled in the order of decreasing $size_j$. A task is scheduled such that it ends at the common deadline C. Steps of the stair-like schedule are filled one by one and no sooner is the less loaded step used than the more loaded one is completely full. On the last occupied step the piece of the task is shifted to the left in order to preserve the stair-like profile. This method has complexity $O(n^2)$, and results in $O(n^2)$ preemptions. Assuming unit granularity of the schedule length an optimal schedule can be obtained in time $O(n^2(\log n + \log \max_j\{p_j\}))$.

A different approach to scheduling tasks has been proposed in [81]. When scheduling task T_j steps are not filled consecutively, starting from the least loaded one. At most two neighboring steps are used instead: one which has the biggest available interval $C - f(P_j) \leq p_j$, and possibly the next, less loaded one (cf. Figure 25.3(b)). Note similarity to the rule proposed in [74] to schedule 1-tasks on uniform processors. This results in the processor allocation profile which is *pseudo-stairlike*. Consequently, the

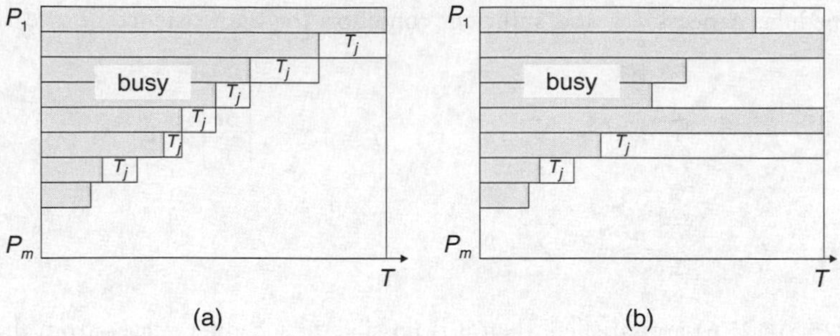

FIGURE 25.3 Scheduling in *cube* model to obtain: (a) Stair-like, (b) Pseudo-stairlike profile.

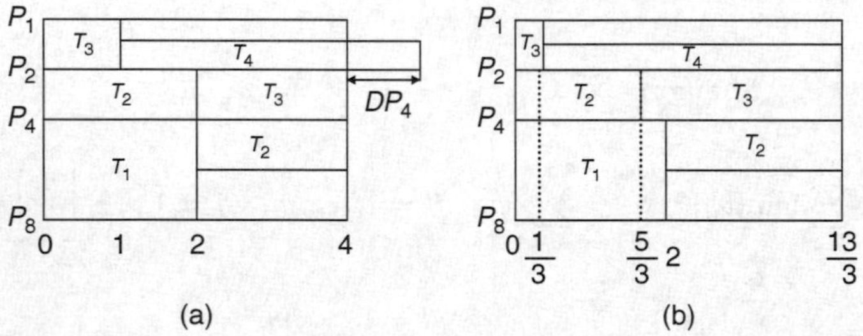

FIGURE 25.4 Example schedule for problem $P \mid cube_j, pmtn \mid C_{\max}$: (a) Partial schedule for $C = 4$, (b) The optimal schedule.

number of preemptions is $O(n)$, and the feasibility testing algorithm runs in $O(n \log n)$ time. A similar approach has been proposed independently in [79].

The concept of stair-like schedules from [80] has been used in [82] to solve problem $P \mid cube_j, pmtn, r_j \mid L_{\max} = 0$, i.e., to decide whether a feasible schedule observing task release times and deadlines exists. The feasibility testing problem has been reduced to a linear program with $O(mn^2)$ variables, and $O(m^2n^2)$ constraints.

In a series of papers [72,76–78] scheduling of multiprocessor tasks on uniform processors, i.e., problem $Q \mid size_j, pmtn \mid C_{\max}$, has been considered. In the first three papers it was assumed that processors form Δ-stripes of equal speed, where $\Delta = \max_j\{size_j\}$. In [76] it was also assumed that $size_j \in \{1, 2\}$, in [77] it was generalized to $size_j \in \{1, \Delta\}$, and finally in [78] to model *cube*. In all the cases a similar method has been used. First, a lower bound C of the schedule length is calculated. Tasks are scheduled according to the order of decreasing size, and the tasks of the same size are considered according to the decreasing processing time. Assume that processors are ordered from the fastest to the slowest one. *Processing capacity* of processor P_i is $PC_i = s_i C$, where s_i is the speed of P_i. Each task is assigned to processors using a rule derived from [74]: A pair of processors (P_i, P_{i+1}) is found such that $PC_i \geq p_j > PC_{i+1}$. Moment x is calculated such that T_j is executed on P_i in interval $[0, x]$, and on P_{i+1} in interval $[x, C]$. A new *composite processor* is created from the remaining interval $[0, x]$ on P_{i+1}, and $[x, C]$ on P_i. After scheduling all tasks of certain size according to the above method, the tasks of the next smaller size are assigned to the processors using the same rule. It has been shown that this method consumes the least processing capacity leaving the biggest possible processing capacity for the smaller size tasks. Unfortunately, a feasible schedule for the lower bound C may not exist in some cases. For example, there is enough processing capacity to execute 1-task T_4 with $p_4 = 4$ in the partial schedule presented in Figure 25.4(a). However, it is available on several processors in parallel. A 1-task may not exploit this parallel processing capacity. In such a situation C must be increased such that conditions (25.5)–(25.6) are satisfied for each task size. Still, not all $m = 8$ processors create processing capacity which can be consumed by T_4. An optimal schedule for the above example is shown in Figure 25.4b. As can be seen processing capacity increased by $m \times \frac{1}{3}$ in total, but only 1 unit can be exploited by T_4. Complex data structures have been utilized in [78] to trace the processors which contribute to the increase of the processing capacity which can be allocated to a task which could not be scheduled feasibly at the initial schedule length C. This procedure is repeated until all tasks are schedulable. Using the above method an algorithm with complexity $O(n \log n + nm)$ has been proposed for problem $Q \mid cube_j, pmtn \mid C_{\max}$ [78]. The algorithm has been tested experimentally. The biggest number of schedule extensions has been observed for $m \approx n$.

Problem $P \mid cube_j, pmtn \mid C_{\max}$, i.e., preemptive scheduling on a hypercube was studied in [83]. A feasibility testing algorithm of [79] was modified to obtain complexity $O(nm)$. As in [72,76–78] tasks are scheduled according to the order of decreasing size and processing time. It was observed that in the optimal schedule at least one task must use all the time remaining to the end of the schedule. This is equivalent to the conclusion that at least one of the conditions (25.5)–(25.6) must be satisfied with equality. In order to trace the flow of processing capacity the algorithm proposed in [83] represents the processing capacity remaining on each l-stripe as a linear function of some common deadline C. The parameters of the function are modified as a result of building partial schedules and consuming the free processing time by the scheduled tasks. Using such a function the optimum schedule length can be calculated in $O(n^2 m^2)$ time for problem $P \mid cube_j, pmtn \mid C_{\max}$. An algorithm based on similar idea was proposed in [84]. Yet, the tasks of the same dimension were not sorted according to the nonincreasing execution time. The computational complexity was $O(n^2 \log^2 n)$ [84].

Problem $P, win \mid cube_j, pmtn \mid C_{\max}$, i.e., scheduling tasks requiring a power-of-two processors on a system with windows of availability has been studied in [72]. A method used to construct an optimal schedule is similar to the one used in [78] to solve $Q \mid cube_j, pmtn \mid C_{\max}$, or in [84] to solve $P \mid cube_j, pmtn \mid C_{\max}$. However, the algorithm presented in [78] required complex data structures to trace accumulation of the processing capacity after extending the schedule length. Analogously, the algorithm in [84] for the same purpose, used a parametric representation of the remaining processing capacity. A conceptually simpler method has been proposed in [72]. When the lower bound C is not adequate to schedule some task T_j, then conditions (25.5)–(25.6) are violated for j. There is some deficiency $DP_j = \sum_{i=a+1}^{a+j} p_i - \sum_{i=1}^{j} PC_i^l$ (cf. Figure 25.4(a)) in the processing capacities, where $a+1$ is the first (and the longest) l-task. Processing capacities grow due to the lengthening the schedule. Yet, only some processors m' contribute to the increase of processing capacity that can be used to feasibly schedule T_j. If we knew the right number of processors m' then the schedule extension could be calculated as DP_j / m'. The actual number of processors m' which produce the necessary processing capacity is in the range $[1, m]$. A binary search over range $[1, m]$ can be applied to find the right number m', and the minimum extension of the schedule length. The binary search is guided by the inequalities (25.5)–(25.6). If conditions (25.5)–(25.6) are satisfied with equality for j then m' is the right one, and the extension is the minimum possible. If conditions (25.5)–(25.6) are satisfied with inequality, then it means that the schedule extension is too long, and m' is too small. Finally, if conditions (25.5)–(25.6) are violated, then the schedule extension is too small, and m' is too big. This method results in an algorithm with complexity $O(n(k + \log m) m \log m)$, where k is the number of time windows of processor availability.

To our best knowledge the first preemptive multiprocessor task scheduling problem ever studied was $P \mid size_j, pmtn \mid C_{\max}$, where $size_j \in \{1, \Delta\}$ [57]. The algorithm for this case first calculates optimum schedule length $C^*_{\max}$, then Δ-tasks are scheduled using McNaughton rule [67] starting from P_1. The Δ-tasks occupy p Δ-stripes in the interval $[0, C^*_{\max}]$, and Δ-stripe $p+1$ in some interval $[0, r]$, where $r \leq C^*_{\max}$. Thus, processors $P_{(i+1)\Delta+1}, \ldots, P_m$ are free in the interval $[0, C^*_{\max}]$. Pieces of the longest 1-tasks that exceed $C^*_{\max} - r$ are scheduled using McNaugthon rule in the interval $[0, C^*_{\max}]$. The rest of 1-tasks are scheduled using the same rule in the remaining available intervals. The crucial element of this method is calculation of the optimum schedule length $C^*_{\max}$. It is calculated according to the formula [57]: $C^*_{\max} = \max\{\frac{Y}{\lfloor \frac{m}{\Delta} \rfloor}, \frac{Z}{m}, p_1, p_{n_\Delta+1}, \max_{j=1}^{n_1}\{\frac{(j-m_1)Y+N_j}{(j-m_1)p+j}\}\}$, where $Y = \sum_{j=1}^{n_\Delta} p_j$, $Z = Y\Delta + \sum_{j=n_\Delta+1}^{n} p_j$, p_1 — is the execution time of the longest Δ-task, $p_{n_\Delta+1}$ — is the execution time of the longest 1-task, $m_1 = m - (p+1)\Delta$ — is the number of processors that can be used by 1-tasks in interval $[0, r]$, $N_j = \sum_{i=n_\Delta+1}^{j} p_i$. The complexity of this algorithm is $O(n)$.

Let us finish this section with an observation that a considerable progress took place over the years in the algorithms for preemptive scheduling of multiprocessor task. The considered problems advanced from $size_j \in \{1, \Delta\}$ [57] to model *cube*, and within model *cube* from feasibility testing algorithms to the algorithms directly constructing optimum schedules.

25.4.2 Minimum Sum Criteria

25.4.2.1 Nonpreemptive Schedules, Arbitrary Processing Times

The nonpreemptive scheduling of the multiprocessor tasks with arbitrary processing times is **NP**-hard already for $m = 2$ (problem $P2 \,|\, size_j \,|\, \sum c_j$) even if $n_2 = 1$ [52]. For the weighted flow time criterion, problem $P2 \,|\, size_j \,|\, \sum w_j c_j$ is s**NP**h [52]. A pseudopolynomial algorithm with complexity $O(nT^{3n_2+1})$ was proposed in [52]. Thus, $P2 \,|\, size_j \,|\, \sum w_j c_j$ is **NP**h in the ordinary sense for any fixed $n_2 \geq 1$.

Subcube allocation and scheduling on hypercubes received a lot of attention (see, e.g., [22,85,86]). In [22] a heuristic and its experimental evaluation were presented for the on-line version of problem $P \,|\, r_j, cube_j \,|\, \sum C_j$. The set of tasks and their parameters were not known in advance. It was observed that sophisticated processor allocation strategies alone could not guarantee good performance. Generally, the allocation algorithms use task sizes to make their decision, while task execution times are not taken into account. One of the shortcomings of a simple task allocator is weak ability to recognize idle subcubes, and inability to compact fragmented subcube areas. A set of *Scan* strategies was proposed which resemble Scan algorithms for controlling movements of the hard disk heads [87]. Tasks with the same $size_j$ are appended to one queue. Queues with different task sizes are scanned in the direction of increasing (or decreasing) size. Tasks from the considered queue are assigned to the processors until the queue is emptied. These scheduling strategies effectively overcome disadvantages of the simple subcube allocators by sequencing the tasks.

25.4.2.2 Nonpreemptive Schedules, Unit Execution Times (UET)

Problem $P \,|\, size_j, p_j = 1 \,|\, \sum c_j$ of unit execution time multiprocessor task scheduling for mean flow time criterion is s**NP**h [88] (reduction from 3-Partition [69]). However, when the number of processors is fixed, that is for problem $Pm \,|\, size_j, p_j = 1 \,|\, \sum w_j c_j$, an algorithm with complexity $O(n^{3m+1})$ has been given in [88]. The algorithm reduces the scheduling problem to the shortest path problem in a graph where nodes represent possible configurations in the executing of the schedule.

As already mentioned a dynamic programming method has been proposed in [66] to solve problem $Pm \,|\, r_j, p_j = 1, size_j \,|\, \sum c_j$.

Problems $Pm \,|\, size_j, p_j = 1 \,|\, \sum w_i U_i$, and $Pm \,|\, size_j, p_j = 1 \,|\, \sum T_i$, that is scheduling for the weighted number of late tasks, and for the weighted tardiness criteria, were identified in [64] to be polynomially solvable for fixed m as an extension of their counterparts in scheduling on dedicated processors [89].

For problem $P \,|\, cube_j, p_j = 1 \,|\, \sum w_j c_j$, that is for the tasks requiring a power-of-two processors, a low order polynomial time algorithm with complexity $O(n(\log n + \log m))$ has been proposed in [88]. This method builds and executes processor feasible sets from the set with biggest weight to the set with the smallest weight. The construction of the processor feasible set with size 2 is based on the observation that 2-task T_j to be elected to the heaviest processor feasible set must compete with a pair T_i, T_l of the highest weight 1-tasks. Analogously, if processor feasible sets with size 2^k are constructed then task T_j with $size_j = 2^k$, competes with two heaviest processor feasible sets of size 2^{k-1}. Thus, processor feasible sets can be constructed recursively from size 1 to size m.

Problem $P2 \,|\, p_j = 1, r_j, size_j \,|\, \sum c_j$, can be solved by a greedy algorithm giving preference to pairs of ready 1-tasks over ready 2-tasks, and to ready 2-tasks over single ready 1-tasks [64]. For problem $P2 \,|\, p_j = 1, size_j \,|\, \sum U_j$, i.e., for the minimum number of late tasks criterion an $O(n \log n)$ algorithm has been proposed in [64]. A general optimality criterion of the type $\gamma = \phi_1(c_1) \odot \phi_2(c_2) \ldots \phi_n(c_n)$, where $\phi_j(c_j)$ is a non-negative nondecreasing cost of completing T_j at c_j, $\odot$ is commutative and associative, and for non-negative numbers x, y, z: $x \leq y \Rightarrow x \odot z \leq y \odot z$, was studied in [64]. This kind of optimality criterion comprises, e.g., criteria $\sum U_j, \sum T_j$. For problems $P2 \mid p_j = 1, size_j \mid \gamma$, a dynamic programming procedure with complexity $O(n^3)$ has been proposed [64].

When precedence constraints are taken into consideration problem $P2 \mid prec, size_j, p_j = 1 \mid \sum c_j$ is already sNPh [64].

25.4.2.3 Preemptive Schedules

It is known that preemptive scheduling of multiprocessor tasks for mean flow time criterion, that is problem $P \mid size_j, pmtn \mid \sum c_j$, is **NP**-hard [88].

25.5 Moldable Tasks

25.5.1 Discrete Execution Time Functions

25.5.1.1 Nonpreemptive Schedules

Problem $P2 \| C_{\max}$ is **NP**h, and $P5 \mid size_j \mid C_{\max}$ is s**NP**h [58]. Therefore, pseudopolynomial time algorithms for moldable tasks may exist only for $m \in \{2, 3, 4\}$ (unless $P = NP$). The issue of strong **NP**-hardness of $P4 \mid any \mid C_{\max}$ remains open. For $P2 \mid any \mid C_{\max}$, and $P3 \mid any \mid C_{\max}$ pseudopolynomial algorithms were proposed in [58]. The algorithms are dynamic programs based on the observation that all schedules for problems $P2 \mid any \mid C_{\max}$, $P3 \mid any \mid C_{\max}$ can be transformed to *canonical schedules* (see Figure 25.5). The dynamic programming methods calculate functions $F(j, x_1, x_2)$ for $m = 2$, and $F(j, x_1, x_2, x_3, x_3, x_5)$ for $m = 3$, which is the smallest execution time of m-tasks among all the schedules for the first j tasks, such that 1-tasks occupy on P_1, P_2, P_3, x_1, x_2, x_3 units of time, respectively, 2-tasks on $\{P_2, P_3\}$ use x_5, and on $\{P_1, P_2\}$ x_5 units of time. For example, for $P2 \mid any \mid C_{\max}$ $F(j, x_1, x_2) = \min\{F(i-1, x_1, x_2) + p_j(2), F(i-1, x_1 - p_j(1), x_2), F(i-1, x_1, x_2 - p_j(1))\}$. The length of the optimum schedule can be calculated as $\min_{x_1, x_2}\{F(n, x_1, x_2) + \max\{x_1, x_2\}\}$. Since $j = 1, \ldots, n$, and $0 \le x_1, \ldots, x_5 \le \sum_{i=1}^{n} p_i = M$ the complexity of this method is $O(nM^2)$ for $m = 2$, and $O(nM^5)$ for $m = 3$.

Scheduling multiprocessor tasks is **NP**h. Therefore, till the end of this section we present several simple heuristics for moldable tasks scheduling. A special case of $P \mid any, \delta_j \mid C_{\max}$ was studied in [90]. A partitionable multiprocessor system consists of m processors, and w controllers, where $n \le w < m$ [90]. A controler is needed to execute a task. Only *parallel schedules* were allowed. In a parallel schedule all tasks, at least in some interval, are executed in parallel. Schedule length can be optimized by partitioning the processors among the tasks. The heuristic starts with assigning one processor to each task. Then, the longest task which is determining $C_{\max}$ is shortened by assigning more processors. This process is repeated until the upper limit δ_j on the sizes of the longest task is achieved, or all m processors are occupied. It was shown that the tight performance ratio of this algorithm is $\min\{n, R/(1 - n/m)\}$, where R is the maximum ratio of two successive acceptable sizes of any task.

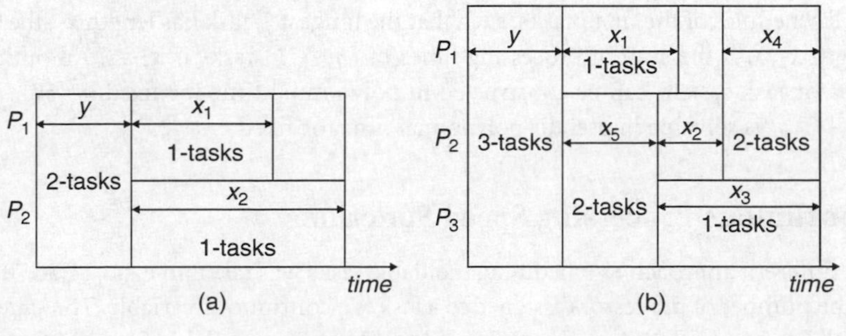

FIGURE 25.5 Canonical schedules for problems: (a) $P2 \mid any \mid C_{\max}$, (b) $P3 \mid any \mid C_{\max}$.

In [91] *earliest completion time* (ECT) heuristic has been proposed for problem $P \mid any, p_j(q) = \frac{p_j}{q}, \delta_j \mid C_{\max}$. ECT is a list scheduling algorithm. In a partial schedule when some tasks are already assigned to the processors, the later a task is started, the more processors are available. The more processors are available, the shorter execution time of the task can be. For a given list of the tasks ECT tries to find for each task the starting time at which the completion time of the task is the earliest. It has been shown that the worst case performance ratio of ECT for $P \mid any, p_j(q) = \frac{p_j}{q}, \delta_j \mid C_{\max}$ is at most $\ln \max_{T_j \in \mathcal{T}}\{\delta_j\} + 2$. ECT heuristic was also analyzed in [92,93].

In [94] a heuristic for problem $P \mid any, sp \mid C_{\max}$ has been proposed, where *sp* denotes series-parallel graphs. Series-parallel graphs are built using two constructions: series union, and parallel union. A single task (node) is a series-parallel graph. In the series union a number of series-parallel graphs are connected sequentially. All the terminal nodes of some preceding graph are connected with all the source nodes of the succeeding graph. In the parallel union a set of series-parallel graphs are independent. In other words the components of the series union must be executed sequentially, and of the parallel union can be executed in parallel. Every series-parallel graph can be represented as a tree of the above two types of unions. The algorithm for $P \mid any, sp \mid C_{\max}$ proposed in [94] uses this tree representation of a series-parallel graph to substitute the tasks (subgraphs) in series/parallel unions by composite equivalent tasks. For each composite task the function of the processing time in the number of assigned processors is calculated using the same type functions of the union components. Using this method, the series-parallel graph is folded to a single equivalent task. To construct the schedule the number of used processors and starting time of each task must be determined. For the given start time of a composite task, and the number of assigned processors, the start time and the number of used processors of the components are computed by unfolding the composite tasks to their initial components.

More about approximate algorithms for scheduling moldable tasks can be found in the next chapter of this book.

25.5.1.2 Preemptive Schedules

Problem $P \mid size_j, pmtn \mid C_{\max}$ is already **NPh** [68]. For moldable tasks problem $P \mid any, pmtn \mid C_{\max}$ is s**NPh**, and problem $P2 \mid any, pmtn \mid C_{\max}$ is **NPh** [58].

For any fixed number of processors a pseudopolynomial algorithm was proposed in [58]. This algorithm is based on the observation that in any schedule for moldable tasks the sizes of the tasks are already selected. Thus, a schedule for moldable tasks is also a schedule for multiprocessor tasks with some selected sizes. Conditions (25.5)–(25.6) must be satisfied for each task size l by any preemptive schedule to be feasible. Thus, an optimum schedule for multiprocessor tasks is determined by the values: $m-1$ longest execution times of 1-tasks, and the sum of all execution times of 1-tasks, $\lfloor \frac{m}{2} \rfloor - 1$ longest execution times of 2-tasks, and the sum of all execution times of 2-tasks, ..., and the sum of execution times of m-tasks. Hence, $H = \sum_{l=1}^{m} \lfloor \frac{m}{l} \rfloor$ numbers determine a schedule for a multiprocessor task system. In order to determine the optimum schedule for moldable tasks at most Z^H multiprocessor task systems must be verified, where Z is an upper bound on the schedule length and H is $O(m \log m)$. The multiprocessor task systems that must be verified are generated using a pseudopolynomial algorithm based on dynamic programming [58]. Namely, function $F(i, x_1, x_2, \ldots, x_{H-1})$ is calculated which is giving the smallest execution time of m-tasks, among all schedules of the first i tasks, such that the longest 1-task has length x_1, the second longest 1-task has length $x_2, \ldots$, the sum of processing times of $(m-1)$-tasks is x_{H-1}. An optimum schedule for multiprocessor task system can be constructed in polynomial time for fixed m [57]. Thus, problem $Pm \mid any, pmtn \mid C_{\max}$ is solvable in pseudo-polynomial time for fixed $m \geq 2$.

25.5.2 Continuous Processing Speed Functions

A qualitatively different approach to scheduling moldable tasks was taken in e.g., [11,56,70,95–99]. It was assumed that the number of processors assigned to a task is a continuous variable. The *state* $x_j(t)$ of a task T_j at time t is the amount of work done on T_j until t. When T_j is available for execution at time r_j, the state of T_j is $x_j(r_j) = 0$. At the completion of T_j at time c_j, the state of T_j is $x_j(c_j) = p_j$. In this case p_j

can be considered as the amount of work to be performed on T_j, or its processing time on one processor. It was assumed that rate $\frac{dx_j}{dt}$ at which the task proceeds towards the completion depends on the amount q_j of the assigned resource, which is the number of the assigned processors q_j, that is $\frac{dx_j}{dt} = f_j(q_j)$. This idea is not new in the scheduling with continuous resources, see for example [99], or other chapters of this book. As the speed of processing $f_j(q)$ does not depend on time, the assignments of the processors are constant over the course of task execution. Under the above assumptions a number of analytical results on the form of the optimum schedules have been obtained. It has been shown [98,99] that for $n < m$ the optimum schedule length can be expressed as:

$$C^*_{\max} = \min\left\{C_{\max} > 0 : \frac{\mathbf{p}}{C_{\max}} \in conv(U)\right\} \tag{25.10}$$

where U is the set of points $\mathbf{v} = [v_1, \ldots, v_n]$ such that $v_j = f_j(q_j)$, for $j = 1, \ldots, n$, and $\sum_{j=1}^{n} q_j = m$, $conv(U)$ is the convex hull of set U, and $\frac{\mathbf{p}}{C_{\max}} = [\frac{p_1}{C_{\max}}, \ldots, \frac{p_n}{C_{\max}}]$ is a straight line in the space containing set U. Using the above result it has been shown that if all functions f_j are convex, then the optimum solution to problem $P|any|C_{\max}$ is to execute all the tasks consecutively using all m processors. This proves optimality of gang scheduling/coscheduling strategies for some types of parallel applications. Note that this assignment is not fractional, and can be applied both to discrete, and to continuous processing speed functions. When all functions f_j are concave, then in the optimum schedule for problem $P|any|C_{\max}$ tasks are executed in parallel. The optimum schedule length $C^*_{\max}$ can be found as the positive root of the equation

$$\sum_{j=1}^{n} f_j^{-1}\left(\frac{p_j}{C_{\max}}\right) = m$$

where f_j^{-1} is the inverse function to f_j, $C_{\max}$ is the independent variable. The above equation can be solved analytically for some types of function f_j. In particular, for $f_j = k_j q^\alpha$, and $\alpha \le 1$ (problem $P|any, p_j(q) = \frac{p_j}{k_j q^\alpha}|C_{\max}$)

$$C^*_{\max} = \left[\frac{1}{m}\sum_{j=1}^{n}\left(\frac{p_j}{k_j}\right)^{\frac{1}{\alpha}}\right]^{\alpha} \tag{25.11}$$

This method has been extended to the case of uniconnected activity networks (uan) form of precedence constraints [11] (problem $P|any, p_j(q) = \frac{p_j}{k_j q^\alpha}|uan|C_{\max}$). In [56] problem $P|any, p_j(q) = \frac{p_j}{k_j q^\alpha}|r_j|L_{\max}$ is considered. Also here processing speed of each task is a continuous function depending on the number (amount) of assigned processors. The problem is reduced to a set of nonlinear equations.

Problem $P|any, p_j(q) = \frac{p_j}{q^\alpha}, prec|C_{\max}$ was examined under the above assumptions in [96,97,100]. Furthermore, it was assumed that speed $\forall_{T_j}\, f_j(q) = q^\alpha$. We describe this case in more detail because under the above conditions the problem of scheduling moldable tasks with precedence constraints has an intuitive physical analogue. It was shown in [96] that the optimal schedule has the following properties: (a) *constant assignment* — the number of processors $q_j > 0$ assigned to T_j is constant during T_j execution, (b) *flow conservation* — when T_j finishes all processors q_j are reallocated to the successors of T_j, (c) *finishing time* — the successors of T_j start immediately after the completion of T_j, (d) *homogeneity* — the optimum solution for $m > 1$ processors can be found by solving the problem for $m = 1$ (note it is a continuous variable), and scaling the processor allocations up to m. It was observed in [97] that processing power, i.e., processors, behave as electric charge, and precedence constraints like wires. The charge, that is processors, pour into a task from its predecessors. After the task's completion the charge is reallocated to the task's successors. This behavior is equivalent to the Kirchoff's current law (KCL). Furthermore, the requirement of the concurrent completion of all the predecessors of some task, and the concurrent start of all the successors of some task can be translated to a set of timing equations formulated in a loop of the precedence constraints graph equivalent to the Kirchoff's voltage law (KVL). The above observations reduce the search for an optimal schedule to solving a set of KCL, KVL constraints. Let us denote by q_{ij}

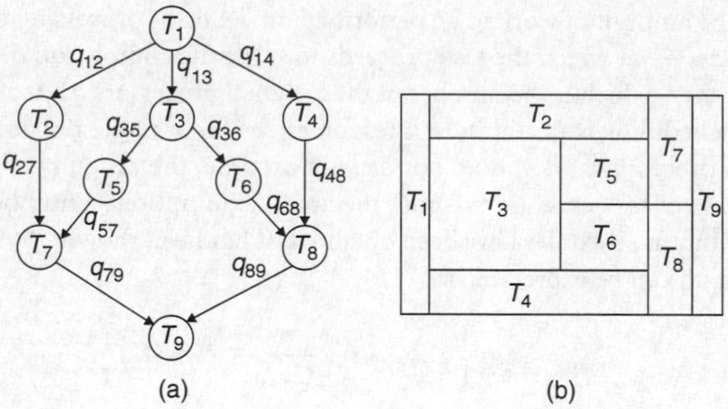

FIGURE 25.6 Example for problem $P\,|\,any, p_j(q) = \frac{p_j}{q^\alpha}, prec\,|\,C_{\max}$ with continuous processor power: (a) Precedence graph, (b) Form of the schedule.

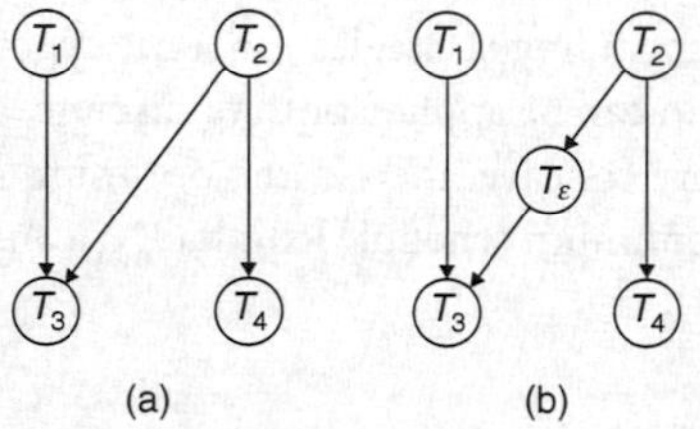

FIGURE 25.7 (a) Example of an edge with nonessential flow, (b) Elimination of nonessential flow.

the amount of processing power, that is continuous processors, which passes from T_i to T_j. For example, for the precedence graph in Figure 25.6(a), the set of constraints is

KCL:

$$q_{12} + q_{13} + q_{14} = m \qquad q_{12} - q_{27} = 0 \qquad q_{13} - q_{35} - q_{36} = 0$$

$$q_{14} - q_{48} = 0 \qquad q_{35} - q_{57} = 0 \qquad q_{36} - q_{68} = 0$$

$$q_{27} + q_{57} - q_{79} = 0 \qquad q_{68} + q_{48} - q_{89} \qquad q_{79} + q_{89} = m$$

KVL:

$$\text{loop } T_1T_2T_7T_5T_3T_1 \qquad \frac{p_2}{q_{12}^\alpha} - \frac{p_5}{q_{35}^\alpha} - \frac{p_3}{q_{13}^\alpha} = 0$$

$$\text{loop } T_1T_3T_6T_8T_4T_1 \qquad \frac{p_3}{q_{13}^\alpha} + \frac{p_6}{q_{36}^\alpha} - \frac{p_4}{q_{14}^\alpha} = 0$$

$$\text{loop } T_1T_2T_7T_9T_8T_4T_1 \qquad \frac{p_2}{q_{12}^\alpha} + \frac{p_7}{(q_{27}+q_{57})^\alpha} - \frac{p_8}{(q_{68}+q_{48})^\alpha} - \frac{p_4}{q_{14}^\alpha} = 0$$

Thus, the problem of optimum scheduling of moldable tasks has been reduced to solving a set E of nonlinear equations, where E is the number of edges in the graph of precedence constraints.

In the above method one difficulty arises when a task may receive charge, i.e., processors, from several predecessors. It is the case when the graph of precedence constraints contains, for example, a subgraph depicted in Figure 25.7(a). The processing power may flow from task T_2 to T_4 only, or from T_2 to T_3 and T_4. In the first case paths $T_1 \rightarrow T_3$, and $T_2 \rightarrow T_4$ are parallel and independent. In the second case, by KVL equations, T_1 and T_2 must be finished simultaneously. In fact, T_1 and T_2 need not finish simultaneously. It is only required that both of them finish before T_3. Consequently, the sum of task execution times along a loop may be satisfied both with equality or with an inequality. The edges causing difficulties of this type were called *nonessential flow edges*. In order to avoid nonessential flow, it is proposed in [97] to insert a tiny task T_ε on each nonessential flow edge, as depicted in Figure 25.7(b).

By using the above method closed form solutions can be derived for series-parallel graphs ($P\,|\,any$, $p_j(q) = \frac{p_j}{q^{\alpha}}, sp\,|\,C_{\max}$). All the series of tasks can be reduced to a task with the sum of processing requirements of the components, and equal assignment of the processing power to the components. A parallel construct which consists of tasks in some set $\mathcal{T}'$ can be reduced to a single task with requirement equal to $p' = (\sum_{T_j \in \mathcal{T}'} p_j^{1/\alpha})^{\alpha}$ (cf. Equation (25.11)). The tasks in $\mathcal{T}'$ are executed in parallel from the beginning till the end, and task $T_j \in \mathcal{T}'$ uses $\frac{p_j^{1/\alpha}}{p'}$ fraction of the processors assigned to execute all the parallel tasks in $\mathcal{T}'$. In this way the whole series-parallel graph of precedence constraints can be folded to a single task. Analogously, by using the homogeneity property, the solution can be unfolded to the components of the graph.

25.6 Malleable Tasks

In this section we consider malleable tasks by which we mean parallel tasks allowing both preemption and changes in the number of used processors. In this section we proceed from simple cases to the more involved ones. We start with work preserving malleable tasks, that is the tasks with linear speedup.

25.6.1 Linear Speedup

Let us note that the problems with no limit on the number of used processors ($\forall_{T_j \in \mathcal{T}} \delta_j \geq m$) are, in a sense, trivial because they reduce to single processor scheduling problems.

The simplest (nontrivial) problem $P\,|\,var, p_j(q) = \frac{p_j}{q}, \delta_j\,|\,C_{\max}$ can be solved in $O(n)$ time by a modified McNaughton's rule [101,102]. The optimum schedule length is

$$C^*_{\max} = \max\left\{\frac{1}{m}\sum_{j=1}^{n} p_j, \max_{T_j \in \mathcal{T}}\left\{\frac{p_j}{\delta_j}\right\}\right\} \tag{25.12}$$

The schedule of the above length can be constructed by the same wrap-around rule as in problem $P\,|\,pmtn\,|\,C_{\max}$. The difference is that the wrap-around may take place more than once for a single task. Consequently, such a task is executed on several processors in parallel. An example instance of problem $P\,|\,var, p_j(q) = \frac{p_j}{q}, \delta_j\,|\,C_{\max}$ and its optimum schedule is shown in Figure 25.8.

Problem $P\,|\,var, p_j(q) = \frac{p_j}{q}, \delta_j, r_j\,|\,C_{\max}$ was studied in [103,104]. It is not difficult to realize that the length of the schedule for this problem is determined by the tasks, or the pieces of the tasks, which remain to be executed after the last ready time. The length of this ending part of the schedule is determined by Equation (25.12). Thus, the optimization algorithm for problem $P\,|\,var, p_j(q) = \frac{p_j}{q}, \delta_j, r_j\,|\,C_{\max}$ must minimize both $\max_{T_j \in \mathcal{T}}\{\frac{p_j}{\delta_j}\}$, and $\sum_{j=1}^{n} p_j$ before reaching the last ready time. Surprisingly, this can be achieved by an adaptation of the algorithm for problems $P2\,|\,prec, pmtn\,|\,C_{\max}$, $P\,|\,tree, pmtn\,|\,C_{\max}$ [105,106]. The algorithm uses two concepts: task height and sharing the processing capability. Task height is the length of the longest path in the precedence constraint graph starting at the considered task and finishing at its furthest successor. Processing time of the task itself is included in the level. Processing capability is a fraction of total processing power m assigned to task T_j in some time interval. The algorithm assigns the

$m = 5$

task	T_1	T_2	T_3	T_4
p_j	2	4	5	9
δ_j	1	2	2	4

(a)

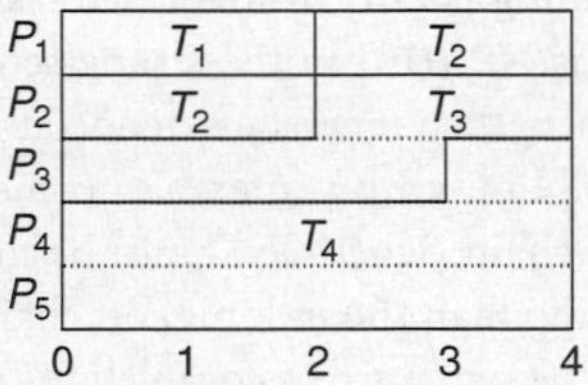

(b)

FIGURE 25.8 (a) Instance of problem $P\,|\,var, p_j(q) = \frac{p_j}{q}, \delta_j\,|\,C_{\max}$, (b) The schedule.

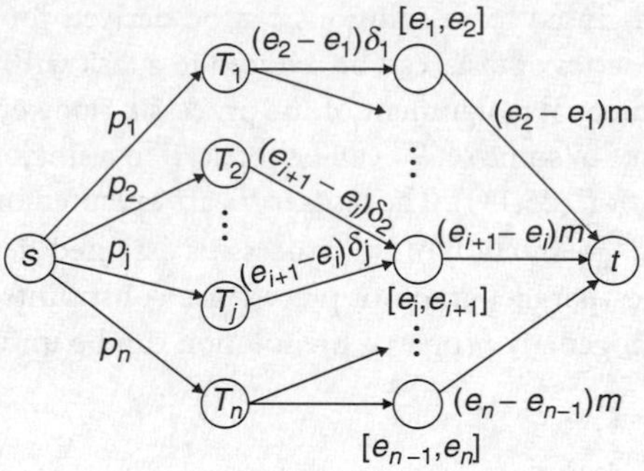

FIGURE 25.9 A network for problem $P\,|\,var, p_j(q) = \frac{p_j}{q}, \delta_j, r_j\,|\,L_{\max}$.

processing capability to the highest tasks first. If there are still some processors free, then they are assigned to the second highest tasks, etc. If two (or more) tasks are equal in the sense of height, then they are assigned equal processing capability. Thus, once two ready tasks become equal with the respect to their heights, they remain equal till the completion of one (or both) of them. For problem $P\,|\,var, p_j(q) = \frac{p_j}{q}, \delta_j, r_j\,|\,C_{\max}$ the height of task T_j is $\frac{p_j}{q}$. The single highest task is assigned δ_j processors. If there are more than one task with equal height in some set $\mathcal{T}'$, and they are able to use together more than the number m' of available processors, then the processing capability assigned to a task $T_j \in \mathcal{T}'$ is $\frac{m'\delta_j}{\sum_{T_i \in \mathcal{T}'} \delta_i}$. The complexity of the algorithm for problem $P\,|\,var, p_j(q) = \frac{p_j}{q}, \delta_j, r_j\,|\,C_{\max}$ is $O(n^2)$ [103,104].

Problem $P\,|\,var, p_j(q) = \frac{p_j}{q}, \delta_j, r_j\,|\,L_{\max}$ was studied in [102]. The method proposed in [102] reduces the problem to a sequence of maximum network flow problems verifying feasibility of the schedule for some test value of $L_{\max}$. This approach is very similar to the algorithm proposed in [107] to solve problem $P\,|\,pmtn, r_j\,|\,L_{\max}$. For a given instance of the problem the events in the task system: ready times $r_1, \ldots, r_n$, and due-dates $d_1, \ldots, d_n$ are sorted increasingly to obtain sequence $e_1, \ldots, e_k$, where $k \le 2n$. The equivalent network is built as follows: For each interval $[e_i, e_{i+1}]$ a vertex is created (cf. Figure 25.9). The vertex representing interval $[e_i, e_{i+1}]$ is connected with a network terminal t by an arc with capacity $(e_{i+1} - e_i)m$. For each task T_j a vertex is created to which the network source s is connected by an arc with capacity p_j. A vertex representing task T_j is connected with the vertices representing time intervals in which T_j can be executed. The capacity of such an arc connecting T_j with $[e_i, e_{i+1}]$ is $(e_{i+1} - e_i)\delta_j$. When the maximum flow saturates all arcs (s, T_j), then a feasible schedule exists. The flow from node T_j to node $[e_i, e_{i+1}]$ determines how much work on T_j must be performed in interval $[e_i, e_{i+1}]$. A partial schedule in each interval $[e_i, e_{i+1}]$ $(i = 1, \ldots, k-1)$ can be obtained using the method for problem $P\,|\,var, p_j(q) = \frac{p_j}{q}, \delta_j\,|\,C_{\max}$, described above. When a feasible schedule does not exist, then the test value of $L_{\max}$ must increase by such an amount that the capacity of the minimum cut in the network is at least equal to $\sum_{j=1}^{n} p_j$. The optimum value of lateness $L^*_{\max}$ can be determined in time $O(n^4 m)$.

The problem $P\,|\,var, chain, p_j(q) = \frac{p_j}{q}, \delta_j\,|\,C_{\max}$ where the precedence constraints are chains of three tasks with the first, and the third tasks sequential $(\delta_j = 1)$, and the second, central task, with unlimited parallelism, was studied in [101]. It was shown that this problem is s**NP**h, which bears a practical consequence that scheduling chains of malleable tasks, that is parallel applications with given parallelism profile (see [23,24]) is **NP**-hard in the strong sense. When an optimum schedule S for at most $m-1$ highest chains is known, then it can be extended to an optimum schedule for all $N > m$ chains using level scheduling algorithm filling empty intervals in schedule S. The height is understood here as the processing required by the given chain. Schedule S must be constructed for at most $m-1$ highest chains demanding strictly more processing than the mth highest chain. S can be found using a linear programming formulation provided that the sequence of completions of the chains' tasks is known. There are at most $\frac{2(m-1)!}{2^{m-1}}$ such sequences. Therefore, problem $Pm\,|\,var, chain, p_j(q) = \frac{p_j}{q}, \delta_j\,|\,C_{\max}$ for a fixed number of processors, and three-task chains can be solved in polynomial time. This result can be extended to chains of any number of tasks bounded by some constant. It can be concluded that the core computational complexity of this problem comes from a set of at most $m-1$ chains, adding more chains increases the complexity only polynomially. Polynomially solvable cases, and approximation algorithms for this problem were also examined in [101].

A problem which resides half-way between the parallel task model considered in this chapter, and the model of scheduling multiprocessor tasks on dedicated processors is studied in [51]. On one hand, each task may be executed on a restricted subset $\mathcal{P}_j$ of the processor set. This makes the problem similar to the case of multiprocessor tasks executed on dedicated processors. On the other hand, a task may use the processors from the given set $\mathcal{P}_j$ as malleable tasks use parallel processors. The problem has been reduced to a sequence of network flow problems [51].

25.6.2 Nonlinear Speedup

In the malleable task scheduling problems considered so far it was assumed that $\forall_{T_j \in \mathcal{T}}\, p_j(q) = \frac{p_j}{q}$. Here we consider other execution speed functions. The methods for solving problem $P \mid any \mid C_{\max}$ with continuous and nonlinear processing speed functions were described in Section 25.5.2. It was assumed that processors, or processing capability, behave like a continuously divisible medium. This assumption is not realistic for small m. Therefore, some method of rounding the continuous processor assignments is needed when m is small.

This problem was studied in [108] for $n \leq m$. The nonlinear functions of processing speed were linearized and substituted by piecewise linear functions. The speed for fractional processor assignment q_j was expressed as a linear combination of the speeds for $\lfloor q_j \rfloor$, and $\lceil q_j \rceil$. We omit the case of convex speed functions, as it was discussed in Section 25.5.2. In the following we consider concave speed functions. The solution of the continuous case was a starting point for construction of integer processor assignments. The solution of the continuous case is based on the observation that the optimum assignments of the resource (processors) to the tasks can be found as a point of intersection of line $\frac{\mathbf{p}}{C_{\max}}$, with a convex hull $conv(U)$ (cf. Equation (25.10)). Calculating this point is easier for piecewise linear speed functions than for arbitrary nonlinear functions, and can be implemented to run in $O(n \max\{m, n \log^2 m\})$ time [108]. In general the point of intersection of line $\frac{\mathbf{p}}{C_{\max}}$ with a facet F of $conv(U)$ defines a fractional processor assignments. This assignment can be expressed as a linear combination $\frac{\mathbf{p}}{C^*_{\max}} = \sum_{i=1}^{l} \lambda_i \mathbf{v_i}$ of the integer assignments defined for the corners $\mathbf{v_1}, \ldots, \mathbf{v_l}$ of facet F, where $\mathbf{v_i} = [v_i^1, \ldots, v_i^n]$ is an n-component vector. Thus, the schedule can be constructed as a sequence of l intervals of length $\lambda_i C^*_{\max}$, for $i = 1, \ldots, l$, with the integer processor assignment $f^{-1}(v_i^j)$ for task T_j, for $j = 1, \ldots, n$. Yet, the computationally hard problem is to find the corners of facet F. For $n \in \{2, 3\}$ it can be done in constant time [108]. For $n \leq m$ the solution of the continuous version of the problem, with assignment q_j for T_j and schedule length $C^*_{\max}$, can be converted to a feasible schedule with the same length and integer assignments without determining corners $\mathbf{v_1}, \ldots, \mathbf{v_l}$ of F. The fractional assignment q_j can be expressed as a linear combination of the assignments $a_j = \lfloor q_j \rfloor$, and $b_j = \lceil q_j \rceil$. The assignment q_j for T_j is converted to two intervals of executing T_j: one using a_j processors and with length $\frac{\lambda_j p_j}{f_j(a_j)}$, and one using b_j processors and with length $\frac{(1-\lambda_j)p_j}{f(b_j)}$, where $\lambda_j = \frac{(b_j - q_j) f_j(a_j)}{f_j(q_j)}$. The schedule for the converted tasks can be found in $O(n)$ time by wrap-around rule, similar to McNaughton's rule, starting with the wider piece (size b_j), and finishing with the narrower piece (size a_j), for each task T_j. We conclude that problem $P \mid var \mid C_{\max}$, for piecewise-linear concave speed functions, and $n \leq m$ can be solved in $O(n \max\{m, n \log^2 m\})$ time [108].

25.7 Conclusions

In this chapter we presented the parallel task scheduling problem. This problem has several special forms depending on the nature of the tasks. The cases of multiprocessor tasks, moldable tasks, and malleable tasks have been distinguished. The complexity and polynomial algorithms for the above problems were discussed. We encourage the interested reader to study the next chapter on approximation algorithms for parallel task scheduling problems.

Acknowledgments

I would like to express my thanks to Philippe Baptiste, Kai Baumgarten, Pierre-François Dutot, Klaus Jansen, Chung-Yee Lee, Ramesh Krishnamurti, Maciej Machowiak, Ceyda Oğuz, Ewa Ratajczyk, and Denis Trystram for their help in preparing this chapter of the handbook.

This research has been partially supported by a grant of Polish State Committee for Scientific Research.

References

[1] Baker, K.R., *Introduction to sequencing and scheduling*, John Wiley & Sons, New York, 1974.

[2] Coffman E.G., Jr., ed., *Scheduling in Computer and Job Shop Systems*, J. Wiley, New York, 1976.

[3] Conway, R.W., Maxwell, W.L., and Miller, L.W., *Theory of Scheduling*, Addison-Wesley, Reading MA, 1967.

[4] Lloyd, E.L., Concurrent task systems, *Operations Research*, 29, 189–201, 1981.

[5] Cirne, W., Using Moldability to Improve the Performance of Supercomputer Jobs, Ph.D. Thesis, Computer Science and Engineering, University of California, San Diego, 2001.

[6] Cirne, W. and Berman, F., A model for moldable supercomputer jobs, *Proceedings of 15th International Parallel & Distributed Processing Symposium*, 2001.

[7] Li, K. and Pan, Y., Probabilistic analysis of scheduling precedence constrained parallel tasks on multicomputers with contiguous processor allocation, *IEEE Transactions on Computers*, 49, 1021–1030, 2000.

[8] Feitelson, D.G., Rudolph, L., Schwiegelshohn, U., Sevcik, K., and Wong, P., Theory and practice in parallel job scheduling, Lecture Notes in Computer Science 1291, Springer-Verlag, Berlin, 1–34, 1997.

[9] Ratajczyk, E., Hybrid algorithms for scheduling multiprocessor tasks (in Polish), Ph.D. Thesis, Institute of Computing Science, Poznań University of Technology, 2001.

[10] Dutot, P.-F., Ordonnancement de chaînes de tâches malléables, Proceedings of RenPar'14, Hammamet, 2002.

[11] Błażewicz, J., Ecker, K., Pesch, E., Schmidt, G., and Węglarz, J., *Scheduling Computer and Manufacturing Processes*, Springer-Verlag, Berlin, 1996.

[12] Brucker, P., *Scheduling Algorithms*, Springer-Verlag, Heidelberg, 1998.

[13] Drozdowski, M., Scheduling multiprocessor tasks — an overview, *European Journal of Operational Research*, 94, 215–230, 1996.

[14] Drozdowski, M., *Selected problems of scheduling tasks in multiprocessor computer systems*, Poznań University of Technology Press, Series: Monographs, No.321, Poznań, 1997. `http://www.cs.put.poznan.pl/~maciejd/txt/h.ps`

[15] Veltman, B., Lageweg, B.J., and Lenstra, J.K., Multiprocessor scheduling with communication delays, *Parallel Computing*, 16, 173–182, 1990.

[16] Feitelson, D.G., The forgotten factor: Facts on performance evaluation and its dependence on workloads, in B. Monien, R. Feldman, (eds.), *Proceedings of Euro-Par 2002*, Lecture Notes in Computer Science 2400, 49–60, 2002.

[17] Parallel workload archive, `http://www.cs.huji.ac.il/labs/parallel/workload`

[18] Coffman, E.G., Jr., Garey, M.R., and Johnson, D.S., Bin packing with divisible item sizes, *Journal of Complexity*, 3, 406–428, 1987.

[19] Feitelson, D.G. and Rudolph, L., Distributed hierarchical control for parallel processing, *IEEE Computer* 23, 65–77, 1990.

[20] Feitelson, D.G. and Rudolph, L., Evaluation of design choices for gang scheduling using distributed hierarchical control, *Journal of Parallel and Distributed Computing*, 35, 18–34, 1996.

[21] Hori, A., Ishikawa, Y., Konaka, H., Maeda, M., and Tomokiyo, T., A scalable time-sharing scheduling for partitionable distributed memory parallel machines, in *Proceedings of 28th Annual Hawaii International Conference on System Sciences (HICSS'95)*, 173–182, 1995.

[22] Krueger, P., Lai, T.-H., and Dixit-Radiya, V.A., Job scheduling is more important than processor allocation for hypercube computers, *IEEE Transactions on Parallel and Distributed Systems*, 5, 488–497, 1994.

[23] Ghosal, D., Serazzi, G., and Tripathi, S.K., The processor working set and its use in scheduling multiprocessor systems, *IEEE Transactions on Software Engineering*, 17, 443–453, 1991.

[24] Kumar, M., Measuring parallelism in computation-intensive scientific/engineering applications, *IEEE Transactions on Computers*, 37, 1088–1098, 1988.

[25] Choundhary, A.N., Narahari, B., Nicol, D.M., and Simha, R., Optimal processor assignment for a class of pipelined computations, *IEEE Transactions on Parallel and Distributed Systems*, 5, 439–445, 1994.

[26] Ercan, M.F., Oguz, C., and Fung, Y.F., Performance evaluation of heuristics for scheduling pipelined multiprocessor tasks, in V.N. Alexandrov et al. (eds.), *Proceedings of ICCS 2001*, Lecture Notes in Computer Science 2073, 61–70, 2001.

[27] Feitelson, D.G. and Rudolph, L., Gang scheduling performance benefits for fine-grain synchronization, *Journal of Parallel and Distributed Computing*, 16, 306–318, 1992.

[28] Gehringer, E.F., Siewiorek, D.P., and Segall, Z., *Parallel Processing: The Cm* Experience*, Digital Press, 1987.

[29] Ousterhout, J.K., Scheduling techniques for concurrent systems, in *Proceedings of Third International Conference on Distributed Computing Systems*, 22–30, 1982.

[30] Zahorjan, J., Lazowska, E.D., and Eager, D.L., The effect of scheduling discipline on spin overhead in shared memory parallel systems, *IEEE Transactions on Parallel and Distributed Systems*, 2, 180–198, 1991.

[31] Amdahl, G.M., Validity of the single processor approach to achieving large scale computing capabilities, in *AFIPS Conference Proceedings (Atlantic City Apr. 18-20, 1967)*, vol. 30, AFIPS, 483–485, 1967.

[32] Downey, A., Using queue time predictions for processor allocation, Lecture Notes in Computer Science 1291, Springer-Verlag, Berlin, 35–57, 1997.

[33] Gustafson, J.L., Reevaluating Amdahl's law, *Communications of the ACM*, 31, 532–533, 1988.

[34] Avizienis, A., Gilley, G.C., Mathur, F.P., Rennels, D.A., Rohr, J.A., and Rubin, D.K., The STAR (self-testing and repairing) computer: An investigation of the theory and practice of fault-tolerant computer design. *IEEE Transactions on Computers*, 20, 1312–1321, 1971.

[35] Hopkins, A.L., Lala, J.M., and Smith, T.B., FTMP — a highly reliable fault-tolerant multiprocessor for aircraft, in *Proceedings of IEEE*, 66, 1221–1239, 1978.

[36] Kopetz, H. and Veríssimo, P., Real time and dependability concepts, in Mullender, S., (ed.), *Distributed Systems*, Addison-Wesley and ACM Press, Reading, MA, 411–446, 1994.

[37] Czarnowski, J., Jędrzejowicz, P., and Ratajczyk, E., Scheduling fault-tolerant programs on multiple processors to maximize schedule reliability, in Felici, M., Kanoun, K. and Pasquini, A., (eds.), *Computer Safety, Reliability and Security*, Lecture Notes in Computer Science 1698, Springer, New York, 385–395, 1999.

[38] Jędrzejowicz, P. and Wierzbowska, I., Scheduling multiple variant programs under hard real-time constraints, *European Journal of Operational Research*, 127, 458–465, 2000.

[39] Drabowski. M., Scheduling tasks in multimicroprocessor systems (in Polish), Ph.D. thesis, Poznań University of Technology, 1985.

[40] Craig, G.L., Kime, C.R., and Saluja, K.K., Test scheduling and control for VLSI built-in self-test. *IEEE Transactions on Computers*, 37, 1099–1109, 1988.

[41] Krawczyk, H. and Kubale, M., An approximation algorithm for diagnostic test scheduling in multicomputer systems, *IEEE Transactions on Computers*, 34, 869–872, 1985.

[42] Bar-Noy, A., Canetti, R., Kutten, S., Mansour, Y., and Schieber, B., Bandwidth allocation with preemption, *SIAM Journal on Computing*, 28, 1806–1828, 1999.

[43] Dharwadkar, P., Siegel, H.J., and Chong, E.K.P., A heuristic for dynamic bandwidth allocation with preemption and degradation for prioritized requests, in *Proceedings of the 21st International Conference on Distributed Computing Systems (ICDCS'01)*, IEEE, 2001.

[44] Krunz, M., Zhao, W., and Matta, I., Scheduling and bandwidth allocation for distribution of archived video in VoD systems, *Journal of Telecommunication Systems, Special Issue on Multimedia*, 9, 1998.

[45] ATM Forum White Paper, ATM service categories: The benefits to the user. ATM Forum, San Francisco.

[46] den Boef, E., Verhaegh, W., and Korst, J., Smoothing streams in an in-home digital network: Optimization of bus and buffer usage, *Proceedings of 5th Workshop on Models and Algorithms for Planning and Scheduling Problems* (MAPSP'01) Aussois, 2001.

[47] Feng, W. and Rexford, J., Performance evaluation of smoothing algorithms for transmitting prerecorded variable-bit-rate video, *IEEE Transactions on Multimedia* 1, 302–313, 1999.

[48] Zhang, Z.-L., Kurose, J., Salehi, J.D., and Towsley, D., Smoothing, statistical multiplexing and call admission control for stored video, *IEEE Journal on Selected Areas in Communication*, 15, 1997.

[49] Coffman, E.G., Jr., Garey, M.R., Johnson, D.S. and Lapaugh, A.S., Scheduling file transfers, *SIAM Journal on Computing*, 14, 744–780, 1985.

[50] Jain, R., Somalwar, K., Werth, J., and Browne, J.C., Scheduling parallel I/O operations in multiple bus systems, *Journal of Parallel and Distributed Computing*, 16, 352–362, 1992.

[51] Serafini, P., Scheduling jobs on several machines with the job splitting property, *Operations Research*, 44, 617–628, 1996.

[52] Lee, C.-Y. and Cai, X., Scheduling one and two-processor tasks on two parallel processors, *IEEE Transactions*, 31, 445–455, 1999.

[53] Coffman, E.G., Jr., An introduction to combinatorial models of dynamic allocation, *SIAM Review*, 25, 311–325, 1983.

[54] Wilson, P., Johnstone, M., Neely, M., and Boles, D., Dynamic storage allocation: A survey and critical review, in Baker, H.G., (ed.), *Memory Management, Proceedings of International Workshop IWMM 95*, Lecture Notes in Computer Science 986, 1–116, 1995.

[55] Herroelen, W., De Reyck, B., and Demeulemeester, E., Resource-constrained project scheduling: A survey of recent developments, *Computers and Operations Research*, 25, 279–302, 1997.

[56] Węglarz, J., Scheduling under continuous performing speed vs. resource amount activity models, in Słowiński, R., and Węglarz, J., (eds.), *Advances in Project Scheduling*, Elsevier, Amsterdam, 273–295, 1989.

[57] Błażewicz, J., Drabowski, M., and Węglarz, J., Scheduling multiprocessor tasks to minimize schedule length, *IEEE Transactions on Computers*, 35, 389–393, 1986.

[58] Du, J. and Leung, J.Y-T., Complexity of scheduling parallel task systems, *SIAM Journal on Discrete Mathematics*, 2, 473–487, 1989.

[59] Błażewicz, J. and Liu, Z., Scheduling multiprocessor tasks with chain constraints, *European Journal of Operational Research*, 94, 231–241, 1996.

[60] Coffman, E.G. Jr. and Graham, R.L., Optimal scheduling for two-processor systems, *Acta Informatica*, 1, 200–213, 1972.

[61] Coppersmith, D. and Winograd, S., Matrix multiplication via arithmetic progressions, *Journal of Symbolic Computation*, 9, 251–280, 1990.

[62] Amoura, A.K., A note on scheduling multiprocessor tasks with precedence constraints on parallel processors, *Information Processing Letters*, 63, 119–122, 1997.

[63] Błażewicz, J. and Liu, Z., Linear and quadratic algorithms for scheduling chains and opposite chains, *European Journal of Operational Research*, 137, 248–264, 2002.

[64] Brucker, P., Knust, S., Roper, D., and Zinder, Y., Scheduling UET task systems with concurrency on two parallel identical processors, *Mathematical Methods of Operations Research* 52, 369–387, 2000.

[65] Baptiste, P. and Schieber, B., A note on scheduling tall/small multiprocessor tasks with unit processing time to minimize maximum tardiness, *Journal of Scheduling*, 6, 395–404, 2003.

[66] Baptiste, P., A note on scheduling multiprocessor tasks with identical processing times, to appear in *Computers and Operations Research*, 30, 2071–2078, 2003.

[67] McNaughton, R., Scheduling with deadlines and loss functions, *Management Science*, 6, 1–12, 1959.

[68] Drozdowski, M., On complexity of multiprocessor tasks scheduling, *Bulletin of the Polish Academy of Sciences, Technical Sciences*, 43, 381–392, 1995.

[69] Garey, M.R. and Johnson, D.S., *Computers and Intractability: A guide to the theory of NP-completeness*, Freeman, San Francisco, 1979.

[70] Błażewicz, J., Cellary, W., Słowiński, R., and Węglarz, J. Scheduling under resource constraints: deterministic models, *Annals of Operations Research*, vol. 7, 1986.

[71] Błażewicz, J., Drozdowski, M., de Werra, D., and Węglarz, J., Deadline scheduling of multiprocessor tasks, *Discrete Applied Mathematics*, 65, 81–96, 1996.

[72] Błażewicz, J., Dell'Olmo, P., Drozdowski, M., and Mączka, P., Scheduling multiprocessor tasks on parallel processors with limited availability, *European Journal of Operational Research*, 149, 377–389, 2003.

[73] Jansen, K. and Porkolab, L., Preemptive parallel task scheduling in $O(n) + \text{Poly}(m)$ time, in Lee, D.T., and Teng, S.H., (eds.), *Proceedings of ISAAC 2000*, Lecture Notes in Computer Science 1969, 398–409, 2000.

[74] Gonzalez, T. and Sahni, S., Preemptive scheduling of uniform processor systems, *Journal of the ACM*, 25, 92–101, 1978.

[75] Horvath, E.C., Lam, S., and Sethi, R., A level algorithm for preemptive scheduling, *Journal of the ACM*, 24, 32–43, 1977.

[76] Błażewicz, J., Drozdowski, M., Schmidt, G., and de Werra, D., Scheduling independent two processor tasks on a uniform duo-processor system, *Discrete Applied Mathematics* 28, 11–20, 1990.

[77] Błażewicz, J., Drozdowski, M., Schmidt, G., and de Werra, D., Scheduling independent multiprocessor tasks on a uniform k-processor system, *Parallel Computing*, 20, 15–28, 1994.

[78] Drozdowski, M., Scheduling multiprocessor tasks on hypercubes, *Bulletin of the Polish Academy of Sciences, Technical Sciences*, 42, 437–445, 1994.

[79] Ahuja, M. and Zhu, Y., An $O(n \log n)$ feasibility algorithm for preemptive scheduling of n independent jobs on a hypercube, *Information Processing Letters*, 35, 7–11, 1990.

[80] Chen, G.-I. and Lai, T.-H., Preemptive scheduling of independent jobs on a hypercube, *Information Processing Letters*, 28, 201–206, 1988.

[81] van Hoesel, C.P.M., Preemptive scheduling on a hypercube, Technical Report 8963/A, Erasmus University, Rotterdam, 1989.

[82] Plehn, J., Preemptive scheduling of independent jobs with release times and deadlines on a hypercube, *Information Processing Letters*, 34, 161–166, 1990.

[83] Shen, X. and Reingold, E.M., Scheduling on a hypercube, *Information Processing Letters*, 40, 323–328, 1991.

[84] Zhu, Y., and Ahuja, M., On job scheduling on a hypercube, *IEEE Transactions on Parallel and Distributed Systems*, 4, 62–69, 1993.

[85] Dutt, S. and Hayes, J.P., Subcube allocation in hypercube computers, *IEEE Transactions on Computers*, 40, 341–352, 1991.

[86] Ghonien, S.A., and Fahny, H.M.A., Job preemption, fast subcube compaction, or waiting in hypercube systems? A selection methodology, *Parallel Computing*, 29, 111–134, 2003.

[87] Silberschatz, A., Peterson, J.L., and Galvin, P.B., *Operating System Concepts*, Addison-Wesley Publishing Co., Reading, MA, 1991.

[88] Drozdowski, M. and Dell'Olmo, P., Scheduling multiprocessor tasks for mean flow time criterion, *Computers and Operations Research*, 27, 571–585, 2000.

[89] Brucker, P. and Krämer, A., Polynomial algorithms for resource constrained and multiprocessor task scheduling problems, *European Journal of Operational Research*, 90, 214–226, 1996.
[90] Krishnamurti, R. and Ma, E., An approximation algorithm for scheduling tasks on varying partition sizes in partitionable multiprocessor systems, *IEEE Transactions on Computers*, 41, 1572–1579, 1992.
[91] Wang, Q. and Cheng, K.H., List scheduling of parallel tasks, *Information Processing Letters*, 37, 291–297, 1991.
[92] Efe, K. and Krishnamoorthy, V., Optimal scheduling of compute-intensive tasks on a network of workstations, *IEEE Transactions on Parallel and Distributed Systems*, 6, 668–673, 1995.
[93] Wang, Q. and Cheng, K.H., A heuristic of scheduling parallel tasks and its analysis, *SIAM Journal on Computing*, 21, 281–294, 1992.
[94] Baumgarten, K., Static scheduling of parallel modules with series-parallel dependence, *Proceedings of New Trends in Scheduling for Parallel and Distributed Systems*, CIRM, Marseille, 2001.
[95] Błażewicz, J., Machowiak, M., Węglarz, J., Mounie, G., and Trystram, D., Scheduling malleable tasks with convex processing speed functions, *Computación y Sistemas*, 4, 158–165, 2000.
[96] Prasanna, G.N.S. and Musicus, B.R., Generalized multiprocessor scheduling using optimal control, in *Proceedings of Third Annual ACM Symposium on Parallel Algorithms and Architectures*, 216–228, 1991.
[97] Prasanna, G.N.S. and Musicus, B.R., Generalized multiprocessor scheduling for directed acyclic graphs, in *Proceedings of Supercomputing 1994*, IEEE Press, 237–246, 1994.
[98] Węglarz, J., Project scheduling with continuously-divisible, doubly constrained resources, *Management Science*, 27, 1040–1052, 1981.
[99] Węglarz, J., Modelling and control of dynamic resource allocation project scheduling systems, in: Tzafestas, S.G., (ed.), *Optimization and Control of Dynamic Operational Research Models*, North-Holland, Amsterdam, 1982.
[100] Prasanna, G.N.S., Agarwal, A. and Musicus, B.R., Hierarchical compilation of macro dataflow graphs for multiprocessors with local memory, *IEEE Transactions on Parallel and Distributed Systems*, 5, 720–736, 1994.
[101] Drozdowski, M. and Kubiak, W., Scheduling parallel tasks with sequential heads and tails, *Annals of Operations Research*, 90, 221–246, 1999.
[102] Vizing, V.G., Minimization of maximum delay in servicing systems with interruption, *U.S.S.R. Computational Mathematics and Mathematical Physics*, 22, 227–233, 1982.
[103] Drozdowski, M., Real-time scheduling of linear speedup parallel tasks, *Information Processing Letters*, 57, 35–40, 1996.
[104] Drozdowski, M., New applications of the Muntz and Coffman algorithm, *Journal of Scheduling*, 4, 209–223, 2001.
[105] Muntz, R.R. and Coffman, E.G., Jr., Optimal preemptive scheduling on two-processor systems, *IEEE Transactions on Computers*, 18, 1014–1020, 1969.
[106] Muntz, R.R. and Coffman, E.G., Jr., Preemptive scheduling of real-time tasks on multiprocessor systems, *Journal of ACM*, 17, 324–338, 1970.
[107] Labetoulle, J., Lawler, E.L., Lenstra, J.K., and Rinnoy Kan, A.H.G., Preemptive scheduling of uniform processors subject to release dates, in: Pulleybank, W.R. (ed.), *Progress in Combinatorial Optimization*, Academic Press, New York, 245–261, 1984.
[108] Machowiak, M., Scheduling malleable tasks in multiprocessor systems (in Polish), Ph.D. thesis, Institute of Computing Science, Poznań University of Technology, 2002.

26

Scheduling Parallel Tasks Approximation Algorithms

Pierre-François Dutot
CNRS

Grégory Mounié
ENSIMAG

Denis Trystram
Institut National Polytechnique

26.1 Introduction: Parallel Tasks in Parallel Processing

26.1.1 Motivation

As it is reflected in this book, scheduling is a very old problem which motivated a lot of researches in many fields. In the parallel processing area, this problem is a crucial issue for determining the starting times of the tasks and the processor locations. Many theoretical studies were conducted [1–3] and some efficient practical tools have been developed, e.g., Pyrros [4] and Hypertool [5].

1-58488-397-9/$0.00+$1.50

Scheduling in modern parallel and distributed systems is much more difficult because of new characteristics of these systems. These last few years, supercomputers have been replaced by collections of large number of standard components, physically far from each other and heterogeneous [6]. The design of efficient algorithms for managing these resources is a crucial issue for a more popular use. Today, the lack of adequate software tools is the main obstacle for using these powerful systems in order to solve large and complex actual applications.

The classical scheduling algorithms that have been developed for parallel machines of the nineties are not well adapted to new execution platforms. The most important factor is the influence of communications. The first attempt that took into account the communications into computational models were to adapt and refine existing models into more realistic ones, e.g., Delay Model with unitary delays [7] and LogP Model [8,9]. However, even the most elementary problems are already intractable [10], especially for large communication delays (the problem of scheduling simple bipartite graphs is already NP-hard [11]).

26.1.2 Discussion about Parallel Tasks

The idea behind parallel tasks (PTs) is to consider an alternative for dealing with communications, especially in the case of large delays. For many applications, the developers or users have a good knowledge of their behavior. This qualitative knowledge is often enough to guide the parallelization.

Informally, a PT is a *task* that gathers elementary operations, typically a numerical routine or a nested loop, which contains itself enough parallelism to be executed by more than one processor. This view is more general than the standard case and contains the sequential tasks as a particular case. Thus, the problems of scheduling PT are at least as difficult to solve. We can distinguish two ways for building PT:

1. PTs as parts of a large parallel application. Usually offline analysis is possible as the time of each routine can be estimated quite precisely (number of operations, volume of communications), with precedence between PT.
2. PTs as independent jobs (applications) in a multiuser context. Usually, new PTs are submitted at any time (online). The time for each PT can be estimated or not (clairvoyant or not) depending on the type of applications.

The PT model is particularly well-adapted to grid and global computing because of the intrinsic characteristics of these new types of supports: large communication delays which are considered implicitly and not explicitly like they are in all standard models, the hierarchical character of the execution support which can be naturally expressed in PT model, and the capacity to react to disturbances or to imprecise values of the input parameters. The heterogeneity of computational units or communication links can also be considered by uniform or unrelated processors for instance.

26.1.3 Typology of Parallel Tasks

There exist several versions of PTs depending on their execution on a parallel and distributed system (see [12] and Drozdowski's chapter of this book):

1. *Rigid* when the number of processors to execute the PT is fixed *a priori*. This number can either be a power of 2 or any integer number. In this case, the PT can be represented as a rectangle in a Gantt chart. The allocation problem corresponds to a strip-packing problem [13].
2. *Moldable* when the number of processors to execute the PT is not fixed but determined before the execution. As in the previous case this number does not change until the completion of the PT.
3. In the most general case, the number of processors may change during the execution (by preemption of the tasks or simply by data redistributions). In this case, the PTs are *malleable*.

Practically, most parallel applications are moldable. An application developer does not know in advance the exact number of processors that will be used at run time. Moreover, this number may vary with the input problem size or number of nodes availability. This is also true for many numerical parallel libraries. The main restriction is the minimum number of processors needed because of time, memory, or storage constraints. Some algorithms are also restricted to particular data sizes and distributions like the FFT algorithm where 2^q processors are needed or Strassen's matrix multiplication with its decomposition into 7^q subproblems [14].

Most parallel programming tools or languages have some malleability support, with dynamic addition of processing nodes support. This is already the case since the beginning for the well-known message passing language PVM, where nodes can be dynamically added or removed. This is also true from MPI-2 libraries. It should be noticed that an even more advanced management support exists when a client/server model is available like in CORBA, RPC (remote procedure call), or even MPI-2 [15]. Modern advanced academic environments, such as Condor, Mosix, Cilk, Satin, Jade, NESL, PM^2, or Athapascan, implement very advanced capabilities, such as resilience, preemption, and migration, or at least the model allows us to implement these features.

Nevertheless, most of the time moldability or malleability must still be taken explicitly into account by the application designers as computing power will appear, move, or be removed. This is easy in a master/worker scheme but may be much more difficult in an SPMD scheme where data must then be redistributed. Environments abstracting nodes may, theoretically, manage these points automatically.

The main restriction in the moldability use is the need for efficient scheduling algorithm to estimate (at least roughly) the parallel execution time as function of the number of processors. The user has this knowledge most of the time but this is an inertia factor against the more systematic use of such models.

Malleability is much more easily usable from the scheduling point of view but requires advanced capabilities from the runtime environment, and thus restricts the use of such environments and their associated programming models. In the near future, moldability and malleability should be used more and more.

26.1.4 Task Graphs

We will discuss briefly in this section how to obtain and handle PTs.

We will consider two types of PTs corresponding respectively to independent jobs and to applications composed of large tasks.

The purpose of this chapter is not to detail and discuss the representation of applications as graphs for the sake of parallelization. It is well known that obtaining a symbolic object from any application implemented in a high-level programming language is difficult. The graph formalism is convenient and may be declined in several ways. Generally, the coding of an application can be represented by a directed acyclic graph where the vertices are the instructions and the edges are the data dependencies [16].

In a typical application, such a graph is composed of tens, hundreds, or thousands of tasks, or even more. Using symbolic representation such as in [17], the graph may be managed at compile time. Otherwise, the graph must be built online, at least partially, at the execution phase. A moldable, or malleable, task graph is a way to gather elementary sequential tasks which will be handled more easily as it is much smaller.

There are two main ways to build a task graph of PTs: either the user has a relatively good knowledge of its application and is able to provide the graph (top–down approach), or the graph is built automatically from a larger graph of sequential tasks generated at runtime (down–top).

26.1.5 Content of the Chapter

In the next section, we introduce all the important definitions and notations that will be used throughout this chapter. The central problem is formally defined and some complexity results are recalled.

We then start with a very simple case, to introduce some of the methods that are used in approximation algorithms. The following six sections are oriented toward some interesting problems covering most of the possible combinations that have been studied in the literature and have been resolved with very different techniques.

The sections are ordered in the following way:

1. The criterion is most important. We start with $C_{\max}$, then $\sum C_i$ and finally both (see next section for definitions).
2. The offline versions are discussed before the online versions.
3. Moldable tasks are studied before malleable tasks.
4. Finally precedence constraints: from the simple case of independent tasks to more complicated versions.

This order is related to the difficulty of the problems. For example offline problems are simpler than online problems.

Finally we conclude with a discussion on how to consider other characteristics of new parallel and distributed systems.

26.2 Formal Definition and Theoretical Analysis

Let us first introduce informally the problem to solve: Given a set of PTs, we want to determine at what time the tasks will start their execution on a set of processors such that at any time no more than m processors are used.

26.2.1 Notations

We will use in this chapter standard notations used in the other chapters of the book. Unless explicitly specified, we consider n tasks executed on m identical processors without preemption.

The execution time is denoted $p_j(q)$ when task j (for $1 \leq j \leq n$) is allocated to q processors. The starting time of task j is $\sigma(j)$, and its completion time is $C_j = \sigma(j) + p_j(q)$. When needed, the number q of processors used by task j will be given by $q = nbproc(j)$.

The *work* of task j on q processors is defined as $w_j(q) = q \times p_j(q)$. It corresponds to the surface of the task on the Gantt chart (time-space diagram).

We will restrict the analysis on PTs that start their execution on all processors simultaneously. In other words, the execution of rigid tasks or moldable tasks corresponds to rectangles. The execution of malleable tasks corresponds to a union of contiguous rectangles.

Figure 26.1 represents the execution of two PTs in the three contexts of rigid, moldable and malleable tasks. The tasks used in this figure have the execution times presented in Table 26.1 for the moldable case. For the malleable case, percentage of these times are taken.

TABLE 26.1 Tasks Used in Figure 26.1

No. of Processors	Big Task	Small Task
1	24	10
2	12	9
3	8	6
4	7	5

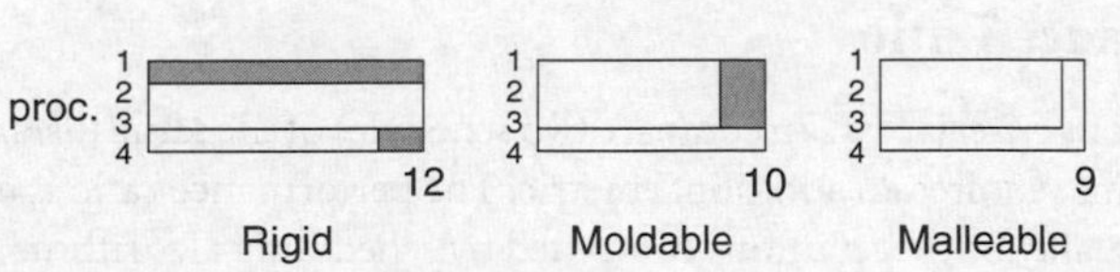

FIGURE 26.1 Comparison of the execution of rigid, moldable, and malleable tasks.

In the rigid case, the big task can only be executed on two processors and the small task on one processor. In the moldable case, the scheduler can choose the allocation but cannot change it during the execution. Finally in the malleable case, the allocation of the small task is on one processor for 80% of its execution time and on all the four processors for the remaining 20%.

As we can see, moldable and malleable characters may improve the rigid execution.

26.2.2 Formulation of the Problem

Let us consider an application represented by a precedence task graph $G(V, E)$. The parallel tasks schedule (PTS) problem is defined as follows:

Instance: A graph $G = (V, E)$ of order n, a set of integers $p_j(q)$ for $1 \leq q \leq m$ and $1 \leq j \leq n$.

Question: Determine a feasible schedule which minimizes the objective function f. We will discuss, in the next section, the different functions used in the literature.

A feasible schedule is a pair of functions $(\sigma, \mathit{nbproc})$ of $V \rightarrow N \times [1, \ldots, m]$, such that

1. the precedence constraints are verified: $\sigma(j) \geq \sigma(i) + p_i(\mathit{nbproc}(i))$ if task j is a successor of i (there is no communication cost),
2. at any time slot no more than m processors are used.

26.2.3 Criteria

The main objective function used historically is the *makespan*. This function measures the ending time of the schedule, i.e., the latest completion time over all the tasks. However, this criterion is valid only if we consider the tasks altogether and from the viewpoint of a single user. If the tasks have been submitted by several users, other criteria can be considered. Let us review briefly the different possible criteria usually used in the literature:

1. Minimization of the *makespan* (completion time $C_{\max} = \max(C_j)$ where C_j is equal to $\sigma(j) + p_j(\mathit{nbproc}(j))$
2. Minimization of the average completion time ($\sum C_i$) [18,19] and its variant weighted completion time ($\sum \omega_i C_i$). Such a weight may allow us to distinguish some tasks from each other (priority for the smallest ones, etc.).
3. Minimization of the mean stretch (defined as the sum of the difference between release times and completion times). In an online context it represents the average response time between the submission and the completion.
4. Minimization of the maximum stretch (i.e., the longest waiting time for a user).
5. Minimization of the tardiness. Each task is associated with an expected due date and the schedule must minimize either the number of late tasks, the sum of the tardiness or the maximum tardiness.
6. Other criteria may include rejection of tasks or normalized versions (with respect to the workload) of the previous ones.

In this chapter, we will focus on the first two criteria, which are the most studied.

26.2.4 Performance Ratio

Considering any previous criteria, we can compare two schedules of the same instance. But the comparison of scheduling algorithms requires an additional metric. The performance ratio is one of the standard tools used to compare the quality of the solutions computed by scheduling algorithms [20].

It is defined as follows: The performance ratio ρ_A of algorithm A is the maximum overall instances $\mathcal{I}$ of the ratio $f(\mathcal{I})/f^*(\mathcal{I})$ where f is any minimization criterion and f^* is the optimal value.

Throughout the text, we will use the same notation for optimal values. The performance ratios are either constant or may depend on some instance input data, such as the number of processors, tasks, or precedence relation.

Most of the time the optimal values can not be computed in reasonable time unless $P = NP$. Sometimes, the worst case instances and values may not be computed either. In order to do the comparison, approximations of these values are used. For correctness, a lower bound of the optimal value and an upper bound of the worst-case value are computed in such cases.

Some studies also use the mean performance ratio, which is better than the worst-case ratio, either with a mathematical analysis or by experiments.

Another important feature for the comparison of algorithms is their complexities. As most scheduling problems are NP-hard, algorithms for practical problems compute approximate solutions. In some contexts, algorithms with larger performance ratio may be preferred, thanks to their lower complexity, instead of algorithms providing better solutions but at a much greater computational cost.

26.2.5 Penalty and Monotony

The idea of using PTs instead of sequential ones was motivated by two reasons, namely to increase the granularity of the tasks in order to obtain a better balance between computations and slow communications, and to hide the complexity of managing explicit communications.

In the PTs model, communications are considered as a global *penalty* factor which reflects the overhead for data distributions, synchronization, preemption, or any extra factors coming from the management of the parallel execution. The penalty factor implicitly takes into account some constraints, when they are unknown or too difficult to estimate formally. It can be determined by empirical or theoretical studies (benchmarking, profiling, performance evaluation through modeling or measuring, etc.).

The penalty factor reflects both influences of the operating system and the algorithmic side of the application to be parallelized.

In some algorithms, we will use the following hypothesis which is common in the parallel application context. Adding more processors usually reduces the execution time, at least until a threshold. But the speedup is not super-linear. From the application point of view, increasing the number of processors also increases the overhead: more communications, more data distributions, longer synchronizations, termination detection, etc.

Hypothesis 26.1 (Monotony) *For all tasks* j, p_j *and* w_j *are monotonic:*

1. $p_j(q)$ *is a decreasing function in* q.
2. $w_j(q)$ *is an increasing function in* q.

More precisely,

$$p_j(q+1) \leq p_j(q)$$

and

$$w_j(q) \leq w_j(q+1) = (q+1)p_j(q+1) \leq \left(1+\frac{1}{q}\right) q\, p_j(q) = \left(1+\frac{1}{q}\right) w_j(q)$$

Figure 26.2 gives a geometric interpretation of this hypothesis.

From the parallel computing point of view, this hypothesis may be interpreted by the Brent's lemma [21]: if the instance size is large enough, a parallel execution should not have super-linear speedup. Sometimes, parallel applications with memory hierarchy cache effect, race condition on flow control, or scheduling anomalies described by Graham [22], may lead to such super-linear speedups. Nevertheless, most parallel applications fulfill this hypothesis as their performances are dominated by communication overhead.

TABLE 26.2 NP-Hard Problems and Associated Reductions

	Problem		
	Indep.	Prec.	Reduction
$C_{\max}$	Offline	—	From 3-partition
	Online clairvoyant	—	From the offline case
	Online nonclairvoyant	—	From the offline case
	—	Offline	From $P \mid p_i = 1, prec \mid C_{\max}$
$\sum \omega_i C_i$	Offline	—	From $P \mid\mid \sum \omega_i C_i$
	Online	—	From the Offline case

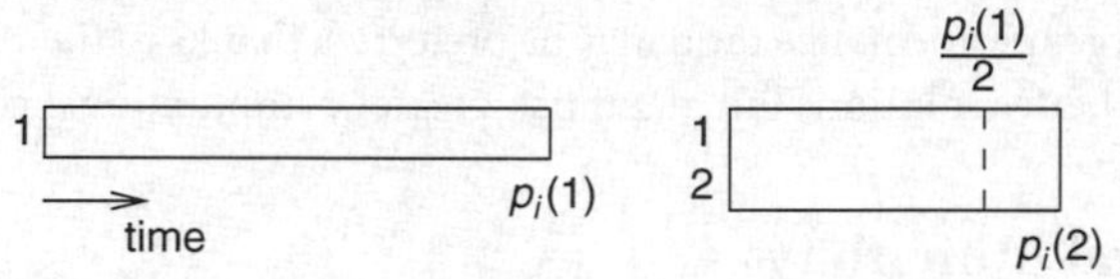

FIGURE 26.2 Geometric interpretation of the penalty on two processors.

Other general hypotheses will be considered over this chapter, unless explicitly stated:

1. A processor executes at most one task at a time.
2. Preemption between PTs is not allowed (but preemption inside PT can be considered, in this case, its cost will be included as part of the penalty). A task cannot be stopped and then resumed, or restarted. Nevertheless the performance ratio is sometimes established in regard to the preemptive optimal solution.

26.2.6 Complexity

Table 26.2 presents a synthetic view of the main complexity results linked with the problems we are considering in this chapter. The rigid case has been deeply studied in the survey [23].

All the complexity proof for the rigid case involving only sequential tasks can be extended to the moldable case and to the malleable case with a penalty factor which does not change the execution time on any number of processors. All the problems of Table 26.2 are NP-hard in the strong sense.

26.3 Preliminary Analysis of a Simplified Case

Let us first detail a specific result on a very simple case. Minimizing $C_{\max}$ for identical moldable tasks is one of the simplest problems involving moldable tasks. This problem has some practical interest, as many applications generate at each step a set of identical tasks to be computed on a parallel platform.

With precedence constraints this problem is as hard as the classical problem of scheduling precedence constrained unit-execution-time (UET) tasks on multiprocessors, as a moldable task can be designed to run with the same execution time on any number of processors.

Even without precedence constraints, there is no known polynomial optimal scheduling algorithm and the complexity is still open. To simplify the problem even more, we introduce a phase constraint. A set of tasks is called a *phase* when all the tasks in the set start at the same time, and no other task starts executing on the parallel platform before the completion of all the tasks in the phase.

This constraint is very practical as a schedule where all the tasks are run in phases is easier to implement on an actual system.

For identical tasks, if we add the restriction that tasks are run in phases, the problem becomes polynomial for simple precedence graph like trees. When the phases algorithm is used for approximating the general problem of independent tasks, the performance ratio is exactly 5/4.

26.3.1 Dominance

When considering such a problem, it is interesting to establish some properties that will restrict the search for an optimal schedule. With the phase constraint, we have one such property.

Proposition 26.1

For a given phase time length, the maximum number of tasks in the phase is reached if all the tasks are alloted the same number of processors, and the number of idle processors is less than this allocation.

The proof is rather simple, let us consider a phase with a given number of tasks. Within these tasks, let us select one of the tasks which are the longest. This task can be chosen among the longest as one with the smallest allocation. There is no task with a smaller allocation than the selected one because the tasks are monotonic.

This task runs in less than the phase length. All other tasks are starting at the same time as this task. If the tasks with a bigger allocation are given the same allocation as the selected one, they will all have their allocation reduced (therefore this transformation is possible). The fact that their running time will probably increase is not a problem here, as we said that within a phase all the tasks are starting simultaneously. Therefore it is possible to change any phase in a phase where all tasks have the same allocation. The maximum number of tasks is reached if there is not enough idle processors to add another task.

26.3.2 Exact Resolution by Dynamic Programming

Finding an optimal phase-by-phase schedule is a matter of splitting the number n of tasks to be scheduled into a set of phases that will be run in any order. As the number of tasks in a phase is an integer, we can solve this problem in polynomial time using integer dynamic programming. The principle of dynamic programming is to say that for one task the optimal schedule is one phase of one task, for two tasks the optimal is either one phase of two tasks or one phase of one task plus the optimal schedule of one task and so on.

The makespan ($C_{\max}(n)$) of the computed schedule for n tasks is

$$C_{\max}(n) = \min_{i=1,\dots,m} \left(C_{\max}(n-i) + p_j\left(\left\lfloor \frac{m}{i} \right\rfloor\right)\right)$$

The complexity of the algorithm is $O(mn)$.

26.3.3 Approximation of the Relaxed Version

We may think that scheduling identical tasks in a phase-by-phase schedule produces the optimal result even for the problem where this phase-by-phase constraint is not imposed. Indeed there is a great number of special cases where this is true. However, there are some counter examples as in Figure 26.3. This example is built on five processors, with moldable tasks running in six units of time on one processor, three units of time on two processors, and two units of time on either three, four, and five processors.

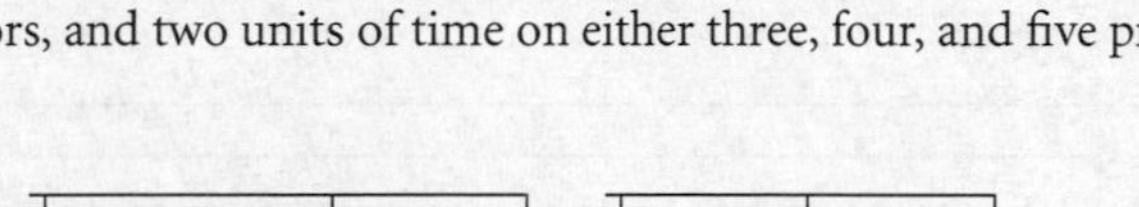

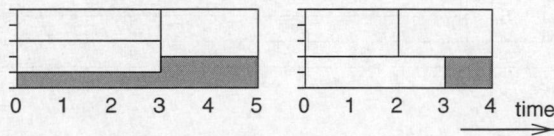

FIGURE 26.3 The optimal schedules with and without phases for three moldable tasks on five processors.

This example shows that the performance ratio reached by the phase algorithm is greater than 5/4. To prove that it is exactly 5/4, we need to make a simple but tedious and technical case analysis on the number of tasks (see [24] for the details).

26.4 Independent Moldable Tasks, C_{max}, Offline

In this section, we focus on the scheduling problem itself. We have chosen to present several results obtained for the same problem using different techniques.

Let us consider the scheduling of a set of n independent moldable tasks on m identical processors for minimizing the makespan. Most of the existing methods for solving this problem have a common geometrical approach by transforming the problem into two-dimensional packing problems. It is natural to decompose the problem in two successive phases: determining the number of processors for executing the tasks, then solve the corresponding problem of scheduling rigid tasks.

The next section will discuss the dominance of the geometrical approach.

26.4.1 Discussion about the Geometrical View of the Problem

We discuss here the optimality of the rectangle packing problem in scheduling moldable tasks. Figure 26.4 shows an example of noncontiguous allocation in the optimal schedule. Moreover, we prove that no contiguous allocation reaches the optimal makespan in this example.

The instance is composed by the eight tasks given in Table 26.3, to be executed on a parallel machine with four processors.

The proof is left to the reader that these tasks verify the monotony assumptions.

The minimum total workload (sum of the first column) divided by the number of processors gives a simple lower bound for the optimal makespan. This optimal value is reached with the schedule presented in Figure 26.4, where task 8 is allocated to processors a, c, and d.

We will now prove that no contiguous allocation can have a makespan of 25. One can easily verify that no permutation of the four processors in Figure 26.4 gives a contiguous allocation.

TABLE 26.3 Execution Times of the Eight Tasks of Figure 26.4

	Processor			
Tasks	1	2	3	4
1	13	13	13	13
2	18	18	18	18
3	20	20	20	20
4	22	22	22	22
5	6	3	2	2
6	6	3	3	3
7	12	6	6	6
8	3	1.5	1	1

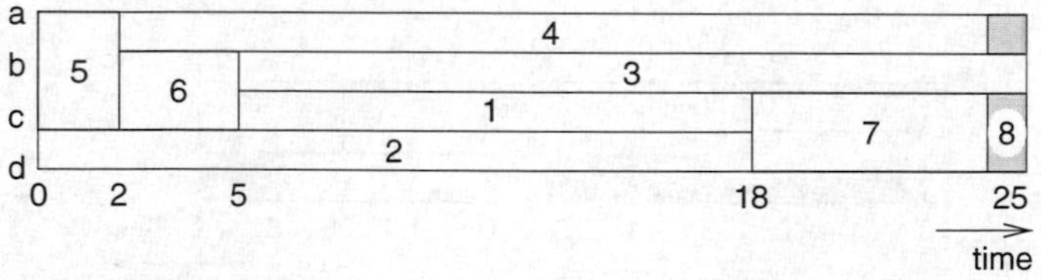

FIGURE 26.4 Optimal noncontiguous allocation.

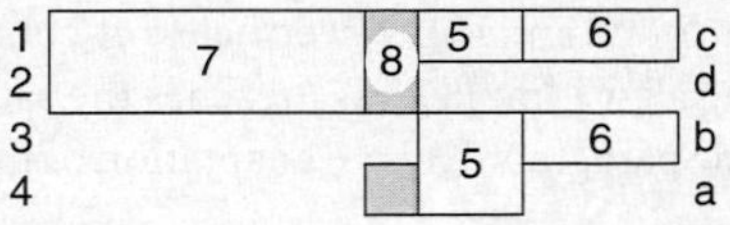

FIGURE 26.5 The resulting allocation.

First, let us look at the possible allocations without considering if a schedule is feasible or not.

Given the sizes of tasks 1–4, we cannot allocate two of these tasks to a processor. Therefore, let us say that task 1 is allocated to processor 1, task 2 to processor 2, and so on. The idle time left on the processors is 12 units of time for processor 1, 7 for processor 2, 5 for processor 3, and 3 for processor 4.

Task 7, being the biggest of the remaining tasks, is a good starting point for a case study. If task 7 is done sequentially, it can only be alloted on processor 1 and leaves no idle time on this processor. In this case, we have processors 2, 3, and 4 with respectively 7, 5, and 3 units of idle time. The only way to fill the idle time of processor 3 is to put task 5 on all three processors and task 6 on two processors. With the allocation of task 5 we have 5, 3, and 1 units of idle time; and with allocation of task 6 we have 2, 0, and 1 units of idle time. Task 8 cannot be allocated to fill two units of time on processor 2 and one on processor 4. Therefore the assumption that task 7 can be done sequentially is wrong.

If task 7 cannot be done sequentially, it has to be done on processor 1 and 2 as these are the only ones with enough idle time. This leaves respectively 6, 1, 5, and 3 units of idle time. The only way to fill processor 2 is to allocate task 8 on three processors, which leaves either 5, 4, and 3; or 5, 5, and 2; or 6, 4, and 2. With only task 5 and 6 remaining, the only possibility to perfectly fit every task in place is to put task 5 on three processors and task 6 on two processors.

The resulting allocation is shown in Figure 26.5. In this figure, no scheduling has been made. The tasks are just represented on the processors they are allotted to, according to the previous discussion. The numbers on the left are the processor indices and the letters on the right show that we can relate each processor to one from Figure 26.4. The only possible allocation is the one used in the schedule of Figure 26.4. As we said that no permutation of the processors can give a contiguous representation of this allotment, there is no contiguous scheduling in 25 units of time.

26.4.2 Two-Phase Approach

We present in this section a first approximation algorithm for scheduling independent moldable tasks using a two-phase approach: first determining the number of processors for executing the tasks, then solving the corresponding rigid problem by a strip packing algorithm.

The idea that has been introduced in [25] is to optimize in the first phase the criterion used to evaluate the performance ratio of the second phase. The authors proposed to realize a trade-off between the maximum execution time (critical path) and the sum of the works.

The following algorithm gives the principle of the first phase. A more complicated and smarter version is given in the original work.

Compute an allocation with minimum work for every task
while $\frac{\sum w_j(nbproc(j))}{m} < \max_j(p_j(nbproc(j)))$ **do**
  Select the task with the largest execution time
  Change its allocation for another one with a strictly smaller execution time and the smallest work
end while

After the allocation has been determined, the rigid scheduling may be achieved by any algorithm with a performance ratio function of the critical path and the sum of the works. For example, a strip-packing algorithm like Steinberg's one [26] fulfills all conditions with an absolute performance ratio of 2. A recent survey of such algorithms may be found in [13]. Nevertheless if contiguity is not mandatory, a simple rigid multiprocessor list scheduling algorithm like [27] reaches the same performance ratio of 2. We will detail this algorithm in the following paragraph. It is an adaptation of the classical Graham's list scheduling under resource constraints.

The basic version of the list scheduling algorithm is to schedule the tasks using a list of priority executing a rigid task as soon as enough resources are available. The important point is to schedule the tasks at the earliest starting time, and if more than one task is candidate, consider first the one with the highest priority. In the original work, each task being executed uses some resources. The total number of resources is fixed and each task requires a specific part of these resources. In the case of rigid independent tasks scheduling, there is only one resource which corresponds to the number of processors allocated to each task and no more than m processors may be used simultaneously. In Graham's paper, there is a proof for obtaining a performance ratio of 2, which can be adapted to our case.

Proposition 26.2

The performance ratio of the previous algorithm is 2.

The main argument of the proof is that the trade-off achieved by this algorithm is the best possible, and thus the algorithm is better in the first phase than the optimal scheduling. The makespan is driven by the performance ratio of the second phase (2 in the case of Steinberg's strip packing).

The advantage of this scheme is its independence in regard to any hypothesis on the execution time function of the tasks, like monotony. The major drawback is the relative difficulty of the rigid scheduling problem which constrains here the moldable scheduling. In the next section, we will take another point of view: put more emphasis on the first phase in order to simplify the rigid scheduling in the second phase.

It should be noticed that none of the strip packing algorithms explicitly use in the scheduling the fact that the processor dimension is discrete. We present such an algorithm in Section 26.4.3 with a better performance ratio of only $3/2 + \epsilon$ for independent moldable tasks with the monotony assumption.

26.4.3 A Better Approximation

The performance ratio of Turek's algorithm is fixed by the corresponding strip packing algorithm (or whatever rigid scheduling algorithm used). As such problems are NP-hard, the only way to obtain better results is to solve different allocation problems which lead to easier scheduling problems.

The idea is to determine the task allocation with great care in order to fit them into a particular packing scheme. We present below a 2-shelves algorithm [28] with an example in Figure 26.6.

This algorithm has a performance ratio of $3/2 + \epsilon$. It is obtained by stacking two shelves of respective sizes λ and $\lambda/2$ where λ is a guess of the optimal value C^*_{max}. This guess is computed by a dual approximation

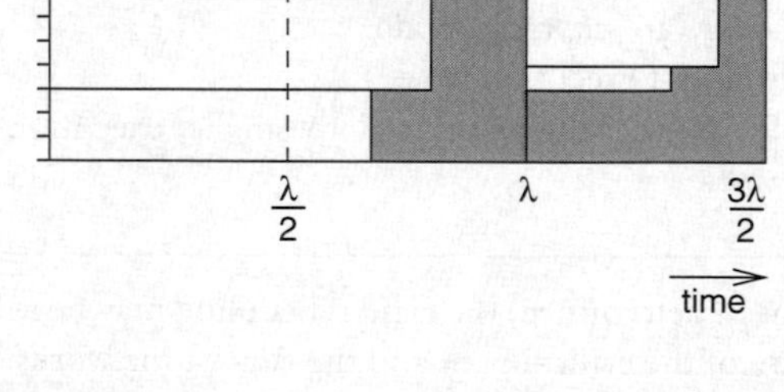

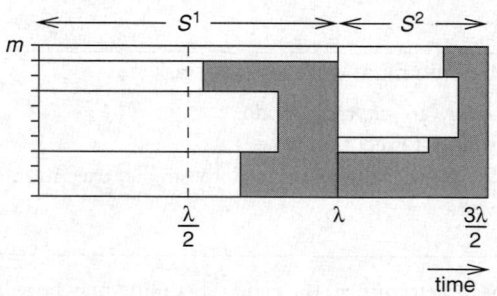

FIGURE 26.6 Principle of the 2-shelf allocation S^1 and S^2.

scheme [29]. Informally, the idea behind dual approximation is to fix a hypothetical value for the guess λ and to check if it is lower than the optimal value $C^*_{\max}$ by running a heuristic with a performance ratio equal to ρ and a value $C_{\max}$. If $\lambda < 1/\rho C_{\max}$, by definition of the performance ratio, λ is underestimated. A binary search allows us to refine the guess with an arbitrary accuracy ϵ.

The guess λ is used to bound some parameters on the tasks. We give below some constraints that are useful for proving the performance ratio. In the optimal solution, assuming $C^*_{\max} = \lambda$

1. $\forall j,\ p_j(nbproc(j)) \leq \lambda$.
2. $\sum w_j(nbproc(j)) \leq \lambda m$.
3. When two tasks share the same processor, the execution of one of these tasks is lower than $\lambda/2$. As there are no more than m processors, less than m processors are used by the tasks with an execution time larger than $\lambda/2$.

We will now detail how to fill the two shelves S^1 and S^2 (Figure 26.6), as well as we can with respect to the sum of the works. Every task in a shelf starts at the beginning of the shelf and is allocated on the minimum number of processors to fit into its shelf. The shelf S^2 (of length lower than $\lambda/2$) may be overfilled, but the sum of processors used for executing the tasks in the first shelf S^1 is imposed to be less than m. Hopefully, the partitioning problem can be solved in polynomial time by dynamic programming with a complexity in $O(nm)$. We detail below the corresponding algorithm.

---

define $\gamma(j,d) :=$ minimum *nbproc(j)* such that $p_j(nbproc(j)) \leq d$
$W_{0,0} = 0;\ W_{\forall j, q<0} = +\infty;$
for $j = 1, \ldots, n$ **do**
  for $q = 1, \ldots, m$ **do**

$$W_{j,q} = \min \begin{pmatrix} W_{j-1,q-\gamma(j,\lambda)} + W_j(\gamma(j,\lambda)) \ \text{// in } S^1 \\ W_{j-1,q} + W_j(\gamma(j,\lambda/2)) \quad \text{// in } S^2 \end{pmatrix}$$

  end for
end for

The sum of the works is smaller than λm (otherwise the λ parameter of the dual approximation scheme is underestimated and the guess must be changed).

Let us now build a feasible schedule. All tasks with an allocation of 1 (sequential tasks) and an execution time smaller than $\lambda/2$ will be put away and considered at the end of the algorithm.

The goal is to ensure that most processors compute for at least λ time units, until all the tasks fit directly into the two shelves.

All tasks scheduled in S^2 are parallel ($nbproc(j) \geq 2$) and, according to the monotony assumption, have an execution time greater than $\lambda/4$. We have to ensure that tasks in S^1 use more than $3\lambda/4$ of processing power. While S^2 is overfilled, we do some technical changes among the following ones:

1. Stack two sequential tasks ($nbproc(j) = 1$) in S^1 with an execution time smaller than $3\lambda/4$, and schedule them sequentially on a single processor.
2. Decrease $nbproc(j)$ by one processor for a PT in S^1 whose execution time is smaller than $3\lambda/4$ and schedule it alone on $nbproc(j) - 1$ processors.
3. Schedule one task from S^2 in S^1 without overfilling it, changing the task allocation to get an execution time smaller than λ.

The first two transformations use particular processors to schedule one or two "large" tasks, and liberate some processors in S^1. All transformations decrease the sum of the works. A surface argument shows that the third transformation occurs at most one time and then the scheduling becomes feasible. Up to this moment, one of the previous transformations is feasible.

The small sequential tasks that have been removed at the beginning fit between the two shelves without increasing $C_{\max}$ more than $3\lambda/2$ because the sum total of the works is smaller than λm (in this case, at least one processor is always used less than λ units of time).

for Dichotomy over $\lambda : \sum w_j(1)/m \leq \lambda \leq \sum w_j(m)/m$ **do**
  $small = tasks\ j\colon p_{j,1} \leq \lambda/2$
  $large = remaining\ tasks$
  knapsack for selecting the tasks in S^1 and S^2
  if $\sum w_j(nbproc(j)) > \lambda m$ **then**
    Failed, increase λ
  else
    Build a feasible schedule for *large*
    insert *small* between the two shelves
    Succeed, decrease λ
  end if
end for

Proposition 26.3

The performance ratio of the 2-shelves algorithm is $\frac{3}{2} + \epsilon$.

The proof is quite technical, but it is closely linked with the construction. It is based on the following surface argument: the total work remains always lower than the guess λm. Details can be found in [30].

26.4.4 Linear Programming Approach

There exists a polynomial-time approximation scheme for scheduling moldable independent tasks [31]. This scheme is not fully polynomial as the problem is NP-hard in the strong sense: the complexity is not polynomial in regard to the chosen performance ratio.

The idea is to schedule only the tasks with a large execution time. All combinations of the tasks with all allocations and all orders are tested. At the end, the remaining small tasks are added. The important point is to keep the number of large tasks small enough in order to keep a polynomial time for the algorithm.

The principle of the algorithm is presented below:

for $j = 1, \ldots, n$ **do**
  $d_j = min_{l=1,\ldots,m} p_j(l)$ (we assume the d_j are decreasingly ordered)
end for
$D = \sum_{j=1}^{n} d_j$
$\mu = \epsilon/2m$
$K = 4m^{m+1}(2m)^{\lceil 1/\mu \rceil + 1}$
Find the smallest $k \leq K$ such that $(d_k + \cdots + d_{2mk+3m^{m+1}-1} \leq \mu D)$
Construct the set of all the relative order schedules involving the k tasks
with the largest d_j (denoted by $\mathcal{L}$)
for $R \in \mathcal{L}$ **do**
  Solve (approximately) R mappings, using linear programming
  Build a feasible schedule including the remaining tasks.
end for
return := the best built schedule.

We give now the corresponding linear program for a particular element of $\mathcal{L}$. A relative order schedule is a list of g snapshots $M(i)$ (not to be mistaken with M the set of available processors). A snapshot is a subset of tasks and the processors where they are executed. $P(i)$ is the set of processors used by snapshot $M(i)$. $\mathcal{F} = \{M \backslash P(i), i = 1, \ldots, g\}$, i.e., the set of free processors in every snapshot. $P_{F,i}$ is one of the n_F partitions of $F \in \mathcal{F}$. For each partition the number of processor sets F_h, with cardinality l is denoted $a_l(F,i)$. A task appears in successive snapshots, from snapshot α_i to snapshot ω_i. D^l is the total processing time for all tasks not in $\mathcal{L}$ (denoted S). Note that solutions are nonintegers, thus the solution is postprocessed in order to build a feasible schedule.

Minimize t_g s.t.

1. $t_0 = 0$
2. $t_i \geq t_{i-1}, i = 1, \ldots, g$
3. $t_{w_j} - t_{\alpha_j - 1} = p_j, \forall T_j \in \mathcal{L}$
4. $\sum_{i:P(i)=M \setminus F} (t_i - t_{i-1}) = e_F, \forall F \in \mathcal{F}$
5. $\sum_{i=1}^{n_F} x_{F,i} \leq e_F, \forall F \in \mathcal{F}$
6. $\sum_{F \in \mathcal{F}} \sum_{i=1}^{n_F} a_l(F, i) x_{F,i} \geq D^l, l = 1, \ldots, m$
7. $x_{F,i} \geq 0, \forall F \in \mathcal{F}, i = 1, \ldots, n_F$
8. $\sum_{T_j \in S} t_j(l) y_{jl} \leq D^l, l = 1, \ldots, m$
9. $\sum_{l=1}^{m} y_{jl} = 1, \forall T_j \in S$
10. $y_{jl} \geq 0, \forall T_j \in S, l = 1, \ldots, m$

where t_i are snapshot end time. The starting time t_0 is 0 and the makespan is t_g. e_F is the time while processors in F are free. $x_{F,i}$ is the total processing time for $P_{F,i} \in \mathcal{P}_F, i = 1, \ldots, n_F, F \in \mathcal{F}$ where only processors of F are executing short tasks and each subset of processors $F_j \in P_{F,i}$ executes at most one short task at each time step in parallel. The last three constraints defined moldable allocation. In the integer linear program, y_{jl} is equal to 1 if task T_j is allocated to l processors, 0 otherwise.

The main problem is to solve a linear program for every $(2^{m+2}k^2)^k$ allocations and orders of the k tasks. This algorithm is of little practical interest, even for small instances and a large performance ratio. Actual implementations would prefer algorithms with a lower complexity like the previous algorithm with a performance ratio of $3/2 + \epsilon$.

26.5 General Moldable Tasks, $C_{\max}$, Offline

26.5.1 Moldable Tasks with Precedence

Scheduling PTs that are linked by precedence relations corresponds to the parallelization of applications composed by large modules (library routines, nested loops, etc.) that can themselves be parallelized.

In this section, we give an approximation algorithm for scheduling any precedence task graph of moldable tasks. We consider again the monotonic hypothesis. We will establish a constant performance ratio in the general case.

The problem of scheduling moldable tasks linked by precedence constraints has been considered under very restricted hypotheses like those presented in Section 26.5.2.

Another way is to use a direct approach like in the case of the two-phase algorithms. The monotony assumption allows us to control the allocation changes. As in the case of independent moldable tasks, we are looking for an allocation which realizes a good trade-off between the sum of the works (denoted by W) and the critical path (denoted by T_∞).

Then, the allocation of tasks is changed in order to simplify the scheduling problem. The idea is to force the tasks to be executed on less than a fraction of m, e.g., $m/2$. A task does not increase its execution time more than the inverse of this fraction, thanks to the monotony assumption. Thus, the critical path of the associated rigid task graph does not increase more than the inverse of this fraction.

With a generalization of the analysis of Graham [22], any list scheduling algorithm will fill more than half (i.e., 1 − the fraction) of the processors at any time, otherwise, at least one task of every path of the graph is being executed. Thus, the cumulative time when less than $m/2$ processors are occupied is smaller than the critical path. As on one hand, the algorithm doubles the value of the critical path, and on the other hand the processors work more than $m/2$ during at most $2W/m$, the overall guarantee is $2W/m + 2T_\infty$, leading to a performance ratio of 4.

Let us explain in more detail how to choose the ratio. With a smart choice [32] a better performance ratio than 4 may be achieved. The idea is to use three types of time intervals, depending on if the processors are used more or less than μ and $m - \mu$ (see I_1, I_2, and I_3 in Figure 26.7. For the sake of clarity, the intervals have been represented as contiguous ones). The intervals I_2 and I_3 where tasks are using less than $m - \mu$

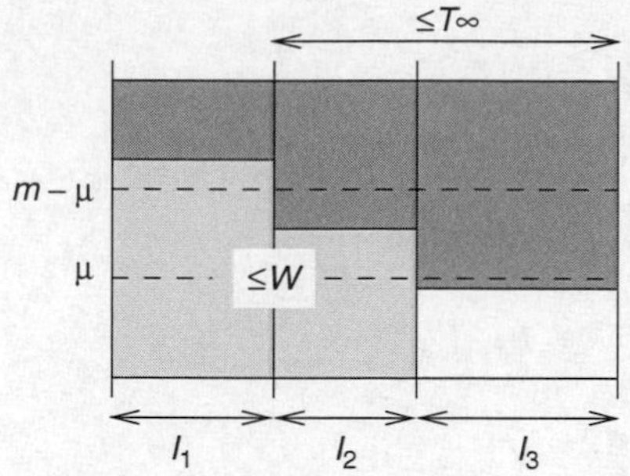

FIGURE 26.7 The different time interval types.

processors are bounded by the value of the critical path, and the sum of the works bounds the surface corresponding to intervals I_1 and I_2 where more than μ processors are used. The best performance ratio is reached for a value of parameter μ depending on m, with $1 \leq \mu \leq m/2 + 1$, such that

$$r(m) = \min_{\mu} \max\left\{\frac{m}{\mu}, \frac{2m - \mu}{m - \mu + 1}\right\}$$

We can now state the main result.

Proposition 26.4

The performance ratio of the previous algorithm is $\frac{3+\sqrt{5}}{2}$ *for series-parallel graphs and trees.*

The reason of the limitation of the performance ratio is the ability to compute an allocation which minimizes the critical path and sum of the works. An optimal allocation may be achieved for structured graphs like trees using dynamic programming with a deterministic and reasonable time. In the general case, it is still possible to choose an allocation with a performance ratio of 2 in regard to the critical path and sum of the works. The overall performance ratio is then doubled (that is $3 + \sqrt{5}$).

26.5.2 Relaxation of Continuous Algorithms

We have presented in Section 26.4 some ways to deal with the problem of scheduling independent moldable tasks for minimizing the makespan. The first two approaches considered direct constructions of algorithms with a small complexity and reasonable performance ratios, and the last one used a relaxation of a continuous linear program with a heavy complexity for a better performance. It is possible to obtain other approximations from a relaxation of continuous resources (i.e., where a PT may be allocated to a fractional number of processors).

Several studies have been done for scheduling precedence task graphs:

1. Prasanna and Musicus [33] studied the scheduling of graphs where all tasks have the same penalty with continuous allocations. The speed-up functions (which are inversely proportional to the penalty factors) are restricted to values of type q^{α}, where q is the fraction of the processors allocated to the tasks and $0 \leq \alpha \leq 1$. This hypothesis is stronger than the monotony assumption and is far from the practical conditions in Parallel Processing.
2. Using restricted shapes of penalty functions (concave and convex), Węglarz provided optimal execution schemes for continuous PTs [34]. For concave penalty factors, we retrieve the classical result of the optimal execution in gang for super-linear speedups.
3. Another related work considered the folding of the execution of rigid tasks on a smaller number of processors than specified [35]. As this folding has a cost, it corresponds in fact to the opposite of the monotony assumption, with super-linear speedup functions. Again, this hypothesis is not practically realistic for most parallel applications.

4. A direct relaxation of continuous strategies has been used for the case of independent moldable tasks [36] under the monotony assumption. This work demonstrated the limitations of such approaches. The continuous and discrete execution times may be far away from each other, e.g., for tasks which require a fractional number of processors lower than 1. If a task requires a continuous allocation between 1 and 2, there is a rounding problem which may multiply the discrete times by a factor of 2. Even if some theoretical approximation bounds can be established, such an approach has intrinsic limitations which did not show any advantage over ad hoc discrete solutions like those described in Section 26.4. However, they may be very simple to implement!

26.6 Independent Moldable Tasks, $C_{\max}$, Online Batch

An important characteristic of the new parallel and distributed systems is the versatility of the resources: at any moment, some processors (or groups of processors) can be added or removed. On another side, the increasing availability of the clusters or collections of clusters involved new kinds of data intensive applications (like data mining) whose characteristics are that the computations depend on the data sets. The scheduling algorithm has to be able to react step by step to arrival of new tasks, and thus, offline strategies cannot be used. Depending on the applications, we distinguish two types of online algorithms, namely, clairvoyant online algorithms when most parameters of the PTs are known as soon as they arrive, and non-clairvoyant ones when only a partial knowledge of these parameters is available. We invite the readers to look at the survey of Sgall [37] or the chapter of the same author in this book.

Most of the studies about online scheduling concern independent tasks, and more precisely the management of parallel resources. In this section, we consider only the clairvoyant case, where a good estimate of the task execution time is known.

We present first a generic result for batch scheduling. In this context, the tasks are gathered into sets (called batches) that are scheduled together. All further arriving tasks are delayed to be considered in the next batch. This is a nice way for dealing with online algorithms by a succession of offline problems. We detail below the result of Shmoys, Wein, and Williamson [38] that proposed how to adapt an algorithm for scheduling independent tasks without release dates (all tasks are available at date 0) with a performance ratio of ρ into a batch scheduling algorithm with unknown release dates with a performance ratio of 2ρ.

Figure 26.8 gives the principle of the batch execution and illustrates the notations used in the proof.

The proof of the performance ratio is simple. First, let us remark that for any instance, the online optimal makespan is greater than the offline optimal makespan. By construction of the algorithm, every batch schedules a subset of the tasks, thus every batch execution time τ_k is smaller than ρ times the optimal offline makespan.

The previous last batch starts before the last release date of a task. Let σ_{n-1} be the starting time of this batch. In addition, all the tasks in the last batch are also scheduled after σ_{n-1} in the optimal. Let τ_n be the execution time of the last batch. As the part of the optimal schedule after the time instant σ_{n-1}

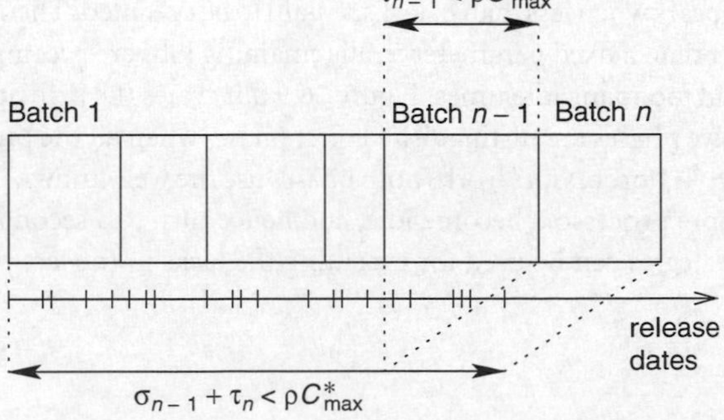

FIGURE 26.8 Online schedule with batches.

contains at least all the tasks of the last batch, the length l of this part times ρ is greater than τ_n. Therefore $\sigma_{n-1} + \tau_n < \sigma_{n-1} + \rho l < \rho C^*_{\max}$.

If we consider the total time of our schedule as the sum of the time of previous last batch (τ_{n-1}) and the time of all other batches ($\sigma_{n-1} + \tau_n$), the makespan is clearly lower than $2\rho C^*_{\max}$.

Now, using the algorithm of Section 26.4.3 with a performance ratio of $3/2+\epsilon$, it is possible to schedule moldable independent tasks with release dates with a performance ratio of $3+\epsilon$ for $C_{\max}$. The algorithm is a batch scheduling algorithm, using the independent tasks algorithm at every phase.

26.7 Independent Malleable Tasks, $C_{\max}$, Online

Even without knowing tasks execution times, when malleable tasks are ideally parallel it is possible to get the optimal competitive ratio of $1 + \phi \approx 2.6180$ with the following deterministic algorithm [35]:

---

if an available task i requests *nbproc*(j) processors and *nbproc*(j) processors are available **then**
  schedule the task on the processors
end if
if less than m/ϕ processors are busy and some task is available **then**
  schedule the task on all available processors, folding its execution
end if

We consider in this section a particular class of nonclairvoyant PTs that is important in practice in the context of exploitation of parallel clusters for some applications [39]: the expected completion times of the PTs is unknown until completion (it depends on the input data), but the qualitative parallel behavior can be estimated. In other words, the p_j are unknown, but the penalty functions are known.

We will present in this section a generic algorithm which has been introduced in [40] and generalized in [41]. The strategy uses a restricted model of malleable tasks which allows two types of execution, namely sequential and rigid. The execution can switch from one mode to the other. This simplified hypothesis allows us to establish some approximation bounds and is a first step towards the general malleable case.

26.7.1 A Generic Execution Scheme

We consider the online execution of a set of independent malleable tasks whose p_j are unknown. The number of processors needed for the execution of j is fixed (it will be denoted by q_j). The tasks may arrive at any time, but they are executed by successive batches. We assume that j can be scheduled either on 1 or q_j processors and the execution can be preempted.

We propose a generic framework based on batch scheduling. The basic idea is simple: when the number of tasks is large, the best way is to allocate the tasks to processors without idle times and communications. When enough tasks have been completed, we switch to a second phase with (rigid) PTs. In the following analysis, we assume that in the first phase each job is assigned to one processor, thus working with full efficiency. Inefficiency appears when less than m tasks remain to be executed. Then, when the number of idle processors becomes larger than a fixed parameter α, all remaining jobs are preempted and another strategy is applied in order to avoid too many idle times. Figure 26.9 illustrates the principle of an execution of this algorithm. Three successive phases are distinguished: first phase, when all the processors are busy; second phase, when at least $m - \alpha + 1$ processors work (both phases use the well-known Graham's list scheduling); final phase, when α or more processors become idle, and hence turn to a second strategy with PTs.

We note that many strategies can be used for executing the tasks in the last phase. We will restrict the analysis to rigid tasks.

26.7.2 Analysis

We provide now a brief analysis of the generic algorithm with idle regulation. More details can be found in [41].

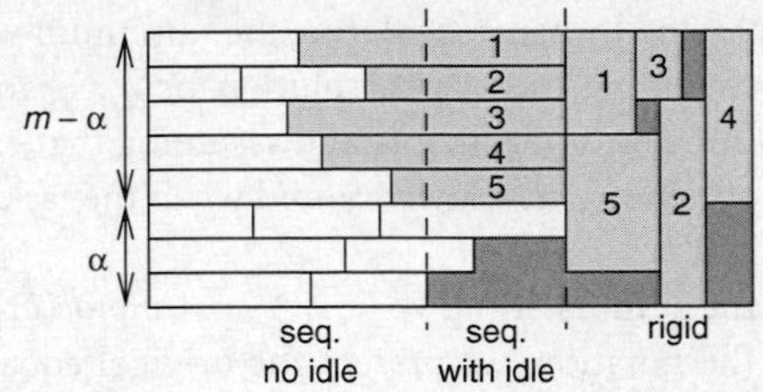

FIGURE 26.9 Principle of the generic scheme for partial malleable tasks.

Proposition 26.5

The performance ratio of the generic scheme is bounded by:

$$\frac{2m - q_{\max}}{m - q_{\max} + 1} - \alpha\left(\frac{1}{m - q_{\max} + 1} - \frac{1}{m}\right)$$

where $q_{\max}$ is the maximum of the q_j.

The proof is obtained by bounding the time in each phase. Let us check now that the previous bound corresponds to existing ones for some specific cases:

1. The case $\alpha = 0$ corresponds to schedule only rigid jobs by a list algorithm. This strategy corresponds to a 2-dimensional packing problem which has already been studied in this chapter.
2. The case $\alpha = 1$ for a specific allocation of processors in gang in the final phase (i.e., where each task is allocated to the full machine: $q_{\max} = m$) has been studied in [40].
3. The case $\alpha = m$ corresponds simply to list scheduling for sequential tasks (the algorithm is restricted to the first phase). As $q_{\max} = 1$, the bound becomes $2 - 1/m$.

It is difficult to provide a theoretical analysis for the general case of malleable tasks (preemption at any time for any number of processors). However, many strategies can be imagined and implemented. For instance, if the penalties are high, we can switch progressively from the sequential execution to two, then three (and so on) processors. If the tasks are very parallel ones, it is better to switch directly from 1 to a large number of processors.

26.8 Independent Moldable Tasks, $\sum C_i$, Offline

In this section, we come back to the original problem of scheduling independent moldable PTs focusing on the minimization of the other criterion, namely, the average completion time.

For a first view of the problem, we present two lower bounds and the principle of the algorithm from Schwiegelshohn et al. [42] that is dedicated to this criterion.

In this section we will use i instead of j for indexing the tasks because of the classical notation of $\sum C_i$.

26.8.1 Lower Bounds

With this criterion, there is a need for new lower bounds instead of the ones generally used with the makespan: critical path and sum of the works.

A first lower bound is obtained when all the tasks start at time 0. The tasks complete no sooner than their execution times which depend on their allotment. Thus $H = \sum p_i(nbproc(i))$ is a lower bound of $\sum C_i$ for a particular allotment.

The second lower bound is obtained when considering the minimum work for executing all the tasks. From classical single processor scheduling, the optimal solution for $\sum C_i$ is obtained by scheduling the tasks by increasing size order. Combining both arguments and assuming that each task may use m processors without increasing its area, we obtained a new lower bound when the tasks are sorted by increasing area: $A = \frac{1}{m} \sum w_i(nbproc(i))(n - i + 1)$.

This last bound is refined by the authors, using $W = \frac{1}{m} \sum w_i(nbproc(i))$ and a continuous integration. Like in the article, to simplify the notation, we present the original equation for rigid allocations. The uncompleted ratio of task i at time t is defined as

$$r_i(t) = \begin{cases} 1 & \text{if } t \leq \sigma(i) \\ 1 - \frac{t - \sigma(i)}{p_i} & \text{if } \sigma(i) \leq t \leq C_i \\ 0 & \text{if } C_i \leq t \end{cases}$$

Thus,

$$\sum_{i=1}^{n} \int_0^{+\infty} r_i(t)\, dt = \sum_{i=1}^{n} \left(\int_0^{C_i} 1\, dt - \int_{\sigma(i)}^{C_i} \frac{t - \sigma(i)}{p_i}\, dt \right)$$

As $C_i = \sigma(i) + p_i$ and $\int_{\sigma(i)}^{C_i} \frac{t-\sigma(i)}{p_i}\, dt = \frac{p_i}{2}$, it can be simplified as

$$\sum_{i=1}^{n} \int_0^{+\infty} r_i(t)dt = \sum_{i=1}^{n} C_i - \frac{1}{2} H$$

This result holds also for a transformation of the instance where the tasks keep the same area but use m processors. For this particular instance, gang scheduling by increasing height is optimal, thus $\sum_{i=1}^{n} C_i' = A$ and $H' = W'$. As A is a lower bound of $\sum_{i=1}^{n} C_i$ and $H > H' = W' = W$:

$$\sum_{i=1}^{n} C_i - \frac{1}{2} H \geq A - \frac{1}{2} W$$

namely,

$$\sum_{i=1}^{n} C_i \geq A + \frac{1}{2} H - \frac{1}{2} W$$

The extension to moldable tasks is simple: these bounds behave like the critical path and the sum of the works for the makespan. When H decreases, $A + \frac{1}{2} H - \frac{1}{2} W$ increases. Thus, there exists an allotment minimizing the maximum of both lower bounds. It can be used for the rigid scheduling.

26.8.2 Scheduling Algorithm

We do not detail too much the algorithm as an algorithm with a better performance ratio is presented in Section 26.9.

The "smart SMART" algorithm of [42] is a shelf algorithm. It has a performance ratio of 8 in the unweighted case and 8.53 in the weighted case ($\sum \omega_i C_i$). All shelves have a height of 2^k (1.65^k in the weighted case). All tasks are bin-packed (first fit, largest area first) into one of the shelves just sufficient to include it. Then all shelves are sorted in order to minimize $\sum C_i$, using a priority of $H_l / \sum_l \omega_i$, where H_l is the height of shelf l.

The basic point of the proof is that the shelves may be partitioned in two sets: a set including exactly one shelf of each size, and another one including the remaining shelves. Their completion times are respectively bounded by H and by A. The combination can be adjusted to get the best performance ratio (leading to the value of 1.65 in the weighted case).

26.9 Independent Moldable Tasks, Bicriterion, Online Batch

Up to now, we only analyzed algorithms with respect to one criterion. We have seen in Section 26.2.3 that several criteria could be used to describe the quality of a schedule. The choice of which criterion to choose depends on the priorities of the users.

However, one could wish to get the advantage of several criteria in a single schedule. With the makespan and the sum of weighted completion times, it is easy to find examples where there is no schedule reaching the optimal value for both criteria. Therefore you cannot have the cake and eat it, but you can still try to find for a schedule how far the solution is from the optimal one for each criterion. In this section, we will look at a generic way to design algorithms with guarantees on two criteria and at a more specific algorithm family for the moldable case.

26.9.1 Two Phases, Two Algorithms ($\mathcal{A}_{\sum C_i}$, $\mathcal{A}_{C_{\max}}$)

Let us use two known algorithms $\mathcal{A}_{\sum C_i}$ and $\mathcal{A}_{C_{\max}}$ with performance ratios respectively $\rho_{\sum C_i}$ and $\rho_{C_{\max}}$ with respect to the sum of completion time and the makespan [43].

Proposition 26.6

It is possible to combine $\mathcal{A}_{\sum C_i}$ and $\mathcal{A}_{C_{\max}}$ in a new algorithm with respective performance ratios of $2\rho_{\sum C_i}$ and $2\rho_{C_{\max}}$ at the same time.

Let us remark that delaying by τ the starting time of the tasks of the schedule given by $\mathcal{A}_{C_{\max}}$ increases the completion time of the tasks with the same delay τ.

The starting point of the new algorithm is the schedule built by $\mathcal{A}_{\sum C_i}$. The tasks ending in this schedule before $\rho_{C_{\max}} C^*_{\max}$ are left unchanged. All tasks ending after $\rho_{C_{\max}} C^*_{\max}$ are removed and rescheduled with $\mathcal{A}_{C_{\max}}$, starting at $\rho_{C_{\max}} C^*_{\max}$ (see Figure 26.10). As $\mathcal{A}_{C_{\max}}$ is able to schedule all tasks in $\rho_{C_{\max}} C^*_{\max}$ and it is always possible to remove tasks from a schedule without increasing its completion time, all these tasks will complete before $2\rho_{C_{\max}} C^*_{\max}$.

Now let us look at the new values of the two criteria. Any task scheduled by $\mathcal{A}_{\sum C_i}$ ending after $\rho_{C_{\max}} C^*_{\max}$ does not increase its completion time by a factor more than 2, thus the new performance ratio is no more than twice $\rho_{\sum C_i}$. On the other hand, the makespan is lower than $2\rho_{C_{\max}} C^*_{\max}$. Thus the performance ratios on the two criteria are the double of the performance ratio of each single algorithm.

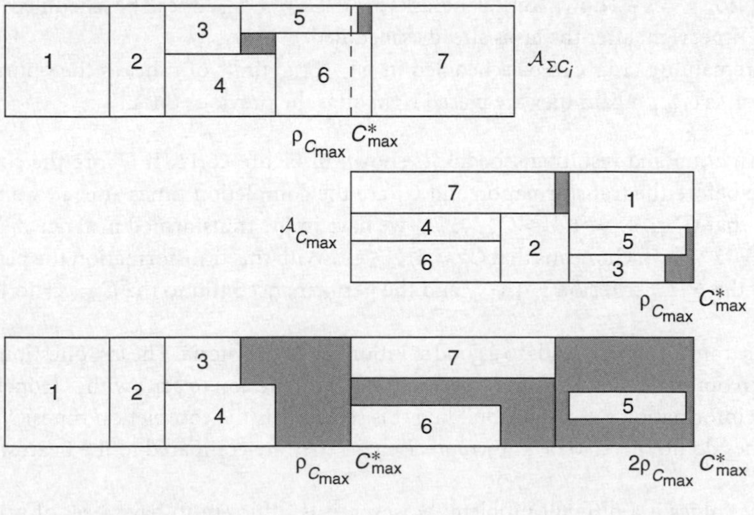

FIGURE 26.10 Bicriterion scheduling combining two algorithms.

We can also remark that in Figure 26.10 the schedule presented has a lot of idle times and the makespan can be greatly improved by just starting every task as soon as possible with the same allocation and order. However, even if this trick can give very good results for practical problems, it does not improve the theoretical bounds proven on the schedules, as it cannot always be precisely defined.

26.9.1.1 Tuning Performance Ratios

It is possible to decrease one performance ratio at the expense of the other. The point is to choose a border proportionally to $\rho_{C_{\max}} C^*_{\max}$, namely, $\lambda * \rho_{C_{\max}} C^*_{\max}$. The performance ratios are a Pareto curve of λ.

Proposition 26.7

It is possible to combine $\mathcal{A}_{\sum C_i}$ and $\mathcal{A}_{C_{\max}}$ in a new algorithm with respective performance ratios of $\frac{1+\lambda}{\lambda}\rho_{\sum C_i}$ and $(1+\lambda)\rho_{C_{\max}}$ at the same time.

Combining the algorithms of Sections 26.4.3 and 26.8 it is possible to schedule independent moldable tasks with a performance ratio of 3 for the makespan and 16 for the sum of the completion time.

26.9.2 Multiple Phases, One Algorithm ($\mathcal{A}_{C_{\max}}$)

The former approach required the use of two algorithms, one per criterion and mixed them in order to design a bicriterion scheduling algorithm. It is also possible to design an efficient bicriterion algorithm just by adapting an algorithm $\mathcal{A}_{C_{\max}}$ designed for the makespan criterion [44].

The main idea is to create a schedule which has a performance ratio on the sum of completion times based on the result of algorithm $\mathcal{A}_{C_{\max}}$ without losing too much on the makespan. To have this performance ratio $\rho_{\sum C_i}$ on the sum of the completion times, we actually try to have the same performance ratio $\rho_{\sum C_i}$ on all the completion times.

We give below a sketch of the proof. Let us now consider that we know one of the optimal schedules for the $\sum C_i$ criterion. We can transform this schedule into a simpler but less efficient schedule as follows:

1. Let $C^*_{\max}$ be the optimal makespan for the instance considered. Let k be the smallest integer such that in the $\sum C_i$ schedule considered, there is no task finishing before $C^*_{\max}/2^k$.
2. All the tasks i with $C_i < C^*_{\max}/2^{k-1}$ can be scheduled in $\rho_{C_{\max}} C^*_{\max}/2^{k-1}$ units of time, as $C^*_{\max}/2^{k-1}$ is the makespan of a feasible schedule for the instance reduced to these tasks, therefore bigger than the optimal makespan for the reduced instance.
3. Similarly for $j = k-2$ down to 1, all the tasks i with $C_i < C^*_{\max}/2^j$ can be scheduled in $\rho_{C_{\max}} C^*_{\max}/2^j$ units of time, right after the tasks already scheduled.
4. All the remaining tasks can be scheduled in $\rho_{C_{\max}} C^*_{\max}$ units of time, as the optimal value of the makespan is $C^*_{\max}$. Again they are placed right after the previous ones.

The transformation and resulting schedule is shown in Figure 26.11. If $C_i^{\$}$ are the completion times in the schedule before the transformation and $C_i^{\dagger}$ are the completion times after it, we can say that for all tasks i such that $C^*_{\max}/2^j < C_i^{\$} \le C^*_{\max}/2^{j-1}$ we have in the transformed instance $\rho_{C_{\max}} C^*_{\max}/2^{j-1} < C_i^{\dagger} \le \rho_{C_{\max}} C^*_{\max}/2^{j-2}$, which means that $C_i^{\dagger} < 4\rho_{C_{\max}} C_i^{\$}$. With this transformation the performance ratio with respect to the $\sum C_i$ criterion is $4\rho_{C_{\max}}$ and the performance ratio to the $C_{\max}$ criterion increased to $2\rho_{C_{\max}}$.

The previous transformation leads to a good solution for both criteria. The last question is "do we really need to know an optimal schedule with respect to the $\sum C_i$ criterion to start with?" Hopefully the answer is no. The only information needed for building this schedule is the completion times $C_i^{\$}$. Actually these completion times do not need to be known precisely, as they are compared to the nearest lower rounded values $C^*_{\max}/2^j$.

Getting these values is a difficult problem; however, it is sufficient to have a set of values such as the schedule given in Figure 26.11 which is feasible and the sum of completion times is a minimum. As the

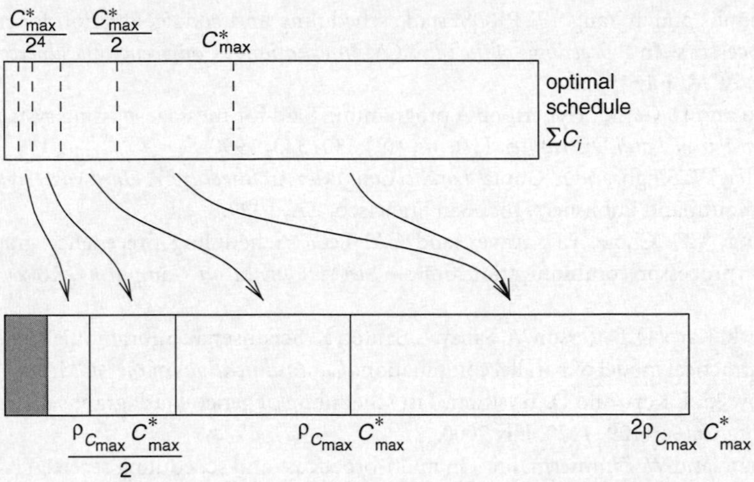

FIGURE 26.11 Transformation of an optimal schedule for $\sum C_i$ in a bicriterion schedule (with $k = 4$).

performance ratio for $\sum C_i$ refers to a value smaller than the optimal one, the bound is still valid for the optimal.

The last problem is to find a partition of the tasks into k sets where all the tasks within set S_j can be run in $\rho_{C_{\max}} C^*_{\max}/2^{j-1}$ (with algorithm $\mathcal{A}_{C_{\max}}$) and where $\sum_j |S_j| C^*_{\max}/2^j$ is a minimum. Filling the batches from left to right with as much weight as possible actually gives the best solution.

26.10 Conclusion

In this chapter, we have presented an attractive model for scheduling efficiently applications on parallel and distributed systems based on PTs. It is a nice alternative to conventional computational models particularly for large communication delays and new hierarchical systems. We have shown how to obtain good approximation scheduling algorithms for the different types of PTs (namely, rigid, moldable, and malleable) for two criteria ($C_{\max}$ and $\sum C_i$) for both offline and online cases. All these cases correspond to systems where the communications are rather slow, and versatile (some machines may be added or removed at some times). Most studies were conducted on independent PTs, except for minimizing the makespan of any task graphs in the context of offline moldable tasks.

Most of the algorithms have a small complexity and thus, may be implemented in actual parallel programming environments. For the moment, most of them do not use the moldable or malleable character of the tasks, but it should be more and more the case. We did not discuss in this chapter how to adapt this model to the other features of the new parallel and distributed systems; it is very natural to deal with hierarchical systems (see a first study in [45]). The heterogeneous character is more complicated because most of the methods assumed the monotony of the PTs. In the heterogeneous case, the execution time does not depend on the number of processors alloted to it, but on the set of processors as all the processors might be different.

References

[1] J. Błażewicz, K. Ecker, E. Pesch, G. Schmidt, and J. Węglarz. *Scheduling in Computer and Manufacturing Systems.* Springer-Verlag, Heidelberg, 1996.

[2] M. Pinedo. *Scheduling: Theory, Algorithms, and Systems.* Prentice-Hall, Englewood Cliffs, 1995.

[3] P. Brucker. *Scheduling.* Akademische Verlagsgesellschaft, Wiesbaden, 1981.

[4] A. Gerasoulis and T. Yang. PYRROS: static scheduling and code generation for message passing multiprocessors. In *Proceedings of the 6th ACM International Conference on Supercomputing*, pages 428–437, ACM, July 1992.

[5] M.-Y. Wu and D. Gajski. Hypertool: A programming aid for message-passing systems. *IEEE Transactions on Parallel and Distributed Systems*, 1(3):330–343, 1990.

[6] D. E. Culler, J. P. Singh, and A. Gupta. *Parallel Computer Architecture: A Hardware/Software Approach*. Morgan Kaufmann Publishers, Inc., San Francisco, CA, 1999.

[7] J.J. Hwang, Y.C. Chow, F.D. Anger, and C.Y. Lee. Scheduling precedence graphs in systems with interprocessor communication times. *SIAM Journal on Computing*, 18(2):244–257, April 1989.

[8] D. Culler, R. Karp, D. Patterson, A. Sahay, E. Santos, K. Schauser, R. Subramonian, and T. von Eicken. LogP: A practical model of parallel computation. *Communications of the ACM*, 39(11):78–85, 1996.

[9] T. Kalinowski, I. Kort, and D. Trystram. List scheduling of general task graphs under LogP. *Parallel Computing*, 26(9):1109–1128, July 2000.

[10] D. Trystram and W. Zimmermann. On multi-broadcast and scheduling receive-graphs under logp with long messages. In S. Jaehnichen and X. Zhou, editors, *The Fourth International Workshop on Advanced Parallel Processing Technologies — APPT 01*, pages 37–48, Ilmenau, Germany, September 2001.

[11] E. Bampis, A. Giannakos, and J.-C. König. On the complexity of scheduling with large communication delays. *European Journal of Operational Research*, 94(2):252–260, 1996.

[12] D. G. Feitelson. Scheduling parallel jobs on clusters. In Rajkumar Buyya, editor, *High Performance Cluster Computing*, Vol. 1, Chap. 21, Architectures and Systems, pages 519–533. Prentice Hall, Upper Saddle River, NJ, 1999.

[13] A. Lodi, S. Martello, and M. Monaci. Two-dimensional packing problems: A survey. *European Journal of Operational Research*, 141(2):241–252, 2002.

[14] I. Foster. *Designing and Building Parallel Programs: Concepts and Tools for Parallel Software Engineering*. Addison-Wesley, Reading, MA, 1995.

[15] A. S. Tanenbaum and M. van Steen. *Distributed Systems: Principles and Paradigms*. Prentice Hall, Upper Saddle River, NJ, 2002.

[16] E.G. Coffman and P.J. Denning. *Operating System Theory*. Prentice Hall, New York 1972.

[17] E. Jeannot. *Allocation de graphes de tâches paramétrés et génération de code*. PhD thesis, Ecole Normale Supérieure de Lyon et Ecole Doctorale Informatique de Lyon, 1999.

[18] H. Shachnai and J. Turek. Multiresource malleable task scheduling to minimize response time. *Information Processing Letters*, 70:47–52, 1999.

[19] F. Afrati, E. Bampis, A. V. Fishkin, K. Jansen, and C. Kenyon. Scheduling to minimize the average completion time of dedicated tasks. *Lecture Notes in Computer Science*, 1974, 2000.

[20] D. Hochbaum, editor. *Approximation Algorithms for NP-hard Problems*. PWS, September 1996.

[21] R.P. Brent. The parallel evaluation of general arithmetic expressions. *Journal of the ACM*, 21(2):201–206, July 1974.

[22] R.L. Graham. Bounds on multiprocessing timing anomalies. *SIAM Journal on Applied Mathematics*, 17(2):416–429, March 1969.

[23] M. Drozdowski. Scheduling multiprocessor tasks — an overview. *European Journal of Operational Research*, 94(2):215–230, 1996.

[24] B. Monien, T. Decker, and T. Lücking. A 5/4-approximation algorithm for scheduling identical malleable tasks. Technical Report tr-rsfb-02-071, University of Paderborn, 2002.

[25] J. Turek, J. Wolf, and P. Yu. Approximate algorithms for scheduling parallelizable tasks. In *4th Annual ACM Symposium on Parallel Algorithms and Architectures*, pages 323–332, 1992.

[26] A. Steinberg. A strip-packing algorithm with absolute performance bound 2. *SIAM Journal on Computing*, 26(2):401–409, 1997.

[27] M. R. Garey and R. L. Graham. Bounds on multiprocessor scheduling with resource constraints. *SIAM Journal on Computing*, 4:187–200, 1975.

[28] G. Mounié. *Ordonnancement efficace d'application parallèles: les tâches malléables monotones.* PhD thesis, INP Grenoble, juin 2000.

[29] D.S. Hochbaum and D.B. Shmoys. Using dual approximation algorithms for scheduling problems: theoretical and practical results. *Journal of the ACM*, 34:144–162, 1987.

[30] G. Mounié, C. Rapine, and D. Trystram. A 3/2-dual approximation algorithm for scheduling independent monotonic malleable tasks. Technical report, ID-IMAG Laboratory, 2000. http://www-id.imag.fr/~trystram/publis_malleable.

[31] K. Jansen and L. Porkolab. Linear-time approximation schemes for scheduling malleable parallel tasks. *Algorithmica*, 32(3):507, 2002.

[32] R. Lepère, D. Trystram, and G.J. Woeginger. Approximation scheduling for malleable tasks under precedence constraints. In *9th Annual European Symposium on Algorithms — ESA 2001*, number 2161 in LNCS, pages 146–157, Springer-Verlag, Heidelberg, 2001.

[33] G. N. S. Prasanna and B. R. Musicus. Generalised multiprocessor scheduling using optimal control. In *3rd Annual ACM Symposium on Parallel Algorithms and Architectures*, pages 216–228, ACM, 1991.

[34] J. Węglarz. Modelling and control of dynamic resource allocation project scheduling systems, in *Optimization and Control of Dynamic Operational Research Models.* North-Holland, Amsterdam, 1982.

[35] A. Feldmann, M-Y. Kao, and J. Sgall. Optimal on-line scheduling of parallel jobs with dependencies. In 25*th Annual ACM Symposium on Theory of Computing*, pages 642–651, San Diego, CA, 1993. url: http://www.ncstrl.org, CS-92-189.

[36] J. Błażewicz, M. Machowiak, G. Mounié, and D. Trystram. Approximation algorithms for scheduling independent malleable tasks. In *Europar 2001*, number 2150 in LNCS, pages 191–196. Springer-Verlag, Heidelberg, 2001.

[37] J. Sgall. On-line scheduling. *Lecture Notes in Computer Science*, 1442:196–231, 1998, chap. 9.

[38] D.B. Shmoys, J. Wein, and D.P. Williamson. Scheduling parallel machine on-line. *SIAM Journal on Computing*, 24(6):1313–1331, 1995.

[39] J. Błażewicz, K. Ecker, B. Plateau, and D. Trystram. *Handbook on Parallel and Distributed Processing.* International handbooks on information systems. Springer, New York, 2000.

[40] C. Rapine, I. Scherson, and D. Trystram. On-line scheduling of parallelizable jobs. In *Proceedings of EUROPAR'98*, number 1470 in LNCS, pages 322–327, Springer Verlag, Heidelberg, 1998.

[41] A. Tchernykh and D. Trystram. On-line scheduling of multiprocessor jobs with idle regulation. In *Proceedings of PPAM'03*, 2003 (Unpublished).

[42] U. Schwiegelshohn, W. Ludwig, J. Wolf, J. Turek, and P. Yu. Smart SMART bounds for weighted response time scheduling. *SIAM Journal on Computing*, 28, 1998.

[43] C. A. Phillips, C. Stein, E. Torng, and J. Wein. Optimal time-critical scheduling via resource augmentation (extended abstract). In *Proceedings of the Twenty-Ninth Annual ACM Symposium on Theory of Computing*, pages 140–149, El Paso, TX, 1997.

[44] L. A. Hall, A. S. Schulz, D. B. Shmoys, and J. Wein. Scheduling to minimize average completion time: Off-line and on-line approximation algorithms. *Mathematics of Operations Research*, 22:513–544, 1997.

[45] P.-F. Dutot and D. Trystram. Scheduling on hierarchical clusters using malleable tasks. In *Proceedings of the 13th annual ACM symposium on Parallel Algorithms and Architectures — SPAA 2001*, pages 199–208, SIGACT/SIGARCH and EATCS, ACM Press, Crete Island, July 2001.

[46] D.G. Feitelson and L. Rudolph. Parallel job scheduling: Issues and approaches. *Lecture Notes in Computer Science*, 949:1–18, 1995.

IV

Real-Time Scheduling

27

The Pinwheel: A Real-Time Scheduling Problem

Deji Chen
Fisher-Rosemount Systems, Inc.

Aloysius Mok
University of Texas at Austin

27.1 Introduction

The *pinwheel* problem is motivated by the performance requirements of a ground station that processes data from a number of satellites. The ground station needs to receive messages from n satellites. Each satellite has a different *repetition interval* associated with it. A satellite having repetition interval a repeatedly sends the same message to the ground station for exactly a successive time slots regardless of whether or not the ground station receives the message. Afterwards the satellite sends a new message for another a successive slots. This continues forever. At each time slot, the ground station will receive a message from a single satellite. In order not to miss messages coming from all the satellites, the ground station needs to come up with a schedule for receiving messages which ensures that a satellite with repetition interval a is allocated at least one slot in *any* interval of a slots.

The pinwheel is a formalization of this problem. Given a multiset of integers $A = \{a_1, a_2, \ldots, a_n\}$ (an instance), a sucessful schedule S is an infinite sequence "$j_1, j_2, \ldots$" over $\{1, 2, \ldots, n\}$ such that any subsequence of $a_i (1 \leq i \leq n)$ consecutive entries (slots) contains at least one item i. Without loss of generosity, we assume $a_1 \leq a_2 \leq \cdots \leq a_n$. The interpretation is that during the kth time slot, the ground station is servicing satellite j_k. For example, "$S = 1\ 2\ 1\ 2 \ldots$" is a schedule for $A = \{2, 3\}$. Notice that the 1st(2nd) satellite(item) is serviced at least once within any interval consisting of 2(3) slots. The ground station receives message all the time; some satellites are handled more frequently than required. In other words, the schedule does not contain any unassigned slots.

1-58488-397-9/$0.00+$1.50

The *pinwheel decision problem* concerns whether such a schedule exists; the *pinwheel scheduling problem* involves producing a representation of the corresponding schedule. In general, the pinwheel decision problem is in PSPACE. It turns out that if a schedule exists for an instance, there exists a cyclic schedule of cycle length no greater than $\prod_{i=1}^{n} a_i$. The term *pinwheel* comes from this cyclic nature of the schedule.

The *density* ρ of an instance A is defined as $\rho(A) = \sum_{i=1}^{n} \frac{1}{a_i}$. The justification for the name density is that in a cyclic schedule, i occupies at least $\frac{1}{a_i}$ of the slots. We call $\frac{1}{a_i}$ the density of the ith item. Instances with $\rho = 1.0$ are called *dense* instances; instances with $\rho < 1.0$ are called *nondense* instances. The best we can show for general dense instances is that the decision problem is in NP but is not known to be NP-hard.

A density of over 1.0 are impossible to schedule, always requiring more slots than exist in any given cycle length. Unfortunately, condition $\rho(A) \leq 1.0$ is only necessary but not sufficient. Consider the instances $\{2, 3, x\}$ for any arbitrary $x \geq 6$; their total density is $\frac{1}{2} + \frac{1}{3} + \frac{1}{x} \leq 1.0$, but they cannot be scheduled. It is conjectured that any instances with $\rho \leq \frac{5}{6}$ can be scheduled. Until now, no one could prove or disprove this conjecture, despite strong evidence indicating its correctness.

A scheduling algorithm X is said to have *density threshold* d_X if X can schedule all pinwheel instances with density $\leq d_X$. Later we shall show such algorithms. For example, **Sx** has a density threshold of $\frac{13}{20}$ and **Sxy** has a density threshold of 0.7.

Typically, many of the satellites that must be monitored will be identical. This leads to instances with small number of distinct integers. Instances with one distinct integer is trivial, which are always schedulable if $\rho \leq 1.0$. Instances with two distinct integers and $\rho \leq 1.0$ also can always be scheduled. For dense instances with three distinct integers, the decision problem is PTIME and a global greatest common divisor greater than 1 is a necessary but not a sufficient condition for schedulability. For dense instances with four or more distinct integers, a greatest common divisor greater than 1 is neither necessary nor sufficient.

The schedule cycle length could be exponential to the problem size. The minimum length for dense instances that can be scheduled is exactly the Least Common Multiple (LCM) of all intervals. We are concerned with finding *useful* representations of the schedules — providing, of course, that they exist. What is really needed is not the schedule itself but a Fast OnLine Scheduler (FOLS) — a program that generates the scheduling sequence in constant time per item generated. Such programs would be suitable to act as drivers for the respective ground stations. From a particular cyclic schedule one can easily build corresponding FOLS as a finite state machine. Unfortunately, since the cyclic schedules are, in general, of exponential length, this seems to be an impractical or unrealistic approach. A better approach would be to construct polynomial time program generators (PTPGs) that take as input an instance of the pinwheel problem and produce as output a corresponding FOLS. We do not expect a PTPG exists that will produce FOLS for all schedulable instances. However, it might exist for classes of instances where the pinwheel decision problem is in PTIME. FOLSs can be constructed in polynomial time for all schedulable dense instances of up to three distinct numbers and many instances with density no more than the threshold.

The standard form for a FOLS program P is the following:

$P : \alpha;$
do
$\beta;$
forever

α is an initialization code segment and β is a simple segment of straight-line code that can generate one symbol of the sequence in one time unit. On each iteration of the **do**-loop β selects an item for a fixed slot. Thus, P generates the scheduling sequence in constant time per item generated. The existence of PTPGs that will produce FOLSs for a class of scheduling problem defines a complexity class. We will concentrate on the class *S-P-C*, for scheduling-polynomial-constant. That means there exists a program that runs in polynomial time and determines whether a schedule exists, and if so generates a scheduler that runs in constant time per item scheduled. This constitutes our working definition of a *useful* representation of a schedule.

An instance can have *compact representation* in which identical integers are represented with (a, x) — a pair of a value a and a count x. For example, the representation of two distinct integers is defined as $\{(a_1, x_1), (a_2, x_2)\}$ where a_1 and a_2, called *frequencies*, are the distinct numbers of the multiset representation, and x_1 and x_2, called *cardinalities*, specify the number of occurrences of each. Any compact representation $\{(a_1, x_1), \ldots, (a_n x_n)\}$ can be transformed to the original model in a straightforward manner by simply representing each pair (a_i, x_i) as a sequence of x_i distinct a_i's. However, such a transformation will result in a multiset of size $\sum_{i=1}^{n} x_i$ integers; in general, this represents an exponential increase in the size of the representation. The complexity of the decision problem given the compact representation is NP-hard.

Give two instances $A = \{a_1, a_2, \ldots, a_n\}$ and $A' = \{a'_1, a'_2, \ldots, a'_n\}$. Assume $a_i \geq a'_i, 1 \leq i \leq n$. By definition, any sucessful schedule of A' is a schedule of A. This technique is called integer reduction and proves to be very useful for general pinwheel decision problems. For example, we show that instances with $\rho \leq 0.5$ may always be scheduled by reduction to instances consisting solely of power of 2 and $\rho \leq 1.0$. The latter instances are schedulable because instances consisting solely of multiples (i.e., $i < j \Rightarrow a_i \mid a_j$) with density ≤ 1.0 are schedulable via a simple greedy strategy.

In the next section, we shall dive into more details of the pinwheel problem just discussed. Section 27.3 explains other hard real-time problems that are related to the pinwheel problem. Section 27.4 summarizes. In each section we shall reference the publications from which the content is derived.

27.2 Pinwheel Problem Exclusive

In this section, we try to present all the known properties of the pinwheel problem. However, due to space limit, we could not provide every detail, especially for the proofs. Instead, we organize them in a systematic fashion. The references at the end are good resources for further reading.

Subsection 27.2.1 draws its content mainly from [1] and talks about pinwheel in general and some results for dense instances; Subsection 27.2.2 discusses instances with two distinct integers [2]; Subsection 27.2.3 discusses instances with three distinct integers [1,3]; Subsection 27.2.4 uses *integer reduction* to find the density thresholds of general instances [4]; Subsection 27.2.5 uses *double-integer reduction* to find better density thresholds of general instances [4,5].

27.2.1 Some General Results and the Dense Instances

Theorem 27.1

If A has a schedule then A has a cyclic schedule whose cycle length is no greater than $\prod_{i=1}^{n} a_i$.

Proof

Let $m = \prod_{i=1}^{n} a_i$. Let $S = j_0 j_1 \ldots$ be an infinite schedule for A. Consider the prefix of S, $S' = j_0 j_1, \ldots, j_{2m}$. Each slot in S' can be represented by a vector $\langle c_1, \ldots, c_n \rangle$ where $c_j, 1 \leq j \leq n$, denotes the number slots since the last occurence of j. For the slots indexed from m to $2m$ these vectors are well defined, i.e., for each such vector each position $j, 1 \leq j \leq n$, in the vector contains a value between 0 and $a_j - 1$. Now there are $\prod_{i=1}^{n} a_i$ such unique vectors. Hence, there exist indices s and t, $m \leq s < t \leq 2m$ such that slots s and t are represented by the same vector. It is now easy to see that "$j_s, \ldots, j_{t-1}$" could be a cycle of a cyclic schedule for A whose length is no greater than $\prod_{i=1}^{n} a_i$. □

The next corollary follows directly from the proof of Theorem 27.1.

Corollary 27.1

The pinwheel decision problem is in PSPACE.

Schedules for dense instances exhibit certain regularities that make them easier to reason about. For instance, the minimum schedule length for instances that can be scheduled is the LCM of all the integer intervals; slots assigned to item i must occur exactly a_i slots apart. These and related properties do not hold for nondense instances. The methods used in dense pinwheel instances involve the use of divisibility and number theory. Nondense instances require additional machineries, notably concerned with the properties of floor and ceiling functions.

Lemma 27.1

Let $A = \{a_1, \ldots, a_n\}$ be a dense instance. Let S be a schedule cycle for A of length m. Then each i must occur exactly $\frac{m}{a_i}$ times in S. Furthermore, successive occurrences of i in S must be exactly a_i slots apart. Lastly, m must be a multiple of LCM$(a_1, \ldots, a_n)$.

Proof

Clearly each $i, 1 \leq i \leq n$, must occur at least $\lceil \frac{m}{a_i} \rceil$ times in S. Because each i must occur at least $\lceil \frac{m}{a_i} \rceil$ times in S we get m to be $\geq \sum_{i=1}^{n} \lceil \frac{m}{a_i} \rceil$. But $m = m \sum_{i=1}^{n} \frac{1}{a_i} = \sum_{i=1}^{n} \frac{m}{a_i}$. Hence, we get a contradiction if for any i, $\lceil \frac{m}{a_i} \rceil > \frac{m}{a_i}$. Thus each i must occur exactly $\frac{m}{a_i}$ times in S. As a result, we see that $\frac{m}{a_i}$ must be integral for every i and hence LCM$(a_1, \ldots, a_n)$ must divide m. □

Theorem 27.2

If $A = \{a_1, \ldots, a_n\}$ is dense and schedulable then A has a cyclic schedule whose cycle length is LCM$(a_1, \ldots, a_n)$. Furthermore, shorter schedule cycles for A are not possible.

Proof

Let S be a schedule cycle for A of length m. From Lemma 27.1 it follows that $m = r\text{LCM}(a_1, \ldots, a_n)$. Hence, shorter schedules for A are not possible. Now divide S into r identical segments, each of size LCM$(a_1, \ldots, a_n)$. Consider an arbitrary slot k in the first segment. Suppose this slot contains i. Since the kth slot of the jth segment is the $(k + a_i(j\text{LCM}(a_1, \ldots, a_n)/a_i))$th slot of S, it follows from Lemma 27.1 that this slot also contains i. Hence, S consists of r identical segments, each of which by itself constitutes a valid schedule cycle for A. □

Theorem 27.3

The pinwheel decision problem for the class of dense instances is in NP.

Proof

Let $A = \{a_1, \ldots, a_n\}$ be dense. Consider the construction of an infinite schedule for A. We need to select starting slots for each index that will insure the avoidance of collisions. By Lemma 27.1, we know that if integer i first appears in slot p_i then it will occupy slots $p_i + ja_i$, for every integer $j \geq 0$. Starting slots permissible for item i are $0, \ldots, a_i - 1$. Hence, A is schedulable if and only if $\exists p_1, \ldots, p_n, 0 \leq p_k < a_k$, $1 \leq k \leq n$, such that $\forall ij, 1 \leq i, j \leq n, \neg[\exists a, b \geq 0$ such that $p_i + aa_i = p_j + ba_j]$.

The question "$\exists a, b \geq 0$ such that $p_i + aa_i = p_j + ba_j$?" is an instance of integer linear programming in two variables and hence is solvable in deterministic polynomial time. A nondeterministic polynomial time algorithm should now be apparent. □

The pinwheel decision problem for the class of dense instances is NP-hard if the instances are in compact representation forms.

27.2.2 Instances with Two Distinct Integers

In this subsection we show that all instances with two distinct integers and density ≤ 1.0 are schedulable, and there also exist FOLSs. A full discussion can be found in [2]. For any instance, we define a schedule cycle length and a method that produces a successful schedule with the cycle length for the instance. An FOLS algorithm is then given.

We use compact representation forms. For instance with two distinct integers $A = \{(a_1, x_1), (a_2, x_2)\}$, the following properties hold: (1) $x_1 > 0$, (2) $x_2 > 0$, (3) $\frac{a_1}{x_1} > 1$, and (4) $\frac{a_2}{x_2} > 1$.

The method makes use of a pair of partition functions that generates two sets of integers that never overlap and yet jointly define all the non-negative integers. For example, $2i$ and $2i+1, i \geq 0$ generate such sets of integers {0 2 4...} and {1 3 5...}. In the following definition, *Place1* and *Place2* define such a pair.

$$H(A) = \frac{a_2 \mathrm{LCM}(x_1, a_2 - x_2)}{a_2 - x_2}$$

$$Place1(i) = i + \left\lceil \frac{ix_2}{a_2 - x_2} \right\rceil, \qquad 0 \leq i < \frac{H(A)}{a_2}(a_2 - x_2)$$

$$Place2(i) = \left\lfloor \frac{ia_2}{x_2} \right\rfloor + 1, \qquad 0 \leq i < \frac{H(A)}{a_2} x_2$$

The scheduling method creates a cyclic schedule of cycle length $H(A)$. In the cycle, all items of frequency a_1 are scheduled repeatedly in sequence at the slots produced by *Place1*; all items of frequency a_2 are scheduled repeatedly in sequence at the slots produced by *Place2*. This scheduling method amounts to distributing slots for items of frequency a_1 as evenly as possible over the first a_2 slots. This creates a *skeleton* of length a_2 that will be repeated as many times as necessary to reach length $H(A)$, which is a multiple of a_2. *Place1* selects slots by skipping slots in the ratio of $\frac{x_2}{a_2 - x_2}$, which is the ratio of slots not needed for a_1 to slots needed for a_1. The ceiling term represents slots skipped—that is, when incrementing i causes the ceiling term to increase, a slot is skipped. Finally, the limiting formula $\frac{H(A)}{a_2}(a_2 - x_2)$ is derived from the fraction of slots used for $a_1 (\frac{a_2}{a_2 - x_2})$.

We now give an example of this scheduling method. The instance $A = \{(15, 7), (6, 3)\}$ yields $H(A) = 42$. Interleaving the contents lists as specified by *Place1* and *Place2* yields the schedule cycle "1,8,2,9,3,10,4, 8,5,9,6,10,7,8,1,9,2,10, 3,8,4,9,5,10,6,8,7,9,1,10,2,8,3,9,4,10,5,8,6,9,7,10."

It could be shown that this method has seven properties:

1. *Place1* and *Place2* never select the same slot.
2. There are enough slots for the items of frequency a_1.
3. There are enough slots for the items of frequency a_2.
4. For any item of frequency a_1, the distance between consecutive slot assignments within the cycle is no more than a_1.
5. For any item of frequency a_2, the distance between consecutive slot assignments within the cycle is no more than a_2.
6. The schedule cycles correctly for items of frequency a_1.
7. The schedule cycles correctly for items of frequency a_2.

With these properties, we have:

Theorem 27.4

The method given above results in a successful cyclic schedule for an instance $A = \{(a_1, x_1), (a_2, x_2)\}$.

Armed with this theorem, we define the program to generate the scheduler.

Algorithm 27.1

$P_1 := 0$; $P_2 := 0$; *slot* $:= 0$; $I_1 := 1$; $I_2 := x_1 + 1$;
do
if $Place1(P_1) = slot$ **then begin**
  $Output(I_1)$;
  $I_1 := I_1 + 1$; **if** $I_1 > x_1$ **then** $I_1 := 1$;
  $P_1 := P_1 + 1$;
if $Place1(P_1) \geq H(A)$ **then** $P_1 := 0$;
end

```
if Place2(P_2) = slot then begin
  Output(I_2);
  I_2 := I_2 + 1; if I_2 > x_1 + x_2 then I_2 := x_1 + 1;
  P_2 := P_2 + 1;
  if Place2(P_2) ≥ H(A) then P_2 := 0;
end
slot := (slot + 1) mod n;
forever
```

Note that the above program meets all the requirements to be an FOLS, and that it can be generated in deterministic polynomial time. Thus we have the following theorem.

Theorem 27.5

The pinwheel scheduling problem restricted to instances with only two distinct integers is in S-P-C.

We conclude this subsection with some comments. We could switch a_1 and a_2, x_1 and x_2 to get a different $H(A)$ value and successful scheduling algorithm. $H(A)$ may not be the minimum cycle of any successful schedule. To derive the minimum cycle, we define function $M(A, n)$ that gives the difference between the potential cyclic schedule length n and the minimum number of slots that must be available in a schedule cycle of that length. It turns out that the minimum possible n satisfying $M(A, n) = 0$ is the minimum successful cycle. We define it to be $LM(A)$. [6] proves that $LM(A)$ can be computed in linear time. Apply the same method and algorithm above with $LM(A)$ and the following two partition functions, we could get a successful schedule of minimum cycle length.

$$Place1(i) = i + \left\lfloor i \frac{x_2 \left\lceil \frac{LM(A)}{a_2} \right\rceil}{x_1 \left\lceil \frac{LM(A)}{a_1} \right\rceil} \right\rfloor, \quad 0 \le i < x_1 \left\lceil \frac{LM(A)}{a_1} \right\rceil$$

$$Place2(i) = i + \left\lfloor i \frac{x_1 \left\lceil \frac{LM(A)}{a_1} \right\rceil}{x_2 \left\lceil \frac{LM(A)}{a_2} \right\rceil} \right\rfloor + 1, \quad 0 \le i < x_2 \left\lceil \frac{LM(A)}{a_2} \right\rceil$$

27.2.3 Instances with Three Distinct Integers

Instances with two distinct numbers are always schedulable. This is no longer the case for three distinct numbers. As mentioned earlier, $A = \{2, 3, x\}, x \ge 6$ is not schedulable. Here we give an algorithm **Pinwheel3** that schedules any instances with $\rho \le \frac{5}{6}$. Again, we represent an instance in the compact form $A = \{(a_1, x_1), (a_2, x_2), (a_3, x_3)\}$. Recall in the last subsection we devised a pair of functions that partition the integer set. **Pinwheel3** further partitions one of the subset into two more disjoint sets. With each set for an integer number, we get an algorithm that can successfully schedule any instance with $\rho \le \frac{5}{6}$. The three partition functions are.

$$Place1(i, \beta) = \lceil i\beta \rceil - 1, \quad i \ge 1$$
$$Place2(i, y, \alpha) = \lfloor \lfloor iy \rfloor \alpha \rfloor, \quad i \ge 1$$
$$Place3(i, z, \alpha) = \lfloor (\lceil iz \rceil - 1)\alpha \rfloor, \quad i \ge 1$$

Note the functions are parameterized with α, β, x, and y. In the following algorithm, they are calculated according to the input instance. Without loss of generality, we assume $\frac{x_1}{a_1} \ge \frac{x_2}{a_2} \ge \frac{x_3}{a_3}$.

Algorithm 27.2 (Pinwheel3)

1 $y := \lfloor a_2(1 - \frac{x_1}{a_1}) \rfloor / x_2$;
2 **if** $y \leq 1$ **then begin**
3 *swap the values of* a_1 *and* a_2; *swap the values of* x_1 *and* x_2;
4 $y := \lfloor a_2(1 - \frac{x_1}{a_1}) \rfloor / x_2$;
5 **end**
6 **if** $y \leq 1$ **then** *output "not schedulable"*
7 **else begin**
8 $\beta := \frac{a_1}{x_1}; \alpha := \frac{1}{1-1/\beta}; z := \frac{1}{1-1/y}; d := \lceil \lceil x_3 z \rceil \alpha \rceil$;
9 **if** $d > a_3$ **then** *output "not schedulable"*
10 **else begin**
11 *generate schedule with slots from Place1, Place2, Place3 assigned to items from frequency* a_1, a_2, *and* a_3 *respectively.*
11 **end**
12 **end**

Theorem 27.6

The schedule produced by **Pinwheel3** *is a successful one. Its running time is* $\bigcirc(1)$.

The above theorem can be proved by showing that (i) *Place1*, *Place2*, and *Place3* disjointly partition the integer set, and (ii) The schedule satisfies the distance requirement of each item.

Theorem 27.7

Any pinwheel instance with three distinct integers and with density $\leq \frac{5}{6}$ *can always be scheduled by* **Pinwheel3**.

Proof

A close look at **Pinwheel3** shows that if $y > 1$ in line 6 and $d \leq a_3$ in line 9, then **Pinwheel3** will always generate successful schedule. So, in the following, we prove that if $\rho(A) \leq \frac{5}{6}$, then y will be greater than 1 in line 6 and d will be less than or equal to a_3 in line 9. In line 1, $y = \lfloor a_2(1 - \frac{x_1}{a_1}) \rfloor / x_2$. If $y > 1$, it will never need to run lines 3–4. Otherwise, it will swap the values of first two integers and recompute y in line 4. First, we analyze the properties of y.

Let $g = (1 - \frac{x_1}{a_1})\frac{a_2}{x_2}$. Since $x_2 \geq 1$ and $y = \lfloor a_2(1 - \frac{x_1}{a_1}) \rfloor / x_2 = \lfloor x_2 g \rfloor / x_2, \lfloor g \rfloor \leq y \leq g$. Now $\frac{a_3}{x_3} \geq \frac{a_2}{x_2} = \frac{g}{1-x_1/a_1} \geq \frac{y}{1-x_1/a_1}$. So $a_3 \geq \frac{x_3 y}{1-x_1/a_1}$. Because a_3 is an integer, $a_3 \geq \lceil \frac{x_3 y}{1-x_1/a_1} \rceil$. According to the values of g and $\frac{x_1}{a_1}$, there are four cases to analyze:

CASE I: $g \geq 2$
CASE II: $1.5 \leq g < 2$ and $\frac{x_1}{a_1} < 0.5$
CASE III: $1.5 \leq g < 2$ and $\frac{x_1}{a_1} \geq 0.5$
CASE IV: $g < 1.5$

All of the above cases can be proved to have the same properties that the value of y in line 6 of **Pinwheel3** will be greater than 1 and d will be less than or equal to a_3 in line 9 if $\rho(A) \leq \frac{5}{6}$. Therefore, **Pinwheel3** will produce a successful schedule. □

Note an instance could still be scheduled by **Pinwheel3** if $\rho > \frac{5}{6}$. For dense instances with three distinct integers, an algorithm exists to decide its schedulability in PTIME and the class of dense instances with three distinct integers is in S-P-C.

27.2.4 Instances with Any Integers — Single Integer Reduction

In this subsection and the next, we shall provide algorithms for general instances using *integer reduction* technique. We shall also derive density thresholds for these algorithms.

The integer reduction technique is based on the observations in the following two theorems.

Theorem 27.8

Let instances $A = \{a_1, a_2, \ldots, a_n\}$, $B = \{b_1, b_2, \ldots, b_n\}$, and $a_i \geq b_i, 1 \leq i \leq n$. If B is schedulable, then so is A.

Proof

Use the schedule for B as the schedule for A. In this schedule, successive i's appear no more than b_i slots apart. Since $a_i \geq b_i$, clearly, successive i's appear no more than a_i slots apart. □

We say A is reduced to B.

Theorem 27.9

If $A = \{a_1, a_2, \ldots, a_n\}$, $\rho(A) \leq 1.0$, and $a_i \mid a_j$ for $1 \leq i < j \leq n$, then A is schedulable.

Proof

The following **SpecialSingle**(A) algorithm successfully generates a scheduling cycle for A. □

SpecialSingle(A) is a driver to a process called parallelized simple-greedy (PSG).

Algorithm 27.3 SpecialSingle(A)

/* *Assume $\rho(A) \leq 1.0$, and $a_i \mid a_j$ for all $1 \leq i < j \leq n$.* */
/* *send(P, message): send a message to process P.* */
/* *receive(Q, message): wait for a message from process Q.* */
*send(***PSG***, A);*
do
  *send(***PSG***, null); receive(***PSG***,i);*
  schedule i in the current slot;
  go to the next slot;
forever

PSG divides logically into two phases:

1. A phase that generates a subschedule S for time slots 0 to $T - 1$ where
   - T is the length of the cycle, unknown until the end of this phase,
   - symbol n is sent for the very first time at time $T - 1$, and
   - two successive i's are sent exactly a_i time slots apart.
2. A phase that repeats the subschedule S over and over again.

Algorithm 27.4 PSG

/* *Assume $\rho(A) \leq 1.0$, and $a_i \mid a_j$ for all $1 \leq i < j \leq n$.* */
/* *$t \equiv$ time slot* */
/* *$k \equiv$ next new symbol (a new symbol is one that has not been sent)* */
/* *$c_i \equiv$ the time at which symbol i is first sent* */
/* *$T \equiv$ length of cycle* */
/* *OutputFlag $\equiv$ flag indicating whether a new symbol can be sent* */
  /* *phase(1): subschedule S spanning time slot 0 to $T - 1$* */
  receive(id, A), $t := 0, k := 1$;
  repeat
    receive(id,null);
    OutputFlag := 0;

```
    for i := 1 to k − 1 do if t mod a_i = c_i then OutputFlag := 1, send(id,i)
    if OutputFlag = 0 then send(id,k), c_k := t, k := k + 1;
    t := t + 1;
until (k > n)
/* phase(2): repeating schedule S */
T := t;
do
    receive(id,null);
    for i := 1 to n do if (t mod T) mod a_i = c_i then send(id,i)
    t := t + 1;
forever
```

Corollary 27.2

SpecialSingle *is an FOLS using* $\bigcirc(n)$ *hardware without concurrent read/write operations.*

Now we make use of above two theorems to derive schedules for general instances.

Definition 27.1

Let $x \geq 2$ *be an integer, we say that* A *is special with respect to* x *if* $\forall a \in A, a = x2^j, j \geq 0$.

According to Theorem 27.9, any instance that is special to an integer and $\rho \leq 1.0$ is schedulable.

The *single-integer reduction* transforms an instance to an instance that is *special* with respect to a single integer.

Corollary 27.3

If $\rho(A) \leq 0.5$, *then* A *is always schedulable.*

Proof

A can be reduced to $B = \{b_1, \ldots, b_n\}$, where $b_i = 2^{\lfloor \log_2 a_i \rfloor} \leq 2^{\log_2 a_i} = a_i$. B is special with respect to 2 and $\rho(B) = \sum_{i=1}^{n} \frac{1}{2^{\lfloor \log_2 a_i \rfloor}} \leq \sum_{i=1}^{n} \frac{1}{2^{\log_2 a_i - 1}} = 2\sum_{i=1}^{n} \frac{1}{a_i} = 2\rho(A) \leq 1.0$. Hence B is schedulable and then A is also schedulable. □

A more general scheduler than that in above proof is called **Sa** and can be formally stated as follows:

Algorithm 27.5 Sa(A)

```
B :=specialize A with respect to a_1;
if ρ(B) ≤ 1 then SpecialSingle(B)
else report failure
```

Theorem 27.10

The density threshold

$$d_{\mathbf{Sa}} = \frac{1 + \frac{1}{a_1} - \frac{1}{a_1 a_n}}{2 - \frac{1}{a_n}}.$$

The theorem can be proved from the fact that $\rho(B) \leq 1.0$ if $\rho(A) \leq d_{\mathbf{Sa}}$.

The following algorithm specializes A with respect to a value in $[\frac{a_1}{2}, a_1]$ and has a better result.

Algorithm 27.6 $\mathbf{Sx}(A)$

$x := a_1$;
repeat
  $B :=$ *specialize* A *with respect to* x;
  $x := x - 1$;
until $((x \le \frac{a_1}{2}) \vee (\rho(B) \le 1))$
if $\rho(B) \le 1$ **then SpecialSingle**(B)
else report *failure*

The complexity of Algorithm 27.6 hinges on the loop to find x and B. With some improvement on this, **Sx** can determine whether or not it can schedule any instance in $\bigcirc(n)$ time.

Theorem 27.11

If a_1 *is even, the density threshold* $d_{\mathbf{Sx}} = \frac{1}{a_1} + \frac{1}{a_1+2} + \sum_{k=\lfloor \frac{a_1}{2} \rfloor+2}^{a_1} \frac{1}{k}$; *if* a_1 *is odd, the density threshold* $d_{\mathbf{Sx}} = \frac{1}{a_1} + \sum_{k=\lfloor \frac{a_1}{2} \rfloor+2}^{a_1} \frac{1}{k}$;

Corollary 27.4

$0.65 \le d_{\mathbf{Sx}} < \ln 2$. *Moreover,* $\lim_{a_i \to \infty} d_{\mathbf{Sx}} = \ln 2$.

27.2.5 Instances with Any Integers — Double Integer Reduction

The above subsection makes use of instances special to a single integer. Recall in Subsection 27.2.2 that any instances with two distinct integers are schedulable. One could wonder if we could make use of instances special to two integers. An instance is special to two integers if it can be divided into two subsets, each of which is special to an integer. This subsection looks at this approach and gives several algorithms with better density thresholds.

Let instance $A = B \bigcup C$ where B is special to integers b and C is special to integer c. Let $p = \lceil b\rho(B) \rceil, q = \lceil c\rho(C) \rceil$. Let **Schedule2** be a scheduling algorithm for $D = \{(b, p), (c, q)\}$ with two distinct integers. From the schedule for D we could arrive at a schedule for A: in the schedule for D, we reschedule all slots allocated to b's for B using **PSG** introduced in the last subsection; we reschedule all slots allocated to c's for C the same way. The algorithm is called **SpecialDouble**:

Algorithm 27.7 SpecialDouble(A)

/* $A = B \bigcup C$. */
/* $B = \{a_{u_1}, \ldots, a_{u_r}\}$ *is special with respect to* b. */
/* $C = \{a_{v_1}, \ldots, a_{v_s}\}$ *is special with respect to* c. */
$p := \lceil b\rho(B) \rceil, q := \lceil c\rho(C) \rceil$
$B' := \{a_{u_1} p/b, \ldots, a_{u_r} p/b\}$
$C' := \{a_{v_1} q/c, \ldots, a_{v_s} q/c\}$
$D := \{(b, p), (c, q)\}$
*send(***Schedule2**, *D); send(***PSG_B**, *B'); send(***PSG_C**, *C');*
do
  *send(***Schedule2**, *null); receive(***Schedule2**, *i);*
  if $1 \le i \le p$ **then** *send(***PSG_B**, *null), receive(***PSG_B**, *j); schedule* u_j
  else *send(***PSG_C**, *null), receive(***PSG_C**, *j); schedule* v_j
  go to the next slot;
forever

Theorem 27.12
If $\frac{\lceil b\rho(B)\rceil}{b} + \frac{\lceil c\rho(C)\rceil}{c} \leq 1$, *then* **SpecialDouble** *will produce a schedule for A.*

Proof
This proof depends on the validity of **Schedule2** and **PSG**. Without loss of generality, let us consider the items of A in B as a similar argument holds for the items of A in C. **Schedule2** guarantees that each symbol $i, 1 \leq i \leq p$, appears no more than b slots apart and there are at least $p = \lceil b\rho(B)\rceil$ slots allocated for B's items (B-slots) in any interval of b slots. After converting B to B', $\rho(B') = \frac{b\rho(B)}{p} \leq 1$ since $\rho(B) \leq \frac{p}{b}$, and B' is special to p. Thus, the **PSG_B** when given B' will schedule item $b2^j$ in B (item $p2^j$ in B') at most $p2^j$ B-slots apart (in the schedule produced by **Schedule2**). This is equivalent to having the element scheduled at most $b2^j$ slots apart. Thus, the theorem is proved. □

Armed with Algorithm 27.7 and its property, we now introduce scheduler **S23** that reduces an instance to another instance special to integers 2 and 3. It reduces any integer between $2(2^j)$ and $3(2^j)$ to $2(2^j)$; it reduces any integer between $3(2^j)$ and $2(2^{j+1})$ to $3(2^j)$. In case the resulting instance is not schedulable, we try to **NORMALIZE** the two subsets by re-reducing some of the items from one subset to another.

Algorithm 27.8 S23(A)
(B, C) :=*specialize A with respect to 2 and 3;*
(B', C') := **NORMALIZE**$(B, C, 2, 3)$;
if $\frac{\lceil 2\rho(B')\rceil}{2} + \frac{\lceil 3\rho(C')\rceil}{3} \leq 1$ **then SpecialDouble**$(B' \bigcup C')$;
else report *failure*

Algorithm 27.9 NORMALIZE(B, C, b, c)
/* **EXCESS**(X, x) *returns* Y: $\lceil x\rho(X)\rceil = y, Y \subseteq X, x\rho(X - Y) = y - 1$. */
P := **EXCESS**(B, b)
Q := **EXCESS**(C, c)
P' := *specialize P to c;*
Q' := *specialize Q to b;*
case $(\frac{2b\rho(P)}{c} + \rho(Q) \leq \frac{1}{c})$: $B' := B - P, C' := C \bigcup P'$
case $(\frac{1}{c} < \frac{2b\rho(P)}{c} + \rho(Q) \leq \frac{2}{c}) \wedge (\rho(P) + \frac{c\rho(Q)}{b} \leq \frac{1}{b})$: $B' := B \bigcup Q', C' := C - Q$
case $(\frac{1}{c} < \frac{2b\rho(P)}{c} + \rho(Q) \leq \frac{2}{c}) \wedge (\rho(P) + \frac{c\rho(Q)}{b} > \frac{1}{b})$: $B' := B - P, C' := C \bigcup P'$
case $(\frac{2b\rho(P)}{c} + \rho(Q) > \frac{2}{c})$: $B' := B, C' := C$

There are many such schedulers depending on how the two integers are selected.

S23 Specialize to integers 2 and 3
Sbc Specialize to a_1 and $\lfloor\sqrt{2}a_1\rfloor$ or $\lceil\sqrt{2}a_1\rceil$
Sby Specialize to a_1 and $y, a_1 \leq y < 2a_1$
Sxy Specialize to $x, \frac{a_1}{2} \leq x < a_1$ and $y, x < y \leq a_1$

Theorem 27.13
S23, Sbc, Sby, *and* **Sxy** *are all FOLSs and their density thresholds are* $\frac{7}{12}$, $\frac{2}{3}$, *0.6964, and 0.7 respectively.*

27.3 Pinwheel Problem Extended

The pinwheel is a special case of real-time scheduling problem which is concerned with the scheduling of n periodic, real-time processes on a single processor [7,8], and is a special case of *latency scheduling* problems for *sporadic processes* [9]. In this section, we shall address the relationship of pinwheel with other real-time problems. Subsection 27.3.1 introduces *Distance-Constraint Task System* (DCTS) [10,11];

Subsection 27.3.2 introduces the concept of *proportionate fairness* [12,13]; Subsection 27.3.3 introduces *Periodic Maintenance Problem* (PMP) [14]; finally Subsection 27.3.4 talks about other problems.

27.3.1 Pinwheel Problem with Non-Integer Values — Distance Constrained Tasks

During single-integer reduction, we reduce an instance A to an instance B special to an integer b. As long as $\rho(B) \leq 1.0$, A is schedulable. So one of the objective is to make $\delta = \rho(B) - \rho(A)$ as small as possible. The algorithms introduced in the last section choose different b to minimize δ. One might wonder if δ will be further reduced if b is not limited to integers. This idea, surprisingly, could be applied to scheduling DCTS.

In a DCTS task set $T = \{T_1, T_2, \ldots, T_n\}$, every task T_i consists of infinite sequences of jobs $J_{i1}, J_{i2}, J_{i3}, \ldots$. Task T_i has execution time e_i and distance constraint c_i. The distance between two consecutive jobs of a task is defined to be the difference of the finishing times of these two jobs. Let f_{ij} denote the finishing time of job $J_{ij}, 1 \leq i \leq n, j \geq 1$. The distance between J_{ij} and $J_{i,j+1}$ is defined to be $J_{i,j+1} - J_{ij}$. The distance constraint requires that $f_{i1} \leq c_i$ and $J_{i,j+1} - J_{ij} \leq c_i, j \geq 1$.

For comparison, we shall call a DCTS $T = \{T_1, T_2, \ldots, T_n\}$ also an instance, and define $T_i = (c_i, e_i)$, $1 \leq i \leq n$, density $\rho(T_i) = \frac{e_i}{c_i}, \rho(T) = \sum_{i=1}^{n} \rho(T_i)$.

The pinwheel problem can be viewed as a special case of the discrete version of the DCTS problem since the pinwheel requirement that "there is at least one item i within any subsequences of a_i consecutive items" is equivalent to the distance constraint that "the temporal distance between two consecutive occurrences of symbol i must be less than or equal to a_i." A scheduling problem for a DCTS instance $\{(c_i, e_i) \mid 1 \leq i \leq n\}$ with all execution times $e_i = 1$ and all instance constraints c_i being integers is exactly the same as that for the pinwheel instance $\{a_i = c_i \mid 1 \leq i \leq n\}$.

On the other hand, the schedulers designed for the pinwheel problem can also be used to schedule DCTSs. Each DCTS task T_i is transformed into a task with unit execution time and distance constraint $a_i = \lfloor \frac{c_i}{\lceil e_i \rceil} \rfloor$. The DCTS problem is then transformed to the pinwheel problem with instance $A = \{a_1, a_2, \ldots, a_n\}$. We can imagine that every $\lceil e_i \rceil$ consecutive slots allocated to the ith item are actually allocated to one job request of task T_i. In this way, we can use the algorithms designed for the pinwheel problem to solve the DTCS problem. Note that the distance constraints are satisfied since $\lceil e_i \rceil \lfloor \frac{c_i}{\lceil e_i \rceil} \rfloor \leq c_i$.

The potential benefit of adopting the pinwheel model is that system schedules can be produced in polynomial time. Now we introduce a more involved adoption. We shall first reduce a DCTS instance into another one that is special to a real value (not necessarily integer), then schedule the resulting instance with *Distance-Constraint Monotonic* (DCM) algorithm.

Similar to the *Rate-Monotonic* (RM) algorithm, the DCM algorithm assigns priorities to tasks before run time in such a way that tasks with smaller distance constraints get higher priorities (ties are broken arbitrarily). At run-time, the system always executes the task with the highest priority among all active tasks. To prevent a DCTS task from being over-executed, we define *separation constraint* s_i: a task T_i only becomes ready again for execution s_i time after current request is finished.

Theorem 27.14

For a DCTS task set A, if $c_i \mid c_j$ *for all* $1 \leq i < j \leq n$ *and* $\rho(A) \leq 1$, *let* $s_i = c_i - f_{i1}$. *A can be scheduled by DCM.*

The proof could be found by extending the method in the proof of Theorem 27.9.

For a DCTS instance $A = \{(c_1, e_1), (c_2, e_2), \ldots, (c_n, e_n)\}$, we define **Sc** to reduce it to $B = \{(b_1, e_1), (b_2, e_2), \ldots, (b_n, e_n)\}$ that is special to a value c such that the largest task density will not increase. In other words, $\frac{c_1}{2} < c = c_m / 2^{\lceil \log_2 \frac{c_m}{c_1} \rceil}$ where c_m is from the task with the biggest ρ.

Theorem 27.15

Given a DCTS task set A, **Sc** *can schedule A if* $\rho(A) \leq \frac{1}{2} + \frac{e_m}{2c_m}$.

Proof

We only prove the density threshold. The correctness of **Sc** follows from the correctness of **Sa** and Theorem 27.14. It is easy to see that $\frac{c_i}{b_i} < \frac{2b_i}{b_i} = 2$, for $1 \leq i \leq n$ and $i \neq m$, and $\frac{c_m}{b_m} = 1$. This means $\rho(B) = \sum_{i=1}^{n} \frac{e_i}{b_i} = \frac{e_m}{b_m} + \sum_{i=1, i \neq m}^{n} \frac{e_i}{b_i} < \sum_{i=1}^{n} \frac{2e_i}{c_i} - \frac{e_m}{c_m}$. If $\rho(A) = \sum_{i=1}^{n} \frac{e_i}{c_i} \leq \frac{1}{2} + \frac{e_m}{2c_m}$, then $\rho(B) < 1$. □

Like in the last section, we could also improve **Sc** to schedule more instances.

Sc Special to $c = c_m/2^{\lceil \log_2 \frac{c_m}{c_1} \rceil}$.
Sr Special to $c = c_i/2^{\lceil \log_2 \frac{c_i}{c_1} \rceil}$ for i that minimizes $\delta = \rho(B) - \rho(A)$.
Sr^g Like **Sr** but use g as the base instead of 2.
Sr^b Like **Sr**^{g} but choose the best g from 2 to $\lfloor \frac{c_n}{c_1} \rfloor$.

Theorem 27.16

The density thresholds for **Sr**, **Sr**^{g}, *and* **Sr**^{b} *are* $n(2^{1/n} - 1)$, $\frac{n}{g-1}(g^{1/n} - 1)$, *and* $n(2^{1/n} - 1)$ *respectively.*

27.3.2 Pinwheel Problem and Fairness — Pfair Scheduling

The issue of *fairness* in resource-allocation and scheduling has recently attracted considerable attention. It means that a task is allocated resource in proportion to its utilization(density) as time progresses. Let a task $T = (a, x)$ where a is the period and x is the execution requirement. The weight is $w = \frac{x}{a}$. For any schedule S, we define **allocated**$_T(t)$ to be $\left| \{t' \mid t' \in [0, t) \text{ and } T \text{ is scheduled at } t'\} \right|$, which is the time allocated to T from 0 to t; we define **lag**(S, T, t) to be $wt -$ **allocated**$_T(t)$, which is the difference between what should be allocated and what is actually allocated. S is *pfair* if and only if $\forall T, t : -1 <$ **lag**$(S, T, t) < 1$.

The following theorem has been proved in [13], and an algorithm **PF** produces such a pfair schedule.

Theorem 27.17

Every uniprocessor system of periodic tasks for which $\sum w \leq 1$ *has a pfair schedule.*

In a pfair schedule, the jth request of T is scheduled between **earliest**$(T, j) = \lfloor \frac{j}{w} \rfloor$ and **latest**$(T, j) = \lfloor \frac{j+1}{w} \rfloor - 1$.

Applying this result, we get **Pinfair** that schedules pinwheel instances in compact representation $A = \{(a_1, x_1), \ldots, (a_n x_n)\}$.

Algorithm 27.10 Pinfair

Step 1: For each $(a_i, x_i), 1 \leq i \leq n$, *define weight* $w_i = (x_i + 1)/a_i$.
Step 2: If $\sum_{i=1}^{n} w_i > 1$ *return failure.*
Step 3: Schedule A using Algorithm **PF**.

Theorem 27.18

Any pinwheel instance satisfying $\sum_{i=1}^{n} w_i < 1$ *can be successfully scheduled by* **PF**.

Proof

Observe first that, if $\sum_{i=1}^{n} w_i < 1$ is satisfied, then **Pinfair** reduces to **PF** with task weights as defined in Step 1. In the remainder of this proof, we therefore study the behavior of **PF** on a task system in which the task weights are as defined in Step 1. From the perspective of meeting the pinwheel condition, the worst case that can occur during the scheduling of task system using **PF** is for task $T = (a, x)$ to get scheduled for the jth time at the earliest possible slot and for the next x allocations to T to occur as late as possible. In that case, there will be exactly x allocations to T over the interval between time instants (**earliest**$(T, j) + 1$) and (**latest**$(T, j + x) + 1$). In order to ensure that **PF** schedules task T for at least x of every a consecutive slots, it suffices to ensure that the size of this interval never exceeds a, for any j. That is, for all j, (**latest**$(T, j + x) + 1$) − (**earliest**$(T, j) + 1$) $\leq x$. This turns out to be correct by definition. Hence, **Pinfair** will schedule task T at least x times in any interval of size a. □

Note $w_i = \frac{x_i+1}{a_i} > \frac{x_i}{a_i} = \rho(T_i)$.

Theorem 27.19

Algorithm **Pinfair** *successfully schedules any pinwheel instance A with density at most* $\frac{a_1}{a_1+1}$.

Proof

$\sum_{i=1}^{n} w_i = \sum_{i=1}^{n} \frac{x_i+1}{a_i} = \sum_{i=1}^{n} \frac{x_i(1+\frac{1}{x_i})}{a_i} \leq \sum_{i=1}^{n} \frac{x_i(1+\frac{1}{x_i})}{a_i} = (1 + \frac{1}{a_1})\rho$. From Theorem 27.18, it follows that, in order for **Pinfair** to schedule A, it is sufficient that $((1 + \frac{1}{a_1})\rho \leq 1) \equiv (\rho \leq \frac{1}{1+\frac{1}{a_1}}) \equiv (\rho \leq \frac{a_1}{a_1+1})$. □

Note $\lim_{a_1 \to \infty} \frac{a_1}{a_1+1} = 1$. **PF** was initially defined in the context of multiprocessor scheduling of periodic task systems. Using **PF**, and techniques essentially identical to ones used here, a multiprocessor variant of **Pinfair** can be defined, which schedules any pinwheel task system on m processors, provided $\sum_{i=1}^{n} w_i \leq m$. This is a really nice by-product.

27.3.3 Pinwheel Problem and Periodic Maintenance Problem

The closest relative to pinwheel problem is the *Periodic Maintenance Problem*(PMP) of [15] introduced before the pinwheel problem. The PMP is motivated by the need to schedule a mechanic's time to perform periodic maintenance. Recast into our terminology this problem requires item i to be scheduled *exactly* every a_i slots. That is, if item i is scheduled into slot k, it must also be scheduled into slot $k + ja_i$ for all nonnegative integers j. This is indeed the case for our dense instances, so such pinwheels are also instances of the PMP. The difference appears in the case of nondense instances. The PMP does not allow an item to be scheduled early; pinwheel does not allow empty slots in a schedule. Thus, the pinwheel problem is concerned with scheduling the server's time as tightly as possible, while the PMP is concerned with minimizing the downtime of the machines being served.

PMP was initially defined for multiple servers. If there are k servers, each time slot can be assigned up to k different items. We briefly mentioned a neat result by the end of last subsection about pinwheel problem with multiple processors.

For every instance of the single-server PMP there is a corresponding instance of the pinwheel problem, and a schedule for the former may be translated into a schedule for the latter. This may be done by *padding* the periodic maintenance instance with new items whose frequency is the LCM of the given items, yielding a dense pinwheel instance. The pinwheel instance has a schedule if and only if the original periodic maintenance instance does. A cyclic schedule for the pinwheel instance can then be transformed into a cyclic schedule for the periodic maintenance instance by changing to *blank* all those slots allocated to the new items. Thus, the pinwheel instances may be exponentially longer than the corresponding periodic maintenance instances; thus complexity results may not transfer. Finally, the inclusion is proper, so pinwheel schedules do not generally imply periodic maintenance problem schedules. For example, $A = \{2, 3\}$ has a pinwheel schedule but not a periodic maintenance schedule.

It turns out that the schedulability problem for MPM of $k \geq 1$ servers is NP-complete in the strong sense. Define single-server class C_M in which each maintenance interval is a multiple of all smaller intervals and C_2 in which there are only two distinct maintenance intervals. Applying similar method developed for pinwheel, we could prove that, for C_M and C_2, the decision problem can be solved in PTIME and an FOLS can be constructed in deterministic polynomial time. That is, both classes are in the complexity class S-P-C. The results for C_M can easily be generalized to multiple servers.

27.3.4 Pinwheel Problem and Other Problems

The pinwheel is one of a growing family of hard-real-time scheduling problems. Since its introduction, the pinwheel model has been used to model the requirements of a wide variety of real-time systems. Besides what's mentioned in above subsections, [16] extended this research to distributed systems; [17] used pinwheel scheduling to construct static schedules for sporadic task systems. The issues of maintaining pinwheel schedules in a dynamic network environment has also been studied [18]. [19,20] applied pinwheel techniques to real-time network scheduling, [21] modeled fault-tolerance and real-time requirements of broadcast disks by generalizing the pinwheel model. Extensions of the pinwheel scheduler for sporadic tasks and resource sharing have been proposed [22].

Let's compare the pinwheel with the classical periodic task model. In the process model of real-time systems (e.g., [7,8,23]), each process is parameterized by an ordered pair (c, p), where c and p are respectively the computation time and period of the process. The scheduling problem is to ensure that every process is executed in every period, i.e., the process (c, p) requests service at time $= i * p, i \geq 0$, and must complete execution by time $= (i + 1) * p$. the satellites in the pinwheel problem may be modeled as processes. However, a process in this case may request service at any time, subject to the condition that two consecutive requests from the same process must be at least p time units apart.

To conclude, we shall reprove Corollary 27.3 with the classical periodic task model.

Proof

For any pinwheel instance $A = \{a_1, \ldots, a_n\}$, we define a periodic task $S = \{(c_1, p_1), \ldots, (c_n, p_n)\}$ where $c_i = 1, p_i = \frac{a_i}{2}, 1 \leq i \leq n$. The utilization of S is $U = \sum_{i=1}^{n} \frac{C_1}{P_1} = \sum_{i=1}^{n} \frac{1}{a_i/2} = 2\sum_{i=1}^{n} \frac{1}{a_i} = 2\rho(A) \leq 2 \times 0.5 = 1.0$. So S is schedulable by *earliest deadline first* (EDF) Algorithm. In this schedule, the longest possible distance between two consecutive allocations of task T_i is when the first request is allocated right at the request time and the second is allocated at the end of the period, which is $2p_i - 1 = 2\frac{a_i}{2} - 1 < a_i$. This implies that for any length of a_i or longer T_i is scheduled at least once. If we remove all the empty slots in the EDF schedule and ignore the fact that T may be scheduled before its *request time*, we get a successful schedule for A. □

27.4 Conclusion

One of the applications motivating the pinwheel problem is the real-time scheduling of satellite communications by a ground station. The study of the pinwheel problem generated fruitful results beyond the pinwheel problem itself. Here we condensed the research literature on this topic and presented the results for both pinwheel scheduling and its application to other problems. We showed that the schedulability problem in general is in NP. For dense instances, FOLS could be found for instances with up to three distinct integers and general instances with density no more than a density threshold. Nondense instances with up to two distinct integers are always schedulable. The density threshold for instances with three distinct numbers is $\frac{5}{6}$. For any instance, the best density threshold is 0.7. We also presented other real-time problems associated with pinwheel. Especially we talked about distance-constrained task system, proportional fairness, and periodic maintenance problem.

References

[1] Holte, R., Mok, A., Rosier, L., Tulchinsky, I., and Varvel, D.: The Pinwheel: A Real-Time Scheduling Problem. *22nd Hawaii International Conference on System Sciences.* January (1989)

[2] Holte, R., Rosier, L., Tulchinsky, I., and Varvel, D.: Pinwheel Scheduling with Two Distinct Numbers. *Theoretical Computer Science.* **100** June (1992) 105–135

[3] Lin, S.S and Lin, K.J.: A Pinwheel Scheduler for Three Distinct Numbers with a Tight Schedulability Bound. *Algorithmica.* **019(04)** (1997) 411–426

[4] Chan, M.Y. and Chin, F.: Schedulers for Larger Classes of Pinwheel Instances. *Algorithmica.* **9** (1993) 425–462
[5] Chan, M.Y. and Chin, F.: General Schedulers for the Pinwheel Problem Based on Double-Integer Reduction. *IEEE Transactions on Computers.* **41(6)** July (1992) 755–768
[6] Romer, T.H. and Rosier, L.E.: An Algorithm Reminiscent of Euclidean-gcd for Computing a Function Related to Pinwheel Scheduling. *Algorithmica.* **17** (1997) 1–10
[7] Leung, J. Y-T. and Merrill, M.L.: A Note on Preemptive Scheduling of Periodic, Real-Time Tasks. *Information Processing Letters.* **11(3)** November (1980) 115–118
[8] Liu, C.L. and Layland, J.W.: Scheduling Algorithms for Multiprogramming in a Hard-Real-Time Environment. *Journal of ACM.* **20(1)** (1973)
[9] Mok, A.: Fundamental Design Problems of Distributed Systems for the Hard-Real-Time Environment. Ph.D. thesis. Massachusetts Institute of Technology, Cambridge, MA. (1983)
[10] Han, C.C. and Lin, K.J.: Scheduling Distance-Constraint Real-time Tasks. *IEEE Real-Time Systems Symposium.* December (1992) 814–826
[11] Han, C.C., Lin, K.J. and Hou, C.J.: Distance-Constrained Scheduling and its Applications to Real-Time Systems. *IEEE Transactions on Computers.* **45(7)** July (1996) 814–826
[12] Baruah, S.: Fairness in Periodic Real-Time Scheduling. *IEEE Real-Time Systems Symposium.* (1995) 200–209
[13] Baruah, S., Cohen, N. K., Plaxton, C. G., and Varvel, D. A.: Proportionate Progress: A Notion of Fairness in Resource Allocation. *Algorithmica.* **15** (1996) 600–625
[14] Baruah, S., Mok, A., Rosier, L., Tulchinsky, I., and Varvel, D.: The Complexity of Periodic Maintenance. *International Computer Symposium.* (1990)
[15] Wei, W. and Liu, C.: On a Periodic Maintenance Problem. *Operations Research Letters.* **2(2)** (1983) 90–93
[16] Hsueh, C.W. and Lin, K.J.: Scheduling Real-time Systems with End-to-End Timing Constraints Using the Distributed Pinwheel Model. *IEEE Transactions on Computers.* **50(1)** January (2001) 51–66
[17] Baruah, S., Rosier, L., and Varvel, D.: Static and Dynamic Scheduling of Sporadic Tasks for Single-Processor Systems. *3rd Euromicro Workshop Real-Time Systems.* (1991)
[18] Hou, C.J. and Tsoi, K.S.: Dynamic Real-Time Channel Setup and Tear-Down in DQDB Networks. *IEEE Real-Time Systems Symposium.* December (1995) 232–241
[19] Han, C.C. and Shin, K.G.: A Polynomial-Time Optimal Synchronous Bandwidth Allocation Scheme for the Timed-Token MAC Protocol. *INFOCOM'95.* **2** (1995) 875–882
[20] Han, C.C. and Shin, K.G.: Real-Time Communication in FieldBus Multiaccess Networks. *IEEE Real-Time Technology and Applications Symposium.* (1995) 86–95
[21] Baruah, S. and Bestavros, A.: Pinwheel Scheduling for Fault-Tolerant Broadcast Disks in Real-Time Database Systems. *IEEE International Conference on Data Engineering.* April (1997) 543–551
[22] Lin, K.J. and Herkert, A.: Jitter Control in Time-Triggered Systems. *29th Hawaii Conference on System Sciences.* (1996)
[23] Mok, A. and Sutanthavibul, S.: Modeling and Scheduling of Dataflow Real-Time Systems. *IEEE Real-Time Systems Symposium.* December (1985)
[24] Baruah, S. and Lin, S.S.: Pfair Scheduling of Generalized Pinwheel Task Systems. *IEEE Transactions on Computers.* **47(7)** July (1998) 812–816
[25] Chen, D., Mok, A., and Baruah, S.: On Modeling Real-time Task Systems. Lecture Notes in Computer Science — Lectures on Embedded Systems. **1494** October (1997).
[26] Dertouzos, M.: Control Robotics: The Procedural Control of Physical Processes. *Proceedings of the IFIP Congress.* (1974) 807–813
[27] Garfinkel, R. S. and Plotnicki, W. J.: A Solvable Cyclic Scheduling Problem with Serial Precedence Structure. *Operations Research.* **28(5)** (1980) 1236–1240
[28] Hsueh, C.W. and Lin, K.J.: An Optimal Pinwheel Scheduler Using the Single-Number Reduction Technique. *IEEE Real-Time Systems Symposium.* (1996) 196–205

[29] Hsueh, C.W. and Lin, K.J.: On-line Schedulers for Pinwheel Tasks Using the Time-driven Approach. *10th Euromicro Workshop on Real-Time Systems.* (1998) 180–187
[30] Mok, A., Rosier, L., Tulchinsky, I. and Varvel, D.: Algorithms and Complexity of the Periodic Maintenance Problem. *Microprocessing and Microprogramming.* **27** (1989) 657–664
[31] Ward, S.A.: An Approach to Real-Time Computation. *7th Texas Conference on Computing Systems — Computing System for Real-Time Applications.* (1978) 5.26–5.34

28

Scheduling Real-Time Tasks: Algorithms and Complexity*

Sanjoy Baruah
The University of North Carolina at Chapel Hill

Joël Goossens
Université Libre de Bruxelles

28.1 Introduction

The use of computers to control safety-critical real-time functions has increased rapidly over the past few years. As a consequence, real-time systems — computer systems where the correctness of a computation is dependent on both the logical results of the computation and the time at which these results are produced — have become the focus of much study.

Since the concept of "time" is of such importance in real-time application systems, and since these systems typically involve the sharing of one or more resources among various contending processes, the concept of scheduling is integral to real-time system design and analysis. Scheduling theory as it pertains to a finite set of requests for resources is a well-researched topic. However, requests in real-time environments are often of a recurring nature. Such systems are typically modeled as finite collections of simple, highly repetitive tasks, each of which generates jobs in a very predictable manner. These jobs have upper bounds upon their worst-case execution requirements and associated deadlines. For example, in a *periodic* task system [1–6] each task makes a resource request at regular periodic intervals. The processing time and the time elapsed between the request and the deadline are always the same for each request of a particular task; they may, however, be different for different tasks. A *sporadic* task system [5,7–12] is similar, except that the

* Supported in part by the National Science Foundation (Grant Nos. CCR-9988327, ITR-0082866, and CCR-0204312).

1-58488-397-9/$0.00+$1.50

request times are not known beforehand; thus, sporadic tasks may be used to model event-driven systems. Real-time scheduling theory has traditionally focused upon the development of algorithms for *feasibility analysis* (determining whether all jobs can complete execution by their deadlines) and *run-time scheduling* (generating schedules at run-time for systems that are deemed to be feasible) of such systems.

28.2 Definitions and Terminology

In hard-real-time systems, there are certain basic units of work known as jobs, which need to be executed by the system. Each such job has an associated deadline, and it is imperative for the correctness of the system that all such jobs complete by their deadlines. In this survey article, we restrict our attention to a *preemptive* model of scheduling — a job executing on the processor may be interrupted, and its execution resumed at a later point in time. There is no penalty associated with such preemption. For results concerning real-time scheduling in nonpreemptive environments, see, for example, [10,13,14].

For our purposes, each real-time job is characterized by three parameters — a *release time*, an *execution requirement*, and a *deadline* — with the interpretation that the job needs to be executed for an amount equal to its execution requirement between its release time and its deadline. We will assume throughout this chapter that *all job execution requirements are normalized to processor computing capacity*, i.e., that job execution-requirements are expressed in units that result in the processor's computing capacity being one unit of work per unit time.

Definition 28.1 (Jobs and Instances)

A real-time ***job*** $j = (a, e, d)$ *is characterized by three parameters — an arrival time* a*, an execution requirement* e*, and a deadline* d *— with the interpretation that this job must receive* e *units of execution over the interval* $[a, d)$*. A real-time* ***instance*** J *is a finite or infinite collection of jobs:* $J = \{j_1, j_2, \ldots\}$*.*

Real-time instances may be generated by collections of periodic or sporadic tasks (see Section 28.2.1 below).

Definition 28.2 (Schedule)

For any collection of jobs J*, a (uniprocessor) schedule* **S** *is a mapping from the Cartesian product of the real numbers and the collection of jobs to* $\{0, 1\}$*:*

$$S\colon \mathbb{R} \times J \longrightarrow \{0, 1\}$$

with $S(t, j)$ *equal to one if schedule* S *assigns the processor to job* j *at time-instant* t*, and zero otherwise.* [1]

Definition 28.3 (Active Job)

A job $j = (a, e, d)$ *in* J *is defined to be active in some schedule* S *of* J *at time instant* t *if*

1. $a \leq t$;
2. $t \leq d$; and
3. $\left(\int_{t'=a}^{t} S(j, t)\right) < e$

That is, an active job is one that has arrived, has not yet executed for an amount equal to its execution requirement, and has not yet had its deadline elapse.

[1] Hence, for all t, there is at most one $j \in J$ for which $S(t, j) = 1$.

28.2.1 Periodic and Sporadic Tasks

In the **periodic** model of real-time tasks, a task T_i is completely characterized by a four-tuple (a_i, e_i, d_i, p_i), where

1. the **offset** a_i denotes the instant at which the first job generated by this task becomes available for execution;
2. the **execution requirement** e_i specifies an upper limit on the execution requirement of each job generated by this task;
3. the **relative deadline** d_i denotes the temporal separation between each job's arrival time and deadline — a job generated by this task arriving at time-instant t has a deadline at time-instant $(t + d_i)$; and
4. the **period** p_i denotes the temporal separation between the arrival times of successive jobs generated by the task.

That is, $T_i = (a_i, e_i, d_i, p_i)$ generates an infinite succession of jobs, each with execution-requirement e_i, at each instant $(a_i + k \cdot p_i)$ for all integer $k \geq 0$, and the job generated at instant $(a_i + k \cdot p_i)$ has a deadline at instant $(a_i + k \cdot p_i + d_i)$.

Sporadic tasks are similar to periodic tasks, except that the parameter p_i denotes the *minimum*, rather than exact, separation between successive jobs of the same task. A sporadic task is usually characterized by three parameters rather than four: $T_i = (e_i, d_i, p_i)$ with the interpretation that T_i generates an infinite succession of jobs each with an execution requirement equal to e_i and a deadline d_i time-units after its arrival time, and with the arrival-instants of successive jobs being separated by at least p_i time units.

A **periodic task system** comprises a finite collection of periodic tasks, while a **sporadic task system** comprises a finite collection of sporadic tasks.

The **utilization** $U(T_i)$ of a periodic or sporadic task T_i is defined to be the ratio of its execution requirement to its period: $U(T_i) \stackrel{\text{def}}{=} e_i / p_i$. The utilization $U(\tau)$ of a periodic, or sporadic task system τ is defined to be the sum of the utilizations of all tasks in τ: $U(\tau) \stackrel{\text{def}}{=} \sum_{T_i \in \tau} U(T_i)$.

28.2.1.1 Special Kinds of Task Systems

Real-time researchers have often found it convenient to study periodic and sporadic task models that are more restrictive than the general models described above. Some of these restricted models include

Implicit-deadline task systems. Periodic and sporadic task systems in which every task has its deadline parameter equal to its period (i.e., $d_i = p_i$ for all tasks T_i).

Constrained-deadline task systems. Periodic and sporadic task systems in which every task has its deadline parameter no larger than its period (i.e., $d_i \leq p_i$ for all tasks T_i).

Synchronous task systems. Periodic task systems in which the offset of all tasks is equal (i.e., $a_i = a_j$ for all tasks T_i, T_j; without loss of generality, a_i can then be considered equal to zero for all T_i). Tasks comprising a synchronous period task system are typically specified by three parameters rather than four — the offset parameter is left unspecified, and assumed equal to zero.

28.2.2 Static- and Dynamic-Priority Algorithms

Most uniprocessor scheduling algorithms operate as follows: at each instant, each active job is assigned a distinct priority, and the scheduling algorithm chooses for execution the currently active job with the highest priority.

Some scheduling algorithms permit that periodic/sporadic tasks T_i and T_j both have active jobs at times t and t' such that at time t, T_i's job has higher priority than T_j's job while at time t', T_j's job has higher priority than T_i's. Algorithms that permit such "switching" of priorities between tasks are known as **dynamic** priority algorithms. An example of dynamic priority scheduling algorithm is the earliest deadline first (EDF) scheduling algorithm [1,15].

By contrast, **static** priority algorithms satisfy the property that for every pair of tasks T_i and T_j, whenever T_i and T_j both have active jobs, it is always the case that the *same* task's job has higher priority. An example of a static-priority scheduling algorithm for periodic scheduling is the rate-monotonic scheduling algorithm [1].

28.2.3 Real-Time Scheduling Problems

As stated in Section 28.1, real-time scheduling theory is primarily concerned with obtaining solutions to the following two problems:

1. **The feasibility-analysis problem:** *Given* the specifications of a task system, and constraints on the scheduling environment (e.g., whether dynamic-priority algorithms are acceptable or static-priority algorithms are required, whether preemption is permitted, etc.), *determine* whether there exists a schedule for the task system that will meet all deadlines.

   Any periodic task system generates exactly one collection of jobs, and the feasibility problem is concerned with determining whether this collection of jobs can be scheduled to meet all deadlines. Since the sporadic task model constrains the *minimum*, rather than the exact, interarrival separation between successive jobs of the same task, a sporadic task system, on the other hand, is legally permitted to generate infinitely many distinct collections of real-time jobs. The feasibility-analysis question for a sporadic task system is thus as follows: given the specifications of a sporadic task system determine whether all legal collections of jobs that could be generated by this task system can be scheduled to meet all deadlines.
2. **The run-time scheduling problem:** *Given* a task system (and associated environmental constraints) that is known to be feasible, *determine* a scheduling algorithm that schedules the system to meet all deadlines.

   For sporadic task systems, the requirement once again is that all deadlines be met for all possible collections of jobs that could legally be generated by the sporadic task system under consideration.

28.3 Dynamic-Priority Scheduling

Below, we first summarize the results concerning dynamic-priority scheduling of periodic and sporadic task systems. Then in Sections 28.3.1–28.3.6, we delve into further details about these results.

Dynamic-priority scheduling algorithms place no restrictions upon the manner in which priorities are assigned to individual jobs. Within the context of preemptive uniprocessor scheduling, it has been shown [1,15] that the EDF scheduling algorithm, which at each instant in time chooses for execution the currently-active job with the smallest deadline (with ties broken arbitrarily), is an *optimal* scheduling algorithm for scheduling arbitrary collections of independent real-time jobs in the following sense: If it is possible to preemptively schedule a given collection of independent jobs such that all the jobs meet their deadlines, then the EDF-generated schedule for this collection of jobs will meet all deadlines as well.

Observe that EDF is a dynamic-priority scheduling algorithm. As a consequence of the optimality of EDF for preemptive uniprocessor scheduling, the run-time scheduling problem for preemptive uniprocessor dynamic-priority scheduling is essentially solved (the absence of additional constraints) — EDF is the algorithm of choice, since any feasible task system is guaranteed to be successfully scheduled using EDF.

The feasibility-analysis problem, however, turns out to be somewhat less straightforward. Specifically,

1. Determining whether an arbitrary periodic task system τ is feasible has been shown to be intractable — co-NP-complete in the strong sense. This intractability result holds even if the utilization $U(\tau)$ of the task system τ is known to be bounded from above by an arbitrarily small constant.

2. An exponential-time feasibility test is known, which consists essentially of simulating the behavior of EDF upon the periodic task system for a sufficiently long interval.
3. The special case of *implicit-deadline systems* (recall that these are periodic or sporadic task systems in which ($d_i = p_i$) for all tasks T_i) is, however, tractable: a necessary and sufficient condition for any implicit-deadline system τ to be feasible upon a unit-capacity processor is that $U(\tau) \leq 1$.
4. The special case of *synchronous systems* is also not quite as difficult as the general problem. The computational complexity of the feasibility-analysis problem for synchronous systems is, to our knowledge, still open; for synchronous periodic task systems τ with $U(\tau)$ bounded from above by a constant strictly less than 1, however, a pseudopolynomial time feasibility-analysis algorithm is known.
5. For sporadic task systems, it turns out that system τ is feasible if and only if the synchronous periodic task system τ' comprises exactly as many tasks as τ does, and with each task in τ' having execution-requirement, deadline, and period exactly the same as those of the corresponding task in τ, is feasible. Thus, the feasibility-analysis problem for sporadic task systems turns out to be equivalent to the feasibility-analysis problem for synchronous periodic task systems. That is, the computational complexity of the general problem remains open; for sporadic task systems τ that have $U(\tau)$ bounded from above by a constant strictly less than 1, a pseudopolynomial time feasibility-analysis algorithm is known.

28.3.1 The Optimality of EDF

It has been shown [1,15] that the EDF scheduling algorithm is an optimal preemptive uniprocessor run-time scheduling algorithm not just for scheduling periodic and sporadic task systems, but rather for scheduling arbitrary real-time instances (recall from Definition 28.1 that a real-time instance J is a finite or infinite collection of jobs: $J = \{j_1, j_2, \ldots\}$).

Definition 28.4 (EDF)

The **earliest-deadline first** (EDF) *scheduling algorithm is defined as follows: At each time-instant t schedule the job j active at time-instant t whose deadline parameter is the smallest (ties broken arbitrarily).*

In the remainder of this chapter, let EDF.J denote the schedule generated by EDF upon a real-time collection of jobs J. The following theorem is the formal statement of the optimality of EDF from the perspective of meeting deadlines.

Theorem 28.1

If a real-time instance J can be scheduled to meet all deadlines upon a preemptive uniprocessor, then EDF.J *meets all deadlines of J upon the preemptive uniprocessor.*

Proof

This result is easily proved by induction. Let Δ denote an arbitrarily small positive number. Consider a schedule S for J in which all deadlines are met, and let $[t_o, t_o + \Delta)$ denote the first time interval over which this schedule makes a scheduling decision different from the one made by EDF.J. Suppose that job $j_1 = (a_1, e_1, d_1)$ is scheduled in S over this interval, while job $j_2 = (a_2, e_2, d_2)$ is scheduled in EDF.J. Since S meets all deadlines, it is the case that S schedules j_2 to completion by time-instant d_2. In particular, this implies that S schedules j_2 for an amount equal to Δ prior to d_2. But by definition of the EDF scheduling discipline, $d_2 \leq d_1$; hence, S schedules j_2 for an amount equal to Δ prior to d_1 as well. Now the new schedule S' obtained from S by swapping the executions of j_1 and j_2 of length Δ each would agree with EDF.J over $[0, t + \Delta)$. The proof of optimality of EDF.J now follows by induction on time, with S' playing the role of S in the above argument. □

28.3.2 The Intractability of Feasibility Analysis

Our next result concerns the intractability of feasibility analysis for arbitrary periodic task systems. To obtain this result, we will reduce the *simultaneous congruences problem* (SCP), which has been shown [4,6] to be NP-complete in the strong sense, to the feasibility-analysis problem. The SCP is defined as follows:

Definition 28.5 (Simultaneous Congruences Problem (SCP))

Given *a set* $A = \{(x_1, y_1), (x_2, y_2), \ldots, (x_n, y_n)\}$ *of ordered pairs of positive integers, and an integer* k, $1 \leq k \leq n$. *Determine whether there is a subset* $A' \subseteq A$ *of* k *ordered pairs and a positive integer* z *such that for all* $(x_i, y_i) \in A'$, $z \equiv x_i \bmod y_i$.

Theorem 28.2

The feasibility-analysis problem for arbitrary periodic task systems is co-NP-hard in the strong sense.

Proof

Leung and Merrill [2] reduced SCP to the (complement of the) feasibility-analysis problem for periodic task systems, as follows.

Let $\sigma \stackrel{\text{def}}{=} \langle\{(x_1, y_1), \ldots, (x_n, y_n)\}, k\rangle$ denote an instance of SCP. Consider the periodic task system τ comprised of n tasks $T_1, T_2, \ldots, T_n$, with $T_i = (a_i, e_i, d_i, p_i)$ for all i, $1 \leq i \leq n$. For $1 \leq i \leq n$, let

- $a_i = x_i$,
- $e_i = \frac{1}{k-1}$,
- $d_i = 1$, and
- $p_i = y_i$.

Suppose that $\sigma \in$ SCP. Then there is a positive integer z such that at least k of the ordered pairs (x_i, y_i) "collide" on z: i.e., $z \equiv x_i \bmod y_i$ for at least k distinct i. This implies that the k corresponding periodic tasks will each have a job arrive at time-instant z; since each job's deadline is one unit removed from its arrival time while its execution requirement is $\frac{1}{k-1}$, not all these jobs can meet their deadlines.

Suppose now that τ is *in*feasible, i.e., some deadline is missed in the EDF-generated schedule of the jobs of τ. Let t_o denote the first time-instant at which this happens. Since all task parameters (except the execution requirements) are integers, it must be the case that t_o is an integer. For t_o to represent a deadline miss, there must be at least k jobs arriving at time-instant $t_o - 1$. It must be the case that the ordered pairs (x_i, y_i) corresponding to the tasks generating these k jobs all collide at $(t_o - 1)$, i.e., $(t_o - 1) \equiv x_i \bmod y_i$ for each of these k ordered pairs. Hence, $\sigma \in$ SCP. □

The utilization $U(\tau)$ of the periodic task system constructed in the proof of Theorem 28.2 above is equal to $1/((k-1)\sum_{i=1}^{n} y_i)$. However, observe that the proof would go through essentially unchanged if $d_i = \epsilon$ and $e_i = \frac{\epsilon}{k-1}$, for any positive $\epsilon \leq 1$, and such a task system would have utilization $\epsilon/((k-1)\sum_{i=1}^{n} y_i)$. By choosing ϵ arbitrarily small, this utilization can be made as small as desired, yielding the result that the intractability result of Theorem 28.2 holds even if the utlization of the task system being analyzed is known to be bounded from above by an arbitrarily small constant:

Corollary 28.1

For any positive constant c, *the feasibility analysis problem for periodic task systems* τ *satisfying the constraint that* $U(\tau) < c$ *is co-NP-hard in the strong sense.*

Observe that every periodic task T_i constructed in the proof of Theorem 28.2 above has its deadline parameter d_i no larger than its period parameter p_i. Therefore, the result of Theorem 28.2 holds for the more restricted *constrained-deadline* periodic task systems as well:

Corollary 28.2

The feasibility-analysis problem for constrained-deadline periodic task systems is co-NP-hard in the strong sense.

28.3.3 Feasibility-Analysis Algorithms

The intractability results of Section 28.3.2 above make it unlikely that we will be able to perform feasibility analysis on arbitrary (or even constrained-deadline) periodic task systems in polynomial time. In Section 28.3.3.1 below, we present an exponential-time algorithm for presenting exact feasibility analysis; in Section 28.3.3.2, we outline a *sufficient*, rather than exact, feasibility-analysis test that has a better run-time complexity for periodic task systems with utilization bounded from above — by Corollary 28.1, exact feasibility analysis of such systems is also intractable.

28.3.3.1 An Exponential-Time Feasibility-Analysis Algorithm

In this section we briefly derive an exponential-time algorithm for feasibility analysis of arbitrary periodic task systems; the intractability result of Theorem 28.2 makes it unlikely that we will be able to do much better.

The approach adopted in developing the feasibility-analysis algorithm is as follows:

1. In Definition 28.6 below, we define the **demand** $\eta_\tau(t_1, t_2)$ of a periodic task system τ over an interval $[t_1, t_2)$ to be the cumulative execution requirement by jobs of tasks in the system over the specified interval.
2. In Theorem 28.3 ahead, we reduce the feasibility determination question to an equivalent one concerning the demand function; specifically we prove that τ is infeasible if and only if $\eta_\tau(t_1, t_2)$ exceeds the length of the interval $[t_1, t_2)$ for some t_1 and t_2.
3. In Lemma 28.1 ahead, we obtain a closed-form expression for computing $\eta_\tau(t_1, t_2)$ for any τ, t_1 and t_2. This expression is easily seen to be computable in time linear in the number of tasks in τ.
4. In Lemmas 28.2–28.4 ahead, we derive a series of results that, for any infeasible τ, bound from above the value of *some* t_2 such that $\eta_\tau(t_1, t_2) > (t_2 - t_1)$. This upper bound $B(\tau)$ (given in Equation 28.4 ahead) is linear in the offset and deadline parameters of the tasks in τ as well as the least common multiple of the periods of the tasks in τ, and is hence at most exponential in both the size of the representation of periodic task system τ, and the values of the parameters of τ. The immediate consequence of this is that to determine whether τ is infeasible, we need only check whether $\eta_\tau(t_1, t_2) > (t_2 - t_1)$ for t_1 and t_2 satisfying $0 \le t_1 < t_2 <$ this upper bound.
5. From the optimality of EDF as a preemptive uniprocessor scheduling algorithm, it follows that periodic task system τ is infeasible if and only if EDF misses some deadline while scheduling τ. We extend this observation to show (Lemma 28.5 ahead) that τ is infeasible if and only if the EDF schedule for τ would miss a deadline at or below the bound $B(\tau)$ of Equation (28.4). Since EDF can be implemented to run in polynomial time per scheduling decision, this immediately yields an exponential-time algorithm for feasibility analysis of an arbitrary periodic task system τ: generate the EDF schedule for τ out until the bound $B(\tau)$, and declare τ feasible if and only if no deadlines are missed in this schedule.

Definition 28.6 (demand)

For a periodic task system τ and any two real numbers t_1 and t_2, $t_1 \le t_2$, the **demand** *$\eta_\tau(t_1, t_2)$ of τ over the interval $[t_1, t_2)$ is defined to be cumulative execution requirement by jobs generated by tasks in τ that have arrival times at or after time-instant t_1, and deadlines at or before time-instant t_2.*

Theorem 28.3

Periodic task system τ is feasible if and only if for all t_1, t_2, $t_1 < t_2$, it is the case that $\eta_\tau(t_1, t_2) \le (t_2 - t_1)$.

Proof

if: Suppose that τ is infeasible, and let t_f denote the earliest time-instant at which the EDF schedule for τ misses a deadline. Let $t_b < t_f$ denote the latest time-instant prior to t_f at which this EDF schedule is not executing jobs with deadlines $\leq t_f$, since 0^- is a time-instant prior to t_f at which this EDF schedule is not executing jobs with deadlines $\leq t_f$, time-instant t_b exists and is well-defined. Thus over $[t_b, t_f)$, the EDF schedule for τ executes only jobs arriving at or after t_b, and with deadlines at or before t_f; these are precisely the jobs whose execution requirements contribute to $\eta_\tau(t_b, t_f)$. Since some job nevertheless misses its deadline at t_f, it follows that $\eta_\tau(t_b, t_f) > t_f - t_b$.

Only if: Suppose that τ is feasible, i.e., all jobs of τ complete by their deadlines in some optimal schedule S. For all $t_1, t_2, t_1 < t_2$, it is that case that all jobs generated by τ that both arrive in, and have their deadlines within, the interval $[t_1, t_2)$ are scheduled within this interval (along with, perhaps, parts of other jobs). It therefore follows that $\eta_\tau(t_1, t_2) \leq (t_2 - t_1)$. □

For the remainder of this section let τ denote a periodic task system comprised of n tasks: $\tau = \{T_1, T_2, \ldots, T_n\}$, with $T_i = (a_i, e_i, d_i, p_i)$ for all i, $1 \leq i \leq n$. Let P denote the least common multiple of the periods of the tasks in τ: $P \stackrel{\text{def}}{=} \text{lcm}\{p_1, p_2, \ldots, p_n\}$.

Lemma 28.1

For any t_1, t_2 satisfying $t_1 < t_2$,

$$\eta_\tau(t_1, t_2) = \sum_{i=1}^{n} e_i \max\left(0, \left\lfloor \frac{t_2 - a_i - d_i}{p_i} \right\rfloor - \max\left\{0, \left\lceil \frac{t_1 - a_i}{p_i} \right\rceil\right\} + 1\right) \tag{28.1}$$

Proof

The number of jobs of T_i that lie within the interval $[t_1, t_2)$ is the number of non-negative integers k satisfying the inequalities

$$t_1 \leq a_i + kp_i$$

and

$$a_i + kp_i + d_i \leq t_2$$

The Lemma follows, since there are exactly

$$\max\left(0, \left\lfloor \frac{t_2 - a_i - d_i}{p_i} \right\rfloor - \max\left\{0, \left\lceil \frac{t_1 - a_i}{p_i} \right\rceil\right\} + 1\right)$$

such k's. □

Lemma 28.2

For any t_1, t_2, satisfying $\max_{1 \leq i \leq n}\{a_i\} \leq t_1 < t_2$,

$$\eta_\tau(P + t_1, P + t_2) = \eta_\tau(t_1, t_2) \tag{28.2}$$

Proof

$$\eta_\tau(P + t_1, P + t_2) = \max\left\{0, \left\lfloor \frac{P + t_2 - a_i - d_i}{p_i} \right\rfloor - \max\left\{0, \left\lceil \frac{P + t_1 - a_i}{p_i} \right\rceil\right\} + 1\right\}$$

Since $t_1 \geq a_i$, we may replace $\max\{0, \lceil \frac{P+t_1-a_i}{p_i} \rceil\}$ by $\lceil \frac{t_1+P-a_i}{p_i} \rceil$, and since p_i divides P, we may extract P/p_i from the floor and ceiling terms. Thus,

$$\begin{aligned}
\eta_\tau(P+t_1, P+t_2) &= \max\left\{0, \frac{P}{p_i} + \left\lfloor \frac{t_2 - a_i - d_i}{p_i} \right\rfloor - \frac{P}{p_i} - \left\lceil \frac{t_1 - a_i}{p_i} \right\rceil + 1\right\} \\
&= \max\left\{0, \left\lfloor \frac{t_2 - a_i - d_i}{p_i} \right\rfloor - \left\lceil \frac{t_1 - a_i}{p_i} \right\rceil + 1\right\} \\
&= \eta_\tau(t_1, t_2)
\end{aligned}$$

□

Lemma 28.3

Let $U(\tau) \leq 1$, $t_1 \geq \max_{1 \leq i \leq n}\{a_i\}$, and $t_2 \geq (t_1 + \max_{1 \leq i \leq n}\{d_i\})$. Then

$$(\eta_\tau(t_1, t_2 + P) > t_2 + P - t_1) \Rightarrow (\eta_\tau(t_1, t_2) > t_2 - t_1) \tag{28.3}$$

Proof

$$\begin{aligned}
\eta_\tau(t_1, t_2 + P) &= \sum_{i=1}^{n} e_i \max\left\{0, \left\lfloor \frac{t_2 + P - a_i - d_i}{p_i} \right\rfloor - \max\left\{0, \left\lceil \frac{t_1 - a_i}{p_i} \right\rceil\right\} + 1\right\} \\
&= \sum_{i=1}^{n} e_i \max\left\{0, \frac{P}{p_i} + \left\lfloor \frac{t_2 - a_i - d_i}{p_i} \right\rfloor - \left\lceil \frac{t_1 - a_i}{p_i} \right\rceil + 1\right\}
\end{aligned}$$

as in the proof of Lemma 28.2. Since $P/p_i \geq 0$, we now have

$$\begin{aligned}
\eta_\tau(t_1, t_2 + P) &\leq \sum_{i=1}^{n} \left(e_i \cdot \frac{P}{p_i} + e_i \max\left\{0, \left\lfloor \frac{t_2 - a_i - d_i}{p_i} \right\rfloor - \left\lceil \frac{t_1 - a_i}{p_i} \right\rceil + 1\right\}\right) \\
&\leq P\left(\sum_{i=1}^{n} \frac{e_i}{p_i}\right) + \eta_\tau(t_1, t_2) \\
&\leq P + \eta_\tau(t_1, t_2)
\end{aligned}$$

(The last step follows because $\sum_{i=1}^{n} \frac{e_i}{p_i} \leq 1$.) If we now suppose $\eta_\tau(t_1, t_2 + P) > t_2 + P - t_1$, we have

$$\begin{aligned}
\eta_\tau(t_1, t_2) &\geq \eta_\tau(t_1, t_2 + P) - P \\
&> t_2 + P - t_1 - P \\
&= t_2 - t_1
\end{aligned}$$

□

Lemma 28.4

Suppose that τ is infeasible and $U(\tau) \leq 1$. There exist t_1 and t_2, $0 \leq t_1 < t_2 < B(\tau)$ such that $\eta_\tau(t_1, t_2) > (t_2 - t_1)$, where

$$B(\tau) \stackrel{\text{def}}{=} 2P + \max_{1 \leq i \leq n}\{d_i\} + \max_{1 \leq i \leq n}\{a_i\} \tag{28.4}$$

Proof

Follows directly from the above two lemmas, and Theorem 28.3. □

Lemma 28.5

Suppose that τ is infeasible and $U(\tau) \leq 1$. Let t_1, t_2 be as specified in the statement of Lemma 28.4. The EDF *schedule for τ will miss a deadline at or before t_2.*

Proof

Immediate. □

Given the above lemma as well as the optimality of EDF, we have the following theorem, and the consequent fact that feasibility analysis of an arbitrary periodic task system τ can be performed in exponential time, by generating the EDF schedule of the system over the interval $[0, B(\tau))$.

Theorem 28.4

Periodic task system τ is infeasible if and only if the EDF *schedule for τ misses a deadline at or before time-instant $B(\tau)$, where $B(\tau)$ is as defined in Equation 28.4. Hence, feasibility analysis of periodic task systems can be performed in exponential time.*

28.3.3.2 A More Efficient Sufficient Feasibility-Analysis Algorithm

By Corollary 28.1, exact feasibility analysis is intractable even for arbitrary periodic task systems τ satisfying the constraint that $U(\tau) < c$ for some constant c. However, it turns out that a *sufficient* feasibility test for such systems can be devised, that runs in time pseudopolynomial in the representation of τ. This algorithm is based upon the following lemma, which we state without proof; this lemma relates the feasibility of an arbitrary periodic task system to the feasibility of a (different) *synchronous* periodic task system:

Lemma 28.6

Let $\tau = \{T_1 = (a_1, e_1, d_1, p_1), T_2 = (a_2, e_2, d_2, p_2), \ldots, T_i = (a_i, e_i, d_i, p_i), \ldots, T_n = (a_n, e_n, d_n, p_n)\}$ denote an arbitrary periodic task system. Task system τ is feasible if the synchronous periodic task system $\tau' = \{T_1' = (e_1, d_1, p_1), T_2' = (e_2, d_2, p_2), \ldots, T_i' = (e_i, d_i, p_i), \ldots, T_n' = (e_n, d_n, p_n)\}$ is feasible.

In Section 28.3.5, a pseudopolynomial time algorithm for feasibility analysis of synchronous periodic task systems is presented; this algorithm, in conjunction with the above lemma, immediately yields a sufficient feasibility-analysis algorithm for arbitrary periodic task systems τ satisfying the constraint that $U(\tau) < c$ for some constant c.

28.3.4 Implicit-Deadline Systems

In implicit-deadline periodic and sporadic task systems, all tasks $T = (a, e, d, p)$ have their deadline parameter d equal to their period parameter p.

In the special case of implicit-deadline periodic task systems, the feasibility-analysis problem is quite tractable:

Theorem 28.5

A necessary and sufficient condition for implicit-deadline periodic task system τ to be feasible is $U(\tau) \leq 1$.

Proof

Let τ consist of n implicit-deadline periodic tasks $T_1, T_2, \ldots, T_n$. Consider a "processor-sharing" schedule S obtained by partitioning the time-line into arbitrarily small intervals, and scheduling each task T_i for a fraction $U(T_i)$ of the interval — since $U(\tau) \leq 1$, such a schedule can indeed be constructed. This schedule S schedules each job of T_i for $U(T_i) \times \frac{e_i}{p_i} = e_i$ units between its arrival and its deadline. □

A minor generalization to Theorem 28.5 is possible: $U(\tau) \leq 1$ is a necessary and sufficient feasibility condition for any task system τ in which *all* tasks $T = (a, e, d, p)$ satisfy the condition that $d \geq p$.

28.3.5 Synchronous Periodic Task Systems

In synchronous periodic task systems, the offset of all tasks is equal (i.e., $a_i = a_j$ for all tasks T_i, T_j; without loss of generality, a_i can then be considered equal to zero for all T_i).

Observe that the NP-hardness reduction in the proof of Theorem 28.2 critically depends upon being permitted to assign different values to the offset parameters of the periodic task system being reduced to; hence, this proof does not extend to the special case of synchronous periodic task systems. To our knowledge, the computational complexity of feasibility-determination for synchronous periodic task systems remains open. However, it can be shown that the result of Corollary 28.1 does not hold for synchronous periodic task systems — if the utilization $U(\tau)$ of synchronous periodic task system τ is bounded from above by a positive constant $c < 1$, then the feasibility-analysis problem for τ can be solved in pseudopolynomial time.

The crucial observation concerning the feasibility analysis of synchronous periodic task systems is in the following lemma:

Lemma 28.7

For any synchronous periodic task system τ

$$\forall\, t_1, t_2, \quad 0 \leq t_1 < t_2, \quad (\eta_\tau(t_1, t_2) \leq \eta_\tau(0, t_2 - t_1))$$

That is, the demand by jobs of synchronous task system τ over any interval of a given length is maximized when the interval starts at time-instant zero.

Proof

We omit a formal proof of this lemma, and instead provide some intuition as to why it should be true. It is easy to observe that the demand — i.e., the cumulative execution requirement — of any particular periodic task over an interval of length ℓ is maximized if a job of the task arrives at the start of the interval, since this (intuitively speaking) permits the largest number of future jobs of the task to have their deadlines within the interval. In a synchronous task system, however, time-instant zero is one at which all the tasks have a job arrive; i.e., the interval $[0, \ell)$ is the one of length ℓ with maximum demand for each individual task, and consequently the one with maximum demand for task system τ. □

As a consequence of Lemma 28.7, we may "specialize" Theorem 28.3 for synchronous periodic task systems, by restricting t_1 to always be equal to zero:

Theorem 28.6 (Theorem 28.3, for synchronous task systems)

Synchronous periodic task system τ is feasible if and only if for all t_o, $0 < t_o$, it is the case that $\eta_\tau(0, t_o) \leq t_o$.

Proof

By Theorem 28.3, τ is feasible if and only if for all $t_1, t_2, t_1 < t_2$, it is the case that $\eta_\tau(t_1, t_2) \leq (t_2 - t_1)$. But by Lemma 28.7 above, $\eta_\tau(t_1, t_2) < \eta_\tau(0, t_2 - t_1)$. The theorem follows, by setting $t_o = t_2 - t_1$. □

Despite the statement of Theorem 28.6 (which may seem to indicate that feasibility analysis of synchronous systems may not be as intractable as for arbitrary systems) there is no synchronous feasibility-analysis algorithm known that guarantees significantly better performance than the exponential-time algorithm implied by Lemma 28.5. For the special case when the utilization of a synchronous periodic task system is *a priori* known to be bounded from above by a positive constant strictly less than one, however, Lemma 28.8 below indicates that pseudopolynomial time feasibility anaysis is possible (contrast this to the case of arbitrary periodic task systems, where Corollary 28.1 rules out the existence of pseudopolynomial time feasibility analysis unless $P = NP$).

Lemma 28.8

Let c be a fixed constant, $0 < c < 1$. Suppose that τ is infeasible and $U(\tau) \leq c$. There exists a t_2 satisfying

$$0 < t_2 < \frac{c}{1-c} \max\{p_i - d_i\}$$

such that $\eta_\tau(0, t_2) > t_2$.

Proof

Suppose that $\eta_\tau(0, t_2) > t_2$. Let $I = \{i \mid d_i \leq t_2\}$. Substituting $a_i = 0$ for all i, $1 \leq i \leq n$, and $t_1 = 0$ in Lemma 28.1, we have

$$\begin{aligned} t_2 &< \eta_\tau(0, t_2) \\ &= \sum_{i=1}^{n} e_i \max\left\{0, \left\lfloor \frac{t_2 - d_i}{p_i} \right\rfloor + 1\right\} \\ &= \sum_{i \in I} \left(\left\lfloor \frac{t_2 - d_i}{p_i} \right\rfloor + 1 \right) e_i \\ &\leq \sum_{i \in I} \frac{t_2 - d_i + p_i}{p_i} e_i \\ &= \sum_{i \in I} \left(\frac{t_2 e_i}{p_i} + \frac{(p_i - d_i)\, e_i}{p_i} \right) \\ &\leq c\, t_2 + c \max\{p_i - d_i\} \end{aligned}$$

Solving for t_2, we get

$$t_2 < \frac{c}{1-c} \max\{p_i - d_i\}$$

□

For constant c, observe that $\frac{c}{1-c} \max\{p_i - d_i\}$ is pseudopolynomial in the parameters of the task system τ. A pseudopolynomial time feasibility analysis algorithm for any synchronous periodic task system τ with bounded utilization now suggests itself, in the manner of the algorithm implied by Lemma 28.5. Simply generate the EDF schedule for τ out until $\frac{c}{1-c} \max\{p_i - d_i\}$, and declare τ feasible if and only if no deadlines are missed in this schedule.

Theorem 28.7

Feasibility analysis of synchronous periodic task systems with utilization bounded by a constant strictly less than one can be performed in time pseudopolynomial in the representation of the system.

28.3.6 Sporadic Task Systems

Each periodic task system generates exactly one (infinite) collection of real-time jobs. Since the sporadic task model constrains the *minimum*, rather than the exact, interarrival separation between successive jobs of the same task, a sporadic task system, on the other hand, is legally permitted to generate infinitely many distinct collections of real-time jobs. This is formalized in the notion of a *legal sequence of job arrivals* of sporadic task systems:

Definition 28.7 (Legal sequence of job arrivals)

*Let τ denote a sporadic task system comprised of n sporadic tasks: $\tau = \{T_1, T_2, \ldots, T_n\}$, with $t_i = (e_i, d_i, p_i)$ for all $i, 1 \leq i \leq n$. A job arrival of τ is an ordered pair (i, t), where $1 \leq i \leq n$ and t is a positive real number, indicating that a job of task T_i arrives at time-instant t (and has execution requirement e_i and deadline at $t + d_i$). A **legal sequence of job arrivals of** τ R_τ is a (possibly infinite) list of job arrivals of τ satisfying the following property: if (i, t_1) and (i, t_2) both belong to R_τ, then $|t_2 - t_1| \geq p_i$.*

The feasibility-analysis question for a sporadic task system is as follows: given the specifications of a sporadic task system τ, determine whether there is any legal sequence of job arrivals of τ which cannot be scheduled to meet all deadlines. (Constrast this to the periodic case, in which *feasibility for a periodic task system* meant exactly the same thing as *feasibility* — the ability to schedule to meet all deadlines — *for the* unique *legal sequence of job arrivals* that the periodic task system generates. In a sense, therefore, the term "feasibility" for a periodic task system is merely shorthand for feasibility of a specific infinite sequence of jobs, while "feasibility" for a sporadic task system means something more — feasibility for infinitely many distinct infinite sequences of jobs.)

For periodic task systems the set of jobs to be scheduled is known *a priori*, during feasibility analysis. For sporadic task systems, however, there is no *a priori* knowledge about which set of jobs will be generated by the task system during run-time. In analyzing sporadic task systems, therefore, every conceivable sequence of possible requests must be considered. Fortunately, it turns out that, at least in the context of dynamic-priority preemptive uniprocessor scheduling, it is relatively easy to identify a unique "worst-case" legal sequence of job arrivals, such that all legal sequences of job arrivals can be scheduled to meet all deadlines if and only if this worst-case legal sequence can. And, this particular worst-case legal sequence of job arrivals is exactly the unique legal sequence of job arrivals generated by the synchronous periodic task system with the exact same parameters as the sporadic task system:

Lemma 28.9

Let $\tau = \{T_1 = (e_1, d_1, p_1), T_2 = (e_2, d_2, p_2), \ldots, T_i = (e_i, d_i, p_i), \ldots, T_n = (e_n, d_n, p_n)\}$ denote an arbitrary sporadic task system. Every legal sequence of jobs' arrivals R_τ of τ can be scheduled to meet all deadlines if and only if the synchronous periodic task system $\tau' = \{T'_1 = (e_1, d_1, p_1), T'_2 = (e_2, d_2, p_2), \ldots, T'_i = (e_i, d_i, p_i), \ldots, T'_n = (e_n, d_n, p_n)\}$ is feasible.

Proof

We will not prove this lemma formally here; the interested reader is referred to [12] for a complete proof. The main ideas behind the proof are the following:

1. Sporadic task system τ is infeasible if and only if there is some legal sequence of job arrivals R_τ and some interval $[t, t + t_o)$ such that the cumulative execution requirement of job arrivals in R_τ that both have their execution requirement and deadlines within the interval exceeds t_o, the length of the interval.
2. The cumulative execution requirement by jobs generated by sporadic task system τ over an interval of length t_o is maximized if each task in τ generates a job at the start of the interval, and then generates successive jobs as rapidly as legal (i.e., each task T_i generates jobs exactly p_i time-units apart).
3. But this is exactly the sequence of jobs that would be generated by the synchronous periodic task system τ' defined in the statement of the lemma.

□

Lemma 28.9 above reduces the feasibility-analysis problem for sporadic task systems to the feasibility-analysis problem for synchronous periodic task systems. The following two results immediately follow, from Theorem 28.4 and Theorem 28.7, respectively:

Theorem 28.8

Sporadic task system τ is infeasible if and only if the EDF *schedule for τ misses a deadline at or before time-instant $2P$, where P denotes the least common multiple of the periods of the tasks in τ.*[2] *Hence, feasibility analysis of sporadic task systems can be performed in exponential time.*

Theorem 28.9

Feasibility analysis of sporadic task systems with utilization bounded by a constant strictly less than one can be performed in time pseudopolynomial in the representation of the system.

28.4 Static-Priority Scheduling

Below, we first summarize the main results concerning dynamic-priority scheduling of periodic and sporadic task systems. Then in Sections 28.4.2.1–28.4.4, we provide further details about some of these results.

Recall that the *run-time scheduling problem* — the problem of choosing an appropriate scheduling algorithm — was rendered trivial for all the task models we had considered in the dynamic-priority case due to the proven optimality of EDF as a dynamic-priority run-time scheduling algorithm. Unfortunately, there is no static-priority result analogous to this result concerning the optimality of EDF; hence, the run-time scheduling problem is quite nontrivial for static-priority scheduling.

In the static-priority scheduling of periodic and sporadic task systems, all the jobs generated by an individual task are required to be assigned the same priority, which should be different from the priorities assigned to jobs generated by other tasks in the system. Hence, the run-time scheduling problem essentially reduces to the problem of associating a unique priority with each task in the system. The specific results known are as follows:

1. For *implicit-deadline* sporadic and synchronous periodic task systems, the rate monotonic (RM) priority assignment algorithm, which assigns priorities to tasks in inverse proportion to their period parameters with ties broken arbitrarily, is an optimal priority assignment. That is, if there is any static priority assignment that would result in such a task system always meeting all deadlines, then the RM priority assignment for this task system, which assigns higher priorities to jobs generated by tasks with smaller values of the period parameter, will also result in all deadlines always being met.
2. For implicit-deadline periodic task systems that are not synchronous, however, RM is provably not an optimal priority-assignment scheme (Section 28.4.3.2).
3. For *constrained-deadline* sporadic and synchronous periodic task systems, the deadline monotonic (DM) priority assignment algorithm, which assigns priorities to tasks in inverse proportion to their deadline parameters with ties broken arbitrarily, is an optimal priority assignment. (Observe that RM priority assignment is a special case of DM priority assignment.)
4. For constrained-deadline (and hence also arbitrary) periodic task systems which are not necessarily synchronous, however, the computational complexity of determining an optimal priority assignment remains open. That is, while it is known (see below) that determining whether a constrained-deadline periodic task system is static-priority feasible is co-NP-complete in the strong sense, it is unknown whether this computational complexity is due to the process of assigning priorities, or merely to validating whether a given priority-assignment results in all deadlines being met. In other words, the computational complexity of the following question remains open [4, p. 247]: *Given a constrained-deadline periodic task system τ that is known to be static-priority feasible, determine an optimal priority assignment for the tasks in τ.*

---

[2] Although this does not follow directly from Theorem 28.4, this $2P$ bound can in fact be improved to P.

Feasibility Analysis

Determining whether an arbitrary periodic task system τ is feasible has been shown to be intractable — co-NP-complete in the strong sense. This intractability result holds even if $U(\tau)$ is known to be bounded from above by an arbitrarily small constant.

Utilization-Based Feasibility Analysis

For the special case of *implicit-deadline* periodic and sporadic task systems (recall from above that rate-monotonic priority assignment is an optimal priority-assignment scheme for such task systems), a simple sufficient utilization-based feasibility test is known: an implicit-deadline periodic or sporadic task system τ is static-priority feasible if its utilization $U(\tau)$ is at most $n(2^{1/n} - 1)$, where n denotes the number of tasks in τ. Since $n(2^{1/n} - 1)$ monotonically decreases with increasing n and approaches $\ln 2$ as $n \to \infty$, it follows that any implicit-deadline periodic or sporadic task system τ satisfying $U(\tau) \leq \ln 2$ is static-priority feasible upon a preemptive uniprocessor, and hence can be scheduled using rate-monotonic priority assignment.[3]

This utilization bound is a sufficient, rather than exact, feasibility-analysis test: It is quite possible that an implicit-deadline task system τ with $U(\tau)$ exceeding the bound above be static-priority feasible (as a special case of some interest, it is known that any implicit-deadline periodic or sporadic task system in which the periods are harmonic — i.e., for every pair of periods p_i and p_j in the task system it is either the case that p_i is an integer multiple of p_j or p_j is an integer multiple of p_i — is static-priority feasible if and only if its utilization is at most one).

Nevertheless, this is the best possible test using the utilization of the task system, and the number of tasks in the system, as the sole determinants of feasibility. That is, it has been shown that $n(2^{1/n} - 1)$ is the best possible utilization bound for feasibility-analysis of implicit-deadline periodic and sporadic task systems, in the following sense: For all $n \geq 1$, there is an implicit-deadline periodic task system τ with $U(\tau) = n(2^{1/n} - 1) + \epsilon$ that is not static-priority feasible, for ϵ an arbitrarily small positive real number.

28.4.1 Some Preliminary Results

In this section, we present some technical results concerning static-priority scheduling that will be used later, primarily in Sections 28.4.3 and 28.4.4 when the rate-monotonic and deadline-monotonic priority assignments are discussed.

For the remainder of this section, we will consider the static-priority scheduling of a periodic/sporadic task system τ comprised of n tasks: $\tau = \{T_1, T_2, \ldots, T_n\}$. We use the notation $T_i \succ T_j$ to indicate that task T_i is assigned a higher priority than task T_j in the (static) priority-assignment scheme under consideration.

The **response time** of a job in a particular schedule is defined to be the amount of time that has elapsed between the arrival of the job and its completion; clearly, in order for a schedule to meet all deadines it is necessary that the response time of each job not exceed the relative deadline of the task that generates the job. The following definition is from [16, p. 131]

Definition 28.8 (critical instant)

A critical instant of a task T_i is a time-instant which is such that

1. *a job of T_i released at the instant has a maximum response time of all jobs of T_i, if the response time of every job of T_i is less than or equal to the relative deadline of T_i, and*
2. *the response time of the job of T_i released at the instant is greater than the relative deadline if the response time of some job of T_i exceeds the relative deadline.*

The response time of a job of T_i is maximized when it is released at its critical instant.

[3]Here, ln 2 denotes the natural logarithm of 2 (approximately 0.6931).

The following lemma asserts that for synchronous task systems in which the deadline of all tasks are no larger than their periods, a critical instant for all tasks $T_i \in \tau$ occurs at time-instant zero (i.e., when each task in τ simultaneously releases a job). While this theorem is intuitively appealing — it is reasonable that a job will be delayed the most when it arrives simultaneous with a job from each higher-priority task, and each such higher-priority task generates successive jobs as rapidly as permitted — the proof turns out to be quite nontrivial and long. We will not present the proof here; the interested reader is referred to a good text-book on real-time systems (e.g., [16,17]) for details.

Lemma 28.10

Let $\tau = \{T_1, T_2, \ldots, T_n\}$ be a synchronous periodic task system with constrained deadlines (i.e., with $d_i \leq p_i$ for all $i, 1 \leq i \leq n$). When scheduled using a static-priority scheduler under static priority assignment $T_1 \succ T_2 \succ \cdots \succ T_n$, the response time of the first job of task T_i is the largest among all the jobs of task T_i.

It was proven [18] that the restriction that τ be comprised of *constrained-deadline* tasks is necessary to the correctness of Lemma 28.10, i.e., Lemma 28.10 does not extend to sporadic or synchronous periodic task systems, in which the deadline parameter of tasks may exceed their period parameter.

A schedule is said to be *work-conserving* if it never idles the processor while there is an active job awaiting execution.

Lemma 28.11

Let τ denote a periodic task system, and S_1 and S_2 denote work-conserving schedules for τ. Schedule S_1 idles the processor at time-instant t if and only if schedule S_2 idles the processor at time-instant t, for all $t \geq 0$.

Proof

The proof is by induction: we assume that schedules S_1 and S_2 both idle the processor at time-instant t_o, and that they both idle the processor at the same time-instants at all time-instants prior to t_o. The base case has $t_o = 0$.

For the inductive step, let t_1 denote the first instant after t_o at which either schedule idles the processor. Assume without loss of generality that schedule S_1 idles the processor over $[t_1, t_2)$. Since S_1 is work-conserving, this implies that all jobs that arrived prior to t_2 have completed over $[t_1, t_2)$, i.e., the cumulative execution requirement of jobs of τ arriving prior to t_2 is equal to $(t_1 - t_o)$. But since S_2 is also work-conserving, this would imply that S_2 also idles the processor over $[t_1, t_2)$. □

28.4.2 The Feasibility-Analysis Problem

In Section 28.4.2.1 below, we show that the static-priority feasibility-analysis problem is intractable for arbitrary periodic task systems. We also show that the problem of determining whether a *specific* priority assignment results in all deadlines being met is intractable, even for the special case of implicit-deadline periodic task systems. However, all these intractability results require asynchronicity: for *synchronous* task systems, we will see (Section 28.4.2.2) that static-priority feasibility analysis is no longer quite as computationally expensive.

28.4.2.1 The Intractability of Feasibility Analysis

Theorem 28.10

The static-priority feasibility-analysis problem for arbitrary periodic task systems is co-NP-hard in the strong sense.

Proof

Leung and Whitehead [4] reduced SCP to the complement of the static-priority feasibility-analysis problem for periodic task systems, as follows. (This transformation is identical to the one used in the proof of Theorem 28.2.)

Let $\sigma \stackrel{\text{def}}{=} \langle\{(x_1, y_1), \ldots, (x_n, y_n)\}, k\rangle$ denote an instance of SCP. Consider the periodic task system τ comprised of n tasks $T_1, T_2, \ldots, T_n$, with $T_i = (a_i, e_i, d_i, p_i)$ for all $i, 1 \leq i \leq n$. For $1 \leq i \leq n$, let

- $a_i = x_i$,
- $e_i = \frac{1}{k-1}$,
- $d_i = 1$, and
- $p_i = y_i$.

Suppose that $\sigma \in$ SCP. Then there is a positive integer z such that at least k of the ordered pairs (x_i, y_i) "collide" on z: i.e., $z \equiv x_i \bmod y_i$ for at least k distinct i. This implies that the k corresponding periodic tasks will each have a job arrive at time-instant z; since each job's deadline is one unit removed from its arrival time while its execution requirement is $\frac{1}{k-1}$, not all these jobs can meet their deadlines regardless of the priority assignment.

Suppose now that $\sigma \notin$ SCP. That is, for no positive integer w is it the case that k or more of the ordered pairs (x_i, y_i) collide on w. This implies that at no time-instant will k periodic tasks each have a job arrive at that instant; since each job's deadline is one unit removed from its arrival time while its execution requirement is $1/(k-1)$, all deadlines will be met with any of the $n$possible priority assignments. □

Observe that every periodic task T_i constructed in the proof of Theorem 28.10 above has its deadline parameter d_i no larger than its period parameter p_i. Therefore, the result of Theorem 28.10 holds for constrained-deadline periodic task systems as well:

Corollary 28.3

The static-priority feasibility-analysis problem for constrained-deadline periodic task systems is co-NP-hard in the strong sense.

Theorem 28.10 above does not address the question of whether the computational complexity of static-priority feasibility-analysis problem for arbitrary periodic task systems arises due to the complexity of (i) determining a suitable priority assignment, or (ii) determining whether this priority-assignment results in all deadlines being met. The following result asserts that the second question above is in itself intractable; however, the computational complexity of the first question above remains open.

Theorem 28.11

Given an implicit-deadline periodic task system τ and a priority assignment on the tasks in τ, it is co-NP-hard in the strong sense to determine whether the schedule generated by a static-priority scheduler using these priority assignments meets all deadlines. (Since constrained-deadline and arbitrary periodic task systems are generalizations of implicit-deadline periodic task systems, this hardness result holds for constrained-deadline and arbitrary periodic task systems as well.)

Proof

This proof, too, is from [4].

Let $\sigma \stackrel{\text{def}}{=} \langle\{(x_1, y_1), \ldots, (x_n, y_n)\}, k\rangle$ denote an instance of SCP. Consider the periodic task system τ comprised of $n+1$ tasks $T_1, T_2, \ldots, T_n, T_{n+1}$, with $T_i = (a_i, e_i, d_i, p_i)$ for all $i, 1 \leq i \leq n+1$. For $1 \leq i \leq n$, let

- $a_i = x_i$,
- $e_i = \frac{1}{k}$, and
- $d_i = p_i = 1$.

Let $T_{n+1} = (0, \frac{1}{k}, 1, 1)$. The priority-assignment is according to task indices, i.e., $T_i \succ T_{i+1}$ for all $i, 1 \leq i \leq n$. Specifically, T_{n+1} has the lowest priority. We leave it to the reader to verify that all jobs of task T_{n+1} meet their deadlines if and only if $\sigma \notin$ SCP. □

28.4.2.2 More Tractable Special Cases

Observe that the NP-hardness reduction in the proof of Theorem 28.10 critically depends upon the fact that different periodic tasks are permitted to have different offsets; hence, this proof does not hold for sporadic or for synchronous periodic task systems. In fact, feasibility analysis is known to be more tractable for sporadic and synchronous periodic task systems in which all tasks have their deadline parameters no larger than their periods (i.e., deadline-constrained and implicit-deadline task systems).

Theorem 28.12

The static-priority feasibility-analysis problem for synchronous constrained-deadline (and implicit-deadline) periodic task systems can be solved in time pseudopolynomial in the representation of the task system.

Proof

In Sections 28.4.3 and 28.4.4, we will see that the *priority-assignment problem* — determining an assignment of priorities to the tasks of a static-priority feasible task system such that all deadlines is met — has efficient solutions for implicit-deadline and constrained-deadline task systems. By Lemma 28.10, we can verify that such a synchronous periodic task system is feasible by ensuring that each task meets its first deadline; i.e., by generating the static-priority schedule under the "optimal" priority assignment out until the largest deadline of any task. □

28.4.3 The Rate-Monotonic Scheduler

The rate-monotonic priority assignment was defined by Liu and Layland [1] and Serlin [19], for sporadic and synchronous periodic implicit-deadline task systems. That is, each task T_i in task system τ is assumed characterized of two parameters: execution requirement e_i and period p_i. The RM priority-assignment scheme assigns tasks priorities in inverse proportion to their period parameter (equivalently, in direct proportion to their *rate* parameter — hence the name), with ties broken arbitrarily.

Computing the priorities of a set of n tasks for the rate monotonic priority rule amounts to ordering the task set according to their periods. Hence the time complexity of the rate-monotonic priority assignment is the time complexity of a sorting algorithm, i.e., $\mathcal{O}(n \log n)$.

28.4.3.1 Optimality for Sporadic and Synchronous Periodic Implicit-Deadline Task Systems

Theorem 28.13

Rate-monotonic priority assignment is optimal for sporadic and synchronous periodic task systems with implicit deadlines.

Proof

Let $\tau = \{T_1, T_2, \ldots, T_n\}$ denote a sporadic or synchronous periodic task system with implicit deadlines. We must prove that if a static priority assignment would result in a schedule for τ in which all deadlines are met, then a rate-monotonic priority assignment for τ would also result in a schedule for τ in which all deadlines are met.

Suppose that priority assignment ($T_1 \succ T_2 \succ \cdots \succ T_n$) results in such a schedule. Let T_i and T_{i+1} denote two tasks of adjacent priorities with $p_i \geq p_{i+1}$. Let us exchange the priorities of T_i and T_{i+1}: if the priority assignment obtained after this exchange results in all deadlines being met, then we may conclude that any rate-monotonic priority assignment will also result in all deadlines being met since any rate-monotonic priority assignment can be obtained from any priority ordering by a sequence of such priority exchanges.

To see that the priority assignment obtained after this exchange results in all deadlines being met, observe the following:

1. The priority exchange does not modify the schedulability of the tasks with a higher priority than T_i (i.e., $T_1, \ldots, T_{i-1}$).
2. The task T_{i+1} remains schedulable after the priority exchange, since its jobs may use all the free time-slots left by $\{T_1, T_2, \ldots, T_{i-1}\}$ instead of only those left by $\{T_1, T_2, \ldots, T_{i-1}, T_i\}$.
3. Assuming that the jobs of T_i remain schedulable (this will be proved below), by Lemma 28.11 above we may conclude that the scheduling of each task T_k, for $k = i+2, i+3, \ldots, n$ is not altered since the idle periods left by higher priority tasks are the same.
4. Hence we need only verify that T_i also remains schedulable. From Lemma 28.10 we can restrict our attention to the first job of task T_i. Let r_{i+1} denote the response time of the first job of T_{i+1} before the priority exchange: the feasibility implies $r_{i+1} \leq p_{i+1}$. During the interval $[0, r_{i+1})$ the processor (when left free by higher priority tasks) is assigned first to the (first) job of T_i and then to the (first) job of T_{i+1}; the latter is not interrupted by subsequent jobs of T_i since $p_i > p_{i+1} \geq r_{i+1}$. Hence, after the priority exchange, the processor allocation is exchanged between T_i and T_{i+1}, and it follows that T_i ends its computation at time r_{i+1} and meets its deadline since $r_{i+1} \leq p_{i+1} \leq p_i$. □

As a consequence of Theorem 28.13 and Lemma 28.10, it follows that synchronous implicit-deadline periodic task system τ is static-priority feasible if and only if the first job of each task in τ meets its deadline when priorities are assigned in rate-monotonic order. Since this can be determined by simulating the schedule out until the largest period of any task in τ, we have the following corollary:

Corollary 28.4

Feasibility analysis of sporadic and synchronous periodic implicit-deadline task systems can be done in pseudopolynomial time.

28.4.3.2 Nonoptimality for Asynchronous Periodic Implicit-Deadline Task Systems

If all the implicit-deadline periodic tasks are not required to have the same initial offset, however, RM priority assignment is no longer an optimal priority-assignment scheme. This can be seen by the following example task system [20]:

$$\tau = \{T_1 = (0, 7, 10), T_2 = (4, 3, 15), T_3 = (0, 1, 16)\}$$

The RM priority assignment ($T_1 \succ T_2 \succ T_3$) results in the first deadline of T_3 being missed, at time-instant 16. However, the priority assignment ($T_1 \succ T_3 \succ T_2$) does not result in any missed deadlines: this has been verified [20] by constructing the schedule over the time-interval $[0, 484)$, during which no deadlines are missed, and observing that the state of the system at time-instant 4 is identical to the state at time-instant 484.[4]

[4] Leung and Whitehead [4, Theorem 3.5] proved that any constrained-deadline periodic task system τ meets all deadlines in a schedule constructed under a given priority assignment if and only if it meets all deadlines over the interval $(a, a + 2P]$, where a denotes the largest offset of any task in τ, and P the least common multiple of the periods of all the tasks in τ. Hence, it suffices to test this schedule out until $4 + 240 \times 2 = 484$.

28.4.3.3 Utilization Bound

In this section, we restrict our attention to sporadic or synchronous periodic implicit-deadline periodic task systems — hence unless explicitly stated otherwise, all task systems are either sporadic or synchronous periodic, and implicit-deadline.

Definition 28.9

Within the context of a particular scheduling algorithm, task system $\tau = \{T_1, \ldots, T_n\}$ is said to fully utilize the processor if all deadlines of τ are met when τ is scheduled using this scheduling algorithm, and an increase in the execution requirement of any T_i $(1 \leq i \leq n)$ results in some deadline being missed.

Theorem 28.14

When scheduled using the rate-monotonic scheduling algorithm, task system $\tau = \{T_1, \ldots, T_n\}$ fully utilizes the processor if and only if the task system $\{T_1, \ldots, T_{n-1}\}$

1. *meets all deadlines, and*
2. *idles the processor for exactly e_n time units over the interval $[0, p_n)$,*

when scheduled using the rate-monotonic scheduling algorithm.

Proof

Immediately follows from Lemma 28.10. □

Let b_n denote the lower bound of $U(\tau)$ among all task systems τ comprised of exactly n tasks which fully utilize the processor under rate-monotonic scheduling.

Lemma 28.12

For the subclass of task systems satisfying the constraint the the ratio between the periods of any two tasks is less than 2, $b_n = n(\sqrt[n]{2} - 1)$.

Proof

Let $\tau = \{T_1, \ldots, T_n\}$ denote an n-task task system, and assume that $p_1 \leq p_2 \leq \cdots \leq p_n$. We proceed in several stages.

Stage 1. We first show that, in the computation of b_n, we may restrict our attention to task systems τ fully utilizing the processor such that $\forall i < n: \ e_i \leq p_{i+1} - p_i$.

Suppose that τ is a task system that fully utilizes the processor and has the smallest utilization from among all task systems that fully utilize the processor. Consider first the case of e_1 and suppose that $e_1 = p_2 - p_1 + \Delta$ ($\Delta > 0$; notice that we must have that $p_2 < 2p_1$, otherwise $e_1 > p_1$ and the task set is not schedulable). Notice that the task system $\tau' = \{T_1', \ldots, T_n'\}$ with $p_i' = p_i \ \forall i$ and $e_1' = p_2 - p_1$, $e_2' = e_2 + \Delta$, $e_3' = e_3, \ldots, e_n' = e_n$ also fully utilizes the processor. Furthermore, $U(\tau) - U(\tau') = \frac{\Delta}{p_1} - \frac{\Delta}{p_2} > 0$, contradicting our hypothesis that the utilization of $T_1, \ldots, T_n$ is minimal.

The above argument can now be repeated for $e_2, e_3, \ldots, e_{n-1}$; in each case, it may be concluded that $e_i \leq p_{i+1} - p_i$.

Stage 2. Next, we show that, in the computation of b_n, we may restrict our attention to task systems τ fully utilizing the processor such that $\forall i < n: \ e_i = p_{i+1} - p_i$.

It follows from Theorem 28.14, the fact that each task T_i with $i < n$ releases and completes exactly two jobs prior to time-instant p_n (this is a consequence of our previously derived constraint that $e_i \leq p_{i+1} - p_i$ for all $i < n$) that

$$e_n = p_n - 2\sum_{i=1}^{n-1} e_i$$

this is since the first $n-1$ tasks meet all deadlines, and over the interval $[0, p_n)$ they together use $\sum_{i=1}^{n-1} e_i$ time units, with $\sum_{i=1}^{n-1} e_i \leq p_n - p_1 < p_1$).

Consider first the case of e_1 and suppose that $e_1 = p_2 - p_1 - \Delta$ $(\Delta > 0)$. Notice that the task system $\tau'' = \{T_1'', \ldots, T_n''\}$ with $p_i'' = p_i\ \forall i$ and $e_1'' = e_1 + \Delta = p_2 - p_1, e_n'' = e_n - 2\Delta, e_i'' = e_i$ for $i = 2, 3, \ldots, n-1$, also fully utilizes the processor. Furthermore, $U(\tau) - U(\tau'') = -\frac{\Delta}{p_1} + \frac{2\Delta}{p_n} > 0$, contradicting our hypothesis that the utilization of $T_1, \ldots, T_n$ is minimal.

The above argument can now be repeated for $e_2, e_3, \ldots, e_{n-1}$; in each case, it may be concluded that $e_i = p_{i+1} - p_i$.

Stage 3. Finally, let $g_i \stackrel{\text{def}}{=} \frac{p_n - p_i}{p_i}$ $(i = 1, \ldots, n-1)$; we get

$$U(\tau) = \sum_{i=1}^{n} \frac{e_i}{p_i} = 1 + g_1\left(\frac{g_1 - 1}{g_1 + 1}\right) + \sum_{i=2}^{n-1} g_i\left(\frac{g_i - g_{i-1}}{g_i + 1}\right)$$

This expression must be minimzed; hence

$$\frac{\partial U(\tau)}{\partial g_j} = \frac{g_j^2 + 2g_j - g_{j_1}}{g_j + 1^2} - \frac{g_{j+1}}{g_{j+1} + 1} = 0, \quad \text{for } j = 1, \ldots, n-1$$

The general solution for this can be shown to be $g_j = 2^{\frac{n-j}{n}} - 1 (j = 1, \ldots, n-1)$, from which it follows that $b_n = n(\sqrt[n]{2} - 1)$. □

The restriction that the ratio between task periods is less than 2 can now be relaxed, yielding the desired utilization bound.

Theorem 28.15 [1]

Any implicit-deadline sporadic or synchronous periodic task system τ comprising of n tasks is successfully scheduled using static-priority scheduling with the rate-monotonic priority assignment, provided

$$U(\tau) \leq n(\sqrt[n]{2} - 1)$$

Proof

Let τ denote a system of n tasks that fully utilizes the processor. Suppose that for some i, $\lfloor \frac{p_n}{p_i} \rfloor > 1$, i.e., there exists an integer $q > 1$ such that $p_n = q \cdot p_i + r, r \geq 0$. Let us obtain task system τ' from task system τ by replacing the task T_i in τ by a task T_i' such that $p_i' = q \cdot p_i$ and $e_i' = e_i$, and increase e_n by the amount needed to again fully utilize the processor. This increase is at most $e_i(q-1)$, the time within the execution of T_n occupied by T_i but not by T_i' (it may be less than e_i if some slots left by T_i' are used by some T_j with $i < j < n$). We have

$$U(\tau') \leq U(\tau) - \frac{e_i}{p_i} + \frac{e_i}{p_i'} + \left[(q-1)\frac{e_i}{p_n}\right]$$

i.e.,

$$U(\tau') \leq U(\tau) + e_i(q-1)\left[\frac{1}{qp_i + r} - \frac{1}{qp_i}\right]$$

Since $q - 1 > 0$ and

$$\frac{1}{q \cdot p_i + r} - \frac{1}{q \cdot p_i} \leq 0, \quad U(\tau') \leq U(\tau)$$

By repeated applications of the above argument, we can obtain a τ'' in which no two task periods have a ratio greater than two, such that τ'' fully utilizes the processor and has utilization no greater than $U(\tau)$. That is, the bound b_n derived in Lemma 28.12 represents a lower bound on $U(\tau)$ among *all* task systems τ comprised of exactly n tasks which fully utilize the processor under rate-monotonic scheduling, and not just those task systems in which the ratio of any two periods is less than 2.

To complete the proof of the theorem, it remains to show that if a system of n tasks has an utilization factor less than the upper bound b_n, then the system is schedulable. This immediately follows from the above arguments, and the fact that b_n is strictly decreasing with increasing n. □

28.4.4 The Deadline-Monotonic Scheduler

Leung and Whitehead [4] have defined the deadline-monotonic priority assignment (also termed the *inverse-deadline* priority assignment): priorities assigned to tasks are inversely proportional to the deadline. It may be noticed that in the special case where $d_i = p_i (1 \leq i \leq n)$, the deadline-monotonic assignment is equivalent to the rate-monotonic priority assignment.

28.4.4.1 Optimality for Sporadic and Synchronous Periodic Constrained-Deadline Task Systems

Theorem 28.16

The deadline-monotonic priority assignment is optimal for sporadic and synchronous periodic task systems with constrained deadlines.

Proof

Let $\tau = \{T_1, T_2, \ldots, T_n\}$ denote a sporadic or synchronous periodic task system with constrained deadlines. We must prove that if a static priority assignment would result in a schedule for τ in which all deadlines are met, then a deadline-monotonic priority assignment for τ would also result in a schedule for τ in which all deadlines are met.

Suppose that priority assignment $(T_1 \succ T_2 \succ \cdots \succ T_n)$ results in such a schedule. Let T_i and T_{i+1} denote two tasks of adjacent priorities with $d_i \geq d_{i+1}$. Let us exchange the priorities of T_i and T_{i+1}: if the priority assignment obtained after this exchange results in all deadlines being met, then we may conclude that any deadline monotonic priority assignment will also result in all deadlines being met since any deadline-monotonic priority assignment can be obtained from any priority ordering by a sequence of such priority exchanges. To see that the priority assignment obtained after this exchange results in all deadlines being met, observe the following:

1. The priority exchange does not modify the schedulability of the tasks with a higher priority than T_i (i.e., $T_1, \ldots, T_{i-1}$).
2. The task T_{i+1} remains schedulable after the priority exchange, since its jobs may use all the free time-slots left by $\{T_1, T_2, \ldots, T_{i-1}\}$ instead of merely those left by $\{T_1, T_2, \ldots, T_{i-1}, T_i\}$.
3. Assuming that the jobs of T_i remain schedulable (this will be proved below), by Lemma 28.11 above we may conclude that the scheduling of each task T_k, for $k = i+2, i+3, \ldots, n$ is not altered since the idle periods left by higher priority tasks are the same.
4. Hence we need only verify that T_i also remains schedulable. From Lemma 28.10 we can restrict our attention to the first job of task T_i. Let r_{i+1} denote the response time of the first job of T_{i+1} before the priority exchange: the feasibility implies $r_{i+1} \leq d_{i+1}$. During the interval $[0, r_{i+1})$ the processor (when left free by higher priority tasks) is assigned first to the (first) job of T_i and then to the (first) job of T_{i+1}; the latter is not interrupted by subsequent jobs of T_i since $p_i \geq d_i \geq d_{i+1} \geq r_{i+1}$.

Hence after the priority exchange, the processor allocation is exchanged between T_i and T_{i+1}, and it follows that T_i ends its computation at time r_{i+1} and meets its deadline since $r_{i+1} \le d_{i+1} \le d_i$. □

As a consequence of Theorem 28.16 and Lemma 28.10, it follows that synchronous periodic task system τ with constrained deadlines is static-priority feasible if and only if the first job of each task in τ meets its deadline when priorities are assigned in deadline-monotonic order. Since this can be determined by simulating the schedule out until the largest deadline of any task in τ, we have the following corollary:

Corollary 28.5

Feasibility analysis of sporadic and synchronous periodic constrained-deadline task systems can be done in pseudopolynomial time.

28.4.4.2 Nonoptimality for Sporadic and Synchronous Periodic Arbitrary-Deadline Task Systems

We have already stated that the computational complexity of optimal priority-assignment for constrained-deadline periodic task systems that are not necessarily synchronous is currently unknown; in particular, DM is *not* an optimal priority-assignment scheme for such systems.

Even for *synchronous* periodic (as well as sporadic) task systems which are not constrained deadline (i.e., in which individual tasks T_i may have $d_i > p_i$), however, the deadline-monotonic rule is no longer optimal. This is illustrated by the following example given by Lehoczky [18].

Example 28.1

Let $\tau = \{T_1 = (e_1 = 52, d_1 = 110, p_1 = 100), T_2 = (e_2 = 52, d_2 = 154, p_2 = 140,)\}$. *Both the RM and the DM priority-assignment schemes would assign higher priority to* T_1*:* $T_1 \succ T_2$*. With this priority assignment, however, the first job of* T_2 *misses its deadline at time* 154 *with the priority assignment* $T_2 \succ T_1$*, it may be verified that all deadlines are met.*

28.5 Conclusions and Further Reading

In this chapter, we have presented what we consider to be some of the more important results that have been published over the past 30 years, on the preemptive uniprocessor scheduling of recurring real-time tasks. Without a doubt, our decisions regarding which results to present have been driven by our own limited knowledge, our preferences and biases, and the page limits and time constraints that we have worked under. Below, we list some other ideas that we consider very important to a further understanding of preemptive uniprocessor scheduling of recurring tasks that we have chosen to not cover in this chapter.

More efficient feasibility tests. We chose to restrict ourselves to showing the computational complexity of the various feasibility-analysis problems we discussed here. Much work has been done on obtaining more efficient implementations of some of the algorithms described here; although these more efficient implementations have about the same worst-case computational complexity as the algorithms described here, they typically perform quite a bit better on average. Important examples of such algorithms include the work of Ripoll, Crespo, and Mok [21] and the recent sufficient feasibility test of Devi [22] on dynamic-priority scheduling, and the large body of work on *response time analysis* [23–25] on static-priority scheduling.

More general task models. Recent research has focused upon obtaining more general models for recurring tasks than the simple periodic and sporadic models discussed in this chapter. Some interesting new models include the *multiframe* model [26–28], which was proposed to accurately model multimedia traffic, and the *recurring real-time task* model [29], which allows for the modeling of conditional real-time code. (Another interesting recurring task model — the *pinwheel* model — is the subject of another chapter [30] in this handbook.)

Resource sharing. Real-time systems typically require the sharing of some other resources as well as the processor; often, a process or task needs exclusive access to such resources. Issues that arise when such resources must be shared by recurring tasks have been extensively studied, and form the subject of another chapter [31] of this handbook.

References

[1] C. Liu and J. Layland. Scheduling algorithms for multiprogramming in a hard real-time environment. *Journal of the ACM*, 20(1):46–61, 1973.

[2] J. Leung and M. Merrill. A note on the preemptive scheduling of periodic, real-time tasks. *Information Processing Letters*, 11:115–118, 1980.

[3] E. Lawler and M. Martel. Scheduling periodically occurring tasks on multiple processors. *Information Processing Letters*, 12:9–12, 1981.

[4] J. Leung and J. Whitehead. On the complexity of fixed-priority scheduling of periodic, real-time tasks. *Performance Evaluation*, 2:237–250, 1982.

[5] A. K. Mok. Fundamental Design Problems of Distributed Systems for The Hard-Real-Time Environment. PhD thesis, Laboratory for Computer Science, Massachusetts Institute of Technology, Cambridge, MA, 1983. Available as Technical Report No. MIT/LCS/TR-297.

[6] S. Baruah, R. Howell, and L. Rosier. Algorithms and complexity concerning the preemptive scheduling of periodic, real-time tasks on one processor. *Real-Time Systems: The International Journal of Time-Critical Computing*, 2:301–324, 1990.

[7] J. Lehoczky, L. Sha, and J. Stronider. Enhanced aperiodic responsiveness in hard real-time environments. In *Proceedings of the Real-Time Systems Symposium*, pages 261–270, San Jose, CA, December 1987. IEEE.

[8] K. Hong and J. Leung. On-line scheduling of real-time tasks. *IEEE Transactions on Computers*, 41:1326–1331, 1992.

[9] B. Sprunt, J. Lehoczky, and L. Sha. Exploiting unused periodic time for aperiodic service using the extended priority exchange algorithm. In *Proceedings of the Real-Time Systems Symposium*, pages 251–258, Huntsville, AL, December 1988. IEEE.

[10] K. Jeffay, D. Stanat, and C. Martel. On non-preemptive scheduling of periodic and sporadic tasks. In *Proceedings of the 12th Real-Time Systems Symposium*, pages 129–139, San Antonio, TX, December 1991. IEEE.

[11] B. Sprunt, L. Sha, and J. Lehoczky. Scheduling sporadic and aperiodic events in a hard real-time system. Technical Report ESD-TR-89-19, Carnegie Mellon University, 1989.

[12] S. Baruah, A. Mok, and L. Rosier. The preemptive scheduling of sporadic, real-time tasks on one processor. In *Proceedings of the 11th Real-Time Systems Symposium*, pages 182–190, Orlando, FL, 1990. IEEE.

[13] K. Jeffay. Analysis of a synchronization and scheduling discipline for real-time tasks with preemption constraints. In *Proceedings of the 10th Real-Time Systems Symposium*, Santa Monica, CA, December 1989. IEEE.

[14] R. Howell and M. Venkatrao. On non-preemptive scheduling of recurring tasks using inserted idle times. Technical Report TR-CS-91-6, Kansas State University, Department of Computer and Information Sciences, 1991.

[15] M. Dertouzos. Control robotics : the procedural control of physical processors. In *Proceedings of the IFIP Congress*, pages 807–813, 1974.

[16] J. W. S. Liu. *Real-Time Systems*. Prentice-Hall, Upper Saddle River, NJ, 2000.

[17] G. C. Buttazzo. *Hard Real-Time Computing Systems: Predictable Scheduling Algorithms and Applications*. Kluwer, Norwell, MA, 1997.

[18] J. P. Lehoczky. Fixed priority scheduling of periodic tasks with arbitrary deadlines. In *IEEE Real-Time Systems Symposium*, pages 201–209, Dec. 1990.

[19] O. Serlin. Scheduling of time critical processes. In *Proceedings of the 1972 Spring Joint Computer Conference*, volume 40 of *AFIPS Conference Proceedings*, 1972.

[20] J. Goossens and R. Devillers. The non-optimality of the monotonic priority assignments for hard real-time offset free systems. *Real-Time Systems: The International Journal of Time-Critical Computing*, 13(2):107–126, 1997.

[21] I. Ripoll, A. Crespo, and A. K. Mok. Improvement in feasibility testing for real-time tasks. *Real-Time Systems: The International Journal of Time-Critical Computing*, 11:19–39, 1996.

[22] U. Devi. An improved schedulability test for uniprocessor periodic task systems. In *Proceedings of the Euromicro Conference on Real-time Systems*, Porto, Portugal, 2003. IEEE.

[23] M. Joseph and P. Pandya. Finding response times in a real-time system. *The Computer Journal*, 29(5):390–395, 1986.

[24] M. Joseph, editor. *Real-Time Systems: Specification, Verification and Analysis*. Prentice Hall International Series in Computer Science. Prentice Hall, New York, 1996.

[25] N. C. Audsley, A. Burns, R. I. Davis, K. W. Tindell, and A. J. Wellings. Fixed priority preemptive scheduling: An historical perspective. *Real-Time Systems*, 8:173–198, 1995.

[26] A. K. Mok and D. Chen. A multiframe model for real-time tasks. In *Proceedings of the 17th Real-Time Systems Symposium*, Washington, DC, 1996. IEEE.

[27] A. K. Mok and D. Chen. A multiframe model for real-time tasks. *IEEE Transactions on Software Engineering*, 23(10):635–645, 1997.

[28] S. Baruah, D. Chen, S. Gorinsky, and A. Mok. Generalized multiframe tasks. *Real-Time Systems: The International Journal of Time-Critical Computing*, 17(1):5–22, 1999.

[29] S. Baruah. Dynamic- and static-priority scheduling of recurring real-time tasks. *Real-Time Systems: The International Journal of Time-Critical Computing*, 24(1):99–128, 2003.

[30] D. Chen and A. Mok. The pinwheel: A real-time scheduling problem. In this volume.

[31] L. Sha. Synchronization issues in real-time task systems. In this volume.

29

Real-Time Synchronization Protocols

Lui Sha
University of Illinois

Marco Caccamo
University of Illinois

29.1 Introduction

An important problem that arises in the context of real-time systems is the duration for which high-priority tasks are blocked by the execution of low-priority jobs. A milestone in this area was the discovery, analysis, and solutions of the unbounded priority inversion problem in the 1980s.

Ideally, when two jobs are competing for the CPU, the higher-priority job should win. In practice, there are exceptions. For example, the higher-priority job will be delayed when it wants to lock a semaphore that has been locked by the lower-priority job. Priority inversion is the phenomenon where a higher-priority job J is blocked by lower-priority jobs. A common situation arises when two jobs attempt to access shared data. Intuitively, it seems that the duration of priority inversion should be a function of the length of critical sections of lower-priority tasks that share resource with J. Unfortunately, this was not the case in the older generation of operating systems using the classical semaphores.

Example 29.1

Suppose that J_1, J_2, and J_3 are three jobs arranged in descending order of priority with J_1 having the highest one. We assume that jobs J_1 and J_3 share a data structure guarded by a binary semaphore S. Suppose that at time t_1, job J_3 locks the semaphore S and executes its critical section. During the execution of job J_3's critical section, the high priority job J_1 is initiated, preempts J_3, and later attempts to use the shared data. However, job J_1 will be blocked on the semaphore S. We would expect that J_1, being the

1-58488-397-9/$0.00+$1.50

highest priority job, will be blocked no longer than the time for job J_3 to complete its critical section. However, the duration of blocking is, in fact, unpredictable. This is because job J_3 can be preempted by the intermediate priority job J_2. The blocking of J_3, and hence that of J_1, will continue until J_2 and any other pending intermediate jobs are completed.

The blocking period in Example 29.1 is not bounded by the duration of critical sections, since it includes the execution time of J_2; hence the name "unbounded priority inversions." Unbounded priority inversion can be avoided if a job in its critical section will be executed at the highest priority level, so that it cannot be preempted. However, this solution is only appropriate for very short critical sections, because it creates unnecessary blocking. For instance, once a low-priority job enters a long critical section, a high-priority job, which does not access the shared data structure, may be needlessly blocked. An improved solution was proposed by Lampson et al. [1] in 1980. The idea is that a monitor should be executed at the highest priority of any task that will ever use this monitor.

During the late 80s, Sha et al. found that Ada tasking rules suffer unbounded priority inversion problem. They coined the phrase "unbounded priority inversion," and proposed the basic priority inheritance protocol (PIP) [2], which has since been adopted by most commercially available operating systems, because of its simplicity. Later, Sha et al. developed the priority ceiling protocol (PCP) and analyzed the properties of PIP and PCP [3,4] for static priority scheduling. PCP prevents deadlocks and ensures that a job can be blocked by lower priority jobs at most once, no matter how many locks are used.

The priority inheritance approach has since been extended for both fixed priority systems and dynamic priority systems, and adopted by POSIX real-time extensions, Ada 95, and is being adopted by the vast majority of commercially available operating systems.

In view of the existing books that have also covered this topic [5–7], we will cover the material as follows. First, we will focus on only the key results so that readers can quickly master the essentials. Second, we will present the latest research on hybrid task sets that has not been covered in existing books. Third, we will provide a new conceptual framework to describe all the key results. Readers who are interested in many practical application examples may consult the book by Klein et al. [5]. Readers who want to see an extensive review of all the extensions prior to 1999 may consult [6].

29.1.1 Basic Concepts and Properties of the Priority Inheritance Approach

We use τ_i to represent a periodic task and J_i to represent a job, i.e., an instance of τ_i. In addition, we use π_i to represent the priority of job J_i. In static priority scheduling such as rate monotonic scheduling (RMS) (or deadline monotonic scheduling), every job of a task has the same priority. In dynamic priority scheduling such as earliest deadline first (EDF) scheduling, different jobs in a task have different priorities. Also, in our notation for a pair of jobs $\{J_i, J_j, i < j\}$, we assume that job J_i has higher priority. We now describe three basic assumptions used by the priority inheritance approach.

Assumption 29.1 *A fixed priority is assigned to each job.*

Note that this assumption holds for both jobs scheduled by RMS, and by EDF assuming that a job's deadline will not change.

Assumption 29.2 *A job will use its assigned priority unless it is in a critical section.*

Assumption 29.3 *When a job executes inside a critical section, its priority could be changed by pairs of actions that raise and lower this job's priority. However, a job's priority will never be lowered below its assigned priority.*

When a job J is delayed by a job J_L with a lower assigned priority, we say that job J is blocked by J_L. When a higher priority job J takes away the processor from a lower priority job J_L that has already started execution, we say that job J preempts job J_L. We say that J_L waits for the execution of J, if J delays the start of execution of J_L.

Lemma 29.1

A job J can be blocked by a lower priority job J_L, only if when J becomes ready, J_L is inside its critical section.

Proof

If J_L is not within a critical section, it will be executed with its assigned low priority. By Assumptions 29.1, 29.2, and 29.3, the priority used by J is greater than that of J_L, i.e., $\pi > \pi_L$. By the definition of priority-based preemptive scheduling, J_L has to wait for J's execution. □

That is, J can only be blocked by the execution of lower priority jobs' critical sections.

Lemma 29.2

When there is no deadlock, a job J can be blocked by a lower-priority job J_L for at most the duration of one outermost nested critical section of J_L, regardless of the number of semaphores J and J_L share.

Proof

By Lemma 29.1, for J_L to block J, J_L must be currently executing a critical section z_L. Once J_L exits z_L, it will use its assigned priority $\pi_L < \pi$. It follows that J will preempt J_L. By Assumptions 29.1, 29.2, and 29.3, J's priority will always be higher than J_L's assigned priority π_L. The lemma follows. □

29.2 Static Priority Real-Time Synchronization Protocols

In the next two subsections, we will review the basic priority inheritance and the priority ceiling protocols [3].

29.2.1 Basic Priority Inheritance Protocol

The basic idea of the PIP is that when a job J blocks one or more higher-priority jobs, it executes its critical section at the highest priority level among all the jobs that it blocks. After exiting its critical section, job J returns to its assigned priority π. Priority inheritance is transitive. For example, there are three jobs, J_H, J_M, and J_L, with priorities high, medium, and low, respectively. Suppose that J_L blocks J_M and J_M blocks J_H, then J_L inherits the inherited priority of J_H of J_M.

To illustrate this idea, we apply PIP to Example 29.1. Suppose that job J_1 is blocked by job J_3. The priority inheritance protocol requires that job J_3 executes its critical section at job J_1's priority. As a result, job J_2 will be unable to preempt job J_3 when J_3 is inside its critical section. That is, job J_2 now must wait for the critical section of the lower priority job J_3 to complete, because job J_3 inherits the priority of job J_1.

It is important to note that under PIP, a job J can be blocked by a lower priority job J_L, even if J and J_L do not share any resource. In Example 29.1, job J_2 can be blocked by the execution of the critical section of J_3, when J_3 inherits the priority of J_1. We now examine the property of this protocol.

PIP does not prevent deadlocks. When deadlock is prevented by some algorithm, e.g., acquiring locks in a given order, an upper bound on the blocking time for jobs under PIP is as follows.

Theorem 29.1

A job J can be blocked by each of the lower priority jobs that may use a semaphore at most once.

Proof

By Lemma 29.2, each of the n lower-priority jobs can block job J_0 for at most the duration of a single outermost critical section. The theorem follows. □

Theorem 29.1 tells us that basic priority inheritance protocol solves the unbounded priority inversion problem, since the blocking time is now the sum of the duration of lower-priority jobs' critical sections.

29.2.2 Priority Ceiling Protocol

Priority ceiling protocol is the basic priority inheritance protocol with an additional priority ceiling rule. Under the priority ceiling protocol, the priority ceiling of a semaphore S is defined as the maximum priority of all the jobs that may ever use this semaphore. That is,

$$ceil(S) = \max_i \{\pi_i \mid \tau_i \text{ may lock } S\} \tag{29.1}$$

moreover, a *system ceiling* for a job J_i, $\Pi(J_i)$, is defined as the highest priority ceiling among all semaphores currently locked by other jobs.

$$\Pi(J_i) = \max_k \{ceil(S_k) \mid \text{semaphore } S_k \text{ is currently locked by all other jobs } J_j, j \neq i\} \tag{29.2}$$

Under preemptive priority scheduling rule, a job cannot start to execute unless it has the highest priority among all the jobs that are ready to execute. The priority ceiling protocol imposes another constraint on a job J's execution when it attempts to lock a semaphore S. J cannot lock S even if S is not locked, unless its priority is strictly higher than the priority ceiling of all the semaphores currently locked by all the other jobs, i.e., $\pi > \Pi(J)$.

The priority ceiling protocol has two nice properties. First, it prevents deadlocks. Second, a job can be blocked by lower priority jobs at most once. We first illustrate the deadlock avoidance property.

Example 29.2

Suppose that we have two jobs J_1 and J_2 in the system. In addition, there are two shared data structures protected by the binary semaphores S_1 and S_2, respectively. We define the priority ceiling of a semaphore as the priority of the highest priority job that may lock this semaphore. Suppose that these two jobs try to lock S_1 and S_2 in opposite order.

$$J_1 = \{\ldots, P(S_1), P(S_2), V(S_2), V(S_1), \ldots\}$$

$$J_2 = \{\ldots, P(S_2), P(S_1), V(S_1), V(S_2), \ldots\}$$

Recall that the priority of job J_1 is assumed to be higher than that of job J_2. Thus, the priority ceilings of both semaphores S_1 and S_2 are equal to the priority of job J_1, π_1.

The sequence of events described below is depicted in Figure 29.1. A line at a low level indicates that the corresponding job is being blocked by, or being preempted by or waiting for a higher-priority job. A line raised to a higher level indicates that the job is executing. The absence of a line indicates that the job has not yet been initiated or has completed. Shaded portions indicate execution of critical sections.

Suppose that at time t_1, J_2 is initiated and it begins execution and then locks semaphore S_2. At time t_2, job J_2 locks semaphore S_2. At t_3, job J_1 starts to execute and preempts job J_2. At t_4, job J_1 attempts to lock the unlocked semaphore S_1. However, it cannot do so since its priority is just equal to the priority ceiling of semaphore S_2 locked by job J_2. Should J_1 be allowed to lock S_1, it would lead to a deadlock.

J_1 is blocked by J_2 under the rule of priority ceiling protocol. Now, J_2 inherits J_1's priority and resumes the execution of its critical section. It then locks S_1, unlocks $S1$ and finally unlocks S_2 at time t_5. At this instant, J_2 returns to its original assigned priority and is preempted by job J_1. J_1 now locks S_1 and then S_2. Next, J_2 unlocks S_2 and S_1 at time t_6. J_1 continues its execution of noncritical section code. After J_1 finishes, J_2 resumes its execution and finally finishes at t_7. We now illustrate the "blocked at most once" property.

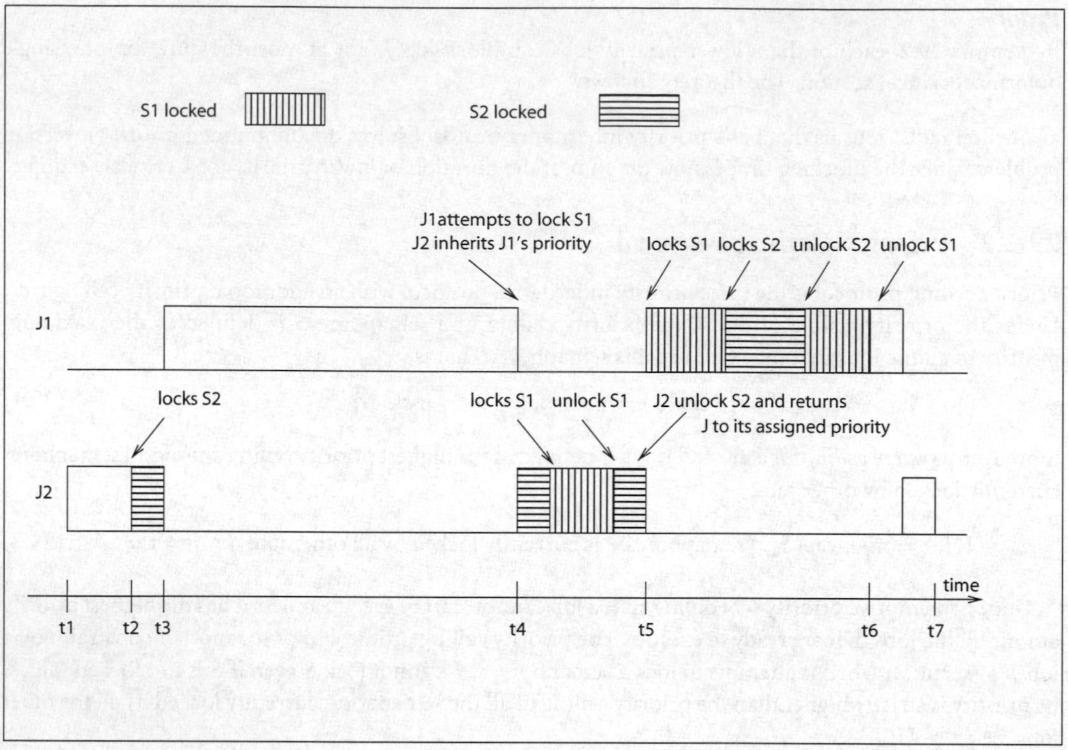

FIGURE 29.1 Deadlock avoidance.

Example 29.3

Consider three jobs in the system (J_1, J_2, and J_3 as in Figure 29.2) and suppose that job J_1 needs to access S_1 and S_2 sequentially, while J_2 accesses S_1 and J_3 accesses S_2. Hence, the priority ceilings of semaphores S_1 and S_2 are equal to the highest priority job J_1.

$$J_1 = \{\ldots, P(S_1), P(S_2), V(S_2), V(S_1), \ldots\}$$

$$J_2 = \{\ldots, P(S_1), V(S_1), \ldots\}$$

$$J_3 = \{\ldots, P(S_2), V(S_2), \ldots\}$$

Note that the priority ceilings of both semaphores are equal to π_1, the priority of job J_1. We now attempt to create multiple blockings for job J_1 by starting jobs in the order of J_3, J_2, and J_1.

At time t_0, job J_3 starts to execute and at t_1, it locks S_2. At t_2, J_2 preempts J_3 and starts to execute. At t_3, J_2 attempts to lock S_1. However, under the rule of PCP, J_2 cannot lock S_1 since its priority is not higher than the system ceiling, which is the priority ceiling of S_2. Should it be allowed to lock S_1, J_1 could be blocked by both J_2 and J_3.

Now, J_3 inherits the priority of J_2 and resumes its execution. At t_4, job J_1 preempts J_3 and starts to execute. At t_5, J_1 attempts to lock S_1. However, under PCP, J_1 is not allowed to lock S_1. J_3 now inherits the priority of J_1 and resumes execution, since it blocks J_1 via the priority ceiling of semaphore S_2. At t_6, J_3 unlocks S_2 and its priority returns to its originally assigned one. J_1 immediately preempts J_3 and resumes execution. It locks S_1 and S_2 and then unlocks S_2 and S_1 at t_7. It goes on to complete its execution at t_8. J_2 now preempts J_3, locks S_1, and resumes execution. It later unlocks S_1 and finishes at t_9. Finally, J_3 resumes its execution and finishes at t_{10}.

We now prove the properties of the priority ceiling protocol.

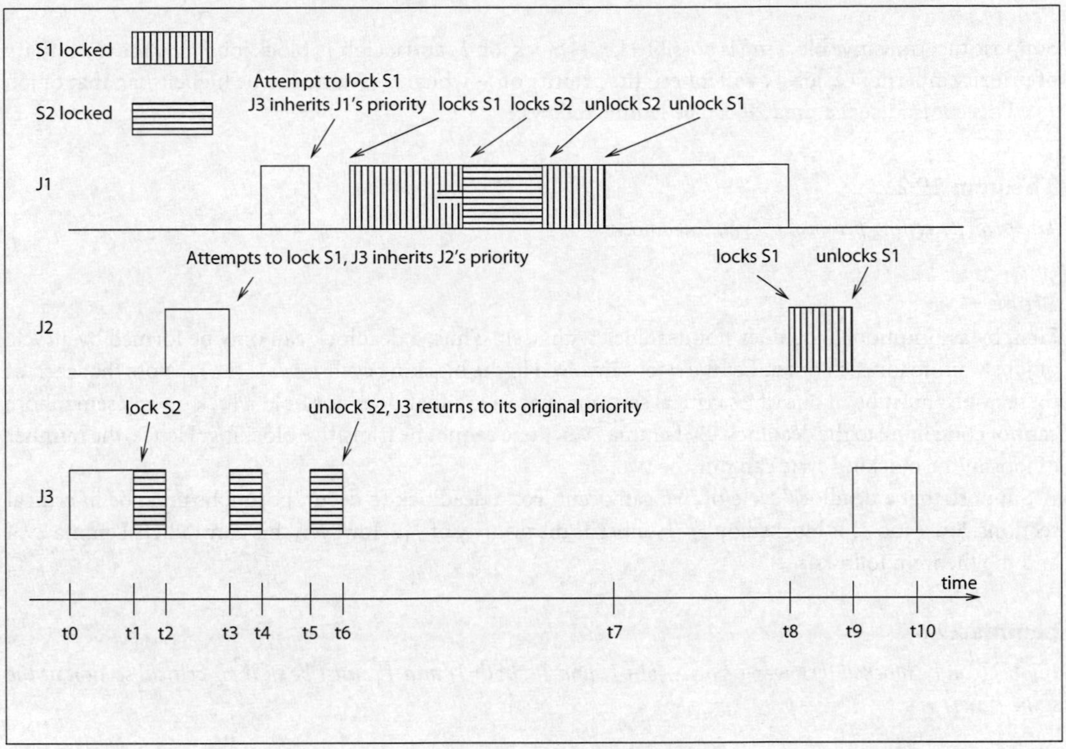

FIGURE 29.2 Blocked at most once.

Lemma 29.3

A job J can be blocked by a lower-priority job J_L, only if the priority of job J is not higher than the system priority ceiling $\Pi(J)$ for job J.

Proof

Under PCP, the higher-priority job J cannot be delayed by J_L if J's priority is higher than the system ceiling for J. □

Lemma 29.4

Suppose that a critical section z_j of job J_j is preempted by job J_i's critical section z_i. Under PCP, job J_j cannot inherit a priority level that is higher than or equal to π_i, the priority of job J_i.

Proof

Suppose that this lemma is false. That is, job J_j is able to inherit a priority higher than or equal to π_i. Under PCP, job J_j is able to inherit the higher priority of a job J, only if J is blocked by J_j. Since J_i is able to enter its critical section under PCP, we know that $\pi_i > \Pi(J_i)$. Since J's priority $\pi > \pi_i$, it follows that $\pi > \Pi(J_i)$. Hence, under PCP neither J nor J_i can be blocked by J_j. By contradiction, the lemma follows. □

Transitive blocking is said to occur if a job J is blocked by J_i, which is, in turn, blocked by another job J_j.

Lemma 29.5

The priority ceiling protocol prevents transitive blocking.

Proof

Suppose that transitive blocking is possible. Let J_3 block job J_2 and let job J_2 block job J_1. By the transitivity of priority inheritance, job J_3 will inherit the priority of J_1 which is assumed to be higher than that of job J_2. This contradicts Lemma 29.4. The lemma follows. □

Theorem 29.2

The priority ceiling protocol prevents deadlocks.

Proof

First, by assumption, a job does not deadlock with itself. Thus, a deadlock can only be formed by a cycle of jobs waiting for each other. Let the n jobs involved in the blocking cycle be $J_1, \ldots, J_n$. Note that each of these n jobs must be in one of its critical sections, since a job that does not hold a lock on any semaphore cannot contribute to the deadlock. By Lemma 29.5, there cannot be transitive blocking. Hence, the number of jobs in the blocking cycle can only be two.

Suppose that a deadlock cycle of two can occur. For a deadlock to occur, both jobs must be in critical sections. Since job J_1 is blocked by J_2, J_2 inherits the priority of J_1. However, this contradicts Lemma 29.4 and the theorem follows. □

Lemma 29.6

If job J can be blocked by lower-priority jobs J_i and J_j, both J_i and J_j must be in their critical section at the same time.

Proof

By Lemma 29.1, if J_i (or J_j) is not in its critical section when J starts, it cannot block J. The lemma follows. □

Theorem 29.3

A job J can be blocked for at most the duration of an outermost critical section from lower-priority jobs.

Proof

Suppose that job J can be blocked by $n > 1$ critical sections. By Lemma 29.2, the only possibility is that job J is blocked by n different lower-priority jobs. Consider any two lower-priority jobs J_i and J_j that block J. By Lemma 29.6, when J starts, both J_i and J_j must be in their critical sections. Since both J_i and J_j block J, they inherit the priority of J, π. We now show the contradiction.

- Suppose that J_j enters its critical section first. By Lemma 29.3, if job J is blocked by J_j, J's priority π is no higher than the priority ceiling of some semaphore S_j locked by J_j. That is, $\pi \leq ceil(S_j)$.
- Since J_i enters its critical section after J_j locks S_j, we have $\pi_i > ceil(S_j)$. Since J_j blocks J, we have $\pi > \pi_i$ by the definition of blocking. Since $\pi_i > ceil(S_j)$, we have $\pi > ceil(S_j)$. □

PCP has since been generalized in the context of database concurrency control [8,9] and multiprocessors [10]. Rajkumar gave a comprehensive treatment of real-time synchronization research up to early 1991 [4]. An important refinement is the stack-based priority ceiling protocol developed by Baker [11]. First, it extends PCP to the EDF scheduling. Second, it allows multithreaded embedded devices sharing a single runtime stack. Finally, Caccamo et al. extend PCP to include the synchronization between periodic and aperiodic tasks which may run out of budgets and forced to be suspended [12].

In the context of static priority scheduling, the stack protocol makes a simple modification to the priority ceiling rule as follows. Under the original PCP rule, a job J cannot enter a critical section unless its priority is higher than all the priority ceilings of semaphores currently locked by other jobs. Under the stack protocol, a job J cannot start execution unless its priority is higher than all the priority ceilings of semaphores currently locked by other jobs. This early blocking feature is pessimistic if job J does not use any semaphore. Otherwise, it saves a context switch. The major advantage of this early blocking feature is that it allows all the tasks share a single runtime stack. Note that under this modified rule, once a task starts execution it cannot be blocked by lower-priority tasks. This permits the system to use a single stack shared by many tasks and is especially useful in small devices.

29.2.3 Schedulability Analysis

First, we note that the Liu and Layland bound can be extended by incorporating the effect of blocking time.

Theorem 29.4

A set of n periodic tasks using the priority inheritance protocols can be scheduled by the rate-monotonic algorithm if the following conditions are satisfied [3]:

$$\frac{C_1}{T_1} + \cdots + \frac{C_n}{T_n} + \max\left\{\frac{B_1}{T_1}, \ldots, \frac{B_{n-1}}{T_{n-1}}\right\} \leq n(2^{1/n} - 1) \tag{29.3}$$

where B_i is the worst case blocking time of task τ_i under either PIP or PCP.

Finally, if we want to use the exact schedulability test for a job, all we need to do is to move the job's deadline from T_i to $(T_i - B_i)$, where B_i is the worst case blocking time of job J under either basic priority inheritance or priority ceiling protocol. We now illustrate their use.

Example 29.4

Consider the case of two periodic tasks. Task τ_1: $C_1 = 40$; $T_1 = 100$; $B_1 = 20$; task τ_2: $C_2 = 40$; $T_2 = 150$.

For task τ_1 we apply the scheduling bound:
$(C_1 + B_1)/T_1 = 0.6 <= 1.0$. Hence, task τ_1 is schedulable.
For task τ_2, we use exact test:
$a_0 = C1 + C2 = 80$
$a_1 = \lceil (a_0/100) * 40 + 40 = 80 = a_0 \rceil$
$a_1 < (150 - 30) = 120$
Hence task τ_2 is also schedulable.

29.3 Stack Resource Policy

The stack resource policy (SRP) is a concurrency control protocol proposed by Baker [11] to bound the priority inversion phenomenon in static as well as dynamic priority systems. In this work, Baker made the insightful observation that under EDF, only jobs from tasks with long relative deadline cannot preempt jobs from tasks with short relative deadline [11]. The relative deadline of a job is the duration between a job's deadline and its arrival time. This is illustrated in Figure 29.3.

As we can see, at time t_0, the first job $J_{1,1}$ of periodic task τ_1 is ready. So is the first job $J_{2,1}$ of periodic task τ_2. Since $J_{1,1}$'s deadline is at t_2 while $J_{2,1}$'s deadline is at t_4, job $J_{2,1}$ must wait for job $J_{1,1}$. Should $J_{2,1}$ start earlier, it will be preempted by $J_{1,1}$. At time t_1, job $J_{1,1}$ finishes and $J_{2,1}$ starts. At t_2, job $J_{1,2}$ becomes ready. However, $J_{1,2}$ has deadline at t_5 and therefore it must wait for $J_{2,1}$.

The key observation here is that when $J_{2,1}$ becomes the job with an earlier deadline at t_4 in Figure 29.3, $J_{1,2}$ cannot start its execution until $J_{2,1}$ finishes. Since $J_{1,2}$ cannot start, it cannot enter its critical section.

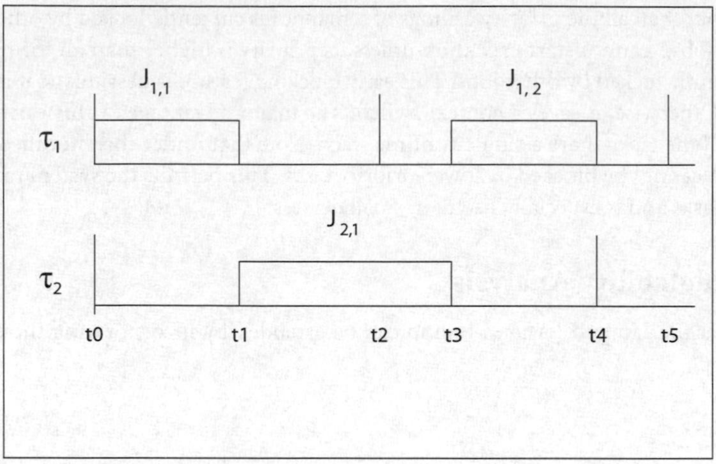

FIGURE 29.3 Condition for preemption.

Since it cannot enter its critical section, it cannot block higher-priority job $J_{2,1}$. This observation is summarized by Lemma 29.7.

Lemma 29.7

Let job J_i have relative deadline D_i, J_j have relative deadline D_j, and $D_i < D_j$. Under EDF, job J_i can never block the execution of J_j.

Proof

There are three cases.

- *Case 1:* J_i has an earlier absolute deadline, that is, $d_i < d_j$. In this case, J_i has higher priority. By definition, a higher-priority job does not block a lower-priority job.
- *Case 2:* J_i has absolute deadline equal to that of J_j, that is, $d_i = d_j$. There are two cases:

  a. *Case 2.1:* We break the tie in favor of J_i, there is no blocking by definition.

  b. *Case 2.2:* We break the tie in favor of J_j. Since $D_j > D_i$ and J_j has higher priority, when J_i becomes ready, it must wait for J_j to continue its execution until it finishes.
- *Case 3:* J_i has a deadline later than that of J_j, that is, $d_i > d_j$. Since $D_j > D_i$ and J_j has higher priority, when J_i becomes ready, it must wait for J_j to continue its execution until it finishes.

□

Lemma 29.7 tells us that under EDF, a job cannot block another job with longer relative deadline. Thus, in the study of blocking under EDF, we only need to consider the case where jobs with longer relative deadlines block jobs with shorter relative deadlines.

When the tasks are periodic, jobs from a periodic task with a shorter period have shorter relative deadlines. Since jobs from a task can only be blocked by jobs from tasks with longer periods, blocking behaviors under EDF and RMS are logically the same.

This observation allows Baker to define preemption levels separately from priority levels under EDF. For periodic tasks, the preemption level under stack protocol can be assigned rate monotonically: the higher the rate, the higher is the preemption level. It follows that the semaphore preemption ceiling under SRP can be defined in the same way as priority ceiling under PCP. This mapping allows SRP to have the same property of blocked at most once and freedom from deadlocks. Like the static priority case, a job J cannot start its execution unless its preemption level is higher than its system ceiling $\Pi(J)$.

29.3.1 The Algorithm

Under SRP, with the EDF scheduling algorithm, each job of τ_i is assigned a priority p_i according to its absolute deadline d_i and a static *preemption level* $\pi_i = \frac{1}{D_i}$. Each shared resource[1] is assigned a ceiling which is the maximum preemption level of all the tasks that will lock this resource. More formally, every resource R_k is assigned a static[2] *ceiling* defined as

$$ceil(R_k) = \max_i\{\pi_i \mid \tau_i \text{ needs } R_k\} \tag{29.4}$$

Moreover, a *system ceiling* for a job J_i, $\Pi(J_i)$, is defined as the highest ceiling of all resources currently locked by other jobs.

$$\Pi(J_i) = \max_k\{ceil(R_k) \mid \text{resource } R_k \text{ is currently locked by all other jobs } J_j, j \neq i\}. \tag{29.5}$$

Finally, SRP scheduling rule requires that

> *a job J is not allowed to start executing until its priority is the highest among the active tasks and its preemption level is greater than its system ceiling $\Pi(J)$.*

Stack resource policy ensures that once a job of a periodic task is started, it will never block until completion under the assumption that there is no self-suspension; it can only be preempted by higher priority tasks. This protocol has several interesting properties. For example, it applies to both static and dynamic scheduling algorithms, prevents deadlocks, bounds the maximum blocking times of tasks, reduces the number of context switches, can be easily extended to multiunit resources,[3] allows tasks to share stack-based resources, and its implementation is straightforward.

Under the SRP there is no need to implement waiting queues. In fact, a task never blocks during execution: it simply cannot start executing if its preemption level is not high enough. As a consequence, the blocking time B_i considered in the schedulability analysis refers to the time for which task τ_i is kept in the ready queue by the preemption test. Although the task never blocks, B_i is considered as a "blocking time" because it is caused by tasks having lower preemption levels.

In general, the maximum blocking time for a task τ_i is bounded by the duration of the longest critical section among those that can block τ_i. Assuming relative deadlines equal to periods, the maximum blocking time for each task τ_i can be computed as the longest critical section among those with a ceiling greater than or equal to the preemption level of τ_i:

$$B_i = \max\{z_{jh} \mid (D_i < D_j) \ \wedge \ \pi_i \leq ceil(\rho_{jh})\} \tag{29.6}$$

where z_{jh} is the worst-case execution time of the hth critical section of task τ_j, D_j is its relative deadline and, ρ_{jh} is the resource accessed by the critical section z_{jh}. Given these definitions, the feasibility of a task set with resource constraints (when only periodic and sporadic tasks are considered), can be tested by the following sufficient condition [11]:

$$\forall i, \ 1 \leq i \leq n \quad \sum_{k=1}^{i} \frac{C_k}{T_k} + \frac{B_i}{T_i} \leq 1 \tag{29.7}$$

which assumes that all the tasks are sorted by decreasing preemption levels, so that $\pi_i \geq \pi_j$ only if $i < j$.

---

[1]Since each shared resource R_k has a unique semaphore S_k associated with itself, semaphores and resources can be interchanged even though they are not conceptually the same.

[2]In the case of multiunit resources, the ceiling of each resource is dynamic as it depends on the number of units actually free.

[3]The definition of SRP in the presence of multiunit resources can be found in [11] and [7].

29.3.2 Theoretical Validation of the Model

In this section we formally describe and prove the properties of SRP. By exploiting these properties, the correctness of the schedulability bound will be finally proved. Before starting to analyze the SRP properties, the notions of *interval of continuous utilization* and *processor demand* for EDF are introduced according to the definitions given by Buttazzo [7].

Definition 29.1

The processor demand of task τ_i in any interval $[t_1, t_2]$, denoted as $D_i(t_1, t_2)$, is the sum of the computation times of all the instances of τ_i with arrival time greater than or equal to t_1 and deadline less than or equal to t_2. Moreover, the total demand $\mathcal{D}(t_1, t_2)$ of a task set $\Gamma(\tau_1, \ldots, \tau_n)$ in any interval $[t_1, t_2]$ is equal to

$$\mathcal{D}(t_1, t_2) = \sum_i D_i(t_1, t_2)$$

Definition 29.2

Let us consider an EDF schedule with a time overflow, and let t_{ov} be the instant at which the time overflow occurs. The interval of continuous utilization *$[t, t_{ov}]$ is the longest interval of busy CPU, before the time overflow, such that either instances (say the set of jobs Γ_{cont}) with deadline less than or equal to t_{ov} are executed within $[t, t_{ov}]$ or a critical section $\bar{s}$ of a job $\bar{J}$ with deadline greater than t_{ov} is executed within $[t, t_{ov}]$ blocking a job belonging to the set Γ_{cont} (see example of Figure 29.4).*

Notice that t must be the release time of some job J with deadline less than or equal to t_{ov} (J belongs to the set Γ_{cont}).

Lemma 29.8

If the preemption level of a job J is greater than its current system ceiling, then there are sufficient available resources to meet

- *the resource requirements of job J;*
- *the resource requirements of every job that can preempt J.*

Proof

Case A. Suppose that $\pi(J) > \Pi(J)$ at time t, but one (say R^*) of the resources needed by job J is busy. However, according to the definition of system ceiling, if resource R^* (needed by J) is currently busy, the system ceiling $\Pi(J)$ must be greater than or equal to the ceiling of R^*. As a consequence, it follows that $\Pi(J) \geq ceil(R^*) \geq \pi(J)$, which is a contradiction.

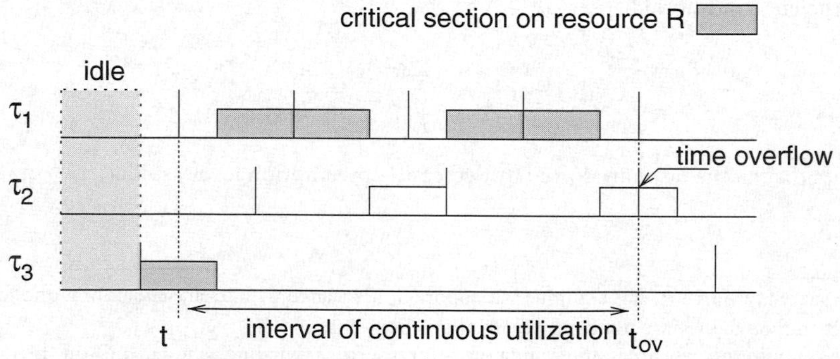

FIGURE 29.4 Interval of continuous utilization.

Case B. Suppose that $\pi(J) > \Pi(J)$ at time t, and assume that a resource R^* needed by job J_h is busy and J_h preempted J. By using the same argument, the system ceiling $\Pi(J)$ must be greater than or equal to the ceiling of resource R^*. As a consequence, it follows that $\Pi(J) \geq ceil(R^*) \geq \pi(J_h) > \pi(J)$, which is a contradiction. □

Theorem 29.5

If no job J is permitted to start until its preemption level becomes greater than its current system ceiling ($\pi(J) > \Pi(J)$), then none of the jobs can be blocked after it starts.

Proof

The theorem follows immediately from Lemma 29.8. In fact, if each job J is allowed to start only if its preemption level is greater than its system ceiling, there must be always enough available resources to satisfy the requirement of job J and all the jobs that can preempt J. □

Theorem 29.6

The Stack Resource Policy (SRP) prevents deadlocks.

Proof

According to Theorem 29.5, a job cannot be blocked after it starts. As a consequence, a job cannot be blocked while holding a resource: it follows that deadlock cannot ever occur. □

Lemma 29.9

If a task set Γ is scheduled according to EDF and SRP and a task $\bar{\tau}$ misses its deadline at time t_{ov}, where $[t, t_{ov}]$ is the interval of continuous utilization before the deadline miss, then no more than a single outermost critical section $z_{l,h}$ of task τ_l with deadline $d_l > t_{ov}$ can execute within interval $[t, t_{ov}]$.

Proof

Let $\bar{\Gamma}$ be the set of jobs with deadline less than or equal to t_{ov}, whose execution is within $[t, t_{ov}]$. Suppose that a second task τ^* with deadline $d^* > t_{ov}$ executes within interval $[t, t_{ov}]$. According to EDF, τ^* has to arrive before t and has to hold a resource R^* at time t that blocks a job belonging to $\bar{\Gamma}$. Moreover, τ_l has to preempt τ^* while it is already holding R^* or vice-versa. Let us suppose that τ_l preempts τ^* (the other case is analogous); according to SRP, $\pi(J_l) > \Pi(J_l) \geq ceil(R^*)$. However, if resource R^* can block job $\bar{J}$ belonging to $\bar{\Gamma}$, we have $ceil(R^*) \geq \pi(\bar{J}) > \pi(J_l)$. This immediately yields a contradiction. □

Theorem 29.7

Under SRP, a job J can be blocked for at most the duration of one critical section.

Proof

The theorem follows immediately from Lemma 29.9 by considering the interval $[r_j, d_j]$ (r_j is J's arrival time and d_j is J's absolute deadline) instead of considering the interval of continuous utilization. □

Theorem 29.8

Let Γ be a task set composed of n hard periodic tasks, supposing tasks are ordered by decreasing preemption level (so that $\pi_i \geq \pi_j$ only if $i < j$), the hard tasks are schedulable by EDF + SRP if

$$\forall i,\ 1 \leq i \leq n \quad \sum_{j=1}^{i} \frac{C_j}{T_j} + \frac{B_i}{T_i} \leq 1 \tag{29.8}$$

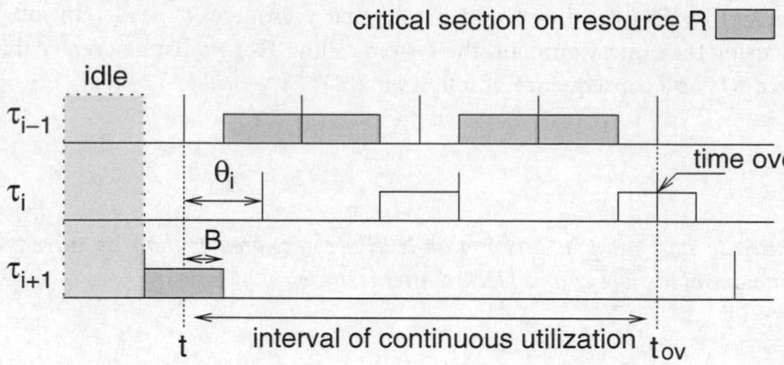

FIGURE 29.5 Task τ_i misses its deadline suffering a blocking time due to task τ_{i+1}.

where C_j is the worst case execution time of task τ_j, T_j is its period, and B_i is the maximum blocking time of task τ_i.

Proof

Suppose Equation (29.8) is satisfied for each τ_i, but task τ_j misses its deadline at time t_{ov}. We have to analyze two cases:

Case A. Task τ_i misses its deadline at time t_{ov} because it is blocked by task τ_{i+1}, where $[t, t_{ov}]$ is the interval of continuous utilization before the deadline miss (see Figure 29.5). According to Lemma 29.9 and Definition 29.2, it follows that no task τ_k with $k > (i+1)$ can execute within interval $[t, t_{ov}]$. Since there is a deadline miss at time t_{ov}, the total demand $\mathcal{D}(t, t_{ov})$ within interval $[t, t_{ov}]$ along with the blocking time B (suffered by jobs running within $[t, t_{ov}]$ and having absolute deadline less than or equal to t_{ov}) must be greater than the interval itself. Hence, it follows that

$$\mathcal{D}(t, t_{ov}) + B = \sum_{j=1}^{i} \left\lfloor \frac{t_{ov} - t - \theta_j}{T_j} \right\rfloor C_j + B > t_{ov} - t \tag{29.9}$$

$$\sum_{j=1}^{i} \frac{t_{ov} - t}{T_j} C_j + B > t_{ov} - t$$

$$\sum_{j=1}^{i} \frac{C_j}{T_j} + \frac{B}{t_{ov} - t} > 1$$

Noticing that $B_i \geq B$ and $(t_{ov} - t) \geq T_i$, it follows that

$$\sum_{j=1}^{i} \frac{C_j}{T_j} + \frac{B_i}{T_i} > 1$$

hence, Equation (29.8) is not satisfied by task τ_i, which is a contradiction.

Case B. Task τ_i misses its deadline at time t_{ov} because it is blocked by task τ_k, where $k > i+1$ and $[t, t_{ov}]$ is the interval of continuous utilization before the deadline miss. According to Lemma 29.9 and Definition 29.2, it follows that no task τ_h with $h > k$ can execute within interval $[t, t_{ov}]$. Since there is a deadline miss at time t_{ov}, the total demand $\mathcal{D}(t, t_{ov})$ within interval $[t, t_{ov}]$ along with the blocking time B (suffered by

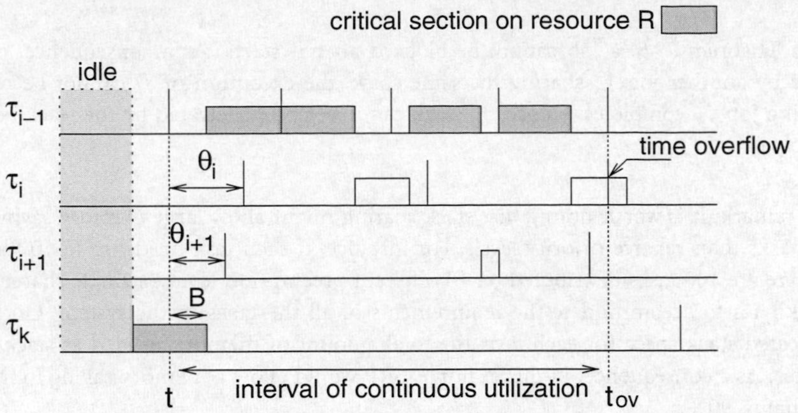

FIGURE 29.6 Task τ_i misses its deadline suffering a blocking time due to task τ_k, where $k > i + 1$.

jobs running within $[t, t_{ov}]$ and having absolute deadline less than or equal to t_{ov}) must be greater than the interval itself. Hence, it follows that

$$\mathcal{D}(t, t_{ov}) + B = \sum_{j=1}^{k-1} \left\lfloor \frac{t_{ov} - t - \theta_j}{T_j} \right\rfloor C_j + B > t_{ov} - t \qquad (29.10)$$

It is worth noting that the index of the summation in Equation (29.10) is up to $k - 1 > i$ instead of i unlike in Equation (29.9); the additional terms due to tasks $\{\tau_{i+1}, \ldots, \tau_{k-1}\}$ represent the interference of those tasks having longer relative deadlines than τ_i, but preempting τ_i since they have earlier absolute deadlines (according to EDF). This case is depicted by Figure 29.6. In addition, Equation (29.10) can be rewritten as

$$\sum_{j=1}^{k-1} \frac{t_{ov} - t}{T_j} C_j + B > t_{ov} - t$$

$$\sum_{j=1}^{k-1} \frac{C_j}{T_j} + \frac{B}{t_{ov} - t} > 1$$

Since task τ_{k-1} executes within interval $[t, t_{ov}]$, it follows that $(t_{ov} - t) \geq T_{k-1}$:

$$\sum_{j=1}^{k-1} \frac{C_j}{T_j} + \frac{B}{T_{k-1}} > 1$$

Moreover, since the blocking time B (due to task τ_k) can block any task τ_z with $z \leq (k - 1)$, the term B can be a blocking time for τ_{k-1}. Finally, noticing that B_{k-1} is the maximum blocking time of τ_{k-1}, it follows that $B_{k-1} \geq B$:

$$\sum_{j=1}^{k-1} \frac{C_j}{T_j} + \frac{B_{k-1}}{T_{k-1}} > 1$$

Hence, Equation (29.8) is not satisfied by task τ_{k-1}, which is a contradiction. □

Theorem 29.9

The Stack Resource Policy (SRP) allows tasks to share a common run-time stack.

Proof

According to Theorem 29.5, a job cannot be blocked after it starts. As a consequence, once a job J_l is preempted by another job J_h sharing the same stack, the execution of J_l cannot be resumed until the preempting job J_h completes. Hence, J_h stack can never be penetrated by the stack belonging to a preempted job J_l. □

As a final remark, it is worth noting that stack sharing might allow large memory saving if there are many more tasks than relative priority levels. For instance, if each task needs up to 10 kbytes of stack space and there are 100 tasks distributed on 10 different preemption levels, a single shared stack would require 100 kbytes to accommodate the requirements of all the tasks in the system. On the contrary, using a dedicated stack space for each task, the total amount of memory needed as stack space would be 1000 kbytes. As a consequence, in this example, SRP would allow a memory saving in the amount of 900 kbytes; that is, 90%.

29.4 Resource Sharing Among Hybrid Tasks

Most current real-time applications require the execution of real-time activities with different levels of criticality. Often, such tasks need to share common resources in exclusive mode, so the system has to prevent unbounded blocking (that may arise from priority inversion phenomena) to ensure predictable execution of hard tasks.

Classic solutions proposed in the literature for handling resource constraints did not consider hybrid (hard and soft) task sets, but developed the analysis only for homogeneous sets consisting of hard periodic tasks [3,11]. Section 29.2 described the PIP and PCP protocols, while Section 29.3 described SRP: all of them make the assumption of a homogeneous task set composed of hard periodic tasks. A method for analyzing the schedulability of hybrid task sets where hard tasks may share resources with soft tasks handled by dynamic aperiodic servers was presented by Ghazalie and Baker [13]. Their approach is based on reserving an extra budget to the aperiodic server for synchronization purpose and uses the utilization-based test [14] for verifying the feasibility of the schedule. Lipari and Buttazzo [15] extended the analysis to a total bandwidth server.

In the following section, we will describe an efficient method based on the work of Caccamo and Sha [12] for integrating hard and soft activities under resource constraints. This approach improves the results achieved earlier [13,15] since it can efficiently handle soft real-time requests with a variable or unknown execution behavior within a hard real-time environment under the EDF [14] scheduling policy.

To avoid unpredictable delays on hard tasks, soft tasks are isolated through a bandwidth reservation mechanism (capacity-based server), according to which each soft task is assigned a fraction of the CPU (a bandwidth) and it is scheduled in such a way that it will never demand more than its reserved bandwidth, independently of its actual requests. For the sake of simplicity, we will refer to a specific aperiodic server, the constant bandwidth server (CBS) [16]. However, the described protocol can be used with any capacity-based aperiodic server.

Introducing resource sharing among hard periodic and soft aperiodic tasks using the CBS cannot be done in a straightforward manner. The key challenge in the integration of CBS and SRP is that the preemption levels were developed under the assumption that relative deadlines are static and thus the resulting preemption levels. Unfortunately, handling resource constraints along with a server like CBS, we have to deal with varying server relative deadlines and the resulting dynamic preemption levels.

Another challenge, due to using a capacity-based server, is to avoid having an aperiodic task suspend its execution inside a critical section. In fact, whenever an aperiodic task runs out of budget while inside a critical section, we should allow the task to continue executing until it leaves the critical section. Such an additional execution time is an overrun whose effects need to be taken into account in the feasibility test. SRP for hybrid task sets consists of two distinct parts:

- an extension of SRP, based on the notion of dynamic preemption levels;
- a compliant capacity-based server (it has to satisfy specific requirements in order to make the idea of hybrid resource sharing work).

Combining these two elements, maintains all basic SRP properties (including the guarantee test) in spite of unknown aperiodic relative deadlines. The approach is completely transparent with respect to hard periodic tasks whose scheduling decisions follow classical EDF and SRP.

29.4.1 Assumptions and Terminology

We consider a hybrid task set consisting of hard periodic tasks and soft aperiodic tasks that have to be executed on a uniprocessor system. Tasks are preemptable and are scheduled using the earliest deadline first (EDF) algorithm. Hard and soft tasks may share a set $\mathcal{R}$ of mutually exclusive resources accessed using the SRP.

Each task τ_i consists of an infinite sequence of identical activities, called instances or jobs. A hard periodic task τ_i is characterized by a pair (C_i, T_i), where C_i is the worst-case computation time (WCET) of the task, and T_i is the job activation period. Each periodic task has a relative deadline D_i equal to its period.

Aperiodic tasks have irregular job activation and job arrival times are not known *a priori*. Soft tasks do not have deadlines of their own, but they are assigned a fictitious deadline by the CBS algorithm for scheduling purposes. To achieve isolation among soft tasks, we assume that each soft aperiodic task is handled by a dedicated CBS, which keeps the pending job requests in a queue served by a given discipline. For the sake of simplicity, we assume jobs are served in a FIFO order, but any arbitrary (preemptive or nonpreemptive) discipline can be used instead.

Each CBS server is characterized by a pair (Q_s, T_s), where Q_s is the maximum budget and T_s is the server period. The ratio $U_s = Q_s / T_s$ is denoted as the server bandwidth. A CBS also maintains two dynamic variables: the current budget c_s and the current deadline d_s assigned to the served job. Basic CBS rules[4] can be summarized as follows:

1. Each server S_i is characterized by a budget c_i and by an ordered pair (Q_i, T_i), where Q_i is the maximum budget and T_i is the period of the server. At each instant, a fixed deadline $d_{i,k}$ is associated with the server. At the beginning $\forall i,\ d_{i,0} = 0$.
2. When a task instance $\tau_{i,j}$ arrives and the server is idle, the server generates a new deadline $d_{i,k} = \max(r_{i,j}, d_{i,k-1}) + T_i$ and c_i is recharged at the maximum value Q_i.
3. Whenever a served job executes, the budget c_i is decreased by the same amount.
4. When the server is active and c_i becomes equal to zero, the server budget is recharged at the maximum value Q_i and a new server deadline is generated as $d_{i,k} = d_{i,k-1} + T_i$.

Throughout this section, whenever a deadline postponement occurs, d_s^{new} denotes the new postponed server deadline and d_s^{old} denotes the previous one. To simplify the notation in the examples, the jobs handled by a server will be denoted as $\{J_1, J_2, \ldots, J_i, \ldots\}$. Due to the CBS bandwidth reservation mechanism, each job may be assigned several fictitious deadlines during its execution. As a consequence, a job J_i can be seen as a sequence of chunks $H_{i,j}$ $(j = 1, 2, \ldots)$, each having an arrival time $a_{i,j}$ and a fixed deadline $d_{i,j}$. If job J_i becomes eligible at time e_i, the first chunk $H_{i,1}$ has arrival time $a_{i,1} = e_i$ and deadline $d_{i,1} = d_s$ equal to the current server deadline. Whenever a new deadline d_s^{new} is assigned to the job at time $\bar{t}$, then the current chunk $H_{i,j}$ is terminated and a new chunk $H_{i,j+1}$ is generated at time $a_{i,j+1} = \bar{t}$ with deadline $d_{i,j+1} = d_s^{\text{new}}$.

[4]This is a simplified version of the CBS algorithm. Its complete formulation can be found in [16].

TABLE 29.1 Parameters of the Task Set

Task	Type	Q_s or C	T_s or T	R_a	R_b
J_1	Soft aperiodic	4	10	—	3
τ_1	Hard periodic	2	12	2	—
τ_2	Hard periodic	6	24	4	2

29.4.2 Preventing Budget Exhaustion Inside Critical Sections

When shared resources are accessed in mutual exclusion by tasks handled by a capacity-based server, problems arise if the server exhausts its budget when a task is inside a critical section. In order to prevent long blocking delays due to the budget replenishment rule, a job which exhausts its budget should be allowed to continue executing with the same deadline, using extra budget until it leaves the critical section. At this time, the budget can be replenished at its full value and the deadline postponed.

The maximum interference created by the budget overrun mechanism occurs when the server exhausts its budget immediately after the job entered its longest critical section. Thus, if ξ is the duration of the longest critical section of task τ handled by server S, the bandwidth demanded by the server becomes $\frac{Q_s+\xi}{T_s}$. This approach inflates the server utilization.

Alternatively, a job can perform a budget check before entering a critical section at time t. If the current budget c_s is not sufficient to complete the job's critical section, the budget is replenished and the server deadline postponed. The remaining part of the job follows the same procedure until the job completes.

This approach dynamically partitions a job into *chunks*. Each chunk has execution time such that the consumed bandwidth is always less than or equal to the available server bandwidth U_s. By construction, a chunk has the property that it will never suspend inside a critical section. The following example illustrates two different solutions using the CBS server with the SRP protocol still maintaining static preemption levels:

Example 29.5

The task set consists of an aperiodic job J_1, handled by a CBS with $Q_s = 4$ and $T_s = 10$, and two periodic tasks, τ_1 and τ_2, sharing two resources R_a and R_b. In particular, J_1 and τ_2 share resource R_b, whereas τ_1 and τ_2 share resource R_a. The task set parameters are shown in Table 29.1.

A simple-minded solution could maintain a fixed relative deadline whenever the budget must be replenished and the deadline postponed. The advantage of having a fixed relative deadline is to keep the SRP policy unchanged for handling resource sharing between soft and hard tasks. In this way, the budget is recharged by a variable amount according to the formula: $c_s = c_s + (d_s^{\text{new}} - d_s^{\text{old}})U_s$, where d_s^{new} is the postponed server deadline and d_s^{old} is the previous server deadline.

A possible solution produced by CBS+SRP is shown in Figure 29.7. Notice that the ceiling of resource R_a is $ceil(R_a) = 1/12$, and the ceiling of R_b is $ceil(R_b) = 1/10$. When job J_1 arrives at time $t = 2$, its first chunk $H_{1,1}$ receives a deadline $d_{1,1} = a_{1,1} + T_s = 12$ according to the CBS algorithm. At that time, τ_2 is already inside a critical section on resource R_a, however $H_{1,1}$ of job J_1 is able to preempt, having its preemption level $\pi_1 = 1/10 > \Pi(J_1)$. At time $t = 5$, J_1 tries to access a critical section, however its residual budget is equal to 1 and is not sufficient to complete the whole critical section. As a consequence, a new chunk $H_{1,2}$ is generated with an arrival time $a_{1,2} = 5$ and a deadline $d_{1,2} = a_{1,2} + T_s = 15$ (the relative deadline is fixed). The budget is replenished according to the available server bandwidth; hence, it follows that $c_s = c_s + (d_s^{\text{new}} - d_s^{\text{old}})U_s = 1 + 1.2$. Unfortunately, the current budget is not sufficient to complete the critical section and an extra budget[5] equal to 0.8 is needed. Hence, we have to inflate the

[5]The extra budget is generated at the cost of reserving some extra bandwidth for the CBS server: it has to be taken into account during the schedulability analysis.

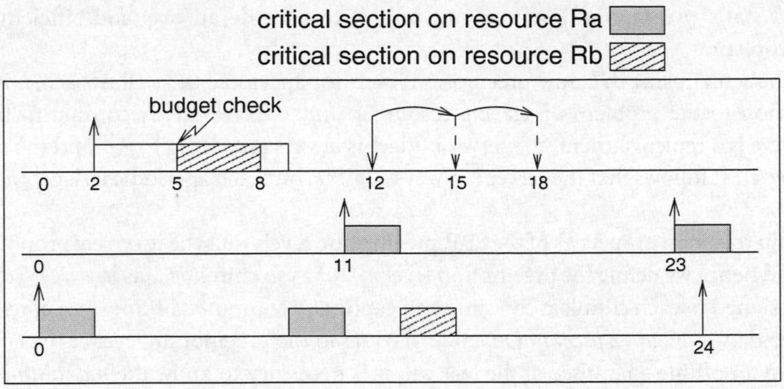

FIGURE 29.7 CBS+SRP with static preemption levels.

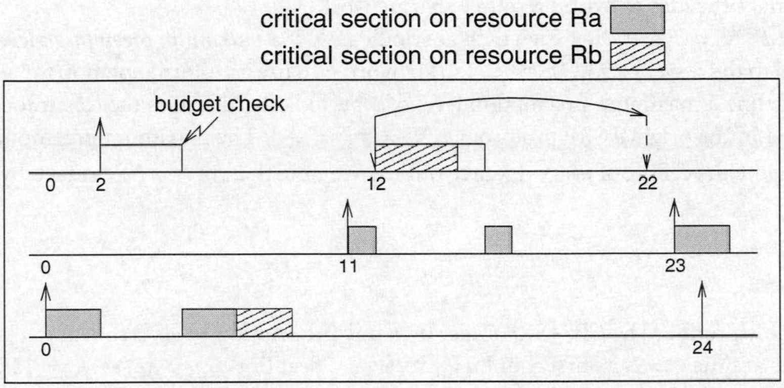

FIGURE 29.8 CBS+SRP with static preemption levels and job suspension.

budget wasting bandwidth. The remaining part of the job follows the same procedure maintaining a *fixed relative deadline* ($d_{1,3} = a_{1,3} + T_s = 8 + 10 = 18$) until the job completes. This approach has two main drawbacks: an extra budget still needs to be reserved, and jobs are cut in too many chunks, so increasing the algorithm overhead.

Another simple-minded solution could suspend a job whenever its budget is exhausted until the current server deadline. Only at that time, the job would become again eligible, and a new chunk would be ready to execute with the budget recharged at its maximum value ($c_s = Q_s$), and the deadline postponed by a server period.

The schedule produced using this approach on the previous example is shown in Figure 29.8. When job J_1 arrives at time $t = 2$, its first chunk $H_{1,1}$ receives a deadline $d_{1,1} = a_{1,1} + T_s = 12$ according to the CBS algorithm. As previously shown, at time $t = 5$, J_1 tries to access a critical section, however its residual budget is equal to one and is not sufficient to complete the whole critical section. As a consequence, J_1 is temporarily suspended and a new chunk is released at time $t = 12$, with deadline $d_{1,2} = 22$ and the budget replenished ($c_s = Q_s = 4$). This approach also has two main drawbacks: it increases the response time of aperiodic tasks and, whenever the budget is recharged, the residual budget amount (if any) is wasted due to job suspension.

29.4.3 Dynamic Preemption Levels

The two methods described in the previous section show that, although the introduction of budget check can prevent budget exhaustion inside a critical section without inflating the server size, fixed relative deadline and static preemption levels do not permit to provide an easy and efficient solution to the addressed problem.

We now show that using dynamic preemption levels for aperiodic tasks allows achieving a simpler and elegant solution to the problem of sharing resources under CBS+SRP. According to the new method, whenever there is a replenishment, the server budget is always recharged by Q_s and the server deadline is postponed by T_s. It follows that the server is always eligible, but each aperiodic task gets a dynamic relative deadline.

To maintain the main properties of the SRP, preemption levels must be inversely proportional to relative deadlines, and hence we define the preemption level $\pi_{i,j}$ of a job chunk $H_{i,j}$ as $\pi_{i,j} = 1/(d_{i,j} - a_{i,j})$. Notice that $\pi_{i,j}$ is assigned to each chunk at run time and cannot be computed offline. As a consequence, a job J_i is characterized by a *dynamic preemption level* π_i^d equal to the preemption level of the current chunk.

To perform an offline guarantee of the task set, it is necessary to know the *maximum preemption level* that can be assigned to each job J_i by the server. Therefore, the deadline assignment rule of the server must be compliant in order to guarantee that each chunk has a minimum relative deadline $D_i^{\min}$ equal to its server period (in Section 29.4.5, we identify the specific requirements the aperiodic server needs to meet in order to make the idea of hybrid resource sharing work).

By setting $D_i^{\min} = T_s$, we can assign each aperiodic task τ_i a *maximum preemption level* $\pi_i^{\max}$ inversely proportional to the server period ($\pi_i^{\max} = 1/T_s$). In order to use a uniform notation for all the tasks in the system, we define a maximum preemption level of a periodic hard task as the classical preemption level typically used in the original SRP protocol ($\pi_i^{\max} = \pi_i = \frac{1}{D_i}$). The maximum preemption levels will be used to compute the ceiling of every resource offline. We note that $\pi_i^d \leq \pi_i^{\max}$, in fact, by definition,

$$\forall i, \quad j\ \pi_{i,j} = \frac{1}{d_{i,j} - a_{i,j}} = \pi_i^d \leq \frac{1}{D_i^{\min}} = \frac{1}{T_s} = \pi_i^{\max} \tag{29.11}$$

The schedule produced by CBS+SRP under dynamic preemption levels is shown in Figure 29.9. When job J_1 arrives at time $t = 2$, its first chunk $H_{1,1}$ receives a deadline $d_{1,1} = a_{1,1} + T_s = 12$ according to the CBS algorithm. At that time, τ_2 is already inside a critical section on resource R_a; however, $H_{1,1}$ of job J_1 is able to preempt, having a preemption level $\pi_{1,1} = 1/10 > \Pi(H_{1,1})$. At time $t = 5$, J_1 tries to access a critical section; however, its residual budget is equal to one and is not sufficient to complete the whole critical section. As a consequence, the deadline is postponed and the budget replenished ($c_s = c_s + Q_s = 1 + 4$). Hence, the next chunk $H_{1,2}$ of J_1 starts at time $a_{1,2} = 5$ with deadline $d_{1,2} = d_{1,1} + T_s = 22$ and budget $c_s = 5$. However, chunk $H_{1,2}$ cannot start because its preemption level $\pi_{1,2} = 1/17 < \Pi(H_{1,2})$. It follows that τ_2 executes until the end of its critical section. When its system ceiling becomes zero, J_1 is able to preempt τ_2. We note that the bandwidth consumed by any chunk is no greater than U_s, since whenever the budget is refilled by Q_s, the absolute deadline is postponed by T_s.

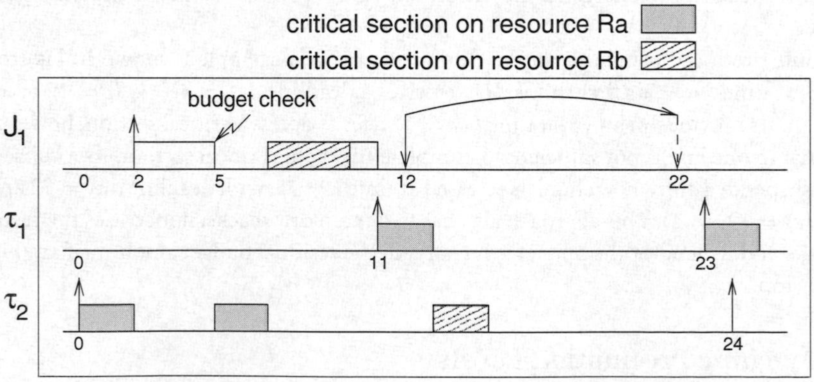

FIGURE 29.9 CBS+SRP with dynamic preemption levels.

The main advantage of the proposed approach is that it does not need to reserve extra budget for synchronization purposes and does not waste the residual budget (if any) left by the previous chunk. However, we need to determine the effects that dynamic preemption levels have on the properties of the SRP protocol.

It can be noted that, since each chunk is scheduled by a fixed deadline assigned by the CBS, each chunk inherits the SRP properties. In particular, each chunk can be blocked for at most the duration of one critical section by the preemption test and, once started, it will never be blocked for resource contention. However, since a soft aperiodic job may consist of many chunks, it can be blocked more than once. The behavior of hard tasks remains unchanged, permitting resource sharing between hard and soft tasks without jeopardizing the hard tasks' guarantee. The details of the proposed technique are described in the next section.

29.4.4 The Algorithm

In this section we first define the rules governing SRP with dynamic preemption levels that have been informally introduced in the previous section. We then prove its properties. To comply with the classical SRP rules, a chunk $H_{i,j}$ starts its execution only if its priority is the highest among the active tasks and its *preemption level* $\pi_{i,j} = 1/(d_{i,j} - a_{i,j})$ is greater than its system ceiling. In order for the SRP protocol to be correct, every resource R_i is assigned a static[6] ceiling $ceil(R_i)$ (we assume binary semaphores) equal to the highest maximum preemption level of the tasks that could be blocked on R_i when the resource is busy. Hence, $ceil(R_i)$ can be computed as follows:

$$ceil(R_i) = \max_k\{\pi_k^{\max} \mid \tau_k \text{ needs } R_i\} \tag{29.12}$$

It is easy to see that the ceiling of a resource computed by Equation (29.12) is greater than or equal to the one computed using the dynamic preemption level of each task. In fact, as shown by Equation (29.11), the maximum preemption level of each aperiodic task represents an upper-bound of its dynamic value.

Finally, in computing the blocking time for a periodic/aperiodic task, we need to take into account the duration of the critical section of an aperiodic task without considering its relative deadline. In fact, the actual relative deadline of a chunk belonging to an aperiodic task is assigned online and it is not known in advance.

To simplify our formulation, we assume that each hard periodic task is handled by a dedicated CBS server with $Q_{s_i} \geq C_i$ and $T_{s_i} = T_i$. With such a parameters assignment, hard tasks do not really need a server in order to be scheduled; we prefer to use a server for hard tasks also, since it gives us the opportunity to treat periodic and aperiodic tasks in the same way.

The blocking times can be computed as a function of the minimum relative deadline of each aperiodic task, as follows:

$$B_i = \max\{z_{j,h} \mid (T_{s_i} < T_{s_j}) \wedge \pi_i^{\max} \leq ceil(\rho_{j,h})\} \tag{29.13}$$

where $z_{j,h}$ is the worst-case execution time of the hth critical section of task τ_j, $\rho_{j,h}$ is the resource accessed by the critical section $z_{j,h}$, and T_{s_i} is the period of the dedicated server. The parameter B_i, computed by Equation (29.13), is the blocking time experienced by a hard or soft task. In fact, $T_{s_i} = D_i^{\min}$ for a soft task and $T_{s_i} = D_i$ for a hard periodic task.

The correctness of the described approach will be formally proved in Section 29.4.7. It will be shown that the modifications introduced in the SRP protocol do not change its properties and permit keeping a static ceiling for the resources even though the relative deadline of each chunk is dynamically assigned at run time by the aperiodic server.

[6]In the case of multiunits resources, the ceiling of each resource is dynamic as it depends on the number of units actually free.

29.4.5 Requirements for the Aperiodic Server

Some additional requirements have to be introduced to handle resource constraints. In particular, the correctness of the proposed technique relies on the following rules:

- Each job chunk must have a minimum relative deadline known *a priori*.
- A task must never exhaust its budget when it is inside a critical section.

The CBS, according to its original formulation [16], is not compliant with both of the requirements above. As a consequence, to make it work when a hybrid task set shares common resources in exclusive mode, CBS needs to be modified as follows:

1. Each server S_i is characterized by a budget c_i and by an ordered pair (Q_i, T_i), where Q_i is the maximum budget and T_i is the period of the server. At each instant, a fixed deadline $d_{i,k}$ is associated with the server. At the beginning $\forall i,\ d_{i,0} = 0$.
2. When a job $\tau_{i,j}$ arrives and the server is idle, the server generates a new deadline $d_{i,k} = \max(r_{i,j}, d_{i,k-1}) + T_i$ and c_i is recharged at the maximum value Q_i.
3. Whenever a served job executes, the budget c_i is decreased by the same amount.
4. When the server is active and c_i becomes equal to zero, the server budget is recharged at the maximum value Q_i and a new server deadline is generated as $d_{i,k} = d_{i,k-1} + T_i$.
5. Whenever a served job $\tau_{i,j}$ tries to access a critical section, if $c_i < \xi_i$ (where ξ_i is the duration of the longest critical section of job $\tau_{i,j}$ such that $\xi_i < Q_i$), a budget replenishment occurs, that is $c_i = c_i + Q_i$ and a new server deadline is generated as $d_{i,k} = d_{i,k-1} + T_i$.

It is worth noting that, with respect to the original definition given in [16], we modified rule 2 and introduced rule 5. Rule 2 has been modified in order to guarantee that each job chunk has a minimum relative deadline equal to the server period. In fact, whenever a job J_i arrives and the server is idle, the job gets an absolute deadline greater than or equal to the arrival time plus the server period. The budget is recharged in such a way that the consumed bandwidth is always no greater than the reserved bandwidth $U_s = Q_s/T_s$.

Rule 5 has been added to prevent a task from exhausting its budget when it is using a shared resource. This is done by performing a budget check before entering a critical section. If the current budget is not sufficient to complete a critical section, the budget is replenished and the deadline postponed.

These two minor changes allow the CBS server to become compliant with the proposed approach without modifying its global behavior.

29.4.6 An Example

The following example illustrates the usage of the CBS server in the presence of resource constraints. The task set consists of an aperiodic job J_1, handled by a CBS with maximum budget $Q_s = 4$ and server period $T_s = 8$ and two periodic tasks τ_1 and τ_2, which share two resources R_a and R_b; in particular, J_1 and τ_1 share resource R_b, while τ_1 and τ_2 share resource R_a. The task set parameters are shown in Table 29.2.

The schedule produced by CBS+SRP is shown in Figure 29.10. When job J_1 arrives at time $t = 3$, its first chunk $H_{1,1}$ receives a deadline $d_{1,1} = a_{1,1} + T_s = 11$ according to the CBS algorithm. At that time, τ_2 is already inside a critical section on resource R_a; however, $H_{1,1}$ of job J_1 is able to preempt it, because its preemption level is $\pi_{1,1} = 1/8 > \Pi(H_{1,1})$. At time $t = 6$, J_1 tries to access a critical section;

TABLE 29.2 Parameters of the Task Set

Task	Type	Q_s or C	T_s or T	R_a	R_b
J_1	Soft aperiodic	4	8	—	3
τ_1	Hard periodic	2	10	1	1
τ_2	Hard periodic	6	24	3	—

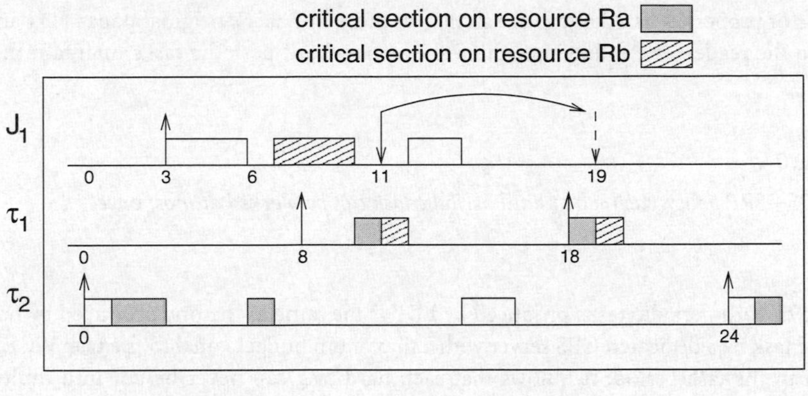

FIGURE 29.10 Schedule produced by CBS+SRP.

however, its residual budget is equal to one and is not sufficient to complete the whole critical section. As a consequence, the deadline is postponed and the budget replenished. Hence, the next chunk $H_{1,2}$ of J_1 starts at time $a_{1,2} = 6$ with deadline $d_{1,2} = 19$. The chunk $H_{1,2}$ of J_1 cannot start because its preemption level $\pi_{1,2} = 1/13 < \Pi(H_{1,2})$. It follows that τ_2 executes until the end of its critical section. When its system ceiling becomes zero, J_1 is able to preempt τ_2. When J_1 frees resource R_b, τ_1 starts executing. It is worth noting that each chunk can be blocked for at most the duration of one critical section by the preemption test and, once it is started, it will never be blocked for resource contention.

In the next section, the SRP properties are formally proved and the validity of the guarantee test is analyzed.

29.4.7 Theoretical Validation of the Model

In this section we prove the properties of SRP for a hybrid task set. In particular, we show that all SRP properties are preserved for hard periodic tasks and for each chunk of soft aperiodic tasks. Finally, we provide a sufficient guarantee test for verifying the schedulability of hybrid task sets consisting of hard and soft tasks.

Since a preemption level is always inversely proportional to the relative deadline of each chunk, the following properties can be derived in a straightforward fashion:

Property 1

A chunk $H_{i,h}$ is not allowed to preempt a chunk $H_{j,k}$, unless $\pi_{i,h} > \pi_{j,k}$.

Property 2

If the preemption level of a chunk $H_{i,j}$ is greater than its current system ceiling, then there are sufficient resources available to meet the requirement of $H_{i,j}$ and the requirement of every chunk that can preempt $H_{i,j}$.

Property 3

If no chunk $H_{i,j}$ is permitted to start until $\pi_{i,j} > \Pi(H_{i,j})$, then no chunk can be blocked after it starts.

Property 4

Under the CBS+SRP policy, a chunk $H_{i,j}$ can be blocked for at most the duration of one critical section.

Property 5

CBS+SRP prevents deadlocks.

The proofs of properties listed above are similar to those in Baker's original paper [11] and are left as an exercise to the reader. The following lemma shows how hard periodic tasks maintain their behavior unchanged:

Lemma 29.10

Under the CBS+SRP policy, each job of hard periodic task can be blocked at most once.

Proof

The schedule of hard periodic tasks produced by EDF is the same as the one produced by handling each hard periodic task by a dedicated CBS server with a maximum budget equal to the task WCET and server period equal to the task period; it follows that each hard task can never be cut into multiple chunks. Hence, using Property 4, it follows that each instance of a hard periodic task can be blocked for at most the duration of one critical section. □

The following theorem provides a simple sufficient condition to guarantee the feasibility of hard tasks when they share resources with soft tasks under the CBS+SRP algorithm.

Theorem 29.10

Let Γ be a task set composed of n hard periodic tasks and m soft aperiodic tasks, each one (soft and hard) scheduled by a dedicated server. Supposing tasks are ordered by decreasing maximum preemption level (so that $\pi_i^{\max} \geq \pi_j^{\max}$ only if $i < j$), then the hard tasks are schedulable by CBS+SRP if

$$\forall i, \ 1 \leq i \leq n+m \quad \sum_{j=1}^{i} \frac{Q_j}{T_j} + \frac{B_i}{T_i} \leq 1, \tag{29.14}$$

where Q_j is the maximum budget of the dedicated server, T_j is the server period, and B_i is the maximum blocking time of task τ_i.

Proof

Suppose Equation (29.14) is satisfied for each τ_i. Notice that aperiodic tasks get dynamic relative deadlines due to the deadline assignment rule of the CBS algorithm; hence, it follows that each task chunk has a relative deadline greater than or equal to its server period. Therefore, we have to analyze two cases:

Case A. Task τ_i has a relative deadline $D_i = T_i$. Using Baker's guarantee test (see Equation (29.7) in Section 29.3), it follows that the task set Γ is schedulable if

$$\forall i, \ 1 \leq i \leq n+m \quad \sum_{j=1}^{i-1} \frac{Q_j}{D_j} + \frac{Q_i}{T_i} + \frac{B_i^{\text{new}}}{T_i} \leq 1 \tag{29.15}$$

where D_j ($D_j \geq T_j$) is the relative deadline of task τ_j and B_i^{new} is the blocking time τ_i might experience when each τ_j has a relative deadline equal to D_j. Notice that a task τ_j can block as well as preempt τ_i varying its relative deadline D_j; however, τ_j cannot block and preempt τ_i simultaneously. In fact, if the current instance of τ_j preempts τ_i, its absolute deadline must be before τ_i's deadline; hence, the same instance of τ_j cannot also block τ_i, otherwise it should have its deadline after τ_i's deadline. From considerations above, the worst-case scenario happens when τ_i experiences the maximum number of preemptions: it occurs by shortening as much as possible the relative deadline of each task τ_j, that is, setting $D_j = T_j$. Hence, it follows that:

$$\forall i, \; 1 \le i \le n+m \quad \sum_{j=1}^{i-1} \frac{Q_j}{D_j} + \frac{Q_i}{T_i} + \frac{B_i^{\text{new}}}{T_i} \le \sum_{j=1}^{i-1} \frac{Q_j}{T_j} + \frac{Q_i}{T_i} + \frac{B_i}{T_i}$$

is worth noting that, even though B_i might be less than B_i^{new}, the last inequality holds since τ_j's interference due to preemption is always greater than or equal to its blocking effect, that is, $\sum_{j=1}^{i-1}(Q_j/T_j - Q_j/D_j) \ge (B_i^{\text{new}} - B_i)/T_i$. Finally,

$$\sum_{j=1}^{i-1} \frac{Q_j}{T_j} + \frac{Q_i}{T_i} + \frac{B_i}{T_i} \le 1$$

Notice that the last inequality holds for the theorem hypothesis; hence, Equation (29.15) is satisfied and the task set is schedulable.

Case B. Task τ_i has a relative deadline $D_i > T_i$. As in Case A, the task set Γ is schedulable if

$$\forall i, \; 1 \le i \le n+m \quad \sum_{j=1}^{i-1} \frac{Q_j}{D_j} + \frac{Q_i}{D_i} + \frac{B_i^{\text{new}}}{D_i} \le 1 \qquad (29.16)$$

From the considerations above, it follows that the worst-case scenario also occurs when $(\forall j, D_j = T_j)$, hence

$$\forall i, \; 1 \le i \le n+m \quad \sum_{j=1}^{i-1} \frac{Q_j}{D_j} + \frac{Q_i}{D_i} + \frac{B_i^{\text{new}}}{D_i} \le \sum_{j=1}^{i-1} \frac{Q_j}{T_j} + \frac{Q_i}{D_i} + \frac{B_i^{\text{new}}}{D_i}$$

Notice that tasks are sorted in decreasing order of maximum preemption levels and each task τ_j has the relative deadline set as $D_j = T_j$, except task τ_i whose relative deadline is $D_i > T_i$. Since τ_i has an unknown relative deadline whose value changes dynamically, Equation (29.16) has to be checked for each D_i, where D_i is greater than T_i. Hence, from Equation (29.6) (see Section 29.3) we derive that the blocking time B_i^{new} of task τ_i is a function of the actual relative deadline D_i as follows:

$$T_i \le D_i < T_{i+1} \Rightarrow B_i^{\text{new}} = B_i$$

$$T_{i+1} \le D_i < T_{i+2} \Rightarrow B_i^{\text{new}} = B_{i+1}$$

$$\vdots$$

$$T_{n+m-1} \le D_i < T_{n+m} \Rightarrow B_i^{\text{new}} = B_{n+m-1}$$

$$T_{n+m} \le D_i \Rightarrow B_i^{\text{new}} = B_{n+m} = 0$$

It is worth noting that the terms $B_i, B_{i+1}, \ldots, B_{n+m}$ are the blocking times computed by Equation (29.13) and are experienced by hard or soft tasks if the relative deadline of each task is set equal to the period of its dedicated server. Finally, a $k \ge i$ will exist such that

$$T_k \le D_i < T_{k+1} \Rightarrow B_i^{\text{new}} = B_k$$

so, it follows that

$$\sum_{j=1}^{i-1}\frac{Q_j}{T_j}+\frac{Q_i}{D_i}+\frac{B_i^{\text{new}}}{D_i}=\sum_{j=1}^{i-1}\frac{Q_j}{T_j}+\frac{Q_i}{D_i}+\frac{B_k}{D_i}$$
$$\le\sum_{j=1}^{i-1}\frac{Q_j}{T_j}+\frac{Q_i}{T_i}+\frac{B_k}{T_k}$$
$$\le\sum_{j=1}^{i}\frac{Q_j}{T_j}+\sum_{h=i+1}^{k}\frac{Q_h}{T_h}+\frac{B_k}{T_k}$$

The last inequality holds because k must be greater than or equal to i and $D_i \ge T_k \ge T_i$. Finally

$$\sum_{j=1}^{i}\frac{Q_j}{T_j}+\sum_{h=i+1}^{k}\frac{Q_h}{T_h}+\frac{B_k}{T_k}=\sum_{j=1}^{k}\frac{Q_j}{T_j}+\frac{B_k}{T_k}\le 1$$

The above inequality holds for the theorem hypothesis; hence, Equation (29.16) is satisfied and the task set is schedulable. □

References

[1] B.W. Lampson and D.D. Redell, Experiences with Processes and Monitors in Mesa, *Communications of the ACM*, 23(2), 105–117, 1980.

[2] D. Cornhill and L. Sha, Priority Inversion in Ada, *Ada Letters*, No. 7, 30–32, November 1987.

[3] L. Sha, L.R. Rajkumar, and J.P. Lehoczky, Priority Inheritance Protocols: An Approach to Real-Time Synchronization, *IEEE Transactions on Computers*, 39(9), 1990.

[4] R. Rajkumar, *Synchronization in Real-Time Systems: A Priority Inheritance Approach*, Kluwer Academic Publishers, Dordrecht, 1991.

[5] M. Klein, T. Ralya, B. Pollak, R. Obenza, and M.G. Harbour, *A Practitioner's Handbook for Real-Time Analysis*, Kluwer Academic Publishers, Dordrecht, 1994.

[6] J. Liu, *Real-Time Systems*, Prentice Hall, New York, 2000.

[7] G.C. Buttazzo, *Hard Real-Time Computing Systems: Predictable Scheduling Algorithms and Applications*, Kluwer Academic Publishers, Boston, 1997.

[8] L. Sha, R. Rajkumar, S. Son, and C.H. Chang, A Real-Time Locking Protocol, *IEEE Transactions on Computers*, 40(7), 793–800, 1991.

[9] M.I. Chen, Schedulability Analysis of Resource Access Control Protocols in Real-Time Systems, Ph.D. Thesis, Technical Report: UIUCDCS-R-91-1705, CS, UIUC, 1991.

[10] R. Rajkumar, L. Sha, and J.P. Lehoczky, Real-Time Synchronization Protocols for Multiprocessors, *Proceedings of the IEEE Real-Time System Symposium*, pp. 259–269, 1988.

[11] T.P. Baker, Stack-Based Scheduling of Real-Time Processes, *The Journal of Real-Time Systems*, 3(1), 67–100, 1991.

[12] M. Caccamo and L. Sha, Aperiodic Servers with Resource Constraints, *Proceedings of IEEE Real-Time Systems Symposium*, London, December 2001.

[13] T.M. Ghazalie and T.P. Baker, Aperiodic Servers in a Deadline Scheduling Environment, *The Journal of Real-Time Systems*, 9, 21–36, 1995.

[14] C.L. Liu and J.W. Layland, Scheduling Algorithms for Multiprogramming in a Hard Real-Time Environment, *Journal of the ACM* 20(1), 40–61, 1973.

[15] G. Lipari and G.C. Buttazzo, Schedulability Analysis of Periodic and Aperiodic Tasks with Resource Constraints, *Journal of Systems Architecture*, 46(4), 327–338, 2000.

[16] L. Abeni and G. Buttazzo, Integrating Multimedia Applications in Hard Real-Time Systems, *Proceedings of the IEEE Real-Time Systems Symposium*, Madrid, Spain, December 1998.

30

A Categorization of Real-Time Multiprocessor Scheduling Problems and Algorithms

John Carpenter
University of North Carolina at Chapel Hill

Shelby Funk
University of North Carolina at Chapel Hill

Philip Holman
University of North Carolina at Chapel Hill

Anand Srinivasan
Microsoft Corporation

James Anderson
University of North Carolina at Chapel Hill

Sanjoy Baruah
University of North Carolina at Chapel Hill

30.1 Introduction

Real-time multiprocessor systems are now commonplace. Designs range from single-chip architectures, with a modest number of processors, to large-scale signal-processing systems, such as synthetic-aperture radar systems. For uniprocessor systems, the problem of ensuring that deadline constraints are met has been widely studied: effective scheduling algorithms that take into account the many complexities that arise in real systems (e.g., synchronization costs, system overheads, etc.) are well understood. In contrast, researchers are just beginning to understand the trade-offs that exist in multiprocessor systems. In this chapter we analyze the trade-offs involved in scheduling independent, periodic real-time tasks on a multiprocessor.

1-58488-397-9/$0.00+$1.50

Research on real-time scheduling has largely focused on the problem of scheduling of recurring processes, or *tasks*. The periodic task model of Liu and Layland is the simplest model of a recurring process [1,2]. In this model, a task T is characterized by two parameters: a worst-case *execution requirement* e and a *period* p. Such a task is invoked at each nonnegative integer multiple of p. (Task invocations are also called *job releases* or *job arrivals*.) Each invocation requires at most e units of processor time and must complete its execution within p time units. (The latter requirement ensures that each job is completed before the next job is released.) A collection of periodic tasks is referred to as a *periodic task system* and is denoted by τ.

We say that a task system τ is *schedulable* by an algorithm A if A ensures that the timing constraints of all tasks in τ are met. τ is said to be *feasible* under a class C of scheduling algorithms if τ is schedulable by some algorithm $A \in C$. An algorithm A is said to be *optimal* with respect to class C if $A \in C$ and A correctly schedules every task system that is feasible under C. When the class C is not specified, it should be assumed to include all possible scheduling algorithms.

30.1.1 Classification of Scheduling Approaches on Multiprocessors

Traditionally, there have been two approaches for scheduling periodic task systems on multiprocessors: *partitioning* and *global scheduling*. In global scheduling, all eligible tasks are stored in a single priority-ordered queue; the global scheduler selects for execution the highest priority tasks from this queue. Unfortunately, using this approach with optimal uniprocessor scheduling algorithms, such as the rate-monotonic (RM) and earliest-deadline-first (EDF) algorithms, may result in arbitrarily low processor utilization in multiprocessor systems [3]. However, recent research on *proportionate fair* (Pfair) scheduling has shown considerable promise in that it has produced the only known optimal method for scheduling periodic tasks on multiprocessors [4–8].

In partitioning, each task is assigned to a single processor, on which each of its jobs will execute, and processors are scheduled independently. The main advantage of partitioning approaches is that they reduce a *multiprocessor* scheduling problem to a set of *uniprocessor* ones. Unfortunately, partitioning has two negative consequences. First, finding an optimal assignment of tasks to processors is a bin-packing problem, which is NP-hard in the strong sense. Thus, tasks are usually partitioned using suboptimal heuristics. Second, as shown later, task systems exist that are schedulable if and only if tasks are not partitioned. Still, partitioning approaches are widely used by system designers.

In addition to the above approaches, we consider a new "middle" approach in which each job is assigned to a single processor, while a task is allowed to migrate. In other words, interprocessor task migration is permitted only at job boundaries. We believe that migration is eschewed in the design of multiprocessor real-time systems because its true cost in terms of the final system produced is not well understood. As a step toward understanding this cost, we present a new taxonomy that ranks scheduling schemes along the following two dimensions:

1. **The complexity of the priority scheme.** Along this dimension, scheduling disciplines are categorized according to whether task priorities are **(i)** static, **(ii)** dynamic but fixed within a job, or **(iii)** fully dynamic. Common examples of each type include (i) RM [2], (ii) EDF [2], and (iii) least-laxity-first (LLF) [9] scheduling.
2. **The degree of migration allowed.** Along this dimension, disciplines are ranked as follows: **(i)** no migration (i.e., task partitioning), **(ii)** migration allowed, but only at job boundaries (i.e., dynamic partitioning at the job level), and **(iii)** unrestricted migration (i.e., jobs are also allowed to migrate).

Because scheduling algorithms typically execute upon the same processor(s) as the task system being scheduled, it is important for such algorithms to be relatively simple and efficient. Most known real-time scheduling algorithms are work-conserving (see below) and operate as follows: at each instant, a *priority* is associated with each active job, and the highest-priority jobs that are eligible to execute are selected for execution upon the available processors. (A job is said to be *active* at time instant t in a given schedule if **(i)** it has arrived at or prior to time t; **(ii)** its deadline occurs after time t; and **(iii)** it has not yet completed execution.) In *work-conserving* algorithms, a processor is never left idle while an active job exists (unless migration constraints prevent the task from executing on the idle processor). Because the runtime overheads of such algorithms tend to be less than those of non-work-conserving algorithms, scheduling algorithms that make scheduling decisions online tend to be work-conserving. In this chapter, we limit our attention to work-conserving algorithms for this reason.[1]

To alleviate the runtime overhead associated with job scheduling (e.g., the time required to compute job priorities, to preempt executing jobs, to migrate jobs, etc.), designers can place constraints upon the manner in which priorities are determined and on the amount of task migration. However, the impact of these restrictions on the schedulability of the system must also be considered. Hence, the effectiveness of a scheduling algorithm depends on not only its runtime overhead, but also its ability to schedule feasible task systems.

The primary motivation of this work is to provide a better understanding of the trade-offs involved when restricting the form of a system's scheduling algorithm. If an algorithm is to be restricted in one or both of the above-mentioned dimensions for the sake of reducing runtime overhead, then it would be helpful to know the impact of the restrictions on the schedulability of the task system. Such knowledge would serve as a guide to system designers for selecting an appropriate scheduling algorithm.

30.1.2 Overview

The rest of this chapter is organized as follows. Section 30.2 describes our taxonomy and some scheduling approaches based on this taxonomy. In Section 30.3, we compare the various classes of scheduling algorithms in the taxonomy. Section 30.4 presents new and known scheduling algorithms and feasibility tests for each of the defined categories. Section 30.5 summarizes our results.

30.2 Taxonomy of Scheduling Algorithms

In this section we define our classification scheme. We assume that job preemption is permitted. We classify scheduling algorithms into three categories based upon the available degree of interprocessor migration. We also distinguish among three different categories of algorithms based upon the freedom with which priorities may be assigned. These two axes of classification are orthogonal to one another in the sense that restricting an algorithm along one axis does not restrict freedom along the other. Thus, there are $3 \times 3 = 9$ different classes of scheduling algorithms in this taxonomy.

30.2.1 Migration-Based Classification

Interprocessor migration has traditionally been forbidden in real-time systems for the following reasons:

- In many systems, the cost associated with each migration — the cost of transferring a job's context from one processor to another — can be prohibitive.
- Until recently, traditional real-time scheduling theory lacked the techniques, tools, and results to permit a detailed analysis of systems that allow migration. Hence, partitioning has been the preferred approach due largely to the nonexistence of viable alternative approaches.

Recent developments in computer architecture, including single-chip multiprocessors and very fast interconnection networks over small areas, have resulted in the first of these concerns becoming less of an issue. Thus, system designers need no longer rule out interprocessor migration solely due to implementation considerations, especially in tightly coupled systems. (However, it may still be desirable to restrict migration in order to reduce runtime overhead.) In addition, results of recent experiments demonstrate

[1]Pfair scheduling algorithms, mentioned earlier, that meet the Pfairness constraint as originally defined [6] are not work-conserving. However, work conserving variants of these algorithms have been devised in recent work [4,8].

that scheduling algorithms that allow migration are competitive in terms of schedulability with those that do not migrate, even after incorporating migration overheads [10]. This is due to the fact that systems exist that can be successfully scheduled only if interprocessor migration is allowed (refer to Lemmas 30.3 and 30.4 in Section 30.3).

In differentiating among multiprocessor scheduling algorithms according to the degree of migration allowed, we consider the following three categories:

1. No migration (partitioned) — In partitioned scheduling algorithms, the set of tasks is partitioned into as many disjoint subsets as there are processors available, and each such subset is associated with a unique processor. All jobs generated by the tasks in a subset must execute only upon the corresponding processor.
2. Restricted migration — In this category of scheduling algorithms, each job must execute entirely upon a single processor. However, different jobs of the same task may execute upon different processors. Thus, the runtime context of each job needs to be maintained upon only one processor; however, the task-level context may be migrated.
3. Full migration — No restrictions are placed upon interprocessor migration.

30.2.2 Priority-Based Classification

In differentiating among scheduling algorithms according to the complexity of the priority scheme, we again consider three categories.

1. Static priorities — A unique priority is associated with each task, and all jobs generated by a task have the priority associated with that task. Thus, if task T_1 has higher priority than task T_2, then whenever both have active jobs, T_1's job will have priority over T_2's job. An example of a scheduling algorithm in this class is the RM algorithm [2].
2. Job-level dynamic priorities — For every pair of jobs J_i and J_j, if J_i has higher priority than J_j at some instant in time, then J_i *always* has higher priority than J_j. An example of a scheduling algorithm that is in this class, but not the previous class, is EDF [2,11].
3. Unrestricted dynamic priorities — No restrictions are placed on the priorities that may be assigned to jobs, and the relative priority of two jobs may change at any time. An example scheduling algorithm that is in this class, but not the previous two classes, is the LLF algorithm [9].

By definition, unrestricted dynamic-priority algorithms are a generalization of job-level dynamic-priority algorithms, which are in turn a generalization of static-priority algorithms. In uniprocessor scheduling, the distinction between job-level and unrestricted dynamic-priority algorithms is rarely emphasized because EDF, a job-level dynamic-priority algorithm, is optimal [2]. In the *multiprocessor* case, however, unrestricted dynamic-priority scheduling algorithms are strictly more powerful than job-level dynamic-priority algorithms, as we will see shortly.

By considering all pairs of restrictions on migrations and priorities, we can divide the design space into $3 \times 3 = 9$ classes of scheduling algorithms. Before discussing these nine classes further, we introduce some convenient notation.

Definition 30.1

A scheduling algorithm is **(*x*,*y*)-restricted** *for* $x \in \{1, 2, 3\}$ *and* $y \in \{1, 2, 3\}$, *if it is in priority class x and migration class y (here, x and y correspond to the labels defined above).*

For example, a $(2, 1)$-restricted algorithm uses job-level dynamic priorities (i.e., level-2 priorities) and partitioning (i.e., level-1 migration), while a $(1, 3)$-restricted algorithm uses only static priorities (i.e., level-1 priorities) but allows unrestricted migration (i.e., level-3 migration). The nine categories of scheduling algorithms are summarized in Table 30.1. It is natural to associate classes of scheduling algorithms with the sets of task systems that they can schedule.

TABLE 30.1 A Classification of Algorithms for Scheduling Periodic Task Systems Upon Multiprocessor Platforms

	1. Static	2. Job-level Dynamic	3. Unrestricted Dynamic
3: Full migration	$(1,3)$-restricted	$(2,3)$-restricted	$(3,3)$-restricted
2: Restricted migration	$(1,2)$-restricted	$(2,2)$-restricted	$(3,2)$-restricted
1: Partitioned	$(1,1)$-restricted	$(2,1)$-restricted	$(3,1)$-restricted

Priority-assignment constraints are on the x-axis, and migration constraints are on the y-axis. In general, increasing distance from the origin may imply greater generality.

Definition 30.2

An ordered pair denoted by $\langle \boldsymbol{x,y} \rangle$ *denotes the set of task systems that are feasible under* (x,y)*-restricted scheduling.*

Of these nine classes, $(1,1)$-, $(2,1)$-, and $(3,3)$-restricted algorithms have received the most attention. For example, $(1,1)$-restricted algorithms have been studied in [3,12–14], while $(2,1)$-restricted algorithms (and equivalently, $(3,1)$-restricted algorithms) have been studied in [3,15,16]. The class of $(3,3)$-restricted algorithms has been studied in [1,4,6,8]. In addition to these, $(1,3)$- and $(2,3)$-restricted algorithms were recently considered in [17] and [18], respectively.

30.3 Schedulability Relationships

We now consider the problem of establishing relationships among the various classes of scheduling algorithms in Table 30.1. (Later, in Section 30.4, we explore the design of efficient algorithms in each class and present corresponding feasibility results.)

As stated in Section 30.1, our goal is to study the trade-offs involved in using a particular class of scheduling algorithms. It is generally true that the runtime overhead is higher for more-general models than for less-general ones: the runtime overhead of a (w,x)-restricted algorithm is at most that of a (y,z)-restricted algorithm if $y \geq w \wedge z \geq x$. However, in terms of schedulability, the relationships are not as straightforward. There are three possible relationships between (w,x)- and (y,z)-restricted scheduling classes, which we elaborate below. It is often the case that we discover some partial understanding of a relationship in one of the following two forms: $\langle w,x \rangle \subseteq \langle y,z \rangle$ and $\langle w,x \rangle \not\subseteq \langle y,z \rangle$, meaning "any task system in $\langle w,x \rangle$ is also in $\langle y,z \rangle$" and "there exists a task system that is in $\langle w,x \rangle$ but not in $\langle y,z \rangle$," respectively.

- The class of (w,x)-restricted algorithms is *strictly more powerful* than the class of (y,z)-restricted algorithms. That is, any task system that is feasible under the (y,z)-restricted class is also feasible under the (w,x)-restricted class. Furthermore, there exists at least one task system that is feasible under the (w,x)-restricted class but not under the (y,z)-restricted class. Formally, $\langle y,z \rangle \subset \langle w,x \rangle$ (where $\subset$ means proper subset). Of course, $\langle y,z \rangle \subset \langle w,x \rangle$ can be shown by proving that $\langle y,z \rangle \subseteq \langle w,x \rangle \wedge \langle w,x \rangle \not\subseteq \langle y,z \rangle$.
- The class of (w,x)-restricted algorithms and the class of (y,z)-restricted algorithms are equivalent. That is, a task system is feasible under the (w,x)-restricted class if and only if it is feasible under the (y,z)-restricted class. Formally, $\langle w,x \rangle = \langle y,z \rangle$, which can be shown by proving that $\langle w,x \rangle \subseteq \langle y,z \rangle \wedge \langle y,z \rangle \subseteq \langle w,x \rangle$.
- The class of (w,x)-restricted algorithms and the class of (y,z)-restricted algorithms are *incomparable*. That is, there exists at least one task system that is feasible under the (w,x)-restricted class but not under the (y,z)-restricted class, and vice versa. Formally, $\langle w,x \rangle \otimes \langle y,z \rangle$, which is defined as $\langle w,x \rangle \not\subseteq \langle y,z \rangle \wedge \langle y,z \rangle \not\subseteq \langle w,x \rangle$.

These potential relationships are summarized in Table 30.2.

TABLE 30.2 Possible Relationships Between the (w, x)- and (y, z)-Restricted Algorithm Classes

Notation	Semantically	Proof Obligation
$\langle w, x\rangle = \langle y, z\rangle$	(w, x)- and (y, z)-restricted classes are equivalent	$\langle w, x\rangle \subseteq \langle y, z\rangle \wedge \langle y, z\rangle \subseteq \langle w, x\rangle$
$\langle w, x\rangle \otimes \langle y, z\rangle$	(w, x)- and (y, z)-restricted classes are incomparable	$\langle w, x\rangle \not\subseteq \langle y, z\rangle \wedge \langle y, z\rangle \not\subseteq \langle w, x\rangle$
$\langle w, x\rangle \subset \langle y, z\rangle$	(y, z)-restricted class dominates (w, x)-restricted class	$\langle w, x\rangle \subseteq \langle y, z\rangle \wedge \langle y, z\rangle \not\subseteq \langle w, x\rangle$

Among the nine classes of scheduling algorithms identified in Table 30.1, it is intuitively clear (and borne out by formal analysis) that the class of $(3, 3)$-restricted algorithms is the most general in the sense that any task system that is feasible under the (x, y)-restricted class is also feasible under the $(3, 3)$-restricted class, for all x and y. Unfortunately, the runtime overhead of $(3, 3)$-restricted algorithms may prove unacceptably high for some applications, in terms of runtime complexity, preemption frequency, and migration frequency.

At first glance, it may seem that the class of $(1, 1)$-restricted algorithms is the least general class, in the sense that any task system that is feasible under the $(1, 1)$-restricted class is also feasible under the (x, y)-restricted class, for all x and y. However, Leung and Whitehead [19] have shown that $\langle 1, 1\rangle \otimes \langle 1, 3\rangle$. We have discovered that several other class pairs are similarly incomparable.

Some class relations are easily derived: since every static-priority algorithm is, by definition, a job-level dynamic-priority algorithm, and every job-level dynamic-priority algorithm is an unrestricted dynamic-priority algorithm, Theorem 30.1 (shown below) trivially holds.

Theorem 30.1

The following relationships hold across the rows of Table 30.1.

- $\langle 1, 1\rangle \subseteq \langle 2, 1\rangle \subseteq \langle 3, 1\rangle$
- $\langle 1, 2\rangle \subseteq \langle 2, 2\rangle \subseteq \langle 3, 2\rangle$
- $\langle 1, 3\rangle \subseteq \langle 2, 3\rangle \subseteq \langle 3, 3\rangle$

Similarly, as stated in the previous section, the optimality of EDF (a job-level dynamic algorithm) on uniprocessors implies the following relationship.

Theorem 30.2

$\langle 2, 1\rangle = \langle 3, 1\rangle$.

However, some of the relationships are not quite that straightforward to decipher, as the result of Leung and Whitehead [19] mentioned above and formally stated below in Theorem 30.3 shows.

Theorem 30.3

The $(1, 1)$-restricted and $(1, 3)$-restricted classes are incomparable, i.e., $\langle 1, 1\rangle \otimes \langle 1, 3\rangle$.

Below is a list of task systems that will be used to further separate the algorithm classes. Each task is written as an ordered pair (e, p), where e is its execution requirement and p is its period. The number of available processors is denoted by M.

$\mathcal{A}$: $T_1 = (1, 2)$, $T_2 = (2, 3)$, $T_3 = (2, 3)$; $M = 2$
$\mathcal{B}$: $T_1 = (2, 3)$, $T_2 = (2, 3)$, $T_3 = (2, 3)$; $M = 2$
$\mathcal{C}$: $T_1 = (12, 12)$, $T_2 = (2, 4)$, $T_3 = (3, 6)$; $M = 2$

$\mathcal{D}$: $T_1 = (3, 6)$, $T_2 = (3, 6)$, $T_3 = (6, 7)$; $M = 2$
$\mathcal{E}$: $T_1 = (3, 4)$, $T_2 = (5, 7)$, $T_3 = (3, 7)$; $M = 2$
$\mathcal{F}$: $T_1 = (4, 6)$, $T_2 = (7, 12)$, $T_3 = (4, 12)$, $T_4 = (10, 24)$; $M = 2$
$\mathcal{G}$: $T_1 = (7, 8)$, $T_2 = (10, 12)$, $T_3 = (6, 24)$; $M = 2$
$\mathcal{H}$: $T_1 = (4, 6)$, $T_2 = (4, 6)$, $T_3 = (2, 3)$; $M = 2$
$\mathcal{I}$: $T_1 = (2, 3)$, $T_2 = (3, 4)$, $T_3 = (5, 15)$, $T_4 = (5, 20)$; $M = 2$

In several of the following lemmas, we make implicit use of Theorem 30.1 to establish certain results. (For instance, in Lemma 30.1 below, the implications in the statement of the Lemma follow directly from Theorem 30.1.)

Lemma 30.1

$\mathcal{A} \in \langle 1, 2\rangle$ (which implies that $\mathcal{A} \in \langle 2, 2\rangle$ and $\mathcal{A} \in \langle 3, 2\rangle$).

Proof

Consider the following $(1, 2)$-restricted algorithm: T_2 has higher priority than T_1, which has higher priority than T_3. Restricted migration is permitted, i.e., each job must execute on only one processor, but different jobs may execute upon different processors. The resulting schedule, depicted in Figure 30.1, shows that $\mathcal{A} \in \langle 1, 2\rangle$. (Only the schedule in $[0, 6)$ is shown since 6 is the least common multiple (LCM) of all the task periods, and the schedule starts repeating after time 6.) □

Lemma 30.2

$\mathcal{A} \notin \langle 1, 1\rangle$ and $\mathcal{A} \notin \langle 2, 1\rangle$ and $\mathcal{A} \notin \langle 3, 1\rangle$.

Proof

The tasks cannot be divided into two sets so that each set has utilization at most one. □

Lemma 30.3

$\mathcal{B} \in \langle 3, 3\rangle$.

Proof

Figure 30.2 depicts a $(3, 3)$-restricted schedule. In this schedule, T_1 and T_2 execute over the interval $[0, 1)$, T_1 and T_3 execute over the interval $[1, 2)$, and T_2 and T_3 execute over the interval $[2, 3)$. Thus, over the interval $[0, 3)$, each task receives two units of processor time. □

We now prove that task system $\mathcal{B}$ is only feasible under the $(3, 3)$-restricted class.

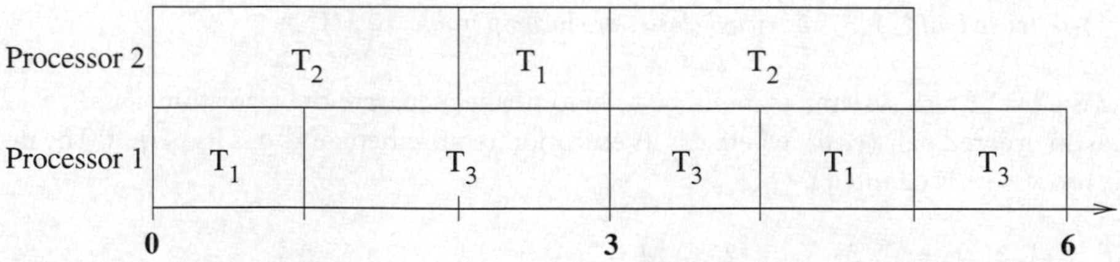

FIGURE 30.1 A $(1, 2)$-restricted schedule for task system $\mathcal{A}$.

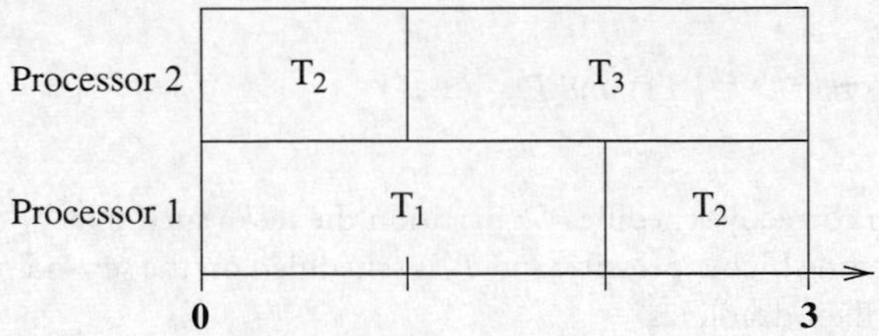

FIGURE 30.2 A (3, 3)-restricted schedule for task system $\mathcal{B}$.

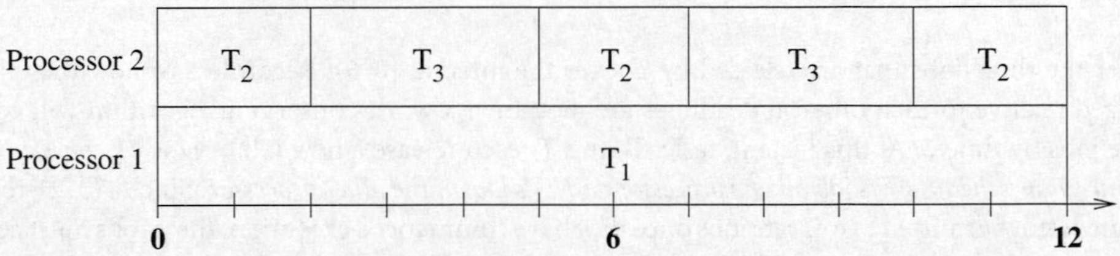

FIGURE 30.3 A (2, 1)-restricted schedule (which is also (2, 2)- and (2, 3)-restricted) for task system $\mathcal{C}$.

Lemma 30.4

$((x \neq 3) \vee (y \neq 3)) \Rightarrow \mathcal{B} \notin \langle x, y \rangle$.

Proof

Consider the first job of each task in $\mathcal{B}$. If $((x \neq 3) \vee (y \neq 3))$, then either these jobs cannot migrate or their relative prioritization is fixed. Note that all three jobs are released at time 0 and must finish by time 3. If these jobs are not allowed to migrate, then two jobs must completely execute on one processor, which is not possible since each requires two units of processor time. Similarly, if the prioritization is fixed, then the lowest-priority job cannot start execution before time 2 and hence will miss its deadline at time 3. Thus, $\mathcal{B} \notin \langle x, y \rangle$. □

Lemma 30.5

$\mathcal{C} \in \langle 2, 1 \rangle$ (which implies that $\mathcal{C} \in \langle 3, 1 \rangle$), $\mathcal{C} \in \langle 2, 2 \rangle$ (which implies that $\mathcal{C} \in \langle 3, 2 \rangle$), and $\mathcal{C} \in \langle 2, 3 \rangle$ (which implies that $\mathcal{C} \in \langle 3, 3 \rangle$).

Proof

The following algorithm correctly schedules $\mathcal{C}$: all jobs of task T_1 are given the highest priority, while jobs of T_2 and T_3 are prioritized using an EDF policy. Note that this is a job-level dynamic priority algorithm.

Since T_1 has a utilization of 1, it will execute solely on one of the processors. Thus, this algorithm will produce the same schedule as if the tasks were partitioned into the sets $\{T_1\}$ and $\{T_2, T_3\}$. Furthermore, correctness follows from the optimality of EDF on uniprocessors: T_2 and T_3 can be correctly scheduled by EDF on a uniprocessor. Figure 30.3 shows the resulting schedule. □

Lemma 30.6

$\mathcal{C} \notin \langle 1, 1 \rangle$, $\mathcal{C} \notin \langle 1, 2 \rangle$, and $\mathcal{C} \notin \langle 1, 3 \rangle$.

Proof

This task set is not feasible when using static priorities because no static-priority scheme can schedule the task system comprised of T_2 and T_3 upon a single processor. Clearly, T_1 must be executed solely on one processor. Regardless of how T_2 and T_1 are statically prioritized, the lowest priority task will miss a deadline. □

Lemma 30.7

$\mathcal{D} \in \langle 1, 1\rangle$ *(which implies that* $\mathcal{D} \in \langle 2, 1\rangle$ *and* $\mathcal{D} \in \langle 3, 1\rangle$*).*

Proof

The following algorithm correctly schedules $\mathcal{D}$: partition the tasks such that T_1 and T_2 are scheduled on one processor with T_1 getting higher priority, and T_3 is scheduled on the second processor. It is easy to see that all three tasks meet their deadlines. □

Lemma 30.8

$\mathcal{D} \notin \langle 3, 2\rangle$ *(which implies that* $\mathcal{D} \notin \langle 2, 2\rangle$ *and* $\mathcal{D} \notin \langle 1, 2\rangle$*).*

Proof

Consider the three jobs that are released by $\mathcal{D}$ over the interval $[0, 6)$. Regardless of how these jobs are prioritized relative to each other, if deadlines are met, then a work-conserving algorithm will complete all three jobs by time 6. At this instant, tasks T_1 and T_2 each release a new job. *Any work conserving algorithm must now schedule* T_1*'s job on one processor and* T_2*'s job on the other processor.* Since a $(3, 2)$-restricted algorithm is not permitted to migrate jobs once they have commenced execution, these jobs must complete execution on the processors upon which they begin. When the second job of T_3 is released at time 7, both T_1 and T_2 have two units of execution left. It is easy to see that whichever processor T_3's job is assigned to, a deadline miss will result.

□

Lemma 30.9

$\mathcal{D} \in \langle 1, 3\rangle$ *(which implies that* $\mathcal{D} \in \langle 2, 3\rangle$ *and* $\mathcal{D} \in \langle 3, 3\rangle$*).*

Proof

Consider an algorithm that assigns T_3 the highest priority, and T_1 the lowest priority. Over any interval $[6k, 6k + 6)$, where k is any integer, T_3 cannot execute for more than 6 units. Since jobs may freely migrate between the processors, there are 6 consecutive units of processor time available for T_1 and T_2 to execute over every such interval; therefore, they will meet their deadlines. Figure 30.5 depicts the resulting schedule.

□

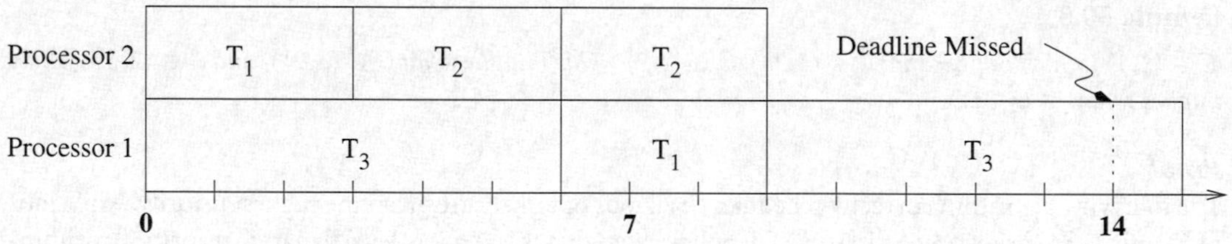

FIGURE 30.4 Deadline miss in a $(3, 2)$-restricted schedule for task system $\mathcal{D}$.

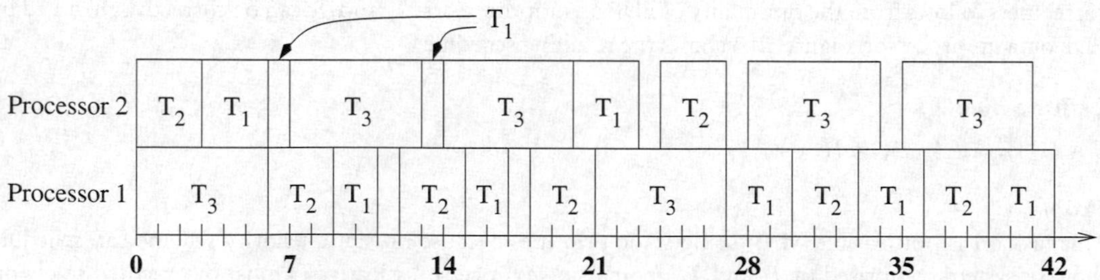

FIGURE 30.5 A $(1, 3)$-restricted schedule for task system $\mathcal{D}$.

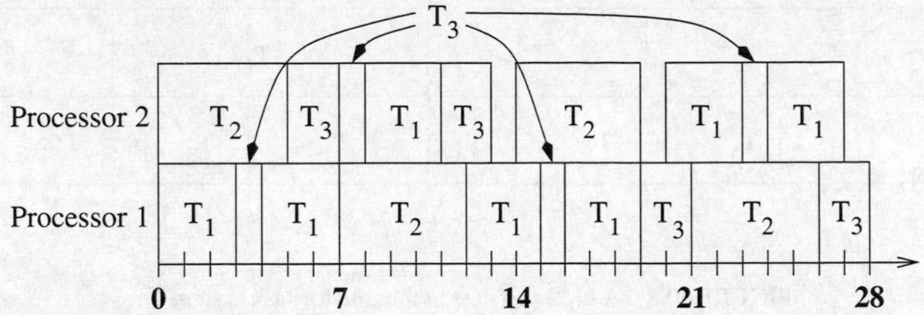

FIGURE 30.6 A (1, 3)-restricted schedule for task system $\mathcal{E}$.

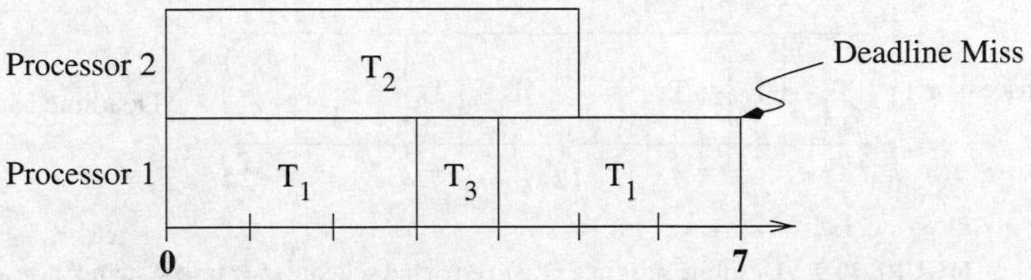

FIGURE 30.7 Deadline miss in a (1, 2)-restricted schedule for task system $\mathcal{E}$.

Lemma 30.10

$\mathcal{E} \in \langle 1, 3\rangle$ *(which implies that* $\mathcal{E} \in \langle 2, 3\rangle$ *and* $\mathcal{E} \in \langle 3, 3\rangle$*).*

Proof

As shown in Figure 30.6, if the priority order (highest to lowest) is T_1, T_2, T_3, then all jobs complete by their deadlines when full migration is permitted. □

Lemma 30.11

$\mathcal{E} \notin \langle 1, 2\rangle$.

Proof

If only restricted migration is allowed, it may be verified that for each of the $3! = 6$ possible static-priority assignments, some task in $\mathcal{E}$ misses a deadline. Figure 30.7 illustrates the schedule assuming the priority order T_1, T_2, T_3. □

The following two lemmas are somewhat counterintuitive in that they imply that *job-level migration sometimes is better than full migration* when static priorities are used.

Lemma 30.12

$\mathcal{F} \in \langle 1, 2\rangle$ *(which implies that* $\mathcal{F} \in \langle 2, 2\rangle$ *and* $\mathcal{F} \in \langle 3, 2\rangle$*).*

Proof

If the priority order (highest to lowest) is T_1, T_2, T_3, T_4 then all jobs will meet their deadlines. The resulting schedule is depicted in Figure 30.8. A region of interest in this schedule occurs over the time interval $[7, 10)$ — since jobs cannot migrate, T_3 does not preempt T_4 during this interval despite having greater priority. This allows the job of T_4 to execute to completion. □

Lemma 30.13

$\mathcal{F} \notin \langle 1, 3\rangle$.

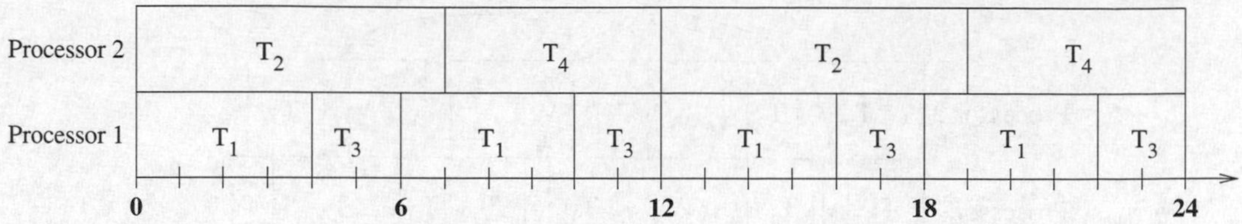

FIGURE 30.8 A (1,2)-restricted schedule for task system $\mathcal{F}$.

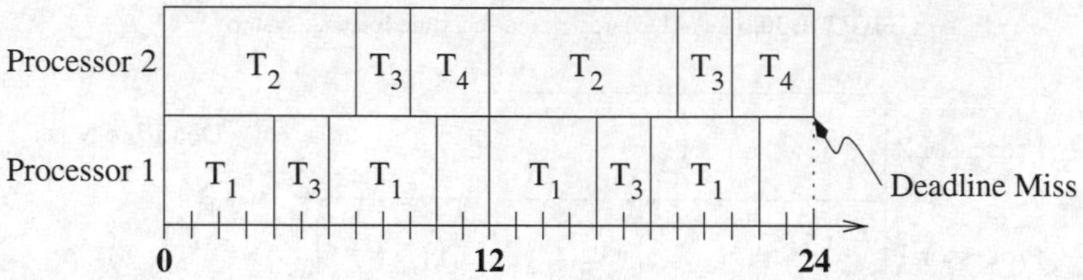

FIGURE 30.9 Deadline miss in a $(1, 3)$-restricted schedule for task system $\mathcal{F}$.

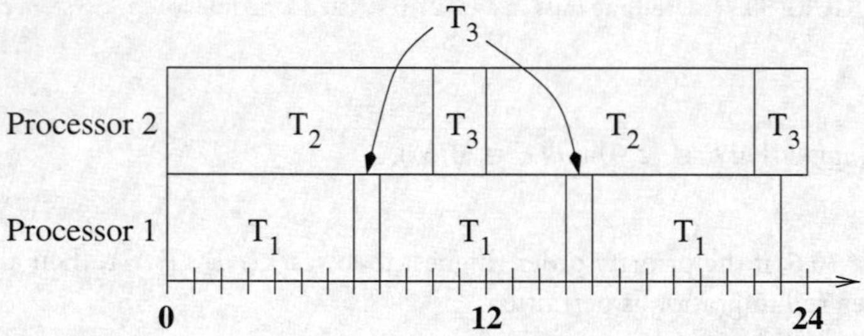

FIGURE 30.10 A $(1, 3)$-restricted schedule for task system $\mathcal{G}$.

Proof

We have verified that all $4! = 24$ possible priority assignments result in a deadline miss. Figure 30.9 illustrates the schedule assuming the priority order T_1, T_2, T_3, T_4. □

Lemma 30.14

$\mathcal{G} \in \langle 1, 3\rangle$ (*which implies that* $\mathcal{G} \in \langle 2, 3\rangle$ *and* $\mathcal{G} \in \langle 3, 3\rangle$).

Proof

If the priority order (highest to lowest) is T_1, T_2, T_3, then all jobs will meet their deadlines if full migration is allowed. Figure 30.10 shows the resulting schedule. □

Lemma 30.15

$\mathcal{G} \notin \langle 3, 2\rangle$ (*which implies that* $\mathcal{G} \notin \langle 2, 2\rangle$ *and* $\mathcal{G} \notin \langle 1, 2\rangle$).

Proof

Over the interval $[0, 24)$, the tasks in $\mathcal{G}$ release six jobs with execution requirements of 7, 7, 7, 10, 10, and 6, respectively. Since jobs cannot migrate, in order to complete all jobs before time 24, these jobs must

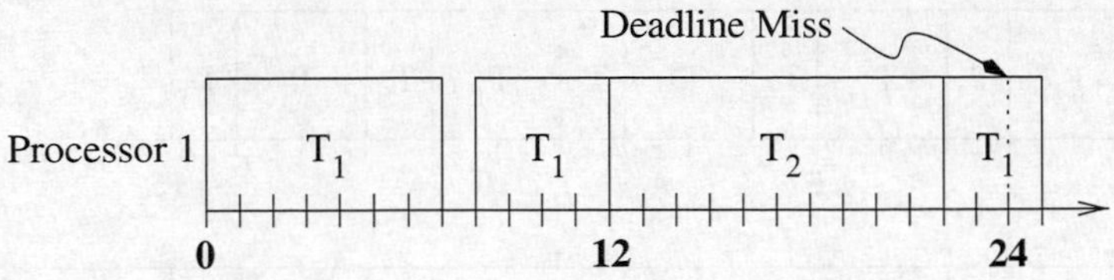

(a)

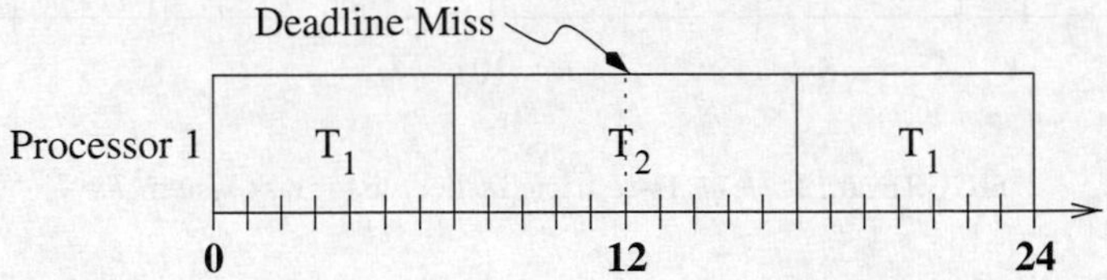

(b)

FIGURE 30.11 Deadline miss in a (3, 2)-restricted schedule for task system $\mathcal{G}$.

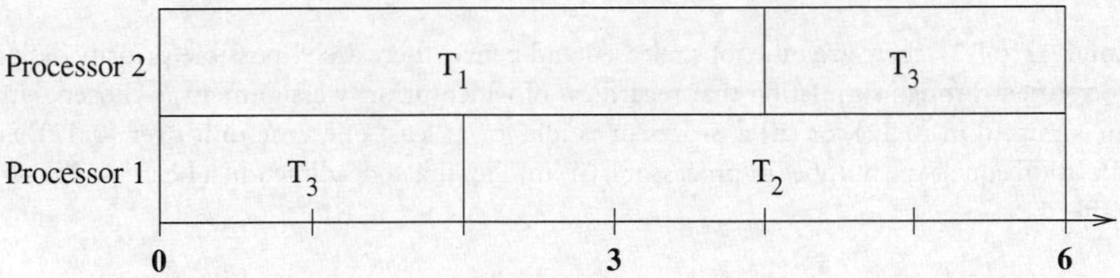

FIGURE 30.12 A (3, 2)-restricted schedule for task system $\mathcal{H}$.

be partitioned into two groups such that the sum of the execution requirements in each group does not exceed 24. The only such partition is into 7, 7, 10 and 6, 7, 10.

Consider the processor that must run jobs from the first group, which have execution requirements 7, 7, and 10, respectively. The job with execution requirement 10 executes either over the interval $[0, 12)$ or over $[12, 24)$. If the interval is $[12, 24)$, then only 7 units over the interval $[0, 8)$ can be utilized, since both jobs with execution requirement 7 belong to task T_1. Therefore, the processor must be idle for one slot, implying that one of the other two jobs misses its deadline. This is illustrated in Figure 30.11(a). On the other hand, if the interval is $[0, 12)$, then a job with execution requirement 7 must execute in either $[0, 8)$ or $[8, 16)$. Regardless of which, the demand over $[0, 16)$ is 17, and thus, a deadline is missed at time 12 or 16, as illustrated in Figure 30.11(b). □

Lemma 30.16

$\mathcal{H} \notin \langle 3, 1\rangle$ (which implies that $\mathcal{H} \notin \langle 2, 1\rangle$ and $\mathcal{H} \notin \langle 1, 1\rangle$).

Proof

$\mathcal{H}$ cannot be partitioned into two sets, each of which has utilization at most one. □

Lemma 30.17

$\mathcal{H} \in \langle 3, 2\rangle$ and $\mathcal{H} \in \langle 2, 3\rangle$.

Proof

The schedule in Figure 30.12 shows that $\mathcal{H} \in \langle 3, 2\rangle$. This schedule can also be produced by the following priority order: T_1's job, T_3's first job, T_2's job, T_3's second job. Hence, $\mathcal{H} \in \langle 2, 3\rangle$. □

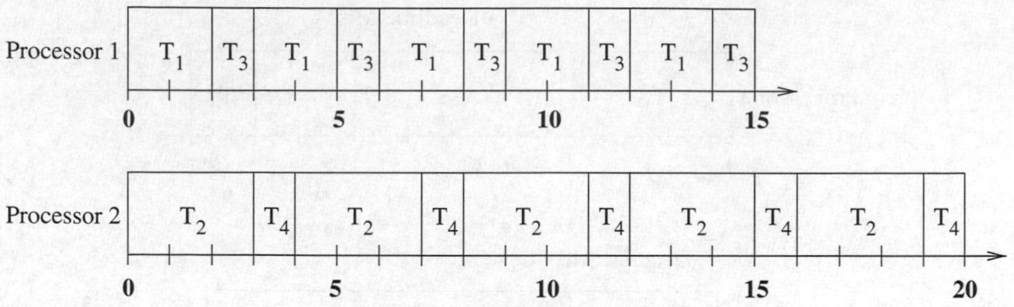

FIGURE 30.13 A (1, 1)-restricted schedule for task system $\mathcal{I}$.

Lemma 30.18

$\mathcal{I} \notin \langle 2, 3 \rangle$.

Proof

In the interval $[0, 12)$, there are nine jobs released and hence, there are 9! possible priority assignments. We have verified through simulation that regardless of which priority assignment is chosen, either **(i)** a deadline is missed in $[0, 12)$, or **(ii)** a processor is idle for at least one time unit over $[0, 12)$. Since the total utilization equals the number of processors, (ii) implies that a deadline must be missed at some time instant after 12. □

Lemma 30.19

$\mathcal{I} \in \langle 1, 1 \rangle$ *(which implies that* $\mathcal{I} \in \langle 2, 1 \rangle$ *and* $\mathcal{I} \in \langle 3, 1 \rangle$*).*

Proof

The following algorithm correctly schedules $\mathcal{I}$: partition the tasks into the sets $\{T_1, T_3\}$ and $\{T_2, T_4\}$. Each set can be correctly scheduled using the RM algorithm on a single processor as shown in Figure 30.13. Figure 30.13 shows the schedules on each processor up to the LCM of the periods of all the tasks assigned to that processor. □

Lemma 30.20

$\mathcal{I} \in \langle 3, 2 \rangle$.

Proof

Consider the schedule suggested in the proof of Lemma 30.19. Such a schedule can be accomplished by an algorithm in $\langle 3, 2 \rangle$ simply by setting the appropriate jobs to the highest priority at each instant. (Note that this idea does not work in the proof of Lemma 30.8. In particular at time 2, T_3's job has higher priority than T_4's job, while at time 3, T_4's job has higher priority. Thus this schedule cannot be produced by any algorithm that assigns fixed priorities to jobs.) □

Using the above lemmas, it is easy to derive many of the relationships among the classes of scheduling algorithms. These relationships are summarized in Table 30.3.

30.3.1 Discussion

It is easy to see from Table 30.3 that several of the scheduling classes are incomparable. In particular, we can observe the following:

TABLE 30.3 Relationships Among the Various Classes

	$\langle 1,1\rangle$	$\langle 2,1\rangle$	$\langle 3,1\rangle$	$\langle 1,2\rangle$	$\langle 2,2\rangle$	$\langle 3,2\rangle$	$\langle 1,3\rangle$	$\langle 2,3\rangle$	$\langle 3,3\rangle$
$\langle 1,1\rangle$	$=$	$\subset$	$\subset$	$\otimes$	$\otimes$	$\otimes$	$\otimes$	$\otimes$	$\subset$
$\langle 2,1\rangle$	$\supset$	$=$	$=$	$\otimes$	$\otimes$	$\otimes$	$\otimes$	$\otimes$	$\subset$
$\langle 3,1\rangle$	$\supset$	$=$	$=$	$\otimes$	$\otimes$	$\otimes$	$\otimes$	$\otimes$	$\subset$
$\langle 1,2\rangle$	$\otimes$	$\otimes$	$\otimes$	$=$	$\subset$	$\subset$	$\otimes$	??	$\subset$
$\langle 2,2\rangle$	$\otimes$	$\otimes$	$\otimes$	$\supset$	$=$	$\subseteq$	$\otimes$	??	$\subset$
$\langle 3,2\rangle$	$\otimes$	$\otimes$	$\otimes$	$\supset$	$\supseteq$	$=$	$\otimes$	$\otimes$	$\subset$
$\langle 1,3\rangle$	$\otimes$	$\otimes$	$\otimes$	$\otimes$	$\otimes$	$\otimes$	$=$	$\subset$	$\subset$
$\langle 2,3\rangle$	$\otimes$	$\otimes$	$\otimes$	??	??	$\otimes$	$\supset$	$=$	$\subset$
$\langle 3,3\rangle$	$\supset$	$\supset$	$\supset$	$\supset$	$\supset$	$\supset$	$\supset$	$\supset$	$=$

Observation 30.1 *Under static priorities, all three migration classes are incomparable.*

As can be seen from the table, there are two relationships that are still open. It is also not known at this time whether $\langle 2,2\rangle$ is proper subset of $\langle 3,2\rangle$.

30.4 Algorithm Design and Feasibility Analysis

In this section we discuss feasibility analysis and online scheduling of periodic task systems. We will use the following notation: $\tau = \{T_1, T_2, \ldots, T_n\}$ denotes the system of periodic tasks to be scheduled on M processors. Let $U(T) = T.e/T.p$, where $T.e$ denotes T's execution requirement and $T.p$ denotes its period. Also, let $U(\tau) = \sum_{T\in\tau} U(T)$.

An (x, y)-restricted feasibility test accepts as input the specifications of τ and M, and determines whether some (x, y)-restricted algorithm can successfully schedule τ upon M processors. Such a test is *sufficient* if any task system satisfying it is guaranteed to be successfully scheduled by some (x, y)-restricted algorithm, and *exact* if it is both sufficient and *necessary* — no task system failing the test can be scheduled by any (x, y)-restricted algorithm. Feasibility tests are often stated as utilization bounds; such a bound is the largest value $U_{\max}$ such that every task system τ satisfying $U(\tau) \leq U_{\max}$ is guaranteed to be feasible.

For several of the classes identified in Table 30.1, exact feasibility analysis is provably intractable: a transformation from the 3-Partition problem can be used to show that feasibility analysis is NP-hard in the strong sense. The class of $(3, 3)$-restricted algorithms is the most general class defined in Section 30.2. The following result is well-known, and immediately yields an efficient and exact $(3, 3)$-restricted feasibility test.

Theorem 30.4

A periodic task system τ can be scheduled upon M processors using some $(3, 3)$-restricted algorithm if and only if $U(\tau) \leq M$.

$(3, 3)$-restricted scheduling algorithms have been the subject of several papers [4–6,20–22]. For example, Leung [22] studied the use of global LLF scheduling on multiprocessors. Since LLF adopts a processor-sharing approach, ensuring that at most one task executes upon a processor at each instant in time may introduce an arbitrarily large number of preemptions and migrations. The Pfair scheduling approach, introduced by Baruah et al. [6] and extended by Anderson et al. [4,5,20,21], reduces the number of preemptions and migrations by scheduling for discrete time units or "quanta." To summarize this work, the current state of the art concerning $(3, 3)$-restricted scheduling is as follows: there is no schedulability penalty (Theorem 30.4) and efficient runtime implementations are known, but the number of preemptions and interprocessor migrations may be high. Pfair scheduling is discussed in detail in the chapter titled "Fair Scheduling of Real-time Tasks on Multiprocessors" in this volume. We now briefly describe results relating to the remaining classes of scheduling algorithms.

30.4.1 Partitioning Approaches

There has been a considerable amount of research on $(x, 1)$-restricted scheduling algorithms, i.e., partitioning approaches [3,12,14,16,23,24]. Recall that, under partitioning, each task is assigned to a processor on which it will exclusively execute. Finding an optimal assignment of tasks to processors is equivalent to a bin-packing problem, which is known to be NP-hard in the strong sense. Several polynomial-time heuristics have been proposed for solving this problem. Examples include first fit (FF) and best fit (BF). In FF, each task is assigned to the first (i.e., lowest-indexed) processor that can accept it (based on the feasibility test corresponding to the uniprocessor scheduling algorithm being used). On the other hand, in BF, each task is assigned to a processor that **(i)** can accept the task, and **(ii)** will have minimal remaining spare capacity after its addition.

Surprisingly, the *worst-case achievable utilization* on M processors for all of the above-mentioned heuristics (and also for an optimal partitioning algorithm) is only $(M+1)/2$, even when an optimal uniprocessor scheduling algorithm such as EDF is used. In other words, there exist task systems with utilization slightly greater than $(M + 1)/2$ that cannot be correctly scheduled by any partitioning approach. To see why, note that $M + 1$ tasks, each with execution requirement $1 + \epsilon$ and period 2, cannot be partitioned on M processors, regardless of the partitioning heuristic and the scheduling algorithm.

Theorem 30.5

No partitioned-based scheduling algorithm can successfully schedule all task systems τ with $U(\tau) \leq B$ on M processors, where $B > \frac{1}{2}(M + 1)$.

Lopez et al. [23] showed that EDF with FF (or BF) can successfully schedule any task system with utilization at most $(\beta M + 1)/(\beta + 1)$, where $\beta = \lfloor 1/\alpha \rfloor$ and α satisfies $\alpha \geq U(T)$ for all $T \in \tau$. A (2,1)-restricted, sufficient feasibility test immediately follows.

Theorem 30.6

If $U(\tau) \leq (\beta M + 1)/(\beta + 1)$, where $\beta = \lfloor 1/\alpha \rfloor$ and α satisfies $\alpha \geq U(T)$ for all $T \in \tau$, then τ is feasible on M processors under the $(2, 1)$-restricted class.

We obtain the following result as a corollary of Theorem 30.6 by letting $\alpha = 1$ (and hence $\beta = 1$).

Corollary 30.1

If $U(\tau) \leq \frac{1}{2}(M + 1)$, then τ is feasible on M processors under the $(2, 1)$-restricted class.

Theorem 30.5 and Corollary 30.1 imply that using EDF with FF or BF is an optimal partitioning approach with respect to utilization bounds.

The worst-case achievable utilization is much smaller for RM-scheduled systems since RM is not an optimal uniprocessor scheduling algorithm. Let $U_{RM,FF}$ denote the worst-case achievable utilization under RM with FF (RM-FF). Oh and Baker proved the following bounds on $U_{RM,FF}$ [13]:

Theorem 30.7

$(\sqrt{2} - 1) \times M \leq U_{RM,FF} \leq (M + 1)/(1 + 2^{\frac{1}{M+1}})$.

Thus, task systems in which the total utilization does not exceed $(\sqrt{2} - 1) \times M$ ($\approx 0.41 \times M$) are schedulable using RM-FF. Though this value is significantly small, RM is still popular because of its simplicity and predictability under overload. Several researchers have proposed partitioning heuristics that improve upon FF and BF. Oh and Son proposed an improved variant of the FF heuristic called first fit decreasing utilization (FFDU) [14]. They showed that for RM-scheduled systems, the number of processors required by FFDU is at most 5/3 the optimal number of processors. (Dhall and Liu had shown earlier that the number of processors required by FF and BF is at most twice the optimal number [3].)

Burchard et al. proposed new sufficient feasibility tests for RM-scheduled uniprocessor systems that perform better when task periods satisfy certain relationships [12]. They also proposed new heuristics that try to assign tasks satisfying those relationships to the same processor, thus leading to better overall utilization. Lauzac et al. also proposed similar schedulability tests and heuristics, in which tasks are initially sorted in order of increasing periods [25]. One disadvantage of these heuristics is that they can lead to unacceptable overhead when used online due to the sorting overhead; when scheduling online, FF and BF are preferred.

30.4.2 Other Classes

An upper bound on the worst-case achievable utilization of any algorithm that is not $(3, 3)$-restricted is easily obtained. Consider the same example used in the proof of Theorem 30.5: a system of $M + 1$ tasks, each with a period of 2 and an execution requirement $(1 + \epsilon)$, to be scheduled on M processors. Consider the set of jobs released by each task at time 0. If job migration is not allowed, then over the interval $[0, 2)$, two of these jobs must be executed on the same processor, implying that one of them will miss its deadline. Furthermore, this task system cannot be scheduled by a $(1, 3)$- or $(2, 3)$-restricted algorithm because when the $M+1$ jobs are released at time 0, the lowest-priority job will miss its deadline. As $\epsilon \to 0$, the utilization approaches $(M + 1)/2$. Thus, by Theorem 30.5, we have the following.

Theorem 30.8

Unless $x = 3$ and $y = 3$, no (x, y)-restricted algorithm can successfully schedule all task systems τ with $U(\tau) \leq B$ on M processors, where $B > \frac{1}{2}(M + 1)$.

We now present some results involving the other classes of scheduling algorithms.

30.4.2.1 (2, 3)-Restricted Algorithms

These algorithms associate a fixed priority with each job but permit jobs to migrate among processors arbitrarily often. (Notice that global scheduling with EDF belongs to this class.) In these algorithms, a preemption, and hence a migration, can only be caused by a job release or (another) migration. Hence, the total number of preemptions and migrations can be bounded at an amortized number of one per job by designing the scheduling algorithm appropriately. Thus, while such algorithms do incur some preemption and migration overhead, this cost can be bounded, and may be acceptable for certain applications. Furthermore, some algorithms in this category, particularly EDF, have very efficient implementations.

As stated above, EDF is a $(2, 3)$-restricted scheduling algorithm. Although EDF is a very popular algorithm in uniprocessor real-time systems, studies (e.g., [9,22]) have suggested that it tends to miss many deadlines in the multiprocessor case. Srinivasan and Baruah [18] recently presented a new $(2, 3)$-restricted algorithm based upon EDF. In their algorithm, tasks that have utilizations at least $M/(2M - 1)$ are statically assigned the highest priority in the system, while the remaining tasks are prioritized on an EDF basis. They proved that this new algorithm can schedule all task systems τ with $U(\tau) \leq \frac{M^2}{2M-1}$ (which simplifies to $(\frac{M}{2} + \frac{M}{4M-2})$) upon M processors. Thus, we have the following theorem.

Theorem 30.9

A task system τ is feasible under the $(2, 3)$-restricted class if $U(\tau) \leq \frac{M^2}{2M-1}$.

30.4.2.2 (1, 3)-Restricted Algorithms

Andersson et al. [17] developed an algorithm similar to that of Srinivasan and Baruah, but based upon RM rather than EDF: tasks that have utilizations at least $M/(3M-2)$ are statically assigned the highest priority in the system, while the remaining tasks are prioritized on a RM basis. They proved that their algorithm can schedule all task systems τ with $U(\tau) \leq \frac{M^2}{3M-2}$ upon M processors. Hence, we have the following theorem.

Theorem 30.10

A task system τ is feasible under the $(1,3)$*-restricted class if* $U(\tau) \leq \frac{M^2}{3M-2}$.

It can be shown that if task periods are harmonic, then the schedule produced by RM is a valid EDF schedule. Therefore, we obtain the following result.

Theorem 30.11

A task system τ, in which all task periods are harmonic, is feasible under the $(1,3)$*-restricted class if* $U(\tau) \leq \frac{M^2}{2M-1}$.

30.4.2.3 (2, 2)-Restricted Algorithms

These algorithms associate a fixed priority with each job, and restrict each job to execute exclusively on a single processor; however, different jobs of the same task may execute upon different processors. Such algorithms are particularly appropriate for scheduling task systems in which each job has a considerable amount of state (as a consequence, it is not desirable to migrate a job between processors), but not much state is carried over from one job to the next. Baruah and Carpenter [26] have designed a $(2,2)$-restricted algorithm which successfully schedules any periodic task system τ satisfying $U(\tau) \leq M - \alpha(M-1)$, where α is as defined earlier. Hence, we have the following result.

Theorem 30.12

If $U(\tau) \leq M - \alpha(M-1)$, where α satisfies $\alpha \geq U(T)$ for all $T \in \tau$, then τ is feasible on M processors under the $(2,2)$*-restricted class.*

The results in this section are summarized in Table 30.4. The exact utilization bound for $(3,3)$-restricted algorithms follows from Theorem 30.4. The bounds on worst-case achievable utilization for partitioned algorithms follow from Theorems 30.5–30.7. (The bounds for $(3,1)$-restricted algorithms follow from those for $(2,1)$-restricted algorithms because $\langle 3,1\rangle = \langle 2,1\rangle$. Refer to Table 30.3.) The upper bounds for the rest of the classes follow from Theorem 30.8. The lower bounds on worst-case achievable utilization for $(1,3)$-, $(2,3)$-, and $(2,2)$-restricted algorithms follow from Theorems 30.10, 30.9, and 30.12, respectively. (The lower bound for $(3,2)$-restricted algorithms follows because $\langle 2,2\rangle \subseteq \langle 3,2\rangle$. Refer to Table 30.3.)

TABLE 30.4 Known Bounds on Worst-Case Achievable Utilization (denoted U) for the Different Classes of Scheduling Algorithms

	1: Static	2: Job-level dynamic	3: Unrestricted dynamic
3: Full migration	$\frac{M^2}{3M-2} \leq U \leq \frac{M+1}{2}$	$\frac{M^2}{2M-1} \leq U \leq \frac{M+1}{2}$	$U = M$
2: Restricted migration	$U \leq \frac{M+1}{2}$	$M - \alpha(M-1) \leq U \leq \frac{M+1}{2}$	$M - \alpha(M-1) \leq U \leq \frac{M+1}{2}$
1: Partitioned	$(\sqrt{2}-1)M \leq U \leq \frac{M+1}{1+2^{\frac{1}{M+1}}}$	$U = \frac{M+1}{2}$	$U = \frac{M+1}{2}$

30.5 Summary

In this chapter we presented a new taxonomy of scheduling algorithms for scheduling preemptive real-time tasks on multiprocessors. We described some new classes of scheduling algorithms and considered the relationship of these classes to the existing well-studied classes. We also described known scheduling algorithms that fall under these classes and presented sufficient feasibility conditions for these algorithms.

References

[1] C. Liu. Scheduling algorithms for multiprocessors in a hard real-time environment. *JPL Space Programs Summary*, 37–60(II):28–31, 1969.

[2] C. L. Liu and J. W. Layland. Scheduling algorithms for multiprogramming in a hard-real-time environment. *Journal of the ACM*, 20(1):46–61, 1973.

[3] S. Dhall and C. Liu. On a real-time scheduling problem. *Operations Research*, 26:127–140, 1978.

[4] J. Anderson and A. Srinivasan. Early-release fair scheduling. In *Proceedings of the 12th Euromicro Conference on Real-time Systems*, pp. 35–43, June 2000.

[5] J. Anderson and A. Srinivasan. Mixed Pfair/ERfair scheduling of asynchronous periodic tasks. In *Proceedings of the 13th Euromicro Conference on Real-time Systems*, pp. 76–85, June 2001.

[6] S. Baruah, N. Cohen, C.G. Plaxton, and D. Varvel. Proportionate progress: A notion of fairness in resource allocation. *Algorithmica*, 15:600–625, 1996.

[7] M. Moir and S. Ramamurthy. Pfair scheduling of fixed and migrating periodic tasks on multiple resources. In *Proceedings of the 20th IEEE Real-time Systems Symposium*, pp. 294–303, December 1999.

[8] A. Srinivasan and J. Anderson. Optimal rate-based scheduling on multiprocessors. In *Proceedings of the 34th ACM Symposium on Theory of Computing*, pp. 189–198, May 2002.

[9] A. Mok. Fundamental Design Problems of Distributed Systems for Hard Real-time Environments. PhD thesis, Massachusetts Institute of Technology, Cambridge, MA, 1983.

[10] A. Srinivasan, P. Holman, J. Anderson, and S. Baruah. The case for fair multiprocessor scheduling. In *Proceedings of the 11th International Workshop on Parallel and Distributed Real-time Systems*, April 2003.

[11] M. Dertouzos. Control robotics: The procedural control of physical processors. In *Proceedings of the IFIP Congress*, pp. 807–813, 1974.

[12] A. Burchard, J. Liebeherr, Y. Oh, and S. Son. Assigning real-time tasks to homogeneous multiprocessor systems. *IEEE Transactions on Computers*, 44(12):1429–1442, 1995.

[13] D. Oh and T. Baker. Utilization bounds for n-processor rate monotone scheduling with static processor assignment. *Real-time Systems*, 15(2):183–192, 1998.

[14] Y. Oh and S. Son. Allocating fixed-priority periodic tasks on multiprocessor systems. *Real-time Systems*, 9(3):207–239, 1995.

[15] S. Davari and S. Dhall. On a real-time task allocation problem. In *Proceedings of the 19th Hawaii International Conference on System Science*, February 1986.

[16] S. Davari and S. Dhall. An on-line algorithm for real-time tasks allocation. In *Proceedings of the Seventh IEEE Real-time Systems Symposium*, pp. 194–200, 1986.

[17] B. Andersson, S. Baruah, and J. Jansson. Static-priority scheduling on multiprocessors. In *Proceedings of the 22nd IEEE Real-time Systems Symposium*, pp. 193–202, December 2001.

[18] A. Srinivasan and S. Baruah. Deadline-based scheduling of periodic task systems on multiprocessors. *Information Processing Letters*, 84(2):93–98, 2002.

[19] J. Leung and J. Whitehead. On the complexity of fixed-priority scheduling of periodic, real-time tasks. *Performance Evaluation*, 2(4):237–250, 1982.

[20] J. Anderson and A. Srinivasan. Pfair scheduling: Beyond periodic task systems. In *Proceedings of the 7th International Conference on Real-time Computing Systems and Applications*, pp. 297–306, December 2000.

[21] P. Holman and J. Anderson. Guaranteeing Pfair supertasks by reweighting. In *Proceedings of the 22nd IEEE Real-time Systems Symposium*, pp. 203–212, December 2001.

[22] J. Leung. A new algorithm for scheduling periodic real-time tasks. *Algorithmica*, 4:209–219, 1989.

[23] J. Lopez, M. Garcia, J. Diaz, and D. Garcia. Worst-case utilization bound for edf scheduling on real-time multiprocessor systems. In *Proceedings of the 12th Euromicro Conference on Real-time Systems*, pp. 25–33, June 2000.

[24] S. Saez, J. Vila, and A. Crespo. Using exact feasibility tests for allocating real-time tasks in multiprocessor systems. In *Proceedings of the 10th Euromicro Workshop on Real-time Systems*, pp. 53–60, June 1998.

[25] S. Lauzac, R. Melhem, and D. Mosse. An efficient RMS admission control and its application to multiprocessor scheduling. In *Proceedings of the 12th International Symposium on Parallel Processing*, pp. 511–518, April 1998.

[26] S. Baruah and J. Carpenter. Multiprocessor fixed-priority scheduling with restricted interprocessor migrations. In *Proceedings of the 15th Euromicro Conference on Real-time Systems*, pp. 195–202, July 2003.

31

Fair Scheduling of Real-Time Tasks on Multiprocessors

James Anderson
University of North Carolina at Chapel Hill

Philip Holman
University of North Carolina at Chapel Hill

Anand Srinivasan
Microsoft Corporation

31.1 Introduction

There has been much recent interest in fair scheduling algorithms for real-time multiprocessor systems. The roots of much of the research on this topic can be traced back to the seminal work of Baruah et al. on *Proportionate fairness* (*Pfairness*) [1]. This work proved that the problem of optimally scheduling periodic tasks[1] on multiprocessors could be solved online in polynomial time by using Pfair scheduling algorithms. Pfair scheduling differs from more conventional real-time scheduling approaches in that tasks are explicitly required to execute at steady rates. In most real-time scheduling disciplines, the notion of a rate is implicit. For example, in a periodic schedule, a task T executes at a rate defined by its required utilization ($T.e/T.p$) over large intervals. However, T's execution rate over short intervals, e.g., individual periods, may vary significantly. Hence, the notion of a rate under the periodic task model is a bit inexact.

[1]A periodic task T is characterized by a *phase* $T.\phi$, an *execution requirement* $T.e$, and a *period* $T.p$: a job release (i.e., task invocation) occurs at time $T.\phi + (k-1) \cdot T.p$ for each integer $k \geq 1$ and the kth job must receive $T.e$ units of processor time by the next release (at time $T.\phi + k \cdot T.p$). A periodic task system is *synchronous* if each task in the system has a phase of 0.

1-58488-397-9/$0.00+$1.50

Under Pfair scheduling, each task is executed at an approximately uniform rate by breaking it into a series of quantum-length subtasks. Time is then subdivided into a sequence of (potentially overlapping) subintervals of approximately equal lengths, called *windows*. To satisfy the Pfairness rate constraint, each subtask must execute within its associated window. Different subtasks of a task are allowed to execute on different processors (i.e., interprocessor migration is permitted), but may not execute simultaneously (i.e., parallelism is prohibited).

By breaking tasks into uniform-sized subtasks, Pfair scheduling circumvents many of the bin-packing-like problems that lie at the heart of intractability results that pertain to multiprocessor scheduling. Indeed, Pfair scheduling is presently the only known approach for optimally scheduling periodic tasks on multiprocessors. Three Pfair scheduling algorithms have been proven optimal: PF [1], PD [2], and PD^2 [3]. Several suboptimal algorithms have also been proposed, including the earliest-pseudo-deadline-first (EPDF) algorithm [4,5], the weight-monotonic (WM) algorithm [6,7], and the deadline-fair-scheduling (DFS) algorithm [8].

In this chapter we present an overview of Pfair scheduling and many of the extensions to it that have been proposed. In Section 31.2 we formally describe Pfairness and its relationship to periodic task scheduling. In Section 31.3 we present the *intrasporadic fairness* (ISfairness) [9,10] constraint, which extends the Pfairness constraint to systems in which the execution of subtasks may be delayed. In Section 31.4 we present results relating to dynamic task systems, i.e., systems in which tasks are allowed to leave and join [11]. Section 31.5 describes prior work on using hierarchical scheduling under a global Pfair scheduler [4,12]. We then present resource-sharing protocols [16,17] designed for Pfair-scheduled systems in Section 31.6. Section 31.7 summarizes the chapter.

31.2 Periodic Task Systems

In this section we consider the problem of scheduling a set τ of synchronous periodic tasks on a multiprocessor. (Recall that $T.\phi = 0$ for synchronous tasks.) We assume that processor time is allocated in discrete time units, or *quanta*, and refer to the time interval $[t, t+1)$, where t is a nonnegative integer, as *slot* t. We further assume that all task parameters are expressed as integer multiples of the quantum size. As mentioned in the footnote in Section 31.1, each periodic task T is characterized by a period $T.p$, a per-job execution requirement $T.e$, and a phase $T.\phi$. We refer to the ratio of $T.e/T.p$ as the weight of task T, denoted by $wt(T)$. Informally, $wt(T)$ is the rate at which T should be executed, relative to the speed of a single processor. A task with weight less than 1/2 is called a *light* task, while a task with weight at least 1/2 is called a *heavy* task.

The sequence of scheduling decisions over time defines a *schedule*. Formally, a schedule S is a mapping $S : \tau \times \mathcal{Z} \mapsto \{0, 1\}$, where τ is a set of periodic tasks and $\mathcal{Z}$ is the set of nonnegative integers. If $S(T, t) = 1$, then we say that *task* T *is scheduled in slot* t.

31.2.1 Pfair Scheduling

In a perfectly fair (ideal) schedule for a synchronous task system, every task T would receive $wt(T) \cdot t$ quanta over the interval $[0, t)$ (which implies that all deadlines are met). However, such idealized sharing is not possible in a quantum-based schedule. Instead, Pfair scheduling algorithms strive to closely track the allocation of processor time in the ideal schedule. This tracking is formalized in the notion of per-task *lag*, which is the difference between each task's allocation in the Pfair schedule and the allocation it would receive in an ideal schedule. Formally, the *lag of task* T *at time* t, denoted $lag(T, t)$,[2] is defined as follows:

$$lag(T, t) = wt(T) \cdot t - \sum_{u=0}^{t-1} S(T, u) \tag{31.1}$$

[2]For brevity, we leave the schedule implicit and use $lag(T, t)$ instead of $lag(T, t, S)$.

A schedule is *Pfair* if and only if

$$(\forall T, t :: -1 < lag(T, t) < 1) \tag{31.2}$$

Informally, Equation (31.2) requires each task T's allocation error to be less than one quantum at all times, which implies that T must receive either $\lfloor wt(T) \cdot t \rfloor$ or $\lceil wt(T) \cdot t \rceil$ quanta by time t.

It is straightforward to show that the Pfairness constraint subsumes the periodic scheduling constraint. For example, in the case of a synchronous system, a periodic task T must be granted $T.e$ quanta in each interval $[k \cdot T.p, (k+1) \cdot T.p)$, where $k \geq 0$. At $t = k \cdot T.p$, $wt(T) \cdot t = (T.e/T.p) \cdot (k \cdot T.p) = k \cdot T.e$, which is an integer. By Equation (31.2), each task's allocation in a Pfair schedule will match that of the ideal schedule at period boundaries. Since all deadlines are met in the ideal schedule, they must also be met in each possible Pfair schedule.

31.2.1.1 Windows

Under Pfair scheduling, each task T is effectively divided into an infinite sequence of quantum-length *subtasks*. We denote the ith subtask of task T as T_i, where $i \geq 1$. The lag bounds in Equation (31.2) constrain each subtask T_i to execute in an associated *window*, denoted $w(T_i)$. $w(T_i)$ extends from T_i's *pseudorelease*, denoted $r(T_i)$, to its *pseudo-deadline*, denoted $d(T_i)$. (For brevity, we often drop the "pseudo-" prefix.) $r(T_i)$ and $d(T_i)$ are formally defined as shown below.

$$r(T_i) = \left\lfloor \frac{i-1}{wt(T)} \right\rfloor \tag{31.3}$$

$$d(T_i) = \left\lceil \frac{i}{wt(T)} \right\rceil \tag{31.4}$$

As an example, consider a task T with weight $wt(T) = 8/11$. Each job of this task consists of eight windows, one for each of its subtasks. Using Equations (31.3) and (31.4), it is easy to show that the windows within each job of T are as depicted in Figure 31.1.

31.2.1.2 Feasibility

A Pfair schedule on M processors exists for τ if and only if

$$\sum_{T \in \tau} \frac{T.e}{T.p} \leq M \tag{31.5}$$

This result was proved by Baruah et al. by means of a network-flow construction [1]. Let L denote the least common multiple of $\{T.p \mid T \in \tau\}$. By restricting attention to subtasks that fall within the first hyperperiod of τ (i.e., in the interval $[0, L)$), the sufficiency of Equation (31.5) can be established by constructing a flow graph with integral edge capacities and by then applying the Ford-Fulkerson result [13]

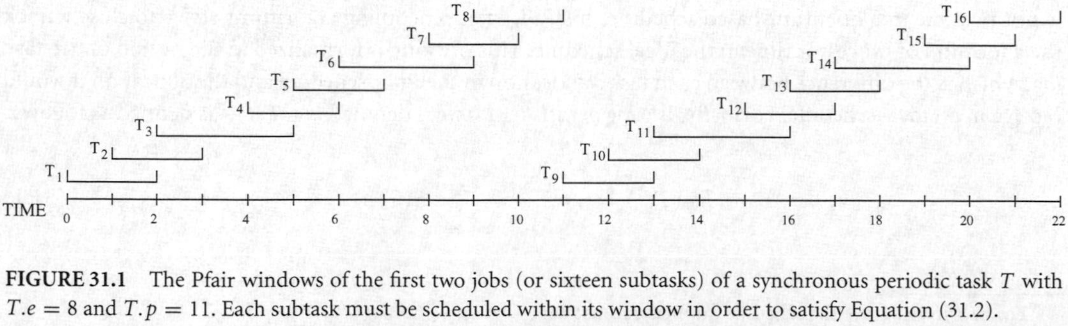

FIGURE 31.1 The Pfair windows of the first two jobs (or sixteen subtasks) of a synchronous periodic task T with $T.e = 8$ and $T.p = 11$. Each subtask must be scheduled within its window in order to satisfy Equation (31.2).

to prove the existence of an integer-valued maximum flow for that graph. This integral flow defines a correct schedule over $[0, L)$. (Interested readers are referred to [1] for a detailed proof of this result).

31.2.2 Optimal Pfair Algorithms

At present, three Pfair scheduling algorithms have been proven optimal for scheduling synchronous periodic tasks on multiprocessors: PF [1], PD [2], and PD^2 [3]. These algorithms prioritize subtasks on an EPDF basis, but differ in the choice of tie-breaking rules. Breaking ties appropriately turns out to be the key concern when designing optimal Pfair algorithms. One tie-break parameter that is common to all three algorithms is the *successor bit*, which is defined as follows:

$$b(T_i) = \left\lceil \frac{i}{wt(T)} \right\rceil - \left\lfloor \frac{i-1}{wt(T)} \right\rfloor \tag{31.6}$$

Informally, $b(T_i)$ denotes the number of slots by which T_i's window overlaps T_{i+1}'s window (see (31.3) and (31.4)). For example, in Figure 31.1, $b(T_i) = 1$ for $1 \leq i \leq 7$ and $b(T_8) = 0$. (Note that the last subtask of a job of a periodic task has a successor bit of 0.) We now briefly describe each of the three algorithms.

31.2.2.1 The PF Algorithm

Under the PF algorithm, subtasks are prioritized as follows: at time t, if subtasks T_i and U_j are both ready to execute, then T_i has higher priority than U_j, denoted $T_i \succ U_j$, if one of the following holds:

(i) $d(T_i) < d(U_j)$,
(ii) $d(T_i) = d(U_j)$ and $b(T_i) > b(U_j)$,
(iii) $d(T_i) = d(U_j)$, $b(T_i) = b(U_j) = 1$, and $T_{i+1} \succ U_{j+1}$.

If neither subtask has priority over the other, then the tie can be broken arbitrarily. Given the PF priority definition, the description of the PF algorithm is simple: at the start of each slot, the M highest priority subtasks (if that many eligible subtasks exist) are selected to execute in that slot.

As shown in Rule (ii), when comparing two subtasks with equal pseudo-deadlines, PF favors a subtask T_i with $b(T_i) = 1$, i.e., if its window overlaps that of its successor. The intuition behind this rule is that executing T_i early prevents it from being scheduled in its last slot. Avoiding the latter possibility is important because scheduling T_i in its last slot effectively shortens $w(T_{i+1})$, which makes it more difficult to schedule T_{i+1} by its pseudo-deadline. If two subtasks have equal pseudo-deadlines and successor bits of 1, then according to Rule (iii), their successor subtasks are recursively checked. This recursion will halt within $\min(T.e, U.e)$ steps, because the last subtask of each job has a successor bit of 0.

In [1], PF was proven optimal using an inductive swapping argument. The crux of the argument is to show that, if there exists a Pfair schedule S such that all decisions in S before slot t are in accordance with PF priorities, then there exists a Pfair schedule S' such that all the scheduling decisions in S' before slot $t+1$ are in accordance with PF priorities. To prove the existence of S', the scheduling decisions in slot t of S are systematically changed so that they respect the PF priority rules, while maintaining the correctness of the schedule. We briefly summarize the swapping arguments used to transform S.

Suppose that T_i and U_j are both eligible to execute in slot t and $T_i \succ U_j$. Furthermore, suppose that, contrary to the PF priority rules, U_j is scheduled in slot t in S, while T_i is scheduled in a later slot t' in S. There are three possibilities.

- $T_i \succ U_j$ by Rule (i). Because T_i's pseudo-deadline is less than U_j's pseudo-deadline, and because the windows of consecutive subtasks overlap by at most one slot, U_{j+1} is scheduled at a later slot than T_i. Therefore, T_i and U_j can be directly swapped, as shown in Figure 31.2(a).
- $T_i \succ U_j$ by Rule (ii). By Rule (ii), $b(U_j) = 0$, which implies that U_{j+1}'s window does not overlap that of U_j. Hence, T_i and U_j can be directly swapped without affecting the scheduling of U_{j+1}, as shown in Figure 31.2(b).

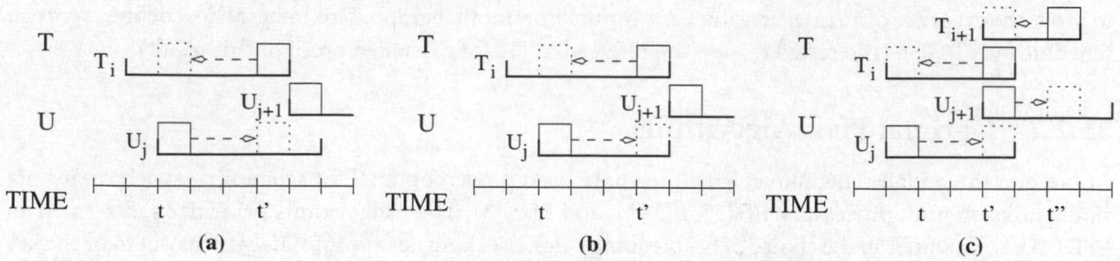

FIGURE 31.2 Correctness proof for PF. Dashed arrows indicate the movement of quanta when swapping allocations. Insets (a–c) illustrate the three cases considered in the proof.

- $T_i \succ U_j$ by Rule (iii). In this case, it may not be possible to directly swap T_i and U_j because U_{j+1} may be scheduled in the same slot as T_i (i.e., swapping T_i and U_j would result in U_j and U_{j+1} being scheduled in the same slot). If U_{j+1} is indeed scheduled in slot t', then it is necessary to first swap T_{i+1} and U_{j+1}, as shown in Figure 31.2(c), which may in turn necessitate the swapping of later subtasks.

From this inductive proof, we obtain the following result:

Theorem 31.1 [1]

PF *is optimal for scheduling periodic tasks on multiprocessors.*

31.2.2.2 The PD and PD^2 Algorithms

Though optimal, PF is inefficient due to the recursion in Rule (iii). In PD, Rule (iii) is replaced by three additional rules, each of which involves only a simple calculation. (Hence, PD uses four rules to break pseudo-deadline ties.) On the other hand, PD^2 replaces Rule (iii) with a single rule and requires only two tie-break parameters. Since PD^2 is a simplified version of PD, we discuss PD^2 first.

Under Rule (iii) of the PD^2 priority definition, group deadlines of competing subtasks are compared. The group deadline of a task T is only important when the task is heavy but does not have unit weight, i.e., when $1/2 \leq wt(T) < 1$. If a task does not satisfy this criterion, then its group deadline is 0.

To motivate the definition of the group deadline, consider a sequence $T_i, \ldots, T_j$ of subtasks such that $b(T_k) = 1 \wedge |w(T_{k+1})| = 2$ for all $i \leq k < j$. Note that scheduling T_i in its last slot forces the other subtasks in this sequence to be scheduled in their last slots. For example, in Figure 31.1, scheduling T_3 in slot 4 forces T_4 and T_5 to be scheduled in slots 5 and 6, respectively. The group deadline of a subtask T_i, denoted $D(T_i)$, is the earliest time by which such a cascade must end. Formally, it is the earliest time t, where $t \geq d(T_i)$, such that either $(t = d(T_k) \wedge b(T_k) = 0)$ or $(t + 1 = d(T_k) \wedge |w(T_k)| = 3)$ for some subtask T_k. (Intuitively, if we imagine a job of T in which each subtask is scheduled in the first slot of its window, then the slots that remain empty exactly correspond to the group deadlines of T within that job.) For example, in Figure 31.1, $D(T_3) = d(T_6) - 1 = 8$ and $D(T_7) = d(T_8) = 11$.

Using this definition, Rule (iii) of the PD^2 priority rules is defined as follows.

(iii) $d(T_i) = d(U_j), b(T_i) = b(U_j) = 1$, and $D(T_i) > D(U_j)$

(Rules (i) and (ii) remain unchanged, and ties not resolved by all three rules can still be broken arbitrarily.) As shown, PD^2 favors subtasks with later group deadlines. The intuition behind this prioritization is that scheduling these subtasks early prevents (or at least reduces the extent of) cascades. Cascades are undesirable since they constrain the scheduling of future slots.

The priority definition used in the PD algorithm adds two additional tie-break parameters (and rules) to those used by PD^2. The first of these is a task's weight. The second is a bit that distinguishes between the two different types of group deadline. For example, in Figure 31.1, $D(T_1)$ is a type-1 group deadline because its placement is determined by the $(t+1 = d(T_k) \wedge |w(T_k)| = 3)$ condition, while $D(T_6)$ is a type-0 group deadline. (Effectively, the value of this bit is the value of $b(T_k)$, where T_k is the subtask that defines the group deadline.) The optimality of PD^2 shows that these two additional tie-break parameters are not needed.

PD was proved optimal by a simulation argument that showed that PD closely tracks the behavior of the PF algorithm; the optimality of PF was then used to infer the optimality of PD [2]. On the other hand, PD^2 was proved optimal through a swapping technique, similar to that used in the optimality proof of PF [3]. We omit these proofs because they are quite lengthy.

Theorem 31.2 [2]

PD *is optimal for scheduling periodic tasks on multiprocessors.*

Theorem 31.3 [3]

PD^2 *is optimal for scheduling periodic tasks on multiprocessors.*

31.2.2.3 Implementation

We now describe an implementation strategy that results in a scheduler with $O(M \log N)$ time complexity, where M is the number of processors, and N the number of tasks [2]. First, assume that eligible subtasks are stored in a priority-ordered "ready queue" R and that ineligible subtasks that will become eligible at time t are stored in the "release queue" Q_t. At the beginning of slot t, Q_t is merged into R. The $\min(M, |R|)$ highest-priority subtasks in R are then extracted from R and selected for execution. For each selected subtask T_i, its successor, T_{i+1}, is initialized and inserted into the appropriate release queue. Using binomial heaps (which are capable of performing all basic heap operations in $O(\log N)$ time) to implement the various queues yields the desired $O(M \log N)$ time complexity.

31.2.3 ERfair Scheduling

One undesirable characteristic of Pfair scheduling is that jobs can be ineligible according to the Pfairness constraint, despite being ready. Consequently, processors may idle while ready but unscheduled jobs exist. *Early-release fairness* (ERfairness) [3] was proposed to address this problem. Under ERfairness, the -1 lag constraint is dropped from Equation (31.2). Instead, each subtask T_i is assumed to have an *eligibility time* $e(T_i) \le r(T_i)$, which is the time at which T_i becomes eligible to execute. (The scheduling of T_i must still respect precedence constraints, i.e., T_i cannot become eligible until T_{i-1} is scheduled, regardless of $e(T_i)$.) In [3], PD^2 was shown to correctly schedule any ERfair task set satisfying Equation (31.5). Note that a Pfair schedule will always respect ERfairness, but not vice-versa. Figure 31.3 illustrates how one job of the task in Figure 31.1 might be scheduled under ERfair scheduling.

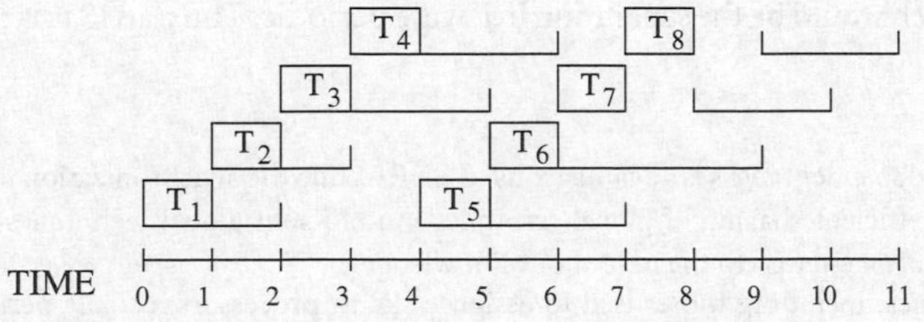

FIGURE 31.3 The Pfair windows of the first jobs of a task T with weight 8/11 are shown. The schedule shown is a valid ERfair schedule, but not a valid Pfair schedule.

31.2.3.1 Servicing Aperiodic Jobs

One important application of ERfair scheduling is to permit mixed scheduling of Pfair tasks and aperiodic jobs. An aperiodic job is a one-shot job that is not associated with any task. Such jobs typically represent the execution of service routines, including those used for interrupt handling. The response times of aperiodic jobs can be improved by allowing server tasks to early release their subtasks [14]. This improves responsiveness while also ensuring that the schedulability of periodic tasks is not compromised.

31.2.3.2 Implementation

PF, PD, and PD^2 can be easily adapted to allow early releases. In particular, if T_i is selected for execution, and if its successor T_{i+1} is eligible (perhaps due to an early release), then T_{i+1} can be inserted immediately into the ready queue. Hence, ERfair variants tend to be more efficient than their counterparts since fewer queue-merge operations are needed.

31.2.4 Practicality

Because of the quantum-based nature of Pfair scheduling, the frequency of preemptions and migrations is a potential concern. A recent experimental comparison conducted by Baruah and us [15] showed that PD^2 has comparable performance (in terms of schedulability) to a task-partitioning approach in which each task is statically assigned to a processor and the well-known earliest-deadline-first (EDF) scheduling algorithm is used on each processor. In this study, real system overheads such as context-switching costs were considered. Moreover, the study was biased against Pfair scheduling in that only static systems with independent[3] tasks of low weight[4] were considered. In spite of the frequent context-switching and cache-related overheads, PD^2 performs competitively because the schedulability loss due to these overheads is offset by the fact that PD^2 provides much better analytical bounds than partitioning.

To improve the practicality of Pfair scheduling, Holman and Anderson investigated many techniques for reducing overhead and improving performance in Pfair-scheduled systems [12,16–19]. These techniques have targeted many aspects of the system, including the frequency of context switching and migrations, scheduling overhead, cache performance, and contention for shared hardware and software resources.

Chandra, Adler, and Shenoy investigated the use of Pfair scheduling in general-purpose operating systems [8]. The goal of their work was to determine the efficacy of using Pfair scheduling to provide quality-of-service guarantees to multimedia applications. Consequently, no formal analysis of their approach was presented. Despite this, the experimental evaluation of their work convincingly demonstrates the practicality of Pfair scheduling.

31.3 Intrasporadic Task Systems

The *intrasporadic* (IS) task model was proposed as an extension of the well-studied periodic and sporadic models [9,10]. The sporadic model generalizes the periodic model by allowing *jobs* to be released late, i.e., the separation between consecutive job releases of a task is allowed to be more than the task's period. The IS model generalizes this by allowing *subtasks* to be released late, as illustrated in Figure 31.4. More specifically, the separation between subtask releases $r(T_i)$ and $r(T_{i+1})$ is allowed to be more than $\lfloor i/wt(T) \rfloor - \lfloor (i-1)/wt(T) \rfloor$, which would be the separation if T were periodic. Thus, an IS task is obtained by allowing

[3]Considering only independent tasks is advantageous to EDF. While the synchronization techniques described later in Section 31.6 permit efficient sharing of global resources, no efficient global techniques have been proposed for partitioned EDF-scheduled systems (to the best of our knowledge).

[4]Bin-packing heuristics, including those used to assign tasks to processors, usually perform better with smaller items. As the mean task utilization increases, the performance of these heuristics tends to degrade, resulting in more schedulability loss. Although heavy tasks are rare, some techniques, such as hierarchical scheduling, can introduce heavy "server" tasks. Hence, it is unrealistic to assume that only light tasks will occur in practice.

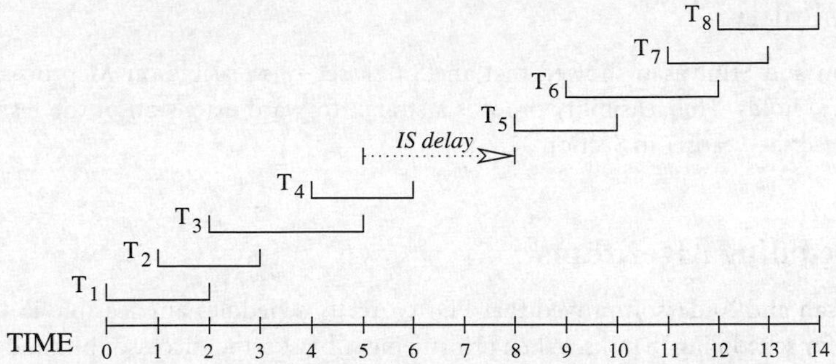

FIGURE 31.4 The PF-windows of the first eight subtasks of an IS task T with weight 8/11. Subtask T_5 is released three units late causing all later subtask releases to be delayed by three time units.

a task's windows to be right-shifted from where they would appear if the task were periodic. Figure 31.4 illustrates this.

Under the IS model, each subtask T_i has an *offset*, denoted $\theta(T_i)$, that gives the amount by which its window has been right-shifted. Hence, Equations (31.3) and (31.4) can be expressed as follows:

$$r(T_i) = \theta(T_i) + \left\lfloor \frac{i-1}{wt(T)} \right\rfloor \tag{31.7}$$

$$d(T_i) = \theta(T_i) + \left\lceil \frac{i}{wt(T)} \right\rceil \tag{31.8}$$

The offsets are constrained so that the separation between any pair of subtask releases is at least the separation between those releases if the task were periodic. Formally, the offsets must satisfy

$$k \geq i \Rightarrow \theta(T_k) \geq \theta(T_i) \tag{31.9}$$

Under the IS model, a subtask T_i is permitted to execute before the beginning of its Pfair window. That is, each subtask T_i has an eligibility time $e(T_i)$, as in ERfair scheduling, and $e(T_i)$ is allowed to be less than $r(T_i)$. The interval $[r(T_i), d(T_i))$ is said to be T_i's *PF-window*, while the interval $[e(T_i), d(T_i))$ is said to be its *IS-window*.

Using the definitions above, it is easy to show that sporadic and periodic tasks are special cases of IS tasks. In particular, using the expression $J(T_i) = \lfloor \frac{i-1}{T.e} \rfloor + 1$, which maps a subtask to the index (starting at 1) of the associated job, a periodic task T can be expressed as an IS task with $e(T_i) = T.\phi + (J(T_i) - 1) \cdot T.p$, i.e., each grouping of $T.e$ subtasks become eligible simultaneously when the associated job is released. Similarly, all subtasks associated with a job under the sporadic model will become eligible when the associated job is released. All subtasks within a single job of a sporadic task have the same offset, i.e., only the window of the first of these subtasks may be separated from that of its predecessor by an amount exceeding that in a periodic system.

The IS model allows the instantaneous rate of subtask releases to differ significantly from the average rate (given by a task's weight). Hence, it is more suitable than the periodic model for many applications, particularly those in networked systems. Examples include web servers that provide quality-of-service guarantees, packet scheduling in networks, and the scheduling of packet-processing activities in routers [20]. Due to network congestion and other factors, packets may arrive late or in bursts. The IS model treats these possibilities as first-class concepts and handles them more seamlessly. In particular, a late packet arrival corresponds to an IS delay. On the other hand, if a packet arrives early (as part of a bursty sequence), then its eligibility time will be less than its Pfair release time. Note that its Pfair release time determines its deadline. Thus, in effect, an early packet arrival is handled by postponing its deadline to where it would have been had the packet arrived on time.

31.3.1 Feasibility

In [9], Anderson and Srinivasan showed that an IS task set τ is feasible on M processors if and only if Equation (31.5) holds. This feasibility proof is a straightforward extension of the feasibility proof for periodic tasks described earlier in Section 31.2.1.

31.3.2 Scheduling Algorithms

In [10], Srinivasan and Anderson proved that PD^2 correctly schedules any feasible IS task system, and thus, is optimal for scheduling IS tasks. When prioritizing subtasks, the successor bits and group deadlines are calculated as if no IS delays occur in the future. For example, consider subtask T_5 in Figures 31.1 and 31.4. In Figure 31.1, T_5's group deadline is at time 8, while in Figure 31.4, its group deadline is at time 11. However, T_3's group deadline is at time 8 in both figures because T_3 is prioritized as if no IS delays occur in the future. In both figures, each of T_3 and T_5 has a b-bit of 1.

Theorem 31.4 [10]

PD^2 *is optimal for scheduling intrasporadic tasks on multiprocessors.*

Srinivasan and Anderson also showed that the EPDF algorithm, which uses only Rule (i) to prioritize subtasks, is optimal for scheduling IS tasks on M ($>$1) processors if the weight of each task is at most $\frac{1}{M-1}$. (This result is fairly tight. In particular, if the weight of a task is allowed to be at least $\frac{1}{M-1} + \frac{1}{(M-1)^2}$, then EPDF can miss deadlines.)

Although the periodic model represents the worst-case behavior of an IS task, it turns out that the optimality proofs mentioned previously do not extend directly to the IS case. The primary reason for this is that the proofs are based on swapping arguments, which require *a priori* knowledge of the positions of future windows. To prove the optimality of PD^2, Srinivasan and Anderson developed a novel and effective lag-based argument that is more flexible than swapping-based arguments. (Interested readers are referred to [10] for the complete proof.) This same proof technique was also used to obtain the EPDF results mentioned above, and also Theorem 31.7 in the next section [11].

31.4 Dynamic Task Systems

In many real-time systems, the set of runnable tasks may change dynamically. For example, in an embedded system, different modes of operation may need to be supported; a mode change may require adding new tasks and removing existing tasks. Another example is a desktop system that supports real-time applications such as multimedia and collaborative-support systems, which may be initiated at arbitrary times. When considering dynamic task systems, a key issue is that of determining when tasks may join and leave the system without compromising schedulability.

31.4.1 Join and Leave Conditions

For IS task systems, a join condition follows directly from Equation (31.5), i.e., a task can join if the total weight of all tasks will be at most M after its admission. It remains to determine when a task may leave the system safely. (Here, we are referring to the time at which the task's share of the system can be reclaimed. The task may actually be allowed to leave the system earlier.) As shown in [21,22], if a task with negative lag is allowed to leave, then it can rejoin immediately and effectively execute at a rate higher than its specified rate, which may cause other tasks to miss their deadlines. Based on this observation, the conditions shown below are at least necessary, if not sufficient.

(C1) *Join condition*: A task T can join at time t if and only if Equation (31.5) continues to hold after joining.[5]
Leave condition: A task T can leave at time t if and only if $t \geq d(T_i)$, where T_i is the last-scheduled subtask of T.

The condition $t \geq d(T_i)$ is equivalent to $lag(T, t) \geq 0$. To see why, note that since $t \geq d(T_i)$, task T receives at least i units of processor time in the ideal schedule by time t. Because T_i is the last-scheduled subtask of T in the actual schedule, T receives at most i quanta by time t. Hence, $lag(T, t) \geq 0$. As a straightforward extension of the feasibility proof for IS task systems [9], it can be easily shown that a feasible schedule exists for a set of dynamic tasks if and only if condition (C1) is satisfied. Indeed, condition (C1) has been shown to ensure schedulability when using *proportional-share* scheduling on a uniprocessor [21,22]. However, as shown below, (C1) is not sufficient when using a priority-based Pfair algorithm (such as PF, PD, or PD^2) on a multiprocessor.

The theorem below applies to any "weight-consistent" Pfair scheduling algorithm. An algorithm is *weight-consistent* if, given two tasks T and U of equal weight with eligible subtasks T_i and U_j, respectively, where $i = j$ and $r(T_i) = r(U_j)$ (and hence, $d(T_i) = d(U_j)$), T_i has priority over a third subtask V_k if and only if U_j does. All known Pfair scheduling algorithms are weight-consistent.

Theorem 31.5 [11]

No weight-consistent Pfair scheduler can guarantee all deadlines on multiprocessors under (C1).

Proof

Consider task systems consisting of only two weight classes: class X with weight $w_1 = 2/5$ and class Y with weight $w_2 = 3/8$. Let $X_f = \{T_1 \mid T \in X\}$ and $Y_f = \{T_1 \mid T \in Y\}$. We construct a counterexample based upon which task weight is favored by the scheduler. (In each of our counterexamples, no subtask is eligible before its PF-window.) We say that X_f *is favored* (analogously for Y_f) if, whenever subtasks in X_f and Y_f are released at the same time, those in X_f are given priority over those in Y_f.

Case 1. X_f is favored. Consider a 15-processor system containing the following sets of tasks.

Set A: 8 tasks of weight w_2 that join at time 0.
Set B: 30 tasks of weight w_1 that join at time 0 and leave at time 3.
Set C: 30 tasks of weight w_1 that join at time 3.

Because $30w_1 + 8w_2 = 15$, this task system is feasible and the join condition in (C1) is satisfied. Furthermore, since $d(T_1) = \lceil \frac{5}{2} \rceil = 3$ for every $T \in B$, the leave condition in (C1) is also satisfied.

Since subtasks in X_f are favored, tasks in Set B are favored over those in Set A at times 0 and 1. Hence, the schedule for [0,3) will be as shown in Figure 31.5(a). Consider the interval $[3, 8)$. Each task in Sets A and C has two subtasks remaining for execution, which implies that Set A requires 16 quanta and Set C requires 60 quanta by time 8. However, the total number of quanta in $[3, 8)$ is $15 \times (8 - 3) = 75$. Thus, one subtask will miss its deadline at or before time 8.

Case 2: Y_f is favored. Consider an 8-processor system containing the following sets of tasks.

Set A: 5 tasks of weight w_1 that join at time 0.
Set B: 16 tasks of weight w_2 that join at time 0 and leave at time 3.
Set C: 16 tasks of weight w_2 that join at time 3.

Because $5w_1 + 16w_2 = 8$, this task system is feasible and the join condition in (C1) is satisfied. Furthermore, since $d(T_1) = \lceil \frac{8}{3} \rceil = 3$ for every $T \in B$, the leave condition in (C1) is also satisfied.

---

[5]If T joins at time t, then $\theta(T_1) = t$. A task that rejoins after having left is viewed as a new task.

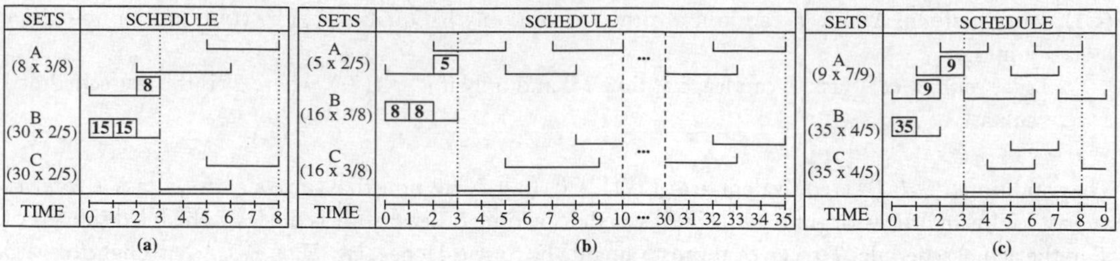

FIGURE 31.5 Counterexamples demonstrating insufficiency of (C1) and tightness of (C2). A boxed integer value n in slot t means that n of the subtasks from the corresponding task set are scheduled in that slot. The dotted vertical lines depict intervals with excess demand. (a) Theorem 31.5. Case 1: Tasks of weight 2/5 are favored at times 0 and 1, (b) Theorem 31.5. Case 2: Tasks of weight 3/8 are favored at times 0 and 1, (c) Theorem 31.6. The tasks of weight 4/5 are allowed to leave at time 3 and rejoin immediately.

Since subtasks in Y_f are favored, tasks in Set B are favored over those in Set A at times 0 and 1. Hence, the schedule for $[0, 3)$ will be as shown in Figure 31.5(b). Consider the interval $[3, 35)$. In this interval, each task in Set A requires $35 \times 2/5 - 1 = 13$ quanta. Similarly, each task in Set C requires $(35-3) \times 3/8 = 12$ quanta. However, the total requirement is $5 \times 13 + 16 \times 12 = 257$, whereas $[3, 35)$ contains only $(35-3) \times 8 = 256$ quanta. Thus, a deadline miss will occur at or before time 35. □

The problem illustrated by the preceding proof is that subtasks are prioritized according to the requirements of their successors. When a subtask is the last subtask that a task will release, then it has no successors and hence should be given a successor bit of 0 and a trivial group deadline (i.e., 0). However, without *a priori* knowledge of task departures, the scheduler must assign parameter values according to the normal priority rules, which are designed for persistent tasks. Indeed, Theorem 31.5 can be circumvented if the scheduler has such knowledge. For example, in Figure 31.5(a), if the scheduler had known that the subtasks in Set B had no successors, then it would have given those subtasks lower priority than those in Set A (by setting their successor bits to 0). However, in general, such *a priori* knowledge of task departures may not be available to the scheduler.

The examples in Figure 31.5(a, b) show that allowing a light task T to leave at $d(T_i)$ when $b(T_i) = 1$ can lead to deadline misses. We now consider heavy tasks.

Theorem 31.6 [11]

If a heavy task T is allowed to leave before $D(T_i)$, where T_i is the last-released subtask of T, then there exist task systems that miss a deadline under PD^2.

Proof

Consider a 35-processor system containing the following sets of tasks (where $2 \le t \le 4$).

Set A: 9 tasks of weight 7/9 that join at time 0.
Set B: 35 tasks of weight 4/5 that join at time 0, release a subtask, and leave at time t.
Set C: 35 tasks of weight 4/5 that join at time t.

All tasks in Sets A and B have the same PD^2 priority at time 0, because each has a deadline at time 2, a successor bit of 1, and a group deadline at time 5. Hence, the tasks in Set B may be given higher priority.[6] Assuming this, Figure 31.5(c) depicts the schedule for the case of $t = 3$.

[6]This counterexample also works for PF and PD since both will give higher priority to the tasks in Set B at time 0.

Consider the interval $[t, t+5)$. Each task in Sets A and C has four subtasks with deadlines in $[t, t+5)$ (see Figure 31.5(c)). Thus, $9 \times 4 + 35 \times 4 = 35 \times 5 + 1$ subtasks must be executed in $[t, t+5)$. Since 35×5 quanta are available in $[t, t+5)$, one subtask will miss its deadline. □

The cases considered in the proofs of Theorems 31.5 and 31.6 motivate the new condition (C2), shown below. In [11], this condition is shown to be sufficient for ensuring the schedulability of dynamic IS task sets scheduled by PD^2. By Theorems 31.5 and 31.6, this condition is tight.

(C2) *Join condition*: A task T can join at time t if and only if Equation (31.5) continues to hold after joining.
Leave condition: A task T can leave at time t if and only if $t \geq \max(D(T_i), d(T_i) + b(T_i))$, where T_i is the last-scheduled subtask of T.

Theorem 31.7 [11]

PD^2 *correctly schedules any dynamic IS task set that satisfies* (C2).

Observe that (C2) guarantees that periodic and sporadic tasks can always leave the system at period boundaries. This follows from the fact that the last subtask T_i in each job will have a successor bit of 0, which implies that $\max(D(T_i), d(T_i) + b(T_i)) = \max(d(T_i), d(T_i) + 0) = d(T_i)$.

31.4.2 QRfair Scheduling

In dynamic task systems, spare processing capacity may become available that can be consumed by the tasks present in the system. One way for a task T to consume such spare capacity is by early releasing its subtasks. However, by doing this, it uses up its future subtask deadlines. As a result, if the system load later increases, then T will be competing with other tasks using deadlines far into the future. Effectively, T is penalized when the load changes for having used spare capacity in the past.

In recent work, Anderson, Block, and Srinivasan proposed an alternative form of early-release scheduling called *quick-release fair* (*QRfair*) scheduling [23]. QRfair scheduling algorithms avoid penalizing tasks for using spare capacity by allowing them to shift their windows forward in time when an idle processor is detected. The benefits of doing this strongly resemble those provided by uniprocessor fair scheduling schemes based on the concept of *virtual time* [22].

31.5 Supertasking

In [4], Moir and Ramamurthy observed that the migration assumptions underlying Pfair scheduling may be problematic. Specifically, tasks that communicate with external devices may need to execute on specific processors and hence cannot be migrated. They further noted that statically binding tasks to processors may significantly reduce migration overhead in Pfair-scheduled systems.

To support nonmigratory tasks, they proposed the use of *supertasks*. In their approach, a supertask $\mathcal{S}_p$ replaces the set of tasks that are bound to processor p, which become the *component tasks* of $\mathcal{S}_p$. (We use $\mathcal{S}_p$ to denote both the supertask and its set of component tasks.) Each supertask $\mathcal{S}_p$ then competes with a weight equal to the cumulative weight of its component tasks, as shown below.

$$wt(\mathcal{S}_p) = \sum_{T \in \mathcal{S}_p} wt(T)$$

Whenever a supertask is scheduled, one of its component tasks is selected to execute according to an internal scheduling algorithm. Although Moir and Ramamurthy proved that EPDF is optimal for scheduling component tasks, they also demonstrated that component-task deadline misses may occur when using each of PF, PD, and PD^2 as the global scheduling algorithm.

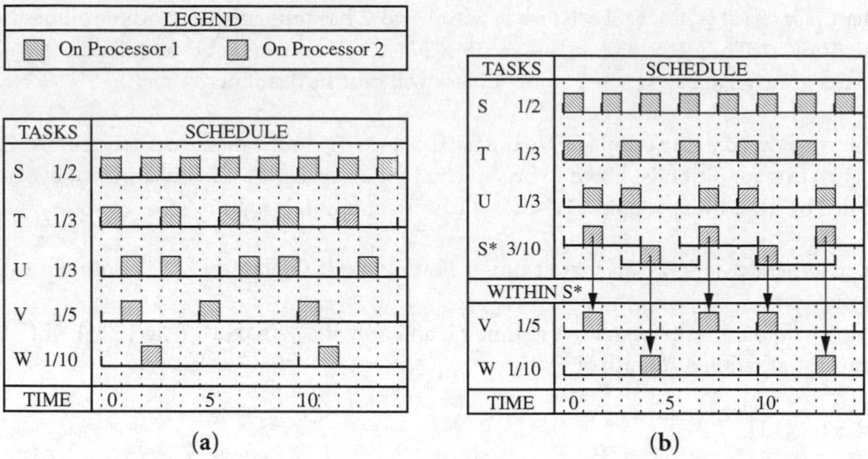

FIGURE 31.6 Sample Pfair schedules for a task set consisting of five tasks with weights 1/2, 1/3, 1/3, 1/5, and 1/10, respectively. (**a**) Normal schedule produced when no supertasks are used, (**b**) Schedule produced when tasks V and W are combined into the supertask S^*, which competes with weight 3/10 and is bound to processor 2.

Figures 31.6(a, b) illustrate supertasking. Figure 31.6(a) shows a PD^2 schedule in which two processors are shared among five tasks, labeled $S, \ldots, W$, that are assigned weights 1/2, 1/3, 1/3, 1/5, and 1/10, respectively. Vertical dashed lines mark the slot boundaries and boxes show when each task is scheduled. Figure 31.6(b) is derived from Figure 31.6(a) by placing tasks V and W into a supertask S^* that competes with weight $wt(S^*) = 3/10 = wt(V) + wt(W)$ and is bound to processor 2. The arrows show S^* passing its allocation (upper schedule) to one of its component tasks (lower schedule) based upon an EPDF prioritization. Although no component-task deadlines are missed in this example, such a weight assignment is not sufficient, in general, to guarantee that all deadlines are met when using the PD^2 algorithm [4].

31.5.1 Supertasking as a Hierarchical-Scheduling Mechanism

In [12], Holman and Anderson considered supertasking more generally as a mechanism for achieving hierarchical scheduling in Pfair-scheduled systems. Hierarchal scheduling is particularly interesting (and desirable) under Pfair scheduling because many common task behaviors (e.g., blocking and self-suspension) can disrupt the fair allocation of processor time to tasks, making scheduling more difficult. One way to compensate for these behaviors is to schedule a set of problematic tasks as a single entity and then use a second-level (component-task) scheduler to allocate the group's processor time to the member (component) tasks. In this approach, fairness is somewhat relaxed in that the group is required to make progress at a steady rate rather than each individual task within the group. (Since fair scheduling was introduced primarily to improve schedulability, weakening the fairness guarantee in order to improve schedulability should be an acceptable trade-off in many cases.)

One immediate advantage of this relaxed fairness is that the component-task scheduler need not be a fair scheduler. Indeed, using an unfair scheduler can result in substantial improvements, as discussed in [12] and as demonstrated in [18]. In addition, using supertasking to achieve hierarchical scheduling ensures that only one component task is executing at each instant, i.e., scheduling within a supertask is a single-resource problem. This characteristic provides two advantages. First, selectively preventing tasks that share resources from executing in parallel can reduce contention, and hence, improve the schedulability of the system, as demonstrated in [16,17]. (This advantage is discussed in more detail in the next section.) Second, component-task schedulability can be analyzed using demand-based arguments and other simple uniprocessor techniques. Informally, a demand-based argument states that a deadline miss must be preceded by an interval of time during which the total time required to service all pending jobs (in priority order) exceeded the available processor time. Such analysis is desirable due both to its simplicity and also to the ease with which it can be understood and adapted to new situations. Unfortunately, demand-based arguments tend to be ineffective on multiprocessors because they do not account for parallel execution.

In the rest of this section we provide an overview of problems and results relating to supertasking. We begin by discussing the scheduling problems that must be addressed. We conclude by briefly describing an open problem relating to supertasking.

31.5.2 Scheduling within a Supertask

Since a supertask is effectively a single resource, an inductive swapping argument can be used to prove that both EPDF and EDF scheduling algorithms can schedule any periodic component-task set that is feasible under a given supertask allocation, provided that all component tasks are independent. We briefly sketch the proof for the EDF case below.

First, suppose, to the contrary of the claim, that a schedule S exists in which all job deadlines are met, but the schedule produced by EDF, S^{EDF}, does not meet all deadlines. Also, let t be the earliest time (slot) at which the scheduling decisions differ. Specifically, select x and y so that $x \in S_t$, $x \notin S_t^{\mathrm{EDF}}$, $y \notin S_t$, and $y \in S_t^{\mathrm{EDF}}$. By the EDF prioritization, y must have a deadline that is at most the deadline of x. Furthermore, since $y \notin S_t$ and both schedules are identical up to time t, it follows that y must be scheduled next at some time t' in S, where $t' > t$. By the initial assumptions, no deadlines are missed in schedule S, which implies that y's (and hence x's) deadline must be at or after $t' + 1$. (Recall that job deadlines come at the end of slots while scheduling decisions are made at the start of each slot.) It follows that x and y can be swapped in schedule S without introducing a deadline miss. This process can then be repeated until all scheduling decisions at time t are identical in both schedules. By then inducting over time, all decisions in schedule S can be made to agree with those in S^{EDF} without introducing a deadline miss. However, this contradicts the claim that S^{EDF} contains a deadline miss, which completes the proof. Hence, EDF is an optimal policy when scheduling within a supertask. The proof is identical for EPDF, with the only exception being that subtask deadlines are considered instead of job deadlines.

31.5.3 Scheduling the Supertask

The main problem when using supertasks is to schedule a supertask so that its component-task set is feasible. There are two basic approaches to solving this problem. First, the global scheduler could be modified to handle supertasks as a special class. When invoked, the global scheduler would first need to determine when each supertask should be scheduled to guarantee the schedulability of its component tasks and then schedule the remaining tasks around the supertasks. Unfortunately, it seems unlikely that supertasks could be granted such special treatment while maintaining the optimality of the global scheduler. In addition, this approach is unappealing in that it would likely increase scheduling overhead and would not provide a clean separation between global and component-task scheduling.

An alternative approach, which Holman and Anderson investigated in [12,18], is to assign to each supertask a weight that is sufficient to guarantee component-task feasibility under *any* Pfair schedule. We refer to this weight-selection problem as the reweighting problem. Unfortunately, Moir and Ramamurthy's work implies that this approach will necessarily result in some schedulability loss, at least in some cases. In [12], Holman and Anderson presented rules for selecting a supertask weight. These rules are based upon demand-based analysis and stem from the theorem shown below, which bounds the amount of processor time available to a supertask over any interval of a given length.

Theorem 31.8 (Holman and Anderson)

A Pfair-scheduled supertask with weight w will receive at least $\lfloor wL \rfloor - 1$ quanta of processor time over any interval spanning L slots.

The theorem stated below follows directly from the rules presented in [12] and highlights some of the interesting practical ramifications of supertasking.

Theorem 31.9

All component tasks of a Pfair-scheduled supertask $\mathcal{S}$ will meet their deadlines if

$$wt(\mathcal{S}) = \min\left(1, \sum_{T \in \mathcal{S}} wt(T) + \frac{1}{L}\right)$$

provided that either (i) component tasks are scheduled using EPDF and L is the shortest window length *of any component task, or (ii) component tasks are scheduled using EDF and L is the* smallest period *of any component task.*

The first implication of the above theorem is that EDF is less costly than EPDF when scheduling component tasks. Although this would seem to contradict the earlier claim that both algorithms are optimal in this context, it actually highlights the fundamental difference between fair and unfair scheduling. Specifically, EDF is optimal for scheduling when *job* deadlines must be met (i.e., for periodic-task scheduling), but is not optimal when *subtask* deadlines must be met (i.e., for Pfair scheduling). Since the Pfairness constraint is stricter than the periodic-task constraint, it is more costly to provide this guarantee when scheduling component tasks. Hence, EDF should be used for scheduling component tasks whenever fairness is not essential to the correctness of the system.

The second implication is that the schedulability loss that stems from reweighting a supertask is bounded (by $\frac{1}{L}$) and should often be small in practice. For instance, if the quantum size is 5 msec and the smallest component-task period is 250 msec, the inflation will be at most 0.02. In addition, this penalty can be reduced when using EDF by increasing component-task periods.

31.5.4 Feasibility

Although sufficient schedulability conditions have been derived for hierarchical scheduling with supertasks (in [12,18]), the corresponding feasibility problem (i.e., the problem of devising necessary and sufficient schedulability conditions) remains open. In [4], Moir and Ramamurthy proved that a periodic task system containing nonmigratory tasks is feasible if and only if it satisfies

$$\sum_{T \in \tau} wt(T) \leq M \wedge \left(\forall p : 1 \leq p \leq M : \sum_{T \in \mathcal{S}_p} wt(T) \leq 1\right)$$

This proof is based on a flow-graph construction similar to that used by Baruah et al. [1] to prove the feasibility condition (31.5) for periodic task systems. Unfortunately, this proof does not appear to extend to the supertasking approach. Consequently, there is no known feasibility condition for systems in which periodic tasks are scheduled within Pfair-scheduled supertasks.

31.6 Resource Sharing

In this section we discuss techniques for supporting task synchronization under Pfair scheduling, including techniques for lock-free [16] and lock-based [17] synchronization.

31.6.1 Lock-Free Synchronization

In [16], Holman and Anderson considered Pfair-scheduled systems in which lock-free shared objects are used. Lock-free algorithms synchronize access to shared software objects while ensuring system-wide liveness, even when faced with process halting failures. Specifically, lock-free algorithms guarantee that

```
typedef Qtype:
    record data: valtype; next: pointer to Qtype

shared var
    Head, Tail: pointer to Qtype

private var
    old, new: pointer to Qtype; input: valtype;
    addr: pointer to pointer to Qtype

procedure Enqueue(input)
    *new := (input, nil);
    do old := Tail;
        if old ≠ nil then addr := &((*old).next)
        else addr := &Head fi
    while ¬CAS2(&Tail, addr, old, nil, new, new)
```

FIGURE 31.7 This figure shows a lock-free enqueue operation using the CAS2 primitive. The first two parameters of CAS2 specify addresses of two shared variables, the next two parameters are values to which these variables are compared, and the last two parameters are new values to assign to the variables if both comparisons succeed. Although CAS2 is uncommon, it makes for a simple example here.

some task is always capable of making progress. Lock-based algorithms cannot satisfy this property since a task can prevent all other tasks from making progress by halting while holding a lock.

Lock-free algorithms typically avoid locking through the use of "retry loops" and strong synchronization primitives. Figure 31.7 depicts a lock-free enqueue operation. An item is enqueued in this implementation by using a *two-word compare-and-swap* (CAS2) instruction to atomically update a tail pointer and either the next pointer of the last item in the queue or a head pointer, depending on whether the queue is empty. This loop is executed repeatedly until the CAS2 instruction succeeds. An important property of lock-free implementations such as this is that operations may interfere with each other. In this example, an interference results when a successful CAS2 by one task causes another task's CAS2 to fail.

Blocking due to lock requests complicates Pfair scheduling considerably, as explained below. Because of this, the lock-free approach will likely perform better when it is applicable.

31.6.1.1 Problems with Lock-Based Synchronization

Locking is problematic in Pfair-scheduled systems for three reasons. First, Pfair scheduling can lead to long delays. Specifically, low-weight tasks are scheduled infrequently, as shown in Figure 31.8. In general, the worst-case delay experienced by lock-requesting tasks due to a lock-holding Pfair-scheduled task T is inversely-proportional to $T.w$. Second, critical sections can span multiple subtasks. Hence, a task's priority may change *during* one of its critical sections. Most existing locking protocols implicitly assume that each task's priority is static, at least until its critical section completes. Third, Pfair weights are a form of *reservation* mechanism, which complicates blocking-time accounting.

31.6.1.2 Accounting for Lock-Free Overhead

Lock-free algorithms are usually viewed as impractical for real-time multiprocessor systems because bounding the worst-case number of retries is difficult. However, the tight synchrony provided by Pfair

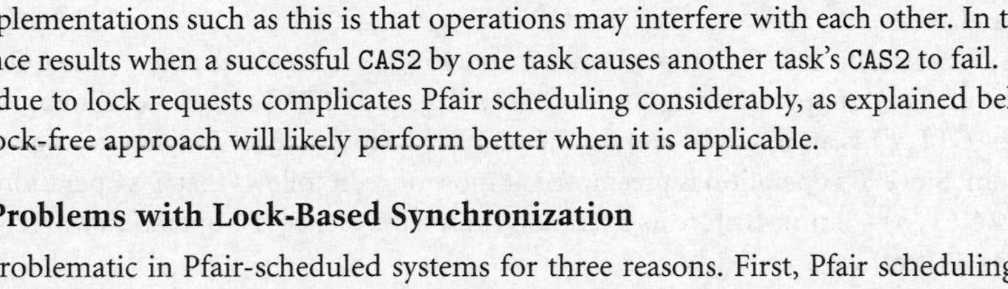

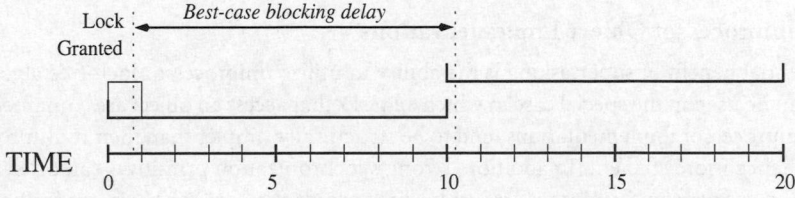

FIGURE 31.8 This figure illustrates the blocking delay bounds implied by Pfair scheduling for a task T of weight 1/10. In this example, task T obtains a lock within slot 0 and is preempted before releasing it. Assuming the lock will be released shortly after the start of the next quantum allocated to T, the shared resource will still be unavailable for an interval of at least 9 slots.

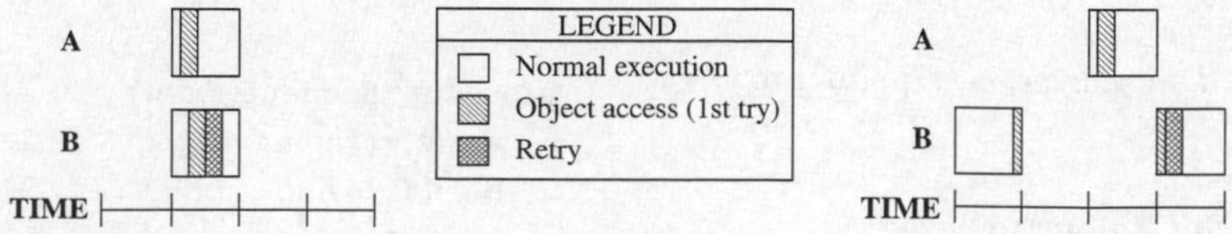

FIGURE 31.9 This figure shows scenarios under which task A causes task B's lock-free operation to fail, and hence be retried. (a) A is scheduled in parallel with B, (b) B is preempted during the operation.

scheduling facilitates worst-case analysis. Specifically, Pfair scheduling guarantees that quantum boundaries align across processors and that at most M tasks execute within each time slot. Hence, a scheduled task faces interference from no more than $M - 1$ other tasks in each time slot in which it executes. In addition, Pfair's tight synchrony implies that interference takes one of two possible forms: interference within a quantum and interference across multiple quanta. Both forms are illustrated in Figure 31.9.

Bounding the number of interferences across multiple quanta is trivial in most cases due to practical considerations. In experiments conducted by Ramamurthy [24] on a 66 MHz processor, operation durations for a variety of common objects (e.g., queues, linked lists, etc.) were found to be in the range of tens of microseconds. On modern processors, these operations will likely require no more than a few microseconds. Since quantum sizes typically range from hundreds of microseconds to milliseconds, it is unlikely that any shared-object operation will be preempted more than once. Hence, we only consider using lock-free techniques when all operations are guaranteed to be preempted at most once, which is expected often to be the case. Therefore, no more than one retry is needed due to interference across multiple quanta.

Now, consider a specific task T that performs an operation on a lock-free object ρ. Let $Q(T, \rho)$ denote the worst-case number of operations applied to ρ by any group of $M - 1$ tasks, excluding T, within one quantum, i.e., $Q(T, \rho)$ is an upper bound on the number of interferences experienced by T within any single quantum. Since T's operation is preempted at most once, it follows that T's operation is interfered with at most $2Q(T, \rho) + 1$ times before its operation completes. Hence, T's operation will require no more than $2Q(T, \rho) + 1$ retries.

31.6.1.3 Reducing Retry Overhead

Recall that interference within a quantum can only occur when two or more tasks that share a common object are scheduled in parallel. One simple technique for reducing such interferences is to place tasks that share common objects into the same supertask, thereby reducing the value of $Q(T, \rho)$. For instance, consider a 4-processor system in which four tasks, denoted $A, \ldots, D$, share an object. Further, suppose that $A, \ldots, D$ make at most 1, 4, 3, and 6 accesses to the object within a single quantum. Without supertasking, task A can experience up to $6 + 4 + 3 = 13$ interferences within each quantum due to tasks $B, \ldots, D$. The worst-case scenario occurs when $A, \ldots, D$ are all scheduled in the same time slot. However, if $B, \ldots, D$ are placed within a supertask, task A can experience no more than $\max(6, 4, 3) = 6$ interferences within each quantum since A can be scheduled together with at most one task from the supertask.

31.6.1.4 Uniprocessor Object Implementations

Another potential benefit of supertasking is the ability to utilize uniprocessor lock-free algorithms. Such algorithms can be used in the special case in which all tasks that access an object are contained in the same supertask. Uniprocessor implementations tend to be structurally simpler than their multiprocessor counterparts and hence more efficient. In addition, strong synchronization primitives can often be efficiently implemented in software on a uniprocessor [25], which avoids the need for hardware support.

31.6.1.5 Evaluating the Supertasking Trade-off

Unfortunately, supertasking's benefits may be counterbalanced by the schedulability loss due to reweighting. To determine whether supertasking is a viable tool for reducing lock-free overhead, Holman and

Anderson conducted a series of experiments for 2-, 4-, 8-, and 16-processor systems [16]. In each experiment, a task set that shares lock-free objects was randomly generated and then the cumulative weight of the system was computed both with and without supertasks. (The heuristic used to assign tasks to supertasks and the experimental setup are described in detail in [16].) For 2-processor systems, schedulability loss due to reweighting was found to outweigh the reduction in lock-free overhead in the vast majority (91%) of the cases considered. This is because interference within a quantum is already limited to only a single interfering task per time slot. Hence, the use of supertasking provides only marginal benefits. On the other hand, supertasking improved schedulability in 82.5, 98.8, and 99.9% of the task sets considered for the 4-, 8-, and 16-processor systems, respectively.

31.6.2 Locking Synchronization

In [17], Holman and Anderson presented two approaches for supporting (non-nested) critical sections under Pfair scheduling. In this section we summarize the concepts underlying these approaches. The analysis of systems using these approaches is straightforward; details can be found in [17].

31.6.2.1 Limitations of Lock-Free Synchronization

Although lock-free algorithms avoid many problems that come with locking, they have many limitations. As previously mentioned, lock-free algorithms often require strong synchronization primitives. Such primitives may not be available in some multiprocessor systems. In addition, time and space overheads do not scale well with respect to object complexity. Complex objects can usually be implemented much more efficiently using locking. Finally, lock-free techniques can only be applied to objects implemented in software. Locks are still needed to synchronize access to external devices.

31.6.2.2 Approach 1: Timer-Based

In the first approach presented by Holman and Anderson, the durations of all critical sections guarded by the lock in question are assumed to be much shorter than the length of a scheduling quantum. (For reasons already explained, this approach is expected to be widely applicable.) For cases in which this assumption holds, a per-lock timer signal, called FREEZE (see the diagram below), can be used that occurs towards the end of each time slot. Once this signal is received, lock granting is disabled. By placing this signal appropriately in time, a task that risks being preempted while holding the lock can be prevented from obtaining it. Hence, no locks using this implementation can be held across a quantum boundary, which avoids many of the previously mentioned problems associated with locking. Effectively, the blocking overhead caused by the preemption of a lock-holding task is traded for blocking overhead caused by the frozen intervals. We discuss the effects of this trade-off at the end of this section.

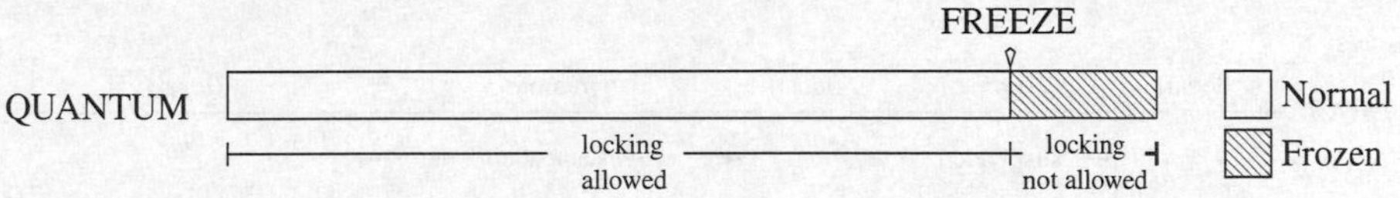

31.6.2.3 Approach 2: Server-Based

This approach can be used to implement any lock, i.e., no assumptions are made about critical-section durations. In this approach, a server executes all critical sections guarded by a specific lock in place of the requesting task. The primary alternative to a server-based approach is the use of some form of inheritance mechanism. Unfortunately, inheritance-based protocols are problematic under Pfair scheduling as explained below.

Under an inheritance-based protocol, a lock-holding task is allowed to inherit some of the scheduling parameters of a lock-requesting task that it blocks. For example, under the priority inheritance protocol [26], a lock-holding task T inherits the priority of a higher-priority task that it blocks. Under Pfair scheduling, a task T that inherits another task U's scheduling parameters is temporarily bound to U's state. Thus, an explicit distinction is made between scheduling states and the tasks that are bound to them. Usually a task T is bound to its own state, but when inheritance occurs, this binding may change. Specifically, T may become bound to another task U's state, or even to *both* its own state *and* U's state, leaving U temporarily bound to no state.

The problem with using inheritance-based protocols in Pfair-scheduled systems is that blocking durations are both determined by task weights and compensated for by changing task weights. Specifically, the blocking experienced by a task T, which is determined by the weights of tasks that share resources with T, is compensated for by adjusting T's weight. However, changing T's weight changes its Pfair window layout, which may either increase or decrease the worst-case delay before a lock held by T is released. This change in delay may then require compensatory changes in other tasks' weights that share locks with T. Unfortunately, when these other weights change, the worst-case blocking experienced by T will also change, requiring T to change weight again. Identifying an optimal weight assignment is a nontrivial task under inheritance-based protocols. Furthermore, changing the weights of tasks may also change the worst-case blocking scenario, which further complicates the weight assignment process. Ultimately, the interdependence of weights caused by inheritance-based protocols makes the effect of a weight change difficult to predict, which may be undesirable even if the weight-selection algorithm's overhead is reasonable.

The protocol proposed by Holman and Anderson, referred to as the static-weight server protocol (SWSP), uses statically-weighted lock servers to execute critical sections in place of the requesting task [17]. Effectively, critical sections guarded by a common lock are implemented as remote procedure calls serviced by a common server. Delays caused by critical-section requests are easily modelled and accounted for by using the IS task model, as illustrated in Figure 31.10. In addition, the use of statically-defined server weights makes blocking-term computations straightforward and the responsiveness of the server more predictable.

31.6.2.4 Comparison

Holman and Anderson conducted breakdown utilization tests for systems with 2, 4, 8, 16, and 32 processors to evaluate the schedulability loss under each of these protocols [17]. As expected, they found that the timer-based approach performs and scales very well when critical sections are much shorter than the quantum's duration. However, breakdown utilizations dropped off quickly once critical section durations longer than approximately one-quarter of the quantum's duration were permitted. The SWSP, on the other hand, had typical breakdown utilizations ranging from 25% to 75% of the system's total utilization. However, in the context of long critical sections, the SWSP did outperform the timer-based approach in many cases.

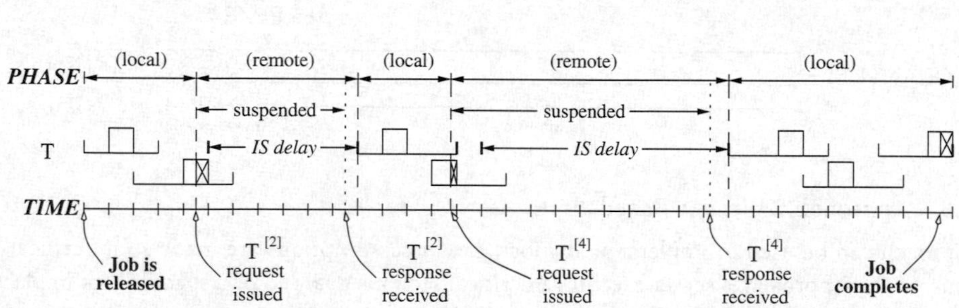

FIGURE 31.10 Behavior of a lock-requesting task T with weight 7/19 under the SWSP. The figure shows the release pattern for T's windows, where T has five phases per job. The first, third, and fifth phases consist of local work, while the second and fourth phases require locks and are hence executed remotely by the servers. Phase transitions are shown across the top of the figure and time is shown across the bottom. Arrows show when each RPC begins and ends. Unshaded boxes show where T executes locally while boxes containing X's denote unutilized processor time.

31.7 Summary

In this chapter we discussed the basic concepts behind Pfair scheduling and the various extensions that have been proposed in the literature. We described the three optimal Pfair algorithms: PF, PD, and PD^2. The PD^2 algorithm is optimal for scheduling tasks under the IS model, which generalizes both the periodic and sporadic models. We also discussed extensions that support dynamic task sets, hierarchical scheduling, and resource sharing.

References

[1] S. Baruah, N. Cohen, C.G. Plaxton, and D. Varvel. Proportionate progress: A notion of fairness in resource allocation. *Algorithmica*, 15:600–625, 1996.

[2] S. Baruah, J. Gehrke, and C.G. Plaxton. Fast scheduling of periodic tasks on multiple resources. In *Proceedings of the 9th International Parallel Processing Symposium*, pp. 280–288, April 1995.

[3] J. Anderson and A. Srinivasan. Early-release fair scheduling. In *Proceedings of the 12th Euromicro Conference on Real-Time Systems*, pp. 35–43, June 2000.

[4] M. Moir and S. Ramamurthy. Pfair scheduling of fixed and migrating periodic tasks on multiple resources. In *Proceedings of the 20th IEEE Real-Time Systems Symposium*, pp. 294–303, December 1999.

[5] A. Srinivasan and J. Anderson. Efficient scheduling of soft real-time applications on multiprocessors. In *Proceedings of the 15th Euromicro Conference on Real-Time Systems*, pp. 51–59, July 2003.

[6] S. Baruah. Fairness in periodic real-time scheduling. In *Proceedings of the 16th IEEE Real-Time Systems Symposium*, pp. 200–209, December 1995.

[7] S. Ramamurthy and M. Moir. Static-priority periodic scheduling of multiprocessors. In *Proceedings of the 21st IEEE Real-Time Systems Symposium*, pp. 69–78, December 2000.

[8] A. Chandra, M. Adler, and P. Shenoy. Deadline fair scheduling: Bridging the theory and practice of proportionate-fair scheduling in multiprocessor servers. In *Proceedings of the 7th IEEE Real-Time Technology and Applications Symposium*, pp. 3–14, May 2001.

[9] J. Anderson and A. Srinivasan. Pfair scheduling: Beyond periodic task systems. In *Proceedings of the 7th International Conference on Real-Time Computing Systems and Applications*, pp. 297–306, December 2000.

[10] A. Srinivasan and J. Anderson. Optimal rate-based scheduling on multiprocessors. In *Proceedings of the 34th Annual ACM Symposium on Theory of Computing*, pp. 189–198, May 2002.

[11] A. Srinivasan and J. Anderson. Fair scheduling of dynamic task systems on multiprocessors. *Journal of Systems and Software* (to appear).

[12] P. Holman and J. Anderson. Guaranteeing Pfair supertasks by reweighting. In *Proceedings of the 22nd IEEE Real-time Systems Symposium*, pp. 203–212, December 2001.

[13] L.R. Ford and D.R. Fulkerson. *Flows in Networks.* Princeton University Press, Princeton, NJ, 1962.

[14] A. Srinivasan, P. Holman, and J. Anderson. Integrating aperiodic and recurrent tasks on fair-scheduled multiprocessors. In *Proceedings of the 14th Euromicro Conference on Real-Time Systems*, pp. 19–28, June 2002.

[15] A. Srinivasan, P. Holman, J. Anderson, and S. Baruah. The case for fair multiprocessor scheduling. In *Proceedings of the 11th International Workshop on Parallel and Distributed Real-Time Systems*, April 2003.

[16] P. Holman and J. Anderson. Object sharing in Pfair-scheduled multiprocessor systems. In *Proceedings of the 13th Euromicro Conference on Real-Time Systems*, pp. 111–122, June 2002.

[17] P. Holman and J. Anderson. Locking in Pfair-scheduled multiprocessor systems. In *Proceedings of the 23rd IEEE Real-time Systems Symposium*, pp. 149–158, December 2002.

[18] P. Holman and J. Anderson. Using hierarchal scheduling to improve resource utilization in multiprocessor real-time systems. In *Proceedings of the 15th Euromicro Conference on Real-Time Systems*, pp. 41–50, July 2003.
[19] P. Holman and J. Anderson. The staggered model: Improving the practicality of Pfair scheduling. In Proceedings of the 24th IEEE Real-time Systems Symposium Work-in-Progress Session, pp. 125–128, December 2003.
[20] A. Srinivasan, P. Holman, J. Anderson, S. Baruah, and J. Kaur. Multiprocessor scheduling in processor-based router platforms: Issues and ideas. In *Proceedings of the 2nd Workshop on Network Processors*, pp. 48–62, February 2002.
[21] S. Baruah, J. Gehrke, C.G. Plaxton, I. Stoica, H. Abdel-Wahab, and K. Jeffay. Fair on-line scheduling of a dynamic set of tasks on a single resource. *Information Processing Letters*, 26(1):43–51, January 1998.
[22] I. Stoica, H. Abdel-Wahab, K. Jeffay, S. Baruah, J. Gehrke, and C.G. Plaxton. A proportional share resource allocation algorithm for real-time, time-shared systems. In *Proceedings of the 17th IEEE Real-Time Systems Symposium*, pp. 288–299, December 1996.
[23] J. Anderson, A. Block, and A. Srinivasan. Quick-release fair scheduling. In Proceedings of the 24th Real-time Systems Symposium, pp. 130–141, December 2003.
[24] S. Ramamurthy. A lock-free approach to object sharing in real-time systems. Dissertation, University of North Carolina at Chapel Hill. 1997.
[25] J. Anderson and S. Ramamurthy. A framework for implementing objects and scheduling tasks in lock-free real-time systems. In *Proceedings of the 17th IEEE Real-Time Systems Symposium*, pp. 92–105, December 1996.
[26] R. Rajkumar. *Synchronization in real-time systems — A priority inheritance approach*. Kluwer Academic Publishers, Boston, 1991.

32

Approximation Algorithms for Scheduling Time-Critical Jobs on Multiprocessor Systems

Sudarshan K. Dhall
University of Oklahoma

32.1 Introduction

Today computers are being widely used for many control and monitoring applications. Many of the jobs arising in these applications are activated in response to external signals that arise at regular intervals. Each activation of these jobs has a prespecified deadline which failing to meet may cause irreparable loss. These jobs are called periodic time critical (PTC) jobs. The main goal of scheduling a set of periodic time-critical jobs is to produce a feasible schedule — a schedule that guarantees that all deadlines are honored. As the number and the processing demand of these jobs increase, the use of multiprocessors may be needed to meet the increased computational requirements. For scheduling periodic time-critical jobs on multiprocessor systems, two approaches have been considered in literature. In one approach different activations of a job may be executed on any available processor (nonpartitioning or the global approach). In the other approach, each job is executed on a single designated processor and does not migrate to any other processor

1-58488-397-9/$0.00+$1.50

during its lifetime (partitioning approach). This chapter considers approximation algorithms for allocating a set of periodic time-critical jobs on a multiprocessor system and analyzes their worst-case behavior.

32.1.1 The Model

We assume that the jobs under consideration possess the following properties.

1. Each job is activated periodically, with a constant interval between successive activations.
2. The processing time of all activations of a job has the same constant value. The processing time is the time that a job needs for completion without being interrupted.
3. The processing of each activation of a job must be completed before its next activation.
4. Jobs are independent of each other; that is, activation of a job does not depend on the activation or completion of any other job.

With these requirements, it turns out that a periodic time-critical job can be completely described by the pair (c, p), where c is the processing time and p is the period of the job. The ratio $1/p$ is the **activation rate** and c/p is the **utilization factor** or **load factor** of the job. Note that the utilization factor of a job is the fraction of the time the processor is busy working on the job. The utilization factor of a set of jobs is the sum of the utilization factor of all the jobs in the set.

32.1.2 Scheduling on a Single Processor

The problem of scheduling a set of periodic time-critical jobs on a single processor was first studied by Liu and Layland [1], and Serlin [2]. An important requirement of any scheduling algorithm for time-critical jobs is that it finds a feasible schedule for a given set of jobs, that is, the schedule ensures that every activation of every job is completed on or before its deadline. The scheduling algorithms for a single processor proposed in these studies are priority based and are broadly categorized as dynamic or static priority algorithms. In static priority algorithms priorities are assigned to jobs with the result that an activation of a job has the same priority as the priority of the job. In dynamic priority algorithms priority is assigned to each activation of a job, with the result that priorities of the jobs change with their activations. A job with a higher priority preempts an executing job with a lower priority. An example of a dynamic priority algorithm is the earliest deadline first (EDF) algorithm that assigns the highest priority to the activation of a job whose deadline is the earliest. An example of a static priority algorithm is the rate-monotonic (RM) algorithm. According to the RM algorithm, priorities to jobs are assigned according to nonincreasing order of their activation rate (or according to the nondecreasing order of their period). That is, a job with the highest activation rate or the shortest period is assigned the highest priority. Some of the important results that will be referred to later in this chapter about these algorithms are summarized below.

Theorem 32.1 [1]

The necessary and sufficient condition for a set of jobs to be feasibly scheduled on a single processor according to EDF algorithm is that its utilization factor is less than or equal to 1.

Theorem 32.2 [1]

Among all static priority algorithms, the RM algorithm is the best in the sense that if a set of jobs can be feasibly scheduled by any static priority algorithm, then it can also be feasibly scheduled using the RM algorithm.

Theorem 32.3 [1]

A set of n jobs can be feasibly scheduled on a single processor if the utilization factor of the set of jobs is less than or equal to $n(2^{1/2} - 1)$. As $n \to \infty$, this value tends to $\ln 2$.

The condition of Theorem 32.3 is not a necessary, but only a sufficient condition. That is, for every n, we can find sets of jobs with utilization factor greater than $n(2^{1/2} - 1)$ that can be feasibly scheduled and that cannot be feasibly scheduled according to RM algorithm.

Recently, Devillers and Goossens [3] pointed out that though the result of Theorem 32.3 is correct, the proof as given in the paper by Liu and Layland [1] has some weaknesses, and presented a corrected version of the proof.

The following schedulability condition due to Burchard et al. [4] is stronger than the one in Theorem 32.3. The better bound is obtained by taking the periods of jobs into account.

Theorem 32.4

Let $\{J_i = (c_i, p_i), i = 1 \text{ to } n\}$, be a set of n periodic jobs. Let $S_i = \log p_i - \lfloor \log p_i \rfloor, i = 1, 2, \ldots, n$, and let

$$\beta = \max_{1 \le i \le n} S_i - \min_{1 \le i \le n} S_i$$

This job set can be feasibly scheduled by the RM algorithm if

$$\sum_{i=1}^{n} \frac{c_i}{p_i} \le \begin{cases} (n-1)(2^{\beta/(n-1)} - 1) + 2^{1-\beta} = 1 & \text{if } \beta < 1 - 1/n \\ n(2^{1/n} - 1) & \text{if } \beta \ge 1 - 1/n \end{cases} \tag{32.1}$$

Though, in some cases the bound on utilization given in Theorem 32.4 is better than the one in Theorem 32.3, this is again only a sufficient condition.

32.1.3 Scheduling on Multiprocessor Systems

As mentioned above, the rate-monotonic algorithm is optimal among all fixed-priority scheduling algorithms on a single processor in the sense that if any set of jobs can be feasibly scheduled by any fixed-priority scheduling algorithm, it can also be feasibly scheduled by the rate-monotonic scheduling algorithm. Similarly, the earliest-deadline scheduling algorithm is optimal for a single processor since it can schedule all job sets with total utilization less than or equal to one. It is natural to assume that extending these algorithms to multiprocessor systems is likely to produce satisfactory results. Unfortunately, as the following two examples illustrate, both these algorithms may perform very poorly when extended in a natural way to multiprocessor systems. In both these examples, the total utilization of a set of $n + 1$ jobs is close to 1, yet both these algorithms fail to produce a feasible schedule for these jobs on n identical processors.

Example 32.1

Let $J_i = (\eta, 1)$ for $i = 1$ to n, and $J_{n+1} = (1, 1 + \eta)$, where η is infinitesimally small. According to rate-monotonic algorithm, the first n jobs will be scheduled on the n processors occupying the first η units of time on each processor. At time 1, all these jobs will make their second request. Since all these jobs have priority over job J_{n+1}, they will occupy all the processors during the time interval $[1, 1 + \eta]$. This leaves only $1 - \eta$ units of time on any single processor to execute job J_{n+1} up to time $1 + \eta$. Thus, according to rate-monotonic algorithm, this set of jobs cannot be feasibly scheduled on n processors.

Example 32.2

Let $J_i = (2\eta, 1)$ for $i = 1$ to n, and $J_{n+1} = (1, 1 + \eta)$, where η is infinitesimally small. According to earliest-deadline algorithm, each of the first n jobs will be scheduled on the n processors during the interval $[0, 2\eta]$ on each processor, respectively. Job J_{n+1} can be scheduled on any one of the n processors starting at time 2η. This leaves only $1 - \eta$ units of time available on any processor for job J_{n+1} up to its deadline $1 + \eta$, making it impossible to meet the deadline of its first request.

Note that in both of the above examples, if we dedicate one processor to job J_{n+1}, then all jobs can be feasibly scheduled on n processors.

An algorithm given by Dertouzos and Mok [5] guarantees that a set of n jobs with total utilization less than or equal to m can be feasibly scheduled on m processors, if $\sum_{i=1}^{n} c_i/p_i \leq m$, and

$$t = GCD\left(T, T \times \frac{c_1}{p_1}, \cdots, T \times \frac{c_n}{p_n}\right)$$

is an integer, where

$$T = GCD(p_1, p_2, \ldots, p_n)$$

If this condition is satisfied, the schedule is produced by assigning $T \times \frac{c_i}{p_i}$ units of time to job J_i during every interval of T units. Note that for any job J_i, the schedule provides $\frac{p_i}{T} \times T \times \frac{c_i}{p_i} = c_i$ units of time during every interval of length p_i.

The above algorithm is applicable only in very special cases. There is no satisfactory algorithm for the general case. Another algorithm, called the Slack-Time algorithm, both for single and multiprocessor systems was proposed by Leung [6]. However, it has been shown that the problem of deciding whether a given set of jobs can be feasibly scheduled by a static- or dynamic-priority algorithm is NP-hard, for each fixed m, the number of processors [6,7]. In light of this, researchers have resorted to development of algorithms that are not optimal, yet provide satisfactory performance at a low computational cost.

At the end of Examples 32.1 and 32.2, it was observed that dedicating processors to execute certain jobs may produce better results for multiprocessor systems. This approach has been adopted by a number of researchers [4,8–11,14] to develop efficient algorithms for scheduling PTC jobs on multiprocessor systems. In this approach, jobs are partitioned into subsets, such that each subset can be feasibly scheduled on a single processor using any suitable algorithm for a single processor. Obviously, the number of processors needed to execute a set of jobs will be the number of subsets in the division. A natural goal of this division is to minimize the number of processors needed to execute a set of jobs. The division itself will depend on the type of algorithm to be used for each processor. Recall that the earliest deadline algorithm ensures feasibility for a set of jobs whose total utilization is less than or equal to one (Theorem 32.1). Thus, if the target algorithm is the EDF algorithm for each processor, this problem reduces to the bin-packing problem [15–17], where given a set of packets, with size of each packet in the range $(0, 1]$, it is required to pack all packets in bins, each of size 1, using minimum number of bins. For the PTC job problem, the job utilization is equivalent to the size of the packet, and the achievable utilization (which is 1 for EDF) on each processor is equivalent to the size of the bin. Thus, all bin-packing algorithms are directly applicable in this case. On the other hand, if the intended algorithm is a fixed priority algorithm, e.g., rate-monotonic, then, though in principle, the two problems are similar in nature, the job allocation problem is much more complicated than the bin-packing problem. This is mainly because the achievable utilization on a single processor depends on the number and type of jobs to be scheduled on that processor. Thus, in this case, the capacity of a bin (processor) is not fixed at 1, but varies as $n(2^{1/n} - 1)$ or as the bound given in (32.1), where n is the number of packets (jobs) allocated to the bin (processor). This is further complicated by the fact that both these bounds on utilization factor are only a sufficient but not a necessary condition for n jobs to be feasibly scheduled on a single processor. Depending on the set of jobs, this value can be as large as 1. Since fixed-priority algorithms have less overhead than dynamic-priority algorithms like the EDF, in the following, it is assumed that the algorithm to be used for each processor is the rate-monotonic algorithm. Two types of algorithms — offline and online algorithms are presented in this chapter. In offline algorithms, information about all jobs is available before the allocation process begins. In online algorithms, the allocation decision about a job is made based on the current information without taking into account future arrival of jobs.

Since the allocation problem of periodic time-critical jobs is very similar to the bin-packing problem, the design and analysis of these algorithms is very much patterned after the bin-packing algorithm. For example, the Rate-Monotonic Next-Fit, the Rate-Monotonic First-Fit (RMFF) algorithms of Dhall [12] and Dhall and Liu [13], the First-Fit Decreasing-Utilization of Davari and Dhall [8], follow the pattern of Next-Fit, First-Fit, and First-Fit Decreasing algorithms for bin-packing [15,16]. The Next-Fit-M algorithm of Davari and Dhall [11] is similar to the Next-Fit algorithm of Lee and Lee [17]. Similarly, the RM-FF algorithm, the Rate-Monotonic Best-Fit algorithm and the Refined Rate-Monotonic First-Fit (RRM-FF) algorithms of Oh and Son [14] also follow the similar patterns of bin-packing algorithms [15,16]. What complicates matter for allocation of real-time jobs is the variable amount of utilization factor that can be allocated to a processor.

32.1.4 Performance Evaluation

To evaluate the performance of various algorithms considered in this chapter, we consider their worst-case behavior. For any set S of jobs, let $N_0(S)$ be the minimum number of processors required using an optimal algorithm, and let $N_A(S)$ represent the number of processors required by algorithm A. Let $R_A(S) \equiv N_A(S)/N_0(S)$. Then, the asymptotic worst-case performance ratio of algorithm A, denoted by R_A^∞, is defined as [18]:

$$R_A^\infty = \inf\{r \geq 1 : \text{ for some } N^* > 0, R_A(S) \leq r \text{ for all sets S with } N_0(S) \geq N^*\}$$

For large problems, therefore, the algorithm A may be R_A^∞ times as expensive as an optimal algorithm (if one exists), but no worse.

The rest of the chapter is organized as follows. Section 32.2 deals with offline algorithms and Section 32.3 deals with online algorithms. Some concluding remarks are given in the last section.

32.2 Offline Algorithms

In this section we consider algorithms that require that all information about jobs be available before application of the algorithm. Algorithms in this category work in two stages. In the first stage, jobs are rearranged in a certain order. The actual allocation is performed in the second stage of the algorithm. Before starting with the discussion of algorithms, a few sufficiency conditions used in these algorithms to decide whether a job should or should not be allocated to the processor under consideration are given below. One of these sufficient conditions has already been stated above as Theorem 32.2.

Theorem 32.5 [1]
A critical instant for any job occurs whenever the job is activated simultaneously with activation of all higher priority jobs.

Since in the job model considered in this chapter, the first activation of each job occurs simultaneously at time 0, the above theorem provides a necessary and sufficient condition for a set of jobs to be feasibly scheduled on a single processor using the RM algorithm. The theorem states that a set of jobs can be feasibly scheduled using the RM algorithm if and only if activations of all jobs up to the deadline of the first activation of the job with the longest period are satisfied. Thus, it is sufficient to test the schedule up to the end of the first longest period. This theorem was restated as follows by Lehoczky, Sha, and Ding [19].

Theorem 32.6 [19]
Let $J_i = (c_i, p_i), i = 1, 2, \ldots, n$, be a set of n jobs, with $p_1 \leq p_2 \leq \cdots \leq p_n$. Job $J_i, i = 1, 2, \ldots, n$, can be feasibly scheduled with jobs $J_1, J_2, \ldots, J_{i-1}$ using RM algorithm if and only if $L_i \leq 1$, where

$$L_i = \min_{\{0 \leq t \leq p_i\}} \frac{1}{t} \sum_{j=1}^{i} c_j \left\lceil \frac{t}{p_j} \right\rceil$$

Though the theorem statement requires minimization over the continuous interval $[0, p_i]$, in practice, one needs to check only a finite number of points for each i.

Theorem 32.7 [11]

Let $J_i = (c_i, p_i), i = 1, 2, \ldots, n$, be a set of n jobs with periods $p_1 \leq p_2 \leq \cdots \leq p_n$. Let $u = \sum_{i=1}^{n-1} c_i/p_i \leq (n-1)(2^{1/(n-1)} - 1)$. Then, if $c_n/p_n \leq 2[1 + u/(n-1)]^{-(n-1)} - 1$, then the set of n jobs can be feasibly scheduled using the RM algorithm. As $n \to \infty$ the right-hand side of the above inequality becomes $2e^{-u} - 1$.

The above theorem can be used for testing the feasibility of a job with other jobs provided period of this job is larger than the period of any other job already scheduled on a processor. However, when only two jobs are involved, this condition on the periods is not necessary. The following theorem, on the other hand, tests feasibility of a job with other jobs based solely on its utilization factor.

Theorem 32.8 [14]

Let $J_i = (c_i, p_i), i = 1, 2, \ldots, n-1$, be a set of $n-1$ jobs, and let $u_i = c_i/p_i$, $i = 1, 2, \ldots, n-1$, be their utilization factors. Assume that these jobs are feasibly scheduled on a single processor. Then, another job with utilization $u_n = c_n/p_n$ can be feasibly scheduled using the RM algorithm if

$$u_n \leq 2\left[\prod_{i=1}^{n-1}(1+u_i)^{-1}\right] - 1$$

The feasibility tests of Theorems 32.2, 32.7, and 32.8 are only sufficiency tests and require a simple calculation. The feasibility test of Theorem 32.6 is both necessary and sufficient. However, it is much more computation intensive than the other tests. Therefore, to keep computation cost reasonably low, almost all allocation algorithms use simpler sufficiency tests. The use of Theorem 32.6 is restricted to the case of testing the feasibility of two or three jobs. For testing feasibility of two jobs, this test translates as follows.

Theorem 32.9

Let (c_1, p_1) and (c_2, p_2) with $p_1 \leq p_2$ and let $m = \lfloor p_2/p_1 \rfloor$. The two jobs can be feasibly scheduled on a single processor using RM algorithm if and only if, $c_2 \leq m(p_1 - c_1) + \max\{0, p_2 - m * p_1 - c_1\}$.

The advantage of this test over the other three tests can be easily seen from the following example.

Example 32.3

Consider the two jobs $(0.4, 1)$, and $(1.0, 1.8)$ with utilizations 0.4 and 0.55, respectively. If the feasibility test of any of the Theorems 32.2, 32.7, or 32.8 is used, these two jobs will not be allocated to the same processor. However, the two jobs can be feasibly scheduled on a single processor according to Theorem 32.9.

32.2.1 Rate-Monotonic Next-Fit (RMNF) Algorithm

This algorithm first sorts jobs in nondecreasing order of their periods. It allocates as many jobs to the first processor as can be feasibly scheduled on a single processor. Once it encounters a job that cannot be allocated to the processor, it starts a new processor. The algorithm is described in Figure 32.1. Time requirement of Step 1 of the algorithm is the time for sorting m objects that takes $O(m \log m)$ time. During allocation, we just check whether the job can be scheduled on the processor by simply comparing its utilization with the remaining capacity of the processor according to Theorem 32.7. Thus, the total allocation time is $O(m)$. Therefore, the complexity of the algorithm is $O(m \log m)$.

The final value of n given by the algorithm is the number of processors needed when the RMNF algorithm is used.

Theorem 32.10

$2.4 \leq R^{\infty}_{\text{RMNF}} \leq 2.67$.

Let $J_i = (c_i, p_i), i = 1, \ldots, m$ be a set of m jobs. Let $U_n = \sum_{i=1}^{k_n} c_{ji}/p_{ni}$ be the utilization of the k_n jobs allocated to processor n.

1. Rearrange all jobs in nondecreasing order of their periods.
2. Let $n = 1, i = 1$.
3. While $i \le m$ do {
   If $c_i/p_i \le 2[1 + U_n/k_n]^{-k_n} - 1$
   Assign J_i to processor n.
   Else set $n = n + 1$, assign J_i to processor n
   Set $i = i + 1$.
   }

FIGURE 32.1 The Rate-Monotonic Next-Fit algorithm.

Proof

In order to establish the lower bound, we show that given $\varepsilon > 0$, there exists a set of jobs for which $\frac{N}{N_0} > 2.4 - \varepsilon$, where N is the number of processors used by the algorithm and N_0 is the minimum number of processors required for the same set of jobs. Let $n = 12k$, for some large integer k. Consider the set of jobs:

$$(1,2), (\eta, 2), (1,3), (\eta, 3), (2,4), (\eta, 4), (2,6), (\eta, 6), \ldots$$
$$(2^{\frac{n}{2}-1}, 2^{\frac{n}{2}}), (\eta, 2^{\frac{n}{2}}), (2^{\frac{n}{2}-1}, 3 \times 2^{\frac{n}{2}-1}), (\eta, 3 \times 2^{\frac{n}{2}-1})$$

The Next-Fit algorithm will divide this set of jobs into n partitions, placing jobs $(2^{i-1}, 2^i)$ and $(\eta, 2^i)$ to processor i, and jobs $(2^{i-1}, 3 \times 2^{i-1})$ and $(\eta, 3 \times 2^{i-1})$ to processor $i + 1$, for $i = 1$to $\frac{n}{2}$. Thus, $N = n$.

Now divide this set of jobs into $k + 1$ subsets as follows. For $i = 1, 2, \ldots, k$, set i contains the following jobs:

$$(2^{6(i-1)}, 2 \times 2^{6(i-1)}), (2^{6(i-1)}, 3 \times 2^{6(i-1)}), (2^{6(i-1)+1}, 2 \times 2^{6(i-1)+1}),$$
$$(2^{6(i-1)+1}, 3 \times 2^{6(i-1)+1}), \ldots, (2^{6(i-1)+5}, 2 \times 2^{6(i-1)+5}), (2^{6(i-1)+5}, 3 \times 2^{6(i-1)+5}),$$

All jobs with computation time η form set number $k+1$. We can choose η appropriately small to ensure that all jobs in set $k + 1$ are feasibly scheduled on a single processor. Jobs in each of the first k subsets can be divided into five groups as follows such that jobs in each group can be feasibly scheduled on a single processor:

Group 1: $(2^{6(i-1)}, 2 \times 2^{6(i-1)})$ and $(2^{6(i-1)+1}, 2 \times 2^{6(i-1)+1})$
Group 2: $(2^{6(i-1)+2}, 2 \times 2^{6(i-1)+2})$ and $(2^{6(i-1)+3}, 2 \times 2^{6(i-1)+3})$
Group 3: $(2^{6(i-1)+4}, 2 \times 2^{6(i-1)+4})$ and $(2^{6(i-1)+5}, 2 \times 2^{6(i-1)+5})$
Group 4: $(2^{6(i-1)}, 3 \times 2^{6(i-1)}), (2^{6(i-1)+1}, 3 \times 2^{6(i-1)+1})$, and $(2^{6(i-1)+2}, 3 \times 2^{6(i-1)+2})$
Group 5: $(2^{6(i-1)+3}, 3 \times 2^{6(i-1)+3}), (2^{6(i-1)+4}, 3 \times 2^{6(i-1)+4})$, and $(2^{6(i-1)+5}, 3 \times 2^{6(i-1)+5})$

Thus, all jobs can be feasibly scheduled using $5k + 1 = \frac{5}{12}n + 1$ processors. Thus, $N_0 \le \frac{5}{12}n + 1$. Thus, $\frac{N}{N_0} \ge \frac{12n}{(5n+12)}$. By selecting n to be sufficiently large, this ratio can be made to approach the limit 2.4. In other words, given $\varepsilon > 0$, there is an integer N, such that

$$\frac{N}{N_0} > 2.4 - \varepsilon \tag{32.2}$$

To establish the upper bound, define the following function:

$$f(u) = \begin{cases} 2.67u, & 0 \le u < 0.75 \\ 2.0, & u \ge 0.75 \end{cases}$$

where u is the utilization factor of a job. Consider a set of m jobs. Let N_0 be the number of processors required in any optimal allocation. Then, for any processor P_i, $0 \le i \le N_0$,

$$\sum_{J \text{ on } P_i} f(u(J)) \le 2.67 \sum_{J \text{ on } P_i} u(J) \le 2.67$$

since sum of utilization of all jobs scheduled on any processor is less than or equal to 1. Summing up over all processors,

$$\sum_{i=1}^{m} f(u(J_i)) \le 2.67 \tag{32.3}$$

Now consider the jobs allocated to two successive processors, say, P_i and P_{i+1} according to the RMNF algorithm. Then, either

$$\sum_{J \text{ on } P_i} f(u(J)) \ge 1 \tag{32.4}$$

or

$$\sum_{J \text{ on } P_i} f(u(J)) + f(u(J')) \ge 2 \tag{32.5}$$

where J' is the job that could not be allocated to processor P_i according to RMNF algorithm and was bumped to processor P_{i+1}. If (32.4) does not hold, then the utilization of any job allocated to P_i is strictly less than 0.75, and $\sum_{J \text{ on } P_i} f(u(J)) < 2.67 \sum_{J \text{ on } P_i} u(J) < 1$. Therefore,

$$\sum_{J \text{ on } P_i} u(J) < \frac{1}{2.67}$$

Thus, according to Theorem 32.7, $u(J') > 2e^{-\sum_{J \text{ on } P_i} u(J)} - 1$. If $u(J') \ge 0.75$, then (32.5) holds. Otherwise, $f(x) + f(u(J')) \ge 2.67(x + 2e^{-x} - 1) \ge 2$, where x is the sum of utilization of all jobs allocated to processor P_i.

In view of the above result

$$\sum_{i=1}^{m} f(u(J_i)) = \sum_{i=1}^{N} \sum_{J \text{ on } P_i} f(u(j)) \ge N - 1$$

This together with (32.3) gives

$$N - 1 \le \sum_{i=1}^{m} f(u(J_i)) \le 2.67 N_0$$

or

$$\frac{N}{N_0} \le 2.67 + \frac{1}{N_0} \tag{32.6}$$

The result follows from Equations (32.2) and (32.6).

In allocating jobs to processors, the RMNF algorithm only checks the last processor under consideration to see if the job can be allocated to this processor. It is very likely that the job not assigned to the processor under consideration may well be feasibly scheduled on one of the earlier processors. The next algorithm just attempts to do that. Of course, if earlier processors are to be considered for allocations, it is necessary

Let $\{J_i = (c_i, p_i), i = 1, \ldots, m\}$ be a set of m jobs.

1. Arrange all jobs in nondecreasing order of their periods and renumber the jobs in this order.

2. Let $i = 1$.

3. While $(i \leq m)$ do {
   While (job J_i is not assigned) do {
   Let $j = 1$.
   If J_i can be feasibly scheduled on P_j with already allocated jobs, then
   Allocate J_i to P_j, and set $i = i + 1$.
   else set $j = j + 1$.
   }
   }

FIGURE 32.2 Rate-Monotonic First-Fit algorithm.

to keep track of all the processors to which allocations have been made so far. This requires the algorithm to maintain extra information, thus needing space, but is likely to reduce the total number of processors needed. □

32.2.2 Rate-Monotonic First-Fit Scheduling (RMFF) Algorithm

As in the previous algorithm, this algorithm also first sorts the jobs in nondecreasing order of their period. However, if a job does not fit on the current processor, this processor is not set aside, but is considered for allocation of later jobs. The algorithm is given in Figure 32.2.

This algorithm will use j processors, where j is the largest index used by the allocation procedure. Before analyzing the asymptotic worst-case performance of the algorithm, let us consider the complexity of the algorithm. The first step is sorting jobs in nondecreasing order of their periods. This requires $O(m \log m)$, where m is the number of jobs in the set. The next step is to allocate jobs to various processors. If at any time p processors have already been allocated at least one job, in the worst case we need to test all p processors to test whether the job under consideration for allocation can be allocated to one of these processors. How much time will a test for allocation take? Any one of a number of feasibility tests can be used. Using utilization of all the jobs, and that of the job under consideration, takes a simple test whose computation time is independent of the number of jobs already allocated to the processor. If we use this test, we will not be able to reap advantage of the jobs being already sorted in nondecreasing order of their periods. According to Theorem 32.6 we know that a job set can be feasibly scheduled on a single processor if requests of all jobs up to the end of the first longest period are satisfied. If we use this test, we can save the schedule of the jobs already allocated to each processor up to the longest period so far. Since the period of the current job is greater than or equal to the longest period of jobs allocated to any processor, we will simply have to extend the schedule only up to the period of the job under consideration. Unfortunately, the computation cost of this test is a function of the number of jobs already allocated to the processor, and may lead to an exponential time complexity for the algorithm. To keep the cost of test reasonable, we use the utilization test of Theorem 32.7 if the number of jobs already allocated to a processor is more than one. Otherwise, we use the test of Theorem 32.9. Thus, the test cost is constant. Hence, the algorithm has a worst-case complexity of $O(m \log m) + O(N^2)$, where N is the number of processors used by the algorithm. Since N is a function of m, the overall complexity can be be written as $O(m^2)$.

Note that the First-Fit algorithm for the bin-packing problem can be implemented in time $O(m \log m)$ [16]. Though the description of the two algorithms is the same, the type of tests used here to achieve a lower cap on the number of processors needed for allocation poses a difficulty in implementing this algorithm with the same complexity. The problem stems from the test that is used for processors that have been allocated a single job. This test prevents us from placing a logical ordering on the processors that have already been used. For example, assume P_1 is allocated the job $(0.4, 1)$ and P_2 is allocated the job $(0.7, 1.4)$. Depending on the period, the range of utilization of the job that can be allocated to P_1 is $[3/7, 3/5]$ and for P_2 the range is $[1/3, 1/2]$. Though the two ranges overlap, if the next job were $(0.8, 1.6)$ with utilization $1/2$, it will be allocated to P_1, whereas job $(0.66, 1.45)$ with utilization that lies in the range of P_1 cannot be allocated to P_1, but will be allocated to P_2. Thus, the range information is not sufficient to permit change in the order of processors to be tested to achieve overall reduction in complexity.

We now begin to analyze the performance of this algorithm. Towards this end, we first establish a few preliminary results.

Lemma 32.1

If a set of three jobs cannot be feasibly scheduled on two processors according to the RMFF algorithm, then the utilization factor of the set of jobs is greater than $\frac{3}{1+2^{1/3}}$.

The proof of this lemma can be found in [12].

Lemma 32.2

If a set of m, *jobs* $m \geq 3$, *cannot be feasibly scheduled on* $(m-1)$ *processors according to the Rate-Monotonic First-Fit algorithm, then the utilization factor of the set of jobs must be greater than* $\frac{m}{1+2^{1/3}}$.

Proof

Since m jobs cannot be feasibly scheduled on $(m-1)$ processors, no three of these jobs can be scheduled on two processors. Hence, the utilization factor of any set of three jobs is greater than $\frac{3}{1+2^{1/3}}$. Therefore, considering all possible subsets of three jobs, we have

$$\binom{m-1}{2}(u_1 + u_2 + \cdots + u_m) > \binom{m}{3}\frac{3}{1+2^{1/3}}$$

or

$$\sum_{i=1}^{m} u_i > \frac{m}{1+2^{1/3}}$$

□

Theorem 32.11

$$2 \leq R_{\text{RMFF}}^{\infty} \leq \frac{4 \times 2^{1/3}}{1+2^{1/3}}.$$

Proof

Define the function $f : [0, 1] \to [0, 1]$, as follows:

$$f(u) = \begin{cases} 2u, & 0 \leq u \leq \frac{1}{2} \\ 1, & \frac{1}{2} \leq u \leq 1 \end{cases}$$

It is claimed that when jobs are assigned to processors according to RMFF, amongst all processors to each of which two jobs are assigned, there is at most one processor for which the total utilization factor of assigned jobs is less than $1/2$. If not, let P_i and P_j, $i < j$, be two such processors. Let J_{i1} and J_{i2} be the jobs assigned to P_i and J_{j1} and J_{j2} be the jobs assigned to P_j.

Case 1: Jobs to P_j are assigned after they are assigned to P_i. Then,

$$u_{i1} + u_{i2} + u_{j1} > 3(2^{1/3} - 1)$$

and

$$u_{i1} + u_{i2} + u_{j2} > 3(2^{1/3} - 1)$$

or

$$u_{j1} + u_{j2} > 6(2^{1/3} - 1) - 2(u_{i1} + u_{i2}) > 6(2^{1/3} - 1) - 1 > 1/2$$

which contradicts the assumption that the utilization of jobs on P_j is less than 1/2.

Case 2: Both jobs to P_j were assigned after the first job to P_i had been assigned. In this case,

$$u_{i1} + u_{j1} > 2(2^{1/2} - 1)$$
$$u_{i1} + u_{j2} > 2(2^{1/2} - 1)$$

Hence,

$$u_{j1} + u_{j2} > 4(2^{1/2} - 1) - 2u_{i1} > 4(2^{1/2} - 1) - 1 > 1/2$$

again, contradicting the assumption.

Case 3: Jobs were assigned alternately to the two processors. Here,

$$u_{i1} + u_{j1} > 2(2^{1/2} - 1)$$
$$u_{i1} + u_{i2} + u_{j2} > 3(2^{1/3} - 1)$$

from which, we again get the contradiction

$$u_{j1} + u_{j2} > 1/2$$

Define deficiency δ for a processor that is assigned k jobs as follows:

$$\delta = \begin{cases} 0, & \text{if } U(k) \geq k(2^{1/k} - 1) \\ 2(1 + \frac{U(k)}{k})^{-k} - 1, & \text{otherwise} \end{cases}$$

where $U(k)$ is the utilization of the k jobs assigned to the processor. Define **coarseness** α_i of processor P_i as follows:

$$\alpha_i = \begin{cases} 0, & \text{if } i = 1 \\ \max_{1 \leq j \leq i-1} \delta_j, & \text{for } i > 1 \end{cases}$$

□

Lemma 32.3

Let $k_i > 2$ be the number of jobs assigned to processor P_i with coarseness $\alpha < 1/6$. If

$$\sum_{j=1}^{k_i} u_{ij} \geq \ln 2 - \alpha_i$$

then

$$\sum_{j=1}^{k_i} f(u_{ij}) \geq 1$$

Proof

If any of the assigned job has a utilization $\geq 1/2$, then the result is immediate. Otherwise,

$$\sum_{j=1}^{k_i} f(u_{ij}) = 2\sum_{i=1}^{k_i} u_{ij} \geq 2(\ln 2 - \alpha_i) > 2(\ln 2 - 1/6) > 1$$

□

Lemma 32.4

Assume processor P_i with coarseness α_i is allocated k_i jobs for which

$$\sum_{j=1}^{k_i} f(u_{ij}) = 1 - \beta \tag{32.7}$$

with $\beta > 0$. Then, either

$$k_i = 1 \quad \textit{and} \quad u_{i1} < 1/2$$

or

$$k_i = 2 \quad \textit{and} \quad u_{i1} + u_{i2} < 1/2$$

or

$$\sum_{j=1}^{k_i} \leq \ln 2 - \alpha_i - \beta/2 \tag{32.8}$$

Proof

Obviously, if $k_i \leq 2$, one of the first two conclusions holds. So let $k_i > 2$. Let $J_{ij} = (c_{ij}, p_{ij})$, for $j = 1, \ldots, k_i$. Then by Theorem 32.3, we have

$$U = \sum_{j=1}^{k_i} u_{ij} < \ln 2 - \alpha_i$$

Let

$$U = \ln 2 - \alpha_i - \lambda \tag{32.9}$$

where $\lambda > 0$. Consider a set of $J'_{ij} = (c'_{ij}, p_{ij})$ of k_i jobs, in which $c'_{ij} \geq c_{ij}, c'_{ij}/p_{ij} < 1/2$, for $j = 1$ to k_i, and

$$\sum_{j=1}^{j=k_i} u'_{ij} = \sum_{j=1}^{j=k_i} u_{ij} + \lambda$$

Since the total utilization factor of this new set of jobs is equal to $\ln 2 - \alpha_i$, this set of jobs can be feasibly scheduled on a single processor. Also, because no job has a utilization factor greater than 1/2,

$$\sum_{j=1}^{j=k_i} f(u'_{ij}) = \sum_{j=1}^{j=k_i} f(u_{ij}) + f(\lambda) \geq 1$$

Hence,

$$\sum_{j=1}^{j=k_i} f(u_{ij}) \geq 1 - f(\lambda) = 1 - 2\lambda \tag{32.10}$$

Combining (32.7), (32.9), and (32.10), we get (32.8). □

Proof of the main theorem. Let N_0 be the minimum number of processors required to schedule the given set of m jobs. Let k_i be the number of jobs allocated to processor $P_i, i = 1, \ldots, N_0$. Then

$$\sum_{i=1}^{m} f(u_i) = \sum_{i=1}^{N_0}\sum_{j=1}^{k_i} f(u_{ij}) \leq \sum_{i=1}^{N_0} 2 \sum_{j=1}^{k_i} u_{ij} \leq 2N_0 \tag{32.11}$$

Now assume RMFF uses N processors for the given set of jobs. Let $P_1, P_2, \ldots, P_s$, after relabelling, if necessary, be all the processors for each of which $\sum f(u)$, where summation runs over all jobs assigned to the processor, is less than 1. Let k_i be the number of jobs allocated to processor $P_i, i, \ldots, s$, and u_{ij} be the utilization factor of the jth job on the ith processor. Then,

$$\sum_{j=1}^{k_i} f(u_{ij}) = 1 - \beta_i, \quad \text{where } \beta_i > 0, i = 1, 2, \ldots, s$$

We divide these s processors into three partitions as follows:

1. Processors with only 1 job assigned. Suppose there are p processors in this set.
   Note that each of these jobs has utilization less than 1/2 and the total utilization of all jobs assigned to these p processors $\geq p/(1 + 2^{1/3})$. Moreover,
   $$\sum f(u) \geq \frac{2p}{1 + 2^{1/3}}$$
   Further, note that no job in this set can have utilization less than 1/3, because a job with utilization factor less than 1/3 can be feasibly scheduled on a single processor together with a job of utilization factor less than 1/2. Hence, in any optimal partition of the set of jobs, no more than two of these jobs can be scheduled on a single processor. Therefore, $N_0 \geq p/2$.
2. Processors with only two jobs assigned. Number of such processors q is 0 or 1.
3. Processors with more than two jobs. Note that coarseness $\alpha_i < 1/6$ for each processor i in this set. Let the number of such processors be r.

Clearly, $s = p + q + r$. For each of the r processors in partition 3, we have

$$U_i = \sum_{j=1}^{k_i} u_{ij} \leq \ln 2 - \alpha_i - \beta_i/2$$

Also, note that deficiency δ_i, for $i = 1, 2, \ldots, r - 1$, satisfies the following relationship:

$$\alpha_{i+1} \geq \delta_i \geq \ln 2 - U_i$$

Thus,

$$\alpha_i + \beta_i/2 \leq \ln 2 - U_i \leq \alpha_{i+1} \quad \text{for } i = 1, 2, \ldots, r - 1$$

Hence,

$$\frac{1}{2}\sum_{i=1}^{r-1} \beta_i \leq \alpha_r - \alpha_1 < 1/6$$

Thus, for the first $r - 1$ processors in partition 3,

$$f(U) = \sum_{i=1}^{r-1} f(U_i) = r - 1 - \sum_{i=1}^{r-1} \beta_i \geq r - 1 - 1/3$$

Thus,

$$\begin{aligned}\sum_i f(u_i) &\geq (N-s) + (r-4/3) + 2p/(1+2^{1/3}) \\ &= N - (p+q+r) + (r-4/3) + 2p/(1+2^{1/3}) \\ &= N - p\left[1 - \frac{2}{1+2^{1/3}}\right] - 4/3 - q \end{aligned} \tag{32.12}$$

From (32.11) and (32.12), we get

$$N \leq 2N_0 + \frac{p(2^{1/3}-1)}{1+2^{1/3}} + \frac{4}{3} + q \tag{32.13}$$

Or,

$$\begin{aligned}\frac{N}{N_0} &\leq 2 + \frac{p(2^{1/3}-1)}{N_0(1+2^{1/3})} + \frac{4/3+q}{N_0} \\ &\leq 2 + 2\frac{2^{1/3}-1}{2^{1/3}+1} + \frac{7}{3N_0}\end{aligned} \tag{32.14}$$

which leads to the desired result.

For the lower bound, consider the following set of N jobs:

$$\begin{aligned} J_1 &= (1, 1+2^{1/N}) \\ J_2 &= (2^{1/N}+\eta, 2^{1/N}(1+2^{1/N})) \\ J_3 &= (2^{2/N}+\eta, 2^{2/N}(1+2^{1/N})) \\ &\vdots \\ J_N &= (2^{(N-1)/N}+\eta, 2^{(N-1)/N}(1+2^{1/N})) \end{aligned} \tag{32.15}$$

where η is such that no job has utilization factor greater than 1/2. First-Fit algorithm will require N processors, because no two of these jobs can be scheduled on a single processor. However, any pair of these jobs can be scheduled on a single processor using the EDF algorithm. Therefore, $N_0 = \lceil \frac{N}{2} \rceil$, and so, $N/N_0 \geq N/\lceil N/2 \rceil$. Taking N sufficiently large, we can make the ratio $N/N_0 > 2 - \varepsilon$.

Sorting of jobs in the above algorithms is based on the job periods. The next algorithm sort jobs based on their utilization factor. □

32.2.3 First-Fit Decreasing-Utilization (FFDU) Algorithm

In this case jobs are first sorted in the order of nonincreasing utilization factor. Then, jobs are allocated to processors in this order. The algorithm can be stated as in Figure 32.3. The value n returned by the algorithm is the number of processors used by this algorithm. The following theorem describes the behavior of this algorithm.

Theorem 32.12

$R^{\infty}_{\text{FFDU}} = 2$.

Proof

Let $f : [0,1] \to [0,1]$ be as follows:

$$f(u) = \begin{cases} 2u, & 0 \leq u \leq \frac{1}{2} \\ 1, & \frac{1}{2} \leq u \leq 1 \end{cases}$$

Given a set $\{J_1, J_2, \ldots, J_m\}$ of m jobs.

1. Sort and renumber jobs in nonincreasing order of their utilization.
2. $i = 1, n = 1$.
3. while $(i \leq m)$ {
    $k = 1$; *assigned* = *false*;
    while (not *assigned*) {
        If J_i is feasible along with other assigned jobs on P_k {
            assign J_i to P_k;
            set $i = i + 1; n = \max\{n, k\}$;
            assigned = *true*;
        }
        else
            $k = k + 1$;
    }
}

FIGURE 32.3 First-Fit Decreasing algorithm.

Let $u_1, u_2, \ldots, u_m$ be the utilization factors of the m jobs. Let k_i be the number of jobs assigned to processor i, $1 \leq i \leq N_0$, in an optimal assignment. Then,

$$\sum_{i=1}^{m} f(u_i) = \sum_{j=1}^{N_0}\sum_{i=1}^{k_j} f(u_{ji})$$
$$\leq \sum_{j=1}^{N_0} 2, \quad \text{since} \quad \sum_{i=1}^{k_j} u_{ji} \leq 1$$
$$= 2N_0 \tag{32.16}$$

Now consider the allocation of this set of jobs according to FFDU algorithm. Assume the allocation uses N processors and let $\{J_{i1}, J_{i2}, \ldots, J_{ik_i}\}$ be the set of jobs assigned to processor P_i, $1 \leq i \leq N$, and let α_i be the utilization factor of the first job assigned to processor P_i. Since jobs are assigned in nonincreasing order of utilization, α_i is the largest utilization of any job assigned to processor P_i. Let U_i denote the total utilization of all jobs assigned to processor P_i, $1 \leq i \leq N$.

In the following, we attempt to find $\sum_{i=1}^{m} f(u_i)$ by applying the function f to jobs assigned to each processor. We consider the following cases.

Case 1: Assume $\alpha_N \leq \ln 2 - 0.5$. Since a job is assigned to the last processor only if it cannot be assigned to the first $N - 1$ processors, and since any set of jobs with utilization factor less than or equal to ln 2 can be feasibly scheduled on a single processor using the RM algorithm (Theorem 32.3), for each processor P_i, $1 \leq i \leq N - 1$,

$$U_i + \alpha_N > \ln 2$$

or

$$U_i > \ln 2 - \alpha_N \geq 0.5$$

Therefore,

$$\sum_{j=1}^{k_i} f(u_{ij}) \geq 1, \quad \text{for } 1 \leq i \leq N - 1 \tag{32.17}$$

Case 2: Assume $\ln 2 - 0.5 < \alpha_N \leq 2^{1/3} - 1$. Since jobs are assigned in the nonincreasing order of their utilization factor, when the job with utilization α_N is assigned to processor P_N, no job with a utilization factor less than or equal to $\ln 2 - 0.5$ has yet been considered for assignment. We divide the first $N - 1$ processors into two classes — (a) Processors for which the utilization factor of the first job assigned is greater than $2^{1/3} - 1$, and (b) all the other processors. For a processor i in class (a) with $\alpha_i \geq 0.5$, we have

$$\sum_{j=1}^{k_i} f(u_{ij}) \geq 1 \tag{32.18}$$

For processors in class (a) with $2^{1/3} - 1 < \alpha_i < 0.5$, which has three or more jobs assigned before the job with utilization factor α_N was considered for allocation on processor P_N, we have

$$\sum_{j=1}^{k_i} f(u_{ij}) \geq 2(2(\ln 2 - 0.5) + 2^{1/3} - 1) > 1 \tag{32.19}$$

Now consider the set of processors in class (a) for which $2^{1/3} - 1 < \alpha_i < 0.5$ and each such processor is assigned one or two jobs before the job with utilization α_N was assigned to P_N. By Theorem 32.3 two jobs with utilization factor up to $2(2^{1/2} - 1)$, and three jobs with total utilization up to $3(2^{1/3} - 1)$ can be feasibly scheduled on a single processor. Since $\alpha_N + 0.5 < 3(2^{1/3} - 1) < 2(2^{1/2} - 1)$, the utilization factor of the jobs assigned to any such processor must have been greater than 0.5. Therefore, for each such processor, we have

$$\sum_{j=1}^{k_i} f(u_{ij}) \geq 1 \tag{32.20}$$

Now consider class (b) processors. Since utilization factor of the first job assigned to any of these processors is less than $2^{1/3} - 1$, utilization factor of all jobs assigned to it is less than $(2^{1/3} - 1)$. Also since $\alpha_N \leq 2^{1/3} - 1$, each such processor must have at least three jobs assigned to it before the job with utilization factor α_N was considered for assignment; otherwise this job should have been assigned to any one of these processors. Further, since $\alpha_N > \ln 2 - 0.5$, for each such processor,

$$\sum_{j=1}^{k_i} f(u_{ij}) \geq 2 * 3 * (\ln 2 - 0.5) \geq 1 \tag{32.21}$$

Combining (32.18)–(32.21), in this case, we get

$$\sum_{j=1}^{k_i} f(u_{ij}) \geq 1, \quad \text{for } 1 \leq i \leq N - 1 \tag{32.22}$$

Case 3: $2^{1/3} - 1 < \alpha_N \leq 1/3$. In this case, for all processors, $1 \leq i \leq N - 1$, $\alpha_i > 2^{1/3} - 1$. Again, we divide all processors into two classes — (a) $\alpha_i \geq 0.5$ and (b) $2^{1/3} - 1 < \alpha_i < 0.5$. For each processor in class (a) we obviously have

$$\sum_{j=1}^{k_i} f(u_{ij}) \geq 1 \tag{32.23}$$

For processors in class (b), each processor must have assigned two jobs to it before the first job on P_N with utilization factor α_N was assigned. This is because $\alpha_i < 0.5$, and according to Theorem 32.7, a job with utilization $\geq \frac{1-0.5}{1+0.5} = \frac{1}{3} \geq \alpha_N$, could have been feasibly scheduled on this processor when the job with utilization α_N was considered for assignment. Therefore, for any processor in this category,

$$\sum_{j=1}^{k_i} f(u_{ij}) \geq 2 * 2 * (2^{1/3} - 1) > 1 \tag{32.24}$$

Combining all these results, if $\alpha_N \leq 1/3$, then

$$\sum_{i=1}^{m} f(u_i) \geq \sum_{j=1}^{N-1} \sum_{i=1}^{k_i} f(u_{ji}) \geq \sum_{i=1}^{N-1} 1 = N - 1 \tag{32.25}$$

Combining Equations (32.16) and (32.25), when $\alpha_N \leq 1/3$, we get

$$N \leq \sum_{i=1}^{m} f(u_i) + 1 \leq 2N_0 + 1 \tag{32.26}$$

or

$$\frac{N}{N_0} \leq 2 + \frac{1}{N_0} \tag{32.27}$$

Now, if $\alpha_N > 1/3$, then, since the jobs are assigned in nonincreasing order of their utilization, each of the N processors is assigned at least one job with utilization $>1/3$. Thus, the number of jobs with utilization $>1/3$ is greater than or equal to N. Since, in any optimal assignment, no more than two such jobs can be assigned to any processor, $N_0 \geq \lceil N/2 \rceil$. Thus, in this case, also

$$\frac{N}{N_0} \leq 2 \tag{32.28}$$

Combining equations (32.27) and (32.28), we get $R^{\infty}_{\text{FFDU}} \leq 2$.

To see that this bound is tight, consider the set of jobs given in (32.15). Using the FFDU algorithm will need N processors. However, any pair of these jobs can be feasibly scheduled on a single processor according to the EDF algorithm. Thus, all these jobs can be feasibly scheduled on $\lceil \frac{N}{2} \rceil$ processors. Thus, $N_0 = \lceil \frac{N}{2} \rceil$, which gives

$$\frac{N}{N_0} = \frac{N}{\lceil N/2 \rceil}$$

Taking N sufficiently large, $\frac{N}{N_0} > 2 - \varepsilon$, for any $\varepsilon > 0$. □

To consider time complexity of this algorithm, the first step of the algorithm sorts all jobs in nonincreasing order of their utilization. This step will take $O(m \log m)$ time. In this case, since the test is based only on the utilization of the next job under consideration, it can be implemented in a similar fashion as the First-Fit algorithm for bin-packing, and hence, takes $O(m \log m)$ time. Thus, the overall complexity of this algorithm is $O(m \log m)$.

32.2.4 Rate-Monotonic General-Task (RMGT) Algorithm

This algorithm [4] has a better performance than any of the algorithms described so far. This is achieved partly by improving the bounds for the feasibility test of a set of jobs over the bounds given in Theorem 32.3. This improvement allows, in some cases, higher utilization for processors. The major factor, however, that contributed to lowering the upper bound is the development of a better estimate (originally conjectured in [12,13] on the utilization of processors that are allocated only one job. This also allows a better estimate on the minimum number of processors required in an optimal allocation.

The algorithm divides jobs into two groups and allocates these jobs to two different groups of processor. Group 1 contains jobs with utilization in the range $(0, \alpha]$ — called the small jobs group — and group 2

1. Reorder jobs in nondecreasing order of their S value (S as defined in Theorem 32.9). That is, $0 \leq S_1 \leq S_2 \leq \cdots \leq S_m$. Let $S_{m+1} = S_1 + 1$.
2. Let $i = 1$, $j = 1$.
3. Assign job J_i to processor P_j. Let $U_j = u_i$, $S = S_i$. $i = i + 1$.
4. Let $\beta_j = S_i - S$. If $U_j + u_i \leq \max[\ln 2, 1 - \beta_j \ln 2]$ then
   assign J_i to P_j, and set $U_j = U_j + u_i$;
   else $j = j + 1$. Assign J_i to P_j, $U_j = u_i$, $S = S_i$.
5. $i = i + 1$. If $i \leq n$ go to Step 4.
6. Set $\beta_j = S_{n+1} - S$.

FIGURE 32.4 Algorithm RMST.

contains jobs with utilization greater than α. Partitioning of n jobs into two groups requires $O(n)$ computation time.

First, consider allocation of jobs in group 1. This algorithm, known as RMST is given in Figure 32.4. The following theorem gives the performance bounds for this algorithm.

Theorem 32.13 [4]
If $\alpha \leq 1/2$ is the maximum utilization of any job, then $R^{\infty}_{\text{RMST}} \leq \frac{1}{1-\alpha}$.

Proof
First note that in Step 4 of the algorithm, allocation to a processor is made only if the feasibility condition of Theorem 32.4 is satisfied. Assume the algorithm allocates N processors. From the description of the algorithm in Figure 32.4, one can easily see that

$$U_j = \sum_{J_r \text{ allocated to processor } P_j} u_r, \quad 1 \leq j \leq N$$

Let i_j be the index of the first job allocated to processor j, according to the algorithm RMST. Then,

$$\beta_j = S_{i_{j+1}} - S_{i_j}, \quad 1 \leq j \leq N \tag{32.29}$$

From Step 4 of the algorithm, it is clear that the job with index i_{j+1} is not assigned to processor $P_j, 1 \leq j < N$, because $U_j + u_{i_{j+1}} > 1 - \beta_j \ln 2$. Therefore,

$$U_j > 1 - \beta_j \ln 2 - u_{i_{j+1}} > 1 - \beta_j \ln 2 - \alpha, \quad \text{for } 1 \leq j < N, \text{ since } u_i \leq \alpha, 1 \leq i \leq m \tag{32.30}$$

and since J_{i_N} is not allocated to processor $P_{i_{N-1}}$,

$$U_{N-1} + U_N > U_{N-1} + U_{i_N} > 1 - \beta_{N-1} \ln 2 \tag{32.31}$$

Let N_0 be the number of processors in an optimal allocation. Then, from Equations (32.29)–(32.31), we get

$$\begin{aligned} N_0 \geq U = \sum_{i=1}^{m} u_i &= \sum_{j=1}^{N} U_j \\ &> \sum_{j=1}^{N-2} (1 - \beta_j \ln 2 - \alpha) + 1 - \beta_{N-1} \ln 2 \\ &\geq (N-2)(1-\alpha) + 1 - \ln 2 \left(\text{since} \sum_{j=1}^{N} \beta_j = 1 \right) \end{aligned} \tag{32.32}$$

Hence,

$$\frac{N}{N_0} \leq \frac{1}{1-\alpha} + \frac{2}{N_0}$$

which gives the required result. □

For the general algorithm, choose $\alpha = 1/3$. Jobs in the first group are allocated using the RMST algorithm. Jobs in the other group are allocated using the First Fit algorithm, similar to the RMFF algorithm, except that this algorithm does not perform the sort operation. The main advantage for this algorithm comes from the fact that a processor in this group can be allocated at most two jobs. As such the exact feasibility condition of Theorem 32.9 is used that has a fixed cost when only two jobs are involved. As pointed out in RMFF algorithm above, the complexity of this algorithm increases to $O(m^2)$. However, this algorithm gives a much better upper bound on the number of processors used by the algorithm. The following theorem gives the performance bound of the RMGT algorithm.

Theorem 32.14

$R^{\infty}_{\text{RMGT}} \leq \frac{7}{4}$.

We need a few preliminary results before taking up the proof of the main theorem.

Lemma 32.5

Assume a set of m arbitrary nonnegative numbers $y_1, \ldots, y_k$ has the property that for $1 \leq j \leq m$ there exists an index i such that

$$y_j = k_i = \min\{2y_1, \ldots, 2y_{i-1}, y_{i+1}, \ldots, y_m\} \quad \textit{or} \quad 2y_j = k_i$$

For such a set, either

$$y_1 \leq \cdots \leq y_m \leq 2y_1 \tag{32.33}$$

or there exists an index $1 \leq j \leq m$ and numbers $a > b \geq 0$ such that

$$y_i = \begin{cases} a & \textit{if } i < j \\ b & \textit{if } i = j \\ 2a & \textit{if } i > j \end{cases} \tag{32.34}$$

Proof

For a proof of this lemma, refer to [4]. □

Lemma 32.6

Consider the set $\{(c_i, p_i) | i = 1, \ldots, m\}$ of m jobs with $p_1 \leq \cdots \leq p_m \leq 2p_1$. Then, these jobs can be scheduled on $m - 1$ processors using Rate-Monotonic algorithm if $U = \sum_{i=1}^{m} c_i / p_i \leq \frac{m}{1+2^{1/m}}$.

Proof

The only way the set of jobs cannot be scheduled on $m - 1$ processors is that no two of these jobs can be scheduled on a single processor. This means that (Theorem 32.9) $c_i + c_j > p_i$ if $i < j$ or $c_i + 2c_j > p_i$ if $i > j$. Consider the problem of minimizing U subject to the conditions:

$$\begin{aligned} &c_i + c_j \geq p_i && \text{if } i < j \\ &c_i + 2c_j \geq p_i && \text{if } i > j \\ &p_1 \leq \cdots \leq p_m \leq 2p_1 && \\ &0 \leq c_i \leq p_i && \text{for } i = 1, 2, \ldots, m \end{aligned} \tag{32.35}$$

First assume c_i's are fixed. Since

$$\frac{\partial U}{\partial p_i} = -\frac{c_i}{p_i^2} < 0, \quad \text{for } i = 1, 2, \ldots, m$$

U is minimized by taking each p_i as large as possible subject to conditions (32.35). To satisfy the first two of these conditions, let

$$p_i = c_i + k_i \quad \text{where } k_i = \min[2c_1, \ldots, 2c_{i-1}, c_{i+1}, \ldots, c_m], \text{ for } i = 1, \ldots, m \tag{32.36}$$

Then,

$$U[c_1, \ldots, c_m] = \sum_{i=1}^{m} \frac{c_i}{c_i + k_i}$$

and

$$\frac{dU}{dc_i} = \frac{k_i}{(c_i + k_i)^2}, \quad \text{for } i = 1, \ldots, m \tag{32.37}$$

Now assume that the c_i's are such that (32.33) is not satisfied, i.e., there is some index j corresponding to which neither there is $i < j$ such that $k_i = c_j$, nor $i > j$ such that $k_i = 2c_j$. Since from (32.37) $\frac{dU}{dc_j}$ is positive, U is decreased by decreasing values for c_j. Thus, condition (32.34) is satisfied at any critical point of U. Therefore, from Lemma 32.5, either

$$c_1 \leq c_2 \leq \cdots \leq c_m \leq 2c_1 \tag{32.38}$$

or, there exists an index $j \in [1, m]$ and numbers $a > b \geq 0$, such that

$$c_i = \begin{cases} a & \text{if } i < j \\ b & \text{if } i = j \\ 2a & \text{if } i > j \end{cases} \tag{32.39}$$

If Equation (32.38) does not hold then values of c_i's are given by Equation (32.39). With these values of c_i's, we get

$$p_i = \begin{cases} a + b & \text{if } i < j \\ 2a + b & \text{if } i = j \\ 2a + 2b & \text{if } i > j \end{cases}$$

and

$$U = (m - 1)\frac{a}{a + b} + \frac{b}{2a + b}$$

Writing $t = b/a$, U as a function of t becomes

$$U(t) = (m-1)\frac{1}{1+t} + \frac{t}{2+t}$$

For $m \geq 2$, and $0 \leq t \leq 1$, $U(t)$ is nonincreasing. Therefore, $U(t)$ has the minimum value when $t = 1$. But then, $a = b$, which contradicts the assumption that $a > b$. Hence, (32.38) holds. This, combined with (32.36) gives $p_i = c_i + c_{i+1}$, for $i = 1, \ldots, m-1$, and $p_m = c_m + 2c_1$. Now the problem of minimizing U reduces to minimizing

$$U[c_1, \ldots, c_m] = \sum_{i=1}^{m-1} \frac{c_i}{c_{i+1}} + \frac{c_m}{c_m + 2c_1}$$

subject to $c_i \geq 0$, for $i = 1, \ldots, m$.

Writing x_i for $\log_2 \frac{c_{i+1}}{c_i}$, $1 \leq i < m$, and x_m for $\log_2 \frac{2c_1}{c_m}$, the problem reduces to minimizing

$$U[x_1, \ldots, x_m] = \sum_{i=1}^{m} \frac{1}{1+2^{x_i}}$$

subject to the conditions

$$\sum_{i=1}^{m} x_i = 1 \text{ and } x_i \geq 0, \quad \text{for } i = 1, \ldots, m$$

This is a strictly convex minimization problem in which the objective function U and the boundary conditions are symmetric with respect to the indices of the variable x. Therefore, the solution to the problem should also be symmetric under permutation of the indices. This observation together with the restriction $\sum_{i=1}^{m} x_i = 1$ suggests the solution $x_i = 1/m$ for $i = 1, \ldots, m$, which gives

$$U = \frac{m}{1+2^{1/m}}$$

This shows that if the set of m jobs cannot be scheduled on $m-1$ processors, its utilization must be greater than U as given above. □

This result can be extended to the general case where the job periods are not restricted as in the above Lemma. For details, refer to [4].

Proof

Assume the number of processors used by group 1 and group 2 jobs is N_1 and N_2, respectively. Divide the processors used for group 2 into two subgroups — N_{21} processors that are allocated only one job and N_{22} processors that are allocated two jobs. Then, if U_1, U_{21}, and U_{22} are utilizations of all jobs allocated to N_1, N_{21}, and N_{22} processors,

$$U_1 > \frac{2}{3}(N_1 - 2) + 1 - \ln 2 > \frac{N_1}{2} - \ln 2 + \frac{1}{6} \tag{32.40}$$

$$U_{21} > \frac{N_{21}}{1 + 2^{\frac{1}{N_{21}}}} > \frac{N_{21}}{2} - \frac{\ln 2}{4} \quad \text{assuming } N_{21} \geq 2 \tag{32.41}$$

$$U_{22} > \frac{2N_{22}}{3} > \frac{N_{22}}{2} \tag{32.42}$$

Let N_0 be the number of processors used by an optimal allocation algorithm. Then, $N_0 \geq U_1 + U_{21} + U_{22}$. Also, since the RMGT algorithm uses necessary and sufficient condition for allocation of jobs to processors allocated only one job, $N_0 \geq N_{22}$. Combining all this,

$$\begin{aligned} N_0 &> \frac{2}{3}(N_1 + N_{22}) + \frac{N_{21}}{2} - \frac{1}{3} - \ln 2 + \frac{\ln 2}{4} \\ &> \frac{2}{3}(N_1 + N_{21} + N_{22}) - \frac{1}{6}N_{21} - \frac{1}{3} - \frac{5}{4}\ln 2 \\ &> \frac{2}{3}N - \frac{1}{6}N_0 - \frac{1}{3} - \frac{5}{4}\ln_2 \end{aligned} \tag{32.43}$$

Or,

$$\frac{N}{N_0} \leq \frac{7}{4} + \frac{\frac{1}{3} + \frac{5}{4}\ln_2}{N_0} \tag{32.44}$$

from which the claim follows. □

Remark 32.1

Lemma 32.6 used in the proof of the above theorem has a great impact on the upper bound of the RMFF algorithm. In [13], it was conjectured that if a set of m jobs $\{(c_i, p_i), i = 1, 2, \ldots, m\}$, *with* $p_i \leq p_{i+1}, i = 1, 2, \ldots, m-1$, *cannot be scheduled on* $m-1$ *processors according to the RMFF algorithm, then* $\sum_{i=1}^{m} c_i/p_i > \frac{m}{1+2^{1/m}}$. *The consequence of Lemma 32.6 is that the term* $\frac{2^{1/3}-1}{2^{1/3}+1}$ *in (32.13) can be replaced with* $\frac{2^{1/p}-1}{2^{1/p}+1}$. *The term* $\frac{p(2^{1/p}-1)}{N_0(1+2^{1/p})}$ *tends to 0 as* $N_0 \to \infty$. *As a result* $R^{\infty}_{\text{RMFF}} \leq 2$.

32.3 Online Algorithms

The previous section described algorithms that require that all jobs be known in advance. These algorithms reorder jobs in some fashion before starting job allocation. This section presents two online algorithms. Online algorithms do not require prior knowledge of parameters of all the jobs. The allocation decision on each job is based solely on the parameter of that job and the information about the current state of the system. No assumptions are made about future jobs.

32.3.1 Next-Fit-M

This algorithm categorizes jobs into $M \geq 2$ different classes based on their utilization factor. Similarly, a processor that is allocated a class-k, $k = 1$ to M, job is designated as a class-k processor. All jobs with utilization factor in the range $(2^{1/(k+1)} - 1, 2^{1/k} - 1]$ are placed in class k, $k = 1$ to $M - 1$. All jobs with utilization factor in the range $(0, 2^{1/M} - 1]$ belong to class M. Jobs are processed in the order they arrive. Once a job of class-k is allocated to a class-k processor it becomes an **active** processor. The processor remains active till no more jobs of that class can be allocated to the processor. At that time the processor becomes **filled** and it is set aside. According to Theorem 32.3 exactly k jobs of class-k, $k = 1$ to $M - 1$, can be feasibly scheduled on a single processor using the RM algorithm. Class-M jobs are allocated to a class-M processor till no more jobs can be allocated to it. Class-M jobs are allocated to a class-M processor so long as the sum of the utilization factor of all allocated jobs to a processor is less than or equal to ln 2. Once a job of class-M is encountered that cannot be allocated to an active class-M processor, a new class-M processor is started. By the very nature of the job allocation procedure it is clear that the utilization factor of jobs assigned to all but at most one class-k processor, $1 \leq k < M$, is greater than $k(2^{\frac{1}{k+1}} - 1)$, and that for class-M processors is greater than $\ln 2 - (2^{\frac{1}{M}} - 1)$. Also note that for any set of jobs the number of class-k processors, $1 \leq k < M$, used by the algorithm is the same, irrespective of the order of their arrival. In other words, except for class-M processors, any permutation of jobs in the original list will result in the same number of processors used by Next-Fit-M.

The NFM algorithm first determines the class of a job and then assigns it to a processor of the appropriate class. The determination of the class of a job takes $O(\log M)$ time and the allocation of a job to the processor takes a constant time. Thus, the total time taken by the algorithm is $O(m \log M)$, where m is the number of jobs. Since M is fixed at the start of the algorithm, the overall complexity of the algorithm is $O(m)$. Also, since once a processor is filled, it is not considered for further assignment, we need to keep only M processors active. Therefore, the space complexity of the algorithm is $O(1)$.

Theorem 32.15

$R^{\infty}_{\text{NFM}} \leq S_M$. *($S_M$ is defined later.)*

Proof

Let N denote the total number of processors used by the algorithm, and let N_k, $1 \leq k \leq M$, denote the number of class-k processors used, and let n_k, $1 \leq k \leq M$, denote the number of jobs of class-k. Then,

$$N \leq \sum_{k=1}^{M} N_k = n_1 + \left\lceil \frac{n_2}{2} \right\rceil + \left\lceil \frac{n_3}{3} \right\rceil + \cdots + \left\lceil \frac{n_{M-1}}{M-1} \right\rceil + N_M \tag{32.45}$$

Let U_M denote the total utilization factor of all class-M jobs in the list of jobs to be scheduled. Since the total utilization factor of jobs assigned to each filled class-M processor is greater than $\ln 2 - (2^{1/M} - 1)$, we have,

$$N_M \leq \frac{U_M}{\ln 2 - (2^{1/M} - 1)} + 1$$

Thus,

$$N = n_1 + \frac{n_2}{2} + \frac{n_3}{3} + \cdots + \frac{n_{M-1}}{M-1} + \frac{U_M}{\ln 2 - (2^{1/M} - 1)} + M$$

Let u_i denote the utilization factor of job J_i. Define the following cost function:

$$f(u) = \begin{cases} \frac{1}{k} & \text{if } u \in (2^{\frac{1}{k+1}} - 1, 2^{\frac{1}{k}} - 1] \\ \frac{u}{\ln 2 - (2^{1M} - 1)} & \text{if } u \leq 2^{1M} - 1 \end{cases}$$

Equation (32.45) can then be written as

$$N < \sum_{i=1}^{m} f(u_i) + M \tag{32.46}$$

Further, if u_k is the utilization of a class-k processor, $1 \leq k < M$, then $(2^{\frac{1}{k+1}} - 1) < u_k$, and, therefore,

$$\frac{f(u_k)}{u_k} = \frac{1}{ku_k} < \frac{1}{k(2^{\frac{1}{k+1}} - 1)}$$

Since $k(2^{\frac{1}{k+1}} - 1)$ is a monotonically increasing function of k, we have

$$\frac{f(u_k)}{u_k} < \frac{1}{k(2^{\frac{1}{k+1}} - 1)}, \quad \text{for } 2^{\frac{1}{k+1}} - 1 \leq u_k \leq 2^{\frac{1}{k}} - 1 \quad \text{and} \quad 1 \leq k < M \tag{32.47}$$

Let N_0 be the number of processors used by an optimal algorithm. Let t_i be the number of jobs on processor i, $1 \leq i \leq N_0$. Then,

$$\sum_{j=1}^{t_i} u_{ij} \leq 1, \quad 1 \leq i \leq N_0$$

and

$$F_{i,M} = \sum_{j=1}^{t_i} f(u_{ij}), \quad \text{for } 1 \leq i \leq N_0$$

Consider a set of t, $t \geq 1$, jobs, with utilization factors $u'_1, \ldots, u'_t$, having the following properties:

1. $u'_k > 0, 1 \leq k \leq t$.
2. $\sum_{k=1}^{t} u'_k \leq 1$.
3. $F_M = \sum_{k=1}^{t} f(u'_k) \geq F_{i,M}$ for $1 \leq i \leq N_0$.

Then, we can rewrite (32.46) as:

$$\begin{aligned} N &\leq \sum_{i=1}^{n} f(u_i) + M \\ &= \sum_{i=1}^{N_0} \sum_{j=1}^{t_i} f(u_{ij}) + M \\ &\leq N_0 \times F_M + M \end{aligned}$$

This gives

$$\frac{N}{N_0} \leq F_M + \frac{M}{N_0}$$

Thus, it is sufficient to find the upper bound on the number F_M, where

$$F_M = \sum_{j=1}^{t} f(u'_j), \quad \text{subject to} \sum_{j=1}^{t} u'_j \leq 1$$

To this end, define a sequence of integers K_i, as follows:

$$K_1 = K_2 = 1$$

For $i > 2$, K_i is the smallest integer K such that the following inequality is satisfied:

$$\left(2^{\frac{1}{K+1}} - 1\right) < 1 - \sum_{j=1}^{i-1} \left(2^{\frac{1}{K_j+1}} - 1\right) \tag{32.48}$$

Thus, by definition, for $i > 2$,

$$\left(2^{\frac{1}{K_i}} - 1\right) \geq 1 - \sum_{j=1}^{i-1} \left(2^{\frac{1}{K_j+1}} - 1\right) \tag{32.49}$$

It can be checked that $K_3 = 4$, $K_4 = 30$, $K_5 = 2635$. For $M \geq 3$ and $K_i < M \leq K_{i+1}$ define

$$S_M = \sum_{r=1}^{i} \frac{1}{K_r} + \frac{1 - \sum_{r=1}^{i} \left(2^{\frac{1}{K_r+1}} - 1\right)}{\ln 2 - \left(2^{\frac{1}{M}} - 1\right)}$$

We show that $F_M = \sum_{j=1}^{t} f(u'_j) < S_M$. Without loss of generality, assume that $u_1 \geq u_2 \geq \cdots \geq u_t$. Since $\sum_{i=1}^{t} u_i \leq 1$, this set can have at most two class-1 jobs. So consider the following three cases:

Case 1: The set has no class-1 job. Then, by Equation (32.47), we have

$$F_M = \sum_{j=1}^{t} f(u'_j) \leq \frac{\sum_{j=1}^{t} u'_j}{2(2^{1/3} - 1)} \leq \frac{1}{2(2^{1/3} - 1)} < 1.93$$

Case 2: There is only one class-1 job, that is $u'_1 > 2^{1/2} - 1$. All other jobs are class-k ($k \geq 2$) jobs. Therefore,

$$F_M = \sum_{j=1}^{t} f(u'_j) = f(u'_1) + \sum_{j=2}^{t} f(u'_j) \leq 1 + \frac{\sum_{j=2}^{t} u'_j}{2(2^{1/2} - 1)}$$

$$\leq 1 + \frac{1 - (2^{1/2} - 1)}{2(2^{1/3} - 1)} < 2.1269$$

Case 3: Both jobs 1 and 2 are class-1 jobs. First, consider the subcase in which job j, $1 \leq j \leq i$, is a class-K_j job. Since $K_i < M < K_{i+1}$, job j belongs to class-M, for $i + 1 \leq j \leq t$. Then,

$$F_M = \sum_{j=1}^{t} f(u_j) < \sum_{j=1}^{i} \frac{1}{K_j} + \frac{1 - \sum_{j=1}^{i} (2^{1/(K_j+1)} - 1)}{\ln 2 - (2^{\frac{1}{M}} - 1)} = S_M$$

Note that in this case, job 1 and job 2 belong to class-K_1 and class-K_2, respectively. What is the effect on the value of F_M if job i, $i > 2$, does not belong to class K_i? Then, $u_j \leq (2^{\frac{1}{K_j+1}} - 1)$, for $i \leq j \leq t$. Therefore, value of F_M will change by the quantity

$$\frac{2^{\frac{1}{K_i+1}} - 1}{(K_i + 1)(2^{\frac{1}{K_i+2}} - 1)} - \frac{1}{K_i}$$

Since

$$\frac{2^{\frac{1}{K_i+1}} - 1}{(K_i + 1)(2^{\frac{1}{K_i+2}} - 1)} < \frac{1}{K_i}$$

the value of F_M will decrease. It can similarly be shown that if job j is not in class K_j, for $j = i - 1$, then the value of F_M will decrease further. This argument can be repeated for jobs $j = i - 2, i - 3, \ldots, 3$, showing that F_M has its maximum value when job j is in class K_j, for $1 \leq j \leq i$, in which case its maximum value is S_M. Thus,

$$\frac{N}{N_0} \leq S_M + \frac{M}{N_0}$$

from which the result follows.

It can be verified that $S_3 = 2.3960$, $S_4 = 2.3404$, $S_{12} = 2.2860$, $S_{30} = 2.2841$, $S_{31} = 2.2837$, $S_\infty = 2.2837$. From this it is clear that there is no significant difference between S_{12} and S_∞. Thus, in practice $M = 12$ is a good choice.

As the following example shows the upper bound of this algorithm is very tight. Let $M = K_{i+1}$. Choose an integer N that is divisible by M. For each j, $j = 1, \ldots, i$, choose N class-K_j jobs with utilization $u_j = (2^{1/(K_j+1)} - 1) + \varepsilon$. Choose N jobs with utilization $u_{i+1} = (1 - \sum_{j=1}^{i} -i \times \varepsilon)$. By definition of K_i, $(2^{1/(M+1)} - 1) < u_{i+1} < (2^{1/M} - 1)$. Thus, algorithm NFM will allocate M jobs with utilization u_{i+1} to each class-M processors. Let $N(\text{NFM})$ denote the number of processors used by NFM. Then,

$$N(\text{NFM}) = \sum_{j=1}^{i} \frac{N}{K_j} + \frac{N}{M}$$

Since $\sum_{j=1}^{i+1} u_j = 1$, the number of processors needed by an optimal allocation is N. Thus,

$$\frac{N(\text{NFM})}{N_0} = \sum_{j=1}^{i} \frac{1}{K_j} + \frac{1}{M} = \sum_{j=1}^{i+1} \frac{1}{K_j}$$

Given a set of m jobs with utilizations $u_1, u_2, \ldots, u_m$, respectively. In the following $R_j = 2[\prod_{i=1}^{k_j}(1 + u_{ji})]^{-1} - 1$ represents the remaining capacity of processor P_j that has been allocated k_j jobs.

1. $i = 1$.
2. while ($i \leq m$) do {

    $j = 1$; *assigned* = *false*;

    while (not assigned) {

    if $u_i \leq R_j$, { allocate J_i to P_j; $i = i + 1$; *assigned* = *true*;}

    else $j = j + 1$.

    }

}

FIGURE 32.5 The Rate-Monotonic First-Fit (RM-FF) algorithm.

For $M = 4$, $M = 30$, and $M = \infty$, the right-hand side is approximately 2.25, 2.2833, and 2.2837, compared with the corresponding S_M values of 2.3404, 2.2841, and 2.2837. It can be shown that if in a given set of jobs, no job has utilization in the range $(2^{1/2} - 1, 1/2]$, then $R_{\text{NFM}}^{\infty}(S) \leq 1.913$ [11]. □

32.3.2 Rate-Monotonic First-Fit (RM-FF) Algorithm

As opposed to the RMFF algorithm of Section 32.2, the RM-FF algorithm [14] does not sort jobs before the allocation process begins. The allocation decision is based on the utilization factors of jobs without regard to their periods. The rest of the algorithm is the same as before, except that it uses the feasibility test of Theorem 32.8. Since no sorting is done before allocation begins, the algorithm is online. The algorithm is described in Figure 32.5.

The largest index j used in Step 2 is the number of processors used by the algorithm. The algorithm starts a new processor only if the current job cannot be allocated to any of the processors currently in use. Before analyzing the performance of this algorithm, a few preliminary results are derived below.

Lemma 32.7 [14]

If m jobs cannot be feasibly scheduled on $m - 1$ processors according to RM-FF, then the total utilization of the m jobs is greater than $\frac{m}{1+2^{1/2}}$.

Proof

According to Theorem 32.3, if two jobs cannot be scheduled on a single processor, their utilization is greater than $2(2^{1/2} - 1) = \frac{2}{1+2^{1/2}}$. Thus, the result is true when $m = 2$.

Now assume the result holds for all job sets of size m, $m = 2, \ldots, k$, for some $k \geq 2$. Assume there exists a set of $k + 1$ jobs that cannot be scheduled on k processors. Then the first k of these jobs cannot be scheduled on $k - 1$ processors, because otherwise the given set would be feasible on k processors. Therefore,

$$\sum_{i=1}^{k} u_i > \frac{k}{1 + 2^{1/2}} \tag{32.50}$$

Consider the $(k+1)$th job. This job cannot be scheduled with any of the first k jobs on a single processor. Therefore,

$$u_i + u_{k+1} > \frac{2}{1+2^{1/2}} \quad \text{for } i = 1, \ldots, k \tag{32.51}$$

Summing up these k equations together with $(k-1)$ times (32.50), and performing some simplification gives

$$\sum_{i=1}^{k+1} u_i > \frac{k+1}{1+2^{1/2}} \tag{32.52}$$

Thus, by induction, the result holds for all m. □

Lemma 32.8 [14]

Consider the set of all processors to which $n \geq k \geq 1$ jobs have been assigned according to the algorithm RM-FF. Among all the processors in this set, there is at most one processor for which the utilization of all allocated jobs is less than or equal to $k(2^{1/(k+1)} - 1)$.

Proof

Assume there are two or more processors for which the total utilization is less than or equal to $k(2^{1/(k+1)} - 1)$. Let P_i and P_j, with $i < j$, be two such processors with $n_i \geq k$ and $n_j \geq k$ jobs, respectively. Since P_j has $n_j \geq k$ jobs and its total utilization is less than or equal to $k(2^{1/(k+1)} - 1)$, at least one of the jobs on P_j must have utilization less than or equal to $2^{1/(k+1)} - 1$. Now, if the utilization of this job is added to that of processor P_i, the total utilization of jobs is less than or equal to $(k+1)(2^{1/(k+1)} - 1)$. Therefore, according to RM-FF this job should have been allocated to P_i. Hence, the lemma is true. □

Theorem 32.16 [14]

$R^{\infty}_{\text{RM-FF}} \leq 2 + (3 - 2^{3/2})/(2^{4/3} - 2) \approx 2.33$.

Proof

Let N and N_0 be the number of processors needed to allocate a given set of jobs according to RM-FF and according to an optimal allocation, respectively. Let $a = 2(2^{1/3} - 1)$, and define f: $(0, 1] \rightarrow (0, 1]$ as follows:

$$f(u) = \begin{cases} u/a, & 0 < u \leq a \\ 1, & a \leq u \leq 1 \end{cases}$$

Let $u = u_1 + u_2 + \cdots + u_k \leq 1$ be the utilization of the allocated jobs to a processor. If none of the jobs has utilization greater than a, then $\sum_{i=1}^{k} f(u_i) = u/a < 1/a$. Otherwise, assume, without loss of generality, that $u_1 \geq a$. Then, $f(u) = \sum_{i=1}^{k} f(u_i) = f(u_1) + \sum_{i=2}^{k} u_i \leq 1 + (1-a)/a = 1/a$. Thus, in all cases, $\sum_{i=1}^{k} f(u_i) \leq 1/a$. Therefore, for an optimal allocation,

$$\sum_{i=1}^{m} f(u_i) = \sum_{j=1}^{N_0} \sum_{k=1}^{k_j} f(u_{jk}) \leq \sum_{j=1}^{N_0} 1/a = N_0/a \tag{32.53}$$

Divide all allocated processors into two categories.

1. Sum of utilization of all jobs allocated to the processor is greater than or equal to a. For any such processor j with k_j jobs, $\sum_{i=1}^{k_j} f(u_{ji}) \geq 1$.
2. Sum of utilization of all jobs allocated to the processor is less than a. For each such processor j with k_j jobs, $\sum_{i=1}^{k_j} f(u_{ji}) = \sum_{i=1}^{k_j} u_{ji}/a < 1$.
   a. Let N_1 be the number of processors in this category that have only one job allocated. Then, by Lemma 32.7, if U is the sum of utilizations of all these jobs, then $U > N_1(2^{1/2} - 1)$. Also, since for each of these processors $f(u_i) < 1$, the utilization of each job in this set is less than a. Hence,

$$f(U) > N_1(2^{1/2} - 1)/a$$

   Further, by Lemma 32.8, there is at most one processor for which $u < (2^{1/2} - 1)$. Thus, in any optimal allocation, no more than two such jobs can be allocated. This means that $N_0 \geq N_1/2$.
   b. Let N_2 be the number of processors each of which has two or more jobs. By Lemma 32.8, there is at most one such processor, giving $N_2 \leq 1$.

From the above discussion, it follows that

$$\begin{aligned}\sum_{i=1}^{m} f(u_i) &\geq N - (N_1 + N_2) + N_1(2^{1/2} - 1)/a \\ &= N - N_1(1 - (2^{1/2} - 1)/a) - N_2 \\ &\geq N - 2N_0[1 - (2^{1/2} - 1)/a] - N_2 \end{aligned} \tag{32.54}$$

Combining this with (32.53), and the fact that $N_2 \leq 1$, we get

$$N_0 > \sum_{i}^{n} f(u_i) \geq N - 2N_0[1 - 2^{1/2} - 1)/a] - N_2$$

or

$$\frac{N}{N_0} \leq [2 + (3 - 2^{3/2})/2(2^{1/3} - 1)] - \frac{1}{N_0}$$

which gives

$$R_{\text{RM-FF}}^{\infty} \leq [2 + (3 - 2^{3/2})/(2^{4/3} - 2)] \approx 2.33$$

To establish the lower bound, consider the following sequence of jobs with $k \geq 1$:

1. First $2640k$ jobs, each with utilization $2^{2641} - 1 + \varepsilon$,
2. Next $2640k$ jobs, each with utilization $2^{1/31} - 1 + \varepsilon$,
3. Next $2640k$ jobs each with utilization $2^{1/5} - 1 + \varepsilon$,
4. Next $5280k$ jobs, each with utilization $2^{1/2} - 1 + \varepsilon$.

Choose ε appropriately small. This set of jobs will require $k + 88k + 660k + 5280k = 6029k$ processors using RM-FF. An optimal allocation will require 2640k processor, with each processor getting two jobs of the last type and 1 job each of the other three types. Hence, in this case $N/N_0 \geq 2.2837$. Thus, $R(\text{RM-FF})^{\infty} \geq 2.2837$.

The implementation of this algorithm is similar to the First-Fit algorithm for bin-packing; as such its computation time is $O(m \log m)$. □

In special cases when there are no jobs in the set with utilization factor greater than or equal to $2(2^{1/2} - 1)$, the algorithm provides much better asymptotic upper bounds. In particular, the following result was proved in [14].

Theorem 32.17

When RM-FF algorithm is applied to schedule a set of jobs in which the maximum utilization factor of a job is $\alpha \leq 2^{1/(1+c)} - 1$, *then*

$$R^{\infty}_{\text{RM-FF}}(\alpha) \leq \frac{1}{(c+1)(2^{1/(2+c)}-1)}, \quad \textit{for } c = 0, 1, 2, \ldots$$

As $c \to \infty$, $R^{\infty}_{\text{RM-FF}}(\alpha) \leq 1/(\ln 2)$.

In view of this theorem together with the result in Lemma 32.6, Oh and Son [14] also proposed a refinement of the above algorithm as RRM-FF algorithm to reduce the upper bound to 2. As in the RMGT algorithm, RRM-FF also allocates jobs with utilization greater than 1/3 and those with utilization $\leq 1/3$ to different groups of processors. However, the computational complexity of this algorithm increases to $O(m^2)$ (though the authors claimed no increase in its computational complexity).

32.4 Conclusion

In this chapter we attempted to provide almost a complete picture of the state of the art in approximation algorithms for allocating periodic time-critical jobs in a multiprocessor system together with their asymptotic worst-case behavior. One of the algorithms not described herein because of space limitation is the Rate-Monotonic Best-Fit algorithm and its refinement, whose asymptotic bounds are similar to the RM-FF algorithm [14]. The upper bound for all the algorithms has been greater than or equal to 2, till the paper by Burchard et. al. [4]. This significant improvement has been due to two factors — an efficient schedulability test for a single processor that takes not only the utilization factor but also the periods of the jobs into account, and better estimate on the number of optimal processors. For the estimate of the minimum number of processors needed for a set of jobs hitherto, almost all algorithms have relied on the total utilization factor of the given set of jobs. However, the improvement in this estimate comes at a cost of increasing the computational complexity of the algorithm from $O(m \log m)$ to $O(m^2)$. The main reason for this is that it is difficult to order the remaining available capacity of a processor, because it depends not only on the allocated capacity but also on the period of the job under consideration for allocation. In view of the result of Oh and Baker [20] that worst-case achievable utilization for $n (n \geq 2)$ lies between $n(2^{1/2} - 1)$ and $(n+1)/(1+2^{1/(n+1)})$, the performance of the algorithms presented in this chapter is very good. In fact, simulation results reported in [4] indicate that on the average for many of the algorithms the ratio N/U is much less than 1.5, where N is the number of processors used by the algorithm and U is the total utilization of a given set of jobs.

References

[1] Liu, C.L. and Layland, J.W. (1973). Scheduling Algorithms for Multiprogramming in a Hard-Real-Time Environment. *Journal of the ACM,* 20, pp. 46–61.

[2] Serlin, P. (1972). Scheduling of Time-Critical Processes, *Proceedings of the Spring Joint Computers Conference*, 40, pp. 925–932.

[3] Devillers, R. and Goossens, J. (2000). Liu and Layland's Schedulability Test Revisited. *Information Processing Letters*, 73(5-6), pp.157–161.

[4] Burchard, A., Liebeherr, J., Oh, Y., and Son, S.H. (1995). New Strategies for Assigning Real-Time Tasks to Multiprocessor Systems. *IEEE Transactions on Computers*, 44(12), pp. 1429–1442.

[5] Dertouzos, Michael L. and Mok, Aloysius K. (1989). Multiprocessor On-line Scheduling of Hard-Real-Time Tasks. *IEEE Transactions on Software Engineering*, 15(12), pp. 1497–1506.

[6] Leung, J.Y.T. (1989). A New Algorithm for Scheduling Periodic Real-Time Tasks. *Algorithmica*, 4, pp. 209–219.

[7] Leung, J.Y.T. and Whitehead, J. (1982). On the Complexity of Fixed-Priority Scheduling of Periodic Real-Time Tasks. *Performance Evaluation*, 2, pp. 237–250.
[8] Davari, S. and Dhall, S.K. (1986). On a Real-Time Task Allocation Problem. *Proceedings of the Nineteenth Annual Hawaii International Conference on Information Sciences and Systems*, pp. 133–141.
[9] Davari, S. and Dhall, S.K. (1986). A Simple On-line Algorithm for Real-Time Tasks Allocation, *Proceedings of the Twentieth Conference on Information Sciences and Systems*, Princeton, NJ, pp 178–182.
[10] Davari, S. and Dhall, S.K. (1986). An On-line Algorithm for Real-Time Tasks Allocation, *Proceedings of the Real-Time Systems Symposium*, pp. 194–200.
[11] Davari, S. and Dhall, S.K. (1995). On-line Algorithms for Allocating Periodic-time-critical Tasks on Multiprocessor Systems. *Informatica*, 19, pp. 83–96.
[12] Dhall, S. K. (1977). Scheduling Periodic Time-Critical Jobs on Single and Multiprocessor Systems. University of Illinois Technical Report No. UIUCDCS-R-77–859.
[13] Dhall, S.K. and Liu, C.L. (1978). On a Real-Time Scheduling Problem. *Operations Research*, 26(1), pp. 127–140.
[14] Oh, Y. and Son, S.H. (1994). Allocating Fixed-Priority Periodic Tasks on Multiprocessor Systems. *Journal of Real-Time Systems*, 5, pp. 1–33.
[15] Johnson, D.S. (1974). Fast Algorithms for Bin-Packing. *Journal of Computer System Science* 8, pp. 272–314.
[16] Johnson, D.S., Demers, A., Ullman, J.D, Garey, M.R., and Graham, R.L. (1974). Worst-Case Performance Bound for Simple One-dimensional Packing Algorithms. *SIAM Journal of Computing*, 3, pp. 299–325.
[17] Lee, C.C. and Lee, D.T. (1985). A Simple On-line Bin-Packing Algorithm. *Journal of the ACM*. 32(3), pp. 562–572.
[18] Garey, Michael R. and Johnson, David S. (1979) *Computers and Intractability A Guide to the Theory of NP-Completeness*, W.H. Freeman, San Francisco.
[19] Lehoczky, J.P., Sha, L., and Ding, Y. (1989). The Rate-Monotonic Scheduling Algorithm: Exact Characterization and Average Case Behavior. *IEEE Real-Time Symposium*, pp. 166–171.
[20] Oh, Dong-Ik and Baker, T.P. (1998). Utilization Bounds for N-Processor Rate Monotone Scheduling with Static Processor Assignment. *Real-Time Systems*, 15, 183–192.

33

Scheduling Overloaded Real-Time Systems with Competitive/ Worst-Case Guarantees

Gilad Koren
Bar-Ilan University

Dennis Shasha
New York University

33.1 Introduction

Once limited to exotic applications, real-time systems now can be found in applications from cars to video-on-demand to space stations. Real-time systems control production and safety in power plants, factories, labs, and perhaps soon in our homes.

An overloaded real-time system is one for which even a clairvoyant scheduler cannot meet all deadlines with the available computational resources. Overload can arise either as the result of failures of some computational resources or as a transient condition. For example, an overloaded multimedia network following an exciting sports event may cause some packets to be dropped. Arbitrage opportunities may come in a cluster to a financial analyst who must then pick and choose among them.

Formally, a system is *underloaded* if there exists a schedule that will meet the deadline of every task and is *overloaded* otherwise. Scheduling underloaded systems is a well-studied topic, and several online

1-58488-397-9/$0.00+$1.50

algorithms have been proposed for the optimal scheduling of these systems on a uniprocessor [1,2]. Examples of such algorithms include *earliest deadline first* (EDF) and *least slack time*[1] *(LST) first*. However, none of these classical algorithms make performance guarantees during times when the system is overloaded. In fact, Locke [3] has experimentally demonstrated that these algorithms perform quite poorly when the system is overloaded.

Practical systems are prone to intermittent overloading caused by a cascading of exceptional situations. A good scheduling algorithm, therefore, should give a performance guarantee in overloaded as well as underloaded circumstances. Obtaining a guarantee becomes especially interesting and challenging in the *online* setting when the exact nature (deadlines and resource requirements) of the incoming tasks and their arrival pattern are unknown to the scheduler.

A common approach to handle overloaded situations is to add an acceptance test module which is executed at task release and checks whether the newly arriving task can be scheduled with the existing tasks without causing overload. If not, the new task is rejected or some existing task(s) may be aborted according to value considerations [4–7]. We describe here algorithms that follow this approach and offer performance guarantees.

For uniprocessor systems, researchers and designers of real-time systems have devised online heuristics to handle overloaded situations [8,9]. Locke proposed several clever heuristics as part of the CMU Archons project [3].

Multiprocessor real-time scheduling is an active field of research. Both shared-memory [10,11] and distributed-memory [2,12–14] architectures have been studied. *Static binding* of tasks to processors (i.e., no migration) is assumed in [2,11,14] while dynamic binding is assumed in [10,15]. For a survey of scheduling issues for uniprocessor and multiprocessor systems see [16,17].

Mok and Dertouzos [15] showed that in a multiprocessor environment an optimal algorithm must have an *a priori* knowledge of release times. Hence, no online optimal algorithm exists even when the system is underloaded. Locke [3, pp. 124–134] also presented heuristics for the multiprocessor environment. Ramamritham and Stankovic [12] studied the question of scheduling firm deadline tasks in a distributed environment. They proposed a scheduler that assumes, at the design phase, that the system is underloaded for *critical* tasks. The noncritical tasks are scheduled dynamically and heuristically using any surplus processing power. The *MOCA algorithm* described below could be used for the noncritical tasks in their environment.

33.2 Competitive Analysis

Suppose a Computer Science graduate student receives a small fellowship during his/her program. As it happens, he/she gets offers, from time to time, for short term consulting assignments from different high technology companies. Each assignment is at a high daily consulting wage. These assignments are so demanding that taking even one of them means quitting graduate school immediately. Returning to graduate school with a fellowship, after quitting, is impossible. There is no way to estimate the flow of consulting projects over time. If, however, the flow does continue, the cumulative value is much higher than the fellowship (we are ignoring the benefit of the education itself).

Each time a new consulting assignment is offered, the student is facing a real-time scheduling problem — should he/she quit graduate school or not? Note that the overall objective is to maximize the value obtained over the entire period of the fellowship no matter what the flow of consulting offers.

In general there is no way to ensure obtaining the best possible outcome in online problems but one would like to eliminate the possibility of a fiasco — a terribly low value compared to the best possible value. This is the goal of competitive analysis is to protect against such disasters.

[1] The *slack time* of a task is the time till its deadline minus its remaining computation time. Hence, the slack time of a task is a measure of its urgency — a task with zero slack time would have to be scheduled immediately for it to hope to meet its deadline.

Competitive analysis is the study of strategies that give a performance guarantee no matter what the external circumstances. The goal is a guaranteed fraction of the best possible value in all possible developments. This is to be compared with heuristics that, in general, may perform well in many cases but in some cases may give poor results. For an algorithm that achieves the best possible competitive guarantee, by contrast, no other algorithm could possibly obtain a higher value in all cases.

33.3 Competitive Analysis: Application for RT Scheduling

In real-time computing systems, correctness may depend on the completion time of tasks as much as on their logical input/output behavior. Tasks in real-time systems have deadlines. If the deadline for a task is met, then the task is said to have *succeeded*. Otherwise it is said to have *failed*.

Real-time systems may be categorized by how they react when a task fails. In a *hard real-time system*, a task failure is considered intolerable. The underlying assumption is that a task failure would result in a disaster, e.g., a fly-by-wire aircraft may crash if the altimeter is read a few milliseconds too late.

A less stringent class of systems is denoted as *soft real-time systems*. In such systems, each task has a positive value. The goal of the system is to obtain as much value as possible. If a task succeeds, then the system acquires its value. If a task fails, then the system gains less value from the task [2]. In a special case of soft real-time systems, called a *firm* real-time system [5], there is no value for a task that has missed its deadline, but there is no penalty either.

An *online* scheduling algorithm is one that is given no information about a task before its release time. Different task models can differ in the kind of information (and its accuracy) given upon release. Intuitively, we assume the following: when a task is released its value, required computation time, and deadline are known precisely. Also, preemption is allowed and task switching takes no time.

The *value density* of a task is its value divided by its computation time. The *importance ratio* of a collection of tasks is the ratio of the largest value density to the smallest value density. When the importance ratio is 1, the collection is said to have *uniform value density*, i.e., a task's value equals its computation time. We will denote the importance ratio of a collection by k.

As in [18–20] we quantify the performance guarantee of an online scheduler by comparing it with a *clairvoyant* [2, p. 39] scheduling algorithm (also called an offline scheduler). A clairvoyant scheduler has complete *a priori* knowledge of all the parameters of all the tasks. A clairvoyant scheduler can choose a "scheduling sequence" that will obtain the maximum possible value achievable by any scheduler.[2]

As in [18–20] we say that an online algorithm has a *competitive factor* r, $0 < r \leq 1$, *if and only if* it is guaranteed to achieve a cumulative value of at least r times the cumulative value achievable by a clairvoyant algorithm on *any* set of tasks. For convenience of notation, we use *competitive multiplier* as the figure of merit. The *competitive multiplier* is defined to be "one over the competitive factor." The smaller the competitive multiplier is, the better the guarantee is. Our goal is to devise online algorithms whose guarantee is the best possible.

Unlike the models described here, other work assumes a slightly different model. For example, the online competitive guarantee is limited to a restrictive set of inputs (for example, all tasks have some minimal laxity upon release [22,23]) or to a model in which the online scheduler gets some additional strength (for example, a faster processor). Kalyanasundaram and Pruhs [22] applied the weak adversary model of *resource augmentation* to analyze the uniprocessor scheduling problem. In resource augmentation analysis the online scheduler is given more resources, say faster processor or more processors, than the clairvoyant adversary. They describe an online algorithm that with a $(1 + \epsilon)$ faster processor achieves a constant (i.e., independent of importance ratio) competitive multiplier which approaches 1 as ϵ increases. Baruah and Harita [23] show that in the uniform value density case doubling the processor speed lets the online scheduler acquire the entire value obtained by a clairvoyant scheduler even in cases of overload.

---

[2]Finding the maximum achievable value for such a scheduler, even in the uniprocessor case, is reducible from the knapsack problem [21], hence is NP-hard.

33.4 Notation and Assumptions

We shall make the following assumptions:

1. *Task Model.* A task may enter the system at any time. Its value, computation time, and deadline are known precisely at the time of arrival (the assumption of exact knowledge is weakened in [24]). Nothing is known about a task before it appears. (We relax this in section 33.7 that studies an offline model).
2. *Upper Bound on Importance.* We assume that an upper bound on the possible importance ratio is known *a priori* and can be used by the online scheduler (this bound is denoted by k). For the uniprocessor case, this assumption can be relaxed [25], but this is unknown for the multiprocessor case. Recall that the importance ratio is the ratio of value densities.
3. *Normalized Importance.* Without loss of generality, assume that the smallest value density of a task is 1. Hence if a collection of tasks has *importance ratio* of k, the highest possible value density of a task in that collection is k.
4. *Task Switching Takes No Time.* A task can be preempted and another one scheduled for execution instantly. This simplifying assumption holds in situations in which all process states are held in memory and time slices are long compared to switch times.

In uniprocessor environments we add the following assumption:

No Overloaded Periods of Infinite Duration. We assume that overloaded periods of infinite duration will not occur. This is a realistic assumption since overload is normally the result of a temporary emergency or failure.

Indeed, in the uniprocessor case, Baruah et al. [26] showed that there is no competitive online algorithm when overloaded periods of infinite duration are possible.[3] Note that the number of tasks may be infinite provided no infinitely long overload period is generated.[4]

In multiprocessor environments we add the following assumption:

Identical Processors. All processors have the same speed and all tasks can be scheduled on any of the processors.[5]

33.5 Uniprocessor Environments

33.5.1 Lower Bound Results

Baruah et al. [18,26] demonstrated, using an adversary argument that, in the uniform value density setting, there can be *no* online scheduling algorithm with a competitive multiplier smaller than 4. They further showed, for environments with an importance ratio k, a lower bound of $(1 + \sqrt{k})^2$ on the possible competitive multiplier of an online scheduler. These results and their extensions are described in the rest of this section.

In proving bounds such as those mentioned above, one usually refers to the online algorithm under consideration as the player. The bounds are best described as a game between a player and an adversary

[3] Intuitively, the adversary can generate a sequence of tasks with ever growing values. This will force any competitive scheduler to abandon the current task in favor of the next one and so on. If the competitive scheduler attempts to complete a task in favor of a new larger one, then the adversary completes the larger one. In either case, the online schedule will result in a small value compared with an arbitrarily large value for a clairvoyant scheduler.
This problem does not arise in the multiprocessor case.

[4] For the definition of overloaded periods, see [27, Section 3.3].

[5] Leung [28] studied the case of *uniform processors* (processors of different speed), See Section 33.6.

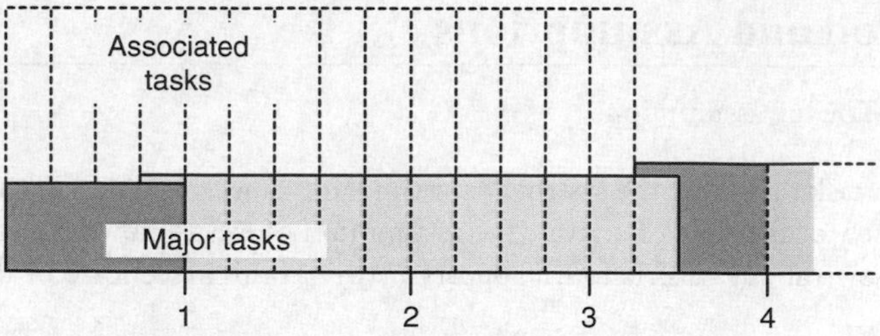

FIGURE 33.1 The lower bound game. Here the online player decided to preempt the first major task (of length 1) in favor of the second major task (of length 3). However, the player did not preempt the second major task in favor of the third (of length 8) so the adversary offers no more associated tasks after the completion of the second major task. In this game the online algorithm achieved a value of 3, the clairvoyant algorithm scheduled to completion all associated tasks and the third major task to obtain a total value of approximately 12.

who makes up part of the task set, observes the player's response to it, and then extends the history by creating a new task (see Figure 33.1).[6]

This process is repeated until the entire task set is complete. At the end of this process, the adversary indicates its schedule, the optimal offline schedule.

For the game in this case, the tasks created by the adversary are of two kinds:

1. **Major Tasks**, which are relatively long in duration and have no slack time.
2. **Associated Tasks**, which are very short in duration and hence have low value and no slack time.

The adversary uses the following device to force the hand of the player. For a major task S_i of length L_i, with deadline d_i, the adversary may also create a sequence of associated tasks of length ϵ, each one being released at the deadline of the previous one of the sequence and before the deadline of S_i. Clearly any algorithm that schedules even one task of this associated task sequence cannot schedule S_i. This sequence of associated tasks stops the moment the player chooses to abandon S_i in favor of a task of the sequence. Otherwise, the sequence continues until d_i is reached. If the player chose to abandon S_i in favor of a task of the sequence, the value obtained by the player is ϵ rather than L_i. The adversary chooses ϵ to be arbitrarily small compared to L_i. A major task S_i that has associated tasks as above is called a *bait*. Otherwise it is simply called *normal*.

Time is divided into *epochs*. In each epoch, the adversary starts off by first creating a major task T_0 of length $t_0 = 1$. In general, after releasing major task T_i of length t_i, the adversary releases a major task T_{i+1} of length t_{i+1}, at time ϵ before the deadline of T_i. If the player ever chooses to schedule an associated task, then no more tasks are released. If the player chooses to abandon T_i in favor of T_{i+1}, this process continues. If the player sticks with T_i, no tasks are released after T_{i+1} is released. In the above description, all tasks except T_{i+1} are baits; T_{i+1} is normal. At any rate, no epoch continues beyond the release of T_m, where m is a finite positive integer.

We note that the player never abandons a bait for one of its associated tasks, since in doing so the value obtained by the player during the epoch is negligible, i.e., ϵ. Thus, during an epoch the player either schedules only $T_i, i < m$ to completion, or the player schedules only T_m to completion.

Theorem 33.1

Given uniform value density, there does not exist an online scheduling algorithm with a competitive multiplier smaller than 4 (*i.e., a competitive factor greater than* 1/4).

[6]Hence, we assume an *adaptive offline* adversary [29]. The adversary is allowed to see the previous responses of the online scheduler before releasing any new tasks.

Proof

For this proof the length of task T_{i+1} is computed according to

$$t_{i+1} = c \cdot t_i - \sum_{j=0}^{i} t_j$$

where c is a constant whose exact value will be specified later in this proof. If the player scheduled only T_i, $i < m$ to completion, the player's value is t_i, whereas the adversary obtains value arbitrarily close to $\sum_{j=0}^{i+1} t_j$ (by performing the associated tasks for $T_0, \ldots, T_i$ and then performing T_{i+1}). In this case the player's value is $1/c$ times the adversary's value. If the player scheduled only T_m to completion, the player's value is t_m, while the adversary's value is arbitrarily close to $\sum_{j=0}^{m} t_j$. If c and m can be chosen such that the ratio $t_m / \sum_{j=0}^{m} t_j$ is no larger than $1/c$, then in either case the player obtains no more than $1/c$ times the adversary's value. In attempting to provide the tightest bound on the competitive factor of an online algorithm, therefore, our attempt is to find the smallest $1/c$ (equivalently, the largest c) such that the series defined by the recurrence relation

$$t_0 = 1 \quad \text{and} \quad t_{i+1} = c \cdot t_i - \sum_{j=0}^{i} t_j$$

satisfies the property

$$\exists m\colon m \geq 0 : \frac{t_m}{\sum_{j=0}^{m} t_j} \leq \frac{1}{c}$$

Standard techniques from the theory of difference equations can be used to show that the property is satisfied when $c < 4$, and that the property is not satisfied when $c \geq 4$. It therefore follows that 4 is a lower bound on the competitive multiplier that can be archived by any online scheduling algorithm in an overloaded environment. □

The next question to address is — how does the importance ratio k affect the best possible guarantee? Unfortunately, the guarantees that can be made by an online algorithm gets weaker as k grows. The lower bound proof for environments with $k > 1$, follows the proof technique of Theorem 33.1. The game works as follows: the major tasks have the lowest possible value density and the associated tasks have the highest possible value density (the graduate student conundrum).

Theorem 33.2

For an environment with importance ratio k, no online scheduling algorithm operating in this environment can have a competitive multiplier smaller than $(1 + \sqrt{k})^2$.

It is interesting to ask how the above lower bound may be refined for special environments. For example, what is the bound when all tasks necessarily have some amount of slack time upon their release? What is the bound for environments of lighter load than the load presented in the lower bound proof?[7]

A partial answer to these questions was given in Baruah et al. [18,26] using the notion of *loading factor*. According to their definition, an environment has a *loading factor b* iff it is guaranteed that there will be no interval of time $[t_x, t_y)$ such that the sums of the execution times of all task-requests making requests and having deadlines within this interval is greater than $b(t_y - t_x)$. For example, the loading factor of the task set in the game above equals 2. When the load factor is smaller or equal to 1, the system is underloaded

[7]For environments of higher load, the bound cannot be lowered because it is matched by the D^{over} algorithm for all environments, and D^{over} is probably optimal.

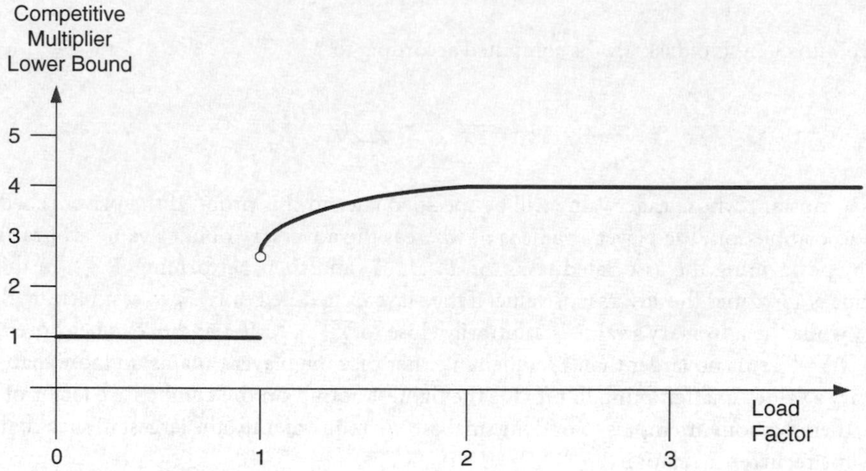

FIGURE 33.2 Lower bound on the competitive multiplier as function of load factor.

and a competitive multiplier of 1 is achievable. For the uniform value density case, as the load exceeds 1, the lower bound on the competitive multiplier rises to 2.597, and as the load increases from one to two, it rises to 4. Please see [18,26] for more information. (see Figure 33.2)

33.5.2 D^{over} Algorithm

In [30,31] we presented an online scheduling algorithm called D^{over} that has an optimal competitive multiplier of $(1 + \sqrt{k})^2$ for environments with importance ratio k. Hence, the lower bound is tight for all k. D^{over} also gives 100% of the value obtainable for periods in which the system is underloaded [27]. D^{over} can be implemented using balanced search trees, and runs at an *amortized* cost of $O(\log n)$ time per task, where n is the maximum over all time instants of the number of concurrent tasks in the system.

In the algorithm described below, there are three kinds of *events* (each causing an associated interrupt) considered:

1. Task Completion: Successful termination of a task. This event has the highest priority.
2. Task Release: Arrival of a new task. This event has low priority.
3. Latest-Start-Time Interrupt: The indication that a task must immediately be scheduled in order to complete by its deadline, i.e., the task's remaining computation time is equal to the time remaining until its deadline. This event has also low priority (the same as task release).

If several interrupts happen simultaneously, they are handled according to their priorities. A task completion interrupt is handled before the task release and latest-start-time (*LST*) interrupts, which are handled in random order. It may happen that a task completion event suppresses a lower priority interrupt, e.g., if the task completion handler schedules a task that has just reached its *LST* then the task scheduling will obviate the need for the latest-start-time interrupt.

At any given moment, the ready tasks that are not currently running are partitioned into two disjoint sets: *privileged* tasks and *waiting* tasks. Whenever a task is preempted it becomes a *privileged* task. However, whenever some task is scheduled as the result of a latest-start-time interrupt *all* the ready tasks (whether preempted or never scheduled) become *waiting* tasks.

D^{over}'s code is presented in Figure 33.3. The following is an intuitive description of the algorithm: as long as no overload is detected (i.e., there is no *LST* interrupt), D^{over} schedules in the same way as EDF. Tasks that are preempted during this phase in favor of a task with an earlier deadline become privileged tasks.

D^{over} maintains a special quantity called availtime. Suppose a new task is released into the system and its deadline is the earliest among all ready tasks. The value of availtime is the maximum computation time that can be taken by such a task without causing the current task or any of the privileged tasks to miss their deadlines.

D^{over} requires three data structures, called Q_privileged, Q_waiting, and Qlst. Each entry in these data structures corresponds to a task in the system. Q_privileged contains exactly the privileged tasks and Q_waiting contains the *waiting* tasks. These two structures are ordered by the tasks' deadlines. In addition, the third structure, Qlst, contains all tasks (again, not including the current task) ordered by their latest-start-times (*LST*).

These data structures support Insert, Delete, Min and Dequeue operations.

- The Min operation for Q_privileged or Q_waiting returns the entry corresponding to the task with the earliest deadline among all tasks in Q_privileged or Q_waiting. For Qlst the Min operation returns the entry corresponding to the task with the earliest *LST* among all tasks in the queue. The Min operation does not modify the queue.
- A Dequeue operation on Q_privileged (or Q_waiting) deletes from the queue the element returned by Min, in addition it deletes this element from Qlst. Likewise a Dequeue operation on Qlst will delete the corresponding element from either Q_privileged, if the element is a *privileged* task or from Q_waiting, if the element is a *waiting* task.

An entry of Q_waiting and Qlst consists of a single task, whereas an entry of Q_privileged is a 3-tuple (T, Previous-time, Previous-avail) where T is a task that was previously preempted at time Previous-time. Previous-avail is the value of the variable availtime at time Previous-time. All of these data structures are implemented as balanced trees (e.g., 2–3 trees).

In the following code,

- *Now*() is a function that returns the current time.
- *Schedule*(T) is a function that gives the processor to task T.
- *Laxity*(T) is a function that returns the amount of time the task has left until its deadline less its remaining computation time. That is, $laxity(T) = deadline(T) - (now() + remaining\_computation\_time(T))$.
- ϕ denotes the empty set.

This code includes lines manipulating *intervals*. The notion of an interval is needed for purpose of analysis only, so these lines are commented.

```
recentval    := 0      (* This will be the running value of privileged tasks. *)
availtime    := ∞
                       (* Availtime will be the maximum computation time that
                       can be taken by a new task without causing the current task
                       or the privileged tasks to miss their deadlines. *)
Qlst         := φ      (* All ready tasks, ordered according to their latest start time. *)
Q_privileged := φ      (* The privileged tasks ordered by deadline order *)
Q_waiting    := φ      (* All the waiting tasks ordered by their deadlines. *)
idle         := true   (* In the beginning the processor is idle *)

loop
```

FIGURE 33.3 D^{over}.

9 task completion :

10 if (both Q_privileged and Q_waiting are not empty) then

11 (* *Both queues are not empty and contain together all the ready tasks. The ready task with the earliest deadline will be scheduled unless it is a task of* Q_waiting *and it cannot be scheduled with* <u>all</u> *the privileged tasks. The first element in each queue is probed by the Min operation.* *)

12

13 $(T_{Q_privileged}, t_{prev}, avail_{prev}) := \mathrm{Min}(\mathrm{Q_privileged});$

14

15 (* *Next, compute the current value of* availtime*. This is the correct value because* $T_{Q_privileged}$ *is the task last inserted of those tasks currently in* Q_privileged; t_{prev} *is the time when* $T_{Q_privileged}$ *was preempted; and the available computation time has decreased by the time elapsed since this element was inserted to the queue.* *)

16

17 availtime := $avail_{prev} - (now() - t_{prev})$;

18 (* *Probe the first element of* Q_waiting *and check which of the two tasks should be scheduled.* *)

19 $T_{Q_waiting} := \mathrm{Min}(\mathrm{Q_waiting});$

20 if $d_{Q_waiting} < d_{Q_privileged}$ and

 availtime\geq remaining_computation_time($T_{Q_waiting}$) then

21 (* *Schedule the task from* Q_waiting. *)

22 Dequeue(Q_waiting);

23 availtime:= availtime − remaining_computation_time($T_{Q_waiting}$);

24 availtime:= min(availtime, laxity($T_{Q_waiting}$));

25 Schedule $T_{Q_waiting}$;

26 else

27 (* *Schedule the task from* Q_privileged. *)

28 Dequeue(Q_privileged);

29 recentval := recentval − value($T_{Q_privileged}$);

30 Schedule $T_{Q_privileged}$;

31 endif (* *which task to schedule.* *)

32 else if (Q_waiting is not empty) then

33 (* Q_privileged *is empty. The current interval is closed here,* $t_{close} = now()$. *The first task in* Q_waiting *is scheduled* *)

34

35 $T_{current} := \mathrm{Dequeue}(\mathrm{Q_waiting});$

36 availtime:= laxity($T_{current}$);

37 (* *A new interval is created with* $t_{begin} = now()$.*)

38

39 Schedule $T_{current}$;

40 else if (Q_privileged is not empty)

41 (* Q_waiting *is empty. The first task in* Q_privileged *is scheduled* *)

42

43 $(T_{current}, t_{prev}, avail_{prev}) := \mathrm{Dequeue}(\mathrm{Q_privileged});$

44 recentval := recentval − value($T_{current}$);

45 availtime := $avail_{prev} - (now() - t_{prev})$;

46 Schedule $T_{current}$;

47 else

48 (* *Both queues are empty. The interval is closed here,* $t_{close} = now()$. *)

FIGURE 33.3 (Continued).

```
            idle := true;
            availtime:= ∞;
        endif
    end(*task completion *)
    task release : (* T_arrival is released. *)
        if (idle ) then
            Schedule T_arrival;
            availtime:= laxity(T_arrival);
            idle := false;
            (* A new interval is created with t_begin = now().*)
        else (*T_current is executing *)
            if d_arrival < d_current and
            availtime≥ computation_time(T_arrival) then
              (* No overload is detected, so the running task is preempted. *)
              Insert T_current into Qlst;
              Insert (T_current, now(), availtime) into Q_privileged;
              (* The inserted task will be, by construction, the task with the earliest deadline
              in Q_privileged*)
              availtime:= availtime − remaining_computation_time(T_arrival);
              availtime:= min(availtime, laxity(T_arrival))
              recentval := recentval + value(T_current);
              Schedule T_arrival;
            else (* T_arrival has later deadline or availtime is not big enough.*)
              (* T_arrival is to wait in Q_waiting *)
              Insert T_arrival into Qlst and Q_waiting;
            endif
        endif(*idle *)
    end(*release *)

    latest-start-time interrupt :
        (* The processor is not idle and the current time is the latest start time of the first
        task in Qlst. *)

        T_next = Dequeue(Qlst);
        if (v_next > (1 + sqrt(k)) (v_current + recentval)) then
            (*v_next is big enough; it is scheduled. *)
            Insert T_current into Qlst and Q_waiting;
            Remove all privileged tasks from
            Q_privileged and insert them into Qlst and Q_waiting;
            (* Q_privileged = φ *)
            recentval := 0;
            availtime:= 0
            Schedule T_next;
        else (*v_next is not big enough; it is abandoned. *)
            Abandon T_next;
        endif
    end(*LST *)
    end{loop }
```

FIGURE 33.3 (Continued).

The task with the earliest deadline (either a newly released task or a *waiting* task) will be scheduled provided it does not cause overload when added to the privileged tasks. This proviso is always met in situations of underload. During overload, when a *waiting* task reaches its *LST*, it will cause a latest-start-time interrupt. This means that some task must be abandoned: either the task that reached its *LST* or some of the privileged tasks. The latest-start-time interrupt routine compares the value of that task against the *sum* of the values of all the privileged tasks. If its value is greater than $(1+\sqrt{k})$ times that sum, then this task will execute on the processor while all the privileged tasks will lose their privileged status to become *waiting* tasks (these tasks might later be successfully rescheduled). Otherwise the task reaching its *LST* is abandoned. A task T that was scheduled by a latest-start-time interrupt can be abandoned in favor of another task T' that reaches its *LST* but only if T' has at least $(1+\sqrt{k})$ times more value than T. D^{over} returns to schedule according to EDF when some task scheduled by its latest-start-time interrupt completes.

33.5.2.1 Analyzing D^{over} Performance

The reader may be curious to know why D^{over} compares *values* rather than *value densities* and why the values are compared using the magic factor of $(1+\sqrt{k})$. The lower bound proof [18,26] shows why value density cannot be a good criterion for choosing which task to abandon.[8] The factor of $(1+\sqrt{k})$ happened to be the one that gave the desired result since it yields the correct ratio between the minimal value gained by D^{over} and the maximal value that might have been missed.

For the purposes of analyzing D^{over}, the collection of tasks is partitioned according to the question of whether the task, under D^{over} scheduling, had completed exactly at its deadline or before its deadline or failed.

1. Let F denote the set of tasks that were abandoned.
2. Let S^p for successful with positive time before the deadline denote the set of tasks that completed successfully and that ended some positive time before their deadlines.
3. Let S^0 denote the set of tasks that completed successfully but ended exactly at their deadlines.

The proof strategy is to partition time into time intervals following D^{over} scheduling decisions. Each interval is a period in which the processor is non-idle. An interval starts when the processor becomes non-idle and ends when *Q_waiting* becomes empty. Each interval is associated with the tasks that were successfully completed or abandoned during it. The value obtained by D^{over} is therefore the value of the tasks successfully completed in all of the intervals.

A bound on the best possible value obtained by the clairvoyant scheduler on each interval is devised. The clairvoyant scheduler has possibly some freedom in deciding where to schedule tasks of S^p. On the other hand, tasks of F and S^0 are constrained to start executing during their respective intervals and possibly continue executing beyond the interval endpoint. The scheduling rules of D^{over} guarantee that the length of this "overflow" is bounded compared to the value obtained by D^{over} in the interval. Hence, by bounding the length of the interval and the extent of possible overflow we get a bound on the best possible value obtained by the clairvoyant from $S^0 \cup F$, summed over all intervals,

$$\text{Value of } (S^0 \cup F) \leq (1+\sqrt{k})^2 \times (\text{value achieved by } D^{over}) - (\text{value of all } S^p \text{ tasks})$$

In the best possible scenario for the clairvoyant it can also schedule to completion all tasks of S^p (in whatever interval possible). Hence

$$\text{Value of } (S^0 \cup F \cup S^p) \leq (1+\sqrt{k})^2 \times (\text{value achieved by } D^{over})$$

See [27,30] for complete analysis.

---

[8] In that proof, going after high value density tasks (the short *teasers*) will give the online scheduler minuscule value compared to the clairvoyant scheduler that will schedule a low value density task that has long computation time and hence big value. This is like our graduate student in Section 33.2.

33.6 Multiprocessor Environments

We describe here algorithms and lower bound results for multiprocessor scheduling of overloaded real-time systems. We consider two memory models: a shared memory model where thread *migration* is cheap, and a distributed memory model where thread migration is impractical. In both cases, we assume a centralized scheduler. In the first model, tasks can *migrate* cheaply (and quickly) from one processor to another. Hence, if a task starts to execute on one processor it can later continue on any other processor (and migration takes no time). In the second model (the *fixed* model), once a task starts to execute on one processor it cannot execute on any other processor. For both models, we assume that preemption within a processor takes no time. Main results are given below:

1. Inherent Bound on the Best Possible Competitive Multiplier

   In [32] we have shown that, for a system with n processors and maximal value density of $k > 1$, there is no online scheduling algorithm with competitive multiplier smaller than $\frac{k}{(k-1)}n(k^{1/n} - 1)$.

   When n tends to infinity this lower bound tends to $\frac{k}{(k-1)} \ln k$.

   This result holds even when migration is allowed. In fact, migration can help only when tasks have slack time — since all tasks of the lower bound game have no slack time, migration cannot help.
2. The MOCA Algorithm

   In [32] we presented an algorithm that does not use migration called the *MOCA algorithm: Multiprocessor Online Competitive Algorithm.* For a system with $2n$ processors and a maximum importance ratio of $k > 1$, this algorithm has a competitive guarantee of at most

$$1 + 2n \min_{(0\leq\omega<n;n=\omega+\psi)} \left\{ \max_{1\leq i\leq\psi} \frac{k^{i/\psi}}{\omega + \frac{(k^{i/\psi}-1)}{(k^{1/\psi}-1)}} \right\}$$

   When n tends to infinity this bound is at most $2 \ln k + 3$, which is within a small multiplicative constant factor from the lower bound for the same system.

   In [27,33] we presented an algorithm called the *Safe-Risky* algorithm, for two-processor systems with uniform value density (i.e., $n = 2$ and $k = 1$) that achieves the best possible competitive multiplier of 2 even when tasks may have slack time but migration is allowed.[9] For the no-migration model, a variant of this algorithm, called the *Safe-Risky-(fixed)*, achieves a competitive multiplier of 3.

   Recently, Leung [28] extended the above result to obtain a modified algorithm called *Safe-Risky** that achieves in the no-migration model an improved competitive multiplier of $2(1 + \frac{1}{\sqrt{6}}) \approx 2.82$.

   Leung also showed that the competitive multiplier for the migration model can be lowered by combining a fast processor with a slow processor as compared with using two parallel and identical processors having the same aggregate power. Specifically, for a two-processor system where one processor has speed $s > 1$ and the other has a speed of 1, an extension of the *Safe-Risky* algorithm achieves a competitive multiplier of $1 + \frac{1}{s}$.

In the rest of this section we give some insight into the lower bound result and the MOCA Algorithm.

33.6.1 The Lower Bound

We would first like to show that every online algorithm has a competitive multiplier of at least $\frac{k}{(k-1)}n(k^{1/n} - 1)$ for a system with n processors and importance ratio of k. As exemplified above for the uniprocessor case, we assume that a game is played between an adversary and the online scheduler.

[9] This was already known when tasks have no slack time [26,34].

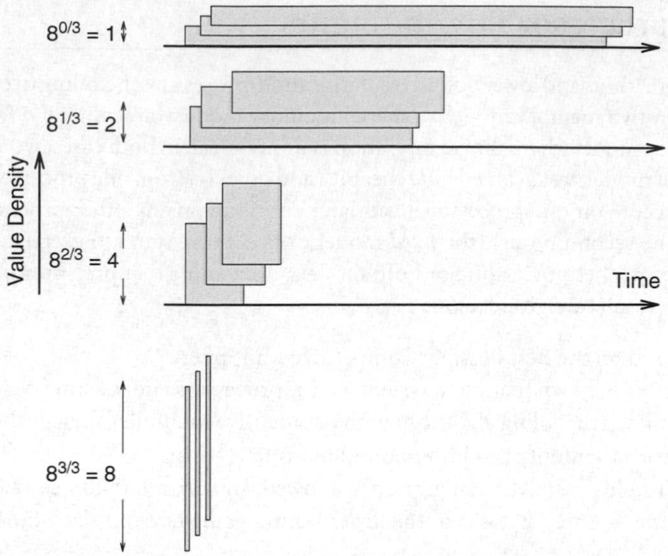

FIGURE 33.4 A complete set for $n = 3$ and $k = 8$. Note that there is one more level than processors and the levels correspond to value densities k^0, $k^{1/3}$, $k^{2/3}$, and $k^{3/3}$. Each level has as many tasks as there are processors. Finally, each lower value density task T is so much longer (some imagination is needed to see that in the drawing) than any higher value density task T' that the value of T is much greater than the value of T'.

We consider $n+1$ possible levels of value density $1, k^{1/n}, k^{2/n}, \ldots, k^{n/n} = k$, call them levels $0, 1, \ldots, n$. With each level we associate a period. A task of some value density level will have a computation time and deadline equal to the corresponding period. Hence, the value of a task of level i equals the length of the ith period times the ith value density. The length of the 0th level's period is set to 1. We choose all other periods in such a way that the value of an $(i + 1)$th level task is only a small fraction of the ith level task's value. In fact, we choose it so the $(i + 1)$th task's *"effective value density"* taken over the ith period is arbitrarily small (say ϵ for some small ϵ). A collection of tasks that has n identical tasks for each level, where all are released at the same time is called a *complete set*.[10] Figure 33.4 shows a complete set for a system with 3 processors and importance ratio of 8.

The adversary controls the release of tasks, making decisions after observing the actions (schedule) of the online algorithm so far. In the following we describe the game played by the adversary and the online scheduler.

The game is played in stages, the first one beginning at time 0. At the beginning of each stage the adversary releases a complete set of tasks. The adversary releases tasks only in complete sets and only at the beginning of a stage. The behavior of the online scheduler dictates *when* the adversary releases the next complete set (i.e, the beginning of the next stage). Denote by t_l the beginning of the lth stage. At time t_l (in particular at time 0), the online algorithm has to schedule a new complete set and possibly some previously released tasks. The number of possible scheduling decisions is vast. However, since the number of processors is smaller than the number of levels, at least one level is not represented in the online schedule (at time t_l). Let i_0 be an index of *some* level (to be specified later) that is not represented. Then, t_{l+1} is set to be the end of the current i_0th level period. This means that up to that time the adversary will release no new tasks. We will say that the stage starting at t_l is associated with level i_0. The game goes on in that manner for a large enough number of stages.

[10]Hence, a complete set has $n(n + 1)$ tasks.

Suppose that the stage starting at t_l is associated with level i_0, then what can the clairvoyant scheduler do? One possibility is to execute n tasks of level i_0 to completion between t_l and t_{l+1}. In this scheme, the clairvoyant scheduler schedules all the processors in the same way, no processor is ever idle, and all current tasks complete immediately before a new set is released.

The idea behind the lower bound game is that while the clairvoyant scheduler gets a value density of $k^{i_0/n}$ for the duration of the entire stage on *all* the processors. The online scheduler utilizes its processors either on lower value density tasks or on higher value density tasks that have very short duration (hence have little value). After the completion of these short high value density tasks, the associated processors will be left idle because no more tasks are released before the end of the stage.

The following theorem is proven using the game described above. The full proof can be found in [27,32].

Theorem 33.3

For a system with n processors and maximal value density of k, there is no online scheduling algorithm with competitive multiplier smaller than $\frac{k}{(k-1)}n(k^{1/n}-1)$.

Corollary 33.1

As the number of processors n tends to infinity, no online algorithm can have a competitive multiplier smaller than $\ln k$ *(natural logarithm).*

Remark 33.1

For $n = 1$ the lower bound is k which is not as good as the already known tight lower bound of $(1+\sqrt{k})^2$. For $k = 1$ a different treatment is needed. Bar-Noy et al. [35] study the case of $k = 1$ but in their model tasks have no slack time. They give a lower bound of $\approx 1/0.66 = 1.5$ on the best possible competitive multiplier. Bar-Noy et al. [35,36] devised online scheduling algorithms for this case with competitive multipliers of $4n/2n-1$ for an even number of processors n, and 2 for an odd number of processors.

33.6.2 Algorithmic Guarantees

Having proved the lower bound on the best possible competitive multiplier, we would like to have an online scheduler that achieves this bound. In the following we describe an algorithm that does so (up to a small multiplicative factor) in many cases.

The algorithm breaks the processors into *bands* (of two processors each) and one *central pool.* The main idea of the algorithm is to attempt to assign a task, upon its release, to the band corresponding to its value density. Tasks that are assigned to a band are guaranteed to complete and can all complete on a single processor. This means that they constitute a uniprocessor underloaded system and can be scheduled according to the EDF. Suppose the new task cannot be added to the band that corresponds to its value density (because it will cause overload at that band). Then the scheduler will determine whether the new task can be scheduled on the next band below (i.e, a band corresponding to lower value density). If the band below cannot accept the new task, the task will continue to *cascade* downwards. If a task cascades to the lowest band but still cannot be scheduled there, it can go into the central pool.

If a newly released task is accepted by one of the bands or by the central pool, it is guaranteed to complete before its deadline (these tasks are called *privileged*). If it is not, it awaits its *LST*, at which time it tries again to be scheduled in a way we explain below.

Throughout this section we assume a system with $2n$ processors. We break the processors into two disjoint groups: 2ψ processors will constitute a *band structure* and the other 2ω processors will constitute a *central pool* as described below ($n = \psi + \omega$ and $n > \omega \geq 0$).

We consider ψ intervals (*levels*) of value density $[1 \cdots k^{1/\psi})$, $[k^{1/\psi} \cdots k^{2/\psi})$, ..., $[k^{(\psi-1)/\psi} \cdots k]$, call these levels $1, \ldots, \psi$ respectively. The ith band is said to be "lower" than the $(i+1)$th band. Suppose the entire set of tasks to be scheduled is Γ. We partition this set according to the value density of the tasks: $\Gamma = \Gamma_1 \cup \Gamma_2 \cdots \cup \Gamma_\psi$ where Γ_i contains all tasks with value density in the range $[k^{(i-1)/\psi}, k^{i/\psi})$. We allocate two processors (*a band*) for each of the ψ value density levels. The remaining 2ω processors are allocated as a *central pool* that will be used by tasks of all levels.

The algorithm has three major components:

1. Upon task release, assign a task to a band if possible (possibly after cascading) making that task privileged.
2. At *LST* (of a nonprivileged task), decide whether and where the task should be scheduled or maybe abandoned.
3. The method used in scheduling each band (and the central pool).

Different choices for these three components would create different variants of the algorithm. One specific variant is presented in [32]. This variant is called the MOCA algorithm. In this variant, the central pool is also broken into bands of two processors each.[11] The MOCA algorithm schedules according to the following rules:

1. At each moment, every band has one of its processors designated as the *safe processor* (SP) and the other as the *risky processor* (RP). Each band has its own queue called *Q_privileged*, the tasks in *Q_privileged* are guaranteed to complete. In addition to the local *Q_privileged* queues there is one global queue called *Q_waiting*. This queue contains all the ready tasks that are not privileged.
2. When a new task T is released, it is assigned to a band as follows:
   a. It is added to the *Q_privileged* of its own band if this does not create overload (i.e, all tasks including the new task can complete on SP). Otherwise, T cascades downward.
   b. If T was not accepted by any band (including all the bands in the central pool) it enters *Q_waiting* where it waits until its *LST* occurs.

   So, at release time only the SPs are examined. A task might not be scheduled even if an RP is idle.[12]
3. A task T that reached its *LST* is assigned to a processor as follows:
   a. If there is any idle RP among all the lower level bands (including T's own level) then schedule T on one of these processors.[13]
   b. If there is no idle RP among lower level bands, we might abandon a task executing on one of these RPs in order to schedule T, based on the following rule:

      Let T^* be the task with the earliest deadline among all the tasks executing on these RPs.

      If T has a later deadline than T^* then abandon T^* and schedule T in its place; otherwise, abandon T.[14]
4. If, at a task completion event, the SP of a band becomes idle then the two processors should switch roles; the *safe-processor* becomes the *risky-processor* and vice versa.[15] Relabelling processors in this way does not require task migration.

Figure 33.5 is a schematic description of MOCA algorithm. The bands structure as described above gives priority to high value density tasks over low value density tasks. Higher value density tasks start their

[11]The bands of the central pool are ordered so that a task that reaches the pool starts with the first band in the pool and if not accepted it cascades to the second band and onwards. If the task is not accepted by the last band in the pool it awaits its *LST*.

[12]Using idle RPs and scheduling tasks of *Q_waiting* before they reach their *LST* is a heuristic that can improve the average case behavior of the scheduler.

[13]Heuristics can be used to choose the processor in case more than one RP is idle.

[14]If T is to be abandoned while there is an idle processor (above T's own band), scheduling T on an idle processor (with or without guaranteeing its completion) can only improve the average case behavior of the scheduler.

[15]The current task on SP (that was RP) becomes privileged.

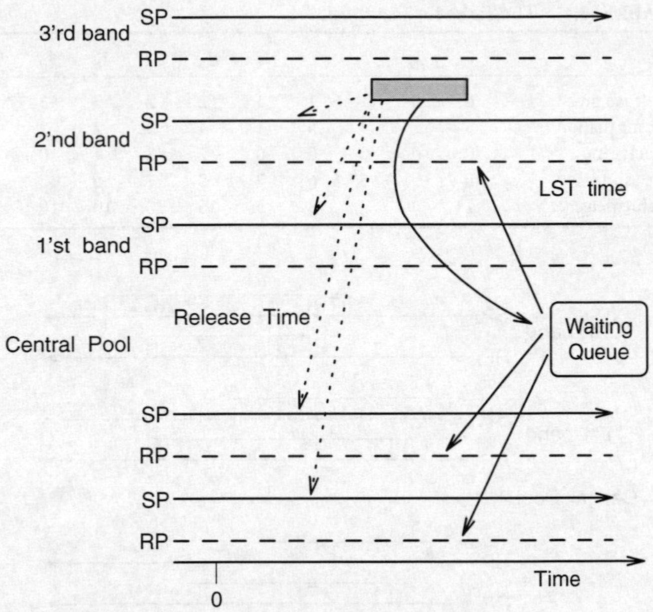

FIGURE 33.5 A schematic description of MOCA algorithm.
In this figure, the system has 10 processors divided into three bands and a central pool. At release time a task tries to be scheduled on one of the SPs starting with its own value density band. If unsuccessful, it awaits its *LST* in *Q_waiting*. At *LST* the task tries to be scheduled on one of the RPs, again starting from its own value density band.

cascading at a higher point and cascading is possible in only one direction: downwards.[16] An algorithm that uses the "pure" bands structure (i.e., with no central pool) can be crippled when the task set consists of mostly low value density tasks since all the higher bands will be left idle. In order to minimize the loss of such cases we add the central pool to the bands structure. If all tasks are of low value density then all high bands would still be left idle but the bands in the central pool would be utilized.

A big enough central pool will offset the damage caused by higher idle bands. However, making the central pool too big can cause another problem — weakening the advantages of the higher value density tasks. We conclude that choosing the right size of the central pool is a delicate and important aspect of the MOCA algorithm. An analogy to everyday life is to a well-balanced investment strategy: some money might be placed on high risk/high value stocks whereas most money should not.

Example 33.1

The following is a small example of MOCA's scheduling. Assume that the highest possible value density is 16, number of processors is 6 of which two are allocated as a central pool and the rest constitute two bands (i.e., $k = 16, 2n = 6, \psi = 2$, and $\omega = 1$). The first band will be for tasks with value density below 4 and the second for tasks with value density of 4 and above. For this example, consider the tasks depicted in Table 33.1. Figure 33.6 shows the schedule created by MOCA algorithm for these tasks.

The first two tasks to be released are scheduled on the SP of the first band and the central pool (T_2 cascades into the central pool). When T_3 is released it cannot be scheduled on an SP, so it is inserted into *Q_waiting* only to create an *LST* interrupt immediately. Then, it is scheduled on the RP of the first band. In the same way T_4 is scheduled on the RP of the central pool. But when T_5 arrives it can be scheduled neither on any of the SPs nor on any of the RPs, hence is abandoned (in the *LST* routine). Note that T_5

[16]Hence, higher value density tasks have more bands that can possibly accommodate them.

TABLE 33.1 The Tasks for Example 33.1

Task	T_1	T_2	T_3	T_4	T_5	T_6	T_7	T_8	T_9	T_{10}
Release time	0	0	1	1	1	2	2	3	3	6
Computation time	5	5	5	5	1	4	3	2	2	1
Slack time	0	0	0	0	0	2	2	1	0	1
Deadline	5	5	6	6	2	8	7	6	5	8
Value density	1	2	3	3	3	16	10	10	10	16

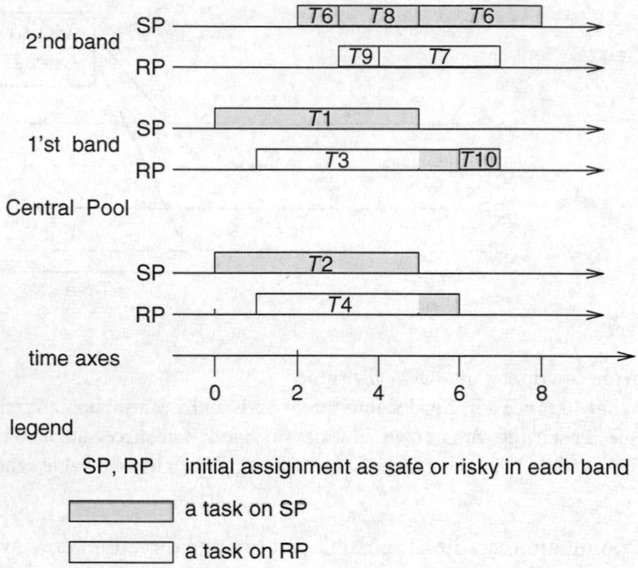

FIGURE 33.6 The MOCA algorithm scheduling for Example 33.1.

is abandoned even though the second band is idle (a task can cascade only downwards unless we invoke heuristics).

All the remaining tasks have value density high enough to be scheduled on the second band. T_6 is scheduled on the SP. T_7 cannot be scheduled on any of the SPs and it enters *Q_waiting* (with *LST* at 4). T_8 can be added to the SP of the second band preempting T_6 (which has a later deadline). T_9 cannot be scheduled on any of the SPs; it reaches its *LST* and is scheduled on the RP of the second band, but at time 4 it is abandoned in favor of T_7 which has reached its *LST* and has a later deadline.

At time 5, the SP of the first band becomes idle, which creates a switch of roles between the SP and RP of that band. Later at time 6, T_{10} is released; it cannot be scheduled on its own band's SP but after cascading it is scheduled on the new SP of the first band.

All in all, the MOCA algorithm completed all the tasks but T_5 and T_9. A clairvoyant scheduler could schedule all the tasks (T_5 can be scheduled on the idle SP and T_9 can be scheduled before its *LST* on the same processor). As we have pointed out in the notes above, heuristics should be used to increase the performance in practice. We have downplayed these heuristics because they do not improve the worst case guarantees.

33.6.2.1 The Algorithm's Competitive Analysis

In this section we would like to present the competitive multiplier of the MOCA algorithm. The idea behind the analysis is to bound the amount of value gained by the clairvoyant algorithm whenever the MOCA algorithm has to abandon a task. The bands structure and the scheduling strategy are such that whenever a task is abandoned then all the lower bands (and the pool bands) are "productive" which leads to a bound on the ratio between the best possible clairvoyant value and the value obtained by the MOCA algorithm. This bound is formalized in the theorem below [32].

Theorem 33.4

For a system with $2n$ processors and maximal value density of $k > 1$ the MOCA algorithm has a competitive multiplier of at most

$$1 + 2n \min_{(0 \leq \psi \leq n; n = \omega + \psi)} \left\{ \max_{1 \leq i \leq \psi} \frac{k^{i/\psi}}{\omega + \frac{(k^{i/\psi} - 1)}{(k^{1/\psi} - 1)}} \right\} \tag{33.1}$$

Remark 33.2

1. Note that the MOCA algorithm does not use migration (i.e., tasks cannot be moved once they are released), hence the previous result holds both whether migration is allowed or not.
2. When $k = 1$, there is no need for the bands structure, hence the central pool consists of all the processors (in our notation $\omega = n - 1$ and $\psi = 1$). This leads to a competitive multiplier of $2 + 1$ (when some tasks may have slack time). For $n = 2$ this corresponds to the results for two processor systems mentioned in the beginning of this section [27,33].
3. When the number of processor is odd, a similar result can be obtained. For a system with $2n + 1$ processors, create bands and pool from the first $2n$ processors. The leftover processor can be used, for example, as a second SP for one of the bands. This leads to a bound of

$$1 + (2n + 1) \min_{(0 \leq \psi \leq n; n = \omega + \psi)} \left\{ \max_{1 \leq i \leq \psi} \frac{k^{i/\psi}}{\omega + \frac{(k^{i/\psi} - 1)}{(k^{1/\psi} - 1)}} \right\}$$

   However, this result does not specialize to a uniprocessor system because at least two processors are needed to create a band.

33.6.2.2 Setting ψ

One can estimate the complex expression for the upper bound on the competitive multiplier given by Theorem 33.4 by setting[17] $\psi = n \frac{\ln k}{\ln k + 1}$ (hence $\omega = n \frac{1}{\ln k + 1} = \frac{\psi}{\ln k}$). This produces the following lemma:

Lemma 33.1

For a system with $2n$ processors and maximal value density of $k > 1$ the MOCA algorithm has a competitive multiplier of at most

$$1 + 2n(k^{1/\psi} - 1) \tag{33.2}$$

where $\psi = n \frac{\ln k}{\ln k + 1}$ (recall that the lower bound is bigger than $2n(k^{1/2n} - 1)$)

---

[17] ln is the natural logarithm.

33.7 Power of Migration in Offline Multiprocessor Real-Time Scheduling

In this section, we are concerned with the offline scheduling of multiprocessor real-time systems. A task is said to *migrate* if it can run on a different processor from the one it was preempted in. We would like to investigate the power of migration. There are both practical and theoretical reasons for such an investigation. From a practical perspective, systems without migration are less centralized and less complicated and incur smaller overhead. However, migration gives more power in processor utilization, in versatility, and in fault handling. It may also be easier to design algorithms and prove bounds on systems with migration.

In an attempt to understand the power of migration in real-time systems Koren, Dar, and Amir [37] posed the following question: *Does a clairvoyant scheduler without migration have the same power as a clairvoyant scheduler with migration?* If we can pinpoint the *performance ratio*, ω_n, (as defined below) between offline schedulers with migration and without, any c approximation algorithm for the optimal nonmigratory schedule becomes a $c\omega_n$ approximation algorithm for the optimal migratory schedule. Likewise, we will have gone a step forward in obtaining online schedulers as well. We will be able to limit ourselves to either the case of migration or no-migration. Any competitive online scheduling algorithm that will be obtained, will guarantee a competitive online scheduling algorithm for both cases.

Definition 33.1

A set of tasks S, in a system with n processors, has a performance ratio *of* $\omega_n(S) = \omega_n$ *if*

$$\omega_n = \frac{\{\textit{value obtained by an optimal migratory schedule for } S\}}{\{\textit{value obtained by an optimal no-migratory schedule for } S\}}$$

An offline no-migratory scheduling algorithm $\mathcal{A}$ *in a system with n processors, has a* performance ratio *of* $\omega_n(\mathcal{A}) = \omega_n$ *if for all S*

$$\omega_n \geq \frac{\{\textit{value obtained by an optimal migratory schedule for } S\}}{\{\textit{value obtained when } \mathcal{A} \textit{ schedules } S\}}$$

Hence, a smaller performance ratio implies better performance for $\mathcal{A}$.

33.7.1 Bounds on the Performance Ratio

The MOCA algorithm presented above does not use migration, but its competitive guarantee holds even when migration is allowed. It follows that the competitive guarantees of MOCA translates into upper bounds on the performance ratio (we are using MOCA as an offline algorithm setting aside the fact that it is an online algorithm). For the uniform value density case, MOCA has a performance ratio of 3, for the general case of $k > 1$ see Theorem 33.4. Note that when tasks have no slack time there is no advantage in migration.

For the *Uniform Value Density* case, [37] provided both an upper and lower bounds on the performance ratio of offline schedulers in a multiprocessor real-time system with n processors (Figure 33.7). They proved that

$$\frac{2n-1}{n} \leq \omega_n \leq \frac{1}{1-\left(1-\frac{1}{2n}\right)^n} < 3$$

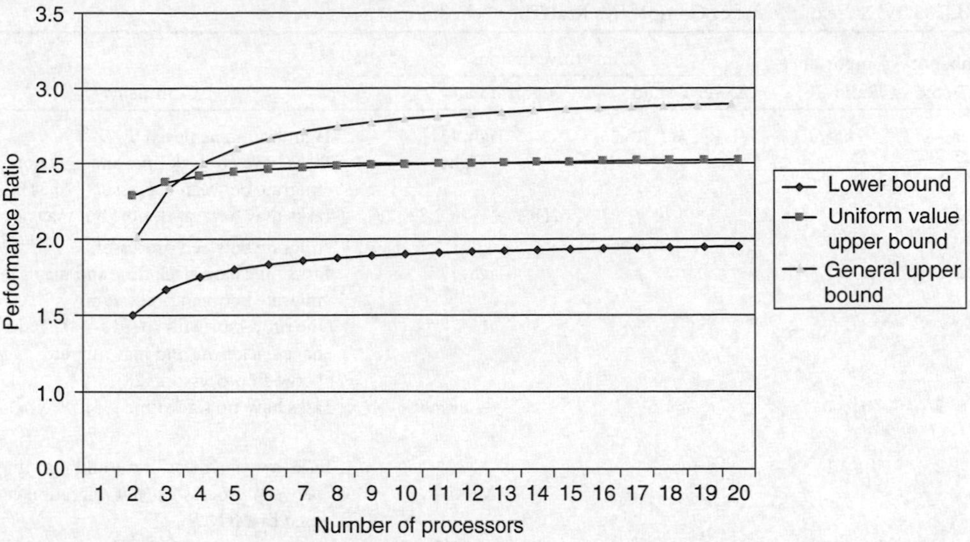

FIGURE 33.7 Lower and upper bounds. This graph shows a lower bound on the performance ratio, the upper bound for the uniform value density case and the upper bound for the general case.

Asymptotically (as $n \to \infty$) the lower bound approaches 2 and the upper bound approaches $\frac{\sqrt{e}}{\sqrt{e}-1} \sim 2.5414$.

Kalyanasundram and Pruhs [38] studied this problem from a different and a very interesting angle: suppose a set of tasks can be scheduled with migration on n processors, how many processors are needed to schedule these tasks without migration? Their result is that $3n - 2$ processors will do in all cases. This result is not limited to the uniform value density case and as such gives an upper bound of $\omega_n \leq \frac{n}{3n-2}$ for the nonuniform value density case.

33.8 Conclusion

As real-time algorithms proliferate in consumer products, the hard deadline model (i.e., missing a deadline is catastrophic) will give way to the firm deadline model (i.e., missing a deadline denies you some gain). Heuristic algorithms will make their best efforts to meet deadlines. However, it is reassuring to be able to *guarantee* that an algorithm can do as well as a clairvoyant algorithm within some small multiplicative factor. Such a guarantee is analogous to a certification in software engineering or a seal of approval from Consumer Reports. Here we summarize the results known to date. Many open problems remain. For example, if missing a deadline not only denies gain, but extracts a penalty, the algorithms described here do not directly help. We believe the ideas are similar, but that remains to be seen. We look forward to seeing far more work in this field.

Table 33.2 summarizes the current state of the art of competitive real-time scheduling. Here, n is the number of processors in the system; k is the *importance ratio*, that is the highest possible value per unit of computation time that any task can possibly obtain (normalizing the lowest to 1). The bounds are expressed in terms of *competitive multipliers*.

Table 33.3 summarizes the power of migration in offline real-time scheduling in terms of *performance ratio*.

TABLE 33.2 State of the Art of Competitive Real-Time Scheduling

Number of Processors	Importance Ratio	Competitive Bounds: Lower Bound	Competitive Bounds: Algorithmic	Comments
1	$k \geq 1$	$(1+\sqrt{k})^2$ [18,26]	Tight [31]	Tight bound achieved by D^{over}
2	1	2	Tight	Tasks have *no* slack time and may not migrate between processors [26,34].
2	1	2	$2(1+\frac{1}{\sqrt{6}}) \approx 2.82$ [28]	Tasks may have slack time but may not migrate between processors.
2	1	2	Tight [27]	Tasks may have slack time and may migrate between processors.
2	1		$1+\frac{1}{s}$	One processor with speed $s > 1$. Tasks may have slack time and may migrate between processors [28].
$n > 2$	1	$\approx$1.5	$\frac{4n}{2n-1}$ even n 2 odd n	Tasks have no slack time [35].
$n \geq 2$	1		3 [32] MOCA	Tasks may have slack time and may migrate between processors (but migration is not used by MOCA).
$n \geq 2$	$k > 1$	$\frac{k}{(k-1)} n(k^{1/n}-1)$ 33	$1+n(k^{1/\psi}-1)$ where $\psi = \frac{n}{2}\,\frac{\ln k}{\ln k+1}$	Exact algorithmic guarantee can be seen in Theorem 33.4 [32].
$n \gg 2$	$k > 1$	$\frac{k}{(k-1)} \ln k$	$2\ln k + 3$	Asymptotic behavior [32]

TABLE 33.3 Performance Ratio Results

Importance Ratio	Performance Ratio: Lower Bound	Performance Ratio: Algorithmic	Comments
1	$\frac{2n-1}{n}$	$\frac{1}{1-\left(1-\frac{1}{2n}\right)^n}$	[37]
1	2	$\frac{\sqrt{e}}{\sqrt{e}-1} \sim 2.5414$	Asymptotic behavior
$k > 1$		$\frac{n}{3n-2}$	[38]

References

[1] M. L. Dertouzos, Control robotics: the procedural control of physical processes, *Proceedings IFIF Congress*, 1974, pp. 807–813.

[2] A. K.-L. Mok, Fundamental design problems of distributed systems for the hard real-time environment, Ph.D. thesis, Massachutsetts Institute of Technology, Boston, MA, May 1983.

[3] C. D. Locke, Best-effort decision making for real-time scheduling, Ph.D. thesis, Carnegie-Mellon University, Pittsburgh, PA, 1986.

[4] K. Ramamritham and J. A. Stankovic, Dynamic task scheduling in hard real-time distributed systems, *IEEE Software* (1984), 65–75.

[5] J. R. Haritsa, M. J. Carey, and M. Livny, On being optimistic about real-time constraints, *Proceedings of the PODS Conference* (Nashville, TN), ACM, April 1990, pp. 331–343.

[6] K. Schwan and H. Zhou, Dynamic scheduling of hard real-time tasks and real-time threads, *IEEE Transactions on Software Engineering* **18** (1992), no. 8, 736–748.

[7] G.C. Buttazzo and J. Stankovic, Adding robustness in dynamic preemptive scheduling, In D.S. Fussel and M. Malek (editors), *Responsive Computer Systems: Steps Toward Fault-Tolerant Real-Time Systems*, Kluwer, Dordrecht, 1995.

[8] T. P. Baker and Alan Shaw, The cyclic executive model and ada, *The Journal of Real-Time Systems* **1** (1989), no. 1, 7–25.

[9] L. Sha, J. P. Lehoczky, and R. Rajkumar, Solutions for some practical problems in prioritized preemptive scheduling, *Proceedings of the 7th Real-Time Systems Symposium* (New Orleans, LA), IEEE, December 1986, pp. 181–191.

[10] E. L. Lawler and C. U. Martel, Scheduling periodically occurring tasks on multiple processors, *Information Processing Letters* **12** (1981), no. 1, 9–12.

[11] R. Rajkumar, L. Sha, and J. P. Lehoczky, Real-time synchronization protocols for multiprocessors, *Proceedings of the 9th Real-Time Systems Symposium*, IEEE, December 1988, pp. 159–269.

[12] J. A. Stankovic and K. Ramamritham, The spring kernel: A new paradigm for real-time systems, *IEEE Software* (1991), 62–72.

[13] H. Zhou, K. Schwan, and I. F. Akyildiz, Performance effects of information sharing in a distributed multiprocessor real-time scheduler, *Proceedings of the 13th Real-Time Systems Symposium* (Phoenix, Arizona), IEEE, December 1992, pp. 46–55.

[14] L. Sha and S. S. Sathaye, A systematic approach to designing distributed real-time systems, *IEEE Computer* (1993), 68–78.

[15] M. L. Dertouzos and A. K.-L. Mok, Multiprocessor on-line scheduling of hard-real-time tasks, *IEEE Transactions on Software Engineering* **15** (1989), no. 12, 1497–1506.

[16] N. Audsley and A. Burns, Real time system scheduling, Computer Science Department Technical Report No. YCS134, York University, UK, January 1990.

[17] S.-C. Cheng, J. A. Stankovic, and K. Ramamritham, Scheduling algorithms for hard real-time systems: A brief survey, J. A. Stankovic and K. Ramamritham, (editors), *Hard Real-Time Systems: Tutorial*, IEEE, 1988, pp. 150–173.

[18] S. Baruah, G. Koren, B. Mishra, A. Raghunathan, L. Rosier, and D. Shasha, On-line scheduling in the presence of overload, *Proceedings of the 32nd Annual Symposium on the Foundations of Computer Science* (San Juan, Puerto Rico), IEEE, October 1991, pp. 101–110.

[19] A. Karlin, M. Manasse, L. Rudolph, and D. Sleator, Competitive snoopy caching, *Algorithmica* **3** (1988), no. 1, 79–119.

[20] D. Sleator and R. Tarjan, Amortized efficiency of list update and paging rules, *Communications of the ACM* **28** (1985), 202–208.

[21] M. R. Garey and D. S. Johnson, *Computers and intractability: a guide to the theory of np-completeness*, W. H. Freeman, New York, 1979.

[22] B. Kalyanasundaram and K. Pruhs, Speed is as powerful as clairvoyance, *Journal of the ACM* **47** (2000), no. 4, 617–643.

[23] S. Baruah and J. Harita, Scheduling for overloaded in real time systems, *IEEE Transactions on Computers* (1997), no. 46, 1034–1039.

[24] G. Koren and D. Shasha, D-over: An optimal on-line scheduling algorithm for overloaded real-time systems, Computer Science Department Technical Report No. 594, Courant Institute, NYU, New York, January 1992; Technical Report No. 138, INRIA, Rocquencourt, France, Feb. 1992.

[25] B. Schieber, Private communication, I.B.M, T.J. Watson Research Center, Yorktown Heights, NY, 1992.

[26] S. Baruah, G. Koren, D. Mao, B. Mishra, A. Raghunathan, L. Rosier, D. Shasha, and F. Wang, On the competitiveness of on-line task real-time task scheduling, *The Journal of Real-Time Systems* **4** (1992), no. 2, 124–144; *Proceedings of the 12th Real-Time Systems Symposium* (San Antonio, TX), Dec. 1991, pp. 106–115, .

[27] G. Koren, Competitive on-line scheduling for overloaded real-time systems, Ph.D. thesis, Computer Science Department, Courant Institute, NYU, New York, September 1993.

[28] J. Y-T Leung, Improved competitive algorithms for two-processor real-time systems, (Unpublished).

[29] S. Ben-David, A. Borodin, R. Karp, G. Tardos, and A. Wigderson, On the power of randomization in on-line algorithms, *Proceedings of the 22nd Annual ACM Symposium on Theory of Computing* (Baltimore, MD), ACM, MAY 1990, pp. 379–386.

[30] Gilad Koren and Dennis Shasha, D-over: An optimal on-line scheduling algorithm for overloaded real-time systems, *SIAM Journal on Computing* **24** (1995), no. 2, 318–339.

[31] Gilad Koren and Dennis Shasha, D-over: An optimal on-line scheduling algorithm for overloaded real-time systems, *Proceedings of the 13th Real-Time Systems Symposium* (Phoenix, Arizona), IEEE, December 1992, pp. 290–299.

[32] Gilad Koren and Dennis Shasha, MOCA: A multiprocessor on-line competitive algorithm for real-time system scheduling, *Theoretical Computer Science* (1994), no. 128, 75–97. (Special Issue on Dependable Parallel Computing)

[33] Gilad Koren and Dennis Shasha, Competitive algorithms and lower bounds for on-line scheduling of multiprocessor real-time systems, Computer Science Department Technical Report No. 639, Courant Institute, NYU, New York, June 1993.

[34] F. Wang and D. Mao, Worst case analysis for on-line scheduling in real-time systems, Department of Computer and Information Science Technical Report No. 91-54, University of Massachusetts, Amherst, MA, June 1991.

[35] A. Bar-Noy, R. Canetti, S. Kutten, Y. Mansour, and B. Schieber, Bandwidth allocation with preemption, *27th ACM Symposium on Theory of Computing*, 1995, pp. 616–625.

[36] A. Bar-Noy, Y. Mansour, and B. Schieber, Private communication, I.B.M, T.J. Watson Research Center, Yorktown Heights, NY, 1992.

[37] G. Koren, A. Amir, and E. Dar, The power of migration in multi-processor scheduling of real-time systems, *SIAM Journal on Computing* **30** (2000), no. 2, 511–527.

[38] B. Kalyanasundaram and K. Pruhs, Eliminating migration in multi-processor scheduling, SODA: *ACM-SIAM Symposium on Discrete Algorithms* (A Conference on Theoretical and Experimental Analysis of Discrete Algorithms), 1999.

[39] G. C. Buttazzo, *Hard real-time computing systems, predictable scheduling algorithms and applications*, Kluwer Academic Publishers, Dordrecht, 1997.

[40] G. Koren, D. Shasha, and S.-C. Huang, MOCA: A multiprocessor on-line competitive algorithm for real-time system scheduling, *Proceedings of the 14th Real-Time Systems Symposium* (Raleigh-Durham, NC), IEEE, December 1993, pp. 172–181.

[41] C. L. Liu and J. Layland, Scheduling algorithms for multiprogramming in a hard real-time environment, *Journal of the ACM* **20** (1973), no. 1, 46–61.

34

Minimizing Total Weighted Error for Imprecise Computation Tasks and Related Problems

Joseph Y-T. Leung
New Jersey Institute of Technology

34.1 Introduction

Scheduling problems with due date related objectives are usually concerned with penalties such as the weighted number of late jobs (i.e., $\sum w_j U_j$), or the weighted amount of time between the completion time of the late job and its due date (i.e., $\sum w_j T_j$). In some applications, however, it is more meaningful to consider penalties involving the weighted number of tardy units (i.e., the weighted number of time units that are late), regardless of how late these units are. This is the case, for example, in a computerized control system, where data are collected and processed periodically. Any data that are not processed before the arrival of the next batch will be lost, and the lost data will have a negative effect on the accuracies of the calculations that are used to control the real-time process. Another example can be found in processing perishable goods, such as harvesting. In this case, jobs represent different stretches of land that need to be harvested. Because of differences in climate and soil conditions and crop culture, the different stretches need to be harvested during different time periods. Crops will perish after its due date, which will cause financial loss. In this application, minimizing the weighted number of tardy units is more meaningful than the other objectives.

Blazewicz [1] was the first to study this problem. He formulated the problem as follows. We are given m parallel processors and n jobs. Each job j has a ready time r_j, due date d_j, processing time p_j, and weight w_j. With respect to a schedule, a job is said to be *late* if it is completed after its due date; otherwise, it is said to be *on-time*. The number of tardy units of job k is the amount of processing of job k done after its due date, and is denoted by Y_k. The problem is to find a schedule such that the total weighted number of tardy units (i.e., $\sum w_j Y_j$) is minimized. As an extension of the $\alpha \mid \beta \mid \gamma$ notation, we denote the nonpreemptive

1-58488-397-9/$0.00+$1.50

version of this problem as $P \mid r_j \mid \sum w_j Y_j$ and the preemptive version as $P \mid pmtn, r_j \mid \sum w_j Y_j$. For the unweighted case, the above two problems will be denoted by $P \mid r_j \mid \sum Y_j$ and $P \mid pmtn, r_j \mid \sum Y_j$, respectively.

The Imprecise Computation Model [2–5] was introduced in the real-time systems community to allow for the trade-off of the accuracy of computations in favor of meeting the deadline constraints of jobs. In this model, each job is logically composed of two subjobs, mandatory and optional. The optional subjob cannot start until the mandatory subjob has finished execution. Mandatory subjobs are required to complete by their deadlines, while optional subjobs can be left unfinished. If a job has an unfinished optional subjob, it incurs an error equal to the processing time of its unfinished portion. The goal is to find a schedule that minimizes the total weighted error. The Imprecise Computation Model is particularly suitable to model iterative algorithms, where the mandatory subjob corresponds to the work needed to set up an initial solution and the optional subjob corresponds to the iterations used to improve the quality of the solution. In this model, it is clear that mandatory subjobs must finish by its deadline, while optional subjobs need not.

Each imprecise computation job j is represented by two subjobs: mandatory (M_j) and optional (O_j). Both jobs have ready times r_j, due date d_j, and weight w_j. M_j has processing time m_j while O_j has processing time o_j. Let $p_j = m_j + o_j$.

The problem of minimizing the total weighted number of tardy units is a special case of imprecise computation in which the processing times of the mandatory subjobs are zero. For a single processor, an algorithm for the total weighted number of tardy units can be used to solve the imprecise computation problem. This can be done by setting the weight of each mandatory subjob to be higher than any optional subjob. Because the mandatory subjobs have higher weights than any optional subjob, the mandatory subjobs are guaranteed to complete (if there is a feasible schedule to complete all the mandatory subjobs). Note that this method cannot be used in multiprocessor systems since a job's mandatory subjob and optional subjob cannot be executed in parallel on different processors.

Blazewicz [1] developed a linear programming solution for the problem $P \mid pmtn, r_j \mid \sum w_j Y_j$, thereby establishing polynomial time complexity of the problem. He also extended the linear programming approach to solve $Qm \mid pmtn, r_j \mid w_j Y_j$. However, the algorithm for uniform processors is only polynomial time for each fixed m. Later, Blazewicz and Finke [6] formulated a minimum-cost-maximum-flow solution to solve both $P \mid pmtn, r_j \mid \sum w_j Y_j$ and $Q \mid pmtn, r_j \mid \sum w_j Y_j$. Using Orlin's $O(|A| \log |V| (|A| + |V| \log |V|))$-time algorithm for the minimum-cost-maximum-flow problem [7], where A and V denote the edge set and vertex set, respectively, $P \mid pmtn, r_j \mid \sum w_j Y_j$ and $Q \mid pmtn, r_j \mid \sum w_j Y_j$ can be solved in $O(n^4 \log n)$ and $O(m^2 n^4 \log mn + m^2 n^3 \log^2 mn)$ times, respectively. These algorithms will be described in the next section.

For a single processor, Hochbaum and Shamir [8] gave an $O(n \log n)$-time algorithm for $1 \mid pmtn, r_j \mid \sum Y_j$ and an $O(n^2)$-time algorithm for $1 \mid pmtn, r_j \mid \sum w_j Y_j$. Later, Leung et al. [9] gave an even faster algorithm for $1 \mid pmtn, r_j \mid \sum w_j Y_j$ that runs in $O(n \log n + kn)$ time, where k is the number of distinct weights. We shall describe this algorithm in the next section.

Potts and van Wassenhove [10] gave an $O(n \log n)$-time algorithm for $1 \mid pmtn \mid \sum Y_j$ and showed that $1 \mid\mid \sum Y_j$ is NP-hard in the ordinary sense. They also gave a pseudo-polynomial time algorithm for $1 \mid\mid \sum Y_j$. Based on the pseudo-polynomial time algorithm, they later gave two fully polynomial approximation schemes for this problem [11].

In the real-time community, Chung et al. [12] gave a network flow approach to solve the total error problem for imprecise computation; their algorithm runs in $O(n^2 \log^2 n)$ time. For a single processor, Shih et al. [13] gave an $O(n^2 \log n)$-time algorithm for the weighted case, and an $O(n \log n)$-time algorithm for the unweighted case.

In [13], Shih et al. proposed an added constraint (called the 0/1-constraint) to be put on the Imprecise Computation Model, where each optional subjob is either fully executed or entirely discarded. This added constraint is motivated by some applications. For example, many jobs can be solved by either a fast or a slow algorithm, with the slow algorithm producing better quality results than the fast one. Due to deadline constraints, it might not be possible to execute the slow algorithm for every job. The problem of scheduling jobs with primary (slow algorithm) and alternate (fast algorithm) versions can be transformed into one of scheduling with 0/1-constraint [14]. The processing time of the mandatory subjob is the processing time of the fast algorithm, while the processing time of the optional subjob is the difference between the processing times of the slow algorithm and the fast one.

With the 0/1-constraint, two problems were proposed in [13]: (1) minimize the total error and (2) minimize the number of *imprecisely scheduled* jobs (i.e., jobs whose optional subjobs are discarded). For a single processor, Shih et al. [13] showed that minimizing the total error is NP-hard and minimizing the number of imprecisely scheduled jobs is polynomial-time solvable if the optional subjobs have identical processing times. Ho et al. [15] later showed that minimizing the total error is solvable in pseudo-polynomial time, while minimizing the number of imprecisely scheduled jobs can be solved in $O(n^9)$ time.

Motivated by the computational complexity of the problems, Ho et al. [15] proposed two approximation algorithms, one for minimizing the total error and the other for minimizing the number of imprecisely scheduled jobs. Both algorithms have time complexity $O(n^2)$. The one for minimizing the total error has a worst-case bound of 3, which is tight. The one for minimizing the number of imprecisely scheduled jobs has a worst-case bound of 2, which is also tight. Interestingly, the number of precisely scheduled jobs in an optimal schedule is also at most twice the number produced by the algorithm. Both algorithms will be described in Section 34.3.

The problem of minimizing the total weighted number of tardy units has been extended to flow shops and open shops [16–18]. Blazewicz et al. [17] considered the problem $F2 \mid d_i = d \mid \sum w_j Y_j$; see also [18]. They showed that the problem is NP-hard in the ordinary sense and gave a pseudo-polynomial time algorithm for it. Blazewicz et al. [16] also considered the open shop problem. They showed that $O \mid pmtn, r_j \mid \sum w_j Y_j$ and $O2 \mid d_j = d \mid \sum Y_j$ are both polynomial-time solvable, while $O2 \mid d_j = d \mid \sum w_j Y_j$ is NP-hard in the ordinary sense. We shall describe these results in Section 34.4.

Minimizing the maximum weighted number of tardy units has also been studied. Ho et al. [19] gave an $O(n^2)$-time algorithm for a single processor and an $O(n^3 \log^2 n)$-time algorithm for multiprocessors. They also considered other dual criteria optimization problems [19]. These results are described in the next chapter.

34.2 Total Weighted Tardy Units

In this section we shall concentrate on identical and uniform processors. We first show that $1 \mid\mid \sum Y_j$ is NP-hard in the ordinary sense; see also [10]. We then give an $O(n \log n + kn)$-time algorithm for $1 \mid pmtn \mid \sum w_j Y_j$, where k is the number of distinct weights; see also [9]. Finally, we describe a network flow approach to solve $P \mid pmtn \mid \sum w_j Y_j$ and $Q \mid pmtn \mid \sum w_j Y_j$ as well as for imprecise computation jobs; see also [6,12].

Theorem 34.1

$1 \mid\mid \sum Y_j$ is NP-hard in the ordinary sense.

Proof

We shall reduce the Partition problem (see Chapter 2 for a definition) to $1 \mid\mid \sum Y_j$. Given an instance $A = (a_1, a_2, \ldots, a_n)$ of the Partition problem, we construct an instance of $1 \mid\mid \sum Y_j$ as follows. There are $n + 1$ jobs. Job j, $1 \leq j \leq n$, has processing time a_j and due date $B = \frac{1}{2} \sum a_j$, while job $n + 1$ has processing time B and due date $2B$. The threshold for $\sum Y_j$ is B.

Since the first n jobs have (common) due date B and job $n+1$ has due date $2B$, by a simple interchange argument, we may assume that the completion time of job $n + 1$ in an optimal schedule is $2B$ or later. The total tardy units from the first n jobs is at least B, since their common due date is B and their total processing times is $2B$. Thus, if we have a schedule with $\sum Y_j \leq B$, then there must be no tardy units from job $n + 1$ which implies that it completes at time $2B$. This means that the total tardy units from the first n jobs is exactly B, which implies that a partition exists. Conversely, if there is a partition, it is easy to see that there is a schedule with $\sum Y_j \leq B$. □

Potts and Van Wassenhove [10] have given a pseudo-polynomial time algorithm for $1 \,||\, \sum Y_j$, which forms the basis of two fully polynomial approximation schemes [11].

We now present an $O(n \log n + kn)$-time algorithm for $1 \,|\, pmtn, r_j \,|\, \sum w_j Y_j$. The algorithm given below solves the problem $1 \,|\, pmtn, r_j \,|\, \sum Y_j$, which forms the basis for the weighted case. The algorithm only schedules the nontardy units of a job, assuming that the tardy units are either not scheduled or scheduled at the end. Let the jobs be ordered in decreasing order of release times. If several jobs have identical release times, they are ordered in decreasing order of due dates. Further ties can be broken arbitrarily. Let $0 = u_0 < u_1 < \cdots < u_p = \max\{d_j\}$ be the $p + 1$ distinct integers from the multiset $\{r_1, \ldots, r_n, d_1, \ldots, d_n\}$. These $p + 1$ integers divide the time frame into p segments: $[u_0, u_1]$, $[u_1, u_2]$, ..., $[u_{p-1}, u_p]$. The algorithm uses an $n \times p$ matrix S to represent a schedule, where $S(i, j)$ contains the number of time units job i is scheduled in segment j (i.e., $[u_{j-1}, u_j]$). (As we shall see below, we can also use a doubly linked list to represent a schedule.)

Algorithm NTU

(1) **For** $j = 1, \ldots, p$ **do:** $l_j \leftarrow u_j - u_{j-1}$.
(2) **For** $i = 1, \ldots, n$ **do:**
  Find a satisfying $u_a = d_i$ and b satisfying $u_b = r_i$.
  For $j = a, a - 1, \ldots, b + 1$ **do:**
    $\delta \leftarrow \min\{l_j, p_i\}$.
    $S(i, j) \leftarrow \delta, l_j \leftarrow l_j - \delta, p_i \leftarrow p_i - \delta$.
  repeat
  repeat

Algorithm NTU schedules jobs in decreasing order of release times. When a job is scheduled, it is assigned from the latest segment $[u_{a-1}, u_a]$ until the earliest segment $[u_b, u_{b+1}]$, with the maximum number of time units assigned in each segment.

Let us examine the time complexity of algorithm NTU. The time it takes to sort the jobs into decreasing order of release times as well as obtaining the integers $u_0 < u_1 < \cdots < u_p$ is $O(n \log n)$. Step 1 takes linear time and a straightforward implementation of Step 2 takes $O(n^2)$ time. Thus, it appears that the running time of the algorithm is $O(n^2)$. However, observe that whenever a value of some $S(i, j)$ is increased, either all the units of a job are scheduled or a segment is saturated (or both). Hence, at most $n + p - 1 = O(n)$ values of $S(i, j)$ will be positive in the solution. If we can avoid scanning all those pairs (i, j) for which $S(i, j) = 0$, then Step 2 will only take linear time and hence the overall running time of the algorithm is $O(n \log n)$. As it turns out, this can be done by the special UNION-FIND algorithm due to Gabow and Tarjan [20].

A schedule produced by algorithm NTU will be denoted as NTU-schedule. Define a *block* as a maximal time interval in which there is only one job assigned (job block) or the processor is idle (idle block). Without any increase in time complexity, the algorithm can be modified to produce a schedule represented by a doubly linked list of blocks. The proof of the following lemma can be found in [9].

Lemma 34.1

The number of blocks in a NTU-schedule is no more than $2n + 1$.

We can use algorithm NTU to solve the weighted case as follows. Sort the jobs in nonincreasing order of their weights. Let π_i, $1 \leq i \leq n$, denote the number of nontardy units of job i in an optimal schedule. Once the values of π_i are known, an optimal schedule can be obtained in $O(n \log n)$ time by the earliest due date rule. We determine these values in phases as follows. After j phases, it has already determined the values $\pi_1, \pi_2, \ldots, \pi_j$. In the $(j + 1)$th phase, it uses algorithm NTU to solve the unweighted problem for jobs $1, 2, \ldots, j+1$, where the processing times of the first j jobs are $\pi_1, \ldots, \pi_j$, and the processing time of the $(j+1)$th job is p_{j+1}. Let x be the number of tardy units in the NTU-schedule obtained. π_{j+1} is then set to be $p_{j+1} - x$, and the algorithm proceeds to the next phase.

The above algorithm makes n calls to algorithm NTU. Since algorithm NTU takes linear time after the initial sorting, the running time of the algorithm becomes $O(n^2)$. Note that we can assign x tardy units to job $j+1$ because the first j jobs can be scheduled on time.

The drawback of the above approach is that jobs are scheduled one by one, which slows down the algorithm. To speed up the algorithm, we need to schedule several jobs simultaneously, say all jobs with identical weights. However, if we schedule several jobs together, then it is not clear how to assign tardy units to the jobs. The trick here is to find a way to allocate tardy units to the new jobs so that the previously scheduled jobs remain intact.

Let there be k different weights, $w_1 > w_2 > \cdots > w_k$, and let T_j, $1 \le j \le k$, be the set of all jobs with weight w_j. We use T to store the jobs (and their processing times) whose nontardy units have already been determined; initially, T is an empty set. Let S be an empty schedule. The jobs are scheduled in phases. At the end of the jth phase, the algorithm has already determined the nontardy units of the jobs in $T_1, \ldots, T_j$. These jobs and their processing times are stored in T. In the $(j+1)$th phase, the algorithm uses algorithm NTU to construct a schedule S_{j+1} for $T \cup T_{j+1}$. It then goes through an adjustment step (described below), transforming S_{j+1} into S'_{j+1} with S as a template. The adjustment step is needed to allocate nontardy units to the new jobs (i.e., jobs in T_{j+1}) in such a way that the jobs previously scheduled (i.e., jobs in T) remain intact. We now set T to be $T \cup T_{j+1}$ and the processing times of the jobs in T are set to be the same as the nontardy units in S'_{j+1}. Finally, we apply algorithm NTU to T to obtain the schedule S before we go to the next phase. We repeat this process until the kth phase is finished.

The adjustment step proceeds as follows. Let there be q blocks in S: $V_i = [v_{i-1}, v_i]$, $1 \le i \le q$. S_{j+1} is transformed block by block from V_1 to V_q. Transformation is applied only to the job blocks in S, but not the idle blocks. Let V_i be a job block in S and let job l be scheduled in the block. Let $N(l)$ (resp. $N_{j+1}(l)$) denote the number of time units job l has executed in S (respectively S_{j+1}) from the beginning until time v_i. If $N(l) > N_{j+1}(l)$, then assign $(N(l) - N_{j+1}(l))$ more time units to job l within V_i in S_{j+1}, by removing any job, except job l, that was originally assigned in V_i. (Note that this reassignment can always be done.) Otherwise, no adjustment is needed.

Figure 34.1 gives a set of jobs with two distinct weights. The schedule S after the first phase is shown in Figure 34.1(a). S_2 and S'_2 are shown in Figures 34.1(b) and 34.1(c), respectively. Finally, the schedule S after the second phase is shown in Figure 34.1(d), which is an optimal schedule for the set of jobs.

Algorithm WNTU

(1) Let there be k distinct weights, $w_1 > \cdots > w_k$, and let T_j be the set of jobs with weight w_j.
(2) Let S be an empty schedule and T be an empty set.
(3) **For** $j = 1, \ldots, k$ **do:**

```
    S_j ← schedule obtained by algorithm NTU for T ∪ T_j.
    Begin (Adjustment Step)
        Let there be q blocks in S: V_i = [v_{i-1}, v_i], 1 ≤ i ≤ q.
        For i = 1,...,q do:
            If V_i is a job block in S, then
                Let job l be executed within V_i in S. Let N(l) (respectively N_j(l)) be the
                number of time units job l has executed in S (respectively S_j) from the
                beginning until time v_i.
                If N(l) > N_j(l), then
                    assign (N(l) − N_j(l)) more time units to job l within V_i in S_j, by
                    replacing any job, except job l, that was originally assigned within V_i.
                endif
            endif
        repeat
```

J_i	r_i	d_i	p_i	w_i
J_1	6	15	5	5
J_2	4	8	3	5
J_3	2	6	3	5
J_4	0	7	3	5
J_5	10	14	3	2
J_6	4	5	1	2
J_7	3	10	6	2
J_8	0	17	3	2
J_9	0	5	4	2

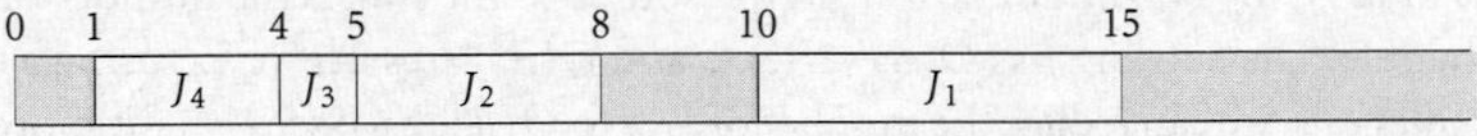

(a) Schedule S obtained after the first phase;

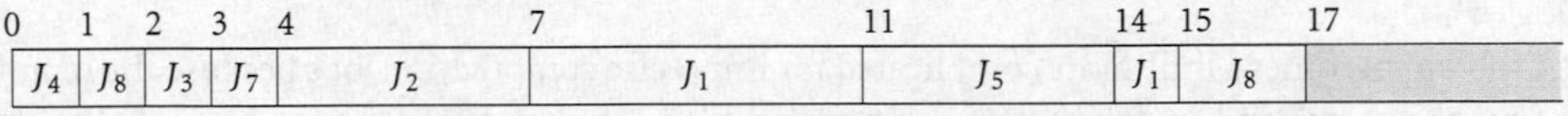

(b) Schedule S_2 obtained in the second phase;

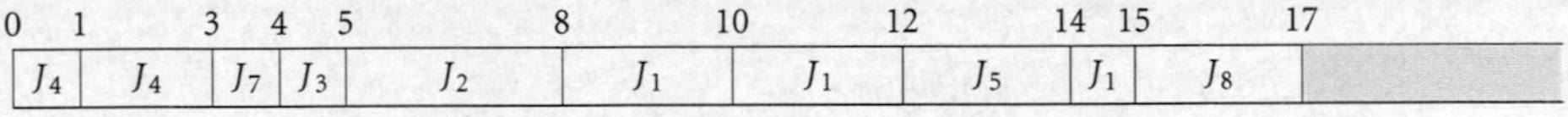

(c) Schedule S_2' obtained in the second phase;

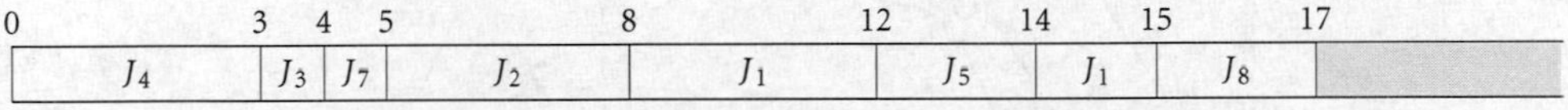

(d) Final schedule obtained by algorithm WNTU;

FIGURE 34.1 An example illustrating algorithm WNTU.

End
$S_j' \leftarrow S_j$.
Set the processing time of each job in T_j to be the number of nontardy units in S_j'.
$T \leftarrow T \cup T_j$.
$S \leftarrow$ schedule obtained by algorithm NTU for T.
repeat

Let us examine the time complexity of algorithm WNTU. Observe that algorithm WNTU utilizes algorithm NTU to construct schedules for various subsets of the jobs. Algorithm NTU requires that the release times and due dates of the jobs be ordered. With an initial sort of the release times and due dates of the jobs, we can obtain in linear time an ordering of the release times and due dates for various subsets of jobs. Once the jobs are ordered, algorithm NTU requires only linear time to construct a schedule.

Steps (1) and (2) of algorithm WNTU takes $O(n \log n)$ time and Step (3) is iterated k times. If we can show that each iteration of Step (3) takes linear time (after an initial sort), then the overall running time of algorithm WNTU becomes $O(n \log n + kn)$. It is clear that every substep in Step (3), with the possible exception of the adjustment step, takes linear time. We now show that the adjustment step can be implemented in linear time. As mentioned before, algorithm NTU can be implemented, with no increase in time complexity, to produce a schedule represented by a doubly linked list of blocks. Thus, we may assume that S and S_j are in this representation. The adjustment process is performed by traversing the two linked lists, modifying S_j if necessary, as the lists are traversed. By Lemma 34.1, the number of blocks is linear. The values $N(l)$ and $N_j(l)$ can be obtained with the help of two one-dimensional arrays L and L': $L(l)$ (respectively $L'(l)$) contains the number of time units job l has executed in S (respectively S_j) since the beginning. L and L' initially have zero in each entry, and they are updated as the linked lists are traversed. Thus, the adjustment process takes linear time.

Theorem 34.2

There is an $O(n \log n + kn)$ time implementation of algorithm WNTU.

We now consider multiprocessor systems. The problem $P \mid pmtn, r_j \mid \sum Y_j$ can be solved by the network flow approach as in [21]; the maximum flow from the source to the sink is the number of nontardy units in an optimal schedule. The network has $O(n)$ vertexes and $O(n^2)$ arcs. However, using a balanced binary tree to represent intervals, we can reduce the number of arcs to $O(n \log n)$; see [12] for more details. We can use the algorithm described in [22] to find the maximum flow with running time $O(|V||A| \log |V|)$, where V and A denote the vertex and arc sets, respectively. Thus, the running time of the algorithm is $O(n^2 \log^2 n)$.

The problem becomes a little more complicated for imprecise computation jobs because the mandatory subjobs must be executed in full and the optional subjobs cannot be executed in parallel with their corresponding mandatory subjob. Nonetheless, it can still be solved by a network flow approach. Figure 34.2

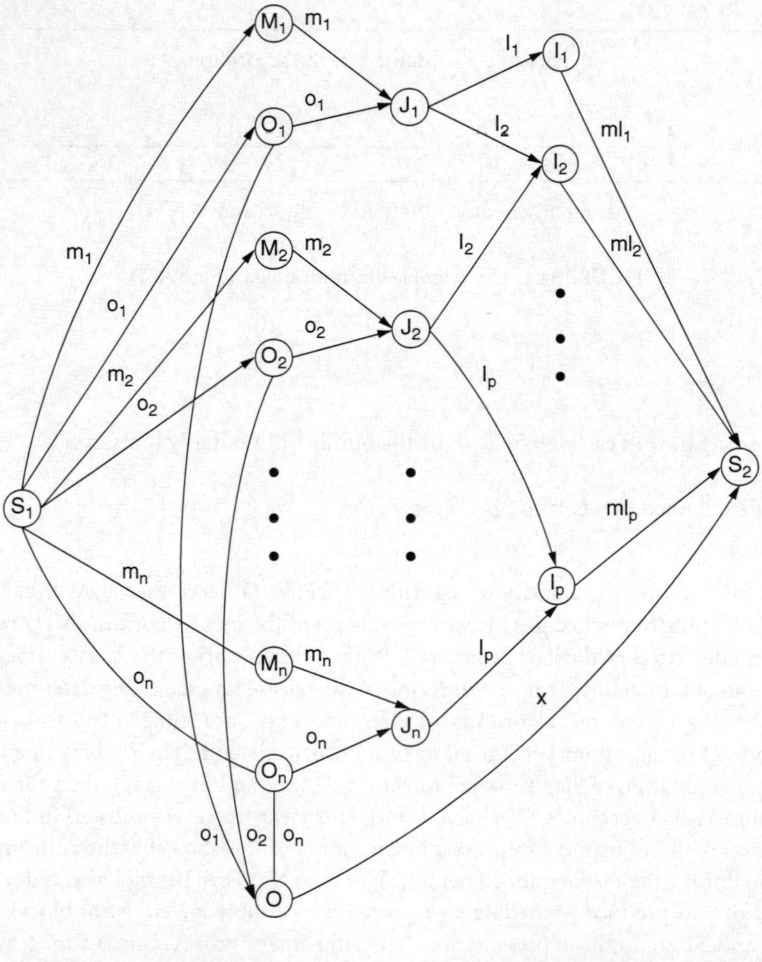

FIGURE 34.2 Network for minimizing total error for imprecise computation jobs.

shows such a network which we will denote by $G(x)$; x is the flow capacity on the arc from vertex O to the sink S_2. The source and sink are represented by S_1 and S_2, respectively. Each job k has three vertexes — M_k, O_k, and J_k. There are arcs from the source to M_k and O_k, $1 \leq k \leq n$, with flow capacities set to m_k and o_k, respectively. Both vertexes M_k and O_k have an arc to J_k, with flow capacities set to m_k and o_k, respectively. In addition, each vertex O_k has an arc to a new vertex O and the flow capacity on the arc is o_k. Each time interval $[u_{i-1}, u_i]$, $1 \leq i \leq p$, is represented by an interval vertex I_i. There is an arc from each J_k to each interval vertex in which it can execute; the flow capacity on the arc is the length of the interval. Finally, each interval vertex has an arc to the sink and the flow capacity on the arc is m times the length of the interval. Vertex O has an arc to the sink with flow capacities x which is a parameter that we can set.

Consider the network $G(0)$. Suppose we replace the flow capacity on each arc that has a flow capacity o_j, $1 \leq j \leq n$, by zero. Let this network be denoted by $G'(0)$. It is clear that $G'(0)$ is the network for mandatory subjobs only. First, we find the maximum flow in $G'(0)$. If the maximum flow is less than $\sum m_j$, then the set of jobs is infeasible. So we may assume that the maximum flow is exactly $\sum m_j$. Next, we find the maximum flow in $G(0)$, say u. If $u = \sum p_j$, then there is a schedule with no error; otherwise, the total error is $\epsilon = \sum p_j - u$. We now set x to be ϵ. The maximum flow in $G(\epsilon)$ is $\sum p_j$ and the flow pattern gives a schedule with total error ϵ. The network has $O(n)$ vertexes and $O(n^2)$ arcs; again, we can reduce the number of arcs to $O(n \log n)$ by using a balanced binary tree to represent the time intervals. Thus, the running time is again $O(n^2 \log^2 n)$.

The problem $P \mid pmtn, r_j \mid \sum w_j Y_j$ can be solved by a minimum-cost-maximum-flow approach. Consider a network constructed as follows. Let the source and sink be represented by S_1 and S_2, respectively. Each job k will be represented by a job vertex J_k. There is an arc from S_1 to each J_k with the maximum flow on the arc equal to p_k and the cost of the flow equal to zero. For each segment $[u_{i-1}, u_i]$, $1 \leq i \leq p$, we create an interval vertex I_i that corresponds to the segment. Each job vertex J_k has an arc to each of the interval vertexes in which job k can execute. The maximum flow on each of these arcs is the length of the interval (i.e., $l_i = u_i - u_{i-1}$ for the interval vertex I_i) and the cost of the flow is zero. In addition, each job vertex J_k has an arc to S_2 with the maximum flow equal to p_k and the cost equal to w_k. Finally, each interval vertex I_i has an arc to S_2 with the maximum flow equal to ml_i and the cost equal to zero, where m is the number of processors. The maximum flow in this network is $\sum p_j$. The minimum-cost-maximum-flow yields a schedule with the minimum weighted number of tardy units.

Figure 34.3 shows the network described above. There are $O(n)$ vertexes and $O(n^2)$ arcs in the network; again, the number of arcs can be reduced to $O(n \log n)$. Using Orlin's minimum-cost-maximum-flow algorithm, $P \mid pmtn, r_j \mid \sum w_j Y_j$ can be solved in $O(n^2 \log^3 n)$ time.

We can combine the ideas in the networks in Figures 34.2 and 34.3 to solve the total weighted error problem for imprecise computation jobs. The running time of the algorithm is again $O(n^2 \log^3 n)$.

The minimum-cost-maximum-flow approach can be extended to solve the uniform processors case, by using the idea in [23] to handle the different speeds of the processors. Assume that processing speeds are ordered in such a way that $s_1 > s_2 > \cdots > s_m$ (the case of identical speeds for some processors can also be handled [23]). In each interval I_i, no job can obtain more than $s_1 l_i$ units of processing, two jobs not more than $(s_1 + s_2)l_i$ units, and so on. The network is the same as in Figure 34.3, except that each interval vertex I_i is expanded into m vertexes, $(I_i, 1), (I_i, 2), \ldots, (I_i, m)$. If job k can execute in the segment $[u_{i-1}, u_i]$, then job vertex J_k has an arc to each vertex interval (I_i, j), $1 \leq j \leq m$, with maximum flow equal to $(s_j - s_{j+1})l_i$ and cost equal to zero. We assume that $s_{m+1} = 0$. Finally, each interval vertex (I_i, j) has an arc to the sink with maximum flow equal to $j(s_j - s_{j+1})l_i$ and cost equal to zero.

Figure 34.4 shows the network described above. There are $O(mn)$ vertexes and $O(mn^2)$ arcs; it is not clear if we can reduce the number of arcs in this case. Using Orlin's minimum-cost-maximum-flow algorithm, $Q \mid pmtn, r_j \mid \sum w_j Y_j$ can be solved in $O(m^2 n^4 \log mn)$ time.

We can also combine the ideas in the networks in Figures 34.2 and 34.4 to solve the total weighted error problem on uniform processors. The running time is again $O(m^2 n^4 \log mn)$.

Finally, we note that $Q \mid pmtn, r_j \mid \sum Y_j$ can be solved in $O(m^2 n^3 \log mn)$ time. This is because for the unweighted case, we can simply find the maximum flow in a network as in the case of $P \mid pmtn, r_j \mid \sum Y_j$. The time saving is due to the fact that we do not need to find a minimum cost flow.

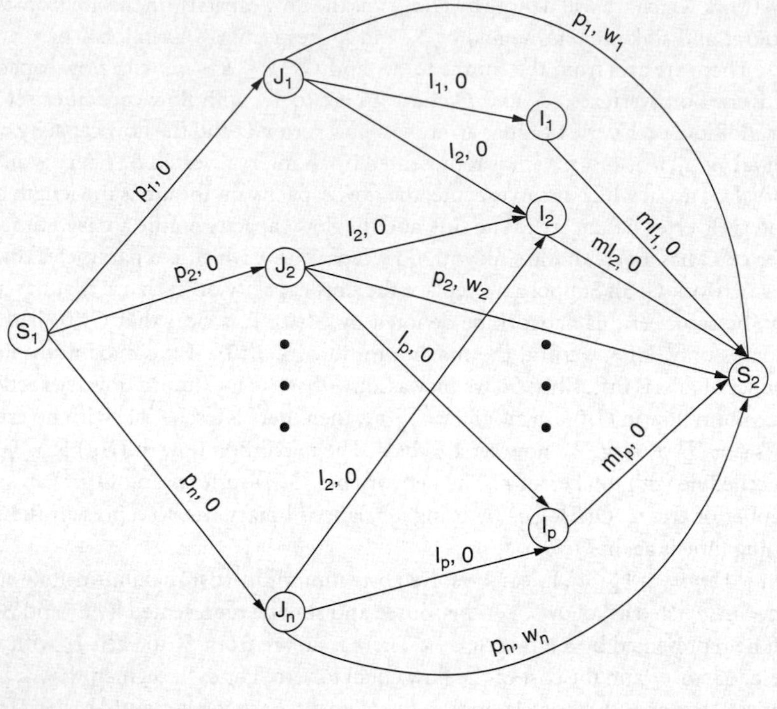

FIGURE 34.3 Network for minimizing the weighted number of tardy units on identical processors.

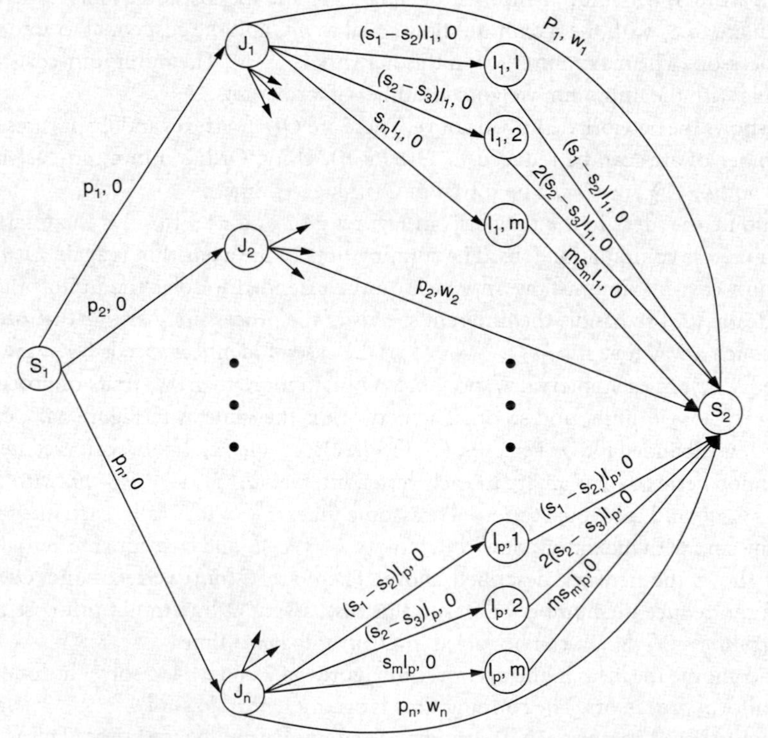

FIGURE 34.4 Network for minimizing the weighted number of tardy units on uniform processors.

Theorem 34.3

$P \mid pmtn, r_j \mid \sum Y_j$ and $Q \mid pmtn, r_j \mid \sum Y_j$ can be solved in $O(n^2 \log^2 n)$ and $O(m^2 n^3 \log mn)$ times, respectively, and $P \mid pmtn, r_j \mid \sum w_j Y_j$ and $Q \mid pmtn, r_j \mid \sum w_j Y_j$ can be solved in $O(n^2 \log^3 n)$ and $O(m^2 n^4 \log mn)$ times, respectively. Furthermore, the corresponding problems for imprecise computation jobs can also be solved with the same time complexity.

34.3 Constraints

In this section we consider scheduling imprecise computation jobs on a single processor with the added constraint that either an optional subjob is executed fully by its deadline or discarded entirely. We consider two performance metrics: (1) the total error and (2) the number of imprecisely scheduled jobs (i.e., the number of jobs whose optional subjob is discarded); see also [13,25].

Lawler [24] has given an $O(n^3 W^3)$-time algorithm for $1 \mid pmtn, r_j \mid \sum w_j U_j$, where n is the number of jobs and $W = \sum w_j$ is the total weight of all the jobs. We can use Lawler's algorithm to solve the total error problem. For each job j, we create two jobs M_j and O_j. The processing time of O_j is o_j and its weight is o_j, The processing time of M_j is m_j and its weight is $\sigma + 1$, where $\sigma = \sum o_j + 1$. It is clear that a schedule that minimizes the weighted number of late jobs is also a feasible schedule that minimizes the total error. Thus, the problem of minimizing the total error can be solved in $O(n^6 \sigma^6)$ time, and hence it is pseudo-polynomial time solvable. This is probably the best one can hope for since the problem is NP-hard in the ordinary sense.

Lawler's algorithm can also be used to solve the problem of minimizing the number of imprecisely scheduled jobs. For each job j, we create two jobs M_j and O_j. The processing time of M_j is m_j and its weight is $n + 1$. The processing time of O_j is o_j and its weight is 1. It is clear that a schedule that minimizes the weighted number of late jobs is also a feasible schedule that minimizes the number of imprecisely scheduled jobs. The running time of the algorithm becomes $O(n^9)$.

Motivated by the computational complexity of the problem, we consider two approximation algorithms, one for minimizing the total error and the other for minimizing the number of imprecisely scheduled jobs. Both algorithms have time complexity $O(n^2)$, and both schedule jobs in succession from an ordered list. The one for the total error, called largest processing time first (LPTF), ordered the jobs in descending order of optional processing times, while the one for the number of imprecisely scheduled jobs, called shortest processing time first (SPTF), ordered the jobs in ascending order of optional processing times. The LPTF has a worst-case bound of 3 and the SPTF has a worst-case bound of 2; both bounds are tight.

Let T be a set of n imprecise computation jobs. Let $T' \subseteq T$ contains all the jobs with nonzero optional processing time, and let $n' = \mid T' \mid$. If H is a scheduling algorithm, we let $E_H(T')$ and $L_H(T')$ denote the set of precisely scheduled and imprecisely scheduled jobs, respectively, in the schedule produced by H. We use OPT to denote an optimal algorithm. Let $\epsilon_H(T)$ denote the total error in the schedule produced by H. With a slight abuse of notation, we use $\epsilon(S)$ to denote the total error in the schedule S.

Algorithm LPTF

(1) Let $T' = \{1, 2, \ldots, n'\}$ be sorted in descending order of the optional processing times. $E_{\text{LPTF}}(T') \leftarrow \emptyset$. Let $T" = \{M_1, M_2, \ldots, M_n\}$.
(2) **For** $i = 1, 2, \ldots, n'$ **do:**
Use algorithm NTU to construct a schedule S for $T" \cup \{O_i\}$.
If $\epsilon(S) = 0$, **then** $E_{\text{LPTF}}(T) \leftarrow E_{\text{LPTF}}(T) \cup \{i\}$ and $T" \leftarrow T" \cup \{O_i\}$.
repeat

The time complexity of algorithm LPTF is $O(n^2)$, since algorithm NTU takes linear time after an initial sort.

It is interesting to note that the worst-case performance of algorithm LPTF would be unbounded if the jobs were sorted in ascending order of optional processing times. Consider two jobs: job 1 has $r_1 = 0$,

$m_1 = 0$, $o_1 = 1$, and $d_1 = 1$; job 2 has $r_2 = 0$, $m_2 = 0$, $o_2 = x$, and $d_2 = x$. If jobs were sorted in ascending order of optional processing times, job 1 would be scheduled first and the total error produced would be x. Algorithm LPTF would schedule job 2 first and produce an error of 1. Thus, the bound can be made arbitrarily large by taking x large enough.

In the following we will show that the worst-case bound of algorithm LPTF is 3, which is tight. To show that the bound is achievable, consider the following four jobs: job 1 has $r_1 = x - \delta$, $d_1 = 2x - \delta$, $m_1 = 0$, and $o_1 = x$; job 2 has $r_2 = 0$, $d_2 = x$, $m_2 = 0$, and $o_2 = x$; job 3 has $r_3 = 2x - 2\delta$, $d_3 = 3x - 2\delta$, $m_3 = 0$, and $o_3 = x$; job 4 has $r_4 = x$, $d_4 = 2x - 2\delta$, $m_4 = 0$, and $o_4 = x - 2\delta$, where $x > \delta > 0$. Algorithm LPTF will schedule job 1 precisely and the remaining jobs imprecisely, yielding a total error of $3x - 2\delta$. The optimal algorithm will schedule job 1 imprecisely and the remaining jobs precisely, yielding a total error of x. The bound approaches 3 as δ approaches 0.

We shall prove the upper bound by contradiction. Let T be the smallest set of jobs, in terms of $|T'|$, that violates the bound. A job j is said to be *inappropriate* if it is precisely scheduled by algorithm LPTF, but imprecisely scheduled by OPT. The following lemma is instrumental to the proof of the upper bound; its proof can be found in [15] and will be omitted.

Lemma 34.2

Let s be an inappropriate job in T and let $\hat{T}$ be obtained from T by setting o_s to be zero. Then we have $\epsilon_{\mathrm{LPTF}}(\hat{T}) > \epsilon_{\mathrm{LPTF}}(T) - 3o_s$.

Lemma 34.3

$E_{\mathrm{LPTF}}(T') \cap E_{\mathrm{OPT}}(T') = \emptyset$ and $E_{\mathrm{LPTF}}(T') \cup E_{\mathrm{OPT}}(T') = T'$.

Proof

We shall prove the lemma by contradiction. Suppose $E_{\mathrm{LPTF}}(T') \cap E_{\mathrm{OPT}}(T') \neq \emptyset$. Let job j be in both $E_{\mathrm{LPTF}}(T')$ and $E_{\mathrm{OPT}}(T')$. Consider the set of jobs $\hat{T}$ obtained from T by setting the mandatory and optional processing times of job j to be $m_j + o_j$ and 0, respectively. Clearly, $|\hat{T}'| < |T'|$, $\epsilon_{\mathrm{LPTF}}(\hat{T}) = \epsilon_{\mathrm{LPTF}}(T)$, and $\epsilon_{\mathrm{OPT}}(\hat{T}) = \epsilon_{\mathrm{OPT}}(T)$. Thus, $\hat{T}$ is a smaller set of jobs violating the bound, contradicting our assumption that T is the smallest set.

If $E_{\mathrm{LPTF}}(T') \cup E_{\mathrm{OPT}}(T') \neq T'$, let j be the job that is not in $E_{\mathrm{LPTF}}(T') \cup E_{\mathrm{OPT}}(T')$. Consider the set of jobs $\hat{T}$ obtained from T by setting the optional execution time of job j to be zero. Clearly, $|\hat{T}'| < |T'|$, $\epsilon_{\mathrm{LPTF}}(\hat{T}) = \epsilon_{\mathrm{LPTF}}(T) - o_j$, and $\epsilon_{\mathrm{OPT}}(\hat{T}) = \epsilon_{\mathrm{OPT}}(T) - o_j$. Thus, $\epsilon_{\mathrm{LPTF}}(\hat{T}) = \epsilon_{\mathrm{LPTF}}(T) - o_j > 3\epsilon_{\mathrm{OPT}}(T) - o_j = 3\epsilon_{\mathrm{OPT}}(\hat{T}) + 2o_j > 3\epsilon_{\mathrm{OPT}}(\hat{T})$. Hence, $\hat{T}$ is a smaller set of jobs violating the bound. □

Lemma 34.4

$|E_{\mathrm{LPTF}}(T')| = 0$.

Proof

Suppose $|E_{\mathrm{LPTF}}(T')| \neq 0$. Let job j be in $E_{\mathrm{LPTF}}(T')$. By Lemma 34.3, job j is an inappropriate job in T. Consider the set of jobs $\hat{T}$ obtained from T by setting the optional execution time of job j to be zero. Clearly, $|\hat{T}'| < |T'|$. By Lemma 34.2, we have

$$\epsilon_{\mathrm{LPTF}}(\hat{T}) > \epsilon_{\mathrm{LPTF}}(T) - 3o_j.$$

Since $\epsilon_{\mathrm{LPTF}}(T) > 3\epsilon_{\mathrm{OPT}}(T)$, we have

$$\epsilon_{\mathrm{LPTF}}(\hat{T}) > 3(\epsilon_{\mathrm{OPT}}(T) - o_j).$$

Since job j is not in $E_{\mathrm{OPT}}(T)$, we have

$$\epsilon_{\mathrm{OPT}}(\hat{T}) = \epsilon_{\mathrm{OPT}}(T) - o_j.$$

Thus, $\epsilon_{\mathrm{LPTF}}(\hat{T}) > 3\epsilon_{\mathrm{OPT}}(\hat{T})$, contradicting our assumption that T is the smallest set violating the bound. □

Theorem 34.4

For any set of jobs T, we have $\epsilon_{\mathrm{LPTF}}(T) \leq 3\epsilon_{\mathrm{OPT}}(T)$. Moreover, the bound can be achieved asymptotically.

Proof

Lemma 34.4 implies that algorithm LPTF cannot schedule any jobs in T' precisely. Lemma 3 implies that the optimal algorithm schedules every job precisely. These two facts lead to an impossibility. □

We now consider the problem of minimizing the number of imprecised scheduled jobs. The following algorithm operates exactly like algorithm LPTF, except that jobs are ordered in ascending order of the optional processing times.

Algorithm SPTF

(1) Let $T' = \{1, 2, \ldots, n'\}$ be sorted in ascending order of the optional processing times. $E_{\mathrm{SPTF}}(T'') \leftarrow \emptyset$.
Let $T" = \{M_1, M_2, \ldots, M_n\}$.
(2) **For** $i = 1, 2, \ldots, n'$ **do:**
Use algorithm NTU to construct a schedule S for $T" \cup \{O_i\}$.
If $\epsilon(S) = 0$, **then** $E_{\mathrm{SPTF}}(T) \leftarrow E_{\mathrm{SPTF}}(T) \cup \{i\}$ and $T" \leftarrow T" \cup \{O_i\}$.
repeat

It is clear that algorithm SPTF produces a feasible schedule satisfying the 0/1-constraint. The time complexity of algorithm SPTF is the same as algorithm LPTF, $O(n^2)$.

In the above we have shown that if algorithm SPTF were used for minimizing the total error, it would give an unbounded worst-case performance. As it turns out, if algorithm LPTF were used for minimizing the number of imprecisely scheduled jobs, it would also give an unbounded worst-case performance. Consider the following $n+1$ jobs: job i, $1 \leq i \leq n$, has $r_i = (i-1)x$, $d_i = ix$, $m_i = 0$, and $o_i = x$; job $n+1$ has $r_{n+1} = 0$, $d_{n+1} = nx$, $m_{n+1} = 0$, and $o_{n+1} = nx$. Algorithm LPTF schedules job $n+1$ precisely and the remaining jobs imprecisely, while the optimal algorithm schedules job $n+1$ imprecisely and the remaining jobs precisely. Thus, the bound can be made arbitrarily large by taking n large enough. Fortunately, algorithm SPTF gives a bound of 2, which is tight.

Before we prove the upper bound, we will show that the bound of 2 can be achieved. Consider the following three jobs: job 1 has $r_1 = x - \delta$, $d_1 = 2x - \delta$, $m_1 = 0$, and $o_1 = x$; job 2 has $r_2 = 0$, $d_2 = x$, $m_2 = 0$, and $o_2 = x$; job 3 has $r_3 = 2(x - \delta)$, $d_3 = 3x - 2\delta$, $m_3 = 0$, and $o_3 = x$. Algorithm SPTF schedules job 1 precisely and the remaining jobs imprecisely, while the optimal algorithm schedules job 1 imprecisely and the remaining jobs precisely.

In the following we shall prove the upper bound by contradiction. Let T be the smallest set of jobs, in terms of $|T'|$, that violates the bound. Call a job *inappropriate* if it is precisely scheduled by algorithm SPTF but not by the optimal algorithm. The following lemma is instrumental to the proof of the upper bound. We will omit the long proof here; it can be found in [15].

Lemma 34.5

Let s be an inappropriate job in T. Let $\hat{T}$ be obtained from T by setting o_s to be zero. Then we have $|L_{\mathrm{SPTF}}(\hat{T})| \geq |L_{\mathrm{SPTF}}(T)| - 2$.

The proofs of the following two lemmas are very similar to those of Lemmas 34.3 and 34.4; we will omit the proofs here.

Lemma 34.6

$E_{\text{SPTF}}(T') \cap E_{\text{OPT}}(T') = \emptyset$ *and* $E_{\text{SPTF}}(T') \cup E_{\text{OPT}}(T') = T'$.

Lemma 34.7

$|\, E_{\text{SPTF}}(T')\,| = 0$.

Theorem 34.5

For any set of jobs T*, we have* $|L_{\text{SPTF}}(T)\,| \leq 2|L_{\text{OPT}}(T)|$. *Moreover, the bound is tight.*

Proof

Lemma 34.7 implies that algorithm SPTF cannot schedule *any* jobs in T' precisely. Lemma 34.6 implies that the optimal algorithm schedules *every* job in T' precisely. These two facts lead to an impossibility. □

While Theorem 34.5 gives a relationship between the number of imprecisely scheduled jobs, it does not give a meaningful relationship between the number of precisely scheduled jobs. The next theorem shows that they are also related by the same multiplicative factor.

Theorem 34.6

For any set of jobs T*, we have* $|\, E_{\text{OPT}}(T)\,| \leq 2\,|\, E_{\text{SPTF}}(T)\,|$. *Moreover, the bound is tight.*

Proof

The set of jobs used to prove Theorem 34.5 shows that $|\, E_{\text{OPT}}(T)\,| = 2\,|\, E_{\text{SPTF}}(T)\,|$. The upper bound is proved by contradiction. Let T be the smallest set of jobs, in terms of $|\, T'\,|$, that violates the bound. It is easy to verify that Lemma 34.6 also holds for T'. Thus, $E_{\text{SPTF}}(T') = L_{\text{OPT}}(T')$ and $E_{\text{OPT}}(T') = L_{\text{SPTF}}(T')$. Hence,

$$|\, E_{\text{OPT}}(T')\,| = |\, L_{\text{SPTF}}(T')\,| \leq 2|L_{\text{OPT}}(T')\,| = 2|E_{\text{SPTF}}(T')\,|$$

contradicting our assumption that T violates the bound. □

From Theorem 34.5 and the fact that

$$|\, E_{\text{SPTF}}(T')\,| + |\, L_{\text{SPTF}}(T')\,| = |\, E_{\text{OPT}}(T')\,| + |\, L_{\text{OPT}}(T')\,| = n'$$

we have

$$|\, E_{\text{SPTF}}(T')\,| = n' - |\, L_{\text{SPTF}}(T')\,| \geq n' - 2|L_{\text{OPT}}(T')\,| = 2|E_{\text{OPT}}(T')\,| - n'$$

From Theorem 34.6, we have

$$|\, E_{\text{SPTF}}(T')\,| \geq \frac{|E_{\text{OPT}}(T')|}{2}$$

Now, $2\,|\, E_{\text{OPT}}(T')\,| - n' \geq \frac{|\, E_{\text{OPT}}(T')\,|}{2}$ if and only if $|\, E_{\text{OPT}}(T')\,| \geq \frac{2n'}{3}$. Thus, when $|\, E_{\text{OPT}}(T')\,| \geq \frac{2n'}{3}$, Theorem 34.5 gives a better bound for $|\, E_{\text{SPTF}}(T)\,|$; otherwise, Theorem 34.6 gives a better bound.

34.4 Open Shops and Flow Shops

Blazewicz et al. [16,17] extended the weighted number of tardy units problem to flow shops and open shops. For flow shops, they [17] showed that $F2\,|\, d_j = d\,|\sum w_j Y_j$ is NP-hard in the ordinary sense and gave a pseudo-polynomial time algorithm for it. For open shops, they [16] showed that $O\,|\, pmtn, r_j\,|\sum w_j Y_j$ and $O2\,|\, d_j = d\,|\sum Y_j$ are solvable in polynomial time, while $O2\,|\, d_j = d\,|\sum w_j Y_j$ is ordinary NP-hard.

Recall that in a two-processor flow shop, each job j has two operations, one on the first processor and the other on the second processor with processing times $p_{j,1}$ and $p_{j,2}$, respectively. Each job has to finish its first operation before it can start its second operation. The number of tardy units of a job is the amount of work done after its due date, including both the first operation and the second operation. The next result is due to Blazewicz et al. [17].

Theorem 34.7

$F2 \mid d_j = d \mid \sum w_j Y_j$ is NP-hard in the ordinary sense.

Proof

We shall reduce the Partition problem to $F2 \mid d_j = d \mid \sum w_j Y_j$. Given an instance $A = (a_1, a_2, \ldots, a_n)$, we construct an instance of $F2 \mid d_j = d \mid \sum w_j Y_j$ as follows. There are $n+1$ jobs. Job j, $1 \leq j \leq n$, has processing times $p_{j,1} = 0$ and $p_{j,2} = a_j$, and weight $w_j = 1$. Job $n+1$ has processing times $p_{n+1,1} = B$ and $p_{n+1,2} = 1$, and weight $w_{n+1} = B+1$, where $B = \frac{1}{2}\sum a_j$. The common due date is $d = B+1$. The threshold for $\sum w_j Y_j$ is B. It is easy to see that a partition exists if and only if there is a schedule with $\sum w_j Y_j \leq B$. □

The complexity of $F2 \mid d_j = d \mid \sum Y_j$ is still open. However, if we allow two different due dates, then the problem becomes NP-hard. If we allow arbitrary number of due dates, then the problem becomes NP-hard in the strong sense.

Theorem 34.8

$F2 \mid d_j \in \{d_1, d_2\} \mid \sum Y_j$ is NP-hard in the ordinary sense. $F2 \mid\mid \sum Y_j$ is NP-hard in the strong sense.

Proof

We use the same reduction as above. If we set the due dates of the first n jobs to be B, the due date of job $n+1$ to be $B+1$, and the weight of job $n+1$ to be 1, then the reduction will work for $\sum Y_j$. For arbitrary number of due dates, we can give a reduction from the 3-Partition problem (see a definition of 3-Partition in Chapter 2) to $F2 \mid\mid \sum Y_j$. □

Blazewicz et al. [17] also gave a dynamic program to solve $F2 \mid d_j = d \mid \sum w_j Y_j$. The basic idea is to determine the first late job and the set of on-time jobs. The on-time jobs are scheduled by Johnson's rule [25] and the late jobs are scheduled in any order. We refer the readers to [17,18] for a full description of the algorithm.

Open shops are like flow shops, with each job having one operation per processor. However, unlike flow shops, the order of execution on the different processors is immaterial (i.e., it does not need to follow a strict order).

The algorithm for $O \mid pmtn, r_j \mid \sum w_j Y_j$ starts by defining the p intervals $[u_0, u_1], [u_1, u_2], \ldots, [u_{p-1}, u_p]$ obtained from the release times and due dates of the jobs; see Section 34.2 for a definition of the intervals. It then uses linear programming to determine the lengths of the early parts of each job executed in each of these intervals. Finally, it uses the algorithm of Gonzalez and Sahni [26] to schedule the early parts. The readers are referred to [16,18] for more details.

The algorithm for $O_2 \mid d_j = d \mid \sum Y_j$ is based on the algorithm of Gonzalez and Sahni [26] for $O_2 \mid\mid C_{\max}$. A schedule is first constructed by the algorithm of Gonzalez and Sahni. The schedule is then modified by shifting some jobs in order to minimize idle times before the common due date. Full details of the algorithm can be found in [16,18].

While minimizing $\sum Y_j$ is polynomially solvable for one common deadline, it loses polynomial-time solvability with an additional deadline.

Theorem 34.9

$O2 \mid d_j \in \{d_1, d_2\} \mid \sum Y_j$ is NP-hard in the ordinary sense.

Proof

The same reduction used in Theorem 34.8 can also be used for this theorem. □

Blazewicz et al. [16] gave a pseudo-polynomial time algorithm for $O_2 \mid d_j = d \mid \sum w_j Y_j$. The idea of the algorithm is to use dynamic programming to calculate the early parts of each job that are executed before the common due date such that $\sum w_j Y_j$ is minimized. They also showed that the problem is NP-hard in the ordinary sense.

Theorem 34.10

$O_2 \mid d_j = d \mid \sum w_j Y_j$ is NP-hard in the ordinary sense.

Proof

We again use the Partition problem for reduction. Given an instance $A = (a_1, a_2, \ldots, a_n)$ of the Partition problem, we create $n + 1$ jobs. Job j, $1 \leq j \leq n$, has $p_{j,1} = p_{j,2} = a_j$ and $w_j = 1$. Job $n + 1$ has $p_{j,1} = p_{j,2} = B$ and $w_{n+1} = 2B + 1$, where $B = \frac{1}{2} \sum a_j$. The common deadline is $2B$ and the threshold for $\sum w_j Y_j$ is also $2B$. It is easy to see that a partition exists if and only if there is a schedule with $\sum w_j Y_j \leq 2B$. □

34.5 Conclusions

In this chapter we have surveyed algorithms and complexity results for the total weighted number of tardy units problem. We have also reviewed algorithms for minimizing the total weighted error for imprecise computation jobs. These two problems are closely related in terms of algorithms and time complexity. Scheduling imprecise computation jobs is also related to power-aware scheduling (see [27]), while power-aware scheduling is related to scheduling jobs with controllable processing times (see [28–31]). Due to extensive applications in many different fields, these scheduling problems promise to receive a lot of attentions in the near future.

Acknowledgment

This work is supported in part by the NSF Grant DMI-0300156.

References

[1] J. Blazewicz, Scheduling preemptible tasks on parallel processors with information loss, *Technique et Science Informatiques*, 3(6), 415–420, 1984.

[2] K-J. Lin, S. Natarajan, and J. W. S. Liu, Concord: a distributed system making use of imprecise results, In *Proceedings of COMPSAC '87*, Tokyo, Japan, 1987.

[3] K-J. Lin, S. Natarajan, and J. W. S. Liu, Imprecise results: utilizing partial computations in real-time systems, In *Proceedings of the 8th Real-Time Systems Symposium*, San Francisco, CA, 1987.

[4] K-J. Lin, S. Natarajan, and J. W. S. Liu, Scheduling real-time, periodic job using imprecise results, In *Proceedings of the 8th Real-Time Systems Symposium*, San Francisco, CA, 1987.

[5] J. W. S. Liu, K-J. Lin, S. W. Shih, A. C. Yu, J-Y. Chung, and W. Zhao, Algorithms for scheduling imprecise computations, *IEEE Computers*, 24, 58–68, 1991.

[6] J. Blazewicz and G. Finke, Minimizing mean weighted execution time loss on identical and uniform processors, *Information Processing Letters*, 24, 259–263, 1987.

[7] J. B. Orlin, A faster strongly polynomial minimum cost flow algorithm, In *Proceedings of 20th ACM Symposium on Theory of Computing*, 377–387, 1988.

[8] D. S. Hochbaum and R. Shamir, Minimizing the number of tardy job unit under release time constraints, *Discrete Applied Mathematics*, 28, 45–57, 1990.

[9] J. Y-T. Leung, V. K. M. Yu, and W-D. Wei, Minimizing the weighted number of tardy task units, *Discrete Applied Mathematics*, 51, 307–316, 1994.

[10] C. N. Potts and L. N. Van Wassenhove, Single machine scheduling to minimize total late work, *Operations Research*, 40(3), 586–595, 1991.

[11] C. N. Potts and L. N. Van Wassenhove, Approximation algorithms for scheduling a single machine to minimize total late work, *Operations Research Letters*, 11, 261–266, 1992.

[12] J. Y. Chung, W. K. Shih, J. W. S. Liu, and D. W. Gillies, Scheduling imprecise computations to minimize total error, *Microprocessing and Microprogramming*, 27, 767–774, 1989.

[13] W. K. Shih, J. W. S. Liu, and J. Y. Chung, Algorithms for scheduling imprecise computations with timing constraints, *SIAM Journal on Computing*, 20, 537–552, 1991.

[14] A. L. Liestman and R. H. Campbell, A fault-tolerant scheduling problem, *IEEE Transactions on Software Engineering*, 12, 1089–1095, 1986.

[15] K. Ho, J. Y-T. Leung, and W-D. Wei, Scheduling imprecise computation tasks with 0/1-constraint, *Discrete Applied Mathematics*, 78, 117–132, 1997.

[16] J. Blazewicz, E. Pesch, M. Sterna, and F. Werner, Open shop scheduling problems with late work criteria, *Discrete Applied Mathematics*, (to appear).

[17] J. Blazewicz, E. Pesch, M. Sterna, and F. Werner, The two-machine flow-shop problem with weighted late work criterion and common due date, *European Journal of Operational Research* (to appear).

[18] M. Sterna, *Problems and Algorithms in Non-Classical Shop Scheduling*, Scientific Publishers of the Polish Academy of Sciences, Poznan, 2000.

[19] K. Ho, J. Y-T. Leung, and W-D. Wei, Minimizing maximum weighted error for imprecise computation tasks, *Journal of Algorithms*, 16, 431–452, 1994.

[20] H. N. Gabow and R. E. Tarjan, A linear-time algorithm for a special case of disjoint set union, *Journal of Computing System and Science* 30, 209–221, 1985.

[21] W. A. Horn, Some simple scheduling algorithms, *Naval Research Logistics Quarterly*, 21, 177–185, 1974.

[22] R. E. Tarjan, *Data Structures and Network Algorithms*, Society for Industrial and Applied Mathematics, Philadelphia, PA 1983.

[23] A. Federgruen and H. Groenevelt, Preemptive scheduling of uniform machines by ordinary network flow techniques, *Management Science*, 32, 341–349, 1986.

[24] E. L. Lawler, A dynamic programming algorithm for preemptive scheduling of a single machine to minimize the number of late jobs, *Annals of Operations Research*, 26, 125–133, 1990.

[25] S. M. Johnson, Optimal two- and three-stage production schedules, *Naval Research Logistics Quarterly*, 1, 61–68, 1954.

[26] T. Gonzalez and S. Sahni, Open shop scheduling to minimize finish time, *Journal of ACM*, 23, 665–679, 1976.

[27] H. Aydin, R. Melhem, and D. Mosse, Periodic reward-based scheduling and its application to power-aware real-time systems, chap. 36, this volume.

[28] H. Hoogeveen and G. J. Woeginger, Some comments on sequencing with controllable processing times, *Computing*, 68, 181–192, 2002.

[29] E. Nowicki and S. Zdrzalka, A survey of results for sequencing problems with controllable processing times, *Discrete Applied Mathematics*, 26, 271–287, 1990.

[30] N. V. Shakhlevich and V. A. Strusevich, Single machine scheduling with controllable release and processing parameters, Technical Report, School of Computing, University of Leeds, Leeds LS2 9JT, U.K., 2003.

[31] V. A. Strusevich, Two machine flow shop scheduling problems with no-wait in process: controllable machine speeds, *Discrete Applied Mathematics*, 59, 75–86, 1995.

35

Dual Criteria Optimization Problems for Imprecise Computation Tasks

Kevin I-J. Ho
Chung-Shan Medical University

35.1 Introduction

In real-time scheduling problems, one of the paramount criteria is meeting the deadlines of all the tasks. Unfortunately, it is impossible to satisfy this criterion all the time, especially in heavily loaded real-time systems. To cope with this situation, one can completely discard some less critical tasks in favor of meeting the deadlines of more important ones. Or, one can schedule tasks partially, i.e., executing tasks not completely, so as to make more tasks receive minimum required execution time and meet their deadlines. The latter approach is applicable to situations such as the numerical analysis and the curing process of composite products. To model this kind of task system, Lin et al. [1–3] proposed the *Imprecise Computation Model*, where the accuracy of the results can be traded for meeting the deadlines of the tasks. For details of the model, the readers are referred to the previous chapter and [4]. In this model, each task can be logically decomposed into two subtasks, mandatory and optional. For each task, the mandatory subtask is required to be completely executed by its deadline to obtain acceptable result, while the optional subtask is to enhance the result generated by the mandatory subtask and hence can be left incomplete.

1-58488-397-9/$0.00+$1.50

If the optional subtask of a task is left incomplete, an error is incurred, which is defined as the execution time of the unfinished portion. Task systems in the traditional scheduling model can be treated as a special case where the execution time of each task's optional subtask is zero.

In the Imprecise Computation Model, each task T_i in the task system TS is symbolized by the quadruple (r_i, d_i, m_i, o_i), where r_i, d_i, m_i, and o_i denote its release time, deadline, mandatory subtask's execution time and optional subtask's execution time, respectively. Let e_i denote the total execution time of task T_i, i.e., $e_i = m_i + o_i$. In certain circumstances, the importance, regarding the criteria, of tasks may be different. To represent the difference among tasks, each task may be assigned one or more weights.

In this model, a *legitimate* schedule of a task system satisfies the following requirements: (1) no task is assigned to more than one processor and (2) at most one task is assigned to a processor at any moment of time. A legitimate schedule is said to be *feasible* if each task starts no earlier than its release time and finishes no later than its deadline, and the mandatory subtask is completely executed. A task system is said to be feasible if there exists at least one feasible schedule for it. The feasibility of a task system can be determined in $O(n^2 \log^2 n)$ time for parallel and identical processors [5] and $O(n \log n)$ time for single processor [6]. Without loss of generality, all the task systems discussed in this chapter are assumed to be feasible, since the time complexity of all the algorithms introduced in this chapter is higher than that of the feasibility test. Additionally, we assume that all the task parameters are rational, all the tasks are independent, and all the processors are identical. Furthermore, in this chapter, only preemptive schedules are considered.

Before we proceed, we define general terminology and notations to facilitate the reading. Let S be a feasible schedule for a task system TS with n tasks in the Imprecise Computation Model. For each task T_i in TS, let $\alpha(T_i, S)$ denote the amount of processor time assigned to T_i in S. The error of T_i in S, denoted by $\varepsilon(T_i, S)$, is defined as $e_i - \alpha(T_i, S)$. The total error of S, denoted by $\varepsilon(S)$, is defined as $\sum_{i=1}^{n} \varepsilon(T_i, S)$. The minimum total error of TS, denoted by $\varepsilon(TS)$, is defined as $\min\{\varepsilon(S) : S \text{ is a feasible schedule for } TS\}$. If the importance of the tasks with respect to a given criterion are not the same, then a positive value w_i is assigned to each task T_i. And the product of $\varepsilon(T_i, S)$ and w_i is called the w-*weighted* error of T_i in S, denoted by $\varepsilon_w(T_i, S)$. The total w-*weighted* error of S and minimum total w-*weighted* error of TS are defined analogously, and are denoted by $\varepsilon_w(S)$ and $\varepsilon_w(TS)$, respectively.

In the past, several researchers have studied the total unweighted and total weighted error problems in the Imprecise Computation Model; see details in Section 35.2. But, most of the studies are only concerned with the total unweighted or weighted error, without considering the distribution of errors among tasks. Therefore, the obtained schedules might result in very uneven error distribution among the tasks. Even worse, some important tasks have large errors, while less important tasks have small errors. This kind of schedule might not be suitable for some task systems.

In this chapter we will discuss, based on the research results done by Ho et at. [7,8], the problems related to the error distribution. We will focus on three dual-criteria optimization problems of preemptively scheduling a set of imprecise computation tasks on $m \geq 1$ identical processors. The criterion of the first optimization problem is the maximum weighted error of a given task system. The second problem is to minimize the total weighted error subject to the constraint that the maximum weighted error is minimized. The third one is to minimize the maximum weighted error subject to the constraint that the total weighted error is minimized. In these problems, each task may have two different weights, one for computing the total weighted error and the other for the maximum weighted error. In Section 35.2, we will briefly review some works related to the problems of minimizing total (weighted) error and the problems of error distribution. Then, the description of the problems on which this chapter focuses is given. In Section 35.3, we will study the problem of minimizing the maximum weighted error. Section 35.4 will focus on the problem of minimizing the total weighted error subject to the constraint that the maximum weighted error is minimized. The problem of minimizing the maximum weighted error under the constraint that the total weighted error is minimized will be discussed in Section 35.5. In the last section, we will make some concluding remarks for this chapter and discuss the relationship among the problems studied in Sections 35.4 and 35.5.

35.2 Related Works

In this section we will review some research results related to the topics discussed in this chapter. Blazewicz [9] was the pioneer of studying the problem of minimizing the total weighted error for the special case where each task has an optional subtask only, i.e., $m_i = 0$ for each $1 \le i \le n$. He reduced the problems of scheduling tasks on parallel and identical processors as well as uniform processors to minimum-cost-maximum-flow problems, which can then be transformed to linear programming ones. Then, by using the minimum-cost-maximum-flow approach again, Blazewicz and Finke [10] gave faster algorithms for both problems. Potts and van Wassenhove [11] proposed an $O(n \log n)$-time algorithm for the same problem on single processor, with the added constraint that all tasks have identical release times and weights. Moreover, they proved that the problem becomes NP-hard for the nonpreemptive case, and present a pseudo-polynomial time algorithm for the nonpreemptive case. In [12], Potts and van Wassenhove gave a polynomial approximation scheme and two fully polynomial approximation schemes based on the pseudo-polynomial time algorithm.

Now, consider the general case in which each task has mandatory, optional, or both subtasks. For parallel and identical processors, Shih et al. [5] proposed an $O(n^2 \log^2 n)$-time algorithm to minimize the total error. And they also showed that the weighted version problem can be transformed to a minimum-cost-maximum-flow problem. Using Orlin's $O(|A| \log |V| (|A| + |V| \log |V|))$-time algorithm for the minimum-cost-maximum-flow problem [13], where A and V denote the arc set and vertex set, respectively, the problem of minimizing the total weighted error on parallel and identical processors can be solved in $O(n^2 \, log^3 n)$ time, since in the network shown in [5], $|A|$ is equal to $O(n \log n)$ and $|V|$ is equal to $O(n)$. Regarding the single processor case, Shih et al [14] gave an algorithm that runs in $O(n \log n)$ time for the unweighted case and $O(n^2 \log n)$ time for the weighted case. Later, Leung et al. [15] gave a faster algorithm that runs in $O(n \log n + kn)$ time, where k denotes the number of distinct weights.

In 1991, Shih et al. [14] proposed a new constraint, the so-called *0/1-constraint*, to be added on the Imprecise Computation Model, where each optional subtask is either fully executed or completely discarded. The 0/1-constraint is motivated by some real world problems. For example, an algorithm for solving a task in the real world might have two different versions. One runs slower than the other, but produces a result with better quality. To meet the deadline constraints of tasks, some tasks might be forced to execute the fast version of the algorithm. By treating the time difference between the fast and slow versions as the execution time of the optional subtask, one can easily transform this type of problem into the problem of scheduling with 0/1-constraint [16]. For a single processor, Shih et al. [14] showed that the problem of minimizing the total error is NP-hard. Later, Ho et al. [17] showed that the problem can be reduced to the problem of minimizing the weighted number of late tasks in classical scheduling theory (denoted as the $1 \mid pmtn, r_j \mid \sum w_j U_j$ problem in the classification scheme of [18]). Lawler [19] proposed an $O(n^3 W^2)$-time algorithm for the $1 \mid pmtn, r_j \mid \sum w_j U_j$ problem, where W is the total weight of all the n tasks. Therefore, the problem of minimizing the total error on a single processor with 0/1-constraint can be solved in pseudo-polynomial time $O(n^5 \sigma^2)$, where σ denotes the total execution time of all the optional subtasks. Motivated by the computational complexity of the problem, Ho et al. also developed an $O(n^2)$-time approximation algorithm for it, with a worst-case performance bound of 3, which is tight.

All the research results stated above were only concerned with minimizing the total (weighted) error, without any regard to the distribution of errors among tasks. Thus, the error distribution of tasks might be biased in the schedules generated by all the algorithms mentioned above. In other words, some tasks have very large (weighted) errors while others have very small (weighted) errors, which might not be acceptable under certain circumstances. Motivated by this consideration, Shih and Liu [20] studied the problem of preemptively scheduling a task system TS on single processor to minimize the maximum *normalized* error, defined as $\max\{\varepsilon(T_i, S)/o_i : T_i \in TS\}$, under the constraint that the total error is minimum. An $O(n^3 \, log n)$-time algorithm was given to solve this problem.

To generalize the problem proposed by Shih and Liu [20], Ho et al. [7] defined a *doubly weighted* task system, in which each task T_i is weighted with two weights, w_i^T and w_i^M, for two different criteria, total w^T-*weighted* error and maximum w^M-*weighted* error, respectively. (Note that the normalized error defined in [20] can be interpreted as each task T_i having two weights, $w_i^T = 1$ and $w_i^M = 1/o_i$. And then, the problem studied in [20] is to minimize the maximum w^M-*weighted* error subject to the constraint that the total w^T-*weighted* error is minimized.) In [7], they considered the problems of preemptively scheduling a doubly weighted task system with two objectives: (1) minimizing the maximum w^M-*weighted* error and (2) minimizing the total w^T-*weighted* error under the constraint that the maximum w^M-*weighted* error is minimized. They also presented two algorithms for both objectives that run in $O(n^3 \log^2 n)$ for parallel and identical processors and $O(n^2)$ for single processor.

To continue the study of preemptively scheduling a doubly weighted task system, Ho et al. [8] considered the problem of minimizing the maximum w^M-*weighted* error under the constraint that the total w^T-*weighted* error is minimized. An algorithm was presented to solve this problem, with time complexity $O(kn^3 \log^2 n)$ for the parallel and identical processors case and $O(kn^2)$ for the single processor case, where k is the number of distinct w^T-*weights*. It is easy to verify that the problem studied by Shih and Liu [20] is a special case of the problem by letting $w_i^T = 1$ and $w_i^M = 1/o_i$ for each task T_i.

35.3 Minimizing Maximum Weighted Error

In this section we will focus on the problem of minimizing the maximum w^M-*weighted* error of a given doubly weighted task system. At the beginning we will formally define the *Doubly Weighted Task Systems*, and then the related terminology and notations used throughout this section. Let $TS = (\{T_i\}, \{r_i\}, \{d_i\}, \{m_i\}, \{o_i\}, \{w_i^T\}, \{w_i^M\})$, $1 \leq i \leq n$, be a doubly weighted task system consisting of n tasks, where w_i^T and w_i^M denote the two positive weights for computing the total weighted error and the maximum weighted error, respectively. Let S be an arbitrary feasible schedule for TS on $m \geq 1$ parallel and identical processors. The symbol $E_{w^M}(S)$ denotes the maximum w^M-*weighted* error of S (i.e., $E_{w^M}(S) = \max_{1\leq i \leq n} \{\varepsilon_{w^M}(T_i, S)\}$) and the symbol $E_{w^M}(TS)$ denotes the minimum maximum w^M-*weighted* error of TS (i.e., $E_{w^M}(TS) = \min \{E_{w^M}(S) : S$ is a feasible schedule for $TS\}$). The symbol $\varepsilon_{w^T}^{w^M}(TS)$ denotes the minimum total w^T-*weighted* error under the constraint that the maximum w^M-*weighted* error is minimized, i.e., $\varepsilon_{w^T}^{w^M}(TS) = \min \{\varepsilon_{w^T}(S)$: S is a feasible schedule for TS and $E_{w^M}(S) = E_{w^M}(TS)\}$.

Let $\min_{1\leq i\leq n}\{r_i\} = t_0 < t_1 < \cdots < t_p = \max_{1\leq i \leq n}\{d_i\}$ be all the distinct release times and deadlines of all tasks in TS. These $p+1$ distinct values divide the time frame into p intervals: $[t_0, t_1]$, $[t_1, t_2]$,. . ., $[t_{p-1}, t_p]$, denoted by $I_1, I_2, \ldots, I_p$. The length of the interval I_j, denoted by L_j, is equal to $t_j - t_{j-1}$. By using McNaughton's rule [21], a schedule S for TS can be described by a $n \times p$ matrix SM_S such that $SM_S(i, j)$ represents the amount of processor time assigned to T_i in I_j. If S is a feasible schedule, SM_S must satisfy the following inequalities: (1) $SM_S(i, j) \leq L_j$ for $1 \leq i \leq n$ and $1 \leq j \leq p$; (2) $\sum_{i=1}^{n} SM_S(i, j) \leq m \times L_j$ for $1 \leq j \leq p$. In the remainder of this chapter, we will represent schedules using the matrix forms to facilitate the reading, if necessary.

An interval I_j is said to be *unsaturated* if $\sum_{i=1}^{n} SM_S(i, j) < m \times L_j$; otherwise, it is *saturated*. A task T_i is said to be *available* in time interval $I_j = [t_{j-1}, t_j]$ if $r_i \leq t_{j-1} < t_j \leq d_i$. A task T_i is said to be *fully scheduled* in I_j if $SM_S(i, j) = L_j$; otherwise, it is said to be *partially scheduled*. A task is said to be *precisely scheduled* in S if $\varepsilon(T_i, S) = 0$; otherwise, it is said to be *imprecisely scheduled*.

The remainder of this section is organized as follows. A polynomial-time preemptive scheduling algorithm, called Algorithm *MME*, for the problem will be given in Section 35.3.1. The algorithm runs in $O(n^3 \log^2 n)$ time for multiprocessor environment and in $O(n^2)$ time for a single processor environment. Then, we will analyze Algorithm *MME*, including the time complexity and the correctness, based on three properties of a specific type of tasks shown in Section 35.3.2. The correctness of these properties will be proved in the last subsection.

35.3.1 Algorithm for Minimizing Maximum Weighted Error

Before we show Algorithm *MME*, we need to define some terminology. For a given task system TS, a task T_r is said to be *removable* if $\varepsilon(TS) = \varepsilon(TS - \{T_r\})$; otherwise, *irremovable*. We will state two properties of removable tasks by Properties 35.1 and 35.3 in Section35.3.2. Using these two properties, one can identify removable tasks from a feasible schedule for *TS* with minimum total error. Also, task T_s is said to be *reducible* if $o_s > 0$; otherwise, *irreducible*. Intuitively, only irremovable and irreducible tasks might contribute errors to the minimum total error of a task system. The correctness of the intuition will be shown in Section 35.3.3. The idea behind Algorithm *MME* is to distribute the minimum total error of the given task system to all the reducible tasks so as to make the incurred w^M-*weighted* errors of all the reducible tasks to be as equal as possible.

Algorithm *MME* proceeds iteratively. Starting with the given task system *TS*, a new task system is constructed in each iteration based on the task system generated in the previous iteration. (The given task system is treated as the one generated by the zeroth iteration.) The task system created in the ith iteration, $i \geq 1$, denoted by $TS(i)$, is obtained by first initializing it to be $TS(i-1)$. Then, the execution times of the optional subtasks of all the reducible tasks will be reduced by a certain amount and some (not necessarily all) removable tasks may also be removed from $TS(i)$, before the algorithm proceeds to the next iteration.

In the remainder of this section, $T_j(i)$ denotes T_j in $TS(i)$. The execution times of the mandatory and optional subtasks of $T_j(i)$ are denoted by $m_j(i)$ and $o_j(i)$, respectively. Since the weights of each task in each $TS(i)$ is invariant, we use the same notations for the task systems generated by the algorithm. Initially, $TS(0)$ is initialized to *TS* and $\eta(0)$ (for storing $E_{w^M}(TS(0))$) is initialized to be zero. Then, a schedule $S(0)$ with minimum total error is constructed for $TS(0)$. If $\varepsilon(S(0)) = 0$, the algorithm stops; otherwise, it proceeds to the next iteration. (In the following, we will call the initialization iteration the zeroth iteration.) In the ith iteration, the algorithm first initializes $TS(i)$ to $TS(i-1)$. Then, it will apportion $\varepsilon(TS(i-1))$ amount of error to all the reducible tasks in $TS(i)$ in such a way that the maximum w^M-*weighted* error is minimized. It is easy to verify that the maximum w^M-*weighted* error would be minimized if the execution time of the optional subtask of each reducible task $T_j(i)$ in $TS(i)$ is reduced by $\varepsilon(TS(i-1))/(\Delta \times w_j^M)$ amount (and hence incurs the same amount of w^M-*weighted* error), where $\Delta = \sum_{T_k(i) \in TS(i) \text{ and reducible}} 1/w_k^M$. Thus, for each reducible task $T_j(i)$ in $TS(i)$, $o_j(i)$ is reduced by $\varepsilon(TS(i-1))/(\Delta \times w_j^M)$ amount (if it can), and $T_j(i)$ is marked as irreducible if $o_j(i)$ becomes zero. Now, $\eta(i)$ is set to be $\eta(i-1) + \varepsilon(TS(i-1))/\Delta$. A schedule $S(i)$ with minimum total error is then constructed for $TS(i)$. If $\varepsilon(S(i)) = 0$, the algorithm stops and $\eta(i)$ gives the $E_{w^M}(TS)$. Otherwise, it will remove some (not necessarily all) removable tasks in $TS(i)$ before proceeding to the next iteration. From Properties 35.3, removable tasks can be found by locating an unsaturated interval I_m (if any) in $S(i)$ in which a nonempty set of tasks is partially scheduled; these (removable) tasks will then be removed from $TS(i)$. A formal description of the algorithm is given below.

Algorithm *MME*

Input: *A doubly weighted task system* $TS = (\{T_i\}, \{r_i\}, \{d_i\}, \{m_i\}, \{o_i\}, \{w_i^T\}, \{w_i^M\})$ *with n tasks and* $m \geq 1$ *parallel and identical processors.*

Output: $E_{w^M}(TS)$.

Step 1: Initialize $\eta(0)$, $TS(0)$, and loop index i to be 0, *TS*, and 0, respectively.

Step 2: Construct a schedule $S(0)$ with minimum total error for $TS(0)$.

Step 3: While $\varepsilon(S(i)) \neq 0$, perform Steps 4–8.

Step 4: Increase i by 1; Set $TS(i)$ to $TS(i-1)$; Set Δ to $\sum_{T_k \in TS(i) \text{ and reducible}} 1/w_k^M$.

Step 5: For each reducible task $T_j(i) \in TS(i)$ do:

- Set $y_j(i)$ to $\varepsilon(S(i-1))/(\Delta \times w_j^M)$.
- Set $o_j(i)$ to $\max\{0, o_j(i) - y_j\}$.
- If $o_j(i) = 0$, mark $T_j(i)$ as irreducible.

Step 6: Set $\eta(i)$ to $\eta(i-1) + \varepsilon(S(i-1))/\Delta$.

Step 7: Construct a schedule $S(i)$ with minimum total error for $TS(i)$.

Step 8: Find an unsaturated interval I_m in $S(i)$ in which a nonempty set of tasks is partially scheduled and delete these tasks from $TS(i)$.

Step 9: Output $\eta(i)$ and terminate.

T_i	T_1	T_2	T_3	T_4	T_5	T_6	T_7	T_8
r_i	0	0	0	6	6	2	6	6
d_i	6	6	5	9	9	10	9	10
m_i	1	2	0	1	1	2	0	1
o_i	7	6	4	5	7	5	5	3
w_i^T	1	1	1	1	1	1	1	1
w_i^M	4	2	1	1	2	2	4	1

(a) An example task system, $TS(0)$.

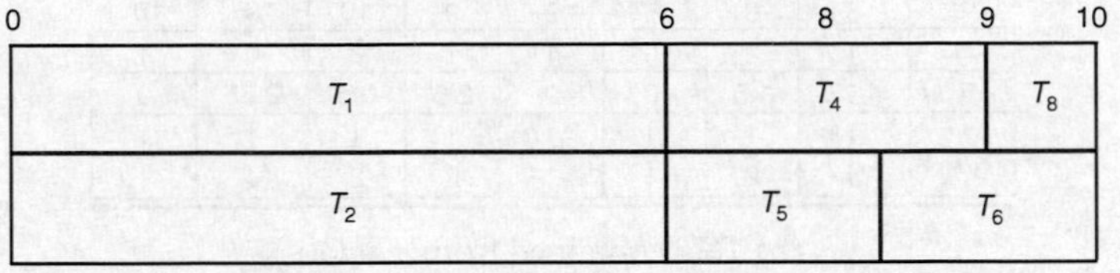

(b) Schedule $S(0)$ with minimum total error for $TS(0)$.

FIGURE 35.1 Initial stage of algorithm *MME*.

T_i	T_1	T_2	T_3	T_4	T_5	T_6	T_7	T_8
y_i	1.5	3	6	6	3	3	1.5	6

(a) Amount of time to be reduced for each task.

T_i	T_1	T_2	T_3^*	T_4^*	T_5	T_6	T_7	T_8^*
r_i	0	0	0	6	6	2	6	6
d_i	6	6	5	9	9	10	9	10
m_i	1	2	0	1	1	2	0	1
o_i	5.5	3	0	0	4	2	3.5	0
w_i^T	1	1	1	1	1	1	1	1
w_i^M	4	2	1	1	2	2	4	1

(b) $TS(1)$ obtained from $TS(0)$.

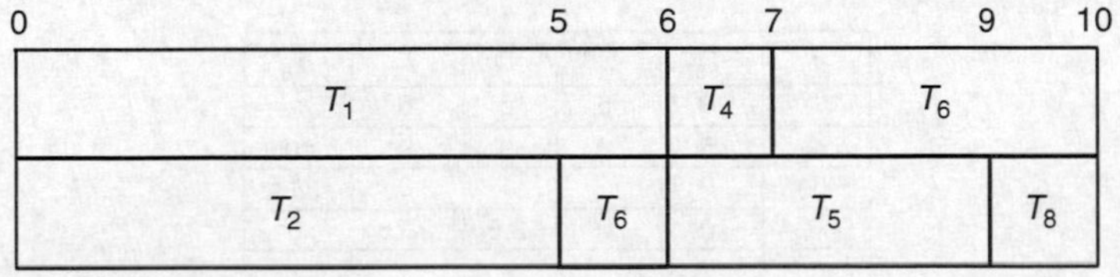

(c) Schedule S(1) with minimum total error for $TS(1)$.

FIGURE 35.2 First iteration of algorithm *MME*.

After formally describing Algorithm *MME*, we illustrate the algorithm by the task system given in Figure 35.1(a). Figure 35.1(b) is the schedule $S(0)$ for $TS(0)$ with minimum total error. Since $\varepsilon(S(0)) = 30 > 0$, the algorithm proceeds to the first iteration. The amount of time to be reduced for each task in $TS(0)$ is computed at the beginning of this iteration and shown in Figure 35.2(a). Figure 35.2(b) gives the task system $TS(1)$ obtained from $TS(0)$ by reducing the execution time of the optional subtask of each task; the irreducible tasks are marked by an asterisk in the figure. A schedule for $TS(1)$ with minimum total error is shown in Figure 35.2(c). Since there is no unsaturated interval in this schedule, we can conclude that no task can be removed.

The total error of the schedule in Figure 35.2(c) is 6, which is greater than 0, so Algorithm *MME* continues with the next iteration. The amount of time to be reduced for each task in $TS(1)$ is given in Figure 35.3(a) and the task system $TS(2)$ is shown in Figure 35.3(b). It is easy to verify that no new irreducible task is generated in this iteration. A schedule, $S(2)$, with minimum total error for $TS(2)$ is shown in Figure 35.3(c). Since time interval [5,6] is unsaturated in $S(2)$, all the tasks partially scheduled

T_i	T_1	T_2	T_3	T_4	T_5	T_6	T_7	T_8
y_i	0.75	1.5	—	—	1.5	1.5	0.75	—

(a) Amount of time to be reduced for each task.

T_i	T_1	T_2	T_3^*	T_4^*	T_5	T_6	T_7	T_8^*
r_i	0	0	0	6	6	2	6	6
d_i	6	6	5	9	9	10	9	10
m_i	1	2	0	1	1	2	0	1
o_i	4.75	1.5	0	0	2.5	0.5	2.75	0
w_i^T	1	1	1	1	1	1	1	1
w_i^M	4	2	1	1	2	2	4	1

(b) $TS(2)$ obtained from $TS(1)$.

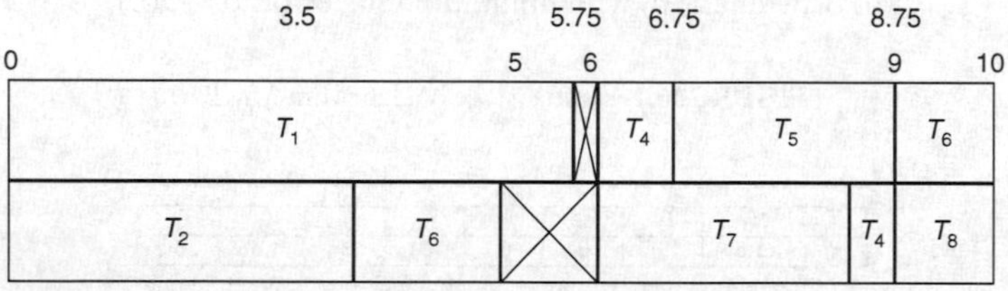

(c) Schedule S(2) with minimum total error for $TS(2)$.

FIGURE 35.3 Second iteration of algorithm *MME*.

T_i	T_3	T_4	T_5	T_7	T_8
y_i	—	—	5/6	5/12	—

(a) Shrunk amount of each task.

T_i	T_3^*	T_4^*	T_5	T_7	T_8^*
r_i	0	6	6	6	6
d_i	5	9	9	9	10
m_i	0	1	1	0	1
o_i	0	0	2.5	2.75	0
w_i^T	1	1	1	1	1
w_i^M	1	1	2	4	1

(b) $TS(3)$ obtained from $TS(2)$.

FIGURE 35.4 Third iteration of algorithm *MME*.

in this interval must be removable, which include T_1, T_2, and T_6. Therefore, the algorithm removes these three tasks from $TS(2)$ and proceeds to the next iteration (because of $\varepsilon(S(2)) > 0$).

Figure 35.4 shows the results, except the schedule, generated in the third iteration. From Figure 35.4(b), we may construct a schedule with zero total error, so Algorithm *MME* terminates after this iteration and outputs the minimum maximum w^M-*weighted* error of $TS(0)$ (which is $10\frac{2}{3}$).

35.3.2 Algorithm Analysis

In this subsection we will show the correctness and the time complexity of Algorithm *MME*, based on three important properties of the removable tasks. The correctness of these properties will be given in the next subsection. The properties are stated as follows.

Properties 35.1

Let T_r be a removable task in TS. Then, T_r must be precisely scheduled in any schedule S for TS with minimum total error, i.e., $\varepsilon(S) = \varepsilon(TS)$.

Properties 35.2

Let T_r be a removable task in TS. Then, $E_{w^M}(TS) = E_{w^M}(TS - \{T_r\})$.

Properties 35.3

Let S be a schedule for TS with the minimum total error, i.e., $\varepsilon(S) = \varepsilon(TS)$. If T_r is partially scheduled in an unsaturated interval in S, T_r must be removable.

To facilitate the analysis of Algorithm *MME,* we need to introduce some more notation. Recall that in Algorithm *MME,* the task system $TS(i)$ is obtained by first initializing it to be $TS(i-1)$. Then, the execution times of the optional subtasks of all the reducible tasks are reduced by a given amount computed based on $\varepsilon(TS)$, and finally some removable tasks, if they exist, are removed from $TS(i)$. In the sequel, $TS^+(i)$ is used to represent the task system of $TS(i)$ right after the reducing operation, i.e., Step 5 of Algorithm *MME.*

To show the time complexity of Algorithm *MME,* we need to prove an important property of the algorithm as stated in the following lemma.

Lemma 35.1

If Algorithm MME does not terminate in the ith iteration, i.e., $\varepsilon(S(i)) > 0, i \geq 1$, then, either a task is marked as irreducible in Step 5 or at least one task is removed in Step 8 in the ith iteration.

Proof

The lemma will be proved by contradiction. We assume that no task is marked as irreducible and also no task is removed in the ith iteration, for some $i \geq 1$. With the assumption, we can show that there must exist a task in $TS^+(i)$ such that this task is partially scheduled in an unsaturated interval in $S(i)$. Then, from Properties 35.3, the task must be a removable task that will be removed in the ith iteration, contradicting the original assumption.

Let $\widehat{TS}$ be the task set in that each task is reducible and irremovable with respect to $TS(i-1)$, i.e., $\widehat{TS} = \{T_k\colon o_k(i-1) > 0 \text{ and } T_k \in TS^+(i-1) \cap TS^+(i)\}$. Since no task is marked as irreducible in the ith iteration, each task T_k in $\widehat{TS}$ is reduced by the amount $y_k(i) = \varepsilon(S(i-1))/(w_k^M \times \Delta)$ in this iteration, where $\Delta = \sum_{T_m \in \widehat{TS}} 1/w_m^M$. Furthermore, $o_k(i) = o_k(i-1) - y_k(i) > 0$ for each task $T_k \in \widehat{TS}$. Let $TS^* = TS^+(i-1) - TS^+(i)$, i.e., all the tasks in TS^* are removable in the $(i-1)$th iteration. The following equality holds.

$$\sum_{T_z(i-1) \in TS^+(i-1)} e_z(i-1) = \sum_{T_z(i) \in TS^+(i)} e_z(i) + \sum_{T_k \in \widehat{TS}} y_k(i) + \sum_{T_x \in TS^*} e_x(i-1) \tag{35.1}$$

From the definition of the total error of a schedule, we have the following two equalities.

$$\varepsilon(S(i-1)) = \sum_{T_z(i-1) \in TS^+(i-1)} e_z(i-1) - \sum_{T_z(i-1) \in TS^+(i-1)} \alpha(T_z(i-1), S(i-1)) \tag{35.2}$$

$$\varepsilon(S(i)) = \sum_{T_z(i) \in TS^+(i)} e_z(i) - \sum_{T_z(i) \in TS^+(i)} \alpha(T_z(i), S(i)) \tag{35.3}$$

Substituting (35.2) and (35.3) into (35.1), the following equality holds.

$$\sum_{T_z(i-1)\in TS^+(i-1)} \alpha(T_z(i-1), S(i-1)) + \varepsilon(S(i-1)) = \sum_{T_z(i)\in TS^+(i)} \alpha(T_z(i), S(i)) + \varepsilon(S(i)) + \sum_{T_k\in \widehat{TS}} y_k(i) + \sum_{T_x\in TS^*} e_x(i-1) \qquad (35.4)$$

From the definitions of $\widehat{TS}$ and TS^*, $\widehat{TS} \bigcup TS^* = TS^+(i-1)$. Moreover, since all the tasks in TS^* are removable in $TS^+(i-1)$, we know that no task in TS^* contributes error to $\varepsilon(S(i-1))$ from Properties 35.1. By assumption, none of the tasks in $TS^+(i)$ is marked as irreducible, so $o_z(i) \geq y_z(i)$ for each $T_z(i)$ in $TS^+(i)$. Thus, we have

$$\varepsilon(S(i-1)) = \sum_{T_k\in \widehat{TS}} y_k(i) \qquad (35.5)$$

Substituting (35.5) into (35.4), we obtain

$$\sum_{T_z(i-1)\in TS^+(i-1)} \alpha(T_z(i-1), S(i-1)) = \sum_{T_z(i)\in TS^+(i)} \alpha(T_z(i), S(i)) + \varepsilon(S(i)) + \sum_{T_x\in TS^*} e_x(i-1) \qquad (35.6)$$

Since the algorithm does not terminate at the end of the ith iteration, $\varepsilon(S(i)) > 0$. So, we have

$$\sum_{T_z(i-1)\in TS^+(i-1)} \alpha(T_z(i-1), S(i-1)) - \sum_{T_x\in TS^*} e_x(i-1) > \sum_{T_z(i)\in TS^+(i)} \alpha(T_z(i), S(i)) \qquad (35.7)$$

(35.7) can be rewritten in matrix form as follows:

$$\sum_{1\leq j\leq p} \sum_{T_z(i-1)\in TS^+(i-1)} SM_{S(i-1)}(z, j) - \sum_{T_x\in TS^*} e_x(i-1) > \sum_{1\leq j\leq p} \sum_{T_z(i)\in TS^+(i)} SM_{S(i)}(z, j) \qquad (35.8)$$

Since only the removable tasks can be removed from $TS^+(i-1)$ by Algorithm *MME*, all the tasks removed at the end of the $(i-1)$th iteration, i.e. all the tasks in TS^*, must be precisely scheduled in $S(i-1)$ by Properties 35.1. That is,

$$\sum_{1\leq j\leq p} \sum_{T_x\in TS^*} SM_{S(i-1)}(x, j) = \sum_{T_x\in TS^*} e_x(i-1) \qquad (35.9)$$

Substituting (35.9) into (35.8), we have

$$\sum_{1\leq j\leq p} \sum_{T_z(i)\in TS^+(i)} SM_{S(i-1)}(z, j) > \sum_{1\leq j\leq p} \sum_{T_z(i)\in TS^+(i)} SM_{S(i)}(z, j) \qquad (35.10)$$

From (35.10), there must exist an interval I_a such that

$$\sum_{T_z(i)\in TS^+(i)} SM_{S(i-1)}(z, a) > \sum_{T_z(i)\in TS^+(i)} SM_{S(i)}(z, a) \qquad (35.11)$$

Because $S(i-1)$ is a feasible and legitimate schedule for $TS(i-1)$, we have the following inequality from (35.11) and the constraints of $SM_{S(i-1)}$:

$$\begin{aligned} m \times L_a &\geq \sum_{T_z(i-1)\in TS^+(i-1)} SM_{S(i-1)}(z, a) \\ &\geq \sum_{T_z(i)\in TS^+(i)} SM_{S(i-1)}(z, a) \\ &> \sum_{T_z(i)\in TS^+(i)} SM_{S(i)}(z, a) \end{aligned}$$

So, we can conclude that I_a is an unsaturated interval in $S(i-1)$.

From (35.11), there must also exist a task $T_x(i)$ in $TS^+(i)$ such that $SM_{S(i-1)}(x,a) > SM_{S(i)}(x,a)$. Since $SM_{S(i)}(x,a) \geq 0$, we have $SM_{S(i-1)}(x,a) > 0$. Therefore, T_x is available in I_a. Moreover, $SM_{S(i)}(x,a) < L_a$, since $SM_{S(i-1)}(x,a) \leq L_a$. So, T_x is partially scheduled in I_a in $S(i)$. From Properties 35.3, T_x is removable. □

With Lemma 35.1, we can show the time complexity of Algorithm *MME*.

Theorem 35.1

The time complexity of Algorithm MME is $O(n^3 \log^2 n)$ for m parallel and identical processors and $O(n^2)$ for single processor.

Proof

We will show that Algorithm *MME* must terminate by the $(2n-1)$th iteration. From Lemma 35.1, a task is either marked irreducible or removed in each iteration. Thus, in the worst case, a task is marked as irreducible in one iteration and then removed in a subsequent iteration. Consequently, there will be at most one task left after the $(2n-2)$th iteration. Clearly, the remaining task will either be marked as irreducible or removed in the $(2n-1)$th iteration and hence Algorithm *MME* will terminate.

It is clear that in each iteration, every step, except Steps 7 and 8, takes at most $O(n)$ time. For multiprocessors, Step 7 can be done in $O(n^2 \log^2 n)$ time by applying the algorithm proposed by Shih et al. [5]. Step 8 can be done in $O(n^2)$ time, since there are at most $O(n)$ intervals that need to be examined and in each interval there exist at most n available tasks. Thus, Algorithm *MME* runs $O(n^3 \log^2 n)$ time for m parallel and identical processors, where the number of processors does not affect the complexity.

In a single processor environment, the algorithm presented by Leung et al. [15] generates a schedule with minimum total error in $O(n)$ time if tasks are sorted in nondecreasing order of the release times. Thus, with a preprocessing step for sorting, Step 7 of Algorithm *MME* can be done in $O(n)$ time. We will show that Step 8 can also be done in $O(n)$ time by using a preprocessing step. By $O(n^2)$ time, we can build doubly linked lists according to tasks and intervals to represent the sets of tasks in all intervals. As tasks are removed in one iteration, they will be deleted from the doubly linked lists. At Step 8, we can locate the first unsaturated interval in that there is a nonempty set of available tasks, and then remove all the partially scheduled tasks in this task set based on Properties 35.3. Thus, Step 8 can be done in $O(n)$ time. Hence, the time complexity of the proposed algorithm is $O(n^2)$. □

Now, we proceed to show the correctness of Algorithm *MME*. Before that, we need to prove the following three lemmas. The first lemma gives the lower bound of the minimum maximum w^M-*weighted* error of a given doubly weighted task system *TS*.

Lemma 35.2

For a given doubly weighted task system TS, $E_{w^M}(TS) \geq \varepsilon(TS)/\Delta$, where $\Delta = \sum_{T_i \in TS \text{ and reducible}} 1/w_i^M$.

Proof

By definition, only reducible tasks can contribute to the total error and minimum maximum w^M-*weighted* error of *TS*. It is easy to see that the lowest possible maximum error of any task system *TS* would occur when the w^M-*weighted* errors of all the reducible tasks are equal and the total error of all the reducible tasks is equal to $\varepsilon(TS)$. Hence, the lower bound of $E_{w^M}(TS)$ can be computed by using the following system of equations, where n' denotes the number of reducible tasks, y_{i_h} the error of the reducible task T_{i_h}, and K the maximum w^M-*weighted* error.

$$y_{i_1} + y_{i_2} + \cdots + y_{i_{n'}} = \varepsilon(TS)$$

$$y_{i_1} \times w_{i_1}^M = y_{i_2} \times w_{i_2}^M = \cdots = y_{i_{n'}} \times w_{i_{n'}}^M = K$$

Clearly, $E_{w^M}(TS) \geq K = \varepsilon(TS)/\Delta$, where $\Delta = \sum_{T_i \in TS \text{ and reducible}} 1/w_i^M$. □

In the next lemma, we will show a relationship between the maximum w^M-*weighted* errors of two task systems when one is obtained from the other by reducing the execution times of the optional subtasks of the reducible tasks by some amount.

Lemma 35.3

For an arbitrary task system TS, let TS be obtained from TS by resetting the execution time of the optional subtask of each task T_i to be* $\max(0, o_i - x/w_i^M)$, *where* $0 \le x \le E_{w^M}(TS)$. *Then,* $E_{w^M}(TS^*) + x = E_{w^M}(TS)$.

Proof

Without loss of generality, we may assume that $E_{w^M}(TS^*) > 0$; otherwise, the lemma holds vacuously since $x \le E_{w^M}(TS)$. The lemma will be proved by contradiction. Suppose that

$$E_{w^M}(TS^*) + x > E_{w^M}(TS) \tag{35.12}$$

We will show that there is a feasible schedule S^* for TS^* such that $E_{w^M}(S^*) < E_{w^M}(TS^*)$, contradicting the definition of $E_{w^M}(TS^*)$.

Let S be a feasible schedule for TS with minimum maximum w^M-*weighted* error. By the definition of $E_{w^M}(S)$, we know that, for each task $T_i \in TS$, $\varepsilon_{w^M}(T_i, S) \le E_{w^M}(S) = E_{w^M}(TS)$. Thus, we have the following inequality:

$$\alpha(T_i, S) \begin{cases} \ge m_i + o_i - \frac{E_{w^M}(TS)}{w_i^M} & \text{if } o_i > \frac{E_{w^M}(TS)}{w_i^M} \\ \ge m_i & \text{otherwise} \end{cases} \tag{35.13}$$

From (35.12) and (35.13), we can get the following inequality:

$$\alpha(T_i, S) \begin{cases} > m_i + o_i - \frac{E_{w^M}(TS^*)}{w_i^M} - \frac{x}{w_i^M} & \text{if } o_i > \frac{E_{w^M}(TS)}{w_i^M} \\ \ge m_i & \text{otherwise} \end{cases} \tag{35.14}$$

The schedule S^* is constructed from S as follows. (Note that T_i^* and T_i denote corresponding tasks belonging to TS^* and TS, respectively.)

Schedule T_i^* in S^* as T_i in S, except that the total processor time assigned to T_i^* is restricted to be $\min\{m_i + o_i^*, \alpha(T_i, S)\}$.

In fact, schedule S^* can be obtained from S by simply converting some of the processor time assigned to T_i into idle time whenever $\alpha(T_i, S) > m_i + o_i^*$.

From the way that S^* is constructed, it is clear that $\alpha(T_i^*, S^*) \ge m_i$ for each T_i^* in TS^*. Thus, S^* is a feasible schedule for TS^*. We will show that $E_{w^M}(S^*) < E_{w^M}(TS^*)$ by showing that $\varepsilon_{w^M}(T_i^*, S^*) < E_{w^M}(TS^*)$. The tasks in TS^* can be partitioned into three groups: Γ_1, Γ_2, and Γ_3 that are defined as follows:

Γ_1: $\{T_i^* : \alpha(T_i^*, S^*) = m_i + o_i^*\}$
Γ_2: $\{T_i^* : \alpha(T_i^*, S^*) = \alpha(T_i, S) \text{ and } o_i > E_{w^M}(TS)/w_i^M\}$
Γ_3: $\{T_i^* : \alpha(T_i^*, S^*) = \alpha(T_i, S) \text{ and } o_i \le E_{w^M}(TS)/w_i^M\}$

Each task in Γ_1 is precisely scheduled in S^*. Thus, $\varepsilon_{w^M}(T_i^*, S^*) = 0 < E_{w^M}(TS^*)$ if T_i^* is in Γ_1. For each task T_i^* in Γ_2, we have $o_i^* = \max\{0, o_i - x/w_i^M\} = o_i - \frac{x}{w_i^M}$, since $o_i > E_{w^M}(TS)/w_i^M$. From the definition of Γ_2 and (35.14), we have $\alpha(T_i^*, S^*) = \alpha(T_i, S) > m_i + o_i - E_{w^M}(TS^*)/w_i^M - x/w_i^M$. So, $\alpha(T_i^*, S^*) > m_i + o_i^* - E_{w^M}(TS^*)/w_i^M$. Hence $\varepsilon_{w^M}(T_i^*, S^*) < E_{w^M}(TS^*)$ for each task in Γ_2. Finally, for each task T_i^* in Γ_3, we have $o_i \le E_{w^M}(TS)/w_i^M$. Thus, $o_i^* = \max\{0, o_i - x/w_i^M\} \le \max\{0, E_{w^M}(TS)/w_i^M - x/w_i^M\} < E_{w^M}(TS^*)/w_i^M$ from (35.12) and the fact that $E_{w^M}(TS^*)/w_i^M > 0$. Therefore, we can conclude that $\varepsilon_{w^M}(T_i^*, S^*) \le E_{w^M}(TS^*)$ for each task T_i^* in TS^*. □

Using Properties 35.2 and 35.3 and Lemmas 35.2 and 35.3, we can show another important property of Algorithm *MME* in the following lemma.

Lemma 35.4

Suppose that Algorithm MME terminates at the kth iteration, $k \geq 1$, for a given task system TS. Then, $\eta(i) + E_{w^M}(TS^+(i)) \leq E_{w^M}(TS)$ for each $0 \leq i \leq k$.

Proof

The lemma will be proved by induction on i. The lemma holds obviously for $i = 0$, since $\eta(i) = 0$ and $E_{w^M}(TS^+(0)) \leq E_{w^M}(TS(0)) = E_{w^M}(TS)$. By assuming that the lemma is true for $i - 1$, we will show that it is true for i.

Let TS^* be the set of tasks removed in the $(i-1)$th iteration, and let $\widehat{TS} = TS^+(i-1) - TS^*$. From Properties 35.1 , 35.2, and 35.3, the tasks in TS^* will not affect the maximum w^M*-weighted* error as well as the total error. Thus, $E_{w^M}(\widehat{TS}) = E_{w^M}(TS^+(i-1))$ and $\varepsilon(\widehat{TS}) = \varepsilon(TS^+(i-1))$. Furthermore, $TS^+(i)$ is obtained from $\widehat{TS}$ by reseeting the execution time of the optional subtask of each task $T_j(i)$ by $\max\{0, o_j(i-1) - x/w_j^M\}$, where $x = \varepsilon(TS^+(i-1))/\Delta$. From Lemma 35.2, we have $x = \varepsilon(TS^+(i-1))/\Delta \leq E_{w^M}(TS^+(i-1))$. Thus, applying Lemma 35.3 to $\widehat{TS}$ and $TS^+(i)$, we have

$$\frac{\varepsilon(TS^+(i-1))}{\Delta} + E_{w^M}(TS^+(i)) \leq E_{w^M}(\widehat{TS}) = E_{w^M}(TS^+(i-1)).$$

On the other hand, $\eta(i) = \eta(i-1) + \varepsilon(TS^+(i-1))/\Delta$. Thus, we have

$$\begin{aligned}\eta(i) + E_{w^M}(TS^+(i)) &= \eta(i-1) + \frac{\varepsilon(TS^+(i-1))}{\Delta} + E_{w^M}(TS^+(i)) \\ &\leq \eta(i-1) + E_{w^M}(TS^+(i-1)) \\ &\leq E_{w^M}(TS).\end{aligned}$$

□

Now, we show the correctness of Algorithm *MME* in the following theorem.

Theorem 35.2

For any task system TS, Algorithm MME correctly computes $E_{w^M}(TS)$.

Proof

Without loss of generality, we assume that Algorithm *MME* terminates at the end of the kth, $k \geq 1$, iteration for TS. We will show that $\eta(k) = E_{w^M}(TS)$. From the nature of the algorithm, both the total error and maximum w^M*-weighted* error of $TS^+(k)$ are zero. Also, $\eta(k) \leq E_{w^M}(TS)$ from Lemma 35.4. Since $\varepsilon(TS^+(k)) = 0$, we can conclude that if the execution time of the optional subtask of each task T_i in TS was reset to be $\max\{0, o_i - \eta(k)/w_i^M\}$, there exists a feasible schedule S in which every task is precisely scheduled. Clearly, S is a feasible schedule for TS with $E_{w^M}(S) \leq \eta(k)$. Thus, we have $\eta(k) \geq E_{w^M}(S) \geq E_{w^M}(TS)$, and consequently, $\eta(k) = E_{w^M}(TS)$. □

35.3.3 Characteristics of Specific Tasks

Since the time complexity and the correctness of Algorithm *MME* are based on the properties of removable tasks stated in the previous subsection, we need to show the correctness of these properties to complete the analysis of the algorithm. We begin by proving Properties 35.1 in the following theorem.

Theorem 35.3

Let T_r be a removable task in TS. Then, T_r must be precisely scheduled in any schedule S for TS with minimum total error, i.e., $\varepsilon(S) = \varepsilon(TS)$.

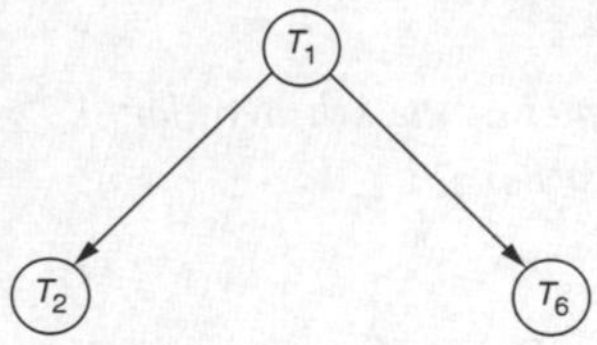

FIGURE 35.5 Directed tree G constructed based on the schedule in Figure 35.3

Proof

Let T_r be a removable task in *TS* and *S* be an arbitrary schedule for *TS* with minimum total error. Consider the schedule S^* obtained from S by deleting T_r from S. Clearly, S^* is a feasible schedule for $TS - \{T_r\}$. Furthermore, $\varepsilon(S^*) = \varepsilon(S) - \varepsilon(T_r, S)$. If T_r is imprecisely scheduled in S, we have $\varepsilon(T_r, S) > 0$, and hence $\varepsilon(S) > \varepsilon(S^*)$. Since $\varepsilon(S^*) \geq \varepsilon(TS - \{T_r\})$, we have $\varepsilon(TS) = \varepsilon(S) > \varepsilon(S^*) \geq \varepsilon(TS - \{T_r\})$, contradicting the fact that T_r is a removable task. Thus, T_r must be precisely scheduled in S. □

To facilitate the proofs of Properties 35.2 and 35.3, we introduce a directed tree $G = (V, A)$ constructed based on the triple (*TS*, *S*, Γ), where *TS* is a task system, *S* is a schedule for *TS* with minimum total error, and Γ is a subset of *TS*. Each vertex in V represents a task in *TS*. Note that a task is represented by at most one vertex (and possibly none) in V. A directed edge from T_i to T_j, $<T_i, T_j>$, implies that, in S, T_j is partially scheduled in an interval in which T_i is scheduled. G is constructed as follows: Initially, V consists of only the tasks in Γ and A is an empty set. Then, the unvisited tasks in V are visited in a breadth-first manner. Suppose we are visiting the task T_i. Let $\Phi(i)$ denote the set of intervals in which T_i is scheduled in S. Expand G by adding the tasks $\{T_j : T_j \notin V$ and T_j is partially scheduled in one of the interval in $\Phi(i)\}$ to V and the directed edges $\{<T_i, T_j> : T_j \notin V$ and T_j is partially scheduled in one of the interval in $\Phi(i)\}$ to A. The above process is iterated until all the tasks in V have been visited. From the way that G is expanded, a task is represented by at most one vertex in G. Thus, the number of vertexes in G is finite and the expansion of G will stop within a finite number of steps. The next lemma gives an important property of the tasks in G. Then, by using the lemma, Properties 35.2 will be proved in Theorem 35.4.

To illustrate the construction of the directed tree, we show the constructed directed tree G in Figure 35.5, where *TS* and *S* are based on the ones given in Figure 35.3 and Γ consists of only T_1. From Figure 35.1(a), we know that T_1 is available in three intervals, [0,2], [2,5], and [5,6]. From the schedule shown in Figure 35.3(c), T_1 is scheduled in interval [5,6] and the tasks T_2 and T_6 are partially scheduled in this interval. Therefore, T_2 and T_6 are added into V and the directed edge $<T_1, T_2>$ and $<T_1, T_6>$ are added to A. Note that if we choose interval [0,2] instead of [5,6], we can only add T_6 into V, since T_2 is fully scheduled in this interval. But, if we choose interval [2,5], we can get the same result as choosing [5,6], since both T_2 and T_6 are partially scheduled in [2,5].

Lemma 35.5

Let G be the directed tree constructed by the above process for (TS, S, $\{T_r\}$), where S is a feasible schedule for TS with minimum total error. If T_r is a removable task in TS, all the tasks in G are precisely scheduled in S.

Proof

By Theorem 35.3, T_r is precisely scheduled in S. We may assume that $e_r > 0$; otherwise, the lemma holds obviously, since T_r is the only task in G.

Suppose that there is a task $T_m \in G$ and $m \neq r$ such that T_m is imprecisely scheduled in S. Let $P = (T_r = T_{m_1}, T_{m_2}, \ldots, T_{m_z} = T_m)$ be the unique path from T_r to T_m in G. From the way that G is constructed, there is a sequence of intervals $(I_{n_1}, I_{n_2}, \ldots, I_{n_{z-1}})$ such that T_{m_h} is scheduled in I_{n_h} and $T_{m_{h+1}}$ is partially scheduled in I_{n_h} for $1 \leq h < z$. Let ρ be the minimum of the following three values:

1. $\min_{1\leq h<z}\{SM_S(m_h, n_h)\}$;
2. $\min_{1\leq h<z}\{L_{m_h} - SM_S(m_{h+1}, n_h)\}$;
3. $\varepsilon(T_m, S)$.

Consider the schedule S^* obtained from S as follows:

Delete T_r from S. Then, shift ρ amount of processor time assigned to $T_{m_{h+1}}$ from $I_{n_{h+1}}$ into I_{n_h} for $1 \leq h < z-1$, and assign ρ amount more processing time to T_m in $I_{n_{z-1}}$.

From the definition of ρ, it is easy to verify that S^* is also a feasible schedule for the task system $TS-\{T_r\}$. As a result, $\varepsilon(S^*) \geq \varepsilon(TS - \{T_r\})$. From the definition of the removable task, we have $\varepsilon(TS - \{T_r\}) = \varepsilon(TS)$. Thus, $\varepsilon(S^*) \geq \varepsilon(S)$. Also from the way that S^* is constructed, we have $\varepsilon(S^*) = \varepsilon(S) - \rho$. So, $\varepsilon(S) - \rho \geq \varepsilon(S)$. Therefore, showing that $\rho > 0$ is sufficient to prove the lemma by contradiction.

Since T_{m_h} is scheduled in I_{n_h} for $1 \leq h < z$, we have $\min_{1\leq h<z}\{SM_S(m_h, n_h)\} > 0$. Since $T_{m_{h+1}}$ is partially scheduled in I_{n_h} for each $1 \leq h < z$, we have $\min_{1\leq h<z}\{L_{m_h} - SM_S(m_{h+1}, n_h)\} > 0$. Finally, since T_m is assumed to be imprecisely scheduled in S, we have $\varepsilon(T_m, S) > 0$. Thus, we have $\rho > 0$. □

Now, we proceed to prove Properties 35.2 in the following theorem.

Theorem 35.4

Let T_r be a removable task in TS. Then, $E_{w^M}(TS) = E_{w^M}(TS - \{T_r\})$.

Proof

We may assume that $e_r > 0$; otherwise, the theorem holds vacuously. It is clear that $E_{w^M}(TS) \geq E_{w^M}(TS - \{T_r\})$. To complete the proof, we only need to show that $E_{w^M}(TS) \leq E_{w^M}(TS - \{T_r\})$.

Let S_T be a feasible schedule for *TS* with minimum total error, and S_M be a feasible schedule for $TS - \{T_r\}$ with minimum maximum w^M-*weighted* error. From S_T and S_M, we will construct a feasible schedule $\hat{S}$ for *TS* such that

$$E_{w^M}(\hat{S}) \leq E_{w^M}(S_M) \tag{35.15}$$

Since $\hat{S}$ is a feasible schedule for *TS*, we have

$$E_{w^M}(TS) \leq E_{w^M}(\hat{S}) \tag{35.16}$$

From (35.15), (35.16) and the definition of S_M, we have

$$E_{w^M}(TS) \leq E_{w^M}(\hat{S}) \leq E_{w^M}(S_M) = E_{w^M}(TS - \{T_r\}),$$

and hence the desired result.

First, we construct the directed tree G based on the triple $(TS, S_T, \{T_r\})$. Then, construct $\hat{S}$ from S_T and S_M such that, for $1 \leq j \leq p$,

$$SM_{\hat{S}}(i, j) = \begin{cases} SM_{S_T}(i, j) & \text{if } T_i \in G; \\ SM_{S_M}(i, j) & \text{otherwise.} \end{cases}$$

In other words, if $T_i \in G$, schedule T_i in $\hat{S}$ as in S_T; otherwise, schedule T_i as in S_M.

By Lemma 35.5, all the tasks in G are precisely scheduled in S_T. Since all the tasks in G are scheduled in $\hat{S}$ as in S_T, they are also precisely scheduled in $\hat{S}$. Thus, the maximum w^M-*weighted* error of these tasks is zero in $\hat{S}$. Moreover, all the tasks not in G are scheduled in $\hat{S}$ as in S_M, so the maximum w^M-*weighted* error of these tasks is equal to $E_{w^M}(S_M)$. Therefore, we can conclude that $\hat{S}$ must satisfy (35.15).

It is easy to see that $\alpha(T_i, \hat{S}) \geq m_i$ for all $T_i \in TS$, so it remains to be shown that $\hat{S}$ is a legitimate schedule. That is, for each interval I_j, $1 \leq j \leq p$, we have

$$SM_{\hat{S}}(i, j) \leq L_j, \quad \text{for each } 1 \leq i \leq n. \tag{35.17}$$

and

$$\sum_{i=1}^{n} SM_{\hat{S}}(i,j) \leq m \times L_j \tag{35.18}$$

From the way that $\hat{S}$ is constructed, each task T_i satisfies (35.17) in each interval I_j, whether it is scheduled as in S_T or S_M. We now show that (35.18) also holds for each interval I_j. The intervals in $\hat{S}$ can be partitioned into two groups. The first group consists of all those intervals in that either the tasks in G are scheduled, or the tasks not in G are scheduled, but not both. The second group consists of all those intervals in that both the tasks in G and not in G are scheduled. Obviously, all the intervals in the first group satisfy (35.18), as they are identical to either S_T or S_M. To complete the proof, we need to show that each interval in the second group also satisfies (35.18). Let I_j be an interval in the second group. Since there is a task in G scheduled in I_j, every task not in G but available in I_j must be fully scheduled in I_j in S_T; otherwise, it would be in G as well. As a result, $\sum_{T_i \notin G} SM_{\hat{S}}(i,j) = \sum_{T_i \notin G} SM_{S_M}(i,j) \leq \sum_{T_i \notin G} SM_{S_T}(i,j)$. Thus, we have

$$\begin{aligned}\sum_{i=1}^{n} SM_{\hat{S}}(i,j) &= \sum_{T_i \in G} SM_{S_T}(i,j) + \sum_{T_i \notin G} SM_{S_M}(i,j) \\ &\leq \sum_{T_i \in G} SM_{S_T}(i,j) + \sum_{T_i \notin G} SM_{S_T}(i,j) \\ &= \sum_{i=1}^{n} SM_{S_T}(i,j) \\ &\leq m \times L_j.\end{aligned}$$

□

We now proceed to prove Properties 35.3. We first construct the directed tree $G = (V, A)$ based on the triple $(TS, S, \{T_x\})$, where TS is a task system, S is a feasible schedule for TS with minimum total error, and T_x is an arbitrary task in TS. To prove Properties 35.3, we need to show another important characteristic of G in the following lemma.

Lemma 35.6

If T_x is partially scheduled in an unsaturated interval I_j in S, all the tasks in G must be precisely scheduled in S.

Proof

We may assume that $e_x > 0$; otherwise, T_x is the only task in G and the lemma holds vacuously. It is easy to see that T_x is precisely scheduled in S. Otherwise, we can assign T_x more processor time in I_j to obtain a schedule with smaller total error than S, contradicting the fact that $\varepsilon(S) = \varepsilon(TS)$.

We now show that all the tasks in G, other than T_x, are precisely scheduled in S. Suppose that there exists a task T_y in G and $x \neq y$ such that T_y is imprecisely scheduled in S. Let $P = (T_x = T_{m_1}, T_{m_2}, \ldots, T_{m_z} = T_y)$ be the unique path from T_x to T_y. From the way that G is constructed, there is a sequence of intervals $(I_{n_1}, I_{n_2}, \ldots, I_{n_{z-1}})$ such that T_{m_h} is scheduled in I_{n_h} and $T_{m_{h+1}}$ is partially scheduled in I_{n_h} for $1 \leq h < z$. Consider the schedule $\hat{S}$ obtained from S as follows.

Let ρ be the minimum of the following five values:

1. $m \times L_j - \sum_{i=1}^{n} SM_S(i,j)$,
2. $L_j - SM_S(x,j)$,
3. $\min_{1 \leq h < z}\{SM_S(m_h, n_h)\}$,
4. $\min_{1 \leq h < z}\{L_{n_h} - SM_S(m_{h+1}, n_h)\}$,
5. $\varepsilon(T_y, S)$.

If $n_1 \neq j$, shift ρ amount of processor time assigned to T_x from I_{n_1} to I_j and leave ρ amount of extra idle time in I_{n_1} for accommodating T_{n_2}. Otherwise, there must have at least ρ amount of idle time in

$I_{n_1} = I_j$ for accommodating T_{n_2}, from the definition of ρ and the given fact that I_j is unsaturated. Next, shift ρ amount of processor time assigned to $T_{n_{h+1}}$ from $I_{n_{h+1}}$ into I_{n_h} for $1 \leq h < z - 1$. Finally, assign ρ amount more processor time in $I_{n_{z-1}}$ to T_y.

From the way that $\hat{S}$ is obtained from S, every task, except T_y, in G receives the same amount of processor time in $\hat{S}$ and S, and T_y receives ρ amount more in $\hat{S}$ than that in S. Thus, $\hat{S}$ is a feasible schedule for TS, and $\varepsilon(\hat{S}) = \varepsilon(S) - \rho$. So, if $\rho > 0$, we would have $\varepsilon(\hat{S}) < \varepsilon(S)$, contradicting the fact that $\varepsilon(S) = \varepsilon(TS)$.

We will show that ρ is greater than zero by showing that the above five values are all positive. Since I_j is unsaturated and T_x is partially scheduled in it, the first two values must be positive. From the way that G is constructed, we know that the third and the fourth values are positive too. Since T_y is assumed to be imprecisely scheduled in S, the last value must also be positive. Thus, ρ is positive and hence the desired contradiction. □

Now, we show the correctness of Properties 35.3 in Theorem 35.5 by using the result of Lemma 35.6.

Theorem 35.5

Let S be a feasible schedule for TS with minimum total error. If T_x is partially scheduled in an unsaturated interval I_j in S, T_x must be a removable task.

Proof

We may assume the e_x is positive; otherwise, $\varepsilon(TS) = \varepsilon(TS - \{T_x\})$, so the theorem holds obviously. The theorem will be proved by contradiction. It is clear that $\varepsilon(TS) \geq \varepsilon(TS - \{T_x\})$. Therefore, proving $\varepsilon(TS) \leq \varepsilon(TS - \{T_x\})$ is sufficient to show the correctness of the theorem.

Let S^* be a feasible schedule for $TS - \{T_x\}$ with minimum total error. From S and S^*, we can construct a feasible schedule $\hat{S}$ for TS such that

$$\varepsilon(\hat{S}) \leq \varepsilon(S^*) \tag{35.19}$$

Since $\hat{S}$ is a feasible schedule for TS, we have

$$\varepsilon(TS) \leq \varepsilon(\hat{S}) \tag{35.20}$$

By (35.19) and (35.20), we have

$$\varepsilon(TS) \leq \varepsilon(\hat{S}) \leq \varepsilon(S^*) = \varepsilon(TS - \{T_x\})$$

and hence the desired result.

We first construct the directed tree G based on the triple $(TS, S, \{T_x\})$. By Lemma 35.6, all the tasks in G are precisely scheduled in S. Now, construct $\hat{S}$ as follows. If $T_i \in G$, schedule T_i as in S; otherwise, schedule it as in S^*. Since all the tasks in G are scheduled in $\hat{S}$ as in S, they are precisely scheduled in $\hat{S}$. Moreover, all the tasks not in G are scheduled in $\hat{S}$ as in S^*, so $\hat{S}$ must satisfy (35.19). Since T_x is in G and both S and S^* are feasible for TS and $TS - \{T_x\}$, respectively, $\hat{S}$ must be feasible for TS based on a similar argument used in Theorem 35.4. □

35.4 Constrained Total Weighted Error

In this section we continue the study in the previous section. For a given *doubly weighted* task system, we will focus on the problem of minimizing the total w^T-*weighted* error of TS under the constraint that the maximum w^M-*weighted* error is minimized. Note that, in this problem, minimizing maximum weighted error is the primary criterion, and minimizing total weighted error is the secondary one. We will present an algorithm, Algorithm *cTWE*, based on the algorithm given by Ho et al. [7], to solve this problem.

To meet the primary criterion, we need to force each task T_i to be scheduled for at least $\overline{e}_i = m_i + \max\{0, o_i - E_{w^M}(TS)/w_i^M\}$ amount of time. To guarantee that each T_i in TS is assigned at least $\overline{e}_i$ amount of time, the simplest way is to modify TS by setting the execution times of mandatory and optional subtasks of T_i to be $\overline{e}_i$ and $e_i - \overline{e}_i$, respectively. Therefore, Algorithm *cTWE* first invokes Algorithm *MME*, which is given in the previous section, to obtain the minimum maximum w^M-*weighted* error of TS, $E_{w^M}(TS)$. Then, it constructs a new task system $\overline{TS}$ from TS by resetting the execution times of mandatory and optional subtasks as described above. Finally, Algorithm *cTWE* applies the algorithm of Shih et al. [5] for the multiprocessor case or the algorithm of Leung et al. [15] for the single processor case to construct a feasible schedule for $\overline{TS}$ with minimum total w^M-*weighted* error. The formal description is given as follows.

Algorithm *cTWE*

Input: *A doubly weighted task system* $TS = (\{T_i\}, \{r_i\}, \{d_i\}, \{m_i\}, \{o_i\}, \{w_i^T\}, \{w_i^M\})$ *with n tasks and* $m \geq 1$ *parallel and identical processors.*

Output: *A feasible schedule* S_{cTWE} *for TS with minimum total* w^M*-weighted error subject to that the maximum* w^M*-weighted error is minimized.*

Step 1: Invoke Algorithm *MME* to compute $E_{w^M}(TS)$.

Step 2: Construct a new task system $\overline{TS}$ from TS as follows:

For each task T_i in TS, create a corresponding task $\overline{T}_i$ in $\overline{TS}$ such that $\overline{r}_i = r_i$, $\overline{d}_i = d_i$, $\overline{w}_i^T = w_i^T$, $\overline{w}_i^M = w_i^M$, and

$$\overline{m}_i = \begin{cases} m_i + o_i - E_{w^M}(TS)/w_i^M & \text{if } o_i \times w_i^M > E_{w^M}(TS); \\ m_i & \text{otherwise} \end{cases}$$

$$\overline{o}_i = \begin{cases} E_{w^M}(TS)/w_i^M & \text{if } o_i \times w_i^M > E_{w^M}(TS); \\ o_i & \text{otherwise} \end{cases}$$

Step 3: If $m > 1$, invoke the algorithm of Shih et al. [5] for $\overline{TS}$ to generate S_{cTWE}; otherwise, invoke the algorithm of Leung et al. [15] for $\overline{TS}$ to generate S_{cTWE}.

It is easy to verify that Algorithm *cTWE* runs in $O(n^3 \log^2 n)$ time for $m > 1$ identical and parallel processors and $O(n^2)$ time for a single processor environment. Now, we proceed to prove the correctness of Algorithm *cTWE*. First, we show a relationship between TS and $\overline{TS}$ in the following lemma. Then, the correctness of the algorithm can be easily proved based on the lemma. Let $FS(\overline{TS})$ denote the set of feasible schedules for $\overline{TS}$ and $CFS(TS)$ denote the set of feasible schedules for TS with minimum maximum w^M-*weighted* error.

Lemma 35.7

$FS(\overline{TS}) = CFS(TS)$.

Proof

First, we show that $FS(\overline{TS}) \subseteq CFS(TS)$, and then $FS(\overline{TS}) \supseteq CFS(TS)$. Let $\overline{S}$ be an arbitrary schedule in $FS(\overline{TS})$. From the way that $\overline{TS}$ is constructed, $\overline{m}_i \geq m_i$ for each task $\overline{T}_i$. Since $\overline{S}$ is a feasible schedule for $\overline{TS}$, it must also be a feasible schedule for TS. We also know that $\overline{o}_i \leq E_{w^M}(TS)/w_i^M$ for each $\overline{T}_i$ and hence $E_{w^M}(\overline{S}) \leq E_{w^M}(TS)$. Since $E_{w^M}(\overline{S}) \geq E_{w^M}(TS)$, based on the definition of $E_{w^M}(TS)$ and the fact that $\overline{S}$ is feasible for TS, $E_{w^M}(\overline{S}) = E_{w^M}(TS)$. So, we can conclude that $\overline{S}$ is a member of $CFS(TS)$.

After showing $FS(\overline{TS}) \subseteq CFS(TS)$, we now show that $FS(\overline{TS}) \supseteq CFS(TS)$. Let S be an arbitrary member of $CFS(TS)$. Since $E_{w^M}(S) = E_{w^M}(TS)$, $\varepsilon_{w^M}(T_i, S) \leq E_{w^M}(TS)$ for each task T_i in TS. So, we have the following inequality.

$$e_i - \alpha(T_i, S) \begin{cases} \leq o_i & \text{if } o_i \times w_i^M \leq E_{w^M}(TS) \\ \leq \frac{E_{w^M}(TS)}{w_i^M} & \text{otherwise} \end{cases} \tag{35.21}$$

From (35.21), we have

$$\alpha(T_i, S) \begin{cases} \geq m_i & \text{if } o_i \times w_i^M \leq E_{w^M}(TS) \\ \geq m_i + o_i - \frac{E_{w^M}(TS)}{w_i^M} & \text{otherwise} \end{cases} \tag{35.22}$$

From (35.22) and the definition of $\overline{m}_i$, we know that $\alpha(T_i, S) \geq \overline{m}_i$. So, S is a feasible schedule for $\overline{TS}$, i.e. $FS(\overline{TS}) \supseteq CFS(TS)$. □

From Step 3 of Algorithm *cTWE*, S_{cTWE} is a feasible schedule with minimum total w^M-*weighted* error for $\overline{TS}$. Also, from Lemma 35.7, we know that every feasible schedule for $\overline{TS}$ is a feasible schedule with minimum maximum w^M-*weighted* error for *TS*. Thus, we can conclude that S_{cTWE} is a feasible schedule for *TS*, with minimum total w^T-*weighted* error under the constraint that the maximum w^M-*weighted* error is minimized.

Theorem 35.6

Let TS be an arbitrarily given doubly weighted task system. Algorithm cTWE generates a feasible schedule that has the minimum total w^T*-weighted error among all feasible schedules with the minimum maximum* w^M*-weighted error.*

35.5 Constrained Maximum Weighted Error

In this section we will study the problem of preemptively scheduling a doubly weighted task system, with the objective of minimizing the maximum w^M-*weighted* error under the constraint that the total w^T-*weighted* error is minimized. A polynomial-time algorithm, called *cMME* in the following, will be shown for solving this problem, which runs in $O(kn^3 \log^2 n)$ time for parallel and identical processors and in $O(kn^2)$ time for a single processor, where k is the number of distinct w^T-*weights*. It is easy to verify that the problem studied by Shih and Liu [20] is a special case of this problem by simply letting $w_i^T = 1$ and $w_i^M = 1/o_i$ for each task T_i.

This section is organized as follows. In the next subsection we will present Algorithm *cMME*. In Section 35.5.2, we prove an important property that forms the basis of the algorithm.

Before we proceed, we shall define more terminologies and notations that will be needed throughout this section. Let S be a feasible schedule for a doubly weighted task system *TS*. The symbol $E_{w^M}^{w^T}(TS)$ denotes the minimum maximum w^M-*weighted* error under the constraint that the total w^T-*weighted* error is minimized; i.e., $E_{w^M}^{w^T}(TS) = \min\{E_{w^M}(S) : S \text{ is a feasible schedule for } TS \text{ and } \varepsilon_{w^T}(S) = \varepsilon_{w^T}(TS)\}$. Likewise, $\varepsilon_{w^T}^{w^M}(TS)$ is defined to be the minimum total w^T-*weighted* error under the constraint that the maximum w^M-*weighted* error is minimized.

35.5.1 Algorithm for Constrained Maximum Weighted Error

In this subsection we will present Algorithm *cMME* for generating a preemptive schedule that minimizes the maximum w^M-*weighted* error under the constraint that the total w^T-*weighted* error is minimized.

One way to find a schedule with the minimum total w^T-*weighted* error is as follows; see [15] for more details. Let $W_1^T > W_2^T > \cdots > W_k^T$ be the k distinct w^T-*weights* of a doubly weighted task system *TS*. A task whose w^T-*weight* is W_j^T will be called a W_j^T-*weight* task. We iterate the following process k times, from $j = 1$ until k. In the jth iteration, the optional execution time of every task, except the W_j^T-*weight* tasks, are set to zero. Then a schedule S with the minimum total error is obtained, using the algorithms in [15] for the single processor case and the algorithm in [5] for the multiprocessor case. (Note that the tasks with nonzero optional execution times have identical weights, namely, W_j^T.) We then adjust the mandatory execution time of every W_j^T-*weight* task, say T_l, to be $\alpha(T_l, S)$, before we proceed to the next iteration. At the end of the kth iteration, the mandatory execution time of every task is the amount of processor time assigned to each task in the final schedule. The basic idea is that it is always better (for the purpose of minimizing the total w^T-*weighted* error) to execute more of the tasks with larger w^T-*weights* than the ones with smaller w^T-*weights*.

As it turns out, if the w^T-*weights* are identical, then there is a schedule that simultaneously minimizes the total w^T-*weighted* error and the maximum w^M-*weighted* error. This property is stated below as Properties 35.4 and will be proved in Section 35.5.2.

Properties 35.4

For an arbitrary doubly weighted task system TS, if the w^T-weights are identical, then there is a feasible schedule S such that $E_{w^M}(S) = E_{w^M}(TS)$ and $\varepsilon_{w^T}(S) = \varepsilon_{w^T}(TS)$. In other words, $E_{w^M}^{w^T}(TS) = E_{w^M}(TS)$ and $\varepsilon_{w^T}^{w^M}(TS) = \varepsilon_{w^T}(TS)$.

Due to Properties 35.4, we can modify the method mentioned at the beginning of this subsection for finding the minimum total w^T-*weighted* error to solve our problem. At each iteration, instead of finding a schedule with the minimum total error, we use Algorithm *cTWE*, given in Section 35.4, to find a schedule with the minimum total w^T-*weighted* error, subject to the constraint that the maximum w^M-*weighted* error is minimized. Since the tasks with nonzero optional execution times have identical w^T weights at each iteration, by Properties 35.4, the obtained schedule also minimizes the maximum w^M-*weighted* error.

For a given task system *TS*, Algorithm *cMME* iteratively constructs a sequence of k task systems, denoted by $TS(j) = (\{T_i(j)\}, \{r_i(j)\}, \{d_i(j)\}, \{m_i(j)\}, \{o_i(j)\}, \{w_i^T(j)\}, \{w^M{}_i(j)\})$ for $1 \le j \le k$, each of which consists of n tasks. For each task T_i in *TS*, there is a corresponding task $T_i(j)$ in $TS(j)$ such that $r_i(j) = r_i, d_i(j) = d_i, w_i^T(j) = w_i^T$ and $w_i^M(j) = w_i^M$. For each $1 \le j \le k$, if $T_i(j)$ is a W_l^T-*weight* task, $1 \le l \le j$, then $m_i(j)$ is set to be the total processor time assigned to T_i in the schedule for $TS(j)$ with minimum total w^T-*weighted* error under the constraint that the maximum w^M-*weighted* error is minimized and $o_i(j)$ is set to zero. Otherwise, they are set to m_i and 0, respectively. Finally, $m_i(k), 1 \le i \le n$, gives the total processor time assigned to T_i in the final schedule for *TS*.

Initially, $TS(0)$ is set to be the same as *TS*, except that $o_i(0)$ is set to zero for each $1 \le i \le n$. Algorithm *cMME* proceeds in k iterations. In the jth iteration, $1 \le j \le k$, the algorithm sets $TS(j) = (\{T_i(j)\}, \{r_i(j)\}, \{d_i(j)\}, \{m_i(j)\}, \{o_i(j)\}, \{w_i^T(j)\}, \{w_i^M(j)\})$ to be the same as $TS(j-1)$, except that $o_i(j)$ is set to o_i for each W_j^T-*weight* task $T_i(j)$. Algorithm *cTWE* given in Section 35.4 is then invoked to generate a schedule $S(j)$ for $TS(j)$. The variable λ for storing the total w^T-*weighted* error of *TS* (which is initialized to be zero) is increased by the total w^T-*weighted* error of $S(j)$. Also, the variable ρ for storing the maximum w^M-*weighted* error of *TS* (which is initialized to be zero) is set to be the larger of $E_{w^M}(S(j))$ and the current value of ρ. Finally, for each W_j^T-*weight* task $T_i(j)$, the algorithm sets $m_i(j)$ to $\alpha(T_i(j), S(j))$ and $o_i(j)$ to zero. The algorithm then repeats the above process in the next iteration. (Notice that the mandatory and optional execution times of a W_j^T-*weight* task might be different at the beginning of the jth iteration than at the end of the jth iteration.) Algorithm *cMME* is formally represented as follows.

Algorithm *cMME*

Input: *A doubly weighted task system* $TS = (\{T_i\}, \{r_i\}, \{d_i\}, \{m_i\}, \{o_i\}, \{w_i^T\}, \{w_i^M\}), 1 \le i \le n$, *and* $m \ge 1$ *parallel and identical processors.*

Output: *A feasible schedule* S_{cMME} *for TS.*

Step 1: Sort all tasks in nonincreasing order of the w^T-*weights* and let $W_1^T > W_2^T > \cdots > W_k^T$ be the k distinct w^T-*weights* in *TS*;
Step 2: Initialize λ and ρ to zero;
Step 3: Initialize $TS(0)$ to be *TS*, except that $o_i(0)$ is set to zero for each $1 \le i \le n$;
Step 4: For $j = 1, \ldots, k$ perform Steps 5-9.

Step 5: Let each task $T_i(j) \in TS(j)$ be the same as the corresponding one in $TS(j-1)$, except that $o_i(j)$ is set to o_i if $T_i(j)$ is a W_j^T*-weight* task;
Step 6: Invoke Algorithm *cTWE* to construct a schedule $S(j)$ for $TS(j)$;
Step 7: Set λ to be $\lambda + \varepsilon_{w^T}(S(j))$;
Step 8: Set $\rho = \max\{E_{w^M}(S(j)), \rho\}$;
Step 9: For each W_j^T*-weight* task $T_i(j)$, set $m_i(j)$ to be $\alpha(T_i(j), S(j))$ and $o_i(j)$ be zero.
Step 10: Output $S(k)$ and stop;

Theorem 35.7

Algorithm cMME constructs a schedule that minimizes the maximum w^M*-weighted error, subject to the constraint that the total* w^T*-weighted error is minimized Its time complexity is* $O(kn^3 \log^2 n)$ *for parallel and identical processors and* $O(kn^2)$ *for a single processor, where k is the number of distinct* w^T*-weights.*

Proof

The correctness of Algorithm *cMME* is based on Properties 35.4 (which will be proved in the following subsection) and the result of [15] (as stated at the beginning of Section 35.5.1). Algorithm *cMME* calls Algorithm *cTWE* which is given in Section 35.4, k times. Its time complexity immediately follows from the time complexity of Algorithm *cTWE*. □

35.5.2 Proofs of Properties

In this subsection we will prove Properties 35.4 to complete the proof of the correctness of Algorithm *cMME*. We first show, in Theorem 35.8, that if all the w^T*-weights* are identical, then $\varepsilon_{w^T}^{w^M}(TS) = \varepsilon_{w^T}(TS)$. We then show, in Theorem 35.9, that $E_{w^M}^{w^T}(TS) = E_{w^M}(TS)$ for the same condition. From these two theorems, Properties 35.4 is proved.

To prove the correctness of Properties 35.4, we need to use the directed tree $G = (V, A)$, introduced in Section 35.3.3, based on the triple (TS, S, Γ), where TS is a doubly weighted task system, S is a feasible schedule for TS with minimum total error, and Γ is a subset of tasks in TS. Readers may refer the construction of G in Section 35.3.3.

Theorem 35.8

Let TS be an arbitrary doubly weighted task system. If the w^T*-weights are identical, then* $\varepsilon_{w^T}^{w^M}(TS) = \varepsilon_{w^T}(TS)$.

Proof

Let S^* be a feasible schedule for TS with minimum total w^T*-weighted* error, i.e. $\varepsilon_{w^T}(S^*) = \varepsilon_{w^T}(TS)$. Also, let $\hat{S}$ be a feasible schedule for TS with the minimum total w^T*-weighted* error under the constraint that the maximum w^M*-weighted* error is minimized, i.e. $\varepsilon_{w^T}(\hat{S}) = \varepsilon_{w^T}^{w^M}(TS)$ and $E_{w^M}(\hat{S}) = E_{w^M}(TS)$. Since S^* is less restricted than $\hat{S}$, $\varepsilon_{w^T}(S^*) \leq \varepsilon_{w^T}(\hat{S})$. The theorem will be proved using contradiction, so we assume

$$\varepsilon_{w^T}(S^*) < \varepsilon_{w^T}(\hat{S}) \tag{35.23}$$

We will construct a feasible schedule $\overline{S}$ for TS satisfying the following two inequalities.

$$\varepsilon_{w^T}(\hat{S}) \leq \varepsilon_{w^T}(\overline{S}) \tag{35.24}$$

$$\varepsilon_{w^T}(\overline{S}) \leq \varepsilon_{w^T}(S^*) \tag{35.25}$$

From (35.24) and (35.25), we have $\varepsilon_{w^T}(\hat{S}) \leq \varepsilon_{w^T}(S^*)$, contradicting (35.23).

Let W be the common w^T-*weights* of the tasks in TS. By the definition of total w^M-*weighted* error of a schedule, the assumption stated in (35.23) can be rewritten as

$$W \times \left(\sum_{i=1}^{n} (e_i - \alpha(T_i, S^*)) \right) < W \times \left(\sum_{i=1}^{n} (e_i - \alpha(T_i, \hat{S})) \right)$$

which implies that

$$\sum_{i=1}^{n} \alpha(T_i, \hat{S}) < \sum_{i=1}^{n} \alpha(T_i, S^*) \tag{35.26}$$

(35.26) can be rewritten in the matrix form as follows.

$$\sum_{j=1}^{p} \sum_{i=1}^{n} SM_{\hat{S}}(i, j) < \sum_{j=1}^{p} \sum_{i=1}^{n} SM_{S^*}(i, j) \tag{35.27}$$

(35.27) implies that there exists at least one interval I_a in $\hat{S}$ such that

$$\sum_{i=1}^{n} SM_{\hat{S}}(i, a) < \sum_{i=1}^{n} SM_{S^*}(i, a) \leq m \times L_a$$

which means that I_a is an unsaturated interval in $\hat{S}$. Let $\Phi^* = \{I_j : \sum_{i=1}^{n} SM_{\hat{S}}(i, j) < \sum_{i=1}^{n} SM_{S^*}(i, j)\}$. It is clear that Φ^* is not empty. Let $\Gamma = \{T_i : T_i$ is partially scheduled in $I_j \in \Phi^*$ in $\hat{S}\}$. Note that for each $I_j \in \Phi^*$, there is at least one task partially scheduled in I_j in $\hat{S}$. Thus, $\Gamma \neq \emptyset$.

Now, construct the directed graph $G = (V, A)$ based on the triple $(TS, \hat{S}, \Gamma)$. To prove the theorem, we need the following proposition, which will be proved in Lemma 35.8.

Proposition 35.1

All the tasks in G are precisely scheduled in $\hat{S}$.

The schedule $\overline{S}$ is constructed as follows: If $T_i \in G$, then schedule T_i in $\overline{S}$ as in $\hat{S}$; otherwise, schedule T_i in $\overline{S}$ as in S^*. We have the following proposition, which will be proved in Lemma 35.9, for the constructed schedule $\overline{S}$.

Proposition 35.2

$\overline{S}$ is a feasible schedule for TS.

Using the above two propositions, we now prove (35.24) and (35.25) to complete the proof.

First, we show that (35.25) holds, since the proof is simpler. Since a task in G is scheduled in $\overline{S}$ as in $\hat{S}$, it is precisely scheduled in $\overline{S}$, by Proposition 35.1. Thus, we have $\sum_{T_i \in G} \alpha(T_i, \overline{S}) = \sum_{T_i \in G} \alpha(T_i, \hat{S}) \geq \sum_{T_i \in G} \alpha(T_i, S^*)$ and hence

$$\begin{aligned}
\sum_{T_i \in TS} \alpha(T_i, \overline{S}) &= \sum_{T_i \in G} \alpha(T_i, \overline{S}) + \sum_{T_i \notin G} \alpha(T_i, \overline{S}) \\
&= \sum_{T_i \in G} \alpha(T_i, \hat{S}) + \sum_{T_i \notin G} \alpha(T_i, S^*) \\
&\geq \sum_{T_i \in G} \alpha(T_i, S^*) + \sum_{T_i \notin G} \alpha(T_i, S^*) \\
&= \sum_{T_i \in TS} \alpha(T_i, S^*)
\end{aligned} \tag{35.28}$$

Since the w^T-*weights* are identical, (35.25) follows from (35.28).

We now show that (35.24) also holds. The intervals in $\overline{S}$ can be partitioned into three groups. The first group, Φ_1, consists of all those intervals in which only the tasks in G are scheduled. Let I_x be an interval in Φ_1. Since the tasks in G are scheduled in $\overline{S}$ as in $\hat{S}$, we have

$$\sum_{i=1}^{n} SM_{\hat{S}}(i,x) = \sum_{i=1}^{n} SM_{\overline{S}}(i,x) \tag{35.29}$$

The second group, Φ_2, consists of all those intervals in which only the tasks not in G are executed. Let I_y be an interval in Φ_2. We wish to show that

$$\sum_{i=1}^{n} SM_{\hat{S}}(i,y) \geq \sum_{i=1}^{n} SM_{\overline{S}}(i,y) \tag{35.30}$$

Clearly, $\sum_{i=1}^{n} SM_{\overline{S}}(i,y) = \sum_{T_i \notin G} SM_{S^*}(i,y)$. Thus, showing that $\sum_{i=1}^{n} SM_{\hat{S}}(i,y) \geq \sum_{T_i \notin G} SM_{S^*}(i,y)$ is sufficient to prove (35.30). There are two cases to be considered.

Case 1: All the tasks available in I_y are not in G.
Case 2: Some tasks available in I_y are in G.

For Case 1, since all the tasks available in I_y are not in G, we have

$$\sum_{i=1}^{n} SM_{\hat{S}}(i,y) = \sum_{T_i \notin G} SM_{\hat{S}}(i,y) \tag{35.31}$$

and

$$\sum_{i=1}^{n} SM_{S^*}(i,y) = \sum_{T_i \notin G} SM_{S^*}(i,y)$$

Assume that

$$\sum_{T_i \notin G} SM_{S^*}(i,y) > \sum_{T_i \notin G} SM_{\hat{S}}(i,y) \tag{35.32}$$

Then, we have

$$\sum_{i=1}^{n} SM_{S^*}(i,y) = \sum_{T_i \notin G} SM_{S^*}(i,y) > \sum_{T_i \notin G} SM_{\hat{S}}(i,y) \tag{35.33}$$

From (35.31) and (35.33), we obtain

$$\sum_{i=1}^{n} SM_{S^*}(i,y) > \sum_{i=1}^{n} SM_{\hat{S}}(i,y)$$

which implies that I_y is in Φ^* from the definition of Φ^*. Moreover, there exists at least one task T_i partially scheduled in I_y in $\hat{S}$. Thus, T_i would be in Γ from the definition of Γ and hence in G. This contradicts the definition of Φ_2. So, (35.32) is not true. In other words, $\sum_{i=1}^{n} SM_{\hat{S}}(i,y) \geq \sum_{T_i \notin G} SM_{S^*}(i,y)$.

Now we consider Case 2. Since there is no task in G scheduled in I_y in $\overline{S}$ from the definition of Φ_2, all those tasks in G that are available in I_y must not be scheduled in I_y in $\hat{S}$. Therefore, (35.31) also holds in this case. We again assume that (35.32) holds in this case. Using the same argument as in Case 1, we can conclude that I_y is in Φ^*. For each task T_i available in I_y but not in G, it must be fully scheduled in I_y in $\hat{S}$; otherwise, it would be in Γ and hence in G. Thus, we have

$$SM_{S^*}(i,y) \leq SM_{\hat{S}}(i,y)$$

which implies that (35.32) does not hold. Thus, we have the same result as in Case 1.

The third group, Φ_3, consists of all those intervals in which both the tasks in G and the tasks not in G are executed. Let I_z be an interval in Φ_3. Since there is a task in G executed in I_z, every task not in G that are available in I_z must be fully scheduled in I_z in $\hat{S}$; otherwise, it would be in G as well. Consequently, we have $\sum_{T_i \notin G} SM_{S^*}(i,z) \leq \sum_{T_i \notin G} SM_{\hat{S}}(i,z)$. Thus, for each I_z in Φ_3, we have

$$\begin{aligned}\sum_{i=1}^{n} SM_{\hat{S}}(i,z) &= \sum_{T_i \in G} SM_{\hat{S}}(i,z) + \sum_{T_i \notin G} SM_{\hat{S}}(i,z) \\ &\geq \sum_{T_i \in G} SM_{\hat{S}}(i,z) + \sum_{T_i \notin G} SM_{S^*}(i,z) \\ &= \sum_{i=1}^{n} SM_{\overline{S}}(i,z) \end{aligned} \tag{35.34}$$

From (35.29), (35.30) and (35.34), we conclude that

$$\sum_{i=1}^{n} \alpha(T_i, \hat{S}) \geq \sum_{i=1}^{n} \alpha(T_i, \overline{S})$$

Since the w^T-*weights* are identical, we conclude that (35.24) holds. □

To complete the proof of Theorem 35.8, we will prove Proposition 35.1 and 35.2 in the following two lemmas.

Lemma 35.8

Let $G = (V, A)$ be the directed graph based on the triple $(TS, \hat{S}, \Gamma)$, where $\hat{S}$ and Γ are defined as in the proof of Theorem 35.8. Then, all tasks in G are precisely scheduled in $\hat{S}$.

Proof

We will prove the lemma by contradiction. Assume that there is a task in G that is imprecisely scheduled in $\hat{S}$. We could construct a feasible schedule $\tilde{S}$ for TS with the following two inequalities.

$$E_{w^M}(\tilde{S}) \leq E_{w^M}(\hat{S}) \tag{35.35}$$

$$\varepsilon_{w^T}(\tilde{S}) < \varepsilon_{w^T}(\hat{S}) \tag{35.36}$$

These two inequalities contradict the given properties of $\hat{S}$: $\varepsilon_{w^T}(\hat{S}) = \varepsilon_{w^T}^{w^M}(TS)$ and $E_{w^M}(\hat{S}) = E_{w^M}(TS)$. Thus, the lemma is proved.

First, we consider the tasks in Γ. Let T_x be a task in Γ that is imprecisely scheduled in $\hat{S}$. From the definition of Γ, T_x is partially scheduled in an unsaturated interval I_q in $\hat{S}$. Construct $\tilde{S}$ from $\hat{S}$ by scheduling more of T_x in I_q. It is obvious that $\tilde{S}$ satisfies both (35.35) and (35.36).

We next consider the tasks in G but not in Γ. We can use the same argument for proving Lemma 35.6 to generate a schedule $\tilde{S}$ and to show that $\tilde{S}$ satisfies (35.35) and (35.36). □

Lemma 35.9

$\overline{S}$ is a feasible schedule for TS.

Proof

Every task in TS is scheduled in $\overline{S}$ either according to S^* or according to $\hat{S}$. Since both S^* and $\hat{S}$ are feasible schedules, the mandatory subtask of each task must be completely scheduled in $\overline{S}$. It remains to be shown that for each interval I_q, $1 \leq q \leq p$, we have the following two inequalities.

$$SM_{\overline{S}}(i,q) \leq L_q, \quad \text{for each } 1 \leq i \leq n \tag{35.37}$$

$$\sum_{i=1}^{n} SM_{\overline{S}}(i,q) \leq m \times L_q \tag{35.38}$$

From the way that $\overline{S}$ is constructed, it is obvious that every task satisfies (35.37) in each interval I_q. We now show that (35.38) also holds for each interval I_q. The intervals in $\overline{S}$ can be partitioned into two groups. The first group consists of all those intervals in which either the tasks in G are scheduled, or the tasks not in G are scheduled, but not both. If I_q is an interval in the first group, then (35.38) holds since tasks are scheduled in I_q either according to S^* or according to $\hat{S}$. The second group consists of all those intervals in which both the tasks in G and the tasks not in G are scheduled. Let I_q be an interval in the second group. Let T_x be a task in G and T_y be a task not in G and both of them are scheduled in I_q in $\overline{S}$. Since T_x is in G and scheduled in I_q in $\hat{S}$, T_y must be fully scheduled in I_q in G; otherwise, T_y would be in G as well from the way that G is constructed. Thus, we have

$$\begin{aligned}\sum_{i=1}^{n} SM_{\overline{S}}(i,q) &= \sum_{T_i \in G} SM_{\overline{S}}(i,q) + \sum_{T_i \notin G} SM_{\overline{S}}(i,q)\\ &= \sum_{T_i \in G} SM_{\hat{S}}(i,q) + \sum_{T_i \notin G} SM_{S^*}(i,q)\\ &\leq \sum_{T_i \in G} SM_{\hat{S}}(i,q) + \sum_{T_i \notin G} SM_{\hat{S}}(i,q)\\ &= \sum_{i=1}^{n} SM_{\hat{S}}(i,q)\\ &\leq m \times L_q. \end{aligned} \tag{35.39}$$

Using Theorem 35.8, we can prove a similar relationship between $E_{w^M}^{w^T}(TS)$ and $E_{w^M}(TS)$. □

Theorem 35.9

For any doubly weighted task system TS, if all the w^T-weights are identical, then $E_{w^M}^{w^T}(TS) = E_{w^M}(TS)$.

Proof

Let S be a feasible schedule for TS such that $E_{w^M}(S) = E_{w^M}(TS)$ and $\varepsilon_{w^T}(S) = \varepsilon_{w^T}^{w^M}(TS)$. By Theorem 35.8, we have $\varepsilon_{w^T}(S) = \varepsilon_{w^T}(TS)$. Thus, $E_{w^M}^{w^T}(TS) \leq E_{w^M}(S) = E_{w^M}(TS)$. Since $E_{w^M}^{w^T}(TS) \geq E_{w^M}(TS)$, we have $E_{w^M}^{w^T}(TS) = E_{w^M}(TS)$. □

Theorem 35.8 and 35.9 immediately imply the following theorem and Properties 35.4 is proved.

Theorem 35.10

For any doubly weighted task system TS, if the w^T-weights are identical, then there is a feasible schedule S for TS such that $E_{w^M}(S) = E_{w^M}(TS)$ and $\varepsilon_{w^T}(S) = \varepsilon_{w^T}(TS)$.

35.6 Conclusion

In this chapter we study the preemptive scheduling problems in *Imprecise Computation Model*. Most of the research results are not concerned with the error distribution among tasks. Therefore, the schedules generated by the algorithms of these studies might result in very biased error distribution among tasks. In other words, some important tasks might have large errors, while some less important tasks have small errors. This kind of schedule might not be acceptable in certain environments. To cope with this situation, Shih and Liu [20] first studied the problem concerned with the error distribution, where the subject is minimizing the maximum normalized error. To generalize the problem proposed by Shih and Liu [20], Ho et al. [7,8] defined a *doubly weighted task systems* in *Imprecise Computation Model*, where each task is associated with two different weights: one (denoted by w^T) for total weighted error and the other (denoted by w^M) for maximum weighted error. In [7] and [8], Ho et al. considered the problems with the following three different subjects.

1. Minimizing the maximum w^M-*weighted* error.
2. Minimizing the total w^T-*weighted* error under the constraint that the maximum w^M-*weighted* error is minimized.
3. Minimizing the maximum w^M-*weighted* error under the constraint that the total w^T-*weighted* error is minimized.

We focus on these three subjects, and the content of this chapter is based on their research results. In Section 35.3, we present a polynomial-time algorithm, Algorithm *MME*, for the problem of minimizing the maximum w^M-*weighted* error and show that the algorithm runs in $O(n^3 \log^2 n)$ and $O(n^2)$ time for multiprocessor and single processor case, respectively. The second subject is introduced in Section 35.4. An algorithm, Algorithm *cTWE*, with the same time complexity as Algorithm *MME* is shown in this section. In Section 35.5, we consider the problem of minimizing the maximum w^M-*weighted* error subject to the total w^T-*weighted* error being minimized. We give a polynomial-time algorithm, Algorithm *cMME*, that runs in $O(kn^3 \log^2 n)$ time for parallel and identical processors and in $O(kn^2)$ time for a single processor. Also, an important relationship between the last two subjects is also shown.

From Properties 35.4, we know that if the w^T-*weights* are all equal, i.e. the weights for computing the total weighted error, then there exists a schedule that simultaneously minimizes the total w^T-*weighted* error and the maximum w^M-*weighted* error. One wonders whether the converse is also true. Ho et al. [8] gave a counter example as follows to show that it is not true even for two distinct w^T-*weights*.

Example 35.1

A task system *TS* with $m+1$ tasks, where m is the number of processors. In *TS*, $T_1 = (0,3,0,4,4,1)$, $T_2 = (0,3,0,3,1,1)$, and $T_i = (0,3,3,0,1,1)$ for $3 \le i \le m+1$. It is easy to verify that $\varepsilon_{w^T}(TS) = 7 \neq 10 = \varepsilon_{w^T}^{w^M}(TS)$ and $E_{w^M}(TS) = 2 \neq 3 = E_{w^M}^{w^T}$.

But, if the parameters of T_2 are changed to $(3,6,0,3,1,1)$, then we can find a schedule that simultaneously minimizes the total w^T-*weighted* error and the w^M-*weighted* error.

References

[1] K-J. Lin, S. Natarajan, and J. W-S. Liu. Concord: A distributed system making use of imprecise results. In *Proceedings of the COMPSAC'87*, Tokyo, Japan, October 1987.

[2] K-J. Lin, S. Natarajan, and J. W-S. Liu. Scheduling real-time, periodic job using imprecise results. In *Proceedings of the 8th Real-Time Systems Symposium*, San Francisco, CA, December 1987.

[3] K-J. Lin, S. Natarajan, and J. W-S. Liu. Imprecise results: Utilizing partial computations in real-time systems. In *Proceedings of the 8th Real-Time Systems Symposium*, San Francisco, CA, December 1987.

[4] J.W-S. Liu, K-J. Lin, S.W. Shih, A.C. Yu, J.Y. Chung, and W. Zhao. Algorithms for scheduling imprecise computations. *IEEE Computer*, 24:58–68, 1991.

[5] W-K. Shih, J. W. S. Liu, J-Y. Chung, and D. W. Gillies. Scheduling tasks with ready times and deadlines to minimize average error. *ACM Operating Systems Review*, July 1989.

[6] W. A. Horn. Some simple scheduling algorithms. *Naval Research Logistics Quarterly*, 21:177–185, 1974.

[7] K. I-J. Ho, J. Y-T. Leung, and W-D. Wei. Minimizing maximum weighted error for imprecise computation tasks. *Journal of Algorithms*, 16:431–452, 1994.

[8] K. I-J. Ho, and J. Y-T. Leung. Dual criteria preemptive scheduling problem for minimax error of imprecise computation tasks. (in press)

[9] J. Blazewicz. Scheduling preemptible tasks on parallel processors with information loss. *Technique et Science Informatiques*, 3:415–420, 1984.

[10] J. Blazewicz, and G. Finke. Minimizing weighted execution time loss on identical and uniform processors. *Information Processing Letters*, 24:259–263, 1987.
[11] C. N. Potts and L. N. van Wassenhove. Single machine scheduling to minimize total late work. *Operations Research*, 40:586–595, 1992.
[12] C. N. Potts and L. N. van Wassenhove. Approximation algorithms for scheduling a single machine to minimize total late work. *Operations Research*, 40:586–595, 1992.
[13] J. B. Orlin. A faster strongly polynomial minimum cost flow algorithm. In *Proceedings of the 20th ACM Symposium on the Theory of Computing*, 1988.
[14] W-K. Shih, J. W. S. Liu, and J-Y. Chung. Algorithms for scheduling imprecise computations with timing constraints. *SIAM Journal on Computing*, 20:537–552, 1991.
[15] J. Y-T. Leung, V. K. M. Yu, and W-D. Wei. Minimizing the weighted number of tardy task units. *Discrete Applied Mathematics*, 51:307–316, 1994.
[16] A. L. Liestman and R. H. Campbell. A fault-tolerant scheduling problem. *IEEE Transactions on Software Engineering*, 12:1089–1095, 1986.
[17] K. I-J. Ho, J. Y-T. Leung, and W-D. Wei. Scheduling imprecise computation tasks with 0/1 constraint. *Discrete Applied Mathematics*, 78:117–132, 1997.
[18] R.L. Graham, E.L. Lawler, L.K. Lenstra, and A.H.G. Rinnooy Kan. Optimization and approximation in deterministic sequencing and scheduling: A survey. *Annals of Discrete Mathematics*, 5:287–326, 1979.
[19] E. L. Lawler. New and improved algorithm for scheduling a single machine to minimize the weighted number of late jobs. *Discrete Applied Mathematics*, 78:117–132, 1989.
[20] W-K. Shih and J. W. S. Liu. Algorithms for scheduling imprecise computation with timing constraints to minimize maximum error. *IEEE Transactions on Computers*, 44:466–470, 1995.
[21] R. McNaughton. Scheduling with deadlines and loss functions. *Management Science*, 6:1–12, 1959.

36

Periodic Reward-Based Scheduling and Its Application to Power-Aware Real-Time Systems

Hakan Aydin
George Mason University

Rami Melhem
University of Pittsburgh

Daniel Mossé
University of Pittsburgh

36.1 Introduction

Hard real-time systems are characterized by the utmost importance of satisfying the timing constraints at run-time. Consequences of missing the deadlines in a hard real-time system can be very serious, even catastrophic. Examples of real-time systems can be found in control systems of nuclear power plants, air traffic systems, and command-and-control applications. Thus, a vast collection of research works in *hard* real-time scheduling theory study the problem of guaranteeing the timely completion of tasks under various task and system models, namely the *feasibility* problem (see [1]). These studies necessarily consider worst-case scenarios when analyzing the problem, such as worst-case task execution times and minimum task inter-arrival times.

A tacit assumption of the hard real-time scheduling framework is that a task's output is of no value if it is not completed by the deadline. However, in a large number of emerging applications a *partial*

1-58488-397-9/$0.00+$1.50

or *approximate* result is acceptable as long as it is produced in a *timely* manner. Such applications can be found in the areas of multimedia, image and speech processing, time-dependent planning, robot control/navigation systems, medical decision making, information gathering, real-time heuristic search, and database query processing. For example, given a short amount of time, an approximate (or fuzzy) image can be produced and transmitted by a multimedia system, or a radar tracking system can compute the approximate location of the target.

Reward-based (RB) scheduling is based on the idea of trading precision for timeliness when the available resources are not sufficient to provide worst-case guarantees. Such a situation can also occur as a result of transient overload and/or faults. In this framework, each task is logically decomposed into two subtasks: the *mandatory* part produces a result of acceptable quality, and the *optional* part refines the result within the limits of available computing capacity. A nondecreasing *reward function*, associated with the execution of the optional part, captures its contribution to the overall system utility. The primary objective of RB scheduling is thus assuring the timely completion of mandatory parts, while maximizing a performance metric — usually the total reward accrued by the optional parts.

The apparent complexity of RB scheduling is partly due to the fact that it is a general framework encompassing both hard and soft real-time scheduling theories. In fact, a task with no optional part corresponds to a traditional hard real-time task, and a task with no mandatory part can model a soft real-time task.

36.2 Reward-Based Task Model, Problem Formulation, and Notation

A RB task T_i comprises a mandatory part M_i and an optional part O_i. Throughout the chapter, we will denote the worst-case execution times of M_i and O_i by m_i and o_i, respectively. The mandatory part M_i of a task becomes ready at task's release time r_i. The optional part O_i becomes ready for execution only when the mandatory part M_i completes. The mandatory part must complete successfully by the task's deadline, whereas the optional part may be left uncompleted by the deadline if more important/urgent objectives require so. In other words, no mandatory or optional execution can take place beyond the task's deadline d_i.

During its execution, O_i *refines* the approximate result produced by M_i. To quantify the "utility" (or "accuracy") of the refinement process, we associate a reward function $R_i(t_i)$ with the execution of the optional part O_i, where t_i is the amount of *optional* service time T_i receives beyond the mandatory part.

All studies in RB scheduling [2–7] assume that the reward functions are nondecreasing, which is based on the observation that the utility of a task does not decrease by allowing it to run longer in almost every practical application.

A schedule of RB tasks is *feasible* if all the mandatory parts complete in a timely fashion. A feasible schedule is also *valid* if additional constraints that may be imposed by the problem definition, such as precedence constraints or execution without preemption, are satisfied. Furthermore, a feasible and valid schedule is *optimal* if the reward accrued maximizes a performance metric. The performance metric most often considered is to maximize the (weighted) total reward of RB tasks [2–4,6,8].

An interesting question concerns the types of reward functions which represent realistic application areas. Figure 36.1 shows the form of the most common reward function types. A *linear* reward function [3,6,8–11] models the case where the benefit to the overall system increases *uniformly* during the optional execution. The simplest case is to have *identical* linear functions, while *nonidentical* (or weighted) linear functions allow us to distinguish between the importance of optional parts of different tasks.

Concave reward functions [4,5,7,12] go further by addressing the case where the greatest increase/refinement in the output quality is obtained during the first portions of the optional execution. A function $f(x)$ is concave if and only if for all x, y and $0 \leq \alpha \leq 1$, $f(\alpha x + [1-\alpha]y) \geq \alpha f(x) + (1-\alpha) f(y)$. Geometrically, this condition means that the line joining any two points of a concave curve *may not be above the curve.* Examples of concave functions are linear functions $(kx + c)$, logarithmic functions $(\ln[kx + c])$, some exponential functions $(c \cdot [1 - e^{-kx}])$, and kth root functions ($x^{1/k}$, where $k \geq 1$). Note that the first derivative of a nondecreasing concave function is nonincreasing (diminishing returns).

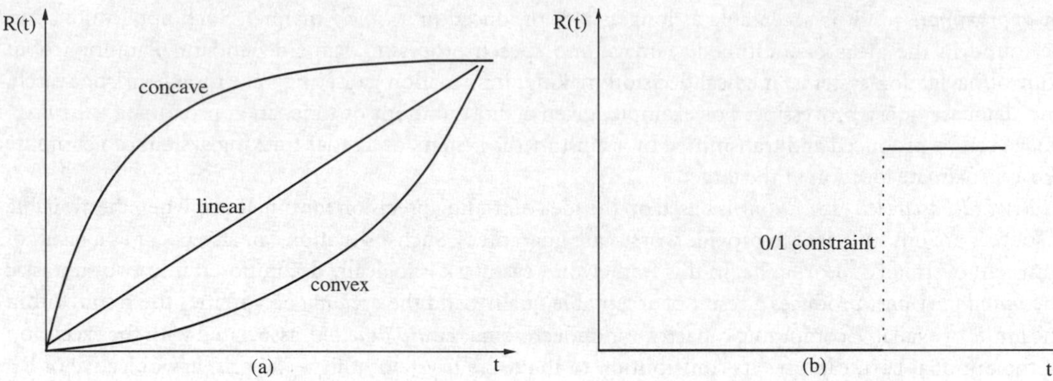

FIGURE 36.1 Reward functions: (a) continuous concave, convex, and linear; (b) 0-1 constraint.

Having nondecreasing concave reward functions means that although the reward monotonically increases beyond M_i, its *rate of increase* decreases or remains constant with time. Hence, linear and concave reward functions successfully model the applications where earlier slots of optional executions are *at least* as valuable as the later ones. In fact, a large number of typical applications reported in the literature can be modeled by concave functions since they have nonincreasing marginal returns. As mentioned above, these areas include image and speech processing, multimedia applications, time-dependent planning, robot control/navigation systems, real-time heuristic search, information gathering, and database query processing.

Convex reward functions (a function f is convex if the function $-f$ is concave) represent the case where a considerable benefit is obtained only later during the optional execution. It should be noted that reward functions with 0-1 constraints, where no reward is accrued unless the *entire* optional part is executed, have also received interest in the literature [6,8,13]. Unfortunately, scheduling with 0-1 constraints has been shown to be NP-Complete in [8].

Note that the reward functions must also take into account the worst-case execution time of optional parts: no reward can be accrued by O_i beyond the upper bound o_i. Throughout the chapter we will assume that the reward function $R_i(t_i)$ of a task T_i is given by

$$R_i(t_i) = \begin{cases} f_i(t_i) & \text{if } 0 \leq t_i \leq o_i \\ f_i(o_i) & \text{if } t_i > o_i \end{cases} \tag{36.1}$$

where t_i is the amount of CPU time allocated to the *optional* part O_i (recall that the mandatory part M_i must be executed fully). We will suppose that each f_i above is a nondecreasing, concave and continuously differentiable function over nonnegative real numbers, unless stated otherwise.

We are now ready to define the general RB Scheduling Problem.

Reward-Based Scheduling Problem: Consider a RB real-time task set $\mathbf{T} = \{T_1, \ldots, T_n\}$, where each task T_i is decomposed to a mandatory part M_i and an optional part O_i. A nondecreasing and concave reward function $R_i(t_i)$ is associated with the execution of each optional part O_i. Given a time point Z, determine the *optimal schedule* $\mathcal{OPT}$ in the interval $[0, Z]$, such that $\mathcal{OPT}$ is feasible and valid, and each optional part O_i receives service for $t_i \leq o_i$ units of time so as to maximize the total system reward $\sum_i R_i(t_i)$.

Determining the optimal schedule for RB tasks clearly involves the computation of optimal optional service times (the t_i values). Noting that the reward accrued by each optional part O_i does not increase beyond the upper bound o_i, this computation can be expressed as an optimization problem where the objective is to find t_i values[1] so as to

$$\text{Maximize} \quad \sum_{i=1}^{n} R_i(t_i) \tag{36.2}$$

$$\text{subject to} \quad 0 \leq t_i \leq o_i, \; i = 1, \ldots, n \tag{36.3}$$

$$\text{There exists a feasible and valid schedule with } \{m_i\} \text{ and } \{t_i\} \text{ values} \tag{36.4}$$

We reiterate that the "validity" condition in the constraint (36.4) may capture any requirements imposed by the task and system model (such as nonpreemptive scheduling or the existence of precedence constraints).

36.3 Related Scheduling Frameworks

A number of research studies that appeared in real-time scheduling literature have common traits with the RB scheduling framework introduced in Section 36.2. Below we briefly review these models and major research studies therein.

36.3.1 Imprecise Computation

A large body of work in RB scheduling originated from *Imprecise Computation* model, where the study in [9] can be considered as a starting point. In this work, the mandatory/optional semantic distinction and the imprecise computation model were introduced. The objective of the scheduling problem was defined as guaranteeing the timeliness of mandatory parts, while minimizing the total error. The *error* of a task is the amount of optional work left uncompleted. Later, the concept was generalized to weighted and general error functions. As a task's optional part executes, the *(precision) error* decreases, according to the specified error function. Notice that the error functions in this context are analogous to (dual of) the reward functions of our framework: a concave reward function corresponds to a convex error function, and vice versa.

In [14], optimal preemptive algorithms for Imprecise Computation model were first proposed using network-flow formulation. Faster algorithms were devised later in [8] to optimally schedule n independent tasks with identical linear reward functions (in time $O(n \log n)$) and linear reward functions with different weights (in time $O(n^2)$). Subsequently, Leung, Yu, and Wei [15] gave an $O(n \log n + kn)$-time algorithm for a single processor, where k is the number of distinct weights. Minimizing the total error with 0-1 constraints have been shown to be NP-hard even with identical weights in [8]. Ho, Leung, and Wei [13] proposed a $O(n^2)$ heuristic for this last problem, along with polynomial-time algorithms for maximizing the *number* of optional tasks that are entirely executed under 0-1 constraints. Note that the imprecise computation problem with precedence constraints is not computationally harder than the one in the independent task model, since we can first modify the ready times/deadlines of tasks to "reflect" precedence constraints and then solve that problem [6].

More recently, the *extended imprecise computation model* was introduced by Feng and Liu in [16]. This study describes several heuristics regarding the problem of assigning service times to tasks within chains. All the tasks in a chain share a common, single temporal constraint. Each chain is composed of imprecise computation tasks with linear precedence constraints and the input quality of each task depends upon the output quality of its predecessor. This work assumes linear error functions.

The problem of imprecise computations for *periodic* tasks was first addressed in [10,17]. A more detailed study appeared in [3] where the possible application areas are classified as "error noncumulative" and

[1] When considering the periodic task model, the execution time of each task instance (t_{ij}) should be considered as a separate unknown (see Section 36.4).

"error cumulative." In "error cumulative" applications, such as radar tracking, an optional instance must be executed completely at every (predetermined) k invocations. The authors further proved that the case of error cumulative jobs is an NP-Complete problem. On the other hand, in "error noncumulative" applications, errors (or optional parts left unexecuted) have no effect on the future instances of the same task. Well-known examples of this category are image/speech processing, information retrieval, and display tasks. For these jobs, the authors showed how to guarantee a feasible schedule by constantly favoring mandatory parts and proposed a class of heuristics to minimize the total error through *Mandatory-First* approach (see Section 36.4.1). All heuristics, except one, operate independently from the nature of error functions.

Online scheduling of imprecise computation tasks has been examined in [11]. The algorithms address separately the cases where the workload consists solely of online tasks and of a mix of online and offline tasks. Only identical linear error functions are considered. However, Baruah and Hickey later proved [18] that, in general, an online algorithm for imprecise computation tasks may perform arbitrarily badly when compared to a clairvoyant scheduler.

Performance metrics other than maximizing the (weighted) total error have been also studied within the imprecise computation model. In [19], two polynomial-time algorithms are presented for unit-weight error functions to obtain a "balanced" distribution of error among imprecise computation tasks. In other words, the aim is to minimize the maximum error. For the problem of minimizing maximum error, Ho, Leung, and Wei [20] gave an $O(n^2)$-time algorithm for a single processor and an $O(n^3 log^2 n)$-time algorithm for multiprocessors.

Leung and Wong investigated the problem of minimizing the number of tardy tasks for a given maximum, tolerable error; and showed it to be NP-hard [21]. In [22], minimizing the average response time, while keeping the total error less than a threshold has been shown to be NP-hard for almost every reasonable model.

36.3.2 Increased-Reward-with-Increased-Service Model

The Increased-Reward-with-Increased-Service (IRIS) framework allows tasks to get increasing reward with increasing service, without an upper bound on the execution times of the tasks and without the separation between mandatory and optional parts [4]. A task executes for as long as the scheduler allows. Typically, a nondecreasing *concave* reward function is associated with each task's execution time. In [4,5] the problem of maximizing the total reward in a system of independent aperiodic tasks is explored. An optimal polynomial-time solution with static task sets (identical ready times) is presented, as well as two extensions that include mandatory parts and online policies for dynamic task arrivals.

36.3.3 Quality-of-Service-Based Resource Allocation Model

A QoS-based resource allocation model (Q-RAM) was proposed in [7,12,23,24]. In that study, the problem is to optimally allocate multiple resources to the various applications such that they simultaneously meet their minimum requirements along multiple QoS dimensions and the total system utility is maximized. In one aspect, this can be viewed as a generalization of optimal CPU allocation problem in RB scheduling, to multiple resources and multiple reward metrics (quality dimensions). Further, dependent and independent quality dimensions are separately addressed for the first time in this work.

However, a fundamental assumption of Q-RAM model is that the resource allocation is made in terms of *utilization of resources*, usually expressed as a fraction (percentage) of the available resource capacity. In the CPU allocation case for the periodic real-time tasks, this translates to identical service time allocation assumption for each instance of a given task. The periodic RB scheduling framework we consider in this chapter assumes that the reward accrued has to be computed separately over *each* instance for the periodic case. In other words, we are not making the assumption that all the instances of a periodic task will have the same CPU allocation time, which can lead to suboptimal results for some settings.

In [7], Rajkumar et al. addressed the case of a single resource and multiple QoS (reward) dimensions and provided a pseudopolynomial time algorithm for continuous concave utility functions. In [12], the problem of a single resource and two or more QoS dimensions (with one dimension being discrete and dependent on another) is explored. The problem is proven to be NP-hard, and an approximation algorithm based on Concave Majorant Optimization is presented. Then, the authors show that the problem of multiple resources and single QoS dimension can be reduced to a mixed integer programming problem; they also report the run-time of example problems when solved by commercial software packages.

Lee et al. further considered discrete QoS dimensions in [23], as opposed to the continuous QoS dimension assumption of their previous research in [7]. They proposed two approximation algorithms for the case of single resource and multiple quality dimensions. The first algorithm yields a solution that is within a known bounded distance from the optimal solution, while the second is capable of returning an allocation whose distance from the optimal solution can be reduced by giving more time to the allocation routine.

Finally, the more general case of multiple resources and QoS dimensions was addressed in [24]. Due to the intractable nature of the problem, a mixed integer programming formulation is developed to yield near-optimal results, followed by an approximation algorithm based on local search techniques. The execution time and quality distance to the optimal solution of these algorithms are experimentally evaluated.

36.4 Periodic Reward-Based Scheduling

Many real-time tasks are *periodic* in nature: Assuming that it is ready for execution at $time = 0$, a periodic real-time task T_i with period P_i will generate a new *task instance (job)* every P_i time units. Specifically, the jth instance of T_i (denoted by T_{ij}) is invoked at $t = (j-1) \cdot P_i$ and it has to be completed by the time of next invocation at $t = j \cdot P_i$ (in other words, we will assume that the relative deadline of each instance is equal to its period).

A large number of periodic real-time applications can be modeled by the RB scheduling framework. Well-known examples are periodic tasks that receive, process, and transmit video, audio, or compressed images and information retrieval tasks. With the advance of multimedia and networking technologies, this class of applications are likely to become more and more widespread. Moreover, these applications readily admit a mandatory/optional semantic distinction: the mandatory part produces an image, frame, or output of acceptable quality and the optional part improves on it.

The extension of the basic RB task model to the periodic execution settings is relatively simple: a schedule of periodic RB tasks is **feasible** if mandatory parts meet their deadlines at every task invocation. Note that the optional execution times of different instances belonging to a given periodic task T_i can be potentially different.

The common performance metric [2,3] for the feasible periodic RB task schedules is **the average cumulative reward**, given by

$$REW_{CUM} = \frac{P_i}{P} \sum_{i=1}^{n} \sum_{j=1}^{P/P_i} R_i(t_{ij}) \tag{36.5}$$

where P is the *hyperperiod*, that is, the least common multiple of $P_1, P_2, \ldots, P_n$ and t_{ij} is the amount of CPU time assigned to the jth instance of optional part of task T_i (i.e., O_{ij}). Observe that it is necessary and sufficient to generate a schedule that maximizes the total reward during the hyperperiod P, since the schedule will repeat itself every P time units. Also note that, in the above expression, the average reward of T_i is computed over the number of its invocations during the hyperperiod P.

Thus, the solution to the periodic RB scheduling problem must

1. ensure that the mandatory part of each periodic task instance meets its deadline, and
2. determine the optional execution times of all the periodic task instances to maximize the performance metric given by Equation (36.5).

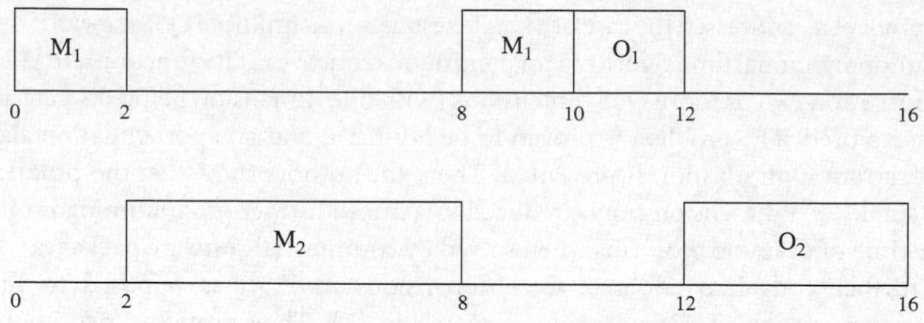

FIGURE 36.2 The solution produced by a mandatory-first algorithm.

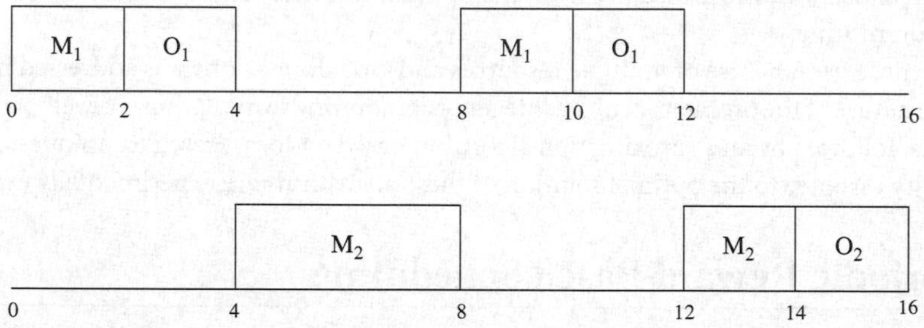

FIGURE 36.3 An optimal schedule.

36.4.1 Mandatory-First Solutions

In one of the earliest attempts attacking the periodic RB scheduling problem, Chung, Liu, and Lin [3] gave priority to guaranteeing the feasibility constraint. To achieve this, they proposed statically assigning higher scheduling priorities to all the mandatory parts with respect to the optional parts. In other words, in this *Mandatory-First* strategy, no optional part is scheduled if there is a ready mandatory part in the system. In their original proposal, the relative scheduling priorities of mandatory parts are determined according to the Rate Monotonic Scheduling policy [25], but it is clear that other policies can be also used. For the priority assignment to the optional parts, the authors proposed a number of heuristics, such as considering the deadlines, periods, or best incremental rewards. As an example, consider an RB task set consisting of two tasks, where $P_1 = 8, m_1 = 2, o_1 = 2, P_2 = 16, m_2 = 6, o_2 = 10$. Let the reward functions of T_1 and T_2 be given by $f_1(t_1) = 8 \cdot t_1$ and $f_2(t_2) = 2 \cdot t_2$. Then, an "intelligent" Mandatory-First algorithm, considering that the reward associated by O_1 is much higher than that of O_2, would produce the schedule shown in Figure 36.2 yielding a total cumulative reward of 12.

Yet, it is easy to see that the optimal solution for this specific problem instance involves delaying the execution of M_2 to the favor of "valuable" O_1, giving a total cumulative reward of 18 (see Figure 36.3).

We can see that the reward performance of the mandatory-first schemes is inherently limited: In the example above, the best mandatory-first scheme could yield only 2/3 of the optimal reward. In fact, in [2] it is proven that, in the worst-case, the relative performance of the mandatory-first schemes can be arbitrarily bad when compared to the optimal policy (i.e., the reward ratio of the best mandatory-first scheme to the reward of the optimal policy can be arbitrarily close to 0). It is clear that the optimal solution must be based on principles fundamentally different from Mandatory-First heuristic.

36.4.2 Optimal Solution

The difficulty involved in the periodic RB scheduling problem stems from the fact that the feasibility and reward maximization objectives must be achieved simultaneously. Moreover, even if we temporarily disregard the feasibility constraint above, the optimization problem itself remains challenging: the number of unknowns to be determined (t_{ij}'s in the previous expression (36.5)) can be potentially *exponential* in the number of tasks, n. Consequently, any *efficient* solution to the problem hinges in the first place on being able to reduce the number of unknowns without sacrificing optimality.

The theorem below establishes that for the most common (i.e., concave and linear) reward functions, the number of unknowns is only n (the number of tasks):

Theorem 36.1 (from [2])

If all the reward functions are concave, then the periodic RB scheduling problem has an optimal solution in which a given task T_i receives the same *optional service time at* every *instance, that is, $t_{ij} = t_{ik} = t_i \ \forall j, k$.*

The same study [2] also shows that, once $t_1, \ldots, t_n$ are determined, then *any periodic real-time scheduling policy that can achieve 100% CPU utilization* may be used to obtain a feasible schedule with these assignments. Examples of such policies are Earliest-Deadline-First (EDF) [25] and Least-Laxity-First (LLF) [26].

Interestingly, Theorem 36.1 extends also to the *identical multiprocessor* settings [2]. However, in general, EDF and LLF cannot be used to achieve full utilization on multiprocessors. An example policy that can be adopted for this purpose is proposed by Bertossi and Mancini in [27].

Theorem 36.1 is crucial in reducing the number of unknowns to n, but it is based on an existence proof and the problem of computing $t_1, \ldots, t_n$ efficiently (in polynomial-time) is equally important. One can obtain the following nonlinear optimization problem formulation to compute the optimal optional service times $t_1, \ldots, t_n$:

$$\text{Maximize} \quad \sum_{i=1}^{n} f_i(t_i) \tag{36.6}$$

$$\text{subject to} \quad \sum_{i=1}^{n} \frac{P}{P_i} t_i \leq P - \sum_{i=1}^{n} \frac{P}{P_i} m_i \tag{36.7}$$

$$0 \leq t_i \leq o_i, \quad i = 1, \ldots, n \tag{36.8}$$

The constraint set (36.8) reflects that negative service times do not have a physical interpretation and executing an optional part O_i beyond its worst-execution time o_i does not increase the reward. The constraint (36.7) encodes the fact that the sum of optional and mandatory service times cannot exceed the total available time P (the length of the hyperperiod).

The work in [2] presents a polynomial-time algorithm to solve the problem above. Observe that if the available time can accommodote all the optional parts in their entirety, that is, if $\sum_{i=1}^{n} \frac{P}{P_i}(m_i + o_i) \leq P$, then setting $t_i = o_i \ \forall i$ clearly maximizes the objective function. Otherwise, because of the nondecreasing nature of reward functions, we should fully use the timeline and consider the constraint (36.7) as an equality constraint. Thus, the algorithm in [2] temporarily ignores the inequality constraints (36.8) and first produces a solution that only satisfies the equality constraint through *Lagrange multipliers* technique. In case that the constraint set (36.8) is violated, then the solution is iteratively improved using the *Kuhn-Tucker optimality conditions* [28] for nonlinear optimization problems. In [2], it is formally proven that this iterative algorithm is guaranteed to converge to the optimal solution in time $O(n^2 \log n)$.

Though the solution above addresses the most common task and reward function models, it is worth investigating the possible extensions to different models. Clearly, Theorem 36.1 is the key, eliminating a potentially exponential number of unknowns (the t_{ij} values) and, arguably, it is very hard to solve the problem without such a property. In fact, [2] also established that the result on the optimality of identical execution times no longer holds if we assume that

1. the relative deadline of a task instance can be smaller than its period, or
2. a static-priority assignment (such as Rate-Monotonic Priority Assignment) is to be used as opposed to EDF/LLF, or
3. the reward functions are not concave.

The solution to the periodic RB scheduling problem with any of the above assumptions is still open as of this writing. Further, it was formally proven in [2] that the concavity assumption is absolutely necessary for computational tractability, *even for tasks sharing a common period and deadline*:

Theorem 36.2 (from [2])

The RB scheduling problem with convex *reward functions is NP-hard.*

36.5 Power-Aware Scheduling

The RB scheduling framework is used when attempting to assign CPU cycles to various applications to maximize the reward (user-perceived utility). Many of these applications do also run on embedded, mobile, and wireless computing devices that rely on *battery power*. Thus, power management has been recently elevated to a major research area in computer science and engineering. Different components of computer systems, such as CPU, memory, disk subsystems, and network interfaces have been subject to various energy-efficiency studies. The reader is referred to [29,30] for comprehensive surveys.

One powerful and increasingly popular technique, known as Variable Voltage Scheduling (*VVS*) or Dynamic Voltage Scaling (*DVS*), is based on reducing CPU energy consumption through the adjustment of the CPU speed by varying both the supply voltage and the clock frequency. The technique is motivated by the observation that the dynamic power dissipation P_d of an on-chip system is given by the following formula:

$$P_d = C_f \cdot V_s^2 \cdot f$$

where C_f is the effective switched capacitance, V_s is the supply voltage, and f is the frequency of the clock. Hence, in principle, it is possible to obtain significant power savings by simultaneously reducing the supply voltage and the clock frequency. On the other hand, clearly, the response time will increase in a linear fashion when we reduce the clock frequency. Thus, trading the speed for energy savings is the main idea in Power-Aware Scheduling. The exact form of power/speed relation depends on the specific technology in use, but as a rule, it is a strictly increasing convex function, specifically a polynomial of at least the second degree. The prospects of reducing the energy consumption through DVS have been materialized in recently-announced processors such as Transmeta's Crusoe processor and Intel Xscale architecture.

Note that in its simplest form, the power/speed relation can be exploited by using the maximum CPU speed when executing tasks and shutting down the CPU whenever the system is (likely to stay) idle for a reasonably long period. However, this *predictive shutdown* technique remains inefficient and suboptimal even for a scheduler with perfect knowledge of idle intervals, due to the convex dependence between the speed and the power consumption: a better approach is to *reduce the CPU speed* while executing tasks.

Early works on DVS focused on task systems with no explicit deadlines. Weiser et al. adopted an approach where time is divided into 10–50 msec intervals, and the CPU clock speed (and the supply voltage) is adjusted using predictions based on processor utilization in recent intervals [31]. Govil, Chan, and Wasserman proposed and evaluated several predictive and nonpredictive approaches for voltage changes [32].

In real-time systems, where tasks have to meet their timing constraints, one cannot arbitrarily reduce the CPU speed. Thus, the real-time power-aware scheduling (RT-PAS) problem[2] can be stated as follows: **Determine the CPU speed to minimize the total energy consumption while still meeting all the deadlines.** Observe that, the solution to the RT-PAS problem depends on the task model under consideration

[2]Alternative names that appeared in the literature for the same framework include *real-time dynamic voltage scaling* (RT-DVS) and *real-time variable voltage scheduling* (RT-VVS).

(aperiodic/periodic/hybrid task sets or preemptive/nonpreemptive scheduling). Further, clearly, the optimal CPU speed in the solution may be *a function of the time and/or identity of the running task.*

RT-PAS research can be traced back to the important work of Yao, Demers, and Shenker [33], where the authors provided a polynomial-time algorithm to compute the optimal speed assignments assuming *aperiodic tasks* and *worst-case execution times.* Other works in this area include heuristics for extended task and system models, such as online scheduling of aperiodic and periodic tasks [34], nonpreemptive RT-PAS [35], and RT-PAS with upper bound on the voltage change rates [36].

36.5.1 Modeling Real-Time Workload and Energy on a Variable Speed CPU

Traditional real-time scheduling theory mandates the consideration of the worst-case workload, which translates to *worst-case execution times* under a *fixed* CPU speed. On a variable speed CPU, it is clear that the execution time of a task is dependent on both the number of CPU cycles required, and the CPU speed. Consequently, the *worst-case number of CPU cycles* is adopted as the measure of the worst-case workload.

On a variable speed CPU, the processor speed S (expressed in the number of CPU cycles per time unit) can be changed between a lower bound $S_{\min}$ and an upper bound $S_{\max}$. We will assume that the speed can be varied continuously over the range $[S_{\min}, S_{\max}]$ and that the supply voltage is also adjusted in accordance with the speed. We will normalize the speed values with respect to $S_{\max}$; that is, at any time, $0 \leq S_{\min} \leq S \leq S_{\max} = 1$.

We will denote by C_i the number of CPU cycles required by the task T_i. If a task executes with constant speed S, then its worst-case execution time is given by C_i/S. Observe that, since $S_{\max} = 1.0$, the worst-case execution time of the task T_i under the maximum speed is C_i time units.

In current processor arcitectures implemented in CMOS technology, the CPU power consumption of task T_i executing at the speed S is given by a strictly increasing and convex function $g_i(S)$. In general, $g_i(S)$ can be represented by a polynomial of the second or the third degree [33,36]. If the task T_i occupies the processor during the interval $[t_1, t_2]$ and if the processor speed changes according to a function $S(t)$, then the *energy consumed* in this interval is $E(t_1, t_2) = \int_{t_1}^{t_2} g_i(S(t))dt$. Note that the last expression simply becomes $E(t_1, t_2) = g_i(S) \cdot (t_2 - t_1)$ if the speed S is constant during the interval.

36.5.2 Correlating Real-Time Power-Aware Scheduling to Reward-Based Scheduling

Before addressing the specific case of *periodic* tasks, in this section, we correlate the *general* RB scheduling problem given by (36.2), (36.3), and (36.4) in Section 36.2 to *general* RT-PAS problem. We underline that the relation we prove is preserved regardless of the specific task model (preemptive/nonpreemptive scheduling, independent/dependent task sets) *or* the number of the processors as long as we make the same assumptions for both RB scheduling and RT-PAS problems. First, we state the general RT-PAS problem.

Real-Time Power-Aware Scheduling Problem: Consider a CPU with variable voltage/speed $S(S_{\min} \leq S \leq S_{\max})$ facility, and a set $\mathbf{T} = \{T_1, \ldots, T_n\}$ of hard real-time tasks, in which each task T_i is subject to a worst-case workload of C_i expressed in the number of required CPU cycles. The power consumption of the task T_i is given by a strictly increasing convex function $g_i(S)$, which is a polynomial of at least the second degree. Given a time point Z, determine *the energy-optimal schedule* ($\mathcal{EOS}$) and *the processor speed* $S(t)$ in the interval $[0, Z]$, such that $\mathcal{EOS}$ is feasible and valid, and the total energy consumption $E(0, Z) = \int_0^Z g_i(S(t))dt$ is minimized.

A major difficulty with the *real-time power-aware scheduling* (RT-PAS) problem lies in the *possibility* of having a *nonconstant* speed in the optimal solution: determining the exact form of $S(t)$ for every point in the interval $[0, Z]$ may be a serious challenge. Fortunately, the *convexity* assumption about $g_i(S)$ makes possible to prove the following helpful property (see [37] for a formal proof).

Proposition 36.1

A task T_i requiring C_i CPU cycles can be executed with a constant speed S_i, *without increasing the energy consumption or the completion time.*

That is, we can assume that the CPU speed will not change while executing a given task (task instance for the periodic task model), without compromising the optimality or feasibility. However, note that the optimal speed *may* be different for different tasks. We can now formulate the general RT-PAS as an optimization problem where the aim is to determine the speeds $S_1, \ldots, S_n$ so as to

$$\text{Minimize} \quad \sum_{i=1}^{n} \frac{C_i}{S_i} \cdot g_i(S_i) \tag{36.9}$$

$$\text{subject to} \quad S_{\min} \leq S_i \leq S_{\max}, \quad i = 1, \ldots, n \tag{36.10}$$

$$\text{There exists a feasible and valid schedule with } \{S_i\} \text{ values} \tag{36.11}$$

The expression $\frac{C_i}{S_i} \cdot g_i(S_i)$ in the objective function (36.9) quantifies the energy consumption of task T_i when run at the speed S_i. The constraint set (36.10) encodes the lower and upper bounds on the speed, whereas the constraint (36.11) enforces the feasibility and validity.

Proposition 36.2

Solving an instance of RT-PAS problem is equivalent to solving an instance of RB scheduling problem with concave reward functions.

Proof

Observe that determining S_i is effectively equivalent to determining *the CPU time allocation of* T_i, which will be denoted by X_i ($X_i = C_i/S_i$). In addition, the minimum and maximum speed bounds impose explicit bounds on X_i. Specifically, for any task T_i, the CPU allocation X_i should lie in the interval $[X_{\min}, X_{\max}]$, where $X_{\min} = \frac{C_i}{S_{\max}}$ and $X_{\max} = \frac{C_i}{S_{\min}}$. Thus, we can rewrite RT-PAS as to determine the CPU time allocations $\{X_i\}$ so as to

$$\text{Minimize} \quad \sum_{i=1}^{n} X_i \cdot g_i\left(\frac{C_i}{X_i}\right) \tag{36.12}$$

$$\text{subject to} \quad \frac{C_i}{S_{\max}} \leq X_i \leq \frac{C_i}{S_{\min}} \quad i = 1, \ldots, n \tag{36.13}$$

$$\text{There exists a feasible and valid schedule with } \{X_i\} \text{ values} \tag{36.14}$$

That is, T_i must receive service for at least $X_{\min}$ units, which can be considered as a *mandatory* execution. Depending on the specific speed assignment S_i, T_i can receive an additional service of up to $X_{\max} - X_{\min}$, which can be seen as an *optional* execution. Moreover, the lower the speed S_i, the longer the "optional" execution beyond $X_{\min}$ and the lower the energy consumption (see Figure 36.4). In addition, minimizing the energy consumption is equivalent to *maximizing energy savings*. Thus, consider the following variable substitutions:

$$m_i = \frac{C_i}{S_{\max}}$$

$$t_i = X_i - m_i$$

$$o_i = \frac{C_i}{S_{\min}} - \frac{C_i}{S_{\max}}$$

$$R_i(t_i) = -\left(t_i + \frac{C_i}{S_{\max}}\right) g_i\left(\frac{C_i}{t_i + \frac{C_i}{S_{\max}}}\right)$$

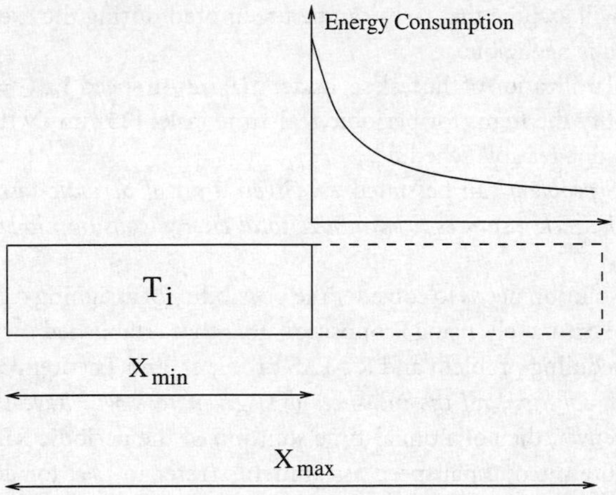

FIGURE 36.4 Energy consumption as a function of CPU allocation.

Using this transformation, we obtain a new form for the RT-PAS problem:

$$\text{Maximize} \quad \sum_{i=1}^{n} R_i(t_i) \tag{36.15}$$

$$\text{subject to} \quad 0 \leq t_i \leq o_i \quad i = 1, \ldots, n \tag{36.16}$$

$$\text{There exists a feasible and valid schedule with } \{m_i\} \text{ and } \{t_i\} \text{ values} \tag{36.17}$$

The final formulation is an instance of the general RB scheduling problem defined in Section 36.2 by Equations (36.2), (36.3), and (36.4). Further, the reward function $R_i(t_i)$ above is concave, since

$$\left(t_i + \frac{C_i}{S_{\max}}\right) g_i \left(\frac{C_i}{t_i + \frac{C_i}{S_{\max}}}\right)$$

is convex. To see this, we can use the result from [28] stating that if a and b are both convex functions and if a is increasing, then $a(b(x))$ is also convex. Thus, by setting

$$h(t_i) = \frac{C_i}{t_i + \frac{C_i}{S_{\max}}}$$

and observing that the multiplication by $(t_i + \frac{C_i}{S_{\max}})$ does not affect the convexity, we justify the concavity of $R_i(t_i) = -g_i(h(t_i))$. □

36.5.3 Determining Optimal Speed Assignments for Periodic Real-Time Tasks

In this section, we focus on the most common type of real-time task models, namely, *periodic task sets*. Thus, we consider a set $\mathcal{T} = \{T_1, \ldots, T_n\}$ of n periodic real-time tasks. As in Section 36.4, the period of T_i is denoted by P_i, which is also equal to the deadline of the current invocation. All tasks are assumed to be independent and ready at $t = 0$. The worst-case number of CPU cycles required by each instance[3] of

[3]Note that we restrict our power-aware scheduling discussion to hard real-time tasks that must be executed in their entirety. That is, we do not address the case where each task can be decomposed to a mandatory and optional part. The reader is referred to a recent study [38] for the power-aware scheduling of RB tasks.

T_i is denoted by C_i. We will assume that tasks can be preempted during the execution and the overhead of changing voltage/speed is negligible.

We define U as the total utilization of the task set under maximum speed $S_{\max} = 1$, that is, $U = \sum_{i=1}^{n} \frac{C_i}{P_i}$. Note that the schedulability theorems for periodic real-time tasks [25] imply that $U \leq 1$ is a necessary condition to have at least one feasible schedule.

Thus, *periodic RT-PAS problem* can be stated as: *Given a set of periodic tasks, determine the optimal speed assignments of all the task instances to minimize total energy consumption while still meeting all the deadlines.*

Note that the optimal solution needs to consider the possibility of assigning different speeds to different instances of a given task. Fortunately, using Proposition 36.2 that established the equivalence between RB (but constant-speed) scheduling problem and RT-PAS problem, and Theorem 36.1, we can immediately assert that *in the optimal solution, all the instances of a given task will have the same CPU allocation, thus the same speed.* Moreover, the polynomial-time solution of the periodic RB scheduling problem [2] can be adopted to compute the optimal speed assignments (refer to [39] for details). Finally, thanks to Theorem 36.1, we can state that EDF and LLF policies may be used to obtain a feasible schedule with these speed assignments.

Moreover, if the power consumption functions of all the tasks can be assumed to be identical (a realistic assumption in many systems), that is, if $g_i(S) = g_j(S) = g(S)\, \forall i, j$, a stronger result can be proven [40]:

Proposition 36.3

If task power consumption functions are identical, then the optimal CPU speed to minimize the total energy consumption while meeting all the deadlines of the periodic hard real-time task set is constant and equal to $S_{\text{opt}} = \max\{S_{\min}, U\}$*, where* $U = \sum_{i=1}^{n} \frac{C_i}{P_i}$.

In other words, the optimal speed with identical power consumption functions is equal to the utilization of the task set under maximum speed (namely, U), subject to the $S_{\min}$ constraint. Observe that this speed choice results in an *effective* utilization of $\sum_{i=1}^{n} \frac{C_i}{S_{\text{opt}} \cdot P_i} = 1.0$ (a fully utilized timeline), assuming that the lower bound on the CPU speed is not violated. Thus, the optimal solution consists in "stretching out" the CPU allocations of all task instances in equal proportions.

Note that the solution presented above is *static* in the sense that it assumes a worst-case workload for each periodic task instance. This is necessary because any solution for hard real-time systems must provision for the worst-case scenarios. However, in practice, many real-time applications complete before presenting their worst-case workload [41]. Thus, at run-time, the static optimal CPU speed can be further reduced when early completions do occur. A class of *dynamic* scheduling techniques to adjust the speed to the *actual* workload were investigated in depth in recent years (see [40,42,43] as representative studies). A crucial consideration in dynamic power-aware scheduling is to avoid deadline misses when reducing the speed and reclaiming unused computation times (*slack*). Thus, the proposed dynamic techniques differ in their ability to exploit slack while preserving the feasibility.

If the actual workload is likely to differ from the worst-case, then additional energy gains can be achieved through a *speculative* strategy. In this approach, the CPU speed is *aggressively* reduced in *anticipation* of early completions. Clearly, such a speculative strategy risks violating the timing constraints if the actual workload of tasks turns out to be higher than predicted. Consequently, the scheduler must guarantee that all the deadlines will be met by adopting a *high* speed in the later parts of the schedule, should there be a need. The idea of speculative speed reduction was first introduced in [44] for tasks sharing a common period. The works in [40,42,45] provide representative research on speculative scheduling policies for general periodic real-time tasks. Finally, the study in [46] presents a detailed performance comparison of state-of-the-art dynamic and speculative dynamic voltage scaling techniques.

36.5.4 Practical Considerations

In this section, we look at two issues that must be carefully considered in any practical real-time system that uses Dynamic Voltage Scaling. The first one is the *delay overhead* involved in performing the speed and supply voltage reductions. This overhead may be particularly significant especially for the *dynamic* scheduling schemes that change the CPU speed at run-time upon detecting early completions.

While Namgoong, Yu, and Meg reported the time taken to reach steady state at the new voltage level as under 6 ms/V as of 1996 [47], Burd et al. indicate that this delay is limited by 70 μs in the new ARM V4 processor based system, including frequency transition [48]. For the Low-Power lpARM processor, clock frequency transitions take approximately 25 μs [49]. In a study based on StrongARM SA-1100 processor operating under constant supply voltage, this delay is reported to be less than 150 μs [50]. When modifying the supply voltage and the clock frequency of the StrongARM SA-1100 processor, Pouwelse, Langendoen, and Sips have found that the voltage and speed increase is rapidly handled (40 μs), but the decrease can take up to 5.5 ms [51]. A full voltage transition can be performed in less than 300 μs in the Transmeta Crusoe 5400 processor [52]. Thus, the technology trends are toward even more efficient DVS-enabled processors and it can be claimed that the overhead involved can be ignored in many cases, especially when the speed changes occur only at context switch time. In case that this overhead is not negligible, it can be incorporated into the worst-case workload of each task [53].

The second issue is related to the fact that, unlike the assumptions of this section's framework where the CPU speed can be changed continuously in the interval $[S_{\min}, S_{\max}]$, current technologies support only a finite number of speed levels [52], though the number of available speed levels is likely to increase in the future. The solutions obtained for the continuous speed model can be always adopted to these settings by choosing the lowest speed level that is equal to or greater than the value suggested by the algorithms. Our preliminary experimental results indicate that this simple approach results in an energy overhead of 15–17% with respect to continuous speed settings, when the number of available speed levels is only 5. When the number of speed levels exceeds 30, the difference reduces to 3%. More comprehensive recent studies that address this issue and other overhead sources in dynamic voltage scaling can be found in [43,54].

36.6 Conclusion

In this chapter, we introduced the RB and the RT-PA scheduling frameworks, and we established the relationship between these two models. RB scheduling is a general framework which unifies hard and soft real-time scheduling theories by addressing feasibility and maximum utility issues simultaneously. Assuring timely completion of mandatory parts guarantees an acceptable overall performance, while optimal scheduling of optional parts maximizes the total user-perceived utility (reward). RB scheduling promotes graceful degradation and resource utilization where worst-case guarantees cannot be given due to faults, overload conditions and for applications that admit an intratask mandatory/optional division. However, simultaneously achieving feasibility and optimality is not a trivial problem, since the reward functions associated with the optional parts bring another dimension to the traditional scheduling framework.

After introducing various research works and models that can be classified as part of the RT-RB scheduling, we focused on RB scheduling for periodic real-time tasks. According to a main result due to [2], when the reward functions are linear and/or concave (the most common case), it is always possible to find an optimal schedule where the service time of a given task remains constant from instance to instance. This, in turn, implies that well-known scheduling policies such as EDF and LLF can be used to meet all the timing constraints with these optimal assignments. Moreover, it is possible to compute these optimal service times in polynomial time by solving iteratively a non-linear optimization problem. Finally, we briefly commented on the existing difficulties when one attempts to extend these results to different models.

RT-PAS has recently become a major research area with ever increasing emphasis on energy saving strategies for portable and embedded devices that rely on battery power. Thanks to Dynamic Voltage

Scaling technology, it is possible to obtain considerable energy savings by simultaneously reducing the CPU speed and the supply voltage. However, this is accompanied by a linear increase in response time, which may compromise the correctness of hard real-time systems. Thus, RT-PAS solutions focus on minimizing the total energy consumption (through speed reduction) while still meeting all the deadlines.

We have shown that the RT-PAS problem can be solved within the framework of RB scheduling, through an appropriate transformation. For the most common, periodic real-time task model, it turns out that the optimal speed to meet all the deadlines while minimizing the energy consumption is constant and equal to the utilization (load) of the task set under maximum speed, if the power consumption of CPU is independent of the running task.

References

[1] G. Buttazzo. *Hard Real-Time Computing Systems: Predictable Scheduling Algorithms and Applications.* Kluwer Academic Publishers, Boston, 1997.

[2] H. Aydin, R. Melhem, D. Mossé, and P.M. Alvarez. Optimal reward-based scheduling for periodic real-time tasks. *IEEE Transactions on Computers,* 50(2): 111–130, 2001.

[3] J.-Y. Chung, J. W.-S. Liu, and K.-J. Lin. Scheduling periodic jobs that allow imprecise results. *IEEE Transactions on Computers,* 19(9): 1156–1173, 1990.

[4] J. K. Dey, J. Kurose, D. Towsley, C.M. Krishna, and M. Girkar. Efficient on-line processor scheduling for a class of IRIS (increasing reward with increasing service) real-time tasks. In *Proceedings of ACM SIGMETRICS Conference on Measurement and Modeling of Computer Systems,* May 1993.

[5] J. K. Dey, J. Kurose, and D. Towsley. On-line scheduling policies for a class of IRIS (increasing reward with increasing service) real-time tasks. *IEEE Transactions on Computers,* 45(7):802–813, 1996.

[6] J. W.-S. Liu, K.-J. Lin, W.-K. Shih, A. C.-S. Yu, C. Chung, J. Yao, and W. Zhao. Algorithms for scheduling imprecise computations. *IEEE Computer,* 24(5): 58–68, 1991.

[7] R. Rajkumar, C. Lee, J. P. Lehoczky, and D. P. Siewiorek. A resource allocation model for QoS management. In *Proceedings of 18th IEEE Real-Time Systems Symposium,* December 1997.

[8] W.-K. Shih, J. W.-S. Liu, and J.-Y. Chung. Algorithms for scheduling imprecise computations to minimize total error. *SIAM Journal on Computing,* 20(3), 1991.

[9] K.-J. Lin, S. Natarajan, and J. W.-S. Liu. Imprecise results: utilizing partial computations in real-time systems. In *Proceedings of 8th IEEE Real-Time Systems Symposium,* December 1987.

[10] J. W.-S. Liu, K.-J. Lin, and S. Natarajan. Scheduling real-time, periodic jobs using imprecise results. In *Proceedings of 8th IEEE Real-Time Systems Symposium,* December 1987.

[11] W.-K. Shih and J. W.-S. Liu. On-line scheduling of imprecise computations to minimize error. *SIAM Journal on Computing,* October 1996.

[12] C. Lee, R. Rajkumar, J. P. Lehoczky, and D. P. Siewiorek. Practical solutions for QoS-based resource allocation problems. In *Proceedings of 19th IEEE Real-Time Systems Symposium,* December 1998.

[13] K.I.J. Ho, J.Y.T. Leung, and W.D. Wei. Scheduling imprecise computations with 0/1 constraint. *Discrete Applied Math.,* 78, 117–132, 1997.

[14] W.-K. Shih, J. W.-S. Liu, J.-Y. Chung, and D.W. Gillies. Scheduling tasks with ready times and deadlines to minimize average error. *ACM Operating Systems Review,* July 1989.

[15] J. Y. T. Leung, V. K. M. Yu, and W. D. Wei. Minimizing the weighted number of tardy task units. *Discrete Applied Math.,* 51, 307–316, 1994.

[16] W. Feng and J.W.-S. Liu. Algorithms for scheduling real-time tasks with input error and end-to-end deadlines. *IEEE Transactions on Software Engineering,* 23(2): 93–106, 1997.

[17] J.-Y. Chung and J. W.-S. Liu. Algorithms for scheduling periodic jobs to minimize average error. In *Proceedings of 9th IEEE Real-Time Systems Symposium,* December 1988.

[18] S. K. Baruah and M. E. Hickey. Competitive on-line scheduling of imprecise computations. *IEEE Transactions on Computers,* 47(7): 1027–1033, 1998.

[19] W.-K. Shih and J. W.-S. Liu. Algorithms for scheduling imprecise computations with timing constraints to minimize maximum error. *IEEE Transactions on Computers,* 44(3):466–471, 1995.

[20] K. I. J. Ho, J. Y. T. Leung, and W. D. Wei. Minimizing maximum weighted error for imprecise computation tasks. *Journal of Algorithms,* 16, 431–452, 1994.

[21] J. Y. T. Leung and C. S. Wong. Minimizing the number of late tasks with error constraint. *Information and Computation,* 106, 83–108, 1993.

[22] J. Y. T. Leung, T. W. Tam, C. S. Wong, and G. H. Young. Minimizing mean flow time with error constraint. *Algorithmica,* 20, 101–118, 1998.

[23] C. Lee, R. Rajkumar, J. P. Lehoczky, and D. P. Siewiorek. On quality of service optimization with discrete QoS options. In *Proceedings of the IEEE Real-Time Technology and Applications Symposium,* June 1998.

[24] C. Lee, J. P. Lehoczky, D. P. Siewiorek, R. Rajkumar, and J. Hansen. A scalable solution to the multi-resource QoS problem. In *Proceedings of 20th IEEE Real-Time Systems Symposium,* December 1999.

[25] C. L. Liu and J. W. Layland. Scheduling algorithms for multiprogramming in hard real-time environment. *Journal of ACM* 20(1), 1973.

[26] A. K. Mok. Fundamental Design Problems of Distributed Systems for the Hard Real-Time Environment. Ph.D. Dissertation, Massachusetts Institue of Technology, Cambridge, MA, 1983.

[27] A. Bertossi and L. Mancini. Scheduling algorithms for fault-tolerance in hard real-time systems. *Real Time Systems Journal,* 7, 1994.

[28] D. Luenberger. *Linear and Nonlinear Programming.* Addison-Wesley, Reading, MA, 1984.

[29] R. GrayBill and R. Melhem (editors). *Power Aware Computing,* Series in Computer Science. Kluwer Academic/Plenum Publishers, New York, May 2002.

[30] P. J. M. Havinga and G. J. M. Smith. Design techniques for low-power systems. *Journal of Systems Architecture,* 46(1): 1–21, 2000.

[31] M. Weiser, B. Welch, A. Demers, and S. Shenker. Scheduling for reduced CPU energy. In *USENIX Symposium on Operating Systems Design and Implementation,* 13–23, 1994.

[32] K. Govil, E. Chan, and H. Wasserman. Comparing algorithms for dynamic speed-setting of a low-power CPU. In *ACM International Conference on Mobile Computing and Networking,* 13–25, 1995.

[33] F. Yao, A. Demers, and S. Shenker. A scheduling model for reduced CPU energy. *IEEE Annual Foundations of Computer Science,* 374–382, 1995.

[34] I. Hong, M. Potkonjak, and M. B. Srivastava. On-line scheduling of hard real-time tasks on variable voltage processor. In *Computer-Aided Design, ICCAD'98,* 653–656, 1998.

[35] I. Hong, D. Kirovski, G. Qu, M. Potkonjak, and M. Srivastava. Power optimization of variable voltage core-based systems. In *Design Automation Conference,* 1998.

[36] I. Hong, G. Qu, M. Potkonjak, and M. Srivastava. Synthesis techniques for low-power hard real-time systems on variable voltage processors. In *Proceedings of 19th IEEE Real-Time Systems Symposium, RTSS'98,* Madrid, December 1998.

[37] H. Aydin. Enhancing Performance and Fault Tolerance in Reward-Based Scheduling. Ph.D. Dissertation, University of Pittsburgh, August 2001.

[38] C. Rusu, R. Melhem, and D. Mossé. Maximizing rewards for real-time applications with energy constraints. *ACM Transactions on Embedded Computing Systems.* (Unpublished)

[39] H. Aydin, R. Melhem, D. Mossé, and P. M. Alvarez. Determining optimal processor speeds for periodic real-time tasks with different power characteristics. In *Proceedings of the 13th EuroMicro Conference on Real-Time Systems, ECRTS'01,* June 2001.

[40] H. Aydin, R. Melhem, D. Mossé, and P. M. Alvarez. Dynamic and aggressive power-aware scheduling techniques for real-time systems. In *Proceedings of the 22nd IEEE Real-time Systems Symposium, RTSS'01,* December 2001.

[41] R. Ernst and W. Ye. Embedded program timing analysis based on path clustering and architecture classification. In *Computer-Aided Design, ICCAD'97,* 598–604, 1997.

[42] P. Pillai and K. G. Shin. Real-time dynamic voltage scaling for low power embedded operating systems. In *Proceedings of the 18th Symposium on Operating System Principles,* October 2001.

[43] S. Saewong and R. Rajkumar. Practical voltage-scaling for fixed-priority real-time systems. In *Proceedings of the 9th IEEE Real-Time and Embedded Technology and Applications Symposium, RTAS'03,* Washington D.C., May 2003.

[44] D. Mossé, H. Aydin, B. Childers, and R. Melhem. Compiler-assisted dynamic power-aware scheduling for real-time applications. *Workshop on Compilers and Operating Systems for Low-Power, COLP'00,* Philadelphia, PA, October 2000.

[45] F. Gruian. Hard real-time scheduling for low-energy using stochastic data and DVS processors. In *International Symposium on Low Power Electronics and Design 2001,* August 6–7, 2001.

[46] W. Kim, D. Shin, H. S. Yun, J. Kim, and S. L. Min. Performance comparison of dynamic voltage scaling algorithms for hard real-time systems. In *Proceedings of the 8th Real-time Technology and Applications Symposium,* San Jose, CA, 2002.

[47] W. Namgoong, M. Yu, and T. Meg. A high efficiency variable-voltage CMOS dynamic DC–DC switching regulator. *IEEE International Solid-State Circuits Conference,* 380–391, 1996.

[48] T. D. Burd, T. A. Pering, A. J. Stratakos, and R. W. Brodersen. A dynamic voltage scaled microprocessor system. *IEEE Journal of Solid-State Circuits,* 35(11): 1571–1580, 2000.

[49] T. Pering, T. Burd and R. Brodersen. Voltage scaling in the IpARM microprocessor system. In *Proceedings of the 2000 International Symposium on Low power Electronics and design,* 2000.

[50] R. Min, T. Furrer, and A. Chandrakasan. Dynamic voltage scaling techniques for distributed microsensor networks. In *Proceedings of the IEEE Computer Society Annual Workshop on VLSI, WVLSI '00,* April 2000.

[51] J. Pouwelse, K. Langendoen, and H. Sips. Dynamic voltage scaling on a low-power microprocessor. In *Proceedings of Mobile Computing Conference, MOBICOM,* 2001.

[52] http://www.transmeta.com

[53] N. AbouGhazaleh, D. Mossé, B. Childers, and R. Melhem. Toward the placement of power management points in real time applications. In *Workshop on Compilers and Operating Systems for Low Power, COLP'01,* Barcelona, Spain, 2001.

[54] B. Mochocki, X. Hu, and G. Quan. A realistic variable voltage scheduling model for real-time applications. In *Proceedings of ICCAD 2002.*

37 Routing Real-Time Messages on Networks

37.1 Introduction

37

Routing Real-Time Messages on Networks

G. Young
California State Polytechnic University

37.1 Introduction

In a distributed system, processes residing at different nodes in the network communicate by passing messages. For a distributed real-time system, the problem of determining whether a set of messages can be sent on time becomes an important issue. The purpose of this chapter is to study the complexity of this message routing problem. Formally, we represent a network by a directed graph $G = (V, E)$, where each vertex in V represents a node of the network and each directed edge in E represents a communication link. If $(u, v) \in E$, then there is a *transmitter* in node u and a *receiver* in node v dedicated to the communication link (u, v). Thus a node can simultaneously send and receive several messages, provided that they are transmitted on different communication links.

A set of n messages $M = (M_1, \ldots, M_n)$ needs to be routed through the network. Each message M_i is represented by the quintuple $(s_i, e_i, l_i, r_i, d_i)$, where s_i denotes the *origin node* (i.e., M_i originates from node s_i), e_i denotes the *destination node* (i.e., M_i is to be sent to node e_i), l_i denotes the *length* (i.e., M_i consists of l_i packets of information), r_i denotes the *release time* (i.e., M_i originates from s_i at time r_i), and d_i denotes the *deadline* (i.e., M_i must reach e_i by time d_i).

An instance of the message routing problem consists of a network G and a set of messages M. Our goal is to determine if the messages in M can be routed through the network G such that each message M_i is sent from node s_i to node e_i in the time interval $[r_i, d_i]$.

We make several assumptions in our model. First, it takes one time unit to send a packet on a communication link. Thus it takes l_i time units for M_i to traverse on any communication link. Second, there is a central controller to construct a route for the messages that will be broadcast to each node. The central controller has complete information on the topology of the network and the characteristics of the messages. (As it turns out, the polynomially solvable cases shown in this chapter can all be done by an

1-58488-397-9/$0.00+$1.50

Online algorithm.) Third, each message must be completely received by a node before it is forwarded to the next one.

We will be considering both nonpreemptive and preemptive transmissions. In nonpreemptive transmission a message once transmitted on a communication link (u, v) must continue until the entire message is received by node v. By contrast, transmission can be interrupted and later resumed in preemptive transmission. We assume, however, that preemption can only occur at integral time units (i.e., at packet boundary) and that there is no time loss in preemption.

In the last two decades, real-time message routing has been studied extensively on different types of networks [1–35]. In this chapter, we discuss, based on the research results done by Leung et al. [25,26], the message routing problem with release time and deadline constraints. Related research results in the literature can be found in [6,7,11,13,14,21,26–29] The message routing problem studied in this chapter is related to the File Transfer problem studied in [12,32], the Data Transfer problem studied in [8-10,31], the Satellite Communication problem studied in [4,19], and the Link Scheduling problem studied in [20,23]. However, there are several fundamental differences between our model and those studied in [4,8–10,12,19,20,23,31,32]. First, the models studied there all assume that the network is a complete graph; i.e., any node can send a message to any other node without going through intermediate nodes. Second, each node in their models is assumed to be equipped with a number of dual-purpose ports that can be used as both receiver and transmitter. Third, the messages are assumed to be released at the same time. Fourth, their primary objective is to minimize the overall finishing time. The major results obtained in [4,8–10,12,19,20,23,31,32] are primarily concerned with showing various subproblems being solvable in polynomial time while others are NP-hard. Furthermore, approximation algorithms are presented with their worst-case performance bounds analyzed.

The purpose of this chapter is to study the complexity of the message routing problem under various restrictions of four parameters — origin node, destination node, release time, and deadline. We look into both the online and offline versions of the stated problem. For online routing, each node in the network routes messages without any knowledge of the messages in the other nodes or the release times of the messages. By contrast, for offline routing, each node in the network routes messages with knowledge of all messages in advance. In Section 37.2, we introduce notations and definitions that are commonly used in the chapter. We also show that if the network is an arbitrary directed graph, then the problem of offline routing of variable-length messages is NP-hard even when all four parameters are fixed; i.e., the messages have identical values in all four parameters. Motivated by the complexity of the problem, in the remainder of this chapter, we concentrate our effort only on simple networks. In Section 37.3, we study the offline routing of variable-length messages. In Section 37.4, we study the online routing of variable-length messages. In Section 37.5, we study the online routing of equal-length messages. Finally, we draw some conclusions in the last section.

37.2 The Model and Preliminaries

In this section we formally define the model used in this chapter and derive some preliminary results. Let $MRNS = (G, M)$ be a message-routing network system, where $G = (V, E)$ is a network and $M = (M_1, M_2, \ldots, M_n)$ is a set of n messages to be routed through G. A *configuration* q is a quintuple (u, v, m, t_1, t_2), where $0 \leq t_1 < t_2$ and $(u, v) \in E$, which denotes that message m is transmitted on the edge (u, v) in the time interval $[t_1, t_2]$. Let $Q_i = \{q_1, q_2, \ldots, q_k\}$ be a set of configurations for M_i and $TT(u, v, \hat{t}_1, \hat{t}_2, Q_i)$ denote the total time that M_i is transmitted on the edge (u, v) in the time interval $[\hat{t}_1, \hat{t}_2]$; i.e., $TT(u, v, \hat{t}_1, \hat{t}_2, Q_i) = \sum_{q \in Q'} (t_2 - t_1)$, where $Q' = \{q = (u, v, M_i, t_1, t_2) \mid q \in Q_i \text{ and } \hat{t}_1 \leq t_1 < t_2 \leq \hat{t}_2\}$. Q_i is a *route* for $M_i = (s_i, e_i, l_i, r_i, d_i)$ if M_i reaches a sequence of vertexes $s_i = x_1, x_2, \ldots, x_p = e_i$ at times $\hat{t}_1 < \hat{t}_2 < \cdots < \hat{t}_p$, respectively, such that the following three conditions are satisfied:

1. $(x_1, x_2, \ldots, x_p)$ is a path in G.
2. M_i must be completely received by vertex x_j before it can be sent to vertex x_{j+1}; i.e., $TT(x_j, x_{j+1}, \hat{t}_j, \hat{t}_{j+1}, Q_i) = l_i$ for each $1 \leq j < p$.

3. Q_i consists of only the configurations of M_i sending along the path $(x_1, x_2, \ldots, x_p)$ at appropriate times; i.e., for each $(u, v, M_i, t_1, t_2) \in Q_i$, there exist $x_j, x_{j+1}, \hat{t}_j$, and $\hat{t}_{j+1}$ such that $\hat{t}_j \leq t_1 < t_2 \leq \hat{t}_{j+1}, u = x_j$, and $v = x_{j+1}$.

A *transmission S* for *MRNS* $= (G, M)$ is defined to be $S = \bigcup_{i=1}^{n} Q_i$, where each Q_i is a route for M_i such that the following two conditions are satisfied:

1. M_i is transmitted only after its release time r_i; i.e., for each $(u, v, M_i, t_1, t_2) \in Q_i, t_1 \geq r_i$.
2. At most one message is transmitted on an edge at a time; i.e., for all (u, v, M_i, t_1, t_2) and $(u, v, M_j, \hat{t}_1, \hat{t}_2) \in S$ such that $i \neq j$, either $\hat{t}_1 \geq t_2$ or $t_1 \geq \hat{t}_2$.

A *transmission S* for *MRNS* is said to be *feasible* if the deadlines of all the messages are met; i.e., for all $(u, v, M_i, t_1, t_2) \in Q_i, t_2 \leq d_i$. A *nonpreemptive* transmission, denoted by S_{NP} , is a transmission such that for each $(u.v, M_i, t_1, t_2) \in S_{NP}, t_2 - t_1 = l_i$. A *preemptive* transmission, denoted by S_P, is a transmission defined exactly as before. A *MRNS* is *feasible* with respect to nonpreemptive (preemptive) transmission if there is a feasible nonpreemptive (preemptive) transmission for *MRNS*.

Example 37.1

Let *MRNS* $= (G, M)$, where G is as shown in Figure 37.1, $M = \{M_1, M_2, M_3, M_4\}$, $M_1 = (1, 4, 3, 0, 6)$, $M_2 = (1, 4, 2, 2, 6)$, $M_3 = (3, 5, 2, 0, 6)$, and $M_4 = (4, 5, 4, 0, 4)$. Consider the transmission $S_{NP} = \{(1, 2, M_1, 0, 3),\ (2, 4, M_1, 3, 6),\ (1, 3, M_2, 2, 4),\ (3, 4, M_2, 4, 6),\ (3, 4, M_3, 0, 2),\ (4, 5, M_3, 4, 6),\ (4, 5, M_4, 0, 4)\}$, as shown in Figure 37.1(a). Clearly, S_{NP} is a feasible nonpreemptive transmission. Thus *MRNS* is feasible with respect to nonpreemptive transmission.

Consider a new system *MRNS′* obtained from *MRNS* by modifying M_4 to be $(4, 5, 2, 3, 5)$. Then *MRNS′* is not feasible with respect to nonpreemptive transmission. However, the transmission $S'_P = \{(1, 2, M_1, 0, 3),\ (2, 4, M_1, 3, 6),\ (1, 3, M_2, 2, 4),\ (3, 4, M_2, 4, 6),\ (3, 4, M_3, 0, 2),\ (4, 5, M_3, 2, 3),\ (4, 5, M_3, 5, 6),$

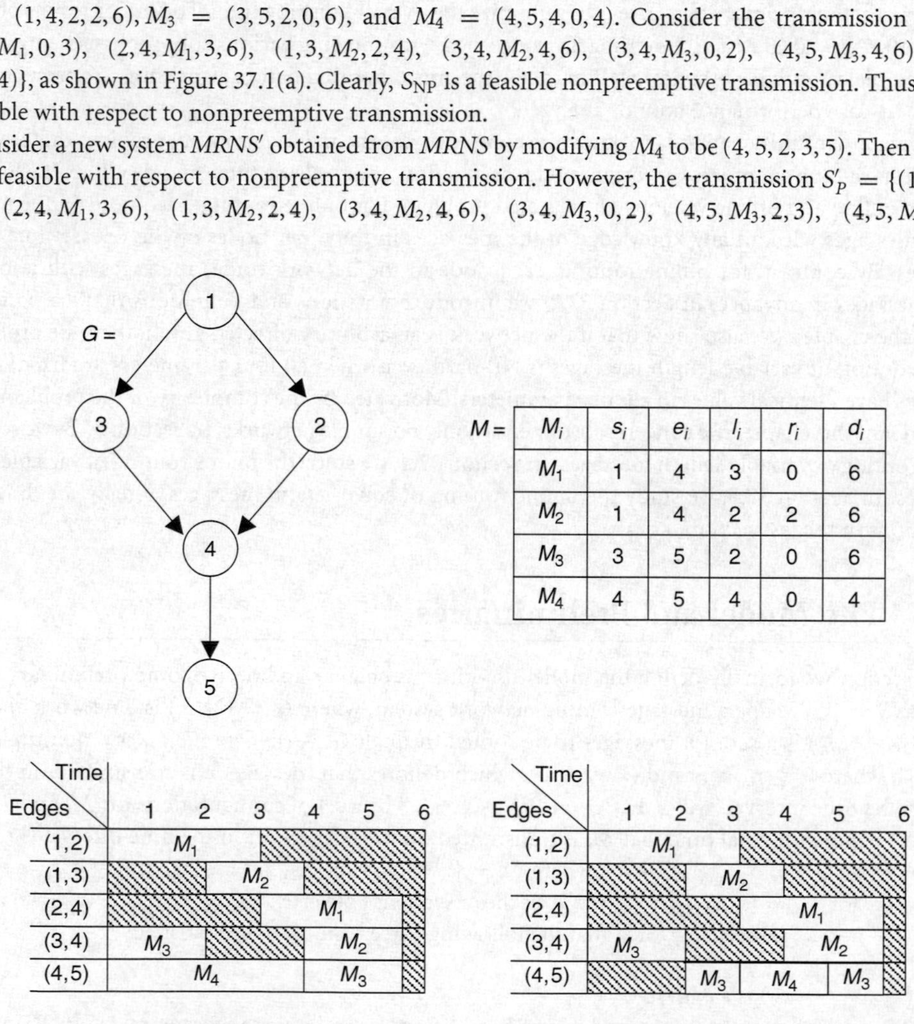

M_i	s_i	e_i	l_i	r_i	d_i
M_1	1	4	3	0	6
M_2	1	4	2	2	6
M_3	3	5	2	0	6
M_4	4	5	4	0	4

FIGURE 37.1 Illustrating Example 37.1: (a) nonpreemptive transmission and (b) preemptive transmission.

$(4, 5, M_4, 3, 5)\}$, as shown in Figure 37.1(b), is a feasible preemptive transmission. Thus *MRNS'* is feasible with respect to preemptive transmission, but not nonpreemptive transmission.

We now show that if the network is an arbitrary directed graph, the problem of determining whether there is a feasible transmission (preemptive or nonpreemptive) for a message-routing network system is NP-complete, even when all four parameters are fixed. Our reduction is from the 3-Partition problem, which is known to be strongly NP-complete [36].

3-Partition. Given a list, $A = (a_1, a_2, \ldots, a_{3z})$ of $3z$ integers such that $\sum_{i=1}^{3z} a_i = zB$ and $B/4 < a_i < B/2$ for each $1 \leq i \leq 3z$, can $I = \{1.2, \ldots, 3z\}$ be partitioned into $I_1, I_2, \ldots, I_z$ such that $\sum_{i \in I_j} a_i = B$ for each $1 \leq j \leq z$?

Theorem 37.1

Given MRNS $= (G, M)$, *where G is an arbitrary directed graph and M is a set of messages with identical origin nodes, destination nodes, release times, and deadlines, the problem of determining whether MRNS is feasible with respect to preemptive transmission is NP-complete.*

Proof

Preemption is necessary only when a new message arrives at a node. Thus the number of preemptions is bounded by a polynomial function of the size of the input, and hence we can guess a preemptive transmission in polynomial time and verify that the transmission is feasible. This shows that the decision problem is in NP. To complete the proof, we reduce the 3-Partition problem to it as follows. Let $A = (a_1, a_2, \ldots, a_{3z})$ be an instance of the 3-Partition problem. We construct an instance of the message routing problem as follows: *MRNS*$= (G, M)$, where $G = (\{0, 1, 2, \ldots, z+1\}, \{(0, 1), (0, 2) \ldots, (0, z), (1, z+1), (2, z+1), \ldots, (z, z+1)\}), M = \{M_1, M_2, \ldots, M_{4z}), $ and $M_i = (0, z+1, a_i, 0, 5B)$ for each $1 \leq i \leq 3z$ and $M_i = (0, z+1, 2B, 0, 5B)$ for each $3z+1 \leq i \leq 4z$. We call the first $3z$ messages the *partition* messages and the remaining z messages the *enforcer* messages. Clearly, all messages in M have identical origin nodes, destination nodes, release times, and deadlines. It is easy to see that the construction can be done in polynomial time. For the reduced instance, there are z distinct paths from vertex 0 to vertex $z+1$ in G. In any feasible transmission for *MRNS*, each enforcer message must be routed along a distinct path from vertex 0 to vertex $z+1$. The remaining partition messages must be distributed among these z paths. We now show that above *MRNS* is feasible with respect to preemptive transmission if and only if A has a 3-Partition.

Suppose $I_1, I_2, \ldots, I_z$ is a solution to the 3-Partition problem. We can construct a feasible nonpreemptive (and hence preemptive) transmission as follows. For each $1 \leq j \leq z$, the messages M_{3z+j} and M_i, where $i \in I_j$, are transmitted from node 0 to node $z+1$ through the edges $(0, j)$ and $(j, z+1)$. M_{3z+j} is transmitted through the edges $(0, j)$ and $(j, z+1)$ in the time intervals $[0, 2B]$ and $[2B, 4B]$, respectively. The messages M_i, where $i \in I_j$, are nonpreemptively transmitted through the edges $(0, j)$ and $(j, z+1)$ in the time intervals $[2B, 3B]$ and $[4B, 5B]$, respectively. Clearly, this is a feasible nonpreemptive transmission. Therefore, *MRNS* is feasible with respect to preemptive transmission.

Now, suppose S is a feasible preemptive transmission for *MRNS*. Observe that there are exactly z distinct paths from node 0 to node $z+1$; namely, $P_1 = (0, 1, z+1)$, $P_2 = (0, 2, z+1), \ldots,$ and $P_z = (0, z, z+1)$. Therefore, each message must be transmitted through one of these z paths. If two enforcer messages are transmitted through the same path in S, then one of them will arrive node $z+1$ at time $6B$ or later, and hence it will miss its deadline. Thus, we may assume that no two enforcer messages are transmitted through the same path in S. For each $1 \leq j \leq z$, let $I_j = \{i \mid 1 \leq i \leq 3z$ and M_i is transmitted through P_j in $S\}$. We claim that $I_1, I_2, \ldots, I_z$ is a solution to the 3-Partition problem. If the claim is not true, then there is an index k such that $\sum_{i \in I_k} a_i > B$. Let $X \cup \{M'\}$ be the set of messages transmitted through P_k in S, where $X = \{M_i \mid i \in I_k\}$ and M' is the enforcer message. Let t' be the time instant at which M' is completely transmitted to node k, and let X_1 and X_2 be a partition of X such that X_1 and X_2 contain all the messages in X that are completely transmitted to node k before and after time t', respectively.

Clearly, the total time taken to transmit all the messages in $X \cup \{M'\}$ is at least

$$t' + 2B + \sum_{M_i \in X_2} l_i \geq \sum_{M_i \in X_1} l_i + 2B + 2B + \sum_{M_i \in X_2} l_i = 4B + \sum_{M_i \in X} l_i > 5B$$

Thus, at least one of the messages in $X \cup M'$ has its deadline missed, contradicting our assumption that S is a feasible transmission. □

Corollary 37.1

Given MRNS $= (G, M)$, *where G is an arbitrary directed graph and M is a set of messages with identical origin nodes, destination nodes, release times, and deadlines, the problem of determining whether MRNS is feasible with respect to nonpreemptive transmission is NP-complete.*

Proof

The same reduction works for nonpreemptive transmission as well. □

Theorem 37.1 and Corollary 37.1 show that the message routing problem is a computationally hard problem if the network is allowed to be arbitrary. Thus we restrict our attention to a simple class of networks, with the hope that the problem can be solved efficiently. We have chosen the class of unidirectional rings because of its simplicity and popularity. A unidirectional ring G is defined to be $G = (V, E)$, where $V = \{1, 2, \ldots, m\}$ for some integer $m > 1$ and $E = \{(1, 2), (2, 3), \ldots, (m-1, m), (m, 1)\}$. In the remainder of this chapter, we will assume that the network is a unidirectional ring with m nodes, unless stated otherwise.

In the remainder of this section, we will give some fundamental results that will be used later in this chapter. First, we need to introduce the following notations. Let S be a transmission for $MRNS = (G, M)$, where G is a unidirectional ring with m nodes and M is a set of n messages. We use $f(M_i, S)$ to denote the finishing time of M_i in S; i.e., $f(M_i, S) = \max_{(u,v,M_i,t_1,t_2) \in S} t_2$. The *makespan* of S, denoted by $MS(S)$ is defined to be $\max_{1 \leq i \leq n} f(M_i, S)$. For a set of messages with identical origin nodes, destination nodes, release times, and deadlines, the next lemma gives a lower bound for the makespan of any feasible transmission (preemptive or nonpreemptive).

Lemma 37.1

Let S_P *be a feasible preemptive transmission for a set of messages with the same origin node, destination node, release time, and deadline. If the common origin node and destination node of all the messages are 1 and e, respectively, then we have* $MS(S_P) \geq \sum_{i=1}^{n} l_i + (e-2)(\max_{1 \leq i \leq n} l_i)$.

Proof

If the lemma is not true, then there is a feasible preemptive transmission S_P such that $MS(S_P) < \sum_{i=1}^{n} l_i + (e-2)(\max_{1 \leq i \leq n} l_i)$. Without loss of generality, we may assume that the common release time of the messages is time 0. Let M_P be the longest message; i.e., $l_p \geq l_i$ for each $1 \leq i \leq n$. For each $2 \leq j \leq e$, let t_j be the time instant at which M_P is completely transmitted to node j on the edge $(j-1, j)$ in S_P. Thus, $t_e = f(M_P, S_P)$. For each $1 \leq i \leq n$ and $2 \leq j \leq e$, let $X(i, j)$ and $Y(i, j)$ be the total amounts of M_i transmitted on the edge $(j-1, j)$ in S_P in the time intervals $[0, t_j]$ and $[t_j, MS(S_P)]$, respectively. Clearly, we have $X(i, j) + Y(i, j) = l_i$ for each $1 \leq i \leq n$ and $2 \leq j \leq e$. Let $A = \{M_i \mid M_i$ is completely transmitted to node 2 before t_2 in $S_p\}$, $B = \{M_i \mid M_i$ is completely transmitted to node 2 after t_2 in S_p, and M_i is completely transmitted to node k before t_k for some node $k, 3 \leq k \leq e\}$, and $C = \{M_i \mid M_i$ is completely transmitted to node j after t_j in S_p for each $2 \leq j \leq e\}$. By our definitions, the sets A, B, C, and $\{M_P\}$ are pairwise disjoint, and the set $A \cup B \cup C \cup \{M_P\}$ contains all the messages.

We now compute a lower bound for $f(M_P, S_P)$ (or equivalently, t_e). Each $M_i \in A$ is completely transmitted to node 2 before M_p does, and hence it delays the finishing time of M_p by at least l_i amount.

For each $M_i \in B$, there is a node $k, 2 \leq k < e$, such that M_i is completely transmitted to node k after M_p does, but it is completely transmitted to node $k+1$ before M_p does. Thus, it again delays the finishing time of M_p by at least l_i amount. Each $M_i \in C$ is completely transmitted to node $j, 2 \leq j \leq e$, after M_p does. However, the total amount it is transmitted on the edge $(e-1, e)$ by time t_e is $X(i, e)$, and hence it delays the finishing time of M_p by at least $X(i, e)$ amount. Thus, we have

$$f(M_P, S_P) = t_e \geq \sum_{M_i \in A} l_i + (e-1)l_p + \sum_{M_i \in B} l_i + \sum_{M_i \in C} X(i, e)$$

Now,

$$\begin{aligned} MS(S_p) &\geq f(M_P, S_P) + \sum_{M_i \in C} Y(i, e) \\ &\geq \sum_{M_i \in A} l_i + (e-1)l_p + \sum_{M_i \in B} l_i + \sum_{M_i \in C} X(i, e) + \sum_{M_i \in C} Y(i, e) \\ &= \sum_{M_i \in A} l_i + (e-1)l_p + \sum_{M_i \in B} l_i + \sum_{M_i \in C} l_i \\ &= \sum_{i=1}^{n} l_i + (e-2)(\max_{i=1}^{n} l_i), \end{aligned}$$

a contradiction. □

Lemma 37.2

Let S_{NP} be a feasible nonpreemptive transmission for a set of messages with the same origin node, destination node, release time, and deadline. If the common origin node and the common destination node of all the messages are 1 and e, respectively, then we have $MS(S_{\text{NP}}) \geq \sum_{i=1}^{n} l_i + (e-2)(\max_{1 \leq i \leq n} l)$.

Proof

Let S_1 be a feasible nonpreemptive transmission for the set of messages such that it has the minimum makespan among all feasible nonpreemptive transmissions. Let S_2 be a feasible preemptive transmission for the set of messages such that it has the minimum makespan among all feasible preemptive transmissions. Since $MS(S_{\text{NP}}) \geq MS(S_1) \geq MS(S_2)$, the lemma follows immediately from Lemma 37.1. □

Let $MRNS = (G, M)$ be a message-routing network system, where $G = (\{1, 2, \ldots, m\}, \{(1, 2), (2, 3), \ldots, (m-1, m), (m, 1)\})$ is an unidirectional ring with m nodes and $M = (M_1, M_2, \ldots, M_n)$ is a set of n messages. The inverse of *MRNS*, denoted by *MRNS*$'$, is defined as follows: $MRNS' = (G', M')$, where $G' = (\{1, 2, \ldots, m\}, \{(2, 1), (3, 2), \ldots, (m, m-1), (1, m)\})$ and $M' = \{M_i' = (e_i, s_i, l_i, d - d_i, d - r_i) \mid (s_i, e_i, l_i . r_i, d_i) \in M$ and d is the largest deadline among all the messages in $M\}$. The next lemma shows that *MRNS* is feasible with respect to nonpremptive (preemptive) transmission if and only if *MRNS*$'$ is feasible with respect to nonpreemptive (preemptive) transmission.

Lemma 37.3

Let $MRNS = (G, M)$ be a message-routing network system with an unidirectional ring. Then MRNS is feasible with respect to nonpreemptive (preemptive) transmission if and only if the inverse of MRNS, $MRNS' = (G', M')$, is feasible with respect to nonpreemptive (preemptive) transmission.

Proof

If S is a feasible nonpreemptive (preemptive) transmission for *MRNS*, we can construct a nonpreemptive (preemptive) transmissions S' for *MRNS*$'$ as follows. For each configuration $q = (u, v, M_i . t_1, t_2) \in S$, add the configuration $q' = (v, u, M_i', d - t_2, d - t_1)$ to S', where d is the largest deadline among all messages in

M. It is easy to see that S' is a feasible nonpreemptive (preemptive) transmission for *MRNS'*. Conversely, if S' is a feasible nonpreemptive (preemptive) transmission for *MRNS'*, then we construct a nonpreemptive (preemptive) transmission S for *MRNS* as follows. For each configuration $q' = (u, v, M_i', t_1, t_2)$ to S', add the configuration $q = (v, u, M_i, d' - t_2, d' - t_1)$ to S, where d' is the largest deadline among all messages in M'. Again, it is easy to see that S is a feasible nonpreemptive (preemptive) transmission for *MRNS*. □

37.3 Offline Routing of Variable-Length Messages

In this section we consider offline routing of variable-length messages on a unidirectional ring. For offline routing, each node in the network routes messages with knowledge of all messages in advance. For nonpreemptive transmission, we show that the problem is solvable in polynomial time if three of the four parameters are fixed, but becomes NP-complete if only two of them are fixed. The same kind of complexity results hold for preemptive transmission, except for the following two cases: (1) identical origin nodes and release times and (2) identical destination nodes and deadlines. The complexity of these two cases remains open.

We first consider nonpreemptive transmission on a unidirectional ring. We first show that the problem is solvable in polynomial time if three of the four parameters are fixed. We then show that the problem becomes NP-complete if only two of them are fixed.

Now we give several polynomial-time algorithms, and show that they solve various restricted cases of the problem. We then show that these algorithms can be combined into a single algorithm. In the following, we say that a message is ready at a node if it has been received completely by the node and is ready to be sent to another one. By renaming nodes if necessary, we may assume that node 1 is the smallest-indexed origin node and node e is the largest-indexed destination node among all the messages.

Algorithm A *(Earliest Available Message Strategy).*Whenever a node is free for transmission, send that ready message which is available at the node the earliest. Ties can be broken arbitrarily.

Algorithm B *(Earliest Available Message and Farthest Destination Strategy).*Whenever a node is free for transmission, send that ready message which is available at the node the earliest. If there is a tie, choose the one with the farthest destination. Further ties can be broken arbitrarily.

Algorithm C *(Earliest Available Message and Earliest Deadline Strategy).*Whenever a node is free for transmission, send that ready message which is available at the node the earliest. If there is a tie, choose the one with the earliest deadline. Further ties can be broken arbitrarily.

We first show that Algorithm A solves the message routing problem if all four parameters are fixed. First, we need to show the following lemma.

Lemma 37.4

Let S_A be the transmission generated by Algorithm A for a set of n messages with identical origin nodes, destination nodes, release times, and deadlines. Let the order of the messages transmitted on the edge $(1, 2)$ be $M_{a_1}, M_{a_2}, \ldots, M_{a_n}$. Then, for each $1 \le i \le n$, M_{a_i} is transmitted on the edge $(j-1, j)$ in the time interval $[b_{i,j}, c_{i,j}]$, where $b_{i,j} \sum_{k=1}^{i-1} la_k + (j-2)(\max_{l \le k \le i} l_{a_k})$ and $c_{i,j} = \sum_{k=1}^{i} la_k + (j-2)(\max_{1 \le k \le i} la_k)$ for each $1 < j \le e$.

Proof

By Algorithm A, the order of messages transmitted on each edge $(j-1, j)$ is $M_{a_1}, M_{a_2}, \ldots, M_{a_n}$. We will prove the lemma by induction on i. For $i = 1$, it is easy to see that M_{a_1} is transmitted on the edge $(j-1, j)$ in the time interval $[(j-2)l_{a_1}, (j-1)l_{a_i}]$ for each $2 \le j \le e$. Assume the lemma is true for all $1 \le i \le p$. where $1 \le p \le n-1$, we will show that it is true for $i = p + 1$. We consider the following two cases.

Case I: *$l_{a_{p+1}} \leq l_{a_x}$ for some $1 \leq x \leq p$*

We will show by induction on j that $M_{a_{p+1}}$ is transmitted on the edge $(j-1, j)$ in the time interval $[b_{a_{p+1}}j, c_{a_{p+1}}j]$. Clearly, $M_{a_{p+1}}$ is transmitted on the edge $(1, 2)$ in the time interval $[\sum_{k=1}^{p} l_{a_k}, \sum_{k=l}^{p+1} l_{a_k}]$. Assume it is true for all $2 \leq j \leq q$, where $2 \leq q \leq e-1$, we will show that it is true for $j = q+1$. By the induction hypothesis, $M_{a_{p+1}}$ is ready at node q at time $c_{a_{p+1}}, q = \sum_{k=1}^{p+1} l_{a_k} + (q-2)(\max_{k=1}^{p+1} l_{a_k})$, and M_{a_p} finishes transmission on the edge $(q, q+1)$ at time $c_{a_p,q+1} = \sum_{k=1}^{p} l_{a_k} + (q-1)(\max_{k=1}^{p} l_{a_k})$. Since $l_{a_{p+1}} \leq l_a$ for some $1 \leq x \leq p$, we have $c_{a_{p+1}}, q \leq c_{a_p,q+1}$, and hence $M_{a_{p+1}}$ is transmitted on the edge $(q, q+1)$ starting at time $c_{a_{p+1},q+1} = b_{a_{p+1},q+1}$. Thus, the claim is true for $j = q+1$.

Case II: *$l_{a_{p+1}} > l_{a_x}$ for some $1 \leq x \leq p$*

Again, we will prove it by induction on j. The basis case, $j = 2$, is obviously true. Assume it is true for all $2 \leq j \leq q$, where $2 \leq q \leq e-1$, we will show that it is true for $j = q+1$. By the induction hypothesis, $M_{a_{p+1}}$ is ready at node q at time $c_{a_{p+1},q}$ and M_{a_p} finishes transmission on the edge $(q, q+1)$ at time $c_{a_p,q+1}$. Since $l_{a_{p+1}} > l_{a_x}$ for all $1 \leq x \leq q+1$, we have $c_{a_{p+1}}, q > c_{a_p,q+1}$, and hence $M_{a_{p+1}}$ is transmitted on the edge $(q, q+1)$ starting at time $c_{a_{p+1},q} = b_{a_{p+1},q+l}$. Thus, the claim is true for $j = q+1$.

By induction, the lemma is true for all $1 \leq i \leq n$ and $2 \leq j \leq e$. □

Theorem 37.2

A set of messages with identical origin nodes, destination nodes, release timers and deadlines is feasible with respect to nonpreemptive transmission if and only if the transmission generated by Algorithm A is feasible.

Proof

The if part is obvious. To prove the only if part, all we need to show is that Algorithm A always generates a minimum makespan transmission. By Lemma 37.1, the transmission S_A generated by Algorithm A has $MS(S_A) = \sum_{i=1}^{n} l_i + (e-2)(\max_{1 \leq i \leq n} l_i)$. By Lemma 37.2, S_A *is a minimum makespan transmission.* □

If all the messages have identical origin nodes, release times, and deadlines, Algorithm B can be used to determine feasibility, as the next theorem shows.

Theorem 37.3

A set of messages with identical origin nodes, release times, and deadlines is feasible with respect to nonpreemptive transmission if and only if the transmission generated by Algorithm B is feasible.

Proof

The if part is obvious. We will prove the only if part by contradiction. Let M be a set of n messages such that the transmission S_B generated by Algorithm B is infeasible, yet there is a feasible nonpreemptive transmission S_{NP} for M. Let the order of the messages transmitted on the edge $(1, 2)$ in S_B be $M_{a_1}, M_{a_2}, \ldots, M_{a_n}$. By the nature of Algorithm B, $e_{a_1} \geq e_{a_2} \geq \cdots \geq e_{a_n}$, and the order of the messages transmitted on all the other edges are identical to that on the edge $(1, 2)$. Let M_{a_k} miss its deadline in S_B. Consider the new set of messages $M' = \{M'_{a_i} \mid 1 \leq i \leq k$ and M'_{a_i} is obtained from M_{a_i} by changing its destination node to $e_{a_k}\}$. Clearly, M' is a set of messages with identical origin nodes, destination nodes, release times, and deadlines. Let S'_B be the transmission obtained by restricting S_B to the messages $M_{a_i}, 1 \leq i \leq k$, and the edges $(j-1, j), 2 \leq j \leq e_{a_k}$. It is easy to see that S'_B is identical to the transmission by Algorithm A for M' with the order of the messages transmitted on the edge $(1, 2)$ being $M'_{a_1}, M'_{a_2}, \ldots, M'_{a_k}$. Thus the transmission generated by Algorithm A for M' is infeasible. Let S'_{NP} be the transmission obtained by restricting S_{NP} to the messages$M_{a_i}, 1 \leq i \leq k$, and the edges $(j-1, j), 2 \leq j \leq e_{a_k}$. Clearly, S'_{NP} is a feasible nonpreemptive transmission for M'. This is a contradiction to Theorem 37.2, since M' is feasible with respect to nonpreemptive transmission, but not Algorithm A. □

If all the messages have identical destination nodes, release times, and deadlines, we can use Algorithm A to determine feasibility, as the next theorem shows.

Theorem 37.4

A set of messages with identical destination nodes, release times, and deadlines is feasible with respect to nonpreemptive transmission if and only if the transmission generated by Algorithm A is feasible.

Proof

The if part is obvious. We will use the inverse operation defined in Section 37.2 to prove the only if part. Let *MNRS*$= (G, M)$ be feasible with respect to nonpreemptive transmission, where M is a set of n messages with identical destination nodes, release times, and deadlines. Let e, 0, and d be the common destination node, release time, and deadline, respectively, of all the messages in M. Let S_A be the transmission generated by Algorithm A for *MNRS*, and let $M_{a_1}, M_{a_2}, \ldots, M_{a_n}$ be the order of the messages arriving at node e. We will show that S_A is a feasible transmission. By the nature of Algorithm A, $s_{a_1} \geq s_{a_2} \geq \cdots \geq s_{a_n}$; i.e., the origin node of M_{a_1} is the closest to node e. Since *MRNS* is feasible with respect to nonpreemptive transmission, the inverse of *MRNS*, denoted by *MRNS*$'= (G', M')$, is also feasible with respect to nonpreemptive transmission, by Lemma 37.3. Observe that M' is a set of n messages with identical origin nodes, release times, and deadlines. Clearly, e, 0, and d are the common origin node, release time, and deadline, respectively, of all the messages in M'. By Theorem 37.3, the transmission generated by Algorithm B for *MRNS*$'$ is feasible. Let S_B be the transmission generated by Algorithm B for *MRNS*$'$ such that the order of the messages sent from node e is $M_{a_n}, M_{a_{n-1}}, \ldots, M_{a_1}$. It is easy to verify that S_B is a transmission consistent with the rules in Algorithm B. An example is shown in Figure 37.2, where Figure 37.2(a) shows the transmission S_A generated by Algorithm A for *MRNS* and Figure 37.2(b) shows the transmission S_B generated by Algorithm B for *MRNS*$'$.

Let $\hat{S}_B$ be the transmission obtained by *right-justifying* the transmissions in S_B. $\hat{S}_B$ is constructed from S_B as follows: If M_x is transmitted in S_B on the edge (i, j), then delay its transmission as late as possible until one of the three events occurs: (1) M_x will miss its deadline, (2) M_x will be passed by the message transmitted after it on the same edge, and (3) M_x will transmit on the edge (i, j) at the same time it is transmitting on the edge (j, k). $\hat{S}_B$ is the final transmission obtained by delaying the transmissions in S_B by the above process until no further delay can be obtained. Figure 37.2(c) shows the transmission $\hat{S}_B$ obtained from the transmission S_B shown in Figure 37.2(b). Clearly, $\hat{S}_B$ is a feasible transmission for *MRNS*$'$. From $\hat{S}_B$, we construct $\hat{S}'_B$ as follows: For each configuration $q = (v, u, M_i, t_1, t_2) \in \hat{S}_B$, add the configuration $q' = (v, u, M_i, d - t_2, d - t_1)$ to $\hat{S}'_B$. Figure 37.2(d) shows the transmission $\hat{S}'_B$ obtained from the transmission $\hat{S}_B$ shown in Figure 37.2(c). It is easy to see that $\hat{S}'_B$ is a feasible transmission for the inverse of *MRNS*$'$, namely, *MRNS*, and that $\hat{S}'_B$ is exactly S_A. Thus S_A is a feasible transmission for *MRNS*. □

If all the messages have identical origin nodes, destination need, and release times, Algorithm C can be used to determine feasibility, as the next theorem shows.

Theorem 37.5

A set of messages with identical origin nodes, destination nodes, and release times is feasible with respect to nonpreemptive transmission if and only if the transmission generated by Algorithm C is feasible.

Proof

The if part is obvious. We will prove the only if part by contradiction. Let M be a set of n messages such that the transmission S_C generated by Algorithm C is infeasible, and yet there is a feasible nonpreemptive transmission S_{NP} for M. Let the order of the messages transmitted on the edge (1, 2) in S_C be $M_{a_1}, M_{a_2}, \ldots, M_{a_n}$. By the nature of Algorithm C, $d_{a_1} \leq d_{a_2} \leq \cdots \leq d_{a_n}$, and the order of the messages transmitted on all the other edges are the same as that on the edge (1, 2). Let M_{a_k} misses its deadline in S_C; i.e., $f(M_{a_k}, S_C) > d_{a_k}$. Consider the new set of messages $M' = \{M'_{a_i} | 1 \leq i \leq k$ and M'_{a_i} is obtained from M_{a_i} by changing its deadline to $d_{a_k}\}$. Clearly, M' is a set of messages with the same origin node, destination node, release time, and deadline. Let S'_C be the transmission obtained by restricting S_C to the

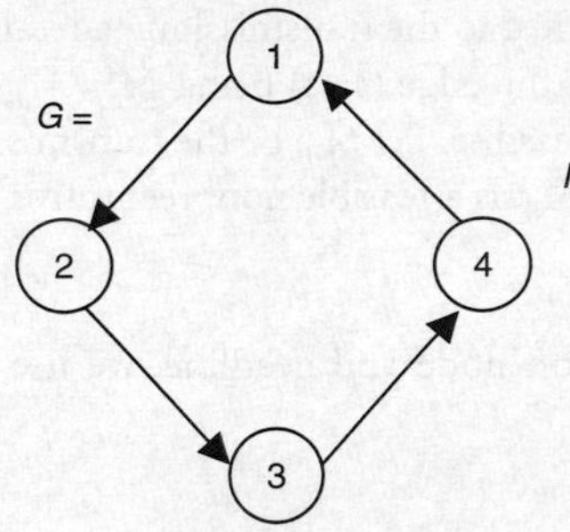

M =

M_i	s_i	e_i	l_i	r_i	d_i
M_1	2	4	2	0	12
M_2	2	4	3	0	12
M_3	1	4	2	0	12
M_4	1	4	1	0	12

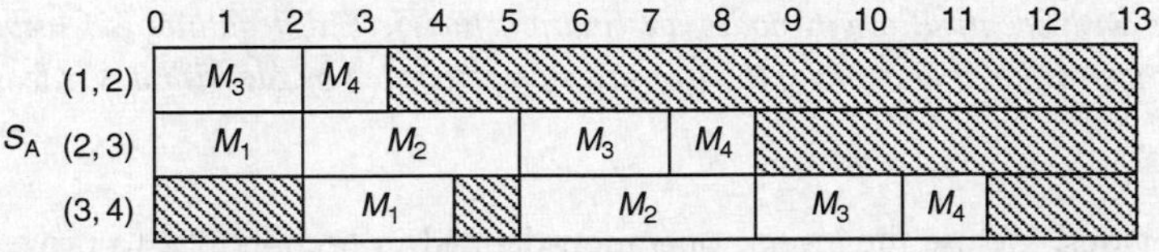

(a)

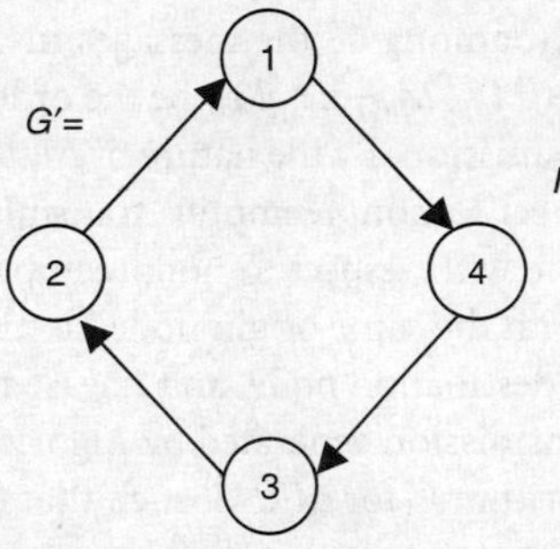

M′ =

M'_i	s'_i	e'_i	l'_i	r'_i	d'_i
M'_1	4	2	2	0	12
M'_2	4	2	3	0	12
M'_3	4	1	2	0	12
M'_4	4	1	1	0	12

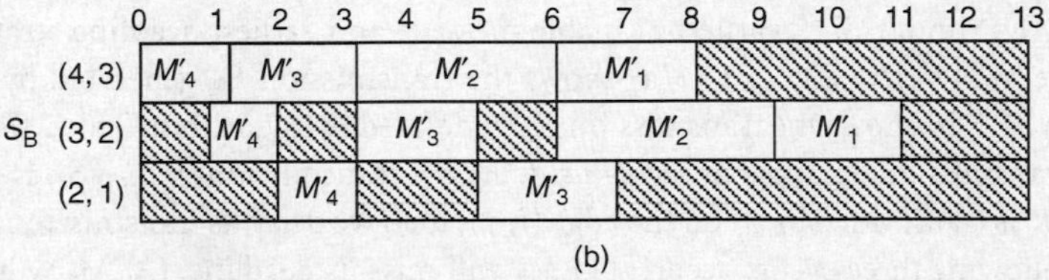

(b)

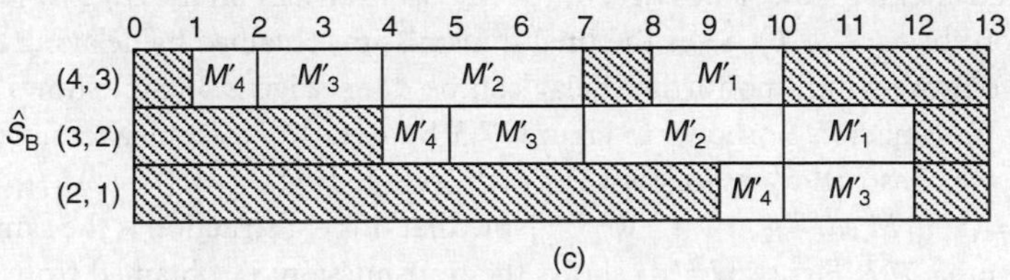

(c)

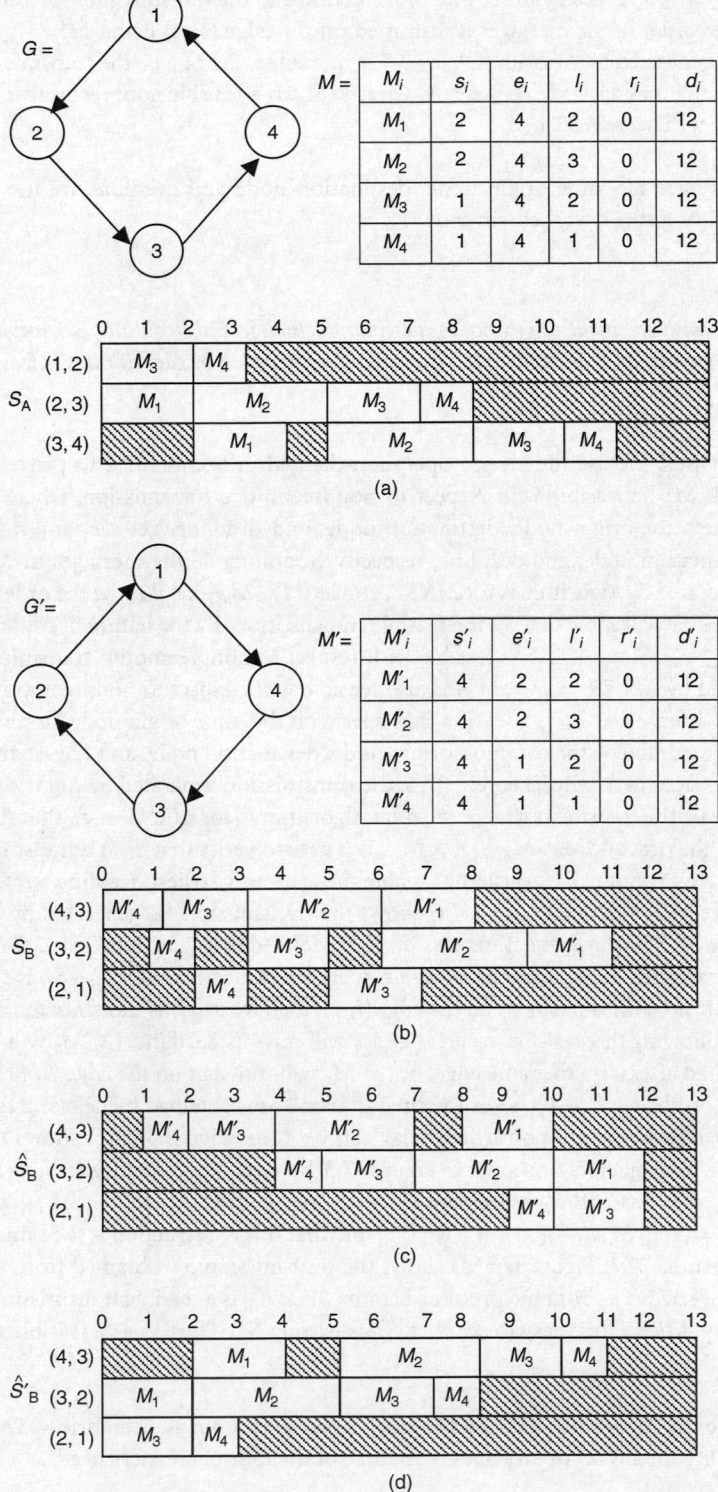

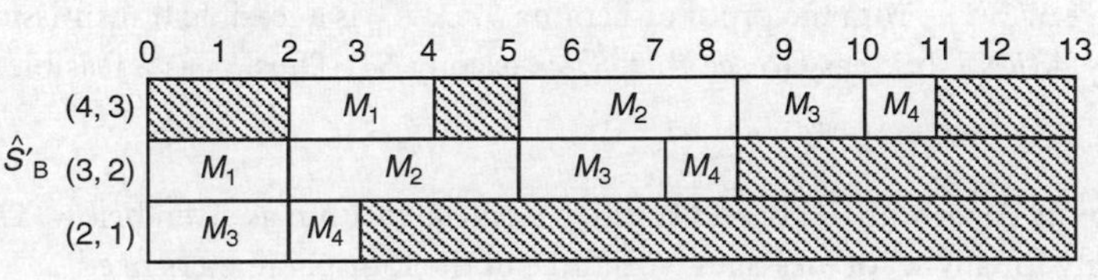

(d)

FIGURE 37.2 Illustrating the Proof of Theorem 37.4: (a) transmission generated by Algorithm A for $MRNS = (G, M)$, (b) transmission generated by Algorithm B for $MRNS' = (G', M')$, (c) transmission obtained by right-justifying S_B, and (d) transmission obtained by taking the inverse of $\hat{S}_B$.

messages $M_{a_i}, 1 \leq i \leq k$. It is easy to see that S'_C is identical to the transmission generated by Algorithm A for M' with the order of the messages transmitted on the edge (1, 2) being $M'_{a_1}, M'_{a_2}, \ldots, M'_{a_k}$. Thus, the transmission generated by Algorithm A for M' is infeasible. Let S'_{NP} be the transmission obtained by restricting S_{NP} to the messages $M_{a_i}, 1 \leq i \leq k$. Clearly, S'_{NP} is a feasible nonpreemptive transmission for M'. This contradicts Theorem 37.1. □

If all messages have the same origin node, destination node and deadline, we use Algorithm A to determine feasibility, as the next theorem shows.

Theorem 37.6

A set of messages with identical origin nodes, destination nodes, and deadlines is feasible with respect to nonpreemptive transmission if and only if the transmission generated by Algorithm A is feasible.

Proof

The if part is obvious. We use the inverse operation defined in Section 37.2 to prove the only if part. Let *MNRS* $= (G, M)$ be feasible with respect to nonpreemptive transmission, where M is a set of n messages with the same origin node, destination node, and deadline. Let s, e, and d be the common origin node, destination node, and deadline, respectively, among all the messages in M. Let S_A be the transmission generated by Algorithm A for *MNRS*, and let $M_{a_1}, M_{a_2}, \ldots, M_{a_n}$ be the order of the messages arriving at node e. We will show that S_A is a feasible transmission. By the nature of Algorithm A, we have $r_{a_1} \leq r_{a_2} \leq \cdots \leq r_{a_n}$. Since *MRNS* is feasible with respect to nonpreemptive transmission, the inverse of *MRNS*, denoted by *MRNS'* $= (G', M')$, is also feasible with respect to nonpreemptive transmission, by Lemma 37.3. Observe that M' is a set of n messages with the same origin node, destination node, and release time. Let e, s and 0 be the common origin node, destination node, and release time, respectively, among all the messages in M'. By Theorem 37.5, the transmission generated by Algorithm C for *MRNS'* is feasible. Let S_C be the transmission generated by Algorithm C for *MRNS'* such that the messages sent from node e is in the order of $M_{a_n}, M_{a_{n-1}}, \ldots, M_{a_t}$. It is easy to verify that S_C is a transmission consistent with the rules in Algorithm C; i.e., earliest available message and earliest deadline strategy. An example is shown in Figure 37.3, where Figure 37.3(a) shows the transmission S_A generated by Algorithm A for *MRNS* and Figure 37.3(b) shows the transmission S_C generated by Algorithm C for *MRNS'*.

Let $\hat{S}_C$ be the transmission obtained by *right-justifying* the transmissions in S_C. $\hat{S}_C$ is constructed from S_C as follows: If M_x is transmitted in S_C on the edge (i, j), then we delay its transmission as late as possible until one of the following three events occurs: (1) M_x will miss its deadline, (2) M_x will be passed by the message transmitted after it on the same edge, or (3) M_x will transmit on the edge (i, j) at the same time it is transmitting on the edge (j, k). $\hat{S}_C$ is the final transmission obtained by delaying the transmissions in S_C by the above process until no further delay can be done. Figure 37.3(c) shows the transmission obtained from the transmission S_C shown in Figure 37.3(b). Clearly, $\hat{S}_C$ is still a feasible transmission for *MRNS'*. From $\hat{S}_C$, we construct $\hat{S}'_C$ as follows: For each configuration $q = (u, v, M_i, t_1, t_2) \in \hat{S}_c$, add the configuration $q' = (v, u, M_i, d - t_2, d - t_1)$ to $\hat{S}'_C$. Note that this construction is the same as the one used in the proof of Lemma 37.3. Figure 37.3(d) shows the transmission $\hat{S}'_C$ obtained from the transmission $\hat{S}_C$ shown in Figure 37.3(c). From the proof of Lemma 37.3, $\hat{S}'_C$ is a feasible transmission for the inverse of *MRNS'*, namely, *MRNS*. It is easy to see that $\hat{S}'_C$ is exactly S_A. Thus S_A is a feasible transmission for *MRNS*. □

We can incorporate Algorithms A, B, and C into a single algorithm as given below. This algorithm can determine feasibility for any set of messages with three of the four parameters fixed.

Algorithm D *(Earliest Available Message, Earliest Deadline, and Farthest Destination Strategy).*Whenever a node is free for transmission, send that ready message which is available at the node the earliest. If there is a tie, choose the one with the earliest deadline. If again there is a tie, choose the one with the farthest destination. Any further ties can be broken arbitrarily.

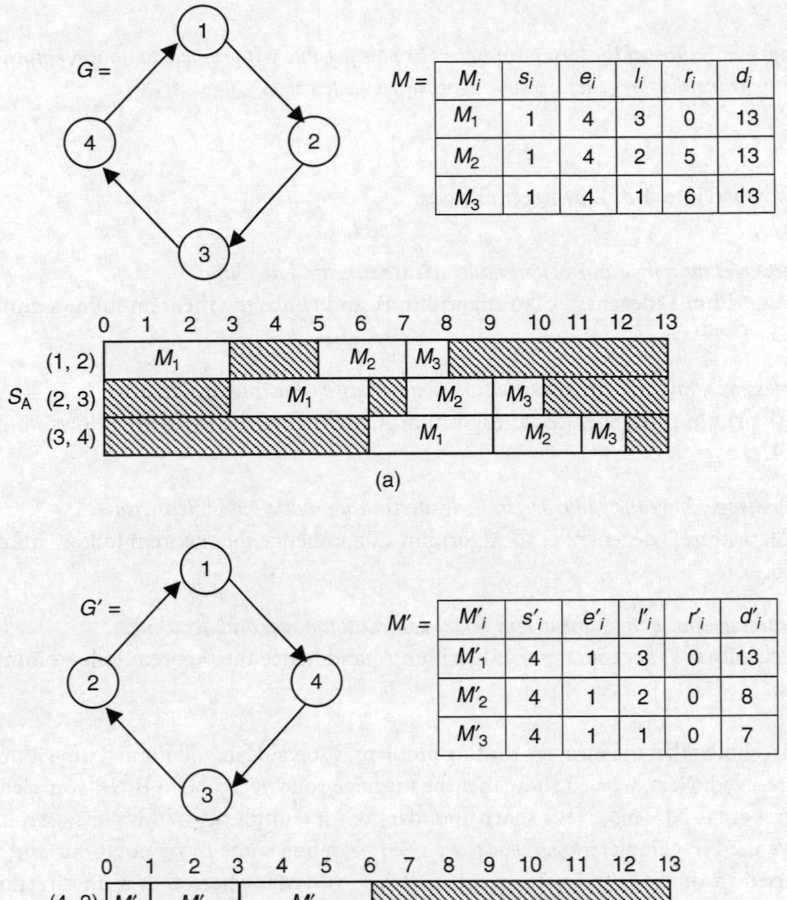

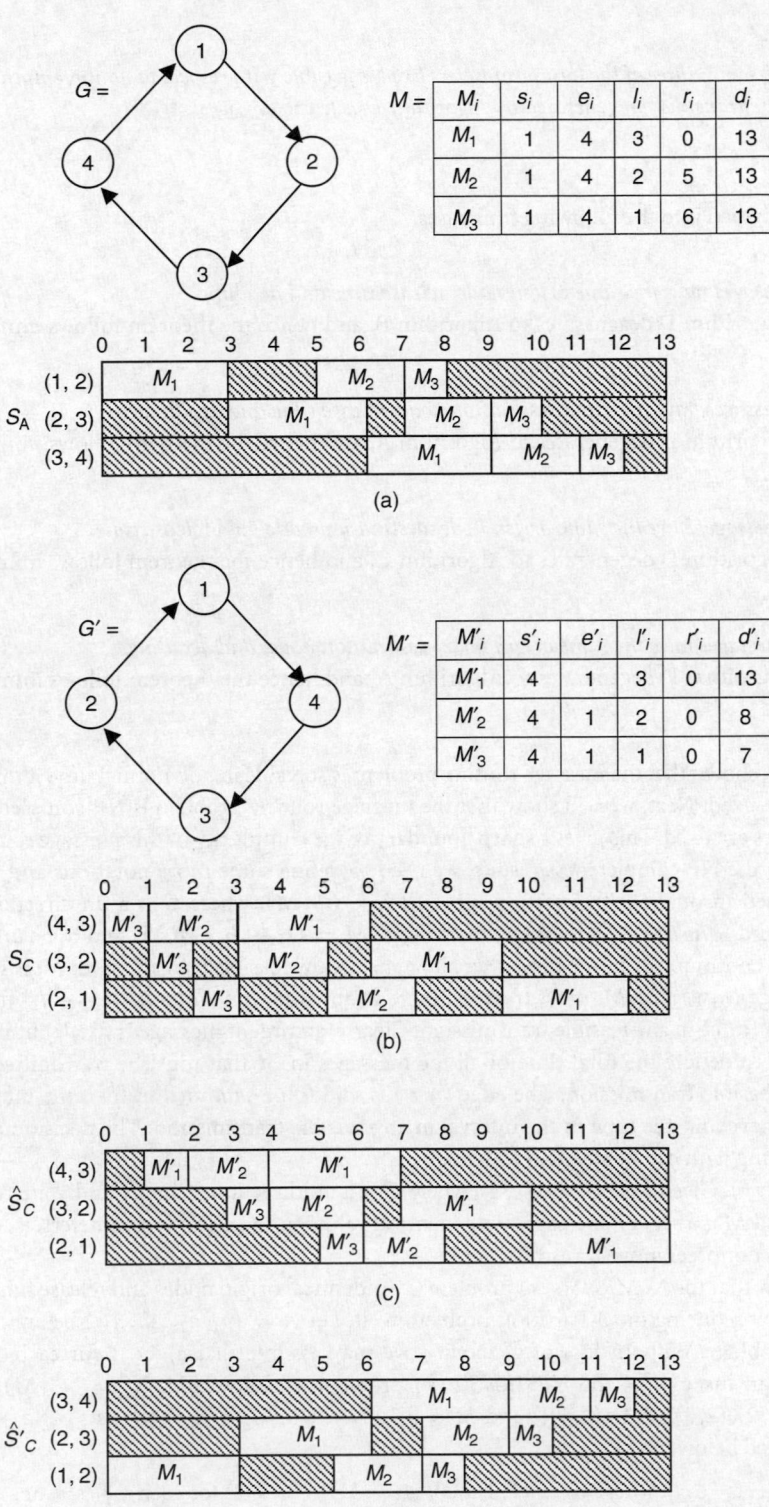

FIGURE 37.3 Illustrating the Proof of Theorem 37.6: (a) transmission generated by Algorithm C for $MRNS = (G, M)$, (b) transmission generated by Algorithm C for $MRNS' = (G', M')$, (c) transmission obtained by right-justifying S_C, and (d) transmission obtained by taking the inverse of $\hat{S}_C$.

Theorem 37.7

A set of messages with three of the four parameters fixed is feasible with respect to nonpreemptive transmission if and only if the transmission generated by Algorithm D is feasible.

Proof

The proof is divided into the following four cases.

Case I: *All messages have the same origin node, release time and deadline.*
In this case Algorithm D degenerates to Algorithm B, and hence the theorem follows immediately from Theorem 37.3.

Case II: *All messages have the same destination node, release time and deadline.*
In this case Algorithm D degenerates to Algorithm A, and hence the theorem follows immediately from Theorem 37.4.

Case III: *All messages have the same origin node, destination node, and release time.*
In this case Algorithm D degenerates to Algorithm C, and hence the theorem follows immediately from Theorem 37.5.

Case IV: *All messages have the same origin node, destination node, and deadline.*
In this case Algorithm D degenerates to Algorithm A, and hence the theorem follows immediately from Theorem 37.6. □

We have just shown that the message routing problem is solvable in polynomial time if three of the four parameters are fixed. Next, we will show that the message routing problem is NP-complete if two of the four parameters are fixed. This gives a sharp boundary of the complexity of this message routing problem. Before we give the NP-completeness results, we need to define some more notations and terminologies that will be used throughout this section. Let *MRNS*= (G, M), where G is a unidirectional ring with $m > 1$ nodes and M is a set of n messages. A message $M_i = (s_i, e_i, l_i, r_i, d_i)$ is said to be *urgent* if it must be transmitted continuously in the time interval $[r_i, d_i]$ in any feasible transmission. M_i is said to be *fixed* if for every edge (u, v) that M_i must traverse, there is only one time interval $[t, t + l_i]$ that M_i can be transmitted on (u, v) in any feasible transmission. Clearly, an urgent message is fixed, but not conversely. Let *TOTAL*(u, v) denote the total time of all the messages in M that must be transmitted on the edge (u, v) in any feasible transmission. The edge (u, v) is said to be *saturated* in the time interval $[t_1, t_2]$ if (u, v) cannot have any idle time in the interval in any feasible transmission. The decision problems that we will be dealing with can be stated as follows.

NPMRNS(x, y). Given *MRNS* $= (G, M)$, where G is a unidirectional ring, x and y are two of the four parameters, and M is a set of n messages with identical values in the x and y parameters, is *MRNS* feasible with respect to nonpreemptive transmission?

We first show that the *NPMRNS*(s, r) problem (i.e., identical origin nodes and release times) is strongly NP-complete by reducing the 3-Partition problem to it. Let $A = (a_1, a_2, \ldots, a_{3z})$ be an instance of the 3-Partition problem. Without loss of generality, we may assume that $a_i > 1$ for each $1 \leq i \leq 3z$. We construct an instance of the *NPMRNS*(s, r) problem as follows: *MRNS* $= (G, M)$, where $G = (\{1, 2, 3, 4\}, \{(1, 2), (2, 3), (3, 4), (4, 1)\})$ and $M = L \cup N \cup U \cup (\bigcup_{i=1}^{z-1} X_i)$ consists of $4z + Bz - B + 1$ messages defined below.

$L = \{L_1, L_2, \ldots, L_{3z}\}$, where $L_i = (1, 4, a_i, 0, (4z - 3)B + z + 2)$ for each $1 \leq i \leq 3z$.
$N = \{N_1, N_2\}$ where $N_1 = (1, 4, (z - 1)B + 1, 0, 3(z - 1)B + 3)$ and $N_2 = (1, 2, (z - 1)B, 0, (3z - 2)B + 1)$
$U = (U_1, U_2, \ldots, U_{z-1})$, where $U_i = (1, 4, 1, 0, (3z - 3 + i)B + 3 + i)$ for each $1 \leq i \leq z - 1$.
$X_i = \{X_{i,1}, X_{i,2}, \ldots, X_{i,B}\}$ for each $1 \leq i \leq z - 1$, where $X_{i,j} = (1, 3, 1, 0, (3z - 3 + i)B + 2 + i + j)$ for each $1 \leq j \leq B$.

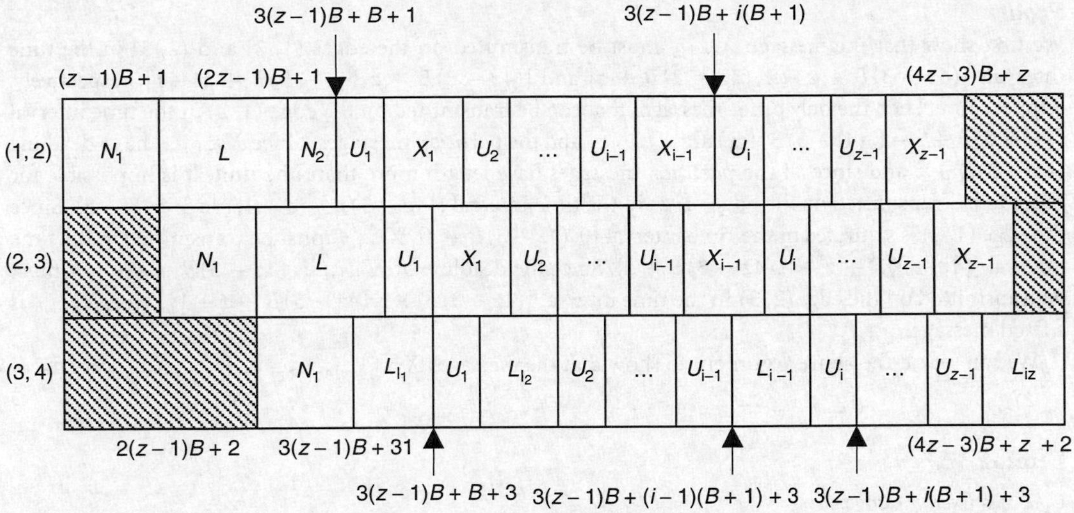

FIGURE 37.4 Illustrating the Proof of Lemma 37.5.

Clearly, all the messages in M have identical origin nodes and release times. It is easy to see that the construction can be done in polynomial time. In the *MRNS* defined above, we call the messages in L the partition messages, the messages in N the enforcer messages, and the messages in U and X_i the unit messages. The next lemma shows that a solution to A implies a solution to the constructed *MRNS*.

Lemma 37.5

If A has a solution, the constructed MRNS also has a solution.

Proof

Let $I_1, I_2, \ldots, I_z$ be a solution to A. We can construct a nonpreemptive transmission S_{NP} for the constructed *MRNS*, as shown in Figure 37.4. It is easy to see that S_{NP} is a feasible transmission for the constructed *MRNS*. □

We next show that a solution to the constructed *MRNS* implies a solution to A. The idea of the proof is to show that the enforcer and unit messages are fixed, which must be transmitted on the edges in the time intervals as shown in Figure 37.4. This will leave z disjoint time intervals on the edge (3, 4), each of length B, for the transmission of the partition messages. Thus a solution to the constructed *MRNS* implies a solution to A. Observe that N_1 is an urgent message and hence must be transmitted continuously from node 1 to node 4, starting at time 0. Therefore, the edges (2, 3) and (3, 4) must be idle in the time intervals $[0, (z-1)B+1]$ and $[0, 2(z-1)B+2]$, respectively. Since $TOTAL(1, 2) = (4z-3)B+z$ and since no message can be transmitted on the edge (1, 2) later than time $(4z-3)B+z$, the edge (1, 2) must be saturated in the time interval $[0, (4z-3)B+z]$. Since no message can be transmitted on the edge (2, 3) before time $(z-1)B+1$ and after time $(4z-3)B+z+1$, and since $TOTAL(2, 3) = (3z-2)B+z$, the edge (2, 3) must be saturated in the time interval $[(z-1)B+1, (4z-3)B+z+1]$. Finally, since no message can be transmitted on the edge (3, 4) before time $2(z-1)B+2$ and after time $(4z-3)B+z+2$, and since $TOTAL(3, 4) = (2z-1)B+z$, the edge (3, 4) must be saturated in the time interval $[2(z-1)B+2, (4z-3)B+z+2]$. The next four lemmas show that the *enforcer* and *unit* messages are fixed.

Lemma 37.6

The messages in X_{z-1} are fixed.

Proof

We first show that the message $X_{z-1,B}$ must be transmitted on the edges $(1, 2)$ and $(2, 3)$ in the time intervals $[(4z-3)B+z-1, (4z-3)B+z]$ and $[(4z-3)B+z, (4z-3)B+z+1]$, respectively. Suppose not. Then the only other messages that can be transmitted on the edge $(1, 2)$ in the time interval $[(4z-3)B+z-1, (4z-3)B+z)$ are $X_{z-1,B-1}$ and the *partition* messages. Since $X_{z-1,B-1}$ has a deadline $(4z-3)B+z$ and since all the *partition* messages have length more than one unit, it is impossible for them to be transmitted on the edge $(1, 2)$ in the time interval $[(4z-3)B+z-1, (4z-3)B+z]$. Since the edge $(1, 2)$ is saturated in the time interval $[0, (4z-3)B+z]$, $X_{z-1,B}$ must be transmitted in the time interval $[(4z-3)B+z-1, (4z-3)B+z]$. Since the deadline of $X_{z-1,B}$ is $(4z-3)B+z+1$, it must be transmitted on the edge $(2, 3)$ in the time interval $[(4z-3)B+z, (4z-3)B+z+1]$. Thus $X_{z-1,B}$ is a fixed message.

We can repeat the above argument to show that the messages $X_{z-1,B-1}, X_{z-1,B-2}, \ldots, X_{z-1,1}$ are fixed. □

Lemma 37.7

U_{z-l} is a fixed message.

Proof

We claim that U_{z-1} must be transmitted on the edges $(1, 2)$, $(2, 3)$ and $(3, 4)$ in the time intervals $[(4z-4)B+z-1, (4z-4)B+z]$, $[(4z-4)B+z, (4z-4)B+z+1]$, and $[(4z-4)B+z+1, (4z-4)B+z+2]$, respectively. Suppose not. Then the only other messages that can be transmitted on the edge $(1, 2)$ in the time interval $[(4z-4)B+z-1, (4z-4)B+z]$ are $X_{z-2,B}$, the *partition* messages, and the messages in X_{z-1}. By Lemma 37.6, the messages in X_{z-1} are fixed. Thus $X_{z-2,B}$ and the *partition* messages are the only possibilities. Since $X_{z-2,B}$ has a deadline $(4z-4)B+z$ and all the *partition* messages have length more than one unit, it is impossible for them to be transmitted on the edge $(1, 2)$ in the time interval $[(4z-4)B+z-1, (4z-4)B+z]$. Since the edge $(1, 2)$ is saturated in the time interval $[0, (4z-3)B+z]$, U_{z-l} must be transmitted on the edge $(1, 2)$ in the time interval $[(4z-4)B+z-1, (4z-4)B+z]$. Since the deadline of U_{z-1} is $(4z-4)B+z+2$, it must be transmitted on the edges $(2, 3)$ and $(3, 4)$ in the time intervals $[(4z-4)B+z, (4z-4)B+z+1]$ and $[(4z-4)B+z+1, (4z-4)B+z+2]$, respectively. Thus U_{z-1} is a fixed message. □

Lemma 37.8

The messages in $U \cup (\bigcup_{i=1}^{z-1} X_i)$ are fixed.

Proof

By Lemmas 37.6 and 37.7, U_{z-1} and the messages in X_{z-1} are fixed. We can repeat the above argument for all the other messages in the order of $X_{z-2}, U_{z-2}, X_{z-3}, \ldots, X_1$ and U_1. Thus the messages in $U \cup (\bigcup_{i=1}^{z-1} X_i)$ are fixed. In particular, the message U_i must be transmitted on the edge $(3, 4)$ in the time interval $[(3z-3+i)B+2+i, (3z-3+i)B+3+i]$ for each $1 \leq i \leq z-1$. □

Lemma 37.9

N_1 and N_2 are fixed messages.

Proof

By the definition of *MRNS*, N_1 is an urgent message and hence it is fixed. By Lemma 37.8, the edge $(1, 2)$ is used by the *unit* messages in the time interval $[(3z-2)B+1, (4z-3)B+z)]$. We claim that N_2 must be transmitted on the edge $(1, 2)$ in the time interval $[(2z-1)B+1, (3z-2)B+1]$. Suppose not. Then only the *partition* messages can be transmitted in this time interval. Consider the time interval $[(3z-2)B, (3z-2)B+1]$. Since all the *partition* messages have length more than one unit, it is impossible for them to be transmitted in this time interval and still meet their deadlines. Since the edge (1, 2) is saturated in the time interval $[0, (4z-3)B+z]$, N_2 must be transmitted in the time interval $[(2z-1)B+1, (3z-2)B+1]$. Thus N_2 is a fixed message. □

We are now ready to show that a solution to the constructed *MRNS* implies a solution to *A*.

Lemma 37.10

If there is a solution to the constructed MRNS, there is a solution to A.

Proof

Let S_{NP} be a feasible nonpreemptive transmission for *MRNS*. By Lemmas 37.8 and 37.9, the *unit* and *enforcer* messages are fixed. The transmission of these fixed messages will leave the time intervals $[(z-1)B+1, (2z-1)B+1]$ and $[(2z-2)B+2, (3z-2)B+2]$ on the edges (1, 2) and (2, 3), respectively, for the transmission of the *partition* messages. There are exactly z time intervals, each of length B, left on the edge (3, 4) for the transmission of the partition messages. Since S_{NP} is a feasible nonpreemptive transmission, each *partition* message must be completely transmitted in one of these z intervals. For each $1 \leq j \leq z$, let $I_j = \{a_i \mid L_i$ is transmitted in the jth interval on the edge (3, 4)$\}$. Clearly, $I_1, I_2, \ldots, I_z$ is a solution to A. □

Theorem 37.8

The NPMRNS(s, r) problem is strongly NP-complete.

Proof

The *NPMRNS*(s, r) problem is clearly in NP. To complete the proof, we reduce the 3-Partition problem to the *NPMRNS*(s, r) problem as given above. By Lemmas 37.5 and 37.10, the instance of 3-Partition problem has a solution if and only if the constructed instance of the *NPMRNS*(s, r) problem has a solution. □

We next show the NP-completeness of the *NPMRNS*(e, r) problem (i.e., identical destination nodes and release times).

Theorem 37.9

The NPMRNS(e, r) problem is strongly NP-complete.

Proof

The *NPMRNS*(e, r) problem is clearly in NP. To complete the proof, we reduce the 3-Partition problem to it as follows. Let $A = (a_1, a_2, \ldots, a_{3z})$ be an instance of the 3-Partition problem. We construct an instance of the *NPMRNS*(e, r) problem as follows: *MRNS* $= (G, M)$, where $G = (\{1, 2, \ldots, 2z+2\}, \{(1,2), (2,3), \ldots, (2z+1, 2z+2), (2z+2, 1)\})$ and $M = \{M_1, M_2, \ldots, M_{4z+1}\}$. For each $1 \leq i \leq 3z$, $M_i = (2z, 2z+2, a_i, 0, 2zB)$, and for each of $3z+1 \leq i \leq 4z+1$, $M_i = (2(i-3z-1)+1, 2z+2, B, 0, (2(4z-i)+3)B)$. Clearly, all the messages in M have identical destination nodes and release times. It is easy to see that the construction can be done in polynomial time.

In the *MRNS* defined above, we call the first $3B$ messages in M the *partition* messages and the last $z+1$ messages the *enforcer* messages. Notice that the *enforcer* messages are urgent and hence must be sent at time 0. The transmission of the *enforcer* messages will leave exactly z disjoint time intervals, each of length B, on the edges $(2z, 2z+1)$ and $(2z+1, 2z+2)$ for the transmission of the *partition* messages.

If $I_1, I_2, \ldots, I_z$ is a solution to A, we can construct a nonpreemptive transmission S_{NP}, as shown in Figure 37.5, for *MRNS*. It is easy to verify that S_{NP} is feasible. Conversely, if S_{NP} is a feasible nonpreemptive transmission for *MRNS*, then the messages $M_{3z+1}, M_{3z+2}, \ldots, M_{4z+1}$ must be transmitted in S_{NP} exactly like that shown in Figure 37.5, since they are all urgent messages. This will leave exactly z disjoint time intervals, each of length B, on the edges $(2z, 2z+1)$ and $(2z+1, 2z+2)$ for the transmission of the messages

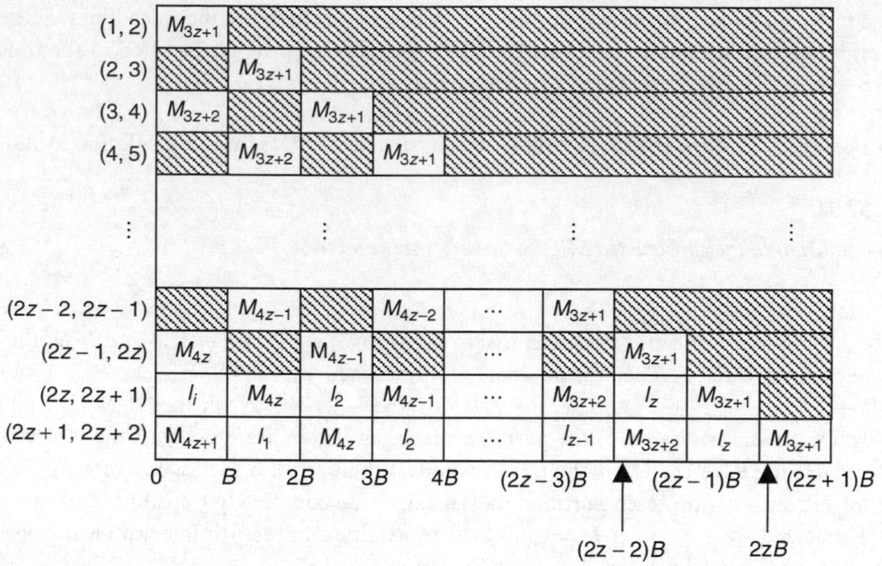

FIGURE 37.5 Illustrating the Proof of Theorem 37.9.

$M_1, M_2, \ldots, M_{3z}$. Let $I_j = \{a_i \mid M_i$ is transmitted in the jth time interval on the edge $(2z, 2z+1)\}$. It is easy to see that $I_1, I_2, \ldots, I_z$ is a solution to A. □

We now consider the NP-completeness of the *NPMRNS*(r, d) problem (i.e., identical release times and deadlines).

Theorem 37.10

The NPMRNS(r, d) problem is strongly NP-complete.

Proof

The *NPMRNS*(r, d) problem is clearly in NP. To complete the proof, we reduce the 3-Partition problem to it as follows. Let $A = (a_1, a_2, \ldots, a_{3z})$ be an instance of the 3-Partition problem. We construct an instance of the *NPMRNS*(r, d) problem as follows: *MRNS*$= (G, M)$, where $G = (\{1, 2, \ldots, 4z+2\}, \{(1,2), (2,3), \ldots, (4z+1, 4z+2), (4z+2, 1)\})$ and $M = \{M_1, M_2, \ldots, M_{4z+1}\}$. For each $1 \leq i \leq 3z$, $M_i = (2z, 2z+2, a_i, 0, 2zB)$, and for each $3z+1 \leq i \leq 4z+1$, $M_i = (2(i-3z-1)+1, 2(i-2z)-1, B, 0, 2zB)$. Clearly, all the messages in M have identical release times and deadlines. It is easy to see that the construction can be done in polynomial time.

If $I_1, I_2, \ldots, I_z$ is a solution to A, we can construct a nonpreemptive transmission S_{NP}, as shown in Figure 37.6, for *MRNS*. It is easy to verify that S_{NP} is feasible. Conversely, if S_{NP} is a feasible nonpreemptive transmission for *MRNS*, then the messages $M_{3z+1}, M_{3z+2}, \ldots, M_{4z+1}$ must be transmitted in S_{NP} exactly like that shown in Figure 37.6, since they are all urgent messages. This will leave exactly z disjoint time intervals, each with length B, on the edges $(2z, 2z+1)$ and $(2z+1, 2z+2)$ for the transmission of the messages $M_1, M_2, \ldots, M_{3z}$. Let $I_j = \{a_j \mid M_i$ is tranmitted in the jth time interval on the edge $(2z, 2z+1)\}$. It is easy to see that $I_1, I_2, \ldots, I_z$ is a solution to A. □

We next consider the NP-completeness of the *NPMRNS*(s, e) problem (i.e., identical origin nodes and destination nodes).

Theorem 37.11

The NPMRNS(s, e) problem is strongly NP-complete.

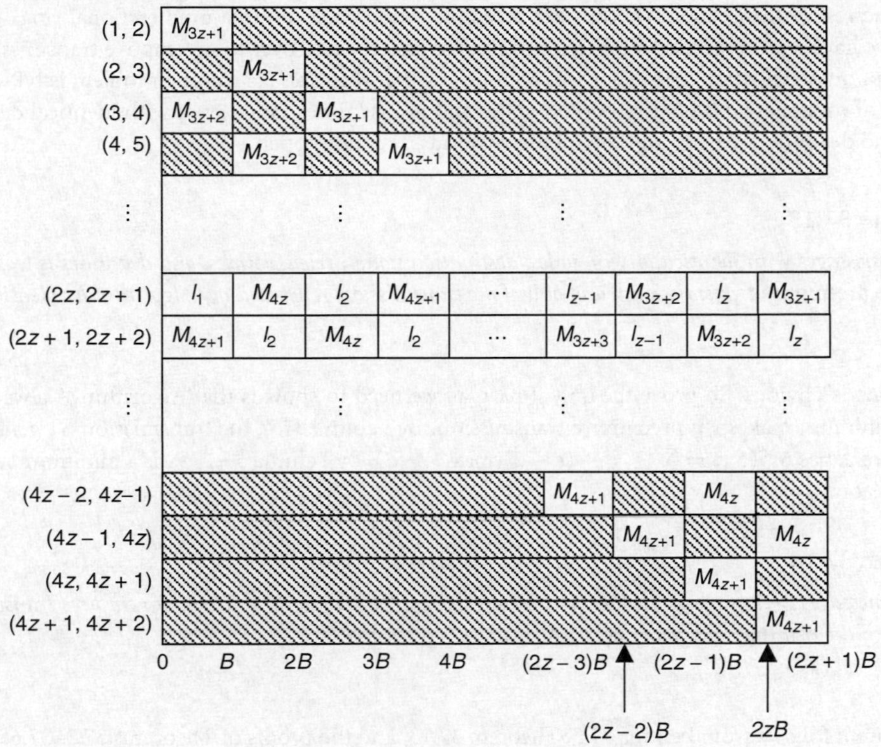

FIGURE 37.6 Illustrating the Proof of Theorem 37.10.

Proof

The *NPMRNS*(s, e) problem is clearly in NP. We reduce the 3-Partition problem to it as follows. Let $A = (a_1, a_2, \ldots, a_{3z})$ be an instance of the 3-Partition problem. We construct an instance of the *NPMRNS*(s, e) problem as follows: $MRNS = (G, M)$, where $G = (\{1, 2, 3\}, \{(1, 2), (2, 3), (3, 1)\})$ and $M = \{M_1, M_2, \ldots, M_{4_{z+1}}\}$. For each $1 \leq i \leq 3z$, $M_i = (1, 3, a_i, 0, (2z + 1)B)$, and for each $3z + 1 \leq i \leq 4z + 1$, $M_i = (1, 3, B, 2(i - 3z - 1)B, 2(i - 3z)B)$. Clearly, all the messages in M have identical origin nodes and destination nodes. It is easy to see that the construction can be done in polynomial time. We leave to the reader the routine proof that the constructed *MRNS* has a solution if and only if A has a solution. □

The NP-completeness of the *NPMRNS*(s, d) problem (i.e., identical origin nodes and deadlines) and the *NPMRNS(e, d)* problem (i.e., identical destination nodes and deadlines) follow from Lemma 37.3, Theorem 37.8 and 37.9.

Theorem 37.12

The NPMRNS(s, d) *and NPMRNS*(e, d) *problems are strongly NP-complete.*

Proof

Given an instance of the *NPMRNS*(s, d) problem, we can construct the inverse of *MRNS*, *MRNS*′, which is an instance of the *NPMRNS*(e, r) problem. By Lemma 37.3 and Theorem 37.9, the *NPMRNS*(s, d) problem is strongly NP-complete. Similarly, given an instance of the *NPMRNS*(e, d) problem, we can construct the inverse of *MRNS*, *MRNS*′, which is an instance of the *NPMRNS*(s, r) problem. By Lemma 37.2 and Theorem 37.8, the *NPMRNS*(e, d) problem is strongly NP-complete. □

For the rest of this section, we consider preemptive transmission on a unidirectional ring. All of the results we have with respect to nonpreemptive transmission carry over to preemptive transmission, with the exception of Theorem 37.8. Thus we cannot show that the message routing problem is NP-complete for a set of messages with identical origin nodes and release times (and hence with identical destination nodes and deadlines). The complexity of these two cases remains open.

Theorem 37.13

A set of messages with identical origin nodes, destination nodes, release times, and deadlines is feasible with respect to preemptive transmission if and only if the transmission generated my Algorithm A is feasible.

Proof

The if part is obvious. To prove the only if part, all we need to show is that Algorithm A always generates a minimum makespan preemptive transmission. By Lemma 37.4, the transmission S_A generated by Algorithm A has $MS(S_A) = \sum_{i=1}^{n} l_i + (e-2)(\max_{1 \leq i \leq n} l_i)$. By Lemma 37.1, S_A is a minimum makespan transmission. □

Theorem 37.14

A set of messages with three of the four parameters fixed is feasible with respect to preemptive transmission if and only if the transmission generated by Algorithm D is feasible.

Proof

The theorem follows from Lemma 37.3, Theorem 37.13, and the proofs of Theorems 37.3–37.6. □

We next consider the NP-completeness of the message routing problem with respect to preemptive transmission. The decision problems that we will be dealing with can be stated as follows.

PMRNS(x, y). Given *MRNS* $= (G, M)$, where G is a unidirectional ring, x and y are two of the four parameters, and M is a set of n messages with identical values in the x and y parameters, is *MRNS* feasible with respect to preemptive transmission?

Theorem 37.15

The PMRNS(e, r), PMRNS(r, d), PMRNS(s, e), and PMRNS(s, d) problems are strongly NP-complete.

Proof

The NP-completeness of *PMRNS*(e, r), *PMRNS*(r, d), and *PMRNS*(s, e) follow directly from the reductions in Theorems 37.9, 37.10, and 37.11, respectively. The NP-completeness of *PMRNS*(s, d) follows from the NP-completeness of *PMRNS*(e, r) and the argument used in the proof of Theorem 37.12. □

37.4 Online Routing of Variable-Length Messages

Algorithm D given in Section 37.3 is an online algorithm. That is, each node in the network routes messages without any knowledge of the messages in the other nodes or the release times of the messages. The study of online routing is meaningful only when the release times of the messages are arbitrary. Thus we will assume this throughout this section and Section 37.5. In this section we continue to work on routing variable-length messages, and we study the issue of whether it is possible to have an optimal online algorithm. An optimal nonpreemptive (preemptive) online algorithm is one that produces a feasible nonpreemptive (preemptive) transmission, whenever one exists. We show that no such optimal nonpreemptive (preemptive) online algorithm can exist unless all of the remaining three parameters are fixed.

Theorem 37.16

For a set of messages with identical origin nodes, destination nodes, and deadlines, Algorithm D is an optimal nonpreemptive (preemptive) online algorithm.

Proof

From Theorems 37.7 and 37.14, we see that Algorithm D is an optimal nonpreemptive (preemptive) online algorithm for a set of messages with identical origin nodes, destination nodes, and deadlines. □

While Theorem 37.16 shows that it is possible to have an optimal online algorithm if all three parameters (origin node, destination node, and deadline) are fixed, we will show in the remainder of this section that no such algorithm can exist if only two of them are fixed.

Theorem 37.17

No optimal nonpreemptive (preemptive) online algorithm can exist for a set of messages with identical origin nodes and destination nodes.

Proof

We will prove the theorem for preemptive transmission only. It will readily be seen that the proof is also applicable to nonpreemptive transmission. Suppose X is an optimal online algorithm for a set of messages with identical origin nodes and destination nodes, Consider the messaging-routing network system $MRNS = (G, M)$, where $G = (\{1, 2, 3\}, \{(1, 2), (2, 3), (3, 1)\})$ and $M = (M_1, M_2, M_3)$. Suppose at time 0, M_1 and M_2 are released from node 1 with $M_1 = (1, 3, 1, 0, 11)$ and $M_2 = (1, 3, 2, 0, 11)$. There are two cases to consider.

Case I: *M_1 is transmitted at time 0 on the edge (1, 2).*
In this case let $M_3 = (1, 3, 4, 2, 10)$ be released from node 1 at time 2. Since M_3 is an urgent message, it is easy to see that one of the messages in M will miss its deadline. However, as shown in Figure 37.7(a), there is a feasible transmission for M. Thus X cannot be optimal.

Case II: *M_2 is transmitted at time 0 on the edge (1, 2).*
In this case let $M_3 = (1, 3, 4, 1, 9)$ be released from node 1 at time 1. Since M_3 is an urgent message, it is easy to see that one of the messages in M will miss its deadline. However, as shown in Figure 37.7(b), there is a feasible transmission for M. Thus, X cannot be optimal.

In both cases X does not produce a feasible transmission for a set of messages with identical origin nodes and destination nodes, even though one exists. Thus X is not an optimal online algorithm. □

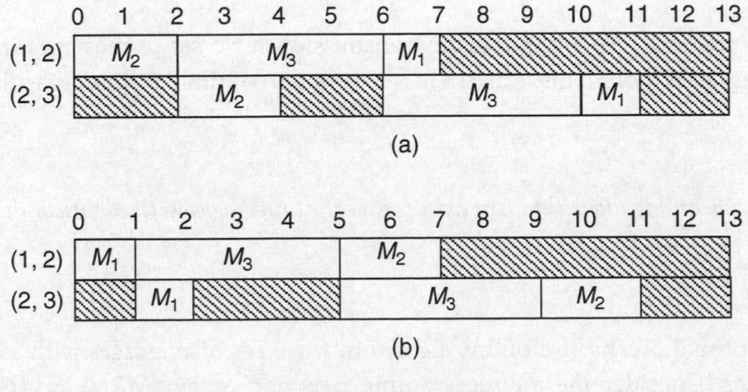

FIGURE 37.7 Illustrating the Proof of Theorem 37.17: (a) case I and (b) case II.

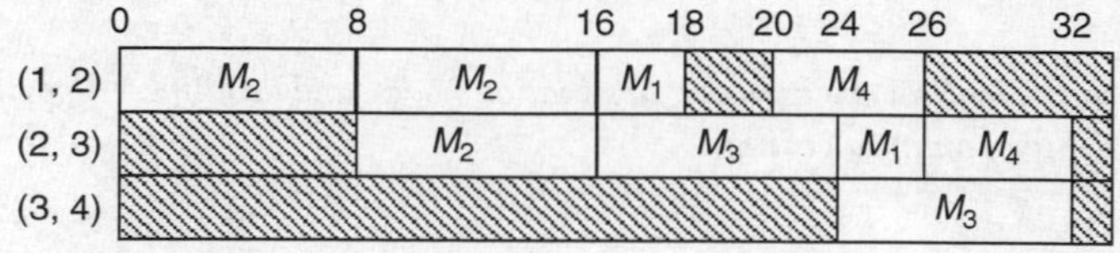

(a)

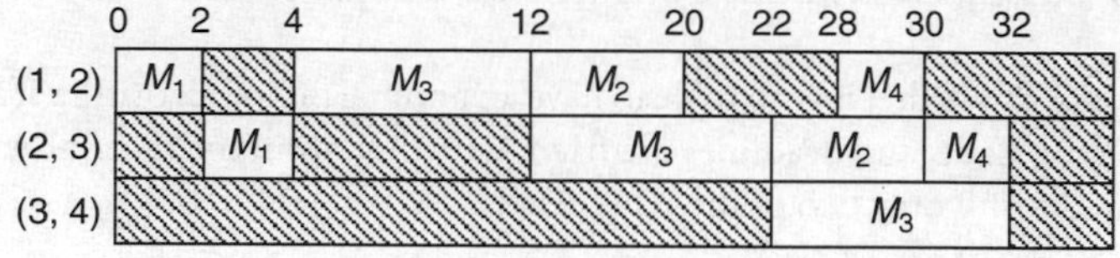

(b)

FIGURE 37.8 Illustrating the Proof of Theorem 37.18: (a) case I and (b) case II.

Theorem 37.18

No optimal nonpreemptive (preemptive) online algorithm can exist for a set of messages with identical origin nodes and deadlines.

Proof

We will prove the theorem for preemptive transmission only. It will readily be seen that the proof is also applicable to nonpreemptive transmission. Suppose X is an optimal online algorithm for a set of messages with identical origin nodes and deadlines. Consider $MRNS = (G, M)$, where $G = (\{1, 2, 3, 4\}, \{(1, 2), (2, 3), (3, 4), (4, 1)\})$ and $M = \{M_1, M_2, M_3, M_4)$. Suppose at time 0, M_1 and M_2 are released from node 1 with $M_1 = (1, 3, 2, 0, 32)$ and $M_2 = (1, 3, 8, 0, 32)$. There are two cases to consider, depending on whether M_1 is transmitted at time 0 on the edge (1, 2).

Case I: *M_1 is transmitted at time 0 on the edge (1, 2).*
In this case let $M_3 = (1, 4, 8, 8, 32)$ and $M_4 = (1, 3, 6, 20, 32)$ be released from node 1 at times 8 and 20, respectively. Since M_3 and M_4 are urgent messages, it is easy to see that one of the messages in $\{M_2, M_3, M_4\}$ will miss its deadline. However, as shown in Figure 37.8(a), there is a feasible transmission for M. Thus X cannot be optimal.

Case II: *M_1 is not transmitted at time 0 on the edge (1, 2).*
In this case let $M_3 = (1, 4, 10, 2, 32)$ and $M_4 = (1, 3, 2, 28, 32)$ be released from node 1 at times 2 and 28, respectively. Since M_3 and M_4 are urgent messages, it is easy to see that one of the messages in M will miss its deadline. However, as shown in Figure 37.8(b), there is a feasible transmission for M. Thus X cannot be optimal.

In both cases X does not produce a feasible transmission for a set of messages with identical origin nodes and deadlines, even though one exists. Thus X is not an optimal online algorithm. □

Theorem 37.19

No optimal preemptive online algorithm can exist for a set of messages with identical destination nodes and deadlines.

Proof

Suppose X is an optimal preemptive online algorithm for a set of messages with identical destination nodes and deadlines. Consider the message-routing network system $MRNS = (G, M)$, where $G = (\{1, 2, 3, 4\}, \{(1, 2), (2, 3), (3, 4), (4, 1)\})$ and $M = \{M_1, M_2, M_3\}$. Suppose at time 0, M_1 and M_2 are released

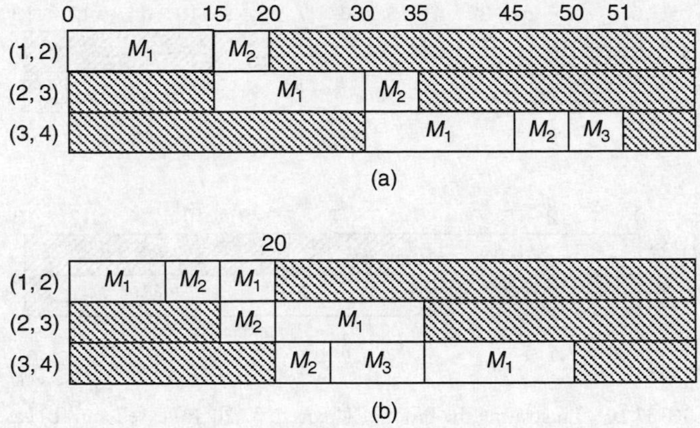

FIGURE 37.9 Illustrating the Proof of Theorem 37.19: (a) case I and (b) case II.

from node 1 with $M_1 = (1, 4, 15, 0, 51)$ and $M_2 = (1, 4, 5, 11, 51)$. It is easy to see that in any transmission, M_1 cannot be completely transmitted to its destination node by time 26, but M_2 can. We consider the following two cases.

Case I: *M_2 is completely transmitted to its destination node by time* 26.
In this case let $M_3 = (3, 4, 1, 50, 51)$ be released from node 3 at time 50. Clearly, either M_1 or M_3 will miss its deadline. However, as shown in Figure 37.9(a), there is a feasible transmission for M. Thus X cannot be optimal.

Case II: *M_2 is not completely transmitted to its destination node by time* 26.
In this case let $M_3 = (3, 4, 10, 26, 51)$ be released from node 3 at time 26. Clearly, one of the messages in M will miss its deadline. However, as shown in Figure 37.9b, there is a feasible transmission for M. Thus X cannot be optimal.

In both cases X does not produce a feasible transmission for a set of messages with identical destination nodes and deadlines, even though one exists. Thus, X is not an optimal online algorithm. □

Theorem 37.20

No optimal nonpreemptive online algorithm can exist for a set of messages with identical destination nodes and deadlines.

Proof

Suppose X is an optimal nonpreemptive online algorithm for a set of messages with identical destination nodes and deadlines. Consider the message-routing network system $MRNS = (G, M)$, where $G = (\{1, 2, 3\}, \{(1, 2), (2, 3), (3, 1)\})$ and $M = \{M_1, M_2, M_3\}$. Suppose at time 0, M_1 is released from node 1 with $M_1 = (1, 3, 5, 0, 12)$. There are two cases to consider.

Case I: *M_1 is transmitted at time* 0.
In this case let $M_2 = (1, 3, 1, 1, 12)$ be released from node 1 at time 1, and $M_3 = (2, 3, 4, 3, 12)$ be released from node 2 at time 3. Clearly, one of the messages in M will miss its deadline. However, as shown in Figure 37.10(a), there is a feasible transmission for M. Thus X cannot be optimal.

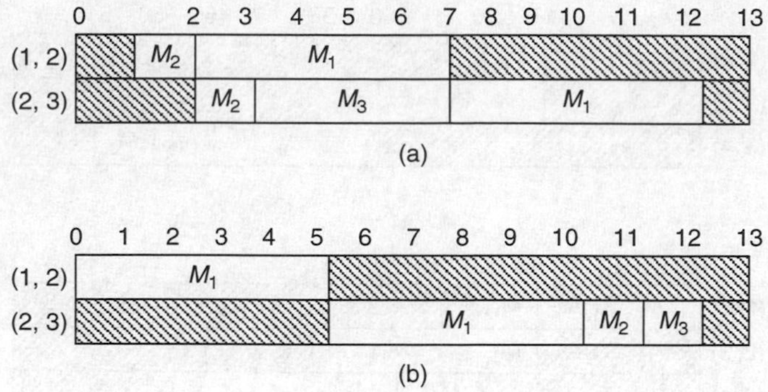

FIGURE 37.10 Illustrating the Proof of Theorem 37.20: (a) case I and (b) case II.

Case II: *M_1 is not transmitted at time* 0.
In this case let $M_2 = (2, 3, 1, 10, 12)$ and $M_3 = (2, 3, 1, 10, 12)$ be released from node 2 at time 10. Clearly, one of the messages in M will miss its deadline. However, as shown in Figure 37.10(b), there is a feasible transmission for M. Thus X cannot be optimal.

In both cases X does not produce a feasible transmission for a set of messages with *identical destination nodes and deadlines*, even though one exists. Thus, X is not an optimal online algorithm. □

37.5 Online Routing of Equal-Length Messages

In Section 37.3 and Section 37.4 we have studied the message routing problem for a set of variable-length messages, and in this section we consider only equal-length messages. Specifically, we study the issue whether it is possible to have an optimal online algorithm for the following networks — unidirectional ring, out-tree, in-tree, bidirectional tree, and bidirectional ring. The definition of optimal online routing algorithm is given in Section 37.4. The problem is considered under various restrictions of the four parameters — origin node, destination node, release time, and deadline.

Instead of using the quintuple notation, $(s_i, e_i, l_i, r_i, d_i)$, used in the previous sections for a variable-length message, we use a shorter quadruple notation, (s_i, e_i, r_i, d_i), in this section as lengths of messages, l_i, are equal. Formally, a set of n unit-length messages $M = \{M_1, \ldots, M_n\}$ needs to be routed through the network. Each message M_i is represented by the quadruple (s_i, e_i, r_i, d_i), where s_i denotes the *origin node* (i.e., M_i originates from s_i), e_i denotes the *destination node* (i.e., M_i is to be sent to e_i), r_i denotes the *release time* (i.e., M_i originates from s_i at time r_i), and d_i denotes the *deadline* (i.e., M_i must reach e_i by time d_i). Throughout this section, we assume that release times and deadlines are integers, while the length of each message is one unit.

An instance of the message routing problem consists of a network $G$and a set of unit-length messages M. The ordered pair $MRNS = (G, M)$ will be called a message routing network system. Given $MRNS = (G, M)$, our goal is to route the messages in M such that each M_i is sent from the node s_i to the node e_i in the time interval $[r_i, d_i]$.

Let $MRNS = (G, M)$, where $G = (V, E)$ and $M = \{M_1, \ldots, M_n\}$. A *configuration* q is a quadruple (u, v, m, t), where $t > 0$ and $(u, v) \in E$, which denotes that message m reaches vertex v from vertex u at time t through the edge (u, v). Let $Q_i = \{q_1, \ldots, q_p\}$ be a set of p configurations. Q_i is called a *route* for $M_i = (s_i, e_i, r_i, d_i)$ if M_i reaches a sequence of vertices $s_i = x_0, x_1, \ldots, x_p = e_i$ at times $t_0 < t_1 < \cdots < t_p$, respectively, such that the following two conditions are satisfied:

1. $(x_0, x_1, \ldots, x_p)$ is a directed path in G.
2. $q_j = (x_{j-1}, x_j, M_i, t_j)$ for each $1 \leq j \leq p$.

A *transmission* S for $MRNS = (G, M)$ is defined to be $S = \bigcup_{i=1}^{n} Q_i$, where Q_i is a route for M_i, such that the following two conditions are satisfied:

1. Each message is transmitted after its release time; i.e., for each $(u, v, M_i, t) \in S$, we have $t > r_i$.
2. At most one message is transmitted on an edge at a time; i.e., for all (u, v, M_i, t) and $(u', v', M_j, t') \in S$ such that $i \neq j, u = u'$, and $v = v'$, we have $t \neq t'$.

S is said to be a *feasible* transmission for *MRNS* if the deadlines of all the messages are met; i.e., for all $(u, v, M_i, t) \in S, t \leq d_i$. *MRNS* is said to be *feasible* if there is a feasible transmission for *MRNS*. Since release times and deadlines are integers and since the length of each message is one unit, it is never beneficial to delay the transmission of a message whenever a communication link is free. Thus we will assume that all feasible transmissions are such that a communication link remains idle only when there is no ready message.

In subsection 37.5.1, we consider three unidirectional networks — unidirectional ring, out-tree, and in-tree. In subsection 37.5.2, we consider two bidirectional networks — bidirectional tree and bidirectional ring.

37.5.1 Unidirectional Networks

In this subsection we consider online routing on unidirectional ring, out-tree, and in-tree, under various restrictions of the four parameters. We first give three online algorithms, and then discuss the results for unidirectional ring, out-tree, and in-tree.

For unidirectional ring, out-tree, and in-tree, any message will be routed along the unique simple path from its origin node to its destination node. Before we introduce the algorithms, we need to define the following notations and terminologies. The *distance* between two nodes u and v is the number of edges on the unique simple path from u to v. If M_i resides in node u at time t, $D(M_i, t)$ denotes the distance between u and e_i. The *slack time* of M_i at time t, denoted by $SLT(M_i, t)$, is defined to be $SLT(M_i, t) = d_i - t - D(M_i, t)$, i.e., the maximum time span that M_i can be delayed for transmission without missing its deadline. The three online algorithms are given below.

Earliest Deadline (ED) Algorithm. Whenever a communication link is free for transmission, send that ready message with the earliest deadline. Ties can be broken arbitrarily.

Farthest Away (FA) Algorithm. Whenever a communication link is free for transmission, send that ready message whose destination node is the farthest away. Ties can be broken arbitrarily.

Smallest Slack Time (SST) Algorithm. Whenever a communication link is free for transmission, send that ready message with the smallest slack time. Ties can be broken arbitrarily.

We next give the results for a unidirectional ring. We first show that the ED, FA, and SST algorithms are optimal when one of the four parameters is fixed (Theorems 37.21–37.24), and then show that no optimal algorithm can exist if all four parameters are arbitrary (Theorem 37.25).

Theorem 37.21

The ED algorithm is optimal for a set of messages with identical destination nodes.

Proof

Let S be a feasible transmission for $MRNS = (G, M)$, where G is a unidirectional ring and M is a set of messages with identical destination nodes. We will show that S can be transformed, without violating any feasibility, into the same transmission as produced by the ED algorithm. Let t be the earliest time at which S differs from the transmission produced by the ED algorithm. Then there must be two messages, say M_a and M_b, residing in node v_k at time t with $d_a \leq d_b$, such that M_b is transmitted before M_a in S, while the opposite is true for the ED algorithm. Since M_a and M_b have a common destination node, $D(M_a, t) = D(M_b, t) = l$.

Let $(v_k, v_{k+1}, \ldots, v_{k+l})$ be the simple path from v_k to the common destination node of M_a and M_b. Let PR_a and PR_b be the partial routes in S after time t for M_a and M_b, respectively, where $PR_a = \{(v_k, v_{k+1}, M_a, t_1), (v_{k+1}, v_{k+2}, M_a, t_2), \ldots, (v_{k+l-1}, v_{k+l}, M_a, t_l)\}$ and $PR_b = \{(v_k, v_{k+1}, M_b, \hat{t}_1), (v_{k+1}, v_{k+2}, M_b, \hat{t}_2), \ldots, (v_{k+l-1}, v_{k+l}, M_b, \hat{t}_l)\}$. Clearly, $\hat{t}_1 < t_1$. Since S is a feasible transmission, $t < t_1 < t_2 < \cdots < t_l \leq d_a$ and $t < \hat{t}_1 < \hat{t}_2 < \cdots < \hat{t}_l \leq d_b$. Let p be the largest index, $1 \leq p \leq l$, such that $\hat{t}_i < t_i$ for each $1 \leq i \leq p$.

Consider the partial routes $\overline{PR}_a$ and $\overline{PR}_b$ obtained by interchanging the first p configurations of PR_a with PR_b. That is, $\overline{PR}_a = \{(v_k, v_{k+1}, M_a, \hat{t}_1), (v_{k+1}, v_{k+2}, M_a, \hat{t}_2), \ldots, (v_{k+p-1}, v_{k+p}, M_a, \hat{t}_p), (v_{k+p}, v_{k+p+1}, M_a, t_{p+1}), \ldots, (v_{k+l-1}, v_{k+l}, M_a, t_l)\}$ and $\overline{PR}_b = \{(v_k, v_{k+1}, M_b, t_1), (v_{k+1}, v_{k+2}, M_b, t_2), \ldots, (v_{k+p-1}, v_{k+p}, M_b, t_p), (v_{k+p}, v_{k+p+1}, M_b, \hat{t}_{p+1}), \ldots, (v_{k+l-1}, v_{k+l}, M_b, \hat{t}_l)\}$. Since $\hat{t}_p < t_p < t_{p+1} < \hat{t}_{p+1}$, $\overline{PR}_a$ and $\overline{PR}_b$ must be valid partial routes for M_a and M_b, respectively. Let $\overline{S}$ be the transmission obtained from S by replacing PR_a and PR_b by $\overline{PR}_a$ and $\overline{PR}_b$, respectively.

We will show that M_a and M_b reach their common destination node by their deadlines in $\overline{S}$, establishing the feasibility of $\overline{S}$. There are two cases to consider, depending on whether $p = l$ or not. If $p = l$, M_a and M_b reach their common destination node at times $\hat{t}_l$ and t_l, respectively. Since $\hat{t}_l < t_l \leq d_a \leq d_b$, their deadlines are met. On the other hand, if $p < l$, M_a and M_b reach their common destination node at the same times as in S. Thus their deadlines must be met.

Repeating the above transformation, we can transform S, without violating any feasibility, into the same transmission as produced by the ED algorithm. □

Theorem 37.22

The FA algorithm is optimal for a set of messages with identical deadlines.

We will omit the proof of Theorem 37.22, as it is very similar to that of Theorem 37.21; it can be found in [32].

Theorem 37.23

The SST algorithm is optimal for a set of messages with identical origin nodes.

Proof

Let S be a feasible transmission for $MRNS = (G, M)$, where G is a unidirectional ring and M is a set of messages with identical origin nodes. Let v_0 be the common origin node of the messages in M. Since G is a unidirectional ring and since all messages in M originate from v_0, all nodes, except v_0, have at most one ready message to be sent to the next node at any time. We will show that S can be transformed, without violating any feasibility, into the same transmission as produced by the SST algorithm. Let t be the first time at which S differs from the transmission produced by the SST algorithm. Then there must be two messages, say M_a and M_b, residing in node v_0 at time t, with $SLT(M_a, t) \leq SLT(M_b, t)$, such that M_b is transmitted before M_a in S, while the opposite is true for the SST algorithm. Let M_a and M_b be transmitted from v_0 in S at times t_1 and $\hat{t}_1$, respectively, where $\hat{t}_1 < t_1$. We consider two cases, depending on whether $D(M_b, t) \leq D(M_a, t)$ or not.

Case I. $D(M_b, t) \leq D(M_a, t)$

Let $D(M_b, t) = l$ and $D(M_a, t) = l + q$, where $q \geq 0$. Let $(v_0, v_1, \ldots, v_l, v_{l+1}, \ldots, v_{l+q})$ be the simple path from v_0 to v_{l+q}, where v_l and v_{l+q} are the destination nodes of M_b and M_a, respectively. Let PR_a and PR_b be the partial routes in S after time t for M_a and M_b, respectively. Clearly, $PR_a = \{(v_0, v_1, M_a, t_1 + 1), (v_1, v_2, M_a, t_1 + 2), \ldots, (v_{l-1}, v_l, M_a, t_1 + l), (v_l, v_{l+1}, M_a, t_1 + l + 1), \ldots, (v_{l+q-1}, v_{l+q}, M_a, t_1 + l + q)\}$ and $PR_b = \{(v_0, v_1, M_b, \hat{t}_1 + 1), (v_1, v_2, M_b, \hat{t}_1 + 2), \ldots, (v_{l-1}, v_l, M_b, \hat{t}_1 + l)\}$. Since S is a feasible transmission, $t_1 + l + q \leq d_a$ and $\hat{t}_1 + l \leq d_b$.

Consider the partial routes $\overline{PR}_a$ and $\overline{PR}_b$ obtained by interchanging the first l configurations of PR_a with PR_b. That is, $\overline{PR}_a = \{(v_0, v_1, M_a, \hat{t}_1 + 1), (v_1, v_2, M_a, \hat{t}_1 + 2), \ldots, (v_{l-1}, v_l, M_a, \hat{t}_1 + l), (v_l, v_{l+1}, M_a, t_1 + l + 1), \ldots, (v_{l+q-1}, v_{l+q}, M_a, t_1 + l + q)\}$ and $\overline{PR}_b = \{(v_0, v_1, M_b, t_1 + 1), (v_1, v_2,$

$M_b, t_1 + 2), \ldots, (v_{l-1}, v_l, M_b, t_1 + l)\}$. Since $\hat{t}_1 < t_1, \overline{PR}_a$ and $\overline{PR}_b$ are valid partial routes for M_a and M_b, respectively. Let $\overline{S}$ be the transformation obtained from S by replacing PR_a and PR_b by $\overline{PR}_a$ and $\overline{PR}_b$, respectively.

We will show that M_a and M_b reach their destination nodes by their deadlines in $\overline{S}$, establishing the feasibility of $\overline{S}$. Since $SLT(M_a, t) \leq SLT(M_b, t), d_a - t - (l + q) \leq d_b - t - l$, and hence $d_a - q \leq d_b$. Since $t_1 + l + q \leq d_a, t_1 + l \leq d_a - q \leq d_b$. Since M_b reaches its destination node at time $t_1 + l$, its deadline is met. M_a reaches its destination node at time $\hat{t}_1 + l$ or $t_1 + l + q$, depending on whether $q = 0$ or not. In both cases, its deadline is met.

Case II. $D(M_b, t) > D(M_a, t)$.
Let $D(M_a, t) = l$ and $D(M_b, t) = l + q$, where $q > 0$. Let $(v_0, v_1, \ldots, v_l, v_{l+1}, \ldots, v_{l+q})$ be the simple path from v_0 to v_{l+q}, where v_l and v_{l+q} are the destination nodes of M_a and M_b, respectively. Let PR_a and PR_b be the partial routes in S after time t for M_a and M_b, respectively. Clearly, $PR_a = \{(v_0, v_1, M_a, t_1 + 1), (v_1, v_2, M_a, t_1 + 2), \ldots, (v_{l-1}, v_l, M_a, t_1 + l)\}$ and $PR_b = \{(v_0, v_1, M_b, \hat{t}_1 + 1), (v_1, v_2, M_b, \hat{t}_1 + 2), \ldots, (v_{l-1}, v_l, M_b, \hat{t}_1 + l), (v_l, v_{l+1}, M_b, \hat{t}_1 + l + 1), \ldots, (v_{l+q-1}, v_{l+q}, M_b, \hat{t}_1 + l + q)\}$. Since S is a feasible transmission, $t_1 + l \leq d_a$ and $\hat{t}_1 + l + q \leq d_b$.

Consider the partial routes $\overline{PR}_a$ and $\overline{PR}_b$ obtained by interchanging the first l configurations of PR_a with PR_b, and modifying the last q configurations of PR_b as follows: $\overline{PR}_a = \{(v_0, v_1, M_a, \hat{t}_1 + 1), (v_1, v_2, M_a, \hat{t}_1 + 2), \ldots, (v_{l-1}, v_l, M_a, \hat{t}_1 + l)\}$ and $\overline{PR}_b = \overline{PR}_{b,1} \cup \overline{PR}_{b,2} = \{(v_0, v_1, M_b, t_1 + 1), (v_1, v_2, M_b, t_1 + 2), \ldots, (v_{l-1}, v_l, M_b, t_1 + l)\} \cup \{(v_l, v_{l+1}, M_b, t_1 + l + 1), \ldots, (v_{l+q-1}, v_{l+q}, M_b, t_1 + l + q)\}$. Clearly, $\overline{PR}_a$ is a valid route for M_a. The first portion of $\overline{PR}_b$ (i.e.,$\overline{PR}_{b,1}$) is a valid partial route for M_b. Since no message is transmitted in S on the edge (v_{l+i-1}, v_{l+i}) in the time interval $[t_1 + l + i - 1, t_1 + l + i]$ for each $1 \leq i \leq q$, the second portion of $\overline{PR}_b$ (i.e.,$\overline{PR}_{b,2}$) is a valid partial route for M_b. Thus $\overline{PR}_b$ is also a valid route for M_b. Let $\overline{S}$ be the transmission obtained from S by replacing PR_a and PR_b by $\overline{PR}_a$ and $\overline{PR}_b$, respectively.

We will show that M_a and M_b reach their destination nodes by their deadlines in $\overline{S}$, establishing the feasibility of $\overline{S}$. Since $\hat{t}_1 < t_1, \hat{t}_1 + l < t_1 + l \leq d_a$. Since M_a reaches its destination node at time $\hat{t}_1 + l$, its deadline is met. Since $SLT(M_a, t) \leq SLT(M_b, t), d_a - t - l \leq d_b - t - (l + q)$, and hence $d_a + q \leq d_b$. Since $t_1 + l \leq d_a, t_1 + l + q \leq d_a + q \leq d_b$. Since M_b reaches its deadline at time $t_1 + l + q$, its deadline is also met.

Repeating the above transformations, we can transform S, without violating any feasibility, into the same transformation as produced by the SST algorithm. □

Theorem 37.24

The SST algorithm is optimal for a set of messages with identical release times.

We will omit the proof of Theorem 37.24, as it is similar to that of Theorem 37.23; it can be found in [22]. In the next theorem we will show by adversary argument that no optimal algorithm can exist for a set of messages with all four parameters arbitrary.

Theorem 37.25

No optimal algorithm can exist for a set of messages with all four parameters arbitrary.

Proof

We will prove the theorem by contradiction. Suppose X is an optimal algorithm for a set of messages with all four parameters arbitrary. Consider $MRNS = (G, M)$, where $G = (\{1, 2, 3, 4\}, \{(1, 2), (2, 3), (3, 4), (4, 1)\})$ and $M = \{M_1, M_2, M_3, M_4, M_5\}$. Let M_1 and M_2 be released from node 1 at time 0, with $M_1 = (1, 3, 0, 4)$ and $M_2 = (1, 4, 0, 6)$. There are two cases to consider, depending on whether M_1 is transmitted by X on the edge (1, 2) at time 0 or not.

If M_1 is transmitted by X on the edge (1, 2) at time 0, let $M_3 = (3, 4, 3, 4)$, $M_4 = (3, 4, 4, 5)$, and $M_5 = (3, 4, 5, 6)$. It is easy to see that one of the messages in M will miss its deadline, even though M is feasible. Thus X cannot be optimal. On the other hand, if M_1 is not transmitted by X on the edge (1, 2) at time 0, let $M_3 = (2, 3, 2, 3)$, $M_4 = (2, 3, 3, 4)$, and $M_5 = (1, 2, 2, 3)$. It is easy to see that one of the messages in M will miss its deadline, even though M is feasible. Thus X cannot be optimal. □

We next give the results, Theorems 37.26–37.29, for an out-tree, which is a directed tree in which every node, except the root, has in-degree one. We first show that the ED and SST Algorithms are optimal for a set of messages with one of the three parameters — origin node, destination node, and release time — fixed, and then show that no optimal algorithm can exist if all three parameters are arbitrary.

Theorem 37.26

The ED algorithm is optimal for a set of messages with identical destination nodes.

Theorem 37.27

The SST algorithm is optimal for a set of messages with identical origin nodes.

Theorem 37.28

The SST algorithm is optimal for a set of messages with identical release times.

The proofs of Theorems 37.26, 37.27, and 37.28 are very similar to those of Theorems 37.21, 37.23, and 37.24, respectively, and hence will be omitted; they can be found in [32].

Theorem 37.29

No optimal algorithm can exist for a set of messages with identical deadlines and all three parameters — origin node, destination node, and release time — arbitrary.

Proof

We will prove the theorem by contradiction. Suppose X is an optimal algorithm for a set of messages with all three parameters arbitrary. Consider $MRNS = (G, M)$, where $G = (\{1, 2, 3, 4\}, \{(1, 2), (2, 3), (2, 4)\})$ and $M = \{M_1, M_2, M_3\}$. Suppose M_1 and M_2 are released from node 1 at time 0, with $M_1 = (1, 3, 0, 3)$ and $M_2 = (1, 4, 0, 3)$. There are two cases to consider, depending on whether M_1 is transmitted by X on the edge (1, 2) at time 0 or not.

If M_1 is transmitted by X on the edge (1, 2) at time 0, let $M_3 = (2, 4, 2, 3)$. It is easy to see that one of the messages in M will miss its deadline, even though M is feasible. Thus X cannot be optimal. On the other hand, if M_1 is not transmitted by X on the edge (1, 2) at time 0, let $M_3 = (2, 3, 2, 3)$. It is easy to see that one of the messages in M will miss its deadline, even though M is feasible. Thus X cannot be optimal. □

We next give the results, Theorems 37.30–37.33, for an in-tree, which is a directed tree in which every node, except the root, has out-degree one. We first show that the ED, FA, and SST algorithms are optimal for a set of messages with one of the three parameters — origin node, destination node, and deadline — fixed, and then show that no optimal algorithm can exist if all three parameters are arbitrary.

Theorem 37.30

The ED algorithm is optimal for a set of messages with identical destination nodes.

Theorem 37.31

The FA algorithm is optimal for a set of messages with identical deadlines.

Theorem 37.32

The SST algorithm is optimal for a set of messages with identical origin nodes.

The proofs of Theorems 37.30, 37.31, and 37.32 are very similar to those of Theorems 37.21, 37.22, and 37.23, respectively, and hence will be omitted; they can be found in [32].

Theorem 37.33

No optimal algorithm can exist for a set of messages with identical release times and all three parameters — origin node, destination node, and deadline — arbitrary.

Proof

We will prove the theorem by contradiction. Suppose X is an optimal algorithm for a set of messages with all three parameters arbitrary. Consider $MRNS = (G, M)$, where $G = (\{1, 2, 3, 4, 5, 6, 7\}, \{(1, 2), (2, 3), (3, 4), (7, 6), (6, 5), (5, 3)\})$ and $M = \{M_1, M_2, M_3, M_4\}$. Suppose M_1 and M_2 are released from node 2 at time 0, with $M_1 = (2, 3, 0, 2)$ and $M_2 = (2, 4, 0, 4)$. There are two cases to consider, depending on whether M_1 is transmitted by X on the edge (2, 3) at time 0 or not.

If M_1 is transmitted by X on the edge (2, 3) at time 0, let $M_3 = (6, 4, 0, 3)$ and $M_4 = (7, 4, 0, 4)$. It is easy to see that one of the messages in M will miss its deadline, even though M is feasible. Thus X cannot be optimal. On the other hand, if M_1 is not transmitted by X on the edge (2, 3) at time 0, let $M_3 = (1, 3, 0, 2)$ and $M_4 = (3, 4, 0, 1)$. It is easy to see that one of the messages in M will miss its deadline, even though there is a feasible transmission for M. Thus X cannot be optimal. □

37.5.2 Bidirectional Networks

In this subsection we consider online routing on a bidirectional tree and a bidirectional ring, under various restrictions of the four parameters. For any network, a route with cycles can be transformed, without violating any feasibility, into another one with no cycle. Without loss of generality, we may assume that all transmissions are cycle-free.

We next give the results, Theorems 37.34–37.36, for a bidirectional tree. We first show that the ED and SST algorithms are optimal for a set of messages with the origin node or destination node fixed, and then show that no optimal algorithm can exist if both the origin node and the destination node are arbitrary.

Theorem 37.34

The SST algorithm is optimal for a set of messages with identical origin nodes.

Proof

Since the messages have identical origin nodes, the route for each message is a unique simple path. This becomes the same case as the out-tree for a set of messages with identical origin nodes, and hence the theorem follows from Theorem 37.27. □

Theorem 37.35

The ED algorithm is optimal for a set of messages with identical destination nodes.

Proof

Since the messages have identical destination nodes, the route for each message is a unique simple path. This becomes the same case as the in-tree for a set of messages with identical destination nodes, and hence the theorem follows from Theorem 37.30. □

Theorem 37.36

No optimal algorithm can exist for a set of messages with identical release times and deadlines, but arbitrary origin nodes and destination nodes.

Proof

We will prove the theorem by contradiction. Suppose X is an optimal algorithm for a set of messages with identical origin nodes and destination nodes, but arbitrary release times and deadlines. Consider $MRNS = (G, M)$, where $G = (\{1, 2, 3, 4, 5, 6\}, \{(1, 2), (2, 3), (2, 5), (3, 4), (5, 6), (2, 1), (3, 2), (5, 2), (4, 3), (6, 5)\})$ and $M = \{M_1, M_2, M_3\}$. Suppose M_1 and M_2 are released from node 1 at time 0, with $M_1 = (1, 3, 0, 3)$ and $M_2 = (1, 5, 0, 3)$. There are two cases to consider, depending on whether M_1 is transmitted by X on the edge (1, 2) at time 0 or not.

If M_1 is transmitted by X on the edge (1, 2) at time 0, let $M_3 = (4, 5, 0, 3)$. It is easy to see that one of the messages in M will miss its deadline, even though M is feasible. Thus X cannot be optimal. On the other hand, if M_1 is not transmitted by X on the edge $(1, 2)$ at time 0, let $M_3 = (6, 3, 0, 3)$. It is easy to see that one of the messages in M will miss its deadline, even though M is feasible. Thus X cannot be optimal. □

We next give the results, Theorems 37.37–37.39, for a bidirectional ring. It is clear that if the messages have identical origin nodes and release times, we can compute a feasible transmission (possibly with exponential time), whenever one exists, in the common origin node of the messages. Thus there is an optimal algorithm for this case.

For a set of messages with identical origin nodes and destination nodes, each message will be sent along one of the two simple paths from the origin node to the destination node. Let P_s and P_l denote the shorter and the longer of these two paths, respectively. It is clear that all nodes, except the origin node and destination node, have at most one message to be sent to the next node at any time. We call these nodes *passing* nodes. The destination node acts as a receiver, while the origin node schedules the messages to be sent along P_s or P_l. The last online routing algorithm in this chapter is given below.

Earliest Deadline Shortest Path (EDSP)Algorithm. At any moment of time t, the origin node, destination node, and passing nodes operate as follows:

1. Among all the ready messages in the origin node, send that message with the earliest deadline, say M_i, along the path P_s. Among all the ready messages that can be sent along the path P_l without violating any feasibility, send that message with the earliest deadline, say $M_j (j \neq i)$, along P_l. Ties can be broken arbitrarily.
2. The destination node acts as a receiver.
3. The passing nodes pass the message, if any, to the next node along the simple path $P_s (P_l)$ if the node is on $P_s (P_l)$.

Theorem 37.37

The EDSP algorithm is optimal for a set of messages with identical origin nodes and destination nodes.

Proof

Let S be a feasible transmission for $MRNS = (G, M)$, where G is a bidirectional ring and M is a set of messages with identical origin nodes and destination nodes. Without loss of generality, we may assume that the messages are sent along P_s or P_l in S. We will show that S can be transformed, without violating any feasibility, into the same transmission as produced by the EDSP algorithm. Let t be the earliest time at which S differs from the transmission produced by the EDSP algorithm. Since the EDSP algorithm also sends messages along the paths P_s and P_l, the two transmissions can only differ at the common origin node. We consider two cases, depending on whether the EDSP algorithm sends one or two messages from the origin node at time t.

Case I: *The EDSP algorithm sends one message from the origin node at time t.*

Let M_i be the message sent from the origin node at time t by the EDSP algorithm. Clearly, M_i is sent along P_s. Since only one message was sent by the EDSP algorithm at time t, there is also only one message, say M_a, sent from the origin node at time t in S. From the nature of the EDSP algorithm, $d_i \leq d_a$. Since M_a

cannot be sent along P_l without violating any feasibility (otherwise, the EDSP algorithm would have sent M_a along P_l), it must be sent along P_s in S. Let M_i be sent from the origin node at time t' in S, where $t' > t$. Consider the transmission $\overline{S}$ obtained from S by interchanging the route of M_a with M_i. Since $d_i \leq d_a$, both messages reach their common destination node by their deadlines in $\overline{S}$.

Case II: *The EDSP algorithm sends two messages from the origin node at time t.*
Let M_i and M_j be the two messages sent by the EDSP algorithm at time t along the paths P_s and P_l, respectively. From the nature of the EDSP algorithm, $d_i \leq d_j$. Let M_a and M_b be the two messages sent in S at time t along P_s and P_l, respectively. (Note that M_b can be a "null" message; we assume a null message has a very large deadline.) If $a = j$, we can interchange the route of M_a with M_b in S without violating any feasibility. Similarly, if $b = i$, we can interchange the route of M_a with M_b in S. Since M_b (or equivalently, M_i) has the earliest deadline among all the ready messages, the new transmission must also be feasible. Thus we may assume that $a \neq j$ and $b \neq i$, and hence we have the following three cases to consider:

1. $a = i$ and $b \neq j$. In this case $d_j \leq d_b$. Let M_j be sent in S at time t', where $t' > t$. Consider the transmission $\overline{S}$ obtained from S by interchanging the route of M_b with M_j. Since $d_j \leq d_b$, both messages reach their common destination node by their deadlines in $\overline{S}$.
2. $a \neq i$ and $b = j$. In this case $d_i \leq d_a$. Let M_i be sent in S at time t', where $t' > t$. Consider the transmission $\overline{S}$ obtained from S by interchanging the route of M_a with M_i. Since $d_i \leq d_a$, both messages reach their common destination node by their deadlines in $\overline{S}$.
3. $a \neq i$ and $b \neq j$. In this case $d_i \leq d_a$ and $d_j \leq d_b$. Let M_i and M_j be sent in S at times t' and t'', respectively, where $t' > t$ and $t'' > t$. Consider the transmission $\overline{S}$ obtained from S by interchanging the routes of M_a and M_b with M_i and M_j, respectively. Since $d_i \leq d_a$ and $d_j \leq d_b$, all four messages reach their common destination node by their deadlines in $\overline{S}$.

Repeating the above transformations, we can transform S, without violating any feasibility, into the same transformation as produced by the EDSP algorithm. □

Theorem 37.38

No optimal algorithm can exist for a set of messages with the origin node arbitrary and the other three parameters fixed.

Proof
We will prove the theorem by contradiction. Suppose X is an optimal algorithm for a set of messages with the origin node arbitrary and the other three parameters fixed. Consider $MRNS = (G, M)$, where $G = (\{1, 2, 3, 4\}, \{(1, 2), (2, 3), (3, 4), (4, 1), (1, 4), (4, 3), (3, 2), (2, 1)\})$ and $M = \{M_1, M_2, M_3\}$. Suppose M_1 is released from node 1 at time 0, with $M_1 = (1, 3, 0, 2)$. There are two cases to consider, depending on whether M_1 is transmitted by X on the edge $(1, 2)$ at time 0 or not.

If M_1 is transmitted by Xon the edge $(1, 2)$ at time 0, let $M_2 = (2, 3, 0, 2)$ and $M_3 = (2, 3, 0, 2)$. It is easy to see that one of these messages will miss its deadline, even though M is feasible. Thus X cannot be optimal. On the other hand, if M_1 is not transmitted by X on the edge $(1, 2)$ at time 0, let $M_2 = (4, 3, 0, 2)$ and $M_3 = (4, 3, 0, 2)$. It is easy to see that one of the messages in M will miss its deadline, even though M is feasible. Thus X cannot be optimal. □

Theorem 37.39

No optimal algorithm can exist for a set of messages with identical origin nodes and deadlines, but arbitrary destination nodes and release times.

Proof
We will prove the theorem by contradiction. Suppose X is optimal for a set of messages with identical origin nodes and deadlines, but arbitrary destination nodes and release times. Consider $MRNS = (G, M)$,

where $G = (\{1, 2, 3, 4\}, \{(1, 2), (2, 3), (3, 4), (4, 1), (1, 4), (4, 3), (3, 2), (2, 1)\})$ and $M = \{M_1, M_2, M_3, M_4, M_5, M_6\}$. Suppose M_1, M_2, M_3, and M_4 are released from node 1 at time 0, with $M_1 = (1, 2, 0, 3)$, $M_2 = (1, 2, 0, 3)$, $M_3 = (1, 4, 0, 3)$, and $M_4 = (1, 4, 0, 3)$. There are three cases to consider.

Case I

M_1 and M_2 are transmitted by X on the edges (1, 2) and (1, 4) at time 0, respectively. In this case, let $M_5 = (1, 4, 1, 3)$ and $M_6 = (1, 2, 2, 3)$. It is easy to see that one of these messages will miss its deadline, even though M is feasible. Thus X cannot be optimal.

Case II

M_3 and M_4 are transmitted by X on the edges (1, 2) and (1, 4) at time 0, respectively. In this case, let $M_5 = (1, 2, 1, 3)$ and $M_6 = (1, 4, 2, 3)$. It is easy to see that one of these messages will miss its deadline, even though M is feasible. Thus X cannot be optimal.

Case III

Either M_1 or M_2 is transmitted on the edge $(1, 2)$ and either M_3 or M_4 is transmitted on the edge $(1, 4)$ at time 0. In this case, let $M_5 = (1, 4, 1, 3)$ and $M_6 = (1, 4, 1, 3)$. It is easy to see that one of these messages will miss its deadline, even though M is feasible. Thus X cannot be optimal. □

37.6 Conclusions

In this chapter we have considered the problem of routing real-time messages on networks. For offline routing and variable-length messages, we showed that if the network is an arbitrary directed graph, the problem is NP-complete even when all four parameters are fixed. In Section 37.3 we then considered a simple network — a unidirectional ring under various restrictions of the four parameters — origin node, destination node, release time, and deadline. For nonpreemptive transmission we showed that the problem is solvable in polynomial time if three of the four parameters are fixed, but becomes NP-complete when only two of them are fixed. The same complexity results hold for preemptive transmission, except for the following two cases: (1) identical origin nodes and release times and (2) identical destination nodes and deadlines. The complexity of the above two cases remains open. For online routing and variable-length messages, the issue of whether there is an optimal online algorithm is also considered for both preemptive and nonpreemptive transmissions. In Section 37.4 we showed that no such algorithm can exist unless all of the remaining three parameters are fixed. In Section 37.5 we considered the problem of routing equal-length messages on several networks — unidirectional ring, out-tree, in-tree, bidirectional tree, and bidirectional ring. For online routing and equal-length messages, again we have considered the issue whether it is possible to have an optimal online routing algorithm under various restrictions of the four parameters — origin node, destination node, release time, and deadline. We showed that (1) for a unidirectional ring, no such algorithm can exist unless one of the four parameters is fixed (i.e., all messages have identical values for that parameter); (2) for an out-tree, no such algorithm can exist unless one of the three parameters — origin node, destination node, and release time — is fixed; (3) for an in-tree, no such algorithm can exist unless one of the three parameters — origin node, destination node, and deadline — is fixed; (4) for a bidirectional tree, no such algorithm can exist unless the origin node or the destination node is fixed; (5) for a bidirectional ring, no such algorithm can exist unless the origin node and either the destination node or the release time are fixed. Our results give a sharp boundary delineating those instances for which an optimal algorithm exists and those for which no such algorithm can exist.

For future research, it is desirable to settle the complexity of the unsolved cases. There are other variations of our model that are worthwhile to investigate. For example, we can consider preemptive transmission in which a message can be partially sent along different paths of the network, or a node can forward messages that have only been partially received (rather than waiting until the entire message has been received). Also it will be interesting to study online routing for other classes of network.

References

[1] Aspnes, J. et al., On-line Routing of Virtual Circuits with Applications to Load Balancing and Machine Scheduling, *Journal of ACM*, 44, pp. 486–504, 1997.

[2] Baldi, M. and Ofek, Y., A Comparison of Ring and Tree Embedding for Real-Time Group Multicast, *IEEE/ACM Transactions on Networking*, 11, pp. 451–464, 2003.

[3] Bejerano, Y., Cidon, I., and Naor, J., Efficient Handoff Rerouting Algorithms: A Competitive On-line Algorithmic Approach, *IEEE/ACM Transactions on Networking*, 10, pp. 749–760, 2002.

[4] Bongiovanni, G., Tang, D. T., and Wong, C. K., A General Multibeam Satellite Switching Algorithm, *IEEE Transactions on Communications*, 29, pp. 1025–1036, 1981.

[5] Chang, Y. L. and Hsu, C. C., Routing in Wireless/Mobile Ad-hoc Networks Via Dynamic Group Construction, *Mobile Networks and Applications*, 5, pp. 27–37, 2000.

[6] Chen, B. et al., Connection-Oriented Communications for Real-Time Applications in FDDI-ATM-FDDI Heterogeneous Networks, *Proceedings of the 17th International Conference on Distributed Computing Systems*, 1997.

[7] Choi, B. K., Bettati, R., Efficient Resource Management for Hard Real-Time Communication over Differentiated Services Architectures, *Proceedings of the 7th International Conference on Real-Time Systems and Applications*, 2000.

[8] Choi, H.-A. and Hakimi, S. L., Scheduling File Transfers for Trees and Odd Cycles, *SIAM Journal of Computing*, 16, pp. 162–168, 1987.

[9] Choi, H.-A. and Hakimi, S. L., Data Transfers in Networks, *Algorithmica*, 3, pp. 223–245, 1988.

[10] Choi, H.-A. and Hakimi, S. L., Data Transfers in Networks with Transceivers, *Networks*, 18, pp. 223–251, 1988.

[11] Choi, S. and Shin K., A Cellular Wireless Local Area Network with QoS Guarantees for Heterogeneous Traffic, *ACM Mobile Networks and Applications (MONET)*, 1997.

[12] Coffman, E. G. et al., Scheduling File Transfers, *SIAM Journal of Computing*, 14, pp. 744–780, 1985.

[13] Devalla, B. et al., Connection-Oriented Real-Time Communication For Mission Critical Applications — An Introduction to NETEX: A Portable and Efficient Tool Kit, *Proceedings of NAECON '97*, pp. 698–707, 1997.

[14] Feng F., Kumar, A., and Zhao, W., Bounding Application-to-Application Delays for Multimedia Traffic in FDDI-Based Communication Systems, *Multimedia Computing and Networking*, pp. 174, 1996.

[15] Fu, H. and Knightly, E., A Simple Model of Real-Time Flow Aggregation, *IEEE/ACM Transactions on Networking*, 11, pp. 422–435, 2003.

[16] Whitehead, J., The Complexity of File Transfer Scheduling with Forwarding, *SIAM Journal of Computing*, 19, pp. 222–245, 1990.

[17] Garey, M. R. and Johnson, D. S., *Computers and Intractability: A Guide to the Theory of NP-Completeness*, Freeman, San Francisco, 1979.

[18] Goel, A, Meyerson, A., and Plotkin, S., Combining Fairness With Throughput: On-line Routing With Multiple Objectives, *Proceedings of The Thirty-Second Annual ACM Symposium on Theory of Computing*, 1999.

[19] Gopal, I. S. and Wong, C. K., Minimizing The Number of Switchings in an SS/TDMA System, *IEEE Transactions on Communications*, 33, pp. 497–501, 1985.

[20] Hajek, B., Link Schedules, Flows, and The Multichromatic Index of Graphs, *Proceedings of Information Science and System*, Princeton, NJ, 1984.

[21] Hajek, B. and Sasaki, G., Link Scheduling in Polynomial Time, Preprint, Department of Electrical and Computer Engineering, University of Illinois, Urbana–Champaign, IL, 1985.

[22] Jiang, L., Real-Time Message Transmission Over the Scalable Coherent Interface (SCI), Department of Computer Science and Engineering, University of Nebraska-Lincoln, Master Thesis, 1997.

[23] Lee, S., Real-Time Wormhole Channels, *Journal of Parallel and Distributed Computing*, 63, pp. 299–311, 2003.

[24] Leighton, F. T., *Introduction to Parallel Algorithms and Architecture*, Morgan Kaufmann, San Mateo, CA, 1991.
[25] Leung, J. Y-T., Tam, T., Wong, C., and Young, G., Routing Messages with Release Time and Deadline Constraints, *Journal of Parallel and Distributed Computing*, 31, pp. 65–76, 1995.
[26] Leung, J. Y-T., Tam, T., Wong, C., and Young, G., On-line Routing of Real-Time Messages, *Journal of Parallel and Distributed Computing*, 34, pp. 211–217, 1996.
[27] Malcolm, N. and Zhao. W., Hard Real-Time Communication in Multiple-Access Networks, *Journal of Real-Time Systems*, 8, pp. 35–77, 1995.
[28] Manimaran, G., Resource Management with Dynamic Scheduling In Parallel and Distributed Real-Time Systems, Ph.D. Thesis, Department of Computer Science and Engineering, Indian Institute of Technology, Madras, 1998.
[29] Norden, S. et al., Deterministic Protocols for Real-Time Communication in Multiple Access Networks, *Computer Communications*, 22, pp. 128–136, 1999.
[30] Raha, A., Kamat S., and Zhao, W., Using Traffic Regulation to Meet End-to-End Deadlines in ATM Networks, *IEEE Transactions on Computers*, 1995.
[31] Reeves, D. and Salama, H. A., Distributed Algorithm for Delay-Constrained Unicast Routing, *IEEE/ACM Transactions on Networking*, 8, pp. 239–250, 2000.
[32] Rivera-Vega, P. I., Varadarajan, R., and Navathe, S. B., Scheduling Data Redistribution in Distributed Databases, *Proceedings of the 6th International Conference on Data Engineering*, pp. 166–173, 1990.
[33] Rivera-Vega, P. I., Varadarajan, R., and Navathe, S. B., Scheduling Data Transfers in Fully Connected Networks, *Networks*, 22, pp. 563–588, 1992.
[34] Tam, T. W., Message Routing in Distributed Real-Time Systems, Ph.D. Thesis, University of Texas, Dallas, 1990.
[35] Varadarajan, R. and Rivera-Vega, P. I., An Efficient Approximation Algorithm for The File Redistribution Scheduling Problem in Fully Connected Networks, *Congr. Numer.*, 91, pp. 129–139, 1992.
[36] Varadarajan, R., Rivera-Vega, P. I., and Navathe, S. B., Data Redistribution Scheduling in Fully Connected Network, *Proceedings of 27th Annual Allerton Conference on Communication, Control, and Computing*, 1989.

V

Stochastic Scheduling and Queueing Networks

38

Offline Deterministic Scheduling, Stochastic Scheduling, and Online Deterministic Scheduling: A Comparative Overview

Michael Pinedo
New York University

1-58488-397-9/$0.00+$1.50

38.1 Introduction

An enormous amount of research has been done on scheduling problems under various different sets of assumptions. A distinction can be made between

1. offline deterministic scheduling,
2. stochastic scheduling, and
3. online deterministic scheduling.

In offline deterministic scheduling, all information is known *a priori*. That is, all the data with regard to the problem, including the number of jobs, their release dates, due dates, weights, etc. are known at the outset. The decision maker has a combinatorial optimization problem of minimizing or maximizing some given objective function. Most of the research in this area has focused on the development of polynomial time algorithms, complexity proofs, exact (usually enumerative) algorithms, heuristics, and worst case analyses.

In stochastic scheduling a different framework is used. The number of jobs is fixed and known in advance. The processing time of a job is not known in advance, but it is known to be a random draw from a given probability distribution. The actual processing time only becomes known when the processing has been completed. Different jobs may have different processing time distributions. The release dates and due dates may also be random variables from known distributions. If the release dates or due dates are random variables, then an actual release date or due date only becomes known when the release date or due date occurs. (One can also imagine models in which the occurence of a release also fixes a corresponding due date.) The decision maker has to optimize (usually minimize) a given objective function, knowing only the probability laws governing the random scheduling data. Since the problem the decision maker faces is a stochastic optimization problem, he has to determine the policy that minimizes the objective in some stochastic sense, most often in expectation. There are various classes of policies, e.g.,

1. the class of static policies,
2. the class of nonpreemptive dynamic policies, and
3. the class of preemptive dynamic policies.

If the decision maker has to adopt a static policy, then he has to specify at the outset all actions to be taken during the evolution of the process and he is not allowed to deviate from those actions when more information becomes available. If the decision maker is allowed to adopt a dynamic policy, then he can make his decisions at any time as a function of all the information that has become available up to that point. However, in the class of nonpreemptive dynamic policies he is not allowed to preempt.

In online scheduling even less information is known *a priori*. The information is released gradually to the decision maker. The decision maker does not even know in advance the number of jobs that are going to be released. He knows nothing about release dates or processing times, whether or not they are deterministic or random. He does not even know the distribution of the random variables. There are two types of online scheduling models: in the first type, the decision maker is given a job's exact processing time the moment it has been released. In the second type, the decision maker knows the processing time of a job only when the job has been completed. In both types of online models, a certain objective function has to be minimized and the decision maker has to determine the best action to take every time a new job is released. Clearly, the optimal policy of the decision maker in this framework results in a value of the objective function that is higher than the minimum value that can be obtained in the offline case, i.e., when the decision maker has all the information at $t = 0$. The quality of an online algorithm is often measured by how close it comes to the minimum in the perfect information case. An online algorithm is said to be *c-competitive* if the resulting value of the objective function is at most c times the minimum value for all possible schedules of the given input instance.

So, offline deterministic scheduling deals with perfect information. Stochastic scheduling deals with some information that may be perfect and other information that is only distributional. Online deterministic scheduling deals with the least amount of information, i.e., not even the distributions of the variables are given; however, in some models the exact processing time of a job may be given upon its arrival. In this chapter we may refer to offline scheduling as the perfect information case, to stochastic scheduling as the distributional information case, and to online scheduling as the no-information case.

Offline deterministic scheduling is, in a sense, an extreme case of stochastic scheduling. All distributions involved in the problem are basically deterministic with zero variance. Stochastic scheduling can also be compared with online scheduling. Consider a stochastic scheduling problem with a random number of jobs with the distribution of this number not known. The processing times of the jobs also have an unknown (arbitrary) distribution. So, in online scheduling even the probability laws governing the job characteristics are not known in advance.

The goals of this chapter are threefold: first, we give a brief overview of the past research directions in the three scheduling areas; second, we compare the results that have been obtained in the three different scheduling areas with one another and discuss their relationships; third, we describe new research directions that lie at the interfaces between current research areas.

38.2 Offline Deterministic Scheduling: The Case With Perfect Information

38.2.1 Overview

An enormous amount of work has been done on offline deterministic scheduling over the years. To cover most of the problems that fall within the standard theory, the well-known $\alpha \mid \beta \mid \gamma$ framework was introduced by Graham, Lawler, Lenstra, and Rinnooy Kan (1978). The α field determines the machine environment (e.g., single machine, flow shop, and so on), the β field determines the processing characteristics and constraints (e.g., preemptions, setup times, release dates, and so on), and the γ field specifies the objective function (e.g., makespan, total weighted tardiness, and so on).

Since most of these scheduling problems are combinatorial optimization problems, a lot of the research in the past has focused on determining the computational complexity of the problems and trying to find out where the borderline lies between polynomial time solvable problems and NP-hard problems.

Since so many scheduling problems are NP-hard, a significant amount of research has focused on the development of integer programming formulations and branch-and-bound methods. A certain amount of research has focused also on the development of heuristics; such research often included worst case analyses. In the remaining part of this section we focus on those results that have appeared in the offline deterministic scheduling literature that also have an impact on stochastic scheduling and on online deterministic scheduling.

38.2.2 Complexity Hierarchies

The more general a scheduling problem is, the higher the level of its complexity is in the complexity hierarchy. For example, a problem with precedence constraints is of higher level of complexity than the same problem without precedence constraints. However, if the problem is a single machine problem, then the precedence constraints may actually reduce the number of computations that are necessary to find the optimal schedule.

Typically, but not always, a preemptive version of a problem is easier than its nonpreemptive counterpart. Allowing preemptions in a scheduling problem can make the problem significantly easier and actually polynomial time solvable. For example, the nonpreemptive problem to minimize the makespan on m parallel machines is well-known to be NP-hard, whereas the same problem with preemptions allowed can be solved in polynomial time (the preemptive rule that assigns at each point in time the m jobs with the longest remaining processing time to the m machines minimizes the makespan).

When in a nonpreemptive environment the jobs have different release dates, then it may be advantageous to keep a machine idle even when there are jobs waiting for processing. Keeping a machine idle while there are jobs waiting for processing may be referred to as *unforced idleness.* In some scheduling problems it may be assumed that unforced idleness is allowed, whereas in other scheduling problems it may be assumed that it is not allowed. Problems that do not allow unforced idleness tend to be easier than problems that do allow unforced idleness.

38.2.3 Bounds on Objective Functions under Given Scheduling Rules

This subsection focuses on the worst-case analyses of four scheduling rules. Each one of these four worst-case examples has a relationship to results in either stochastic scheduling or in online scheduling.

1. Consider the $Pm \,||\, C_{\max}$ problem. There are m identical machines in parallel and n jobs. Job j has a processing time p_j. The goal is to assign each one of the n jobs to one machine in such a way that the makespan is minimized. This problem is well-known to be NP-hard. Suppose that at time $t = 0$ these jobs are presented to the decision maker in a certain list. The decision maker has to schedule these jobs one by one in the order in which they appear on the list; when the decision maker puts a job on a machine in a certain time period, he does not know the processing times of the jobs that appear further down the list. The LIST schedule instructs the decision maker, when he takes the next job of the list, to schedule that job as early as possible on the machine that is freed first. This rule is referred to as the LIST rule. This rule has led to one of the first examples of a worst-case analysis. Graham (1966) showed that

$$\frac{C_{\max}(\text{LIST})}{C_{\max}(\text{OPT})} \leq 2 - \frac{1}{m}$$

   This rule implies that not considering all the processing times in advance results in a schedule with a makespan that may be close to twice the minimum makespan under an optimal schedule that is achieved when all processing times are given at time 0. This worst-case analysis of the LIST rule has been one of the cornerstones of online scheduling.
2. Consider again the $Pm \,||\, C_{\max}$ problem. The objective is again the minimization of the makespan. One very popular heuristic selects, each time a machine is freed, among the remaining jobs the job with the longest processing time. This rule is referred to as the *longest processing time* (LPT) first rule. This rule also has led to a fine example of a worst-case analysis. Graham (1969) showed that

$$\frac{C_{\max}(\text{LPT})}{C_{\max}(\text{OPT})} \leq \frac{4}{3} - \frac{1}{3m}$$

   This rule implies that knowing the processing times in advance results in a schedule that is at worst 1/3 higher than the best schedule. This is in contrast to results in online scheduling that indicate that when the processing times are not known in advance the schedule may be significantly worse.
3. Consider the $Pm \mid prmp \mid \sum C_j$ problem. There are m identical machines in parallel and n jobs. Job j has a processing time p_j. The objective is to minimize the total completion time while preemptions are allowed. The following so-called Round-Robin algorithm cycles through the list of jobs, giving each job τ units of processing time in turn. For τ small enough, a schedule that is constructed using this rule has the property that at any point in time each uncompleted job has received the same amount of processing. If $\tau \to 0$, then the rule is at times referred to as *processor sharing* (PS). It has been shown that with regard to the Round-Robin (RR) algorithm the following inequality holds:

$$\frac{\sum C_j(\text{RR})}{\sum C_j(\text{OPT})} \leq 2 - \frac{2m}{(n+m)}$$

   This worst case bound is attained when the processing times of all the n jobs are equal to one another.
4. Consider the $Pm \,||\, \sum w_j C_j$ problem. There are m identical machines in parallel and n jobs. Job j has a processing time p_j and a weight w_j. The goal is to assign each one of the n jobs to one machine

in such a way that the total weighted completion time is minimized. This problem is well-known to be NP-hard. One heuristic would select, each time a machine is freed, among the remaining jobs the job with the highest w_j/p_j ratio. This rule is referred to as the *weighted shortest processing time* (WSPT) first rule. This rule has led to the following worst-case bound. Kawaguchi and Kyan (1986) showed that

$$\frac{\sum w_j C_j(\text{WSPT})}{\sum w_j C_j(\text{OPT})} \leq \frac{1}{2}(1+\sqrt{2}) = 1.2071$$

It is easy to show that in an example that attains the worst-case bound, the n jobs must have w_j/p_j ratios that are very close to one another, i.e.,

$$\frac{w_1}{p_1} \approx \frac{w_2}{p_2} \approx \ldots \approx \frac{w_n}{p_n}$$

Consider, for example, the following instance with two machines and three jobs. Job 1 has $w_1 = p_1 = 1$; job 2 has $w_2 = p_2 = 1$; job 3 has $w_3 = p_3 = 2$. All three ratios are 1, so every schedule is WSPT. In this simple example, which is not the worst case,

$$\frac{\sum w_j C_j(\text{LPT})}{\sum w_j C_j(\text{SPT})} = \frac{8}{7} = 1.14$$

It turns out that the LPT rule in this case is better than the SPT rule. By making some very minor adjustments in the processing times, an example can be constructed with LPT being equivalent to the WSPT rule and SPT being the optimal rule. In a subsequent section we consider a stochastic counterpart of this same instance.

38.3 Stochastic Scheduling: The Case with Distributional Information

38.3.1 Overview

There is an extensive literature on stochastic scheduling. The most common assumptions in stochastic scheduling are the following: Usually, the number of jobs (n) is fixed and the weights of the jobs (w_j) are fixed as well. However, the processing times are assumed to be random variables of which the distributions are known in advance. The data with regard to the release dates and/or due dates may be either fixed (deterministic) or random. Many of the results obtained and properties shown in the stochastic scheduling literature also hold when the number of jobs is random. Having random weights for the jobs usually does not add any complexity to the problem (a problem typically does not change if a random weight is replaced by a fixed weight that is equal to the expected weight).

Most of the results in the stochastic scheduling literature has been obtained for models with a fixed number of jobs (n). However, the results obtained for the models usually tends to be of a structural nature, e.g., the optimality of a monotone priority policy such as the *longest expected processing time* (LEPT) first rule or the *shortest expected processing time* (SEPT) first rule.

No framework or classification scheme has ever been introduced for stochastic scheduling problems. It is clearly more difficult to develop such a scheme for stochastic scheduling problems than for deterministic scheduling problems. Providing a full characterization of a stochastic scheduling problem requires many specifications. For example, the distributions of the processing times have to be specified as well as the distributions of the due dates (which may be different). It has to be specified whether the processing times of the n jobs are independent or correlated (e.g., processing times may be independent draws from a given distribution or they may be equal to the same draw of that distribution) and also which class of policies is being considered. The release processes of the jobs may be subject to various forms of randomness. For example, a release process may be

1. deterministic,
2. according to a Poisson process with exponential interarrival times,
3. according to an arbitrary arrival process with given distributional assumptions,
4. according to an arrival process without any distributional assumptions.

Because of all these possible modeling alternatives, no problem framework has yet been established for stochastic scheduling.

38.3.2 Reductions of Stochastic Problems to Deterministic Problems

Some stochastic scheduling problems can be reduced to a corresponding deterministic scheduling problem. In this subsection we describe four ways in which this can happen.

1. A single machine problem with processing times that have arbitrary distributions may be transformed into a deterministic problem with processing times that are equal to the means of the distributions in the stochastic problem. The optimal schedule in the deterministic counterpart is equivalent to the optimal policy for the original stochastic problem in the class of static policies as well as in the class of nonpreemptive dynamic policies. This is, for example, the case with the stochastic counterpart of $1 \,||\, \sum w_j C_j$ when the processing times have arbitrary distributions. Such a connection between a stochastic scheduling problem and its deterministic counterpart is extremely strong. The optimal schedule for its deterministic counterpart immediately translates into an optimal policy for the original stochastic problem.
2. A problem with exponential processing times, release dates, and due dates assuming either preemptive dynamic policies or nonpreemptive dynamic policies can often be formulated as a *markov decision process* (MDP). An MDP can be formulated as a linear program. However, the fact that such a problem can be formulated as a linear program does not imply that the stochastic scheduling problem can be solved in polynomial time; the reason is that formulating the scheduling problem as a MDP and then transforming it in a linear program may increase the size of the problem exponentially. The number of possible states in the state space of the MDP may be an exponential function of the number of jobs. For example, consider the stochastic counterpart of the problem $Pm \,||\, \sum w_j U_j$ with n processing times and n due dates. If all the processing times and due dates are exponentially distributed, then the possible states in which the process can end up include all possible combinations of jobs already completed and due dates already occurred. It is clear that such a relationship between a stochastic scheduling problem and the equivalent deterministic problem is much weaker here than in the previous case.
3. If the density functions of any two random variables in a stochastic scheduling problem do not *overlap*, then the results that have been obtained for offline deterministic counterparts of a stochastic problem may be extendable to the stochastic setting. The density functions of two random variables X_1 and X_2 do not overlap, when $P(X_1 \geq X_2) = 1$, i.e., X_1 is larger than or equal to X_2 with probability 1. Consider, for example, the stochastic counterpart of $Pm \,||\, C_{\max}$ with n processing times that have nonoverlapping density functions, i.e., the n processing times of n jobs can be ordered in such a way that $X_1 \leq X_2 \leq \cdots \leq X_n$ with probability 1. Assume that the decision maker is allowed to adopt a nonpreemptive dynamic policy. The worst-case bound for the deterministic LPT rule can be extended to this stochastic case and it can be shown that

$$\frac{E(C_{\max}(\text{LEPT}))}{E(C_{\max}(\text{OPT}))} \leq \frac{4}{3} - \frac{1}{3m}$$

This result is actually not quite immediate. Note that the LEPT rule is easy to implement as a nonpreemptive dynamic policy; however, the optimal policy in this stochastic setting is even harder to determine than the optimal policy in the deterministic setting, since the decision maker does not know the exact processing times *a priori*. In order to show that the inequality above holds condition on the n processing times, and assume they are $x_1, \ldots, x_n$. After conditioning on the processing times the problem basically becomes a deterministic problem and for this instance with processing times $x_1, \ldots, x_n$

$$C_{\max}(\text{LPT}) \leq \left(\frac{4}{3} - \frac{1}{3m}\right) C_{\max}(\text{OPT})$$

The deterministic LPT rule (which can be used after conditioning on the processing times) results in exactly the same schedule that would have been generated in the stochastic setting when the processing times would not have been known in advance, i.e., the LPT rule knowing all processing times *a priori* results in exactly the same realization of the makespan as the LEPT rule would have not knowing all the processing times *a priori* (this follows from the fact that the density functions are not overlapping). However, the optimal schedule in the setting that is conditional on the processing times may not be the same one as the one generated in the stochastic setting when the processing times are not known in advance (following an optimal nonpreemptive dynamic policy). Using the optimal nonpreemptive dynamic policy in the stochastic setting may (with a positive probability) result in a makespan that is strictly *larger* than the optimal makespan knowing all the processing times in advance. So in a stochastic setting with nonpreemptive dynamic policies

$$E(C_{\max}(\text{LEPT})) \leq \left(\frac{4}{3} - \frac{1}{3m}\right) E(C_{\max}(\text{OPT}))$$

Of course, there are also many examples of stochastic scheduling problems with non-overlapping density functions that are not easy extensions of their offline deterministic counterparts.

4. There are a number of special cases of stochastic scheduling problems that are equivalent to deterministic problems. Consider a stochastic version of the two machine flow shop problem $F2 \mid block \mid C_{\max}$. The blocking mechanism in between the two machines implies that whenever machine 1 has completed the processing of a job and machine 2 has not yet completed the processing of its job, then machine 1 cannot release its job; the job remains on machine 1, thereby preventing the next job from starting on the machine. The processing time of job j on machine 1 (2) is the random variable X_{1j} (X_{2j}) from an arbitrary distribution F_{1j} (F_{2j}). The objective is find the job sequence that minimizes the expected makespan $E(C_{\max})$. It has been shown that this problem is equivalent to a deterministic *travelling salesman problem* (TSP).

38.3.3 Bounds on Expected Performance Measures

The research literature has focused on various other relationships between stochastic scheduling and offline deterministic scheduling. Some of this research has focused on bounds. Consider a stochastic scheduling environment with a fixed schedule, i.e., it has been predetermined whenever a machine has been freed which job has to be processed next. Assume the n processing times are the random variables $X_1, \ldots, X_n$. If the objective function under consideration can be expressed as a function of $X_1, \ldots, X_n$ which only uses additions of random variables or the "max" operator, then bounds can be established for these objective functions based on the classes of distributions under consideration.

It turns out that in any single machine environment as well as in any permutation flow shop (i.e., a flow shop in which the job sequence does not change from one machine to the next), a lower bound can be obtained for the expected total completion time objective by replacing all the random processing times by their respective means and turning the stochastic scheduling problem into a deterministic scheduling problem, see Pinedo and Wie (1986). However, it is not possible to obtain such a lower bound for a parallel machine environment.

38.4 Comparisons of Stochastic Scheduling Problems with Offline Deterministic Counterparts

38.4.1 Stochastic Scheduling Problems that are Easier than Their Offline Deterministic Counterparts

Research in stochastic scheduling has shown that there are a number of stochastic scheduling problems that are tractable while their deterministic counterparts are NP-hard. Usually, the random processing times in these stochastic scheduling problems are exponentially distributed. The exponential distribution has a host of very nice properties, including the memoryless property, a coefficient of variation equal to 1, a completion rate (also referred to at times as failure rate or hazard rate) that is a constant independent of time. Four examples of NP-hard deterministic problems that have stochastic counterparts with a very nice structure are

1. $1 \mid r_j, prmp \mid \sum w_j C_j$,
2. $1 \mid d_j = d \mid \sum w_j U_j$,
3. $1 \mid d_j = d \mid \sum w_j T_j$,
4. $Pm \mid\mid C_{\max}$.

The first problem allows for a nice solution when the processing times are exponential and the release dates are arbitrarily distributed. The optimal policy is then the preemptive weighted shortest expected processing time (WSEPT) rule. When the processing time distributions are anything but exponential, it appears that the preemptive WSEPT rule is not necessarily optimal. The stochastic counterparts of the second and third problem also lead to the WSEPT rule when the processing time distributions are exponential and the jobs have a common due date which is arbitrarily distributed. Also here if the processing times are anything but exponential, the optimal rule is not necessarily WSEPT.

The stochastic counterparts of $Pm \mid\mid C_{\max}$ are slightly different. When the processing times are exponential, the LEPT rule minimizes the expected makespan in all classes of policies. However, this holds for other distributions also. If the processing times have a *decreasing completion rate* (DCR), e.g., they are hyperexponentially distributed, and satisfy a fairly strong form of stochastic dominance, then the LEPT rule remains optimal. Note that when preemptions are allowed, and the processing times are DCR, the nonpreemptive LEPT rule remains optimal.

38.4.2 Stochastic Scheduling Problems that are Comparable to Their Offline Deterministic Counterparts

There are many problems of which the stochastic versions exhibit very strong similarities to their deterministic counterparts. Examples of such problems are

1. $1 \mid r_j, prmp \mid L_{\max}$,
2. $1 \mid prec \mid h_{\max}$,
3. $F2 \mid\mid C_{\max}$,
4. $J2 \mid\mid C_{\max}$.

It can be shown that the preemptive EDD rule is optimal for the deterministic problem $1 \mid r_j, prmp \mid L_{\max}$ and that it remains optimal when the processing times are random variables that are arbitrarily distributed. The objective $h_{\max}$ is a generalization of the objective $L_{\max}$, that is

$$h_{\max} = \max(h_1(C_1), h_2(C_2), \ldots, h_n(C_n))$$

where $h_j(C_j)$ denotes a nondecreasing penalty function that corresponds to job j. It can be shown that the algorithm for the stochastic counterpart of $1 \mid prec \mid h_{\max}$ is very similar to the algorithm for the deterministic version. The same can be said with regard to $F2 \,||\, C_{\max}$ and $J2 \,||\, C_{\max}$ when the processing times are exponential.

38.4.3 Stochastic Scheduling Problems that are Harder than their Offline Deterministic Counterparts

Of course, there are also problems of which the deterministic versions are easy and the versions with exponential processing times are hard. Examples of such problems are

1. $Pm \mid p_j = 1, tree \mid C_{\max}$,
2. $O2 \,||\, C_{\max}$.

For the deterministic problem $Pm \mid p_j = 1, tree \mid C_{\max}$ the *critical path* (CP) rule is optimal. According to this rule the decision maker selects, whenever a machine is freed, among the remaining jobs the one that is at the head of the longest string of jobs in the tree. For the version of the same problem with all processing times i.i.d. exponential the optimal policy is not known and may depend on the form of the tree. For the $O2 \,||\, C_{\max}$ problem the *longest alternate processing time* (LAPT) rule is optimal, i.e., whenever one of the machines is freed, select among the jobs that still need processing on the other machine that job that has the longest processing time on the other machine. When the processing times are exponential the problem appears to be very hard.

38.5 Online Deterministic Scheduling: The Case with No-Information

38.5.1 Overview

There are two types of online scheduling problems. In both types of online scheduling problems there is not any information with regard to future events (not even distributional).

However, in a problem of the first type it is assumed that whenever a job has entered the system, its processing time is known immediately after it arrives (this type of online scheduling is at times referred to as clairvoyant scheduling). In a problem of the second type it is assumed that whenever a job has entered the system, its processing time remains unknown; its processing time only becomes known when its processing has been completed (this type of online scheduling is at times referred to as nonclairvoyant scheduling). This second type of problem may in a sense be regarded as an extreme case of the stochastic scheduling paradigm.

It is of interest to compare both classes of online scheduling problems with stochastic scheduling problems considering the level of information that is available to the decision maker.

An online scheduling problem of the first type has in one sense more information available than the corresponding stochastic scheduling problem and in another sense less information. With regard to the arrival process of the jobs, there is less information in the online scheduling problem. Usually, in a stochastic scheduling problem there are some distributional assumptions with regard to future job arrivals, whereas in an online scheduling problem there is no information with regard to future arrivals. However, with regard to the processing times of the jobs, there is more information in the online scheduling problem: immediately after a job has entered the system, its processing time is known with certainty (i.e., its processing time is a deterministic distribution). When a job enters the system in a stochastic scheduling problem, its processing time is a random variable of which only the distribution is known.

In an online scheduling problem of the second type there is in every aspect less information than in either a corresponding stochastic scheduling problem or a corresponding online problem of the first type. There is no information with regard to the arrival process of the jobs and there is no information with regard to the processing times of the jobs (i.e., when a job enters the system its processing time is a random variable from an unknown distribution).

38.5.2 Bounds and Competitive Ratios

The bounds that have been obtained for this second type of online scheduling problem (nonclairvoyant) also provides bounds (which are admittedly weak) for a fairly large class of stochastic scheduling problems. In what follows we present two examples.

1. Consider an online scheduling problem of the second type with m machines in parallel. Jobs are presented to the decision maker one at a time. It is of interest to determine the competitive ratio of the online rule which, whenever a new job is presented to the decision maker, puts the new job as early as possible on that machine that becomes available first. The competitive ratio of this rule turns out to be the same as the worst-case bound of the LIST scheduling rule determined by Graham (1966,1969).

   The worst case analysis of the LIST rule implies that the makespan under the LIST rule is at most twice the optimal makespan. More precisely,

$$\frac{C_{\max}(\text{LIST})}{C_{\max}(\text{OPT})} \leq 2 - \frac{1}{m}$$

   This implies that in a stochastic scheduling problem with m machines and n jobs of which the processing time distributions are given (i.e., $G_1, \ldots, G_n$), the expected makespan under an arbitrary policy is at most twice the expected makespan under the optimal policy. This can be shown as follows: Condition on the realization of the n processing times. For any realization of the processing times $x_1, \ldots, x_n$, the $C_{\max}(\text{OPT})$ (obtained by using the policy that is optimal for the stochastic scheduling problem) is larger than or equal to the lower bound on the makespan $LB(C_{\max})$ that can be obtained by finding the optimal schedule with deterministic processing times $x_1, \ldots, x_n$. Also, for any realization of the processing times $x_1, \ldots, x_n$, the $C_{\max}(\text{OPT})$ (obtained by using the policy that is optimal for the stochastic scheduling problem) is smaller than or equal to the upper bound on the makespan $UB(C_{\max})$ that can be obtained by finding the schedule with the worst possible makespan when the processing times are deterministic $x_1, \ldots, x_n$. So, after conditioning on the processing times,

$$LB(C_{\max}) \leq C_{\max}(\text{OPT}) \leq UB(C_{\max}) \leq \left(2 - \frac{1}{m}\right) \times LB(C_{\max})$$

   This chain of inequalities holds for any realization of processing times $x_1, \ldots, x_n$. Unconditioning yields that the expected makespan under an arbitrary policy π in a stochastic setting never can be larger than twice the expected makespan under an optimal policy, i.e.,

$$\frac{E(C_{\max})(\pi)}{E(C_{\max})(\text{OPT})} \leq 2 - \frac{1}{m}$$

2. Consider again an online scheduling problem of the second type with m identical machines in parallel. The objective is to minimize the sum of the completion times of the jobs and preemptions are allowed. A job's processing time becomes known upon its completion. The following so-called RR algorithm cycles through the list of jobs, giving each job τ units of processing time in turn. For τ small enough, a schedule that is constructed using this rule has the property that at any point in time each uncompleted job has received the same amount of processing. If $\tau \to 0$, then the rule is at times referred to as PS. It can be shown that the competitive ratio of the RR algorithm is

$$\frac{\sum C_j(\text{RR})}{\sum C_j(\text{OPT})} \leq 2 - \frac{2m}{n+m}$$

   This result has also implications for the stochastic scheduling problem with m machines and n jobs with arbitrary processing time distributions $G_1, \ldots, G_n$. The objective is to minimize

$E(\sum C_j)$. Condition again on the processing times, and assume the processing times are $x_1, \ldots, x_n$. From the fact that the above bound applies to each realization of the processing times, it follows that

$$\frac{E(\sum C_j)(\text{RR})}{E(\sum C_j)(\text{OPT})} \leq 2 - \frac{2m}{n+m}$$

for arbitrary processing time distributions $G_1, \ldots, G_n$.

38.6 New Research Directions

There are several areas and subareas in the general scheduling domain that have received very little attention. For example, very little work has been done on worst-case analyses in stochastic scheduling (which is an area that lies somewhere in between offline deterministic scheduling and stochastic scheduling) and stochastic online scheduling (which is an area that lies somewhere in between stochastic scheduling and online deterministic scheduling).

38.6.1 Stochastic Worst-Case Analyses

The second section of this chapter described the worst-case behavior of the WSPT rule in a parallel machine environment with deterministic processing times. The following stochastic counterpart is of interest: assume that all the processing times are independent and exponentially distributed with different means. What is the worst-case behavior of the WSEPT first rule? We conjecture that the worst-case behavior of a stochastic problem in general is not as bad as the worst case behavior of its deterministic counterpart.

Consider the following stochastic counterpart of the example described at the end of Section 38.2. The processing times of jobs 1 and 2 are both exponentially distributed with mean 1 and the processing time of job 3 is exponentially distributed with mean 2 (i.e., with rate 0.5). The weights are 1, 1, and 2, respectively. There are two different nonpreemptive dynamic policies. Both of which are according to the WSEPT rule. However, one of them is according to the LEPT first, while the other is according to the SEPT first. It can be shown easily that

$$\frac{E(\sum w_j C_j(\text{SEPT}))}{E(\sum w_j C_j(\text{LEPT}))} = 21/20 = 1.05$$

This value is closer to 1 than the value for its deterministic counterpart which was 1.14.

Recall that the coefficient of variation of the deterministic distribution is 0, whereas the coefficient of variation of the exponential distribution is 1. Consider the Erlang(k, λ) distribution which is a convolution of k independent exponentially distributed random variables with rate λ. The exponential distribution is one extreme in this class of distributions and the deterministic is the other extreme. The coefficient of variation of the Erlang distribution is in between 0 and 1. Consider again the two machine example and three jobs with mean processing times 1, 1, and 2 and weights 1, 1, and 2. Assume that the processing time distributions of the three jobs are Erlang$(2, \lambda)$. It can be shown easily that in this case

$$\frac{E(\sum w_j C_j(\text{SEPT}))}{E(\sum w_j C_j(\text{LEPT}))} = 1.06385$$

which lies indeed in between the worst case ratios corresponding to the exponential case and to the deterministic case.

One may conjecture that if the processing time distributions have a coefficient of variation that is higher than 1, the worst case bound may come even closer to 1. However, this may not always be the case. Consider the following instance with three jobs, again with means 1, 1, 2 and weights 1, 1, 2. The distributions of all three jobs are mixtures of two fixed exponentials with means 0 and 4. The mixing probabilities of the first two jobs are 0.75 and 0.25 (yielding a mean of 1) and the mixing probabilities of the third job are 0.5 and 0.5 (yielding a mean of 2). It turns out that for this case, somewhat surprisingly,

$$\frac{E(\sum w_j C_j(\text{SEPT}))}{E(\sum w_j C_j(\text{LEPT}))} = 1$$

Consider now another instance with three jobs, again with means 1, 1, 2 and with weights 1, 1, 2. The distributions of all three jobs are now mixtures of two fixed exponentials with means 1 and 3. The mixing probabilities of the first two jobs are 1 and 0 (yielding a mean of 1) and the mixing probabilities of the third job are 0.5 and 0.5 (yielding a mean of 2). It turns out that for this case

$$\frac{E(\sum w_j C_j(\text{SEPT}))}{E(\sum w_j C_j(\text{LEPT}))} = 1.0566$$

which is higher than the 1.05 when the processing time distributions are pure exponentials.

38.6.2 Stochastic Online Scheduling

There is another area in scheduling that has not been explored yet. This area is the area of stochastic online scheduling. Consider the following framework: the type of distribution of the processing times is known in advance; however, the values of the parameters of the distribution for each one of the processing times are not known in advance. Immediately after the decision has been made to process the job and the job has been put on the machine the parameter becomes known. In a nonpreemptive setting, it usually does not help much if the parameter becomes known after the job has begun with its processing. In what follows we consider three examples of stochastic online scheduling.

1. Consider the following stochastic counterpart of the deterministic online scheduling problem with m machines in parallel and the makespan objective (with the competitive ratio being $2 - 1/m$): in the stochastic counterpart there are also m machines in parallel and jobs are presented to the decision maker one by one. However, the decision maker does not receive all the information with regard to the processing time of the job; he only knows that the jobs have exponential processing time distributions but does not know the means (or rates) of the exponential distributions. His objective is to schedule the jobs in such a way that he minimizes the expected makespan. We would like the competitive ratio that compares the expected makespan obtained by scheduling the jobs without knowing the means versus the expected makespan assuming the decision maker would have known the means of the exponential processing times beforehand. It turns out that this competitive ratio is smaller than the competitive ratio of the classical online scheduling problem which assumes that all the processing time distributions are deterministic.

   Consider the special case with two machines. In the deterministic setting the worst case result for a LIST schedule with $m = 2$ is 1.5. Consider the following stochastic counterpart with $m = 2$. There are n exponential jobs with mean $1/n$, where n is a very large number. The sum of these n processing times is equal to 1 with a very small variance (i.e., the sum of these n processing times is like a deterministic distribution). There is one additional exponential job with rate λ and the mean of this job is significantly larger than the mean of one of the small jobs. If the means of the distributions are known in advance then the optimal policy is LEPT, which assigns the long job at time $t = 0$ immediately to one of the machines and the small jobs to the other machine. The expected makespan can be computed easily and is

$$E(C_{\max}) = \frac{1}{2}\left(1 + \frac{1}{\lambda} + \frac{1}{\lambda}e^{-\lambda}\right)$$

The worst possible schedule not knowing the means *a priori* puts the longest job last. The expected makespan is then

$$E(C_{\max}) = \frac{1}{2} + \frac{1}{\lambda}$$

So the ratio that divides the best by the worst is

$$\frac{1 + 2x}{1 + x + xe^{-\frac{1}{x}}}$$

where $x = 1/\lambda$. In order to determine the x that maximizes this ratio, the derivative with respect to x has to be set equal to 0. After some elementary computations a worst-case ratio can be obtained for $m = 2$, i.e.,

$$\frac{E(C_{\max})(\text{LIST})}{E(C_{\max})(\text{OPT})} \approx 1.28$$

This value is significantly closer to 1, than the value for its deterministic counterpart (1.5), see Krasik and Pinedo (2003).

Straightforward computations yield also the worst-case ratio when the number of machines m becomes very large, i.e., if $m \to \infty$, then the worst-case ratio approaches 1.466 (this is in contrast to the value of 2 for the deterministic counterpart).

2. Consider the same setting as in the previous example. However, now the processing time distributions belong to a different class of distributions, namely the class of Erlang$(2, \lambda)$ distributions. The processing time of job j is a convolution of two exponentials, both having rate λ. It can be shown that when we have two machines in parallel, the ratio that divides the worst by the best schedule is

$$\frac{1 + 4x}{1 + 2x + (1 + 2x)e^{-1/x}}$$

where $x = 1/\lambda$. The maximum value of this ratio is 1.3358.

3. Consider now the stochastic counterpart of the RR algorithm for the preemptive online scheduling problem with m machines in parallel and the total completion time as the objective to be minimized. The competitive ratio presented in the previous section for this algorithm is attained when all the processing times are equal. Consider now the case when all the processing times are exponentially distributed with the same mean. In contrast to the deterministic result, it turns out that the RR algorithm is actually *optimal* when the processing times are independent and distributed according to the same exponential. The RR algorithm is also optimal when the processing times of the jobs are independent and identically distributed and the distribution has a *decreasing completion rate.*

38.7 Discussion

Several remarks are in order.

1. There is a distributional connection between the offline deterministic scheduling models and the stochastic models with exponential processing time distributions. Consider the Erlang(k, λ) class of distributions. Such a distribution is a k fold convolution of an exponential distribution with rate λ and mean $1/\lambda$. So the mean of this distribution is k/λ. If the mean is 1 and $k = 1$ then this distribution is the exponential distribution with rate 1 and if the mean is 1 and $k = \infty$ then this distribution is the deterministic. When k moves from 1 to ∞ (while k/λ is kept fixed at 1), the distribution moves from the exponential to the deterministic.

   Consider the $Pm \mid prmp \mid C_{\max}$ problem and its stochastic counterparts. When $k = 1$ the *longest expected remaining processing time* (LERPT) rule is optimal, when $k = \infty$ the LRPT rule is optimal. What happens for k in between? Is it always LERPT?

Consider the $Pm \mid\mid C_{max}$ problem and its stochastic counterparts. When $k = 1$ the LEPT rule is optimal, when $k = \infty$ the problem is NP-hard. What happens for k in between?

2. Various conclusions can be drawn from the comparisons in Section 38.4 between offline deterministic scheduling problems and their stochastic scheduling counterparts. Problems with all variables independent exponentially distributed may often be similar or easier than their offline deterministic counterparts. Problems with all processing times deterministic and equal to 1 tend to be easier than problems will all processing times exponentially distributed with mean 1. Problems with precedence constraints tend to be easier in an offline deterministic setting than in a stochastic setting.
3. Comparing the worst case analyses done in offline deterministic scheduling with some of the preliminary results obtained so far for worst case analyses of stochastic scheduling problems, it seems that the worst cases in offline deterministic scheduling are "worse" than the worst cases in stochastic scheduling, i.e., the worst-case ratios are higher in a deterministic setting than in a stochastic setting. However, this observation may only be true in cases when the distributions of the processing times satisfy certain stochastic orderings. When they do not satisfy certain stochastic orderings the worst cases in stochastic settings may behave worse than the worst cases of their deterministic counterparts.
4. The values of the performance measures in an offline deterministic problem can be computed easily given the schedule. The expected value of a performance measure in a stochastic scheduling problem cannot be computed that easily; it has to be approximated.

References

R.L. Graham (1966). Bounds for Certain Multiprocessing Anomalies, *Bell System Technical Journal,* Vol. 45, pp. 1563–1581.

R.L. Graham (1969). Bounds on Multiprocessing Timing Anomalies, *SIAM Journal of Applied Mathematics,* Vol. 17, pp. 263–269.

R.L. Graham, E.L. Lawler, J.K. Lenstra, and A.H.G. Rinnooy Kan (1979). Optimization and Approximation in Deterministic Sequencing and Scheduling: A Survey, *Annals of Discrete Mathematics,* Vol. 5, pp. 287–326.

T. Kawaguchi and S. Kyan (1986). Worst Case Bound of an LRF Schedule for the Mean Weighted Flow Time Problem, *SIAM Journal of Computing,* Vol. 15, pp. 1119–1129.

V. Krasik and M. Pinedo (2003). On Competitive Ratios in Stochastic On-line Scheduling, Working Paper, IOMS Department, Stern School of Business, New York University.

E.L. Lawler, J.K. Lenstra, and A.H.G. Rinnooy Kan (1982). Recent Developments in Deterministic Sequencing and Scheduling: A Survey, in *Deterministic and Stochastic Scheduling,* Dempster, Lenstra, and Rinnooy Kan (eds.), pp. 35–74, D. Reidel, Dordrecht.

R. Motwani, S. Phillips, and E. Torng (1994). Nonclairvoyant Scheduling, *Theoretical Computer Science,* Vol. 130, pp. 17–47.

M. Pinedo and S.-H. Wie (1986). Inequalities for Stochastic Flow Shops and Job Shops, *Applied Stochastic Models and Data Analysis,* Vol. 2, pp. 61–69.

M. Pinedo (2002). *Scheduling — Theory, Algorithms and Systems,* 2nd ed., Prentice-Hall, Upper Saddle River, New Jersey.

K. Pruhs, J. Sgall, and E. Torng (2004). On-line Scheduling Algorithms, in *Handbook of Scheduling: Algorithms, Models, and Performance Analysis,* J. Leung (ed.), CRC Press, Boca Raton, FL.

R. Righter (1994). Stochastic Scheduling, Chapter 13 in *Stochastic Orders,* M. Shaked and G. Shanthikumar (eds.), Academic Press, San Diego, California.

D.B. Shmoys, J. Wein, and D.P. Williamson (1995). Scheduling Parallel Machines On-line, *SIAM Journal of Computing,* Vol. 24, pp. 1313–1331.

G. Weiss (1982). Multi-Server Stochastic Scheduling, in *Deterministic and Stochastic Scheduling,* Dempster, Lenstra and Rinnooy Kan (eds.), D. Reidel, Dordrecht, pp. 157–179.

39

Stochastic Scheduling with Earliness and Tardiness Penalties

Xiaoqiang Cai
The Chinese University of Hong Kong

Xian Zhou
The Hong Kong Polytechnic University

39.1 Introduction

Scheduling to minimize both earliness and tardiness costs has emerged to be a main thrust of research in the scheduling field. Investigation of these problems has largely been motivated by the adoption of the Just-In-Time concept in manufacturing industry, which aims to complete jobs exactly at their due dates, not earlier and not later. There are, however, many other applications that espouse the concept of earliness-tardiness (E/T) minimization. An example is the harvest of crop products, which should be conducted around the ripe time of the crop. Another example is the production and delivery of perishable products like fresh meat or milk. Such a product should not be finished too early in order to avoid its possible decay, and should not be completed too late in order not to miss the delivery.

Deterministic E/T scheduling problems have been extensively studied by many authors in the past two decades. See, for example, reviews of Baker and Scudder (1990), Chen, Potts, and Woeginger (1998), Gordon, Proth, and Chu (2002), and Kanet and Sridharan (2002). Study of stochastic E/T scheduling has started, however, only in recent years, although scheduling problems in the real-world involve, in essence, randomness and uncertainty. In fact, a place where scheduling can play a critical role is to provide the best policies/strategies regarding the timing to perform the required work under randomness and uncertainty, so as to minimize the possible loss.

1-58488-397-9/\$0.00+\$1.50

In this chapter we will provide a review of stochastic E/T scheduling. We will introduce models that consider random processing times, random due dates or deadlines, and stochastic machine breakdowns. We will describe some main results that have been achieved on squared E/T functions, piecewise linear E/T functions, and other E/T functions. In particular, we will show that, although an optimal dynamic policy is usually quite difficult to obtain for E/T scheduling problems, there is an interesting model which exhibits optimal analytical solutions in both classes of static and dynamic policies, even when job preemptions are allowed. In summary, despite the difficulties in dealing with randomness because of the combinatorial nature of sequencing problems, which is compounded further by the presence of uncertain factors, and the nonregular feature of E/T objective functions, some significant advancements have been achieved on stochastic E/T scheduling. We will also point out some open questions for future investigations.

We organize this chapter as follows. In Section 39.2 we describe a basic model for E/T problems and introduce notations and elementary concepts. E/T problems with quadratic and linear E/T functions are introduced in Sections 39.3 and 39.4, respectively. In Section 39.5, we consider stochastic E/T scheduling under an uncertain delivery time, where issues on dynamic policies will be discussed. Finally, some concluding remarks are given in Section 39.6, which includes open issues and opportunities for further research.

39.2 A Basic Model

We describe a basic model in this section. Models to be considered in the following sections can be regarded as variants of the basic model, carrying different features and assumptions to capture certain decision-making concerns or to facilitate the development of certain results.

Suppose that a set $\{1, 2, \ldots, n\}$ of independent jobs are to be processed by a single machine. All jobs are ready for processing at time $t = 0$. The amount of time required to complete each job i, referred to as the *processing time* of job i and denoted by P_i, $i = 1, 2, \ldots, n$, is a random variable with a probability distribution F_i. Denote the mean and variance of P_i by μ_i and σ_i^2, respectively. The processing times are nonnegative and mutually independent.

Each job i has a due date D_i, which can be a real due date that is prescribed and known in advance, or some other deadlines (such as the random arrival of a storm before which a crop should be gathered in order to avoid the damage, or the uncertain departure time of a transporter to deliver the finished jobs). Assume that the due dates of jobs $\{D_i\}$ are random variables independent of the processing times $\{P_i\}$. The problem with deterministic due dates is a special case of that with random due dates and is therefore also included in the model here.

The machine is subject to stochastic breakdowns, with the breakdown process being characterized by a sequence of finite-valued, positive random vectors $\{(Y_k, Z_k)\}_{k=1}^{\infty}$, where Y_k and Z_k are the durations of the kth uptime and the kth downtime, respectively. Assume that the uptimes of the machine are independent of its downtimes and that the downtimes are mutually independent and identically distributed with mean ν and variance σ^2. For the stochastic sequence $\{Y_k\}_{k=1}^{\infty}$, define a counting process $\{N(t) : t \geq 0\}$ by

$$N(t) = \sup\{k \geq 0 : S_k \leq t\} \tag{39.1}$$

where S_k denotes the total uptime of the machine before its kth breakdown, i.e.,

$$S_k = \sum_{i=1}^{k} Y_i, \quad k \geq 1 \tag{39.2}$$

For convenience we define $S_0 = 0$.

Consider the situation where $N(t)$ is a Poisson process with rate κ. In addition, consider the so-called *preempt-resume* model (cf. Birge et al., 1990) regarding the processing of a job on the machine. Under the preempt-resume model, it is assumed that, if a machine breakdown occurs during the processing of a job, the work done on the job prior to the breakdown is not lost, and the processing of the disrupted job can be continued from where it was interrupted. Thus, the processing times P_i, $i = 1, \ldots, n$, are independent of any machine breakdowns. In contrast with the preempt-resume model, the *preempt-repeat* model assumes that the work done prior to a breakdown is lost, and the processing of the job will have to be restarted as soon as the machine resumes operable. This is a substantially more difficult model (see Cai, Sun, and Zhou, 2003), on which little result has been found in the E/T scheduling literature. Therefore, the following sections will be restricted to the *preempt-resume* breakdown pattern.

In a stochastic environment, there are two classes of policies that a decision maker may be allowed to adopt (cf. Pinedo, 1995):

The class of static policies — A decision cannot be altered as soon as it is determined and executed. In other words, the decision will be maintained throughout the whole process of operation. This models the situations where alteration of any decision may be very expensive or prohibited.

The class of dynamic policies — A decision can be adjusted at any time after it is determined and applied, by taking into account the new information that becomes available up to the current time. In other words, the decision maker may observe the occurrences of random events and the realizations of the random variables, to adjust his/her decision dynamically.

In most of the models to be introduced in Sections 39.3–39.5 below, job preemption is not allowed; that is, once the processing of a job starts, it must continue without interruption until the job is completed. In Subsection 39.5.3, however, we consider an E/T scheduling model that allows preemption; that is, a job may be preempted by another job, if this is found to be beneficial based on the information available at the time of making the decision.

In the rest of this section we will limit our description of the problem to the class of static, nonpreemptive policies $\zeta = (\lambda, r)$, where

1. λ is a sequence to process the jobs, denoted by $\lambda = (i_1, i_2, \dots, i_n)$, which is a permutation of $(1, 2, \dots, n)$, with $i = i_k$ if job i is the kth to be started and processed by the machine, and
2. r is the amount of time that the machine is kept idle before it starts to process its first job.

Many studies in the E/T scheduling literature implicitly assume that the machine will start to process its first job at time zero, that is, $r = 0$. Such an assumption, however, is not appropriate for E/T problems (unless the problem itself does not allow the machine to be idle at any time), since a main concern of E/T scheduling is how to optimally delay the processing of jobs so as to reduce the earliness penalty. In fact, a more desirable policy should determine the optimal idle time inserted before each and every job. Nevertheless, E/T problems considering such idle times are often extremely difficult. The only known result in stochastic E/T scheduling that includes job-dependent idle times is Cai and Zhou (1999), which will be reviewed in Section 39.5. In other sections of this chapter, we will consider only a deterministic r before all jobs.

For any ζ, the completion time of job i can be expressed as

$$C_i(\zeta) = R_i(\zeta) + \sum_{k=0}^{N(R_i(\zeta))} Z_k \tag{39.3}$$

where $\mathcal{B}_i(\zeta)$ is the set of jobs sequenced no later than job i under ζ, and

$$R_i(\zeta) = r + \sum_{k \in \mathcal{B}_i(\zeta)} P_k \tag{39.4}$$

represents the total uptime of the machine before job i is finished and $Z_0 \equiv 0$. Note that the completion time $C_i(\zeta)$ depends on ζ, but we will use the notation C_i when there is no ambiguity.

Completing a job either earlier or later than its due date will incur a penalty (cost). Let the earliness and tardiness costs for job i be denoted as follows, respectively:

$$E_i(\zeta) = g_i(D_i - C_i) \quad \text{if } C_i < D_i \tag{39.5}$$

$$T_i(\zeta) = h_i(C_i - D_i) \quad \text{if } C_i > D_i \tag{39.6}$$

where $g_i(x)$ and $h_i(x)$ are functions of the magnitudes of earliness and tardiness, respectively.

Remark 39.1

1. If $g_i(D_i - C_i) = \alpha_i(D_i - C_i)$ and $h_i(C_i - D_i) = \beta_i(C_i - D_i)$, where α_i and β_i are constants, representing respectively the unit earliness and tardiness costs, then the problem is said to have linear E/T cost functions. In particular, if $\alpha_i = \beta_i = w_i$ for all i, then the problem is said to have symmetric E/T weights w_i.
2. If $g_i(D_i - C_i) = \alpha_i(D_i - C_i)^2$ and $h_i(C_i - D_i) = \beta_i(C_i - D_i)^2$, the problem is said to have squared (quadratic) E/T cost functions. In particular, if $g_i(D_i - C_i) = (C_i - \bar{C})^2$ and also $h_i(C_i - D_i) = (C_i - \bar{C})^2$, where

$$\bar{C} = \frac{1}{n}\sum_{i=1}^{n} C_i \tag{39.7}$$

   is the mean completion time under ζ, then the problem reduces to the well-known completion time variance (CTV) problem.
3. A performance measure with earliness and tardiness (E/T) costs is nonregular, in the sense that it is *not monotone* in the completion time.

To summarize, the problem of stochastic E/T scheduling we consider in this chapter is to determine an optimal ζ^* so as to minimize the *expected total cost* defined by

$$TC(\zeta) = E\left[\sum_{C_i < D_i} g_i(D_i - C_i) + \sum_{C_i > D_i} h_i(C_i - D_i)\right] \tag{39.8}$$

where $E(X)$ denotes the expectation of a random variable X.

39.3 Quadratic Cost Functions

We examine in this section models with quadratic cost functions, which are relatively easy to analyze because of the differentiability of a squared function. Models with linear E/T functions will be discussed in Section 39.4, which are substantially harder, due to the nondifferentiability of a piecewise linear function.

39.3.1 The Symmetric Case with a Common Due Date

Consider the problem with $g_i(D_i - C_i) = (d - C_i)^2$ and $h_i(C_i - D_i) = (C_i - d)^2$, for $i = 1, 2, \ldots, n$, where $d = D_i$ for all $i = 1, 2, \ldots, n$ is called the common due date of the jobs, and is assumed to be a deterministic number. The processing times P_i are assumed to be random variables, following general probability distributions. The counting process $N(t)$ for machine breakdowns is assumed, in this subsection, to be a generalized Poisson process. Thus, it follows from Parzen (p. 130, 1962) that $E[N(t)] = \kappa t E(Y)$ and $\text{Var}(N(t)) = \kappa t E(Y^2)$, where $\kappa > 0$, Y is a nonnegative integer-valued random variable, and $\text{Var}(X)$ denotes the variance of a random variable X. When $Y = 1$ with probability 1, $N(t)$ reduces to a Poisson process, and thus $E[N(t)] = \kappa t$ and $\text{Var}(N(t)) = \kappa t$.

In this case, an optimal solution $\zeta^* = (\lambda^*, r^*)$ is to be determined so as to minimize

$$TC(\lambda, r) = E\left[\sum_{i=1}^{n}(C_i - d)^2\right] \tag{39.9}$$

Denote this problem as Model 1. Note that it involves both random processing times and random machine breakdowns. Mittenthal and Raghavachari (1993) examine a similar model, but with deterministic processing times.

Without loss of generality, we can assume that

$$\mu_1 \le \mu_2 \le \cdots \le \mu_n \tag{39.10}$$

39.3.1.1 Deterministic Equivalent

Cai and Tu (1996) have developed the following deterministic equivalent, where Λ is defined as the set of the $n!$ permutations of the integers $1, 2, \ldots, n$. The deterministic equivalent facilitates the application of approaches/techniques available for deterministic problems.

Theorem 39.1

Model 1 is equivalent to

$$\min_{r\in[0,d],\ \lambda\in\Lambda} \quad TC(\lambda, r) = \sum_{i=1}^{n} (\hat{\psi}_i - \hat{d})^2 + \sum_{i=1}^{n} \hat{\phi}_i \tag{39.11}$$

where

$$\hat{\psi}_i = \hat{r} + \sum_{j\in\mathcal{B}_i(\lambda)} (\hat{a}\mu_j), \quad \hat{r} = \hat{a}r, \quad \hat{d} = d - \frac{\hat{b}}{2\hat{a}}, \quad \hat{\phi}_i = \sum_{j\in\mathcal{B}_i(\lambda)} (\hat{a}^2\sigma_j^2) \tag{39.12}$$

$$\hat{a} = 1 + \mu\kappa E(Y) \quad \textit{and} \quad \hat{b} = \kappa(\sigma E(Y) - \mu^2 E(Y) + \mu^2 E(Y^2)). \tag{39.13}$$

39.3.1.2 Optimality Properties

V-shaped sequence is an important concept for E/T scheduling, which was firstly introduced by Eilon and Chowdhury (1977), who find that an optimal sequence for the waiting time variance (WTV) problem is V-shaped with respect to processing times. In the deterministic case, a sequence is said to be V-shaped with respect to processing times if the jobs before (after) the shortest job follow nonincreasing (nondecreasing) order of processing times.

Note that the $TC(\lambda, r)$ in (39.11) consists of two sums. Given any r, it can be shown that the first sum is minimized by a V-shaped sequence with respect to mean processing times $\{\mu_i\}$, whereas the second sum is minimized by a SVPT (shortest variance of processing time first) sequence, which can be regarded as a special V-shaped sequence with respect to the variances of processing times. Although the two sums are each minimized by a V-shaped sequence, the orders of the jobs in the two V-shaped sequences are likely to differ, and therefore the optimal sequence for the combination of the two sums is not necessarily V-shaped. In the following we provide a sufficient condition, under which a V-shaped sequence exists which minimizes the overall objective function $TC(\lambda, r)$.

Theorem 39.2 (Cai and Tu, 1996)

Given any $r \in [0, d]$, there exists a sequence λ^* minimizing $TC(\lambda, r)$ which is V-shaped with respect to mean processing times $\{\mu_i\}$, if

$$\left|\frac{\sigma_i^2}{\mu_i} - \frac{\sigma_j^2}{\mu_j}\right| \le |\mu_i - \mu_j|, \quad \forall \mu_i \ne \mu_j \tag{39.14}$$

Condition (39.14) to ensure an optimal V-shaped sequence is quite elegant. We can easily verify that the condition is satisfied when the processing times P_i follow many well-known distributions, including *Exponential:* $\sigma_i^2 = \mu_i^2$; *Uniform over* $[0, b_i]$: $\sigma_i^2 = \mu_i^2/3$; *Gamma:* $\sigma_i^2 = \mu_i^2/b$ (with constant $b > 0$); *Erlang:* $\sigma_i^2 = \mu_i^2/K$ (where K is a positive integer); *Rayleigh:* $\sigma_i^2 = \frac{4-\pi}{\pi}\mu_i^2$; and *Chi-square:* $\sigma_i^2 = 2\mu_i$.

It is, however, a sufficient condition only. Cai and Tu (1996) have shown, by devising a counterexample, that a sequence may be optimal even though (39.14) is not satisfied.

The following theorem narrows down the possible range for the optimal r^*.

Theorem 39.3 (Cai and Tu, 1996)

The optimal r^* is bounded by

$$\max\{0, r_L\} \le r^* \le \max\{0, r_S\} \tag{39.15}$$

where

$$r_L = \frac{1}{\hat{a}}d - \frac{\hat{b}}{2\hat{a}^2} - \frac{1}{n}\sum_{i=1}^{n} i\mu_i \tag{39.16}$$

$$r_S = \frac{1}{\hat{a}}d - \frac{\hat{b}}{2\hat{a}^2} - \frac{1}{n}\sum_{i=1}^{n} (n-i+1)\mu_i \tag{39.17}$$

Let λ_{SVPT} denote a SVPT sequence (shortest variance of processing time first), and λ_{SEPT} a SEPT sequence (shortest expected processing time first). Cai and Tu have further established the following optimality properties.

Proposition 39.1

$\lambda^* = \lambda_{\text{SVPT}}$ *and* $r^* = \max\{0, \frac{1}{\hat{a}}d - \frac{\hat{b}}{2\hat{a}^2} - \frac{1}{2}(n+1)\mu_c\}$ *constitute an optimal solution, if* $\mu_i = \mu_c, \forall i$.

Proposition 39.2

λ^* *is V-shaped with respect to* $\{\mu_i\}$ *if* $\sigma_i^2 = \sigma_c^2, \forall i$.

Proposition 39.3

When $d \le \frac{\hat{b}}{2\hat{a}} + \frac{\hat{a}}{2}(\mu_1 + \mu_2)$, $\lambda^* = \lambda_{\text{SEPT}}$ *and* $r^* = 0$ *constitute an optimal solution under the agreeable condition that* $\mu_i < \mu_j$ *implies* $\sigma_i^2 \le \sigma_j^2, \forall i, j$.

Proposition 39.4

$r^* = 0$ *if* $d \le \frac{\hat{b}}{2\hat{a}} + \frac{\hat{a}}{n}\sum_{i=1}^{n}(n-i+1)\mu_i$.

39.3.1.3 Algorithm

Here we assume that μ_i and r take integer values. We also assume that the sufficient condition (39.14) is satisfied.

Cai and Tu (1996) developed a dynamic programming (DP) algorithm to generate the best solution among all V-shaped solutions for Model 1. We describe the algorithm in detail here, since it exhibits the key idea of a class of dynamic programs to derive the best V-shaped solutions for E/T scheduling. Clearly, if condition (39.14) is satisfied, then the solution generated by the algorithm will be optimal. Otherwise, the algorithm can still be applied. If the problem happens to have a V-shaped optimum, then the best V-shaped solution found will be optimal, otherwise the solution may serve as an approximate solution.

Let $r_{\min} = \max\{0, \lfloor r_L \rfloor\}$ and $r_{\max} = \max\{0, \lfloor r_S \rfloor\}$, where r_L and r_S are given by (39.16) and (39.17), respectively, and $\lfloor x \rfloor$ represents the largest integer less than x. It follows from Theorem 39.2 that the optimal r^* can only take values in the set

$$\mathcal{R} = \{r_{\min}, r_{\min} + 1, r_{\min} + 2, \ldots, r_{\max}\} \tag{39.18}$$

Moreover, let $\mathcal{N}_i = \{1, 2, \ldots, i\}$, and let θ_i be the earliest starting time of the jobs in $\mathcal{N}_i$. Clearly θ_i must be contained in the set

$$\mathcal{H}_i = \{r_{\min}, r_{\min} + 1, r_{\min} + 2, \ldots, r_{\max} + \Theta_n - \Theta_i\} \tag{39.19}$$

where $\Theta_i = \sum_{k=1}^{i} \mu_k$, for $i \in \mathcal{N}_n$

Define

$$f_i(\theta_i) = \min_{\lambda_i \in \mathcal{V}_i} \left\{ \sum_{j \in \mathcal{N}_i} [\hat{a}\theta_i + \hat{a} S_j(\lambda_i) - \hat{d}]^2 + \hat{a}^2 \sum_{j \in \mathcal{N}_i} T_j(\lambda_i) \right\} \tag{39.20}$$

where λ_i denotes a sequence for all jobs in $\mathcal{N}_i$, $\mathcal{V}_i$ is the set of those V-shaped λ_i, $S_j(\lambda_i)$ $(T_j(\lambda_i))$ is the sum of μ_k (σ_k^2) of those jobs in $\mathcal{N}_i$ that are sequenced no later than job j under λ_i. Clearly, for a given r, the optimal value of $TC(\lambda, r)$ equals $f_n(r)$, provided that the problem with the given r has an optimum λ which is V-shaped.

Note that we have indexed the jobs according to (39.10). In a V-shaped sequence λ_i, job i has only two possible locations, either the first or the last job in the sequence. If job i is the first, it follows from (39.20) that $f_i(\theta_i)$ equals:

$$\begin{aligned} f_i^b(\theta_i) &= [\hat{a}(\theta_i + \mu_i) - \hat{d}]^2 + \hat{a}^2\sigma_i^2 + \sum_{j \in \mathcal{N}_{i-1}} [\hat{a}(\theta_i + \mu_i) + \hat{a} S_j(\lambda_{i-1}) - \hat{d}]^2 \\ &\quad + \hat{a}^2 \sum_{j \in \mathcal{N}_{i-1}} \left[\sigma_i^2 + T_j(\lambda_{i-1})\right] \\ &= [\hat{a}(\theta_i + \mu_i) - \hat{d}]^2 + i\hat{a}^2\sigma_i^2 + f_{i-1}(\theta_i + \mu_i) \end{aligned} \tag{39.21}$$

Otherwise, if job i is the last, then $f_i(\theta_i)$ equals

$$\begin{aligned} f_i^a(\theta_i) &= [\hat{a}(\theta_i + \Theta_i) - \hat{d}]^2 + \hat{a}^2\Omega_i + \sum_{j \in \mathcal{N}_{i-1}} [\hat{a}\theta_i + \hat{a} S_j(\lambda_{i-1}) - \hat{d}]^2 + \hat{a}^2 \sum_{j \in \mathcal{N}_{i-1}} T_j(\lambda_{i-1}) \\ &= [\hat{a}(\theta_i + \Theta_i) - \hat{d}]^2 + \hat{a}^2\Omega_i + f_{i-1}(\theta_i) \end{aligned} \tag{39.22}$$

where $\Omega_i = \sum_{k=1}^{i} \sigma_k^2$, for $i \in \mathcal{N}$. It follows from the principle of optimality of DP that

$$f_i(\theta_i) = \min\left\{ f_i^b(\theta_i), f_i^a(\theta_i) \right\} \tag{39.23}$$

subject to the boundary conditions:

$$f_0(\theta) = 0, \ \forall\theta \quad \text{and} \quad f_i(\theta) = +\infty, \quad \forall\theta \notin \mathcal{H}_i, \ i \geq 1 \tag{39.24}$$

We can now describe the following algorithm.

Algorithm 39.1

1. *For* $i = 1, 2, \ldots, n$, *compute* $f_i(\theta_i)$ *for all* $\theta_i \in \mathcal{H}_i$, *according to* (39.21)–(39.24).
2. *Let* $TC(\lambda^*, r^*) = f_n(r^*) = \min_{r \in \mathcal{R}}\{f_n(r)\}$.
3. *Construct the sequence* λ^* *that achieves* $f_n(r^*)$ *by a backward tracking procedure.*

Theorem 39.4 (Cai and Tu, 1996)

Algorithm 39.1 can find the solution (λ^*, r^*) *in time* $O(n\Theta_n)$, *which is optimal for Model 1 if condition* (39.14) *is satisfied.*

39.3.2 The Symmetric Weighted case

We now consider the problem with $g_i(D_i - C_i) = w_i(D_i - C_i)^2$ and $h_i(C_i - D_i) = w_i(C_i - D_i)^2$, for $i = 1, \ldots, n$, where w_i is called the *weight* of job i, $i = 1, \ldots, n$, and $D_1, \ldots, D_n$ are random due dates independent of the processing times. We assume that the counting process $N(t)$ for machine breakdowns is a Poisson process with rate κ. Furthermore, we need the assumption that the processing times P_i follow exponential distributions, with means μ_i, $i = 1, \ldots, n$.

An optimal solution $\zeta^* = (\lambda^*, r^*)$ is to be determined so as to minimize

$$TC(\lambda, r) = E\left[\sum_{i=1}^{n} w_i (C_i - D_i)^2\right] \tag{39.25}$$

The deterministic case with d_0 as a common due date usually requires $r \leq d_0$. This restriction is natural, since it is obviously unrealistic if $r > d_0$, namely, the machine is still kept idle while the due date of the jobs has passed away. Analogous to the deterministic case, we here restrict r within $[0, d]$, where $d = E[D_1]$ is the common expected due date.

Denote the problem above as Model 2. Without loss of generality, we can assume that

$$\mu_1/w_i \leq \mu_2/w_2 \leq \cdots \leq \mu_n/w_n \tag{39.26}$$

Cai and Zhou (2000) have investigated Model 2, with random due dates D_i following a common distribution and $r = 0$. It is not difficult to allow $r > 0$ by an analysis similar to the derivation of equation (13) in Cai and Zhou (2000), which leads to the following deterministic equivalent of Model 2.

Theorem 39.5

Let $W = \sum_{i=1}^{n} w_i$. Then Model 2 is equivalent to

$$\min_{r\in[0,d],\lambda\in\Lambda} \quad TC(\lambda, r) = \sum_{i=1}^{n} w_i(1+\nu\kappa)^2\left\{\sum_{k\in\mathcal{B}_i}\mu_k^2 + \left(\sum_{k\in\mathcal{B}_i}\mu_k\right)^2\right\} + E[(D_1 - r)^2]W$$
$$+\sum_{i=1}^{n} w_i\{\kappa(\nu^2+\sigma^2) - 2(d-r)(1+\nu\kappa)\}\sum_{k\in\mathcal{B}_i}\mu_k \tag{39.27}$$

The following result reveals the V-shape property for Model 2.

Theorem 39.6 (Cai and Zhou, 2000)

Given any $r \in [0, d]$, an optimal sequence λ^* that minimizes $TC(\zeta)$ of Model 2 is V-shaped with respect to $\{\mu_i/w_i\}$.

It is interesting to note that the V-shape property does not rely on the parameters of the machine breakdowns.

Based on the V-shape property of Theorem 39.6, we can develop a dynamic programming algorithm to find the optimal solution for Model 2. The algorithm developed here also assumes that d and all μ_i, $i = 1, 2, \ldots, n$, are integers. Also assume that r takes an integer value. Then the feasible set for r can be given as follows:

$$\mathcal{R} = \{0, 1, \ldots, d\} \tag{39.28}$$

Let $\Theta = \sum_{i=1}^{n} \mu_i$ and $\Phi = \sum_{i=1}^{n} \mu_i^2$.

For any given $r \in \mathcal{R}$, we consider a set of jobs $\mathcal{N}_i = \{1, 2, \ldots, i\}$. In a V-shaped sequence with respect to $\{\mu_i/w_i\}$, job i will be sequenced either the first or the last among all jobs in $\mathcal{N}_i$, because the jobs have been numbered according to (39.26). Assuming that λ^* is the best V-shaped sequence and $\bar{\mathcal{N}_i}$ is the set of jobs sequenced before all jobs in $\mathcal{N}_i$ under λ^*, we denote

$$\Theta_i = \sum_{j\in\bar{\mathcal{N}}_i}\mu_j \quad \text{and} \quad \Phi_i = \sum_{j\in\bar{\mathcal{N}}_i}\mu_j^2 \tag{39.29}$$

Define $g_i(\Theta_i, \Phi_i)$ as the contribution of all jobs in $\mathcal{N}_i$ to the cost function (39.27), given Θ_i and Φ_i. Then, it is clear that, if job i is sequenced as the first among all jobs in the set $\mathcal{N}_i$, then

$$g_i(\Theta_i, \Phi_i) = g_i^a(\Theta_i, \Phi_i) = g_{i-1}\left(\Theta_i + \mu_i, \Phi_i + \mu_i^2\right) + w_i(1+\nu\kappa)^2\left\{\Phi_i + \mu_i^2 + (\Theta_i + \mu_i)^2\right\}$$
$$+ w_i\{\kappa(\nu^2 + \sigma^2) - 2(d-r)(1+\nu\kappa)\}(\Theta_i + \mu_i) \quad (39.30)$$

If job i is sequenced as the last among all jobs in the set $\mathcal{N}_i$, then

$$g_i(\Theta_i, \Phi_i) = g_i^b(\Theta_i, \Phi_i) = g_{i-1}(\Theta_i, \Phi_i) + w_i(1+\nu\kappa)^2\{\Phi_i + \bar{\Phi}_i + (\Theta_i + \bar{\Theta}_i)^2\}$$
$$+ w_i\{\kappa(\nu^2 + \sigma^2) - 2(d-r)(1+\nu\kappa)\}(\Theta_i + \bar{\Theta}_i) \quad (39.31)$$

where

$$\bar{\Theta}_i = \sum_{j \in \mathcal{N}_i} \mu_j \quad \text{and} \quad \bar{\Phi}_i = \sum_{j \in \mathcal{N}_i} \mu_j^2 \quad (39.32)$$

Then, λ^* must sequence this job such that

$$g_i(\Theta_i, \Phi_i) = \min\left\{g_i^a(\Theta_i, \Phi_i), g_i^b(\Theta_i, \Phi_i)\right\}$$
$$\text{for} \quad \Theta_i = 0, 1, \ldots, \Theta - \bar{\Theta}_i \quad \text{and} \quad \Phi_i = 0, 1, \ldots, \Phi - \bar{\Phi}_i \quad (39.33)$$

subject to the boundary conditions

$$g_0(x_1, x_2) = \begin{cases} (d-r)^2 W & \text{if } x_1 = 0, 1, \ldots, \Theta \text{ and } x_2 = 0, 1, \ldots, \Phi \\ +\infty & \text{otherwise} \end{cases} \quad (39.34)$$

The following algorithm solves Model 2.

Algorithm 39.2

1. *For each* $r \in \mathcal{R}$, *go to steps 2–3.*
2. *For* $i = 1, 2, \ldots, n$, *compute* $g_i(\Theta_i, \Phi_i)$ *for all* $\Theta_i = 0, 1, \ldots, \Theta - \bar{\Theta}_i$ *and* $\Phi_i = 0, 1, \ldots, \Phi - \bar{\Phi}_i$, *according to* (39.30)–(39.34).
3. *Let* $G_n^*(r) = \min_{\Theta_n, \Phi_n}\{g_n(\Theta_n, \Phi_n)\}$.
4. *Search for the optimal* r^* *that achieves* $G_n^* = \min_{r \in \mathcal{R}} G_n^*(r)$.
5. *Construct, by a backward tracking process, the sequence* λ^* *that achieves* G_n^*.

Theorem 39.7

Algorithm 39.2 can find the optimal solution ζ^* *for Model 2 in* $O(nd\Theta\Phi)$ *time.*

39.3.3 The Weighted, Asymmetric Case

We now consider a model that is similar to Model 2, except that $g_i(D_i - C_i) = \alpha w_i(D_i - C_i)^2$ and $h_i(C_i - D_i) = \beta w_i(C_i - D_i)^2$, for $i = 1, 2, \ldots, n$, where α and β represent unit costs for earliness and tardiness penalties respectively. Moreover, we assume that $D_1, \ldots, D_n$ are exponentially distributed with a common mean $1/\delta$. Again, Cai and Zhou (2000) consider the case of $r = 0$ only, but it is not difficult to incorporate a positive r into the model.

An optimal solution ζ^* is to be determined so as to minimize

$$TC(\zeta) = E\left[\alpha \sum_{C_i < D_i} w_i(D_i - C_i)^2 + \beta \sum_{C_i > D_i} w_i(C_i - D_i)^2\right]$$
$$= \sum_{i=1}^{n} w_i\left\{E\left[\alpha(D_i - C_i)^2 I_{\{C_i < D_i\}}\right] + \beta E\left[(C_i - D_i)^2 I_{\{C_i > D_i\}}\right]\right\} \quad (39.35)$$

where I_A is the indicator of an event A which takes value 1 if A occurs and 0 otherwise.

Denote the problem above as Model 3. Similar to the arguments in Cai and Zhou (2000) we can show the V-shape of the optimal sequence.

Theorem 39.8

For each $r \in [0, d]$, *an optimal sequence* λ^* *which minimizes* $TC(\zeta)$ *in (39.35) is V-shaped with respect to* $\{\mu_i/w_i\}$.

A dynamic programming algorithm similar to Algorithm 39.2 can also be developed, based on Theorem 39.8, to derive the optimal solution for Model 3. For details, see Cai and Zhou (2000).

39.3.4 Additional References and Comments

Chakravarthy (1986) examines the completion time variance (CTV) problem with random processing times, which can be shown as a special case of Model 1 we discussed in Section 39.2 (Remark 39.1). He indicates that an optimal sequence is V-shaped with respect to the mean processing times, if all processing times have an equal mean, or have an equal variance, or follow exponential distributions.

Cai (1996) further derives a sufficient condition for an optimal sequence for the CTV problem to be V-shaped with respect to the mean processing times. Zhou and Cai (1996) reveal an equivalent relation between the CTV problem and the WTV (waiting time variance) problem when processing times are random. Cai and Zhou (1997a) study the stochastic CTV problem, where processing times are random variables and the weights are job dependent. They show that the problem is NP-complete, establish a W-shaped property for optimal sequences, and develop a pseudopolynomial time algorithm. Their work also generalizes what is studied in Vani and Raghavachari (1987), where a CTV problem without job-dependent weights is considered, and some optimality conditions are derived.

Luh, Chen, and Thakur (1999) consider a stochastic scheduling problem with uncertain arrival times, processing times, due dates, and weights (priorities). The machines are configured as a job shop, and assumed to be deterministically and continuously available. The problem is to find the schedule to process all the operations of the jobs so as to minimize the expected total cost: $E[\sum_{i=1}^{n} \alpha_i(C_i - D_i)^2 + \beta_i(C_i - D_i)^2]$. A solution approach based on combined Lagrangian relaxation and stochastic dynamic programming is developed.

Qi, Yin, and Birge (2000b) study a relevant but different stochastic scheduling problem with quadratic E/T functions, where processing times can be compressed, at a compression cost. Specifically, $P_i = \hat{p}_i - Q_i$, where p_i is the base processing time for job i (which is known and fixed), and Q_i is the amount of compression. They assume that Q_i, $i = 1, 2, \ldots, n$ are independently and identically distributed (i.i.d.) random variables. Stochastic machine breakdowns are considered. The objective is to minimize the sum of squared deviations of the job completion times from their due dates and the compression costs.

39.4 Linear E/T Costs

When the cost functions $g_i(x)$ and $h_i(x)$ of (39.5)–(39.6) are linear, (39.8) reduces to

$$TC(\zeta) = E\left[\sum_{C_i \leq D_i} \alpha_i(D_i - C_i) + \sum_{C_i > D_i} \beta_i(C_i - D_i)\right] \tag{39.36}$$

where $\alpha_i \geq 0$ and $\beta_i \geq 0$ are, respectively, the unit earliness and tardiness penalties.

While (39.36) looks simple because the cost functions involved are all linear, the stochastic E/T problem is in fact a very complicated one, much harder than the case with nonlinear, quadratic cost functions considered in the above section. The main reason is the nondifferentiability exhibited in (39.36). For example, when $\alpha_i = \beta_i = 1$, for all $i = 1, 2, \ldots, n$, (39.36) reduces to

$$TC(\zeta) = E\left[\sum_{i=1}^{n} |C_i - D_i|\right] \tag{39.37}$$

which involves a nondifferentiable absolute function.

Soroush and Fredendall (1994) have considered a problem with linear E/T functions, where processing times are random variables and due dates are deterministic constants. They obtain a deterministic equivalent, when the processing times are normally distributed. Nevertheless, the deterministic equivalent has become a complicated, highly nonlinear function, which lacks any additive or multiplicative form. Observing this inherent difficulty, they point out that it is unlikely to apply a dynamic programming kind of methodology to solve the problem, and turn to constructing heuristics to find approximate solutions.

In fact, due to the inherent complications with nondifferentiability, the literature on stochastic E/T scheduling with linear cost functions is scarce, although in the deterministic setting, this is the class of E/T problems that have been studied most extensively.

In this Section, we will mainly introduce the results obtained by Cai and Zhou (1997b). The problem is to determine an optimal sequence λ^* and an optimal machine starting time r^* so as to minimize the expectation of a weighted combination of the earliness penalty, the tardiness penalty, and the flow time penalty, namely,

$$\min_{\lambda, r} TC(\lambda, r) = E\left[\sum_{C_i \le D_i} \alpha |D_i - C_i| + \sum_{C_i > D_i} \beta |C_i - D_i| + \sum_{i=1}^{n} \gamma C_i\right] \tag{39.38}$$

where $\alpha \ge 0$ is the unit earliness penalty, $\beta \ge 0$ is the unit tardiness penalty, and $\gamma \ge 0$ is the unit flow time penalty. Machine breakdowns are not considered, although the results described below may be extended to a breakdown model as given in Section 39.2.

The processing times P_i, $i = 1, 2, \ldots, n$, are assumed to follow normal distributions with known means μ_i and known variances σ_i^2. Moreover, it is assumed that the variances are proportional to the means, namely, $\sigma_i^2 = a\mu_i$, $i \in N$. This relationship holds when each job consists of a large number of independent elementary tasks and all the elementary tasks are independent and identically distributed. It also approximately holds under some mild conditions, even if the tasks follow different distributions (see Cai and Zhou, 1997b).

As usual, we assume that the parameters μ_i are positive integers. Since the processing times are nonnegative, a should be restricted such that the probability of a processing time being negative is negligible. Technically, we assume that $a \le \mu_i/4$ for all i, which ensures that $\Pr(P_i < 0) = \Phi(-\mu_i/\sigma_i) = \Phi(-\sqrt{\mu_i/a}) < 0.025$ for all i.

The due dates D_i, $i = 1, 2, \ldots, n$, are nonnegative random variables, independent of $\{P_i\}$, with a common distribution. Let D be a representative of D_i's.

To generalize the constraint $r \le d_0$ for a deterministic common due date d_0 to the stochastic due dates, here we restrict r to an interval $[0, \bar{r}]$, where $\bar{r}$ is an upper bound on r, which is to be determined so that the probability that the due date D occurs before the starting time r would not be greater than that after r. More precisely, we assume

$$\Pr(r - x \le D < r) \le \Pr(r \le D \le r + x) \quad \forall x \ge 0 \text{ and } r \in [0, \bar{r}] \tag{39.39}$$

If D has a symmetric distribution, it is easy to see that (39.39) is satisfied with exact equality and $\bar{r} = E(D)$. In particular, if D is a deterministic due date d_0, this reduces to $\bar{r} = d_0$.

Denote the problem described above as Model 4.

39.4.1 Deterministic Equivalent

Define the following functions on $(0, \infty)$:

$$F(x) = (\alpha + \beta)\phi(x) + x[(\alpha + \beta)\Phi(x) - \alpha] \tag{39.40}$$

$$f(x) = x \int_0^\infty F\left(\frac{x}{a} - \frac{t - r}{x}\right) d\Pr(D \le t) + \gamma\left(r + \frac{x^2}{a}\right) \tag{39.41}$$

and

$$g(x) = f(\sqrt{ax}) \tag{39.42}$$

where

$$\phi(x) = \frac{1}{\sqrt{2\pi}} e^{-\frac{1}{2}x^2} \quad \text{and} \quad \Phi(x) = \frac{1}{\sqrt{2\pi}} \int_{-\infty}^{x} e^{-\frac{1}{2}y^2} dy$$

are the probability density and the cumulative distribution function respectively of the standard normal distribution.

Theorem 39.9 (Cai and Zhou, 1997b)
Let $\theta_i = \sum_{j \in B_i(\lambda)} \mu_j$. Then the objective function (39.3) is equivalent to

$$TC(\lambda, r) = \sum_{i=1}^{n} \left\{ \sqrt{a\theta_i} \int_0^\infty F\left(\frac{\theta_i + r - t}{\sqrt{a\theta_i}}\right) d\Pr(D \le t) + \gamma(r + \theta_i) \right\} \tag{39.43}$$

or

$$TC(\lambda, r) = \sum_{i=1}^{n} f(\sqrt{a\theta_i}) = \sum_{i=1}^{n} g(\theta_i) \tag{39.44}$$

39.4.2 Optimality Properties

A function is said to be V-shaped on an interval (a, b) if there exists a $\delta \in [a, b]$ such that the function is decreasing on (a, δ) and increasing on (δ, b). By showing the function $f(x)$ of (39.41) is V-shaped on the interval $(\sqrt{a\mu_1}, \infty)$, Cai and Zhou (1997b) obtain the following result.

Theorem 39.10
Given any starting time r, an optimal sequence λ^* which minimizes $EET(\lambda, r)$ must be V-shaped with respect to $\{\mu_i\}$.

Define

$$H(\lambda, r) = \sum_{i=1}^{n} \int_0^\infty \Phi\left(\frac{\theta_i + r - t}{\sqrt{a\theta_i}}\right) d\Pr(D \le t) \tag{39.45}$$

Cai and Zhou (1997b) also obtained the following result regarding the possible range of r.

Theorem 39.11

Let λ_S and λ_L denote, respectively, the sequence in order of shortest expected processing time first and the sequence in order of longest expected processing time first sequence, and r_S, r_L satisfy

$$H(\lambda_S, r_S) = \frac{n(\alpha - \gamma)}{\alpha + \beta} \qquad H(\lambda_L, r_L) = \frac{n(\alpha - \gamma)}{\alpha + \beta} \tag{39.46}$$

If (λ^*, r^*) is the optimal solution, then

$$\max\{r_L, 0\} \leq r^* \leq \min\{r_S, \bar{r}\} \tag{39.47}$$

In particular, if $r_S \leq 0$, or equivalently if

$$H(\lambda_S, 0) \geq \frac{n(\alpha - \gamma)}{\alpha + \beta} \tag{39.48}$$

then $r^* = 0$.

The following properties show that the SEPT sequence λ_S is optimal under certain conditions.

Proposition 39.5

Given any $r \in [0, \bar{r}]$, if

$$\int_0^\infty \Phi\left(\frac{\mu_1 + r - t}{\sqrt{a\mu_1}}\right) d\Pr(D \leq t) \geq \frac{\alpha - \gamma}{\alpha + \beta} \tag{39.49}$$

then λ_S is optimal.

Proposition 39.6

If $\alpha \leq \gamma$, then λ_S is optimal and the corresponding $r_S = 0$.

Proposition 39.7

If the due dates equal a deterministic common value d_0, then λ_S is optimal for

$$d_0 \leq \mu_1 + \sqrt{a\mu_1}\Phi^{-1}\left(\frac{\beta + \gamma}{\alpha + \beta}\right) \tag{39.50}$$

Proposition 39.8

When $\alpha \leq \beta + 2\gamma$, λ_S is optimal if the due dates equal a deterministic common value $d_0 \leq \mu_1$.

39.4.3 Algorithms

Without loss of generality, we assume that

$$\mu_1 \leq \mu_2 \leq \cdots \leq \mu_n \tag{39.51}$$

Let $\mathcal{N}_i = \{1, 2, \ldots, i\}$. Then, according to Theorem 39.10, job i should be sequenced either the first or the last among all jobs in $\mathcal{N}_i$. Let $h_i(\theta, r)$ be the contribution of the jobs in $\mathcal{N}_i$ toward the overall objective function (39.43), given that θ is the sum of mean processing times of the jobs sequenced before the jobs in $\mathcal{N}_i$. The following dynamic programming algorithm can find an optimal solution for Model 4, where $\Theta_i = \sum_{j \in \mathcal{N}_i} \mu_j$.

Algorithm 39.3 (Cai and Zhou, 1997b)

1. *Evaluate r_L and r_S according to (39.46)–(39.48). Let $r_{\min} = \max\{r_L, 0\}$, $r_{\max} = \min\{r_S, \bar{r}\}$, and $\mathcal{R}$ be the set of integers contained in $[r_{\min}, r_{\max}]$.*
2. *For each $r \in \mathcal{R}$ and $i = 1, 2, \ldots, n$, compute:*

$$h_i(\theta, r) = \min\{g(\theta + \mu_i) + h_{i-1}(\theta + \mu_i, r);\ g(\theta + \Theta_i) + h_{i-1}(\theta, r)\} \tag{39.52}$$

for $\theta = 0, 1, \ldots, \Theta_n - \Theta_i$, subject to the boundary condition:

$$h_0(\theta, r) = 0, \quad \forall \theta \tag{39.53}$$

3. *Optimal r^* is the one that satisfies $h_n(0, r^*) \le h_n(0, r)$, $\forall r \in \mathcal{R}$, and optimal λ^* is the sequence constructed by a backtracking procedure that achieves $h_n(0, r^*)$.*

The following theorem gives the time complexity of Algorithm 39.3.

Theorem 39.12

Algorithm 39.3 can find the optimal solution $\zeta^* = (\lambda^*, r^*)$ for Model 4 in $O(n\Theta_n|\mathcal{R}|)$ time.

The time complexity of Algorithm 39.3 reduces to $O(n\Theta_n)$ when (39.48) is satisfied. In general, it is bounded above by $O(n\Theta_n\bar{r})$ since $|\mathcal{R}| \le \bar{r}$.

Given any r, we can find accordingly an optimal sequence λ that minimizes $TC(\lambda, r)$ of (39.43). Thus, we can let $G(r) = \min_\lambda TC(\lambda, r)$, and the problem becomes one of finding an optimal r^* to minimize $G(r)$. Cai and Zhou (1997b) suggest to use the Fibonnaci method to search for r^*. Clearly the algorithm with such a Fibonnaci search would be an optimal one if $G(r)$ is a unimodal function of r. Otherwise, the solution $\zeta = (\lambda, r)$ thus obtained would be approximate. Cai and Zhou carried out computational experiments to evaluate the approximate algorithm and found that in many cases the approximate algorithm produced good near-optimal solutions.

A Fibonnaci search over the range $\mathcal{R}$ needs $O(\log|\mathcal{R}|)$ time. Consequently, the time complexity of the approximate algorithm reduces to $O(n\Theta_n \log|\mathcal{R}|)$.

39.4.4 Extensions

The algorithms discussed in Section 39.4.3 are based on the V-shape property of Theorem 39.10, which depends on the assumption that the processing times are normally distributed with variances proportional to means. Cai and Zhou (1997b) suggest to extend the algorithms to problems without such assumptions, by the following approach.

Consider a problem with processing times that follow certain probability distributions $J_i(x)$ with means ν_i and variances ς_i^2, $i = 1, 2, \ldots, n$. It is proposed that the distribution $J_i(x)$ be approximated by a normal distribution with mean $\mu_i = \nu_i$ and variance $\sigma_i^2 = a\mu_i = a\nu_i$, where a is a parameter to be determined so that σ_i^2, $i = 1, 2, \ldots, n$, are the best approximations to the original variances ς_i^2, under the following criterion:

$$\min_a \sum_{i=1}^{n} \left[a\nu_i - \varsigma_i^2\right]^2 \tag{39.54}$$

Solving (39.54), we obtain

$$a = \frac{\sum_{i=1}^{n} v_i^2}{\sum_{i=1}^{n} v_i \varsigma_i^2} \tag{39.55}$$

Then, using a as given by (39.55) and $\mu_i = v_i$ for $i = 1, 2, \ldots, n$, we may apply Algorithm 39.3 or the approximate algorithm using Fibonnaci search to find an approximate solution to the problem with the distributions $J_i(x)$.

Cai and Zhou conducted several computational experiments, and found that the algorithms generated very good solutions for a number of processing time distributions, including *General normal* distributions without the assumption of variances being proportional to means, *Uniform* distributions, *Laplace* distributions, and *Exponential* distributions.

39.4.5 Additional References and Comments

Generally speaking, stochastic scheduling with linear E/T cost functions is difficult. What we have presented above focus on a class of normal distributions for processing times. Hussain and Sastry (1999) consider a similar situation where processing times follow normal distributions with given means and variances. They, however, apply a genetic algorithm in order to tackle the problem. Other relevant results are introduced below.

Jia (2001) has studied a problem with symmetric E/T weights, where the objective is to minimize $TC(\lambda) = E[\sum_{i=1}^{n} w_i |C_i - D|]$. Under the assumptions that both the processing times P_i, $i = 1, 2, \ldots, n$, and the common due date D are exponential, he shows that an optimal sequence should be V-shaped with respect to $\{E(P_i)/w_i\}$. He considers machine breakdowns, under both preempt-resume and preempt-repeat patterns (which are equivalent in his settings due to the memoryless property of the exponential processing times). He also assumes that the machine always starts to process the first job at time zero (i.e., $r = 0$).

Soroush (1999) studies a problem of simultaneous due-date determination and sequencing of a set of n jobs with random processing times. Each job i has a deterministic and distinct due date d_i. Consequently, given a job sequence, it is possible that the due date is matched so as to minimize the E/T costs. In some sense, this problem is relatively easy. To see this, consider the case where all processing times are deterministic and the objective is to minimize the total E/T cost. One can see that an optimal solution is simply to assign the due date of a job equal to its completion time (this will make the E/T costs exactly equal to zero). The problem becomes meaningful when the processing times are random, or the objective also involves other cost such as weighted flowtime, or there are other operational constraints. Cheng (1986, 1991) studies the problem with random processing times where the selection of due dates is restricted by some prespecified rules. Cheng (1986) examines the TWK (total work content) method, where due dates are assumed to be proportional to the processing times. Cheng (1991) considers the SLK (slack) method, which assumes that the due date of a job is equal to the processing time of the job plus a constant waiting allowance.

Al-Turki, Mittenthal, and Raghavachari (1996) consider a single-machine problem with random job completion times, which arise from either machine breakdowns or other random fluctuations. The objective is to minimize an expected weighted combination of due dates, completion times, earliness, and tardiness costs. For a given schedule, they consider the determination of optimal distinct due dates or optimal common due date. The scheduling problem for a common due date is also considered when random completion times are caused by machine breakdowns.

Qi, Yin, and Birge (2000a) study a problem with linear E/T functions, which is similar to (39.38) except that the last term is a cost incurred by the due date instead of the total flow time. Random processing times are considered. Sufficient conditions to ensure an optimal SEPT sequence are developed. Some further results are also obtained when processing times follow exponential and normal distributions. Qi, Yin, and Birge (2002) study another problem with compressible processing times. They assume that the amounts of compressions for the n jobs are i.i.d. random variables. An objective consisting of linear earliness and tardiness costs and compression costs are examined. Some results on optimal V-shaped sequence are obtained.

39.5 E/T Scheduling Under an Uncertain Delivery Time

Here we consider the problem where the finished jobs have to be delivered by a transporter, but the arrival time of the transporter is delayed due to certain unexpected factors such as poor weather conditions, operational delays, or industrial action. Such a problem often occurs when a public transportation service (such as a cargo flight) is used. It is cheap, but the manufacturing company has no control over the schedule of the transporter, and therefore will have to plan carefully to minimize the possible loss when it is known that the transporter is delayed but the actual departure time becomes uncertain. Cai and Zhou (1999) provided two examples of this nature. The model considered and the results introduced in this section are organized based on Cai and Zhou (1999).

Let D denote the departure time of the transporter, which is uncertain and thus is assumed to be a random variable following an exponential distribution with parameter δ. If job i is completed before D, there is a waiting (inventory) cost $g_i(D - C_i)$. If it is completed after D, it will incur a tardiness cost w_i. The tardiness cost represents, for example, the extra cost to arrange an alternative transporter to deliver the job to its destination if it misses the original transporter.

The machine is subject to stochastic breakdowns, and the breakdown pattern is as modeled in Section 39.2. It is assumed that $N(t)$ is a Poisson process with rate κ.

Idle time before the machine starts processing each job is considered. Let r_i denote the amount of time that the machine is kept idle immediately before job i is started, and let $\mathbf{r} = \{r_1, r_2, \ldots, r_n\}$.

The definition (39.4) should now be rewritten as:

$$R_i(\zeta) = \sum_{k \in \mathcal{B}_i(\lambda)} (r_k + P_k) \tag{39.56}$$

The completion time $C_i = C_i(\zeta)$ can still be expressed by (39.3), with $Z_0 \equiv 0$.

The problem is to determine an optimal policy $\zeta^* = (\lambda^*, \mathbf{r}^*)$, from the class of (static or dynamic) policies that can be adopted, to minimize the expected total cost:

$$TC(\zeta) = E\left[\sum_{C_i \le D} g_i(D - C_i) + \sum_{C_i > D} w_i\right] \tag{39.57}$$

Denote the problem above as Model 5.

39.5.1 Deterministic Equivalent

When the policy to be adopted is static, Cai and Zhou (1999) obtain the following result. Let:

$$\alpha_i = \int_0^{+\infty} g_i(y)\delta e^{-\delta y} dy \tag{39.58}$$

$$f_k = E[e^{-\eta P_k}] \tag{39.59}$$

$$\eta = \delta + \kappa \Pr(D \le Z) \tag{39.60}$$

where Z is a random variable whose distribution is the same as the common distribution of the downtimes Z_k.

Theorem 39.13

When ζ is a static policy, the expected total cost is equivalent to

$$TC(\zeta) = \sum_{i=1}^{n} (\alpha_i - w_i) \prod_{k \in \mathcal{B}_i(\lambda)} e^{-\eta r_k} f_k + \sum_{i=1}^{n} w_i \tag{39.61}$$

It can be shown that (39.56) *is also equivalent to*

$$TC(\zeta) = \sum_{i=1}^{n} \{\alpha_i \Pr(C_i < D) + w_i \Pr(C_i > D)\} \tag{39.62}$$

Remark 39.2

The deterministic counterpart of Model 5 is NP-hard even in the special case with $g_i(\cdot) \equiv 0$ for all i (which is equivalent to the Knapsack problem).

39.5.2 Optimal Static Policy

Let $\mathcal{I}$ and $\mathcal{I}^c$ be sets of jobs defined such that $\alpha_i < w_i$ if $i \in \mathcal{I}$ and $\alpha_i \geq w_i$ if $i \in \mathcal{I}^c$.

Theorem 39.14 (Cai and Zhou, 1999)

An optimal static policy for Model 5 processes the jobs according to the following rules:

(a) *Jobs in $\mathcal{I}$ are sequenced in nonincreasing order of $(w_i - \alpha_i) f_i/(1 - f_i)$ and processed from time zero with no idle time inserted between any two consecutive jobs;*
(b) *All jobs in $\mathcal{I}^c$ are processed in an arbitrary order, starting as soon as D has occurred or the last job in $\mathcal{I}$ has been completed, whichever comes later.*

Remark 39.3

1. Under the policy of Theorem 39.14, there may be an idle time between the last completed job in $\mathcal{I}$ and the first started job in $\mathcal{I}^c$, which is a random variable as it depends upon D.
2. Note that Model 5 has an implicit assumption that all jobs must be processed. If the scheduling problem does not require that all jobs must be completed, then Rule (b) in Theorem 39.14 can be replaced by
(b') Jobs in $\mathcal{I}^c$ are not processed at all.

39.5.3 Optimal Dynamic Policy

We now address the situations where a policy can be revised dynamically. The problem is to determine, at any time t, a job for processing, by taking into account all information available up to t.

Consider first the case where preemptions are not allowed, namely, a new decision can be applied only when a job is completed. The following theorem can be proven by using mathematical induction (Cai and Zhou, 1999).

Theorem 39.15

When preemptions are forbidden, the optimal static policy as specified by Theorem 39.14 is optimal in the class of dynamic policies for Model 5.

We now turn to the case where preemptions are allowed. In this case, the processing of a job may be interrupted, if necessary, before its completion so that another job is started.

When ζ is a preemptive dynamic policy, we can show that

$$TC(\zeta) = \sum_{i=1}^{n} (\alpha_i - w_i) E[e^{-\eta R_i(\zeta)}] + \sum_{i=1}^{n} w_i \tag{39.63}$$

where α_i and η are given by (39.58) and (39.60), respectively, and $R_i(\zeta)$ is defined by (39.56).

By (39.27) one can see that, to minimize $ETC(\zeta)$, $R_i(\zeta)$ for $i \in \mathcal{I}$ (i.e., $\alpha_i < w_i$) should be made as small as possible whereas $R_i(\zeta)$ for $i \in \mathcal{I}^c$ should be made as large as possible. This implies that the idle time before any job in the set $\mathcal{I}$ must be zero; namely, $r_i = 0$ for $i \in \mathcal{I}$. On the other hand, all jobs in the set $\mathcal{I}^c$ should be delayed as late as possible.

Consequently, with respect to the jobs in $\mathcal{I}$, the problem is to find an optimal policy ζ^* that minimizes

$$TC_{\mathcal{I}}(\zeta) = E\left(\sum_{i\in\mathcal{I}}(\alpha_i - w_i)e^{-\eta R_i(\zeta)}\right) \tag{39.64}$$

in the class of preemptive dynamic policies, where no idle times should be inserted before any jobs in $\mathcal{I}$. This problem can be optimally solved by Gittins theory (Gittins, 1989). Define:

$$G_i(T_i(t)) = \sup_{\kappa > T_i(t)} \frac{(w_i - \alpha_i)\int_{(T_i(t),\kappa]} e^{-\eta s}\, dQ_i(s)}{\int_{(T_i(t),\kappa]}(1 - Q_i(s))e^{-\eta s}\, ds}, \quad i \in \mathcal{I} \tag{39.65}$$

where $Q_i(s)$ is the cumulative distribution function of processing time P_i and $T_i(t)$ denotes the realization of the processing time on job i up to the moment t.

In the discrete case where P_i takes integer values, the Gittins Index reduces to:

$$G_i(T_i(t)) = \max_{\kappa > T_i(t)} \frac{(w_i - \alpha_i)\sum_{s=T_i(t)+1}^{\kappa} e^{-\eta s}\Pr(P_i = s)}{\sum_{s=T_i(t)+1}^{\kappa} e^{-\eta s}\Pr(P_i \geq s)}, \quad i \in \mathcal{I} \tag{39.66}$$

To minimize $TC_{\mathcal{I}}(\zeta)$ of (39.64), a dynamic policy ζ^* should, at any time t, process the job i^* that has the maximum Gittins Index among all the unfinished jobs at time t. This leads to the following result.

Theorem 39.16

When preemption is allowed, an optimal dynamic policy for Model 5 will process the jobs according to the following rules:

(a) *At any time $t < D$, choose the job i^* with*

$$G_{i^*}(T_{i^*}(t)) = \max_{i\in\mathcal{I}_u(t)} G_i(T_i(t)) \tag{39.67}$$

to be the present job to process, where $\mathcal{I}_u(t)$ denotes the set of jobs which belong to $\mathcal{I}$ and have not been completed at time t.

(b) *At any time $t \geq D$, process all unfinished jobs in an arbitrary order with no preemptions.*

Some results can be further obtained with the concept of *hazard rate* of P_i, define as $\rho_i(s) = q_i(s)/(1 - Q_i(s))$, where $q_i(s)$ represents the density function of P_i if P_i is continuous, or the probability mass function if P_i is discrete. One can show that, if $\rho_i(s)$ is a nondecreasing function in s within the support of $q_i(s)$, then $G_i(x)$ is a nondecreasing function in x. In this case, it follows from Theorem 39.16 that job i will never be preempted if it has been selected to start, although preemption is allowed. Consequently, the problem reduces to one with no preemptions as addressed by Theorem 39.15. Thus, we have the following:

Corollary 39.1

The static policy of Theorem 39.14 remains optimal in the class of dynamic policies even though preemptions are allowed, if $\rho_i(s)$, for all $i \in \mathcal{I}$, are nondecreasing functions in s.

Note that the condition of Corollary 39.1 is satisfied when P_i are deterministic constants p_i. This leads to the following result.

Corollary 39.2

If processing times are deterministic, the nonpreemptive policy of Theorem 39.14, where $f_i = e^{-\eta p_i}$ *for all* i*, is optimal in the class of dynamic policies, even though preemptions are allowed.*

39.5.4 Additional References and Comments

The multimachine case of Model 5 has been studied in Cai and Zhou (1999), where the weights α_i and w_i need to be job independent.

A class of open-loop dynamic policies are studied in Cai and Zhou (1998, 1999), which minimize, at any time $t \geq 0$, the expected cost starting from the current time t by taking into account the new measurements (realizations of random variables such as processing times) received up to t, under the assumption that a policy adopted now would not be changed later. The computational experiments in Cai and Zhou (1998) show that, at time $t > 0$ when new information is known, if the optimal static policy for the remaining jobs is computed again, then r can be nonzero. This means that inserted idle time before each job can be more desirable.

Federgruen and Mosheiov (1997) consider single machine scheduling problems with a common due-date, arbitrary monotone earliness and tardiness costs and arbitrary breakdown and repair processes. The processing times are deterministic. They show a V-shaped (static) schedule without idle times is optimal, if the deterministic equivalent cost function is quasi-convex. They also derive conditions under which the equivalent cost function is quasi-convex. A pseudopolynomial algorithm is developed to compute an optimal V-shaped schedule. A main assumption of Federgruen and Mosheiov (1997) is that the weights for all jobs must be equal. This is essential to ensure that a V-shaped optimal sequence is possible.

Cai and Zhou (2003) consider a stochastic harvest scheduling where there is a random rainfall to arrive that may create a severe damage to the crops to be harvested, whereas the crop products that have been gathered should be delivered to one of two possible markets. Optimal solutions in the classes of static and dynamic policies are derived.

39.6 Concluding Remarks

Among the many interesting results that have been obtained on the characteristics of optimal solutions for stochastic E/T scheduling, it appears that the most important one is a type of V-shape property of optimal sequences. The V-shape property has enabled efficient algorithms to be obtained based on dynamic programming. There are, however, two main problems related to the V-shape property.

1. The existence of a V-shaped optimal schedule often requires that the earliness weights α_i and the tardiness weights β_i be common for all jobs (or meet certain compatibility conditions). Such a requirement may not make sense in many practical situations, as the costs for different jobs are bound to be different. How could we overcome this difficulty?
2. The V-shaped structure seems to hinder the development of optimal dynamic policy. So far the only result available on the optimal policy in the class of dynamic policies is Cai and Zhou (1999), for Model 5 above. Because of some coincidental combination of the problem components, the derivation of an optimal dynamic policy in Cai and Zhou (1999) is actually equivalent to one based on a regular cost measure (see (39.64) above). Could we find optimal dynamic policies for other stochastic E/T problems?

The above two questions represent two interesting lines of further research.

We should emphasize that an important decision that should not be overlooked under an E/T criterion is the *idle time* inserted before processing of each job. The essence of E/T scheduling is, in fact, the optimal delay of certain job completions so as to reduce the earliness cost. As we have reviewed above, there have been some studies considering the idle time before the machine processes its first job. In the deterministic case with a common due date for all jobs, it is easy to show that an optimal solution needs only this idle time, because idle times inserted between any two consecutive jobs should be zero. For stochastic E/T problems, a nonzero idle time may be needed, however, as the previous job might be completed earlier than expected. Another interesting line of further research is to investigate the modelling and determination of such optimal idle times.

Acknowledgments

This research was partially supported by the Research Grants Council of Hong Kong under Earmarked Grants CUHK 4418/99E, CUHK 4166/01E, and Poly U1542/01E.

References

[1] Al-Turck, U.M., J. Mittenthal, and M. Raghavachari (1996). The single-machine absolute-deviation early-tardy problem with random completion times. *Naval Research Logistics*. **43**, 573–587.

[2] Baker, K.R. and G.D. Scudder (1990). Sequencing with earliness and tardiness penalties: a review. *Operations Research*. **38**, 22–36.

[3] Birge, J., J.B.G. Frenk, J. Mittenthal, and A.H.G. Rinnooy Kan (1990). Single-machine scheduling subject to stochastic breakdowns. *Naval Research Logistics*. **37**, 661–677.

[4] Cai, X. (1996). V-Shape property for job sequences that minimize the expected completion time variance. *European Journal of Operational Research*. **91**, 118–123.

[5] Cai, X. and F.S. Tu (1996). Scheduling jobs with random processing times on a single machine subject to stochastic breakdowns to minimize early-tardy penalties. *Naval Research Logistics*. **43**, 1127–1146.

[6] Cai, X. and X. Zhou (1997a). Sequencing jobs with random processing times to minimize weighted completion time variance. *Annals of Operations Research*, Special Issue on Scheduling Theory and Applications. **70**, 241–260.

[7] Cai, X. and X. Zhou (1997b). Scheduling stochastic jobs with asymmetric earliness and tardiness penalties. *Naval Research Logistics*. **44**, 531–557.

[8] Cai, X. and X. Zhou (1999). Stochastic scheduling on parallel machines subject to random breakdowns to minimize expected costs for earliness and tardy jobs. *Operations Research*. **47**, 422–437.

[9] Cai, X. and X. Zhou (2000). Asymmetric earliness and tardiness scheduling with exponential processing times on an unreliable machine. *Annals of Operations Research*. **98**, 313–331.

[10] Cai, X., X.Q. Sun, and X. Zhou (2003). Stochastic scheduling with preemptive-repeat machine breakdowns to minimize the expected weighted flowtime. *Probability in the Engineering and Informational Sciences* **17**, 467–485.

[11] Cai, X. and X. Zhou (2003). Stochastic harvest scheduling of fresh products to supply two potential markets. (Submitted)

[12] Chakravarthy, S. (1986). A single machine scheduling problem with random processing times. *Naval Research Logistics Quarterly*. **33**, 391–397.

[13] Chen, B., C.N. Potts, and G.J. Woeginger (1998). A review of machine scheduling: complexity, algorithms, and approximability. In: D.-Z. Du and P. Pardalos, (eds.). *Handbook of Combinatorial Optimization* Kluwer, Dordrecht.

[14] Cheng, T.C.E. (1986). Optimal due-date assignment for a single machine sequencing problem with random processing times. *International Journal of Systems Science*. **17**, 1139–1144.

[15] Cheng, T.C.E. (1991). Optimal assignment of slack due dates and sequencing of jobs with random processing times on a single-machine. *European Journal of Operational Research*. **51**, 348–353.

[16] Eilon, S. and I.G. Chowdhury (1977). Minimizing waiting time variance in the single machine problem. *Management Science*. **23**, 567–575.

[17] Federgruen A. and G. Mosheiov (1997). Single machine scheduling problems with general breakdowns, earliness and tardiness costs. *Operations Research*. **45**, 66–71.

[18] Gittins, J.C. (1989). *Multi-armed Bandit Allocation Indices*. John Wiley & Sons, Chichester.

[19] Gordon, V.S., J.M. Proth, and C.B. Chu (2002). Due date assignment and scheduling: SLK, TWK and other due date assignment models. *Production Planning & Control.* **13**, 117–132.
[20] Hussain, S.A. and V.U.K. Sastry (1999). Application of genetic algorithm to stochastic single machine scheduling problem with earliness and tardiness costs. *International Journal of Computer Mathematics.* **70**, 383–391.
[21] Jia, C.F. (2001). Stochastic single machine scheduling with an exponentially distributed due date. *Operations Research Letters.* **28**, 199–203.
[22] Kanet, J.J. and V. Sridharan (2002). Scheduling with inserted idle time: Problem taxonomy and literature review. *Operations Research.* **48**, 99–110.
[23] Luh, P.B., D. Chen, and L.S. Thakur (1999). An effective approach for job-shop scheduling with uncertain processing requirements. *IEEE Transactions on Robotics and Automation.* **15**, 328–339.
[24] Mittenthal, J. and M. Raghavachari (1993). Stochastic single machine scheduling with quadratic early-tardy penalties. *Operations Research.* **41**, 786–796.
[25] Parzen, E. (1962). *Stochastic Processes*, Holden-Day, San Francisco.
[26] Pinedo, M. (1995). *Scheduling: Theory, Algorithms, and Systems.* Prentice Hall, Englewood Cliffs, NJ.
[27] Qi, X.D., G. Yin, J.R. Birge (2000a). Scheduling problems with random processing times under expected earliness/tardiness costs. *Stochastic Analysis and Applications.* **18**, 453–473.
[28] Qi, X.D., G. Yin, J.R. Birge (2000b). Single-machine scheduling with random machine breakdowns and randomly compressible processing times. *Stochastic Analysis and Applications.* **18**, 635–653.
[29] Qi, X.D., G. Yin, J.R. Birge (2002). Single machine scheduling with randomly compressible processing times. *Stochastic Analysis and Applications.* **20**, 591–613.
[30] Soroush, H.M. (1999). Sequencing and due-date determination in the stochastic single machine problem with earliness and tardiness costs. *European Journal of Operational Research.* **113**, 450–468.
[31] Soroush, H.M. and L.D. Fredendall (1994). The stochastic single machine scheduling problem with earliness and tardiness costs. *European Journal of Operational Research.* **77**, 287–302.
[32] Vani, V. and M. Raghavachari (1987). Deterministic and random single machine sequencing with variance minimization. *Operations Research.* **35**, 111–120.
[33] Zhou, S. and X. Cai (1996). Variance minimization — relationship between completion time variance and waiting time variance. *Journal of the Australian Mathematical Society, Series B.* **38**, 126–139.

40

Developments in Queueing Networks with Tractable Solutions

Xiuli Chao
North Carolina State University

40.1 Introduction

A queueing network is a system consisting of a finite number of *stations* that provide services to *jobs*. The processing stations in the network are typically referred to as *nodes*. Examples of queueing networks include computer systems, manufacturing systems, job shops, airport terminals, railway or highway systems, and telephone systems. In these settings, jobs (data, parts or subassemblies, customers, planes, vehicles, phone calls, etc.) arrive at the system, and require some form of service (operation executions, assembly processes, machining, airplane take-offs, bridge or toll booth passings, phone conversations, etc.). Queueing network models have been successfully applied in the performance evaluation and optimization of computer systems, communication systems, manufacturing systems, and logistic systems. Typical performance measures of practical interest are sojourn time, congestion level, blocking probability, and throughput; and system design and optimization issues include dynamic routing control of jobs (packets), trunk designs, resource allocation, load balancing, or throughput maximization.

A number of methods have been developed for analyzing queueing networks, each with its own limitations. For example, rapid advances in computers have made it possible to perform large-scale simulations.

1-58488-397-9/$0.00+$1.50

Still, in addition to the high costs because of the extensive use of computer time and memory, the results of large-scale simulations tend to be application-specific and are not always useful for detecting general trends in performance measures. Another method for evaluating a queueing system is approximation. However, a theoretical basis is often required to guarantee that an approximation is not far from the real solution. A theoretical analysis of a queueing model is therefore not only important for its own sake; it is also important to complement simulation results and approximations. Such a theoretical analysis involves determining the stationary distribution of the network states, e.g., the number of jobs at each node, from which various performance measures can be derived.

Clearly, a closed form solution for the stationary distribution, if obtainable, is the most preferred. Of the networks with tractable solutions, networks with product form stationary distributions are the ones most researchers have focused on and most applications are based on. Networks with product form solutions have many properties that facilitate their analysis. In this class of networks, in spite of the high level of interaction between the nodes, the joint distribution of all the nodes is the product of the marginal distributions of the individual nodes. Roughly speaking, it implies that the stationary distribution of the network can be obtained by multiplying the stationary distributions of the individual nodes assuming that each node is in isolation and subject to Poisson arrivals. Due to this property, the analysis of a queueing network reduces to the analysis of single node queues, simplifying the applications tremendously. Nevertheless, we shall see that this area is, from a theoretical as well as from a practical point of view, not as narrow as it appears. As a matter of fact, were it not because of Jackson's celebrated product form result and its extensions, applications of queueing networks would most likely not have been as widespread as they are today.

The study of queueing network starts with the celebrated papers of Jackson (1957) (1963). Other work in the 1960s includes Whittle (1968) and Gordon and Newell (1967). These models focus on networks with exponential processing times. Significant breakthrough of queueing network research appeared in mid 1970s with the work of Baskett, et al. (1975), and Kelly (1975) (1976), that extend the product form results to networks with arbitrary processing time distributions and multiple classes of jobs. During the 1980s and 1990s researchers extend the theory of queueing networks with batch movements and networks with instantaneous movements and signals, and the representative works in this area are Henderson and Taylor (1990), Gelenbe (1991), Chao and Pinedo (1993), and Chao and Miyazawa (2000). See also the recent books by Chao et al. (1999) and Serfozo (1999).

In this chapter we present an overview of the latest developments in queueing networks with tractable solutions. To present the result in a general format we shall start with an abstract framework. We first develop sufficient conditions for the queueing network to possess product form stationary distribution, and for the case with no instantaneous movements we also present the necessary and sufficient conditions for the network to possess product form solution. Numerous examples are given that are covered by the result of the chapter as special cases.

This chapter consists of nine sections. In the following two sections we present the definition of quasi-reversibility for both nodes without triggering and with triggering. In Sections 40.4 and 40.5 we introduce networks with quasi-reversible nodes, without and with triggering respectively. Section 40.6 presents a special class of queueing networks called networks with positive and negative signals as well as their solution. Section 40.7 addresses the following question: What is the necessary and sufficient condition for a network to possess a product form stationary distribution, and a complete answer is given to this question for the class of networks that involve simultaneous transitions of at most two nodes. Quasi-reversibility is revisited in Section 40.8 under the framework of Section 40.7, and several classes of networks are investigated for which quasi-reversibility is not only a sufficient, but also a necessary condition for product form. We conclude with a brief discussion in Section 40.9.

40.2 Quasi-Reversibility of Queues

Quasi-reversibility is an input-output property of queues. It implies that when the system is in stochastic equilibrium, the future arrival processes, the current state of the system, and the past departure processes are independent.

In conventional queueing models, a job arrives at a system to receive service, and leaves the system after its service is completed. The networks discussed in this chapter include, in addition to conventional jobs, other entities that carry along commands and induce actions at the nodes where they arrive. These entities are, in case they do not trigger instantaneous departures, still referred to as *jobs*. If, however, an arrival has a positive probability of triggering a departure, it is called a *signal*. Thus, the cascading effects of signals may generate throughout the network an arbitrary number of arrivals and departures simultaneously.

It is useful to make a distinction between different classes of jobs and signals. Jobs of different classes may have different characteristics with respect to their processing requirements, their routings through the network, etc. Signals of different classes may carry different messages and may have different effects on the system. When there is no need to make a distinction between them, jobs as well as signals are referred to as *entities*; they may be viewed as different classes of entities.

This section and the next section present a formal theory of quasi-reversibility for queues with and without signals. We first modify the definition of continuous time Markov chains by including a transition rate from each state back to itself, i.e., $q(x,x)$, which is a non-negative number and is *not* equal to $-\sum_y q(x,y)$ as defined conventionally. Such a transition is usually excluded in the theory of continous time Markov chains, but we find it convenient for queueing applications.

In some stochastic systems, such as Example 40.1 below, both the arrival and the service completion of a regular job result in a transition from $n+1$ to n. Hence, different events may result in the same transition from x to x'. For this reason, we introduce the following notation. For each pair of states (x,x'), we decompose the transition rate function $q(x,x')$ of the queue into three types of rates, namely,

$$q_u^{\mathrm{A}}(x,x'), \quad u \in T$$

$$q_v^{\mathrm{D}}(x,x'), \quad v \in T$$

$$q^{\mathrm{I}}(x,x')$$

where T is the set of the classes of arrivals and departures, which is countable. Even though in many queueing systems the classes of arrivals are different from the classes of departures, we use a single index set T because we can take T as the union of both arrival and departure classes. Thus the transition rate of the queue can be written as

$$q(x,x') = \sum_{u \in T} q_u^{\mathrm{A}}(x,x') + \sum_{v \in T} q_v^{\mathrm{D}}(x,x') + q^{\mathrm{I}}(x,x'), \quad x,x' \in \mathcal{S} \tag{40.1}$$

These thinned transition rate functions q_u^{A}, q_v^{D}, and q^{I} generate the embedded point processes corresponding to class u arrivals, class v departures and the internal transitions, respectively. The first two embedded point processes are often referred to as the *arrival process of class u entities* and the *departure process of class v entities*. The superscripts A, D, and I stand for arrival, departure, and internal. The internal transition typically represents a change of status of the jobs such as a decrease of their remaining processing times, or, in the case the node contains multiple processing stations, the movements of jobs among the different stations of the node.

If the supports of the rate functions in (40.1) are disjoint, the decomposition above would only have one term. However, we do not make any restriction with regard to their supports. They are distinguished only by the probabilities, i.e., rate decomposition. It should be noted that, even though $q(x,x')$ is said to be decomposed into three types of components (arrival, departure, and internal transition rates) that result in the same transition from x to x', it is the opposite in applications. One is usually given the arrival, departure and internal rates that result in the same transition, and they have to be added up in order to obtain $q(x,x')$.

Example 40.1

Consider an $M/M/1$ queue with two classes of arrivals. Let $T = \{c, c^-\}$ with c and c^- representing regular and negative customers. If a processing completion is classified as class c, then it follows from the assumption of exponential processing times with mean $1/\mu$ that

$$q_c^{\mathrm{D}}(n, n-1) = \mu, \quad n = 1, 2, \ldots.$$

Customers arrive according to a Poisson process with rate α, so

$$q_c^{\mathrm{A}}(n, n+1) = \alpha, \quad n = 0, 1, \ldots.$$

Suppose negative customers arrive according to a Poisson process with rate α^- and reduce the number of customers by 1, we have

$$q_{c^-}^{\mathrm{A}}(n, n-1) = \alpha^-, \quad n = 1, 2, \ldots.$$

Finally, a negative customer that arrives at an empty node simply disappears, thus

$$q_{c^-}^{\mathrm{A}}(0, 0) = \alpha^-$$

Let $q(n, n')$ denote the transition rate of the queue, then its nonzero transition rates are

$$\begin{aligned} q(n, n+1) &= q_c^{\mathrm{A}}(n, n+1), & n \geq 0 \\ q(n, n-1) &= q_c^{\mathrm{D}}(n, n-1) + q_{c^-}^{\mathrm{A}}(n, n-1), & n \geq 1 \\ q(0, 0) &= q_{c^-}^{\mathrm{A}}(0, 0) \end{aligned}$$

Definition 40.1

The continuous time Markov chain with transition rate q is called quasi-reversible with respect to $\{q_u^{\mathrm{A}}(x, x'); u \in T\}$, $\{q_u^{\mathrm{D}}(x, x'); u \in T\}$ and $q^{\mathrm{I}}(x, x')$ if there exist two sets of nonnegative numbers $\{\alpha_u; u \in T\}$ and $\{\beta_u; u \in T\}$ such that

$$\sum_{x' \in \mathcal{S}} q_u^{\mathrm{A}}(x, x') = \alpha_u, \quad x \in \mathcal{S}, u \in T \tag{40.2}$$

$$\sum_{x' \in \mathcal{S}} \pi(x') q_u^{\mathrm{D}}(x', x) = \beta_u \pi(x), \quad x \in \mathcal{S}, u \in T \tag{40.3}$$

where π is the stationary distribution of the Markov chain q.

The nonnegative numbers α_u and β_u are often called the arrival rate and departure rate of class u entities.

Quasi-reversibility is a property concerning the arrival and departure processes. In fact, it is often useful to study this property with regard to only a portion of the arrival and departure processes. This is particularly true in networks of queues where only some of the departures from a node join another node, while the rest are either absorbed at that node or exit the network. These cases, however, are included in the definition above since one can classify the nonrouted arrivals (or departures) as internal transitions.

An alternative definition for quasi-reversibility is the following.

Definition 40.2

A stationary continuous time Markov chain $\{X(t); t \geq 0\}$ with transition rate q of (40.1) is quasi-reversible if the following two conditions hold.

(i) *The $X(t)$ is independent of the arrival process of class u entities subsequent to time t for all $u \in T$.*
(ii) *The $X(t)$ is independent of the departure process of class u entities prior to time t for $u \in T$.*

Quasi-reversibility is closely related to Poisson flows, as is shown in the following theorem.

Theorem 40.1

Definitions 1 and 2 are equivalent, and each of them implies that

(a) *the arrival process of class $u \in T$ entities are Poisson and the arrival processes of different classes of entities are independent,*
(b) *the departure process of class $u \in T$ entities are Poisson and the departure processes of different classes of entities are independent.*

Proof
The first quasi-reversibility condition (40.2) implies condition (i) of Definition 40.2 and part (a) of the theorem. On the other hand, condition (i) implies that the left hand side of condition (40.2) is a constant. Denoting this constant by α_u, we obtain (40.2). We next show that condition (40.3) of the first definition implies condition (ii) of the second definition and part (b) of the theorem. To this end, consider the reversed process $X(-t)$. Define $\tilde{q}_u^{\mathrm{A}}$, $\tilde{q}_u^{\mathrm{D}}$, and $\tilde{q}^{\mathrm{I}}$ as

$$\tilde{q}_u^{\mathrm{A}}(x,x') = \frac{\pi(x')}{\pi(x)} q_u^{\mathrm{D}}(x',x)$$

$$\tilde{q}_u^{\mathrm{D}}(x,x') = \frac{\pi(x')}{\pi(x)} q_u^{\mathrm{A}}(x',x)$$

$$\tilde{q}^{\mathrm{I}}(x,x') = \frac{\pi(x')}{\pi(x)} q^{\mathrm{I}}(x',x)$$

Let $\tilde{q}$ be the transition rate function of $X(-t)$. Its transition rates are

$$\tilde{q}(x,x') = \sum_u \tilde{q}_u^{\mathrm{A}}(x,x') + \sum_u \tilde{q}_u^{\mathrm{D}}(x,x') + \tilde{q}^{\mathrm{I}}(x,x')$$

Thus, the time reversed process also represents a queueing model with thinned transitions $\tilde{q}_v^{\mathrm{A}}$, $\tilde{q}_u^{\mathrm{D}}$, and $\tilde{q}^{\mathrm{I}}$. From condition (40.3), we have

$$\sum_{x'} \tilde{q}_u^{\mathrm{A}}(x,x') = \frac{1}{\pi(x)} \sum_{x'} \pi(x') q_u^{\mathrm{D}}(x',x) = \beta_u$$

This implies that the class u arrival process in $X(-t)$ is Poisson with rate β_u. However, from the definition of $\tilde{q}_u^{\mathrm{A}}$ and the detailed Kelly lemma (see Chao et al., 1999), a class u arrival in the reversed process $X(-t)$ corresponds to a class u departure in process $X(t)$. Thus we obtain condition (ii) of the second definition. Since a Poisson process reversed in time is Poisson with the same rate, we have (b). Finally, condition (ii) implies that the arrival epochs in the reversed process $X(-t)$ are generated at a constant rate independent of the current state, which gives (40.3). □

40.3 Quasi-Reversibility of Queues with Triggered Departures

The quasi-reversibility defined in the last section is concerned with arrival and departure epochs. One prominent feature is that an arrival cannot occur at the same time as a departure. In queueing systems with signals, however, the arrival of a signal may immediately trigger a departure. Thus the definition of quasi-reversibility is not applicable in these cases. In this section we extend the notion of quasi-reversibility to include such simultaneous events.

As before, let q be the transition rate of the node and let it be decomposed into the components $\{q_u^{\mathrm{A}}; u \in T\}$, $\{q_u^{\mathrm{D}}; u \in T\}$, and q^{I} of (40.1). Assume that q admits the stationary distribution π. Furthermore, assume that when a class u entity arrives and induces the state of the node to change from x to x', it instantaneously triggers a class v departure with triggering probability $f_{u,v}(x,x')$, where

$$\sum_{v\in T} f_{u,v}(x,x') \leq 1, \quad u \in T,\ x,x' \in \mathcal{S}$$

With probability

$$1 - \sum_{v\in T} f_{u,v}(x,x')$$

the class u arrival does not trigger any departure.

Note that when $\sum_{v\in T} f_{u,v}(x,x') \equiv 0$ for all x and u, the system reduces to that of the previous section with no instantaneous movements.

Definition 40.3

If there exist two sets of nonnegative numbers $\{\alpha_u; u \in T\}$ *and* $\{\beta_u; u \in T\}$ *such that*

$$\sum_{x'\in\mathcal{S}} q_u^{\mathrm{A}}(x,x') = \alpha_u, \quad x \in \mathcal{S},\ u \in T \tag{40.4}$$

$$\sum_{x'\in\mathcal{S}} \pi(x') \left(q_u^{\mathrm{D}}(x',x) + \sum_{v\in T} q_v^{\mathrm{A}}(x',x) f_{v,u}(x',x) \right) = \beta_u \pi(x), \quad x \in \mathcal{S},\ u \in T \tag{40.5}$$

then the queue with signals is said to be quasi-reversible with respect to $\{q_u^{\mathrm{A}}, f_{u,v}; u \in T, v \in T\}$, $\{q_u^{\mathrm{D}}; u \in T\}$, *and* q^{I}.

As in the last section, quasi-reversibility for queues with signals implies that the arrivals of the different classes of entities form independent Poisson processes, and the departures of different classes of entities, including both triggered and nontriggered departures, also form independent Poisson processes. Moreover, future arrivals and past departures are independent of the current state of the system.

In many applications, triggered and nontriggered departures belong to different classes, i.e.,

$$q_v^{\mathrm{A}}(x,x') f_{v,u}(x,x') q_u^{\mathrm{D}}(x,x') = 0 \quad \text{for all } x, x' \text{ and } u, v \in T$$

Let T' and T'' be the sets of the nontriggered and triggered departure classes, respectively, such that

$$T = T' \cup T'' \quad \text{and} \quad T' \cap T'' = \emptyset$$

Then (40.5) is reduced to

$$\sum_{x'\in\mathcal{S}} \pi(x') q_u^{\mathrm{D}}(x',x) = \beta_u \pi(x), \quad x \in \mathcal{S},\ u \in T'$$

$$\sum_{x'\in\mathcal{S}} \pi(x') \sum_{v\in T'} q_v^{\mathrm{A}}(x',x) f_{v,u}(x',x) = \beta_u \pi(x), \quad x \in \mathcal{S},\ u \in T''$$

This is equivalent to saying that both the triggered and nontriggered departure processes are independent Poisson with rate β_u for class $u \in T$.

The triggering arrivals and triggered departures are referred to as signals since they pass through a node and change its state instantaneously. The following example illustrates this.

Example 40.2

Consider an $M/M/1$ queue with two classes of arrivals, denoted by c and s. Class c refers to the regular jobs, and class s refers to signals. When a signal arrives at the node, it triggers a job to depart immediately as a class s departure, provided the queue is not empty upon its arrival. If a signal arrives at an empty queue, nothing occurs and no departure is triggered. The job departures generated by regular processing completions are still classified as class c departures. The decomposed transition rates are

$$q_c^{\mathrm{A}}(n, n+1) = \alpha, \quad n \geq 0,$$
$$q_s^{\mathrm{A}}(n, n-1) = \alpha^-, \quad n \geq 1,$$
$$q_s^{\mathrm{A}}(0, 0) = \alpha^-,$$
$$q_c^{\mathrm{D}}(n, n-1) = \mu, \quad n \geq 1$$

All other transition rates are zero. By the triggering mechanism, we have

$$f_{c,c}(n, n') = f_{c,s}(n, n') = 0, \quad n, n' \geq 0$$
$$f_{s,s}(n, n-1) = 1, \quad n \geq 1$$

Since the dynamics of this queue is the same as that of a regular $M/M/1$ queue with arrival rate α and service rate $\mu + \alpha^-$, its stationary distribution π is given by

$$\pi(n) = \left(1 - \frac{\alpha}{\mu + \alpha^-}\right)\left(\frac{\alpha}{\mu + \alpha^-}\right)^n, \quad n \geq 0$$

If we set

$$\beta = \frac{\alpha\mu}{\mu + \alpha^-},$$
$$\beta^- = \frac{\alpha\alpha^-}{\mu + \alpha^-}$$

then this system is quasi-reversible with departure rates β and β^-, since,

$$\sum_{n'} q_c^{\mathrm{A}}(n, n') = \alpha, \quad n \geq 0$$
$$\sum_{n'} q_s^{\mathrm{A}}(n, n') = \alpha^-, \quad n \geq 0$$
$$\sum_{n'} \pi(n')\left(q_c^{\mathrm{D}}(n', n) + \sum_{u=c,s} q_u^{\mathrm{A}}(n', n) f_{u,c}(n', n)\right) = \sum_{n'} \pi(n') q_c^{\mathrm{D}}(n', n) = \beta\pi(n), \quad n \geq 0$$
$$\sum_{n'} \pi(n')\left(q_s^{\mathrm{D}}(n', n) + \sum_{u=c,s} q_u^{\mathrm{A}}(n', n) f_{u,s}(n', n)\right) = \sum_{n'} \pi(n') q_s^{\mathrm{A}}(n', n) f_{s,s}(n', n) = \beta^-\pi(n), \quad n \geq 0$$

This is a very simple system, but many queueing networks with negative signals are generated by this model.

40.4 Networks of Quasi-Reversible Nodes

In this section we connect N quasi-reversible nodes into a queueing network with Markovian routing mechanisms. The main result is that such a network has a product form solution, i.e., the stationary distribution of the network factorizes into the product of the marginal distributions of the individual nodes. The omitted proofs of this and the next section can be found in Chao and Miyazawa (2000a).

We consider a queueing network with an arbitrary Markovian routing mechanism and multiple classes of entities. As discussed earlier, entities include both jobs and signals, and their effects on the nodes can be quite general. For instance, the arrival of an entity may decrease the number of jobs or trigger other actions before instantaneously moving to another node. In this section we consider a network structure without signals, i.e., an arrival does not trigger any instantaneous departure. This enables us to give an explicit expression for the network transition rates. The model in this section forms the basis for the network with signals that will be discussed in Section 40.5. However, when signals are present, the model becomes more involved, and a mathematical expression for the network transition rates becomes complicated without the use of matrix operators.

Suppose the network has N nodes. Each node represents a single processing station, or a cluster of stations (subnetwork). In addition to these nodes we have node 0, which represents the outside world. In this section, the state space $\mathcal{S}_0$ of node 0 is the singleton, i.e., $\mathcal{S}_0 = \{0\}$. Node 0 is a Poisson source, i.e., exogenous entities arrive at the network according to a Poisson process. Even though our main concern is an open network, the arguments can be applied to closed networks as well by simply removing node 0. However, we keep node 0 for consistency. For node $j = 0, 1, \ldots, N$, let T_j denote the class of arrival and departure entities at node j. As discussed earlier, we do not make any distinction between arrival and departure classes, even though they may be different. In case they are different we simply let T_j be the union of the arrival and departure classes. For instance, in Example 40.1, $T_j = \{c, c^-\}$, even though the arrival classes are $\{c, c^-\}$ and the departure class is $\{c\}$.

Let x_j be the state of node j with state space $\mathcal{S}_j$. For node 0, apparently $x_j \equiv 0$. When node j contains a single station, x_j may represent, for instance, the number of each class of jobs as well as their positions in the queue. Since node j may also be a subnetwork, x_j can be more general, e.g., it may represent the number of each class of jobs present at each station as well as their positions at the stations within the node. It may also include the remaining processing times at the node when the processing times are not exponentially distributed. Furthermore, since each node may be a subnetwork, there may be internal transitions within x_j, e.g., job movements between different stations within the same node, or between positions in the same station of node j.

What is the necessary information to construct a queueing network model? A little reflection reveals that we need two types of information: Node (or local) information, i.e., how does each node operate and react to arrivals from other nodes? and internode (or global) information, i.e., how are the nodes interconnected?

With regard to node information, we first note that the arrival process at each node of the network is not known before the network is put together. Therefore the arrival transition rate of each node, i.e., q_j^{A}, is not known, nor is it needed for the construction of the queueing network. What we do have to know is what would happen with the node when an arrival occurs. Thus, in order to construct the network, we need for each node the following information.

(i) *Arrival effects:* The rules according to which the node changes state with the arrival of an entity.
(ii) *Departure transition rates:* The rate at which the state of the node changes and it may induce the state of another node to change.
(iii) *Internal transition rates:* The rate at which the state of the node changes and it does not affect the states of other nodes.

For these reasons, we specify each node by a transition *probability* function that describes the changes of state upon arrivals and transition *rate* functions that describe changes of state due to departures and internal transitions. Thus, for node j and an entity of class u, we introduce functions p_{ju}^{A}, q_{ju}^{D}, and q_j^{I} on state space $\mathcal{S}_j$.

$p_{ju}^{\mathrm{A}}(x_j, x'_j) =$ the probability that a class u arrival at node j changes the state from x_j to x'_j, where it is assumed that

$$\sum_{x'_j \in \mathcal{S}_j} p_{ju}^{\mathrm{A}}(x_j, x'_j) = 1, \quad x_j \in \mathcal{S}_j$$

$q_{ju}^{\mathrm{D}}(x_j, x'_j) =$ the rate at which class u departures change the state of node j from x_j to x'_j.

$q_j^{\mathrm{I}}(x_j, x'_j) =$ the rate at which internal transitions change the state of node j from x_j to x'_j.

For node 0, we set $p_{0,u}^{\mathrm{A}}(0,0) = 1$, $q_{0,u}^{\mathrm{D}}(0,0) = \beta_{0u}$, and $q_j^{\mathrm{I}}(0,0) = 0$. This implies that exogenous class u entities, i.e., class u departures from node 0, arrive at the network from the outside according to a

Poisson process with rate β_{0u}. Note that $p^{\mathrm{A}}_{ju}(x_j, x_j)$ may be positive, i.e., an arrival may not cause a change of state with a positive probability. We refer to p^{A}_{ju} as the *arrival effect function.*

We describe each queue by the three components q^{A}_u, q^{D}_u, and q^{I}. If a queue in the network is initially characterized by q^{A}_u, q^{D}_u, and q^{I}, then the arrival effect function may be defined as

$$p^{\mathrm{A}}_u(x, x') = \frac{q^{\mathrm{A}}_u(x, x')}{\sum_y q^{\mathrm{A}}_u(x, y)} \tag{40.6}$$

and q^{D}_u, q^{I} are the departure and internal transition functions. However, unless a node is a separate queue, we assume that it is characterized by p^{A}_u, q^{D}_u, and q^{I} because, as discussed earlier, the arrival process at a node of a network depends on the structure of the entire network.

Example 40.3

Assume that node j of the network is a queue with negative customers, i.e., Example 40.1. It has two classes of arrivals and a single class of departures, $T_j = \{c, c^-\}$, and is characterized by the following arrival and departure functions:

$$p^{\mathrm{A}}_{jc}(n_j, n'_j) = \begin{cases} 1, & n'_j = n_j + 1 \\ 0, & \text{otherwise} \end{cases}$$

$$p^{\mathrm{A}}_{jc^-}(n_j, n'_j) = \begin{cases} 1, & n'_j = n_j - 1 \geq 0 \\ 0, & \text{otherwise} \end{cases}$$

$$p^{\mathrm{A}}_{jc^-}(0, 0) = 1,$$

$$q^{\mathrm{D}}_{jc}(n_j, n'_j) = \begin{cases} \mu_j, & n'_j = n_j - 1, n_j \geq 1 \\ 0, & \text{otherwise} \end{cases}$$

There are no class c^- departures and there are no internal transitions, so

$$q^{\mathrm{D}}_{jc^-}(n_j, n'_j) \equiv 0, \quad n_j, n'_j \geq 0$$

$$q^{\mathrm{I}}_j(n_j, n'_j) \equiv 0, \quad n_j, n'_j \geq 0$$

Node j has a single server with service rate μ_j. When a customer arrives, the number of customers in the node increases by 1, when a negative customer arrives, the number of customer decreases by 1, provided the node is not empty. When a negative customer arrives at an empty node, the state does not change. Note that the arrival process at node j, which is denoted by q^{A}_{ju}, is not given. It is characterized by p^{A}_{ju}, which describes what happens when a class u ($u = c, c^-$) entity arrives at the node when it is in state n_j. Also note that p^{A}_{ju} and q^{A}_{ju} in Example 40.1 satisfy (40.6).

The interactions between the nodes are defined as follows. A class u departure from node j enters node k as a class v arrival with probability $r_{ju,kv}$, and an exogenous class u arrival is routed to node k as a class v arrival with probability $r_{0u,kv}$. It is assumed that

$$\sum_{k=0}^{N} \sum_{v \in T_k} r_{ju,kv} = 1, \quad j = 0, 1, \ldots, N, \ u \in T_j \tag{40.7}$$

Note that class u departures from node j leave the network with probability $\sum_{v \in T_0} r_{ju,0v}$. This probability is often denoted by $r_{ju,0}$. In this way, we associate the departures from one node with the arrivals at another.

Let

$$\mathcal{S} = \mathcal{S}_1 \times \mathcal{S}_2 \times \cdots \times \mathcal{S}_N$$

be the product state space. Then,

$$\boldsymbol{x} = (x_1, x_2, \ldots, x_N) \in \mathcal{S}$$

is the state of the network. This network is a continuous time Markov chain with state space $\mathcal{S}$ and transition rate function q, where

$$q(\boldsymbol{x}, \boldsymbol{x}') = \sum_{j=0}^{N} \sum_{k=0}^{N} \sum_{u \in T_j} \sum_{v \in T_k} q_{ju}^{\mathrm{D}}(x_j, x_j')\, r_{ju,kv}\, p_{kv}^{\mathrm{A}}(x_k, x_k')\, 1[x_\ell = x_\ell' \text{ for all } \ell \neq j, k]$$
$$+ \sum_{j=0}^{N} q_j^{\mathrm{I}}(x_j, x_j')\, 1[x_\ell = x_\ell' \text{ for all } \ell \neq j] \tag{40.8}$$

for $\boldsymbol{x} = (x_1, x_2, \ldots, x_N) \in \mathcal{S}$ and $\boldsymbol{x}' = (x_1', x_2', \ldots, x_N') \in \mathcal{S}$. The first summation on the right-hand side of (40.8) represents the state changes due to job transfers from one node to another, and the second summation represents internal state changes. If

$$q_{ju}^{\mathrm{D}}(x_j, x_j) = p_{ju}^{\mathrm{A}}(x_j, x_j) = q_j^{\mathrm{I}}(x_j, x_j) = 0$$

then the transition rate function (40.8) can be partitioned into disjoint sets

$$q(\boldsymbol{x}, \boldsymbol{x}') = \begin{cases} \sum_{u \in T} \sum_{v \in T} q_{ju}^{\mathrm{D}}(x_j, x_j')\, r_{ju,kv}\, p_{kv}^{\mathrm{A}}(x_k, x_k'), & x_\ell = x_\ell' \quad \text{for all } \ell \neq j, k \\ q_j^{\mathrm{I}}(x_j, x_j'), & x_\ell = x_\ell' \quad \text{for all } \ell \neq j \\ 0 & \text{otherwise} \end{cases}$$

This is a typical situation in a conventional queueing network such as Jackson network. However, it is not true in general. For instance, if a departing job transforms itself into a negative signal and there is no job present at the node where it arrives, then the signal does not have any effect. In this case only the state of the node from which the job departs changes. That is, the transition caused by a signal may result in a change of state that is similar to an internal transition.

We now derive the stationary distribution for the queueing network just constructed. Assuming that each node in isolation is quasi-reversible, we show that the stationary distribution of the network process has product form.

Consider for each node j the following auxiliary process:

$$q_j^{(\alpha_j)}(x_j, x_j') = \sum_{u \in T_j} \left(\alpha_{ju} p_{ju}^{\mathrm{A}}(x_j, x_j') + q_{ju}^{\mathrm{D}}(x_j, x_j') \right) + q_j^{\mathrm{I}}(x_j, x_j'), \quad x_j, x_j' \in \mathcal{S}_j \tag{40.9}$$

Clearly, $q_j^{(\boldsymbol{\alpha}_j)}(x_j, x_j')$ can be viewed as node j being in isolation, with class $u \in T_j$ entities arriving according to a Poisson process with rate α_{ju}. In general, $p_{ju}^{\mathrm{A}}, q_{ju}^{\mathrm{D}}$, and q_j^{I} are allowed to be functions of $\boldsymbol{\alpha}_j = \{\alpha_{ju}; u \in T_j\}$. However, this dependency of $\boldsymbol{\alpha}_j$ is made implicit for simplicity.

Suppose $q_j^{(\boldsymbol{\alpha}_j)}$ has a stationary distribution $\pi_j^{(\boldsymbol{\alpha}_j)}$, i.e.,

$$\pi_j^{(\alpha_j)}(x_j) \left(\sum_{u \in T_j} \left(\alpha_{ju} + \sum_{x_j' \in \mathcal{S}_j} q_{ju}^{\mathrm{D}}(x_j, x_j') \right) + \sum_{x_j' \in \mathcal{S}_j} q_j^{\mathrm{I}}(x_j, x_j') \right)$$
$$= \sum_{u \in T_j} \sum_{x_j' \in \mathcal{S}_j} \pi_j^{(\alpha_j)}(x_j') \left(\alpha_{ju} p_{ju}^{\mathrm{A}}(x_j', x_j) + q_{ju}^{\mathrm{D}}(x_j', x_j) \right) + \sum_{x_j' \in \mathcal{S}_j} \pi_j^{(\alpha_j)}(x_j') q_j^{\mathrm{I}}(x_j', x_j),$$
$$x_j, x_j' \in \mathcal{S}_j \tag{40.10}$$

This $\pi_j^{(\alpha_j)}$ is expected to be the marginal distribution of node j for some parameters $\boldsymbol{\alpha}_j = (\alpha_{ju}; u \in T_j)$. However, the exact values of $\boldsymbol{\alpha}_1, \ldots, \boldsymbol{\alpha}_N$ are not yet known. Thus, for the time being, the $\boldsymbol{\alpha}_j$ may be regarded as dummy parameters, and their values will be determined later by the traffic equations.

First, note that we always have

$$\sum_{x'_j \in S_j} \alpha_{ju} p^{\mathrm{A}}_{ju}(x_j, x'_j) = \alpha_{ju}, \quad u \in T_j$$

Hence, quasi-reversibility is equivalent to the property that there exists a set of nonnegative numbers $\{\beta_{ju}; u \in T_j\}$ such that

$$\sum_{x'_j \in \mathcal{S}_j} \pi_j^{(\alpha_j)}(x'_j) q^{\mathrm{D}}_{ju}(x'_j, x_j) = \beta_{ju} \pi_j^{(\alpha_j)}(x_j), \quad x_j \in \mathcal{S}_j \tag{40.11}$$

for all $j = 1, 2, \ldots, N$ and $v \in T_j$. By (40.11), β_{ju} is determined by

$$\beta_{ju} = \sum_{x_j, x'_j \in \mathcal{S}_j} \pi_j^{(\alpha_j)}(x_j) q^{\mathrm{D}}_{ju}(x_j, x'_j) \tag{40.12}$$

Consider now the network generated by linking nodes $1, 2, \ldots, N$ through the routing probability matrix $R = \{r_{ju,kv}\}$, where node j is defined by p^{A}_{ju}, q^{D}_{ju}, and q^{I}_j. Assume that class $u \in T_0$ entities arrive at the network from the outside (node 0) at rate β_{0u}, which is given, and that each entity joins node k as a class v entity with probability $r_{0u,kv}$. Let β_{kv} be the average departure rate of class v entities from node k. The average arrival rate of class u entities at node j satisfies

$$\alpha_{ju} = \sum_{k=0}^{N} \sum_{v \in T_k} \beta_{kv} r_{kv,ju}, \quad j = 0, 1, \ldots, N, \ u \in T_j \tag{40.13}$$

These equations are referred to as the *traffic equations*. Note that β_{jv} is a nonlinear function of $\boldsymbol{\alpha}_j$, which is determined by (40.12), so the traffic equations are, in general, nonlinear in the α_{ju}'s. Finding solutions of (40.13) can be considered a fixed point problem concerning the vector $\boldsymbol{\alpha} = \{\alpha_{ju}; j = 0, 1, \ldots, N, u \in T_j\}$.

Theorem 40.2

For $(\boldsymbol{\alpha}_0, \boldsymbol{\alpha}_1, \ldots, \boldsymbol{\alpha}_N)$ satisfying (40.11) and (40.13), if each node of the network with transition rate $q_j^{(\alpha_j)}$ is quasi-reversible then the stationary distribution of the network is

$$\pi(\boldsymbol{x}) = \prod_{j=1}^{N} \pi_j^{(\alpha_j)}(x_j), \quad \boldsymbol{x} \equiv (x_1, x_2, \ldots, x_N) \in \mathcal{S}. \tag{40.14}$$

In many queueing systems (40.11) holds for a range of $\boldsymbol{\alpha}_j$. If this is the case the queue is called uniformly quasi-reversible.

Definition 40.4

Node j, characterized by $\{p^{\mathrm{A}}_{ju}; u \in T_j\}$, $\{q^{\mathrm{A}}_{ju}; u \in T_j\}$ and q^{I}_j, is called uniformly quasi-reversible *if it is quasi-reversible for all $\boldsymbol{\alpha}_j$ for which the stationary distribution $\pi_j^{(\alpha_j)}$ exists.*

If node j is uniformly quasi-reversible, then the departure rate β_{ju} is well defined on the range of $\boldsymbol{\alpha}_j$ for which $\pi_j^{(\alpha_j)}$ exists. Thus it can be considered a function of $\boldsymbol{\alpha}_j$, and the determination of $\boldsymbol{\alpha}_j$, for which the marginal distribution $\pi_j^{(\alpha_j)}$ of node j is computed, requires the solution of the nonlinear traffic equations. Thus uniform quasi-reversibility is important in computing the stationary distribution of the network.

The remaining problem is whether the traffic equations have a solution and how they can be determined. This is a fixed point problem as we stated before on which there exists an extensive literature, thus we will not discuss it further here.

We refer to a network of quasi-reversible nodes as a *quasi-reversible network*. Indeed, as seen from the following corollary, when the entire network is viewed as a system, it is also quasi-reversible.

Corollary 40.1

In quasi-reversible queueing networks, the class u departure process from node j to the outside is Poisson with rate $\beta_{ju} \sum_{v \in T_0} r_{ju,0v}$. *The network is quasi-reversible with respect to arrivals from the outside and departures to the outside.*

However, this does not imply that the departure and arrival processes at each node of the network are Poisson. Actually, it can be shown that the flow on a link is Poisson if and only if it is not part of a cycle. For instance, the flows in between any two nodes in a feedforward network are Poisson.

The following example illustrates Theorem 40.2.

Example 40.4

Consider a network with N single-server nodes. Each node has exponentially distributed processing times and two classes of entities: regular customers and negative customers, i.e., the type of node discussed in Example 40.1. Regular as well as negative customers arrive at node j from the outside according to independent Poisson processes with rates λ_j and λ_j^-. Upon a processing completion at node j, a customer joins node k as a regular customer with probability $r_{jc,kc}$ and as a negative customer with probability r_{jc,kc^-}, $k = 0, 1, \ldots, N$. The arrival of negative customer removes a customer from the node. The state of the network is represented by a vector $\boldsymbol{n} = (n_1, \ldots, n_N)$, where n_j is the number of regular customers at node j, $n_j \in \mathcal{S}_j = \{0, 1, \ldots\}$.

Node j, characterized by (40.9), is subject to Poisson arrivals of regular and negative customers with rates α_j and α_j^-. From Example 40.1, it follows that the stationary distribution π_j of node j is of a geometric form, and the node is uniformly quasi-reversible, and the π_j exists if and only if

$$\alpha_j < \mu_j + \alpha_j^- \tag{40.15}$$

The departure rate from node j is

$$\beta_j = \frac{\alpha_j \mu_j}{\mu_j + \alpha_j^-}$$

The traffic equations are

$$\alpha_j = \lambda_j + \sum_{k=1}^{N} \frac{\alpha_k \mu_k}{\mu_k + \alpha_k^-} r_{kc,jc}, \quad j = 1, \ldots, N$$

$$\alpha_j^- = \lambda_j^- + \sum_{k=1}^{N} \frac{\alpha_k \mu_k}{\mu_k + \alpha_k^-} r_{kc,jc^-}, \quad j = 1, \ldots, N$$

Thus, if the stability condition (40.15) is satisfied, then, by Theorem 40.2, the stationary distribution of the network, π, is the product of the π_j, i.e.,

$$\pi(\boldsymbol{n}) = \prod_{j=1}^{N} \left(1 - \frac{\alpha_j}{\mu_j + \alpha_j^-}\right) \left(\frac{\alpha_j}{\mu_j + \alpha_j^-}\right)^{n_j}$$

This is the network first studied by Gelenbe (1991).

40.5 Networks with Signals and Triggered Movements

This section extends the results of the last section to networks with instantaneous movements. Since instantaneous movements are triggered by signals, these networks are often referred to as networks with signals.

Consider a network with N nodes. Each node is a quasi-reversible queue with signals as described in Section 40.3. Let $\mathcal{S}_j$ be the state space of node j, and let T_j be the set of arrival and departure entity classes, $j = 1, 2, \ldots, N$. As discussed in Section 40.2, we need to specify for each node the transition *probability* functions that describe state changes due to arrivals, and the transition *rate* functions that describe departures and internal state changes. For these, we use the same notation as in last section, i.e., p^{A}_{ju}, q^{D}_{ju} and q^{I}_{j}. However, since there are instantaneous movements when there is an arrival at a node, we also have to specify the probability functions for the arrivals to induce departures, i.e., the *triggering probability* functions $f_{ju,v}(x_j, x'_j)$. When a class u entity arrives at node j and the state changes from x_j to x'_j, it simultaneously induces a class v departure with triggering probability $f_{ju,v}(x_j, x'_j)$. These probabilities satisfy

$$\sum_{v \in T_j} f_{ju,v}(x_j, x'_j) \leq 1, \quad u \in T_j, \quad x_j, x'_j \in \mathcal{S}_j, \quad j = 1, 2, \ldots, N$$

We allow p^{A}_{ju}, $f_{ju,v}$, q^{D}_{ju}, and q^{I}_{j} to be functions of a nonnegative vector $\boldsymbol{\alpha}_j = \{\alpha_{ju}; u \in T_j\}$, even though this dependency is made implicit for convenience. Also, $p^{\mathrm{A}}_{ju}(x_j, x_j)$ may be positive, i.e., an arrival may, with a positive probability, cause no change of state.

The dynamics of the network is described as follows. Class $u \in T_j$ entities from the outside, i.e., class u departures from node 0, arrive at the network according to a Poisson process with rate β_{0u}, and each is routed to node j as a class v entity with probability $r_{0u,kv}$. A class u departure from node j, either triggered or nontriggered, joins node k as a class v arrival with probability $r_{ju,kv}$, $k = 0, 1, \ldots, N$, where

$$\sum_{k=0}^{N} \sum_{v \in T_k} r_{ju,kv} = 1, \quad j = 0, 1, \ldots, N, \ u \in T_j$$

Furthermore, whenever there is a class u arrival at node j, either from the outside or from other nodes, it causes the state of the node to change from x_j to x'_j with probability $p^{\mathrm{A}}_{ju}(x_j, x'_j)$, it also triggers a class v departure with probability $f_{ju,v}(x_j, x'_j)$, and it triggers no departure from node j with probability

$$1 - \sum_{v \in T_j} f_{ju,v}(x_j, x'_j)$$

In this way, we associate the departures, both regular departures and triggered departures, from one node with the arrivals at another.

A distinctive feature of this network is that there are simultaneous arrivals and departures. For instance, if, for nodes $j_1, j_2, \ldots, j_k$ and classes $u_\ell, u'_\ell \in T_\ell$, $\ell = j_1, j_2, \ldots, j_k$,

$$\begin{aligned} &p^{\mathrm{A}}_{j_1 u_1}(x_{j_1}, x'_{j_1}) f_{j_1 u_1, u'_1}(x_{j_1}, x'_{j_1}) r_{j_1 u'_1, j_2 u_2} \times \cdots \times p^{\mathrm{A}}_{j_{k-1} u_{k-1}}(x_{j_{k-1}}, x'_{j_{k-1}}) \\ &\quad \times f_{j_{k-1} u_{k-1}, u'_{k-1}}(x_{j_{k-1}}, x'_{j_{k-1}}) r_{j_{k-1} u'_{k-1}, j_k u_k} p^{\mathrm{A}}_{j_k u_k}(x_{j_k}, x'_{j_k}) > 0 \end{aligned}$$

then a class u_1 arrival at node j_1 will simultaneously create arrivals at nodes $j_1, j_2, \ldots, j_k$, and change the states of these nodes to $x'_{j_1}, x'_{j_2}, \ldots, x'_{j_k}$ with a positive probability. Note that the same node may be visited several times on this route, and as a result, the state of the node may change a number of times at one point in time.

Let

$$\boldsymbol{X}(t) = (X_1(t), X_2(t), \ldots, X_N(t))$$

denote the state of the network at time t, with $X_j(t)$ being the state of node j. Then $\boldsymbol{X}(t)$ is a Markov process on the state space $\mathcal{S}$. Let q denote the transition rate function of the Markov process $\boldsymbol{X}(t)$.

As in the earlier section, we need to first consider each individual node j with an auxiliary transition rate $q_j^{(\boldsymbol{\alpha}_j)}$ to compute the stationary distribution of the network, where

$$q_j^{(\alpha_j)}(x_j, x_j') = \sum_{u \in T_j} \left(\alpha_{ju} p_{ju}^{\mathrm{A}}(x_j, x_j') + q_{ju}^{\mathrm{D}}(x_j, x_j') \right) + q_j^{\mathrm{I}}(x_j, x_j'), \quad x_j, x_j' \in \mathcal{S}_j \tag{40.16}$$

The $\boldsymbol{\alpha}_j = (\alpha_{ju}; u \in T_j)$ are considered dummy parameters and their values are to be determined by the traffic equations. Assume $q_j^{(\boldsymbol{\alpha}_j)}$ has a stationary distribution $\pi_j^{(\boldsymbol{\alpha}_j)}$, $j = 1, 2, \ldots, N$. We now require that the nodes with signals be quasi-reversible. Note that $q_{ju}^{\mathrm{A}} \equiv \alpha_{ju} p_{ju}^{\mathrm{A}}$ satisfies condition (40.5) automatically. So the quasi-reversibility is equivalent to the existence of nonnegative numbers $\{\beta_{iu}; u \in T_j\}$ for all $j = 1, 2, \ldots, N$ such that

$$\sum_{x_j' \in \mathcal{S}_j} \pi_j^{(\alpha_j)}(x_j') \left(q_{ju}^{\mathrm{D}}(x_j', x_j) + \sum_{v \in T_j} \alpha_{jv} p_{jv}^{\mathrm{A}}(x_j', x_j) f_{jv,u}(x_j', x_j) \right) = \beta_{ju} \pi_j^{(\alpha_j)}(x_j) \quad u \in T_j,\ x_j \in \mathcal{S}_j \tag{40.17}$$

Since α_{ju} and β_{iu} are the arrival and departure rates of class u entities at node j, the traffic equations

$$\alpha_{ju} = \sum_{k=0}^{N} \sum_{v \in T_k} \beta_{kv} r_{kv,ju}, \quad j = 0, 1, \ldots, N,\ u \in T_j$$

have to be satisfied. We need the following condition to ensure that the network process is regular:

$$\sum_{j=1}^{N} \sum_{x_j \in \mathcal{S}_j} \pi_j^{(\alpha_j)}(x_j) \sum_{x_j' \in \mathcal{S}_j} q_i^{(\alpha_j)}(x_j)(x_j, x_j') < \infty \tag{40.18}$$

A simple sufficient condition for (40.18) is

$$\sum_{u \in T_j} \left(\alpha_{ju} + \beta_{ju} + \sum_{x_j' \in \mathcal{S}_j} q_j^{\mathrm{I}}(x_j)(x_j, x_j') \right) < \infty, \qquad \text{for all } j = 1, \ldots, N$$

which is satisfied by all the examples in this chapter.

The following result for networks with signals is an extension of Theorem 40.2 for networks without instantaneous movements.

Theorem 40.3

If each node of the network is a quasi-reversible queue with signals, i.e., Equation (40.17) *is satisfied, and if* $\boldsymbol{\alpha}_j$, $j = 1, \ldots, N$ *are the solutions of the traffic Equations* (40.13), *then the queueing network with signals has the product form stationary distribution*

$$\pi(\boldsymbol{x}) = \prod_{j=1}^{N} \pi_j^{(\alpha_j)}(x_j), \quad \boldsymbol{x} \equiv (x_1, x_2, \ldots, x_N) \in \mathcal{S} \tag{40.19}$$

The following example illustrates Theorem 40.3.

Example 40.5

Consider a queueing network with jobs and negative signals, as described in Example 40.2. When a job completes its processing at node j, it goes to node k as a regular job with probability $r_{jc,kc}$, and as a negative signal with probability $r_{jc,ks}$, $k = 0, 1, \ldots, N$. When a negative signal arrives at node j, it induces a job, if there is one present, to depart. The job then joins node k as a regular job with probability $r_{js,kc}$, and as a negative signal with probability $r_{js,ks}$, $k = 0, 1, \ldots, N$. Let λ_j and λ_j^- be the exogenous arrival rates of jobs and signals at node j. The state of the network is given by vector $\boldsymbol{n} = (n_1, \ldots, n_N)$, where n_j is the number of jobs at node j.

Node j, defined by (40.16), is a queue discussed in Example 40.2, with service rate μ_j, job arrival rate α_j, and signal arrival rate α_j^-. It is quasi-reversible with departure rates of jobs and signals given by

$$\beta_j = \frac{\alpha_j \mu_j}{\mu_j + \alpha_j^-}, \quad \beta_j^- = \frac{\alpha_j \alpha_j^-}{\mu_j + \alpha_j^-}$$

Thus the traffic equations are

$$\alpha_j = \lambda_j + \sum_{k=1}^{N} \frac{\alpha_k \mu_k}{\mu_k + \alpha_k^-} r_{kc,jc} + \sum_{k=1}^{N} \frac{\alpha_k \alpha_k^-}{\mu_k + \alpha_k^-} r_{ks,jc}, \quad j = 1, 2, \ldots, N \tag{40.20}$$

$$\alpha_j^- = \lambda_j^- + \sum_{k=1}^{N} \frac{\alpha_k \mu_k}{\mu_k + \alpha_k^-} r_{kc,js} + \sum_{k=1}^{N} \frac{\alpha_k \alpha_k^-}{\mu_k + \alpha_k^-} r_{ks,js}, \quad j = 1, 2, \ldots, N \tag{40.21}$$

Suppose these traffic equations have positive solutions α_j, α_j^- such that

$$\frac{\alpha_j}{\mu_j + \alpha_j^-} < 1, \quad j = 1, 2, \ldots, N$$

Since each node is uniformly quasi-reversible, applying Theorem 40.3 yields the stationary distribution

$$\pi(\boldsymbol{n}) = \prod_{j=1}^{N} \left(1 - \frac{\alpha_j}{\mu_j + \alpha_j^-}\right) \left(\frac{\alpha_j}{\mu_j + \alpha_j^-}\right)^{n_j}$$

40.6 Networks with Positive and Negative Signals

In this section we apply Theorem 40.3 to a queueing network with two types of signals: positive signals and negative signals. The first subsection considers the case of a single class of positive signals and a single class of negative signals, and the second subsection considers multiple classes of positive and negative signals. These models include most networks with batch movements as special cases.

40.6.1 Single Class of Positive and Negative Signals

Consider a network with N nodes and a single server at each node. There are three classes of entities: *jobs, positive signals*, and *negative signals*, denoted by $T = \{c, s^+, s^-\}$. Class c refers to the regular jobs; their arrivals do not trigger any instantaneous movements. Positive and negative signals, however, represent signaling mechanisms that induce immediate transitions at the nodes where they arrive. Assume that the state of the network is $\boldsymbol{n} = (n_1, n_2, \ldots, n_N)$, where n_j is the number of jobs at node j, $j = 1, 2, \ldots, N$. The arrival of a positive signal at a node increases the number of jobs at that node by 1, and then leaves immediately for another node. The arrival of a negative signal at a node triggers a job to depart, provided the node is not empty upon its arrival. A negative signal disappears when it arrives at an empty node.

Assume that jobs arrive from the outside at node j according to a Poisson process with rate λ_j, and positive and negative signals arrive from the outside at node j according to Poisson processes with rates λ_j^+ and λ_j^-. Node j, $j = 1, \ldots, N$, has exponential processing times with rate μ_j.

Upon a processing completion at node j, a job leaves for node k as a regular job with probability $r_{jc,kc}$, as a positive signal with probability r_{jc,ks^+}, as a negative signal with probability r_{jc,ks^-}, and it leaves the network with probability $r_{jc,0}$, where

$$\sum_{k=1}^{N}(r_{jc,kc} + r_{jc,ks^+} + r_{jc,ks^-}) + r_{jc,0} = 1, \quad j = 1, \ldots, N$$

When a positive signal arrives at node j, either from the outside or from another node, it adds one job and then leaves immediately for node k as a regular job with probability $r_{js^+,kc}$, as a positive signal with probability r_{js^+,ks^+}, as a negative signal with probability r_{js^+,ks^-}, and it leaves the network with probability $r_{js^+,0}$, where

$$\sum_{k=1}^{N}(r_{js^+,kc} + r_{js^+,ks^+} + r_{js^+,ks^-}) + r_{js^+,0} = 1, \quad j = 1, \ldots, N$$

Finally, when a negative signal arrives at node j, either from the outside or from another node, it triggers a job, if any, to depart. The departing job goes to node k as a regular job with probability $r_{js^-,kc}$, as a positive signal with probability r_{js^-,ks^+}, as a negative signal with probability r_{js^-,ks^-}, and it leaves the network with probability $r_{js^-,0}$, where again

$$\sum_{k=1}^{N}(r_{js^-,kc} + r_{js^-,ks^+} + r_{js^-,ks^-}) + r_{js^-,0} = 1, \quad j = 1, \ldots, N$$

As indicated earlier, a negative signal that arrives at an empty node is assumed to be lost. We refer to this model as *a network with positive and negative signals.*

In this network there can be any number of job additions or deletions at various nodes of the network at the same point in time. For instance, if there is a sequence $j_1, j_2, \ldots, j_k$ such that

$$r_{j_1s^+,j_2s^+} r_{j_2s^+,j_3s^+} \cdots r_{j_{k-1}s^+,j_ks^+} > 0$$

then the arrival of a positive signal at node j_1 would, with positive probability, add one job at each one of nodes $j_1, j_2, \ldots, j_k$. There can also be batch arrivals at node j if $r_{js^+,js^+} > 0$, and the arrival of a positive signal can add a batch of random size at a number of nodes. Unlike in networks with negative signals, in which a signal may be interrupted on its route once it hits an empty node, a positive signal in this model will never be interrupted. It disappears only when it is transformed into another class of entity or when it leaves the network.

To ensure that the network is stable, i.e., it will not be overloaded, we have to exclude the case that a positive signal from any node generates an infinite number of jobs in the network at one point in time. Hence we make the following technical assumption in order to ensure that the stochastic process is regular: The Markov chain with state space $\{0, 1, \ldots, N\}$ and transition probabilities $p_{\cdot,\cdot}$ given by

$$p_{j,k} = r_{js^+,ks^+}, \quad j, k = 1, \ldots, N$$

$$p_{j,0} = 1 - \sum_{k=1}^{N} r_{js^+,ks^+} - \sum_{k=0}^{N} r_{js^+,kc}, \quad j = 1, \ldots, N$$

$$p_{0,0} = 1$$

has only one recurrent state 0.

We are interested in the stationary probability of this network. However, it is known that such networks do not have closed form solutions. In the following theorem, we modify the network process so as to obtain a product form solution for the network. This modification may appear artificial, and it is introduced purely to obtain the uniform quasi-reversibility of each node so that Theorem 40.3 can be applied. However, under some conditions the product form solution serves as a stochastic upper bound for the original network.

Suppose the following traffic equations have a nonnegative solution $\{\alpha_j; j = 1,\ldots,N\}$, $\{\alpha_j^+; j = 1,\ldots,N\}$, and $\{\alpha_j^-; j = 1,\ldots,N\}$:

$$\alpha_j = \lambda_j + \sum_{k=1}^{N} \rho_k \mu_k r_{kc,jc} + \sum_{k=1}^{N} \rho_k \alpha_k^- r_{ks^-,jc} + \sum_{k=1}^{N} \rho_k^{-1} \alpha_k^+ r_{ks^+,jc} \tag{40.22}$$

$$\alpha_j^+ = \lambda_j^+ + \sum_{k=1}^{N} \rho_k \mu_k r_{kc,js^+} + \sum_{k=1}^{N} \rho_k \alpha_k^- r_{ks^-,js^+} + \sum_{k=1}^{N} \rho_k^{-1} \alpha_k^+ r_{ks^+,js^+} \tag{40.23}$$

$$\alpha_j^- = \lambda_j^- + \sum_{k=1}^{N} \rho_k \mu_k r_{kc,js^-} + \sum_{k=1}^{N} \rho_k \alpha_k^- r_{ks^-,js^-} + \sum_{k=1}^{N} \rho_k^{-1} \alpha_k^+ r_{ks^+,js^-} \tag{40.24}$$

for $j = 1,\ldots,N$, where

$$\rho_j = \frac{\alpha_j + \alpha_j^+}{\mu_j + \alpha_j^-}$$

Now modify the network with positive and negative signals such that whenever node j is empty, a Poisson departure process of positive signals is activated with rate $\rho_j^{-1}\alpha_j^+$.

The following result can be proved by verifying the conditions of Theorem 40.3.

Theorem 40.4

If the solution of the traffic equations satisfy $\rho_j < 1$ for $j = 1,\ldots,N$, the modified network described above has the product form stationary distribution

$$\pi(n_1,\ldots,n_N) = \prod_{j=1}^{N} (1-\rho_j)\,\rho_j^{n_j} \tag{40.25}$$

Since the modified network has additional departures of positive signals, the network process stochastically dominates the corresponding process without the additional departures if a positive signal cannot be transformed into a negative signal. This can be easily proved using sample path stochastic comparison by constructing the two processes and coupling the numbers of jobs in the two networks. Note that, if the modified network has a stationary distribution, and if it stochastically dominates the original network, then the original network must also have a stationary distribution. Thus we obtain the following result.

Corollary 40.2

Let π^0 be the stationary distribution of the network without the additional departure processes. If $r_{js^+,ks^-} = 0$ and $r_{jc,ks^-} = 0$ for all $j,k = 1,\ldots,N$, then π^0 is stochastically dominated by the product form geometric distribution obtained in Theorem 40.4, i.e.,

$$\sum_{k_j \ge n_j, j=1,\ldots,N} \pi^0(k_1,\ldots,k_N) \le \prod_{j=1}^{N} \left(\frac{\alpha_j + \alpha_j^+}{\mu_j + \alpha_j^-}\right)^{n_j} \tag{40.26}$$

40.6.2 Multiple Classes of Positive and Negative Signals

We extend the results of the last subsection to networks with multiple classes of positive and negative signals. Suppose there is a single class of jobs denoted by c, I^+ classes of positive signals denoted by $\{u^+; u = 1, 2, \ldots, I^+\}$, and I^- classes of negative signals denoted by $\{u^-; u = 1, 2, \ldots, I^-\}$, where I^+ and I^- may be infinity. There is a single server at node j, and the processing times at node j are exponentially distributed with rate μ_j. Jobs arrive at node j from the outside according to a Poisson process with rate λ_j. Class u^+ positive signals, $u = 1, 2, \ldots, I^+$, arrive at node j from the outside according to a Poisson process with rate λ^+_{ju}, and class u^- negative signals, $u = 1, 2, \ldots, I^-$, arrive at node j from the outside according to a Poisson process with rate λ^-_{ju}.

The effects of positive and negative signals at a node are the same as before. That is, the arrival of a class u^+ positive signal at node j adds one job at node j and then departs, whereas the arrival of a class u^- negative signal at node j triggers one job, if any one is present, to depart. If a negative signal arrives at an empty node, nothing happens and the signal disappears. Thus, node j is characterized by

$$\begin{aligned} p^{\mathrm{A}}_{jc}(n_j, n_j + 1) &= 1, \quad n_j \geq 0 \\ p^{\mathrm{A}}_{ju^+}(n_j, n_j + 1) &= 1, \quad n_j \geq 0, u = 1, 2, \ldots, I^+ \\ p^{\mathrm{A}}_{ju^-}(n_j, n_j - 1) &= 1, \quad n_j \geq 1, u = 1, 2, \ldots, I^- \\ p^{\mathrm{A}}_{ju^-}(0, 0) &= 1, \quad u = 1, 2, \ldots, I^- \\ q^{\mathrm{D}}_{jc}(n_j, n_j - 1) &= \mu_j, \quad n_j \geq 1 \end{aligned}$$

The triggering probabilities are

$$\begin{aligned} f_{jc,w}(n_j, n'_j) &= 0, \quad w = c, u^+, v^-, \text{ and } n_j, n'_j \geq 0 \\ f_{ju^+,u^+}(n_j, n_j + 1) &= 1, \quad n_j \geq 0, \ u = 1, 2, \ldots, I^+ \\ f_{ju^-,u^-}(n_j, n_j - 1) &= 1, \quad n_j > 0, \ u = 1, 2, \ldots, I^- \end{aligned}$$

The routing probabilities are defined as follows. Upon a processing completion at node j, a job goes to node k as a regular job with probability $r_{jc,kc}$, as a class v^+ positive signal with probability r_{jc,kv^+}, as a class v^- negative signal with probability r_{jc,kv^-}, and it leaves the network with probability $r_{jc,0}$, where

$$\sum_{k=1}^{N} \left(r_{jc,kc} + \sum_{v=1}^{I^+} r_{jc,kv^+} + \sum_{v=1}^{I^-} r_{jc,kv^-} \right) + r_{jc,0} = 1 \quad \text{for } j = 1, \ldots, N$$

The arrival of a class u^+ positive signal at node j, either from the outside or from another node, adds one job to node j, and the signal leaves immediately for node k as a job with probability $r_{ju^+,kc}$, as a class v^+ positive signal with probability r_{ju^+,kv^+}, as a class v^- negative signal with probability r_{ju^+,kv^-}, and it leaves the network with probability $r_{ju^+,0}$, where

$$\sum_{k=1}^{N} \left(r_{ju^+,kc} + \sum_{v=1}^{I^+} r_{ju^+,kv^+} + \sum_{v=1}^{I^-} r_{ju^+,kv^-} \right) + r_{ju^+,0} = 1 \quad \text{for } j = 1, \ldots, N, \text{ and } u = 1, 2, \ldots, I^+$$

Finally, the arrival of a class u^- negative signal at node j, either from the outside or from another node, triggers one job from the node to depart, provided the queue is not empty upon its arrival. The triggered job then goes to node k as a job with probability $r_{ju^-,kc}$, as a class v^+ positive signal with probability

r_{ju^-,kv^+}, as a class v^- negative signal with probability r_{ju^-,kv^-}, and it leaves the network with probability $r_{ju^-,0}$, where

$$\sum_{k=1}^{N}\left(r_{ju^-,kc}+\sum_{v=1}^{I^+}r_{ju^-,kv^+}+\sum_{v=1}^{I^-}r_{ju^-,kv^-}\right)+r_{ju^-,0}=1,\quad \text{for } j=1,\dots,N, \text{ and } u=1,2,\dots,I^-$$

The arrival of a negative signal at an empty node does not have any effect and disappears.

Remark 40.1

The only additional feature of the network with multiple classes of positive and negative signals is the class-dependent routing. However, the class-dependent routing is very useful and it is general enough to include most queueing networks with batch arrivals and batch processings as special cases. See example 40.6.

Let α_{jc}, $\{\alpha^+_{ju}; j=1,\dots,N, u=1,\dots,I^+\}$, and $\{\alpha^-_{ju}; j=1,\dots,N, u=1,\dots,I^-\}$ denote the average arrival rates of jobs, positive signals, and negative signals at node j. They are determined by the traffic equations

$$\alpha_j=\lambda_{jc}+\sum_{k=1}^{N}\rho_k\mu_k r_{kc,jc}+\sum_{k=1}^{N}\sum_{v=1}^{I^-}\rho_k\alpha^-_{kv}r_{kv^-,jc}+\sum_{k=1}^{N}\sum_{v=1}^{I^+}\rho_k^{-1}\alpha^+_{kv}r_{kv^+,jc}$$
$$j=1,\dots,N \tag{40.27}$$

$$\alpha^+_{ju}=\lambda^+_{ju}+\sum_{k=1}^{N}\rho_k\mu_k r_{kc,ju^+}+\sum_{k=1}^{N}\sum_{v=1}^{I^-}\rho_k\alpha^-_{kv}r_{kv^-,ju^+}+\sum_{k=1}^{N}\sum_{v=1}^{I^+}\rho_k^{-1}\alpha^+_{kv}r_{kv^+,ju^+}$$
$$j=1,\dots,N,\ u=1,\dots,I^+ \tag{40.28}$$

$$\alpha^-_{ju}=\lambda^-_{ju}+\sum_{k=1}^{N}\rho_k\mu_k r_{kc,ju^-}+\sum_{k=1}^{N}\sum_{v=1}^{I^-}\rho_k\alpha^-_{kv}r_{kv^-,ju^-}+\sum_{k=1}^{N}\sum_{v=1}^{I^+}\rho_k^{-1}\alpha^+_{kv}r_{kv^+,ju^-}$$
$$j=1,\dots,N,\ u=1,2,\dots,I^- \tag{40.29}$$

where

$$\rho_j=\frac{\alpha_j+\alpha_j^+}{\mu_j+\alpha_j^-},\quad j=1,\dots,N$$

and α_j^+ and α_j^- are defined as

$$\alpha_j^+=\sum_{v=1}^{I^+}\alpha^+_{jv},\quad j=1,\dots,N$$

$$\alpha_j^-=\sum_{v=1}^{I^-}\alpha^-_{jv},\quad j=1,\dots,N$$

Clearly, α_j, α_j^+, and α_j^- are the average arrival rates of jobs, positive signals and negative signals at node j. Also, the total average arrival rate of jobs, including regular job and those added by positive signals, is $\alpha_j+\alpha_j^+$.

Theorem 40.5

Suppose the traffic Equations (40.27), (40.28), *and* (40.29) *have nonnegative solutions such that*

$$\rho_j \equiv \frac{\alpha_j + \alpha_j^+}{\mu_j + \alpha_j^-} < 1, \quad \textit{for all } j = 1, \ldots, N$$

If the network is modified so that whenever node j is empty, there is an additional departure process of class u^+ positive signals with rate

$$\frac{\mu_j + \alpha_j^-}{\alpha_j + \alpha_j^+} \alpha_{ju}^+$$

then the stationary probability of the network is

$$\pi(\boldsymbol{n}) = \prod_{j=1}^{N} (1 - \rho_j) \rho_j^{n_j} \tag{40.30}$$

Corollary 40.3

Let π^0 be the stationary distribution of the network without the additional departures of positive signals. If $r_{ju^+,kv^-} = 0$ and $r_{jc,kr^-} = 0$ for all j, k, u and v, then π^0 is stochastically dominated by the geometric product form π of (40.30).

The following example illustrates how the multiple classes of signals can be used to model batch movements.

Example 40.6

Consider a network of N single-server nodes. Jobs arrive at node j from the outside according to a Poisson process with rate λ_j, $j = 1, 2, \ldots, N$. The jobs are served in batches of a fixed size K_j, and the processing time of a batch is exponentially distributed with rate μ_j. Upon a processing completion at node j, the K_j jobs coalesce into a single job, and this single job goes to node k with probability r_{jk}, $k = 0, 1, \ldots, N$, where 0 is the outside world. In case there are less than K_j jobs in node j upon a processing completion at the node, these jobs coalesce into a partial batch and are removed from the system. When a job arrives at node j when the number of jobs at the node is less than K_j, it joins the batch currently being served; otherwise it waits in queue.

This model is a special case of the network with multiple classes of negative signals. To see this, consider a network with a single class of jobs and $K_j - 1$ classes of negative signals at node j, denoted by u^- for $u = 1, 2, \ldots, K_j - 1$. Jobs arrive at node j from the outside according to a Poisson process with rate λ_j. The jobs are served one at a time and the service rate at node j is μ_j. The routing probabilities, denoted by r^*, are defined as

$$\begin{aligned} r^*_{jc,j(K_j-1)^-} &= 1, \quad j = 1, \ldots, N \\ r^*_{ju^-,j(u-1)^-} &= 1, \quad u = 2, 3, \ldots, K_j - 1, \quad j = 1, \ldots, N \\ r^*_{j1^-,kc} &= r_{jk}, \quad k = 0, 1, \ldots, N, \quad j = 1, \ldots, N \end{aligned}$$

That is, upon a processing completion at node j a job goes back to node j as a class $(K_j - 1)^-$ negative signal with probability 1; the arrival of a class u^- $(u = 2, 3, \ldots, K_j - 1)$ negative signal at node j removes one job and then immediately goes to node j as a class $(u-1)^-$ signal with probability 1; a class 1^- negative signal arrives at node j and reduces the number of jobs by 1, then goes to node k as a regular job with probability r_{jk}. This implies that a regular processing completion at node j instantaneously removes K_j jobs from node j, provided there are at least K_j jobs present, and then goes to node k as a regular job with probability r_{jk}. That a negative signal that arrives at an empty queue disappears translates to the fact that, when there are less than K_j jobs at node j upon a processing completion, the entire batch is removed from the network. This is exactly the network under consideration.

By Theorem 40.5, this network has a geometric product form stationary distribution. Since there are no positive signals, no additional departure process is required.

40.7 Necessary and Sufficient Conditions for Product Form

In the preceding sections we discussed various network models that possess product form stationary distributions. Most of the results are obtained through quasi-reversibility. A natural question is whether quasi-reversibility is also a necessary condition for product form. The answer is negative. This section presents the necessary and sufficient conditions for product form for the class of networks whose transitions involve at most two nodes, i.e., there is no instantaneous triggering. Such a characterization yields a general procedure for verifying whether a network has a product form solution and obtaining it when it exists. Furthermore, the network has a product form stationary distribution and is *biased locally balanced* if and only if the network is quasi-reversible and certain traffic equations are satisfied. We also consider various scenarios in which quasi-reversibility is a necessary condition for product form. The omitted proofs in this and the next section can be found in Chao et al. (1998).

The network consists of N nodes, indexed 1 to N, and the outside is labeled as node 0. However, unlike the formulation in earlier sections, we here assume that the outside, i.e., node 0, has multiple states. Since departures from node 0 are arrivals to the network, such a formulation allows the arrival process to the network from the outside to be arbitrary. The state of the network is a vector of the states of the individual nodes and the outside world. Its state space is

$$\mathcal{S} = \mathcal{S}_0 \times \mathcal{S}_1 \times \cdots \times \mathcal{S}_N$$

For convenience we only consider the case of single class of transitions. Extension to multiple classes of transitions is straightforward. As in earlier sections, node j, $j = 0, 1, \ldots, N$, is subject to three types of state transitions, referred to as arrival, departure and internal transitions, denoted by

$$\left\{p_j^{\mathrm{A}}(x_j, y_j); x_j, y_j \in \mathcal{S}_j\right\}$$
$$\left\{q_j^{\mathrm{D}}(x_j, y_j); x_j, y_j \in \mathcal{S}_j\right\}$$
$$\left\{q_j^{\mathrm{I}}(x_j, y_j); x_j, y_j \in \mathcal{S}_j\right\}$$

They represent, respectively, the transition probabilities due to arrivals, the transition rate due to departures, and the internal transition rate. Thus we must have

$$\sum_{y_j \in \mathcal{S}_j} p_j^{\mathrm{A}}(x_j, y_j) = 1, \quad x_j \in \mathcal{S}_j$$

The network process is characterized by the following system dynamics:

(i) When node j is in state x_j, the departure transition rate that changes the state from x_j to y_j is $q_j^{\mathrm{D}}(x_j, y_j)$, $y_j \in S_j$.
(ii) A departure from node j is transferred to node k as an arrival with probability r_{jk}, $k = 0, 1, \ldots, N$ (recall that node 0 represents the outside).
(iii) An arrival at node k changes its state from x_k to y_k with probability $p_k^{\mathrm{A}}(x_k, y_k)$, $y_k \in \mathcal{S}_k$.
(iv) The internal transition rate at node j is $q_j^{\mathrm{I}}(x_j, y_j)$ when its state is x_j. We here redefine the internal transition so as to represent all transitions that do not trigger state changes at other nodes, i.e., it includes case (ii) with $j = k$. Denote this new transition rate by $q_j^{\mathrm{I}*}$, i.e.,

$$q_j^{\mathrm{I}*}(x_j, y_j) = q_j^{\mathrm{I}}(x_j, y_j) + \sum_{x'_j} q_j^{\mathrm{D}}(x_j, x'_j) r_{jj} p_j^{\mathrm{A}}(x'_j, y_j)$$

The network process has the transition rates

$$q(\mathbf{x}, \mathbf{x}') = \sum_{j,k} q_{jk}(\mathbf{x}, \mathbf{x}'), \quad \mathbf{x}, \mathbf{x}' \in \mathcal{S}$$

where

$$q_{jk}(\mathbf{x}, \mathbf{x}') = \begin{cases} q_j^{\mathrm{D}}(x_j, x_j') r_{jk} p_k^{\mathrm{A}}(x_k, x_k') 1[y_\ell = x_\ell, \ell \neq j, k] & \text{if } j \neq k \\ q_j^{\mathrm{I}*}(x_j, x_j') 1[x_\ell' = x_\ell, \ell \neq j] & \text{if } j = k \end{cases}$$

Our objective is to find the necessary and sufficient conditions for the network to have a product form stationary distribution.

The following notation will be used in our analysis. For a probability distribution π_j on $\mathcal{S}_j$, define

$$q_j^{\mathrm{D}}(x_j) = \sum_{y_j} q_j^{\mathrm{D}}(x_j, y_j)$$

$$q_j^{\mathrm{I}*}(x_j) = \sum_{y_j} q_j^{\mathrm{I}*}(x_j, y_j)$$

$$\tilde{p}_j^{\mathrm{A}}(x_j) = \frac{\sum_{y_j} \pi_j(y_j) p_j^{\mathrm{A}}(y_j, x_j)}{\pi_j(x_j)}$$

$$\tilde{q}_j^{\mathrm{D}}(x_j) = \frac{\sum_{y_j} \pi_j(y_j) q_j^{\mathrm{D}}(y_j, x_j)}{\pi_j(x_j)}$$

$$\tilde{q}_j^{\mathrm{I}*}(x_j) = \frac{\sum_{y_j} \pi_j(y_j) q_j^{\mathrm{I}*}(y_j, x_j)}{\pi_j(x_j)}$$

$$\beta_j = \sum_{x_j} \sum_{y_j} \pi_j(x_j) q_j^{\mathrm{D}}(x_j, y_j)$$

$$\nu_j = \sum_{x_j} \sum_{y_j} \pi_j(x_j) q_j^{\mathrm{I}*}(x_j, y_j)$$

Notice that $q_j^{\mathrm{D}}(x_j)$ ($q_j^{\mathrm{I}*}(x_j)$) is different from the transition rate function $q_j^{\mathrm{D}}(x_j, y_j)$ ($q_j^{\mathrm{I}*}(x_j, y_j)$). They are distinguished only by their arguments. When they are used without arguments (e.g., q_j^{D}), they represent the transition rate functions, e.g., $q_j^{\mathrm{D}}(x_j, y_j)$. Assume that β_j and ν_j are finite. Keep in mind that $\tilde{p}_j^{\mathrm{A}}(x_j), \tilde{q}_j^{\mathrm{D}}(x_j), \tilde{q}_j^{\mathrm{I}*}(x_j)$ as well as β_j, ν_j are functions of π_j. The following relationships can be easily verified:

$$\sum_{x_j} \pi_j(x_j) q_j^{\mathrm{D}}(x_j) = \sum_{x_j} \pi_j(x_j) \tilde{q}_j^{\mathrm{D}}(x_j) = \beta_j \tag{40.31}$$

$$\sum_{x_j} \pi_j(x_j) q_j^{\mathrm{I}*}(x_j) = \sum_{x_j} \pi_j(x_j) \tilde{q}_j^{\mathrm{I}*}(x_j) = \nu_j \tag{40.32}$$

We first consider the possible forms of the marginal distributions when the network process has a product form stationary distribution. Define the transition rate q_j for each node j by

$$q_j(x_j, y_j) = \alpha_j p_j^{\mathrm{A}}(x_j, y_j) + (1 - r_{jj}) q_j^{\mathrm{D}}(x_j, y_j) + q_j^{\mathrm{I}*}(x_j, y_j), \quad x_j, y_j \in \mathcal{S}_j \tag{40.33}$$

where α_j is a parameter to be determined. Consider this process as node j operating in isolation. The first term in the summation indicates that this isolated node has Poisson arrivals with rate α_j. The second and third terms are transition rates associated respectively with departures from node j and internal transitions at node j, where an internal transition may be a departure that returns to the same node (see (iv)).

Theorem 40.6

If the network process has the product form stationary distribution

$$\pi(\boldsymbol{x}) = \prod_{j=0}^{N} \pi_j(x_j)$$

then each π_j *is the stationary distribution for the* q_j *defined by* (40.33) *in which coefficients* α_j *are the solution to traffic equations*

$$\alpha_j = \sum_{k \neq j} \beta_k(\alpha_k) r_{kj}, \quad j = 0, 1, \ldots, N \tag{40.34}$$

where $\beta_j(\alpha_j)$ *denotes the* β_j *of* (40.31) *which depends on* α_j *through* π_j.

The next theorem provides the necessary and sufficient conditions for the network process to have a product form distribution.

Theorem 40.7

The network has the product form stationary distribution

$$\pi(\boldsymbol{x}) = \prod_{j=0}^{N} \pi_j(x_j), \quad \boldsymbol{x} \in \mathcal{S}$$

if and only if each π_j *is the stationary distribution of* q_j *with coefficients* α_j *satisfying the traffic Equations* (40.34) *and*

$$\left(\tilde{q}_j^{\mathrm{D}}(x_j) - \beta_j\right) r_{jk} \left(\tilde{p}_k^{\mathrm{A}}(x_k) - 1\right) + \left(\tilde{q}_k^{\mathrm{D}}(x_k) - \beta_k\right) r_{kj} \left(\tilde{p}_j^{\mathrm{A}}(x_j) - 1\right) = 0 \tag{40.35}$$

for all $j \neq k$ *and* $x_j \in \mathcal{S}_j$, $x_k \in \mathcal{S}_k$.

Clearly, the following are three sufficient conditions for (40.35), so they are sufficient for the stationary distribution of the network to be product form.

(a) Both nodes j and k are quasi-reversible. Recall that node j is quasi-reversible if $\tilde{q}_j^{\mathrm{D}}(x_j)$ is independent of x_j; in this case it must equal to β_j.
(b) Both nodes j and k are noneffective with respect to arrivals. Node j is said to be noneffective with respect to arrivals if $\tilde{p}_j^{\mathrm{A}}(x_j) = 1$ for all $x_j \in \mathcal{S}_j$.
(c) Either node j or node k is quasi-reversible and noneffective with respect to arrivals.

These sufficient conditions are further weakened if $r_{jk} = 0$ or $r_{kj} = 0$. These and other special cases will be discussed in the next section. Note that when the outside (node 0) is a Poisson source, then node 0 has only one state (say 0) which is noneffective with respect to arrivals, i.e., the state of the outside source is not changed when a job departs the network. On the other hand, the Poisson source is clearly quasi-reversible, so it belongs to case (c). Therefore, when a network is subject to Poisson arrivals from the outside, condition (40.35) only has to be verified for nodes other than 0. In this case the product form stationary distribution $\prod_{j=0}^{N} \pi_j(x_j)$ can be written as $\prod_{j=1}^{N} \pi_j(x_j)$.

Theorem 40.7 yields the following procedure for establishing the existence of a product form stationary distribution for the network process and obtaining the distribution when it exists.

Step 1. For the dummy parameter α_j compute the stationary distribution π_j of node j defined by q_j of (40.33).
Step 2. Compute β_j using (40.34), which is a function of α_j since π_j is. So write it as $\beta_j(\alpha_j)$.
Step 3. Solve the traffic Equations (40.34).
Step 4. Check condition (40.35) for each pair j, k and all x_j, x_k.

If this four-step procedure is successful, then $\pi(\boldsymbol{x}) = \prod_{j=0}^{N} \pi_j(x_j)$ is the stationary distribution of the network process.

Finding vector $\boldsymbol{\alpha} = (\alpha_0, \alpha_1, \ldots, \alpha_N)$ that satisfies the traffic Equations (40.34) is a fixed point problem whose solution is usually established by Brouwer's fixed point theorem. It follows from Theorem 40.7 that such a fixed point always exists when the network has a product form stationary distribution.

Theorem 40.8

The network has a product form stationary distribution if and only if there exists a solution to the traffic Equations (40.34) and it satisfies the condition of Step 4.

Therefore, the procedure above, in principle, applies to any queueing networks with product solutions. If the procedure is successful it gives the product form stationary distribution of the network; otherwise, i.e., if it does not lead to a solution that satisfies the condition of Step 4, then the procedure concludes that the network does not have a product form solution. For a particular application, one may be able to construct an algorithm to compute the fixed point, e.g., an iterative algorithm.

40.8 Quasi-Reversibility Revisited

As we observed earlier, quasi-reversibility is a sufficient condition but not a necessary condition for product form. This section is concerned with how much stronger than necessary this condition is. It turns out that quasi-reversibility is equivalent to a product form that satisfies the biased local balance equations.

A Markov chain with transition rate q is said to satisfy *biased local balance* with respect to a positive probability measure π on $\mathcal{S}$ and real numbers $\gamma = \{\gamma_j; j = 0, 1, \ldots, N\}$ if $\sum_j \gamma_j = 0$ and

$$\pi(\boldsymbol{x})\left(\sum_k \sum_{\boldsymbol{y}} q_{jk}(\boldsymbol{x}, \boldsymbol{y}) + \gamma_j\right) = \sum_k \sum_{\boldsymbol{y}} \pi(\boldsymbol{y}) q_{kj}(\boldsymbol{y}, \boldsymbol{x}), \quad \boldsymbol{x} \in \mathcal{S},\ j = 0, 1, \ldots, N \tag{40.36}$$

The π must be the stationary distribution for the Markov chain q since the global balance equations are the sum of these biased local balance equations over j. Also, we say that q is *locally balanced* with respect to π when all the γ_j's are 0.

Theorem 40.9

The following statements are equivalent:

(i) *The network satisfies biased local balance with respect to a product form distribution $\pi(\boldsymbol{x}) = \prod_{j=0}^{N} \pi_j(x_j)$ and $\gamma = \{\gamma_j; j = 0, 1, \ldots, N\}$*

(ii) *Each node q_j is quasi-reversible with respect to π_j for some α_j that satisfies*

$$\alpha_j = \sum_{k \neq j} \beta_k r_{kj}, \quad j = 0, 1, \ldots, N \tag{40.37}$$

If these statements hold, then

$$\gamma_j = \alpha_j - (1 - r_{jj})\beta_j, \quad j = 0, 1, \ldots, N \tag{40.38}$$

The remaining part of this section explores scenarios under which quasi-reversibility is also a necessary condition for product form.

Corollary 40.4

If in a queueing network there are no immediate turn around loops, i.e., $r_{jk} \neq 0$ *implies* $r_{kj} = 0$, *and* $p_k^{\mathrm{A}}(x_k)$ *is not identically 1 for all* x_k, *i.e.,*

$$\tilde{p}_k^{\mathrm{A}}(x_k) \neq 1 \quad \textit{for at least one } x_k \in \mathcal{S}_k \tag{40.39}$$

then the product form implies that node j *is quasi-reversible. In particular, if the discrete-time Markov chain on* $\mathcal{S}_k$ *with transition probability* $\{p_k^{\mathrm{A}}(x_k, y_k); x_k, y_k \in \mathcal{S}_k\}$ *is transient, then* (40.39) *is satisfied.*

Proof

Under the assumptions, Equation (40.35) is reduced to

$$\left(\tilde{q}_j^{\mathrm{D}}(x_j) - \beta_j\right) r_{jk} \left(\tilde{p}_k^{\mathrm{A}}(x_k) - 1\right) = 0 \tag{40.40}$$

Fix an x_k that satisfies (40.39). Since (40.40) holds for every x_j and $r_{jk}(\tilde{p}_k^{\mathrm{A}}(x_k) - 1) \neq 0$, we conclude that $\tilde{q}_j^{\mathrm{D}}(x_j) = \beta_j$ for all x_j, i.e., node j is quasi-reversible.

To prove the second part, first note that $\tilde{p}_k^{\mathrm{A}}(x_k) = 1$ for all x_k is, by the definition of $\tilde{p}_k^{\mathrm{A}}(x_k)$, equivalent to π_k being a positive stationary measure for the Markov chain with transition probability $\{p_k^{\mathrm{A}}(x_k, y_k); x_k, y_k \in \mathcal{S}_k\}$; this cannot be true if p_k^{A} is transient. Thus there must be at least one x_k such that (40.39) is satisfied. □

In job based queues with no signals, arrivals do not decrease the number of jobs at the node, so p_k^{A} is clearly transient. This is not true, however, in networks with negative signals, in which the arrival of a negative signal reduces the number of regular jobs.

Definition 40.5

Node j *is said to be* nonterminal *if* $1 - r_{jj} - r_{j0} > 0$, *i.e., a departure from node* j *arrives at other nodes in the network with a positive probability.*

Feedforward networks clearly satisfy the first condition of Corollary 40.4 for all nodes that are nonterminal. Thus we obtain the following result.

Corollary 40.5

A job based feedforward queueing network has a product form stationary distribution if and only if all the nonterminal nodes are quasi-reversible.

Of course, there are many queueing networks with feedback that satisfy the conditions of Corollary 40.4.

Example 40.7

Consider the job based network with four nodes. Jobs arrive at nodes 1 and 2 according to Poisson processes. Departures from nodes 1 and 2 join node 3, and departures from node 4 either join node 1, node 2, or leave the network. Clearly, this network satisfies the conditions of Corollary 40.5, so quasi-reversibility of each node is both necessary and sufficient for the stationary distribution of the network to be product form.

To present the next result, we need to introduce two new concepts. Node j is called a *conventional queue* if it has an empty state, denoted by 0, from which there can be no departures or internal transitions, and state 0 cannot be reached via arrival or internal transitions. That is,

$$\tilde{p}_j^{\mathrm{A}}(0) = q_j^{\mathrm{D}}(0) = 0$$
$$q_j^{\mathrm{I}*}(x_j, x_j') = 0 \quad \text{if either } x_j = 0 \text{ or } x_j' = 0$$

Clearly, if a network has an outside Poisson source, then node 0 is not conventional. A queueing network is called conventional if its outside source is Poisson and all other nodes are conventional. Let α_j^+ denote the average arrival rate at node j including the feedback, i.e.,

$$\alpha_j^+ = \sum_{k=0}^{N} \beta_k r_{kj} = \alpha_j + r_{jj}\beta_j$$

Node j is said to be *internally balanced,* if $\alpha_j^+ = \beta_j$, i.e., the average arrival rate equals the average departure rate.

Theorem 40.10

Suppose a queueing network has conventional nodes and the outside source, i.e., node 0, is noneffective with respect to arrivals. If the network has a product form stationary distribution, then a nonterminal node k is quasi-reversible if and only if either one of the following two conditions holds.

(a) *Node k has a path connecting it to an internally balanced node.*
(b) *Node k is directly connected to some node $j \neq 0$, but node j is not directly connected to node k, i.e, $r_{kj} > 0$ and $r_{jk} = 0$.*

In these cases, all the nonterminal nodes are internally balanced. Note that condition (b) is satisfied if the destination node k is a terminal node.

From sufficient condition (c) of Section 40.7 for product form stationary distribution it follows that if the outside source is Poisson, then no condition is required on the terminal nodes for the product form to hold.

The next result follows immediately from Theorem 40.10.

Corollary 40.6

Consider a queueing network with all nodes being conventional and all nonterminal nodes satisfy either (a) or (b) of Theorem 40.10. The network has a product form stationary distribution if and only if all the nonterminal nodes are quasi-reversible.

Considering an even more special case we obtain the following result.

Corollary 40.7

Consider a conventional queueing network with Poisson arrivals from the outside and each node is internally balanced, i.e., the departure rate of each node is equal to its arrival rate. The network has a product form stationary distribution if and only if all nonterminal nodes are quasi-reversible. If the outside source node is also a conventional queue, then the network has a product form stationary distribution if and only if all nodes are quasi-reversible.

Note the difference between Corollary 40.5 and Corollary 40.7. In Corollary 40.5 the nonterminal nodes do not need to be conventional or internally balanced, but the topological structure of the network is restricted. On the other hand, in Corollary 40.7, the network topology may be arbitrary, but each nonterminal node has to be conventional and internally balanced.

40.9 Conclusion

In this chapter we reviewed some latest developments on queueing networks with tractable stationary distributions. Clearly, if possible it is always preferred to find the closed form analytical solution for a network problem, and only when this is not possible will one resort to approximation methods. Furthermore, we note that even when analytical solution is not available, the necessary and sufficient condition often can help obtain bounds and approximations for the nonproduct form networks. This is because of the fact that necessary and sufficient condition reveals the additional conditions that need to be imposed for the network problem to yield a product form solution. In many cases, the network after imposing additional conditions, that has a product form solution, gives rise to a stochastic bound for the original problem. Moreover, it is clear that, if the additional conditions only have minor impact on the performance of the original problem, then the product form solution obtained can be used as a good approximation for the original problem.

This chapter focused on queueing network models with exponentially processing times. For models with arbitrary processing time distributions, state-dependent transition rates (such as multiple-server queues, etc.), and discrete time models, the reader is referred to Chao et al. (1999).

Acknowledgment

This research was partially supported by NSF under DMI-0196084 and DMI-0200306, and a grant from the National Natural Science Foundation of China under No. 70228001.

References

[1] N. Asaddathorn and X. Chao, (1999), A decomposition approximation for assembly-disassembly types of queueing networks. *Annals of Operations Research*, 87, 247–261.

[2] F. Baskett, K.M. Chandy, R.R. Muntz and F.G. Palacios, (1975), Open, closed and mixed networks of queues with different classes of customers. *Journal of the Association for Computing Machinery*, 22, 248–260.

[3] R. Boucherie and X. Chao, (2001), Queueing networks with string transitions of mixed vector additions and vector removals. *Journal Systems Science and Complexity*, 14, 337–355.

[4] R. Boucherie, X. Chao, and M. Miyazawa, (2003), Arrival first queueing networks with applications in Kanban production systems. *Performance Evaluation*, 51, 83–102.

[5] Chandy, K.M., Howard, J.H. Jr. and Towsley, D.F., (1977), Product form and local balance in queueing networks. *Journal of the Association for Computing Machinery*, 24, 250–263.

[6] X. Chao, (1994), A note on networks of queues with signals and random triggering times. *Probability in the Engineering and Informational Sciences*, 8, 213–219.

[7] X. Chao, (1995), On networks of queues with arbitrary processing time distributions. *Operations Research*, 43, 537–544.

[8] X. Chao, (1995), A queueing network model with catastrophe and product form solution. *Operations Research Letters*, 18, 75–79.

[9] X. Chao, (1997), Partial balances in batch arrival, batch service and assemble-transfer queueing networks. *Journal of Applied Probability*, 34, 745–752.

[10] X. Chao, W. Henderson and P. Taylor, (2001), State-dependent coupling of general queueing networks. *Queueing Systems: Theory and Application*, 39, 337–348.

[11] X. Chao and M. Miyazawa, (1998) On quasi-reversibility and partial balance: An unified approach to product form results. *Operations Research*, 46, 927–933.

[12] X. Chao and M. Miyazawa, (2000a), Queueing networks with instantaneous movements: A unified approach by quasi-reversibility. *Advances in Applied Probability*, 32, 284–313.

[13] X. Chao and M. Miyazawa, (2000b), On truncation properties for finite buffer queues and queueing networks. *Probability in the Engineering and Informational Sciences*, 14, 409–423.

[14] X. Chao, M. Miyazawa, and M. Pinedo, (1999), *Queueing Networks: Customers, Signals, and Product Form Solutions*, John Wiley & Sons, Chichester.

[15] X. Chao, M. Miyazawa, R. Serfozo, and H. Takada, (1998), Markov network processes with product form stationary distribution. *Queueing Systems: Theory and Application*, 28, 377–401.

[16] X. Chao and M. Pinedo, (1995), Queueing networks with signals and stage dependent routing. *Probability in the Engineering and Informational Sciences*, 9, 341–354.

[17] X. Chao and M. Pinedo, (1995), On networks of queues with batch services, signals, and product form solution. *Operations Research Letters*, 237–242.
[18] X. Chao, M. Pinedo, and D. Shaw, (1996), An assembly network of queues with product form solution. *Journal of Applied Probability*, 33, 858–869.
[19] X. Chao and S. Zheng, (1998), A result on networks of queues with customer coalescence and state dependent signaling. *Journal of Applied Probability*, 35, 151–164.
[20] X. Chao and S. Zheng, (2000), Triggered concurrent batch arrivals and batch departures in queueing networks. *Discrete Event Dynamic Systems*, 10, 115–129.
[21] E. Gelenbe, (1991), Product-form queueing networks with negative and positive customers. *Journal of Applied Probability*, 28, 656–663.
[22] W.P. Glynn, (1990), Diffusion approximation, in *Handbook in OR & MS: Stochastic Models*, D.P. Heyman and M.J. Sobel, (eds.), Vol. 2, pp. 145–198.
[23] W.J. Gordon and G.F. Newell (1967), Closed queueing systems with exponential servers. *Operations Research*, 15, 254–265.
[24] J.M. Harrison and R.J. Williams (1987), Brownian models of open queueing networks with homogeneous customer populations. *Stochast.* 22, 77–115.
[25] J.M. Harrison and R.J. Williams (1992), Brownian models of feedforward queueing networks: Quasireversibility and product form solutions. *Annals Applied Probability*, 2, 263–293.
[26] W. Henderson, W. Pearce, P.K. Pollett, and P.G. Taylor, (1992), Connecting internally balanced quasireversible Markov processes. *Advances in Applied Probability*, 24, 934–959.
[27] J.R. Jackson, (1957), Networks of waiting lines. *Operations Research*, 5, 516–523.
[28] J.R. Jackson, (1963), Jobshop-like queueing systems. *Management Science*, 10, 131–142.
[29] F.P. Kelly, (1975), Networks of queues with customers of different types. *Journal of Applied Probability*, 12, 542–554,
[30] F.P. Kelly, (1976), Networks of queues. *Advances in Applied Probability*, 8, 416–432.
[31] F.P. Kelly, (1979), *Reversibility and Stochastic Networks*, John Wiley & Sons, New York.
[32] F.P. Kelly, (1982), Networks of quasi-reversible nodes. In *Applied Probability-Computer Science: The Interface*, R.L. Disney and T.J. Ott, (eds.), Vol. I, pp. 3–26.
[33] J.F.C. Kingman, (1969), Markov population processes. *Journal of Applied Probability*, 6, 1–18.
[34] Y.V. Malinkovsky, (1990), A criterion for pointwise independence of states of units in an open stationary Markov queueing network with one class of customers. *Theory of Probability and Applications*, 35(4), 797–802.
[35] R.R. Muntz, (1972), Poisson Departure Processes and Queueing Networks. IBM Research Report RC4145. In *Proceedings of the Seventh Annual Conference on Information Science and Systems*, Princeton, NJ, 435–440.
[36] M. Neuts, (1981), *Structured Stochastic Matrices of M/G/1 Type and Their Applications.* Marcel Dekker, New York.
[37] P.K. Pollett, (1986), Connecting reversible Markov processes. *Advances in Applied Probability*, 18, 880–1986.
[38] R.F. Serfozo, (1989), Poisson functionals of Markov processes and queueing networks. *Advances in Applied Probability*, 21, 595–611.
[39] R.F. Serfozo, (1999), *Introduction to Stochastic Networks*, Springer-Verlag, New York.
[40] N. van Dijk, (1993), *Queueing Networks and Product Forms: A Systems Approach*, Wiley & Sons, New York.
[41] J. Walrand, (1988), *An Introduction to Queueing Networks*, Prentice Hall, New York.
[42] P. Whittle, (1968), Equilibrium distributions for an open migration process. *Journal of Applied Probability*, 5, 567–571.
[43] P. Whittle, (1986), *Systems in Stochastic Equilibrium*, Wiley & Sons, New York.

41

Scheduling in Secondary Storage Systems

Alexander Thomasian*

New Jersey Institute of Technology

41.1 Introduction

The memory hierarchy in a computer system consists of one to three levels of CPU caches: caches for file and database systems in main memory, the cache associated with a disk array controller; an onboard disk cache, disks; and automated tape libraries, which are themselves cached. We are concerned here with accesses to database and file systems, which may be satisfied by the buffer in main memory, but in the case

*The author acknowledges the support of NSF through Grant 0105485 in Computer System Architecture.

1-58488-397-9/$0.00+$1.50

of a cache miss, an I/O request is generated for a data block residing on disk. This I/O request is intercepted by the disk array controller, which checks its cache in order to satisfy the I/O request. In case of a cache miss an I/O request is issued to the disk and the disk controller checks its cache to satisfy the request. A disk access is initiated if we had misses at all levels.

The role of the storage hierarchy is to take advantage of *spatial and temporal locality* to improve performance, such that small caches at a higher level of the memory hierarchy result in a significant filtering out of requests, so that a few requests propagate to lower levels.

Main measures of disk performance are (i) the *disk transfer rate* which is important for large data transfers, since it determines the time to complete the operation; (ii) the *access bandwidth*, which is the number of accesses a disk can handle without saturation; (iii) the *kilobytes accessed per second* (KAPS) performance measure combines the first two measures into one [1]. The access bandwidth is determined by the spatial distribution of the blocks being accessed. Measures (ii) and (iii) are affected by the sizes of the blocks being transferred.

Disk access bandwidth is the inverse of disk access time, which is the sum of *controller overhead time*, *seek time*, *rotational latency time*, and *transfer time*. The transfer time is determined by the size of the block being transferred and the transfer rate, which depends on the characteristics of the disk. The sum of seek time and rotational latency is called *positioning time*.

Disk characteristics have had a profound effect on computer system performance. Multiprogramming was introduced to hide disk latencies and increase CPU (or processor) utilization and system throughput. When an application program makes a disk request the operating system switches to a ready-to-run program, if any.

Online transaction processing (OLTP) applications are a challenging workload from the viewpoint of the disk subsystem, for the following three reasons: (i) transaction throughputs are rather high and each transaction generates multiple disk accesses, (ii) the cache hit rate for random disk accesses is expected to be small, (iii) disks are subjected to accesses to randomly placed data blocks, resulting in rather high mean disk access time.

Mean transaction response time (R_{txn}) in OLTP applications is dominated by the mean number of disk accesses per transaction due to cache misses ($\overline{n}_{da}$), since CPU processing time is small. Given a transaction arrival rate λ, which may be in the thousands of transactions per second, the rate of disk requests is $\Lambda_{disk} = \overline{n}_{da}\Lambda$.

Given a balanced disk load, which can be attained via striping (see Section 41.6.2), each one of the N disks in the system will be subjected to $\lambda_{disk} = \Lambda_{disk}/N$ requests per second. N should be chosen such that the utilization of each disk $\rho_{disk} = \lambda_{disk}\overline{x}_{disk} < 1$ and furthermore the mean response time of each disk (r_{disk}) is acceptably low (say below 100 milliseconds). If a FCFS policy is used r_{disk} can be estimated approximately using the M/G/1 queuing model [2,3]: $r_{disk} = \overline{x}_{disk} + \lambda_{disk}\overline{x^2}_{disk}/(2(1-\rho_{disk}))$. The number of required disks is then $N_{disks} = \Lambda_{disk}/\lambda_{disk}$.

A pragmatic approach to improve OLTP performance is caching, since it will result in a reduction in the number of disk accesses. In the case of a record accessed via a B+ tree index, which usually has three or four levels, the number of disk accesses can be reduced by two by ensuring that the highest two levels of the index are held in the database buffer. In benchmarking studies the mean transaction throughput is determined at the point when $R_{txn} \leq 2$ seconds, for example. So every attempt should be made to minimize R_{txn}. Interactive users of shared servers also benefit from subsecond response times, since it has been shown to result in faster user responses and a synergistic improvement in user productivity.

Disk requests in the aforementioned discussion are considered *discrete requests*, since they tend to be independent from each other. *Continuous requests* are to successive blocks of the same file. Such requests tend to be associated with multimedia applications. A brief discussion of the processing of a mixture of such requests appears in Section 41.4.4.

Disk scheduling policies are mainly concerned with minimizing positioning time, since disk transfer time is fixed, but depends on the disk zone to which the file is assigned (see Section 41.3). The goal of disk scheduling is reducing disk response time, while ensuring the variance (or its square root: the standard deviation) remains small. Percentiles of disk response time can be expressed as the sum of the mean and multiples of the standard deviation. In the case of the M/G/1 queuing model, the mean and standard deviation of response time can be estimated rather easily, but its percentiles require in the general case the numerical inversion of the Laplace-Stieltjes Transform of the response time [2,3].

We discuss basic disk scheduling methods in Section 41.4.1 and a set of hybrid disk scheduling methods in Section 41.4.2. Our simulation results and those of others have shown that the *shortest positioning time first* (SPTF) has the "best" performance. Several variations of SPTF to alleviate this shortcoming are discussed in Section 41.4.3. We finally discuss disk arm prepositioning as a means of improving performance.

Besides disk arm scheduling, there are several other techniques that can be used to improve performance. Three methods are discussed in Section 41.5: (i) minimizing disk access time by allocating frequently accessed files at the middle disk cylinders, (ii) log-structured file systems, which minimizes disk update overhead, and (iii) utilizing the disk arm while the disk is not processing user requests and/or opportunistically processing low priority requests.

We next discuss *redundant arrays of independent disks* (RAID) as follows: motivation for RAID and basic RAID concepts in Section 41.6, mirrored disks in Section 41.7, and RAID level 5 (RAID5) rotated parity disk arrays in Section 41.8. This discussion is important since more and more disks are sold as disk arrays, rather than individual disks. Performance evaluation studies of secondary storage systems are discussed in Section 41.9, which is followed by Conclusions.

41.2 Scheduling in Computer Systems

With dropping costs, less attention is being paid to scheduling computing resources than before. This is especially true for personal computers and workstations, which are idle most of the time. In fact there is a lot of computing power available in idle workstations, which can be exploited by tools like Condor [4]. More recently there have been proposals to exploit the computational power of disk drive controllers [5].

Resource scheduling in highly parallel computer systems remains an important area, since such computational resources are expensive to purchase and operate. Resources can be shared spatially, as when a subset of the processing elements in an array are dedicated to a job, but finer grained sharing is available through multithreading, i.e., another process can proceed with the computation when one process has encountered a cache miss. Another form of sharing the CPU is multiprogramming, where the operating system switches to another job when the current job encounters an I/O request and is put to sleep.

There have been numerous simulation and analytic studies of CPU scheduling in single processor systems and in *shared memory multiprocessors* (SMPs), see, e.g., [6,7]. To simplify the discussion we consider CPU scheduling in isolation. Our interest is in finding parallels between CPU and disk scheduling methods.

FCFS scheduling of the CPU is known to provide poor performance when CPU processing times are highly variable. Consider a Poisson stream with rate Λ and $1 \leq i \leq I$ streams. The fractions of requests which belong to the ith stream is f_i. Requests in the ith stream have a mean service time $\overline{x_i}$, and a second moment $\overline{x_i^2}$. The overall mean and second moment are: $\bar{x} = \sum_{i=1}^{I} f_i \overline{x_i}$ and $\overline{x_i^2} = \sum_{i=1}^{I} f_i \overline{x_i^2}$. The utilization factor of the CPU is $\rho = \Lambda \overline{x}$.

The mean response time for the ith stream is $R_i = \overline{x_i} + W$, where the mean waiting time according to the M/G/1 queuing model is $W = \Lambda \overline{x^2}/2(1-\rho)$. It is easy to see that R_i is affected by W, which is a measure of unfinished work in the system due to all classes. The overall mean response time is $R = \sum_{i=1}^{I} R_i f_i = \overline{x} + W$.

Head-of-the-line (HOL) priority queuing discipline can be utilized to prioritize important requests. If the first class has the highest priority then $R_1 = \overline{x_1} + \lambda \overline{x^2}/2(1-\rho_1)$, where $\rho_1 = \Lambda f_1 \overline{x_1}$ is the utilization factor due to class-one requests only, i.e., high priority requests are affected by the server utilization in their own class only.

Requests submitted to a computer system have very different processing times, so that the processing of long requests concurrently with short requests may result in a rather poor performance for the latter. The *round-robin* (RR) policy is an improvement over the FCFS policy as follows. Each request in the queue receives a processing quantum q per CPU visit, until the request is completed. *Processor sharing*

(PS) is a limiting case of RR, where each one of n requests in the queue is processed $1/n$th of the time. The mean response time in this case is strictly a function of the processing time of a request (x), so that that $R(x) = x/(1-\rho)$, where ρ is the utilization factor over all classes. It follows that the overall mean response time is $R = \overline{x}/(1-\rho)$.

The multilevel feedback or *foreground/background* (FB) policy is similar to the RR, but requests after receiving a quantum of CPU processing are routed to a lower priority queue than the one from which they originated. In effect this is a combination of PS and HOL priority queuing. The FB policy has the advantage that long, lower priority requests do not compete with short requests for the CPU.

If the processing time of requests is known *a priori* and preemption is not allowed then the *shortest job first* (SJF) or SPTF policy minimizes the number of requests in the system and hence also the mean response time [7]. If the processing distribution is known then without preemption *shortest expected processing time* (SEPT) first and with preemption *shortest expected remaining processing time* (SERPT) first are appropriate [7].

While there are similarities between CPU and disk scheduling, the similarities are limited due to the following reasons: (i) disk requests tend to be nonpreemptable, especially during the seek time, (ii) CPU processing allows a preemptive resume policy with very small overhead, to save the state of the preempted process, (iii) the processing time of disk requests depends on the processing order, and (iv) the processing time of a disk request can be estimated rather accurately by the disk controller, while this is not true for CPU requests.

In our discussion we consider a modification of the SPTF policy, which allows the conditional prioritization of one class of requests with respect to another. Some degree of preemption can be introduced by treating the reading of a disk as a sequence of accesses, rather than an interruptible sequence. This permits the system to occasionally check whether other higher priority requests are enqueued. The processing of multiple long requests from a disk is multiplexed in the case of video-streaming to allow glitch-free viewing and to reduce buffering requirements.

While disk requests are considered to be nonpreemptable, a *split-seek* option is considered in [8], which allows a disk request to be preempted at the point it has completed its seek. The requests being preempted are track-read requests as part of the rebuild process for RAID5 (see Section 41.8.3), but more aggressive preemptive policies for track reads and destage-write requests (see Section 41.8.1) are considered in [9]. Preemptions are allowed during latency for writes and even during the transfer phase for track reads. Note that in this case we have an approximation to the preemptive repeat discipline with resampling [7]. If all of the sectors of a track are not read in the first access, then multiple accesses will be required until all the sectors are read.

41.3 Disk Organization

Computer disks consist of *platters* attached to a *spindle*, which is rotated at a *constant angular velocity* specified in *rotations per minute* (RPM). Disk RPMs have increased gradually over the years and 15,000 RPM disk drives with a 4-ms rotation time are available. *Constant linear velocity* is used in compact discs (CDs), which store the data in a helix.

Platters are coated on both sides with magnetic material on which data is recorded in concentric circles called *tracks*. The number of disk platters might be varied in the same disk family to achieve different capacities. Disks in the same family have similar characteristics except that their power consumption increases with the number of platters. Tracks with the same radius on different platters constitute a *cylinder*.

Adaptive spindown policies have been proposed to conserve energy in mobile battery-backed systems [10]. Such power saving strategies are regularly applied to CRTs and have the additional advantage of increasing the longevity of the screen. While the spinning down of a disk, which has not been in use for a while, seems to be a winning proposition, the following disadvantages need to be considered: (i) spin-up from a stationary position takes more energy, (ii) spin-up time causes user dissatisfaction, and is unacceptable for applications such as critical patient care, (iii) reduces disk reliability.

A large portion of the power budget in server environments goes into the I/O subsystem [11]. This served as the motivation for a simulation study of several workloads on server-class disks with dynamic speed control for power management, i.e., the disks are rotated slower when the disk utilization is low. This study shows significant energy saving with little impact on performance, but requires the disk to be able to operate at two speeds.

Data is recorded in fixed length 512 byte blocks called *sectors*, which is the smallest unit of data transfer. Each sector is protected by an error correcting code (ECC) and there is a block identifying its address and an alternate address if the sector is faulty. The *noID* technology holds this information in a semiconductor memory, allowing a significant increase in sectors per track and disk capacity.

The disk arm has a set of *read-write* (R/W) heads, where each R/W head can access only one track. The set of R/W heads attached to a disk arm can access all tracks on a disk cylinder, but only one of the R/W heads is activated at any time and the time to activate another head is termed *head switching time*. Disk blocks are accessed by moving the arm to the appropriate cylinder and activating the R/W head on the track where the data resides. Given the beginning sector number, the disk determines cylinder and track numbers.

Data is written sequentially on consecutive sector numbers, starting with the lowest numbered track (top track on outermost cylinder) continuing from track to track and from cylinder to cylinder. The reason for allocating data in this manner is that it allows efficient sequential access to data files. *Track skew* and *cylinder skew* are introduced to ensure that after switching to the next head or the next cylinder, the head will land on the next consecutive sector, so that an unnecessary disk rotation is not incurred. Logically consecutive sectors may not be mapped onto physically consecutive sectors, since there are faulty sectors, faulty tracks, and spare cylinders to hold faulty sectors and tracks. The mapping from physical to logical block numbers is carried out by the disk controller and the noID feature allows the alternative location of a data block to be determined without accessing the disk.

In early disk drives the number of bits and hence the number of sectors per track was determined by the highest permissible linear recording density and the circumference of the innermost track. This results in wasted-disk capacity, since outer disk cylinders can potentially store more sectors than inner ones. *Zoned bit recording* or *zoning* maintains the linear recording density almost constant at all tracks, so that the ratio $1 + \alpha$ of the number of sectors on outermost and innermost tracks is roughly proportional to their circumference. Zoning in this case entails in a $100\alpha/2$ percent increase in disk capacity. For easy bookkeeping the disk is subdivided into *zones*, where the number of tracks per zone is fixed. Zoned recording results in variable data transfer rates from disk tracks. The tracks corresponding to the outermost zone have the highest data transfer rate, since they can transfer the most sectors per disk rotation.

Disk service time is the sum of *disk controller time*, *seek time*, *rotational latency*, and *transfer time*, which varies from zone to zone for the same block size. In fact only disk controller may be involved in the processing of a disk request, since the disk request may be satisfied by the *onboard disk cache*, which is a multimegabyte DRAM that is part of the disk assembly.

The onboard cache is mainly used for prefetching, which is initiated by the disk controller upon observing sequentiality in the disk access pattern. The remaining or all sectors of a track from which a data block was accessed are transferred into the cache. The onboard cache is also used to match the disk transfer rate to the transfer rate of the I/O bus, which is higher. When data is being read from disk it is first transferred to the cache, but the bus transfer can be started even before the block transfer is completed from disk.

The number of devices (disks) attached to a *small computer system interface* (SCSI) bus depends on its bandwidth, e.g., higher 320 MB/s buses can handle 16 devices. The onboard cache obviates a problem in early disk drives, which required a circuit-switched path to carry out a successful data transfer. A *rotational position sensing* (RPS) misses occurred when one or more disk drives were ready to transmit, while another bus transfer was in progress. The disk encountering the blocking had to wait for a full rotation before attempting the transfer again.

The seek time characteristic $t(d)$ for a seek distance d (expressed in number of cylinders) is a convex (nondecreasing) function of d. $t(d)$ is measured based on the average of repeated seeks of distance d. The maximum seek time $t(C - 1)$ is for a *full-stroke seek*. The seek time characteristic can be summarized by curve-fitting to experimental data. A representative equation is $t(d) = a + b\sqrt{d}$ for $d \leq d_0$ and

$t(d) = e + fd$ for $d_0 \le d \le C - 1$, where d_0 is the beginning of the linear portion of $t(d)$. The average seek time $\bar{x}_{seek} = \sum_{d=1}^{C-1} P[d]t(d)$, where $P[d], 1 \le d \le C - 1$ is the probability of a seek of distance d.

Disk requests can be categorized into *discrete* and *continuous*. Discrete requests, which are generated by *infinite* sources, tend to be to random disk blocks. When there is no zoning, requests will tend to be unform over cylinders, since all cylinders have the same number of records. It can be shown in this case that $P[0] = 1/C$ and $P[d] = 2(C - d)/C^2, 1 \le d \le C - 1$, where the latter drops linearly with the seek distance (note that only two full-stroke seeks are possible). A similar analysis can be used to obtain $P[d]$ in a disk with zoning, which tend to favor shorter seeks for uniform block accesses, since there is a heavier concentration of data blocks on outer tracks.

Rotational latency is uniformly distributed over disk rotation time (T_{rot}), so that $\bar{x}_{latency} = T_{rot}/2$. Disks with *zero-latency* or *roll mode* capability can start a read or write access with an intermediate sector in the block being accessed. In fact $T_{rot}/2$ is still a good approximation for the average latency in this case, as long as the blocks being accessed are small, i.e., the number of sectors in the block is a small fraction of the sectors of the innermost track. When a full track is being transferred, $\overline{x}_{latency} = T_{sector}/2$, where T_{sector} is the time to transfer a sector.

The sum of seek and rotational latency, referred to as *positioning time*, constitutes the overhead portion of disk access time, while disk transfer time is considered "useful work". In the case of a prefetching transfer only the fraction of transferred data, which is accessed by user applications is considered useful work, but this fraction is difficult to determine.

There is an inherent access intensity associated with the volume of data stored on disk, which of course varies drastically with the type of data. Increased data volumes are expected with increasing disk capacities, which have experienced dramatic increases in size, but this also depends on the utilization of disk capacity. It might be that older disks are replaced by new disks not because their capacity was exhausted, but rather due to lower performance with respect to new disks.

A partial explanation why the access rate of disks is not increasing as rapidly as would be expected is the *five-minute rule*: keep a data item in electronic memory if its access frequency is five minutes or higher [1]. The time parameter of this rule is an increasing function of time as larger DRAM memories become more and more affordable. The break-even reference interval in seconds is given as the ratio of pages per megabyte (MB) of DRAM (128 eight-KB pages) divided by its price per MB ($15) and accesses per second per disk (64) divided by the price per disk drive ($2000) [1]. Using the 1997 parameters yields 266 seconds or roughly 5 minutes.

There are many competing technologies with magnetic disks. The longevity of disk storage with respect to DRAM is attributable to the lower cost of disk storage per MB by a factor of 1000 to one, the higher recording density, and also nonvolatility. The nonvolatility feature is quite important and micro-electro-mechanical system (MEMS)-based storage, which is nonvolatile and an order of magnitude faster than disks is a technology contending to replace disks, although it is currently ten times more expensive per MB [12]. Another interesting technology is *Magnetic RAM* (MRAM), which has approximately DRAM speeds, but is less costly, stores hundreds of megabytes per chip and is nonvolatile. It will take time for these technologies to replace disks, so given the current dominance of magnetic disks and disk arrays the current chapter will solely concern itself with this technology.

41.4 Disk Arm Scheduling

A disk is at its best when it sequentially transfers large blocks of data, so that it is operating close to its maximum transfer rate. Large block sizes are usually associated with *synchronous requests*, which are based on processes that generate requests to successive blocks of data. Synchronous requests commonly occur in multimedia applications, such as video-on-demand (VOD). Requests have to be completed at regular time intervals to allow glitch-free video viewing. Disk scheduling is usually round-based, i.e., a set of streams are processed periodically in a fixed order. An *admission control policy* may be used to ensure that processing of a new stream is possible with satisfactory performance for all streams and that this is done by taking into account buffer requirements.

Discrete requests originate from an infinite number of sources and the arrival process in this case is usually assumed to be Poisson (with rate λ). In an OLTP system sources correspond to transactions executing at a high level of concurrency, which generate I/O requests during their execution. A disk access is only required when the requested data block is not cached at a level of the memory hierarchy preceding the disk.

OLTP applications which tend to have stringent response time requirements generate accesses to small blocks of data, e.g., 96% to 4 KB and 4% to 24 KB [13].

In this section we first discuss basic scheduling policies for discrete requests to small randomly placed data blocks. Given the importance of the *shortest access time first* (SATF) policy, we discuss its variations in a separate section. We next proceed to discuss some hybrid disk scheduling methods, and this is followed by the scheduling of mixed requests, i.e., discrete and random requests. Anticipatory disk scheduling is discussed last.

41.4.1 Basic Disk Scheduling Methods

The default FCFS scheduling policy provides a rather poor performance in the case of accesses to small randomly placed data. Excluding the time spent at the disk controller, the mean service time is the sum of a mean (random) seek time, the mean latency, and transfer time. The transfer time of small (4 or 8 KB) blocks tends to be negligibly small with respect to positioning time (sum of seek time and rotational latency).

The observation that disk queue-lengths are short was used as an argument to discourage disk scheduling studies [14], but evidence to the contrary, that queue-lengths can be significant is given in Figure 1 in [15]. It is important to note that with the advent of the SCSI protocol the scheduling of disk requests are enqueued at the disk controller, which carries out request scheduling, rather than the operating system.

The high ratio of the maximum seek and rotation time for some early disk drives [15] served as the motivation for the *shortest seek time first* (SSTF) and (SCAN) policies, which are aimed at reducing the seek time [16].

SSTF schedules enqueued requests according to the seek distance from the currently completed request. SSTF is a greedy policy, so that requests in some disk areas may encounter starvation and long delays.

The SCAN (or elevator) scheduling policy moves the disk arm in alternate directions, stopping at cylinders to serve all pending requests and is thus less prone to starvation than SSTF.

Cyclical SCAN (CSCAN) returns the disk arm to one end after each scan, in order to alleviate the bias in serving requests on middle disk cylinders twice. LOOK and CLOOK are a minor variation of SCAN and CSCAN, reversing the direction of the scan as soon as there are no more requests in that direction, rather than reaching the innermost and/or outermost cylinder if there are no requests in that direction. We refer to SCAN and CSCAN while more strictly we mean LOOK and CLOOK.

The SATF or SPTF scheduling policy gives priority to requests which when processed next (i.e., after the completion time of the current request) will minimize positioning time [15]. This policy is desirable for modern disks, where seek time is comparable to rotational latency. SATF is similar to SJF policy, which is applicable when the mean service times of requests are known *a priori*, but is different from SJF in that disk service times are determined by the order in which they are processed.

Several simulation experiments have shown that SPTF, which minimizes positioning time, outperforms SSTF and SCAN in minimizing the mean response time [17,18]. Methods to improve SPTF performance even further and methods to bound percentiles of its response time are discussed in the next subsection.

41.4.2 Variations of the SATF Policy

Prioritizing the processing of one category of requests with respect to another works when there is a distinct category of requests whose performance needs to be improved, The *HOL priority queuing discipline* [7] serves requests (in FCFS order) from the higher priority queue until it is empty, at which point it switches to a lower priority queue.

The SATF policy can be modified by giving conditional priority to read requests with respect to write requests (any other two or more request types can be handled in this manner) [18]. SATF winner read requests are processed unconditionally, while the service time of an SATF winner write request is compared to the best read request and the write request is processed only if $\bar{x}^{best}_{write}/\bar{x}^{best}_{read} < t$, where the threshold $0 \leq t \leq 1$. In effect $t = 1$ corresponds to pure SATF, while $t = 0$ is a reader priority discipline, which is similar to HOL except that the FCFS policy is replaced by the SATF policy. Simulation results show that an intermediate value of t can be selected, which improves read response time at the cost of a small reduction in throughput.

SATF performance can be improved by applying lookahead [18]. Consider an SATF winner request A, whose processing according to SATF will be followed by request X. There might be some other request B, which when followed by Y (also according to SATF) yields a total processing time $T_B + T_Y < T_A + T_X$. With n requests in the queue, the cost of the algorithm increases from $O(n)$ to $O(n^2)$. In fact the second requests (X or Y) may not be processed at all, so that in comparing the sum of the processing times, the processing time of the second request is multiplied by a discount factor $0 \leq \alpha \leq 1$. The improvement in performance due to lookahead is more significant if requests demonstrate locality.

SATF in its basic form is prone to starvation. The *weighted shortest time first* (WSTF) also considered in this study computes the effective positioning time as $T^{eff}_{pos} = T_{pos}(1 - w/W_M)$, where w is the waiting time and W_M the maximum waiting time [19]. The difficulty in implementing this method is selecting the value of W_M. Note the similarity of this method to *time-dependent priorities method* in [7], where the priority of a job increases as function of its waiting time.

Two other proposals maintain a sliding window on the number of considered requests and the arrival time of requests. The goal is to improve the standard deviation of response time, hopefully at a small sacrifice to its mean value. One such proposal maintains a sliding window so that only the first n requests in a FCFS queue are considered for SATF scheduling [15]. The window-based SATF method only considers requests, which have arrived within a time window T, since the arrival of the oldest request in the queue [20]. The difficulty with both methods is finding the optimum n and T.

Additional SATF-based methods are described below.

41.4.3 Hybrid Disk Scheduling Methods

There have been many proposals for hybrid disk scheduling methods, which are combinations of other well-known methods. We discuss two variations of SCAN and two policies which combine SCAN with SSTF and SPTF policies in order to reduce the variance of response time.

In *N-step SCAN* the request queue is segmented into subqueues of length N and requests are served in FCFS order from each subqueue [21]. When $N = 1$ we have the FCFS policy, otherwise when N is very large N-step SCAN amounts to the SCAN policy.

FSCAN is another variation of SCAN with $N = 2$ queues [22]. Requests are served from one queue, while the other queue is being filled with requests. The SCAN policy only serves requests, which were there at the beginning of the SCAN, but no new requests are served. Note that it is possible some of these new requests can be served at no cost (opportunistically during the latency phase of another request) or little additional cost (e.g., following a processed request on the same track). When $N = 1$ we have the SCAN policy and when $N \to \infty$ the FCFS policy.

The $V(R)$ disk scheduling algorithm ranges from V(0) = SSTF to V(1) = SCAN, so that it provides a continuum of algorithms. To bias the arm to continue its movement in the direction of the current scan, $d_{bias} = R \times C$, where C denotes the number of disk cylinders, is subtracted from the seek distance in that direction. The value of R is varied in simulation results to determine the value that minimizes the sum of the mean response time and a constant k times its standard deviation. This sum is tantamount to a percentile of response time. It is shown that for lower arrival rates SSTF is best, i.e., $R = 0$, while at higher arrival rates $R = 0.2$ provides the best performance.

Grouped shortest time first (GSTF) is a combination of SCAN and STF (i.e., SATF) policies [19]. A disk is divided into groups of consecutive cylinders and the disk arm completes the processing of requests in the current group according to SATF, before proceeding to the next group. When there is only one group we have SATF and with as many groups as cylinders we have SCAN.

41.4.4 Disk Scheduling for Continuous Data Requests

Continuous data requests have an implicit deadline associated with the delivery of the next data block, e.g., video segment in a video stream for glitch-free viewing.

The *earliest deadline first* (EDF) scheduling policy [23] is a natural choice in that it attempts to minimize the number of missed deadlines. On the other hand it incurs high positioning overhead, which is manifested by a reduction in the number of video streams that can be supported. SCAN-EDF improves performance by using SCAN, while servicing requests with the same deadline.

Scheduling in *rounds* is a popular scheduling paradigm, so that the successive blocks of all active requests need to be completed by the end of the round. The size of the blocks being read and the duration of the rounds should be chosen carefully to allow glitch-free viewing. RR, SCAN, or *group sweeping scheduling* (GSS) [24] policies have been proposed for this purpose.

In addition to continuous requests (C-requests), media servers also serve discrete requests (D-requests). One method to serve requests is to divide a round to subrounds, which are used to serve requests of different types. More sophisticated scheduling methods are described in [25], two of which are described here.

One scheduling method serves C-requests according to SCAN and intervening D-requests according to either SPTF or OPT(N). The latter determines the optimal schedule after enumerating all $N!$ schedules ($N = 6$ is used in the paper).

The *FAir MIxed-scan ScHeduling* (FAMISH) *method* ensures that all C-requests are served in the current round and that D-requests are served in FCFS order. More specifically this method constructs the SCAN schedule for C-requests, but also incorporates the D-requests in FCFS order in the proper position in the SCAN queue, up to the point where no C-request misses its deadline. D-requests can also be selected according to SPTF. The relative performance of various methods is studied using simulation and reported in [25].

There are two categories of timing guarantees for I/O requests for continuous media, also referred to as *service levels*. *Deterministic* and statistical service levels guarantees that all and 99% of requests will be completed on time, respectively. The latter category is meaningful for *variable-bit-rate* (VBR) rather than *constant-bit-rate* (CBR) streams, since the required I/O bandwidth for VBR varies and provisioning for worst case behavior will degrade performance inordinately.

Scheduling of I/O requests in the context of scheduling real-time transactions is discussed in [26]. Transactions are served in FCFS order or assigned priorities based on EDF or *least slack time* (LST) policies [27]. LST is different from EDF in that the scheduler takes into account the processing acquired by the transaction which is being executed according to a *preemptive resume policy* [7]. I/O requests may be assigned the same priority as the transaction which initiated them, so that in effect disk requests are processed in a random order, with no performance improvement over FCFS scheduling, i.e., a poor disk response time is to be expected.

Scheduling with *quality-of-service* (QOS) guarantees requires the investigation of the following issues [28]: controlling and limiting the allocation of resources for different levels, scheduling of resources to meet performance goals, determining system design and its parameters, investigating the tradeoffs between statistical, and deterministic guarantees on throughput.

In spite of the randomness of disk service times, the scheduling problem in the deterministic case can be approached as follows: requests are grouped into rounds and served according to a seek optimizing policy, so that all requests will meet their deadline (with a worst-case assumption for latency). A two level queuing model first proposed in [29] and adopted in [28] for supporting multiple QOS levels, where the first level queue acts as a scheduler, as well as an admission controller to a *pool of requests* from which requests are scheduled on disks.

There are three request categories: (i) periodic requests, which are CBR or VBR and require service at regular intervals of time (deterministic or statistical guarantee), (ii) interactive requests require quick service as for playing a video-game (best effort mean response time), and (iii) aperiodic requests correspond to regular file transfers (guarantee of a minimum level of service or bandwidth, but this is of course possible if the disks are not 100% utilized).

To deal with bursts of interactive requests *leaky bucket control* is utilized in [30], e.g., a limited number of such requests are admitted per second into a pool from which requests are scheduled on disk. While the proposed techniques improve throughput significantly over other methods, further improvements are considered possible by dynamic allocation and adaptive performance guarantees. That remains an area requiring further research.

41.4.5 Disk Arm Prepositioning

The service time of disk accesses can be reduced by prepositioning the disk arm, e.g., positioning the disk arm at the middle disk cylinder will reduce the seek distance from $C/3$ to $C/4$ for random requests [31]. With mirrored disks (see Section 41.7) it is best to position one arm at $C/4$ and another arm at $3C/4$, when all requests are reads, so that the positioning time is reduced to $C/8$ [31]. It is shown in [31] that when the fraction of write requests is w then both arms should be placed at $C/2$ when $w > 0.5$ and otherwise at $1/2 \pm s$, where $s_{opt} = 0.25(1 - 2w)/(1 - w)$.

Given the reduction in mean seek time, there will also be an improvement in mean response time when the arrival rate of user requests (λ) is very low (both for single and mirrored disks). As λ is increased there is a point beyond which user requests are delayed by prepositioning requests and their response time is degraded.

The *anticipatory disk scheduling* method described in [32] defers the processing of a new request, until it has ascertained that there are no further pending requests in the area of disk at which the R/W head resides. The improvement in performance due to this method is explored in the context of Apache file servers, Andrews filesystem, and the TPC-B benchmark. This is in effect a demonstration that a *nonwork-conserving scheduler*, which introduce *forced idleness* [7], can improve performance in some cases.

41.5 Techniques to Improve Disk Performance

The following methods to improve disk performance are discussed in this section: (i) disk reorganization and defragmentation; (ii) log-structured file systems; (iii) active disks and free-block disk scheduling.

41.5.1 Disk Reorganization and Defragmentation

The simplest form of disk reorganization is defragmentation, which attempts to place the *fragments* of a file into one contiguous region. This will obviously improve file system performance, because the overhead of positioning the R/W head at the beginning of the file needs to be incurred only once, in case the whole file has to be transferred sequentially. Defragmentation programs are packaged with some operating systems and carry out their task when the system is otherwise idle.

A more sophisticated method for improving the efficiency of access of files in a disk is the *organ pipe arrangement*, which places the most popular file on the middle disk cylinder and other files are placed next to it, alternating from one side to the other [33]. More formally we have n records denoted by $R_1, \ldots, R_n$, which are accessed with probability $p_1, \ldots, p_n$, which sum to one. Since the records have equal lengths, they can be postulated to represent dots which are allocated on a real line at points $0, \pm 1, \pm 2, \ldots$.

The objective function of the allocation is to minimize the expected head travel distance: $D = \sum_{\{i,j\}} p_i p_j d(i, j)$ and given that i and j are located at I and J, we have $d(i, j) = |I - J|$, which is the Euclidean distance of the points. Given that $p_1 \geq p_2 \geq \cdots \geq p_n$, then D is minizimized with the following arrangement: $(\ldots R_4, R_2, R_0, R_1, R_3, \ldots)$. The distance for a random allocation is $n/3 + O(1)$, an optimal static allocation yields $7n/30 + O(1)$, while a dynamic allocation yields $7n/30 + \alpha n + o(1)$

(the details of this allocation are not discussed here for the sake of brevity). In the case of a zoned disk the optimal starting point is between the outermost and the middle disk cylinders.

The optimal solution is NP-hard in the static case when the records have variable lengths [34]. Two heuristics to allocate records in this case to minimize the expected seek time are presented in [34].

In mass storage systems, which is a form of tertiary storage, tape cartridges are placed in two dimensions and there is a R/W head that can move horizontally on a bar parallel to the x axis, which itself can move up or down the y axis. The access time is the maximum of the two durations. The placement of cartridges when their access frequencies are known is discussed in [33].

File access frequencies, which vary over time, can be determined by monitoring file accesses. This information can then be used to adaptively reorganize the files to approximate an organ pipe organization [35].

The average seek distance can be reduced further by allocating disk files by taking into account the frequency with which files are accessed together. Clustering algorithms utilizing the file access pattern have been proposed for this purpose, see, e.g., [36].

Log-structured arrays (LSA), which are described in Section 41.8.1 follow a combination of the two approaches [37], i.e., combine frequently accessed blocks into segments, which are then written onto the middle disk cylinders.

41.5.2 Log-Structured File Systems

The *log-structured file systems* (LFS) paradigm is useful in an environment where read requests are not common, while there are many writes. This is so in a system with a large cache, where most read requests are satisfied by cache accesses and the disks are mainly used for writing, since there is usually a timeout for modified data in the cache to be destaged onto disk. Another operational model is that the data is cached at client's workstation or PC. The user is continuously involved in modifying the locally cached file, so that this operation does not require disk I/O, but saving the file occasionally does. If there are many such users, a relatively heavy write load will be generated at the disks of the centralized server.

LFS accumulates modified files in a large cache, which is destaged to disk in large segments, e.g., the size of a cylinder, to prorate arm positioning overhead [38]. Space previously allocated to altered files is designated as free space and a background garbage collection process is used to consolidate segments with free space to create almost full and absolutely empty segments for future destaging.

The write cost in LFS is a steeply increasing function of the utilization of disk capacity (u), which is the fraction of live data in segments (see Figure 3 in [38]) and a crossover point with respect to improved UNIX's FFS occurs at $u = 0.5$. The following segment cleaning policies are investigated: (i) when should the segment cleaner be executed? (ii) how many segments should it clean? (iii) the choice of the segment to be cleaned, and (iv) the grouping of live blocks into segments. The reader is referred to [38] for more details.

41.5.3 Active Disks and Free-Block Scheduling

Given the high value of disk arm utilization there have been recent proposals for *freeblock scheduling*, which utilizes the disk arm opportunistically in processing regular background, low-priority requests [39]. Freeblock scheduling can be used in conjunction with LFS for garbage collection, nearest-neighbor queries for multimedia search, and data mining [5,40,41].

Modern disks are equipped with microprocessors, so that even if these microprocessors are slower than those used in servers, the overall processing capacity at the disks may easily exceed that of the server. The off-loading of CPU processing to the disk controller works well for computations that can be localized to a disk. Otherwise activities at several disks have to be coordinated at the server. Another advantage of application downloading is the reduction of the volume of data to be transmitted.

41.6 Raid Concepts

The RAID paradigm was introduced in [42], which also introduced the classification of RAID level 1 through level 5. Two more RAID levels (RAID0 and RAID6) were added later [43]. In this section we start our discussion with the motivation for RAID, followed by two concepts that RAID levels have in common: data *striping* and *fault-tolerance* in the form of check disks.

41.6.1 Motivation for RAID Systems

The motivation for RAID was to replace *single large expensive disks* (SLEDs) with an array of inexpensive disks. This requires the rewriting of the OS to handle the so-called *fixed block architecture* (FBA) disks (with fixed size sectors), but an alternative solution turned out to be the winner.

In the past the operating systems, such as IBM's MVS (later OS/390 and now z/OS) used in conjunction with mainframe computers, was involved intimately in the processing of a disk request: initiating a seek and waiting for its completion; initiating a search for the required data block; and finally it was interrupted after the data was transferred via the *direct memory access* (DMA) unit.

Currently, the old-fashioned I/O requests to nonexisting disks (say IBM 3380 or 3390) with variable block size *extended count-key-data* (ECKD) formats are received by a RAID controller, which generate requests to FBA disks, which are the only disks that are available today. In effect SLEDs with ECKD format disks are simulated on inexpensive FBA disks.

One of the early advantages of RAID was the potential for higher parallelism, since files on SLEDs were partitioned into stripe units and allocated in a RR manner across multiple small inexpensive disks (striping is discussed in more detail in the next section). Today all disks are inexpensive and have a much higher capacity than (the early) SLEDs, but due to striping the data on a SLED can be allocated over multiple disks in the array.

Since the OS carries out request queuing at the level of nonexisting SLEDs, there might be cases when such queuing is unnecessary, since the data actually resides on two different disks. *Parallel access volume* (PAV) in z/OS obviates this serialization in request processing by providing multiple queues for original disks [44], but this improvement may be only available to IBM's disk arrays as a competitive advantage. This is an example of a trivial scheduling decision that can improve performance in practice.

Disks with the SCSI interface have requests queuing capability, but unfortunately the scheduling policies of disk controllers are not disclosed. A further complicating factor is the onboard disk cache, which is used for prefetching and also stores data to be written to disk. By treating a disk as a black box and subjecting it to a stream of requests it should be possible to conjecture its scheduling and caching policies. Similar approaches have been used to extract disk parameters [45].

41.6.2 Striping

Data access skew was a common occurrence in large mainframe computer installations, i.e., disk utilizations were not balanced and only a few disks had a high utilization, while the majority of disks were almost idle. Tools were developed to optimize the allocation of files to minimize the average response time for accessing files, or some other performance measure. Such tools take into account file access frequencies, the characteristics of heterogeneous disks, and the bandwidth of the associated I/O paths. Since a complete reallocation of files is costly, local perturbations to the current allocation are usually considered to be sufficient [46].

Striping has been used as a cure to the access skew problem. Files are partitioned into *striping units*, which are allocated in a RR manner at all disks, so that the striping units in a row constitute a *stripe*. It is argued that by intermixing data from all files at all disks will result in a balanced disk load.

Parity striping maintains a regular data layout, but allows enough space on each disk for parity protection [47]. A counter argument to striping is that without striping the user has direct control over data allocation so as to ensure that two hot files are not placed on the same disk, but this is not guaranteed with striping.

Another disadvantage of striping is that distributing the segments of a file among several disks makes backup and recovery quite difficult. For example, if more than two disks fail in a RAID5 system, we need to reconstruct all disks, rather than just the two disks that failed, but parities need to be recomputed in any case.

Randomized data allocation schemes are a variation to striping. A precomputed distribution to implement the random allocation in a multimedia storage server is presented in [48] and compared with data striping.

The striping unit size has been an area of investigation, since RAID5 was proposed. The striping unit should be large enough to satisfy the most commonly occurring accesses, e.g., 8 KB for transaction processing and 64 kilobyte for decision support applications. Otherwise multiple accesses to several disks would be required to satisfy one logical request. There are the following considerations: (i) in a lightly loaded system the parallelism is beneficial and will reduce the response time of the requests, (ii) in a heavily loaded system the overhead associated with positioning time will result in an increase in the mean overall response time, (iii) data access skew may result from larger stripe unit sizes, but very large stripe units are required for this to happen, and (iv) larger stripe units allow sequential accesses to large files, but disallow parallel accesses.

A study for maximizing performance in striped disk arrays [49] estimates the size of the optimal striping unit size in KB as the product of a factor (equal to 1/4), the average positioning time (in milliseconds), data transfer rate (in MB per second), and the degree of concurrency minus one, plus 0.5 KB.

If disks are used exclusively for parallel processing, synchronization delay for seek times can be minimized by moving the disk arms together. *Spindle synchronization*, a functionality available to SCSI disk drives, can be used to minimize synchronization time due to latency.

Skew across disk arrays is also possible, since the files allocated to different arrays have different access rates. One solution to this problem is to stripe across arrays.

41.6.3 RAID Fault-Tolerance and Classification

Fault-tolerance through coding techniques was proposed for early RAID designs to attain a reliability level comparable to SLEDs [42,50]. The most simple scheme utilized in RAID3 and RAID4 is based on an extra (parity) disk to protect N disks against the failure of a single disk. Note that the identity of this disk is known to the system. Let $d_n, 1 \leq n \leq N$ denote N data blocks at N RAID5 disks, then the parity block at the $(N+1)$th disk is given by $p \leftarrow d_1 \oplus d_2 \cdots \oplus d_N$, where $\oplus$ is the exclusive-OR operation.

RAID3 and RAID4 are similar in that they utilize a dedicated parity disk. RAID3 is used in parallel processing so that all disk drives are accessed together, while RAID4 utilizes a larger stripe unit than RAID3 and its disks may be accessed independently.

A weakness of RAID4 is that the parity disk may become a bottleneck for write-intensive applications, but this effect is alleviated by rotating the parity at the stripe unit level, so that all disks have the same number of parity blocks. The left symmetric (NE to SW diagonal) parity organization, has been shown to have several desirable properties [50].

RAID1 or mirrored disks is a degenerate case of RAID4 with one disk protected by the other. It is discussed in more detail in the next section.

The RAID6 classification was introduced to classify disk arrays, which have two check disks and can tolerate two disk failures. The first check disk (P) is parity and the second check disk (Q) is computed according to the Reed-Solomon code, which requires computations across finite fields. The EVENODD scheme has the same redundancy level as RAID6, which is the minimal level of redundancy, but only uses the parity operation [51].

The practice of using parity to protect against single disk failures continues, although the previous dichotomy of more reliable SLEDs vs. less reliable and inexpensive high volume disks has disappeared. The underlying reasons are as follows: (i) there is no data loss when a disk failure occurs and the system continues its operation by recreating requested blocks on demand, although performance is degraded and (ii) the system can automatically rebuild the contents of the failed disk on a spare disk. A *hot spare* is required, since most systems do not allow *hot swapping*, replacing a broken disk while the system is running.

41.7 RAID1 or Mirrored Disks

This section is organized as follows. We first provide a categorization of routing and disk scheduling policies in mirrored disks. We consider two cases where each disk has an independent queue and when the disks have a shared queue.

We next consider the scheduling of requests when an NVS cache is provided (in the shared queue) to hold write requests. This allows read requests to be processed at a higher priority than writes. Furthermore, writes are processed in batches to reduce the completion time of write requests by taking advantage of disk geometry.

41.7.1 Request Scheduling with Mirrored Disks

In addition to higher reliability, mirrored disks provide *more than twice* the access bandwidth of single disks for processing read requests, which tend to dominate write requests in many applications. This effect was first noted in [52] assuming seeks which are uniform over (0,1) and also quantified in [53] assuming uniform requests over C disk cylinders. Routing read requests to the disk providing the shorter seek distance results in its reduction (on the average) from $C/3$ to $C/5$ in the discrete and 5/24 in the continuous case. The expected seek distance for writes is the expected value of the maximum of seek distances at the two disks, which is $7C/15$ in the discrete and $11C/24$ in the continuous case.

Mirrored disks have two configurations as far as their queuing structure is concerned [54]: (i) *Independent Queues* (IQ): Read requests are immediately routed to one of the disks according to some routing policy, while write requests are sent to both disks, (ii) *Shared Queue* (SQ): Read requests are held in this queue, so that there are more opportunities for improving performance. Many variations are possible, e.g., deferred forwarding of requests from the shared queue (at the router) to independent queues.

Request routing in IQ can be classified as *static* or *dynamic*. Uniform and RR are examples of static routing, where the latter is easier to implement and improves performance by making the interarrival times *more regular*, i.e., Erlang with a coefficient of variation equal to $1/\sqrt{2}$, rather than exponential with a coefficient of variation equal to one. That this is so can be shown with the assumption that disk service times are exponentially distributed [54]. In fact both uniform and RR policies balance the queue-lengths.

The router, in addition to checking whether a request is a read or a write, can determine other request attributes, e.g., the address of the data being accessed. In the case of a read request, it can be routed to the disk providing the shortest positioning time, but this would be only possible if the disk layout and the current position of the arm is known. An approximation to the SSTF can be implemented based on the block numbers.

If a small set of files are being accessed at a high rate then it makes sense that each disk be assigned the processing of a subset of the files. This can be carried out indirectly without the knowledge of file identities, but checking the range of block numbers being accessed (assumed to be the same for both disks). Such *affinity based routing* will improve the hit ratio at the onboard buffer at each disk.

Dynamic policies take into account the number of requests at each disk, the composition of requests at a disk, etc. *Join the shortest queue* (JSQ) policy can be used in the first case, but this policy does not improve performance when requests have a high service time variability [55]. Simulation studies have shown that the routing policy has a negligible effect for random requests, so that performance is dominated by the local scheduling policy [54].

SQ provides more opportunities than IQ to improve performance, since more requests are available for minimizing the positioning time. For example, the SATF policy with SQ attains a better performance than IQ, since the shared queue is approximately twice the length of individual queues [54].

41.7.2 Scheduling of Write Requests with an NVS Cache

The performance of a mirrored disk system without an *nonvolatile storage* (NVS) cache can be improved by using a *write anywhere policy* on one disk (to minimize disk arm utilization and susceptibility to data loss), while the data is written in place later on the primary disk. Writing in place allows efficient sequential accesses, while a write anywhere policy can write data very quickly at the first available slot. A special directory is required to keep track of the individual blocks, however. The *distorted mirrors* method described in [56] operates in this manner.

Caching of write requests in NVS can be used to improve the performance of mirrored disks. Prioritizing the processing of read requests yields a significant improvement in response time, especially if the fraction of write requests is high. We can process write requests more efficiently by scheduling them in batches optimized with respect to the data layout on disk.

The scheme proposed in [57] runs mirrored disks in two phases, while one disk is processing read requests, the other disk is processing writes in a batch mode using CSCAN. The performance of the above method can be improved by [58]: (i) eliminating the forced idleness in processing write requests individually; (ii) using an exhaustive enumeration (for sufficiently small batch sizes, say 10) or SATF, instead of CSCAN, to find an optimal destaging order; and (iii) introducing a threshold for the number of read requests, which when exceeded, defers the processing of write batches.

41.8 RAID5 Disk Arrays

We describe the operation of a RAID5 system in normal, degraded, and rebuild modes with emphasis on the scheduling of disk requests.

41.8.1 RAID5 Operation in Normal Mode

The updating of a single data block from d_{old} to d_{new} requires the updating of the corresponding parity block p_{old}: $p_{new} \leftarrow d_{old} \oplus d_{new} \oplus p_{old}$. When d_{old} and p_{old} are not cached the writing of a single block requires four disk accesses, which is referred to as the *small write penalty*.

This penalty can be reduced by carrying out the reading and writing of data and parity blocks as *read-modify-writes* (RMW) accesses, so that extra seeks are eliminated. The data and parity are read, modified, and written back after one disk rotation. On the other hand it may be possible to process other requests opportunistically, before it is time to write after a disk rotation.

Simultaneously starting the RMW for data and parity blocks may result in the parity disk being ready for writing *before* the data block has been read. One way to deal with this problem is to incur additional rotations at the parity disk until the necessary data becomes available. A more efficient way is to start the RMW for parity only after the data block has been read. Such precedence relationships can be represented as a *task graph* [6] or *directed acyclic graphs* (DAGS) introduced in [59].

A RMW request based on the above discussion can be succinctly specified as the following two graphs, where the former graph provides more parallelism and may minimize completion time:

$$\{(\text{read old data})\ , (\text{read old parity}) < (\text{write new data})\ , (\text{write new parity})\}$$

$$\{(\text{read old data}) < (\text{read old parity}), (\text{write old data}) < (\text{write new parity})\}$$

Several techniques have been proposed to reduce the write overhead in RAID5. One method provides extra space for parity blocks, so that their writing incurs less rotational latency, i.e., the *write anywhere policy*. These techniques are reviewed in [60], which proposes another technique based on batch updating of parity blocks. The system logs the *difference blocks* (exclusive-OR of the old and new data blocks) on a dedicated disk. The blocks are then sorted in batches, according to their disk locations, so as to reduce the cost of updating the parity blocks.

Modified data blocks can be first written into a duplexed NVS cache, which has the same reliability level as disks. Destaging of modified blocks due to write requests and the updating of associated parity blocks can be deferred and carried out at a lower priority than reads.

Several policies for disk array caches are discussed in [61]. Destage is initiated when NVS cache utilization reaches a certain threshold (high-mark) and it is stopped when it reaches a low-mark. An alternative method based on the *rate* at which the cache is filled is proposed in [62] and shown to be an improvement over the previous method using a trace-driven simulation study.

The small write penalty can be eliminated if we only have a *full stripe write*. A *reconstruct write* is preferable when $n > N/2$ of the stripe units are being written. In this case we simply read the remaining $N + 1 - n$ stripe units to compute the parity. Reconstruct and full stripe writes are rare, especially if the stripe unit is large, unless we have a batch application.

41.8.1.1 Log-Structured Arrays

Log-structured array extends the LFS paradigm to RAID5 disk arrays [37], but this scheme seems to have been implemented in StorageTek's Iceberg disk array [43], which is in fact also a RAID6 and led to this classification in [43]. LSA stores a stripe's worth of data in the cache, before writing it out, to make full-stripe writes possible. The previous version of updated blocks is designated free space and garbage collection is carried out by a background process, to create new empty stripes. There are many interesting issues associated with the selection of the stripes to be garbage collected, which are beyond the scope of this discussion.

The processing required for space reorganization is a function of the utilization of the capacity of the disk. High and low thresholds can be used to initiate and stop the garbage collection process.

If we assume that disk reorganization is started when disks are idle, then we may model garbage collection as a vacationing server [3]. When each disk in the array is not busy with processing user requests, it reads a stripe unit from a stripe selected for garbage collection. The reading of each stripe unit corresponds to one vacation and the server repeats taking vacations until a user request arrives, at which point it resumes the processing of user requests after completing the current vacation. The $K \geq 2$ stripes read into the disk array controller memory can be compressed into K_f almost full stripes, creating $K_e = K - K_f$ empty stripes, where at best $K_f = 1$. Writing can be continued without a glitch as long as empty stripes are available. A preliminary analysis of LFS on a single disk is reported in [63].

41.8.2 RAID5 Operation in Degraded Mode

A RAID5 disk array can operate in *degraded mode* with one failed disk. Blocks on the failed disks can be reconstructed on demand by accessing all of the corresponding blocks on surviving disks according to a *fork-join request* and XORing them to recreate the block. For example, to access a block which resides on disk one: $d_1 \leftarrow d_2 \oplus d_3 \oplus \ldots d_{N+1}$, where one of the N blocks is the parity block. The time taken by a fork-join request is the maximum of the response times at all disks.

As far as writes are concerned, if the disk at which the parity block resides is broken then we simply write the data block. If say the $(N + 1)$th disk at which the data resided is broken and the parity resides on disk one then: $p_1 \leftarrow d_2 \oplus d_3 \ldots d_{N+1}$, which requires N reads and one write.

Given that each one of the surviving disks has to process its own requests, in addition to the fork-join requests, results in a doubling of disk loads when all requests are reads (we have assumed that disk loads were balanced due to striping). This requires the disk utilization to be below 50% in normal mode, when all requests are reads, but higher initial disk utilizations are possible when a fraction of disk requests are writes.

The *clustered RAID* (also called *parity declustering*) organization is proposed in [64] to reduce the increase in disk load in degraded mode. This is accomplished by selecting the *parity group size* to be smaller than the number of disks, so that a less intrusive reconstruction is possible, because only a fraction of disks are involved in the rebuild process. *Balanced incomplete block designs* (BIBDs) have been proposed to balance disk loads on finite capacity disks [65,66]. BIBD layouts are available for certain parameters only and *nearly random permutations* is a more flexible approach, but is more expensive computationally [67].

Properties for ideal layouts are given in [66]: (i) Single failure correcting: the stripe units of the same stripe are mapped to different disks. (ii) Balanced load due to parity: all disks have the same number of parity stripes mapped onto them. (iii) Balanced load in failed mode: the reconstruction workload should be balanced across all disks. (iv) Large write optimization: each stripe should contain $N - 1$ contiguous stripe units, where N is the parity group size. (v) Maximal read parallelism: reading $n \leq N$ disk blocks entails in accessing n disks. (vi) Efficient mapping: the function that maps physical to logical addresses is easily computable. The *permutation development data layout* (PDDL) is a mapping function described in [68], which has excellent properties and good performance in light and heavy loads.

41.8.3 RAID5 Operation in Rebuild Mode

In addition to performance degradation, a RAID5 system operating in degraded mode is vulnerable to data loss if a second disk failure occurs. If a hot spare is provided, the rebuild process is initiated as soon as possible when a disk fails. A hot spare is required in systems where a failed disk cannot be replaced without bringing the system down, i.e., disrupting system operation. Systems with hot swapping require (error-prone) operator intervention, while otherwise failed disks may be replaced periodically.

Instead of wasting the bandwidth of the hot spare, which does not contribute to disk performance, the *distributed sparing* scheme distributes a capacity equivalent to the spare disk as spare areas over $N + 2$ disks [69]. Distributed sparing is also used in PDDL [68]. The drawback of hot sparing is that a second round of data movement is required to return the system to its original configuration. The *parity sparing* method utilizes two parity groups, one with $N + 1$ and the other with $N' + 1$ disks, which are combined to form a single group with $N + N' + 1$ disks after one disk fails [70].

The two performance metrics in rebuild processing are the mean response times of user requests (especially read requests) when rebuild is in progress and the rebuild time. In fact the performance degradation with respect to degraded mode with disk-oriented rebuild policy (described below) is equal to the mean residual service time [2] of the rebuild unit, e.g., a track, which is roughly equal to half a disk rotation if several tracks are read each time.

Read redirection can be used to improve performance as rebuild progresses [64]. When a data block has been rebuilt on the spare disk it can simply be retrieved from that disk, so that invoking an expensive fork-join request is unnecessary. Of course blocks already rebuilt on the spare disk need to be updated.

The rate at which the spare disk can be rebuilt is higher than at which such data becomes available in ordinary RAID5 systems. This is because the volume of data to be read for rebuilding equals the volume at which it is being written, since the load on the spare disk approaches that of surviving disks only at time when the rebuild process nears its end. This is not so in clustered RAID, since the volume of data being written exceeds the volume of the data being read (per disk). To ensure that the rebuild process is not slowed down by the spare disk, a control policy is introduced in [64] which reduces the load in the spare disk by regulating the fraction of redirected reads. This scheme is less relevant if the workload is dominated by writes.

Three variations of RAID5 rebuild have been proposed in the research literature.

The *disk-oriented rebuild* scheme reads successive *rebuild units* from all surviving disks, while the disks are idle. When all the corresponding rebuild units (say disk tracks) have been read they are XORed to obtain the track on the failed disk, which is then written onto a spare disk. The performance of this method is studied via simulation in [66] and analytically in [8,69] using an M/G/1 queuing model with server vacations.

The *stripe-oriented rebuild* scheme synchronizes at the level of the rebuild unit. In other words we complete the reading of a rebuild unit from all disks, before proceeding to the next one. The reading of rebuild units may be carried out when the disk is idle, so as to reduce interference with user requests, although this is not strictly necessary. The performance of this method is evaluated via simulation in [66] and shown to be inferior compared to the former rebuild method. This is because for each rebuild unit we incur the synchronization overhead.

A third technique is to allow one rebuild request at a time and process them at the same priority as user generated requests [67]. When a rebuild request is completed at a disk, a new request is immediately substituted in its place. This method can be analyzed as a *queuing system with permanent customers* [71], but its relative performance with respect to others remains to be quantified. The processing of rebuild requests at the same priority as user requests is expected to degrade the response times of user requests, as compared to the case when rebuild processing is only carried out when the disk system is otherwise idle. On the other hand prioritizing the processing of user requests is expected to result in a reduction in the *mean time to repair* (*MTTR*).

It has been shown that the *mean time to data loss* (*MTDL*) in a RAID5 with $N + 1$ disks is: $MTDL = MTTF^2/(N(N+1)MTTR)$, where *MTTF*, the mean time to disk failure, is in the millions of hours. This implies that even a factor of two reduction in MTTR will not have much affect on the RAID5 *availability* given as $MTDL/(MTTR + MTDL) \approx 1$.

Disks indicate problems in reading data blocks and other failures to the disk array controller, which may predict that a disk is about to fail. The contents of such a disk are copied onto a spare disk, with the rebuild process being invoked sparingly, only for damaged sectors and tracks. Algorithms to predict disk failures have been implemented in operational systems and it has been observed that disk failures are correlated, so that simple reliability models, such as the one above, tend to overestimate the MTDL.

41.9 Performance Evaluation of Scheduling Methods

Simulation and analytic studies have been used extensively in evaluating the performance of disk scheduling policies. We first discuss the performance evaluation of drums and disks, followed by the performance evaluation of disk arrays.

41.9.1 Performance Analysis of Drums

Drums preceded disks because of their simpler construction. In drums data is recorded on parallel tracks on its surface and they differ from disks in that there is a fixed R/W head per track. The drum is rotated at a fixed velocity to make data accessible to the R/W heads. The R/W heads in drums are shared among a fewer data blocks than in disks, so that drums were more expensive and were replaced by disks.

There are two types of drums: *paging drums* and *file drums*. The paging drum, used as a backup device for virtual memory systems, stores fixed sized blocks or pages on fixed boundaries, referred to as sectors. File drums store variable sized blocks starting at *arbitrary points* on the circumference of a track, which in fact correspond to word boundaries. *Sectored drums* store variable size blocks starting on sector boundaries.

Drum performance with respect to the FCFS policy can be improved if the instantaneous position of the drum is known by adopting the *shortest latency time first* (SLTF) policy. It is a greedy policy which simply selects the next request to be served to be the one which minimizes latency (note similarity to the SPTF policy for disks). Given a fixed number of randomly placed variable sized requests it is shown that [99] *an SLTF schedule for a collection of records is never as much as one revolution longer than an optimal schedule.*

The *minimal total processing time* (MTPT) algorithm determines the minimal processing time for N requests in $O(N \log N)$ time, but improved and more expensive variations to MTPT are available [72]. Simulation results show that SLTF outperforms MTPT for the majority of considered workloads.

Analytic solutions of drums with FCFS and SLTF policies are relatively simple, since drum service time has only two components, latency and transfer time. If we assume Poisson arrival drums can be analyzed using an M/G/1 queuing model, enhanced with vacationing servers in some cases [6,72,73].

41.9.2 Simulation and Analytic Studies of Disks

Disk simulation studies can be classified into trace- and random-number driven simulations. Trace-driven simulation studies have shown that an extremely detailed model is required to achieve high accuracy in estimating disk performance [17,74].

One of the problems associated with simulating disk drives is having access to their detailed characteristics. The DIXTRAC tool automates the extraction of disk parameters [45], which can be used by the DiskSim simulation tool [75], which is a successor to RAIDframe [59].

An I/O trace, in addition to being used in trace-driven simulations, can be analyzed to characterize the I/O request pattern, which can then be used in a random-number driven simulation study. For example, the analysis of an I/O trace for an OLTP environment showed that 96% of requests are to 4 KB blocks and 4% to 24 KB blocks [13].

In dealing with discrete requests most analytic and even simulation studies assume Poisson arrivals. With the further assumption that disk service times are independent, the mean disk response time with FCFS scheduling can be obtained using the *M/G/1 queuing model* [3,76]. There is a dependence among successive disk requests as far as seek times are concerned [77], but simulation studies have shown that this queuing model yields fairly accurate results.

The analysis is quite complicated for other scheduling policies [78], even when simplifying assumptions are introduced to make the analysis tractable. The analysis of the SCAN policy in [79] assumes (i) Poisson arrivals to each cylinder; (ii) the disk arm seeks cylinder-to-cylinder, even visiting cylinders not holding any requests; and (iii) satisfying a request takes the same time at all cylinders. Clearly this analysis cannot be used to predict the performance of the SCAN policy in a realistic environment.

Most early analytic and simulation studies were concerned with the relative performance of various disk scheduling methods [21], e.g., SSTF has a better performance than FCFS [77], at high arrival rates SCAN outperforms SSTF [78]. Several studies propose a new scheduling policy and use simulation to compare its performance with respect to standard policies [19,80]. Two relatively recent studies compare the performance of several scheduling methods [17,18].

A theoretical analysis of disk scheduling is reported in [81]. The authors consider the offline scheduling of a fixed number (n) of disk requests to minimize access time. Disk service time is specified by a convex (nondecreasing) reachability function ($f(.)$) from one request to another, which are represented as points on the surface of the disk. The disk is represented by a $2\pi \times 1$ rectangle, where request i is specified by its coordinates $(r_i, \theta_i), 0 \leq r_i \leq 1, 0 \leq \theta_i \leq 2\pi$. The distance between two requests R_i and R_j is then $d(R_i, R_j) = \min\{k : f(\theta_j - \theta_i + 2k\pi) \geq |R_j - R_i|\}$.

In other words, the distance between two points is defined by the angular distance plus k rotations required to move the arm radially (k is set to an integer k to facilitate the proof). The NP-hardness of the problem is shown via a reduction to a restricted version of the Hamiltonian Circle Problem (a variant of *traveling salesman problem* (TSP)). An approximation algorithm which takes $(3/2)T_{opt} + a$ disk rotations is developed, where T_{opt} is the number of disk rotations for the optimal algorithm and a is a constant independent of n.

41.9.3 RAID Performance

There have been numerous analytical and simulation studies of disk arrays. In fact early work on RAID was concerned with RAID reliability. Experimental results showed that the disk failures follow an exponential distribution [50], so that Markov chain models can be utilized for reliability modeling [76]. All components, rather than just disks should be taken into account in reliability modeling. The cross-hatch disk array connects $(N + 1) \times (N + 1)$ disks to a horizontal and a vertical bus, but parity groups are assigned diagonally, so that a higher reliability is achieved [82].

Markovian models (with Poisson arrivals and exponential service times) have been used successfully in investigating the relative performance of variations of RAID5 disk arrays [83]. Fork-join queuing delays required for the analysis in degraded mode are based on an approximation developed in [84].

Performance evaluation studies of RAID5 disk arrays based on the M/G/1 queuing model are more accurate than Markovian models, since disk service time is not exponential. Most notably RAID5 performance is compared with parity striping [47] in [85] and with mirrored disks in [86]. An analysis of clustered RAID in all three operating modes is given in [67]. An analysis of RAID5 disk arrays in normal, degraded, and rebuild mode is given in [8] and [69], where the former (resp. latter) study deals with RAID5 systems with dedicated (resp. distributed) sparing. The analysis in [8] and [69] is presented in a unified manner in [87], which also reviews other analytical studies of disk arrays.

A simulation study of clustered RAID is reported in [66], which compares the effect of disk-oriented and stripe-oriented rebuild. The effect of parity sparing on performance is investigated in [70]. The performance of a RAID system tolerating two disk failure is investigated in [88] via simulation, while an M/G/1-based model is used to evaluate and compare the performance of RAID0, RAID5, RAID6, and EVENODD organizations in [89].

Simulation study of mirrored disks with independent and shared queues is reported in [54]. It is difficult to evaluate the performance of disk arrays via a trace-driven simulation, because in effect we have a very large logical disk, whose capacity equals the sum of the capacities of several disks. This difficulty is overcome in [12], which utilizes a trace-driven simulation study to investigate the performance of a heterogeneous RAID1 system: MEMS-based storage backed-up by a magnetic disk is reported in [12].

As far as MEMS-based storage is concerned, it moves the R/W heads arranged in a two-dimensional array in x-y coordinates to align them with the data to be read. Several simulation and analytic tools have been developed to evaluate the performance of MEMS-based storage, see, e.g., [90].

41.10 Conclusions

The emphasis of this chapter is on the scheduling of discrete requests to randomly placed data blocks, but we also consider the scheduling of a mixture of continuous and discrete requests. The SATF policy yields the best performance, which is the reason why we present variations of this policy to reduce the variance of response time.

We considered two configurations for mirrored disks: independent queues and a shared queue. The routing of requests in the former case can be static or dynamic, i.e., make use of information about the current state of the disk. Static routing may interpret the request to achieve a performance advantage. With a shared queue better performance is attainable due to resource sharing.

In RAID5 and RAID6 disk arrays single user requests may be translatable into dags providing a time-space trade-off. Data allocation is an important issue in disk arrays, since disk loads should be balanced in degraded as well as normal modes. Parity declustering was proposed to ensure that in the worst case (with all read requests) a disk array can process requests at more than 50% of its maximum disk utilization in normal mode.

Disks and disk arrays can also be analyzed under favorable assumptions, when the FCFS or head-of-the-line priority discipline is in effect, but a realistic analysis of disk scheduling policies, especially the SATF policy remains a challenging problem.

Two important topics beyond the scope of this discussion are (i) design tools for automating the data allocation and configuration of large storage systems [91] and (ii) tools for managing storage [92], since the cost of managing storage is much higher than its cost. Finally *direct attached storage*, which is the usual paradigm of attaching disks to servers, is being replaced by *network attached storage*, with disk arrays accessible by the SCSI protocol, which runs on top of TCP/IP. This introduces many new and challenging problems, see [93].

References

[1] J. Gray and G. Graefe. The five-minute rule ten years later and other computer storage rules of thumb, *ACM SIGMOD Rec. 26*(4): 63–68 (1997).

[2] L. Kleinrock. *Queuing Systems Vol. I: Theory*, Wiley-Interscience, 1975.

[3] H. Takagi. *Queueing Analysis. Vol. 1: Vacation and Priority Systems*, North-Holland, 1991.

[4] M. Litzkow, M. Livny, and M. Mutka. Condor — A hunter of idle workstations, *Proc. 8th ICDCS*, 1988, pp. 104–111.

[5] E. Riedel, C. Faloutsos, G. R. Ganger, and D. F. Nagle. Active disks for large scale data processing, *IEEE Comput. 34*(6): 68–74 (2001).

[6] E. G. Coffman, Jr. and P. J. Denning. *Operating Systems Theory*, Prentice-Hall, 1972.

[7] L. Kleinrock. *Queueing Systems Vol. II: Computer Applications*, Wiley-Interscience, 1976.

[8] A. Thomasian and J. Menon. Performance analysis of RAID5 disk arrays with a vacationing server model, *Proc. 10th ICDE Conf.*, 1994, pp. 111–119.

[9] A. Thomasian. Priority queueing in RAID disk arrays, *IBM Research Report RC 19734*, 1994.

[10] F. Douglis, P. Krishnan, and B. N. Bershad. Adaptive disk spin-down policies for mobile computers, *Symp. on Mobile and Location Ind. Comput.*, USENIX, 1995, pp. 121–137; *Comput. Syst. 8*(4): 381–413 (1995).

[11] S. Gurumurthi, A. Sivasubramaniam, M. Kandemir, and H. Franke. DRPM: Dynamic speed control for power management in server-class disks, *Proc. 30th ISCA*, 2003, pp. 169–181.

[12] M. Uysal, A. Merchant, and G. A. Alvarez. Using MEMS-based storage in disk arrays, *Proc. 2nd Conf. File Storage Technol. — FAST '03*, USENIX, 2003.

[13] K. K. Ramakrishnan, P. Biswas, and R. Karedla. Analysis of file I/O traces in commercial computing environments, *Proc. Joint ACM SIGMETRICS/Perform. '92 Conf.*, 1992, pp. 78–90.

[14] W. C. Lynch. Do disk arms move? *Perform. Eval. Rev. 1*(4): 3–16 (Dec. 1972).

[15] D. Jacobson and J. Wilkes. Disk scheduling algorithms based on rotational position, *HP Tech. Rpt. HPL-CSP-91-7rev*, 1991.

[16] P. J Denning. Effects of scheduling in file memory operations, *Proc. AFIPS Spring Joint Comput. Conf.*, 1967, pp. 9–21.

[17] B. L. Worthington, G. R. Ganger, and Y. L. Patt. Scheduling for modern disk drivers and nonrandom workloads, *Proc. ACM SIGMETRICS Conf. 1994*, pp. 241–251.

[18] A. Thomasian and C. Liu. Some new disk scheduling policies and their performance, *Proc. ACM SIGMETRICS Conf.*, 2002, pp. 266–267; *Perform. Eval. Rev. 30*(2): 31–40 (Sept. 2002).

[19] M. I. Seltzer, P. M. Chen, and J. K. Ousterhout. Disk scheduling revisited, *Proc. 1990 USENIX Summer Tech. Conf.*, 1990, pp. 307–326.

[20] A. Riska, E. Riedel, and S. Iren. Managing overload via adaptive scheduling, *Proc. 1st Workshop Algorithms Arch. Self-Managing Systems*, 2003.

[21] T. J. Teorey and T. B. Pinkerton. A comparative analysis of disk scheduling policies, *Commun. ACM 15*(3): 177–184 (1972).

[22] E. G. Coffman, Jr., E. G. Klimko, and B. Ryan. Analyzing of scanning policies for reducing disk seek times, *SIAM J. Comput. 1*(3): 269–279 (1972).

[23] C. L. Liu and J. W. Layland. Scheduling algorithms for multiprogramming in hard read-time environment, *J. ACM 20*(1): 46–61 (Jan. 1973).

[24] M. S. Chen, D. D. Kandlur, and P. S. Yu. Optimization of the grouped sweeping scheduling for heterogeneous disks, *Proc. 1st ACM Int'l Conf. Multimedia*, 1993, pp. 235–242.

[25] E. Balafoutis et al. Clustered scheduling algorithms for mixed media disk workloads in a multimedia server, *Cluster Comput. 6*(1): 75–86 (2003).

[26] R. K. Abbott and H. Garcia-Molina. Scheduling real-time transactions: A performance evaluation, *ACM Trans. Database Syst. 17*(3): 513–560 (Sept. 1992).

[27] J. W. S. Liu. *Real-Time Scheduling*, Prentice-Hall, 2000.

[28] R. Wijayaratne and A. L. Narasimha Reddy. Providing QOS guarantees for disk I/O, *Multimedia Syst. 8:* 57–68 (2000).

[29] P. J. Shenoy and H. M. Vin. Cello: A disk scheduling framework for next generation operating systems, *Proc. ACM SIGMETRICS Conf.*, 1998, pp. 44–55.

[30] ATM Forum. *ATM User Network Interface Specification*, Prentice-Hall, 1993.

[31] R. P. King. Disk arm movement in anticipation of future requests, *ACM Trans. Comput. Syst. 8*(3): 214–229 (1990).

[32] S. Iyer and P. Druschel. Anticipatory scheduling: A disk scheduling framework to overcome deceptive idleness in synchronous I/O, *Proc. 17th SOSP*, 2001, pp. 117–130.

[33] C. K. Wong. Minimizing head movement in one dimensional and two dimensional mass storage systems, *ACM Comput. Surv. 12*(2): 167–178 (1980).

[34] U. I. Gupta, D. T. Lee, J. Y.-T. Leung, J. W. Pruitt, and C. K. Wong. Record allocation for minimizing seek delay, *Theor. Comput. Sci. 16*: 307–319 (1981).

[35] S. Akyurek and K. Salem. Adaptive block rearrangement, *ACM Trans. Comput. 13*(2): 89–121 (May 1995).

[36] B. T. Bennett and P. A. Franaszek. Permutation clustering: An approach to online storage reorganization, *IBM J. Res. Dev. 21*(6): 528–533 (1977).

[37] J. Menon. A performance comparison of RAID5 and log-structured arrays, *Proc. 4th IEEE HPDC*, 1995, pp. 167–178.

[38] M. Rosenblum and J. K. Ousterhout. The design and implementation of a log-structured file system, *ACM Trans. Comput. Syst. 10*(1): 26–52 (Feb. 1992).

[39] C. R. Lumb, J. Schindler, G. R. Ganger, and D. F. Nagle. Toward higher disk head utilization: Extracting free bandwidth from busy disk drives, *Proc. 4th OSDI Symp.* USENIX, 2000, pp. 87–102.

[40] A. Acharya, M. Uysal, and J. H. Saltz. Active disks: Programming model, algorithms, and evaluation, *Proc. ASPLOS VIII*, 1998, pp. 81–91.

[41] E. Riedel, C. Faloutsos, G. R. Ganger, and D. F. Nagle. Data mining in an OLTP system (nearly) for free, *Proc. ACM SIGMOD Int'l Conf.*, 2000, pp. 13–21.

[42] D. A. Patterson, G. A. Gibson, and R. H. Katz. A case study for redundant arrays of inexpensive disks, *Proc. ACM SIGMOD Int'l Conf.*, 1988, pp. 109–116.

[43] P. M. Chen, E. K. Lee, G. A. Gibson, R. H. Katz, and D. A. Patterson. RAID: High-performance, reliable secondary storage, *ACM Comput. Surv. 26*(2): 145–185 (1994).

[44] A. S. Meritt et al. z/OS support for IBM TotalStorage enterprise storage server, *IBM Syst. J. 42*(2): 280–301 (2003).

[45] J. Schindler and G. R. Ganger. Automated disk drive characterization, *CMU SCS Technical Report CMU-CS-99-176*, 1999.

[46] J. L. Wolf. The placement optimization program: A practical solution to the disk file assignment problem, *Proc. ACM SIGMETRICS Conf.*, 1989, pp. 1–10.

[47] J. Gray, B. Horst, and M. Walker. Parity striping of disk arrays: Low-cost reliable storage with acceptable throughput, *Proc. 16th Int'l VLDB Conf.*, 1990, 148–161.

[48] J. R. Santos, R. R. Muntz, and B. Ribeiro-Neto. Comparing random data allocation and data striping in multimedia servers, *Proc. ACM SIGMETRICS Conf.* 2000, pp. 44–55.

[49] P. M. Chen and D. A. Patterson. Maximizing performance on a striped disk array, *Proc. 17th ISCA*, 1990, pp. 322–331.

[50] G. A. Gibson. *Redundant Disk Arrays: Reliable, Parallel Secondary Storage*, The MIT Press, 1992.

[51] M. Blaum, J. Brady, J. Bruck, and J. Menon. EVENODD: An optimal scheme for tolerating disk failure in RAID architectures, *IEEE Trans. Comput. 44*(2): 192–202 (Feb. 1995).

[52] S. W. Ng. Reliability, availability, and performance analysis of duplex disk systems, *Reliability and Quality Control*, M. H. Hamza (editor), Acta Press, 1987, pp. 5–9.

[53] D. Bitton and J. Gray. Disk shadowing, *Proc. 14th Int'l VLDB Conf.*, 1988, pp. 331–338.

[54] A. Thomasian et al. Mirrored disk scheduling, *Proc. Symp. Perform. Eval. Comput. Telecomm. Syst. — SPECTS '03*, 2003.

[55] W. Whitt. Deciding which queue to join: Some counterexamples, *Oper. Res. 34*(1): 226–244 (Jan. 1986).

[56] J. A. Solworth and C. U. Orji. Distorted mirrors, *Proc. 1st Int'l Conf. Parallel Distr. Inf. Syst. — PDIS*, 1991, pp. 10–17.

[57] C. Polyzois, A. Bhide, and D. M. Dias. Disk mirroring with alternating deferred updates, *Proc. 19th Int'l VLDB Conf.*, 1993, 604–617.

[58] A. Thomasian and C. Liu. Performance of mirrored disks with a shared NVS cache". (Unpublished)

[59] W. V. Courtright II et al. RAIDframe: A rapid prototyping tool for raid systems, http://www.pdl.cmu.edu/RAIDframe/raidframebook.pdf.

[60] D. Stodolsky, M. Holland, W. C. Courtright II, and G. A. Gibson. Parity logging disk arrays, *ACM Trans. Comput. Syst.* 12(3): 206–325 (1994).

[61] K. Treiber and J. Menon. Simulation study of cached RAID5 designs, *Proc. 1st IEEE Symp. High Perform. Comput. Arch. — HPCA*, 1995, pp. 186–197.

[62] A. Varma and Q. Jacobson. Destage algorithms for disk arrays with nonvolatile caches, *IEEE Trans. Comput. 47*(2): 228–235 (1998).

[63] J. T. Robinson and P. A. Franaszek. Analysis of reorganization overhead in log-structured file systems, *Proc. 10th Int'l Conf. Data Eng. — ICDE*, 1994, pp. 102–110.

[64] R. Muntz and J. C. S. Lui. Performance analysis of disk arrays under failure, *Proc. 16th Int'l VLDB Conf.*, 1990, pp. 162–173.

[65] S. W. Ng and R. L. Mattson. Uniform parity distribution in disk arrays with multiple failures, *IEEE Trans. Comput. 43*(4): 501–506 (1994).

[66] M. C. Holland, G. A. Gibson, and D. P. Siewiorek. Architectures and algorithms for on-line failure recovery in redundant disk arrays, *Distributed and Parallel Databases 11*(3): 295–335 (1994).

[67] A. Merchant and P. S. Yu. Analytic modeling of clustered RAID with mapping based on nearly random permutation, *IEEE Trans. Comput. 45*(3): 367–373 (1996).

[68] T. J. E. Schwarz, J. Steinberg, and W. A. Burkhard. Permutation development data layout (PDDL) disk array declustering, *Proc. 5th IEEE Symp. High Perform. Comput. Arch. — HPCA*, 1999, pp. 214–217.

[69] A. Thomasian and J. Menon. RAID5 performance with distributed sparing, *IEEE Trans. Parallel Distributed Syst. 8*(6): 640–657 (June 1997).

[70] J. Chandy and A. L. Narasimha Reddy. Failure evaluation of disk array organizations, *Proc. 13th Int'l Conf. Distributed Comput. Syst. — ICDCS*, 1993, pp. 319–326.

[71] O. J. Boxma and J. W. Cohen. The M/G/1 queue with permanent customers, *IEEE J. Selected Topics Commun. 9*(2): 179–184 (1991).

[72] S. H. Fuller. *Analysis of Drum and Disk Storage Units — LNCS 31*, Springer, 1975.

[73] S. H. Fuller and F. Baskett. An analysis of drum storage units, *J. ACM 22*(1): 83–105 (1975).

[74] C. Ruemmler and J. Wilkes. An introduction to disk drive modeling, *IEEE Comput. 27*(3): 17–28 (March 1994).

[75] http://www.pdl.cmu.edu/DiskSim/

[76] K. S. Trivedi. *Probability and Statistics with Reliability, Queueing and Computer Science Applications*, 2nd ed., Wiley-Interscience, 2002.

[77] M. Hofri. Disk scheduling: FCFS vs. SSTF revisited, *Commun. ACM 23*(11): 645–653 (Nov. 1980), (Corrigendum *24*(11): 772).

[78] E. G. Coffman, Jr. and M. Hofri. Queueing models of secondary storage devices, In *Stochastic Analysis of Computer and Communication Systems*, H. Takagi (editor), North-Holland, 1990, pp. 549–588.

[79] E. G. Coffman, Jr. and M. Hofri. On the expected performance of scanning disks, *SIAM J. Comput. 11*(1): 60–70 (1982).

[80] R. M. Geist and S. Daniel. A continuum of disk scheduling algorithm, *ACM Trans. Comput. Syst. 5*(1): 77–92 (1987).

[81] M. Andrews, M. A. Bender, and L. Zhang. New algorithms for disk scheduling, *Algorithmica 32*: 277–301 (2002).

[82] S. W. Ng. Crosshatch disk array for improved reliability and performance, *Proc. 21st ISCA*, 1994, pp. 255–264.

[83] J. Menon. Performance of RAID5 disk arrays with read and write caching, *Distributed and Parallel Databases 11*(3): 261–293 (1994).

[84] R. Nelson and A. Tantawi. Approximate analysis of fork-join synchronization in parallel queues, *IEEE Trans. Comput. 37*(6): 739–743 (1988).

[85] S.-Z. Chen and D. F. Towsley. The design and evaluation of RAID5 and parity striping disk array architectures, *J. Parallel and Distributed Comput. 10*(1/2): 41–57 (1993).

[86] S.-Z. Chen and D. F. Towsley. A performance evaluation of RAID architectures, *IEEE Trans. Comput. 45*(10): 1116–1130 (1996).

[87] A. Thomasian. RAID5 disk arrays and their performance evaluation, In *Recovery Mechanisms in Database Systems*, V. Kumar and M. Hsu (editors), pp. 807–846.

[88] G. A. Alvarez, W. A. Burkhard, and F. Cristian. Tolerating multiple failures in RAID architectures with optimal storage and uniform declustering, *Proc. 24th ISCA*, 1997, pp. 62–72.

[89] C. Han and A. Thomasian. Performance of two-disk failure tolerant disk arrays, *Proc. Symp. Perform. Eval. Comput. Telecomm. Syst. — SPECTS '03*, 2003.

[90] J. L. Griffin, S. W. Shlosser, G. R. Ganger, and D. F. Nagle. Modeling and performance of MEMS-based storage devices, *Proc. ACM SIGMETRICS 2000*, pp. 56–65.

[91] G. A. Alvarez et al. Minerva: an automated resource provisioning tool for large-scale storage systems, *ACM Trans. Comput. Syst. 19*(4): 483–518 (2001).

[92] J. Wilkes. Data services — from data to containers, *2nd Conf. File Storage Technol. — FAST '03*, USENIX. (Keynote Speech)

[93] K. A. Salzwedel. Algorithmic approaches for storage networks, *Algorithms for Memory Hierarchies, Advanced Lectures, Lecture Notes in Computer Science 2625*, Springer 2003, pp. 251–272.

[94] E. Anderson et al. Ergastulum: quickly finding near-optimal storage system designs, http://www.hpl.hp.com/research/ssp/papers/ergastulum-paper.pdf

[95] J. Gray and P. J. Shenoy. Rules of thumb in data engineering, *Proc. 16th ICDE*, 2000, pp. 3–16.

[96] E. D. Lazowska, J. Zahorjan, G. S. Graham, and K. C. Sevcik. *Quantitative System Performance*, Prentice-Hall 1984.

[97] E. K. Lee and R. H. Katz. The performance of parity placements in disk arrays, *IEEE Trans. Comput. 42*(6): 651–664 (June 1993).

[98] S. W. Ng. Improving disk performance via latency reduction, *IEEE Trans. Comput. 40*(1): 22–30 (Jan. 1991).

[99] H. S. Stone and S. H. Fuller. On the near-optimality of the shortest-latency time first drum scheduling discipline, *Commun. ACM 16*(6): 352–353 (1973).

42

Selfish Routing on the Internet

Artur Czumaj
New Jersey Institute of Technology

42.1 Introduction

In recent years, we are more and more frequently dealing with large networks, like the Internet, that are not only computationally very complex, but they are also following highly nonstandard mechanisms and rules. The complex structure of such large networks is caused by their size and growth, their almost spontaneous emergence, their open architecture, and their lack of any authority that monitors and regulates network operations. Classical analyses of such networks have made assumptions either about some global regulation mechanisms or were designed with cooperative users in minds; in large networks nowadays both of these assumptions seem to be not very realistic. This initiated in recent years a growing interest in the analysis of large, noncooperative networks, where all the users decide by themselves on their own behavior. In such networks, each user is typically interested only in maximizing its own benefits at the lowest possible costs, and its choice of action will depend on those of other network's users. The behavior of users in such an environment has been traditionally addressed in the framework of *game theory* that provides systematic tools to study and understand their socio-economic behavior, see, e.g., [1–8] and the references therein. A fundamental notion that arises in game theory is that of *Nash equilibrium*, which, informally, is the state where no user finds it beneficial to change it. Game theoretical models have been employed in networking in the context of traffic and bandwidth allocation, flow control, routing, etc. These studies

1-58488-397-9/$0.00+$1.50

typically investigate the properties of Nash equilibria and supply the insight into the nature of networking under decentralized and noncooperative control.

It is well known that Nash equilibria do not always optimize the overall performance of the system and exhibit suboptimal network performance. A classical example is the so-called "Prisoner's Dilemma" (see, e.g., [4,5]), which describes a 2-players game in which the optimal solution for both players may have the cost significantly better than the optimal strategy of each user independently, that is, in the Nash equilibrium. In order to understand the phenomenon of noncooperative systems, Koutsoupias and Papadimitriou [3] initiated investigations of the *coordination ratio*, which is the ratio between the cost of the *worst possible Nash equilibrium* and that of the *social* (i.e., overall) *optimum*. In other words, this analysis seeks the price of uncoordinated selfish decisions ("the price of anarchy"). Koutsoupias and Papadimitriou [3,7] proposed to investigate the coordination ratio for routing problems in which a set of agents is sending traffic along a set of parallel links with linear cost functions. This model has been later extended to more realistic models, most notably to more general cost functions (including in particular, queuing functions), and in which the underlying network is arbitrary. This research initialized also the study of properties of Nash equilibria vs. those of social optimum. And so, how to modify Nash equilibria to improve their cost, how to modify the underlying networks to ensure that the coordination ratio will be low, and how to converge to "good" Nash equilibria.

In this survey we will discuss some of the aspects that arose in that research, and our main focus is on the analysis of the performance in traffic routing in large noncoordinated networks.

42.1.1 The Models

Our main goal is to study the coordination ratio of traffic modeling in networks. The traffic modeling problem can be described as the problem of sending a set of packets (tasks) through a network to their destinations. The performance of the network depends on two main parameters: performance of the packet on the route from the source to the destination and the service time (or waiting time) observed at the destination. These two performance parameters are important to distinguish between two main models that we study in this survey: a *task allocation model* and a *routing flow model*. To some extent, one can see the task allocation model as the traffic modeling problem in which we concentrate attention only on the destination congestion parameter, while in the routing flow model we concentrate on the scenario in which the traffic can be modeled by a network flow. The differences between these two models are quite subtle and in some cases they even disappear.

In the *task allocation model* [3], we essentially consider the problem of allocating scheduling tasks to machines according to certain game theoretical assumptions. The important feature of these allocations is that each task is considered as a single entity and cannot be split to be assigned in a part to different machines. One can see this model as the routing model in which the underlying network consists of two nodes, a source and a sink, and a set of parallel links from the source to the sink [3]. In this model, the main parameter to be investigated is that of the link congestion, that is, what is the maximum cost associated with any machine (or any link). Equivalently, one can consider this model as that of focusing on the service time only; we study the performance of the system only in terms of the load observed at the destinations.

In the *routing flow model* [8,9], we assume that the *tasks* are to be routed to the servers (machines) and the focus of the problem is on efficient routing of the tasks. Additionally, in the routing flow model, unless clearly stated otherwise, we assume that there is an infinite number of tasks and each task corresponds to a negligible fraction of the overall network traffic and therefore it can be modeled as a *flow* (and hence it can be split).

42.2 Task Allocation Model

In this section we describe a basic routing model that is essentially a rather nonstandard variant of a classical makespan minimization scheduling problem. This model and the problem statement have been formally introduced by Koutsoupias and Papadimitriou [3] and we follow here their description.

42.2.1 Task Allocation Model with Linear Cost Functions

We consider a scheduling problem with m independent *machines* with *speeds* $s_1, \ldots, s_m$ and n independent *tasks* with *weights* $w_1, \ldots, w_n$. If all tasks have identical weights then we will talk about *identical tasks*; if all speeds are identical, then we will talk about *identical machines*. The goal is to allocate the tasks to the machines to minimize the maximum load (makespan) of the machines in the system.

We assume that each task can be assigned to a single machine but the decision to which machine it is to be assigned is determined by each user's strategy (here, and throughout the entire survey, by a user or an agent we mean the entity that decides about the allocation of a given task). If a task i is allocated deterministically to a machine in $[m]$ then we say it follows a *pure strategy*; if it is allocated according to some probability distribution on $[m]$, then we say it follows a *mixed strategy*. (Throughout the paper, we use notation $[N] = \{1, \ldots, N\}$.)

Let $(j_1, \ldots, j_n) \in [m]^n$ be a sequence of pure strategies, one for each task, its *cost* for task i is

$$\frac{1}{s_{j_i}} \cdot \sum_{j_k = j_i} w_k$$

which is the time needed for machine j_i chosen by task i to complete all tasks allocated to that machine[1].

Similarly, for a sequence of pure strategies $(j_1, \ldots, j_n) \in [m]^n$, the *load* of machine j is defined as

$$\frac{1}{s_j} \cdot \sum_{j_k = j} w_k$$

Given n tasks of length $w_1, \ldots, w_n$ and m machines of speed $s_1, \ldots, s_m$, let *opt* denote the *social optimum*, that is, the minimum cost of a pure strategy:

$$opt = \min_{(j_1,\ldots,j_n)\in[m]^n} \max_{j\in[m]} \frac{1}{s_j} \cdot \sum_{i: j_i = j} w_i \qquad (42.1)$$

For example, if all machines have the same unit speed ($s_j = 1$ for every $j \in [m]$) and all weights are the same ($w_i = 1$ for every $i \in [n]$), then the social optimum is $\lceil \frac{n}{m} \rceil$. Furthermore, it is easy to see that in any system

$$opt \geq \frac{\max_i w_i}{\max_j s_j} \qquad (42.2)$$

We notice also that computing the social optimum is $\mathcal{NP}$-hard even for identical speeds (because then, as observed in [3], it is essentially the partition problem).

Let p_i^j denote the probability that an agent $i \in [n]$ sends the entire traffic w_i to a machine $j \in [m]$. Let ℓ_j denote the *expected load* on a machine $j \in [m]$, that is,

$$\ell_j = \frac{1}{s_j} \cdot \sum_{i \in [n]} w_i p_i^j$$

For a task i, the *expected cost of task i on machine j* (or its *finish time* when its load w_i is allocated to machine j [3]) is equal to

$$c_i^j = \frac{w_i}{s_j} + \frac{1}{s_j} \cdot \sum_{t \neq i} w_t p_t^j = \ell_j + (1 - p_i^j) \frac{w_i}{s_j}$$

With the notation defined above, we are now ready to define the notion of Nash equilibria.

[1]In the original formulation of Koutsoupias and Papadimitriou [3], an additional additive "initialization" term L^{j_i} was used. However, since in all papers we are aware of all analyses assumed $L^{j_i} = 0$, we skipped that term in our presentation. We want to point out, however, that our analyses are not affected by these additive terms.

Definition 42.1 (Nash equilibrium)

The probabilities $(p_i^j)_{i\in[n],j\in[m]}$ *define a* Nash equilibrium *if and only if any task* i *will assign nonzero probabilities only to machines that minimize* c_i^j, *that is,* $p_i^j > 0$ *implies* $c_i^j \le c_i^q$, *for every* $q \in [m]$.

In other words, a Nash equilibrium is characterized by the property that there is no incentive for any task to change its strategy.

As an example, we observe that in the system considered above in which all machines have the same unit speed and all weights are the same, the uniform probabilities $p_i^j = \frac{1}{m}$ for all $j \in [m]$ and $i \in [n]$ define a Nash equilibrium. As another example, for identical tasks and machines, we consider a balanced deterministic allocation in which we allocate task i to machine $(i \bmod m) + 1$ (that is, $p_i^j = 1$ for $j = (i \bmod m) + 1$ and is zero otherwise); also in that case the system is in a Nash equilibrium.

Let us begin with a simple claim describing a basic property of Nash equilibria.

Claim 42.1 *In an arbitrary Nash equilibrium, if* $p_i^q > 0$ *for certain* $i \in [n]$ *and* $q \in [m]$, *then* $\ell_j + \frac{w_i}{s_j} \ge \ell_q$ *for every* $j \in [m]$. *In particular, if* $\ell_j + 1 < \ell_q$ *then* $w_i > s_j$.

Furthermore, if $p_i^j > 0$ *and* $p_i^q > 0$, *then* $|\ell_j - \ell_q| \le opt$.

Proof

First, let us notice that $c_i^j \le \ell_j + \frac{w_i}{s_j}$ and $c_i^q = \ell_q + (1 - p_i^q)\frac{w_i}{s_q} \ge \ell_q$. Therefore, since $p_i^q > 0$, the definition of Nash equilibria implies that $c_i^q \le c_i^j$ and hence, $\ell_q \le \ell_j + \frac{w_i}{s_j}$.

The second claim follows trivially from the first one. The third claim follows from the first one and from the observation that $\frac{w_i}{s_j} \le opt$, by (42.2). □

Fix an arbitrary Nash equilibrium, that is, fix probabilities $(p_i^j)_{i\in[n],j\in[m]}$ that define a Nash equilibrium. Consider randomized allocation strategies in which each task i is allocated to a single machine chosen independently at random according to the probabilities p_i^j, that is, task i is allocated to machine j with probability p_i^j. Let C_j, $j \in [m]$, be the random variable indicating the *load of machine* j in our random experiment. We observe that C_j is the weighted sum of independent 0-1 random variables J_i^j, $\mathbf{Pr}[J_i^j = 1] = p_i^j$, such that

$$C_j = \frac{1}{s_j}\sum_{i=1}^{n} w_i \cdot J_i^j \tag{42.3}$$

Let $\mathfrak{c}$ denote the *maximum expected load* over all machines, that is,

$$\mathfrak{c} = \max_{j\in[m]} \ell_j \tag{42.4}$$

Notice that $\mathbf{E}[C_j] = \ell_j$, and therefore $\mathfrak{c} = \max_{j\in[m]} \mathbf{E}[C_j]$.

Finally, we define the *social cost* $\mathfrak{C}$ to be the expected maximum load (instead of maximum expected load), that is,

$$\mathfrak{C} = \mathbf{E}[\max_{j\in[m]} C_j] \tag{42.5}$$

Observe that $\mathfrak{c} \le \mathfrak{C}$ and possibly $\mathfrak{c} \ll \mathfrak{C}$. Recall that *opt* denotes the *social optimum* (i.e., the minimum cost of a pure strategy). In this section our main focus is on estimating the *coordination ratio* which is the worst-case ratio

$$\mathfrak{R} = \max \frac{\mathfrak{C}}{opt} \tag{42.6}$$

where the maximum is over all Nash equilibria.

Example

Let us consider a simple scenario of two machines, $m = 2$. If the machines are identical, then a "bad" Nash scenario can be achieved already for two identical tasks: if each task is allocated uniformly at random, that is, if $p_i^j = \frac{1}{2}$ for $i, j \in [10]$. One can check that this scenario gives rise to a Nash equilibrium, the expected maximum load is $\mathfrak{C} = \frac{3}{2}$, and the social optimum is $opt = 1$. This shows that already for identical tasks and identical two machines, the worst-case coordination ratio is larger than 1, namely it is $\frac{3}{2}$.

Next, we show a more complex example in which this bound is increased by assuming machines of different speeds. Consider the scenario with $n = m = 2$ and with nonidentical tasks and machines. Let $s_1 = 1$ and $s_2 \geq s_1$, and set the weights $w_1 = s_1$ and $w_2 = s_2$. We consider a scenario with the distribution $p_1^2 = \frac{s_2^2}{s_1(s_1+s_2)}$, $p_1^1 = 1 - p_1^2$, and $p_2^1 = \frac{s_1^2}{s_2(s_1+s_2)}$, $p_2^1 = 1 - p_2^2$. Notice that if $s_2 > \phi$, where $\phi = (1+\sqrt{5})/2$, then these probabilities are outside the interval $[0, 1]$, and therefore we assume that $s_2 \leq \phi$.

One can show that such defined probabilities define a Nash equilibrium [3, Theorem 3]. Furthermore, one can compute the expected maximum load to be

$$\mathfrak{C} = \left(\frac{p_1^1 \cdot p_2^1}{s_1} + \frac{p_1^2 \cdot p_2^2}{s_2}\right)(w_1 + w_2) + \left(\frac{p_1^1 \cdot p_2^1}{s_1} + \frac{p_1^2 \cdot p_2^2}{s_2}\right) w_2 = \frac{s_1 + 2s_2}{s_1 + s_2} = \frac{1 + 2s_2}{1 + s_2}$$

and we clearly have the social optimum $opt = 1$. This implies that the worst-case coordination ratio is $\mathfrak{R} = \frac{s_1+2s_2}{s_1+s_2} = \frac{1+2s_2}{1+s_2}$. Since this value is maximized when s_2/s_1 is maximized, and since $s_2/s_1 \leq \phi$, we can conclude that the worst-case coordination ratio for two (not necessarily identical) machines is at least $1 + \frac{\phi}{1+\phi} = \phi \approx 1.618$.

42.2.2 Basic Results for the Coordination Ratio for Linear Cost Functions

Koutsoupias and Papadimitriou [3] initiated the study of the worst-case coordination ratio and showed some of its basic properties in the task allocation model.

- For two identical machines the worst-case coordination ratio is exactly $\frac{3}{2}$.
- For two machines (not necessarily identical, that is, with possibly different speeds) the worst-case coordination ratio is at least $\phi = \frac{1+\sqrt{5}}{2} \approx 1.618$ (see example above).
- For m identical machines the worst-case coordination ratio is $\Omega(\frac{\log m}{\log\log m})$ and it is at most $3 + \sqrt{4\, m \ln m}$.
- The worst-case coordination ratio for any number of tasks and m (not necessarily identical) machines is $\mathcal{O}\left(\sqrt{\frac{s_1}{s_m} \sum_{j=1}^{m} \frac{s_j}{s_m}} \sqrt{\log m}\right)$, where s_j is the speed of machine j, and $s_1 \geq s_2 \geq \cdots \geq s_m$.

These bounds give almost a complete characterization of the problem for two machines. For a larger number of machines, they show only that one can make an upper bound for the coordination ratio to be independent of the number of tasks and that it is at least $\Omega(\frac{\log m}{\log\log m})$. Koutsoupias and Papadimitriou [3] conjectured also that for m identical machines the worst-case coordination ratio is $\Theta(\frac{\log m}{\log\log m})$.

Mavronicolas and Spirakis [11] greatly extended some of the bounds above in the so-called *fully-mixed model*. The *fully-mixed model* is a special class of Nash equilibria in which all p_i^j are nonzero. Mavronicolas and Spirakis [11] showed that for *m identical machines* in the fully-mixed Nash equilibrium the worst-case coordination ratio is $\Theta(\frac{\log m}{\log\log m})$. Similarly, they showed that for m (not necessarily identical) machines and *n identical tasks* in the fully-mixed Nash equilibrium, if $m \leq n$, then the worst-case coordination ratio is $\mathcal{O}(\frac{\log n}{\log\log n})$.

These results seemed to confirm the conjecture of Koutsoupias and Papadimitriou [3] that for m identical machines the worst-case coordination ratio is $\Theta(\frac{\log m}{\log\log m})$, but the complete proof of that result has remained elusive. Soon after, Czumaj and Vöcking [12] gave an elegant analysis of the task allocation model and obtained ultimate results for the general case. We will state their results in details in Section 42.2.3.3, but we first begin with their analysis of the simplest case, when all machines are identical, resolving the conjecture of Koutsoupias and Papadimitriou [3]. Because their analysis is relatively simple, we present their details here.

42.2.3 Analysis for Identical Machines

In this section we present a proof of the basic result about the system with identical machines. After the paper of Koutsoupias and Papadimitriou [3], the main research effort was to show that for m identical machines the worst-case coordination ratio is essentially obtained when all tasks are assigned at random, when it achieves the value of $\Theta(\log m/ \log\log m)$. This problem has been resolved independently by Koutsoupias et al. [13] and by Czumaj and Vöcking [12] with a significantly simpler proof (for $n = \mathcal{O}(m)$, this bound was proven independently also in [14]). We present here the simple proof from [12]; a tighter bound is proven in [12] that shows that for m identical machines the worst-case coordination ratio is exactly[2] $\Gamma^{(-1)}(m) + \Theta(1)$ (Theorem 42.3), where the bound is *tight up to an additive constant term.*

42.2.3.1 Lower Bound

One of the standard results about the balls-into-bins processes shows that if one throws n balls into m bins independently and uniformly at random, then the load distribution is very similar to the Poission distribution with the parameter n/m, and in particular, with high probability there exists a bin having the load of[3] $\Omega(\frac{n}{m} + \frac{\log m}{\log((m/n)\log m)})$. With this classical result, we can immediately show the following theorem, stated first in [3].

Theorem 42.1

For m identical machines the worst-case coordination ratio is greater than or equal to $\Gamma^{(-1)}(m) - \frac{3}{2} + o(1) = \Theta(\log m/ \log\log m)$.

Proof

We consider the scenario with $n = m$, all tasks having identical weights equal to 1, and all speeds equal to 1. We analyze the scenario in which each task chooses its machine independently and uniformly at random, that is, $p_i^j = \frac{1}{m}$ for every $i, j \in [m]$.

Let us verify that this setting of probabilities defines a Nash equilibrium. We first observe that for every $j \in [m]$, the expected load on machine j is

$$\ell_j = \sum_{i\in[m]} \frac{w_i\, p_i^j}{s_j} = \sum_{i\in[m]} \frac{1}{m} = 1$$

Similarly, for every task $i \in [m]$ and every machine $j \in [m]$, the expected cost of task i on machine j is

$$c_i^j = \ell_j + (1 - p_i^j)\frac{w_i}{s_j} = 2 - \frac{1}{m}$$

Therefore, since all values of c_i^j are identical, the definition of Nash equilibrium, Definition 42.1, implies that this strategy defines a Nash equilibrium.

---

[2]We use standard notation to denote by $\Gamma(N)$ the *Gamma (factorial) function*, which for any natural number N is defined by $\Gamma(N + 1) = N!$ and for an arbitrary real number $x > 0$ is defined as $\Gamma(x) = \int_0^\infty t^{x-1} e^{-t}\, dt$. The *inverse of the Gamma function* is denoted by $\Gamma^{(-1)}(N)$. It is well known that $\Gamma^{(-1)}(N) = \frac{\log N}{\log\log N}(1 + o(1))$.

[3]To simplify the notation, throughout the entire survey, for any nonnegative real x we use $\log x$ to denote $\log x = \max\{\log_2 x, 1\}$.

Now, it is easy to see that the social optimum *opt* is equal to 1, which is obtained by assigning every task i to machine i. To analyze the social cost $\mathfrak{C}$, we use the aforementioned analyses of random processes allocating balls into bins. In particular, it is a standard result that $\mathfrak{C} = \Theta(\log m/\log\log m)$, and in fact, Gonnet [15] obtained a very precise bound and showed that in that case $\mathfrak{C} = \Gamma^{(-1)}(m) - \frac{3}{2} + o(1)$. This clearly implies that the coordination ratio is at least $\mathfrak{C} = \Gamma^{(-1)}(m) - \frac{3}{2} + o(1) = \frac{\log m}{\log\log m}(1 + o(1))$, which concludes the analysis. □

42.2.3.2 Upper Bound

In this section we present the proof from [12] of the following result.

Theorem 42.2

For m identical machines the worst-case coordination ratio is $\mathcal{O}(\log m/\log\log m)$.

Proof

Let us first rescale the weights and speeds in the problem and assume, without loss of generality, that all speeds are equal to 1 and that the social optimum is *opt* = 1.

Since *opt* = 1, then, by (42.2), we must have

$$w_i \le 1 \quad \text{for all } i \in [n]$$

Furthermore, by Claim 42.1, the assumption that the system is in a Nash equilibrium implies that

$$\mathfrak{c} = \max_{j\in[m]} \ell_j < 2 \quad \text{for all } j \in [m]$$

Next, to estimate the load C_j of any machine $j \in [m]$ we apply to (42.3) a standard concentration inequality due to Hoeffding[4] to obtain the following bound that holds for any $t > 0$:

$$\mathbf{Pr}[C_j \ge t] \le (e \cdot \mathbf{E}[C_j]/t)^t \le (2e/t)^t$$

where the last inequality follows from the fact that $\mathbf{E}[C_j] = \ell_j$ and $\ell_j < 2$.

Therefore, if we pick $t \ge 3 \ln m/\ln\ln m$, then $\mathbf{Pr}[C_j \ge t] \ll 1/m$ and, intuitively, this should yield $\mathfrak{C} = \mathbf{E}[\max_{j\in[m]} C_j] = \mathcal{O}(t)$. The following inequalities prove this more formally.

$$\mathfrak{C} = \mathbf{E}[\max_{j\in[m]} C_j] \le t + \sum_{\tau=t}^{\infty} \mathbf{Pr}[\exists_{j\in[m]} C_j \ge \tau] \le t + \sum_{\tau=t}^{\infty} m \cdot (2e/\tau)^{\tau} \le t + \sum_{\tau=t}^{\infty} 2^{-\tau} \le t + 1$$

□

Actually, Czumaj and Vöcking showed [12] how to tighten this bound to match exactly the bound from Theorem 42.1 and proved the following theorem.

Theorem 42.3

For m identical machines the worst-case coordination ratio is exactly

$$\Gamma^{(-1)}(m) + \Theta(1) = \frac{\log m}{\log\log m}(1 + o(1))$$

[4] We use the following standard version of Hoeffding bound [16]: Let $X_1, \ldots, X_N$ *be independent random variables with values in the interval* $[0, z]$ *for some* $z > 0$*, and let* $X = \sum_{i=1}^{N} X_i$*, then for any t it holds that* $\mathbf{Ar}\sum_{i=1}^{N} X_i \ge t \le (e \cdot t[X]\,/\,t)^{t/z}$.

42.2.3.3 Ultimate Analysis for the General Case

In this section we present the complete characterization of the worst-case coordination ratio for the task allocation model obtained by Czumaj and Vöcking [12]. All the bounds listed here are asymptotically tight.

The main result is an ultimate and tight bound for the worst-case coordination ratio.

Theorem 42.4

The coordination ratio for m machines is

$$\Theta\left(\min\left\{\frac{\log m}{\log\log\log m}, \frac{\log m}{\log\left(\frac{\log m}{\log(s_1/s_m)}\right)}\right\}\right)$$

where it is assumed that the speeds satisfy $s_1 \geq \cdots \geq s_m$.

In particular, the worst-case coordination ratio for m machines is

$$\Theta\left(\frac{\log m}{\log\log\log m}\right)$$

The bound obtained in this theorem is rather surprising not only because of the fact that the worst-case coordination ratio is a rather unusual function $\mathcal{O}(\frac{\log m}{\log\log\log m})$, but also that it shows that even if the machines may have arbitrary speeds, the coordination ratio is just a little larger than that for identical machines proved in Theorem 42.3. Furthermore, this bound shows that the coordination ratio is upper bounded by a function of m only, and thus independent of the number of tasks, of their weights, and of the machine speeds; no such bound has been known before.

The proof of the upper bound in Theorem 42.4 follows directly from the following two lemmas, that are interesting by themselves.

Lemma 42.1

The maximum expected load $\mathfrak{c}$ *satisfies*

$$\mathfrak{c} \leq \min\left\{opt \cdot (\Gamma^{(-1)}(m) + 1), opt \cdot (2\log(s_1/s_m) + \mathcal{O}(1))\right\}$$

where it is assumed that the speeds satisfy $s_1 \geq \cdots \geq s_m$.

In particular,

$$\mathfrak{c} = opt \cdot \mathcal{O}\left(\min\left\{\frac{\log m}{\log\log m}, \log\left(\frac{s_1}{s_m}\right)\right\}\right)$$

Lemma 42.2

The social cost $\mathfrak{C}$ *satisfies*

$$\mathfrak{C} = opt \cdot \mathcal{O}\left(\frac{\log m}{\log\left(\frac{opt \cdot \log m}{\mathfrak{C}}\right)} + 1\right)$$

Lemmas 42.1 and 42.2 give some interesting results for a few special cases. We first state their applications for systems in which all users follow only *pure strategies* (that is, $p_i^j \in \{0, 1\}$ for every i, j). Since in that case, $\ell_j = C_j$ for every $j \in [m]$, we also have $\mathfrak{c} = \mathfrak{C}$. Therefore Lemma 42.1 gives immediately the following result.

Corollary 42.1

For pure strategies the worst-case coordination ratio for m machines is bounded from above by

$$\min\{\Gamma^{(-1)}(m) + 1, 2\log(s_1/s_m) + \mathcal{O}(1)\}$$

where it is assumed that the speeds satisfy $s_1 \geq \cdots \geq s_m$.

Notice also that this theorem matches the bound from Theorem 42.2 for identical machines.

The lower bound in Theorem 42.4 follows from a construction, analog to the upper bound, that, for every positive integer m, for every positive real r and every $S \geq 1$, there exists a set of n tasks and a set of m machines with certain speeds $s_1 \geq s_2 \geq \cdots \geq s_m$ with $\frac{s_1}{s_m} = S$, being in a Nash equilibrium and satisfying

1. $opt = r$
2. $\mathfrak{C} = opt \cdot \Omega\left(\min\left\{\frac{\log m}{\log\log m}, \log\left(\frac{s_1}{s_m}\right)\right\}\right)$
3. $\mathfrak{C} = opt \cdot \Omega\left(\frac{\log m}{\log\left(\frac{opt \cdot \log m}{c}\right)}\right)$

42.2.4 Extensions

The work discussed in the previous section has been extended in some aspects. One such extension, in which more general classes of cost functions has been considered, will be the theme of Section 42.2.5; three others are discussed here.

42.2.4.1 Restricted Assignments in the Task Allocation Model

Awerbuch et al. [17] discussed the problem from the previous section in the restricted assignment model, that is, in which each task can be assigned to a subset of the machines. Using similar techniques as in [12], Awerbuch et al. [17] showed that essentially, the results from Corollary 42.1 and Theorem 42.4 are asymptotically tight in such a scenario as well. In particular, they showed that the worst-case coordination ratio is $\Theta\left(\frac{\log m}{\log\log m}\right)$ for systems of pure strategies and $\Theta\left(\frac{\log m}{\log\log\log m}\right)$ for systems of mixed strategies (and arbitrary speeds and weights).

42.2.4.2 Analysis of the Task Allocation Model for Average-Cost Functions

Very recently, Berenbrink et al. [18] analyzed the coordination ratio between the social cost incurred by selfish behavior and the optimal social cost in terms of the average-cost functions — the average delay over tasks. They consider essentially the same model as that described in the previous section with the only differences in Equations (42.1), (42.4), and (42.5) that define the cost of the allocation. Berenbrink et al. [18] use the following definitions:

$$opt = \min_{(j_1,\ldots,j_n)\in[m]^n} \sum_{j\in[m]} \sum_{i:j_i=j} \frac{w_i}{s_j}$$

$$\mathfrak{c} = \sum_{j\in[m]} \ell_j$$

$$\mathfrak{C} = \mathbf{E}\left[\sum_{j\in[m]} C_j\right]$$

Berenbrink et al. [18] showed that under such a cost measure, even if the problem is trivial if all tasks and machines are identical, in general, the coordination ratio satisfies significantly different properties than those analyzed in Section 42.2.2. They analyzed two main scenarios, one in which all machines are identical and another in which all tasks are identical.

For identical machines (identical speeds), Berenbrink et al. [18] showed that the ratio between even the best Nash equilibrium and the social optimum can be arbitrarily large, but on the other hand, they proved that the coordination ratio is upper bounded by $\mathcal{O}(\max_i w_i / \min_i w_i)$. They also showed that the ratio between the cost of the worst Nash equilibrium and the best Nash equilibrium is upper bounded by a constant.

For identical tasks (identical weights), Berenbrink et al. [18] showed that there exist instances for which the ratio between the cost of the best Nash equilibrium and that of the social optimum is at least $\frac{5}{4}$, while at the same time, the ratio between the cost of the worst Nash equilibrium and the best Nash equilibrium is upper bounded by 3. They also described an $\mathcal{O}(nm)$-time algorithm that for identical weights finds a lower-cost Nash allocation and an optimal allocation.

42.2.4.3 Existence of Nash Equilibria and the Complexity of Their Finding

In the previous discussion of the task allocation model, we have neglected the discussion of the existence of Nash equilibria. Actually, it is quite easy to show that there exists always at least one *mixed* Nash equilibrium, and Rosenthal proved that actually there exists always at least one *pure* Nash equilibrium (see also Melhorn) [19, Theorem 1].

Once we established an existence of one pure Nash equilibrium, it is natural to ask how to find one. In [19], it is shown that it is $\mathcal{NP}$-hard to find the best and the worst pure Nash equilibria, but on the other hand, there exists a simple polynomial-time algorithm for finding *a* pure Nash equilibrium. The later result was strengthened by Feldmann et al. [20], who designed a polynomial-time algorithm that for any given pure task allocation computes a Nash equilibrium with no larger cost. An interesting consequence of this result is the existence of a PTAS (polynomial-time approximation scheme) for the problem of computing a Nash equilibrium with minimum social cost. Indeed, we can first apply a PTAS for min-makespan (due to Hochbaum and Shmoys [21]) that computes a $(1 + \epsilon)$-approximation of the social optimum (minimum cost allocation of the task to the machines) and then apply the algorithm above to obtain a Nash equilibrium whose social cost is not larger than $(1 + \epsilon)$ times the social optimum.

An undesired property of the algorithms discussed above is that they are off-line and centralized. Recently, we have seen some research on noncentralized algorithms for finding Nash equilibria, most notably, on the issues of convergence to a Nash equilibrium. This problem has received significant attention in the game theory literature (e.g., see [22]), but only recently we have seen quantitative analyses for the task allocation model.

The problem can be described as a load balancing process. We assume there is some initial task allocation in the task allocation model. This allocation does not have to be social minimum and does not have to be a Nash equilibrium; it is arbitrary. However, if the system is not in a Nash equilibrium, then there is at least one task for which it is beneficial to reallocate to another machines. Therefore, we consider various "strategies" that reallocate the tasks according to the selfish rule: a task migrating to a new machine must reduce the observed cost. We consider only scenarios in which only one task is reallocated in a single step.

Even-Dar et al. [23] proved that if the system performs a sequence of the migration steps according to the selfish rule, then after sufficiently many steps, the system will arrive in a Nash equilibrium. Feldmann et al. [20] analyzed the convergence time of such strategies and showed that even for identical machines, if we have $m = \Theta(\sqrt{n})$, then there is a sequence of migration moves that requires $2^{\Omega(\sqrt{n})}$ migration steps to reach a Nash equilibrium. On the other hand, they showed that for identical machines any sequence of $2^n - 1$ migration steps will reach a Nash equilibrium [20]. Even-Dar et al. [23] studied a more general situation of nonidentical machines. Let K be the number of different weights, $W = \sum_{i \in [n]} w_i$ be the total weight of all the tasks, and $w_{\max} = \max_{i \in [n]} w_i$ be the maximum weight of a task. The first result proven by Even-Dar et al. [23] is that independently of the initial allocation, any sequence of $\min\{m^n, (\mathcal{O}(1+\frac{n}{Km}))^{Km}\}$ migration steps will reach a Nash equilibrium, and if all weights are integer, then $\mathcal{O}(4^W)$ will suffice. Even-Dar et al. [23] obtained also some further bounds for special cases and analyzed some specific (global) strategies of task migration. For example, for identical machines, if one first moves the maximum weight task to a machine in which its observed load is minimal, then a Nash equilibrium is reached in at most n steps. A related model was also investigated in [24].

42.2.5 Task Allocation Model with Arbitrary Cost Functions

After a fairly complete analysis of the task allocation model with linear cost function has been obtained in [12], it has been natural to consider more general classes of cost function. Already Papadimitriou [3,6] raised that question and his main interest has been in the analysis of the classes of cost functions that arise in Internet and networking applications, most notably cost functions that correspond to the expected delay in queueing systems. Czumaj et al. [25] presented a thorough study of this model for general, monotone families of cost functions and for cost functions from queueing theory. Unfortunately, unlike in the case of linear cost functions, the coordination ratios for other cost functions are typically significantly worse, and often even unbounded. In this section we will discuss the results from [25].

42.2.5.1 Web Server Farms

The main motivation behind our study comes from a specific example in which our model is the natural choice: we investigate the effects of selfish behavior on a *Web server farm*. Consider a scenario in which some companies maintain sets of servers distributed all over the world and they compete to offer content providers to store data for them. Such servers could store large embedded files or pictures and movies, for which the uploading cost is usually significant. The request streams that would normally go to the servers of the content provider will be now redirected to these new Web servers. To achieve a high quality of service, in the allocation process to which server each stream will be assigned one has to perform some load balancing to ensure that the service cost in each server will not be too large. Important aspects that have to be taken into account in this allocation are that different streams might have different characteristics, e.g., caused by different file lengths, and that different servers may have different service cost. For practical studies that investigate the reasons and impacts of this variability in traffic see, e.g., [26–30].

In nowadays server farms the mapping of data streams to servers is typically done by a centralized or distributed algorithm that is under control of the provider of the server farm. We can imagine, however, that such a service can be offered in a completely different way without global control. For example, each stream of requests is managed by a selfish agent (e.g., the content provider) that decides to which server the stream is directed. In this case, every agent would aim to minimize its own cost, e.g., the expected latency experienced by the requests in the stream or the fraction of requests that are rejected.

42.2.5.2 Details of the Model

To present our framework in its full generality and to fit the Web server farm model, we slightly modify the model discussed in Section 42.2.1. We assume that the tasks are arriving continuously as streams, and thus, we consider our problem as that of assigning n data (task) streams to m machines (servers). The set of streams is denoted by $[n]$ and the set of machines is denoted by $[m]$. The data streams shall be mapped to the machines such that a cost function (describing, e.g., waiting or service times) is minimized. We aim at comparing the assignment obtained by selfish users with a min-max optimal assignment.

Following the framework of server farms discussed in [25], we assume that the machines use identical policy to serve the tasks; different machines, however, may have different *bandwidths* (corresponding to the speed in the model from Section 42.2.1, larger bandwidth is a larger speed). Let b_j denote the bandwidth of machine j.

Data streams are infinite sequences of requests for service. These sequences are assumed to be of a stochastic nature. For simplicity, we make a standard assumption that requests are issued by a large number of independent users and hence they arrive with *Poisson* distribution. Let r_i denote the *injection rate* of data stream i. For the lengths of the sessions (execution time) we allow general probability distributions and we assume that the *session length* of stream i is determined by an arbitrary probability distribution $\mathcal{D}_i$. We define the *weight* of stream i to be λ_i,

$$\lambda_i = r_i \cdot \mathbf{E}[\text{session length with respect to } \mathcal{D}_i]$$

We distinguish between *fractional* and *integral* assignments of data streams to the machines. In an *integral assignment*, every stream must be assigned to exactly one machine. The mapping is described by an assignment matrix $\mathcal{X}$,

$$\mathcal{X} = \left(x_i^j\right)_{i\in[n],j\in[m]}$$

where x_i^j is an indicator variable with $x_i^j = 1$ if stream i is assigned to machine j and 0 otherwise. In a *fractional assignment* the variables x_i^j can take arbitrary real values from $[0,1]$, subject to the constraint $\sum_{j\in[m]} x_i^j = 1$, for every $i \in [n]$.

The cost occurring at the machines under some fixed assignment is defined by *families of cost functions* $\mathcal{F}_B = \{f_b : b \in B\}$, where B denotes the domain of possible bandwidth values and f_b describes the cost function for machines with bandwidth $b \in B$. For example, a collection of identical machines with some specified bandwidth b is formally described by a family of cost functions $\mathcal{F}_B$ with $B = \{b\}$.

The *load* of a machine j under an assignment $\mathcal{X}$ is defined by w_j,

$$w_j = \sum_{i=1}^{n} \lambda_i \cdot x_i^j$$

and the *cost* of machine j is defined by C_j,

$$C_j = f_{b_j}\left(x_1^j, \ldots, x_n^j\right)$$

Unless otherwise stated, we consider the routing problem with respect to the *min-max objective*. That is, we assume an *optimal assignment* minimizes the maximum cost over all machines:

$$opt = \min_{\mathcal{X}} \max_{j\in[m]} \quad f_{b_j}\left(x_1^j, \ldots, x_n^j\right)$$

We note here, that in the definition of *opt* the minimization is over all matrices $\mathcal{X}$ that are either integral (in the case of integral assignments) or fractional (in the case of fractional assignments). This distinction will be clear from the context.

42.2.5.3 Integral Assignments and Nash Equilibria

We assume the decision about the assignment of a data stream $i \in [n]$ to a machine is performed by an agent i who uses certain strategy to assign its data stream. Similarly as in Section 42.2.1, we distinguish between mixed and pure strategies. The set of *pure strategies* for agent $i \in [n]$ is $[m]$, that is, a pure strategy maps every stream to exactly one machine and hence can be described by an integral assignment matrix $\mathcal{X}$. A *mixed strategy* is defined to be a probability distribution over pure strategies. In particular, the probability that agent i maps its stream to machine j is denoted by p_i^j.

Observe that under these assumptions the load w_j and the cost C_j of machine j are random variables depending on the probabilities $p_i^j, i \in [n]$. Let ℓ_j denote the *expected cost on machine* j, that is,

$$\ell_j = \mathbf{E}[C_j]$$

For a stream i, we define the *expected cost of stream* i *on machine* j by c_i^j,

$$c_i^j = \mathbf{E}\left[C_j \mid x_i^j = 1\right]$$

Our objective is to minimize the maximum cost over all machines and therefore we define the *social cost* of a mixed assignment by

$$\mathfrak{C} = \mathbf{E}[\max_{j\in[m]} C_j]$$

If selfish players aim to minimize their individual cost, then the resulting (possible mixed) assignment is in *Nash equilibrium*, that is, $p_i^j > 0$ implies $c_i^j \leq c_i^q$, for every $i \in [n]$ and $j, q \in [m]$. In other words, a Nash equilibrium is characterized by the property that there is no incentive for any agent to change its strategy.

42.2.5.3.1 *Class of Cost Functions Investigated in Section 42.2.1*

Let us notice that the model of Koutsoupias and Papadimitriou [3] described in Section 42.2.1 is essentially the integral assignment model with linear cost functions, that is, with cost functions of the form $f_b(x_1, \ldots, x_n) = \sum_{i=1}^{n} \lambda_i x_i / b$ (where w_i from Section 42.2.1 is replaced by λ_i now).

42.2.5.4 Coordination Ratios: Bounded vs. Unbounded

The *worst-case coordination ratio for a fixed set of machines and data streams* is defined by max $\frac{\mathfrak{C}}{opt}$, where the maximum is over all Nash equilibria. The *worst-case coordination ratio* $\mathfrak{R}$ *over a family of cost functions* $\mathcal{F}_B$ is defined to be the maximum worst-case coordination ratio over all possible sets of streams and machines with cost functions from $\mathcal{F}_B$. Typically $\mathfrak{R}$ is described by an asymptotic function in m.

A cost function f is called *simple* if it depends only on the injected load, that is, if the cost of a machine is a function of the sum of the weights mapped to the machine but does not depend on other characteristics like, e.g., the session length distribution. A simple cost function is called *monotone* if it is nonnegative, continuous, and nondecreasing. For an ordered set B, a family of simple cost functions $\mathcal{F}_B$ is called *monotone* if (i) f_b is monotone for every $b \in B$ and (ii) the cost functions are nonincreasing in b, i.e., $f_b(\lambda) \geq f_{b'}(\lambda)$, for every $\lambda \geq 0$ and $0 < b \leq b'$.

We say a worst-case coordination ratio $\mathfrak{R}$ over a family of cost functions $\mathcal{F}_B$ is *bounded* if for every m and every server farm with m servers with cost functions from $\mathcal{F}_B$ there exists $\Gamma > 0$ such that for every set of streams the value of the worst-case Nash equilibrium is at most $\Gamma \cdot opt$. (Observe that Γ might depend on m, and thus bounded means bounded by a function in m.) Otherwise, the worst-case coordination ratio is *unbounded*. Czumaj et al. [25] gave a precise characterization of those monotone families of cost functions for which the worst-case coordination ratio is bounded.

Theorem 42.5

The worst-case coordination ratio over a monotone family $\mathcal{F}_B$ of cost functions is bounded if and only if

$$\exists \alpha \geq 1 \quad \forall b \in B \quad \forall \lambda > 0 \quad f_b(2\lambda) \leq \alpha \cdot f_b(\lambda)$$

Actually, the analysis from [25] shows that for every fixed family of cost functions the worst-case coordination ratio is either unbounded or it is polynomially bounded in the number of machines, m.

Let us illustrate the use of this theorem on some basic examples. First, we consider families over *polynomial cost functions*, i.e., functions of the form $\sum_{r=0}^{k} a_r \cdot \lambda^r$, for a fixed $k \geq 0$. For these families we can pick $\alpha = a_k \cdot 2^k$ to conclude that here the coordination ratio is bounded. In contrast, there is no such α for *exponential cost functions*, i.e., cost functions for which an additive increase in the load leads to a multiplicative increase in the cost.

42.2.5.5 Fractional Assignments and Selfish Flow

The motivation to consider fractional assignments is to assume that every stream consists of infinitely many units each carrying an infinitesimal (and thus negligible) amount of flow (traffic). Each such unit behaves in a selfish way. Intuitively, we expect each unit to be assigned (selfishly) to a machine promising minimum cost, taking into account the behavior of other units of flow. Assuming infinitesimal small units of flow, we came to a fractional variant of the integral assignment model. (This fractional model has been frequently considered in the literature and will be discussed in more details in Section 42.3.)

The fractional model does not distinguish between mixed and pure strategies. There are several equivalent ways to define a Nash equilibrium in this model. We use the characterization of Wardrop [31], see also [9]: a fractional assignment is in *Nash equilibrium* if $x_i^j > 0$ implies $C_j \leq C_q$, for every $i \in [n]$ and $j, q \in [m]$. The *worst-case coordination ratio* is defined analogously to the integral assignment model, and the *worst-case coordination ratio* over a family of cost functions is denoted by $\mathfrak{R}^*$.

Czumaj et al. [25] show that unlike in integral assignments, all monotone cost functions behave very well under *fractional assignments*, and in particular, if cost functions are described by monotone cost functions, then $\mathfrak{R}^* = 1$.

42.2.5.6 Allocating Streams with Negligible Weights

By comparing the bounds for the coordination ratio for fractional assignments with Theorem 42.5, we see that integrality can lead to a dramatic performance degradation. As mentioned before, the fractional flow model is assumed to be a simplification that aims to model the situation in which each stream carries only a negligible fraction of the total load. Therefore, Czumaj et al. [25] investigated the relationship between fractional flow and integral assignments of streams with tiny weights more closely. They show that for any monotone family of cost functions $\mathcal{F}_B$, informally, even if we consider only streams of negligible weights to be allocated, the coordination ratio is bounded if and only if the worst-case coordination ratio without this restriction is bounded.

A motivation for considering fractional flow instead of integral assignments is that these two models are sometimes assumed to be "essentially equivalent," see, e.g., Remark 2.3 in [9]. The result above disproves this equivalence for general cost functions. Moreover, the instances proving the characterization of unbounded worst-case coordination ratios use only pure strategies. Hence, even pure assignments with negligible weights are different from fractional flow.

42.2.5.7 Integral Assignments under Bicriteria Measures

Since we identified several instances of cost functions for which $\mathfrak{R}$ is unbounded, in order to analyze the performance of the system for such cost functions, we investigate *bicriteria* characteristics of the system. Let opt_Γ denote the value of an optimal solution over pure strategies assuming that all the injection rates r_i are increased by a factor of Γ. Then, the *bicriteria ratio* $\overline{\mathfrak{R}}$ is defined to be the smallest Γ satisfying $C \leq opt_\Gamma$ over all Nash equilibria. In other words, the bicriteria ratio describes how many times the injected rates (the amount of traffic in the system) must be decreased so that the worst-case cost in Nash equilibrium cannot exceed the optimal cost for the original rates.

It is not surprising that selfish routing can lead to a dramatic cost increase when the cost function has an ∞-pole. In principle, bicriteria measures can be much more informative as they in some sense filter out the extreme behavior of such cost functions at the pole. However, as shown in [25], if the worst-case coordination ratio $\mathfrak{R}$ is unbounded, then the bicriteria ratio $\overline{\mathfrak{R}}$ has value at least m, even if all streams are restricted to be tiny.

The example proving this bad ratio is a server farm of identical machines. In fact, for the case of identical machines one can easily show that a bicriteria ratio of m is the worst possible. This is because a Nash equilibrium cannot be worse than mapping all streams to the same machine, and the cost of this extremely unbalanced solution is bounded above by an optimal assignment for an instance with all weights blown up by a factor of m.

42.2.5.8 Average-Cost Objective Function vs. Min-Max Objective One

Besides the min-max objective function investigated above, we can also study the *average-cost (or total latency) objective function* that will be discussed in more detail in Section 42.3 (see also Section 42.2.4.2). This objective function aims at minimizing the expected weighted average cost over all streams. Formally, the cost under this objective function is defined by

$$\mathfrak{C}_{ave} = \frac{1}{\lambda} \sum_{j \in [m]} \mathbf{E}[w_j \cdot C_j]$$

and the social optimum is defined by

$$opt_{ave} = \min_{\mathcal{X}} \left(\frac{1}{\lambda} \sum_{j \in [m]} C_j \cdot f_{b_j}\left(x_1^j, \ldots, x_n^j\right) \right)$$

where $\lambda = \sum_{i \in [n]} \lambda_i$ is the total injected weight and the minimum is taken over all integral assignment matrices. These definitions are equivalent to the respective definitions from Section 42.3.

We can consider various coordination ratios for the average-cost objective function similarly as for the min-max objective function. These average-cost coordination ratios are defined in the same way as for the min-max model considered before, the only difference is that now one compares the average cost in Nash equilibrium with the average-cost optimum. Czumaj et al. [25] showed that in the case of integral assignments the average-cost objective leads to exactly the same characterizations for coordination and bicriteria ratios as those given above for the min-max objective.

42.2.5.9 Cost Functions from Queueing Theory

A typical example of a monotone family of cost functions that is derived from the formula for the expected system time (delay) on an M/M/1 machine (server) with injection rate λ and service rate b, namely, $\frac{1}{b-\min\{b,\lambda\}}$. Already Koutsoupias and Papadimitriou, in their seminal work [3], asked for the price of selfish routing under cost functions of this form. The characterization of bounded and unbounded coordination ratios given in Theorem 42.5 immediately implies that the integral coordination ratios for this family of functions are unbounded.

The paper [25] contains a detailed discussion of further properties of this and other cost functions from queueing theory, and we refer for more details to that paper. In particular, they considered a generic class of "monotone queueing functions" (which contains most of cost functions from queueing theory) for which they show that $\mathfrak{R}^* = 1$, $\mathfrak{R} = \infty$, and $\overline{\mathfrak{R}} \geq m$, even under the restriction that all streams have negligible weights.

42.2.5.10 Queueing under Heterogeneous Traffic

Until now we implicitly assumed homogeneous traffic, i.e., all streams have the same (general) session length distribution. However, several practical studies show that Internet traffic is far away from being homogeneous, see [26,28,29]. Following these studies, one has to take into account different session lengths distributions.

Czumaj et al. [25] studied a scenario in which the cost functions are defined using the *Pollaczek-Khinchin* (P-K) formula (see [32]) describing expected waiting time in M/G/1 queues, that is, the expected delay of requests on sequential machines under *heterogeneous* traffic with arbitrary service time distributions. They show that not only that $\mathfrak{R} = \infty$, $\overline{\mathfrak{R}} \geq m$ for the P-K cost function family, which follows easily from our prior discussion. They proved that in the fractional flow model, unlike under homogeneous traffic, where $\mathfrak{R}^* = 1$, the coordination ratio $\mathfrak{R}^*$ for the P-K cost function family under heterogeneous traffic is unbounded.

An important conclusion made in [25] is that the optimality of fractional flow in Nash equilibrium is a special property of homogeneous traffic in the allocation model, and hence one must take into account the heterogeneous nature of Web traffic when studying the price of selfish routing in the Internet.

42.2.5.11 Machines with Parallel Channels and Rejection

Until now, we assumed that all requests are served, regardless of how long they have to wait for service. In practice, however, Web servers reject requests when they are overloaded. For simplicity, let us assume that a server rejects requests whenever all service channels are occupied and then these requests disappear from the system. In this case, the fraction of rejected requests is completely independent of the service time distribution. In other words, there is no difference between homogeneous and heterogeneous traffic under this service model. In fact, the fraction of rejected requests can be derived from the *Erlang loss formula*, see [33,34]. We obtain the following cost function family $\mathcal{F}_{\mathbb{N}_{>0}}$ for machines that can open up to b channels simultaneously:

$$f_b(x_1, \ldots, x_n) = \frac{\lambda^b/b!}{\sum_{k=0}^{b} \lambda^k/k!} \quad \text{with } \lambda = \sum_{i=1}^{n} \lambda_i\, x_i$$

One can show that for the family $\mathcal{F}_{\mathbb{N}_{>0}}$ of Erlang loss cost functions, $\mathfrak{R}^* = 1, \mathfrak{R} = \infty$ and $\overline{\mathfrak{R}} \geq m$ [25]. However, unlike in previous examples, even if the coordination ratios are large, the absolute cost of selfish routing under the Erlang loss cost function family is small, which is confirmed by the following theorem.

Theorem 42.6 [25]

Let δ satisfy $\delta \geq 2/\log_2 m$. Consider a server farm of m machines with bandwidths $b_1 \geq \cdots \geq b_m$ and cost functions from the family $\mathcal{F}_{\mathbb{N}_{>0}}$ of Erlang loss cost functions. Suppose $\sum_{i\in[n]} \lambda_i \leq \frac{1}{6e}\sum_{j\in[m]} b_j$ and $\max_{i\in[n]} \lambda_i \leq \frac{b_m}{3\delta \log_2 m}$. Then, any Nash equilibrium has social cost at most $m^{-\delta+1} + m \cdot 2^{-b_m/4}$.

Hence, if the total injected load is at most a constant fraction of the total bandwidth and every stream has not too large weight, then the fraction of rejected requests is at most $m^{-\delta+1} + m\,2^{-b_m/4}$, assuming constant δ. Under the same conditions, an optimal assignment would reject a fraction of $2^{-\Theta(b_1)}$ packets. Taking into account that typical Web servers can open several hundred TCP connections simultaneously, so that b_m can be assumed to be quite large, we conclude that the cost of selfish routing is very small in absolute terms, even though the coordination ratio comparing this cost with the optimal cost is unbounded.

42.2.5.12 Final Remarks about Web Server Farms

The results from [25] listed above have some important algorithmic consequences in the Web server farms model. They show that the choice of the queueing discipline should take into account the possible performance degradation due to selfish and uncoordinated behavior of network users. It is shown that the worst-case coordination ratio for queueing systems without rejection is unbounded. The same is true for server farms that reject requests in case of overload. However, there is a fundamental difference between these two kinds of queueing policies: the delay under selfish routing in the queuing systems without rejection is in general unbounded, and in fact, the selfish routing in such queueing systems can lead to an arbitrary large delay even when the total injected load can potentially be served by a single machine. In contrast, the fraction of rejected requests under selfish routing can be bounded above by a function that is exponentially small in the number of TCP connections that can be opened simultaneously.

Czumaj et al. [25] concluded that server farms that serve all requests, regardless of how long requests have to wait, cannot give any reasonable guarantee on the quality of service when selfish agents manage the traffic. However, if requests are allowed to be rejected, then it is possible to guarantee a high quality of service for every individual request stream. Thus, the typical practice of rejecting requests in case of overload is a necessary condition to ensure efficient service under game theoretic measures.

42.3 Routing in a Flow Model

In the second part of the survey, we consider another model of costs functions. For the purpose of this analysis, we assume that each network user controls a negligible fraction of the overall traffic and therefore it can be modeled as a flow. For example, each agent could represent a car in a highway system, or a single packet in a communication network.

We consider a directed network $\mathcal{N} = (V, E)$ with node set V, edge set E, and k source-destination node pairs $\{s_1, t_1\}, \{s_2, t_s\}, \ldots, \{s_k, t_k\}$. We denote the set of (simple) s_i–t_i paths by $\mathcal{P}_i$ and define $\mathcal{P} = \bigcup_{i=1}^{k} \mathcal{P}_i$. A *flow* is a function $f : \mathcal{P} \to \mathbb{R}_{\geq 0}$, and for a fixed flow f and for any $e \in E$, we define $f_e = \sum_{\pi\in\mathcal{P}:e\in\mathcal{P}} f_\pi$. With each pair $\{s_i, t_i\}$, we associate a positive *traffic rate* r_i that denotes the amount of flow to be sent from node s_i to node t_i. We say a flow f is *feasible* if for all $i \in [k]$, $\sum_{\pi\in\mathcal{P}_i} f_\pi = r_i$.

We assume that each edge $e \in E$ is given a load-dependent *latency function* denoted by $\ell_e(\cdot)$. We will later make some more precise assumptions about the latency functions, but for now, one can assume that every ℓ_e is nonnegative, differentiable, and nondecreasing. The latency of a path π with respect to a flow f is defined as the sum of the latencies of the edges in the path, denoted by $\ell_\pi(f)$,

$$\ell_\pi(f) = \sum_{e\in\pi} \ell_e(f_e)$$

Finally, we define the cost $\mathfrak{C}(f)$ of a flow f in $\mathcal{N}$ as the total latency incurred by f, that is,

$$\mathfrak{C}(f) = \sum_{\pi \in P} \ell_\pi(f) f_\pi$$

Notice also that by summing over the edges in a path π and reversing the order of summation, we may also write $\mathfrak{C}(f) = \sum_{e \in E} \ell_e(f_e) f_e$.

Following Roughgarden and Tardos [9], we call the triple $(\mathcal{N}, r, \ell)$ an *instance*.

42.3.1 Flows at Nash Equilibria

Since we assume that all agents carry a negligible fraction of the overall traffic, we can informally assume that each agent chooses a single path of the network (see [35] for a more detailed and formal discussion). Therefore, we focus our discussion on pure strategies and define Nash equilibria for *pure strategies* only.

A flow f feasible for instance $(\mathcal{N}, r, \ell)$ is at *Nash equilibrium* if for all $i \in [k]$, $\pi_1, \pi_2 \in \mathcal{P}_i$ with $f_{\pi_1} > 0$, and $\delta \in (0, f_{\pi_1})$, we have

$$\ell_{\pi_1}(f) \leq \ell_{\pi_2}(\tilde{f})$$

where

$$\tilde{f}_\pi = \begin{cases} f_\pi - \delta & \text{if } \pi = \pi_1 \\ f_\pi + \delta & \text{if } \pi = \pi_2 \\ f_\pi & \text{if } \pi \notin \{\pi_1, \pi_2\} \end{cases}$$

In other words, a flow is at Nash equilibrium if no network user has an incentive to switch paths.

If we let δ tend to 0, continuity and monotonicity of the edge latency function yields the following useful characterization of a flow at Nash equilibrium, called in this context also a *Wardrop equilibrium* [31].

Theorem 42.7

A flow f feasible for instance $(\mathcal{N}, r, \ell)$ is at Nash equilibrium if and only if for every $i \in [k]$, and $\pi_1, \pi_2 \in \mathcal{P}_i$ with $f_{\pi_1} > 0$, it holds that $\ell_{\pi_1}(f) \leq \ell_{\pi_2}(f)$.

An important consequence of this characterization is that if f is Nash equilibrium then all s_i–t_i paths to which f assigns a positive amount of flow have *equal latency*, that we will denote by $L_i(f)$. Therefore, we can express the cost of a flow f at Nash equilibrium in the following useful form.

Theorem 42.8

If f is a flow at Nash equilibrium for instance $(\mathcal{N}, r, \ell)$, then

$$\mathfrak{C}(f) = \sum_{i=1}^{k} L_i(f) r_i$$

42.3.2 Optimal Flows vs. Flows at Nash Equilibrium

Beckmann et al. [36] noticed a similarity between the characterizations of optimal solutions and of flows at Nash equilibrium that leads to a very elegant characterization of an optimal flow as a flow at Nash equilibrium with a different set of edge latency functions. For any edge $e \in E$, let us denote by ℓ_e^* the *marginal cost* of increasing flow on edge e,

$$\ell_e^*(x) = \frac{d}{dy}(y \cdot \ell_e(y))(x) = \ell_e(x) + x \cdot \ell_e'(x)$$

Then, one can show the following theorem (see Beckmann et al. [36], Dafarmos and Sparrow [37], Roughgarden and Tardos [9]).

Theorem 42.9

Let $(\mathcal{N}, r, \ell)$ be an instance in which $x \cdot \ell_e(x)$ is a convex function for each edge and let ℓ^* be the marginal cost function as defined above. Then a flow f feasible for $(\mathcal{N}, r, \ell)$ is optimal if and only if f is at Nash equilibrium for the instance $(\mathcal{N}, r, \ell^*)$.

This precise and simple characterization of optimal flows plays a crucial role in the analysis of the flow model and in the understanding of that model. From now on, we will denote the optimal flow by f^* and the marginal cost function by ℓ^*.

We notice that the convexity of $x \cdot \ell_e(x)$ is necessary in Theorem 42.9. Therefore, in this survey we shall mostly focus on standard latency functions, where a latency function ℓ is *standard* if it is differentiable and if $x \cdot \ell_e(x)$ is convex on $[0, \infty)$. Many but not all of interesting latency functions are standard. Typical examples of standard latency functions include differentiable convex functions, and examples of nonstandard latency functions are differentiable approximations of step functions.

An important consequence of using standard latency functions is that in network with standard latency functions the optimal flow (up to an arbitrarily small additive error term) can be found in polynomial time using convex programming, see [8,9] for more details.

Using the similarity between the optimal flow and a flow at Nash equilibrium, Beckmann et al. [36] proved the existence and essential uniqueness of Nash equilibria (see also [9, Lemma 2.9]).

Theorem 42.10 [36]

Any instance $(\mathcal{N}, r, \ell)$ with continuous, nondecreasing latency functions admits a feasible flow at Nash equilibrium. Moreover, if f and $\tilde{f}$ are flows at Nash equilibrium, then $\mathfrak{C}(f) = \mathfrak{C}(\tilde{f})$.

We notice that actually, any such an instance admits a feasible acyclic flow at Nash equilibrium, see [8, Section 2.6].

Finally, following our definition from the previous sections, we define a *coordination ratio* (which has been also frequently called in this context a *price of anarchy*) as the ratio between the cost of a flow at Nash equilibrium in an instance $(\mathcal{N}, r, \ell)$ over the minimum cost of the optimal flow for that instance,

$$\frac{\mathfrak{C}(f)}{\mathfrak{C}(f^*)}$$

where f is a flow at Nash equilibrium and f^* denotes the optimal flow. Notice that by Theorrem 42.10, all Nash equilibria have the same cost and thus for any instance $(\mathcal{N}, r, \ell)$ the coordination ratio is the same for all flows at Nash equilibrium.

42.3.3 Unbounded Coordination Ratio

In this section we show that without making any assumption about the latency functions the coordination ratio can be arbitrary large.

Let us consider the example from Figure 42.1 (this is essentially an example of Pigou [38]). Let p be any positive integer. In that example, we are given a very simple network with two nodes, a source s and a sink t, two links, each from the source to the sink, and with the latency functions $\ell_1(x) = 1$ and $\ell_2(x) = x^p$. Let us consider the instance $(\mathcal{N}, r, \ell)$ with such defined network $\mathcal{N}$ and the latency functions ℓ, and with the rate $r = 1$. Notice that if we allocate flow q on the 1st link and $1 - q$ on the 2nd link, then the cost of that flow is $q + (1 - q)(1 - q)^p = q + (1 - q)^{p+1}$.

(a) Latency functions (b) Marginal cost functions

FIGURE 42.1 Pigou's example with the source node s, the sink node t, and the latency functions $\ell_1(x) = 1$ and $\ell_2(x) = x^p$.

Let us first notice that the flow at Nash equilibrium allocates the entire flow to the 2nd link, with the cost $\mathfrak{C}(f) = 1$. Indeed, if any positive amount of flow has been allocated to the 1st link, then by Theorem 42.7, we had $\ell_2(f) \geq \ell_1(f) = 1$, which means that $\ell_2(f) \geq 1$, contradicting the assumption that not the entire flow was allocated to the 2nd link.

To see what is the optimal flow, we apply Theorem 42.9. We have the marginal cost function $\ell_1^*(x) = 1$ and $\ell_2^*(x) = (p+1)x^p$, and hence, optimal flow f^* is at Nash equilibrium for the instance $(\mathcal{N}, 1, \ell^*)$. From that, we observe that at Nash equilibrium we will allocate positive flow to both links, and hence, by our comments after Theorem 42.7, this amount of flow will have to make $\ell_1^*(f^*) = \ell_2^*(f^*)$. Therefore, in the optimal flow we allocate $1 - (1+p)^{-1/p}$ to the 1st link and $(1+p)^{-1/p}$ to the 2nd link. (Indeed, notice that $\ell_1(1 - (1+p)^{-1/p}) = 1$ and $\ell_2^*((1+p)^{-1/p}) = (p+1)((1+p)^{-1/p})^p = 1$.) Therefore, the minimum cost of the optimal flow for instance $(\mathcal{N}, 1, \ell)$ is $\mathfrak{C}(f^*) = 1 - \frac{p}{1+p}(1+p)^{-1/p}$.

From that, we obtain the coordination ratio

$$\frac{1}{1 - \frac{p}{1+p}(1+p)^{-1/p}} = \frac{(1+p)\sqrt[p]{1+p}}{(1+p)\sqrt[p]{1+p} - p} = \Theta\left(\frac{p}{\ln p}\right)$$

which tends to ∞ with increasing p. Therefore, this simple example shows that the coordination ratio can be arbitrary large even for polynomial latency functions.

42.3.4 Routing in Arbitrary Networks: Latency Functions Are Hardest for Parallel Links

In Section 42.3.3 we have seen that for general (standard) latency functions the coordination ratio can be unbounded even in the simplest case of a two-node network. However, we are still interested in the coordination ratio for typical (or natural) classes of latency functions. The first step in this direction was obtained by Roughgarden and Tardos [9], who showed that for the class of linear latency functions, that is, latency functions of the form $\ell(x) = a\,x + b$ for $a, b \geq 0$, the coordination ratio is always bounded by a small constant, namely by $\frac{4}{3}$.

Let us construct (after [9]) an example showing that this bound cannot be improved. We consider again the network from Figure 42.1 with $p = 1$. By our analysis in Section 42.3.3, we know that in Nash equilibrium we allocate the entire flow to the 2nd link and in the optimal flow, we allocate $\frac{1}{2}$ flow to the 1st link and the same amount of flow to the 2nd link. Therefore, the flow at Nash equilibrium has the cost 1 and the optimal flow has the cost $\mathfrak{C}(f^*) = \frac{1}{2} \cdot 1 + \frac{1}{2} \cdot \frac{1}{2} = \frac{3}{4}$. In this case, we obtain the coordination ratio $\frac{4}{3}$.

Even if the analysis of Roughgarden and Tardos [9] for linear latency functions was tight and nice, a possible extension of this analysis to other latency functions seems to be quite tedious and actually, this analysis does not seem to be generalizable to arbitrary functions, for example for queuing functions. Recently, Roughgarden [39] came up with a more elegant approach that greatly simplified that analysis. He showed that essentially (and very informally) among all the networks, the worst coordination ratio is achieved by networks consisting of two nodes connected by parallel links, sometimes even with two parallel links. This result shows an importance of the model of parallel links and it significantly simplifies the analysis of the coordination ratio in networks.

Let us call a class $\mathcal{L}$ of latency functions *standard* if it contains a nonzero function and each function $\ell \in \mathcal{L}$ is standard. By analyzing the scenario of the Pigou's network (see Figure 42.1) with various standard latency functions, Roughgarden [39] observed that in many cases the coordination ratio for that example leads to the worst possible coordination ratio. This motivates us to define the anarchy value to be the worst-case coordination ratio in Pigou's example from Figure 42.1 that use the latency function ℓ on the 2nd link. More formally, if ℓ is a nonzero standard latency function, then we define the *anarchy value* $\alpha(\ell)$ of ℓ as

$$\alpha(\ell) = \sup_{r>0:\ell(r)>0} \frac{1}{\lambda\mu + (1-\lambda)}$$

where $\lambda \in [0, 1]$ is a solution of $\ell^*(\lambda r) = \ell(r)$ and $\mu = \ell(\lambda r)/\ell(r)$.

We further extend this notion to classes of functions, and define the *anarchy value of a standard class* $\mathcal{L}$ *of latency functions*, by

$$\alpha(\mathcal{L}) = \sup_{0 \neq \ell \in \mathcal{L}} \alpha(\ell)$$

With these definitions, we can state now the first main result from [39], which gives a precise upper bound for the coordination ratios.

Theorem 42.11

Let $\mathcal{L}$ be a standard class of latency functions with anarchy value $\alpha(\mathcal{L})$. Let $(\mathcal{N}, r, \ell)$ denote an instance with latency functions drawn from $\mathcal{L}$. Then, the coordination ratio is upper bounded by $\alpha(\mathcal{L})$.

The bound in Theorem 42.11 is tight, but actually, the class of networks for which it matches lower bounds can be even simplified further if some additional conditions are satisfied. Let us define two new classes of latency functions: a class $\mathcal{L}$ is *diverse* if for each positive constant $c > 0$ there is a latency function $\ell \in \mathcal{L}$ with $\ell(0) = c$, and a class $\mathcal{L}$ is *nondegenerate* if $\ell(0) \neq 0$ for certain $\ell \in \mathcal{L}$. A network is called a *union of paths* if it can be obtained from a network of parallel links by repeated subdivisions.

In the following theorem, the term worst-case coordination ratio is with respect to the worst-case choice of the underlying instances.

Theorem 42.12

Let $\mathcal{N}_m$ denote the network with one source node, one sink node, and m parallel links directed from source to sink.

- *If $\mathcal{L}$ is a standard class of latency functions containing the constant functions, then the worst-case coordination ratio for all instances $(\mathcal{N}, r, \ell)$ with latency functions in $\mathcal{L}$ is identical to the worst-case coordination ratio for all instances $(\mathcal{N}_2, r, \ell)$ with latency functions in $\mathcal{L}$.*
- *If $\mathcal{L}$ is a standard and diverse class of latency functions, then the worst-case coordination ratio for all instances $(\mathcal{N}, r, \ell)$ with latency functions in $\mathcal{L}$ is identical to the worst-case coordination ratio for all instances $(\mathcal{N}_m, r, \ell)$ with latency functions in $\mathcal{L}$.*
- *If $\mathcal{L}$ is a standard and nondegenerate class of latency functions, then the worst-case coordination ratio for all instances $(\mathcal{N}, r, \ell)$ with latency functions in $\mathcal{L}$ is identical to the worst-case coordination ratio for all instances $(\mathcal{N}^*, r, \ell)$ with latency functions in $\mathcal{L}$ and with the underlying networks $\mathcal{N}^*$ that are union of paths.*

TABLE 42.1 The Coordination Ratio for Basic Classes of Latency Cost Functions Determined in [39]

Description	Typical Representative	Worst-case Coordination Ratio
Linear	$ax + b$	$\frac{4}{3} \approx 1.333$
Quadratic	$ax^2 + bx + c$	$\frac{3\sqrt{3}}{3\sqrt{3}-2} \approx 1.626$
Cubic	$ax^3 + bx^2 + cx + d$	$\frac{4\sqrt[3]{4}}{4\sqrt[3]{4}-3} \approx 1.896$
Polynomials of degree $\leq p$	$\sum_{i=0}^{p} a_i x^i$	$\frac{(1+p)\sqrt[p]{1+p}}{(1+p)\sqrt[p]{1+p}-p} = \Theta(p/\ln p)$
M/M/1 delay functions	$(u - x)^{-1}$	$\frac{1}{2}\left(1 + \sqrt{\frac{u_{\min}}{u_{\min}-R_{\max}}}\right)$
M/G/1 delay functions	$\frac{1}{u} + \frac{x(1+\sigma^2 u^2)}{2u(u-x)}$	$\sup_{\ell}\left(1 + \sqrt{\frac{u_\ell}{u_\ell - R_{\max}}}\right) \frac{2u_\ell + R_{\max}(\sigma_\ell^2 u_\ell^2 - 1)}{4u_\ell + (u_\ell + R_{\max} - \sqrt{u_\ell(u_\ell - R_{\max})})(\sigma_\ell^2 u_\ell^2 - 1)}$

Polynomial coefficients are assumed to be nonnegative. The parameters u and σ are the expectation and standard deviation of the associated queue service rate distribution, $R_{\max}$ denotes the maximum allowable amount of network traffic, and $u_{\min}$ is the minimum allowable edge service rate (capacity). In M/G/1 case, the parameters u_ℓ and σ_ℓ denote the expectation and standard deviation of the service rate distribution associated with latency ℓ, see [39] for more details.

42.3.4.1 Applications of Theorem 42.11

Theorem 42.12 can be immediately applied to the analysis of the coordination ratio for large classes of latency functions. We list in Table 42.1 the results proven by Roughgarden using the method discussed in Theorem 42.12.

42.3.5 Bicriteria Analysis: Bandwidth Increase to Compensate Losses in Selfish Routing

In the previous section we have seen that for many very natural latency cost functions the coordination ratio may be either large or even unbounded. In order to overcome these negative results, Roughgarden and Tardos [9] proposed to study bicriteria scenarios. In this study, one asks how much the *bandwidth on all links must be increased* to compensate the losses due to selfish routing. The following theorem due to Roughgarden and Tardos [9] characterizes the relation between the bandwidth increase and the cost of selfish routing.

Theorem 42.13

If f is a flow at Nash equilibrium for $(\mathcal{N}, r, \ell)$ and f^* is feasible for $(\mathcal{N}, 2r, \ell)$, then $\mathfrak{C}(f) \leq \mathfrak{C}(f^*)$. Moreover, for any $\gamma > 0$, if f is a flow at Nash equilibrium for $(\mathcal{N}, r, \ell)$ and f^* is feasible for $(\mathcal{N}, (1+\gamma)r, \ell)$, then $\mathfrak{C}(f) \leq \frac{1}{\gamma}\mathfrak{C}(f^*)$.

The first result should be interpreted that the losses due to selfish routing can be completely compensated for by doubling the bandwidth on all links.

By analyzing the example from Figure 42.1 (with $r = 1$), one can observe that the bound in Theorem 42.13 is essentially tight, in the sense that by taking p sufficiently large, one can obtain an instance admitting an optimal flow for a traffic rate arbitrarily close to $(1+\gamma)$ with cost strictly less than γ and an optimal flow feasible for rate $1+\gamma$ with cost arbitrarily close to γ.

42.3.6 Flows at Approximate Nash Equilibrium

Roughgarden and Tardos [9] discussed also the flows at *approximate* Nash equilibria. This analysis is motivated by a realistic assumption that an agent can only distinguish between paths that differ in their latency by more than a $(1+\epsilon)$ factor for some small $\epsilon > 0$. We extend the definition of Nash equilibria to obtain the following analogue of Theorem 42.7 (for more details, see [9]).

Lemma 42.3

A flow f feasible for instance $(\mathcal{N}, r, \ell)$ is at ϵ-approximate Nash equilibrium if and only if for every $i \in [k]$, and $\pi_1, \pi_2 \in \mathcal{P}_i$ with $f_{\pi_1} > 0$, it holds that $\ell_{\pi_1}(f) \leq (1+\epsilon)\ell_{\pi_2}(f)$.

Roughgarden and Tardos [9] showed that essentially most of the results mentioned in the previous section can carry on for approximate Nash equilibria if the bounds are multiplied by some function of $(1 \pm \epsilon)$. And so, for example, they show an analog of Theorem 42.13 that if f is a flow at ϵ-approximate Nash equilibrium for $(\mathcal{N}, r, \ell)$ and f^* is feasible for $(\mathcal{N}, 2r, \ell)$, then $\mathfrak{C}(f) \leq \frac{1+\epsilon}{1-\epsilon}\,\mathfrak{C}(f^*)$.

42.3.7 Unsplittable Flows

In the previous discussion on flows at Nash equilibria, we were making an often unrealistic assumption that there are an infinite number of noncooperative agents, each controlling a negligible fraction of the overall traffic. In this section, we study the scenario in which the number of agents is finite and each agent must select a single path for routing.

An important consequence of the analysis due to Roughgarden and Tardos [9] is that in the model of unsplittable flows not only the coordination ratio may be unbounded but also the bicriteria bound. In particular, Roughgarden and Tardos [9] (see also [8]) showed that a flow at Nash equilibrium for $(\mathcal{N}, r, \ell)$ can be arbitrarily more costly than an optimal instance obtained from $(\mathcal{N}, r, \ell)$ by increasing the amount of flow each agent must route by an arbitrary factor. They gave an example of a network with latency functions with unbounded derivatives for which they showed that routing a strictly positive amount of additional flow on an edge may increase the latency on that edge by an arbitrarily large amount. This bound also holds for such important classes of latency functions as the delay functions of M/M/1 queues (latency of the form $\ell(x) = 1/(b - x)$, where b is the service rate) that are of particular interest in networking. On a positive side, Roughgarden and Tardos proved that if the largest possible change in edge latency resulting from a single agent rerouting of its flow is not too large, some bicriteria result is possible, see [9, Theorem 5.5].

42.3.8 How Fair (or Unfair) Is Optimal Routing

In the prior discussion we concentrated our interest on the analysis of how much can we loose by allowing noncooperative traffic allocations in networks. Very recently, Roughgarden [40] studied a somehow reverse question of how unfair can an optimal routing be, or in other words, how much worse off can network users be in an optimal assignment than in an assignment that is at Nash equilibrium. This study has begun from an observation that optimal flow allocation strategies can force some users of the network to behave in a very noneconomic way so that some number of users "sacrifice" their own performance to improve the overall cost.

Roughgarden [40] defined the *unfairness* of instance $(\mathcal{N}, r, \ell)$, denoted by $\mathfrak{u}(\mathcal{N}, r, \ell)$, as the maximum ratio between the latency of a flow path of an optimal flow for $(\mathcal{N}, r, \ell)$ and flow path at Nash equilibrium for $(\mathcal{N}, r, \ell)$. Similarly as for the coordination ratio $\mathfrak{R}$ discussed in Section 42.3.3, one can construct problem instances for which $\mathfrak{u}(\mathcal{N}, r, \ell)$ can be arbitrarily large if no additional restriction on the latency functions is put. (Take the Pigou's example from Figure 42.1, with the modification that $\ell_1(x) = (1+p)(1-\epsilon)$; in that case, $\mathfrak{u}(\mathcal{N}, r, \ell) = (1+p)(1-\epsilon)$.)

For standard latency functions a better bound is achievable. Let $\mathcal{L}$ be a class of standard latency functions. For any $\ell \in \mathcal{L}$, define $\gamma(\ell)$ by $\gamma(\ell) = \sup_{x>0} \frac{\ell^*(x)}{\ell(x)}$, where $\gamma(\ell)$ can be interpreted as the biggest discrepancy between how optimal and Nash flows evaluate the per-unit cost of using an edge. If $\gamma(\mathcal{L}) = \sup_{\ell \in \mathcal{L}} \gamma(\ell)$ and $(\mathcal{N}, r, \ell)$ is an instance with latency functions drawn from $\mathcal{L}$, then Roughgarden [40] proved that $\mathfrak{u}(\mathcal{N}, r, \ell) \leq \gamma(\mathcal{L})$. As an example demonstrating this result, we observe that if $\mathcal{L}$ contains some polynomials with nonnegative coefficients of degree at most p, then the unfairness is at most $1 + p$ (and it is actually exactly $1 + p$).

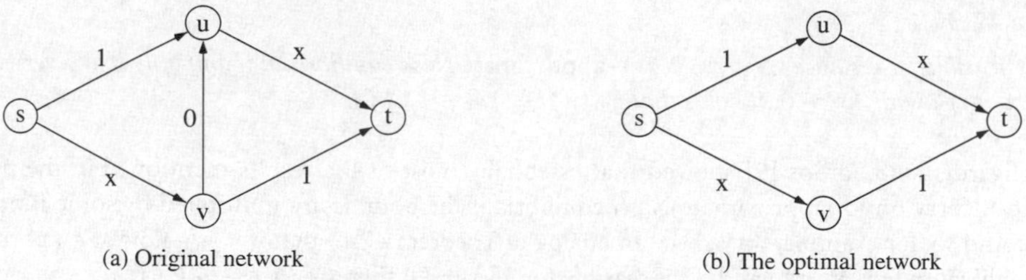

FIGURE 42.2 Braess's paradox.

42.3.9 Constructing "Nash-"Efficient Networks Is Hard

Since the prior discussion shows that the coordination ratio may be very large or even unbounded, it is interesting to study the question of how to design networks to ensure that the coordination ratio is small, that is, for a given class of latency functions, how to modify a given network to obtain the best possible flow at Nash equilibrium. It is well-known that removing the edges from a network may improve its performance. This phenomenon was first shown by Braess [41] (see also [42]) and is commonly known as *Braess's Paradox*, see Figure 42.2. Roughgarden [43] gave optimal inapproximability results and approximation algorithms for several basic network design problems of the following type: given a network with n nodes, edge latency functions, a single-source pair, and a rate of traffic, construct a subnetwork minimizing the flow cost at Nash equilibrium. For networks with continuous, nonnegative, nondecreasing latency functions, Roughgarden showed that on one hand, there is no $(n/2 - \epsilon)$-approximation polynomial-time algorithm unless $\mathcal{P} = \mathcal{NP}$, and on the other hand, there is an $n/2$-approximation polynomial-time algorithm for this problem; for linear latency cost functions, the bound of $n/2$ should be replaced by $\frac{4}{3}$. Moreover, Roughgarden [43] showed that essentially the following trivial algorithms achieve the optimal bounds: given a network of candidate edges, build the entire network. For more discussion about this problem we refer the reader to [43]; research on a related model has been recently reported in [10].

42.3.10 Further Extensions

There has been much more research on game theoretical aspects of the traffic flow in networks that we could cover not in this survey because of space limitations. Below, we just list some of other exciting topics discussed in the literature.

Schulz and Stier Moses [44] investigated the flow model in *capacitated networks*. They show that for general latency functions the capacities can significantly change the behavior of the system if we compare with the analyses from [9,39]. On the other hand, for a standard class of latency functions, they showed that the worst coordination ratio is achieved by uncapacitated networks.

Roughgarden [45] investigated the scenario in which some portion of the traffic is globally controlled (regulated) by a centralized authority that aims at optimizing the overall system performance (such scenarios are known as *Stackelberg games* in game theory). He studied the most basic model of a network consisting of two nodes (source and sink) that are connected by parallel links. Roughgarden showed that by employing a carefully chosen scheduling strategy the centralized authority can greatly reduce the inefficiency caused by selfish users in a system of shared machines possessing load-dependent latencies. In particular, Roughgarden [45] proved that if an α fraction of the overall traffic can be globally regulated, then for an arbitrary class of latency functions, there is a strategy that ensures that the coordination ratio is at most $1/\alpha$. Stackelberg games in a similar setting have been also investigated, e.g., in [1,46–48].

In other works, recently, Cole et al. [49,50] studied the use of economic means ("taxes") on influencing the negative effects of selfish behavior in congested networks. Friedman [33] analyzed the flow model discussed in this section and showed that in any network, for most of traffic rates the coordination ratio is much smaller than the worst-case value. In other words, even if the losses due to selfish routing can be large in general, one should expect them to be small on average.

42.4 Further References

Despite our best efforts, it is totally impossible to present the full picture of the area of selfish routing in one short survey. In particular, we omitted a lot of research in networking literature, game theory literature, and complexity and algorithms literature related to our theme. A reader interested in more references and in further advances in this field is recommended to check the most recent editions of proceedings of major conferences in theoretical computer science, networking, and game theory, in particular, *STOC* (Annual ACM Symposium on Theory of Computing), *FOCS* (IEEE Symposium on Foundations of Computer Science), *SODA* (Annual ACM-SIAM Symposium on Discrete Algorithms), *INFOCOM* (Annual Joint Conference of the IEEE Computer and Communications Societies), and *ACM Conference on E-commerce*; most of the research discussed in this survey has been published there.

Acknowledgment

This research was supported partially by NSF grant CCR-0105701.

References

[1] Y. A. Korilis, A. A. Lazar, and A. Orda. Achieving network optima using stackelberg routing strategies. *IEEE/ACM Transactions on Networking*, 5(1):161–173, February 1997.

[2] Y. A. Korilis, A. A. Lazar, and A. Orda. Capacity allocation under noncooperative routing. *IEEE Transactions on Automatic Control*, 42(3):309–325, 1997.

[3] E. Koutsoupias and C. H. Papadimitriou. Worst-case equilibria. In C. Meinel and S. Tison, (eds.), *Proceedings of the 16th Annual Symposium on Theoretical Aspects of Computer Science (STACS)*, volume 1563 of *Lecture Notes in Computer Science*, pp. 404–413, Trier, Germany, March 4–6, 1999. Springer-Verlag, Berlin.

[4] R. B. Myerson. *Game Theory: Analysis of Conflict*. Harvard University Press, Cambridge, MA, 1991.

[5] G. Owen. *Game Theory*. Academic Press, Orlando, FL, 3rd ed., 1995.

[6] C. H. Papadimitriou. Algorithms, games, and the Internet. In *Proceedings of the 33rd Annual ACM Symposium on Theory of Computing (STOC)*, pp. 749–753, Hersonissos, Crete, Greece, July 6–8, 2001. ACM Press, New York, NY.

[7] C. H. Papadimitriou. Game theory and mathematical economics: A theoretical computer scientist's introduction (Tutorial). In *Proceedings of the 42th IEEE Symposium on Foundations of Computer Science (FOCS)*, pp. 4–8, Las Vegas, NV, October 14–17, 2001. IEEE Computer Society Press, Los Alamitos, CA.

[8] T. Roughgarden. *Selfish Routing*. PhD thesis, Cornell University, Department of Computer Science, May 2002.

[9] T. Roughgarden and É. Tardos. How bad is selfish routing? *Journal of the ACM*, 49(2):236–259, March 2002. In *Proceedings of the 41st IEEE Symposium on Foundations of Computer Science (FOCS)*, pp. 93–102, Redondo Beach, CA, November 12–14, 2000. IEEE Computer Society Press, Los Alamitos, CA.

[10] E. Anshelevich, A. Dasgupta, É. Tardos, and T. Wexler. Near-optimal network design with selfish agents. In *Proceedings of the 35th Annual ACM Symposium on Theory of Computing (STOC)*, San Diego, CA, June 9–11, 2003. ACM Press, New York, NY.

[11] M. Mavronicolas and P. Spirakis. The price of selfish routing. In *Proceedings of the 33rd Annual ACM Symposium on Theory of Computing (STOC)*, pp. 510–519, Hersonissos, Crete, Greece, July 6–8, 2001. ACM Press, New York, NY.

[12] A. Czumaj and B. Vöcking. Tight bounds for the worst-case equilibria. In *Proceedings of the 13th Annual ACM-SIAM Symposium on Discrete Algorithms (SODA)*, pp. 413–420, San Francisco, CA, January 6–8, 2002. SIAM, Philadelphia, PA.

[13] E. Koutsoupias, M. Mavronicolas, and P. Spirakis. Approximate equilibria and ball fusion. In *Proceedings of the 9th International Colloquium on Structural Information and Communication Complexity (SIROCCO)*, Andros, Greece, June 10–12, 2002. Carleton Scientific, Ottawa, Canada.

[14] P. G. Bradford. On worst-case equilibria for the Internet. Manuscript, 2001.

[15] G. Gonnet. Expected length of the longest probe sequence in hash code searching. *Journal of the ACM*, 28(2):289–304, 1981.

[16] W. Hoeffding. Probability inequalities for sums of bounded random variables. *Journal of the American Statistical Association*, 58(301):13–30, March 1963.

[17] B. Awerbuch, Y. Azar, Y. Richter, and D. Tsur. Tradeoffs in worst-case equilibiria. In *Proceedings of the 1st Workshop on Approximation and Online Algorithms (WAOA)*, Budapest, Hungary, September 15–20, 2003.

[18] P. Berenbrink, L. A. Goldberg, P. Goldberg, and R. Martin. Utilitarian resource assignment. Manuscript, March 2003.

[19] D. Fotakis, S. Kontogiannis, E. Koutsoupias, M. Mavronicolas, and P. Spirakis. The structure and complexity of Nash equilibria for a selfish routing game. In P. Widmayer, F. Triguero, R. Morales, M. Hennessy, S. Eidenbenz, and R. Conejo, (eds.), *Proceedings of the 29th Annual International Colloquium on Automata, Languages and Programming (ICALP)*, volume 2380 of Lecture Notes in Computer Science, pp. 123–134, Málaga, Spain, July 8–13, 2002. Springer-Verlag, Berlin.

[20] R. Feldmann, M. Gairing, T. Lücking, B. Monien, and M. Rode. Nashification and the coordination ratio for a selfish routing game. In *Proceedings of the 30th Annual International Colloquium on Automata, Languages and Programming (ICALP)*, pp. 514–526, Lecture Notes in Computer Science, Eindhoven, The Netherlands, June 30 – July 4, 2003. Springer-Verlag, Berlin.

[21] D. Hochbaum and D. Shmoys. A polynomial approximation scheme for scheduling on uniform processors: Using the dual approximation approach. *SIAM Journal on Computing*, 17(3):539–551, 1988.

[22] D. Fudenberg and D. Levine. *The Theory of Learning Games*. MIT Press, 1998.

[23] E. Even-Dar, A. Kesselman, and Y. Mansour. Convergence time to Nash equilibria. In *Proceedings of the 30th Annual International Colloquium on Automata, Languages and Programming (ICALP)*, pp. 502–513, Lecture Notes in Computer Science, Eindhoven, The Netherlands, June 30–July 4, 2003. Springer-Verlag, Berlin.

[24] I. Milchtaich. Congestion games with player-specific payoff functions. *Games and Economic Behavior*, 13:111–124, 1996.

[25] A. Czumaj, P. Krysta, and B. Vöcking. Selfish traffic allocation for server farms. In *Proceedings of the 34th Annual ACM Symposium on Theory of Computing (STOC)*, pp. 287–296, Montréal, Québec, Canada, May 19–21, 2002. ACM Press, New York, NY.

[26] M. E. Crovella and A. Bestavros. Self-similarity in World Wide Web traffic: Evidence and possible causes. *IEEE/ACM Transactions on Networking*, 5(6):835–846, 1997. In *Proceedings of the ACM SIGMETRICS, the International Conference on Measurement and Modeling of Computer Systems*, pp. 160–169, Philadelphia, PA, May 23–26, 1996. ACM Press, New York, NY.

[27] F. Douglis, A. Feldmann, B. Krishnamurthy, and J. Mogul. Rate of change and other metrics: A live study of the World Wide Web. In *Proceedings of the 1st USENIX Symposium on Internet Technologies and Systems*, pp. 147–158, Monterey, CA, December 8–11, 1997.

[28] A. Feldmann, A. C. Gilbert, P. Huang, and W. Willinger. Dynamics of IP traffic: A study of the role of variability and the impact of control. In *Proceedings of the ACM SIGCOMM, Conference on

Applications, Technologies, Architectures, and Protocols for Computer Communication, pp. 301–313, Cambridge, MA, August 30–September 3, 1999. ACM Press, New York, NY.

[29] K. Park, G. Kim, and M. E. Crovella. On the relationship between file sizes, transport protocols, and self-similar network traffic. In *Proceedings of the 4th International Conference on Network Protocols (ICNP)*, pp. 171–180, Columbus, OH, October 29–November 1, 1996.

[30] V. Paxson and S. Floyd. Wide area traffic: The failure of Poisson modeling. *IEEE/ACM Transactions on Networking*, 3:226–244, 1995.

[31] J. G. Wardrop. Some theoretic aspects of road traffic research. In *Proceedings of the Institution of Civil Engineers, Part II*, volume 1, pp. 325–362, 1952.

[32] L. Kleinrock. *Queueing Systems*, volume I: Theory. John Wiley & Sons, New York, NY, 1975.

[33] E. Friedman. Genericity and congestion control in selfish routing. Manuscript, May 2003. Available from `http://www.orie.cornell.edu/%7Efriedman/papers.htm`.

[34] D. Gross and C. M. Harris. *Queueing Theory*, volume I: Theory. John Wiley & Sons, New York, NY, third edition, 1998.

[35] A. Haurie and P. Marcotte. On the relationship between Nash-Cournot and Wardrop equlibria. *Networks*, 15:295–308, 1985.

[36] M. Beckmann, C. B. McGuire, and C. B. Winston. *Studies in the Economimices of Transportation*. Yale University Press, 1956.

[37] S. C. Dafermos and F. T. Sparrow. The traffic assignment problem for a general network. *Journal of Research of the National Bureau Standards, Series B*, 73B(2):91–118, 1969.

[38] A. C. Pigou. *The Economics of Welfare*. Macmillan, London, UK, 1920.

[39] T. Roughgarden. The price of anarchy is independent of the network topology. In *Proceedings of the 34th Annual ACM Symposium on Theory of Computing (STOC)*, pp. 428–437, Montréal, Québec, Canada, May 19–21, 2002. ACM Press, New York, NY.

[40] T. Roughgarden. How unfair is optimal routing? In *Proceedings of the 13th Annual ACM-SIAM Symposium on Discrete Algorithms (SODA)*, pp. 203–204, San Francisco, CA, January 6–8, 2002. SIAM, Philadelphia, PA.

[41] D. Braess. Über ein Paradoxon der Verkehrsplanung. *Unternehmensforschung*, 12:258–268, 1968. Available from `http://homepage.ruhr-uni-bochum.de/Dietrich.Braess/paradox.pdf`.

[42] F. P. Kelly. Network routing. *Philosophical Transactions of the Royal Society, London A*, 337:343–367, 1991.

[43] T. Roughgarden. Designing networks for selfish users is hard. In *Proceedings of the 42nd IEEE Symposium on Foundations of Computer Science (FOCS)*, pp. 472–481, Las Vegas, NV, October 14–17, 2001. IEEE Computer Society Press, Los Alamitos, CA.

[44] A. S. Schulz and N. Stier Moses. On the performance of user equilibria in traffic networks. In *Proceedings of the 14th Annual ACM-SIAM Symposium on Discrete Algorithms (SODA)*, pp. 86–87, Baltimore, MD, January 12–14, 2003. SIAM, Philadelphia, PA.

[45] T. Roughgarden. Stackelberg scheduling strategies. In *Proceedings of the 33rd Annual ACM Symposium on Theory of Computing (STOC)*, pp. 104–113, Hersonissos, Crete, Greece, July 6–8, 2001. ACM Press, New York, NY.

[46] V. S. Anil Kumar and M. V. Marathe. Improved results for Stackelberg scheduling strategies. In P. Widmayer, F. Triguero, R. Morales, M. Hennessy, S. Eidenbenz, and R. Conejo (eds.), *Proceedings of the 29th Annual International Colloquium on Automata, Languages and Programming (ICALP)*, volume 2380 of *Lecture Notes in Computer Science*, pages 776–787, Málaga, Spain, July 8–13, 2002. Springer-Verlag, Berlin.

[47] T. Başar and R. Srikant. A Stackelberg network game with a large number of followers. *Journal Optimization Theory and Applications*, 115(3):479–490, December 2002.

[48] Y. A. Korilis, A. A. Lazar, and A. Orda. The role of the manager in a noncooperative network. In *Proceedings of the 15th IEEE INFOCOM, Annual Joint Conference of the IEEE Computer and Communications Societies*, pp. 1285–1293, San Francisco, CA, March 24–28, 1996. IEEE Computer Society Press, Los Alamitos, CA.

[49] R. Cole, Y. Dodis, and T. Roughgarden. How much can taxes help selfish routing? In *Proceedings of the 4th ACM Conference on E-commerce (EC)*, pp. 98–107, San Diego, CA, June 9–12, 2003. ACM Press, New York, NY.
[50] R. Cole, Y. Dodis, and T. Roughgarden. Pricing network edges for heterogeneous selfish users. In *Proceedings of the 35th Annual ACM Symposium on Theory of Computing (STOC)*, pp. 521–530, San Diego, CA, June 9–11, 2003. ACM Press, New York, NY.
[51] P. Krysta, P. Sanders, and B. Vöcking. Scheduling and traffic allocation for tasks with bounded splittability. In *Proceedings of the 28th International Symposium on Mathematical Foundations of Computer Science*, pp. 500–510, Bratislava, Slovak Republic, August 25–29, 2003. Springer-Verlag, Berlin.

VI

Applications

43

Scheduling of Flexible Resources in Professional Service Firms

Yalçin Akçay
Koc University

Anantaram Balakrishnan
The University of Texas at Austin

Susan H. Xu
Penn State University

43.1 Introduction

Flexibility is the ability of an organization to effectively cope with uncertainties and changes in the market by employing resources that can process different types of jobs. In today's ever-changing business environment, driven by the challenges of quick response, increasing customization, shorter product life cycles, and intense competition, flexibility has become vital for both manufacturing and service operations to hedge these systems against uncertainty while controlling costs. Decision makers well understand the strategic significance of flexibility to compete effectively (Jones and Ostroy, 1984). However, these strategic decisions must be integrated with operational policies and resource usage decisions to fully realize the benefits of flexibility. At the strategic level, practitioners and researchers focus on the role of flexibility in handling uncertainties, translating flexibility requirements into capacity and performance objectives. This area has been a popular research field due to the emergence of related flexible manufacturing technologies (e.g., Fine and Freund, 1990; Jordan and Graves, 1995; Mieghem, 1998). Operational level issues, on the other hand, deal with designing specific methods for *deployment* of flexibility. Given the available level of flexibility in resources, the decision maker faces the problem of effectively allocating these flexible resources to various operations in order to achieve the strategic goals. In this chapter we investigate operational policies for deployment of flexible resources in professional service firms such as consulting, legal, or training firms. In particular, the work is motivated by resource assignment decisions within the *workplace learning industry*.

1-58488-397-9/$0.00+$1.50

In this environment, since personnel costs accounts for a large portion of total costs, scheduling the available resources (people) is one of the most important managerial activities (Maister, 1997). Different project types undertaken by these firms require different skill sets. In this context, a particular individual (resource) is flexible if this person is proficient in multiple skills, and so can be assigned to different project types. Observe that, here, the scheduling function is mainly concerned with decisions regarding which resource to assign to each project or job. As Maister (1997) notes, although service firms often consider scheduling as routine and tend to underestimate its impact compared to other activities, the frequency of these operational level decisions and their implications in terms of being able to accept new business opportunities as they arise make the scheduling function critical to the overall success of the firm.

The motivating context for this chapter is firms in the *workplace learning industry*. Rapid technological change and the emergence of global markets and competition have profoundly changed the way people and organizations work. In turn, these dramatic changes require continuously upgrading the knowledge and skills of employees to adapt to the new environment. Consequently, workplace learning has become a strategic necessity to maintain the competitive advantage of both individuals and employers. Workplace learning companies provide training, typically on-site, for the employees within a corporation by offering educational modules on topics relevant to different managerial levels, such as supervision, quality, global management, and so on. Being a significant supplier of education and skill development for businesses, the workplace learning industry has grown rapidly in response to increasing corporate training needs. Indeed, according to the 2002 International Comparisons Report of the American Society for Training Development (ASTD), the percentage of employees who received training, in more than 550 organizations from 42 countries, was 76.7% in 2000, compared to 74.9% in 1999 and 67.6% in 1998. Similarly, the training expenditures of organizations as a percentage of their total annual payrolls increased from 1.8% in 1997 to 2.5% in 2000. Organizations spent an average of $630 per employee on training in 2000, and this amount is projected to continue its upward trend in upcoming years. On a global scale, organizations devoted an average of 26.2% of their total training expenditures to outsourced training in 2000. These statistics indicate that the workplace learning industry is large and growing.

As the setting for our decision problem, we consider a workplace learning provider (WLP) who offers various types of training programs, such as programs on leadership, change management, and quality, to its clients. These clients are corporations seeking outsourced training for their employees. The WLP utilizes instructor-led classroom as its method of training delivery. Despite attempts to promote e-learning, the face-to-face delivery approach is still the most dominant form of training according to the 2002 report of the ASTD. Instructors employed by the WLP have different capabilities based on their qualifications — some are specialized to teach only one particular type of training program, whereas others are experienced enough to teach multiple types of programs. This characteristic introduces resource flexibility to the problem context. In other words, for any particular client request, specifying the type of program needed and date of delivery, the WLP manager has the freedom to assign any available instructor who has the required capabilities. Client requests arrive dynamically, and the firm must decide whether to accept the request, and if so which instructor to assign to the project. The goal of the decision maker is to maximize the WLP's expected revenues by intelligently utilizing its limited resources to meet the stochastic demand for workplace training courses. At the strategic level, the firm's leadership must decide what courses to offer, how to price these courses, how many instructors to hire and with what skills. The operational level accept/reject decisions of client requests and instructor assignment decisions, made online (immediately when the request arrives), constitute our *scheduling* problem. If a current request is accepted, the decision maker gives up the opportunity to use the assigned instructor for a potentially more profitable future request. On the other hand, if the request is rejected, there is a risk of having to later assign the instructor to a less profitable request or even being left with an idle instructor at the end of the decision horizon. The dynamic resource assignment (scheduling) decision must account for this tradeoff between the current opportunity and future prospects.

Our flexible resource scheduling problem shares a number of common features with revenue management applications (e.g., airlines, lodging and hospitality, and automobile rentals; see McGill and VanRyzin (1999) for a recent survey of revenue management research). In particular, in terms of Weatherford and

Bodily's (1992) unified taxonomy of revenue management problems, resources become extinct (or perish) after a certain deadline, inventory of available resources during the decision horizon is fixed (or almost impossible to replenish within the duration of the decision horizon), demand for these resources is stochastic, and different resources can be used to satisfy a given customer demand. Revenue management has become a very powerful managerial tool to exploit the revenue-enhancement potential in many businesses (Cross, 1997; Belobaba, 1989; Smith et al., 1992). Our problem context differs from the typical revenue management applications such as airline seat allocation in two important aspects. First, in the airline, hotel, or car rental applications, the different types of demand typically correspond to different customer classes, and the focus is on assigning or reserving capacity from a common pool of resources (e.g., economy class seats) to these different demand types. In contrast, the resource flexibility structure in professional service firms is much richer and complex. Resources are not identical, neither are the job requirements. Thus, not all resources can perform every job. Moreover, resource flexibility is mixed, i.e., some resources are highly specialized, others have some flexibility, and still others are very versatile, and the available resource assignment options may not have any special structure. Second, in the airline and other contexts, prices can be adjusted dynamically (daily or even hourly); so, pricing is an important lever to manage the demand arrival process. In contrast, for professional services, pricing decisions are more strategic and long-term, and cannot be coupled with resource assignment decisions as a demand management tactic. Rather, we assume that prices are prespecified, and focus instead on accepting or not accepting incoming jobs, and assigning resources to accepted jobs. In summary, although the frequency of resource assignment decisions for professional service firms is not as high as in the airline or other conventional revenue management applications, the complexity of the resource flexibility structure makes the decision problem quite complex.

The professional service resource assignment problem entails dynamic decision-making, as jobs arrive, and therefore can be naturally formulated as a profit-maximizing stochastic dynamic program. However, since the state space is very large, finding the optimal policy by solving the associated Bellman equations is impractical. Therefore, like other researchers, we focus on identifying some structural properties of the optimal policy, analyzing some special cases, and developing approximate solution methods based on these properties and special cases. In particular, we consider three different types of heuristic approaches — based respectively on a threshold policy, a newsvendor-type capacity reservation policy, and a rollout policy with a certainty equivalence approximation — and benchmark the performance of these methods against both a naïve (first-come, first-served) approach and an upper bound on the maximum expected profit. Our computational tests indicate that the methods are very effective, providing near-optimal solutions with minimal computational effort.

The rest of this chapter is organized as follows. In Section 43.2, we formally define the flexible resource assignment problem, formulate it as a stochastic dynamic program, and discuss some related literature. We analyze a special case and also present two upper bounds for our problem in Section 43.3. In Section 43.4, we propose approximate solution methods, motivated by analysis of special cases. Section 43.5 reports our computational results for a wide range of problem instances, and Section 43.6 presents concluding remarks.

43.2 Problem Definition and Formulation

We consider a WLP who offers m different types of training programs to its clients, and employs l different types of instructors. Instructors of the same type have similar skills and can teach the same set of training programs. We refer to instructors who can only teach one type of training program as *specialized* instructors, and those who can teach multiple programs as *flexible* instructors. We distinguish instructors who can teach *all* programs by referring to them as *versatile* instructors. We index the program type by $i = 1, 2, \ldots, m$, and let π_i denote the return or revenue (net of any direct costs) for a type-i training program; we assume, for convenience, that this revenue does not depend on the type of instructor assigned to teach the program. We focus on the demand and resource allocation for training programs scheduled on a particular day. Client requests for training on this day arrive randomly, starting T periods before this date.

Given the number of instructors of each type available at the start of the T-period decision horizon and the probability distribution of customer arrivals (for each program type) in each period of the horizon, the manager must decide resource assignments in each period with the objective of maximizing the expected total revenue over the decision horizon. Note that each *period* in this T-period decision horizon corresponds to a decision period for resource assignment (i.e., the firm accumulates client requests during the period, say, one day and allocates resources at the end of the period); by making the period length small enough we can represent instantaneous decision-making as each customer request arrives. The manager must make two major decisions for each client request — *acceptance* and *resource assignment.* The manager can either accept or reject an incoming client request. For instance, the manager might reject a client request, and thus forego the associated revenue, if none of the currently available (unassigned) resources can handle the requested program type, or if the manager feels that available resources must be conserved for anticipated future requests for more profitable projects. We refer to this accept/reject decision as the *acceptance decision*. For each accepted request, the manager must then decide which instructor to assign to this engagement; we refer to this decision as the (resource) *assignment decision.* In making these decisions, the manager must consider the future demand, the availability of instructors and their flexibility, and relative monetary returns for different types of training programs. To keep our subsequent discussions general, we use the term *resources* instead of instructors, and *jobs* instead of training programs or clients. We do not assume any particular demand distributions (demand can even be nonstationary), but require demands across time periods to be independent. We also allow batch arrivals to occur in each period, and as we noted earlier we assume prespecified revenues or prices for each job type. Moreover, a client cannot cancel a program request, and the firm cannot recall or reassign a resource that has already been assigned to a particular job.

We can model this job acceptance and resource assignment problem as a stochastic dynamic program in which stages correspond to periods of the finite decision horizon. The state of the system at each stage is determined by the number of available (*unassigned*) resources and the realized demand in that period for each job type. We can then write the maximum expected total revenue as a recursive function, with the number of each type of resource assigned to incoming jobs during that period as the decision variables. The stochastic dynamic program must incorporate constraints to ensure that each accepted job is assigned a resource that is capable of performing that job, and the number of resources of each type that are assigned to accepted jobs does not exceed the currently available resources of that type.

To formulate the problem as a dynamic program, we adopt the following notation and conventions. Recall that the m job types are indexed from $i = 1$ to $i = m$. For convenience, we index the job types in decreasing order of revenue π_i, i.e., $\pi_1 \geq \pi_2 \cdots \geq \pi_m$. Let $j = 1, 2, \ldots, l$, denote the index of resource types. Each resource type has distinctive capabilities in terms of the types of jobs it can process. To capture these capabilities, for $i = 1, 2, \ldots, m$ and $j = 1, 2, \ldots, l$, we define the indicator parameter a_{ij} which is 1 if a type-j resource can process a type-i job, and 0 otherwise. The matrix $\mathbf{A} = \{a_{ij}\}$, therefore, represents the firm's *resource flexibility structure*. If all the resource types are specialized, i.e., every resource type can only perform one unique job type, then $\mathbf{A}$ is an identity matrix. At the other extreme, if all resources are versatile, i.e., each resource can perform all job types, then $\mathbf{A}$ consists of a single column of ones. Let t denote the index of time period in the decision horizon; we index time periods backwards, starting with $t = 0$ as the date on which the accepted jobs must be performed to $t = T$ at the start of the decision horizon. Let n_{jt}, for $j = 1, 2, \ldots, l$ and $t = 1, 2, \ldots, T$, be the number of available (unassigned) type-j resources at time t; $\mathbf{n}_t = \{n_{1t}, n_{2t}, \ldots, n_{lt}\}$ is the *resource availability* vector at time t. In terms of this notation, the firm's resource *capacity*, comprising the number of available resources at the start of the decision horizon, is $\mathbf{n}_T = \{n_{1T}, n_{2T}, \ldots, n_{lT}\}$.

The *demand* for each job type, expressed in terms of the number of customer requests for that job type, in every period is uncertain. Let D_{it}, for $i = 1, 2, \ldots, m$ and $t = 1, 2, \ldots, T$, be the random variable representing the demand for type-i jobs in period t. The corresponding period-t demand vector is $\mathbf{D}_t = \{D_{1t}, D_{2t}, \ldots, D_{mt}\}$. We use lower-case letters to denote a particular realization of the random variable. Thus, d_{it} and $\mathbf{d}_t = \{d_{1t}, d_{2t}, \ldots, d_{mt}\}$ represent, respectively, the actual demand for type-i jobs and the actual demand vector in period t. Let $P(\mathbf{D}_t = \mathbf{d}_t)$ denote the probability that the demand in period t equals $\mathbf{d}_t$.

The stochastic dynamic programming formulation of the flexible resource assignment problem consists of T stages, one for each period $t = 1, 2, \ldots, T$. The state of the system in any period is described by the available resources and realized demand in that period. Thus, the vector pair $(\mathbf{n}_t, \mathbf{d}_t)$ characterizes the system state in period t. To formulate the dynamic program, define $u_t(\mathbf{n}_t, \mathbf{d}_t)$ as the *maximum expected revenue-to-go* at stage t given the current state $(\mathbf{n}_t, \mathbf{d}_t)$. For $i = 1, 2, \ldots, m$ and $j = 1, 2, \ldots, l$, the stage-t decision variable x_{ijt} is the number of type-$j$resources assigned to type-$i$jobs in period t. Let $\mathbf{e}_j$ denote the jth unit vector. Then, for all periods $t = 1, 2, \ldots, T$, all possible resource availability vectors satisfying $\mathbf{0} \leq \mathbf{n}_t \leq \mathbf{n}_T$, and possible demand vectors $\mathbf{d}_t$, we can express the maximum revenue-to-go function $u_t(\mathbf{n}_t, \mathbf{d}_t)$ in terms of the revenue function for the next period (i.e., period $t-1$) as follows:

$$u_t(\mathbf{n}_t, \mathbf{d}_t) = \max_{x_{ijt}:\forall(i,j)} \left\{ \sum_{i=1}^{m} \sum_{j=1}^{\ell} \pi_i a_{ij} x_{ijt} \right\} + \sum_{\forall \mathbf{d}_{t-1}} \left(u_{t-1} \left(\mathbf{n}_t - \sum_{i=1}^{m} \sum_{j=1}^{\ell} \mathbf{e}_j a_{ij} x_{ijt}, \mathbf{d}_{t-1} \right) P(\mathbf{D}_{t-1} = \mathbf{d}_{t-1}) \right) \tag{43.1}$$

subject to

$$\sum_{j=1}^{\ell} a_{ij} x_{ijt} \leq d_{it} \quad \text{for } i = 1, 2, \ldots, m \tag{43.2}$$

$$\sum_{i=1}^{m} a_{ij} x_{ijt} \leq n_{jt} \quad \text{for } j = 1, 2, \ldots, l \tag{43.3}$$

$$x_{ijt} \geq 0 \text{ and integer for } i = 1, 2, \ldots, m \text{ and } j = 1, 2, \ldots, l \tag{43.4}$$

The boundary condition, at $t = 0$, is

$$u_0(\mathbf{n}_0, \mathbf{d}_0) = 0 \quad \text{for all } \mathbf{n}_0 \leq \mathbf{n}_T \tag{43.5}$$

where, without loss of generality, we assume $\mathbf{d}_0 = \mathbf{0}$.

Constraint (43.4) states that the decision variables x_{ijt} must be nonnegative and integer-valued. The recursive Equation (43.1) expresses the revenue-to-go $u_t(\mathbf{n}_t, \mathbf{d}_t)$ in period t as the sum of the *rewards* collected in period t (the first term in (43.1)) and the maximum expected revenue over the remaining time horizon. Constraints (43.2) and (43.3) are the demand and supply restrictions on the decision values, specifying respectively that the number of resources assigned to a particular job type must not exceed period t's (actual) demand for that job type and the number of assigned resources of each type must not exceed the availability of that type. The boundary condition (43.5) simply states that the value of any resources left over at the end of the horizon is zero, reflecting the fact that capacity cannot be inventoried in service settings. The optimal values of x_{ijt} specify the optimal policy to follow at each stage and state of the system, and the value

$$\sum_{\forall \mathbf{d}_T} u_T(\mathbf{n}_T, \mathbf{d}_T) P(\mathbf{D}_T = \mathbf{d}_T)$$

is the maximum expected revenue over the decision horizon given the firm's resource capacity.

The dynamic program (43.1) to (43.5) shares some similarities, but is also more complex, than the *dynamic and stochastic knapsack problem* (DSKP). Kleywegt and Papastavrou (1998) define the DSKP as the problem of allocating the capacity of a knapsack (a finite-capacity resource) to items with random arrival times, resource requirements (size), and rewards; an item's resource requirement and reward are revealed upon its arrival. As in our problem, the decision maker must accept/reject decisions sequentially as items arrive. If an item is accepted, it is loaded into the knapsack, occupying the required amount of capacity. The DSKP considers only a single versatile resource type (the knapsack) that can handle all jobs, whereas our flexible resource assignment problem is more general because it permits multiple resource types with varying levels of flexibility.

To find the optimal policy for the flexible resource allocation problem, we must solve the recursive Equations (43.1). However, because of the exponential state space, finding the optimal policy requires enormous computational effort even for problems with moderate number of resources and limited decision horizon. Therefore, we focus on developing good approximation (heuristic) procedures that are based on an understanding of the optimal policy structure and analysis of special cases. We next discuss these properties, and later present our solution procedures and discuss their computational performance.

43.3 Special Case and Upper Bounds

In this section we first focus on a special case of the problem with two job types, and identify some key structural properties of the optimal online policy. We show that a threshold-type policy is optimal for this case. This result is consistent with the findings in other discrete and continuous-time dynamic programming models applied to different contexts, which often show that the optimal policy for such a problem can be described by a critical number or a monotone switching curve (Puterman, 1994). We use this result as a building block for our approximate algorithms in the next section. We also study deterministic versions of our problem and establish two upper bounds on the expected value of the optimal online solution. These bounds use network flow models and replace stochastic demands with their certainty equivalents (expected values).

43.3.1 Special Case: Two Job Types with Unit Job Arrivals

Consider a system with two types of jobs, types 1 and 2, and three types of resources, types 1, 2, and 3. Type-1 and type-2 resources are specialized resources that can only process type-1 and type-2 jobs, respectively. Type-3 resources are flexible (versatile) resources that can handle either job type. In each period t, we assume that no more than one job arrives (effectively, this assumption, reasonable for the WLP setting, precludes batch arrivals, i.e., customer requests for multiple programs). In particular, either a single unit of a type-i job arrives, with probability p_i, for $i = 1, 2$, or no demand arrives at all, with probability $(1-p_1-p_2)$. We wish to identify structural properties of the optimal online policy for this two-job case, and use the insights to guide the development of the approximate methods in Section 43.4. We summarize our results without proof; Akcay et al. (2003) provide more details.

We define the state of the system in any period by the number of available resources $\mathbf{n} = (n_1, n_2, n_3)$ and the type of demand on hand, $\mathbf{e}_i, i = 1, 2$, where $\mathbf{e}_1 = (1, 0)$ and $\mathbf{e}_2 = (0, 1)$. Let $u_t(\mathbf{n}; \mathbf{e}_i)$ be the maximum expected revenue-to-go, given that the state of the system is $(\mathbf{n}; \mathbf{e}_i)$ in period t. The following observations are intuitively true and can also be proved formally using interchange arguments.

- The optimal online policy must accept as many type-1 jobs as possible, utilizing the type-1 resources before the type-3 resources.
- The optimal online policy must accept a type-2 job whenever type-2 resources are available.

The problem is more challenging when a type-2 job arrives and only type-1 and type-3 resources are available. To analyze this case, we define $\Delta_t(n_1, n_3; \mathbf{e}_2)$ as the increment in the optimal solution value if we add an extra flexible resource to the system when a type-2 job is on hand and no type-2 resources are available, i.e., $\Delta_t(n_1, n_3; \mathbf{e}_2) = u_t(n_1, 0, n_3+1; \mathbf{e}_2) - u_t(n_1, 0, n_3; \mathbf{e}_2)$. Thus, $\Delta_t(n_1, n_3; \mathbf{e}_2)$ is the opportunity cost of a flexible resource in the current state. The following lemma, which we can prove by induction on t, develops some monotonicity properties of the opportunity cost function.

Lemma 43.1

The opportunity cost function $\Delta_t(n_1, n_3; \mathbf{e}_2)$ *is*

1. *nonincreasing in* n_1 *and* n_3, *for any* $t = 1, 2, \ldots, T$
2. *nondecreasing in* t *for any* $n_1 \geq 0$ *and* $n_3 \geq 0$

Part 1 of the lemma states that the marginal value of an extra unit of the flexible resource decreases or stays the same as the number of available type-1 or type-3 resources increases (equivalently the value of a flexible resource increases as the capacity of the system becomes tighter). Similarly, part 2 states that, as the end of decision horizon approaches, the flexible resources become more valuable. Therefore, it would make sense to reserve the flexible resources for jobs arriving toward the end of the decision horizon. Based on these preliminary results, we propose the following theorem for the optimal online policy.

Theorem 43.1

Suppose a type-2 job arrives in period t, and let $(n_1, 0, n_3; \mathbf{e}_2)$ *be the state of the system in this period.*

1. *There exists a nonincreasing function* $F_t(n_1)$ *such that the optimal online policy accepts the type-2 job and assigns a type-3 resource to it if and only if* $n_3 \geq F_t(n_1)$.
2. $F_t(n_1)$ *is a nondecreasing function of t for any given* n_1.

For a given period t, the nonincreasing function $F_t(n_1)$ specifies the threshold value that the number of type-3 resources should exceed in order for the optimal policy to assign this resource to a type-2 job in state $(n_1, 0, n_3; \mathbf{e}_2)$. Part 2 shows that, as the end of the decision horizon approaches, the optimal policy becomes more lenient toward accepting a type-2 job for the same number of type-1 resources.

Unfortunately, determining the precise form of the threshold function $F_t(n_1)$ is not always easy. However, if $p_1 + p_2 = 1$, that is, a type-1 or type-2 job always arrives in each period (and hence the total demand for the two job types is known), then the threshold function takes the simple form $F_t(n_1) = t - n_1$, where t, the number of periods remaining until the end of the decision horizon, is also the total remaining demand. Thus, in this case, the optimal online policy accepts an incoming type-2 job and uses a flexible resource only if $n_3 \geq F_t(n_1) = t - n_1$. As expected, this threshold function satisfies the properties of Theorem 43.1.

43.3.2 Network Flow Model: Full Information Solution and Upper Bound

Next, we consider the general problem with multiple (more than two) job types and develop two upper bounds on the expected value (total revenue) of the optimal online solution, based on a network flow model. Suppose the decision maker has *full* information about the total future demand of every job type over the remaining t periods. (Alternatively, suppose the decision maker can wait until the end of the decision horizon before making acceptance and resource assignment decision.) In this situation, the decision maker faces a *deterministic* (single-period) resource allocation problem that we can formulate as a network flow model. For this setting, we let $\tilde{\mathbf{D}}_t = \{\tilde{D}_{1t}, \tilde{D}_{2t}, \ldots, \tilde{D}_{mt}\}$ denote the vector of random variables $\tilde{D}_{it}$ for the *total demand* for type-i jobs from period t until the end of the decision horizon. Let $\tilde{\mathbf{d}}_t = \{\tilde{d}_{1t}, \tilde{d}_{2t}, \ldots, \tilde{d}_{mt}\}$ be a particular realization of total demand from period t; our full information model assumes that this demand is known. Then, using the decision variables x_{ij} to denote the number of type-j resources assigned to type-i jobs, for $i = 1, \ldots, m$ and $j = 1, \ldots, l$, the following integer program maximizes total revenue for the known demand $\tilde{\mathbf{d}}_t$ and set of available resources $\mathbf{n}_t$.

$$\tilde{u}_t(\mathbf{n}_t, \tilde{\mathbf{d}}_t) = \max \sum_{i=1}^{m} \sum_{j=1}^{l} \pi_i a_{ij} x_{ij} \tag{43.6}$$

$$\textit{Demand}\text{ constraints:} \quad \sum_{j=1}^{l} a_{ij} x_{ij} \leq \tilde{d}_{it}, \quad i = 1, 2, \ldots, m \tag{43.7}$$

$$\textit{Supply}\text{ constraints:} \quad \sum_{i=1}^{m} a_{ij} x_{ij} \leq n_{jt}, \quad j = 1, 2, \ldots, l \tag{43.8}$$

$$x_{ij} \geq 0 \text{ and integer}, i = 1, 2, \ldots, m \text{ and } j = 1, 2, \ldots, l \tag{43.9}$$

As before, the demand and supply constraints (43.7) and (43.8) ensure that the number of resources assigned to a particular job type does not exceed the demand for that job type, and the total number of assigned resources of each type does not exceed the availability of that resource. We can represent the optimization problem (43.6)–(43.9) as a transportation problem, with sources representing resource nodes, one for each resource type, and destinations corresponding to job nodes. The available supply at each resource node is n_{jt}, for $j = 1, 2, \ldots, l$, and the maximum requirement (or demand) at job node i is $\tilde{d}_{it}$, for $j = 1, 2, \ldots, l$. A type-j resource node is connected to a type-i job node only if $a_{ij} = 1$, i.e., only if the type-j resource can perform the type-i job. The profit per unit of flow into the type-i job node is π_i (or equivalently, the cost is $-\pi_i$). If all the demand and resource availability values are integers, the linear programming relaxation of this problem has an optimal integer solution. We will refer to this network flow model as the *full information* (*omniscient*) model, and next discuss two upper bounds we can obtain using this model.

The full information model determines the optimal resource allocation and maximum revenue for a particular demand scenario $\tilde{\mathbf{d}}_t$. Since the total demand for each job type in the remaining t periods is a random variable, the maximum revenue is also a random variable that depends on the stochastic total demand $\tilde{\mathbf{D}}_t$. By taking the weighted average of $\tilde{u}_t(\mathbf{n}_t, \tilde{\mathbf{d}}_t)$ over all possible demand realizations $\tilde{\mathbf{d}}_t$, we obtain the expected value $E[\tilde{u}_t(\mathbf{n}_t, \tilde{\mathbf{D}}_t)]$ of the full information solution which is an upper bound on the expected revenue using any online resource assignment policy.

What if we replace the actual total demand $\tilde{d}_{it}$ in the demand constraints (43.7) with the expected total demand? Let $\tilde{\mu}_{it} = E(\tilde{D}_{it})$ denote the expected total demand for type-i jobs from period t until the end of the decision horizon. We refer to the modified model, obtained by replacing $\tilde{d}_{it}$ by $\tilde{\mu}_{it}$ in constraints (43.7), for all $i = 1, 2, \ldots, m$, and relaxing the integrality requirements in constraints (43.9), as the *certainty equivalent* model, and denote its optimal value as $\tilde{u}_t^{LP}(\mathbf{n}_t, \tilde{\boldsymbol{\mu}}_t)$. The following theorem (proved in Akcay et al., 2003) states that this optimal value is an upper bound on the expected revenue of the optimal full information solution which in turn is an upper bound on the optimal online solution value.

Theorem 43.2

The optimal online solution value $E[u_t(\mathbf{n}_t, \mathbf{D}_t)]$ is bounded from above as follows:

$$\tilde{u}_t^{LP}(\mathbf{n}_t, \tilde{\boldsymbol{\mu}}_t) \geq E[\tilde{u}_t(\mathbf{n}_t, \tilde{\mathbf{D}}_t)] \geq E[u_t(\mathbf{n}_t, \mathbf{D}_t)] \tag{43.10}$$

43.4 Approximate Solution Methods: Online Policies

In Section 43.2 we modeled the job acceptance and resource assignment problem as a stochastic dynamic program. Since finding the optimal online policy is computationally intractable even for moderate-size problems, we next present three approximate policies based on our previous insights on the optimal policy structure. In particular, for the two job-type special case, we proved in Section 43.3 that a threshold-type policy is optimal. In this section, we first propose a *basic threshold policy* for the general problem with multiple jobs; in this policy, we use a rather simplistic approach to determine the acceptance threshold value. Subsequently, we propose another threshold-type policy, which we call the *capacity reservation policy*, that relies on the analysis of a newsvendor subproblem to determine the threshold function. Finally, we introduce the *rollout policy* which approximates the stochastic dynamic program by using the bounds discussed in Section 43.3. In the next section we present the results of our computational study and demonstrate the effectiveness of these online policies.

43.4.1 Basic Threshold Policy

For two job-type problems, Theorem 43.1 showed that a flexible resource (type-3 resource) should be assigned to the less valuable job (type-2 job) if and only if the number of flexible resource exceeds a threshold value which depends on the number of available specialized resources for the more valuable job (type-1 job), i.e., $n_3 \geq F_t(n_1)$. Based on this observation, we propose the following threshold policy, consisting of an acceptance rule followed by a resource assignment rule for the accepted job. For this method, recall that we index the job types in nonincreasing order of unit revenues.

Suppose a type-i job arrives in period t. Then, the basic threshold policy uses the following rules:

1. Accept a type-i job if and only if
   (a) the total number of resources that can only be assigned to either the type-i jobs or any job type of lesser value (job types with indices larger than i) is greater than zero, or
   (b) the total number of resources that can be assigned to job types that are more valuable than type-i jobs (i.e., jobs with indices smaller than i) exceeds the total expected demand for these job types over the remaining time horizon.
2. If the type-i job is accepted in step 1, assign a resource which is expected to generate the smallest return among all resources that can be assigned to the type-i job.

In part (a) of step 1, the policy accepts the new type-i job if there are any resources for which type-i jobs are the most profitable among all job types that this resource can perform. If not, the policy applies a threshold-type criterion in part (b) to decide whether to accept the job. For this basic policy, we set the threshold equal to the total expected remaining demand for job types that are more valuable than type-i jobs. If the total number of resources that can be assigned to these job types exceeds this threshold value, the policy accepts the type-i job. Otherwise, the policy rejects the current request, anticipating that the flexible resources will later be required for more valuable jobs. Once a job is accepted, it is assigned to the least profitable resource type with the smallest flexibility index that can process this job. We measure the profitability or expected revenue of a resource type as the weighted average revenue over all the job types that this resource can process (the weight for the revenue of each job type is the relative expected total future demand for this job type).

The basic threshold policy is simple and easy to implement, but the acceptance criterion does not consider demand variability and also does not differentiate the relative reward differences of the more valuable jobs. Therefore, we expect its performance to deteriorate when demand has a large variance and the rewards of different job types differ significantly.

Next, we demonstrate the application of the basic threshold policy using a simple example. We will later use the same example to illustrate other online policies. Consider a system with three job types, $m = 3$, and seven resource types, $l = 7$. Resource types 1, 2, and 3 are *specialized* resources that can process job types 1, 2, and 3, respectively. Resource types 4, 5, and 6 are *flexible* resources that can process job types 1 and 2, 1 and 3, and 2 and 3, respectively. Type-7 resources are *versatile*, capable of handling all three types of jobs. Suppose the average demand rate is $\lambda_1 = \lambda_2 = \lambda_3 = 1$ arrival per period, and let $\pi_1 = 4$, $\pi_2 = 2$, and $\pi_3 = 1$. Consider the stage at $t = 4$, with three remaining future periods, and suppose the number of available resources in this stage is $n_1 = n_2 = n_3 = n_6 = 0$ and $n_4 = n_5 = n_7 = 1$. Since $E(\tilde{D}_1 + \tilde{D}_2) = 3 \times (1 + 1) = 6$ is greater than $n_4 + n_5 + n_6 + n_7 = 3$, the basic threshold policy would reject a type-3 job at this stage. The policy anticipates that the system would need six flexible resources to satisfy future type-1 and type-2 demands (which are more valuable than the type-3 job), whereas there are only three available flexible resources that can be assigned to these job types. Hence, the policy reserves these limited flexible resources and rejects the type-3 job. Similarly, since $E(\tilde{D}_1) = 3 \times 1 = 3$ is not less than $n_4 + n_5 + n_7 = 3$, the policy would also reject a type-2 job at this stage.

43.4.2 Capacity Reservation Policy

The newsvendor model addresses the core underlying problem of choosing the quantity of a perishable item to stock, given the item's cost, selling price, salvage values, and the probability distribution of demand. The newsvendor principle applies naturally to the simple two-fare seat allocation problem with sequential arrivals. For instance, Belobaba (1989) introduced the expected marginal seat revenue (EMSR) technique to set booking limits for this single resource revenue management problem. EMSR determines the booking limits using a newsvendor model, and treats these limits as thresholds that the available capacity should exceed for a seat to be sold to a lower-class passenger. Since the resource flexibility structure in our problem is significantly more complex, we propose an alternative newsvendor-based threshold policy which we call the *capacity reservation* policy.

In the capacity reservation policy, we first determine the total number of resources that should be reserved for future arrivals of jobs with greater returns than the type-i job on hand. If Q denotes this quantity, the standard newsvendor model selects its optimal value, say Q^*, such that

$$F(Q^*) = \frac{c_u}{c_u + c_o}$$

where c_u is the underage cost, c_o is the overage cost, and $F(.)$ is the cumulative demand distribution. For our problem, the overage cost is the lost revenue (π_i) if we do not accept the type-i job and the flexible resource remains unutilized at the end of the finite time horizon. And, if $\bar{\pi}_i$ denotes the expected return assuming that the flexible resource will be assigned to a more valuable job (than type-i jobs) in the remaining periods, then the underage cost ($\bar{\pi}_i - \pi_i$) is the lost revenue if we accept the type-i job, and we forego the opportunity of using it later for a more valuable job. Unfortunately, calculating the value of $\bar{\pi}_i$ is not straightforward. To approximate the value of $\bar{\pi}_i$, we first estimate the conditional expected total demand of each job type running out of its respective specialized resource (and, therefore, requiring flexible resources to satisfy its excess demand). We call this quantity the overflow demand of the job type. We then compute $\bar{\pi}_i$ as the weighted average revenue over all the job types that are more valuable than type-i, where the weight for each job type's revenue is the relative expected overflow demand of the job type. We define $F(.)$ as the distribution of the total overflow demand of all job types that are more valuable than type-i, and compute the probability that the total overflow demand will exceed the total number of flexible resources that can process type-i jobs as well as more valuable jobs. Using these principles, the capacity reservation policy entails applying the following procedure:

1. Accept a type-i job if and only if
   (a) the total number of resources that can only be assigned to either type-i jobs or any job of lesser value (jobs with indices larger than i) is greater than zero, or
   (b) the probability that the overflow demand of job types that are more valuable than the type-i job is greater than or equal to the total number of flexible resources that can process type-i or more valuable jobs is less than the threshold value of $\pi_i/\bar{\pi}_i$.
2. If the type-i job is accepted in the first step, assign a resource which is expected to generate the smallest return among all resources that can be assigned to the type-i job.

Clearly, we would expect the capacity reservation policy to outperform the basic threshold policy since it considers the relative returns from different types of jobs as well as the actual demand distributions. However, it requires more computation effort compared to the basic threshold policy.

Let us return to the numerical example from Section 43.4.1 to illustrate our capacity reservation policy. If a flexible resource is not assigned to a type-3 job, but rather reserved for a future arrival of a more valuable job (either a type-1 or a type-2 job), the expected return from the resource would be $\bar{\pi}_3 = (\pi_1 + \pi_2)/2 = (4+2)/2 = 3$ (because the system is symmetric). The capacity reservation policy would reject a type-3 job since $P(\tilde{D}_1 + \tilde{D}_2 > n_4 + n_5 + n_6 + n_7) = 0.849 > \pi_3/\bar{\pi}_3 = 1/3$. On the other hand, if a flexible resource is not assigned to a type-2 job, but saved for a future arrival of a type-1 job, then $\bar{\pi}_2 = \pi_1 = 4$. The capacity reservation policy would accept a type-2 job since $P(\tilde{D}_1 > n_4 + n_5 + n_7) = 0.353$ is less than the critical ratio of $\pi_2/\bar{\pi}_2 = 2/4$ (recall that the basic threshold policy would reject a type-2 job). In this case, the policy would assign a type-4 resource to the type-2 job since the semiflexible type-4 resources are expected to generate smaller revenues than the versatile type-7 resource.

43.4.3 Rollout Policy

For difficult stochastic dynamic programming problems (such as our flexible resource allocation problem), Bertsekas and Tsitsiklis (1998) propose a general class of approximate methods, which they refer to as *neuro-dynamic programming*. When making a decision at the current stage, the method uses an approximate, instead of optimal, future reward function in the Bellman equation. The method used to construct the approximate reward function is a key issue in this approach. Bertsekas and Tsitsiklis' *rollout heuristic* simulates the system under a selected policy, called the *base policy*, beginning from the next stage, and applies the policy for all possible states of the problem. Then, the current decision that leads to the state with the maximum heuristic reward is selected. Bertsekas et al. (1997) and Bertsekas and Castanon (1999) show that the rollout heuristic significantly improves the performance of the base policy. Instead of simulating the system, Bertsekas and Tsitsiklis (1998) also suggest approximating the future reward function using the *certainty equivalence* approximation; this method computes the reward of the future stages by simply assuming fixed values for the problem unknowns. Variations of this heuristic have been successfully applied to many areas, ranging from Markov games (Chang and Marcus, 2001) to scheduling of straight-line code compilers (McGovern and Moss, 1998).

The rollout heuristic with the certainty equivalence approximation as the base policy is equivalent to the bid price control (BPC) policy discussed in the revenue management literature. Simpson (1989) and Williamson (1992) propose BPC to solve the multi-leg version of the airline revenue management problem, which they model as a static network. Talluri and van Ryzin (1999) prove that BPC is not optimal for general dynamic networks and investigate why it fails to produce optimal decisions. Bertsimas and Popescu (2003) introduce a certainty equivalent control (CEC) heuristic, based on an approximate dynamic programming model, and report that this method is superior to BPC in terms of revenue generation and robustness.

We apply the rollout heuristic to our problem setting by using the certainty equivalence approach to approximate the future reward function. Specifically, for the next stage problem, we solve the network flow model with the expected values of the remaining total demands. As we discussed in Section 43.3.2, the optimal value of this model is an upper bound on the optimal expected future revenue. Based on this approximation, we define the *opportunity cost* of a type-j resource that is capable of processing type-i jobs as $\omega_{jt} = \tilde{u}_{t-1}^{LP}(\mathbf{n}_{t-1}, \tilde{\boldsymbol{\mu}}_{t-1}) - \tilde{u}_{t-1}^{LP}(\mathbf{n}_{t-1} - e_j, \tilde{\boldsymbol{\mu}}_{t-1})$. Thus, the opportunity cost ω_{jt} is the change in the value of objective function in the LP relaxation of the network flow model for a unit change in the value of n_{jt}; this value is the same as the shadow price associated with the supply constraint for type-j resources. Using this principle, our rollout heuristic for the online assignment problem consists of the following steps:

1. Accept a type-i job if and only if its reward π_i is greater than or equal to the minimum opportunity cost ω_{jt} among the available resources that can process type-i jobs.
2. If the type-i job is accepted, then assign to it the available resource with the minimum opportunity cost.

In the rollout policy, we check whether an available resource capable of processing type-i jobs has a shadow price that is less than or equal to the revenue from the type-i job. If so, we assign the type-i job to the resource that has the minimum shadow price. Unlike the basic threshold policy, the rollout policy is an optimization-based heuristic and requires solving the network flow model in each stage. The rollout policy uses the shadow prices and rewards to differentiate among job types, but does not consider demand variation. Hence, we expect its performance to deteriorate when demand variability is large.

To illustrate the rollout policy, consider again our previous three job-type example. The rollout policy first solves the network flow problem using the expected demands of $\tilde{\mu}_1 = \tilde{\mu}_2 = \tilde{\mu}_3 = 3$. As the following argument demonstrates, the shadow price of the type-4 resource is equal to 2, i.e., $\omega_4 = 2$. The current optimal solution can only accept one type-2 job (whereas all three type-1 jobs are accepted). An extra unit of a type-4 resource can, therefore, be used to accept the second type-2 job, increasing the optimal solution value by $\pi_2 = 2$, which is the shadow price for the type-4 resource. Similarly the shadow prices of type-5 and type-7 resources are $\omega_5 = \omega_7 = 2$. Since $\min\{\omega_5, \omega_7\} = 2$ is larger than $\pi_3 = 1$, the rollout policy would not accept a type-3 job at $t = 4$. On the other hand, the policy would accept a type-2 job since $\min\{\omega_4, \omega_7\} = 2$ is equal to $\pi_2 = 2$; in this case, the policy would assign either a type-4 or type-7 resource since they have equal shadow prices.

43.5 Computational Results

In this section we summarize our computational results using the online resource assignment policies that we developed in Section 43.4. The primary goal of our study is to assess the effectiveness of these policies under various operating scenarios such as different levels of capacity availability and resource flexibility. We measure each policy's performance by comparing its revenue to an upper bound on the revenue from the optimal online policy and to the revenue obtained using a first-come first-served policy. Since we focus on operational level *scheduling* decisions, we assume as given the strategic decisions such as pricing, resource capacities, and flexibility levels. For our computational tests, the following parameters characterize these system design decisions.

- **Job types and resource types:** We study a system with three job types ($m = 3$) and seven resources types ($l = 7$). Resource types 1, 2, and 3 are *specialized* resources that can process job types 1, 2, and 3, respectively. Resource types 4, 5, and 6 are *semiflexible* resources that can process job types 1 and 2, 1 and 3, and 2 and 3, respectively. Type-7 resources are *versatile*, capable of handling all three types of jobs.
- **Level of resource flexibility:** We define the *flexibility index* of a type-j resource, denoted by F_j, as the total number of different job types that the resource can perform. So, for our test problems, $F_1 = F_2 = F_3 = 1$, $F_4 = F_5 = F_6 = 2$, and $F_7 = 3$. We then compute the flexibility index of the given *system* as the weighted average flexibility over all resource types, where the weights are the number of each type of resource available. That is, the system flexibility index F is

$$F = \frac{\sum_{j=1}^{\ell} F_j n_{jT}}{\sum_{j=1}^{\ell} n_{jT}}$$

- **Level of resource availability:** We define the *availability index* A of the system as the ratio of the total number of available resources to the expected total demand for all job types over the entire decision horizon, i.e.,

$$A = \frac{\sum_{j=1}^{\ell} n_{jT}}{\sum_{i=1}^{m} \bar{\mu}_{iT}}$$

  Clearly, A is a measure of system capacity; as its value increases, we expect to meet a larger proportion of the incoming demand.
- **Job rewards:** For convenience, we assume a fixed percentage difference between the revenue for one job type and the next type, i.e., $\pi_i/\pi_{i+1} = r$ for $i = 1, 2, \ldots, m-1$ and $\pi_m = 1$. We refer to the constant r as the *reward ratio*.
- **Demand process:** We assume that the demand for each job type follows a Poisson distribution in each period.

$$P(D_{it} = d) = \frac{\lambda_i^d e^{-\lambda_i}}{d!}$$

  where $E(D_{it}) = \lambda_i$ and $\text{Var}(D_{it}) = \lambda_i$.

In order to reduce the number of control parameters for our computational study, we assume that the system is *symmetric*. That is, we assume that all three job types have identical, but independent, demand distributions, and the system contains the same number of resources for all semiflexible resources and all specialized resources, i.e., $n_{1T} = n_{2T} = n_{3T}$ and $n_{4T} = n_{5T} = n_{6T}$.

We consider a *base case* system with $T = 10$ periods, reward ratio $r = 2$, Poisson demand rates $\lambda_1 = \lambda_2 = \lambda_3 = 2$ per period, resource availability index $A = 0.8$ and resource flexibility index $F = 2.0$

(obtained by setting initial resource capacities as $n_1 = n_2 = n_3 = n_7 = 0$ and $n_4 = n_5 = n_6 = 16$). We generate other scenarios with varying levels of availability and flexibility by changing the number of resources (while preserving symmetry); we also vary the length of the decision horizon and reward rate. We implemented the algorithms using the C programming language, using CPLEX 7.5 for the optimization subroutines (e.g., upper bounding procedure).

For each set of problem parameters or *scenario*, we simulate the system for 100 runs, i.e., we generate 100 demand streams or problem *instances*. The certainty equivalent solution, obtained by solving the network flow model with expected demand values, provides an upper bound (UB) on the revenue for each scenario. For each problem instance within a particular scenario, we apply each of the three online policies, and also determine the full information (omniscient) solution (i.e., network flow solution with the actual demand realization for that instance). We then compute, for each scenario and every problem instance:

- the revenue for the full information solution (OMN)
- the revenues for the capacity reservation policy (CRP), rollout policy (ROP), basic threshold policy (TP), and first-come first-served (FCFS) policy

We define the percentage revenue gap of a resource assignment policy (CRP, ROP, TP, or FCFS) for a particular problem instance as the difference in revenues for that policy and the OMN solution as a fraction of the OMN solution value. For each scenario, we average the percentage gap values over all the instances for that scenario. The percentage revenue gap between the average revenue for the OMN solution and the certainty equivalent UB, expressed as a percentage of the average OMN value, measures the quality of the upper-bounding procedure.

Figure 43.1 depicts the average value of the OMN solution for various levels of resource flexibility and availability. Although the OMN value is an upper bound on the revenue from the optimal online solution, we use it to assess the impact of flexibility and availability on the optimal revenues.

Figure 43.1 shows that increasing the resource flexibility improves the system performance regardless of the level of resource availability. In other words, for a given number of resources, more flexible systems can attain higher revenues. However, the impact of resource flexibility is most significant when system capacity is tight. For smaller values of the availability index, the slope of the curves are steeper compared to the slopes at higher values of the availability index. Therefore, little flexibility is sufficient to achieve the maximum solution value for systems with moderate capacities, whereas larger levels of flexibility are required for systems with tighter capacities. In our system, even with modest capacity $A = 0.6$, the system can achieve the benefit of maximum flexibility with a flexibility index F of just 1.5.

Figure 43.2 depicts the percentage gap between the average value of the OMN solution and the certainty equivalent solution (UB) for different levels of resource flexibility and availability. The percentage difference

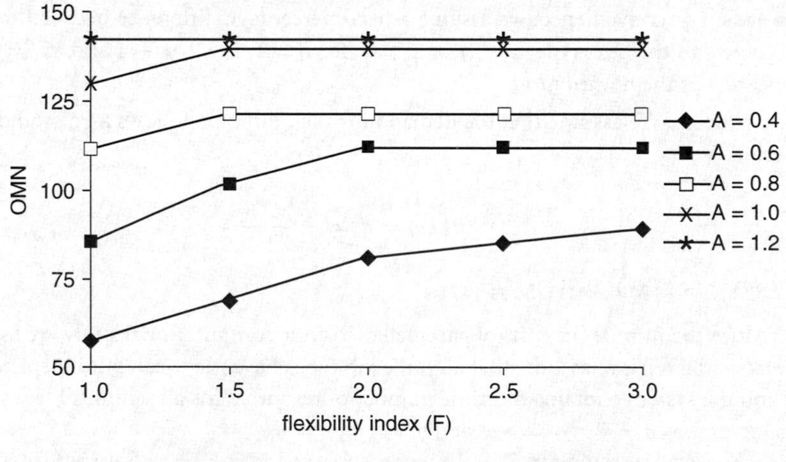

FIGURE 43.1 Average value of OMN solution at different levels of resource availability and flexibility.

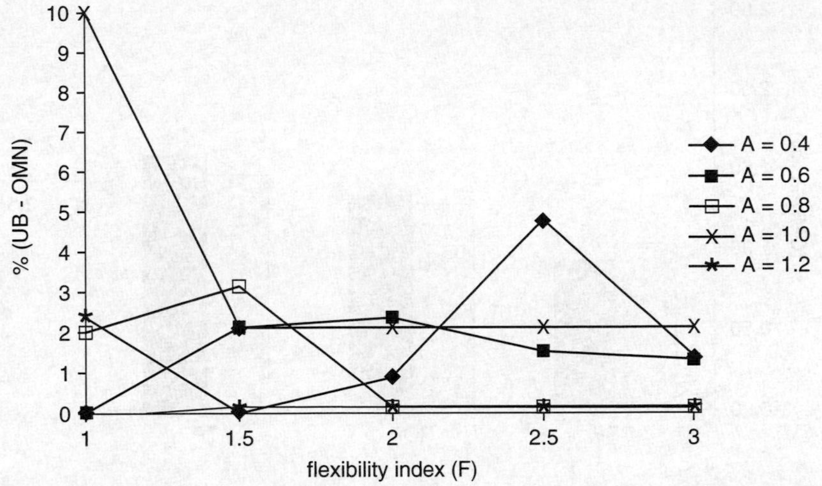

FIGURE 43.2 Percentage gap between average OMN value and certainty equivalent solution (UB) at different levels of resource flexibility and availability.

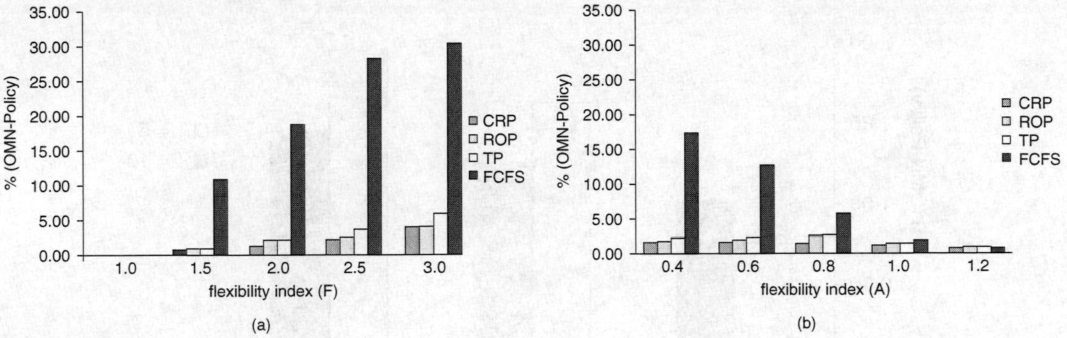

FIGURE 43.3 Average percentage gap between OMN and online policy revenues at different levels of resource flexibility and availability.

between these two upper bounds is often less than 3%, but it can be as large as 10%. We also observe from Figure 43.2 that, as the flexibility index increases, the gap between the two bounds tends to increase for the small availability index ($A \leq 0.6$), but decreases for the large availability index ($A \geq 0.8$). This behavior suggests that the certainty equivalent model, which does not consider the stochasticity of demand, can significantly overestimate the achievable capacity utilization and revenue for the underlying system with stochastic demand, especially when resource availability is low (high) and flexibility is high (low).

Next, we evaluate the performance of our online policies. In Figure 43.3, we present the average (over all problem instances for a scenario) percentage gap between the online policy solution values and the OMN solution value under various levels of resource availability and flexibility.

As the flexibility index increases, the difference between OMN and online policy solutions also increases. This behavior indicates that, as the system becomes more flexible, either the performance of our policies deteriorates, or the OMN solution becomes a poorer approximation of the optimal online solution, or both. We cannot precisely determine the relative impact of these two causes since the optimal online policy is difficult to compute. Nevertheless, all policies perform remarkably well compared to the FCPS decision rule at different flexibility and availability levels. The policies achieve near-optimal results when the availability index is either small (only accept the valuable jobs) or large (accept all jobs). The gap between the best online policy and OMN is highest when resource availability is moderate, making the

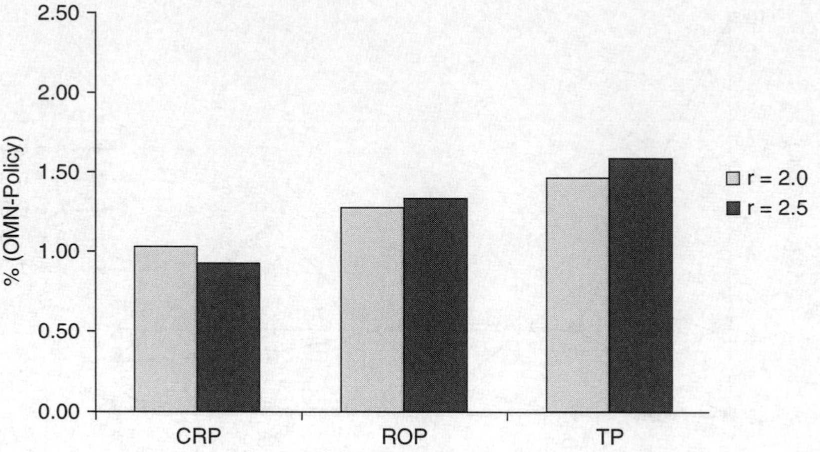

FIGURE 43.4 Average percentage gap between OMN solution and online policies for different reward rates.

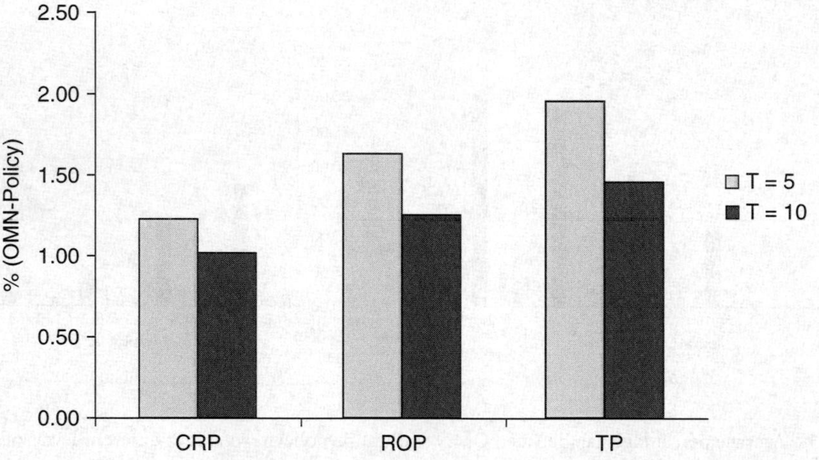

FIGURE 43.5 Percentage gap between average value of the full information solution (OMN) and online policies for different decision horizons.

acceptance and assignment decisions more critical. Again, the quality of the OMN solution value as an upper bound on the value of the optimal online policy might also contribute to this higher gap.

Observe that the CRP rule consistently performs best among all other rules, followed by the ROP and TP methods. The FCFS policy performs very poorly except when system capacity is large. Over all the problem instances and scenarios, the CRP deviates from OMN by an average of 1.01%, whereas the ROP, TP, and FCFS deviates by 1.24%, 1.44%, and 8.24%, respectively. The maximum deviation from OMN for the three online policies are 3.38%, 3.84%, 4.97%, and 31.72% for CRP, ROP, TP, and FCFS, respectively.

Figure 43.4 compares the effect of reward ratio on the algorithms' performance. Increasing the reward ratio adversely affects the performance of ROP and TP rule, albeit marginally, but favorably affects the performance of CRP. This difference in behavior arises because CRP is more judicious in using flexible resources for less valuable jobs, compared to ROP and TP. Therefore, reserving a flexible resource for a potential future demand of a more valuable job pays off significantly for CRP when the jobs differ significantly in their relative revenues.

Finally, in Figure 43.5 we study the effect of the length of the decision horizon on the performance of the online policies. Observe that the performance of the policies deteriorates as the decision horizon becomes shorter. This deterioration occurs due to two reasons. First, as the decision horizon gets longer, the algorithms have a greater chance of recovering from poor decisions in earlier periods, and make up for them in the remaining periods. However, if the horizon is shorter, the number of decisions made is proportionally smaller, and any mistake will significantly affect the overall performance of the algorithm. Second, with other things being equal, a system with a long horizon benefits from the pooling effect of demand variance over time, and thus experiences less stochasticity of demand, thereby improving performance (i.e., having smaller percentage gaps with respect to OMN). We have found this effect to hold even for decision horizons longer than ten periods.

43.6 Concluding Remarks

Motivated by the important operational decisions facing professional service firms in the workplace learning industry, we have proposed and studied the flexible resource assignment problem. Firms in this rapidly growing industry face the problem of dynamically allocating their instructors with different skill sets to various types of incoming projects. In contrast to typical revenue management applications such as airline seat allocation, the resource flexibility structure in professional service firms is much richer and complex, which makes the task of scheduling workforce quite challenging.

We formulate the flexible resource assignment problem as a finite horizon stochastic dynamic program. Because of the exponential state space of the problem, standard solution techniques of dynamic programming are not practical. Therefore, our focus is on developing approximate online policies which give high quality solutions and are also computationally efficient. We first study special cases of the flexible resource assignment problem, and identify some structural properties of the optimal policy. We then propose three heuristic approaches, namely, the basic threshold policy, capacity reservation policy, and rollout policy, based on these properties and special cases. Our test problems show how the system performance is affected by various factors such as resource flexibility, resource availability, demand distributions, and reward differences of various job types. In particular, we observe that little flexibility is sufficient to achieve the maximum reward for the systems with moderate capacity, whereas more flexibility is required for the system with tighter capacity. Our computational tests also demonstrate that our methods significantly outperform the naïve benchmark method (first-come first-served), and provide near-optimal solutions in most problem instances. While all three heuristics perform remarkably well, the capacity reservation policy consistently dominates the other solution approaches, since it incorporates more problem parameters (rewards, demand distributions) into its decision process.

References

Akcay Y., Balakrishnan A., and Xu. S. H. (2003), On-line assignment of flexible resources, Working Paper. Department of Supply Chain and Information Systems, Penn State University.

Belobaba, P. P. (1989), Application of a probabilistic decision model to airline seat inventory control, *Operations Research* 37(2), 183–197.

Bertsekas, D. P. (1997), Differential training of rollout policies, in Proceedings of the 35th Allerton Conference on Communication, Control, and Computing, Allerton Park, IL.

Bertsekas, D. P., and Castanon, D. A. (1999), Rollout algorithms for stochastic scheduling problems, *Journal of Heuristics* 5(1), 89–108.

Bertsekas, D. P., and Tsitsiklis, J. N. (1998), Neuro-Dynamic Programming, Athena Scientific, Belmont, MA.

Bertsekas, D. P., Tsitsiklis, J. N., and Wu, C. (1997), Rollout algorithms for combinatorial optimization, *Journal of Heuristics* 3(3), 245–262.

Bertsimas, D., and Popescu, I. (2003), Revenue management in a dynamic network environment, *Transportation Science* 37(3), 257–277.

Chang, H. S., and Marcus, S. I. (2001), Markov games: Receding horizon approach, Technical Research Report TR 2001-48, Institute for Systems Research, University of Maryland, College Park, MD.

Cross, R. G. (1997), Revenue Management: Hard-Core Tactics for Market Domination, Broadway Books, New York, NY.

Fine, C., and Freund, R. (1990), Optimal investment in product-flexible manufacturing capacity, *Management Science* 36(4), 449–466.

Jones, R. A., and Ostroy, J. M. (1984), Flexibility and uncertainty, *Review of Economic Studies* 51, 13–32.

Jordan, W. C., and Graves, S. C. (1995), Principles on the benefits of manufacturing process flexibility, *Management Science* 41(4), 577–594.

Kleywegt, A. J., and Papastavrou, J. D. (1998), The dynamic and stochastic knapsack problem, *Operations Research* 46, 17–35.

Maister, D. H. (1997), *Managing the Professional Service Firm*, Simon and Schuster Adult Publishing.

McGill, J. I., and Ryzin, G. J. V. (1999), Revenue management: Research overview and prospects, *Transportation Science* 33(2), 233–256.

McGovern, A., and Moss, E. (1998), Scheduling straight-line code using reinforcement learning and rollouts, in Proceedings of the 11th Neural Information Processing Systems Conference (NIPS '98), pp. 903–909.

Mieghem, J. A. V. (1998), Investment strategies for flexible resources, *Management Science* 44(8), 1071–1078.

Puterman, M. L. (1994), Markov Decision Processes. John Wiley & Sons, Inc.

Simpson, R. W. (1989), Using network flow techniques to find shadow prices for market and seat inventory control, Memorandum M89-1, MIT Flight Transportation Laboratory, Cambridge, MA.

Smith, B. C., Leimkuhler, J. F., and Darrow, R. M. (1992), Yield management at American Airlines, *Interfaces* 22(1), 8–31.

Talluri, K. T., and Ryzin, G. J. V. (1999), An analysis of bid-price controls for network revenue management, *Management Science* 44, 1577–1593.

Weatherford, L. R., and Bodily, S. E. (1992), A taxonomy and research overview of perishable-asset revenue management: yield management, *Operations Research* 40, 831–844.

Williamson, E. L. (1992), Airline Network Seat Control, PhD thesis, MIT.

44

Novel Metaheuristic Approaches to Nurse Rostering Problems in Belgian Hospitals

Edmund Kieran Burke
University of Nottingham

Patrick De Causmaecker
KaHo Sint-Lieven

Greet Vanden Berghe
KaHo Sint-Lieven

44.1 Introduction

There is an increasing pressure of work in healthcare organizations in Belgium. This continues to remain a serious problem in spite of recent significant technological advances. One potential way of easing this pressure is to develop better nurse rostering decision support systems that can help to produce rosters which employ resources more efficiently. However, there is more than just one goal when generating personnel rosters in hospitals. Resource efficiency is important but so is the satisfaction level of patients. Personnel rosters also affect the organizational structure of the hospital and they directly influence the private lives of the staff. It is therefore important to provide an interactive system that generates high quality scheduling solutions within a reasonable amount of computing time. Such schedules should cover the hospital requirements while avoiding patterns that are detrimental to the nurses' and patients' priorities.

The presented approach concentrates on the short-term problem of assigning specific tasks to a sufficient number of qualified nurses. This problem is often referred to as *rostering*, *scheduling*, or *timetabling*. A part of the problem data, such as the number of personnel in a ward, the required qualifications, the definition of shift types, etc., is determined at the strategic level. Although these settings are not usually considered

1-58488-397-9/$0.00+$1.50

to be part of the nurse rostering problem, some longer term strategic decisions can affect the solution strategies. The model described in this chapter therefore provides several possibilities for flexible problem setting. For example, shift types can be divided over several nurses, personnel demands can be expressed in terms of shorter intervals than shift length, night shifts can be assigned to a special category of night nurses, possibilities exist for creating part time work, people can temporarily be assigned to different wards in order to address emergencies, personnel members can express certain preferences for particular times in the planning period, etc.

The problem of finding a high quality solution for the personnel timetabling problem in a hospital ward has been addressed by many scientists, personnel managers, and schedulers over a number of years. In recent years, the emergence of larger and more constrained problems has presented a real challenge for researchers. A flexible planning system should incorporate as much knowledge as possible to relieve the personnel manager or head nurse from the unrewarding task of setting up objective schedules that attempt to satisfy a range of conflicting objectives.

This chapter presents each of the different steps in developing the system: a comparison with related publications (Section 44.2), the modeling of the nurse rostering problem (Section 44.3), the setting up of a solution framework (Section 44.4), and the development of appropriate search techniques (Section 44.5). We also investigate the applicability of a multi-criteria approach for solving the nurse rostering problem (Section 44.6). The possibility of assigning weights to certain criteria or conditions guides the search through a different set of solutions and produces interesting results of a very good quality. We compare the results of the developed algorithms, summarize their benefits and drawbacks, and end with a general discussion in Section 44.7.

44.2 Related Literature

Employee scheduling covers staffing, budgeting, and short-term scheduling problems. Although these fields have variable time horizons, they are strongly interrelated. Scheduling of hospital personnel is particularly challenging because of different staffing needs on different days and shifts. Unlike most other facilities, healthcare institutions work around the clock.

Until recently, nearly all personnel scheduling problems in Belgian hospitals were solved manually. Planners had no automatic tool to test the quality of a constructed schedule. They made use of very straightforward constraints on working time and idle time in a recurring process.

44.2.1 Categories of Nurse Scheduling

We distinguish between different categories of nurse or hospital scheduling. Hospital *staffing* involves determining the number of personnel of the required skills in order to meet predicted requirements [1–7]. Factors that make this task complex are the organizational structure and characteristics, personnel recruitment, skill classes of the personnel, working preferences, patient needs, circumstances in particular nursing units, etc. Another significant staffing decision is to define work agreements for part time workers, to decide whether substitution of skill classes is allowed and for which people.

Two major advantages of *centralized scheduling* [2,4,5,8] are fairness to employees through consistent, objective, and impartial application of policies and opportunities for cost containment through better use of resources. We refer to *unit scheduling* when head nurses or unit managers are given the responsibility of generating the schedules locally [9–14]. It is sometimes considered to be an advantage because nurses get more personalized attention. Conversely, it can be the case that personnel members see their schedule as a punishment or suspect the head nurse to be giving preferential treatment to certain people.

Self-Scheduling is a manual process [15,16]. The technique is more time consuming than automatic scheduling but it has the advantage that the nurses cooperate and are asked for advice. Generally it is performed by the personnel members themselves and coordinated by the head nurse of a ward. Self-scheduling is so common that complete automation is not recommended.

Cyclical Scheduling

Cyclical scheduling concerns organizations in which each person works a cycle of a number of weeks [17–19]. Staff know their schedule a long time in advance and the same schedule patterns are used over and over again. There are significant benefits [5] but cyclical schedules unfortunately lack generality. They cannot address (without major changes) many of the flexible features and personal preferences that are part of the modern problem.

44.2.2 Complexity Factors

When only considering short term scheduling, we distinguish between two main goals: meeting the coverage requirements [12,20–22] and satisfying time related constraints [23,24]. Approaches that solve all time related constraints generally consider a low number of such constraints.

Many models allow limited violations of coverage constraints and consider them in the evaluation function [13,25]. These violations are not allowed in our model, unless they are explicitly approved of as a relaxation [26]. The approach provides different planning options for precisely defining the desired personnel coverage.

The personnel scheduling literature has examples with strictly separated skill categories [27–31]. Other approaches apply hierarchically substitutable qualifications (in which higher skilled people can replace less experienced colleagues) [9,11]. User definable substitution [20,21,24,32] is particularly well suited to real world practice.

In simplified research examples, problem definitions have the same constraints for all the personnel members [25,27,31]. The assignment of schedules to people is then very arbitrary. More realistic examples take part time contracts into account and provide flexibility to define personal work agreements [13,21,22,33]. Some approaches generate schedules that consist of days off and days on [15,29]. The assignment of actual shifts to people, is then left for a head nurse to carry out manually. Algorithms developed for use in practical healthcare environments do not usually work with three strictly distinct shift types (like in [9,23,34]). The activities in hospitals are so diverse that a large number of user-definable shifts is often necessary [12,13,22,31,33,35].

The personnel requirements are nearly always expressed as a number of people required per shift type or even per day [9,11,20,21]. In a more flexible approach [26,36], the number of possible shift types is higher than in other problems. However, the idea of composing a schedule with different combinations of shift types, through time interval requirements, has not been well studied in the nurse rostering literature.

We have compared a large number of time related soft constraints with those that are implemented in our model [37]. Some models apply set values for the constraints [9,35,38] whereas they are user definable in more advanced approaches [20,21,23].

44.2.3 Nurse Rostering Approaches

Since the 1960s, publications on various aspects of healthcare personnel scheduling appear regularly. We categorize approaches into optimization, heuristic, and artificial intelligence techniques.

Most mathematical scheduling approaches make use of an objective function, which is optimized subject to certain constraints. Nearly all of the earlier papers in the literature (e.g., [24,32,34,39]) mention mathematical optimizing techniques for their linear models. Simplifications of the real data are unavoidable. Real world problems are too complex to be *optimized* and many authors employ heuristics (e.g., [40–43]). It is also common to define more than one objective [23,25,27,29,30,38].

Although cyclical schedules are generally considered to be less difficult to generate, most of them are constructed with heuristic [12,40–42] techniques. In the 1980s and later, artificial intelligence techniques for nurse scheduling (declarative approaches [44], constraint programming [12,13,35,45], expert systems [25,30,42], case-based reasoning [46,47], etc.) were investigated. Some of these approaches are still relevant to today's research issues [13,33,45]. Many of the most recent papers (the 1990s and later) tackle the problem with metaheuristic approaches such as simulated annealing [48,49], tabu search [11,23], and evolutionary algorithms [9,20,28,50,51].

There are many advanced models for practical personnel scheduling, but none of them is suited for the problem that we discovered in Belgian hospitals. Strong arguments for the importance of the model and solution methods presented in this chapter are the flexibility of the approach, the applicability in practice, and the generic problem formulation.

44.3 The Nurse Rostering Problem

44.3.1 Problem Dimensions

This research arose from the need for automated rostering assistance in Belgian healthcare organizations. We developed a general model for the nurse rostering problems and refer to it as *advanced nurse rostering model* (ANROM). A software package based on the model and the solution framework was first implemented in a hospital in 1995 but the system is still evolving to cope with the new and more complex real-world problems that keep appearing. So far, over 40 hospitals in Belgium, of which some have about 100 wards, have replaced their time consuming manual scheduling by this system.

Although the problem is user-defined to a large extent, the software has to be efficient in different settings. Every specific hospital ward should be able to formulate its problem within the restrictions of the model described in the following sections. The main goal of the system is to create a schedule by assigning shift types to skilled personnel members, in order to meet the requirements in a certain planning period. Personnel have work regulations limiting their assignments.

Hospitals are organized in wards with fixed activities, usually a settled location, and, for the most part, they have a permanent *team of nurses*. Although practical situations often allow people to be moved to another ward whenever a personnel shortage is unsolvable, this personnel rostering problem concerns a group of personnel belonging to the same ward.

Personnel members in a ward belong to *skill categories*. The division into categories is based upon the particular level of qualification, responsibility, job description, and experience of the personnel. Rather than employing strictly disjoint skill categories or hierarchical substitutability, we opted for a solution that is closer to the reality in hospitals. The problem of replacing people is solved in ANROM by assigning *alternative* skill categories to certain people. People with more experience, or who have taken some exams, can be substitutes for higher skill categories.

Hospital personnel have *work regulations* or contracts with their employer. Several job descriptions such as part time work, night nurses' contracts and weekend work are possible. The regulations involve different constraints but they can make the schedules much more flexible. Moreover, very personal arrangements like *free Wednesday afternoons* or *refresher courses* at regular points in time, can easily be set. It is not unlikely to have personalized contracts for the majority of personnel members in Belgian hospitals.

A *shift type* is a predefined period with a fixed start and end time in which personnel members can be on or off duty. Different part time contracts require a large variation in start and end times and in duration. Table 44.1 presents a simplified example of a set of shift types. It is common that hospital schedulers define shift types according to their needs.

Planning periods for nurse rostering vary from a couple of days to a few months. Since cyclical rosters are not common at all, it is important for individual employees to know their schedule some time in advance. Long term scheduling, on the other hand, should not be too detailed because the personnel requirements and preferences fluctuate and are not predictable in the long term.

TABLE 44.1 Example of Shift Types

Code	Shift	Start	End
M	Morning	06:45	14:45
L	Late	14:30	22:00
N	Night	22:00	07:00

	Mon	Tue	Wed	Thu	Fri	Sat	Sun
P1	M	M	L	L	N		
P2	N		N	L	L		
P3	M	M	M	M	M	M	M
P4	M		L	N	N	N	
P5	M L	L	L	L			

FIGURE 44.1 Roster example for 5 people (P1, . . . , P5) and 1 week; M, L, and N being the shift types introduced in Table 44.1.

	Schedule Example																				
P1	*	-	-	*	-	-	-	*	-	-	*	-	-	-	*	-	-	-	-	-	-
P2	-	-	*	-	-	-	-	-	*	-	*	-	-	*	-	-	-	-	-	-	-
P3	*	-	-	*	-	-	*	-	-	*	-	-	*	-	-	*	-	-	*	-	-
P4	*	-	-	-	-	-	-	*	-	-	-	*	-	-	*	-	-	*	-	-	-
P5	*	*	-	-	*	-	-	*	-	-	*	-	-	-	-	-	-	-	-	-	-

FIGURE 44.2 Schedule corresponding to the roster in Figure 44.1: '*' denotes that there is an assignment in the schedule, '-' denotes that the schedule is free.

Short planning periods enable the search algorithms to find good quality results, much faster than longer planning periods. However, guaranteeing fairness among personnel members is restricted when the planning period is short.

The roster, in which the shift assignments are stored, is called the *schedule*. We define assignment units as entities of minimum allocation in a schedule. They are mainly introduced to express and evaluate the soft constraints on the personnel's schedules. Each shift type corresponds to an assignment unit in practice.

We illustrate the meaning of assignment units with a simple example. A fragment of a possible personnel roster is presented in Figure 44.1. We notice that there are five people in the ward, and the shift types correspond to Table 44.1. Figure 44.2 presents the *schedule* that corresponds to the roster of Figure 44.1. Each column in the schedule represents an assignment unit. For each day of the planning period there are three columns, one for each shift type. The assignment units are ordered according to the start times of the shift types that they represent. When two shift types have the same start time, the first assignment unit will match the shift type with the earliest end time.

44.3.2 Hard and Soft Constraints

Hard constraints are those that must be satisfied at all costs. *Soft constraints* are those which are desirable but which may need to be violated in order to generate a workable solution. We call a *feasible* solution one that satisfies the following hard constraints:

1. All the shift types specified in the personnel requirements have to be assigned to a personnel member.
2. One person cannot be assigned twice to the same shift on the same day.
3. Shifts can only be assigned to people of the right skill category.

The real-world situation addressed in this research incorporates a high number of soft constraints on the personal schedules. The soft constraints will preferably be satisfied, but violations can be accepted to a certain extent. It is highly exceptional in practice to find a schedule that satisfies all the soft constraints. The aim of the search algorithms is to minimize the penalties due to violations of these constraints.

The model includes exceptions for the evaluation in addition to certain corrections which are required in holiday periods or periods of illness absence. Boundary conditions at the beginning and end of the planning period have an important impact on the evaluation. No penalty is generated when a violated constraint can still be satisfied by scheduling suitable shifts in the next planning period.

The list of soft constraints can be divided into three categories:

1. Certain constraints hold for the entire hospital
   - Minimum time between two assignments
   - Use of an alternative skill category in certain situations
2. Another set of soft constraints is the same for all the people with the same contract (full-time, half-time, night nurses, etc.) Values are set by the users
   - Maximum number of assignments in the planning period
   - Minimum/Maximum number of consecutive days
   - Minimum/Maximum number of hours worked
   - Minimum/Maximum number of consecutive free days
   - Maximum number of assignments per day of the week
   - Maximum number of assignments for each shift type
   - Maximum number of a shift type per week
   - Number of consecutive shift types
   - Assign two free days after night shifts
   - Assign complete weekends
   - Assign identical shift types during the weekend
   - Maximum number of consecutive working weekends
   - Maximum number of working weekends in a 4-week period
   - Maximum number of assignments on bank holidays
   - Restriction on the succession of shift types
   - Patterns enabling specific cyclic constraints (e.g., a free Wednesday afternoon every 2 weeks)
   - Balancing the workload among personnel
3. When individual personnel members have an agreement with the personnel manager or head nurse, then certain constraints can be implemented
   - Day off; shifts off
   - Requested assignments
   - Tutorship (people not allowed to work alone)
   - People not allowed to work together

For more details about these soft constraints, we refer to [37].

44.4 Solution Framework

44.4.1 Evaluation of Solutions

The search heuristics for solving the ANROM model are driven by an evaluation function that estimates the quality of schedules. Since hard constraints have to be satisfied at all costs, only soft constraint violations contribute to that quality. In this section, we briefly introduce the method that was developed to model and evaluate the complex time-related constraints. It obviously takes care of boundary constraints imposed by previous planning periods.

The evaluation makes use of a simple algorithm and requires very little memory and computation time. This is useful for evaluating intermediate solutions while exploring the large search space. Since hard constraints have to be satisfied at all costs, only soft constraint violations contribute to the quality of a solution. The procedure tackles the characteristics of the soft constraints (cost parameters, tolerable deviations, and restrictions on consecutiveness) in a modular way. It is easily extendible and provides a very

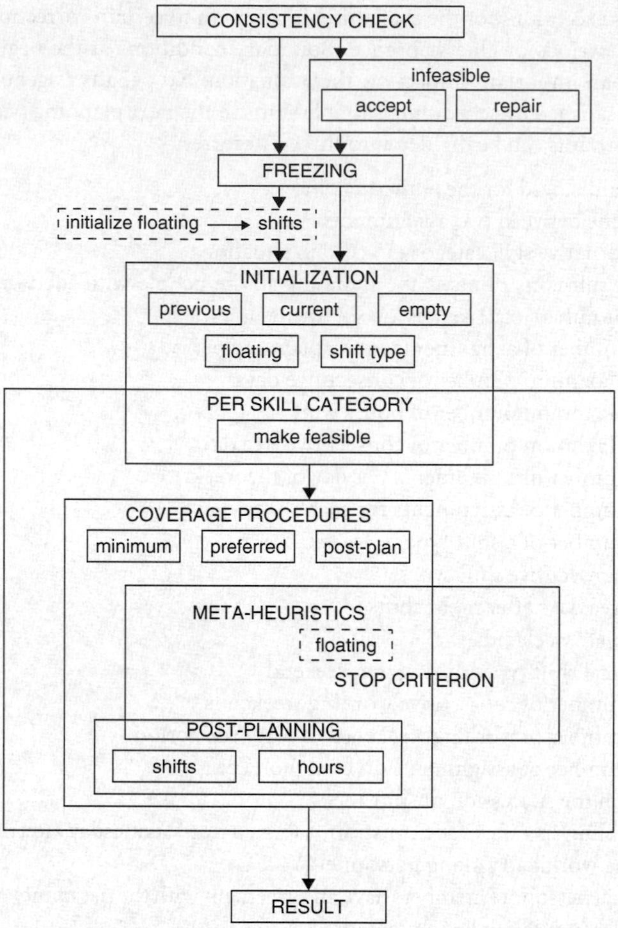

FIGURE 44.3 Overview of the solution framework.

structural technique for incorporating new constraints that appear in real world problems. The modular nature of the approach allows the system to provide some feedback. This functionality assists the user with the interpretation of the quality of the result.

44.4.2 Planning Procedures

The general metaheuristics for ANROM are supplied with a few planning options for varying objectives. They have been developed as separate procedures in order to split typical hospital goals from general search algorithms. Figure 44.3 schematically demonstrates the order in which these planning procedures (presented in bold) appear in the total planning process.

Boxes that are on the same level indicate alternatives. There is one general box for metaheuristics because every algorithm that respects the hard constraints can be plugged into the framework.

44.4.2.1 Consistency Check on Available People

When the hard constraints are so strict that no feasible solution exists, planners can opt to relax them. The planning system is not developed for handling infeasible problems. In most cases, the hard constraints are so strong that it is obvious, after a preliminary check, that some of the soft constraints cannot be satisfied.

The procedure that we developed can handle shift type as well as time interval requirements. Apart from an obvious check on the hard constraints, the users demanded an extra check on some *precedence* soft constraints, namely on patterns, personal requests for days and shifts off, and requested shift assignments. Either the user can accept the remedy by relaxing the personnel requirements when necessary or he/she can deliberately choose to violate the soft constraints that were checked. Both options are demonstrated in the overview of Figure 44.3.

44.4.2.2 Freezing Parts of the Schedule

The search space for rostering algorithms can be restricted for several reasons. We distinguish between three arguments to prevent algorithms from modifying certain parts of the schedule:

1. Some hospitals prefer interaction and they schedule the timetables of certain nurses manually.
2. The start and end date of personal contracts do not necessarily coincide with the start and end times of the planning period. Also, people can temporarily be assigned to other wards.
3. In case urgent rescheduling is required, it is recommended not to alter already existing personal schedules completely but to freeze certain periods in time.

No matter which parts of the schedule are frozen, the evaluation procedure considers the entire schedule.

44.4.2.3 Initialization

The initialization of the scheduling algorithm consists of two phases for constructing a feasible initial solution. In the first phase, the input is loaded, after a coverage procedure has been selected. Planners can make a choice among a random initial schedule and an initial schedule that is copied from the previous or current planning period (if that exists).

The second phase makes the schedule feasible by randomly adding or removing assignments until the hard constraints are satisfied [26].

44.4.2.4 Planning Order of Skill Categories

Each skill category is scheduled separately in this model. Dividing the rostering into subproblems reduces the search space. The number of personnel belonging to each skill category is often considerably smaller than the entire staff in the ward. The number of shifts to be assigned (whether or not translated from floating requirements) is also lower.

In the case when people have permission to carry out shifts for alternative skill categories, there are a few difficulties. After the planning for a skill category has stopped, the algorithm moves on to the next skill category and temporarily freezes the already assigned shifts.

44.4.2.5 Time Interval Requirements

Time interval requirements provide an alternative way of expressing the coverage needs. This idea was put forward by users of the initial version of the model. It is now part of the solution framework and it can be combined with any of the algorithms.

Hospitals define a large number of shift types that match different activities and enable many kinds of part time work. Personal schedules are always set up with shift types although it is quite complex for planners to express the daily personnel coverage in terms of shift types. With this new option, we concentrate on an advanced formulation of requirements in terms of intervals of personnel coverage. It often meets the habits and needs that occur in practice more appropriately, it reduces nonproductive attendance at work and it results in much more flexible timetables by splitting and recombining shift types. Details of this model for formulating coverage constraints can be found in [36].

44.4.2.6 Coverage Procedures

In practice, the number of required personnel on a certain day is not completely strict. Experienced planners know very well when it is reasonable to plan more or less personnel than is supposedly *required.* However, there exist no clear rules for decisions like this. Planners using ANROM can optionally choose among different coverage strategies. We call some of them postplanning algorithms. Examples include:

1. **Minimum-preferred requirements.** Each of these two kinds of requirement can be set as hard constraints for the search algorithms.
2. **Plan toward preferred requirements.** Instead of strictly setting the hard constraints, this option allows a range in which the hard constraints are considered to be satisfied. The algorithm to organize this option first takes the minimum requirements as hard constraints. After a result has been calculated by the scheduling algorithms, the system searches possibilities for adding shifts to the schedule whenever this does not involve an extra violation of soft constraints.

   When personnel requirements are expressed as a number of personnel needed per time interval (and not per shift type, as is mostly the case), a slightly different approach is required [36]. Instead of searching the best candidate to assign an extra shift or an extra pair of shifts for, we have to add switches from a shift to a longer shift, from a pair of shifts to a single shift with a longer entire duration or from a single shift to a pair of shifts that last longer (see [36]).
3. **Adding hours.** In the personal schedules that do not have enough hours assigned, this planning option assigns extra shifts. The only condition is that the preferred requirements for the new shift types are not zero and that the extra assignment does not generate additional violations.

44.5 Metaheuristics and Hybrids

44.5.1 Variable Neighborhood Search

In this section, we present a method that applies the problem characteristics to dynamically modify the environments of the search heuristics. The exploration of the search space can be improved by combining short sighted search with greedy search in a wider environment. The main concepts of the variable neighborhood approach are explained in [52].

44.5.1.1 Environments for Personnel Rostering

The metaheuristics that we developed to solve nurse rostering problems change neighborhoods when they cannot find better solutions during a number of iterations. All the solutions in an environment must satisfy the hard constraints. Although the cost function is the driving force for the heuristics, it may remain blind for improvements if it cannot interpret certain problem characteristics. A few of the environments were designed only for the purpose of finding such improvements. Examples include:

1. **Single shift-day neighborhood.** The simplest environment contains all solutions that differ from the current solution by moving one assignment to another person's schedule. Figure 44.4 illustrates this neighborhood for an example with four nurses and four different shift types. Nurse A can

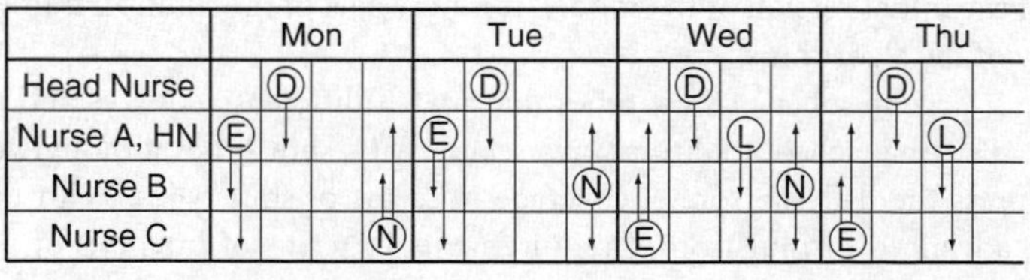

FIGURE 44.4 Possible moves in a single shift-day neighborhood. Each day is divided into columns that represent the shifts. Shifts are early (E), day (D), late (L), and night (N).

replace the head nurse (HN), the other two regular nurses cannot. The arrows in the figure show the only possible moves that do not violate the hard constraints. Moving the head nurse's D shift to nurse B does not result in a feasible solution. Neither does moving nurse A's E shift to the schedule of the head nurse.

2. **Soft constraint related neighborhoods.** Planners with practical experience were often more concerned about violations of certain soft constraints than about the overall solution quality. This resulted in a particular set of environments that consider the search space in a completely different way. They are blind for the global effect of a move but attempt to ameliorate the quality of a roster by only considering the soft constraints that are being addressed. Of course, this approach is completely opposed to the idea of the general evaluation function that abstracts from individual constraints. Therefore, the entire search procedure should not end immediately after exploring this neighborhood.

   Examples of such environments look at weekend constraints, overtime, alternative qualifications, personal preferences, etc. A special neighborhood only looks at the most violated constraint (no matter which constraint that is).
3. **Swapping large sections of personal schedules.** Unlike the previous two, in which neighboring solutions only differ in the position of one single shift type, this set of neighborhoods looks at schedules that differ substantially from the current solution.

   The *shuffle* environment considers switches from a part of the worst personal schedule, in terms of the evaluation function, with any other schedule. Instead of moving single shift types, all the assignments in a longer period (that can be one day or a number of days that equals half the planning period) are switched. A similar environment is called the *greedy shuffle*. It consists of all the possible shuffles between any set of two people.

   Reallocating larger groups of shifts at once is often less harmful for the quality of a schedule than moving shifts around. The drawback of applying this category of neighborhoods is that the number of neighboring solutions is very large, and iterations thus require a large amount of computation time.

44.5.1.2 Algorithms and Search Order

Different scenarios are possible in variable neighborhood search. For the experiments that we carried out in [53], we only applied steepest descent and tabu search as local search algorithms. The results mainly provided insight into the characteristics of the search space and into the importance of the order in which to explore neighborhoods. We discovered that the most efficient application of different neighborhoods is to search them in order of increasing size (i.e., starting from the single shift-day environment and ending with swapping large sections). Each time a better solution is found, the search procedure restarts from the smallest size neighborhood.

Some large size environments take the search procedure to a kind of end state. The resulting roster is (in general) very satisfactory, especially since it is nearly impossible for experienced planners to make manual improvements. This explains why it is not useful to modify such end states by applying another small size neighborhood afterward.

44.5.2 Hybrid Tabu Search

44.5.2.1 Tabu Search for Personnel Scheduling

Hybrid tabu search algorithms are integrated in the software package that is based on ANROM (see [21,53] for more details). One of the major benefits of tabu search is its ability to escape from local optima, by preventing the search from going through the same areas. As long as there are improvements, tabu search behaves like a steepest descent algorithm. There are aspiration criteria to prevent better solutions in a *tabu* part of the search space from being ignored. The efficiency of the search can be increased by applying diversification.

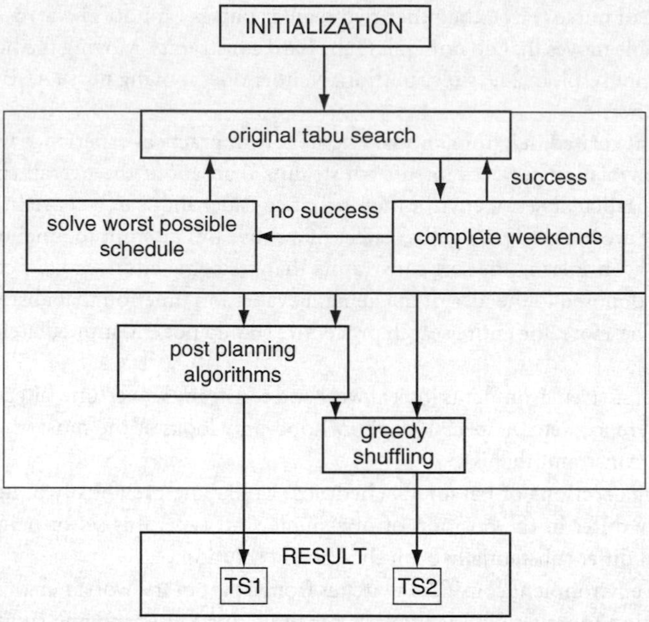

FIGURE 44.5 Diagram of the hybrid tabu search algorithms for the nurse rostering problem, plug-in for Figure 44.3.

The basic algorithm is the combination of tabu search with the single shift-day neighborhood that was explained in Section 44.5.1. Certain characteristics of encountered solutions (and similar solutions in the neighborhood) are put into the tabu list in order not to circle round.

Instead of carrying out random diversifications when the basic algorithm cannot generate better solutions during a number of iterations, we opted for implementing more problem specific steps. The different neighborhoods that we introduced above are useful in this respect. They are particularly useful when considering the following:

1. The soft constraint on full weekends
2. Improving the worst personal roster (shuffle)
3. All the large improvements between two personal rosters (the greedy shuffle environment)

44.5.2.2 Hybrid Algorithms for Practical Use

With the two following algorithms, we offer planners a choice between a fast algorithm that generates schedules of acceptable quality in a very short amount of time, and a more thorough one that is deployed when a final solution is required. The course of both options is illustrated in Figure 44.5. That diagram fits into the metaheuristics box of Figure 44.3.

We consider two versions of hybrid *tabu search* (TS) (see [21,53] for more details):

TS1: Fast planning. When the basic tabu search algorithm does not find improvements, the diversification consists of searching in the weekend environment, and after that in the shuffle environment. This process is iteratively repeated until the global stop criterion is reached.

TS2: Thorough planning (TS1 + greedy shuffle). This combination requires more time but the results are of a considerably higher quality. Especially since the greedy step is carried out at the end of the calculations (and not iteratively in the process), the computation time is kept under control. The confidence of the users is high because all the manual modifications make the obtained solution worse.

Hybrid algorithms (compared to single TS) are much better suited for the personnel rostering problems that we tackle. We refer to [21] for test results on real world problems. Application in practice reveals that the increased quality of solutions generated with TS2 compensates for the large amount of computation time required.

44.5.3 Memetic Algorithms

Population based techniques can overcome problems that occur when a single solution *evolves*. We briefly introduce a set of memetic algorithms for personnel rostering, that make use of crossover (in order to copy good characteristics of previous generations) and mutation (for random diversity). Memetic algorithms apply local search on every individual of a population. The computation time is large but the results demonstrate that memetic algorithms offer more to rostering than the sum of the component algorithms (see [20] for more details).

44.5.3.1 Evolutionary Approaches for Personnel Rostering

It is quite complex to apply crossover operators to ANROM because the standard operators do not maintain feasibility. Moreover, the identification of good roster features is particularly difficult since the nature of the constraints hardly allows any partial evaluations. Combining parts of nurse schedules seldom results in good quality. Figure 44.6 demonstrates how the memetic algorithms for ANROM are composed. The diagram needs to be plugged into Figure 44.3 in order to see the full process.

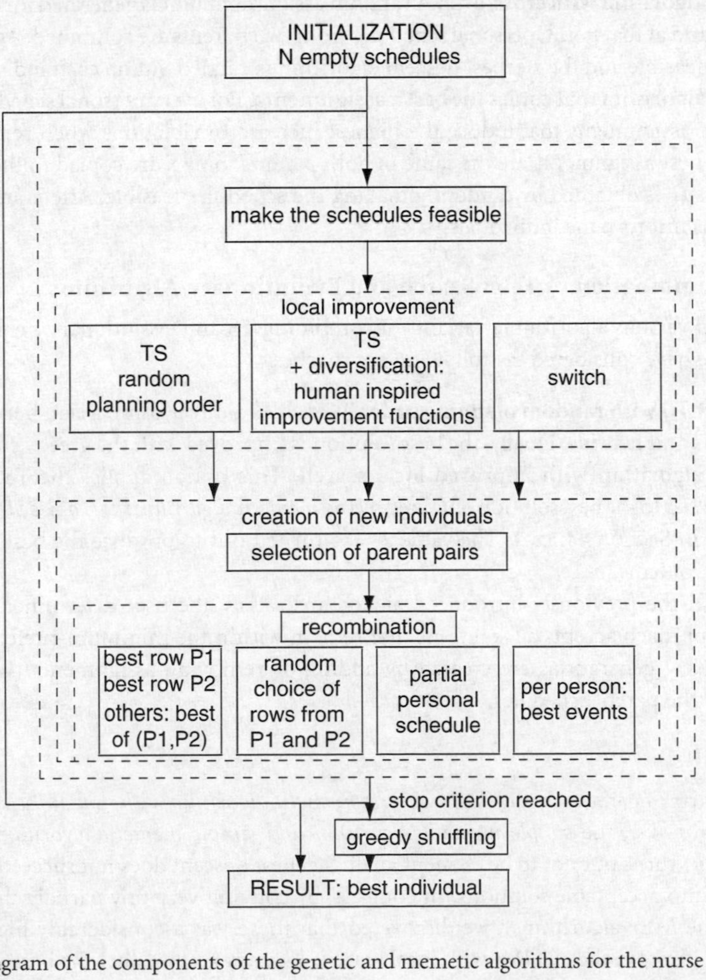

FIGURE 44.6 Diagram of the components of the genetic and memetic algorithms for the nurse rostering problem, plug-in for Figure 44.3.

An initial population consists of N individual schedules that match the hard constraints. The quality of a roster is determined by the sum of the qualities of all individual rosters. Therefore, some crossover operators will copy full personal rosters to the next generations. Others will only consider good assignments. In any case, nearly all the operators require repair functions that make the newly created individual solutions feasible. The procedure that was introduced in Section 44.4.2 to make initial schedules feasible, is also applicable here. We briefly discuss a few algorithmic variants:

1. **Simple memetic algorithm.** It applies steepest descent (in the single shift-day neighborhood from Section 44.5.1) on each newly created individual. The planning order of qualifications (Section 44.4.2) remains unchanged. A simple tournament selects the best parents. The first child is obtained by copying the best personal schedule from the first parent, together with the best one (for another member of staff) from the other parent. A personal schedule is simply the work schedule for one particular person. All the other personal schedules are copied randomly from both parents. For the second child, we start from the best solution of the other parent. Both new individuals are normally infeasible but they are repaired by randomly adding or removing assignments until the coverage constraints are met.
2. **Diverse memetic algorithm.** This algorithm is based on the *simple memetic algorithm* but the planning order of skill categories is randomly chosen in the local search step.
3. **Diverse memetic algorithm with random selection.** It applies the general features of the *diverse memetic algorithm* but all the personal schedules are randomly selected from both parent pairs.
4. **Memetic algorithm with cross-over.** A random assignment unit is generated for each person in the schedule and at that point, personal schedules from two parents are combined. Again, the schedules are made feasible and the steepest descent algorithm is carried out on each individual.
5. **Memetic algorithm that copies the best x assignments.** For every personal schedule, the algorithm selects the assignments that induce the highest increase in violations when removed. In the case when the best assignments are the same in both parents, only x are copied to the next generation. The diversity is obtained by randomly making the schedules feasible. Afterward, the local search algorithms improve the individuals.

44.5.3.2 A Combination of Tabu Search and Evolutionary Algorithms

We have explored various algorithmic variants of combining TS and evolutionary methods (see [20] for more details). We have considered the following approaches:

1. **Multistart TS with random planning order.** It enables a direct comparison between TS, memetic algorithms and hybrids. Finally, the best solution is improved with the greedy shuffle.
2. **Memetic algorithm with improved local search.** This option applies the TS2 algorithm from Section 44.5.2 to the best solution obtained by the *memetic algorithm that copies the best 4 assignments* described in Section 44.5.3.1. The value $x = 4$ turned out to produce the best results for a wide range of problems.
3. **Switch.** All the previously introduced algorithms satisfy the user determined hard constraints but this approach accepts all solutions that remain within the minimum-preferred requirements interval. New generations are created by adding or removing assignments (within the feasible region) in the parent schedules.

44.5.3.3 Results

Copying entire parts of parent schedules (the *diverse memetic algorithm with random selection*, the *memetic algorithm with crossover*, the *simple memetic algorithm and simple memetic algorithm*, and the *diverse memetic algorithm*) turns out not to be efficient at all. Steepest descent does not succeed in modifying the new individuals into acceptable solutions. In contrast, by copying very tiny partial schedules with good qualities (as in the MEx algorithms), we discovered that there was a considerably higher diversity and more freedom to improve the *children* with local search. We may assume that the best x assignments of a schedule influence the rest so much that solutions evolve toward good quality schedules.

Multistart TS with random planning order cannot compete with the real memetic algorithms. This proves that longer computation time and variance in initial solutions does not lead to the quality that a hybrid approach can reach. The memetic-tabu hybrid *memetic algorithm that copies the best x assignments* behaves extremely well for very complex problems (see [20] for a thorough discussion of the test results). It outperforms all the other algorithms, except *Switch,* and clearly demonstrates the benefits of hybrid approaches. *Switch*, which is a little bit less strict about the selected hard constraints, provides very interesting results. It demonstrates that relaxing the hard constraints, to a controlled extent, is often not as harmful as expected [20].

44.6 Multicriteria Approach

Users of the planning software that we have described [21,26,36] are confronted with the difficult task of expressing their needs in terms of the set of constraints and corresponding cost parameters. We have developed a multicriteria approach, which attempts to overcome some practical difficulties that planners deal with [54].

The evaluation function sums violations of constraints that are completely different. Yet, it is not desirable that hospital planners have to compare all these different measures. The lack of insight into the characteristics of the solution space often leads to very poor parameter settings. Experiments teach us, for example, that higher values of cost parameters can give rise to high barriers in the search space and make it even more difficult to find better solutions.

Multicriteria modeling allows users to express the importance of criteria according to their own preferences. The approach bears more resemblance to the customs in hospitals than the cost function approach, and it supports better control of compensation between constraints.

Nurse rostering problems are generally unsolvable in terms of satisfying all the constraints that planners initially state are hard — they usually have to *soften* some of these constraints. There is considerable scope for the uptake of multi-objective methods which promote a reasonable approach for searching for a compromise between conflicting objectives.

44.6.1 A Multicriteria Model for Nurse Rostering

We implemented a multicriteria approach for nurse rostering in which the violations of one soft constraint are measured with one criterion. Constraints of a very different nature, that are expressed in different measures, can now be treated simultaneously (see [54] for more details).

We apply *compromise programming,* which is based on the concept of the distance to an ideal point. Each schedule is represented as a point in the criteria space. The number of soft constraints determines the dimensions of the criteria space. For real world problems, solutions that correspond to the ideal point do not usually exist. The anti-ideal point is represented by a schedule with the worst value for all the criteria. In order to treat all the criteria in dimensionless units, a preference space is created (Figure 44.7). The

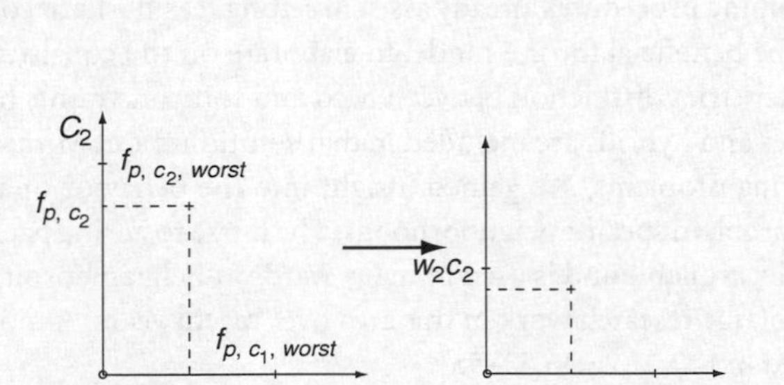

FIGURE 44.7 Mapping from the criteria space into the preference space for person p's schedule.

model requires a best and a worst value for each of the criteria. It is acceptable to state that the value 0 (denoting no violations of the corresponding soft constraint), matches the best value, even if that is not feasible. Estimating the worst value of criteria is more complex. We opted for generating the most extreme values instead of determining realistic worst values in the feasible region. The obtained worst value for each criterion is mapped to the relative priority or the weight (w in Figure 44.7) assigned to the constraint. A smaller distance to the ideal point in the preference space denotes a *better* schedule.

Each personal roster is evaluated separately by measuring its distance to the ideal point. We refer to [54] for more details about the metrics (called L_p) that we used. Small values of p allow for compensation between criteria. Good values for a criterion can make up for a worse evaluation of another one. By increasing the distance measure p to ∞ only the value of the largest coordinate contributes to the quality.

The sum of the distances of all the personnel members determines the quality of the entire schedule. For the experiments that we undertook, we applied the previously described metaheuristics to generate better solutions. The evaluation function is replaced by the measurement of the distance to the ideal point. The metaheuristics ensure that the search does not leave the feasible part of the search space.

44.6.2 Results

The proposed multicriteria approach is very promising for nurse rostering. Conclusions are drawn from experiments on real world test problems and are reported in [54]. The approach is particularly appropriate for treating completely different criteria. Users determine the weights for each of the soft constraints and thus set relative priorities. They can control compensation of these constraints by manipulating the distance measure.

In general, this multicriteria approach has the potential to more accurately reflect the daily real world situation in hospitals than the methods that employ a single evaluation function.

44.7 Conclusions

Nurse rostering is a very complex combinatorial problem, for which software assistance is not common (at least in Belgium). During our work it became clear that, especially in Belgian hospitals, any assistance for the head nurses or ward managers to automatically generate their monthly rosters could save a lot of time and effort. Several levels of decision making can be distinguished in nurse scheduling but the problem dealt with in this model is situated at the short-term timetabling level. Its main objective is to understand and automatically generate comfortable shift schedules for personnel members in order to meet the staff coverage. We captured an extensive set of realistic constraints, and integrated them, together with explicit and implicit objectives, in a general, flexible model.

The development of the solution framework, with modifiable evaluation tools and a large set of heuristics targeting specific objectives, constitutes a major improvement for hospital schedulers. However, the model often requires more insight into the characteristics of specific data than planners in practice can be expected to have. Some of the planning procedures already assist in setting feasible hard constraints or relaxing them when necessary. It will be beneficial for the model to elaborate on the consistency check and to take the idea of relaxing the rather strict distinction between hard and soft constraints further.

A set of metaheuristics and hybrids are included in that solution framework, as the central search force for solving nurse rostering problems. We gained insight into the behavior of applying heuristics and in making use of different problem specific neighborhoods. The nurse rostering package based upon ANROM has become commercially available and is used in many wards in Belgian hospitals. This overview chapter has summarized much of our research work in this area over recent years. A more detailed analysis of our approaches can be found in [20,21,26,36,53–55].

Future research will certainly build upon the promising early findings of testing our multicriteria approach on nurse rostering. It opens perspectives for releasing the planners from setting the cost parameters. It is more realistic and increases the flexibility in setting the weights and thus modifying the relative priority of the constraints.

By automating the nurse rostering problem for Belgian hospitals, the scheduling effort and time are reduced considerably as compared to the manual approach that was previously used. The time for automatic schedule generation can be tailored to suit the time available by selecting appropriate search heuristics. The proposed solution method provides an unbiased way of generating the schedules for all the personnel members. It enables simple verification of the constraints, helps redefine unrealistic hard constraints, and thus leads to an overall higher satisfaction among the personnel, as is manifest in many applications.

Although the nurse rostering model was developed explicitly to address hospital personnel rostering, the techniques and methods developed as a result of this research are certainly adaptable to other personnel scheduling problems. Of course the presented algorithms deal with an extensive set of soft constraints, of which many are only valid in healthcare. Moreover, other sectors require the evaluation of constraints on locations, equipment, etc., that are irrelevant in nurse rostering.

References

[1] G. de Vries: Nursing workload measurement as management information, *European Journal of Operational Research*, Vol. 29, 1987, 199–208.

[2] F. Easton, D. Rossin, W. Borders: Analysis of alternative scheduling policies for hospital nurses, *Production and Operations Management*, Vol.1, No. 2, 1992, 159–174.

[3] D. Schneider, K. Kilpatrick: An optimum manpower utilization model for health maintenance organisations, *Operations Research*, Vol. 23, No. 5, 1975, 869–889.

[4] V. Smith-Daniels, S. Schweikhart, D. Smith-Daniels: Capacity management in health care services: Review and future research directions, *Decision Sciences*, Vol. 19, 1988, 891–919.

[5] M. Warner: Nurse staffing, scheduling, and reallocation in the hospital, *Hospital & Health Services Administration*, 1976, 77–90.

[6] H. Wolfe, J.P. Young: Staffing the nursing unit: Part I, *Nursing Research*, Vol. 14, No. 3, 1965, 236–234.

[7] H. Wolfe, J.P. Young: Staffing the nursing unit: Part II, *Nursing Research*, Vol. 14, No. 4, 1965, 299–303.

[8] S.P. Siferd, W.C. Benton: Workforce staffing and scheduling: Hospital nursing specific models, *European Journal of Operational Research*, Vol. 60, 1992, 233-246.

[9] U. Aickelin, K. Dowsland: Exploiting problem structure in a genetic algorithm approach to a nurse rostering problem, *Journal of Scheduling*, Vol. 3, No. 3, 2000, 139–153.

[10] D. Bradley, J. Martin: Continuous personnel scheduling algorithms: A literature review, *Journal of the Society of Health Systems*, Vol. 2, 1990, 8–23.

[11] K. Dowsland: Nurse scheduling with tabu search and strategic oscillation. *European Journal of Operations Research*, Vol. 106, 1998, 393–407.

[12] A. Meisels, E. Gudes, G. Solotorevski: Employee timetabling, constraint networks and knowledge-based rules: A mixed approach, In E.K. Burke, P. Ross (Editors), *Practice and Theory of Automated Timetabling; First International Conference Edinburgh*, Springer, 1995, 93–105.

[13] H. Meyer auf'm Hofe: ConPlan/SIEDAplan: Personnel assignment as a problem of hierarchical constraint satisfaction, *Proceedings of the Third International Conference on the Practical Application of Constraint Technology*, London, 1997, 257–271.

[14] D. Sitompul, S. Randhawa: Nurse scheduling models: A state-of-the-art review, *Journal of the Society of Health Systems*, Vol. 2, 1990, 62–72.

[15] H.E. Miller: Implementing self scheduling, *The Journal of Nursing Administration*, Vol. 14, March 1984, 33–36.

[16] K.K. Ringl, L. Dotson: Self-scheduling for professional nurses, *Nursing Management*, Vol. 20, 1989, 42–44.

[17] S.E. Bechtold, L.W. Jacobs: Implicit modeling of flexible break assignments in optimal shift scheduling, *Management Science*, Vol. 36, 1990, 1339–1351.

[18] R. Hung: A cyclical schedule of 10-hour, four-day workweeks, *Nursing Management*, Vol. 22, No. 5, 1991, 30–33.

[19] N. Muslija, J. Gaertner, W. Slany: Efficient generation of rotating workforce schedules, In E.K. Burke, W. Erben (Editors), *Proceedings of the 3rd International Conference on the Practice and Theory of Automated Timetabling*, ISBN 3-00-003866-3, 2000, 314–332.
[20] E.K. Burke, P. Cowling, P. De Causmaecker, G. Vanden Berghe: A memetic approach to the nurse rostering problem, *Applied Intelligence Special Issue on Simulated Evolution and Learning*, Vol. 15, No. 3, Springer, 2001, 199–214.
[21] E.K. Burke, P. De Causmaecker, G. Vanden Berghe: A hybrid tabu search algorithm for the nurse rostering problem, B. McKay et al. (Editors), *Simulated Evolution and Learning, 1998*; Artificial Intelligence, Vol. 1585, Springer, 1999, 187–194.
[22] A. Schaerf, A. Meisels: Solving employee timetabling problems by generalised local search, *Proceedings Italian AI*, 1999, 493–502.
[23] I. Berrada, J. Ferland, P. Michelon: A multi-objective approach to nurse scheduling with both hard and soft constraints, *Socio-Economic Planning Science*, 30, 1996, 183–193.
[24] H.E. Miller, W. Pierskalla, G. Rath: Nurse scheduling using mathematical programming, *Operations Research*, Vol. 24, 1976, 857–870.
[25] J.-G. Chen, T. Yeung: Hybrid expert system approach to nurse scheduling, *Computers in Nursing*, 1993, 183–192.
[26] P. De Causmaecker, G. Vanden Berghe: Relaxation of coverage constraints in hospital personnel rostering, E.K. Burke, P. De Causmaecker (Editors), *Selected Revised Papers of 4th International Conference on Practice and Theory of Automated Timetabling*, LNCS 2740, 129–147.
[27] J. Arthur, A. Ravindran: A multiple objective nurse scheduling model, *AIIE Transactions*, Vol. 13, No. 1, 1981, 55–60.
[28] H. Kawanaka, K. Yamamoto, T. Yoshikawa, T. Shinogi, S. Tsuruoka: Genetic algorithm with the constraints for nurse scheduling problem, *Proceedings of Congress on Evolutionary Computation*, Seoul, IEEE Press, 2001, 1123–1130.
[29] A. Musa, U. Saxena: Scheduling nurses using goal-programming techniques, *IEEE*, 1984, 216–221.
[30] I. Ozkarahan: A disaggregation model of a flexible nurse scheduling support system, *Socio-Economical Planning Science*, Vol. 25, No. 1, 1991, 9–26.
[31] G. Weil, K. Heus, P. Francois et al.: Constraint programming for nurse scheduling, *IEEE Engineering in Medicine and Biology*, 1995, 417–422.
[32] M. Warner, J. Prawda: A mathematical programming model for scheduling nursing personnel in a hospital, *Management Science*, Vol. 19, 1972, 411–422.
[33] M. Chiarandini, A. Schaerf, F. Tiozzo: Solving employee timetabling problems with flexible workload using tabu search, In E.K. Burke, W. Erben (Editors), *Proceedings of the 3rd International Conference on the Practice and Theory of Automated Timetabling*, ISBN 3-00-003866-3, 2000, 298–302.
[34] V.M. Trivedi, M. Warner: A branch and bound algorithm for optimum allocation of float nurses, *Management Science*, Vol. 22, No. 9, 1976, 972–981.
[35] M. Okada: An approach to the generalised nurse scheduling problem — generation of a declarative program to represent institution-specific knowledge, *Computers and Biomedical Research*, Vol. 25, 1992, 417–434.
[36] E.K. Burke, P. De Causmaecker, S. Petrovic, G. Vanden Berghe: Floating Personnel Requirements in a Shift Based Timetable, working paper KaHo Sint-Lieven, 2003.
[37] G. Vanden Berghe: An advanced model and novel metaheuristic solution methods to personnel scheduling in healthcare, *PhD dissertation*, University of Gent, 2002.
[38] A. Jaszkiewicz: A metaheuristic approach to multiple objective nurse scheduling, *Foundations of Computing and Decision Sciences*, Vol. 22, No. 3, 1997, 169–184.
[39] M. Warner: Scheduling nursing personnel according to nursing preference: A mathematical programming approach, *Operations Research*, Vol. 24, 1976, 842–856.
[40] R. Blau: Multishift personnel scheduling with a microcomputer, *Personnel Administrator*, Vol. 20, No. 7, 1985, 43–58.

[41] M. Kostreva, K. Jennings: Nurse scheduling on a microcomputer, *Computers and Operations Research*, 18, 1991, 731–739.

[42] L.D. Smith, D. Bird, A. Wiggins: A computerised system to schedule nurses that recognises staff preferences, *Hospital & Health Service Administration*, 1979, 19–35.

[43] L.D. Smith, A. Wiggins: A computer-based nurse scheduling system, *Computers and Operations Research*, Vol. 4, No. 3, 1977, 195–212.

[44] S.J. Darmoni, A. Fajner, N. Mahe, A. Leforestier, O. Stelian, M. Vondracek, M. Baldenweck: Horoplan: computer-assisted nurse scheduling using constraint based programming, *Journal of the Society for Health Systems*, Vol. 5, 1995, 41–54.

[45] P. Chan, G. Weil: Cyclical staff scheduling using constraint logic programming, In E.K. Burke, W. Erben (Editors), *Practice and Theory of Automated Timetabling; Third International Conference*, Konstanz, Springer, 2000, 159–175.

[46] G.R. Beddoe, S. Petrovic, G. Vanden Berghe: Case-based reasoning in employee rostering: learning repair strategies from domain experts, University of Nottingham. (Unpublished).

[47] S. Petrovic, G. Beddoe, G. Vanden Berghe: Storing and adapting repair experiences in personnel rostering, In E.K. Burke, P. De Causmaecker (Editors), *Selected Revised Papers of the 4th International Conference on Practice and Theory of Automated Timetabling*, LNCS 2740, 148–168.

[48] M.J. Brusco, L.W. Jacobs: Cost analysis of alternative formulations for personnel scheduling in continuously operating organizations, *European Journal of Operational Research*, Vol. 86, 1995, 249–261.

[49] M. Isken, W. Hancock: A heuristic approach to nurse scheduling in hospital units with non-stationary, urgent demand, and a fixed staff size, *Journal of the Society for Health Systems*, Vol. 2, No. 2, 1990, 24–41.

[50] F. Easton, N. Mansour: A distributed genetic algorithm for employee staffing and scheduling problems, *Conference on Genetic Algorithms*, San Mateo, 1993, 360–367.

[51] J. Tanomaru: Staff scheduling by a genetic algorithm with heuristic operators, *Proceedings of the IEEE Conference on Evolutionary Computation*, 1995, 456–461.

[52] N. Mladenović, P. Hansen: Variable neighborhood Search, *Computers & Operations Research*, Vol. 24, 1997, 1097–1100.

[53] E.K. Burke, P. De Causmaecker, S. Petrovic, G. Vanden Berghe: Variable neighborhood search for nurse rostering problems, In Mauricio G.C. Resende, Jorge Pinho de Sousa (editors), *Metaheuristics: Computer Decision-Making*, Chapter 7, Kluwer, 2003, 153–172.

[54] E.K. Burke, P. De Causmaecker, S. Petrovic, G. Vanden Berghe: A multi criteria metaheuristic approach to nurse rostering, *Proceedings of Congress on Evolutionary Computation*, CEC2002, Honolulu, IEEE Press, 2002, 1197–1202.

[55] E.K. Burke, P. De Causmaecker, S. Petrovic, G. Vanden Berghe: Fitness evaluation for nurse scheduling problems, *Proceedings of Congress on Evolutionary Computation*, CEC2001, Seoul, IEEE Press, 2001, 1139–1146.

[56] I. Ozkarahan: An integrated nurse scheduling model, *Journal of the Society for Health Systems*, Vol. 3, No. 2, 1991, 79–101.

45

University Timetabling

Sanja Petrovic
University of Nottingham

Edmund Burke
University of Nottingham

45.1 Introduction

Timetabling can be considered to be a certain type of *scheduling* problem. In 1996, Wren described timetabling as the problem of placing certain resources, subject to constraints, into a limited number of time slots and places with the aim being to satisfy a set of stated objectives to the highest possible extent (Wren, 1996). This general representation is a broadly accepted description of the class of timetabling problems. Such problems arise in a wide variety of domains including education (e.g., university and school timetabling), healthcare institutions (e.g., nurse and surgeon timetabling), transport (e.g., train and bus timetabling) and sport (e.g., timetabling of matches between pairs of teams). Some views on the terms *scheduling* and *timetabling* can be found in the literature. For example, Wren (1996) emphasises that scheduling often aims to minimize the total cost of resources used, while timetabling often tries to achieve the desirable objectives as nearly as possible. On the other hand, Carter (2001) points out that timetabling decides upon the *time* when events will take place, but does not usually involve the allocation of resources in the way that scheduling often does. For example, the process of generating a university course timetable does not usually involve specifying which teachers will be allocated to which particular course.

1-58488-397-9/$0.00+$1.50

This information is usually decided upon well before the timetable is actually constructed. However, in our view, the subtle differences between differing views of what is meant by the terms *timetabling* and *scheduling* are not really important. The important point is that scientific progress in understanding and solving real world timetabling problems from across the application areas outlined above will be made by drawing on the state of the art in other scheduling problems and *vice versa*.

The purpose of this chapter is to highlight some recent research advances that have been made by the authors and co-workers. We aim to overview our recent developments and to point the interested reader in the right directions to find out more about this crucial scheduling application area.

Educational timetabling is a major administrative activity at most universities. It can be classified into two main categories: course and exam timetabling. In course (exam) timetabling problems a number of courses (exams) are allocated into a number of available classrooms and a number of timeslots, subject to constraints. Usually, we distinguish between two types of constraints: hard and soft. Hard constraints are rigidly enforced by the university and, therefore, have to be satisfied. Solutions that do not violate hard constraints are called *feasible solutions*. Common hard constraints are as follows:

- No person can be allocated to be in more than one place at any one time.
- The total resources required in each time period must not be greater than the resources that are available.

Soft constraints are those that it is desirable to satisfy, but they are not essential. In real-world university timetabling problems it is usually impossible to satisfy all of the soft constraints. The quality of a feasible timetable can be assessed on the basis of how well the soft constraints are satisfied. However, some problems are so complex that it is difficult to find even a feasible solution.

The sets of hard and soft constraints differ significantly from university to university. More details about different constraints that are imposed by universities can be seen by consulting Carter and Laporte (1998, 1996) and Burke et al. (1996a).

There are significant differences between course and exam timetabling. This can be illustrated by noting that in many situations there is exactly one exam for each course, while a single course may need to be held multiple times per week. Another example which illustrates the difference between course and exam timetabling is obtained by observing that a number of exams can often be scheduled into one room or an exam may be split across several rooms. This is, of course, not usually acceptable in course timetabling where it is most typically the case that one course has to be scheduled into exactly one room.

In many universities, the large number of events to be scheduled (this can be several hundreds or even thousands of courses/exams) and the wide variety of constraints which have to be taken into consideration makes the construction of a timetable an extremely difficult task. Its manual solution usually requires a significant amount of effort. The development of automated methods for university timetabling is a challenging research issue which has attracted the attention of the scientific community from a number of disciplines (especially from Operational Research and Artificial Intelligence) for over 40 years.

A wide variety of approaches to timetabling problems have been described in the literature and tested on real world university course and examination data. These approaches can be classified with respect to different criteria. Here we extend the classification presented by Carter and Laporte (1996, 1998), which draws upon the employed methodology. Our extended classification includes approaches that have been developed since the above papers were published.

1. **Approaches based on Mathematical programming.** An example of an early integer linear programming formulation of university timetabling (in which binary variables represent the assignment of events) was presented by Akkoyunly (1973). The high computational demand of these methods makes them less attractive for large timetabling problems, but there is still research work into these methods and their hybridization with other methods. For example, see Dimopoulou and Miliotis (2001).
2. **Approaches based on Graph coloring.** Timetabling problems can be represented by graphs where vertices denote events (courses/exams), while edges denote conflicts between the events (such as

when students have to attend both events). For example, see de Werra (1985). The construction of a conflict-free timetable can be modelled as a graph coloring problem. The task is to color the vertices of a graph so that no two adjacent vertices are colored by the same color. Each color corresponds to a time period in the timetable. A variety of heuristics for constructing conflict-free timetables based upon graph coloring is available in the literature. Usually, these heuristics order the events based on an estimation of how difficult it is to schedule them. The rationale behind this is that the events that are more difficult to schedule should be considered first. Examples of common graph coloring/timetabling heuristics are as follows:

- *Largest degree first*: prioritizes the events that have a large number of conflicts with other events.
- *Largest weighted degree*: takes into account the conflicts of events with other events weighted by the number of students involved.
- *Saturation degree*: schedules first those events that have the smallest number of valid periods available for scheduling in the timetable constructed so far.
- *Color degree*: orders the events by the number of conflicts that an event has with the events that have already been scheduled.

One of the main reasons that these heuristics have been widely employed in timetabling is that they are easy to implement. However, they do lack the power of more modern intensive search methods. Nevertheless, their hybridization with other search methods still plays a key role in modern timetabling research. For example, see Burke, de Werra, and Kingston, (2003c). Indeed, we employ saturation degree in one of the methods described below. For a more detailed discussion of graph coloring in timetabling and the associated heuristics, see (Burke, de Werra, and Kingston, 2003c) and (Carter, 1986).

3. **Cluster methods.** These methods first split the set of events into groups so that events within a group satisfy the hard constraints. Then the groups are assigned to time periods to fulfil the soft constraints. For example, see White and Chan (1979), Lotfi and Cerveny (1991). However, these approaches can result in a poor quality timetable.
4. **Constraint based approaches.** In these methods, the events of a timetabling problem are modelled by a set of variables to which values (i.e., resources such as rooms and time periods) have to be assigned subject to a number of constraints. For example, see Brailsford et al. (1999), White (2000). When the propagation of the assigned values leads to an infeasible solution, a backtracking process enables the reassignment of value(s) until a solution that satisfies all of the constraints is found.
5. **Meta-heuristic methods**. The last two decades have seen an increased research interest in the development of meta-heuristic approaches to university timetabling. Such approaches include simulated annealing, tabu search, genetic algorithms and their hybridization. Examples of all of these methods can be found in the four "Practice and Theory of Automated Timetabling" books (Burke and Ross, 1996b; Burke and Carter, 1998; Burke and Erben, 2001b; Burke and De Causmaecker, 2003a). Meta-heuristic methods begin with one or more initial solutions and employ search strategies that try to avoid getting stuck in local optima. An advantage of these search algorithms is that they can produce high quality solutions. However, they often require significant tuning of parameters and also often have a considerable computational cost.
6. **Multicriteria approaches**. In these approaches, each criterion measures the violation of a corresponding constraint. The criteria have different levels of importance in different situations and for different institutions. A multiphase exam scheduling package has been developed by Arany and Lotfi (1989) and Lotfi and Cerveny (1991), where each phase attempts to minimize the violation of a certain constraint. A multicriteria approach that is based on Compromise Programming and which can handle all the criteria simultaneously was described in Burke et al. (2001a). A multicriteria extension of the Great Deluge algorithm which drives a trajectory based search of the criteria space by dynamic change of criteria weights was presented in Petrovic and Bykov (2003). These latter two approaches are described in more detail below.

7. **Case-based reasoning approaches**. These are relatively new approaches that utilize past solutions as building blocks to construct a solution to new timetabling problems (Burke et al., 2003b, 2002a, 2002b, 2001c, 2001d; Petovic and Qu, 2002). The issue of defining similarity measures between two timetabling problems plays a crucial role in these approaches. These approaches are described in more detail below.

A number of comprehensive surveys of existing timetabling methods and applications can be found in the literature (deWerra, 1985; Carter, 1986; Carter and Laporte, 1996, 1998; Bardadym, 1996; Burke et al., 1997; Schaerf, 1999; Burke and Petrovic, 2002a). Also, a number of real-world university examination timetabling problems have been collected and used as benchmark problems. They can be downloaded from ftp://ftp.mie.utoronto.ca/pub/carter/testprob/ and ftp://ftp.cs.nott.ac.uk/ttp/Data/. At present a set of benchmark course timetabling problems is under construction. The current collection of data sets can be seen at http://www.asap.cs.nott.ac.uk/themes/timetabling. In addition, the EU Meta-heuristics network maintains a collection of test problems that it used in a course timetabling competition in 2002. See http://www.idsia.ch/Files/ttcomp2002/ for more details.

As mentioned before, this chapter aims to highlight a key class of important scheduling problems (university timetabling problems) and it aims to highlight a number of different approaches to both course and exam timetabling that have been developed by the authors. We focus on four main streams of research, which in our opinion represent particularly promising areas of timetabling research. The first theme is concerned with developing methods that are less reliant upon parameter tuning than the current state of the art in meta-heuristic development. The goal here is not only to reduce the reliance upon parameter tuning but also to develop approaches which employ parameters that are more relevant to real world users than the current state of the art. The second theme concerns the integration of meta-heuristic and multicriteria approaches to timetabling. Here the goal is to develop special purpose timetabling methods which can produce high quality solutions in real world environments. The third research theme is to explore the employment of case based reasoning in timetabling. The fourth stream of research is concerned with developing approaches that attempt to raise the level of generality of timetabling systems. Such approaches aim to facilitate the development of intelligent systems that can choose an appropriate heuristic/method for solving a timetabling problem at hand. Such systems are applicable to wider range of timetabling problems with different constraints and requirements. Of course, these four broad themes of timetabling research are not disjoint. For example, we will describe (see below) a successful investigation of case based reasoning as a timetabling heuristic selector (in an attempt to raise the level of generality of timetabling systems). The chapter is organized as follows. The next section gives a formal description of course and exam timetabling problems. Section 45.3 presents the modified Great Deluge algorithm and discusses its performance on both examination and course timetabling. Section 45.4 describes two algorithms for multicriteria examination timetabling: an approach based on Compromise programming and a modified Great Deluge algorithm with dynamic weights. Section 45.5 discusses two case-based reasoning systems which utilize the memorized timetabling problems and their solutions in solving new timetabling problems. Section 45.6 is devoted to adaptive systems and hyper-heuristics and is followed by some brief concluding comments in Section 45.7.

45.2 Course and Examination Timetabling Problem Statement

As we have discussed above, during the construction of a timetable we need to take different constraints into account simultaneously. These constraints may be conflicting in the sense that an attempt to satisfy one of the constraints can lead to the violation of another. For example, suppose we have a situation where we are trying to schedule two exams which need to be scheduled immediately before/after each other. Suppose also, that we have a constraint which requires an even spread of exams. In such a situation we can see that satisfaction of the first constraint will probably lead to the violation of the second. In addition, the violations of different constraints are measured by different units with different scales (e.g., the number of students involved in the violation, the number of exams, etc.).

Such observations motivated a multicriteria statement of timetabling problems in which criteria measure the violations of the corresponding constraints. A university may assess the quality of a timetable from very different points of view (expressed by departments, students, a timetable officer, etc.). The university often assigns different importances to the constraints that are imposed on the timetabling problem. For example, a department may have strong demands for a specific classroom for certain examinations, while students are usually concerned with the order in which exams are scheduled and their proximity, etc.

Before we present a multicriteria statement of course timetabling problems, we introduce the following notation:

- N is the number of courses.
- P is the number of time periods available.
- l_n is the number of lectures of the course n, $n = 1, \ldots, N$, that have to take place during P time periods.
- r_p is the number of rooms available at period p, $p = 1, \ldots, P$.
- $C = [c_{nm}]_{N \times N}$ is the symmetrical matrix which represents the conflicts of courses.
- c_{nm}, is the number of students taking both courses n and m where $n, m = 1, \ldots, N$.
- K is the number of criteria.
- $T = [t_{np}]_{N \times P}$ is the matrix which represents assignments of the courses into time periods.
- $t_{np} = \begin{cases} 1, & \text{if course } n \text{ is scheduled in time period } p, n = 1, \ldots, N, p = 1, \ldots, P. \\ 0, & \text{otherwise} \end{cases}$
- $f_k(T)$ is the value of criterion C_k, $k = 1, \ldots, K$.
- w_k is the weight of criterion C_k, $k = 1, \ldots, K$.

The multicriteria course timetabling problem can be stated in the following way.

Given the number of time periods P, the number of courses N with the required number of lectures, determine the timetable T which makes all the components of the vector $\boldsymbol{WF} = (w_1 f_1(T), \ldots, w_k f_k(T), \ldots, w_K f_K(T))$ as small as possible, subject to

$$\sum_{p=1}^{P} t_{np} = l_n, \quad n = 1, \ldots, N \tag{45.1}$$

$$\sum_{n=1}^{N} t_{np} \leq r_p, \quad p = 1, \ldots, P \tag{45.2}$$

$$\sum_{n=1}^{N-1} \sum_{m=n+1}^{N} \sum_{p=1}^{P} t_{np} t_{mp} c_{nm} = 0 \tag{45.3}$$

Constraints (45.1), (45.2) and (45.3) ensure that each course has the required number of lectures, that lectures can be accommodated in a given number of rooms, and that courses which have students in common are not scheduled in the same time period, respectively.

Similarly, we define exam timetabling problems using the following notation:

- N is the number of exams.
- P is the number of time periods available.
- $C = [c_{nm}]_{N \times N}$ is the symmetrical matrix which represents the conflicts of exams.
- c_{nm}, is the number of students taking both exams n and m, $n, m = 1, \ldots, N$.
- K is the number of criteria.
- $T = [t_{np}]_{N \times P}$ is the matrix which represents assignments of the exams into time periods.
- $t_{np} = \begin{cases} 1, & \text{if exam } n \text{ is scheduled in time period } p, n = 1, \ldots, N, p = 1, \ldots, P. \\ 0, & \text{otherwise} \end{cases}$

- $f_k(T)$ is the value of criterion $C_k, k = 1, \ldots, K$.
- w_k is the weight of criterion $C_k, k = 1, \ldots, K$.

Multicriteria exam timetabling problems can be defined as follows.

Given the number of exams N and the number of time periods P, determine the timetable T which makes all the components of the vector $\boldsymbol{WF} = (w_1 f_1(T), \ldots, w_k f_k(T), \ldots, w_K f_K(T))$ as small as possible, subject to

$$\sum_{p=1}^{P} t_{np} = 1, \quad n = 1, \ldots, N \tag{45.4}$$

$$\sum_{n=1}^{N-1} \sum_{m=n+1}^{N} \sum_{p=1}^{P} t_{np} t_{mp} c_{nm} = 0 \tag{45.5}$$

Constraints (45.4) and (45.5) require that each exam is scheduled only once, and that no two conflicting exams are scheduled in the same time period, respectively.

Many approaches reported in the timetabling literature on course and exam timetabling are reduced to single criterion problems where the criterion to be minimized is the weighted sum of components of the corresponding vector ***WF***.

45.3 A Modified Great Deluge Algorithm for Timetabling

One of the major drawbacks of many meta-heuristic approaches for timetabling (and other scheduling problems) is that they can be very dependent upon a range of parameters. The effectiveness of the approach can often be very sensitive to the parameter settings and determining the correct settings can create significant difficulties. These difficulties are exacerbated if the parameters have to be set by nonexperts in meta-heuristics (such as most university timetabling officers). These observations led us to investigate meta-heuristic approaches that are not as dependent upon parameters. In fact, they led us to explore not only the reduction of such parameters but also to employ parameters that can be understood by nonexpert users. For example, it is unlikely that a timetabling officer will have much feel for the cooling schedule for simulated annealing or the mutation probability in a genetic algorithm. However, most real world users will be quite happy to consider *computational time* (amount of time the computer is going to spend to try and solve the problem in hand) as a parameter. With this goal in mind we developed a timetabling method which was based on the Great Deluge Algorithm that was introduced by Dueck in 1993 (Dueck, 1993). This approach (for examination timetabling) is presented, discussed, and evaluated in Burke et al. (2001e). An exploration of the approach for course timetabling is presented in Burke et al. (2003d). The approach discussed in these two papers actually requires two parameters. In addition to *computational time* (already mentioned) the approach also needs an *estimate* of the value of the evaluation function that is required. Note that this parameter (like *computational time*) is something that real world users can easily get a feel for. They can easily relate to the *quality* of a potential solution (in terms of the evaluation function, i.e., a *penalty* value) and the time that they want the computer to spend to try and get as close to that *penalty* value as possible. Of course, the difficulty in the design of the algorithm is in ensuring that the algorithm uses all the time it is given in effectively exploring the search space. In this chapter, we will briefly outline the basic method and then discuss the results of applying that basic method to examination timetabling and course timetabling.

45.3.1 The Basic Great Deluge Timetabling Algorithm

This meta-heuristic always accepts better solutions than the one that it is currently considering. It will accept *worse* solutions if the evaluation function value is less than (or equal) to a particular upper limit (which Dueck called the *level*). This is lowered during the run (according to a *decay rate*). The decay rate is a simple function of the two parameters. It is defined as the evaluation value of the initial solution subtracted by the estimate of the desired evaluation value, all divided by the desired number of moves (computational expense). The algorithm can be outlined by the following pseudo code, which is taken from Burke et al. (2001e).

```
Specify the initial solution, s
Input the two parameters described above
The initial level, L is defined to be f(s) where f is the evaluation function
Calculate the decay rate ΔL (as described above)
While further improvement is impossible
          Define the Neighborhood N(s)
          Randomly select the candidate solution s' from N(s)
          Calculate f(s')
          If f(s') ≤ f(s)
          Then accept s'
          Else if f(s') ≤ L
                Then accept s'
          Lower the level (set L to L − ΔL)
```

For more detail and discussion about this algorithm see Burke et al. (2001e).

45.3.2 Employing the Great Deluge Algorithm on Exam Timetabling Problems

The above method was investigated for examination timetabling in Burke et al. (2001e) and was evaluated on a range of standard benchmark exam timetabling problems (see Section 45.1). The algorithm was applied to the 13 benchmark problems at ftp://ftp.mie.utoronto.ca/pub/carter/testprob. It produced the best results for several of the 13 problems. Another algorithm (which we called time predefined simulated annealing) is described in the same paper. This approach modifies the acceptance condition of simulated annealing. One of the conclusions of Burke et al. (2001e) is that the Great Deluge Exam Timetabling Algorithm significantly outperforms time predefined simulated annealing because it is far less sensitive to initial parameters. It also produces superior best results in 9 of the 13 benchmark problems and superior average results in 11 of the 13 problems. The superiority of the great deluge algorithm over the time predefined simulated annealing algorithm was underlined when we applied the two algorithms to three benchmark problems at ftp://ftp.cs.nott.ac.uk/ttp/Data. These problems are more complex (in terms of the constraints). The great deluge algorithm produced better results than the time predefined simulated annealing algorithm and indeed it produced better results than any other published method (albeit by employing more time). The time that the algorithm used for the three problems was between 345 and 1268 seconds. On one particular problem (Nottingham University examinations — 1994), we let the algorithm run for a significant amount of time to see how good the results could be. The previous lowest penalty for this problem was 490 (Burke and Newall, 1999). The Great Deluge Algorithm produced a cost of 384 in 612 seconds. In 2.5 hours it could lower the penalty cost to 256 and in 67 hours, the cost was reduced to 225. Note that although 67 hours looks like a lot of computing time it corresponds approximately to one weekend (Friday afternoon to Monday morning). Exam timetables are not produced very frequently (usually once or twice a year). Leaving a computer running over a weekend to produce a good quality solution is an entirely feasible prospect in most universities. One potential drawback of the Great Deluge algorithm is that it needs to work with feasible solutions. Some methods (e.g., Burke and Newall, 1999) can simply assign a high penalty to the breaking of a hard constraint. This option is not open when employing the Great Deluge approach.

45.3.3 Employing the Great Deluge Algorithm on Course Timetabling Problems

In 2002 the EU Meta-heuristics network (see http://iridia.ulb.ac.be/~meta/newsite/index.php) ran a competition on course timetabling. The problem sets are available from http://www.idsia.ch/Files/ttcomp2002/. The great deluge algorithm (although originally developed for exam timetabling) was applied to these problems and entered into the competition. The approach came third and the results are discussed in Burke et al. (2003d). It is interesting to see that the great deluge approach works very well indeed on some of the problems. For example, it produces the best results on 8 of the 23 problem instances. Indeed, the approach was the only one able to produce a penalty of 0 for one of the problems. The winner of the competition (P. Kostuch) produced the best results for 14 of the 23 problems. The second place candidate (B. Jaumard et al.) produced results which were best on 1 of the 23 problems. It is interesting to note that the great deluge approach was second in terms of the number of best results. However, the competition measured average performance and for some of the problem instances, the great deluge algorithm performed very poorly indeed (which affected the averages). For two of the problems, it gets the very worse results among the six leading participants. An investigation into why the algorithm should vary between performing extremely well and extremely poorly for instances of the course timetabling problem is currently underway.

45.4 Multicriteria Approaches to Timetabling

In the majority of timetabling algorithms, a solution is evaluated using only a subset of constraints. A single cost function, which penalizes the violations of the constraints, is usually additive (as we observed in Section 45.2). This requires the timetable officer to set appropriate weights to compensate for a variety of units with different scales that violations of the constraints are measured by. In addition, the additive cost function is not flexible enough in the sense that it allows the compensation of a heavily violated constraint by one that is lightly violated. This is a feature that the timetable officer may not always be satisfied with.

This motivated the development of multicriteria approaches to timetabling which aim to offer a more flexible way of handling different types of constraints simultaneously. Two multicriteria approaches are presented in this section: an approach based on Compromise programming and a modified Great Deluge algorithm with dynamic weights.

45.4.1 A Multicriteria Approach Based on Compromise Programming

A multicriteria approach was developed for examination timetabling in which each criterion measures the number of violations of the corresponding constraint (Burke et al., 2001a). It is based on the principal idea of Compromise programming — a multicriteria decision making method that uses the distance between a solution and an *ideal point* to assess the quality of the solution (Zeleny, 1974). An ideal point is a solution that optimizes all the criteria simultaneously, i.e., in the context of timetabling it is a timetable which does not violate any of the constraints. Usually, such a solution to a real-world complex, highly constrained timetabling problem does not exist. The notion of an *anti-ideal point* is used to denote the solution that has the worst possible values of all the criteria.

Usually in multicriteria decision making each solution is represented as a point in the criteria space. The dimensionality of this space is equal to the number of criteria. In order to overcome the problem of comparison of different criteria values, a new *preference space* is introduced (Petrovic, 1995). A linear mapping is defined which maps the criteria space into the preference space. Figure 45.1 depicts the mapping when there are two criteria (i.e., $K = 2$). Namely, for each criterion $C_k, k = 1, \dots, K$, a linear function is defined which maps each value of the criterion f_k to the value $s_k \in [0, w_k]$, where w_k is the relative importance of the criterion (the notation presented in Section 45.2 is used). The linear function (s_k) is

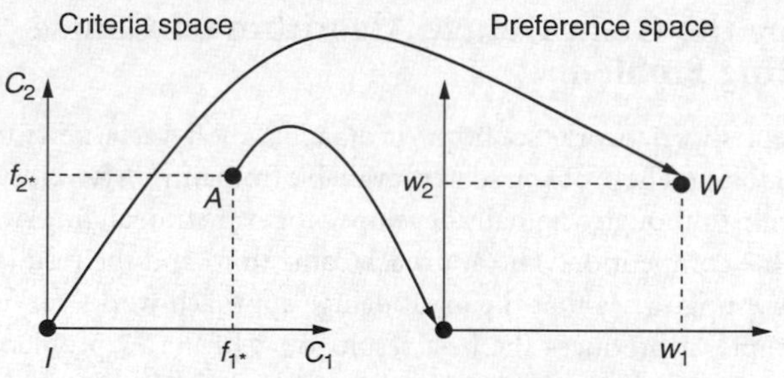

FIGURE 45.1 Mapping of the criteria space into the preference space when $K = 2$.

defined so that the worst value of the criterion, denoted by f_{k^*}, is mapped onto the value 0, while the best value of the criterion, denoted by f_k^*, is mapped onto w_k. It can be formulated as follows:

$$s_k = w_k \frac{f_{k^*} - f_k}{f_{k^*} - f_k^*} \tag{45.7}$$

where f_k and s_k denote the values of the criterion C_k in the criteria space and the preference space, respectively.

According to formula (45.7) the ideal point $I = (f_1^*, \ldots, f_K^*)$, $f_k^* = 0, k = 1, \ldots, K$, is mapped into the point $W = (w_1, \ldots, w_K)$, while the anti-ideal point $A = (f_{1^*}, \ldots, f_{K^*})$ is mapped to the point 0 in the preference space. It is usually impossible to find exact worst values of criteria in a complex timetabling problem and, consequently, an anti-ideal point can be only calculated approximately. A detailed example of the calculation of worst values of a criterion is given in Burke et al. (2001a).

A new algorithm for heuristic search of the preference space was developed. It consists of two phases.

1. The goal of the first phase is to find initial solutions that are good with respect to a single criterion. It is carried out by using a *Saturation degree* heuristic which orders the exams and then the exam selected for scheduling is placed in the time period that causes the lowest increase of the criterion value.
2. In the second phase, the algorithm iteratively searches the neighborhood of each of the initial solutions aiming to get closer to the ideal point. Two operators are used during the search:
   - Hill-Climbing directs the search toward the local optima. Namely, it randomly chooses an exam to reschedule and places it in the valid time period (so that it does not cause violation of any hard constraint) which yields a maximum decrease in the distance from the ideal point.
   - A mutation operator directs the search away from local optima toward new areas of the preference space. It reschedules the exams that contribute the most to the distance from the ideal point.

The search stops when the multiple application of the Hill-Climbing operator followed by the mutation operator have not decreased the distance between the solution and the ideal point for a predefined number of steps.

Therefore, the search of the neighborhoods of each initial solution results in one timetable. The final timetable is the solution which is the closest to the ideal point.

A family of L_p metrics are used to evaluate the distance between the solution $S = (s_1, s_2, \ldots, s_K)$ and the ideal point $W = (w_1, w_2, \ldots, w_K)$ in the preference space. This family is calculated by the following formula:

$$L_p(S, W) = \left\{ \sum_{k=1}^{K} [s_k - w_k]^p \right\}^{1/p}, \quad 1 \leq p \leq \infty \tag{45.8}$$

The value of parameter p enables the timetable officer to express his/her preference toward the solutions. Three values of parameter p are usually of particular interest:

- $p = 1$: This enables highly satisfied constraints in the timetable to compensate for less satisfied ones.
- $p = \infty$ (it can be proved that $L_\infty(S, W) = \max_k[w_k - s_k]$): This prevents trade-offs between the criteria values, i.e., only the solutions which satisfy all the constraints reasonably well are of interest.
- $p = 2$: This gives the solution geometrically closest to the ideal point in the preference space.

The approach described in Burke et al. (2001a) included a higher number of constraints than other benchmark problems do. In total, nine constraints were considered. These constraints referred to room capacities, proximity of the exams, and time and order of exams. C_1 represents the number of times that students cannot be seated in the timetable. C_2 represents the number of conflicts where students have exams in adjacent periods on the same day. C_3 represents the number of conflicts where students have two or more exams in the same day. C_4 represents the number of conflicts where students have exams in adjacent days. C_5 represents the number of conflicts where students have exams in overnight adjacent periods. C_6 represents the number of times that a student has an exam that is not scheduled in a time period of the proper duration. C_7 represents the number of times that a student has an exam that is not scheduled in the required time period. C_8 represents the number of times that a student has an exam that is not scheduled before or after another specified exam and C_9 represents the number of times that a student has an exam that is not scheduled immediately before or after another specified exam.

The approach cannot be compared with other approaches based on a single cost function, because the measures used for timetable evaluations are incomparable and they operate with different constraints. However, this approach gives more flexibility to the timetable officer to express his/her preference. It does so by enabling the assignment of relative importances to each of the criteria (the timetable officer does not have to worry about how easy it is to satisfy the corresponding constraint). The approach also allows the timetable officer to choose the distance measure which reflects his/her attitude toward compensation among the criteria values.

45.4.2 A Modified Great Deluge Algorithm with Dynamic Weights

The main idea behind the Great Deluge algorithm is to direct the search of the solution space along a trajectory by changing the acceptance level of the cost function values. This has been implemented in a multicriteria environment (Petrovic and Bykov, 2003). The method requires the timetable officer to specify a reference solution that reflects his/her preference toward the criteria values. It can be obtained either by some other method (not necessarily a multicriteria one) or the timetable officer sets the values of the criteria that he/she aims at in the solution. The method conducts the search along the trajectory that is drawn in the criteria space between the reference solution and the origin, aiming to find a solution that is better than the reference one. The direction of the search is toward the origin because it represents the ideal solution. As mentioned, the ideal solution is usually not achievable in real-world timetabling problems. However, the search does not start from the reference point itself, because it may either lay in the local optimum (thus the search will be stuck), or if the point was set manually it may not be possible to find a solution that corresponds to this point in the criteria space. The search starts from a solution that is in the vicinity of the reference point. Therefore, the search is divided into two phases. The aim of the first phase is to find a solution which is close to the reference one. The search starts from a randomly chosen initial solution toward the reference point and stops when there is no improvement in the cost function in a predefined number of iterations. In the second phase, the search is conducted along the trajectory drawn between the reference point and the origin.

The Great Deluge algorithm is employed with the cost function defined as the weighted sum of the criteria values. It can be proved that lowering the acceptance value of the cost function values is equivalent

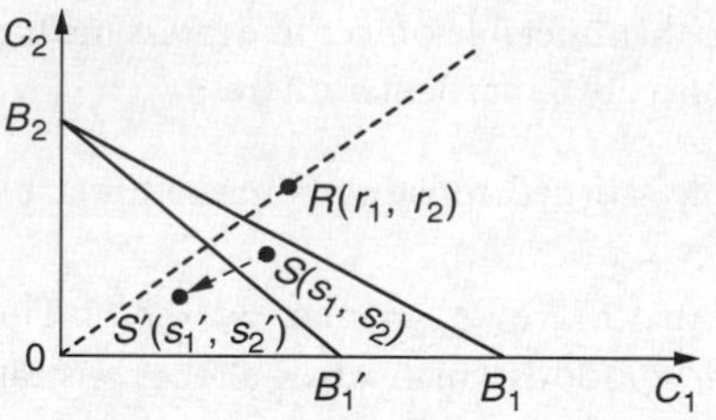

FIGURE 45.2 Change of criteria weights enables the search along the trajectory.

to changing the weights of criteria in the cost function (Petrovic and Bykov, 2003). Namely, formula (45.9) can be transformed into (45.10) when $\Delta W = \Delta B/(B - \Delta B)$. The two formulae are presented below:

$$w_1 f_1 + w_2 f_2 \le B - \Delta B \qquad (45.9)$$

$$w_1 f_1(1 + \Delta W) + w_2 f_2(1 + \Delta W) \le B \qquad (45.10)$$

However, in order to drive the search along the trajectory instead of changing the weights of all criteria equally, the weight of a single criterion is changed. The choice of the criterion whose weight is to be changed depends upon the position of the current solution relative to the trajectory. The idea of the algorithm is illustrated in Figure 45.2, when the number of criteria is equal to two ($K = 2$). Let us suppose that the current solution $S(s_1, s_2)$ is placed below the trajectory *RO*. The border line B_1B_2 determines the acceptance level of the cost function values of the solutions generated in the next iteration of the search. To direct the search toward the trajectory, the solutions which improve the criterion C_1 should be accepted, which means that the weight of the criterion has to be increased. It is equivalent to the change of the acceptance level which is depicted as the rotation of the border line from B_1B_2 to $B_{1'}B_2$.

The algorithm requires two items of data to be specified in advance: the initial weights of criteria $(w_1^0, \ldots, w_K^0)$ and the increase of weights ΔW in each iteration. The initial values of the weights determine the initial border line B_1B_2, and thus affect the performance of algorithm. The experimental studies have shown that when $w_1^0 = s_1^0/r_1$ where $S(s_1^0, \ldots, s_K^0)$ and $R(r_1, \ldots, r_K)$ were the initial and reference solutions, respectively, then the algorithm performed well. The value of ΔW affects the time required to conduct the search. The algorithm does not necessarily converge in a given number of iterations. However, our experiments, performed on real-world timetabling data, showed that the algorithm succeeded in obtaining solutions that were better than the reference one as they dominated it by all criteria.

45.5 Case-Based Reasoning Approaches to Timetabling

The nature of the human reasoning process is based on *experience* rather than solely on a set of general guidelines or first principles. Case-based reasoning (CBR) attempts to rely on previous knowledge and experience rather than on formal models or rules [RIE89], [KOL93], [LEA96]. In CBR, the description of the actual problems and their solutions are memorized as cases and organized in a case base. Given the description of a new problem, the case(s) that is (are) the most appropriate for the new problem is (are) retrieved from the case base and employed in solving the new problem. CBR has been successfully applied to various types of scheduling problems including the planning and scheduling of large-scale airlift operations (Koton, 1989), dynamic job-shop scheduling (Bezirgan, 1993), repair problems in job-shop scheduling (Miyashita and Sycara, 1995), travelling salesman problems (Cunningham and Smyth, 1997), production planing and control problems (Schmidt, 1998), nurse rostering (Petrovic et al., 2003), etc. These applications indicated to us that CBR could be a valuable tool in solving timetabling problems.

A CBR approach can have a twofold role in solving timetabling problems:

1. As a *solution reuse* technique. In this situation, the system finds a *case(s)* that is the most similar to the problem in hand. The solution of the previously solved problem(s) is adapted to meet the requirements of the new problem.
2. As a *methodology reuse* technique. Here, the system proposes a method (heuristic, meta-heuristic or a hybrid method) which is the most appropriate for the new problem. The choice of method is based upon what worked well on *similar* problems in the case base. Of course, the development of appropriate *similarity* measures is a challenging research issue.

The remainder of this section gives a more detailed look at the CBR approaches developed for university timetabling that utilize both solution and methodology reuse.

45.5.1 Case-Based Solution Reuse in Timetabling

Following the basic idea of CBR, a number of timetabling problems (cases), are stored in a case base. Each case stores a timetabling problem and the solution for it. When a new problem is to be solved, a case which contains the timetabling problem which is most similar to the new problem (with respect to the structure of the problem), is retrieved from the case base. The idea is that the solution of the retrieved case is a potentially useful starting point in the solution of the new problem. However, the retrieved solution needs to be adapted to take into consideration the differences between the new problem and the retrieved one. The development of CBR approaches for timetabling was motivated by the construction of timetables in practice where the established custom is often to start from *last year's timetable*, and to make changes to it to fit the new problem requirements. The usual differences between new timetabling problems and old ones are in student demand patterns, the number of courses, etc.

We present here an overview of a CBR system developed for course timetabling problems which is presented in far more detail in (Burke et al., 2000; Burke et al., 2001c; Burke et al., 2001d; Rong, 2002). The following issues will be addressed: the representation of complex course timetabling problems, the organization of timetabling problems into a case base, the definition of a similarity measure which drives the retrieval process and the adaptation of the solution of a retrieved case.

45.5.1.1 Representation of the Cases

A list of (feature, value) pairs, which is often employed in case based reasoning systems to represent cases (Leake, 1996), is not a rich enough structure for representing complex timetabling problems with various constraints imposed between courses and on the courses themselves. Instead, attribute graphs are used to represent timetabling problems structurally. The vertices of the graph represent courses, while edges represent constraints between pairs of courses. In addition, attributes are assigned to both vertices and edges to represent constraints on courses and on pairs of courses, respectively. For example, attributes assigned to vertices may denote that the course should be held once per week, or n times per week or that the course should be scheduled in a particular time period, etc. The attributes of the edges may denote that two courses are in conflict with each other (i.e., they have common students), that one course should be held before or after the other, that two courses should be consecutive to each other, etc. An attribute graph is represented by an adjacency matrix whose diagonal elements contain the attributes of the vertices, while the other elements denote the attributes of the edges between the corresponding vertices. The order of courses in a new problem and the timetabling problems stored in the case base is important in the retrieval process. Therefore, for each timetabling problem in the case base, all the possible permutations of the courses are stored, represented by corresponding adjacency matrices.

For illustration purposes, an example of an attribute graph for a simple 3-course timetabling problem is given in Figure 45.3. The vertex for *Maths* is labelled by 1 with value 2 to denote that this is a multiple course, held twice a week. Label 0 for *Physics* and *Spanish* means that these two courses are standard courses in the sense that they are held just once a week. *Physics* should be held before *Maths* which is labelled by 4.

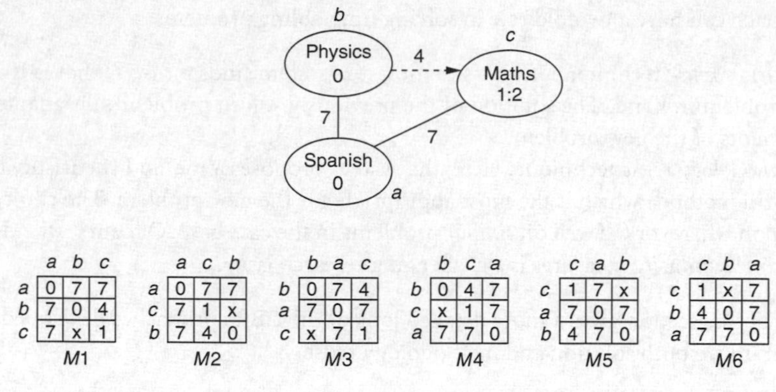

FIGURE 45.3 An attribute graph for a 3-course timetabling problem and its adjacency matrices.

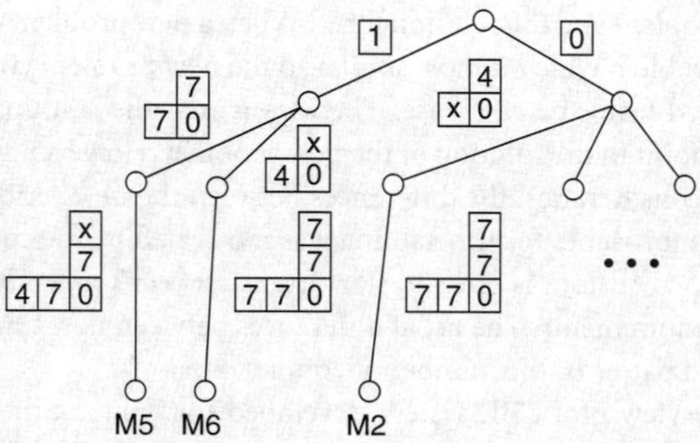

FIGURE 45.4 A decision tree which stores adjacency matrices.

Label 7 on the edges between *Spanish* and both *Physics* and *Maths*, denotes that *Spanish* should not be scheduled simultaneously with these two courses because they have students in common. For this 3-course timetabling problem, a total of six adjacency matrices is needed (denoted by $M1, \ldots, M6$) to store all the possible permutations of the courses. The element x in the adjacency matrix denotes that there is no edge between two vertices.

45.5.1.2 Organization of the Case Base

In order to enable an efficient search of the case base, cases are stored hierarchically in a decision tree. Each level of the decision tree stores additional elements of the adjacency matrices. For example, the nodes of level 1, 2, and 3 store elements a_{11}, $a_{21}a_{22}a_{12}$, $a_{31}a_{32}a_{33}a_{23}a_{13}$ of the adjacency matrix, respectively. For example, the 3-course problem given in Figure 45.3, which requires six permutations of courses to be stored will lead to the decision tree presented in Figure 45.4.

45.5.1.3 The Similarity Measure and Retrieval Process

The goal of the retrieval process is to find a case from the case base which is structurally the most similar to the new timetabling problem. The matching process between two timetabling problems represented by attribute graphs is based on graph isomorphism. An algorithm for graph isomorphism presented in Messmer (1995) has been employed. However, instead of using an exact match between the attributes of the new timetabling problem and the attributes stored in the decision tree, a partial match has been introduced which is based on the penalties defined for the pairs of attributes. If the penalty is below a given threshold, the match is considered to be successful.

The search stops when none of the nodes on the level below the current one match successfully with the new problem. In such a way, the new timetabling problem is classified to a node in the decision tree. The node contains a number of full or partial timetabling problems that give information about the possible isomorphism between the problems stored with it and a new one. The similarity measure between the new timetabling problem and the timetabling problem retrieved from the case base is defined carefully taking into consideration the adaptation necessary to perform on the retrieved case to match the new problem. It was observed that not only problems which are graph isomorphic to the new problem can be useful, but also graphs of similar structure with some differences. The costs of the substitution, deletions and insertions of the vertices and edges that have to be carried out on the retrieved case are defined empirically and included in the similarity measure. For example, the substitution costs are lower than the deletion and insertion costs, while the deletion costs are lower than the insertion ones.

The similarity measure between the timetabling problem from the case base (also called the source case) T_S and the new timetabling problem T_N can be defined in the following way:

$$S(T_S, T_N) = 1 - \frac{\sum_{i=0}^{n} p_i + \sum_{k=0}^{m} a_k + \sum_{l=0}^{k} d_l}{P + A + D} \tag{45.11}$$

where

- n is the number of matched attributes between the cases T_S and T_N.
- p_i is the cost assigned for substituting a vertex or edge of the retrieved case T_S with a vertex or edge of T_N.
- m and k are the total numbers of the vertices or edges that have to be inserted into and deleted from the retrieved case T_S, respectively, to obtain a match with T_N.
- a_k, d_l are the penalties assigned for inserting and deleting a vertex or edge k and l into and from the retrieved case T_S, respectively, to obtain a match with T_N.
- P is the sum of the penalties for substitution of each possible pair of vertices or edges in the retrieved case T_S to match T_N.
- A and D are the sums of the costs of inserting and deleting all of the vertices or edges into and from the retrieved case T_S respectively, to match T_N.

45.5.1.4 The Multiple Retrieval Process

The drawback of the CBR approach described above, which requires all the permutations of the courses to be memorized, is that a limited number of small timetabling problems can be stored in the case base. However, a large number of experiments indicated that the system worked well when new problems were of similar or smaller size than the cases from the case base. Not surprisingly, timetabling problems of small size could not provide good solutions for new problems of larger size. On the other hand, it was reported in the literature, and experienced in practice, that algorithms for timetabling can cope better with the large size of problems by decomposing them into smaller subproblems (Carter, 1983; Robert and Hertz, 1995; Burke and Newell, 1999).

In order to enable the CBR system to handle timetabling problems of large size, a multiple retrieval system has been developed which retrieves, in each iteration, a solution for the part of the new timetabling problem for which the best match has been found in the case base. In such an approach, the number of subproblems and courses assigned to them are not fixed in advance, as in other approaches based on decomposition, but they change dynamically as they depend on the current contents of the case base. In each iteration, the following steps are performed on the attribute graph of the new problem.

1. The vertices in the new problem that match the vertices of the retrieved case are replaced by a super vertex. The attribute of the super vertex is determined following the predefined priorities of the vertex attributes.

2. The super vertex in the new attribute graph keeps the edges of all of its vertices. If there is more than one edge between the super vertex and some other vertex, then the attribute of the edge is determined following the predefined priorities of the edge attributes. For example, in order to preserve the feasibility of the final solution, the label 7 which denotes the hard constraint that two courses have common students has the highest priority and overwrites the other labels.

45.5.1.5 Adaptation

The final solution of the new timetabling problem consists of super vertices, which hold the solutions of the subproblems, and the vertices for which no match could be found. According to the isomorphisms found, the courses in the solution of the retrieved cases are replaced with matching courses from the new problem. It is to be expected that the solutions of the retrieved case(s) violate some of the constraints of the new timetabling problem. Therefore, they have to be adapted to fulfil the constraints of the new problem. However, as the retrieval process compares the structures of the timetabling cases and the similarity measure takes into consideration how difficult it is to adapt the solution of the retrieved case, the adaptation process usually does not require a great effort.

The adaptation process performs the following steps:

1. All the courses of the retrieved cases that match the courses of the new problem are kept, while the ones that do not match are removed and placed into the *Unscheduled list.*
2. All the courses that violate the soft constraints are removed and placed into the *Unscheduled list*.
3. All the remaining courses are allocated in the smallest available room.
4. All the courses in the *Unscheduled list* are rescheduled using the *Largest degree first* heuristics with tournament selection. The course selected for scheduling is put into the first available time slot that does not violate any of the constraints. If there is no such slot, the course is scheduled into the timeslot which causes the least number of violations of soft constraints (measured by the penalty function which is explained below).
5. The course is left unscheduled if it cannot be scheduled without violating hard constraints or no room is available.

45.5.1.6 Penalty Function

The penalty function which is used to evaluate the quality of the proposed timetables takes into consideration the number of unscheduled courses (U) and soft constraints (S). It can be defined as follows:

$$P(t) = 100U(t) + 5S(t) \tag{45.12}$$

Solutions with unscheduled courses are penalized highly to ensure that partial timetables are less acceptable.

We have carried out a large number of experiments and have concluded that a case base with small complex cases enables good performance of the multiple retrieval process by providing small scheduling blocks for timetables of large size. Small complex cases here refer to timetabling problems with the number of courses between 6 and 10, while each course has between 1 and 4 constraints with other courses. The number of cases in the case base in the experiments performed varied from 5 to 15.

45.5.2 Case-Based Methodology Reuse in Timetabling

While various techniques have been satisfactorily applied to a wide range of university timetabling problems, a natural question arises concerning the robustness of a particular implementation when a problem's specifications change. The disadvantage of employing specially tailored heuristics for the given problem is that usually they work well only in the environments that they are developed (tailored) for. The issue of how to increase the level of generality of existing heuristic methods has recently attracted the attention of scheduling/optimization researchers (Burke et al., 2003e). The idea is to develop a system which selects the right method for the given problem in an intelligent way. Therefore, instead of designing a heuristic which works directly on a specific problem, the system should consider information about the problem which helps in the intelligent selection of an appropriate method for that particular problem.

We present here two case-based reasoning approaches to the intelligent selection of appropariate heuristics for university timetabling problems. The first approach proposes heuristics which are appropriate for solving the *whole* timetabling problem, i.e., a heuristic is chosen and then that approach is used on the particular timetable problem in hand. The heuristics do not change during the construction of the timetable. It was developed for course timetabling and is discussed in Petrovic and Qu (2002), Burke et al. (2003b). However, it was noticed that a dynamic change of heuristics during the timetable construction leads to solutions of higher quality. This observation resulted in the development of a CBR approach that dynamically proposes an appropriate heuristic at each step of the timetable construction (Burke et al., 2002b). This second CBR approach addresses examination timetabling. We will first address the issues that are common to both approaches and then give specific details of each approach.

45.5.2.1 The Cases

Each case consists of two parts, namely, the problem description and two best heuristics for them. We decide to store two heuristics, because, very often, the timetables constructed by two different heuristics are of similar quality. If for some cases all, or a majority of the heuristics perform the same, then the case is not representative and thus is not stored in the case base.

45.5.2.2 Case Representation

A case is represented by a list of (feature, value) pairs. However, the features of cases to be used in the retrieval process are not predefined. Knowledge discovery (Fayyad et al., 1996) is used to identify those features of timetabling problems which are potentially useful in choosing heuristics for solving new timetabling problems.

45.5.2.3 The Similarity Measure

A nearest neighbor approach is used which calculates the weighted sum of differences between each pair of features of the two cases being compared. It can be formulated as follows:

$$S(C_S, C_N) = \frac{1}{\sum_i w_i (f_{S_i} - f_{N_i})^2 + 1} \tag{45.9}$$

where

- C_S and C_N are the source case and the new problem, respectively.
- f_{Si} and f_{Ni} are values of the feature i (for the source and new problem, respectively).
- w_i is the importance of the feature i.

The higher the value of $S(C_S, C_N)$ the higher the similarity between the cases C_S and C_N.

45.5.2.4 The Retrieval Process

The case with the highest similarity with the new problem is retrieved from the case base and its best heuristic for the new problem is suggested.

45.5.2.5 Training Cases

A set of training cases is defined and used in the knowledge discovery process to define the features of the cases. The heuristic obtained as the best for the training case is compared with the two heuristics proposed by the retrieved case. If it is equal to one of them then the retrieval is considered to be successful.

45.5.2.6 System Performance

The percentage of successful retrievals for all the training cases determines the system performance.

45.5.2.7 The Discovery of Cases Feature List

In the feature discovery process, the search space consists of all possible enumerations of features. Tabu search is employed (Glover, 1998) to search for the best features to be used in the case representation. Each solution presents a set of selected features. Three operators are performed on the solutions. They are:

- Change of a randomly selected feature
- Addition of a new feature into the solution
- Removal of the randomly chosen feature

The system performance serves as an objective function. Namely, for each solution the objective function calculates the number of successful retrievals for all the training cases. The size of the tabu list is set to be 9. That is a value within the range of values recommended in the literature (Reeves, 1996), and proved to be good in our experiments. The stopping condition for the tabu search is a predefined number of iterations.

45.5.2.8 The Size of the Case Base

The cases stored in a case base significantly affect the system performance. In order to refine the case base, knowledge discovery is employed again. This time it is used to select cases which are useful for solving new timetabling problems and to remove the irrelevant ones which contain redundant or even nonrepresentative information. A *Leave-One-Out* strategy is used, which removes (one by one) source cases from the case base and then checks the system performance on training cases. If the system performance is decreased, the source case is removed from the case base, otherwise it is retained. As a result of the *Leave-One-Out* strategy, the case base will contain only cases which are relevant and useful for heuristic selection.

45.5.3 A Case-Based Single Heuristic Selector

A case-based heuristic selector was developed for course timetabling. A case provides the description of the course timetabling problem and two heuristics that are the best to solve it with respect to the defined cost function. The features that are initially considered to be important in the heuristic selection process for course timetabling are as follows:

- f_1 — number of courses
- f_2 — number of time periods
- f_3 — number of constraints
- f_4 — number of rooms
- f_5 (f_6) — number of hard (soft) constraints
- f_7 (f_8) — number of courses that should (should not be scheduled to a fixed time period)
- f_9 (f_{10}) — number of courses that should (should not be scheduled to consecutive time periods)

The ratio of these features may also be decisive for the heuristic selection. The knowledge discovery process identified only a small number of features to be relevant for heuristic selection. It also reveals that if more complex features are included in the case representation, then the weights of these features do not affect the selection process (here complex features are ratios between two simple features). The knowledge discovery process included 100 training cases, small and big case bases (with 45 and 90 cases respectively) and 5 heuristics (*Large degree first*, *Largest degree with tournament selection*, *Color degree*, *Saturation degree* and *Hill-climbing*). The features that led to the highest system performance and thus were identified to be important for the heuristic selection are: f_3/f_2, f_{10}/f_2, f_3/f_8, f_2/f_1, f_3.

The quality of the produced timetables was evaluated taking into consideration the number of violations of the soft constraints.

The refinement of the case base demonstrated that only 6 out of the original 45 source cases (8 out of 90 when the large case base was used) are relevant enough for course timetable selection and had even better performance than the original cases. Obviously, nonrelevant cases in the case base significantly deteriorate the performance of the CBR system.

45.5.4 Case-Based Multiple Heuristics Selector

A heuristic selector has been developed for exam timetabling which dynamically suggests the most appropriate heuristics during the construction of the exam timetable (Burke et al., 2002b). The cases represent partial solutions obtained during the timetabling problem solving process and, for each of them, two best heuristics which cause the least increment of the penalty function are stored. The penalty function P takes into consideration the proximity of exams in the timetable t. It can be defined as follows:

$$P(t) = \sum_{p=1,2,3} w_p S_p \tag{45.13}$$

where

- S_p is the number of occurrences where students have exams scheduled p time periods apart.
- w_p is the importance of the correspondent constraint. (The occurrences where students have to sit exams in consecutive periods are penalized more than if there are free time slots between the exams and therefore $w_1 = 16, w_2 = 8, w_3 = 4$.)

The features that were initially found to be useful in the multiple heuristics selection for exam timetabling are: the number of exams, the number of time periods, the number of hard constraints, the number of rooms, the density of the conflict matrix, the number of exams that are already scheduled in the partial solution, the number of times that exams with common students are scheduled in consecutive time periods in the partial solution, the penalty of the partial solution, the cost of scheduling the exam into the partial solution, the number of the most constrained exams and the number of constraints of the most constrained exams.

It was noticed that during the timetable construction, the same heuristics were selected within a number of steps (usually around 10). In order to keep the size of the case base small, and thus easily manageable, each source and training case are sampled every 10 steps and partial solutions and heuristics for them are stored.

In a similar way as for course timetabling, a small number of features are shown to lead to good system performance. For example, experiments performed on a case base with 95 source cases and 95 training cases showed that a small number of features were representative enough (from 3 to 7 either simple features or ratios between simple features). However, in contrast to single heuristic selection, more cases were retained in the case base after the refinement process. A possible explanation is that there is a high number of possible and relevant partial solutions which are worth memorizing and using in future timetabling problems. On the other hand, there is a smaller number of timetabling problems that are representative enough for choosing heuristics for new problems.

Experimental results have also demonstrated that higher quality timetables have been obtained by using a CBR approach to dynamically choose heuristics during the timetable construction than by applying single heuristics for the whole timetable.

45.6 Adaptive Systems and Hyper-Heuristic Approaches

Burke and Petrovic (2002) published a paper which explored some research trends in automated timetabling. In that paper we briefly discussed hyper-heuristics as a promising timetabling research theme. A hyper-heuristic can be described as a "heuristic to choose a heuristic." See Burke et al. (2003e) for a detailed discussion about hyper-heuristics. Ross et al. (1998) argued that a future way forward in timetabling research (in terms of genetic algorithms) might be to investigate genetic algorithms to choose the most appropriate heuristic for a problem rather than to operate directly on the problem.

Indeed, Terashima-Marin et al. (1999) explored a genetic algorithm approach to the evolution of strategies in examination timetabling.

In this section of the chapter we will explore some of our recent work on raising the level of generality at which timetabling systems can operate. This work is concerned with hyper-heuristic methods and with heuristic approaches that can adapt to the particular problem in hand. Such work is motivated by observing that many timetabling systems (and indeed, scheduling systems) are very problem-instance specific. This observation is not new. Carter and Laporte (1996a) pointed this out in terms of examination timetabling. We believe that it is increasingly important to develop methods which have a wider applicability and which would thus be much cheaper to implement than bespoke tailor made timetabling/scheduling systems.

45.6.1 Adapting Heuristic Orderings for Exam Timetabling

A basic (and well studied) approach to exam timetabling problems is to order each exam that is to be scheduled in terms of how difficult it is thought to be (to schedule). Then each exam is taken in turn (the most difficult first) and scheduled (if possible) in the period of least penalty. The measure of difficulty can be given by one of the graph coloring heuristics that were mentioned in Section 45.1. Burke and Newall (2002c) presented and discussed an approach which *adapts* the difficulty measure. Suppose we start with a heuristic H and suppose also that H orders a particular exam e in position i in its ordering of *difficulty*. In addition, assume that when the standard approach (outlined above) comes to place exam e in the schedule that it cannot do so. In this case, the heuristic H did (perhaps) not produce a very good ordering. The ordering (in this case) could be improved by moving e up the ordering and running the algorithm again. This overall idea underpins the adaptive approach to exam timetabling problems which was presented in Burke and Newall (2002c). The method starts with a given heuristic ordering and that ordering is adapted according to the given problem. The approach is shown to work reasonably well even when started off on deliberately bad heuristics. It compares well against published results. It is more general than most of the published methods and is much easier and significantly quicker to implement. Moreover, it produced results which are comparable with the best known published results. These results are presented and discussed in detail in Burke and Newall (2002). In Burke and Newall (2003g), this approach is used to initialize the great deluge exam timetabling algorithm discussed above and in Burke et al. (2001e). It is demonstrated that the great deluge algorithm is improved by combining it with the adaptive method in this way.

45.6.2 A Hyper-Heuristic Applied to Course Timetabling

As mentioned above, one of the motivating goals behind hyper-heuristic research is to raise the level of generality at which timetabling/scheduling systems can operate. Of course, this goal ranges beyond educational timetabling problems. Indeed, a tabu search hyper-heuristic is presented in Burke, Kendall, and Soubiega (2003), which is demonstrated on nurse rostering problems and on course timetabling problems. The overall approach selects from a set of low level heuristics at each decision point in the search. The heuristics can be thought of as competing to be chosen. This competition is based upon reinforcement learning. During the search, a tabu list of heuristics is managed by the hyper-heuristic. The method is compared against bespoke algorithms for nurse rostering and course timetabling. It is shown to be competitive with those algorithms across both of these very different problems.

45.7 Conclusions

This chapter presented an overview of some state-of-the-art university timetabling methods that have been developed by the authors and others from the same research team. It emphasized a number of important issues which can be summarized as follows:

- Multicriteria approaches to timetabling better reflect real-world timetabling problems.
- The usage of parameters which are transparent to the decision maker (timetable officer) can increase the uptake of timetabling systems in the real world. We particularly explored two relevant parameters: the time which the decision maker is willing to spend on the construction of a timetable and an estimation of the quality of the desired timetable.
- The employment of previous experience gained during solving timetabling problems can provide significant useful knowledge when faced with a new problem.
- The learning mechanisms which aim to improve the performance of an algorithm play a crucial role in the development of knowledge based timetabling systems.
- The development of more general approaches to timetabling/scheduling problems is a particularly challenging research goal. Progress in this direction could significantly increase uptake by enabling the implementation of systems which can be successfully applicable to a range of problems.

These issues are discussed and evaluated in the context of university timetabling. However, we hope that the presented methodologies may inspire researchers to model and solve a wide variety of other timetabling/scheduling problems from different domains.

References

Akkoyunly, E.A., A Linear Algorithm for Computing the Optimum University Timetable, *The Computer Journal*, 16(4), 1973, pp. 347–350.

Arani, T. and Lotfi, V., A Three Phased Approach to Final Exam Scheduling, *IEE Transactions* 21(1), 1989, 86–96.

Balakrishnan, N., Lucena, A., and Wong, R.T., Scheduling Examinations to Reduce Second-Order Conflicts, *Computers & Operations Research*, 19, 1992, pp. 353–361.

Bezirgan, A., A Case-Based Approach to Scheduling Constraints, In: Dorn, J. and Froeschl, K.A., (eds.), *Scheduling of Production Processes*, Ellis Horwood Limited, 1993, pp. 48–60.

Brailsford, S.C., Potts, C.N., and Smith, B.M., Constraint Satisfaction Problems: Algorithms and Applications, *European Journal of Operational Research*, 119, 1999, pp. 557–581.

Burke, E. and De Causmaecker, P., (eds.), The Practice and Theory of Automated Timetabling III: *Selected Revised Papers from the 4th International Conference on the Practice and Theory of Automated Timetabling*, KaHo St.-Lieven, Gent, Belgium, 21-23 August 2002, *Springer Lecture Notes in Computer Science Series* Vol. 2740, 2003a.

Burke, E., MacCarthy, B., Petrovic, S., and Qu, R., Knowledge Discovery in a Hyper-Heuristic Using Case-Based Reasoning for Course Timetabling, (Burke, and De Causmaecker, 2003a), (in press).

Burke, E.K., de Werra, D., and Kingston, J., Applications in Timetabling, Section 5.6 of the Handbook of Graph Theory. In: J. Yellen and J. Grossman (eds.), Chapman Hall/CRC Press, (in Press).

Burke, E.K., Bykov, Y., Newall, J.P., and Petrovic, S., A Time-Predefined Approach to Course Timetabling, *Journal of Operational Research* (*YUJOR*), (in press).

Burke, E.K., Hart, E., Kendall, G., Newall, J.P., Ross, P., and Schulenburg, S. (2003), Hyper-Heuristics: An Emerging Direction in Modern Search Technology, Glover and Kochenberger (eds.), Chapter 16, Handbook of Meta-Heuristics, pp. 457–474, Kluwer 2003e.

Burke, E.K., Kendall, G., and Soubiega, S., A Tabu-Search Hyper-Heuristic for Timetabling and Rostering, *Journal of Heuristics*, Vol. 9(5), (in press).

Burke, E.K. and Newall, J.P., Enhancing Timetable Solutions with Local Search Methods, E.K.Burke and P. De Causmaecker (eds.), The Practice and Theory of Timetabling IV, *Selected Revised Papers from the 4th International Conference on the Practice and Theory of Automated Timetabling*, Gent, August 2002, *Springer Lecture Notes in Computer Science* Vol. 2740, Springer (in press).

Burke, E.K. and Petrovic, S., Recent Research Directions in Automated Timetabling, *European Journal of Operational Research — EJOR*, 140(45.2), 2002a, pp. 266–280.

Burke, E., Petrovic, S., and Qu, R., Case-Based Heuristic Selection for Examination Timetabling, In: *Proceedings of the 4th Asia-Pacific Conference on Simulated Evolution and Learning — SEAL'02*, Singapore, November 18-22, 2002b, ISBN: 981-04-7523-3, pp. 277–281.

Burke, E.K. and Newall, J.P., Solving Examination Timetabling Problems through Adaptation of Heuristic Orderings, University of Nottingham, School of Computer Science and IT Technical Report NOTTCS-TR-2002-2, *Annals of Operations Research*, (in press).

Burke, E., Bykov, Y., and Petrovic, S., A Multicriteria Approach to Examination Timetabling, In: *Selected Papers from the 3rd International Conference on the Practice and Theory of Automated Timetabling* (*PATAT 2000*), Konstanz, Germany, 16–18 August 2000, *Springer-Verlag Lecture Notes in Computer Science*, Vol. 2079, E. Burke and W. Erben, (eds.), 2001a, pp. 118–131.

Burke, E. and Erben, W., (eds.), The Practice and Theory of Automated Timetabling III: *Selected Papers from the 3rd International Conference on the Practice and Theory of Automated Timetabling*, University of Applied Sciences, Konstanz, August 16–18, 2000, *Springer Lecture Notes in Computer Science Series*, Vol. 2079, 2001b.

Burke, E., MacCarthy, B., Petrovic, S., and Qu, R., Case-Based Reasoning in Course Timetabling: An Attribute Graph Approach, In: *Case-Based Reasoning Research and Development, 4th. International Conference on Case-Based Reasoning, ICCBR-2001*, Vancouver, Canada, 30 July–2 August 2001, *Springer-Verlag Lecture Notes in Artificial Intelligence*, Vol. 2080, D.W. Aha, I. Watson and Q. Yang, eds., 2001c, pp. 90–104.

Burke, E., MacCarthy, B., Petrovic, S., and Qu, R., Multi-Retrieval in a Structured Case Based Reasoning Approach for Course Timetabling Problems, Technical Report NOTTCS-TR-2001-7, University of Nottingham, School of Computer Science and IT, 2001d.

Burke, E., Bykov, Y., Newall, J., and Petrovic, S., A Time-Predefined Local Search Approach to Exam Timetabling Problems, Nottingham School of Computer Science and IT Technical Report 2001-6, *IIE Transactions on Operations Engineering*, (in press).

Burke, E., MacCarthy, B., Petrovic, S., and Qu, R., Structured Cases in CBR — Re-using and Adapting Cases for Time-Tabling Problems, *Knowledge-Based Systems* 13 (2-3), 2000, pp. 159–165.

Burke, E.K. and Newall, J.P., A Multi-Stage Evolutionary Algorithm for the Timetable Problem, *IEEE Transactions on Evolutionary Computation*, 3(1), 1999, pp. 63–74.

Burke, E. and Carter, M., (eds.), The Practice and Theory of Automated Timetabling II: *Selected Papers from the 2nd International Conference on the Practice and Theory of Automated Timetabling*, University of Toronto, August 20–22, 1997, *Springer Lecture Notes in Computer Science Series*, Vol. 1408, 1998.

Burke, E., Kingston, J., Jackson, K., and Weare, R., Automated University Timetabling: The State of the Art, *The Computer Journal* 40(9) 1997, pp. 565–571.

Burke, E.K., Elliman, D.G., Ford, P., and Weare, R.F., Examination Timetabling in British Universities — A Survey. In (Burke and Ross, 1996b), 1996a, pp. 76–92.

Burke, E. and Ross, P., (eds.), The Practice and Theory of Automated Timetabling: *Selected Papers from the First International Conference on the Practice and Theory of Automated Timetabling*, Napier University, August/September 1995, *Springer Lecture Notes in Computer Science Series*, Vol. 1153, 1996b.

Carter, M.W., A Survey of Practical Applications of Examination Timetabling, *Operations Research*, 34, 1986, pp 193–202.

Carter, M.W., A Decomposition Algorithm for Practical Timetabling Problems, Technical Paper 83-06, Department of Industrial Engineering, University of Toronto, 1983.

Carter, M.W. and Laporte, G., Recent Developments in Practical Examination Timetabling, in (Burke and Ross, 1996b), 1996a, pp. 3–21.

Carter, M.W., Laporte, G., and Lee, S.Y., Examination Timetabling: Algorithmic Strategies and Applications, *Journal of the Operational Research Society* 47(3), 1996b, pp. 373–383.

Carter M.W. and Laporte G., Recent Developments in Practical Course Timetabling, in (Burke and Carter, 1998), 1998, pp. 3–19.

Carter, M., Timetabling, In: Gass, S. and Harris, C., (eds.), *Encyclopedia of Operations Research and Management Science*, Kluwer Academic Publishers, 2001, pp. 833–836.

Cunningham, P. and Smyth, B., Case-Based Reasoning in Scheduling: Reusing Solution Components, *The International Journal of Production Research*, 35, 1997, pp. 2947–2961.

de Werra, D., An Introduction to Timetabling, *European Journal of Operational Research*, 19, 1985, pp. 151–162.

Dimopoulou, M. and Miliotis, P., Implementation of a University Course and Examination Timetabling System, *European Journal of Operational Research*, 130(1), 2001, pp. 202–213.

Dueck, G., New Optimization Heuristics, The Great Deluge Algorithm and the Record to Record Travel, *Journal of Computational Physics*, 104, 1993, pp 86–92.

Fayyad, U., Piatetsky-Shapiro, G., Smyth R., and Uthurusamy (eds.), *Advances in Knowledge Discovery and Data Mining*, AAAI Press, Melo Park, CA, 1996.

Glover, F., *Tabu Search*, Kluwer Academic Publishers, 1998.

Kolodner, J., *Case-Based Reasoning*, Morgan-Kaufmann, 1993.

Koton, P., SMARTPlan: A Case-Based Resource Allocation and Scheduling System, In: *Proceedings: Workshop on Case-Based Reasoning (DARPA)*, Hammond, K., (ed.), Pensacola Beach, Florida, San Mateo, CA:Morgan Kaufmann, 1989, pp. 285–289.

Leake, D. (ed.), *Case-Based Reasoning, Experiences, Lessons & Future Directions*, AAAI Press, 1996.

Lotfi, V. and Cerveny, R., A Final Exam-Scheduling Package, *Journal of the Operational Research Society*, 42(3), 1991, pp. 205–216.

Messmer BT., Efficient Graph Matching Algorithms for Preprocessed Model Graph, PhD thesis, University of Bern, Switzerland, 1995.

Miyashita, K. and Sycara, K., CABINS: A Framework of Knowledge Acquisition and Iterative Revision for Schedule Improvement and Reactive Repair, *Artificial Intelligence*, 76, 1995, pp. 377–426.

Paechter, B., Cumming, A., and Luchian, H., The Use of Local Search Suggestion Lists for Improving the Solution of Timetabling Problems with Evolutionary Algorithms, In: Fogarty, T.C., (ed.), Evolutionary Computing, Timetabling Section, *Springer Lecture Notes in Computer Science Series*, Vol. 993. 1995, pp. 86–102.

Paechter, B., Cumming, A., Norman, M.G., and Luchian, H., Extensions to a Memetic Timetabling System, in (Burke and Ross, 1996b), 1996, pp. 251–266.

Petrovic, S., Beddoe, G., and Vanden Berghe, G., Storing and Adapting Repair Experiences in Personnel Rostering (in press).

Petrovic, S. and Bykov, Y., A Multiobjective Optimisation Technique for Exam Timetabling Based on Trajectories (in press).

Petrovic, S. and Qu, R., Case-Based Reasoning as a Heuristic Selector in a Hyper-Heuristic for Course Timetabling Problems, *Proceedings of the Sixth International Conference on Knowledge-Based Intelligent Information Engineering Systems and Applied Technologies*, KES'02, Vol. 82., Milan, Italy, September 16–18, 2002, pp. 336–340.

Petrovic, S. and Petrovic, R., Eco-Ecodispatch: DSS for multicriteria loading of thermal power generators, *Journal of Decision Systems*, 4(4), 1995, pp. 279–295.

Riesbeck, C. and Schank, R., *Inside Case-Based Reasoning*, Lawrence Erlbaum Associates Publishers, 1989.

Reeves, C., Modern Heuristic Techniques, In: Rayward-Smith V.J., Osman I.H., Reeves, C.R., and Smith G.D., (eds.), *Modern Heuristic Search Methods*, John Wiley & Sons Ltd, 1996.

Robert, V. and Hertz, A., How to Decompose Constrained Course Scheduling Problems into Easier Assignment Type Subproblems, in: (Burke and Ross, 1996b), pp 364–373, 1996.

Rong, Q., Case-Based Reasoning for Course Timetabling Problems, PhD thesis, School of Computer Science and Information Technology, University of Nottingham, 2002.

Schaerf, A., A Survey of Automated Timetabling, *Artificial Intelligence Review*, 13/2, 1999, pp. 87–127.

Schmidt, G., Case-Based Reasoning for Production Scheduling, *International Journal of Production Economics*, 56–57, 1998, pp. 537–546.

White, G.M., Constrained Satisfaction, Not So Constrained Satisfaction and the Timetabling Problem, A Plenary Talk in the *Proceedings of the 3rd International Conference on the Practice and Theory of Automated Timetabling*, University of Applied Sciences, Konstanz, August 16–18, 2000, pp. 32–47.

White, G.M. and Chan, P.W., Toward the Construction of Optimal Examination Timetables, *INFOR* 17, 1979, pp. 219–229.

Wren, A., Scheduling, Timetabling and Rostering — A Special Relationship?, in (Burke and Ross, 1996a), 1996, pp. 46–75.

Zeleny, M., A Concept of Compromise Solutions and the Method of Displaced Ideal, *Computers and Operations Research*, 1(4), 1974, pp. 479–496.

46

Adapting the GATES Architecture to Scheduling Faculty

R. P. Brazile
University of North Texas

K. M. Swigger
University of North Texas

46.1 Introduction

Scheduling problems have been studied in both the operations-research (OR) and artificial-intelligence (AI) literature for many years. In OR literature, the scheduling problem has been described as a mathematical formulation in which the problem solver attempts to maximize or minimize the value of an objective function subject to certain constraints [1]. In contrast, AI researchers have used knowledge-based approaches coupled with a set of heuristics derived from an expert's experience and knowledge to guide the program toward an acceptable solution [2]. Like all debates, the question of whether one technique is better than another is subject to interpretation, depending on whether one is more familiar with OR or expert-system techniques. Regardless of one's research orientation, however, the search for a general framework that will solve all scheduling problems continues to be a Holy Grail for the scheduling community. While the authors do not presume to have found the holy relic, we do offer a first attempt toward providing some basic approaches that might develop into a meta-framework for scheduling problems. Having developed a solution to the problem of scheduling planes at gates, we apply our techniques to scheduling teachers to classes. Each of the problems elucidates notions of constraints (natural extension of constraint definition by a specific preference), preferences, satisfaction degrees, solutions, and degrees

1-58488-397-9/$0.00+$1.50

of consistencies. Therefore, one of our objectives is to suggest general characteristics of constraint-based approaches with preferences in the area of timetabling and present a metaframework for representing such problems.

The following chapter, therefore, presents the metaframework for scheduling problems and reports on our experiences with developing a solution to several real-world problems. After a short introduction, including a description of our particular framework for solving constraint satisfaction problems, we describe how to apply our technique to a second, related problem. Since we intend to give a unifying framework to particular approaches, we describe how the original framework can accommodate the new but related problem. Our goal is to compare different techniques, and develop criteria and a metaframework that can accommodate several problems. In the next section, we provide a brief overview of the original gate assignment scheduling program. After the basic description of the *gate assignment and tracking expert system* (GATES) framework, we continue with its application to a new domain. This is followed by a discussion of other models and frameworks that are used to develop expert systems in this area. In doing so, we characterize some of the advantages and disadvantages of the various models and show why our particular approach can be applied to other similar problems. In our conclusion section, we try to outline some general characteristics of the problem that may indicate why our approach is preferable to other techniques.

46.2 Background

In the summer of 1986, the authors were asked to investigate a problem for *trans world airlines* (TWA). The problem consisted of scheduling incoming and outgoing airline flights to gates at the JFK airport in New York city [3]. This particular problem falls into the more general class of problems, known as constraint satisfaction problems, in which the goal is to allocate resources, given a set of specific constraints. Because of the number and kind of scarce resources, this class of problems often does not have an obvious solution. For example, the gate assignment problem contains a number of different types of aircraft, gates, and flights, all with their own set of constraints. Our approach, therefore, was to duplicate the human expert by developing a program that would produce a reliable schedule for the planes.

The resulting program called GATES was a constraint satisfaction expert system developed to create TWA's monthly and daily gate assignments at JFK and St. Louis airports. Obtained from an experienced ground controller, the domain knowledge was represented by Prolog predicates as well as several rule like data structures, including preference rules (The GATEOK predicates) and denial rules (the Conflict predicates). These two types of rules determined when a set of gates could or could not be assigned to a particular flight. The system operators would modify schedules by retracting rules, adjusting tolerances and deleting information.

The system used the following procedures to produce monthly gate assignments:

1. Consider an unassigned flight that has the most constraints first (a set of FLIGHT rules).
2. Select a particular gate for a particular flight by using a set of GATEOK rules that have been arranged in some priority.
3. Verify that the gate assignment is correct by checking it against a set of CONFLICT rules.
4. Make adjustments by relaxing constraints so that all flights are assigned to gates.
5. After all assignments are made, adjust assignments to maximize gate utilization, minimize personnel workloads, and maximize equipment workload.

The GATES system monitored itself in three passes:

Pass 1. Use of all the constraints: Try to schedule all the planes. With all the constraints in use, usually only about 75% of the planes get scheduled, and the system moves to Pass 2. (If successful in scheduling all planes, the system still goes to Pass 2.)

Pass 2. Relax the constraints so that all of the planes are scheduled, and move to Pass 3.

Pass 3. Put back the constraints in an attempt to improve the schedule.

All passes recurse through passes 1, 2, and 3. Thus, passes 1, 2, and 3 act as a type of subprogram that is called during every phase of the program.

As previously stated, GATES was used in production mode by personnel at JFK and St. Louis airports to schedule airline flights to airport gates after the planes have landed. GATES was also used in a human benchmarking experiment that compared human performance to expert system performance [4]. In this study, human subjects were asked to schedule a preselected number of flights to gates. The scheduling task included a plan to follow to do the task, the constraints (restrictions) to use in doing the task, a worksheet and extra blank paper, and the scheduling task itself. While the human performance varied, it was clear that successful schedulers used similar techniques to solve the problem, and that those techniques were similar to the way the expert system solved the same problem. Similar to the expert system, humans organized the constraints into several different categories to help them in scheduling flights. Humans also applied various procedures to insure that the constraints were satisfied while making the gate assignments.

After completing the human subjects study, we began to explore the notion that the GATES framework could be applied to other engineering and industrial problems where limited resources and weakly defined constraints exist. In most real-life situations, problems are expressed in terms of constraints, possibilities, probabilities, and costs. While some problems require optimal solutions, others are ill-defined and include uncertainties. Most of these problems need to include a set of preferences as a list of facts or as part of the control structures. Furthermore, constraints are often organized into several different categories to assist in the design of the system, and production rules are used to insure that the constraints are satisfied. The remaining part of this chapter illustrates how these same procedures are applied to a second, scheduling problem.

46.3 Teacher Assignment Problem

The teacher assignment problem can be viewed as a special case of scheduling called timetabling [5]; that is, the allocation, subject to constraints, of given resources to objects being placed in space-time, in such a way as to satisfy as nearly as possible a set of desirable objectives [6]. The specific instance of timetabling described in this section focuses on assigning all classes that are scheduled within a given semester to the available faculty members of a specific department, in such a way as to satisfy the constraints on the number of classes that each faculty must teach as well as their preferences. This is unlike the timetabling problem discussed in [7], which is concerned with scheduling faculty as well as the courses themselves. In the problem discussed in this chapter, we are given the schedule of classes and must assign teachers to the classes (or classes to the teachers, as there are more classes than teachers). Thus the problem consists of assigning teachers to courses while maximizing satisfaction of preferences and number.

For this particular problem, we assumed that the standard teaching load for a teacher was two classes each semester. However, individual instructors may teach fewer classes, depending on their status or circumstances. For example, sometimes an instructor may be assigned additional administrative duties, such as graduate coordinator, and will be assigned a fewer number of courses. In this case, the teaching load is normally reduced by one class each semester. On the other hand, some faculty members are designated as teaching faculty and, as a result, are assigned more than two classes per semester.

In addition to determining the number of classes that each faculty member is assigned, the system must also try to satisfy different faculty and departmental preferences. For those faculty members who are given a two-class assignment, the department *prefers* that the instructors teach one graduate and one undergraduate course; although sometimes this is not possible. Furthermore, most instructors prefer to teach their assigned classes on the same days, that is Monday–Wednesday or Tuesday–Thursday. Some faculty *prefer* to teach evening classes, while others *prefer* to teach classes only during the day. Since everyone is allowed to submit a list of preferred classes, ranked according to their preferences, the system can use this information to determine a teacher's class assignments. One of the major goals of the program is to give each person their first choice. If this is not possible, then the system should use a set of priorities to make appropriate assignments.

46.4 Solution Using the GATES Structure

The aim of this section is to present a generalized knowledge-based approach to the teacher assignment problem, which is often derived in an ad hoc manner. Using the basic structure outlined in the above section, and in the original GATES paper [3], it is possible to derive a solution to the teacher assignment problem using the procedures developed for GATES. For example, a set of constraints, together with the logic for these constraints, can be easily defined for the teacher assignment problem. Rules can be developed in Prolog to implement the solution process that mirror many of the same rules developed for GATES. Our experience in constructing the GATES system informs us about the type of constraints that are needed in the teacher assignment problem, and how to organize the rules that will enforce those constraints. We further suggest that the entire process can be completed in multiple passes. The first pass should consist of matching teachers to classes that do not violate any constraint. If the schedule is still incomplete, a second pass can be called in which rules are retracted or tolerances relaxed to assign all previously unassigned classes. This process can continue until all teachers have been assigned their required number of classes.

In the previous work a set of constraints were characterized as either hard, those that cannot be violated, or soft, those that may sometimes be violated. In the GATES system, we divided the set of soft constraints even further by creating a number of subcategories of constraints. These various categories and subcategories serve to help organize the existing set of constraints and allow a system designer to identify any missing constraints that may have been overlooked. A more complete description of the various categories and subcategories of constraints is given below.

46.4.1 Constraints without Exceptions

This type of constraint represents a *hard* rule, meaning that it can never be relaxed or violated. Often these types of constraints characterize some physical property such as size, weight, or height. For example, the following constraint represents the fact that a teacher cannot be in two places at the same time.

C1: A faculty member may not be assigned two classes whose meeting times overlap.

46.4.2 Constraints with Exceptions

These types of constraints represent requirements that should be met, but can be made optional if the program fails to produce a valid schedule. In the GATES system, these types of constraints were retracted after the first pass if planes were left unassigned. The following example illustrates an exception constraint for the teacher assignment problem. The semantics of this constraint indicate that a faculty member should normally be assigned to both a graduate and undergraduate course, unless there are no more courses to assign in that category.

C2: A faculty member should be assigned one graduate and one undergraduate class, if the number to be assigned is two. An exception to this rule is allowed if the department schedule needs it.

46.4.3 Constraints with Changing Tolerances

Some of the constraints contain a number that represents a tolerance level that is maintained while achieving a particular goal or subgoal. A tolerance is defined as a particular *optimal* time interval that may be reduced as the problem proceeds through the search space. The program is given some flexibility in assigning the different levels of tolerance. For example, the tolerances can be set high for the first pass, and then relaxed during the second pass in order to achieve a completed schedule. The following example illustrates a rule with *changing tolerances.*

C3: Assign a person classes with at least one class period between the assigned classes. The tolerance is 1. This may be relaxed to 0, if necessary to make a good schedule.

46.4.4 Guidelines

A guideline is a type of constraint that represents a request for a *better* schedule. Under ideal conditions, the program should produce a schedule that would follow all the program's guidelines. However, this is not always possible. When a guideline cannot be met, it can be ignored. Although similar to constraints with exceptions, guidelines are different in that they can be ignored at any time, as opposed to being relaxed under certain circumstances. An example of a guideline is the faculty preference for having courses scheduled on the same days. This particular constraint would look like the following:

C4: A faculty member assigned two classes should have both classes on the same days.

46.4.5 Convenience Constraints

Convenience constraints specify the program/user's desirable preferences. This is the weakest constraint category in the system. The program tries to satisfy the convenience constraints, but it is not required to meet these requirements to produce an acceptable schedule. The example below illustrates a constraint that captures a user's preference for teaching night classes.

C5: If a person prefers teaching in the evening, assign that person an evening class.

46.5 Production Rules

The original GATES expert system was developed using the Prolog programming language to guide the assignment process. Production rules, in the form of Prolog predicates, checked each assignment so that it did not violate any constraints. Prolog works by checking each predicate in a rule to make sure that it is true before proceeding to the next predicate. Once all predicates in a rule are satisfied (i.e., true), then the rule *fires*. In the case of the original GATES program, a rule that fires means that a plane is assigned to a gate. In the case of the teacher assignment problem, a rule that fires means that a teacher is assigned to a specific course. The rules are designed in such a way that the system continues to assign classes to instructors until all instructors have been assigned the required number of classes, or there are no more assignments that can be made with the current set of constraints and tolerances. This latter condition indicates that the program has completed the first pass over the data. At this point, constraints are retracted or tolerances relaxed, and the system continues to assign classes. This process can be done automatically, or it may be interrupted by live-user who adjusts tolerances or constraints through a user interface. Regardless of which method is used, the system continues to try and assign teachers to courses until it can no longer match teachers to classes.

To make the various course assignments, the system uses two different types of predicates, permissive rules and conflict rules. A discussion of these two types of predicates now follows.

46.5.1 Permissive Rules

Permissive rules, as the name implies, allow assignments to be made. These rules also verify that all constraints currently in effect are met. Permissive rules determine the order in which various types of assignments will be considered, as well as the priority within the different assignment category levels. For example, the following rule can be used to give all teachers their first choice of classes, before all other assignments are tried.

R1: If a person has asked for a class as their first preference and no one else has requested that class, then assign the class to the person.

46.5.2 Conflict Rules

These rules specify when an assignment cannot be made. The conflict rules contain many of the constraints that were described in the previous section. An example of a conflict rule is one that prevents an assignment. For example, the following rule ensures that only one teacher will be assigned to a single course.

R2: Do **not** *assign the same class to two people.*

46.5.3 Priorities

The program can prioritize the way it assigns the teachers to classes by either ordering the rules or the predicates within the rules. For example, if one of the major goals of the system is to give teachers everything they want, then the developer can place the permission rules before the conflict rules. Priorities can also be set by adding explicit rules that force the system to consider certain characteristics before others. For example, we can assign tenured faculty to classes before untenured faculty through the following rule.

R3: Assign permanent tenure track faculty first.

46.6 Course Assignment

The process of assigning teachers to classes is made when no permissive rule, conflict rule, or tolerance level is violated. If the program can accomplish the goal of assigning all teachers to classes within the first pass, then the program terminates. If some teachers have not been assigned classes, then the program should deactivate optional constraints (such as guidelines and convenience constraints), and reduce noncritical tolerance levels and repeat the assignment process. At this point, the program may also need to retract some existing assignments to create a valid schedule. Given a sufficient number of classes, the system will eventually produce a schedule with all teachers assigned to classes.

46.6.1 Data Structures

While the original GATES had only two major predicates, we anticipate the need for at least four predicates for the teacher assignment program. One predicate is needed to denote the available classes; a second to represent the teacher and teacher load; a third to represent the teacher's preferences; and a fourth to store the actual class assignments.

The initial data for the program can be stored in two predicates: *available* and *teacher*. The *available* predicate would contain a list of classes for the semester, the day and meeting time. The *teacher* predicate would contain the teacher name and the number of classes that are associated with a particular teacher. For example, the *available* predicate might be defined as follows:

Available(course name, meeting day, start time, end time).

Very simply, the *available* predicate would be used to describe the classes that need to be assigned. The *course name* would be the variable for the course; *meeting day* would be the sequence of days that the class meets (e.g., mw means Monday and Wednesday); *start time* would refer to the time that the course starts; and *end time* would be the time that the class ends. After a class is assigned, it is removed from the available list.

The *teacher* predicate would contain the names of the various faculty, along with the number of courses that they are required to teach. For example, the teacher predicate can be defined as

teacher(teacher name, required classes).

```
available(5290.001, tth, 1530, 1650).
available(5250.001, tth, 1930, 2050).
available(5330.002, mw, 1600, 1720).
available(5330.003, mw, 1230, 1350).
available(5420.001, mw, 1400, 1520).
available(5450.001, mw, 1230, 1350).
available(5520.001, mw, 1930, 2050).
available(6330.001, mw, 1530, 1650).
available(6330.003, th, 1500, 1750).
available(5350.001, tth, 1800, 1920).
teacher(brazile, 2).
teacher(swigger, 1).
teacher(mikler, 2).
teacher(akl, 2).
preference(brazile, 5350.001, 1).
preference(swigger, 5290.001, 1).
```

FIGURE 46.1 Example of input data.

```
assignall:-teacher(T,X),X>0,available(C,_,_,_),preference(T,C,1),
        nodup(C,1),
        asserta(assigned(T,C)),
        retract(teacher(T,X)),Y=X−1,
        asserta(teacher(T, Y)),
        retract(available(C,_,_,_)),
        asserta(why(T,C,1,"top priority"),
        fail.
```

FIGURE 46.2 Example of an assignment rule.

If the developer decides to give class preferences to faculty members who are tenure track, an additional field can be added to the teacher predicate to represent this particular information. Alternatively, the user might want to define this fact in a separate predicate such as *tenuretrack(teacher name)* to indicate tenured faculty status.

Faculty preferences can be represented through the preference predicate. This predicate can be defined as

preference(teacher name, course name, rank)

where *teacher name* indicates the teacher for the course, *course name* is the course, and *rank* specifies a number that indicates the degree of preference for that particular course (e.g., 1 is high and 5 is low). Faculty members would be encouraged to provide a list of several preferences in case the initial request got denied. Below is an example of the different input predicates, as well as specific information for each of the variables identified in the predicates (ssee Figure 46.1).

Once the system finds a match for a teacher with a course, it stores this information in the *assigned* predicate. The assigned predicate is defined as

assigned(teacher name, course name).

where *teacher name* is the name of the teacher, and *course name* is the name of the course. This predicate stores the program's output, which is the schedule of teacher assignments. The final output is the list of teachers along with their assigned classes. This predicate is asserted into the fact base whenever one of the permissive rules fires. An example of this process (see Figure 46.2).

46.7 Reassignment of Faculty

Once a master schedule is produced, it may require some changes. For example, a professor may get sick and be unable to teach a class. Or, a professor may need to alter a teaching assignment because of a sudden commitment. The program can be modified very easily to accommodate reassignments. For example, if there are an insufficient number of students enrolled in a class, then the professor will need to be reassigned to another class. If this occurs, then the system will need to assign the teacher to a previously unassigned class. The program's initial input is the formerly completed schedule, minus the class (or group of classes) that is no longer offered. The input should also contain the now unassigned teacher (or group of teachers). The system should be able to start with this initial input and reschedule the teacher(s) using only the rules that are associated with the constraints without exceptions, since at this point the unassigned teacher must relinquish any preferences.

46.8 User Interface

The original user interface for the GATES had two major purposes: (i) to query the expert system for specific information, (ii) to ask the system why it derived a specific answer. A similar user interface would be appropriate for the teacher-assignment problem. A user would be allowed to ask for information about both assigned classes and teachers. The user should also be allowed to ask why a specific teacher was assigned a class. If a user is dissatisfied with the solution, then they should be allowed to adjust the tolerances or retract rules and reschedule. An example of a simple interface for the teacher assignment expert system is shown in Figure 46.3.

Most of the menu selections are obvious. The command "c" allows the user to enter a class identifier and receive information about which teacher was assigned a particular class. The command "i" allows the user to enter an instructor's name and find out a teacher's assigned classes. The commands "a" and "u" allow the user to add or delete an instructor's class assignment. The "d" command permits the user to prevent a teacher from being assigned a specific class. The "n" command displays a list of instructors who were not assigned classes. The command "p" prints the complete list of assignments. The "w" command allows the user to ask why a specific assignment was made. That is, it displays a list of rule(s) that permitted the assignment to be made. The "x" command is used to ask why various assignments were **not** made, which displays the list of *conflict* rules that prevented the assignment from being made.

The "t" command would allow the user to change the tolerance values. Generally the user wants to decrease tolerances to their lowest levels so that a full assignment may be obtained. But higher tolerances

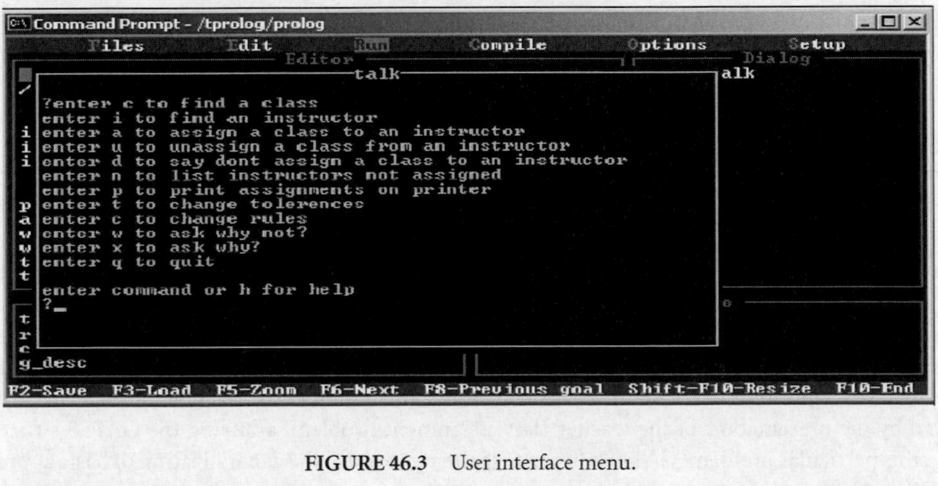

FIGURE 46.3 User interface menu.

can be enforced whenever the goal is to create a *better* schedule. Finally, the "c" command allows the user to retract rules and try to reschedule the unassigned teacher.

46.9 Related Research

Early research on scheduling problems (CSPs) can be traced to the seventies [8–10], but widespread applications did not really begin until the work in expert systems and logic programming[11–13]. Expert systems' programming has been applied to a wide range of scheduling problems including those related to managing and automating processes [14], personnel and military campaigns [15], and robot planning [16]. A special case of scheduling is timetabling, the allocation of resources subject to specific time constraints [6]. Timetabling problems may be solved by different methods inherited from operational research such as graph coloring and artificial intelligence programming such as search procedures, simulated annealing, and genetic algorithms [17]. Our concentration has been on methods related to constraint programming approaches.

Constraint satisfaction together with constraint logic programming has been applied to a wide range of timetabling problems starting with research in the nineties[18–21]. Constraint logic programming is particularly well suited for timetabling problems because it allows for the formulation of the problem in a more direct way than other approaches [22,23]. Suitable representation of the problem by the variables and their domains can greatly influence the way constraints can be formulated and the overall computational effectiveness of the program. Basically, search within the solution space is performed using some type of heuristic or other nonsystematic method. Within the constraint satisfaction programming approach, preferences for selecting feasible solutions are implemented with the help of some heuristics[7,19]. For example, the authors in [7] use constraint logic programming to attack the timetabling problem for universities. They state, "An essential component is an automatic heuristic solution search with an interactive user-intervention facility. The user will, however, only be able to alter a timetable or schedule in such a way that no hard constraints are violated." This approach is similar to what is used in GATES and suggested in this chapter. The automatic heuristic solution portion of the GATES' search is implemented by considering the priority of the various assignments under consideration and then ordering the goal predicates using priority information. Soft constraints are automatically retracted or modified, but the user may intervene and retract or modify their own set of the constraints.

Jackson, Havens, and Dollard [24] and Doan Domingos, and Halevy [25] also define a set of soft constraints, but they include a cost for violating the different constraints in the domain. Final value with minimal cost drives the course selection process. This is similar to the *constraints with tolerances* that are described in this chapter.

Finally, in [5] the author expresses the need to apply some type of *preferences* that have to be included into both declarative and control part of the solution. We have included preferences in both the GATES and teacher-assignment problem as a form of *soft* constraint that should be satisfied whenever it is possible. These preferences can be relaxed or extended during the program's multiple passes through the data.

46.10 Conclusion

Using the GATES architecture, we have proposed a unifying structure for dealing with scheduling problems, particularly timetabling issues. The proposed architecture helps in defining the initial constraints and rules by providing a framework that encourages the recognition of appropriate constraints and preferences. The structure provides a natural extension of constraint definition, preferences, satisfaction, and consistency degrees.

After an initial introduction of GATES and its history, we presented research in the area of scheduling, including a description of particular systems related to timetabling and class scheduling. This was then followed by the presentation of the teacher class assignment problem, applying the GATES structure to solving this particular problem. GATES offers a first attempt toward the generalization of basic approaches through problem relaxation. A number of constraint categories were introduced that were intended to deal with similar types of limitations found within these two domains. More specifically, examples of hard and soft constraints were described, and these were shown to be effective in producing a solution. This led to a discussion of rule types and how these impact the scheduling process. Priorities and tolerances were introduced to help guide the scheduling task and produce better results. The overall problem was then characterized as a set of requirements that help match classes to teachers while taking into account preferential requirements of teachers toward any acceptable solution.

Finally we compared our structure with other approaches and solutions of comparable problems. The proposed structure seems to incorporate many of the features that are being expressed in the current literature on constrained logic programming. We believe that our proposed structure is comparable to other methods while preserving the advantage of being both fast as well as effective in solving a problem.

References

[1] R. Bellman, et al. *Mathematical Aspects of Scheduling and Applications*, Pergamon Press, Elmsford, NY, 1982.

[2] B.R. Fox and K.G. Kempf. A representation for opportunistic scheduling. In *Proceedings of Third International Symposium for Robotic Research*, MIT Press, Cambridge, MA, pp 111–117, 1986.

[3] R.P. Brazile and K.M. Swigger. GATES: An airline gate assignment and tracking expert system. *IEEE Expert*, 33–39, 1988.

[4] H. O'Neil, E. Baker, Y. Ni, A. Jacoby, and K. Swigger. Human benchmarking for the evaluation of expert systems. In H. O'Neil, E.L. Baker (editors), *Technology Assessment in Software Applications*, Lawrence Erlbaum, Hillsdale, NJ, pp. 13–45, 1994.

[5] H. Rudova. Constraint satisfaction with preferences. *PhD Thesis*, Faculty of Informatics, Masaryk University, Brno, Czech Republic, Jan. 2001.

[6] A. Wren. Scheduling, timetabling and rostering — a special relationship? In E. Burke and P. Ross (editors), *Practice and Theory of Automated Timetabling*, Springer-Verlag LNCS 1153, pp. 46–75, 1996.

[7] H.-J. Goltz, G. Küchler, and D. Matzke. Constraint-based timetabling for universities. In: *Proceedings of INAP'98, 11th International Conference on Applications of Prolog*, Tokyo, pp. 75–80, Sep. 1998.

[8] U. Monatanari. Networks of constraints: Fundamental properties and application to picture processing. *Information Sciences*, 7(2):95–132, 1974; Technical Report, Carnegie Mellon University, 1970.

[9] A. K. Mackworth. Consistency in networks of relations. *Artificial Intelligence*, 8(1):99–118, 1977.

[10] E. C. Freuder. Synthesizing constraint expressions. *Communications ACM*, 21(11):958–966, 1978.

[11] R. A. O'Keefe. *The Craft of Prolog*, MIT Press, 1990.

[12] D. Maier and D.S. Warren. *Computing with Logic, Logic Programming with Prolog*, The Benjamin/Cummings Publishing Company, Inc., 1988.

[13] D. M. Gabbay, C.J. Hogger, and J.A. Robinson (editors). *Handbook of Logic in Artificial Intelligence and Logic Programming*, vol. 1, *Logical Foundations*, Clarendon Press, Oxford, 1993.

[14] F.M. Proctor, K.N. Murphy, and R.J. Norcross. Automating robot programming in the cleaning and deburring workstation of the AMRF. In *Proceedings of the SME Deburring and Surface Conditioning*, San Diego, CA, February 1989.

[15] T. Emerson and M. Burstein. Development of a constraint-based airlift scheduler by program synthesis from formal specifications. In *Proceedings of the 1999 Conference on Automated Software Engineering*, Orlando, FL, September, 1999.

[16] T. Fukuda and O. Hasegawa. 3-D image processing and grasping planning expert system for distorted objects. In *Proceedings of IEEE International Conference on Robotics & Automation*, Vol. 4, pp. 726–731, 1989.

[17] A. Schaerf. A survey of automated timetabling. *Artificial Intelligence Review* (13): 87–127, 1999.
[18] P. Boizumault, Y. Delon, and L. Péridy. Solving a real-life planning exam problem using constraint logic programming. In Manfred Meyer (editor), *Constraint Processing: Proceedings of the International Workshop at CSAM'93*; Research Report RR-93-39, DFKI Kaiserslautern, pp. 107–112, August 1993.
[19] F. Azevedo and P. M. Barahona. Timetabling in constraint logic programming. In *Proceedings of the 2nd World Congress on Expert Systems*, 1994.
[20] M. Yoshikawa, K. Kaneko, Y. Nomura, and M. Watanabe. A constraint-based approach to high-school timetabling problems: A case study. In *Proceedings of the Sixteenth National Conference on Artificial Intelligence and the Eleventh Innovative Applications of Artificial Intelligence Conference*, AAAI Press/MIT Press, pp. 1111–1116, 1994.
[21] P. Boizumault, C. Guéret, and N. Jussien. Efficient labelling and constraint relaxation for solving timetabling problems. In P. Lim and J. Jourdan (editors), *Proceeding of the 1994 ILPS Post-Conference Workshop on Constraint Languages/Systems and their Use in Problem Modelling*, vol. 1 (Application and Modelling), pp. 116–130, November 1994.
[22] G. Lajos. Complete university modular timetabling using constraint logic programming. In E. Burke and P. Ross (editors), *Practice and Theory of Automated Timetabling*, Springer-Verlag LNCS 1153, pp. 146–161, 1996.
[23] C. Guéret, N. Jussien, P. Boizumault, and C. Prins. Building university timetables using constraint logic programming. In E. Burke and P. Ross (editors), *Practice and Theory of Automated Timetabling*, Springer-Verlag LNCS 1153, pp. 130–145, 1996.
[24] W.K. Jackson, W.S. Havens, and H. Dollard. Staff scheduling: A simple approach that worked. Technical Report CMPT97-23, School of Computing Science, Simon Fraser University, 1997.
[25] Doan, P. Domingos, A. Halevy. Reconciling schemas of disparate data sources: A machine-learning approach. *SIGMOD*, 2001.
[26] Laborie, Nuijten, "Constraint-Based Scheduling in an A. I. Planning and Scheduling Perspective", AIPS '02 Tutorial.

47

Constraint Programming for Scheduling

John J. Kanet
University of Dayton

Sanjay L. Ahire
University of Dayton

Michael F. Gorman
University of Dayton

47.1 Introduction

Scheduling has been a focal point in operations research (OR) for over fifty years. Traditionally, either special purpose algorithms or integer programming (IP) models have been used. More recently, the computer science field, and in particular, logic programming from artificial intelligence has developed another approach using a declarative style of problem formulation and associated constraint resolution algorithms

1-58488-397-9/$0.00+$1.50

for solving such combinatorial optimization problems [1–3]. This approach, termed as constraint logic programming or CLP (or simply CP), has significant implications for the OR community in general, and for scheduling research in particular.

In this chapter, our goal is to introduce the constraint programming (CP) approach within the context of scheduling. We start with an introduction to CP and its distinct technical vocabulary. We then present and illustrate a general algorithm for solving a CP problem with a simple scheduling example. Next, we review several published studies where CP has been used in scheduling problems so as to provide a feel for its applicability. We discuss the advantages of CP in modeling and solving certain types of scheduling problems. We then provide an illustration of the use of a commercial CP tool (OPL Studio®[1]) in modeling and designing a solution procedure for a classic problem in scheduling. Finally, we conclude with our speculations about the future of scheduling research using this approach.

47.2 What is Constraint Programming (CP)?

We define CP as an approach for formulating and solving discrete variable constraint satisfaction or constrained optimization problems that systematically employs deductive reasoning to reduce the search space and allows for a wide variety of constraints. CP extends the power of logic programming through application of more powerful search strategies and the capability to control their design using problem-specific knowledge [1,3].

CP involves the use of a mathematical/logical modeling language for encoding the formulation, and allows the user to apply a wide range of search strategies, including customized search techniques for finding solutions. CP is very flexible in terms of formulation power and solution approach, but requires skill in declarative-style logic programming and in developing good search strategies.

We begin with a description of the structural features of CP problems and then present a general CP computational framework. In this process, we introduce several technical CP keywords/concepts (denoted in *italics*) that are either not encountered in typical OR research or are used differently in the CP context.

47.2.1 Constraint Satisfaction Problem

At the heart of CP is the constraint satisfaction problem (CSP). Brailsford, Potts, and Smith [4] define CSP as follows: Given a set of discrete variables, together with finite domains, and a set of constraints involving these variables, find a solution that satisfies all the constraints.

In constraint satisfaction problems, variables can assume different types including: Boolean, integer, symbolic, set elements, and subsets of sets. Similarly, a variety of constraints are possible, including:

- mathematical: $C = s + p$ (completion time = start time + processing time)
- *disjunctive*: tasks J and K must be done at different times
- relational: at most five jobs to be allocated at machine R50
- explicit: only jobs A, B, and E can be processed on machine Y50

Finally, a CSP can involve a wide variety of constraint operators, such as: $=, <, >, \geq, \leq, \neq$, subset, superset, union, member, $\vee$ (Boolean OR), $\bullet$ (Boolean AND), $\Rightarrow$ (implies), and $\Leftrightarrow$ (iff). In addition to these constraints, CP researchers have developed special purpose constraints that can implement combinations of the above types of constraints efficiently. For example, the *alldifferent* constraint for a set of variables implements their pairwise inequalities efficiently through logical filtering schemes [5].

A (feasible) *solution* of a CSP is an assignment of a value from its domain to every variable, in such a fashion that every constraint of the problem is satisfied. When tackling a CSP, we may want to identify just one or multiple solutions.

[1] OPL Studio® (ILOG, Inc., Mountain View, CA).

47.3 How Does Constraint Programming Solve a CSP?

47.3.1 A General Algorithm for CSP

Figure 47.1 summarizes a general algorithm for solving a CSP. It starts with CSP formulation including defining variables, their domains, and constraints (block 1). Constraints are stored in a *constraint store.*

In block 2 of the figure, the domains of individual variables are reduced using logic-based *filtering algorithms* for each constraint that systematically reduce the domains of the variables. As the domain of a variable is reduced, each constraint that uses the variable is then activated for application of its associated filtering algorithm. This systematic process is called *constraint propagation* and *domain reduction.*

After constraint propagation and domain reduction, two possibilities arise (block 3), i.e., either a solution is found or not. If a solution is found the algorithm terminates (End1). If all solutions are required, the basic process is repeated. If no solution is found, problem *inconsistency* (the state where the domain of at least one variable has become empty) at the current stage is examined (block 4).

If inconsistency is not proven, then a search is undertaken using some search strategy for *branching* (block 6). Branching divides the main problem into a set of mutually exclusive and collectively exhaustive subproblems by temporarily adding a constraint. Branching selects one of the branches and propagates all constraints again using the filtering algorithms (block 2).

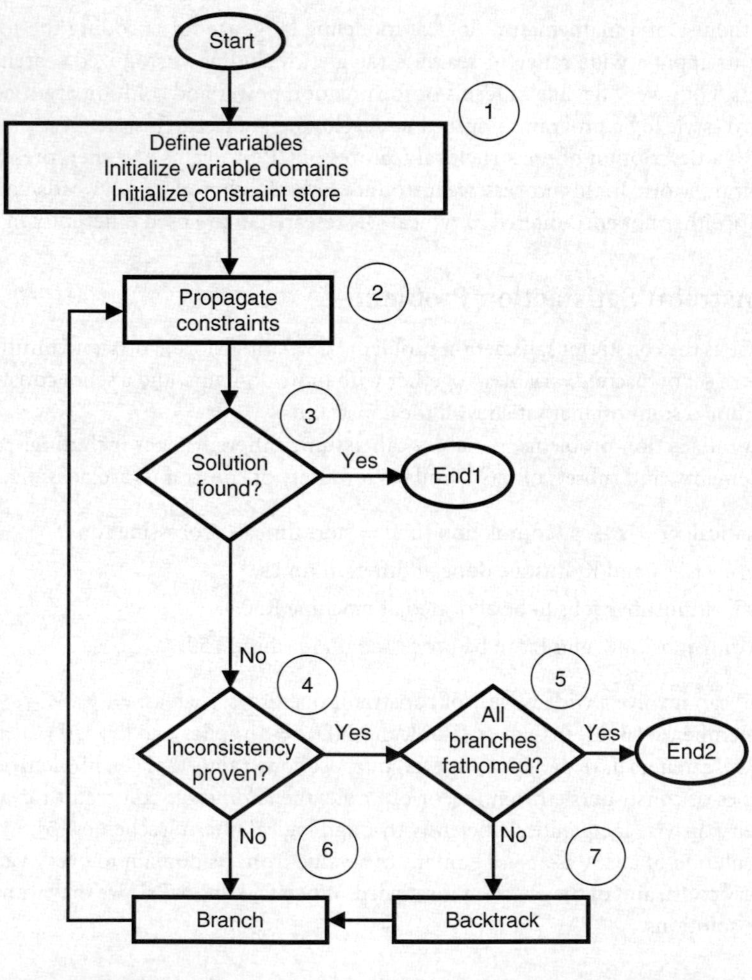

FIGURE 47.1 A general algorithm for constraint satisfaction problems (CSP).

If, in block 4, inconsistency is proven (a *failure*), the search tree is examined in block 5 to check if all subproblems have been explored (fathomed). If all branches have been fathomed, problem inconsistency is proven. If not, the algorithm *backtracks* (block 7) to the previous stage and branches to a different subproblem (block 6).

47.3.2 Constraint Propagation

47.3.2.1 Within-Constraint Domain Reduction

Constraint propagation (Figure 47.1-block 2) uses the concept of logical *arc consistency checking* to communicate information about variable domain reduction within and across constraints involving the specific variables. Figure 47.2 illustrates the concept. In this figure, part (a) shows that job 1 has a processing time of 3 (that may not be interrupted). We start with an initial domain of [0, 1, 2, 3, 4, 5, 6] for the variables C_1 (completion time of job 1) and s_1 (start time of job 1). In part (b), the constraint $C_1 = s_1 + 3$ reduces the domain of C_1 using the domain values of s_1, making arc $C_1 \leftarrow s_1$ consistent. In part (c), the constraint now reduces the domain of s_1 using the updated domain values of C_1, making arc $C_1 \rightarrow s_1$ consistent. This bi-directional arc consistency check ensures full reduction of domains of the two variables involved in this constraint.

47.3.2.2 Between-Constraint Domain Reduction

Constraint propagation also entails communication of information between constraints involving common variables to reduce their domains. Figure 47.3 illustrates the concept of constraint propagation across related constraints. In this case, the goal is to schedule two jobs, job 1 (described in Figure 47.2) and a second job (job 2) with processing time of 2 over the time domain [0, 1, 2, 3, 4, 5, 6] on a single-machine (assuming the disjunctive constraint that the machine can handle only one job at a time). In step 1, the domain of C_1, s_1, C_2, and s_2 are reduced with the help of arc consistency checking of the constraints binding C_1 to s_1 and C_2 to s_2, respectively. Let us say that there is an additional constraint on C_1 due to its due date ($C_1 < 5$). In step 2, this additional constraint reduces the domain of C_1 as shown ($C_1 = [3, 4]$). In step 3, this information about the updated domain of C_1 is communicated to the C_1 to s_1 constraint and the disjunctive constraint. The domain of s_1 is reduced as shown ($s_1 = [0, 1]$), and the domain of C_2 is reduced as shown ($C_2 = [5, 6]$). In step 4, the domain of s_2 is in turn reduced as shown ($s_2 = [3, 4]$). The sequence of steps illustrates the concept of sequential and full communication between the related constraints. Obviously, the actual sequence can be interchanged with the same final result (for example, step 3 and 4).

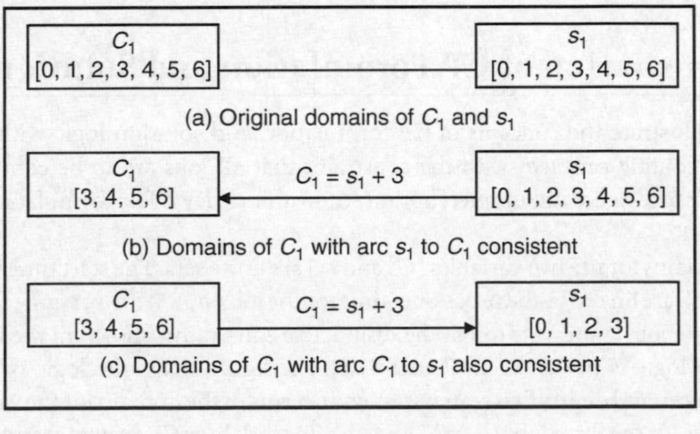

FIGURE 47.2 An illustration of domain reduction and are consistency checking logic.

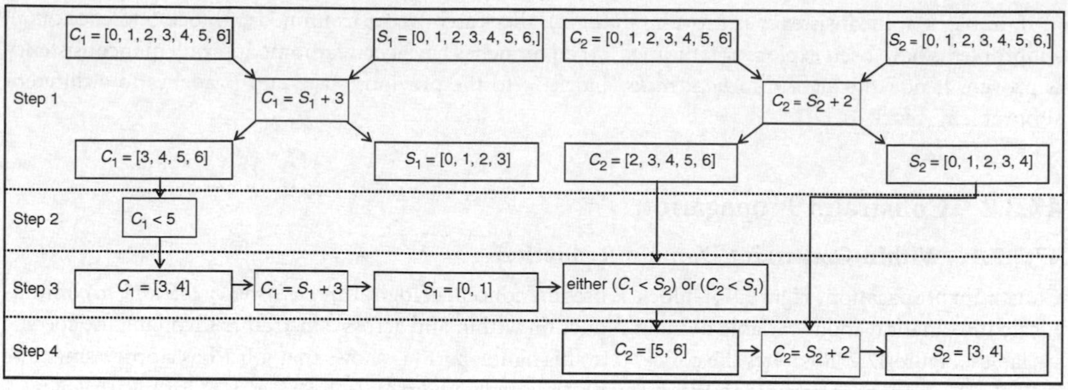

FIGURE 47.3 An illustration of constraint propagation logic.

47.3.3 Branching

Any branching strategy could be deployed in block 6 of Figure 47.1. Typically, a depth-first strategy is deployed but other more sophisticated look-ahead strategies could be used (see [4,6] for a more complete discussion). In any case, the decision of what to branch on is open as well. Often, we branch by first selecting a variable whose domain is not yet *bound* (reduced to a single value). Selection of the variable is often based on a *heuristic* (rule of thumb used to guide efficient search) such as "smallest current domain first." Once a variable is selected, the branch is established by *instantiating* the variable to one of the values in its current domain. Again, this choice could be made according to a heuristic like "smallest value in the current domain first." In fact, decision could be made to branch based on temporary constraint(s) involving one value of a variable, multiple values of the variable, multiple values of multiple variables. In CP jargon, the points at which the search strategy makes an advance along the search tree using one of these several choices are called *choice points.*

47.3.4 Adapting the CSP Algorithm to Constrained Optimization Problems

CSP problems can easily be extended to constrained optimization problems (COP). Suppose our COP has an objective Z to be minimized. If and when a first solution to the original CSP is found, its objective value is calculated (Z'), constraint $Z < Z'$ is added to the constraint store of the solved CSP to form a new CSP, and the CSP algorithm is repeated on this new problem. This process is repeated until inconsistency is found and all branches are fathomed. The last found solution is the optimum.

47.4 An Illustration of CP Formulation and Solution Logic

In this section we illustrate the concepts of CP formulation and solution logic with a simplified 3-job, single-machine scheduling problem with the constraint that all jobs are to be completed on or before their due dates and processed without interruption (time unit = day). The formulation of this problem is presented in Figure 47.4.

Note that the domains for the two variables (C_j and s_j) are finite sets. The start time (s_j) and completion time (C_j) for each job are linked by the processing time for the job (p_j). We express the fact that the machine can perform only one job at a time through the disjunctive constraints shown in the formulation.

The CP solution logic is presented in Figure 47.5. The figure maps the logic of domain reduction, constraint propagation, and depth-first search strategy in terms of the domain for the variables representing the completion times for the three jobs (C_1, C_2, and C_3). To track how CP evaluates and screens the solution space, at each step, the number of unexplored candidate values for C_1, C_2, and C_3 are provided. The values

Parameters: (nonnegative integers)
n = number of jobs
p_j = Processing Time of Job j; $j = 1,\ldots, n$
d_j = Due Date of Job j; $j = 1,\ldots, n$

Variables: (nonnegative integers)
C_j = Completion Time of Job j; $j = 1,\ldots, n$
s_j = Start Time of Job j; $j = 1,\ldots, n$

Objective: To find all schedules (combinations of $C1$, $C2$, and $C3$) that result in zero total tardiness

Variable Domains:
$p = p_1 + \cdots + p_n$
$C_j, s_j \in [0, 1,\ldots, p]$; $j = 1,\ldots, n$

Constraints:
$s_j = C_j - p_j$; $j = 1,\ldots, n$
$C_j \leq d_j$; $j = 1,\ldots, n$
$C_j \leq s_k \vee C_k \leq s_j$; $j, k = 1,\ldots, n$; $j \neq k$

FIGURE 47.4 CP formulation of a single-machine scheduling problem.

of the variables that are rendered inconsistent due to domain reduction induced through constraint propagation are shaded and labeled with the step at which they were rendered inconsistent. Finally, the bound values are shaded black and labeled with the step at which they were made. Note that, to start with, there are 343 ($7 \times 7 \times 7$) candidate assignments.

In step 1, domain reduction occurs on all the three variables individually with the constraints involving the completion time, processing time, start time, and the due date for each job. For example, because job 1 has a processing time of 3 days, it cannot be completed on days 0, 1, and 2. Also, job 1 is due on day 5. So, it cannot be completed on day 6. Note that the initial domain reduction results in 90 unexplored assignments (based on blank cells in Figure 47.5) after pruning 253 potential assignments.

Since no further domain reduction is possible and we do not have a solution, the depth-first search strategy is initiated in step 2. Using the search heuristic of "smallest current domain first", C_1 is chosen for instantiation. As soon as $C_1 = 3$ is executed, constraint propagation occurs with the effects shown on the domains of C_1, C_2 and C_3. C_1 can no longer assume values of 4 and 5. C_2 can no longer assume values of 2, 3, and 4. Finally, $C3$ can no longer assume values of 1, 2, and 3. Step 2 results in reduction of unexplored assignments from 90 to 6.

In step 3, C_2 is chosen to be instantiated again using the "smallest current domain first" heuristic. The first value of $C_2 = 5$ renders $C_2 = 6$ inconsistent, and also renders $C_3 = 4$ and $C_3 = 5$ inconsistent. Thus, in step 3, the search actually encounters the first feasible solution to this problem ($C_1 = 3, C_2 = 5, C_3 = 6$).

If we desire to identify all solutions, the depth-first search strategy now backtracks to step 2 for the next C_2 values (note that for $C_1 = 3, C_2$ could be either 5 or 6), and bounds C_2 to the next value in its current domain, namely, $C_2 = 6$. As soon as it instantiates $C_1 = 3$ and $C_2 = 6$, the propagation process now updates the domain of C_3 to see what values of C_3 might be consistent with these instantiations. It encounters the single consistent value of $C_3 = 4$, and identifies the second solution: $C_1 = 3, C_2 = 6, C_3 = 4$.

Backtracking, step 5 now reverts back to the C_1 level and instantiates it to $C_1 = 4$, and repeating the logic mentioned in steps 1 through 3, it identifies the third solution: $C_1 = 4, C_2 = 6, C_3 = 1$. Finally, instantiating $C_1 = 5$ yields the fourth (and last) solution: $C_1 = 5, C_2 = 2, C_3 = 6$.

Goal: To find combinations of C_j that satisfy all disjunctive constraints and result in zero total tardiness.

Legend:
- [ ] Unexplored candidate assignment
- [t] Inconsistent assignment at step t
- [t] (black) Bounded values at step t

Data:

Job	p_j	d_j
1	3	5
2	2	6
3	1	8

Solution Steps:

Step 1: C_j domain reduction based on processing times, due dates, and target tardiness constraints.

Day	0	1	2	3	4	5	6
Job 1	1	1	1				1
Job 2	1	1					
Job 3	1						

Step 2: C_1 bound to 3.

Day	0	1	2	3	4	5	6
Job 1	1	1	1	2	2	2	1
Job 2	1	1	2	2	2		
Job 3	1	2	2	2			

Step 3: C_1 bound to 3; C_2 bound to 5 (=> $C_3 = 6$).

Day	0	1	2	3	4	5	6
Job 1	1	1	1	2	2	2	1
Job 2	1	1	2	2	2	3	3
Job 3	1	2	2	2	3	3	3

$C_1 = 3$; $C_2 = 5$; $C_3 = 6$ is the first solution.

Step 4: C_1 bound to 3; C_2 bound to 6 (=> $C_3 = 4$).

Day	0	1	2	3	4	5	6
Job 1	1	1	1	2	2	2	1
Job 2	1	1	2	2	3	4	4
Job 3	1	2	2	2	4	4	4

$C_1 = 3$; $C_2 = 6$; $C_3 = 4$ is the second solution.

Step 5: C_1 bound to 4 (=> $C_2 = 6$; $C_3 = 1$).

Day	0	1	2	3	4	5	6
Job 1	1	1	1	5	5	5	1
Job 2	1	1	5	5	5	5	5
Job 3	1	5	5	5	5	5	5

$C_1 = 4$; $C_2 = 6$; $C_3 = 1$ is the third solution.

Step 6: C_1 bound to 5 (=> $C_2 = 2$; $C_3 = 6$).

Day	0	1	2	3	4	5	6
Job 1	1	1	1	6	6	6	1
Job 2	1	1	6	6	6	6	6
Job 3	1	6	6	6	6	6	6

$C_1 = 5$; $C_2 = 2$; $C_3 = 6$ is the fourth solution.

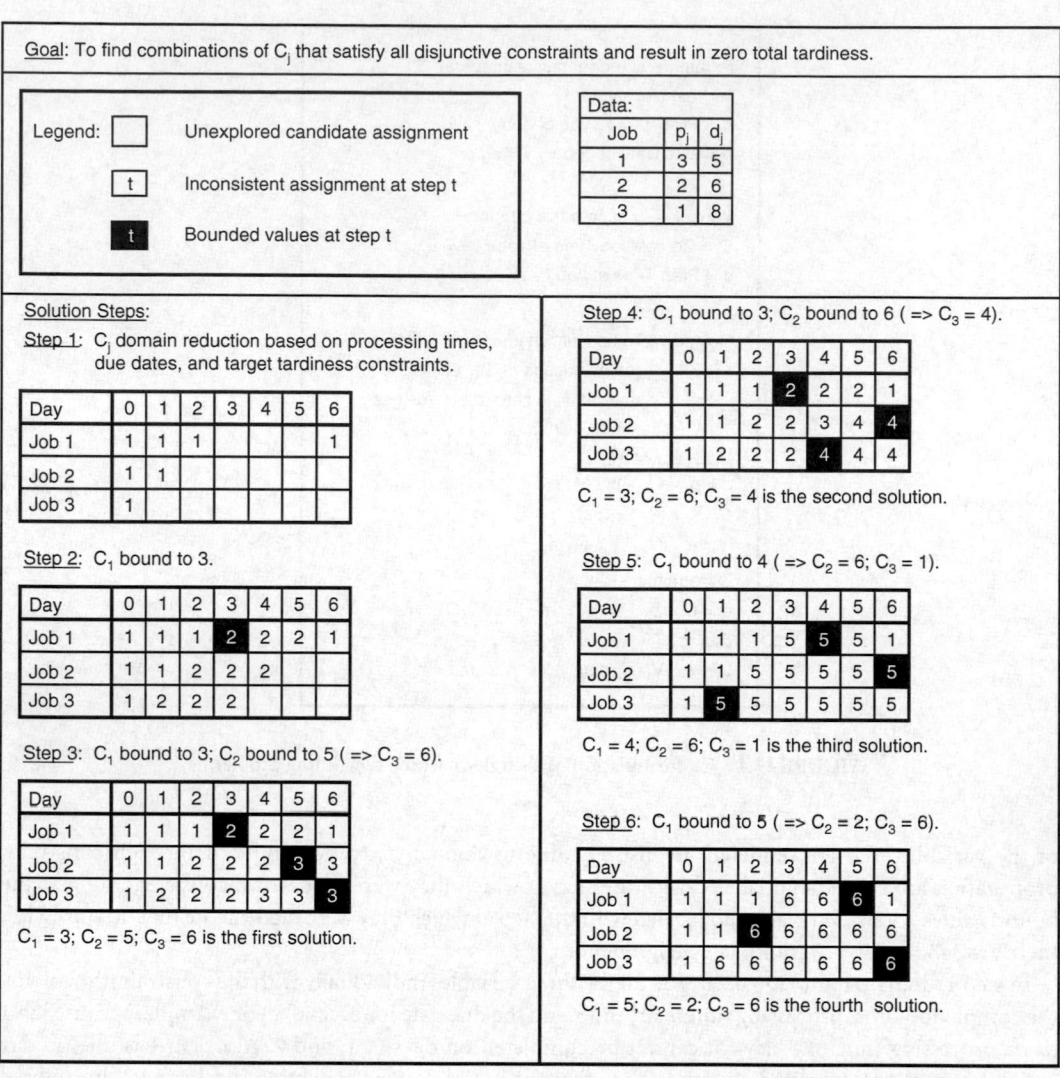

FIGURE 47.5 CP logic for a single-machine, three-job scheduling problem.

47.5 Selected CP Applications in the Scheduling Literature

In this section we describe selected applications of CP to scheduling-related problems from recent studies in the literature. We review applications from a number of specific scheduling subject areas: job shop scheduling, single-machine sequencing, parallel machine scheduling, vehicle routing, and timetabling[2]. Finally, we review recent developments in the integration of the CP and IP paradigms.

[2]Throughout this section, numerous specific software products are referenced as the specific CP or IP software used to implement solutions. These products are:
ILOG Solver®, OPL Studio®, and CPLEX®(ILOG, Inc., Mountain View, CA);
OSL®– Optimization Solutions and Library – (IBM, Inc., Armonk, NY);
CHARME is a retired predecessor language of ILOG Solver;
ECLiPSe, free to researchers through Imperial College, London. http://www.icparc.ic.ac.uk/eclipse;
Friar Tuck, free software available at http://www.friartuck.net/
CPLEX and OSL are MIP software; all others are CP-based software.

47.5.1 Job Shop Scheduling

Brailsford, Potts, and Smith [4] reviewed the application of CP in the job shop scheduling environment. They described a general job shop scheduling problem, in which there are n jobs each with a set of operations which must be conducted in a specific order. The jobs are scheduled on m machines which have a capacity of one operation at any one time and the objective is to minimize the makespan. Brailsford et al. [4] noted that the problem instances of this type with as few as 15 jobs and 10 machines have not been solved with current branch and bound algorithms because these methods suffer from weak lower bounds which inhibit effective pruning of the branch and bound tree. Further, Brailsford et al. [4] concluded that "CP compares favorably with OR techniques in terms of ease of implementation and flexibility to add new constraints."

Nuijten and Aarts [7] addressed the problem described by Brailsford et al. [4] above. They provided a CP-based approach for solving these classic job shop problems and found that their search performed favorably compared with branch and bound algorithms when measured by solution speed. Their study leveraged the flexibility of CP search by incorporating randomization and restart mechanisms. Lustig and Puget [8] have given an instructive example of a simple job shop scheduling formulation in OPL Studio.

Darby-Dowman, Little, Mitra, and Zaffalon [9] applied CP to a generalized job assignment machine scheduling problem, in which different products with unique processing requirements are assigned to machines in order to minimize makespan. The problem is similar to Brailsford et al. [4], but in Darby-Dowman et al. [9], "machine cells" have a capacity greater than one. They found CP to be intuitive and compact in its formulation. They solved their CP problem using ECLiPSe and the IP version in CPLEX. They reported that CP performed more predictably than IP (when measured by search speed). CP's domain reduction approach reduced the search space quickly, contributing to its superior performance.

47.5.2 Single-Machine Sequencing

It is noteworthy that relatively little application of CP has taken place in the job sequencing area. The majority of this literature uses special-purpose codes to take advantage of the special problem structure of the job sequencing problem. A notable exception is Jordan and Drexl [10], which addressed the single-machine batch sequencing problem with sequence-dependent setup times. They assumed every job must be completed before its deadline, and all jobs are available at time zero. They solved this problem under a number of different objectives, including minimize setup costs, minimize earliness penalties, both setup and earliness costs combined, and finally, with no objective function (as a constraint satisfaction problem). They used CHARME to solve their CP formulation and compared computational results of the same model solved using OSL, an IP optimization software. They found that CP worked best when the machine was at high capacity (tight constraints; smaller feasible solution space), and IP worked best when the shop was at low capacity. Solution speeds are comparable for the two methods, but CP showed smaller variability in solution times overall. Both methods performed disappointingly on larger problems.

47.5.3 Parallel Machine Scheduling

Jain and Grossman [11] examined an application for assigning customer orders to a set of nonidentical parallel machines. Their objective was to minimize the processing cost of assignment of jobs to machines which have different costs and processing times for each job. Each job had a specific release (earliest start) and due (latest completion) dates. They found that for their problem, the CP formulation required approximately one-half the number of constraints and two-thirds the number of the variables of IP. They solved the CP problems using ILOG Solver, and the IP problems using CPLEX. They found that CP outperforms IP for smaller problems, and its performance is comparable to IP for larger problems.

47.5.4 Timetabling

Henz [12] utilized a CP methodology to schedule a double round-robin college basketball schedule in a minimal number of dates subject to a number of constraints imposed by the league. Henz [12] was able to model unusual constraints such as "no more than two away games in a row" and "no two final away games" in a direct way with CP. Henz [12] found a dramatic improvement in performance over Nemhauser and Trick [13]. Nemhauser and Trick [13] solved the problem in three phases using OR methods such as pattern generation, set generation and timetable generation in 24 hours of computer time. Henz [12] reported solution time under 1 min using Friar Tuck, a CP software written specifically for addressing tournament scheduling problems.

More recently, Baker, Magazine and Polak [14] and Valouxis and Housos [15] have explored school timetabling using CP. Baker et al. [14] created a multiyear timetable for courses sections in order to maximize contact among student cohorts over a minimum time horizon. They found CP straightforwardly reduces the size of the solution space, thereby facilitating the search for an optimum. Valouxis and Housos [15] used a combination of CP and IP to address a daily high school timetabling problem in which the teachers move to several different class sections during the day and the students remain in their classrooms. Their objective function was to minimize the idle hours between the daily teaching responsibilities of all the teachers while also attempting to satisfy their requests for early or late shift assignments. They developed a hybrid search method for improving the upper bounding which helped reduce the search space and facilitated faster solution speed.

47.5.5 Vehicle Dispatching

The synchronized vehicle dispatching problem of Rousseau, Gendreau and Pesant [16] requires coordinating a number of vehicles such as ambulances with complementary resources such as medics to service demands such as patients. These resources are coordinated across time intervals in order to minimize the travel cost of the vehicles. Their model is a real-time model, with customer orders arriving sporadically even after vehicles have been dispatched (all orders are not known at the beginning of the time horizon). New customers are inserted as constraints into the model which is then resolved based on the new conditions. Rousseau et al. found modeling the complicated nature of the synchronization constraints is simplified in a CP environment. Further, because of the real-time nature of customer arrivals, insertion of additional customer constraints (orders) was found to be easy in the CP paradigm.

47.5.6 Integration of the CP and IP Paradigms

One of the most active and potentially highest-opportunity areas of recent research is in the exploration of hybrid methods for CP and IP search methodologies. It is widely held that CP and IP have complementary strengths (e.g., see [11,17]). CP's strengths arise from its flexibility in modeling, a rich set of operators, and domain reduction in combinatorial problems. IP offers specialized search techniques such as relaxation, cutting planes and duals for specific mathematical problem structures. Hooker [17], Milano, Ottoson, Refalo, and Thorsteinsson [18] and Brailsford et al. [4] have provided general discussions on developing a framework for integrating CP and IP.

There are recent studies which integrate CP and IP for solving specific problems. Darby-Dowman et al. [9] made an early attempt at integrating IP and CP by using CP as a preprocessor to limit the search space for IP, but had limited success. They suggested, but did not develop, using a more integrated combination of IP and CP: CP for generating cuts on the tree and IP for determining search direction. Jain and Grossman [11] presented a successful deeper integration of these methods for the machine scheduling problem based on a relaxation and decomposition approach. Finally, Valouxis and Housos [15] used local search techniques from OR to tighten the constraints for CP and improve the search time. If efforts at integrating these methods are successful, the end state for these methods may be that researchers view IP and CP as special case algorithms for a general problem type.

47.6 The Richness of CP for Modeling Scheduling Problems

As a general framework for developing search strategies, CP has a rich set of operators and variable types, which often allows for a succinct and intuitive formulation of scheduling problems. The following section describes the richness of the CP modeling environment with examples of variable indexing, and constraints such as strict inequality, logical constraints, and global constraints.

47.6.1 Variable Indexing

Just as in other computer programming languages, CP allows "variable indexing", in which one variable can be used as an index into another. This capability creates an economy in the formulation which reduces the number of required decision variables in the scheduling problem formulation. The result is often a more compact and intuitive expression of the problem.

To illustrate the value of variable indexing capability, an adaptation of single-machine batch sequencing problem with sequence-dependent setup costs is taken from Jordan and Drexl [10]. In this problem, the objective is to minimize setup costs and earliness penalties given varying setup costs for adjacent jobs in the sequence. In IP, a binary variable, $y[i, j]$, is used to indicate if job i immediately precedes job j. If n is the total number of jobs in the problem, with this formulation, there are $n(n-1)$ binary indicator variables to express the potential job sequences. If setup cost $[i, j]$ is the cost of setup for job i immediately preceding job j, then the total setup cost for the set of assignments is specified as the sum of setup cost $[i, j]y[i, j]$ for all i and $j, i \neq j$.

To formulate the problem in CP, we create the decision variable, job$[k]$ which takes the label of the job assigned to position k in the sequence. We then use job$[k]$ as an indexing variable to calculate the total setup cost of the assignments as the sum of setup cost[job$[k-1]$, job$[k]$] for all $k > 1$. With this indexing capability, the number of variables in the problem is reduced to the number of jobs, n. An additional benefit of variables indexing is that the model is more expressive of the original problem statement. The solution to the problem is more naturally thought of as "the sequence of jobs on the machine" as in CP, than, "the collection of immediately adjacent jobs on a machine", as is the case in the IP formulation.

Another example of the usefulness of variable indexing in the job shop literature comes from Darby-Dowman et al. [9]. They described a generalized job assignment problem that IP treats with mn decision variables (m machines, n jobs), but CP addresses with only n decision variables (jobs), which take on the value of the machine to which a job is assigned. In this setting as well, the flexibility of CP's variable indexing construct more intuitively and compactly expresses the problem as an assignment of jobs to machines, rather than IP's expression of a machine-job pair.

Generally stated, problems that can be expressed as a matching of one set to another can be more succinctly expressed using an index into one of the sets (as afforded in CP) than using a binary variable to represent every possible combination of the cross product of the two sets (as is the case in IP).

47.6.2 Constraints

CP allows for the use of a number of operators such as set operators and logical conditions which enable it to handle a variety of special constraints easily. In this section, we describe some of these special constraints that are common in scheduling: strict inequality, logical constraints, and global constraints, which, as Williams and Wilson [5] describe, are straight forward in CP, but difficult in IP.

47.6.2.1 Inequalities with Boolean Variables

We start with the simple example of strict inequality constraint for Boolean variables. Let us say in a scheduling application, the assignment of a job to a pair of machines can be expressed as Boolean variables (call them MachineA and MachineB) which take the value of 1 if a job is assigned, and the value of 0 if no job is assigned. Further, let us say that it is desirable for either one of two machines to have a job assigned, but not both, or equivalently that one Boolean variable does not equal another. We express, the inequality directly in CP as

$$\text{MachineA} \neq \text{MachineB}$$

In IP, we specify the same constraint as:

$$\text{MachineA} + \text{MachineB} = 1$$

Both formulations accurately and efficiently impose the condition in a single constraint that either MachineA or MachineB receives a job (equals one), but not both. However, the CP statement is a more intuitive and direct expression of the naturally occurring constraint without the indirection necessary as in the IP formulation to express the constraint as the sum of two variables.

47.6.2.2 Inequalities with Integer Variables

Now we consider another situation in which MachineA and MachineB are integer variables representing the job number assigned to each machine. Further, we assume the same job cannot be assigned to both MachineA and MachineB. In CP, the constraint is again expressed using the simple constraint:

$$\text{MachineA} \neq \text{MachineB}$$

However, in IP, the $\neq$ condition on integer variables must be modeled as two inequalities ($>$, $<$) and an exclusive or (Xor) condition:

$$(\text{MachineA} > \text{MachineB}) \text{ Xor } (\text{MachineB} > \text{MachineA})$$

Further, in IP we are restricted to $\geq$ and $\leq$ constraints. So in order to capture the essence of the strict inequality, we add or subtract some small value, ε, to the integer values:

$$(\text{MachineA} \geq \text{MachineB} + \varepsilon) \text{ Xor } (\text{MachineB} \geq \text{MachineA} + \varepsilon)$$

Finally, in IP the logical Xor is modeled with a Boolean indicator variable, δ for each constraint, and a "BigM" multiplier is used to ensure that at least one but not both constraints hold:

$$\text{MachineA} - \text{MachineB} - \varepsilon + \delta\text{BigM} \geq 0$$
$$\text{MachineB} - \text{MachineA} - \varepsilon + (1 - \delta)\text{ BigM} \geq 0$$

In this example, $\delta = 1$ implies MachineB > MachineA; $\delta = 0$ implies MachineA > MachineB. Familiar and widely-used constructs such as "BigM" and Boolean indicator variables have been borne out of necessity to create formulations which meet the linear problem structure requirements for IP.

47.6.2.3 Logical Constraints

Logical constraints are common in scheduling. The "precedes" condition is a ubiquitous example from sequencing. For example, let us say that the completion of JobA must precede the start of JobB or that the completion of JobB must precede the start of JobA, but not both. This is an example of the exclusive or (Xor) constraint. CP handles constraints of this form with relative ease with constraints of the form:

$$(\text{A.start} > \text{B.end}) \text{ Xor } (\text{B.start} > \text{A.end})$$

In IP, the relationship is stated as follows:

$$\text{A.start} - \text{B.end} - \varepsilon + \delta\text{BigM} \geq 0$$
$$\text{B.start} - \text{A.end} - \varepsilon + (1 - \delta)\text{ BigM} \geq 0$$

where $\delta = 1$ implies job A is first and $\delta = 0$ implies job B is first.

CP expresses the Xor condition directly and naturally. It is interesting to note that because the $\neq$ condition and the Xor condition are logically equivalent, both are modeled in the same way in IP, even though the basic problem statements (precedes and $\neq$) are perceived differently.

47.6.2.4 Global Constraints

Global constraints apply across all, or a large subset of, the variables. For example, imagine we are assigning jobs to m machines, and it is desired to assign different jobs to each machine in an efficient way. We may want to extend the $\neq$ constraint discussed above to apply to all of the machines under consideration. In CP, the alldifferent constraint is used:

alldifferent(Machine).

This constraint assures that no two machines are assigned to the same job. It should be noted that the alldifferent constraint in CP is a single, global constraint, and that a constraint of this form has stronger propagation properties than n inequality constraints (see Hooker, 2002).

In IP, the alldifferent concept is implemented as an inequality constraint (as described above) for every machine pair. For m machines, this would require $m(m-1)$ constraints, and $m(m-1)/2$ indicator variables.

Other global constraint constructs exist in CP. For the preceding example, we may want to assure that each machine gets no more than one job in the final solution. In CP, we use the "distribute(Machine)" constraint — a global constraint restricting the count of jobs on each machine to one. In IP, we create an indicator variable and accompanying constraint for each machine which indicates if it has a job assigned or not, then, create a constraint on the sum of the indicator variables.

The need for the indicator variables and "BigM" constructs is obviated in the more general modeling framework of CP. Because the problem can be directly and succinctly stated, accurate formulations are easily created, maintained, modified and interpreted, and the potential for errant formulations are reduced.

Other special-purpose CP constructs have been developed to aid in model building that are useful to the scheduling community. For examples, ILOG's OPL includes the "cumulative" constraint, which defines the maximum resource availability for some time period, and "reservoir" constraints, which capture resources that can be stockpiled and replenished (such as raw materials, funds, or inventory). These constructs are special-purpose routines that take advantage of the known properties of the scheduling problem in order to represent resources accurately and leverage their properties in order to reduce domains more rapidly.

47.7 Some Insights into the CP Approach

We do not want to encourage discussion on "which is better", CP or IP. Which performs better depends on a specific problem, the data, the problem size, the researcher, the commercial software and the model of the problem (among other things!). In any case, the debate on which is better, CP or IP, may be moot because the two are so different. IP is a well-defined solution methodology based on particular mathematical structures and search algorithms. CP, on the other hand, is a method of modeling using a logic-based programming language (see [8]). Also, as covered previously, the two approaches have complementary strengths that can be combined in many ways (see [11,17]). Despite their orthogonality in approach, in this section we compare and contrast CP and IP as approaches to solving scheduling problems, and provide some insight into when CP may be an appropriate method to consider.

47.7.1 Contrasting the CP and IP Approaches to Problem Solving

When applied to scheduling-related problems, CP and IP both attempt to solve NP-hard combinatorial optimization problems; however, they have different approaches for addressing them. IP leverages the mathematical structure of the problem. In the CP approach, the freedom and responsibility for creating constraints and a search strategy falls primarily to the researcher. This is similar to the requirement of developing a good IP model that fits the linear structure required of an IP solver. CP search methods rely less on particular mathematical structure of the objective function and constraints, but more on the domain knowledge of specific aspects of the problem. Consider the trivial example of the single-machine job sequencing problem to minimize total tardiness. An IP formulation would not likely contain an explicit constraint that states there need be no gaps between jobs on a machine; it is rather a logical outcome of the optimization process. Using the CP paradigm, the researcher who knows, logically, the optimal solution has no such gaps might utilize this knowledge by introducing an explicit constraint to that effect, possibly reducing the solution space and improving performance of CP search.

In general, IP is objective-centric while CP is constraint-centric. IP follows a "generate-and-test strategy" evaluating the objective function for various values of the decision variables. CP focuses more on the constraints in the problem, constantly reevaluating the logical conclusions resulting from the interactions of the constraints and, as a result, reducing the domain of feasible solutions.

Another difference between CP and IP falls in what might be called "modeling philosophy". In IP the modeler is encouraged to specify a minimal set of constraints sufficient to capture the essence of a given problem. Fewer constraints are generally viewed as "better" or more elegant (for examples, see [19, p. 194] and [20, p. 34]). In CP, the modeling philosophy is somewhat different. The modeler is encouraged to add constraints that more carefully capture the nuances of a particular problem. More constraints are better, because taken together they reduce the solution domain. Contrary to the IP approach, even redundant constraints are considered desirable in CP in some cases if they improve constraint propagation and domain reduction.

47.7.2 Appropriateness of CP for Scheduling Problems

There are certain attributes of problems that researchers can be aware of when deciding if CP is an appropriate methodology to employ for solving scheduling problems. First, CP is most appropriate for pure integer, or Boolean decision variable problems; CP methods are not effective for floating point decision variables. CP is more successful reducing domains for variables with finite domains at the outset.

Second, CP is well suited for combinatorial optimization problems that tend to have a large number of logical, global and disjunctive constraints that CP handles well with its rich set of operators and special-purpose constraints. CP more naturally expresses variable relationships in combinatorial optimization problems. It can be useful for quickly formulating these types of problems without significant formulation difficulties.

Third, CP operates more efficiently when there are a large number of interrelated constraints with relatively few variables in each constraint. This characteristic results in better constraint propagation and domain reduction, and better performance of CP algorithms. For example, if a constraint is stated as the sum($x1$ to $x100$) < 1000, and $x1$ is instantiated, not much can be said about $x2$ to $x100$ because there are so many variables in the constraint. On the other hand, if $x1 + x2 < 100$, and $x1$ is bound to 50, then it follows immediately that $x2 < 50$, and the domain of $x2$ is significantly reduced. Further, if $x2$ is in another constraint with only $x3$, then $x2$'s domain reduction propagates immediately to $x3$'s domain reduction. Thus, a problem with a large number of interrelated constraints with few variables in each constraint tends to be well suited for CP.

Finally, in a more general sense, CP applies well to any problem that can be viewed as an optimal mapping of one ordered set to another ordered set, where the relation between variables in each set can be expressed in mathematical terms that are significant to the objective. Scheduling problems fit into this description. An example may be a set of operations with precedence conditions being assigned to machines over time-indexed periods. Both sets are ordered in some way, and the solution is a mapping between them. The decision variables are integer or Boolean, constraints tend to be logical, global or disjunctive in nature, and there is a large number of constraints that often contain few variables. The strengths of CP seem well suited for application in scheduling research.

47.8 Formulating and Solving Scheduling Problems via CP

We now illustrate how CP might be useful in modeling scheduling problems. We will use the well-known single-machine weighted tardiness problem as a specific case in point. In the notation of Pinedo [21] this is described as $1||\Sigma w_j T_j$. The $1||\Sigma w_j T_j$ problem is simple to specify but well known to be NP-hard (see [22]). It well serves the purpose of showing the look and feel of a CP approach without excessive clutter. Typically $1||\Sigma w_j T_j$ is solved by special purpose branch-and-bound algorithms implemented in a high level general purpose programming language such as C++. See for example [23]. Our illustration will show several ways the problem can be formulated via CP. In doing this we will use the commercially available CP modeling language OPL as implemented in ILOG OPL Studio. For details regarding the OPL programming language the reader is directed to Van Hentenryck [24].

An OPL program is commonly comprised of three major program blocks including:

- a Declarations/Initializations block for declaring data structures, initializing, and reading data
- an Optimize/Solve block for specifying, in a declarative mode, the problem specifications (either a constraint satisfaction or an optimization problem)
- an optional Search block for specifying how the search is to be conducted

The absence of any search specification invokes the OPL default search (to be described below).

47.8.1 A Basic Formulation

Figure 47.6 shows a fairly compact encoding of $1||\Sigma w_j T_j$ in OPL. We refer to it hereafter as Model 1.

Lines 2 to 5 in the Declarations/Initializations block define and initialize the number of jobs, processing times, due dates and job weights (n, $p[.]$, $d[.]$, $w[.]$). The use of "int+" and "float+" specify nonnegative integer and floating point data types, respectively. The notation "= . . .;" signifies that data is to be read from an accompanying data file. Lines 6 and 7 define the two nonnegative integer variables $C[.]$, and $s[.]$. Since weighted tardiness is regular, we can ignore schedules with inserted idle time (see, e.g., Baker, 2000, p. 2.4). We take advantage of this fact by limiting the domains for $s[.]$ and $C[.]$ to $\{0, 1, \ldots, \Sigma_{j\in 1,\ldots,n} p[j]\}$.

Lines 9 to 15 of Figure 47.6 constitute the Optimize/Solve block. The presence of the keyword "minimize" signals that this is a minimization problem and that the associated objective follows. The keywords "subject to" introduce the so-called "constraint store." The constraint in line 13 assures the desired relation between $s[.]$ and $C[.]$. We could as well have specified $C[j] - s[j] >= p[j]$. However, the equality specification yields a smaller search space. Lines 13 and 14 in Figure 47.6 together define the disjunctive requirement that for all pairs of jobs j, k that either j precedes k or k precedes j but not both.

```
01 //SINGLE MACHINE WEIGHTED TARDINESS SCHEDULING MODEL 1
02 int + n= ... ;//n is the number of jobs
03 int + p[1 .. n] = ... ;//p[i] is the processing time for job i
04 int + d[1 .. n]= ... ;//d[i] is the due date for job i
05 float + w[1 .. n]= ... ;//w[i] is the weight for job i
06 var int + C[j in 1 .. n] in 0 .. sum(j in 1 .. n) p[j];//C[i] is the completion time for job i
07 var int + s[j in 1 .. n] in 0 .. sum(j in 1 .. n) p[j];//s[i] is the start time for job i
08 //END OF DECLARATIONS/INITIALIZATIONS BLOCK
09 minimize
10 sum(j in 1 .. n) w[j]*maxl(0,C[j]-d[j])
11 subject to{
12 forall(j in 1 .. n){
13 C[j]-s[j] = p[j];
14 forall(k in 1 .. n: k>j) (C[j] <= s[k]) V (C[k] <= s[j]);
15 };
16 };//END OF OPTIMIZE/SOLVE BLOCK
```

FIGURE 47.6 Example OPL program for $1||\Sigma w_j T_j$ (Model 1).

TABLE 47.1 Sample $1||\Sigma w_j T_j$ Problem (Elmaghraby, 1968)

Job	Processing Time	Due Date	Weight
1	3	2	1
2	3	5	3
3	2	6	4
4	1	8	1
5	5	10	2
6	4	15	3
7	4	17	1.5

TABLE 47.2 OPL Solution Using Model 1 on Elmaghraby Problem Data

Optimal Solution with Objective Value: 25.0000	
$s[1] = 19$	$C[1] = 22$
$s[2] = 0$	$C[2] = 3$
$s[3] = 3$	$C[3] = 5$
$s[4] = 5$	$C[4] = 6$
$s[5] = 6$	$C[5] = 11$
$s[6] = 11$	$C[6] = 15$
$s[7] = 15$	$C[7] = 19$

Since Model 1 provides no search specifications the default search in OPL is used. It works as follows. All possible domain reduction is first accomplished. If a solution does not result, then a depth-first search ensues. Variables are chosen for instantiation in the order of smallest domain size first. Values within the domain of a variable are chosen in order of smallest value first. Each instantiation represents a choice point.

As can be observed, Model 1 requires $2n$ variables and $n(n+1)/2$ constraints. We executed Model 1 on the 7-job instance of $1||\Sigma w_j T_j$ found in Elmaghraby [25] and depicted in Table 47.1.

Using OPL Studio we arrived at the solution in Table 47.2 after generating 254 choice points.

47.8.2 A Second Formulation

A second formulation (called Model 2) for $1||\Sigma w_j T_j$ is provided in Figure 47.7. Here the variable $s[.]$ is replaced with the variable position[.] to represent the position of job j in the sequence. The constraint in line 12 assures that for all combinations of job pairs the position numbers are different. Line 13 binds the two variables position[.] and $C[.]$ by specifying an equivalence relation for all permutations of job pairs. Line 14 is required to assure the job j in position 1 is completed at time $t = p[j]$. The formulation is relatively compact ($2n$ variables, $(3n^2 - n)/2$ constraints), but the performance is lackluster. Running this model for the sample Elmaghraby data causes 5391 choice points to be created. The constraint set as specified is sufficient to assure solution, but the filtering algorithms are not strong enough to enable much domain reduction as in Model 1. We begin to see with this example that model building in CP is, to a great degree, a craft. The goal is not to minimally express a problem specification but rather to express as much information in the constraint store so as to foster domain reduction.

47.8.3 Strengthening the Constraint Store

A first improvement to the Model 2 is accomplished by replacing line 12 with the so-called "all different" constraint; i.e., change line 12 to read "alldifferent(position);". This is a good example of a single global constraint as described earlier. The alldifferent constraint operates at once on the entire set of variables,

```
01 //SINGLE MACHINE WEIGHTED TARDINESS SCHEDULING MODEL 2
02 int + n = ... ;//n is the number of jobs
03 int + p[1 .. n] = ... ; //p[i] is the processing time for job i
04 int + d[1 .. n] =... ;//d[i] is the due date for job i
05 float + w[1 .. n] = ... ;//w[i] is the "weight" for job i
06 var int + C[j in 1 .. n] in 0 .. sum (i in 1 .. n)p[i]; //C[i] is the completion time for job i
07 var int + position[j in 1 .. n] in 1 .. n; //position[i] is job i's position in the sequence
08 //END OF DECLARATIONS/INITIALIZATIONS BLOCK
09 minimize
10 sum(j in 1 .. n) w[j]*maxl(0,C[j]-d[j])
11 subject to{
12 forall(j, k in 1 .. n: k>j) position[j] <> position[k];
13 forall (j, k in 1 .. n: j<>k) position[j] > position[k] <=> C[k] <= C[j]-p[j];
14 forall (j in 1 .. n) position[j] = 1 => C[j] = p[j];
15 };//END OF OPTIMIZE/SOLVE BLOCK
```

FIGURE 47.7 Example OPL program for $1||\Sigma w_j T_j$ (Model 2).

and is considered a single constraint. Making this replacement in the code reduces the number of choice points for the sample problem from 5391 to 3953.

We can improve the performance of Model 2 further by adding more detailed information regarding the relationship between position and completion time. The equivalence constraints of line 13 can be made stronger when jobs are adjacent in the schedule, for then the difference in their completion times is exactly the processing time of the first job in the pair. We implement this knowledge by adding the following adjacency constraints to the Model 2 formulation.

$$\text{forall } (j, k \text{ in } 1..n\text{: } j <> k) \text{ position}[j] = \text{position}[k] + 1 <==> C[k] = C[j] - p[j]$$

We applied this additional constraint set to Model 2 and applied the new model to the Elmaghraby sample data. The result was a further reduction in the number of choice points from 3953 to 1969.

47.8.4 A Final Improvement to Model 2

We can improve Model 2 further by adding problem-specific domain knowledge to the formulation. For example, note that when a job occupies position j we can place a lower bound on its completion time, namely the sum of its processing time plus the sum of the remaining $j - 1$ jobs with smallest processing times. A revised Model 2 is provided in Figure 47.8 which includes this idea along with all the other aforementioned revisions to Model 2. The set of constraints bounding the jobs' minimum completion times is implemented in lines 19 to 24. Running this model on the sample problem data drastically reduces the number of choice points further from 1969 to 50. Note that in this case for simplicity we assume the jobs are numbered in order of nondecreasing processing time so the comparison in performance is not 100% fair. Nevertheless, it vividly illustrates the rather craft-like aspect of model building for scheduling using CP.

47.8.5 Utilizing the Special Scheduling Objects of OPL

Up to this point we have not taken advantage of the specialized scheduling objects embedded in the OPL modeling language. Objects like "activity" and "resource" can be used to relieve the modeler from some tedium as well as to take advantage of a number of built-in functions peculiar to scheduling applications. We illustrate with another formulation of $1||\Sigma w_j T_j$ as depicted in Figure 47.9 and refer to it as Model 3. Here line 6 defines the special object "scheduleHorizon" which is used to limit the domain of the search space. Line 7 declares a set of variable structures of type "Activity" with durations $p[.]$. Line 8 defines the variable

```
//SCHEDULING MODEL 2 (revised)
int + n= ... ;//n is the number of jobs
nt + p[1 .. n] = ... ;//p[i] is the processing time for job i
int + d[1 .. n]= ... ;//d[i] is the due date for job i
float + w[1 .. n]= ... ;//w[i] is the "weight" for job i
var int + C[j in 1 .. n] in 0 .. sum (i in 1 .. n)p[i]; //C[i] is the completion time for job i
var int + position[j in 1 .. n] in 1 .. n; //Job position number in sequence
//END OF DECLARATIONS/INITIALIZATIONS BLOCK
minimize
sum(j in 1 .. n) w[j]*maxl(0,C[j]-d[j])
subject to{
forall (j in 1 .. n) position[j] = 1 => C[j] = p[j];
alldifferent(position);
forall (j, k in 1 .. n: j<>k){
position[j] > position[k] <=> C[k] <= C[j]-p[j];
position[j] = position[k]+1 <=> C[k] = C[j]-p[j];
};
//NOTE: The following requires job numbers in non-decreasing processing time order
forall (j in 1 .. n){
forall (k in 1 .. n){
position[j]=k & j<=k => C[j] >= sum(l in 1 .. n: l<=k) p[l];
position[j]=k & j>k  => C[j] >= p[j]+sum(l in 1 .. n: l<k) p[l];
};
};
};//END OF OPTIMIZE/SOLVE BLOCK
```

FIGURE 47.8 Example OPL program for $1||\Sigma w_j T_j$ (Model 2 with all revisions).

```
01 //SINGLE MACHINE WEIGHTED TARDINESS SCHEDULING MODEL 3
02 int + n= ... ;//n is the number of jobs
03 int + p[1 .. n] = ... ;//p[i] is the processing time for job i
04 int + d[1 .. n]= ... ;//d[i] is the due date for job i
05 float + w[1 .. n]= ... ;//w[i] is the weight for job i
06 scheduleHorizon = sum(j in 1 .. n) p[j];
07 Activity job[j in 1 .. n](p[j]);
08 UnaryResource Machine;
09 //END OF DECLARATIONS/INITIALIZATIONS BLOCK
10 minimize sum(j in 1 .. n) w[j]*maxl(0,job[j].end-d[j])
11 subject to{
12 forall(j in 1 .. n) job[j] requires Machine;
13 };//END OF OPTIMIZE/SOLVE BLOCK
```

FIGURE 47.9 Example OPL program for $1||\Sigma w_j T_j$ (Model 3).

"Machine" as a "UnaryResource" (i.e., one that can service only one activity at a time). Line 12 specifies the constraint that each job (Activity) is to be processed on "Machine." Built in to the UnaryResource type along with the special constraint predicate "requires" is the assurance of the disjunctive constraint that only one job can occupy Machine at a given time. Note that, as with the other models presented earlier, no search block is provided so that OPL invokes the default search algorithm as needed. This rather compact formulation when run using the Elmaghraby sample data yields a solution after considering 117 choice points. We can attribute this relatively good performance to using OPL's special scheduling objects. Their use triggers special purpose filtering algorithms leading to more efficient domain reduction. Of course we could now commence as before with attempts to improve the performance by the addition of problem domain specific knowledge (more constraints).

```
14 search{
15   while(not isRanked(Machine)) do
16     select (j in 1.. n: isPossibleFirst(Machine,job[j])
17         ordered by increasing  maxl(d[j]-dmin(job[j].start), p[j])/w[j])
18         tryRankFirst(Machine,job[j]);
19 };
```

FIGURE 47.10 Example OPL search block for $1||\Sigma w_j T_j$ (Model 3).

47.8.6 Controlling the Search

OPL allows the modeler significant freedom in designing a search strategy. We provide a few simple examples to illustrate. In solving minimization problems with objective function Z, OPL has the aforementioned feature of adding the constraint $Z < z'$ to the constraint store where z' is always the objective value of the current best found solution. Suppose for $1||\Sigma w_j T_j$ we wish to steer the search to discover a good trial solution early on. We might be motivated then to make use of a good heuristic. A recent paper by Kanet and Li [26] reports on a heuristic dispatching rule "weighted modified due date" (WMDD) for $1||\Sigma w_j T_j$. At time $t = 0$, it computes for each job j the following value:

$$\text{WMDD}[j] = \max\{d[j] - t, p[j]\}/w[j]$$

It then selects the job k with smallest WMDD to occupy the machine. At time $t = t + p[k]$ the process is repeated on the remaining jobs so that a schedule is constructed from beginning to end. With OPL we can create a search strategy, which first constructs a schedule from beginning to end, plunging to a complete solution in conformance to WMDD, and then backtracks. To illustrate how we might implement this in OPL, we append a search block to Model 3; the OPL specification for which is depicted in Figure 47.10. The functionality of the various OPL keywords is almost self-explanatory. In OPL, a unary resource is said to be ranked when a permutation in which it services activities is completely specified. The keyword "dmin" is a "reflective" function that returns the current minimum value of the domain for its argument. In this case the argument is the variable job$[j]$.start. This is what affords the building of the schedule from start to finish and the intended dynamic recalculation of WMDD. With each execution of line 18 another job is appended to the end of the schedule. Doing so clips the domains of the start times for the remaining unscheduled jobs accordingly so that on the next call to dmin the WMDD calculation in line 17 is dynamic. Running this version of the model on the sample Elmaghraby data produces the desired result; the number of choice points drops from 117 to 37.

From the previous examples we see that it is quite simple using OPL to organize the search either over job completion times or over jobs' positions. Alternatively, we might want to search over the jobs that occupy the different positions. To clarify, consider a variable job$[j]$ to represent the job occupying position j in the sequence. After the appropriate declaration we need only the following two lines in the optimize/solve block.

$$\text{alldifferent(job);}$$
$$\text{forall}(j \text{ in } 1..n)\ \text{C[job}[j]] = \text{sum}(k \text{ in } 1..n : k <= j)p[\text{job}[k]];$$

The second constraint set assures that the completion time for a job occupying position number j is the sum of job completion times through j (and illustrates the use of indexing variables described earlier).

In CP, controlling the organization of the search space is even more flexible than suggested above. As described earlier, instead of branching on the different values of a variable within its domain, we might wish to branch on a specific condition (i.e., insert a constraint or set of constraints). Upon backtracking we introduce its negation. In the aforementioned Model 2, for example, we might wish to define a binary search where at each node in the search tree we choose, for some job pair (j, k), either to have position$[j]$ < position$[k]$ or position$[k]$ < position$[j]$; i.e., at each choice point we first introduce the constraint position[j] < position[k], then upon backtracking introduce position[j] > position[k]. A simple OPL implementation of this would look like the following.

```
search{
    forall (i in 1..n-1)
        try position[i] < position[i+1] } position[i] > position[i+1] endtry;
};
```

In addition to controlling the organization of the search space, OPL offers the modeler several choices for the basic branching strategy. Although depth-first search is the default search strategy, OPL offers a number of other choices including a best-first strategy (commonly used in branch-and-bound codes for scheduling).

47.9 Concluding Remarks

47.9.1 CP and Scheduling Problems

We have argued that CP has good application to problems rife with logical constraints, particularly problems with many constraints involving few variables, because this affords numerous interactions inducing an abundance of domain reduction. Is this an inherent property of scheduling problems? We would so argue. Consider the definition of scheduling offered by Baker ([27], p. 2): "*Scheduling is the allocation of resources over time to perform a collection of tasks.*" Were it not for the little phrase "over time" then scheduling problems might not present the challenge that they do. It is this little phrase that begets the logical connections between the allocations. For example, consider the case of unary resources. Here the implication of the phrase "over time" means that no two jobs can be allocated in the same time interval — a set of two-variable (binary) constraints. For the case of sequence-dependent setup times the phrase "over time" affects pairs of chronologically adjacent allocations (another set of binary constraints). As another example consider the problem of assigning basketball referee crews to a league schedule of games. One constraint that makes this assignment problem a scheduling problem is the obvious requirement that a given crew may not be allocated to two different games that are scheduled to be played at the same time. It is the phrase "over time" which makes scheduling problems rife with logical conditions and constraints with few variables and thus amenable to the CP paradigm.

47.9.2 The Art of Constraint Programming

We have defined CP as a method for formulating and solving discrete variable constraint satisfaction/constrained optimization problems and have highlighted its reliance on logic-based computer programming. As such, CP involves choosing variables and data structures, representing the relations between these entities in a constraint store, and designing the search strategy that may ensue.

Employing CP for scheduling problems is a craft involving several interrelated skills perhaps not so customary to operations researchers. CP involves modeling skills for capturing relations (and including them in the constraint store) about the nature of the problem so as to enhance constraint propagation and domain reduction. Operations researchers have traditionally been trained to believe that, when formulating integer programs, expressing problems with as few variables and constraints as possible is generally better since it often leads to smaller memory requirements, faster solutions, and means for a cleaner, more elegant — more efficient formulation.[3] In CP, a compact formulation is not necessarily a good one. The goal is not to minimally represent the problem in terms of variable and constraint definitions but to use knowledge about the nature of the solution to fortify the constraint store. The discussion in the previous section

[3] However, as Williams (1999) points out with the prevalence of "presolve" algorithms in commercial solvers today, the modeler can afford to be more verbose in his modeling, without losing computational efficiency.

where the adjacency constraints were added to Model 2 illustrates this point. Although these constraints were unnecessary in terms of expressing a correct model specification, their introduction served to boost the deductive power of the constraint store and induce more domain reduction.

A second related skill in the craft of CP for scheduling is the ability to embed scheduling knowledge into the constraint set or into the design of the search strategy. For almost a half century there has been a steady stream of advances in the science of scheduling resulting in a great base of knowledge in the form of theorems and algorithms for specific scheduling problems. For scheduling problems we often know many pieces of information regarding the nature of optimum solutions, i.e., sufficient (but not necessary) conditions for optimality. For example, there are a number of precedence theorems now collected in the scheduling literature for the $1||\Sigma w_j T_j$ problem of the previous section [25,28–30]. Such theorems (scheduling domain knowledge) take the form "*If <condition> then there exists an optimum schedule in which job a precedes job b.*" Similarly there is a wealth of knowledge in scheduling about heuristic rules with empirical evidence to show they provide good results. The WMDD dispatching heuristic for $1||\Sigma w_j T_j$ described and illustrated in the previous section is a good example. Our experience in working with CP is that such scheduling-specific domain knowledge is relatively easy to implement within the CP framework. So we can look to CP as a tool that complements scheduling algorithmic knowledge by serving as a vehicle for its easy implementation. We see this as a trend, a trend that will undoubtedly be nurtured by further development and more widespread availability of CP tools and familiarity of operations researchers to the CP paradigm.

References

[1] Baptiste, P., Le Pape, C., and Nuijten, W., *Constraint-Based Scheduling: Applying Constraint Programming to Scheduling Problems*, Kluwer Academic Publishers, Boston, MA, 2001.

[2] Tsang, E., *Foundations of Constraint Satisfaction*, Wiley, Chichester, England, 1994.

[3] Van Hentenryck, P., *Constraint Satisfaction in Logic Programming*, MIT Press, Cambridge, MA, 1989.

[4] Brailsford, S.C., Potts, C.N., and Smith, B.M., Constraint satisfaction problems: algorithms and applications, *European Journal of Operational Research*, 119, 557, 1999.

[5] Williams, H. and Wilson, J., Connections between integer linear programming and constraint logic programming — an overview and introduction to the cluster of articles, *INFORMS Journal on Computing*, 10, 261, 1998.

[6] Freuder, E.C. and Wallace, M., Constraint satisfaction, in: Glover, F. and Kochenberger, G.A. (eds.) *Handbook of Metaheuristics*, Kluwer Academic Publishers, Boston, MA, 2003, chap. 14.

[7] Nuijten, W. and Aarts, E., A computational study of constraint satisfaction for multiple capacitated job shop scheduling, *European Journal of Operational Research*, 90, 269, 1996.

[8] Lustig, I. and Puget, J., Program does not equal program: constraint programming and its relation to mathematical programming, *Interfaces*, 31, 29, 2001.

[9] Darby-Dowman, K., Little, J., Mitra, G., and Zaffalon, M., Constraint logic programming and integer programming approaches and their collaboration in solving an assignment scheduling problem, *Constraints*, 1, 245, 1997.

[10] Jordan, C. and Drexl, A., A comparison of constraint and mixed integer programming solvers for batch sequencing with sequence dependent setups, *ORSA Journal on Computing*, 7, 160, 1995.

[11] Jain, V. and Grossman, I., Algorithms for hybrid MILP/CP models for a class of optimization problems, *INFORMS Journal on Computing*, 13, 258, 2001.

[12] Henz, M., Scheduling a major college basketball conference — revisited, *Operations Research*, 49, 163, 2001.

[13] Nemhauser, G. and Trick, M., Scheduling a major college basketball conference, *Operations Research*, 46, 1, 1998.

[14] Baker, K., Magazine, M., and Polak, G., Optimal block design models for course timetabling, *Operations Research Letters*, 30, 1, 2002.

[15] Valouxis, C. and Housos, E., Constraint programming approach for school timetabling, *Computers & Operations Research*, 30, 1555, 2003.

[16] Rousseau, L., Gendreau, M., and Pesant, G., The synchronized vehicle dispatching problem, *Working Paper*, Center for Transportation Research, University of Montreal, 2002.

[17] Hooker, J. N., Logic, optimization, and constraint programming, *INFORMS Journal on Computing*, 14, 295, 2002.

[18] Milano, M., Ottoson, G., Refalo, P., and Thorsteinsson, E., The role of integer programming techniques in constraint programming's global constraints, *INFORMS Journal on Computing*, 4, 387, 2002.

[19] Hillier, F.S. and Lieberman, G.J., *Introduction to Operations Research*, Holden-Day, Inc., San Francisco, 1980.

[20] Williams, H.P., *Model Building in Mathematical Programming*, Wiley, New York, 1999.

[21] Pinedo, M., *Scheduling: Theory Algorithms, and Systems*, Prentice Hall, Englewood Cliffs, NJ, 1995.

[22] Du, J. and Leung, J.Y.-T., Minimizing total tardiness on one machine is NP-hard, *Mathematics of Operations Research*, 15, 483, 1990.

[23] Potts, C.N. and Van Wassenhove, L.N., A branch and bound algorithm for the total weighted tardiness problem, *Operations Research*, 33, 363, 1985.

[24] Van Hentenryck, P., *The OPL Optimization Programming Language*, MIT Press, Cambridge, MA, 1999.

[25] Elmaghraby, S.E., The one machine sequencing problem with delay costs, *Journal of Industrial Engineering*, 19, 105, 1968.

[26] Kanet, J.J. and Li, X., A weighted modified due date rule for sequencing to minimize weighted tardiness, *Journal of Scheduling* (forthcoming).

[27] Baker, K. R., *Introduction to Sequencing and Scheduling*, Wiley, New York, 1974.

[28] Akturk, M. S. and Yildirim, M. B., A new dominance rule for the total weighted tardiness problem, *Production Planning & Control*, 10, 138, 1999.

[29] Rachamadugu, R.M.V., A note on the weighted tardiness problem, *Operations Research*, 35, 450, 1987.

[30] Rinnooy Kan, A.H.G., Lageweg, B.J., and Lenstra, J.K., Minimizing total costs in one-machine scheduling, *Operations Research*, 23, 908, 1975.

[31] Baker, K. R., *Elements of Sequencing and Scheduling*, K. R. Baker (ed.), Hanover, NH, 2000.

48

Batch Production Scheduling in the Process Industries

Karsten Gentner
University of Karlsruhe

Klaus Neumann
University of Karlsruhe

Christoph Schwindt
University of Karlsruhe

Norbert Trautmann
University of Karlsruhe

48.1 Introduction

In the process industries (for example, in chemical, pharmaceutical, food, or metal casting industries), final products are generally produced through several successive chemical or physical transformations (called *tasks*) from raw materials. We consider *batch production*, where a *batch* means a task together with the quantity produced. The input of a task is supposed to be consumed at its start and the output is available at its completion. The processing of a batch is called an *operation*. The processing time of an operation is assumed to be independent of the batch size. A task may be carried out more than once resulting in several operations.

To perform the tasks, several types of resources are required: *manpower*, *processing units* which have to be cleaned between successive operations (e.g., reactors, heaters, filters, or agitators), and *storage facilities* for products (e.g., containers or silos). The resources are available only in limited capacity, where the available manpower is not necessarily constant over time. Some tasks can be executed on alternative processing units that may differ in speed or cleaning time. Several identical processing units may form a larger *multi-processing system* (e.g., a multichambered autoclave) whose units can only be run jointly. *Break calendars* specify time intervals (e.g., lunch break) where special tasks cannot be performed. Some tasks may be interrupted (e.g., packaging) and then have to be resumed immediately after the break. Other tasks must not be interrupted (e.g., chemical reactions).

1-58488-397-9/$0.00+$1.50

Given the primary requirements for all final products, *short-term production planning* generally strives for processing all batches in a minimum amount of time or for minimally exceeding given due dates, that is, the *makespan* or *maximum lateness*, respectively, is to be minimized. Short-term planning can be decomposed into batching and batch scheduling. The *batching problem* provides the number and sizes of batches for all tasks observing inventory balance and storage capacity constraints. This problem can be formulated as a nonlinear mixed-integer program and transformed into a linear mixed-binary program of moderate size (cf. Neumann et al. [1,2]). In practice, batch sizes are often predetermined by technology, regulations, or as output at a higher planning level. Thus, in what follows, we will only deal with the *batch scheduling problem*, which given the set of all operations, seeks to schedule all operations on the processing units where scarce resources have to be taken into account.

Most solution approaches to short-term planning in process industries from literature are based on mixed-integer formulations, cf. e.g., Kondili et al. [3], Pinto and Grossmann [4], or Blömer and Günther [5,6]. Grunow et al. [7] and Timpe [8] have proposed hybrid methods combining mixed-integer and constraint programming. The drawback of mixed-integer programs is that the number of decision variables does not grow polynomially in the number of operations. The approach discussed in what follows models the batch scheduling problem as a *resource-constrained project scheduling problem*, where, basically, the number of decision variables equals the number of operations plus the number of alternative processing units. For the latter problem, efficient solution methods developed recently permit us to solve large problem instances from practice.

The rest of the chapter, where we sometimes follow Neumann et al. [1,2], is organized as follows. Section 48.2 presents a case study, which contains most of the features typical of batch production in process industries. Section 48.3 shows how to model the batch scheduling problem as a resource-constrained project scheduling problem. Section 48.4 sketches a solution procedure for the latter problem. In Section 48.5, a decomposition method is proposed for approximately solving large problem instances with more than thousand operations. Section 48.6 provides a solution to the case study from Section 48.2.

48.2 Case Study

In this section, we describe an extended version of a case study presented by Westenberger and Kallrath [9], which is based on an existing plant. To represent the batch production scheduling problem, we use a modification of the state-task network (STN) concept introduced by Kondili et al. [3]. An STN is a directed graph which includes three types of elements (cf. Figure 48.1):

1. *State nodes* represent the raw materials, intermediates, and final products. They are drawn as ellipses labelled with the respective state number, and the initial, minimum, and maximum stocks of the corresponding product. Some of the intermediate products cannot be stocked, which is indicated by the label "ns". The value ∞ for the initial or maximum stock means that there is sufficient initial stock or unlimited storage capacity, respectively.

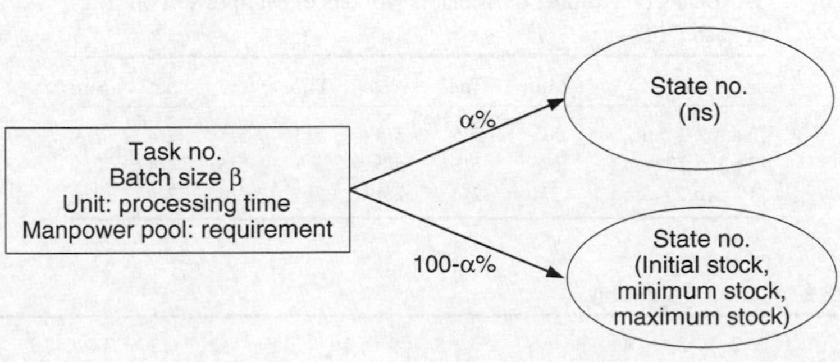

FIGURE 48.1 Elements of state-task networks.

2. *Task nodes* stand for the chemical or physical transformations of materials from one or more input states into one or more output states. Task nodes are represented by rectangles indicating the task number, the batch size, the required processing units, the processing time, and the requirement for each manpower pool. A manpower pool contains all workers with the same skills. If a task can be executed on alternative processing units, the corresponding processing times are listed for each of the units.
3. *Arcs* indicate flow of material. If more than one input product is consumed or output product is produced, the fractions of input or output products are given on the arcs.

Figure 48.2 shows the STN for the batch production process under study with 19 products, 17 tasks, and 9 processing units. The dark-shaded areas represent the processing units. Each area groups the nodes of the tasks that can be processed on the corresponding unit. The identical processing units U_6 and U_7 form a multiprocessing system (MPS), i.e., the starts and completions of overlapping operations carried out on units U_6 and U_7 have to be synchronized. Alternative processing units are available for executing tasks 10 to 14, 16, and 17. In order to guarantee purity of products, every processing unit has to be cleaned when passing to an operation with higher task index (tasks are numbered according to increasing quality requirements). The time needed for cleaning a processing unit equals one half of the processing time of the preceding operation.

As indicated in Figure 48.2, workers of different skills are needed for processing tasks. According to their skills, the workers are grouped in two manpower pools MP_1 and MP_2. Each of the tasks 1 to 12 requires one worker of the first pool, and tasks 13 to 17 take up two or three workers of the second pool. The workers operate in two day shifts and one night shift, which differ in the number of available workers (cf. Table 48.1). The planning period is one week starting on Monday at 6 a.m., and the time unit is one hour.

There is a meal break for the staff of pool MP_2 during the 6th hour of each day shift. We distinguish between tasks which may be interrupted by these production breaks (tasks being performed manually, e.g., packaging) and noninterruptible tasks (chemical reactions or tasks using expensive equipment). In our example, tasks 15 and 17 can be interrupted by production breaks, whereas the remaining tasks must be processed without any interruption.

The primary requirements for the final products corresponding to the states 15 to 19 within the planning horizon are given by the demand vector $(30, 30, 40, 20, 40)$. There is no initial stock of final products available. Table 48.2 shows the number of batches for each task (batching solution). Altogether there are 76 batches or operations. For what follows, we assume that the operations are numbered consecutively from 1 to 76 according to increasing task numbers.

The problem consists in assigning a processing unit and a start time to each operation such that:

(i) the makespan is minimized
(ii) a sufficient amount of each input product is available at the start of an operation
(iii) there is enough storage space available for the intermediate products to be stored

TABLE 48.1 Number of Available Workers of Manpower Pool 1/2 in Each Shift

	Mon	Tue	Wed	Thu	Fri	Sat	Sun
[6 a.m., 2 p.m.[	5/5	5/5	5/5	5/5	5/5	2/4	2/4
[2 p.m., 10 p.m.[	5/5	5/5	5/5	5/5	5/5	2/4	2/4
[10 p.m., 6 a.m.[	2/4	2/4	2/4	2/4	2/4	2/4	2/4

TABLE 48.2 Batching Solution

Task	1	2	3	4	5	6	7	8	9	10	11	12	13	14	15	16	17
# Batches	11	8	10	4	2	2	4	3	3	3	3	6	3	3	4	3	4

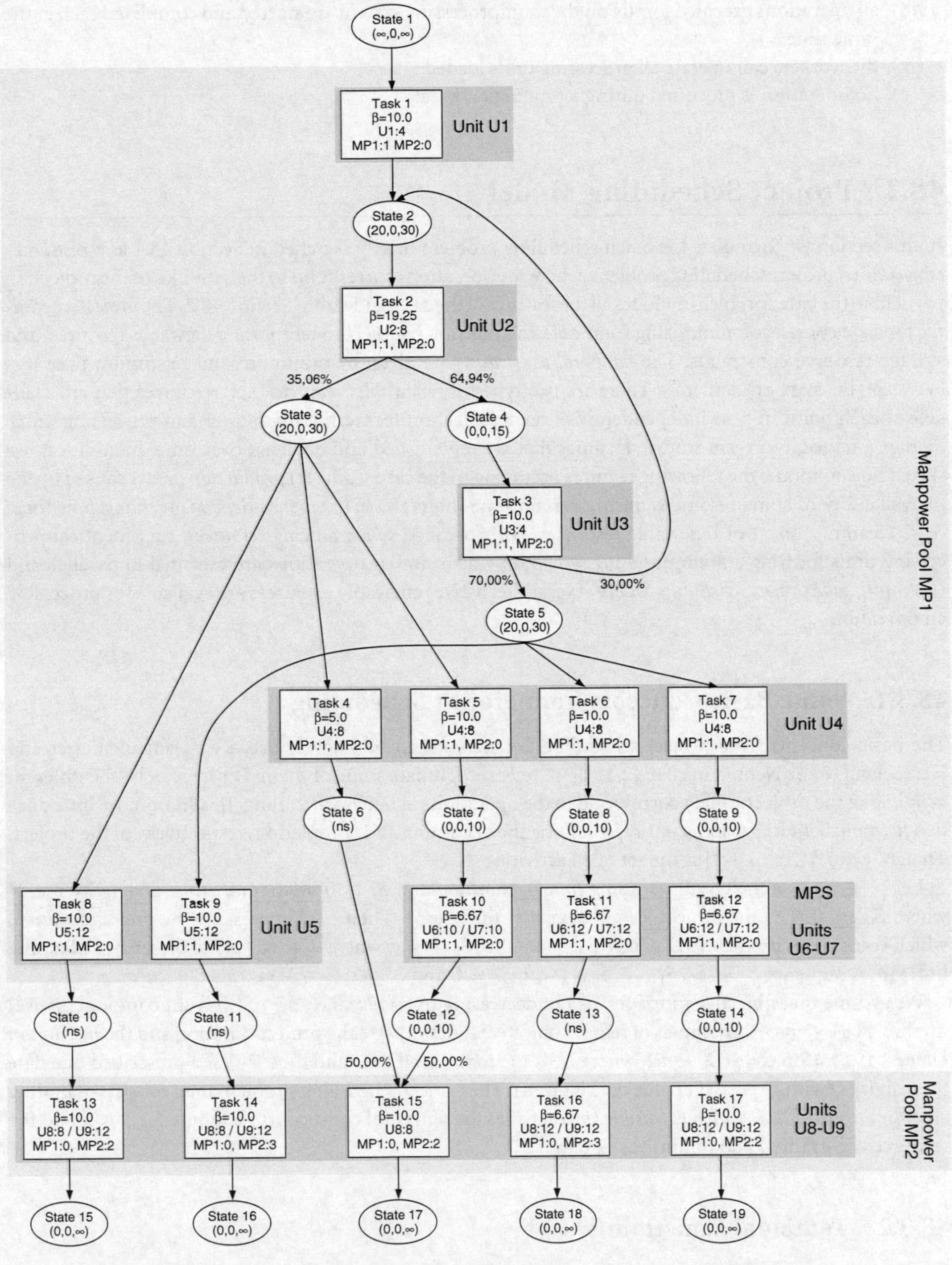

FIGURE 48.2 State-task network of case study.

(iv) the intermediate products which cannot be stored are consumed immediately
(v) each operation which can be executed on alternative processing units is assigned to exactly one of them
(vi) each processing unit performs at most one operation at a time
(vii) no processing unit carries out an operation during a cleaning time

(viii) all operations executed jointly on the multiprocessing system are started and completed each at the same time
(ix) the workers can operate all processing units loaded
(x) no operation is processed during a production break

48.3 Project Scheduling Model

In this section we formulate the batch scheduling problem briefly sketched in Section 48.1 as a resource-constrained project scheduling problem, where the operations correspond to the activities of the project. In particular, the latter problem includes all the features of the case study from Section 48.2. The *project scheduling problem* consists of minimizing some *objective function* f subject to temporal, renewable-resource, and storage-resource constraints. The *temporal constraints* are given by minimum and maximum time lags between (the start of) activities. There are two types of *renewable resources* (i.e., resources that are available at each point in time independently of their utilization formerly): manpower and processing units. *Storage resources* represent storage facilities that are replenished and depleted over time. In addition, we sketch how to model the following features occurring in the case study: (i) *production breaks* caused by the unavailability of some resources during certain time intervals, (ii) *sequence-dependent changeover times* (e.g., cleaning times) of renewable resources, (iii) so-called *synchronizing resources*, e.g., identical processing units forming a *multiprocessing system* on which similar operations are executed in parallel, and (iv) *multimode project scheduling* where several alternative renewable resources are available for processing an operation.

48.3.1 Some Basic Concepts from Project Scheduling

The basic concepts from project scheduling are assumed to be known (see, e.g., Demeulemester and Herroelen [10] or Neumann et al. [2,11]), we only recapitulate some of them. Let $1,\ldots,n$ be the different *activities* of the project, which correspond to the operations in batch scheduling. In addition, we introduce two *fictitious activities* 0 and $n+1$ representing the beginning and completion, respectively, of the project. Then $V = \{0, 1, \ldots, n+1\}$ is the set of all activities.

Let $p_i \in \mathbb{Z}_{\geq 0}$ be the given *processing time* or *duration* and $S_i \geq 0$ be the *start time* of activity $i \in V$ where $S_0 := 0$ (i.e., the project always begins at time zero). Then S_{n+1} represents the *project duration*, which coincides with the makespan. We assume that $S_{n+1} \leq \overline{d}$ where $\overline{d} \in \mathbb{Z}_{\geq 0}$ is the prescribed planning horizon. A sequence $S = (S_0, S_1, \ldots, S_{n+1})$ with $S_0 = 0$ and $S_i \geq 0$ $(i \in V)$ is called a *schedule*.

We assume that objective function f is nondecreasing or *regular*, i.e., $S \leq S'$ (meant componentwise) implies $f(S) \leq f(S')$. Examples of regular objective functions are the project duration and the *maximum lateness* $L_{\max} = \max_{i \in V}(C_i - d_i)$, where C_i is the *completion time* and $d_i \in \mathbb{Z}_{\geq 0}$ is a prescribed due date for activity i (with $d_i := \overline{d}$ if no due date is given). The project scheduling problem then consists of finding an *optimal schedule* S, which is *feasible* (i.e., satisfies the temporal constraints including $S_{n+1} \leq \overline{d}$ and the resource constraints) and minimizes function f.

48.3.2 Temporal Constraints

A *minimum time lag* $d_{ij}^{\min} \in \mathbb{Z}_{\geq 0}$ or a *maximum time lag* $d_{ij}^{\max} \in \mathbb{Z}_{\geq 0}$ can be prescribed between the start of different activities or operations $i, j \in V$, i.e., $S_j - S_i \geq d_{ij}^{\min}$ or $S_j - S_i \leq d_{ij}^{\max}$, respectively. For example, if operations i and j are carried out in succession on the same processing unit that has to be cleaned after executing operation i, then $d_{ij}^{\min} = p_i + c_i$ where $c_i \in \mathbb{Z}_{\geq 0}$ is the cleaning time (sequence-dependent changeover times will be discussed in Subsection 48.3.6). When operation i provides a perishable product that has to be consumed without any delay by some operation j, we have $d_{ij}^{\min} = d_{ij}^{\max} = p_i$. To satisfy the condition $S_{n+1} \leq \overline{d}$, we set $d_{0,n+1}^{\max} := \overline{d}$. If for two activities $i, j \in V$, there is a minimum time lag $d_{ij}^{\min}$, we put $\delta_{ij} := d_{ij}^{\min}$, and if there is a maximum time lag $d_{ij}^{\max}$, we set $\delta_{ji} := -d_{ij}^{\max}$. Let E be the set of activity pairs (i, j) for which there is a time lag δ_{ij}, then

$$S_j - S_i \geq \delta_{ij} \quad ((i, j) \in E) \tag{48.1}$$

is the set of *temporal constraints*. A schedule S that satisfies (48.1) is called *time-feasible*.

It is well-known that an *activity-on-arc project network* N can be uniquely assigned to the underlying project (cf. e.g., Neumann et al. [2]). The nodes of N are identified with the activities $i \in V$, and the arcs $\langle i, j\rangle$ of N correspond to the activity pairs $(i, j) \in E$ where δ_{ij} is the weight of arc $\langle i, j\rangle$. Since there are maximum time lags in addition to minimum ones, network N contains cycles.

48.3.3 Renewable Resources

Each processing unit or worker represents a renewable resource of capacity one. R identical processing units or R workers with the same skills correspond to a renewable resource of capacity R. In the case study from Section 48.2, the workers form two manpower pools representing two renewable resources of capacity five each. The identical processing units U_6 and U_7 are combined in a pool of capacity two, whereas processing units U_1 to U_5, U_8, and U_9 are renewable resources of capacity one each.

Let $\mathcal{R}^\rho$ be the set of all renewable resources, $R_k \in \mathbb{N}$ be the capacity of resource $k \in \mathcal{R}^\rho$, and $r_{ik} \in \{0, 1, \ldots, R_k\}$ be the amount of resource k used by activity $i \in V$. In the case study, for manpower pool k, r_{ik} is the number of workers required for carrying out operation i (one worker of pool 1 for each operation belonging to tasks 1 to 12, and two or three workers of pool 2 for each operation from tasks 13 to 17). If operation i has to be executed on processing unit (pool) k, the corresponding resource requirement is $r_{ik} = 1$.

Given schedule S, $\mathcal{A}(S, t) := \{i \in V \mid S_i \leq t < S_i + p_i\}$ is the set of activities in progress at time t, and $r_k^\rho(S, t) := \sum_{i \in \mathcal{A}(S,t)} r_{ik}$ is the amount of resource $k \in \mathcal{R}^\rho$ used at time t. Then the *renewable-resource constraints* are

$$r_k^\rho(S, t) \leq R_k \quad (k \in \mathcal{R}^\rho, \ 0 \leq t \leq \bar{d}) \tag{48.2}$$

A schedule S that satisfies (48.2) is termed *resource-feasible*.

In the case study, there is a reduced size of work force during night shifts and weekends, that is, the capacity of the corresponding resource is no longer constant over time. The latter case can be modelled as follows. Let $a_k(t)$ be the capacity of renewable resource k at time t where $a_k(\cdot)$ is a (right-continuous) step function, and let $R_k := \max_{0 \leq t \leq \bar{d}} a_k(t)$. For each (half-open) interval $[t', t''[$ where $a_k(\cdot)$ is constant, we introduce an auxiliary activity i' with duration $p_{i'} := t'' - t'$, start time $S_{i'} := t'$, and resource requirement $r_{i'k} := R_k - a_k(t')$. To enforce the fixed start time $S_{i'} = t'$, we additionally introduce the time lags $d_{0i'}^{\min} := d_{0i'}^{\max} := t'$.

48.3.4 Storage Resources

Nonperishable products can be stocked in storage facilities, which are depleted and replenished over time and represent storage resources. Let $\mathcal{R}^\sigma$ be the set of all storage resources, and let $\underline{R}_k \geq 0$ and $\overline{R}_k \geq \underline{R}_k$ be the given minimum inventory (safety stock) and maximum inventory (capacity), respectively, of storage resource $k \in \mathcal{R}^\sigma$. Moreover, let r_{ik} be the (possibly negative) increase in the inventory level of storage resource k caused by operation or activity i. If $r_{ik} < 0$, then activity i depletes resource k by $-r_{ik}$ units at its start time S_i. If $r_{ik} > 0$, activity i replenishes resource k by r_{ik} units at its completion time $S_i + p_i$. In the case study, r_{ik} is determined as follows. The initial stock of storage resource k corresponds to r_{0k}. Let $\tau(i)$ be the task belonging to operation i. If task $\tau(i)$ consumes a product P, then the depletion $-r_{ik} > 0$ equals the batch size of $\tau(i)$ multiplied by the corresponding input proportion of P. For example, task 15 with a batch size of 10 consists of four batches representing operations 66 to 69 (cf. Table 48.2). Fifty percent of the input of task 15 represent nonperishable product P_{12}, which is supposed to be stocked in storage facility 12. Then $-r_{i,12} = 10 \times 0.5 = 5$ for $i = 66, \ldots, 69$. Analogously, if task $\tau(i)$ produces product P, the replenishment $r_{ik} > 0$ equals the batch size of $\tau(i)$ multiplied by the output proportion of P.

Let $V_k^- := \{i \in V \mid r_{ik} < 0\}$ and $V_k^+ := \{i \in V \mid r_{ik} > 0\}$ be the sets of all activities depleting and replenishing, respectively, resource $k \in \mathcal{R}^\sigma$, where, for simplicity, we assume that $V_k^- \cap V_k^+ = \emptyset$. Given schedule S, $\mathcal{A}_k(S,t) := \{i \in V_k^- \mid S_i \le t\} \cup \{i \in V_k^+ \mid S_i + p_i \le t\}$ is the set of activities depleting or replenishing resource $k \in \mathcal{R}^\sigma$ by time t, and $r_k^\sigma(S,t) := \sum_{i \in \mathcal{A}_k(S,t)} r_{ik}$ is the inventory level of storage resource k at time t. Then the *storage-resource* or *inventory constraints* are

$$\underline{R}_k \le r_k^\sigma(S,t) \le \overline{R}_k \quad (k \in \mathcal{R}^\sigma, 0 \le t \le \overline{d}) \tag{48.3}$$

A schedule S that satisfies (48.3) is called *inventory-feasible.*

In summary, our *basic project scheduling problem* reads as follows

$$\begin{aligned} \text{Minimize} \quad & f(S) \\ \text{subject to} \quad & (48.1), (48.2), \text{and } (48.3) \\ & S_0 = 0, \ S_i \ge 0 \quad (i \in V) \end{aligned} \tag{48.4}$$

In other words, we seek to determine a schedule S which is *feasible*, i.e., time-, resource-, and inventory-feasible, and minimizes function f. As already mentioned, such a schedule S is said to be *optimal.*

Next, we briefly sketch how to model the special features of our case study already mentioned: (i) production breaks, (ii) sequence-dependent changeover times, (iii) synchronizing resources, and (iv) multimode project scheduling. For more details we refer to Neumann et al. [2].

48.3.5 Break Calendars

In the case study, there are meal breaks for the staff of pool 2 in the day shifts during which the operations belonging to tasks 13 to 17 cannot be executed. In practice, production breaks where some renewable resources are not available may also arise from weekends or holidays. A break can be represented by a *break calendar*, i.e., a (right-continuous) step function $b{:}[0, \overline{d}] \to \{0, 1\}$, where $b(t){:} = 0$ if t falls into a break and $b(t){:} = 1$, otherwise. $\int_{t'}^{t''} b(\tau)d\tau$ is the *total working time* in interval $[t', t''[$. Different resources may have different calendars. Thus, we assign an *activity calendar* b_i to each activity $i \in V$, where $b_i(t){:} = 0$ if there is some resource used by activity i which is not available at time t and $b_i(t){:} = 1$, otherwise.

In the case study, the operations from tasks 15 and 17 may be interrupted during a production break (and are resumed at the end of the break), whereas the operations from tasks 13, 14, and 16 must be processed without any interruption. In general, let V^{int} be the set of activities that may be interrupted and let $V^{nint} := V \setminus V^{int}$ be the set of activities that must not be interrupted. For $i \in V^{int}$, we assume that the length of a working time interval between two successive breaks is at least ε_i, where $\varepsilon_i > 0$ is a given minimum execution time, e.g., one unit of time. For $i \in V^{nint}$, we set $\varepsilon_i := p_i$. Hence,

$$b_i(t) := 1 \quad \text{for } S_i \le t < S_i + \varepsilon_i \quad (i \in V) \tag{48.5}$$

The completion time of activity i generally depends on activity calendar b_i:

$$C_i(S_i) = \begin{cases} \min\left\{t \ge S_i + p_i \mid \int_{S_i}^{t} b_i(\tau)d\tau = p_i\right\}_j & \text{for } i \in V^{int} \\ S_i + p_i, \quad \text{for } i \in V^{nint} \end{cases}$$

Similarly, the time lags may depend on so-called time lag calendars b_{ij} and are now denoted by $\Delta_{ij}(S_i)$. Thus,

$$S_j - S_i \ge \Delta_{ij}(S_i) \quad ((i,j) \in E) \tag{48.6}$$

and the temporal constraints (48.1) have to be replaced by the *calendar constraints* (48.5) and (48.6).

48.3.6 Sequence-Dependent Changeover Times

In practice, there is often a *changeover time* $\vartheta_{ij}^k \in \mathbb{Z}_{\geq 0}$ between the execution of two successive activities i and j on one and the same unit of renewable resource $k \in \mathcal{R}^\rho$, e.g., for cleaning that unit. During the changeover, that resource unit is not available for processing an activity. The changeover time generally depends on the sequence of activities i, j, i.e., $\vartheta_{ij}^k \neq \vartheta_{ji}^k$. In the case study, if operation j executed on processing unit (pool) k produces a product whose quality is higher than that of the output product(s) of operation i, then ϑ_{ij}^k equals the cleaning time of resource k; otherwise, $\vartheta_{ij}^k = 0$. Also, if i is the last operation on resource k, there is a positive cleaning time $\vartheta_{i,n+1}^k$.

We assume that the so-called *weak triangle inequality*

$$\vartheta_{hi}^k + p_i + \vartheta_{ij}^k \geq \vartheta_{hj}^k \tag{48.7}$$

is satisfied for all triples of different activities h, i, and j that use resource $k \in \mathcal{R}^\rho$. If (48.7) were not satisfied, it would be possible to save changeover time by processing additional activities. Moreover, for notational convenience we assume that $\vartheta_{0i}^k = \vartheta_{i,n+1}^k = 0$ and $\vartheta_{ii}^k = \infty$ for all $k \in \mathcal{R}^\rho$ and $i \in V_k$, where $V_k := \{i \in V \mid r_{ik} > 0\}$ is the set of all activities using resource $k \in \mathcal{R}^\rho$.

The condition that changeover times are to be observed and resource demands by activities and changeovers do not exceed the capacities R_k of resources $k \in \mathcal{R}^\rho$ can be modelled as follows. For $k \in \mathcal{R}^\rho$ and $i \in V_k$, let $X_k(i)$ with $|X_k(i)| = r_{ik}$ be the set of units of resource k allocated to activity i. A schedule S is called *changeover-feasible* if for each resource $k \in \mathcal{R}^\rho$, mapping $X_k : V_k \rightarrow 2^{\mathbb{N}}$ can be chosen such that

$$\left.\begin{array}{l} X_k(i) \cap X_k(j) \neq \emptyset \text{ implies} \\ S_j \geq S_i + p_i + \vartheta_{ij}^k \text{ or } S_i \geq S_j + p_j + \vartheta_{ji}^k \end{array}\right\} \quad (i, j \in V_k, i \neq j) \tag{48.8}$$

and

$$X_k(i) \subseteq \{1, \ldots, R_k\} \quad (i \in V_k) \tag{48.9}$$

Condition (48.8) says that activities i and j processed on one and the same unit of resource $k \in \mathcal{R}^\rho$ must not overlap where the changeover time is taken into account. Condition (48.9) ensures that the capacity R_k of resource k is not exceeded. Note that a changeover-feasible schedule is resource-feasible as well.

48.3.7 Synchronizing Resources

Certain renewable resources may process several activities sharing similar properties (and thus forming a so-called *activity family*) batchwise, i.e., simultaneously, where their execution is begun at the same time. We then speak of *synchronizing resources*, also called *batching machines* in machine scheduling (cf. e.g., Potts and Kovalyov [12]). In the case study, the multiprocessing system consisting of identical processing units U_6 and U_7 represents one synchronizing resource. The operations are grouped in activity families in such a way that each family contains operations with identical processing times.

Let $\mathcal{R}^\beta$ be the set of all synchronizing or batching resources. Since $\mathcal{R}^\beta \subseteq \mathcal{R}^\rho$, the resources $k \in \mathcal{R}^\beta$ have to satisfy the renewable-resource constraints (48.2). Let φ_i denote the activity family that contains activity i. Moreover, let again $\mathcal{A}(S, t)$ be the set of activities in progress at time t and V_k be the set of activities using resource k. Then the so-called *homogeneity constraints*

$$\varphi_i = \varphi_j \quad (i, j \in V_k \cap \mathcal{A}(S,t) \text{ for some } k \in \mathcal{R}^\beta \text{ and some } t \in [0, \overline{d}]) \tag{48.10}$$

guarantee that only activities from one and the same family can overlap in time on a synchronizing resource. The *simultaneity constraints*

$$S_i = S_j \quad (i, j \in V_k \cap \mathcal{A}(S,t) \text{ for some } k \in \mathcal{R}^\beta \text{ and some } t \in [0, \overline{d}]) \tag{48.11}$$

ensure that activities being executed simultaneously on a synchronizing resource have to be begun at the same time. A schedule S that satisfies (48.2) with $\mathcal{R}^{\beta}$ instead of $\mathcal{R}^{\rho}$, (48.10), and (48.11) is called *synchronization-feasible.*

48.3.8 Multimode Project Scheduling

In the case study, tasks 13, 14, 16, and 17 can be carried out on alternative processing units belonging to different pools. For each pool of identical processing units an operation can be executed on, we introduce an *execution mode* of that operation, and e.g., the processing times and resource requirements refer to the individual execution modes instead of the operations. We then speak of a *multimode project scheduling problem*, which decomposes into two interdependent subproblems, a *mode assignment problem* and a *single-mode project scheduling problem.*

In general, let $\mathcal{M}_i$ be the set of alternative execution modes for activity $i \in V$. Then a *mode assignment* is a vector $x = (x_{im})_{i \in V,\, m \in \mathcal{M}_i}$ with

$$x_{im} \in \{0, 1\} \quad (i \in V,\ m \in \mathcal{M}_i) \tag{48.12}$$

where $x_{im} := 1$ if activity i is carried out in mode m and $x_{im} := 0$, otherwise, and

$$\sum_{m \in \mathcal{M}_i} x_{im} = 1 \quad (i \in V) \tag{48.13}$$

(i.e., each activity has to be executed in exactly one mode).

Each mode assignment x defines a corresponding single-mode project scheduling problem whose input data depend on x. Let p_{im} be the duration of activity i if carried out in mode m. Then $p_i(x) := \sum_{m \in \mathcal{M}_i} p_{im} x_{im}$ is the duration of activity i given x. Analogously, let $\delta_{imjm'}$ be the time lag corresponding to activity pair (i, j) where activities i and j are carried out in modes $m \in \mathcal{M}_i$ and $m' \in \mathcal{M}_j$, respectively. Then

$$\delta_{ij}(x) = \sum_{m \in \mathcal{M}_i} \sum_{m' \in \mathcal{M}_j} \delta_{imjm'} x_{im} x_{jm'}$$

is the time lag for activity pair (i, j) given x, and the temporal constraints are

$$S_j - S_i \geq \delta_{ij}(x) \quad ((i, j) \in E) \tag{48.14}$$

In the same way we obtain the sequence-dependent changeover time

$$\vartheta^k_{ij}(x) = \sum_{m \in \mathcal{M}_i} \sum_{m' \in \mathcal{M}_j} \vartheta^k_{imjm'} x_{im} x_{jm'}$$

from activity i to activity j on renewable resource $k \in \mathcal{R}^{\rho}$ given x, where $\vartheta^k_{imjm'}$ denotes the changeover time for execution modes $m \in \mathcal{M}_i$ and $m' \in \mathcal{M}_j$. Let r_{ikm} be the amount of renewable resource $k \in \mathcal{R}^{\rho}$ used by activity i if carried out in mode m. Given assignment x, $r_{ik}(x) := \sum_{m \in \mathcal{M}_i} r_{ikm} x_{im}$ is the requirement of activity i for resource $k \in \mathcal{R}^{\rho}$. Given schedule S in addition, $\mathcal{A}(S, t, x) := \{i \in V \mid S_i \leq t < S_i + p_i(x)\}$ is the set of activities in progress at time t, and $r^{\rho}_k(S, t, x) := \sum_{i \in \mathcal{A}(S,t,x)} r_{ik}(x)$ is the requirement for resource $k \in \mathcal{R}^{\rho}$ at time t. Then the renewable-resource constraints read

$$r^{\rho}_k(S, t, x) \leq R_k \quad (k \in \mathcal{R}^{\rho},\ 0 \leq t \leq \overline{d}) \tag{48.15}$$

Assuming for simplicity that only renewable resources are needed and that none of the special features discussed in Subsections 48.3.5, 48.3.6, or 48.3.7 are present, we are then looking for an assignment x and a schedule S which satisfy the constraints (48.12) to (48.15) and minimize objective function f.

48.4 Solution Procedure

In Section 48.3 we have shown how the batch scheduling problem can be modelled as a multimode project scheduling problem with temporal constraints, calendars, as well as renewable, storage, and synchronizing resources. In this section we describe a solution procedure of type branch-and-bound for the project scheduling problem. The enumeration scheme of the algorithm, which is discussed in Subsection 48.4.1, is based on relaxing the mode assignment constraints and the different types of resource constraints. A solution to such a relaxation is generally infeasible due to two reasons. First, there exist activities for which no execution mode has been assigned and second, the corresponding schedule does not satisfy all resource constraints. In the course of the algorithm, the relaxation is then stepwise refined by iteratively selecting execution modes for activities and introducing precedence relationships between activities whose joint resource requirements lead to a conflict on some resource. In Subsection 48.4.2, we explain how to find appropriate sets of conflicting activities. Depending on the type of resource constraint violated, the precedence relationships introduced give rise to additional minimum and maximum time lags between activities, which are the subject of Subsection 48.4.3. In Subsection 48.4.4, we are concerned with implementation issues of the algorithm. In particular, we outline the main principles of a filtered beam search procedure that is based on the branch-and-bound algorithm.

48.4.1 Enumeration Scheme

Let (PSP) denote the project scheduling problem to be solved. As we have seen in Subsection 48.3.8, problem (PSP) decomposes into a mode assignment problem and a single-mode project scheduling problem for given mode assignment. A feasible solution to (PSP) consists in a *schedule-assignment pair* (S, x), where x is a mode assignment and S is a feasible schedule for x. Basically, (PSP) can be solved by enumerating alternative mode assignments and computing an optimal schedule for each mode assignment. This *sequential approach* has been used by De Reyck and Herroelen [13] for a tabu search procedure solving a multimode project scheduling problem with renewable resources. In what follows we describe an *integrated approach* devised by Heilmann [14], where the mode assignment and project scheduling problems are considered simultaneously.

The constraints that make (PSP) a hard problem are the mode assignment constraints (48.13) and the different types of resource constraints introduced in Subsections 48.3.3, 48.3.4, 48.3.6, and 48.3.7. By relaxing the mode assignment constraints we allow for *partial mode assignments* $\underline{x} = (\underline{x}_{im})_{i \in V, m \in \mathcal{M}_i}$ where $\underline{x}_{im} \in \{0, 1\}$ for all $i \in V$, $m \in \mathcal{M}_i$ and $\sum_{m \in \mathcal{M}_i} \underline{x}_{im} \leq 1$. Analogously to (full) mode assignments, we can associate a single-mode project scheduling problem (PSP($\underline{x}$)) to each partial mode assignment $\underline{x}$. We construct (PSP($\underline{x}$)) in such a way that for each full mode assignment $x \geq \underline{x}$, problem (PSP($\underline{x}$)) represents a relaxation of problem (PSP(x)). To this end, consider the sets

$$\mathcal{M}_i(\underline{x}) := \begin{cases} \mathcal{M}_i, & \text{if } \sum_{m \in \mathcal{M}_i} \underline{x}_{im} = 0 \\ \{m\} \text{ with } \underline{x}_{im} = 1, & \text{otherwise} \end{cases}$$

of modes that can be assigned to activity $i \in V$ in any full mode assignment $x \geq \underline{x}$. We define the durations, time lags, resource requirements, and changeover times for problem (PSP($\underline{x}$)) to be $p_i(\underline{x}) := \min_{m \in \mathcal{M}_i(\underline{x})} p_{im}$, $\delta_{ij}(\underline{x}) := \min_{m \in \mathcal{M}_i, m' \in \mathcal{M}_j} \delta_{imjm'}$, $r_{ik}(\underline{x}) := \min_{m \in \mathcal{M}_i(\underline{x})} r_{ikm}$, and $\vartheta_{ij}^k(\underline{x}) := \min_{m \in \mathcal{M}_i, m' \in \mathcal{M}_j} \vartheta_{imjm'}^k$ for all $i, j \in V$ and all $k \in \mathcal{R}^\rho$. Obviously, $p_i(\underline{x})$, $\delta_{ij}(\underline{x})$, $r_{ik}(\underline{x})$, and $\vartheta_{ij}^k(\underline{x})$ represent lower bounds on the respective quantities in all full mode assignments $x \geq \underline{x}$. We say that a schedule S is *feasible for* $\underline{x}$ if S satisfies all calendar and resource constraints of (PSP($\underline{x}$)).

To obtain a problem that can be solved efficiently, we additionally relax the resource constraints of (PSP($\underline{x}$)). The resulting problem is referred to as a *temporal scheduling problem* (TSP($N, \underline{x}$)) with calendars in project network N. Problem (TSP($N, \underline{x}$)) consists in finding a time-feasible schedule S satisfying the calendar constraints (48.5) and (48.6) and minimizing regular objective function f. Such a schedule can be determined by using a polynomial-time label-correcting algorithm which starting with $S = 0$ iteratively delays activities $i \in V$ until all calendar constraints are met (see Franck et al. [15]). It can be shown that the resulting time-feasible schedule is componentwise minimal and hence solves (TSP($N, \underline{x}$)) to optimality independently of the objective function f under consideration.

Since we have omitted the resource constraints, the schedule S arising from temporal scheduling is generally not feasible for $\underline{x}$. In this case, S causes so-called *resource conflicts* at certain points in time $t \in [0, \overline{d}]$. We distinguish between the following types of resource conflicts:

1. *capacity overflow* on a renewable resource $k \in \mathcal{R}^\rho$, i.e., $r_k^\rho(S,t) > R_k$ if k does not require changeovers, and $X_k(i) \not\subseteq \{1, \ldots, R_k\}$, otherwise,
2. *inventory shortage* for a storage resource $k \in \mathcal{R}^\sigma$, i.e., $r_k^\sigma(S,t) < \underline{R}_k$,
3. *inventory excess* for a storage resource $k \in \mathcal{R}^\sigma$, i.e., $r_k^\sigma(S,t) > \overline{R}_k$,
4. *homogeneity conflict* for a synchronizing resource $k \in \mathcal{R}^\beta$, i.e., $\varphi_i \neq \varphi_j$ for some $i, j \in V_k \cap \mathcal{A}(S,t)$, or
5. *simultaneity conflict* for a synchronizing resource $k \in \mathcal{R}^\beta$, i.e., $S_i \neq S_j$ for some $i, j \in V_k \cap \mathcal{A}(S,t)$.

Each of those resource conflicts can be resolved stepwise by introducing precedence relationships $i \prec j$ among conflicting activities i, j. Depending on the type of resource conflict, the precedence relationships $i \prec j$ can be formulated as specific minimum or maximum time lags $d_{ij}^{\min}$ or $d_{ji}^{\max}$, respectively, which in turn can be represented as arcs $\langle i, j \rangle$ with weights δ_{ij} in an (expanded) project network $\overline{N}$. In general, one and the same resource conflict can be removed in different ways, which correspond to alternative sets of precedence relationships. Those alternatives are enumerated in the course of the branch-and-bound procedure.

In principle, the integrative approach to solving project scheduling problem (PSP) is as follows. We start with the empty partial mode assignment $\underline{x} = 0$ where each activity i may be executed in any execution mode $m \in \mathcal{M}_i$ and we initialize the expanded project network $\overline{N}$ with network N. In each iteration of the algorithm, we then solve the corresponding temporal scheduling problem (TSP($\overline{N}, \underline{x}$)) by applying the label-correcting algorithm. If the resulting schedule S is feasible for $\underline{x}$, we check whether all activities have been assigned to an execution mode. In the affirmative case, $\underline{x}$ is a (full) mode assignment and schedule-assignment pair $(S, \underline{x})$ is a feasible solution to (PSP). If in addition $f(S)$ is smaller than the objective function value f^* of the best feasible solution found so far, we store $(S, \underline{x})$ and update f^*. If S is feasible for $\underline{x}$ but partial mode assignment $\underline{x}$ is still incomplete, we select some activity $i \in V$ with $\sum_{m \in \mathcal{M}_i} \underline{x}_{im} = 0$ and branch over all alternative assignments of an execution mode $m \in \mathcal{M}_i$ to i. In the case where the schedule S obtained from the temporal scheduling computations is not feasible for $\underline{x}$, we determine a resource conflict for S and branch over alternative precedence relationships $i \prec j$ between two activities i and j which are involved in the conflict. For each $i \prec j$, we add an arc $\langle i, j \rangle$ with weight δ_{ij} to expanded project network $\overline{N}$. Arc weight δ_{ij} is equal to the time lag to be introduced between activities i and j, which depends on the type of resource conflict at hand. By selecting an execution mode for some activity or adding an arc to $\overline{N}$ we obtain a refined temporal scheduling problem, which is then solved again. The steps performing temporal scheduling, assigning execution modes to activities, and adding arcs to $\overline{N}$ are reiterated until all alternative mode assignments and all alternative ways of resolving emerging resource conflicts have been investigated.

Algorithm 48.1 shows a depth-first search implementation of this procedure, where Q is a stack and temporal scheduling yields schedule $S^\infty := (\infty, \ldots, \infty)$ if problem (TSP($\overline{N}, \underline{x}$)) is not solvable. In the latter case, the precedence relationships introduced contradict the original temporal constraints of problem (PSP). Provided that there exists a feasible solution to (PSP), the algorithm yields an optimal solution (S^*, x^*) in a finite number of iterations. In Subsections 48.4.3 and 48.4.4 we explain in more detail how to detect and to remove resource conflicts.

Algorithm 48.1 Enumeration scheme for solving (PSP)

Input: A multimode project scheduling problem (PSP).
Output: An optimal solution (S^*, x^*) to (PSP).
  put $\overline{N} := N$, $\underline{x} := 0$, $Q := \{(\overline{N}, \underline{x})\}$, and $f^* := \infty$;
  while $Q \neq \emptyset$ **do**
    remove some $(\overline{N}, \underline{x})$ from Q;
    determine schedule S by solving temporal scheduling problem (TSP$(\overline{N}, \underline{x})$);
    if S is feasible for $\underline{x}$ **then**
      if $\underline{x}$ is full mode assignment **then**
        if $f(S) < f^*$ **then** put $(S^*, x^*) := (S, \underline{x})$ and $f^* := f(S)$;
      else
        determine some $i \in V$ with $\sum_{m \in \mathcal{M}_i} \underline{x}_{im} = 0$;
        for all $m \in \mathcal{M}_i$ **do**
          put $\underline{x}' := \underline{x}$, $\underline{x}'_{im} := 1$, and add $(\overline{N}, \underline{x}')$ to Q;
    elsif $S < S^\infty$ **then**
      determine resource conflict and list L of precedence relationships $i \prec j$;
      for all precedence relationships $i \prec j$ on list L **do**
        put $\overline{N}' := \overline{N}$, add arc $\langle i, j \rangle$ with weight δ_{ij} to $\overline{N}'$, and add $(\overline{N}', \underline{x})$ to Q;
  if $f^* < \infty$ **then return** optimal solution (S^*, x^*);

48.4.2 Detecting Resource Conflicts

We consider a time-feasible schedule S which is obtained by performing the temporal scheduling step in some iteration of the enumeration scheme from Subsection 48.4.1. In what follows, we show how resource conflicts induced by S can be identified efficiently and how for a given resource conflict, an appropriate set of conflicting activities can be constructed. For simplicity, in what follows we omit designator $\underline{x}$ in the notations of durations, time lags, resource requirements, and changeover times belonging to project scheduling problem (PSP$(\underline{x})$) in question.

From the definition of sets $\mathcal{A}(S, t)$ it follows that for a renewable resource $k \in \mathcal{R}^\rho$ with vanishing changeover times, the maximum requirement $r_k^\rho(S, t)$ for k always occurs at the start time of some activity $i \in V$. Moreover, for a storage resource $k \in \mathcal{R}^\sigma$ the definition of sets $\mathcal{A}_k(S, t)$ implies that the inventory of k always attains its minimum at the start time of an activity $i \in V_k^-$. Symmetrically, it holds that there always exists an activity $i \in V_k^+$ such that the inventory of k is maximum at the completion time $S_i + p_i$ of i. Now consider some synchronizing resource $k \in \mathcal{R}^\beta$. The homogeneity constraints for k say that any two activities overlapping in time on k must belong to the same activity family. This condition is met precisely if at the start time of any activity $i \in V_k$, no activity $j \in V_k$ from a different family is executed on k. Similarly, the simultaneity constraints for k, which require the simultaneous start of activities overlapping on k, are satisfied exactly if at the start time of any activity $i \in V_k$, no activity $j \in V_k$ started earlier is still in progress. In summary, the resource constraints for renewable resources without changeover times, for storage resources, and for synchronizing resources can be checked by building up and scanning a sorted list of activity start and completion times for each resource, which altogether can be done in $O(n \log n + mn)$ time where m denotes the number of resources under consideration.

We now turn to the renewable resources $k \in \mathcal{R}^\rho$ that necessitate a sequence-dependent changeover time between the execution of successive activities. Here, for given schedule S and resource k the problem arises to allocate the units of resource k to the activities $i \in V_k$ such that the time interval between the execution of two activities sharing common resource units is sufficiently large to perform the changeover (see condition (48.8) from Subsection 48.3.6). The changeover-feasibility of schedule S requires that the number of resource units needed for such an allocation does not exceed the capacity R_k of resource k (cf. condition (48.9) and recall that this condition also includes the renewable-resource constraints (48.2)).

Next, we describe an efficient way proposed in Neumann et al. [2] for verifying the changeover-feasibility of S by applying classical network flow techniques.

Let $P_k(S) := \{(i, j) \in \overline{V}_k \times \overline{V}_k \mid S_j \geq S_i + p_i + \vartheta^k_{ij}\}$ where $\overline{V}_k := V_k \cup \{0, n+1\}$ be the set of all activity pairs (i, j) for which schedule S enables a changeover of resource $k \in \mathcal{R}^\rho$ between the completion of activity i and the start of activity j. For each pair $(i, j) \in P_k(S)$ we want to determine the number $\Phi^k_{ij} \geq 0$ of resource units to be changed over from activity i to activity j. Then condition (48.8) can be stated equivalently as

$$\sum_{(h,i)\in P_k(S)} \Phi^k_{hi} = \sum_{(i,j)\in P_k(S)} \Phi^k_{ij} = r_{ik} \quad (i \in V_k) \tag{48.16}$$

Equation (48.16) says that for each activity $i \in V_k$, r_{ik} units of resource k are taken from preceding activities $h \in \overline{V}_k$ and r_{ik} resource units are passed from i to some succeeding activity $j \in \overline{V}_k$. $\sum_{i\in V_k} \Phi^k_{0i}$ corresponds to the number of units of resource k that have to be set up (i.e., changed over from project beginning 0 to some activity $i \in V_k$). Thus, in conjunction with Equation (48.16), condition (48.9) can be rewritten as

$$\sum_{i\in V_k} \Phi^k_{0i} \leq R_k \tag{48.17}$$

Vector $\Phi^k := (\Phi^k_{ij})_{(i,j)\in P_k(S)}$ can be regarded as a *flow* in the network G_k with nodes $i \in \overline{V}_k$ and arcs $\langle i, j\rangle$ for $(i, j) \in P_k(S)$, where the left Equation in (48.16) is the flow conservation condition at node i. The right Equation in (48.16) fixes the outflow $\sum_{(i,j)\in P_k(S)} \Phi^k_{ij}$ of node i to be equal to r_{ik}. In the terms of network flow theory, this corresponds to the definition of a lower and an equally large upper node capacity for i. Inequality (48.17) means that the value $\sum_{i\in V_k} \Phi^k_{0i}$ of flow Φ^k must be less than or equal to resource capacity R_k. Hence, the changeover-feasibility of schedule S can be verified by computing for each resource $k \in \mathcal{R}^\rho$ a flow of minimum value in network G_k. Such a minimum-flow problem can be solved in $O(n^3)$ time by two applications of a so-called preflow-push algorithm for the maximum-flow problem (see Ahuja et al. [16]). Aside from a minimum flow, the algorithm also provides a $0-(n+1)$ node cut of maximum capacity in G_k, i.e., a node set $\mathcal{A}_k(S)$ such that removing the nodes $i \in \mathcal{A}_k(S)$ from G_k separates $n+1$ from 0 and $\sum_{i\in\mathcal{A}_k(S)} r_{ik}$ is maximum. It can be shown (see Neumann et al. [2]) that $\mathcal{A}_k(S)$ coincides with a set of activities that must be assigned to different resource units and have maximum joint requirements for k.

Now suppose that we have found a resource conflict on a resource k at some time $t \in [0, \overline{d}]$. Then depending on the type of conflict we obtain the following list L of alternative precedence relationships $i \prec j$ to be investigated in Algorithm 48.1:

1. in case of a capacity overflow on a renewable resource $k \in \mathcal{R}^\rho$ without (or with) changeover times, L contains $i \prec j$ for all $i, j \in \mathcal{A}(S, t)$ (or $i, j \in \mathcal{A}_k(S)$, respectively) with $i \neq j$
2. in case of an inventory shortage for a storage resource $k \in \mathcal{R}^\sigma$, L contains $i \prec j$ for all $i \in V^+_k \setminus \mathcal{A}_k(S, t)$ and all $j \in V^-_k \cap \mathcal{A}_k(S, t)$
3. in case of an inventory excess for a storage resource $k \in \mathcal{R}^\sigma$, L contains $i \prec j$ for all $i \in V^-_k \setminus \mathcal{A}_k(S, t)$ and all $j \in V^+_k \cap \mathcal{A}_k(S, t)$
4. in case of a homogeneity conflict for a synchronizing resource $k \in \mathcal{R}^\beta$, L contains $i \prec j$ for all $i, j \in \mathcal{A}(S, t)$ with $\varphi_i \neq \varphi_j$
5. in case of a simultaneity conflict for a synchronizing resource $k \in \mathcal{R}^\beta$, L contains $i \prec j$ for all $i, j \in \mathcal{A}(S, t)$ with $S_i \neq S_j$

48.4.3 Removing Resource Conflicts

It remains to associate a time lag δ_{ij} (or respectively arc weight δ_{ij} in network $\overline{N}$) with each precedence relationship $i \prec j$ enumerated in the course of Algorithm 48.1. Figure 48.3 shows the case of a capacity overflow on a renewable resource $k \in \mathcal{R}^\rho$ without changeover times. For two activities $i, j \in \mathcal{A}(S, t)$, the two precedence relationships $i \prec j$ and $j \prec i$ on list L correspond to the time lags $\delta_{ij} = p_i$ and $\delta_{ji} = p_j$,

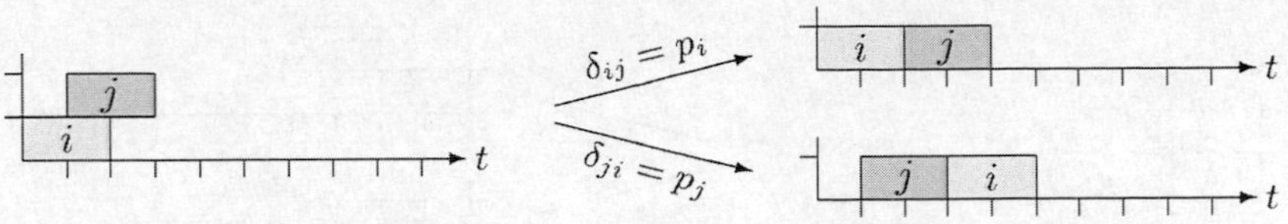

FIGURE 48.3 Removing a capacity overflow: case of vanishing changeover times.

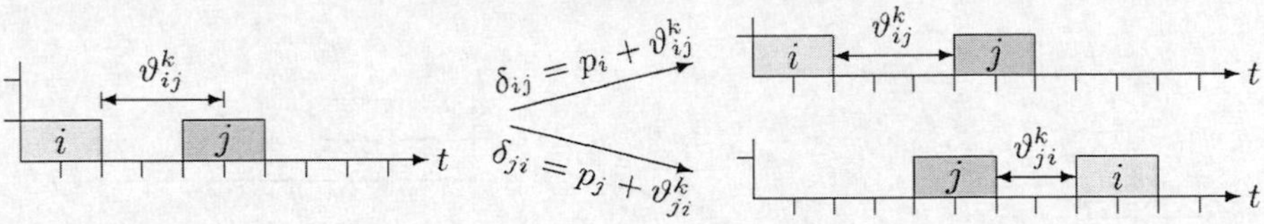

FIGURE 48.4 Removing a capacity overflow: case of positive changeover times.

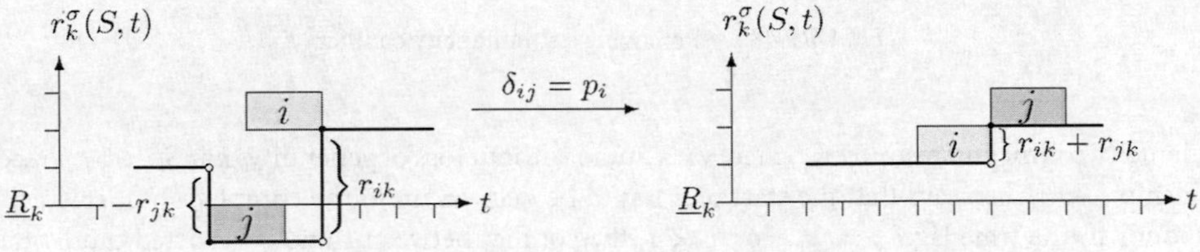

FIGURE 48.5 Removing an inventory shortage.

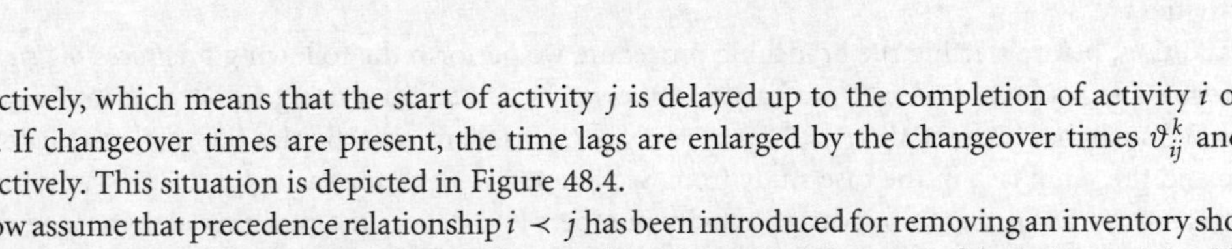

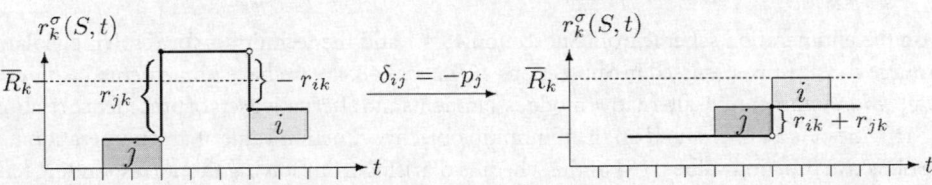

FIGURE 48.6 Removing an inventory excess.

respectively, which means that the start of activity j is delayed up to the completion of activity i or *vice versa*. If changeover times are present, the time lags are enlarged by the changeover times ϑ^k_{ij} and ϑ^k_{ji}, respectively. This situation is depicted in Figure 48.4.

Now assume that precedence relationship $i \prec j$ has been introduced for removing an inventory shortage for some storage resource $k \in \mathcal{R}^\sigma$. As it is illustrated in Figure 48.5, the start of depleting activity j has to be shifted behind the replenishment of resource k occurring at the completion of activity i, i.e., in analogy to the case of a capacity overflow on a renewable resource, we obtain time lag $\delta_{ij} = p_i$. Figure 48.6 displays the case of an inventory excess for storage resource k. Here, precedence relationship $i \prec j$ means that the replenishment at the completion time of activity j must not occur before activity has depleted resource k at its start, i.e., $S_j + p_j \geq S_i$. The latter condition can be formulated as time lag $\delta_{ij} = -p_j$, which corresponds to a maximum time lag $d^{\max}_{ji} = p_j$ between the start of activities j and i.

Eventually, we consider the case where a resource conflict occurs on some synchronizing resource $k \in \mathcal{R}^\beta$. If there is a homogeneity conflict on k, which is illustrated in Figure 48.7, the meaning of precedence relationships $i \prec j$ and $j \prec i$ is the same as for a renewable resource without changeover times, i.e., we introduce the time lags $\delta_{ij} = p_i$ and $\delta_{ji} = p_j$, respectively. Figure 48.8 deals with the case of

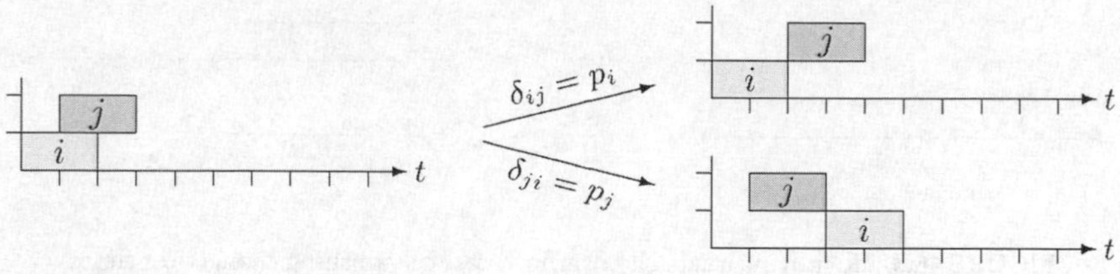

FIGURE 48.7 Removing a homogeneity conflict.

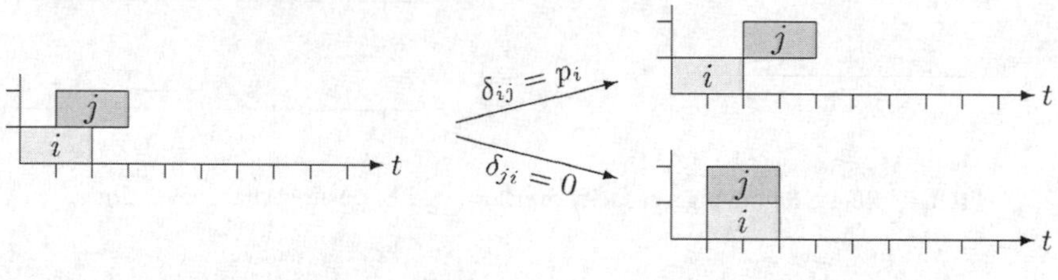

FIGURE 48.8 Removing a simultaneity conflict.

a simultaneity conflict on resource k, where we assume without loss of generality that $S_i < S_j$. Precedence relationship $i \prec j$ then says that the start of j has to be delayed up to the completion of i, which again corresponds to the time lag $\delta_{ij} = p_i$. For $j \prec i$, the conflict between i and j is sorted out by delaying activity i such that both activities may start simultaneously, i.e., $\delta_{ji} = 0$.

48.4.4 Implementation Issues

Based on the enumeration scheme from Subsection 48.4.1 and the techniques for resolving violations of the resource constraints discussed in Subsections 48.4.2 and 48.4.3, we have implemented a *branch-and-bound procedure* enumerating alternative mode assignments and alternative sets of precedence relationships among activities. As a *lower bound* on the minimum objective function value at an enumeration node we use the objective function value $f(S)$ of the schedule S arising from solving the corresponding temporal scheduling problem (TSP($\overline{N}, \underline{x}$)). $f(S)$ constitutes a valid lower bound because (TSP($\overline{N}, \underline{x}$)) is a relaxation of the multimode project scheduling problem with project network $\overline{N}$ and sets of execution modes $\mathcal{M}_i(\underline{x})$ for activities $i \in V$.

In addition, before starting the branching procedure we perform the following *preprocessing step*. For each renewable resource $k \in \mathcal{R}^\rho$, we partition the set V_k of activities using resource k into s subsets $V_k^1, \ldots, V_k^s$ such that each set V_k^r $(r = 1, \ldots, s)$ only contains identical activities (i.e., activities belonging to one and the same task in the case study from Section 48.2). Next, for each $r = 1, \ldots, s$ we order the activities $i \in V_k^r$ according to increasing activity numbers and denote the resulting activity sequence by $(i_1, \ldots, i_\nu)$. Then, we introduce time lags $\delta_{ij} = 0$ between any two consecutive activities $i = i_\mu$ and $j = i_{\mu+1}$ in that ordering $(\mu = 1, \ldots, \nu - 1)$. Moreover, let $a = \lfloor R_k / r_{ik} \rfloor$ with $i \in V_k^r$ be the maximum number of activities from set V_k^r which can be processed in parallel on resource $k \in \mathcal{R}^\rho$. For all $\mu = 1, \ldots, \nu - a$, we additionally introduce the time lag $\delta_{ij} = p_i$ between activities $i = i_\mu$ and $j = i_{\mu+a}$. In the special case where $a = 1$, those time lags imply that we put all identical activities from set V_k^r one after another before we move to the enumeration part of the algorithm. In this way we considerably reduce the number of feasible solutions to be enumerated without any loss of generality.

Aside from the case of small problem instances with less than 50 activities, applying the (exact) branch-and-bound algorithm to problem (PSP) would be too time-consuming. That is why we truncate the branch-and-bound algorithm to a heuristic procedure of type *filtered beam search* as follows (cf. e.g., Neumann et al. [2] or Pinedo [17]). By φ and $1 \leq \beta < \varphi$ we denote the integers corresponding to the *filter width* and the *beam width*, respectively. After the generation of all offsprings of the current enumeration node, we solve the corresponding temporal scheduling problems and order the offspring nodes according to some filter criterion. Subsequently, the first φ nodes are evaluated on the basis of a beam criterion, and the best β nodes are added to the enumeration tree. The remaining offspring nodes are excluded from further consideration. For hard problem instances, where even a beam width of $\beta = 2$ is generally too large, we use realizations β of a random variable $\widetilde{\beta}$ with an expected value $E(\widetilde{\beta}) < 2$ as beam widths.

48.5 Decomposition of Large-Scale Problem Instances

In this section we discuss a heuristic *decomposition algorithm* for multimode project scheduling problem (PSP), which decomposes the original scheduling problem into several smaller *subproblems*. Those subproblems are solved by the filtered beam search procedure from Section 48.4, and the (partial) solutions of the subproblems are integrated to obtain a (complete) solution to (PSP). The decomposition algorithm is used when coping with large-scale problem instances with more than thousand activities, for which the computational requirements of filtered beam search become prohibitively large when addressing the entire problem as a whole.

In the field of machine scheduling, decomposition algorithms have been proposed for a large variety of specific problem settings. A comprehensive overview of those methods can be found in Ovacik and Uszoy [18]. Decomposition algorithms for batch scheduling in the process industries have been treated by Basset et al. [19], Elkamel et al. [20], Kuriyan and Reklaitis [21], and Mauderli and Rippin [22]. Those algorithms mainly differ in the *decomposition scheme*, which defines the way in which the original problem is partitioned into subproblems. Roughly speaking, a scheduling problem can be decomposed with respect to time, resources, or operations (see Ovacik and Uszoy [18] for a general classification of decomposition schemes). Our algorithm is based on the latter type of decomposition, where the operations or activities of a subproblem are determined by solving an integer linear program. The main idea of this approach is to generate the subsets of activities in such a way that in the solution to the corresponding subproblems, bottleneck resources are evenly loaded and changeover times on those resources can be avoided to the greatest possible extent. A similar decomposition scheme has been used by Mauderli and Rippin [22] for campaign planning in the process industries, which also includes solving the batching problem. In Subsection 48.5.1 we explain the basic principle of our decomposition algorithm. Subsections 48.5.2 and 48.5.3 provide details on the decomposition scheme and the modelling of the subproblems.

48.5.1 Basic Principle

A decomposition algorithm generally consists of the four building blocks *problem decomposition*, *modeling of subproblems* for given decomposition, *solution of subproblems*, and *integration of subproblem solutions*. In our algorithm, the decomposition and the modeling and solution of the subproblems are performed in an iterative way. After having solved all subproblems, the solution to the original problem is constructed by concatenating the partial solutions obtained one after another.

The procedure is illustrated in Figure 48.9. In each iteration, we compute an appropriate subset $U \subset V$ of activities to be scheduled next (decomposition step). Set U arises from solving an integer linear program, which refers to the activities not yet scheduled and to the final inventories from the preceding iteration. Aside from set U, the solution to the integer linear program also provides a mode assignment y for the activities $i \in U$. Next, the subproblem belonging to activity set U is modelled (modelling step), where the execution modes for the activities processed on bottleneck resources are fixed in advance according to y. In addition to the activities from U, the subproblem also contains activities of a *frozen zone* in the schedule S' generated in the preceding iteration. The frozen zone consists of all activities which according to the latter schedule, are completed after the earliest point in time t^* where some resource $k \in \mathcal{R}^\rho$ has processed its entire workload. The initial inventories r_{0k} for storage resources $k \in \mathcal{R}^\sigma$ are set equal to the corresponding inventory levels $r_k^\sigma(S', t^*)$ at time t^* from the preceding iteration. To account for time lags δ_{ij} and changeover times ϑ_{ij}^k between activities i already scheduled (and not contained in the frozen zone)

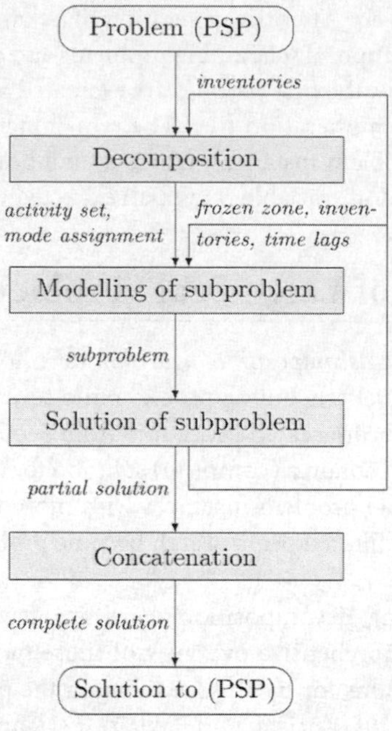

FIGURE 48.9 Decomposition algorithm.

and activities $j \in U$, new time lags δ_{0j} are introduced. The resulting subproblem is then solved by applying filtered beam search (solution step), and the procedure returns to the decomposition step. Finally, when all activities $i \in V$ have been scheduled, a complete solution to (PSP) is constructed by joining the partial solutions of consecutive iterations at the corresponding synchronization time points t^* (concatenation step).

48.5.2 Decomposition Scheme

In this subsection we consider the decomposition step. Let $\widetilde{\mathcal{R}}^\rho \subseteq \mathcal{R}^\rho$ be a set of *bottleneck resources* whose allocation over time may strongly affect the solution quality. For example, resources $k \in \widetilde{\mathcal{R}}^\rho$ can be selected based on the ratio of the workload to be processed on k and resource capacity R_k or on the mean changeover times required. By $\widetilde{V} \subset V$ we denote the set of *critical activities* using some bottleneck resource, i.e., $\widetilde{V} = \bigcup_{k \in \widetilde{\mathcal{R}}^\rho} V_k$. Clearly, the critical activities should be scheduled in such a way that the bottleneck resources are loaded evenly and changeover times are avoided. More precisely, we want to determine an $\subseteq$-maximal set U such that:

(i) no bottleneck resource $k \in \widetilde{\mathcal{R}}^\rho$ processes more than a prescribed maximum workload W
(ii) the critical activities $i \in \widetilde{V} \cap U$ can be executed on the bottleneck resources $k \in \widetilde{\mathcal{R}}^\rho$ with minimum changeover time in between
(iii) the inventories of the storage resources $k \in \mathcal{R}^\sigma$ after the execution of all activities $i \in U$
  (a) do not fall below the safety stocks $\underline{R_k}$
  (b) do not exceed the storage capacities $\overline{R}_k$
(iv) the temporal constraints between activities are taken into account, i.e., $j \in U$ implies $i \in U$ for all $(i, j) \in E$ with $(i, j) \neq (n+1, 0)$

By condition (i) we guarantee an (approximately) equal allocation of the workload among the bottleneck resources. Moreover, at the same time upper bound W limits the number of activities in set U and thus the size of the generated subproblem. Condition (iv) guarantees that scheduling the activities $j \in U$ does not decrease the latest start times of activities i to be scheduled in subsequent iterations. For the special case where aside from pair $(n+1, 0)$ referring to the planning horizon $\overline{d}$ of (PSP) we only have minimum time lags $\delta_{ij} \geq 0$, condition (iv) implies that the activities can be scheduled in chronological order.

In the sequel, we formulate the problem of computing set U as an integer linear program. We suppose that we have one execution mode $m \in \mathcal{M}_i$ for each unit of a bottleneck resource on which activity $i \in V$ can be carried out, and we identify execution mode m with that resource unit. Accordingly, we can assume that each bottleneck resource $k \in \widetilde{\mathcal{R}}^{\rho}$ has capacity $R_k = 1$. To avoid changeover times on bottleneck resources, we assign critical activities $i, j \in \widetilde{V}$ belonging to different tasks $\tau(i)$ and $\tau(j)$ to different resource units (i.e., i and j are performed in different modes). Let y_{im} be a binary decision variable which equals one if activity i is included into set U and executed in mode $m \in \mathcal{M}_i$, and zero, otherwise. The activity set we are looking for is then $U = \{i \in V \mid \sum_{m \in \mathcal{M}_i} y_{im} = 1\}$, where for simplicity we suppose that V only contains the project beginning 0 and the activities which have not yet been scheduled in previous iterations (in the first iteration of the decomposition algorithm, V coincides with the original set of all activities). Our problem can now be formulated as follows, with v_k denoting the final inventory of resource $k \in \mathcal{R}^{\sigma}$ resulting from the partial solution obtained in the preceding iteration (in the first iteration, $v_k = 0$ for all $k \in \mathcal{R}^{\sigma}$):

$$
\left.
\begin{array}{lll}
\text{Maximize} \displaystyle\sum_{i \in V} \sum_{m \in \mathcal{M}_i} y_{im} & & \\
\text{subject to} \displaystyle\sum_{i \in V_k} \sum_{m \in \mathcal{M}_i} p_{im} y_{im} \leq W \quad (k \in \widetilde{\mathcal{R}}^{\rho}) & & \text{(i)} \\
y_{im} + y_{jm} \leq 1 \quad (i, j \in \widetilde{V} : \tau(i) \neq \tau(j),\ m \in \mathcal{M}_i \cap \mathcal{M}_j) & & \text{(ii)} \\
\displaystyle\sum_{i \in V} \sum_{m \in \mathcal{M}_i} r_{ik}\, y_{im} \geq \underline{R}_k - v_k \quad (k \in \mathcal{R}^{\sigma}) & & \text{(iiia)} \\
\displaystyle\sum_{i \in V} \sum_{m \in \mathcal{M}_i} r_{ik}\, y_{im} \leq \overline{R}_k - v_k \quad (k \in \mathcal{R}^{\sigma}) & & \text{(iiib)} \\
\displaystyle\sum_{m \in \mathcal{M}_i} y_{im} \geq \sum_{m \in \mathcal{M}_j} y_{jm} \quad ((i, j) \in E \setminus \{(n+1, 0)\}) & & \text{(iv)} \\
\displaystyle\sum_{m \in \mathcal{M}_i} y_{im} \leq 1 \quad (i \in V) & & \text{(v)} \\
y_{im} \in \{0, 1\} \quad (i \in V,\ m \in \mathcal{M}_i) & & \text{(vi)}
\end{array}
\right\}
\quad (48.18)
$$

The objective function to be maximized corresponds to the number $|U|$ of selected activities. Inequalities (i) to (iv) correspond to the above conditions (i) to (iv). Inequalities (v) ensure that each activity is assigned to at most one execution mode, and constraints (vi) define the domains of decision variables y_{im}.

48.5.3 Generation of Subproblems

In each iteration of the decomposition algorithm, the solution $y = (y_{im})_{i \in V, m \in \mathcal{M}_i}$ to integer linear program (48.18) provides the set $U = \{i \in V \mid \sum_{m \in \mathcal{M}_i} y_{im} = 1\}$ of activities to be scheduled, along with a mode assignment $(y_{im})_{i \in U, m \in \mathcal{M}_i}$ for the activities $i \in U$. Since we want to minimize the changeover times on the bottleneck resources $k \in \widetilde{\mathcal{R}}^{\rho}$, we fix the execution mode for each critical activity $i \in \widetilde{V} \cap U$ according to y (i.e., we only allow for mode assignments $x \geq y$). Basically, the subproblem (P) to be solved then arises from problem (PSP) by replacing activity set V with $U \cup \{0, n+1\}$ and putting $\mathcal{M}_i := \{m\}$ where $y_{im} = 1$ for all $i \in \widetilde{V} \cap U$.

As the individual subproblems considered in the different iterations of the algorithm are linked by the temporal and resource constraints, we have to take into account those dependencies when formulating the subproblems. This can be done by integrating parts of the solution (S', x') found in the preceding iteration into the subproblem (P) under consideration. Let U' be the set of activities which have been scheduled in the preceding iteration. Subproblem (P) with node set U is then modified in the following way.

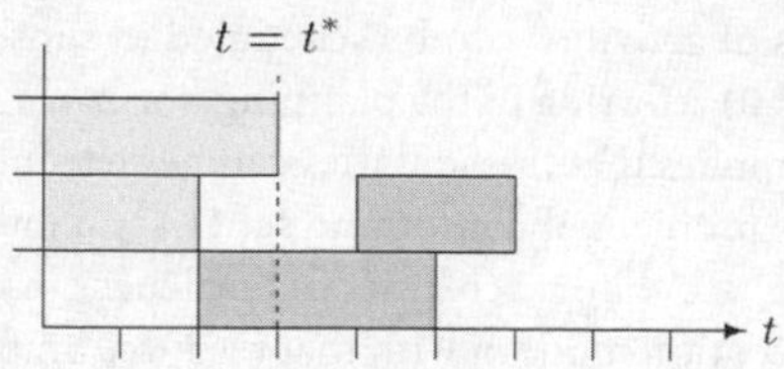

FIGURE 48.10 Frozen zone in schedule S'.

1. We define a frozen zone $F \subseteq U'$ of activities which are added to set U (see schedule S' depicted in Figure 48.10, where activities $i \in F$ are dark-shaded). Let $t^* := \min_{k \in \mathcal{R}^\rho} \max_{i \in U'} \{S_i' + p_i(x') \mid r_{ik}(x') > 0\}$ be the point in time at which the first renewable resource is released after having processed the activities assigned. Time t^* corresponds to the left end of the planning interval (i.e., the project beginning) for subproblem (P). Frozen zone F contains all activities which have not been completed by time t^*, i.e., $F = \{i \in U' \mid S_i' + p_i(x') > t^*\}$. If activity $i \in F$ is started before time t^*, we reduce its duration p_i by $t^* - S_i'$ time units. The start times of activities $i \in F$ are put to $\max(0, S_i' - t^*)$, which is achieved by introducing the time lags $\delta_{0im} = -\delta_{im0} = \max(0, S_i' - t^*)$ for all $m \in \mathcal{M}_i$ (where for simplicity, we omit the index for the execution mode of 0).
2. The initial inventories r_{0k} for storage resources $k \in \mathcal{R}^\sigma$ are set to be equal to the inventory levels $r_k^\sigma(S', x', t^*)$ at time t^* given schedule S' and mode assignment x'.
3. Prescribed time lags δ_{ij} between activities $i \in U' \setminus F$ and $j \in U$ are taken into account by introducing the time lags $\delta_{0jm'} = \max_{i \in U' \setminus F} (S_i' + \delta_{imjm'} - t^*)$ for all $m' \in \mathcal{M}_j$, where $x_{im}' = 1$.
4. Similarly, changeover times on renewable resources between activities $i \in U' \setminus F$ and $j \in U$ lead to time lags $\delta_{0jm'} = \max_{i \in U' \setminus F} \max_{k \in \mathcal{R}^\rho : r_{ikm} r_{jkm'} > 0} (S_i' + p_{im} + \vartheta_{imjm'}^k - t^*)$ for all $m' \in \mathcal{M}_j$, where $x_{im}' = 1$.

The resulting subproblem (P) represents a multimode project scheduling problem of the same type as (PSP), which can be solved by the filtered beam search procedure described in Section 48.4.

48.6 Solution to the Case Study

The filtered beam search procedure from Section 48.4 and the decomposition algorithm from Section 48.5 have been implemented in C++ under MS-Visual Studio 6.0. For the case study described in Section 48.2, we have obtained a makespan of 102 units of time within 30 seconds of CPU time using the filtered beam search procedure (with filter width $\varphi = 2$ and expected beam width $E(\tilde{\beta}) = 1.41$) on a Pentium-800 personal computer with 512 MB memory operating under MS Windows 2000. Figure 48.11 shows the corresponding Gantt chart for the nine processing units, the location of the meal breaks of manpower pool 2, and the manpower requirements for pools 1 and 2 over time. The dark-shaded boxes in the Gantt chart represent cleaning times of processing units. The light-shaded area in the charts of the manpower pools shows the number of available workers. There are two operations ($i = 67, 73$) that are interrupted during a break.

To apply the proposed decomposition algorithm from Section 48.5 to a large problem instance, we have increased 13 times the primary requirements for the final products in the case study from Section 48.2. The planning period is then 3 months or 13 weeks instead of one week, again starting on a Monday at 6 a.m. Table 48.3 shows the corresponding number of batches for each task. Altogether there are now 1111 batches to be scheduled.

The best result has been obtained by choosing processing units U_5 to U_9 as the bottleneck resources and setting workload parameter W equal to 35. We have iteratively computed the subsets U of activities to be scheduled by solving subproblems (48.18) using CPLEX 7.0, where we have stopped the computation after 10 seconds of CPU time if a feasible solution had been found. On a Pentium-1400 personal computer

TABLE 48.3 Batching Solution for Large Problem Instance

Task	1	2	3	4	5	6	7	8	9	10	11	12	13	14	15	16	17
# Batches	203	132	165	52	26	26	52	39	39	39	39	78	39	39	52	39	52

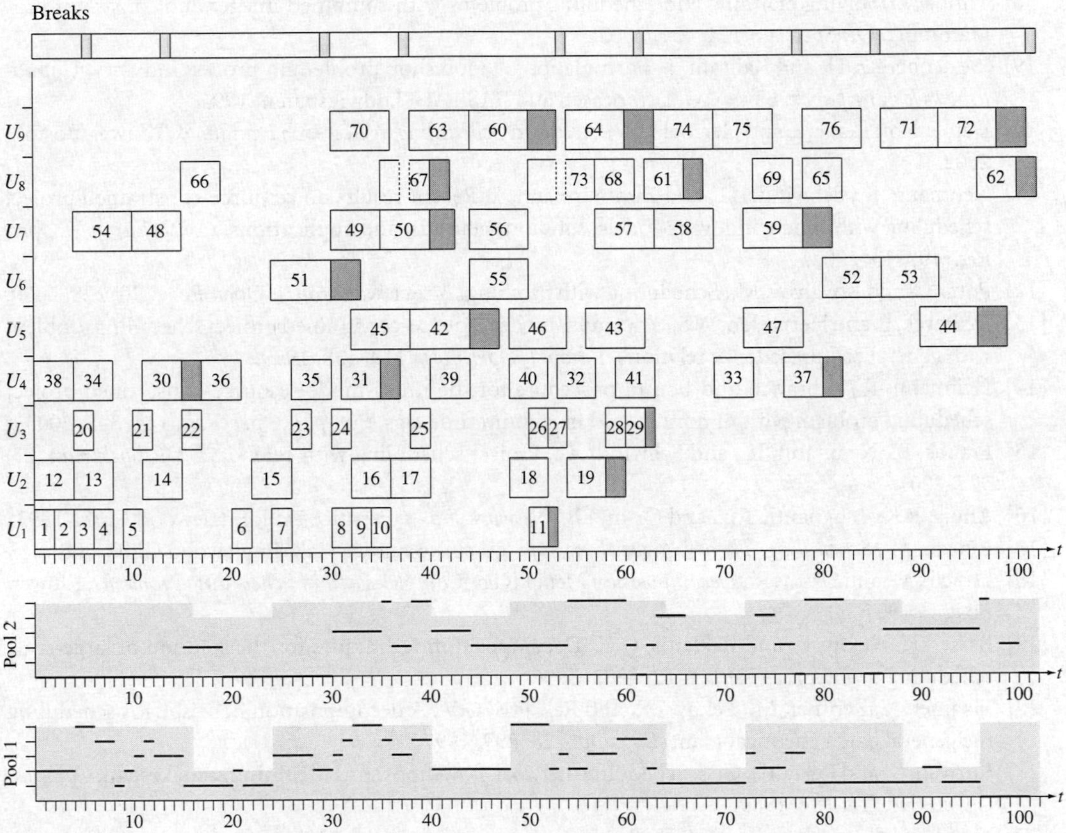

FIGURE 48.11 Charts for feasible solution to case study.

with 640 MB RAM operating under Windows 2000, we have obtained a makespan of 1359 units of time in 500 seconds of CPU time. Without using the decomposition algorithm, no feasible solution has been found within one hour of CPU time.

References

[1] Neumann, K., Schwindt, C., and Trautmann, N., Advanced production scheduling for batch plants in process industries, *OR Spectrum*, 24, 251, 2002.

[2] Neumann, K., Schwindt, C., and Zimmermann, J., *Project Scheduling with Time Windows and Scarce Resources*, 2nd ed., Springer, Berlin, 2003.

[3] Kondili, E., Pantelides, C.C., and Sargent, R.W.H., A general algorithm for short-term scheduling of batch operation — I. MILP formulation, *Comput. Chem. Eng.*, 17, 211, 1993.

[4] Pinto, M. and Grossmann, I.E., A continuous time mixed integer linear programming model for short term scheduling of multistage batch plants, *Ind. Eng. Chem. Res.*, 34, 3037, 1995.

[5] Blömer, F. and Günther, H.O., Scheduling of a multi-product batch process in the chemical industry, *Comput. Ind.*, 36, 245, 1998.
[6] Blömer, F. and Günther, H.O., LP-based heuristics for scheduling chemical batch processes, *Internat. J. Prod. Res.*, 35, 1029, 2000.
[7] Grunow, M., Günther, H.O., and Lehmann, M., Campaign planning for multi-stage batch processes in the chemical industry, *OR Spectrum*, 24, 281, 2002.
[8] Timpe, C., Solving planning and scheduling problems with combined integer and constraint programming, *OR Spectrum*, 24, 431, 2002.
[9] Westenberger, H. and Kallrath, J., Formulation of a job shop problem in process industry, *Unpublished working paper*, Bayer AG, Leverkusen and BASF AG, Ludwigshafen, 1995.
[10] Demeulemester, E.L. and Herroelen, W.S., *Project Scheduling: A Research Handbook*, Kluwer, Boston, 2002.
[11] Neumann, K., Schwindt, C., and Zimmermann, J., Recent results on resource-constrained project scheduling with time windows: Models, solution methods, and applications, *Centr. Europ. J. Oper. Res.*, 10, 113, 2002.
[12] Potts, C. and Kovalyov, M., Scheduling with batching: A review, *Europ. J. Oper. Res.*, 120, 228, 2000.
[13] De Reyck, B. and Herroelen, W.S., The multi-mode resource-constrained project scheduling problem with generalized precedence relations, *Europ. J. Oper. Res.*, 119, 538, 1999.
[14] Heilmann, R., A branch-and-bound procedure for the multi-mode resource-constrained project scheduling problem with minimum and maximum time lags, *Europ. J. Oper. Res.*, 144, 348, 2003.
[15] Franck, B., Neumann, K., and Schwindt, C., Project scheduling with calendars, *OR Spektrum*, 23, 325, 2001.
[16] Ahuja, R.K., Magnanti, T.L., and Orlin, J.B., *Network Flows*, Prentice Hall, Englewood Cliffs, 1993.
[17] Pinedo, M., *Scheduling: Theory, Algorithms, and Systems*, Prentice Hall, Englewood Cliffs, 2001.
[18] Ovacik, M. and Uzsoy, R., *Decomposition Methods For Complex Factory Scheduling Problems*, Kluwer, Boston, 1997.
[19] Basset, H., Pekny, F., and Reklaitis, G.V., Decomposition techniques for the solution of large-scale scheduling problems, *AIChE J.*, 12, 3373, 1996.
[20] Elkamel, A., Zentner, M., Pekny, J.F., and Reklaitis, G.V., A decomposition heuristic for scheduling the general batch chemical plant, *Eng. Opt.*, 28, 299, 1997.
[21] Kuriyan, K. and Reklaitis, G.V., Scheduling network flowshops so as to minimize makespan, *Comput. Chem. Eng.*, 13, 187, 1989.
[22] Mauderli, A. and Rippin, D.W.T., Production planning and scheduling for multi-purpose batch chemical plants, *Comput. Chem. Eng.*, 3, 199, 1979.

49

A Composite Very-Large-Scale Neighborhood Search Algorithm for the Vehicle Routing Problem

Richa Agarwal
Georgia Institute of Technology

Ravindra K. Ahuja
University of Florida

Gilbert Laporte
University of Montreal

Zuo-Jun "Max" Shen
University of Florida

49.1 Introduction

The classical *vehicle routing problem* (VRP) is defined on an undirected graph $G = (N, E)$, where $N = \{0, 1, \dots, n\}$ is a node set and $E = \{(i, j) : i, j \in N\}$ is an edge set. For simplicity (i, j) and (j, i) represent the same edge. Node 0 corresponds to a *depot* at which are based m identical vehicles of capacity C, while the remaining nodes are *customers*. Each customer i has a nonnegative demand q_i. With each edge (i, j) is associated a cost c_{ij} corresponding to a distance or to a travel time. The VRP consists of determining

1-58488-397-9/$0.00+$1.50

vehicle routes of minimum total cost satisfying the following constraints:

1. Each route starts and ends at the depot.
2. Each customer belongs to exactly one route.
3. The total customer demand of any route does not exceed C.
4. The total cost of any route does not exceed a preset limit D.

The VRP lies at the heart of distribution management. This problem underlies several applications including the distribution of goods to a set of customers (see, e.g., Golden, Assad, and Wasil, 2002), dial-a-ride operations (Cordeau and Laporte, 2003), and pickup and delivery problems arising, for example, in express courier companies (Mitrovi'c-Mimi'c et al., 2004). These problems must often be solved dynamically in real-time. Several of these applications contain side constraints such as time windows. The VRP has significant economic importance. Owoc and Sargious (1992) estimate that transportation costs account for a proportion of 11 to 13% of the total production cost of goods.

The VRP is NP-hard and can rarely be solved exactly for values of n in excess of 100. The best exact algorithms are based on branch-and-cut (Naddef and Rinaldi, 2002; Baldacci, Hadjiconstantinou, and Mingozzi, 2004). In practice, the VRP is solved by means of heuristics. Several *classical* heuristics have been proposed over the past forty years (Laporte and Semet, 2002). One of the most famous is the Clarke and Wright (1964) savings algorithm. Initially, it constructs n back and forth routes from the depot and then gradually merges these routes using a cost saving criterion. Wark and Holt (1994) have proposed a modified scheme in which the best set of merges is determined at each step through the solution of a matching problem. In recent years, metaheuristics based on local search (e.g., simulated annealing and tabu search) and on population search (Prins, 2004) have been proposed. For surveys and assessments of such methods, see Gendreau et al. (2002) and Cordeau et al. (2002).

The aim of this chapter is to present a new local search algorithm for the VRP called *very large-scale neighborhood search* (VLSN Search). Local search methods attempt to improve an initial solution by performing an exploration of the search space. At each iteration of a local search method, a move is made from the current solution S to a solution S' in the neighborhood of S. Basically, these algorithms differ in the way neighborhoods are defined and on what constitutes an acceptable solution S'. Two basic neighborhood types exist in a solution. The first is obtained by permuting customers belonging to the same route. This amounts to applying any *traveling salesman problem* (TSP) improvement strategy to a given route. Common examples are 2-opt and 3-opt moves. The second involves exchanges between several routes, such as moving some customers to a different route, swapping customers between two routes, performing customer exchanges or transfers involving several routes, etc. As a rule, more complicated mechanisms yield richer neighborhoods but require more computational effort. Several rules have been proposed for the definition of an acceptable neighbor. Classical or simple local search uses a *descent mechanism*, i.e., in order to be accepted as the next point, solution S' must have a lesser cost than S. If no cost-reducing neighbor can be identified, then the search comes to an end. Metaheuristics typically allow worsening moves. In simulated annealing, such moves are accepted with some probability whose value decreases as the search progresses. Tabu search proceeds to the best neighbor but, to avoid cycling, it excludes from consideration a set of so-called tabu solutions. Some tabu search mechanisms allow the exploration of intermediate infeasible solutions (see, e.g., Gendreau et al., 1994).

The VLSN Search method proposed in this chapter works with very large cardinality neighborhoods involving the displacement of customers between several routes. The best move is identified through the solution of a related network flow problem on an auxiliary graph. Contrary to a recent trend in metaheuristics, VLSN Search applies a descent mechanism. As will be seen, it yields high quality results competitive with those of the best available metaheuristics.

The remainder of the chapter is organized as follows. Section 49.2 contains a survey of neighborhood structures of the VRP. Our VLSN search algorithm is contained in Sections 49.3 and 49.4. Computational results are presented in Section 49.5, followed by conclusions in Section 49.6.

49.2 Survey of Neighborhood Structures for the Vehicle Routing Problem

In this section, we present a survey of several neighborhood structures that have been proposed for the VRP by researchers.

49.2.1 λ-Opt Neighborhood

Since any feasible solution to the VRP is a set of routes, any solution algorithm for the TSP can be applied to each route separately to obtain *within route optimality*. One of the most important neighborhood search algorithm for TSP is λ-opt algorithm of Lin (1965). For a given initial feasible solution S, λ edges are removed from S, which creates λ segments that are disconnected, then λ edges are added and the segments are reconnected so that we get a feasible tour with lower total costs. The algorithm terminates when no further improvements can be found. An example of this neighborhood is given in Figure 49.1. A 2-opt neighborhood of a tour T consists of all tours T' which can be obtained by deleting two arcs, say $(i, i+1)$ and $(j-1, j)$, and adding two new arcs $(i, j-1)$ and $(i+1, j)$ that are not in T.

Or (1976) proposed a general form of the 3-opt interchanges that move strings of three, two, or one consecutive vertices to other locations. Renaud et al. (1996) extended the above 3-opt interchanges to 4-opt interchanges. Finally, Lin and Kernighan (1973) proposed an algorithm that updates λ dynamically throughout the search. These algorithms only focus on *within route optimality*. To get a larger neighborhood, the neighborhood search algorithms need to exchange edges among different routes. Some neighborhood structures that involve two or more routes are reviewed next in this section.

49.2.2 λ-Interchange Neighborhood

Osman (1993) defined a *λ-interchange neighborhood*, which consists of exchanging up to λ customers between two routes. Given an initial feasible solution S to VRP, which is represented by a set of routes $S = (I_1, I_2, \ldots, I_p, I_q, \ldots, I_m)$, a λ-interchange between a pair of routes I_p and I_q is an exchange of a subset of customers between the two routes. More specifically, let $R_1 \subseteq I_p$ and $R_2 \subseteq I_q$ represent subsets from the two routes, respectively, and $|R_1| \leq \lambda, |R_2| \leq \lambda$. The interchange creates two new routes $I'_p = (I_p \backslash R_1) \cup R_2$ and $I'_q = (I_q \backslash R_2) \cup R_1$, and a new solution $S' = (I_1, \ldots, I'_p, I'_q, \ldots, I_m)$. The neighborhood $N_\lambda(S)$ of a given solution S is the set of all neighbors S' generated this way for a given value of λ. If either R_1 or R_2 is empty, then it is called a *customer move*; and if both are nonempty, then it is called a *customer swap*.

In Osman (1993), λ is set to 2 and the neighbors of a current solution are generated by allowing a mix of single and double customer moves, and single and double customer swaps. Osman embedded this approach in a tabu-search framework and tested two strategies for selecting a neighbor solution. The first one, called *best admissible*, selects the best solution; and the second one, called *first admissible*, selects the first admissible improving solution.

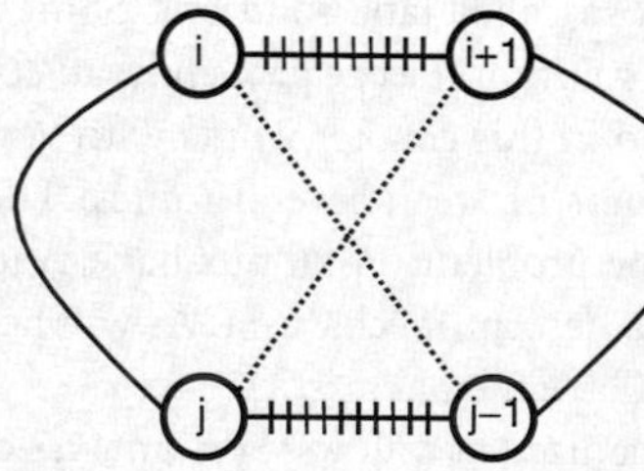

FIGURE 49.1 An example of a 2-opt neighborhood.

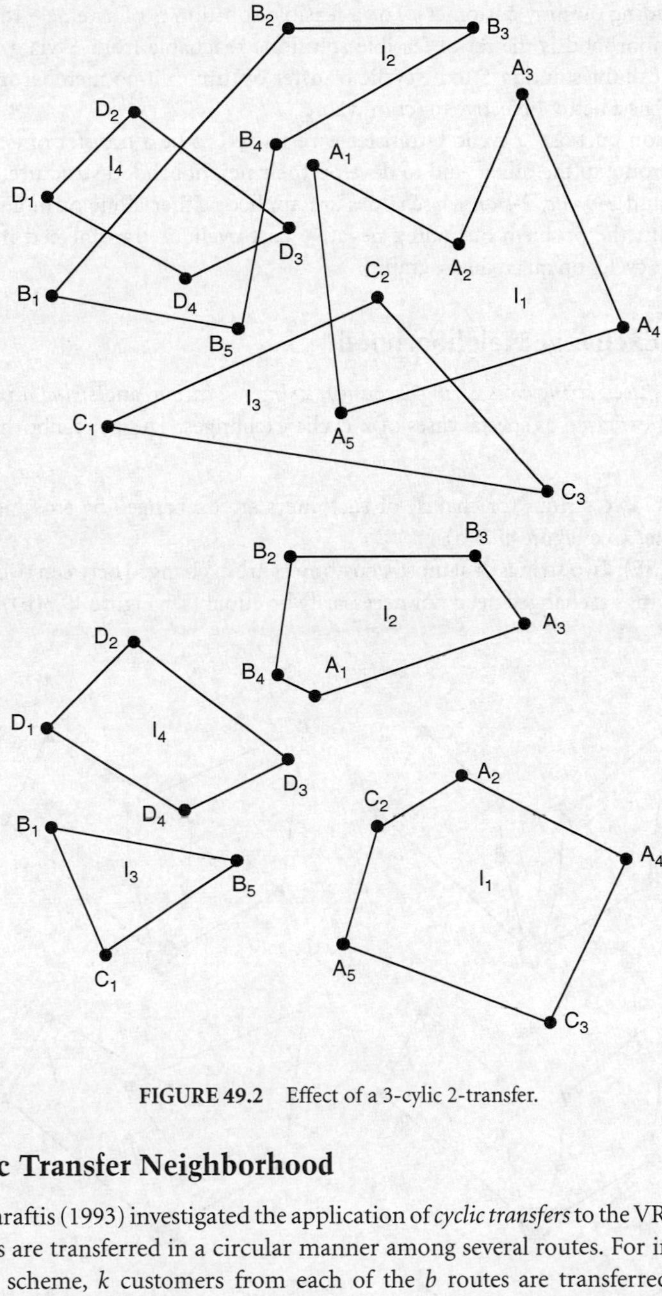

FIGURE 49.2 Effect of a 3-cylic 2-transfer.

49.2.3 Cyclic Transfer Neighborhood

Thompson and Psaraftis (1993) investigated the application of *cyclic transfers* to the VRP. In cyclic transfers, customer demands are transferred in a circular manner among several routes. For instance, in a general *b-cyclic, k-transfer* scheme, k customers from each of the b routes are transferred to the next route. Figure 49.2 shows a detailed example of a 3-cyclic 2-transfer. There are four routes, I_1, I_2, I_3, and I_4 in this example, the customers being served by each route are: $I_1 = \{A_1, A_2, A_3, A_4, A_5\}$, $I_2 = \{B_1, B_2, B_3, B_4, B_5\}$, $I_3 = \{C_1, C_2, C_3\}$ and $I_4 = \{D_1, D_2, D_3, D_4\}$. In a 3-cyclic 2-transfer that involves routes I_1, I_2, and I_3, two customers from route I_1, $\{A_1, A_3\}$, are transferred to route I_2; two customers from route I_2, $\{B_1, B_5\}$, are moved to route I_3; and finally, two customers from $I_3, \{C_2, C_3\}$, are transferred to I_1; I_4 is left unchanged. The cost of a cyclic transfer is the change in optimal objective function value caused by the cyclic transfer. Let $S = (I_1, \ldots, I_p, I_q, \ldots, I_m)$ and $S' = (I'_1, \ldots, I'_p, I'_q, \ldots, I'_m)$ represents sets of m routes before and after a cyclic transfer occurs, and let $f(I)$ represent the optimal cost of route I. Then the cost of the cyclic transfers is $\sum_{i=1}^{m} [f(I'_i) - f(I_i)]$.

Thompson and Psaraftis also considered *path transfers*, which allow the transfer of customers among permutations (rather than cyclic permutations) of routes. A path transfer can be converted to a cyclic transfer easily by adding dummy customers. For a feasible solution S of a vehicle routing problem, the cyclic transfer neighborhood is the set of feasible solutions reachable from S via a cyclic transfer or a path transfer. They call the solution S to be cyclic transfer optimum if no member in the cyclic transfer neighborhood of S has a better objective function value.

For implementation purpose, a cyclic k-transfer is restricted to be a transfer of exactly k consecutive demands from one route to the other. And to develop their neighborhood structure, *2-cyclic, 1-transfer, 3-cyclic, 1-transfer*, and *2-cyclic, 2-transfer* schemes are applied, either alone or in a group with various permutations. Finally, the problem of finding negative cost cyclic is transformed into the problem of finding negative cost cycles on an auxiliary graph.

49.2.4 String Exchange Neighborhood

Van Breedam (1994) used *string cross, string exchange, string relocation*, and *string mix* neighborhood for VRP, which can all be viewed as special cases of 2-cyclic exchanges. These neighborhoods are described next:

String cross (SC). Two strings (or chains) of customers are exchanged by crossing two edges of two different routes (see Figure 49.3(a)).

String exchange (SE). Two strings of at most k customers are exchanged between two routes. The length of the two strings exchanged need not necessarily be equal (see Figure 49.3(b)).

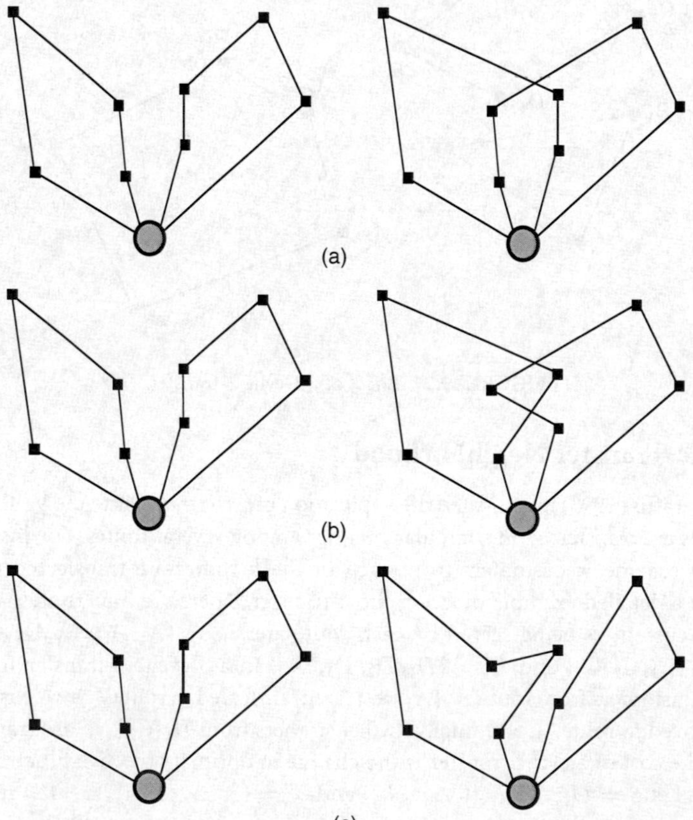

FIGURE 49.3 Examples of string exchange neighborhood. (a) An example of string cross; (b) An example of string exchange; (c) An example of string relocation.

String relocation (SR). A string of at most k customers is moved from one route to another, typically with $k = 1$ or 2. This type of move can potentially reduce the number of routes (see Figure 49.3(c)).

String mix (SM). The best move between SE and SR is selected.

Two local improvement strategies are used to choose an improving neighbor. The first, FI, implements the first move that improves the objective function. The second, BI, evaluates all possible moves and implements the best one. Many parameters that can affect the performance of the algorithm, such as string length (k), types of moves (SE, SR, SM), the selection strategy (FI, BI), and initial solution (poor, good), are fine tuned in the algorithm.

49.2.5 Tabu Route Neighborhood

The tabu-route algorithm presented in Gendreau et al. (1994) contains several innovative features and has produced many best known solutions for the standard test bed of the vehicle routing problem available in the OR library (http://graph.ms.ic.ac.uk/info.html). They define the neighborhood of a current solution S as the set of all solutions reachable from S by removing a customer v from its current route I_p and inserting it in another route I_q containing one of its closest neighbors. This insertion is performed by means of GENI procedure. GENI, developed by Gendreau et al. (1992) for TSP, is a generalized insertion routine. Every insertion is executed simultaneously with a local reoptimization of the current tour. To limit the size of the neighborhood, in each iteration only a randomly selected subset of customers is considered for reinserting into other routes. A tabu-search framework is used in which reinserting v in the route I_p is declared tabu for a specified number of iterations. Intermediate infeasible solutions are also considered and penalized through the use of penalty parameters.

49.2.6 Ejection Chain Neighborhood

An ejection chain method is presented in Rego (1998), which identifies a *reference structure* to guide the generation of alternating paths and cycles from the current solution. An intermediate *flower structure* is defined as consisting of one path called stem that emanates from the depot, and several blossoms (or routes) attached to the depot (see Figure 49.4). Depot is known as the *core* vertex (c) of the flower and the end of the stem as the *root* vertex (r) of the flower.

To define neighbors of the current solution, first a flower structure is created from the solution based on either of the following rules:

Rule 1. Delete an edge (v_c, v_j) that belongs to the set of edges in a route, thus transforming a route into a stem (see Figure 49.5(a)), or

Rule 2. Divide a route into one blossom and one stem by suitably inserting and deleting edges (see Figure 49.5(b)).

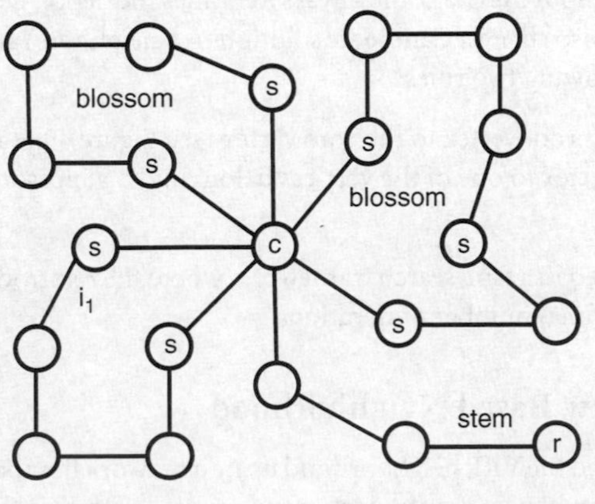

FIGURE 49.4 The flower reference structure.

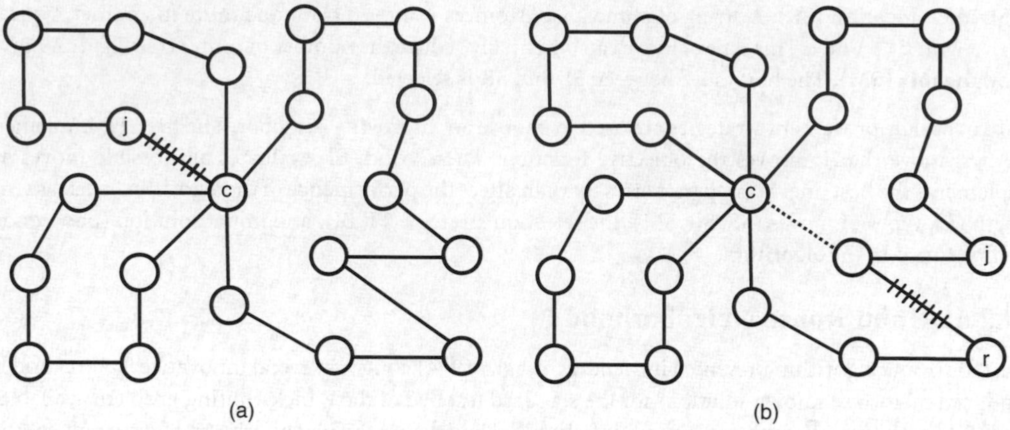

FIGURE 49.5 Rules for creating a flower structure.

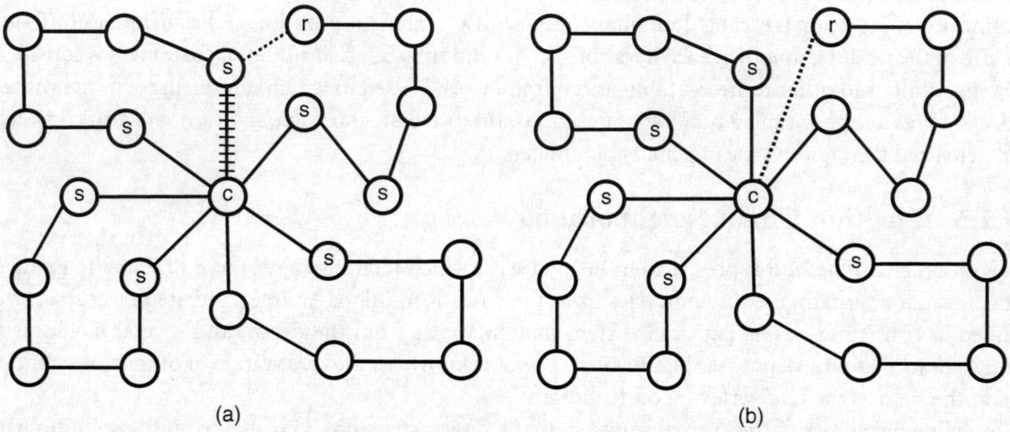

FIGURE 49.6 Recovering the solution from a flower.

Rules are then defined to produce ejection moves that would allow the transition from one flower structure to another. The resulting flower structure transformations successively transform a blossom and a stem into another blossom and stem, thus rearranging the flower structure. The core vertex is maintained constant throughout the chain and always identifies the depot. During this process the number of active vehicle routes may also change. Candidate solutions (or neighbors) are recovered from these flower structures based on the following two rules:

Rule 1. Relink the current root vertex to the core vertex (see Figure 49.6(a)), or

Rule 2. Relink the root vertex to one of the vertices belonging to a blossom and cut a suitable edge (see Figure 49.6(b)).

This approach is embedded in a tabu search framework, where the reintroduction of an edge just deleted is declared tabu for a predefined number of iterations.

49.2.7 Network Flow Based Neighborhood

Xu and Kelly (1996) generated the VRP neighborhood using a network flow based approach. This approach provides a mechanism for exchanging customers among routes such that the distance and feasibility are simultaneously considered. We give an example of the underlying network in Figure 49.7.

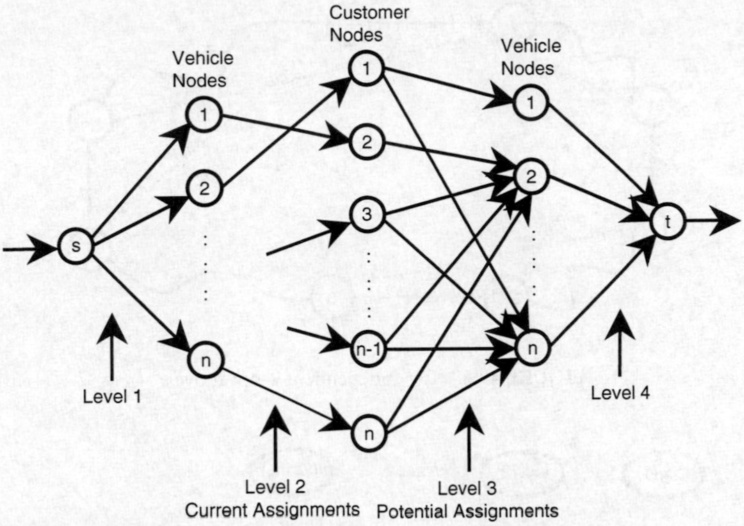

FIGURE 49.7 A network flow model to generate neighborhoods.

The flow on level 1 arcs represents the number of customers that are removed from the current route, and the input flow c specifies the total number of such moves. Level 2 and level 3 arcs have a flow of 1 or 0 to indicate whether a customer is removed from a route (level 2) or reinserted in a route (level 3). On level 4, the flow represents the number of customers added to the routes. The ejection moves are determined by minimizing the flow on this network. This is a very generic model. A bound on the arcs in level 1 can be imposed to specify the maximum or minimum number of customers that are allowed to be removed from that vehicle route. The model represents the best ejection/insertion move by fixing c to 1. Nonpermissible ejections/insertions are modeled by assigning large prohibitive costs to corresponding arcs.

Besides the above network flow based moves, customer swaps between two routes are executed after a certain number of iterations. Intermediate infeasible solutions with respect to the capacity constraint are allowed by introducing a dynamic penalty system. The 3-opt and 2-opt methods are also used in order to find the best sequence of customers clustered into a route. This approach is embedded into a tabu search framework that involves a large number of parameters, adjusted dynamically throughout the execution of the algorithm.

49.2.8 Path Based Neighborhood

Ergun (2001) defined compounded independent moves neighborhood for the vehicle routing problem. This method accommodates the compounding of different types of independent simple moves such as 2-opts, customer swaps, and insertions in creating neighborhood. Two 2-opt moves are considered to be independent if the subtours that they affect do not overlap. Specifically, two single 2-opt moves in Figure 49.8 involving $(i, i+1)$ and $(j-1, j)$, and $(s, s+1)$ and $(t-1, t)$ are independent moves if either of the following relations hold:

$$\text{Min}\{i, j-1\} > \text{Max}\{s+1, t\}$$

$$\text{Max}\{i+1, j\} < \text{Min}\{s, t-1\}$$

The compounded independent 2-opt neighborhood for tour T is the set of all tours T' which can be obtained by applying a collection of independent 2-opt moves to T. Compounded swaps and insertions are also defined in the same fashion.

The swap moves are enhanced by exchanging consecutive sets of customers instead of single customer swaps and replacing single customer insertions by multiple customer insertions. Node sets of sizes

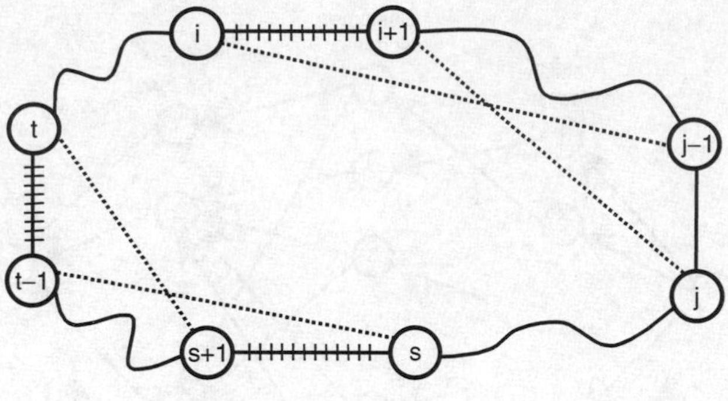

FIGURE 49.8 Two independent 2-opt moves.

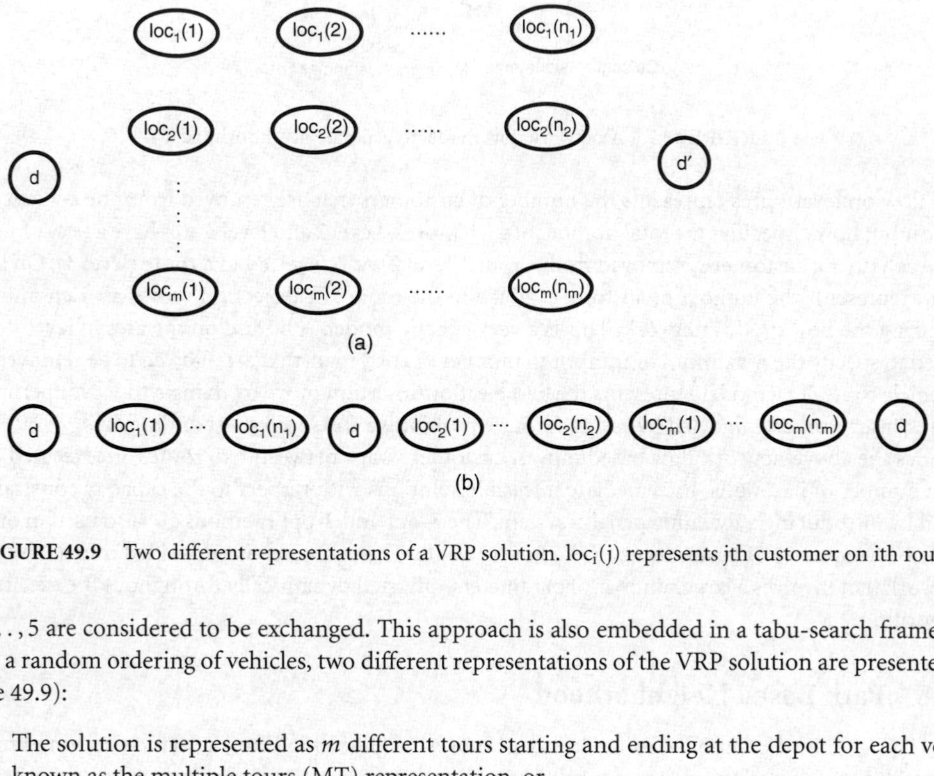

FIGURE 49.9 Two different representations of a VRP solution. $loc_i(j)$ represents jth customer on ith route.

$1, 2, \ldots, 5$ are considered to be exchanged. This approach is also embedded in a tabu-search framework. Given a random ordering of vehicles, two different representations of the VRP solution are presented (see Figure 49.9):

1. The solution is represented as m different tours starting and ending at the depot for each vehicle, known as the multiple tours (MT) representation, or
2. The solution is represented as one big tour, which is obtained by concatenating the m tours in an arbitrary order, known as the single tour (ST) representation.

Based on the two representations, two neighborhood structures are considered. With the MT representation, cyclic exchange neighborhood framework of Thompson and Psaraftis (1993) is used with the aforementioned compounded swaps and insertions. With the ST representation, VRP is solved as a TSP, and compounded 2-opt, insertions, and swaps are considered.

49.3 The Composite VLSN Algorithm for the VRP

Our neighborhood search algorithm for the vehicle routing problem starts with a feasible solution S, it then defines the *neighbors* of S by performing composite multiexchanges, identifies a suitable *neighbor* using the concept of improvement graph, and replaces S with it. This process is repeated until a termination criterion is reached. In this section, we describe the basic concepts used in developing our algorithm including neighborhood structure and improvement graph.

49.3.1 Composite Cyclic Exchange

Our technique generalizes the cyclic transfer neighborhood structure described in Section 49.2 in the sense that a set of $k \leq K$ customers can be transferred from one route to another, where K is a user-defined parameter and gives an upper bound on the maximum number of customers that can be transferred from one route to another. However, rather than considering every possible set of customers, which would make the neighborhood too large and too time-consuming to enumerate, we consider only those sets of customers that are served consecutively by a vehicle in the current solution. Consider a feasible solution S with r routes $I_1, I_2, I_3, \ldots, I_r$. For a particular customer i in route $I_l, 1 \leq l \leq r$, we define the following notations:

T_{i1}: customer i in route I_l

T_{i2}: customer i and the customer next to i in the same route, i.e., route I_l

T_{i3}: customer i and the next two customers in route I_l

...

T_{ik}: customer i and the next $k-1$ customers in route I_l

A *composite cyclic exchange* is defined by a sequence of customers $T_{a_1q_1} - T_{a_2q_2} - \cdots - T_{a_rq_r} - T_{a_1q_1}$, where customer a_l belongs to route $I_l, 1 \leq l \leq r$, and $1 \leq q_1, q_2, \ldots q_r \leq K$. The composite cyclic exchange consists of the following moves: the customer a_1 and the next $q_1 - 1$ customers (represented by the notation $T_{a_1q_1}$) move from route I_1 to route I_2, the customer a_2 and the next $q_2 - 1$ customers (represented by $T_{a_2q_2}$) move from route I_2 to route I_3, and so on, and finally the customer a_r and the next $q_r - 1$ customers (represented by $T_{a_rq_r}$) move from route I_r to route I_1 to complete the cyclic exchange. Thus, comparing with the simple cyclic exchanges, the composite cyclic exchanges allow the flexibility of moving more than one customer from one route to another; see Figure 49.10 for a comparison of simple cyclic exchange and composite cyclic exchange.

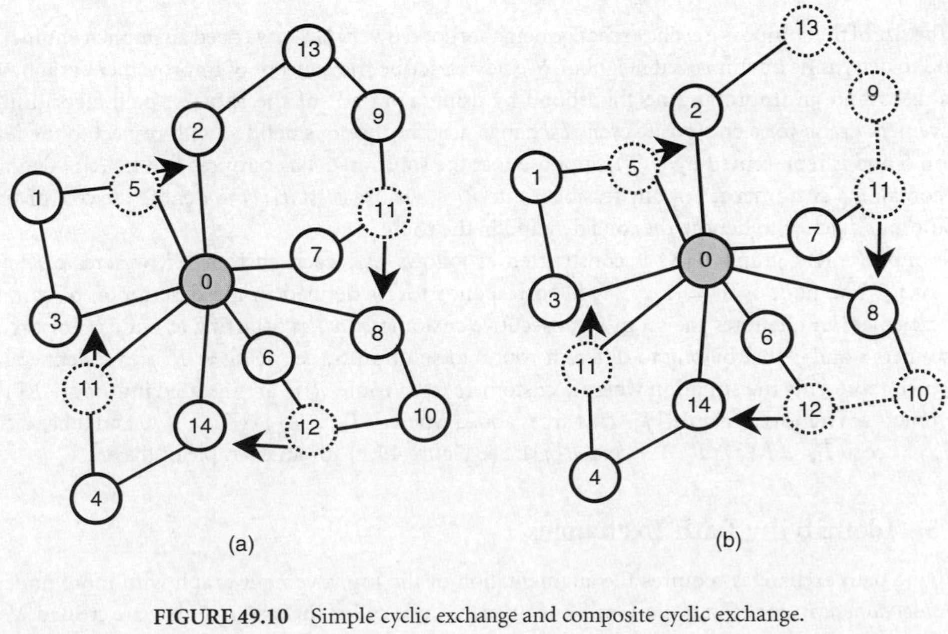

FIGURE 49.10 Simple cyclic exchange and composite cyclic exchange.

49.3.2 Composite Path Exchange

A composite path exchange of customers can be defined similar to the composite cyclic exchange; the difference is that the nodes in the last route do not get transferred to the first route. In other words, a composite path exchange $T_{a_1q_1} - T_{a_2q_2} - \cdots - T_{a_{r-1}q_{r-1}} - T_{a_rq_r}$ represents the following change: the customers in the set $T_{a_1q_1}$ move from route I_1 to route I_2, the customers in the set $T_{a_2q_2}$ move from route I_2 to route I_3 and so on, and finally the customers in the set $T_{a_{r-1}q_{r-1}}$ move from route I_{r-1} to route I_r. A composite path exchange can be transformed to a composite cyclic exchange with the definition of *improvement graph* (as described later); hence only composite cyclic exchanges need to be considered and discussed.

49.3.3 A Comparison of Different Neighborhood Sizes

The composite cyclic exchange neighborhood generalizes the two-exchange and the simple cyclic exchange neighborhoods. In the following discussion, we compare the neighborhood sizes of these three neighborhoods. Consider an instance of VRP with n customers and m vehicles. For simplicity, assume n is a multiple of m and each route has exactly n/m customers. For simplicity, we will ignore the vehicle capacity and maximum route length constraints. Then, a two-exchange neighborhood has $n(n-n/m) = \Omega(n^2)$ neighbors and a simple cyclic exchange neighborhood has $m!(n/m)^m = \Omega(n^m)$ neighbors (since there are n/m ways a customer can be selected in a route, and if one customer is selected from each route then $m!$ cyclic exchanges can be defined). The composite cyclic exchange $T_{a_1q_1} - T_{a_2q_2} - \cdots - T_{a_{r-1}q_{r-1}} - T_{a_rq_r}$ allows up to K possible moves for each customer in the cyclic exchange. Thus, for every simple cyclic exchange, there are K^m possibilities for the corresponding composite cyclic exchange, the number of composite cyclic exchanges can be as large as K^m times the number of simple cyclic exchanges. For instance, if $n = 100$ and $m = 10$, then the two-exchange neighborhood has about 9000 neighbors, simple cyclic exchange neighborhood has about 6300 trillion neighbors and composite multiexchange cyclic neighborhood has 3.8×10^{11} trillion neighbors (assuming $K = 5$). The above analysis shows that the composite cyclic exchange neighborhood is much larger than those of the two-exchange neighborhood and the simple cyclic exchange neighborhood.

49.3.4 Improvement Graph

Since the size of the composite cyclic exchange neighborhood is very large, we need an implicit enumeration method to determine an improved neighbor. We next describe the concept of improvement graph, which allows us to find an improving neighborhood by using a variant of the shortest path algorithm. The improvement graph for a composite cyclic exchange neighborhood is defined with respect to the feasible solution S and is represented by $G(S)$. Suppose that the solution S has r routes and let $R[i]$ denote the route containing customer i. For any feasible route $R[i]$, we use $cost(R[i])$ to denote the cost of serving the route and $d(R[i])$ to denote the total demand in the route.

The improvement graph $G(S)$ is constructed as follows. For each customer i, we create K nodes, $i_1, i_2, \ldots, i_K$. The node $i_k, k = 1, 2, \ldots, K$ corresponds to T_{ik} defined in the description of composite cyclic exchange, and denotes the set of k consecutive customers in $R[i]$ starting at node i. For each pair of customers i and j that belong to different routes in solution S, i.e., $R[i] \neq R[j]$, we define arc $(i_q, j_{q'})$, which represents the situation where q customers from route $R[i]$ are inserted into route $R[j]$ and q' customers are removed from $R[j]$. This arc is added when $d(T_{iq} \cup R[j] \backslash T_{jq'}) \leq C$ and it has a cost of $c(i_q, j_{q'}) = \text{cost}(T_{iq} \cup R[j] \backslash T_{jq'}) - \text{cost}(R[j])$. See Figure 49.11 for an example of the arc.

49.3.5 Identifying Path Exchanges

Identifying path exchanges requires the augmentation of the improvement graph with more nodes and arcs. A *pseudonode* h_i for each route I_i in S and an extra node called the origin node v are created. We add

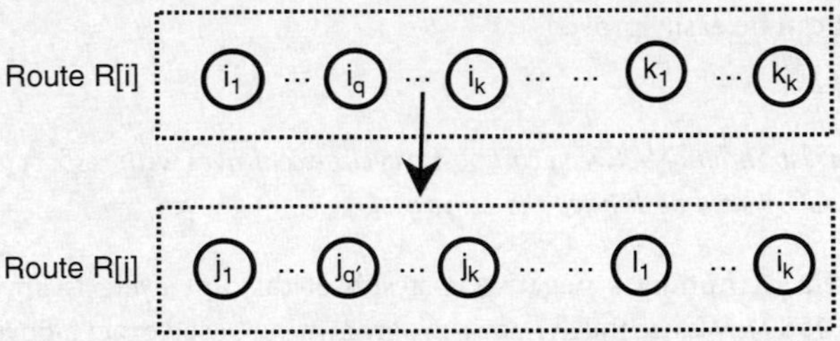

FIGURE 49.11 Arcs in an improvement graph.

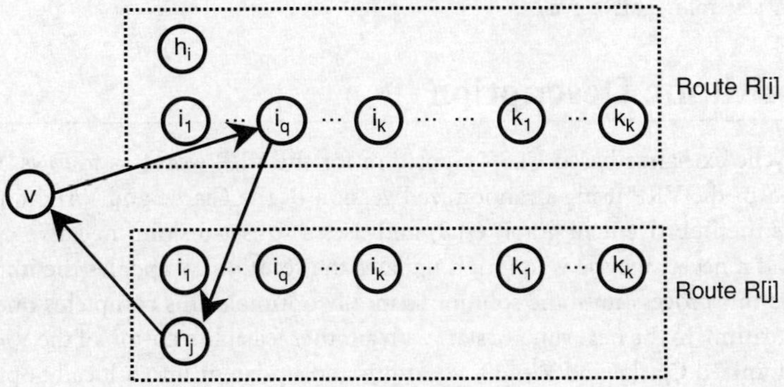

FIGURE 49.12 Augmentation of the improvement graph.

a pseudonode h_i into subset $S[i]$ that corresponds to route I_i. For notational convenience, all the nodes except the newly added pseudonodes and the origin node are referred to as *regular nodes*. The augmentation of the improvement graph also includes the following arcs (see Figure 49.12):

1. The origin node v is connected with each regular node i_q. Arc (v, i_q) has a cost of
$$c(v, i_q) = \text{cost}(R[i]\backslash\{i_q\}) - \text{cost}(R[i])$$
These arcs signify the move where customers T_{iq} are removed from route $R[i]$.
2. Each regular node i_q is connected to each pseudonode h_j if i_q and h_j belong to different subsets and the capacity constraint is satisfied (i.e., $d(T_{iq} \cup R[j]) \leq C$). This arc represents the move where T_{iq} is transferred from route $R[i]$ to $R[j]$, but no customer moves out of $R[j]$. The arc has a cost of $c(i_q, h_j) = \text{cost}(R[j] \cup \{i_q\}) - cost(R[j])$.
3. Each pseudonode h_i is connected to the origin node v. Arc (h_i, v) has a cost of $c(h_i, v) = 0$. There is no actual movement of customers associated with this arc.

The cycle in Figure 49.12 represents a path cyclic exchange, in which we move T_{iq} from $R[i]$ to $R[j]$, but no customer is moved out of $R[j]$. This explains why we only need to consider composite cyclic exchanges. $G(S)$ is a directed graph partitioned into r disjoint subsets $S[1], S[2], \ldots, S[r]$, where subset $S[i]$ corresponds to route I_i . In solution S, each customer i belongs to exactly one route $R[i]$ and thus in $G(S)$, each node belongs to exactly one subset. It is clear that nodes $i_1, i_2, \ldots, i_k$ all belong to the same subset.

Let a_i denote any node in subset $S[i], i = 1, 2, \ldots r$, then we say a directed cycle $a_1 - a_2, \ldots, a_r$ in the improvement graph is a subset-disjoint cycle if $a_1, a_2, \ldots, a_r$ all belong to different subsets. The improvement graph converts each possible composite cyclic exchange with respect to solution S into a *subset-disjoint* cycle in the improvement graph. Thus, the problem of finding a profitable composite cyclic exchange becomes the problem of finding a subset-disjoint cycle with negative cost.

The following result can be easily proved.

Lemma 49.1

There is a one-to-one correspondence between composite cyclic exchanges with respect to solution S and directed subset-disjoint cycles in G(S), and both have the same cost.

However, the problem of finding a negative cost subset-disjoint cycle is an NP-complete problem (Thompson and Orlin, 1989). We use the dynamic programming based exact algorithm to identify subset-disjoint cycles in the improvement graph as proposed by Ahuja et al. (2003). This algorithm provides flexibility of choosing the appropriate neighborhood size based on both computational time and quality of the local optimal solution desired. Furthermore, it has the ability to determine not just one negative cycle, but possibly several negative cycles.

49.4 Algorithmic Description

The composite cyclic exchange based VLSN algorithm for the VRP works as follows. First, it obtains a feasible solution S of the VRP using a randomized version of the Clarke and Wright algorithm (1964). It then constructs the improvement graph $G(S)$ and uses a subset-disjoint negative cycle algorithm to find such a cycle. If a negative cycle is found, it updates the solution S, updates the improvement graph $G(S)$, and repeats this process until the solution is locally optimal. This completes one run of the local improvement algorithm. In the next run, we start with another feasible solution of the VRP obtained again through the randomized Clarke and Wright algorithm and update it into a locally optimal solution by performing a sequence of profitable composite cyclic exchanges. We continue in this manner until either we have performed enough runs or exceeded a specified running time. See the following pseudocode for details (Figure 49.13).

Now we discuss some implementation details of the algorithm.

To allow the possibility of increasing the number of routes when the number of routes obtained from Clarke-Wright algorithm are not optimal, one empty route is added to the initial solution. This is necessary because during the cyclic exchanges, the number of routes is not going to change.

Computing the cost of an arc in $G(S)$ involves solving an NP-hard problem because it involves adding and removing customers from a route and then finding the optimal sequencing of customers in the route, which is equivalent to solving a TSP in the route. To circumvent the problem, the following polynomial-time approximation of computing the arc cost is used: For each $c(i_q, j_{q'}), q'$ customers are removed from

```
algorithm composite VLSN;
begin
   obtain a feasible solution S' using the randomized Clarke and Wright algorithm;
   apply 3-opt on S' to obtain S;
   construct the improvement graph G(S);
   while G(S) contains negative cycles do
   begin
      obtain a subset-disjoint negative cycle W in G(S) using dynamic programming approach;
      use the composite cyclic exchange corresponding to W to update the solution S.
      apply 3-opt on S as a post optimization step.
      update G(S);
   end;
end.
```

FIGURE 49.13 The composite VLSN search algorithm for VRP.

$R[j]$ and then one by one the minimum cost insertion point is found for each of the q customers removed from route $R[i]$ without resequencing.

After a cycle is implemented and the solution is updated, 3-opt is applied to maintain within route optimality. Now, the improvement graph needs to be regenerated based on the updated solution. The practice of only updating the improvement graph rather than reconstructing it all over again is evaluated. Although it would save some time, it was insignificant comparing with the implementation complications it would invoke. Since our graph is quite big and many routes are involved in a single iteration, updating instead of reconstructing it is not very beneficial. Furthermore, since the computational time spent on constructing the improvement graph is of the same order as the time spent on finding a negative cycle, it would probably not help much even if the time on generating the improvement graph can be cut down, unless a better algorithm is also used for negative cycle detection. For problems of smaller sizes, where the number of routes that are affected in any particular iteration are much smaller than the total number of routes, updating the improvement graph rather than reconstructing it can save significant amount of time.

Since the improvement can become very large, the idea of selectively generating the improvement graph is also tested, that is, we only generate part of the improvement graph. This practice significantly reduces the computation time requirement both for the improvement graph generation and for finding the negative cycle. The rules used to reduce the size of the improvement graph are as follows:

1. Generate an arc between two nodes (i, j) only if the distance between them is shorter than a prespecified value.
2. Generate an arc between two nodes (i, j) only if the angle formed by the two arcs connecting i, j to the depot is smaller than a prespecified value.
3. A combination of the above two techniques, i.e., an arc will be generated only if both the conditions in 1 and 2 are satisfied.

49.5 Computational Results

In this section, we present the computational results of the composite cyclic exchange based VLSN search algorithm for the VRP. We compare the VLSN algorithm with the best existing simple local search algorithm, which is due to Wark and Holt (1994), and metaheuristics such as tabu search and simulated annealing. In general, metaheuristics produce better solutions than those from simple local search algorithms, but they are difficult to implement as several parameters need to be fine-tuned. Furthermore, the computational time requirements for metaheuristics are generally high compared to the simple local search algorithms.

Test environment. The VLSN search algorithm for VRP is tested on the fourteen test problems described in Christofides et al. (1979). The problems were obtained from Beasley (1990). These problems contain between 50 and 199 customers in addition to the depot. Problems 1 to 5 and 11, 12 have capacity restrictions only. Problems 6 to 10 and 13, 14 are the same as 1 to 5 and 11, 12, respectively, except that they also have a maximal route length constraint. In Problems 1 to 10, cities are randomly distributed in the plane, while in Problems 11 to 14, cities appear in clusters. The VLSN search algorithm is implemented in C++ on a Sun Ultra-80 workstation. Computational times are reported in seconds.

Overall Performance. In Table 49.1, the first three columns report the characteristics of the problem sets. Specifically, the second column contains the numbers of customers; the third column explains the additional constraints on the problem, where "C" indicates constraint on vehicle capacity, and "D" indicates a limit on the route length. The fourth column reports the current best solutions obtained from simple local search algorithms, which is due to Wark and Holt (1994), and the fifth column reports the best solutions in the literature (mostly obtained by metaheuristics of Rochat and Taillard, 1995). In Table 49.1, we report the best value from 100 runs for each instance. It can be observed that the VLSN algorithm improves the best simple local search solution in 12 out of

TABLE 49.1 Comparison of the VLSN Algorithm with the Best Known Solutions

Problem Data					VLSN Search Algorithm		
Number	Size	Type	Best Known Solution Value	Best Known Local Search	Solution Value	Average Time per Run	Error % against Best Known
1	50	C	524.61	524.6	524.61	1.03	0.0000
2	75	C	835.26	835.8	835.67	9.92	0.0495
3	100	C	826.14	830.7	830.56	18.54	0.5345
4	150	C	1028.42	1038.5	1038.87	126.07	1.0156
5	199	C	1291.45	1321.3	1311.41	201.71	1.5453
6	50	C, D	555.43	555.4	555.43	7.31	0.0000
7	75	C, D	909.68	911.8	909.67	248.87	0.0000
8	100	C, D	865.94	878.0	865.94	244.06	0.0005
9	150	C, D	1162.55	1176.5	1164.31	1134.63	0.1513
10	199	C, D	1395.85	1418.3	1408.43	504.27	0.9012
11	120	C	1042.11	1043.4	1042.24	41.06	0.0127
12	100	C	819.56	819.6	819.56	12.70	0.0000
13	120	C, D	1541.14	1548.3	1543.45	302.40	0.1499
14	100	C, D	866.37	866.4	866.37	224.20	0.0000
						Average % error	0.3100

14 cases; and that the solutions from the VLSN algorithm are on average only 0.31% higher than the best-known solutions obtained from metaheuristic approaches.

To better understand the behavior of the VLSN algorithm, we perform additional tests using different parameter values. We analyze their impacts on the solution quality and solution time. We describe these results next.

Size of the neighborhood. We examine the effect of different neighborhood sizes by varying the value of parameter K, the maximum number of customers that can be transferred in a particular move. For every problem, the algorithm is tested with three different K values. We run the program 100 times and report in Table 49.2 the average value and best value from the 100 runs for each instance. From Table 49.2, it can be observed that the solution quality improves significantly with the increase of the neighborhood size. This result is intuitive since the increase of neighborhood size means a larger search space and more opportunities for improving the solution. However, for K values larger than 5, although the computational times increased dramatically, no better solutions could be found for the cases we tested. Based on this fact, the maximum number of consecutive customers removed from a route are limited to 5 in the following tests.

Length of the negative cycle. We fix the K value at 5 and examine the impact of cycle length in this set of tests. From Table 49.3 we observe that the solution quality improves with the increase of cycle length at the expenses of increased solution time. We also observe a decreasing improvement in the solution quality with the increase of cycle length. In order to achieve a balance between computational time and solution quality, we use the following rule to select negative cycles: Among all the possible negative cycles with lengths at most L, we choose the one with the most negative cost.

Composite neighborhood structure. Table 49.4 illustrates the advantages of applying composite neighborhood structures. The cycle length is fixed at two, thus if $K = 1$, the algorithm is the simple 2-exchange and if $K > 1$, the algorithm is a composite 2-exchange. Figure 49.14 shows that better solutions are obtained with composite 2-exchange than with simple 2-exchange.

Reducing running time requirements. The advantage of VLSN search algorithm is that it searches through more neighbors in each iteration and picks the best among them, thus provides better results than those from pair exchange or other elementary moves. However, the search in a larger neighborhood leads to higher computational cost and it would be helpful to understand how fast the neighborhood size grows with different parameters settings. Table 49.5 shows, for different values of K, the average number of arcs and nodes generated in the improvement graph and the corresponding best solutions obtained with 100 runs of the algorithm. It can be concluded that the size of the

TABLE 49.2 Impact of Neighborhood Size on the Running Time and Quality of the Solution

		K = 1			K = 3			K = 5		
Problem Number	Best Solution Value	Average Solution	Best Solution	Computational Time per Run	Average Solution	Best Solution	Computational Time per Run	Average Solution	Best Solution	Computational Time per Run
1	524.61	560.17	533.68	0.13	551.35	524.61	0.42	543.14	524.61	1.03
2	835.26	864.61	844.12	0.52	857.19	835.67	3.11	855.66	835.67	9.92
3	826.14	861.89	838.77	1.21	850.64	835.04	7.25	847.43	830.56	18.54
4	1028.42	1082.40	1053.94	6.01	1066.05	1044.49	43.00	1061.50	1038.87	126.07
5	1291.45	1357.55	1335.82	13.65	1342.15	1314.07	91.76	1338.16	1311.41	201.71
6	555.43	580.17	555.43	0.24	566.62	555.43	1.62	564.80	555.43	7.31
7	909.68	946.76	921.40	0.76	930.23	909.67	17.46	928.73	909.67	248.87
8	865.94	912.11	880.36	1.95	887.97	866.74	38.74	881.03	865.94	244.06
9	1162.55	1225.69	1181.39	8.61	1191.30	1165.51	60.01	1189.56	1164.31	1134.63
10	1395.85	1470.99	1430.61	15.15	1442.29	1418.44	86.12	1439.02	1408.43	504.27
11	1042.11	1067.71	1043.89	1.94	1059.89	1042.11	15.49	1057.51	1042.24	41.06
12	819.56	830.86	819.56	0.81	827.84	819.56	4.44	821.61	819.56	12.70
13	1541.14	1565.45	1548.23	3.01	1550.50	1543.45	57.59	1559.50	1543.45	302.40
14	866.37	888.73	866.37	1.90	884.31	866.37	32.01	870.20	866.37	224.20

TABLE 49.3 Impact of the Cycle Length on the Running Time and Quality of the Solution

Problem Number	Best Solution Value	Cycle Length = 2			Cycle Length = 4			Cycle Length = 8		
		Average Solution	Best Solution	Computational Time per Run	Average Solution	Best Solution	Computational Time per Run	Average Solution	Best Solution	Computational Time per Run
1	524.61	558.07	526.13	0.59	546.18	524.61	0.72	543.14	524.61	1.03
2	835.26	874.79	857.66	3.37	863.06	838.87	3.69	855.66	835.67	9.92
3	826.14	859.82	835.67	11.88	850.84	835.15	12.43	847.43	830.56	18.54
4	1028.42	1078.31	1052.58	83.80	1065.12	1045.10	100.47	1061.50	1038.87	126.07
5	1291.45	1355.12	1321.12	120.20	1342.01	1317.71	160.10	1338.16	1311.41	201.71
6	555.43	581.60	560.29	1.87	568.03	555.43	2.80	564.80	555.43	7.31
7	909.68	953.22	929.96	5.29	938.20	914.83	11.47	928.73	909.67	248.87
8	865.94	908.25	877.18	30.24	874.03	865.94	50.51	881.03	865.94	244.06
9	1162.55	1230.32	1204.38	144.76	1221.90	1169.89	520.00	1189.56	1164.31	1134.63
10	1395.85	1472.23	1451.39	190.00	1450.23	1438.20	390.99	1439.02	1408.43	504.27
11	1042.11	1063.91	1044.70	25.45	1059.41	1042.24	29.89	1057.51	1042.24	41.06
12	819.56	830.58	820.32	4.95	821.61	819.56	15.53	821.61	819.56	12.70
13	1541.14	1562.80	1547.17	66.98	1554.43	1545.00	154.26	1559.50	1543.45	302.40
14	866.37	891.98	866.79	19.50	870.60	866.36	83.63	870.20	866.37	224.20

TABLE 49.4 Comparison of Simple 2-Exchange with Composite 2-Exchange

		K = 1			K = 3			K = 5		
Problem Number	Best Solution Value	Average Solution	Best Solution	Computational Time per Run	Average Solution	Best Solution	Computational Time per Run	Average Solution	Best Solution	Computational Time per Run
1	524.61	568.19	545.04	0.10	563.09	526.84	0.30	558.07	526.13	0.59
2	835.26	882.52	858.60	0.42	875.99	857.66	2.04	874.79	857.66	3.37
3	826.14	873.38	849.94	0.97	862.81	835.67	5.54	859.82	835.67	11.88
4	1028.42	1101.46	1052.58	5.15	1084.30	1052.58	30.9	1078.31	1052.58	83.80
5	1291.45	1373.62	1338.84	10.71	1360.16	1329.00	59.00	1355.12	1321.12	120.20
6	555.43	597.67	572.34	0.17	584.91	562.65	0.82	581.60	560.29	1.87
7	909.68	969.73	935.44	0.42	957.75	935.44	2.23	953.22	929.96	5.29
8	865.94	943.28	903.92	1.16	916.47	877.18	11.09	908.25	877.18	30.24
9	1162.55	1268.43	1212.37	4.26	1239.72	1199.91	45.43	1230.32	1204.38	144.76
10	1395.85	1501.88	1466.42	5.31	1480.16	1451.39	72.70	1472.23	1451.39	190.00
11	1042.11	1073.56	1044.70	1.56	1064.95	1044.70	10.50	1063.91	1044.70	25.45
12	819.56	832.53	820.32	0.58	830.86	820.32	2.56	830.58	820.32	4.95
13	1541.14	1571.71	1549.59	2.13	1564.40	1547.80	21.39	1562.80	1547.17	66.98
14	866.37	899.98	866.786	1.04	893.43	866.79	7.91	891.98	866.79	19.50

TABLE 49.5 Impact of Parameter K on the Size of the Neighborhood

	K = 1			K = 3			K = 5		
Problem Number	Avg. No. of Arcs	Avg. No. of Nodes	Best Solution	Avg. No. of Arcs	Avg. No. of Nodes	Best Solution	Avg. No. of Arcs	Avg. No. of Nodes	Best Solution
1	1630.48	57.63	533.68	5057.03	157.63	524.61	8551.57	257.63	524.61
2	4052.85	87.99	844.12	14104.51	237.99	835.67	23488.49	387.99	835.67
3	6628.52	110.02	838.77	25759.40	310.02	835.04	48806.80	510.02	830.56
4	15630.10	164.08	1053.94	66768.77	464.08	1044.49	136588.90	764.08	1038.87
5	26218.99	218.02	1335.82	122281.00	616.02	1314.07	260988.70	1014.00	1311.41
6	920.70	58.33	555.43	5348.90	158.33	555.43	11502.00	258.33	555.43
7	1777.77	89.32	921.40	10563.39	239.32	909.67	21397.84	389.32	909.67
8	3416.31	111.66	880.36	23465.86	311.66	866.74	55261.47	511.66	865.94
9	6801.54	167.53	1181.39	46477.03	467.00	1165.51	110484.40	767.53	1164.31
10	9559.63	220.16	1430.61	72209.35	618.16	1418.44	219011.20	1121.00	1408.43
11	9049.72	129.20	1043.89	48052.76	369.20	1042.11	102018.20	609.20	1042.24
12	8264.15	112.00	819.56	45348.20	312.00	819.56	91930.18	512.00	819.56
13	6361.34	133.22	1548.23	41283.95	373.05	1543.45	96398.20	613.05	1543.45
14	8118.45	113.37	866.37	6489.60	313.00	866.37	38246.80	513.00	866.37

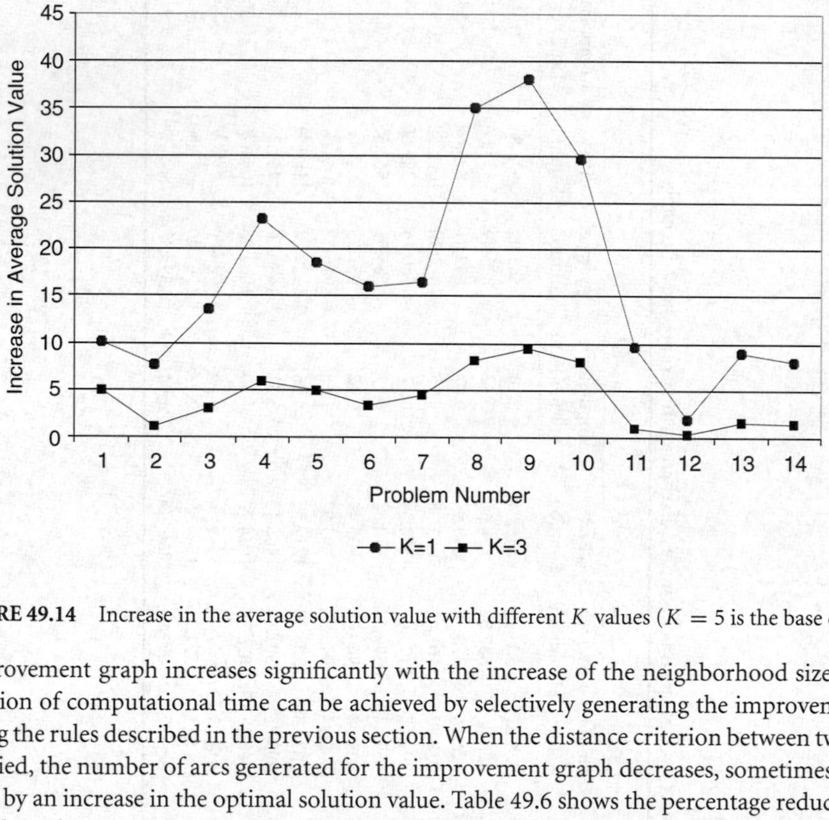

FIGURE 49.14 Increase in the average solution value with different K values ($K = 5$ is the base case).

improvement graph increases significantly with the increase of the neighborhood size. Some reduction of computational time can be achieved by selectively generating the improvement graph using the rules described in the previous section. When the distance criterion between two nodes is applied, the number of arcs generated for the improvement graph decreases, sometimes accompanied by an increase in the optimal solution value. Table 49.6 shows the percentage reduction in the number of arcs and percentage increase in optimal solution value using different distance criteria, denoted by MaxR. Note that for cases with route length constraint, MaxR is set to be the maximum route length allowed by the constraint, and that for other cases, it is set to a value determined based on the initial solution from the Clarke and Wright algorithm. Results shown in Table 49.6 indicate that selectively generating the improvement graph is a good option for reducing computational times, while maintaining the quality of solution obtained. With the problem instances tested, distance criterion set to MaxR/4 appears to be the best candidate for the balance between solution quality and computational cost.

TABLE 49.6 Comparison of Different Criteria for Selective Improvement Graph Generation

	MaxR/2		MaxR/4		MaxR/6	
Problem Number	Reduction in the Number of Arcs (%)	Increase in the Solution Value (%)	Reduction in the Number of Arcs (%)	Increase in the Solution Value (%)	Reduction in the Number of Arcs (%)	Increase in the Solution Value (%)
6	26.1	0	42.2	0	48	8
7	39.9	0	40.1	2.1	52.1	10.1
8	17.3	0	39.7	3	42.1	12.8

TABLE 49.7 Computational Time Taken by the VLSN Search Algorithm

Problem		The VLSN Search Algorithm			
No.	Size	Total Time per Run	Negative Cycle	Improvement Graph	CW Solution+3-opt
1	50	1.03	0.10	0.16	0.05
2	75	9.92	1.57	1.87	0.16
3	100	18.54	1.42	2.83	0.41
4	150	126.07	6.18	11.44	1.14
5	199	201.71	9.12	14.20	2.01
6	50	7.31	1.30	1.52	0.06
7	75	248.87	20.93	21.48	0.17
8	100	244.06	28.84	30.75	0.44
9	150	1134.63	72.47	89.09	1.13
10	199	504.27	90.90	121.00	1.62
11	120	41.06	5.98	9.08	0.73
12	100	12.70	0.82	3.47	0.43
13	120	302.40	40.00	69.09	0.79
14	100	224.20	8.00	11.90	0.52

Table 49.7 shows the computational time required for three major operations in the algorithm, which include generating initial solution using Clarke and Wright algorithm, generating the improvement graph, and finding negative cycles. The time reported in the *negative cycle* column is the time spent on finding the best cycle (up to a certain length, L) in the graph; the one in *improvement graph* column is the time spent on reconstructing the improvement graph after the solution is updated.

We observe that the time spent on finding a negative cycle in the graph is of the same order as the time spent on reconstructing the improvement graph. The values reported for Clarke and Wright algorithm show that the time spent on generating the initial solutions is insignificant. Another observation is that the computational time increases monotonically with the increase of problem size, which is an intuitive result.

Solving large size VRPs. The algorithm is also tested on some large instances of VRP as in Xu and Kelly (1996). Fifty runs of each problem instance are performed and the best results are reported in Table 49.8, which also shows the results obtained using other algorithms. It is observed that the results obtained by the composite neighborhood structures are comparable to those by the sophisticated tabu search algorithm, and in half of the cases the solutions are improved. This shows that the algorithm is more effective on large-scale problems.

49.6 Conclusions

In this chapter, we propose a new local search algorithm for VRP. Computational results obtained on a series of benchmark problems clearly indicate that the composite cyclic exchange based VLSN algorithm performs very well. It outperforms the best existing local search algorithms and produces solutions that are within 0.3% of the best solutions that are obtained by metaheuristics. Furthermore, the VLSN algorithm

TABLE 49.8 Computational Results for Large-Scale VRP Instances

Problem Description			Results from Different Algorithms			
Size	Maximum Capacity	Maximum Route Length	C-W*	X-K†	RTR§	VLSN
240	550	650	5974.23	5646.46	5834.6	5802.02
320	700	900	9257.23	8570.28	9002.26	8837.03
400	900	1200	12282.9	11880.37	11879.95	11644.95
200	900	1800	7230.47	7361.29	6702.73	6613.45
280	900	1500	9372.33	9088.66	9016.93	8828.02
360	900	1300	11901.54	11411.85	11213.31	10797.71
440	1000	1200	12982.65	12825	12514.2	12548.81
323	1000	Inf	838.92	746.56	749.15	756.67
252	1000	Inf	952.74	881.07	881.04	877.83
320	1000	Inf	1221.69	1118.09	1103.69	1119.05
240	200	Inf	771.70	666.84	720.44	717.04

*Clarke and Wright [1964] savings algorithm.
†Xu and Kelly [1996] tabu search algorithm.
§Golden et al. [1998] record-to-record travel algorithm.

does not need much problem specific tailoring and parameter setting and performs very well both in terms of solution values and computational costs. The algorithm is a good candidate for many practical situations where good quality solution needs to be obtained quickly.

We believe there are two main factors that contribute to the success of our VLSN algorithm. The first one is the fact that a much larger neighborhood is considered for each solution, which provides better opportunities; the second one is the implicit enumeration of the neighborhood with the implementation of the improvement graph technique, since the explicit enumeration of the type of neighborhoods considered in VLSN is computationally prohibitive.

The algorithm is also very flexible. Additional constraints such as assigning particular cities to specific vehicles within certain time windows can easily be incorporated. It should also be pointed out that the algorithm could be further improved by introducing some additional features, such as using the tabu-search framework instead of the local-search framework to guide the search process. Other directions for future research include algorithm for VRP with dynamic pick up and dispatching, and algorithm that considers time window constraints, which can be incorporated during the generation of the improvement graph in a similar fashion as the route length and capacity constraints.

Acknowledgments

The research of the first and second authors is supported by the NSF grants: DMI-9900087 and DMI-0217359. The research of third author is supported by Canadian Natural Sciences and Engineering Research Council grant: NSERC OGP00039682. The fourth author received funding from the NSF grant: DMI-0223323.

References

Ahuja, R.K., J.B. Orlin, and D. Sharma. 2003. A composite neighborhood search algorithm for the capacitated minimum spanning tree problem. *Operations Research Letters* **31**, 185–194.

Baldacci, R., E. Hadjiconstantinou, and A. Mingozzi. 2004. An exact algorithm for the capacitated vehicle routing problem based on a two-commodity network flow formulation. *Operations Research.* (Unpublished).

Beasley, J.E., (1990). The Operations Research Library, Imperial College Management School, London U.K.; http://www.ms.ic.ac.uk/info.html, Date Accessed: 07/22/2002.

Christofides, N., A. Mingozzi, and P. Toth. 1979. The vehicle routing problem. In N. Christofides, A. Mingozzi, P. Toth, and C. Sandi (editors) *Combinatorial Optimization*. Wiley, Chichester.

Clarke, G. and J.W. Wright. 1964. Scheduling of vehicles from a central depot to a number of delivery points. *Operations Research* **12,** 568–581.

Cordeau, J.F. and G. Laporte. 2003. A tabu search heuristic for the static multi-vehicle dial-a ride problem. *Transportation Research* **B37,** 579–594.

Cordeau, J.F., M. Gendreau, G. Laporte, J.-Y. Potvin, and F. Semet. 2002. A guide to vehicle routing heuristics. *Journal of the Operational Research Society* **53,** 512–522.

Ergun, O. 2001. *New Neighborhood Search Algorithms Based on Exponentially Large Neighborhoods.* Ph.D. Thesis, Operations Research Center, Massachusetts Institute of Technology, Cambridge, MA.

Gendreau, M., A. Hertz, and G. Laporte. 1992. New insertion and postoptimization procedures for the traveling salesman problem. *Operations Research* **40,** 1086–1094.

Gendreau, M., A. Hertz, and G. Laporte. 1994. A tabu search heuristic for the vehicle routing problem. *Management Science* **40,** 1276–1290.

Gendreau, M., G. Laporte, and J.-Y. Potvin. 2002. Metaheuristics for the capacitated VRP. In P. Toth and D. Vigo (editors) *The Vehicle Routing Problem.* SIAM Monographs on Discrete Mathematics and Applications, Philadelphia.

Golden, B.L., A.A. Assad, and E.A. Wasil. 2002. Routing vehicles in the real world: Applications in the solid waste, beverage, food, dairy and newspaper industries. In P. Toth and D. Vigo (editors) *The Vehicle Routing Problem.* SIAM Monographs on Discrete Mathematics and Applications, Philadelphia.

Golden, B.L., E.A. Wasil, J.P. Kelly, and I-M. Chao. 1998. Metaheuristics in vehicle routing. In T.G. Crainic and G. Laporte (editors), *Fleet Management and Logistics*, Kluwer, Boston, pp. 33–56.

Laporte, G. and F. Semet. 2002. Classical heuristics for the capacitated VRP. In P. Toth and D. Vigo (editors) *The Vehicle Routing Problem.* SIAM Monographs on Discrete Mathematics and Applications, Philadelphia.

Lin, S. 1965. Computer solutions of the traveling salesman problem. *Bell System Technical Journal* **44,** 1447–1459.

Lin, S. and B. Kernighan. 1973. An effective heuristic algorithm for the traveling salesman problem. *Operations Research* **21,** 498–516.

Mitrović-Minić, S., R. Krishnamurti and G. Laporte. 2004. Double-horizon based heuristics for the dynamic pickup and delivery problem with time windows. *Transportation Research* B. (In press).

Naddef, D. and A. Rinaldi. 2002. Branch-and-cut algorithms for the capacitated VRP. In P. Toth and D. Vigo (editors) *The Vehicle Routing Problem.* SIAM Monographs on Discrete Mathematics and Applications, Philadelphia.

Or, I. 1976. *Traveling Salesman Type Combinatorial Problems and Their Relation to the Logistics of Blood Banking*. Ph.D. Thesis, Department of Industrial Engineering and Management Sciences, Northwestern University, Evanston, IL.

Osman, I.H. 1993. Metastrategy simulated annealing and tabu search algorithms for the vehicle routing problem. *Annals of Operation Research* **41,** 421–451.

Owoc, M. and M.A. Sargious. 1992. The role of transportation in free trade competition. In N. Waters (editor) *Canadian Transportation: Competing in a Global Context*. Banff, Alberta.

Prins, C. 2004. A simple and effective evolutionary algorithm for the vehicle routing problem. *Computers and Operations Research.* (Unpublished).

Renaud, J., F.F. Boctor, and G. Laporte. 1996. An improved petal heuristic for the vehicle routing problem. *Journal of the Operations Research Society* **47,** 329–336.

Rego, C. 1998. A Subpath ejection method for the vehicle routing problem. *Management Science* **44,** 1447–1459.

Rochat, Y. and E.D. Taillard. 1995. Probabilistic diversifications and intensifications in local search for vehicle routing. *Journal of Heuristics* **1,** 147–167.

Thompson, P.M. and J.B. Orlin. 1989. The theory of cyclic transfers. Working Paper Number: *OR 200-89*. Operations Research Center, MIT, Cambridge, MA.

Thompson, P.M. and H.N. Psaraftis. 1993. Cyclic transfer algorithms for multivehicle routing and scheduling problems. *Operations Research* **41,** 935–946.

Van Breedam, A. 1994. *An Analysis of the Behavior of Heuristics for the Vehicle Routing Problem for a Selection of Problems with Vehicle-Related, Customer-Related and Time-Related Constraints.* Ph.D. Thesis, University of Antwerp.

Xu, J. and J. Kelly. 1996. A network flow-based tabu search heuristic for the vehicle routing problem. *Transportation Science* **30,** 379–393.

Wark, P. and J. Holt. 1994. A repeated matching heuristic for the vehicle routing problem. *Operational Research Society* **45,** 1156–1167.

50

Scheduling Problems in the Airline Industry

Xiangtong Qi
Hong Kong University of Science and Technology

Jian Yang
New Jersey Institute of Technology

Gang Yu
The University of Texas at Austin

50.1 Introduction

Being in a time-sensitive and mission-critical business, the airline industry bumps from left to right into all sorts of scheduling problems. To just name a few, it faces the challenges posed by its aircraft scheduling, crew scheduling, scheduling disruption management, and other long-term business planning and real-time operational problems that possess scheduling features.

The scheduling problems faced by the airline industry are even more complicated than the traditional machine scheduling problems. A machine scheduling problem more or less deals with the sequencing and scheduling of a set of jobs to be processed on one or a few machines. On the other hand, an airline has to deal with a set of interwoven complex problems. It has to assign different fleets to thousands of flights under various connectivity and compatibility constraints, then find the route for each individual aircraft so that the just-obtained assignment is fulfilled while the maintenance requirements of the aircraft are met; and at the same time, it has to assign crews to the various flights so that not only the connectivity and compatibility constraints are satisfied but also the rest requirements of the crews and other regulatory constraints are satisfied.

For traditional machine scheduling problems, which can undoubtedly be very difficult, people have at least built up a basic framework under which problems can be described using simple and common terms

1-58488-397-9/$0.00+$1.50

that reflect the underlying problem settings and assumptions. The research on airline scheduling, on the other hand, is much less standardized. The innate complexity and versatility of airline operations and the relatively young age of the research field make it virtually impossible for researchers, at least at this stage, not to deal with airline scheduling problems in a more piecemeal and ad hoc fashion. Having said this, we nevertheless do observe the emergence of unifying trends in the field as more and more common traits are realized for different problems. At this stage, although no unified notational system has appeared, we are already able to classify most problems in the field using several major categories. This chapter aims to give an overview to the categories we deem the most important.

Before going on any further, we make a note on our terminology. Here, an aircraft refers to a physical airplane, a fleet refers to a type of aircraft such as Boeing 737 or MD 80, a flight or flight segment refers to a direct connection from one airport A at a departure time to another airport B at an arrival time, and a crew refers to a person who serves on an aircraft, either a pilot or a flight attendant.

The purpose of this chapter is to provide a review of the up-to-date research on important scheduling problems in the airline industry. Due to the limited space here, we have to strike a balance between the breadth of our coverage and the depth of our analysis. We strive to cover all the major areas of airline scheduling, and emphasize more on modeling and general solution approaches than the in-depth analysis of special-purpose algorithms. For specific technical details, please refer to some of the references listed at the end of the chapter.

There are four major categories of scheduling problems that an airline has to deal with. The first is the aircraft scheduling problem. It is concerned with the assignment of individual aircraft to a predetermined network of flights to be operated by the airline. Given that different assignments gain different amounts of profit or incur different amounts of cost, the objective is to maximize total profit or minimize total cost under constraints that reflect physical feasibilities and governmental regulations such as flight connectivity, aircraft capacity, and maintenance requirements.

The second is the crew scheduling problem. Its goal is to determine the allocation of crew members, namely pilots and flight attendants, to individual flights. Similar to the aircraft scheduling problem, the crew scheduling problem aims to minimize the total allocation cost under various constraints. Moreover, this problem is even more difficult since it involves much more constraints. For example, the problem has to take into account that pilots usually have their preferences to specific flights and all these preferences have to be considered with regard to their seniorities.

The third is the disruption management problem, also called the real-time irregular operations scheduling problem. In daily operations, none of the flight schedule, aircraft schedule, and the crew schedule is likely to be executed without interruptions, due largely to the often occurrences of disruptive events such as bad weather, mechanical failure, and crew sickness. It is therefore imperative to decide in a real-time fashion the best plan to be carried out after the disruptive event has changed the operational environment. At the same time, the airline has also to take into account that the deviation of the new plan from the original plan can be costly and that the new plan must converge back to the original plan after a certain amount of time.

The last category is a combination of airline scheduling problems that can be modeled as traditional machine scheduling problems. The category contains the aircraft landing sequencing problem, the pilot training class scheduling problem, the aircraft scheduling problem with the ground delay program, and the workforce scheduling problem for baggage delivery.

We shall point out that the above four categories have by no means exhausted all airline scheduling problems. For a review of more of these problems, the reader is referred to Yu and Thengvall [1] and Yu and Yang [2].

We organize this chapter as follows: In Section 50.2, we briefly review some mathematical models that are used in many airline scheduling problems; discuss aircraft scheduling in Section 50.3 and crew scheduling in Section 50.4; then address the disruption management problem for airline scheduling in Section 50.5; and finally, we discuss airline scheduling problems that can be modeled as traditional machine scheduling problems in Section 50.6.

50.2 Background on Three Formulations

In this section, we briefly review three combinatorial optimization problems that are extensively used for modeling airline scheduling problems: the set partitioning (covering) problem, the multicommodity network flow problem, and the Euler tour problem. The first two problems are well known NP-hard problems, while the last one is solvable in polynomial time [3].

50.2.1 Set Partitioning Problem

The set partitioning problem can be described as follows. There are a base set $S = \{e_i \mid i = 1, 2, \ldots, m\}$ and a collection $\mathcal{S} = \{S_j \mid j = 1, 2, \ldots, n\}$ of subsets of S. With each S_j there is associated a cost c_j. For any subcollection $\mathcal{P}$ of $\mathcal{S}$, we call it a partition of S when (i) $\cup_{S_j \in \mathcal{P}} S_j = S$, and (ii) $S_j \cap S_l = \emptyset$ for any two $S_j, S_l \in \mathcal{P}$. For a partition $\mathcal{P}$, we define its cost to be $\sum_{S_j \in \mathcal{P}} c_j$, the total cost associated with all the individual subsets in $\mathcal{P}$. The set partitioning problem is concerned with finding the least costly partition of S. When we relax the second constraint of exclusiveness in the definition for a partition, the resulting subcollection is called a cover of S and the corresponding problem is called the set covering problem.

There is a natural integer-programming formulation for the set partitioning problem. For any $i = 1, 2, \ldots, m$; $j = 1, 2, \ldots, n$, we may use a constant a_{ij} to denote whether or not element e_i is contained in subset S_j: 1 when it is and 0 when it is not. For any $j = 1, 2, \ldots, n$, we may use a binary variable x_j to denote whether or not S_j belongs to the solution subcollection $\mathcal{P}$: 1 when it is and 0 when it is not. Then, the following integer programming formulation exactly describes the set partitioning problem.

$$\min \sum_{j=1}^{n} c_j x_j \tag{50.1}$$

subject to

$$\sum_{j=1}^{n} a_{ij} x_j = 1, \quad \text{for } i = 1, 2, \ldots, m \tag{50.2}$$

$$x_j \in \{0, 1\}, \quad \text{for } j = 1, 2, \ldots, n \tag{50.3}$$

To obtain the formulation for the set covering problem, we only need to change (50.2) to

$$\sum_{j=1}^{n} a_{ij} x_j \geq 1, \quad \text{for } i = 1, 2, \ldots, m \tag{50.4}$$

With respect to airline scheduling problems, usually the base set S is used to model the set of all required flight segments and each subset S_j is used to model a particular subset of connected flight segments that can be consecutively served by an aircraft or a crew. The set partitioning (covering) problem then naturally represents the problem of finding the least costly way to feasibly cover all flight segments with aircraft or crew. Certainly modifications such as the introduction of more constraints will be made when the above formulations are applied to real situations.

Exact methods for solving the set partitioning (covering) problem are usually based on branch and bound algorithms, which need to repeatedly solve various *linear programming* (LP) relaxation problems. When the LP problems are small, most integer programming solvers will be able to handle the overall problem with ease. When the LP problems become large, however, other techniques need to be employed. One technique people often use is column generation. The technique works as follows. In the beginning, a scaled-down version of the problem with only part of all the columns in the original problem is solved.

Then, based on evaluations made on the just-solved problem, some columns are removed from the current problem and some other columns in the original problem are added so that a new LP problem is formed. Next, the new LP problem is solved. The procedure is thus repeated until the current solution is proven to be optimal for the original LP problem.

50.2.2 Multicommodity Network Flow Problem

The network flow model used in airline schedule revolves around the so-called time-space network. In this network, every node stands for a time-location pair representing the departure or landing time and origin or destination airport of a flight. Two types of basic arcs exist in this network: the ground arcs and the flight arcs. A ground arc connects two nodes associated with the same airport and successive time points and usually represents an aircraft or a crew staying at the airport during the corresponding time interval. A flight arc connects two nodes respectively associated with the origin and destination of a flight. We can regard the movements of aircraft or crew members with the various flights as flows in the time-space network. Thus, the problem of finding the cheapest assignment of aircraft or crew members to flights becomes that of finding the minimum-cost integral flows in the network with the observation of constraints such as flow conservation, arc capacity, and others. When we are considering multiple types of aircraft that have different suitabilities with regard to flights, we have to treat the flows corresponding to these different types differentially. Thus, we obviously have to use the multicommodity version of the network flow problem.

The integer programming formulation of the basic multicommodity network flow problem can be described as follows. Let there be K commodity types, and the underlying network be $G = (V, I)$, where V is the set of nodes and I the set of arcs. We use (v, v') to denote the arc pointing from node v to node v'. For every $v \in V$, let $IN_v = \{i \mid i = (v', v) \in I \text{ where } v' \in V\}$ be the set of arcs that enter into v and $OUT_v = \{i \mid i = (v, v') \in I \text{ where } v' \in V\}$ be the arcs that leave from v. For $k = 1, 2, \ldots, K$ and $i \in I$, let c_{ki} be the cost of sending a unit flow of commodity type k through arc i. For $i \in I$, let u_i be the capacity of arc i. When we use x_{ki} to denote the flow of commodity type k on arc i, our formulation takes the following shape.

$$\min \sum_{k=1}^{K} \sum_{i \in I} c_{ki} x_{ki} \tag{50.5}$$

subject to

$$\sum_{i \in IN_v} x_{ki} = \sum_{i \in OUT_v} x_{ki}, \quad \forall k = 1, 2, \ldots, K, \ v \in V \tag{50.6}$$

$$\sum_{k=1}^{K} x_{ki} \leq u_i, \quad \forall i \in I \tag{50.7}$$

$$x_{ki} \in \{0, 1, 2, \ldots\}, \quad \forall k = 1, 2, \ldots, K, \ i \in I \tag{50.8}$$

In the above, (50.6) enforces flow conservation for each commodity type k at any node v, (50.7) expresses the capacity constraint for each arc i, and (50.8) states the integrality constraint for the flow of every commodity on every arc. Note that it is (50.7) that binds the multiple flow types into one problem: we would only need to solve separate network flow problems were this constraint not present.

The sizes of the problems faced by the airline industry render even the LP relaxations of the corresponding multicommodity network flow problems difficult to solve. To overcome this difficulty, researchers have developed special tools, such as Lagrangian relaxation, column generation, and Dantzig-Wolfe decomposition. The reader may find details in Ahuja, Magnanti, and Orlin [4]. These tools are used for solving the LP relaxations during the branch-and-bound or branch-and-cut processes that eventually solve the original problems to optimality.

50.2.3 Euler Tour Problem

A Euler tour in a directed graph is a closed tour along the arcs such that each arc is traversed exactly once, even though some nodes may be traversed multiple times. Given a directed graph, the Euler tour problem seeks such a tour. The aircraft maintenance routing problem where maintenance is always carried out overnight can be modeled as a Euler tour problem. To do so, each node should model an airport and each arc a daily route for a single aircraft, linking its starting airport to its ending airport. Among all airports, some can conduct overnight maintenance checks. When a Euler tour exists, all aircraft can repeatedly experience the same sequence of daily routes though in any given day, they are all assigned to different daily routes.

The sufficient and necessary condition for a directed graph to have a Euler tour is that (i) the graph is connected and (ii) each node has the same out-degree as its in-degree. For an m-arc directed graph satisfying the above two conditions, an $O(m)$-time algorithm exists for the Euler tour problem. The basic idea of the algorithm is to repeatedly merge arc-disjoint cycles. We start from any node v to traverse previously-untraversed arcs until coming back to v and denote the obtained tour as T. This is achievable due to the two conditions. If all arcs have been traversed, we let T be the Euler tour. Otherwise, we remove from the graph all the arcs in T. There must be a node v' that connects T with the remaining graph. We then as before construct a new tour T' in the remaining graph that starts and ends at v'. Obviously, T and T' can be combined into a single tour that starts and ends at node v. We repeat this process till all arcs are traversed.

50.3 Aircraft Scheduling

In this section, we introduce the scheduling problem for aircraft, i.e., the assignment of individual aircraft to flights. In practice, the aircraft schedule is determined by a sequence of decision processes. To better understand the problem, we need some knowledge about the processes.

First, the airline must design its time-space flight network comprising the flight segments that it will serve. Each flight segment in the flight network is represented by two nodes and an arc pointing from one of the nodes to the other, with the first node being at the departure time and origin airport of the flight and the second node being at the arrival time and destination airport of the flight. For each airport, the flight network also contains ground arcs that link all neighboring nodes of the same airport in the time-forward fashion. In designing the flight network, many factors have to be considered. These factors include the forecast of market demands, the capacities of the airline's fleets, the competition from other airlines, etc. The main goal is for the network to achieve the maximum profit for the airline. To know how the profit from flying each flight segment is estimated, the reader may refer to Dobson and Lederer [5].

Very often an airline owns many different aircraft types such as Boeing 737, Boeing 777, DC10, etc. Due to the commonality among aircraft of the same type and the differences in aircraft of different types, airlines usually treat all their same-type aircraft as fleets. The next decision the airline needs to make is then fleet assignment, that of assigning fleet types to the flight segments in its flight network. After this step, the aircraft scheduling problem can be decomposed into separate subproblems under individual fleet types.

In the last step, individual aircraft within each fleet are assigned to the flight segments reserved for the fleet in the fleet assignment stage. This is called the aircraft maintenance routing problem since at this stage, the main concern is to construct a flight schedule for each aircraft so that the aircraft is able to pass the maintenance bases at the frequency mandated by the *federal aviation administration* (FAA).

In the following, we give an overview of the fleet assignment and aircraft maintenance routing problems. To view an earlier survey, the reader may refer to Gopalan and Talluri [6], which reviewed fleet assignment and aircraft routing along with other topics like traffic forecasting.

50.3.1 Fleet Assignment

A typical airline conducts its fleet assignment in a periodic fashion, mostly on a daily basis. The factors that influence the assignment of fleet types to flights include passenger demands, seating capacities, operational costs, and various technical and FAA requirements. The objective is either to maximize the total profit or equivalently, minimize the total cost.

50.3.1.1 Network Flow Model

The nature of fleet assignment lends itself to a multicommodity network flow model, where fleet types correspond to commodities, the flight network serves as the underlying network, and assignments correspond to flows. Here, we introduce the multicommodity network flow formulation of the fleet assignment problem.

Indices:

k, index for fleet types
v, index for nodes
i, index for arcs
h, index for airports

Input parameters:

K, number of fleet types
V, set of all nodes
I, set of all arcs
I_F, set of flight arcs
n_k, number of available aircraft for fleet k
H, set of all airports
s_h, the node associated with airport h and the beginning of the day
t_h, the node associated with airport h and the end of the day
c_{ki}, cost of assigning fleet k to flight arc i

Decision variables:

x_{ki}, indicating whether fleet type k is assigned to arc i

Model:

$$\min \sum_{k=1}^{K} \sum_{i \in I_F} c_{ki} x_{ki} \tag{50.9}$$

subject to

$$\sum_{i \in IN_v} x_{ki} = \sum_{i \in OUT_v} x_{ki}, \quad \forall k = 1, 2, \ldots, K, \quad v \in V \tag{50.10}$$

$$\sum_{k=1}^{K} x_{ki} = 1, \quad \forall i \in I_F \tag{50.11}$$

$$\sum_{h \in H} \sum_{i \in OUT_{s_h}} x_{ki} \leq n_k, \quad k = 1, 2, \ldots, K \tag{50.12}$$

$$\sum_{i \in OUT_{s_h}} x_{ki} = \sum_{i \in IN_{t_h}} x_{ki}, \quad \forall k = 1, 2, \ldots, K, \quad h \in H \tag{50.13}$$

$$x_{ki} \in \{0, 1\}, \quad \forall k = 1, 2, \ldots, K, \quad i \in I_F \tag{50.14}$$

$$x_{ki} \geq 0, \quad \forall k = 1, 2, \ldots, K, \quad i \in I \tag{50.15}$$

In the model, (50.9) is the objective function which minimizes the total cost of fleet assignment; (50.10) states the conservation of flows at each node for each fleet type; (50.11) makes sure that each flight segment is flown by one and only one aircraft; (50.12) enforces the resource constraint for each fleet type; (50.13) is the aircraft balance constraint: the number of aircraft of any particular fleet flying out from any particular airport in the beginning of the day must be the same as the number of aircraft of that fleet flying back to the same airport at the end of the day; (50.14) is the integral 0-1 constraint on each flight arc; and (50.15) is the nonnegative constraint on each arc.

The above model describes the basic requirement for fleet assignment. To apply the model in practice, however, more details need to be discussed.

Note that it is far from straightforward to estimate each c_{ki}, the cost of assigning fleet type k to flight or ground arc i. In practice, the direct operational costs including fuel costs, crew costs, and landing fees are easy to obtain. The difficulty lies in estimating the indirect costs due to mismatches between the capacities of fleet types and the passenger demands on the flight segments: assigning a bigger aircraft than is needed to a flight induces unnecessarily high direct cost, while assigning a smaller aircraft than is needed to a flight causes potential passengers to be spilled over to overcrowd other flights or be captured by rival airlines or alternative transportation means. Many researchers have tried to find accurate ways to estimate the indirect costs. Barnhart, Kniker, and Lohatepanont [7] proposed the so-called itinerary-based airline fleet assignment, which can capture the assignment costs in more details by combining the basic fleet assignment model and a passenger mix model. Yan and Tsing [8] made similar effort by combining fleet flows and passenger flows in one time-space network.

One important issue in using the above basic model is about the flight connection requirement. An aircraft cannot depart from an airport immediately after it arrives. It has to stay on the ground for some time, say at least 40 min. The simplest way to accommodate this requirement is to make a flight arc end in a node at the time when it is ready for its next departure. In practice there are other complex rules regarding flight connection. For example, the minimum connection time is a function of both incoming and outgoing flight arcs, and hence many connections by the same flight are prohibited. To capture such constraints, the ground arcs have to be refined, for example, by explicitly identifying all possible connection possibilities instead of simply using identical ground arcs. The reader may refer to Rushmeier and Kontogiorgis [9] for more details about this.

In the basic model, the flight segments are given fixed departure and arrival times, which were generated in the network design stage. This, of course, may lead to suboptimal assignments since these times were generated without taking into account the differences in speeds of different fleet types. So it is worthwhile to consider the case where these times can be adjusted at the fleet assignment stage. Actually, we can assume that each flight segment is associated with a flexible departure time window, while the exact departure time within the time window is to be determined during the assignment. We can discretize the time axis to solve the fleet assignment problem with departure time windows: for each discrete time point in the time window of a flight, we create a flight arc departing at that time, with the understanding that one and only one of these flight arcs can have a unit flow. An implementation of such a model may be found in Rexing et al. [10]. Later on, we will show that such a technique is also used in disruption management for flight-aircraft and crew scheduling.

Fleet assignment is mostly done on a daily basis. A weekly schedule can be generated by a repetition of the daily schedule for the weekdays and making proper reductions to the daily schedule for the weekend. A more precise way is to generate a weekly schedule directly while considering different daily requirements. The problem can be modeled in the same way as the daily problem, although it is of a much larger size. When departure times are variable within time windows, additional constraints have to be included to ensure that all flights with the same flight number will depart at the same time each day. Such a model was proposed by Ioachim et al. [11].

50.3.1.2 Set-Partitioning Model

Researchers have also formulated fleet assignment problems as set-partitioning type of problems. We introduce a typical model below (see, e.g., Desaulniers et al. [12]).

Let K be the number of fleets, I_F the set of flight segments, and n_k the number of aircraft of fleet $k = 1, 2, \ldots, K$. For each fleet k, let $\mathcal{S}_k$ be the collection of subsets of flight segments in I_F that form feasible daily schedules for an aircraft in fleet k. Note that two different $\mathcal{S}_k$'s may have a nonempty intersection. Let all the subsets in $\mathcal{S}_k$ be $I_{k1}, I_{k2}, \ldots, I_{k|\mathcal{S}_k|}$. For any subset I_{kj} in any $\mathcal{S}_k$, and any flight i in I_F, let a_{ikj} be the binary constant that denotes whether flight i is in subset I_{kj}. Let c_{kj} be the cost of having fleet k cover the subset I_{kj} of flight segments. Also, let x_{kj} be the binary decision variable that designates whether subset I_{kj} in $\mathcal{S}_k$ has been chosen in the fleet assignment solution. In set partitioning terminology, the fleet assignment problem is about choosing subsets from the various $\mathcal{S}_k$'s to nonredundantly cover all the flights.

Besides the classical set partitioning constraints, we need additional constraints to fully describe our problem. First, the total number of subsets chosen from $\mathcal{S}_k$ cannot exceed the available number n_k of aircraft in fleet k. Secondly, the number of aircraft of any particular fleet flying out from any particular airport in the beginning of the day must be the same as the number of aircraft of that fleet flying back to the same airport at the end of the day. To deal with this constraint, let H be the set of airports, constant o_{hkj} be 1 if subset I_{kj} starts from airport $h \in H$ and 0 if not, and constant d_{hkj} be 1 if subset I_{kj} ends at h and 0 if not. Now, the fleet assignment problem can be expressed as follows.

$$\min \sum_{k=1}^{K} \sum_{j=1}^{|\mathcal{S}_k|} c_{kj} x_{kj} \tag{50.16}$$

subject to

$$\sum_{k=1}^{K} \sum_{j=1}^{|\mathcal{S}_k|} a_{ikj} x_{kj} = 1, \quad \forall i \in I_F \tag{50.17}$$

$$\sum_{j=1}^{|\mathcal{S}_K|} x_{kj} \le n_k, \quad \forall k = 1, 2, \ldots, K \tag{50.18}$$

$$\sum_{j=1}^{|\mathcal{S}_k|} o_{hkj} x_{kj} = \sum_{j=1}^{|\mathcal{S}_k|} d_{hkj} x_{kj}, \quad \forall k = 1, 2, \ldots, K, \ h \in H \tag{50.19}$$

$$x_{kj} \in \{0, 1\}, \quad \forall k = 1, 2, \ldots, K, \ j = 1, 2, \ldots, |\ \mathcal{S}_k\ | \tag{50.20}$$

50.3.2 Aircraft Maintenance Routing

After fleet types are assigned to flight segments, the airline then needs to decide how to assign individual aircraft within each fleet to the flight segments to be flown by that fleet. The main concern at this stage is that each aircraft should be guaranteed of sufficiently frequent maintenance checks. Thus the problem is called aircraft maintenance routing.

The relationship between fleet assignment and aircraft maintenance routing can be best illustrated by the following example. Suppose a Boeing 777 fleet has four aircraft, I, II, III, and IV, and that the daily flight segments assigned to this fleet are as shown in Table 50.1.

For complete coverage, each of Flights 3(CA) and 6(ED) needs an aircraft; while Flights 1(AB), 2(BC), 4(DB), and 5(BE) together need two aircraft, and there are different possibilities to cover these four flights. One such possibility is to use one aircraft to fly Flights 1 and 2 (ABC) and another to fly Flights 4 and 5 (DBE). Thus we have a set of daily routes for the four aircraft, as might be denoted by $O_1 = \{12(ABC), 3(CA), 45(DBE), 6(ED)\}$. Under the same notation, $O_2 = \{15(ABE), 3(CA), 42(DBC), 6(ED)\}$ is then another feasible set of daily routes.

TABLE 50.1 Example of Maintenance Routing

Flight No.	Departure Airport	Departure Time	Arrival Airport	Arrival Time
1	A	8:00 a.m.	B	12:00 p.m.
2	B	2:00 p.m.	C	6:00 p.m.
3	C	10:00 a.m.	A	3:00 p.m.
4	D	9:00 a.m.	B	12:30 p.m.
5	B	2:30 p.m.	E	6:00 p.m.
6	E	11:00 a.m.	D	5:00 p.m.

Each set of daily routes corresponds to a directed graph where nodes stand for airports and each arc represents a daily route from the starting airport of the day to the ending airport of the day. When a Euler tour or a number of Euler tours (when the graph is disconnected) can be found for the graph, a periodic schedule for the fleet can be derived from the set. For instance, the graph corresponding to O_1 has two Euler tours *ACA* and *DED*. We may use Aircraft I (II) to cover route 12(*ABC*) on odd (even) days and route 3(*CA*) on even (odd) days, and use Aircraft III (IV) to cover route 45(*DBE*) on odd (even) days and route 6(*ED*) on even (odd) days. On the other hand, the graph corresponding to O_2 has one Euler tour *AEDCA*. On any given day, each aircraft covers one route in O_2; and on the next day, each aircraft is to cover the route that starts from the airport at which its current route ends.

Suppose for every aircraft, an overnight maintenance check is needed once every four days and the check can only be done at airport A. Then in the example, O_1 is clearly not a feasible routing while O_2 is. The aircraft maintenance routing problem for a given fleet is exactly the problem of finding a feasible set of daily routes so that maintenance requirements are met while the flights to be flown by the fleet are already determined.

In reality, there are four major types (Type A, B, C, and D) of maintenance checks mandated by the FAA. These checks vary in scope, duration, and the required frequency. In many cases, however, only Type A check, a routine inspection of all major systems of an aircraft including the landing gear, engines and control surfaces, is taken into account when aircraft routing is concerned. This is mainly because Type A check's required frequency, once every 65 flight hours as mandated by the FAA, is much higher than those of all other types of checks. In practice, many airlines take more stringent policies to the effect that this type of checks are called for once every 3 or 4 days. Correspondingly, the aircraft routing problems are referred to as the 3-day or 4-day maintenance routing problems.

50.3.2.1 Two-Step Approach

The two-step approach applies to the common case where maintenance checks are done overnight. The first step generates a set of daily routes that take care of all flight segments assigned to the given fleet, such as O_1 or O_2 in the preceding example; and the second step generates periodic schedules for individual aircraft that meet all maintenance requirements. These two steps are run repeatedly until the set of daily routes generated by the first step leads to feasible maintenance routes for each aircraft in the second step. The first step usually uses a simple rule like connecting the flights into and out of any airport in a first-in-first-out (FIFO) fashion, which incidentally produces O_1 in the preceding example. The second step, however, can be more involved.

Given the set of airports and the set of daily routes generated by the first step, we can generate a directed graph $G = (N, A)$, such that N corresponds to the airport set and A the route set, with each arc emanating from the starting airport and entering the ending airport of the corresponding route. There is a subset M of N corresponding to airports that can perform maintenance checks. Apparently, $|A|$ aircraft are needed in the fleet to cover all the daily routes. When there is a Euler tour in G (a set of Euler tours in case G has more than one connected components) without any k nodes in $N \setminus M$ in succession, we can construct feasible periodic routes for all $|A|$ aircraft that meet the maintenance requirements: on any given day, each aircraft covers a route; and on the next day, each aircraft is to cover the route that succeeds its current route on its Euler tour.

For any connected component of a graph, we know that a Euler tour can be found in polynomial time if it exists. The problem now is how to deal with the additional k-node constraint. When $k = 3$, the problem can be solved in polynomial time through a conversion into a Euler tour problem without the k-node constraint. When $k \geq 4$, however, the problem becomes NP-complete. The reader interested in the details can consult Gopalan and Talluri [13] and Talluri [14].

Note that the Euler tour model only deals with the feasibility issue and ignores the cost-optimality issues. In reality, different airports conduct maintenance checks at different prices. If our goal is to find the least expensive aircraft maintenance routing, we should again resort to the multicommodity network flow model. In the model, each commodity represents one aircraft, and the underlying network is almost the same time-space network studied previously with the exception that, now each pair of airport and calendar day makes up one node and each arc corresponds to one daily route. The problem is to find a path of unit flows for each commodity so that the k-node constraint is satisfied and the total maintenance cost is minimized. This model has the advantage that it provides more opportunity for analyzing maintenance costs and helps the allocation of maintenance bases (see Feo and Bard [15]).

50.3.2.2 Other Approaches

The two-step approach cannot guarantee the optimality or even feasibility of the solution, because the first step does not consider the maintenance requirements and the second step cannot directly change flight segments within a daily route. It is better if the daily routes can be generated with the maintenance requirements being taken into account. Clarke et al. [16] modeled such a problem as an integer programming problem and solved it using Lagrangian relaxation techniques.

As we have already known, aircraft maintenance routing is usually done fleet by fleet after the occurrence of fleet assignment. Naturally, a more ambitious approach would be to conduct these two activities simultaneously. Barnhart et al. [17] presented such a unified model and solved it using branch-and-bound and column generation.

There are also researchers who consider more than one type of maintenance checks simultaneously. For example, Sriram and Haghani [18] considered both Type A and Type B checks, where the latter needs to be performed every 300 to 600 flight hours. The problem was solved by a hybrid heuristic of random and depth-first searches.

50.4 Crew Scheduling

Crew scheduling deals with the problem of assigning individual crew members to prescheduled flights. Like aircraft scheduling, crew scheduling is sequentially divided into two separate stages: crew pairing and monthly crew assignment. We will elaborate on what each of these two stages is about later on. Here, a crew can be either a pilot or a flight attendant. Since each pilot is fleet-associated, pilot scheduling can be decomposed into problems in individual fleets. On the other hand, each flight attendant often can work on any fleet. So flight attendant scheduling involves normally all fleets and is a much larger problem. At the same time, pilot scheduling is subject to more strict governmental and contractual regulations, while flight attendant scheduling has to cope with relatively fewer constraints. Besides these differences, similar approaches are usually applied to crew scheduling for both types of crew.

50.4.1 Crew Pairing

In the crew pairing problem, we are given a set of prescheduled flight segments over a period of time. Among all the involved airports, some of them are crew bases where crew members can start and end with over the time period. A legal crew pairing is a sequence of connected flight segments beginning and ending at a crew base that satisfy all legality constraints. The crew pairing problem is that of finding the minimum-cost set of legal crew pairings that cover all the given flight segments.

Both crew pairing and fleet assignment are about covering a given set of flight segments. The crew pairing problem tends to be more difficult mainly because the former involves flights over more than one day while the latter involves flights over one day. Also, crew pairing is usually involved with more constraints due to governmental and contractual restrictions such as maximum daily working hours, minimum overnight rest period, maximum number of flight legs, and maximum time away from a crew base. In addition, a legal crew pairing must start and end at the same crew base, a constraint not to be worried about in fleet assignment.

Much like fleet assignment, existing models for crew pairing can be classified into two main categories, the set partitioning (covering) type and the network flow type. While in fleet assignment, most works are based on the multicommodity network flow model, crew pairing is more often modeled as a set partitioning (covering) problem. This is mainly due to the difficulty of handling many complex constraints inherent in crew pairing using the network flow model.

50.4.1.1 Set Partitioning Model

Using a set partitioning model, each legal pairing is modeled as a subset of flight segments associated with its proper operational cost. The main component of the cost is the so-called pay and credit, defined as the difference between the hours actually needed to cover the subset of flights and the guaranteed hours of pay. The objective of crew pairing is to minimize the total cost of all selected pairings. Let there be m flights and n subsets of legal pairings. Let binary constant a_{ij} indicate whether flight i belongs to subset j, and c_j be the cost associated with subset j. Then, the crew pairing problem can be exactly formulated as a set partitioning problem as described by (50.1), (50.2), and (50.3). If a crew can be on a flight as a passenger, then the corresponding model is of the set covering type, where (50.2) is replaced by (50.4).

We can add additional constraints to the above set partitioning or covering model to accommodate more restrictions. For example, suppose airports $1, 2, \ldots, B$ are crew bases, there are d_b crew members in airport b, and binary constant f_{bj} indicates whether pairing j starts and ends at airport b. Then, the following constraint addresses the crew availability issue.

$$\sum_{j=1}^{n} f_{bj} x_j \leq d_b, \quad b = 1, \ldots, B \tag{50.21}$$

The major drawback with the set partitioning (covering) model is its huge size. There can be billions of potential legal pairings in a medium size problem. It is infeasible in practice to solve such an integer programming problem to optimality. Naturally, people have been developing various algorithms to achieve near-optimal solutions over the past decades. Most of these algorithms contain primarily two modules, the pairing generation module and the pairing optimization module. The pairing generation module generates legal pairings that can potentially be included in the final solution, and the pairing optimization module solves the current problem based on pairings selected by the pairing generation module. A good solution is supposed to be obtained by running these two modules iteratively.

For crew pairing generation, people often first use a set of randomly generated pairings which is of probably a very high cost. In later iterations, some of the pairings used in past iterations still remain. At the same time, new pairings generated randomly (e.g., Anbil et al. [19] and Klabjan, Johnson, and Nemhauser [20]) or through evaluations (e.g., Graves et al. [21]) enter the pool of generated pairings.

Each pairing optimization problem in the second module is itself a set partitioning problem with less subsets than the original pairing problem. Various methods have been tried on this problem, including the column generation method in Lavoie, Minous, and Odier [22], the branch and price method in Barnhart et al. [23], the Lagrangian relaxation method in Anbil et al. [19], the branch and cut method in Hoffman and Padberg [24], etc. All the aforementioned methods have been reported to be successfully implemented and used in real systems.

50.4.1.2 Network Flow Model

Network-flow-based models are also used extensively in crew pairing, though still not as much as set-partitioning-based models. Desaulniers et al. [25] formulated the problem as an integer nonlinear multi-commodity network flow problem with additional resource variables. In the model, a commodity models a crew, nodes model airports at different time points, and arcs model various crew activities, such as operational flight segments, deadhead flight segments, connections, and rests. Additional resource variables are used to model different crew pairing constraints and regulations. A branch-and-bound algorithm based on an extension of the Dantzig-Wolfe decomposition principle was used to solve this problem. The model has been adopted by Air France. Barnhart and Shenoi [26] arrived to a similar network flow model. They obtained the solution for a relaxed model first and then developed the solution for the original model based on that solution.

In some special cases, the problems may become easy. For example, Yan and Tu [27] proposed a pure network flow model for crew pairing in Taiwan's China Airlines where much fewer constraints are needed. They used the network simplex algorithm to solve the problem. There are also some other approaches that do not belong to conventional mathematical programming techniques, such as the simulated annealing algorithm proposed by Emden-Weinert and Proksch [28] and the genetic algorithms proposed by Levine [29] and Ozdemir and Mohan [30].

50.4.2 Monthly Crew Assignment

The task following crew pairing is the monthly crew assignment. It assigns individual crew members into trips, i.e., sequences of crew pairings over a certain period of time, usually one month. In practice, two approaches have been taken to tackle the problem. The approach relying on the crew rostering problem tries to find an assignment scheme covering all pairings that is *fair* to all crew members, after taking into account their requirements and preferences. In the other approach called bidline system, each crew member has a different seniority, and the person with a higher seniority has a higher priority to satisfy his/her preference. Crew rostering is more frequently used in Europe, and bidline procedures are more common in North America.

50.4.2.1 Crew Rostering

The crew rostering problem is often modeled as a set partitioning problem. Here, the pairings form the elements and all possible monthly sequences constitute all the subsets. Ryan [31] used a generalized set partitioning model where each element may need to be covered multiple times to generate a feasible crew rostering with the objective to maximize the total satisfaction of all crew members. The LP relaxation of the problem was first solved by the primal simplex algorithm, then a branch-and-bound approach was used to generate integer solutions.

Gamache et al. [32] introduced a problem somewhat *dual to* the set partitioning problem, where the number of subsets is fixed and the total cost of the uncovered elements is to be minimized. The problem models the situation where there are a fixed number of crew members and the total cost of the uncovered pairings are to be minimized. They solved the LP relaxation of the problem using column generation, and used heuristics to generate good integer solutions. Local search algorithms are also used in crew rostering problems. Examples include the simulated annealing algorithm by Lucic and Teodorovic [33], and the genetic algorithm by El Moudani et al. [34].

Constraint programming, originally invented in the field of Artificial Intelligence, is widely used for solving crew rostering problems. Dawid, Konig, and Strauss [35] proposed an extended set partitioning model and solved it using a recursive implicit enumeration approach incorporating elements of constraint programming. Caprara et al. [36] and Sellmann et al. [37] used constraint programming as their primary solution techniques.

Some researchers are not content with using one single criterion to measure the fairness of the crew assignment in terms of how satisfied crew members are given their preferences. Instead, they try multi-criterion approaches. Two typical measures are as follows. Given the total flying hours required by flight schedules and the total number of crew members, there is an average flying hour per crew. So one measure of fairness is the deviation of the actual flying time of the trips from the average flying time. Another measure is the deviation of the weekend out-of-home time.

Lucic and Teodorovic [33] studied a multicriterion problem, and transformed it into a single-criterion one by assigning weights to individual criteria. Teodorovic and Lucic [38] used fuzzy logic models to handle the multiple criteria. El Moudani et al. [34] treated the total cost of the trips as a primary measure and the satisfaction of pilot preferences as a secondary measure. They first solved the problem minimizing the primary measure while ignoring the secondary objective, and then solved the problem minimizing the secondary measure under the constraint that the primary measure be not worse off than the optimal level obtained previously by a certain percentage.

50.4.2.2 Bidline Systems

In a bidline system, first trips are generated, and then crew members choose their preferred trips in the order of their seniority. It is natural that some trips are preferable to others due to differences in difficulties, lengths of night flying, time zone crosses, etc. As a consequence, senior crew members always get their ideal duties while junior ones often get unwanted duties. It is therefore up to the trip generation stage to ensure that fairness can still be expected from the system.

Jones [39] was one of the earliest to build a fair bidline system using an algorithmic rather than manual approach. The system uses techniques in expert-system design and other heuristics. Jarrah and Diamond [40] used a set partitioning model to generate trips and solve the problem using column generation techniques. Christou et al. [41] developed a two-phase trip-generation algorithm. The first phase constructs as many high-quality trips as possible, and the second phase uses a genetic algorithm to select the best trips from the pool of trips.

The preferential bidding system combines features of both the crew rostering and bidline approaches. In such a system, crew members with higher seniority still have higher priority to satisfy their preferences. But the preferences are satisfied under the condition that the remaining uncovered trips can still be fully covered by other junior crew members. Gamache, Soumis, and Villeneuve [42] proposed an iterative approach to solve this problem. They determined each crew member's trip one by one from the most senior to the most junior, trying to maximize his/her preference while keeping the remaining problem feasible.

50.5 Disruption Management for Airline Scheduling

Flight, aircraft, and crew schedules are all generated in advance at an airline's planning stage. When these schedules are being executed, however, various disruptions may occur that render the schedules unexecutable if unchanged. Possible disruptions stem from equipment failure, crew sickness, bad weather, air traffic control restrictions, and so on. Due to their frequent occurrences, the ability to dynamically revise the original schedules to suit the newly changed operational environment after disruptions is very important to the airline. Disruption management refers to this process of plan adjustment.

50.5.1 An Overview of Disruption Management

We feel the need for an overview of disruption management because this is yet a relatively new area. We shall first base our discussion on a general context rather than the particular airline scheduling setting. Suppose we have an operational plan that is optimal or near optimal under the most expected environment. When the plan is being executed, disruptions may occur that change the environment abruptly from time to time. As a consequence, the original operational plan may not remain optimal or even feasible. After the occurrence of a disruption, therefore, we need to revise the original plan to make it suitable for the new environment.

In addition to the suitability requirement of the revised new plan for the new operational environment, the new plan should also not be too far away from the original one, since the deviation incurs its own cost in real life. For instance, in the airline scheduling setting, the new flight schedule after a major storm should deviate from the published schedule as less as possible even though a certain degree of deviation is inevitable, because any deviation causes confusion and inconvenience to passengers and other operations, real extra operational expenditures, and other penalties. Moreover, in cases where the original plan is to be repeatedly executed, the new plan should gradually converge back to the original plan over a period of time during which no new disruption strikes. Also, the new plan should be generated in a short period of time since it is immediately needed after the occurrence of a disruption.

The study of disruption management originates from airline scheduling and has many applications in other fields such as production planning and scheduling, telecommunications, and public sectors (see, e.g., Clausen et al. [43]). Research in airline operations disruption management has brought about its own new concepts and philosophies on handling uncertainties, among which partial solutions and multiple solutions are two important ones.

In the airline setting, it is sometimes impossible to find a new high-quality plan that covers the whole planning horizon and takes all resources and commitments into consideration in a timely fashion. When this is the case, the airline can first settle with a quickly-found partial solution addressing only the most immediate and important decisions, and during the execution of the partial solution, look for a more considerate and longer-term plan that can take over from the partial solution.

The ability of a disruption management system to provide multiple solutions is also important, since some issues arising in real-time cannot be addressed by the underlying mathematical model without human intervention. When multiple solutions are presented to the human decision maker, however, he/she will have a better chance to find one that addresses the issues.

50.5.2 Flight-Aircraft Rescheduling

When a disruption occurs, flights, aircraft, and crew members may all need to be rescheduled. In practice, this is done sequentially, with flights and aircraft being together rescheduled first. If no satisfactory new crew schedule can be generated after the new flight-aircraft schedule is generated, then the whole process will have to be repeated. We discuss disruption management for flight-aircraft scheduling first.

Two important issues of the flight-aircraft rescheduling problem are related to the original flight-aircraft schedule. First, the disruption management problem often has a time window. It is required that the new schedule must converge back to the original schedule after a predetermined time window to mitigate the long-term effect of the disruption. Specifically, all aircraft must be at correct airports by the end of the time window. Second, within the time window, it is also preferred that the new schedule deviate from the original schedule as minimally as possible. We will explain how to measure and achieve this goal later.

To reschedule the flights and aircraft together, there are several options with different deviation costs, i.e., cost differences between the new and original actions. These options include delaying some flights by certain amounts of time (still to be covered by the same aircraft as originally planned), canceling some flights, using allowable types of aircraft other than the original types to cover flights, and ferrying in aircraft from other airports to cover flights. The objective of the disruption management problem is to find a new flight-aircraft schedule with the minimum total deviation cost that converge back to the original schedule within the time window.

The problem can be modeled as a multicommodity network flow problem revolving around an underlying time-space network similar to the one used in fleet assignment. Each commodity represents a type of aircraft. The network represents the original flight schedule, where ground arcs represent aircraft waiting on the ground and flight arcs represent flights. Aircraft-flight assignments are modeled as flows on the arcs. The time horizon of the network spans from the moment right after the occurrence of the disruption to the time when the time window ends. Every node representing an airport at the beginning of the time horizon has a given integer amount of inflow. This puts the number of aircraft available at the airport right after the disruption into the model. Also, every node representing an airport at the end of the time horizon is required to have a certain integer amount of inflow. This forces the aircraft to reach the positions for resuming the original schedule after the time window. Multiple delay arcs parallel to each flight arc, arcs that connect the same set of airports at delayed time points, are generated, so that having a unit flow at one and only one of these arcs can represent the flight being delayed. Having a zero flow on a flight arc and all its delay arcs naturally represent the flight being canceled. Ferry arcs, very similar to flight arcs, can be added to the network so that a unit flow on a ferry arcs represent an aircraft being ferried from one airport to another.

Besides flight delays and cancellations, another source of huge deviation cost comes from having to use different aircraft to cover different flights that are originally covered by a single aircraft. Passengers have strong preferences to stay in the same aircraft. A model that penalizes this kind of deviation can use a modeling apparatus called protection arcs. For two connected flights that originally use the same aircraft, a protection arc starts from the starting node of the first flight and ends at the terminal node of the second flight. A unit flow on the protection arc means that both flights are covered by the same aircraft. When this arc is assigned a cost that is less than the total costs of the two flight arcs, there will be incentive to use the same aircraft for both flights.

Teodorovic and Guberinic [44] used the network flow model to study the flight-aircraft rescheduling problem. But they only considered the option of delaying flights. Teodorovic and Stojkovic [45] extended the preceding work to include flight cancellations. Jarrah et al. [46] presented two special cases based on the network flow model, one considering only flight delays and the other considering only flight cancellations. While delays and cancellations were not addressed simultaneously, this approach was already considered practical enough to be implemented by United Airlines (see Rakshit, Krishnamurthy, and Yu [47]). Yan and Yang [48] were the first to incorporate flight delays, cancellations, and ferryings in a single model. Yan and Lin [49] extended the above model to handle airport closures. Yan and Tu [50] extended the same model to tackle multiple fleet substitutions. Other related works include Arguello, Bard, and Yu [51,52] and Cao and Kanafani [53,54].

Thengvall, Bard, and Yu [55] first introduced the protection arcs to reduce the deviation cost due to passenger unsatisfaction over having to change aircraft in connection airports. Thengvall, Yu, and Bard [56] considered probably the most complete model so far. They considered flight delays, cancellations, the above peculiar deviation, aircraft ferrying, fleet substitution, and hub closures. Recently, Stojkovic et al. [57] studied a special case of the problem where the disruption is small enough for the airline to be able to keep the original aircraft itineraries. The problem is interesting in that it can be modeled by a pure network flow model and thus solved in polynomial time.

50.5.3 Crew Rescheduling

The next step after flight-aircraft rescheduling is to revise the crew schedule with respect to the new flight-aircraft schedule. Recall that in crew scheduling, crew pairings lasting 2 to 5 days are generated to cover all flights, and crew members are assigned to the pairings through bidding or rostering. When the flight schedule is changed, some original crew pairings are broken. Some disruptions, such as crew sickness and emergency leaves, impair the execution of the original crew schedule directly rather than through a changed flight-aircraft schedule.

The goal of crew rescheduling is to repair the broken pairings so that the entire system can return to the original schedule efficiently within a given time window. The options to be used for the repairing include forming new crew pairings, breaking up old crew pairings, using reserved crew members, and crew deadhead (a crew travels as a passenger to an airport and joins a pairing or back to the home base).

Relative to that of flight-aircraft rescheduling, the crew rescheduling literature is rather sparse. Teodorovic and Stojkovic [58] used a FIFO rule to assign crew members to new flight-aircraft schedules. Wei, Yu, and Song [59] developed a heuristic-based search algorithm for the problem. The solution approach was successfully implemented for Continental Airlines (see Yu et al. [60]). Lettovsky, Johnson, and Nemhauser [61] proposed an integer programming model for the problem, and used a primal-dual subproblem simplex method to solve its LP relaxations. Computational results showed that medium size problems can be solved to optimality within minutes.

Traditionally, people dealt with the flight-aircraft and crew rescheduling problems sequentially. Recently, however, Stojkovic and Soumis [62] made an attempt to consider the two problems simultaneously under a single framework. In the model, changes to the existing flight and crew schedules are considered simultaneously, while the planned aircraft itineraries are kept intact. The objective is to minimize the total cost due to flight delays and cancellations and crew schedule changes. The problem was formulated as an integer nonlinear multicommodity network flow problem with time windows and additional constraints. The solution approach used Dantzig-Wolfe decomposition combined with branch-and-bound. The approach was tested on several input data sets. All of them were successfully solved very quickly.

50.6 Problems Modeled as Machine Scheduling Problems

In this section, we introduce several airline scheduling problems that can be modeled as machine scheduling problems. These problems can be described by the machine scheduling terminology in terms of jobs, machines, job processing times, job completion times, job due dates, etc. Currently, these problems have not received enough attention from the machine scheduling research community, and yet, they are inspired by real applications and pose new opportunities.

50.6.1 Scheduling of Aircraft Landings

We start with the problem of scheduling aircraft landings at an airport. There are n aircraft flying toward an airport during a planning cycle. Without other competitors for the same runway, each aircraft has an ideal landing time d_i which may be its published landing time or the landing time resulting from following the most fuel-efficient flying speed. On the other hand, each aircraft i has its earliest possible landing time a_i resulting from flying at its top speed, and its latest possible landing time b_i resulting from consuming all the fuel it carries before landing. Aircraft i must land within the time window $[a_i, b_i]$. There is also a corresponding cost associated with the aircraft's real landing time C_i: an α_i per unit time penalty for being earlier than d_i and a β_i per unit time penalty for being later than d_i. When all the n aircraft are competing for the same runway, they still have to observe a certain forbidding-period rule: right after aircraft i has landed, there is a p_i amount of time during which no aircraft can land. The job of the aircraft landing scheduling problem is to find a landing time for each of the n aircraft within its time window such that no aircraft lands within a preceding aircraft's forbidding period and the total cost associated with these time spots is the minimum possible.

A single machine scheduling problem with time windows and the earliness-tardiness objective can be used to model this problem. Each aircraft can be modeled as a job and the runway the single machine. The ideal landing time d_i now corresponds to job i's due date, and $[a_i, b_i]$ the job's time window into which its completion time must fall. Aircraft i's forbidding period p_i is now the processing time required by job i. Job i's completion time C_i in turn corresponds to aircraft i's landing time. Let binary variable x_{ki} indicate whether the kth job in the schedule is job i. Then, the machine scheduling problem that exactly addresses concerns of the aircraft landing scheduling problem can be formulated as follows:

$$\min \sum_{i=1}^{n} (\alpha_i \max\{d_i - C_i, 0\} + \beta_i \max\{C_i - d_i, 0\}) \tag{50.22}$$

subject to

$$\sum_{i=1}^{n} x_{ki} = 1, \quad \forall k = 1, 2, \ldots, n \tag{50.23}$$

$$\sum_{k=1}^{n} x_{ki} = 1, \quad \forall i = 1, 2, \ldots, n \tag{50.24}$$

$$\sum_{i=1}^{n} x_{k+1,i} C_i \geq \sum_{i=1}^{n} x_{ki}(C_i + p_i), \quad \forall k = 1, 2, \ldots, n-1 \tag{50.25}$$

$$a_i \leq C_i \leq b_i, \quad \forall i = 1, 2, \ldots, n \tag{50.26}$$

In the formulation, the objective (50.22) is clearly to minimize the total earliness and tardiness cost; (50.23) states that one and only one job occupies each position and (50.24) states that each job occupies one and only one position; (50.25) expresses each job's required processing time; and (50.26) enforces the hard time window for each job.

While earliness-tardiness scheduling problems have been extensively studied in the past two decades, to our knowledge no research has been conducted on the above problem in the machine scheduling context. In airline scheduling, people have been using mathematical programming techniques to address this kind of problem. Beasley et al. [63] presented a mixed-integer zero-one formulation for the problem and solved it using an LP relaxation tree search algorithm and a heuristic algorithm. Ernst, Krishnamurthy, and Storer [64] worked on the same model and developed a specialized simplex algorithm for the problem.

50.6.2 Class Scheduling for Pilot Training

This section introduces the class scheduling problem at the training center of a major airline. The airline's pilots are often awarded with new positions which require them to receive additional training on new skills at the airline's training center. A major problem emerging from this need for pilot training is the so-called training class scheduling problem: with limited training device availabilities, how to determine the daily activities of all pilots in the training center so that they can complete the training in as short amount of time as possible (Yu, Dugan, and Arguello [65]).

Specifically, we are given n classes to be scheduled. Each class has a sequential schedule to have the occasion to use certain training devices. Also, some devices can substitute for other devices. On the other hand, the supply of the training devices is limited and the classes have to use the commonly demanded devices on different days. An unlimited number of free days can be inserted into the schedule of each class when the devices needed are being used by other classes. The inserted days certainly lengthen the duration of each class, which is undesirable since each pilot being trained in a class remains unproductive for its entire duration. The objective of the class scheduling problem is therefore to dispatch the training devices to classes in such a way that the total weighted completion time of the classes is minimized.

The problem can again be modeled as a machine scheduling problem. Here, each training device is modeled as a machine, each class as a job, and any day's activity of a class as a unit-processing-time operation. In the machine scheduling literature, there are some existing works that handle some aspects of the class scheduling problem. For instance, Linn and Zhang [66] considered a hybrid flow shop problem, where a hybrid flow shop has several stations each of which has multiple machines. A job needs to be processed sequentially through these stations and needs only one of the machines at any station. Brucker, Jurisch, and Kramer [67] studied a multipurpose machine scheduling problem, where different machines are partially substitutable, i.e., an operation of a job can be done by any one of a set of candidate machines. Lee [68] considered a scheduling problem with machine unavailability, where a machine may have variable capability over time. On the other hand, none of the existing work in machine scheduling describes all aspect of the class scheduling problem.

Qi, Bard, and Yu [69] gave a full account of the class scheduling problem and solved it using a branch and bound algorithm. In the method, each node D in the branching tree that is t levels down from the root of the tree corresponds to a partial schedule for all classes from days 1 to t, which describes for each class its activities during this period. Each of node D's successor node in the tree corresponds to a $(t + 1)$-day partial schedule for all the classes whose activities during the first t days coincide with those corresponding to D. Various elimination rules and effective lower bounds were used to cut unpromising branches to accelerate the searching speed.

As the problem size grows, however, the running time of the branch-and-bound algorithm increases fairly rapidly. A faster rolling horizon heuristic was then developed. In this approach, classes are ordered in a series as Class $1, 2, \ldots, n$, and the problem is solved iteratively with each iteration a problem with a smaller size of h classes being solved. In the first iteration, Classes $1, 2, \ldots, h$ are scheduled to their completions; in the second iteration, Class 1's schedule is held unchanged, and Classes $2, 3, \ldots, (h + 1)$ are scheduled to their completions, with the new schedules for Classes $2, 3, \ldots, h$ possibly being different from the ones found in the last iteration. This process keeps on going until in the $(n - h + 1)$th iteration, Classes $(n - h + 1), (n - h + 2), \ldots, n$ are all scheduled to their completions. Each iteration is in turn solved by the branch-and-bound algorithm.

50.6.3 Scheduling with the Ground Delay Program

The *ground delay program* (GDP) is one of the several programs that the FAA is currently administering for more efficient and equitable use of the airspace and airports. When bad weather develops around an airport and reduces its operational capacity, the FAA may initiate the GDP which restricts any given airline's landing to the airport to a few time slots. These time slots are usually later than the planned landing times for the airline's flights into the airport. So it is important for the airline to reschedule their flights so that a certain measure of delay can be minimized. Note that this problem also falls into the category of disruption management. We present it here because of its use of machine scheduling modeling.

The problem faced by an airline at an airport with the GDP is stated as follows. There is a set I of in-flights and a set K of out-flights at the airport. An out-flight needs resources, such as crew members and connecting passengers, from some of the in-flights. So the out-flight cannot depart until the corresponding in-flights have landed. Therefore, how much each out-flight is to be delayed depends on how the time slots are assigned to the in-flights. Given a planned departure schedule of all out-flights, the scheduling problem with the GDP is to assign the time slots to in-flights so that a certain measure of lateness related to the out-flight departures is minimized. For certain measures of lateness, we can model the airport as a single machine with a limited set of time slots and model the in-flights as jobs with due dates. Thus the problem becomes a due-date-related machine scheduling problem.

When the maximum lateness of the out-flights is to be minimized, we can in the machine scheduling model define the due date of a job corresponding to an in-flight to be the earliest planned departure time of the out-flight that needs resources from the in-flight. Using a pairwise exchange argument, it can be shown that the *earliest due date* (EDD) rule solves the problem to optimality. When the number of late out-flights is to be minimized, we can still define the job due dates in the same fashion. This time, however, the corresponding machine scheduling problem is strongly NP-hard. A heuristic has been proposed for the problem. For detailed discussions about scheduling with the GDP, the reader is referred to Vasquez-Marquez [70] and Luo and Yu [71].

50.6.4 Scheduling with Varying Machine Speeds

Amaddeo, Nawijn, and van Harten [72] studied a baggage handling problem where the handling speed varies under the varying worker availabilities and different baggages are allowed to be handled simultaneously. The problem can be modeled as a single-machine scheduling problem in which the machine speed varies over time, several jobs might occupy the machine simultaneously, and the goal is to minimize the total weighted completion time of the jobs. More specifically, there are n jobs to be processed on a machine; the total amount of work required by job i is q_i and its weight is w_i; the machine speed at time t is given at $m(t)$; the machine is to dedicate a partial speed $m_i(t)$ to job i at time t, under the constraint that $m(t) = \sum_{i=1}^{n} m_i(t)$; the completion time C_i of job i is determined by the equation $q_i = \int_0^{C_i} m_i(t)dt$; and the goal is to find the machine allocation schedule $\{m_i(t) \mid i = 1, 2, \ldots, n, t \geq 0\}$ that minimizes the total weighted completion time $\sum_{i=1}^{n} w_i C_i$ of the jobs.

It has been shown that there exists an optimal schedule in which no jobs are processed concurrently and no preemption ever occurs. Based on this property, a conventional branch-and-bound algorithm was used to solve the problem in which a node in the searching tree represents a partial schedule where only some of the jobs are scheduled. The problem itself has been proved to be NP-hard. A similar problem has also been studied by Surkis and Dogramaci [73], who proposed a simple heuristic algorithm. Note that the above problem is different from the variable-speed machine scheduling problem studied by Trick [74]. In the latter, the machine speeds can be adjusted at varying prices.

References

[1] Yu, G. and Thengvall, B., Optimization in the airline industry, In Pardalos, P.M. and Resende, M.G.C., (editors), *Handbook of Applied Optimization*, Oxford University Press, New York, 2002.

[2] Yu, G. and Yang, J., Optimization applications in the airline industry, In Du, D.-Z. and Pardalos, P.M., (editors), *A Handbook for Combinatorial Optimization*, Kluwer Academic Publishers, 635–726, 1999.

[3] Garey, M.R. and Johnson, D.S., *Computers and Intractability*, W.H. Freeman, San Francisco, 1979.

[4] Ahuja, R.K., Magnanti, T.L., and Orlin, J.B., *Network Flows: Theory, Algorithms, and Applications*, Prentice Hall, New Jersey, 1993.

[5] Dobson, G. and Lederer, P.J., Airline scheduling and routing in a hub-and-spoke system, *Transportation Science*, 27, 281–297, 1993.

[6] Gopalan, R. and Talluri, K.T., Mathematical models in airline schedule planning: A survey, *Annals of Operations Research*, 76, 155–185, 1998.

[7] Barnhart, C., Kniker, T.S., and Lohatepanont, M., Itinerary-based airline fleet assignment, *Transportation Science*, 36, 199–217, 2002.

[8] Yan, S. and Tsing, C.-H., A passenger demand model for airline scheduling and routing, *Computers & Operations Research*, 29, 1559–1581, 2002.

[9] Rushmeier, R.A. and Kontogiorgis, S.A., Advances in the optimization of airline fleet assignment, *Transportation Science*, 31, 159–169, 1997.

[10] Rexing, B. et al., Airline fleet assignment with time windows, *Transportation Science*, 34, 1–20, 2000.

[11] Ioachim, I. et al., Fleet assignment and routing with schedule synchronization constraints, *European Journal of Operational Research*, 119, 75–90, 1999.

[12] Desaulniers, G. et al., Daily aircraft routing and scheduling, *Management Science*, 43, 841–855, 1997.

[13] Gopalan, R. and Talluri, K.T., The aircraft maintenance routing problem, *Operations Research*, 46, 260–271, 1998.

[14] Talluri, K.T., The four-day aircraft maintenance routing problem, *Transportation Science*, 32, 43–53, 1998.

[15] Feo, T.A. and Bard, J.F., Flight scheduling and maintenance base planning, *Management Science*, 35, 1415–1432, 1989.

[16] Clarke, L. et al., The aircraft rotation problem, *Annals of Operations Research*, 69, 33–46, 1997.

[17] Barnhart, C. et al., Flight string models for aircraft fleeting and routing, *Transportation Science*, 32, 208–220, 1998.

[18] Sriram, C. and Haghani, A., An optimization model for aircraft maintenance scheduling and reassignment, *Transportation Research Part A: Policy and Practice*, 37, 29–48, 2003.

[19] Anbil, R. et al., Recent advances in crew-pairing optimization at American Airlines, *Interfaces*, 21(1), 62–74, 1991.

[20] Klabjan, D., Johnson, E.L., and Nemhauser, G.L., Solving large airline crew scheduling problems: Random pairing generation and strong branching, *Computational Optimization and Applications*, 20, 73–91, 2001.

[21] Graves, G.W. et al., Flight crew scheduling, *Management Science*, 39, 736–745, 1993.

[22] Lavoie, S., Minous, M., and Odier, E., A new approach for crew pairing problem by column generation with an application to air transportation, *European Journal of Operational Research*, 35, 45–58, 1988.

[23] Barnhart, C. et al., Branch-and-price: Column generation for solving huge integer programs, *Operations Research*, 46, 316–329, 1998.

[24] Hoffman, K.L. and Padberg, M., Solving airline crew scheduling problems by branch-and-cut, *Management Science*, 39, 657–682, 1993.

[25] Desaulniers, G. et al., Crew pairing at Air France, *European Journal of Operational Research*, 97, 245–259, 1997.

[26] Barnhart, C. and Shenoi, R.G., An approximate model and solution approach for the long-haul crew pairing problem, *Transportation Science*, 32, 221–231, 1998.

[27] Yan, S. and Tu, Y.-P., A network model for airline cabin crew scheduling, *European Journal of Operational Research*, 140, 531–540, 2002.

[28] Emden-Weinert, T. and Proksch, M., Best practice simulated annealing for the airline crew scheduling problem, *Journal of Heuristics*, 5, 419–436, 1999.

[29] Levine, D., Application of a hybrid genetic algorithm to airline crew scheduling, *Computers & Operations Research*, 23, 547–558, 1996.

[30] Ozdemir, H.T. and Mohan, C.K., Flight graph based genetic algorithm for crew scheduling in airlines, *Information Science*, 133, 165–173, 2001.

[31] Ryan, D.M., The solution of massive generalized set partitioning problems in aircrew rostering, *The Journal of the Operational Research Society*, 43, 459–467, 1992.

[32] Gamache, M. et al., A column generation approach for large-scale aircrew rostering problems, *Operations Research*, 47, 247–263, 1999.

[33] Lucic, P. and Teodorovic, D., Simulated annealing for the multi-objective aircrew rostering problem, *Transportation Research Part A: Policy and Practice*, 33, 19–45, 1999.

[34] El Moudani, W. et al., A bi-criterion approach for the airlines crew rostering problem, *Lecture Notes in Computer Science*, 1993, 486–500, 2001.

[35] Dawid, H., Konig, J., and Strauss, C., An enhanced rostering model for airline crews, *Computers & Operations Research*, 28, 671–688, 2001.

[36] Caprara, A. et al., Integrating constraint logic programming and operations research techniques for the crew rostering problem, *Software-Practice* & *Experience*, 28, 49–76, 1998.

[37] Sellmann, M. et al., Crew assignment via constraint programming: Integrating column generation and heuristic tree search, *Annals of Operations Research*, 115, 207–225, 2002.

[38] Teodorovic, D. and Lucic, P., A fuzzy set theory approach to the aircrew rostering problem, *Fuzzy Sets and Systems*, 95, 261–271, 1998.

[39] Jones, R.D., Development of an automated airline crew bid generation system, *Interfaces*, 19(4), 44–51, 1989.

[40] Jarrah, A.I.Z. and Diamond, J.T., The problem of generating crew bidlines, *Interfaces*, 27(4), 49–64, 1997.

[41] Christou, I.T. et al., A two-phase genetic algorithm for large-scale bidline-generation problems at Delta Airlines, *Interfaces*, 29(5), 51–65, 1999.

[42] Gamache, M., Soumis, F., and Villeneuve, D., The preferential bidding system at Air Canada, *Transportation Science*, 32, 246–255, 1998.

[43] Clausen, J. et al., Disruption management, *ORMS Today*, 28, 40–43, 2001.

[44] Teodorovic, D. and Guberinic, S., Optimal dispatching strategy on an airline network after a schedule perturbation, *European Journal of Operational Research*, 15, 178–182, 1984.

[45] Teodorovic, D. and Stojkovic, G., Model for operational daily airline scheduling, *Transportation Planning and Technology*, 14, 273–285, 1990.

[46] Jarrah, A.I.Z. et al., A decision support framework for airline flight cancellations and delays, *Transportation Science*, 27, 266–280.

[47] Rakshit, A., Krishnamurthy, N., and Yu, G., Systems operations advisor: A real-time decision support system for managing airline operations at United Airlines, *Interfaces*, 26, 50–58, 1996.

[48] Yan, S. and Yang, D., A decision support framework for handling schedule perturbations, *Transportation Research Part B: Methodology*, 30, 405–419, 1996.

[49] Yan, S. and Lin, C., Airline scheduling for the temporary closure of airports, *Transportation Science*, 31, 72–82, 1997.

[50] Yan, S. and Tu, Y.-P., Multifleet routing and multistop flight scheduling for schedule perturbation, *European Journal of Operational Research*, 103, 155–169, 1997.

[51] Arguello, M.F., Bard, J.F., and Yu, G., A GRASP for aircraft routing in response to groundings and delays, *Journal of Combinatorial Optimization*, 5, 211–228, 1997.

[52] Arguello, M.F., Bard, J.F., and Yu, G., Models and methods for managing airline irregular operations aircraft routing. In Yu, G. (editor), *Operations Research in the Airline Industry*, Kluwer Academic Publishers, Boston, 1–45, 1997.

[53] Cao, J. and Kanafani, A., Real-time decision support for integration of airline flight cancellations and delays, part I: mathematical formulations, *Transportation Planning and Technology*, 20, 183–199, 1997.

[54] Cao, J. and Kanafani, A., Real-time decision support for integration of airline flight cancellations and delays, part II: algorithms and computational experiments, *Transportation Planning and Technology*, 20, 201–217, 1997.

[55] Thengvall, B.G., Bard, J.F., and Yu, G., Balancing user preferences for aircraft schedule recovery during irregular operations, *IIE Transactions on Operations Engineering*, 32, 181–193, 2000.

[56] Thengvall, B.G., Yu, G., and Bard, J.F., Multiple fleet aircraft schedule recovery following hub closures, *Transportation Research Part A: Policy and Practice*, 35, 289–308, 2001.

[57] Stojkovic, G. et al., An optimization model for a real-time flight scheduling problem, *Transportation Research Part A: Policy and Practice*, 36, 779–788, 2002.

[58] Teodorovic, D. and Stojkovic, G., Model to reduce airline schedule disturbances, *Journal of Transportation Engineering*, 121, 324–331, 1995.

[59] Wei, G., Yu, G., and Song, M., Optimization model and algorithm for crew management during airline irregular operations, *Journal of Combinatorial Optimization*, 1, 305–321, 1997.

[60] Yu, G. et al., A new era for crew recovery at Continental Airlines, *Interfaces*, 33(1), 5–22, 2003.

[61] Lettovsky, L., Johnson, E.L., and Nemhauser, G.L., Airline crew recovery, *Transportation Science*, 34, 337–348, 2000.

[62] Stojkovic, G. and Soumis, F., An optimization model for the simultaneous operational flight and pilot scheduling problem, *Management Science*, 47, 1290–1305, 2001.

[63] Beasley, J.E. et al., Scheduling aircraft landings — The static case, *Transportation Science*, 34, 180–197, 2000.

[64] Ernst, A.T., Krishnamurthy, M., and Storer, R.H., Heuristic and exact algorithms for scheduling aircraft landings, *Networks*, 34, 229–241, 1999.

[65] Yu, G., Dugan, S., and Arguello, M., Moving toward an integrated decision support system for manpower planning at Continental Airlines: Optimization of pilot training assignments, In Yu, G., (editors), *Industrial Applications of Combinatorial Optimization*, Kluwer Academic Publishers, Boston, 1–24, 1998.

[66] Linn, R. and Zhang, W., Hybrid flow shop scheduling: A survey, *Computers and Industrial Engineering*, 37, 57–61, 1999.

[67] Brucker, P., Jurisch, B., and Kramer, A., Complexity of scheduling problems with multi-purpose machines, *Annals of Operations Research*, 70, 57–73, 1997.

[68] Lee, C.-Y., Two-machine flowshop scheduling with availability constraints, *European Journal of Operational Research*, 114, 420–429, 1999.

[69] Qi, X., Bard, J.F., and Yu, G., Class scheduling for pilot training, *Operations Research*, (Unpublished) 2003.

[70] Vasquez-Marquez, A., American-Airlines arrival slot allocation system (ASAS), *Interfaces*, 21(1), 42–61, 1991.

[71] Luo, S.J. and Yu, G., On the airline schedule perturbation problem caused by the ground delay program, *Transportation Science*, 31, 298–311, 1997.

[72] Amaddeo, H.F., Nawijn, W.M., and van Harten, A., One-machine job-scheduling with nonconstant capacity — Minimizing weighted completion times, *European Journal of Operational Research*, 102, 502–512, 1997.

[73] Surkis, J. and Dogramaci, A., Minimizing the sum of weighted completion times of n-independent jobs when resource availability varies over time: Performance of simple priority rule, *Naval Research Logistics Quarterly*, 35, 35–47, 1988.

[74] Trick, M.A., Scheduling multiple variable-speed machines, *Operations Research*, 42, 234–248, 1994.

51

Bus and Train Driver Scheduling

Raymond S. K. Kwan
University of Leeds

51.1 Introduction

Bus and train driver scheduling is a process that determines the composition of a set of driver shifts for a day's transport operation requiring coverage by drivers. Driver wages are a big percentage, e.g., about 45% for the bus sector in the UK [1], of the running costs of transport operations. Therefore, efficient scheduling of the drivers is vitally important to transport operators. There are often marked differences between different countries in how their transport services are operated. Therefore although the general principles of bus and train driver scheduling may be applicable to most countries, there will be some local variations. This chapter will be based mainly on the author's experience in UK operations.

The aim of Section 51.1 is to introduce the bus and train driver scheduling problem and the contexts under which the scheduling process is performed. A brief outline of the problem defining some terminology will be first presented. Then, the bus and train driver scheduling problem will be discussed under three different contexts: the timescale of when scheduling is performed in relation to the actual running of the transport services; the wider scope of operations planning; and the variety of transport services that have to be scheduled.

Section 51.2 gives a brief overview of the relevant literature. Section 51.3 presents the "Generate and Select" (GaS) approach and some algorithms based on it. The generation part of GaS, which so far has received less attention in literature than it deserves, will be discussed. The selection part of GaS belongs to the class of set covering problems, which have been proven to be NP-complete [2]. Some solution methods based on integer linear programming (ILP) and evolutionary computation for the selection part of GaS and for tackling very large and complex problem instances will be described. Section 51.4 presents an alternative constructive approach using tabu search, which iteratively reforms a single schedule for

1-58488-397-9/$0.00+$1.50

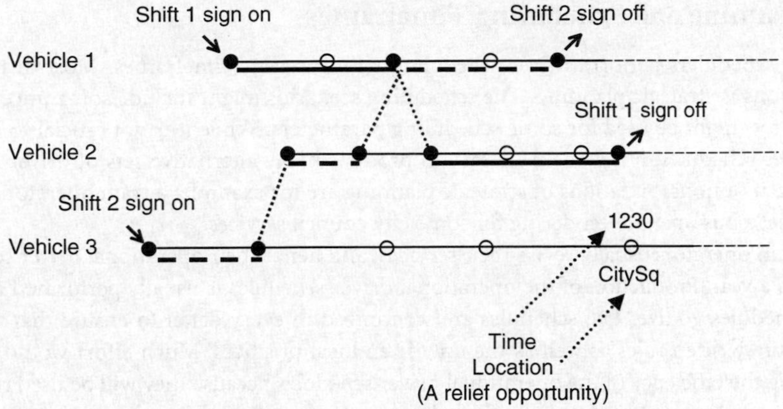

FIGURE 51.1 Vehicle work and driver shifts.

improvements. Finally, Section 51.5 gives some remarks on the approaches and algorithms discussed, and outlines some directions in bus and train driver scheduling research.

51.1.1 Outline of the Problem

A bus or a train may be used continuously all day long, but a driver usually only works up to about 10 h each day. Furthermore, after a few hours work on a vehicle, the driver must be relieved by another driver and takes a break. However only some locations, called *relief points*, in the transport network would be suitable for changing drivers. Figure 51.1 illustrates the work of some vehicles graphically represented on a timescale. The times when a vehicle goes past the relief points are called *relief opportunities* (ROs), and they are circled on the vehicle work graph. It should be emphasized that relief opportunities are only candidates, and usually only a small number of them will actually be used. Obviously, the work of a vehicle must begin and end at ROs that will be used in the driver schedule. The vehicle work between two successive ROs is called a *work piece*. Work pieces are usually the basic building blocks in forming driver shifts, and checking that all work pieces are covered by driver shifts is fundamental to driver scheduling.

The structure of a driver shift generally includes some reporting procedures at the beginning and end called *sign-on* and *sign-off*, respectively. Sign-on and sign-off usually take place at the same crew depot known as the *home depot* of the shift. The total duration from sign-on to sign-off is called the *spreadover* of a shift. Many of the scheduling rules are dependent on ranges of spreadover lengths. Often, scheduling methods would simply cost a shift by its spreadover since actual wage cost is usually proportional to spreadover. From sign-on to sign-off of a shift are alternating driving work and breaks. Each period of driving work would be on the same vehicle and is called a *spell*. Hence a spell is a contiguous sequence of work pieces on the same vehicle. In order to optimize how the vehicle work is carved out between the driver shifts, some of the breaks are only for the sake of switching the driver on to another vehicle and may be relatively short. Such short breaks are called *join-ups*. Some breaks are longer designed for drivers to have a rest, and they are called *mealbreaks*. Figure 51.1 shows examples of a two-spell shift with one mealbreak and a three-spell shift with a join-up and a mealbreak.

Clearly, there would be numerous ways of forming spells and numerous possible combinations of spells in forming driver shifts. The main objective is usually to minimize the total number of shifts needed to cover all the work. However, this objective may not be appropriate when short part-time shifts are present in the schedule. Sometimes a schedule may also include some quite long 4-day-week driver shifts. Therefore, another widely used objective is to minimize the total cost of the schedule.

The terms defined above are by no means universal. Some interesting local terms used in the rail industry are for example: a driver shift is called a *diagram*, hence driver scheduling is called *driver diagramming*; a mealbreak is called a PNB (stands for "Physical Needs Break"!).

51.1.2 Planning and Scheduling Timeframes

Planners and schedulers perform driver scheduling within several timeframes. Most distant from the actual operations is strategic planning. The scheduling scenarios might include some uncertainties, and ball-park figures might be used for some scheduling parameters. While it is not crucial to produce fully operable driver schedules, it may be necessary to produce many alternative sets of driver schedules for different what-if options. Situations of strategic planning are for example: a train operator bidding for a 15-yr franchise; a bus operator tendering for some city council services.

Bus and train operators usually revise their services, and hence their operational driver schedules, two or three times a year. Production of the operational driver schedules is usually performed a few months before the schedules go live. The schedules are scrutinized in every detail to ensure that they are legal according to the written rules as well as the unwritten local practices. Much effort would be expended on maximizing the efficiency of the operational driver schedules because they will be used regularly until the next revision. Also because of the significant lifespan of a set of operational schedules, optimization is more important than minimizing changes from the previous set of schedules, and therefore it is generally acceptable to compile a driver schedule completely afresh.

Driver schedules may be revised for short-term temporary implementation. This may have arisen from engineering work requirements, special holiday operations, and so on. For these scheduling scenarios, it is more important to maintain as much of the regular schedules as possible so that there is less chance of anything going wrong with the services. The short-term nature of the schedules means that it is usually not worthwhile to spend too much effort in optimization.

The above types of scheduling could be described as offline driver scheduling because the process is well in advance of the actual operations. The process does not need to consider actual driver personnel, it only determines how many notional shifts and how they together cover all the work. The notional shifts have yet to be assigned in some other follow-on processes. This chapter will only consider offline driver scheduling for regular operations. The same algorithms would also be appropriate for strategic planning, although the degree of certainty in the input may be lower. With some slight extension of the algorithms, mainly to achieve the minimum change objective, it is relatively easy to cater also for short-term scheduling scenarios.

In contrast to offline driver scheduling, online driver scheduling refers to rescheduling of actual driver personnel when the regular schedules are not running according to plan. The process is a lot more complex than its offline counterpart. It has to take into account the real-time status of many aspects of the operations. The process actually has to extrapolate and predict the operational status when the revised operations can be started, since it would be impossible to implement changes instantly. Little time would be available for the online rescheduling process, and inevitably it is for this kind of fire-fighting that optimization is usually impractical and not expected. Online rescheduling is so different from offline scheduling that it will be excluded from this chapter.

51.1.3 Relationship with Other Planning and Scheduling Tasks

In the wider scope of operations planning and scheduling, there are other tasks preceding and following that of driver scheduling. Since the central aim of all these tasks is to enable accurate and efficient running of the transport operations, it can be argued that all these tasks are entwined in one large optimization problem. Indeed, planners and schedulers do make changes back and forth among these tasks to achieve a more globally optimized operation. However, it is practically convenient to isolate these tasks so that each is not too large and complex to tackle.

Figure 51.2 shows the typical stages of bus and train operations planning and scheduling tasks. It begins with routes and timetable planning, the output of which is what the public perceives as the transport services offered to them. For most operators, especially those in UK, routes and timetable planning is followed by vehicle scheduling and then driver scheduling. The logic is that we first decide how the fleet of vehicles is deployed before we put drivers on these vehicles. The last stage, called rostering, is to package the daily notional driver shifts together with rest days. Often, each roster package is the work for a driver

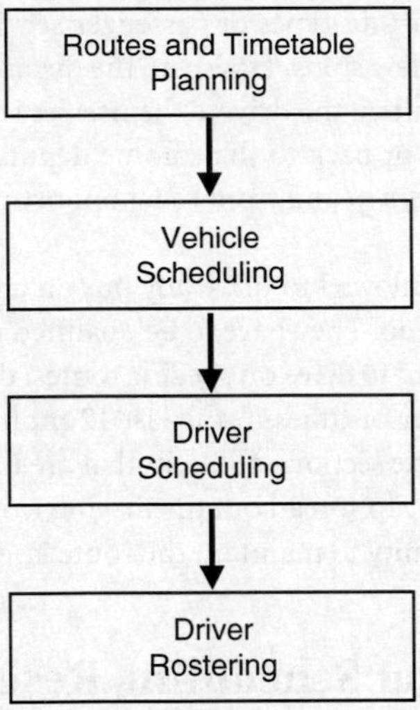

FIGURE 51.2 Stages of transport operations planning and scheduling.

in one week. For fairness, the weekly rosters are often assigned to drivers in rotation, i.e., the driver on roster number 1 this week will be on roster number 2 next week and so on. After rostering, the whole plan is ready for day-to-day actual operations. The management of drivers signing on and off, absences, on call spare drivers, etc., is usually part of the depot management system.

The above division of the planning and scheduling stages helps trace the information flow into and out of the driver scheduling process. This knowledge may be useful because sometimes the key to a better schedule might not be a perfect optimizer, but a vision that more favorable scheduling conditions could be achieved from a preceding stage. For example, a driver might be forced to take a break after a short spell of work because the bus or train is terminating. The best optimizer might not be able to alleviate this problem, but a simple modification of the vehicle schedule might redistribute work between the vehicles so that the vehicle concerned would terminate a bit later, thereby significantly improving the driver schedule.

51.1.4 Transport Service Types

Transport services may differ significantly in terms of service frequency, trip duration and number of stops. These factors are mainly related to the geography of the areas served. The complexity of the driver scheduling problem would be correspondingly different between different types of service.

Urban commuting services are present in both bus and train problems. These services usually have a high frequency and many stops suitable for changing driver en-route. The corresponding driver scheduling problem would be large due to the large number of relief opportunities at short intervals. For maximum optimality, it is necessary to consider the use of short spells in forming shifts. However, lowering the minimum spell length parameter would cause an exponential increase in the number of shifts generated.

Also common to both bus and train problems are suburban and rural services. The service frequencies are smaller and there are fewer stops. Such services therefore might be easier to schedule. However, the boundaries between urban, sub-urban and rural services are not always distinct, and they might be mixed for driver scheduling. The problem then becomes complex. There may be several sets of labor rules involving several crew depots. Whereas some of the work may be only suitable for drivers from a particular depot with a specific set of labor rules, some work may be suitable for more than one depot and hence multiple sets of labor rules. The large number of possibilities in matching such work with different sets of depot specific work makes scheduling rather complex.

In train problems, there are two further types of passenger services, namely, intercity and provincial. They also usually only have relatively few stops. However, the distances covered are relatively much larger than the other types of service. Very often, the drivers themselves rely on the train services to bring them to the starting location of their work or back to their home depots at the end of their shifts. Accurately timed means of transporting the drivers to and from relief opportunities becomes important. Also, these problems often involve many depots.

Whereas bus drivers normally are allowed to drive any bus on any route, train drivers are much more restrictive in what they are allowed to do. They have to be qualified to drive specific types of rolling stock. Not only are they required to be trained to drive on specific routes, the drivers also must have driven along a section of track a minimum number of times in the last 12 months to maintain his/her qualification. This is usually not a big problem for the sections of tracks that are used by many services. For the less well used route sections, it is often desirable to spread out the driving work to as many shifts as possible so that more drivers would have the opportunity to maintain the route knowledge.

51.2 Overview of Driver Scheduling Research

The emergence of automatic driver scheduling algorithms started around the early 1970s, when computers were beginning to find practical applications. In 1975, the first international workshop on computer-aided public transport scheduling was held in Chicago [3], which has since been established as a series of major international conferences held roughly every three years on the subject [4–10]. Driver scheduling is a major research area reported in this conference series. The conference proceedings include two reviews, by Wren in 1981 [11] and by Wren and Rousseau in 1995 [12], specifically on bus driver scheduling.

From the technical perspective, researches in driver scheduling can be broadly divided into three eras. From mid 1960s to early 1980s, the approaches taken were mainly heuristic, e.g., [13–17], trying to mimic how human schedulers perform driver scheduling manually. Although the early heuristic methods did achieve some very impressive results, they were not easily and readily usable for different companies and adaptable for changing problem scenarios, and therefore their adoption was often limited and short-lived. As affordable computing power was increasing, mathematical approaches became more feasible from the late 1970s. Work on mathematical approaches for driver scheduling was most intense during the 1980s, e.g., [18–23]. Mathematical approaches are still the most successful, judging from uptake of such systems by industry, and researches on improving them are still ongoing nowadays. From the late 1980s, there has been an upsurge of general interests in meta-heuristics, local search methods and constraint-based methods, e.g., [24–32]. These modern heuristic approaches have been tried for the driver scheduling problem, and it seems much research is set to continue in the foreseeable future.

From the application perspective, research in driver scheduling has been mainly based on bus operations [11]. Work on train driver scheduling started to be active only since the early 1990s, e.g., [33–36]. Although the uptake of automatic driver scheduling systems by train operators is still very scarce, the research has greatly benefited bus company users. This is because the train problem is like a superset encompassing the bus problem. It is generally more complex and more difficult to solve, and hence when the solution methods designed for the train problem are applied to the bus problem, the task becomes a lot easier.

51.3 The Generate and Select Approach (GaS)

Figure 51.3 highlights the GaS approach. A limiting factor for the size of the candidate shift set is the capability of the solution methods for the Selection phase, which is increasing quite rapidly in the last twenty years owing to computer hardware getting ever more powerful. Advances in linear programming techniques and meta-heuristics have also helped. Nowadays, it is common to build two to four hundred thousand candidate shifts in the Generation phase. For large and complex problem instances, it may be necessary to build up to about a million candidate shifts in order to obtain a good solution.

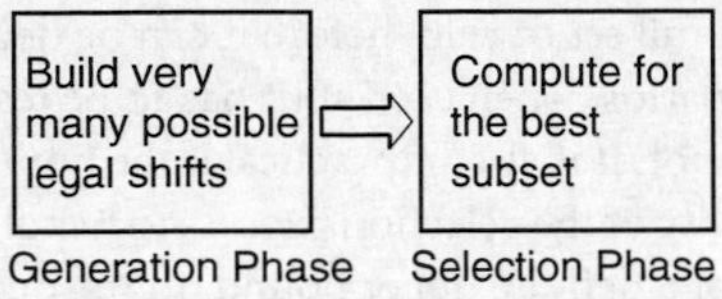

FIGURE 51.3 The Generate and Select approach.

The Selection process naturally fits the well-known set covering model, which can be represented by the following ILP:

$$\text{Minimize} \quad W_1 \sum_{j=1}^{n} c_j x_j + W_2 \sum_{j=1}^{n} x_j$$

$$\text{Subject to} \quad \sum_{j=1}^{n} a_{ij} x_j \geq 1, \quad i = 1, 2, \ldots, m \tag{51.1}$$

$$x_j = 0 \text{ or } 1, \quad j = 1, 2, \ldots, n \tag{51.2}$$

where

n = number of candidate shifts

m = number of work pieces

x_j = shift variable, $x_j = \begin{cases} 1 & \text{if shift } j \text{ is selected} \\ 0 & \text{otherwise} \end{cases}$

c_j = cost of shift j

$a_{ij} = \begin{cases} 1 & \text{if work piece } i \text{ is covered by shift } j \\ 0 & \text{otherwise} \end{cases}$

W_1 and W_2 are weight constants

The above ILP may be augmented by side constraints such as setting bounds on the number of shifts, restricting the number of a certain type of shift, etc. The objective function has two components, the total cost and the total number of shifts, to be minimized. The bi-objectives are weighted dependent on the scheduler's priority.

The biggest advantage of GaS is that the required domain specific knowledge, i.e., the scheduling rules, can be dealt with almost entirely within the Generation phase. Since the Generation process does not have to consider how the candidate shifts fit among themselves, application of the scheduling rules is straightforward and independent for each candidate shift. There is relatively little need for sophisticated optimization in the Generation phase than in the Selection phase. However, the importance of the Generation phase cannot be overstated because no matter how powerful the Selection algorithm is the results are doomed if the Generation phase has not captured the essence of the real scheduling scenario and reflected it in the set of candidate shifts. Driver scheduling knowledge is so diverse and changing that the process of its generalization and incorporation is a continuous incremental process. Fortunately, the GaS approach is particularly suitable for such continuous incremental updates.

51.3.1 Generation of Candidate Shifts

The main aim of the Generation phase is to produce a large pool of legal candidate shifts without missing any shifts that are vital to the formation of the most efficient schedule. Driver schedules are like jigsaw puzzles that it may not be easy to decide which pieces would fit together. Unlike an ordinary jigsaw puzzle, which has a finite number of pieces, we are at liberty to generate as many candidate shifts as is practical, and that is the usual strategy to minimize the chance of missing out critical shifts. It is usually regarded as impractical if the generation process takes many hours to complete. This may happen because every shift generated has to be vigorously tested to ensure that it is legal, and that is time consuming. Some illegal shifts might be revealed just after the first few rules are tested, but in the worst scenario a shift might have to be tested against the full set of rules before it can be declared legal or illegal. The problem is compounded by multidepot scenarios, when each shift has to be tested against the rule set using each depot in turn as the shift's home depot. It is also impractical if the number of candidate shifts generated is so large that it is beyond the capability of the Selection process to handle or to yield an acceptable solution within reasonable time. There is not a definite upper bound on the number of candidate shifts, although it is usually easy to tell by experience when the Generation process has breached such a bound. In the following subsections, some aspects of shift generation will be discussed.

51.3.1.1 Framework for Shift Generation

There are two main approaches for shift generation. The first approach is analytical of the vehicle work patterns, and the generation process is guided by target kinds of shift to be built. The IMPACS system described by Parker and Smith [15] and Smith and Wren [22] is an example using this approach. The advantage of this approach is that the generation process has tight control over the characteristics of shifts it is constructing, and therefore a good schedule might be obtainable from a relatively small set of candidate shifts. This was important before the 1990s, when affordable computer hardware was barely powerful enough to support mathematical solvers in the Selection phase. Indeed, in the IMPACS system before 1990, the ILP solver had a maximum limit of only 5000 candidate shifts. A heuristic process is therefore often needed to throw away some of the candidate shifts constructed. A big disadvantage is that such a heuristic is usually controlled by parameters that have to be set by trial and error, and the quality of the reduced set of candidate shifts might not be satisfactory.

In the second approach, driver shifts are constructed in a uniform fashion, modelled on some generalized structures of driver shifts, and then classified into different types [33,37]. There is less tuning needed over the mechanics of the generation process. In comparison with the first approach, both the classification of shifts and the number of shift types are more flexible. Since the generation process is more straightforward, the second approach also has the advantage that it is easier to incorporate new scheduling rules. Therefore, the discussion will be focussed on the second approach exemplified by the TRACS II system [38–40] hereafter.

With respect to shift construction, the smallest unit of work is a *spell*, which is defined as some continuous work on one vehicle, such that the start of a spell is when the driver either starts work on a vehicle after signing on duty or resumes work after a break, and the end of a spell is when the driver either finishes work on a vehicle before signing off duty or starts a break. A *half-shift* consists of one or two spells of work. There are three main types of full shift namely, *straight shift*, *split shift*, and *part shift*. A straight shift consists of two half-shifts joined by a break of normal duration. At least one of the breaks in a straight shift must be suitable as a mealbreak. A split shift consists of two half-shifts joined by a mealbreak plus a long period, up to a few hours, of inactivity. A part shift consists of one or two half-shifts without requiring a mealbreak. Part shifts are usually for part-time drivers, and therefore their lengths are normally quite short. A full shift may also be referred to by its number of spells. For example, a two-part shift has two spells and so on. The more spells in a shift means the more changeovers of drivers giving rise to higher risks of delay. Drivers generally do not like changing vehicles many times either. Generally, the most favoured are two-part shifts and it is very rarely necessary to have shifts of more than four spells.

TRACS II first constructs all the legal spells of work. All combinations of the spells are then tried in forming valid half shifts. The half shifts are then combined to form straight shifts and split shifts. All two-part shifts will be formed first followed by three-part and four-part shifts. Finally, part shifts are formed. After the driving work of a full shift is defined, a suitable home depot is assigned to the shift. Then, depending on the home depot and where work starts and finishes, appropriate events of signing on and off duty are determined completing the construction of the shift.

51.3.1.2 Driver Scheduling Rules

No matter how intelligent a driver scheduling system is, some rules have to be specified by the human scheduler. Some of the rules may be specific to the local problem instance. For example, in a UK train network, some trains stop at a very busy station so that changing drivers there is undesirable. These trains also stop at a small station just a few minutes away. Therefore if the train goes through both stations in its journey, the big station would be suppressed from being a driver relief point. Drivers travelling as passengers are allowed to get on/off at the big station though. However, if the train terminates at the big station, it does not make sense to interrupt the driving work at the small station and hence the small station is suppressed. Furthermore, the small station is closed late at night and early in the morning, when it has to be suppressed as well. Specific local rules may have to be dealt with by ad hoc means; sometimes customized program codes are inevitable.

Luckily the majority of driver scheduling rules are commonly applicable across the bus and train companies, and generalized facilities for specifying the rules are expected in reasonable scheduling systems. Many rules can be expressed in terms of time durations, e.g., maximum working time without a mealbreak is 4 h 30 min, or clock-face time values, e.g., a particular type of shift must finish before 7 p.m. Some of these rules are hard rules that the transport operator is strictly bound by law or by union agreement. Some are soft rules used for specifying preferred characteristics of the shifts, and may help control the size of the candidate shift set generated. For example, minimum spell length is a soft rule; theoretically no spell is too small, but too many shifts would be generated if a small parameter is used. The general driver scheduling rules can be broadly divided into three categories:

1. Shift structure related, e.g., maximum spreadover length
2. Time allowances, e.g., travel time from relief point to canteen, signing on/off
3. Route and traction (or vehicle type) knowledge

Travel times between locations can often be approximated and expressed as a matrix of fixed allowances. These allowances may have been calculated assuming the use of certain modes of transport such as walking or using taxis. This is often adequate for urban bus problems because the geographical distances are relatively small. For train problems and some bus problems, the transportation arrangements may have to be more precise. For example, it would be necessary to know if and when there will be a train service at a relief point to bring the relieved driver back to his/her home depot late at night. A trip planning algorithm would be needed, but some practical simplification could be made. For example, such trips may be restricted to include no more than two legs on trains.

Although some rules are soft rules, the rule parameters may have to be set on a trial and error basis involving several runs of the scheduling algorithm. Alternatively, some shift feature preferences may be achieved by adding penalty weights to the shift costs. A positive weight would discourage the feature whereas a negative weight would make the feature favorable in the final schedule. The penalty weight method may also be used to implement a utility whereby patterns of previous schedules are recognized and preserved if it is not making the schedule too costly.

51.3.2 Algorithms for the Selection Phase

It is easy to design a very fast greedy heuristic to select shifts from the set of candidate shifts to form a schedule. For example, a greedy heuristic could be: starting with an empty schedule, for each remaining uncovered work piece, randomly select one of the candidate shifts that covers it. However, the resulting schedule is obviously unlikely to be efficient. In practice, inefficiency of using one or two shifts more than optimal would be regarded as significant. Greedy heuristics alone are likely to result in schedules using many more shifts than optimal. Therefore more sophisticated methods are needed. In the next section, an integer linear programming (ILP) method will be described. The ILP method for driver scheduling has proved to be the most successful in the last twenty years, and systems based on it are adopted by many transport operators. In recent years, meta-heuristics are widely used for combinatorial optimization problems. In Sections 51.3.2.2 and 51.3.2.3, two evolutionary algorithms are described. They work in conjunction with a greedy schedule construction heuristic, and are showing promising results.

51.3.2.1 Integer Linear Programming

An example of a successful solver for the set covering ILP discussed at the beginning of Section 51.3 will be described in this section. It is a solver embedded in a series of systems developed at Leeds University from the late 1970s [22,33,34,38–44] which is still evolving. The basic strategy of the algorithm is first to ignore the zero-one integer constraints on the shift variables. The relaxed LP is solved to optimality, and then an all-integer solution is sought using branch-and-bound. The Revised Simplex Method is applied incorporating the steepest-edge algorithm by Goldfarb and Reid [45] and Forrest and Goldfarb [46]. Unless the candidate shift set is very small, a column generation technique described by Fores et al. [41], [42] is also used in the LP relaxation phase. Under this framework, specialization for the driver scheduling problem is made at various stages in order to maximize the performance of the algorithm.

The relaxed LP is solved starting with an initial basic feasible solution, which is then iteratively improved by swapping variables between the basic and nonbasic variable sets to reduce the objective function cost. The initial basic feasible solution could be constructed straightforwardly using commonly used techniques for solving general LPs, e.g., select the slack variables associated with the constraints. However, it is relatively easy to produce a much better, i.e., at lower objective function cost, initial basic feasible solution heuristically. For example, a heuristic can effectively be constructing a schedule by choosing one candidate shift at a time to cover the remaining uncovered work, checking that no work piece or side constraints are violated at each choice, and finally the starting basis is completed by updating the slack and surplus variables according to whether there are any work pieces still uncovered and by how much some work pieces are overcovered.

The number of shifts involved in the initial basic solution above can be treated as a rough upper bound U. Based on Sherali's work on converting multiobjective models into single objective models [47], Willers has shown [43] that there are computational advantages in replacing the objective function by

$$\text{Minimize} \sum_{j=1}^{n} D_j x_j$$

where

$$D_j = \left(1 + \sum_{k=1}^{U} M_k\right) + c_j$$

$M_1, M_2, \ldots, M_n$ are costs of the candidate shifts in descending order.

The relaxed LP is solved in stages designed to narrow down the search space. Prior to the Selection phase of GaS, each candidate shift has been given a rating to indicate its likely goodness, assessed based on its work content and cost, in contributing to an efficient schedule. In the first stage, the candidate shift set is reduced by banning all the shifts that have low goodness ratings. Three-spell and four-spell shifts are also deemed to be of low desirability and banned as well. After the LP is solved for the reduced shift set, the banned shifts are restored and the LP is reoptimized.

In the next stage, the algorithm automatically adds two side constraints and the scheduler may also add or change the side constraints. If the scheduler would like to make certain shifts more favorable or unfavorable, penalty costs on the shifts are also applied at this stage. The first side constraint added automatically is that the sum of the slack variables of the work piece constraints must be less than or equal to zero, so that no uncovered work pieces are allowed. The second side constraint added automatically imposes a tight lower bound on the sum of the shift variables based on the optimal continuous solution found so far. The optimal continuous solution generally involves many times more shift variables, many of which have fractional values, than found in the optimal integer solution. However, practical experience has demonstrated that the sum of the shift variables in the optimal continuous solution is a very good estimation of the number of shifts in the optimal integer solution. Suppose the sum value has an integer part I and a fractional part f. The target number of integer shifts T to be used is computed by:

$$T = \begin{cases} I & \text{if } f < \varepsilon \\ I + 1 & \text{otherwise} \end{cases}$$

where ε is a small constant, and a tight lower bound side constraint is

$$\sum_{j=1}^{n} x_j \geq T$$

After the relaxed LP is reoptimized, unless the continuous solution is already also an integer solution, the branch-and-bound phase is entered to search for integer solutions. Because of the huge number of shift variables, the general variable branching strategy would not be efficient. Two specialized branching strategies are used. They are namely, *relief opportunity branching* and *constraint branching*. Relief opportunity branching is tried first; if the schedule is still fractional, constraint branching is tried; and failing that, the algorithm resorts to variable branching.

The idea of relief opportunity branching comes from the basic principle that a relief opportunity may be either used or unused in the final schedule. Given a fractional solution, the variables for the shifts using a particular relief opportunity may have fractional values. If at a relief opportunity, the sum of the corresponding fractional solution values is close to either 0 or 1, it would be worth trying to force it to be exactly 0 or 1. The candidate shifts that include work spells containing the selected relief opportunity are identified and then divided into two groups depending on whether the selected relief opportunity lies within a work spell or is at either end of a work spell. The relief opportunity branching strategy is to form two branches at a search node, such that each branch corresponds to one of the two groups of shifts being banned.

The constraint branching strategy was originally developed by Ryan and Foster [48,49]. Two work piece constraints are first selected. On one branch, the solution is forced to cover both work pieces by the same shift. On the other branch, the solution is forced to cover the two work pieces by different shifts.

The branch-and-bound process would be considerably quickened if the set of candidate shifts could be reduced in size. A heuristic by Smith and Wren [22] is optionally used whereby those candidate shifts that do not use the relief opportunities used by the optimal continuous solution of the relaxed LP are banned. Practical experience has shown that this heuristic performs well. The run time is much reduced and the solutions obtained are usually nearly as good as those obtained without applying the heuristic. The good performance may be explained from the observation by Kwan, Kwan, and Wren [50] that usually a large proportion (in some cases up to 97%) of the shifts used in the final integer solution are also present in the relaxed LP solution.

51.3.2.2 Evolutionary Algorithms

Evolutionary algorithms have been popular in the last two decades with researchers seeking practical near-optimal solutions to NP-hard problems. Evolutionary algorithms are usually simple to understand and to code, and they often converge very quickly. A naive genetic algorithm for the Selection phase of GaS would be obvious: use the binary string chromosome representation such that each gene determines if a candidate shift is selected or not, and apply the usual genetic operators for reproduction and management of the population. However, since the chromosomes would be exceedingly long, same number of genes as the number of candidate shifts, it would be very difficult for standard crossover and mutation methods to converge in the enormous solution search space. Simple genetic operators also cannot guarantee that the offspring would represent feasible schedules. For example, a single point crossover is likely to result in the wrong mix of shifts such that some work would become uncovered. Hence early researches were aimed at overcoming these drawbacks. Wren and Wren [25], Clement and Wren [24], and Gonzalez, Hernandes, and Corne [27] transform the problem into that of assigning a candidate shift to cover each work piece. Simple genetic algorithms are still employed. Genes in a chromosome record the assignments of shifts to work pieces. The chromosomes are therefore of reasonable lengths. The crossover and mutation operators are adapted so that new offspring are adjusted heuristically to ensure that they represent feasible schedules. Gonzalez, Hernandes, and Corne sub-divided the work pieces into groups, each constituting a separately evolved species. Individuals from each species are chosen and combined to form an overall solution. Only small problem instances had been tested by these early researchers, and the quality of the schedules produced were not very satisfactory.

In the following sub-sections, two recent approaches are described, which have yielded promising results. The first approach is a hybrid scheme, in which a genetic algorithm supplies an evolved partial solution to a heuristic component to build a complete solution. The genetic algorithm also exploits domain specific knowledge obtained from other components in the hybrid scheme. The second approach applies a kind of hypermutation scheme to evolve a single schedule, which also features the use of fuzzy evaluation.

51.3.2.2.1 Genetic Algorithm with Combinatorial Traits (GACT)

Simple genetic algorithms do not work well for driver scheduling probably because they have not utilized much domain knowledge. Kwan, Kwan, and Wren investigated a hybrid genetic algorithm [30,37,50] that exploits the properties that bind shifts together into a good schedule. Patterns of work piece and relief combinations that are keys to fitting the shifts seamlessly to achieve an efficient schedule are called *combinatorial traits*. Combinatorial traits are difficult to be identified and proved, however an evolutionary framework may offer good tolerance in the accuracy of diagnosing combinatorial traits because false combinatorial traits would render offspring unfit and perish.

The optimal continuous solution of the relaxed LP discussed in Section 51.3.2.1 would be a good source for combinatorial traits. Kwan, Kwan and Wren have shown that on average 79% of the shifts in an integer solution found by branch-and-bound are contained within the continuous solution [50]. Furthermore, there are generally many similarly good integer solutions that involve different combinations of the shifts used in the continuous solution. Hence, the continuous relaxed LP solution is used as the starting point, and a binary string chromosome is used to represent a selection from the shifts in the continuous solution. The selection is an implicit combinatorial trait because it is not diagnosed by analysis of structures or patterns in the schedule. Explicitly diagnosed combinatorial traits are also used and will be discussed later. The fitness of the chromosome, and hence the goodness of the implicit combinatorial trait, is evaluated by using a greedy schedule construction heuristic to complete the schedule based on the selection. The better the completed schedule the fitter the chromosome is deemed to be. Simple roulette wheel selection, single-point crossover and adaptive mutation are used to evolve a fixed size population of implicit combinatorial traits. As a by-product of the evaluation, good schedules constructed are stored and finally the best schedule is returned.

Combinatorial traits may be derived explicitly from the schedules constructed. Figure 51.4 shows a *relief chain* as an example of an explicit combinatorial trait. In the example, the timings of some of the shifts in the schedule fit neatly such that the mealbreaks cascade down the work of several vehicles like a chain. If the timings did not fit well, some of the vehicles may have to be taken over by extra new drivers instead of ones who can resume work after their mealbreaks. Explicit combinatorial traits are not subject to crossover or mutation but simply passed on to offspring as some private learning properties. In GACT, when an offspring is created, it will learn from the explicit combinatorial traits passed on from one of its parents before it is evaluated. For example, the shifts involved in a relief chain are added to the set of shifts selected by the offspring's chromosome, i.e., learned, before the schedule construction heuristic is applied.

Figure 51.5 illustrates the overall structure of the GACT hybrid evolutionary algorithm. GACT was tested and the results were compared with those obtained by the ILP-based TRACS II system. Twenty data

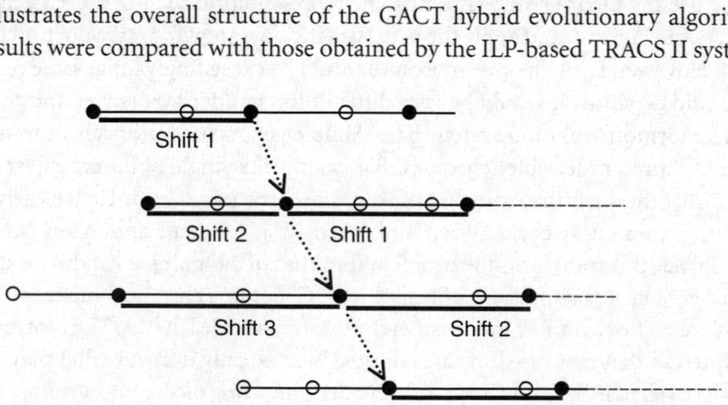

FIGURE 51.4 Relief chain combinatorial trait.

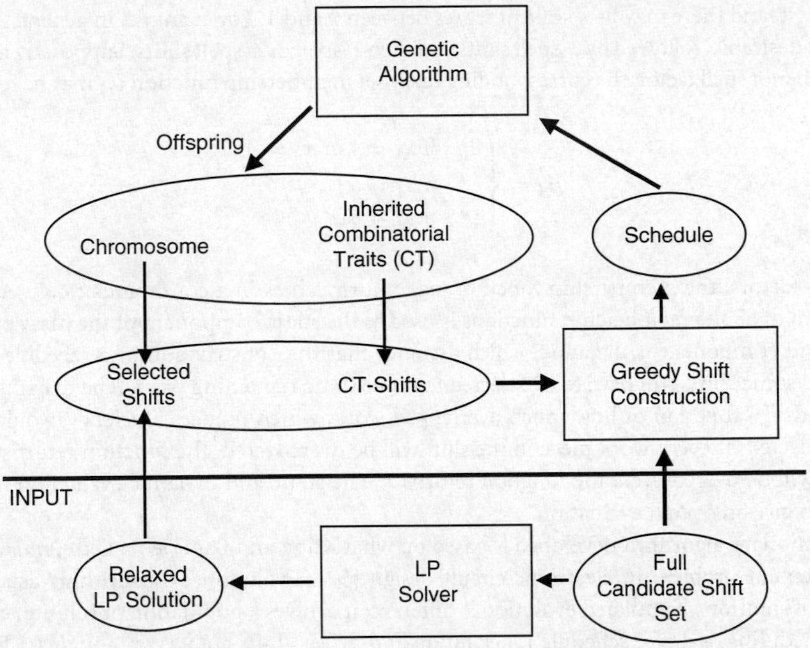

FIGURE 51.5 The GACT hybrid evolutionary algorithm.

sets from bus, tram, and rail were used. GACT obtained better schedules in five instances, equalled in seven instances, and was worse in eight instances. In the instances that GACT excelled, the problems were large. Then, GACT was further tested with multiple runs using different random number seeds. On average, about 62% of the runs yielded the same or better results than those used in the comparison with TRACS II. GACT is therefore quite robust. The further runs also improved its comparison with TRACS II, of the 20 instances, it was better in five instances, equalled in eight, and worse in seven instances.

51.3.2.2.2 Fuzzy Simulated Evolution

When a schedule is formed, its cost is attributable to well-defined costs of its component shifts. When alternative schedules are compared, it might even be possible to infer that one schedule is worse because some large cost shifts are present. However, while solutions are being sought and the final schedule is not yet known, it would obviously be inappropriate to discriminate against shifts entirely based on their costs. The evaluation of a shift at those stages ought to be considering how different aspects of the shift may affect its ability to fit well within an efficient schedule. However, it may not be possible to formulate a satisfactory deterministic shift evaluation function on such a basis. Li and Kwan therefore applied fuzzy set theory [51] on shift evaluation in developing evolutionary algorithms for the selection process of GaS [52–56].

The fuzzy evaluation function consists of two components. The first component is based on some static characteristics of the shift being evaluated, for example:

1. total worked time
2. spreadover
3. ratio of total worked time to spreadover
4. number of spells
5. relaxed LP solution value

The fifth factor above refers to the continuous solution of the relaxed LP discussed in Section 51.3.2.1, in which the value of a shift variable may range between 0 and 1. Fuzzy set membership functions are designed for each factor such that a value of 1 indicates that the shift scores highest in that factor, a value of

0 scores lowest, and there may be a scale of scores between 0 and 1. For example, in general 2-spells shifts are the most desirable, followed by 3-spell shifts and then 1-spell or 4-spell shifts. Suppose x_4 is the variable for the number of spell factor, the corresponding fuzzy set membership function μ_4 may be formulated as:

$$\mu_4 = \begin{cases} 0 & \text{if } x_4 = 1 \text{ or } x_4 = 4 \\ \frac{1}{2} & \text{if } x_4 = 3 \\ 1 & \text{if } x_4 = 2 \end{cases}$$

For other factors, the membership function may return values in a continuous range. A normalized weighted sum of all the membership functions is used as the static component of the fuzzy evaluation.

The second component is dynamic, which assumes that the construction of a schedule is in partial progress and some shifts still have to be selected to cover some remaining work. The dynamic evaluation is formulated as a function of how much overlapped work, which reduces efficiency, would arise if the shift were selected. If every work piece in the shift will be overcovered, the function returns 0, and if no work piece will be overcovered, the function returns 1. The static and dynamic evaluation functions are multiplied to give an overall evaluation.

The evolutionary algorithm developed is based on what Kling and Banerjee called *simulated evolution* for the placement problem in electronic circuit design [57]. Evolutionary algorithms usually evolve a population of solutions. Simulated evolution is different; it evolves a population of solution components. In driver scheduling, a driver schedule is a solution and driver shifts are its components. The algorithm iteratively operates on a single schedule starting with an initial schedule. In each iteration, individuals in the population of shifts are ranked using fuzzy evaluation discussed above. Those shifts below a predefined threshold level of fitness will be marked for deletion from the schedule. Some markings may be randomly reversed as a kind of mutation. After a portion of the schedule is duly deleted, a greedy selection heuristic similar to that in GACT discussed in Section 51.3.2.2.1 is used to repair the schedule using the large pool of candidate shifts. Fuzzy evaluation is also applied on the candidate shifts in the greedy selection heuristic. Hence, each iteration mimics a generation of evolution, the process of which is simulated in the sense that the state of the population is transformed in one go rather than through a series of repeated genetic operations such as crossover and mutation.

Setting the weights for the fuzzy sets above is critical in this approach. Each weight can vary in a continuous range between 0 and 1. Its range is first approximated by a range of several discrete values. A simple genetic algorithm is used to evolve configurations of discretized weights, which are used in short runs of the simulated evolution algorithm for evaluating its fitness inferred from the quality of the best schedule produced. The aim is to calibrate a reasonable set of weights for a fuller run of the simulated evolution algorithm. The simple GA has later been replaced by a technique [58] used in the design of experiments for parameters setting. Basically, the simulated evolution algorithm is experimented a fixed number of times. Each experiment uses one of a range of weight configurations predefined by the experiments design technique, which has a sound statistical basis.

The test results have shown that the simulated evolution with fuzzy evaluation algorithm yields slightly better schedules than GACT. Comparing with the ILP method, it has yielded some slightly better schedules and some slightly worse schedules for problem instances that are not very large; however, its results are significantly better for some very large problem instances. The fuzzy simulated evolution algorithm at present depends on very little domain knowledge. Therefore, it can be quite easily adapted for the general set covering problem, which has wider applications. On the other hand, further research incorporating more domain knowledge may lead to more optimized driver schedules.

51.3.3 Tackling Very Large and Complex Problems

Bus and train driver scheduling problem instances may be difficult because either the data set is large or the scheduling conditions are complex. Sometimes problem instances may also be both large and complex. The largeness of a problem instance is generally proportional to the number of relief opportunities in the data. As a rough guide, instances with over 800 ROs are large and those with over 1500 ROs are very large. Apart from the desire to schedule a large network of services together, large instances often arise from scheduling intensive urban services which are frequent and have many suitable locations for relieving drivers. Although some problem instances are large, they may not be having very complex scheduling rules. Complex problems typically arise when an operating company runs mixed types of service, e.g., urban, suburban and rural, with several depots. The scheduling rules and conditions vary, but overlap, between different types of service. Because of the grey boundaries between different types of service, it would not be optimal if each piece of work is classified into only one particular service type. The problem instance has to be scheduled as a whole, and the interaction between the different sets of scheduling rules makes the instance complex. Often complex problem instances are also large.

Various methods have been attempted in combating large and complex instances because they do appear often in real life, but the methods are usually ad hoc and experimental and therefore not much has been published. In the late 1980s, there was some work at the University of Leeds in a scheme of decomposing a large problem into several subproblems. The vehicle work is analyzed intensively in order to derive the subproblems intelligently. The first subproblem is made larger than the rest, so that after the best schedule for the subproblem has been found, the less efficient shifts in the schedule is taken out and carried over to the next subproblem. The process is repeated and the carry-over of inefficiently scheduled work may go round back to the first subproblem. The results were generally not very satisfactory probably because how the problem is subdivided is critical and it is not easy to derive a good and general algorithm for this problem subdivision task.

Kwan and Wren [29] used a simple genetic algorithm to select ROs to be banned thereby reducing the problem. The method relies on the availability of an accurate estimation heuristic that can quickly predict the minimum number of shifts needed based on the reduced set of ROs. Zhao, Wren, and Kwan have developed a good driver shift estimator [59], but the rationale behind the algorithm is based on normal data one would expect in real life. However, the reduction of ROs by genetic algorithms can make the patterns of remaining ROs rather abnormal. Hence, the results have not been satisfactory.

More recently, a new method called *clone and reduce*, as illustrated in Figure 51.6, has been developed for the BusTRACS [40] system, which is based on GaS. Rather than banning some ROs from the beginning to reduce the problem size, the problem instance is solved by means of tightening the scheduling parameters, only using two-part shifts and reducing the search depths in building the candidate shifts. The problem instance is cloned and then reduced by banning all the ROs that are not in the rough schedule produced. The new schedule would be no worse because all the ROs used by the original schedule are still available. Furthermore, the big reduction in ROs makes it feasible to relax the scheduling parameters and assumptions made earlier, some banned ROs may also be restored. The relaxation could be applied to just a focussed part/aspect of the problem instance, and the clone and reduce steps are iterated switching the focus in each iteration until the full problem is effectively tackled.

In an example big problem instance from a major UK city bus network, there was over 2700 vehicle hours of work and 2339 ROs. The labor agreement was well-defined and not complex, and there was only

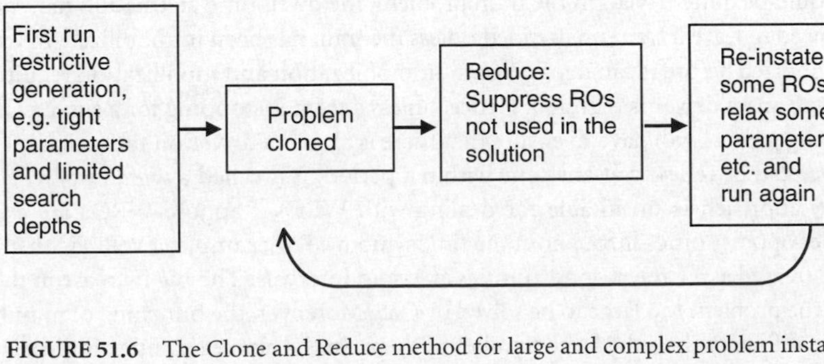

FIGURE 51.6 The Clone and Reduce method for large and complex problem instances.

TABLE 51.1 A Large and Complex Problem Instance

Subproblems	ROs	Vehicle Hours	Shifts	Cost
Park and Ride	284	169:10	22	224:35
City	906	702:10	85	912:22
Country	416	386:36	44	478:22
Total	1606	1261.56	151	1615:19
BusTRACS with Reduce and Clone			142	1603:00
Manual schedule			146	1589:19

one depot. A normal BusTRACS run would be impractical, generating many millions of candidate shifts. Clone and reduce was applied. The reduced runs had about 1000 to 1300 ROs, which were within the capability of BusTRACS to tackle. The first run used 335 shifts with a cost of 2805 h. After 20 iterations of clone and reduce, the final schedule had 306 shifts costing 2704 h, which was better than the company's manual schedule with 308 shifts costing 2707 h.

In another example, also from a major UK bus network, the problem instance was large and complex. There were three types of services, namely, Park and Ride, City, and Country. It would be desirable to keep the drivers to work on only one type of service during a shift, but mixed work was allowed if the schedule would then be cheaper. As shown in Table 51.1, the problem instance was first divided into three subproblems and solved separately. The combined schedule was worse than the existing manual schedule by five shifts and was much more expensive. The City and Country sets were then merged and the clone and reduce method was applied to the merged problem. After three iterations, four shifts were saved. Finally, the merged City and Country set is further merged with the Park and Ride set, and the clone and reduce method was applied. There was a stronger preference to keep the Park and Ride work together, and therefore a minimum change from the previous Park and Ride schedule constraint was added. The final schedule had 142 shifts costing 1603 h, which had saved four shifts at an increased cost compared with the manual schedule. Note that the scheduling exercise also had to deal with some other constraints such as the proportions of longer 4-day-week shifts and 5-day-week shifts, and the resulting schedule was also better than the manual schedule from the perspectives of these other constraints.

51.4 Constructive Approach Using Tabu Search

The driver scheduling problem as depicted in Figure 51.1 has a major simplification: a relief opportunity is assumed to occur at one particular time. But often, the vehicle may stop at a location for some time and it may be allowed to change driver any time within that duration. In the bus problem, a bus may be (but not always) allowed to be left unattended, i.e., without a driver, while it is idle before its next departure. In that case, the relief opportunity can be expanded into two, one at arrival and one at departure. The shift generation process has to be adapted to not having to use a driver to cover the unattended duration, but the adaptation would be quite trivial. In the train problem, the dwell time at a station may vary a lot. It is generally not allowed to leave a train unattended unless the train has been immobilized, in which case the train has to be mobilized before it can depart again. Immobilization and mobilization require some time, e.g., 10 min in total, and a driver is required. Hence, unless a train is stopping for a significant amount of time, the scheduling process will have to ensure that there is always a driver on the train.

When the driver can be relieved at any time within a period, it is called a *window of relief opportunity* (WRO). The GaS approach is unsuitable for dealing with WROs. Suppose WROs are expanded into minute-apart relief opportunities throughout the time window. For example, a WRO with a 10-min time window could be treated as 11 relief opportunities at 1-min intervals. The big increase in the number of ROs might make the problem too large to be solved by GaS. Moreover, the bunching of minute-apart ROs would cause more difficulty than the large number of them. In the above example, if the starting time of the WRO is suitable as the end of a work spell, it is likely that all the 11 expanded ROs would be suitable. In problems without WROs, it is almost certain that a work spell cannot be extended to end at any of the next 11 ROs, which would be farther apart in time. Compounded by the combinations of work spells in forming full shifts, WROs would cause an astronomical number of candidate shifts to be generated.

Shen and Kwan therefore investigated an alternative constructive approach [32, 60] instead of GaS. The algorithm developed is called heuristics for automatic crew scheduling (HACS). HACS constructs an initial schedule using a simple and quick heuristic. It assumes that the average spell length is half the maximum driving work allowed in a shift in the labor agreement. Each vehicle is then partitioned, using its ROs or WROs, into spells of roughly the average spell length. The spells are then coupled to form two-spell shifts. The spells are considered to be the first spells of shifts in ascending order of starting times. For each first spell of a shift, the process will try finding the earliest feasible spell to be coupled with it. If that is not possible, the spell that starts closest to being time feasible, i.e., the driver would have enough time for a mealbreak and travel after the first spell, will be used. Hence, invalid shifts are allowed in the initial solution. A one-spell shift is formed if there is an odd number of spells. The total number of shifts in the initial schedule may be much higher than optimal.

HACS then performs trials of reforming the shifts, two at a time, for improvement. A scheme of dividing the iterations into phases of reducing infeasibility and costs is used. Furthermore, tabu search with multineighborhood moves is employed. If a shift is considered to be some spells linked together, the multineighborhood moves are different ways of swapping the spells and links between the pair of shifts being considered, which include systematic exploration of any WROs involved in achieving a successful swap. Figure 51.7 illustrates two examples of simple swaps and Figure 51.8 illustrates an example of

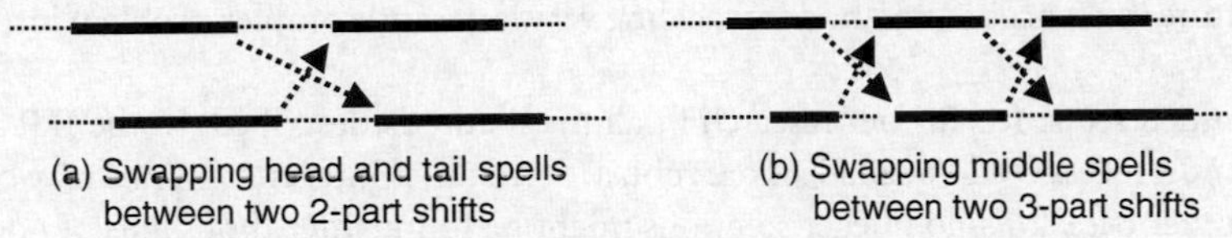

(a) Swapping head and tail spells between two 2-part shifts
(b) Swapping middle spells between two 3-part shifts

FIGURE 51.7 Swapping of work spells.

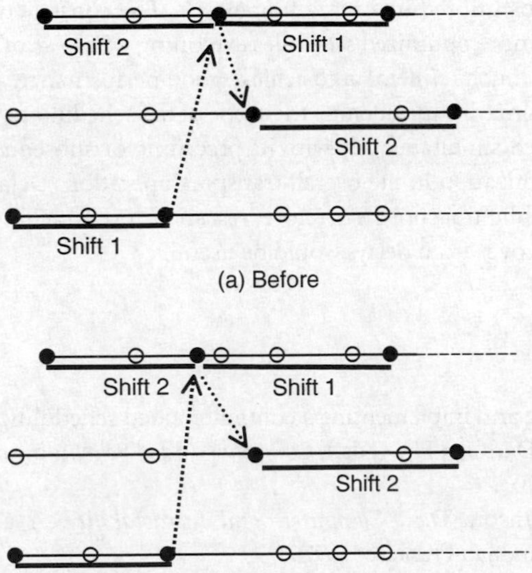

FIGURE 51.8 Reselection from neighboring relief opportunities.

redistributing work between two shifts that requires selecting a neighboring relief opportunity to be used instead.

HACS has successfully produced reasonable schedules for test problems that include WROs. Using TRACS II [39], which is based on GaS, and expanding WROs into a range of ROs at 1-min intervals, it is not possible to produce a solution for even a very small problem. However, when HACS was used for problems without WROs, the schedules produced were generally worse than those produced by TRACS II, although the run times of HACS were only fractions of those of TRACS II. Experiments have also been performed in which TRACS II was first used to solve the problem with each WRO reduced to just the arrival. WROs were then restored and the TRACS II solution was used as the initial solution for HACS. HACS was able to take advantage of WROs and produced improved solutions.

Handling WROs is a step forward in truly solving the real scheduling problem. HACS has demonstrated some promise of the constructive heuristic approach. The heuristic nature of the approach also makes it attractive as a basis for extension to produce vehicle and driver schedules simultaneously, thereby achieving further efficiency as a whole. Further research on HACS may include generating better initial solutions and designing a better scheme in driving the multineighborhood structures.

51.5 Remarks and New Research Directions

In this chapter the bus and train driver scheduling problem has been extensively discussed. The Generate and Select approach and the constructive approach have been described, and some algorithms based on these approaches have been outlined. For detailed tabulation and analysis of results, the reader should refer to the literature referenced. A method for tackling very large and complex problem instances has also been described.

There is still much scope for further research in all the methods described. In the ILP approach, the branch-and-bound process could be made more robust. Sometimes integer solutions may be found quite quickly, but on other occasions no integer solutions might be found after thousands of nodes have been searched. Large problems may still be difficult to solve. The evolutionary algorithms have demonstrated that it helps to incorporate strong domain knowledge for them to be effective. Further research may enable them to attain performance closer or even surpassing that of the ILP approach. More combinatorial traits may be utilized, and perhaps utilized in a fuzzy framework. The constructive approach has opened up the possibility of achieving more optimized schedules exploiting windows of relief opportunity. The tabu search algorithm still needs much refinement to achieve good performance.

Other research challenges may lie in widening the scope of the scheduling problem. The driver scheduling problem could be tackled simultaneously with its preceding or subsequent scheduling tasks thereby achieving more global optimization in the overall transport operation. Delays and service reliability are very important issues in public transport. Therefore, research in producing robust driver schedules and rescheduling methods to recover from delays would be useful.

References

[1] Meilton, M., Selecting and implementing a computer aided scheduling system for a large bus company, in: Voss, S. and Daduna, J.R., (eds.), *Computer-Aided Scheduling of Public Transport*, Springer-Verlag, Berlin, 2001, 203.

[2] Garey, M.R. and Johnson, D.S., *Computers and Intractability: A Guide to the Theory of NP-Completeness*, Freeman, San Francisco, 1979.

[3] Preprints of the International Workshop on Automated Techniques for Scheduling of Vehicle Operators for Urban Public Transportation Services, Chicago, 1975.

[4] Wren, A., (ed.), Computer scheduling of public transport, *Proceedings of the Second International Workshop on Computer-Aided Scheduling of Public Transport*, North-Holland, 1981.

[5] Rousseau, J.-M., (ed.), Computer scheduling of public transport 2, *Proceedings of the Third International Workshop on Computer-Aided Scheduling of Public Transport*, North-Holland, 1985.

[6] Daduna, J.R. and Wren, A., (eds.), Computer-aided transit scheduling, *Proceedings of the Fourth International Workshop on Computer-Aided Scheduling of Public Transport*, Springer-Verlag, 1988.

[7] Desrochers, M. and Rousseau, J.-M., (eds.), Computer-aided transit scheduling, *Proceedings of the Fifth International Workshop on Computer-Aided Scheduling of Public Transport*, Springer-Verlag, 1992.

[8] Daduna, J.R., Branco, I., and Paixáo, J.M.P, (eds.), Computer-aided transit scheduling, *Proceedings of the Sixth International Workshop on Computer-Aided Scheduling of Public Transport*, Springer-Verlag, 1995.

[9] Wilson, N.H.M., (ed.), Computer-aided transit scheduling, *Proceedings of the Seventh International Workshop on Computer-Aided Scheduling of Public Transport*, Springer-Verlag, 1999.

[10] Voss, S. and Daduna J.R., (eds.), Computer-aided scheduling of public transport, *Proceedings of the Eighth International Conference on Computer-Aided Scheduling of Public Transport (CASPT 2000)*, Springer-Verlag, Berlin, 2001.

[11] Wren, A., General review of the use of computers in scheduling buses and their crews, in: Wren, A., (ed.), *Computer Scheduling of Public Transport*, North-Holland, 1981, 3.

[12] Wren, A. and Rousseau, J.-M., Bus driver scheduling — an overview, in: Daduna, J.R., Branco, I., and Paixao, J.M.P., (eds.) Computer-aided transit scheduling, Springer-Verlag, 1995, 173.

[13] Elias, S.E.G., The use of digital computers in the economic scheduling for both man and machine in public transportation, Kansas State University Bulletin, Special Report number 49, 1964.

[14] Manington, B. and Wren, A., A general computer method for bus crew scheduling, in: *Preprints of the International Workshop on Automated Techniques for Scheduling of Vehicle Operators for Urban Public Transportation Services*, Chicago, 1975.

[15] Parker, M.E. and Smith, B.M., Two approaches to computer crew scheduling, in: Wren, A., (ed.), Computer scheduling of public transport, North-Holland, 1981, 193.

[16] Wilhelm, E.B., Overview of the RUCUS package driver run cutting program (RUNS), in: *Preprints of the International Workshop on Automated Techniques for Scheduling of Vehicle Operators for Urban Public Transportation Services*, Chicago, 1975.

[17] Luedtke, L.K., RUCUS II: A review of system capabilities, in: Rousseau, J.-M., (ed.), *Computer Scheduling of Public Transport* 2, North-Holland, 1985, 61.

[18] Rousseau, J.-M., Lessard, R., and Blais, J.-Y., Enhancements to the HASTUS crew scheduling algorithm, in: Rousseau, J.-M., (ed.), *Computer Scheduling of Public Transport* 2, North-Holland, 1985, 295.

[19] Desrochers, M. and Soumis, F., CREW-OPT: Crew scheduling by column generation, in: Daduna, J.R. and Wren, A., (eds.), *Computer-Aided Transit Scheduling*, Springer-Verlag, 1988, 83.

[20] Desrochers, M. et al., CREW-OPT: Subproblem modeling in a column generation approach to urban crew scheduling, in: Desrochers, M. and Rousseau, J.-M., (eds.), *Computer-Aided Transit Scheduling*, Springer Verlag, 1992, 395.

[21] Falkner, J.C. and Ryan, D.M., EXPRESS: Set partitioning for bus crew scheduling in Christchurch, in: Desrochers, M. and Rousseau, J.-M., (eds.), *Computer-Aided Transit Scheduling*, Springer Verlag, 1992, 359.

[22] Smith, B.M. and Wren, A., A bus crew scheduling system using a set covering formulation, in: *Transportation Research*, 22A, 97, 1988.

[23] Daduna, J.R. and Mojsilovic, M., Computer-aided vehicle and duty scheduling using the HOT program system, in: Daduna, J.R. and Wren, A., (eds.), *Computer-Aided Transit Scheduling*, Springer-Verlag, 1988, 133.

[24] Clement, R. and Wren, A., Greedy genetic algorithms, optimizing mutations and bus driver scheduling, in: Daduna, J.R., Branco, I., and Paixao, J.M.P., (eds.), *Computer-Aided Transit Scheduling*, Springer-Verlag, 1995, 213.

[25] Wren, A. and Wren, D.O., A genetic algorithm for public transport driver scheduling, *Computers and Operations Research*, 22, 101, 1995.

[26] Curtis, S.D., Smith, B.M., and Wren, A., Forming bus driver schedules using constraint programming, in: *Proceedings of First International Conference on the Practical Applications of Constraint Technologies and Logic Programming*, The Practical Application Company, Blackpool, 1999, 239.

[27] Gonzalez Hernandez, L.F. and Corne, D.W., Evolutionary divide and conquer of the set covering problem, in: Fogarty, T.C., (ed.), *Evolutionary Computing*, Springer, Berlin, 1996, 198.

[28] Guerinik, N. and Caneghem, M.V., Solving crew scheduling problems by constraint programming, in: Montanari, U. and Rossi, F., (eds.), *Principles and Practice of Constraint Programming CP'95*, Springer, Berlin, 1995, 481.

[29] Kwan, R.S.K. and Wren, A., Hybrid genetic algorithms for bus driver scheduling, in: Bianco, L. and Toth, P., (eds.), *Advanced Methods in Transportation Analysis*, Springer, Berlin, 1996, 609.

[30] Kwan, R.S.K., Kwan, A.S.K., and Wren, A., Evolutionary driver scheduling with relief chains, *Evolutionary Computation*, 9, 445, 2001.

[31] Layfield, C.J., Smith, B.M., and Wren, A., Bus relief opportunity selection using constraint programming, in: *Proceedings of First International Conference on the Practical Applications of Constraint Technologies and Logic Programming*, The Practical Application Company, Blackpool, 1999, 537.

[32] Shen, Y. and Kwan, R.S.K., Tabu search for driver scheduling, in: Voss, S. and Daduna, J.R., (eds.), *Computer-Aided Scheduling of Public Transport*, Springer-Verlag, Berlin, 2001, 121.

[33] Kwan, A.S.K., Kwan, R.S.K., Parker, M.E, and Wren, A., Producing train driver shifts by computer, in: Allan, J. et al., (eds.), *Computer in Railways V, Vol. 1: Railway Systems and Management*, Computational Mechanics Publications, 1996, 421.

[34] Kwan, A.S.K., Kwan, R.S.K., Parker, M.E, and Wren, A., Producing train driver schedules under differing operating strategies, in: Wilson, N.H.M., ed., *Computer-Aided Transit Scheduling*, Springer-Verlag, 1999, 129.

[35] Caprara, A. et al., Solution of large-scale railway crew planning problems: the Italian experience, in: Wilson, N.H.M., (ed.), *Computer-Aided Transit Scheduling*, Springer-Verlag, 1999, 1.

[36] Ernst, A. et al., Rail crew scheduling and rostering optimization algorithms, in: Voss, S. and Daduna, J.R., (eds.), *Computer-Aided Scheduling of Public Transport*, Springer-Verlag, Berlin, 2001, 53.

[37] Kwan, A.S.K., Train driver scheduling, PhD thesis, University of Leeds, Leeds, UK, 1999.

[38] Fores, S., Proll, L., and Wren, A., TRACS II: A hybrid IP/heuristic driver scheduling system for public transport, *J. of OR Society*, 53, 1093, 2002.

[39] Wren, A., Fores, S., Kwan, A.S.K., Kwan, R.S.K., Parker, M.E., and Proll, L., A flexible system for scheduling drivers, *Journal of Scheduling*, 6, 437, 2003.

[40] Kwan, A.S.K., Parker, M.E., Kwan, R.S.K., Fores, S., Proll, L., and Wren, A., Recent advancement in an established driver scheduling system, in: Kendall, G., Burke E., Petrovic, S., (eds.), *Proceedings of the First Multidisciplinary International Conference on Scheduling: Theory and Applications (MISTA)*, Nottingham University, 2003, 704.

[41] Fores, S., Proll, L., and Wren, A., An improved ILP system for driver scheduling, in: Wilson, N.H.M., (ed.), *Computer-Aided Transit Scheduling*, Springer-Verlag, 1999, 43.

[42] Fores, S., Column generation approaches to bus driver scheduling, PhD thesis, University of Leeds, Leeds, UK, 1996.

[43] Willers, W.P., Improved algorithms for bus crew scheduling, PhD thesis, University of Leeds, Leeds, UK, 1995.

[44] Smith, B.M., Bus crew scheduling using mathematical programming, PhD thesis, University of Leeds, Leeds, UK, 1986.

[45] Goldfarb, D. and Reid, J.K., A practical steepest-edge simplex algorithm, *Mathematical Programming*, 12, 361, 1977.

[46] Forrest, J.J. and Goldfarb, D., Steepest-edge simplex algorithms for linear programming, *Mathematical Programming*, 57, 341, 1992.

[47] Sherali, H.D., Equivalent weights for lexicographic multi-objective programs: Characterization and computations, *European Journal of Operational Research*, 18, 57, 1982.
[48] Ryan, D.M., ZIP - a zero one integer programming package for scheduling, Technical report CSS-85, AERE, Computer Science and Systems Division, Harwell, UK, 1980.
[49] Ryan, D.M. and Foster, B.A., An integer programming approach to scheduling, in: Wren, A., (ed.), *Computer Scheduling of Public Transport*, North-Holland, 1981, 269.
[50] Kwan, A.S.K., Kwan, R.S.K., and Wren, A., Driver scheduling using genetic algorithms with embedded combinatorial traits, in: Wilson, N.H.M., (ed.), *Computer-Aided Transit Scheduling*, Springer-Verlag, 1999, 81.
[51] Zadeh, L.A., Fuzzy sets, *Information and Control*, 8, 338, 1965.
[52] Li, J. and Kwan, R.S.K., A fuzzy genetic algorithm for driver scheduling, *European Journal of Operational Research*, 147, 334, 2003.
[53] Li, J. and Kwan, R.S.K., A fuzzy evolutionary approach with Taguchi parameter setting for the set covering problem, in: *Proceedings of the 2002 IEEE Congress on Evolutionary Computation*, IEEE Press, 2002, 1203.
[54] Li, J. and Kwan, R.S.K., A fuzzy simulated evolution algorithm for the driver scheduling problem, in *Proceedings of the 2001 IEEE Congress on Evolutionary Computation*, IEEE Press, 2001, 1115.
[55] Li, J. and Kwan, R.S.K., A fuzzy theory based evolutionary approach for driver scheduling in: Spector, L. et al. (eds.), *Proceedings of the Genetic and Evolutionary Computation Conference*, Morgan Kaufman, 2001, 1152.
[56] Li, J, Fuzzy evolutionary approaches for bus and rail driver scheduling, PhD thesis, University of Leeds, Leeds, UK, 2002.
[57] Kling, R.M. and Banerjee, P., ESP: a new standard cell placement package using Simulated Evolution, in: *Proceedings of the 24th ACM/IEEE Design Automation Conference*, 1987, 60.
[58] Taguchi, G. (1987) *System of Experimental Design*, Vols. 1 and 2, UNIPUB/Kraus International Publications, New York.
[59] Zhao, L., Wren, A., and Kwan, R. S. K., Development of a driver duty estimator, *J. of OR Society*, 46, 1102, 1995.
[60] Shen, Y., Tabu search for bus and train driver scheduling with time windows, PhD thesis, University of Leeds, Leeds, UK, 2001.

52

Sports Scheduling

Kelly Easton
Kansas State University

George Nemhauser
Georgia Institute of Technology

Michael Trick
Carnegie Mellon University

52.1 Introduction

This chapter is a survey of the current body of sports scheduling literature covering a period of time from the early 1970s to the present day. Most of this literature concerns tournament scheduling problems, particularly the *single round robin tournament problem* (SRRTP), in which each team plays each other team exactly once, and the *double round robin tournament problem* (DRRTP), in which each team plays each other team exactly twice (usually once at each venue).

Round robin scheduling is interesting from both a practical and theoretical standpoint. There are numerous sports leagues, including many college basketball, football and baseball leagues and European soccer and cricket leagues, that play single or double round robin tournaments as their regular season schedules. Additionally, there are many theoretical questions that can be asked regarding round robin tournaments. One question treated in the current body of literature pertains to finding the minimum number of breaks, or times that a team has consecutive home or away games, in an n-team round robin tournament. Another question, currently still open, involves finding sufficient conditions for a home-away pattern set to yield a feasible round robin schedule.

While round robin scheduling is the most common problem treated in the sports scheduling literature, there are two other tournament scheduling problems that have received some attention. As indicated above, in round robin tournament scheduling, games are usually played at one of the opponents'

1-58488-397-9/$0.00+$1.50

home venues. In the *balanced tournament design problem* (BTDP), however, there are common facilities which host all the games. The BTDP is similar to the SRRTP in that every team plays every team exactly once, but it requires the set of games played by a given team to be equally (as nearly as possible) distributed over the possible facilities. An alternative statement of the BTDP has teams competing at a single facility at different times during the day. Then the set of games played by a given team must be equally (as nearly as possible) distributed over the possible times.

Secondly, in the *bipartite tournament problem* (BTP), there are two teams with n players each. Each player must compete exactly once against every player on the opposing team. A generalization of the BTP involves assigning athletes to compete in different events during an athletic competition.

Before continuing it will be useful to introduce a few terms particular to the sports scheduling literature. We have defined *single round robin tournament* (SRRT), *double round robin tournament* (DRRT), *balanced tournament design* (BTD), and *bipartite tournament* (BT) above. A *schedule* is a mapping of games to *slots*, or time periods, such that each team plays at most once in each slot. A double round robin tournament schedule may be *mirrored*, meaning that the games played in the first half of the schedule are repeated in the same order in the second half of the schedule with the venues reversed, or *partially mirrored*, meaning that all slots in the schedule are paired such that one is the mirror of the other, but the slots are not necessarily divided into the two halves of the schedule or ordered in any particular fashion. A schedule is said to be *compact* if it includes the minimum possible number of slots. (In this chapter, we will assume that a schedule is compact unless otherwise noted.) A *pattern* is a vector of *home* (H), *away* (A), or *bye* (B) designations for a single team over the slots in the schedule. (A team is assigned a *bye* if it does not play in a particular slot.) Two patterns are said to be *complementary* if in every slot one pattern has a home and the other has an away. A *pattern set* is a collection of patterns, one for each team. A *tour* is the schedule for a single team in the tournament. A *trip* is a series of consecutive away games, and a *home stand* is a series of consecutive home games.

We have chosen to organize the literature considered in this survey according to solution methodology. General solution methods used in sports scheduling include *graph algorithms, integer programming* (IP), *constraint programming* (CP), metaheuristics like *simulated annealing* (SA), and *tabu search* (TS), *problem-specific heuristics*, and *enumeration*. Of course, some research employs more than one solution methodology.

Before delving into these categories, however, we give a few general results on the various tournament scheduling problems.

52.2 Round Robin Tournaments

An SRRT for n teams, n even, requires $n-1$ slots, and each team plays in every slot. If n is odd, n slots are required, and each team has one bye. The SRRTP may be formally stated as follows:

Input: A set of n teams $T = \{1, \ldots, n\}$.
Output: A mapping of the games in the set $G = \{g_{ij} : i, j \in T, i < j\}$, to the slots in the set $S = \{s_k, k = 1, \ldots, n-1$ if n even and $k = 1, \ldots, n$ if n odd$\}$ such that no more than one game including i is mapped to any given slot for all $i \in T$.

The construction of an SRRT falls under the general category of *block designs*. The study of block designs is said to have begun with a paper by the statistician F. Yates in which he considered arrangements of subsets with certain balance properties [1]. A (v, k, λ) configuration describes a collection of k-element subsets (the blocks) of a v-element set S such that each pair of elements in S occurs in exactly λ of the blocks.

A block design is *resolvable* if the blocks can be arranged into r disjoint groups so that the $(b/r) = (v/k)$ blocks of each group are disjoint and contain in their union each element exactly once. An SRRT on n teams, n even, is an $(n, 2, 1)$ resolvable design. The origin of the *circle method* for solving resolvable designs is not known, but it can be used to construct an SRRT. The following description of the circle method is from [1].

Algorithm SRRTP-C

Label the teams $\infty, 1, 2, \ldots, n-1$. On day i, play i vs. ∞, $(i-1)$ vs. $(i+1)$, $(i-2)$ vs. $(i+2), \ldots, (i-(n/2-1))$ vs. $(i+(n/2+1))$, each integer being reduced (mod $n-1$) to lie in the interval $[1, n-1]$.

Proof

Team h will play against team k on day i where $h \equiv i-j, k \equiv i+j \pmod{n-1}$ for some j. But $h \equiv i-j$ and $k \equiv i+j$ if and only if $2i \equiv h+k$ and $2j \equiv k-h \pmod{n-1}$. These congruences have unique solutions since $n-1$ is odd. □

In other words, in the first slot play ∞ vs. 1, 2 vs. $(n-1)$, 3 vs. $(n-2), \ldots, n$ vs. $(n/2+1)$. In successive slots, obtain the opponents by adding 1 (mod $n-1$) to each number except ∞. (Obviously, replace ∞ with n in the final schedule.)

The circle method does not include a method for selecting venues. Theoretically, we can simply assign j to be the home team in any game i vs. j with $i < j$. Practically, however, it is necessary for a schedule to be somewhat balanced with regard to the home and away games played by the various teams. De Werra [2] provides a venue-assignment algorithm for the circle method, shown below.

Algorithm SRRTP-V

For each slot $i, i = 1, \ldots, n-1$, let $i+k$ play at $i-k$ if k is odd or $i-k$ at $i+k$ if k is even. For the games i vs. ∞, assign i at n if i is even and n at i if i is odd.

The DRRTP may be formally stated as follows:

Input: A set of n teams $T = \{1, \ldots, n\}$.
Output: A mapping of the games in the set $G = \{g_{ij} : i, j \in T, i \neq j\}$, to the slots in the set $S = \{s_k, k = 1, \ldots, 2(n-1)$ if n even and $k = 1, \ldots, 2n$ if n odd$\}$ such that no more than one game including i is mapped to any given slot for all $i \in T$.

A canonical DRRT can easily be created by combining two SRRTs, the second a mirrored version of the first. In other words, schedule an SRRT in the first half of the DRRT schedule. Then, for the second half of the schedule, repeat the games in order, reversing the venues. De Werra [3] pointed out that this will create an unfair pattern for some teams in the schedule due to sequences of three consecutive away games. However, this imbalance can be eliminated by a slight modification to Algorithm SRRTP-V. Rather than assigning i at n if i is even and n at i if i is odd for the games i vs. n, instead assign i at n if $i \leq n-5$ is odd or if $i = n-2$; otherwise, assign n at i.

52.2.1 Round Robin Tournaments and Latin Squares

A *latin square* of order n is an $n \times n$ array with each cell containing an element from the set $\psi = \{1, 2, \ldots, n\}$. Each row and column of the array contains each element in ψ exactly once. A latin square is quite simple to generate. One way is to fix the first row with any permutation of the integers $1, \ldots, n$. Then enter the $n-1$ cyclic permutations of the n integers in the remaining rows. This method produces a subset of the very large set of possible latin squares. Specifically, for each n, there are at least $n!(n-1)! \ldots (2!)1$ admissible configurations [4].

There is a direct correspondence between latin squares and SRRTs. This is important not only because algorithms designed to solve latin squares can be used to generate SRRTs, but there is an extensive body of research concerning latin squares, while less comprehensive research exists for the SRRTP and DRRTP. Colbourn has compiled a review of latin square results [5]. A number of these results can be translated to obtain results regarding single and thus (mirrored) double round robin tournaments.

We are specifically interested in even, symmetric latin squares with the restriction that the elements down the principle diagonal (cells in row i and column i for $i = 1, 2, \ldots, n$) are all identical. Every even, symmetric latin square with a single element along the principle diagonal generates a single round robin

6	2	3	4	5	1
2	6	4	5	1	3
3	4	6	1	2	5
4	5	1	6	3	2
5	1	2	3	6	4
1	3	5	2	4	6

FIGURE 52.1 A Symmetric 6 × 6 Latin Square.

Slot 1:	Team 1 vs. Team 6	Team 2 vs. Team 5	Team 3 vs. Team 4
Slot 2:	Team 1 vs. Team 2	Team 3 vs. Team 5	Team 4 vs. Team 6
Slot 3:	Team 1 vs. Team 3	Team 2 vs. Team 6	Team 4 vs. Team 5
Slot 4:	Team 1 vs. Team 4	Team 2 vs. Team 3	Team 5 vs. Team 6
Slot 5:	Team 1 vs. Team 5	Team 2 vs. Team 4	Team 3 vs. Team 6

FIGURE 52.2 A corresponding single round robin tournament for six teams.

tournament, and every single round robin tournament for an even number of teams generates n even, symmetric latin squares, each with a distinct single element along the principle diagonal. Figures 52.1 and 52.2 illustrate this correspondence.

In practical round robin scheduling problems, one of the most common constraints involves fixed games. The concept of a tournament with fixed games is analogous to a *partially completed, symmetric latin square.* A partially completed latin square is a latin square that has some cells filled with admissible numbers. Numerous complexity results concerning completing partially completed latin squares are known. Colbourn showed that completing a symmetric latin square is $\mathcal{NP}$-complete [6]. More recently, Easton and Parker showed that completing a partially-completed latin square with at most three open cells in each row or column is $\mathcal{NP}$-complete even if only the numbers 1, 2, and 3 are missing from any row or column [4]. This is the strongest result attainable since there is a simple polynomial time algorithm if all but two cells are filled in any row or column [4]. Easton and Parker also give a result for sparsely filled latin squares. Namely, completing a partially completed latin square is $\mathcal{NP}$-complete even if on average only three elements are filled in any row or column.

These latin square results can be translated into results for the SRRTP. Thus, we have a polynomial time algorithm for completing a schedule if all but two games for each team are fixed and a polynomial time algorithm for completing a schedule in which any number of at most one team's games are fixed. We know that the SRRTP with all but three slots completely scheduled and every team with at most three unscheduled games is $\mathcal{NP}$-complete and that the SRRTP for n teams is $\mathcal{NP}$-complete even if on average each team has two games scheduled. For full reductions of these latin square results to the SRRTP results, see [7].

52.2.2 Round Robin Pattern Sets

Thus far we have been discussing the creation of a round robin tournament schedule by first assigning team match-ups to slots, then determining the venue for each game. Actually, there is much research to suggest that reversing the order of these tasks is an effective approach in many cases. Later in this survey we will discuss a number of solution methodologies that decompose the round robin scheduling problem, first generating a pattern set then pairing compatible teams in games. Here we consider results that regard the generation of a *good* pattern set as more of a problem unto itself.

For decomposition schemes such as the one just described, the run time of the algorithm depends not only on the speed with which it generates pattern sets but the number of pattern sets generated and the ratio of feasible to infeasible pattern sets that pass through to the game generation subproblem. (Here *feasible* describes a pattern set that gives a round robin tournament and *infeasible* a pattern set that does not yield a round robin tournament.) Clearly, if there were a known set of necessary and sufficient conditions for feasible pattern sets, this would favorably affect the ratio of feasible to infeasible pattern sets, the number of pattern sets generated, and quite possibly the run time of the pattern set generation algorithm.

While the sufficient conditions for general pattern set feasibility are unknown at this point, we do have a collection of necessary conditions that have proven useful in practice. For example, in a feasible pattern set for an SRRT, every pair of patterns must differ in at least one slot. This means that any given pattern can appear at most once in a pattern set. The condition for DRRTs is a bit stronger: in a feasible pattern set for a DRRT, for every pair of patterns i, j such that $1 \leq i < j \leq n$, there must be at least one slot in which i is home and j is away and at least one slot in which j is at home and i is away.

A second necessary condition requires every slot in the pattern set to include an equal number of home and away games.

For SRRTPs in which the number of breaks is restricted to at most one for each team, a feasible pattern set must have the complementary property [8]. In other words, it must be possible to group the set of $n = 2m$ patterns into m disjoint pairs of complementary patterns. This result sounds surprising at first, but it is really quite obvious. Notice that for the condition of equal number of Hs and As to hold, the breaks in this type of pattern set must occur in pairs of one AA break and one HH break. Since no pattern can have more than one break, these pairs must be complementary.

Miyashiro, Iwasaki, and Matsui [9] recently published a more complex necessary condition for feasible pattern sets that can be checked in polynomial time for a pattern set with the minimum number of breaks. Let P represent the patterns in a given pattern set; let Q be a subset of P; let S represent the complete set of slots. Let $|H_Q^s|$ be the number of home games in slot s summed over the patterns in Q. Similarly, let $|A_Q^s|$ be the number of away games in slot s summed over the patterns in Q. Then, for a given pattern set to be feasible, it must have

$$\sum_{s \in S} \min\{|H_Q^s|, |A_Q^s|\} \geq \binom{|Q|}{2}$$

for all $Q \subset P$.

Miyashiro et al. show that this sum is the same for the complement of the subset Q. Along with certain properties of pattern sets with the minimum number of breaks, this allows Miyashiro et al. to give an $O(n^4)$ algorithm for checking this condition for any minimum break pattern set.

52.3 Balanced Tournament Designs

The Balanced Tournament Design Problem may be formally stated as follows:

Input: A set of n teams $T = \{1, \ldots, n\}$ and a number of facilities F.
Output: A mapping of the games in the set $G = \{g_{ij} : i, j \in T, i < j\}$, to the slots available at each facility described by the set $S = \{s_{fk}, f = 1, \ldots, F, k = 1, \ldots, n-1 \text{ if } n \text{ even and } k = 1, \ldots, n \text{ if } n \text{ odd}\}$ such that no more than one game involving team i is assigned to a particular slot and the difference between the number of appearances of team i at two separate facilities is no more than 1.

A common version of the BTDP has $2m$ teams and m facilities. We will denote this problem by BTDP($2m$). It is known that there exist solutions to this problem for all $m \neq 2$. The first constructions for $n \equiv 0$ or $n \equiv 1 (\text{mod} 3)$ teams were provided by Haselgrove and Leech [10], and the full result was established by Schellenberg, van Rees, and Vanstone [11] a short time later. Lamken and Vanstone [12] published a simplified proof in 1985.

Blest and Fitzgerald [13] show a simple construction for a BTD with $2m + 1$ teams and m facilities.

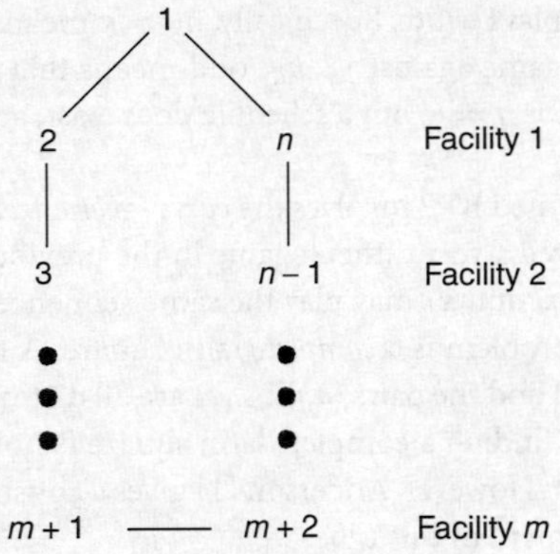

FIGURE 52.3 Bracelet for algorithm BTDP-NO.

Algorithm BTDP-NO

1. Arrange the teams 1 through $2m+1$ in an elongated pentagon or "bracelet" as shown in Figure 52.3. Indicate a facility associated with each row containing two teams as shown.
2. For each slot $k = 1, \ldots, 2m+1$, give the team at the top of the pentagon the bye. For each row with two teams i, j associated with facility f assign g_{ij} to s_{kf}. Then shift the teams around the pentagon one position in a clockwise direction.

Of course, it is easy to see that this algorithm uses the same principle as the circle method described above. Blest and Fitzgerald also give a construction for $n \equiv 2 \pmod 4$.

52.4 Bipartite Tournaments

The BTP may be formally stated as follows:

Input: Two teams with n players $T_1 = \{x_1, \ldots, x_n\}$ and $T_2 = \{y_1, \ldots, y_n\}$.
Output: A mapping of the games in the set $G = \{g_{ij} : i \in T_1, j \in T_2\}$, to the slots in the set $S = \{s_k, k = 1, \ldots, n\}$ such that exactly one game including t is mapped to any given slot for all $t \in T_1 \cup T_2$.

Bipartite tournaments are equivalent to latin squares. To see this, first notice that we can write the opponents for x_i in the ith column of an n by n matrix and obtain a latin square. Conversely, notice that any $n \times n$ latin square will uniquely determine a bipartite tournament. We could also write down a latin square by letting entry l_{ij} = the slot in which x_i plays y_j for $i, j \in 1, \ldots, n$ [1].

We have discussed several latin square results that can be applied to SRRTs. Similarly, these results can be applied to BTs. Latin square methods can also be used to solve several special cases of the BTP. Anderson describes three possibilities [1]. For example, if the n games in each slot are played at n different facilities, it may be desirable to construct a balanced BT in which each team plays exactly once at each facility. The solution to this problem is the join of two *mutually orthogonal latin squares* (MOLS) on $\{1, \ldots, n\}$ in which the rows correspond to slots and the columns correspond to facilities. The join includes ordered pairs (i, j), which correspond to the games x_i vs. y_j. If this type of tournament exists, then the MOLS can be read directly from it; therefore, it exists if and only if $n \neq 2, 6$ because there are no MOLS for these parameters.

Another balanced BTP arises in chess. The player using the white pieces has an advantage because white gets to move first; therefore, the fairest possible tournament would have each player using white and black equally often. Additionally, in each slot, both teams should have exactly one half of their players using white and the other half using black. In this case, we must assume n even. Again, the join of two MOLS A, B gives a solution to this problem. The entries in A give the slots in which games are assigned and the entries in B indicate which team plays white. Specifically, $a_{ij} = k$ means that x_i plays y_j in slot k, $b_{i,j}$ even means that x_i plays white in the game against y_j, $b_{i,j}$ odd means that x_i plays black in the game against y_j. This method cannot be used for $n = 6$, but a schedule does exist, and Anderson gives its construction in [1].

The third example of a constrained BTP involves the *carry-over effect.* A team may gain an advantage if its current opponent has just played a very difficult game in the previous slot. In order to fairly distribute these advantages, no two players x_i and x_j may play the same sequence of opponents y_p followed immediately by y_q. A solution to this problem is a *complete* latin square. A latin square $A = a_{ij}$ is complete if the pairs (a_{ij}, a_{ij+1}) are all distinct and the pairs (a_{ij}, a_{i+1j}) are all distinct. The entry $a_{k,j} = i$ corresponds to x_i playing y_j in slot k. Simply finding a complete latin square is not enough because the rows do not represent the players in T_2 in order. However, Anderson [1] gives a construction for a complete latin square that does yield a bipartite tournament of this type.

Russell [14] showed that if the number of teams is a power of 2, then it is possible to evenly apply the carry-over effect. For other sizes, he formulated an integer program to minimize the variance of carry-over effect that found optimal values for six-team schedules, and heuristic values for slightly larger leagues.

52.5 Graph Algorithms

The graph theoretical model for a single round robin tournament (SRRT) among $2m$ teams is the complete graph K_{2m}. The edge $[i, j]$ represents the game between team i and team j. Then any SRRT can be presented as a 1-factorization $(F_1, \ldots, F_{2m-1})$ of K_{2m} where each 1-factor includes the games scheduled in one slot. A 1-factorization of K_{2m} may also be understood as a coloring of K_{2m} or a decomposition of K_{2m} into perfect matchings.

To completely define the tournament schedule, the 1-factorization must be oriented. The arc (i, j) represents the game *team i at team j*, and the arc (j, i) represents the game *team j at team i*. A 1-factorization of K_{2m} together with an orientation is called an *oriented coloring* of the graph K_{2m}.

To model a DRRT, K_{2m} is replaced by the oriented graph G_{2m} in which each pair of nodes i, j is joined by two arcs: $a_{i,j}$ and $a_{j,i}$. Then a schedule is represented by a decomposition of the arc set that is equivalent to the union of two oriented colorings of K_{2m}.

52.5.1 Minimum Breaks Problem

De Werra, who has done much work in sports scheduling, has largely focused on the SRRTP requiring minimization of the number of breaks (SRRTP-MB). Thus, many of the results described in the current section concern this particular case of the SRRTP.

In an oriented coloring of K_{2m}, there are at least $2m - 2$ breaks [3]. To see this, note that there are only two possible patterns for a round robin tournament with $2m$ teams that have no breaks, one beginning with a home game and one beginning with an away game. As discussed above, no pattern may be repeated in a feasible pattern set; thus, all but two patterns must have breaks. De Werra's algorithm for assigning venues for the games scheduled by the circle method (described above) generates a schedule with exactly $2m - 2$ breaks.

This result for the SRRTP-MB can be extended to the DRRTP requiring a minimization of the number of breaks (DRRTP-MB). Any decomposition of G_{2m} has at least $6m - 6$ breaks [3].

Although nearly every team in an SRRT on an even number of teams must have a break, this is not the case for SRRTs on an odd number of teams. The inclusion of byes in the pattern set increases the number of possible patterns with no breaks. In fact, for each slot s, there are two patterns with a bye in s and no breaks. We need only one of these patterns to form a pattern set, and thus it is possible to form an SRRT with no breaks. Formally, the complete graph on an odd number of nodes K_{2m+1} has an oriented factorization with no breaks [3].

Later De Werra [8] generalized these results: Let G be a d-regular factorizable graph on $2m$ nodes; in every oriented coloring of G, there are at least $2(m - \alpha(G))$ breaks where $\alpha(G)$ is the maximum size of an independent set in G. Also, any d-regular graph G has an oriented coloring with no breaks if and only if G is bipartite.

De Werra [15] also expanded on the circle method by showing how the canonical factorization it gives can be permuted to obtain a set of m canonically feasible factorizations, meaning that each member of the set is a permutation of the original and has $2m - 2$ breaks. We have discussed the fact that an SRRT with at most one break per team is composed of pairs of teams with complementary patterns. It is also clear that there are at most two breaks per slot. Therefore, we can completely characterize the pattern set of an SRRT with $2m - 2$ breaks, up to a permutation of the rows, with a sequence $\{b_1,\dots,b_{m-1}\}, 2 \leq b_1 < b_2 < \cdots < b_{m-1} \leq 2m - 1$ that describes the slots in which breaks occur. Similarly, we can completely characterize the pattern set of an *equitable* SRRT, i.e., one with exactly one break for each team, up to a permutation of the rows, with a sequence $\{b_1,\dots,b_m\}$.

Of course, not all sequences B yield canonically feasible schedules. De Werra showed that a sequence $B = \{b_1,\dots,b_{m-1}\}$ is canonically feasible if and only if $b_1 \leq 3$; $b_{i+1} - b_i \leq 2 (i = 1,\dots,m-2)$; $b_{n-1} \geq 2m - 2$. For equitable schedules, a sequence $B = \{b_1,\dots,b_m\}$ is canonically feasible if and only if $b_1 = 2$; $b_{i+1} - b_i \leq 2 (i = 1,\dots,m-1)$; $b_m = 2m - 1$. From this characterization, it directly follows that there are m canonically feasible sequences for schedules with $2m - 2$ breaks and $m - 1$ canonically feasible sequences for equitable schedules.

There are numerous special cases of the SRRTP-MB. We will consider two here: an SRRTP-MB in which no team may play consecutive away games in the same geographical location, and a SRRTP-MB in which $2m$ teams are divided into p divisions of q teams each with constraints on the positioning of in-division and out-of-division games.

52.5.2 Geographical Location

Some sports leagues include pairs of teams that play in the same geographic location. A well-known example of this is major league baseball (MLB) in the United States. There are four pairs of MLB teams that play in the same small geographic area: the Chicago Cubs and the Chicago White Sox, the New York Yankees and the New York Mets, the San Francisco Giants and the Oakland A's play on either side of the Bay Bridge in Northern California, and the Los Angeles Dodgers and the Anaheim Angels play within an hour of each other in Southern California. An important constraint in the MLB scheduling problem requires complementary patterns for these pairs of teams. This scheme is designed to maximize the attendance at the home games in the regions.

De Werra [15] observed that if fans are attending home games at both teams in the region in consecutive slots, they do not want to see the same visiting team. We have discussed the fact that an SRRT with at most one break per team is composed of pairs of teams with complementary patterns. Assuming that two teams located in the same geographic area are associated with one of these pairs, Algorithm SRRTP-V can be modified to give an oriented factorization for the schedule generated by Algorithm SRRT-C with the property that no team plays consecutive away games at two teams with complementary patterns. Rather than scheduling $2m$ at $2m - 2$ and $2m - 1$ at $2m$, schedule $2m - 2$ at $2m$ and $2m$ at $2m - 1$. Then the complementary pairs are $1, 2m - 1, 2, 3,\dots, 2m - 4, 2m - 3, 2m - 2, 2m$, and it is easy to check that no team visits these teams in consecutive slots.

52.5.3 Divisions

There are many examples of sports leagues that organize themselves in divisions. Again, MLB provides a good example of this. The 30 MLB teams are divided into the National League and the American League. Within each league, the teams are further subdivided into east, central, and west divisions. Here we first consider the problem of a league that has two divisions of $2m$ teams, organized geographically. In other words, teams within a division are grouped relatively close together, but two teams in different divisions might be located a long distance from one another. For this reason, the league requires teams to play divisional games on the weekdays and games between divisions on the weekends. Since each team must play $2m$ weekend games and $2m-1$ weekday games, a schedule with two games per week, one on a weekday and one on a weekend, begins and ends with a weekend game.

De Werra showed how to construct two oriented factorizations of the complete graph with $2m$ nodes, the union of which covers the weekday slots and a third oriented factorization of the bipartite graph K_{XY} where $|X| = |Y| = 2m$ to cover the weekend slots. These are then merged to give the full schedule. For the exact construction, see [15]. In total, this schedule has $4m - 2$ breaks. De Werra noted that this solution depends upon having an equal, even number of teams in each division. Without either one of these conditions, there is no known factorization for the weekdays. This result represents a significant departure from the results above in that it does not depend on a rearrangement of the canonical factorization. In fact, the canonical factorization cannot be used in this case because the union of three canonical factorizations contains a triangle [3].

This result can be generalized to apply to $2m$ teams divided into p divisions of q teams each. Obviously, q must be even, and since $q = 2$ is trivial, $q \geq 4$. Assuming that $p = 2^i$ with $i \geq 1$, De Werra [16] showed that there exists an oriented factorization such that each team has at most three breaks, no team has consecutive breaks, in each division and each slot half the teams have home games, and the total number of breaks is $(5/2)pq - 2(p + q)$.

A closely-related problem requires all games between divisions to be played first and all divisional games to be played at the end of the schedule. Clearly, the factorization described above could be permuted to provide a solution for this problem. De Werra [16] showed that this arrangement also has $(5/2)pq-2(p+q)$ breaks. Note that this is not necessarily the minimum number of breaks. De Werra also proved a lower bound of $2pq - q - 2p$ breaks for this problem, which is attained for $p = 2$, but is not generally not attained for $p > 2$.

52.5.4 Minimum Irregularities Problem

While simply minimizing the breaks in a schedule is desirable in some practical scheduling problems, there is a variation of SRRTP-MB that is sometimes more applicable: the SRRTP in which the number of slots with breaks is minimized (SRRTP-MS). Using MLB as an example again, all teams are required to change pattern across the *all-star break*. In other words, no break is allowed for the slot immediately following the all-star break.

To construct schedules with a minimum number of irregular slots, K_{2m} must be partitioned into a minimum number of bipartite factors. This minimum is $\log_2 m$ [17]. De Werra [18] showed that for a league of $2m$ teams, there exists an SRRT with a minimum number of irregular slots and at most $2\lfloor m/2 \rfloor \lceil \log_2 m \rceil$ breaks.

52.5.5 Practical Applications

Several papers in the literature [19–22] describe using a canonical factorization as the basis for solving a practical scheduling problem.

Schreuder [22] solved the DRRTP for the *Dutch Professional Football League* by first selecting an SRRT canonical schedule with minimum breaks and mirroring it to create a DRRT schedule, and secondly assigning actual teams to the patterns. Schreuder is considered to have pioneered this two-phase decomposition which was later used by Russell and Leung [20], Schaerf [21], and Bartsch, Drexl, and Kroger [19], and incorporated into Nemhauser and Trick's three-phase method [23].

Russell and Leung [20] solved the scheduling problem for the *Texas Baseball League* which can be represented by three consecutive DRRTPs. Their method follows Schreuder's: the first phase involves generating a series of three canonical DRRT timetables, and in the second phase, enumeration is used to find the minimum cost assignment of teams to the patterns in the timetable.

Bartsch, Drexl, and Kroger (BDK) [19] solved the scheduling problems of two European soccer leagues, one from Germany and the other from Austria. The former requires a DRRT schedule made up of two separate SRRTS with complementary pattern sets, and the latter requires four separate SRRTs of which the second and third have complementary patterns as do the first and fourth. For the German league, in the first phase of the algorithm, a canonical minimum break schedule is selected, and rounds swapped in order to generate a timetable that yields a feasible schedule. In phase two, teams are assigned to patterns with a *smart* enumeration or *truncated* branch-and-bound algorithm. The German league's noncompact schedule requires a third phase in which the games in each round are assigned to days of the week. This is accomplished in a greedy fashion with a limited neighborhood search when an infeasibility is encountered. The solution method for the compact schedule of the Austrian league follows the method of Nemhauser and Trick which will be discussed in the section on integer programming.

52.6 Integer Programming

An IP for the SRRTP is easy to formulate. Here we let the binary variables $x_{ijk} = 1$ if i plays j in slot k, and 0 otherwise, for $i < j \in \{1,\ldots,n\}$ and $k \in \{1,\ldots,n-1\}$. (We assume n is even for simplicity.) Since there is no objective in the basic SRRTP, the IP is the feasibility problem of finding x_{ijk} that satisfies the following constraints.

$$\sum_{j:j>i} x_{ijk} = 1 \quad \forall i \in 1,\ldots,n, \;\; k \in 1,\ldots,n-1$$

$$\sum_{k} x_{ijk} = 1 \forall i, \quad j \in 1,\ldots,n, \;\; i < j$$

$$x_{ijk} \quad \text{binary} \; \forall i, j \in \{1,\ldots,n\}, \;\; i < j, \;\; k \in 1,\ldots,n-1$$

The first constraint guarantees that every team plays exactly once in each slot. The second ensures that each team plays every opponent exactly once.

Trick [24] called this formulation the *basic-IP* formulation. He also suggested a *strong-IP* formulation that includes odd-set constraints to strengthen the one-factor constraint which provides exactly one game for each team in each slot. For the SRRTP, the odd-set constraints take the form

$$\sum_{i \in S, j \notin S} x_{ijk} \geq 1$$

where S is a subset of teams, $|S|$ is odd and k is a particular slot in $\{1,\ldots,n-1\}$. The odd-set constraints are too numerous to add them all to the model, but small sets can be selected for addition to the model in a branch-and-cut IP algorithm.

While simple to express, this IP may not be simple to solve when problem-specific constraints are added. Trick benchmarked the Base-IP formulation for several constrained SRRTPs and DRRTPs. We will discuss his results in the section comparing IP and CP below.

Rather than solving a single IP, Nemhauser and Trick [23] broke the DRRTP for the *Atlantic Coast Conference* (ACC) basketball teams into three phases: (i) generate pattern sets, (ii) generate timetables, and (iii) assign actual teams to patterns. Here we use the term *timetable* to denote a schedule with games between patterns rather than actual teams. Phases 2 and 3 are similar to the steps Schreuder used in solving the Dutch professional football league problem [22]. In the Nemhauser and Trick approach, the first two subproblems are solved using IP; in the third phase, all possible solutions are enumerated.

The ACC DRRTP includes a number of problem-specific constraints falling into two broad categories: pattern set constraints and team-specific constraints. Pattern set constraints include infeasible pattern sequences such as three consecutive home or away games, and equally distributed home weekend, away weekend, home weekday and away weekday games. Team-specific constraints include fixed home and away patterns, fixed games and opponent ordering constraints. These types of constraints are typical of the practical sports scheduling problems in the literature. The objective in the ACC problem is to maximize *good* weekend slots, i.e., weekend slots with popular match-ups, later in the season. In order to maximize the number of slots between games i at j and j at i and, possibly more importantly, to simplify the problem, Nemhauser and Trick adopted a partial mirroring scheme.

To generate pattern sets, Nemhauser and Trick first enumerated all feasible patterns. Then they used the following IP to generate pattern sets. Let P be the set of patterns, and S be the set of slots. The binary variable x_i is 1 if pattern i is selected and 0 otherwise. The coefficient h_{ik} is 1 if pattern i has a home in slot k and 0 otherwise. Similarly, the coefficient a_{ik} is 1 if pattern i has an away in slot k and 0 otherwise. The objective coefficient b_i indicates the extent to which pattern i has any bad pattern sequences like AA in the first two slots.

$$\text{Minimize} \sum_{i \in P} \sum_{j>i} \sum_{k} x_{ijk}$$

subject to

$$\sum_{i \in P} h_{ik} x_i = 4 \quad \forall k \in S$$

$$\sum_{i \in P} a_{ik} x_i = 4 \quad \forall k \in S$$

$$x_i \text{ binary } \forall i \in P$$

In a technical note (http://mat.gsia.cmu.edu/accmod.html), Nemhauser and Trick suggested strengthening this IP by adding a constraint to ensure that each pair of patterns has at least one slot with H-A and one with A-H.

After optimizing the IP, a cut is added to prevent the current solution from being regenerated. Then the IP is reoptimized. This process continues until all pattern sets with an objective value less than 2 are found. Given a pattern set and the partial-mirroring scheme, the timetables are generated using a model very similar to the base-IP for the SRRTP described above. The final phase, assigning actual teams to patterns, is accomplished using total enumeration.

52.7 Constraint Programming

Constraint programming is a combinatorial approach to solving hard optimization problems that grew out of logic programming. For a summary of constraint programming approaches, see, for instance, the monograph by Dechter [25].

Henz chose to use CP models to solve the ACC problem specified by Nemhauser and Trick (see [23] for the original problem description and [26] for the CP approach). Henz's CP approach for generating pattern sets is very similar to the IP approach described above. In the pattern generation model, the binary variables are used to represent the home, away, or bye assignments in each slot. Henz reported that the most effective search strategy seemed to be first enumerating the variables associated with the first slot and progressing to those associated with the last slot.

In combining patterns into pattern sets, Henz used a CP formulation very similar to Nemhauser and Trick's strengthened IP. A CP algorithm naturally searches for all feasible solutions, so there is no need to add cuts and restart an optimization process as with IP.

To generate complete schedules from pattern sets, Henz used an approach outlined by Cain [27] in 1977: first assign teams to patterns, then match teams in games. This represents a reversal of the order in which these steps were taken by Nemhauser and Trick. A single CP model is able to accomplish these two steps, although the set of variables matching teams and patterns is completely instantiated before any choices are made for the variables that pair teams into games.

Using a CP model rather than enumeration and reversing the order of the phases resulted in a great deal of run time savings. Henz's total run time was on the order of minutes, a substantial improvement over the 24-hour run time reported by Nemhauser and Trick.

In a more recent paper, Henz, Müller, and Thiel [28] examined the empirical performance of several different propagation algorithms on the two constraints in a number of theoretical SRRTPs and DRRTPs on n teams, n even. The SRRT formulation of Henz, Müller, and Thiel is given below (using the OPL modeling language).

```
int n = ...;
range Teams [1..n];
range Slots [1..n-1];
var Teams opponent[Teams,Slots];

solve {

  forall (i in Teams, k in Slots) opponent[i,t]<>i;
  forall (i in Teams) alldifferent(all (k in Slots) opponent[i,k]);
  forall (k in Slots) onefactor(all (i in Teams) opponent[i,k]);

};
```

An assignment

```
opponent[i,k]=j
```

is interpreted as *i plays j in slot k*. The all-different constraint ensures that every team plays exactly once in each slot. The one-factor constraint guarantees that each team plays every opponent exactly once.

An important consideration in the implementation of a CP algorithm is the trade-off between the strength of a propagation scheme and the computational time required to affect the propagation. For the all-different constraint, two propagation algorithms were considered:

1. representation of the constraint with $n(n-1)$ inequality constraints of the form

   `opponent[i,k] <> opponent[j,k]`

2. arc-consistent propagation with respect to the constraint itself

A CSP like the CP model for the SRRTP is arc-consistent with respect to a constraint if for each affected variable and each value in its domain, there is at least one instantiation of all the other affected variables which satisfies the constraint. So arc-consistent propagation with respect to a constraint is propagation that makes the CSP arc-consistent with respect to that constraint.

For the one-factor constraint, three propagation algorithms were considered:

1. Arc-consistent propagation with respect to the constraints

   `opponent[i,k] <> i` and `opponent[opponent[i,k],k]=i`

2. The propagation described in (1), plus arc-consistent propagation with respect to the redundant constraint

   `alldifferent(all (i in Teams) opponent[i,k])`

3. Arc-consistent propagation with respect to the constraint itself

Henz, Müller, and Thiel tested these propagation algorithms on the basic SRRTP, both the SRRTP and DRRTP with randomly generated forbidden game assignments (so many, in fact, that the problems have few or no solutions), an SRRTP with an objective of minimizing carry-over, a mirrored DRRTP with the objective of minimizing breaks, a DRRTP with fixed pattern set, and the ACC problem. In general, Henz, Müller, and Thiel found that the relatively unconstrained problems were better solved by the arc-constraint propagation of the constraints themselves but that the computational effort was not justified in the case of the more constrained problems.

In addition to Henz and his colleagues, Schaerf has also focused on applying CP methods to tournament scheduling problems. In [21], Schaerf solved a DRRTP with added constraints typical of practical sports scheduling problems. His method follows Schreuder's two-phase decomposition framework, using a canonical factorization for the timetable and then solving a CP to assign the actual teams to positions in the timetable. Added constraints include team complementarity, stadium availability, forbidden patterns for sets of teams, and constraints on the positioning of top games. Schaerf selected a canonical factorization with minimum breaks and used the venue assignment algorithm proposed by De Werra [3] for improving the fairness across patterns in a DRRT.

In the team assignment CP, teams are assigned patterns in a ranked manner: teams involved in the top games are assigned first followed by the rest of the teams according to the degree that they are involved in the hard and soft constraints. The last teams to be assigned patterns are the *free* teams, i.e., the teams involved in neither hard nor soft constraints.

Schaerf tested versions of his DRRT with different subsets of constraints *softened*, i.e., moved from the constraint set to the objective. He reported that his team assignment CP solved all versions of the problem for $n \leq 20$ but was only effective on large problem with $n = 30$ or 40 when almost all of the constraints were hard.

52.8 Comparing IP and CP

Obviously, the ACC problem provides some grounds for comparing the two methods; however, the comparison rings rather false for two reasons: (i) column generation/enumeration may be a better IP approach, and (ii) some of the run time savings found by Henz can be attributed to the reversal of phases 2 and 3.

As a keynote address at Practice and Theory of Automated Timetabling-2002 (PATAT 2002), Trick [24] presented a paper benchmarking the two methods on various types of SRRTPs and DRRTPs. Trick tested the SRRTPs and DRRTPs with randomly generated forbidden game assignments created by Henz, Müller, and Thiel [28]. He compared the performance of the Basic-IP and the standard CP formulation shown in Section 52.7. In the CP algorithm, arc-consistent propagation is used for the all-different constraint (Henz, Müller, and Thiel Scheme 2) while the one-factor constraint is propagated using the redundant all-different constraint (again, Henz, Müller, and Thiel Scheme 2). Trick found that basic-IP is competitive with the CP and is quicker to prove infeasibility for the larger n (in these cases, branching was not necessary).

Trick also tested a divisional SRRTP in which the n teams are divided into two divisions and it is desirable to first play games between divisions then finish with divisional play. This is easy to do for n divisible by 4 but cannot be done in a compact schedule for $n \equiv 2 \pmod 4$. Using $n \equiv 2 \pmod 4$, Trick fixed the first $n/2 - 1$ slots with games between the divisions and put the remaining two intra-divisional games in slot $n/2$. Completing this SRRT with $n - 1$ slots is infeasible; however, the basic-IP and the CP algorithm described above could not prove this for $n > 10$ in 30 min of computational time. Strong-IP was able to prove infeasibility without branching for all the problems tested, up to $n = 22$.

The third test included a randomly generated linear objective function. No additional constraints were included in the SRRTP. Not surprisingly, the IP performed better on these instances; Trick suggested that in order for the CP to compete in these tests, a specialized search strategy would be necessary.

Finally, Trick tested a DRRTP with pattern set constraints. First limiting the number of consecutive home or away games, he found that the constraint program was more effective in generating solutions for these instance, although the IP was able to produce a solution in a reasonable amount of time. Upon the addition of a constraint preventing singleton series, however, the CP was unable to produce solutions for most instances within 30 min. The IP produced solutions in all cases, though slowly. The addition of fixed games stymied all algorithms.

52.9 Combining IP and CP

The *traveling tournament problem* (TTP) was introduced by Michael Trick at the Spring, 1999 INFORMS meeting. The TTP was created as a simple problem (easy to state, minimal data requirements) embodying the difficult nature of any scheduling problem that combines travel and pattern set issues. The TTP was formally introduced in a paper by Easton, Nemhauser, and Trick [29] and is defined as follows:

Input: A set of teams $T = \{1, \ldots, n\}$; D an $n \times n$ integer distance matrix with elements d_{ij}; l, u integer parameters.

Output: A double round robin tournament on the teams in T such that

1. the length of every home stand and road trip is between l and u inclusive
2. the total distance traveled by the teams is minimized

For $u = n-1$, the maximum value for u, a team may visit every opponent without returning home. This is equivalent to a traveling salesman tour. For small u, a team must return home often, and consequently, its travel distance increases. For $u = 1$, the objective is constant, and the problem is solely one of feasibility. The benchmark instances on the web page http://mat.gsia.cmu.edu/TOURN/ have $l = 1$ and $u = 3$, which represents a common constraint in practice. These will be the parameters assumed for the remainder of this section. Additionally, the benchmark instances include a constraint on repeaters, i.e., i may not play at j and host j at home in consecutive slots.

There is an obvious lower bound for the TTP that factors into many of its solution methodologies. A team's optimal tour minimizes the travel distance for that team exclusive of the other teams in the schedule. The sum over n teams of the distances associated with their optimal tours provides a simple but strong lower bound on the TTP. We will refer to this bound as the *independent lower bound* (ILB).

Solutions have been generated using various methods. The two most successful general approaches include hybrid IP–CP algorithms and heuristics such as SA and TS. Here we discuss two of the former approaches; we discuss an SA approach in the following section.

Easton, Nemhauser, and Trick [30] published a solution methodology for the TTP based on a parallel branch-and-price (column generation) algorithm in which individual team tours are the columns. Specifically, a column is given by a vector specifying whether the slot is home or away and if it is away, by the opponent in that slot. A constraint programming model is used to solve the pricing problem. Additionally, a CP model for a full problem is used as a primal heuristic, run on one processor during the execution of the parallel algorithm.

The master IP simply selects one tour for each team in the tournament. Rather than branching on a single tour, however, Easton, Nemhauser, and Trick used higher order branching variables: the patterns indexed by team and slot. In other words, at any given node in which the master LP value is less than the current best solution, the solution space is divided into schedules in which team t is home in slot s and schedules in which team t is away in slot s. In order to select a branching variable, Easton, Nemhauser, and Trick used a strategy known as strong branching in which a small set of candidate variables are selected and tested with a limited number of simplex iterations along each branch.

A column pool is maintained so that it is not necessary to run the pricing problem every time the master prices out. When the pricing subproblem is called, it is run once for each team, generating all possible negative reduced cost tours within a given time increment.

The final component of the Easton, Nemhauser, and Trick solution methodology is a primal heuristic that makes use of an expanded version of the constraint program for the pricing problem, i.e., the primal heuristic works on the whole schedule, as opposed to just one tour. Simply running this CP does not produce *good* solutions quickly; however, by fixing certain elements it is able to generate solutions within an n-dependant percent of the ILB in a reasonable amount of time. Specifically, $n/2$ teams are selected to play one of their optimal tours. For a given set of *fixed* teams, the primal heuristic is run for a specific time interval, outputing improved solutions as they are obtained.

Easton, Nemhauser, and Trick were able to solve to optimality multiple instances of the TTP with up to $n = 8$ teams within a reasonable amount of time.

Benoist, Laburthe, and Rottembourg [31] also published a hybrid IP-CP solution methodology for the TTP. In this approach, CP is the basic algorithm and a specialized TSP code is used to solve the subproblems. The basic formulation is similar to that given above: columns are individual team tours. In this case, however, the specific visiting teams are included. The effect of this interpretation is that each tour carries more information, but the total number of possible tours to consider is substantially increased. Benoist, Laburthe, and Rottembourg found that this is the stronger formulation in terms of their framework.

The main CP model includes a representation of the DRRT constraints plus the two TTP specific constraints. In addition, a global constraint is used to compute a lower bound by Lagrangian relaxation.

Benoist, Laburthe, and Rottembourg compared three strategies in their computational tests using different degrees of lower bounding within the main CP. Their most sophisticated algorithm was able to achieve optimality for $n = 4$ and $n = 6$. Its best solutions for $n = 8$ and $n = 10$ were 3.4 and 17 percent over the ILB, respectively.

A hybrid IP–CP algorithm is a natural choice for the TTP since it has feasibility elements (the home and away pattern) and optimality elements (the travel distance). Prior to introducing the TTP, however, Trick [32] proposed a combined algorithm for the minimum break problem.

In his approach, Trick decomposed the SRRTP-MB into two phases: (i) matching teams in games, and (ii) assigning venues. A CP model was used to accomplish the first task. Game assignment was easy for large n in the absence of any specialized constraints. Additionally, the problem remained easy with the addition of fixed match-ups. Adding a carry-over constraint had an adverse effect on run time to the extent that the algorithm took on the order of a day to find a solution for instances with $n = 10$.

The second phase was more difficult to solve than the first. Trick used an IP model. The variables were

```
start[i]=1 if team i starts at home and 0 otherwise;
to_home[i,k]=1 if team i goes home after slot k and 0 otherwise;
to_away[i,k]=1 if team i goes away after slot k and 0 otherwise.
```

Auxiliary variables were defined to represent whether a team is home or away in slot k. For example,

```
at_home[i,k]=home[i]+sum_{k'<k} (to_home[i,k']-to_away[i,k'])
```

is 1 if i is home in slot k and 0 otherwise.

The basic IP formulation for this phase was strengthened by several addition constraints proposed by Regin [33] in the CP context. For example, it was assumed, without loss of generality, that the first team begins at home. Trick reported only marginal improvement in run time with the addition of this constraint. The model was also strengthened by the addition of the constraints

```
at_home[i,t]+to_home[i,t]<=1;
at_home[i,t]-to_away[i,t]>=0.
```

The resulting improvement was substantial for this addition. Finally, a triangle constraint was added to provide the following: if we have teams i, j, and k playing in slots $s_{ij} < s_{jk} < s_{ik}$, then either:

- i has a break between s_{ij} and s_{ik}, or
- j has a break between s_{ij} and s_{jk}, or
- k has a break between s_{jk} and s_{ik}.

This addition was very effective and allowed instances with up to $n = 22$ to be solved quickly. A true computational comparison with Regin's algorithm was not possible, but the computational times from Trick's CP/IP model appeared to be growing more slowly with instance size.

52.10 Metaheuristics: Simulated Annealing and Tabu Search

In the most recent development on the TTP front, Anagnostopoulos et al. [34] designed an advanced SA algorithm (TTSA) to search for heuristic solutions to the TTP. To date, this approach has yielded the lowest distance solutions for the benchmark TTP instances with $n = 10$ to 16.

Anagnostopoulos et al. divided the TTP constraints into hard and soft categories. The DRRT constraints are always satisfied while the constraints on repeaters and consecutive H/A games are relaxed. In this way, TTSA searches both the feasible region of the TTP, in which both the hard and soft constraints are satisfied, and an infeasible region in which the soft constraints are not satisfied to some extent.

The objective in the Anagnostopoulos et al. model includes a component to penalize solutions from the infeasible region. This penalty is dynamically adjusted to prevent the algorithm from spending too much time in a space where the soft constraints are not satisfied. This is somewhat similar to the strategic oscillation approach used in tabu search algorithms for the generalized assignment problem.

TTSA uses five different moves to explore the neighborhood of a current schedule:

1. Swapping the venues for a particular pair of games, i.e., i at j in slot s and j at i in slot s' becomes i at j in slot s' and j at i in slot s
2. Swapping the positions of two slots of games
3. Swapping the schedules of two teams (except for the games in which they play each other)
4. Partially swapping the games in two slots, i.e., swapping the games for team i, then swapping as few additional games between the two slots as possible to satisfy the DRRT constraints
5. Partially swapping team schedules, i.e., swapping the games in a single slot, then swapping as few additional games between the teams as possible to satisfy to DRRT constraints

As is standard in SA algorithms, the probability of accepting a nonimproving solution decreases over time (this is known as decreasing the temperature). After reaching a low temperature, however, TTSA is able to *reheat* in order to escape a local minimum.

Anagnostopoulos et al. reported that the most successful parameter settings for TTSA include a very slow cooling system, a large number of phases, and long phase duration. The TTSA was able to improve the current best solution (as of January, 2003) to the TTP benchmark instances for $n = 10$ through $n = 16$ by 2 to 5%.

Approximately a decade prior to the development of TTSA, two papers were published describing metaheuristic approaches for scheduling cricket matches. Willis and Terrill [35] used an SA approach to solve the scheduling problem for the *Australian state cricket league.* Their algorithm also treats some constraints as hard and others as soft. Violating the soft constraints is penalized in the objective. Hard constraints include limitations on the scheduling of one-day (vs. four-day) matches, fixed games and fixed venues for one-day matches. Soft constraints result from television requests, pattern set requirements, conflicts with international matches, and the effects of weather in certain regions. The algorithm produced a solution with nine violated soft constraints which was, for the most part, acceptable for the *Australian cricket board.*

Wright [36] used a TS algorithm to schedule the SRRT for an *English cricket league*. A compact schedule is not required by the league; only the last slot in the tournament needs to have the maximum number of games possible. So the set of hard constraints includes those in the basic SRRTP with the equality constraint for the number of times a team plays in a slot relaxed to an inequality. The only other hard constraint in the model is that most of the teams need to have one or two home games in each of the first and last three slots. Soft constraints include team requests for certain games and venues in particular slots, H/A balance, and complementary pattern requirements. In addition to the penalties associated with the soft constraints, the objective also includes a travel component mainly used to minimize total travel distance but also to penalize a solution for giving any team more than one long trip.

Wright's algorithm is initialized with an incomplete solution. All the moves used to explore the neighborhood around a current solution involve moving a game from slot s to slot s'. The algorithm considers sequences of up to four moves. In other words, the game g_{ij} might be moved from slot s to slot s' in which the game g_{ik} is already scheduled forcing g_{ik} to be moved. After three of these moves, the last game to be moved is placed in a slot that does not already contain any games for either of its two opponents. Wright ran several variations of the TS algorithm with different parameter settings. A moderate set of parameters, somewhere between standard tabu search and an extreme variant, produced the best solutions which were, for the most part, acceptable to the league.

52.11 Problem-Specific Heuristics

Campbell and Chen [37] in 1976, Ball and Webster [38] in 1977, and Bean and Birge [39] in 1980, all described heuristics with a similar framework: determine *good* trips, then use these as a basis for building the full schedule.

Campbell and Chen solved a DRRTP with a limit of two consecutive H/A games for a ten-team college basketball league in the United States. In the first phase of their heuristic, optimal trip pairings are generated with a *smart* enumeration algorithm. (One set of pairings gives the optimal tour for every team; the opponent paired with the traveling team is the one single trip.) In the second phase, the pairings are used to establish travel partners and the symmetric latin squares representing all possible SRRTs are enumerated. Home and aways are assigned in order to minimize breaks. The minimum distance SRRT is selected, mirrored, and expanded to give a solution to the DRRTP. Note that this heuristic generates a noncompact schedule since the expansion from the five-team travel partners schedule to the ten-team schedule includes byes (the teams in the pairing assigned the bye in the travel partners schedule play each other and have one bye in the expanded schedule).

Ball and Webster also considered a DRRTP with a limit of two consecutive H/A games. The SEC and Big 10 conference's basketball leagues serve as examples. In the first phase of the heuristic, optimal tours for all teams are found by enumeration (leading to the constant pairings result for n even). In the second phase, venue-specific games are assigned to slots based on the constant pairings from the first phase. After scheduling the first half of the tournament, the second half is completed as a mirror or partial mirror of the first. The exact algorithms for both even and odd n can be found in [38].

Bean and Birge focused on the scheduling problem for the *National Basketball Association* (NBA) in the United States. The teams in the NBA (numbering 22 in 1980) play a noncompact schedule; the most limiting constraints in the model, according to the authors, are the required rest day between games on opposite coasts and the stadium availability which averages only 30% of the total days in the season.

Bean and Birge rejected graph theoretical and IP approaches to the problem because of its size. They then developed the following heuristics:

1. Greedily generate *good* trips with up to 5 away venues.
2. Order the trips from greatest to smallest distance.
3. Select the trip at the top of the queue. If possible, schedule it in the slots with the least stadium availability for the traveling team. If it is not possible to schedule the tour, break it into two or more subtrips and insert these at the end of the queue.
4. Repeat step 3 until only a small number of trips remain in the queue.
5. Use an IP to complete the schedule minimizing travel distance.

Bean and Birge reported that their heuristic was successfully used to schedule two seasons for the NBA.

52.12 Conclusion

Sports scheduling provides an interesting combination of practical application and theoretical challenge. As shown in this review, there are a large number of problem types in this area, and each has been attacked by a variety of solution methods. For most problems, even relatively small instances remain computationally challenging, making this a fruitful area for more research.

References

[1] Anderson, I., 1997. *Combinatorial Designs and Tournament*, Clarendon Press, Oxford, 160–161.

[2] de Werra, D., L. Jacot-Descombes, and P. Masson, 1990. A Constrained Sports Scheduling Problem, *Discrete Applied Mathematics* **26**, 41–49.

[3] de Werra, D., 1981. Scheduling in Sports, *Annals of Discrete Mathematics* **11**, 381–395.

[4] Easton, T. and R.G. Parker, 2001. On Completing Latin Squares, *Discrete Applied Mathematics* **113**, 167–181.

[5] Colbourn, C. and J.H. Dinitz, 1996. *The CRC Handbook of Combinatorial Designs*, CRC Press, Boca Raton, FL.

[6] Colbourn, C., 1983. Embedding Partial Steiner Triple Systems is NP-Complete, *Journal of Combinatorial Theory* (A) **35**, 100–105.

[7] Easton, K., 2003. Solving Sports Scheduling Problems Using Integer Programming and Constraint Programming, Ph.D. thesis, Georgia Institute of Technology.

[8] de Werra, D., 1988. Some Models of Graphs for Scheduling Sports Competitions, *Discrete Applied Mathematics* **21**, 47–65.

[9] Miyashiro, R., H. Iwasaki, and T. Matsui, 2002. Characterizing Feasible Pattern Sets with a Minimum Number of Breaks. (Unpublished)

[10] Haselgrove, J. and J. Leech, 1977. A Tournament Design Problem, *American Mathematical Monthly* **84**, 198–201.

[11] Schellenberg, P.J., G.H.J. van Rees, and S.A. Vanstone, 1977. The existence of Balanced Tournament Designs, *Ars Combinatoria* **3**, 303–318.

[12] Lamken, E.R. and S.A. Vanstone, 1985. The Existence of Factored Balanced Tournament Designs, *Ars Combinatoria* **19**, 157–160.

[13] Blest, D.C. and D.G. Fitzgerald, 1988. Scheduling Sports Competitions with a Given Distribution of Times, *Discrete Applied Mathematics* **22**, 9–19.

[14] Russell, K.G., 1980. Balancing Carry-Over Effects in Round Robin Tournaments, *Biometrica* **67**, 127–131.

[15] de Werra, D., 1980. Geography, Graphs and Games, *Discrete Applied Mathematics* **2**, 327–337.

[16] de Werra, D., 1985. On the Multiplication of Divisions: The Use of Graphs for Sports Scheduling, *Networks* **15**, 125–136.

[17] Kotzig, A., 1972. Decompositions of Complete Graphs into Regular Bichromatic Factors, *Discrete Mathematics* **2**, 383–387.

[18] de Werra, D., 1982. Minimizing Irregularities in Sports Scheduling Using Graph Theory, *Discrete Applied Mathematics* **4**, 217–226.

[19] Bartsch, T., A. Drexl, and S. Kroger, 2002. Scheduling European Soccer Leagues: Models, Methods and Applications, *PATAT 2002*.

[20] Russell, R.A. and J.M.Y. Leung, 1994. Devising a Cost Effective Schedule for a Baseball League, *Operations Research* **42**, 614–625.

[21] Schaerf, A., 1999. Scheduling Sport Tournaments Using Constraint Logic Programming, *Constraints* **4**, 43–65.

[22] Schreuder, J.A.M., 1992. Combinatorial Aspects of Construction of Competition Dutch Professional Football League, *Discrete Applied Mathematics* **35**, 301–312.

[23] Nemhauser, G.L. and M.A. Trick, 1988. Scheduling a Major College Basketball Conference, *Operations Research* **46**, 1–8.

[24] Trick, M.A., 2002. Integer and Constraint Programming Approaches for Round Robin Tournament Scheduling, Keynote Address at *PATAT 2002*.

[25] Dechter, Rina, 2003. *Constraint Processing*, Morgan Kaufmann, New York, NY.

[26] Henz, M., 1999. Constraint-Based Round Robin Tournament Planning, *Proceedings of the 1999 International Conference on Logic Programming*, Las Cruces, NM.

[27] Cain, W.O., Jr., 1977. A Computer-Assisted Heuristic Approach Used to Schedule the Major League Baseball Clubs, In S.P. Ladany and R.E. Machol, (editors), *Optimal Strategies in Sports*, North-Holland Publishing Company, New York, 32–41.

[28] Henz, M., T. Müller, and S. Theil, 2002. Global Constraints for Round Robin Tournament Scheduling, *European Journal of Operations Research* **153**(1), 92–101.

[29] Easton, K., G.L. Nemhauser, and M.A. Trick, 2001. The Traveling Tournament Problem: Description and Benchmarks, *Principles and Practice of Constraint Programming — CP 2001*, Springer Lecture Notes in Computer Science **2239**, 580–585.

[30] Easton, K., G.L. Nemhauser, and M.A. Trick, 2002. Solving the Traveling Tournament Problem: A Combined Integer Programming and Constraint Programming Approach, *PATAT 2002*.

[31] Benoist, T., F. Laburthe, and B. Rottembourg, 2001. Lagrange Relaxation and Constraint Programming Collaborative Schemes for Traveling Tournament Problems, *Proceedings of CP-AI-OR '01*, Wye College, UK, 15–26.

[32] Trick, M.A., 2001. A Schedule-and-Break Approach to Sports Scheduling, In E. Burke and W. Erben (editors), *PATAT 2000*, Springer Lecture Notes in Computer Science **2079**, 242–253.

[33] Regin, J.C., 2000. Modeling with Constraint Programming, *Dagstuhl Seminar on Constraint and Integer Programming*.

[34] Anagnostopolous, A., L. Michel, P. Van Hentenryck, and Y. Vergados, 2003. A Simulated Annealing Approach to the Traveling Tournament Problem. (Unpublished)

[35] Willis, R.J. and B.J. Terrill, 1994. Scheduling the Australian State Cricket Season Using Simulated Annealing, *Journal of the Operations Research Society* **45**, 276–280.

[36] Wright, M., 1994. Timetabling County Cricket Fixtures Using a Form of Tabu Search, *Journal of the Operations Research Society* **45**, 758–770.

[37] Campbell, R.T. and D.S. Chen, 1976. A Minimum Distance Basketball Scheduling Problem, In S.P. Ladany, R.E. Machol, and D.G. Morrison, (editors), *Managament Science in Sports*, North-Holland Publishing Company, New York, 15–26.

[38] Ball, B.C. and D.B. Webster, 1977. Optimal Scheduling for Even-Numbered Team Athletic Conferences, *AIIE Transactions* **9**, 161–169.

[39] Bean, J.C. and J.R. Birge, 1980. Reducing Travelling Costs and Player Fatigue in the National Basketball Association, *INTERFACES* **10**, 98–102.

[40] Armstrong, J. and R.J. Willis, 1993. Scheduling the Cricket World Cup — a Case Study, *Journal of the Operations Research Society* **44**, 1067–1072.

[41] Ferland, J.A. and C. Fleurent, 1991. Computer Aided Scheduling for a Sports League, *INFOR* **29**, 14–25.

[42] Fleurent, C. and J.A. Ferland, 1993. Allocating Games for the NHL Using Integer Programming, *Operations Research* **41**, 649–654.

[43] Henz, M., 2001. Scheduling a Major College Basketball Conference — Revisited, *Operations Research* **49**, 163–168.

[44] Regin, J.C., 1999. Minimization of the Number of Breaks in Sports Scheduling Problems Using Constraint Programming, *DIMACS Workshop on Constraint Programming and Large Scale Discrete Optimization*.

[45] Schreuder, J.A.M., 1980. Constructing Timetables for Sports Competitions, *Mathematical Programming Study* **13**, 58–67.

Index

A

B

C

D

E

F

G

H

I

J

K

L

M

N

O

P

Q

R

S

T

U

V

W

Z

* 9 7 8 1 5 8 4 8 8 3 9 7 5 *

T - #1044 - 101024 - C0 - 254/178/49 [51] - CB - 9781584883975 - Gloss Lamination